QUANTITATIVE HUMAN PHYSIOLOGY

QUANTITATIVE HUMAN PHYSIOLOGY
AN INTRODUCTION

Second Edition

Joseph Feher

Department of Physiology and Biophysics,
Virginia Commonwealth University School of Medicine

AMSTERDAM • BOSTON • HEIDELBERG • LONDON • NEW YORK • OXFORD • PARIS
SAN DIEGO • SAN FRANCISCO • SINGAPORE • SYDNEY • TOKYO

Academic Press is an imprint of Elsevier

Academic Press is an imprint of Elsevier
125 London Wall, London EC2Y 5AS, United Kingdom
525 B Street, Suite 1800, San Diego, CA 92101-4495, United States
50 Hampshire Street, 5th Floor, Cambridge, MA 02139, United States
The Boulevard, Langford Lane, Kidlington, Oxford OX5 1GB, United Kingdom

Copyright © 2012, 2017 Elsevier Inc. All rights reserved.

No part of this publication may be reproduced or transmitted in any form or by any means, electronic or mechanical, including photocopying, recording, or any information storage and retrieval system, without permission in writing from the publisher. Details on how to seek permission, further information about the Publisher's permissions policies and our arrangements with organizations such as the Copyright Clearance Center and the Copyright Licensing Agency, can be found at our website: www.elsevier.com/permissions.

This book and the individual contributions contained in it are protected under copyright by the Publisher (other than as may be noted herein).

Notices
Knowledge and best practice in this field are constantly changing. As new research and experience broaden our understanding, changes in research methods, professional practices, or medical treatment may become necessary.

Practitioners and researchers must always rely on their own experience and knowledge in evaluating and using any information, methods, compounds, or experiments described herein. In using such information or methods they should be mindful of their own safety and the safety of others, including parties for whom they have a professional responsibility.

To the fullest extent of the law, neither the Publisher nor the authors, contributors, or editors, assume any liability for any injury and/or damage to persons or property as a matter of products liability, negligence or otherwise, or from any use or operation of any methods, products, instructions, or ideas contained in the material herein.

British Library Cataloguing-in-Publication Data
A catalogue record for this book is available from the British Library

Library of Congress Cataloging-in-Publication Data
A catalog record for this book is available from the Library of Congress

ISBN: 978-0-12-800883-6

For Information on all Academic Press publications
visit our website at https://www.elsevier.com

Working together to grow libraries in developing countries

www.elsevier.com • www.bookaid.org

Publisher: Katey Birtcher
Acquisition Editor: Steve Merken
Editorial Project Manager: Nate McFadden
Production Project Manager: Stalin Viswanathan
Designer: Maria Ines Cruz

Typeset by MPS Limited, Chennai, India

Contents

Preface ix

Acknowledgments xiii

UNIT 1 PHYSICAL AND CHEMICAL FOUNDATIONS OF PHYSIOLOGY 1

Chapter 1.1 The Core Principles of Physiology 3

Chapter 1.2 Physical Foundations of Physiology I: Pressure-Driven Flow 15

Chapter 1.3 Physical Foundations of Physiology II: Electrical Force, Potential, Capacitance, and Current 31

Problem Set 1.1 Physical Foundations: Pressure and Electrical Forces and Flows 43

Chapter 1.4 Chemical Foundations of Physiology I: Chemical Energy and Intermolecular Forces 46

Chapter 1.5 Chemical Foundations of Physiology II: Concentration and Kinetics 59

Chapter 1.6 Diffusion 75

Chapter 1.7 Electrochemical Potential and Free Energy 87

Problem Set 1.2 Kinetics and Diffusion 94

UNIT 2 MEMBRANES, TRANSPORT, AND METABOLISM 99

Chapter 2.1 Cell Structure 101

Chapter 2.2 DNA and Protein Synthesis 120

Chapter 2.3 Protein Structure 130

Chapter 2.4 Biological Membranes 142

Problem Set 2.1 Surface Tension, Membrane Surface Tension, Membrane Structure, Microscopic Resolution, and Cell Fractionation 156

Chapter 2.5 Passive Transport and Facilitated Diffusion 161

Chapter 2.6 Active Transport: Pumps and Exchangers 170

Chapter 2.7 Osmosis and Osmotic Pressure 182

Problem Set 2.2 Membrane Transport 199

Chapter 2.8 Cell Signaling 205

Chapter 2.9 ATP Production I: Glycolysis 218

Chapter 2.10 ATP Production II: The TCA Cycle and Oxidative Phosphorylation 227

Chapter 2.11 ATP Production III: Fatty Acid Oxidation and Amino Acid Oxidation 241

UNIT 3 PHYSIOLOGY OF EXCITABLE CELLS 253

Chapter 3.1 The Origin of the Resting Membrane Potential 255

Chapter 3.2 The Action Potential 265

Chapter 3.3 Propagation of the Action Potential 280

Problem Set 3.1 Membrane Potential, Action Potential, and Nerve Conduction 289

Chapter 3.4 Skeletal Muscle Mechanics 292

Chapter 3.5 Contractile Mechanisms in Skeletal Muscle 305

Chapter 3.6 The Neuromuscular Junction and Excitation–Contraction Coupling 318

Chapter 3.7 Muscle Energetics, Fatigue, and Training 334

Problem Set 3.2 Neuromuscular Transmission, Muscle Force, and Energetics 349

Chapter 3.8 Smooth Muscle 351

UNIT 4 THE NERVOUS SYSTEM 363

Chapter 4.1 Organization of the Nervous System 365

Chapter 4.2 Cells, Synapses, and Neurotransmitters 375

Chapter 4.3 Cutaneous Sensory Systems 389
Chapter 4.4 Spinal Reflexes 400
Chapter 4.5 Balance and Control of Movement 409
Problem Set 4.1 Nerve Conduction 422
Chapter 4.6 The Chemical Senses 427
Chapter 4.7 Hearing 440
Chapter 4.8 Vision 456
Problem Set 4.2 Sensory Transduction 471
Chapter 4.9 Autonomic Nervous System 473

UNIT 5 THE CARDIOVASCULAR SYSTEM 487

Chapter 5.1 Overview of the Cardiovascular System and the Blood 489
Chapter 5.2 Plasma and Red Blood Cells 498
Chapter 5.3 White Blood Cells and Inflammation 507
Chapter 5.4 The Heart as a Pump 516
Problem Set 5.1 Blood 525
Chapter 5.5 The Cardiac Action Potential 528
Chapter 5.6 The Electrocardiogram 537
Chapter 5.7 The Cellular Basis of Cardiac Contractility 547
Chapter 5.8 The Cardiac Function Curve 556
Problem Set 5.2 Cardiac Work 565
Chapter 5.9 Vascular Function: Hemodynamics 568
Chapter 5.10 The Microcirculation and Solute Exchange 578
Chapter 5.11 Regulation of Perfusion 589
Chapter 5.12 Integration of Cardiac Output and Venous Return 599
Chapter 5.13 Regulation of Arterial Pressure 608
Problem Set 5.3 Hemodynamics and Microcirculation 619

UNIT 6 RESPIRATORY PHYSIOLOGY 621

Chapter 6.1 The Mechanics of Breathing 623
Chapter 6.2 Lung Volumes and Airway Resistance 633
Chapter 6.3 Gas Exchange in the Lungs 642
Problem Set 6.1 Airway Resistance and Alveolar Gas Exchange 653
Chapter 6.4 Oxygen and Carbon Dioxide Transport 656
Chapter 6.5 Acid−Base Physiology I: The Bicarbonate Buffer System and Respiratory Compensation 665
Chapter 6.6 Control of Ventilation 672
Problem Set 6.2 Gas Transport and pH Disturbances 682

UNIT 7 RENAL PHYSIOLOGY 687

Chapter 7.1 Body Fluid Compartments 689
Chapter 7.2 Functional Anatomy of the Kidneys and Overview of Kidney Function 698
Chapter 7.3 Glomerular Filtration 705
Problem Set 7.1 Fluid Volumes, Glomerular Filtration, and Clearance 715
Chapter 7.4 Tubular Reabsorption and Secretion 719
Chapter 7.5 Mechanism of Concentration and Dilution of Urine 730
Chapter 7.6 Regulation of Fluid and Electrolyte Balance 740
Chapter 7.7 Renal Component of Acid−Base Balance 752
Problem Set 7.2 Fluid and Electrolyte Balance and Acid−Base Balance 765

UNIT 8 GASTROINTESTINAL PHYSIOLOGY 769

Chapter 8.1 Mouth and Esophagus 771
Chapter 8.2 The Stomach 785
Chapter 8.3 Intestinal and Colonic Chemoreception and Motility 796
Chapter 8.4 Pancreatic and Biliary Secretion 810
Chapter 8.5 Digestion and Absorption of the Macronutrients 821
Chapter 8.6 Energy Balance and Regulation of Food Intake 834
Problem Set 8.1 Energy Balance 847

UNIT 9 ENDOCRINE PHYSIOLOGY 851

Chapter 9.1 General Principles of Endocrinology 853
Chapter 9.2 Hypothalamus and Pituitary Gland 870
Chapter 9.3 The Thyroid Gland 883
Chapter 9.4 The Endocrine Pancreas and Control of Blood Glucose 895
Chapter 9.5 The Adrenal Cortex 906

Chapter 9.6 The Adrenal Medulla and Integration of Metabolic Control 916

Chapter 9.7 Calcium and Phosphorus Homeostasis I: The Calcitropic Hormones 924

Chapter 9.8 Calcium and Phosphorus Homeostasis II: Target Tissues and Integrated Control 933

Chapter 9.9 Female Reproductive Physiology 946

Chapter 9.10 Male Reproductive Physiology 956

Problem Set 9.1 Ligand Binding 967

Appendix I Important Equations 969

Appendix II Important Physical Constants for Physiology 981

Index 983

Preface

Welcome to the second edition of *Quantitative Human Physiology*! This new edition has been updated with numerous enhancements, many of which were suggested or inspired by instructors and students who used the first edition. These important changes include (but are not limited to):

- Substantial updating of the text throughout to reflect the latest research results, with many sections expanded to include relevant and important information
- Enhanced, updated, and improved figures for better understanding and clarification of challenging topics
- Addition of several new appendices covering statistics, nomenclature of transport carriers, and structural biology of important items such as the neuromuscular junction and calcium release unit
- Addition of new problems within the problem sets and example calculations in the text
- Addition of some Clinical Applications such as dual energy X-ray bone densitometry
- Addition of commentary to power point presentations.

The goal of this new edition was to make important improvements while retaining the features that make this text uniquely suited to the needs of undergraduate bioengineering students. While it is sometimes very tempting to make drastic and sweeping changes in an attempt satisfy everyone, this new edition focuses on refinements and updates that, we hope, most instructors and students will find helpful, informative, and instructionally sound.

THIS TEXT IS A PHYSIOLOGY TEXT FIRST, AND QUANTITATIVE SECOND

The second edition of this text remains faithful to its primary goals: it remains, first and foremost, a physiology text. It is explicitly designed for a certain class of students, those majoring in Biomedical Engineering at Virginia Commonwealth University, and their suggestions, limitations, and desires have shaped the text from the outset. Specifically, the text is designed for students who have never been exposed to Physiology, students who know neither the language nor the concepts of the subject. The text contains all elements of physiology in nine units: physical and chemical foundations; cell physiology; excitable tissue physiology; neurophysiology; cardiovascular physiology; respiratory physiology; renal physiology; gastrointestinal physiology; and endocrinology. The course is best taught in the order of the text but it is possible to present the material in other sequences.

Secondly, the text affirms its aim to be quantitative. Being quantitative has two aspects. The first is about knowing the numerical value for the ranges of crucial aspects of physiology, such as the flows or forces within the body. The second is about discovering the relations between physiological parameters. For many aspects of physiology, there currently is not enough information to make a detailed quantitative analysis, or the analysis at that depth is beyond the scope of this text. In these cases, this text is not very different from more traditional texts. Where possible, the text takes an analytical and quantitative approach.

THE TEXT USES MATHEMATICS EXTENSIVELY

Carl Frederick Gauss famously said, "Mathematics is the queen of sciences." Mathematics is not just about the numerical value of something, such as the magnitude of the arterial blood pressure or the rate of salivary secretion, although that is what many people think of when they think of quantitation. Rather, mathematics is about the relationship between things that can vary with time or position. These relationships cannot be fully understood with words alone. They require the language of mathematics. Students should be able to articulate these relationships with words, but this text demands more. Mathematical statements—equations—are simply logical sentences. You can read an equation in words. But the mathematical statement uses an economy of words. Mathematics also has specific rules for the manipulation of the logical parts of the sentences, so that rearrangement or combination of equations leads to new insight.

NOT ALL THINGS WORTH KNOWING ARE WORTH KNOWING WELL

This text uses lots of mathematics at the level of the calculus and elementary differential equations. These mathematical tools are used to encode the physics or chemistry of processes into mathematical symbolism, and mathematical manipulation leads to useful equations.

The point of the derivations is the useful equation, not the derivations themselves. The derivations are included, sometimes as an appendix, to satisfy the students that the final equations are not magical, but come from the application of mathematics to physical and chemical principles that apply to the body. The point is to be able to apply the final equations. Physics, chemistry, and math at the level of calculus are used to get the equations, but generally algebra is all that is required to apply them. In my view, the derivations are important to teach students the process of encoding the physics and chemistry and deriving the relationship between variables that constitute the useful equations, but rote memorization of derivations is boring and useless.

PERFECT IS THE ENEMY OF GOOD: EQUATIONS AREN'T PERFECT, BUT THEY'RE OFTEN GOOD ENOUGH

The text does not say what I tell my students about the applicability of equations in general. I tell my students in lecture that all of these equations are wrong. They are wrong in something of the same sense that Newtonian mechanics is wrong. Relativity and quantum mechanics supplant Newtonian mechanics (even in the macroscopic world) but Newtonian mechanics will still get the rocket ship to the moon. So I tell them that these equations are theoretically wrong but they give satisfactorily correct answers, in much the same way that Newtonian mechanics still does, even though theoretically incomplete. Fick's Law of Diffusion depends on continuum mathematics for an inherently discrete process. But the discreteness is so finely divided that the distinction is unimportant.

EXAMPLES AND PROBLEM SETS ALLOW APPLICATION OF THE USEFUL EQUATIONS

There are several aids in the text to foster a quantitative and analytical understanding. First, there are some worked calculations in the text that are set apart in text boxes as Examples. These aim to show the students how to apply some of the ideas presented in the text. Second, there are a total of 17 problem sets scattered throughout the text. All units have at least one, most have two, and the cardiovascular section has three. These are meant to cover about three chapters each, so that a problem set can be assigned approximately once a week, for three lectures per week. Students have repeatedly told me that they want a solution set to the problems to see how they can do them, but this makes them useless as a graded assignment. There is no better teacher than wrestling with a problem. The second edition has expanded on these problem sets with new problems.

The problems themselves are generally meant to cover some new idea or concept that could not be, or was not, effectively covered in the text itself, and to expand on the student's understanding of the material. They are not merely busy work or "plug and chug" exercises. The alert student should not merely do the problem, but think about what the result means. In many cases, the problems are written as a sequence of questions, each of which sets up the student for further questions or insights. These illustrate the process necessary for answering a larger question, breaking it up into parts of the answer that must come first. This method aids the students in thinking about larger problems: break it down into its simpler components. Some subject matter unfortunately does not lend itself easily to such problems, but I have attempted to find problems that students can do. The students in my classes find the problems challenging. I allow them to work on them collaboratively, because they are meant to be part of the instruction and less of the evaluation, but the problem sets are graded and contribute significantly to the final grades. Of course, such policies are set at the discretion of the instructor.

LEARNING OBJECTIVES, SUMMARIES, AND REVIEW QUESTIONS GUIDE STUDENT LEARNING

The Learning Objectives are meant as a guide to the construction of examination questions, either directly or indirectly. These learning objectives are fairly broad and together they cover the breadth of physiology. This is my advice to students: read the Learning Objectives first, the chapter Summary second, and then read the text. Next, attempt to answer the Review Questions and return to the Learning Objectives. If you are mystified by any of the Review Questions or Learning Objectives, read the pertinent part of the text again.

CLINICAL APPLICATIONS PIQUE INTEREST

Pathological situations often illuminate normal physiology. Clinical Applications are scattered throughout the text. Because it takes a lot of background material to understand these clinical applications, Clinical Applications are less prevalent in the early parts of the text, which are foundational, than in the latter parts of the text. There are more than 50 such Clinical Applications, with several new additions in the second edition. The students like them because these Clinical Applications tell them that there is a reason for learning what might otherwise appear arcane.

HOW INSTRUCTORS CAN USE THIS TEXT

This text is meant for anyone with a fairly modest level of mathematical skill, at the level of the calculus with elementary differential equations. Students without this level of training will find this book too difficult. I have developed this book specifically with undergraduate Biomedical

Engineering students in mind, and have taught this material for 16 years. Each Chapter is intended to be a single lecture, and the length of the chapter is meant to be readable in a single sitting. There are 77 chapters, so it is most useful in a two-semester sequence. I cover all chapters in that time, and all problem sets. The text would also be useful for instructors with less time by adjusting the breadth and depth of coverage. Units 1, 8, and 9, (Physical and Chemical Foundations, Gastrointestinal Physiology and Endocrinology, respectively) for example, could be eliminated or covered more superficially. Unit 1 could be eliminated because it is a review, Units 8 and 9 because they are relatively peripheral to Biomedical Engineering. Unit 2, Cell Physiology, could be covered more superficially or eliminated if students previously have had Cell Biology. Alternatively, it is possible to cover the breadth of physiology if the depth is reduced. This can be done by focusing on the chapter summaries, which give a broach picture of what is happening with less detail, and combining lectures or omitting some topics. In this regard, the text is useful similarly to a cafeteria: the instructor is free to choose those sections of most interest and to downplay those of lesser interest. Some instructors may feel that knowing how to think about problems is the most important thing, and that the details of the physiology are secondary. Such an instructor may want to focus more on the appendices than on the chapter matter themselves, and more on the problems and how to solve them.

ANCILLARY MATERIALS FOR INSTRUCTORS

For instructors adopting this text for use in a course, the following ancillaries are available: Power Point lecture slides, electronic images from the text, solutions manual for the problem sets, laboratory manuals with example data, and examination questions with answers. The Power Point slides have been updated to include the new figures and more commentary as appropriate for the lectures. Please visit http://textbooks.elsevier.com/9780128008836.

HOW STUDENTS CAN USE THIS TEXT

A student's goal ought to be to learn as much physiology as possible within the constraints of available time. This text is written with considerable detail. The Learning Objectives and Review Questions set the stage for the kinds of things you should be able to do, and the kinds of questions you should be able to answer. The Chapter Summaries encapsulate each chapter in an economy of words and detail. Start with the Learning Objectives, read the Summary, and then look at the Review Questions. Then read the chapter and repeat Learning Objectives, Summary, and Review Questions. What you cannot answer at that point requires you to re-read the pertinent sections a second time.

The approach to the problem sets is different. The first job is: find a bright fellow student to work with. Second, read the question for understanding of what it is asking. Then ask yourself, what is needed to answer this question? How can you find out what is needed? If it is a physical constant, where can you find it? Do you know a relationship or equation that relates what is being asked to what is given? Write it down and rearrange it to give the desired answer. Plug and chug what is given and what else you have looked up to get a numerical answer. It is very simple in principle but sometimes very difficult in execution.

ANCILLARY MATERIALS FOR STUDENTS

Student resources available with this text include a set of online flashcards, a selection of animations based on the figures in the text, and online quizzes for self-study. Please visit http://booksite.elsevier.com/9780128008836.

STUDENT FEEDBACK

Students who have completed the course regularly tell me that it is both one of the most challenging and one of the most rewarding courses of their undergraduate career. This is generally the case: what you get out of an educational enterprise is proportionate to what you put in. Physiology is an integrative science. There is great satisfaction in understanding how a system works from the cellular and subcellular level all the way up to the organism level. Human beings do not come with an owner's manual. The idea that this text is a first approximation to an owner's manual resonates with the students.

Joseph Feher

Acknowledgments

I would like to thank all the reviewers of the proposal and drafts of this project. Their feedback was very helpful in improving the final version:

Brett BuSha	The College of New Jersey
Jingjiao Guan	Florida State University
Nuran M. Kumbaraci	Stevens Institute of Technology
Marie Luby	University of Connecticut
Ken Yoshida	Indiana University—Purdue University Indianapolis
Ryan Zurakowski	University of Delaware

I would also like to acknowledge the help and support of my colleagues at the VCU School of Medicine, especially Clive Baumgarten and Ray Witorsch, for their criticism of early drafts, and Margaret Biber, who expressed the confidence in publishing this text that kept me going. I would like to thank George Ford, VCU School of Medicine, who helped develop the early outlines of the cardiovascular section; Scott Walsh, who provided the basis for some of the endocrine chapters; and Andy Anderson, who helped critique multiple teaching efforts of mine. I especially acknowledge the feedback from many years of sophomore and junior BME students who pointed out errors and difficulties in the material and suggested improvements to help their learning experience, which is really what this text is all about. Special thanks go to students Woon Chow, Yasha Mohajer, Matthew Caldwell, Matthew Painter, Mary Beth Bird, Richard Boe, Linda Scheider, William Eggleston, Alex Sherwood, Ross Pollansch, Kate Proffitt, and Roshan George. Teaching these students and many more like them, and getting to know them, has been tremendously rewarding.

I would like to thank the publishing team at Elsevier including Katey Birtcher, Publisher; Steve Merken, Senior Acquisitions Editor; Nate McFadden, Senior Developmental Editor; Maria Inês Cruz, Cover Designer; and Stalin Viswanathan, Production Project Manager.

Much more indirectly, I wish to thank a long line of scientific mentors who instilled in me the academic integrity and the desire to be thorough and correct. Included in this list are Lemuel Wright, and Donald B. McCormick, of Cornell University, who advised me on my master's thesis; Robert Hall of Upstate Medical Center, with whom I worked for a short but meaningful time and who provided the epiphany for my pursuit of an academic career; Robert Wasserman, of Cornell University, who first inspired me to apply mathematics to transport processes; Norm Briggs, of VCU School of Medicine, who first thought I would be a good bet for a tenure track position; Don Mikulecky, who introduced me to many ideas of theoretical biology; and Margaret Biber, who first presented me with the daunting task of developing a year-long physiology sequence for BME students, with laboratories, that formed the text.

Although I desire to be thorough and correct, often I am mistaken. I acknowledge the help of all those listed above, but reserve to myself the blame for any errors in the text. If the reader finds any errors of fact or analysis, I would appreciate them letting me know so that I can evaluate and correct it.

Lastly, I express my deep appreciation for my wife, Lee, who put up with innumerable late nights and weekends with a husband glued to the chair in front of the computer screen. Her support and encouragement made the work possible. The first edition of this text was dedicated to the memory of Karen Esterline Feher, who died tragically during the writing of the first edition. The second edition was written shortly after the death of Mildred Hastings Feher, who was a most gracious human being. Most of it was also written while my daughter, Teresa, gave birth to three boys. These four ladies in my life have made me understand more fully the circle of life and death, and have taught me to value all parts of it. They are the bringers of life and I can only hope to preserve and protect it, and to finally let it go. I dedicate this effort to Mildred, Lee, Karen and Teresa.

UNIT 1

PHYSICAL AND CHEMICAL FOUNDATIONS OF PHYSIOLOGY

The Core Principles of Physiology   1.1

Learning Objectives

- Define the discipline of physiology
- Describe in general terms how each organ system contributes to homeostasis
- Define reductionism and compare it to holism
- Describe what is meant by emergent properties
- Define homeostasis
- List the four Aristotelian causes and define teleology
- Define mechanism
- Describe how evolution is a cause of human form and function
- Write equations for the conservation of mass and energy for the body
- Give an example of signaling at the organ or cellular level of organization
- List the core principles of physiology
- Contrast feed-forward or anticipatory control to negative feedback control

HUMAN PHYSIOLOGY IS THE INTEGRATED STUDY OF THE NORMAL FUNCTION OF THE HUMAN BODY

ORGAN SYSTEMS WORK TOGETHER TO PRODUCE OVERALL BODY BEHAVIOR

Almost any explanation of something begins with a description of that something's parts. The human body consists of many parts. We consider an assortment of parts that usually relate to each other in defined ways to be a **system**. In physiology, a system is usually considered to be a group of organs that serve some well-defined function of the body. The parts of these systems can be described separately, but they *work together* to produce the overall system behavior. That is, the individual behavior of the parts is *integrated* to produce overall behavior. The various organ systems and their functions in the body are summarized in Table 1.1.1. These systems are further integrated to produce overall bodily function. Physiology is the study of the integrated normal function of the human body.

Each of these organ systems is essential to the survival of the organism, the living human being. It is possible in an artificial environment to survive with a single compromised system—such as persons with failed kidneys or failed immunological systems—but these persons could not survive in natural ecosystems.

REDUCTIONISM EXPLAINS SOMETHING ON THE BASIS OF ITS PARTS

The process of explaining something on the basis of its parts is called **reductionism**. Thus the behavior of the body can be explained by the coordinated behavior of its component organ systems. In turn, each organ system can be explained in terms of the behavior of the component organs. In this reduction recursion, the behavior of the component organs can be explained by their components, the individual cells that make up the organ. These cells, in turn, can be explained by the behavior of their component subcellular organelles; the subcellular organelles can be explained by the macrochemicals and biochemicals that make up these organelles; the biochemicals can be explained by their component atoms; the atoms can be explained by their component subatomic particles; the subatomic particles can be explained by fundamental particles. According to this reductionist recursion, we might anticipate that the final explanation of our own bodies lies in the physics of the fundamental particles. Beyond being impractical, there is a growing realization that it is theoretically impossible to describe complex and complicated living beings solely on this basis of fundamental physics, because at each step in the process some information is lost.

This situation should not trouble us too much. In physics, we speak of the force of friction even though friction is not a fundamental force. It results from tremendous numbers of electromagnetic interactions between particles on two surfaces so that friction really is just a trace of all of those microscopic electromagnetic interactions onto the macroscopic bodies. The details of the surfaces produce those forces and we can experimentally reduce the force simply by polishing the surfaces that interact, making the tiny bumps and valleys on the surfaces smaller. We don't know the details of the surface and so we ignore the reality and lump all of those interactions together and call it the "force" of friction. We have lost some information and abandon the idea of recursing reductionism down to the level of fundamental physics.

QUANTITATIVE HUMAN PHYSIOLOGY: AN INTRODUCTION

TABLE 1.1.1 Organ Systems of the Body

Organ System	Function
Nervous system/ endocrine system	Sensory input and integration; command and control
Musculoskeletal system	Support and movement
Cardiovascular system	Transportation between tissues and environmental interfaces
Gastrointestinal system	Digestion of food and absorption of nutrients
Respiratory system	Regulation of blood gases and exchange of gas with the air
Renal system	Regulation of volume and composition of body fluids
Integumentary system (skin)	Protection from microbial invasion, water vapor barrier, and temperature control
Reproductive system	Pass life on to the next generation
Immune system	Removal of microbes and other foreign materials

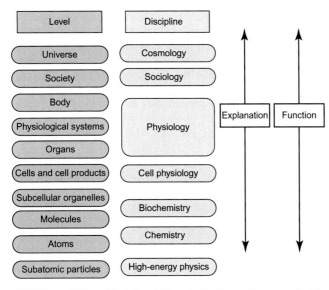

FIGURE 1.1.1 Hierarchical description of physical reality as applied to physiological systems. We attempt to "explain" something in terms of its component parts and describe a function for a part in terms of its role in the "higher" organizational entity.

PHYSIOLOGICAL SYSTEMS ARE PART OF A HIERARCHY OF LEVELS OF ORGANIZATION

The recursion of explanation described above for reductionism involves various levels of complexity in a hierarchical description of living beings, as shown in Figure 1.1.1. Understanding any particular level entails relating that level to the one immediately above it and the one immediately below it. For example, scientists studying a particular subcellular organelle can be said to have mastered it when they can explain how the function of the organelle derives from the activities of its parts, the molecules that make it up, and how the organelle's function is regulated by and contributes to the function of the cell.

REDUCTIONISM IS AN EXPERIMENTAL PROCEDURE; RECONSTITUTION IS A THEORETICAL PROCEDURE

The processes used in going "down" or "up" in this hierarchy are not the same. We use reductionism to explain the function of the whole in terms of its parts, by going "down" in the levels of organization. We describe the function of the parts at one level by showing how they contribute to the behavior of the larger level of organization, going "up" in Figure 1.1.1. These processes are fundamentally different. Reductionism involves actually breaking the system into its parts and studying the parts' behavior in isolation under controlled conditions. For example, we can take a sample of tissue and disrupt its cells so that the cell membranes are ruptured. We can then isolate various subcellular organelles and study their behavior. This procedure characterizes the behavior of the subcellular organelle. Knowing the behavior of the individual parts and paying close attention to how these parts are connected, it is possible to predict system behavior from the parts' behavior using simulation or other techniques. Because it is impossible, except in rare and limited cases, to reassemble broken systems (we cannot unscramble the egg!), we must test our ideas of subcellular function theoretically.

HOLISM PROPOSES THAT THE BEHAVIOR OF THE PARTS IS ALTERED BY THEIR CONTEXT IN THE WHOLE

Holism conveys the idea that the parts of an organism are interconnected and that each part affects others. The parts cannot be studied in isolation because important aspects of their behavior depend solely on their interaction with other parts. Reductionism seems to imply that the whole is the sum of the parts, whereas holism suggests that the whole is greater than the sum of the parts. The system takes on new properties, **emergent properties**, that arise from complex interactions among the parts. Examples of emergent properties include **self-replication**. The ability of cells to form daughter cells is a system property that does not belong to any one part but belongs to the entire system. Consciousness is also an emergent property that arises from neuronal function but at a much higher level of organization.

PHYSIOLOGICAL SYSTEMS OPERATE AT MULTIPLE LEVELS OF ORGANIZATION SIMULTANEOUSLY

It should be clear from Figure 1.1.1 that all the levels of organization simultaneously operate in the living human being. Processes occur at the molecular, subcellular, cellular, organ, and system levels simultaneously and dynamically.

THE BODY CONSISTS OF *CAUSAL MECHANISMS* THAT OBEY THE LAWS OF PHYSICS AND CHEMISTRY

When we say that we explain something, usually we mean that we can trace the output of the system—its behavior—to some cause. That is, we can seamlessly trace cause and effect from some starting point to some ending point. Aristotle (384–322 BC) posited four different kinds of causality:

1. Material Cause
   A house is a house **because** of the boards, nails, shingles, and so on that make it up. We are what we are **because** of the cells and the cell products that make us up.
2. Efficient Cause
   A house is a house **because** of the laborers who assembled the materials to make the house. We are what we are because of the developmental processes that produced us and **because** of all of the experiences we have had that alter us.
3. Formal Cause
   A house is a house **because** of the blueprint that directed the laborers to assemble the materials in a particular way. We are what we are **because** of the DNA that directs our cells to make some proteins and not others, and because of epigenetic effects—those effects resulting from the environment interacting with the genome.
4. Final Cause
   A house is a house **because** someone needed shelter. We are what we are **because**…
   The final cause for humans has a variety of possible answers. This is the only cause that addresses the idea of a purpose. We make a house for a purpose: to provide someone with shelter. What is the purpose of human beings? This cause asks the question of why rather than how.

TELEOLOGY IS AN EXPLANATION IN REFERENCE TO A FINAL CAUSE

A description or explanation of a system based on the reference to the final cause is called a teleological explanation. Teleology has long been ridiculed by scientists because it appears to reverse the scientific notion of cause and effect. In normal usage, cause-and-effect linkages describe only the efficient cause. When a force acts on something, that something reacts in a predictable way. Its predictability is encoded in physical law. Teleology describes the behavior in terms of its final purpose, and not its driving force, which reverses the cause-and-effect link.

HOW? QUESTIONS ADDRESS CAUSAL LINKAGES. WHY? QUESTIONS ADDRESS FUNCTION

To clarify this process, let us ask some questions about blood pressure. How does your body regulate the average aterial blood pressure? This answer takes some time to develop fully (see Unit 5) but we can simplify the answer by saying: by increasing or decreasing the caliber of the arteries, by increasing or decreasing the output of the heart, and by increasing or decreasing the volume of fluid in the arteries. All of these actions—and more—interact to determine the arterial blood pressure. All of these parts to the answer involve actions, and actions are efficient causes. For example, increasing or decreasing the caliber of the arteries depends on a complicated network of signaling pathways and biochemical reactions, but each starts with a cause and ends with an effect. Now we ask: why does your body regulate average aterial pressure at the level it does? The answer is again not so simple but we can simplify it as: to assure perfusion of the tissues. This addresses a purpose. In this sense, it is teleological, but it also makes sense to us. It is not that the cardiovascular system *knows* what it is doing, but its operation has been selected to produce the desired output. Within the framework of the body, organs do serve a purpose and that purpose is their function. All of the functions listed in Table 1.1.1 are final causes for the organ systems.

HUMAN BEINGS ARE NOT MACHINES BUT STILL OBEY PHYSICAL LAW

Aristotle had a different idea about the mind. He posited that the mind was not a material entity, much like an idea is not a material thing. He posited that the mind was what perceived, imagined, thought, emoted, desired, felt pain, reasoned, remembered, and controlled the body. This philosophy in which the mind controls behavior is called **mentalism**. These ideas went largely unchallenged until the Renaissance. René Descartes (1596–1650) wrote a book, *Treatise on Man*, in which he tried to explain how the nonmaterial mind might interact with the material body. In this process, he constructed mechanical analogues to explain sensation and command of movement. He thought that the operation of all things, however complex, could be explained by some **mechanism**. Each mechanism consists of a sequence of events that link an initial causal input to an effect that becomes the cause of the next step in the mechanism. In this view, human beings are very complex machines that obey natural law. In the late 1800s, W.O. Atwater (1844–1907) built a calorimeter to study heat production, gas exchange, and fuel consumption in humans. He found that the energy output of humans matched the chemical energy of the food consumed, within narrow experimental error. This result confirmed that the law of conservation of energy held for the transformation of energy by the human body as well as for inanimate transformations.

IS THERE A GHOST IN THE MACHINE?

The core principle of physiology that states the human body is a mechanism strikes at the heart of the concept of **vitalism**. Vitalism states that living things cannot be described in mechanistic terms alone, and that some organizing force or vital principle forever distinguishes living things from nonliving things. For human beings, we could call this the soul and be in reasonable agreement with the vitalists. So far, science has found no reliable, scientific verification that the human body violates any physical law. The existence of emergent properties gives credence to the idea of vitalism. Emergent properties are new properties that arise from complex systems because of their complexity and topology—the way that subsystems are connected. These emergent properties do not appear to be predictable. That is, we cannot see how we would predict their emergence given the fundamental laws of physics. Examples of these emergent properties are life itself and consciousness. These properties appear to arise from interactions of parts that appear to obey physical law alone, but how these properties arise remains mysterious. These emergent properties are system properties, not mechanism properties. Because of overwhelming evidence that specific deficits in brain function produce specific deficits in mental function, we have come to believe that the brain somehow produces the mystical thing that is conscious and self-aware. This thing is not material in the ordinary sense of the word, just like an idea is not a material thing. The new mind–body problem is the inverse of Descarte's mind–body problem: how can a material thing (the brain) produce the nonmaterial thing that we identify as *self*. Is this relationship a one-way street, or does the mind have a reciprocal effect on the brain? Although these are extremely important questions, most physiologists take a narrower aim of explaining only those phenomena for which we have satisfactory mechanistic models and attempt to extend the range to include all physiologic phenomena, including consciousness.

UNDERSTANDING A SYSTEM IS EQUIVALENT TO MAKING A MODEL OF IT

What we have been discussing so far is what it means to say that we understand something. The something we are trying to understand we will label "the real system." To understand this system, we make a model of it in a process that we will call **encoding** (B in Figure 1.1.2). The model does not have to remain solely in our minds. It could be written down as a set of equations or as a computer algorithm, for example. In the real system, perturbations cause changes in the real system. This **causal link** (A in Figure 1.1.2) between perturbation and effect is what we desire to predict or explain using the model. In the model, the cause is also encoded, and its predicted effect we will call an **inference** (C in Figure 1.1.2). That is, some perturbation of the model is predicted to cause some effect in the model. When we **decode** this effect (D in Figure 1.1.2), the inference of the model is translated as our prediction of the behavior of the real system. A correct model is one that correctly predicts the behavior of the real system. That is

[1.1.1] $$A = B + C + D$$

THE CORE PRINCIPLES OF PHYSIOLOGY

The rest of this chapter will discuss several Core Principles of Physiology. These are:

- Cells are the organizational unit of Life.
- Homeostasis is a central theme of physiology.
- We have evolved from prior life forms and our pedigree is revealed in our genome.
- Physiological systems transform matter and energy while obeying the conservation laws.
- Coordinated command and control requires signaling at all levels of organization.
- Control systems using negative feedback, positive feedback, anticipatory and threshold mechanisms.

CELLS ARE THE ORGANIZATIONAL UNIT OF LIFE

THE CELL THEORY IS A UNIFYING PRINCIPLE OF BIOLOGY

The cell theory states that all biological organisms are composed of cells; cells are the unit of life and all life come from preexisting life. The cell theory is so established today that it forms one of the unifying principles of biology.

The word *cell* was first used by Robert Hooke (1635–1703) when he looked at cork with a simple microscope and found what appeared to be blocks of material making up the cork. The term today describes a microscopic unit of life that separates itself from its surroundings by a thin partition, the cell membrane.

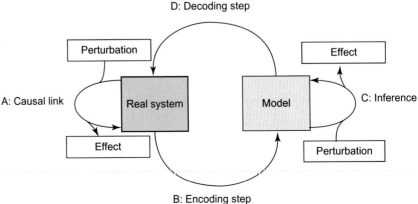

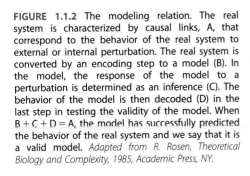

FIGURE 1.1.2 The modeling relation. The real system is characterized by causal links, A, that correspond to the behavior of the real system to external or internal perturbation. The real system is converted by an encoding step to a model (B). In the model, the response of the model to a perturbation is determined as an inference (C). The behavior of the model is then decoded (D) in the last step in testing the validity of the model. When B + C + D = A, the model has successfully predicted the behavior of the real system and we say that it is a valid model. *Adapted from R. Rosen, Theoretical Biology and Complexity, 1985, Academic Press, NY.*

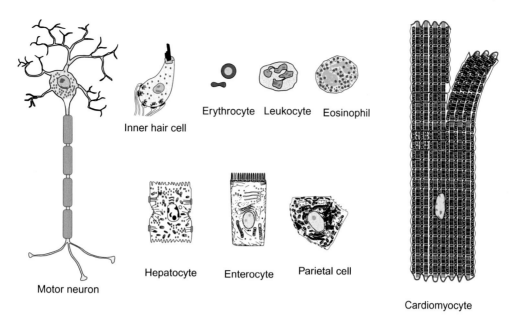

FIGURE 1.1.3 Examples of the different cells that populate the human body. Motor neurons such as the one illustrated are found in the ventral horn of the spinal cord. The inner hair cells are found in the cochlea and form part of our response to sound. Erythrocytes, leukocytes, and eosinophils are all found in the blood. Cells typically are not colored but may be seen in color by their adsorption of histological stains. Hepatocytes in the liver help package nutrients, form bile, and detoxify foreign chemicals. The enterocytes line the small intestine and absorb nutrients from the food into the blood. The parietal cells secrete HCl in the stomach. The cardiomyocyte shown is a ventricular cell whose contraction contributes to the pumping action of the heart. This is a small sampling of the diversity of cell forms in the human body.

Most biologists believe that life arose spontaneously from inanimate matter, but the details of how this could have happened remain unknown, and the time scale was long. Rudolf Virchow, a German pathologist (1821–1902), famously wrote "omnis cellula e cellula"—all cells come from other cells—meaning that spontaneous generation of living things from inanimate matter does not occur over periods as short as our lifetimes.

CELLS WITHIN THE BODY SHOW A MULTITUDE OF FORMS

Large multicellular organisms such as ourselves consist of a vast number of different cells that share some features but vary in size, structure, biochemical makeup, and functions. A sampling of the spectrum of cells that make up the body is illustrated in Figure 1.1.3. Almost all cells in the body have a **cell membrane**, also called the **plasma membrane**, and most contain a nucleus. The simplest cell in the human is probably the **erythrocyte**, which is the only cell in the body that lacks a nucleus.

THE DIVERSITY OF CELLS IN THE BODY DERIVES FROM DIFFERENTIAL EXPRESSION OF THE GENOME

The outward appearance and behavior of an organism define its **phenotype**, which is related to but not identical to the organism's genetic material, its **genotype**. The genotype consists of the set of alternate forms of genes, called **alleles**, that the organism has, and these alternate forms of genes are further defined by the sequence of nucleotides in their DNA. DNA is the genetic material that is passed on through the generations. It determines the kind of proteins that cells can produce, and these materials make up the phenotype. The **genome** is the entirety of the hereditary information, including all of the genes and regions of the DNA that are not involved in producing proteins. Nearly all cells in the body contain the entire genome. The exceptions include the erythrocytes and the reproductive cells. Those cells that are not reproductive cells are called **somatic** cells (from the Greek *soma*, meaning *body*). Thus the great majority of body cells are somatic cells, and they all contain the same amount and kind of DNA. The astounding diversity of the types of human cells derives from their expression of different parts of the genome. Here **expression** means using DNA to produce proteins.

THE CONCEPT OF *HOMEOSTASIS* IS A CENTRAL THEME OF PHYSIOLOGY

EXTRACELLULAR FLUID SURROUNDS ALL SOMATIC CELLS

As described above, each cell in the body is surrounded by a cell membrane that defines the limits of the cell and separates the interior of the cell from its exterior. The interior consists of a number of subcellular organelles suspended in a fluid, the **intracellular fluid**. The exterior consists of an extracellular matrix that holds things in place and an **extracellular fluid**. The extracellular fluid has two components: the plasma and the interstitial fluid. The plasma is that part of the extracellular fluid that is contained in the blood vessels. The interstitial fluid is that part of the extracellular fluid between the cells and the walls of the vasculature. Nearly all cells of the body come in intimate contact

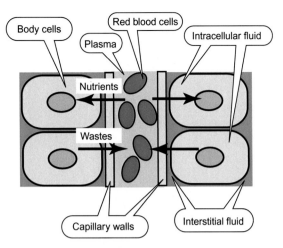

FIGURE 1.1.4 Relationship between cells and the extracellular fluid. All cells of the body are surrounded by a thin layer of extracellular fluid from which they immediately derive nutrients such as amino acids, sugars, and oxygen, and to which they discharge wastes such as carbon dioxide and other end products of metabolism. Nutrients are delivered to the cells and waste products are removed through the circulation, which does not make direct contact with the interstitial fluid, but is separated from it by the walls of the vascular system.

with the extracellular fluid. The last step in the delivery of nutrients and the first step in removal of wastes is achieved through the extracellular fluid (see Figure 1.1.4). The extracellular fluid was called the *milieu interieur*, or the internal environment, by the great French physiologist, Claude Bernard (1813–1878). Survival of the cells depends on the maintenance of a constant internal environment. **The maintenance of a constant internal environment is called homeostasis**, which is literally translated as *same standing*. Contributing to the maintenance of a constant internal environment appears to the "goal" (or final cause or purpose) of many physiological systems, and this homeostasis is the central theme of physiology.

EVOLUTION IS AN EFFICIENT CAUSE OF THE HUMAN BODY WORKING OVER LONGTIME SCALES

EVOLUTION WAS POSTULATED TO EXPLAIN THE DIVERSITY OF LIFE FORMS

Charles Darwin (1809–1892) wrote *On The Origin of Species* in 1858 as his attempt to explain the origin of the tremendous variety of animals and plants in today's ecosystems. He noted that any one species consists of a population of individuals that are capable of breeding among themselves but not with members of other species. The similarity among members of a species defines the species; the differences between them define the individual. These outward appearances constitute the phenotype, as described earlier, which arises from the response of the genotype to the environment. Some of the individual members of a species are better suited to their environment than others and produce more offspring as a consequence. With sufficient time, the frequency of genotypes represented in the population would shift to those better suited to the environment. New variations in the genotype arise by mutation. Over geological time, such **natural selection** gradually changes the population. Darwin believed that such slow changes in the genetic makeup of populations could eventually produce new species, and he termed this slow formation of new species **evolution**.

EVOLUTION RESULTS FROM CAUSE AND EFFECT SUMMED OVER LONGTIME PERIODS

Evolution is like a higher level on the hierarchy of cause-and-effect relationships. As an example, consider a mutation that alters the structure of a critical protein located in a selected group of cells in the body that enhances the function of these cells. The mutation causes an altered protein, which in turn causes enhanced behavior of the organism. This altered behavior of the organism causes greater success in reproduction. Over time, greater success in reproduction replaces the less-fit genotype with the mutated, superior genotype. Thus evolution results from thousands of independent cause-and-effect linkages played out over a population of individuals, over longtime periods.

EVOLUTION WORKS ON PREEXISTING FORMS: COMPARATIVE GENOMICS REVEALS PEDIGREE

At some time in the distant past, there was no life on earth. The origin of life is unknown and, in some sense, how it arose is not a scientific question because we cannot test any hypothesis of events in the past. We can, however, search for the trace of past events in the world today, much like a detective searches for clues to determine what happened earlier. This search has some of the character of an experiment. In this way, the fossil record illuminates the march of evolution to the present day. Similarly, we carry traces of our evolution in our own genome in the form of "fossil genes." Because evolution works on preexisting forms, and because the multicellular organism plan entails the same challenges to homeostasis, and the same problems of cell maintenance, the genomes for many diverse animals and plants share profound similarities. For this reason, similarities in the genome can be used to trace the evolution of the proteins and shed light on the pedigree of species.

EVOLUTION TAILORS THE PHENOTYPE TO THE ECOSYSTEM

Humans live and reproduce within the context of an ecosystem. Our evolution has occurred because of our fit, or lack of it, with a specific environment. This explains some of the diversity of human forms within our species. Skin color and overall body shape, for example, are adaptations that arose to better fit the different levels of sunlight and air temperatures at different latitudes. Evolution has prepared us to meet the challenges of our environment but has not prepared us for unusual challenges. For example, we are adapted to survive short periods without water or longer periods without food, but we cannot do without air even for short periods.

ROBUSTNESS MEASURES THE ABILITY OF THE BODY TO RESPOND TO ENVIRONMENTAL CHALLENGES

A robust system is one that continues to function even when faced with difficult challenges. A fragile system fails easily. Engineered systems aim for a degree of robustness and usually achieve that end by adding redundant or back-up control systems and by building in safety factors. These systems are robust for some kinds of failure while remaining vulnerable to others. For example, autopilot systems in aircraft use multiple computers with different programs so that failure of one does not cause failure of the entire system. But these systems remain vulnerable to general power failure. We also have redundant or reserve function for several physiological systems. We have two kidneys, yet generally we can survive with only one. Liver, intestine, brain, and heart have more functional capacity than generally used so that we can survive if part of these organs fails. Strokes and heart attacks damage parts of the brain or the heart, respectively. Persons who suffer these cardiovascular accidents often recover much of their function, depending on the degree of damage and its location. If the damage is severe or involves a critical area, the victim may be permanently impaired or they may die. History is replete with astonishing stories of the tenacity of humans for life under amazingly harsh conditions. On the other hand, history also tells of the crumbling of civilizations when exposed to unfamiliar pathogens. Thus the human body may surprise us either because of its robustness or its fragility.

REGULATION OF THE GENOME MAY EXPLAIN THE FAST PACE OF EVOLUTION

There is a growing realization among evolutionary biologists that mutations in the genes that encode for somatic proteins—the ones that make us up—are only a small part of the story and cannot account for the rapid pace of evolution. Instead, much of evolution is accounted for in the genes that regulate the expression of other genes. Many modern birds, for example, do not have teeth. Yet it is possible experimentally to induce birds to make teeth, because they retain the genes for making teeth but also have genes that suppress the expression of the genes for teeth. Many diverse groups of animals share most of the genes involved in body building but differ in how and when these genes are used. The result is the differing body forms that are found in the animal kingdom.

EVOLUTION HELPS LITTLE IN EXPLAINING THE NORMAL FUNCTION OF THE BODY

Although evolution is one of the few unifying principles of biology and is one cause of the structure and function of the human body, it answers the question of the efficient cause of the body on a different time scale than the normal operating time scale of the body. Evolution does not aid us very much when trying to explain how our bodies work on a minute-to-minute basis. Instead, we look to control theory to explain the normal functioning of the human body.

LIVING BEINGS TRANSFORM ENERGY AND MATTER

Repair, maintenance, growth, activity, and reproduction all require input of energy and matter in the form of food. The gastrointestinal system breaks down the food, which is absorbed into the blood and distributed among the tissues according to need. The available building blocks must be transformed into cellular or extracellular components, and all of this metabolism requires energy. The energy comes from the oxidation of food and subsequent production of wastes. In addition to this conversion of chemical energy of one compound to another, we also transform chemical energy into other forms of energy, including electrical and mechanical energy. These processes obey physical and chemical laws that govern the transformation of energy and matter. In particular, we can write two equations that describe overall mass and energy balance in the body:

$$M_{in} = M_{out} + \Delta M_{body}$$

$$M_{food} + M_{drink} + M_{inspiredair} = M_{feces} + M_{urine} + M_{expiredair} + M_{exfoliation} + \Delta M_{body}$$

[1.1.2]

where M indicates mass and the subscripts indicate the origin of the mass (in = input; out = output; most of these are self-explanatory) and ΔM_{body} indicates change in body mass. These equations describe mass balance and simply indicate that all of the mass that enters the body must either stay there (ΔM_{body}) or exit the body through one of several routes. A similar equation can be written for energy balance:

$$E_{in} = E_{out} + \Delta E_{body}$$

$$E_{food} + E_{drink} + E_{inspiredair} = E_{feces} + E_{urine} + E_{expiredair} + E_{exfoliation} + \Delta E_{body} + E_{heat} + E_{work}$$

[1.1.3]

The overall mass and energy balance is shown schematically in Figure 1.1.5.

FUNCTION FOLLOWS FORM

Almost all processes carried out by the body, at all levels of organization, depend on the three-dimensional structure of some component. The structure both

QUANTITATIVE HUMAN PHYSIOLOGY: AN INTRODUCTION

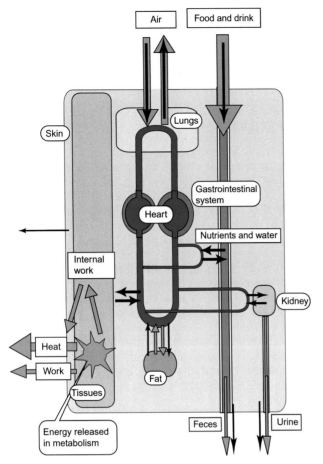

FIGURE 1.1.5 Overall mass and energy balance in the body. The *lighter arrows* indicate energy transfer. The *black arrows* denote transfer of mass. The chemical energy of the ingested food is released by oxidation in metabolism, as indicated by the starburst in the tissues. This released chemical energy is used for internal work, which usually eventually degrades to heat, for external work and for storage in the chemical energy of body components such as glycogen and fat. Growth also entails a form of storage of ingested mass and chemical energy. The laws of conservation of mass and energy (in ordinary chemical reactions) require that the matter and energy that enter the body must be equal to the matter and energy that leave the body plus any change in the matter and energy content of the body.

enables function and constrains it, by determining what can be done and how fast it can be accomplished. These structural considerations apply at the molecular level, at which the three-dimensional shape of protein surfaces determines what binds to the protein, how it is chemically altered, and how it interacts with other surfaces. These structural considerations apply at the subcellular level, at which the organelles themselves can compartmentalize chemicals and so determine or limit rates of reactions by regulating transfer between the compartments. Structural considerations are also important at the tissue level, at which the topology or spatial distribution of cellular processes allows countercurrent flows, for example, that are crucial in clearing metabolites from the blood or concentrating the urine. Structural considerations are important at the organ system level at which the structure and arrangement of nerves and tissues is vital for the proper coordination of activity such as the heart beat or gastrointestinal motility. As another example, both the lungs and the gastrointestinal tract involve transfer of gas or nutrients from the environment to the blood. Both lungs and intestine have enormous surface areas and thin barriers—consequences of their structure—to maximize the rate of transport.

COORDINATED COMMAND AND CONTROL REQUIRES SIGNALING AT ALL LEVELS OF ORGANIZATION

Success of an organism requires adaptive responses to change in the environment. This, in turn, requires sensory apparatus that senses both the external environment (**exteroreceptors**) and the interior environment (**interoreceptors**). These originate signals that pass either to nearby cells or to the central nervous system either for specific, reflex responses or for global responses. These signals are important at all levels of organization. At the subcellular level, these signals regulate the activities of subcellular components such as the expression of specific genes or the regulation of the rates of energy transformation. At the tissue level, local signals can regulate smooth muscle contraction to regulate blood flow within the organ or secretion into ducts; at the organ system level, signals traveling through the blood (hormones) or over nerves can coordinate activity of the system. At the whole organism level, signals at all levels must be used to adapt to whole-body responses such as running to avoid predators. Coordinating command and control for muscle contraction using sensory information from the environment (exteroreceptors) and from the muscle (interoreceptors) is illustrated in Figure 1.1.6. Signaling at the cellular level is illustrated schematically in Figure 1.1.7.

MANY CONTROL SYSTEMS OF THE BODY USE NEGATIVE FEEDBACK LOOPS

One of the main themes of physiological control systems comprises a negative feedback loop. This consists of controlled parameters such as plasma calcium concentration, body temperature, plasma glucose concentration, and plasma pH, a sensor for that parameter, a comparator, and an effector. For many physiologically controlled parameters, there is a set point or reference (see Figure 1.1.8). This is the desired value for the controlled parameter. Its value can change under some circumstances. When the controlled parameter varies from its set point, the variation is detected by the sensor and comparator. The comparator then engages some effector or actuator mechanism to correct the departure of the parameter from its normal, set-point value. An example of this is core body temperature, whose normal set-point value is about 37°C. Whenever heat loss exceeds heat production, body temperature falls below the set point and the person shivers. In this case, the sensor

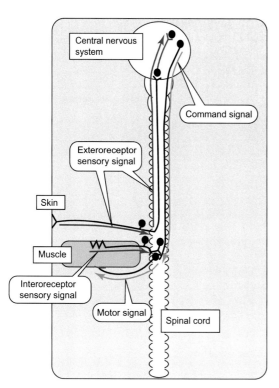

FIGURE 1.1.6 Neural signaling in the control of muscle. Neurons consist of cell bodies (*dark circles*) that have long processes (*black lines*) that bring signals into the central nervous system (*dark lines with arrows*) or take signals out toward the periphery (*light lines with arrows*). Branches at the ends of the long processes signify the junction of one neuron with another or with muscle. Neural signal transmission across these junctions is discussed in Chapter 4.2. Muscles are controlled by motor neurons whose cell bodies lie in the spinal cord. These can be activated in reflexes initiated by exteroreceptors that sense perturbations on the skin and send signals to the spinal cord and eventually activate the motor neurons by a simple reflex involving just a few interneurons in the spinal cord. Muscles can also be activated by another reflex involving a stretch receptor internal to the muscle (interoreceptor). In a third pathway, motor neurons can be activated by command signals originating in the brain. Nervous control of muscle is considered in Chapters 4.4 and 4.5.

The Core Principles of Physiology — 11

detects the temperature, the comparator determines that it has fallen below the set point, and it engages the skeletal muscles as an effector to produce heat by shivering to help raise the temperature back to the set point. In a fever, the set point is elevated and the individual feels chilled even when the temperature is elevated to, say, 40°C. When the fever "breaks," the set point is reset back to 37°C and the person perspires because now the body temperature is elevated above the set point. When the controlled parameter varies from its set point, the variation is detected by the comparator. The comparator then engages some actuator mechanism to correct the departure of the parameter from its normal, set-point value. The controlled parameter can vary from its set point by the action of some physiological disturbance. These control systems are called negative feedback loops because the causality forms a loop ($y \Rightarrow x \Rightarrow u \Rightarrow a \Rightarrow y$) and the adjustment is typically the opposite sign, or the negative, of the disturbance. If d adds to y, the adjustment a subtracts from y; if d subtracts from y, a adds to it. There are many systems that operate through negative feedback loops.

POSITIVE FEEDBACK CONTROL SYSTEMS HAVE DIFFERENT SIGNS FOR THE ADJUSTMENT TO PERTURBATIONS

Positive feedback systems exist for the mechanism of blood clotting, parts of the menstrual cycle, aspects of the action potential, and for parturition (child birth). In these cases, the disturbance is followed by an adjustment in the same direction of the disturbance, so that there is a rapid increase in some component. These positive feedback systems generally are self-limiting and, after the rapid increase in some component, there is a gradual return to baseline levels.

ANTICIPATORY OR FEED-FORWARD CONTROL AVOIDS WIDE SWINGS IN CONTROLLED PARAMETERS

Sometimes rapid changes in a controlled parameter can outstrip the physiological mechanisms for reacting to these changes, resulting in potentially catastrophic changes in the internal environment. To avoid this, some physiological systems anticipate changes in controlled parameters and begin to do something about it even before the parameter changes. Most wide swings in controlled parameters have to do with behavior. Eating, for example, is followed by an influx of nutrients into the blood. The nervous system prepares the gastrointestinal tract for a meal by using sensory cues—the sight, aroma, and taste of food—to induce the secretion of gastrointestinal fluids even before food is swallowed. In another example, controlled parameters including blood pH, P_{CO_2} (the partial pressure of CO_2 in the blood—a measure of CO_2 concentration) and blood P_{O_2} help regulate the depth and frequency of breathing. Negative feedback mechanisms keep these controlled parameters within narrow ranges during normal activity. During strenuous activity, there appears to be little or no error in these controlled parameters, though the depth and frequency of breathing is markedly increased. This occurs through an anticipatory response of the central nervous system in which the depth and frequency of breathing is activated simultaneously with activity.

DEVELOPMENTAL AND THRESHOLD CONTROL MECHANISMS REGULATE NONCYCLICAL AND CYCLICAL PHYSIOLOGICAL SYSTEMS

Although negative feedback control is a major theme in physiology, it does not account for a variety of important physiological events. Developmental events include the onset of puberty and menopause. Pregnancy, parturition (birth), and cyclical events such as the menstrual cycle and the sleep/wake cycle are episodic events that do not obey negative feedback mechanisms and may involve positive feedback mechanisms.

WE ARE NOT ALONE: THE MICROBIOTA

In natural ecosystems, human beings are literally covered with tiny contaminating organisms. We typically have 10 times more bacteria than body cells, but each

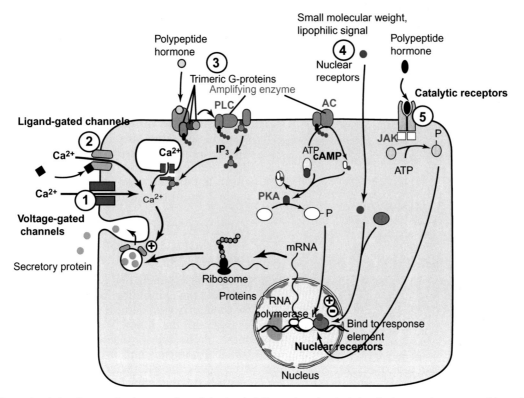

FIGURE 1.1.7 Synopsis of signaling mechanisms on the cellular level. Cells receive electrical signals that can be converted into chemical signals through voltage-gated channels (1). The voltage-gated calcium channel is shown. Chemical signals released from nearby cells can also open ion channels, producing electrical signals in the cell (2). Cells receive chemical signals in the form of polypeptide hormones that cannot penetrate the cell membrane. These can affect the cell by coupling to heterotrimeric G-proteins (3) or to catalytic receptors on the surface of the cell (5). These are coupled to amplifying enzymes or to kinases that phosphorylate intracellular proteins. Small molecular weight, permeant chemical messengers (4) can enter the cell and bind to receptors in the nucleus, which then alter the kind or amount of specific proteins made by the cell. These signaling mechanisms are discussed in detail in Chapter 2.8.

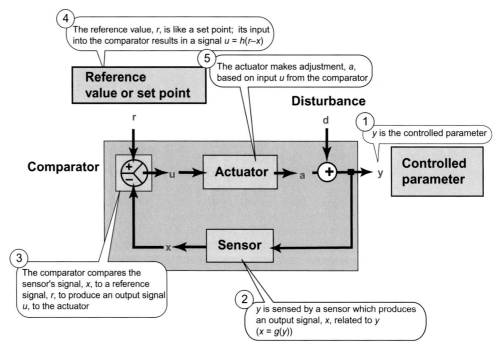

FIGURE 1.1.8 Component parts of a negative feedback loop. A controlled parameter, y, is sensed by a sensor that releases a signal that related to the value of y [$x = g(y)$]. This signal is fed into a comparator, which compares the signal x to some reference or set-point value, causing the comparator to produce a signal u that is a function of the error of x from its reference [$u = h(r - x)$, where u is the signal from the comparator, $r - x$ is the error and $h(r - x)$ gives the functional dependence of u on the error]. The signal u turns on an actuator that makes an adjustment, a, to the value of y. In negative feedback, the value of a reduces the error $(r - x)$ so that the value of y returns towards its normal, set-point level. Disturbances, d, can alter the value of y and the negative feedback loop is engaged to minimize the departure of y from its set-point level.

of these is small, and so the total mass of these bacteria amounts to 0.5–1.5 kg. These organisms are present as surface contaminants, but they are the usual surface contaminants and so they can be considered to be part of us. They live on the surface of the body, and invasion within the body constitutes infection, and must be fought off by our immune systems. They live on the skin, in the oral cavity, in the airways, and in the gastrointestinal system and on the terminal parts of the reproductive system where it melds with the skin. Most of these are within the gastrointestinal system, and fully half of the feces is estimated to consist of bacteria. In addition to the bacteria, many natural ecosystems provide us with a load of other organisms: tapeworms, flukes, helminths (hookworms, pinworms, and roundworms), leeches, fleas, various fungal organisms, and a host of viruses. Although in natural ecosystems, infections with some of the multicellular parasites may be unusual, the load of bacterial contaminants is unavoidable. The aggregate of these hangers-on is called the **microbiota**. The microbiota engage signaling systems of the body and thereby alter our physiological states. This is true of the rhinoviruses that make us sneeze, thereby spreading them around to other individual hosts, and intestinal bugs that induce diarrhea, using the host signaling systems, to likewise spread them to other hosts. Evidence is accruing that even those "benevolent" strains that do not make us frankly sick still manipulate our physiology to their advantage. Thus the microbiota become part of our physiology.

PHYSIOLOGY IS A QUANTITATIVE SCIENCE

As described earlier, homeostasis refers to the maintenance of a constant internal environment, where the internal environment refers to the extracellular fluid that surrounds the cells. This internal environment is characterized by the concentration of a host of materials, and each of these concentrations has a unit and a numerical value. Many of these materials are metabolized by the tissues so that maintaining constant values requires matching supply to consumption. The rates of supply and consumption also have units and numerical values. As Figure 1.1.5 shows, the circulatory system unites all organs of the body by virtue of their perfusion with a common fluid, the blood. Maintenance of this flow requires pressure differences that also have units and magnitudes. Understanding the flows and forces that keep the blood moving and keep its composition relatively constant requires a quantitative approach.

SUMMARY

Physiology is the integrated study of the normal function of the human body. Like many complicated things, the body can be viewed as a set of subcomponents that interact by linking the output of one component to the input of another. These subcomponents are the organ systems. These include the cardiovascular system, the respiratory system, the renal system, the gastrointestinal system, the neuroendocrine system, the musculoskeletal system, the integument, and the reproductive system. Understanding how the body works as a whole requires us to make a model, either implicit or explicit, that explains the integration of the structures that make up organ systems, and the integration of the organ systems that produces the overall system behavior. Explanation requires that cause and effect in the model faithfully predicts cause and effect in the real system. Understanding can occur on different hierarchical levels of integration: the systems level, the organ level, the cell level, and the subcellular level. Each level seeks to explain behavior at that level on the basis of the components that make up that level. This is reductionism, the explanation of the behavior of a complicated object on the basis of its parts. We say that we understand something when we can explain function in terms of the parts one level below and we can show how behavior at that level contributes to behavior one level above.

Aristotle identified four classes of causality: the material cause, the efficient cause, the formal cause, and the final cause. Explanation of something on the basis of the final cause is called teleology. Although science presumes that each component of living things obeys physical law alone, systems produce emergent properties that we seem to be unable to predict. These emergent properties belong to the system as a whole rather than to individual parts within it.

Evolution is one cause of the human form and function, but it aids us only a little in understanding how physiological systems work.

In the hierarchy of levels of organization, cells are the fundamental unit of life. The various cells of the body show a remarkable diversity of form and function, but they all carry the complete genome, with the exception of erythrocytes and reproductive cells. The diversity of form arises from the use of only parts of the genome for each type of cell.

The overriding principle of human physiology is homeostasis, meaning the maintenance of a constant internal environment. Our internal environment is the extracellular fluid that bathes all cells in the body. A combination of internal control systems and external behavior maintains homeostasis. In the final analysis, our life depends on inputs from the environment and so our behavior (feeding, drinking, and temperature control) is crucial to our survival. Internal regulation of the internal environment relies primarily on negative feedback control, in which controlled parameters feed into a sensor that compares the value of the controlled parameter to the desired set point or reference levels. Variation from the set point results in the activation of effector mechanisms that increase or decrease the controlled parameter so that it more closely approximates the set-point level. Anticipatory control also contributes to homeostasis without allowing wide swings in the values of controlled parameters. Some important physiological events, such as puberty, menstruation, pregnancy, parturition, menopause, and the sleep/wake cycle, are not homeostatic. Instead, they incorporate switches between one physiological state and another.

These control systems contribute to the robust control of bodily functions that enable homeostasis under harsh environmental conditions.

REVIEW QUESTIONS

1. What argument does holism make against reductionism?
2. Give some examples of emergent properties
3. What is a cell? Why is it considered to be a fundamental unit of organization of life?
4. Contrast genotype with phenotype
5. What is homeostasis? Why is it central to physiology?
6. What is the internal environment in large multicellular animals such as ourselves?
7. What would constitute proof for the theory of evolution? Do you think science has provided it? Why or why not?
8. From Einstein's equation $E = mc^2$, you have learned in physics that mass and energy are interconvertible. Why can we say that mass and energy are conserved in physiological systems? What does it mean to say that living things "transform" matter and energy?
9. Describe the hierarchical organization of the body.
10. Give examples of signaling at the organism level and cellular level.

Physical Foundations of Physiology I: Pressure-Driven Flow 1.2

Learning Objectives

- Define intensive and extensive variables
- Define flow and flux
- Describe the driving principle for heat flow, electrical current, diffusive flow, and volume flow
- Explain what is meant by fluxes moving downhill
- Write a continuity equation and describe its meaning
- Explain why steady-state flux requires a linear gradient of T, Ψ, C, or P
- List four capacitances commonly encountered in physiology
- Define pressure and be able to convert pressure between atm, mmHg, and Pa
- Write Poiseuille's law, state its assumptions, and be able to calculate flow using it
- Write the Law of Laplace for cylindrical tubes and for spheres

FORCES PRODUCE FLOWS

EXTERNAL AND INTERNAL MOVEMENT IS A HALLMARK OF HUMAN LIFE

For humans in their natural environment, movement is essential for survival. This movement refers to translation of the body from one location to another, and movement of the limbs relative to one another. In addition to this movement of body parts with respect to the external world, movement of materials within the body is also essential. Most important among these internal movements are the movement of the blood, movement of the air in and out of the lungs, movement of food and fecal material along the gastrointestinal tract, and movement of the urine from its formation to elimination. In addition, the body transports materials across barriers such as the gastrointestinal tract lining, lungs, and kidney tubule. Transport also occurs within cells. All of these movements require the continued application of force to overcome inertia and friction.

TRANSPORT OF MATERIAL IS DESCRIBED AS A FLOW OR A FLUX

The transport of material is quantitatively expressed as a **flow**, which we will symbolize by the variable Q. The flow can be expressed as

- volume of material or fluid transported per unit time;
- mass of material transported per unit time;
- number of particles or moles transported per unit time;
- number of ions or unit charges transported per unit time (electrical current).

FLOW DEPENDS ON THE AREA; FLUX IS FLOW PER UNIT AREA

The total flow of volume or solute is an **extensive** variable: the flow depends on the extent or the amount of the system that gives rise to the flow. In the case of two compartments separated by a membrane, doubling the area or extent of the membrane would produce twice as much flow between the two compartments. Dividing the flows by the area normalizes the flows. The normalized flow is the flux, and the flux is an **intensive** variable whose value is independent of the extent of the system. Flux is *defined* as

[1.2.1]
$$J_V = \frac{Q_V}{A}$$
$$J_S = \frac{Q_S}{A}$$

where Q_V is the volume flow, J_V is the volume flux, A is the cross-sectional area through which flow occurs, oriented at right angles to the direction of flow, Q_S is the amount of material (solute) transported per unit time, J_S is the solute flux, and A is the area. The units of flux are amount or volume or mass or charge per unit time per unit area.

Strictly speaking, fluxes and flows are vectors, consisting of the magnitude of the flux or flow and its direction. Unless otherwise noted, we will consider flux or flow only in one direction and therefore we will suppress the vector nature of flux and flow.

FLUX DEPENDS LINEARLY ON ITS CONJUGATE FORCE

For a variety of forces and fluxes, the flux that results from a net force varies linearly with the force:

[1.2.2] $$J_x = LF_x$$

where J_x is the flux of something, L is a phenomenological coefficient, and F_x is the net force that drives the flux. This generic equation holds for a variety of kinds of fluxes. The flux of heat energy, electrical flux (the current density), diffusion of solute and pressure-driven flow all obey this general phenomenological law. In each of these cases, the net force is proportional to the **gradient** of an intensive variable. Strictly speaking, the gradient is a vector quantity, but we use it here to denote the slope of these intensive variables along one dimension:

[1.2.3]
$$\text{Fourier's law of heat conduction: } J_H = -\lambda \frac{dT}{dx}$$
$$\text{Ohm's current law: } J_e = -\sigma \frac{d\psi}{dx}$$
$$\text{Fick's first law of diffusion: } J_S = -D \frac{dC}{dx}$$
$$\text{Pressure-driven flow: } J_V = -L_P \frac{dP}{dx}$$

where J_H is the flux of heat energy, dT/dx is the temperature gradient, λ is the coefficient of thermal conductivity, J_e is the electrical current flux, $d\psi/dx$ is the voltage gradient, σ is the electrical conductivity, J_S is the solute flux, dC/dx is the concentration gradient, D is the diffusion coefficient, J_V is the volume flux, dP/dx is the pressure gradient, and L_P is the hydraulic conductivity. All of these phenomenological equations find application in physiological systems. They are all analogues of Ohm's law.

These equations are true only if the only driving force is the one specified. For example, diffusion of electrolytes, charged solutes, is influenced by electric fields. If a voltage gradient is also present along with a concentration gradient, Fick's first law of diffusion would need to be modified to reflect that influence. A pressure difference that produces a volume flow in the presence of a diffusion gradient will also modify the flux of solute. In general, flows produced by multiple flow processes are not independent. If there are two forces driving flows, we write

[1.2.4]
$$J_1 = L_{11}F_1 + L_{12}F_2$$
$$J_2 = L_{21}F_1 + L_{22}F_2$$

where L_{11} is the coefficient relating flux 1 to its primary driving force 1 and L_{22} relates flux 2 to its primary driving force 2, and L_{12} and L_{21} are the coupling coefficients that describe how secondary forces affect the flows. An example of this is a bimetallic junction. When two unlike metals are joined together, passing a current through the junction causes it to either heat up or cool, and this is called the *Peltier effect*. The coupling coefficient implies that if you heat up or cool the junction, a current will flow. This is the basis of the *thermocouple*. In this case, the two fluxes are heat and current and the two forces are temperature gradient and voltage gradient. Because of a principle called microscopic reversibility, it turns out that the cross-coupling coefficients are equal: $L_{12} = L_{21}$. This is called **Onsager reciprocity**, in honor of Lars Onsager (1903–1976) who earned the Nobel Prize in Chemistry in 1968 for this discovery.

FLUX MOVES DOWNHILL

The relations in Eqn [1.2.3] describe fluxes in one dimension. Both fluxes and gradients are actually vectors, but we consider a single direction here for simplicity. Consider Fick's law of diffusion for solutes. If the gradient of concentration is constant, we may write

[1.2.5] $$J_S = -D \frac{(C_1 - C_2)}{(x_1 - x_2)}$$

for two points (C_1, x_1) and (C_2, x_2). If $C_1 > C_2$ and $x_1 < x_2$, the slope is negative and the flux is positive. If $C_1 < C_2$ and $x_1 < x_2$, then the slope is positive and the flux is negative. Thus, the flux always goes from regions of high concentration to regions of low concentration (see Figure 1.2.1). This is true for all the intensive variables for the fluxes in Eqn [1.2.3]. These fluxes always move downhill, unless acted upon by additional forces.

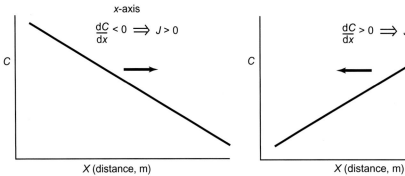

FIGURE 1.2.1 Flux moves downhill. Consider one-dimensional flux, with positive flux defined as in the direction of the x-axis. In this case, we consider diffusion that is driven only by a concentration gradient. If the gradient is negative (higher concentrations at lower values of x), then by Eqn [1.2.3], the flux is positive and directed to the right (middle panel). If the gradient is positive, then the flux is negative and directed to the left. In each case, the flux of solute is from the region of high concentration toward the region of lower concentration.

CONSERVATION OF MATTER OR ENERGY LEADS TO THE *CONTINUITY EQUATION*

Here we consider that a concentration gradient exists and produces a solute flux as a consequence. We consider a cylindrical tube as shown in Figure 1.2.2, having a cross-sectional area A, which is intersected at right angles by planes at $x=x$ and $x=x+\Delta x$. so that the tube is cut into three compartments, the left, middle, and right compartments.

The concentration may vary with time and distance. We define the concentration in any volume element as

[1.2.6] $$C(x,t) = \frac{N(x,t)}{V}$$

where $N(x, t)$ is the number of solute particles in the volume element and V is the volume element. We define $J(x)$ as the *net* number of solute particles crossing the plane at $x = x$ per unit time per unit area, with positive being directed along the x-axis, to the right. The number of particles entering the middle compartment from the left in time Δt is $AJ(x)\Delta t$ and the number leaving the middle compartment by crossing the plane at $x = x + \Delta x$ is $AJ(x + \Delta x)\Delta t$. Here the parenthesis means "function of" and not multiplication. If there is no chemical transformation of the solute particles, their number is conserved and we can write

[1.2.7] $$\Delta N = AJ(x)\Delta t - AJ(x+\Delta x)\Delta t$$
$$\frac{\Delta N}{\Delta t} = AJ(x)t - AJ(x+\Delta x)$$

Dividing by the volume element $V = A\Delta x$ and rearranging, we have

[1.2.8] $$\frac{1}{A\Delta x}\frac{\Delta N}{\Delta t} = \frac{AJ(x) - AJ(x+\Delta x)}{A\Delta x}$$
$$\frac{\Delta \frac{N}{V}}{\Delta t} = -\frac{AJ(x+\Delta x) - AJ(x)}{A\Delta x}$$

In the limit of $\Delta t \to 0$ and $\Delta x \to 0$, this becomes

[1.2.9] $$\frac{\partial C(x,t)}{\partial t} = -\frac{\partial J(x)}{\partial x}$$

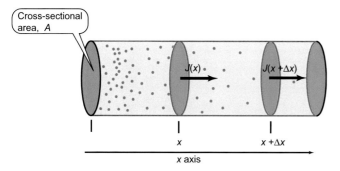

FIGURE 1.2.2 Fluxes as a function of distance in the presence of a concentration gradient.

This equation is called the **continuity equation**. What this equation says is that if the flux of solute is not the same everywhere, then the amount of solute must be building up or becoming depleted somewhere, and this buildup or depletion changes the concentration of solute. It is a straightforward consequence of the conservation of material. This equation is not true if the diffusing chemical undergoes chemical transformation. In this case, it is not conserved.

Similar continuity equations can be written for the flux of heat energy, charge, and volume. Their form is given in Eqn [1.2.10]:

[1.2.10]
$$\text{Heat: } \rho C_P \frac{\partial T}{\partial t} = -\frac{\partial J_H}{\partial x}$$
$$\text{Charge: } \frac{C}{V}\frac{\partial \psi}{\partial t} = -\frac{\partial J_e}{\partial x}$$
$$\text{Solute: } \frac{\partial C_S}{\partial t} = -\frac{\partial J_S}{\partial x}$$
$$\text{Volume: } \frac{C}{V}\frac{\partial P}{\partial t} = -\frac{\partial J_V}{\partial x}$$

where ρ is the density of matter through which heat flows and C_p is its specific heat capacity. In the next equation, C is the electrical capacitance, V is the volume. Next, C_s is the concentration. In the last line C is the compliance, and V is the volume. Here we have the unfortunate situation in which single variables denote different quantities: C stands for electrical capacitance (in farads), concentration (usually in molar) and compliance ($=\Delta V/\Delta P$). Generally the meaning of the variable is clear from its context.

STEADY-STATE FLOWS REQUIRE LINEAR GRADIENTS

In homeostasis, there is a steady supply of nutrients and removal of wastes and a steady withdrawal of nutrients and a steady production of wastes by the tissues. This steady state in which all flows are constant is more easily amenable to mathematical analysis. What "steady state" means is that each of the variables on the left-hand side of Eqn [1.2.10] is zero because at the steady state there are no changes in temperature, charge, concentration, or pressure with time:

[1.2.11]
$$\text{Heat: } \rho C_P \frac{\partial T}{\partial t} = -\frac{\partial J_H}{\partial x} = 0$$
$$\text{Charge: } \frac{C}{V}\frac{\partial \psi}{\partial t} = -\frac{\partial J_e}{\partial x} = 0$$
$$\text{Solute: } \frac{\partial C_S}{\partial t} = -\frac{\partial J_S}{\partial x} = 0$$
$$\text{Volume: } \frac{C}{V}\frac{\partial P}{\partial t} = -\frac{\partial J_V}{\partial x} = 0$$

QUANTITATIVE HUMAN PHYSIOLOGY: AN INTRODUCTION

Substituting in from Eqn [1.2.3] for J_H, J_e, J_S, and J_V, we have

[1.2.12]
$$\frac{\partial^2 T}{\partial x^2} = 0$$
$$\frac{\partial^2 \psi}{\partial x^2} = 0$$
$$\frac{\partial^2 C}{\partial x^2} = 0$$
$$\frac{\partial^2 P}{\partial x^2} = 0$$

This condition is met only if the gradient of T, ψ, C, or P is constant; thus, the slope of T, ψ, C, and P at steady state is constant, and each of these intensive variables varies linearly with distance.

HEAT, CHARGE, SOLUTE, AND VOLUME CAN BE STORED: ANALOGUES OF CAPACITANCE

The steady state is often approximated in the body but rarely achieved. At rest heat production balances heat exhausted to the environment. When we begin exercising, heat production rises rapidly and the temperature of the body rises accordingly until, once again, heat production matches heat transfer to the environment, achieved by using other forces besides the conduction described in Fourier's law. This new steady state of temperature during exercise is achieved at different operating conditions than at rest. In another example, transport of blood through the cardiovascular system is pulsatile, because the pressure that drives transport comes from the heart, and the heart produces force rhythmically. Each of the main four variables we have been discussing, heat, charge, amount of chemicals, and volume, can be temporarily stored or depleted.

Electrical charge can be stored in capacitors. The constitutive relation between charge, voltage, and capacitance is given as

[1.2.13]
$$Q = CV$$

where Q here stands for charge, C is the capacitance, and V is the voltage. Here we are victims of the use of the same variables to denote entirely different quantities. We will use Q most often to signify a flow, but here it signifies charge, in coulombs. In physiology, we often use C to denote concentration, but here it means capacitance, in farads ($=CV^{-1}$); in physiology, V usually signifies volume, but here it means electrical potential, in volts. Electrical capacitance is an important concept for physiologists as well, because membrane potential derives from a separation of electrical charges across the membrane, and the membrane itself acts like a tiny capacitor with two conducting plates, separated by a dielectric. We will discuss this further in the sections on membrane potential, action potential, and the cable properties of nerves (Chapters 3.1–3.3). The other relationships completely analogous to the relation between charge, capacitance, and voltage, are

[1.2.14]
$$H = C_P MT: \text{Heat energy} = \text{heat capacity} \times \text{mass} \times \text{temperature}$$
$$M = VC: \text{Amount} = \text{volume} \times \text{concentration}$$
$$V = CP: \text{Volume} = \text{compliance} \times \text{pressure}$$

where the capacitance-like elements include electrical capacitance (C in Eqn [1.2.13]), thermal mass ($C_p M$, the specific heat capacity times the mass), volume (V), and compliance (C). Note again the multiplicity of uses of a single notation. C variously stands for capacitance, heat capacity, concentration, or compliance.

The capacitances are all expressed as the ratio of an extensive variable and an intensive variable and are all themselves extensive variables. Table 1.2.1 summarizes the four kinds of capacitances.

PRESSURE DRIVES FLUID FLOW

In the case of fluid or air flow, pressure differences drive the flow. The SI unit of pressure is the pascal, Pa, equal to 1 N m^{-2}. However, physiologists still use other units, notably the atmosphere and mmHg. The atmospheric pressure is the weight of a column of air equal to the height of the atmosphere in the earth's gravitational field per unit area of the earth's surface. The actual pressure in the atmosphere decreases as you ascend, but the unit of 1 atm is defined for a standard condition of the air and standard altitude at sea level. The conversion between atmospheres and mmHg is an observed phenomenon. Atmospheric pressure can be measured in units of mmHg as described in Figure 1.2.3.

Figure 1.2.3 illustrates that atmospheric pressure supports a column of 760 mmHg high. The pressure of this column of Hg is equal to atmospheric pressure, and the pressure of the column is simply its weight divided by its area. The weight is the force of gravity acting on the column, and is given as

[1.2.15]
$$W = F = mg$$
$$= \rho V g$$
$$= \rho A h g$$

TABLE 1.2.1 Four Kinds of Capacitances

Capacitance	Expression	Units	Application
Electrical **capacitance**, C	Q/V	Farads = coulombs/volt	Nerve conduction, membrane potential
Thermal capacitance, $C_p M$	H/T	JK^{-1} = joules/temperature	Heat production/loss temperature regulation
Chemical capacitance, V, **volume**	M/C	Volume = moles/molarity, L	Metabolism, filtration
Mechanical capacitance, C, **compliance**	V/P	Compliance = volume/pressure, $m^3 \text{ Pa}^{-1}$	Blood pressure, breathing

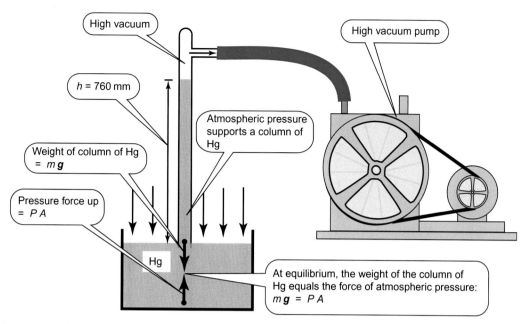

FIGURE 1.2.3 Measurement of atmospheric pressure. A closed vertical tube is connected to a high vacuum pump and evacuated air. When inverted into a dish of mercury, the atmospheric pressure forces mercury up the tube until mechanical equilibrium is achieved when the weight of the column of mercury exerts a pressure equal to the atmospheric pressure. At sea level in dry air, 1 atm will support a column 760 mmHg high.

TABLE 1.2.2 Conversion Between Pressure Units

atm	mmHg	Pa
1	760	1.013×10^5
0.00132	1	133.29
9.87×10^{-6}	0.00750	1

The pressure is just the force per unit area. Dividing Eqn [1.2.15] by the area, we get

$$[1.2.16] \qquad P = \frac{F}{A} = \rho g h$$

Thus, the height of the column of mercury in equilibrium with the atmospheric pressure is independent of its area. We need to specify only the height of the column of mercury. Thus, at sea level, the atmospheric pressure supports a column of 760 mmHg high and we say that 760 mmHg = 1 atm.

The value of atmospheric pressure in pascals = N m^{-2} can be calculated from 760 mmHg by using the density of Hg (13.59 g cm^{-3}) and the acceleration due to gravity (9.81 m s^{-2}). Inserting these values into Eqn [1.2.16], we get

$$P = 13.59 \text{ g cm}^{-3} \times 9.81 \text{ m s}^{-2} \times 0.76 \text{ m}$$
$$= 13.59 \text{ g cm}^{-3} \times (100 \text{ cm m}^{-1})^3 \times 10^{-3} \text{ kg g}^{-1}$$
$$\times 9.81 \text{ m s}^{-2} \times 0.76 \text{ m}$$
$$= 13.59 \times 10^3 \text{ kg m}^{-3} \times 9.81 \text{ m s}^{-2} \times 0.76 \text{ m}$$
$$1 \text{ atm} = 101.3 \times 10^3 \text{ kg m s}^{-2} \text{ m}^{-2} = 1.013$$
$$\times 10^5 \text{ N m}^{-2} = 1.013 \times 10^5 \text{ Pa}$$

We can therefore complete a conversion table for pressure units (Table 1.2.2).

POISEUILLE'S LAW GOVERNS STEADY-STATE LAMINAR FLOW IN NARROW TUBES

In 1835, Jean Leonard Marie Poiseuille experimentally established the relationship between flow through narrow pipes and the pressure that drives the flow. The relationship is

$$[1.2.17] \qquad Q_V = \frac{\pi a^4}{8\eta} \left(\frac{\Delta P}{\Delta x} \right)$$

where Q_V is the flow, in units of volume per unit time, π is the geometric ratio, a is the radius of the pipe, η is the viscosity, ΔP is pressure difference between the beginning and end of the pipe, and Δx is the length of the pipe. This equation describes the relationship between flow and pressure difference only for laminar flow. Laminar flow is steady, streamlined flow, and it is distinguished from turbulent or chaotic flow. This equation is often applied to problems in physiology even though the conditions for its valid application are missing. Its application requires us to understand **viscosity**.

Consider two parallel plates separated by a fluid, as shown in Figure 1.2.4. The top plate can be moved at a constant velocity relative to the stationary bottom plate only if the plate is subjected to a force that continuously overcomes the frictional resistance on the plate caused by its contact with the adjacent fluid.

The viscosity is the resistance of a fluid to shear forces. It is defined by the equation

$$[1.2.18] \qquad \frac{F}{A} = \eta \frac{dv}{dy}$$

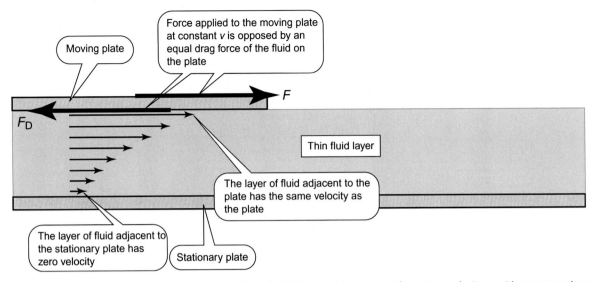

FIGURE 1.2.4 Definition of viscosity. Two plates are separated by a fluid. The top plate moves with constant velocity, v, with respect to the stationary bottom plate. The fluid adheres to the plates and a thin layer of fluid immediately adjacent to the plates has the same velocity as the plates. This results in a velocity profile in the fluid. The steepness of this velocity profile, dv/dy, is the velocity gradient.

where F is the shear force, A is the area, v is the velocity, and y is the dimension perpendicular to the plate. The ratio F/A is called the **shear stress** and the quantity dv/dy is called the **velocity gradient**. F/A in this equation has the units of pressure, $Pa = N\,m^{-2}$ and dv/dy has the units of $m\,s^{-1}\,m^{-1} = s^{-1}$, so the units of η in SI are Pa s. Thus, the viscosity is the ratio of the shear stress, F/A, divided by the velocity gradient, dv/dy. In older texts viscosity is sometimes given in units of poise = 1 dyne cm^{-2} s. These can be converted to Pa s by using the definition of $Pa = 1\,N\,m^{-2}$, $1\,N = 1\,kg\,m\,s^{-2}$ and 1 dyne = 1 g cm s^{-2}: Thus, $1\,N = 10^5$ dyne.

$1\,Pa\,s = 1\,N\,m^{-2}\,s = 1\,kg\,m\,s^{-2}\,m^{-2}\,s = 1\,kg$
$\times 10^3\,g\,kg^{-1}\,m \times 100\,cm\,m^{-1} \times s^{-2} \times m^{-2}$
$\times (0.01\,cm\,m^{-1})^2\,s = 10^3\,g \times 100\,cm \times s^{-2}$
$\times 10^{-4}\,cm^{-2}\,s = 10\,g\,cm\,s^{-2} - cm^{-2}\,s =$
$10\,dyne\,cm^{-2}\,s = 10\,poise$
So 1 Pa s = 10 poise

EXAMPLE 1.2.1 Ultrafiltration in the Kidney

The kidney has a structure called the glomerulus that consists of combined layers of cells and extracellular matrix—bundles of fibers in the extracellular space—that together form an ultrafilter (see Chapter 6.2 for further description). It is called an ultrafilter because the combined layers retain proteins while letting most small solutes pass into the ultrafiltrate. We model the membrane here as a flat membrane that is pierced by many identical right cylindrical holes, or pores. Assume that the radius of the pores is 3.5 nm and the pore length is 50 nm. The viscosity of the fluid is taken to be the same as plasma, 0.02 poise. The aggregate area of the pores makes up 5% of the total surface area of the membrane. The total pressure on the input side, the side of the blood, averages 60 mmHg and on the ultrafiltrate side the total pressure averages 45 mmHg (see Chapter 6.2 for a discussion of the origin of these pressures). The total available area of the membrane is 1.5 m². What is the filtration rate in cm³ min⁻¹?

The situation is depicted schematically in Figure 1.2.5. We use Poiseuille's equation here. The total flow, Q_V, is the sum of the flow through all of the pores:

$$Q_v = Nq_v = N\pi a^4/8\eta\,(\Delta P/\Delta x)$$

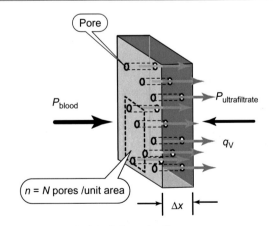

FIGURE 1.2.5 Model of the kidney ultrafilter.

where N is the number of pores and q_v denotes the flow through a single pore. Here a is the radius, $a = 3.5 \times 10^{-9}$ m, η is the viscosity ($\eta = 0.02$ poise $\times$ 1 Pa s/10 poise = 0.002 Pa s), ΔP is the pressure difference = 15 mmHg $\times$ 133.3 Pa mmHg^{-1} = 1999.5 Pa, and $\Delta x = 60 \times 10^{-9}$ m. Now that all the units are compatible, we plug them into the equation and get:

$q_V = \pi \times (3.5 \times 10^{-9} \text{ m})^4 / 8 \times 2 \times 10^{-3} \text{ Pa s} \times (1999.5 \text{ Pa}/60 \times 10^{-9} \text{ m})$

$= 471.44 \times 10^{-36} \text{ m}^4 / 16 \times 10^{-3} \text{ Pa s} \times 33.33 \times 10^9 \text{ Pa} \times \text{m}^{-1}$

$= 982.1 \times 10^{-24} \text{ m}^3 \times \text{s}^{-1} \times (100 \text{ cm} \times \text{m}^{-1})^3 \times 60 \text{ s} \times \text{min}^{-1}$

$= 5.89 \times 10^{-14} \text{ cm}^3 \times \text{min}^{-1}$

This is the flow through a single pore. We need to know how many of them are there. If their aggregate area is 5% of the total, then the number of pores can be calculated from

$N \times \pi a^2 = 0.05 \times 1.5 \text{ m}^2$

Knowing that $a = 3.5 \times 10^{-9}$ m, we solve for N:

$N = 0.075 \text{ m}^2 / \pi \times (3.5 \times 10^{-9} \text{ m})^2 = 1.95 \times 10^{15}$

which is a lot of pores! Multiplying $N \times q_V$, we get

$Q_V = N \times q_V = 1.9 \times 10^{15} \text{ pores} \times 5.89 \times 10^{-14} \text{ cm}^3 \text{ min}^{-1} \text{ pore}^{-1}$

$= \mathbf{111.9 \text{ cm}^3 \text{ min}^{-1}}$

This is a reasonable approximation to the filtration rate in an adult human.

THE *LAW OF LAPLACE* RELATES PRESSURE TO TENSION IN HOLLOW ORGANS

The blood vessels maintain a pressure difference across their walls. These vessels approximate hollow cylinders. The gallbladder, urinary bladder, heart, and lung alveoli also maintain a pressure difference, and approximate hollow spheres. The Law of Laplace can be derived for these ideal geometries by considering mechanical equilibrium. Consider first a circular cylinder of radius r subjected to an external pressure P_o and internal pressure P_i, as shown in Figure 1.2.6.

In Figure 1.2.6, the internal pressure acting on the upper half of the cylinder by the lower half cylinder produces a force on the upper half that is given as $P_i \times 2r \times L$. The external pressure acting over the surface of the upper half of the cylinder produces a force that is equal to $P_o \times 2r \times L$, as can be deduced from the situation where $P_o = P_i$ in a mechanically stable fluid with no walls. The balance of the internal and external pressures produces a net force of $\Delta P \times 2r \times L$ where $\Delta P = P_i - P_o$. This is balanced by the force of the walls of the lower half acting on the upper half. These forces are the wall tension, in units of $\text{N} \times \text{m}^{-1}$, acting along the length of the wall. These forces are directed downward, as shown. For mechanical equilibrium to occur

[1.2.19] $\qquad F_{net} = 0 = T2L = \Delta P 2rL$

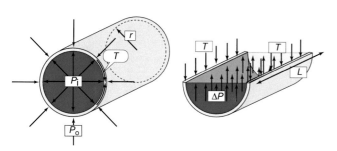

FIGURE 1.2.6 Right circular cylinder of radius r and length L is subjected to a transmural pressure difference. For mechanical equilibrium, the net force on any part of the cylinder must be zero. We split the cylinder in half down its axis. The sum of forces on the upper half must sum to zero.

Rearrangement of this equation gives the Law of Laplace for a cylinder:

[1.2.20] $\qquad \Delta P = \dfrac{T}{r}$

An exactly analogous argument can be made for a sphere. In this case, the pressure is distributed over the cross-sectional area of the hemisphere, and the wall tension is distributed around the circumference. The result gives

[1.2.21] $\qquad F_{net} = 0 = T 2\pi r = \Delta P \pi r^2$

which is easily rearranged to

[1.2.22] $\qquad \Delta P = \dfrac{2T}{r}$

SUMMARY

Flow consists of the movement of something from one place to another in the body. We identify four classes of flow of major importance in physiology: flow of heat energy, electric current, solute, and volume (gas or body fluids). Flow is an extensive variable, depending on the extent of the system. Flux is the flow divided by the area through which the flow occurs. Fluxes are driven by forces. Heat flux is driven by a temperature gradient, electrical current by a potential gradient, solute flux by a concentration gradient, and volume flux by a pressure gradient. Gradients are vector quantities. Here we consider one-dimensional flow and flux and suppress the vector nature of these gradients. In each of these cases, flux occurs "downhill," meaning from regions of high temperature, potential, concentration or pressure, to regions of low temperature, potential, concentration, or pressure. Each flux has the form

$$J_x = -K \dfrac{d\psi_x}{dx}$$

The continuity equation for each of these fluxes states that a gradient of flux produces a time dependence of the driving force. If the flux is not the same everywhere, then temperature, voltage, concentration, or pressure varies with time. The continuity equation is a consequence of conservation of energy, electric charge, chemical species, or volume.

Steady-state flows require linear gradients of temperature, potential, concentration, or pressure.

Heat energy, electric charge, concentration, and volume can all be stored in the body. The ability to store these things is quantified by capacitances. These include thermal capacitance, electrical capacitance, chemical capacitance, and mechanical capacitance. The thermal capacitance is MC_p, where C_p is the specific heat capacity; the units of MC_p are in $J\ K^{-1}$; electrical capacitance is Q/V, in F; chemical capacitance is V, in L; mechanical capacitance is compliance, in $L\ Pa^{-1}$.

Pressure drives flow. Pressure is measured in pascals $= 1\ N\ m^{-2}$. This unit is equal to 9.87×10^{-6} atm or 0.0075 mmHg. Steady-state pressure-driven flow through narrow pipes is described by Poiseuille's law:

$$Q_V = \frac{\pi a^4}{8\eta}\left(\frac{\Delta P}{\Delta x}\right)$$

where a is the radius of the pipe, η is the viscosity, in Pa s, π is the geometric ratio of circumference to diameter of a circle, ΔP is the pressure difference in the pipe, and Δx is the length of the pipe. Viscosity is the resistance of a fluid to shear forces and is defined mathematically as the ratio of the shear stress to the velocity gradient.

A transmural pressure within hollow organs or tubes requires tension in the walls of these organs. So far as these organs approximate thin-walled cylinders or spheres, the wall tension obeys the Law of Laplace:

$$\text{Cylinder:}\quad \Delta P = \frac{T}{r}$$

$$\text{Sphere:}\quad \Delta P = \frac{2T}{r}$$

where ΔP is the transmural pressure difference, T is the wall tension, in $N\ m^{-1}$, and r is the radius of the cylinder or sphere.

REVIEW QUESTIONS

1. Is density an intensive or extensive variable? Is temperature intensive or extensive?
2. What drives current flow? What drives solute flow? If solute is charged, would its movement make a current?
3. What does it mean to say that fluxes "move downhill"?
4. What does capacitance mean for charge? Solute? Heat? Volume?
5. How is Ohm's law like Fick's law of diffusion?
6. What are the units of viscosity?
7. What are the assumptions in the derivation of Poiseuille's law? Are these reasonable?
8. What is the relationship between velocity and flux in fluid flow?
9. What is the relationship between capacitance, charge, and voltage?
10. What is the relationship between volume, amount, and concentration?
11. What is the relationship between compliance, volume, and pressure?
12. What is the relationship between heat capacity, heat energy, and temperature?
13. What is the relationship between wall tension and pressure in a sphere? In a cylinder?

APPENDIX 1.2.A1 DERIVATION OF POISEUILLE'S LAW

POISEUILLE'S LAW DESCRIBES PRESSURE-DRIVEN FLOW THROUGH A CYLINDRICAL PIPE

Consider two fluid compartments which are joined by a right cylindrical pipe, as shown below, of area A. Since pressure is defined as a force per unit area, the total force acting on the fluid in the pipe on the left-hand surface is just $P_L A$, and the total force acting on the fluid in the pipe on the right-hand side is $P_R A$. If P_L and P_R are not equal, then the fluid within the pipe will be subject to a net force and therefore this volume of fluid will be accelerated. The result will be a movement of fluid in the direction of the net force. What we wish to establish is the quantitative relationship between the resulting flow and the pressure difference which drives this flow.

SHEAR STRESS IS THE VISCOSITY TIMES THE VELOCITY GRADIENT

The movement of fluid through the pipe shown in Figure 1.2.A1.1 encounters resistance along the cylindrical surfaces of the pipe. This resistance is key to deriving an expression for the flow as a function of pressure. It is due to the viscosity of the fluid, as shown in Figure 1.2.4. We consider that the flow of fluid occurs in layers, or laminae, and that each layer exerts a viscous drag on the layer immediately above and below. The fluid remains in contact with the walls on either side of the tube, and so has a velocity $v = 0$ at the walls.

In order to achieve a constant velocity, we need to apply a continual force, F, in order to overcome the viscous drag of the fluid. The **shear stress** is the force exerted by one lamina on an adjacent one, and is given by

[1.2.A1.1]
$$\frac{F}{A} = \eta\frac{dv}{dy}$$

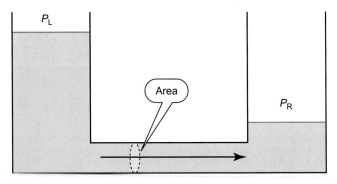

FIGURE 1.2.A1.1 Flow is driven by pressure differences.

where F is the force, A is the area, η is the coefficient of viscosity of the fluid, and dv/dy is the gradient of velocity. This equation makes good common sense: the force necessary to keep the lamina moving depends on how much area is exposed to the resistance, it depends on how sticky the fluid is, and it depends on how fast it is going.

FLOW THROUGH A PIPE

We consider here the flow through a tube or pipe of radius a and length Δx. The pressure at the left end of the pipe is $P(x)$ and the pressure at the right end is $P(x + \Delta x)$. We consider here a constant flow which therefore has a constant velocity. This means that the fluid is not subjected to any net force. Consider the forces experienced by a hollow shell of inner radius r and outer radius $r + dr$, as shown in Figure 1.2.A1.2.

The fluid in contact with the walls of the pipe does not move, so that $v = 0$ at $r = a$. The velocity of the fluid increases as one approaches the center of the tube. The actual velocity profile will be solved on the way to deriving an expression for the total flow through the pipe.

Because the fluid flows through the pipe at a constant velocity along the pipe (but which depends on the distance from the walls of the pipe), the sum of the forces on the hollow shell shown in Figure 1.2.A1.2 must be zero. The forces to the right include the force of the pressure on the left and the drag force of the inner layer of fluid, and the forces to the left include the pressure on the right and the drag force of the outer fluid. Thus, we have

[1.2.A1.2] $\quad 0 = P(x)A + F_2 - P(x + \Delta x)A - F_1$

The area of the hollow sphere on which the pressures act is $2\pi r dr$. The drag forces F_1 and F_2 are given by Eqn [1.2.A1.1] as

[1.2.A1.3]
$$F_1 = 2\pi(r + dr)\Delta x \eta \left(\frac{dv}{dr}\right)_{r+dr}$$
$$F_2 = 2\pi r \Delta x \eta \left(\frac{dv}{dr}\right)_r$$

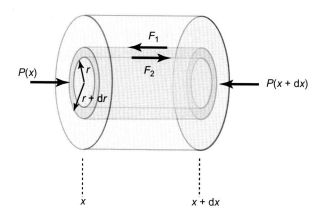

FIGURE 1.2.A1.2 Balance of forces on a hollow cylinder of inner radius r and outer radius $r + dr$. F_1 is the drag force of the lamina with inner radius $r + dr$ on the hollow cylinder; F_2 is the drag force on the hollow cylinder by the lamina immediately inside, with outer radius r.

Insertion of this result into Eqn [1.10.A1.2] gives

[1.2.A1.4]
$$0 = (P(x) - P(x + \Delta x))2\pi r\, dr$$
$$+ 2\pi r \Delta x \eta \left(\frac{dv}{dr}\right)_r$$
$$+ 2\pi (r + dr)\Delta x \eta \left(\frac{dv}{dr}\right)_{r+dr}$$

which can be re-written as

$$0 = -(\Delta P)2\pi r\, dr$$
$$+ 2\pi r \Delta x \eta \left(\frac{dv}{dr}\right)_r$$
$$- 2\pi r \Delta x \eta \left(\frac{dv}{dr}\right)_{r+dr} - 2\pi\, dr \Delta x \eta \left(\frac{dv}{dr}\right)_{r+dr}$$

[1.2.A1.5]

where $\Delta P = P(x + \Delta x) - P(x)$. We next approximate the gradient of velocity at $(r + dr)$ by the first two terms of a Taylor's series:

[1.2.A1.6] $\quad \left(\dfrac{dv}{dr}\right)_{r+dr} \simeq \left(\dfrac{dv}{dr}\right)_r + dr\left(\dfrac{d^2v}{dr^2}\right)_r$

Substitution of Eqn [1.2.A1.6] into Eqn [1.2.A1.5] gives

$$0 = -\left(\frac{\Delta P}{\Delta x}\right)2\pi r\, dr \Delta x$$
$$+ 2\pi r \Delta x \eta \left(\frac{dv}{dr}\right)_r$$
$$- 2\pi r \Delta x \eta \left(\frac{dv}{dr}\right)_r - 2\pi\, dr\, \Delta x \eta \left(\frac{dv}{dr}\right)_r$$
$$- 2\pi r\, dr \Delta x \eta \left(\frac{d^2v}{dr^2}\right)_r - 2\pi r\, dr^2 \Delta x \eta \left(\frac{d^2v}{dr^2}\right)_r$$

[1.2.A1.7]

Canceling out the like terms, and factoring out the term $2\pi r dr\, \Delta x$, we obtain

[1.2.A1.8] $\quad 0 = -\left(\dfrac{\Delta P}{\Delta x}\right) - \dfrac{\eta}{r}\left(\dfrac{dv}{dr}\right)_r - \eta\left(\dfrac{d^2v}{dr^2}\right)_r - dr\dfrac{\eta}{r}\left(\dfrac{d^2v}{dr^2}\right)_r$

In the limit as $dr \to 0$, the last term in Eqn [1.2.A1.8] vanishes. Multiplying through by -1, we are left with the differential equation

[1.2.A1.9] $\quad 0 = \dfrac{\Delta P}{\Delta x} - \dfrac{\eta}{r}\left(\dfrac{dv}{dr}\right) + \eta\left(\dfrac{d^2v}{dr^2}\right)$

This second-order differential equation can be solved by converting it to a first order differential equation with the substitution $y = dv/dr$, and then multiplying through by an integrating factor, ρ. Re-arranging Eqn [1.2.A1.9], we have

[1.2.A1.10]
$$\rho\left(\frac{dy}{dr}\right) + \frac{\rho}{r}y + \frac{\rho}{\eta}\frac{\Delta P}{\Delta x} = 0$$

We choose ρ so that the first two terms are an exact differential. This is true if $\rho = r$. Thus we have

[1.2.A1.11]
$$r\left(\frac{dy}{dr}\right) + y = \frac{d(ry)}{dr} = -\frac{r}{\eta}\left(\frac{\Delta P}{\Delta x}\right)$$

Integrating Eqn [1.2.A1.11], we obtain

[1.2.A1.12]
$$\int d(ry) = \int -\frac{r}{\eta}\left(\frac{\Delta P}{\Delta x}\right)dr$$
$$ry = -\frac{1}{2\eta}\left(\frac{\Delta P}{\Delta x}\right)r^2$$

Canceling the r factor on both sides of Eqn [1.2.A1.12] and recalling that $y = dv/dr$, we integrate Eqn [1.2.A1.12] again:

[1.2.A1.13]
$$\frac{dv}{dr} = -\frac{1}{2\eta}\left(\frac{\Delta P}{\Delta x}\right)r$$
$$\int dv = \int -\frac{1}{2\eta}\left(\frac{\Delta P}{\Delta x}\right)r\,dr$$
$$v = -\frac{1}{4\eta}\left(\frac{\Delta P}{\Delta x}\right)r^2 + C$$

where C is a constant of integration which can be evaluated from the boundary conditions. The boundary conditions are that $v = 0$ when $r = a$. That is, the velocity of the fluid immediately adjacent to the walls of the pipe is zero. Insertion of $v = 0$ and $r = a$ into Eqn [1.2.A1.13] gives

[1.2.A1.14]
$$C = \frac{a^2}{4\eta}\left(\frac{\Delta P}{\Delta x}\right)$$
$$v = \frac{1}{4\eta}\left(\frac{\Delta P}{\Delta x}\right)(a^2 - r^2)$$

This equation gives the velocity of fluid flow as a function of the radial distance from the center ($r = 0$) to the edge ($r = a$) of the pipe. This equation says that the velocity profile is parabolic.

What we wanted to do at the outset was to calculate the total flow through the pipe, Q_V. This is the volume of fluid which crosses the total cross-section of the pipe per unit time. The volume flux, J_V, is the volume moving through a small increment of unit area per unit time. In fact J_V is the velocity of fluid movement. To see this, consider a block of fluid moving at constant velocity, v. The block has a cross-sectional area, A. In time t the block moves a distance $v\Delta t$. The volume of fluid moving in this time is $Av\Delta t$. The volume flux, J_V, is the volume of fluid moved per unit area per unit time. This is

[1.2.A1.15]
$$J_V = \frac{Av\Delta t}{A\Delta t} = v$$

Thus, the volume flux is the velocity of fluid movement. To find the total fluid flow, we integrate the flow as a function of distance from the center of the pipe. Thus

[1.2.A1.16]
$$Q_V = \int_0^a dQ_V = \int_0^a J_V\,dA = \int_0^a J_V 2\pi r\,dr$$
$$= \int_0^a v 2\pi r\,dr$$
$$= \int_0^a \frac{1}{4\eta}\left(\frac{\Delta P}{\Delta x}\right)(a^2 - r^2)2\pi r\,dr$$
$$= \frac{2\pi}{4\eta}\left(\frac{\Delta P}{\Delta x}\right)\int_0^a (a^2 - r^2)r\,dr$$

The last integral can be evaluated by making the substitution $u = (a^2 - r^2)$, so $du = -2r\,dr$, to obtain

[1.2.A1.17]
$$\int_{r=0}^{r=a}(a^2 - r^2)r\,dr = \int_{r=0}^{r=a} u\left(-\frac{du}{2}\right)$$
$$= -\frac{1}{2}\int_{u=a^2}^{u=0} u\,du$$
$$= \frac{a^4}{4}$$

Inserting this result of the integration into Eqn [1.2.A1.16], we obtain

[1.2.A1.18]
$$Q_V = \frac{\pi a^4}{8\eta}\left(\frac{\Delta P}{\Delta x}\right)$$

This last equation gives the total flow through the pipe, Q_V. This equation is called Poiseuille's Law, in honor of the French physician Jean Leonard Marie Poiseuille, who experimentally established the law in 1835. The relation shows that the total flow is linearly dependent on the driving force for the flow, the pressure difference between the left and right ends of the pipe. Further, it is inversely related to the length of the pipe, Δx, and to the viscosity, η. Most importantly, the flow is proportional to the fourth power of the radius of the pipe. One of the points of the derivation given above was to show the mathematical origin of this very steep dependence on the size of the pipe.

APPENDIX 1.2.A2 INTRODUCTORY STATISTICS AND LINEAR REGRESSSION

INTRODUCTION

Statistics have two functions: (1) to describe the variation in some data and (2) perform tests on sets of data to determine the cause of the variation in that data. As an example, we know that people of the same age are not the same height or weight. The set of heights for a population can be described by a statistic that describes the center (mean, median or mode) and by a statistic that describes the variation around that center. Typically the mean is used to describe the center and the standard deviation is used to describe the variation, or spread around that center. These are examples of descriptive statistics. Suppose we wanted to know the causes of variation in human height. Possible causes might include diet, genetics, sleep patterns. It is extremely difficult to determine the cause of variation in humans because these possible causes are not controlled. However, there are populations that differ in a variety of ways. We may ask the question, do two different populations have different average heights? These kinds of questions involve statistical tests.

THE MEAN

Suppose we have a set of variables $\{X_1, X_2, \ldots, X_i, \ldots, X_n\}$. The mean is *defined* as

$$[1.2.A2.1] \quad \overline{X} = \mu_X = \frac{\sum_{i=1}^{n} X_i}{n}$$

THE POPULATION VARIANCE

The variance is a measure of the spread of a population. It is the average of the squared deviations from the mean. The square is taken so that all deviations from the mean contribute to the variance; otherwise, negative deviations would cancel positive deviations: the average deviation from the mean is zero. For a population the variance is given by

$$[1.2.A2.2] \quad \sigma^2 = \frac{\sum_{i=1}^{n} (X_i - \overline{X})^2}{n}$$

THE SAMPLE VARIANCE

The sample variance is a measure of the spread of a population when some sub-set of the population is used so that the mean is estimated rather than known. Its formula is similar to that of the population variance, except that one "degree of freedom" is used for the estimation of the mean. The sample variance is calculated as

$$[1.2.A2.3] \quad s^2 = \frac{\sum_{i=1}^{n} (X_i - \overline{X})^2}{n - 1}$$

here s^2 is used as a symbol for the sample variance, whereas σ^2 is used as a symbol for the population variance. Note that the formula for s^2 is very similar to the formula for the variance except that $n - 1$ is used as the divisor instead of n, the number of measurements used to estimate the mean and variance of the sample. The term $(n - 1)$ is the **degrees of freedom** that refers to the number of independent pieces of information that goes into the estimation of the statistic. Because the mean is used in the calculations, each measurement contributes to the mean and, if you know the mean, the last value can be calculated from all of the other values. Thus the degrees of freedom when the data is used to estimate the mean is $n - 1$. Expanding the square term in Eqn [1.2.A2.3], we have

$$[1.2.A2.4] \quad s^2 = \frac{\sum_{i=1}^{n} (X_i^2 - 2\overline{X} X_i + \overline{X}^2)}{n - 1}$$

This is rewritten as

$$[1.2.A2.5] \quad s^2 = \frac{\sum_{i=1}^{n} X_i^2 - 2\overline{X} \sum_{i=1}^{n} X_i + n \overline{X}^2}{n - 1}$$

Recognizing from Eqn [1.2.A2.1] that

$$[1.2.A2.6] \quad \sum_{i=1}^{n} X_i = n \overline{X}$$

Inserting this into Eqn [1.2.A2.5] gives

$$[1.2.A2.7] \quad s^2 = \frac{\sum_{i=1}^{n} X_i^2 - 2n\overline{X}^2 + n\overline{X}^2)}{n - 1} = \frac{\sum_{i=1}^{n} X_i^2 - n\overline{X}^2}{n - 1}$$

Multiplying both numerator and denominator by n, we reach

$$[1.2.A2.8] \quad s^2 = \frac{n \sum_{i=1}^{n} X_i^2 - n^2 \overline{X}^2}{n(n - 1)}$$

Again making use of Eqn [1.2.A2.6] we arrive at the computational formula for the sample variance:

$$[1.2.A2.9] \quad s^2 = \frac{n \sum_{i=1}^{n} X_i^2 - \left(\sum_{i=1}^{n} X_i\right)^2}{n(n - 1)}$$

THE SAMPLE STANDARD DEVIATION

The sample standard deviation, denoted by the statistic s, is defined to be the positive square root of the sample variance:

$$[1.2.A2.10] \quad s = \sqrt{\frac{n \sum_{i=1}^{n} X_i^2 - \left(\sum_{i=1}^{n} X_i\right)^2}{n(n - 1)}}$$

THE STANDARD ERROR OF THE MEAN

The standard deviation of a sampling distribution of a statistic is called the **standard error** of that statistic. A sampling distribution of a statistic is its probability distribution. What this means is that if we estimate the mean of a population several times using different random samples, we will not get identical answers. There will be some distribution of the mean. In the same way, there will be some variation in our sample variance calculated from the different samples. The standard deviation of the distribution of means is the standard error of the mean. Note that the sample standard deviation does not change with the sample size, because each additional member of the sample contributes both to n and to the squared deviation from the mean. On the other hand, we ought to expect that our estimate of the mean improves with the number of observations in the sample. Thus, the standard error of the mean gets smaller as the number of observations in the sample increases. The formula for the standard error of the mean is

$$[1.2.A2.11] \quad SEM = \frac{s}{\sqrt{n}}$$

PROBABILITY

Probability theory began with the analysis of games of chance. The probability of an event happening, such as drawing a particular card, or a card that belongs to a particular suit, or the appearance of a particular face on a die after a roll, is defined as the ways in which the event can happen compared to the total possible number of outcomes. For example, there are six faces on a die and the likelihood that any particular face ends upright is one out six, or $1/6 = 0.1625$. We can write this as

$$P(1) = 0.1625;\ P(2) = 0.1625;\ P(3) = 0.1625;$$
$$P(4) = 0.1625; P(5) = 0.1625;\ P(6) = 0.1625$$

where $P(i)$ refers to the probability of the face with i dots (pips) on it. If we have two dice, the probability of having a total number of dots depends on how many outcomes produce that number of dots. For example, we can find 12 dots only by having 6 on one die and 6 on the other. There are a total of 36 possible outcomes as shown in Table 1.2.A2.1 below.

Since there are 36 possible outcomes, the probability of getting a 12 is $1/36 = 0.0278$. The probability of getting a total of 7, however, is higher. We can get a 7 from the following combinations: $\{(4,3),\ (3,4),\ (5,2),\ (2,5),\ (6,1),\ \text{and}\ (1,6)\}$. The probability of getting a 7 is thus $6/36 = 0.1625$. Note the probability of getting some outcome is 1.0, and this is the sum of each of the probabilities for the individual outcomes:

[1.2.A2.12]
$$\sum_{i=1}^{i=n} P(i) = 1.0$$

where there are a total of n distinguishable outcomes. The probability of observing a particular outcome for a roll of the dice is the product of the individual outcomes for each die. Thus, the probability of a $\{4,3\}$ outcome is given as

$$P(4,3) = P(4) \times P(3) = 1/6 \times 1/6 = 1/36$$

Here we are making the distinction of which die shows 4 and which shows 3. For example, if the dice were colored this would correspond to a red die with a 4 and a white die with a 3. If they are not distinguishable, then there are two ways to get (4,3), without regard to order. In this case $P(4,3) = P(3,4)$ and the probability of getting a 7 with a 4 and a 3 is $P\{4,3\} = P(4,3) + P(3,4) = 2/36$.

The validity of these conclusions depends on **fair** dice that are **independent**. A die is fair if and only if the probabilities of all six outcomes are the same, and they are independent if and only if the outcome of one die does not influence the outcome of the other.

HYPOTHESIS TESTING

Often in scientific investigations or engineering it is necessary to obtain data to test whether or not some treatment had an effect, or whether or not the data fit some equation whose derivation is based on some theory. The derivation of the equation may have required some assumptions that limit the conditions under which the equation is valid. An excellent fit of the data to the predicted values based on the theory gives us confidence that the theory is valid and the assumptions used were met. There is no guarantee that this is so. The philosophy of science tells us that all we can do is disprove hypotheses. If the data do not agree with the theory, one of two things must be true: either there is something "wrong" with the data or there is something wrong with the theory. The data must be taken at face value but perhaps the data is not what you think it is. For example, suppose you want to investigate the relationship between pressure and flow at steady-state and you make measurements of the flow and pressure in a system. You make some measurements, but do not realize that the system is not at steady-state. Thus the data cannot be analyzed using an equation that assumes steady-state. The data are not "wrong," but you are wrong in thinking that the data represent the steady-state condition. Failure of the equation to match your data is not a problem with the data or the validity of the data, it is a problem of trying to fit data to conditions that do not pertain to how the data was actually obtained. Other possible problems exist. For example, perhaps you measured the pressure with an incorrectly calibrated meter, so the pressures that you report are not the true pressures. There is generally far more ways of making mistakes than there are ways of making the measurements correctly.

TABLE 1.2.A2.1 Possible Outcomes from a Roll of Two Dice

Outcome of Second Die		Outcome of First Die					
		1	2	3	4	5	6
	1	(1,1)	(2,1)	(3,1)	(4,1)	(5,1)	(6,1)
	2	(1,2)	(2,2)	(3.2)	(4,2)	(5,2)	(6,2)
	3	(1,3)	(2,3)	(3,3)	(4,3)	(5,3)	(6,3)
	4	(1,4)	(2,4)	(3,4)	(4,4)	(5,4)	(6,4)
	5	(1,5)	(2,5)	(3,5)	(4,5)	(5,5)	(6,5)
	6	(1,6)	(2,6)	(3,6)	(4,6)	(5,6)	(6,6)

TABLE 1.2.A2.2 Type I and Type II Errors

Inference	Null Hypothesis	
	True	False
Reject H_0	Type I error	Correct inference
Accept H_0	Correct Inference	Type II error

So one kind of hypothesis is this: the derived equation is a good fit for the data. We can call this the "null hypothesis," H_0: there is no difference between the data and what is predicted from the equation. We suppose that the data is not a perfect fit. There are differences between the data and the predicted value for the data. If the differences are small, we may think that the fit is good and we are likely to accept the null hypothesis as being true. If the differences are large, we begin to think something is "wrong" with either the data or the equation, and we reject the null hypothesis as being false. It is possible to make two distinct types of errors in this testing of hypotheses. **The Type I error is the incorrect rejection of a true null hypothesis. The Type II error is the failure to reject a false null hypothesis.** These types of errors are clarified in Table 1.2.A2.2.

Another way of describing these is that the Type I error is a false positive: you conclude there is an effect when there is none present. A Type II error is a false negative. You fail to detect an effect when it is actually present.

The probability of making a Type I error is designated α. The probability of making a Type II error is designated as β. α is generally accepted as the level of significance of a statistical test of H_0. This is to say, we have to agree on how different the data can be from what is predicted from the null hypothesis before we can reject the null hypothesis. There is some probability that the differences that you obtained arose by random chance, and this is the probability of making a Type I error.

In other types of experiments we may want to know if some treatment or variable has some effect in a population of individuals. Suppose we do an experiment in which we measure some variable X for a set of individuals who are not treated, the control group, and we measure the same variable for a set of individuals who receive some treatment. Did the treatment affect the value of X? What we generally have is a set of measurements of variable X from the control group $G_0 = \{X_{01}, \ldots, X_{0n}\}$ and a second set of measurements from the experimental group $G_e = \{X_{E1}, \ldots, X_{En}\}$. The mean for G_0 is designated μ_0 and the mean for the experimental group is μ_E. The null hypothesis is H_0: $\mu_0 = \mu_E$. Generally the means will not be exactly equal. There is some probability that a given difference $\mu_0 - \mu_E$ will arise by chance. The t-statistic for comparison of two means is given as

$$[1.2.A2.13] \qquad t = \frac{\mu_0 - \mu_E}{\sqrt{\text{SEM}_0^2 + \text{SEM}_E^2}}$$

Clearly, the larger the value of t the more likely that the difference in the means does not arise by chance. If the difference arises by chance, then we would be wrong to reject the null hypothesis. The probability that the difference arises by chance, and we reject the null hypothesis, is the probability of the Type I error, and this is called the **level of significance** of the test. Typically it is accepted that $\alpha = 0.05$, which means there is a 5% chance that we will reject the null hypothesis even though it is true. Values of t for given values of α are published, and these depend on the **degrees of freedom** for the t-statistic. The degrees of freedom are the number of independent pieces of information that contribute to the estimation of the statistic, which is the t-value in this case. In a two-sample comparison such as the one described here, the degrees of freedom (symbolized as df or ν) is given as $n_0 + n_E - 2$. We subtract 2 because the means are derived from the data and so the means and the values themselves are not independent, and we have used 2 means in the calculation of the t-statistic.

We could apply these or other types of statistical test to groups of people to compare them and to make statistical statements about them, such as "men are more prone to heart attacks than women", or "black american males have a higher incidence of hypertension than white american males," or "obese white females with multiple children are most prone to gallstones." These statistical statements mean nothing when you are faced with a single incidence of the population: population averages and trends tells you nothing about any specific member of the group. You don't know, without making some measurements, how far away from average this specific person will be. However, we can make general statements about the populations that may guide us in understanding or treating a condition or setting public health policy.

THE NORMAL PROBABILITY DENSITY FUNCTION

A probability density function is one in which the probability of an outcome being in some interval is the product of the density function and the interval. Mathematically, it is given as

$$[1.2.A2.14] \qquad P(a \leq x \leq b) = \int_a^b p(x) dx$$

where $P(a \leq x \leq b)$ is the probability of x being between a and b and $p(x)$ is the probability density function. There are several kinds of probability distribution functions. We will derive the normal or gaussian probability density distribution by considering a random throw of a dart aimed at the origin of a Cartesian plane. We make some basic assumptions of the probability of the dart landing in a particular area. These are:

- the distribution is symmetrical: the probability of being high by some distance is the same as being low by the same distance, and this is equal to the probability of being right by the same distance, or by being left by the same distance. In fact, the probability of being off-center by some distance, r, is independent of the angle θ from the origin.

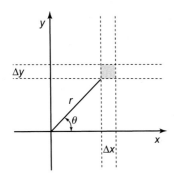

FIGURE 1.2.A2.1 Cartesian coordinate and polar coordinate systems for the analysis of the normal probability distribution function. The probability of a dart hitting the gray area, dxdy, is the probability of it landing in the vertical stripe of width Δx times the probability of it landing in the horizontal stripe with width Δy.

- errors in particular directions are independent. That is, the probability of being off-center by x in the horizontal direction does not alter the probability of being off-center by y in the vertical direction.
- large errors are less likely than small errors.
- our aim is unbiased. That is, on average, the distance from the origin, taking x and y as positive and negative, is zero.

Consider the Cartesian coordinate system shown in Fig. 1.2.A2.1. The probability of the dart landing in the vertical stripe between x and $x + \Delta x$ is given by $p(x)\Delta x$, where $p(x)$ is the probability density function. Similarly, the probability of the dart landing in the horizontal stripe between y and $y + \Delta y$ is $p(y)\Delta y$. What we want to do is determine the mathematical form of the function p.

From the independence assumption, the probability of the dart landing in the shaded region is the product of the probabilties of landing in the horizontal or vertical stripes: $p(x)\Delta x p(y)\Delta y$. Because of symmetry, the probability of the dart landing in the area the size of $\Delta x \Delta y$ located r distance from the origin does not depend on θ and we can write

[1.2.A2.15] $$g(r)\Delta x \Delta y = p(x)\Delta x\, p(y)\Delta y$$

which gives

[1.2.A2.16] $$g(r) = p(x)p(y)$$

Differentiating with respect to θ, we obtain

[1.2.A2.17] $$\frac{dg(r)}{d\theta} = 0 = p(x)\frac{dp(y)}{d\theta} + p(y)\frac{dp(x)}{d\theta}$$

$dg(r)/d\theta = 0$ because $g(r)$ is independent of θ. Inserting $x = r\cos\theta$ and $y = r\sin\theta$, we can re-write Eqn [1.2.A2.17] as

[1.2.A2.18] $$0 = p(x)\frac{dp(y)}{dy}\frac{dr\sin\theta}{d\theta} + p(y)\frac{dp(x)}{dx}\frac{dr\cos\theta}{d\theta}$$

$$0 = p(x)\frac{dp(y)}{dy}r\cos\theta - p(y)\frac{dpx}{dx}r\sin\theta$$

[1.2.A2.19] $$\frac{\frac{dp(x)}{dx}}{p(x)x} = \frac{\frac{dp(y)}{dy}}{p(y)y}$$

This differential equation is true for all x and y, and x and y are independent. This can happen only if the ratio defined on each side of the equation is constant. We let

$$\frac{\frac{dp(x)}{dx}}{p(x)x} = C$$

[1.2.A2.20] $$\frac{dp(x)}{p(x)} = Cx\,dx$$

$$\int \frac{dp(x)}{px} = \int Cx\,dx$$

$$\ln p(x) = C\frac{x^2}{2} + c$$

Taking the exponent of e of both sides of the last equation gives

[1.2.A2.21] $$p(x) = Ae^{\frac{k}{2}x^2}$$

Because large x is less likely than small x, we know that C must be negative and we can re-write this equation as

[1.2.A2.22] $$p(x) = Ae^{-\frac{k}{2}x^2}$$

We can evaluate A by the requirement that the total probability for all outcomes is 1.0. This is expressed mathematically as

[1.2.A2.23] $$\int_{-\infty}^{\infty} Ae^{-\frac{k}{2}x^2}\,dx = 1.0$$

Since A is a constant, it can be removed from the integrand. Due to the symmetry of the problem, this integral is twice the integral from zero to infinity. We write this as

[1.2.A2.24] $$\int_0^{\infty} e^{-\frac{k}{2}x^2}\,dx = \frac{1}{2A}$$

We can combine this with the distribution about y, as these are symmetrical:

[1.2.A2.25] $$\int_0^{\infty} e^{-\frac{k}{2}x^2}\,dx \cdot \int_0^{\infty} e^{-\frac{k}{2}y^2}\,dy = \frac{1}{4A^2}$$

Since x and y are independent, we can re-write this product of integrals as the integral of their products:

[1.2.A2.26] $$\int_0^{\infty}\int_0^{\infty} e^{-\frac{k}{2}(x^2+y^2)}\,dx\,dy = \frac{1}{4A^2}$$

This can be converted to polar coordinates, recognizing that $(x^2 + y^2) = r^2$ and $dxdy = rdrd\theta$. We obtain

[1.2.A2.27] $$\int_0^{\frac{\pi}{2}}\int_0^{\infty} e^{-\frac{k}{2}r^2} r\,dr\,d\theta = \frac{1}{4A^2}$$

We can evaluate the integrals by making the substitution $u = -kr^2/2$. It follows that $du = -krdr$, and thus $-du/k = rdr$. Making these substitutions in the interior integral, we derive

[1.2.A2.28] $$\int_0^{\frac{\pi}{2}} \frac{-1}{k}\left[\int_0^{-\infty} e^u \, du\right] d\theta = \frac{1}{4A^2}$$

Evaluation of the interior integral is -1. Eqn [1.2.A2.28] thus becomes

[1.2.A2.29] $$\int_0^{\frac{\pi}{2}} \frac{1}{k} d\theta = \frac{\pi}{2k} = \frac{1}{4A^2}$$

The value of A is thus given as

[1.2.A2.30] $$A = \sqrt{\frac{k}{2\pi}}$$

The probability distribution in Eqn [1.2.A2.22] becomes

[1.2.A2.31] $$p(x) = \sqrt{\frac{k}{2\pi}} e^{-\frac{k}{2}x^2}$$

We can evaluate k from the variance of the probability distribution. The average value for x is given as

[1.2.A2.32] $$\mu = \int_{-\infty}^{\infty} x \, p(x) \, dx$$

The variance, similar to our earlier definition in Eqn [1.2.A2.2] is given as

[1.2.A2.33] $$\sigma^2 = \int_{-\infty}^{\infty} (x - \mu)^2 \, p(x) \, dx$$

Because of the symmetry of the coordinates, and the fact that $p(x) = p(-x)$, we know that the mean, μ, is zero. With Eqn [1.2.A2.31], and $\mu = 0$, the variance is given as

[1.2.A2.34] $$\sigma^2 = 2\sqrt{\frac{k}{2\pi}} \int_0^{\infty} x^2 e^{-\frac{k}{2}x^2} dx$$

We use symmetry as we did before to integrate from 0 to ∞:

[1.2.A2.35] $$\sigma^2 = 2\sqrt{\frac{k}{2\pi}} \int_0^{\infty} x^2 e^{-\frac{k}{2}x^2} dx$$

We can integrate this by parts by identifying

[1.2.A2.36] $$u = x \quad dv = x e^{-\frac{k}{2}x^2} dx \quad v = -\frac{1}{k} e^{-\frac{k}{2}x^2}$$

Substituting these into Eqn [1.2.A2.35] we obtain

$$\sigma^2 = 2\sqrt{\frac{k}{2\pi}} \int_0^{\infty} x^2 e^{-\frac{k}{2}x^2} dx$$

$$\sigma^2 = 2\sqrt{\frac{k}{2\pi}} \int_0^{\infty} u \, dv$$

$$\sigma^2 = 2\sqrt{\frac{k}{2\pi}} \left[uv \Big|_0^{\infty} - \int_0^{\infty} v \, du \right]$$

$$\sigma^2 = 2\sqrt{\frac{k}{2\pi}} \left[-\frac{x}{k} e^{-\frac{k}{2}x^2} \Big|_0^{\infty} + \int_0^{\infty} \frac{1}{k} e^{-\frac{k}{2}x^2} dx \right]$$

[1.2.A2.37]

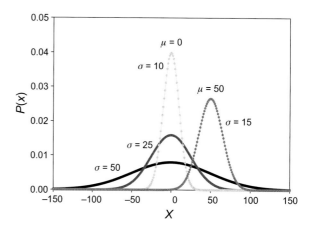

FIGURE 1.2.A2.2 Gaussian probability distribution functions for varying values of σ and μ. Different values of μ move the distribution to the right or to the left; different values of σ influence the shape of the curve, with smaller values resulting in sharper curves and larger values creating more spread out curves.

The first term in brackets is evaluated from $x = 0$ to $x = M$ in the limit as $M \to \infty$, and it is zero. The second term in the brackets has already been done in this derivation (Eqn [1.2.A2.24] and Eqn [1.2.A2.30]). This last equation in Eqn [1.2.A2.37] becomes

[1.2.A2.38] $$\sigma^2 = 2\sqrt{\frac{k}{2\pi}} \left[0 + \frac{1}{k} \frac{\sqrt{2\pi}}{2\sqrt{k}} \right] = \frac{1}{k}$$

This last equation gives $k = 1/\sigma^2$. Inserting this into the probability density function (Eqn. [1.2.A2.31]) gives

[1.2.A2.39] $$p(x) = \frac{1}{\sigma^2 \sqrt{2\pi}} e^{-2\sigma^2}$$

This is the normal probability distribution centered at $\mu = 0$. The general equation with mean μ is achieved by a horizontal shift in x:

[1.2.A2.40] $$p(x) = \frac{1}{\sigma^2 \sqrt{2\pi}} e^{-\frac{(x-\mu)^2}{2\sigma^2}}$$

Calculated Gaussian or normal probability distribution functions are shown in Figure 1.2.A2.2. Note that the value of σ determines the spread of the distribution. Smaller values of σ result in a sharper distribution.

LINEAR REGRESSION

For linear regression, the data we're interested in is generally numerical and comes in sets of ordered pairs: $\{(x_1,y_1), \ldots, (x_n,y_n)\}$. What we desire to know is what is the best linear fit to the ordered pairs? There are many different ways of doing this. What we are going to go over is the **least squares linear regression**. The first thing we are going to do is to make a few assumptions. These are:

1. there is no error in the values of X_i. All of the error is in the values of Y_i
2. the values of Y_i are distributed normally. That is, they obey the normal probability distribution. This is the same as the Gaussian probability distribution and is commonly referred to as the "bell curve". This curve has been described in Eqn [1.2.A2.40]

We introduce a model equation

[1.2.A2.41] $$\hat{Y} = mX = b$$

where $\hat{Y}$ is the predicted value of Y by this model equation. Since we have assumed that the values of X are determined perfectly, with no error, all of the error is in the set of Y_i and how it differs from the predicted values, $\hat{Y}_i$. What we want to do is minimize the squared error between the set of Y_i and $\hat{Y}_i$. The total squared error is

[1.2.A2.42] $$\text{Error} = \sum_i (Y_i - \hat{Y}_i)^2$$

Substituting in from Eqn [1.2.A2.41] we have

[1.2.A2.43] $$\text{Error} = \sum_i (Y_i - mX_i - b)^2$$

The error estimated as the square of the deviations from the observed values, the set of Y_i, from the predicted values, $mX_i + b$, varies with m and b used in the calculations. We require this sum of square errors to be minimized in order to have the best fit line. We achieve this by looking for minima in the square errors as we vary m and b. That is, we look for the m and b that has the least squared error. This occurs when the partial derivatives of the error with respect to m and b are minima:

[1.2.A2.44]
$$\frac{\partial \sum_i (Y_i - mX_i - b)^2}{\partial m} = 0$$
$$\frac{\partial \sum_i (Y_i - mX_i - b)^2}{\partial b} = 0$$

These two form a pair of simultaneous equations in m and b; all of the pairs of (x_i, y_i) are known. Expanding the square term, we get

$$\frac{\partial \sum_i (Y_i^2 - 2mY_iX_i - 2Y_ib + m^2X_i^2 + 2mbX_i + b^2)}{\partial m} = 0$$

$$\frac{\partial \sum_i (Y_i^2 - 2mY_iX_i - 2Y_ib + m^2X_i^2 + 2mbX_i + b^2)}{\partial b} = 0$$

[1.2.A2.45]

Performing the partial differentiation gives

[1.2.A2.46]
$$2m \sum_i X_i^2 + 2b \sum_i X_i - 2 \sum_i Y_iX_i = 0$$
$$2m \sum_i X_i + 2nb - 2 \sum_i Y_i = 0$$

canceling out the common factors of 2, this is recognized as a system of two simultaneous equations in two unknowns:

[1.2.A2.47]
$$\sum_i X_i^2 m + \sum_i X_i b = \sum_i Y_iX_i$$
$$\sum_i X_i m + nb = \sum_i Y_i$$

This system of simultaneous equations in two unknowns (m and b) can be solved by application of Cramer's Rule:

[1.2.A2.48] $$\frac{\begin{vmatrix} \sum_i X_iY_i & \sum_i X_i \\ \sum_i Y_i & n \end{vmatrix}}{\begin{vmatrix} \sum_i X_i^2 & \sum_i X_i \\ \sum_i X_i & n \end{vmatrix}} = m$$

and the expression for b is

[1.2.A2.49] $$\frac{\begin{vmatrix} \sum_i X_i^2 & \sum_i X_iY_i \\ \sum_i X_i & \sum_i Y_i \end{vmatrix}}{\begin{vmatrix} \sum_i X_i^2 & \sum_i X_i \\ \sum_i X_i & n \end{vmatrix}} = b$$

Evaluating the determinants in the numerator and denominator for m in Eqn [1.2.A2.48], we get the computational formula for the slope of the least-squares best fit line:

[1.2.A2.50] $$m = \frac{n \sum_i X_iY_i - \sum_i X_i \sum_i Y_i}{n \sum_i X_i^2 - \sum_i X_i \sum_i X_i}$$

Evaluating the determinants from Eqn [1.2.A2.49] we get

[1.2.A2.51] $$b = \frac{\sum_i X_i^2 \sum_i Y_i - \sum_i X_i \sum_i X_iY_i}{n \sum_i X_i^2 - \sum_i X_i \sum_i X_i}$$

By algebraic manipulation, it can be shown that this last equation for the intercept of the least-squares line is given also as

[1.2.A2.52] $$b = \bar{Y} - m\bar{X}$$

Physical Foundations of Physiology II: Electrical Force, Potential, Capacitance, and Current   1.3

Learning Objectives

- Write Coulomb's law for electrostatic forces
- Define electrical potential at x as the work done moving a unit positive charge from infinity to x
- Write three equivalent but different descriptions of a conservative force
- Define electric field as the electric force per unit charge
- Describe the electric field as the negative gradient of the potential
- Recognize Gauss's law
- Write the formula for capacitance in terms of charge and voltage
- Write the formula for capacitance in terms of area, dielectric constant, and plate separation
- Describe how capacitance varies with area, dielectric, and plate separation
- Be able to calculate the capacitance of biological membranes given k, δ, and physical dimensions
- Be able to calculate electric field intensity and force on a charged particle given $V(x, y, z)$
- Write Kirchhoff's Current Law and Kirchhoff's Voltage Law
- Be able to calculate the time constant for a simple RC circuit.

COULOMB'S LAW DESCRIBES ELECTRICAL FORCES

Electric charge is a fundamental property of some subatomic particles. Electrons have negative charge and protons have positive charge. These designations of positive and negative are arbitrary but rigidly accepted by convention. Separated electrical charges in a vacuum (see Figure 1.3.1) experience a force that is described by Coulomb's law:

$$[1.3.1] \quad F = \frac{q_1 q_2}{4\pi\varepsilon_0 r^2}$$

$$\mathbf{F}_{1 \text{ on } 2} = \frac{q_1 q_2 \mathbf{r}_{12}}{4\pi\varepsilon_0 r^3} = -\mathbf{F}_{2 \text{ on } 1}$$

The bold face symbols signify vector quantities. $\mathbf{F}$ is the force; q_1 and q_2 are electrical point charges, in coulombs, that are separated by the distance r, in meter; ε_0 is a constant, the electrical permittivity of space, which has the value of $8.85 \times 10^{-12} \, \text{C}^2 \, \text{N}^{-1} \, \text{m}^{-2}$. The lower equation signifies that the direction of the force is along the line between the two charges. The magnitude of the force is proportional to the product of the charges and inversely proportional to the square of their separation. Its sign depends on the signs of the point charges. Two positive charges, or two negative charges, result in a positive force which is repulsive and directed away from their center as indicated in Figure 1.3.1. Two charges of opposite sign experience a negative force which is attractive along a line connecting them.

If the intervening space is not vacuum, but some medium, the equation is altered slightly by the inclusion of a **dielectric constant**, κ, whose value depends on the medium:

$$[1.3.2] \quad \mathbf{F}_{1 \text{ on } 2} = \frac{q_1 q_2 \mathbf{r}_{12}}{\kappa 4\pi\varepsilon_0 r^3}$$

The dielectric constant for the vacuum is 1.0. For all other materials, $\kappa > 1.0$. The reduced force in the presence of a dielectric material is due to charges present in the material that reorient themselves in the presence of the external point charges and thereby screen the charges from each other. Materials with asymmetric charge distributions within their materials typically have large dielectric constants.

THE *ELECTRIC POTENTIAL* IS THE WORK PER UNIT CHARGE

Suppose there is a positive charge of magnitude q_{fixed} fixed in space at some location. We have another charge, a unit positive charge, located infinitely far away so that the force between the charges initially is effectively zero. If we bring the unit positive charge q_{test} toward the fixed charge at a constant velocity, then its kinetic energy does not change. As we approach the fixed charge, the repulsive force becomes larger and larger and we must apply an external force to keep the q_{test} at constant velocity. Because our applied external force, $\mathbf{F}_{\text{ext}}$, has moved through a distance, we have performed work on the body, given as

$$[1.3.3] \quad \text{Work}_{i \Rightarrow f} = \int_i^f \mathbf{F}_{\text{ext}} \cdot d\mathbf{s}$$

This work is the amount of energy we have expended in moving the positive q_{test} toward the positive q_{fixed}. Where did that energy go? If we release q_{test}, we find that it moves away from q_{fixed} and gains kinetic energy which is exactly equal to the energy we used to move

© 2017 Elsevier Inc. All rights reserved.
DOI: http://dx.doi.org/10.1016/B978-0-12-800883-6.00003-3

q_{test} toward q_{fixed}. We say that the energy we used to move q_{test} was stored as **potential energy**, U. We define the potential at any place A in space as being the work done to bring a positive unit charge from infinite separation to point A:

[1.3.4] $$U_A = \frac{\text{Work}_{\infty \Rightarrow A}}{q_{test}} = \int_\infty^A \frac{\mathbf{F}_{ext} \cdot \mathbf{ds}}{q_{test}}$$

The unit of potential here is joules coulomb^{-1} = volts. The usefulness of the potential is that the work can be determined easily by multiplying the potential times the charge. $\mathbf{F}_{ext}$ in this equation is the external force required to move the positive test charge with no change in velocity. It is exactly equal to and opposite in sign to $\mathbf{F}_{int}$, the interacting electrostatic force. This is, in turn, given by Coulomb's law (see Eqn [1.3.1]). Since it is directed along $\mathbf{ds}$, we can write

[1.3.5] $$U_A = -\int_\infty^A \frac{\mathbf{F}_{int} \cdot \mathbf{ds}}{q_{test}} = -\int_\infty^A \frac{qq_{test}}{4\pi\varepsilon_0 r^2 q_{test}} dr$$

$$U_A = \frac{1}{4\pi\varepsilon_0} \frac{q}{r_A}$$

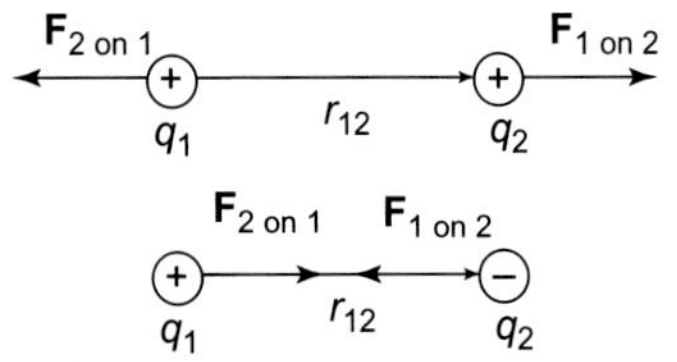

FIGURE 1.3.1 Electrical forces between separated point charges. q indicates charge. Like charges repel, so that the force of q_1 on q_2 is directed away from q_1 on a line connecting their centers. The force of q_1 on q_2 is exactly opposite to the force of q_2 on q_1. Unlike charges attract with forces opposite but in line with the vector connecting their centers.

From this definition of the potential, it should be clear that the potential surrounding a positive charge is positive: it takes work to bring a positive charge toward it. The potential surrounding a negative charge is negative, as we can get energy out of bringing a positive charge toward it. These lead us to an important conclusion: **a separation of charge produces an electric potential**. The potential defined in this way is a scalar quantity, having magnitude but not a direction, whereas the electrical force is a vector. This comes about from integrating the dot product of $\mathbf{F}_{int}$ with $\mathbf{ds}$ where $\mathbf{ds}$ is the distance increment that points along the pathway taken from infinite separation to point A. The dot product means that we add only those components of the force that are directed on the line connecting the centers of the charges. These conclusions are illustrated in Figure 1.3.2.

THE IDEA OF POTENTIAL IS LIMITED TO *CONSERVATIVE* FORCES

THE CONSERVATION OF ENERGY THEOREM STATES THAT ENERGY MAY BE CONVERTED BUT NOT DESTROYED

The First Law of Thermodynamics is the conservation of energy theorem. It states that in ordinary mechanical events, the total energy is constant. It is written in differential form as

[1.3.6] $$dE = dq - dw$$

where E is the total energy, q in this case is the heat energy, and w is the work. This is another unfortunate case where variables are used to denote completely different quantities. In thermodynamics, q symbolizes

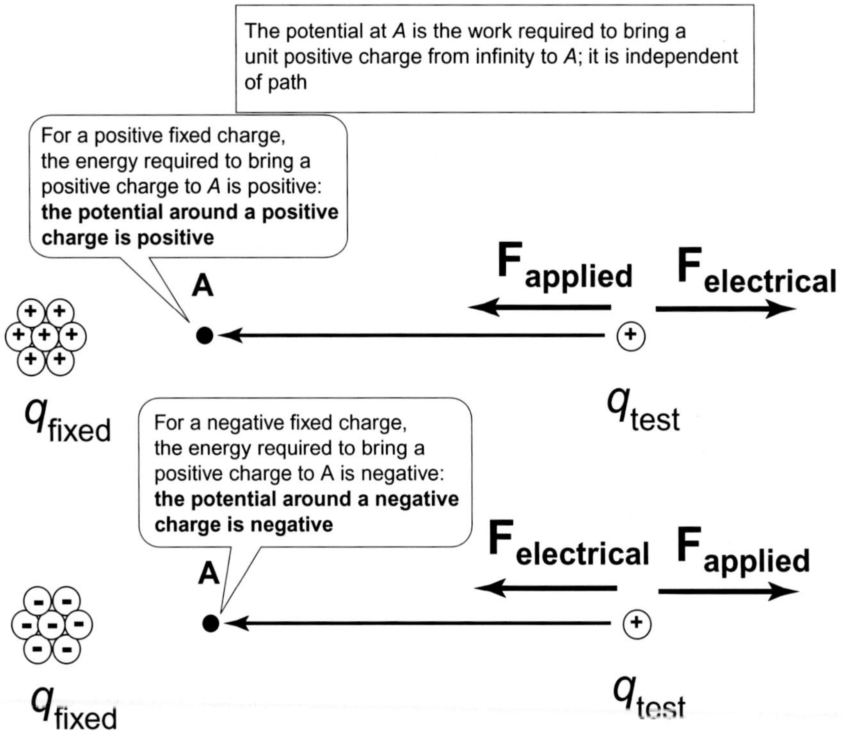

FIGURE 1.3.2 Definition of the electrical potential. The potential at a point A is defined as the work required to bring a unit positive charge (q_{test}) from infinite separation to point A. If there is a fixed positive charge near A, it takes work to bring q_{test} to A (we must apply a force to overcome the repulsive force and we move that force through a distance) and the potential is positive. If there is a fixed negative charge near A, then q_{test} is attracted to it and it takes energy (work) to slow q_{test}—the work is negative because the applied force is opposite to the direction of movement.

heat, and in electrostatics it symbolizes charge. The appearance of heat in this equation is extremely important, because it turns out that there are theoretical limits in the conversion of heat energy to useful work, and this gives rise to the concepts of entropy and free energy.

This equation assumes that heat and work are alternate forms of energy. We take the equivalence of mechanical, thermal, chemical, and electrical energies for granted but historically this idea took some time to develop. In writing that any energy change in the system is the balance between work output and heat input, it is assumed that work is equivalent to heat. The equality of mechanical work and heat was established in 1845 by Joule.

The signs of dq and dw in this equation are important, and are consequences of the definitions of heat and work. The quantity dq is *defined* as the heat absorbed *by* the system from its surroundings, and dw is *defined* as the work done *by* the system on its surroundings (see Figure 1.3.3). The "system" here is anything we have drawn a conceptual line around, usually in agreement with some physical boundary, that sets part of the universe off from the rest of it. In the case of electrostatics, the system is the set of charges distributed in space.

THE WORK DONE BY A CONSERVATIVE FORCE IS PATH INDEPENDENT

The work done by moving q_{test} toward q_{fixed} is the work done by the surroundings (us) on the system of interacting charged particles. It is equal but opposite in sign to the work done by the interacting force. There may be heat generated by the necessity to apply more force than the interacting force in order to overcome friction, if the charges move through some medium, but this is separate from the interacting force (the coulombic or electrostatic force) itself. The coulombic force itself generates no heat at all. It belongs to a class of forces called **conservative forces** that do not dissipate energy as heat. Conservative forces are characterized by three equivalent statements:

1. The work done by a conservative force depends only on the initial and final positions, and not on the path (see Figure 1.3.4).
2. The potential difference between two points depends only on the end points and not the path.
3. The total work done by a conservative force acting around a closed loop is zero.

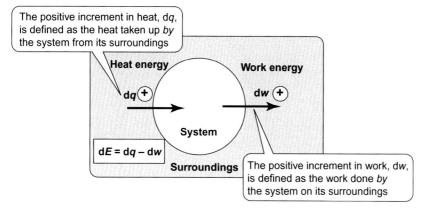

FIGURE 1.3.3 Theorem of the Conservation of Energy. The system is any part of the universe that we have enclosed by some boundary, real or imagined. Positive heat flow is defined as heat energy that is absorbed by the system from its surroundings. Similarly, positive work is defined as work that is done by the system on its surroundings. By these definitions, conservation of energy means that dE = dq − dw. In order to write this equation, it is assumed that heat and work have the same units, that of energy. Here work can be electrical, mechanical, or chemical.

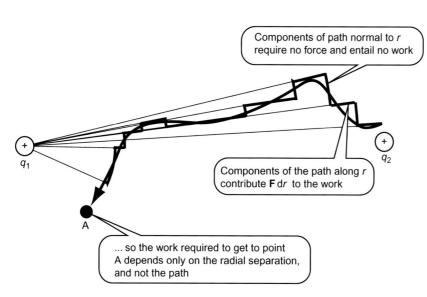

FIGURE 1.3.4 Potential is independent of the path. Any path from start to finish can be successively approximated by a series of paths oriented either parallel to the vector connecting the charges or perpendicular to it. Those components of the path perpendicular to the vector require no force and therefore contribute nothing to the potential at point A, which is defined as the work necessary to bring a unit positive charge (here shown as q_2) from infinite separation to point A. Components of the path oriented parallel to the vector connecting the point charges contribute **F** dr to the force. Therefore, the total work (potential) moving the charge depends only on the radial separation and not the path taken.

POTENTIAL DIFFERENCE DEPENDS ONLY ON THE INITIAL AND FINAL STATES

If the potential depends only on the position, then it is a state function, one that is independent of path and dependent only on the **state** of the system. Thus we can associate a potential with any position, A and B:

[1.3.7]
$$U_A = -\int_\infty^A \frac{\mathbf{F}_{int} \cdot \mathbf{ds}}{q_{test}}$$
$$U_B = -\int_\infty^B \frac{\mathbf{F}_{int} \cdot \mathbf{ds}}{q_{test}}$$
$$U_{A \Rightarrow B} = -\int_A^B \frac{\mathbf{F}_{int} \cdot \mathbf{ds}}{q_{test}} = U_B - U_A$$

If A = B, where A is the initial state and B is the final state, we can write

[1.3.8]
$$U_{initial \Rightarrow final} = 0 = -\oint_{initial}^{final} \mathbf{F}_{int} \cdot \mathbf{ds}$$

This is the mathematical statement that the work performed by the system around any closed loop is zero. This turns out to be equivalent to the statement that the potential is a function of position (state) only and not of the path used to get to that position.

THE *ELECTRIC FIELD* IS THE NEGATIVE GRADIENT OF THE POTENTIAL

The **electric field intensity** is *defined* as the electric force per unit charge:

[1.3.9]
$$\mathbf{E} = \frac{\mathbf{F}_{int}}{q_{test}}$$

Insertion of this into Eqn [1.3.7] and differentiating, we get

[1.3.10]
$$\mathbf{E} = -\frac{dU}{dS}$$

This equation is not correct as written yet, because we have a vector (the electric field intensity) on one side and a scalar on the other! We need to take a particular kind of derivative, the **gradient**, to convert the scalar potential into a vector force. Equation [1.3.10] is correct as written as long as the axis of **ds** corresponds with the direction of **F**. The full three-dimensional vector equation is

[1.3.11]
$$\mathbf{E} = -\mathbf{i}\frac{\partial U}{\partial x} - \mathbf{j}\frac{\partial U}{\partial y} - \mathbf{k}\frac{\partial U}{\partial z}$$
$$\mathbf{E} = -\nabla U$$

where **i**, **j**, and **k** are unit vectors in the *x*, *y*, and *z* directions. The expression on the right-hand side of Eqn [1.3.11] is called the **gradient** of the function *U*. It is a vector whose components on each axis are the slope of the potential projected onto that axis. Generally, the gradient is a vector that does not align with any axis. Instead, it points in the direction of the steepest slope of the potential surface in three dimensions. The force points down this slope. It is the negative of the gradient of the potential. The last equation shows the gradient written in operator notation. The operator ∇ is called **del** and is defined as

[1.3.12]
$$\nabla = \mathbf{i}\frac{\partial}{\partial x} + \mathbf{j}\frac{\partial}{\partial y} + \mathbf{k}\frac{\partial}{\partial z}$$

FORCE AND ENERGY ARE SIMPLE CONSEQUENCES OF POTENTIAL

The usefulness of potential is that it simplifies the idea of force and energy. The force on a charge *q* is given easily by multiplying it times the electric field, which is $-\text{grad } U$. The energy cost in moving a charge from one potential to another is just $q\Delta U$:

[1.3.13]
$$\mathbf{F} = -q\nabla U$$
$$\Delta \text{Energy} = q\Delta U$$

where ∇ is the del operator and Δ signifies the difference between final and initial states. The electric potential, *U*, here is in units of volts. If charge is in coulombs, the force is in units of coulomb-volt per meter; a volt-coulomb is a joule = 1 N m = 1 kg m^2 s^{-2}; therefore, the force is in units of 1 N m/m = N. The energy is in units of joules. From now on we will abandon use of *U* as a symbol for the potential; physiologists typically use *V*, *E*, or ψ as symbols of potential.

GAUSS'S LAW IS A CONSEQUENCE OF COULOMB'S LAW

Gauss's law is written as

[1.3.14]
$$\oint \mathbf{E} \cdot \mathbf{ds} = \frac{q}{\varepsilon_0}$$

where the integral is taken over *any* closed surface of the dot product of the electric field and the area vector, equal to the area increment **ds** and oriented perpendicular to the surface. What this says is that this dot product, summed over any closed surface, is equal to the charge enclosed by the surface divided by ε_0, the electrical permittivity of space. If there is no enclosed charge, the surface integral is zero. This equation is a variant of Coulomb's law (see Eqn [1.3.1]). To see how this equation works, we consider a spherically symmetrical distribution of positive charges as shown in Figure 1.3.5.

The evaluation of the surface integral is simplified by choosing an appropriate surface. In this case, we choose a sphere centered on the symmetrical charge. By symmetry, the electric field is directed radially outward, pointing along the vector **ds**. Similarly, the electric field is everywhere constant in magnitude at a

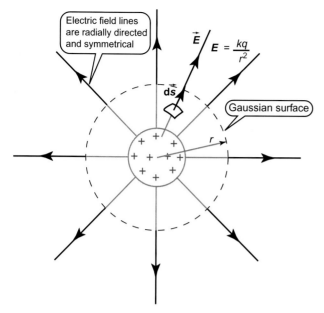

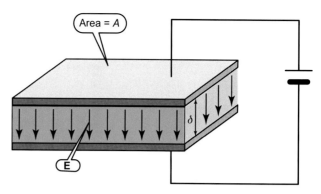

FIGURE 1.3.6 A parallel plate capacitor. Two plates, each of area A, are separated by a distance δ. They are charged by connecting them to a battery that produces a capacitance current until the potential difference between the two plates is equal to that across the two poles of the battery, so that the net potential difference across the entire circuit loop is zero. At this point, there is no more current flow. The separation of charges produces a uniform electric field between the two plates.

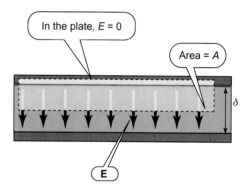

FIGURE 1.3.5 Electric field surrounding a spherically symmetrical distribution of positive charge. d**s** is a vector having a magnitude of the area increment and directed normal to the closed surface. In this case, we take the Gaussian surface, indicated here by a *dashed line*, to be a sphere centered on the symmetrically distributed charge. The electric field vector and the surface normal vector are pointing in the same direction, so that the angle between them, θ, is zero and the dot product of **E** and d**s** is **E** ds, because cos $\theta = 1$.

prescribed distance, r, from the center of the charged body. Thus we can write

[1.3.15]
$$\oint E \cdot ds = E 4\pi r^2 = \frac{q}{\varepsilon_0}$$
$$E = \frac{q}{4\pi r^2 \varepsilon_0}$$

which is the magnitude of the electric field (**E** = **F**/q, the electric force per charge) from Coulomb's law (see Eqn [1.3.1]).

THE *CAPACITANCE* OF A PARALLEL PLATE CAPACITOR DEPENDS ON ITS AREA AND PLATE SEPARATION

As described in Chapter 1.2, the ability to store electric charge is characterized by the capacitance, *defined* as

[1.3.16]
$$C = \frac{Q}{V}$$

where C is the capacitance, Q is the charge, and V is the potential, in volts. We now consider a particular type of device to store charge, a parallel plate capacitor, as described in Figure 1.3.6.

The two charged plates will be attracted to each other and so must be held apart by some dielectric material that insulates the plates and keeps the charges separated. There will be some fringing of the electric field

FIGURE 1.3.7 Parallel plate capacitor with a Gaussian surface. The Gaussian surface is the box indicated by the *dashed lines*. The electric field is constant within the capacitor and oriented as shown. The integral of **E** d**s** in the plate is zero because **E** is zero there. The integral of **E** d**s** in the dielectric between the plates is EA. The integral of **E** d**s** on the sides of the enclosed surface is zero because **E** and d**s** are orthogonal in this region.

around the edges of the plate, which we shall ignore. The resulting electric field within the capacitor is uniform, which can be proved by integrating the Coulomb force over the uniformly distributed charge on a plane, which we will not do here. We draw a rectangular closed surface, one side of which is in the middle of the dielectric and the other in the middle of the plate. Since the plate is a good conductor, the electric field within the plate is zero—the voltage difference in the plate is zero. The closed surface integral is just the constant electric field times the area of the surface in the dielectric. The situation is illustrated in Figure 1.3.7.

Application of Gauss's law according to the description in the legend of Figure 1.3.5 gives

[1.3.17]
$$\oint \mathbf{E} \cdot d\mathbf{s} = \mathrm{E}A = \frac{q}{\varepsilon_0}$$
$$\mathbf{E} = \frac{q}{\varepsilon_0 A}$$

Since the electric field is uniform, its relation to V from Eqn [1.3.10] is given as

$$[1.3.18] \qquad E = -\frac{dV}{dx} = \frac{V_1 - V_2}{\delta}$$

The capacitance is calculated as

$$[1.3.19] \qquad C = \frac{q}{V_1 - V_2}$$

Substituting in for q from Eqn [1.3.17] and for $V_1 - V_2$ from Eqn [1.3.18], we have

$$[1.3.20] \qquad C = \frac{\varepsilon_0 A}{\delta}$$

The presence of a dielectric between the plates reduces E for a given charge, and therefore increases the capacitance. The formula for parallel plates with a dielectric is

$$[1.3.21] \qquad C = \frac{\kappa \varepsilon_0 A}{\delta}$$

where κ is the dielectric constant, a dimensionless ratio.

According to Eqn [1.3.21], the capacitance increases linearly with the area and inversely with the separation between the plates, and is increased by materials with high dielectric constants.

BIOLOGICAL MEMBRANES ARE ELECTRICAL CAPACITORS

Biological membranes share some of the features of parallel plate capacitors and act as electrical capacitors. Their structure is detailed in Chapter 2.4. Briefly, biological membranes consist of an asymmetric bilayer of lipid molecules that assemble to form an interior insulating core. This effectively separates two plates—the surfaces of each bilayer—from each other. The separation distance is typically quite small, on the order of 7 nm. This bilayer structure is shown in Figure 1.3.8. From it, you can see the resemblance of the bilayer to a parallel plate capacitor.

The dielectric constant of some materials is shown in Table 1.3.1. This constant varies with temperature and the chemical make-up of the dielectric. Materials that are polar and mobile, such as water, can orient their partial charges with the electric field, and reduce the field within the dielectric. In this way, more charge can be added to the surfaces of the plates and therefore these dielectrics have a high dielectric constant.

TABLE 1.3.1 Dielectric Constant of Some Materials

Material	Dielectric Constant, κ
Air	1.00059
Water	80
Glycerol	43
Acetic acid	6.2
Benzene	2.3
CCl$_4$	2.2
Oleic acid	2.46
Hexanol	13.3
Hexane	1.89
Stearic acid	2.29
Monopalmitin	5.34

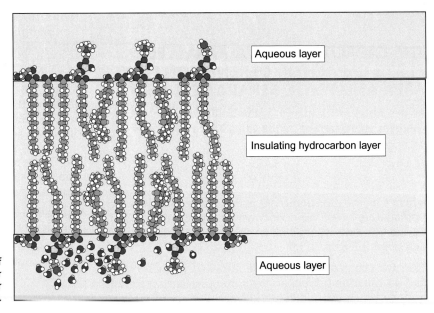

FIGURE 1.3.8 Lipid bilayer membrane consisting of various lipid molecules arranged with their hydrocarbon tails toward the interior of the bilayer and their water-soluble parts facing the water phase.

EXAMPLE 1.3.1 Capacitance of Planar Lipid Bilayers

Dr. Alexandre Fabiato at VCU drilled a narrow and clean hole with a diameter of 250 μm = 0.25 mm into a Lexan partition that separated two electrolyte solutions. He "painted" some phospholipids over the hole using a Teflon stick cut at an angle and dipped into a solution of lipids dissolved in hexane. After "thinning" (passive removal of the hexane through the aqueous phase), the membranes form a planar lipid bilayer (see Figure 1.3.9). Dr. Fabiato measured the capacitance of the membrane using an AC signal connected to electrodes immersed in the solutions. He derived a capacitance of 350 pF (350×10^{-12} F). Calculate the specific capacitance of the membrane Dr. Fabiato made, in F cm^{-2}.

The specific capacitance is just the capacitance per unit area of membrane: $C_m = C/A$. The measured capacitance is 350×10^{-12} F and the area A is πr^2, where $r = 0.0125$ cm; therefore,

$$C_m = 340 \times 10^{-12} \text{ F}/4.9 \times 10^{-4} \text{ cm}^2 = \mathbf{0.71 \ \mu F \ cm^{-2}}$$

What Is the Approximate Thickness of the Bilayer?
Equation [1.3.19] allows us to calculate the thickness as $\delta = \kappa \varepsilon_0 / C_m$, where C_m is the specific capacitance. We do not know κ, but the dielectric constant for lipid-like substances has been determined, as examples shown in Table 1.3.1. Here we use $\varepsilon_0 = 8.85 \times 10^{-12}$ C^2 J^{-1} m^{-1} and $C_m = 0.71 \times 10^{-6}$ C V^{-1} cm^{-2} that we calculated earlier. Using the dielectric constant for n-hexane as an example, we calculate

$$\delta = 1.89 \times 8.85 \times 10^{-12} \text{ C}^2 \text{ J}^{-1} \text{ m}^{-1}/0.71$$
$$\times 10^{-6} \text{ CV}^{-1} \text{ cm}^{-2} \times (100 \text{ cm m}^{-1})^2 = \mathbf{2.36 \times 10^{-9} \text{ m}}$$

The calculated values of δ (Table 1.3.2) are of the same order as expected from electron micrographs of membranes.

If the potential across a membrane is 80 mV, and its thickness is 7 nm, What is the electric field intensity?

The field is uniform inside a capacitor, and so is given by $\mathbf{E} = -\Delta V/\Delta x$, where ΔV is the potential difference and Δx is the separation of the plates. Thus the electric field is

$$\mathbf{E} = -80 \times 10^{-3} \text{ V}/7 \times 10^{-9} \text{ m} = \mathbf{-11.4 \times 10^6 \text{ V m}^{-1}}$$

TABLE 1.3.2 Calculated δ for Various κ

κ	δ (nm)
1.89	2.36
2.29	2.85
2.46	3.07
5.34	6.67

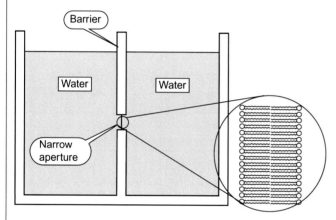

FIGURE 1.3.9 Planar lipid bilayer formed in a narrow hole between two aqueous compartments.

ELECTRIC CHARGES MOVE IN RESPONSE TO ELECTRIC FORCES

As mentioned in the section "Force and Energy Are Simple Consequences of Potential," the usefulness of the concept of potential lies in the ease of calculating the force on a charged particle or the energy needed to move from one region to another. The electrical force on a charged particle is given as

[1.3.22] $\qquad \mathbf{F} = -q\nabla U = q\mathbf{E} = ze\mathbf{E}$

where U is the potential, often written as V, q is the charge, $\mathbf{E}$ is the electric field, z is the valence (+/− integral number of charges per particle), and e is the unit charge of the electron. Thus, a charged particle, of either sign, in an electric field is subjected to an accelerating force. Ions in solution are subjected to these forces and accelerate on account of them. These ions accelerate until they reach a terminal velocity, v, at which point the electrical force is matched by a drag force on the particle by the surrounding solution. Figure 1.3.10 illustrates this situation.

MOVEMENT OF IONS IN RESPONSE TO ELECTRICAL FORCES MAKES A CURRENT AND A SOLUTE FLUX

The drag force on a particle moving through a solution is proportional to its velocity and directed opposite to it. Further, the electrical force given in Eqn [1.3.22] at

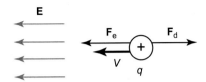

FIGURE 1.3.10 Forces on a charged particle in solution subjected to a constant electric field. The electrical force, F_e, is the product of the charge, q, on the particle and the electric field, E. This electrical force accelerates the charged particle and it moves through the solution. This movement produces a drag force, F_d, which is proportional to the velocity, v. The particle reaches a terminal velocity when the net force on the particle is zero: $F_e + F_d = 0$.

the terminal velocity is equal but opposite to F_d. Therefore, we can write

[1.3.23]
$$F_d = -\beta v$$
$$F_e = -F_d$$
$$F_e = \beta v$$

where β is a drag coefficient or **frictional coefficient**. Thus ions subjected to a constant electrical field will move at a constant terminal velocity, v, and that velocity will be proportional to the electric field. This movement of charged particles constitutes a movement of charge from place to place, and so it is an electrical current. Further, because solute particles carry the charge, the movement also forms a solute flow. The solute flux is related to the velocity by

[1.3.24]
$$J_s = vC$$

where J and v are written as vectors and C is the concentration of the solute (see Figure 1.3.11). Because each solute particle carries the charge ze, the **current density** is

[1.3.25]
$$i = zeJ_s$$

This expression can be converted into Ohm's law by using Eqns [1.3.22]–[1.3.24]:

[1.3.26]
$$i = zeJ_s = zevC$$
$$i = zeC\frac{F_e}{\beta}$$
$$i = \frac{z^2 e^2 C}{\beta}(-\nabla V)$$

The last equation is an analogue of Eqn [1.2.3] for the one-dimensional form of Ohm's law:

[1.2.3]
$$J_e = -\sigma \frac{d\psi}{dx}$$

EXAMPLE 1.3.2 Forces on Charged Particles

Consider the planar lipid bilayer in Example 1.3.1, which has a potential difference of 80 mV across it. What would the electric force be on a Na$^+$ ion in the middle of the bilayer?

The electric force on a charged particle is given as $F_e = qE$. We calculated the electric field intensity, E, in this case to be -11.4×10^6 V m^{-1}. The charge on any ion is ze, where z is the valence or integral number of charges on the particle and e is the charge on the electron. In this case, $z = +1$ and the charge on the electron is given in various units.

The most useful unit here is the coulomb: $1e = 1.6 \times 10^{-19}$ C.

The force is thus given as

$$F_e = 1.6 \times 10^{-19} \text{ C} \times -11.4 \times 10^6 \text{ V m}^{-1}$$
$$= -1.82 \times 10^{-12} \text{ V C m}^{-1}$$
$$= -1.82 \times 10^{-12} \text{ J m}^{-1}$$
$$= -1.82 \times 10^{-12} \text{ N m m}^{-1}$$
$$= \mathbf{-1.82 \times 10^{-12} \text{ N}}$$

THE RELATIONSHIP BETWEEN J AND C DEFINES AN AVERAGE VELOCITY

Consider the right cylindrical tube shown in Figure 1.3.11 that contains solute particles moving to the right at average velocity v. In time Δt, the particles travel a horizontal distance $\Delta x = v\Delta t$. All of the solute particles in the volume $A\Delta x$ will have crossed a cross-sectional plane in the cylinder. Thus the flux will be the total number of solute particles in that volume, per unit area per unit time. The number of solute

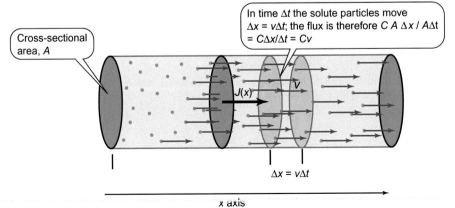

FIGURE 1.3.11 Relationship among J, C, and v. If solutes have an average velocity v, they sweep out a distance $v\Delta t$ in time Δt, and this corresponds to an entire volume of solute, equal to $Av\Delta t$, moving to the right. The number of solute particles in this volume is $CAv\Delta t$. The flux is this number per unit area, per unit time: $J = CAv\Delta t/A\Delta t = Cv$.

particles in the volume $A\Delta x$ is $CA\Delta x$. Thus the flux is given as

[1.3.27]
$$J = \frac{CA\Delta x}{A\Delta t}$$
$$J = Cv$$

From this equation, it is clear that the ratio of J/C defines an average velocity for solute particles.

OHM'S LAW RELATES CURRENT TO POTENTIAL

According to the discussion earlier, a difference in potential produces a force on charged particles, and the force is proportional to the negative gradient of the potential. The movement of charges in response to a potential makes a current, a flow of charges. It is given as

[1.3.28]
$$I = \frac{\Delta q}{\Delta t}$$

The current I is given in amperes = coulomb s^{-1}. **Current is defined as the movement of positive charge**, so that the movement of cations (positive ions) constitutes a current in the direction of the flow; the movement of anions (negative ions) makes a current in the opposite direction to the flow. The current due to an ion is related to its flow by

[1.3.29]
$$I_x = z\mathfrak{F} Q_x$$

where I_x is the current of ion x, in coulombs s^{-1}, Q_x is the flow of ion x in mol s^{-1}, z is the integral charge per ion ($+/- 1, 2, ...$) and $\mathfrak{F}$ is the Faraday (9.649×10^4 coulombs mol^{-1} = 6.02×10^{23} electrons - mol^{-1} $\times$ 1.6×10^{-19} coulombs electron^{-1}); $z\mathfrak{F}$ converts mol to coulombs. Note that current is an extensive variable, while current density is an intensive variable.

The movement of charge through matter-filled space encounters resistance from the matter. Those materials that offer little resistance are called **conductors**. Other materials, such as membrane lipids and the myelin sheath that surrounds nerve axons, offer high resistance and are called **insulators**. The current is greater if the potential driving it is greater and is less according to the resistance of the material through which the current flows. This is Ohm's Law:

[1.3.30]
$$I = \frac{\Delta \psi}{R} = \frac{E}{R}$$

where I is the current, $\Delta \psi$ is the potential difference, often symbolized as E or V, and R is the **resistance**. Resistance has the units of ohms = volts/amps, symbolized as Ω. Ohm's law can also be written as

[1.3.31]
$$I = g \Delta \psi$$

where $g = 1/R$ is the **conductance**. The SI unit for conductance is the siemen = amp/volt.

A battery is a device for using chemical reactions to create a voltage difference. In effect, chemicals trap electrons, with their negative charges, at fixed distances from their positively charged nuclei. Different chemicals will then have different potentials for their electrons. In chemistry these are called **oxidation potentials**. These refer to the energy required to remove the electron from the chemical. The movement of electrons from one chemical to another can then release energy equal to the difference in the oxidation potentials times the number of charges that move. If we hook up a battery in series with a resistance, we can produce a current which is given by Ohm's Law (see Eqn [1.3.30]). This situation is shown schematically in Figure 1.3.12.

KIRCHHOFF'S CURRENT LAW AND KIRCHHOFF'S VOLTAGE LAW

The circuit shown in Figure 1.3.12 is a simple circuit in which a resistor is placed over the terminals of a battery. When the circuit is completed, current flows and this can be measured with an ammeter. **Kirchhoff's Voltage Law states that the total voltage differences around any loop must be zero**. This is a restatement of the conservative nature of the electric force: the work done in any loop is zero. Since the only resistance is between nodes 2 and 4, the voltage differences around the loop are ΔV_{23} and ΔV_{71}, where ΔV_{71} is the voltage across the battery from the negative electrode to the positive electrode. Thus, $\Delta V_{71} = -\Delta V_{17} = -E$, the voltage provided by the battery. The total voltage drop around the loop is given as $0 = \Delta V_{23} + \Delta V_{71} = \Delta V_{23} - E$. Solving this for ΔV_{23}, we find $\Delta V_{23} = E$ and therefore the current through the resistor is $I = E/R$.

Kirchhoff's Current Law states that the sum of current into any node must be zero. Thus if there is a current $I_{23} = E/R$ that enters node 3, there must be a current $I_{34} = I_{23}$ that leaves node 3. The total current into node 3 is $I_{23} - I_{34} = 0$. I_{34} is negative in this last equation because it leaves node 3—it is the opposite direction (with respect to node 3) of the current I_{23}, but it is equal in magnitude.

THE TIME CONSTANT CHARACTERIZES THE CHARGING OF A CAPACITOR IN A SIMPLE RC CIRCUIT

Suppose now that we include a capacitor in the circuit shown in Figure 1.3.12. This expanded circuit is shown in Figure 1.3.13. Initially, with the switch open, there will be no potential across the capacitor, and no current in the circuit. If we flip the switch, current will begin to flow because there will be potential differences in the circuit. But the capacitor is filled with a dielectric that disallows current flow! How can current flow across the

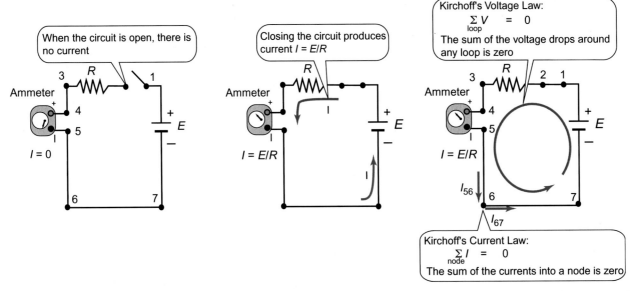

FIGURE 1.3.12 Ohm's Law, Kirchhoff's Current Law, and Kirchhoff's Voltage Law. When the circuit is completed by closing the switch between nodes 1 and 2, current flows as described by Ohm's Law: $I = \Delta V_{23}/R$, where ΔV_{23} is the voltage difference across the resistor (between nodes 2 and 3) and R is the resistance. In this circuit, the only resistance is between nodes 2 and 3 and all other resistances are negligible. Kirchhoff's Current Law (KCL) states that the sum of currents into a node is zero. At node 6, for example, the sum of the currents is $I_{56}-I_{67}$, the negative sign indicating that the current is out of the node. Thus, $I_{56}-I_{67}=0$ or $I_{56}=I_{67}$. Kirchhoff's Voltage Law (KVL) states that the sum of voltage drops around any closed loop is zero. For the circuit shown, the voltage drops are from nodes 2 to 3, where $\Delta V_{23} = I\,R = E$; and from nodes 7 to 1 where $\Delta V_{71} = -\Delta V_{17} = -E$. Thus the voltage drop around the loop is $\Delta V_{23} + \Delta V_{71} = E - E = 0$. All of the other voltage drops between nodes (ΔV_{45}, ΔV_{56}, ΔV_{67}, and ΔV_{12}) are zero because the resistances in this part of the circuit are negligible.

capacitor? The current in this case is a **capacitive current**, not a **resistive current**, like the kind shown in Figure 1.3.12. The flow of positive ions onto the top plate of the capacitor produces an electric field (a force per unit charge) that repels positive charges from the bottom plate. This movement of charges away from the bottom plate is the capacitive current. As the charges move away, there is a separation of charges on the capacitor and it now has a potential difference given by $V = q/C$. Charges continue to move until the potential across the capacitor is equal but opposite to the potential across the battery, E.

To analyze the time course of current and voltage in the circuit, we make use of Kirchhoff's Voltage Law that says the sum of the voltage drops in the circuit must be zero. We write

$$\Delta V_{61} + \Delta V_{23} + \Delta V_{45} = 0$$

[1.3.32]
$$-E + \frac{dq}{dt}R + \frac{q}{C} = 0$$

where q is the charge on the capacitor and dq/dt is the current through the resistor, which is also the current across the capacitor. We can separate variables in Eqn [1.3.32] and integrate to solve this equation for q as a function of t. The rate of charging of the capacitor is given as

[1.3.33]
$$q = EC\left(1 - e^{-\frac{t}{RC}}\right)$$

where t is the time and the combined terms RC is called the **time constant** because it describes the time taken to charge the capacitor. The current can be obtained from differentiation of q to give

[1.3.34]
$$I = \frac{E}{R}e^{-\frac{t}{RC}}$$

The time course of charge and current is shown in Figure 1.3.14. In Eqn [1.3.33], if $t = RC$, then $q = EC(1 - 1/e)$, so the time constant is the time required for the charge to be $1/e = 0.37$ of its final value.

SUMMARY

Some particles in nature either repel or attract other particles, and the force developed between them varies inversely with the square of their separation. These particles are said to be "charged," and there is no more basic description or explanation of their interaction than Coulomb's law that quantifies it. Charges have two types: positive and negative. Like charges repel, unlike charges attract.

This electrostatic force is a conservative force, meaning that the work performed in moving a charge around any closed loop is zero. Conservative forces also mean that the work done in moving a particle around depends only on the initial and final states, and not the path. Equivalently, the potential energy associated with a distributed set of charges depends only on the position and not on the path it takes to get there.

is measured in volts. The charge is measured in coulombs. Energy is measured in joules or volt-coulombs. Because the potential is defined as the integral of the work to move a unit positive charge from infinity to A, the electric force per unit positive charge is the negative derivative of the potential. The potential is a scalar whereas the force is a vector. The derivative here is the **gradient**, which converts the scalar potential into a force vector. The electric force per unit positive charge is the electric field.

The capacitance is defined as: $C = Q/V$, where Q is the charge and V is the potential difference across the capacitor. The capacitance of a parallel plate capacitor depends on several physical characteristics of the capacitor and is given as

$$C = \frac{\kappa \varepsilon_0 A}{\delta}$$

where κ is the dielectric constant that depends on material between the plates, ε_0 is a physical constant, A is the area of the plates, and δ is the spatial separation of the plates. Biological membranes form capacitors with very small δ.

Ions move in response to electrical forces. This movement forms both a solute flux and an electrical current. Movement in response to electrical forces accelerates ions in solution until a terminal velocity is reached. At this point, the net force on the ion is zero and is the balance between the electrical force and the drag or frictional force on the solute particle by the solution. The average terminal velocity is the flux divided by the concentration. Ohm's Law is given as $I = E/R$.

Conservation of charge and the conservative nature of the electric force give rise to Kirchhoff's Current Law and Kirchhoff's Voltage Law. Kirchhoff's Current Law states that the sum of all currents into any node of a circuit must be zero. Kirchhoff's Voltage Law states that the potential differences around any closed loop must sum to zero. When a capacitor is connected in series with a resistor and a voltage source (a battery), the capacitor gradually becomes charged until the potential across it exactly opposes the potential of the battery. The time course of charging depends on RC, the product of the resistance and the capacitance. When the capacitor is fully charged, current in the circuit goes to zero. The value of RC is the time constant for the circuit, which is the time that the charge takes to get to within $1/e$ of its final value.

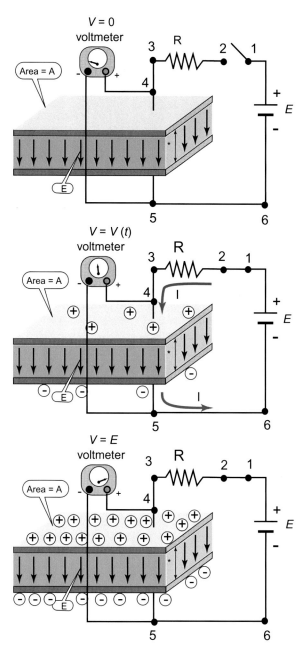

FIGURE 1.3.13 Charging of a capacitor. The capacitor consists of two parallel conducting plates separated by a dielectric, or insulating, material. At the start, top, the circuit is broken by a switch and there is no potential difference across the capacitor. When the switch is closed, middle panel, charge begins to move, making a current. The positive charges on the top plate repel positive charges on the bottom plate, which move back to the battery, completing the circuit for the current. This is a capacitive current, because there is no flow across the dielectric but there is a flow in the circuit. The separation of charges across the capacitor creates a potential difference related to the capacitance of the capacitor: $V = qC$. This builds up as current continues to flow until V is exactly opposite to E. At this point, current flow stops and the capacitor is fully charged.

The electrical potential at a point, A, is defined as the work necessary to bring a unit positive charge from infinite separation to that point. Therefore, positive fixed charges are associated with positive potential and negative fixed charges are associated with negative potential. Separation of charge produces a potential. The potential

REVIEW QUESTIONS

1. What do we mean by "electric potential"?
2. What makes an electrical potential difference between two points?
3. How much energy is gained by charge q over a potential difference V?
4. What is meant by "electric field intensity"?
5. What is the relationship between charge and voltage for a parallel plate capacitor?

QUANTITATIVE HUMAN PHYSIOLOGY: AN INTRODUCTION

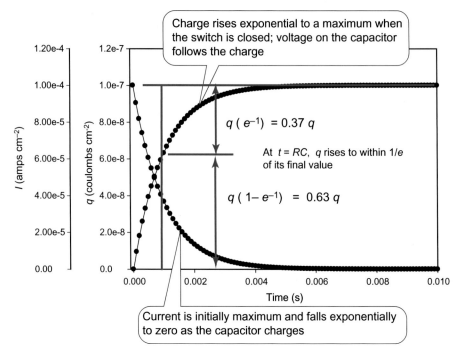

FIGURE 1.3.14 Charging of a capacitor in series with a resistor. Here the specific capacitance was taken as 1 μf cm^{-2} and the specific resistance was taken as 1000 Ω cm^2 and the time constant $RC = 1$ ms. The charge rises exponentially to reach a maximum and the current simultaneously decreases exponentially from an initial value to zero. The time constant is the time required for q to rise to with 1/e of its final value, which is the same as the time required for I to decrease to within 1/e of its final value or for I to decrease to 1/e I_0.

6. Why do biological membranes act as tiny capacitors?
7. How does capacitance depend on membrane thickness and surface area? What is the dielectric constant?
8. What is the relationship between average solute velocity and solute flux?
9. What is the frictional coefficient and how does it relate to velocity and flux?
10. What is Kirchhoff's Current Law? What is Kirchhoff's Voltage Law?
11. What is meant by "resistive current"? "Capacitive current"?
12. Why does current stop in an RC circuit?
13. What is meant by the term "time constant"? What is the time constant in a simple RC circuit?

Problem Set
Physical Foundations: Pressure and Electrical Forces and Flows
1.1

1. Identify whether the following variables are intensive or extensive and explain your reasoning.
   A. Temperature
   B. Heat content
   C. Volume
   D. Density
   E. Mass
   F. Concentration
   G. Moles
   H. Pressure
   I. Area
   J. Flow
   K. Flux
   L. Viscosity
2. Normal systolic blood pressure is about 120 mmHg
   A. Convert this to atmospheres.
   B. Convert this to pascals
3. Normal diastolic blood pressure is about 80 mmHg
   A. Convert this to atmospheres
   B. Convert this to pascals
4. The density of whole blood is typically 1.055 g cm^{-3}. The density of Hg is 13.6 g cm^{-3}.
   A. Derive an equation to express the hydrostatic pressure of a column of blood as the height of Hg that would produce the same hydrostatic pressure as that column of blood
   B. Use the equation to determine the pressure of a 20-cm column of blood, expressed in mmHg
   C. If systolic blood pressure is 120 mmHg, what is the height of blood that this could support (if it was constant)?
5. For dialysis membranes, the L_p was determined to be $6.34 \times 10^{-7} \text{ cm min}^{-1} \text{ mmHg}^{-1}$. A cylindrical hole 1 cm in diameter was cut into two Lexan pieces that were then bolted together with the membrane between. Fluid entered one side through a pump and a pressure transducer was affixed. The flow was adjusted until the pressure (above atmospheric) was 20,000 pascals. The pressure on the opposite side was atmospheric (zero). Assume steady-state flow. What was the flow through the membrane?
6. The viscosity of water at 25 °C is about 0.00089 Pa s. The inner diameter of a PE160 polyethylene tubing is 1.14 mm. Assume steady-state laminar flow.
   A. What pressure is necessary to get a flow of 5 mL min^{-1} through a 20-cm length of this tubing?
   B. What is the velocity of the flow?
   C. What pressure is necessary to get a flow of 5 mL min^{-1} through the same PE160 tubing if plasma is used, with viscosity of 0.002 Pa s?
   D. What pressure would be needed for the same flow of water through a 20-cm length of PE60 tubing with i.d. = 0.76 mm?
7. After the end of a normal inspiration, the volume of air in the lungs is about 2.8 L. Normally quiet inspiration is driven by a pressure difference of about 2 mmHg. The air in the lungs is at 37 °C and after normal expiration it is at atmospheric pressure. Quiet inspiration is driven by the expansion of the chest cavity by contraction of the diaphragm, which expands the air in the lungs. How much is the air expanded to produce an decrease of 2 mmHg in pressure? Use the ideal gas equation, $PV = nRT$, where P is the pressure, V is the volume, n is the number of moles, $R = 0.082 \text{ L atm mol}^{-1} \text{ °K}^{-1}$ is the gas constant, and T is the absolute temperature.
8. A burette is a vertical, right circular cylinder that is open at the top and has a stopcock valve at the bottom to let fluid out. Assume that you have a burette that has an inner diameter of 1 cm and a height of 100 cm.
   A. Assume that you fill the burette with water to some height, h. What is the relation between h and the pressure at the base of the burette?
   B. What is the relation between volume of water in the burette and the pressure?
   C. How does this relation map onto the relation between charge and voltage on a capacitor?
   D. What is the hydraulic analogue of voltage?
   E. What is the hydraulic analogue of charge?

F. Suppose you open the stopcock fully. The fluid will drain out. Assume a constant diameter of the opening of the stopcock that provides a constant resistance, R. Derive an equation that describes the time course of draining the burette.
G. Identify the time constant for the burette emptying.

9. You are given a long vertical tube filled with fluid of viscosity η at sea level where the acceleration due to gravity is $g = 9.81$ m s^{-2}. You have a steel ball or radius r and density ρ_{steel} that you carefully drop into the fluid. Assume that the drag force on the ball obeys Stokes' Law: the drag coefficient is $6\pi r\eta$, where $\mathbf{F}_{drag} = -6\pi r\eta\,\mathbf{v}$. Remember that the steel ball is subjected to a buoyant force equal to the volume of the ball times the density of the fluid, ρ_{fluid}, times the acceleration of gravity.
   A. Derive an expression for the time of approach of the steel ball to terminal velocity.
   B. Derive an expression for the viscosity of the fluid as a function of the terminal velocity. (This is how you can determine the viscosity of a fluid.)

10. The Poiseuille equation that relates flow in narrow tubes to the pressure difference is analogous to Ohm's Law for current flow.
    A. What is the resistance to flow in terms of the parameters of the tube?
    B. What happens to resistance if the radius of the tube is halved?
    C. What happens to resistance if the radius of the tube is doubled?

11. Suppose that you begin to exercise and as a consequence you begin to produce more heat.
    A. What do you suppose will happen to the body temperature if, when environmental temperature does not change, the rate of heat production increases.
    B. How is this change in body temperature described by the continuity equation applied to heat?
    C. How do you suppose the body will shed this excess heat production?

12. You are out camping and it is very cold outside. You are losing heat faster than you can produce it.
    A. What happens to body temperature?
    B. How is this change in body temperature described by the continuity equation applied to heat?
    C. How do you suppose you can prevent hypothermia?

13. Consider a capacitor with a capacitance of 10 μF. You connect it to a variable DC voltage source with a switch in the circuit and a resistor of 1000 Ω in series with the capacitor.
    A. What is the relation between the steady-state voltage across the capacitor and the charge?
    B. Before you close the switch, there is no charge on the capacitor. When you close the switch, current begins to flow from the voltage source at E volts. What is the relation between current and the voltage drop across the resistor in terms of the current and the resistance?
    C. What is the voltage drop across the capacitor in terms of charge and capacitance?
    D. Kirchoff's voltage law says that the voltage drop around any loop must be zero. Write the equation for the voltage drop across the resistor, capacitor, and voltage source.
    E. Solve the equation in part D to derive the time course of charging of the capacitor.
    F. Solve the equation in part C to derive the time course of the current.
    G. Identify the time constant for the charging of the capacitor.

14. An unmyelinated axon can be considered to be a long right circular cylinder. Consider that an axon is 10 cm long with a diameter of 1.0 μ (1 μ = 10^{-6} m).
    A. If the specific capacitance of the membrane is 1 μF cm^{-2}, what is the capacitance of the axon membrane?
    B. How much charge is separated by this membrane to give a potential of 70 mV?

15. A muscle cell approximates a right circular cylinder 10 cm long and 70 μ in diameter. The specific capacitance of the membrane (the capacitance per unit area) is 1 μF cm^{-2}.
    A. What is the capacitance of the muscle membrane?
    B. How much charge is separated by this membrane to give a potential of −85 mV?

16. A bubble is held at a radius of 250 μm. The transmural pressure difference is 2 mm Hg. What is the tension in the wall?

17. The thickness of a single membrane is about 7 nm with a specific capacitance $C_m = 1$ μF cm^{-2}. A myelin sheath consists of multiple membranes produced by coils of Schwann cell or oligodendroglia cells. Suppose a myelin sheath results from 100 membranes stacked on top of each other.
    A. What is the specific capacitance of the myelin?
    B. If the myelin sheath is a right circular cylinder 1 mm long with a radius of 3 m, what is its total capacitance?
    C. How much charge does it take to produce a voltage of −70 mV across this capacitor?
    D. If the myelin were just 1 membrane—i.e., not myelin—what would be its capacitance? How much charge would it take to produce a voltage of −70 mV across the single membrane?

18. The transmural pressure difference across a small vein is 20 mmHg. The radius is 1 mm. What is the wall tension?

19. The heat capacity of the human body is about 3500 joules kg^{-1} °C^{-1}. Suppose that a person weighs 75 kg. How much energy does it take to raise the body temperature from 98°C to 102°C?
20. The charge on the electron is 1.602×10^{-19} coulombs. How much energy, in joules, is gained when an electron is accelerated across a potential difference of 1 v in a vacuum? This amount of energy is called the electron-volt.
21. Suppose that the average diameter of the aorta is 1.1 cm. The flow through the aorta is nearly the entirety of the cardiac output.
    A. If the cardiac output is 5 L/min, what is the average flow through the aorta?
    B. Suppose further that the cardiovascular system is nearly closed to fluid transfer. That is, that on a short-term basis the volume of the blood does not change. This means that ALL of the blood that leaves the heart goes through the arteries, and then capillaries, and then returns to the heart through the veins. Using the continuity equation, what is the flow through the aggregate capillaries?
    C. If the average diameter of the capillary is 4×10^{-4} cm, and flow through the capillary is 0.1 cm s^{-1}, how many capillaries are there?

1.4 Chemical Foundations of Physiology I: Chemical Energy and Intermolecular Forces

Learning Objectives

- List the chemical elements that make up the organic part of the body
- List the major chemical elements found as electrolytes in the body
- List the chemical elements found in trace quantities and used as cofactors for enzymes
- Explain why single CC bonds rotate easily whereas double CC bonds do not
- Define and give examples of structural isomerism, geometric isomerism, and optical isomerism
- Write correct estimates of bond length and energy (within a factor of 2) for covalent bonds
- Define electronegativity
- Describe what is meant by a polar bond
- Distinguish between covalent and ionic bonds
- Define dipole moment and be able to calculate the energy of dipole–dipole interactions
- Describe the hydrogen bond and recognize its typical energy
- Describe what is meant by London dispersion forces
- Draw the Lennard-Jones Potential and label the axes

ATOMS CONTAIN DISTRIBUTED ELECTRICAL CHARGES

In ordinary chemical reactions, atoms are the fundamental particles. The word *atom* derives from the Greek *atomos*, which means *indivisible*. The atoms themselves are composed of simpler subatomic particles, the **neutrons**, **protons**, and **electrons**. These particles are characterized by their rest mass and electrical charge, as shown in Table 1.4.1.

Ernest Rutherford showed that all of the positive charges in an atom are concentrated within a very small volume, called the **nucleus**, and was awarded the 1908 Nobel Prize in Chemistry for the work. The nucleus has dimensions on the order of 10^{-15} m! This requires some new thinking: if positive charges repel each other according to Coulomb's law, how can they be concentrated in the nucleus? The answer is that there are other fundamental forces at work here, the strong nuclear force and weak nuclear force, that have effects only over very short distances ($<10^{-14}$ m) and account for the stability of atomic nuclei.

Each atom has a definite number of neutrons, protons, and electrons. The number of protons in the nucleus is called the **atomic number**, Z, and this number defines the chemical element that describes the atom. In a neutral atom, the number of electrons is equal to Z, and these orbit the nucleus. The behavior of the electrons defines the chemical reactivities of the elements, and the concentrated positive charge of the nucleus, in turn, determines the behavior of the surrounding electrons.

ELECTRON ORBITALS HAVE SPECIFIC, QUANTIZED ENERGIES

Although we refer to electrons as *particles*, in fact they have wave-like characteristics such as constructive and destructive interference. The structure of the atom cannot be explained using classical physics. Instead, it requires quantum mechanics. Quantum mechanics posits that the "orbit" of electrons around the nucleus is described by a wave function, which has been interpreted as being related to the probability of finding the electron in some volume. The wave function has quantum numbers in that it allow electron orbitals to have only specific energy levels, and transitions between them can occur only when the exchange of energy is exactly equal to the difference in the two energy levels. A set of quantum numbers uniquely describes the energy state of each individual electron in an atom. One quantum number describes the electron "shell," a second describes the "orbital" within that shell, and a third describes the spin of the electron. The Pauli exclusion principle states that no two electrons can share the same set of quantum numbers within an atom. These orbitals are generally described as "clouds," indicating the distributed nature of the orbital electrons.

TABLE 1.4.1 Mass and Charge of Subatomic Particles

Particle	Rest Mass (g)	Electrical Charge (C)
Neutron	1.6747×10^{-24}	0
Proton	1.6726×10^{-24}	$+1.602176 \times 10^{-19}$
Electron	9.132×10^{-28}	$-1.602176 \times 10^{-19}$

Chemical Foundations of Physiology I: Chemical Energy and Intermolecular Forces   47

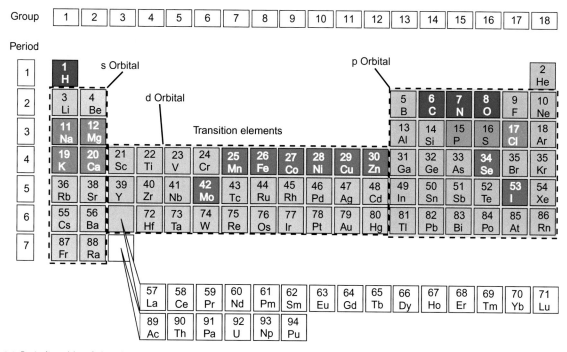

FIGURE 1.4.1 Periodic table of the elements. Each element is distinguished by the atomic number, the number of protons in the nucleus of each atom. Each chemical element is symbolized by a one or two letter abbreviation, as indicated in the figure. Some elements have special biological significance. C, H, O, and N (carbon, hydrogen, oxygen, and nitrogen, respectively) form the basic organic structures of the body. Phosphorus and sulfur (P and S, respectively) are less common parts of the organic substance. Sodium (Na), potassium (K), magnesium (Mg), calcium (Ca), and chlorine (Cl) highlighted in gray make up the electrolytes of the body fluids and are necessary in large amounts. Other elements, particularly transition elements, bind to organic structures and enable their activities. These are required in trace amounts and include iron (Fe), manganese (Mn), cobalt (Co), nickel (Ni), copper, (Cu), zinc (Zn), selenium (Se), molybdenum (Mo), and iodine (I).

HUMAN LIFE REQUIRES RELATIVELY FEW OF THE CHEMICAL ELEMENTS

As noted above, each chemical element consists of atoms whose nuclei contain a definite number of protons and some number of neutrons, which typically is about the same as the number of protons, and an equal number of electrons distributed among the atomic orbitals. There are 94 naturally occurring elements, but relatively few of these are essential to human life, as illustrated in Figure 1.4.1.

ATOMIC ORBITALS EXPLAIN THE PERIODICITY OF CHEMICAL REACTIVITIES

There are eight main "shells," referring to the principal quantum number, $n = (1,2,3,4,5,6,7,8)$ that describes atomic orbitals. There are four major subshells: s, p, d, and f, whose names derive from spectroscopic descriptions of *sharp, principal, diffuse,* and *fundamental*. These orbitals are described by the azimuthal quantum number, $l = (0,1,2,3)$ for (s,p,d,f), respectively. Each subshell has a structure and a capacity for electrons that is described by the magnetic quantum number, m, and the spin quantum number, s. The s subshell is spherically symmetrical and holds only 2 electrons; each set of p orbitals holds 6 electrons, the d orbitals hold 10, and the f orbitals hold 14. The sequential filling of these

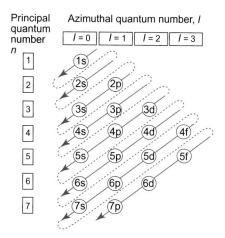

FIGURE 1.4.2 Order of filling of atomic orbitals. Electronic orbits are characterized by a principal quantum number that determines the main shell, an azimuthal quantum number that determines the subshell, a magnetic quantum number that determines the orbital, and the spin quantum number that determines the spin of the electron. There are four subshells: s, p, d, and f. These have 1, 3, 5, and 7 orbitals that each can hold up to two electrons of opposite spin. The order of filling with increasing number of electrons follows the blue diagonal arrows in the diagram: 1s fills first, followed by 2s and 2p; next is 3s followed by 3p and 4s, followed by 3d, 4p, and 5s; next is 4d, 5p, and 6s; then 4f, 5d, 6p, and 7s.

orbitals accounts for the periodic chemical behavior of the elements with their atomic number. This order of filling is shown in Figure 1.4.2. Each subshell (s, p, d, f) is typically filled with the requisite number of electrons before filling the remaining subshells. Each electron has a spin quantum number, s, that is represented as "up" or "down." The orbitals in the subshells are typically filled singly with electrons of parallel spin before double occupancy begins. This is the so-called "bus seat rule," analogous to the filling of a bus where double seats tend to fill with single individuals before double occupancy occurs.

Full orbitals are inherently stable, because they have low energy, and atoms having full orbitals are chemically unreactive. These correspond to the noble gases, helium (He), neon (Ne), argon (Ar), krypton (Kr), xenon (Xe), and radon (Rn). The electronic structure of some of these stable atoms is shown in Figure 1.4.3. All of the other elements can react with other atoms, in order to become more stable by attempting to fill their orbitals. They do this by sharing electrons, a process that constitutes chemical bonding. This sharing can be equal or very unequal, corresponding to the extremes of covalent bonding and ionic bonding.

ATOMS BIND TOGETHER IN DEFINITE PROPORTIONS TO FORM MOLECULES

Two or more elements can combine to form a compound, and the resulting character of the compound is entirely different from the two elements themselves. A classic example is water. Two volumes of hydrogen gas will combine with one volume of oxygen gas to produce water, which at the same temperature is a liquid and behaves altogether differently from either the hydrogen or oxygen gas. Such combinations of elements are called **compounds**, and the fundamental unit of them is the **molecule**. Molecules consist of atoms that are bonded together through the sharing of electrons in their outer atomic orbitals. The electrostatic shielding and energy involved in the orbital electrons overcome the repulsive forces between the positively charged nuclei. The resulting molecule is generally more stable than the starting materials. In order to break apart the molecule, energy must be supplied. This energy is called the **bond energy**, and its magnitude depends on the compound.

COMPOUNDS HAVE CHARACTERISTIC GEOMETRIES AND SURFACES

Carbon has an atomic number of 6. Its electronic structure is $1s^2\ 2s^2\ 2p^2$: there are two electrons in the 1s orbital, two in the 2s orbital, and two in the 2p orbitals, as shown in Figure 1.4.4. Carbon can achieve the stable neon configuration of $1s^2\ 2s^2\ 2p^6$ by sharing electrons with four hydrogen atoms. The resulting

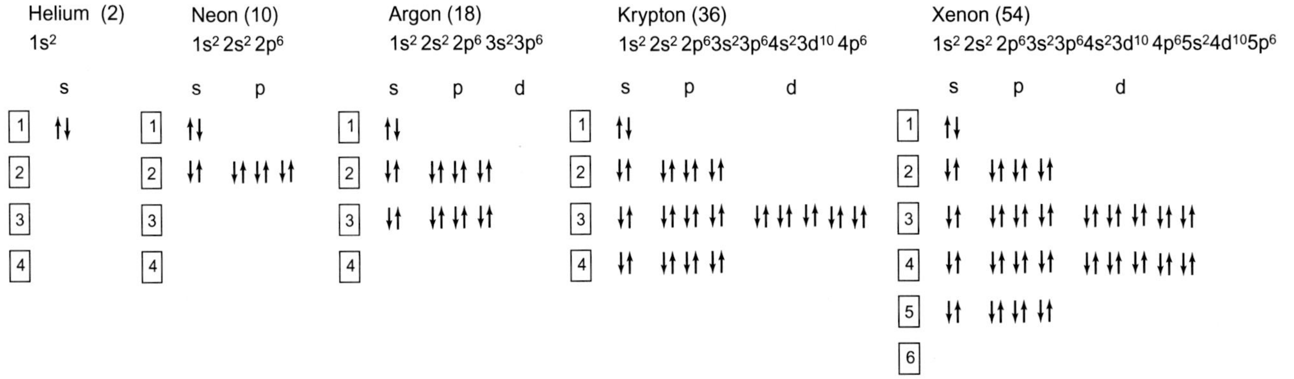

FIGURE 1.4.3 Electronic structure of the inert gases. These inert gases are chemically unreactive because their orbitals are already filled. Helium, with $n = 2$ protons in its nucleus, fills the 1s orbital with 2 electrons of opposite spin. Spin is indicated in the drawing by an arrow pointed upward or downward. Neon ($n = 10$) fills the 2s and 2p orbitals with a total of 8 electrons. Each orbital in the subshells carries at most two electrons. The order of filling of the orbitals corresponds to that shown in Figure 1.4.2.

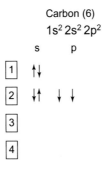

FIGURE 1.4.4 Electronic structure of carbon. Carbon has 6 protons in its nucleus and 6 electrons that occupy the orbitals, 2 in the 1s orbital, 2 in the 2s orbital, and 2 more in the p orbitals. Only two of the three p orbitals are occupied by electrons. Carbon requires four more electrons to reach the stable configuration of Neon.

compound is methane, written as CH$_4$ to convey the definite and fixed proportion of 1C for 4H atoms. All molecules in methane have this compositional **stoichiometry**. The close approach of the H nuclei and C nucleus alters the electronic structure of both the carbon and hydrogen. In molecular orbital theory, both the 2s and 2p orbitals of carbon participate in bonding by forming four hybrid molecular orbitals, termed sp^3, meaning the hybrid of the 2s orbital with 3p orbitals (see Figure 1.4.5). The angle between the neighboring CH bonds is 109°28′. The shape-filling model of methane shows the edges of the carbon and hydrogen atoms as if they were hard spheres, but really the orbitals do not have such definite boundaries. The electron orbitals define these soft edges. All compounds are defined by the relative locations of their atomic nuclei and the three-dimensional distribution of their electronic charges. These make up a three-dimensional surface that can interact with other three-dimensional surfaces. The bedrock of all of chemistry and physiology is the interaction of these surfaces.

SINGLE CC BONDS CAN FREELY ROTATE

Carbon can also form bonds with other carbon atoms. Ethane has the compositional stoichiometry of C$_2$H$_6$ (see Figure 1.4.6). It is two methane molecules in which two CH bonds are replaced by a single CC bond. In the single CC bond, the sp^3 hybrid orbitals overlap along their axis and form a circularly symmetric sigma bond. There is relatively free rotation around the axis of symmetry of this single bond, with three dips of about 12 kJ mol^{-1} for each rotation when the H atom from one methyl group aligns with the space between the H atoms of the opposite methyl group. These ideas are shown schematically in Figure 1.4.6.

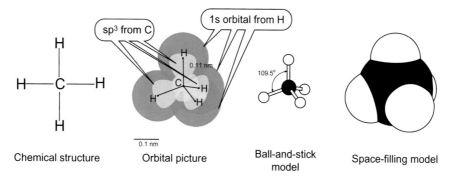

FIGURE 1.4.5 Structure of methane. The compositional stoichiometry of methane is CH$_4$—one carbon atom bonded to four hydrogen atoms. It arises from the sharing of the 1s electron of H with the 2s and 2p electrons of carbon. The bonding arises from overlap of the 1s H electron with electrons with hybrid C orbitals called sp^3—formed from one s orbital and three p orbitals.

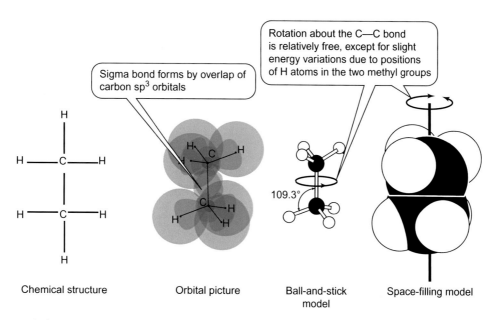

FIGURE 1.4.6 Structure of ethane. Here 1C atom binds to another C and 3H atoms. The CC bond forms by overlap of the sp^3 orbital along its axis to form a sigma bond that has circular symmetry. This bond can rotate about its axis, with some resulting configurations having just a little more stability than others. The most stable arrangement is shown, with the H atom of one methyl group aligned with the space between the H atoms in the opposing methyl group.

DOUBLE CC BONDS PROHIBIT FREE ROTATION

Carbon can form double bonds by altering its molecular orbitals. Ethylene is ethane in which another pair of CH bonds converts to a second CC bond (see Figure 1.4.7). Instead of combining $2s^2$ with $2px^1$, $2py^0$, and $2pz^1$ to form the $4sp^3$ orbitals, it can arrange the electrons in a planar trigonal geometry by hybridizing the $2s^2$ with the $2px^1$ and $2py^0$ orbitals to form three sp^2 orbitals with one electron each, and another pz^1 orbital that can form a second bond, a pi bond, out of the plane of the sp^2 orbitals. This bond resists twisting and a 90° twist breaks the overlap of the p orbitals, and hence breaks the second bond. Thus double bonds such as that in ethylene, shown in Figure 1.4.7, produce a somewhat rigid plane in any molecule in which they are found.

CHEMICAL BONDS HAVE BOND ENERGIES, BOND LENGTHS, AND BOND ANGLES

So far we have discussed CH bonds, CC single bonds, and CC double bonds. There are a host of other kinds of bonds that connect atoms together to make molecules, and each of these bonds has definite average bond lengths, bond angles, and bond energies. Table 1.4.2 summarizes approximate values for some of these bonds. These are approximate because atoms bound to other parts of the molecule can influence the molecular orbitals some distance away, thereby altering the angle, or length, or energy of any particular bond.

BOND ENERGY IS EXPRESSED AS *ENTHALPY* CHANGES

Earlier we wrote the conservation of energy theorem as

[1.3.6] $$dE = dq - dW$$

where dE was the incremental change in the energy of a system, dq is defined as the heat absorbed *by* the system from its surroundings, and dw is defined as the work performed *by* the system on its surroundings. The total energy content of the system, E, does not depend on the path taken to get to its configuration. It is a state variable. If we conduct a change of state at constant pressure, then Eqn [1.3.6] can be written as

[1.4.1] $$dE = dq_p - P\, dV$$

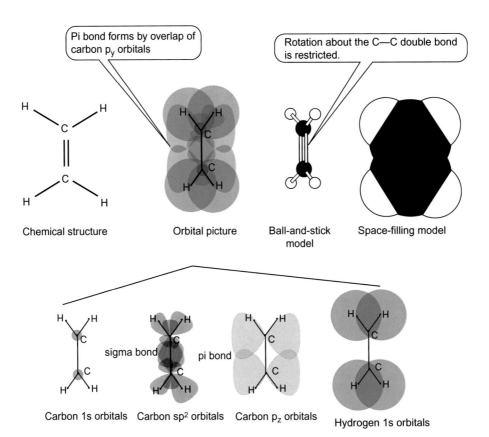

FIGURE 1.4.7 Structure of ethylene. The double CC bond is stronger than the single bond (it requires more energy to break) and locks the C atoms and all of the bonded groups into a single plane. This plane resists twisting because twist along the CC axis rotates the p orbitals away from overlap, breaking the pi bond. The space-filling model is not oriented the same way as the ball-and-stick model; the plane of the space-filling model is parallel to the plane of the paper. The lower part of the diagram shows the component orbitals. The 1s orbitals from carbon do not participate in molecular bonding. The sp^2 orbitals bond to the H 1s orbitals through sigma bonds and form the C—C sigma bond. The p_z orbitals perpendicular to the plane of the sp^2 orbitals form a second C—C bond, a pi bond.

TABLE 1.4.2 Typical Bond Length and Bond Energy for Some Bonds Important in Physiology

Bond	Length (pm)	Energy (kJ mol^{-1})	Bond	Length (pm)	Energy (kJ mol^{-1})	Bond	Length (pm)	Energy (kJ mol^{-1})
CC	154	346	CC	134	602	CC	120	835
CN	147	308						
CO	143	360	CO	120	799			
CS	182	272						
HC	109	411						
HN	101	386						
HO	96	459						
HvS	134	363						
PO	163	335	PO	150	544			
SS	205	226						

Bond lengths are given in pm ($=10^{-12}$ m) and bond energy is reported in kilojoule per mole (kJ mol^{-1}). Many tables of bond energies report them in kcal mol^{-1}. The calorie is defined as the amount of heat energy required to raise the temperature of 1 g of water from 14.5°C to 15.5°C. The calorie is readily converted to the joule using the conversion factor 1 cal = 4.184 J. *Calorie* derives from the Latin *calor*, meaning *to heat*.

if the only work is pressure–volume work. Note here that

$$[1.4.2] \quad F\,dx = \frac{F}{A} A\,dx = P\,dV$$

clarifies that a pressure moving a volume has the same work units as a force moving through a distance. Integrating Eqn [1.4.1] between two states, we obtain

$$[1.4.3] \quad \begin{aligned} \int_i^f dE &= \int_i^f dq_p - \int_{V_i}^{V_f} P\,dV \\ E_f - E_i &= q_p - P(V_f - V_i) \\ (E_f + PV_f) - (E_i + PV_i) &= q_p \\ H_f - H_i &= q_p \\ \Delta H &= q_p \end{aligned}$$

Here we make the definition

$$[1.4.4] \quad H = E + PV$$

where H is the **enthalpy**. Since E, P, and V are all state variables, depending only on the state and not the path taken to that state, enthalpy is also a state variable. The **bond energies are differences in the enthalpy of formation of the products and reactants in any chemical reaction.**

THE MULTIPLICITY OF CX BONDS PRODUCES *ISOMERISM*

Some chemical compounds with identical compositional stoichiometry behave differently. Their different behaviors can be obvious and large, or quite subtle, depending on how the compounds differ. These compounds with identical composition differ in the way the atoms are arranged in the molecule. They are called **isomers**, meaning *same weight*. There are three major classes of isomers, structural, geometric, and optical, which are described below and shown schematically in Figures 1.4.8–1.4.10:

- **Structural isomers** differ in the connectivity of the atoms in the molecule.
- **Stereoisomers** have identical connectivity but the atoms are arranged differently in space. These consist of two classes:
    - **Geometric isomers**, involving a double CC bond that does not allow free rotation
    - **Optical isomers**, existing in two types.
- **Enantiomers** are mirror images of each other. They have identical physical characteristics such as melting point and density but are not superimposable. This requires an asymmetric carbon atom in which four nonidentical groups are bonded. They differ in their ability to rotate the plane of polarized light.
- **Diasteriomers** are optical isomers that are not enantiomers. They typically differ in the spatial distribution about one or more asymmetric carbons, while not being mirror images.

UNEQUAL SHARING MAKES POLAR COVALENT BONDS

The electrons that are shared in covalent bonds distribute themselves according to the charges on the nuclei within the molecule and the other electrons in the unshared orbitals that shield that charge from the electron. In almost all cases, the electrons are not shared equally but tend to spend more time near one or the other of the nuclei involved in the bond. The ability of an atom in a molecule to attract shared electrons is called its **electronegativity**. It generally increases going up the periodic table and going to the right, so that F has the highest electronegativity and Fr has the lowest. The electronegativity of some common elements is shown in Table 1.4.3 in arbitrary *Pauling units* scaled to F at 4.0.

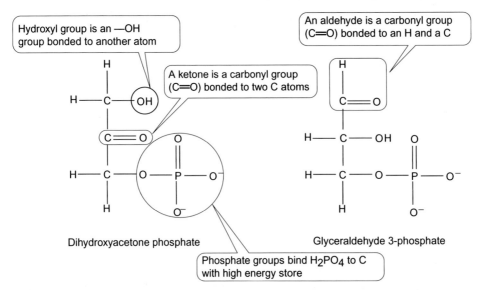

FIGURE 1.4.8 An example of structural isomerism. Both glyceraldehyde 3-phosphate and dihydroxyacetone phosphate contain the same number of each type of atom. However, the connectivity of the atoms differs. Glyceraldehyde 3-phosphate contains an aldehyde group, which is defined as a C atom with a double bond to O and a single bond to H. Dihydroxyacetone phosphate contains a ketone group, which is a C atom with a double bond to O and the remaining bonds to C atoms. Conversion of the two chemicals requires breaking and reforming chemical bonds.

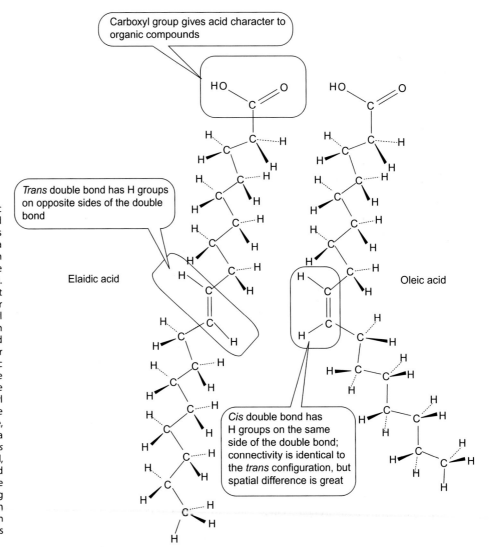

FIGURE 1.4.9 An example of geometric isomerism. Both elaidic acid and oleic acid belong to a class of organic compounds called fatty acids, characterized by a carboxyl group at one end of an unbranched hydrocarbon chain. In these molecules, the chains are 18 carbons long. Both have one double bond beginning at the ninth carbon from the carboxyl end or the ninth carbon from the terminal methyl end. The spatial arrangement of the carbon chain and hydrogens at the double bond can be achieved in two ways: either *cis* or *trans*. In the *cis* arrangement, found in oleic acid, the two hydrogens are on the same side of the double bond, meaning that the hydrocarbon chain toward the carboxyl end and the hydrocarbon chain toward the methyl end are also on the same side, opposite to the hydrogens. This produces a kink in the hydrocarbon chain. In the *trans* arrangement, in the case of elaidic acid, the two H atoms on the doubly bonded C atoms are on opposite sides of the double bond. This has the effect of keeping the hydrocarbon chain straighter. Both compounds have the same composition but different physical properties due to this geometric isomerism.

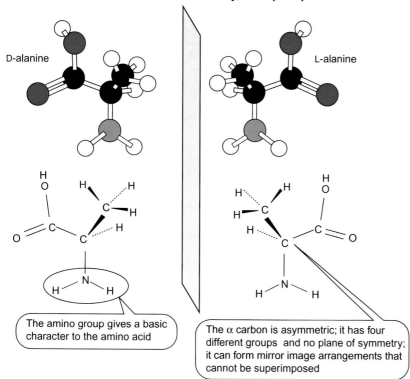

FIGURE 1.4.10 An example of optical isomerism. Amino acids consist of a backbone of a carboxyl group bonded to a central C atom, called the α carbon, and an amino group bonded to the other side of the α carbon. Typically the α carbon has two more bonds, one to an H atom and the other to a variable group, called R for residue, whose composition determines the kind of amino acid. There are 20 different R groups and 20 different amino acids. The R group for alanine, shown here, is a methyl group. Because the α C atom is asymmetric—there is no plane of symmetry—the four groups can be arranged in two nonequivalent ways. Our bodies use only the L-amino acids. In the L-amino acids, starting from the carboxyl group and moving toward the amino group, the R group is to the left. This is very important, as D-amino acids in proteins would have their R groups pointing the wrong way.

TABLE 1.4.3 Electronegativity of Common Atoms

Atoms	H	C	N	O	F	Na	Mg	P	S	Cl
Electronegativity	2.2	2.5	3.0	3.5	4.0	0.9	1.2	2.1	2.5	3.0

Source: L. Pauling, *The Nature of the Chemical Bond*. Cornell University Press, 1960.

Atoms with similar electronegativities will share bonding electrons equally and will produce nonpolar bonds. Examples include the CC bond, CH bond, and HS bond. Bonds such as OH will be polar bonds.

IONIC BONDS RESULT FROM ATOMS WITH HIGHLY UNEQUAL ELECTRONEGATIVITIES

If the difference in electronegativity is too great, an ionic bond will form in which the strongly electronegative atom strips an electron from the weakly electronegative atom. Examples include NaCl, KCl, $CaCl_2$, $MgCl_2$, and a host of other physiologically relevant compounds. These are noted for their dissociation in water to form ions, Na^+, Cl^-, K^+, Ca^{2+}, and Mg^{2+}, for example. These isolated ions are stabilized by their interaction with the polar water molecule, which presents a negative side towards the positive ions (**cations**) and a positive side towards the negative ions (**anions**).

WATER PROVIDES AN EXAMPLE OF A POLAR BOND

The bond angle defined by HOH is 104.5°, which is close to the tetrahedral angle. In this case, O forms 3 orbitals by hybridization of the $2s^2$ orbitals with the $2px$ and $2py$ orbitals. This leaves 2 O electrons in an sp^2 orbital that are unshared and 2 electrons in a pz^2 orbital that are also unshared. These form the **lone electron pairs** of water that participate in yet another kind of bonding, the **hydrogen bond**, discussed later. The electronegative O atom attracts the electrons away from the H nuclei, forming a partial separation of charge in the molecule itself. Thus the bond is said to be **polar** (see Figure 1.4.11).

The estimated charge separation is about -0.67 on the O atom and about $+0.33$ on each of the H atoms. The **dipole moment** is defined as

$$[1.4.5] \qquad \mathbf{p} = q\mathbf{d}$$

where **p** is a vector, the dipole moment, q, is the charge divided into equal q_- and q_+, and **d** is the vector pointing from q_- to q_+. The dipole moment of a single water molecule is 1.855 debye (1 debye = 3.33564×10^{-30} C m), but the dipole moment of water varies with the size of the water cluster, because nearby water molecules rearrange themselves in the presence of an electric field. Dipoles themselves produce an electric field and therefore interact with electric charges in its vicinity. These electrostatic interactions are part of the forces that govern the interaction of surfaces. In an electric field, the dipole experiences a torque given by

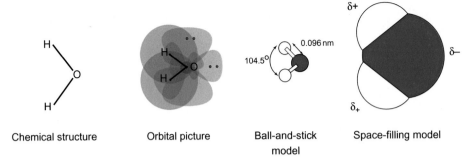

Chemical structure   Orbital picture   Ball-and-stick model   Space-filling model

FIGURE 1.4.11 Structure of water. The HOH angle is 104.5° and the bond length is about 96 pm. The O atom is far more electronegative than the H atoms, thereby causing an asymmetric redistribution of the electrical charges away from the H atoms and surrounding the O atom. These partial separations of charges produce a polar compound and a net dipole moment.

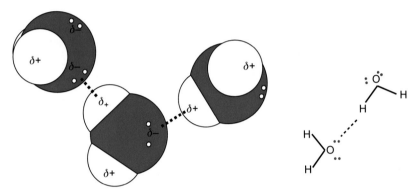

FIGURE 1.4.12 The hydrogen bond in water. The polar OH bond involves a separation of charge. The partial positive charge on the H atom in water is attracted to the lone electron pairs on the opposite side of the molecule on adjacent molecules. This forms a weak bond that is easy to form and easy to break. In many situations, the number of hydrogen bonds significantly stabilizes large structures. In water, there are many hydrogen bonds because each water molecule has the potential of participating in four of them: two because of the H atoms and two more because of the two lone electron pairs.

$$[1.4.6] \quad \boldsymbol{\tau} = \mathbf{p} \times \mathbf{E}$$

where $\boldsymbol{\tau}$ is the torque, $\mathbf{p}$ is the dipole moment, and $\mathbf{E}$ is the electric field strength.

INTERMOLECULAR FORCES ARISE FROM ELECTROSTATIC INTERACTIONS

Chemical bonds join atoms together to form molecules. Molecules can also be attracted to each other through a variety of intermolecular forces that include:

- Hydrogen bonding
- Dipole–dipole interactions
- London dispersion forces.

HYDROGEN BONDING OCCURS BETWEEN TWO ELECTRONEGATIVE CENTERS

The hydrogen bond involves the sharing of the positive charge of hydrogen between two electronegative centers such as oxygen and nitrogen. It requires proximity and proper orientation of the two electronegative centers. In the case of water, the bond is strongest when the OH bond on one water molecule aligns with the lone electron pair orbital of the adjacent water molecule, as shown in Figure 1.4.12. The hydrogen bond requires only 8–40 kJ mol^{-1} to break, compared to much higher values for covalent bonds such as CC (346 kJ mol^{-1}) or CH (411 kJ mol^{-1}). This low bond energy makes it useful, because it means that the bond can form and break under the influence of normal thermal agitation. At the same time, a large number of hydrogen bonds can stabilize structures such as proteins and DNA.

DIPOLE–DIPOLE INTERACTIONS ARE EFFECTIVE ONLY OVER SHORT DISTANCES

By virtue of their spatial separation of charges, dipoles produce an electric field surrounding them whose magnitude is given by

$$[1.4.7] \quad U(r, \theta) = \frac{p \cos \theta}{4 \pi \varepsilon_0 \kappa r^2}$$

An elementary proof of this result is given in Appendix 1.4.A1. The angle θ is defined as the angle between the point at which potential is given and the midpoint between the two separated charges within the dipole. Note that this is a potential, not the force. In Coulomb's law, the potential varied with $1/r$ (see Eqn [1.3.5]) but here it varies with $1/r^2$. Replacing one ion with a dipole causes the interaction to be shorter range. At short range, a nearby charge "sees" both charges; at longer range the charge "sees" two opposing dipole charges that tend to neutralize each other; the interaction becomes weaker at longer distances.

Because they both produce an electric field, pairs of dipoles interact with each other. The energy of interaction is given by

[1.4.8] $$U(r) = -\frac{2}{3kT}\left(\frac{p_A p_B}{4\pi\varepsilon_0 \kappa}\right)^2 \frac{1}{r^6}$$

where k is a new physical constant, Boltzmann's constant, which is the gas constant per molecule. Its value is 1.38×10^{-23} J K^{-1}; T is the temperature in K, p_A and p_B are the dipole moments of the two dipoles, ε_0 is the electrical permittivity of space, given earlier as 8.85×10^{-12} C^2 N^{-1} m^{-2}, and r is the separation of the centers of the dipoles.

LONDON DISPERSION FORCES INVOLVE INDUCED DIPOLES

Water has a permanent dipole moment. Symmetrical compounds such as methane and H_2 have no permanent dipole moment, but these can be induced to form a dipole by the presence of an externally applied electric field. A **polarizable** atom redistributes its internal charge in response to an electric field to form a dipole moment aligned with the applied field. For small electric field strength, the induced dipole is approximately proportional to the applied field:

[1.4.9] $$\mathbf{p}_{ind} = \alpha \mathbf{E}$$

where $\mathbf{p}_{ind}$ is the **induced dipole moment**, $\mathbf{E}$ is the electric field, and α is the **polarizability**. The SI unit for α is C m^2 V^{-1}, but it is often converted to units of volume, cm^3, by multiplying by $1/4\pi\varepsilon_0 \times 10^{-6}$, where ε_0 is the electrical permittivity of space. This effect, shown in Figure 1.4.13, results in the attraction of a neutral molecule to a charged molecule.

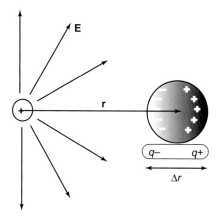

FIGURE 1.4.13 Electrical polarizability. The presence of a charge establishes a local electrical field. Electrons within nearby molecules can respond to this field by redistribution of charge, causing an induced dipole. Because of the separation of charge, the induced dipole experiences unequal forces from the fixed charge. The result is a net movement of the induced dipole.

Even molecules that do not have a permanent dipole moment can transiently produce dipole moments that result in their attraction. We imagine that electrons orbit their nucleus in an "electron cloud" but this picture is an average distribution. At any instant the electrons are separated from their nucleus, producing a transient dipole. When nearby atoms synchronize the distribution of electrons in their clouds, they can produce attractive forces first described by F. London in 1937 and called London dispersion forces.

The totality of forces between atoms or molecules due to dipole–dipole, dipole-induced dipole, and instantaneously induced dipoles (London dispersion force) is called the **van der Waals force**. It excludes the interaction due to covalent bonds or electrostatic interactions.

CLOSE APPROACH OF MOLECULES RESULTS IN A REPULSIVE FORCE THAT COMBINES WITH THE VAN DER WAALS FORCES IN THE LENNARD–JONES POTENTIAL

Imagine two atoms or molecules separated by a large distance. Because of the large distance, their interaction is very small—there is little force between them. Even if they have very little dipole moment, as they approach one another they will experience attractive forces due to London dispersion forces, and these are attractive. As the distance between their atomic nuclei shortens, they begin to experience repulsive forces due to the interpenetration of their atomic or molecular orbitals. This repulsive force varies quite steeply with separation. The overall potential energy for the interaction of nonbonding particles has been mathematically approximated by the **Lennard–Jones potential**. It is given as

$$U(r) = 4\varepsilon\left[\left(\frac{\sigma}{r}\right)^{12} - \left(\frac{\sigma}{r}\right)^6\right] = \varepsilon\left[\left(\frac{r_m}{r}\right)^{12} - 2\left(\frac{r_m}{r}\right)^6\right]$$

[1.4.10]

where $U(r)$ is the potential energy of the interaction, ε is depth of the potential well (a measure of the strength of the attractive forces), and σ is the distance at which the intermolecular potential is zero; r_m is the distance of separation when the potential is $-\varepsilon$ (see Figure 1.4.14).

ATOMS WITHIN MOLECULES WIGGLE AND JIGGLE, AND BONDS STRETCH AND BEND

The bond lengths, angles, and energies listed in Table 1.4.2 are averages. Two atoms involved in a bond actually oscillate back and forth around the average bond length. In addition, the angles defined by, say, HCH are not fixed, but the three atoms oscillate around the average bond angle. The molecules also translate through the solution or gas, and rotate. Some of these motions affect others, as rapid rotation about an axis perpendicular to a bond tends to stretch the bond. All of these modes of action within molecules store kinetic and potential energy on a sub-molecular scale. When a moving body stops because of friction, the kinetic energy does not disappear; it is converted from the macromolecular to the molecular scale—the kinetic energy appears as heat and the temperature of the surfaces increases. This conversion is a one-way street: we can never fully recover that molecular energy back into macromolecular action. The temperature is, in one sense, a measure of this dynamic motion. These various modes of thermal motion are shown in Figure 1.4.15.

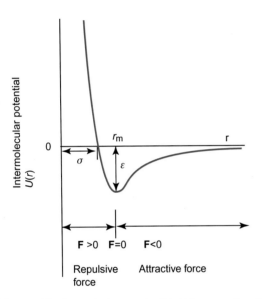

FIGURE 1.4.14 The Lennard–Jones potential. At far separation, there is little force and no potential. Because of the inverse 6th power dependence on the separation, this force becomes larger as particles approach one another, reaching a minimum at $U(r) = \varepsilon$ and $r = r_m$. The separation at which $U(r) = 0$ is $r = \sigma$. The equilibrium for the particles occurs at the minimum potential.

SUMMARY

Atoms are the fundamental units of the elements and are defined by the number of protons in the atomic nucleus. Only relatively few of the chemical elements have active roles in the body. The electrons orbit the nucleus in electron orbitals that have definite energy levels. These electron orbitals describe the distributed nature of the electrons within their atoms. Electrons populate the orbitals in a defined sequence that gives rise to the periodic behavior of the chemical elements within the periodic table.

The chemical elements can combine to form compounds in fixed ratios to each other, often by sharing the outer orbital electrons. These bonds form by combination of their atomic orbitals or by the formation of molecular orbitals in which the electrons are distributed among the nuclei that make up the molecule. Most organic compounds form covalent bonds in which the electrons are shared approximately equally among the nuclei. These bonds have energies in the range 200–500 kJ mol^{-1} and have bond lengths on the order of 0.1–0.2 nm. The bond angles are determined by the kind of bonds that form. Many of the biochemicals within the human body are organic compounds that have CC, CC, CN, COH, CO, and CH bonds. The single bonds typically freely rotate about the axis connecting the two nuclei. Double bonds do not permit free rotation because the rotation would break the second bond and this requires energy.

The rigid form of the CC bond gives rise to geometric isomerism. The different arrangement of the same nuclei around the CC bond can cause large differences in the overall shape of molecules. Isomerism also arises from the different spatial arrangement of chemical groups around an asymmetric C atom. Such isomers can be mirror images of each other (enantiomers) or not. All of the amino acids (except glycine) exist in enantiomeric forms, but only the L-type of amino acids are used to make proteins.

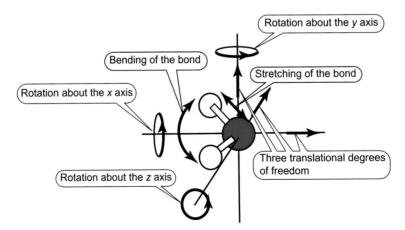

FIGURE 1.4.15 Degrees of freedom of motion in water. The HOH bond can bend, changing its angle; the OH bond can oscillate, stretch, and compress its bond length; the entire molecule can rotate around several independent axes; the molecule can translate in three independent directions. Each of these modes of movement is independent and each carries some kinetic energy. The energy distributed amongst the various modes of motion increases with increase in temperature.

Polar bonds involve unequal sharing of bonding electrons between the two nuclei involved in the bond. A good example is the OH bond in which the oxygen atom is more electronegative. Electronegativity refers to the ability of a nucleus to attract shared electrons. Because O is more electronegative than H, the resulting OH bond is polar, with more negative charge around the O atom than the H atom. This partial separation of charge produces a dipole which is described by its dipole moment, equal to the charge times its separation directed from the − to the + charge. The dipole moment produces an electric field that can interact with other nearby atoms. The OH group is very important because it forms hydrogen bonds with other electronegative atoms. This low energy bond is easy to form and easy to break and can stabilize biological structures.

Ionic bonds occur when two atoms or chemical groups differ greatly in their electronegativities. The more electronegative atom or group effectively "steals" an electron from the less electronegative atom. When dissolved in water, these types of compounds typically dissociate into a cation and anion.

Molecules can interact with each other through ion–ion interactions, described by Coulomb's law, hydrogen bonding, dipole–dipole interactions, and London dispersion forces. The London dispersion forces arise from transient dipoles in atoms that induce transient dipoles in nearby atoms, producing an attractive force. The Lennard–Jones potential describes the overall interaction between nonbonded particles.

Atoms and molecules form surfaces that interact with other surfaces. Almost all of physiology is about how these surfaces interact to produce catalysis or tight binding or loose binding or specificity of binding.

REVIEW QUESTIONS

1. What is the fundamental unit of an element? What is the fundamental unit of a chemical compound?
2. Why do atoms form compounds in definite and fixed proportions?
3. What forms the surfaces of atoms and molecules?
4. Why are the CH bonds in methane arranged to point toward the vertices of a tetrahedron?
5. What is the tetrahedral angle?
6. What are typical energies for CH and CC bonds?
7. What are typical bond distances for CH and CC bonds?
8. What is meant by structural isomerism?
9. What is meant by geometric isomerism?
10. What is optical isomerism?
11. Why does water form polar covalent bonds?
12. What is hydrogen bonding? How much energy is in a hydrogen bond?
13. What is an electric dipole?
14. What are dipole–dipole interactions?
15. What are London dispersion forces?

APPENDIX 1.4.A1 DIPOLE MOMENT

A DIPOLE CONSISTS OF TWO EQUAL CHARGES SEPARATED BY A DISTANCE, D, AND IS DESCRIBED BY ITS ELECTRIC DIPOLE MOMENT

An electric dipole consists of two equal charges, q_+ and q_-, separated by a distance **d**, as shown in Figure 1.4.A1.1. These are typically molecules whose separation distance is small compared to the distance at which electrical effects are noted. As we shall see, cardiomyocytes can also act as electric dipoles. The electric dipole moment is *defined* as

[1.4.A1.1] $$\mathbf{p} = q_+ \mathbf{d}$$

where **d** is a vector pointing from q_- to q_+, as shown in Figure 1.4.A1.1. The net force on a unit positive test charge at any point surrounding the dipole can be determined by the vectorial sum of the component forces from q_+ and q_-, as shown in Figure 1.4.A1.1. We can determine a set of points surrounding the dipole that has the same magnitude of force, but in varying directions. This set forms a curve. The family of lines for a set of force magnitudes, shown in Figure 1.4.A1.2, represents the **electric field** surrounding the dipole. Moving a positive unit charge from a large distance away (∞) to any point within the electric field entails

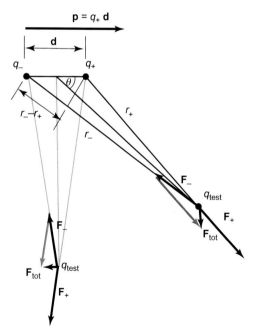

FIGURE 1.4.A1.1 Origin of the electrical forces near an electric dipole determined at two different locations. The electric dipole consists of two equal charges (q_+ and q_-) separated by a distance, **d**. Charges placed nearby experience a force as a result of the electric dipole. The net force is the vector sum of the forces exerted by the two charges, as shown for two different positions. Because the force declines as $1/r^2$, the direction of the force changes with distance and angle, θ, made between the line joining the center of the dipole and the test charge, and the electric dipole moment.

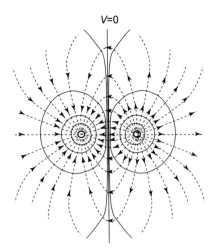

FIGURE 1.4.A1.2 Electric field and electrical potential surrounding a dipole. The electric field lines are shown in *dashed lines with arrows*. They represent the lines of equal force that would be felt by a positive unit test charge as it approaches the dipole. The electric field is a vector which, when magnified by the size of a charge, gives the magnitude and direction of the force. The electric potential contour lines are shown in *solid*. The negative gradient of these contour lines are the electric field lines. Thus the electric field is the steepest slope down the potential surface. The lines of equal potential intersect the electric field lines at right angles, much like the steepest descent off a hill intersects its altitude contour lines at right angles.

expending energy that is stored as potential energy, which is identified as the electrical potential. The potential at any point is the scalar sum of the potential energy associated with each charge. That is, the potential at any point is just the sum of the potential energy of q_+ and that of q_-. Contour lines of equal potential are shown in Figure 1.4.A1.2. These lines intersect the lines of electric force at right angles.

The electric potential surrounding a dipole is just the sum of the potentials associated with each charge. Thus we write:

[1.4.A1.2]
$$\begin{aligned} V_{\text{total}} &= V_+ + V_- \\ &= \frac{q_+}{4\pi\varepsilon_0 r_+} + \frac{q_+}{4\pi\varepsilon_0 r_-} \\ &= \frac{q_+}{4\pi\varepsilon_0}\left[\frac{1}{r_+} - \frac{1}{r_-}\right] \\ &= \frac{q_+}{4\pi\varepsilon_0}\left[\frac{r_- - r_+}{r_+ r_-}\right] \\ &\approx \frac{q_+ \mathbf{d}\cos\theta}{4\pi\varepsilon_0 r^2} = \frac{\mathbf{p}\cos\theta}{4\pi\varepsilon_0 r^2} \end{aligned}$$

where the quantity $q_+\mathbf{d}$ appears. We have identified this as the electric dipole moment, $\mathbf{p} = q_+\mathbf{d}$. Here $\mathbf{d}\cos\theta$ enters the equation as an approximation of $r_- - r_+$, as suggested by the geometry of Figure 1.4.A1.1. This assumption is generally made when $r \gg d$. Thus the voltage at any point is inversely proportional to the square of the distance and varies with the relative position of the point with respect to $\mathbf{p}$, the electric dipole moment. The point of these calculations is to show that the electric dipole creates an electric field and a potential field that can be measured some distance away. The value of the potential depends on the distance and the relative position compared to the dipole.

Appendix I
Important Equations

1. Definition of Flux and Flow:

The flux is given as:

[1.2.1]
$$J_V = \frac{Q_V}{A}$$
$$J_S = \frac{Q_S}{A}$$

where the volume flux, J_V, is defined as the volume flow, Q_V, per unit cross-sectional area (the area element being normal to the direction of flow), and the solute flux, J_S, is the flow of solute per unit cross-sectional area. The volume flow is the volume per unit time; the solute flow is the amount of solute (expressed as grams or moles) per unit time.

2. The Continuity Equation:

The Continuity Equation is given as:

[1.2.9]
$$\frac{\partial C(x,t)}{\partial t} = -\frac{\partial J(x)}{\partial x}$$

where C is the concentration and J is the flux. What this equation says is that if the flux is not the same everywhere, concentration must be changing—either building up or being depleted.

3. Pressure of a Column of Fluid:

The pressure of a column of fluid on the surface of the earth is given by:

[1.2.16]
$$P = \frac{F}{A} = \rho g h$$

where P is the pressure, ρ is the density of the fluid, g is the acceleration due to gravity, and h is the height of the column.

4. Poiseuille's Law:

Poiseuille's Law states:

[1.2.17]
$$Q_V = \frac{\pi a^4}{8\eta}\left(\frac{\Delta P}{\Delta x}\right)$$

The equation describes the relationship between pressure and laminar flow through a narrow and long tube. Q_V is the flow through the tube, a is the inner radius of the tube, π is the geometric ratio of the diameter to the circumference of a circle, ΔP is the pressure difference between the ends of the tube, Δx is the length of the tube, and η is the viscosity of the medium.

5. Definition of Viscosity:

Viscosity is defined from:

[1.2.18]
$$\frac{F}{A} = \eta \frac{dv}{dy}$$

where F is the force, A is the area parallel to the force (F/A is the shear stress), η is the viscosity (usually in Pa s), and dv/dy is the transverse gradient of velocity.

6. Law of Laplace:

The Law of Laplace for a cylinder is:

[1.2.20]
$$\Delta P = \frac{T}{r}$$

where ΔP is the pressure across the wall, T is the wall tension, and r is the radius. For a sphere, this is

[1.2.22]
$$\Delta P = \frac{2T}{r}$$

7. Coulomb's Law:

[1.3.1]
$$\mathbf{F} = \frac{q_1 q_2}{4\pi\varepsilon_0 r^2}$$
$$\mathbf{F}_{1 \text{ on } 2} = \frac{q_1 q_2 \mathbf{r}_{12}}{4\pi\varepsilon_0 r^3} = -\mathbf{F}_{2 \text{ on } 1}$$

where $\mathbf{F}$ is the electrostatic force between two point charges of magnitude q_1 and q_2, r is the distance separating the two charges, π is the geometric ratio, and ε_0 is a physical constant, the electrical permittivity of space. The top equation gives the magnitude of the force; the bottom equation gives its direction.

8. Definition of Electrical Potential:

[1.3.5]
$$U_A = -\int_{\infty}^{A} \frac{\mathbf{F}_{\text{int}} \cdot \mathbf{ds}}{q_{\text{test}}}$$

where U_A is the potential at point A, $\mathbf{F}_{\text{int}}$ is the interaction force acting on a positive test charge of magnitude q_{test}, and $\mathbf{ds}$ is the distance increment. Electrical potential is expressed in volts.

9. Definition of Electric Field Intensity:

The electric field intensity **E** is given by:

[1.3.9][1.3.11] $$\mathbf{E} = \frac{\mathbf{F}_{int}}{q_{test}} = -\nabla U$$

where **F** is the electrical interaction force on a charge of magnitude q_{test}, U is the potential, and ∇ is the gradient operator that converts the scalar function U into a vector with components equal to the partial derivative of U along each axis.

10. Electrical Capacitance:

Electrical capacitance is defined as:

[1.3.16] $$C = \frac{Q}{V}$$

where C is the capacitance, Q is the charge, and V is the voltage difference across the capacitor.

11. Capacitance of Two Parallel Plates:

The capacitance between two parallel plates is given as:

[1.3.21] $$C = \frac{\kappa \varepsilon_0 A}{\delta}$$

where C is the capacitance (in farads = C V^{-1}), κ is the dielectric constant characteristic of the material between the plates (a dimensionless ratio), ε_0 is the electrical permittivity of the vacuum = 8.85×10^{-12} C^2 J^{-1} m^{-1}, A is the area (in m^2), and δ is the distance between the plates (m).

12. Drag Force and Velocity:

[1.3.23] $$\mathbf{F}_d = -\beta v$$

where $\mathbf{F}_d$ is the drag force on a particle moving at constant velocity, v, and β is the drag coefficient.

13. Relation Between Flux and Velocity:

When there is a flux of material, the average velocity of the particles is determined by:

[1.3.24] $$\mathbf{J}_s = vC$$

where J_s is the flux of solute, v is the average velocity, and C is the concentration.

14. The Dipole Moment:

The dipole moment is defined as:

[1.4.5] $$\mathbf{p} = q\mathbf{d}$$

where **p** is the dipole moment, q is the charge separated in the dipole, and **d** is the distance vector pointing from q_- to q_+.

15. Fick's Dilution Principle:

$$\text{Volume} = \frac{\text{amount}}{\text{concentration}}$$

[1.5.3] $$V = \frac{m}{C}$$

where V is the volume of distribution, m is the amount (either in mass units or in mole units), and C is the concentration, either in mass per unit volume or moles per unit volume.

16. The Arrhenius Equation:

The Arrhenius equation describes the dependence of a rate constant on the temperature. It is given as:

[1.5.16] $$k = A\, e^{-E_a/RT}$$

where k is the rate constant, A is the preexponential factor, E_a is the activation energy, R is the ideal gas constant, and T is the absolute temperature. The equation in Chapter 1.5 takes the natural logarithm of the rate constant.

17. The Michaelis–Menten Equation:

The Michaelis–Menten equation describes a particular class of enzymatic reactions that obey simple saturation behavior. Not all enzyme reactions have kinetics that can be described in this way. The equation is:

[1.5.26] $$J = \frac{J_{max}[S]}{K_m + [S]}$$

where J is the enzyme reaction rate, J_{max} is the maximum reaction rate, $[S]$ is the substrate concentration, and K_m is the Michaelis–Menten constant, which is equal to the substrate concentration when the enzyme is one-half maximal.

18. Fick's First Law of Diffusion:

Fick's First Law of Diffusion is given as:

[1.6.3] $$J_s = -D\frac{\partial C}{\partial x}$$

where J_s is the solute flux, D is the diffusion coefficient, C is the concentration, and x refers to distance along the line parallel to the flux. This equation is valid only when there is no other force on the diffusing solute such as an electric field if the solute is charged, or solvent drag. Strictly speaking, J is a vector and so is the gradient. In three dimensions, the equation can be written as:

$$\mathbf{J} = -D\nabla C = -D\left(\frac{\partial C}{\partial x}\mathbf{i} + \frac{\partial C}{\partial y}\mathbf{j} + \frac{\partial C}{\partial z}\mathbf{k}\right)$$

19. Fick's Second Law of Diffusion:

Fick's Second Law of Diffusion is derived from the First Law and the Continuity Equation. It is

[1.6.6] $$\frac{\partial C}{\partial t} = D\frac{\partial^2 C}{\partial x^2}$$

20. Time of Diffusion:

The time for one-dimensional diffusion is estimated as:

[1.6.33] $$x = \sqrt{\overline{x^2}} = \sqrt{2D\Delta t}$$

where x is the distance, D is the diffusion coefficient, and Δt is the time.

21. Concentration Profile for an Initially Sharp Concentration:

For an initially narrow distribution, the distribution after time t and as a function of distance away from the initial source is given as:

$$[1.6.34] \quad C(x,t) = C_0 \sqrt{\frac{1}{4\pi Dt}} e^{\frac{-x^2}{4Dt}}$$

where $C(x,t)$ is the concentration at x after time t, C_0 is the initial concentration at position $x = 0$, D is the diffusion coefficient, and t is the elapsed time.

22. Convection–Diffusion Equation:

Diffusive flux can be augmented by the movement of solvent in solvent drag. The solute flux is given by where the first term on the right-hand side is due to diffusion and the second term is solvent drag.

$$[1.6.36] \quad J_s(x,t) = -D\frac{\partial C(x,t)}{\partial x} + J_v(t)C(x,t)$$

23. Fick's First Law of Diffusion with an External Force:

When a solute is subject to a force, f, per particle, along one dimension, the one-dimensional Fick's equation is modified to:

$$[1.6.42] \quad J = -D\frac{\partial C}{\partial x} + \frac{D}{kT}fC$$

where J is the flux, D is the diffusion coefficient, k is Boltzmann's constant, T is the absolute temperature, f is the force per particle, and C is the concentration.

24. Drag Force and Einstein's Frictional Coefficient:

Particles subject to an external force are accelerated until they reach a terminal velocity. At this time the drag force, $\mathbf{F_d}$, balances the external force, f. It is given as:

$$[1.6.37] \quad \mathbf{F_d} = -f = -\beta v$$

where v is the terminal velocity and β is the frictional coefficient. It is given as:

$$[1.6.38] \quad \beta = \frac{kT}{D}$$

where k is Boltzmann's constant, T is the absolute temperature, and D is the diffusion coefficient.

25. Stokes–Einstein Equation:

For a spherical particle, Stokes determined the frictional coefficient as a function of particle size and medium viscosity. Combining it with the Einstein's frictional coefficient, we derive

$$[1.6.45] \quad 6\pi\eta a_s = \frac{kT}{D}$$

where D is the diffusion coefficient, k is Boltzmann's constant, T is the absolute temperature, π is the geometric ratio, η is the fluid viscosity, and a_s is the radius of the spherical particle.

26. The Electrochemical Potential:

The electrochemical potential of a charged species is given by:

$$[1.7.13] \quad \mu = \mu^0 + RT \ln C + z\mathfrak{F}\Psi$$

where μ^0 sets the zero of the function at some standard state; the second term, $RT \ln C$, is the work of concentrating the material to the concentration, C; the last term is the work necessary to bring the charged species from zero potential to Ψ. The electrochemical potential can be expanded to include other work terms such as pressure–volume work.

27. Gibbs Free Energy:

There are several thermodynamic functions called "free energy." The Gibbs free energy is the free energy under conditions of constant temperature and pressure. You need to know two things about the Gibbs free energy. First, it depends on the composition of a mixture:

$$[1.7.24] \quad G = \sum_i \mu_i n_i$$

where G is the Gibbs free energy, μ_i is the chemical potential of species i in the mixture, and n_i is the number of moles of species i. Second, all spontaneous reactions occur with a decrease in free energy. Thus some reactions with $\Delta G > 0$ can occur by being linked to other reactions so that the overall $\Delta G < 0$.

$\Delta G < 0 \rightarrow$ spontaneous reaction
$\Delta G = 0 \rightarrow$ reaction is at equilibrium
$\Delta G > 0 \rightarrow$ opposite reaction occurs spontaneously
[1.7.25]

28. Gibbs Free Energy of ATP Hydrolysis:

The free energy of ATP hydrolysis is given as:

$$[1.7.30] \quad \Delta G = \Delta G^0 + RT \ln\left(\frac{[ADP][Pi]}{[ATP]}\right)$$

where ΔG^0 is the standard free energy change under standard conditions of 1 M [ADP], [Pi], and [ATP], 250°C, and 1 atm P.

29. Relation Between Gibbs Free Energy and K_{eq}:

Equilibrium occurs when $\Delta G = 0$. This fact can be used in Eqn [1.7.30] to find

$$[1.7.32] \quad \Delta G^0 = -RT \ln K_{eq}$$

where K_{eq} is the equilibrium constant. Note that K_{eq} depends on the absolute temperature, T.

30. Absorbance:

Absorbance in spectrophotometry is defined as:

$$A = \log\frac{I_0}{I}$$

where A is the absorbance, I_0 is the intensity of incident light, and I is the intensity of transmitted light.

31. The Beer–Lambert Law:

This relates absorbance to concentration:

$$A = \varepsilon C d$$

where A is the absorbance, ε is the molar extinction coefficient that depends on the wavelength of the incident light and the chemical nature of the absorbing material, C is the concentration of the absorbing chemical, and d is the path length. The molar extinction coefficient is usually determined with a path length $d = 1$ cm. Often the equation is given with d omitted, with the assumption that it is 1 cm and the molar extinction coefficient is in units of M^{-1}.

32. Microscopic Resolution:

The resolution of a light microscope is given by:

[2.1.A1.1] $$\text{Resolution} = \frac{0.61 \lambda}{\eta \sin \alpha}$$

where λ is the wavelength of the light, η is the refractive index of the medium, and α is the angle of the cone of light collected by the objective lens. The combined terms $\eta \sin \alpha$ are collectively known as the **numerical aperture**. Note that the resolution has the units of distance and refers to the separation of just discernable circular disks. Thus high resolution is associated with small distances.

33. Relative Centrifugal Force:

Relative centrifugal force is given by the formula:

[2.1.A1.15] $$\text{RCF} = \frac{\omega^2 r}{g}$$

where RCF is given in multiples of the acceleration due to gravity at the surface of the earth, ω is the angular velocity ($= 2\pi \times$ revolutions per second), and r is the distance from the center of rotation.

34. Rate of Centrifugal Sedimentation:

The rate of centrifugal sedimentation is given by:

[2.1.A1.17] $$m(1 - \bar{v}\rho) = \beta \left[\frac{dr/dt}{\omega^2 r}\right]$$

where m is the mass of the sedimenting particle, $\bar{V}$ is the partial specific volume, ρ is the density of the medium through which the particle is sedimenting, β is the drag or frictional coefficient, dr/dt is the radial velocity of sedimentation, and $\omega^2 r$ is the magnitude of the centripetal acceleration.

35. Svedberg:

The **Svedberg** is a unit of sedimentation per unit centrifugal force and is defined as:

[2.1.A1.18] $$s = \left[\frac{dr/dt}{\omega^2 r}\right]$$

where s is the Svedberg, dr/dt is the radial velocity of sedimentation, and $\omega^2 r$ is the magnitude of centripetal acceleration.

36. Stokes' Equation:

This gives the frictional or drag coefficient for a spherical particle as:

[2.1.A1.21] $$\beta = 6\pi \eta a_s$$

where η is the viscosity of the medium and a_s is the radius of the sphere.

37. The Partition Coefficient:

The partition coefficient describes the distribution of a material between the watery phase and an immiscible organic phase. It is defined as:

[2.3.1] $$k_s = \frac{[X]_{\text{organic phase}}}{[X]_{\text{water phase}}}$$

where k_s is the partition coefficient and $[x]$ is the concentration of substance X either in the organic phase or in the aqueous phase.

38. Surface Tension:

It is defined by:

[2.4.1] $$dG = \gamma\, dA$$

where G is the Gibbs free energy, A is the area, and γ is the surface tension. Because G is in units of J, or N m, and A is in units of m^2, it follows that γ is in units of $N\, m^{-1}$.

39. Permeability:

The permeability refers to the ability of materials to penetrate membranes. It is operationally defined as:

[2.5.9] $$p = \frac{J_s}{\Delta C}$$

where p is the permeability, J_s is the solute flux in the absence of J_v, and ΔC is the concentration difference across the membrane that drives the passive solute flux.

The permeability depends on the characteristics of the membrane–solute pair. For a solute that penetrates the membrane through microscopic pores, the permeability is given as:

[2.5.10] $$p = \frac{n \pi a^2 D}{\delta}$$

where p is the permeability, n is the number of pores per unit area, π is the geometric ratio, a is the radius of the pore, D is the diffusion coefficient of the solute within the pore, and δ is the thickness of the membrane. For a solute that penetrates through the membrane by dissolving in the lipid phase of the membrane, the permeability is given as:

[2.5.14] $$p = \frac{k_s D_{s,\text{lipid}}}{\delta}$$

where p is the permeability, k_s is the partition coefficient of the solute between the aqueous and lipid phases, D is the diffusion coefficient of the solute within the lipid phase, and δ is the thickness of the membrane.

40. Hydraulic Conductivity:

Pressure-driven laminar flow through a membrane is described phenomenologically as:

[2.7.22] $$J_v = L_p \Delta P$$

where J_v is the volume flux, in mL cm^{-2} s^{-1}, ΔP is the pressure difference that drives bulk flow, and L_p is the hydraulic conductivity, filtration coefficient, or hydraulic permeability.

41. van't Hoff's Law:

The osmotic pressure of an ideal, dilute solution is given as:

[2.7.15] $$\pi = RTC_s$$

where π is the osmotic pressure, R is the gas constant, T is the absolute temperature, and C is the concentration of impermeant solute, in molar. This is a limiting law which is approximately true only for dilute ideal solutions.

42. Osmotic Coefficient:

The osmotic coefficient is defined as:

[2.7.16] $$\varphi = \frac{\pi_{observed}}{RTC_s}$$

It corrects for the nonideality of solutions.

43. Reflection Coefficient:

The reflection coefficient is a characteristic of a membrane and is defined as:

[2.7.24] $$\sigma = \frac{\pi_{eff}}{\varphi RTC}$$

where π_{eff} is the effective osmotic pressure observed with a real membrane which is not perfectly semipermeable.

44. Volume Flux in the Presence of ΔP and $\Delta \pi$:

The volume flux across a real membrane which is not perfectly semipermeable to solute is given as:

[2.7.42] $$J_v = L_p[\Delta P - \sigma \Delta \pi]$$

where L_p is the hydraulic permeability, ΔP is the pressure difference across the membrane, σ is the reflection coefficient, and $\Delta \pi$ is the osmotic pressure difference across the membrane.

45. Volume Response of Cells to Osmotic Changes:

When a cell having an initial volume V_0 is in contact with a solution of osmolarity $\pi_{isotonic}$ and then the bathing solution is changed to one of osmolarity π, it alters its volume according to:

[2.7.48] $$\frac{V_c}{V_0} = \left(1 - \frac{V_b}{V_0}\right)\frac{\pi_{isotonic}}{\pi} + \frac{V_b}{V_0}$$

where V_c is the new cell volume at equilibrium and V_b is a volume taken to be the osmotically inactive volume of the cell.

46. Proton Motive Force:

The proton motive force is the electrochemical potential difference of H$^+$ across the inner mitochondrial membrane. It is given as:

[2.10.10] $$\Delta \mu_{out \to in} = RT \ln \frac{[H^+]_{in}}{[H^+]_{out}} + \mathfrak{F}(\psi_{in} - \psi_{out})$$

Here it can be seen that the concentration contribution can be expressed in volts.

47. The Nernst Equation:

At equilibrium, the membrane potential across a membrane permeable to only one ion is given as:

[3.1.5] $$\frac{RT}{z\mathfrak{F}} \ln \frac{[X^+]_o}{[X^+]_i} = \psi_i - \psi_o$$

where R is the gas constant ($=8.314$ J mol^{-1} K^{-1}), T is the absolute temperature, z is the valence ($\pm$ integer), $\mathfrak{F}$ is the Faraday $= 96,500$ C mol^{-1}; ln is the natural logarithm, [X] is the concentration of ion X, ψ is the membrane potential inside or outside of the membrane. Subscript i indicates inside the membrane, o indicates outside of the membrane.

48. Goldman–Hodgkin–Katz Current Equation:

The current across a membrane that is carried by a single ion is given as:

[3.1.14] $$I_{ion} = \frac{(D/\delta)[(z^2\mathfrak{F}^2/RT)E_m][C_i - C_o e^{(-z\mathfrak{F}/RT)E_m}]}{[1 - e^{(-z\mathfrak{F}/RT)E_m}]}$$

where z is the valence of the ion, E_m is the membrane potential, $\mathfrak{F}$ is the Faraday ($=96,500$ C mol^{-1}), R is the gas constant ($=8.314$ J mol^{-1} K^{-1}), and T is the absolute temperature; C_i and C_o are the inside and outside concentrations of the ion, respectively.

49. Goldman–Hodgkin–Katz Equation:

The potential across a membrane which is permeable to a variety of ions is given as:

[3.1.16] $$E_m = \frac{RT}{\mathfrak{F}} \ln \left[\frac{P_K[K^+]_o + P_{Na}[Na^+]_o + P_{Cl}[Cl^-]_i}{P_K[K^+]_i + P_{Na}[Na^+]_i + P_{Cl}[Cl^-]_o}\right]$$

where E_m is the electrical potential across the membrane, R is the gas constant (8.314 J mol^{-1} K^{-1}), $\mathfrak{F}$ is the Faraday (9.65×10^4 C mol^{-1}), P_i is the permeability of the i-th ion.

50. Chord Conductance:

The chord conductance is defined as:

[3.1.18] $$I_i = g_i(E_m - E_i)$$

where g_i is the chord conductance of ion i, E_m is the membrane potential, E_i is the equilibrium potential of ion i, and I_i is the current carried by the ion.

51. Slope Conductance:

The slope conductance is defined as:

[3.1.19] $$g_i = \frac{dI_i}{dE_m}$$

where g_i is the conductance of ion i, I_i is the current carried by the ion, and E_m is the membrane potential.

52. The Chord Conductance Equation:

The membrane potential at steady-state or resting conditions is given as:

$$E_m = \frac{g_{Na}}{(g_{Na} + g_K + g_{Cl})} E_{Na} + \frac{g_K}{(g_{Na} + g_K + g_{Cl})} E_K$$
$$+ \frac{g_{Cl}}{(g_{Na} + g_K + g_{Cl})} E_{Cl}$$

[3.1.22]

where E_m is the resting membrane potential and g_i is the chord conductance of ion i.

53. Transmembrane Resistance:

The transmembrane resistance is given as:

[3.3.4] $$R = \frac{R_m}{A}$$

where R_m is the transmembrane resistance of a unit area of membrane and A is the area of the membrane. R is in Ω; R_m is in units of $\Omega\ cm^2$.

54. Axoplasmic Resistance:

The axoplasmic resistance is given by:

[3.3.5] $$R_i = \frac{\rho_i d}{A}$$

where R_i is the internal resistance, d is the distance down the axon, A is the cross-sectional area, and ρ_i is the specific resistance in $\Omega\ cm$.

55. Electrotonus:

Injection of current in an axon causes the membrane potential to vary with distance and time from the point of current injection. At infinite time, a steady state is reached, electrotonus. The voltage depends on the distance from the point of current injection as:

[3.3.17] $$V = (V_0 - V_r)e^{(-x/\lambda)} + V_r$$

where V is the voltage at point x, V_0 is the voltage at the point of current injection, V_r is the voltage infinitely far away (the resting potential), x is the distance from the point of current injection, and λ is the space constant, in units of distance.

56. Space Constant (or length constant) for an Axon:

[3.3.15] $$\lambda = \sqrt{\frac{R_m \pi a^2}{\rho_i 2\pi a}} = \sqrt{\frac{R_m}{\rho_i} \frac{a}{2}}$$

where λ is the space constant, R_m is the specific transmembrane resistance, a is the radius of the axon, and ρ_i is the specific resistance of the axoplasm.

57. Decay Under Space Clamp:

If the internal resistance is made small (=space clamp), temporal decay of small departures from the resting membrane potential is given by:

[3.3.22] $$V = (V_0 - V_r)e^{-(t/\tau)} + V_r$$

where V is the voltage at time t, V_0 is the voltage at the beginning of the decay, and V_r is the voltage at infinite time, the resting membrane potential; t is the time and τ is the time constant, in units of time.

58. Time Constant for an Axon:

[3.3.15] $$\tau = R_m C_m$$

where τ is the time constant in s^{-1}, R_m is the specific resistance of the axon membrane (in $\Omega\ cm^2$), and C_m is the specific capacitance (in $F\ cm^{-2}$).

59. Weber–Fechner Law:

The Weber–Fechner Law attempts to relate sensory perception to stimulus intensity. It is written as:

[4.3.1] $$\Delta S = k \frac{\Delta W}{W}$$

where k is a constant, W is the stimulus strength, and S is the sensory perception.

60. Stevens' Power Law:

Steven's Power Law attempts to relate sensory perception to stimulus intensity. It is written as:

[4.3.3] $$\Psi = k\Phi^n$$

where Ψ is the psychological sensation, k is a constant, n is a constant, and Φ is the stimulus intensity.

61. Decibel:

The Decibel is defined as:

[4.7.1] $$\text{Loudness (decibels)} = 10 \log \frac{I_{sound}}{I_{ref}}$$

[4.7.2] $$\text{Loudness (dB)} = 20 \log \frac{\Delta P_{sound}}{\Delta P_{ref}}$$

where I is the intensity in $W\ cm^{-2}$ and ΔP is the increment in pressure over ambient pressure. The reference intensity that physicists use is $10^{-16}\ W\ cm^{-2} = 10^{-12}\ W\ m^{-2}$.

62. Sound Intensity:

The intensity of sound, in $W\ m^{-2}$, is related to the pressure increment according to:

[4.7.A1.32] $$I = \frac{\Delta P_0^2}{2\rho c}$$

where I is the intensity, ΔP_0 is the pressure amplitude, ρ is the density of the air, and c is the velocity of sound.

63. Snell's Law of Refraction:

Snell's law of refraction describes the relationship between the angle of incidence of incoming light, θ_1, and the angle of refraction, θ_2, as being given by:

[4.8.A1.4] $$n_i \sin \theta_i = n_r \sin \theta_r$$

where n_1 is the index of refraction in medium 1 and n_2 is the index of refraction in medium 2.

64. The Thin Lens Formula:

The thin lens formula gives the relationship between the distance to the object from the nodal point, the distance to its image, and the focal length. It is

[4.8.A1.12] $$\frac{1}{O} + \frac{1}{I} = \frac{1}{f}$$

where O is the distance (in m) to the object, I is the distance (in m) to the image, and f is the focal length.

65. Pressure-Driven Flow:

Pressure drives flow with the following relation:

[5.1.3] $$Q = \frac{\Delta P}{R}$$

where Q is the flow, ΔP is the pressure difference that drives the flow, and R is the resistance to the flow. This is the hydraulic analogy to Ohm's law.

66. Mechanical Compliance:

Mechanical compliance is defined as:

[5.1.4] $$C = \frac{\Delta V}{\Delta P}$$

where C is the compliance, ΔV is a volume increment, and ΔP is the pressure increment accompanying ΔV.

67. Hematocrit:

The hematocrit is defined as

$$Hct = \frac{\text{packed cell volume}}{\text{total blood volume}} \times 100$$

which is a number between 0 and 100, or it is defined as the hematocrit ratio, which is the same ratio but without multiplying it by 100 to convert it to percent.

68. Addition of Cardiac Leads I, II, and III:

The three Einthoven leads are defined as:

Left arm − right arm = LEAD I
Left leg − right arm = LEAD II
Left leg − left arm = LEAD III

Because of their definition, Kirchhoff's voltage law is applied as:

[5.6.3] $$I + III = II$$

69. Definition of Stroke Volume:

The stroke volume is the difference between end diastolic volume and end systolic volume:

[5.8.1] $$SV = EDV - ESV$$

where SV is the stroke volume, EDV is the end diastolic volume, and ESV is the end systolic volume.

70. Calculation of Cardiac Output:

Cardiac output can be most easily calculated as:

[5.8.2] $$CO = SV \times HR$$

where CO is the cardiac output, SV is the stroke volume, and HR is the heart rate.

71. Fick Principle of Estimation of Cardiac Output:

The cardiac output can be measured from the total rate of oxygen consumption and the difference in arterial and venous blood total oxygen concentration:

[5.8.7] $$Q_a = \frac{Q_{O_2}}{[O_2]_a - [O_2]_v}$$

where Q_a is the cardiac output, Q_{O_2} is the rate of oxygen consumption, $[O_2]_a$ is the total arterial concentration of oxygen, and $[O_2]_v$ is the total venous concentration of oxygen.

72. Estimation of Cardiac Output by the Indicator Dilution Method:

Cardiac output can be estimated by injecting m moles of indicator into the venous side of the circulation and measuring the concentration of indicator on the opposite side of the heart. The equation gives the cardiac output

[5.8.12] $$Q_a = \frac{m}{\int_0^t C_m \, dt}$$

where Q_a is the cardiac output, m is the moles of injected indicator, and C_m is the concentration of the indicator in the arterial side of the circulation; t is the time.

73. Total Energy of a Fluid and Equivalent Pressure:

The total energy of a moving fluid has components of pressure, kinetic energy, and gravitational potential energy:

[5.9.2] $$E = E_P + E_K + E_G = PV + 1/2 \rho V v^2 + \rho g h V$$

The equivalent pressure is obtained by dividing by the volume, V:

[5.9.3] $$P' = P + 1/2 \rho v^2 + \rho g h$$

where P' is the total equivalent pressure, P is the pressure due to PV work, ρ is the density, v is the velocity, g is the acceleration due to gravity, and h is the height above an arbitrary zero reference point.

74. Definition of Pulse Pressure:

The pulse pressure is defined as:

[5.9.7] $$\Delta P_{pulse} = P_{systolic} - P_{diastolic}$$

This is the difference between the systolic pressure and the diastolic pressure.

75. Mean Arterial Pressure:

The mean arterial pressure is given by:

$$[5.9.8] \qquad P_A = \frac{\int_{t_1}^{t_2} P \, dt}{t_2 - t_1}$$

where P_A is the mean arterial pressure, P is the pressure that varies in the arteries with time, and t is the time. This equation is often approximated as:

$$[5.9.9] \qquad P_A = P_{diastolic} + \frac{\Delta P_{pulse}}{3}$$

where $P_{diastolic}$ is the diastolic pressure and ΔP_{pulse} is the pulse pressure.

76. The Reynolds Number:

The Reynolds number is:

$$[5.9.12] \qquad Re = \frac{2a <V> \rho}{\eta}$$

where Re is the Reynolds number, a is the radius of the tube, $<V>$ is the average velocity in the tube, ρ is the density of the fluid, and η is the viscosity. This is taken as an indicator of laminar versus chaotic or turbulent flow.

77. Venous Concentration of Solute:

The venous concentration of solute is given as:

$$[5.10.10] \qquad C_v = C_i + (C_a - C_i) e^{(-Sp/Q_v)}$$

where C_v is the venous concentration, C_i is the interstitial fluid concentration, presumably constant, C_a is the arterial concentration, S is the surface area of the capillaries through which exchange occurs, p is the permeability, and Q_V is the flow through the capillaries.

78. Interstitial Fluid Concentration of Solute:

The interstitial fluid concentration of solute is given as:

$$[5.10.15] \qquad C_i = \frac{C_a + Q_m}{Q_V E}$$

where C_i is the interstitial fluid concentration, C_a is the arterial concentration, Q_m is the rate of metabolism, which could be positive if the substance is produced by the tissue, or negative if the substance is consumed; Q_V is the flow through the capillaries and E is the extraction. The extraction is given as:

$$[5.10.14] \qquad [1 - e^{-(Sp/Q_V)}] = E$$

79. Starling Forces for Filtration of Fluid Out of the Microcirculation:

The net filtration of fluid out of the microcirculation is given as:

$$[5.10.20] \qquad Q_V = A k_f [(P_C - P_i) - \sigma(\pi_C - \pi_i)]$$

where Q_V is the flow out of the capillary (the filtration flow), A is the area through which the flow occurs (the surface area of the capillary), k_f is the filtration coefficient, P_C is the hydrostatic pressure within the capillary, and P_i is the hydrostatic pressure in the interstitial fluid; π_C is the oncotic pressure within the capillary and π_i is the oncotic pressure in the interstitial fluid; σ is the reflection coefficient of those substances contributing to the oncotic pressure.

80. Capillary Hydrostatic Pressure:

The capillary hydrostatic pressure is given as:

$$[5.11.5] \qquad P_C = P_A \left[\frac{R_V}{R_A + R_V} \right] + P_V \left[\frac{R_A}{R_A + R_V} \right]$$

where P_C is the capillary hydrostatic pressure, P_A is the arterial hydrostatic pressure, P_V is the venous hydrostatic pressure, R_A is the resistance of the arteries to flow, and R_V is the resistance of the veins to flow.

81. Definition of Total Peripheral Resistance:

The total peripheral resistance can be determined from

$$[5.12.2] \qquad Q_{veins} = \frac{P_A - P_{RA}}{TPR}$$

where Q_{veins} is the flow through the veins, which at steady state is the cardiac output, P_A is the arterial pressure, P_{RA} is the right atrial pressure, and TPR is the total peripheral resistance.

82. Vascular Function Curve:

The linear portion of the vascular function curve is given as:

$$[5.12.13] \qquad Q_{veins} = -\frac{[1 + (C_V/C_A)]}{TPR} [P_{RA} - P_{MS}]$$

where Q_{veins} is the flow through the vasculature, C_V is the compliance of the veins, C_A is the compliance of the arteries, P_{RA} is the right atrial pressure, P_{MS} is the mean systemic pressure, and TPR is the total peripheral resistance. This relationship describes only the linear portion of the vascular function curve and fails at low pressures when the veins collapse.

83. The Ideal Gas Law:

Ideal gases obey

$$[6.1.3] \qquad PV = nRT$$

where P is the pressure, V is the volume, R is the gas constant ($=0.082$ L atm mol^{-1} K^{-1}), and T is the absolute temperature.

84. Surface Tension:

The surface tension relates the free energy of the surface to its area according to:

$$[6.1.5] \qquad dG = \gamma dA$$

where dG is the change in surface free energy, γ is the surface tension (energy per unit area or force per unit length), and dA is the change in surface area.

85. Definition of Respiratory Quotient:

The respiratory quotient is defined as:

$$[6.3.1] \qquad R = \frac{Q_{CO_2}}{Q_{O_2}}$$

where R is the respiratory quotient, Q_{CO_2} is the rate of CO_2 production by the body, and Q_{O_2} is the rate of oxygen consumption by the body.

86. Definition of Partial Pressure:

The partial pressure of a gas in a mixture of gases is given by:

$$[6.3.6] \qquad P_A = \frac{n_A}{n} P_B = f_A P_B$$

where P_A is the partial pressure of gas A, n_A is the number of moles of gas A, n is the total number of moles of gas in the mixture, f_A is the mole fraction, and P_B is the barometric or total pressure of the mixture of gases.

87. Partial Pressure in Moist Air:

The partial pressure of a gas in moist tracheal air is given by:

$$[6.3.8] \qquad P_A = f_A (P_B - P_{H_2O})$$

where P_A is the partial pressure in the moist air, f_A is the mole fraction in dry air, P_B is the total pressure of the gas mixture, and P_{H_2O} is the partial pressure of water in the moist air. At body temperature, $P_{H_2O} = 47$ mmHg.

88. Henry's Law:

Henry's Law relates the concentration of gas dissolved in a fluid to the partial pressure of the gas in the air in equilibrium with the fluid:

$$[6.3.9] \qquad x_A = \beta_A P_A$$

where x_A is the mole fraction of gas A in the fluid, β_A is the solubility of the gas in the fluid, and P_A is the partial pressure of the gas in the air. This can be converted, approximately, to:

$$[6.3.11] \qquad [A] = \alpha_A P_A$$

where $[A]$ is the concentration of gas in fluid phase, P_A is the partial pressure of the gas in the air in equilibrium with the fluid, and α_A is the solubility in different units from β.

89. Diffusing Capacity:

The diffusing capacity of a gas across the lungs is given as:

$$[6.3.15] \qquad Q_s = D_L \Delta P$$

where Q_s is the flow of gas, D_L is the diffusing capacity, and ΔP is the partial pressure difference that drives diffusion.

90. Alveolar Ventilation:

Alveolar ventilation can be calculated as:

$$[6.3.16] \qquad Q_A = \nu_R (V_T - V_D)$$

where Q_A is the alveolar ventilation (in L min^{-1}), ν_R is the respiratory rate (in breaths min^{-1}), V_T is the tidal volume (in L), and V_D is the dead space volume, part of the tidal volume that does not exchange gas because it does not reach the alveoli (in L).

91. Physiological Dead Space:

The physiological dead space can be calculated from:

$$[6.3.19] \qquad V_D = V_T \left[1 - \frac{f_{E_{CO_2}}}{f_{A_{CO_2}}} \right]$$

where V_D is the dead space volume, V_T is the tidal volume, $f_{E_{CO_2}}$ is the mole fraction of CO_2 in expired air, and $f_{A_{CO_2}}$ is the mole fraction of CO_2 in the alveolar air.

92. Alveolar Ventilation Equation:

The alveolar ventilation can be determined by

$$[6.3.21] \qquad Q_A = \frac{Q_{CO_2}}{P_{A_{CO_2}}} (P_B - 47)$$

where Q_A is the alveolar ventilation, Q_{CO_2} is the rate of CO_2 production, $P_{A_{CO_2}}$ is the partial pressure of CO_2 in alveolar air, P_B is the barometric or atmospheric pressure, in mmHg, and 47 is P_{H_2O} at body temperature, in mmHg.

93. Alveolar Gas Equation:

The partial pressure of O_2 in the alveolar can be calculated from:

$$[6.3.A1.15] \qquad P_{A_{O_2}} = P_{I_{O_2}} - \frac{1}{R} P_{A_{CO_2}} + f_{I_{O_2}} \left(\frac{(1-R)}{R} \right) P_{A_{CO_2}}$$

where $P_{A_{O_2}}$ is the partial pressure of O_2 in the alveoli, $P_{I_{O_2}}$ is the partial pressure of O_2 in the inspired air, R is the respiratory quotient, $f_{I_{O_2}}$ is the mole fraction of O_2 in dry inspired air, and $P_{A_{CO_2}}$ is the partial pressure of CO_2 in alveolar air.

94. Conversion of V_{BTPS} to V_{STPD}:

$$[6.3.A2.3] \qquad V_{BTPS} = \frac{310}{273} \frac{760}{(760-47)} V_{STPD}$$

$$= 1.2104 V_{STPD}$$

where V_{BTPS} is the volume at body temperature and pressure, saturated with water vapor, and V_{STPD} is the volume of gas at standard temperature and pressure, dry.

95. The Hill Equation for O_2 Binding to Hemoglobin:

The Hill equation for O_2 binding to hemoglobin is given as:

$$[6.4.2] \qquad [Hb \cdot O_2] = \frac{[O_2]^h}{K + [O_2]^h} [Hb \cdot O_2]_{max}$$

where $[Hb \cdot O_2]$ is the concentration of hemoglobin with oxygen bound, $[O_2]$ is the oxygen concentration (or partial pressure), K is the affinity constant, h is the exponent of the O_2 concentration, and $[Hb \cdot O_2]_{max}$ is the maximum O_2 binding capacity of hemoglobin.

APPENDIX I: IMPORTANT EQUATIONS

96. Determination of Oxygen Consumption:

Oxygen consumption can be calculated from the cardiac output and A–V difference in total $[O_2]$:

[6.4.4] $$Q_a([O_2]_a - [O_2]_v) = Q_{O_2}$$

where Q_a is the flow through the arteries (assumed equal to the flow through the veins, equal to the cardiac output); $[O_2]_a$ is the total arterial content of O_2, in mL of O_2 at STPD per 100 mL of blood; $[O_2]_v$ is the total venous content of O_2 at STPD per 100 mL of blood; Q_{O_2} is the rate of oxygen consumption, given as a positive value as mL O_2 at STPD per min.

Oxygen consumption can also be calculated from the respiratory gases:

[6.4.5] $$Q_T^* f_{I_{O_2}} = Q_T f_{E_{O_2}} + Q_{O_2}$$

where Q_T^* is the flow of inspired air, $f_{I_{O_2}}$ is the mole fraction of O_2 in the inspired air, Q_T is the flow of expired air, $f_{E_{O_2}}$ is the mole fraction of O_2 in the expired air, and Q_{O_2} is the consumption of O_2, in mL O_2 at STPD per min. Q_T^* and Q_T should be in volumes at STPD per min.

97. The Henderson–Hasselbalch Equation:

For the dissociation of any acid HA to H^+ and the conjugate base, A^-, the Henderson–Hasselbalch Equation is written as:

[6.5.12] $$pH = pK + \log \frac{[A^-]}{[HA]}$$

where $pH = -\log[H^+]$, and $pK = -\log K$, where K is the acid dissociation constant of the acid.

98. The Henderson–Hasselbalch Equation for the HCO_3^- Buffer System:

The Henderson–Hasselbalch Equation for the bicarbonate buffer system is written as:

[6.5.23] $$pH = 6.10 + \log \frac{[HCO_3^-]}{0.0308 \, P_{CO_2}}$$

here $[HCO_3^-]$ must be in units of mM and P_{CO_2} must be in units of mmHg.

99. Calculation of ICF Volume:

The intracellular fluid compartment can be calculated by:

[7.1.2] $$ICF = TBW - ECF$$

where ICF is the intracellular fluid volume, TBW is the total body water, and ECF is the extracellular fluid volume.

100. Calculation of Interstitial Fluid Volume:

The interstitial fluid volume can be calculated as:

[7.1.3] $$ISF = ECF - plasma$$

where ISF is the interstitial fluid volume, ECF is the extracellular fluid volume, and plasma is the plasma fluid volume.

101. Lean Body Mass:

The lean body mass, LBM, can be calculated from the total body water, TBW, by:

[7.1.4] $$\text{Lean body mass} = \frac{TBW}{0.73}$$

102. Macroscopic Electroneutrality:

Because of the strength of electric forces, solutions are neutral, and so

[7.1.6] $$\sum_i C_i^+ = \sum_i C_i^-$$

where C_i^+ refers to the concentration of the i-th positively charged ion (cation) and C_i^- refers to the concentration of the i-th negatively charged ion (anion).

103. Gibbs–Donnan Equilibrium:

At equilibrium, a membrane that separates a solution of NaCl from NaP will produce a membrane potential given by the Nernst Equation and a distribution of Na and Cl ions such that

[7.1.12] $$\frac{[Na^+]_o}{[Na^+]_i} = \frac{[Cl^-]_i}{[Cl^-]_o} = r$$

Here r is called the Gibbs–Donnan ratio; $[Na^+]_o$ is the concentration of Na^+ in the compartment without impermeant anion; $[Na^+]_i$ is the concentration of Na^+ in the compartment with the impermeant anion.

104. Material Balance Equation for the Kidney:

[7.3.3] Rate of filtration + rate of secretion = rate of excretion + rate of reabsorption

105. GFR as Inulin Clearance:

The glomerular filtration rate can be calculated as the inulin clearance as:

[7.3.6] $$GFR = \frac{Q_u \, U_{inulin}}{P_{inulin}}$$

where GFR is the glomerular filtration rate, usually in mL min^{-1}, Q_u is the rate of urine flow, in mL min^{-1}, U_{inulin} is the urinary concentration of inulin in a timed sample, and P_{inulin} is the plasma concentration of inulin at the time of the urinary sample.

106. Renal Plasma Flow as PAH Clearance:

The renal plasma flow can be calculated as:

[7.3.10] $$RPF = \frac{Q_u \, U_{PAH}}{RA_{PAH} - RV_{PAH}}$$

where RPF is renal plasma flow, typically in mL min^{-1}, Q_u is the rate of urine flow, in mL min^{-1}, U_{PAH} is the urinary concentration of PAH (para aminohippuric acid); RA_{PAH} is the concentration of PAH in the renal artery; RV_{PAH} is the concentration of PAH in the renal vein.

107. Renal Clearance:

The renal clearance of any substance, x, is defined as:

$$[7.3.12] \qquad C_x = \frac{Q_u U_x}{P_x}$$

where C_x is the clearance of substance x; Q_u is the urine flow rate, in mL min^{-1}; U_x is the urinary concentration of x; P_x is the plasma concentration of x.

108. Definition of Filtration Fraction:

The filtration fraction is defined as:

$$[7.3.13] \qquad \text{Filtration fraction} = \frac{\text{GFR}}{\text{RPF}}$$

where GFR is the glomerular filtration rate and RPF is the renal plasma flow.

109. Definition of the Sieving Coefficient:

The sieving coefficient for any substance is defined as:

$$[7.3.14] \qquad \text{Sieving coefficient} = \Theta = \frac{C_B}{C_P}$$

where Θ is the sieving coefficient of a substance, C_B is the concentration of the substance in Bowman's space, and C_P is the concentration of the substance in the plasma.

110. Forces Producing Ultrafiltration:

The forces producing glomerular filtration are written as:

$$[7.3.18] \qquad \text{GFR} = K_f[P_{GC} - P_{BS} - \pi_{GC}]$$

where GFR is the glomerular filtration rate, K_f is the filtration coefficient of the renal corpuscle, P_{GC} is the hydrostatic pressure within the glomerular capillaries, P_{BS} is the hydrostatic pressure within Bowman's space, and π_{GC} is the oncotic pressure (=colloid osmotic pressure) of the blood in the glomerular capillaries.

111. Filtered Load:

The filtered load of any substance is calculated by:

$$[7.4.1] \qquad \text{Filtered load}_x = \text{GFR}\,\Theta_x P_x$$

where GFR is the glomerular filtration rate, Θ_x is the sieving coefficient for substance x, and P_x is the plasma concentration of substance x.

112. Fraction of Water Reabsorbed:

The fraction of water reabsorbed from the glomerulus to any segment of the nephron can be estimated by:

$$[7.4.15] \qquad \frac{V_R}{V_T} = 1 - \frac{1}{(\text{TF}/P)_{\text{inulin}}}$$

where V_R is the volume of fluid reabsorbed, V_T is the volume of fluid filtered, $(\text{TF}/P)_{\text{inulin}}$ is the ratio of the tubular fluid inulin concentration to its concentration in plasma.

113. Fraction of Filtered Load Remaining: The Double Ratio:

The fraction of filtered load of substance x remaining at any locus in the nephron can be calculated by:

$$[7.4.18] \qquad \text{Fraction remaining} = \frac{(\text{TF}/P)_x}{(\text{TF}/P)_{\text{inulin}}}$$

where $(\text{TF}/P)_x$ is the ratio of the tubular fluid concentration of x to its plasma concentration, and $(\text{TF}/P)_{\text{inulin}}$ is the ratio of the tubular fluid concentration of inulin to its plasma concentration.

114. Net Gain of HCO_3^- from the Kidney:

The net gain (or loss) of HCO_3^- from renal processes is calculated as:

$$[7.7.8] \qquad \text{Net } HCO_3^- \text{ gain} = \text{titratable acid} + NH_4^+ - HCO_3^-$$

where titratable acid is the amount of base necessary to titrate the urine to pH 7.4; NH_4^+ is the amount of ammonium excreted in a known amount of time; and HCO_3^- is the amount of bicarbonate in the urine.

115. Energetic Equivalence of Oxygen:

The rate of energy expenditure can be estimated from the rate of oxygen consumption as:

$$[8.6.3] \qquad \begin{aligned} M &= 4.85 \text{ kcal L}^{-1} \times Q_{O_2} \\ &= 20.3 \text{ kJ L}^{-1} \times Q_{O_2} \end{aligned}$$

where M is the total metabolic rate (in energy per unit time) and Q_{O_2} is the rate of oxygen consumption in L (STPD) per min.

116. Gas Consumption and Production as a Function of Fuel Use:

The amount of O_2 consumed or CO_2 produced can be calculated from the amount of fats, carbohydrates, and proteins that are used for fuel, using the following:

$$[8.6.7] \qquad \begin{aligned} V_{O_2} &= 0.746c + 2.02f + 1.01p \\ V_{CO_2} &= 0.746c + 1.43f + 0.844p \end{aligned}$$

here V_{O_2} is the volume of O_2, in L at STPD, consumed; V_{CO_2} is the volume of CO_2 in L at STPD; c is the amount of carbohydrate burned, in g; f is the amount of fats burned, in g; and p is the amount of protein consumed, in g.

117. Metabolic Clearance Rate:

The metabolic clearance rate is the amount of hormone removed per unit time divided by its plasma concentration:

$$[9.1.6] \qquad \text{MCR} = \frac{kH}{H/V_d} = kV_d$$

where MCR is the metabolic clearance rate, in mL min^{-1} or L min^{-1}, k is the first-order rate constant for hormone degradation, in min^{-1}, H is the amount of hormone, in moles, and V_d is the volume of distribution of hormone, in mL or L.

118. The Scatchard Equation:

The basis for the Scatchard plot is written as:

$$[9.1.A1.13] \qquad \frac{B}{[L]} = K(B_{max} - B)$$

where B is the amount of ligand bound, $[L]$ is the free ligand concentration, K is the intrinsic association constant, B_{max} is the maximum amount of ligand that can be bound.

119. The Hill Plot:

The Hill Equation is given as:

$$[9.1.A1.21] \qquad \nu = \frac{[P \cdot L_h]}{[P] + [P \cdot L_h]} = \frac{K[L]^h}{1 + K[L]^h}$$

where ν is the fractional saturation of binding (restricted to $0 \leq \nu \leq 1$), K is the association constant, $[L]$ is the free ligand concentration, and h is the Hill coefficient. This can be rearranged to:

$$[9.1.A1.22] \qquad \log\left[\frac{\nu}{1-\nu}\right] = \log K + h \log[L]$$

which suggests the Hill plot of $\log[\nu/(1-\nu)]$ against $\log[L]$.

Appendix II
Important Physical Constants for Physiology

Avogadro's Number:

$$N = 6.02 \times 10^{23} \text{ particles mol}^{-1}$$

The Faraday:

$$\mathfrak{F} = 96,489 \text{ C mol}^{-1}$$
$$1 \text{ C} = 6.24 \times 10^{18} \text{ electrons}$$
$$1 \text{ electron} = 1.6 \times 10^{-19} \text{ C}$$

Electrical Permittivity of the Vacuum:

$$\varepsilon_0 = 8.85 \times 10^{-12} \text{ C}^2 \text{ J}^{-1} \text{ m}^{-1}$$

Energy Units, Interconversions:

$$1 \text{ J} = 0.239 \text{ cal}$$
$$1 \text{ cal} = 4.186 \text{ cal}$$
$$1 \text{ J} = 1 \text{ N m}$$
$$1 \text{ J} = 1 \text{ V} \times \text{C}$$

Pressure Units, Interconversions:

$$1 \text{ atm} = 760 \text{ mmHg}$$
$$1 \text{ atm} = 1.013 \times 10^5 \text{ Pa}$$
$$1 \text{ Pa} = 1 \text{ N m}^{-2}$$
$$1 \text{ mmHg} = 133.3 \text{ Pa}$$

Gas Constant, Values:

$$R = 0.082 \text{ L atm mol}^{-1} \text{ K}^{-1}$$
$$R = 8.314 \text{ J mol}^{-1} \text{ K}^{-1}$$
$$R = 1.987 \text{ cal mol}^{-1} \text{ K}^{-1}$$

Boltzmann's Constant:

$$k = R/N = 1.38 \times 10^{-23} \text{ J K}^{-1} \text{ molecule}^{-1}$$

Viscosity Units:

Viscosity is in units of Pa s, where $1 \text{ Pa} = 1 \text{ N m}^{-2}$. An alternate archaic unit is the poise $= 1 \text{ dyne cm}^{-2}$ s. The interconversion is 1 Pa s = 10 poise.

Planck's Constant:

$$h = 6.625 \times 10^{-34} \text{ J s}$$

It relates the energy of a photon of light to its frequency: $E = h\nu$.

Index

Note: Page numbers followed by "b," "f," and "t" refer to boxes, figures, and tables, respectively.

A

A-bands: muscle, 305
ABC transporters, 174, 181, 816, 816–817
Abdomen
　abdominal aorta, 698, 699
　abdominal cavity, 625
　forceful expiration, 626
　prevertebral ganglia, 475
Absolute refractory periods, 268
Accessory muscles: breathing mechanics, 626
ACE inhibitors, 617
Acetoacetate, 245
Acetone, 245
Acetyl Coenzyme A, 218, 227
Acetylcholine, 168
　autonomic nervous system, 479
　degradation, 322–323
　muscarinic receptors, 382, 479–480
　muscle excitation, 321–322
　nicotinic receptor binding, 382
　pacemaker potential, 530
　recycling, 322–323
　stomach, 791
Acetylcholine receptor (AchR), 331
Acetylcholine signal
　muscle fiber membrane, 331
Acetylcholinesterase (AChE), 331
Achalasia, 783b
AChE. *See* Acetylcholinesterase (AChE)
AchR. *See* Acetylcholine receptor (AchR)
Acid reflux, 783
Acid secretion, 727, 791–794, 794
Acid–base balance, 752, 765
Acid–base physiology, 665
Acidemia, 665
Acidic amino acids, 130
Acidosis, 665, 669
　ammonium, 762
　hydrogen carbonate, 759
　hyperventilation, 668
　hypoventilation, 668
　kidneys, 759
　lungs, 760
　renal tubular acidosis, 763b
　respiratory acidosis, 668
　ventilatory drive, 678–679
Acids: bile, 814–816
Acinar cells, 810
Acinar secretion, mechanism of, 812f
ACTH
　pituitary–hypothalamus axis, 907–908
　steroid secretion, 908–909
　zona glomerulosa, 911
Actin
　cardiac contractility, 548
　cell structure, 105, 107
　myosin ATPase, 549
　skeletal muscle, 309–310
　smooth muscle, 356

Action potentials, 265
　axons, 280, 319
　cardiac action potentials, 528
　channels, 269–270
　cutaneous sensory systems, 390–391
　motor neuron synaptic inputs, 319
　muscle fiber intracellular calcium, 324
　muscle membranes, 323
　origin, 319
　problem sets, 289
　propagation, 280
　　axons, 280
　　cable properties, 283–285
　　current, 280
　　electrotonic spread, 280–281
　　nerve cable properties, 280–281
　　nerve conduction velocity, 280
　　nodal movement, 286
　　saltatory conduction, 286
　　space constants, 283–285
　　time constants, 283–285
Activation energies
　enzyme effects, 64f
　path, 70
　rates of chemical reactions and, 64
　transition state theory, 67–70
Activation gates: action potentials, 270
Active transport
　antiports, 175–176
　cell structure, 102
　electrochemical potentials, 170–172
　energetics, 170–172
　exchangers, 170
　ion permeation, 170–172
　material transport, 172
　metabolic energy, 172
　phosphorylated intermediates, 173
　pumps, 170
　sodium, potassium-ATPase, 172–173
　sodium, potassium-ATPase as electrogenic, 173–174
　sodium–calcium exchangers, 174–175
　symports, 176
Active zones, 329–330
Actomyosin cross-bridge cycling, 549
Acyl chains, 143–145
Adaptation to stimuli, 392
Adaptive functions: autonomic nervous system, 473
Addison's disease, 914b
Adenine, 120
Adenosine, 594
Adenosine diphosphate (ADP): equilibrium concentrations, 92
Adenosine triphosphate (ATP)
　as energy currency of the cell, 218–219, 220f
　enzymes, 108, 549, 794
　equilibrium concentrations, 92
　exercise duration/intensity, 334–335
　hydrolysis, 90–92

　muscular activity, 312–313, 334
　production
　　amino acid oxidation, 241
　　fatty acid oxidation, 241
　　glycolysis, 218
　　oxidative phosphorylation, 227
　　TCA cycle, 227
　regeneration, 336
Adenylyl cyclase, 591
ADH. *See* Antidiuretic hormone (ADH)
Adherens junctions, 111, 547
Adipocyte lipolysis, 242f
Adipose triglyceride lipase (ATGL), 242, 242f
Adiposity signals, 843–844
ADP. *See* Adenosine diphosphate (ADP)
Adrenal cortex, 702, 906
Adrenal medulla, 475, 596, 916
Adrenaline, 475
Adrenergic receptors
　catecholamine action, 919–920
　norepinephrine release, 480–482
　pharmacology-based classification, 383
Adrenergic stimulation
　smooth muscles, 358
Afferent arterioles, 699, 700–701
　glomerular hydrostatic pressure, 740–741
　renin release, 701–702, 748
Afferent fibers: gut extrinsic innervation, 799
Afferent sensory neurons, 419
Affinity of a chemical for electrons, 229
Afterload: cardiovascular system, 556, 558–559, 599
Ageusia, 437
Agglutinins, 503
Agglutinogens, 503
Agouti-related protein (AgRP), 842–843
Agranular white blood cells, 507
Agrin, 331
AgRP. *See* Agouti-related protein (AgRP)
A-intercalated cells, 757
　acid secretion by, 761
Air molecules: speed of sound, 451–452
Air movement: lung volumes, 625
Air pressure waves: hearing, 441–442
Air transport: breathing, 623–624
Airway mechanoreceptors, 679–680
Airway resistance, 630–631, 633, 653
Airy disk, 113–114
Albumin, 498
Aldosterone, 702, 912. *See also* Renin–angiotensin–aldosterone (RAA) system
　perfusion regulation, 596–597
　zona glomerulosa, 911
Alkalemia, 665
Alkaline phosphatase, 939
Alkaline solution secretion, 811–812
Alkaloids, 506
Alkalosis, 665, 669
　hydrogen carbonate, 760

983

Alkalosis (*Continued*)
   hyperventilation, 668
   kidneys, 760, 761
   lungs, 760
   respiratory compensation, 668
Alleles, 956
Allergies, 509–510
α dystroglycan, 314–315
α melanocyte stimulating hormone (αMSH), 842–843
α-actinin, 105, 310
α-helix, 134–135, 136*f*
α-Klotho (KL), 931
α$_1$-antitrypsin, 496
Alpha$_2$ adrenergic stimulation, 209–210
Alveoli
   anatomic dead space, 647
   carbon dioxide production rates, 648
   duct mechanics, 624
   gas equations, 648, 651–652
   gas exchange, 645–646, 651, 653
   Law of Laplace, 627
   pressure, 627*b*
   surfactants, 627–628
Amacrine cells, 463
Ambient temperature and pressure, saturated (ATPS), 645
Amiloride, 434–435, 733
Amino acids
   ammonium origin, 755–756
   antidiuretic hormone, 870
   ATP production, 241
   basolateral membranes, 824–825, 825
   brush border membranes, 824–825
   insulin secretion, 898
   oxidation, 241
   oxytocin, 870
   protein structure, 139
      peptide bonds, 131–133
      primary protein structure, 134
      secondary protein structure, 134–136
      tertiary protein structure, 136–137
   proximal convoluted tubule, 725
   structure, 139
   transfer RNA, 123
Aminoacyl (A) binding sites, 123
Aminopeptidase A, 823–824
Aminopeptidase N, 823–824
Aminopeptidase P, 823–824
Aminopeptidase W, 823–824
Ammonium, 754, 755–756, 762
Amphipathic molecules, 147–149, 149
   lowering of the surface energy by, 150*f*
AMPK (AMP-activated protein kinase), 761
Amylase, 811
Amylolytic enzymes, 810
Anaerobic metabolism, 222
   lactic acid
      glycolytic flux, 338
Anaerobic thresholds: lactic acid, 340
Anaphase, 957–958
Anatomic dead space, 647
Anemias, 501, 505
Angiotensin, 702. *See also* Renin–angiotensin–aldosterone (RAA) system
   effects, 911–912
   perfusion regulation, 596–597
   zona glomerulosa, 911
Angiotensin converting enzyme (ACE), 748
Angiotensin II, 748
Angiotensin-converting enzyme (ACE) 1 and 2, 823–824
Angiotensinogen, 498–499
Anionic amino acid system, 825

Anions, 53
Ankyrin, 314–315, 501*f*
Annulus fibrosus, 516, 520
ANP. *See* Atrial natriuretic peptide (ANP)
ANS. *See* Autonomic nervous system (ANS)
Anterior olfactory nucleus (AON), 433*f*
Anterior pituitary, 857, 857, 874, 961–962, 962–963
Anterior pyriform cortex (APC), 433*f*
Anterolateral tract, 395
Antibodies, 499, 514–515
Anticipatory control systems, 11
Anticoagulant therapy, 496*b*
Antidiuretic hormone (ADH)
   amino acids, 870
   blood pressure, 616–617
   blood volume, 873
   collecting duct, 733
   diabetes insipidus, 739*b*
   distal nephron permeability, 738, 745
   feedback loops, 745–746
   hyperosmolarity, 744–745
   hypothalamus, 870
   hypovolemia, 744–745
   inner medullary collecting duct, 733
   late distal tubule, 733
   perfusion regulation, 596–597
   plasmaosmolarity, 873
   urine concentration, 730
   water balance, 744
Antigens, 499*f*, 503
Antiparallel beta sheets, 135
Antiports, 175–176
Antithrombin III, 496
Antithromboplastin, 496
Antrum, 952
AON. *See* Anterior olfactory nucleus (AON)
Aorta, 519–520
Aortic arch: arterial pressure regulation, 609–610
Aortic bodies: ventilation control, 676
Aortic regurgitation, 520, 563
Aortic valve, 519–520
APC. *See* Anterior pyriform cortex (APC)
Apex beat, 516
Apical membranes, 725, 779*f*
Aplastic anemia, 506
Apnea, 680*b*
Apolipoproteins, 498, 829–831
Aquaporins, 168, 181, 198, 727, 731, 751
Arcuate arteries, 700–701, 700*f*
Arcuate arterioles, 700–701
Arcuate veins, 701
Arginine vasopressin (AVP), 730. *See also* Antidiuretic hormone (ADH)
Aristotle, 5
Aromatic amino acids, 825
Arrhythmias, 544*b*, 563
Arterial tree: hemodynamics, 570, 572
Arteries
   arcuate arteries, 699, 700–701
   circulatory system overview, 489
   compliance, 570–571
   disease, 563
   interlobar arteries, 699, 700–701
   pressure, 573–574, 608
   stroke volume, 570–571
   vascular function curves, 602–603
   ventilation control, 676
Arterioles
   afferent arterioles, 699, 700–701, 740–741, 748
   arcuate arterioles, 700–701
   branching, 573–574
   circulatory system overview, 489

   efferent arterioles, 701, 740–741
   perfusion regulation, 589, 593
   pressure drops, 573–574
   solute exchange, 578
   terminal arterioles, 578
   vascular function, 604
Arteriosclerosis, 571
Aspartate, 383
Aspiration reflexes, 680
Association constants, 666
Asthma, 650*b*, 669
ATGL. *See* Adipose triglyceride lipase (ATGL)
Atherosclerosis, 571
Atomic mass, 59
Atomic number, 46
Atomic orbitals, 47–48
Atoms
   electrical charges, 46
   molecule formation, 48
   movement, 56*f*
ATP synthase, 233
ATP synthetase, 108
ATPase, 549, 550, 794
   ABC transporters, 174
   F-type ATPases, 174
   P-type ATPases, 174
   V-type ATPases, 174
ATP-driven ion pumps, 181
ATPS. *See* Ambient temperature and pressure, saturated (ATPS)
Atresia, 948
Atria, 518, 519. *See also* Sinoatrial (SA) node
   cardiac action potential, 528
   electrocardiograms, 540, 544
   vascular function curves, 603–604
   ventricular function curves, 558
Atrial natriuretic peptide (ANP)
   blood pressure, 617
   hypervolemia, 749
   perfusion regulation, 596–597
Atrioventricular (AV) conduction blocks, 544
Atrioventricular (AV) impulse transmission, 523
Atropine nerve gas antidote, 484
Atropine poison, 485
Auditory cortex, 446–447
Auditory systems. *See* Hearing
Augmented unipolar limb leads, 543
Auscultation, 520
Autonomic nervous system (ANS), 473
   acetylcholine, 479
   adaptive functions, 473
   cardiac action potential, 529–530
   efferent functions, 473–474, 478–479
   emotional state, 473–474
   homeostasis, 473
   micturition, 483–485
   nerve terminals, 482
   neurotransmitters, 479
   norepinephrine, 479
   reflexes, 473
   target cell receptors, 482
Autoregulation: GFR/RGF, 742
Autosomes, 956
AV. *See* Atrioventricular (AV)...
Avogadro's number, 59
AVP. *See* Arginine vasopressin (AVP)
Axons
   action potentials, 265, 280, 319
   cervical ganglia, 476
   neuropeptides, 384–385
   olfactory receptor cells, 427–428, 429
Axoplasmic resistance: action potentials, 282

B

Back pressure: bleeding restriction, 494
Balance of movement, 409
Baroreceptors, 609–610
Baroreflex, 610–611
Barrett's esophagus, 783
Barrier function, 110
Bartter syndrome, 732
Barttin, 732
Basal ganglia, 372, 413–415
Basal metabolic rate (BMR), 837–838
Base-acid balance, 752, 765
Base-acid physiology, 665
Basement membranes, 708
Basic amino acids, 130
Basolateral membrane, 779f
Basolateral membrane (BLM), 825
   amino acid absorption, 825
Basolateral membranes, 725, 824–825
Basophils, 507, 508–509
Bellini ducts. See Ducts of Bellini
Bending of light, 467–468
Beta adrenergic stimulation, 212, 213f
   cell signaling, 209
Beta blockers: hypertension, 617
β dystroglycan, 314–315
Beta oxidation: fatty acids, 243–245
Beta sheet, 135, 136f
β-hydroxybutyric acid, 245
Between brain. See Diencephalon
Bicarbonate absorption: acid secretion, 727
Bicarbonate buffer systems, 665
Bicarbonate reabsorbtion, 756–757
Bile, 773, 817–819, 833
Bile acids, 814–815, 815, 815–816
   enterohepatic circulation of, 817f
Biliary secretion, 810
Bilirubin, 502–503
Biliverdin, 502–503
B-intercalated cells, 757
Bioassays: hormones, 861–862
Biological membranes. See Membranes
Bipolar cells, 462–463
Bipolar electrodes, 542–543
Bitter taste, 432, 436
Bladder, 698
BLM. See Basolateral membrane (BLM)
Blood
   acid excretion, 753
   ATP production, 221–222
   back pressure, 494
   bleeding disorders, 496b
   blood substitutes, 663b
   blood–brain barrier, 370, 678
   carbon dioxide transport, 661
   coagulation, 494–495
      activation, 495–496
      anticoagulant therapy, 496b
      inhibition, 495–496
   concentration, 61b
   dissolved oxygen content, 656
   ejection by heart, 570
   endocrine glands, 853
   exercise intensity, 338
   gas exchange, 642, 647
   glucocorticoids, 909–910
   glucose, 895
   heart chambers, 518–520
   hemoglobin, 656–657
   hormones, 858–859
   hydrogen carbonate (HCO_3), 661–662
   inflammation, 511
   kidneys, 698
   lungs, 648
   overview, 487
   oxygen
      carrying capacity, 657b
      consumption, 657
      delivery, 661
      diffusion, 659
      hemoglobin, 656–657
   plasma, 494, 498
   pressure
      arterial pressure regulation, 609, 613–615
      hormonal regulation, 615–617
      long-term blood volume regulation, 613–615
      sphygmomanometers, 572–573
   pressure-driven flow, 493
   problem sets, 525
   red cells, 498
   solute exchange, 585
   type classification, 503–504
   vasoconstriction, 494
   vessels
      branching, 573–574
      compliance, 493–494, 570–571
      disease, 563
      overview, 489
      perfusion regulation, 589–590, 592, 593, 596–597
      pressure, 572–573, 608
      solute exchange, 578, 584–585
      stroke volume, 570–571
      vascular function, 569–570, 602–603, 604
      ventilation control, 676
   volume
      antidiuretic hormone, 873
      integrated response, 749
      long-term blood pressure regulation, 613–615
      vascular function curves, 605
   white cells, 507
B-lymphocytes, 499–500, 510
Body fluid compartments, 687
Body temperature and pressure, saturated (BTPS), 644, 645, 652
Bohr effect, 660–661
Boltzmann's constant, 85
Bonds
   amino acids, 123
   angles, 50
   energies, 50
   enthalpy, 50–51
   hydrogen bonds, 53–54, 54
      deoxyribonucleic acid, 121, 121
      secondary protein structure, 134–136
      water, 54
   isomerism, 51
   length, 50
   movement, 56f
   protein structure, 131–133
   rotation, 51
   unequal sharing, 51–53
   water, 53–54
Bone
   erythrocyte formation, 502
   hormones, 935–937
   hypocalcemia, 927
   osteoblasts, 933, 934–935
   osteoclasts, 934, 935, 935
   osteocytes, 933–934
   remodeling, 934
   vitamin D mineralization, 931
Bone cells
   fibroblast growth factor 23 (FGF23), 931
Bovine spongiform encephalopathy (BSE), 139
Bowman's capsule, 698–699, 700–701
Bowman's glands, 428
Bowman's space, 701, 705
Bradycardia, 610
Bradykinin: paracrine secretions, 595
Brain, 370
   adrenal cortex, 702, 906
   adrenal medulla, 475, 596, 916
   anterior pituitary, 857, 874, 961–962, 962–963
   auditory cortex, 446–447
   blood–brain barrier, 370, 678
   brain stem
      auditory cortex, 446–447
      function, 373
      ventilation control, 677
   cerebellum
      control of movement, 413–415
      function, 373
      movement accuracy, 415
   cerebral cortex, 372, 411
   cerebrospinal fluid, 691
      brain surface features, 367
      cushioning effect, 369–370
      ventilation control, 678
      ventricles, 369–370
   consciousness, 366
   cortex
      hearing, 446–447
      kidney function, 698–699
      olfactory output, 429–431
      taste receptors, 438
      vision, 464–465
   diencephalon function, 373
   feeding behavior, 839
   flavor in, 438
   forebrain function, 372
   gray matter organization, 371
   hearing, 442–444, 446–447
   hindbrain function, 373
   hyperosmolarity, 744–745
   hypothalamus, 870
      diabetes insipidus, 739b
      feeding behavior, 839
      function, 373
      sensory afferents, 611–612
      testicular function, 961–963
      thyroid stimulating hormone, 885
   hypovolemia, 744–745
   medulla
      adrenal medulla, 475, 596, 916
      baroreflex, 611–612
      function, 373
      kidney function, 698–699
      reticulospinal tract, 406
      salivary nuclei, 777–778
      swallowing centers, 780
      urea, 733, 736
      vestibular nuclei, 419
   motor cortex, 411
   orbitofrontal cortex, 431
   pituitary gland, 870
   pons, 373
   posterior pituitary, 857, 870
   sensory cortex, 395
   somatosensory information, 394–395
   strength gains, 343
   surface features, 367–369
   testicular function, 961–963
   thalamus
      function, 373
      olfactory output, 431
      taste receptors, 438
      ventilation control, 678
      vision, 464
Breasts: myoepithelial cell contraction, 871
Breathing. See Respiratory systems
Bronchi, 624

INDEX

Bronchiectasis, 640
Bronchioles, 624
Bronchitis, 640
Bronsted–Lowry theory, 665–666
Brush border membranes
　amino acids, 824–825
　　protein digestion, 822–824
　　starch digestion, 825–826
Bruxism, 775
BSE. See Bovine spongiform encephalopathy (BSE)
BTPS. See Body temperature and pressure, saturated (BTPS)
Buffers, 500, 665, 752–753. See also Bicarbonate buffer systems
Bundles of His, 523

C

C peptide, 895
CA. See Carbonic anhydrase
Ca-ATPase pump, 550
Calcitonin (CT), 935, 937f, 941–942
Calcitropic hormones, 924
Calcium. See also Calcitropic hormones
　actomyosin cross-bridge cycling, 549
　ATP production, 239
　bone mineralization, 942
　cardiac action potential, 528, 532–533
　cardiac contraction
　　calsequestrin, 551–552
　　glycosides, 553–554
　　induced, 552–553
　cardiac relaxation, 550–551
　cell signaling, 206–207, 211
　chemical neurotransmission initiation, 379
　cross-bridge cycling, 325–326
　excitation-contraction coupling, 324
　homeostasis, 924, 933
　hypertension channel blockers, 617
　induced cardiac contraction, 552–553
　intestine, 939
　mitochondria, 551
　muscle fiber action potentials, 324
　negative feedback loops, 942–943
　1,25(OH)$_2$-D hormone, 937–939
　osteoclasts, 934
　parathyroid hormone, 941
　perfusion regulation, 591
　presynaptic cells, 322, 380
　regulation, via membrane potential, 354
　　by altering BK$_{Ca}$ channels, 354
　repetitive stimulation effects, 327
　sarcoplasmic reticulum, 324
　smooth muscle, 354–356
　　contraction, 353–354
　　force production, 356–357
　　relaxation, 358
　urinary excretion, 940–941
Calcium Release Unit
　multiple proteins on T-tubules and sarcoplasmic reticulum membranes, 331–333
Calcium-induced calcium release (CICR), 549–550
Calmodulin, 332
Calmodulin-dependent activation of enzymes, 206
Calmodulin-dependent protein kinase (CAM kinase II), 332
Calorimeters, 834, 836–837, 837
Calsequestrin (CASQ), 332, 551–552

cAMP. See Cyclic AMP (cAMP)
Cancellous bone, 933
Capacitance, 31, 281–282
　action potential propagation, 281–282
　coaxial cables, 288
　coaxial capacitors, 287
　Gauss' law, 287
　parallel plate capacitors, 35–36
　pressure-driven flow, 18
Capacitive current, 39–40
Capacitors, 36, 282
　charging of, time constant, 39–40
Capillaries
　branching, 573–574
　circulatory system overview, 489
　inflammation, 511
　perfusion regulation, vasoconstriction, 589–590
　solute exchange, 578
　　bulk fluid movement, 584
　　lymphatics, 585
　　net filtration pressure, 584–585
　　passive mechanisms, 579
　　transcytosis, 584
　types, 578–579
Carbamates, 662
Carbaminohemoglobin, 662
Carbohydrates
　biological membranes, 142
　dietary fiber, 825
　digestion, 825
　glucose, 220–221
Carbon
　bond rotation, 50
　geometries, 48–49
　rotation, 50
　surface, 48–49
Carbon dioxide (CO_2)
　acid secretion, 759
　alveolar ventilation calculations, 648
　ATP production, 227–229
　breathing mechanics, 623
　carbaminohemoglobin, 662
　hydrogen carbonate reabsorption, 757–759
　hydrogen ion secretion, 757–759
　metabolic acidosis, 760
　metabolic alkalosis, 760–761
　pH regulation, 668
　respiratory system, 623
　transport, 656
　ventilation control, 677
Carbon monoxide (CO) poisoning, 662b
Carbonic acid (H_2CO_3), 667–668
Carbonic anhydrase (CA), 662, 667, 727, 752, 753
Carbonic anhydrase on the apical membrane (CAIV), 727
Carboxyhemoglobin, 662
Carboxypeptidase A, 822–823
Carboxypeptidase P, 823–824
Cardiac action potential, 528
Cardiac contractility, 516–518
　cardiac cycle, 520–522
　cardiac glycosides, 553–554
　cellular basis, 547
　myofibrils, 548–549
　parasympathetic stimulation, 553
　regulation, 552
　stretch-based modulation, 554–555
Cardiac cycle
　contractile events, 520–522
　electrocardiograms, 544–545
Cardiac function curves, 556
　cardiac output, 599
　steady-state operating points, 604, 605

Cardiac glycosides: cardiac contraction, 553–554
Cardiac muscle
　calcium-induced calcium release, 549–550
　contraction, 516–518
　features, 547
Cardiac myofibrils: contractility, 548–549
Cardiac output (CO), 556, 599
　distribution by vascular system, 568
　Fick's principle, 560–561
　indicator dilution method, 561–563
Cardiolipin: biological membranes, 146–147
Cardiomyocyte coupling, 547
Cardiomyopathy, 563
Cardiovascular system. See also Blood; Heart; Inflammation; Vascular system
　hemodynamics, 568
　overview, 487
Carnitine carrier substances, 243
Carotid bodies: ventilation control, 676
Carotid chemoreceptors: ventilation control, 676
Carotid sinus: arterial pressure regulation, 609
Carrier classifications, 179
CART. See Cocaine–amphetamine regulated transcript (CART)
CASQ. See Calsequestrin (CASQ)
Catalysts
　cell structure, 101
　posttranslational modification, 137–138
　tricarboxylic acid cycle, 230–231
Cataracts, 466b
Catecholamines, 382
　adrenergic receptor types, 919–920
　degradation, 918–919
　fight or flight, 920–921
　sympathetic stimulation, 916–917
Cationic amino acid system, 825
Cations, 53
Caudal ventrolateral medulla (CVLM), 611–612
Causal link, 6
Caveolae, 153f
　and integral proteins, 153
Caveolae vesicles, 578–579
Caveolin, 578–579
Cavins, 153
CCK. See Cholecystokinin (CCK)
Celiac ganglia, 475, 799
Cell bodies: motor neurons, 265
Cell cortex, 105
Cells
　action potentials, 265
　blood, 494, 498, 507
　body fluid compartments, 692
　cardiac action potential, 528–529
　cardiac contractility, 528–529, 547
　cell theory, 6–7
　complement systems, 514f
　core principles, 6–7
　　cell theory, 6–7
　　diversity, 7
　　forms, 7
　　genome, 7
　cortisol, 910–911
　diffusion coefficients, 79–80
　follicular development, 949–953
　fractionation problem sets, 156
　gametes, 946, 947, 956, 958, 960–961
　glucagon release, 899–900
　glycolysis, 218, 222
　gut, 782
　insulin synthesis, 895
　membranes
　　cell forms, 7

cell structure, 101–102
cell theory, 6–7
smooth muscle contractile filaments, 352
motor neurons, 265, 318, 318
myoepithelial cells, 777, 871
nervous system, 318, 375
osmotic pressure, 192–193
ovarian steroidogenesis, 953
oxygen diffusion, 659
red blood cells, 498
signaling, 205
  cellular response classes, 206
  channels, 206–207, 207–209
  chemical signaling, 205, 207–209, 212
  effectors, 212
  electrical signals, 205, 206
  endocrine signals, 206
  event transduction, 205
  gene expression, 213, 215
  heterotrimeric G-protein-coupled receptors, 209
  ligand-gated ion channels, 207–209
  mechanical signals, 205
  membrane-bound enzymes, 212
  neurotransmitters, 206
  nuclear receptors, 213
  receptors, 212, 213
    gene transcription, 213
    histone acetylase, 214
    transcription factors, 214–215
  voltage-gated ion channels, 206
skeletal muscle contractile mechanisms, 306
smooth muscle coupling, 352–353
solute exchange, 581–582
structure, 99
  attachment, 110–111
  cell membranes, 101–102
  centrifugation separation analyses, 116–117
  cytoskeleton, 103
  cytosol, 103
  electron microscopy, 114
  endoplasmic reticulum, 108
  form, 101
  function, 101
  Golgi apparatus, 108
  lipid processing/synthesis, 108
  lysosomes, 109
  microscopy, 113
  mitochondria, 108–109
  nuclei, 107–108
  organelles, 101, 115
  peroxisomes, 109
  proteasomes, 110
  protein processing/synthesis, 108
  protein synthesis, 108
  ribosomes, 108
  study methods, 113
  subcellular fractionation, 115
thyroglobulin precurosor, 883, 885
types, 7, 125–126
Cellular pacemakers, 675
Cellular respiration, 623
Central chemosensors, 677–678
Central nervous system (CNS). *See also* Brain; Spinal cord
  cutaneous sensory systems, 389
  depression, 669
  feeding behavior, 838–839
  integrative centers, behavioral response, 366
  major area functions, 372–373
  odor adaptation, 431
  organization, 367
  preganglionic neurons, 474
  serotonin, 384
  thoracolumbar spinal cord, 475–477

Central pattern generator (CPG), 773–774
Central sleep disorders, 680
Central venous pressure. *See* Preload
Centrifugation: cell structure, 115–116
Centrioles, 104–105
Centripetal acceleration, 117
Cephalic acid secretion phases, 791–794
Cephalic enzyme secretion phases, 812–813
Ceramide, 146
Cerebellum
  control of movement, 413–415
  function, 373
  movement accuracy, 415
Cerebral cortex
  function, 372
  purposeful movement, 411
Cerebrosides, 146, 146
Cerebrospinal fluid (CSF), 691
  brain surface features, 367
  cushioning of brain, 369–370
  ventilation control, 678
  ventricles, 369–370
Cervical ganglia, 476
CFTR. *See* Cystic fibrosis transmembrane conductance regulator (CFTR)
Chaperones: protein structure, 137
Chaperonins: protein structure, 137
Charge
  action potential propagation, 282
  atoms, 46
  electrical potential, 31–32
  light, 467
  movement, 37
  pressure-driven flow, 18
  vision, 467
Chemical potentials: osmotic pressure, 184
Chemical reactions, rates of
  and activation energy, 64
Chemical signals, 205
  circulatory system overview, 489–490
  ligand-gated ion channels, 207–209
  membrane-bound enzymes, 212
  smooth muscle, 354–356
Chemical species, aggregate of, 72
Chemical sysnapses. *See* Synapses
Chemical trigger zones (CTZ), 804
Chemiosmotic coupling, 233
Chemokines, 512
Chemoreceptors, 676f, 677, 791–793, 797–798
Chemosensors, 676, 677–678, 840–841
Chemotaxic compounds, 513–514
Chenodeoxycholic acid, 814–815
Chest wall, 626, 628–630
Chewing, 773
Chloride ion secretion
  acid–base balance, 757
Cholecystectomy, 819
Cholecystokinin (CCK), 788–789, 812–813, 841
Cholesterol
  bile, 814, 816–817
  biological membranes, 146–147
  steroid hormones, 857, 906–907
Cholic acid, 814–815
Chord conductance, 259–260, 269
Chordae tendinae, 518
Chorionic gonadotropin (hCG), 963
Choroid plexuses, 678
Chromaffin cells, 475, 596
Chromatin, 122
Chromatography, 862
Chromosomes, 120, 122, 956, 958–959
Chronaxie, 268
Chronic acidosis, 762

Chronic bronchitis, 640
Chronic Kidney Disease Epidemiology Collaboration (CKD-EPI) equation, 724
Chronic obstructive pulmonary disease (COPD), 640b
Chronotropic effects: cardiac action potential, 529–530
Chylomicrons, 585, 829–831
Chyme, 800–801
Chymotrypsin, 822–823
CICR. *See* Calcium-induced calcium release (CICR)
Cilia. *See also* Follicles
  cell structure, 104
Ciliary body: pupillary light reflex, 482–483
Circular esophageal smooth muscle, 781
Circulatory pressure, 601
Circulatory systems. *See also* Cardiovascular system; Respiratory systems
  diapedesis, 513–514
Circus arrhythmia, 544
Cisternae, 108
Citric acid cycle. *See* tricarboxylic acid (TCA) cycle
CJD. *See* CREUTZFELDT–Jakob disease (CJD)
Clathrin-coated pits, 153f
  and integral proteins, 153
Clitoris, 946
Clotting
  inflammation, 511
  plasmin, 495
  retraction, 495
CO. *See* Carbon monoxide (CO) poisoningCardiac output (CO)
Coagulation, 494–495
  activation, 495–496
  anticoagulant therapy, 496b
  inhibition, 495–496
Coaxial capacitors, 287
Cobalamin, 506
Cocaine–amphetamine regulated transcript (CART), 842–843
Cochlea
  cochlear microphonic, 448–450
  hair cells, 444–445
  implants, 449b
  sound frequency tonotopic mapping, 445–446
Codons, 123
Cold receptors, 393
Collagen, 627, 710
Collecting ducts
  acid–base balance, 757
  kidney function, 698–699
  urea transport, 733
  water permeability, 733
Colligative properties
  osmotic pressure, 185
Colloid osmotic pressure. *See* Oncotic pressure
Colon
  content reflux, 802
  diverticular disease, 808b
  motility, 796
Command systems, 10
Compartmental analysis, 72–74
  chemical species, aggregate of, 72
  turnover, 72–74
Compensatory pause, 544
Competitive inhibition
  facilitated diffusion, 165
  ligand binding, 868–869
Compliance
  circulatory systems, 601–602
  lung expansion, 626, 627

Compliance (*Continued*)
   pulse pressure, 570–571
   vascular function, 568, 569–570, 602–603
Compton effect, 114
Concave lenses, 468
Concentration, 59
   definition, 60
   facilitated diffusion, 164–165
   osmotic pressure, 182–184
   urine, 730
Concentric contractions, 300, 300
Conductance, 39
Conductance potentials, 258–259, 269–270, 271
Conductors, 39
Confocal microscopy, 119
Connecting ducts: salt, 732–733
Consensual pupillary light reflex, 482–483
Conservation of energy
   electrical potential, 32–33
   pressure-driven flow, 17
Conservation of matter, 17
Conservation of solute calculations, 61
Conservative forces
   electrical potential, 32–33
   work, 33
Constipation, opioid-induced, 807b
Continence, 483, 803
Continuity equations, 17, 83–84
   diffusion, 75
   pressure-driven flow, 17
Continuous capillaries, 578, 578–579
Continuous positive airway pressure (CPAP), 640
Contractile behavior
   muscle fibers, 297–298
Contractile mechanisms
   airway resistance, 639
   colonic motility, 802–803
   excitation-contraction coupling, 206, 318
   frequency codes, 409
   heart, 516–518, 547
      action potentials, 528–529
      cardiac cycle, 520–522
      cardiac glycosides, 553–554
      cellular basis, 547
      cycle, 520–522
      electrical systems, 522–523
      myofibrils, 548–549
      parasympathetic stimulation, 553
      regulation, 552
      stretch-based modulation, 554–555
   perfusion regulation, 592–593
   phasically-based, 351
   population codes, 409
   skeletal muscle, 293, 300, 305
   slow wave activity, 799
   smooth muscle, 351, 352
   stomach, 787–788
   stroke volume, 556
Control systems
   cell structure, 101
   core principles, 10
      anticipatory control systems, 11
      cyclical systems, 11
      developmental control systems, 11
      feed-forward control systems, 11
      negative feedback loops, 10–11
      noncyclical systems, 11
      threshold control systems, 11
   movement, 409
Convection–diffusion equations, 81
Convergence of light, 468
Convex lenses, 468
Cooperativity of oxygen binding, 656–657

Cooperativity plots, 867–868
Coordinated command and control, 10
COPD. *See* Chronic obstructive pulmonary disease (COPD)
Cori cycle, 223
Coronary artery disease, 563
Coronary sinus, 518
Corpus albicans, 949
Corpus cavernosa, 965
Corpus luteum, 949, 954–955
Cortex. *See also* Cerebral cortex
   hearing, 446–447
   kidney function, 698–699
   olfactory output, 429–431
   taste receptors, 438
   vision, 464–465
Cortical masticatory area (CMA), 773–774
Cortical nephrons, 701
Cortical neurons, 396–397
Corticospinal tract, 405
Cortisol, 909–910, 911
Costameres, 314–315
Cotransporters. *See* Symports
Cough reflexes, 680
Coulomb's law, 31, 34–35
Countercurrent exchangers, 736
Counter-transporters. *See* Exchangers
Covalent bonds, 51–53, 123
COX. *See* Cyclooxygenase (COX) enzyme
CPAP. *See* Continuous positive airway pressure (CPAP)
Cranial nerve VII, 773
Cranial nerves, 373, 477–478
Creatinine, 723
Creutzfeldt–Jakob disease (CJD), 139
Cribriform areas, 427–428, 698–699
Cristae, 109f
Cross-bridge cycle, 312
Cross-bridges, 312–313, 325, 549
Crossed-extensor reflexes, 400–401
Crossing-over: genetics, 958–959
Cross-striations: muscle, 548–549
Crystallins, 466
Crystalloids, 584
CSF. *See* Cerebrospinal fluid (CSF)
CT polypeptide hormone
   hypercalcemia, 928
   plasma ion concentration, 928
CTZ. *See* Chemical trigger zones (CTZ)
Current, 31, 39
   action potentials, 265, 280, 282, 286
   ion movement, 37–38
   Ohm's law, 39
   resting membrane potential, 260–261
Current–voltage relationship, 271–272
Cushing's disease/syndrome, 913b
Cutaneous sensory systems, 389
   action potentials, 390–391
   anatomical connection, 391
   anterolateral tract, 395
   central nervous system pathways, 390
   dorsal column pathways, 394–395
   exteroreceptors, 389
   interoreceptors, 389
   long receptors, 390–391
   pain information, 395
   perception, 390
   receptive fields, 392, 396–397
   receptor types, 392–393
   sensation disorders, 395–396
   sense organs, 390
   sensory cortex, 395
   sensory neurons, 392
   sensory receptors, 390
   sensory stimuli, 391

   short receptors, 390–391
   somatosensory information, 394–395, 396–397
   temperature information, 395
   Weber–Fechner law of psychology, 391–392
C–X bonds: isomerism, 51
Cyclic AMP (cAMP), 529–530
Cyclooxygenase (COX) enzyme, 593, 596, 795
Cysteine aminopeptidase, 739
Cystic fibrosis, 623–624
Cystic fibrosis transmembrane conductance regulator (CFTR), 624, 812
Cytokines, 502, 510, 512
Cytoplasm
   actomyosin cross-bridge cycling, 549
   cell structure, 103
Cytosine, 120
Cytoskeletal units
   tensegrity, 106
Cytoskeleton
   cell structure, 103
      actin, 105
      intermediate filaments, 105–106
      microtubules, 103–105
      myosin, 107
   skeletal muscle contractile mechanisms, 314–315
Cytosol
   ATP production, 235–236
   cell structure, 103
   smooth muscle, 354, 358
   taste receptor cells, 435
Cytosolic carbonic anhydrase (CAII), 727
Cytotoxic T cells, 510

D

D (diffuse) subshells, 47–48
Da. *See* Dalton (Da)
Dalton (Da), 59
Darrow–Yannet diagrams, 694–696
Darwin, Charles, 8
Davenport diagrams, 669
Deamination, 247–251
Decoding, 6
Defecation, 803–804
   nervous control of, 805f
Degrees of freedom, 25
Dehydration, 740
Dehydrogenation reactions, 247
Deiodinase type I, 886–888
Delayed rectifying potassium ion channels, 532, 533
Dendrites, 265
Density gradient centrifugation, 116, 116
Deoxycholic acid, 815
Deoxyribonucleic acid (DNA)
   cell diversity, 7
   cell signaling, 214
   genetic code, 123–125
   genomes, 120
   nucleotides, 121–122
   protein synthesis, 120
   replication, 122
   transcription regulation, 125–126
Dephosphorylation, 221–222
Depolarization
   action potentials, 269
   electrocardiograms, 541–542, 544
Depolarizing current, 265
Depolarizing stimuli, 266
Depot fat, 241–242
Desmosomes, 111, 547
Detergents, 153
Detrusor muscles, 483

Deuterium oxide, 689
Developmental control systems, 11
DGC. See Dystrophin–glycoprotein complex (DGC)
DHPR. See Dihydropyridine receptors (DHPR)
Di- and tri-peptide transport systems, 825
Diabetes insipidus, 739b
Diabetes mellitus, 176, 668, 721, 904b
Diapedesis, 513–514
Diaphragm: breathing mechanics, 625
Diaphysis, 933
Diarrhea, 668
Diasteriomers, 51
Diastole relaxation, 516
Diastolic pressure, 520–522, 556–557, 570, 571–572
Diastolic volume, 556–557
Dicrotic pressure, 570
Dictyosomes, 108
Dielectric constants, 281–282
Diencephalon, 373
Dietary fiber, 825, 831b
Differential centrifugation, 115–116
Differential expression of the genome, 7
Differential interference microscopy, 119
Differentiation: cell structure, 101
Diffusion, 75, 579, 580, 645–646, 755–756
 alveolar membranes, 645–646
 electrochemical potential, 87–88
 equilibrium potentials, 255–257
 muscle, 353, 659–660
 oxygen, 659–660, 661
 problem sets, 94
 solute exchange, 581–582
Diffusional permeability
 by dissolution–diffusion mechanism, 197–198
 of microporous membranes, 195–196
Digastric muscle, 773
Digestion, 773, 821. See also Gastrointestinal (GI) system
 bile, 817
 chewing, 773
 exocrine pancreas, 810
 saliva, 775–776
Digoxin, 749
Digoxin drugs, 553–554
Dihydropyridine receptors (DHPR), 331, 549
Dihydroxyacetone phosphate, 242–243
Dilator muscles, 482–483
Dilution
 conservation of solute calculations, 61
 fluid volume calculations, 62
 urine, 730
Dipeptidyl peptidase 4, 823–824
Dipeptidylcarboxypeptidases, 823–824
2,3-diphosphoglycerate (DPG), 660
Dipole moment, 57–58
Dipoles
 dipole–dipole interactions, 54–55
 electrocardiograms, 537, 541
Direct pupillary reflex, 482–483
Discontinuous capillaries, 578
Dissociation constants, 666
Dissolution
 cytosol, 103
 lipid bilayers, 163–164
Dissolution–diffusion mechanism
 diffusional permeability by, 197–198
Distal nephrons
 acid–base balance, 757
 antidiuretic hormone, 738, 745
 osmotic diuresis, 738
 urinary excretion, 940–941

Distal tubules, 701, 732–733
Disulfide exchange, 137
Disulfide isomerase, 137
Diuretics, 617
Divergence of light, 468
Diverticular disease, 808b
Diving reflexes, 679
Dixon plots, 869
DNA. See Deoxyribonucleic acid (DNA)
DNA methylation, 128
DOK7, 329
Dominant ovarian follicle, 949
Donors: universal, 504
Dopamine, 382–383
Dorsal column pathway, 394–395
Dorsal motor nuclei, 612
Dorsal respiratory group (DRG), 672
Dose–response curves: hormones, 859–860
Double carbon–carbon bonds, 50
Double helical structure, 122
Double-reciprocal plots, 66
Downregulated dose–response curves: hormones, 860–861
DPG. See 2,3-Diphosphoglycerate (DPG)
Drag force, 85
DRG. See Dorsal respiratory group (DRG)
Duchenne muscular dystrophy, 314–315
Duct cells: pancreas, 811–812, 817–819
Ductal cells, 810
Ducts of Bellini, 698–699
Duodenum, 788, 788, 796
Duty cycle, 335
Dyads, 549
Dye: body fluid compartments, 689
Dynamic compression: lungs, 639
Dynamic pressure: breathing mechanics, 630–631
Dynamic response phenomena: arterial pressure regulation, 610
Dynamin, 153
Dystrophin, 314–315
Dystrophin–glycoprotein complex (DGC), 314–315

E

E neurons, 673–674
Ear parts. See also Hearing
 inner ear, 441–444
 middle ear, 441–444
  air pressure waves, 441–442
  fluid pressure waves, 441–442
  sound channeling, 441
 outer ear, 441–444
Eccentric contractions: skeletal muscle mechanics, 300
ECF. See Extracellular fluid (ECF)
ECG. See Electrocardiograms (ECG)
ECL. See Enterochromaffin-like (ECL) cellsEssential light chain (ECL)
Ectopic beat, 544
Ectopic focus, 544
Ectopic pregnancy, 948
Edema, 508–509
Edinger–Westphal nuclei, 482–483
EDP. See End-diastolic pressure (EDP)
EDV. See End-diastolic volume (EDV)
Effective renal plasma flow (ERPF), 707
Effectors
 behavior, 367
 cell signaling, 209, 212
Efferent arterioles, 701, 740–741
Efferent autonomic nervous system, 473–474
Efferent fibers, 367
Efferent stimulation: inner hair cell tuning, 445

Einstein's frictional coefficient, derivation of from momentum transfer in solution, 82–86
Einthoven's triangle, 537–539, 543–544
Ejaculation, 965
Ejection cardiac cycle phase, 522
Elastic fibers: lung expansion, 627
Elastin fibers, 627
Electric dipoles. See Dipoles
Electric fields, 34
Electrical activities: smooth muscle, 351–352
Electrical axis definition, 541–542
Electrical capacitors, 36, 282
Electrical charge. See Charge
Electrical force. See Force
Electrical polarizability, 55f
Electrical potentials, 31
 conservative forces, 32–33
 electric fields, 34
 energy, 34
 force, 34
 work per unit charge, 31–32
Electrical signals, 205, 206
 heart contraction, 522–523
 voltage-gated ion channels, 206
Electrical synapses, 378–379
Electrically coupled cardiomyocytes, 547
Electrically coupled smooth muscle cells, 352–353
Electrocardiograms (ECG), 537
Electrochemical gradients: ATP production, 234–235
Electrochemical potentials, 87–88
 active transport, 170–172
 definition, 88
 diffusive forces, 87–88
 electrical forces, 87–88
 flux, 88
 force, 88
 Gibbs free energy, 89
 negative gradients, 88
Electrodes: electrocardiograms, 542–543
Electrodiffusion equations, 257–258
Electrogenic
 sodium, potassium-ATPase as, 173–174
Electrolytes
 bile, 814
 plasma blood cells, 498–500
 problem sets, 765
Electromagnetic radiation, 467
Electromechanical coupling, 590–591
Electron transport chain (ETC), 108, 231
Electronegativity
 ionic bonds, 53
Electronic orbits, 47f
Electrons
 affinity of a chemical for, 229
 atomic structure, 46
 ATP production, 233
 electronegativity, 51, 54
 microscopy, 114
 quantized energies, 46
Electrophoretic uniport, 239, 551
Electrostatic interactions, 54
 dipole–dipole interactions, 54–55
 hydrogen bonding, 54
 induced dipoles, 55
 London dispersion forces, 55
 secondary protein structure, 136
Electrotonic spread, 280–281
Electrotonus, 284
Emesis, 804
Emotional states: autonomic nervous system, 473–474
Emphysema, 640, 669
Emulsification, 828

Enantiomers, 51
Encoding, 6
End-diastolic pressure (EDP), 520–522
End-diastolic volume (EDV), 520–522
   cardiac output, 556
      pressure–volume loops, 556–557
Endergonic reactions, 90, 91f
Endocardium: electrocardiograms, 540–541
Endochondral bone formation, 879f
Endocrine and autocrine signals
   and muscle size, 345–346
Endocrine pancreas, 810, 895
Endocrinology, 851. See also Hormones
Endocytosis, 578–579
   cell structure, 102
   proximal tubules, 728
Endogenous creatinine clearance, 723
Endogenous digitalis-like substances, 749
Endogenous ligands: taste, 437b
Endometrium, 947–948
Endoplasmic reticulum, 116
Endoplasmic reticulum (ER), 108, 154
Endosteum, 933
Endosymbiotic hypothesis, 108–109
Endothelial cells, 510–511
   glomerular filtration, 708
   perfusion regulation, 593
Endothelin, 593
Endothermic reactions, 250–251
End-systolic volume (ESV), 556, 561
Endurance training, 343–345
Energy. See also Adenosine triphosphate (ATP)
   active transport, 170–172
      antiports, 175–176
      electrochemical potentials, 170–172
      exchangers, 170
      material transport, 172
      phosphorylated intermediates, 173
      sodium, potassium-ATPase, 172–173
      sodium, potassium-ATPase as electrogenic, 173–174
      symports, 176
   balance, 834, 847
   conservation of, 17, 32–33
   core principles, 9
   electrical potential, 34
   enthalpy definition, 51
   free energy, 87
   hydrogen ion pumps, 231–232
   hydrolysis, 90–92
   kinetics, 59
      heart total work, 557–558
      Michaelis–Menten formulations, 65–66
      problem sets, 94
   muscle, 334, 349
   pressure-driven flow, conservation, 17
   transition state theory, 67–70
Energy currency of the cell
   adenosine triphosphate (ATP) as, 218–219, 220f
Enhancers of gene expression, 125–126
Enteric nervous system, 473, 797, 799
Enterochromaffin-like (ECL) cells, 791
Enterocytes, 937
Enteroendocrine cells (EECs), 797–798
   chemoreception by, 798f
   in gastrointestinal tract, 800t
Enterogastric inhibitory reflex, 788, 804
Enthalpy: definition, 51
Environmental challenges, robustness and, 9
Enzymes. See also Renin–angiotensin–aldosterone (RAA) system
   acetylcholine recycling, 322–323
   acid–base balance, 752, 753

activation energy, 64f
ATP production, 222, 227, 243, 246
biliary secretion, 810
calmodulin-dependent activation, 206
carbonic anhydrase, 662, 667, 752, 753
cell signaling, 207, 212, 214
cell structure, 109
deoxyribonucleic acid, 120, 121–122
diabetes insipidus, 739b
exocrine pancreas, 810
flippase enzymes, 152
histone code, 126
hormone immunosorbent assays, 863
kidneys, 701–702
lipid bilayers, 152–153
lipolytic enzymes, 810, 811, 828–829
messenger RNA, 122
Michaelis–Menten formulations, 65–66
pancreatic secretion, 810
posttranslational modification, 137–138
ribosomal RNA, 122
single nephron GFR, 743
speeding up reactions, 64
starch digestion, 825–826
stomach, 790, 794
transfer RNA, 123
tyrosine kinases, 212, 899
Eosinophils, 507, 509–510
Ependymal cells, 678
Epicardium, 540–541
Epinephrine, 221, 242, 242, 242f, 382, 475, 480–481
   adrenal medulla, 916
   cardiac action potential, 534
   metabolic control, 922
   perfusion regulation, 596–597
Epiphysis, 933
Epithelial cell attachments, 111
EPO. See Erythropoietin (EPO)
EPSP. See Excitory postsynaptic potentials (EPSP)
Equilibrium centrifugation, 116
Equilibrium concentrations, 92
Equilibrium constants, 63
Equilibrium potentials, 255–257, 269–270
Equipartition theorem, 85
ER. See Endoplasmic reticulum (ER)
Erection of penis, 964–965
ERV. See Expiratory reserve volume (ERV)
Erythroblastosis fetalis, 505b
Erythrocytes
   abundance, 500
   cell forms, 7
   destruction of, 502–503
   formation, 502
   hemoglobin, 500
Erythropoietin (EPO), 498–499, 502, 507, 679, 703
Esophageal manometry, 783
Esophagitis, 783
Esophagus, 769
Essential light chain (ECL), 356
Estrogen reducing adult height, 879–880
ESV. See End-systolic volume (ESV)
ETC. See Electron transport chain (ETC)
Ethane, 50
Ethanolamine, 143
Ethylene, 50
Euchromatin, 214
Evan's blue dye, 689
Evolution
   core principles, 8–9
   cause and effect, 8
   comparative genomics, 8
   diversity, 8

   genome, 9
   phenotypes, 8
   preexisting forms, 8
   robustness, 9
Exchangers, 736
   active transport, 170
   ATP production, 239
   cardiac action potential, 533–534
Excitation-contraction coupling, 206, 318
Excitatory neurotransmitters
   aspartate, 383
   glutamate, 383
Excitory postsynaptic potentials (EPSP), 318–319, 327–328
Exercise, 635
   duration/intensity
      aggregate rates, 334–335
      ATP consumption, 334–335
      blood lactate levels, 338
      metabolic pathways, 336
   glucose, 340, 902
   maximum voluntary ventilation, 635
   metabolism, 838
   oxygen delivery, 660–661
   ventilation, 635, 680
Exergonic reactions, 90
Exhalation, 625
Exit (E) binding sites, 125f
Exocrine pancreas, 810, 895
Exocytosis, 102
Exons, 126f
Exothermic reactions, 250–251
Expiration, 625
Expiration (E) neurons, 673–674
Expiratory reserve volume (ERV), 633
Expired air: steady-state gas exchange equations, 651
Expired oxygen: oxygen consumption calculations, 658–659
Exponential decay
   first-order rate equations and, 63
External intercostal muscles, 672
External sphincter muscles, 484
Exteroreceptors, 10, 389
Extracellular fluid (ECF), 614–615
   body fluid compartments, 690, 694–695, 694, 695
   homeostastis, 7–8
Extracellular resistance: action potential propagation, 282–283
Extracellular volume: body fluid compartments, 689
Extracorporeal shockwave lithotripsy, 819
Extrasystole beat, 544
Extravasation, 511
Extrinsic nerves: stomach, 786
Extrinsic nervous innervation: ileal motility, 803
Extrinsic proteins, 151
Exudation: inflammation, 511
Eyes. See also Vision
   shape, vitreous body, 458
   structure, 456–458

F

Facial nerve, 477–478, 773
Facilitated diffusion, 161
   competitive inhibition, 165
   membrane-bound carriers, 164
   proximal convoluted tubules, 725
   saturation, 164–165
   specificity, 165
F-actin. See Filamentous actin (F-actin)
$FADH_2$: ATP production, 235
Fallopian tubes, 946, 947

Far-sightedness, 459–460
Fasting, 789, 801–802
Fatigue, 303, 334, 341–343, 342, 342–343
Fats: ATP production, 219–220, 241
Fatty acids
  ATP production, 241
  beta oxidation, 243–245
  biological membranes, 143
  depot fat breaking down, 241–242
  mitochondria, 243
  oxidation, 241
  peroxisomes, 243
Fatty acyl chains, 143–145
Fatty alcohols
  plasmanyl phospholipis and plasmenyl phospholipids using, 146
Feedback loops, 10–11
  antidiuretic hormone-renal system, 745–746
  growth hormone, 876
  plasma calcium ions, 942–943
  plasma PI, 942–943
  ventilation control, 677, 677f, 679f
Feedforward mechanisms, 11, 680
Feeding center, 839
Female reproductive systems, 946
Fence function, 110–111
Fenestrae, 709
Fenestrated capillaries, 578, 579
Ferguson reflex, 871
Ferritin, 503
Fetus, 948
Fiber, 825, 831b
Fibrillation, 544
Fibrinogen, 494–495
Fibroblast growth factor 23 (FGF23) and hyperphosphatemia, 942
Fick's dilution principle
  body fluid compartments, 689–690
  fluid volume calculations, 62
Fick's law: solute exchange, 580
Fick's law of diffusion, 75, 75
Fick's principle: cardiac output, 560–561
Fight or flight, 920–921
Filamentous actin (F-actin), 105, 548–549
Filamin, 105
Filling pressure, 558
Filtered loads, 719, 725
Filtering of air: breathing, 623–624
Filtration coefficient, 183f, 186–187
Filtration permeability, 196
Filtration pressure: solute exchange, 584–585
Fimbriae, 947
Fimbrin, 105
First degree heart blocks, 544
First Korotkoff sound, 572–573
First Law of Thermodynamics, 32–33
First pain production by nociceptors, 393
First-order rate equations and exponential decay, 63
FKBP-12 (FK506 binding protein), 332
Flat bones, 933
Flatulation, 803
Flavor in the brain, 438
Flippase enzymes, 152
Flow
  arterial pressure regulation, 608–609
  pressure-driven flow, 15
  problem sets, 43
  solute exchange, 581–582
Flow through a pipe
  Poiseuille's law, 23–24
Flow waves: arterial tree, 572
Flow-induced vasodilation: perfusion regulation, 593

Flow–volume loops: airway resistance, 635–636
Fluid balance, 740, 765
Fluid mosaic model, 153
Fluid pressure waves: hearing, 441–442
Fluid volumes, 62, 715
Fluorescence microscopy, 119
Fluorophore, 119
Flux
  alveolar membranes, 646
  diffusion, 80–81
  electrochemical potentials, 88
  glucose flux, 902
  pressure-driven flow, 15
    conjugate force, 16
    flow per unit area, 15
    movement, 16
FMRI. See Functional magnetic resonance imaging (fMRI)
Focal adhesion complex, 314–315
Focal lengths: vision, 468, 469–470
Focal points: vision, 468
Folic acid, 506
Follicle stimulating hormone (FSH), 948–949, 954, 961–962, 963
Follicles
  cellular development, 949–953
  corpus luteum formation, 954–955
  hormones, 948–949, 954, 961–962, 963
  ovarian development, 949
  ovulation, 947
  thyroid gland, 883
    thyroglobulin precurosor, 883, 885
    thyroxine, 883
    triiodothyronine, 883
Food, 771, 834. See also Gastrointestinal (GI) system
  thermic effect of, 838, 838f
Force
  actin/myosin interactions, 107, 356, 548–549
  cardiac contraction
    force–frequency relations, 552
    stretch-based modulation, 554–555
    sympathetic stimulation, 552–553
  conservative forces, 33–34
  diffusive flux, 80–81
  electric charge movement, 37
  electrical force, 31
    Coulomb's law, 31
    current, 37–38
    electric charge movement, 37
    electrochemical potential, 87–88
    equilibrium potentials, 255–257
    problem sets, 43
    solute flux, 37–38
  electrical potential, 34
  electrochemical potential, 88
  frequency codes, 409
  hair cells, 419
  heart beat, 552
  intermolecular forces, 46
  inverse myotatic reflexes, 403
  London dispersion forces, 55
  myosin/actin interactions, 107, 356, 548–549
  population codes, 409
  pressure-driven flow, 15–16
  repulsive force, 55
  skeletal muscle, 292, 293–295, 295–296, 296–297, 308–309
    architecture, 300–302
    eccentric contractions, 300
    fatigue, 303
    force–velocity curves, 313–314

    isometric force, 293
    motor neuron firing, 295–296
    motor unit activation, 293–295
    muscle length, 296–297
    muscle velocity, 298–299
  smooth muscle, 351–352
    actin–myosin interaction, 356
    calcium ions, 356–357
    myosin light chain phosphorylation, 356
  van der Waals force, 55
"Force" of friction, 3
Forced vital capacity (FVC), 635
Force–frequency relations, 552
Forceful expiration, 626, 639
Forceful inspiration, 626
Force–velocity curves, 313–314
Forebrain, 372
Foreign chemicals: taste, 437b
Forward rate constants, 62–63
Fossil genes, 8
Fourier's law of heat transfer, 75
Frank–Starling law, 558
FRC. See Functional residual capacity (FRC)
Free diffusion coefficients, 79–80
Free energy, 87
Free water clearance, 746–748
Frequency codes, 409
Frequency vibrations, 448–450
Frictional coefficient, 85
Fructose, 827
FSH. See Follicle stimulating hormone (FSH)
F-type ATPases, 174
Functional hyperemia, 594
Functional magnetic resonance imaging (fMRI), 594
Functional residual capacity (FRC), 630–631, 633, 649–650
Fundamental (f) subshells, 47–48
Fusiform muscle fibers, 301
FVC. See Forced vital capacity (FVC)

G

GABA. See Gamma amino butyric acid (GABA)
G-actin. See Globular actin (G-actin)
Gag reflexes, 804
Galactose, 827
Gallbladder, 814, 817
Gallstones, 818b
Gametes, 946, 947, 956, 958, 960–961
Gamma amino butyric acid (GABA), 383–384
Gamma motor systems, 403
γ-tubulin ring complex (γ-TuRC), 104–105
Ganglia
  adrenal medulla, 475
  cervix, 476
  postganglionic neurons, 474
  vision, 462–463, 464
Ganglionic plexuses of nerve cells: gut, 782
Gangliosides, 146, 146
Gap junctions, 111, 534, 547
Gases
  breathing mechanics, 624
  exchange, 623, 642, 653
  Henry's law, 644–645
  problem sets, 653, 682
Gastric accommodation, 787
Gastric acid secretion phases, 791–794
Gastric emptying, 788, 804
Gastric lipase, 790–791
Gastric phases, 812–813, 822–824
Gastric slow waves, 785–786
Gastrin, 787, 789, 791, 808
Gastrin-releasing peptide (GRP), 791
Gastrocolic reflex, 803, 804

Gastroesophageal reflux disease (GERD), 783b
Gastroileal reflex, 803, 804
Gastrointestinal (GI) system, 691, 857, 928
  biliary secretion, 810
  body fluid compartments, 697f
  colonic motility, 796
  esophagus, 769
  growth, 771
  ileum, 796, 802, 833
  intestinal motility, 796
  mouth, 769
  pancreas, 810
  small intestine, 785, 796, 825, 828–829
  stomach, 785
Gastroparesis, 789b
Gating. Ligand-gated ion channels LGICs;. See Voltage-gated ion channels
Gaussian surfaces, 35f, 35f
Gauss's law, 34–35, 287
Genes
  definition, 956
  DNA, 120
  expression
    cell signaling, 213, 215
    thyroid hormones, 888
  genetic code, 123–125
  transcription
    cell signaling, 213
    histone code, 126–127
Genome, 958
  aldosterone, 912
  deoxyribonucleic acid, 120
  differential expression, 7
  evolution, 9
  immune response, 511–512
Genomics, comparative
  revealing pedigree, 8
Genotypes, 956
  blood groups, 503
  cell diversity, 7
  DNA, 120
Geometric isomers, 51
GERD. See Gastroesophageal reflux disease (GERD)
Germ cells. See Gametes
GFR. See Glomeruli—filtration
Ghrelin, 787, 789, 842, 877
GHRH. See Growth hormone releasing hormone (GHRH)
GI. See Gastrointestinal (GI) system
Gibbs free energy, 89
Gibbs–Donnan equilibria, 692–693
GIP. See Glucose-dependent insulinotropic peptide (GIP)
Glaucoma, 457b
Glial cells, 375–376
Globin components: erythrocytes, 500
Globular actin (G-actin), 309
  cardiac contractility, 548–549
  cell structure, 105
Globulins, 496, 498–499
Glomeruli, 429, 698–699
  blood flow, 742–743
  filtration, 705
    autoregulation, 742
    creatinine concentration, 723–724
    endogenous creatinine clearance, 723
    nephron adjustment, 743–744
    oncotic pressure, 740
    problem sets, 715
    tubuloglomerular feedback, 742
    urine output, 740–741
  hydrostatic pressure, 740, 740–741
Glomerulotubular balance, 743–744, 758–759
Glossopharyngeal nerve, 478, 673, 676

Glottis, 625
GLP-1. See Glucagon-like peptide (GLP-1)
Glucagon, 897
  exocrine pancreas, 810
  islets of Langerhans, 899–900
  liver glycogenolysis, 900
  metabolic control, 921
Glucagon-like peptide (GLP-1), 842, 897
Glucocorticoids, 909–910, 921–922
Glucogenic amino acids, 247
Gluconeogenesis, 221, 224, 226f, 703, 900–902
Glucose. See also Adenosine triphosphate (ATP)—production
  absorption, 827
  blood, 895
  diabetes mellitus, 721
  flux, 902
  high intensity exercise, 336–338
  kidneys, 703
  muscle sarcolemma, 340
  renal titration, 720–721
Glucose-dependent insulinotropic peptide (GIP), 897
Glucostatic hypothesis, 841–842
Glucosuria, 721
Glutamate, 383
Glycerol
  biological membranes, 143–145
  depot fat, 241–242
  glycolysis intermediates, 242–243
  metabolism, 242–243
  phosphate shuttles, 222–223, 235
Glycerol kinase, 242–243
Glycerolipids, 146
Glycerophosphates, 242–243
Glycocalyx, 710
Glycogen, 219–220, 338
Glycogenolysis, 900–902
Glycogenoses, 343
Glycolysis, 218, 242–243
Glycolytic flux
  lactic acid, anaerobic metabolism, 338
Glycophorin, 501f
Glycoproteins. See Follicle stimulating hormone (FSH)Luteinizing hormone (LH)
Glycosides, 553–554
GnRH secretion, 962–963
Goblet cells, 623–624
Goitrogens, 892
Goldman–Hodgkin–Katz (GHK) equation, 257–258, 262–264
Golgi apparatus, 108
Golgi stack, 108
Gonadotropin-releasing hormone, 961–962
Gonadotropins, 952–953, 963. See also Follicle stimulating hormone (FSH); Luteinizing hormone (LH)
Gonads, 947
GPCRs. See G-protein-coupled receptors (GPCRs)
G-proteincoupled receptors (GPCR family), 797–798
G-protein-coupled receptors (GPCRs), 381–382
  bitter taste, 436
  cell signaling, 209
  sweet taste, 436
  umami taste, 436
G-proteins
  ATP production, 221
  perfusion regulation, 591
Graafian follicle, 949, 952–953
Gradient, 41
Granular white blood cells, 507. See also Basophils; Eosinophils; Neutrophils

Granule cells
  renin release, 701–702, 748
  single nephron GFR, 743
Granulocytes, 511
Grave's disease, 893–894
Gravity
  hair cells, 419
  heart total work, 557–558
  lung blood flow, 649
Gray matter, 371
Growth
  gastrointestinal system, 771
  thyroid hormone, 888
Growth factors: white blood cell formation, 507
Growth hormone (GH), 875, 881b, 921
  mechanisms of action of, 878f
Growth hormone releasing hormone (GHRH), 875
Growth plates, 877–879
  skeletal growth at, 877–879
GRP. See Gastrin-releasing peptide (GRP)
GS mechanisms, 591
GTPases, 211–212
Guanine, 120
Guanylate cyclase, 357, 591–592
Guanylyl cyclase receptors, 212
Gut. See Stomach

H

Hair cells. See also Follicles
  balance, 415–419
  hearing, 444–445
  inner hair cell tuning, 445
  saccules, 419
  utricles, 419
Haldane effect, 662
Half-life: hormones, 861
hCG. See Human chorionic gonadotropin (hCG)
HCV. See Hepatitis C virus (HCV)
HDL. See High-density lipoproteins (HDL)
Head rotation: vestibular apparatuses, 419
Hearing, 440
  auditory cortex, 446–447
  delay differences, 441
  ear infections, 444b
  ear parts, 441–444
  ear tubes, 444b
  intensity, 440–441
  pitch perception, 447–448
  sound physics, 451–455
  sources of noise, 441
  tests for newborns, 449b
  timbre, 440–441
  tone, 440–441
Heart, 516
  action potential, 528
  arterial pressure, 570, 608
  attack, 563
  calcium-induced calcium release, 549–550
  cardiac cycle, 520–522, 544–545
  cardiomyocyte coupling, 547
  cardiomyopathy, 563
  chambers, 518–520
  circulatory system overview, 489
  contraction, 516–518, 547
    cardiac glycosides, 553–554
    cellular basis, 547
    cycle, 520–522
    electrical systems, 522–523
    myofibrils, 548–549
    parasympathetic stimulation, 553
    regulation, 552
    stretch-based modulation, 554–555

coronary artery disease, 563
coronary sinus, 518
ejection of blood, 570
electrocardiograms, 537
failure, 563b
faulty valves, 563
force–frequency relations, 552
function curves, 556, 599, 599–600, 604, 605
glycosides, 553–554
heartburn, 783
location, 516
murmurs, 520
muscle features, 547
myofibrils, 548–549
output, 556, 599
  distribution by vascular system, 568
  Fick's principle, 560–561
  indicator dilution method, 561–563
rate
  cardiac output, 556
  nucleus tractus solitarius, 611–612
  parasympathetic withdrawal, 612
  respiratory sinus arrhythmia, 612–613
  rostral ventrolateral medulla, 612
total work
  gravitational terms, 557–558
  kinetic terms, 557–558
  pressure, 557–558
valves, 520
venous return, 599
work problem sets, 565
Heat
  circulatory system overview, 489–490
  of combustion, 834–835
  diffusion, 75
  pressure-driven flow, 18
Helper T cells, 510
Hematocrit, 708
Hematocrits, 501, 656
Hematopoietic growth factors, 507
Heme groups, 500–501
Heme recycling, 503
Hemodynamics, 568, 619
Hemoglobin, 61b, 656–657
  erythrocytes, 500
  heme groups, 500–501
  oxygen delivery, 659
  oxygen dissociation curves, 659–660
  polypeptide chains, 500–501
Hemophilia, 496
Hemorrhage, 605, 610
Hemostasis: vascular volume, 494–495
Henderson–Hasselbalch equations, 666
  carbonic acid, 667–668
  isohydric principle, 667
  pH, 669, 753
Henle loops. See Loops of Henle
Henry's law, 644–645
Heparin, 496, 508–509
Hepatitis C virus (HCV), 663
Hepatocytes, 221, 814
Hereditary spherocytosis, 506
Hering–Breuer inflation reflex, 680
Hernia, 783b
Heterochromatin, 214
Heterotrimeric G-proteins
  ATP production, 221
  cell signaling, 209
Hiatal hernia, 783b
High-density lipoproteins (HDL), 498
Hill equation, 656–657
Hill plots, 859, 867–868
Hindbrain, 373
Hippuric acid, 707–708

Histamine
  basophils, 508–509
  perfusion regulation, 595–596
  stomach, 791
Histone acetylase, 214
Histone code, 126–127
Histones, 122, 214
HIV. See Human immunodeficiency virus (HIV)
Hodgkin–Huxley model, 276
Homeostasis, 924
  autonomic nervous system, 473
  calcium, 933
  core principles, 7–8, 13–14, 13
    extracellular fluid, 7–8
    somatic cells, 7–8
  phosphorus, 933
  weight regulation, 838
Homologous chromosomes, 958–959
Horizontal cells: vision, 463
Hormones. See also Antidiuretic hormone (ADH)
  adrenal cortex, 906
  adrenal medulla, 916
  anterior pituitary, 857, 874
  arterial pressure regulation, 609
  ATP production, 221, 242
  blood pressure, 615–617
  blood transport, 858–859
  bone, 935–937
  calcitropic hormones, 924
  cell signaling, 213
  classification, 854–857
  definitions, 853, 853
  dose–response curves, 859–860
  endocrine hormones, 796
  endocrine pancreas, 810, 895
  exocrine pancreas, 810, 895
  follicle stimulating hormone, 948–949, 954, 961–962, 963
  gastrointestinal hormones, 857
  half-life, 861
  hypothalamus, 870
  kidneys, 703
  level measurements, 861–863
  luteinizing hormone, 948–949, 954, 961–962, 963, 963
  menstrual cycle, 953–954
  metabolic clearance rates, 861
  muscle size, 343
  ovarian steroidogenesis, 953
  pancreas, 810, 895
  paracrine secretions, 595–596
  perfusion regulation, 596–597
  pituitary gland, 857, 870
  polypeptide hormones, 854, 857–858
  posterior pituitary, 857, 870
  receptors, 859
  smooth muscle, 353
  steroid hormones, 854, 857, 858
  stomach, 788–789, 791
  target cells, 859, 861
  testicles, 960–961, 961–962
  testosterone, 960–961
  thyroid gland, 883, 921, 963
  urine concentration, 730
Hormone-sensitive lipase (HSL), 242, 242f
Hot taste, 436–437
How? questions, 5
HSL. See Hormone-sensitive lipase (HSL)
HUGO. See Human Gene Organization (HUGO)
Human chorionic gonadotropin (hCG), 963
Human Gene Organization (HUGO), 179, 179
Human immunodeficiency virus (HIV), 663
Humidifying of air: breathing, 623–624

Hyaline membrane disease, 630
Hydraulic conductivity, 183f, 186–187
Hydraulic permeability, 183f, 186–187
Hydrochloric acid secretion, 790, 794
Hydrogen bonds, 53, 54
  deoxyribonucleic acid, 121
  secondary protein structure, 134–136
  water, 54
Hydrogen carbonate, 661–662, 758–759
  acid–base balance, 753–754, 757–759
  bile duct cells, 817–819
  exocrine pancreas, 810
  metabolic acidosis, 760
  metabolic alkalosis, 760–761
  proximal tubules, 727
  respiratory acidosis, 759
  respiratory alkalosis, 760
  ventilation control, 678
Hydrogen ions, 757–759
  acid–base balance, 757–759
  electron transport chain, 231–232
  isohydric principle, 667
  pH definition, 665
  stomach, 791–793, 794
Hydrogen/sodium exchangers: ATP production, 239
Hydrolysis, 90–92, 108, 829–832
Hydrophilic groups, 136, 143–145
Hydrophobic groups, 136, 143
Hydrophobic interactions, 130–131
Hydroxylation, 929
Hygiene hypothesis, 650
Hyperbaric oxygen, 662
Hypercalcemia, 928
Hypercapnia, 677, 678
Hyperemia, 594
Hyperosmolarity, 744–745
Hyperphosphatemia
  fibroblast growth factor 23 (FGF23) and, 942
Hyperpolarization, 266
Hypertension, 563, 617b
Hyperthyroidism, 893–894
Hypertonic solutions, 191, 192f
Hypertrophy, 343, 542
Hyperventilation, 668, 760
Hypervolemia, 749
Hypocalcemia, 926–927
Hypochromic anemias, 502, 506
Hypokalemic metabolic alkalosis, 762
Hypopnea, 680
Hypothalamus, 870
  diabetes insipidus, 739b
  feeding behavior, 839
  function, 373
  sensory afferents, 611–612
  testicular function, 961–963
  thyroid stimulating hormone, 885
Hypothesis testing, 26–27
Hypothyroidism, 889–890
Hypotonic solutions, 191, 192f
Hypoventilation, 668, 760
Hypovolemia, 744–745
Hypoxia, 502, 678–679
Hypoxic vasoconstriction, 650
H-zone: muscle, 306

I
I neurons, 673–674
I-bands: muscle, 305
IC. See Inspiratory capacity (IC)
ICCs. See Interstitial cells of Cajal (ICCs)
ICF. See Intracellular fluid (ICF)
Ideal gas equations, 642–643
IGF. See Insulin-like growth factor (IGF)

IGF-I. See Insulin-like growth factor (IGF-I)
IgG antibodies, 514
IgM antibodies, 514
Ileal brake, 804
Ileal motility, 803
Ileocecal sphincter, 802, 803
Ileogastric reflex, 804
Ileum, 796, 802, 833
Immune response. See InflammationWhite blood cells
Immunoglobulins, 499
Impermeant ions, 692–693
Inactivation gates, 270–271, 270
Inappropriate antidiuretic hormone secretion (SIADH), 739b
Incisura pressure, 570
Incontinence, 803
Incus, 441–442
Indicator dilution method, 561–563
Indigestible carbohydrates, 825
Induced dipoles, 55
Infant respiratory distress syndrome, 630
Inference, 6
Inferior cervical ganglia, 476
Inferior mesenteric ganglia, 475, 799
Inferior vena cava, 518, 698
Inflammation, 507, 595–596
Inflation reflexes, 680
Infundibulum, 947
Inhalation, 625
Inhibin, 948
Inhibitory postsynaptic potentials (IPSP), 318–319, 327–328
Initial segment: action potential origin, 319
Initial training gains, 343
Inner ear, 441–444
Inner hair cells: ear, 445, 448–450
Inner medulla osmotic gradients, 736
Inner medullary collecting duct, 733
Inner mitochondrial membrane, 108
Innervation ratio, 320
  of motor units, 298
Inositol, 143
Inositol triphosphate (IP_3) receptors, 354–356
Inotropic agents: cardiac function curves, 559
Inotropic effects: drugs, 553–554
Inscriptions: muscle, 301
Inspiration, 625
Inspiration (I) neurons, 673–674
Inspiratory capacity (IC), 633
Inspiratory (Insp) neurons, 675
Inspiratory motor neurons, 673
Inspiratory reserve volume (IRV), 633
Inspired air: steady-state gas exchange equations, 651
Inspired oxygen: oxygen consumption calculations, 658–659
Insulators, 39
Insulin, 222, 242, 242
  amino acids, 898
  β cells, 895
  exocrine pancreas, 810
  glucagon-like peptide, 897
  glucose-dependent insulinotropic peptide, 897
  metabolic control, 921
  parasympathetic stimulation, 897–898
  plasma glucose, 895–897
  pulsatile release, 898–899
  somatostatin, 897
  sulfonylurease, 898
  sympathetic stimulation, 897–898
  tyrosine kinases, 899
Insulin-like growth factor (IGF), 876–877, 877
Insulin-like growth factor (IGF-I), 345

Integral proteins, 151
  caveolae and clathrin-coated pits, 153
Integrin, 314–315
Integrins, 314–315
Intercalated cells, 733, 757, 761
Intercalated disks, 523, 534, 547
Intercostal muscles, 625, 672
Interlobar arteries, 699
Interlobar veins, 699, 701
Intermediate filaments, 105–106
  classification of, 107t
  structure of, 106f
Intermolecular forces, 46
Internal intercostal muscles, 672
Interoreceptors, 10, 389
Interphase, 956–957
Interstitial cells of Cajal (ICCs), 785–786, 796, 796–798. See also Pacemakers
Interstitial fluid (ISF)
  body fluid compartments, 690, 692
  circulatory system overview, 489
  oxygen diffusion, 659
  solute exchange, 582–583
Interstitial space clotting, 511
Intestine
  acid secretion phases, 791–794
  calcium ions, 939
  colon
    content reflux, 802
    diverticular disease, 808b
    motility, 796
  lining renewal, 821
  migrating motility complex, 789
  motility, 796
  $1,25(OH)_2$-D hormone, 937–939
  pancreatic enzyme secretion, 813
  PI, 939
  postprandial pancreatic enzyme secretion, 812–813
  protein digestion phases, 822–824
  small intestine, 785
    carbohydrate digestion, 825
    lipolytic activity, 828–829
  stomach acid secretion, 794
  surface area, 821
Intracellular calcium, 206–207
  action potentials, 324
  cell signaling, 211
  smooth muscle contraction, 353–354
Intracellular fluid (ICF): body fluid compartments, 690–691, 694, 695
Intracellular organelles: biological membranes, 142
Intracellular protein digestion phases, 822–824
Intrafusal fiber tension: gamma motor systems, 403
Intraocular fluid, 691
Intrapleural pressure: breathing mechanics, 628–630, 630
Intrapleural space: breathing mechanics, 630
Intrinsic factor, 506, 790–791
Intrinsic nerves: stomach, 786
Intrinsic proteins, 151
Introns, 126f
Inulin, 690, 706–707
  body fluid compartments, 689, 690
  filtered loads, 725
  renal titration curves, 719–720
  water reabsorption, 724–725
Inverse myotatic reflexes, 403
Involuntary ventilation control, 672
Inward ion current, 532–533
Inward rectifying potassium ion channels, 528–529, 532
Iodine deficiency disorders, 893b

Ionic bonds, 53
Ionophores: ion transport, 166–167
Ionotropic receptors, 381–382
Ions, 757. See also Acid–base balance; Action potentials; Electrical potentials; Electrochemical potentials; Resting membrane potentials
  active transport, 170–172
  actomyosin cross-bridge cycling, 549
  aldosterone, 912
  body fluid compartments, 692–693
  bone mineralization, 942
  cardiac
    contraction, 551f
    relaxation, 550–551
  cell signaling, 206–207, 207–209
  chemical neurotransmission initiation, 379
  cross-bridge cycling, 325–326
  current, 37–38
  delayed rectifying potassium ion channels, 532, 533
  excitation-contraction coupling, 324
  hypervolemia, 749
  induced cardiac contraction, 552–553
  intestine, 939
  inward recifying potassium ion channels, 528–529, 532
  ligand-gated channels, 167, 207–209, 381–382
  negative feedback loops, 942–943
  $1,25(OH)_2$-D hormone, 937–939
  osteoclasts, 934
  parathyroid hormone, 941–942
  plasma blood cells, pH, 500
  presynaptic cells, 322, 380
  relaxation, 358
  release, 324
  repetitive stimulation effects, 327
  reuptake, 324–325
  sensitization, 591
  signals, 324
  smooth muscle, 356–357
    contraction, 353–354
  solute flux, 37–38
  stomach, 791–793
  transport, 166–167
  uptake, 324, 551
  urinary excretion, 940–941
  ventricular cardiomyocyte action potential, 530–534
  voltage-gated ion channels, 167, 206
  zona glomerulosa, 911
IP_3. See Inositol triphosphate (IP_3) receptors
IPSP. See Inhibitory postsynaptic potentials (IPSP)
Iris: pupillary light reflex, 482–483
Iron: heme recycling, 503
Iron-deficiency anemia, 506
Irregular bones, 933
IRV. See Inspiratory reserve volume (IRV)
ISF. See Interstitial fluid (ISF)
Islets of Langerhans, 810, 899–900
Isobars, 669
Isoforms, 111
Isohydric principle, 667
Isomerism, 51
Isometric contractions, 293, 300
Isometric force, 293
Isosmotic solution reabsorbtion, 731
Isotonic fluid, 776–777
Isotonic muscle contractions, 298–299
Isotonic saline infusion, 696
Isotopes, 59
Isovolumetric heart contraction, 522
Isovolumetric heart relaxation, 522

J

Jejunum, 796
JGA. *See* Juxtaglomerular apparatus (JGA)
Juxtaglomerular apparatus (JGA), 701–703, 743, 748
Juxtaglomerular cells, 748
Juxtamedullary nephrons, 701

K

Ketogenic amino acids, 247
Ketone bodies, 245
Ketosis, 245–246
Kidneys, 699f, 700–701
   acid–base balance, 752
   adrenal cortex, 906
   anatomy, 698
   body fluid compartments, 696
   concentration mechanisms, 730
   deiodinase type I, 886–888
   dilution mechanisms, 730
   failure, 703b
   fluid balance regulation, 740
   free water clearance, 746–748
   function, 698
   glomerular filtration, 705
   hypocalcemia, 927
   nephrons
      acid–base balance, 757
      antidiuretic hormone, 738, 745
      glomerular filtration, 705
      osmotic diuresis, 738
      osmotic gradients, 733–736
      salt reabsorption, 743–744
      splay, 720
      tubular transport mechanisms, 731–736
      tubuloglomerular feedback, 742
      urinary excretion, 940–941
      water reabsorption, 724, 743–744
   osmotic gradients, 733–736
   pressure-driven flow, 20
   vitamin D, 929
Kinase activation, 590
Kinetics, 59
   heart total work, 557–558
   problem sets, 94
Kirchhoff's Current Law, 39
Kirchhoff's Voltage Law, 39
Korotkoff sound, 572–573
KREBS cycle. *See* Tricarboxylic acid (TCA) cycle
Kuru disease, 139–140

L

L system, 825
Labia majora, 946
Labia minora, 946
Lactate dehydrogenase (LDH), 223
Lactate shuttling
   liver, 339
   mitochondria, 339
   oxidative fibers, 339
Lacteals, 829–831
Lactic acid
   anaerobic metabolism
      glycolytic flux, 338
   anaerobic thresholds, 340
   NADH oxidation carrier systems, 339
Lactose intolerance, 830b
Lagging DNA strands, 123f
Lamellar bone, 934
Laminar flow: airway resistance, 636–637, 638
Laminin, 710
Langmuir adsorption isotherms, 864
Langmuir troughs, 149

Language processing, 447
Laplace law. *See* Law of Laplace
Large intestine. *See* Colon
Late distal tubule, 733. *See also* Connecting ducts
Late proximal tubule
   salt absorption, 728
   water absorption, 728
Latency: action potentials, 267
Lateral corticospinal tract, 405
Lateral hypothalamic area (LHA), 839
Lateral vestibulospinal tract, 405–406
Law of Laplace, 21, 516–517, 627
LBM. *See* Lean body mass (LBM)
LDH. *See* Lactate dehydrogenase (LDH)
LDL. *See* Low-density lipoproteins (LDL)
Le Chatelier's principle, 753
Leading DNA strands, 123f
Lean body mass (LBM), 691
Learning: ventilation control, 680
Least squares linear regression, 29–30
Left electrical axis deviations, 542
Length–tension curves, 296, 306–308
Lennard–Jones potential, 55, 56f
   repulsive force
      van der Waals forces, 55
Lenses
   focal lengths/object distance relationships, 469–470
   image formation/refractive power, 468–469
   light convergence, 468
   light divergence, 468
   near object focus, 459
   thin lens formula, 467–470
Leptin, 843
Leukocytes. *See* White blood cells
Level of significance of the test, 27
Leydig cells, 961, 963
LGICs. *See* Ligand-gated ion channels (LGICs)
LH. *See* Luteinizing hormone (LH)
LHA. *See* Lateral hypothalamic area (LHA)
Ligament of Treitz, 796
Ligand binding, 864, 967
Ligand-gated ion channels (LGICs), 167, 207–209, 381–382
Ligandins, 815–816
Light chain kinase activation, 590
Light convergence, 468
Light divergence, 468
Light focus, 456–458, 467–470
Limbic system, 372
Liminal length, 273
Limiting pH, 757–758
Linear regression, 29–30
Lineweaver–Burk plots, 66
Lipase, 241–242
Lipases, 242
Lipids. Fats;. *See also* Fatty acids
   bilayers/biological membranes, 36, 142
      dynamic motion, 152–153
      extraction, 142
      lipid rafts, 153
      liposome formation, 150
      passive transport, 163–164
      structure, 36f
   cell structure, 108
   digestion, 828
   hydrolysis products, 829–832
Lipolysis, 242
Lipolytic enzymes, 810, 811, 828–829
Lipopolysaccharide (LPS) receptors, 512
Lipoproteins, 241–242, 829–832
Liposome formation, 150
Lipostatic hypothesis, 843
Lithocholic acid, 815

Liver
   absorption of nutrients, 773
   ATP production, 221, 247
   bile, 814, 814–816, 816
   deiodinase type I, 886–888
   digestion, 773
   glycogenolysis, 900
   lactate shuttling, 339
   substrate packaging, 245–246
   vitamin D, 929
   xenobiotics, 816
Local nervous innervation, 803
Local testosterone effects, 963
Locomotor pattern generators, 404–405
London dispersion forces, 55
Lone electron pairs, 53
Long bones, 933
Long receptors, 390–391
Longitudinal esophageal smooth muscle, 781
Longitudinal sarcoplasmic reticulum, 306
Loops of Henle, 701
   permeability, 730, 732
   vasa recta, 736–738
Low-density lipoproteins (LDL), 499
Lower esophageal sphincter (LES), 772–773, 782, 785
Lower pons, 672
LPS. *See* Lipopolysaccharide (LPS) receptors
LRP4, 331
L-type enteroendocrine cells, 798
L-type ion channels, 528–529, 532–533
   epinephrine, 534
Luminal protein digestion phases, 822–824
Lungs
   body fluid compartments, 697f
   breathing patterns, 679–680
   circulatory system overview, 491–493
   collapse, 630
   compliance, 626, 627
   gas equilibration, 649f
   gas exchange, 642
   metabolic acidosis, 760
   pleura, 626
   ventilation pressure, 628–630
   volume, 633
      air movement, 625
      airway resistance, 638–639
      anatomic dead space, 647
      body size, 633
      capacity, 633–634
      spirometry, 633
Lusitropic effect phenomena, 553
Luteinizing hormone (LH), 948–949, 954, 961–962, 963, 963
Lymphatic system, 585, 585–586
Lymphedema, 585
Lymphocytes, 499–500, 507, 510
Lymphoid progenitors, 507
Lysis, 514f
Lysosomes, 109, 116

M

M2 receptors: pacemaker potential, 530
Macrocytic anemias, 501
Macroglobulin, 496, 498–499
Macronutrients. *See* Nutrients
Macrophages, 510–511
   chemokine secretion, 512f
   cytokine secretion, 512f
   erythrocyte destruction, 502
   monocytes, 508
Macroscopic electroneutrality, 692
Macula densa cells, 743
MAG. *See* Monoacylglycerol (MAG)

Magnocellular pathway, 463
Malaria, 506
Malate/aspartate shuttles, 222–223, 235
Male reproductive systems, 956
Malleus, 441–442
Mannan-binding lectin (MBL), 514
Mannose receptors, 512
Manometry, 783
MAPs. *See* Microtubule-associated proteins (MAPs)
Marker enzymes, 116
Markers: body fluid compartments, 689, 690
Mass spectrometry, 862
Matrices: mitochondria, 108–109
Maximum effort exercise, 335–336
Maximum voluntary ventilation, 635
MBL. *See* Mannan-binding lectin (MBL)
McArdle's disease, 346
MCR. *See* Metabolic clearance rates (MCR)
Mean, 25
   standard error of, 25
Mean arterial pressure, 571–572
Mean circulatory pressure, 601
Mean electrical axis definition, 541–542
Mean systemic pressure, 601, 602, 609
Mechanical signals, 205
Mechanically coupled smooth muscle cells, 352–353
Mechanoreceptors, 392–393, 679–680, 840–841
Medial pterygoideus, 773
Medial vestibulospinal tract, 406
Medulla
   adrenal medulla, 475, 596, 916
   baroreflex, 611–612
   function, 373
   kidney function, 698–699
   reticulospinal tract, 406
   salivary nuclei, 777–778
   swallowing centers, 780
   urea, 733, 736
   ventilation control, 672
   vestibular nuclei, 419
Meiosis, 956, 958
Melanopsin photodectors, 482–483
Membrane potential
   problem sets, 289
Membrane potential, calcium regulation via, 354
   by altering BK_{Ca} channels, 354
Membrane proteins, 150–151
Membrane surface tension
   problem sets, 156
Membranes, 142
   action potentials, 266, 323
      chord conductance equation, 269
      depolarization, 269
      threshold, 267–268
      transmembrane resistance, 282
   active transport, 170
   amphiphatic molecules, 149
   attack complexes, 515
   cardiolipin, 146–147
   cell signaling, 213
   cell theory, 6–7
   cholesterol, 146–147
   dynamic motion, 152–153
   electrical capacitors, 36
   facilitated diffusion, 164
   fatty acyl chains, 143–145
   gas diffusion, 645–646
   glycerol, 143–145
   hydrophilic groups, 143–145
   intracellular organelles, 142
   ion transport, 166–167
   lipid rafts, 153
   liposome formation, 150
   organic solvents, 142
   passive transport, 161–163
   phosphate, 143–145
   phospholipids, 142–143, 143–145, 146–147, 149–150
   plasmanyl phospholipids, 146
   plasmenyl phospholipids, 146
   problem sets, 156, 199
   protein bonding affinities, 151–152
   resting potentials, 253
   secreted proteins, 154
   smooth muscle contractile filaments, 352
   sphingolipids, 146, 146–147
   surface pressure, 149
   surface tension, 149
   transport mechanisms, 161
Menarche, 948
Menses, 948
Menstrual cycle, 946–947, 948–949, 953–954
Mentalism, 5
Mesangial cells, 740, 743
Mesenteric ganglia, 475, 799
Messenger ribonucleic acid (mRNA), 122
Metabolic acidosis, 678–679, 760
Metabolic alkalosis, 760–761, 762
Metabolic clearance rates (MCR), 861
Metabolic control, 916
Metabolic hyperemia: perfusion regulation, 594
Metabolism, 336. *See also* Adenosine triphosphate (ATP)—production
   active transport, 172
   endurance training, 343–345
   energy expenditure, 837
   exercise duration/intensity, 336
   food intake, 838–839
   hormones, 861
   muscle fibers, 338
   perfusion regulation, 594–595
   thyroid hormone, 888
Metabotropic receptors, 381–382
Metalloproteinases, 783
Metaphase, 957
Metastable complexes, 68
Methane, 48–49
Methemoglobin, 502f
MGL. *See* Monoacylglycerol lipase (MGL)
Michaelis–Menten formulations, 65–66
Microbiota, 11–13
Microcirculation, 578, 619
Microcytic anemias, 501, 506
Micronutrients
   absorption, 821
   digestion, 821
Microporous membrane
   diffusional permeability of, 195–196
   mechanism of osmosis for, 189–190
   microscopic characteristics, 188
   physical origin of osmotic pressure across, 194–195
Micropuncture studies
   glomerular filtration, 705
   proximal tubule, 725
   ultrafiltrate reabsorption, 725
Microscopic characteristics
   microporous membrane, 188
Microscopy
   cell structure, 113
   problem sets, 156
Microsomal triglyceride transfer protein (MTP), 829–831
Microtubule organizing centers (MTOC), 104–105
Microtubule-associated proteins (MAPs), 103–104
Microtubules
   cell structure, 103–105
   motor proteins, 103–104
Microvilli. *See* Villi
Micturition, 483–485
Midbrain, 373
Middle cervical ganglia, 476
Middle ear, 441–444
   air pressure waves, 441–442
   fluid pressure waves, 441–442
   sound channeling, 441
Midmyocardium: electrocardiograms, 540–541
Migrating motility complex (MMC), 789
Migrating motor complex (MMC), 801–802
Migrating myoelectric complex (MMC), 801–802
Mineralization: vitamin D, 931
Minerolocorticoids. *See* Aldosterone
Mitochondria, 116
   calcium ion uptake, 551
   cell structure, 108–109
   fatty acid metabolism, 243
   hydrogen ion pumps, 231–232
   lactate shuttling, 339
   NADH oxidation carrier systems, 339
   oxygen diffusion, 659
   TCA cycle, 227
   transport mechanisms, 238–239
Mitosis, 956
Mitotic spindles, 104–105
Mitral valve, 519
Mixed sleep disorders, 680
MLCK. *See* Myosin light chain kinase (MLCK)
MLCP. *See* Myosin light chain phosphatase (MLCP)
M-line/M-band
   myomesin joins thick filaments at, 310–311
   muscle, 306
MMC. *See* Migrating motility complex (MMC)
   migrating motor complex (MMC)
   migrating myoelectric complex (MMC)
Modeling relation, 6, 6f
Modification of Diet in Renal Disease (MDRD) Study equation, 724
Modified amino acids, 854
Molecules
   atom binding, 48
   atom movement, 56f
   bond movement, 56f
   bonding of atoms, 48
   potential energy/separation graphs, 67
   size
      diffusion coefficients, 81–82
      Stokes–Einstein equation, 81–82
Moles
   Avogadro's number, 59
   concentration calculations, 60b
   definition, 59
   electrochemical potential, 89
   gas partial pressure, 642–643
Momentum transfer in solution
   derivation of Einstein's frictional coefficient from, 82–86
Monoacylglycerol lipase (MGL), 242f
Monoacylglycerol (MAG), 242, 242f
Monocytes, 507, 510–511
   diapedesis, 513–514
   inflammation, 511
   tissue macrophages formation, 508
Monoglycerol lipase, 242
Monomeric GTPases: cell signaling, 211–212
Monooxygenases, 816
Monosynaptic reflexes, 402–403

Mono-unsaturated fatty acid, 144f
Motilin, 787, 789
Motion sickness, 804
Motoneuron activity, 409–410
Motor activity
   premotor areas, 411–412
   sensory areas, 411–412
Motor control
   hierarchical control, 412–413
   serial control, 412–413
   spinal reflexes, 410–411
Motor cortex: somatotopic organization, 411
Motor efferents: autonomic reflexes, 478–479
Motor end plate, 320
Motor external behavioral responses, 366–367
Motor nerves: myotomes, 410
Motor neurons
   action potentials, 265
     synaptic inputs, 319
   muscle excitation, 321–322
   respiratory muscle control, 672, 673
   skeletal muscle, 295–296, 318
   spinal reflexes
     descending tracks, 405–406
     integrated response, 407
     myotatic reflexes, 402–403
   ventilation control, 673
Motor proteins: cell structure, 104f
Motor units
   innervation ratio of, 298
   muscle fiber, 298
Motor units: skeletal muscle mechanics, 293–295
Mouth, 769
Movement
   balance/control, 409
   cell structure, 101
   gastrointestinal system, 771
mRNA. *See* Messenger ribonucleic acid (mRNA)
MS. *See* Multiple sclerosis (MS)
MTOC. *See* Microtubule organizing centers (MTOC)
MTP. *See* Microsomal triglyceride transfer protein (MTP)
Mucin, 814
Mucus, 775, 790–791
Multiple sclerosis (MS), 386b
Munc-18, 329–330
Muscarine poisoning, 485b
Muscarinic receptors, 382, 479–480
Muscle, sphincters. *See also* Heart; Skeletal muscle; Smooth muscle
   action potentials, 323, 528
   airway resistance, 639
   autonomic nervous systems, 483
   breathing mechanics, 625–626
   defecation, 803–804
   dystrophy, 316b
   endurance training, 343–345
   energetics, 334, 349
   excitation
     acetylcholine, 321–322
     contraction coupling, 318
     motor neurons, 321–322
   exercise
     duration/intensity, 336
     glucose, 336–338
     glycogen, 336–338
   fibers
     action potentials, 324
     metabolic properties, 338
     muscle spindle, 401–402
     types, 346–347
   gamma motor systems, 403
   heart, 516
   hypertrophy, 343
   inverse myotatic reflexes, 403
   micturition, 484
   myosin light chain kinase activation, 590
   of Oddi, 773
   oxygen, 659–660, 660–661
   problem sets, 349
   pupillary light reflex, 482–483
   relaxation, 803
   size, 345–346
   solute exchange, 585–586
   sphincters, 773, 802
     micturition, 484
     pupillary light reflex, 482–483
     relaxation, 804
     upper esophageal, 772–773
   spinal cord circuitry, 410
   training, 334
   twitch waveforms, 326–327
   ventilation control, 672
Muscle fibers, 297–298
   and metabolic ability, 297–298
   motor units, 298
Muscle size
   endocrine and autocrine signals regulating, 345–346
Muscle spindle, 401–402
Muscle strength
   depending on muscle size, 343
Muscular activity, 334
Muscular dystrophy
   clinical applications, 316b
Music processing, 447
MUSK (Muscle-specific kinase), 331
Myelinated axons, 265
Myelinated baroreceptor A fibers, 610
Myeloid progenitors, 507
Myenteric plexus, 782, 796–798
Myocardial contractility, 610
Myoepithelial cells, 777, 871
Myofibrils, 305, 548
Myogenic mechanisms, 592–593, 742, 742–743
Myoglobin, 659–660
Myomesin, 310–311
Myoplasmic calcium ions, 325–326, 326
Myosin
   actomyosin cross-bridge cycling, 549
   ATPase, 549
   cardiac contractility, 548
   cell structure, 105, 107
   skeletal muscle, 308–309
   smooth muscle, 356
Myosin light chain kinase (MLCK)
   perfusion regulation, 590–591
   vascular smooth muscle, 590
Myosin light chain phosphatase (MLCP), 356, 590–591
Myosin light chain phosphorylation, 356
Myostatin, 345–346
Myotatic reflexes, 401, 402–403
Myotomes, 410

N

NAD^+: ATP production, 222–224
NADH. *See* Nictoninamide adenine dinucleotide (NADH)
NANC. *See* Nonadrenergic noncholinergic (NANC) neurons
Narrow tubes: pressure-driven flow, 19–20
Nasal cavities, 427
Natriuretic peptide, 242, 242f
Natural killer cells, 510
Natural selection, 8

Nausea, 804–806
Near object focus, 459
Near-sightedness, 459–460
Nebulin, 309–310
Negative feedback loops, 10–11
   antidiuretic hormone-renal system, 745–746
   growth hormone, 877
   plasma calcium ions, 942–943
   plasma PI, 942–943
   ventilation control, 677
Negative free energies, 90–92
Negative gradients
   electric fields, 34
   electrochemical potentials, 88
Negative intrapleural pressure, 628–630
Nephrogenic diabetes insipidus, 739
Nephrons, 698–699, 699
   acid–base balance, 757
   antidiuretic hormone, 738, 745
   glomerular filtration, 706
   osmotic diuresis, 738
   osmotic gradients, 733–736
   salt reabsorption, 743–744
   splay, 720
   tubular transport mechanisms, 731–736
   tubuloglomerular feedback, 743
   urinary excretion, 940–941
   water reabsorption, 724, 743–744
Nernst–Planck electrodiffusion equation, 257–258
Nerve conduction
   problem sets, 289
Nerve gases, 484b
Nerves
   action potentials, 268, 280, 280, 280, 280–281
   autonomic nervous system, 474, 482
   gut, 782
   hearing, 442–444
   problem sets, 422
   smooth muscle, 353
   stomach, 782, 785, 788–789
   ventilation control, 672, 673
Nervous innervation: ileal motility, 803
Nervous system
   autonomic nervous system, 473
   balance of movement, 409
   cells, 375
   central nervous system
     cutaneous sensory systems, 389
     depression, 669
     feeding behavior, 838–839
     integrative centers, 366
     major area functions, 372–373
     odor adaptation, 431
     organization, 367
     preganglionic neurons, 474
     serotonin, 384
     thoracolumbar spinal cord, 475–477
   chemical senses, 427
   control of movement, 409
   cutaneous sensory systems, 389
   enteric, 797
   gut extrinsic innervation, 799
   neurotransmitters, 375
   organization, 363
     behavior, 365–366
     blood-brain barrier, 370
     brain internal structures, 370–371
     brain surface features, 367–369
     central nervous system divisions, 367
     cerebral spinal fluid, 369–370
     component cells, 371–372
     consciousness, 366
     external behavioral responses, 366–367

Nervous system (*Continued*)
　　gray matter, 371
　　major area functions, 372–373
　　neuroendocrine system, 365
　　peripheral nervous system divisions, 367
　parasympathetic nervous system, 473
　　baroreflex, 611–612
　　cardiac contraction, 553
　　gut extrinsic innervation, 799
　　heart rate, 612
　　insulin secretion, 897–898
　　micturition, 484
　　muscarinic receptors, 479–480
　　origin, 477–478
　　pacemaker potential, 530
　　pupillary light reflex, 482–483
　　target tissue stimulation effects, 482
　somatic nervous system, 474f
　spinal reflexes, 400
　sympathetic nervous system, 473
　　adrenal medulla, 916
　　baroreflex, 611–612
　　calcium ion induced cardiac contraction, 552–553
　　gut extrinsic innervation, 799
　　insulin secretion, 897–898
　　micturition, 484
　　origin, 475–477
　　pacemaker potential, 529–530
　　rostral ventrolateral medulla, 612
　　target tissue stimulation effects, 482
　　vascular system control, 596
　synapses, 375
Nervous tissue
　neurons, 375
　supporting cells, 375
Net filtration pressure, 584–585
Net fluxes
　unidirectional fluxes and, 70–72
Neuroendocrine system. *See* HormonesNervous system
Neuromuscular junction, 318
　complex array of interacting proteins, 329–331
　molecular machinery of, 329–331
　multiple enlargements connected by axon segments, 320
　muscle fiber membrane, 331
Neuromuscular transmission problem sets, 349
Neurons
　action potentials, 265
　autonomic nervous system, 474
　cutaneous sensory systems, 392, 396–397
　gamma amino butyric acid, 383–384
　nervous tissue, 375
　receptive fields, 396–397
　shape, 376–377
　size, 376–377
　ventilation control, 673
　vestibular nuclei, 419
Neuropathic pain, 397
Neuropeptide Y (NPY), 842–843
Neuropeptides, 384–385
Neurotransmission, 320–321, 379
Neurotransmitters, 205, 375
　acetylcholine, 168
　　autonomic nervous system, 479
　　degradation, 322–323
　　muscarinic receptors, 382, 479–480
　　muscle excitation, 321–322
　　nicotinic receptor binding, 382
　　pacemaker potential, 530
　　recycling, 322–323
　　stomach, 791

aspartate, 383
autonomic nervous system, 479
catecholamines, 382
cell signaling, 206
destruction of, 380–381
glutamate, 383
recycling, 322–323, 381
removal of, 380–381
serotonin, 384, 494
　central nervous system, 384
　perfusion regulation, 595
　peripheral nervous system, 384
smooth muscle, 353
Neutral amino acid system, 825
Neutrons, 46
Neutrophils, 507, 508, 513–514
Nicotinic receptor binding, 382
Nictoninamide adenine dinucleotide (NADH), 218
　ATP production, 227–229, 234–235, 235–236
　lactic acid, 339
Nidogen, 710, 710
Nitric oxide (NO)
　perfusion regulation, 591–592, 593
　smooth muscle relaxation, 357
Nitrogen, 837
NO. *See* Nitric oxide (NO)
Noble gases, 48
Nociceptors, 393
Nodes of Ranvier, 265
Nonadrenergic noncholinergic (NANC) neurons, 482
Noncyclical control systems, 11
Nonpolar amino acids, 130
Nonshivering thermogenesis, 250b
Nonspontaneous spontaneous process coupling, 90f
Nonsteroidal anti-inflammatory drugs (NSAIDs), 596, 794
Norepinephrine, 382, 479, 480–482
Normal probability density function, 27–29
Normochromic anemias, 502
Nose
　odors
　　binding proteins, 428–429
　　central nervous system, 431
　　detection limits, 431
　　recognition, 428
　　trigeminal nerve, 431–432
　olfactory bulb, 429
　olfactory epithelium, 427
　olfactory output, 431
　olfactory receptors, 427–428, 429
NPY. *See* Neuropeptide Y (NPY)
NSAIDS. *See* Nonsteroidal anti-inflammatory drugs (NSAIDs)
NTS. *See* Nucleus tractus solitarius (NTS)
Nuclear domain, 346
Nuclear envelopes, 107–108
Nuclear pores, 107–108
Nuclear receptors, 213
Nucleation sites, 105
Nuclei
　atomic structure, 46
　cell structure, 107–108
Nuclei separation, 67–68
Nucleolus
　cell structure, 107–108
　ribosomal RNA assembly, 122–123
Nucleolytic enzymes, 810, 811
Nucleosomes, 126, 214
Nucleotides, 121–122
Nucleus ambiguus, 611–612

Nucleus tractus solitarius (NTS), 611–612, 840–841
Numerical aperture, 113–114
Nutrients
　absorption, 772–773, 821
　catabolism, 837
　circulatory system overview, 489–490
　digestion, 821
　filtered load, 719
　gastrointestinal system, 771, 771–772
　glomerular filtration, 705
　intestine, 794
　pancreatic enzyme secretion, 813

O

Obscurin, 312
Obstructive sleep apnea, 681, 681
Oculomotor nerve, 477
Odorant binding proteins, 428
Odors
　binding proteins, 428–429
　central nervous system, 431
　detection limits, 431
　recognition, 428
　trigeminal nerve, 431–432
Off-center bipolar cells, 462
Off-center ganglion cells, 462–463
1,25(OH)$_2$-D hormone, 937
25(OH)$_2$-D vitamin D, 930–931
Ohm's law, 39
Okazaki fragments, 123f
Olfaction, 429
Olfactory bulb, 429
Olfactory epithelium, 427
Olfactory output, 431
Olfactory receptors, 427
　cribriform plate, 427–428
　glomeruli, 429
　second-order neurons, 429
Olfactory sensory neurons (OSN), 427, 433f
On-center bipolar cells, 462
On-center ganglion cells, 462–463
Oncotic pressure
　plasma proteins, 500
　solute exchange, 584
One-dimensional diffusion, 79
One-dimensional random walk, 76–79
Onsager reciprocity, 16
Oogenesis, 948
Opioid-induced constipation, 807b
Optical isomers, 51
Orad stomach, 787
Oral dissolution therapy, 819
Oral rehydration therapy (ORT), 177
Orbitals, 46, 47–48
Orbitofrontal cortex, 431
Organ systems of the body, 4t
Organelles, 116. *See also* Mitochondria
　biological membranes, 142
　calcium ion uptake, 551
　cell structure, 101, 109, 115
　fatty acid metabolism, 243
　hydrogen ion pumps, 231–232
　lactate shuttling, 339
　NADH oxidation carrier systems, 339
　transport mechanisms, 238–239
　tricarboxylic acid cycle, 227
Organic anions: sodium cotransporters, 725–727
Organic solvents: biological membranes, 142
ORT. *See* Oral rehydration therapy (ORT)
Orthostasis, 610
OSA. *See* Obstructive sleep apnea
Osmolarity, 184
Osmolytes, 184

Osmometers, cells behaving like, 192–193
Osmoreceptor cells: antidiuretic hormone secretion, 744
Osmosis, 182, 191–192, 194–198. *See also* Osmotic...
   aquaporins, 198
   dissolution–diffusion mechanism, diffusional permeability by, 197–198
   filtration permeability, 196
   for microporous membranes, 189–190
   microporous membranes, diffusional permeability of, 195–196
   physical origin of osmotic pressure across a microporous membrane, 194–195
   pressure- and osmosis-driven flow, 196–197
Osmotic and hydrostatic pressures, equivalence of, 186–187
Osmotic diuresis: distal nephron, 738
Osmotic gradients
   inner medulla, 736
   urine concentration, 730–731
Osmotic pressure, 182, 187, 191–192, 727–728
   cells, 192–193
   colligative properties, 185
   hypertonic solutions, 191, 192*f*
   hypotonic solutions, 191, 192*f*
   microporous membranes, 194–195
      mechanism of osmosis for, 189–190
      microscopic characteristics, 188
   nonideal solutions, 185
   osmometers, cells behaving like, 192–193
   osmotic and hydrostatic pressures, equivalence of, 186–187
   osmotic coefficients, 185
   permeable solutes, 187
   physical origin of, across a microporous membrane, 194–195
   rational osmotic coefficient, 186
   reflection coefficients, 187
   regulatory volume decrease (RVD), 193
   regulatory volume increase (RVI), 193
   solute exchange, 584, 584
   solutions, 185
   units for the calculation of, 187*t*
   van't Hoff equations, 182–184
Osmotically inactive, 194
OSN. *See* Olfactory sensory neurons (OSN)
Osteoblasts, 933, 934–935
Osteocalcin, 934
Osteoclastic osteolysis, 935–936
Osteoclasts, 934, 935
Osteocytes, 933–934
Osteocytic osteolysis, 935
Osteoid, 933, 934–935
Osteoprogenitor cells, 934
Osteoprotegerin (OPG), 935
Otitis media, 444
Ouabain, 749
Ouabain drugs, 553–554
Outer ear, 441–444
Outer hair cells: ear structure, 445
Outer mitochondrial membrane, 109*f*
Outward potassium ion current, 532
Ova, 948, 958
Ovaries, 946, 947, 949
Overpressure-induced renal damage, 742–743
Overshoot: action potentials, 268
Oviducts. *See* Fallopian tubes
Ovulation, 947, 948, 949, 954–955
Oxidation
   anaerobic thresholds, 340
   ATP production, 236–238, 241
      acetyl CoA, 218
      pyruvate, 227
   carrier systems, 339

Oxidation potentials, 39
Oxidative fibers: lactate shuttling, 339
Oxidative muscle: oxygen storage, 659–660
Oxidative phosphorylation: ATP production, 227
Oxygen
   consumption, 657
      calculations, 659*b*
      cardiac output, 560–561
   delivery
      in exercise, 661
      to tissue, 659
   dissociation curves, 656, 656–657, 660–661
   electron transport chain, 233
   hemoglobin, 656–657
   myoglobin, 659–660
   respiratory system, 623
   saturation curves, 656
   transport, 623, 656
   ventilation control, 677
Oxytocin, 872–873
   amino acids, 870
   hypothalamus, 870
   myoepithelial cell contraction, 871
   uterine contraction, 871

P

P (principle) subshells, 47–48
P wave atrial depolarization, 540
Pacemakers, 522, 528–529, 675, 675–676, 785–786
Pain production by nociceptors, 393
Pain receptors: taste, 436–437
Painful stimuli, 400
Pancreas, 810, 895
   absorption of nutrients, 773
   digestion, 773
   endocrine pancreas, 810, 895
   exocrine pancreas, 810, 895
   secretory cells, 109*f*
Pancreatic acinar secretion, mechanism of, 812*f*
Papillary muscles, 518
Para-amino hippuric (PAH) acid, 707
   clearance, 723
   renal titration, 722–723
Paracellular pathways: intestinal calcium ions, 937
Paracrine secretions, 595–596
Parallel beta sheets, 135
Parallel muscle fibers, 301
Parallel plate capacitors, 35–36
Parasitic infections, 509–510
Parasympathetic nervous system, 473
   baroreflex, 611–612
   cardiac contraction, 553
   gut extrinsic innervation, 799
   heart rate, 612
   insulin secretion, 897–898
   micturition, 484
   muscarinic receptors, 479–480
   origin, 477–478
   pacemaker potential, 530
   pupillary light reflex, 482–483
   target tissue stimulation effects, 482
Parathyroid hormone (PTH), 703
   derivative control, 927
   destruction of, 927
   goals, 941–942
   hypocalcemia, 926–927
   osteoclastic osteolysis, 935–936
   osteocytic osteolysis, 937
   plasma phosphate regulation, 722
   vitamin D, 929–930

Paraventricular nuclei, 744
Paravertebral ganglia, 475
Parietal lobes: vision, 465
Parietal pleura, 625*f*
Parkinson's disease, 420
Parotid glands, 775
Paroxysmal superventricular tachycardia (PSVT), 544
Partial molar volume, 184
Partial pressures, 642–643, 652
   alveolar air and blood, 657, 658*f*
   diffusing capacity, 646–647
Partial volumes, 652
Particle number in moles, 59
Partition coefficients, 130–131
Parturition, 947–948
Parvocellular pathway, 462–463
Passive mechanisms, 161. *See also* Diffusion
   cell structure, 102
   reabsorbtion of urea, 728
   resting expiration, 626
   solute exchange, 579
   water, 168
Patch clamp experiments, 271
Pauling units, 51
PBC. *See* Pre-Botzinger complex (PBC)
Peak expiratory flow rate (PEFR), 635–636
Pedigrees: evolution, 8
PEEP (positive end-expiratory pressure), 640
PEFR. *See* Peak expiratory flow rate (PEFR)
Peltier effect, 16
Pelvic nerve, 803
Pendred's syndrome, 892*b*
Penis: erection, 964–965
Pepsinogen secretion, 790–791
Peptic ulcers, 794*b*
Peptide bonds, 131–133
Peptidyl (P) binding sites, 125*f*
Perception, 390
Perfusion, 589. *See also* Ventilation/perfusion ratios
   pulmonary circulation, 650
   solute exchange, 583–584
Pericardium, 516
Pericytes, 578–579
Perikaryons, 265
Perilipin, 242
Periodic table, 47*f*
Periodicity, 47–48
Periosteum, 933
Peripheral chemosensors, 677
Peripheral nervous system (PNS)
   connection to CNS, 373
   organization, 367
   serotonin, 384
Peripheral proteins, 151
Peristalsis, 698, 780–781
   intestinal motility, 800–802
   vomiting, 809
Peritoneum, 698
Peritubular capillaries, 701
Permeable solutes: osmotic pressure, 187
Pernicious anemia, 506
Peroxisomes, 109, 116, 243
Perturbations
   positive feedback control systems and, 11
pH. *See also* Acid–base balance
   chemical buffers, 665
   definition, 665
   Henderson–Hasselbach equations, 667–668
   oxygen dissociation curves, 660, 660
   plasma blood cells, 500
   problem sets, 682
   renal system, 665
   respiratory system, 665, 668

pH (*Continued*)
  taste receptor cells, 435
  ventilation control, 676
Phagocytes, 502–503, 508
Phagocytosis, 102
Pharmacomechanical coupling, 590–591
Pharyngeal swallowing phases, 780–781
Phase locking: pitch perception, 447–448
Phase-contrast microscopy, 119
Phasically-based contractile activity, 351
Phenotypes, 956
  cell diversity, 7
  DNA, 120
  evolution, 8
Phenoxybenzamine, 481
Pheochromocytoma, 922b
pH–hydrogen carbonate diagrams, 669–670
Phonocardiograms, 520
Phosphate
  biological membranes, 143–145
  proximal convoluted tubules, 725–727
Phosphatidylinositol 3-kinase (PI-3K), 345
Phospholamban
  cardiac relaxation, 550
  smooth muscle, 354
Phospholamban (PLN), 333
Phospholipase C, 211
Phospholipids
  bile, 814
  biological membranes, 142–143, 143–145, 146–147, 149–150
  water squeezing out, 149f
Phosphorus homeostasis, 924, 933
Phosphorylated intermediates: active transport, 173
Phosphorylation: smooth muscle, 354, 356
Photoreceptor cells, 460
Phrenic nerve, 625, 672
Physiologic dead space, 647–648
Physiology, 66
Physiology, core principles of, 6
PI
  bone mineralization, 942
  equilibrium concentrations, 92
  intestine, 937–939
  negative feedback loops, 942–943
  $1,25(OH)_2$-D hormone, 937–939
  osteoclasts, 933–934
  urinary excretion, 940–941
Pigments: bile, 814
PI-3K. *See* Phosphatidylinositol 3-kinase (PI-3K)
Pinnate muscle fibers, 301
Pinocytosis, 102
Pitch perception, 447–448
Pituitary gland, 857, 870
Pituitary–hypothalamus axis, 907–908
PKA. *See* Protein kinase A (PKA)
PKG. *See* Protein kinase G (PKG)
Placenta, 948
Planar lipid bilayer capacitance, 37b
Plasma, 494, 498
  antidiuretic hormone, 873
  body fluid compartments, 689, 690–691, 692, 693–694, 696
  buffer lines, 669
  calcitropic hormones, 924–925, 928
  creatinine concentration, 723–724
  dehydration, 740
  electrolytes, 498–500
  glomerular filtration, 707–708
  glucose
    absorption, 900–902
    excretion, 720
    glycogenolysis, 900–902
    insulin secretion, 895–897

hepatocytes, 814
kidneys, 697f, 702
membrane cell forms, 7
parathyroid hormone, 941
pH, 500, 665
phosphate, 722
thyroid hormone, 885–886
volume estimations, 62b
water, 498–500
Plasma membrane, 116
Plasmalogens, 146
Plasmanyl glycerol phospholipids, 146, 146f
Plasmanyl phospholipids
  using fatty alcohols, 146
Plasmenyl glyercol phospholipids, 146, 146f
Plasmenyl phospholipids
  using fatty alcohols, 146
Plasmin, 495
Plasminogen, 495
Platelets, 507
  pluripotent stem cells, 507
  vascular hole sealing, 494
Pleura, 626
PLN. *See* Phospholamban (PLN)
Pluripotent stem cells, 502, 507–508
Pneumothorax condition, 630
Podocytes, 708, 709f
Poiseuille equation, 637–638
Poiseuille's law, 19–20, 189, 574–576
  derivation of, 22–24
    pressure-driven flow through a cylindrical pipe, 22
  flow through a pipe, 23–24
  shear stress, 22–23
Polar amino acids, 130
Polar bonds, 53, 53–54, 57
Polar groups, 143
Polarizable atoms, 55
Polycythemia, 501, 679
Polydipsia, 721
Polypeptide chains, 500–501
Polypeptide hormones, 854, 857–858
Polyuria, 721, 738
POMC. *See* Proopiomelanocortin (POMC)
Pons, 373, 672
Pontine respiratory group (PRG), 672, 674f
Pontine reticulospinal tract, 406
Population codes: contractile force, 409
Population variance, 25
Positive chronotropy, 552
Positive feedback control systems, 11
Positive ionotropic agents, 559
Positive pressure breathing, 640b
Positive staircase phenomena, 552
Posterior pituitary, 857, 870
Postganglionic neurons, 474
Postganglionic sympathetic nerves, 799
Postganglionic sympathetic neurons, 480–482
Postprandial pancreatic enzyme secretion, 812–813
Postsynaptic potentials
  decay over time, 319
  electronical spread, 319
  excitory types, 318–319
  graded, 319
  inhibitory types, 318–319
Posttranslational modification, 126, 137–138
Potassium
  acid–base balance, 757, 762
  action potentials, 269–270, 528–529, 532, 533
  aldosterone, 749, 912
  equilibrium potentials, 257
  stomach, 794
  zona glomerulosa, 911

Potential, current and
  Ohm's law, 39
Pre-Botzinger complex (PBC), 673
Preganglionic axons, 477
Preganglionic fibers, 475, 476
Preganglionic neurons, 474
Pregnancy, 948
"Preinspiratory" (Pre-I) neurons, 675
Preload: cardiovascular system, 556, 558–559, 599
Premature depolarization, 544
Premotor areas, 411–412
Preosteoblasts, 934
Pressure
  alveoli, 627b
  arteries, 571–572, 573–574
  blood flow, 493, 494
  breathing mechanics, 624–625, 625, 628–630
  circulatory, 601
  heart
    muscle contraction, 516–518
    total work, 557–558
    valve closure, 520
  mole fractions, 642–643
  osmotic, 182, 727–728
  partial, 642–643, 646–647, 652, 657, 658f
  perfusion regulation, 589–590
  pH regulation, 668
  plasma proteins, 500
  problem sets, 43
  solute exchange, 584–585
  sound, 454–455
  systemic, 601, 602, 609
  vapor pressure, 643, 644
  vascular flow, 568–569
  vascular function, 569–570, 603–604
  ventilation control, 677, 677
Pressure- and osmosis-driven flow, 196–197
Pressure-driven flow, 15
  blood, 493
  capacitance, 18
  charge, 18
  conservation of energy, 17
  conservation of matter, 17
  continuity equations, 17
  through a cylindrical pipe
    Poiseuille's law, 22
  force, 15–16
  heat, 18
  Laplace Law, 21
  Poiseuille's law, 19–20
  solute, 18
  steady-state flow
    linear gradients, 17–18
    Poiseuille's law, 19–20
  volume, 18
Pressure–volume (PV) loops, 556–557
Pre-synaptic cells, 380
  calcium ion efflux mechanisms, 322
Pre-synaptic terminals, 381
Prevertebral ganglia, 475
PRG. *See* Pontine respiratory group (PRG)
Primary active transport, 172–173
Primary auditory cortex, 447
Primary bile acids, 814–815
Primary motor cortex, 411
Primary peristalsis, 780
Primary polydipsia, 739
Primary protein structure, 134
Primary spermatocytes, 960–961
Primordial follicle, 949
Principal cells, 733
Probability, 26
Problem sets

acid–base balance, 765
action potential, 289
airway resistance, 653
alveolar gas exchange, 653
blood, 525
cardiac work, 565
cell fractionation, 156
clearance, 715
diffusion, 94
electrical force, 43
electrolyte balance, 765
energy balance, 847
fluid balance, 765
fluid volumes, 715
gas exchange, 653
gas transport, 682
glomerular filtration, 715
hemodynamics, 619
kinetics, 94
ligand binding, 967
membrane potential, 289
membrane structure, 156
membrane surface tension, 156
membrane transport, 199
microcirculation, 619
microscopic resolution, 156
muscle energetics, 349
muscle force, 349
nerve conduction, 289, 422
neuromuscular transmission, 349
pH, 682
pressure forces and flows, 43
sensory transduction, 471
surface tension, 156
Proliferation: cell structure, 101
Proline and glycine transport system, 825
Proline transport system, 825
Prometaphase, 957
Promoters of gene expression, 125
Proopiomelanocortin (POMC), 842–843
Propanolol, 481
Prophase, 956–957
Propulsion of stomach contents, 787–788
Prostacyclin, 593, 596
Prostaglandins, 595
Prostate specific atigen (PSA), 965
Proteasome, structure of, 110f
Proteasomes, 110
Protein. *See also* Follicle stimulating hormone (FSH); Luteinizing hormone (LH)
　aquaporins, 168, 727, 731, 751
　biliary secretion, 815–816
　biological membranes, 154
　　bonding affinities, 151–152
　　endoplasmic reticulum, 154
　　lipid rafts, 153
　bitter taste, 436
　breathing mechanics, 624
　calcium ion release/reuptake, 324
　cardiac action potential, 529–530
　cell signaling, 209
　cell structure, 104f, 105, 110, 111
　　endoplasmic reticulum, 108
　　proteasomes, 110
　　ribosomes, 108
　cortisol, 909–910
　dietary energy content, 241
　digestion, 822–824
　distal nephron permeability, 745
　endocytosis, 728
　erythrocytes, hemoglobin, 500
　folding diseases, 139b
　glomerular filtration, 710
　G-proteins, 209, 221, 381–382, 436, 591
　mannan-binding lectin, 514

odorant response, 428–429
plasma blood cells, 498–500
　oncotic pressure, 500
　pH, 500
reabsorption mechanisms, 728
semen, 965
skeletal muscle contractile mechanisms, 308–309, 314–315
structure, 130
　activation, 139
　amino acids, 130, 134–136
　four levels of structure, 134–137
　hydrophobic interactions, 130–131
　inactivation, 139
　number of molecules, 139
　partition coefficients, 130–131
　peptide bonds, 131–133
　posttranslational modification, 137–138
　primary protein, structure, 134
　quaternary protein, structure, 137
　reactive surface formation, 133–134
　reversible activation/inactivation, 139
　secondary protein structure, 134–136
　tertiary protein structure, 136–137
sweet taste, 436
synthesis
　deoxyribonucleic acid, 120
　messenger RNA, 122
thyroid hormones, 885–886
umami taste, 436
xenobiotics, 816
Protein kinase A (PKA), 242, 332
Protein kinase G (PKG), 332
Protein phosphatases, 332
Protein YY(3-36), 843
Proteoglycans, 710, 710
Proteolytic cleavage, 137, 514
Proteolytic enzymes, 810–811
Prothrombin, 498–499
Proton electrochemical gradients, 234–235, 234
Proton motive force, 234
Proton pumps, 233, 794
Protons, 46
Proximal tubules, 701, 725–728
　acid–base balance, 753
　ammonium, 755–756
　endocytosis, 728
　isosmotic solution reabsorbtion, 731
　ultrafiltrate reabsorption, 725
PSA. *See* Prostate specific atigen (PSA)
P-selectin protein, 513–514
PSVT. *See* Paroxysmal superventricular tachycardia (PSVT)
PTH. *See* Parathyroid hormone (PTH)
P-type ATPases, 174
Puberty, 948
Pudendal nerve, 803
Pulmonary artery, 518–519
Pulmonary chemoreflex, 680
Pulmonary surfactants, 627–628
Pulmonary valve, 518–519
Pulmonary ventilation, 623, 635. *See also* Alveoli; Respiratory systems
　exercise, 635
Pulse pressure
　artery compliance/stroke volume, 570–571
　mean arterial pressure, 571–572
Pupillary light reflex, 482–483
Purkinje fibers, 523
Pursed-lip breathing, 639–641, 641f
PV. *See* Pressure–volume (PV) loops
Pyloric sphincter, 772–773, 785
Pyruvate, 227

Q

QRS complexes, 540
Quantized energies: electron orbitals, 46
Quaternary protein structure, 137
Quinine alkaloids, 506

R

RAA. *See* Renin–angiotensin–aldosterone (RAA) system
Radial arteries, 699, 700–701
Radial veins, 699, 701
Radioimmunoassays, 862–863
Random coils, 135
Rapsyn, 331
Rate constants: transition state theory, 68–70
Rational osmotic coefficient, 186
RBF. *See* Renal blood flow (RBF)
RDS. *See* Respiratory distress syndrome (RDS)
Reaction rates: transition state theory, 67–70
Reactive oxygen species (ROS), 508
Receptors, 859, 860
　adrenergic receptors, 383
　arterial pressure regulation, 609–610
　autonomic nervous system, 482
　baroreceptors, 609–610
　catecholamine action, 919–920
　cell signaling, 209, 212, 213
　　gene transcription, 213
　　histone acetylase, 214
　　transcription factors, 214–215
　colonic contraction, 803
　cutaneous sensory systems, 389, 390–391, 392–393
　　receptive fields, 392
　　types, 392–393
　G-proteins, 209, 221, 381–382, 436, 591
　guanylyl cyclase, 212
　immune response, 511–512
　metabotropic receptors, 381–382
　muscarinic receptors, 382, 479–480
　nicotinic receptor binding, 382
　passive mechanisms
　　cell structure, 102
　　reabsorbtion of urea, 728
　　resting expiration, 626
　　solute exchange, 579
　　water, 168
　smell, 427, 429
　smooth muscle, 354, 358
　steroid hormones, 125–126
　stomach, 791–793
　stretch receptors
　　antidiuretic hormone secretion, 745
　　colonic contraction, 803
　　stomach, 791–793
　taste, 433–434, 438
　thermoreceptors, 393
　tyrosine kinase, 212
　warm receptors, 393
Recipients: universal, 504
Recoil tendency of lungs, 627, 630
Recruitment
　muscle force, 297
　and muscle force, 297
Rectoanal inhibitory reflex, 804
Red blood cells, 494, 498
Reduction potential, 229, 229–230
Reentry arrhythmia, 544
Referred pain, 478–479
Reflection coefficients, 190–191
　osmotic pressure, 187
Reflexes
　aspiration, 680
　autonomic nervous system, 473, 478–479

Reflexes (*Continued*)
  diving, 679
  Hering–Breuer inflation, 680
  ileogastric, 804
  inflation, 680
  myotatic, 401, 402–403
  spinal, 400
  stomach distension, 803
  swallowing, 680
  ventilation control, 679–680
Reflux, 783, 802
Refraction
  light focusing, 458–459
    angles of incidence, 468
    refractive indices, 468
    speed of light, 467–468
    thin lens formulae, 467–470
  thin lens formulae, 467–470
Refractive index, 113–114
Refractory periods, 268
Regulatory light chain (RLC): smooth muscle, 356
Regulatory T cells, 510
Regulatory volume decrease (RVD), 193
  osmotic pressure, 193
Regulatory volume increase (RVI), 193
  osmotic pressure, 193
Relative humidity, 644
Relative refractory periods, 268
Relaxation volume–pressure curves, 628–630
Renal arteries, 698, 699, 700–701
Renal blood flow (RBF): autoregulation, 741–743
Renal capsule, 698–699
Renal compensation
  acid–base balance, 759
Renal corpuscles, 705
Renal damage, 742–743
Renal function curves, 614–615
Renal plasma flow, 707–708
Renal pyramids, 698–699
Renal system. *See also* Kidneys
  pH regulation, 665, 752–753
Renal titration, 719–720, 722–723
Renal tubular acidosis, 668, 763b
Renal veins, 698, 699, 701
Renin, 701–702, 743, 748
Renin–angiotensin–aldosterone (RAA) system, 749
  blood volume, 749
  hormonal regulation, blood pressure, 615–616
  perfusion regulation, 596–597
  sodium balance, 748–749
Replication forks, 123f
Repressors of gene expression, 125–126
Reproductive systems, 946, 956
Repulsive force
  van der Waals forces
    Lennard–Jones potential, 55
Residual volume (RV), 633, 634t
Resistive current, 39–40
Resolution: cells structure/microscopy, 113–114
Respiratory distress syndrome (RDS), 630
Respiratory rate (RR), 635
  exercise, 635
  pulmonary ventilation, 635
Respiratory systems, 621. *See also* Alveoli; Lungs
  acidosis, 669, 759
  airway resistance, 630–631, 633, 653
  alkalosis, 669, 760
  breathing mechanics, 621
  carbon dioxide transport, 656
  compensation, 665

  exchange ratios, 642
  gas exchange, 642
  muscle control, 672
  oxygen transport, 656
  pH regulation, 665, 668, 752–753
  quotients, 642
  sinus arrhythmia, 612–613
  ventilation control, 672
Resting expiration, 626
Resting membrane potentials, 253
  cardiac cells, 528
    contractile cells, 528–529
    sinoatrial node, 528–529
    ventricular cardiomyocyte action potentials, 530–532
Rest time: intensity of exercise, 335–336
Reticuloendothelial system, 510–511
  erythrocyte destruction, 502–503
  monocytes, 510–511
  tissue macrophages, 510–511
Reticulospinal tract, 406, 484
Retina, 456–458, 460–462
  far-sightedness, 459–460
  near-sightedness, 459–460
  photoreceptor cells, 460
  pupillary light reflex, 482–483
  visual signal processing, 460–462
Retropulsion of stomach contents, 787–788
Reverse peristalsis, 800–802, 804–806
Reverse rate constants: chemical reactions, 62–63
Reynolds number, 637
Rheobase, 268
RhoA G-proteins, 591
Ribonucleic acid (RNA)
  cell structure, 108
  messenger RNA, 122
  ribosomal RNA, 108, 122–123
  transfer RNA, 123
Ribosomal ribonucleic acid (rRNA), 108, 122–123
Ribosomes: cell structure, 108
RNA. *See* Ribonucleic acid (RNA)
Robustness and environmental challenges, 9
ROS. *See* Reactive oxygen species (ROS)
Rostral ventrolateral medulla (RVLM), 612
Rough endoplasmic reticulum, 108
RR. *See* Respiratory rate (RR)
RRNA. *See* Ribosomal ribonucleic acid (rRNA)
Rubrospinal tract, 405
RV. *See* Residual volume (RV)
RVD. *See* Regulatory volume decrease (RVD)
RVI. *See* Regulatory volume increase (RVI)
RVLM. *See* Rostral ventrolateral medulla (RVLM)
Ryanodine receptors (RyR), 331, 332, 332, 549
RyR. *See* Ryanodine receptors

S

s (sharp) subshells, 47–48
SA. *See* Sinoatrial (SA) node
Saccules, 419
Sacral nerves, 477–478
Saliva, 775–776, 778–779
Saliva production, 779–780, 779f, 780
Salivary glands, 775
Salivary nuclei, 777–778
Salivation control, 777–778
Salt
  body fluid compartments, 696b
  distal tubule, 732–733
  late proximal tubule, 728
  loop of Henle, 731, 732
  nephron adjustment, 743–744
  taste, 432, 434–435

Saltatory conduction: action potential propagation, 286
Sample standard deviation, 25
Sample variance, 25
Sarcolemma (SL), 306, 340, 550–551, 590–591
Sarcomeres, 310, 311–312, 548
Sarcoplasmic reticulum membranes, multiple proteins on
  Calcium Release Unit, 331–333
Sarcoplasmic reticulum (SR), 306
  calcium ions
    excitation-contraction coupling, 324
    induced cardiac contraction, 549, 550, 551f
    induced cardiac relaxation, 550–551
    release, 324
    reuptake, 324–325
  phospholamban phosphorylation, 354
  repetitive stimulation, 327
  summation, 326
  tetany, 326
Sarin nerve gas, 484
Satiety center, 839
Satiety signals, 840–843
Saturated fatty acid, 144f
Saturation
  facilitated diffusion, 164–165
  glucose, 720–721
  splay, 721
Scatchard plots, 859, 864–866
Scavenger receptors: immune response, 512
Schwann cells, 265
Scramblase enzymes, 152
Scrapies, 139
Scrotum, 960
Second degree heart blocks, 544
Second pain production by nociceptors, 393
Secondary active amino acid transporters, 725
Secondary active transport
  antiports, 175–176
  sodium–calcium exchangers, 174–175
  symports, 176
Secondary bile acids, 815
Secondary follicle, 951–952
Secondary peristalsis, 780–781
Secondary protein structure, 134–136
Secondary sex characteristics development, 948
Secondary spermatocytes, 960–961
Second-order neurons, 429
Secretin, 794, 808, 813–814
Secretory cells, 108, 221
Secretory materials: cell structure, 108
Sedimentation, 115–116
Sedimentation coefficient, 118
Segmentations: intestinal motility, 800–802
Self-identification, 5–6
Self-replication, 4
Semen, 965
Semenogelin, 965
Semi-lunar valves, 519–520
Seminiferous tubules, 960–961
Sensation disorders, 395–396
Sense organs, 390
Senses, 389, 427
  environmental condition appraisal, 366
  hearing, 440
  smell, 427
  taste, 427
  vision, 456
Sensors: external behavioral responses, 366–367
Sensory afferents: nucleus tractus solitarius, 611–612
Sensory areas: motor activity, 411–412
Sensory cortex: cutaneous sensory systems, 395

Sensory input: autonomic reflexes, 478–479
Sensory neurons: vestibular nuclei, 419
Sensory receptors, 390
Sensory systems: cutaneous sensory systems, 389
Sensory transduction: problem sets, 471
Septicemia, 585
Septum, 518–519
SERCA (smooth endoplasmic reticulum Ca-ATPase), 333
Serine, 143
Serine/threonine receptors, 212
Serotonin, 384, 494
   central nervous system, 384
   perfusion regulation, 595
   peripheral nervous system, 384
Serous glands, 775
Sertoli cells, 961, 963
Sex hormones, 213, 963
Sexual reproduction. *See* Reproductive systems
SGLT. *See* Sodium–glucose linked contransporters (SGLT)
Shear stress
   Poiseuille's law, 22–23
Short bones, 933
Short receptors, 390–391
SIADH. *See* Syndrome of inappropriate antidiuretic hormone secretion (SIADH)
Sickle cell anemia, 501
Sickle cell disease, 506
Sieving coefficients: glomerular filtration, 710
Siggaard–Anderson nomograms, 669
Sight. *See* Vision
Signal function, 111
Single carbon–carbon bonds, 49
Sinoatrial (SA) node
   action potentials, 528–529, 530, 534
   heart beat, 522
Sinus arrhythmia, 544
Sinus bradycardia, 544
Sinus tachycardia, 544
Sinusoidal capillaries. *See* Discontinuous capillaries
Sister chromatids, 956–957
Skeletal growth at growth plates, 877–879
Skeletal muscle, 292
   anatomical arrangement, 292–293
   architecture, 300–302
   breathing mechanics, 625–626
   classification, 292–293
   concentric contractions, 300
   contractile mechanisms, 300, 305
     ATP splitting, 312–313
     cell structure, 305–306
     cross-bridges, 312–313
     cytoskeleton, 314–315
     fiber type dependence, 313–314
     force, 308–309
     force–velocity curves, 313–314
     muscular dystrophy, 316b
     proteins, 314–315
     sliding filament hypotheses, 306–308
     thick filaments, 308–309
     thin filaments, 309–310
     transmembrane proteins, 314–315
   eccentric contractions, 300
   fine structure, 292–293
   force, 292, 293–295, 295–296, 296–297, 308–309
     eccentric contractions, 300
     fatigue, 303
     innervation ratio of motor units, 298
     muscle length, 296–297
     muscle velocity, 298–299
     recruitment, 297
   function, 292
   isometric contractions, 300
   isometric force, 293
   motor neurons, 295–296, 318
   motor units, 293–295
   neural control, 292–293
   power, 299–300
   shortening, 292
   size principles, 295
   velocity, 298–299, 300–302
   voluntary activation, 293
Skeleton. *See also* Bone
   form/support, 933
Skin: vitamin D, 928–929
SL. *See* Sarcolemma (SL)
Sleep apnea, 640, 680b
Sleep disorders, 680b
Sliding filament hypotheses, 306–308
Slit diaphragms, 710
Slit membranes, 710
Slope conductance: resting membrane potential, 258–259
SM proteins, 329–330
Small intestine, 785
   carbohydrate digestion, 825
   lipolytic activity, 828–829
Small nuclear RNAs (snRNA), 126f
Smell. *See* Nose
Smooth endoplasmic reticulum, 108
Smooth muscle, 292, 351
   adrenergic stimulation, 358
   airway resistance, 639
   beta-adrenergic receptors, 358
   cell coupling, 352–353
   chemical signals, 354–356
   contractile filaments, 352
   cross-striations, 351
   cytosolic calcium ions, 354, 358
   electrical activities, 351–352
   esophagus, 780
   force, 351–352
     actin–myosin interaction, 356
     calcium ions, 356–357
     myosin light chain phosphorylation, 356
   guanylate cyclase, 357
   hormones, 353
   intestinal contraction, 799
   intracellular calcium ions, 353–354
   intrinsic activity, 353
   mechanism synopsis, 358–359
   myosin light chain phosphatase, 356
   nerves, 353
   nitric oxide, 357
   perfusion regulation, 591–592
   phasically-based contractile activity, 351
   sphincters, 772–773
   tension, 351
   tonically-based contractile activity, 351
SNAP-25, 329–330
SNARE proteins for SNAP receptor, 329–330
Sneezing, 679
Sodium
   action potentials, 269, 269–270, 532, 533–534
   aldosterone, 749, 912
   ATP production, 239
   filtered load, 720b
   hypervolemia, 749
   integrated response, 749–750
   organic anion cotransporters, 725–727
   renin–angiotensin–aldosterone system, 748–749
Sodium, potassium-ATPase, 172–173
   as electrogenic, 173–174
Sodium chloride. *See also* Salt
   body fluid compartments, 695
   loop of Henle, 732
Sodium–calcium exchangers, 174–175, 533–534
Sodium–glucose linked contransporters (SGLT), 725
Sodium–hydrogen carbonate cotransporters, 753–754
Sodium–hydrogen exchangers, 239
Solitary nuclei, 438
Soluble adenylyl cyclase (sAC), 761
Soluble NSF attachment protein, 329–330
Solute carriers, 179–181
Solute exchange, 578
Solute flux, 37–38
Solute loads, 738
Solute osmotic pressure, 187
Solute pressure-driven flow, 18
Solutions
   dilution calculations, 61b
   making up, 60b
   osmotic pressure, 185, 191–192
Solvents: biological membranes, 142
Soma, 265, 384–385
Somatic cells
   cell diversity, 7
   homeostasis, 7–8
   mitosis, 956
Somatic nervous system, 474f
Somatomedin. *See* Insulin-like growth factor (IGF-I)
Somatosensory cortical neurons: receptive fields, 396–397
Somatosensory information: dorsal column pathway, 394–395
Somatosensory input: cutaneous sensory systems, pain reduction, 396
Somatostatin (SST), 789, 808, 875
   exocrine pancreas, 810
   insulin secretion, 897
   stomach, 791
Somatotopic organization/representation, 394, 411
Sound. *See* Hearing
Sour taste, 432, 435
Space constants: action potential propagation, 283–285
Specific activity, 71
Specificity of facilitated diffusion, 165
Spectrin, 105, 501f
Spectrin-based filament network, 314–315
Speed of light, 467–468
Speed of sound, 451–452
Sperm, 958, 960–961
Spherical lenses, 468–469
Sphincters, 772–773, 802
   micturition, 484
   of Oddi, 773
   pupillary light reflex, 482–483
   relaxation, 803
   upper esophageal, 772–773
Sphingolipids, 146–147
   using sphingosine, 146
Sphingomyelin, 146
Sphingosine, 146
   sphingolipids using, 146
Sphygmomanometers, 572–573
Spinal cord
   muscle control, 410, 672
   nerve function, 373
   respiratory control, 672
   reticulospinal tract, 406, 484
   sympathetic nervous system, 475–477
   ventilation control, 673

Spinal reflexes, 400
   crossed-extensor reflexes, 400–401
   descending tracks, 405–406
      lateral corticospinal tract, 405
      lateral vestibulospinal tract, 405–406
      medial vestibulospinal tract, 406
      medullary reticulospinal tract, 406
      pontine reticulospinal tract, 406
      rubrospinal tract, 405
      tectospinal tract, 406
      ventral corticospinal tract, 405
   gamma motor systems, 403
   integrated response, 407
   intrafusal fiber tension, 403
   inverse myotatic reflexes, 403
   locomotor pattern generators, 404–405
   motor control, 410–411
   muscle spindle, 401–402
   myotatic reflexes, 402–403
   painful stimuli, 400
   reflex definition, 400
   withdrawal reflexes, 400–401
Spirometry
   airway resistance, 635–636
   lung volumes, 633
   maximum voluntary ventilation, 635–636
Splanchnic nerves, 799
Splay, 721
Spliceosomes, 126f
Spongy bone, 933
Spontaneous to nonspontaneous process coupling, 90f
SR. See Sarcoplasmic reticulum (SR)
SST. See Somatostatin (SST)
Standard deviation, sample, 25
Standard error of the mean, 25
Standard Hydrogen Electrode, 229
Standard redox potential, 229
Standard reduction potential, 229
Standard temperature and pressure, dry (STPD), 643t, 645, 652
Stapes, 441–442
Starch, 811, 825–826
Starling forces, 585, 711–713
Starvation and estrogen reducing adult height, 879–880
Static forces of gravity, 419
Static pressure: breathing mechanics, 628
Static response phenomena, 610
Steady-state flows: pressure-driven flow, 17–18, 19–20
Steady-state gas exchange equations, 651–652
Steady-state operating points, 604, 605
Stem cells, 502, 507–508
Stenosis, 563
Stercobilin, 502–503
Stereocilia, 444–445
Stereoisomers, 51
Sternum, 625
Steroid hormones, 125–126, 854, 857, 858, 948
   ACTH, 908–909
   adrenal cortex, 906–907
Steroidogenesis: ovarian, 953
Stethoscopes, 520
STIM1 (stromal interacting molecule), 333
Stimuli awareness, 390
Stimuli perception, 390
Stimulus-secretion coupling, 206
Stock solution dilutions, 61b
Stoichiometry, 48–49
Stokes equation, 119
Stokes–Einstein equation, 81–82
Stomach, 782, 785
   acid secretion, 790

cephalic phases, 791
chemoreceptors, 791–793
colonic motility, 803
contractions, 787–788
emptying, 788, 804
extrinsic innervation, 799
food, 772–773
gastric
   accommodation, 787
   lipase, 790–791
   motility, 785–786, 787
   phases, 791–794, 812–813, 822–824
   slow waves, 785–786
grinding, 787–788
hydrochloric acid secretion stomach, 790
intestinal acid secretion phases, 791–794
intrinsic factor secretion, 790–791
lower esophageal sphincter, 782
migrating motility complex, 789
mucus secretion, 790–791
nerve cell ganglionic plexuses, 782
orad stomach, 787
pacemaker cells, 785–786
pepsinogen secretion, 790–791
postprandial pancreatic enzyme secretion, 812–813
protein digestion, 822–824
regions, 785
small intestine, 785
stretch receptors, 791–793
vomiting, 804
STPD. See Standard temperature and pressure, dry (STPD)
Streamline flow. See Laminar flow
Strength training, 343
Strength–duration relationship, 268
   amount of charge necessary to reach threshold, 274
   threshold depolarization, 273
Stressed volumes, 601–602, 608–609
Stretch receptors
   antidiuretic hormone secretion, 745
   colonic contraction, 803
   stomach, 791–793
Striated muscle, 292, 305, 548
Stroke volume (SV)
   afterload, 556, 558–559
   cardiac output, 556
   contractility, 556
   Frank–Starling law, 558
   preload, 556, 558–559
   pulse pressure, 570–571
Structural isomers, 51
Subcellular fractionation, 115
Subendocardium, 540–541
Subepicardium, 540–541
Sublingual glands, 775
Submandibular glands, 775
Submucosal plexus, 782, 796–798
Substantia nigra, 415
Sulfonylurease, 898
Superior cervical ganglia, 476
Superior mesenteric ganglia, 475, 799
Superior vena cava, 518
Supporting cells: nervous tissue, 375
Suppressor T cells, 510
Supraoptic nuclei, 744
Surface membranes: smooth muscle, 354–356
Surface pressure: biological membranes, 149
Surface recognition: cell structure, 102
Surface tension
   biological membranes, 149
   lung expansion, 627
   problem sets, 156
   pulmonary surfactants, 627–628

Svedberg units, 118
Swallowing, 680, 780
Sweet taste, 432, 436
Swelling, 508–509
Sympathetic nervous system, 473
   adrenal medulla, 916
   baroreflex, 611–612
   calcium ion induced cardiac contraction, 552–553
   gut extrinsic innervation, 799
   insulin secretion, 897–898
   micturition, 484
   origin, 475–477
   pacemaker potential, 529–530
   rostral ventrolateral medulla, 612
   target tissue stimulation effects, 482
   vascular system control, 596
Sympathetic tetralogy, 615
Symports, 176
Synapses, 318, 375
Synaptobrevin, 329–330
Synaptotagmin, 329–330
Syndrome of inappropriate antidiuretic hormone secretion (SIADH), 739b
Synovial fluid, 691
Syntaxin, 329–330
System, understanding, 6
Systemic circulation, 491–493
Systemic perfusion regulation, 592
Systemic pressure, 601
Systemic testosterone effects, 963
Systole contraction, 516
Systolic pressure, 570, 572–573
Systolic volume, 556, 561

T

T wave subepicardium repolarization, 540–541
TAARs. See Trace amine-associated receptors (TAARs)
Talin, 314–315
Target cells, 221
   autonomic nervous system, 482
   cortisol, 910–911
   hormones, 859, 861
   integrated control, 933
Taste, 427
   bitter taste, 436
   hot taste, 436–437
   pain receptors, 436–437
   salty taste, 434–435
   sour taste, 435
   sweet taste, 436
   taste buds, 433–434
   types, 432
   umami taste, 436
Taste receptor cells, 435
TBW. See Total body water (TBW)
TCA. See Tricarboxylic acid (TCA) cycle
Tectorial membrane, 444–445
Tectospinal tract, 406
Teleology, 5
Telophase, 958
Temperature
   cutaneous sensory systems, 395
   oxygen dissociation curves, 660, 660
   speed of sound, 452–453
Temporal lobes
   olfactory output, 429–431
   vision, 465
Temporalis muscle, 773
Temporomandibular joint syndrome, 775b
Tendon muscle force sensors, 403
Tensegrity
   cytoskeletal units, 106

Terminal arterioles, 578
Terminal bronchioles, 624
Terminal cisternae, 306
Terminal ileum, 833
Tertiary follicle, 952
Tertiary protein structure, 136–137
Testicles, 960–961, 961–963
Testosterone, 963
Tetanus, 296
Tetany, 326
Tetralogy of Fallot heart defects, 523
Thalamus
 function, 373
 olfactory output, 431
 taste receptors, 438
Thecal layer, 951–952
Thermal dilution method, 563
Thermocouples, 16
Thermodynamic derivation: van't Hoff's law, 184–185
Thermogenesis, 250b
Thermoreceptors, 393
Thick filaments
 cardiac contractility, 548–549
 skeletal muscle contractility, 305, 308–309
Thick-walled spheres: cardiac muscle contraction, 517–518
Thin filaments
 cardiac contractility, 548–549
 skeletal muscle contractility, 305, 309–310
Thin lens formulae, 467–470
Thin-walled spheres: cardiac muscle contraction, 516–517
Thiokinase, 243
Thiolase, 246
Third degree heart blocks, 544
Thirst, 596–597
Thoracic cage, 625
Thoracic cavity, 516, 625
Thoracic duct, 585, 829–831
Thoracolumbar spinal cord, 475–477
Thorax, 537–539
Threshold depolarization
 strength–duration relationship, 273
Thrombi, 494
Thrombin, 494, 496
Thrombopoietin (TPO), 507
Thromboxane, 494, 593, 595, 596
Thymine, 120
Thyroglobulin, 883, 885
Thyroid gland, 883, 921, 963
Thyroid hormone: metabolic control, 921
Thyroid stimulating hormone (TSH), 885, 963
Thyroxine, 883, 884–885
Tidal volume (TV), 633
 anatomic dead space, 647
 exercise, 635
 pulmonary ventilation, 635
Timbre: hearing, 440–441
Time constant, 40
 capacitor, charging of, 39–40
Tissue factor pathway inhibitor (TFPI), 496
Tissue plasminogen activator (tPA), 495
Tissues
 cardiac output distribution, 568
 inflammation, 511
 integrated control, 933
 macrophages, 508, 510–511
 neurons, 375
 oxygen delivery, 659
 perfusion regulation, 589
 supporting cells, 375
 thyroid hormones, 886–888
Titin filaments, 548
Titratable acid, 754, 755

Titration curve plots, 859, 860, 867–868
T-lymphocytes, 510
Tone: hearing, 440–441
Tonically-based contractile activity, 351
Tonicity, 191–192
Tonotopic mapping of sound frequency, 445–446
Total body water (TBW), 689, 691
Total lung capacity (TLC), 634
Total peripheral resistance (TPR), 602–603, 604, 610, 614
Total peripheral resistance–cardiac output relationships, 614
Toxins: action potentials, 270
tPA. See Tissue plasminogen activator (tPA)
TPO. See Thrombopoietin (TPO)
TPR. See Total peripheral resistance (TPR)
Trabeculae, 933
Trace amine-associated receptors (TAARs), 428
Trachea: breathing mechanics, 623–624
Tracheae: breathing mechanics, 644
Tracheobronchial tree, 624
Training of muscle, 334
Transamination reactions, 247
Transcellular fluid compartments, 691
Transcription, 123–125
 cell signaling, 213
 cortisol, 910–911
 histone code, 126–127
Transcription factors, 125, 214–215
Transcutaneous electrical neural stimulation, 397
Transcytosis, 578–579, 584
Transduction, 205
Transfer ribonucleic acid (tRNA), 123
Transferrin, 499
Transient diabetes insipidus, 739
Transition state theory, 67–70
Translation, 122, 126f
Translocon proteins, 108
Transmembrane proteins, 151–152, 314–315
Transmembrane resistance: action potential propagation, 282
Transmissible spongiform encephalopathies (TSE), 139
Transport mechanisms. See also Diffusion
 active transport
  antiports, 175–176
  cell structure, 102
  electrochemical potentials, 170–172
  energetics, 170–172
  exchangers, 170
  ion permeation, 170–172
  material transport, 172
  metabolic energy, 172
  phosphorylated intermediates, 173
  pumps, 170
  sodium, potassium-ATPase, 172–173
  sodium, potassium-ATPase as electrogenic, 173–174
  sodium–calcium exchangers, 174–175
  symports, 176
 cell structure, 101, 102
 cytosol, 103
 passive mechanisms, 161
  cell structure, 102
  reabsorbtion of urea, 728
  resting expiration, 626
  solute exchange, 579
  water, 168
 tubular, 731–736
 urea, 733
Transport proteins, nomenclature of, 179
 ABC transporters, 181
 aquaporins, 181

ATP-driven ion pumps, 181
 carrier classifications, 179
 HUGO nomenclature, 179
 solute carriers, 179–181
Transverse (T) tubules, 306, 549, 550
Treppe phenomena, 552
TRIC-A (trimeric intracellular cationselective channel), 332–333
Tricarboxylic acid (TCA) cycle, 227
Tricuspid valve, 518
Trigeminal nerve, 431–432, 773
Triglycerides, 241–242
Triiodothyronine, 883
Tripalmitin, 246b
Tropomyosin, 310, 548–549
Troponin, 310
TSE. See Transmissible spongiform encephalopathies (TSE)
TSH. See Thyroid stimulating hormone (TSH)
T-tubules, multiple proteins on Calcium Release Unit, 331–333
T-type calcium ion channels, 528–529
Tubules. See also Nephrons
 acid excretion, 753
 acidosis, 763b
 pH, 757–759
 reabsorption/secretion, 705, 719
 transport mechanisms, 736
Tubulin, 103f
Tubuloglomerular feedback, 742
Tunic intima, 590
Tunica adventitia, 590
Tunica media, 590
Tuning: pitch perception, 447–448
Turbinates, 623
Turbulent airflow, 636–637, 638
Turnover
 compartmental analysis, 72–74
TV. See Tidal volume (TV)
Twitch waveforms, 326–327
TXA$_2$. See Thromboxane
Tympanostomy, 444
Type I error, 27, 27
Type II error, 27, 27
Type IV collagen, 710
Tyrosine, 212, 382, 916
Tyrosine kinases, 212, 899

U

Ubiquitin, 110
Ubiquitinylation reaction, 110f
Ulcers, 794b
Ultrafiltrate/ultrafiltration
 glomerular filtration, 705, 711–713
 barriers, 713
 pressure-driven flow, 20b
 proximal tubule, 725
Umami taste, 432, 436
Uncompensated metabolic acidosis, 670
Uncompensated metabolic alkylosis, 670
Understanding a system, 6
Unidirectional fluxes, 70–72
 and net fluxes, 70–72
Unipolar leads: electrocardiograms, 542–543
Unitary conductances: action potentials, 271
Universal donors/recipients, 504
Unmyelinated axons, 265
Unmyelinated baroreceptor C fibers, 610
Unstressed volumes, 601–602, 608–609
Upper esophageal sphincter, 772–773
Upregulated dose–response curves: hormones, 860–861
Uracil, 122

INDEX

Urea
  elimination, 247–251
  inner medulla
    collecting duct, 733
    osmotic gradients, 736
  loop of Henle, 732
  production, 247–251
  proximal convoluted tubule, 727
  reabsorption mechanisms, 727
Urease, 794
Ureters, 698, 698–699
Urethra, 698
Urinary system, 698. *See also* Kidneys
  concentration mechanisms, 730
  dilution mechanisms, 730
  excretion
    calcium ions, 940–941
    PI, 940–941
  glomerular filtration, 740–741
  limiting pH, 757–758
  metabolic acidosis, 760
  metabolic alkalosis, 760–761
  nitrogen, 837
Urobilin, 502–503
Ursodeoxycholic acid, 815
Ussing flux ratio equation, derivation of, 178–179
Uterus, 871, 946, 947–948
Utricles, 419

V

Vagal efferents, 612
Vagina, 946
Vagovagal reflexes, 780–781, 787, 813
Vagus, 478
  afferent fibers, 799
  gut parasympathetic innervation, 799
  pacemaker potential, 530
  stomach, 791
  ventilation control, 673, 675–676, 676f
Van der Waals force, 55
  repulsive force
    Lennard–Jones potential, 55
Van't Hoff equations, 182–184, 187
Van't Hoff's law, 182, 184–185
Vapor pressure, 643, 644
Variance, population, 25
Variance, sample, 25
Vasa recta, 701, 736
Vascular system
  blood flow, 493
  circulatory system overview, 489
  function
    arterial compliances, 602–603
    arteriolar resistance, 604
    blood volume, 605
    cardiac function matching, 599–600
    hemodynamics, 568
    positive right atria pressures, 603–604
    steady-state operating points, 604
    venous compliances, 602–603
  hemostasis, 494–495
  paracrine secretions, 595–596
  perfusion regulation, 590, 591–592
  platelet plugs, 494
  resistance, 576
  sympathetic nervous system, 596
Vasoactive intestinal peptide (VIP), 778–779
Vasoconstriction, 650
  arteriolar resistance, 604
  bleeding restriction, 494
  kidneys, 704
  perfusion regulation, 589–590

Vasodilation
  arteriolar resistance, 604
  inflammation, 511
  perfusion regulation, 591–592, 593, 594–595
Vasomotion phenomena, 583–584
Vasopressin. *See* Antidiuretic hormone (ADH)
VC. *See* Vital capacity (VC)
Veins
  arcuate veins, 700f, 701
  circulatory system overview, 489
  interlobar veins, 701
Velocity centrifugation, 116
Venoconstriction, 602
Venous compliances, 602–603
Venous return, 599
Ventilation control, 623, 672
Ventilation/perfusion ratios, 649–650
Ventilatory acclimatization, 679
Ventilatory drive
  carbon dioxide pressure, 678
  hypercapnia, 678, 678
  hypoxia, 679
  metabolic acidosis, 678–679
Ventilatory response: chemoreceptor firing rates, 677
Ventral corticospinal tract, 405
Ventral respiratory group (VRG), 672, 673–674
Ventricles, 518, 519
  action potentials, 528, 530–534
  cerebral spinal fluid, 369–370
  electrocardiograms, 540
  filling, 520–522
  function curves, 558
Ventromedial hypothalamus (VMH), 839
Venules, 489, 578
Very-low density lipoproteins (VLDL), 499
Vesicles
  cytoskeleton, 103
  fusion, 379–380
  pre-synaptic terminals, 381
  transcytosis, 584
Vessels. *See* Blood—vessels
Vestibular apparatuses
  afferent sensory neurons, 419
  balance, 415–419
  head rotation, 419
Vestibular nuclei, 419
Vestibulospinal tract, 405–406
Video esophagraphy, 783
Villi, 105, 797
Villin, 105
Vinculin, 314–315
VIP. *See* Vasoactive intestinal peptide (VIP)
Visceral pleura, 625f
Viscosity, 19–20
Vision, 456
  bipolar cells, 462–463
  central pathway crossover, 463–464
  eye shape, 458
  eye structure, 456–458
  far-sightedness, 459–460
  ganglion cells, 462–463, 464
  near-sightedness, 459–460
  system overview, 456
  visual cortex, 464–465
  vitreous body, 458
Vital capacity (VC), 633–634
Vitalism, 5–6
Vitamin B_{12}, 506
Vitamin D
  activation, 929
  bone mineralization, 931, 941–942
  definition, 929
  forms, 930
  goals, 941–942

  inactivation, 929
  kidneys, 703
  metabolism, 929–930
  25(OH)2-D, 930–931
  synthesis, 928–929
Vitreous body: eye shape, 458
VLDL. *See* Very-low density lipoproteins (VLDL)
VMH. *See* Ventromedial hypothalamus (VMH)
Volatile acids, 752
Voltage-dependent changes: action potentials, 268–269
Voltage-gated ion channels, 167, 206
Volume
  blood
    antidiuretic hormone, 873
    integrated response, 749
    long-term blood pressure regulation, 613–615
    vascular function curves, 605
  body fluid compartments, 689, 690
  concentration, 60
  diastolic volume, 556–557
  end-diastolic volume, 520–522, 556–557
  end-systolic volume, 556, 561
  expiratory reserve volume, 633
  Fick dilution principle, 62
  fluid problem sets, 715
  integrated response, 749–750
  lungs, 633
    air movement, 625
    airway resistance, 638–639
    anatomic dead space, 647
    body size, 634
    capacity, 633–634
    spirometry, 633
  partial volumes, 652
  plasma volume estimations, 62b
  pressure-driven flow, 18
  pressure–volume loops, 556–557
  regulatory volumes, 193
  residual volume, 633, 634t
  spirometry, 633
  stressed volumes, 601–602, 608–609
  stroke
    afterload, 556, 558–559
    cardiac output, 556
    contractility, 556
    Frank–Starling law, 558
    preload, 556, 558–559
    pulse pressure, 570–571
  systolic volume, 556, 561
  tidal volume, 633, 635, 647
  unstressed volumes, 601–602, 608–609
  vascular function, 569–570
Volume contraction, 762
Voluntary ventilation control, 672
Vomiting, 668, 804
Von Willebrand factor (vWF), 496
VRG. *See* Ventral respiratory group (VRG)
V-type ATPases, 174
vWF. *See* von Willebrand factor (vWF)
VX nerve gas, 484

W

Warm receptors, 393
Warming of air: breathing, 623–624
Water
  antidiuretic hormone, 744
  aquaporins, 168, 727, 731, 751
  biological membranes
    amphiphatic molecules, 149
    phospholipids, 146–147
    surface tension, 147
  body fluid compartments, 692, 694–695

bonding of atoms, 48, 54
collecting duct, 733
degrees of freedom, 56f
distal nephron permeability, 745
filtered load, 719
free diffusion coefficients, 79–80
gas dissolution, 644–645
hydrogen bonds, 54
inulin, 725
late distal tubule, 733
late proximal tubule, 728
loop of Henle, 730, 732
nephron adjustment, 743–744
osmotic pressure, 182, 584, 584, 727–728
passive transport, 168
plasma blood cells, 498–500
polar bonds, 53–54
protein structure, 134
reabsorption mechanisms, 725, 727–728, 743–744

secondary protein structure, 134–136
structure, 54f
urine concentration, 730–731
vapor pressure, 643, 644
Wave equations: propagation of sound, 454–455
Wave summation: repetitive stimulation, 327
Weber–Fechner law of psychology, 391–392
Weight regulation, 838
White blood cells, 494, 507
White ramus communicans, 475
Whole-cell current–voltage relationship, 275f
Why? questions, 5
Withdrawal reflexes, 400–401
Womb. *See* Uterus
Work
  conservative forces, 33–34

  electrical potential, 31–32
  fatigue, 341–343
Wound closure, 495
Woven bone, 934

X
Xenobiotics, 816
Xenochemicals, 814

Y
Y^+L system, 825

Z
Zero electrodes, 542–543
Z-lines/disks: muscle, 306, 310
Zona glomerulosa, 911
Zonula adherens, 111
Zonula occludens, 110–111
Zwitterions, 130

Chemical Foundations of Physiology II: Concentration and Kinetics 1.5

Learning Objectives

- Be able to calculate the molar concentration given the mass of solute, formula weight, and volume of solution
- Be able to calculate the number of moles of solute in an aliquot of solution given the volume and concentration
- Be able to determine the gram molecular weight of simple compounds
- Know the prefix notations for scientific notation
- Be able to calculate dilutions of stock solutions to form working solutions
- Be able to calculate the volume of distribution using Fick's dilution principle
- Define rate constant
- Be able to derive a first-order rate equation and calculate its half-life
- Define equilibrium constant
- Describe how an enzyme can change the rate of a reaction without changing its equilibrium
- Write the Michaelis—Menten equation and draw a saturation plot. Identify K_m and V_{max}
- Draw a double-reciprocal plot and identify K_m and V_{max} on this plot

AVOGADRO'S NUMBER COUNTS THE PARTICLES IN A MOLE

As described in Chapter 1.4, each chemical species is composed of a definite number of atoms of each element. This is Dalton's law of fixed proportions, which states that a molecule contains integral numbers of each kind of atom. Under ordinary chemical reactions, these atoms cannot be converted into each other. Each of these atoms contributes a tiny but definite mass to the molecule. The **atomic mass unit is defined as 1/12 of the mass of a carbon-12 atom**, the one containing 6 protons and 6 neutrons in its nucleus. This unit of mass is also called the **dalton, Da**. The atomic weight of a carbon-12 atom is *defined* as being 12 daltons. The atomic weight of most elements, including carbon, is not exactly integral because most elements consist of mixtures of **isotopes** that differ slightly in their atomic mass. These are atoms that possess the characteristic number of protons (the Z number) but differ in their numbers of neutrons.

Each molecule has a definite mass that depends on the atoms that make it up. If we add up the atomic masses of the constituent atoms, we get the **molecular weight** in daltons. If we express this molecular weight in grams, we will have the **gram molecular weight** (just scratch out "daltons" and replace it with "grams"). A pile of molecules whose mass in grams is the gram molecular weight is called a **mole**. Such a pile will consist of not one molecule but very many of them, and the number will be the number of daltons in a gram.

To see this, suppose we take a substance with a molecular weight of Z daltons. How many of these molecules will we have to pile up in order to make up a pile with a mass of Z grams? Let N be the number of molecules in the pile of Z grams. The mass of one molecule is Z daltons. The mass of N of them is just $N \times Z$ daltons. This mass will be the gram molecular weight. Thus, we have

$$Z \text{ daltons} \times N = Z \text{ grams}$$

$$N = \frac{Z \text{ grams}}{Z \text{ daltons}} = \frac{1 \text{ g}}{1 \text{ Da}}$$

The actual mass of 1 Da is 1.66×10^{-24} g. Thus, the number of molecules in a gram molecular weight is

$$N = \frac{1 \text{ g}}{1.66 \times 10^{-24} \text{ g}} = 6.02 \times 10^{23}$$

This is, of course, Avogadro's number. It is the number of molecules per mole. A mole is a pile of molecules of a single substance whose mass in grams is equal to its gram molecular weight and which contains Avogadro's number of molecules. We can turn this definition around and *define* the mole as Avogadro's number of particles. This definition is completely general. We did not specify how big the molecule is in the above calculation. No matter the size of the compound, one mole contains Avogadro's number of molecules.

CONCENTRATION IS THE AMOUNT PER UNIT VOLUME

In Chapter 1.2, we defined extensive and intensive variables as those that depend on the extent of the system and those whose value is independent of the extent of the system, respectively. The amount of a solute, and the volume in which it is distributed, are both extensive variables. If you have twice the volume of solution, you have twice the amount of solute dissolved in it. The concentration, on the other hand, is an intensive variable which is given as the ratio of two extensive variables:

[1.5.1]
$$\text{Concentration} = \frac{\text{amount}}{\text{volume}}$$
$$C = \frac{m}{V}$$

where C is the concentration, m is the amount of solute, and V is the volume in which that amount of solute is dissolved.

SCIENTIFIC PREFIXES INDICATE ORDER OF MAGNITUDE

Often concentrations of physiologically relevant materials in physiological fluids are quite small and must be expressed in scientific notation, as in the example of hemoglobin in Example 1.5.3. It is useful to know the established prefixes for units of volume, mass, or length. The standard units are the liter, L, the gram, g, and the meter, m, respectively. The standard prefixes are shown in Table 1.5.1.

EXAMPLE 1.5.1 Calculating Molar Concentration

Suppose we dissolve 18 g of NaCl in water in a 2-L volumetric flask and then add water to the mark so that the solution volume is 2 L. What is its concentration?

We can report this concentration in mass per liter as 9 g L^{-1}. For many solutes, this is an acceptable way of expressing the concentration. For example, the concentration of hemoglobin in whole blood is 15 g dL^{-1}. Here dL means "deciliter" and is one-tenth of a liter or 0.1 L. Since 1 L = 1000 mL, 1 dL = 1000 mL/10 = 100 mL. Thus, a hemoglobin concentration of 15 g dL^{-1} means that there is 15 g of hemoglobin in every 100 mL of whole blood.

The concentration of many solutes is not reported in these units, however, but in units of moles per liter, or **molar**. We can convert from mass to moles by dividing by the gram molecular weight. In the case of NaCl given above, the atomic weight of Na is 22.99 daltons and the atomic weight of chlorine is 34.45 daltons, giving a molecular weight for NaCl of 22.99 + 35.45 = 58.44 daltons; its gram molecular weight is 58.44 g. Thus 58.44 g of NaCl constitutes 1 mole of NaCl. The number of moles of NaCl in 18 g is calculated as

$$X \text{ moles} = \frac{18 \text{ g NaCl}}{58.44 \text{ g NaCl mol}^{-1}} = 0.308 \text{ mol}$$

Since this amount was dissolved in 2 L of solution, its concentration = amount/volume.

$$\text{Concentration} = \frac{0.308 \text{ mol}}{2 \text{ L}} = 0.154 \text{ M}$$

where M designates **molarity** or moles per liter.

EXAMPLE 1.5.2 How to Make Up a Solution

You need to make up 500 mL of a solution containing 0.3 M urea. How much urea do you need?

Here we rearrange the equation $C = m/V$ to read $m = C \times V$.

In this case, $C = 0.3$ M and $V = 0.5$ L. The number of moles of urea is thus

$$m = 0.3 \text{ M} \times 0.5 \text{ L} = \mathbf{0.15 \text{ mol}}$$

We need to convert this into grams so that we can accurately weigh out the required amount of urea on a good balance. To do this, we need the gram molecular weight of urea. We can find this out several ways. We can look it up in a CRC handbook or similar source. Another way is to write out the formula and add up all of the atomic weights of all the atoms in the molecule times their compositional stoichiometry. This could be tedious for a large molecule. We could look on the bottle, because most chemical companies publish the **formula weight** on the bottle. This formula weight may be different from the gram molecular weight because the chemical might have waters of hydration with it or ions to counterbalance charges on the chemical. ATP, for example, is usually sold as $Na_2ATP \cdot 3H_2O$ and its formula weight is not the gram molecular weight of ATP alone. Looking on the bottle seems like a winner, as we must go find the bottle to weigh out the urea. We find that the formula weight for urea is 60.08 g mol^{-1}. The mass of urea is found by multiplying the moles by the gram molecular weight:

$$m = 0.15 \text{ mol} \times 60.08 \text{ g mol}^{-1} = \mathbf{9.012 \text{ g}}$$

This is the amount you need to weigh on an accurate balance.

EXAMPLE 1.5.3 Blood Concentration of Hemoglobin

The blood content of hemoglobin is 15 g dL^{-1}. The molecular weight of hemoglobin is 66,500 daltons. What is its average molar concentration in the blood?

If the molecular weight is 66,500 daltons, then 1 mole has a mass of 66,500 g. The question asks for the average concentration because hemoglobin is not uniformly distributed within the blood, but is contained within red blood cells only. Therefore, its concentration in the red blood cells exceeds its average concentration in the blood. The concentration in blood is given by

$$[Hb] = \frac{15 \text{ g dL}^{-1} \times 1 \text{ dL}/0.1 \text{ L}}{66,500 \text{ g mol}^{-1}} = 0.00266 \text{ M} = \mathbf{2.66 \times 10^{-3} \text{ M}}$$

These examples should enable you to calculate:

- the number of moles in a given volume of solution of given concentration;
- the mass of material in a given volume of solution of given concentration;
- the concentration of solution containing a known mass of material in a given volume.

EXAMPLE 1.5.4 Make a Dilution of a Stock Solution

Suppose we have a stock solution of 0.1 M MgCl$_2$ and we want to make up 50 mL of solution with a final concentration of 5 mM MgCl$_2$. How much of the stock solution should we add to a 50-mL volumetric flask?

We use Eqn [1.5.3] directly here:

$$0.1 \text{ M} \times X \text{ mL} = 0.005 \text{ M} \times 50 \text{ mL}$$

Solving for X, we find $X = \mathbf{2.5 \text{ mL}}$

TABLE 1.5.1 Prefixes Used with Powers of Ten, in 10³ Ratios

Atto	10^{-18}	Femto	10^{-15}	Pico	10^{-12}	Nano	10^{-9}
Micro	10^{-6}	Milli	10^{-3}	—	10^{0}	Kilo	10^{3}
Mega	10^{6}	Giga	10^{9}	Tera	10^{12}		

Thus we can write:

1 picoliter (1 pL) = 10^{-12} L
1 nanoliter (1 nL) = 10^{-9} L
1 microliter (1 μL) = 10^{-6} L
1 milliliter (1 mL) = 10^{-3} L
1 liter (1 L) = 10^{0} L

1 picogram (1 pg) = 10^{-12} g
1 nanogram (1 ng) = 10^{-9} g
1 microgram (1 μg) = 10^{-6} g
1 milligram (1 mg) = 10^{-3} g
1 gram (1 g) = 10^{0} g
1 meter (1 m) = 10^{0} m

1 picometer (1 pm) = 10^{-12} m
1 nanometer (1 nm) = 10^{-9} m
1 micrometer (1 μm) = 10^{-6} m
1 millimeter (1 mm) = 10^{-3} m
1 kilogram (1 kg) = 10^{3} g
1 kilometer (1 km) = 10^{3} m

The prefixes mega-, giga-, and tera- are not typically used for units of volume, mass, or distance, but often find use with other units such as hertz or watts. There are additional prefixes in the SI. These include:

centi 10^{-2} deci 10^{-1} deca 10^{1} hecto 10^{2}

DILUTION OF SOLUTIONS IS CALCULATED USING CONSERVATION OF SOLUTE

Often it is necessary or easier to prepare solutions from more concentrated stock solutions by removing an **aliquot** (a fraction of the solution) of the more concentrated solution and placing it in a volumetric flask. We then add solvent to bring the volume up to the final solution volume, as shown in Figure 1.5.1.

Let V_1 be the volume of the aliquot of the more highly concentrated solution, with concentration C_1. The amount of solute in this aliquot is given by Eqn [1.5.2]:

[1.5.2] $$m = C_1 \times V_1$$

This amount is also the amount in the final solution with volume V_2 and final concentration C_2. Since the amount in the aliquot is still in the final solution, we can write

[1.5.3] $$C_1 V_1 = C_2 V_2$$

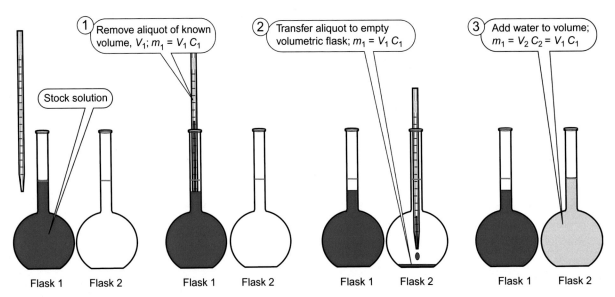

FIGURE 1.5.1 Dilution of a stock solution. Flask 1 contains a stock solution of concentration C_1. We remove a known volume, V_1, with a calibrated pipette and place it in an empty volumetric flask. We then add water, which adds solvent but no solute, to bring the volume up to the final volume. The amount of solute in the second flask is the same amount that we added in the aliquot, V_1, but now it is distributed in the volume V_2.

CALCULATION OF FLUID VOLUMES BY THE FICK DILUTION PRINCIPLE

The equation describing dilution of stock solutions can be used to determine the volume of distribution of materials in the body. Suppose we inject a known amount of a substance, Y, into a person and this substance Y remains in the plasma because it cannot get out of the vascular system and it cannot get into the cells suspended in the blood. We wait a few minutes for Y to become evenly distributed, and then take a sample of plasma and measure Y's concentration. Then we can calculate the volume of the plasma as

[1.5.4]
$$\text{Volume} = \frac{\text{amount}}{\text{concentration}}$$
$$V = \frac{m}{C}$$

This is just a variant of Eqn [1.5.2] in which we solve for V instead of C or m. In our particular case in this example, we write

$$V = m_Y/[Y]$$

where m_Y is the amount of substance Y and $[Y]$ is its concentration.

CHEMICAL REACTIONS HAVE FORWARD AND REVERSE *RATE CONSTANTS*

Suppose that we observe a simple chemical reaction that can be described as

[1.5.5]
$$A + B \rightleftharpoons C$$

where A, B, and C denote different chemicals and the arrows indicate that the reaction proceeds in the direction of the arrow. The reaction actually consists of two separate reactions:

[1.5.6]
$$A + B \rightarrow C$$
$$A + B \leftarrow C$$

The top reaction is the **forward** reaction and the bottom reaction is the **reverse** reaction. Both are characterized by a **rate constant**. The rate constant gives the rate of the reaction when the rate constant is multiplied by the concentration of reactants:

[1.5.7]
$$J_f = k_f[A][B]$$
$$J_r = k_r[C]$$

EXAMPLE 1.5.5 Estimation of Plasma Volume

Suppose that we inject a 50-kg person with 2 mL of a solution of Evans' Blue Dye, at 5 mg mL^{-1}. Evans' Blue Dye is restricted to the plasma because it tightly binds to a plasma protein, albumin, that ordinarily does not escape from the circulation and which only slowly is removed or added to the plasma. We wait 10 min and then obtain a sample of blood and measure the concentration of Evans' Blue Dye in the plasma and find that it is 0.4 mg dL^{-1}. What is the person's plasma volume?

The amount of Evans' Blue Dye that was injected is calculated as

$$m_{inj} = C_{inj}V_{inj} = 5 \text{ mg mL}^{-1} \times 2 \text{ mL} = \mathbf{10 \text{ mg}}$$

The concentration of Evans' Blue Dye after mixing was 0.4 mg dL^{-1}. We calculate its volume of distribution as

$$V = \frac{10 \text{ mg}}{0.4 \text{ mg dL}^{-1}} = 25 \text{ dL} = \mathbf{2.5 \text{ L}}$$

where J_f is the forward reaction rate, J_r is the reverse reaction rate, k_f is the forward reaction rate constant, and k_r is the reverse rate constant, and $[A]$, $[B]$, and $[C]$ are the concentrations of the indicated reactants.

Although this form of the reaction rate is true for elementary reactions, more complicated reactions could have a form that appears to have little to do with the overall reaction as written—its stoichiometric relation or accounting of the number of each kind of molecule that participates in the reaction. This is because complicated reactions occur with intermediary steps that may involve chemicals that do not appear in the overall balanced reaction. Nearly all enzymatic reactions, for example, do not obey Eqn [1.5.7] because the reaction rate is determined largely by the concentration of enzyme. For the moment, we will consider only elementary reactions that obey Eqn [1.5.7].

The rate of a reaction is the number of completed reactions that take place per unit time. You should recognize this as an extensive property. If we had twice the volume of a solution, with the same concentrations of reactants and products, we would have twice the number of reactions taking place per unit time. So we often convert reaction rates into intensive variables by dividing by the volume. The forward rate is thus the number of completed reactions per unit time per unit volume. We express these numbers of reactions in terms of moles, which are related by Avogadro's number to a real number of completed reactions. Thus the forward rate constant has units of $M^{-1} s^{-1}$. By similar reasoning, the reverse rate constant in Eqn [1.5.7] has the units of s^{-1}. The rate of change of reactants A and C can be given as

[1.5.8]
$$\frac{d[A]}{dt} = -k_f[A][B] + k_r[C]$$
$$\frac{d[C]}{dt} = -k_f[A][B] - k_r[C]$$

The negative sign before $J_f(=k_f[A][B])$ in the above equation indicates that this reaction reduces the concentration of reactant A; the positive sign of J_r indicates that this reaction flux adds to the concentration of reactant A. Similar reasoning gives us the rate of change of $[C]$. At equilibrium, the concentrations of A, B, and C are no longer changing. The values of $[A]$, $[B]$, and $[C]$ are altered from their original concentrations so that the forward and reverse rates are equal:

[1.5.9]
$$\frac{d[A]}{dt} = 0 = -k_f[A][B] + k_r[C]$$
$$k_f[A][B] = k_r[C]$$
$$J_f = J_r$$

The rate constants, k_f and k_r, are characteristic of the reaction path involved and the experimental conditions such as ionic strength and temperature of the reactants. They typically do not vary with the concentrations of A, B, and C. Thus, at equilibrium, the middle expression in Eqn [1.5.9] is true for any set of $[A]$, $[B]$, and $[C]$. We rearrange this to get:

[1.5.10]
$$\frac{k_f}{k_r} = \frac{[C]}{[A][B]} = K_{eq}$$

where K_{eq} is the **equilibrium constant**. From the units of k_f as $M^{-1} s^{-1}$ and k_r as s^{-1}, we see that K_{eq} has the units of M^{-1}.

FIRST-ORDER RATE EQUATIONS SHOW EXPONENTIAL DECAY

Some kinds of chemical reactions, such as decomposition reactions, obey the equation

[1.5.11]
$$A \rightarrow B$$

Often this reaction is strongly directed to the right, meaning that $k_f \gg k_r$. The reaction rate is approximated as

[1.5.12]
$$J = -k_f[A]$$

Here we imagine that initially $[B] = 0$ and there is no reverse reaction. Some reactions occur with such completeness that the reverse reaction is negligible. We can rewrite Eqn [1.5.12] as

[1.5.13]
$$\frac{d[A]}{dt} = -k_f[A]$$

This is called a first-order rate equation because the rate of reaction is proportional to the first order or the concentration of reactant—it is proportional to the first power of its concentration. We can separate variables and integrate this equation, as follows:

[1.5.14]
$$\int_0^t \frac{d[A]}{[A]} = \int_0^t -k\,dt$$
$$\ln[A] - \ln[A]_0 = -kt$$
$$[A] = [A]_0 e^{-kt}$$

The last equation describes the concentration of A with time—it decays exponentially. This equation is described as a first-order exponential decay. Many reactions and processes are described by this type of analysis, such as radioactive decay and the disappearance of many different hormones from the circulation. This type of reaction is characterized by its half-life, the time required for $[A]$ to fall from its initial value, $[A]_0$, to one-half of its initial value, $[A]_0/2$. In this case, Eqn [1.5.13] gives

[1.5.15]
$$\frac{[A]_0}{2} = [A]_0 e^{-kt_{1/2}}$$
$$\ln\frac{1}{2} = \ln e^{-kt_{1/2}}$$
$$-\ln 2 = -kt_{1/2}$$
$$t_{1/2} = \frac{\ln 2}{k}$$

Thus the half-life of a first-order reaction is inversely proportional to the first-order rate constant.

RATES OF CHEMICAL REACTIONS DEPEND ON THE *ACTIVATION ENERGY*

The values of k_f and k_r depend on the reaction path taken for the reaction, but K_{eq} is not affected by the reaction path. We shall not prove this result here but will give some rationale for it. A more complete description is given in Appendix 1.5.A1.

The reactants (A and B) and the product, C, can be viewed as possessing some degree of energy. This energy is a potential energy that consists of the potential energy of all of the interactions of their orbital electrons with the positive nuclei. The set of nuclei has some spatial arrangement which changes during the course of the reaction. This is the essence of a chemical reaction, in which the relative positions of the nuclei are altered. The orbital electrons, of course, follow the nuclei so that the energy of the ensemble changes during the course of the reaction. We call this energy the potential energy, and it includes the potential energy of the electrons and the nuclei and the kinetic energy of the electrons, but does not include the kinetic energy of the nuclei. We can plot this potential energy against the "reaction coordinate," which is the actual distance along the minimum energy path from reactants to products. An example of such a graph is shown in Figure 1.5.2.

The rate constants that govern the rates of reaction depend on the energy required to reach the activated complex intermediate between reactants and products. Large activation constants are associated with small rate constants. The relationship is expressed by the Arrhenius equation:

$$[1.5.16] \qquad \ln k = \ln A - \frac{E_a}{RT}$$

where k is the rate constant for the reaction, A is a *preexponential factor* that has to do with the orientation of the reaction and not its temperature dependence, E_a is the activation energy in J mol^{-1}, R is the gas constant (= 8.314 J mol^{-1} K^{-1}), and T is the absolute temperature (K). The equilibrium constant, however, depends only on the initial and final energy levels. Note that the Arrhenius equation means that a larger E_a would cause a lower rate constant, and a smaller E_a would increase the rate constant.

ENZYMES SPEED UP REACTIONS BY LOWERING E_A

It is possible for the reaction to take a different path. For example, A and B could be absorbed onto the surface of an enzyme. The forces that aid in this binding have been discussed in Chapter 1.4: electrostatic interaction between ions in the substrate and enzyme; hydrogen bonding; dipole–dipole interactions; and London dispersion forces. This binding to the enzyme changes the configuration of nuclei and alters the energy of the activated complex. In this way, the enzyme offers an alternative pathway for the reaction that involves far less activation energy. This increases the reaction rates without altering the final energetics of reactants and products. Thus the rates increase without changing the equilibrium constant. Figure 1.5.3 illustrates this idea.

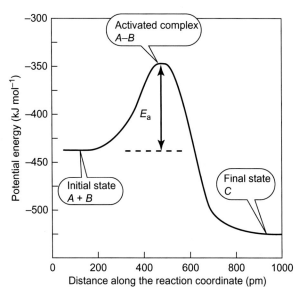

FIGURE 1.5.2 Potential energy along the reaction coordinate for the reaction $A + B \rightarrow C$. Reactants A and B are at low potential energy. The activated complex is a form intermediate between reactants and products, and can be attained by converting kinetic energy into potential energy by a collision between A and B. The difference between the energy of the activated complex and the reactants is the activation energy that must be supplied for the reaction to proceed. The "reaction coordinate" is the distance along the minimum free-energy path from reactants to products.

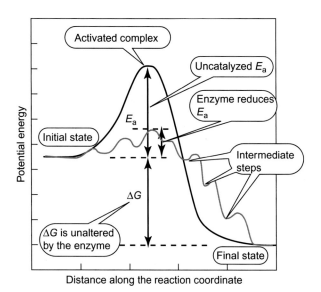

FIGURE 1.5.3 The effect of an enzyme on the overall activation energy for a reaction. The uncatalyzed reaction requires a large activation energy, E_a, and so the reaction occurs slowly. The enzyme offers an alternative path, often by breaking the reaction into a number of small steps, so that the reaction occurs more quickly. Enzymes do not change the overall energetics of the products with respect to the reactants (ΔG in the figure is the change in free energy in the transition between reactants and products) and so the equilibrium constant is unaffected.

THE MICHAELIS–MENTEN FORMULATION OF ENZYME KINETICS

Catalysis is the speeding up of a chemical reaction by a chemical species that does not enter into the stoichiometry of the reaction. Enzymes are catalysts because they speed up reactions without themselves being altered. Enzymes typically bind the reactants and alter their shape (the three-dimensional arrangement of all of the atomic nuclei in the reactants) by virtue of their being attracted to or bound to the surface of the enzyme. The enzyme changes the reaction from a homogeneous reaction in the solution to a heterogeneous reaction occurring on the surface of the enzyme. In this process, the enzyme itself is unchanged. It participates in the reaction but does not show up in an accounting of the reactants and products because it is unchanged.

Figure 1.5.3 shows that enzymes speed up reactions by providing an alternate path for the reaction to take, and that this path requires less activation energy. Michaelis and Menten provided a simple analysis of this reaction path by imagining it to take place in two steps:

$$[1.5.17] \quad E + S \underset{k_2}{\overset{k_1}{\rightleftharpoons}} E-S \overset{k_3}{\rightarrow} E + P$$

where E is the enzyme, S is the substrate, $E-S$ is the substrate–enzyme complex, and P is the product. This is a very simple reaction mechanism. The rate of the enzyme-catalyzed reaction is defined by the rate of product release:

$$[1.5.18] \quad J = k_3[E-S]$$

If we know $[E-S]$ and k_3, we can calculate the reaction rate. Here k_3 is in units of s^{-1}. If we keep $[S]$ and $[P]$ constant, we can define a steady-state rate in which S is converted to P at a constant rate. Under these conditions, we can solve for $[E-S]$. The rate of change of $[E-S]$ is given by

$$[1.5.19] \quad \frac{d[E-S]}{dt} = k_1[E][S] - k_2[E-S] - k_3[E-S]$$

At steady state, $d[E-S]/dt = 0$, so that

$$[1.5.20] \quad k_1[E][S] = (k_2 + k_3)[E-S]$$

Since the enzyme can exist in only two states, E and $E-S$, we have a conservation relation

$$[1.5.21] \quad [E]_{total} = [E] + [E-S]$$

where $[E]_{total}$ is the total concentration of enzyme. Inserting this relation for $[E]$ in Eqn [1.5.20], we get

$$[1.5.22] \quad k_1([E]_{total} - [E-S][S]) = (k_2 + k_3)[E-S]$$

Solving for $[E-S]$, we find

$$[1.5.23] \quad [E-S] = \frac{k_1[E]_{total}[S]}{k_2 + k_3 + k_1[S]}$$

From Eqn [1.5.18], the rate of the reaction is just $k_3[E-S]$. Multiplying both sides of Eqn [1.5.23] by k_3,

and dividing numerator and denominator by k_1, we obtain

$$[1.5.24] \quad J = k_3[E-S] = \frac{k_3[E]_{total}[S]}{(k_2 + k_3)/k_1 + [S]}$$

The maximum velocity occurs when all of the enzyme is present as $[E-S]$. The maximum velocity or rate of the reaction is thus given as

$$[1.5.25] \quad J_{max} = k_3[E]_{total}$$

Inserting this definition into Eqn [1.5.24], we finally obtain

$$[1.5.26] \quad J = \frac{J_{max}[S]}{K_m + [S]}$$

where K_m is a newly defined constant called the Michaelis–Menten constant. It is given as

$$[1.5.27] \quad K_m = \frac{k_2 + k_3}{k_1}$$

The units of k_2 and k_3 are both in s^{-1}, whereas the unit of k_1 is M^{-1} s^{-1}; thus, the unit of K_m is M. If $k_3 \ll k_2$, K_m approximates the value of k_2/k_1, which is the dissociation constant for binding of S to the enzyme. It can be obtained experimentally as the value of the substrate concentration at which the enzyme exhibits one-half maximal velocity. This can be seen from Eqn [1.5.26] by inserting $J = \frac{1}{2}J_{max}$ and finding that $[S] = K_m$ when the rate of the reaction is one-half maximal.

The saturation curve for a Michaelis–Menten type reaction is shown in Figure 1.5.4. In this case, J_{max} was 8 µmol min^{-1} mL^{-1}. At one-half of this maximal velocity, the substrate concentration was 1.5 mM.

The curve shown in Figure 1.5.4 often lacks sufficient points to accurately extrapolate the observed velocity

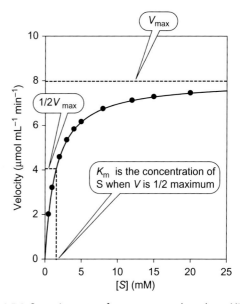

FIGURE 1.5.4 Saturation curve for an enzyme that obeys Michaelis–Menten kinetics. The velocity at low $[S]$ is nearly linear with $[S]$ but quickly levels off. This behavior is called saturation kinetics. The concentration of substrate at one-half maximal velocity (=one-half saturation) is equal to the K_m, the Michaelis–Menten constant.

to the maximal velocity. To aid in this graphical determination of enzyme kinetics, the Michaelis–Menten equation is transformed. If we take the inverse of Eqn [1.5.26], we obtain

[1.5.28]
$$J = \frac{J_{max}[S]}{K_m + [S]}$$
$$\frac{1}{J} = \frac{K_m + [S]}{J_{max}[S]}$$
$$\frac{1}{J} = \frac{1}{J_{max}} + \frac{K_m}{J_{max}}\frac{1}{[S]}$$

This equation suggests that a plot of $1/J$ versus $1/[S]$ should be linear with an intercept on the $1/J$ axis of $1/J_{max}$ and a slope of K_m/J_{max}. The intercept on the $1/[S]$-axis is $-1/K_m$. Such a plot is called a **double-reciprocal plot**, also known as a Lineweaver–Burk plot. An example for the results shown in Figure 1.5.4 is shown in Figure 1.5.5.

Throughout this discussion of enzyme rates, we have used J as the symbol of the rate of product release, or the rate of the reaction. J has the units of moles per unit volume, per unit time. Often this is taken to be the same as the reaction velocity, and the usual symbol in the Michaelis–Menten equation is V, and not J. Sometimes, the reaction velocity in these units is normalized by dividing by the enzyme concentration, so that the reaction velocity is given as a specific activity of the enzyme, in units of moles of reaction per unit enzyme per unit time. These various representations of the velocity all obey the same general form of equations and can be subjected to the same analysis.

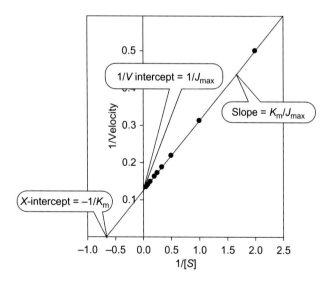

FIGURE 1.5.5 Lineweaver–Burk plot. Values of $1/[S]$ are plotted along the abscissa while values of $1/J$ are plotted on the ordinate. Here J is the velocity, or rate, of the enzyme reaction and $[S]$ is the substrate concentration. The slope of the line is K_m/J_{max}, where J is the enzyme flux ≈ enzyme velocity and J_{max} is the theoretical maximal rate. K_m is the Michaelis–Menten constant. It is also determined as the X-intercept = $-1/K_m$. The Y-intercept is $1/J_{max}$. Deviations from linearity on the Lineweaver–Burk plot suggest that the enzyme does not obey Michaelis–Menten kinetics.

PHYSIOLOGY IS ALL ABOUT SURFACES

As noted above, enzyme catalysis derives from the reaction occurring on the surface of the enzyme rather than in homogeneous solution. Physiology is all about surfaces and their interactions. Proteins have complicated three-dimensional shapes that closely appose the surfaces of some materials and not others—they are specific. Sometimes the fit is very close and therefore the binding strength is very high—the two surfaces form a tight association. Which proteins stick to which other proteins, or which other substrates, or ligands, determines the activity of the proteins, which determines the activities of the cells, and then the organs, and, finally, the organism. Ultimately, physiology is all about what happens on the surfaces of molecules.

SUMMARY

The concentration of a solute in solution is the amount of that solute per unit volume of solution. It can be expressed as the mass of the solute per unit volume or the number of moles of solute per unit volume. A mole of any substance is Avogadro's number of particles. Avogadro's number originates in the ratio of the mass of a carbon-12 atom to 12 g. The dalton is defined as 1/12 of the mass of a carbon-12 atom. The number of daltons in 1 g is Avogadro's number, 6.02×10^{23}.

The concentration in molar is the number of moles per liter of solution. Small concentrations use established prefixes to describe them, in increments of 1000. A millimolar solution is 10^{-3} M, micromolar is 10^{-6} M, nanomolar is 10^{-9} M, and picomolar is 10^{-12} M. These same prefixes are used for units of g, L, and m.

The relationship among concentration, amount of solute, and volume of solution can be used to determine the volume of physiological fluids. Evans' Blue Dye is an example of a solute that can be used to estimate plasma volume, because the dye enters the plasma but cannot leave it easily.

Elementary chemical reactions have forward and reverse rate constants that govern the rate of conversion in either the forward or reverse reaction. The rates of reaction have the units of moles per unit time per unit volume of solution. The ratio of the forward and reverse rate constants is the equilibrium constant. Conversion of reactants to products requires an activation energy, and because kinetic energy increases with temperature, the rate of the reaction also varies with the temperature. The relationship between temperature and rate is described by the Arrhenius equation, which incorporates the activation energy, E_a.

Enzymes speed chemical reactions by altering the path of the reaction by allowing it to proceed on the surface of the enzyme. Thus enzymes convert homogeneous reactions in the fluid phase to heterogeneous reactions on the surface of the enzyme. The alternate path reduces the activation energy for the reaction, thereby allowing it to proceed quicker.

The Michaelis—Menten formulation of enzyme kinetics describes the steady-state rate of enzyme reactions and is derived for a particular kind of enzyme mechanism. Nevertheless, the resulting equation often fits many enzyme-catalyzed reactions. It is given as

$$J = \frac{J_{max}[S]}{K_m + [S]}$$

where J is the reaction flux or velocity, in units of moles of completed reaction per unit time per unit volume, J_{max} is the asymptotic maximal rate achievable extrapolated at infinite concentration, K_m is the Michaelis—Menten constant, equal to the concentration of substrate at half maximal velocity, and $[S]$ is the substrate concentration. This equation describes **saturation kinetics**. It can be analyzed more easily using the inverse of the equation, rearranged, to give

$$\frac{1}{J} = \frac{1}{J_{max}} + \frac{K_m}{J_{max}}\frac{1}{[S]}$$

Plots of $1/J$ against $1/[S]$ yield $1/J_{max}$ as the intercept on the abscissa and $-1/K_m$ as the extrapolated intercept on the ordinate.

REVIEW QUESTIONS

1. What is a dalton? What is meant by molecular weight?
2. What is a mole? What is meant by the gram molecular weight? How would you determine the gram molecular weight of small compounds?
3. Write the relationship among concentration, volume, and amount.
4. What is meant by "first-order reaction"?
5. How do you calculate the half-life of a reaction?
6. In the plot of potential energy against reaction coordinate, what is meant by "reaction coordinate"? What units does it have?
7. What is the activation energy?
8. Why do reaction rates generally depend on the temperature?
9. How could you determine the activation energy for a reaction?
10. In general, how do enzymes speed up biochemical reactions?
11. How would you determine K_m and V_{max} for an enzyme?

APPENDIX 1.5.A1 TRANSITION STATE THEORY EXPLAINS REACTION RATES IN TERMS OF AN ACTIVATION ENERGY

TRANSITION STATE THEORY CALCULATES THE POTENTIAL ENERGY OF REACTANTS AS A FUNCTION OF SEPARATION

Transition state theory gives an expression for the rate constant of a chemical reaction by applying statistical mechanics to quantum mechanical calculations of various configurations of reactant and product. In this procedure, a collection of nuclei with their attendant electrons is treated as a "supermolecule." Quantum mechanical calculations are performed in which the potential energy of the supermolecule is calculated as a function of the relative positions of the nuclei. An exact solution requires solving for the total energy including the kinetic energy terms of all of the nuclei and electrons and the potential energy terms for all electron–electron, electron–nucleus, and nucleus–nucleus pairs. Since this solution is extremely difficult, some simplifying assumptions are made. One is that the electronic motion is extremely rapid compared to translation of the nuclei and that the electrons adjust instantly to any change in the positions of the nuclei. Thus the energy of the electrons and potential energy of the nuclei can be calculated as if the nuclei were at rest. The energy calculated in this way includes everything but the kinetic energy of the nuclei. It is called the potential energy even though it includes the kinetic energy of the electrons.

POTENTIAL ENERGY CAN BE GRAPHED AGAINST SEPARATION IN A SINGLE MOLECULE

The potential energy of a configuration of nuclei with their electrons may be represented as a point on an f-dimensional surface in an $f+1$-dimensional space, where f is the number of independent variables required to specify the relative positions of all nuclei. For a diatomic molecule (like H_2 or HF), $f = 6$ because we need six variables to specify the Cartesian coordinates of the two nuclei. However, three of these can be considered to locate the center of mass of the molecule and another two specify the orientation of the molecular axis. As these do not really concern us, we have only one remaining variable, the internuclear distance, to specify the relative positions of the two nuclei. Since $f = 1$, we have a potential surface which is the one-dimensional potential energy curve in the two-dimensional graph for a diatomic molecule (H_2) as shown in Figure 1.5.A1.1.

POTENTIAL ENERGY CAN BE GRAPHED AGAINST NUCLEI SEPARATION IN A CHEMICAL REACTION

In a configuration of three nuclei, the potential energy surface can be shown as a fixed-angle surface, where the angle is defined as the angle between one bond and the approaching reactant, as shown in Figure 1.5.A1.2. This allows us to plot the potential surface on paper; otherwise we need another physical dimension. The entire potential surface consists of an infinite number of these fixed-angle surfaces, one for each angle of approach.

Such a fixed-angle surface is shown in Figure 1.5.A1.3 for the reaction

$$F + H - H' \rightarrow F - H + H'$$

The surface itself is three-dimensional. What Figure 1.5.A1.3 shows is the two-dimensional projection of the surface onto the plane of the paper. Here every point on a given line has the same potential energy. Thus these lines represent potential energy contours in much

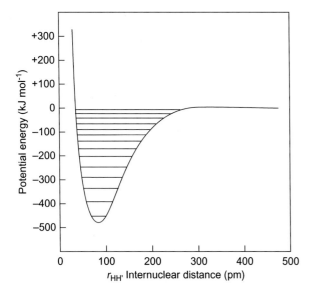

FIGURE 1.5.A1.1 Potential energy diagram for a diatomic molecule, H_2. Potential energy is plotted as a function of the separation of the two H nuclei. The average bond length corresponds to the minimum energy level, which in this case occurs at about 80 pm. The horizontal lines indicate quantum vibrational states. Rotation of the molecule alters the potential energy profile due to a centrifugal effect (rotation stretches the bond length). *Adapted from J.W. Moore and R.G. Pearson,* Kinetics and Mechanism, *John Wiley & Sons, New York, NY, 1981.*

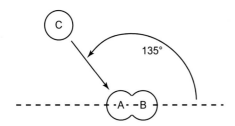

FIGURE 1.5.A1.2 Fixed angle of approach of reactant C to the diatomic molecule AB. The potential energy surface varies with the angle of approach.

the same way as contour lines on a topographic map represent lines of equal altitude.

In the contour map for the reaction $F + H-H' \rightarrow F-H + H'$, several regions are labeled. Region "W" corresponds to separated F, H, and H' nuclei and this is a plateau region where potential energy varies little with the distance between the nuclei. At region "X," F is far removed from $H-H'$ and the potential energy, U, is affected little by changes in r_{FH}, the distance between the F nucleus and the H nucleus. The effect of $r_{HH'}$ on U is that shown earlier for the diatomic molecule, so that a vertical cross-section at "X" would show a deep valley with steep sides. At region "Y," all three nuclei are close together and the potential energy has increased due to van der Waals repulsion. A collision of F with $H-H'$ thus corresponds to the movement of the configuration of nuclei from region "X" to region "Y" where potential energy is increased. If the reaction is completed, then the configuration moves on to region "Z." In this region, r_{FH} is short and $r_{HH'}$ is long, indicating that a bond has formed between $F-H$ and the $H-H$ is broken. The horizontal cross-section at "Z" gives the potential energy dependence of $F-H$ on its bond distance.

THE ACTIVATED COMPLEX IS A METASTABLE COMPLEX OF REACTANTS

The movement of the configuration of nuclei from X to Y to Z in Figure 1.5.A1.3 traces a completed reaction from the initial reactants. At the point Y, there is a saddle point represented by higher potential energy, in this case, than for either reactants or products. The configuration of nuclei at this point is called the **activated complex** which is said to be in the **transition state**. Here all of the nuclei are close together, forming a complex. The path of steepest ascent from reactants to activated complex and on to products is called the **reaction coordinate**. This path is perpendicular to each contour it crosses. It is the minimum energy path from reactants to products.

Recall that the potential energy surface is just that—a potential energy. Collisions between reactants can overcome the potential energy barrier between the reactant and the activated complex if there is sufficient kinetic energy of the molecules that collide. The potential energy can be plotted against the distance along the reaction coordinate. This is the familiar diagram encountered in general chemistry, without any kind of explanation of what the reaction coordinate actually is. For our example reaction, $F + H-H' \rightarrow F-H + H'$, the potential energy versus reaction coordinate plot is shown in Figure 1.5.A1.4. This plot is the conceptual origin of Figure 1.5.2 in the text.

REACTIONS GENERALLY DO NOT FOLLOW A MINIMUM ENERGY PATH

Reactions in general do not follow the reaction coordinate. Two alternative trajectories are shown in Figure 1.5.A1.5. Because the H–H' bond vibrates, the configuration of all three nuclei while F approaches $H-H'$ oscillates. In Figure 1.5.A1.5A, the reaction ascends near to the activated complex but does not go to completion. This trajectory represents a nonreactive, inelastic collision between F and $H-H'$. In this case, there is a transfer of energy from translational kinetic energy to vibrational energy. In Figure 1.5.A1.5B, a completed reaction is shown. Here the oscillations correspond to vibrations of reactant $H-H'$ and product FH.

THE TRANSITION STATE THEORY SAYS THAT THE RATE CONSTANT VARIES WITH THE EXPONENT OF THE ACTIVATION ENERGY

The derivation of a rate equation from the transition state theory is a bit complicated and we will not attempt it in any detail. However, an elementary understanding of it can be provided by thinking of the activated complex as a separate, transient species. Then the rate of completed reaction will be proportional to the amount of activated complex, and the rate per unit volume will be proportional to its concentration. If we imagine that the reactants and

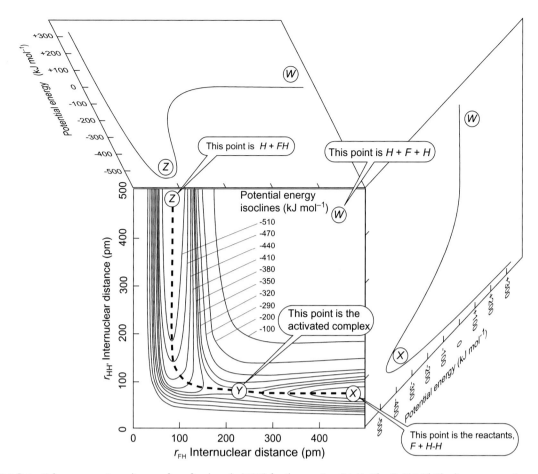

FIGURE 1.5.A1.3 Potential energy contour diagram for a fixed angle (180°) for the reaction $F + H-H' \rightarrow F-H + H'$. The lines represent a set of points of equal potential much like contour lines on a topographic map represent a set of points of equal altitude. The value of the potential energy (in kJ mol^{-1}) for each line is indicated. Point W lies on a plateau of high potential energy corresponding to dissociated F, H, and H' nuclei. X corresponds to the configuration in which F is far from H–H'; it is at a potential energy minimum. Y is the saddle point at which all three nuclei are close; it corresponds to the activated complex. Z corresponds to the configuration in which H' is far from FH. It represents the products of the completed reaction. The graph at the right is a vertical cross-section through the three-dimensional surface at point X. The graph at the top is a vertical cross-section (orthogonal to the one at the right) through the three-dimensional surface at point Z. *Adapted from J.W. Moore and R.G. Pearson,* Kinetics and Mechanism, *John Wiley & Sons, New York, NY, 1981.*

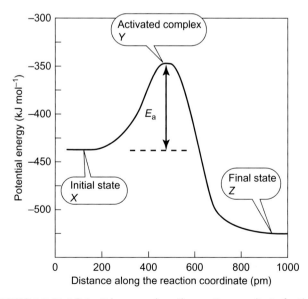

FIGURE 1.5.A1.4 Potential energy along the reaction coordinate for the reaction $F + H-H' \rightarrow F-H + H'$. Reactants $F + H-H'$ are at a low potential energy. The activated complex is at higher potential energy, which can be attained by converting kinetic energy into potential energy through a collision of F and H–H'. The reactants are at lower potential energy. The difference between the energy of the activated complex and the reactants is the activation energy that must be supplied for the reaction to proceed.

the activated complex are in equilibrium, then the rate constant for the overall reaction will be proportional to the equilibrium constant.

$$[1.5.A1.1] \qquad K_r = \kappa \frac{kT}{h} K^*$$

There are a lot of k's in this equation. K_r is the rate constant for the reaction; κ is a **transmission coefficient** that tells us what fraction of activated complexes goes on to complete the reaction; k in kT is Boltzmann's constant, which is equal to the gas constant per molecule, or R/N_o, the gas constant divided by Avogadro's number; h is Planck's constant; and K^* is a constant that has the form of an equilibrium constant for the formation of the activated complex, but strictly speaking it is not an equilibrium constant. The equilibrium constant is related to the change in free energy for the reaction (see Chapter 1.7) according to

$$[1.5.A1.2] \qquad K^* = e^{\frac{-\Delta G^*}{RT}}$$

where ΔG^* is the free-energy change per mole for the formation of the activated complex from the reactants. This energy change is identified with the **activation energy**, E_a, as described in Figure 1.5.A1.4.

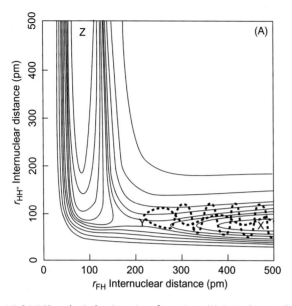 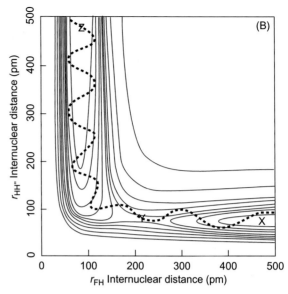

FIGURE 1.5.A1.5 Hypothetical trajectories of reactions. (A) An inelastic collision that does not result in a completed reaction. (B) A completed reaction. The oscillations are due to vibration of the molecule in which bond length oscillates around an average value.

In 1884, van't Hoff proposed that the temperature dependence of equilibrium constants could be described by

$$[1.5.A1.3] \qquad \frac{d \ln K_{eq}}{dT} = \frac{\Delta G^0}{RT^2}$$

It was discovered shortly thereafter that the rates of chemical reactions approximately double for every 10°C increase in temperature. By analogy to the van't Hoff law, Arrhenius proposed in 1889 that the temperature dependence of the rate constant could be given as

$$[1.5.A1.4] \qquad \frac{d \ln K_r}{dT} = \frac{\Delta E^0}{RT^2}$$

The integrated form of this equation is similar to what we see from transition state theory:

$$[1.5.A1.5] \qquad K_r = Ae^{-E_a/RT}$$

where A is the preexponential factor that incorporates the transmission coefficient. This is close to what one would expect by inserting the definition of K^* into Eqn [1.5.A1.1]. The natural log of both sides gives

$$[1.5.A1.6] \qquad \ln K_r = \ln A - \frac{E_a}{RT}$$

Thus the plots of the logarithm of the rate constant against $1/T$ should be linear with a slope $= -E_a/R$. Such a plot is an **Arrhenius plot**. Although transition state theory does not predict a perfectly linear relation between $\ln K_r$ and $1/T$, most reactions obey Eqn [1.5.A1.6] well.

THE ACTIVATION ENERGY DEPENDS ON THE PATH

The potential energy surfaces described in Figure 1.5.A1.3 pertain only to the collisions between the molecules H–H' and F, which is a simple system. Similar potential energy surfaces could be described for more complicated reactions, but the surface becomes much more complicated because of the large number of atomic nuclei involved in typical biochemical reactions. However, the idea of the reaction coordinate and activation energy remains. By absorbing reactants onto their surfaces, enzymes completely alter the potential energy surfaces of a reaction. The net effect is to lower E_a. According to Eqn [1.5.A1.6], lower activation energies mean larger rate constants. Thus enzymes and other catalysts increase the speed of reactions by providing an alternate mechanism such that E_a is reduced. This idea is depicted graphically in Figure 1.5.3 in the text.

APPENDIX 1.5.A2 UNIDIRECTIONAL FLUXES OVER A SERIES OF INTERMEDIATES DEPEND ON ALL OF THE INDIVIDUAL UNIDIRECTIONAL FLUXES

UNIDIRECTIONAL FLUXES DIFFER FROM NET FLUXES

Consider the diagram shown in Figure 1.5.A2.1. Each node, indicated by a dot, represents a state, compartment, or individual chemical species in a series. Movement between nodes indicates either a flux of material between compartments or the transition of one chemical species into another, or the transition between intermediate states of an enzyme, for example.

We will treat all of the transitions between states as being an elementary reaction so that the flux obeys the relation

$$[1.5.A2.1] \qquad J_{ij} = \alpha_{ij} N_i$$

where J_{ij} is the flux from node i to node j, N_i is the population of node i, and α_{ij} is a pseudo-first order rate constant. It is called a pseudo-first order rate constant because it may incorporate the concentration of a ligand if the transition between node i and node j requires it.

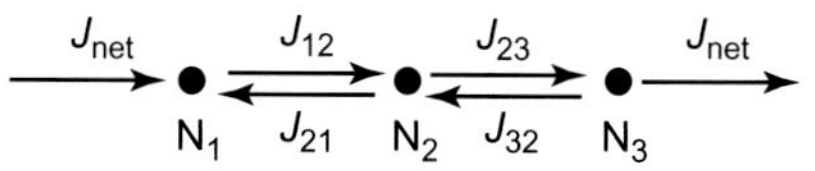

FIGURE 1.5.A2.1 Chemical species can be transformed into various intermediates here represented as three nodes, N1, N2, and N3. The fluxes between the nodes are indexed with the source node first and sink node second. Alternatively, the nodes can represent compartments into which the same chemical is transported, or different states of the same object, such as an enzyme.

At steady state, there could be a throughput of material such that sum flux, J_{net}, of material enters node 1 and the same flux exits the end, at node 3 in this case. At steady state, this requires that the net flux between any two nodes is the same, equal to J_{net}. However, there can be a higher unidirectional flux of material from node 1 to node 2 than this net flux. All that steady state requires is that

[1.5.A2.2] $$J_{ij} - J_{ji} = J_{net}$$

for each ij pair. The unidirectional flux, J_{ij}, can be determined by somehow labeling the population of state i and seeing how fast they are converted to state j. One way this can be done is by adding a pulse of radioactive material of state i, and watching its conversion to state j. We will note the population of radioactive tracer in each state or compartment with an asterisk, which is added into state i alone. We can write the following sets of equations that govern the time change in the population of radioactivity in each state:

[1.5.A2.3]
$$\frac{dN_1^*}{dt} = -J_{12}\frac{N_1^*}{N_1} + J_{21}\frac{N_2^*}{N_2}$$
$$\frac{dN_2^*}{dt} = J_{12}\frac{N_1^*}{N_1} - (J_{21}+J_{23})\frac{N_2^*}{N_2} + J_{32}\frac{N_3^*}{N_3}$$
$$\frac{dN_3^*}{dt} = J_{23}\frac{N_2^*}{N_2} - J_{32}\frac{N_3^*}{N_3}$$

The idea about unidirectional flux is that there is no back flux. Thus, for the transition between state 1 and state 2, we can measure unidirectional flux J_{12} only when N_2^* is negligible. For the top equation in Eqn [1.5.A2.3], we have

[1.5.A2.4]
$$\frac{dN_1^*}{dt} = -J_{12}\frac{N_1^*}{N_1} + J_{21}\frac{N_2^*}{N_2}$$
$$\frac{N_2^*}{N_2} \approx 0$$
$$\frac{dN_1^*}{dt} = -J_{12}\frac{N_1^*}{N_1}$$
$$J_{12} = -\frac{dN_1^*}{dt} \Big/ \frac{N_1^*}{N_1}$$

The last equation in Eqn [1.5.A2.4] defines what is meant by a unidirectional flux obtained by tracer means. In this case, dN_1^*/dt is negative—the amount of tracer in state 1 decreases with time—and thus the flux J_{12} is positive. The factor N_1^*/N_1 in the denominator is called the *specific activity* of the added radioactivity. The units of J_{12} are in units of N per unit time.

Measuring the unidirectional flux from state 1 to state 3 requires a similar condition as the unidirectional flux from state 1 to 2; that is, there must be negligible back flux during the measurement. In this case, this means that N_3^*/N_3 is negligible. We also require a steady state. In this case, steady state means $dN_2^*/dt = 0$. What we are searching for is an overall unidirectional flux that obeys the relation:

[1.5.A2.5] $$\frac{dN_3^*}{dt} = J_{13}\frac{N_1^*}{N_1}$$

Because $N_3^*/N_3 \approx 0$ as a criterion for measuring unidirectional flux, the bottom part of Eqn [1.5.A2.3] becomes

[1.5.A2.6] $$\frac{dN_3^*}{dt} = J_{23}\frac{N_2^*}{N_2}$$

If $dN_2^*/dt = 0$ as a condition of steady state, then by the middle part of Eqn [1.5.A2.3] we can write

[1.5.A2.7] $$\frac{N_1^*}{N_1} = \frac{(J_{21}+J_{23})}{J_{12}}\frac{N_2^*}{N_2}$$

Substituting in for N_2^*/N_2 from Eqn [1.5.A2.3] into Eqn [1.5.A2.7], we obtain

[1.5.A2.8] $$\frac{dN_3^*}{dt} = \frac{J_{12}J_{23}}{(J_{21}+J_{23})}\frac{N_1^*}{N_1}$$

Comparison of Eqn [1.5.A2.5] with Eqn [1.5.A2.8] allows us to identify J_{13} as

[1.5.A2.9] $$J_{13} = \frac{J_{12}J_{23}}{(J_{21}+J_{23})} = J_{12}\frac{J_{23}}{(J_{21}+J_{23})}$$

This result has an intuitive interpretation. It says that the unidirectional flux from node 1 to node 3 is the flux from node 1 to node 2 times the proportion of the flux that goes on to node 3. This proportion is the flux from node 2 to node 3 divided by the total flux away from node 2: the sum of J_{21} and J_{23}.

It can be seen readily that we could just as easily added radioactivity at node 3 and watch its appearance in node 1. From the symmetry, we can write J_{31} as

[1.5.A2.10] $$J_{31} = \frac{J_{32}J_{21}}{(J_{21}+J_{23})} = J_{32}\frac{J_{21}}{(J_{21}+J_{23})}$$

These results are completely general. Suppose that the sequence of nodes, states, or compartments was longer than just three, as shown in Figure 1.5.A2.2. We could use the unidirectional flux to reduce the diagram to one fewer states, and then do so again using the same principle that we discovered here. The results for J_{14} and J_{41} are given as

$$J_{14} = \frac{J_{13}J_{34}}{(J_{31}+J_{34})} = \frac{J_{12}\frac{J_{23}}{(J_{21}+J_{23})}J_{34}}{J_{32}\frac{J_{21}}{J_{21}+J_{23}}+J_{34}} = \frac{J_{12}J_{23}J_{34}}{J_{32}J_{21}+J_{21}J_{34}+J_{23}J_{34}}$$

[1.5.A2.11]

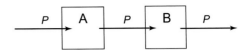

FIGURE 1.5.A2.2 Unidirectional fluxes over a series of intermediate states. The flux between states 1 and 3 can be replaced by two unidirectional fluxes. The process can be repeated over any number of intermediates.

$$J_{41} = \frac{J_{43}J_{31}}{(J_{31}+J_{34})} = \frac{J_{43}\dfrac{J_{21}}{(J_{21}+J_{23})}J_{32}}{J_{32}\dfrac{J_{21}}{J_{21}+J_{23}}+J_{34}} = \frac{J_{43}J_{32}J_{21}}{J_{32}J_{21}+J_{21}J_{34}+J_{23}J_{34}}$$

[1.5.A2.12]

In this way, the unidirectional flux over any number of intermediate states can be calculated from the set of individual unidirectional fluxes.

APPENDIX 1.5.A3 SIMPLE COMPARTMENTAL ANALYSIS

THE AGGREGATE OF A CHEMICAL SPECIES CONSTITUTES ITS BODILY "POOL"

Compounds in the body are normally in a dynamic state in which materials are converted from precursors to products. For example, there is a steady state between the phospholipids in the cytoplasm and in the mitochondria, with continuing interconversion between both. The aggregate of the cytoplasmic phospholipids in the its pool, as is the aggregate of the mitochondrial phospholipids. A "pool" or "compartment" is defined as the set of molecules of a specific compound or group of compounds found in a specific part of the organism. For example, we may speak of "liver glycogen" or "muscle glycogen" or "liver phospholipid" or "plasma phospholipid." In investigating these pools, we are most interested in the identity of the pools, their size, and the rate at which they exchange material with other pools. Radioactive labeling is an effective way of determining these parameters. Generally these pools change only slowly with time. Their maintenance at a fairly steady level is a consequence of homeostasis, the maintenance of a constant internal environment, which is the hallmark of physiological systems. Though blood glucose or blood [Ca^{2+}] is maintained fairly constant, there is always material entering and leaving the plasma pool.

THE TURNOVER DESCRIBES THE RATE OF EXCHANGE OF A POOL

The rate of exchange of material is the rate at which material enters or leaves the pool, and can be given in units of amount per unit time. This is also called the turnover or the rate of renewal of the pool. The fractional turnover rate is the fraction of the total pool which is replaced per unit time. If A is the size of pool (in g or moles) and P g or moles of A are renewed each minute, then the fractional turnover rate is given as

[1.5.A3.1] $$k_1 = \frac{P}{A}$$

where the units of k are reciprocal time. Here we consider pool A of size A which is a precursor of pool B of size B, as shown in Figure 1.5.A3.1. We will assume that the pools are in a steady state (their sizes do not change with time); there is rapid and uniform mixing within the pools and that the rate of transfer does not change with time. Suppose we inject an amount of radioactive compound A, which we denote as A^*, into pool A at time $t = 0$. If the radioactivity mixes well and uniformly, we may describe the change in radioactivity in pool A as

[1.5.A3.2] $$\frac{dA^*}{dt} = -P\frac{A^*}{A}$$

This equation derives from the idea that the radioactive portion of pool A will be turned over in proportion to its concentration in pool A. Thus when P unit of pool A turns over per unit time, PA^*/A unit is radioactive and $P(A-A^*)/A$ unit is not radioactive. The quantity A^*/A is called the specific activity of pool A, denoted as S_A. Equation [1.5.A3.2] can be rewritten as

[1.5.A3.3] $$\frac{1}{A}\frac{dA^*}{dt} = -\frac{P}{A}S_A$$

Since A is constant under steady-state conditions, we may rewrite Eqn [1.5.A3.3] as

[1.5.A3.4] $$\frac{d(A^*/A)}{dt} = -\frac{P}{A}S_A$$

$$\frac{dS_A}{dt} = -k_1 S_A$$

Here we have used the definition of the fractional turnover rate given in Eqn [1.5.A3.1]. The last equation is of the form of first-order decay and is easily integrated to give

[1.5.A3.5] $$S_A = S_{A_0} e^{-k_1 t}$$

where S_{A0} is the specific activity of pool A at time $t = 0$. Plots of $\ln S_A$ against time will allow the calculation of the fractional turnover rate as the negative of the slope. Extrapolation back to time zero gives the pool size: if A^*/A is known, and the amount of radioactivity injected, A^*, is known, then A can be calculated. It is important to note that the determination of the specific activity, $S_A = A^*/A$, does not require knowing A; it does require knowing A^* and A in an aliquot of the pool. Thus the specific activity can be determined even if the pool size A cannot be directly measured. The validity of estimating pool sizes in this way depends on the assumptions of uniform and rapid mixing of the radioactive label with the endogenous material.

The specific activity of pool B is affected by influx from pool A and efflux from pool B. Here we write

[1.5.A3.6] $$\frac{dB^*}{dt} = P\,S_A - P\,S_B$$

We can divide both sides of the equation by B to obtain

$$\frac{1}{B}\frac{dB^*}{dt} = \frac{P}{B}(S_A - S_B)$$

[1.5.A3.7] $$\frac{dB^*/B}{dt} = k_2(S_A - S_B)$$

$$\frac{dS_B}{dt} = k_2(S_A - S_B)$$

We can rewrite the last part of this equation as

[1.5.A3.8] $$\frac{dS_B}{dt} + k_2\,S_B = k_2\,S_A$$

This equation cannot be solved by integration directly, because S_A is a function of time and we don't know S_B as a function of time. We can multiply both sides by an integrating factor, ρ, such that

[1.5.A3.9] $$d(\rho S_B) = \rho\,dS_B + \rho\,k_2\,S_B\,dt$$

We choose ρ such that ρS_B is an exact differential, so that

[1.5.A3.10] $$d(\rho S_B) = \rho\,dS_B + S_B\,d\rho$$

Comparison of Eqns [1.5.A3.9] and [1.5.A3.10] indicates that

[1.5.A3.11] $$d\rho = \rho k_2\,dt$$

Solution of this equation gives

[1.5.A3.12] $$\rho = e^{k_2 t}$$

Multiplying both sides of Eqn [1.5.A3.8] by this multiplication factor gives

[1.5.A3.13] $$\rho\,dS_B + \rho\,k_2\,S_B\,dt = \rho\,k_2\,S_A\,dt$$

We have chosen ρ so that Eqn [1.5.A3.9] is valid, so we can substitute $d(\rho S_B)$ for the left-hand side of Eqn [1.5.A3.13]:

[1.5.A3.14] $$d(\rho\,S_B) = \rho\,k_2 S_A\,dt$$

Integration of this equation is now possible because the left-hand side is a function of ρS_B alone and the right-hand side is a function of t alone. Substitution of ρ from Eqn [1.5.A3.12] and for S_A from Eqn [1.5.A3.5] gives

[1.5.A3.15] $$d(\rho\,S_B) = k_2\,e^{k_2 t} S_{A_0}\,e^{-k_1 t}\,dt$$

Integration from $t = t$ to $t = t$ gives

$$\int_0^t d(\rho\,S_B) = \int_0^t k_2\,e^{k_2 t}\,S_{A_0}\,e^{-k_1 t}\,dt$$

[1.5.A3.16] $$\rho\,S_B \Big|_0^t = \frac{k_2\,S_{A_0}}{k_2 - k_1} e^{(k_2-k_1)t} \Big|_0^t$$

$$S_B e^{k_2 t} = \frac{k_2\,S_{A_0}}{k_2 - k_1}(e^{(k_2-k_1)t} - 1)$$

Dividing both sides by the exponent on the left finally leaves us with

[1.5.A3.16] $$S_B = \frac{k_2}{k_2 - k_1} S_{A_0}(e^{-k_1 t} - e^{-k_2 t})$$

We have now derived equations for S_A and S_B for the conditions shown in Figure 1.5.A3.1. The graph of $\ln S_A$ versus time will give $\ln S_{A_0}$ as the intercept and $-k_1$ as the slope. Eqn [1.5.A3.6] shows that the maximum of the graph of $\ln S_B$ will occur with $S_A = S_B$, because at this point $dB^*/dt = 0$, which means $dS_B/dt = 0$ and therefore $d\ln S_B/dt = 0$. P may be obtained from A and k_1. Figure 1.5.A3.2 shows the results of calculations for Eqns [1.5.A3.5] and [1.5.A3.16] for assumed values of A, A^*, k_1 and k_2. The plot of $\ln S_A$ gives an intercept of 9.2103. Thus $S_{A_0} = 10{,}000$ cpm μmol^{-1}. Since the amount of injected radioactivity was 1×10^8 cpm, we can calculate the pool size, A, as

$$A = A^*/S_{A_0} = 1 \times 10^8 \text{ cpm}/10^4 \text{cpm } \mu\text{mol}^{-1}$$
$$= 10^4\,\mu\text{mol} = 0.02 \text{ mol; thus}$$

$A = 0.02$ mol

The fractional turnover rate is given as the negative of the slope: **$k_1 = 0.02$ min^{-1}**. The turnover of the pool at steady state can be calculated as

$$P = k_1\,A = 4 \times 10^{-4} \text{ mol min}^{-1}; \mathbf{P = 4 \times 10^{-4} \text{ mol min}^{-1}}$$

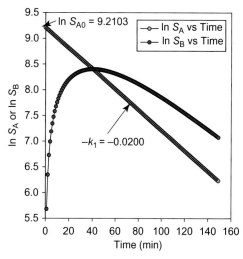

FIGURE 1.5.A3.2 Plot of $\ln S_A$ and $\ln S_B$ against time. The plot of $\ln S_A$ is linear with a Y-intercept of 9.2103 and a slope of -0.02. The plot of $\ln S_B$ intersects that of $\ln S_A$ at the point where $S_A = S_B$.

Note that the plot of $\ln S_B$ peaks at the intersection of the $\ln S_A$ and $\ln S_B$ curve. This occurs at 40.55 minutes and $S_A = S_B = 4444.2$ cpm μmol^{-1}. Eqn [1.5.A3.16] can be reorganized to attempt to solve for k_2. It can be rewritten as

[1.5.A3.17]
$$\frac{(k_2 - k_1)}{k_2} S_B = S_{A_0} e^{-k_1 t} - S_{A_0} e^{-k_2 t}$$

$$\left(1 - \frac{k_1}{k_2}\right) S_B = S_A - S_{A_0} e^{-k_2 t}$$

$$\left(1 - \frac{k_1}{k_2}\right) S_B = S_B - S_B \, e^{k_1 t} \, e^{-k_2 t}$$

$$k_1 \, e^{-k_1 t} = k_2 \, e^{-k_2 t}$$

This result predicts that the point of intersection of the two curves depends only on the two fractional turnover rates and not on the amount of radioactivity injected. This makes intuitive sense. We know k_1 and the time of intersection $t = 40.55$ min. Thus $k_1 \, e^{-k1t} = 0.008888$ min^{-1} and we search for k_2 such that $k_2 \, e^{-k2 \, t} = 0.008888$ min^{-1}. The solution is

$$k_2 = 0.03 \text{ min}^{-1}$$

which can be verified by substitution. This solution can be obtained graphically by the intersection of the line $y = \ln k_2$ with the line $y = 40.55 k_2 - 4.723$.

The pool size B can be calculated from $k_2 = P/B$; we have $k_2 = 0.03$ min^{-1} and $P = 4 \times 10^{-4}$ mol min^{-1}, giving B as

$B = 4 \times 10^{-4}$ mol min^{-1}/0.03 min^{-1} = 0.0133 mol;

$B = 0.0133$ mol

Diffusion 1.6

Learning Objectives

- Define flow and flux
- Describe the meaning of the continuity equation
- Write Fick's First Law of Diffusion
- Recognize Fick's Second Law of Diffusion
- Identify the units of the diffusion coefficient
- Identify the three major assumptions of the one-dimensional random walk
- Describe the diffusion coefficient in terms of the parameters of the one-dimensional random walk
- Describe what is meant by the time of diffusion and its dependence on distance
- Describe the diffusion coefficient in the cytoplasm compared to that in water
- Write the Stokes–Einstein equation and identify the parameters in it

FICK'S FIRST LAW OF DIFFUSION WAS PROPOSED IN ANALOGY TO FOURIER'S LAW OF HEAT TRANSFER

Adolph Fick (1829–1901) was a German physiologist who enunciated what we now call Fick's First Law of Diffusion in 1855. Fick argued from analogy to two well-known laws of physics involving flows and their driving forces. The first was Fourier's Law of Heat Transfer:

[1.6.1] $$J_H = -\lambda \frac{\partial T}{\partial x}$$

where J_H is the rate of heat energy transfer per unit area per unit time and $\partial T/\partial x$ is the temperature gradient. The second law was Ohm's law:

[1.6.2] $$J_e = -\sigma \frac{\partial \psi}{\partial x}$$

where J_e is the electrical flux and $\partial \psi/\partial x$ is the voltage or potential gradient. By analogy, Fick wrote:

[1.6.3] $$J_s = -D \frac{\partial C}{\partial x}$$

This is Fick's First Law of Diffusion in one dimension. This law says that the positive J is in the direction of the negative spatial slope of the concentration. In analogy to the other laws, this law says that solutes move from regions of high concentrations to low concentrations, and that the driving force for such a movement is the concentration **gradient**. The term gradient has a specific meaning that is defined by vector calculus, as described in Chapter 1.3. Fluxes are usually expressed in units of moles per square centimeter per second. Since concentration is in units of moles cm^{-3}, and x is in units of cm, the gradient is in units of moles cm^{-4}. Dividing the units of J by the units of the gradient, we have the units of the diffusion coefficient as $\text{cm}^2 \text{ s}^{-1}$. In free water solutions, most low molecular weight materials have diffusion coefficients on the order of $1 \times 10^{-5} \text{ cm}^2 \text{ s}^{-1}$. Larger materials diffuse more slowly, as we will see.

FICK'S SECOND LAW OF DIFFUSION FOLLOWS FROM THE CONTINUITY EQUATION AND FICK'S FIRST LAW

In Chapter 1.2, we derived the **continuity equation** from the conservation of material flowing along one dimension. It was given as

[1.6.4] $$\frac{\partial C(x,t)}{\partial t} = -\frac{\partial J(x)}{\partial x}$$

If we differentiate Fick's First Law (see Eqn [1.6.3]) with respect to x, we obtain

[1.6.5] $$\frac{\partial J_S}{\partial x} = -D \frac{\partial^2 C}{\partial x^2}$$

By substitution into the continuity equation, we obtain

[1.6.6] $$\frac{\partial C}{\partial t} = D \frac{\partial^2 C}{\partial x^2}$$

This is **Fick's Second Law of Diffusion**. It follows from the First Law of Diffusion by application of the continuity equation. It relates changes in concentration with time with the spatial distribution of solute particles. Given initial conditions of $C(x,0)$ and boundary conditions, solutions of this equation, or its three-dimensional analogue, allow one to determine concentration as a function of time and position ($C(x,t)$).

© 2017 Elsevier Inc. All rights reserved.
DOI: http://dx.doi.org/10.1016/B978-0-12-800883-6.00007-0

FICK'S SECOND LAW CAN BE DERIVED FROM THE ONE-DIMENSIONAL RANDOM WALK

Fick's Second Law of Diffusion can be derived from molecular kinetic theory elucidated by Maxwell and Boltzmann in the latter half of the 19th century. This can be accomplished by analyzing the statistics of the random walk. Application of the continuity equation to the Second Law can then result in the derivation of Fick's First Law. It should become clear, then, that both these laws describing diffusion are a consequence of the random motion of particles. The process of diffusion thus appears to be a statistical result only, there being no literal "driving force" for the diffusive flux. Despite this, we will find that there is energy associated with concentration or dilution of solutes, and that opposition of diffusion, e.g., during active transport, requires real forces.

The random walk we will discuss is a one-dimensional model of real events. We choose to simplify the analysis because otherwise it is intractable and because it turns out to be a good model due to the enormously large number of collisions which occur during diffusion of real solute particles. The main assumptions of the model are:

- Each particle moves in a straight line between collisions.
- The motion consists of steps of length λ taken either to the left or to the right.
- The probability of making a step to the right is equal to the probability of taking a step to the left.

This surprisingly simple model will allow us to derive a one-dimensional form of Fick's Second Law.

If the probability of taking a step to the right is p and the probability of taking a step to the left is q, the third assumption gives $p = q = 1/2$. The question we ask first is this: if a particle starts out at $x = 0$, what is the probability that, after some elapsed time, t, it will be found some distance x away, where x is an integral multiple of λ? We can take λ to be a measure of the **mean free path**. If t_c is the time between collisions, and we wait for the interval t, then there will be t/t_c collisions in this time. These simple assumptions allow us to derive an expression for the probability density that a particle starting at $x = 0$ will be found in a distance interval centered at x at time t.

First we will convert the elapsed time to the number of steps taken in that interval. The size of the steps to the right and to the left is λ, the mean free path between collisions. If the average speed is v, then the average speed times the time between collisions, t_c, will be equal to λ. Thus

[1.6.7] $$t_c = \frac{\lambda}{\langle v \rangle}$$

The number of collisions in the time t is given as the time t divided by the time between collisions:

[1.6.8] $$N = \frac{t}{t_c}$$

Note that we are using N now to count the number of steps a particle of average speed v and mean free path λ will make in the interval t. In this derivation, it does not signify the number of solute molecules. Now the original problem can be reformulated: what is the probability that, after N steps, the particle will be found at a distance $x = m\lambda$, where m is an integer, away from its starting place?

Let R be the number of steps to the right and L be the number of steps to the left. Then

[1.6.9] $$\begin{aligned} N &= R + L \\ R - L &= m \end{aligned}$$

where the total number of steps is N. To travel a distance $m\lambda$ away from the starting point in steps of size λ, $R - L = m$. The desired probability of making R steps to the right and L steps to the left out of a total of N steps is given by the binomial probability distribution:

[1.6.10] $$P_N(R, L) = \left(\frac{1}{2}\right)^N \frac{N!}{R!L!}$$

This is simply the probability that with N trials, R trials will be to the right and L trials will be to the left. From Eqn [1.6.9], we have

[1.6.11] $$\begin{aligned} R &= \frac{N+m}{2} \\ L &= \frac{N-m}{2} \end{aligned}$$

which can be substituted back into Eqn [1.6.10] to give

[1.6.12] $$P_N(R, L) = P_N(m) = \left(\frac{1}{2}\right)^N \frac{N!}{((N+m)/2)!((N-m)/2)!}$$

In this form the probability distribution is a discrete variable, with only integral values of m. This random walk model of diffusion is shown in Figure 1.6.1 where the distribution models diffusion from an initially sharp distribution at $m = 0$ and $P = 1.0$. After 10 steps, the material is distributed between $-10 \leq m \leq 10$; after 20 steps, the material is distributed between $-20 \leq m \leq 20$, but the probability at the extreme ends of the distribution is small. The probability profile loses density in the center and gradually spreads out with the number of steps. Remember here that the number of steps is directly proportional to the time according to Eqn [1.6.8]. This behavior corresponds subjectively to the idea of diffusion: the material gradually diminishes at its source and spreads out with time.

If you look carefully at the binomial probability distribution in Figure 1.6.1, you will notice that its profile appears to be a fairly well-behaved function. As written, it is a discrete probability function, being defined for only integral values of N and m. What we would like is a continuous distribution that describes the envelope of the binomial probability distribution. Now normally the value of N, the number of collisions in the time interval t is extraordinarily large. In a gas at room temperature, there are typically 5×10^9 collisions per second, while

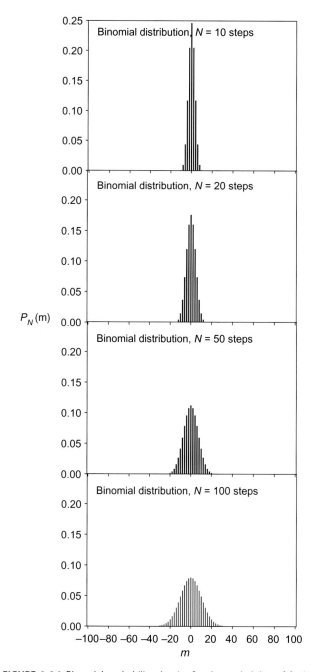

FIGURE 1.6.1 Binomial probability density for the probability of finding a particle that initially was placed at $m = 0$, as a function of the number of steps, $N = 10, 20, 50,$ and 100. Note that material initially present in a narrow band centered at the origin spreads out with increasing number of steps, corresponding to increasing time.

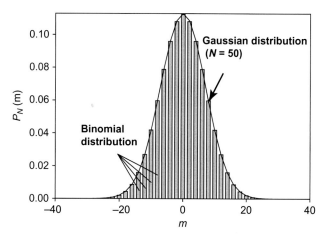

FIGURE 1.6.2 Binomial probability distribution for $N = 50$, with the envelope of the Gaussian distribution with $N = 50$. The binomial probability distribution is discrete, with only integral values of m allowed. The Gaussian distribution is continuous.

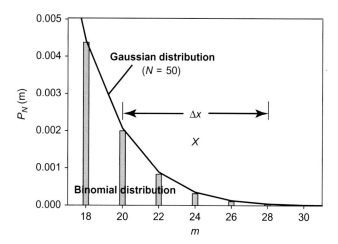

FIGURE 1.6.3 Enlargement from Figure 1.6.2.

Stirling's formula for n factorial is

[1.6.13] $\qquad n! = \sqrt{2\pi n}\, e^{n\,\ln(n-n)}$

This truly amazing formula is the key to converting the discrete probability distribution into a continuous one. Using Stirling's formula in Eqn [1.6.13], plus the approximation that $\ln(1 + \alpha) \approx \alpha$ for small values of α, it is possible to convert Eqn [1.6.12], by a lot of algebraic manipulation, to

[1.6.14] $\qquad P_N(m) = \sqrt{\dfrac{2}{\pi N}}\, e^{-(m^2/2N)}$

This is the **Gaussian** approximation to the probability distribution function $P_N(m)$. This function provides a continuous **envelope** to the discrete probability function given in Eqn [1.6.12], as shown schematically in Figure 1.6.2. The Gaussian probability distribution is discussed in Appendix 1.2.A2.

As mentioned earlier, what we desire is to find the probability that a particle starting at $x = 0$ and $t = 0$ will end up in the interval of Δx centered at x at some time t later. How we can accomplish this can be made clearer if we enlarge part of Figure 1.6.2, as shown in Figure 1.6.3.

the mean free path is typically 10^{-5} cm or so. We cannot count the number of steps accurately nor can we measure the distance accurately, so we resort to a simpler question: What is the probability that a particle starting at $x = 0$ and $t = 0$ will end up in the displacement interval Δx centered at x at time t? We can get this probability from $P_N(m)$ first by letting N get very large to justify the use of Stirling's approximation, and then by counting the number of m's which land the particle in the displacement interval Δx centered at x.

Since each step is of length λ, a net displacement of $R - L = m$ steps leads to the displacement $x = m\lambda$. Thus, $m = x/\lambda$, and we may replace m in Eqn [1.6.14] with x/λ to obtain

$$[1.6.15] \quad P_N(m) = P_N\left(\frac{x}{\lambda}\right) = \sqrt{\frac{2}{\pi N}} e^{-(x^2/2N\lambda^2)}$$

What we desired was the probability that a particle will be in the interval Δx centered at x at some time t after beginning the random walk. Specifying the time t is equivalent to specifying the value of N, as these are related according to Eqn [1.6.8] as $N = t/t_c$, where t_c is the time between collisions. We may substitute this value of N into Eqn [1.6.15] to obtain

$$[1.6.16] \quad P_N(m) = P_N\left(\frac{x}{\lambda}\right) = P(m,t) = P\left(\frac{x}{\lambda},t\right) = \sqrt{\frac{2t_c}{\pi t}} e^{-(x^2 t_c/2\lambda^2 t)}$$

The probability the particle will be in the interval Δx centered at x is the sum of the probabilities that the value of $m\lambda$ will fall in this interval. For a given N (which is the same as for a given time, t) values of $m = R - L$ are either all odd or all even. Thus the possible values of displacement are separated by 2λ. The number of values of m consistent with landing in the interval Δx is thus $\Delta x/2\lambda$. If we take $P(x/\lambda, t)$ given in Eqn [1.6.16] as the **average** probability in the interval Δx, then the probability of finding the particle in the interval is

$$[1.6.17] \quad \sum_{m \in \Delta x} P(m,t) = \frac{\Delta x}{2\lambda} P\left(\frac{x}{\lambda},t\right) = \frac{1}{2\lambda} P\left(\frac{x}{\lambda},t\right)\Delta x = \sqrt{\frac{t_c}{2\pi\lambda^2 t}} e^{-t_c x^2/2\lambda^2 t}\Delta x$$

From this result, we define a **probability density function**:

$$[1.6.18] \quad P(x,t) = \sqrt{\frac{t_c}{2\pi\lambda^2 t}} e^{(-t_c x^2/2\lambda^2 t)}$$

When multiplied by the length of the interval, Δx, this function gives the probability of finding the particle in the interval Δx at time t. The **diffusion coefficient** is defined as

$$[1.6.19] \quad D = \frac{\lambda^2}{2t_c}$$

Note that this definition is consistent with the units of the diffusion coefficient of $cm^2 s^{-1}$ that we obtained from Fick's First Law of Diffusion. Using this definition, Eqn [1.6.18] becomes

$$[1.6.20] \quad P(x,t) = \sqrt{\frac{1}{4\pi Dt}} e^{(-x^2/4Dt)}$$

This is still the probability density function for finding a particle in an interval at some time after beginning the random walk, but all of the parameters of the random walk, the mean free path, and the time between collisions are submerged into a single constant, D. Thus the diffusion coefficient derives its values from microscopic characteristics of the diffusing substance. Since we have defined the average velocity as $\langle v \rangle = \lambda/t_c$ (see Eqn [1.6.7]), we see that the diffusion coefficient is related to the square of the average velocity. Thus thermal agitation should increase the average velocity and thereby increase diffusion.

We can use Eqn [1.6.20] to derive an expression for the concentration of solute particles at position x at time t, which we denote as $C(x,t)$. Ordinarily the concentration is the number of particles per unit volume. In our one-dimensional analogue, it is the number of particles per unit length. At any time, t, the number of particles per unit length, Δx, at position x is the sum of all particles which have random walked into the displacement interval from all other areas. This is the probability density function times the initial concentration summed over all intervals. This can be written as

$$[1.6.21] \quad C(x,t) = \int_{-\infty}^{\infty} C_0(x')P(x-x',t)dx'$$

where $C_0(x')$ is the initial concentration at point x'. Here

$$[1.6.22] \quad P(x-x',t) = \sqrt{\frac{1}{4\pi Dt}} e^{-(x-x')^2/4Dt}$$

Eqn [1.6.21] expresses the concentration at position x and time t in terms of the concentration everywhere at an initial time $t = 0$. The initial time $t = 0$ is chosen arbitrarily. We can write

$$[1.6.23] \quad C(x, t+\Delta t) = \int_{-\infty}^{\infty} C(x',t)P(x-x',\Delta t)dx'$$

Let us substitute in $s = x' - x$, so that $C(x',t) = C(x+s,t)$ and $P(x-x',\Delta t) = P(-s,\Delta t) = P(s,\Delta t)$ (because of symmetry in x in the probability distribution function) and $ds = dx'$. Then Eqn [1.6.23] becomes

$$[1.6.24] \quad C(x, t+\Delta t) = \int_{-\infty}^{\infty} C(x+s,t)P(s,\Delta t)ds$$

We let Δt be small, so that the concentration is altered only by local events. We approximate $C(x+s,t)$ as

$$C(x+s,t) = C(x,t) + s\frac{\partial C(x,t)}{\partial x} + \frac{1}{2}s^2\frac{\partial^2 C(x,t)}{\partial x^2} + \cdots$$

[1.6.25]

which is a Taylor's series expansion of $C(x+s,t)$. Inserting Eqn [1.6.25] back into Eqn [1.6.24], we obtain

$$[1.6.26] \quad \begin{aligned} C(x+t,\Delta t) &= C(x,t)\int_{-\infty}^{\infty} P(s,\Delta t)ds \\ &+ \frac{\partial C(x,t)}{\partial x}\int_{-\infty}^{\infty} sP(s,\Delta t)ds \\ &+ \frac{1}{2}\frac{\partial^2 C(x,t)}{\partial x^2}\int_{-\infty}^{\infty} s^2 P(s,\Delta t)ds \end{aligned}$$

The function $P(s,\Delta t)$ is given from Eqn [1.6.22] and our definition of $s = x' - x$ as

[1.6.27] $$P(s, \Delta t) = \sqrt{\frac{1}{4\pi D \Delta t}} e^{-s^2/4D\Delta t}$$

Evaluation of the integrals gives

[1.6.28] $$\begin{aligned}\int_{-\infty}^{\infty} P(s,\Delta t)ds &= 1 \\ \int_{-\infty}^{\infty} sP(s,\Delta t)ds &= 0 \\ \int_{-\infty}^{\infty} s^2 P(s,\Delta t)ds &= 2D\Delta t\end{aligned}$$

We shall not prove these integration results. You should note their meaning, however. The first integration result is the **normalization** of the probability distribution, which means that the particle must be somewhere with the probability equal to 1. The second result is the **average displacement** about the initial zero displacement s. The zero value of the average displacement means that the distribution is symmetrical: displacement to the right and to the left are equally likely in the random walk. Recall by the definition of $s = x' - x$ that s may be both negative and positive, so that the average is zero. The last integration gives the average square displacement. Both positive and negative values of s contribute to s^2. This integration gives the **variance**, or the average squared displacement, for the distribution. For the case of this one-dimensional model of diffusion, the variance is $2D\Delta t$. Inserting these integration results into Eqn [1.6.26] gives

[1.6.29] $$C(x, t+\Delta t) = C(x,t) + \frac{1}{2}\frac{\partial^2 C(x,t)}{\partial x^2} 2D\Delta t$$

This may be rewritten as

[1.6.30] $$\frac{C(x,t+\Delta t) - C(x,t)}{\Delta t} = D\frac{\partial^2 C(x,t)}{\partial x^2}$$

If we take the limit as $\Delta t \to 0$, we recognize the left-hand side of Eqn [1.6.30] as the partial derivative of the concentration with respect to time. We then have

[1.6.31] $$\frac{\partial C(x,t)}{\partial t} = D\frac{\partial^2 C(x,t)}{\partial x^2}$$

This is Fick's Second Law of Diffusion. From the continuity equation, Fick's First Law can be derived. Although Fick's First Law was originally derived on phenomenological grounds and the Second Law followed it from the continuity equation, this derivation shows that it can be done the other way using a quite simple model which nevertheless embodies the main ideas giving rise to diffusion: there are an enormous number of collisions which give rise to a random motion of particles from one region of space to another.

THE TIME FOR ONE-DIMENSIONAL DIFFUSION INCREASES WITH THE SQUARE OF DISTANCE

How long does it take for a given material to diffuse some distance, x? This question is deceptively simple. It is not asking how long it takes the first molecule to get to x, or how long it takes for the concentration at x to reach a particular value. It is asking about the *population* of molecules that are moving from high concentration to low concentration. From Figure 1.6.1, you can see that an initially sharp distribution gradually broadens with time because of diffusion. What we want is a quantitative measure of the *shape* of the concentration profile. The time of diffusion is usually calculated from the **variance** of the Gaussian distribution:

[1.6.32] $$\bar{x}^2 = \int_{-\infty}^{+\infty} x^2 \sqrt{\frac{1}{4\pi D\Delta t}} e^{-x^2/4D\Delta t}\, dx = 2D\Delta t$$

where the elapsed time of diffusion is Δt. For two-dimensional diffusion, the variance is $4D\Delta t$; for three-dimensional diffusion it is $6D\Delta t$. The elapsed time for diffusion, Δt, is the time taken for the inflection point of the distribution to move from $x = 0$ to $x = x$, given an initially sharp distribution at $x = 0$. The time taken to diffuse a given distance, x, is the mean square displacement divided by $2D$. Alternatively, we can calculate the distance of diffusion in time Δt as

[1.6.33] $$x = \sqrt{\bar{x}^2} = \sqrt{2D\Delta t}$$

For distances smaller than the cell (0–10 μm), diffusion takes less than a ms up to a few ms. For distances on the order of the diameter of muscle cells (40–100 μm), diffusion takes several seconds.

DIFFUSION COEFFICIENTS IN CELLS ARE LESS THAN THE FREE DIFFUSION COEFFICIENT IN WATER

If the initial concentration of a substance is very narrow, then the subsequent distribution some time later due to diffusion will be given by Eqn [1.6.20] as

[1.6.34] $$C(x,t) = C_0 \sqrt{\frac{1}{4\pi Dt}} e^{-x^2/4Dt}$$

This equation can be shown to obey Fick's Second Law of Diffusion, Eqn [1.6.31], which is left as an exercise for the student (see Problem 15 in Problem Set 1.2). Dividing Eqn [1.6.34] by C_0 and taking the logarithm of both sides, we obtain:

[1.6.35] $$\begin{aligned}\ln\frac{C(x,t)}{C_0} &= \ln\sqrt{\frac{1}{4\pi Dt}} + \ln e^{-x^2/4Dt} \\ &= -\frac{x^2}{4Dt} - \ln(2\sqrt{\pi Dt})\end{aligned}$$

Kushmerick and Podolsky (Ionic mobility in cells, *Science* **166**:1297–1298, 1969) microinjected a 3–6 mm segment of muscle fiber from the semitendinosus muscle of the frog with small amounts of tracer materials, and then allowed the materials to diffuse along the fibers for various periods of time. They immersed the fibers in oil to avoid diffusion through the water phase outside the muscle. After the prescribed period, the muscles were dehydrated in acetone, stained, embedded in paraffin,

> ### Example 1.6.1 Time of One-Dimensional Diffusion
>
> The diffusion coefficient for Ca^{2+} in water is $D_{Ca} = 0.8 \times 10^{-5}$ cm^2 s^{-1}; for calbindin, a 9500-Da protein that binds Ca^{2+} with high affinity, $D_P = 0.12 \times 10^{-5}$ cm^2 s^{-1}. How long does it take for Ca^{2+} or calbindin to diffuse $x = \{0.1, 1, 10, 20, 50, 100\ \mu m\}$?
>
> Here we use Eqn [1.6.32] to calculate $\Delta t = x^2/2D$. As a representative calculation, we find Δt for Ca^{2+} for 0.1 μm:
>
> $$\Delta t = \frac{(0.1 \times 10^{-6}\ m \times 10^2\ cm\ m^{-1})^2}{2 \times 0.8 \times 10^{-5} cm^2\ s^{-1}} = 6.25 \times 10^{-6}\ s$$
>
> which is amazingly quick. The other times can be calculated by the same method, and we can fill out Table 1.6.1.
>
> Variation of this calculation was performed by A.V. Hill in 1949 to show that a diffusable substance could not be the activator of muscle cells because the delay between nervous excitation and contraction of the muscle was too fast, only a few ms.
>
> **TABLE 1.6.1** The Calculated Time for One-Dimensional Diffusion of Free Ca^{2+} and Calbindin in Water for Various Distances
>
> | Distance (μm) | Calcium $D_{Ca} = 0.8 \times 10^{-5}$ cm^2 s^{-1} | Calbindin $D_P = 0.12 \times 10^{-5}$ cm^2 s^{-1} |
> |---|---|---|
> | 0.1 | 6.3×10^{-6} | 42×10^{-6} |
> | 1.0 | 0.6×10^{-3} | 4.2×10^{-3} |
> | 10.0 | 62.5×10^{-3} | 420×10^{-3} |
> | 20.0 | 0.25 | 1.67 |
> | 50.0 | 1.56 | 10.4 |
> | 100.0 | 6.25 | 41.7 |

and cut into 25 μm sections, which were then counted. The plots of ln (counts at distance x/total counts) were linear with the square of the distance, as predicted by Eqn [1.6.35], and the diffusion coefficient in the muscle was estimated from the slope. An example of their results for ^{42}K is shown in Figure 1.6.4. Kushmerick and Podolsky found that the diffusion coefficient for most substances they injected was about one-half of the free water diffusion coefficient. These materials included K^+, Na^+, SO_4^{-2}, sorbitol, sucrose, and ATP. For Ca^{2+}, however, the apparent diffusion coefficient was about 50 times less than the free water diffusion. From this they concluded that Ca^{2+} was retarded by interaction with fixed components within the muscle cell. It is important to recognize that this effect is a reduction in the *apparent* diffusion coefficient because the total Ca^{2+} is being partitioned between a freely diffusable form and a bound or fixed form. Thus the apparent diffusion coefficient is reduced because they calculated the diffusion from the total concentration and not just the free concentration.

EXTERNAL FORCES CAN MOVE PARTICLES AND ALTER THE DIFFUSIVE FLUX

The mathematical relations describing the diffusion of nonelectrolytes have been presented to you in the form of Fick's First and Second Laws of Diffusion. These expressions were derived from the one-

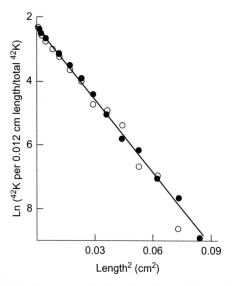

FIGURE 1.6.4 Typical results for the Kushmerick and Podolsky experiment. The longitudinal distribution of ^{42}K$^+$ at 0.02 cm intervals after diffusion for 320 s at 20°C was plotted as a logarithmic transform of the diffusion equation for an infinite slab from an infinitely thin distribution: ln (counts at distance x/ total cts) = $-x^2/4Dt - \ln 2(\pi Dt)^{1/2}$.

dimensional random walk, which considered that the probability of making a step to the right and to the left was the same. Under some circumstances, this is not true. For example, diffusive flux is altered when a bulk flux of fluid occurs simultaneously with the diffusive flux. In this case, we write

$$[1.6.36] \quad J_s(x,t) = -D\frac{\partial C(x,t)}{\partial x} + J_v(t)C(x,t)$$

The first term on the right describes the diffusive flux and the second term describes the flux of solute due to **solvent drag**. Here J_v is the **volume flux**, equal to the volume of fluid moving across an area per unit area per unit time. This volume times the concentration of solute will give the amount of solute moving across that area per unit area per unit time. J_v is equal to the velocity of fluid flow: the volume flux is $V/(A\Delta t) = (A\Delta x)/(A\Delta t) = \Delta x/\Delta t$, the velocity of fluid flow. Eqn [1.6.36] is the **convection–diffusion equation**, because the bulk flow is described as **convection**.

There are other circumstances which alter the flux from that described by Fick's First Law. These circumstances occur when there are external forces applied to the solute particles. Examples of these forces include electrical forces and gravitational forces. Recall in Chapter 1.3 that we considered that electrical forces accelerate ions in solution until they reach a terminal velocity in which the electrical force is balanced by the drag force. The drag force, $\mathbf{F}_d$, is proportional to the terminal velocity. We wrote:

$$[1.6.37] \quad \begin{aligned} \mathbf{F}_e &= ze\mathbf{E} \\ \mathbf{F}_d &= -\beta \mathbf{v} \\ \mathbf{F}_e + \mathbf{F}_d &= 0 \\ \mathbf{F}_e &= \beta \mathbf{v} \end{aligned}$$

where β is the frictional coefficient or drag coefficient. It is given as

$$[1.6.38] \quad \beta = \frac{kT}{D}$$

as originally proposed by Einstein in 1905 (see Appendix 1.6.A1). Here k is Boltzmann's constant, the ideal gas constant divided by Avogadro's number: $k = R/N_0$, and D is the diffusion coefficient. This result makes sense: a large diffusion coefficient is usually associated with small particles, and these would have a small frictional coefficient, encountering less resistance to movement. Incorporating this definition of β into the last equation in Eqn [1.6.37], we have

$$[1.6.39] \quad \mathbf{F}_e = \frac{kT}{D}\mathbf{v}$$

In Chapter 1.3, we also established that the ratio of $\mathbf{J}$ to C defines an average velocity of movement of particles:

$$[1.6.40] \quad \mathbf{J} = \mathbf{v}C$$

Substituting for $\mathbf{v}$ from Eqn [1.6.40] into Eqn [1.6.39] and rearranging, we obtain

$$[1.6.41] \quad \mathbf{J} = \frac{D}{kT}\mathbf{F}_e C$$

What this equation means is that, in the absence of a concentration gradient, an external force will produce a flux that is linearly related to the magnitude of the force per particle and the concentration, with a coefficient related to the diffusion coefficient. We have derived this for an electric force, but the result is completely general for any external force.

In the presence of a concentration gradient, we expect the diffusive flux to add to the flux caused by the application of an external force. From Fick's First Law of Diffusion and Eqn [1.6.41], we obtain, for one dimension, Fick's First Law of Diffusion for solutes subjected to an external force:

$$[1.6.42] \quad J = -D\frac{\partial C}{\partial x} + \frac{D}{kT}fC$$

This can be written in vector notation as

$$[1.6.43] \quad \mathbf{J} = -D\nabla C + \frac{D}{kT}C\mathbf{F}$$

THE STOKES–EINSTEIN EQUATION RELATES THE DIFFUSION COEFFICIENT TO MOLECULAR SIZE

Stokes showed that for a spherical particle, the drag force was related to its size and to the viscosity of the medium:

$$[1.6.44] \quad \mathbf{F}_d = -\beta v = -6\pi\eta a_s \mathbf{v}$$

where η is the **viscosity** of the fluid and a_s is the radius of the sphere. Clearly, Stokes derived an expression for the frictional coefficient. From Eqns [1.6.38] and [1.6.44], we can write

$$[1.6.45] \quad \begin{aligned} 6\pi\eta a_s &= \frac{kT}{D} \\ D &= \frac{kT}{6\pi\eta a_s} \end{aligned}$$

The last expression is the **Stokes–Einstein equation**. It indicates that the diffusion coefficient for a spherical particle should be a function of its radius, the absolute temperature, and the viscosity of the fluid in which it is diffusing.

Recall earlier that Kushmerick and Podolsky found that the apparent diffusion coefficient of readily diffusable substances like K^+, Na^+, and sucrose inside cells was about one-half of their diffusion coefficients in water. From the Stokes–Einstein equation, it would seem that the variable most likely responsible for these decreases in diffusion coefficients is the viscosity of the medium. Thus the cytoplasm appears to be a watery environment, where most small molecular weight materials are free to diffuse but with reduced diffusion coefficients owing to the greater viscosity of the cytoplasm. The **tortuosity** of the path is not included in this analysis. This refers to the blockade of direct diffusion by large structures in the cytoplasm, including organelles and cytoskeleton. Because materials cannot diffuse in a straight line, the diffusion apparently takes longer because the actual path length in the microscopic domain is larger than the apparent path length. In fact, tortuosity and increased viscosity are both partial explanations for the reduced diffusion coefficient inside cells.

SUMMARY

Solutes move by diffusion from regions of high concentration to regions of low concentration. Fick's First Law of Diffusion states that the flux is proportional to the negative of the gradient of C:

$$J_s = -\frac{D\,\partial C}{\partial x}$$

where J_s is the solute flux, D is the diffusion coefficient, and $\partial C/\partial x$ is the one-dimensional gradient. The continuity equation states that changes in concentration with time must be due to changes in flux with distance:

$$\frac{\partial C}{\partial t} = -\frac{\partial J_s}{\partial x}$$

Fick's Second Law of Diffusion derives from his First Law and the Continuity Equation:

$$\frac{\partial C}{\partial t} = \frac{D\,\partial^2 C}{\partial x^2}$$

Fick's Second Law can be derived from a random walk model of diffusion in which molecules take large numbers of small steps. Using Stirling's approximation, the discrete binomial probability distribution can be converted to a continuous one, resulting in a Gaussian probability distribution. For a narrow starting distribution at time $t = 0$, the distribution of solute at time t is given as

$$C(x,t) = C_0\sqrt{1/4\pi Dt}\; e^{-x^2/4Dt}$$

From the random walk, D is identified as $\lambda^2/2t_c$, where λ is the distance between collisions and t_c is the time between collisions. The time of diffusion is typically estimated as the variance of the Gaussian distribution, which gives

$$t = \frac{x^2}{2D}$$

Although diffusion is a statistical result, it is equivalent to a force in that it produces a flow of material. Other forces can also make solutes move. These forces include electrical forces on charged solutes, solvent drag (convection), and gravitational forces. An external force applied to solute particles causes a flux given by

$$J = \frac{D}{kT}fC$$

where f is the force per molecule, k is Boltzmann's constant ($k = R/N_0$), T is the absolute temperature, and C is the concentration. In the presence of a concentration gradient, the total diffusive flux in the presence of an external force is

$$J = -\frac{D\,\partial C}{\partial x} + \frac{DfC}{kT}$$

Stokes derived an equation for the drag force on a spherical object. Einstein combined this with his expression for the drag force in terms of the diffusion constant and gave us the Stokes–Einstein equation:

$$D = \frac{kT}{6\pi\eta a_s}$$

where η is the viscosity of the medium in which diffusion occurs and a_s is the radius of a spherical solute.

REVIEW QUESTIONS

1. Why is diffusive flux proportional to the negative of the gradient and not the gradient?
2. What are the assumptions of the random walk derivation?
3. What are the units of the diffusion coefficient?
4. How does the time of diffusion vary with distance?
5. Why are diffusion coefficients slower in the cytoplasm of cells than in water?
6. What is solvent drag?
7. What is meant by convective flow?
8. How does the diffusion coefficient vary with molecular size? With viscosity of the medium? With temperature? What is Boltzmann's constant?
9. How is Fick's First Law of Diffusion altered in the presence of additional forces acting on the diffusing particles?

APPENDIX 1.6.A1 DERIVATION OF EINSTEIN'S FRICTIONAL COEFFICIENT FROM MOMENTUM TRANSFER IN SOLUTION

Here we consider a right cylindrical volume V of cross-sectional area A and thickness Δx, so that $V = A\Delta x$. We imagine that solute particles may move with velocity $+v$ in the x-direction (to the right) and velocity $-v$ (velocity v to the left), as shown in Figure 1.6.A1.1. Let the number of particles in V with velocity v be $N_+(t)$ and the number with velocity $-v$ be $N_-(t)$. The concentration of particles with velocity $+v$ or $-v$ in V at any time will be given by

[1.6.A1.1]
$$\frac{N_+(t)}{V} = C_+(t)$$
$$\frac{N_-(t)}{V} = C_-(t)$$

The number of particles with a given velocity may change with time. This may happen in four different ways: (1) particles with velocity $+v$ may enter the volume element from the left; (2) particles with velocity $+v$ may leave the volume element at the right; (3) particles with velocity $+v$ within the volume V may convert to particles with velocity $-v$ by colliding with solvent particles; (4) particles with velocity $-v$ could convert to velocity $+v$ by collisions with solvent particles. The entry of particles from the left is given by

[1.6.A1.2]
$$Q_i(x) = v\,A\,C_+(x)$$

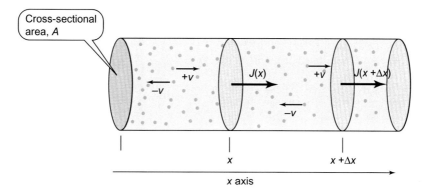

FIGURE 1.6.A1.1 Solutes within a hypothetical volume. Solutes have a velocity $+v$ in the positive x-direction or $-v$ in the opposite direction. All particles have velocity $+v$ or $-v$, although only a few are shown.

where we recognize that vC_+ is the flux of particles with velocity $+v$. The exit of particles at the right is

[1.6.A1.3] $$Q_o(x) = v A C_+(x + \Delta x)$$

The rate of conversion of N_+ to N_- is proportional to N_+ within the volume V, with the proportionality constant having dimensions of reciprocal time. This proportionality constant is $1/t_c$, where t_c is the time between collisions. This conversion reduces N_+ within V, so we may write:

[1.6.A1.4] $$\left(\frac{\partial N_+}{\partial t}\right)_{-\text{collision}} = -\frac{N_+}{t_c} = \frac{V C_+}{t_c}$$

The rate of conversion of N_- to N_+ adds to the value of N_+ and is given by a similar expression:

[1.6.A1.5] $$\left(\frac{\partial N_+}{\partial t}\right)_{+\text{collision}} = \frac{N_-}{t_c} = \frac{V C_-}{t_c}$$

The total net change of N_+ is given by the sum of Eqns [1.6.A1.2, 1.6.A1.3, 1.6.A1.4, 1.6.A1.5]:

$$\left(\frac{\partial N_+}{\partial t}\right) = vA\, C_+(x) - vA\, C_+(x + \Delta x) + \frac{V}{t_c}(C_- - C_+)$$

[1.6.A1.6]

where v means velocity and V means volume. We can approximate $C_+(x + \Delta x)$ by the first two terms of a Taylor's series expansion:

[1.6.A1.7] $$C_+(x + \Delta x) = C_+(x) + \Delta x \left(\frac{\partial C_+}{\partial C}\right) + \cdots$$

Insertion of Eqn [1.6.A1.7] into Eqn [1.6.A1.6] gives

$$\left(\frac{\partial N_+}{\partial t}\right) = vA\left[C_+(x) - C_+(x) - \Delta x\left(\frac{\partial C(x)}{\partial x}\right)\right] + \frac{V}{t_c}(C_- - C_+)$$

$$\left(\frac{\partial N_+}{\partial t}\right) = -vA\Delta x\left(\frac{\partial C_+(x)}{\partial x}\right) - \frac{V}{t_c}(C_+ - C_-)$$

[1.6.A1.8]

Since $A\Delta x = V$, we write

[1.6.A1.9]
$$\frac{1}{V}\frac{\partial N_+}{\partial t} = -v\frac{\partial C_+}{\partial x} - \frac{1}{t_c}(C_+ - C_-)$$

$$\frac{\partial C_+}{\partial t} = -v\frac{\partial C_+}{\partial x} - \frac{1}{t_c}(C_+ - C_-)$$

By completely analogous reasoning, we can determine the rate of change in C_- in V as

[1.6.A1.10] $$\frac{\partial C_-}{\partial t} = v\frac{\partial C_-}{\partial x} + \frac{1}{t_c}(C_+ - C_-)$$

Since the total concentration, $C(x)$ is the sum of C_+ and C_-, then

[1.6.A1.11] $$\frac{\partial C(x)}{\partial t} = \frac{\partial C_+(x)}{\partial t} + \frac{\partial C_-(x)}{\partial t}$$

and substituting into Eqn [1.6.A1.11] from Eqns [1.6.A1.9] and [1.6.A1.10], we obtain

$$\frac{\partial C(x)}{\partial t} = -v\frac{\partial C_+}{\partial x} - \frac{1}{t_c}(C_+ - C_-) + v\frac{\partial C_+}{\partial x} + \frac{1}{t_c}(C_+ - C_-)$$

$$= -v\left(\frac{\partial C_+}{\partial x} - \frac{\partial C_-}{\partial x}\right)$$

$$= -\frac{\partial(v(C_+ - C_-))}{\partial x}$$

[1.6.A1.12]

Now the net flux across any area element within the volume V can be described as the difference between two unidirectional fluxes:

[1.6.A1.13] $$\begin{aligned}J &= J_+ - J_- \\ &= v\,C_+ - v\,C_- \\ &= v\,(C_+ - C_-)\end{aligned}$$

Substitution of this relation into Eqn [1.6.A1.12] gives

[1.6.A1.14] $$\frac{\partial C(x)}{\partial t} = -\frac{\partial J(x)}{\partial x}$$

which is the **continuity equation** we derived earlier (see Eqn [1.2.9]). The purpose of deriving the continuity equation in this way was to familiarize you with this method of accounting for all of the particles and to impress upon you that the collisions with the solvent involving changes in the velocity canceled. These collisions did not affect the total number of particles since a particle contributing to N_+ before collision still contributes to N_- after collision.

Now let us consider what happens to the total momentum, P, of the particles in volume V. Let m be the mass of each solute particle. The total momentum of the particles is

[1.6.A1.15] $$P = (mv)N_+ + (-mv)N_-$$

Because N_+ and N_- change with time, so does P. We write

[1.6.A1.16] $$\frac{\partial P}{\partial t} = mv\left(\frac{\partial N_+}{\partial t}\right) - mv\left(\frac{\partial N_-}{\partial t}\right)$$

Since $N_+ = VC_+$ and $N_- = VC_-$, we have

[1.6.A1.17] $$\frac{\partial P}{\partial t} = mvV\frac{\partial C_+}{\partial t} - mvV\frac{\partial C_-}{\partial t}$$

Inserting our earlier results from Eqns [1.6.A1.9] and [1.6.A1.10], we obtain

[1.6.A1.18] $$\frac{1}{V}\frac{\partial P}{\partial t} = -mv^2\left[\frac{\partial C_+}{\partial x} + \frac{\partial C_-}{\partial x}\right] - \frac{2mv}{t_c}(C_+ - C_-)$$

Recalling that $C = C_+ + C_-$, and using Eqns [1.6.A1.11] and [1.6.A1.13], we obtain

[1.6.A1.19] $$\frac{1}{V}\frac{\partial P}{\partial t} = -mv^2\frac{\partial C}{\partial x} - \frac{2m}{t_c}J$$

The two terms on the right-hand side of Eqn [1.6.A1.19] have specific interpretations. The first term, $-mv^2 \partial C/\partial x\, V$, represents the net flow of momentum carried by particles with velocity $+v$ or $-v$. The second term, $-2m/t_c\, JV$, is the average force exerted by the solvent particles on the solute particles. To see these interpretations, we write the flow of momentum across area A at any point x within the volume V as

$$mv(vC_+(x))A + (-mv)(-vC_-(x))A = mv^2(C_+(x) + C_-(x))A$$
[1.6.A1.20]

The net flow of momentum into volume V across the boundaries between x and $x + \Delta x$ is

$$mv^2(C_+(x) + C_-(x))A - mv^2(C_+(x+\Delta x) + C_-(x+\Delta x))A$$
[1.6.A1.21]

Since $C(x) = C_+(x) + C_-(x)$, the expression in Eqn [1.6.A1.21] becomes

[1.6.A1.22] $$mv^2 A[C(x) - C(x+\Delta x)]$$

Expanding $C(x+\Delta x)$ as $C(x) + \Delta x\, \partial C/\partial x$, the expression in Eqn [1.6.A1.22] becomes

[1.6.A1.23] $$-mv^2\frac{\partial C}{\partial x}\Delta x A = -mv^2\frac{\partial C}{\partial x}V$$

Thus, in Eqn [1.6.A1.19], $-mv^2\, \partial C/\partial x$ is that part of $1/V\, \partial P/\partial t$ which is due to the net momentum change of the particles in V due to what they carried into or out of the volume.

For the second term in Eqn [1.6.A1.19], note that every time a particle moving with velocity $+v$ collides with a solvent obstacle, its velocity becomes $-v$, and thus it experiences a net momentum change of $-2mv$. The total number of such changes per unit time in the volume V is N_+/t_c. So the time-averaged change for the momentum change per unit time due to this collision is $-2mv\, N_+/t_c$. In a similar way, particles traveling with velocity $-v$ experience a momentum change of $+2mv$ upon collision with the solvent and the time-averaged total change in momentum per unit time for this type of collision in the volume V is $+2mv\, N_-/V$. The time-averaged rate of net change in momentum for these types of collisions is just their sum:

$$\frac{\partial P_c}{\partial t} = F_c = -2mv\frac{N_+}{t_c} + 2mv\frac{N_-}{t_c}$$

[1.6.A1.24] $$= F_c = -\frac{2mv}{t_c}V(C_+ - C_-)$$

$$\frac{1}{V}\frac{\partial P_c}{\partial t} = \frac{F_c}{V} = -\frac{2m}{t_c}J$$

where the subscript c denotes that the momentum change and force are due to collisions of the solute molecules with the solvent. Thus the second term in Eqn [1.6.A1.19] is identified as the change in momentum produced by collisions with the solvent. It is equivalent to a force per unit volume exerted by the solvent particles on the solute particles. Equation [1.6.A1.24] can be rewritten as a differential equation in $J(x,t)$. Recall that the total momentum, P, of the solute particles in the volume V is given by Eqn [1.6.A1.15]; substituting in for the definition of the concentration (Eqn [1.6.A1.1] and (J) Eqn [1.6.A1.13]), we have:

$$P = (mv)N_+ + (-mv)N_-$$
[1.6.A1.25] $$= mV\,v(C_+ - C_-)$$
$$P = mV\,J(x,t)$$

Insertion of this result into Eqn [1.6.A1.19] gives

[1.6.A1.26] $$\frac{1}{V}mV\frac{\partial J(x,t)}{\partial t} = -mv^2\frac{\partial C(x,t)}{\partial x} - \frac{2m}{t_c}J(x,t)$$

This can be rearranged to

[1.6.A1.27] $$\frac{\partial J(x,t)}{\partial t} = \frac{2}{t_c}\left[-\frac{t_c v^2}{2}\frac{\partial C(x,t)}{\partial x} - J(x,t)\right]$$

Eqn [1.6.A1.27] describes the buildup of $J(x,t)$ in time. We will pay particular attention to the situation where the concentration gradient, $\partial C(x,t)/\partial x$, is constant and steady-state flux is achieved. Steady-state flux means that $\partial J(x,t)/\partial t = 0$. That is, the flux no longer changes with time. Under these circumstances, by Eqn [1.6.A1.27], we obtain the steady-state flux as

$$[1.6.A1.28] \qquad J = -\frac{t_c v^2}{2}\frac{\partial C}{\partial x}$$

Comparing this to Fick's First Law of Diffusion,

$$[1.6.A1.29] \qquad J = -D\frac{\partial C}{\partial x}$$

we can identify

$$[1.6.A1.30] \qquad D = \frac{t_c v^2}{2}$$

In the one-dimensional random walk model of diffusion, we defined the diffusion coefficient to be

$$[1.6.A1.31] \qquad \begin{aligned} D &= \frac{\lambda^2}{2t_c} \\ &= \frac{(vt_c)^2}{2t_c} \\ &= \frac{t_c v^2}{2} \end{aligned}$$

Thus the derivation performed here is completely consistent with the random walk model of diffusion. Eqn [1.6.A1.24] describes the momentum change of the solute particles that result from the collisions with solvent. It is given per unit volume as

$$[1.6.A1.32] \qquad \frac{1}{V}\frac{\partial P_c}{\partial t} = \frac{F_c}{V} = -\frac{2m}{t_c}J$$

In this equation, F_c is the force per unit volume on the solute particles. This is the same as the drag force on the solute particles when they move through the solution. If we imagine that the particles are subjected to a uniform external force, then the particles will accelerate until they reach a terminal velocity, v. The force F_C will be the sum of all of the drag forces on the particles within the volume, which is the drag force per particle times the number of particles in the volume. The flux, J, will be given as vC. Eqn [1.6.A1.32] can then be rewritten as

$$[1.6.A1.33] \qquad \frac{f_c N}{V} = -\frac{2m}{t_c}vC$$

where f_C is the force on a single particle, N is the number of particles in the volume, V, v is the terminal velocity, and C is the concentration. Since $N/V = C$, the equation is further simplified to

$$[1.6.A1.34] \qquad f_c = -\frac{2m}{t_c}v$$

Here the force f_C is equal to the **drag force** on the particle traveling at velocity v, and the coefficient is the drag or **frictional coefficient**:

$$[1.6.A1.35] \qquad F_D = -\beta v$$

which allows us to identify the drag coefficient as

$$[1.6.A1.36] \qquad \beta = \frac{2m}{t_c}$$

The **equipartition theorem** of thermodynamics gives

$$[1.6.A1.37] \qquad \begin{aligned} \frac{1}{2}mv^2 &= \frac{1}{2}kT \\ v^2 &= \frac{kT}{m} \end{aligned}$$

where k is **Boltzmann's constant** ($=1.38\times 10^{-23}$ J K^{-1}, which is equal to the gas constant, R, divided by Avogadro's number). Insertion of this result into the equation for the diffusion coefficient, Eqn [1.6.A1.31] gives

$$[1.6.A1.38] \qquad \begin{aligned} D &= \frac{t_c kT}{2m} \\ t_c &= \frac{2mD}{kT} \end{aligned}$$

Insertion of this into Eqn [1.6.A1.36] gives

$$[1.6.A1.39] \qquad \beta = \frac{2m}{t_c} = \frac{2m}{\frac{2mD}{kT}} = \frac{kT}{D}$$

This is Einstein's frictional coefficient, given as

$$[1.6.A1.39] \qquad \beta = \frac{kT}{D}$$

Recall Eqn [1.6.A1.19] reproduced here:

$$[1.6.A1.19] \qquad \frac{1}{V}\frac{\partial P}{\partial t} = -mv^2\frac{\partial C}{\partial x} - \frac{2m}{t_c}J$$

substituting in $v^2 = kT/m$ from Eqn [1.6.A1.37] and $t_c = 2mD/kT$ from Eqn [1.6.A1.38], we obtain

$$[1.6.A1.40] \qquad \frac{1}{V}\frac{\partial P}{\partial t} = -kT\frac{\partial C}{\partial x} - \frac{kT}{D}J$$

Recall that this equation describes the rate of momentum change of the **solute** per unit volume V. It has two components: the first is the net momentum carried into the volume by the diffusing solutes, and the second is the momentum change produced upon collisions with the solvent. Note that if $C(x)$ is decreasing with increasing x then the gradient, $\partial C/\partial x$, will be negative and the first term on the right-hand side of Eqn [1.6.A1.40] will be positive. In this case of diffusion of solute towards increasing values of x, J will be positive and the second term, denoting the change in momentum by collisions with solvent, will be negative. This second term denotes the drag of solvent on solute movement. Steady-state diffusion occurs when $\partial P/\partial t$ is zero and $\partial C/\partial x$ is constant. Under these conditions, Eqn [1.6.A1.40] becomes Fick's First Law of Diffusion.

Suppose now that we add an additional force to the solute particles in the volume V. Let the force acting on each particle be f. In the volume element, there are

$N(x) = C(x) V$ particles. The total force acting on these particles is $f C(x) V$. This force clearly contributes to the rate of change of the momentum of the solute particles in the volume. Eqn [1.6.A1.40] becomes

$$[1.6.A1.41] \qquad \frac{1}{V}\frac{\partial P}{\partial t} = -kT\frac{\partial C}{\partial x} - \frac{kT}{D}J + fC$$

To recapitulate, the first term on the right-hand side of Eqn [1.6.A1.41] represents the rate of momentum change by particles entering or leaving the volume V; the second term is due to collisions with the solvent particles; the third term is due to the action of some external force. At steady state, $\partial J/\partial t = 0$, and this implies by Eqn [1.6.A1.25] that $\partial P/\partial t$ also is zero. Under this constraint, Eqn [1.6.A1.41] gives

$$[1.6.A1.42] \qquad J = -D\frac{\partial C}{\partial x} + \frac{D}{kT}fC$$

This is Fick's First Law of Diffusion for solutes subjected to an external force, which is Eqn [1.6.42]. Thus this equation, which forms the basis of the derivation of the electrochemical potential, can be derived using momentum transfer of solutes in a solution and the equipartition theorem of thermodynamics.

Electrochemical Potential and Free Energy 1.7

Learning Objectives

- Write Fick's First Law of Diffusion and explain how a concentration gradient makes a flux
- Describe how an external force such as electrostatic force can make a flux of charged solute
- Write the formula for the electrochemical potential
- Explain the formula for electrochemical potential in terms of the component driving forces
- Explain how the driving forces that produce a flux equal the negative gradient of the electrochemical potential
- Define the term **free energy**
- Describe the relationship between the free energy and the direction of any process
- Be able to calculate the free energy of ATP hydrolysis under specified conditions of temperature and concentrations of reactants ATP, ADP, and Pi
- Write the relationship between the standard free energy change and the equilibrium constant
- Know the approximate value for the free energy of ATP hydrolysis under cellular conditions

DIFFUSIVE AND ELECTRICAL FORCES CAN BE UNIFIED IN THE ELECTROCHEMICAL POTENTIAL

Let us recapitulate what we learned in Chapters 1.3 and 1.6 about the movement of charged particles in solution. First, the relationship between the flux and the concentration defines an average velocity, given as

[1.7.1] $$J = vC$$

where J is the one-dimensional flux, v is the velocity, and C is the concentration. Second, the presence of a concentration gradient, in the absence of any other forces, produces a flux given by Fick's First Law of Diffusion:

[1.7.2] $$J = -D\frac{\partial C}{\partial x}$$

where D is the diffusion coefficient. Here we drop the vector notation because the equation used here is one dimensional and the direction is assumed to be along the x-axis. If we use the gradient of C, instead of the derivative, we retain the three-dimensional vector that describes diffusive flux in three dimensions.

Third, in the absence of a concentration gradient, any external force acting on solute particles causes them to be accelerated until they reach a terminal velocity. At this terminal velocity, the external force is balanced by the drag force on the particle by the solvent. This drag force is proportional to the velocity. Because of this, there is a relationship between the flux and the external force, given as

[1.7.3] $$J = \frac{D}{kT}Cf$$

where f is the force acting on the solute particles, per particle. The concentration gradient has some of the appearances of a force in that it causes a flux. So does an externally applied force. What we seek to do is to combine diffusion and other forces into a single equivalent force. When both are operating, the flux is given as

[1.7.4] $$J = -D\frac{\partial C}{\partial x} + \frac{D}{kT}Cf$$

If the force is an electrical force, its magnitude is given by

[1.7.5] $$F = F_E = ze\mathrm{E}$$

where z is the valence of the particle ($\pm$ integer), e is the charge on the electron, and E is the electric field. Assuming one dimension, we drop the vector notation for force and insert Eqn [1.7.5] into Eqn [1.7.4] to give

[1.7.6] $$J = -D\frac{\partial C}{\partial x} + \frac{D}{kT}Cze\mathrm{E}$$

Thus we have these two equations:

[1.7.7] $$J = \frac{D}{kT}Cf$$
$$J = -D\frac{\partial C}{\partial x} + \frac{D}{kT}Cze\mathrm{E}$$

The top equation says that there is a flux produced by some unknown force. The bottom equation says that now we know what these forces are and we have parceled them out. Part of the flux is caused by diffusion and part of the flux is caused by an electrical force. What we desire is an expression for the total force in the top equation that will produce the flux in the bottom equation. To unite the diffusive force and the electrical force, we solve these two equations for the unknown force, f, and find that

$$[1.7.8] \quad f = -\frac{kT}{C}\frac{\partial C}{\partial x} + ze\mathbf{E}$$

The electric field, $\mathbf{E}$, is the electric force per unit charge. This electric force is the negative gradient of its potential (see Chapter 1.3 for a definition of potential, gradients, and conservative forces). In a three-dimensional model, the electric field is a vector and the potential is a scalar. We will ignore these realities in this one-dimensional model because in one dimension the force and gradient have a single direction. Using Ψ as the symbol for electrical potential energy, we can rewrite Eqn [1.7.8] as

$$[1.7.9] \quad f = -\frac{kT}{C}\frac{\partial C}{\partial x} - ze\frac{\partial \Psi}{\partial x}$$

The units of Ψ are volts. The force in this equation is the force per particle. It is customary to convert this to the force per mole of particles by multiplying by Avogadro's number:

$$[1.7.10] \quad \begin{aligned} F = N_0 f &= -\frac{N_0 kT}{C}\frac{\partial C}{\partial x} - zN_0 e\frac{\partial \Psi}{\partial x} \\ F &= -\frac{RT}{C}\frac{\partial C}{\partial x} - z\Im\frac{\partial \Psi}{\partial x} \end{aligned}$$

where R is the gas constant ($=8.314$ J mol^{-1} K^{-1}) and $\Im$ is the Faraday (9.649×10^4 C mol^{-1} = 6.02×10^{23} electrons mol$^{-1} \times 1.6 \times 10^{-19}$ C electron^{-1}). We *define* an **electrochemical potential** so that the overall force (produced by both diffusion and electrical forces, in this case) is the negative derivative of the electrochemical potential:

$$[1.7.11] \quad F = -\frac{\partial \mu}{\partial x}$$

Combining Eqns [1.7.10] and [1.7.11], we get

$$[1.7.12] \quad -\frac{\partial \mu}{\partial x} = -\frac{RT}{C}\frac{\partial C}{\partial x} - z\Im\frac{\partial \Psi}{\partial x}$$

Integrating, we obtain

$$[1.7.13] \quad \mu = \mu^0 + RT \ln C + z\Im\Psi$$

This is the **electrochemical potential**. We will use it to make calculations about membrane potential and the energetics involved in physiological processes. It is a potential in the same sense as the electrical potential. It has the units of energy per mole. The electrical potential, in volts, is equivalent to joules per coulomb, which can be converted to units of energy per mole. Similarly, the electrochemical potential can be converted into volts.

As can be seen from Eqn [1.7.13], the electrochemical potential has three components. The first component, μ^0, has the sense of a constant of integration. It sets the zero of the electrochemical potential and its reference is a standard state. For solutions, μ^0 refers to the chemical potential of a hypothetical solution of unit molarity and no potential. That is, when $C = 1$ and $\Psi = 0$, $\mu = \mu^0$

by Eqn [1.7.13]. The second term, $RT \ln C$, refers to the work necessary to concentrate the solute, per mole; the third term, $z\Im\Psi$, is an electrical work term. It is the work necessary to bring one mole of particles from zero potential to Ψ. If other kinds of work are involved, we would need to expand Eqn [1.7.13] to include the other work terms.

In a typical physiological solution, several kinds of particles are dissolved. Since their concentrations are nearly independent of one another, they each have an identifiable electrochemical potential. Each solute within a local region, on the other hand, experiences the same electrical potential. Therefore, we can write an electrochemical potential for each material in the solution:

$$[1.7.14] \quad \mu_i = \mu_i^0 + RT \ln C_i + z_i \Im \Psi$$

THE OVERALL FORCE THAT DRIVES FLUX IS THE NEGATIVE GRADIENT OF THE ELECTROCHEMICAL POTENTIAL

The flux produced by a concentration difference and electric potential difference is given by

$$[1.7.3] \quad \mathbf{J} = \frac{D}{kT}C\mathbf{f}$$

where $\mathbf{J}$ and $\mathbf{f}$ are both vectors. $\mathbf{f}$ is the force per molecule. We convert to molar dimensions by multiplying and dividing by Avogadro's number:

$$[1.7.15] \quad \begin{aligned} \mathbf{J} &= \frac{D}{N_0 kT} C N_0 \mathbf{f} \\ \mathbf{J} &= \frac{D}{RT} C \mathbf{F} \end{aligned}$$

This emphasizes that $\mathbf{J}$ is a vector in the same direction as $\mathbf{F}$, the force on the solute particles per mole. The force per mole is, in turn, given by the negative gradient of the electrochemical potential:

$$[1.7.16] \quad \mathbf{F} = -\nabla \mu = -\left[\mathbf{i}\frac{\partial \mu}{\partial x} + \mathbf{j}\frac{\partial \mu}{\partial y} + \mathbf{k}\frac{\partial \mu}{\partial z}\right]$$

Inserting in μ from Eqn [1.7.13] and F into Eqn [1.7.15], we get

$$[1.7.17] \quad \begin{aligned} \mathbf{J} = \frac{D}{RT}C\Bigg[&-\frac{RT}{C}\left(\mathbf{i}\frac{\partial C}{\partial x} + \mathbf{j}\frac{\partial C}{\partial y} + \mathbf{k}\frac{\partial C}{\partial z}\right) \\ &- z\Im\left(\mathbf{i}\frac{\partial \Psi}{\partial x} + \mathbf{j}\frac{\partial \Psi}{\partial y} + \mathbf{k}\frac{\partial \Psi}{\partial z}\right)\Bigg] \end{aligned}$$

which is simplified to

$$[1.7.18] \quad \mathbf{J} = -D\left[\nabla C + \frac{z\Im}{RT}C\nabla\Psi\right]$$

The one-dimensional version of this equation is

$$[1.7.19] \quad J = -D\frac{\partial C}{\partial x} - \frac{D}{RT}z\Im C\frac{\partial \Psi}{\partial x}$$

THE ELECTROCHEMICAL POTENTIAL IS THE *GIBBS FREE ENERGY* PER MOLE

Throughout the derivation of Fick's Laws of Diffusion, we made an unstated assumption that the pressure and temperature of the diffusing particles were constant. Under these conditions, we could derive the electrochemical potential. The constraints of constant temperature and pressure are useful when applied to problems in mammalian physiology, where these conditions are usually met. Under these conditions, it is useful to define a thermodynamic variable called the **Gibbs free energy**:

[1.7.20] $$G = E + PV - TS$$

where E is the internal energy, P is the pressure, V is the volume, T is the temperature, and S is the **entropy**. The internal energy consists of all of the myriads of movements of the particles, including their internal motions. This is the part of the free energy where chemical bonding energy is stored. When compounds undergo chemical transformations, energy is either released or stored, depending on whether the reaction is exothermic or endothermic, respectively. The entropy has a specific thermodynamic definition which has been shown to be related to the number of ways that the particles can be arranged in the system which are consistent with the state of the system (its temperature, pressure, volume, and number of particles). This is related to the probability of finding the state for randomly arranged particles. Thus states with high probabilities have high entropy. Highly organized states, which can be accomplished in only a few ways, have a low probability and a low entropy. For example, a concentrated solution has only a few ways to crowd all the solute particles together compared to a dilute solution, so a concentrated solution has less entropy than a dilute solution.

As mentioned above, the Gibbs free energy is used to describe transformations that occur at constant temperature and under constant pressure. It is a **state variable**, meaning that it depends only on the state of the system and not on the path required to get there. Thus there is a defined difference in the free energy between an initial state and a final state:

[1.7.21] $$\Delta G = G_{final} - G_{initial}$$

The energy involved in a chemical transformation can be harnessed to do useful work. For example, a reaction that gives off heat can be used to expand a gas to move a piston, producing work. The relationship between the free energy and the work that can be accomplished by a chemical transformation is

[1.7.22] $$-\Delta G \geq W$$

where W is the work energy. That is the decrease in free energy in any transformation at constant T and P is equal to the maximum amount of work that can be performed by that transformation. It is for this reason that the function G is called the "free energy"; it is the energy which is available to perform work. The work that can be performed could be mechanical or electrical or chemical or concentration work.

Another variable determining the state of a system is its composition. Suppose that the system is composed of a number of substances, $i, j, k, \ldots$ of mole amounts $n_i, n_j, n_k, \ldots$ If a small amount of substance i is added, then the system will experience a change in its free energy, G. The chemical potential is

[1.7.23] $$\mu_i = \left(\frac{\partial G}{\partial n_i}\right)_{T,P,n_{j \neq i}}$$

Thus the chemical potential is the free energy per mole. The Gibbs free energy is an extensive variable that increases with the volume of the system or the number of moles of materials in the system. The chemical potential is an intensive variable, meaning that it depends on the state of the system and not its extent. However, usually ΔG values are calculated per mole so that G and μ are often used interchangeably. Equation [1.7.23] can be integrated to give

[1.7.24] $$G = \sum_i \mu_i n_i$$

THE SIGN OF ΔG DETERMINES THE DIRECTION OF A REACTION

One of the conclusions of thermodynamics is that natural processes always occur in such a way that the free energy of the universe decreases. That is, in all natural processes, ΔG in Eqn [1.7.21] is negative. A corollary of this conclusion is the condition of equilibrium. Equilibrium occurs when no further change can occur. In this case, the free energy has reached a minimum, and the free energy change is zero with respect to this process. On the other hand, if ΔG is positive, then the reverse reaction will occur spontaneously:

$\Delta G < 0 \Rightarrow$ spontaneous reaction
$\Delta G = 0 \Rightarrow$ reaction is at equilibrium
$\Delta G > 0 \Rightarrow$ opposite reaction occurs spontaneously

[1.7.25]

PROCESSES WITH $\Delta G > 0$ CAN PROCEED ONLY BY LINKING THEM WITH ANOTHER PROCESS WITH $\Delta G < 0$

As described earlier, a natural process can proceed only if $\Delta G < 0$ for that process. Many processes that occur in biological systems require energy, meaning that $\Delta G > 0$. These processes cannot proceed on their own. They can be made to proceed, however, by linking the process for which $\Delta G > 0$ with another process for which $\Delta G < 0$. The combined processes will proceed spontaneously only if the global ΔG for both of them is less than zero. We will consider these kinds of processes in detail when we consider active transport in Chapter 2.6. "Proceed spontaneously" here means that the linked processes will occur with no further input of energy outside the combined processes. It does not mean that the processes will occur rapidly or slowly. Thermodynamics tells us whether or not a process will occur, and with what energy changes, but it does not speak of the kinetics of the processes. In this sense, the term "thermodynamics"

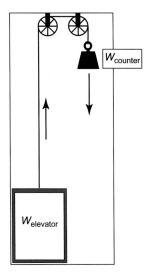

FIGURE 1.7.1 Coupling of a spontaneous process to a nonspontaneous process. An elevator sits on the ground floor. It will not rise spontaneously because the free energy for that process is positive: $\Delta G_{elevator} > 0$. A large counterweight at the top floor will fall spontaneously; the free energy change for this process is negative: $\Delta G_{counter} < 0$. The elevator can be raised if it is coupled to the larger negative free energy of the counterweight's fall: $\Delta G_{elevator} + \Delta G_{counter} < 0$. In this case, the elevator can be raised only once. Repetition of the process would require coupling to yet another process that provides the free energy to raise the counterweight back up again. In a real process, some of the free energy for the counterweight's fall would be dissipated. One can never recover 100% of the energy as useful work.

is something of a misnomer, as really it is about the statics of energy transformation.

An example of coupling is lifting a weight against gravity by using another weight, as shown in Figure 1.7.1.

The coupling of biological process can be visualized using the energy diagrams for chemical reactions such as those shown in Chapter 1.5. Reactions that involve a decrease in free energy ($G_{final} - G_{initial} < 0$ and thus $G_{final} < G_{initial}$) are called **exergonic** reactions, and they proceed spontaneously. A schematic example of an exergonic reaction is shown in Figure 1.7.2A. Reactions that involve an increase in free energy ($G_{final} - G_{initial} > 0$ and thus $G_{final} > G_{initial}$) are called **endergonic** reactions, and they do not proceed spontaneously (see Figure 1.7.2B). If an exergonic reaction can be coupled to an endergonic reaction, and the sum of the ΔG for the two reactions is less than zero, then the combined processes will both proceed spontaneously (see Figure 1.7.2C).

THE LARGE NEGATIVE FREE ENERGY OF ATP HYDROLYSIS POWERS MANY BIOLOGICAL PROCESSES

ATP is adenosine triphosphate and its structure is shown in Figure 1.7.3. It occupies a special position in the cellular flow of energy because of the energy stored in its terminal phosphate bond. It takes a lot of energy to add a phosphate to ADP to form ATP, and that chemical energy becomes available to do work when the bond is split. The cell uses ATP as its energy currency. Energy derived from the oxidation of foodstuffs is stored in this terminal phosphate bond, and it is used in myriad reactions involving synthesis of materials, transport of materials, and the performance of mechanical work. The hydrolysis reaction of ATP to ADP and Pi is shown in Figure 1.7.3.

The free energy change for the ATP hydrolysis reaction is given by a combination of Eqns [1.7.21] and [1.7.24]:

$$\Delta G = G_{final} - G_{initial}$$
$$= n_{final\ ATP}\mu_{ATP} + n_{final\ ADP}\mu_{ADP} + n_{final\ Pi}\mu_{Pi}$$
$$\quad -n_{initial\ ATP}\mu_{ATP} - n_{initial\ ADP}\mu_{ADP} - n_{initial\ Pi}\mu_{Pi}$$
$$= \Delta n_{ATP}\mu_{ATP} + \Delta n_{ADP}\mu_{ADP} + \Delta n_{Pi}\mu_{Pi}$$
$$= \Delta n_{rxn}\mu_{ADP} + \Delta n_{rxn}\mu_{Pi} - \Delta n_{rxn}\mu_{ATP}$$

[1.7.26]

where the subscripts denote the chemical species, μ is the chemical potential, and Δn is the number of moles of each material that participates in the reaction. The last line in Eqn [1.7.26] relates the change in the number of moles of participating reactants to the number of completed reactions. This is just the stoichiometry of the reaction. We may write it as follows:

[1.7.27]
$$1\ ATP = 1\ ADP + 1\ Pi$$
$$-1\ ATP + 1\ ADP + 1\ Pi = 0$$

The coefficients here are the stoichiometry. Here they indicate that, for every completed reaction, the number of molecules of ATP decreases by one and the number of molecules of ADP and Pi increases by one. Thus the last line in Eqn [1.7.26] relates the change in free energy upon completion of a Δn number of reactions.

This number can be specified in moles because, as described in Chapter 1.5, using moles is just another way of counting a large number of things.

We divide the last line of Eqn [1.7.26] by the number of completed reactions to obtain

[1.7.28]
$$\frac{\Delta G}{\Delta n_{rxn}} = \mu_{ADP} + \mu_{Pi} - \mu_{ATP}$$

Substituting in for the chemical potentials, we get

[1.7.29a]
$$\frac{\Delta G}{\Delta n} = \mu^0_{ADP} + RT\ln[ADP] + \mu^0_{Pi} + RT\ln[Pi]$$
$$\quad - \mu^0_{ATP} - RT\ln[ATP]$$

[1.7.29b]
$$\frac{\Delta G}{\Delta n} = \Delta\mu^0 + RT\ln\left(\frac{[ADP][Pi]}{[ATP]}\right)$$

If we will let $\Delta n = 1.0$ to indicate that we are speaking of the free energy change for Avogadro's number of completed reactions, then we have

[1.7.30]
$$\Delta G = \Delta G^0 + RT\ln\left(\frac{[ADP][Pi]}{[ATP]}\right)$$

where ΔG^0 is the **standard free energy change**. It refers to the **free energy change per mole** under standard

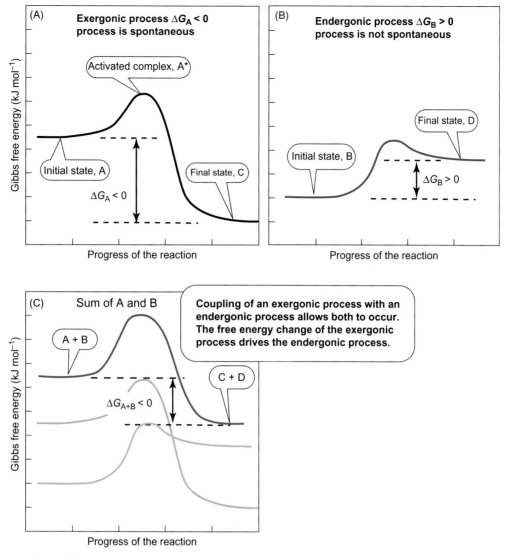

FIGURE 1.7.2 Coupling of an endergonic reaction with an exergonic reaction. Exergonic reactions are those for which $\Delta G_A = G_{final} - G_{initial} < 0$, and so these reactions proceed spontaneously (panel A). Endergonic reactions are those for which $\Delta G_B = G_{final} - G_{initial} > 0$, and these reactions do not proceed spontaneously (panel B). However, an endergonic reaction can be made to proceed if it can obtain a decrease in free energy by linking it to the exergonic reaction. In essence, a coupled reaction involves a completion of both the exergonic and endergonic reactions. If the combined $\Delta G_{A+B} = \Delta G_A + \Delta G_B < 0$, then the combined reaction will occur spontaneously (panel C).

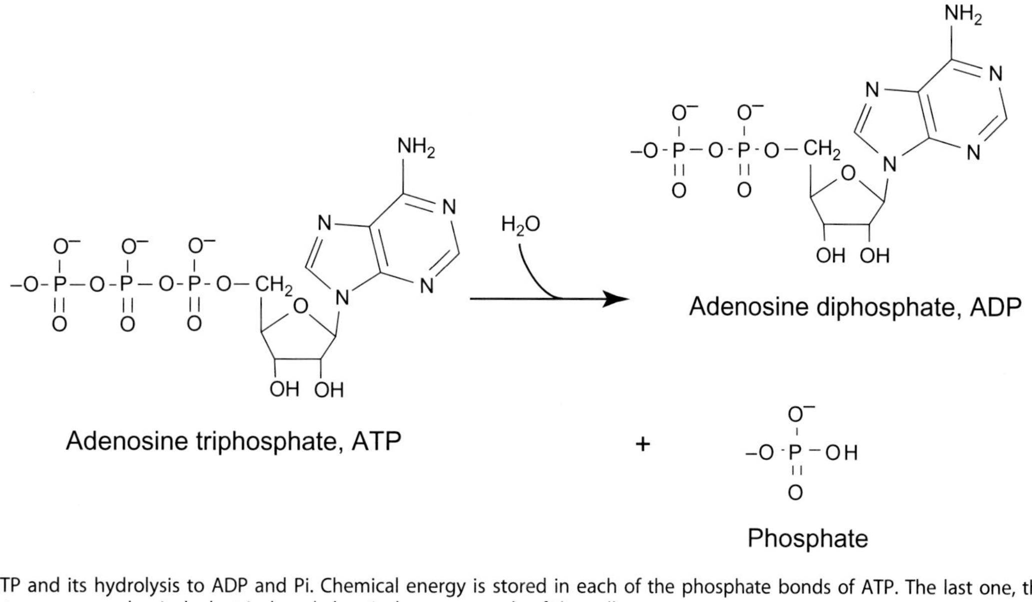

FIGURE 1.7.3 ATP and its hydrolysis to ADP and Pi. Chemical energy is stored in each of the phosphate bonds of ATP. The last one, the γ-phosphate, is typically used to power mechanical, electrical, and chemical energy needs of the cell.

> **EXAMPLE 1.7.1 Free Energy of ATP Hydrolysis Under "Typical" Cell Conditions**
>
> "Typical" values for the concentrations of ATP, ADP, and Pi are about 5 mM for ATP, 5 mM for Pi, and 40 µM for ADP. Calculate the free energy of ATP hydrolysis under these conditions.
>
> Here we use Eqn [1.7.33]. The value of $\Delta G^{0'}$ is given in the text as -7.4 kcal mol^{-1} or -30.9 kJ mol^{-1}. The value of R we use is 8.314 J mol^{-1} K^{-1} and $T = 37°C = 310$ K. **In the calculation, all concentrations must be expressed in M**. We insert these values and calculate:
>
> $$\Delta G = -31.0 \text{ kJ mol}^{-1} + 8.314 \text{ J mol}^{-1} \text{ K}^{-1} \times 310 \text{ K}$$
> $$\times \ln[40 \times 10^{-6} \text{ M} \times 5 \times 10^{-3} \text{ M} / 5 \times 10^{-3} \text{M}]$$
>
> $$= -31.0 \text{ kJ mol}^{-1} + 2.58 \text{ kJ mol}^{-1} \times (-10.12)$$
> $$= \mathbf{-57.1 \text{ kJ mol}^{-1}} = -13.6 \text{ kcal mol}^{-1}$$
>
> This free energy of ATP hydrolysis under cell conditions is the energy available for the various kinds of work undertaken by the cell, including chemical work (synthesis), mechanical work (movement and transport), and electrical work. In the final analysis, nearly all of the work produced by the cells is eventually degraded and appears as heat.

conditions of 25°C, 1 atmosphere pressure and unit concentration. Note that we have said all along that ΔG is an extensive variable, and now it seems that we have transformed it into an intensive variable. Tabulated values of ΔG^0 necessarily report it in units of energy per mole, so these values are actually values of $\Delta \mu^0$.

You can see that if all species were at unit concentration, the second term on the right-hand side of Eqn [1.7.25] would be zero. Under these conditions, $\Delta G = \Delta G^0$. Experimental determination of ΔG^0 takes advantage of the fact that at equilibrium $\Delta G = 0$ (see Eqn [1.7.25]) and the fact that the second term in Eqn [1.7.30] incorporates the equilibrium constant for the reaction:

[1.7.31] $$K_{eq} = \left(\frac{[ADP][Pi]}{[ATP]}\right)$$

So that at equilibrium ($\Delta G = 0$), Eqn [1.7.30] becomes

[1.7.32] $$\Delta G^0 = -RT \ln K_{eq}$$

MEASUREMENT OF THE EQUILIBRIUM CONCENTRATIONS OF ADP, ATP, AND PI ALLOWS US TO CALCULATE ΔG^0

It turns out that the free energy of ATP hydrolysis is a bit more complicated than we have let on here. The chemical species produced have different ionizations at different pH values, and ATP and ADP both bind Mg^{2+} ions. The energy of binding of H^+ and Mg^{2+} ions should be incorporated into the reaction. In addition, we are interested in the hydrolysis of ATP under physiological conditions of 37°C. The free energy of ATP hydrolysis has been determined for these different conditions. Since the details of these conditions differ from cell to cell, no one value can be used. However, a "typical" value is given a special symbol, $\Delta G^{0'}$, and signifies the free energy of ATP hydrolysis under the "typical" cell conditions. **Its value is about -7.4 kcal mol^{-1} or -31.0 kJ mol^{-1}**.

The units of ΔG are those of RT. Values of R usually used here are 1.987 cal mol^{-1} K^{-1} = 8.314 J mol^{-1} K^{-1}.

The conversion between joule and calorie is 1 J = 0.239 cal.

Under cellular conditions, [ATP], [ADP], and [Pi] are not at their equilibrium values, nor are they unit concentrations. The free energy under these conditions is given by

[1.7.33] $$\Delta G = \Delta G^{0'} + RT \ln\left(\frac{[ADP][Pi]}{[ATP]}\right)$$

It is important to realize that the argument of the logarithm ([ADP][Pi]/[ATP]) is generally not equal to the equilibrium constant, but only when the reaction is at equilibrium.

SUMMARY

A difference of concentration produces a flow of material in solution. Application of an electric force to charged particles also produces a flow of material in solution. These two forces can be united in the definition of a single force that is proportional to the negative gradient of a potential. This potential is the electrochemical potential written as

$$\mu = \mu^0 + RT \ln C + z\Im\Psi$$

This electrochemical potential is the Gibbs free energy per mole. The Gibbs free energy, G, is an extensive variable, whereas μ is an intensive variable. G is also a state variable, depending only on the state of a system and not on the path it took to reach that state. The free energy is the maximum energy that can be extracted to do useful work. For all spontaneous reactions, the change in free energy, $\Delta G = G_{final} - G_{initial}$, is negative. Processes that require work occur spontaneously only if they are linked to other processes that lose free energy, so that for the overall process $\Delta G < 0$. For any process at equilibrium, $\Delta G = 0$.

Many cellular processes require energy for the synthesis of materials, transport, or mechanical or electrical work. These occur because energy is supplied by the hydrolysis of ATP. ATP hydrolysis to ADP and Pi has a large negative ΔG under cellular conditions. The free energy of ATP hydrolysis per mole is given as

$$\Delta G_{\text{ATP hydrolysis}} = \Delta G^{0'} + RT \ln\left(\frac{[\text{ADP}][\text{Pi}]}{[\text{ATP}]}\right)$$

Under "typical" cellular conditions, $\Delta G_{\text{ATP hydrolysis}} = -13.6$ kcal mol^{-1} = -57.1 kJ mol^{-1}.

REVIEW QUESTIONS

1. In the formula for the electrochemical potential, what is R? What is T? What are the units of $RT \ln C$? What units must C be in? What is z? What is the Faraday? What is Ψ?
2. What is the Gibbs free energy?
3. What is the relationship between the electrochemical potential and the Gibbs free energy?
4. How does the sign of ΔG determine the direction of a process?
5. Describe thermodynamic coupling.
6. What is the relationship between the free energy change and the equilibrium constant?
7. How would you determine the free energy change for ATP hydrolysis under cellular conditions?
8. What is the approximate free energy change of ATP hydrolysis in cells?

1.2 Problem Set
Kinetics and Diffusion

1. A. The empirical formula of glucose is $C_6H_{12}O_6$. What is its molecular weight?
   B. Isotonic glucose is 5% (w/v) glucose. How much glucose would we need to make 100 mL of isotonic glucose?
2. A. You need to make 250 mL of a stock solution of 0.1 M Na_2 ATP. Its formula weight is 605.2 g mol^{-1}. How much Na_2 ATP should you weigh out?
   B. Your advisor is skeptical of your abilities. He wants you to check out the 0.1 M ATP solution and tells you to do it spectrophotometrically. Spectrophotometry relies on the different abilities of chemicals to absorb light of specific wavelengths. A diagram of a spectrophotometer is shown in Figure 1.PS2.1.

At particular wavelengths, chemicals absorb light according to their chemical structure and their concentration. The law governing the absorption of light is the Beer–Lambert Law:

$$A = \varepsilon C d$$

where A is the absorbance; ε is a constant that depends on the chemical and typically varies with the wavelength of light—it is the **molar extinction coefficient** and is in units of M^{-1}; C is the concentration of the chemical (in M); and d is the path length. The molar extinction coefficient is defined for a path length of 1 cm. The absorbance is defined as

$$A = \log(I_0/I)$$

where I_0 is the incident light intensity and I is the transmitted light intensity. Your advisor tells you that $\varepsilon_{259} = 15.4 \times 10^3 \, M^{-1}$; this is the molar extinction coefficient of ATP at a wavelength of incident light of 259 nm. He tells you to make a dilution of the stock by taking 25 µL of the stock solution and diluting it to 100 mL. What absorbance do you expect of the final diluted solution, if you made it up correctly, at $\lambda = 259$ nm?

3. A. The molecular weight of ryanodine is 493.54 g mol^{-1}. You want to make 10 mL of a 10-mM stock solution. How much ryanodine should you weigh out?
   B. You make a dilution of the 10-mM ryanodine stock by pipetting 10 µL of the stock solution into a 10-mL volumetric flask and adding

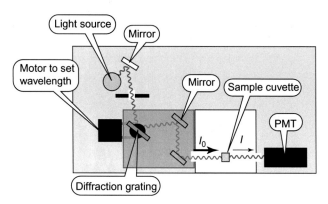

FIGURE 1.PS2.1 Light path in a single beam spectrophotometer. The view is from above. Light from a source is collimated (making a narrow beam) and passed through a monochromator that selects a narrow band of wavelength of light to be passed through the sample. A photomultiplier tube (PMT) detects the light and measures its intensity. Comparison of this intensity, I, to the intensity when the sample is missing, I_0, allows the calculation of the absorbance. Absorbance is recorded with time or as a function of wavelength.

water to the mark. You measure the absorbance as a function of wavelength (against a water blank, using a standard 1 cm path length optical cell) and find a peak at 271 nm with an absorbance of 0.179. What is ε_{271} for ryanodine? (See Problem #2 for a discussion of the Beer–Lambert Law and a definition of the molar extinction coefficient.)

4. A. Magnesium chloride has a formula of $MgCl_2 \cdot 6H_2O$. What is its formula weight?
   B. You desire to make 1 L of 0.1 $MgCl_2$ solution. How much $MgCl_2 \cdot 6H_2O$ should you weigh?
   C. You need to make 25 mL of a 25-mM solution of $MgCl_2$. How much of the 0.1 M stock solution do you add to the 25 mL volumetric flask?
5. The extracellular fluid volume varies with the size of the person. Suppose in an individual we determine that the ECF is 14 L. The average [Na^+] in the ECF is about 143 mM.
   A. What is the total amount of Na^+ in the ECF, in moles? in grams?
   B. Suppose this person works out and sweats 1.5 L with an average [Na^+] of 50 mM. During this time the urine output is 30 mL

with an average [Na$^+$] of 600 mM. How much Na$^+$ is lost during the workout?

C. If the person does not drink fluids at all during the workout, what will be the [Na$^+$] in the plasma at the end of the workout? Assume that all of the fluid in the sweat and urine originated from the ECF.

6. The body normally produces about 2 g of creatinine per day. The amount varies with individuals and is approximately proportional to the muscle mass. It is excreted through the kidneys according to urinary excretion of creatinine = GFR × plasma concentration of creatinine, where GFR is an abbreviation for "glomerular filtration rate." If the GFR is 120 mL min^{-1}, what is the plasma concentration of creatinine at steady state? *Hint*: Assume the body is at steady state with respect to creatinine.

7. Just before noon, your plasma glucose concentration was 100 mg dL^{-1}. This plasma glucose is approximately evenly distributed among 3.5 L of plasma and 10.5 L of interstitial fluid that comprises your 14 L of ECF. Glucose is readily distributed in both compartments. You drink a can of soda that contains 35 g of glucose.
   A. How much would your blood glucose rise if all the glucose in the soda was absorbed and none of it was metabolized?
   B. Given that postprandial (after eating) *increases* in blood glucose amount to maybe 40 mg dL^{-1}, depending on the meal, over a period of an hour, give a crude estimate of the rate of glucose uptake by the peripheral tissues. Assume that the meal contains 100 g of carbohydrates and all of it is absorbed in 1 hour.

8. The association reaction for Ca^{2+} and EGTA (a chemical that binds Ca^{2+}) is written as

$$Ca^{2+} + EGTA \rightleftharpoons Ca \cdot EGTA$$

Under defined and particular conditions of temperature and ionic mixture, the association constant was determined to be $K_A = 2.52 \times 10^6$ M^{-1}. In a chemical mixture, 400 μM total EGTA was included and the free [Ca^{2+}] determined by a Ca^{2+}-selective electrode was found to be 4×10^{-7} M. Assuming that there are no other binding agents for Ca^{2+}, what is the total [Ca^{2+}] in the mixture?

9. 2,4-Dinitrophenyl acetate decomposes in alkaline solution with a pseudo-first-order rate constant of 11.7 s^{-1} at 25°C. It is a "pseudo"-first-order rate constant because it depends on the pH.
   A. If the initial concentration of DNPA is 1 mM, what is its concentration after 15 seconds?
   B. At what time is the concentration reduced to 0.5 mM (i.e., what is the half-life of the reaction)?
   C. After 5 minutes of reaction, what is the concentration of DNPA?

10. The following data were obtained for the rate of the Mg, Ca-ATPase activity of vesicles of cardiac sarcoplasmic reticulum as a function of temperature. What can you tell about the activation energy?

Temperature (°C)	ATPase Rate (μmol min^{-1} mg^{-1})
6.9	0.068
11.5	0.138
15.8	0.300
19.8	0.568
20.2	0.585
25.6	1.236
26.1	1.154
31.0	2.238
34.8	3.030
39.2	4.220

11. Superoxide reduces cytochrome C in the reaction

$$Cyt\ C \cdot Fe^{3+} + O_2^- \Rightarrow Cyt\ C \cdot Fe^{2+} + O_2$$

where Cyt C·Fe^{3+} is the oxidized form and Cyt C·Fe^{2+} is the reduced form of cytochrome C. The reaction can be followed spectrophotometrically at 550 nm. The extinction coefficient for the reduced form of cytochrome C is $\varepsilon_{RED} = 2.99 \times 10^4$ M^{-1} and the extinction coefficient for the oxidized form $\varepsilon_{OX} = 0.89 \times 10^4$ M^{-1} (V. Massey, The microestimation of succinate and the extinction coefficient of cytochrome C. *Biochimica et Biophysica Acta*, 34:255–256, 1959). See Problem #2 for a discussion of extinction coefficients and spectrophotometry. When xanthine oxidase converts xanthine to uric acid, it produces superoxide that can be measured using cytochrome C reduction. The following data were obtained for A_{550}:

Time (min)	A_{550}
0	0.1326
1	0.1478
2	0.1637
3	0.1791
4	0.1941
5	0.2073
6	0.2202

   A. Calculate the rate of cytochrome C reduction.
   B. The xanthine oxidase was added in 75 μL of 6.5 mg XO per mL into a 3-mL reaction mixture. Calculate the specific activity of cytochrome C reduction (moles of cytochrome C reduced per min per mg of XO protein).

12. You suspect you are anemic and your physician orders some tests. He finds that your hemoglobin is 13 g%. The molecular weight of hemoglobin is 66,500 g mol^{-1}.
   A. What is the concentration of hemoglobin in molar in your blood?
   B. Each hemoglobin binds four oxygen molecules. If the hemoglobin is saturated with oxygen, what is the concentration of O_2 bound to Hb, in molar?

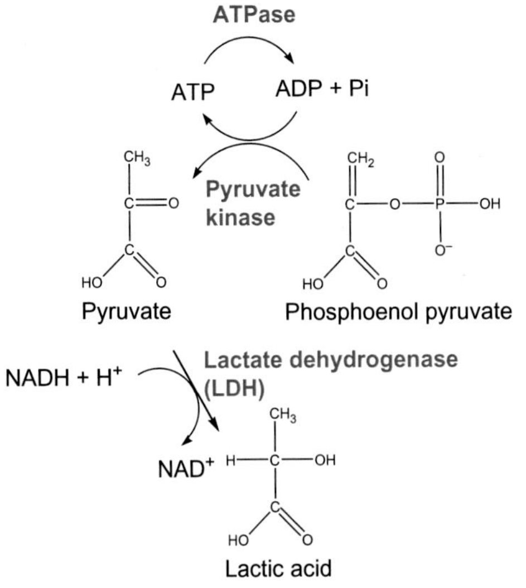

FIGURE 1.PS2.2 ATP hydrolysis by pyruvate kinase converts phosphoenolpyruvate to pyruvic acid. This is coupled by lactate dehydrogenase to the conversion of pyruvic acid to lactic acid and conversion of NADH to NAD^+. The progress of the reaction can be followed spectrophotometrically by the change in absorbance of NADH.

TABLE 1.PS2.1 Diffusion Coefficients and M_r for a Variety of Proteins

Protein	Molecular Weight	$D \times 10^7$ (cm^2 s^{-1})
Milk lipase	6600	14.5
Metallothionein	9700	12.4
Cytochrome C	12,000	12.9
Ribonuclease	12,600	13.1
Myoglobin	16,890	11.3
Chymotrypsinogen	23,200	9.5
Carbonic anhydrase	30,600	10.0
Peroxidase II	44,050	6.8
Albumin	68,500	6.1
Lactoperoxidase	92,620	6.0
Aldolase	149,100	4.6

C. Convert the answer in B to volume using the ideal gas equation, $PV = nRT$, where T is the absolute temperature, $R = 0.082$ L atm mol^{-1} K^{-1}, V is the volume that we seek, and $P = 1$ atm. The conditions for volume of gas are usually STPD—standard temperature and pressure, dry. The standard temperature is 0°C and pressure is 1 atm.

13. The rate of ATP hydrolysis by ATPases can be followed by the coupled enzyme assay shown in Figure 1.PS2.2. The progress of the reaction can be followed by A_{340}. The extinction coefficient of NAD^+ at 340 nm is negligible. The extinction coefficient of NADH at 340 nm is 6.2×10^3 M^{-1}. See Problem #2 for a discussion of extinction coefficients and spectrophotometry. In one reaction, the concentration of Ca-ATPase was 0.22 mg mL^{-1} and A_{340} was 0.65 at $t = 0$ min and 0.455 at $t = 2.0$ min. What is the activity of the Ca-ATPase in units of μmol min^{-1} mg^{-1}?

14. Show by representative calculations that Stirling's formula

$$n! = \sqrt{2\pi n}\, e^{n \ln(n'-n)}$$

is a good approximation for $n!$. Use $n = 1, 2, 3, 4, 5$.

15. Show that the equation

$$C(x,t) = C_0 \sqrt{\frac{1}{4\pi Dt}}\, e^{\frac{-x^2}{4Dt}}$$

obeys Fick's Second Law of Diffusion.

16. The intestinal enterocytes form a covering over the intestinal lining which, to the first approximation, can be considered to be a plane. Assuming no binding or sequestration within the cell, what is the estimated time of diffusion of Ca^{2+} across the intestinal enterocyte? The length of the enterocyte is 20 μm and assume that the effective diffusion coefficient of Ca^{2+} is about 0.4×10^{-5} cm^2 s^{-1}.

17. Table 1.PS2.1 lists the diffusion coefficients and the molecular weight of a variety of proteins. What relationship can you deduce between the size and the diffusion coefficients of these soluble proteins? (Hint: regress $\ln D$ against $\ln M_r$). Is the relationship you found consistent with the Stokes–Einstein equation?

18. The free diffusion coefficient of oxygen in aqueous solutions is about 1.5×10^{-5} cm^2 s^{-1}. If the diffusion distance between air and blood is 0.5 μm, about how long is the diffusion time?

19. Suppose a soluble protein has a molecular weight of 45 kDa and a density of 1.06 g cm^{-3}. Suppose further that the viscosity of the cytoplasm has a viscosity of 0.005 Pa s (about five times that of water—there is debate about the viscosity of cytoplasm with numbers varying from 0.001 to over 0.1 Pa s).
    A. Estimate the diffusion coefficient for the protein in the cytoplasm at 37°C.
    B. If the proteins were synthesized in the cell body, or soma, of a neuron in the spinal cord, about how long would it take to diffuse to the axon terminal 75 cm away?

20. Diffusion coefficients in cytoplasm have been estimated by a technique of photobleaching recovery. In this technique, an area of the cytoplasm is irradiated with light to photobleach a fluorescent probe. Recovery of fluorescence in the region is achieved by diffusion of unbleached probes from adjacent areas of the cytoplasm. The translational diffusion coefficient can be estimated from the half-time of fluorescent recovery. (D. Axelrod et al., Mobility measurements by analysis of fluorescence photobleaching recovery kinetics.

Biophysical Journal **16**:1055–1069, 1976.) This technique was applied to estimate the relative viscosity of cytoplasm and nucleoplasm by microinjecting fluorescein isothiocyanate-labeled dextrans of varying molecular sizes and measuring the fluorescence photobleaching recovery (I. Lang et al., Molecular mobility and nucleoplasmic flux in hepatoma cells. *Journal of Cell Biology* **102**:1183–1190, 1986). These authors obtained the following data:

Probe	Molecular Weight (kD)	Equivalent Radius (nm)	D in Dilute Solution	D in Cytoplasm	D in Nucleoplasm
			D is in units of 10^{-6} cm^2 s^{-1}		
FD20	17.5	3.30	0.651	0.080	–
FD40	41.0	4.64	0.463	0.044	0.069
FD70	62.0	5.51	0.390	0.029	0.056
FD150	156.9	9.07	0.237	0.015	0.036

A. Plot D against $1/a$, where a is the molecular radius, for each of the solutions. From the Stokes–Einstein relation, you would expect the resulting curves to pass through the origin of zero diffusion coefficient with infinite radius. Do the curves extrapolate back in this way? Why or why not?

B. Regardless of the intercept, the slope of the plot from part A ought to be related to the viscosity of the medium. Use the slopes to estimate the relative viscosity of the dilute solution, cytoplasm, and nucleoplasm.

UNIT 2

MEMBRANES, TRANSPORT, AND METABOLISM

Cell Structure 2.1

Learning Objectives

- List the main categories of cellular function
- Describe the general structure, location, and function of the plasma membrane
- Describe the composition, location, and function of the cytosol
- Compare the general structure and function of microtubules, microfilaments, and intermediate filaments
- Describe the general structure, location, and function of the nucleus and its envelope
- Describe the general structure, location, and function of the mitochondria
- Describe the general structure, location, and function of the endoplasmic reticulum
- Describe the general structure, location, and function of the Golgi apparatus
- Describe the general structure, location, and function of lysosomes and peroxisomes
- Describe the general structure, location, and function of proteasomes
- Explain the general purpose of ubiquitinylation of proteins
- List the different types of cell-to-cell junctions

FOR CELLS, FORM FOLLOWS FUNCTION

In Chapter 1.1, we discussed the following points:

- Cells are the organization unit of life.
- Cells come in a multitude of forms, specialized for their function.
- The vast majority of cells are somatic cells, and all of these contain the same set of genetic information, present in DNA and organized into genes.
- The multitude of forms comes from using only specific sets of the genes to make proteins.

Multicellular organisms such as ourselves evolved because multicellular structures can provide an internal environment that is more stable than the natural environment, and thereby enhance the survival of the component cells and of the organism. Free-floating, single-celled organisms live at the mercy of environmental conditions, whereas protected, multicellular organisms can better withstand changes in the environment. Single-celled organisms and cells in multicellular organisms must solve a number of problems in order to survive. These include:

- **Catalysis**: Cells must be able to change one metabolite into another in order to synthesize cellular constituents, degrade them, or provide energy.
- **Transport**: Cells must be able to move things from outside the cell to inside or from one compartment to another within the cell. This includes bulk secretion or uptake of materials from the extracellular fluid.
- **Signal transduction**: Cells must have mechanisms for responding to signals from other cells or from within the cell. These may be chemical signals or electrical signals.
- **Recognition**: Cells attach to other cells and to extracellular structures. They must be able to recognize where they should form attachments.
- **Movement**: At some stage of their development, all cells must be able to move so as to position themselves properly within the cellular matrix that makes us up.
- **Control**: All of the activities of the cell must be coordinated. Control here also means that cells must select the parts of the genome that they will use. "Control" thus implies **differentiation**—the formation of specialized cells uniquely suited to their task.
- **Proliferation**: At appropriate times of development, cells must make new cells. This involves cell division and its control.

ORGANELLES MAKE UP THE CELL LIKE THE ORGANS MAKE UP THE BODY

The cells of the body typically are composed of a relatively small number of **organelles** that carry out specific functions of the cell, much like our organs do for us. These are called organelles because their relation to the cell is like the organs' relation to the body. Table 2.1.1 lists these organelles along with their major function. The disposition of these organelles in a "typical" cell is shown in Figure 2.1.1.

THE *CELL MEMBRANE* MARKS THE LIMITS OF THE CELL

The cell membrane, also known as the plasma membrane, defines the inside and outside of the cell. Like all biological membranes, it consists of two lipid

QUANTITATIVE HUMAN PHYSIOLOGY: AN INTRODUCTION

TABLE 2.1.1 Major Organelles of the Cell and Their Function

Organelles	Function
Plasma membrane	**Customs officer of the cell**: determines what gets into or out of the cell, also **signal transduction** and cell recognition
Cytosol	**Cell sap**: the fluid medium in which soluble biochemicals diffuse and in which the other organelles are suspended
Cytoskeleton	Support, movement, and cell attachment
Nucleus	**Command center**: contains the hereditary material and organizes and controls differentiation
Free ribosomes	**Factory of the cell** for soluble proteins
Rough ER	**Factory of the cell** for membrane proteins and secreted proteins
Smooth ER	Synthesis of lipids and steroids
Golgi apparatus	**Shipping department**: finishing and targeting of proteins to specific locations
Mitochondria	**Powerhouse of the cell**: site of oxidation and energy transfer
Lysosomes	**Garbage disposal**: destruction of worn-out organelles
Proteasomes	Destruction of tagged proteins
Peroxisomes	Oxidation of fatty acids and detoxification of xenobiotics

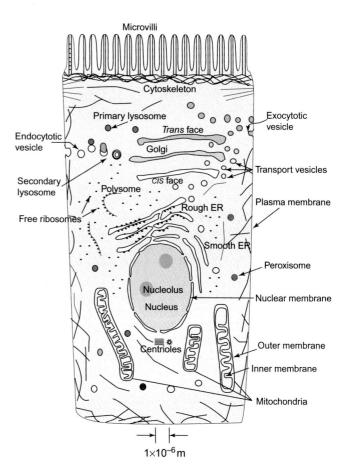

FIGURE 2.1.1 A "typical" human cell showing various subcellular organelles including the plasma or cell membrane, nucleus, nuclear membrane, rough ER, smooth ER, transport vesicles, Golgi apparatus, mitochondria, cytoskeletal elements, lysosomes, peroxisomes, and endocytotic and exocytotic vesicles. Although there is no "typical" cell, most cells contain this set of organelles. Exceptions to this rule include the adult erythrocytes.

layers, or leaflets, in which are embedded a variety of proteins that serve specific functions (see Chapter 2.4 for a discussion of the structure of lipids in biological membranes). Both the inside and outside layers may be coated with carbohydrate units that partly define the function of the lipids. An important function of this membrane is to determine what goes into and what comes out of the cell. The "customs officer" of the cell has several mechanisms that can transport materials into or out of the cell. These include:

- Passive transport
- Active transport
- Exocytosis
- Endocytosis.

We will discuss all of these in more detail in Chapters 2.5 and 2.6. **Passive transport** requires no metabolic energy and may involve diffusion through the lipid layer of the membrane or through water-like channels that are established by proteins that span the membrane. **Active transport** requires the input of cellular metabolic energy and may be primary or secondary. Primary active transport directly couples transport to metabolic energy. Secondary active transport indirectly links transport of a material to metabolic energy. **Exocytosis** and **endocytosis** refer to the movement of materials that are enclosed in **vesicles**. These vesicles are tiny hollow spheres of membrane. Secretory vesicles are filled with some material by the cell. These vesicles can fuse with the plasma membrane, and, in so doing, they dump their contents into the extracellular space. This process is called exocytosis. Endocytosis is similar but occurs in the opposite direction. In this case, parts of the plasma membrane invaginate and pinch off to form endocytotic vesicles inside the cell. Endocytosis of fluid is called **pinocytosis** or "cell drinking"; endocytosis of particulate stuff is called **phagocytosis** or "cell eating."

In addition to these functions, the cell membrane forms the hub of **signal transduction** and **surface recognition**. It must transfer extracellular signals originating from other cells to an intracellular signal inside the cell. This is what is meant by "signal transduction." The cell membrane receives the first messenger in the form of a chemical or electrical signal, and receipt of the first messenger causes the formation of a second messenger inside the cell. Surface recognition occurs through surface proteins that are members of the **major histocompatibility complex** or MHC. These proteins are responsible for beginning transplant rejections by recognizing the transplants as foreign matter.

THE *CYTOSOL* PROVIDES A MEDIUM FOR DISSOLUTION AND TRANSPORT OF MATERIALS

The cell membrane surrounds all the constituents of the cell, which themselves exist in a fluid medium that allows transfer of materials among them. This fluid medium is the **cytoplasm**, which literally means "cell fluid." It includes the **cytosol** and all of the organelles suspended in it. The cytosol itself is the fluid that contains dissolved ions and organic compounds of a bewildering variety. These include amino acids for building proteins, glucose for energy, a tremendous variety of metabolic intermediates, and cytoplasmic enzymes for glycolysis, the first stage in burning carbohydrates for energy. The ionic composition of the cytoplasm varies with different cell types, but Table 2.1.2 gives some reasonable approximate numbers for the "typical" cell. The cytoplasm also serves as a medium for the transmission of control signals from the outer surface of the cell to the interior, and from the nucleus to the rest of the cell. Although we describe it here as a fluid, the cytosol is not like water: ions diffuse through the cytosol slower than they do through water (see Chapters 1.6 and 1.PS2, Problem #20). Cutting a muscle fiber, for example, does not cause its fluid to leak out like water. The cytoplasm is more akin to a gel. Most of this behavior is due to the small volume of fluid and the abundance of cell surfaces. Although bulk water flows, a thin film adheres to any wettable surface. This thin film generally exceeds the thickness of a cell. Thus, at the cell level, surface forces govern much of bulk fluid flow.

THE *CYTOSKELETON* SUPPORTS THE CELL AND FORMS A NETWORK FOR VESICULAR TRANSPORT

Arrays of protein filaments form a network within the cytoplasm. These filaments determine the shape of the cell and provide for the movement of the cell as a whole or for the movement of organelles from one part of the cell to another. There are four major types of filaments comprising the cytoskeleton:

1. Microtubules
2. Intermediate filaments
3. Actin filaments or microfilaments
4. Myosin filaments.

MICROTUBULES ARE THE LARGEST CYTOSKELETAL FILAMENTS

The microtubules are about 25 nm in diameter and are constructed of heterodimers of **tubulin**, a globular polypeptide of 50,000 Da. These dimers are assembled into protofilaments of tubulin dimers with the β-tubulin of one joined to the α-tubulin of the next. The microtubules are assembled from 13 such protofilaments arranged in a cylinder with a hollow core (see Figure 2.1.2).

A number of **microtubule-associated proteins** (MAPs) bind to microtubules. Some of these MAPs are "motor

TABLE 2.1.2 Selected Components of the Cytosol

Component	Concentration
Na^+	14×10^{-3} M
K^+	120×10^{-3} M
Cl^-	10×10^{-3} M
HCO_3^-	10×10^{-3} M
ATP	5×10^{-3} M
ADP	50×10^{-6} M
Mg^{2+}	0.5×10^{-3} M
Ca^{2+}	0.1×10^{-6} M
pH	7.1–7.2
Osmolarity	295 mOsm L^{-1}

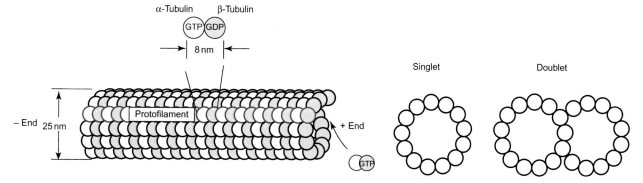

FIGURE 2.1.2 Schematic diagram of the structure of a microtubule. Microtubules consist of protofilaments composed of tubulin dimers, one α-tubulin and one β-tubulin. Both bind GTP, guanosine triphosphate. GTP, like ATP, stores chemical energy in its phosphate bonds. Hydrolysis of GTP to GDP can be used to stabilize protein shapes. The α-tubulin does not hydrolyze its GTP, whereas the β-tubulin hydrolyzes its GTP to GDP; 13 of the protofilaments assemble to form a singlet microtubule. Because of the asymmetrical arrangement of the tubulin monomers, the microtubule has an asymmetry, with a **plus** (+) end and a **minus** (−) end. The plus (+) end is the end pointing away from the origin of the microtubule and is the end to which monomers add to the growing microtubule. These ends differ in the rates of tubulin association and dissociation. Because these rates differ, the microtubule can treadmill—dissolve at one end while lengthening at the other. Tubulin can also form doublet and triplet structures. Cross-sections of a singlet microtubule and a doublet microtubule are shown.

proteins" that can "walk" along the microtubule. Two of these motor proteins are named **kinesin** and **dynein**. Kinesin forms a family of motor proteins with about 40 members. Most of these are " + directed" motors, moving along the microtubules toward the + end. Dyneins comprise a family of − directed motors. These motor proteins can attach vesicles and then move along the microtubules, carrying their vesicles along. In this way, the microtubules can provide a track along which intracellular transport occurs. This is especially important in neurons in which transport must occur down a long narrow process of the cell. Figure 2.1.3 shows a highly schematic cartoon illustrating kinesin and dynein movement along a microtubule.

Microtubules can also form the interior of larger structures called **cilia** that extend out from the cell into the extracellular fluid. These cilia have a special arrangement of nine doublets arrayed circumferentially around two central microtubules. In this case, the microtubules are cross-linked such that the action of dynein causes the cilia to bend. The waving cilia move the extracellular fluid past the fixed cell. In this way, the movement of the cilia causes the movement of extracellular fluid. In the lungs, the cilia move mucus, trapped dust, and foreign material toward the pharynx where they can be expelled. In the Fallopian tubes, the cilia help move the ovum toward the uterus. Figure 2.1.4 shows a schematic diagram of the structure of a cilium. Activation of the dynein arms linking two isolated microtubules will ordinarily cause the two microtubules to slide past each other. In the cilia, linking proteins turn this motion into bending of the cilia.

Microtubules also form the **mitotic spindle** during cell division. Nearly all cells possess two **centrioles** oriented at right angles. These form a microtubule organizing center (MTOC) that provides a scaffold for the assembly of microtubules. Another kind of tubulin, γ-tubulin, binds to accessory proteins to form a γ-tubulin ring complex (γ-TuRC) that acts as a nucleation site for microtubules. The centrosome consists

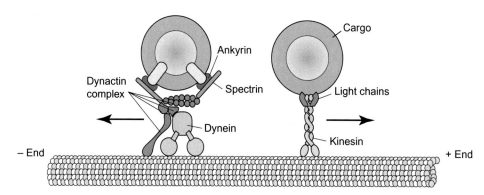

FIGURE 2.1.3 Schematic model of microtubule motor proteins. Kinesins typically have two globular heads and an elongated coiled tail. The tail regions of most kinesins bind cargo, either membrane-enclosed vesicles or microtubules. Dyneins may contain two or three globular heads and a large number of accessory proteins that bind vesicle cargo. Dynein itself is a complex assembly that requires a second complex assembly, dynactin, to transport cargo. A possible arrangement of some of these structures is shown.

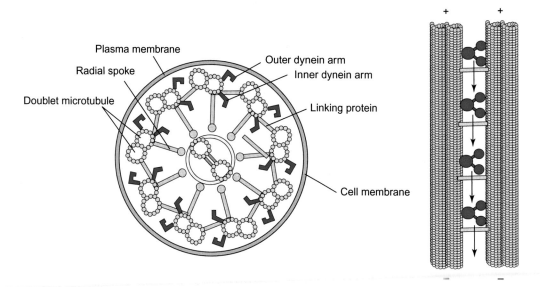

FIGURE 2.1.4 Structure of a cilium. A cilium contains a "9 + 2" arrangement of microtubules. Nine doublet microtubules containing A and B microtubules with 13 and 11 protomers, respectively, surround a central pair of microtubules. Multiple structures link these microtubules. The dynein arms move toward the minus end of the microtubule. In isolated tubules, this would cause sliding of one microtubule past another. The linking proteins turn this motion into a bending of the cilium.

of the two centrioles surrounded by a centrosome matrix containing many copies of the γ-TuRC. The microtubules that grow out of this centrosome complex provide tracks for chromosome movement during cell division.

ACTIN FILAMENTS ARISE FROM NUCLEATION SITES USUALLY IN THE CELL CORTEX

Actin filaments are present in most cells but are especially abundant in muscle cells. The monomer is a globular protein called **G-actin**, with a molecular weight of 41,800 Da. G-actin polymerizes noncovalently into actin filaments, called **F-actin**. Actin filaments consist of two strands of globular molecules twisted into a helix with a repeat distance of about 36 nm. The filament is asymmetric having distinguishable ends that are detectable by the way in which it interacts with **myosin**, another protein that is present in many cell types but is especially abundant in muscle. Thus the actin filament also has a **plus end** (the growing end) and a **minus end** (the nucleation or beginning end). Each individual actin filament is about 3.5 nm across, so that F-actin has a diameter of about 7 nm. Assembly and stabilization of filamentous, or F-actin, is described in Figure 2.1.5.

Actin filaments determine the shape and movement of the cell's surface, including structures such as **microvilli**, which are fingerlike extensions of epithelial cells that line internal structures like the intestinal villi and kidney tubules. The membrane of these cells anchors the actin filaments and extends them into a web of cytoskeletal elements in the main body of the cell. Other proteins can cross-link actin microfilaments together to form bundles of filaments or gel-like networks. These cross-linking proteins include **α-actinin**, **fimbrin**, and **villin**, which bundle actin filaments together. **Spectrin** and **filamin** both have two actin-binding sites. They join two actin filaments together to form a web of supporting filaments.

INTERMEDIATE FILAMENTS ARE DIVERSE

Intermediate filaments were originally named because with diameters between 8 and 10 nm, they are intermediate in size between the microtubules (at 25 nm) and the microfilaments at 7 nm. These intermediate filaments are composed of a number of different proteins. They play some structural or tension-bearing role. They differ from microtubules and microfilaments in that:

- Both microtubules and microfilaments are made by the polymerization of globular monomeric

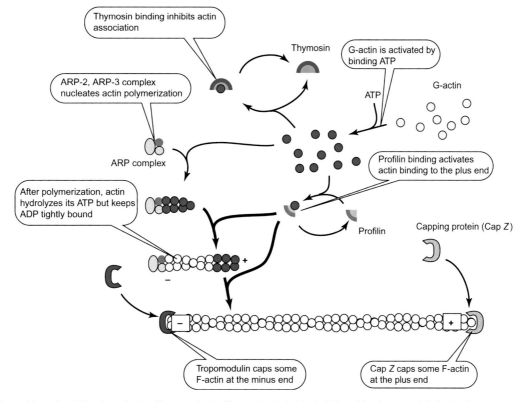

FIGURE 2.1.5 Assembly and stabilization of microfilaments (actin filaments). Actin binds ATP and begins assembly by binding to actin-related proteins (ARPs) that serve as a nucleation site, usually just under the cell membrane in the cortex of the cell. The ARP complex can also bind F-actin on the side of the filament, so it can build a tree-like web from individual actin filaments. After assembly, actin hydrolyzes its bound ATP, but the ADP remains tightly bound. Formation and stabilization of F-actin is regulated by proteins that bind the free monomer. Thymosin binds to the free monomer and inhibits its association with either the minus or plus end of the F-actin. Profilin binds to the free monomer and inhibits its association with the minus end but markedly enhances its association with the plus end. Cap Z binds to the plus end of the F-actin and stabilizes it. The minus end can be stabilized by remaining bound to the ARP complex. In muscle cells, tropomodulin binds to the minus end and stabilizes it.

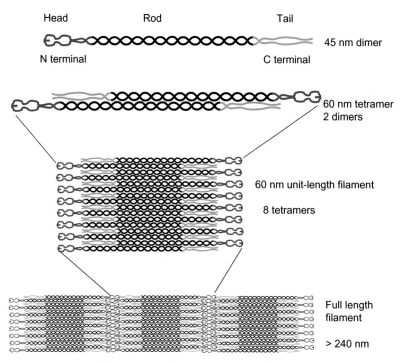

FIGURE 2.1.6 Highly schematic representation of the structure of intermediate filaments. The elementary subunit of the intermediate filaments consists of an elongated rod with an N-terminal "head" and a C-terminal "tail." A variety of these elementary subunits are made by the body. These associated laterally to form a homo- or heterodimer, approximately 45 nm long. These dimers further associate laterally, with an offset, to form a 60-nm-long tetramer. Eight of these tetramers further associate laterally to form a unit length filament. These unit length filaments can associate end to end to form longer filaments. The width of the mature filaments is not eight times the width of the tetramer, as these associate in three dimensions to form a mature filament about 11 nm in diameter. After the initial association into full length filaments, the filaments are radially compacted to form the mature filaments.

proteins, but the intermediate filaments are made of elongated (45 nm) and thin (2–3 nm) rod-like dimers. The intermediate filament units align with their long axis parallel to the filament axis, and filament width is determined by lateral association of the dimers (see Figure 2.1.6).

- Both microtubules and microfilaments are polar, which allows the active movement of motor proteins with their associated cargo along the filaments. Assembled intermediate filaments have no polarity because individual monomers are oriented in both directions along the axis of the filament.
- Intermediate filaments differ from both microtubules and microfilaments in that reversible association and dissociation of intermediate filament dimers can occur all along the length of the filament, whereas association and dissociation of microtubules and microfilaments occur only at their ends. This process is called **dynamic subunit exchange**. However, the exchange occurs much slower than the exchange of subunits in microtubules and microfilaments.
- Unlike tubulin and actin, the subunits of the intermediate filaments do not bind a nucleotide.

The intermediate filaments are diverse; some 73 separate proteins in humans have been identified encoded by over 70 genes. They all consist of three parts: a "head," a long rod-like central part, and a "tail." The members of the IF family have been subdivided into five distinct groups based on their structure, mode of assembly, and developmental expression in different tissues. These groups and their types are summarized in Table 2.1.3. There is considerable variation within types. For example, there are over 50 different varieties of keratin.

CYTOSKELETAL UNITS FORM FREE-FLOATING STRUCTURES BASED ON *TENSEGRITY*

Buckminster Fuller in the 1960s invented the word "tensegrity" as a blend of tension and integrity. He used it to describe architectural structures of remarkable rigidity that were composed of compressive rods and elastic cables. These two elements can be combined to form stable structures. Cells cannot rely on their membranes for structural stability because the membranes themselves are weak. But if you drape the membranes over cytoskeletal elements, structural strength can be achieved. In this view, the microtubules are the rigid rods and intermediate filaments are the tension-bearing elements. The actin and myosin elements allow for the movement of the cytoskeleton and the consequent movement of the attached membrane. In this way, the cell can extend processes or move from one place to another.

TABLE 2.1.3 Classification of the Intermediate Filaments

Types of IFs	Protein	Tissue Distribution	Proposed Function	Associated Diseases
Type I	Acidic keratins	Epithelial tissues	Tissues strength and integrity	Epidermolytic hyperkeratosis
Type II	Basic keratins	Epithelial tissues		
Type III	Desmin GFAP Peripherin Vimentin Syncoilin	Muscle Glial cells	Sarcomere organization	Dilated cardiomyopathy Alexander disease
Type IV	Neurofilaments NF-L, NF-M, and NF-H Nestin Synemin α, β	Neurons	Axon organization	Amyotrophic lateral sclerosis; Parkinson disease
Type V	Nuclear lamins type A, B1, B2, C1,C2	Nucleus	Nuclear organization and signaling	Hutchinson–Gilford progeria; limb–girdle muscle dystrophy; Emery–Dreifuss muscular dystrophy
Type VI	Filensin Phakinin	Lens		Cataracts

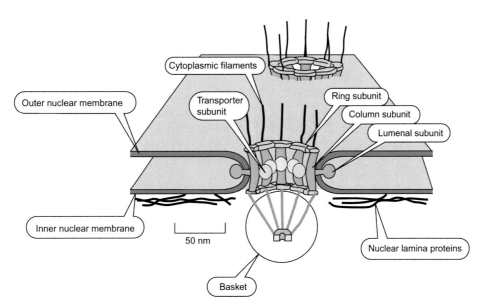

FIGURE 2.1.7 Cartoon of the structure of the nuclear pore. The nuclear pores consist of a complex of more than 50 different proteins that form a complicated structure with octagonal symmetry. Small molecular weight materials (< 20 kD) can pass through these pores in both directions, but the movement of larger materials, such as RNA and ribosomes, is regulated. The outer membrane faces the cytosolic compartment and the inner membrane faces the nuclear compartment. Two rings made of eight subunits each are connected by columnar scaffold subunits. Lumenal subunits anchor the scaffold to the membrane. The ring subunits connect to fibrils that form a basket structure on the nuclear side. Transport subunits in the interior of the pore actively transport materials into or out of the nucleus.

MYOSIN INTERACTS WITH ACTIN TO PRODUCE FORCE OR SHORTENING

Myosin also exists in multiple isoforms. A major form is a protein of about 200 kDa that forms a homodimer with two long tails forming a coil, a hinge region, and a head region that binds actin filaments, hydrolyzes ATP, and "walks" along the actin filament. The interaction of the actin filament relative to the myosin filament causes either shortening of the acto-myosin thread or production of force. This mechanism is responsible for muscle force and also produces movement in nonmuscle cells.

THE *NUCLEUS* IS THE COMMAND CENTER OF THE CELL

Most cells have linear dimensions on the order of 20–50 μm. The nucleus is the largest organelle, with a diameter of about 3–5 μm. The nucleus is bounded by a double membrane, the **nuclear envelope**, that has pores for materials to move between nucleus and cytoplasm, as shown in Figure 2.1.7. The nucleus contains nearly all of the DNA of the cell. As described in Chapter 2.2, this DNA carries the information that allows the synthesis of specific proteins. The nucleus also contains a specialized region called the **nucleolus**. This is a diffuse region that is not delimited by a membrane. The nucleolus is involved in the synthesis of **ribosomes**.

RIBOSOMES ARE THE SITE OF PROTEIN SYNTHESIS

Ribosomes exist either free in the cytoplasm or bound to membranes of the rough endoplasmic reticulum (ER). Both types consist of two main subunits, designated 60S and 40S, where the S refers to Svedbergs and describes how fast the particles sediment during centrifugation (see Appendix 2.1.A1). The larger subunit consists of three different strands of ribonucleic acid (RNA) and about 49 different proteins. The smaller subunit has a single RNA strand and about 33 other proteins. Both are assembled in the nucleolus, a specialized region of the nucleus that is not membrane bound. There, large loops of DNA containing ribosomal RNA genes are transcribed by the enzyme **RNA polymerase I** to ribosomal RNA or **rRNA**. Ribosomal proteins made in the cytoplasm are imported back into the nucleus and assembled, along with the rRNA, into the two subunits. The two subunits join to form a functional ribosome in the cytoplasm that makes a platform on which proteins are synthesized, as described more fully in Chapter 2.2.

THE *ER* IS THE SITE OF PROTEIN AND LIPID SYNTHESIS AND PROCESSING

The ER is a membranous network within the cell fluid, or cytoplasm, that is continuous with both the outer nuclear membrane and the plasma membrane. The membrane forms flattened disks with an enclosed space called **cisternae**. Electron micrographs and density gradient centrifugation reveal two functionally distinct regions of the ER; these are **rough ER** and **smooth ER**. The "rough" ER was given that name because the many ribosomes attached to the membrane give it a granular appearance in electron micrographs. The smooth ER lacks these attached ribosomes.

Special proteins, called **translocons**, span the ER membrane and bind ribosomes to the cytoplasmic face of the rough ER. Protein synthesis occurs on the ribosomes. Some of these are free within the cytoplasm, and the proteins made there are generally soluble proteins that remain in the cytoplasm. The ribosomes on the ER synthesize proteins that pass through the ER membrane as they are being synthesized. These proteins may become embedded in membranes or they may be destined for secretion from the cell. After synthesis, many proteins are further processed within the ER cisternae, preparing them for secretion or targeting them for some location within the cell.

THE *GOLGI APPARATUS* PACKAGES SECRETORY MATERIALS

The Golgi apparatus consists of sets of membrane-delimited smooth-surfaced cisternae. Each set of flattened, disk-shaped cisternae resembles a stack of pancakes. This structure is called a **Golgi stack** or **dictyosome**. It is about 1 μm in diameter and is usually located near the nucleus and near the centrioles that define the cell center. The number of cisternae in a stack varies from 6 to 30, and the number of Golgi stacks in the cell varies enormously with the biochemical activity of the cell.

Golgi stacks are polarized with two distinct faces. The *cis* or **forming face** is nearest a smooth transitional portion of the rough ER. The *trans* or **maturing face** typically faces the plasma membrane. Swarms of small vesicles (about 50 nm in diameter) cluster on the *cis* face of the Golgi stack. A large number of vesicles associate with the sides of the stack near the dilated rims of each cisternae. In electron micrographs, these vesicles sometimes appear to bud off the Golgi cisternae. In secretory cells, larger vesicles containing high concentrations of secreted proteins appear to originate from the *trans* face of a Golgi stack. The Golgi stacks are a processing station for proteins manufactured in the rough ER. Proteins made in the rough ER travel to the Golgi through the small transport vesicles. In the Golgi, the proteins are processed and packaged for delivery to various locations throughout the cell, including packaging for eventual fusion with the plasma membrane and secretion into the extracellular space (see Figure 2.1.8).

THE *MITOCHONDRION* IS THE POWERHOUSE OF THE CELL

Mitochondria produce much of the cell's ATP by coupling the chemical energy of oxidation of metabolites to the synthesis of ATP. Their main structural features are shown schematically in Figure 2.1.9. The matrix contains many different enzymes required for the oxidation of pyruvic acid and fatty acids, including those involved in the **tricarboxylic acid cycle**. The mitochondrial matrix also includes mitochondrial DNA, special mitochondrial ribosomes, tRNAs, and enzymes that are required for the expression of mitochondrial genes. The inner mitochondrial membrane contains a number of important proteins that collectively comprise the **electron transport chain** and another enzyme complex called the **ATP synthetase** that makes ATP from ADP and Pi, the reverse of the ATP hydrolysis reaction discussed in Chapter 1.7. These complexes are discussed further in Chapter 2.10. Briefly, these complexes couple the chemical energy derived from the oxidation of fuels obtained from food to the synthesis of ATP. This is the site of oxygen consumption by aerobic cells.

Lynn Margulis originally postulated that the mitochondria in aerobic cells that contain a nucleus (eukaryotic cells)

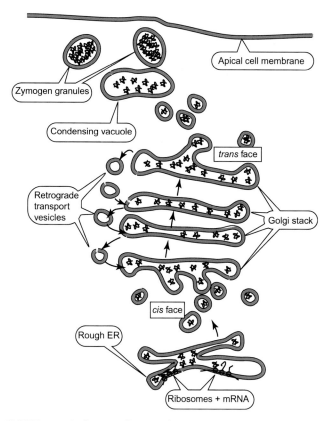

FIGURE 2.1.8 Packaging of secreted proteins in secretory cells of the pancreas. Proteins are synthesized on the membrane of the rough ER and translocated into the lumen of the ER as they are being made. The ER forms transport vesicles that fuse to form the *cis* face of the Golgi stack. These membranes progress through the stack as new layers are added on the *cis* face and taken away on the *trans* face. On the *trans*, or maturing, face of the Golgi stack, the enclosed proteins are collected in secretory vesicles. These then fuse into condensing vacuoles that concentrate the proteins to form zymogen granules. These granules lie in the apical aspect of the secretory cells, adjacent to the plasma membrane. Upon stimulation, these granules fuse with the apical membrane and release their contents of secretory enzymes into the lumen of a duct, or channel, that takes the enzymes into the intestine.

originated from the engulfment of aerobic bacteria by anaerobic single-celled organisms. This hypothesis, called the **endosymbiotic hypothesis**, derives from the similarity of mitochondria to bacteria. Both have circular DNA; both are approximately the same size; both reproduce by dividing into two, asexually; mitochondrial ribosomes resemble bacterial ribosomes rather than eukaryotic ribosomes, and both bacteria and mitochondria share a slightly different genetic code from that in the nucleus.

LYSOSOMES AND *PEROXISOMES* ARE BAGS OF ENZYMES

Lysosomes are membranous bags of **hydrolytic enzymes** including **proteases**, **nucleases**, **glycosidases**, **lipases**, **phospholipases**, and **phosphatases**. These hydrolytic enzymes are acid hydrolases, being optimally active in an acid environment. Lysosomes are typically 0.2–0.3 μm in diameter. They originate from the *trans* face of the Golgi stack and are formed first as **primary lysosomes**. The primary lysosome fuses repeatedly with a variety of membrane-bound substrates including endocytotic vesicles, phagocytotic vesicles, and worn-out intracellular organelles. After fusion, the combined vesicle forms a **secondary lysosome**. Because of its diverse substrate contents, the secondary lysosomes have a diverse morphology. Lysosomes degrade phagocytosed material and worn-out parts of the cell.

The peroxisome is another membrane-bounded vesicle, with a diameter of about 0.5 μm. It contains oxidative enzymes such as catalase, d-amino acid oxidase, and urate oxidase. Like the mitochondria, the peroxisomes are a major site of O_2 utilization. The peroxisome detoxifies foreign chemicals and metabolizes fatty acids. Beta-oxidation, a process in which fatty acids are shortened by two carbons to form acetyl-coenzyme A, occurs in both mitochondria and peroxisomes (see Chapter 2.11).

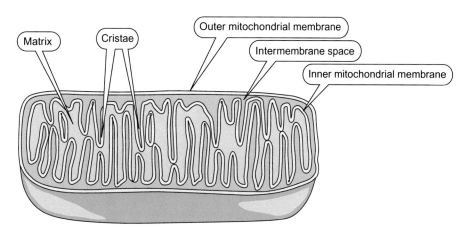

FIGURE 2.1.9 Typical features of a mitochondrion. The mitochondria, shown cut longitudinally, are membrane-delimited structures about 0.5–1.0 μm wide and 1–4 μm long. The **outer mitochondrial membrane** is permeable to many materials with molecular weights below 10 kDa. The **inner mitochondrial membrane** is impermeable to most materials but can transport specific materials. It is folded to form the shelf-like **cristae** and encloses the **matrix**. The intermembrane space lies between the inner and outer membranes.

PROTEASOMES DEGRADE MARKED PROTEINS

Proteasomes are large, multisubunit protein complexes that are scattered throughout the cytoplasm and that degrade cell proteins. Proteins are tagged for degradation by the attachment of a **ubiquitin** molecule to the proteins. Ubiquitin is a protein consisting of 76 amino acids. Proteins can be marked by more than one ubiquitin chain through a complex series of reactions involving several different enzymes. This process is called **ubiquitinylation**, in which a series of enzymes recognizes the proteins to be degraded and adds activated ubiquitin onto them. Different sets of enzymes in this pathway recognize different degradation signals on proteins. In this way, each set of enzymes targets distinct subsets of proteins that bear particular degradation signals. The process of ubiquitinylation is illustrated in Figure 2.1.10.

The proteasome consists of a central hollow cylinder capped at both ends. The central cylinder consists of a stack of four 7-membered rings. The caps recognize the ubiquitin coats of proteins that have been tagged for degradation. Each cap contains ATPase activity, releasing chemical energy in the process. These caps use the energy to unfold the ubiquitinylated proteins and move them into the central cylinder where proteases cleave the protein into its constituent amino acids. The structure of the proteasome is shown in Figure 2.1.11.

CELLS ATTACH TO EACH OTHER THROUGH A VARIETY OF JUNCTIONS

Cells form a variety of attachments to each other or to the extracellular matrix. These include:

- **Zonula occludens**: This "tight" junction joins epithelial cells in an occluding zone at one pole of the epithelium. It has three functions:

1. a *barrier function* that disallows free movement of materials between cells in an epithelial sheet. This is the so-called "tight" part of tight junctions, but in reality the junctions vary considerably in their permeabilities.
2. a *fence function* that prevents the free migration of membrane components from the apical surface

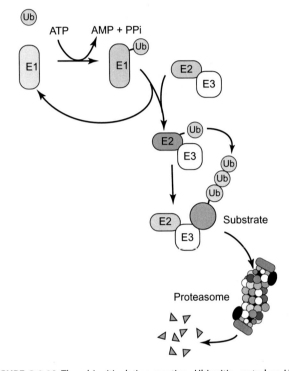

FIGURE 2.1.10 The ubiquitinylation reaction. Ubiquitin, noted as Ub in the figure, is a 76 amino acid protein that is used to "tag" proteins as being ready for degradation. First, Ub is attached to E1 through an ATP-requiring reaction. It is then passed from E1 to E2, and from there to the protein substrate being tagged for demolition. The ubiquitinylated protein is then degraded by a specialized cell structure called the proteasome.

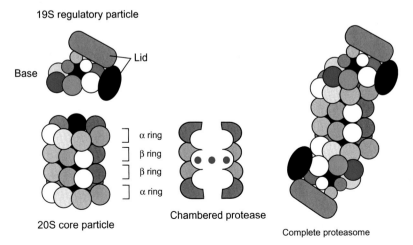

FIGURE 2.1.11 Structure of the proteasome. The overall structure consists of a 20S core particle of 28 subunits capped on one or both ends by a 19S regulatory particle containing at least 19 subunits. The core particle consists of four 7-membered rings that is symmetrical about a plane perpendicular to the long axis of the particle. Proteolytic activity resides in the two middle layers of the core particle.

of the cell to the lateral surface. This effectively partitions the cell membrane into components.
3. a *signal function* that help regulate cell proliferation, differentiation, and polarity.
- **Zonula adherens**: This is a belt of attachment that typically surrounds epithelial cells just below the zonula occludens—meaning toward the basolateral pole of the cell.
- **Desmosomes**: These are "spot welds" between cells and are constructed of different proteins than those that make up zonula adherens.
- **Gap junctions**: These junctions serve to electrically connect cells because they allow small ions to pass from one cell to another, and these ions carry electrical current.

Each of these junctions is made up of complexes of many proteins and a variety of proteins can associate with these. Figure 2.1.12 illustrates the structure of the desmosome, zonula adherens, zonula occludens, and gap junction, and Figure 2.1.13 shows their use in an epithelial sheet in the small intestine, consisting of a lining of cells that separate the ingested food and gastrointestinal secretions from the blood. Many of the proteins that make up these structures are present in the body in various **isoforms**—variants of the protein that serve the same basic function but in different tissues. For example, there are some 20 different types of connexins that associate as hexamers to form the connexons in gap junctions.

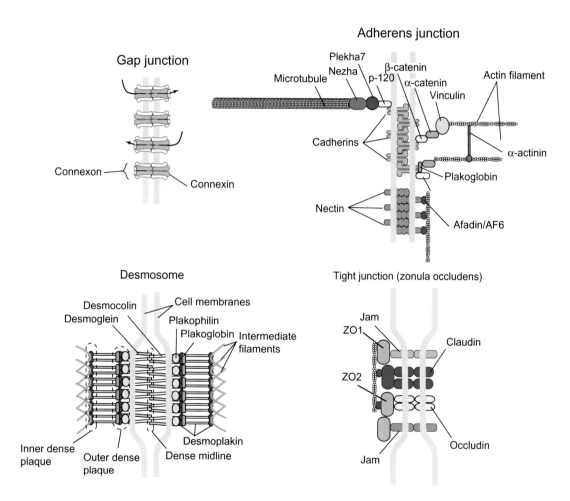

FIGURE 2.1.12 Schematic diagram of the proposed structures of gap junctions, adherens junctions, and desmosomes. Gap junctions form by linking of connexons on opposing membranes. Each connexon consists of a hexamer of connexin units that form a central pore. When two connexons link, the pore of one lines up with the pore of another, forming a watery path between the cells. This allows diffusion of small, soluble materials from one cell to another without crossing the cells' membranes. Adherens junctions form by interaction of extracellular parts of cadherin molecules. These proteins are embedded in the cell membrane and have a short cytoplasmic tail. This cytoplasmic tail binds to a variety of proteins including p-120, plakoglobin, and β-catenin. These in turn bind other proteins that eventually form initiation sites for actin polymerization. Cadherin also directly binds microtubules, although other proteins may stabilize this interaction. Desmosomes adhere two cells together because of the interaction of extracellular domains of desmoglein and desmocolin. These penetrate the membrane and their cytoplasmic domains bind other proteins, plakophilin, plakoglobin, and desmoplakin, that eventually connect to intermediate filaments. Tight junctions consist of binding of claudins and occludins, along with junctional adhesion molecule (JAM). There are multiple isoforms of each of these that produce junctions of varying permeability. These are stabilized by multiple cytoplasmic accessory proteins, including zonula occludens proteins (ZO-1, ZO-2, and ZO-3). These can connect to the cytoskeleton through other proteins such as afadin. In the junction between three cells, a special protein called tricellulin is required to seal the gap between the cells.

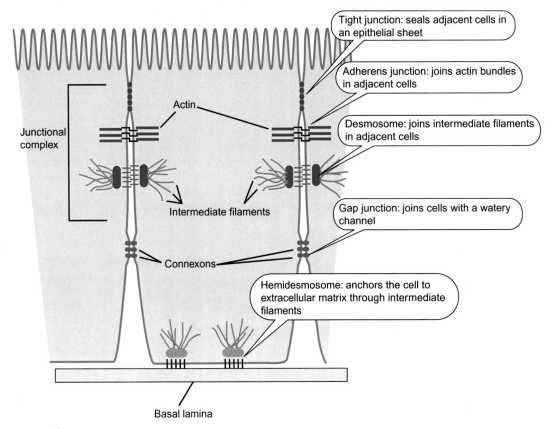

FIGURE 2.1.13 Cell attachments found in an epithelial sheet. The intestinal cells form a layer, referred to as an epithelium, that lines the intestine. The intestinal cells are bound at their apical pole, the side facing the gastrointestinal lumen, by a junctional complex consisting of tight junction, adherens junction, and desmosomes. The tight junction and adherens junction form a belt completely surrounding the cell. The desmosomes are located in spots. The cells are also joined by gap junctions that allow free passage of small molecular weight materials between the cells. The cells attach to the extracellular basal lamina through hemidesmosomes that connect the fibers in the extracellular matrix with the cytoskeletal intermediate filaments.

SUMMARY

The cell is the fundamental organizational unit of the body. Although cells come in many different forms, they share many features. Each cell consists of a number of organelles, so named because they contribute to overall cell function in much the same way as our organs contribute to bodily function.

The cell membrane determines the inside and outside of a cell. This is the "customs officer" of the cell, determining what enters or exits the cell. Materials move through the cytoplasm, the watery cell fluid that transports materials from one place to another, largely by diffusion. A cytoskeleton maintains cell shape and provides movement. Cytoskeletal elements include microtubules, which are hollow rods formed from 13 protomer filaments of a heterodimer of tubulin, actin filaments, composed of a double helix of actin monomers strung together, intermediate filaments of various descriptions, and myosin filaments. These cytoskeletal elements are dynamic and allow the cell to change shape and to transport materials along cytoskeletal tracks. The nucleus is the single largest organelle, and it is the "command post" of the cell. It is enclosed by a double membrane that is pierced by numerous nuclear pores. In response to signals, the nucleus "expresses" select regions of the genome. This means that the nucleus specifically converts some DNA into mRNA, but not all.

The mRNA then makes proteins in the "factory" of the cell located on ribosomes either free in the cytoplasm or bound to the surface of the rough ER, another membranous network in the cytoplasm. The smooth ER makes lipids; the rough ER makes membrane-bound proteins and proteins destined for export from the cell. The rough ER transfers its protein content to the *cis* side of the Golgi apparatus, which is the "shipping and packaging" department of the cell. The materials move from one part of the Golgi apparatus to another in tiny membrane spheres called vesicles. At each stage, the proteins are processed further. The final vesicles leaving the *trans* face of the Golgi stack are ready for export from the cell by exocytosis.

All of these activities of the cell require chemical energy supplied as ATP. The mitochondria make ATP by coupling the chemical energy liberated by oxidation of foodstuffs to the synthesis of ATP. Thus the mitochondrion is the "powerhouse" of the cell. It consists of a double membrane structure. The electron transport chain is on the highly folded inner membrane. This membrane synthesizes ATP from chemical precursors and is the main site of oxygen consumption by the cell.

The cell also contains a variety of other organelles including the lysosome, peroxisome, and proteasome. These structures degrade material engulfed by the cell by endocytosis and also degrade worn-out organelles and cell proteins.

Cells form a variety of attachments to other cells and to the extracellular matrix. These include zona occludens and adherens junctions, gap junctions, and desmosomes. These form by the complex association of a variety of different proteins: connexons on opposite membranes link up to form the gap junction; cadherins link up to form adherens junctions; desmoglein and desmocolin form the desmosomes; claudin and occludin link up in the tight junction. These junctions form by the association of extracellular parts of these transmembrane proteins. The intracellular parts join up with still other proteins that eventually connect the junctions with the cell's cytoskeleton.

REVIEW QUESTIONS

1. What is the plasma membrane? What compartments does it separate? What is endocytosis? Exocytosis? Pinocytosis? Phagocytosis?
2. What is the major cation (positively charged ion that migrates toward the cathode, the negative electrode) of the cytosol? Is the cytosol comparable to water?
3. What makes up the cytoskeleton? What is a microtubule? What is a microfilament? What is an intermediate filament? How do intermediate filaments differ from microtubules and microfilaments?
4. What does the plus (+) end of a microtubule or microfilament mean? What does the minus (−) end mean? What are "motor proteins"? What does it mean to be "+ directed"? Name the family of + directed motor proteins. Name the family of − directed motor proteins. How do motor proteins carry cargo?
5. Describe the structure of a nuclear membrane. Why does the nuclear membrane have an elaborate pore structure? How big is the nucleus? What is the nucleolus?
6. What do ribosomes do? What are they made of? Where are they made? What do "40S" and "60S" descriptions of the major ribosomal subunits mean?
7. What distinguishes "rough ER" from "smooth ER"? What does each make?
8. What is the Golgi apparatus? What is the *cis* face? *Trans* face? What are all those vesicles doing hanging around the Golgi rims and *cis* and *trans* faces? What goes on inside the Golgi cisternae?
9. How big is a mitochondrion? Why do you suppose it has two membranes? Where is the electron transport chain? What do the mitochondria do?
10. What do lysosomes contain? What is the difference between a primary and a secondary lysosome? What goes on inside peroxisomes?
11. What do proteasomes do? How do they know which proteins to degrade?
12. Name four junctions between cells. Which ones involve connexons? Which ones join actin filaments from one cell to another? Which ones join intermediate filaments from one cell to another?

APPENDIX 2.1.A1 SOME METHODS FOR STUDYING CELL STRUCTURE AND FUNCTION

THE MICROSCOPE HAS REVOLUTIONIZED OUR UNDERSTANDING OF BIOLOGY

The invention of the microscope literally opened up a new world view in biology. For the first time, we could look upon the microscopic world and we discovered that it was full of animated objects. Robert Hooke (1635–1703) first used the microscope to study biological material in 1665 and coined the term "cell" from the likeness of the empty cells in cork to the monks' cells within a monastery. Antoine van Leeuwenhoek (1632–1723) was the first to observe live cells with hand-made microscopes that could magnify up to 500 times. He observed an astonishing array of cells from single-celled protists to red blood cells, bacteria, sperm, and muscle fibers. These observations, along with many others, prompted Theodor Schwann to enunciate the Cell Theory in 1839, based in part on conversations with Matthias Schleiden, whom Schwann did not credit. It contained three important elements:

1. The cell is the unit of structure, physiology, and organization in living things.
2. The cell is a distinct entity but is the building block of complicated organisms.
3. Cells form from cell-free material, as in crystallization.

We now know that this last point is incorrect. Its first correction was famously uttered by the German pathologist, Rudolf Virchow, when he said "Omnis cellula e cellula"—all cells come from cells. The entire development of the cell theory was supported by a single advance—the light microscope.

MICROSCOPIC RESOLUTION IS THE ABILITY TO DISTINGUISH BETWEEN TWO SEPARATED OBJECTS

What do we mean when we say that red blood cells, erythrocytes, are invisible to the naked eye? Surely we can see blood. What we mean here is that we cannot make out the individual cells. The ability to distinguish two objects separated by a distance is called the **resolution**.

Light produces a diffraction pattern around all objects. This pattern consists of areas of maximum and minimum intensity of light. The resolution of optical devices is determined by considering the diffraction pattern produced by a circular aperture of radius *a*. This circular aperture is called an **Airy disk**. The convention is that two apertures can be distinguished if the first maximum

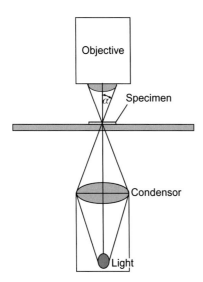

FIGURE 2.1.A1.1 Angle of the cone of light passing through the specimen and incident to the objective.

of the diffraction pattern from one pattern falls on the first minimum of the second. This is called the Rayleigh criterion. The diffraction pattern can be described analytically using Bessel functions, which we will not do here. The result is given by

$$[2.1.A1.1] \qquad \text{Resolution} = \frac{0.61\,\lambda}{\eta \sin \alpha}$$

where λ is the wavelength of light, η is the **refractive index** of the medium between the specimen and the lens, and α is the angle of the cone of light that passes between the specimen and the lens (see Figure 2.1.A1.1). The expression $\eta \sin \alpha$ is called the **numerical aperture** of the lens being used. This description of resolution is the inverse of what is ordinarily meant. Eqn [2.1.A1.1] defines resolution in terms of a distance. However, when we can resolve two objects that are close together, we ordinarily speak of a high resolution. According to Eqn [2.1.A1.1], the resolved distance is smallest (and the resolution is highest) when α is 90° and the refractive index is increased by placing oil between the specimen and lens. The refractive index of air is close to 1.0, whereas that of oil is 1.4–1.5.

THE ELECTRON MICROSCOPE HAS ADVANCED OUR UNDERSTANDING OF CELL STRUCTURE

Until the 1950s, the workhorse of cellular structure was the optical microscope. Our ideas of the structure and composition of cells were further enhanced by the electron microscope. This device allows tremendous magnification of cell components but suffers from the disadvantage that the cells must be fixed and stained before viewing their structures because the electron microscope views specimens in a vacuum. Otherwise, the incident electron beam would be scattered by air molecules. The electron microscope illuminates the structure of cells only after artificial preparation. Without functional studies, the activity going on in the structures remains elusive.

The electron microscope can achieve much higher resolution (it can resolve objects that are separated by a smaller distance) than the optical microscope because the wavelength of electrons is so much shorter than the wavelength of visible light. The fact that electrons have a wavelength at all was an astounding discovery. The first clue to the wavelength of electrons was the **Compton effect**, reported by Arthur H. Compton in 1922. Compton found that incident X-rays were scattered by a carbon target, subsequently shifting the X-rays to a lower wavelength and dislodging an electron from the crystal. The dislodged electron has a momentum. To describe this collision of light with electrons, it was necessary to postulate that the incident light possessed momentum, in order to preserve the conservation of momentum theorem of physics. Compton produced a quantum-mechanical analysis of the scattering that differed from classical explanations, but agreed with the experimental observations.

The photoelectric effect and Compton effects showed that light had distinctive particle-like aspects. It was already known, from interference and diffraction experiments, that light also had distinctive wave-like aspects. Louis de Broglie in 1924 postulated that the wave–particle duality of photons might also apply to particles; if photons have momentum, particles such as the electrons ought to have a wavelength. From the Compton experiments, the wavelength would be given by

$$[2.1.A1.2] \qquad \lambda = \frac{h}{mv}$$

where h is Planck's constant $= 6.625 \times 10^{-34}$ J s. The wave nature of electrons was confirmed in 1927 by Davisson and Germer, who exposed a single crystal of nickel to electrons having 54 electron volts of kinetic energy. They observed an electron diffraction pattern that confirmed de Broglie's relation.

The short wavelength of electrons opened the possibility of a microscope with astounding resolution. The electron microscope, invented in the 1930s, was first applied to living tissues by Albert Claude, Keith Porter, and George Palade in the late 1940s and 1950s.

The resolution of the electron microscope is determined not only by the wavelength of the incident radiation, which in this case is a beam of electrons, but also by the numerical aperture. Although theoretically the resolution of the electron microscope should be close to 0.002 nm, in principle it is much larger than this because the inherent properties of magnetic lenses limit the aperture angle to about 0.5°.

Because biological specimens lack inherent contrast, the practical resolution is further reduced to about 1–2 nm. Nevertheless, this is a marked improvement over the optical microscope. The optical microscope was useful to a magnification of about 1000×; the electron microscope could attain more than a 100-fold better magnification.

SUBCELLULAR FRACTIONATION ALLOWS STUDIES OF ISOLATED ORGANELLE BUT REQUIRES DISRUPTION OF CELL FUNCTION AND STRUCTURE

Although the optical microscope and electron microscope provided keen insights into the structure of living things, the function of the structures could not be directly investigated. In the late 1940s and 1950s, Albert Claude and Christian de Duve developed methods for **subcellular fractionation** for separating cells into their component parts and elucidating the function of the cellular constituents. This method disrupted the cells and separated the parts by **differential centrifugation**.

DIFFERENTIAL CENTRIFUGATION PRODUCES ENRICHED FRACTIONS OF SUBCELLULAR ORGANELLES

The first step in subcellular fractionation is the disruption of the cell into its component subcellular **organelles**. This process usually uses homogenization, and its aim is to break the plasma membrane that delimits the cell, thereby releasing the cellular contents, without damaging those contents. Cells can be homogenized by sonication (exposure to high-frequency sound waves), shearing the cells between two surfaces such as a Teflon mortar and glass pestle, or placing the cells in a high-speed blender. These treatments break the cell membrane and leave the remaining parts of the cells relatively intact. These relatively harsh treatments obviously disrupt some of the normal relationships between parts of the cell, and may scramble normal constituents of cells.

The resulting mixture, the homogenate, can be separated into its component parts on the basis of their size and density (see Figures 2.1.A1.2 and 2.1.A1.3). Following lysis of the cells, the homogenate is placed in a tube in a centrifuge and spun at relatively low speed ($1000 \times$ gravity) for a short time (10–20 minutes). The centrifugation causes materials to move away from the axis of centrifugation. When the particles reach the bottom of the tube, they form a pellet. This process is called **sedimentation**. Particles that sediment quickly reach a terminal velocity in the centrifuge tube. At this terminal velocity, the frictional drag on the particle provides the acceleration necessary to keep the particle in approximate uniform circular motion (see below for an explanation of the forces during sedimentation). The frictional drag depends on the density of the particle, viscosity of the medium, and speed of centrifugation. The first centrifugation, at low speed and short times, sediments unbroken cells, and the nuclei because these

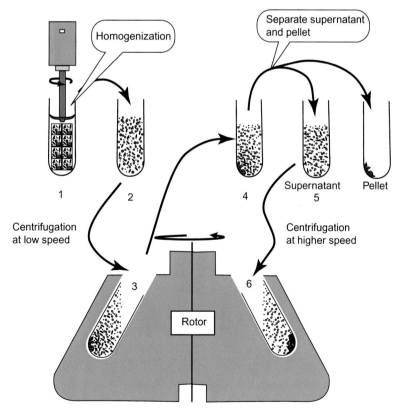

FIGURE 2.1.A1.2 Separation of subcellular organelles by differential centrifugation. Whole tissue is first homogenized, which disrupts cell membranes and releases subcellular organelles. The homogenate is then centrifuged to separate out particles on the basis of their sedimentation. In general, large particles sediment with small centrifugal forces and smaller particles require larger forces. Successive centrifugation at progressively faster revolutions per minute (RPM) separates organelles on the basis of their sedimentation characteristics.

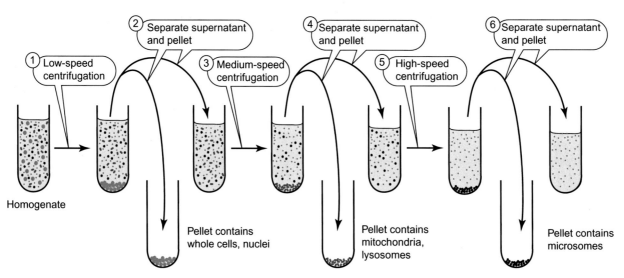

FIGURE 2.1.A1.3 Differential centrifugation. The first slow speed spin causes large and heavy particles to sediment. These are separated from the particles that do not sediment. Successively faster centrifugations cause progressively smaller and lighter particles to sediment.

are the largest and heaviest of subcellular structures. Therefore, a fraction of the homogenate is produced that is enriched in nuclei. The supernatant fraction, the suspension that lies above the first pellet, is cleared of nuclei but still contains other subcellular particles.

Successive centrifugations at higher speeds and longer times produce fractions enriched in the **mitochondria**, **lysosomes**, and **peroxisomes**. Further centrifugation at still higher speeds sediments the broken fragments of the **plasma membrane** and **endoplasmic reticulum**. De Duve identified these subcellular fractions through the use of **marker enzymes**. The basic idea here is that each of the subcellular structures has a unique biochemical composition that enables their unique biochemical function within the cell. Part of their biochemical composition is their component enzymes. By measuring the distribution of an enzyme, the scientist can track the distribution of the particles that contain it. Markers for the mitochondria, for example, include cytochrome C oxidase and succinate dehydrogenase, among others. By measuring cytochrome C oxidase in the various fractions, one can estimate the amount of mitochondria in those fractions. De Duve deduced the existence of the lysosome on the basis of an enzyme that distributed itself differently from all other known markers.

DENSITY GRADIENT CENTRIFUGATION ENHANCES PURITY OF THE FRACTIONS

The centrifuge is a crude instrument for the separation of subcellular fractions because of the way in which it separates the different subcellular organelles. Differential centrifugation can produce fractions enriched in one particle or another, but the fractions are not pure. This is due to the fact that sedimentation occurs over the considerable length of the centrifuge tube. When the tube is spun, particles throughout the tube are subject to centrifugal forces that cause them to sediment. Although heavier particles sediment more quickly, the heavier particles at the top of the tube have much further to travel than those near the bottom. By the time all of the heavy particles sediment, some of the lighter particles near the bottom of the tube, or at the middle, also sediment. Therefore, pure fractions of subcellular particles cannot be achieved easily by simple differential centrifugation. Further purification can be achieved by using **density gradient centrifugation**. In this method, subcellular organelles are separated by centrifugation through a gradient of a dense substance, such as sucrose. In **velocity centrifugation**, the material to be separated is layered on top of a sucrose gradient, and then centrifuged. Particles of different sizes and density sediment through the gradient at different rates moving as discrete bands. At the end of the centrifugation, the different layers consist of purified organelles, and they can be collected for further experiments. In **equilibrium centrifugation**, the density gradient is used to separate particles based on their buoyant density. Instead of being separated by their sedimentation velocity, particles will sediment until they reach a layer with the same density as the particles. At this point, sedimentation stops and the purified organelles can be collected at the equilibrium position. These methods of separating subcellular organelles are illustrated in Figure 2.1.A1.4.

ANALYSIS OF CENTRIFUGATION SEPARATION

CIRCULAR MOTION REQUIRES AN INWARD CENTRIPETAL FORCE

Centrifugation typically involves spinning tubes of material at a constant angular velocity, except for the angular acceleration to that velocity, and the deceleration when the spin stops. The position of any particle in the tube at any time may be represented by the vector **r**, as shown in Figure 2.1.A1.5.

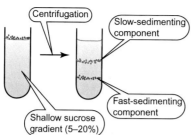
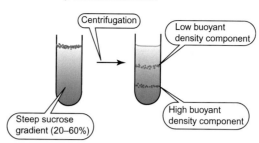

FIGURE 2.1.A1.4 Comparison of velocity sedimentation and equilibrium sedimentation for the separation of subcellular particles. In velocity sedimentation, separation relies on different velocity of sedimentation through a shallow sucrose (or other material) gradient. The sucrose is added to prevent mixing by convection. In equilibrium sedimentation, subcellular particles sediment until they reach a zone of solution that matches their density. At equilibrium, the organelles distribute themselves over a narrow region of the tube corresponding to the density of the organelle. The organelles can then be harvested from a narrow band.

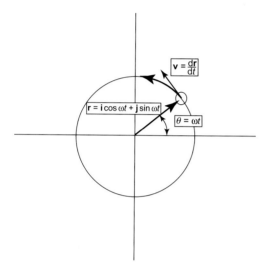

FIGURE 2.1.A1.5 A particle in uniform circular motion around a central pivot point. The angular displacement is a linear function of time.

This vector makes the angle θ from an arbitrary zero reference, the x-axis, and this angle increases with time. The angular velocity is defined as

$$[2.1.A1.3] \qquad \omega = \frac{d\theta}{dt}$$

By integrating this we see easily that:

$$[2.1.A1.4] \qquad \theta = \omega t$$

The position vector indicating the location of the point relative to the center of rotation is thus given as

$$[2.1.A1.5] \qquad \vec{r} = \vec{i}\,\cos\omega t + \vec{j}\,\sin\omega t$$

where **r** is the position vector and **i** and **j** are unit vectors along the x-axis and y-axis, respectively. The velocity vector at any time is the derivative of this position vector:

$$[2.1.A1.6] \qquad \vec{V} = d\frac{\vec{r}}{dt} = -\vec{i}\,\omega\sin\omega t + \vec{j}\,\omega\cos\omega t$$

This velocity vector is orthogonal to the position vector as seen by the dot product: it is given as

$$\vec{r}\cdot\vec{V} = -\omega\sin\omega t\cos\omega t + \omega\sin\omega t\cos\omega t = 0$$
$$[2.1.A1.7]$$

This means that the velocity vector is oriented at 90° to the position vector, as shown in Figure 2.1.A1.5. The acceleration at any time is the derivative of the velocity:

$$[2.1.A1.8] \qquad \vec{a} = d\frac{\vec{V}}{dt} = -\vec{i}\,\omega^2\cos\omega t - \vec{j}\,\omega^2\sin\omega t$$
$$\vec{a} = -\omega^2\vec{r}$$

We see here that the magnitude of the acceleration is $\omega^2 r$ and its direction is directly opposite that of the position vector (the negative sign in Eqn [2.1.A1.8]). (Compare Eqn [2.1.A1.8] to [2.1.A1.5] for the identity of **r** in **a**.) This is the centripetal acceleration. In order for the particle to remain in uniform circular motion, the velocity must be continuously bent toward the center. The acceleration vector is orthogonal to the velocity vector, so that all of the acceleration is used to change the direction of the velocity, and not its magnitude.

CENTRIPETAL FORCE IN A SPINNING TUBE IS PROVIDED BY THE SOLVENT

As described above, particles that are spinning in a rotor at a constant angular velocity, ω, are subjected to a centripetal acceleration given by

$$[2.1.A1.9] \qquad \mathbf{a} = \omega^2 r$$

where **a** is the acceleration, r is the radius, and ω is the angular velocity, in radians per second ($=2\pi \times$ revolutions per second). This **centripetal acceleration** is the acceleration necessary to keep a particle rotating about the axis, at a distance r. If the actual force is less than this, the particle will move away from the axis of rotation. In the centrifuge, the centripetal acceleration is provided by collisions with solvent particles. The net force of these collisions under ordinary conditions (i.e., not in the centrifuge) adds to the force of gravity on the particle. In the centrifuge, the centripetal force necessary to keep the particle rotating at angular velocity ω at distance r from the axis is

$$[2.1.A1.10] \qquad F_c = (m_{particle} - m_{solution})\omega^2 r$$

Here the mass of solution is the mass of the volume of solution which is displaced by the particle and is the origin of the buoyant force. Eqn [2.1.A1.10] can be rewritten as

$$[2.1.A1.11] \quad \begin{aligned} F_c &= (m - V_p\rho)\omega^2 r \\ &= (m - m\bar{v}\rho)\omega^2 r \\ &= m(1 - \bar{v}\rho)\omega^2 r \end{aligned}$$

where m is the mass of the particle, $\bar{V}$ is the partial specific volume, equal to $1/\rho$ for the particle, and ρ is the density of the solution.

As pointed out above, this centripetal force is the force required to maintain an orbit at angular velocity ω at a distance r from the axis of revolution. The source of this force is the collisions with solvent molecules, which have a net direction toward the axis of revolution only when there is a net velocity of the particle in the opposite direction. That is, a particle more dense than the solution will move toward the bottom of the tube (away from the axis of revolution) with some velocity relative to the tube. Because it is moving through the solution, it experiences a drag force that is proportional to the velocity. This drag force on a particle moving outward from the axis of revolution is given by

$$[2.1.A1.12] \quad F_D = -\beta V$$

where β is a frictional coefficient. This equation says that the drag force is proportional to the velocity but opposite in direction. A terminal velocity, v, is reached when there is a balance between F_c and F_D. From Eqns [2.1.A1.11] and [2.1.A1.12], this is

$$[2.1.A1.13] \quad m(1 - \bar{v}\rho)\omega^2 r = \beta v$$

THE MAGNITUDE OF THE CENTRIPETAL FORCE CAN BE EXPRESSED AS RELATIVE CENTRIFUGAL FORCE

The frame of reference for the analysis presented so far is the nonrotating frame of the laboratory. One can also view the situation from the accelerated, rotating frame of reference of the solution within the rotor. In this case, there is an apparent force on every particle, the centrifugal force, which is equal but opposite to the centripetal force, which appears to drive particles heavier than the solution to the bottom of the tube. This motion within the tube is opposed by the frictional force, given above, which is opposite to the direction of motion. The terminal velocity is reached when the centrifugal force is equal to the frictional force, as given by Eqn [2.1.A1.13]. Note that the centrifugal force is a fictional force which must be invented in order to apply Newton's laws in a uniformly accelerated frame of reference. The centrifugal force is just that force given by Eqns [2.1.A1.10] and [2.1.A1.11]. Usually the centrifugal force is given in multiples of g, the acceleration due to gravity:

$$[2.1.A1.14] \quad a = \frac{\omega^2 r}{g} g$$

The relative centrifugal field (RCF) is given as

$$[2.1.A1.15] \quad \text{RCF} = \frac{\omega^2 r}{g}$$

where $g = 981$ cm s^{-2} or 9.81 m s^{-2}.

THE VELOCITY OF SEDIMENTATION IS MEASURED IN SVEDBERGS OR S UNITS

The terminal velocity within the tube can be written as

$$[2.1.A1.16] \quad V = \frac{dr}{dt}$$

where r is the distance from a sedimenting particle and the center of rotation. Insertion of this definition into Eqn [2.1.A1.13] and rearranging, we obtain:

$$[2.1.A1.17] \quad m(1 - \bar{v}\rho) = \beta \left[\frac{\frac{dr}{dt}}{\omega^2 r}\right]$$

The term in the brackets on the right-hand side of the equation defines the **sedimentation coefficient, s. The sedimentation coefficient is the rate of sedimentation per unit of centrifugal force.** Sedimentation coefficients are usually of the order of 10^{-13} s. They are reported in **Svedberg** units, where 1 **Svedberg** = 10^{-13} s. This unit is named after T. Svedberg, an early pioneer in the design of centrifuges and their use in investigations of biological material.

The sedimentation coefficient is obtained experimentally by plotting the logarithm of the radius of the maximum of the concentration profile against the time. To see this, consider again the definition of the sedimentation coefficient:

$$[2.1.A1.18] \quad s = \frac{\frac{dr}{dt}}{\omega^2 r}$$

multiplying through by ω^2.

$$[2.1.A1.19] \quad \omega^2 s = \frac{\frac{dr}{dt}}{r} = \frac{d \ln r}{dt}$$

Thus the slope of a plot of $\ln r$ against t will give $\omega^2 s$.

The sedimentation coefficient varies with the concentration of solute, usually decreasing as the total protein concentration increases. The sedimentation coefficient at infinite dilution, s^0, is usually obtained by extrapolation of plots of $1/s$ against protein concentration to zero protein concentration.

Sedimentation information can sometimes inform us about molecular dimensions. Eqn [2.1.A1.17] can be rearranged to give

$$[2.1.A1.20] \quad \frac{m(1 - \bar{v}\rho)}{s} = \beta$$

where β is the frictional coefficient described in Eqn [1.2.A1.12]. The **Stokes equation** gives the frictional coefficient as

[2.1.A1.21] $$\beta = 6\pi\eta a_s$$

where η is the viscosity of the solution and a_s is the radius of the molecule, assuming spherical geometry. Thus the measurement of s and the knowledge of the molecular mass, m, and its partial molar volume allow estimate of its size.

DENSITY GRADIENT CENTRIFUGATION

Macromolecules or subcellular organelles can be sedimented through gradients of increasing density on a preparative scale to purify them. There are a variety of materials which can be used to prepare such density gradients, including CsCl, D_2O, Ficoll, glycerol, sorbitol, sucrose, and percoll. The gradients usually are formed by mixing two limiting solutions in varying proportions in order to produce the desired gradient. Gradients can be discontinuous, in which solutions of varying densities are layered on top of each other manually, or continuous. Continuous gradients can have a variety of shapes, although linear gradients are most common. Materials will sediment until they reach a solution with the same density as the particles, and then they stop sedimenting.

OTHER OPTICAL METHODS

A number of other optical approaches have proven useful for the modern investigation of cell function. These include **phase-contrast microscopy** and **differential interference microscopy**. Both of these optical techniques use variation in the refractive index of cell structures, rather than variation in light absorption, to produce contrast between the structures. These images can be clarified by using video cameras and computerized image analysis and processing. Other techniques include **fluorescence microscopy** and **confocal microscopy**. The main advantage in these techniques is that intact and living cells can be investigated.

In fluorescence microscopy, specific components of the cells can be labeled by attaching a **fluorophore** to them. The fluorophore is a fluorescent molecule. It absorbs light at one wavelength and emits it at a second, lower wavelength. The location of the fluorescent molecule is achieved by illuminating the specimen at the excitation wavelength and using filters to collect the emitted light. For example, incubating cells with a fluorophore-tagged antibody directed against a specific protein allows the study of the distribution of the protein. This technique can be used with either living or fixed cells. In other experiments, native proteins expressed by cells can be "tagged" with a fluorescent protein, green fluorescent protein (GFP by incorporating the gene for GFP into the gene for the protein. Thus the location of the tagged protein can be followed by fluorescent microscopy.

Confocal microscopy allows for optical sectioning of live cells. Confocal microscopy refers to the idea that both incident and emitted light are in focus. In a bright field, light illuminates the specimen and the objective lens is moved to focus the light. However, out-of-focus light still reaches the detector. In confocal microscopy, an aperture is placed in front of the detector so that out-of-focus light is eliminated. The result is that the image is formed from a narrow plane of in-focus light. By moving the plane up and down, one can obtain a series of optical sections. By computer techniques, three-dimensional reconstructions of the object can be obtained. Because the image is obtained and stored digitally, the resulting image can be viewed from any perspective. Confocal microscopy can be combined with fluorescent probes to achieve outstanding detail of localization.

2.2 DNA and Protein Synthesis

Learning Objectives

- Compare genotype and phenotype
- Describe the components of a nucleotide
- Name the different nucleotides that comprise DNA and RNA
- Describe how the nucleotides are connected to form single-stranded DNA or RNA
- Describe how hydrogen bonding is useful in combining antiparallel nucleotide strands to form the double helix
- Explain why antiparallel strands require special arrangement for DNA replication
- List the ways RNA differs from DNA
- List the various kinds of RNA and their function in the cell
- Know the origin of the RNA polymerases responsible for formation of the various RNA classes
- Describe what is meant by "the genetic code"
- Define and contrast "transcription" and "translation"
- Distinguish among "response elements," "intron," and "exon"
- Describe histones and their postulated function in the structure of chromosomes
- Describe what is meant by the "histone code"
- Describe how DNA is methylated and how this methylation can be passed from parent to daughter cells

DNA MAKES UP THE GENOME

As described in Chapter 1.1, almost all cells in the body have the same amount and kind of DNA. The diversity of human cell forms derives from their **expression** of different parts of the DNA. The total DNA with its division into units, called **genes**, constitutes the **genome**. Expression of a gene means that the DNA that makes up the gene is used to direct the synthesis of a specific protein. As we will see in this chapter, genes associate with a host of proteins that regulate the expression of the genes. The human genome refers to the set of genes that are normally present in humans.

The DNA in human cells is organized into compact units called **chromosomes**, meaning "colored body," which refers to their appearance in fixed and stained preparations. Each chromosome carries a defined set of genes that carries the instructions for making a set of proteins. Because each chromosome is paired, nearly every gene comes in pairs, but the two pairs are usually not identical. Paired genes carry the instructions for the synthesis of analogous materials, but they differ in the details. These alternate forms of the genes in a single person are called **alleles**. The set of alleles of a particular person is called the **genotype**. The set of proteins and other materials that the person actually makes, and which determine their outward appearance and behavior, is called the **phenotype**. Humans have 23 pairs of chromosomes.

Two chromosomes determine the sex of the individual. These are the X and Y chromosomes. Persons with two X chromosomes are genotypic females; having one X and one Y makes a genotypic male. Because the Y chromosome is smaller than the X, some of the genes carried on the X chromosome are not paired. This is the one exception to the rule that all genes are paired.

DNA CONSISTS OF TWO INTERTWINED SEQUENCES OF *NUCLEOTIDES*

DNA IS BUILT FROM NUCLEOTIDES

DNA stands for **deoxyribonucleic acid**. It is located primarily in the nucleus of cells but important parts are also present in the mitochondria. It is composed of a sequence of building blocks called **nucleotides**. These nucleotides come in two different types and four varieties. The types are the purines and pyrimidines. The purines in DNA are **adenine** and **guanine**, and the pyrimidines are **thymine** and **cytosine**. Each of these nucleotides consists of the base (adenine, guanine, thymine, and cytosine) linked to a sugar, **deoxyribose**, and **phosphate**. The chemical structures of the nucleotides are shown in Figure 2.2.1.

NUCLEOTIDES ARE LINKED TOGETHER TO FORM A CHAIN

These four bases are linked together to form a long sequence of nucleotides. The DNA is elongated by reacting a nucleotide triphosphate (with two more phosphates linked to the phosphate shown in Figure 2.2.1) on the 3′ end of an existing chain. This reaction is catalyzed by an enzyme called **DNA polymerase**. This enzyme is involved in the **replication** of DNA, where two complete DNA strands are made from

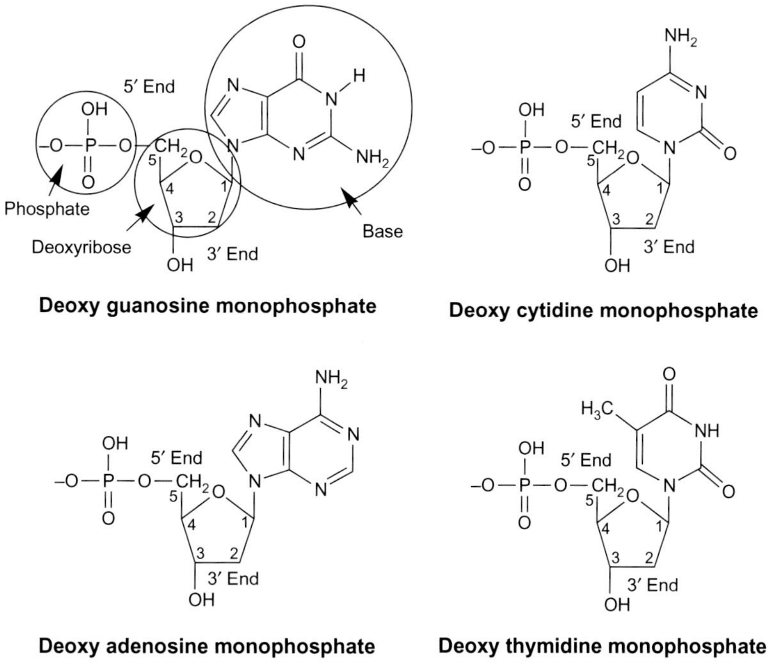

FIGURE 2.2.1 Structures of the nucleoside monophosphates.

one, using the original DNA as a template. The structure of single-stranded DNA is shown in Figure 2.2.2.

The single-stranded DNA described in Figure 2.2.2 is just half of the story. In humans, DNA is normally present as double strands that are held together by **hydrogen bonds**, as shown in Figure 2.2.3. In double-stranded DNA, adenine on one strand pairs with thymine on the opposite strand and guanine on one strand pairs with cytosine on the opposite strand. The two strands have opposite polarity: the 5′ end of one strand is opposite to the 3′ end of the other.

HYDROGEN BONDING ALLOWS FOR DNA STABILITY WITH RAPID DISSOCIATION

As discussed in Chapter 1.4, hydrogen bonds involve sharing of the positive H atom between two electronegative centers. It requires the right spatial separation and orientation of these centers, and has low dissociation energy. This allows the H-bond to form or break rapidly. However, many hydrogen bonds can stabilize large structures like DNA and proteins.

THE DNA TEMPLATE SETS THE SEQUENCE OF NUCLEOTIDES

DNA polymerase adds nucleotides to the 3′ end, using a nucleotide triphosphate as a substrate. The base on the opposite strand determines which nucleotide is incorporated. Thus DNA polymerase replicates DNA on the basis of the DNA already present. The DNA strand unwinds to form two single strands. The DNA polymerase adds nucleotides on both strands to form two complete DNA double strands. The hydrogen bonding between nucleotides is crucial to the ability of DNA to

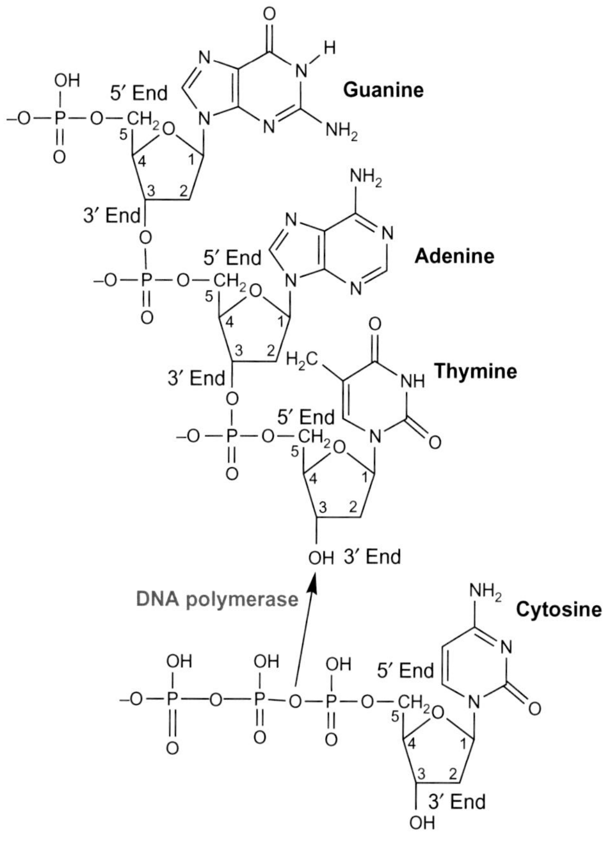

FIGURE 2.2.2 Arrangement of bases in single-stranded DNA. The phosphate–deoxyribose part of the nucleotide triphosphates forms a backbone of alternating phosphate and deoxyribose molecules. Attached to this backbone are the four bases: guanine, adenine, thymine, and cytosine. The 3′ and 5′ ends of the strand derive from the numbering of the ribose carbons.

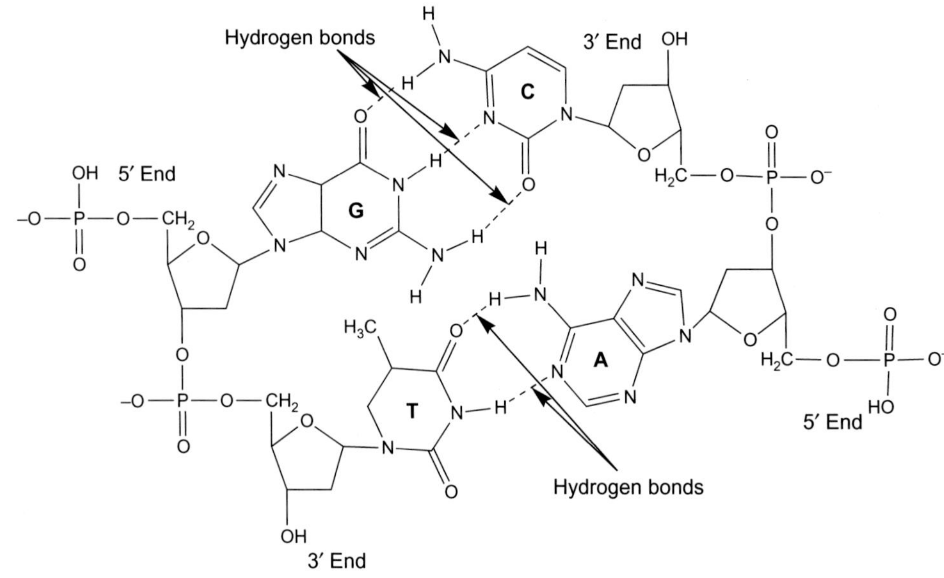

FIGURE 2.2.3 Pairing of bases in double-stranded DNA.

serve as a template for its own replication and for the synthesis of RNA.

THE DOUBLE HELIX IS SUPERCOILED IN CHROMOSOMES

The two DNA strands intertwine around each other to form a double helix. This helix can be further wrapped around proteins called **histones**. The complex of DNA and its associated proteins is called **chromatin**. The complex resembles "beads on a string" that are visible in electron micrographs (see Figure 2.2.9). The entire structure can be further coiled and coiled again to form the chromosomes. All of the DNA becomes condensed like this when the cell divides. Between divisions, the chromosomes partially "unravel" to form a less dense form of chromatin that is the working state of DNA.

THE DOUBLE HELIX POSES SPECIAL CHALLENGES FOR DNA REPLICATION

The replication of DNA by adding nucleotides only to the 3′ end of the growing DNA strand poses a problem for the duplication of double-stranded DNA because the strands are of opposite polarity. This leads to the replication of DNA in spurts, as shown in Figure 2.2.4 and described in the legend.

RNA IS CLOSELY RELATED TO DNA

RNA is **ribonucleic acid**. Structurally, it is very similar to DNA but it differs in several ways. First, the sugar in the backbone in DNA is **deoxyribose**, whereas in RNA it is **ribose**. Second, the nucleotide base thymine in DNA is replaced by **uracil** in RNA. Uracil hydrogen bonds with adenine, taking the place of thymine. Third, RNA in eukaryotic cells is single stranded. Fourth, RNA is not replicated. All of the RNA is produced from DNA using DNA as a template.

There are different kinds of RNA:
- mRNA: "messenger" RNA
- tRNA: "transfer" RNA
- rRNA: "ribosomal" RNA
- snRNA, scRNA: "small nuclear" and "small cytoplasmic" RNA
- Mitochondrial RNA.

MESSENGER RNA CARRIES THE INSTRUCTIONS FOR MAKING PROTEINS

mRNA is "messenger" RNA. mRNA is synthesized in the nucleus using the nucleotide sequence of DNA as a template. This process requires nucleotide triphosphates as substrates and is catalyzed by the enzyme **RNA polymerase II**. The process of making mRNA from DNA is called **transcription**, and it occurs in the nucleus. The mRNA directs the synthesis of proteins, which occurs in the cytoplasm. mRNA formed in the nucleus is transported out of the nucleus and into the cytoplasm where it attaches to the **ribosomes**. Proteins are assembled on the ribosomes using the mRNA nucleotide sequence as a guide. Thus mRNA carries a "message" from the nucleus to the cytoplasm. The message is encoded in the nucleotide sequence of the mRNA, which is complementary to the nucleotide sequence of the DNA that served as a template for synthesizing the mRNA. Making proteins from mRNA is called **translation**.

RIBOSOMAL RNA IS ASSEMBLED IN THE NUCLEOLUS FROM A DNA TEMPLATE

As discussed in Chapter 2.1, ribosomes are complex structures comprised of ribosomal RNA (rRNA) and a number of proteins. **RNA polymerase I** makes rRNA form a large loop of DNA called the **nucleolar organizer** region. The rRNAs then combine with proteins that migrate into the nucleolus from the cytoplasm to form the small and large ribosomal subunits. These ribosomal

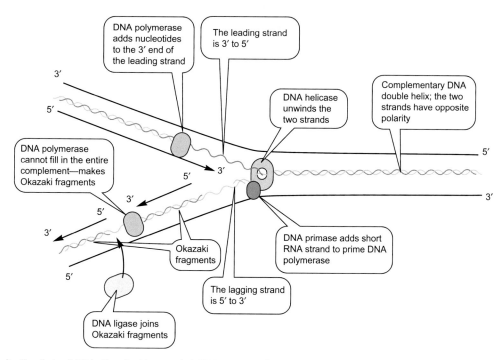

FIGURE 2.2.4 Replication fork of DNA. The double-stranded DNA consists of two strands of opposite polarity. **DNA helicase** unwinds the strands, forming two single-stranded DNAs. The **leading strand** is 3' to 5' so its complementary strand being newly synthesized is 5' to 3' and DNA polymerase adds nucleotides to the 3' end as they become available from the helicase. The **lagging strand** is 5' to 3' and its complementary strand is 3' to 5'. DNA primase adds a short RNA strand that primes the DNA polymerase. DNA polymerase then makes the complement by progressing away from the **replication fork**. While it is making DNA, the helicase unwinds more DNA so another DNA polymerase starts replication nearer the replication fork. In this way, the lagging strand is filled in with **Okazaki fragments** that bind to the lagging strand but are not connected. **DNA ligase** connects the Okazaki fragments to complete replication of the lagging strand.

subunits are then transferred to the cytoplasm where they are fully assembled to form an 80S functional ribosome and become protein factories.

TRANSFER RNA *COVALENTLY BINDS AMINO ACIDS AND RECOGNIZES SPECIFIC REGIONS OF MRNA*

How does mRNA specify the sequence of amino acids in a protein? Which amino acid is to be incorporated into the protein is specified by a sequence of three nucleotides called a **codon**. The mRNA triplets do not directly recognize and specify the amino acids; they do so through the use of another kind of RNA called **transfer RNA** or **tRNA**. These remarkable molecules are adapters that can link with an amino acid and recognize the triplets of nucleotides on the mRNA, the codons. They do this by containing a sequence complementary to the codon: the **anticodon**. The function that maps triplets of nucleotides on the mRNA to specific amino acids is called the **genetic code**. Figure 2.2.5 shows the genetic code in look-up table format.

The tRNA consists of a single strand of RNA from 70 to 90 nucleotides long that is held together by hydrogen bonding within nucleotides on the same chain. One end of the tRNA allows for covalent attachment of an amino acid. Another section of the tRNA contains a sequence of three nucleotides that forms the anticodon. Precursors to the tRNA are transcribed from DNA by **RNA polymerase III**.

Another key in the formation of proteins is the attachment of amino acids to the specific tRNA. Specific enzymes called **aminoacyl-tRNA synthetases** couple the amino acid to the appropriate tRNA. There is a different synthetase for each amino acid. One attaches glycine to tRNA^{Gly}, another attaches alanine to tRNA^{Ala}, and so on. These synthetases must recognize both the amino acid and the tRNA that contain the right anticodon. The overall processing of RNA and protein synthesis is shown in Figure 2.2.6. Translation is shown in Figure 2.2.7.

THE GENETIC CODE IS A SYSTEM PROPERTY

The genetic code shown in Figure 2.2.5 lists an amino acid or other signal (such as STOP) for every triplet nucleotide in mRNA. The code is the function that maps the nucleotide sequence onto instructions for protein synthesis. That is,

[2.2.1] $$f\{N_i : \{A, U, C, G\}\} = \{A_j : \{\text{control steps, amino acids}\}\}$$

This describes the function that turns a set of N_i nucleotides on the mRNA (selected from adenosine, uracil, cytosine, and guanosine nucleotides) into a set of A_j amino acids. Where does this code reside? The code should not be confused with the message or the means of writing the message. Therefore, the code itself is not a

First position 5' end	Second position				Third position 3' end
	U	C	A	G	
U	Phe	Ser	Tyr	Cys	U
	Phe	Ser	Tyr	Cys	C
	Leu	Ser	STOP	STOP	A
	Leu	Ser	STOP	Trp	G
C	Leu	Pro	His	Arg	U
	Leu	Pro	His	Arg	C
	Leu	Pro	Gln	Arg	A
	Leu	Pro	Gln	Arg	G
A	Ile	Thr	Asn	Ser	U
	Ile	Thr	Asn	Ser	C
	Ile	Thr	Lys	Arg	A
	Met	Thr	Lys	Arg	G
G	Val	Ala	Asp	Gly	U
	Val	Ala	Asp	Gly	C
	Val	Ala	Glu	Gly	A
	Val	Ala	Glu	Gly	G

Amino acid	Three-letter code	One-letter code	Codons
Alanine	Ala	A	GCA GCC GCG GCU
Arginine	Arg	R	AGA AGG CGA CGC CGG CGU
Asparagine	Asn	N	AAC AAU
Aspartic acid	Asp	D	GAC GAG
Cysteine	Cys	C	UGC UGU
Glutamic acid	Glu	E	GAA GAG
Glutamine	Gln	Q	CAA CAG
Glycine	Gly	G	GGA GGC GGG GGU
Histidine	His	H	CAC CAU
Isoleucine	Ile	I	AUA AUC AUU
Leucine	Leu	L	UUA UUG CUA CUC CUG CUU
Lysine	Lys	K	AAA AAG
Methionine	Met	M	AUG
Phenylalanine	Phe	F	UUC UUU
Proline	Pro	P	CCA CCC CCG CCU
Serine	Ser	S	AGU AGC UCU UCC UCA UCG
Threonine	Thr	T	ACA ACC ACG ACU
Tryptophan	Trp	W	UGG
Tyrosine	Tyr	Y	UAC UAU
Valine	Val	V	GUA GUC GUG GUU

FIGURE 2.2.5 The genetic code. Each amino acid that is incorporated into a protein is specified by a triplet sequence of nucleotides on the mRNA. These triplets are called codons. Which codons specify which amino acids is shown here in two formats. The left format shows which amino acids correspond to which codons given as a first, second, and third position. Here we use the Biochemists' shorthand for RNA bases and amino acids, where U = uracil, C = cytosine, A = adenine, and G = guanosine, and each amino acid is given by its three-letter shorthand designation, where Phe = phenylalanine, Ser = serine, Tyr = tyrosine, Cys = cysteine, Leu = leucine, Pro = proline, His = histidine, Arg = arginine, Gln = glutamine, Ile = isoleucine, Thr = threonine, Asn = asparagine, Lys = lysine, Met = methionine, Val = valine, Ala = alanine, Asp = aspartic acid, Gly = glycine, and Glu = glutamic acid. The right format lists the amino acids together with their three-letter designation and single-letter designation, with a list of the codons that specify them.

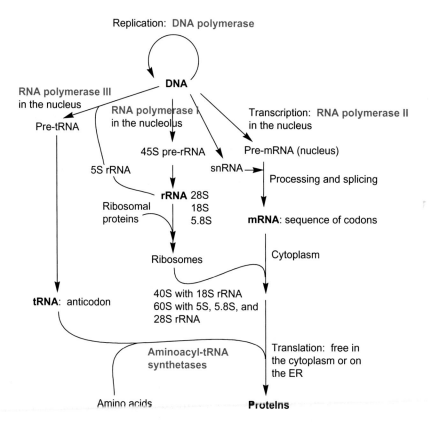

FIGURE 2.2.6 Processing of RNA and DNA. Replication of DNA is accomplished by DNA polymerase using the original DNA as a template. Messenger RNA is synthesized in the nucleus as a precursor that is processed to form the final mRNA. The synthesis of mRNA is called transcription and is accomplished by RNA polymerase II. The final mRNA travels to the cytoplasm where it binds to ribosomes. The ribosomal subunits are formed in the nucleolus from proteins and ribosomal RNA that is made as a precursor and cut into the final rRNA strands. rRNA is made from DNA by RNA polymerase I. The mRNA directs the sequential addition of amino acids to form proteins in a process called translation. This requires tRNA, made from DNA by RNA polymerase III.

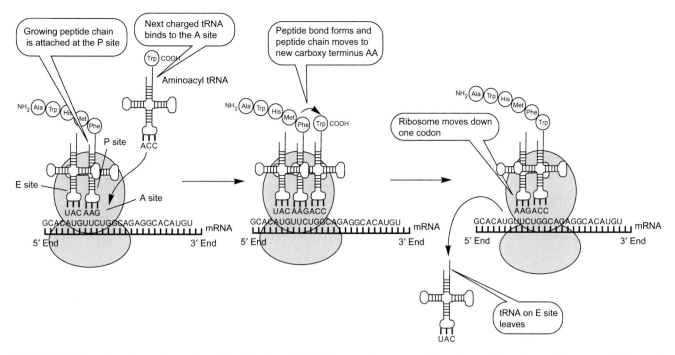

FIGURE 2.2.7 Elongation of a growing polypeptide chain. The mRNA binds to a ribosome consisting of some 82 proteins and 4 separate rRNA strands. There are 3 tRNA binding sites: an **aminoacyl** or A site, a **peptidyl** or P site, and an **exit** or E site. The aminoacyl tRNA is escorted to the ribosome by an **elongation factor** that hydrolyzes GTP. The scheme begins with a short polypeptide bound to a tRNA at the peptidyl site along with the tRNA^{Met} that remains at the exit site. The next aminoacyl tRNA binds to the aminoacyl site; in this case it is tRNA^{Trp} that is charged with its amino acid. The peptide bond is formed between the peptide and its next amino acid on the carboxy terminus. The ribosome shifts over one codon; the tRNA at the exit site leaves and the former occupant of the peptidyl site now occupies the exit site. The peptide now occupies the peptidyl site one codon further along the mRNA.

property of the mRNA. The DNA itself also does not contain the code, as its message has no meaning without the tRNA. The tRNA is synthesized from other parts of the DNA that are not directly transcribed for protein synthesis. Is the code in the tRNA that links the triplets of mRNA to a specific amino acid? Or is the code in the aminoacyl-tRNA synthetases, the enzymes that couple the amino acid to the specific tRNA? If the code is in the aminoacyl-tRNA synthetases, then the code is in some sense also in the DNA because the DNA directs the synthesis of the aminoacyl-tRNA synthetases! But the DNA does not "make" proteins; mRNA does not "make" proteins. The proper proteins are synthesized with DNA as the store of information, mRNA carrying the message, tRNA converting, in a single step, the nucleotide information into the protein information, and preexisting proteins catalyzing the entire series of events. The code itself is an emergent property. The genetic code does not exist in any single component of the mechanism for making proteins. It is not "in" the DNA, or the mRNA or the tRNA or the tRNA synthetases. Rather, it is a system property that emerges from the interactions of all of these parts.

REGULATION OF DNA TRANSCRIPTION DEFINES THE CELL TYPE

The differentiation of cell types produces the wide spectrum of cell types in the body, but our understanding of the process is still rudimentary: we do not know how to change one cell type into another. We do know, however, that the hallmark of differentiation is selective expression of the cell's DNA. Selective expression of DNA as proteins requires selective transcription. Initiation of transcription by RNA polymerase II requires a number of specific proteins called **transcription factors**. These come in two flavors: some are required for activity at all genes and are therefore called general transcription factors. Other transcription factors bind to DNA sequences that control the expression of specific genes. The overall process is shown in Figure 2.2.8.

Most genes contain both transcribed and untranscribed regions. On the 3' end of the DNA gene is a specialized sequence called the **promoter** that helps regulate gene expression. This region contains a TATAA sequence (called a TATA box) some 25–30 nucleotides upstream from the initiation site. Initiation begins when a transcription factor **TFIID** (transcription factor polymerase II) binds to the DNA TATAA sequence. Some promoters do not contain a TATA box, but instead contain an initiator sequence. Nevertheless, TFIID is involved in initiation of transcription even on promoters that lack the TATA box.

In addition to these general transcription factors, there are a number of transcription factors that may act as **enhancers** or **repressors** of gene expression. These factors interact with specific sites on the DNA that are generally further away from the gene than the promoter region and act as regulatory elements for gene transcription. The signals that turn on or turn off the production of transcription factors ultimately determine the phenotypic fate of cells. How these transcription factors work is still being investigated. An example of this is the

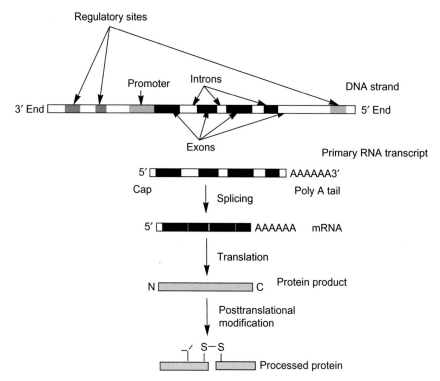

FIGURE 2.2.8 Overall processes of transcription and translation. The gene has a promoter that often contains a TATAA sequence for initiation of transcription by RNA polymerase II, several **exons** (expressed or coding regions of the DNA sequence) and **introns** (intervening or noncoding regions), and regions that bind transcription factors to either enhance or repress expression. The primary gene transcript contains the sequence of bases complementary to the introns and exons. This primary RNA transcript is then further processed on large complexes called **spliceosomes**, composed of protein and **small nuclear RNAs** or **snRNA**. The spliceosomes remove the introns by clipping them out and then splice together the remaining exons to form a sequential mRNA. This mRNA is then translated in the cytosol to a primary protein product which may then require further processing.

steroid hormone receptors. Steroid hormones and similar materials like vitamin D bind to receptor proteins that in turn bind to specific sequences on the DNA called **response elements**. Binding of the receptor proteins then activates gene expression. A number of accessory proteins are required for this process.

THE *HISTONE CODE* PROVIDES ANOTHER LEVEL OF REGULATION OF GENE TRANSCRIPTION

As described earlier, the double helix of DNA winds around specific proteins called histones. These histones form the basic unit of chromatin called the **nucleosome**. A schematic of this structure is shown in Figure 2.2.9. When wrapped up in this way, DNA is inaccessible to RNA polymerase II and so cannot be transcribed to form mRNA. Most cells sequester away large portions of their DNA in this way. In order to express DNA, cells must unwrap it from the chromatin. Determining which sections of DNA should be unwrapped is the first step in regulating gene transcription. This is accomplished through covalent modifications of the histones. Covalent modification of histones after synthesis is called **posttranslational modification**. Histones undergo

- acetylation of lysine and arginine amino acids in the histones;
- methylation of lysine and arginine amino acids;
- phosphorylation of serine and threonine amino acids in the histones;
- ubiquitinylation of lysines.

All of these modifications are accomplished by enzymes that must themselves be regulated. These enzymes possess **histone acetyltransferase** (HAT) activity, **histone deacetylation** (HDAC) activity, **histone methyltransferase** (HMT) activity, and **histone kinase** activity. The function of these enzymes is evident from their names: HAT adds acetyl groups to the histones; HDAC removes them; HMT adds methyl groups; and histone kinase phosphorylates the histones. Figure 2.2.10 summarizes the various known modifications of the core histones.

These posttranslational modifications of the histones have functional consequences. For example, the combination of acetylation at lysine at position 8 on H4 and lysine at position 14 on H3 with phosphorylation of serine at position 10 on H3 is associated with transcription. Trimethylation of lysine at position 9 on H3 with the lack of acetylation of H3 and H4 correlates with transcriptional repression. These observations have led to the hypothesis that gene expression is regulated in part by a "histone code." This hypothesis requires two components:

1. Specific enzymes write the code by adding or removing modifications at specific sites in the histones.
2. Other proteins recognize the histone markers and interact with histones and other factors to mediate functional effects.

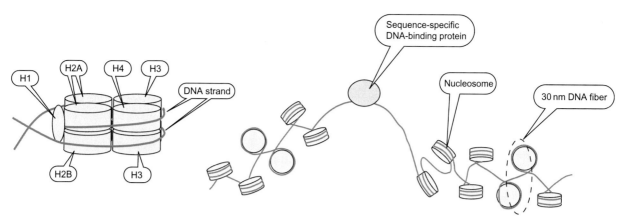

FIGURE 2.2.9 Schematic diagram of nucleosome structure and 30-nm DNA fiber. DNA is wrapped around a complex of eight histone proteins: a tetramer of H3 and H4 and two dimers of H2A and H2B, to form the nucleosome. Strands of DNA bound to nucleosomes resemble beads on a string. The interaction of DNA with the nucleosomes is altered by H1, a linker histone, that enables the nucleosomes to condense to form a 30-nm DNA strand. H1 interacts with both the histones in the core and DNA. The nucleosome contains 147 base pairs of DNA wrapped nearly twice around the core histones. A short "linker" of 10–60 base pairs separates each nucleosome from its neighbor. The linear sequence of nucleosomes forms "beads on a string" in electron micrographs. This becomes more highly condensed to form 30-nm-thick fibers that are stabilized by the H1 histones.

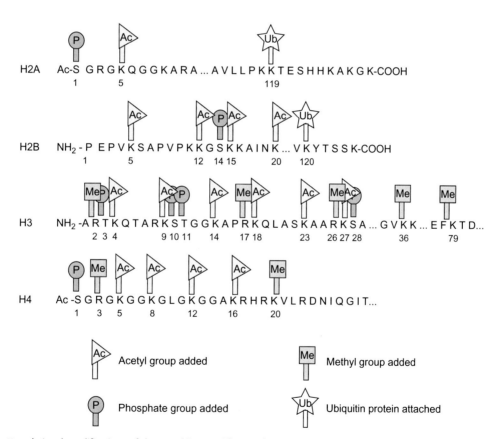

FIGURE 2.2.10 Posttranslational modifications of the core histones. The one-letter amino acid abbreviation follows that in Figure 1.3.5. The numbers below the amino acids refer to the number of amino acids in the sequence, starting from the amino terminus. Source: *From C.L. Peterson and M.-A. Lanie, Histones and histone modifications,* Current Biology *14:R546–R551, 2004.*

The histone code may be a misnomer if it is viewed as being like the genetic code. In the genetic code, given triplets of nucleotides on mRNA *always* produce the same result, independent of which cell or tissue is being analyzed. The term "histone code" implies that a particular combination of histone modifications will *always* produce the same biological result. Evidence suggests that the same pattern of histone modifications can be interpreted differently by different cells depending on the gene and its cellular context.

DNA METHYLATION REPRESSES TRANSCRIPTION

A methyl group can be added to the 5 position of cytosine to form 5 methyl cytosine by the action of DNMT, DNA methyltransferase, as shown in Figure 2.2.11. These enzymes come in two classes. DNMT1 is responsible for maintenance methylation, which methylates the new strand of recently replicated DNA, so that the methylation pattern is passed down from stem cells to daughter cells (see Figure 2.2.12). This explains the unidirectionality of most developmental processes. Stem cells become differentiated cells and the differentiated cells maintain their differentiation, partly through a pattern of DNA methylation. The second class of DNA methylation transferases, represented by DNMT3a and DNMT3b, is responsible for de novo methylation. Both the maintenance methylation enzymes and de novo methylation enzymes methylate cytosine in a CpG sequence in the DNA (see Figure 2.2.11).

The consequence of DNA methylation is typically repression of transcription for the genes that are methylated. The methyl group does two things: it interferes with the binding of transcription factors that eventually recruit RNA polymerase II, and it also allows the binding of a set of proteins that specifically recognize the methylated DNA, and these proteins recruit histone deacetylases that modify the histones associated with the DNA. These actions result in a repressed transcription of the methylated parts of the DNA.

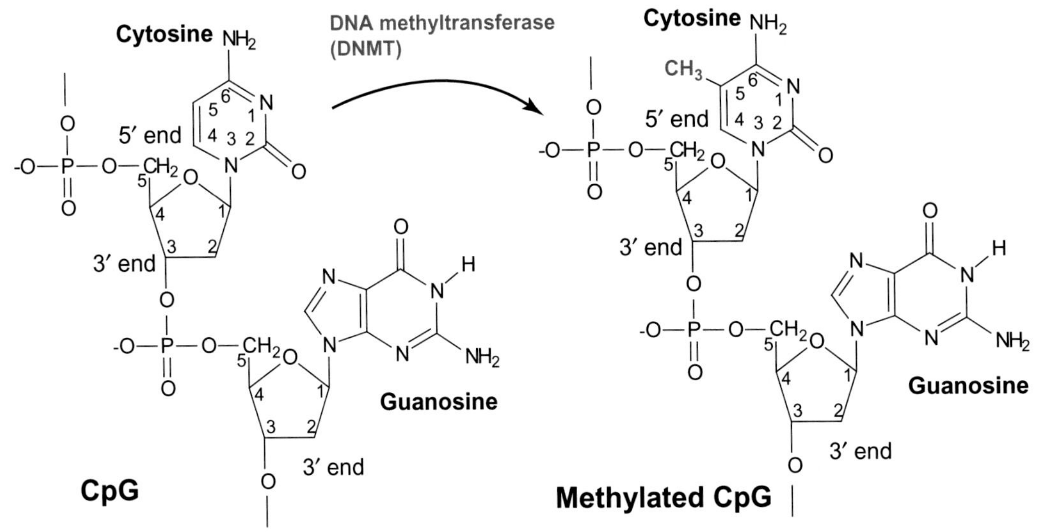

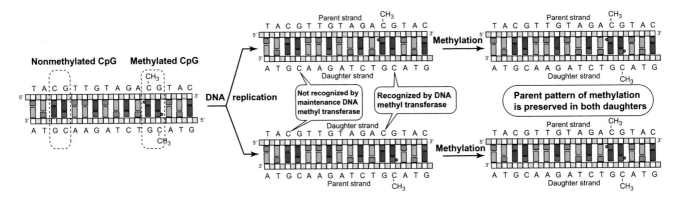

FIGURE 2.2.11 DNA methylation. DNA is methylated at sites with the sequence CpG by DNA methyltransferase. It can also be demethylated, but the process is not simply a reversal of the methylation reaction. Demethylases form hydroxymethyl cytosine, which is then cut out and replaced with cytosine by DNA repair mechanisms.

FIGURE 2.2.12 Maintenance methylation. The parent DNA has some CpG sequences that are methylated, and some that are not. These are determined by de novo methylation reactions catalyzed by DNA methyl transferase 3A and 3B (DNMT3a and DNMT3b). When the DNA is replicated, the daughter strands are not methylated. The maintenance DNA methyl transferase (DNMT1) recognizes the CpG sequence on the daughter strand that corresponds to the methylated CpG sequence on the parent strand, and then methylates the daughter strand cytosine. DNMT1 does not bind to the unmethylated CpG sequences and therefore does not methylated these. The result is that the methylated pattern of the parent strand is replicated in the methylated pattern of the two daughter strands. In this way, the pattern of gene repression in a differentiated cell line is passed onto its descendants.

SUMMARY

Genetic information is stored in DNA in the nucleus and mitochondria of cells. DNA consists of two strands of nucleotides on a phosphodeoxyribose backbone. The two strands form a double helix that is stabilized by the formation of hydrogen bonds between nucleotide bases on the two strands. Replication of DNA is based on making strands complementary to the two strands produced when the double-stranded DNA is unwound. The four bases include two purines (adenine and guanine) and two pyrimidines (cytosine and thymine). They pair up as A:T and G:C, with two hydrogen bonds between A and T and three between G and C.

DNA serves as a template for making a variety of RNA types: RNA polymerase I makes ribosomal RNA (rRNA) from nuclear DNA; RNA polymerase II makes messenger RNA (mRNA); RNA polymerase III makes transfer RNA (tRNA). All of these contribute to the synthesis of proteins and control of gene expression.

Protein synthesis occurs on the ribosomes, using mRNA as a template for tRNA. The ribosomes themselves are large and complex structures consisting of rRNA and a number of proteins. The ribosomes consist of 60S and 40S subunits. The 60S subunit has three rRNAs (28S, 5S, and 5.8S) and 49 other proteins; the 40S subunit has an 18S rRNA and another 33 proteins. The ribosome binds to mRNA and provides a reaction site for the peptide bound to tRNA (P site), a second reaction site for the next amino acid covalently attached to its tRNA (A site), and a third site for the tRNA about to leave the ribosome (E site). The ribosome brings the peptide carboxyl terminal close to the amino terminal of the next amino acid, and then forms the peptide bond, simultaneously shifting the peptide from the P to the A site. The mRNA provides a sequence of nucleotides. Groups of three nucleotides form codons that are recognized by complementary nucleotide sequences on the tRNA—the anticodons. The specific binding of anticodon to codon allows the mRNA to determine the sequence of amino acids in the proteins being made. The attachment of specific amino acids to specific tRNAs, however, is assured by the tRNA synthetases that hook the amino acids onto tRNA.

mRNA directs the synthesis of specific proteins by virtue of its sequence of codons. The set of proteins that are made determines the type of cell, because cell structure and activity derives from the kinds and amounts of proteins expressed by the cell. Cells have developed elaborate methods for determining what parts of the DNA are transcribed into mRNA. These methods involve transcription factors, enhancers and repressors, and the histone code. The DNA in cells is wrapped around a complex of histone proteins, forming a nucleosome. Modification of the histones allows specific sections of the DNA to be either used to make proteins or silenced.

A second method for repression of gene transcription is methylation of cytosines in the DNA in sequences of CpG. These are originally methylated by de novo methylation, but the pattern of methylation can survive DNA replication through a maintenance methyltransferase. The result is a stable pattern of gene repression that survives proliferation. DNA methylation interacts with histone modification to determine which DNA sequences will be silenced.

REVIEW QUESTIONS

1. What is the genotype? What is the phenotype? What is an allele? What is the usefulness in having two copies of each gene?
2. What are the parts of a nucleotide? Name the purine bases. Name the pyrimidine bases.
3. How are nucleotides linked together to form single-stranded DNA or RNA? What distinguishes the 5' end from the 3' end? What enzyme makes DNA from nucleotides?
4. What holds double-stranded DNA together? Why are hydrogen bonds useful? How many hydrogen bonds link guanine to cytosine? How many such bonds link adenine to thymine?
5. During DNA replication, what determines the sequence of DNA in the new strands?
6. Does DNA polymerase add nucleotides at the 5' end or the 3' end of the strand? What problem does this make for DNA replication? What is an Okazaki fragment?
7. How does RNA differ from DNA? What is mRNA? What is tRNA? What is rRNA? What RNA polymerase makes mRNA? Which makes tRNA? rRNA?
8. What is a codon? Is it on mRNA, DNA, or tRNA? What is an anticodon?
9. What couples tRNA with amino acids? Are these enzymes specific for the anticodon?
10. What is a ribosome? What is the A site? What is the P site? What is the E site?
11. What is the genetic code? Where is it located in the cell?
12. What is meant by "transcription"? What is meant by "translation"?
13. How are inactive portions of DNA locked up by the cell? How do they get unlocked? What is a "response element"? What is an "intron"? What is an "exon"?
14. What are histones? What promotes DNA binding? What promotes DNA unraveling from the histones? What is meant by "the histone code"?
15. What is DNA methylation? Where does it occur? What is the consequence of DNA methylation? How does DNA methylation pattern get passed on to daughter cells?

2.3 Protein Structure

Learning Objectives

- Draw the basic structure of an amino acid, specifying the amino group, carboxyl group, alpha carbon, and R group
- For each amino acid, tell whether it is polar, nonpolar, acidic, or basic
- Draw the reaction describing the formation of a peptide bond
- Describe what is meant by the primary structure of a protein
- Describe what is meant by the secondary, tertiary, and quaternary structure of a protein
- Describe the four kinds of noncovalent interactions that stabilize protein structure
- Define posttranslational modification
- List four major classes of posttranslational modification
- List the major kinds of chemical modification of proteins
- List the amino acids involved in glycosylation and its overall function in proteins
- Describe the consequence of gamma carboxylation of proteins and name the vitamin involved
- Describe two distinct ways of varying the activity of proteins in cells
- Describe three distinct ways membrane proteins can be anchored in the membrane by covalent modification

AMINO ACIDS MAKE UP PROTEINS

In Chapter 2.2, we described how proteins are made on ribosomes by linking amino acids together in long chains. Which amino acids make up a specific protein, and in which order, is determined by the sequence of triplet nucleotide codons residing on the mRNA, which in turn is produced on a DNA template in the nucleus. Because these constituent amino acids determine the detailed shape of the protein surface, we should find the key to protein activity in the three-dimensional arrangement of these amino acids.

The general chemical structure of the amino acids is shown in Figure 2.3.1. The amino acids are named for the two functional groups each of them possesses: an **amino group** ($-NH$) and a **carboxylic acid group** ($-COOH$). These two groups have widely different reactivities. The amino group is basic and will be positively charged at neutral pH, forming an ammonium ion ($-NH_3^+$). The carboxyl group is acidic and will form a negatively charged carboxyl ion ($-COO^-$) at neutral pH. Thus at neutral pH, many amino acids will have a positive charge on the amino end and a negative charge on the carboxyl end. Compounds possessing both positive and negative charges simultaneously are called **zwitterions**, as shown in Figure 2.3.2.

Although the amino and carboxyl groups are important for the structure of proteins, the tremendous variety of structure and function of proteins is produced by the R groups (R stands for "residue" and refers to the part of the amino acid other than the amino or carboxyl group covalently bonded to the α carbon). There are 20 different amino acids differing only in their R groups. The R groups confer properties on the amino acids that classify them as **nonpolar**, **polar**, **basic**, or **acidic**. Each of these groups is shown in Figures 2.3.3–2.3.5. Biochemists often use a three-letter abbreviation for the amino acids or a one-letter code for a further shorthand. These abbreviations are shown under each structural formula (see Figure 2.3.6).

As their name implies, the nonpolar amino acids contain side groups that are nonpolar and therefore not attracted to water. Because water is so attracted to itself, these groups naturally are repulsed by the water and are said to be hydrophobic, or water hating. For glycine and alanine, the nonpolar side groups are so small that the amino acid has little preference for either a watery (hydrophilic) or nonwatery (hydrophobic) environment. Phenylalanine and tryptophan are highly hydrophobic and seek environments away from water.

HYDROPHOBIC INTERACTIONS CAN BE ASSESSED FROM THE PARTITION COEFFICIENT

We can estimate the strength of hydrophobic interactions by measuring the partitioning of a material between the water phase and an immiscible solvent such as *N*-octanol or ethyl acetate, whose properties may be regarded as being similar to the interior of protein away from the water phase. The partition coefficient is

[2.3.1] $$k_s = \frac{[x]_{\text{organic phase}}}{[x]_{\text{water phase}}}$$

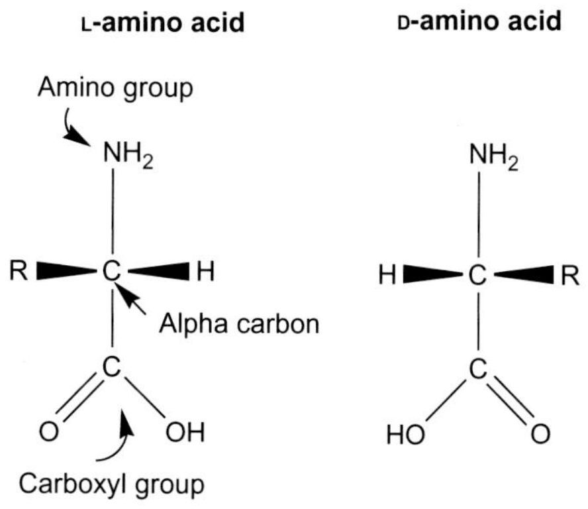

FIGURE 2.3.1 Chemical structure of the amino acids. Amino acids consist of a central carbon, the alpha carbon, to which are bonded a carboxyl group, an amino group, a hydrogen atom, and a variable group, called R. Because the alpha carbon has four different groups bonded to it, it is an asymmetric carbon that is capable of stereoisomerism. All of the amino acids can be made in L- and D-form (L and D originally described the ability of compounds to rotate the plane of polarized light; L for *levo*, meaning "left," and D for *dextro* meaning "right." The symbols L and D for the amino acids do not actually refer to the direction of rotation of the plane of polarized light). All naturally occurring amino acids in higher organisms are the L-form. The L and D forms are mirror images of each other and are not superimposable.

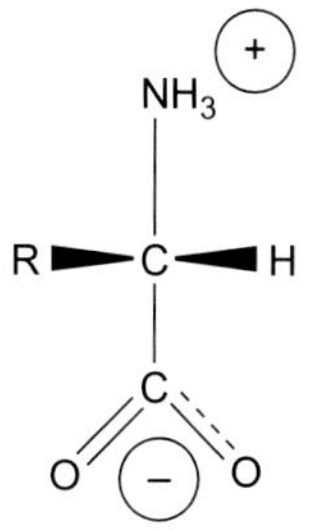

FIGURE 2.3.2 Zwitterion form of an amino acid. At neutral pH (pH = 7), the carboxyl group is dissociated to form the anionic $-COO^-$ group, and the amino group binds a H^+ ion to form $-NH_3^+$. Thus both ends of the amino acid are charged but with opposite polarity.

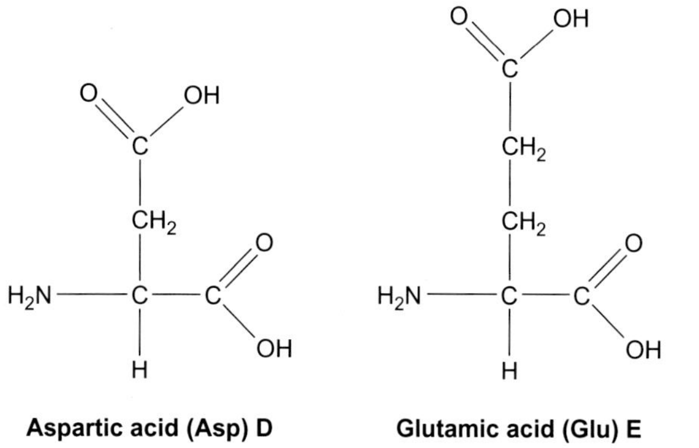

FIGURE 2.3.3 The acidic amino acids aspartic acid and glutamic acid. The R group in both cases contains another carboxylic acid group that dissociates to form an H^+ ion and a negatively charged carboxyl group ($-COO^-$). The presence of this group classifies these amino acids as acidic amino acids. The negatively charged carboxyl group imparts a negative charge to the region of any protein that contains these groups.

where the concentrations are the equilibrium concentrations, as shown in Figure 2.3.7. This equilibrium condition obeys Eqn [1.7.32]:

$$\Delta G^0 = -RT \ln K_{eq}$$

[2.3.2] $$\Delta G_T^0 = -RT \ln \frac{[x]_{\text{organic phase}}}{[x]_{\text{water phase}}}$$

where ΔG_T^0 is the standard Gibbs free energy change for transfer of substance X from water to the organic phase. Although this can represent the strength of hydrophobic interaction of a small molecule, it cannot reliably predict the behavior of a polymer of the material or of the material when it participates in a heteropolymer such as a protein. This is true for amino acids in particular because some of their functional groups that determine the octanol/water partition are altered when the amino acids are linked up to make proteins. Nevertheless, parts of molecules can be described as being hydrophobic or hydrophilic, depending on whether or not they would partition themselves into the organic phase.

Hydrophobicity has two components: the "squeezing out" of water-insoluble components due to the attraction of water for itself and the self-association of nonpolar materials due to dipole–dipole interactions, dipole–induced-dipole, and induced-dipole–induced-dipole interactions (London dispersion forces) discussed in Chapter 1.4. Water repels nonpolar materials because the surface of the nonpolar material cannot form hydrogen bonds with the water, and its shape therefore reduces the number of hydrogen bonds that the water can form at the surface—the water molecules at the surface are in a higher energetic state, having some of their hydrogen bonds broken. Thus it takes energy to insert a nonpolar material into the water phase.

The amino acids can be classified solely on the basis of their hydrophobicity or hydrophilicity. In this case, we have three categories: the hydrophilic amino acids include aspartic acid, glutamic acid, lysine, and arginine; the hydrophobic amino acids include valine, leucine, isoleucine, methionine, phenylalanine, and tryptophan; the "neutral" amino acids, those that are neither strongly hydrophilic nor strongly hydrophobic, include glycine, alanine, serine, histidine, proline, threonine, cysteine, glutamine, and asparagine.

PEPTIDE BONDS LINK AMINO ACIDS TOGETHER IN PROTEINS

Proteins are formed as a linear, unbranched chain of amino acids. The amino acids are covalently linked by a **peptide bond** formed between the amino group of one amino acid and the carboxyl group of the next. The formation of the peptide bond is a dehydration reaction, as shown in Figure 2.3.8.

Cells make proteins by the sequential addition of amino acids to the carboxyl terminus of the growing chain. This is accomplished on the ribosome when specific tRNAs (transfer RNAs) with their specific bound amino

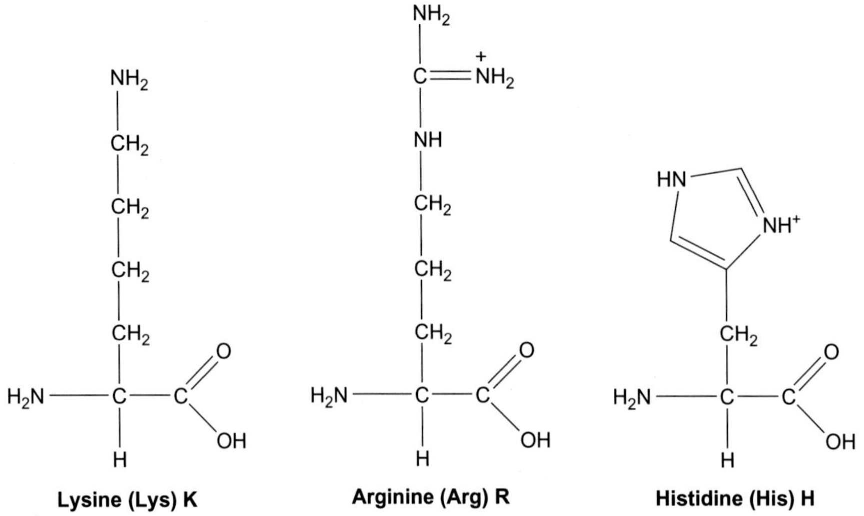

132 QUANTITATIVE HUMAN PHYSIOLOGY: AN INTRODUCTION

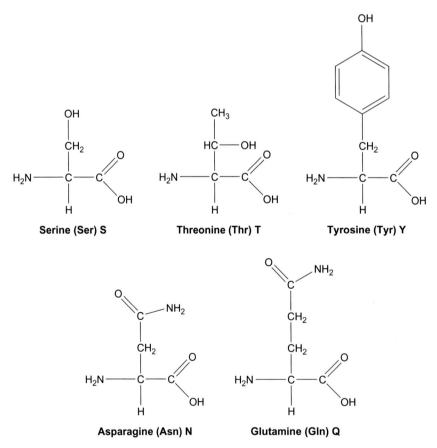

FIGURE 2.3.4 The chemical structures of the basic amino acids, lysine, arginine, and histidine. These contain ionizable NH groups so that at neutral pH, these residues would contribute positive charge to the protein.

FIGURE 2.3.5 The chemical structure of the polar amino acids. The polar amino acids contain groups that can form hydrogen bonds with water. These groups are soluble in water. As a result, those portions of proteins which contain large numbers of polar residues will usually be exposed to water. They help solubilize proteins in solution, so we might expect soluble proteins to be coated with polar, acid, or basic R groups.

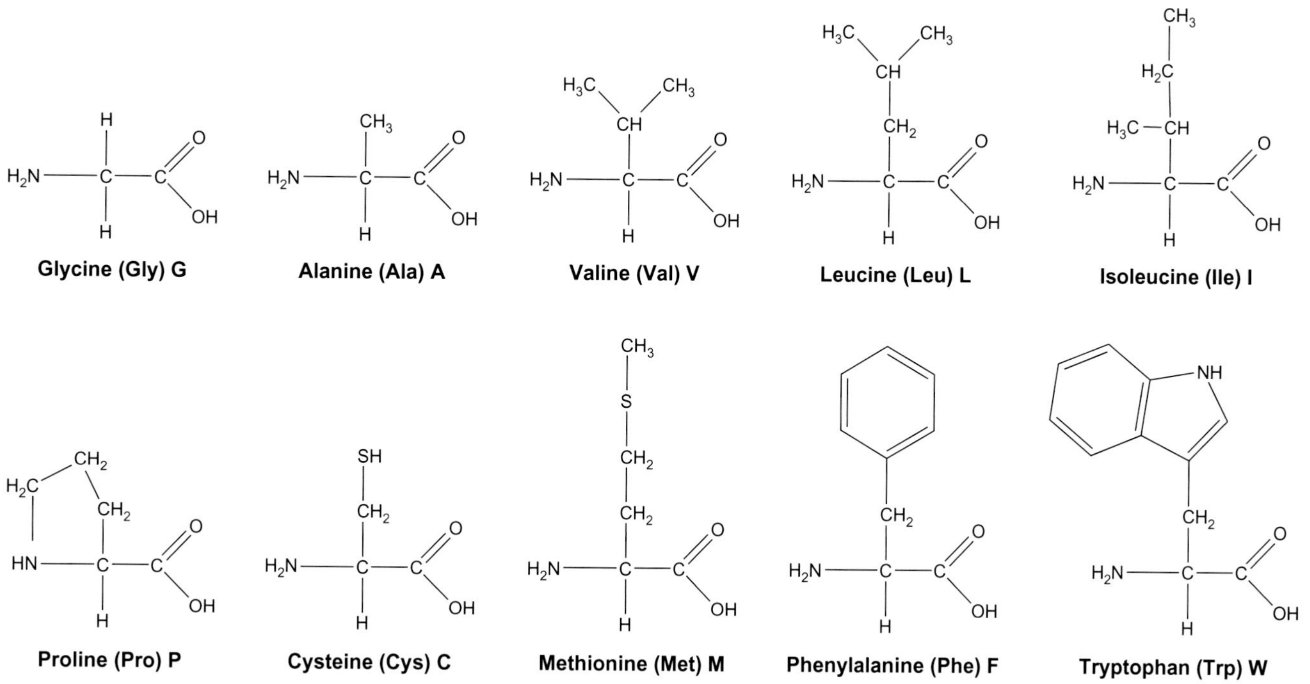

FIGURE 2.3.6 Chemical structures of the nonpolar amino acids.

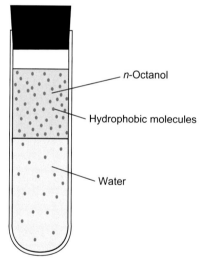

FIGURE 2.3.7 Assessment of hydrophobicity by the partition coefficient between octanol and water. A hydrophobic molecule is dissolved in n-octanol and then an aliquot of the solution is placed in contact with water and shaken. The material will distribute itself between the two phases. The ratio of the concentrations at equilibrium in the organic phase to the water phase defines the partition coefficient, which is an equilibrium constant that can be used to calculate the free energy of transfer from the organic phase to the water phase. Hydrophobic materials will have a higher concentration in the n-octanol phase; hydrophilic materials will have a higher concentration in the water phase.

PROTEIN FUNCTION CENTERS ON THEIR ABILITY TO FORM REACTIVE SURFACES

Proteins perform a variety of functions in cells, and these can be broadly classified as **structural**, **catalytic**, **transport**, or **regulatory**. All of these functions require the surface of the protein to interact with other surfaces.

As discussed in Chapter 1.5, enzymes provide a surface on which biochemical reactions can occur. These reactions can occur in the water phase, but only slowly because of the large activation energy required. Binding of the substrates to the enzyme surface alters the path of the reaction, enabling it to proceed more quickly. This lower energy pathway is produced by the interaction of the protein surface with the substrates of the biochemical reaction. The details of the protein surface enable catalysis, the speeding up of a reaction without appearing in the overall stoichiometry of the reaction. The surface of the protein must closely match the surface of the substrates, which means that it cannot match other substrates well, because the protein surface cannot match two different surfaces. Thus the protein surface simultaneously enables catalysis while it confers **specificity**—the enzyme works only with specific substrates.

In the same way that protein enzymes interact with biochemical substrates on the surfaces of the proteins, structural proteins interact with other components by virtue of their surfaces. The proteins that make up connective tissue are sticky. They bind to themselves and to a variety of other proteins. Their surfaces make them sticky. Closely matching surfaces allow proteins to interact with other proteins, thereby allowing one protein to

acid bind to the appropriate triplet codon of the mRNA. The amino acid is attached to the tRNA via its carboxyl group. When the peptide bond is formed, the entire growing chain is transferred to the free amino terminal of the next amino acid. The ribosome then moves one frame (nucleotide triplet) and the process is repeated (see Chapter 2.2).

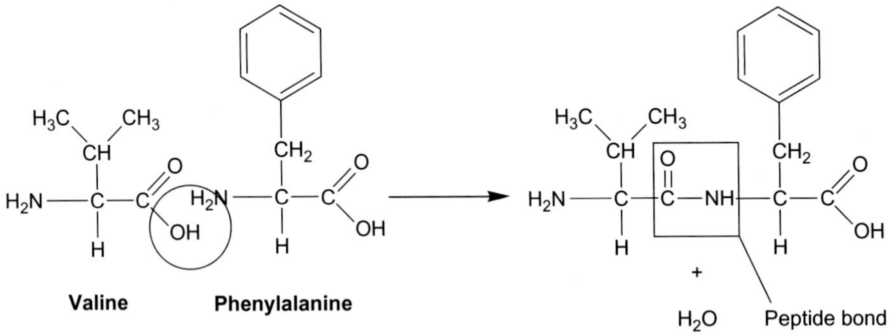

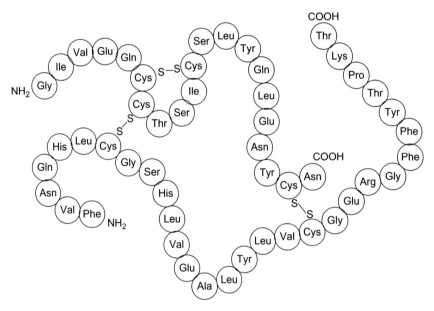

FIGURE 2.3.8 Formation of a peptide bond between two amino acids. The overall reaction is shown. The actual reaction involves many intermediary steps, catalyzed by enzymes.

FIGURE 2.3.9 Primary structure of insulin. Insulin is a protein hormone secreted by the pancreas in response to high blood glucose. It causes peripheral tissues to take up the glucose, thereby lowering the plasma glucose concentration back toward normal. Insulin is synthesized as a larger, single polypeptide chain that is modified by excision of two peptides to form the A chain, with 21 amino acids, and the B chain with 30. This is one example of posttranslational modification that occurs with many proteins.

regulate another through the binding together of matching surfaces.

THERE ARE FOUR LEVELS OF DESCRIPTION FOR PROTEIN STRUCTURE

Proteins take on their remarkably diverse functions because they can fold to form specific shapes and because some of the amino acids that make up the proteins have inherent chemical reactivity. These shapes provide a surface for the binding of materials, and this binding originates all of the functions of proteins. The shape of proteins has four levels of description:

1. Primary
2. Secondary
3. Tertiary
4. Quaternary.

THE PRIMARY STRUCTURE OF A PROTEIN IS ITS AMINO ACID SEQUENCE

The primary structure of a protein refers to its amino acid sequence. It defines the chemical connectivity of the constituent amino acids. In 1953, Sanger determined the amino acid sequence of the A and B chain of insulin. This was the first protein whose entire amino acid sequence was determined and the work was a major milestone in biochemistry. The chemical structure of insulin is shown in Figure 2.3.9.

THE SECONDARY STRUCTURE OF PROTEIN REFERS TO THE FOLDING OF AMINO ACIDS IN ADJACENT SEQUENCES

Proteins have three regular secondary structures: the α-helix, the β-sheet, and the β-turn. These are complicated, three-dimensional structures. The α-helix was first postulated by Linus Pauling, Robert Corey, and Herman

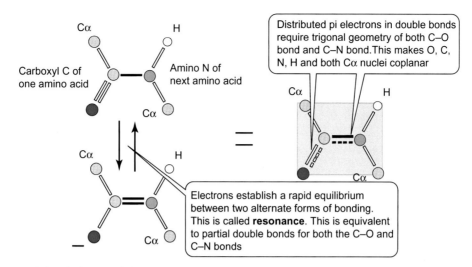

FIGURE 2.3.10 Planar peptide bond. The peptide bond involves a combination of an carboxyl group of one amino acid with the amino group of the next amino acid. The C−N bond and C=O bond **resonate**: the electrons alternate between a single C−N and double C=N bond, and simultaneously between a double C=O bond and a C−O⁻ bond. This is equivalent to a distributed electron density over both bonds. This restricts free rotation about the C−N bond, locking all six nuclei into a single plane.

Branson (The Structure of Proteins: Two Hydrogen-Bonded Helical Configurations of the Polypeptide Chain, *Proc. Natl. Acad. Sci. USA* **37**:205−211, 1951) on the basis of a planar peptide bond and linear hydrogen bond length of 0.272 nm, and no requirement for an integral number of amino acids per turn of the helix. The planar nature of the peptide bond was key. Pauling realized that resonance between the carbonyl oxygen and the C−N bond would produce a partial double bond character to the C−N bond that would prohibit free rotation about its axis. The H bonding would be trigonal and therefore planar. The carboxyl C would be similarly planar, and because both the N and the carboxyl C were connected, all six nuclei connected to the peptide bond would be coplanar. This idea is shown in Figure 2.3.10.

Although the C−N bond cannot freely rotate, due to its partial double bond nature, the N−Cα bond is not so restricted. This Cα nucleus is connected to the next C that participates in a peptide bond, and this Cα−C bond can also rotate. The result is that the next peptide bond, which also forms a plane, is rotated relative to the first. Two dihedral angles are defined to describe this rotation: φ is the dihedral angle about the N−Cα bond and ψ is the dihedral angle about the Cα−C bond. These are shown diagrammatically in Figure 2.3.11.

Pauling and co-workers used the idea of the stiff plane of the peptide bond and the known length of the hydrogen bond to deduce the structure of the alpha helix, shown in Figure 2.3.12. In this structure, the polypeptide backbone traces a right-handed helix (going from the amino terminus to the carboxy terminus). Note that the C=O bonds point towards the carboxy terminus, where they hydrogen bond with the amino hydrogen. The protein forms a rod with the side groups of the amino acid sticking out more or less radially.

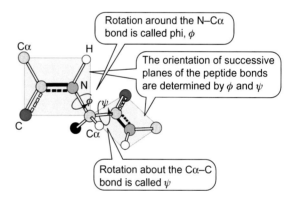

FIGURE 2.3.11 Rotation of successive peptide bonds. Each peptide bond defines a plane. Between each bond there is allowable rotation around the N−Cα bond (defined as φ) and rotation about the Cα−C bond (defined as ψ). The result is a change in direction of the polypeptide chain.

The second major secondary structure in proteins is the **beta sheet**, shown schematically in Figure 2.3.13. These result from sideways hydrogen bonding of a linear chain of amino acids whose peptide plane is bent at the Cα carbon. There are two ways for this to be accomplished: **parallel** beta sheets join segments of the polypeptide chain whose amino to carboxyl direction is going in the same direction. **Antiparallel** beta sheets join polypeptide chains of opposite N to C polarity.

Structures not fitting into these categories are often called "random coils," although they may not be random at all. There are four main principles involved in the formation of these secondary structures.

Hydrogen Bonding Stabilizes Structure

The polar centers of the carbonyl oxygen and the amide nitrogen are hydrogen bonded to other structures in the protein, either another amide bond or a polar side group of an amino acid. These hydrogen bonds stabilize protein structure and are instrumental in forming the α-helix.

Hydrophilic Groups Face Water; Hydrophobic Groups Face Away from Water

The protein will fold so that highly charged groups and polar groups are on the outside of the protein facing water. Here the interaction between these hydrophilic groups and the hydroxyl groups of water stabilizes the structure.

Side Groups Cannot Occupy the Same Space

This is a simple way of saying there is **steric hindrance** in the structure. Atoms in the amino acids exhibit repulsive forces when placed too close together. These forces ultimately arise from the interpenetration of atomic or molecular electron orbitals.

Electrostatic Interactions Stabilize Structures

Groups with opposite electric charge attract each other and this attractive force can stabilize the three-dimensional arrangement of amino acids. Similarly, groups with the same sign electric charge repel each other and this repulsive force can also stabilize the structure by preventing closer movement of the electrically charged groups.

TERTIARY STRUCTURE DEALS WITH THE THREE-DIMENSIONAL ARRANGEMENT OF ALL OF THE AMINO ACIDS

The tertiary structure of proteins deals with how the regional structures are put together in space. For example, the α-helices may be oriented parallel to each other or at right angles. So the tertiary structure refers to the folding of the different segments of helices, sheets, turns, and the remainder of the protein into its native three-dimensional structure. Commonly, membrane proteins are anchored in the membrane by hydrophobic

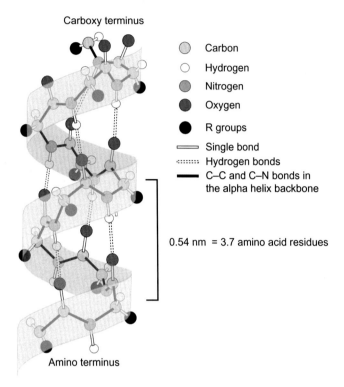

FIGURE 2.3.12 The alpha helix. The dark bonds indicate those in the polypeptide backbone that are directly involved in the peptide bonds. The dashed lines indicate hydrogen bonds that bond successive turns of the helix with each other. These hydrogen bonds connect the carbonyl of one amino acid with the amino hydrogen of the fourth amino acid down the chain.

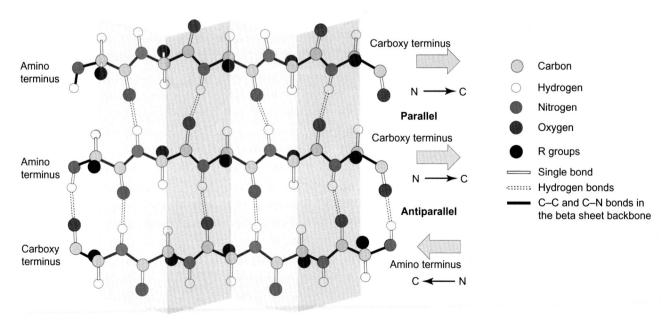

FIGURE 2.3.13 Structure of the beta sheet. Strings of amino acids are bonded laterally through hydrogen bonds. The arrangement can be between strings that have the same amino to carboxy orientation, called parallel, or between strings with opposite orientation, called antiparallel. The planar peptide bonds line up to form a kind of pleated sheet.

alpha helices that may be far removed in the primary sequence but closely apposed in the tertiary structure of the protein.

QUATERNARY STRUCTURE REFERS TO THE INTERACTION OF A PROTEIN WITH OTHER PROTEINS

The quaternary structure refers to the interaction of a protein with other proteins or other components of the cell. This association of the protein with other elements of the cell can alter its three-dimensional shape and its activities because the close proximity of another protein's surface can alter the shape of the protein. Many proteins exist in the cell in complex macromolecular assemblies in which quaternary structure enables or regulates the function of the component proteins.

POSTTRANSLATIONAL MODIFICATION REGULATES AND REFINES PROTEIN STRUCTURE AND FUNCTION

Proteins that come off the ribosome are not yet finished. The cell processes the newly made proteins in several ways, including:

- formation of disulfide bonds
- folding into the functional form
- cleaving the proteins at specific sites
- chemical modification.

PROTEIN DISULFIDE ISOMERASE CATALYZES DISULFIDE EXCHANGE

Proteins often are stabilized by disulfide bonds between cysteine residues in the protein. These cysteines are not necessarily close to each other on the primary sequence but must be close in the tertiary structure of the protein. The **protein disulfide isomerase** catalyzes the interconversion of disulfide bonds until the right ones are formed.

CHAPERONES AND CHAPERONINS HELP PROTEINS FOLD

As they are made on the ribosome, proteins begin to fold up. Sometimes the primary structure of the protein alone can determine the proper final shape, and **denaturing** the protein (adding materials or heat that disrupts its shape) is reversible. In other cases, proteins need help in determining their shape, and they are synthesized on a kind of scaffold that ensures that they fold properly. These scaffolds are generally other proteins called **chaperones**. There are two kinds of chaperones. **Molecular chaperones** bind to unfolded or partially unfolded proteins and stabilize their structure, preventing them from being degraded. **Chaperonins** directly facilitate the folding of proteins. Molecular chaperones are members of the **Hsp70** family of proteins; hsp means "heat shock protein" because these increase after heat stress to an animal that would denature proteins. Complexes of eight **Hsp60** molecules form a barrel-shaped chaperonin that aids protein folding.

PROTEOLYTIC CLEAVAGE

Cells make many proteins in precursor form with longer primary sequences than the finished product. Proteolytic cleavage forms the final protein by chopping off unwanted parts of the protein.

PROTEINS ARE CHEMICALLY MODIFIED IN A VARIETY OF WAYS

Cells chemically modify proteins after translation in a variety of ways, some of which are shown in Figure 2.3.14. These include:

- acetylation
- methylation

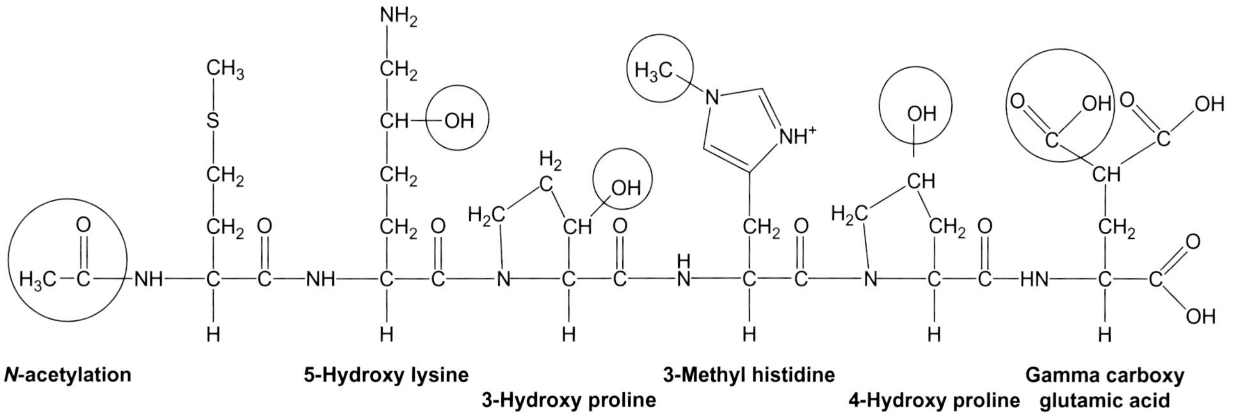

FIGURE 2.3.14 Examples of some chemical posttranslational modifications of proteins. The N-terminal amino acid of many proteins (about 80% of them) is acetylated. Lysine groups can also be acetylated (not shown). Acetylation may regulate the life span of proteins as nonacetylated proteins rapidly degrade. Hydroxylation of lysine in the 5 position occurs in collagen. In collagen, proline is converted to 3 hydroxyproline or 4 hydroxyproline. Histidine, particularly in actin, is methylated. Arginine can also be methylated (not shown) and this modification is part of the histone code (see Chapter 2.2). Glutamic acid residues in prothrombin (a protein involved in clotting of blood) and in bone proteins are carboxylated at the γ-side chain carbon to form gamma carboxy glutamic acid. The close proximity of the two carboxyl groups in gamma carboxy glutamic acid creates a binding site for calcium ions, and this confers Ca^{2+} sensitivity to the coagulation process. The gamma carboxylation reaction requires vitamin K, which is necessary for normal blood coagulation.

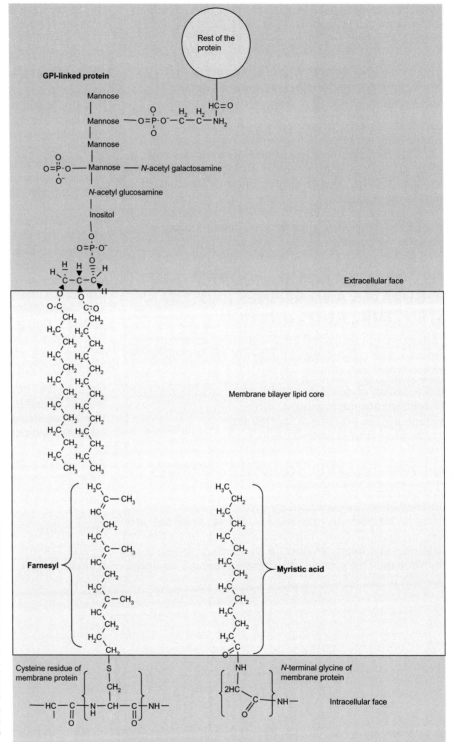

FIGURE 2.3.15 Anchors of membrane proteins. Some membrane proteins are anchored in the membrane by attachment of hydrophobic parts such as myristic acid or palmitic acid. Myristic acid is a 14-carbon hydrocarbon chain with a carboxyl group on one end that can covalently attach to the N-terminal glycine of membrane proteins. Palmitic acid is a 16-carbon saturated monocarboxylic fatty acid and attaches to proteins similar to myristic acid. The process of attaching myristic acid or palmitic acid is called myristoylation or palmitoylation. Farnesyl is a polymer of three 5-carbon units called isoprene. Farnesyl attaches covalently via a thioester bond to cysteine residues somewhere in the middle of the protein and helps anchor some proteins in the membranes. Other proteins link to **GPI** or glycosylphosphatidylinositol, a glycosylated membrane lipid that helps keep proteins in the membrane.

- hydroxylation
- gamma carboxylation
- glycosylation
- myristoylation or palmitoylation
- phosphorylation.

Proteins in the endoplasmic reticulum are often **glycosylated**, meaning that sugars or sugar derivatives are covalently attached to the proteins. N-linked glycosylation occurs when carbohydrate branches are added to the side chain NH_2 of asparagine through N-acetylglucosamine. O-glycosylation occurs on the side chain OH of serine, threonine, or hydroxylysine and the connecting carbohydrate is N-acetylgalactosamine.

Proteins stick in membranes because they are anchored there by a variety of posttranslational modifications, as shown in Figure 2.3.15.

PROTEIN ACTIVITY IS REGULATED BY THE NUMBER OF MOLECULES OR BY REVERSIBLE ACTIVATION/ INACTIVATION

The cell can alter the activity of its component proteins by altering the number of copies of the protein, or by activating or inactivating the proteins that are already present.

Regulating the number of protein molecules requires synthesis of new protein molecules or degradation of existing ones. This takes time and degrading existing proteins wastes energy. Reversible activation or inactivation of proteins can achieve rapid regulation that also conserves energy. Cells use phosphorylation/dephosphorylation of proteins to regulate their activities. Attachment of a phosphate group changes the charge on a local region of the protein, which alters its three-dimensional shape and changes its activity. Serine, threonine, and tyrosine residues are the targets for these phosphorylation reactions (see Figure 2.3.16). An enzyme that phosphorylates proteins is called a **protein kinase**; one that dephosphorylates proteins is called a **protein phosphatase**. Cells contain a variety of protein kinases and protein phosphatases (see Figure 2.3.17).

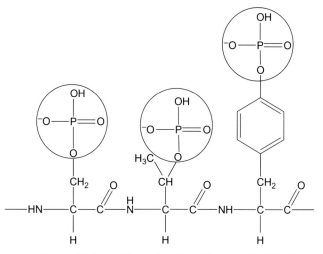

FIGURE 2.3.16 Phosphorylation sites of proteins. Serine, threonine, and tyrosine all have hydroxyl groups that can be esterified with phosphate. This alters the shape and charge of the protein surface in that area, leading to changes in protein activity.

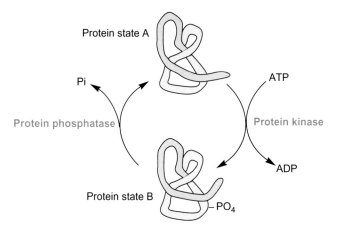

FIGURE 2.3.17 Phosphorylation cycle for protein regulation. Some proteins have phosphorylation sites that can be covalently linked to phosphate from ATP through the action of a variety of protein kinases. This alters the local charge of the protein, which in turn changes its shape and its activity. The protein returns to its dephosphorylated state by the action of protein phosphatases.

Clinical Applications: Protein Folding Diseases

With the advent of the microscope in the mid-1800s, Pasteur, Koch, and many others formulated the germ theory of infectious diseases. We now know that microscopic viral, bacterial, protozoan, or parasitic agents cause a long list of diseases: smallpox, polio, rabies, HIV, yellow fever, anthrax, bubonic plague, syphilis, tuberculosis, cholera, gonorrhea, malaria, sleeping sickness, schistosomiasis, to name but a few. Each of these infectious agents contains DNA or RNA that codes for the organism's own proteins and enables replication of its nucleic acid. Viruses do this by using the host's machinery. Bacteria, protozoans, and parasites are self-contained organisms that use the host's environment for their own reproduction. Investigation of a group of diseases called **transmissible spongiform encephalopathies (TSE)** required revolutionarily new thinking about infection. These diseases include **Creutzfeldt–Jakob disease**, **kuru**, Gerstmann–Straussler syndrome (GSS), and fatal familial insomnia (FFI) in humans, and **scrapies** and **bovine spongiform encephalopathy (BSE)** in animals.

Kuru is a neurological disease among the Fore, a linguistic group of people in Papua New Guinea. In ritual cannabalism, these people ate the bodies of dead relatives. After long incubations, they developed partial paralysis and loss of motor control and eventually died. Because the disease seemed to run in families, a genetic cause was first proposed, but later rejected. Carleton Gajdusek, a pediatrician and virologist, was unable to transmit the disease to any animal, including primates. Igor Klatzo, a neuropathologist, examined tissues sent by Gajdusek and found that kuru was a unique disease without precedent, the closest condition being Creutzfeldt–Jakob disease. In a bizarre twist, William Hadlow, a veterinary neuropathologist, saw some of Klatzo's photomicrographs in an exhibit at the Wellcome Medical Museum in London, and noted a startling resemblance between neurohistological changes in kuru and those in scrapie, a neurological disease in sheep first described in 1732. It was known to be infectious, but the infectious agent was not yet identified. Hadlow wrote to Gajdusek and published a letter in *The Lancet*

(Continued)

Clinical Applications: Protein Folding Diseases (Continued)

in 1959. The similarity of kuru to infectious scrapie prompted Gajdusek to further attempt inoculating animals with kuru. In 1966, he and his co-workers transmitted kuru to chimpanzees and then Creutzfeldt–Jakob disease to chimpanzees in 1968. Gajdusek earned the 1976 Nobel Prize in Medicine.

Although it was initially assumed that the infectious agent must be some virus, investigators failed to identify any virus or any immunological response to one. The agent was not destroyed by UV irradiation or nucleases that typically inactivate nucleic acids. Tikvah Alper and her co-workers in 1966 found that the infectious agent of scrapie was too small to provide a nucleic acid code. In 1967, J.S. Griffith proposed that a protein alone could be the infectious agent of TSE if it was a pathogenic form of a host protein that could convert normal host protein to the pathogenic form. Finally, in 1982, Stanley Prusiner identified the infectious agent as a protein devoid of nucleic acids. He coined the term "**prion**" which stands for proteinaceous infectious particle. Prusiner earned the 1997 Nobel Prize in Medicine.

Prions are infectious proteins. They "reproduce" by causing normal cellular prion protein (PrP^C) to fold up differently, converting it into the pathogenic, or scrapie, isoform (PrP^{Sc}). The native PrP^C appears to have three α-helices and two short β-strands; PrP^{Sc} has two α-helices and much more β-sheet. This transition from α-helix to β-sheet is the fundamental event underlying prion diseases. Proteolysis of PrP^{Sc} produces a smaller, protease-resistant molecule of about 142 amino acids (Prp 27–30) which polymerizes into **amyloid** that presents itself in the disease state.

Ritualistic cannabalism transmitted kuru among the Fore people of New Guinea; industrial cannabalism spread BSE ("mad cow disease") in Europe. There is more than one bad way to fold a protein. Increasing numbers of patients have contracted a new variant of Creutzfeldt–Jakob disease (vCJD) from prion-tainted beef. Because of its long incubation time, we do not yet know the price of mad cow disease. Current thinking that the disease can be eradicated completely by control of infection is wrong. The disease can spontaneously appear without infection.

SUMMARY

The core of amino acids is their asymmetric alpha carbon. Attached to one side is an amino group. On the other side is a carboxyl group. The third group is hydrogen and the fourth group is a variable group, usually referred to as a side chain or R group. These variable groups define the class of the amino acids. Glutamic acid and aspartic acid have a carboxyl group on the side chain. At neutral pH, this group ionizes and therefore has a negative charge. Lysine, arginine, and histidine are basic amino acids because their side chains have a basic chemical character. At neutral pH, these are positively charged. Serine, threonine, and tyrosine have hydroxyl groups that confer polar character. Asparagine and glutamine have an amide group that is also polar. Nonpolar amino acids include glycine, alanine, valine, leucine, isoleucine, proline, cysteine, methionine, tryptophan, and phenylalanine. Some of these are highly hydrophobic, such as tryptophan, isoleucine, leucine, and phenylalanine.

Proteins are made by the formation of peptide bonds between the amino group of one amino acid and the carboxyl group of another. Because of this bond, the basic character of the amino group is neutralized, and the acidic character of the carboxyl group is neutralized, and the character of the chain of amino acids is determined by the sequence of the side chains.

We describe protein structure on four levels: the primary sequence describes the linear sequence of amino acids along the peptide chain backbone, proceeding from the amino terminus to the carboxy terminus. The protein folds into secondary local structures such as α-helices, β-sheets, and β-turns. The arrangement of these secondary structures in three-dimensional space produces the tertiary structure. Combination of proteins with other structures produces macromolecular complexes with quaternary structure. Some proteins spontaneously fold into their "native" shape, whereas others are assembled on a kind of scaffold that helps them fold up properly into their active form. Denaturation of proteins occurs when the protein loses its normal shape. In some cases this is reversible, but usually loss of the proper folding causes irreversible loss of function. Hydrogen bonding, electrostatic interactions, hydrophobic interactions, and steric hindrance all help stabilize proteins in their secondary structures. Disulfide bonds between cysteine side chains help stabilize higher order structure.

Proteins undergo posttranslational modifications after they are synthesized. These include *N*-glycosylation or *O*-glycosylation, proteolytic cleavage, hydroxylation, methylation, acetylation, γ-carboxylation, covalent attachment of hydrophobic molecules that anchor proteins in membranes, and phosphorylation of specific hydroxyl groups on side chains of serine, threonine, and tyrosine.

Protein function depends on the way their surfaces interact with the surfaces of other materials—substrates, structural elements, or other materials to which the proteins bind. Catalytic activity or structural or regulatory roles of proteins depend on the close match of their surface with the surface of the things they bind to. This also determines the specificity of the protein's action.

In general, activity of proteins in the cell can be regulated by altering the amount of protein or by altering its intrinsic activity. Reversible regulation can be achieved by phosphorylation/dephosphorylation of proteins.

REVIEW QUESTIONS

1. Name the two acidic amino acids. What makes them acidic? Name three basic amino acids. What makes them basic?
2. Name five polar amino acids. What makes them polar? Name the nonpolar amino acids.

3. What does "hydrophobic" mean? What does "hydrophilic" mean? What is the partition coefficient? How does it measure hydrophobicity?
4. What is a peptide bond? Where do you find it?
5. What is the primary structure of a protein?
6. What is an α-helix? β-sheet? What interactions among protein side chains stabilize these structures?
7. What is "posttranslational modification"? Name four different kinds of posttranslational modification.
8. What residues are most often acetylated? What residues are hydroxylated? What is "γ-carboxylation"? What amino acid is γ-carboxylated? What is glycosylation? What is myristoylation or palmitoylation?
9. How can proteins be anchored in hydrophobic membranes?
10. Describe the phosphorylation/dephosphorylation cycle for regulating activity of proteins in the cell.

2.4 Biological Membranes

Learning Objectives

- Describe fatty acids and what is meant by saturated and unsaturated fatty acids
- Distinguish between *cis* and *trans* arrangement around a double bond in an unsaturated fatty acid
- Describe the constituents of phosphatidic acid, phosphatidylcholine, phosphatidylethanolamine, phosphatidylserine, and phosphatidylinositol
- Identify hydrophilic and hydrophobic groups in membrane lipids
- Recognize the steroid ring structure of cholesterol
- Recognize the chemical structures of cardiolipin, sphingosine, sphingomyelin, and ceramide
- Define surface tension
- Describe how amphipathic lipids reduce surface tension
- Describe motion in the plane of a lipid bilayer
- Describe the fluid mosaic model of biological membranes
- Distinguish between integral and peripheral proteins
- Describe caveolae and clathrin-coated pits
- Describe how secreted proteins are synthesized on the ER membrane

BIOLOGICAL MEMBRANES SURROUND MOST INTRACELLULAR ORGANELLES

As discussed in Chapter 2.1, cells contain a variety of subcellular organelles and the hallmark of most of them is that they are surrounded by a membrane. These membranes divide the cell into several compartments in which enzymes and substrates are sequestered away from the rest of the cell. This separation into compartments is required for the functioning of these organelles. Maintenance of this compartmentalization requires selective transport of materials across the membranes. Table 2.4.1 lists the various subcellular organelles with their approximate contributions to the cell volume and membrane area. The proportions of cell volume and area represented by the subcellular organelles vary markedly with cell type and activity.

BIOLOGICAL MEMBRANES CONSIST OF A LIPID BILAYER CORE WITH EMBEDDED PROTEINS AND CARBOHYDRATE COATS

The composition of biological membranes varies enormously among the different subcellular organelles, but all biological membranes share a basic structure. The core of the membrane is a lipid bilayer. Embedded in this core are a variety of proteins that carry out many of the activities of the membrane, including selective membrane transport, and some of both the lipids and proteins have carbohydrate coats. This basic structure is shown in Figure 2.4.1. The rest of this chapter expands on this general description.

ORGANIC SOLVENTS CAN EXTRACT LIPIDS FROM MEMBRANES

Gorter and Grendel, in 1925, provided early evidence for the lipid bilayer structure of membranes when they extracted the lipids from erythrocytes with acetone and spread them over the surface of water. They noted that the area occupied by the lipids was about twice the calculated area of the surface of the erythrocytes. Because the only membrane in the erythrocytes is the plasma membrane, they concluded that the membrane was a lipid bilayer. This confirmed earlier observations by Ernst Overton in the 1890s that there is an excellent correlation between the ability of a number of solutes to enter cells and their solubility in olive oil. Overton concluded that the surface of the cell was made up of lipids similar to olive oil.

BIOLOGICAL MEMBRANES CONTAIN MOSTLY *PHOSPHOLIPIDS*

Organelles can be isolated by cellular disruption and differential centrifugation, as described in Appendix 2.1.A1. The lipids in the membranes can be extracted into an organic phase, typically a chloroform/methanol mixture, because these lipids are hydrophobic (see Figure 2.3.7), and then the lipids can be separated into their component classes by chromatography, and the amounts can be measured. Approximate lipid composition of some membranes is given in Table 2.4.2.

TABLE 2.4.1 The Relative Area and Enclosed Volumes of the Major Subcellular Membranes in a Typical Liver Cell

Membrane	Percent of Total Cell Volume Enclosed	Approximate Number per Cell	Percent of Total Cell Membrane
Plasma membrane	100	1	2
Mitochondria	22	1700	39
Rough ER	9	1	35
Smooth ER	2	1	16
Golgi apparatus	4	1	7
Nuclear inner membrane	6	1	0.2
Lysosomes	1	300	0.4
Peroxisomes	1	400	0.4

Values were estimated by quantitative electron microscopy.

However, it should be kept in mind that the lipid composition varies not only with the kind of organelle but also with the kind of cell.

PHOSPHOLIPIDS CONTAIN FATTY ACYL CHAINS, GLYCEROL, PHOSPHATE, AND A HYDROPHILIC GROUP

Table 2.4.2 shows that **phospholipids** comprise the most abundant class of lipids in membranes. Each consists of four parts: a **glycerol** backbone, a **phosphate** esterified to one end of the glycerol molecule, a **polar head group** attached to the phosphate, and two **fatty acids** esterified to the other two hydroxyl groups of the glycerol. The glycerol phospholipids form the major subclass of the phospholipids, and they contain phosphate esterified to the C-3 hydroxyl group of glycerol and two fatty acids esterified to the C-2 and C-1 hydroxyl groups. Fatty acids consist of two distinct regions (see Figure 2.4.2): a long hydrocarbon chain and a carboxylic acid group (—COOH). The most frequent length of the carbon chain is 16 or 18 carbon atoms. As discussed in Chapter 1.4, carbon can form four single bonds with bond angles that approximate those between the center of a tetrahedron and its vertices. In a hydrocarbon chain, two of these bonds link a carbon with adjacent carbons and two remain for other bonds. When hydrogen is covalently bound to all of the two remaining bonding orbitals, the hydrocarbon is **saturated**. **Unsaturated** fatty acids contain one or more double bonds between carbon atoms. Those containing more than one double bond are **polyunsaturated**. The variety of phospholipids is produced from the variety of fatty acids and from the different polar head groups. The most common lipid constituent of membranes is phosphatidylcholine, with other phospholipids making major contributions. The components of the simplest phospholipid, phosphatidic acid, are shown in Figure 2.4.2.

A variety of polar groups can be attached to the phosphate of phospholipids. These groups include **serine**, **inositol**, and **ethanolamine**, which can be methylated to form **choline**. The resulting phospholipids are shown in Figure 2.4.3.

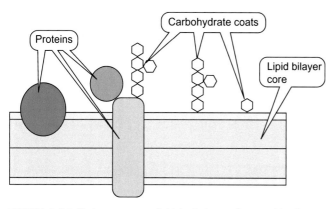

FIGURE 2.4.1 Basic structure of biological membranes. Membranes consist of a lipid bilayer core to which proteins are attached in a variety of ways. In addition, some lipids and proteins have carbohydrate groups attached to them.

TABLE 2.4.2 Approximate Lipid Composition of Different Cell Membranes

Lipids	Percentage of Total Lipids by Weight		
	Plasma Membrane	Mitochondrial Membrane	Endoplasmic Reticulum Membrane
Phosphatidylcholine	24	39	40
Phosphatidylethanolamine	7	35	17
Phosphatidylserine	4	2	5
Cholesterol	17	3	6
Sphingomyelin	19	0	5
Glycolipids	7	0	0
Other	22	21	27

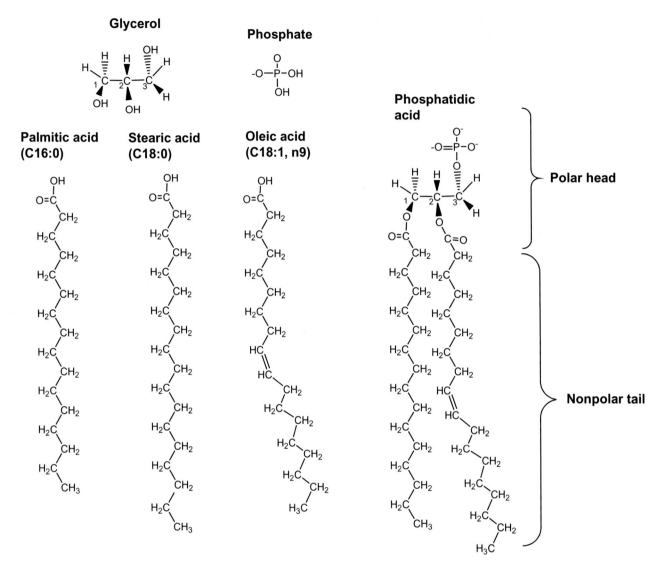

FIGURE 2.4.2 Components of a simple phospholipid, phosphatidic acid. Glycerol forms the backbone. Each of the three carbons in glycerol covalently bonds to a hydroxyl group (−OH). Rotation about the C−C single bonds can position the C1 OH or C3 OH in any orientation. The central carbon, C2, is an asymmetric carbon when C1 and C3 are differently substituted, and will show stereoisomerism when different groups attach to the two ends. To distinguish the 1 and 3 positions, the numbering nomenclature shown in the figure is used. The structure of inorganic phosphate is also shown. Fatty acids form the hydrophobic core of phospholipids. The chemical structure of three common fatty acids: palmitic acid, stearic acid, and oleic acid is shown. The single carbon−carbon bonds allow free rotation about the axis connecting the two carbon nuclei. Each carbon atom forms bonds that nearly line up the centers of two tetrahedrons with their respective apices. Palmitic acid is a 16-carbon **saturated fatty acid**, meaning that every carbon's bonds other than the carboxyl carbon are fully occupied with single bonds to carbon or hydrogen. In the nomenclature of fatty acids, it is designated 16:0, indicating a 16-carbon fatty acid with no double C−C bonds. Stearic acid is a 18-carbon saturated fatty acid, designated as 18:0. Oleic acid is a **mono-unsaturated fatty acid**, meaning that one pair of carbon atoms are joined by a double bond. In this case, the double bond is between carbons 9 and 10, numbering down from the carboxyl carbon. Its designation is 18:1 Δ^9, where the 1 indicates one double bond and the Δ^9 indicates the double bond begins at Carbon 9. There is an alternate number system starting from the terminal methyl group of the fatty acid, instead of starting at the carboxyl group. This nomenclature uses the prefix omega (ω); thus oleic acid is 18:1 $\omega - 9$. Here the 18 stands for the length of the hydrocarbon chain, 1 indicates the number of double bonds, and $\omega - 9$ indicates the position of the double bond numbering from the methyl end. Substitution of the omega with the letter *n* is becoming popularized, with the same meaning. As discussed in Chapter 1.4, double bonds produce kinks in the hydrocarbon chain due to restricted rotation about C=C double bonds. Oleic acid has a *cis* orientation of the H atoms around the double bond, meaning the two hydrogens are on the same side of the double bond. The structure of a phosphatidic acid is shown on the right. This particular phosphatidic acid has an oleic acid molecule esterified to C-2 of glycerol and a palmitic acid molecule esterified to C-1. The fatty acids differ from molecule to molecule, but this is a typical arrangement in which saturated fatty acids occupy the C-1 position and unsaturated or polyunsaturated fatty acids occupy the C-2 position. Phosphatidic acid illustrates a common property of this class of lipids in that it consists of spatially separated water-soluble polar or hydrophilic groups and water-insoluble nonpolar or hydrophobic groups.

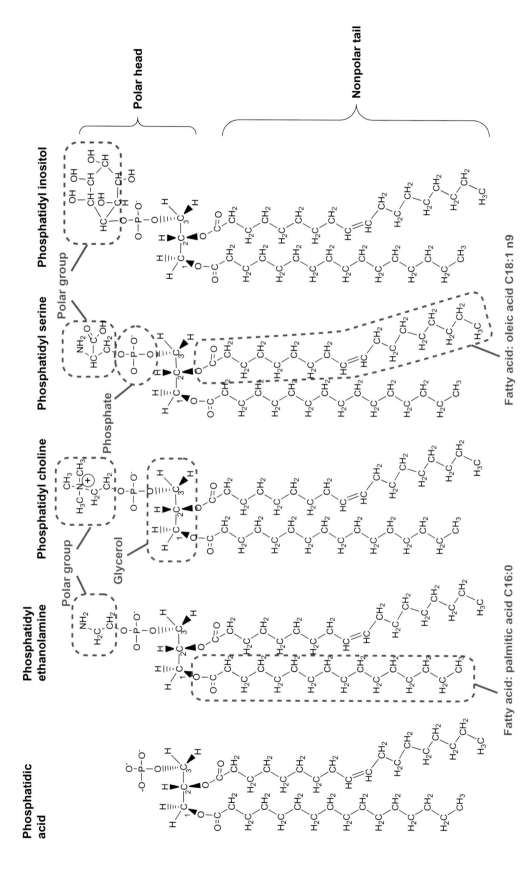

FIGURE 2.4.3 Chemical structures of some common glycerophospholipids. The phosphate group in phosphatidic acid is esterified to the hydroxyl group of several other hydrophilic molecules including ethanolamine, choline, serine, and inositol. These form the lipids shown. Each of these are named for the hydrophilic group and the fatty acids, as in 1-palmitoyl, 2-oleoyl phosphatidylcholine.

PLASMANYL PHOSPHOLIPIDS AND PLASMENYL PHOSPHOLIPIDS USE FATTY ALCOHOLS INSTEAD OF FATTY ACIDS

A major subclass of the phospholipids use fatty alcohols instead of fatty acids, forming an ether linkage with glycerol instead of an ester. These are called plasmanyl glycerol phospholipids. Most of these are modified to contain a vinyl ether linkage, in which the alcohol group is doubly bonded to the rest of the fatty alcohol chain. These are termed plasmenyl glycerol phospholipids or plasmalogens. Their structures are shown in Figure 2.4.4. They make up 15–20% of the total phospholipids of cell membranes.

SPHINGOLIPIDS USE SPHINGOSINE AS A BACKBONE AND ARE PARTICULARLY RICH IN BRAIN AND NERVE TISSUES

Sphingolipids are present in many membranes but they are particularly rich in brain and nerve tissues. There are three classes of sphingolipids: **sphingomyelin**, **cerebrosides**, and **gangliosides**. Sphingomyelin is the only one of these that is a sphingophosphatide. It is analogous to the glycerophosphatides except that it contains sphingosine instead of glycerol as the core structure that links the hydrophilic phosphate and choline to the hydrophobic hydrocarbon chains. Sphingosine is a derivative of the amino acid, serine. The chemical structures of sphingosine and sphingomyelin are shown in Figure 2.4.5.

The fatty acid amide of sphingosine alone is called a **ceramide** (see Figure 2.4.5). Ceramides can be linked through the hydroxyl group to sugar groups to form another class of sphingolipids, the **cerebrosides**. The sugar part contains a number of hydroxyl groups and is hydrophilic. In some cases the sugar part is quite large and branched, forming a **ganglioside**. Both cerebrosides and gangliosides form another class of lipids called glycolipids because they incorporate sugar derivatives.

OTHER LIPID COMPONENTS OF MEMBRANES INCLUDE *CARDIOLIPIN*, *SPHINGOLIPIDS*, AND *CHOLESTEROL*

CARDIOLIPIN IS TWO GLYCEROLIPIDS LINKED BACK TO BACK

The structure of cardiolipin is shown in Figure 2.4.6. It consists of two phosphatidic acid molecules linked through another glycerol. Mitochondrial membranes are particularly rich in cardiolipin.

CHOLESTEROL CONDENSES MEMBRANES

Cholesterol, shown in Figure 2.4.7, is the most abundant steroid in animal tissues. All steroid hormones are derived from cholesterol. It possesses a rigid ring structure that attracts normally flexible phospholipid chains to itself, causing membranes to become more rigid in the vicinity of this molecule.

PHOSPHOLIPIDS IN WATER SELF-ORGANIZE INTO LAYERED STRUCTURES

All of the glycerophospholipids and sphingolipids we have discussed are characterized by the spatial separation of a polar head group, consisting of ionized groups, hydroxyls and carbonyl oxygens, and a long tail consisting mainly of hydrocarbons. The polar head group is capable of interacting with water and each of the materials there individually is water soluble. This

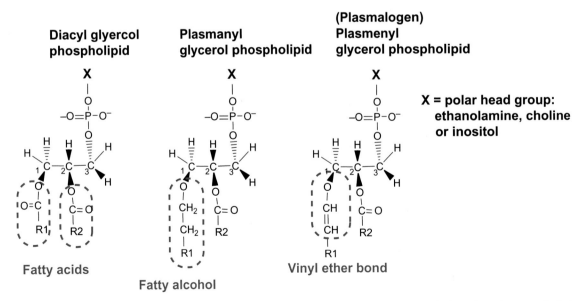

FIGURE 2.4.4 Plasmanyl glycerol phospholipids and plasmenyl glycerol phospholipids. Some long-chain hydrocarbons have an alcohol and not a carboxyl group at their end. These can be joined to glycerol through an ether linkage, forming a plasmanyl glycerol phospholipid which is typically enriched in the sn-2 position with polyunsaturated fatty acids. In most cases, the carbon adjacent to the ether is joined in a double bond, forming a vinyl ether bond that characterizes the plasmenyl glycerol phospholipids, also known as the plasmalogens.

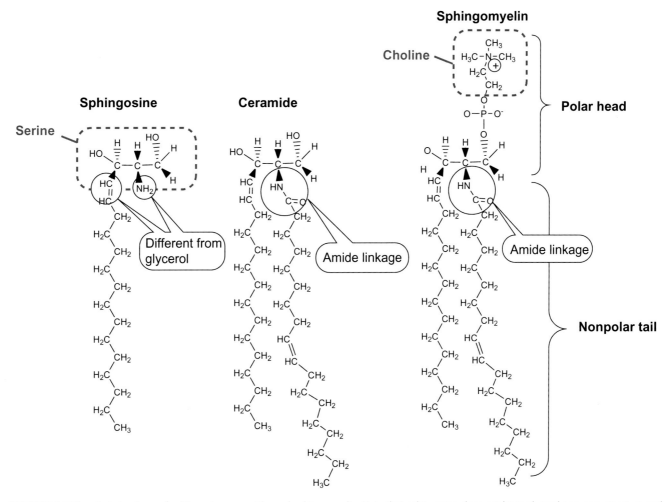

FIGURE 2.4.5 Chemical structures of sphingosine, ceramide, and sphingomyelin. Note that sphingosine does not have glycerol as a core structure and joins a hydrocarbon through an amide linkage instead of an ester to form sphingomyelin.

region of the molecule is **hydrophilic**, meaning *water loving*. The hydrocarbon tail is not soluble in water; it is **hydrophobic** or *water hating*. Hydrophobic parts are also described as **lipophilic** or fat loving. Molecules having both of these separate domains are said to be amphipathic, from the Latin and Greek *amphi* meaning *having two sides*. These two sides are illustrated by the space-filling model in Figure 2.4.8. These two separate domains of phospholipids are crucial to their behavior in cells. When placed in water, the polar heads of these molecules remain associated with water and form hydrogen bonds with it; the long hydrocarbon, nonpolar tails repel the water and associate with each other, forming a self-organized structure, the lipid bilayer. Much of this behavior of the lipids resides in the nature of water. To see this, we need to learn more about the behavior of water at hydrophobic interfaces.

SURFACE TENSION OF WATER RESULTS FROM ASYMMETRIC FORCES AT THE INTERFACE

At the air–water interface, water molecules are subjected to asymmetric forces, as shown in Figure 2.4.9. These molecules have lost some of their bonds connecting them to the bulk phase and are, therefore, in a higher energy state than water in the bulk phase. These molecules are partially evaporated. Thus it takes energy to promote water molecules to the interface and the energy of the surface increases with its area. At constant temperature and pressure, the change in surface energy is the change in the **Gibbs free energy, G**. The change is given by:

[2.4.1] $$dG = \gamma \, dA$$

where dA is the increment in area, dG is the increment in Gibbs free energy, and γ is the **surface tension**. Since the energy has units of force × distance, the tension has units of force per unit length. Typical units of γ are dynes cm^{-1}. In SI units, γ is expressed in N m^{-1} or J m^{-2}. The surface tension is a measure of how much more a water molecule at the surface is attracted to the bulk water phase because of the increase in intermolecular forces compared to the surface.

WATER "SQUEEZES OUT" AMPHIPATHIC MOLECULES

Recall that amphipathic molecules consist of spatially separated water-loving or hydrophilic head group and a

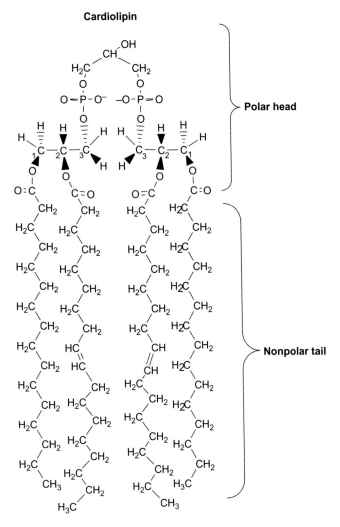

FIGURE 2.4.6 Chemical structure of cardiolipin.

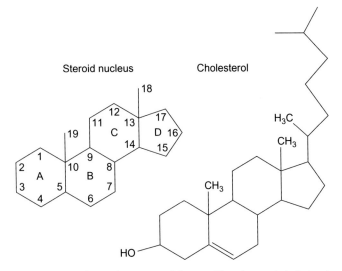

FIGURE 2.4.7 Chemical structure of the steroid nucleus and cholesterol. The steroids are all derived from the steroid nucleus, a perhydrocyclopentanophenanthrene nucleus of four rings: three fused rings in the phenanthrene arrangement and a five-carbon ring attached. Therefore, the nucleus has the name "cyclopentano" to refer to the five-carbon ring; "phenanthrene" refers to the three six-membered rings and "perhydro" indicates that the double bonds of phenanthrene are saturated with hydrogen. The steroid nucleus is numbered as shown.

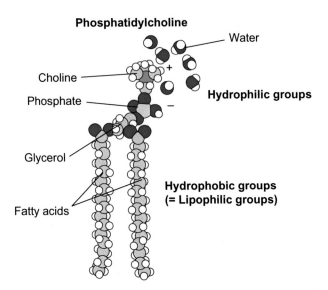

FIGURE 2.4.8 Space-filling model of phosphatidylcholine. Nonpolar surfaces are shown in gray or white. Polar surfaces are dark blue (oxygen), light blue (phosphorus), or intermediate blue (nitrogen). Charged surfaces attract water by dipole–dipole interactions. The hydrophilic or water-loving parts of the molecule are all concentrated at one end. The hydrophobic (water-hating) or lipophilic (fat-loving) groups are located at the opposite end.

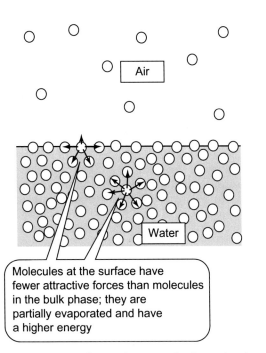

FIGURE 2.4.9 Asymmetric forces of water molecules at the air–water interface. In the bulk phase of liquid water, the intermolecular forces acting on any water molecule are on average equal in all directions. In the air, intermolecular interactions are markedly reduced. In order to evaporate, energy must be added to the water molecules to break the attractive intermolecular forces. At the air–water interface, water molecules are subject to more intermolecular forces than those water molecules in the air phase, and fewer forces than water molecules in the bulk liquid phase. Therefore, water molecules at the surface have more energy than those in the bulk phase. In a sense, they are partially evaporated, having lost some but not all of their intermolecular bonds. If we were to increase the surface area, we would have to put in energy proportional to the area of increase.

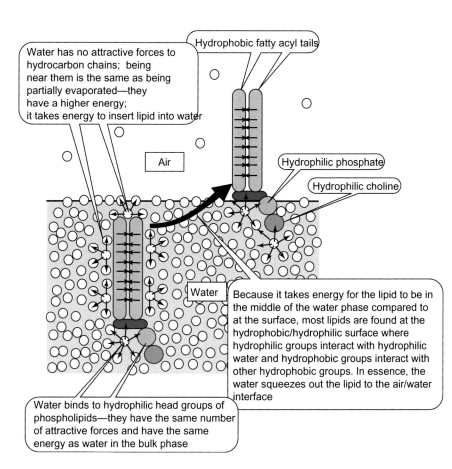

FIGURE 2.4.10 Water "squeezes out" phospholipids. The hydrocarbon parts of a phospholipid molecule within the bulk aqueous phase cannot form hydrogen bonds with the adjacent water molecules, so these water molecules cannot form as many hydrogen bonds as the other water molecules in the bulk phase. The set of water molecules surrounding the hydrocarbon are essentially partially evaporated and have a higher energy than the water molecules in the bulk phase. Therefore, to insert the hydrocarbon in the water takes energy. Phospholipids on the surface of the water, on the other hand, are in a lower energy state. The self-association of water thus "squeezes out" the phospholipid to the surface.

hydrophobic or water-fearing hydrocarbon tail. Dissolving an amphipathic molecule in the bulk water phase disrupts the self-association of water, creating an interface within the bulk phase that requires energy. Because of this, the amphipathic molecule is "squeezed out" of the bulk water phase to the surface of the solution adjacent to the air. In this situation, the water molecules at the surface interact with hydrophilic groups in the amphipathic molecule, which lowers the energy of the surface, and the hydrophobic groups in the amphipathic molecule can interact with other hydrophobic groups through London dispersion forces. These ideas are shown diagrammatically in Figures 2.4.10 and 2.4.11.

The free energy change for the transfer of phospholipid from the bulk phase to the surface is given as

[2.4.2] $\quad \Delta G_{bulk \Rightarrow surface} = G_{surface} - G_{bulk}$

Since $G_{surface}$ is less than G_{bulk}, the free energy change for the transfer is negative, and therefore the transfer occurs spontaneously.

AMPHIPATHIC MOLECULES SPREAD OVER A WATER SURFACE, REDUCE SURFACE TENSION, AND PRODUCE AN APPARENT SURFACE PRESSURE

When amphipathic molecules such as phosphatidylcholine or oleic acid are dissolved in a volatile organic solvent (e.g., hexane, decane) and then layered over water, the organic solvent evaporates and leaves a thin film of the lipid. As shown in Figures 2.4.10 and 2.4.11, these amphipathic molecules are squeezed out to the surface of the water, forming a layer a single molecule thick. The lipids form a **monolayer**. These lower the surface tension according to Eqn [2.4.1] because they lower the energy required to move water molecules to the surface (see Figure 2.4.11). This lowering of the surface tension can be measured using a **Langmuir trough**, as shown in Figure 2.4.12. This device has two barriers on the surface of the water. One barrier is fixed, the other is movable. Since the monolayer decreases the surface tension, the movable barrier feels a net force in the direction of the clean surfaces. By lowering the surface tension, the lipids appear to exert a **surface pressure** defined as

[2.4.3] $\quad \pi = \gamma_0 - \gamma$

where π is the surface pressure, γ_0 is the surface tension of the clean surface, and γ is the surface tension in the presence of the monolayer. Surface pressure has the same units as the surface tension.

PHOSPHOLIPIDS FORM BILAYER MEMBRANES BETWEEN TWO AQUEOUS COMPARTMENTS

Two monolayers can orient themselves back to back to form a bilayer between two aqueous compartments, as shown in Figure 2.4.13. This is the low energy state for this situation for the reasons we have already described.

If the phospholipids were dispersed throughout the solution, the surface area between the water phase and the hydrophobic, hydrocarbon phase would be large. Since it takes energy to produce this surface, the dispersed hydrocarbon is a high-energy state compared to the condensed one. A secondary cause of this spontaneous organization of the lipids is the attractive interactions among the hydrocarbon chains.

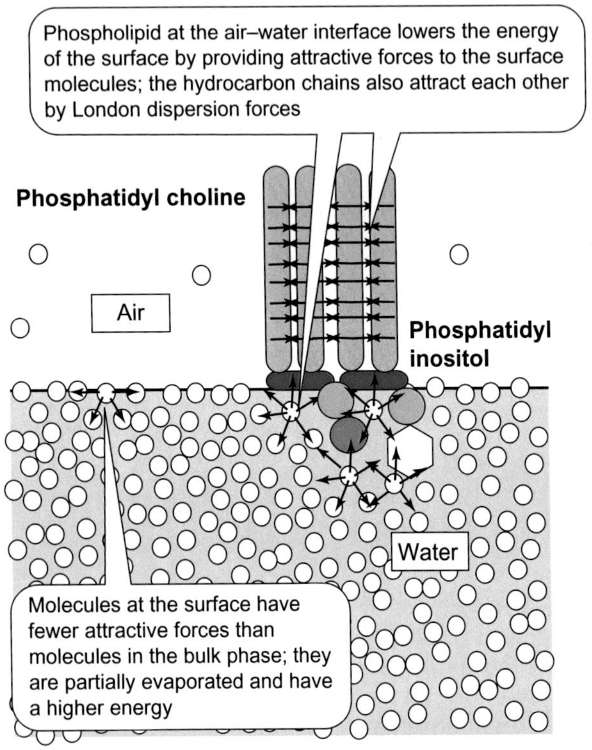

Macroscopic planar bilayer membranes can be formed across a narrow aperture separating two solutions, as shown in Figure 2.4.14. These membranes are useful because they allow the study of single membrane channels incorporated into the membrane.

LIPID BILAYERS CAN ALSO FORM *LIPOSOMES*

The macroscopic lipid bilayer shown in Figure 2.4.14 is unstable to mechanical forces. When disrupted, the membrane breaks apart to form spherical bilayers separating an internal and external watery compartment. These hollow spheres are called **liposomes** (see Figure 2.4.15). These structures can also be generated directly from phospholipids by soaking them in water and adding sonic energy.

ALTHOUGH LIPIDS FORM THE CORE, MEMBRANE PROTEINS CARRY OUT MANY OF THE FUNCTIONS OF MEMBRANES

So far we have been describing membrane lipids that form the core of biological membranes. These contribute to the barrier function of membranes, but many other functions of biological membranes are performed by protein constituents of the membranes. These functions include:

- **Transport**: Cells must be able to move things into and out of the cell.
- **Signal transduction**: Cell must have mechanisms for responding to signals from other cells or from within the cell. These may be chemical, electrical, or mechanical signals.
- **Recognition**: Cells attach to other cells and to extracellular structures. They must be able to recognize where they should form attachments.
- **Attachment**: Cells must anchor themselves to the extracellular matrix or to each other. Often these attachments also provide a signaling pathway.

FIGURE 2.4.11 Lowering of the surface energy by amphipathic molecules. Water at its ordinary surface is in a higher energy state—it takes energy to break the hydrogen bonds that binds the water molecules in the bulk aqueous phase. When phospholipids are present at the surface, water binds to the hydrophilic groups of the phospholipids, thereby reducing the energy needed to form additional surface. Because the energy of the surface is related to its area by the surface tension, reducing the energy of the surface per unit area is the same as reducing the surface tension.

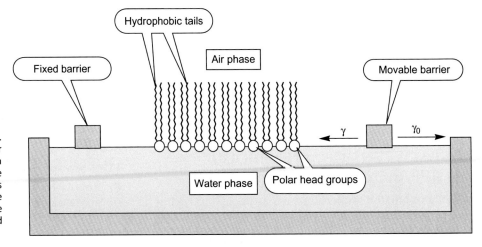

FIGURE 2.4.12 The Langmuir trough. A clean water surface has a fixed barrier and a movable barrier. When only clean water forms the surface, the surface tension is γ_0. Adding a lipid film reduces the surface tension to γ. The movable barrier experiences a net force toward the clean surface without lipid. Thus the lipid appears to exert a surface pressure.

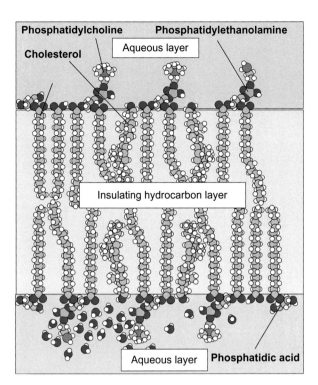

FIGURE 2.4.13 Structure of the lipid bilayer. Only some lipids are shown and in expanded format for clarity. In reality, the lipid bilayer forms a closed surface that approximates a plane. The interior of the lipid bilayer is fluid, consisting of hydrocarbon chains that are saturated (the straight chains in the figure) or unsaturated (bent chains in the figure). The phospholipids form the bulk of the bilayer. Lipids such as cholesterol have a rigid backbone that partially stiffens and solidifies the membrane. Cholesterol may accumulate in heterogeneous patches of membrane called lipid rafts.

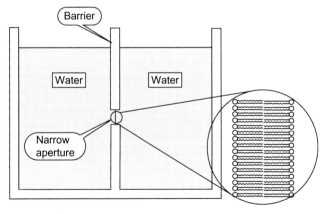

FIGURE 2.4.14 Planar lipid bilayer between two aqueous compartments. Phospholipids were dissolved in hexane and "painted" over a small aperture drilled in a Lexan partition that separated two electrolyte solutions. After thinning, with the passive removal of hexane through the aqueous phase, a lipid bilayer forms between the two compartments.

- **Movement or force production**: Cells often must move or transmit a force from inside the cell to the extracellular matrix. This requires connection of the cell's cytoskeleton through the membrane to the extracellular matrix.

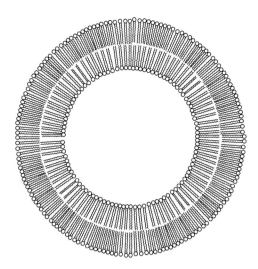

FIGURE 2.4.15 A schematic drawing of a cross-section of a liposome. These are small structures about 50–150 nm across. Because each bilayer is about 7.5–10 nm across, the thickness of the bilayer occupies a considerable portion of the entire liposome volume and the enclosed volume, the lumen, is small. Because of the high curvature and small size, the area on the outside of the liposome is nearly twice the area on the inside, and therefore significantly more lipid faces the outside compared to the inside surface of the liposome. Liposomes may find use someday to deliver drugs to specific locations within the body by incorporating recognition signals into the lipid bilayer.

MEMBRANE PROTEINS BIND TO MEMBRANES WITH VARYING AFFINITY

The proteins that perform the various functions listed above can be loosely classified according to how tightly they are bound to the membrane. Loosely bound proteins are called **peripheral** proteins, also sometimes called **extrinsic** proteins. They can be released from the membrane by relatively gentle procedures such as washing with a salt solution. Other proteins are called **integral** proteins, also sometimes called **intrinsic** proteins. These are tightly bound by the membrane and can be released only by resorting to drastic measures such as dissolving the membrane with a detergent. By coating the hydrophobic parts of the membrane proteins with hydrophilic material, detergents **solubilize** the membrane proteins. Examples of useful detergents in membrane research include the ionic **sodium dodecyl sulfate** (**SDS**) and the nonionic family of **Triton** detergents. Figure 2.4.16 illustrates integral and peripheral proteins.

Proteins are held in membranes by the same kinds of interactions that hold lipids in the bilayer: hydrophobic and hydrophilic interactions. Many proteins have sequences of amino acids that penetrate all the way across the membrane. These are **transmembrane proteins**. In the parts of the protein exposed to lipid, hydrophobic amino acid side chains appose the lipid core: valine, leucine, isoleucine, phenylalanine, tryptophan, and methionine. Those parts of the protein exposed to water have a preponderance of hydrophilic amino acids: aspartic acid, glutamic acid, lysine, and arginine. The neutral amino acids can be present in

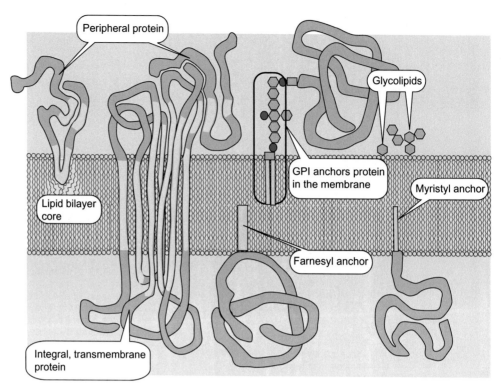

FIGURE 2.4.16 Schematic arrangement of membrane proteins in the lipid bilayer. Darker areas of proteins are predominantly hydrophilic; lighter areas denote hydrophobic areas. Hexagons signify hydrophilic carbohydrates. Proteins are usually anchored by hydrophobic sequences of amino acids; less often they are anchored by covalent attachment to hydrophobic materials such as GPI, farnesyl, palmitic acid, or myristic acid.

either domain. Some proteins have multiple sequences that cross the membrane, with domains facing each of the watery solutions on the two sides of the bilayer. The sequences that cross the membrane are typically arranged as alpha helices, with the hydrophobic amino acids facing the hydrophobic lipid core. Other proteins may bind to the membrane by covalently attached hydrophobic groups. These include **myristic acid** (14:0 fatty acid), **palmitic acid**, (16:0) **farnesyl**, and **glycosylphosphatidylinositol** (GPI) (see Figure 2.4.16).

LIPIDS MAINTAIN DYNAMIC MOTION WITHIN THE BILAYER

Researchers have labeled phospholipids with probes that are sensitive indicators of molecular motion and have tracked the mobility of lipids and proteins within the plane of the bilayer. The results of these experiments show a variety of molecular motions, shown in Figure 2.4.17. This has given rise to the fluid mosaic model of biological membranes. The term fluid mosaic model describes a dynamic system in which the lipids form a plane that gradually curves around to form a closed surface—there are no exposed lipid edges. Lipid motion in the plane of the membrane includes rotation, flexion, stretch, and lateral diffusion in the plane. All of these movements are rapid. The proteins in the membrane can also move, unless they are anchored by binding to other components of the membrane. There is a gradient of fluidity of motion from the polar head group to the center of the hydrophobic interior of the membrane, being most fluid in the center and most anchored at the head group. Lipid movement from one side of the membrane to the other—the "flip-flop"

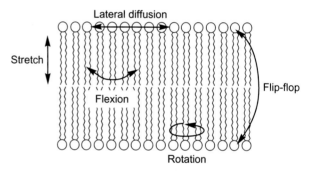

FIGURE 2.4.17 Possible motions of lipids within a bilayer. Lipid molecules can move laterally within the plane of the membrane, rotate about their long axis, flex within the fluid interior of the membrane, or move from one side of the bilayer to the other. Most of these motions are fast, but the "flip-flop" reaction is very slow, occurring less than once in 2 weeks for any individual lipid molecule. Exchange of neighbors occurs very fast, on the order of 10^7 times per second. This rapid exchange gives rise to a rapid lateral movement within each half of the bilayer.

reaction—is slow because it requires lipids to bring their hydrophilic head group through the lipid phase, which is energetically costly. Because of the very slow movement of lipids from one half of the bilayer to another, membranes can maintain an asymmetric composition, but it must actively make and sustain it. In most biological membranes, the two half bilayers differ in their composition. Cells add lipids only to the cytoplasmic side of membranes. An enzyme called a **"scramblase"** flips lipids from the cytoplasmic half to the extracellular half, but this enzyme is not very specific. The different composition of membranes arises from a second enzyme, a **"flippase"**, that flips only some phospholipids from the cytoplasmic half to the extracellular half.

The fluidity of membrane lipids depends on their composition. Saturated fatty acids are known to be very stiff compared to unsaturated fatty acids. Saturated fatty acyl chains can be packed closely together, whereas the kinks produced by *cis* double bonds make it difficult to pack these chains close together. Because of this, unsaturated fatty acids promote fluid movement within the bilayer. Cholesterol is also a very rigid molecule. Cholesterol molecules orient themselves in the membrane with their steroid nucleus adjacent to the hydrophobic tails and the hydroxyl group adjacent to the polar head groups. The steroid nucleus is a flat, plate-like structure that partially immobilizes the nearby fatty acyl chain, thereby reducing the fluidity of the bilayer and increasing the mechanical stability of the bilayer.

The motions of the lipids in the bilayer makes it appear as a two-dimensional fluid. It is fluid within the plane of the membrane, but relatively rigid perpendicular to this plane. The membrane proteins more or less "float" in this lipid see like so many icebergs in the North Atlantic. This combination of fluid lipids and iceberg proteins was the origin of the descriptive term, **fluid mosaic model**.

LIPID RAFTS ARE SPECIAL AREAS OF LIPID AND PROTEIN COMPOSITION

Lipid rafts are microdomains in biological membranes that contain different proportions of lipids and proteins from the rest of the membrane. They were discovered when portions of membranes were found to be less easily solubilized by detergents. Detergents are chemicals that dissolve membranes by providing their constituents with flotation devices: they bind the hydrophobic domains and coat them with water-soluble material. These detergent-resistant areas of membrane accumulate cholesterol and sphingolipids. Sphingolipids generally contain longer and straighter fatty acyl chains. These attract each other more forcefully than do unsaturated fatty acids, because the straight chains can pack more closely without the kinks in their chains. These aggregate into the raft microdomain. Because these chains are straighter, the membranes are also thicker at the rafts. The plasma membrane is thought to have many such rafts about 70 nm in diameter.

CAVEOLAE AND *CLATHRIN-COATED PITS* ARE STABILIZED BY INTEGRAL PROTEINS

The surface of cells forms a variety of specializations that curve inwardly, forming an indentation of the membrane. Caveolae are one of these. Caveolae are a subset of lipid rafts, but not all lipid rafts are caveolae. Caveolae are 60–80 nm pits in the membrane that contain some 140–150 oligomeric caveolin molecules. There are three mammalian caveolins: CAV1, CAV2, and CAV3, all have parts that bind to membranes. Their oligomeric structure is stabilized by a family of cytoplasmic proteins called cavins. Caveolar membranes are enriched in both cholesterol and phosphatidylserine (another phospholipid in which the head group is serine instead of ethanolamine, choline, or inositol). Depletion of the cholesterol or mechanical flattening of caveolae causes dissociation of cavin from caveolin (see Figure 2.4.18). Flattening of the caveolae occurs upon stretch of skeletal muscle, cardiac myocytes, endothelial cells, and fibroblasts. It may be that caveolae are involved in the sensing or response to mechanical stretch.

Membranes can also form clathrin-coated pits that are involved in receptor-mediated endocytosis, in which parts of the membrane invaginate and pinch off, forming an interior vesicle with enclosed extracellular material. **Clathrin** consists of three heavy chain subunits (CHC17 or CHC22) and three light chains (CLC) that trimerize to form a triskelion, the unit of clathrin. These units then associate on membranes to form a clathrate (lattice) structure. The lattice structure consists of a number of pentagons and hexagons. The clathrin protein itself has a curvature to it, and this imparts a curvature to the clathrate and stabilizes the budding part of the membrane. The membrane is then pinched off by another protein complex called dynamin (see Figure 2.4.19).

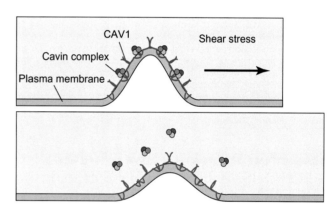

FIGURE 2.4.18 Caveolae response to stretch. Caveolae are indentations or pits in the plasma membrane that are stabilized by a network of integral proteins that include oligomers of caveolin (CAV1). Cytoplasmic proteins called cavins stabilize the caveolin structure. When the membrane is stretched, the caveolae flatten and cavin dissociates from the caveolin. This may be part of how cells sense or respond to stretch.

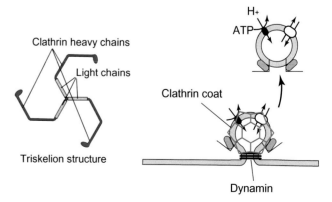

FIGURE 2.4.19 Clathrin-coated pits and endocytosis. Clathrin consists of a trimer of heavy chains each of which binds a light chain. These self-assemble to form a clathrin coat that stabilizes budding membranes. The budding membranes are pinched off through the actions of dynamin.

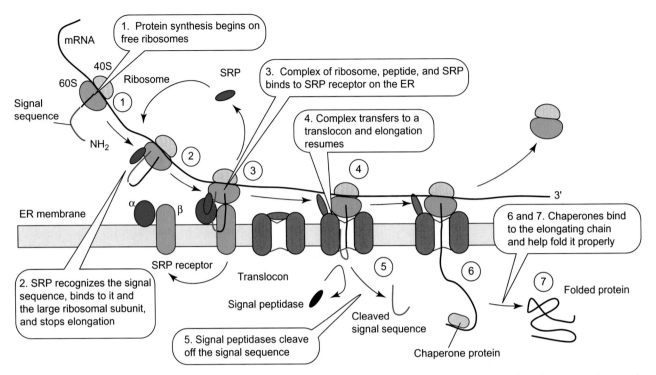

FIGURE 2.4.20 Mechanism of synthesis of secretory proteins into the ER lumen in the pancreas. Synthesis begins on free ribosomes in the cytoplasm (1). The initial N-terminus of the protein contains a signal sequence that is recognized and bound to by cytosolic SRP, signal recognition particle, a nucleoprotein consisting of RNA and a set of six separate proteins (2). Binding of SRP stops elongation. The complex of nascent polypeptide, ribosome, and SRP is bound to the ER by the SRP receptor consisting of a peripheral α-subunit and an integral β-subunit (3). The SRP receptor transfers the ribosome to a translocon, a complex of proteins that spans the ER membrane and provides an aqueous channel for the protein across the membrane (4). The SRP dissociates from the complex in the process. The polypeptide chain grows and translocates across the ER membrane at the same time. The signal sequence is cleaved off (5) by signal peptidases in the ER lumen. The growing polypeptide chain is eventually completed and folds up into its active conformation (7). The folding is assisted by chaperone proteins and enzymes such as Hsp-70, protein disulfide isomerase, and peptidyl prolyl isomerase (6). *Source: Modified from Lodish et al., Molecular Cell Biology, 4th Ed, W.H. Freeman, New York, NY, 2000.*

SECRETED PROTEINS HAVE SPECIAL MECHANISMS FOR GETTING INSIDE THE ENDOPLASMIC RETICULUM

The synthesis of membrane proteins or proteins destined for secretion poses a problem for cells because the ribosomes on which the proteins are made are in the cytosolic compartment but their products are either in the membranes or in the extracellular compartment. This problem is solved by an elaborate mechanism, shown in Figure 2.4.20.

SUMMARY

Lipids readily dissolve in organic solvents such as chloroform while they are sparingly soluble in water. Several classes make up biological lipids including fatty acids, phospholipids, and steroids. They are typically made up of long chains of hydrocarbons or carbon rings. The major constituent of biological membranes are the phospholipids. These consist of a polar head group connected covalently to a nonpolar tail. The polar head group in turn consists of a hydrophilic group like inositol, serine, choline, or ethanolamine linked to a phosphate group, which in turn is esterified to glycerol. All of these are highly water soluble. The other hydroxyls of the glycerol are esterified to two fatty acids, which are highly water insoluble. Thus these phospholipids spatially separate hydrophilic and hydrophobic parts. When placed in water, the hydrophilic parts associate with the water while the hydrophobic parts associate with other hydrophobic molecules. When placed on top of water, these amphipathic molecules form a lipid monolayer.

The surface tension of the water results from asymmetric forces on the surface from the bulk water phase and the air phase. Because the hydrophilic parts of the lipids attract water molecules on the surface, they reduce the asymmetry in forces. Accordingly, lipids reduce the surface tension. Experimentally, this appears as a surface pressure. Folding such a monolayer back on itself produces a bilayer membrane. This consists of a double layer of molecules in which the hydrophilic domain faces the water phase and the hydrophobic domain faces the interior of the membrane, occupied by other hydrophobic parts of lipid molecules. Other lipid aggregates include the liposome. The liposome is a bilayer that forms a hollow sphere.

As in all molecules, chemical bonds in lipids can stretch, rotate, and flex. These motions along long chains produce motion within the hydrophobic interior of membranes. The hydrophobic chains are relatively well anchored at the polar head group, so there is a gradient of fluidity in the membrane, it being most fluid in the center and less at the periphery. Double bonds in the hydrocarbon chains make a kink in the chain that disallows close packing of the chains. Thus double bonds promote fluidity within the bilayer. Saturated fats, those that contain no double bonds, make membranes more rigid. Cholesterol is a rigid molecule composed of a plate-like steroid nucleus and a hydrophobic tail. It generally makes membranes more rigid.

Lipids freely move in the plane of the membrane while motion across the membrane, the "flip-flop" reaction, is slow. Cells take advantage of this slow flip-flop to maintain asymmetric distributions of lipids in the two halves of the bilayers. Proteins embed in the membrane and move around, something like icebergs floating in a lipid sea. Thus the membrane is described as a "fluid mosaic." However, thicker microdomains of the membrane contain concentrations of sphingolipids and cholesterol. These microdomains are called lipid rafts.

Proteins can be loosely associated with membranes or tightly bound. The loosely bound proteins are called peripheral or extrinsic proteins and the tightly bound proteins are called integral or intrinsic proteins. Proteins are held in the membrane either by hydrophilic–hydrophobic interactions between their amino acids and the lipid and water phases or by attachment of hydrophobic groups such as myristic acid, palmitic acid, farnesyl, or GPI. Many proteins and lipids have carbohydrate coats.

The synthesis of secreted proteins requires the synthesis of an endoplasmic reticulum (ER) signal sequence that is recognized by a signal recognition particle (SRP), which then binds to a receptor on the ER membrane. This enables transfer of the cytosolic ribosome to the ER membrane and subsequent simultaneous translation and translocation of protein across the ER membrane. Once inside the ER, the synthesized protein undergoes posttranslational modification.

REVIEW QUESTIONS

1. Name the major membranes in the cell. Which membrane accounts for most of the membranes in the cell?
2. What is a saturated fatty acid? What is an unsaturated fatty acid? What effect does unsaturation have on the structure of the fatty acid? What do fatty acids attach to in phospholipids?
3. What is a phospholipid? What are the major types of phospholipids? Which chemical groups on the phospholipid are hydrophilic? Which groups are hydrophobic? What is the significance of spatial separation of hydrophilic and hydrophobic character in lipids?
4. What is the general structure of cardiolipin? What is sphingosine? What is sphingomyelin? What is a ceramide? How do these differ from phosphatidylcholine?
5. What is surface tension? What are its units? What do amphipathic lipids do to the surface tension? Why?
6. Name the various degrees of freedom of lipid motion in a bilayer. Which is the slowest? Which is the fastest?
7. What is meant by the term "fluid mosaic model"?
8. What is an integral protein? What is a peripheral protein? In what ways can proteins be anchored to membranes?
9. What is a liposome? What is a planar lipid bilayer?
10. What are lipid rafts?
11. What is meant by "caveolae"? What is a clathrin-coated pit?
12. How do secreted proteins get inside secretory vesicles? What is meant by "signal sequence"?

Problem Set 2.1
Surface Tension, Membrane Surface Tension, Membrane Structure, Microscopic Resolution, and Cell Fractionation

1. Consider a soap film stretched over a wire frame, one end of which is movable. Experimentally, one observes that there is a force exerted on the movable member as indicated in Figure 2.PS1.1. Clearly, this force depends on the dimensions of the wire frame. Therefore, we express the force per unit length as γ. Write an expression for the work performed in expanding the film a distance dx. Rewrite this in terms of the area increment, dA, by which the film is expanded.

2. Consider a soap film again in the form of a bubble as shown in Figure 2.PS1.2. The surface tension can be thought of as either the force per unit length or the energy per unit area. The minimal energy form for a soap film is the minimum area for a given volume. This is the sphere. So, in the absence of other effects, including gravity, the soap bubble should be a sphere.
   A. What is the total surface energy of the sphere? Remember that the variable we have been using for surface tension is γ.
   B. If the radius were to decrease by dr, what would be the *change* in the surface energy?
   C. Since shrinking decreases the surface energy, at equilibrium the tendency to shrink must be balanced by a pressure difference across the film, ΔP. At equilibrium, the work against this pressure for an increment in radius dr is exactly equal to the decrease in surface energy. That is, at equilibrium the free energy change is zero. Otherwise, the bubble would not be stable and it would shrink. What is the work that must be done against this pressure difference? *Hint*: Pressure is force per unit area, so the total force must be the area times the pressure. Work is force times distance.
   D. Equate the pressure–volume work in part C to the surface energy decrease in part B. From this equation, derive an expression for ΔP in terms of γ and r. This result is a famous equation, the Law of Laplace, which finds application in respiratory physiology and cardiovascular physiology.

3. When heart cells are exposed to a hypotonic medium, they swell and measurements show that their volume has increased. Measurements of their membrane capacitance, however, do not change. How can this happen?

4. Liposomes form structures 100 nm across their outside diameter. The average density of the lipids used to form the liposomes is 0.89 g cm^{-3}. Assume that the thickness of the bilayer is 8 nm.
   A. What is the volume of the lipid shell? What is its mass?
   B. What is the ratio of the outer surface area to the inner surface area of the liposomes?
   C. What is the enclosed volume of the liposome?

FIGURE 2.PS1.1 Soap film on a wire frame. The soap film exerts a force per unit length on the movable barrier. This force is the surface tension. To expand the film, we must do work.

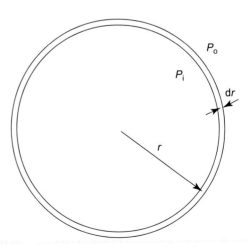

FIGURE 2.PS1.2 A soap bubble of radius r. Because the surface tension results in an inwardly directed force, the bubble will tend to collapse unless there is a pressure difference across the membrane that prevents its collapse.

D. Using the answers to A and C, what is the overall density of the liposomes? Assume that the liposome is filled with water with density = 1 g cm^{-3}.
E. How many liposomes can be derived from 100 mg of lipids?
F. A drug is soluble to 5 mM. If the liposomes were formed in a solution of this drug, and therefore the enclosed volume included the drug to this concentration, how many moles of drug would be contained in the liposomes derived from 100 mg of lipids?

5. Isolated cardiac sarcoplasmic reticulum (SR) vesicles have an average outside diameter of about 150 nm. The membrane itself is about 10 nm thick. The enclosed volume can be estimated by measuring the efflux of passively loaded tracer materials such as mannitol, and the result gives 5 μL mg^{-1} SR protein.
   A. How many vesicles are there per mg of SR protein?
   B. What is the surface area of the vesicles per mg of SR protein?
   C. If the SR Ca-ATPase Ca^{2+} uptake activity is 4 μmol min^{-1} mg^{-1}, what is the uptake activity per unit surface area?

6. The method of measuring the surface tension of a liquid–air interface is the **drop weight method**. In this method, drops are allowed to form at the end of a tube of known radius, and a number of them are collected and then weighed so that the weight per drop can be determined accurately. The weight per drop is given by Tate's Law (1864):

   [2.PS1.1] $W = 2\pi r \gamma$

   where r is the radius of the tube. This equation uses the idea that the surface tension is the force per unit length and that the maximum force that can be used to support the weight of the forming drop is the circumference of the tube times its surface tension. In practice, the weight of the drop is less than that given by Tate's Law because some of the liquid supported by the tube remains after the drop falls. More detailed analysis makes use of a correction factor such that

   [2.PS1.2] $W' = 2\pi r \gamma f$

   where W' is the actual weight per drop and f is the correction factor. It turns out that the correction factor f varies with $rV^{-1/3}$, where V is the volume of the drop. Approximate values of the correction factor are given in Table 2.PS1.1:
   A. Using a tip with an outside diameter of 0.40 cm and an inside diameter of 0.20 cm, 20 drops of an organic liquid weighed 0.80 g. The density of the liquid was 0.95 g cm^{-3}. This liquid wet the tip. (Hint: This goes to determine whether you use the inside or the outside diameter!) Use the appropriate correction factor from Table 2.PS1.1, and the drop weight, to calculate the surface tension of the organic liquid.
   B. Using a tip with and outside diameter of 0.21 cm and an inside diameter of 0.18 cm, 20 drops of a liquid had a mass of 0.766 g. The density of the liquid was 1.00 g cm^{-3}. This liquid wet the tip. (Hint: This goes to determine whether you use the inside or the outside diameter!) Use the appropriate correction factor from Table 2.PS1.1, and the drop weight, to calculate the surface tension of the organic liquid.

7. The surface tension of pure water is 72.0 dyne cm^{-1}. 1 dyne cm^{-1} is equivalent to 1 mN m^{-1}, which is the SI unit for surface tension. When dipalmitoyl lecithin is spread at 50 Å^2 mol^{-1}, the surface pressure is 11 mN m^{-1}. What is the surface tension when dipalmitoyl lecithin is spread on the surface?

8. The tension in a biological membrane can be measured in a variety of ways. One way is called the **pipette aspiration technique**. In this technique, a specially manufactured micropipette is attached to a vesicle by light suction. These pipettes typically have a open diameter of 1–2 μm and have a square end. Application of increasing suction draws the vesicle into the pipette, forming a cylindrical part within the pipette and a spherical part outside of it (see Figure 2.PS1.3).
   A. Assume that the Law of Laplace holds for both the spherical part of the vesicle outside the pipette and the hemisphere within the pipette. Write the two equations relating P_v, P_o to D_v and T, the tension in the vesicle, and P_v, P_p to D_p. There is only one tension in the membrane, which is the same everywhere.
   B. Defining $\Delta P = P_o - P_p$, use the answer in part A to solve for ΔP in terms of T, D_v, and D_p:

TABLE 2.PS1.1 Correction Factors for the Drop Weight Method

$rV^{-1/3}$	f
0.3	0.7256
0.4	0.6828
0.5	0.6515
0.6	0.6250
0.7	0.6093
0.8	0.6000
0.9	0.5998
1.0	0.6098
1.1	0.6280
1.2	0.6535
1.3	0.640
1.4	0.603
1.5	0.567
1.6	0.535

Source: Data from A.W. Adamson, Physical Chemistry of Surfaces, Interscience, New York, NY, 1967.

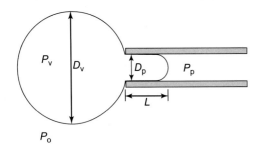

FIGURE 2.PS1.3 Pipette aspiration technique. A vesicle obtained from a "bleb" on a cell when the cell is exposed to hypotonic medium is excised and attached to a micropipette by application of suction. Increasing the suction draws the vesicle into the pipette a distance L and reduces the diameter of the remaining vesicle. By LaPlace's law, this increases the tension in the membrane. Continual decreases in P_p eventually causes the vesicle to rupture. The critical tension for rupture can be determined in this way.

TABLE 2.PS1.2 Data for the Dilatation of Sarcolemma Vesicles Using the Micropipette Aspiration Technique

Tension (mN m^{-1})	$\Delta A/A$	Tension (mN m^{-1})	$\Delta A/A$
1.4	0.0027	9.1	0.0178
2.6	0.0058	10.5	0.0212
3.9	0.0078	11.9	0.0248
5.1	0.0112	13.1	0.0261
6.6	0.0127	14.5	0.0296
7.9	0.0160		

J.A. Nichol and O.F. Hutter, Tensile strength and dilatational elasticity of giant sarcolemmal vesicles shed from rabbit muscle, *J Physiol*. 493:187–198 (1996)

C. Solve part B to express T in terms of ΔP, D_p, and D_v.
D. A giant sarcolemmal vesicle was obtained from a rabbit muscle and subjected to the pipette aspiration technique (J.A. Nichol and O.F. Hutter, Tensile strength and dilatational elasticity of giant sarcolemmal vesicles shed from rabbit muscle, *J Physiol*. 493:187–198 (1996)). The pipette diameter (D_p) was 19 μm and the vesicle diameter was 66 μm. At a pipette suction of 8 cm H$_2$O, what is the tension in the membrane?
E. What is the total area of the membrane in terms of D_v and D_p and the length L?
F. When suction pressure is increased, L increases and D_v decreases. The increased tension stretches the membrane, causing a dilatation. The elastic area expansion modulus is *defined* as

$$K = \frac{T}{\frac{\Delta A}{A}}$$

where ΔA is the expansion of the membrane area due to dilatation and T is the tension. Considering your answer for part E, write an expression for the increase in area attributable to membrane dilatation, ΔA, in terms of an initial condition with D_{vi} and L_i and a final condition with D_{vf} and L_f.
G. In practice, the increase in the length of the projection is not long enough to cause an easily measurable difference in D_{vf} compared to D_{vi}. Assuming that the volume of the vesicle plus projection remains constant, express D_{vf} in terms of D_{vi}, ΔL, and D_p.

9. Using the micropipette aspiration technique described in Problem #8, the following data were obtained for the tension and area dilatation for sarcolemma vesicles obtained from rabbit skeletal muscle (see Table 2.PS1.2):
The vesicle ruptured at the last point recorded. Calculate the elastic area expansion modulus (see Problem 2.PS1 problem #8 for a definition of the elastic area expansion modulus).

10. For light of wavelength 5000 Å (=500 nm), calculate the theoretical maximum resolution of an optical microscope.

11. Typically the energy of the electron beam in an electron microscope is known, because the voltage through which the electrons are accelerated is known. One electron volt is the energy gained by an electron when it is accelerated across a potential of 1 V. One electron has a charge of 1.602×10^{-19} C. So the electron volt is 1.6×10^{-19} V C = 1.6×10^{-19} J (1 J = 1 V C = 1 N m). The rest mass of the electron is 9.109×10^{-31} kg.
    A. Using this information, and assuming that all of the energy are converted to kinetic energy, calculate the momentum of a 150-keV electron (the denominator in Eqn (1.2.A1.2)). (*Hint*: Kinetic energy $E = p^2/2m$.)
    B. Calculate the wavelength of an electron having a kinetic energy of 150 keV.
    C. Using the result of (B), calculate the theoretical resolving power of an electron microscope using a 150-keV electron beam.

12. In a Sorvall T-865 fixed angle rotor, the distance to the axis of rotation is 3.84 cm at the top of the tube and 9.10 cm at the bottom of the tube. Calculate the RCF at 20,000 rpm at the top and bottom of the tube.

13. We are centrifuging a collection of particles with diameter 150 nm and average density of 1.10 g cm^{-3} through a water solution with density 1.0 g cm^{-3} at 20,000 rpm. The viscosity of the water is 1×10^{-3} Pa s, where Pa is the pascal = 1 N m^{-2}.
    A. At a distance of 8.5 cm from the axis of rotation, what is the net force on a particle?
    B. From Stoke's equation, calculate the frictional coefficient.
    C. What is the particle's terminal velocity?
    D. What direction is the net force?
    E. What causes this net force?

14. In eukaryotic cells (cells with a nucleus), ribosomes have two major subunits, a 60s and a 40s. If we assume both are spheres and have the same average density, what are their relative sizes?
15. Intestinal cells make a calcium-binding protein called calbindin. Calbindin has a molecular weight of about 9000 Da. Its synthesis requires the active form of vitamin D, $1,25(OH)_2$ cholecalciferol, to turn on the gene for the protein. When $1,25(OH)_2$ cholecalciferol is given to vitamin-D-deficient people, it takes about 45 minutes for the intestinal cells to make the first complete calbindin.
    A. Identify the major steps that could account for the 45-minute lag in appearance of calbindin.
    B. Eukaryotic cells (cells with a nucleus) attach amino acids to new proteins at the rate of about 2 per second. Is the synthesis of the protein the major part of the lag? (*Hint*: The average molecular weight of an amino acid is about 100 Da. You can estimate how many amino acids are in the protein from this information—you could look it up because its sequence is known, but we are just doing a "back of the envelope" calculation here.)
16. We have a double-stranded DNA segment of 1000 base pairs. Its nucleotide composition is randomly distributed among A, T, C, and G. Assume that each hydrogen bond in the double strand has an energy of 4 kcal mol^{-1} (1 J = 0.239 cal).
    A. How many hydrogen bonds are there in the segment? (*Hint*: Consult Figure 2.2.3 for the numbers of hydrogen bonds for each base pair.)
    B. What is the total energy necessary to pry apart the two strands, assuming that hydrogen bonding is the only force keeping them together? (It is not.)
    C. Assume that hydrogen bonds break when they are stretched 0.2 nm. How much force is necessary to break one?
    D. If all the hydrogen bonds in our DNA segment were to be ruptured all at once, how much force would be necessary?
17. The molecular weight of the protein myosin is 525,000 g mol^{-1}. Its sedimentation coefficient is 6.4S (this S is the Svedberg, not seconds) in water with a density of 1.0 g cm^{-3}, and its partial specific volume is 0.73 cm^3 g^{-1}. Calculate the frictional coefficient, β. This is sometimes called f.
    A. From the molecular weight and the specific volume, calculate the radius myosin would have if it were spherical.
    B. From the radius calculated in (B), determine the frictional coefficient from Stoke's equation. This is called f_0. The viscosity of water at 25 °C is 1×10^{-3} Pa s, where Pa is the pascal = 1 N m^{-2}.
    C. If the protein were spherical we would expect $f = f_0$. Is myosin spherical?
18. Consider the device shown in Figure 2.PS1.4 that consists of two reservoirs, S_1 and S_2, that initially contain the limiting concentrations C_1 and C_2, respectively, where $C_1 < C_2$. A pump removes fluid from S_1 at rate R_1 and places it in reservoir S_2. Therefore, as soon as the pumping starts the concentration in S_2 begins to change. A magnetic stir bar rapidly mixes reservoir S_2, and a second pump withdraws fluid from S_2 at rate R_2 and places it in a centrifuge tube. The total volume of $S_1 = S_2$ and both are one-half of the capacity of the tube, so that when all of the solutions are pumped into the tube, the tube is filled. Show that for $R_2 = 2 \times R_1$, the gradient is linear in volume from C_2 at the bottom of the tube to C_1 at the top.
19. The SR is a specialized endoplasmic reticulum in skeletal, cardiac, and smooth muscle cells. It contains a Ca-ATPase pump that actively pumps Ca^{2+} ions from the cytosol to an enclosed compartment within the SR, its lumen. The activity of the SR can be estimated by the rate of oxalate-supported Ca^{2+} uptake. This activity is useful because it can also be measured in homogenates of the tissue. Evidence suggests that the

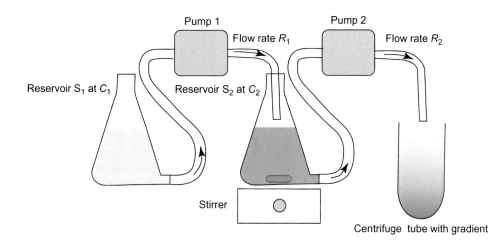

FIGURE 2.PS1.4 One way to make a gradient. Two reservoirs have the limiting concentrations C_1 and C_2. Pump 1 removes fluid from reservoir S_1 at rate R_1 and places it in reservoir S_2, initially at C_2 but then becomes diluted with fluid from S_1. Pump 2 removes fluid from S_2 at rate R_2 and places it in a centrifuge tube. The gradient that is formed depends on the values of R_1 and R_2 and the volumes of the reservoirs. Both S_1 and S_2 begin with identical volumes equal to one-half of the volume delivered to the centrifuge tube.

oxalate-supported Ca^{2+} uptake rate is only due to the SR and other organelles—the surface membrane or mitochondria—do not contribute to it. The left ventricles from a set of dogs were removed under general anesthesia, weighed and homogenized in 3 volumes of buffer (3 mL of buffer for every g wet weight of heart) and homogenate protein, volume and oxalate-supported Ca^{2+} uptake rate were measured. The homogenate was subjected to differential and then sucrose-gradient centrifugation to isolate membrane vesicles of the SR. The following was obtained from an average of 10 preparations: heart homogenate volume: 4.1 mL g^{-1} wet weight of heart, heart homogenate protein: 46.1 mg mL^{-1}, heart homogenate oxalate-supported Ca^{2+} uptake rate: $119 \text{ nmol min}^{-1}$ (mg homogenate protein)$^{-1}$ isolated SR oxalate-supported Ca^{2+} uptake rate: $3.44 \text{ }\mu\text{mol min}^{-1}$ (mg SR protein)$^{-1}$.

A. Calculate the homogenate protein per g wet weight of heart tissue.
B. Calculate the total homogenate Ca^{2+} uptake rate per g wet weight of tissue.
C. Assuming that the SR is 100% pure, how much SR is there, in mg of SR protein, per g of wet weight of heart tissue? (*Hint*: think about the units in the calculation.)

Passive Transport and Facilitated Diffusion 2.5

Learning Objectives

- Describe the microporous membrane as a model of passive membrane transport
- Describe the lipid bilayer model of passive membrane transport
- Define the permeability of a membrane
- Describe how the permeability depends on the microscopic character of the membrane for a porous membrane
- Describe how the permeability depends on the microscopic character of the membrane and solute for a dissolution model of passive transport
- Be able to determine the free energy change for passive transport
- Distinguish between facilitated transport and diffusional transport on the basis of saturability, specificity, rates, and competition
- Write an equation showing the rate of facilitated transport as a function of its solute concentration with zero-*trans* concentration. Identify the variables and describe their meaning
- Distinguish between an ionophore and a channel
- Describe what is meant by channel gating
- Distinguish between voltage-gated channels and ligand-gated channels

MEMBRANES POSSESS A VARIETY OF TRANSPORT MECHANISMS

As described in Chapter 2.4, membranes serve as effective barriers to the free movement of materials, thereby dividing the cell into compartments. This compartmentalization is necessary. In muscle, for example, it allows for the control of contraction by releasing Ca^{2+} ions from a store (the specialized endoplasmic reticulum of the muscle cell) into the cytoplasmic compartment. Relaxation is then brought about by removing Ca^{2+} ions from the cytosol back into the storage compartment. In another example, compartmentalization allows mitochondria to transduce the energy of oxidation of foodstuffs into the chemical energy of ATP. However, compartmentalization does not make sense if material absolutely cannot travel between the compartments. What is necessary is *selective* transport and *regulated* transport. The cell must be able to control what goes across the membranes and how fast.

There are three main mechanisms for transport:

A. Passive transport
   1. Diffusion
   2. Facilitated transport
B. Active transport
   1. Primary active transport
   2. Secondary active transport
C. Osmosis.

In this chapter, we will consider passive transport across two types of hypothetical membranes: a microporous membrane and a lipid bilayer membrane. The mechanisms of passive transport differ considerably between these two models, but the overall form of the equations is similar. In Chapter 2.6, we will consider active transport and then in Chapter 2.7, we will discuss osmosis.

A MICROPOROUS MEMBRANE IS ONE MODEL OF A PASSIVE TRANSPORT MECHANISM

Here we introduce the porous membrane as a model for biological membranes as shown in Figure 2.5.1. We consider here that a microporous membrane separates two solutions of different concentrations but the same pressure. The pores allow solute particles to pass, but the rest of the membrane that lacks pores is impermeable to the solute, and also to solvent water. We assume that the solute particles are small compared to the pores.

First we write the flux, the flow per unit area within a single pore. This is governed by Fick's Laws of Diffusion given as

[2.5.1]
$$j_s = -D\frac{\partial C(x)}{\partial x}$$
$$\frac{\partial C(x)}{\partial t} = D\frac{\partial^2 C(x)}{\partial x^2}$$

where j_s indicates the flux within the pore. We use the lower case "j" purposefully to distinguish it from J_s, which we will use to signify the macroscopic flux across the entire membrane. The top equation is Fick's First Law of Diffusion; the bottom equation is Fick's Second Law of Diffusion. Here we are concerned only with flux across the membrane, in one direction, and the

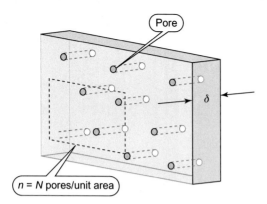

FIGURE 2.5.1 Schematic of the hypothetical microporous membrane. In this model, the membrane is a thin sheet of thickness δ. It is pierced by many cylindrical pores oriented perpendicular to the surface of the membrane. The radius of each pore is a and the number of pores, N, per unit area A is $n = N/A$. This model might not pertain to some cellular membranes, but it may describe some extracellular membranes such as the basement membrane, which supports many cells, especially epithelial cells, or it may represent the filtration membrane present in the kidney.

one-dimensional forms of Fick's laws apply to this situation. Let C_L be the concentration on the left side of the membrane and C_R be the concentration on the right. We can arrange it so that the volumes of the two baths are so large that C_L and C_R are effectively kept constant. Under these conditions, the solute flow will come to a steady state or stationary value. This means that neither the fluxes nor the concentration of solute changes with time. Fick's Second Law of Diffusion becomes

$$[2.5.2] \qquad 0 = D\frac{\partial^2 C(x)}{\partial x^2}$$

Note that this situation cannot be literally true, as solute is moving from one compartment to another, so there must be some changes in $C(x)$ with time. However, $C(x)$ can be so nearly constant that we can ignore the very slight error. The solution to this equation is that $\partial C(x)/\partial x$ is constant. This means that the concentration within the pores is linear with x. We solve this equation by two successive integrations, incorporating the boundary conditions that at $x = 0$, $C(x) = C_L$ and at $x = \delta$, $C(x) = C_R$ (see Figure 2.5.2). The concentration is written as

$$[2.5.3] \qquad C(x) = C_L + \left(\frac{C_L - C_R}{0 - \delta}\right)x$$

and the concentration gradient is

$$[2.5.4] \qquad \frac{\partial C(x)}{\partial x} = -\left(\frac{C_L - C_R}{\delta}\right) = -\frac{\Delta C}{\delta}$$

The flux in the pore is given from Eqns [2.5.1] and [2.5.4] as

$$[2.5.5] \qquad j_s = D\frac{\Delta C}{\delta}$$

The total flow of solute **per pore**, q_s, is given by the area of the pore times the flux within the pore:

$$[2.5.6] \qquad q_s = \pi a^2 j_s$$

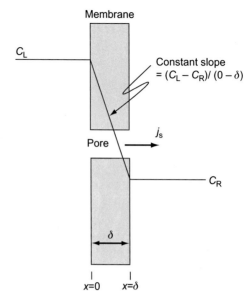

FIGURE 2.5.2 Cross-section of a microporous membrane in the vicinity of a pore. Superimposed on the cross-section is a graph of the concentration gradient. The left compartment has a higher concentration (C_L) than the right compartment with concentration C_R. Under this situation, the flux through the pore is to the right.

The total flow across an area A of the membrane containing N pores is

$$[2.5.7] \qquad Q_s = Nq_s = N\pi a^2 j_s$$

The macroscopically observed flux across the membrane is the total flow of solute (Q_s) divided by the macroscopic area of the membrane.

$$[2.5.8] \qquad \begin{aligned} J_s &= \frac{Q_s}{A} \\ &= \frac{N\pi a^2 j_s}{A} = n\pi a^2 j_s \\ &= \frac{n\pi a^2 D}{\delta}\Delta C \end{aligned}$$

According to this equation, the observed macroscopic solute flux across the membrane is linearly related to the concentration difference by a coefficient that includes the thickness of the membrane (δ), the density of pores in the membrane (n), the radius of the pores (a), and the diffusion coefficient of the solute (D). Often many of these parameters are not known with accuracy and we lump all of these terms together to write

$$[2.5.9] \qquad J_s = p\Delta C$$

where p is the **permeability** of the membrane to the solute. This **phenomenological coefficient** has the units of cm s^{-1} and includes all of the microscopic parameters of the membrane:

$$[2.5.10] \qquad p = \frac{n\pi a^2 D}{\delta}$$

In this model, the permeability increases when the size of the pores increases, when the number of pores per unit area of membrane increases, when the thickness of the membrane decreases, and when the diffusion coefficient of the transported solute increases. Which of these can be regulated? Typically membranes do not regulate their thickness, nor can the diffusion coefficient be altered. **Channels** can act like pores and they can be **gated**. That is, the channels can be opened or shut. This has the effect of controlling the area through which materials can be transported and this is a common way of regulating ion transport. Another way of physiologically regulating passive transport is by controlling the number of pores (or channels) in a membrane.

The distinction between pores and channels lies in the substrate. We think of pores as holes in a substrate that will not collapse, as if we drilled a tiny hole in a thin plastic sheet. Lipid bilayers, however, will not support a watery void in their interior. Channels are proteins embedded in the membrane that line a watery pathway across the membrane and prevent its collapse by providing mechanical support on the sides of the pathway. The watery path across the membrane is made by the proteins that form the channel.

DISSOLUTION IN THE LIPID BILAYER IS ANOTHER MODEL FOR PASSIVE TRANSPORT

Consider now a markedly different model of the membrane. In this case, there are no pores, but we envision that a molecule may penetrate from the left to the right of the membrane by dissolving in the lipid bilayer core of the membrane, diffusing across the lipid, and then being extracted back into the aqueous phase on the other side of the membrane. This model of passive transport is shown schematically in Figure 2.5.3.

The dissolution of the solute in the lipid membrane is described quantitatively by a constant called the **partition coefficient** (see Chapter 2.3):

[2.5.11] $$k_s = \frac{\text{equilibrium C in the lipid phase}}{\text{equilibrium C in the water phase}}$$

If equilibrium is reached quickly at both the left and right surface of the membrane, then diffusion through the lipid phase would limit the rate of transport. Let the concentration on the left side of the membrane be C_L and the concentration on the right side of the membrane be C_R. The concentration immediately inside the membrane on the left, by Eqn [2.5.11], will be $k_s C_L$, and the concentration on the right inside the membrane will be $k_s C_R$. The steady-state flux through the lipid phase is

[2.5.12] $$J_s = D_{s,\text{lipid}} \left(\frac{k_s C_L - k_s C_R}{\delta} \right)$$

where $D_{s,\text{lipid}}$ is the diffusion coefficient of the solute in the lipid phase and δ is the thickness of the lipid phase. We write J_s here because the entire area of the membrane is available for dissolution of the solute and

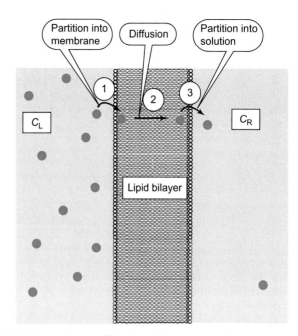

FIGURE 2.5.3 Cartoon of the lipid bilayer model of passive transport. The left and right compartments are separated by a lipid bilayer membrane. Solute, shown here as blue spheres, moves across the membrane in three well-defined steps. In step 1, the particle partitions itself into the lipid phase of the membrane. In step 2, the material diffuses across the lipid bilayer. In step 3, the material partitions itself back into the aqueous phase on the right side of the membrane. Overall transport rates are determined by the rates of steps 1, 2, and 3.

diffusion across the lipid bilayer. This equation can be rewritten as

[2.5.13] $$J_s = \frac{k_s D_{s,\text{lipid}}}{\delta} \Delta C$$

This last equation is identical in form to that derived earlier in the microporous membrane model:

[2.5.9] $$J_s = p \Delta C$$

In the case of the dissolution−diffusion−solution model, we identify the permeability as

[2.5.14] $$p = \frac{k_s D_{s,\text{lipid}}}{\delta}$$

Once again, the permeability is a single phenomenological parameter that relates the flux to the concentration difference. It incorporates all of the microscopic parameters of the membrane−solute pair into a single parameter. In this case, these microscopic parameters are the partition coefficient, the thickness of the membrane, and the diffusion coefficient of the solute in the membrane.

Equation [2.5.14] neatly sums up many experimental observations concerning the permeability of materials through biological membranes. These are briefly summarized in **Overton's rules**:

A. The permeability is proportional to the lipid solubility.
B. The permeability is inversely proportional to molecular size.

Thus we expect ethanol to permeate cell membranes quite easily because it is small and it is lipid soluble. On the other hand, glucose is larger and it is not readily lipid soluble and so it requires another mechanism to enter the cell. This mechanism is the carrier. Overton's rules derive from two main components of Eqn [2.5.14]: the dependence of p on k_s means that lipid solubility is a direct determinant of the permeability and the dependence on D_s means that size is an inverse determinant of permeability.

Lipid solubility here depends on the hydrophobicity or lipophilicity of the solute. Different chemical groups in any molecule confer hydrophilic or hydrophobic character to those parts of the molecule, as described in Chapters 2.3 and 2.4. In particular, electric charge make a solute highly hydrophilic and not lipophilic. Thus charged or ionized solutes are generally highly impermeable by this dissolution mechanism of transport. Solutes enriched in hydroxyl groups, carboxyl groups, and amino groups are generally not easily permeable through this mechanism unless they are also very small.

FACILITATED DIFFUSION USES A MEMBRANE-BOUND CARRIER

For some valuable materials, the membrane permeability is not large enough for the cell's needs and facilitated diffusion is used to carry solute across the membrane. One possible way a **carrier** could operate is shown in Figure 2.5.4. The transformations that occur to allow this transport are not known in detail. It is likely that the carrier provides something like a pore for the solute, but the pore is specifically designed to fit the solute. In this case, the solute molecule never dissolves in the lipid bilayer but is protected from it by a pocket of the protein carrier.

The mechanism involved in facilitated diffusion shown in Figure 2.5.4 involves a sequence of four reactions:

[2.5.15]
$$\begin{aligned} S_L + E_L &\to E - S_L \\ E - S_L &\to E - S_R \\ E - S_R &\to E_R + S_R \\ E_R &\to E_L \\ \hline \text{Sum}: S_L &\to S_R \end{aligned}$$

The sum of these four reactions is the movement of solute from the left side of the membrane to the right side. The carrier concentration does not enter into the overall stoichiometry of the reaction, because its presence on both sides of the reactions cancels itself out. It acts as a catalyst for transport because it determines its rate without being altered by the process.

FACILITATED DIFFUSION *SATURATES* WITH INCREASING SOLUTE CONCENTRATIONS

Facilitated diffusion can be distinguished from a purely diffusional mechanism because facilitated diffusion is **saturable** and it is **specific**. Plots of the flux versus the concentration are not linear, as you would expect from the diffusion mechanisms as shown in Eqns [2.5.8] and [2.5.13]. The rate increases with concentration but only up to a point. This is due to the fact that there are only so many carrier molecules in the membrane. When they are all busy, there can be no further increase in the rate of transport. This is analogous to the ferrying of people across a river that is too deep for most of them to wade. Although some can wade, most must cross only by ferry. When there are not too many people, the ferries can accommodate them easily and the transport rate will increase with each increase in the number of people waiting on shore. When the crowd on the shore gets too great, however, the ferries become full and the rate of transport can be increased further only by the number of brave souls who can wade the river (diffuse across the membrane) or by increasing the number of ferries. Figure 2.5.5 shows the kinetics of a saturable transport mechanism.

These curves often closely resemble the hyperbolic plots characteristic of Michaelis–Menten enzyme kinetics and can be fit to

[2.5.16]
$$Q_{\text{trans}} = \frac{Q_{\max} C}{K_m + C}$$

where Q_{trans} is the flow of transported material across the membrane in moles per unit time, $Q_{\max}$ is the maximum flow, C is the concentration of the transported

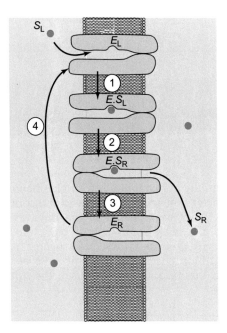

FIGURE 2.5.4 Schematic diagram of carrier-mediated passive transport. The carrier is designated as "E." In this scheme, an integral protein molecule in the membrane binds to a solute molecule on one side of the membrane. The carrier molecule then undergoes a transformation that has the effect of changing the side of the membrane that is accessible to the binding site. The solute molecule then dissociates from the carrier on the opposite side. The carrier then returns to its original shape.

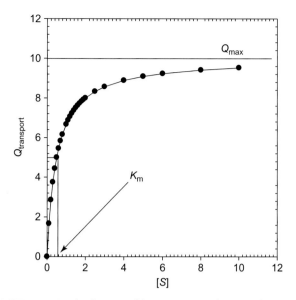

FIGURE 2.5.5 Graph of a saturable transport mechanism. The rate of transport in moles per unit time per unit area is plotted against the concentration of material on the feed side [S]. A maximum transport rate, Q_{max}, can be identified. The substrate concentration at half-maximal transport is used to characterize the affinity of the transport mechanism for its substrate.

solute on the feed side of the membrane, keeping the concentration on the opposite side at zero, and K_m is a constant characteristic of the carrier. The term "K_m" comes from Michaelis–Menten kinetics, and these carriers almost certainly do not have the mechanism first proposed by Michaelis and Menten. Nevertheless, the term "K_m" has come to mean "the concentration of substrate at half-maximal activity." Sometimes this is referred to as K_t, the concentration at half-maximal transport. Eqn [2.5.16] is a simplified version of the exact solution of the kinetics of the scheme shown in Figure 2.5.4.

FACILITATED DIFFUSION SHOWS *SPECIFICITY*

Another distinguishing feature of carrier-mediated facilitated diffusion is its structural **specificity**. The parts of the carriers that bind transported solute are specifically designed for that solute and not others. For example, most cell membranes in the human contain carriers for glucose. They will transport D-glucose but not its **enantiomer** (mirror image compound) L-glucose. The carrier for glucose will not transport amino acids and vice versa.

FACILITATED DIFFUSION SHOWS *COMPETITIVE INHIBITION*

The specificity of carrier-mediated facilitated diffusion also gives rise to **competitive inhibition**. Compounds that closely approximate the shape of the natural substrate may also bind to the carrier and be transported across the membrane. Since the carrier cannot carry both compounds at the same time, the transport of the natural substrate is reduced by the presence of competitive inhibitors. This competitive inhibition is closely related to that observed in enzyme kinetics. In other cases, a compound sharing some similarity with the natural substrate may bind to the carrier but not be transported. If the binding is at the transport site, such a compound might inhibit transport.

PASSIVE TRANSPORT OCCURS SPONTANEOUSLY WITHOUT INPUT OF ENERGY

The chemical reaction for the overall transport is written as

[2.5.17] $$\text{Solute}_L \rightarrow \text{Solute}_R$$

and the free energy change for the reaction is

[2.5.18] $$\Delta G = G_R - G_L$$

Substituting in with the chemical potential, we obtain

[2.5.19] $$\Delta G = n(\Delta \mu^0 + RT \ln C_R) - n(\Delta \mu^0 + RT \ln C_L)$$
$$= n\,RT \ln \frac{C_R}{C_L}$$

where n is the number of moles of solute moving from left to right. Here there is no electrical work term because the charge on the molecule, z, is zero. If $C_R > C_L$, ΔG calculated from Eqn [2.5.19] will be positive. This means that the opposite process will occur. That is, solute will move from the right to the left, opposite to the direction shown in Eqn [2.5.17]. If $\Delta G = 0$, then no net movement occurs and $C_R = C_L$. If $C_R < C_L$, then ΔG calculated according to Eqn [2.5.19] will be negative and the reaction will proceed as written, with solute moving from the left to the right side of the membrane. Thus thermodynamics tells us what process can occur and with what change in free energy, but it does not give us an expression for the permeability or the rate at which the process will occur.

Diffusion through aqueous pores or through the lipid barrier of membranes or by facilitated diffusion is called passive transport because none of these mechanisms requires "outside" energy. These flows occur spontaneously. What this means is that the energy that drives them is contained within the solutions themselves. This does not mean that they occur rapidly, but only that they occur naturally without the addition of any "outside" force. The rate at which they occur depends on the mechanisms of transfer. The analysis of the mechanism gives us additional information, such as what determines and regulates the rate.

The overall ΔG for facilitated diffusion is the ΔG for the sum, which is the same in Eqns [2.5.15] and [2.5.17]. The net $\Delta G = nRT \ln C_L/C_R$. Thus the participation of the carrier, which remains unchanged by the transport reaction, does not alter the reaction *energetics* at all, whereas it does alter the reaction *kinetics*. The carrier is a catalyst. It speeds up the reaction, which in this case is transport, without entering into the stoichiometry of the reaction. This is an example of how thermodynamic

Example 2.5.1 Specificity of Transport

There are a variety of transporters for glucose that are called GLUT (for glucose transporter). The GLUT-1 transporter imports glucose into a variety of cell types. This is an integral membrane protein with a molecular weight of 45 kDa. Its K_m for glucose is 1.5 mM. It will also transport L-glucose with a K_m of 3000 mM. Glucose is typically about 100 mg% in the extracellular fluid. At what fraction of Q_{max} will glucose be transported at this concentration?

The normal plasma [glucose] is given as 100 mg%, which is 100 mg of glucose per 100 mL of plasma. This is $P_g = 100$ mg/100 mL $\times$ 1000 mL L^{-1} = 1 g L^{-1}. Since the molecular weight of glucose is 180 Da, its gram molecular weight is 180 g mol^{-1} and the normal plasma glucose concentration is

$$P_g = 1 \text{ g L}^{-1} / 180 \text{ g mol}^{-1} = 0.0056 \text{ M} = 5.6 \text{ mM}$$

The rate of transport, assuming zero-trans glucose, is given by Eqn [2.5.16] as

$$Q_{trans} = [5.6 \text{ mM}/(1.5 \text{ mM} + 5.6 \text{ mM})]Q_{max} = \mathbf{0.789\ Q_{max}}$$

At plasma [glucose], the transporters are nearly saturated and the rate of transport could be increased mostly by affecting the number of transporters, i.e., increasing Q_{max}.

What would the L-glucose rate of transport be at the same concentration as D-glucose?

$$Q_{trans\ L\text{-}glucose} = [5.6 \text{ mM}/(3000 \text{ mM} + 5.6 \text{ mM})]Q_{max}$$
$$= \mathbf{0.0019\ Q_{max}}$$

D-Glucose is transported almost 400 times more quickly than L-glucose.

Example 2.5.2 Effect of Substrate Concentration on Flux

GLUT-1 glucose transporter is the most ubiquitous form of glucose transporter. It is present in high amounts in erythrocytes and endothelial cells, the blood–brain barrier and in the proximal straight tubule of the nephron. If its K_m for glucose is 1.5 mM, how much would transport increase if plasma glucose were increased from 80 mg% to 120 mg%?

First, we convert the plasma glucose concentrations to mM. 80 mg% means 80 mg per 100 mL or 80 mg/0.1 L = 800 mg/L. The molecular weight of glucose is 180 g, so this concentration is equivalent to 0.8 g/180 g mol^{-1}/L = 4.44 mM. Similarly, 120 mg% is 6.67 mM.

If the K_m for glucose is 1.5 mM, then the transport rate at 80 mg% glucose would be

$$Q = Q_{max} \times 4.4 \text{ mM}/(1.5 \text{ mM} + 4.44 \text{ mM}) = \mathbf{0.75 Q_{max}}$$

And at 120 mg% glucose it would be

$$Q = Q_{max} \times 6.66 \text{ mM}/(1.5 \text{ mM} + 6.66 \text{ mM}) = \mathbf{0.82 Q_{max}}$$

The transport rate increases 9% when the blood glucose increases by 50%. These GLUT1 transporters are insensitive to changes in blood glucose.

GLUT2 is another glucose transporter that is present in beta cells of the islets of Langerhans, in the pancreas, and also in the kidney, intestine, and liver. Its K_m for glucose is much higher, 17 mM. How much would transport increase if plasma glucose were increased from 80 mg% to 120 mg%?

We can use the molar concentrations for glucose that we used before for the GLUT1 calculations. The transport rates are calculated at 80 mg% glucose as

$$Q = Q_{max} \times 4.4 \text{ mM}/(17 \text{ mM} + 4.44 \text{ mM}) = \mathbf{0.207 Q_{max}}$$

and at 120 mg% glucose, it would be

$$Q = Q_{max} \times 6.66 \text{ mM}/(17 \text{ mM} + 6.66 \text{ mM}) = \mathbf{0.282 Q_{max}}$$

Here the transport rate increases 36% when blood glucose increases 50%.

Thus GLUT1 and GLUT2 serve different functions. GLUT1 operates close to maximal rates relatively independently of blood glucose levels. GLUT2 increases transport almost proportionately with blood glucose, so that GLUT1 is used for basal glucose transport into metabolizing tissues, whereas GLUT2 finds use as part of the sensor apparatus for glucose concentrations in blood.

---

analysis of the reaction is independent of the mechanism: it tells us about the energetics without telling us anything about the reaction's path or its rate.

The saturability of carrier-mediated facilitated diffusion distinguishes it from the other passive transport mechanisms that show a linear relationship between flow and the concentration difference across the membrane. Neither mechanism can concentrate solute. Flow of material always occurs from the side with the higher concentration to the side with the lower concentration.

When the concentrations on the two sides of the membrane are equal, no further flow occurs because the two solutions are in equilibrium.

IONS CAN BE PASSIVELY TRANSPORTED ACROSS MEMBRANES BY *IONOPHORES* OR BY *CHANNELS*

So far we have considered passive diffusion of nonelectrolytes. Suppose now that the diffusing species are electrically charged. **Charged species are poorly soluble in the lipid phase**, and so they cannot merely dissolve in the lipid on one side of the membrane, diffuse across, and then enter the compartment on the opposite side of the membrane. They need either carriers or channels to get across.

IONOPHORES CARRY IONS ACROSS MEMBRANES OR FORM CHANNELS

Fungi and bacteria make a class of poison called **ionophores**. These are molecules that allow ions to cross membranes. The fungi and bacteria make these compounds to kill off competition by disrupting the permeability barrier of their competitors' membranes. These ionophores are of two types: carriers and channel formers. Figure 2.5.6 illustrates these two types of ionophores.

An example of a carrier is **A23187**. This material is commercially obtained from *Streptomyces chartreusis* and has weak antibiotic activity against gram-positive bacteria. It is particularly active for divalent cations with a specificity of $Mn^{2+} > Ca^{2+} > Mg^{2+} > Sr^{2+} > Ba^{2+} > Li^+ > Na^+ > K^+$. It is predominantly used as a carrier for Ca^{2+}. Other examples of natural molecules that act as carriers include **valinomycin** (a K^+ ionophore) and **nigericin** (an H^+ ionophore).

An example of a channel former is **gramicidin A**. This is an antibiotic polypeptide containing 15 amino acids that is isolated from the bacterium *Bacillus brevis*. The molecule appears to form a pore by linking two molecules of gramicidin A across the bilayer. The gramicidin pore appears to behave like a water-filled pore. Other examples of pore-forming antibiotics include **amphotericin** and **nystatin**. Amphotericin makes a channel by interacting with cholesterol in cell membranes.

ION CHANNELS

A variety of integral membrane proteins form channels for ions. These **ion channels** exhibit some of the characteristics of carriers in that they are highly selective. These channels exhibit other characteristics such as **gating**. Gating refers to the fact that these channels act as if they have gates that are opened sometimes, allowing ions to cross the membrane, and are closed at other times, preventing ions from moving. The percent of the time the channels are opened is referred to as the open probability, p_o, and can be regulated in various ways. Some channels open when another molecule binds to the channel. These are **ligand-gated** channels. Other channels sense the local potential, probably through the presence of charged groups on the channel, and open or close depending on the potential. These are **voltage-gated** channels. A cartoon of these types of channels is shown in Figure 2.5.7.

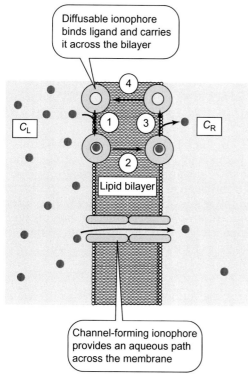

FIGURE 2.5.6 How ionophores work. Some ionophores increase the passive diffusion across a lipid bilayer by providing a hydrophilic pocket that binds a solute and sequesters it away from the hydrophobic lipid interior. These ionophores generally show specificity of transport because the pocket binds some ions better than others. For these types of ionophores, the ionophore–ligand complex is believed to diffuse across the lipid bilayer, carrying the ligand with the ionophore. Other ionophores form an aqueous channel across the lipid bilayer. These channel-forming ionophores are less specific but still show specificity due to the size and shape of the channel.

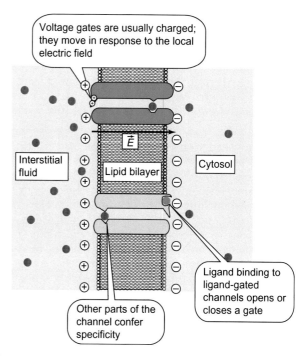

FIGURE 2.5.7 Voltage- and ligand-gated channels. Voltage-gated channels typically have a highly charged part of the protein that responds to the local electrical field produced by the separation of charge on the two sides of the membrane. Changes in this electric field alter the disposition of the gate to either open or close access to a hydrophilic pathway across the membrane. Ligand-gated channels bind a regulatory ligand that alters the shape of the channel so as to open or close its pathway.

Examples of voltage-gated and ligand-gated channels abound, and a full description of them here is premature because we have not yet studied membrane potential or action potentials. **Voltage-gated Na$^+$ and K$^+$ channels** allow ions to flow across the membrane only under specific circumstances. The flow of ions is an electric current, because charged ions are moving. The membrane itself is a tiny capacitor, as we saw in Chapter 1.3. The currents going through these channels can discharge the capacitor, changing the voltage across the membrane, or they can charge it back up again. In this way, opening of the fast Na$^+$ channel in neurons causes a brief, pulse-like change in the voltage across the neuronal cell membrane. The membrane potential is reestablished by a later opening of the K$^+$ channels. These actions produce the nerve impulse, the brief change in nerve cell membrane potential that propagates down the nerve and activates its target—either another neuron, a muscle fiber, or some secretory cell.

Other voltage-gated channels include the **T-, L-, and N-type Ca^{2+} channels**. The designation "T" signifies that this channel opens "transiently"; the "L" stands for "long-lasting"; and the N indicates that this type of channel is "neuronal." The channels generally open upon depolarization of the cell membrane and they specifically transport Ca^{2+}. The consequence is that the [Ca^{2+}] inside the cell increases in the vicinity of the channel, and this Ca^{2+} binds to cellular elements to change their activity.

Ligand-gated channels may be present on the surface membrane and also on interior membranes. The endoplasmic reticulum of many cells contains a large tetrameric protein called the **IP3 receptor**. This receptor forms a channel for Ca^{2+} across the ER membrane and opens in response to IP3 (inositol trisphosphate) that is liberated from the surface membrane as part of signal transduction. Gating by IP3 causes Ca^{2+} release from the ER and the increased cytoplasmic [Ca^{2+}] alters cellular activity.

Many ligand-gated channels are present on the surface membrane and respond to neurotransmitters or hormones. These channels are gated by the binding of a chemical rather than by the voltage difference across the membrane. **Acetylcholine** is the neurotransmitter involved in skeletal muscle neurotransmission. When activated, the motor neuron nerve terminal releases acetylcholine near the skeletal muscle membrane. The acetylcholine binds to **nicotinic acetylcholine receptors**, so named because of their sensitivity to nicotine, on the muscle membrane. Binding of acetylcholine opens a large conductance pathway mainly for Na$^+$. This causes a depolarization of the muscle membrane that propagates along the muscle surface, eventually activating muscle contraction.

WATER MOVES PASSIVELY THROUGH AQUAPORINS

Passive transport of water across biological membranes also occurs through water channels. These are tiny pores formed by proteins called **aquaporins**. There are a variety of aquaporins and they are present on virtually every cell membrane. AQP1 has a molecular weight of 29 kDa and forms a channel by the association of four monomers. In some membranes, the number of aquaporins is physiologically regulated so that water movement through the cell can be regulated. This is particularly important in the kidney, because the kidney has the final job of retaining water when it is scarce and excreting it when it is in excess. Although water obeys Fick's Laws of Diffusion, its movement is dominated by pressure-driven flow.

SUMMARY

Materials cross biological membranes by a variety of mechanisms including passive transport, active transport, and osmosis. Passive transport mechanisms require no input of metabolic energy. Because of this, passive transport always entails the movement of materials from regions of high concentration to regions of low concentration. The free energy change per mole in the reaction $S_L \rightarrow S_R$ is

$$\Delta\mu = RT \ln[C_R/C_L]$$

where C_R and C_L are the concentrations of S on the right- and left-hand sides of the membrane, respectively. If $C_R < C_L$, then $\Delta\mu < 0$ and the reaction proceeds from the high concentration (C_L) to the low concentration (C_R).

Passive diffusion across membranes is characterized by a linear relationship between the rate of transport and the concentration difference across the membrane:

$$J_s = p\Delta C$$

where J_s is the macroscopically observed flux and p is the permeability. This equation holds true if we envision the membrane as a microporous membrane in which diffusion occurs through tiny pores, or if we envision the solute as dissolving in the lipid bilayer and diffusing across it. The dependence of p on the microscopic characteristics of the membrane differs in these two models. For a microporous membrane

$$p = n\pi a^2 D/\delta$$

where n is the number of pores per unit area, a is the radius of the pore, D is the diffusion coefficient of the solute, and δ is the thickness of the membrane. For a solute dissolving in the lipid bilayer

$$p = KD_{lipid}/\delta$$

where K is the partition coefficient of the material in the lipid phase, D_{lipid} is the diffusion coefficient in the lipid bilayer, and δ is the thickness of the bilayer.

Some membrane proteins bind solutes and provide an alternative path across membranes. The alternative path facilitates the diffusion of the solute across the membrane. These proteins are carriers for the solutes. The kinetics of transport shows specificity, saturation, and competition with similar solutes. The overall transport rate often obeys an equation of the form

$$Q_{trans} = Q_{max}C/[K_m + C]$$

where Q_{max} is the maximum transport rate, typically limited by the number of carriers in the membrane, and K_m is a measure of the dissociation constant of the carrier for the solute.

Passive transport mechanisms include lipid dissolution and diffusion, facilitated diffusion, ligand-gated channels, voltage-gated channels, diffusion-mediated ionophores, and pore-forming ionophores.

REVIEW QUESTIONS

1. Why is the gradient for a diffusive process linear at steady state?
2. For a microporous membrane, what effect would increasing the number of pores have on diffusive flux across a membrane? What effect would increasing the size of the pores have? What effect should result from increasing the size of the diffusing solute? Increasing the temperature?
3. For a solute dissolution model, what effect would increasing the partition coefficient have? Increasing the particle size of the diffusing solute? Increasing the temperature?
4. Why should two models as different as the microporous membrane and solute dissolution model have identical relationship between J and ΔC?
5. Under what conditions is the free energy for transfer across a membrane for a solute equal to zero?
6. How does the function relating rate of transport to concentration differ between facilitated diffusion and simple diffusion?
7. Why does simple diffusion not show specificity or competition?
8. What two general mechanisms are used to regulate the open or closed state of channels?
9. What is a channel?
10. How would you determine K_m for facilitated diffusion? Q_{max}?

2.6 Active Transport: Pumps and Exchangers

Learning Objectives

- Write the equation for the electrochemical potential for an ion
- Give the approximate concentration of the major ions inside and outside of a heart cell
- Be able to calculate the free energy change for ion movement across the cell membrane under given conditions of membrane potential and ion concentrations
- Be able to calculate the free energy change for ion transport when coupled to other processes
- Distinguish between active and passive transport
- Distinguish between primary and secondary active transport
- Give an example of a primary and secondary active transport mechanism
- Distinguish between P-type, V-type, F-type, and ABC-type active transporters
- Define the terms: symport, cotransporter, antiport, and exchanger

THE ELECTROCHEMICAL POTENTIAL DIFFERENCE MEASURES THE ENERGETICS OF ION PERMEATION

Here we consider the transport of four ions: Na^+, K^+, Cl^-, and Ca^{2+}, across the surface membrane of heart cells at rest. These ions have different concentrations inside and outside of the cell, and are also subjected to electrical forces because there is a potential difference across the cell membrane and thus an electric field within the membrane. The cardiomyocytes maintain a resting **membrane potential**. We will learn later how this is established, but for now it is enough to know that there is a separation of charge in the outside and inside compartments. At rest, there is an accumulation of negative charges inside the cell and positive charges outside. This separation of charges gives rise to the membrane potential, which is always taken as the difference between the electrical potential inside and outside of the membrane:

[2.6.1] $$\Delta\psi = \psi_i - \psi_o$$

The membrane potential is sometimes identified with the variables ψ, V_m, or E_m. The situation is described in Figure 2.6.1.

Let us first take the case of Na^+. The concentration of Na^+ outside the cell is about 145 mM and inside the cell it is 12 mM. The higher concentration outside of the cell favors a net Na^+ flow from outside to inside, driven by diffusion. Further, the negative potential inside the cell also favors Na^+ movement into the cell. We write the free energy change per mole for the movement of Na^+ from out to in as

[2.6.2] $$\Delta\mu = \mu_i - \mu_o$$

Recall here that the free energy change is the free energy of the final state (inside the cell, in this case) minus the free energy of the initial state (outside the cell in this case). When we are dealing with the free energy per mole, we write μ. This is an intensive property which is defined by the conditions and not by the extent of the cell or its membrane or of the amount of material being transported. The free energy itself is an extensive property that depends on how much material is being transported. We can insert the definition of electrochemical potential that we justified earlier (see Eqn [1.7.14]):

[2.6.3] $$\mu_x = \mu_x^0 + RT \ln C_x + z_x \mathfrak{F} \psi_x$$

where the subscript x denotes substance x to avoid confusion with the subscript i or o which denotes the inside or outside of the cell, respectively. Substituting in for the conditions of the inside and outside of the cell, we find

$$\begin{aligned}\Delta\mu &= \mu_i - \mu_o \\ &= \mu_{Na_i}^0 + RT\ln[Na^+]_i + \mathfrak{F}\psi_i - \mu_{Na_o}^0 - RT\ln[Na^+]_o - \mathfrak{F}\psi_o \\ &= RT\ln\frac{[Na^+]_i}{[Na^+]_o} + \mathfrak{F}(\psi_i - \psi_o)\end{aligned}$$

[2.6.4]

The standard free energy per mole (μ^0) cancels out because it is independent of condition. The μ^0 is the part of the electrochemical potential that incorporates the chemical energy involved in the formation of bonds. Since in this case there is no chemical transformation (the Na^+ ion is not chemically transformed in any way; it is simply transported from one side of the membrane to the other), $\Delta\mu^0 = 0$. Also, the z in the formula for electrochemical potential is the integer charge, which is +1 for the Na^+ ion. Remember here that ln is the natural logarithm, not logarithm base 10.

The calculation of $\Delta\mu$ for the conditions of the cell shows that $\Delta\mu < 0$ (see Example 2.6.1). This means that the reaction as written is spontaneous: it will occur in the direction written **passively**, without additional forces. Our analysis of the energy does not tell us anything about how fast the process will occur, because the thermodynamic analysis is independent of the mechanism, and the mechanism is what determines the rate. What we can say, however, is that a channel for Na^+ that opens in this membrane will cause a rapid inflow of Na^+ from the extracellular fluid into the cell. Net flow would not occur in the opposite direction. What we can also say is that if the membrane has any non-zero permeability to Na^+, the flux will be from the extracellular fluid (the outside) into the cell.

Now let us consider what happens with Ca^{2+}. In a completely analogous way, we write the change in free energy for Ca^{2+} entry into the cell as

$$\Delta\mu = \mu_i - \mu_o$$
$$= RT \ln[Ca^{2+}]_i + 2\mathfrak{F}\psi_i - RT \ln[Ca^{2+}]_o - \mathfrak{F}\psi_o$$
$$= RT \ln\frac{[Ca^{2+}]_i}{[Ca^{2+}]_o} + 2\mathfrak{F}(\psi_i - \psi_o)$$

[2.6.5]

Here the 2 in the equation arises because Ca^{2+} has two positive charges per ion. The two charges correspond to z_x in Eqn [2.6.3]. The electrical force on a Ca^{2+} ion is twice the force on an Na^+ ion in the same electric field and thus movement produces twice the energy.

Note that this process (see Example 2.6.2) has a negative $\Delta\mu$, so it also occurs spontaneously. In this case, the free energy change is much more negative. This is due to the fact that the concentration gradient for Ca^{2+} contains more energy and the electrical energy gain is twice as great because each Ca^{2+} ion has twice the charge of an Na^+ ion. A channel for Ca^{2+} on the cell membrane would let Ca^{2+} into the cell under these conditions.

Now let us calculate the free energy change for K^+ entry into the cell. The formula is:

$$\Delta\mu = \mu_i - \mu_o$$
$$= RT \ln[K^+]_i + \mathfrak{F}\psi_i - RT \ln[K^+]_o - \mathfrak{F}\psi_o$$
$$= RT \ln\frac{[K^+]_i}{[K^+]_o} + \mathfrak{F}(\psi_i - \psi_o)$$

[2.6.6]

In this case, $\Delta\mu$ is positive (see Example 2.6.3). This means that K^+ under these conditions does not passively enter the cell. Rather, the spontaneous process is K^+ *exit* from the cell. If a channel specific for K^+ was to open in the membrane, K^+ would leave the cell.

In this last example, let us calculate the free energy associated with Cl^- entry into the cell. Again we write the difference in the electrochemical potentials:

$$\Delta\mu = \mu_i - \mu_o$$
$$= \mu_{Cl_i}^0 + RT \ln[Cl^-]_i + (-1)\mathfrak{F}\psi_i - \mu_{Cl_o}^0 + RT \ln[Cl^-]_o$$
$$- (-1)\mathfrak{F}\psi_o$$
$$= RT \ln\frac{[Cl^-]_i}{[Cl^-]_o} - \mathfrak{F}(\psi_i - \psi_o)$$

[2.6.7]

Note that the valence on the Cl^- ion is negative and it is entered that way in the equation. According to the calculation (see Example 2.6.4), at rest $[Cl^-]$ is distributed at equilibrium across the cell membrane. That is, there is no free energy change for Cl^- transport across the resting muscle membrane.

These calculations show how the electrochemical potential calculates the energetics of transport, and further

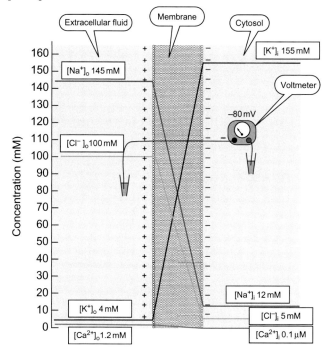

FIGURE 2.6.1 Concentrations of Na^+, K^+, and Ca^{2+} and the resting membrane potential across the resting cardiac muscle cell membrane. The subscript "o" refers to the "outside" of the cell; "i" denotes the "inside" compartment.

EXAMPLE 2.6.1 Free Energy of Na^+ Transport

For the conditions shown in Figure 2.6.1, calculate the free energy of transport of Na^+ from outside to inside the cardiomyocyte.

Inserting the values for the concentrations into Eqn [2.6.4] and using $R = 8.314$ J mol^{-1} K^{-1} (=1.987 cal mol^{-1} K^{-1}; 1 J = 0.239 cal) and $T = 310$ K, and remembering that the faraday ($\mathfrak{F}$) is 9.649×10^4 C mol^{-1}, we get

$\Delta\mu = 8.314$ J mol^{-1} K^{-1} $\times 310$ K $\ln[(12 \times 10^{-3}$ M$)/(145 \times 10^{-3}$ M$)]$

$+ 9.649 \times 10^4$ C mol$^{-1} \times (-0.080$ V$)$

$= -6.42$ kJ mol$^{-1} - 7.72$ kJ mol^{-1}

$= \mathbf{-14.14}$ **kJ mol**$^{-1}$ $= -3.38$ kcal mol^{-1}

EXAMPLE 2.6.2 Free Energy of Ca²⁺ Transport

For the conditions shown in Figure 2.6.1, calculate the free energy of transport of Ca^{2+} from outside to inside the cardiomyocyte.

Inserting the values for the concentrations and membrane potential into Eqn [2.6.5], we get:

$\Delta\mu = 8.314 \text{ J mol}^{-1} \text{ K}^{-1} \times 310 \text{ K ln}[(0.1 \times 10^{-6} \text{ M})/(1.2 \times 10^{-3} \text{ M})]$
$+ 2 \times 9.649 \times 10^4 \text{ C mol}^{-1} \times (-0.080 \text{ V})$
$= -24.2 \text{ kJ mol}^{-1} - 15.4 \text{ kJ mol}^{-1}$
$= \mathbf{-39.6 \text{ kJ mol}^{-1} = -9.46 \text{ kcal mol}^{-1}}$

EXAMPLE 2.6.3 Free Energy of K⁺ Transport

For the conditions shown in Figure 2.6.1, calculate the free energy of transport of K^+ from outside to inside the cardiomyocyte.

We can insert the values for $[K^+]_i = 155 \times 10^{-3}$ M and $[K^+]_o = 4 \times 10^{-3}$ M into Eqn [2.6.6] to calculate $\Delta\mu$:

$\Delta\mu = 8.314 \text{ J mol}^{-1} \text{ K}^{-1} \times 310 \text{ K ln}[(155 \times 10^{-3} \text{ M})/(4 \times 10^{-3} \text{ M})]$
$+ 9.649 \times 10^4 \text{ C mol}^{-1} \times (-0.080 \text{ V})$
$= -9.43 \text{ kJ mol}^{-1} - 7.72 \text{ kJ mol}^{-1}$
$= \mathbf{+1.71 \text{ kJ mol}^{-1} = +0.41 \text{ kcal mol}^{-1}}$

EXAMPLE 2.6.4 Free Energy of Cl⁻ Transport

For the conditions shown in Figure 2.6.1, calculate the free energy of transport of Cl^- from outside to inside the cardiomyocyte.

We can insert the values for $[Cl^-]_i = 5 \times 10^{-3}$ M and $[Cl^-]_o = 100 \times 10^{-3}$ M into Eqn [2.6.7] to calculate $\Delta\mu$:

$\Delta\mu = 8.314 \text{ J mol}^{-1} \text{ K}^{-1} \times 310 \text{ K ln}[(5 \times 10^{-3} \text{ M})/(100 \times 10^{-3} \text{ M})]$
$- 9.649 \times 10^4 \text{ C mol}^{-1} \times (-0.080 \text{ V})$
$= -7.72 \text{ kJ mol}^{-1} + 7.72 \text{ kJ mol}^{-1}$
$= \mathbf{0 \text{ kJ mol}^{-1} = 0 \text{ kcal mol}^{-1}}$

show that both the electrical and diffusive forces enter into the equations that determine the direction of ion flow. Things get a bit more complicated and interesting when ion flow is coupled to other processes.

ACTIVE TRANSPORT MECHANISMS LINK METABOLIC ENERGY TO TRANSPORT OF MATERIALS

All of the ion movements we have discussed so far involve movement "down" the electrochemical gradient: the free energy was higher for the initial condition than for the final condition, and the free energy change $\Delta\mu = \mu_{final} - \mu_{initial}$ is negative ($\Delta\mu < 0$). Passive diffusion through pores or the lipid bilayer, carriers, and channels are all passive. This does not mean that energy is not involved. What it means is that the energy does not derive from metabolism. The energy comes from the solutions themselves. However, cells also concentrate some materials by moving them from a region of low electrochemical potential to a region of higher electrochemical potential. This movement has $\Delta\mu > 0$. It can occur spontaneously only when the positive $\Delta\mu$ for transport is coupled to another process with a more negative $\Delta\mu$. Typically this process is ATP hydrolysis.

Primary active transport moves materials against an electrochemical gradient by the direct involvement of ATP hydrolysis. Examples of molecules that are involved in active transport include the **ion pumps**: Na, K-ATPase, Ca-ATPase, and H-ATPase.

Secondary active transport moves materials against an electrochemical gradient by the indirect involvement of ATP hydrolysis. ATP is used to establish an electrochemical gradient for something, usually Na^+, and the energy stored in the electrochemical gradient for Na^+ is then used to pump material "uphill." Examples of secondary active transport are the **Na–glucose cotransport** in the intestinal epithelium and renal proximal tubule and the **Na–Ca exchange** in the heart surface membrane.

NA,K-ATPASE IS AN EXAMPLE OF PRIMARY ACTIVE TRANSPORT

The analysis of free energy changes on ion movement that we performed earlier in this chapter indicated that

> **EXAMPLE 2.6.5 Calculate the Free Energy for Operation of Na,K-ATPase**
>
> For the conditions of the cell, calculate the free energy for the Na,K-ATPase.
>
> We have already calculated that $\Delta\mu$ for ATP hydrolysis is -57.1 kJ mol^{-1} (see Chapter 1.7).
>
> We have calculated $\Delta\mu$ for Na$_o^+$ going into the cell: $\Delta\mu_{Na_o \to Na_i} = -14.14$ kJ mol^{-1}. This is the opposite process of what the pump does. Thus $\Delta\mu$ for Na$^+$ exit is $\Delta\mu_{Na_i \to Na_o} = +14.14$ kJ mol^{-1}.
>
> We also calculated $\Delta\mu$ for K$^+$ entry as $\Delta\mu_{K_o \to K_i} = +1.71$ kJ mol^{-1}.
>
> Inserting the values for $\Delta\mu$, we get
>
> $$\Delta\mu_{Na,K\text{-ATPase}} = -57.1 \text{ kJ mol}^{-1} + 3 \times 14.14 \text{ kJ mol}^{-1}$$
> $$+ 2 \times 1.71 \text{ kJ mol}^{-1} = \mathbf{-11.26 \text{ kJ mol}^{-1}}$$

the distribution of Na$^+$, K$^+$, and Ca^{2+} is far from equilibrium. This implies that the cell actively maintains these concentrations away from equilibrium, or else the equilibrium distribution would eventually occur. The mechanism responsible for maintaining the resting concentrations of Na$^+$ and K$^+$ is the Na–K pump. This pump moves three Na$^+$ ions out of the cell at the same time that it transports two K$^+$ ions into the cell. The movement of these ions is coupled to the hydrolysis of ATP. The whole process is written as

[2.6.8] $\text{ATP} + 3\text{Na}_i^+ + 2\text{K}_o^+ \Rightarrow \text{ADP} + \text{Pi} + 3\text{Na}_o^+ + 2\text{K}_i^+$

The $\Delta\mu$ for this entire process is equal to the sum of the $\Delta\mu$ for three separate processes:

[2.6.9a] $\text{ATP} \Rightarrow \text{ADP} + \text{Pi}$

[2.6.9b] $3\text{Na}_i^+ \Rightarrow 3\text{Na}_o^+$

[2.6.9c] $2\text{K}_o^+ \Rightarrow 2\text{K}_i^+$

The overall free energy for this coupled process is written as

$\Delta\mu_{Na,K\text{-ATPase}} = \Delta\mu_{\text{ATP} \Rightarrow \text{ADP+Pi}} + 3\Delta\mu_{Na_i \Rightarrow Na_o} + 2\Delta\mu_{K_o \Rightarrow K_i}$
[2.6.10]

Recall that $\Delta\mu$ is the free energy per mole. In this case, it is the free energy *per mole of completed reaction* with the **stoichiometry** given by the overall reaction in Eqn [2.6.8].

$\Delta\mu_{Na,K\text{-ATPase}} = -11.26$ kJ mol^{-1} (see Example 2.6.5) means that the net change in free energy per mole of reaction, not per mole of Na$^+$ or K$^+$ or ATP, is -11.26 kJ. This is excess free energy of ATP hydrolysis beyond that required to transport Na$^+$ and K$^+$. ATP hydrolysis has a total of 57.1 kJ of energy per mole of ATP that can be harnessed to do work, and the Na,K-ATPase uses 45.84 kJ of energy per mole of reaction to do electrochemical work, under cell conditions. According to this result, there is enough energy in ATP hydrolysis to drive the Na,K-ATPase reaction. These calculations hold for the resting cell under the conditions we have investigated. The free energy for the Na,K-ATPase reaction changes with changes in cellular [Na$^+$], [K$^+$], $\Delta\psi$, [ATP], [ADP], or [Pi]. Changes in [ATP], [ADP], or [Pi] alter the energy available to the Na,K-ATPase to do work. Changes in [Na$^+$], [K$^+$], or $\Delta\psi$ alter the energy necessary for transport.

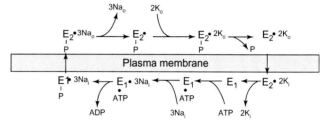

FIGURE 2.6.2 Modified post-Albers scheme for the reaction mechanism of Na,K-ATPase. Follow the reaction scheme in the direction of the arrows and you will see that the net reaction is the hydrolysis of ATP and the transport of three Na$^+$ ions from in to out and two K$^+$ ions from out to in. The processes are coupled in the reaction mechanism of the pump. The pump cannot hydrolyze ATP without binding Na$^+$ and K$^+$ in sequence. Neither can Na$^+$ be transported without K$^+$ transport and ATP hydrolysis. The coupling is made possible in part by the phosphorylation of the enzyme at an aspartic acid residue. This is shown by E–P in the diagram. The formation of a phosphoenzyme is common to the P-type active transport pumps.

The negative $\Delta\mu$ that we have calculated for the Na,K-ATPase indicates that the Na,K-ATPase reaction will occur spontaneously, but it will not tell us at what rate. The rate is a consequence of the mechanism of the pump.

NA,K-ATPASE FORMS A PHOSPHORYLATED INTERMEDIATE

The mechanism of ion pumps is generally complicated. It is useful to think of the enzyme as being characterized by a limited number of conformations to which we can give identifying labels. The reaction mechanism is then viewed as the sequential steps that occur among these conformations to achieve ATP hydrolysis and ion transport. Transformations between these conformations are determined by **rate constants**. Each step has two rate constants, one for the forward and the other for the reverse reaction. A simplified scheme for Na,K-ATPase is shown in Figure 2.6.2.

THE NA,K-ATPASE IS *ELECTROGENIC*

The overall reaction of the Na,K-ATPase shown in Eqn [2.6.8] indicates the stoichiometry of 3 Na$^+$ being transported out of the cell and 2 K$^+$ ions being transported into the cell. Thus the numbers of charges moving in

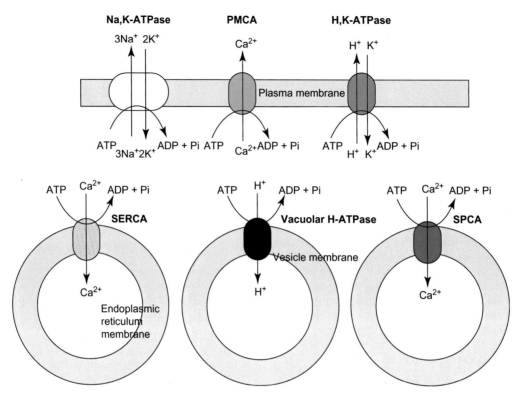

FIGURE 2.6.3 Some types of primary active transport mechanisms. Some are located on the plasma membrane of specific cell types; others, such as the smooth endoplasmic reticulum Ca-ATPase (SERCA), are located on subcellular organelles. All of the primary active transporters hydrolyze ATP. PMCA is the plasma membrane Ca-ATPase; SERCA is the smooth endoplasmic reticulum Ca-ATPase; SPCA is the secretory pathway Ca-ATPase.

the two directions are not equal, and each turnover of the pump corresponds to the movement of one net positive charge out of the cell. Thus the operation of the pump involves a transmembrane current and a net separation of charge. For this reason, the pump is called *electrogenic*—it generates an electric current and produces an electrical potential. As we will see later, its contribution to the resting membrane potential is small, but it is not zero.

THERE ARE MANY DIFFERENT PRIMARY ACTIVE TRANSPORT PUMPS

There are a variety of primary active transport pumps encoded by the human genome. These can be classified into four major groups.

P-type ATPases: These all form phosphorylated intermediates in the pump mechanism, like that shown for the Na,K-ATPase in Figure 2.6.2. Among these are **gastric H⁺-ATPase** that is responsible for acidification of the stomach contents; **Na⁺,K⁺-ATPase** that is responsible for maintaining ionic gradients in most cells; **PMCA** (for **p**lasma **m**embrane **c**alcium **A**TPase) responsible for pumping Ca^{2+} out of cells; the **SERCA** family of pumps, where SERCA stands for **s**mooth **e**ndoplasmic **r**eticulum Ca^{2+}ATPase, which is responsible for removing Ca^{2+} from the cytosol of a variety of cell types and placing it in storage in internal sacs within cells; and **SPCA**, the **s**ecretory **p**athway **C**a-**A**TPase, which pumps Ca^{2+} and other ions, such as Mn^{2+} and Zn^{2+}, into the Golgi to bind to secretory proteins. The SPCA is distinguishable functionally from the SERCA pumps because the SERCA pumps are inhibited by thapsigargin at low concentrations whereas SPCA is not. Figure 2.6.3 shows some of these primary active transporters.

V-type ATPases: Membranes of lysosomes and secretory vesicles contain a vacuolar-type H⁺-ATPase that pumps H⁺ ions from the cytoplasm into the vesicles. This V-type H⁺-ATPase differs from the gastric H⁺-ATPase in that it does not require K⁺. The structure and mechanism of V-type ATPases differs from the P-type active transporters.

F-type ATPases: These are more commonly referred to as ATP-synthetases, because they usually work in the reverse mode to make ATP rather than hydrolyze it for the purpose of transport. The main example in the human is the F_0F_1ATPase of the inner mitochondrial membrane, which is discussed in Chapter 2.10. It uses the electrochemical gradient of H⁺ to make ATP, but it can also hydrolyze ATP.

ABC transporters: The ABC here stands for "**A**TP-**b**inding **c**assette." This is a large family of proteins that engage in the primary active transport of a wide variety of solutes.

THE NA—CA EXCHANGER AS AN EXAMPLE OF SECONDARY ACTIVE TRANSPORT

According to our earlier calculation, $\Delta \mu$ for Ca^{2+} entry into the heart muscle cell was -39.6 kJ mol^{-1}. If left to itself Ca^{2+} would slowly leak into the cell and disturb the resting $[Ca^{2+}]_i$. Heart cells have at least two mechanisms for pumping the Ca^{2+} back out. One of these, mentioned above, is a Ca-ATPase that directly couples ATP hydrolysis to the outward transport of a single Ca^{2+} ion. This is the PMCA Ca^{2+} pump. A second mechanism is called the **Na–Ca exchanger**. Its **stoichiometry**, or relative mole numbers, is 3Na:1Ca. The overall reaction is

[2.6.11] $\qquad 3Na_o^+ + Ca_i^{2+} \rightleftharpoons 3Na_i^+ + Ca_o^{2+}$

The $\Delta\mu$ for this entire process is equal to the sum of the $\Delta\mu$ for two separate processes:

[2.6.12] $\qquad \begin{array}{c} 3Na_o^+ \rightleftharpoons 3Na_i^+ \\ Ca_i^{2+} \rightleftharpoons Ca_o^{2+} \end{array}$

So we write

[2.6.13] $\quad \Delta\mu_{Na,Ca\ exchange} = 3\Delta\mu_{Na_o \rightleftharpoons Na_i} + \Delta\mu_{Ca_i \rightleftharpoons Ca_o}$

According to this result (see Example 2.6.6), there is enough energy in the Na^+ electrochemical gradient to drive Ca^{2+} out of the cell. However, it requires coupling the entry of three Na^+ ions for each Ca^{2+} ion that exits the cell. If the Na^+–Ca^{2+} exchange was to couple the movement of two Na^+ to each Ca^{2+}, there would be insufficient energy to drive Ca^{2+} efflux, and the exchanger would actually work in reverse mode, with the Ca^{2+} gradient driving the efflux of Na^{2+}.

The Ca^{2+} efflux that occurs through the Na–Ca exchanger is "uphill," meaning that it requires energy. Therefore, it is an active transport. The energy, however, does not come from ATP hydrolysis directly. It comes from the energy stored in the electrochemical gradient of Na^+. This energy, in turn, comes from the operation of the Na,K-ATPase that establishes and maintains the Na^+ and K^+ gradients across the cell membrane. Therefore, the Na–Ca exchange is an example of **secondary active transport**. It requires energy from a source outside of the solutes themselves. The energy is supplied by the Na^+ gradient. The Na^+ gradient is established using ATP hydrolysis.

SECONDARY ACTIVE TRANSPORT MECHANISMS ARE SYMPORTS OR ANTIPORTS

The Na–Ca exchanger, NCX, described above is an example of an **antiport**. It has this description because the two materials being transported go in opposite directions. Such a device is also called an **exchanger** or a **counter-transporter**. There are a variety of antiport secondary active transport mechanisms, as summarized in Figure 2.6.4. Thus far we have largely discussed cationic transporters, those that transport the cations. Cations are ions that migrate toward the cathode, the negatively charged electrode, and so cations are positively charged ions. Devices also transport anions or negatively charged ions. An important example of these is the Cl^-–HCO_3^- exchanger. In the red blood cell membrane this exchanger is the AE1 protein, which comprises a large fraction of the integral proteins of the erythrocyte membrane. This anion transporter exchanges Cl^- for HCO_3^- in the ratio of 1:1. This

---

EXAMPLE 2.6.6 Calculate the Free Energy for Operation of the Na–Ca Exchanger

We have already calculated $\Delta\mu$ for Na^+ entry as: $\Delta\mu_{Na_o \to Na_i} = -14.14$ kJ mol^{-1}

We have also calculated $\Delta\mu$ for Ca^{2+} entry as $\Delta\mu_{Ca_o \to Ca_i} = -39.6$ kJ mol^{-1}.

But in this case we are dealing with Ca^{2+} *exit*, which has $\Delta\mu = +39.6$ kJ mol^{-1}. The overall $\Delta\mu$ for the Na,Ca exchanger is thus

$\Delta\mu_{NCX} = 3 \times (-14.14$ kJ mol$^{-1}) + 39.6$ kJ mol$^{-1} = $ **-2.8 kJ mol^{-1}**

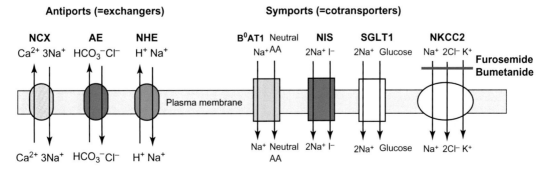

FIGURE 2.6.4 Examples of secondary active transport. NCX means "Na calcium exchanger"; AE means "anion exchanger"; NHE means "Na H exchanger"; B^0 is a specific name of a type of Na–amino acid transporter, of which there are several types; NIS means "Na iodine symport"; SGLT means "sodium glucose linked transporter." Many of these transporters exist in multiple forms or isoforms. NKCC2 means "Na, K two chloride cotransporter".

exchanger is important, as you will see later, in helping the erythrocytes to carry waste CO_2 (see Chapter 6.4).

Other secondary active transport mechanisms transport two materials in the same direction and therefore are categorized as **symports**, also called **cotransporters**. Examples include the Na—glucose transporter in the intestine and kidney membranes that transports glucose in the lumen of these organs into the absorptive cells lining the lumen. These cells also contain a number of Na—amino acid transporters that do the same thing as the glucose transporter: they transport amino acids from the lumen into the cells. The amino acids and glucose transported in this way are then transported into the blood by another mechanism, usually by facilitated diffusion. Some examples of symports are shown in Figure 2.6.4.

Secondary active transporters and facilitated diffusion proteins are classified in the family of solute carriers (SLC). There is a wide variety of these proteins, over 300 of them, and Figure 2.6.4 shows just a sampling of them. Appendix 2.6.A2 describes the nomenclature of these transport proteins.

SUMMARY

The movement of ions across cell membranes involves a change in free energy that depends on the concentration of the ion on both sides of the membrane, the charge on the ion, and the membrane potential. The free energy change can be calculated as

$$\Delta \mu = \mu_{final} - \mu_{initial}$$

where

$$\mu = \mu^0 + RT \ln C + z\Im\psi$$

where R is the gas constant = 8.314 J mol^{-1} K^{-1}, z is the charge on the ion, $\Im$ is the faraday = 9.649×10^4 C mol^{-1}, and T is the temperature in Kelvin.

At rest in muscle cells, the free energy per mole ($\Delta \mu$) for Na$^+$ entry into the cell is negative, meaning that it occurs spontaneously. Similarly, $\Delta \mu$ for K$^+$ exit is negative, whereas $\Delta \mu$ for Cl$^-$ is near zero, implying that Cl$^-$ distribution is near equilibrium. Cells maintain the Na$^+$ and K$^+$ gradients by actively pumping out Na$^+$ ions and pumping in K$^+$ ions. This movement of ions requires free energy that is supplied by the energy in the terminal phosphate bond of ATP. The Na,K-ATPase couples the outward movement of three Na$^+$ ions and the inward movement of two K$^+$ ions to the hydrolysis of 1 ATP molecule. The enzyme mechanism is responsible for this coupling. Under cell conditions, the overall $\Delta \mu$ for the Na,K-ATPase is negative because $\Delta \mu$ for ATP hydrolysis is more negative than the combined positive $\Delta \mu$ for Na$^+$ and K$^+$ transport "uphill."

The Na,K-ATPase is an example of primary active transport in which the transport of ions is directly linked to the hydrolysis of ATP. Other transporters couple the positive $\Delta \mu$ for solute transport with the negative $\Delta \mu$ for Na$^+$ entry into the cell. These are secondary active transporters, because they use energy to concentrate materials but the energy is derived directly from the Na$^+$ gradient and indirectly from ATP hydrolysis. Examples of secondary active transport include the surface membrane Na—Ca exchanger, Na—glucose cotransport, and Na—amino acid cotransport.

Transporters carrying materials in the same direction are called symports or cotransporters. Those carrying materials in opposite directions are called antiports or exchangers. Multiple examples of both classes occur in the body.

Clinical Applications: SGLT2 Inhibitors and Diabetes Mellitus

Diabetes mellitus is characterized by an abnormally high plasma glucose concentration caused by insufficient production of insulin by the pancreas. Insulin is a protein hormone that increases glucose transporters in peripheral tissues (GLUT4) that remove glucose from the circulation. There are two major classifications of persons with diabetes. Those with Type 1 diabetes require insulin injections and have little or no production of insulin, generally caused by a destruction of the beta cells in the pancreas that produce the hormone. Persons with Type 2 diabetes generally produce insulin but the body cells are resistant to the hormone and the circulating levels of insulin are inadequate to lower plasma glucose levels. High blood glucose causes glycosylation of proteins, as measured clinically by HbA$_{1c}$, glycosylated hemoglobin. Long-term control of plasma glucose is monitored by the HbA$_{1c}$ level. Diabetes mellitus is described in more detail in Chapter 9.4.

The kidney produces urine by filtering large volumes of blood and then reabsorbing the desired materials and discarding the rest. Glucose is reabsorbed in the kidney in the proximal tubule of the nephron, the functional unit of the kidney, by SGLT2 present in the apical membrane of the tubule (see Figure 2.6.5). Final transport of glucose into the blood occurs over GLUT2, a facilitated transport mechanism. 90% of glucose reabsorption in the kidney occurs via the SGLT2 glucose uptake mechanism. A relatively new class of drugs inhibits the SGLT2 so that not all of the filtered glucose is reabsorbed, and glucose appears in the urine. These drugs include empagliflozin, canagliflozin, and dapagliflozin. The chemical structures of these drugs are shown in Figure 2.6.6.

The term "diabetes" derives from the Greek meaning "to siphon" and this refers to the increased urinary flow in diabetic persons. This is caused by the inability of the kidney to reabsorb all of the glucose, and the excretion of this glucose in a larger volume of water. The effect of the SGLT2 inhibitors is to (1) lower blood glucose levels, (2) reduce HbA$_{1c}$ levels, (3) generally cause a slight loss of weight, and (4) tend to cause dehydration due to increased urine flow. These drugs are of limited value to persons with dysfunctional kidneys (Whalen K, Miller, S, and Onge E, The role of sodium—glucose co-transporter 2 inhibitors in the treatment of type 2 diabetes, *Clin. Therap.* **37**:1150—1166, 2015).

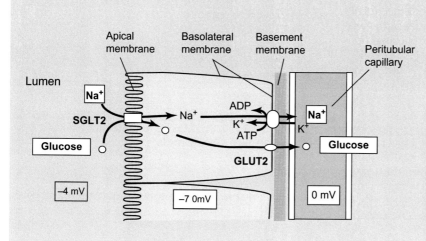

FIGURE 2.6.5 Mechanism of reabsorption of glucose from the ultrafiltrate in the lumen of the nephron proximal tubule. SGLT2 transports 2 Na$^+$ ions into the cell along with glucose. The energy is supplied by the electrochemical gradient for Na$^+$ that is established by pumping Na$^+$ out of the cell at the basolateral membrane. Glucose enters the blood through GLUT2, a facilitated diffusion carrier on the basolateral membrane.

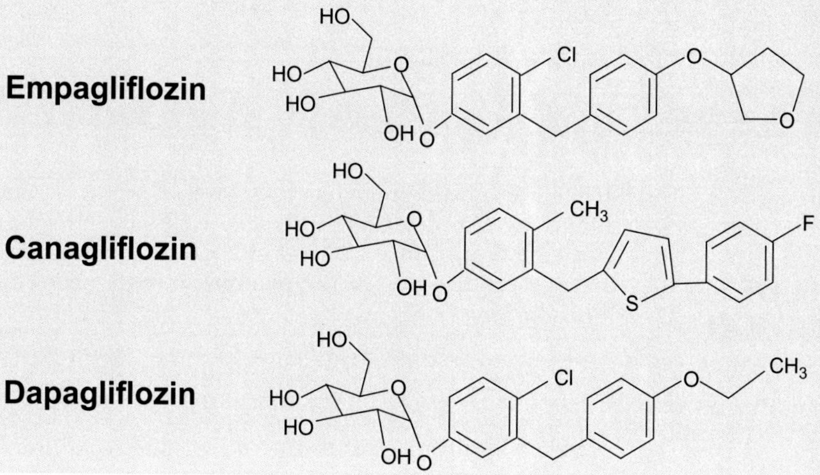

FIGURE 2.6.6 Chemical structure of the SGLT2 inhibitors.

Clinical Applications: Oral Rehydration Therapy

One of most important public health issues in the world is the availability of clean drinking water. Contaminated water supplies carry cholera and other infectious agents that result in diarrhea and vomiting that cause dehydration that can be fatal, especially in children. Prior to the introduction of oral rehydration therapy (ORT), diarrhea was the leading cause of infant mortality in developing nations. ORT is estimated to have reduced world-wide infant deaths from 5 million per year to 3 million per year (2006 figures). However, diarrhea remains the second leading cause of death in children less than 5 years old (18%, after pneumonia at 19%).

The WHO (World Health Organization) and UNICEF (United Nations Children's Fund, shortened from the original United Nationals International Children's Emergency Fund) jointly publish guidelines for the composition of oral rehydration solution (ORS). Its current formulation is: 2.6 g NaCl; 2.9 g trisodium citrate dihydrate ($Na_3C_6H_5O_7 \cdot 2\ H_2O$); 1.5 g KCl; 13.5 g glucose per L of solution.

The molar ratio of sodium to glucose in this ORS is 1.0, and it is slightly hyposmotic. It should be made with clean water, but when clean water is not available other fluids may be substituted—but not sugar-containing fluids like fruit juices.

The effectiveness of ORT relies on the SGLT1 system in the intestine that transports glucose along with 2 Na$^+$ ions into the enterocyte and then into the blood. The transport of nutrients creates an osmotic reabsorption of water along with the solutes. This is enough to counteract the loss of fluid through diarrhea or vomiting. ORT was not used until the 1960s. Before that time, rehydration was accomplished by intravenous fluid administration. In the developing world, IV therapy is not widely available and ORT is a low-technology solution that is much more readily available.

REVIEW QUESTIONS

1. If a Na channel was to open on the surface of a cardiac cell, at rest, in which direction would Na ions travel? If the channels were K-specific, in which direction would K ions travel? If it were Ca^{2+}-specific, which way would Ca^{2+} go? If it was a Cl^- channel, which way would Cl^- ions go?
2. Which part of Eqn [2.6.4] gives the energy due to diffusion? Which part gives the energy due to electrical forces? For Na^+, which is bigger, the part due to concentration differences or the part due to electrical forces? Which part of the total energy, concentration or electrical, is bigger for Ca^{2+}?
3. Where does the energy come from for the movement of materials by passive transport?
4. Where does the energy come from for the movement of materials by primary active transport?
5. Where does the energy come from for the movement of materials by secondary active transport?
6. Do you think that the NCX can go in reverse?
7. Why cannot thermodynamics predict reaction rates?
8. What is a symport? What is an antiport? What is a cotransporter?

APPENDIX 2.6.A1 DERIVATION OF THE USSING FLUX RATIO EQUATION

The passive flux of an ion in a solution is given by Fick's First Law of Diffusion with an external force, which was introduced in Eqn [1.7.19]. The one-dimensional form of this equation is reproduced here:

[2.6.A1.1] $$J_s = -D\frac{\partial C}{\partial x} - \frac{D}{RT}Cz\Im\frac{\partial \psi}{\partial x}$$

where J is the flux, in $mol\ cm^{-2}\ s^{-1}$, D is the diffusion coefficient in $cm^2\ s^{-1}$, C is the concentration (converted to units of $mol\ cm^{-3}$), R is the gas constant ($=8.314\ J\ mol^{-1}\ °K^{-1} = 1.987\ cal\ mol^{-1}\ °K^{-1}$), T is the absolute temperature (°K), z is the integer of the charge on the ion (± 1 or 2, generally), $\Im$ is the Faraday ($=96,489\ coulomb\ mol^{-1}$) and ψ is the electrical potential, in joules ($=$volt-coulomb). The quantity $\partial C/\partial x$ is the magnitude of the gradient of C, and $\partial \psi/\partial x$ is the magnitude of the electric field. The flux can be obtained by integrating this equation. To begin, we multiply both sides of the equation by an integrating factor, ρ, and we choose ρ so that the right-hand side of the equation becomes an exact differential:

[2.6.A1.2] $$J_s\rho = -D\left[\rho\frac{\partial C}{\partial x} + \frac{Cz\Im\rho}{RT}\frac{\partial \psi}{\partial x}\right]$$

We choose ρ so that the terms in brackets are an exact differential. Thus we want

[2.6.A1.3] $$\rho\frac{\partial C}{\partial x} + \frac{Cz\Im\rho}{RT}\frac{\partial \psi}{\partial x} = \frac{d(\rho C)}{dx} = \rho\frac{\partial C}{\partial x} + C\frac{\partial \rho}{\partial x}$$

From comparing the left of Eqn [2.6.A1.3] to the right, we see that multiplication by ρ transforms the equation into an exact differential if

[2.6.A1.4] $$\frac{\partial \rho}{\partial x} = \frac{z\Im\rho}{RT}\frac{\partial \psi}{\partial x}$$

We rearrange this to get

[2.6.A1.5] $$\frac{\partial \rho}{\rho} = \frac{z\Im}{RT}\partial \psi$$

A solution to Eqn [2.6.A1.5] is

[2.6.A1.6] $$\rho = e^{\frac{z\Im}{RT}\psi}$$

We may insert this result back into Eqn [2.6.A1.2], with Eqn [2.6.A1.3], to obtain

[2.6.A1.7] $$J_s e^{\frac{z\Im}{RT}\psi} = -D\frac{d\left(Ce^{\frac{z\Im}{RT}\psi}\right)}{dx}$$

which may be rewritten as

[2.6.A1.8] $$J_s e^{\frac{z\Im}{RT}\psi}dx = -D\,d\left(Ce^{\frac{z\Im}{RT}\psi}\right)$$

If we are considering passive transport across a membrane, we can determine the *passive* flux by integrating this equation from $x = 0$ (one side of the membrane) to $x = \delta$, the other side of the membrane for a membrane with thickness δ:

[2.6.A1.9] $$\int_0^\delta J_s e^{\frac{z\Im}{RT}\psi}dx = -D\int_0^\delta \delta\left(Ce^{\frac{z\Im}{RT}\psi}\right)$$

Here we limit ourselves to the steady-state condition. In this case, J_s does not vary with distance across the membrane—it is constant. Therefore, J_s may be removed from the integral and we get

[2.6.A1.10] $$J_s = \frac{-D\int_0^\delta \delta\left(Ce^{\frac{z\Im}{RT}\psi}\right)}{\int_0^\delta e^{\frac{z\Im}{RT}\psi}dx}$$

The numerator in this equation is the integral of an exact differential and can be immediately evaluated between the boundaries. This gives

[2.6.A1.11] $$J_s = \frac{-D\left[C(\delta)e^{\frac{z\Im}{RT}\psi(\delta)} - C(0)e^{\frac{z\Im}{RT}\psi(0)}\right]}{\int_0^\delta e^{\frac{z\Im}{RT}\psi}dx}$$

The denominator in this equation can be evaluated only if $\psi(x)$ is known. However, generally $\psi(x)$ is unknown. Ussing made the observation that the presence of an active transport mechanism would not obey Eqn [2.6.A1.11], because the flux would not be passive, and this equation describes the passive flux. He further made the observation that we don't need to know how ψ varies with x if we take the ratio of the unidirectional fluxes. The unidirectional flux is the flux that you would observe if the concentration on the other side was zero.

We define here the unidirectional flux $J_{i\to o}$ to be the flux from inside to outside, with $x = 0$ on the inside and $x = \delta$ on the outside. This flux is given from Eqn [2.6.A1.11] by setting $C(\delta) = 0$, and we obtain

$$[2.6.A1.12] \qquad J_{i\to o} = \frac{D\, C(0)\, e^{\frac{z\mathfrak{F}}{RT}\psi(0)}}{\int_0^\delta e^{\frac{z\mathfrak{F}}{RT}\psi}\, dx}$$

We further define the unidirectional flux $J_{o\to i}$ to be the unidirectional flux from outside with $x = \delta$ to the inside, with $x = 0$. This flux is also given from Eqn [2.6.A1.11] by setting $C(0) = 0$; we obtain

$$[2.6.A1.13] \qquad J_{o\to i} = \frac{-D\, C(\delta)\, e^{\frac{z\mathfrak{F}}{RT}\psi(\delta)}}{\int_0^\delta e^{\frac{z\mathfrak{F}}{RT}\psi}\, dx}$$

Here the minus sign conveys a convention that the outward flux is taken as positive. Thus a negative flux simply means that the flux is directed inward, and a positive flux is directed outward. The magnitude of the fluxes is given by the absolute values of the fluxes. If we take the ratio of the unidirectional fluxes, the denominators in each cancel each other out, and we don't need to do the integration that requires knowledge of $\psi(x)$. Taking the ratio of the two unidirectional fluxes, we arrive at

$$[2.6.A1.14] \qquad \frac{J_{i\to o}}{J_{o\to i}} = \frac{D\, C(0)\, e^{\frac{z\mathfrak{F}}{RT}\psi(0)}}{-D\, C(\delta)\, e^{\frac{z\mathfrak{F}}{RT}\psi(\delta)}}$$

Using the notation that $C(0) = C(i)$, the inside concentration, and $C(\delta) = C(o)$, the outside concentration, this equation can be simplified to

$$[2.6.A1.15] \qquad \frac{J_{i\to o}}{J_{o\to i}} = -\frac{C(i)}{C(o)} e^{\frac{z\mathfrak{F}}{RT}\psi(0) - \psi(\delta)}$$

Using the definition that the difference in potential across the membrane, $\psi(0) - \psi(\delta) = E_m$, the membrane potential, this last Eqn [2.6.A1.15] becomes

$$[2.6.A1.16] \qquad \frac{J_{i\to o}}{J_{o\to i}} = -\frac{C(i)}{C(o)} e^{\frac{z\mathfrak{F}}{RT} E_m}$$

This last equation is the **Ussing Flux Ratio Equation**. It describes the expected ratio of the unidirectional fluxes *if the ions are transported passively across the membrane*. Deviations from the flux ratio equation are taken to indicate that the fluxes are not transported passively. That is, deviations from the expected flux ratio can be taken to indicate that an active transport mechanism is present. Ussing considered other possible deviations from the expected flux ratio such as single-file transport.

Hans Ussing (1911–2000) derived his flux ratio equation in the late 1940s, soon after radioactive isotopes became available to measure unidirectional ion fluxes. He was the first to prove the existence of active transport mechanisms, using the frog skin as a model. The Na,K-ATPase was later discovered, in 1957, by Jens Skou, who earned a Nobel Prize for the discovery.

APPENDIX 2.6.A2 NOMENCLATURE OF TRANSPORT PROTEINS

HUGO NOMENCLATURE

Proteins have "trivial" or common names that were generally first provided by their discoverers. These names have often been changed when a new function for the protein was discovered or its relationship to other proteins was discovered. Since these proteins derive from genes, it is now increasingly useful to use the name for the gene to also describe the resulting protein. The **HUGO Gene Nomenclature Committee (HGNC)** provides a unique identifier for each gene. This committee is a part of the **Hu**man **G**ene **O**rganization (HUGO). Full and updated databases are available at www.gene-names.org.

CARRIER CLASSIFICATIONS

The transport proteins as described in Chapters 2.5 and 2.6 can be generally classified as belonging to a small number of types. These are:

Passive Transporters:
 Facilitated transporters
 Ion channels
 Water channels
Active Transporters:
 Secondary active transporters: exchangers and cotransporters
 ATPase pumps (P, V, and F-types)
 ABC (ATP-binding cassette) transporters.

These functional classifications do not map precisely onto the HUGO nomenclature. The HUGO classification lumps facilitated transporters and secondary active transporters into one group, the solute carriers, SLC. Carriers are often grouped into families based on sequence homology rather than functionality. An example of this is the sodium–iodine exchanger located in the thyroid gland, which is part of the sodium–glucose cotransport family.

SOLUTE CARRIERS

The solute carriers include the facilitated diffusional carriers such as GLUT1 and GLUT2 and the secondary active transport carriers such as NCX, NHE, SGLT, NIS, and AE. All members of the SLC superfamily are named according to

SLC n X m

where SLC indicates the superfamily, n is an integer that denotes the family, X is a letter that denotes the subfamily, and m is a second integer that denotes the isoform. As an example, the facilitated glucose carriers are all members of the SLC2A subfamily of which there are 14 members: SLC2A{1…14}. These correspond to their common names, as shown in Table 2.6.A2.1.

There are an enormous number of these SLC genes. There are at least 52 families (n) with a total of 396 genes (Σm_i) that encode for transporter proteins. It is

TABLE 2.6.A2.1 Nomenclature of the Facilitated Glucose Transporters (Augustin, R. "The protein family of glucose transport facilitators; It's not about glucose after all." *Life* **62**:315–333, 2010)

HUGO Name	Common Name	Location and Function
SLC2A1	GLUT1	K_m for glucose = 1.5 mM; ubiquitous but important for erythrocytes and uptake of glucose into the cerebrospinal fluid (CSF)
SLC2A2	GLUT2	K_m for glucose = 17 mM; liver, kidney, intestine, β cells of pancreas; transports glucose out of intestine and kidney cells
SLC2A3	GLUT3	K_m for 2-deoxyglucose = 1.4 mM; present in neurons, spermatozoa, placenta
SLC2A4	GLUT4	K_m for glucose = 5 mM; skeletal and cardiac muscle, adipose tissue; rate-limiting step in insulin-stimulated glucose uptake into tissues
SLC2A5	GLUT5	K_m for fructose = 6 mM; mainly jejunum of small intestine but also found in kidney, brain, muscle and fat; primarily a fructose transporter
SLC2A6	GLUT6	brain, spleen, and leukocytes
SLC2A7	GLUT7	K_m for glucose = 0.3 mM; apical membrane of intestine and colon; also transports fructose
SLC2A8	GLUT8	K_m for glucose = 2 mM; testis, cerebellum, liver, spleen, lung, fat; intracellular without insulin-stimulated translocation to the surface
SLC2A9	GLUT9	K_m for glucose = 0.6 mM but also transports urate; kidney and liver and β cells of pancreas; exchanges urate for glucose or fructose
SLC2A10	GLUT10	Present in heart, lung, brain, liver, muscle, pancreas, placenta, and kidney; may be intracellular
SLC2A11	GLUT11	Three isoforms (GLUT11{A,B,C}); in humans GLUT11 is exclusively expressed in slow-twitch muscle fibers
SLC2A12	GLUT12	
SLC2A13	GLUT13; HMIT	Does not transport sugar; H^+-coupled myoinositol symporter; K_m = 0.1 mM for myoinositol; located intracellularly; highest in brain
SLC2A14	GLUT14	

entirely unreasonable to attempt memorization of all of them. The list of these and other gene families is available in tabular format on the HUGO website.

Note that some designations within a family can be confusing in that the function of the carrier is not related to its stated family! For example, GLUT 9 (SLC2A9) is primarily regarded as a urate transporter rather than a glucose transporter, but it is a member of the facilitated diffusion carriers for glucose.

A second example of this is found in the sodium–glucose transport family. SGLT1 and SGLT2 are part of the SLC5A subfamily that includes 12 members (SLC5A{1…12}. SLC5A1 corresponds to SGLT1 and SLC5A2 corresponds to SGLT2, but SLC5A5 corresponds to the sodium–iodine exchanger (NIS), shown in Figure 2.6.4, that does not transport glucose! The NIS is grouped in this way because of structural similarities in the transport proteins that leads researchers to suppose that they belong to the same subfamily of transporters. SGLT1 is the main transporter in the intestine whereas SGLT2 is the main transporter in the kidney, although SGLT1 is also found in the kidney.

The amino acid transporter shown in Figure 2.6.4 has three members (B^0AT1 3) that correspond to genes SLC6A19, SLC6A15, and SLC6A18, respectively. These all carry neutral amino acids. Because there are so many amino acids and also a variety of neurotransmitters and other solutes that need carrying, there are a large number of these kinds of carriers. They belong to the SLC1, SLC3, SLC6, SLC7, SLC36, SLC38, and SLC43 subfamilies.

The anion exchanger shown in Figure 2.6.4, AE, also has three subtypes (AE1–3) that correspond to SLC4A1–3. These are all electroneutral $Cl^- - HCO_3^-$ exchangers. The SLC4 family contains 10 members (SLC4A{1…5; 7…11} that include Na^+ and HCO_3^- or CO_3^{-2} cotransporters.

The Na^+-Ca^{2+} exchanger (NCX) of Figure 2.6.4 also has three members (NCX1–3) encoded by genes SLC8A1–3, respectively. There is also a mitochondrial Na^+-Ca^{2+} exchanger, NCLX, as a gene product of SLC8B1. There are additional transporters for Ca^{2+} including NCKX ($Na^+-Ca^{2+}-K^+$ exchanger, SLC24 family) and CCX (Ca^{2+}-cation exchanger).

The Na^+-H^+ exchanger (NHE) of Figure 2.6.4 has several members. NHE1–5 are all present on the surface membrane, and NHE3 and NHE5 recycle between the surface and intracellular membranes. They correspond to SLC9A1–5, respectively. NHE6, NHE7, and NHE9 appear to be located in intracellular membranes along the secretory pathway. They correspond to SLC9A6, SLC9A7, and CLC9A9, respectively.

The Na–K–2Cl cotransporter shown in Figure 2.6.4 comes in two forms, NKCC1 and NKCC2, corresponding to genes SLC12A1 and SLC12A2.

There are a large number of other transporters of the SLC gene superfamily that transport neurotransmitters or ions, or vitamins, or metabolites across plasma membranes or internal membranes. A complete listing is available on the HUGO website.

ATP-DRIVEN ION PUMPS

Names for the genes for the ATP-driven ion pumps have the form

ATP n X m

where ATP indicates the superfamily, n is an integer that denotes the family, X is a letter that denotes the subfamily, and m is an integer that denotes the member. Many of the ATP-driven pumps consist of multiple subunits, and these are generally organized by having the same family integer, n, and a different subfamily name. For example, the Na,K-ATPase has an α and a β subunit, and these are indicated as ATP1A and ATP1B. There are four varieties of each, so that the α subunit of the Na,K-ATPase corresponds to ATP1A{1...4} and the β subunit has genes ATP1B{1...4}.

The Ca^{2+} pumps are members of the ATP2 family. ATP2A{1,2,3} correspond to SERCA1, SERCA2, and SERCA3, respectively, that are located on the internal membranes of the cell, the endoplasmic reticulum, or the sarcoplasmic reticulum. The plasma membrane Ca-ATPases, PMCA{1...4} are encoded by the ATP2B {1...4} genes, respectively. SPCA1 and SPCA2 are encoded by ATP2C1 and ATP2C2. The gastric H^+–K^+-ATPase is encoded by ATP4A (the α subunit) and ATP4B (the β subunit}. All of the above-mentioned pumps constitute the P-class of ATP-driven ion pumps.

The F-type and V-type ATPase ion pumps are much more complex, consisting of multiple subunits and multiple copies of some of these subunits. Therefore, there is no one gene that encodes the entire operating complex. The mitochondrial F-type ATPase consists of an F_1 and an F_O subunit that themselves are complexes of additional subunits. These are encoded by the ATP5 family of genes, with subfamilies denoted by the letters A, B, C, D, E, F, G, H, I, J, L, and O.

Somewhat differently, the V-type ATPase that acidifies lysosome contents by pumping in H^+ ions is designated ATP6V0 and ATP6V1 for the V0 and V1 subunits, and letters corresponding to the subunits within V0 and V1.

ABC TRANSPORTERS

ABC transporters hydrolyze ATP to transport a wide variety of substrates. There are 48 transporters classified in 7 families denoted in this case by letters alone:

ABC X m

where ABC denotes "ATP-binding cassette", X is a letter indicating the family, and m is an integer indicating the member.

AQUAPORINS

Aquaporins are proteins that increase water movement across biological membranes. They have a molecular weight of around 30 kDa and associate as tetramers, although each monomer has a water channel. Several of the aquaporins will transport other small molecular weight, electrically neutral substrates such as glycerol or urea. There are a variety of aquaporins named AQP n where n is the member of the family. There are 14 members of the family, named AQP{1...12}. AQP0 has been renamed MIP for "major intrinsic protein" of the lens, and AQP12 has A and B subtypes. In the HUGO classification, aquaporins are considered to be a subtype of ion channels.

2.7 Osmosis and Osmotic Pressure

Learning Objectives

- Define osmosis and osmotic pressure
- Write van't Hoff's limiting law for osmotic pressure dependent on concentration
- List the colligative properties of solutions and explain why vapor pressure depression perfectly predicts osmotic pressure
- Describe the osmotic coefficient for correction of nonideality in solutions
- Be able to calculate the predicted osmotic pressure for a solution
- Define hydraulic conductivity or hydraulic permeability
- Define the reflection coefficient
- Explain how the hydraulic conductivity depends on the microscopic parameters of the membrane
- Write an equation for volume flow across a microporous membrane in the presence of hydraulic and osmotic pressures
- Be able to calculate the hydraulic conductivity and reflection coefficient given appropriate data
- Describe the origin of the osmotic pressure for microporous membranes
- Describe how erythrocyte cell volume changes when placed in contact with solutions of varying [NaCl] or [glucose]
- Contrast the concepts of tonicity and osmolarity
- Describe the behavior of a perfect osmometer
- Explain why cells are not perfect osmometers
- Define RVD and RVI

OSMOSIS IS THE FLOW OF WATER DRIVEN BY SOLUTE CONCENTRATION DIFFERENCES

Probably no concept is more confusing to beginning students than osmosis and osmotic pressure, partly because it is defined backwards, as you will see. We begin with the experiments of Pfeffer in 1877.

Pfeffer made a precipitation membrane in the walls of unglazed porcelain cups by reacting copper salts with potassium ferricyanide. He used the precipitation membrane that resulted to separate a sucrose solution on the inside of the cup from water on the outside. He observed that water flowed from the outside to the inside. This is the primary observation of **osmosis**. **Osmosis refers to the movement of fluid across a membrane in response to differing concentrations of** *solute* **on the two sides of the membrane.** The word "osmosis" originates from the Greek, meaning "thrust" or "impulse."

Pfeffer also observed that the flow was proportional to the sucrose concentration inside the cup. When water was inside the cup, he observed that a pressure applied to the inside compartment would force water out of the cup, and this flow was proportional to the pressure. When sucrose was inside a closed cup, a pressure would develop inside the cup and this pressure was proportional to the sucrose concentration. He recognized this as an equilibrium state in which the outward filtration of water balanced the inward movement caused by osmosis. He *defined* the **osmotic pressure** as the **pressure necessary to stop osmotic flow across a barrier that is impermeable to the solute**. The osmotic pressure historically is given the symbol π (see Figure 2.7.1). This definition is the key: it defines the osmotic pressure as the pressure needed to stop osmotic flow rather than the pressure that drives osmotic flow. It is also defined only for a **semipermeable membrane**, one that is impermeable to solute but permeable to water.

THE VAN'T HOFF EQUATION RELATES OSMOTIC PRESSURE TO CONCENTRATION

Pfeffer's data (see Figure 2.7.2) showed that the osmotic pressure of solutions was linearly related to their concentration. In 1887, van't Hoff argued from Pfeffer's results, and from thought experiments considering gases in equilibrium with solutions, that the osmotic pressure should be given by

$$[2.7.1] \qquad \pi = RTC_s$$

where R is the gas constant (0.082 L atm mol^{-1} K^{-1}), T is the absolute temperature, and C is the molar concentration of solute in the inner compartment. Equation [2.7.1] is known as **van't Hoff's Law**. It gives the osmotic pressure due to the solute, s. In a mixture, the osmotic pressures due to each solute particle add up, much like Dalton's Law of Partial Pressures in gases. The result is that

$$[2.7.2] \qquad \pi = RT \sum C_s$$

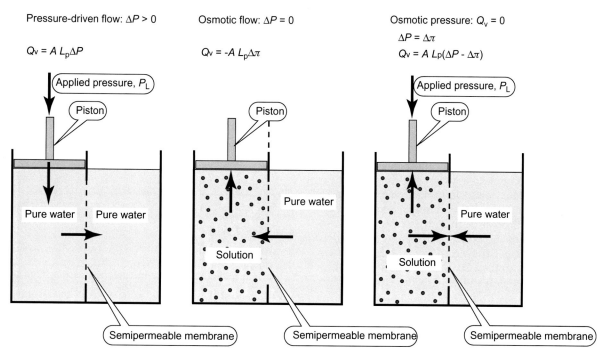

FIGURE 2.7.1 Equivalence of hydrostatic and osmotic pressures in driving fluid flow across a membrane. Left panel: An ideal, semipermeable membrane is freely permeable to water but is impermeable to solute. When the membrane separates pure water on the right from pure water on the left, application of a pressure, P_L, to the left compartment forces water across the semipermeable membrane. The flow is linearly related to the pressure difference by the area of the membrane (A) and a proportionality constant, L_P, that is characteristic of the membrane. This constant is variously called the **hydraulic conductivity, hydraulic permeability,** or **filtration coefficient**. Positive Q_V is taken as flow to the right. Middle panel: The ideal, semipermeable membrane separates a solution on the left from pure water on the right, and water moves to the solution side by osmosis. The flow, Q_V, is linearly related to the difference in osmotic pressure, $\Delta\pi$, by the area of the membrane and the hydraulic conductivity, L_P. The flow causes expansion of the left compartment and movement of the piston, which is assumed here to be weightless. Right panel: Application of a pressure, P_L, to a solution so that $\Delta P = \Delta\pi$ results in no net flow across the membrane. The osmotic pressure of a solution is *defined* as the pressure necessary to stop fluid flow when an ideal semipermeable membrane separates pure water from the solution.

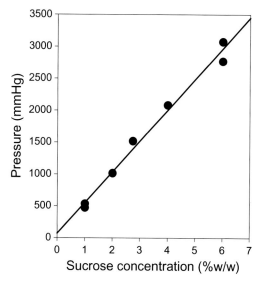

FIGURE 2.7.2 Plot of the data of Pfeffer (1877) for the osmotic pressure of sucrose solutions. A copper ferrocyanide precipitation membrane was formed in the walls of an unglazed porcelain cup. The membrane separated a sucrose solution in the inner chamber from water outside the cup. The inner chamber was then attached to a manometer and sealed. The linear relationship between the pressure measured with this device and the sucrose concentration was the experimental impetus for deriving van't Hoff's Law. [*Data from Pfeffer, W.* Osmotische Untersuchungen Studien zur Zellmechanik, *Leipzig [translated by G.R. Kepner and E.J. Tadelmann:* Osmotic investigations: Studies on cell membranes, *1985, Van Nostrand Reinhold, New York].*

The concentration in van't Hoff's law, ΣC_s, refers to the *concentration of osmotically active solute particles in the solution*. Organic compounds such as glucose typically dissolve to form one particle for each molecule of solute, and for these compounds C_s is the same as the molar concentration. Strong salts, on the other hand, dissociate to form more than one particle for each mole of salt. NaCl, for example, dissociates in solution to form one Na^+ and one Cl^- ion. The concentration of osmotically active particles is twice the concentration of NaCl. Similarly, $CaCl_2$ dissociates nearly completely to form one Ca^{2+} ion and two Cl^- ions, and the total concentration of particles is 3 times the concentration of $CaCl_2$. The **osmolarity** of a solution equals ΣC_s and is expressed in osmoles per liter to indicate clearly that we are referring to the number of osmotically active solute particles, called **osmolytes**, rather than the concentration of the solute.

THERMODYNAMIC DERIVATION OF VAN'T HOFF'S LAW

We have learned that all spontaneous processes (those that occur naturally without any additional forces) are accompanied by a decrease in free energy. We have also learned that the free energy per mole is the chemical potential (see Chapter 1.7). In the case of a solution separated from pure water by a semipermeable membrane, water movement will occur when there is a difference in the chemical potential of *water* on the two sides of the membrane, such that water movement results in a decrease in the free energy. At equilibrium, when the pressure applied to the solution is equal to the osmotic pressure, the chemical potential of water is equal on both sides of the membrane, and no net movement of water occurs.

In the derivation that follows, we consider that a semipermeable membrane separates two compartments. On the right is pure water and on the left is some solution with concentration C_s. We know that there will be flow of water from the pure water to the solution side and that application of pressure, π, to the solution side will stop the flow. What is the relation between π and C_s? We can discover this by looking at the equilibrium condition when osmotic pressure balances hydrostatic pressure. At this point, the free energy change for water across the membrane is zero.

THE CHEMICAL POTENTIAL INCLUDES PRESSURE–VOLUME WORK

The equation we have used for chemical potential of a solute, the free energy per mole, is

[2.7.3] $\quad \mu_s = \mu_s^0 + RT \ln C_s + z\Im\psi$

In this equation, $z\Im\psi$ represents the work done, per mole, in moving the material from the standard state to the condition in which the material is placed. In this case, it is the electrical work. There are other kinds of work such as pressure–volume work. The general equation for the chemical potential is

[2.7.4] $\quad \mu_s = \mu_s^0 + RT \ln C_s + \text{work terms}$

where the work terms include all work (except concentration work, which is included explicitly in $RT \ln C_s$) necessary to bring the material from the standard state to its present state.

THE ACTIVITY CORRECTS THE CHEMICAL POTENTIAL FOR INTERACTIONS BETWEEN SOLUTE PARTICLES

It turns out that Eqn [2.7.4] is an approximation. In our derivation of the general Fick's Law, we did not consider some other kinds of interactions, such as solute molecules bumping into each other. The accurate equation is

[2.7.5] $\quad \mu_s = \mu_s^0 + RT \ln a_s + \text{work terms}$

where a_s is the **activity** of the solute. In the case of osmosis, the work term is the pressure–volume work and there is no electrical work term. At equilibrium, where the pressure across the semipermeable membrane is the osmotic pressure, the chemical potential of water on the two sides of the semipermeable membrane must be equal (because the free energy change at equilibrium is zero). Therefore, we write the equality of chemical potential for water on the left and right sides as

[2.7.6] $\quad \mu_w^0 + RT \ln a_{w,L} + \overline{V}_w P_L = \mu_w^0 + RT \ln a_{w,R} + \overline{V}_w P_R$

where the subscripts L and R refer to the left and right sides of the semipermeable membrane, μ_w^0 is the chemical potential of liquid water in its standard state (pure water at 1 atm pressure), V_w is the volume of water per mole (the **partial molar volume**), P is the pressure, and a_w is the activity of water. For an **ideal solution**, the activity of water is its **mole fraction**:

[2.7.7] $\quad a_w = X_w = \dfrac{n_w}{n_w + n_s}$

where X_w is the usual variable denoting the mole fraction of water, and n_w and n_s are the moles of water and solute, respectively, in any aliquot of the solution. Substituting in for a_w and canceling the μ_w^0 on both sides of Eqn [2.7.6], we come to

[2.7.8] $\quad \overline{V}_w P_L - \overline{V}_w P_R = RT \ln X_{w,R} - RT \ln X_{w,L}$

Since the right-hand solution is pure water, $X_{w,R} = 1.0$ and $\ln X_{w,R} = 0$. Thus we have

[2.7.9] $\quad \overline{V}_w(P_L - P_R) = -RT \ln X_{w,L}$

Now the mole fractions of solute and water in a solution must sum to 1.0. This is expressed as

[2.7.10] $\quad \begin{array}{c} X_{w,L} + X_{s,L} = 1.0 \\ \ln X_{w,L} = \ln(1 - X_{s,L}) \end{array}$

In dilute solutions, $X_{s,L} \ll 1.0$, so we may approximate $\ln(1 - X_{s,L}) \approx -X_{s,L}$. Substitution of this result into Eqn (2.7.9) gives

[2.7.11] $$(P_L - P_R) = \frac{RT}{\overline{V}_w} X_{s,L}$$

The left-hand side of Eqn [2.7.11] is just the osmotic pressure, π, which is equal to the extra pressure that must be applied to the left-hand side in order to establish equality of the chemical potential of water on the two sides of the membrane. From the definition of mole fraction, Eqn [2.7.11] becomes

[2.7.12] $$(P_L - P_R) = \pi = \frac{RT}{\overline{V}_w} \frac{n_s}{n_s + n_w}$$

For a dilute solution, $n_s \ll n_w$, so we approximate this result as

[2.7.13] $$\pi = \frac{RT}{\overline{V}_w} \frac{n_s}{n_w}$$

Again for a dilute solution, $n_w V_w \approx V$, the volume of the solution. Thus Eqn [2.7.13] gives

[2.7.14] $$\pi = RT \frac{n_s}{V}$$

or

[2.7.15] $$\pi = RTC_s$$

This last equation is the van't Hoff equation for the osmotic pressure. This thermodynamic derivation entails two assumptions: the solution is dilute enough to approach ideality and that the solution is incompressible so that the pressure–volume work is $V_w \Delta P$. It is important to recognize that the van't Hoff equation is not exact for physiological solutions. Rather, it is an approximation that is strictly true only for dilute ideal solutions.

OSMOTIC PRESSURE IS A PROPERTY OF SOLUTIONS RELATED TO OTHER *COLLIGATIVE PROPERTIES*

Osmotic pressure is closely related to some other properties of solutions, the **colligative properties**. These include the **freezing point depression**, the **boiling point elevation**, and the **vapor pressure depression**, all caused by dissolving solutes in a solution. The osmolarity is often determined from vapor pressure depression or freezing point depression, rather than from direct osmotic pressure measurements. The osmolarity is the concentration necessary to observe these phenomenon.

To see the connection between osmotic pressure and vapor pressure depression, consider Figure 2.7.3. A solution placed in a sealed container with a source of pure water will gain water because its vapor pressure is lower than that of the water. This situation is formally equivalent to osmosis, where the semipermeable membrane is the intervening air between the two surfaces.

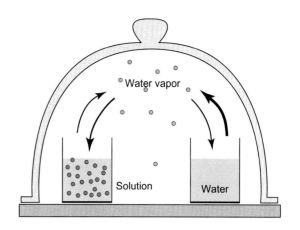

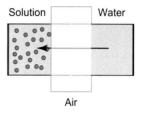

FIGURE 2.7.3 Equivalence of vapor pressure and osmotic pressure. Two beakers containing a solution or pure water are both placed in a single, sealed compartment. The air above the fluids contains air molecules plus water vapor. The partial pressure of water is its contribution to the total pressure, and it is proportional to the water concentration in the gas phase. The **vapor pressure** is *defined* as the **partial pressure of water in equilibrium with the liquid phase**. Water molecules will leave the liquid to moisten a dry gas. At equilibrium, molecules will evaporate from the liquid phase and condense from the gas phase at equal rates, so that a dynamic equilibrium is established. The vapor pressure of pure water is higher than the vapor pressure of the solution. This vapor pressure depression is one of the colligative properties of solutions. Because there is only one vapor pressure, it cannot simultaneously be in equilibrium with the water and with the solution. Thus water from the pure water beaker will continue to evaporate, because its vapor pressure is higher than that in the air, and water will continue to condense into the solution, because the partial pressure of water is higher than the vapor pressure of the solution.

Thus osmotic pressure and vapor pressure depression are perfect predictors of each other because essentially they are the same phenomenon.

THE OSMOTIC COEFFICIENT φ CORRECTS FOR THE ASSUMPTION OF DILUTE SOLUTION AND FOR NONIDEAL BEHAVIOR

As noted above, the van't Hoff equation makes two assumptions: the solution is dilute and it is ideal. The assumption of ideality enters when we equate the activity of water with its mole fraction. The assumption of dilute solutions allows us to identify $\ln(1 - X_s)$ with $-X_s$. We can correct for both assumptions by identifying

[2.7.16] $$\varphi = \frac{\pi_{observed}}{RTC_s}$$

Here φ is the **osmotic coefficient**. The osmotic coefficient can be less than or greater than 1.0.

THE RATIONAL OSMOTIC COEFFICIENT CORRECTS FOR THE ASSUMPTION OF IDEALITY

Equation [2.7.9] gives the osmotic pressure in terms of the mole fraction of water:

$$[2.7.17] \qquad \pi = -\frac{RT}{\overline{V}_w} \ln X_w$$

But the true equation for osmotic pressure is given by the manipulation of Eqn [2.7.6] as

$$[2.7.18] \qquad \pi = -\frac{RT}{\overline{V}_w} \ln a_w$$

We introduce the rational osmotic coefficient, g, to make these two equations give the same result:

$$[2.7.19] \qquad g \ln X_w = \ln a_w$$

The value of g can be calculated from Eqns [2.7.17] and [2.7.19] as

$$[2.7.20] \qquad g = \frac{\pi_{observed}}{-\dfrac{RT}{\overline{V}_w} \ln X_w}$$

Thus the rational osmotic coefficient corrects for the discrepancy between the osmotic pressure and the osmotic pressure calculated from the mole fraction of water. This assumes ideality, in which the activity is equal to the mole fraction, but not dilution. Thus the calculations of osmotic pressure based on Eqn [2.7.17] are better than those calculated using van't Hoff's Law because calculations based on Eqn [2.7.17] are valid even for solutions that are not dilute. However, Eqn [2.7.17] still requires the assumption of ideal solution behavior, or that the activity of water is equal to its mole fraction. Eqn [2.7.18] gives the osmotic pressure without assuming either a dilute solution or ideality. It can be calculated from vapor pressure measurements as

$$[2.7.21] \qquad \pi = -\frac{RT}{\overline{V}_w} \ln a_w = -\frac{RT}{\overline{V}_w} \ln \frac{p}{p^0}$$

Figure 2.7.4 shows the ratio of the observed osmotic pressure to that calculated by Eqn [2.7.15], Eqn [2.7.17], or Eqn [2.7.21]. These give an estimate of φ and g. The approximate values of φ at physiological concentrations for a variety of common solutes are given in Table 2.7.1. Units used in the calculation of osmotic pressure and appropriate values for R are given in Table 2.7.2.

EQUIVALENCE OF OSMOTIC AND HYDROSTATIC PRESSURES

As mentioned earlier, Pfeffer observed a linear relationship between the flow across the membrane and the pressure when water was on both sides of the membrane. This relationship can be described as

$$[2.7.22] \qquad \begin{aligned} Q_V &= A\, L_P(P_L - P_R) = AL_P\, \Delta P \\ J_V &= L_P(P_L - P_R) = L_P\, \Delta P \end{aligned}$$

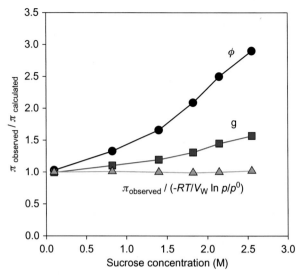

FIGURE 2.7.4 Osmotic coefficients as a function of sucrose concentration. The osmotic coefficient, φ, was calculated according to Eqn [2.7.16] by dividing the observed osmotic pressure by RTC (circles). The rational osmotic coefficient was obtained according to Eqn [2.7.20] by dividing the observed osmotic pressure by $-RT/V_w \ln X_w$ (squares). The correlation of vapor pressure to osmotic pressure was tested by dividing the observed osmotic pressure by $-RT/V_w \ln p/p^0$ (triangles). This figure shows that the van't Hoff equation is good for dilute solutions but fails at high solute concentrations, due to the failure of the assumptions that solutions are dilute and ideal. The osmotic coefficient corrects for these failures. The rational osmotic coefficient deviates significantly at higher sucrose solutions where solution behavior is further from ideal. The ratio of the observed osmotic pressure to that calculated from vapor pressure measurements is close to 1.0 over the entire concentration range. *Data from Glasstone, S.* Textbook of Physical Chemistry. *Princeton: Van Nostrand, 1946.*

TABLE 2.7.1 Approximate Values of the Osmotic Coefficient for Common Solutes Under Physiological Conditions

Solute	Number of Particles Formed upon Solution	Molecular Weight (g mol^{-1})	Osmotic Coefficient (φ)
NaCl	2	58.4	0.93
KCl	2	74.6	0.92
CaCl$_2$	3	111.0	0.85
Na$_2$SO$_4$	3	142.0	0.74
MgCl$_2$	3	95.2	0.89
MgSO$_4$	2	120.4	0.58
NaHCO$_3$	2	84.0	0.96
Alanine	1	89.1	1.00
Mannitol	1	182.2	1.00
Glucose	1	180.2	1.01
Sucrose	1	342.3	1.02

where Q_V is the flow in units of volume per unit time and J_V is the flux, or flow per unit area, in units of velocity. Here the positive flow is taken from left to right and pressure drives the flow. Thus if $P_L > P_R$, then

TABLE 2.7.2 Units for the Calculation of Osmotic Pressure

Pressure Units	1 atm Equivalent	Gas Constant (R)	Solute Osmolyte Concentration (ΣC_S)[a]
atm	1	0.082 L atm mol^{-1} K^{-1}	mol L^{-1}
mm Hg	760	62.36 L mm Hg mol^{-1} K^{-1}	mol L^{-1}
Pa = N m^{-2}	1.013×10^5	8.314 N m mol^{-1} K^{-1}	mol m^{-3} = mol (1000 L)$^{-1}$
dyne cm^{-2}	1.013×10^6	8.314×10^7 dyne cm mol^{-1} K^{-1}	mol cm^{-3}

[a] Osmolarity (osmol L^{-1}) is defined as the concentration of osmotically active particles, osmolytes, in mol L^{-1}. Therefore, the units osmoles and moles cancel in the calculation of osmotic pressure.

Example 2.7.1 Calculate the Osmotic Pressure of 0.9% NaCl

0.9% NaCl means 0.9 g per 100 mL of solution. This is 9 g L^{-1}. We can convert this to molarity by dividing by the molecular weight, 58.4 g mol^{-1}: [NaCl] = 9 g L^{-1}/58.4 g mol^{-1} = 0.154 M.

The effective osmolarity of this solution is 2 osmol mol^{-1} × 0.154 M × 0.93 = 0.2866 osmol L^{-1}.

The osmotic pressure is calculated as $\pi = RT\varphi C$ = 0.082 L atm mol^{-1} K^{-1} × 310 K × 0.2866 mol L^{-1} = **7.29 atm**.

This is an enormous pressure on the physiological scale.

Q_V is positive and flow is to the right. If $P_L < P_R$, then $(P_L - P_R)$ is negative, Q_V is negative and flow is to the left. L_p is a coefficient characteristic of the membrane, variously called the **hydraulic conductivity**, **hydraulic permeability**, or **filtration coefficient**.

When Pfeffer added impermeant solutes to the inner chamber (left chamber), flow was observed into the solution in the absence of any macroscopic hydrostatic pressure differences. That is, when $\Delta P = 0$ there was a negative flow, and that flow was proportional to the osmotic pressure. Thus it appears that the solute caused a *reduction* in the pressure on the solution side because the flow is inward and additional pressure on the solution side, the osmotic pressure, is necessary to stop the osmotic flow. The osmotic pressure, which is characteristic of the solution, is equivalent to a reduction in the hydrostatic pressure. If solution is present on both sides of the membrane, we write

[2.7.23]
$$Q_V = A\,L_P[(P_L - \pi_L) - (P_R - \pi_R)]$$
$$Q_V = A\,L_P[(P_L - P_R) - (\pi_L - \pi_R)]$$
$$Q_V = A\,L_P(\Delta P - \Delta \pi)$$

THE REFLECTION COEFFICIENT CORRECTS VAN'T HOFF'S EQUATION FOR PERMEABLE SOLUTES

When membranes are partially permeable to the solute (leaky membranes), the measured osmotic pressure is less than that predicted by van't Hoff Law. This has led to the definition of another membrane parameter, σ, the **reflection coefficient**, which is *defined* as

[2.7.24]
$$\sigma = \frac{\pi_{eff}}{\varphi RTC}$$

where π_{eff} is the *effective* or observed osmotic pressure and φRTC is the theoretical osmotic pressure that would be observed if the membrane was perfectly semipermeable. The reflection coefficient is different for each pair of solute and membrane. Its range is $0 \leq \sigma \leq 1.0$. Rewriting Eqn [2.7.24], we have

[2.7.25]
$$\pi_{eff} = \sigma\varphi RTC = \sigma\pi$$

It is important to note that **the osmotic pressure is a characteristic of the solution**, and it is defined as the pressure necessary to stop osmotic flow when a solution is separated by an ideal, perfectly semipermeable membrane from pure water. Recall that a semipermeable membrane is defined as one that has zero permeability to solute. Every aqueous solution has an osmotic pressure—it is a concentration measure like molarity is—and it can be measured by any of the colligative properties. When a solution is placed in contact with a membrane that is not semipermeable, its osmotic pressure will be reduced by the membrane's reflection coefficient, which is characteristic not only of the membrane but also of the solute. As a consequence of this, flow across a real membrane that is not semipermeable will be altered. It is governed by the equation

[2.7.26]
$$Q_V = A\,L_P[(P_L - \Sigma_L\sigma_i\pi_i) - (P_R - \Sigma_R\sigma_i\pi_i)]$$
$$Q_V = A\,L_P(\Delta P - \Sigma\Delta\sigma_i\pi_i)$$

where the summation sign means that all osmotically active solutes in the solution contribute to the total osmotic pressure.

We now have three phenomenological coefficients that describe volume and solute flux across a membrane. These are summarized in Table 2.7.3. The question before us now is: what is the physical meaning of L_P and σ?

Example 2.7.2 Calculate the Net Driving Force and Flow Across a Dialysis Membrane.

A semipermeable dialysis membrane has an area of 90.5 cm^2 and $L_p = 6.46 \times 10^{-6}$ cm min^{-1} mmHg^{-1}. Inside the dialysis membrane was a solution of 5% (w/v) sucrose at a pressure of 10 mmHg. Outside was a solution of 2% sucrose at a pressure of 50 mmHg. The entire apparatus was equilibrated to room temperature, 20°C. What is the net pressure across the membrane? What is the net flow across the membrane? Assume φ for sucrose is 1.0. The molecular weight of sucrose is 342 g mol^{-1}.

First, we calculate the osmotic pressures of the solutions. 5% sucrose means 5 g of sucrose per 100 mL of solution or 50 g L^{-1}. We can convert this to molarity by dividing by the molecular weight, 342 g mol^{-1}:

$$[\text{sucrose}]_{in} = 50 \text{ g L}^{-1}/342\text{g mol}^{-1} = \mathbf{0.146 \text{ M}}.$$

The molar concentration of 2% sucrose is similarly calculated as

$$[\text{sucrose}]_{out} = 20 \text{ g L}^{-1}/342\text{g mol}^{-1} = \mathbf{0.058 \text{ M}}$$

The osmotic pressure of the two solutions is calculated as

$$\pi_{in} = \varphi RT \times 0.146 \text{ M}$$
$$= 62.36 \text{ L mmHg mol}^{-1} \text{ K}^{-1} \times 293 \text{ K} \times 0.146 \text{ mol L}^{-1}$$
$$= \mathbf{2668 \text{ mmHg}}$$

$$\pi_{out} = \varphi RT \times 0.058 \text{ M}$$
$$= 62.36 \text{ L mmHg mol}^{-1} \text{ K}^{-1} \times 293 \text{ K} \times 0.058 \text{ mol L}^{-1}$$
$$= \mathbf{1060 \text{ mmHg}}$$

The net driving force is calculated as

$$(P_{in} - P_{out}) - (\pi_{in} - \pi_{out}) = (10 \text{ mmHg} - 50 \text{ mmHg})$$
$$- (2668 \text{ mmHg} - 1060 \text{ mmHg})$$
$$= \mathbf{-1648 \text{ mmHg}}$$

Positive pressure would force fluid outward; negative net pressure means flow is inward. Note that the systematic insertion of the pressure values is essential to obtaining the correct result.

The inward flow is given by

$$Q_V = A\, L_p(\Delta P - \Delta \pi)$$
$$= 90.5 \text{ cm}^2 \times 6.46 \times 10^{-6} \text{cm min}^{-1} \text{ mmHg}^{-1} \times -1648 \text{ mmHg}$$
$$= \mathbf{-0.963 \text{ cm}^3 \text{ min}^{-1}}.$$

TABLE 2.7.3 Phenomenological Membrane Coefficients

Coefficient	Parameter	Calculated as
p	Permeability	$(J_s/\Delta C)_{J_V=0}$
L_p	Hydraulic conductivity	$(J_V/\Delta P)_{\Delta \pi=0}$; $-(J_V/\sigma \Delta \pi)_{\Delta P=0}$
σ	Reflection coefficient	$-(J_V/L_p \Delta \pi)_{\Delta P=0}$

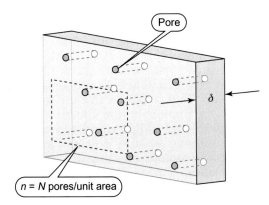

FIGURE 2.7.5 Schematic drawing of the hypothetical microporous membrane. We imagine that this membrane is a thin, flat sheet that is pierced by right cylindrical pores of radius a and depth δ, which is equivalent to the thickness of the membrane. There are $n = N/A$ pores per unit area of the membrane. The membrane separates two aqueous solutions with a solute concentration C_L on the left and C_R on the right. We further suppose that the left chamber is subject to the pressure P_L and the right side to the pressure P_R.

L_P FOR A MICROPOROUS MEMBRANE DEPENDS ON THE MICROSCOPIC CHARACTERISTICS OF THE MEMBRANE

Here we consider fluid flow across a microporous membrane such as that shown in Figure 2.7.5. We consider that the membrane itself is impermeant to water and solute, but both may go through tiny pores in the membrane. We consider three cases:

1. There is a pressure difference across the membrane but either there is water on both sides of the membrane or the solute particles are very small compared to the size of the pore. Pressure-driven flow will occur.
2. There is no pressure difference across the membrane that separates water from solution, but the solutes on one side of the membrane are too large to fit through the pores. This will produce an osmotic pressure difference and an osmotic flow.
3. There is no pressure difference across the membrane that separates water from solution, but the solutes are small enough to fit through the pore, but not as easily as water. This will produce a smaller osmotic pressure and a smaller osmotic flow than in case 2.

CASE 1: THE SOLUTE IS VERY SMALL COMPARED TO THE PORE

The volume flow through the membrane is the number of pores times the volume flow per pore:

[2.7.27] $$Q_v = N\, q_v$$

where Q_V signifies the overall observed macroscopic flow, N is the number of pores, and q_v is the flow through a single pore. If we assume **laminar flow** (see Chapter 1.2), then the flow through each pore will be given by **Poiseuille's Law**, which gives the flow through a right cylindrical pipe of radius a and length δ as

[2.7.28] $$q_v = \frac{\pi a^4}{8\eta\delta}\Delta P$$

where η is the **viscosity** of the fluid. The use of Poiseuille's law in this situation requires the assumptions of laminar flow, Newtonian fluids, and a pore length that is long compared to the entrance length of the pore. The entrance length is the distance it takes for the fluid to establish a parabolic velocity profile within the pore. The observed macroscopic flux, J_v, is the flow per unit area of membrane per unit time. Substituting Poiseuille's law into Eqn [2.7.27] gives the volume flux as

[2.7.29] $$J_v = \frac{Nq_v}{A} = nq_v = \frac{n\pi a^4}{8\eta\delta}\Delta P$$

where $n = N/A$ is the density of pores in the membrane – the number of pores per unit area. Comparing this to Eqn [2.7.22], $J_V = L_P \Delta P$, we can identify

[2.7.30] $$L_p = \frac{n\pi a^4}{8\eta\delta}$$

According to this equation, the observed macroscopic flux is linearly related to the pressure difference by a coefficient which includes the thickness of the membrane, δ, the density of pores in the membrane, n, the radius of the pores, a, and the viscosity of the fluid, η. L_p has units of volume per unit time per unit area per unit pressure.

CASE 2: THE SOLUTE IS LARGER THAN THE PORE: THE MECHANISM OF OSMOSIS FOR MICROPOROUS MEMBRANES

If the solute particles are larger than the pore, then they cannot get through the membrane and the membrane is perfectly semipermeable. Pfeffer has already experimentally determined what happens in this case, and his experimental results permit several conclusions. First, the solute in the water causes the flow because there is no flow when pure water is on both sides of the membrane. Second, the pressure at equilibrium or the flow is directly proportional to the concentration of the solute, if the solutions are sufficiently dilute. Third, the solute causes a reduction in the pressure of the solution because the flow is inward and additional pressure on the solution side is necessary to stop the osmotic flow. Fourth, because the observed osmotic pressure obeys van't Hoff's Law only when the membrane is impermeable to solute, the osmotic pressure and the osmotic flow result from the interaction of the solute with the membrane. If the membrane is highly permeable to solute also, no osmosis and no osmotic pressure are observed. We should look to the interaction of the membrane with the solute to explain the mechanism of osmosis. The derivation for one mechanism of osmotic pressure flow and pressure generation is given in Appendix 2.7.A1. Briefly, this derivation recognizes that the total pressure in the bulk solution results from contributions of both solute and solvent molecules. However, solute molecules cannot enter the pores and so they cannot contribute to the pressure within the pore. Thus there is a pressure deficit on the solution side immediately upon entering the pore from that side. Since pressure results from collisions of molecules, resulting in momentum change in the molecules, there is a momentum deficit within the pores due to restricted entry of solute particles into the pores.

In Case 2, we have the situation where $P_L = P_R$, $C_L = C_L$, and $C_R = 0$ for a membrane which is impermeant to solute. From Eqn [2.7.A1.9], the pressure immediately within the pore, P_L', is given by

[2.7.31] $$P_L' = P_R - RTC_L$$

assuming the validity of van't Hoff's law. A plug of water extending the length of the pore is subjected to a pressure difference, which is given by

[2.7.32] $$P_L' - P_R = -RTC_L$$

because, in this case, $P_L = P_R$. The negative sign indicates the pressure is higher on the right than on the left. Thus fluid flow should occur from right to left according to Poiseuille's law:

[2.7.33] $$q_v = \frac{\pi a^4}{8\eta\delta}(-\pi)$$

where q_v is the flow per pore. In this equation, we see the absurd situation in which one symbol is used to signify two entirely different quantities. The π in the fraction is the geometric ratio, while the π in parenthesis signifies the osmotic pressure. The negative sign indicates flow to the left. The overall volume flux for the membrane is then given as

[2.7.34] $$J_v = N\frac{q_v}{A} = \frac{n\pi a^4}{8\eta\delta}(-\pi)$$

It is clear, then, that J_v in the presence of osmotic flow is given by

[2.7.35] $$J_v = -L_p \pi$$

where L_p is identical to that described earlier for pressure-driven flow (see Eqn [2.7.30]).

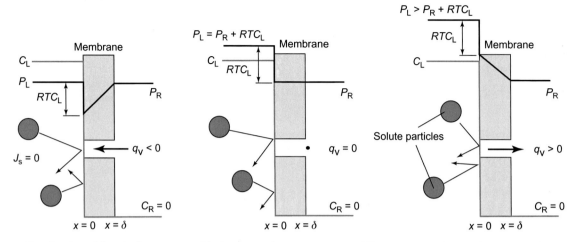

FIGURE 2.7.6 The direction of flow in the presence of hydrostatic and osmotic pressure differences across the membrane. Far left: The hydrostatic pressure is the same on both sides of the membrane but the left side contains a solution of concentration C_L. In the pore, there is a pressure gradient and flow is toward the solution. Middle: The pressure on the left was increased by $RTC_L = \pi$, the osmotic pressure. This is the equilibrium situation and no flow occurs. Right: Pressure on the left is more than RTC_L greater than pressure on the right, and flow is to the right.

It is instructive to consider the pressure profiles within the pore which are established due to the osmotic pressure. Poiseuille's Law is given as

[2.7.36]
$$Q_v = \frac{\pi a^4}{8\eta}\left(\frac{\Delta P}{\Delta X}\right)$$

Since $\partial Q_v/\partial x = 0$ at steady state, it follows that $\Delta P/\Delta x$ must be a constant in x and in t: the pressure gradient is linear in x. This allows us to draw the schematic diagrams of pressure and concentration gradients during osmotic flow as shown in Figure 2.7.6.

CASE 3: THE REFLECTION COEFFICIENT RESULTS FROM PARTIALLY RESTRICTED ENTRY OF SOLUTES INTO THE PORES

When the membrane is freely permeable to the solute (Case 1), there is no osmosis and no osmotic pressure and hydrostatic pressure drives fluid flow. When the membrane is completely impermeable to the solute (Case 2), and there is no hydrostatic pressure, the osmotic pressure is ideally given by van't Hoff's Law and drives fluid flow. Here we consider Case 3, a membrane which distinguishes between solute and solvent but which is not completely impermeable to solute.

Here we consider pores that are large compared to the solvent particles but they *partially* exclude solute particles. The partial exclusion is due to the fact that the effective area of the pore is reduced compared to the area available to the solvent. When solute particles enter the pore, it reduces the effective osmotic pressure in direct proportion to the fraction of solute particles that can enter the pore. Consider that the pores have a radius a and that the solute particles are spherical with a radius a_s. If $a_s < a$, then at least some of the solute particles can get through the pore. Suppose that if a solute

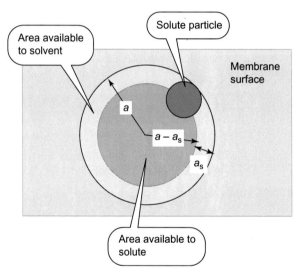

FIGURE 2.7.7 Relative areas of the pore available to solvent water and solute particles. The figure depicts a single pore, looking down along its axis normal or perpendicular to the surface of the membrane. The outer circle is the dimension of the pore that is available to the solvent water. The inner circle represents the dimensions of the pore available to the solute, which is less than that available to solvent.

particle hits the rim of the pore prior to entry, then it is *reflected* back into the bulk solution. The effective pore area will be reduced due to this reflection. The situation is depicted schematically in Figure 2.7.7, looking perpendicular to the membrane along the axis of the pore. The area of the pore which is accessible to solute is given by

[2.7.37]
$$A_s = \pi(a-a_s)^2 = \pi a^2 \left(1-\frac{a_s}{a}\right)^2$$

The relative area available to solute compared to that available to solvent is

[2.7.38]
$$\frac{A_s}{A} = \frac{\pi a^2 \left(1-\frac{a_s}{a}\right)^2}{\pi a^2} = \left(1-\frac{a_s}{a}\right)^2$$

The fraction of solute particles which are reflected by the pore is approximated by the ratio of the area in light blue in Figure 2.7.7 to the total area. This is identified with σ, the **reflection coefficient**:

$$\sigma = \frac{A - A_s}{A} = 1 - \frac{A_s}{A}$$

[2.7.39]

$$\sigma = 1 - \left(1 - \frac{a_s}{a}\right)^2$$

If $\xi = a_s/a$, then

[2.7.40] $$\sigma = 2\xi - \xi^2$$

We may expect the ratio of the concentration in the pore to the concentration in the bulk solution to be the same as the ratio of the pore area available to solute to the area available to solvent, A_s/A. This is given by Eqn [2.7.39] to be

[2.7.41] $$\frac{C'_L}{C_L} = \frac{A_s}{A} = 1 - \sigma$$

In the absence of solvent drag (when $J_v = 0$), the concentration profile may be drawn as shown in Figure 2.7.8.

In this model, some of the solute molecules can enter the pore and therefore they contribute to the pressure in the pore. We expect the pressure deficit within the pore to be due only to those molecules which are reflected. Thus the osmotic pressure should be $\sigma RT\Delta C = \sigma \Delta \pi$. The expression for laminar flow in the pore is

[2.7.42]
$$J_v = \frac{n\pi a^4}{8\eta\delta}[\Delta P - \sigma \Delta \pi]$$

$$J_v = L_p[\Delta P - \sigma \Delta \pi]$$

In the above model, σ is viewed as being due to a hindered entry into the pore. There may be additional hindrance to solute flow through the pore due to interactions with the pore wall, but these are ignored here. According to this view, the reflection coefficient has physical meaning only in the range $0 \leq \sigma \leq 1.0$. If $A_s = A$, then the pore is large enough so that the membrane does not discriminate between solute and solvent, and $\sigma = 0$ according to Eqn [2.7.39]. If $a_s \geq a$, then $A_s = 0$ and $\sigma = 1.0$.

SOLUTIONS MAY BE *HYPERTONIC* OR *HYPOTONIC*

When cells are placed in contact with a solution, they may either swell or shrink as shown in Figures 2.7.9 and 2.7.10. These observations introduce the idea of **tonicity**, which is operationally defined. If we place a cell in a solution and the cell swells, the solution is called hypotonic. If we place a cell in a solution and the cell shrinks, we call that solution hypertonic. If the cell neither swells nor shrinks, the solution is isotonic.

OSMOSIS, OSMOTIC PRESSURE, AND TONICITY ARE RELATED BUT DISTINCT CONCEPTS

Osmotic pressure is a theoretical concept. It is the pressure necessary to stop osmotic flow if a solution is separated from pure water by a semipermeable membrane. A semipermeable membrane is *defined* as a membrane that allows the passage of some molecules but not others. It is freely permeable to water but impermeable

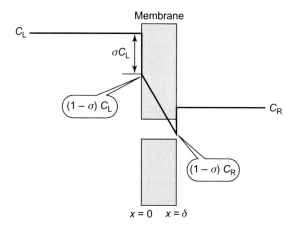

FIGURE 2.7.8 Concentration profile within a pore that partially excludes solute particles. This profile pertains only when solvent drag is zero. Solvent drag washes the concentration gradient in the direction of volume flow and alters J_s.

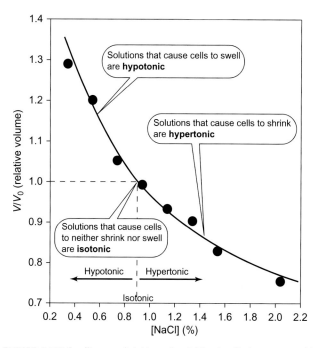

FIGURE 2.7.9 Swelling or shrinking of red blood cells in contact with different concentrations of NaCl. Whole blood was centrifuged to separate the red blood cells from the plasma. Then an aliquot of the plasma was removed and replaced with an equal volume of the indicated concentration of NaCl. When the plasma was replaced with NaCl solution less than about 0.9%, the cells' volume increased relative to that in normal plasma. These solutions are described as being **hypotonic**. When the [NaCl] that replaced plasma was greater than 0.9%, the red blood cells shrank. These solutions are called **hypertonic**. Replacement with 0.9% NaCl caused the cells to neither swell nor shrink: this solution is **isotonic**.

to solute. All solutions have an osmotic pressure. In the approximation of the ideal, dilute solution, the magnitude of the osmotic pressure is given by van't Hoff's Equation (see Eqn 2.7.7), where C is the molar concentration of osmotically active solutes. This C is the **osmolarity**.

Both osmolarity and osmotic pressure are properties of a solution, but tonicity is not. Tonicity refers to the direction of osmotic flow when a particular solution is placed in contact with a particular cell. Tonicity involves real membranes rather than an ideal, semipermeable membrane. The real membrane has a set of reflection coefficients, one for each solute on each side of the membrane. Whether shrinking or swelling occurs depends on the osmotic concentrations and compositions of the two solutions on the two sides of the cell membrane, and also on how these solutes interact with the real biological membrane.

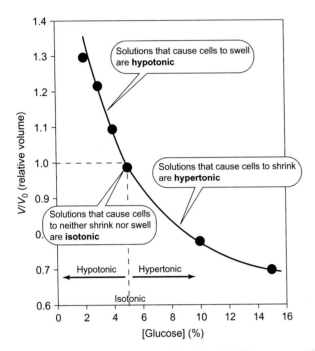

FIGURE 2.7.10 Swelling or shrinking of red blood cells in contact with different concentrations of glucose. Whole blood was centrifuged to separate the red blood cells from the plasma. Then an aliquot of the plasma was removed and replaced with an equal volume of the indicated concentration of glucose. When the plasma was replaced with solutions of glucose less than about 5%, the cells' volume increased relative to that in normal plasma. These solutions are described as being **hypotonic**. When the [glucose] that replaced plasma was greater than 5%, the red blood cells shrank. These solutions are called **hypertonic**. Replacement with 5% glucose caused the cells to neither swell nor shrink: this solution is **isotonic**.

CELLS BEHAVE LIKE OSMOMETERS

Consider a cell with a volume V_0 and a total concentration of osmotically active solutes, ΣC_i, resting in an isotonic medium. Now suppose that we rapidly replace the medium with another with a different osmotic pressure which we will symbolize as π. If the new medium is hypotonic, then by definition the cell will swell to a new volume, at which point the cell's cytoplasmic tonicity will match the medium's. Similarly, if the new medium is hypertonic, then the cell will shrink to a new volume that will match the tonicity of the medium. We assume here that the volume of the medium is so large that water movement into the cell or out of the cell does not appreciably affect the osmolarity of the medium. We further assume that water movement is fast compared to the movement of solutes or ions across the membrane, and that we can measure the cell's volume after water movement has occurred but before solute movement. What is the relationship between the medium osmotic pressure and the final cell volume, V_c?

In this situation, the total amount of osmotically active solutes in the cell is constant—the intracellular solutes do not move. The total intracellular osmolytes are equal to $V_0 \Sigma C_i$, where V_0 is the volume under isotonic conditions and C_i is the concentration of solute i in the cytoplasm in the isotonic condition. Now the total amount of osmotically active solutes at any time is

---

Example 2.7.3 Isosmotic Solutions May Not Be Isotonic

A. Calculate the osmolarity and osmotic pressure of isotonic saline (NaCl):

Isotonic saline is 0.9% NaCl. This is 0.9 g NaCl per dL or 100 mL of solution = 0.9/0.1 L or 9.0 g L^{-1}.

The formula weight for NaCl is 58.44 g mol^{-1}. The *molarity* of NaCl is 9 g L^{-1}/58.44 g mol^{-1} = 0.154 M.

Since NaCl dissociates into two particles per mole, the osmolarity is 2 × 0.154 M = 0.308 OsM.

The osmotic coefficient of NaCl is φ_{NaCl} = 0.93, so that the measured osmolarity of this solution would be $\varphi \times C$ = 0.93 × 0.308 OsM = **0.286 OsM**. The osmotic pressure at 37°C is **7.27 atm**.

B. Calculate the osmolarity and osmotic pressure of 1.8% urea in water:

The formula weight for urea is 60.0 g mol^{-1} and φ_{urea} = 0.95.

Its osmolarity is φC = 0.95 × 18 g L^{-1}/60 g mol^{-1} = **0.285 OsM**. Its osmotic pressure is **7.24 atm**.

C. What happens when red blood cells are placed in 0.9% NaCl? 1.8% urea?

Red blood cells placed in contact with 0.9% NaCl neither shrink nor swell. The solution is isotonic. When placed in 1.8% urea, the cells swell so much that the cells lyse or break open. Although 0.9% NaCl and 1.8% urea are isoosmotic with red blood cell contents, 0.9% NaCl is isotonic and 1.8% urea is not isotonic.

distributed in the volume of the cell, V_c. So the osmolarity at any time is

$$[2.7.43] \qquad \frac{V_0 \sum C_i}{V_c}$$

When the osmotic pressure of the medium equilibrates with that of the cell, with the assumption of no solute movement (which is equivalent to assuming $\sigma_i = 1.0$), we get

$$[2.7.44] \qquad \pi = RT \frac{V_0 \sum C_i}{V_c}$$

This can be rewritten as

$$[2.7.45] \qquad \pi = \frac{V_0 \sum RTC_i}{V_c} = \frac{V_0}{V_c} \pi_{\text{isotonic}}$$

which we can rearrange to obtain

$$[2.7.46] \qquad V_c = V_0 \frac{\pi_{\text{isotonic}}}{\pi}$$

According to Eqn [2.7.46], a cell acting as a perfect osmometer would show a cell volume that was inversely proportional to the osmolarity of the external medium, with an intercept of zero. Typically the cell's volume under isotonic conditions, V_0, is the control for their volume under nonisotonic conditions. Thus we rewrite Eqn [2.7.46] as

$$[2.7.47] \qquad \frac{V_c}{V_0} = \frac{\pi_{\text{isotonic}}}{\pi}$$

Figure 2.7.11 shows the plot of the volume of cardiac cells exposed to various osmolar solutions, with volume normalized to the volume under isotonic conditions. According to Eqn [2.7.47], we expect the data to be linear with an intercept of zero. Actual data from these cardiomyocytes show that the response of relative volume (V_c/V_0) is linear within a considerable range, but the curve does not extrapolate to zero volume. These real cells are not perfect osmometers.

The response of real cells is described by the empirical equation

$$[2.7.48] \qquad \frac{V_c}{V_0} = \left(1 - \frac{V_b}{V_0}\right) \frac{\pi_{\text{isotonic}}}{\pi} + \frac{V_b}{V_0}$$

The intercept of this line on the volume axis is V_b/V_0, which in the case of cardiomyocytes shown in Figure 2.7.11 is 0.34. This is interpreted to mean that there is a fraction of the cell's volume that is **osmotically inactive**. This is partly to be expected. Not all of the volume of the cell is water, and it is the volume of water that dissolves osmotically active solutes and is responsive to changes in the medium osmolarity. Thus the sum of all of the volumes of large particles such as DNA, RNA ribosomes, and membranes contributes to an osmotically inactive volume. Although this is certainly part of the explanation for the osmotically inactive volume, the issue is by no means settled. Other components of the inactive volume might include the volume of a variety of macromolecular assemblies such as the cytoskeleton.

CELLS ACTIVELY REGULATE THEIR VOLUME THROUGH RVDs AND RVIs

Although the presence of a hypotonic or hypertonic solution initiates swelling or shrinking, respectively, often the volume change is not maintained. A cell that initially swells when placed in a hypotonic medium may eventually lose some of its acquired volume: it undergoes a **regulatory volume decrease or RVD**. The swelling of the cell activates compensatory mechanisms that cause transport of osmotically active solutes (osmolytes) out of the cell. Similarly, cell shrinking can activate an influx of osmolytes in some cells leading to a compensatory swelling that is called a **regulatory volume increase** or **RVI**. These RVDs and RVIs are accomplished by altering the cell's contents of osmolytes.

SUMMARY

Osmosis refers to the movement of fluid across a membrane in response to different concentrations of solutes on the two sides of the membrane. The movement of fluid is toward the more concentrated solution. Osmotic pressure is defined as the pressure that must be applied to the solution side to stop fluid movement when a semipermeable membrane separates a solution from pure water. Here, the semipermeable membrane is permeable to water but not to solute. The osmotic pressure for dilute ideal solutions obeys van't Hoff's Law:

$$\pi = RT \sum C_s$$

which can be derived on thermodynamic grounds. Because the solutions are not ideal, the equation is

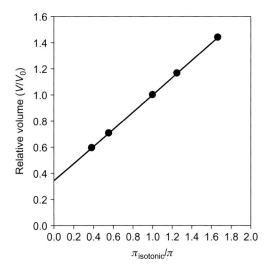

FIGURE 2.7.11 Relative volume of isolated cardiac cells exposed to test solutions of different osmolarities. *Source: Data from Drewnowska and Baumgarten,* American Journal of Physiology **260**:C122−131, 1991.

refined by including an osmotic coefficient, φ_s, characteristic of each solute:

$$\pi_{\text{observed}} = RT\Sigma\, \varphi_s\, C_s$$

Defined in this way, the osmotic pressure is a pressure deficit caused by dissolving solutes. However, membranes that are somewhat permeable to the solute develop a transient osmotic pressure that is less than that predicted by van't Hoff's Law. Thus the actual pressure developed across a membrane that separates a solution from pure water depends on the interaction of the solute with the membrane. The correction for partially permeable membranes requires the reflection coefficient, σ:

$$\pi_{\text{observed}} = RT\Sigma\, \sigma_s\, \varphi_s\, C_s$$

The magnitude of the pressure tells us nothing of the flow. Osmotic pressure and hydrostatic pressure add to drive fluid flow across the membrane, with a proportionality constant, L_p. The phenomenological equation describing fluid flow is

$$Q_v = A\, L_p (\Delta P - \sigma \Delta \pi)$$

Osmosis and osmotic pressure is a thermodynamic concept which exists independently of mechanism. In microporous membranes, osmosis is caused by a momentum deficit within the pores due to the reflection of solute molecules by the membrane. This reduces the pressure on the solution side of the pore by π for a semipermeable membrane.

Thus there are three characteristic parameters for describing passive material transfer across membranes: the permeability, p, the hydraulic conductivity, L_p, and the reflection coefficient, σ.

Osmolarity is a kind of concentration measure, distinct from molarity. It is related to other colligative properties of solutions including freezing point depression, vapor pressure depression, and boiling point elevation. Tonicity is a related concept but involves a real, biological membrane that may not be semipermeable. Tonicity makes reference to a particular cell and its membrane. Solutions may be isoosmotic but not isotonic.

Cells respond to swelling or shrinking according to the empirical relation:

$$V/V_0 = (1 - V_b/V_0)\, \pi_{\text{isotonic}}/\pi + V_b/V_0$$

where V_0 is the volume of the cell under isotonic conditions. V_b is interpreted as the osmotically inactive volume. If $V_b = 0$, the cell would behave like an ideal osmometer.

REVIEW QUESTIONS

1. In which direction does osmotic flow occur? Why?
2. What equation gives the magnitude of the osmotic pressure? What are the limitations of this equation? How would you correct for its errors?
3. What is the reflection coefficient? How does it relate to permeability?
4. The equations in this chapter were derived for a microporous membrane. Would they still hold for a lipid dissolution model of a membrane?
5. How does tonicity differ from osmolarity? Define hypotonic, hypertonic, and isotonic.
6. What is the relationship between volume and osmotic pressure in a perfect osmometer?
7. What is meant by "osmotically inactive" volume?
8. How do cells regulate their volume?

APPENDIX 2.7.A1 MECHANISM OF OSMOSIS: FILTRATION VERSUS DIFFUSION DOWN A CONCENTRATION GRADIENT

PHYSICAL ORIGIN OF THE OSMOTIC PRESSURE ACROSS A MICROPOROUS MEMBRANE

Consider what happens in the vicinity of a single pore, as shown in Figure 2.7.A1.1. Since the pore excludes solute particles, the average concentration must decrease from C_L in the bulk solution to zero in moving along the axis of the pore. Thus there is a concentration gradient near the opening of the pore. By Fick's First Law of Diffusion, we should expect a net diffusion of solute particles into the pore. However, the particles hit the rim of the pore and are reflected back into the bulk solution. On the average, these particles experience a force in the $-x$ direction. The magnitude of this average force can be obtained from the generalized Fick's equation (see Eqn [1.6.42]):

[2.7.A1.1] $$j_s = -D\frac{\partial C(x)}{\partial x} + \frac{D}{RT} F\, C(x)$$

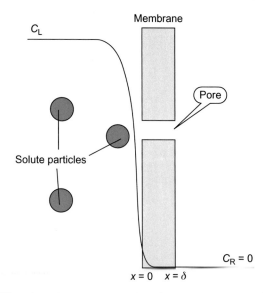

FIGURE 2.7.A1.1 Concentration gradient for solute particles near a pore. In this case, the solute particles are large compared to the pore and cannot penetrate into the membrane.

We write R in place of k and F in place of f in Eqn [2.7.A1.1] because we are speaking of the force per mole rather than the force per molecule, as was done in Eqn [1.6.42].

Since $j_s = 0$ for an impermeant membrane, Eqn [2.7.A1.1] becomes

$$[2.7.A1.2] \qquad FC(x) = RT\frac{\partial C(x)}{\partial x}$$

The force F is an external force that acts on the solute particles. The total force acting on the solute particles is just $Fn(x)$, where n is the number of moles of particles. Dividing by the volume, V, we have $Fn(x)/V = FC(x)$, which is the average force per unit volume. This is the force that the membrane is exerting on the solute bodies, per unit volume, located in the volume of fluid directly in front of the pore. The consequence of this is that there will be a pressure drop at the pore entrance. To see this, we will analyze the forces acting on an element of fluid located directly in front of a pore, as shown in Figure 2.7.A1.2.

Here we consider the forces acting on an element of fluid with an area A from a point x well within the bulk solution to a point $x + \Delta x$ within the pore near its surface. We consider that the volume element is in mechanical equilibrium. The forces acting on this volume are the forces acting on the solute particles and the forces acting on the surfaces in contact with adjacent fluid. The sum of these forces must be zero to meet the condition of mechanical equilibrium. We identify these forces with the "body" forces, F_B, and "contact" forces, F_C. Thus we write:

$$[2.7.A1.3] \qquad 0 = F_B + F_C$$

for the condition of mechanical equilibrium. The net contact force is the balance of the pressure acting on the surface area to the right and left of the volume element:

$$[2.7.A1.4] \qquad F_C = A\,P(x) - A\,P(x + \Delta x)$$

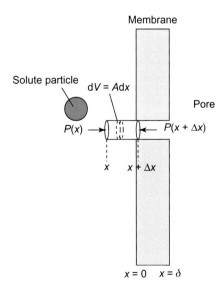

FIGURE 2.7.A1.2 Forces acting on the plug of fluid immediately in front of a pore in a porous membrane.

The total body force is the body force per unit volume integrated over the volume.

$$[2.7.A1.5] \qquad F_B = \int_x^{x+\Delta x} F\,C(x)\,\mathrm{d}V$$

Inserting $\mathrm{d}V = A\mathrm{d}x$ and $FC(x) = RT\,\partial C(x)/\partial x$ from Eqn [2.7.A1.2], we obtain

$$[2.7.A1.6] \qquad \begin{aligned} F_B &= ART\int_x^{x+\Delta x}\frac{\partial C(x)}{\partial x}\,\mathrm{d}x \\ &= ART[C(x + \Delta x) - C(x)] \end{aligned}$$

Since $C(x + \Delta x) = 0$ because solute particles are not in the pore, Eqn [2.7.A1.6] becomes

$$[2.7.A1.7] \qquad F_B = -ART\,C(x)$$

The negative sign in Eqn [2.7.A1.7] indicates that F_B is directed to the left. Inserting Eqns [2.7.A1.4] and [2.7.A1.7] into Eqn [2.7.A1.3], we obtain

$$[2.7.A1.8] \qquad A\,P(x) - A\,P(x + \Delta x) = ART\,C(x)$$

or

$$[2.7.A1.9] \qquad P(x) - P(x + \Delta x) = RTC_L$$

This equation says that the net pressure experienced by the volume of fluid immediately adjacent to the pore decreases as one moves from the left compartment into the pore, and that the drop in pressure is RTC_L. This analysis suggests that the osmotic pressure develops as a consequence of the interactions between the solute particles and the membrane. The solute particles contribute momentum in the solution. When the particles collide with the membrane, that momentum is transferred to the membrane and not to the fluid within the pore. As a consequence, there is a momentum deficit within the pore (compared to the bulk solution). Since the pressure is the average momentum change per unit area experienced by particles colliding within the fluid, this momentum deficit shows up as a pressure deficit. The resulting pressure difference produces a flow that is given as

$$[2.7.A1.9] \qquad J_v = \frac{n\pi a^4}{8\eta\delta}\Delta\pi$$

where we identify

$$[2.7.A1.10] \qquad L_p = \frac{n\pi a^4}{8\eta\delta}$$

DIFFUSIONAL PERMEABILITY OF MICROPOROUS MEMBRANES

In the absence of a hydrostatic or osmotic pressure gradient, water will diffuse across a microporous membrane. If we assume the membrane is impermeable except at the pores, the diffusion through the pores will obey Fick's Laws of diffusion:

[2.7.A1.11]
$$j_w = -D\frac{\partial C}{\partial x}$$
$$\frac{\partial C}{\partial t} = D\frac{\partial C^2}{\partial x^2}$$

where j_w is the flux through a single pore. At steady state, the flux does not change with time, or with distance, and so the gradient is linear in x and we can write

[2.7.A1.12]
$$j_w = \frac{\pi a^2 D_w}{\delta}\Delta C_w$$

where j_w is the flux of water through the pore, a is the radius of the pore, D_W is the diffusion coefficient of water in the pore, δ is the length of the pore (equal to the thickness of the membrane), and ΔC_W is the concentration difference of water. The overall macroscopic flux for a microporous membrane is given as

[2.7.A1.13]
$$J_w = \frac{n\pi a^2 D_w}{\delta}\Delta C_w$$

where n is the density of the pores, $n = N/A$, the number of pores per unit area of the membrane. This is analogous to the passive solute flux discussed in Chapter 2.5. We can similarly define a diffusional permeability for water from the relation:

[2.7.A1.14]
$$J_w = P_d \Delta C_w$$

and we can identify a **diffusional permeability coefficient** for water from Eqns [2.7.A1.13] and [2.7.A1.14] as

[2.7.A1.15]
$$P_d = \frac{n\pi a^2 D_w}{\delta}$$

FILTRATION PERMEABILITY IN THE PRESENCE OF A PRESSURE DIFFERENCE IN A MICROPOROUS MEMBRANE

In Chapter 1.6, we found that diffusional flux can be altered by the application of an additional force. In particular, a flux of solute obeyed the relation:

[2.7.A1.16]
$$J_s = -\frac{D}{RT}C F$$

where J_S is the flux of solute, D is its diffusion coefficient, R is the gas constant, T is the temperature (K), C is the concentration, and F is the total force on the particle per mole. We went on to define the electrochemical potential so that

[2.7.A1.17]
$$F = -\frac{d\mu}{dx}$$

These equations have general validity and are also true for water. We can write for the flux in the pore that

[2.7.A1.18]
$$j_w = \frac{D_w}{RT}C\frac{d\mu_w}{dx}$$

From the thermodynamic derivation of van't Hoff's Law, we write for the chemical potential of water

[2.7.A1.19]
$$\mu_w = \mu_w^0 + RT \ln a_w + \overline{V}_w P$$

insertion of this into Eqn [2.7.A1.18] and at steady-state flow, we obtain for the macroscopic membrane

[2.7.A1.20]
$$J_w = \frac{n\pi^2 D_w}{RT\delta}C_w\overline{V}_w\Delta P$$

V_w is the partial molar volume (in volume per mole, L mol^{-1}), and C_W is the water concentration, in moles L^{-1}, and these cancel: $V_W C_W = 1.0$. The flux given here is in mole per unit area per unit time. To convert to J_V we multiply by V_W:

[2.7.A1.21]
$$J_V = \frac{\overline{V}_w n\pi a^2 D_w}{RT\delta}\Delta P$$

This describes pressure-driven flow across the membrane. We recognize that the same terms for P_d are present in this equation. We recognize Eqn [2.7.A1.21] as $J_V = L_P \Delta P$ and identify L_P as

[2.7.A1.22]
$$L_P = \frac{\overline{V}_w n\pi a^2 D_w}{RT\delta} = \frac{\overline{V}_w P_f}{RT}$$

where P_f is the **filtration permeability for water**. It has the same expression for P_d given in Eqn [2.7.A1.15] but we obtain it experimentally from L_P as

[2.7.A1.23]
$$P_f = L_P\frac{RT}{\overline{V}_w} = \frac{n\pi a^2 D_w}{\delta}$$

Thus comparing Eqn [2.7.A1.15] to Eqn [2.7.A1.23], P_d and P_f should be the same if the mechanism of osmosis is by diffusion.

PRESSURE- AND OSMOSIS-DRIVEN FLOW ACROSS A LIPID BILAYER BY DISSOLUTION–DIFFUSION

Here we consider a membrane that presents all of its area to the solution phase, and water crosses by dissolving in the lipid of the membrane on one side of the membrane, diffusing across the membrane essentially like a vapor, and then coming out of lipid solution back into the aqueous phase on the other side. We imagine that dissolution is rapid (the solution phase and membrane phase are in equilibrium) and that diffusion is comparatively slow. Equilibrium of water in the solution phase with water in the membrane phase is described by equating the chemical potential of water in the two phases:

$$\mu_{w(solution)}^0 + RT \ln X_{w(solution)} + \overline{V}_{w(solution)}P$$
$$= \mu_{w(membrane)}^0 + RT \ln X_{w(membrane)} + \overline{V}_{w(membrane)}P$$

[2.7.A1.24]

where $X_{w(solution)}$ and $X_{w(membrane)}$ are the mole fractions of water in the solution in equilibrium with the membrane and in the membrane phase, respectively. The partition coefficient is defined as

[2.7.A1.25] $$K_w = \frac{X_{w(\text{membrane})}}{X_{w(\text{solution})}}$$

Consider the case where only osmotic pressure drives water flow and the hydrostatic pressure difference across the membrane is zero. Since water generally partitions poorly into hydrocarbon solvents, we may assume that the mole fraction of water in the membrane phase is low. Thus the water concentration is dilute, and we may replace the mole fraction of water with its concentration:

[2.7.A1.26] $$C_{W(\text{membrane})} \approx \frac{X_{w(\text{membrane})}}{\overline{V}_{\text{lipid}}}$$

If equilibrium is rapid, we can combine Eqn [2.7.A1.25] and Eqn [2.7.A1.26] to get

[2.7.A1.27] $$C_{w(\text{membrane})} \approx K_w \frac{X_{w(\text{solution})}}{\overline{V}_{\text{lipid}}}$$

For dilute solutions, $X_{W(\text{solution})}$ can be replaced by $1 - \overline{V}_w C_S$ where C_S is the solute concentration:

[2.7.A1.28] $$C_{w(\text{membrane})} \approx K_w \frac{1 - \overline{V}_w C_S}{\overline{V}_{\text{lipid}}}$$

The concentration of water immediately inside the left-hand side of the membrane is given by Eqn [2.7.A1.28] where C_S is the concentration of solute in the left compartment. A similar expression pertains to the water concentration immediately inside on the right side. The net diffusive flux of water across the membrane is given as

[2.7.A1.29] $$J_w = -D_w^m \frac{C_{w,L} - C_{w,R}}{0 - \delta}$$

where D_w^m is the diffusion coefficient of water in the membrane phase. Substituting in for the concentrations from Eqn [2.7.A1.28], we get

[2.7.A1.30] $$J_w = -\frac{D_w^m K_w \overline{V}_w (C_{S,L} - C_{S,R})}{\overline{V}_{\text{lipid}} \delta} = -\frac{D_w^m K_w \overline{V}_w \Delta C_S}{\overline{V}_{\text{lipid}} \delta}$$

The flux given here is in mole per unit area per unit time. To convert to J_V, we multiply by V_w:

[2.7.A1.31] $$J_V = \overline{V}_w J_w = -\frac{D_w^m K_w \overline{V}_w^2 \Delta C_S}{\overline{V}_{\text{lipid}} \delta} = -\frac{D_w^m K_w \overline{V}_w^2}{\overline{V}_{\text{lipid}} RT \delta} RT \Delta C_S$$

The last term on the far right is the osmotic pressure difference, $\Delta \pi$. Equation [2.7.A1.31] relates the volume flux to the osmotic pressure when the mechanism of water flow is dissolution and diffusion. We recognize it as a form of the phenomenological equation, $J_V = -L_P \Delta \pi$ and therefore we identify L_P as

[2.7.A1.32] $$L_P = \frac{D_w^m K_w \overline{V}_w^2}{\overline{V}_{\text{lipid}} RT \delta}$$

In Eqn [2.7.A1.23], we described P_f, the filtration water permeability, as

[2.7.A1.23] $$P_f = L_P \frac{RT}{\overline{V}_w}$$

Applying this result to Eqn [2.7.A1.32], we obtain

[2.7.A1.33] $$P_f = \frac{D_w^m K_w \overline{V}_w}{\overline{V}_{\text{lipid}} \delta}$$

This equation was derived for the situation in which there was an osmotic gradient ($\Delta C_S > 0$) in the absence of a hydrostatic pressure gradient ($\Delta P = 0$). The expression for the situation where $\Delta C_S = 0$ and $\Delta P > 0$ can be obtained by returning to Eqn [2.7.24] and setting the mole fractions of water to 1.0 while the pressures differ. The result is that exactly the same L_P is derived for pressure-driven flow as for osmotic flow when the mechanism is by rapid dissolution of water followed by slow diffusion of water through the membrane phase.

DIFFUSIONAL PERMEABILITY BY THE DISSOLUTION–DIFFUSION MECHANISM: P_D

The permeability of lipid membranes to a diffusional water flux is expressed by the equation:

[2.7.A1.34] $$J_w = P_d \Delta C_w$$

where P_d is the diffusional permeability and ΔC_w is the difference in water concentration across the membrane. The overall permeation of water takes three steps: dissolution into the membrane phase at the left interface, diffusion across the membrane phase, and reversal of the dissolution at the right interface. If we assume, as we did in the derivation of P_f, that the rate-limiting step is diffusion through the membrane phase, then we may rewrite Eqn [2.7.A1.34] as

[2.7.A1.35] $$J_w = D_w^m \frac{\Delta C_{w(\text{membrane})}}{\delta}$$

Substituting in for $\Delta C_{w(\text{membrane})}$ from Eqn [2.7.A1.27], this is

[2.7.A1.36] $$J_w = D_w^m \frac{K_w \Delta X_{w(\text{solution})}}{\overline{V}_{\text{lipid}} \delta}$$

since $\Delta X_w = V_w \Delta C_w$, this becomes

[2.7.A1.37] $$J_w = D_w^m \frac{K_w \overline{V}_w}{\overline{V}_{\text{lipid}} \delta} \Delta C_w$$

and we can identify P_d by comparing Eqns [2.7.A1.37] and [2.7.A1.34] as

[2.7.A1.38] $$P_d = \frac{D_w^m K_w \overline{V}_w}{\overline{V}_{\text{lipid}} \delta}$$

Comparing the results for P_f (see Eqn [2.7.A1.33]) and for P_d (see Eqn [2.7.A1.38]), the dissolution–diffusion mechanism of water transport indicates that $P_f/P_d = 1.0$.

If we measure the diffusional permeability of a lipid membrane and L_P, and calculate P_f from the L_P according to Eqn [2.7.A1.23], we should expect them to be equal provided that the mechanism is dissolution and diffusion. Our earlier results with microporous membranes also concluded that $P_f/P_d = 1.0$ if the mechanism of transport was by diffusion.

EXPERIMENTS CONFIRM $P_F/P_D = 1.0$ FOR LIPID BILAYERS BUT $P_F/P_D \gg 1.0$ FOR MOST BIOLOGICAL MEMBRANES

The measurement of P_f (from L_P) and P_d in planar lipid bilayers confirm that they are the same, indicating that water transfer across pure lipid bilayer membranes is by diffusion. However, in most biological membranes, P_f/P_d is much greater than 1.0, suggesting that the mechanism is not diffusional, but hydrodynamic. The erythrocyte membrane was used in many of these studies, and it was shown that compounds that interact with protein sulfhydryl groups markedly reduced P_f while leaving P_d relatively unchanged, and causing the ratio of P_f/P_d to become 1.0. These early studies pointed to the existence of proteinaceous pores in the erythrocyte membrane, and subsequently these were identified as AQP1, the first in the family of aquaporins. Peter Agre earned the Nobel Prize in 2003 for his discovery of the aquaporins.

AQUAPORINS ACCOUNT FOR MOST WATER TRANSPORT IN CELLS

Aquaporins are a family of proteins, all about 28 kDa, that are found in a variety of tissues that have high water transport rates such as the intestines and kidneys, salivary glands, pancreas, and more. There are 13 human varieties, labeled AQP0, AQP1–12. The proteins all have six transmembrane domains and associate to form a functional tetramer, although each part has its own water channel. Although water can permeate through lipid membranes, this pathway is much slower than the AQP-mediated pathway. Thus real biological membranes are a mosaic of lipid pathways for diffusional transport of water and pores for pressure-driven transport of water.

Problem Set 2.2
Membrane Transport

1. The GLUT-1 transporter has a K_m for glucose of 1.5 mM. The normal, resting plasma glucose concentration is about 90 mg%. This is a clinical unit that is not part of the ISI but you have to get familiar with it anyway. X mg% means that the solute has that many mg in a deciliter of plasma. One deciliter is 0.1 L = 100 mL. The molecular weight of glucose is 180 g mol^{-1}. From this information, calculate the rate of glucose transport by GLUT-1, as a percent of its maximum, when exposed to normal plasma. Assume saturation kinetics.

2. When C_L was 2.3×10^{-6} M and C_R was zero, the flux across a microporous membrane was found to be 0.234 pmol cm^{-2} s^{-1}. The free diffusion coefficient of the material being measured was 0.8×10^{-5} cm^2 s^{-1}.
   A. What is the permeability of the membrane to the material?
   B. If the thickness of the membrane is 10×10^{-6} m, what is the equivalent relative area available for diffusion of the material?

3. Two membranes, A and B, have permeabilities P_A and P_B for a given solute. These two membranes are joined together to form a single composite, two-layered membrane, as shown in Figure 2.PS2.1. Examples are the successive filtration/diffusion barriers in the kidney glomerulus, successive permeability barriers in the lung, and successive permeability barriers of unstirred layers adjacent to intestinal epithelial membranes.

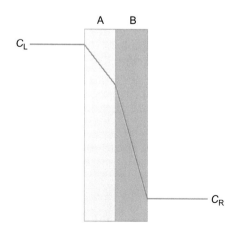

FIGURE 2.PS2.1 Composite membrane formed from the sandwich of two membranes, A and B, with different characteristics.

   A. How would you define the overall permeability P of the two-layered membrane? *Hint*: Think about how you would define permeability for any membrane. Think about what you need to know to calculate P. Use C_L for the left concentration and C_R for the concentration on the right.
   B. Why is the steady-state solute flux through the composite membrane the same through the membrane A layer and the membrane B layer? *Hint*: Think of the continuity equation and what it means.
   C. What is the concentration profile through the composite membrane? *Hint*: Calculate the concentration at the interface of membranes A and B; call it C_m, in terms of C_L and C_R. *Hint*: Equate the flux through the two membranes.
   D. Find an expression for P in terms of P_A and P_B.
   E. Do the permeabilities act like inverse resistances in a series arrangement?

4. When C_L was 2.3×10^{-6} M and C_R was zero, the flux across a microporous membrane was found to be 0.234 pmol cm^{-2} s^{-1}. This flux was determined with vigorous stirring, which virtually eliminated any unstirred layers adjacent to the membrane. When the stirrer was turned off, the flux decreased to 0.157 pmol cm^{-2} s^{-1}. The free diffusion coefficient of the material being measured was 0.8×10^{-5} cm^2 s^{-1}.
   A. What is the permeability of the membrane plus unstirred layer?
   B. What is the permeability of the membrane alone?
   C. Using the information in Problem #3, what is the thickness of the unstirred layer? (Assume that the diffusion coefficient in the unstirred layer is equal to the free diffusion coefficient.)

5. The surface area of the lungs is about 75 m^2, and the thickness of the alveolar diffusion layer is about 0.5 μm. The P_{O_2} in alveolar air is 100 mmHg, while the P_{O_2} of venous blood is 40 mmHg. The diffusion coefficient of O_2 in water is about 1.5×10^{-5} cm^2 s^{-1}. The solubility of O_2 is given by Henry's Law as $[O_2] = 0.024\ P_{O_2}$; here $[O_2]$ is expressed in mL O_2 at STPD (standard temperature and pressure, dry: 0°C and 1 atm pressure) per mL of water and P_{O_2} is in

atmospheres. Only the dissolved O_2 diffuses. What is the initial rate of O_2 diffusion from the aggregate alveoli to the blood? (Initial rate means to pretend that the venous P_{O_2} is clamped at 40 mmHg.) Give the answer in mL O_2 per minute and in mol O_2 per minute using the Ideal Gas Law. Assume $37°C$, $R = 0.082$ L atm mol^{-1} K^{-1}. Remember that 1 atm = 760 mmHg.

6. Heart cells contain a Na–Ca exchanger with a stoichiometry of 3Na:1Ca. The following questions pertain to this transporter.
   A. The free energy of transport of Ca^{2+} across the sarcolemma of the heart cell can be calculated from the following conditions during the rest phase of the heart beat:
   $$[Ca^{2+}]_o = 1.2 \times 10^{-3} \text{ M}$$
   $$[Ca^{2+}]_i = 0.1 \times 10^{-8} \text{ M}$$
   $$E_m = -0.085 \text{ V}$$
   Calculate the free energy for the reaction $Ca_{out} \rightarrow Ca_{in}$. Recall that $R = 8.314$ J mol^{-1} K^{-1}, $T = 310$ K, $\Im = 9.649 \times 10^4$ C mol^{-1}. Remember that Ca^{2+} has two electrical charges per atom.
   B. Calculate the free energy for the reaction $Na_{in} \rightarrow Na_{out}$ for the following conditions during the rest phase of the heart:
   $$[Na^+]_o = 145 \times 10^{-3} \text{ M}$$
   $$[Na^+]_i = 14 \times 10^{-3} \text{ M}$$
   $$E_m = -0.085 \text{ V}$$
   C. Which way does the Na–Ca exchange proceed at rest?

7. Ischemia refers to the condition of no blood flow. When the artery perfusing an area of tissue is blocked, oxygen can no longer be delivered to support energy metabolism. Under these conditions, the ATP concentration falls and ADP and Pi concentrations rise. What will this do to the free energy of ATP hydrolysis (also called the **affinity** of ATP hydrolysis)? What do you think will be the consequence of ischemia on the effectiveness of the ion pumps? (Kammermeir, Schmidt, and Jungling, Free energy change of ATP hydrolysis: a causal factor of early hypoxic failure of the myocardium? *J. Mol. Cell. Cardiol.* 14:267–277, 1982).

8. The membrane of the sarcoplasmic reticulum (an internal membrane in muscle cells) has multiple ion channels, so the membrane potential across this membrane is believed to be zero. This membrane has a Ca-ATPase pump that links two Ca^{2+} atoms to the hydrolysis of ATP. Using a free energy of ATP hydrolysis of -57.1 kJ mol^{-1}, what is the thermodynamic limit of Ca^{2+} accumulation if the free $[Ca^{2+}]$ on the cytosolic face is 1×10^{-7} M? (*Hint*: The thermodynamic limit is when the free energy change for the transport reaction is zero.)

9. Each of the reactions shown in Eqn [2.5.15] has a forward and reverse rate constant. Show that for a passive transport process the product of all the forward rate constants is equal to the product of all the reverse rate constants. (*Hint*: Use a principle called **detailed balance**, which states that at equilibrium all steps in a reaction sequence must also be at equilibrium.)

10. Consider that a membrane separates two compartments, each containing a solution of 10 mL. The left compartment has an initial concentration C_L. The membrane has permeability p to the solute and area A.
    A. Derive an expression for C_L and C_R as a function of time.
    B. Suppose that you obtained experimental values for $C_L(t)$ or $C_R(t)$. What plot of the data could you make to determine p?
    C. What would happen if we doubled both the volume and the surface area to the time course of equilibration of C_L and C_R.

11. The surface area of the lungs is about 75 m^2, and the thickness of the alveolar diffusion layer is about 0.5 μm. The P_{O_2} in alveolar air is 100 mmHg, while the P_{O_2} of venous blood is 40 mmHg. The diffusion coefficient of O_2 in water is about 1.5×10^{-5} cm^2 s^{-1}. The solubility of O_2 is given by Henry's Law as $[O_2] = 0.024 P_{O_2}$; here $[O_2]$ is expressed in mL O_2 per mL of water and P_{O_2} is in atmospheres. Only the dissolved O_2 diffuses. The volume of blood in the lung is 70 mL, and the volume of air at the end of normal inspiration volume is about 2.8 L (at body temperature of $37°C$).
    A. Derive an equation for the time course of oxygen equilibration between blood and air in the lungs.
    B. Using the values given here, estimate the half-time of equilibration.
    C. Most of the oxygen in the blood is not free but is bound to hemoglobin within the red blood cells. Do you think this would accelerate or decelerate the rate of equilibration of blood and air oxygen pressures?

12. Vesicles of the sarcoplasmic reticulum have embedded in their membrane an active Ca-ATPase pump. Several different isoforms of this primary active pump are expressed in different tissues. When exposed to ATP, Mg, and Ca, Ca^{2+} ions are accumulated and eventually reach a steady-state uptake. If the pump is quickly quenched by adding extravesicular EGTA, which complexes activator Ca^{2+} and thereby stops the pump, the accumulated Ca^{2+} will leak back out. Monitoring the intravesicular Ca^{2+} with time allows one to estimate the permeability of the vesicles. Derive an expression for the amount of Ca^{2+} remaining in the vesicle as a function of time and suggest a plot to determine the permeability. What other information might you need to know to determine the permeability?

13. The osmotic coefficient for $CaCl_2$ under physiological conditions is 0.85. Calculate the osmotic pressure of a solution of 10 mM $CaCl_2$ at $37°C$.

Recall here that $R = 0.082$ L atm mol^{-1} K^{-1}. Give the answer in both atm and mmHg (1 atm = 760 mmHg).

14. The kidney filters plasma to produce an **ultrafiltrate**, which is the first step in the formation of urine. This filtrate is called an ultrafiltrate because the kidney can retain even small particles like plasma proteins. The force behind ultrafiltration is the blood pressure. The kidney ultrafiltration occurs at a structure called the **glomerulus**, which is a group of small blood vessels (capillaries) that are closely joined to another structure, **Bowman's capsule**, that forms a double-walled cup for the collection of the ultrafiltrate. The filtration barrier is a combination of the capillary walls and structure in Bowman's capsule.
    A. Calculate the filtration coefficient (L_p) for the basement membrane of kidney glomeruli using the following approximations:
        1. Pore radius = 35 Å
        2. Pore length = 600 A
        3. Fractional pore area = 5%
        4. Blood plasma viscosity = 0.02 poise (dyne s cm^{-2})

        The "fractional pore" is the total area of the pores divided by the total area of the membrane. A dyne is a g cm s^{-2}.
    B. Calculate the glomerular filtration rate (GFR) assuming a total area for both kidneys of 1.5 m^2 and a driving force $\Delta P = 20$ mmHg. The GFR is the total volume of ultrafiltrate produced per minute. Its units should be in cm^3 min^{-1}. Make it so.

15. The observed osmotic pressure of solutions of plasma proteins increases more rapidly than concentration. Empirical fits to the concentration dependence of osmotic pressure are given by Landis and Pappenheimer (*Handbook of Physiology*, vol 2, section 2, pp. 961–1034, 1963):

$$\pi_{albumin} = 2.8C + 0.18C^2 + 0.012C^3$$

where π is in units of mmHg and C is in units of g% (i.e., g of protein per deciliter of plasma).
    A. According to this equation, as C becomes more dilute the relation approaches van't Hoff's Equation. Keeping in mind the units of the variables, what is the molecular weight of albumin? Assume that the temperature is 37°C. (By the way, osmotic measurements were the first measurements of protein molecular weights.)
    B. What is the contribution of albumin to the osmotic pressure of plasma when it contains 4.0 g% of albumin?
    C. The osmotic pressure of plasma proteins and associated ions is called the **oncotic pressure**. If the plasma oncotic pressure is 25 mmHg, how much of the oncotic pressure is contributed by globulins, fibrinogen, and other components?

16. Assume that serum albumin is a sphere of diameter 31 Å. Assume that the glomerular membrane is pierced by pores of equivalent diameter of 35 Å.
    A. Give an estimate of σ for albumin for the glomerular membrane.
    B. Calculate the concentration of albumin in the ultrafiltrate.
    C. If the GFR is 120 mL min^{-1}, calculate the daily filtered load of albumin (how much is filtered every day). How does this compare with the recommended dietary intake of 0.8 g protein per kg body weight per day?

17. The value of L_p for the red blood cell is about 1.8×10^{-11} cm^3 dyne^{-1} s^{-1}. Its surface area is about 1.35×10^{-6} cm^2 (Solomon, *Methods in Enzymology*, pp. 192–222, 1989).
    A. What is the initial osmotic flow if the osmolarity inside is initially 300 mOs M and the osmolarity outside is 275 mOs M? (Assume $\sigma = 1.0$ for all solutes.)
    B. If the volume of the cell is 100×10^{-12} cm^3, how long would it take to double its volume provided that the osmotic pressure and area of the membrane and L_p did not change?
    C. In the case described, how much water would be required to enter the cell to equilibrate the osmotic pressure between inside and outside? Assume that the outside bath is essentially infinite so that its osmotic pressure is kept constant.

18. Osmotic pressure is one of a class of properties of solutions that are called **colligative properties**. The others in this class include **vapor pressure depression**, **boiling point elevation**, and **freezing point depression**. These properties are different expressions of the same phenomenon: the lowering of the activity of water by dissolution of solutes. Various osmometers have been made using one or another or these properties. Table 2.PS2.1 shows several solutions of sucrose and glucose, their water concentrations,

TABLE 2.PS2.1 Solute Concentration, Water Concentration, and Freezing Point Depression in Sucrose and Glucose Solutions

Sucrose Solutions			Glucose Solutions		
[Sucrose] (M)	[Water] (M)	Δ (°C)	[Glucose] (M)	[Water] (M)	Δ (°C)
0	55.45	0	0	55.46	0
0.029	55.12	0.06	0.028	55.28	0.05
0.059	54.77	0.11	0.056	55.11	0.10
0.081	54.42	0.16	0.084	54.94	0.16
0.118	54.07	0.23	0.112	54.76	0.21
0.179	53.72	0.35	0.140	54.59	0.27

Source: Data from Handbook of Chemistry and Physics, CRC Co, Cleveland, OH, 1965.

and their freezing point depression. Because dissolving solutes invariably dilutes the solvent, water, you can see from the table that the water concentration also decreases with increasing solute concentration. Sucrose, however, is twice as large as glucose, so we might expect that dissolution of sucrose would dilute the water further. Therefore, the colligative properties of solutions cannot be proportional to both solute and solvent concentration. Plot separately the freezing point depression against the water concentration and against the solute concentration. Which relationship shows the same dependence? From this result, do you expect osmosis to be dependent on solvent water concentration or on solute concentration?

19. From the data in Table 2.PS2.1, calculate the coefficient relating freezing point depression to molarity in the equation:

[2.PS2.1] $$T_f - T = \Delta = k_f C$$

where T_f is the freezing point of the pure liquid, T is the freezing point of the solution, k_f is the coefficient, and C is the concentration.

20. Chapter 2.7 gives us several expressions for the osmotic pressure. The most complete is derived from Eqn [2.7.6] and is

[2.PS2.2] $$P_L - P_R = \pi = \frac{RT}{\overline{V}_w} \ln \frac{a_{\text{pure water}}}{a_{\text{solution}}}$$

Here the subscripts L and R refer to the left and right sides of the semipermeable membrane, respectively, where pure water is on the right side. If the solution was ideal, we can use the mole fraction for the activity and this equation becomes

[2.PS2.3] $$\pi = -\frac{RT}{\overline{V}_w} \ln X_w$$

In the further approximation of a dilute solution, this equation was transformed to

[2.PS2.4] $$\pi = RT\, C_s$$

The activity of water in Eqn [2.PS2.2] is measured by the vapor pressure. Table 2.PS2.2 tabulates the calculations of the osmotic pressure from the vapor pressure measurements, the mole fraction of water and from the concentration of solute. Plot the ratio of each of the observed osmotic pressures to the calculated osmotic pressure, in columns 3, 4, and 5, against the sucrose concentration. The values you calculate for the ratio of the observed π to the calculated π in column 3 is $\varphi\,(=\pi/RT \ln C_s)$, the osmotic coefficient. The ratio of the observed π to column 4 (i.e., $(\pi/[-RT/\overline{V}_w \ln x_w])$ defines the **rational osmotic coefficient**, g.

A. Why is the equation using the mole fraction of water a better predictor of the osmotic pressure than the van't Hoff Equation?

TABLE 2.PS2.2 Values for the Observed Osmotic Pressure and the Osmotic Pressure Calculated from the van't Hoff Equation, from the Mole Fraction of Water, and from the Measured Vapor Pressure of Solutions

[Sucrose] (M)	Observed π (atm)	$RT\,C_s$ (atm)	$-RT/\overline{V}_w \ln x_w$ (atm)	$RT/\overline{V}_w \ln (P_0/P_s)$ (atm)
0.098	2.47	2.40	2.48	2.50
0.824	27.2	20.4	24.6	26.6
1.399	58.4	35.1	48.8	57.3
1.823	95.2	45.5	72.6	93.2
2.146	139.0	55.7	96.0	135.6
2.55	187.3	64.5	118.9	186.5

Source: Data from Glasstone, Textbook of Physical Chemistry, van Nostrand, 1946.

TABLE 2.PS2.3 Molecular Weight and Concentration of Plasma Proteins.

Protein	Average Molecular Weight	Concentration in Plasma (g dL^{-1})
Albumin	69,000	4.2
Fibrinogen	330,000	0.3
Immunoglobin G	150,000	0.8
Immunoglobin M	750,000	0.3
α_2 Globulins	100,000	0.7
α_1 Globulins	50,000	0.5
β Globulins	100,000	0.8

B. Why is the equation using the vapor pressure a better predictor of the osmotic pressure than the equation using the mole fraction of water?

C. Are these methods good predictors of the osmotic pressure in dilute solutions?

21. Krebs–Henseleit buffer has the following composition: 119 mM NaCl, 25 mM NaHCO$_3$, 3.2 mM KCl, 1.2 mM MgSO$_4$, 1.2 mM KH$_2$PO$_4$, 11 mM glucose, and 1.4 mM CaCl$_2$. Calculate its osmolarity and osmotic pressure, at 37°C, assuming $\varphi = 1$ for all solutes. Would you expect this solution to be hypertonic, hypotonic, or isotonic to mammalian cells?

22. Plasma contains a variety of proteins that exert osmotic pressure. A list of these, with their molecular weights and approximate protein concentrations, is given in Table 2.PS2.3. From these data, calculate the approximate osmotic pressure in plasma that is due just to plasma proteins.

23. A. The device shown in Figure 2.PS2.2 was used to determine the steady-state flow and pressure across a dialysis membrane. The data

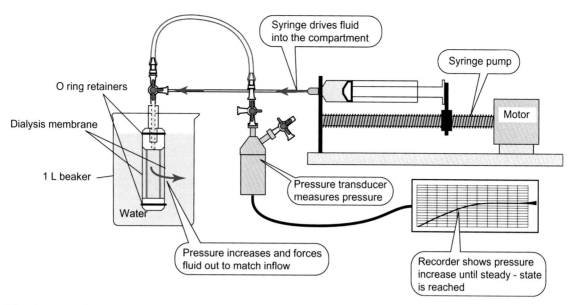

FIGURE 2.PS2.2 Device used to determine L_P. Water was injected into the inner chamber at a known rate using a syringe pump. This increases the pressure within the chamber and forces fluid out through the membrane. The pressure increases as more water is injected and eventually a steady state is reached in which the rate of injection matches the rate of filtration through the membrane. The steady-state pressure is measured continuously by a pressure transducer and recorded.

TABLE 2.PS2.4 Pressure at Steady-State Flow Across a Dialysis Membrane

Flow Rate (cm³ min⁻¹)	Pressure (mmHg)
0	0
0.0097	180
0.0194	360
0.0388	680

that were obtained are given in Table 2.PS2.4. The area of the membrane that was exposed to flow was 90.5 cm². Determine L_P for the membrane.

B. The membrane in part A was used to separate pure water on the outside from a 0.75 M sucrose solution on the inside. The flow across the membrane was measured using the device shown in Figure 2.PS2.3.

24. When vesicles of the cardiac sarcoplasmic reticulum (CSR) are incubated with ATP, Mg^{2+} and Ca^{2+}, they take up Ca^{2+} and reach a pseudo steady state. This is a steady state that changes, but slowly. The uptake of Ca^{2+} is mediated by the SERCA2a Ca-ATPase. The uptake reaction can be quenched by adding EGTA to the external solution, which binds the Ca^{2+} outside of the vesicles, or by adding glucose plus hexokinase, that converts the ATP to ADP and glucose-6 phosphate. When the uptake reaction is stopped, Ca^{2+} that was already taken up by the vesicles leaks out passively.

A. The amount of Ca^{2+} taken up by the vesicles is generally normalized to the amount

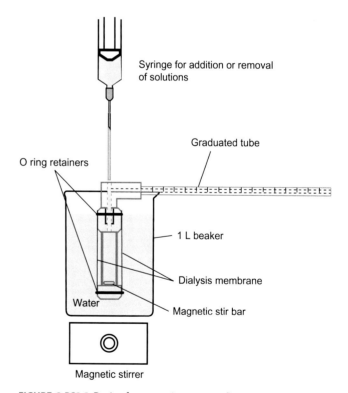

FIGURE 2.PS2.3 Device for measuring osmotic flow at constant $\Delta P = 0$. The inner compartment could be drained and then filled with various experimental solutions. The outer compartment contained pure water. As fluid enters the compartment across the dialysis membrane as a result of the osmotic pressure difference, it forces fluid down the horizontal tube without any increase in the hydrostatic pressure. The rate of fluid flow can be estimated from the rate of fluid movement down the tube.

of CSR protein in mg rather than being expressed as a concentration. A typical steady-state Ca^{2+} uptake is 40 nmol mg⁻¹. In separate experiments, the enclosed volume of the CSR vesicles was determined to be 5 μL mg^{-1}. What is the approximate concentration of Ca^{2+} inside the vesicles at steady state?

B. The average vesicle size determined by electron microscopy is about 150 nm. What is the volume and surface area of a vesicle this size, assuming it is a sphere?

C. Given that the enclosed volume of the aggregate vesicles is 5 μL mg^{-1}, how many vesicles are there per mg of CSR protein? How much surface area is there per mg of CSR protein?

D. The initial passive efflux at a load of 40 nmol mg^{-1} when the pump is stopped is 16 nmol min^{-1} mg^{-1}. Convert this to a flux in units of nmol cm^{-2} s^{-1} by dividing by the surface area per mg of CSR protein and converting min to s.

E. From the information in part D, What is the passive permeability to Ca^{2+} in units of cm s^{-1}?

25. Oral Rehydration Solution used for Oral Rehydration Therapy, as recommended by WHO/UNICEF, has the following composition:
2.6 g NaCl
2.9 g Na$_3$C$_6$H$_5$O$_7$•2 H$_2$O (trisodium citrate dihydrate)
1.5 g KCl
13.5 g glucose
per L of solution.

Calculate the osmolarity of this solution. Trisodium citrate will dissociate fully at neutral pH to 3 Na$^+$ ions and one citrate^{-3} ion. Assume complete dissociation of NaCl and KCl.

Cell Signaling 2.8

Learning Objectives

- Distinguish among autocrine, paracrine, endocrine, neural, and neuroendocrine signaling
- Explain why Ca^{2+} is a special ion with respect to signaling
- Describe in general terms what ligand-gated ion channels do
- List the major steps in turning on and off of heterotrimeric G-protein-coupled receptors
- Distinguish among G_s, G_i, G_q, and $G_{12/13}$ mechanisms
- Describe the steps in a G_s mechanism's activation of PKA and how they differ among tissues
- Describe the steps in a G_q mechanism's activation of CAM kinase
- List the four types of catalytic receptors
- Describe the steps in a JAK—STAT pathway
- Describe in general terms lipophilic signaling molecules' effects on gene transcription

SIGNALING *TRANSDUCES* ONE EVENT INTO ANOTHER

In its broadest context, cell signaling involves the **transduction** of some event into another event. In sensory transduction, a sensory cell is exposed to some external signal that is *transduced* to produce a nervous signal, the action potential. As we will see later in Chapter 3.2, this action potential can move along cell membranes to rapidly convey the signal, the action potential, to remote parts of the sensory neuron. The action potential is then *transduced* to release neurotransmitter at the synapse—the gap between one neuron and another. The neurotransmitter is then *transduced* to form the response of the postsynaptic cell, the one on the other side of the synapse. In the case of cutaneous (skin) senses, the original sensory signal is mechanical—a push or a pull on the nerves in the skin. The mechanical signal is transduced to an electrical signal, and the electrical signal is then transduced to a chemical signal. This simple series of events illustrates the use of **mechanical**, **electrical**, and **chemical** signals in the body (see Figure 2.8.1).

CELL-TO-CELL COMMUNICATION CAN ALSO USE DIRECT MECHANICAL, ELECTRICAL, OR CHEMICAL SIGNALS

Mechanical signals can originate in the external environment, as in the case of sensory transduction, or they can be the signals from another cell. Mechanical signaling requires close contact of cells and generally occurs through cell junctions as discussed in Chapter 2.2. Mechanical force originates on filaments within cells that eventually connect to the extracellular matrix through cytoskeletal elements. Transmission of these forces occurs through the extracellular matrix, but it is also sensed by neighboring cells. All of the pressure sensors in the body are really stretch receptors, in which mechanical stretch is transduced into electrical signals or chemical signals.

Electrical signals are usually used within the cell, as part of a signaling pathway to communicate intracellularly, and most often to move the signal rapidly from one place in the cell to another. Less frequently, direct electrical coupling occurs between cells. Such electrical coupling uses gap junctions, whose structure was discussed in Chapter 2.1. This mechanism is vitally important in coordinating some smooth muscle contraction and cardiac contraction.

Chemical signals do not require close contact and can be classified according to the distances involved, the mechanism of transmission, and the target of the chemical signals. These various classes of inter- and intracellular communication are shown in Figure 2.8.2. In some cases the signaling molecule remains bound to the cell and so transmission of this signal requires contact between the signaling cell and its target. This **contact-dependent signaling** is important in development and in immune responses. In other cases, the signaling cell releases a chemical that either acts locally (a **paracrine** or **autocrine** signal) or travels through the blood to act on remote targets (an **endocrine** signal). Autocrine signals have receptors on the signaling cell itself or others like it. Paracrine signals affect other types of cells located in the neighborhood of the signaling cell. Neurons also release signaling molecules, usually at the end of a long extension of the cell, the axon. When the neural chemical signal enters the blood and acts on distant targets, it is called a **neuroendocrine** signal. When it acts in the vicinity of its release site, it is a **neurotransmitter**.

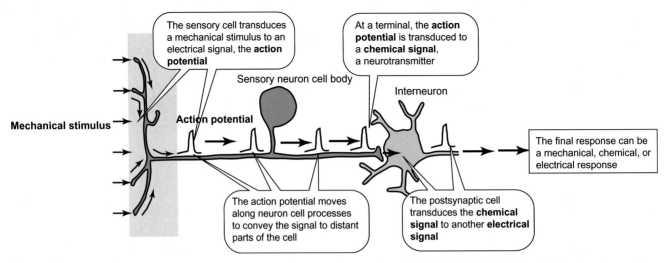

FIGURE 2.8.1 Transduction of signals. Some kinds of sensory cells can transduce mechanical stimuli to electrical signals which can be conveyed along their surface for rapid spatial relay of the signal. At the end of the cell, the electrical signal is transduced to a chemical signal to convey the signal across the gap between the cells. The postsynaptic cell transduces this chemical signal back to an electrical signal.

SIGNALS ELICIT A VARIETY OF CLASSES OF CELLULAR RESPONSES

Intra- and intercellular signals begin a cascade of events that eventually changes cell behavior. The response of cells to signaling events includes altered:

- ion transport;
- metabolism;
- gene expression or differentiation;
- shape, movement, or force production;
- cell growth or cell division;
- apoptosis or programmed cell death.

ELECTRICAL SIGNALS AND NEUROTRANSMITTERS ARE FASTEST; ENDOCRINE SIGNALS ARE SLOWEST

The speed of response to an initial stimulus depends on the mode of delivery of the signal and the mechanism of the response in the target cells. Electrical signals are the fastest way to transmit a signal from one place in the body to another, in milliseconds, but the overall response depends on what happens in the target cell. If the response involves changes in activity of proteins already present in the target cell, the response can be rapid. If the response involves altered gene expression that requires synthesis of new protein, response can take hours. If it involves altered cell growth, it can take days to complete. Neurotransmitter signaling is the fastest response, followed by changes in cell shape or the development of force. Endocrine signals are slowest but last longer.

VOLTAGE-GATED ION CHANNELS CONVEY ELECTRICAL SIGNALS

Ion channels allow ions to cross biological membranes that they otherwise cannot penetrate. Because they are electrically charged, movement of ions makes a current, and the current either charges or discharges the membrane—the current alters the membrane potential. Thus the voltage-gated ion channels are largely responsible for alterations in the membrane potential that is rapidly conveyed through the cell.

VOLTAGE-GATED Ca^{2+} CHANNELS TRANSDUCE AN ELECTRICAL SIGNAL TO AN INTRACELLULAR Ca^{2+} SIGNAL

Cells maintain a very low intracellular $[Ca^{2+}]$ (<100 nM) to avoid Ca^{2+} precipitation with phosphate and organic phosphates (ATP, etc.) present in high concentrations in the cytoplasm. The low cytoplasmic $[Ca^{2+}]$ allows increases in cytoplasmic $[Ca^{2+}]$ to be used as a signal. Multiple types of voltage-gated Ca^{2+} channels (voltage-dependent calcium channel, VDCCs) reside on the surface membrane of many cells. Depolarization of the cell membrane opens these channels, causing Ca^{2+} to move from the ECF, with 1.2 mM $[Ca^{2+}]$, to the intracellular compartment. The cytoplasm contains a number of proteins that bind Ca^{2+} with high affinity and that change shape or activity upon Ca^{2+} binding. The effects of increasing cytoplasmic $[Ca^{2+}]$ include the following:

1. **Stimulus—secretion coupling**: The increased $[Ca^{2+}]$ binds to Ca^{2+} sensors on vesicles, causing the fusion of the secretory vesicles with the plasma membrane and release of secreted products into the ECF.
2. **Excitation—contraction coupling**: The increased $[Ca^{2+}]$ binds to Ca-sensitive elements on contractile filaments or cytoskeletal elements, causing either force development or shortening by muscle cells.
3. **Calmodulin-dependent activation of enzymes**: Calmodulin is a small cytosolic protein that binds four Ca^{2+} molecules and then activates many enzymes such as myosin light chain kinase in smooth muscle.

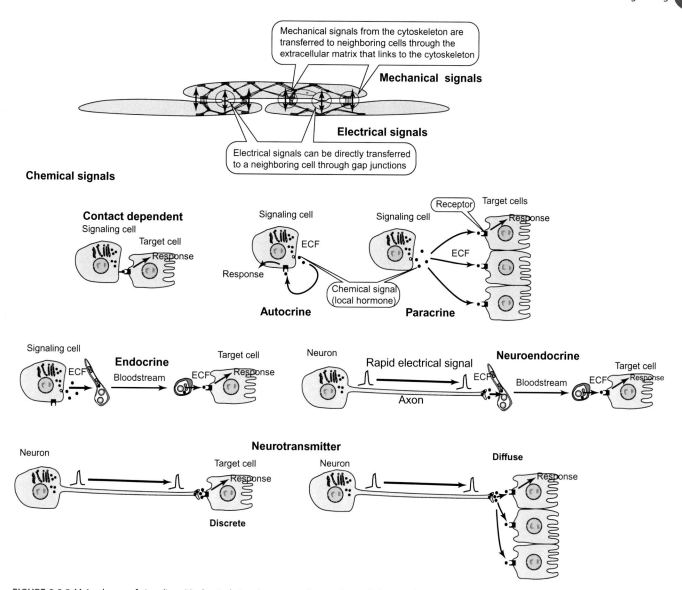

FIGURE 2.8.2 Main classes of signaling. Mechanical signals can pass from cell to cell through filaments in the extracellular matrix attached to membrane-bound proteins in the surfaces of cells, particularly at cell junctions such as desmosomes. Electrical signals can also pass directly from one cell to another through gap junctions. Some signaling molecules remain bound to the surface and so the signal affects the target cell only by direct contact of the signaling cell with the target cell. Cells can release chemical signals that act locally. When they affect the signaling cell, or others like it, they are autocrine signals. When they affect other nearby cells, they are paracrine signals. Signals that are released into the bloodstream to affect distant target cells are endocrine signals. If they are released from long processes by neurons, they are neuroendocrine signals. Nerve cells release a variety of chemical signals at terminals near target cells. These are neurotransmitters. If they are released very close to clustered receptors on target cells, they are discrete neurotransmitters. If they are released into a general area to affect multiple cells, they are diffuse neurotransmitters. Electrical signals are most often used intracellularly to rapidly convey the action potential from one part of the cell to another. This is the fastest movement of a signal in the body.

4. **Direct activation of enzymes**: Ca^{2+} can directly bind to some enzymes, such as PKC, and activate them. Figure 2.8.3 illustrates these aspects of Ca^{2+} signaling in cells.

LIGAND-GATED ION CHANNELS OPEN UPON BINDING WITH CHEMICAL SIGNALS

Fast release of chemical signals by electrical signals in nerve terminals, followed by an electrical signal in the target cell, is the fastest mechanism of signaling used in the body. This is the classic neurotransmitter mechanism: the electrical signal (the action potential) on the presynaptic cell propagates down the axon to the nerve terminal where it causes a local intracellular Ca^{2+} signal that releases chemical neurotransmitter into the ECF immediately adjacent to the postsynaptic cell. The chemical signal then binds to a receptor which is also an ion channel, causing a flux of ions across the postsynaptic cell membrane and a resulting electrical signal in the postsynaptic cell. Three major classes of surface membrane, ligand-gated ion channels (LGICs) have been identified, as shown in Figure 2.8.4, and many of these have multiple subtypes. The major types are distinguished by their structures. Receptors for acetylcholine, serotonin (=5 hydroxy tryptamine), gamma amino butyric acid (GABA), and glycine all consist of five subunits. Each family, such as **nicotinic ACh receptor**, has

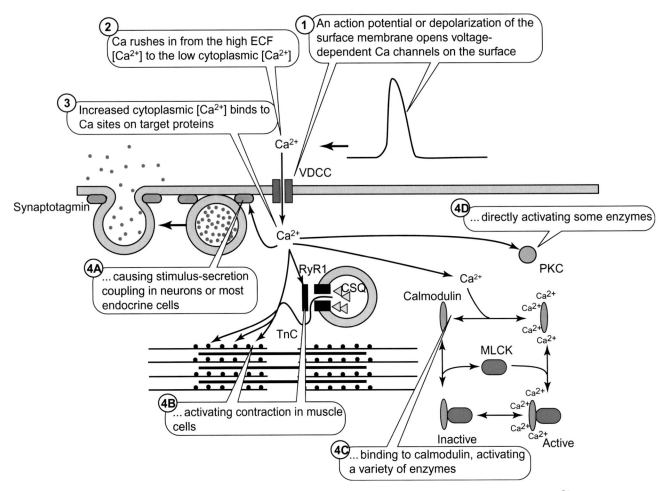

FIGURE 2.8.3 Electrically coupled calcium signaling. Depolarization of the cell membrane opens a calcium channel that lets Ca^{2+} into the cell. At rest, the cytoplasmic $[Ca^{2+}]$ is very low. Upon stimulation, influx of Ca^{2+} raises the $[Ca^{2+}]$ enough for Ca^{2+} to bind to Ca^{2+}-binding sites on specific proteins. Ca^{2+} binding to synaptotagmin causes fusion of secretory vesicles with the plasma membrane. Binding to troponin C (TnC) activates force in skeletal or cardiac muscle. Ca^{2+} binding to calmodulin activates a number of enzymes such as MLCK (myosin light chain kinase) involved in smooth muscle contraction. In other cases, Ca^{2+} directly activates some enzymes; PKC is shown. RyR1 is the ryanodine receptor on the endoplasmic reticulum membrane; CSQ is calsequestrin, a calcium-binding protein in the lumen of some ER membranes.

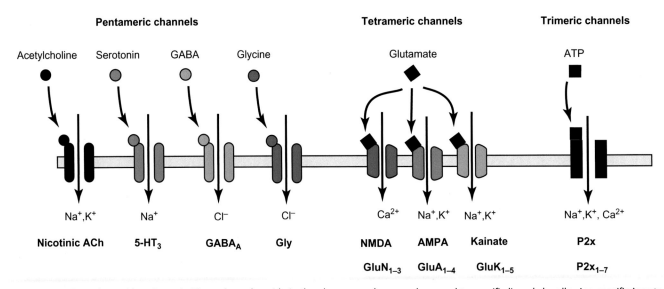

FIGURE 2.8.4 Ligand-gated ion channels. These channels reside in the plasma membrane and respond to specific ligands by allowing specific ions to cross the membrane. The channels are classified according to their structure and agonist or chemical signal that opens the channel. The names of the channels are at the bottom of the figure, and alternate naming conventions have been proposed. Each family of channels has multiple isoforms that depend on the subunit make-up of the channels.

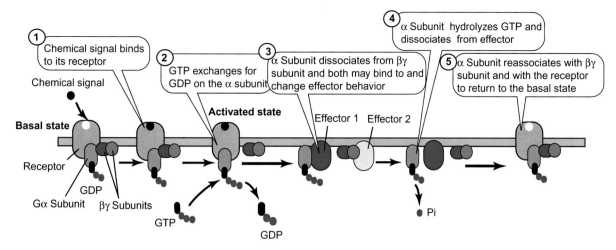

FIGURE 2.8.5 General scheme for heterotrimeric GPCRs. The receptors are membrane-bound proteins that bind chemical signals. The heterotrimeric G-protein consists of an α subunit that binds and hydrolyzes GTP and a βγ subunit that does not dissociate. Binding of the ligand to its receptor triggers the exchange of GTP for GDP and subsequent dissociation of the α subunit and βγ subunit and both are then able to alter the behavior of effector targets in the cell. The α subunit spontaneously hydrolyzes its GTP, and the α subunit reassociates with the βγ subunit to return to the basal, unstimulated state.

multiple subtypes consisting of different subunits, but each member of the subtype responds to one chemical signal, acetylcholine in this case. Receptors for glutamate each have four subunits, and this family of LGIC has further subtypes distinguished by artificial agonists (stimulators of the receptor), NMDA (N-methyl D-aspartic acid), AMPA (α-amino-3-hydroxy-5-methyl-4 isoxazole propionic acid), and kainate.

HETEROTRIMERIC G-PROTEIN-COUPLED RECEPTORS (GPCRS) ARE VERSATILE

G-protein-coupled signaling pathways are versatile because of their modular structure: they consist of **receptors**, **heterotrimeric G-proteins**, and **effectors**. Receptors are membrane-bound proteins that bind signaling molecules on the external surface of cells. Binding alters the conformation of the receptor, and this change is transferred to a heterotrimeric G-protein, consisting of α, β, and γ subunits. In the unstimulated state, the Gα subunit binds GDP. Upon ligand binding to its receptor, the receptor causes the Gα·GDP to exchange GTP for the bound GDP, and the Gα·GTP dissociates from the βγ subunits. The dissociated subunits can then bind to effector molecules, exerting some change in their behavior. The Gα·GTP has inherent GTPase activity, so it reverts back to Gα·GDP and reassociates with the βγ subunit. The overall plan of the G-protein signaling pathway is shown in Figure 2.8.5.

THERE ARE FOUR CLASSES OF G-PROTEINS: $G\alpha_S$, $G\alpha_I/G\alpha_O$, $G\alpha_Q$, AND $G\alpha_{12}/G\alpha_{13}$

β ADRENERGIC STIMULATION IS AN EXAMPLE OF A $G\alpha_S$ MECHANISM

A number of signaling molecules can bind to a receptor that is coupled to a $G\alpha_s$-protein. All $G\alpha_s$-protein's α subunit binds to a membrane-bound enzyme, **adenylyl cyclase**, which converts ATP into 3′,5′ cyclic AMP, or cAMP. Increasing the concentration of cAMP activates **protein kinase A (PKA)**, which phosphorylates (adds a phosphate to) specific target proteins. Phosphorylation of these target proteins alters cell behavior, the final consequence of exposure to the chemical signal. The basic pathway for this signaling mechanism is shown in Figure 2.8.6. This signal is turned on by activation of adenylyl cyclase and then PK and then phosphorylation of proteins. It is turned off by the simultaneous inactivation of all three of these. The α subunit of the G-protein spontaneously hydrolyzes its GTP and reassociates with the βγ subunit. This inactivates adenylyl cyclase and stops its synthesis of cAMP. The cAMP is also broken down by **phosphodiesterase (PDE)** so that the increased [cAMP] is removed. The cAMP dissociates from the regulatory subunits of the PKA, causing it to inactivate so that target proteins are no longer phosphorylated. However, this does not reverse phosphorylation of proteins. The phosphorylated proteins are dephosphorylated by specific enzymes called protein phosphatases. There are four classes of serine/threonine **phosphoprotein phosphatases**: **PP1**, **PP2a**, **PP2b**, and **PP2c**, which dephosphorylate proteins phosphorylated at serine and threonine residues. Fine control of this system is provided by the regulation of both adenylyl cyclase, as described, and by the regulation of both PDE and the protein phosphatases.

α_2 ADRENERGIC STIMULATION IS AN EXAMPLE OF A $G\alpha_I/G\alpha_O$ MECHANISM

Other molecules bind to their G-protein-coupled receptor (GPCR) and release a Gα subunit that inhibits adenylyl cyclase. These are referred to as G_i mechanisms. An example is epinephrine binding to α_2 receptors on neurons and is illustrated in Figure 2.8.7. Other members of this class achieve the same end—reduction in cAMP levels—by activating PDE. Retinal photoreceptor

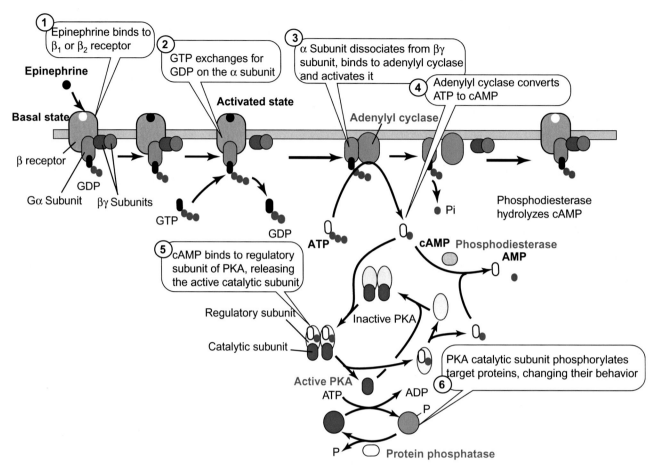

FIGURE 2.8.6 Mechanism of G_s-coupled receptors. Many different kinds of ligands bind to G_s-coupled receptors. Epinephrine is shown, which binds to β_1 and β_2 receptors that are coupled to G_s proteins. Activation follows the general scheme shown in Figure 2.8.5. Here the $G\alpha_s$ subunit binds to adenylyl cyclase, activating it and increasing the cytoplasmic concentration of cAMP. This activates PKA that phosphorylates a variety of target proteins. The signal is turned off by (1) hydrolysis of GTP by the $G\alpha$ subunit and dissociation and removal of activation of adenylyl cyclase; (2) hydrolysis of cAMP by PDE and removal of activation of PKA; (3) dephosphorylation of target proteins by protein phosphatases.

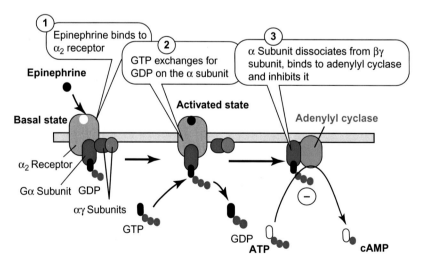

FIGURE 2.8.7 Mechanism of G_i-coupled receptors. Here epinephrine binds to α_2 receptors, which is followed by the inhibition of adenylyl cyclase and a reduction in cytoplasmic [cAMP].

cells, for example, activate cGMP PDE. This mechanism is illustrated in Figure 2.8.8. Still other ligands, such as acetylcholine, bind to M2 receptors that cause inhibition of adenylyl cyclase, and the $\beta\gamma$ subunit activates a K^+ channel (see Figure 2.8.9). Thus this class of GPCR exerts a variety of effects including direct inhibition of adenylyl cyclase, activation of PDE, and direct activation of K^+ channels.

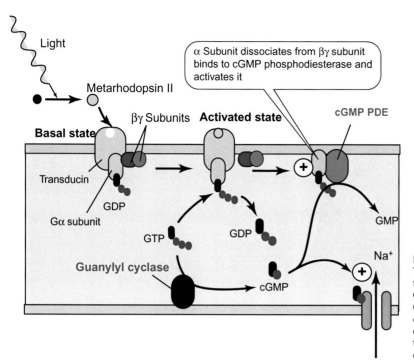

FIGURE 2.8.8 GPCR involved in retinal signal transduction. Through a series of steps, light converts rhodopsin to metarhodopsin II, which binds to a heterotrimeric G-protein called transducin. The Gα subunit exchanges GTP for GDP, dissociates from the βγ subunit, and activates cGMP PDE. Guanylyl cyclase in these cells makes cGMP continuously, and cGMP opens a Na$^+$ channel. Degrading the cGMP reduces [cGMP] and therefore regulates the open state of the channel, producing an electrical signal.

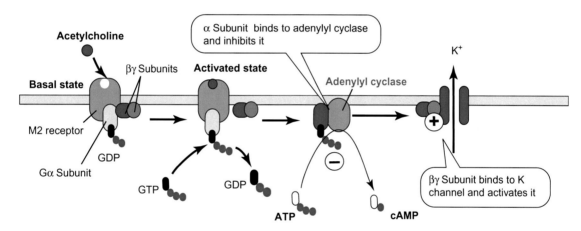

FIGURE 2.8.9 G$_i$ mechanism involved in M2 GPCR. Acetylcholine binds to a variety of receptors. The M2 receptor, present in heart, couples binding of acetylcholine to a Gα$_i$ subunit that inhibits adenylyl cyclase. The βγ subunit released by acetylcholine binding directly activates a K$^+$ channel.

Gα$_Q$/Gα$_{11}$ GPCR ACTIVATES PHOSPHOLIPASE C AND RELEASES CA FROM INTRACELLULAR STORES

A third class of GPCR activates **phospholipase C** on the surface membrane, which cleaves phosphatidyl inositol bisphosphate to produce **diacylglycerol (DAG)** and **inositol triphosphate (IP3)**. The IP3 releases Ca^{2+} from the endoplasmic reticulum, while the DAG activates **protein kinase C (PKC)** (see Figure 2.8.10).

Gα$_{12}$/Gα$_{13}$-COUPLED RECEPTORS ACTIVATE SMALL MONOMERIC GTPASES

The Gα$_{12}$ is the last of the four major families of heterotrimeric G-proteins (G$_s$, G$_i$, G$_q$, and G$_{12}$) that we will discuss. Gα$_{12}$ and Gα$_{13}$ are linked to **GTP exchange factors (GEFs)** that activate small monomeric G-proteins by exchanging their bound GDP with GTP. A second set of modulatory proteins, called **GAPs**, for **GTPase Activating Proteins**, facilitate the inactivation of these small monomeric GTPases. The overall plan is shown in Figure 2.8.11.

The superfamily of small monomeric GTPases is divided into several major subfamilies, including:

Ras, Rho, Rab, Ran, and Arf

These are all small (20–40 kDa) proteins that are membrane bound due to covalent attachment of lipids (such as *N*-myristoylation on Arf proteins). The Rho GTPases regulate the cytoskeleton and play a role in the regulation of smooth muscle contraction. One of its effectors is Rho Kinase, a serine–threonine protein kinase that phosphorylates myosin light chain phosphatase and inactivates it (see Chapter 3.8). Rho is also involved in cell cycle progression and gene expression. The Rab family of monomeric GTPases regulates vesicular traffic as well as modulation of actin

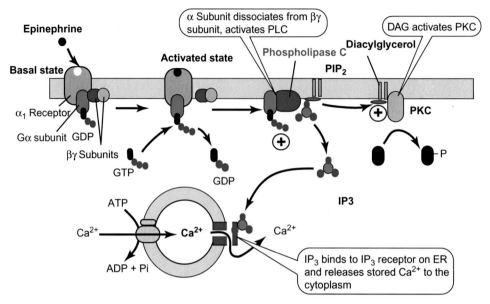

FIGURE 2.8.10 Mechanism of Gα_q signaling. Binding of ligand to its GPCR results in release of the Gα_q and $\beta\gamma$ subunits. The Gα_q subunit activates phospholipase C, which cleaves phosphatidyl inositol bisphosphate in the surface membrane, liberating DAG and IP3. IP3 releases Ca^{2+} stored in the ER and DAG activates PKC that phosphorylates sets of target proteins.

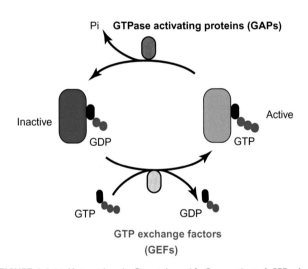

FIGURE 2.8.11 Heterotrimeric G-proteins with Gα_{12}-activated GEFs that promote the exchange of GTP for GDP on small monomeric G-proteins. This converts the small monomeric G-proteins into an active form. Reversion to the inactive form is catalyzed by GAPs, GTPase activating proteins. The GEF binds to the Gα_{12} subunit.

dynamics. Members of the Ran family regulate transport of materials between the cytoplasm and nucleus. Arf stands for "ADP-ribosylation factor." All six mammalian Arfs are located in the Golgi and regulate vesicular transport there.

THE RESPONSE OF A CELL TO A CHEMICAL SIGNAL DEPENDS ON THE RECEPTOR AND ITS EFFECTORS

According to Figures 2.8.6, 2.8.7, and 2.8.10, epinephrine can bind to β GPCRs, with the final activation of adenylyl cyclase, increased [cAMP], and activation of PKA; or it can bind to α_2 receptors with the inhibition of adenylyl cyclase, decreased [cAMP], and no activation of PKA; or it can bind to α_1 receptors with subsequent activation of PLC, release of Ca^{2+} from intracellular stores, and activation of PKC. These different effects typically occur in separate cells. Thus the response of the cell depends on the receptor for the chemical signal and not on the chemical signal alone. Further, some cells respond differently than others because they express entirely different sets of target proteins. In the liver, for example, the primary target for PKA from the stimulation of β adrenergic receptors is phosphorylase kinase, which then phosphorylates phosphorylase, activating it, and glycogen synthetase inactivating it. In the heart, the primary targets of β adrenergic stimulation are the voltage-dependent Ca^{2+} channels in the surface membrane, phospholamban on the sarcoplasmic reticulum, ryanodine receptors (RyR2) on the sarcoplasmic reticulum, and troponin I (TnI). These are illustrated in Figure 2.8.12. Thus the effects of epinephrine using the same type of receptor are glycogenolysis in the liver and increased contractile strength in the heart. These differences indicate that the final effect is a function of (1) the chemical signal, (2) its receptor, (3) the effector, and (4) the targets within the cell.

CHEMICAL SIGNALS CAN BIND TO AND DIRECTLY ACTIVATE MEMBRANE-BOUND ENZYMES

A variety of extracellular chemical signals can bind to receptors on the surface membrane; these are amplifying enzymes or directly activate an amplifying enzyme. They are of four main types: **receptor guanylyl cyclase**, **receptor serine/threonine kinase**, **receptor tyrosine kinase**, and **receptor-associated tyrosine kinase**. These are illustrated in Figure 2.8.13.

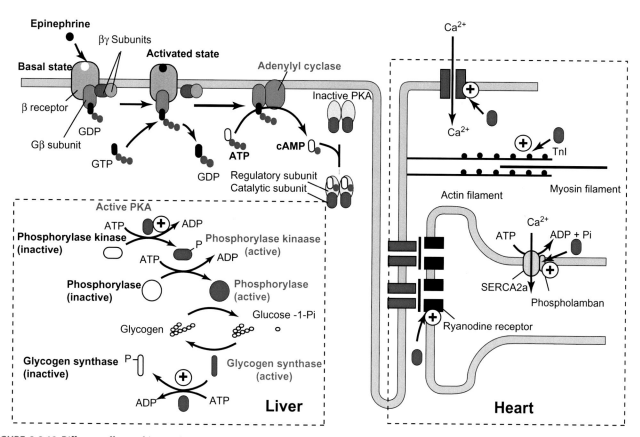

FIGURE 2.8.12 Different effects of beta adrenergic stimulation of liver cells versus heart cells. In the liver, the primary response of beta adrenergic stimulation is activation of glycogenolysis through activation of glycogen phosphorylase through a cascade of protein phosphorylation reactions. In the heart, the primary response of beta adrenergic stimulation is faster activation and relaxation of the muscle through control of cytoplasmic [Ca^{2+}] by phosphorylation of voltage-dependent Ca^{2+} channels on the surface of the cell, phosphorylation of the ryanodine receptor on the surface of the sarcoplasmic reticulum, phosphorylation of TnI on the contractile filaments, and phosphorylation of phospholamban to relieve inhibition of the SERCA2a Ca^{2+} pump on the surface of the SR.

MANY SIGNALS ALTER GENE EXPRESSION

So far we have discussed signaling molecules that cannot penetrate the cell membrane. These bind to receptors on the surface of the cell, and the binding is transduced into a second messenger such as Ca^{2+}, cAMP, cGMP, IP3, and DAG, one of a number of small monomeric GTPases, or causes phosphorylation of intracellular proteins. A number of lipophilic signaling molecules penetrate the cell membrane and bind to receptors either in the cytoplasm or in the nucleus and alter the expression of specific genes in the cell. Hormones that alter gene expression this way include the sex hormones **testosterone**, **estrogen**, and **progesterone**; **corticosteroids**, including **glucocorticoids**, produced by the adrenal cortex; **mineralocorticoids**, produced by the adrenal gland that regulate electrolyte and water balance; **vitamin D**, which regulates calcium and phosphate balance, among other effects; and **thyroid hormone** and **retinoic acid**, which have effects in almost all cells.

NUCLEAR RECEPTORS ALTER GENE TRANSCRIPTION

The nuclear receptors constitute a **superfamily** of proteins that are structurally related and perform similar functions, but they exhibit specificity for binding ligands and specificity of action. These nuclear receptors include the following: estrogen receptors α and β (ERα and ERβ); androgen receptor (AR); progesterone receptor (PR); glucocorticoid receptor (GR); mineralocorticoid receptor (MR); vitamin D receptor (VDR); thyroid receptors α and β (TRα and TRβ); retinoic acid receptor types α, β, and γ (RARα, RARβ, and RARγ); and 9-cis-retinoic acid receptors (RXRα, RXRβ, and RXRγ). These nuclear receptors are restricted to the nucleus with the exception of the mineralocorticoid receptor (MR) and the GR which reside in the cytoplasm. Upon binding with its ligand, MR and GR move into the nucleus where they bind to the regulatory region of a modulated gene. Binding of the lipophilic ligand with its receptor changes the conformation of the receptor–ligand complex, allowing the receptor to bind to specific nucleotide sequences on the DNA, called **response elements**. Binding of these receptors begins a cascade of events that eventually activates the transcription of specific genes. The newly transcribed mRNA is transferred to the cytoplasm where it is translated into a protein. The set of proteins regulated by the hormones confers specific capabilities on the cells in which they are expressed. Thus cell function is regulated by controlling cell concentration of specific active proteins.

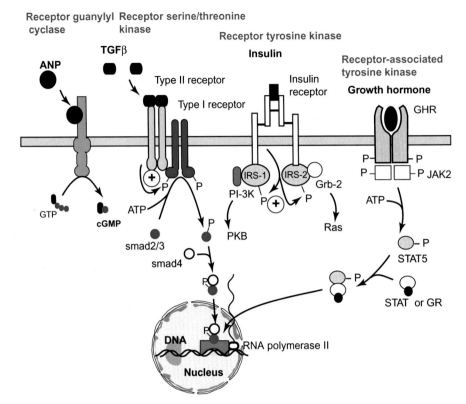

FIGURE 2.8.13 Catalytic receptors. Chemical signals binding to the extracellular surface of the cell are coupled to enzymes that alter intracellular components. **Receptor guanylate cyclases** respond to atrial natriuretic peptide or brain natriuretic peptide, and binding increases the concentration of cGMP in the cell, which increases activity of cGMP-dependent protein kinase. **Receptor serine/threonine kinases**: dimers of a variety of growth factors including transforming growth factor beta (TGF-β), myostatin, and bone morphogenic protein (BMP) bind to an activin type II receptor that recruits an activin type I receptor and phosphorylates it. This active serine/threonine kinase then phosphorylates one of a family of proteins called smads (smad2 or smad3 is shown), which binds to smad4. The complex enters the nucleus and regulates gene expression. **Receptor tyrosine kinases**: Insulin or insulin-like growth factor (e.g., IGF-1 and IGF-2) binds to the insulin receptor. The activated receptor phosphorylates several intracellular substrates. Insulin receptor substrate-1 (IRS-1) is shown. The phosphorylated proteins can activate two main pathways: the PI-3K (phosphatidyl inositol 3 kinase) pathway activates PKB and PKC downstream. The Ras pathway is activated by Grb-2 that binds to IRS and then activates a Ras GTP exchange protein leading eventually to transcription factors. **Receptor-associated tyrosine kinases** are used by growth hormone (shown), prolactin and erythropoietin, and most interleukins. Their binding induces close proximity of two receptors that bind members of the Janus family of tyrosine kinases (JAKs: JAK1-3 and TYK2). These transphosphorylate themselves and their receptor and phosphorylate proteins called STAT (for **s**ignal **t**ransduction and **a**ctivation of **t**ranscription). These form dimers with other transcription factors and regulate gene transcription.

NUCLEAR RECEPTORS RECRUIT HISTONE ACETYLASE TO UNWRAP THE DNA FROM THE HISTONES

Heterochromatin is highly condensed DNA that cannot be transcribed. **Euchromatin** is more easily accessible for the assembly of transcriptional subunits, and DNA in this configuration has a higher rate of transcription. The configuration of chromatin is regulated by the **acetylation** of **histones**. Histones are a family of proteins, described in Chapter 2.2, that form a complex with DNA called a **nucleosome** that is stabilized by the attraction of the negatively charged DNA to the positively charged histones. In this form, the DNA cannot be transcribed. Acetylation reduces the association of the DNA with the histones by reducing the positive charge on the histones. Deacetylation promotes condensation of the DNA into heterochromatin. The enzyme **histone acetyl transferase (HAT)** sticks acetyl groups on the histones, and **histone deacetylase** removes them. Both of these enzymes associate with coactivators and repressors of transcription. A variety of proteins associated with nuclear receptors possess HAT activity.

NUCLEAR RECEPTORS RECRUIT TRANSCRIPTION FACTORS

Once acetylation of the histones has allowed the reorganization of chromatin, several other complexes of proteins bind to the DNA to initiate transcription. The transcription factor TFIID (for transcription factor polymerase II) is a complex of proteins that binds to a TATAA sequence on the DNA some 25—30 nucleotides upstream of the initiation site (see Chapter 1.3). TFIID contains a **TATAA-binding protein (TBP)**, which binds directly to the TATAA sequence, and a series of other factors called **TBP-associated factors (TAFs)**. TBP then binds a second basal transcription factor, **TFIIB**. This allows the binding of RNA polymerase II to the complex, which is then fully activated by the binding of assorted other transcription factors, **TFIIF, TFIIE,** and **TFIIH**. The nuclear receptor influences transcription through specific proteins that interact both with the nuclear receptor on its recognition site and with the RNA polymerase complex on its initiation site. This interaction is possible because the DNA can form loops that closely appose the nuclear receptors and the

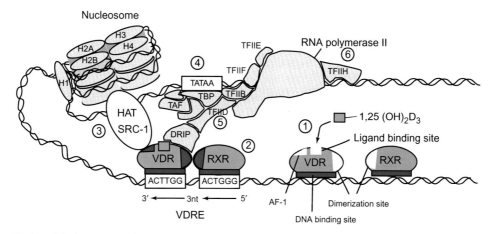

FIGURE 2.8.14 Simplified model of activation of transcription of DNA by vitamin D_3. Not all steps in this process are established, and the figure is meant to convey some of the players and their postulated roles. It is not to be taken too seriously. The active form of vitamin D_3, $1,25(OH)_2D_3$, binds to its receptor, VDR (vitamin D receptor), which is nonspecifically bound to DNA. Binding of the ligand (1) results in dimerization of the VDR with RXR (9-*cis*-retinoic acid receptor) and binding of the dimer to specific vitamin D-responsive elements on the DNA (VDRE) (2). These are similar repeat motifs on the DNA as indicated by the sequences ACTTGG and ACTGGG. Transactivation begins with the recruitment of coactivators with HAT activity (3). One of these is SRC-1 for the steroid receptor coactivator. Acetylation of histones causes chromatin remodeling that facilitates transcription. Actual initiation of transcription requires the binding of TBP to the TATAA box, along with several TAFs (4). This complex is TFIID, for transcription factor for RNA polymerase II. TFIID then binds TFIIB, which forms a bridge to RNA polymerase II. Several proteins, called vitamin D-receptor interacting proteins, or DRIPs, form a bridge to RNA polymerase II, stabilizing the preinitiation complex (5). Transcription is then initiated (6).

preinitiation complex. This process is illustrated for VDR in Figure 2.8.14. In other cases, nuclear receptors can also regulate gene expression by suppressing transcription. For example, glucocorticoids suppress the effects of transcription factor **nuclear factor** κB (**NF-κB**), which stimulates genes as part of the inflammatory response. Glucocorticoids reduce inflammation by this effect (see Chapter 9.5).

OTHER SIGNALING PATHWAYS ALSO REGULATE GENE EXPRESSION

The binding of signaling molecules to nuclear receptors is only one of many routes for the regulation of gene expression. As noted earlier, receptor serine/threonine kinases can alter gene expression through phosphorylation of smads; receptor tyrosine kinases alter gene expression through Ras, and receptor-associated tyrosine kinase alters gene expression through phosphorylation of STATs. Signals that affect 3′,5′ cyclic AMP levels also regulate gene expression. Specific genes possess a regulatory sequence called the cAMP response element or **CRE**. PKA phosphorylates **CREB** (CRE-binding protein), a transcription factor that binds to the CRE. CREB can also be phosphorylated by CaM kinase II and CaM kinase IV. Transcriptional activation by CREB requires coactivators including CREB binding protein (**CBP**) with a molecular weight of 300 kDa. CREB forms homodimers to activate transcription, but it can also form heterodimers with CREM (CRE modifier) that either activate or inhibit transcription. A schematic of CREB's involvement in the regulation of transcription is shown in Figure 2.8.15.

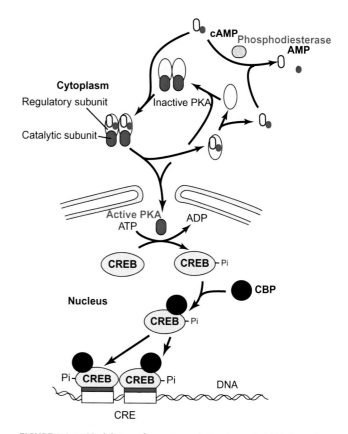

FIGURE 2.8.15 Modulation of gene transcription by cyclic AMP-dependent PKA. Activation of adenylyl cyclase occurs on the surface membrane of the cell through a G_s-coupled receptor for a hormone. The increased cAMP activates PKA by dissociating the regulatory subunit from the catalytic subunit. The catalytic subunit translocates to the nucleus where it phosphorylates CREB, a protein that binds to cyclic AMP responsive elements (CRE) on the DNA strand. CREB binds to CBP and may form homodimers or heterodimers and may activate or inhibit transcription. A variety of cofactors are recruited to complete the mechanism.

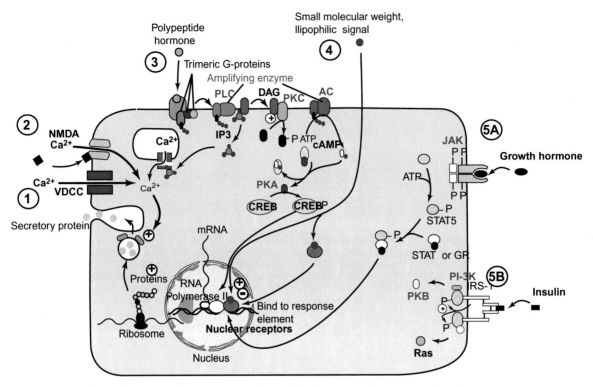

FIGURE 2.8.16 Summary of major signaling pathways. (1) Voltage-dependent channels open to convey electrical responses. (2) LGICs convert chemical signals into electrical signals. (3) Four different classes of heterotrimeric GPCRs convert extracellular chemical signals into intracellular chemical signals. The GPCR coupled to phospholipase C (G_q) and to adenylyl cyclase (G_s) is shown, but other responses also occur. (4) Small lipophilic signaling molecules enter the cell and affect gene transcription and other processes. (5) Extracellular chemical signals, larger proteins, activate enzymes that produce intracellular signals. Growth hormone and insulin receptor types are shown.

SUMMARY OF SIGNALING MECHANISMS

A synopsis of the signaling mechanisms discussed in this chapter is shown in Figure 2.8.16. The main classes of signaling include the following: (1) voltage-gated ion channels, including the fast Na^+ channel and K^+ channels involved in action potential origination and propagation. These channels maintain electrical signaling that is rapidly conveyed over the surface of the cell. (2) LGICs, the mainstay of synaptic transmission between neurons and between nerve and muscle. (3) Heterotrimeric GPCR, including four main subtypes: those that excite adenylyl cyclase (G_s mechanisms); those that inhibit adenylyl cyclase (G_I mechanisms); those that activate phospholipase C, releasing IP3 and DAG (G_q mechanisms); and those that stimulate GEFs to activate one of a family of small GTPase proteins. (4) Extracellular signals, which directly activate enzymes such as guanylyl cyclase, receptor serine/threonine kinase, receptor tyrosine kinase, and receptor-associated tyrosine kinase. (5) Signaling molecules that bind cytoplasmic or nuclear receptors. Other signaling mechanisms, such as those involving sphingosine phosphate, are not shown here.

SUMMARY

Mechanical, electrical, or chemical signals are used by cells to communicate. Mechanical signaling and some chemical signaling require close contact. Electrical signaling is the fastest way to move a signal from one part of a cell to another and to adjacent cells. Chemical signals can be used by a cell to regulate itself (autocrine), its near neighbors (paracrine) or distant cells through the medium of the blood (endocrine). Nerve cells use electrical signals over long cell processes to cause release of chemical signals near their target cells. Neuroendocrine signals are chemical signals released by neurons into the blood.

Voltage-gated ion channels make electrical signals possible. The influx of Ca^{2+} ions carries an electrical and a chemical signal because Ca^{2+} binds to specific receptors inside the cell to initiate secretion, to activate enzymes indirectly through CAM kinase or directly through activation of other enzymes, or to activate contraction.

LGICs convert chemical into electrical signals. This is used in neurotransmission: an action potential on one cell is converted to an electrical signal on another. Many chemical signals have multiple types of receptors, so that the effect in the postsynaptic cell depends on the chemical released by the presynaptic cell and the receptor expressed by the postsynaptic cell.

Many chemical signals bind to receptors on the surface membrane that are linked to heterotrimeric G-proteins (GPCRs). These dissociate upon ligand binding and generally the α subunit activates an amplifying enzyme such as adenylyl cyclase (for G_s mechanisms) or phospholipase C (for G_q mechanisms), which increase the formation of 3′,5′ cyclic AMP or IP3, respectively. Other GPCRs inhibit adenylyl cyclase (G_i) or recruit a number of small monomeric G-proteins such as Ras and Rho ($G_{11/12}$). The βγ subunit can also affect intracellular targets. Increased cytosolic cAMP activates PKA that phosphorylates specific target proteins. IP3 released by G_q-coupled receptors causes Ca^{2+} release from ER stores and activation of CaM kinase.

Other chemical signals bind to surface receptors that are catalytic. The four classes are receptor guanylyl cyclase, receptor serine/threonine kinase, receptor tyrosine kinase, and receptor-associated tyrosine kinase. Examples of these signals include insulin and growth hormone. Insulin binding activates an intrinsic tyrosine kinase that phosphorylates insulin receptor substrates that bind phosphatidyl inositol 3 kinase, PI-3K. This forms PIP3, which activates a phosphoinositide-dependent protein kinase (PDK). Growth hormone activates a receptor that in turn activates a member of the Janus Kinase family of proteins, which then phosphorylates STAT5 (signal transduction and activation of transcription). The phosphorylated STAT molecule turns on specific genes.

Small lipophilic chemical signals such as thyroxine, vitamin D, and the steroid hormones penetrate the cell membrane and bind to receptors either in the cytosol or in the nucleus. These receptors bind to specific regions of DNA called response elements. The receptors recruit a large number of accessory proteins that unravel the DNA and direct the synthesis of mRNA that codes for specific proteins.

The phosphorylation state of a set of regulated proteins depends not only on the activity of the protein kinase but also on the activity of the protein phosphatases. Both the kinases and the phosphatases may be regulated to alter the phosphorylation state of cellular proteins.

REVIEW QUESTIONS

1. What is an autocrine hormone? Paracrine hormone? Endocrine hormone?
2. What is the fastest way to convey a signal from one part of the body to another?
3. How does an electrical signal on the surface of a cell become a Ca^{2+} signal in its interior? List four distinct ways that Ca^{2+} can affect cell function.
4. What is a ligand-gated ion channel? What is the source of the extracellular ligand? Is the ligand a chemical, electrical, or mechanical signal?
5. What is meant by "G-protein-coupled receptor"? A $G\alpha_s$ mechanism couples to what amplifying enzyme? What is the product of this enzyme? What does this product do? What is a $G\alpha_i$ mechanism?
6. What amplifying enzyme is activated by a $G\alpha_q$ mechanism? What are its products? What do these products do?
7. Name the four classes of catalytic receptors.
8. Why do peptide hormones have receptors facing the extracellular space?
9. How do small lipophilic signals affect cell function? What is meant by "nuclear receptor"? What do these receptors do? What is meant by the term "response element"?
10. How does cAMP alter gene expression?

2.9 ATP Production I: Glycolysis

Learning Objectives

- Be able to draw a diagram showing the relationship among glycolysis, tricarboxylic acid cycle, and electron transport chain
- Explain what is meant in describing ATP as the "energy currency" of the cell
- Write the empirical formula for glucose
- List three sources of glucose in the body
- Define glycogenolysis
- Define gluconeogenesis
- Describe how glycogenolysis is regulated in the liver by epinephrine
- Explain why muscle tissue does not contribute to plasma glucose directly
- Describe how glucose gets into cells
- Describe what is meant by substrate-level phosphorylation
- Explain the function of lactate dehydrogenase during rapid glycolysis
- Describe how the rate of gluconeogenesis can be increased

TAKE A GLOBAL VIEW OF METABOLISM

Intermediary metabolism comprises all of the transformations of biological chemicals that allow the cell to produce energy and synthesize materials that make it up. It is a bewildering array of chemicals and their interconnected pathways. Within this, there are processes that are the composite of many of the individual processes. **Glycolysis**, for example, occupies a special place in the metabolic scheme. We ought to have some appreciation of its place without having to recall all of the transformations that occur within it. The same is true of the **citric acid cycle**, also known as the **Kreb's cycle** or the **tricarboxylic acid cycle**. This set of metabolic transformations is central to energy production in cells. We ought to understand the role of the metabolic pathways without necessarily knowing all of the individual transformations that occur within them.

ENERGY PRODUCTION OCCURS IN THREE STAGES: BREAKDOWN INTO UNITS, FORMATION OF ACETYL COA, AND COMPLETE OXIDATION OF ACETYL COA

Figure 2.9.1 shows the overall plan of energy-producing reactions in cells. These occur in three stages. In Stage 1, foodstuffs consisting of proteins, lipids, and carbohydrates are broken down into their constituent subunits. These are the amino acids, simple sugars like glucose, and fatty acids and glycerol.

In the second stage, these simple subunits are broken down to form **acetyl coenzyme A**. Coenzyme A is a chemical that acts as a carrier for the two-carbon acetyl group, but it is not a carrier in the sense of being transported across a membrane. It is being carried forward in a biochemical reaction. The formation of acetyl CoA is accompanied by the incorporation of some of the energy of the food into ATP, and some limited formation of another compound, **NADH**. NADH is **nicotinamide adenine dinucleotide**. It acts as a carrier for **reducing equivalents**. We will learn more about NADH later on in this chapter. These reducing equivalents are later used to produce ATP in the mitochondria.

The third stage of energy production takes place in the mitochondria and involves the complete **oxidation** of acetyl CoA to water and CO_2 and produces the major proportion of reducing equivalents. The energy stored in NADH produced in this stage is converted to energy stored in ATP via the **electron transport chain** which is coupled indirectly to the **ATP synthetase** in the inner mitochondrial membrane.

ATP IS THE ENERGY CURRENCY OF THE CELL

Electricity is a very versatile form of energy that has come to dominate modern society. We use it to operate heavy machinery, melt metal for casting or extrusion, power drills, pumps, and saws; run television, toasters, ovens, and computers—we use it for almost everything. We generate this electric power by burning coal, natural gas, and even public refuse, but we can also "burn" nuclear material. These methods

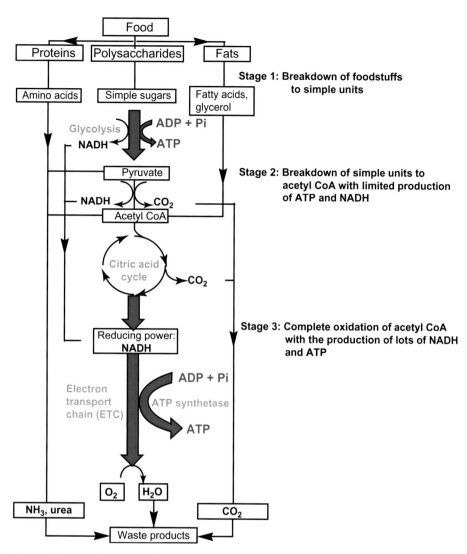

FIGURE 2.9.1 Overall scheme of intermediary metabolism. In the first stage, macronutrients found in food are broken down into their constituent subunits. In the second stage, these are converted to acetyl CoA in a process that produces only a little ATP and NADH. In the third stage, the acetyl CoA is completely oxidized, accompanied by the production of lots of ATP and NADH.

generate power by boiling water to turn a turbine connected to a dynamo. We can also generate electric power by turning a dynamo by moving water or wind. We can also use solar radiation to generate useful electrical power.

Our cells have an analogue of the power plant: the mitochondrion. It does not make electric power, but it does generate chemical energy in the form of ATP. Just like electric power, ATP can be generated from multiple kinds of fuel. Carbohydrates, fats, or amino acids can all be "burned" to produce energy that is stored in the terminal phosphate bond of ATP. Just like societal production of electrical energy, ATP formation has a final common pathway in the mitochondria, the "power house of the cell." Analogous to electricity, ATP can also be produced outside of the mitochondria. Like electricity, this form of chemical energy is very versatile. ATP fuels chemical work such as the synthesis of materials. It fuels mechanical work such as muscle contraction and movement of the cytoskeleton. It fuels electrical work in moving ions across membranes. The extra energy not directly captured by these processes is used to heat the body. All of these activities require ATP to be split into ADP and inorganic phosphate, Pi. The human body continuously splits ATP, and the steady state requires that this continuous splitting is matched to a continuous reformation of ATP from ADP and Pi. This idea is shown in Figure 2.9.2.

FUEL RESERVES ARE STORED IN THE BODY PRIMARILY IN FAT DEPOTS AND GLYCOGEN

Energy that the body uses for movement, biochemical synthesis, and transport all ultimately derives from chemical energy stored in food. However, the body stores some of this energy in its own materials. These include the **fat deposits** in **adipose tissue** and **glycogen** granules stored in the muscles and liver. Energy is not stored as protein deposits, but body proteins can be

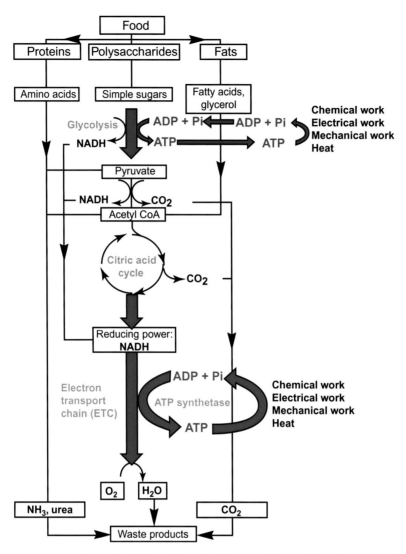

FIGURE 2.9.2 ATP as the energy currency of the cell. ATP is continuously being used for a variety of purposes that include chemical synthesis, production of mechanical force and transport of materials, and movement of ions that constitutes an electrical current. ATP hydrolysis also generates heat. This continuous use of ATP varies with the state of activation. ATP hydrolysis is coupled to resynthesis of ATP in order to maintain a constant supply of energy so that activation that is coupled to increased rates of ATP hydrolysis is simultaneously linked to increased rates of ATP synthesis.

and are continuously used as energy sources. We begin our discussion of energy metabolism with glucose.

GLUCOSE IS A READILY AVAILABLE SOURCE OF ENERGY

Many cells of the body, particularly those in the central nervous system, depend crucially on glucose as an energy source. Glucose is a six-carbon compound with the empirical formula of $C_6H_{12}O_6$. It is called a **carbohydrate** because its chemical formula is close to $C_n(H_2O)_n$, indicating a 1:1 ratio between carbon and water. Thus its empirical formula is equivalent to a hydrated carbon atom. The chemical structure of glucose is shown in Figure 2.9.3. The blood plasma

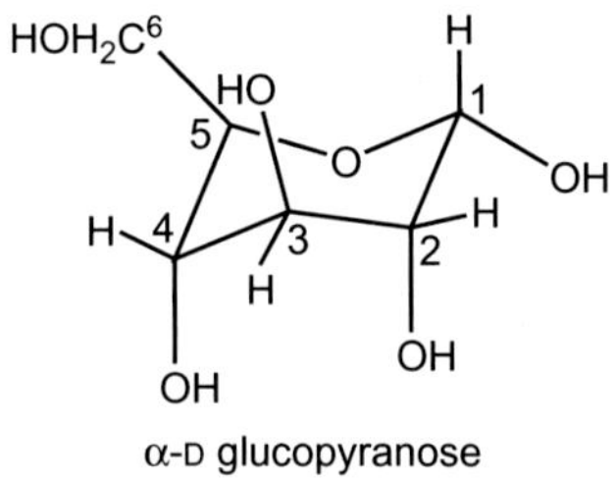

FIGURE 2.9.3 Chemical structure of α-D-glucopyranose. Glucose can exist in several configurations, one of which is shown here. The glucose atoms within the molecule are numbered 1 through 6 as shown in the figure. The pyranose ring forms a six-membered structure that approximates a plane. The hydroxyl side groups project from the plane either up or down. At C-1, C-2, and C-4 it is down and at C-3 it is up. When the hydroxyl group is down it is designated as α; when it is up it is designated as β.

typically contains glucose at levels between 80 and 120 mg%. Recall that mg% is mg of glucose per 100 mL of plasma (=1 dL). Glucose enters the circulation from several sources. The first source is directly from foodstuffs. Plant starches in the food we eat are broken down to glucose which is absorbed from the intestine into the portal blood (blood that flows from intestine to liver) and then into the general circulation.

Another source of glucose is from **glycogen** stored in the liver and in muscle. Glycogen is a polymer of glucose in which the glucose subunits are stuck together end to end. There are two ways of doing this, called an α-1,4 glucosidic bond and an α-1,6 glucosidic bond. This nomenclature merely names the numbers of the carbon atoms that are attached to one another and the α signifies the stereochemistry of how the bond is formed. The chemical structure of glycogen is shown in Figure 2.9.4. Glucose is stored as glycogen in many cells, but in large quantities in the muscles and liver. The glucose in glycogen cannot release its chemical energy while it is bound in the glycogen. It must first be broken down to the constituent subunits, the glucose molecules, by a process called **glycogenolysis**. The root word "lysis" means "break down," so glycogenolysis means "glycogen break down." Liver glycogen can contribute to blood glucose, whereas muscle glycogen is converted to glucose in the muscle fiber and used only for muscle activities.

A third source of glucose is from amino acids. Some amino acids can be used to produce glucose through a process called **gluconeogenesis**. Literally, this means "new glucose formation."

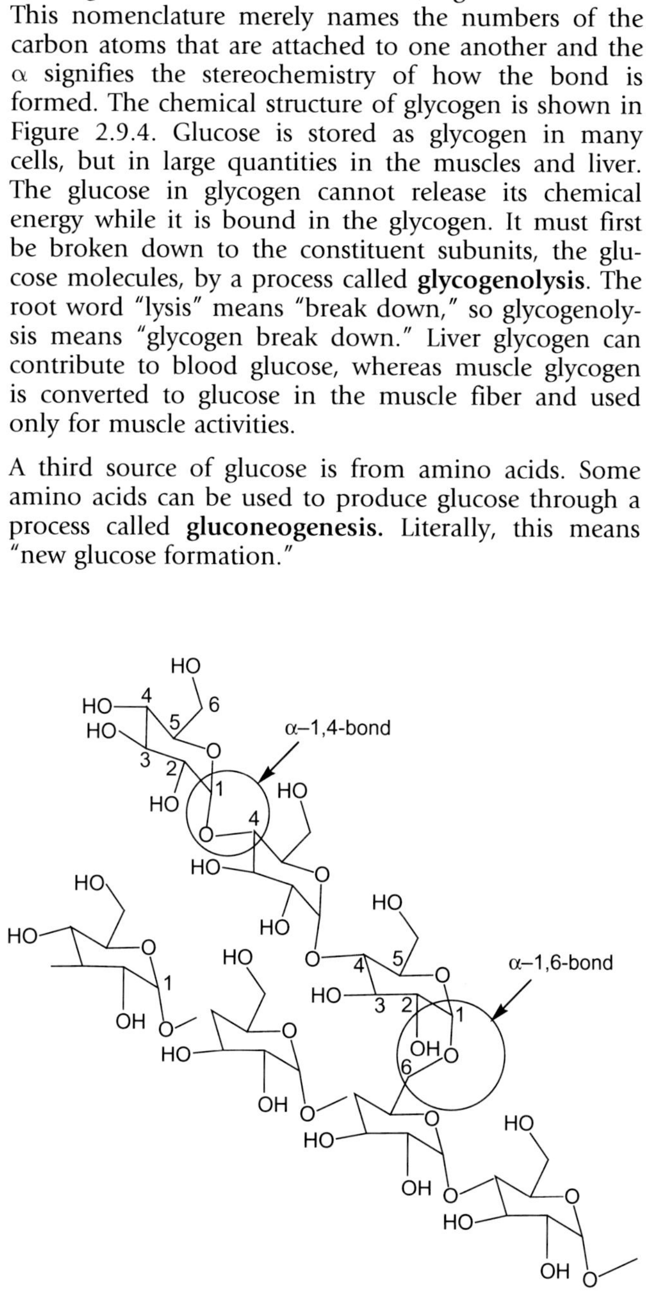

FIGURE 2.9.4 Structure of glycogen. Note that glycogen is a **branched** polymer of glucose. The α-1,4 glucosidic bond connects linear chains of glucose molecules. The α-1,6 glucosidic bond causes the chain to branch.

GLUCOSE RELEASE BY THE LIVER IS CONTROLLED BY HORMONES THROUGH A SECOND MESSENGER SYSTEM

Glycogenolysis in the liver is controlled partly by hormones. A hormone is a material which is released from **secretory cells** in the body that travels through the body via the blood, and has an effect on **target cells** located some distance away (see Chapter 2.8). One of the important hormones regulating glycogenolysis in the liver is **epinephrine**. Epinephrine does not enter the liver cell. It binds to a receptor on the **hepatocyte** (liver cell) surface and a "second messenger" is produced within the cell. The receptor for epinephrine is a G-protein-coupled receptor (**GPCR**), as discussed in Chapter 2.8. The receptor is coupled to a heterotrimeric **G-protein**, a class of protein that binds GTP, guanosine triphosphate. In the case of the epinephrine receptor, the G protein is a Gαs, meaning that the α subunit of the heterotrimeric G-protein stimulates **adenylyl cyclase** to increase the cytosolic concentration of **cyclic AMP** (3′,5′cyclic adenosine monophosphate). The cAMP is the "second messenger" within the hepatocyte.

The cAMP then activates an enzyme, **protein kinase A (PKA)**, in the liver cell. PKA begins a cascade of phosphorylation reactions that shuts down glycogen synthesis and activates glycogen breakdown according to the scheme shown in Figure 2.9.5.

After activation by cAMP, the system returns to its inactivated state in two ways. First, the cAMP produced by adenylyl cyclase is degraded to AMP (adenosine monophosphate) by another enzyme, **cAMP phosphodiesterase**. This turns off the second messenger signal. Second, **protein phosphatases** dephosphorylate the proteins that were phosphorylated during activation of the cascade. There are four classes of serine/threonine phosphoprotein phosphatases: **PP1**, **PP2a**, **PP2b**, and **PP2c**. PP1 dephosphorylates many of the proteins phosphorylated by PKA. The balance between phosphorylated and dephosphorylated proteins is set by the competing activity of the kinases and phosphatases.

THE LIVER EXPORTS GLUCOSE INTO THE BLOOD BECAUSE IT CAN DEPHOSPHORYLATE GLUCOSE-6-P

In the liver, glycogenolysis ends at glucose-1-P. This is converted to glucose-6-P by another enzyme, **phosphoglucomutase**. The glucose-6-P is then converted to glucose by **glucose-6-phosphatase**. This enzyme is extremely important because only the liver, kidney, and intestine have it, allowing them to release glucose into the blood from glucose-6-P; neither glucose-1-P nor glucose-6-P can exit the cell. Muscle cells have glycogen stores that can be broken down to provide energy, but only for the muscle cell because they lack

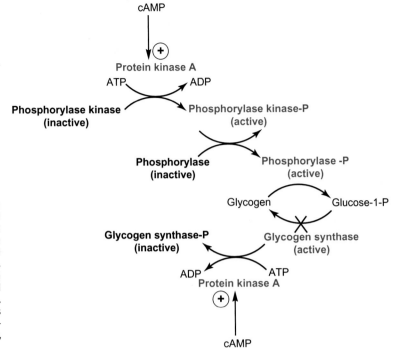

FIGURE 2.9.5 Cascade of activation events to shut down glycogen synthesis and activate glycogenolysis upon stimulation of the liver with epinephrine. Epinephrine binds to a G-Protein-Coupled Receptor on the surface of the hepatocytes which stimulates adenylyl cyclase to increase formation of 3′,5′ cyclic adenosine monophosphate (cAMP). The increased cAMP stimulates protein kinase A, which then phosphorylates the enzyme phosphorylase kinase, so-named because it phosphorylates another enzyme, phosphorylase. Phosphorylase has its name because it phosphorylates glycogen during glycogenolysis to produce glucose-1-phosphate. PKA also phosphorylates glycogen synthase, converting it from its active form to an inactive form.

glucose-6-phosphatase. Muscles cannot contribute glucose to the blood.

A SPECIFIC GLUCOSE CARRIER TAKES GLUCOSE UP INTO CELLS

In muscle cells, glucose can be taken up from the blood by a glucose transporter, GLUT, of which there are multiple isoforms. The one in muscle and fat is GLUT4. The number of these receptors is regulated hormonally, and they exist in a latent form in vesicles stored within the cell. The GLUT4 transporters are particularly sensitive to the hormone **insulin**. Brain, liver, and red blood cells have GLUT transporters that are not regulated by insulin, and therefore these tissues are insensitive to insulin. Muscle cells can also derive glucose from glycogenolysis within the cell. The fate of glucose, whether derived from blood or glycogen, is conversion to pyruvate through the process of **glycolysis**.

GLYCOLYSIS IS A SERIES OF BIOCHEMICAL TRANSFORMATIONS LEADING FROM GLUCOSE TO PYRUVATE

Figures 2.9.6 and 2.9.7 show the reactions of glycolysis that produce pyruvate from glucose. These reactions occur in the cytoplasm. The pyruvate then enters the mitochondria where it is completely oxidized and produces a number of ATP molecules per molecule of pyruvate.

GLYCOLYSIS GENERATES ATP QUICKLY IN THE ABSENCE OF OXYGEN

Glycolysis can generate ATP in the absence of oxygen. This is described as **anaerobic** metabolism. It results from **substrate-level phosphorylation**. This is distinct from **oxidative phosphorylation** that occurs in the mitochondria. Substrate-level phosphorylation refers to the formation of ATP from ADP and a phosphorylated intermediate, rather than from ADP and inorganic phosphate, Pi, as is done in oxidative phosphorylation.

The amount of ATP that is generated by glycolysis is relatively low. Two ATP molecules are required to start glycolysis (from glucose), and four are generated by substrate-level phosphorylation. An additional two NADH molecules are generated, which can be used to generate another three to five ATP molecules through the electron transport chain in the mitochondria. So a net gain of 5−7 moles of ATP can be generated from the conversion of 1 mole of glucose to 2 moles of pyruvate. The total energy in the oxidation of glucose is 2867 kJ mol^{-1}. The energy in 7 moles of ATP is about 7×57.1 kJ mol^{-1} = 399.7 kJ mol^{-1}. This represents capture of only some 14% of the total energy available from glucose oxidation.

GLYCOLYSIS REQUIRES NAD$^+$

Glycolysis occurs in the cytoplasm and it generates some NADH from NAD$^+$. The NAD$^+$ is an obligatory substrate for the reaction of glyceraldehyde-3-phosphate to 1,3-diphosphoglycerate. If NAD$^+$ is not

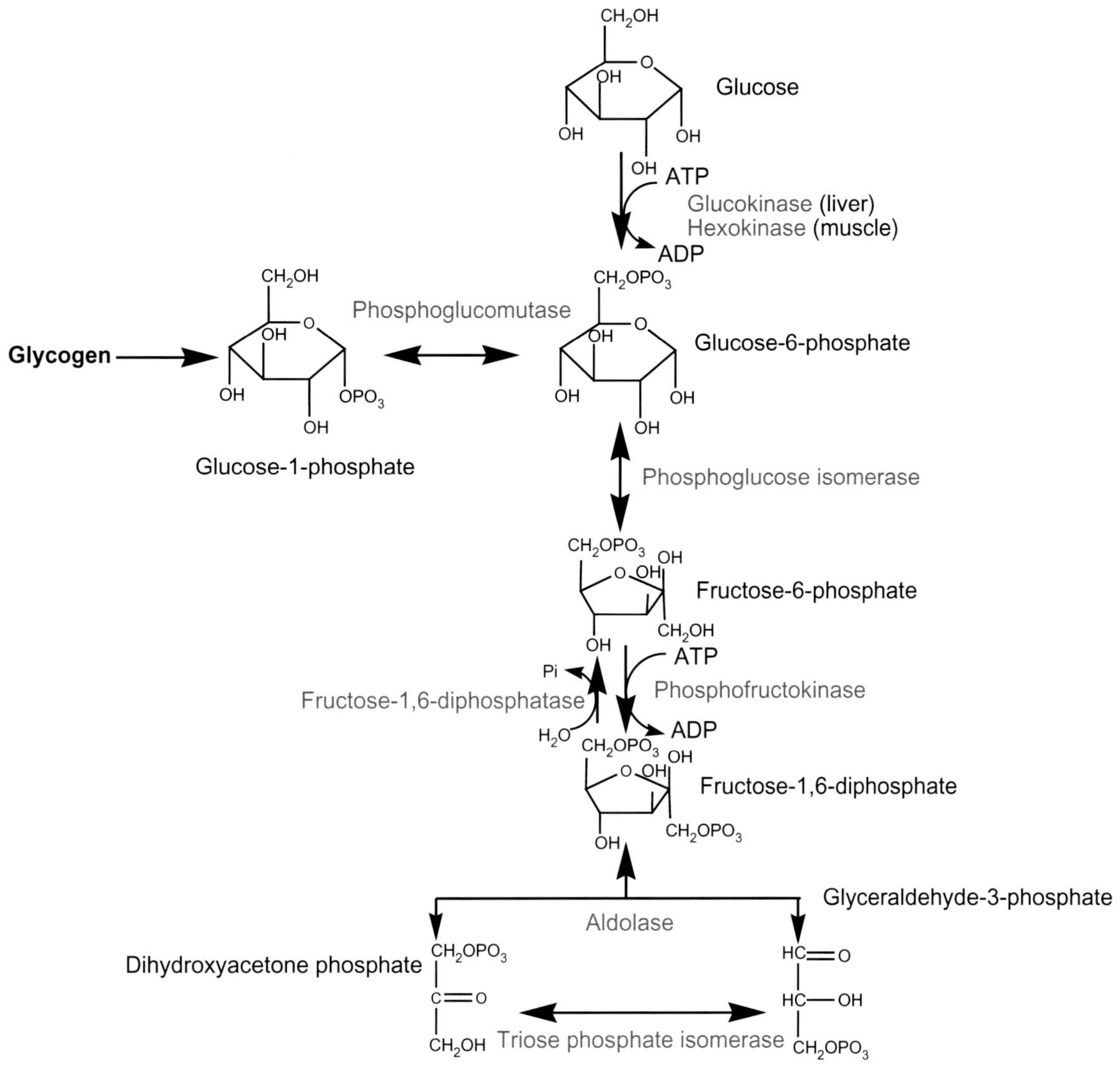

FIGURE 2.9.6 First part of glycolysis, leading from glucose to two three-carbon intermediates that are readily interconvertible. Chemical structures and names of the intermediates are shown in black. The enzymes that participate in the interconversions are shown in blue. Glycolysis begins by phosphorylation of glucose in two successive steps, forming glucose-6-phosphate and then forming fructose-1,6-diphosphate. These steps in glycolysis require ATP to "prime" the process.

regenerated, glycolysis will halt. In the presence of oxygen, NADH is oxidized in the mitochondria to regenerate NAD^+, but NADH itself cannot cross the mitochondrial membrane. Two shuttles transfer the "reducing equivalents" across the mitochondrial membrane. These are the **glycerolphosphate shuttle** and the **malate/aspartate shuttle** (see Chapter 2.10). Figure 2.9.8 illustrates the requirement of glycolysis for NAD^+.

If the glycolytic generation of NADH exceeds the mitochondrial oxidation of cytoplasmic NADH, then cytoplasmic NAD^+ will become depleted and its absence will limit the metabolic flux through glycolysis. Under these conditions, the cell must regenerate NAD^+ from NADH in order to allow glycolysis to continue. This is achieved by making lactic acid from pyruvate through the enzyme **lactate dehydrogenase**,

LDH. Lactic acid production occurs all the time, but increases when glycolysis is going faster than the mitochondria can accommodate the metabolic flux of cytoplasmic NADH, regardless of the state of oxygenation of the tissue.

A good example of this occurring physiologically is in muscle during brief strenuous exercise such as a 200-m sprint. In this case, nearly all of the energy will be supplied by glycolysis. In order for glycolysis to continue, the muscle will produce lactic acid, which will leave the muscle and travel to the liver. The oxygen necessary to oxidize the accumulated lactic acid constitutes part of the "oxygen debt" that must be repaid when oxygen is available.

Provided the liver is adequately oxygenated, the liver will reoxidize the lactic acid to pyruvate, which can then

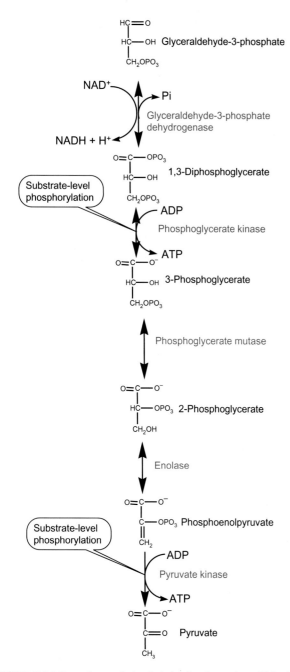

FIGURE 2.9.7 Second part of glycolysis leading from glyceraldehyde-3-P to pyruvate. Chemical structures and names of the intermediates are shown in black. The enzymes that participate in the interconversions are shown in blue. ATP is formed twice in this sequence, once in the conversion of 1,3-diphosphoglycerate to 3-phosphoglycerate and for a second time in the conversion of phosphoenolpyruvate to pyruvate. The formation of ATP directly from phosphorylated intermediary metabolites is called substrate-level phosphorylation. Two molecules of glyceraldehyde-3-phosphate are formed from every molecule of glucose. Thus glycolysis produces 4 ATP per molecule of glucose.

is converted to muscle glucose is called the **Cori cycle** (see Chapter 3.7).

GLUCONEOGENESIS REQUIRES REVERSAL OF GLYCOLYSIS

Energy transduction in cells involves glycolysis, as we have reviewed it, plus the complete oxidation of pyruvate in the mitochondria, plus the oxidation of other fuels such as fats and proteins. Some tissues (liver, intestine, kidney) export glucose into the blood for the muscles to use during exercise. As mentioned earlier, the liver can mobilize glycogen stores for this purpose, but it can also make new glucose from the amino acids derived from proteins. The process of making new glucose from proteins is called **gluconeogenesis**. It involves chemically transforming the hydrocarbon parts of amino acids into intermediates of carbohydrate metabolism, and then running glycolysis backwards to form glucose. How this is accomplished is illustrated in Figure 2.9.9 for the effect of glucagon on liver cells. Briefly, glucagon activates glycogenolysis through means similar to what we have described earlier for epinephrine. This produces glucose-1-phosphate. In the liver, phosphoglucomutase converts glucose-1-phosphate to glucose-6-phosphate. Glucose-6-phosphatase removes the phosphate from glucose-6-phosphate to produce glucose, which is then released into the blood stream. Activated PKA also phosphorylates **CREB**, the **cyclic AMP responsive element binding protein**. This activates its binding to the **CRE, cAMP responsive element**. Activation of CRE increases the transcription of another transcriptional activator that then turns on the synthesis of **PEPCK, phosphoenolpyruvate carboxy kinase**. This enzyme converts **oxaloacetate** to **phosphoenolpyruvate**. The oxaloacetate is a common carbohydrate intermediate formed from the glucogenic amino acids. These are amino acids that form glucose (see Chapter 2.11). PKA also indirectly regulates a key controlling enzyme in glycolysis: **phosphofructokinase 1/fructose biphosphatase 1(PFK1/FBPase1)**. PFK converts fructose-6-phosphate to fructose-1,6-biphosphate; FBPase converts fructose-1,6-biphosphate to fructose-6-phosphate. The FBPase1 activity and PFK1 activities are regulated by cytosolic levels of **fructose-2,6-biphosphate (FBP)**. Frucose-2,6-biphosphate stimulates PFK activity and it inhibits FBPase activity. Fructose-2,6-biphosphate levels are determined by the activity of **phosphofructose kinase 2** and **fructose-2,6-biphosphatase (FBPase2)** which convert fructose-6-phosphate to fructose-2,6-biphosphate. The activities of PFK2/FBPase 2 reside on a single polypeptide chain. PKA phosphorylates PFK2/FBPase2, stimulating the FBPase2 activity and inhibiting the PFK2 activity. This reduces the level of fructose-2,6-biphosphate, which subsequently removes activation of PFK1 and removes inhibition of FBPase1. The net result is an inhibition of PFK1, which thereby slows glycolysis, and activation of FBPase1, which increases gluconeogenesis.

be converted to glucose by gluconeogenesis. The glucose so formed can be released by the liver into the blood for use again by the muscle. The overall process by which muscle glucose becomes blood lactic acid which

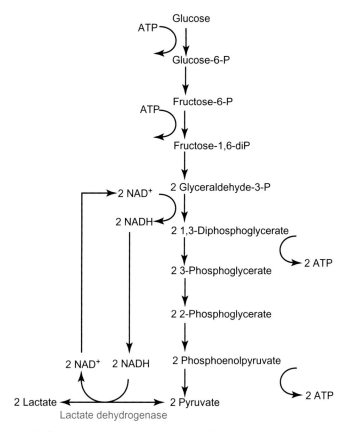

FIGURE 2.9.8 Necessity for regenerating NAD$^+$ during rapid glycolysis. When NADH oxidation by the mitochondria cannot keep pace with glycolysis, [NAD$^+$] falls and [NADH] rises. The oxidation of NADH by lactate dehydrogenase, converting pyruvate to lactate, occurs to regenerate NAD$^+$ so that glycolysis can continue to generate some ATP.

SUMMARY

Cells use chemical energy to power their synthetic, mechanical, and transport work. The chemical energy stored in the terminal phosphate bond of ATP is used as a common energy source for all of these processes. Cells produce ATP by linking the energy of oxidation of foodstuffs to the chemical synthesis of ATP. Oxidation of carbohydrates, fats, and proteins all give rise to ATP as a common energy currency for the cell.

The overall process of energy production occurs in three stages: (1) breakdown of foodstuffs into component units (amino acids for the proteins; fatty acids and glycerol for fats; glucose and fructose for carbohydrates); (2) formation of acetyl CoA with limited formation of ATP and NADH; (3) complete oxidation of acetyl CoA with the production of lots of NADH and ATP through the electron transport chain in the mitochondria.

Carbohydrates provide the most rapid source of ATP. Glucose in the blood can be taken up by tissues through specific glucose transporters in their cell membranes (GLUT1, GLUT4) to provide a ready source of energy. Liver and muscle cells store carbohydrates in a readily usable form called glycogen. Liver can convert glycogen stores to blood glucose but muscle uses its glycogen stores inside the muscle by converting it to glucose-1-phosphate. Glycogen utilization begins with its breakdown into component glucose molecules, a process called glycogenolysis. In liver cells this is regulated by hormones. One important hormone, epinephrine, helps raise blood glucose by mobilizing liver glycogen. It achieves this task by binding to a receptor on the outside surface of hepatocytes. This receptor is coupled to a G-protein, so-named because it binds and then hydrolyzes guanosine triphosphate, GTP. This G-protein consists of three subunits: α, β, and γ. Upon binding of hormone, the α subunit dissociates and activates adenylyl cyclase, which converts intracellular ATP to 3′,5′-cyclic AMP (cAMP). The cAMP then activates protein kinase A (PKA), which phosphorylates a number of target proteins involved in glycogen metabolism. PKA phosphorylates phosphorylase kinase, which then phosphorylates phosphorylase, the enzyme that breaks down glycogen. It also phosphorylates glycogen synthase, inactivating it. In this way, increasing cAMP turns on glycogenolysis and inhibits glycogen synthesis.

The end product of glycogenolysis is glucose-1-phosphate. This is converted to glucose-6-phosphate by phosphoglucomutase. Glucose-6-phosphate can then enter glycolysis, the conversion of glucose to pyruvic acid that occurs in the cytoplasm. Liver cells can convert

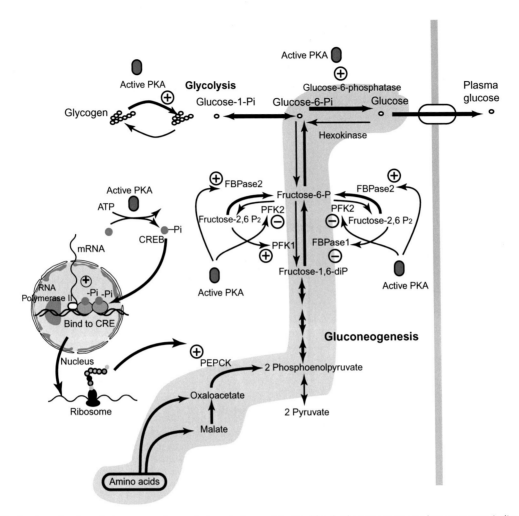

FIGURE 2.9.9 Mechanism of action of glucagon on liver cells to put glucose into the blood. Glucagon increases key processes, indicated by the circled + signs, through increasing cAMP in the hepatocytes. Gluconeogenesis is the synthesis of new glucose from amino acids, indicated by the pathway highlighted in blue. Gluconeogenesis requires the conversion of fructose-1,6 diphosphate to fructose-6-phosphate, the reverse of the reaction that occurs during glycolysis. This is accomplished by inhibiting PFK1 and activating FBPase. Inhibition of PFK1 and activation of FBPase are brought about by decreasing fructose-2,6 diphosphate levels by stimulating FBPase2 and inhibiting PFK2 through phosphorylation mediated by PKA.

glucose-6-phosphate to glucose, which it then exports into the blood. Muscle cells lack glucose-6-phosphatase and so cannot export glucose into the blood.

The first stage in glucose oxidation is glycolysis, in which one molecule of glucose is converted to two molecules of pyruvic acid. Glycolysis generates some ATP by substrate-level phosphorylation (occurring at the level of the phosphorylated intermediates of glycolysis as opposed to synthesis from ADP and Pi). It also requires another compound, nicotinamide adenine dinucleotide (NAD^+). This compound is converted to NADH during glycolysis, and under aerobic conditions is regenerated from NADH by the mitochondria. When glycolysis outstrips the ability of mitochondria to regenerate NAD^+, NADH can be converted to NAD^+ by linking this conversion to the production of lactic acid from pyruvate through the enzyme lactic dehydrogenase. Thus in strenuous activity lactic acid is produced to regenerate NAD^+ so that glycolysis can continue.

REVIEW QUESTIONS

1. What is glucose? Name three processes that supply body cells with glucose.
2. What is glycogen? What is glycolysis? Why can liver convert glycogen to blood glucose but muscle cannot?
3. Why is ATP first consumed in glycolysis instead of being produced? What is substrate-level phosphorylation?
4. Why do muscle cells produce lactic acid during bursts of activity?
5. What is lactate dehydrogenase?
6. What is the Cori cycle?
7. What is gluconeogenesis?

ATP Production II: The TCA Cycle and Oxidative Phosphorylation 2.10

Learning Objectives

- Describe the reaction catalyzed by pyruvate dehydrogenase
- Describe in general terms the function of the water-soluble vitamins
- Describe what it means to say that NADH is a carrier for reducing equivalents
- Define reduction potential and describe how it can be measured
- Be able to calculate the energy released in a reduction–oxidation reaction
- List the number of NADH molecules generated per turn of the TCA cycle
- List the number of $FADH_2$ molecules generated per turn of the TCA cycle
- List the number of ATP molecules (or equivalent) produced by substrate-level phosphorylation per turn of the TCA cycle
- Indicate where CO_2 is released during glucose oxidation
- Describe what is meant by the "electron transport chain"
- Tell the approximate magnitude and sign of the membrane potential across the inner mitochondrial membrane
- Give the stoichiometry of ATP formation from NADH; from $FADH_2$
- Be able to calculate the electrochemical potential difference for H^+ ions across the inner mitochondrial membrane
- Describe in words how the ATP synthase makes ATP
- Describe the chemiosmotic hypothesis for oxidative phosphorylation
- Describe in general terms how cytoplasmic NADH enters the mitochondria
- Describe how ADP and Pi get into the mitochondrion and how ATP leaves it

OXIDATION OF PYRUVATE OCCURS IN THE MITOCHONDRIA VIA THE TCA CYCLE

Pyruvate is the end product of glycolysis. Its metabolism continues in the mitochondria via the "**TCA cycle**," the **tricarboxylic acid cycle**, so named because many of the intermediates have three carboxyl groups. It is also referred to as the **Krebs cycle** in honor of Sir Hans Krebs, who did much of the pioneering work in describing it, and it is also referred to as the **citric acid cycle** because citric acid is formed in it. This series of metabolic transformations occurs in the inner mitochondria of cells. Its fuel source is pyruvic acid derived from glycolysis in the cytosol. The TCA cycle can also be initiated within the mitochondria by the oxidation of fatty acids to form acetyl CoA (see Chapter 2.11).

PYRUVATE ENTERS THE MITOCHONDRIA AND IS CONVERTED TO ACETYL CoA

The mitochondria have two membranes, an outer membrane and an inner membrane. The outer membrane is relatively permeable, whereas the inner membrane is highly impermeable to most materials. Pyruvate produced in the cytosol by glycolysis crosses the inner mitochondrial membrane by facilitated diffusion on its own **pyruvate carrier**. Inside the **matrix** of the mitochondria, pyruvate is converted to **acetyl coenzyme A**. This conversion of pyruvate to **acetyl CoA** requires three different enzymes and five different coenzymes, which are organized into a multienzyme complex called **pyruvate dehydrogenase**. Three of the coenzymes required here are vitamins: thiamine, riboflavin, and niacin. **The water-soluble vitamins all find their use in mammals as part of enzymatic reactions, and most of the B vitamins are involved in carbohydrate metabolism.** The overall reaction is shown in Figure 2.10.1 along with the structure of coenzyme A.

PYRUVATE DEHYDROGENASE RELEASES CO_2 AND MAKES NADH

The production of acetyl CoA from pyruvate is noteworthy because here is the first production of CO_2 from glucose. This gas forms a major waste product that must be eliminated, largely through the lungs. Second, the reaction produces NADH from NAD^+. NADH is a carrier for **reducing equivalents** in the cell. The structures of NAD^+ and NADH are shown in Figure 2.10.2.

The conversion of NAD^+ to NADH is a **reduction reaction**. Oxidation and reduction are two halves of the same process, dealing with the exchange of electrons

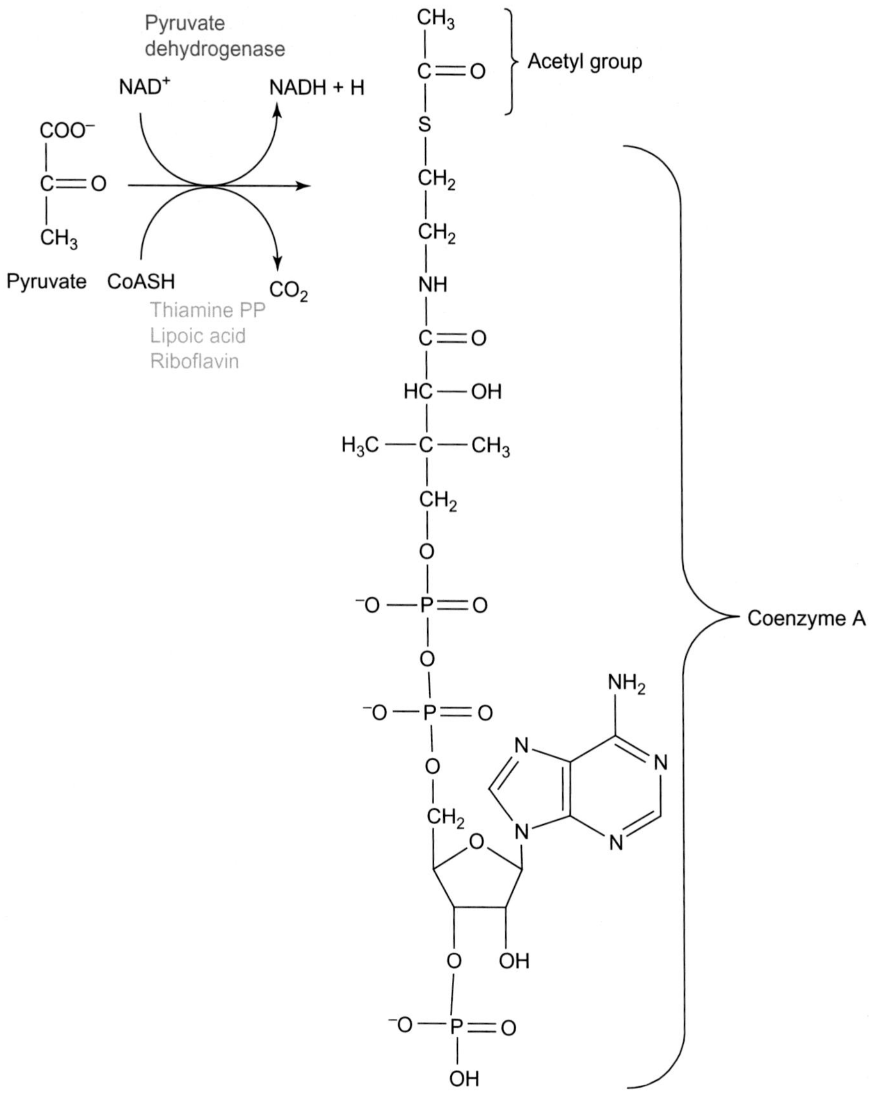

FIGURE 2.10.1 Conversion of pyruvate to acetyl CoA by pyruvate dehydrogenase, indicated in blue. Note that the reaction releases CO_2. Coenzyme A is a complex of ATP with pantothenic acid. It carries the acetyl group on through biochemical reactions in the cell. The reaction requires five different coenzymes (coenzyme A, NAD^+, thiamine pyrophosphate, lipoic acid, and riboflavin). Three of these coenzymes are vitamins (niacin, thiamine, and riboflavin).

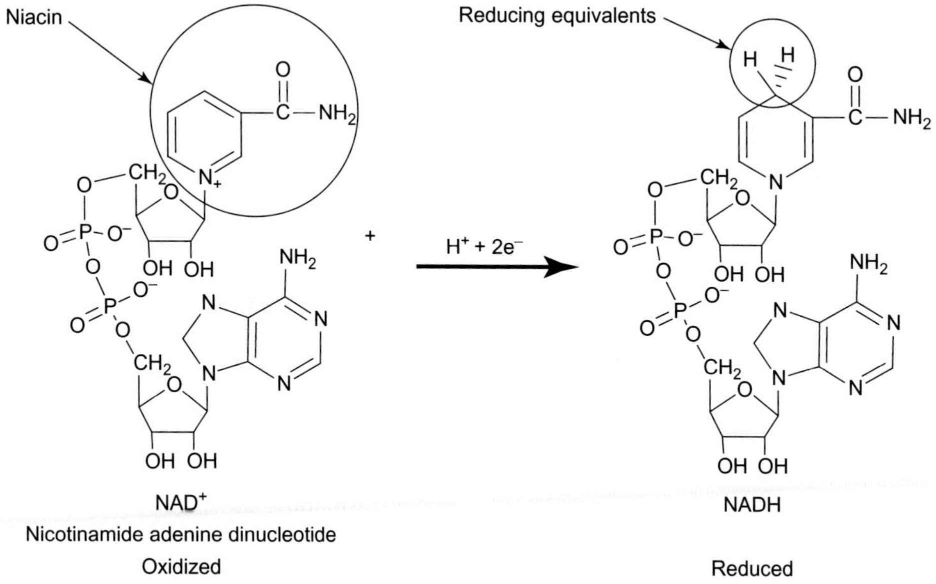

FIGURE 2.10.2 Formation of NADH from NAD^+. NAD^+ is the oxidized form; NADH is the reduced form.

between chemicals. A mnemonic device for oxidation/reduction reactions is

LEO GER

which stands for "Loss of Electrons, Oxidation; Gain of Electrons, Reduction." When NAD^+ gains two electrons in the form of H^-, it is reduced. At the same time, the chemical from which the H^- is extracted is oxidized. In oxidation/reduction reactions, one chemical is reduced while the other is oxidized. Now the electrons in the reduction reaction had to come from someplace, and that someplace is another chemical. So in a reduction–oxidation reaction there are always two redox pairs, one being reduced and the other being oxidized in the process.

THE AFFINITY OF A CHEMICAL FOR ELECTRONS IS MEASURED BY ITS STANDARD REDUCTION POTENTIAL

If compound A binds electrons more tightly than compound B, we expect that A will take electrons from B in the reaction $A + B \rightarrow A^- + B^+$. In this reaction, A is reduced and B is oxidized. This can be written as the sum of two **half-reactions**:

$$A + e^- \rightarrow A^- \text{ and } B \rightarrow B^+ + e^-$$

The relative tendency for a compound to be reduced is called its **reduction potential**. The **standard redox potential** is measured against a standard half-reaction, arbitrarily assigned the reduction potential of zero. This is the reduction/oxidation of hydrogen:

$$2H^+(aq) + 2e^- \rightarrow H_2(g) \quad E_0 = 0.00$$

Here E_0 is referred to as the **standard reduction potential** of hydrogen. Because oxidation is the reverse reaction of reduction, the standard oxidation potential is the negative of the standard reduction potential.

Standard reduction potentials are measured as shown in Figure 2.10.3 against a **Standard Hydrogen Electrode**. It is measured under standard conditions of 1 M concentration of all reactants and 1 atm pressure of H_2 gas. The standard reduction potential is given the symbol E_0.

Recall from Chapter 1.3 that the potential is the work done in bringing a **positive** unit charge from infinite separation to the point at which the potential is to be defined. The work done in moving a charge through a potential is just the charge times the potential. This was expressed in Eqn [1.3.13]:

[2.10.1] $$\Delta \text{Energy} = q\Delta U$$

where U is the potential and q is the total charge. In electrochemistry we generally use E for the potential, as we have done above, in units of volts, and q is in coulombs. This gives the energy change in volt-coulombs or joules. This energy change is the change in free energy, and so Eqn [2.10.1] can be rewritten as

[2.10.2] $$\Delta G = z\Im \Delta E_0'$$

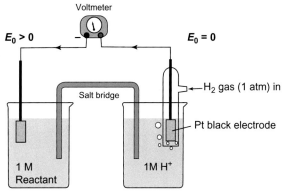

FIGURE 2.10.3 Measurement of the standard electrode potential. One half-cell containing standard concentrations of reactant (1 M) is connected to the standard hydrogen electrode (SHE) with a finely divided Pt black electrode bubbled with 1 atm of H_2 gas and 1 M H^+ in solution. The voltage between the two half-cells is the reduction potential. If electrons flow to the reactant, then the reactant is being reduced and has a positive standard reduction potential—it has a higher affinity for electrons than hydrogen. Note that current is defined as positive charge flow, which is opposite to electron flow. Voltages are typically measured with a potentiometer that finds the voltage necessary to stop current flow, so that the measurement occurs at equilibrium when no current flows.

where $\Delta E_0'$ is the difference in reduction potential between the two half-cells, z is the valence of the carrier, and $\Im$ is the Faraday = 98,500 coulombs mol^{-1}. The Faraday converts the charges to coulombs and normalizes the free energy change to the free energy change per mole. Since the charge carrier is the electron, $z = -1$, this equation becomes

[2.10.3] $$\Delta G = -\Im \Delta E_0'$$

This is the free energy change *per mole of electrons*. If there are n electrons involved per reaction, the free energy change *per mole of reaction* is

[2.10.4] $$\Delta G = -n\Im \Delta E_0'$$

THE REDUCTION POTENTIAL DEPENDS ON THE CONCENTRATION OF OXIDIZED AND REDUCED FORMS, AND THE TEMPERATURE

The standard reduction potential is defined at unit concentrations of all reactants. When the concentrations are not 1 M, the measured reduction potential changes. Consider the reduction of A in contact with a Standard Hydrogen Electrode, as described earlier. We can write the two half-cell reactions as

$$A + n\, H_{\text{test}}^+ + n\, e^- \rightarrow AH_n$$

$$n/2\, H_2 \rightarrow n\, H_{\text{SHE}}^+ + n\, e^-$$

$$A + n/2\, H_2 + n\, H_{\text{test}}^+ \rightarrow AH_n + n\, H_{\text{SHE}}^+$$

where SHE denotes Standard Hydrogen Electrode. The free energy per mole for the overall reaction is calculated as

$$\Delta\mu = \mu^0_{AH_n} + RT \ln[AH_n] + n\,\mu^0_{H_{SHE}} + n\,RT \ln[H^+_{SHE}]$$

$$- \mu^0_A - RT \ln[A] - \frac{n}{2}\mu^0_{H_2} - \frac{n}{2} RT \ln f_{H_2}$$

$$- n\,\mu^0_{H_{test}} - n\,RT \ln[H^+_{test}]$$

[2.10.5]

where f_{H2} is the fugacity of hydrogen, analogous to the activity in aqueous solutions. Since the pressure of H_2 gas at the Standard Hydrogen Electrode is 1 atm, f_{H2} is 1.0. Collecting terms in Eqn [2.10.5], we get

$$\Delta\mu = \mu^0_{AH_n} + n\,\mu^0_{H_{SHE}} - \mu^0_A - \frac{n}{2}\mu^0_{H_2} - n\,\mu^0_{H_{test}} + RT \ln[AH_n]$$

$$- RT \ln[A] + n\,RT \ln[H^+_{SHE}] - n\,RT \ln[H^+_{test}]$$

[2.10.6]

The top line of Eqn [2.10.6] is the free energy per mole for the standard reduction potential for A. This is given by Eqn [2.10.4]. Remembering that $[H^+]$ for the Standard Hydrogen Electrode = 1 M, its term drops out, because $\ln[H^+] = \ln 1 = 0$, and we write

$$\Delta\mu = -n\Im\Delta E_0 + RT \ln\frac{[AH_n]}{[A]} - n\,RT \ln[H^+_{test}]$$

[2.10.7]

The observed potential is related to the free energy change of the reaction through Eqn [2.10.4] and so we have

$$-n\Im\Delta E = -n\Im\Delta E_0 + n\,RT \ln[H^+_{test}] + RT \ln\frac{[AH_n]}{[A]}$$

[2.10.8]

In most situations, the $[H^+]$ in the test half-cell can be kept nearly constant by the use of chemical buffers. In this case, its contribution to the free energy will also be constant, and we can define a practical reduction potential ($\Delta E_0'$) that incorporates the standard reduction potential and the pH term. Doing this, plus dividing both sides by $-n\Im$, we come to

[2.10.9] $$\Delta E = \Delta E_0' + \frac{RT}{n\Im} \ln\frac{[A]}{[AH_n]}$$

The argument of the logarithm is inverted from Eqn [2.10.8] because of multiplying through by −1 to convert the minus sign in $-n\Im$ to positive values. What this equation means is that the actual reduction potential depends on the relative concentrations of the oxidized form ([A]) and reduced form ([AH_n]). The reduction potential as a function of the oxidation state of a redox reaction is shown in Figure 2.10.4 for NADH, Ubiquinone, and Cytochrome C (more on these later). Note that when [A] = [AH], which occurs

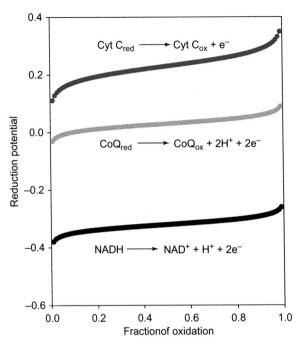

FIGURE 2.10.4 Reduction potential for various redox reactions found in the cell as a function of their oxidation state. The reduced form is AH_n in Eqn [2.10.9] and the oxidized form is A. As AH_n is oxidized by addition of a strong oxidant, it is converted to A and the reduction potential changes according to Eqn [2.10.9]. When the reaction is 50% complete, [A] = [AH_n] and the argument of the logarithm becomes 1.0, and ln 1.0 = 0. At this point, the measured reduction potential is the practical standard reduction potential: $\Delta E = \Delta E_0'$. For NADH this is −0.32 volts; for Coenzyme Q (CoQ) this is +0.030 volts; for cytochrome C this is +0.23 volts. The higher reduction potential means that oxidized CoQ will oxidize NADH by taking electrons from it, and oxidized Cyt C will oxidize CoQ by taking electrons from it.

when the reactant is 50% oxidized, the measured ΔE is equal to $\Delta E_0'$.

THE TCA CYCLE IS A CATALYTIC CYCLE

The biochemical transformations that constitute the TCA cycle are shown in Figure 2.10.5. This is a catalytic cycle in that the intermediates themselves are not altered by the cycle. It starts with oxaloacetate, a 4-carbon dicarboxylic acid, condensing with acetyl CoA to produce citrate. As the cycle continues, NADH is generated in each of three separate reactions (at isocitrate dehydrogenase, α-ketoglutarate dehydrogenase, and malate dehydrogenase) and $FADH_2$ is generated at succinate dehydrogenase. $FADH_2$ is the oxidized form of FAD, flavin adenine dinucleotide. It is another chemical carrier of reducing equivalents whose structure is shown in Figure 2.10.6. GTP is generated at succinyl CoA synthetase, and CO_2 is generated twice, at isocitrate dehydrogenase and α-ketoglutarate dehydrogenase. The GTP generated in the cycle by substrate-level phosphorylation is formally equivalent to ATP as the two high-energy compounds are readily interconverted. The overall TCA cycle is

$$\text{Acetyl CoA} + 2H_2O + GDP + Pi + FAD + 3NAD^+ \rightarrow$$
$$\text{CoASH} + 2\,CO_2 + GTP + 3\,NADH + 3\,H^+ + FADH_2$$

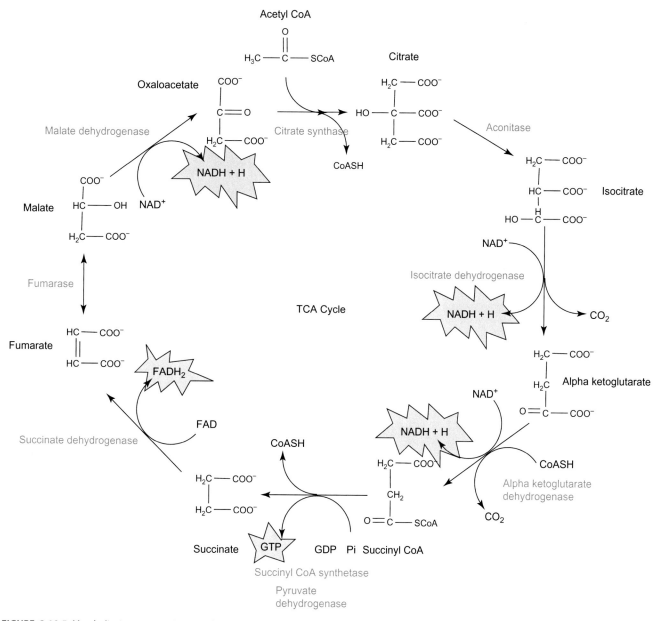

FIGURE 2.10.5 Metabolic interconversions in the TCA cycle. Note that the two-carbon fragment, acetic acid, is carried by CoA to combine with oxaloacetate to form citrate. The oxaloacetate is regenerated by the cycle. Thus the cycle's intermediates are catalysts in that they are not consumed. The two carbons in acetate are converted to CO_2 by the cycle.

The result is that the two carbons in acetate are converted to CO_2 and a bunch of reducing equivalents (8 per 2-carbon acetate). Note that O_2 is not explicitly required for this process, but it is required for the continued operation of this cycle. If we add in the formation of acetyl CoA from pyruvate, the overall reaction is

Pyruvate + 3 H_2O + GDP + Pi + FAD + 4 NAD^+ →

3 CO_2 + GTP + 4 NADH + 4 H^+ + $FADH_2$

The NADH and $FADH_2$ produced during this overall series of reactions must be returned to NAD^+ and FAD, or the process will stop. The reduced NADH and $FADH_2$ are oxidized by a special system called the **electron transport chain (ETC)**. In the process of being oxidized, the energy stored in these compounds enables the synthesis of ATP through a process called **oxidative phosphorylation**.

THE ETC LINKS CHEMICAL ENERGY TO H^+ PUMPING OUT OF THE MITOCHONDRIA

The ETC consists of an array of proteins inserted in the inner mitochondrial membrane. The overall plan is this: NADH delivers two electrons to a series of chemicals that differ in their chemical affinity for these electrons (see Figure 2.10.7). This is expressed in their reduction potential (see above) which is related to

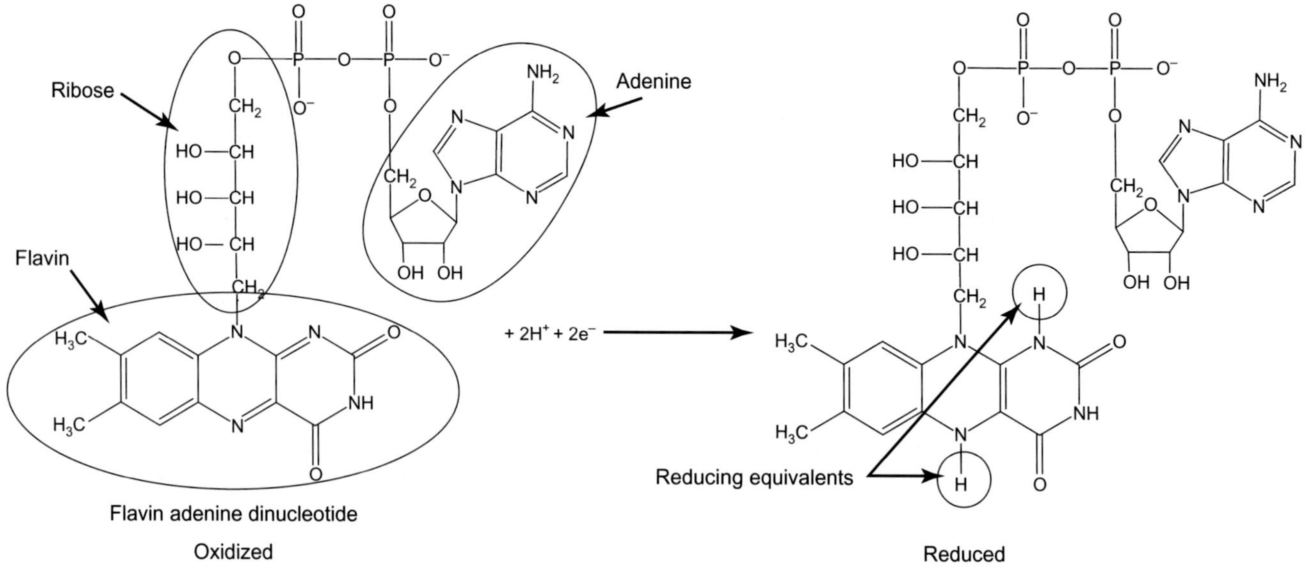

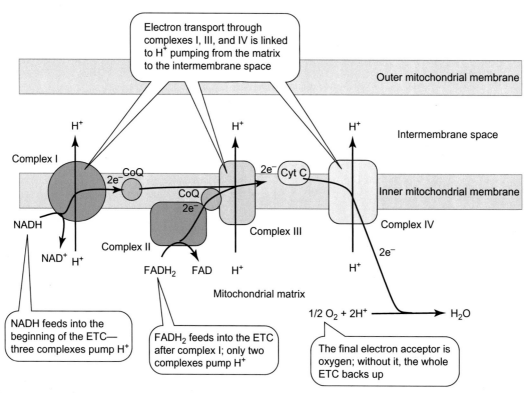

FIGURE 2.10.6 Structure of oxidized and reduced FAD. FAD stands for "flavin adenine dinucleotide." It consists of a flavin part, a ribose part, and an adenine part. The ribose + flavin is better known as riboflavin or vitamin B_2. The binding of reducing equivalents is associated with gain of energy, which can be released on oxidation.

FIGURE 2.10.7 The electron transport chain (ETC). NADH feeds in reducing equivalents at the beginning of the ETC, which hands them on to proteins with progressively higher affinity until at the end of the chain the electrons are combined with oxygen. Complexes I, III, and IV use the chemical energy of oxidation to pump H^+ ions from the mitochondrial matrix to the intermembrane space. This makes an electrical current that separates charge and produces a potential difference across the mitochondrial membrane.

their free energy. The energy is released gradually, in steps, and the ETC complexes use the decrease in free energy to pump hydrogen ions from the matrix space to the intermembrane space between the inner and outer mitochondrial membranes. This pumping of hydrogen ions produces an electrochemical gradient for hydrogen ions and the energy in this gradient is used to generate ATP from ADP and Pi.

OXYGEN ACCEPTS ELECTRONS AT THE END OF THE ETC

Molecular oxygen oxidizes the last step in the ETC. This is the point at which oxygen is consumed by the mitochondria, producing water. Without oxygen to finally oxidize the ETC, the chain itself will remain reduced and no further reducing agents can be fed through it. That is, NADH cannot be converted to NAD^+ (at the far left of the chain) in the absence of oxygen. $FADH_2$ also cannot be converted to FAD in the absence of oxygen. The $\Delta E_0'$ for the redox reaction $\frac{1}{2}O_2 + 2H^+ + 2e^- \rightarrow H_2O$ is $+0.82$ volts. The cascade of electrons down the ETC is shown diagrammatically in Figure 2.10.8. Note that the electrons travel down the cascade towards ever more positive reduction potentials, as these signify increasing affinity for electrons.

PROTON PUMPING AND ELECTRON TRANSPORT ARE TIGHTLY COUPLED

If the proton gradient is at equilibrium with the free energy of electron transport, then the electrons cannot be transported through the ETC. The energy stored in the electrochemical gradient of protons across the inner mitochondrial membrane must also be drained in some way for the ETC to continue operating. **The energy in the proton electrochemical gradient is used to make ATP.** The coupling of the electrochemical gradient of H^+ across the inner mitochondrial membrane with ATP synthesis is called **chemiosmotic coupling** (because there is a concentration difference across the membrane and an electric potential). It was first proposed by Peter Mitchell in 1961, who was awarded the Nobel Prize for the work in 1978.

THE ATP SYNTHASE COUPLES INWARD H^+ FLUX TO ATP SYNTHESIS

The inner mitochondrial membrane contains many copies of a protein called the **F0F1ATPase**. This is also called **ATP synthase**. It consists of two parts: the F_0 component spans the membrane and provides a channel for protons to move into the matrix from the intermembrane space. The F_1 component is a complex of five proteins with the composition $\alpha_3\beta_3\gamma\delta\varepsilon$, with a molecular weight of about 360,000. The F_0 Part of the complex consists of an integral membrane, a subunit, a b dimer, and 8–15 small c-subunits. The structure of the ATP synthase is shown in Figure 2.10.9. This remarkable complex couples the movement of H^+ to the synthesis of ATP through mechanical intermediates. Hydrogen ions from the inner matrix access the c subunit via a channel in the a subunit, causing a rotation of the c subunit turbine. This rotates the γ subunit, which has a cam-like protrusion that deforms the α and β subunits. Each time the cam passes one of the three $\alpha\beta$ complexes, ATP is formed from ADP and Pi bound to the $\alpha\beta$ subunits. Because there are three of these $\alpha\beta$ subunits, each turn of the c-protein turbine produces 3 ATP molecules.

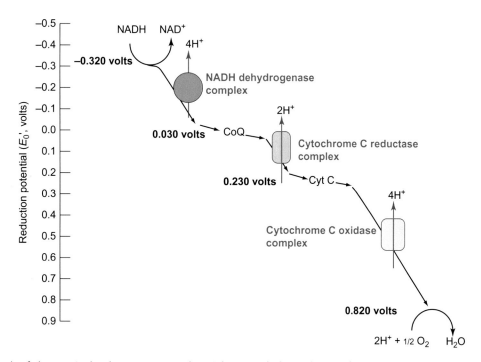

FIGURE 2.10.8 Cascade of electrons in the electron transport chain. It begins with the production of NADH by reactions in glycolysis or TCA cycle. The reduction potential of NADH/NAD^+ is -0.320 volts. Electrons are passed to the NADH dehydrogenase complex that pumps $4H^+$ per 2 e^- out of the mitochondrial matrix to the intermembrane space. Electrons are then passed to Coenzyme Q, with a reduction potential of 0.030 volts. Coenzyme Q passes the electrons to the cytochrome C reductase complex that pumps $2H^+$ per 2 e^-. Electrons then are transferred to cytochrome C with a reduction potential of 0.230 volts. The electrons that are taken by the Cytochrome C oxidase complex that pumps $4H^+$ per 2 e^-. Cytochrome C oxidase is finally oxidized by molecular oxygen, whose reduction potential is 0.820 volts.

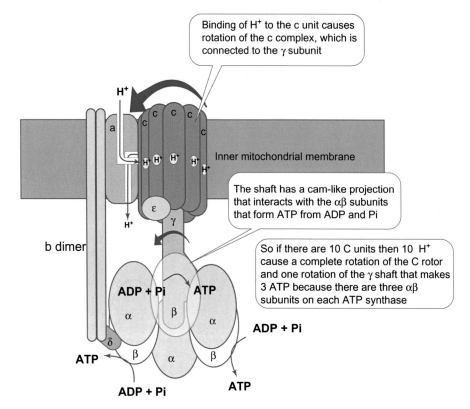

FIGURE 2.10.9 The F_0F_1ATPase or the ATP synthase of mitochondria. It consists of a complex of proteins that make up a tiny H^+-driven turbine. H^+ ions enter from the intermembrane space through protein a. They bind to a c unit in the rotator, which cause a rotation of the c-complex. After a nearly complete rotation, the H^+ ion is removed to the mitochondrial matrix. The c-complex binds the γ subunit and rotates it. The $\alpha\beta$ subunits of the head are kept steady by the stator components, δ and the b dimer. The γ subunit has a projection that interacts with each of the $\alpha\beta$ subunits, and this mechanical interaction is used to synthesize ATP from ADP and Pi. Complete rotation of the c-complex requires as many H^+ as c-subunits. For each complete rotation, the ATP synthase makes 3 ATP molecules.

THE PROTON ELECTROCHEMICAL GRADIENT PROVIDES THE ENERGY FOR ATP SYNTHESIS

The ETC pumps H^+ ions out of the matrix into the intermembrane space. The stoichiometry is about $10H^+$ ions per $2e^-$ when they originate from NADH, and about $6H^+$ when the $2e^-$ originate from $FADH_2$ (see Figure 2.10.7). Because the H^+ ions move without counter ions, this movement is an outward current that separates charge, and therefore there is a potential developed across the inner mitochondrial membrane. This potential varies depending on the state of mitochondrial activity, but a typical value is about 160 mV, negative inside. In addition to the potential, there is a concentration difference in H^+ established across the membrane. The pH of the intermembrane space is about 7.0, whereas the pH of the matrix is about 8.0. Recall that $pH = -\log [H^+]$, so that $pH = 7.0$ implies that $[H^+] = 10^{-7}$ M and $pH = 8$ means $[H^+] = 10^{-8}$ M. Thus there is a 10-fold difference in the $[H^+]$ established by the ETC. When H^+ ions travel from the intermembrane space to the matrix, they release the free energy stored in the electrochemical gradient for H^+, enabling the F_1 subunit to synthesize ATP from ADP and Pi. This **proton electrochemical gradient** is sometimes called the **proton motive force**. The free energy for H^+ transfer from the intermembrane space to the mitochondrial matrix is calculated as

[2.10.10]
$$\begin{aligned}\Delta\mu_{out\to in} &= \mu_{in} - \mu_{out} \\ &= \mu^0 + RT\ln[H^+]_{in} + \Im\psi_{in} - \mu^0 \\ &\quad - RT\ln[H^+]_{out} - \Im\psi_{out} \\ &= RT\ln\frac{[H^+]_{in}}{[H^+]_{out}} + \Im(\psi_{in} - \psi_{out})\end{aligned}$$

The free energy change for ATP synthesis under the conditions of the cell varies from cell to cell and from place to place within the cell because the local concentrations of ADP, Pi, ATP, and ions that bind to them (H^+, Ca^{2+}, and Mg^{2+}) also vary from place to place. Nevertheless, we have already calculated an approximate free energy change for ATP hydrolysis under conditions of the cell to be -57.1 kJ mol^{-1}. The free energy of ATP synthesis should be the opposite of this, $+57.1$ kJ mol^{-1}.

According to the result in Example 2.10.1, there is not enough energy in one H^+ transport to synthesize ATP. If we assume integral stoichiometry, we need at least three of them. The free energy for the reaction

[2.10.11] $\qquad ADP + Pi + 3H^+_{out} \to ATP + 3H^+_{in}$

> **EXAMPLE 2.10.1 Calculate the Free Energy in the Mitochondrial H$^+$ Electrochemical Gradient**
>
> The free energy per mole of H$^+$ is given by Eqn [2.10.1] as
>
> $$\Delta \mu_{H_{out} \to H_{in}} = RT \ln\left(\frac{[H^+]_{in}}{[H^+]_{out}}\right) + \Im(\psi_{in} - \psi_{out})$$
>
> Inserting values of $R = 8.314$ J mol^{-1} K^{-1}, $T = 310$ K, $[H^+]_{in} = 10^{-8}$ M, $[H^+]_{out} = 10^{-7}$ M, we calculate the chemical part as
>
> $$RT \ln\left(\frac{[H^+]_{in}}{[H^+]_{out}}\right) = 8.314 \text{ J mol}^{-1} \text{ K}^{-1} \times 310 \text{ K} \times \ln\left(\frac{10^{-8} \text{ M}}{10^{-7} \text{ M}}\right)$$
> $$= -5.93 \text{ kJ mol}^{-1}$$
>
> Using $\Im = 9.649 \times 10^4$ C mol^{-1} and $\psi_{in} = -0.16$ V (with $\psi_{out} = 0$), the electrical part of the free energy change is
>
> $$\Im(\psi_{in} - \psi_{out}) = 9.649 \times 10^4 \text{ C mol}^{-1} \times (-0.16 \text{ V} - 0)$$
> $$= -15.44 \text{ kJ mol}^{-1}$$
>
> Thus the total free energy change for H$^+$ transfer from the intermembrane space to the matrix, for this condition given here, is **-21.37 kJ mol^{-1}**.

is the sum of the free energy of two processes:

[2.10.12]
$$ADP + Pi \to ATP$$
$$3H^+_{out} \to 3H^+_{in}$$

We add the two to get

[2.10.13]
$$\Delta \mu = \Delta \mu_{ADP+Pi \to ATP} + 3\Delta \mu_{H^+_{out} \to H^+_{in}}$$
$$= 57.1 \text{ kJ mol}^{-1} + 3(-21.37 \text{ kJ mol}^{-1})$$
$$= -7.01 \text{ kJ mol}^{-1}$$

The negative free energy change for this coupled reaction indicates that this process will proceed spontaneously. That is, there is enough energy in the electrochemical gradient of H$^+$ across the inner mitochondrial membrane to synthesize 1 ATP for every 3H$^+$ ions transported. As it turns out, the stoichiometry is not integral.

NADH FORMS ABOUT 2.5 ATP MOLECULES; FADH$_2$ FORMS ABOUT 1.5 ATP MOLECULES

The amount of ATP formed from oxidative phosphorylation has been controversial but a consensus seems to have been reached. Measurements show that electron transport beginning with NADH results in 10H$^+$ ions being transported from matrix to intermembrane space, and with FADH$_2$ the number is 6, because the first complex is bypassed. What happens to these H$^+$ ions? Most are used to drive the ATP synthase as described in Figure 2.10.9, but some are used to bring the phosphate into the mitochondria from the cytosolic compartment (see Figure 2.10.10) and some are used for other transport processes. Mitochondria from the heart of cows has 8 c-subunits in their ATP synthase, suggesting that 8H$^+$ ions are needed for one complete rotation of the rotor and synthesis of 3 ATP molecules. This gives a nonintegral stoichiometry: each ATP requires $8/3 = 2.67$ H$^+$ ions! Our calculation above indicates that the minimum number is 57.1 kJ mol^{-1}/21.87 kJ mol^{-1} = 2.61. Because 1H$^+$ is required to import Pi, the number of ATP produced by NADH becomes $10/(2.67 + 1) = 2.7$ ATP/NADH. If the proton motive force is used to drive other processes, the ATP yield will be lower. Recent studies suggest a relatively constant H$^+$/ATP ratio of 4.0, including transport processes. In this case ATP production from NADH would be 10H$^+$ per NADH/4H$^+$ per ATP = 2.5 ATP per NADH. For FADH$_2$, the ratio would be 6H$^+$ per FADH$_2$/4H$^+$ per ATP = 1.5 ATP per FADH$_2$. These numbers are approximate and tentative. The ATPase from different sources has different c-ring sizes that may cause differences in the H$^+$/ATP ratio and therefore the ATP/NADH ratio.

ATP CAN BE PRODUCED FROM CYTOSOLIC NADH

The NADH produced in the cytosol by glycolysis cannot enter the mitochondrial matrix, yet it must be oxidized back to NAD$^+$ to allow glycolysis to proceed. This can be accomplished by lactate dehydrogenase, as described in Chapter 2.9, but this does not extract the energy of combustion remaining in the lactic acid. Two types of shuttle mechanisms have the effect of bringing cytosolic reducing equivalents into the matrix, without NADH itself actually entering the matrix. These shuttles are the **glycerol phosphate shuttle** and the **malate/aspartate shuttle**.

In the glycerol phosphate shuttle, NADH is oxidized to NAD$^+$ by the cytosolic glycerol-3-phosphate dehydrogenase, while dihydroxyacetone phosphate is simultaneously reduced to glycerol-3-phosphate. Glycerol-3-phosphate then penetrates the mitochondrial outer membrane and reduces FAD to FADH$_2$ by the mitochondrial glycerol-3-phosphate dehydrogenase to form dihydroxyacetone phosphate. In this way, we start with cytoplasmic NADH and dihydroxyacetone phosphate and we end up with mitochondrial FADH$_2$ and cytoplasmic NAD$^+$ and dihydroxyacetone phosphate. So the reducing power of NADH is transferred to FADH$_2$, which then enters the ETC to generate 1.5 ATP molecules. Because complex I is bypassed, only 1.5 ATP are made per molecule of NADH passed on to the mitochondria by the glycerol P shuttle. Figure 2.10.11 illustrates the glycerol phosphate shuttle.

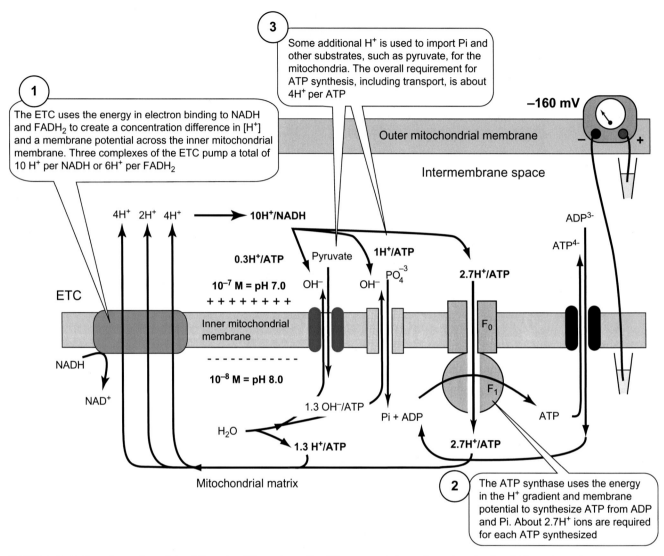

FIGURE 2.10.10 Overall coupling of the ETC to the ATP synthase. The ETC pumps electrons from the mitochondrial matrix to the intermembrane space, creating a [H^+] concentration gradient and an electrical potential difference. The ATP synthase uses the energy in this gradient to link ATP synthesis to H^+ ions going down their electrochemical gradient. Because of its mechanism (see Figure 2.10.9) in heart mitochondria $8H^+$ ions make 3 ATP molecules, for a stoichiometry of 2.7H^+/ATP. However, the H^+ gradient is used for mitochondrial transport as well. For each Pi that enters the mitochondria, 1.0H^+ ion is used—the exit of OH^- is equivalent to the entry of H^+. Small amounts of H^+ flow is also used for other transport processes. The total H^+ required for ATP synthesis, including transport, is about 4.0. Because each NADH causes ETC to pump 10H^+ ions, this means that about 2.5 ATP molecules are formed per NADH. ATP exit and ADP entry do not require energy, as shown.

In the malate/aspartate shuttle, cytoplasmic NADH is used to convert cytoplasmic oxaloacetate to malate, which can be carried across the mitochondrial inner membrane by a **dicarboxylate carrier** by facilitated diffusion with no metabolic energy expenditure. Inside the matrix, the malate is converted back to oxaloacetate, generating the NADH back, which then transfers electrons to the ETC. To complete the cycle, oxaloacetate must get back outside. This is accomplished by converting oxaloacetate to aspartate (using glutamate as a substrate). The aspartate is transported out of the mitochondria where it is converted back to oxaloacetate and glutamate. The cycle is completed when glutamate goes back into the mitochondria. In this shuttle, 2.5 ATP molecules are produced for each cytosolic NADH because it is effectively transferred into the matrix as NADH. Figure 2.10.12 illustrates the malate/aspartate shuttle.

MOST OF THE ATP PRODUCED DURING COMPLETE GLUCOSE OXIDATION COMES FROM OXIDATIVE PHOSPHORYLATION

Figure 2.10.13 shows the production of ATP throughout glycolysis and the TCA cycle. Glycolysis utilizes 2 moles of ATP per mole of glucose and then produces 4 moles ATP per mole by substrate-level phosphorylation. The 2 moles of NADH produced by glyceraldehyde-3-P dehydrogenase can be converted to either 5 moles of

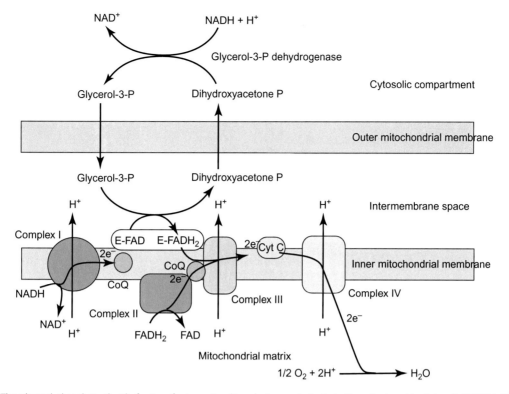

FIGURE 2.10.11 The glycerol phosphate shuttle for transferring cytosolic reducing equivalents to the mitochondria. Cytosolic NADH + H$^+$ is converted to NAD$^+$ by glycerol P dehydrogenase. Glycerol-3-P crosses the outer mitochondrial membrane and is converted back to dihydroxyacetone P by a mitochondrial glycerol-3-P dehydrogenase in the inner mitochondrial membrane. The dihydroxyacetone P goes back into the cytosol. The reduced mitochondrial glycerol-3-P dehydrogenase reduces ubiquinone in the inner membrane, which passes the reducing equivalents on to complex III of the ETC.

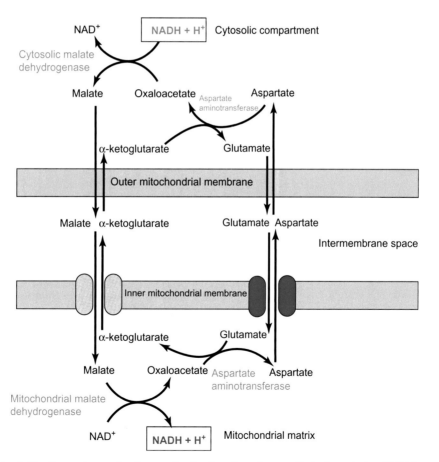

FIGURE 2.10.12 The malate shuttle. Two cycles running in opposite directions have the net effect of transferring NADH from cytosol to mitochondrial matrix. In one cycle, oxaloacetate is converted to malate while NADH is converted to NAD$^+$. Malate crosses the inner mitochondrial membrane where it is converted back to oxaloacetate and NADH. The oxaloacetate is then converted to aspartate, which leaves the mitochondria and passes back into the cytosol where the aspartate is converted back to oxaloacetate. A second cycle runs in the opposite direction. Malate entry into the mitochondrial matrix is accompanied by α-ketoglutarate exit. In the cytosol, the latter is converted to glutamate by aspartate amino transferase, which links the reaction α-ketoglutarate → glutamate to the reaction aspartate → oxaloacetate. The glutamate exchanges with aspartate across the inner mitochondrial membrane.

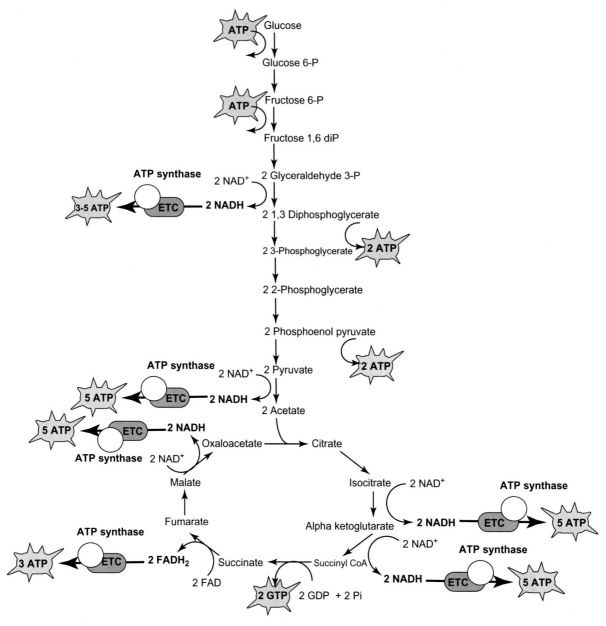

FIGURE 2.10.13 Overall ATP production from glycolysis, TCA cycle, and ETC. Glycolysis produces 4 ATP, but consumes 2 ATP molecules, per mole of glucose. It also produces 2 moles NADH per mole of glucose. These reducing equivalents are cytosolic and can result in 3−5 moles of ATP depending on how the NADH enters the mitochondria. In the mitochondria, a total of 8 moles of NADH are produced per mole of glucose, including 2 in the pyruvate dehydrogenase step and 6 in the TCA cycle. These each produce 2.5 moles of ATP per mole of NADH, so the NADH produces 8 moles NADH/mol glucose × 2.5 moles ATP/mol NADH = 20 moles ATP/mol glucose. The 2 $FADH_2$ produced at succinate dehydrogenase produce 1.5 moles of ATP per mole of $FADH_2$ for a total of 3 moles ATP/mol glucose. Succinyl CoA synthetase produces 2 moles GTP/mol glucose, which is energetically equivalent to 2 moles ATP/mol glucose. Total ATP production is thus 5−7 moles in glycolysis +25 moles in the mitochondria, for a total of 30−32 moles of ATP per mole of glucose.

ATP through the malate/aspartate shuttle or 3 moles of ATP through the glycerol phosphate shuttle.

MITOCHONDRIA HAVE SPECIFIC TRANSPORT MECHANISMS

In order for oxidative phosphorylation to work, the inner mitochondrial membrane must be impermeable to H^+ ions. H^+ ions in solution are usually bound to water as H_3O^+, the hydronium ion, which is very small. Therefore, the inner mitochondrial membrane must be relatively impermeable to most ions in order for the membrane to establish the potential difference and concentration difference in $[H^+]$. At the same time, things have to be able to get in and out. We have already discussed two such carriers—those that operate the shuttles that allow cytosolic reducing equivalents to enter the matrix. Several other carriers are shown schematically in Figure 2.10.14.

ATP produced in the matrix must leave the matrix to power cellular activities. This occurs through facilitated diffusion by the **ATP−ADP translocase**. Since ATP has

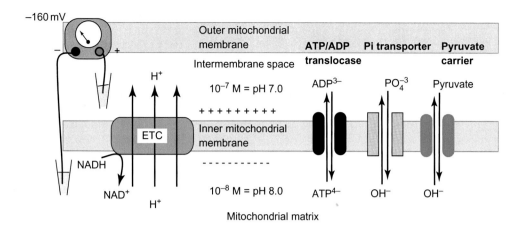

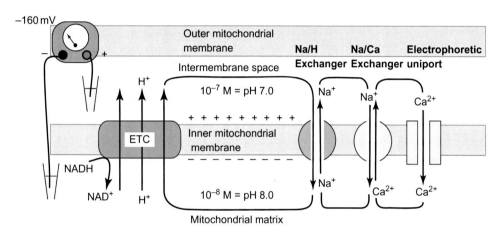

FIGURE 2.10.14 Selected transport mechanisms in mitochondria. See text for details.

more negative charges, and its concentration is higher in the mitochondria where it is produced, the transport is "downhill" in both directions. This translocase is poisoned by **atractyloside**.

Phosphate must also enter the matrix in order to be incorporated into ATP. Since phosphate is highly charged, movement across the membrane against a strong electric force is energetically unfavorable. Phosphate is carried across in exchange for OH^- ions. The outward flow of OH^- is equivalent to an inward flow of H^+. Therefore, some of the energy of the electrochemical H^+ gradient is used to transport phosphate into the mitochondria.

Mitochondria also take up Ca^{2+} because of the large negative potential inside. This uptake occurs through a channel called the **electrophoretic uniport**. This name signifies that the electrical gradient is the driving force for Ca^{2+} movement and that only one ion moves (uniport). Ca^{2+} taken up this way must be able to exit. This is accomplished by a **Na/Ca exchanger** that couples Ca^{2+} exit with Na^+ entry. This is another example of secondary active transport, in which "uphill" transport of Ca^{2+} is linked to "downhill" movement of Na^+. Of course, the Na^+ taken up by this process must also have an exit. Na^+ efflux from the mitochondria is through a **Na/H exchanger**. The Na/H exchanger is also an example of secondary active transport, in which uphill movement of Na^+ is coupled to downhill movement of H^+. In the mitochondria neither Na^+ efflux nor H^+ entry is linked to ATP hydrolysis. They are powered by the ETC pumping H^+ out of the mitochondrial matrix into the intermembrane space, thereby creating a large potential and a concentration gradient for H^+ ions.

SUMMARY

Glycolysis converts glucose into pyruvate, which enters the mitochondria by facilitated diffusion on its own carrier. The mitochondria couples the oxidation of pyruvate to the formation of the high-energy chemical bond in ATP, a process called oxidative phosphorylation. Inside the mitochondrial matrix, pyruvate enters into a series of reactions, beginning with pyruvate dehydrogenase. This series of reactions converts the three-carbon pyruvate molecule into three molecules of CO_2. In the first step, pyruvate dehydrogenase converts pyruvate into acetyl coenzyme A, producing a molecule of CO_2 and a molecule of NADH, nicotinamide adenine dinucleotide. Some of the chemical energy of oxidation of pyruvate is captured in the energy of electrons binding to NADH.

This energy is converted to ATP later on by the ATP synthase that relies on the electrochemical gradient of H^+ that is established by the ETC.

The second reaction combines acetyl CoA with oxaloacetate to form citric acid. This is the first step in the citric acid cycle, or the tricarboxylic acid cycle (TCA cycle), also called the Krebs cycle for Sir Hans Krebs. This cycle is a catalytic cycle in that it regenerates oxaloacetate and all parts of the acetate part of acetyl CoA are converted to CO_2 or H_2O. The cycle captures energy of oxidation by forming NADH or $FADH_2$, flavin adenine dinucleotide, or by forming GTP directly in the conversion of succinyl CoA to succinate by succinyl CoA synthetase. Each turn of the cycle produces 3 NADH molecules, 1 $FADH_2$ molecule, 1 molecule of GTP, and 2 molecules of CO_2.

The mitochondria make additional ATP molecules from NADH and $FADH_2$. These transfer their electrons to the ETC, which uses the energy of binding of the electrons, expressed as the reduction potential, to pump hydrogen ions from the mitochondrial matrix to the intermembrane space. This active pumping produces a concentration difference of H^+ ions and a large electrical potential. The ATP synthase in the inner mitochondrial membrane uses the energy in the electrochemical potential difference for H^+ to make ATP by a process that converts electrochemical energy to mechanical energy and back again to ATP. NADH enters the ETC early and its energy causes H^+ pumping at complexes I, III, and IV of the ETC. A total of $10H^+$ are pumped out per NADH molecule. The ATP synthase requires $4H^+$ ions to make ATP (about 3 for the direct use of the ATP synthase and 1 for transport of reactants into the mitochondria). As a consequence, each NADH produces about 2.5 ATP molecules. $FADH_2$ enters the ETC after complex I, and so it makes only 1.5 ATP molecules.

Thus complete oxidation of pyruvate to CO_2 and H_2O forms 4 NADH, 1 $FADH_2$, and 1 GTP for a total equivalent of $4 \times 2.5 + 1 \times 1.5 + 1 = 12.5$ ATP molecules. Since there are two pyruvate molecules formed from glucose, the TCA cycle accounts for 25 ATP molecules per glucose molecule.

NADH is also generated by glycolysis in the cytoplasm. This NADH is oxidized in the mitochondria, but indirectly because NADH itself cannot cross the inner mitochondrial membrane. Instead, two shuttle systems have the effect of transferring cytosolic NADH to mitochondrial matrix NADH. The glycerol phosphate shuttle converts cytosolic NADH to mitochondrial $FADH_2$, whereas the malate shuttle converts it to mitochondrial NADH. Glycolysis produces a net gain of 2 ATP and 2 NADH per glucose molecule.

REVIEW QUESTIONS

1. How does pyruvate get into the mitochondria? How does cytosolic NADH get into the mitochondria? How do ADP and Pi get into the mitochondria? How do ATP get out of the mitochondria?
2. Is NADH reduced or oxidized? Is FAD reduced or oxidized? Why does NADH make more ATP than $FADH_2$?
3. What is the TCA cycle? How many CO_2 molecules are released per pyruvate? Per acetate? In general, where is CO_2 released?
4. How many ATP molecules are produced during glycolysis per mole of glucose? Does this ATP production require oxygen?
5. What determines the direction of electron flow in the ETC? How do you calculate the energy of a reduction−oxidation reaction?
6. How many ATP molecules are produced in mitochondria during oxidative phosphorylation? Does this ATP production require oxygen?
7. Why is there a membrane potential across the inner mitochondrial membrane? If the ATP synthetase lets in H^+, why does not this current depolarize the mitochondria?
8. What is the chemiosmotic hypothesis? Some materials are proton ionophores. What effect would these have on oxidative phosphorylation?

ATP Production III: Fatty Acid Oxidation and Amino Acid Oxidation 2.11

Learning Objectives

- Describe the chemical structure of a triglyceride
- Describe how adipocyte lipolysis is regulated by catecholamines and insulin
- Describe the main route of glycerol oxidation
- Describe the role of carnitine in the import of fatty acids into mitochondria
- List the number of NADH, FADH$_2$, and acetyl CoA produced for each turn of the beta oxidation cycle
- Account for the number of beta oxidation cycles for palmitic acid
- List the number of NADH, FADH$_2$, and GTP produced for oxidation of acetyl CoA
- Account for the total numbers of ATP molecules produced by oxidation of palmitic acid
- List three chemicals that comprise the ketone bodies
- Distinguish between the terms glucogenic and ketogenic for amino acids
- List the amino acids that are exclusively ketogenic
- Recognize the amino acids that are exclusively glucogenic
- Name the amino acid that is required for feed into the urea cycle

FATS AND PROTEINS CONTRIBUTE 50% OF THE ENERGY CONTENT OF MANY DIETS

In the previous chapters, we saw how carbohydrates are metabolized through glycolysis to form pyruvate, producing some energy in the process. Pyruvate is then converted to acetyl CoA, which enters the TCA cycle to produce reducing equivalents. These reduced compounds, NADH and FADH$_2$, are then oxidized by the respiratory chain of the mitochondria to produce a proton electrochemical gradient that is then used to produce ATP. The typical American diet contains only about 49% of its calories as carbohydrates, with another 35% coming from fat and 16% from protein. Thus the fats and proteins must also be used to generate cellular energy. How are fats and proteins used to make ATP?

DEPOT FAT IS STORED AS TRIGLYCERIDES AND BROKEN DOWN TO GLYCEROL AND FATTY ACIDS FOR ENERGY

Most fatty acids in the body are stored as triglycerides (triacylglycerol, TAG; see Figure 2.11.1), the acyl esters of three fatty acids with a glycerol molecule. Blood carries triglycerides in special structures called **lipoproteins**, in which the water-insoluble lipids are coated with special proteins. These lipoproteins are made solely in the intestines and the liver. All tissues can also store triglycerides in lipid droplets in the cytoplasm,

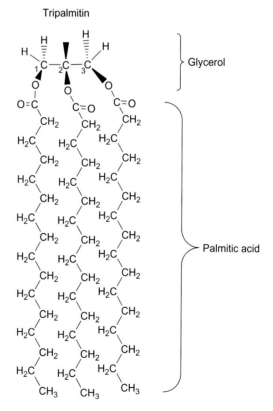

FIGURE 2.11.1 Structure of tripalmitin as an example of a triglyceride. These are stored in adipose tissue and released into the circulation. Lipase breaks down the triglyceride by hydrolyzing the ester bonds between the fatty acids carboxyl group and the glycerol hydroxyl group.

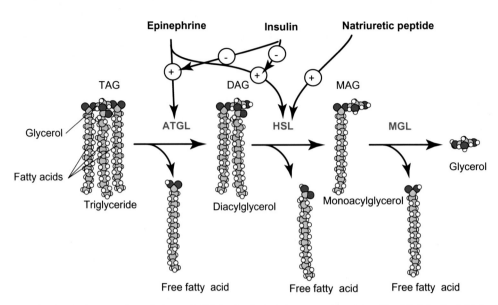

FIGURE 2.11.2 Adipocyte lipolysis. Adipose triglyceride lipase (ATGL) begins lipolysis by converting triglycerides (TAG) in the lipid droplet to diacylglycerol (DAG) and a free fatty acid. Hormone-sensitive lipase (HSL) continues the process by hydrolyzing a second fatty acid, producing monoacylglycerol (MAG). Monoacylglycerol lipase (MGL) then hydrolyzes the last fatty acid ester bond to produce glycerol and a third free fatty acid. Hormones and nerves control lipolysis by activating or inhibiting ATGL and HSL. Epinephrine can activate ATGL and HSL through β receptors and this action is blocked by insulin. Epinephrine can inhibit ATGL and HSL through α_2 receptors. Natriuretic peptides also can activate lipolysis through independent mechanisms.

but adipocytes contain very large lipid droplets that may occupy most of the cell. The adipocytes are specialized for storage and release of lipids for energy. The circulating triglycerides can be hydrolyzed to form glycerol and three fatty acids through the action of lipoprotein **lipase** that both hydrolyzes the lipid and acts as a bridge in lipoprotein uptake. Lipid droplets can also be hydrolyzed for energy within the cell, but the adipocyte alone can export the glycerol and free fatty acids derived from the lipid store.

Lipid droplets are surrounded by a phospholipid monolayer to which is absorbed the protein **perilipin**, which has five members (PLIN1-5) that are distributed in different tissues. In adipose tissue, the major form is PLIN1. PLIN1 stabilizes the lipid droplet against lipolysis. Lipolysis begins by **ATGL, adipose triglyceride lipase**, that converts tricacylglycerides (TAG) into diacylglycerol (DAG) and a free fatty acid (FA). **Hormone-sensitive lipase** (HSL) then converts the DAG into monoacylglycerol (MAG) and a free fatty acid. The MAG is further converted to glycerol and an FA by the enzyme **monoglycerol lipase**. Control of this process is hormonal and neural, as shown in Figures 2.11.2 and 2.11.3. The catecholamines, derived either from circulating epinephrine or from sympathetic nervous stimulation, can either activate or inhibit lipolysis. Insulin inhibits lipolysis, and natriuretic peptide stimulates it.

In Chapter 2.9, we learned that epinephrine mobilizes liver glycogen stores by activating glycogenolysis through a G_s mechanism. Epinephrine can also mobilize lipid stores by activating lipolysis through a G_s mechanism, as shown in Figure 2.11.3. Binding of catecholamines to β_1 or β_2 receptors on the adipocyte is followed by the activation of adenylyl cyclase and an increase in 3',5'-cyclic AMP in the adipocyte cytosol. This activates protein kinase A (PKA) that phosphorylates a number of targets including perilipin, PLIN1, and HSL (hormone-sensitive lipase). PLIN1 binds ABHD5 (α/β hydrolase domain containing 5) that activates ATGL. Phosphorylation of PLIN1 releases the ABHD5 to activate ATGL. Simultaneously, HSL that was previously located in the cytosol binds to the lipid droplet where it participates in lipolysis.

Adipocytes also have α_2 receptors that are coupled to a G_i mechanism that inhibits adenylyl cyclase activity. Occupancy of these receptors has the opposite effect, an inhibition of lipolysis, as occupancy of the β receptors. Insulin inhibits lipolysis by activating cAMP phosphodiesterase, type 3B, by phosphorylation by PKB. This activated PDE-3B decreases the cAMP concentration, leading to inactivation of lipolysis.

Natriuretic peptide increases lipolysis by activation of HSL through phosphorylation by PKG, a cGMP-activated protein kinase.

GLYCEROL IS CONVERTED TO AN INTERMEDIATE OF GLYCOLYSIS

Plasma glycerol is taken up by tissues and converted to α-glycerophosphate by an enzyme called **glycerol kinase**, by phosphorylating the glycerol with the terminal phosphate group of ATP. The glycerophosphate is converted to **dihydroxyacetone phosphate**

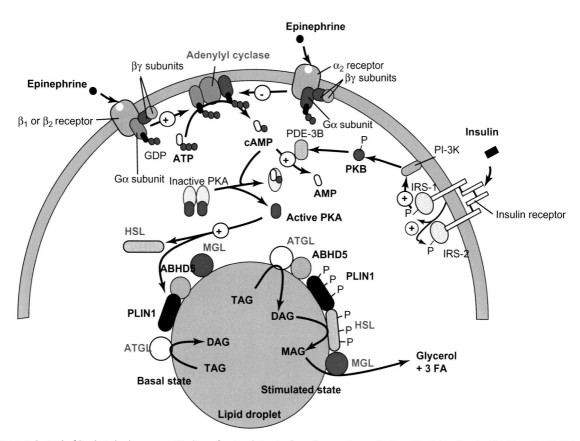

FIGURE 2.11.3 Control of lipolysis by hormones. Binding of epinephrine to β_1 or β_2 receptors activates adenylyl cyclase by the $G_{\alpha s}$ subunit. This increases formation of cAMP and therefore increases the concentration of cAMP in the cell. Increased cAMP activates protein kinase A that phosphorylates PLIN1 and HSL. Phosphorylation of PLIN1 releases ABHD5 that then binds to ATGL and activates it. PKA phosphorylation of HSL results in its translocation from the cytosol to the lipid droplet surface. In this position, the enzymes completely hydrolyze TAG to glycerol and 3 fatty acids, which are exported out of the adipocyte. Lipolysis is inhibited by binding of epinephrine to α_2 receptors that inhibit adenylyl cyclase. Insulin binding to its receptor results in a cascade that activates PKB that phosphorylates cAMP phosphodiesterase. This lowers the cAMP concentration and inhibits lipolysis.

by another enzyme, α-glycerophosphate dehydrogenase, requiring NAD^+, and thus the glycerol can enter into glycolysis and the TCA cycle to be fully oxidized to CO_2 and H_2O to provide energy as ATP to the cell (see Figure 2.11.4).

FATTY ACIDS ARE METABOLIZED IN THE MITOCHONDRIA AND PEROXISOMES

Free fatty acids are formed in the cytoplasm by the action of lipase on stored triglycerides, but the fatty acids themselves are degraded and oxidized only in the mitochondria and peroxisomes. The fatty acids have surface activity (they lower the surface tension) and can impair membrane integrity. Therefore, the fatty acids are carried in solution by **fatty acid binding proteins**. These are low-molecular-weight proteins (about 14,000 Da) that probably have a dual function of decreasing the concentration of the free fatty acids and of enhancing the diffusion through the cytoplasm by carrying the fatty acids.

The first step in the metabolism of free fatty acids is their import into the inner mitochondrial matrix by combining with a carrier substance, **carnitine** (see Figure 2.11.5). First, the fatty acid is combined with coenzyme A by the enzyme **thiokinase**, which hydrolyzes ATP to AMP and PPi. This fatty acyl CoA is then transferred to carnitine by the enzyme **carnitine fatty acyl transferase**. Once the fatty acyl carnitine is transferred to the mitochondrial matrix, it is once again combined with CoA to form fatty acyl CoA. In this form and in this place, the fatty acid can be oxidized in a systematic way to produce energy.

BETA OXIDATION CLEAVES TWO CARBON PIECES OFF FATTY ACIDS

Beta oxidation is the process by which fatty acids are processed progressively to release two-carbon segments in the form of acetyl CoA. This series of reactions is summarized in Figure 2.11.6. These reactions produce acetyl CoA and 1 $FADH_2$ and 1 NADH per turn of the beta oxidation cycle.

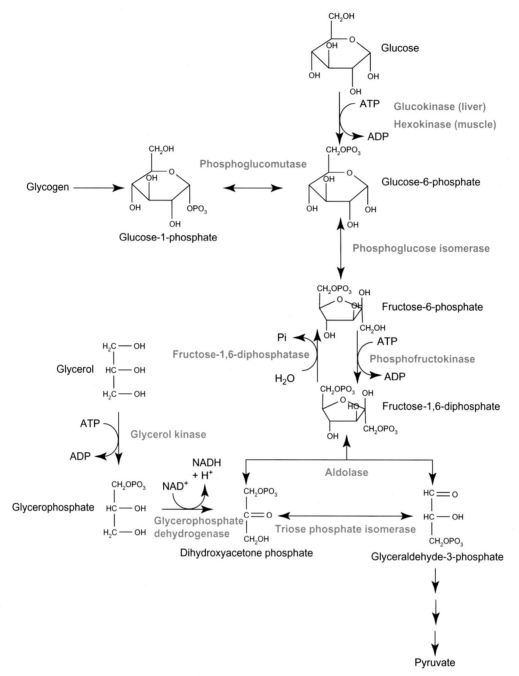

FIGURE 2.11.4 Metabolism of glycerol. Glycerol is phosphorylated and then converted into dihydroxyacetone phosphate, an intermediate in glycolysis.

EXAMPLE 2.11.1 ATP Yield from Glycerol

How much ATP is made from glycerol, under optimal conditions?

Glycerol requires 1 ATP per molecule for the conversion into glycerol phosphate. The glycerol phosphate is then converted to dihydroxyacetone phosphate with the production of 1 molecule of NADH. The dihydroxyacetone phosphate can then be oxidized fully to CO_2 through glycolysis and the TCA cycle as outlined in Chapters 2.9 and 2.10.

In the conversion to pyruvic acid, dihydroxyacetone phosphate generates 1 NADH and 2 ATP molecules per molecule of glycerol. The net effect of converting glycerol to pyruvic acid is -1 ATP + 2 ATP + 2 NADH = 1 ATP + 2 NADH = 6 ATP, assuming that the NADH are both oxidized with the generation of 2.5 ATP per NADH. The complete oxidation of pyruvate produces 4 NADH, 1 $FADH_2$, and 1 GTP. When the reducing equivalents are oxidized through the electron transport chain, this produces

4 NADH × 2.5 ATP/NADH + 1 $FADH_2$ × 1.5ATP/$FADH_2$ + 1 GTP
× 1 ATP/GTP = 12.5 ATP

Each glycerol molecule liberated from a triglyceride thus produces at most 18.5 **molecules ATP per molecule glycerol**. Glycerol has a gram molecular weight of 92 g mol^{-1}. It produces 18.5 moles ATP/92 g mol^{-1} = **0.20 mol g^{-1}**.

This is a little more than the energy derived from glucose: 32 moles ATP/180 g mol^{-1} = 0.18 mol g^{-1}.

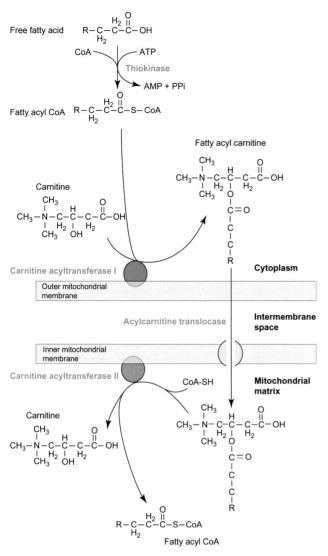

FIGURE 2.11.5 Involvement of carnitine in the entry of free fatty acids into the inner mitochondrial space. Fatty acids are combined with coenzyme A in a reaction that effectively costs 2 ATP molecules per reaction. The fatty acyl CoA is then converted to fatty acyl carnitine, which can penetrate the inner mitochondrial membrane. In the mitochondrial matrix, the carnitine is removed by a second, different carnitine acyl transferase.

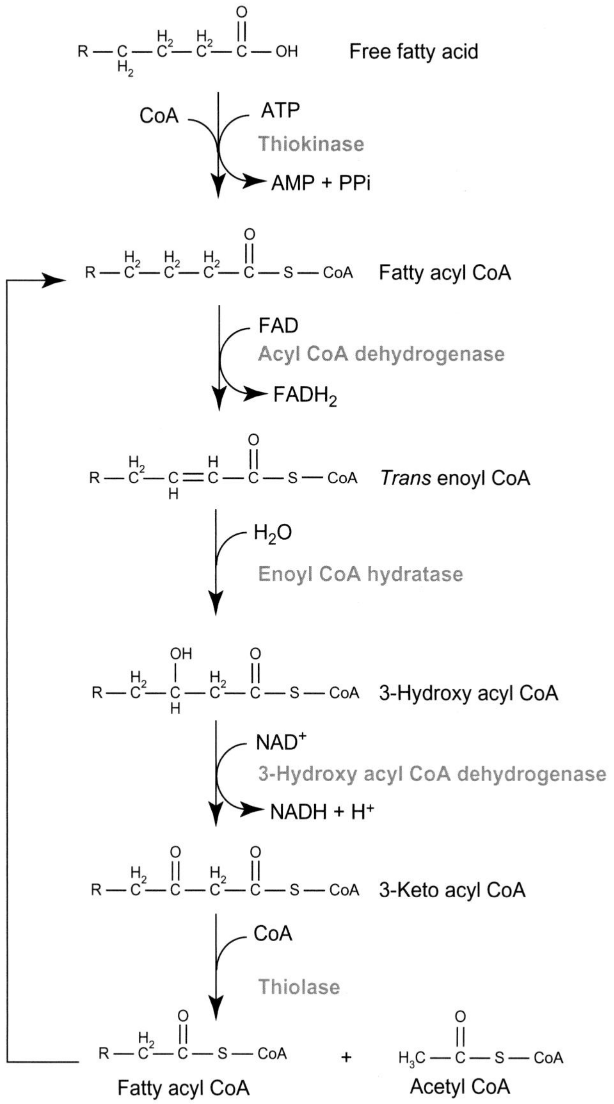

FIGURE 2.11.6 Beta oxidation of free fatty acid.

Once coupled with coenzyme A, the fatty acyl chain is oxidized, producing $FADH_2$. The fatty acyl chain is further oxidized by adding water and then removing two more hydrogens by the 3-hydroxy acyl CoA dehydrogenase, this time producing NADH. The final step in each turn of the beta oxidation cycle is the production of acetyl CoA and the regeneration of fatty acyl CoA, shortened by two carbon atoms. This shorter fatty acyl CoA reenters the beta oxidation cycle until, at the end, all of the chain is converted to acetyl CoA. In this way, palmitic acid, for example, will produce 8 acetyl CoA molecules and 7 $FADH_2$ molecules and 7 NADH molecules (there are only 7 because the last turn of the cycle produces two acetyl CoA molecules and so the last two carbons do not enter the cycle again to produce $FADH_2$ and NADH).

THE LIVER PACKAGES SUBSTRATES FOR ENERGY PRODUCTION BY OTHER TISSUES

During exercise, when fatty oxidation occurs rapidly, the liver packages acetyl CoA in a form that can be readily used by other tissues for the generation of energy. In the liver, two molecules of acetyl CoA combine to form acetoacetate. The acetoacetate that forms can then be converted to β-hydroxybutyric acid and, to a lesser extent, to acetone. All three of these compounds, **acetoacetate**, **β-hydroxybutyric acid**, and **acetone**, are referred to as **ketone bodies** (see Figure 2.11.7). They leave the liver cell and travel in the blood to the peripheral tissues, which take up the compounds and metabolize them for energy.

The combined concentration of the ketone bodies is typically less than about 3 mg%. Despite these low

> **EXAMPLE 2.11.2 ATP Yield from Fatty Acids**
>
> Each turn of the beta oxidation cycle produces NADH and FADH$_2$. For palmitic acid, $C_{16}H_{32}O_2$, a total of eight acetyl CoA molecules are produced from seven turns of the beta oxidation cycle. The NADH and FADH$_2$ feed into the electron transport chain to produce ATP. The total ATP produced from beta oxidation is thus
>
> 7 NADH/palmitate × 2.5 ATP/NADH + 7 FADH$_2$/palmitate
> × 1.5 ATP/FADH$_2$ = 28 ATP/palmitate
>
> Each acetyl CoA produced by beta oxidation enters the TCA cycle where it is further oxidized to CO_2 and produces reducing equivalents that are used by the ETC to make ATP. Each acetyl CoA molecule produces 3 NADH, 1 FADH$_2$, and 1 GTP molecule during a single turn of the TCA cycle. These are eventually used by the ETC to produce
>
> 3 NADH/acetyl CoA × 2.5 ATP/NADH + 1 FADH$_2$/acetylCoA
> × 1.5 ATP/FADH$_2$ + 1 GTP/acetyl × 1 ATP/GTP
> = 10 ATP/acetyl CoA
>
> Since there are eight acetyl CoA molecules per palmitic acid produced from beta oxidation, the total ATP produced from acetyl CoA from palmitic acid is
>
> 10 ATP/acetyl CoA × 8 acetyl CoA/palmitate = 80 ATP/palmitate
>
> We add this to the ATP produced from beta oxidation and subtract the two ATP needed to start beta oxidation from the initial thiokinase reaction to get
>
> 28 ATP from beta oxidation + 80 ATP from TCA cycle
> −2 ATP from thiokinase reaction = **106 ATP/palmitic acid**

> **EXAMPLE 2.11.3 Compare ATP Yield from Glucose to that of Tripalmitin**
>
> In Chapter 2.10, we found that the maximum yield of ATP from the complete oxidation of glucose was 32 ATP per glucose. This corresponds to 7 ATP per glucose from the oxidation of glucose to pyruvate, and 25 ATP from the complete oxidation of pyruvate to CO_2 and H_2O. The gram molecular weight of glucose is 180 g mol^{-1}, so the ATP production is
>
> 32 moles ATP/mol glucose = 32 moles ATP/180 g
> = **0.18 mol ATP per g of glucose**
>
> The gram molecular weight of tripalmitin is 807.3 g mol^{-1}, tripalmitin consisting of three palmitic acid molecules and one glycerol. We have calculated that glycerol gives rise to at most 18.5 ATP per glycerol, and each palmitic acid produces at most 106 ATP per palmitic. The total for tripalmitin is thus
>
> 336.5 moles ATP/mole tripalmitin = 336.5 moles/807.3 g
> = **0.42 ATP per g of tripalmitin**
>
> Thus fat has about 2.3 times as much energy stored per unit weight.

concentrations, the flux of energy to the metabolizing tissues can be great because the ketone bodies are taken up so quickly. The concentrations of the ketone bodies can occasionally rise to very high levels, a condition known as **ketosis**. This condition occurs whenever metabolism of fats is emphasized such as in starvation and in diabetes mellitus. In this case, the urine can contain ketones and the presence of acetone is sometimes detectable by its odor in the exhaled air.

In the peripheral tissues, acetoacetic acid is taken up and converted in the mitochondria to acetoacetyl CoA by the transfer of a CoA moiety from succinyl CoA, an intermediate in the TCA cycle. Since succinyl CoA is usually converted to succinate with the formation of 1 molecule of GTP, conversion of succinyl CoA to succinate in this reaction removes the potential synthesis of 1 molecule of GTP. So, although no energy in the form of ATP is directly involved in this transfer, it has a net cost of 1 molecule of ATP. The acetoacetyl CoA can then be converted to two molecules of acetyl CoA by **thiolase**, the same enzyme involved in the production of acetyl CoA from fatty acyl CoA during the beta oxidation pathway of fatty acids.

AMINO ACIDS CAN BE USED TO GENERATE ATP

Amino acids can be used to build body proteins and they can be broken down to yield energy. In the steady state of the adult, the body store of proteins remains constant and there is a constant throughput of amino acids, equal to the dietary intake, that is converted to metabolic energy. In the typical American diet, about 16% of the calories are provided by dietary protein.

Because the hepatic portal blood leaves the intestines and travels to the liver, the liver has the first opportunity to metabolize all the nutrients, including the amino acids absorbed from digested proteins in the intestinal lumen. The liver does several things: it catabolizes a

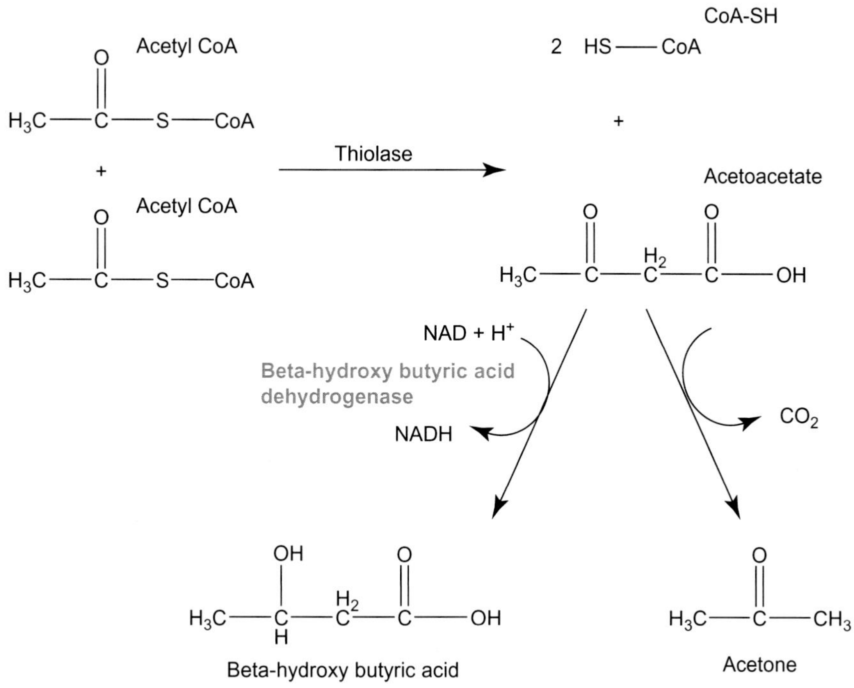

FIGURE 2.11.7 Formation of ketone bodies by the liver. During rapid lipid metabolism, when production of acetyl CoA outstrips the liver's ability to oxidize it, acetyl CoA coalesces to form acetoacetate, beta-hydroxybutyric acid, and acetone.

large fraction of the amino acids (57%), releases some unchanged into the general circulation (23%), and utilizes some 20% for synthesis of proteins that either remain in the liver or are released into the blood.

Catabolism of amino acids can be broadly categorized into two processes: the breakdown of amino acids to carbohydrate precursors and potentially leading to the formation of glucose; and transformations leading to acetyl CoA that result in the potential formation of ketone bodies. Amino acids that break down into carbohydrate precursors are called **glucogenic**; those leading to acetyl CoA are called **ketogenic**.

- **Leucine** and **lysine** are the only exclusively ketogenic amino acids.
- **Isoleucine, threonine, phenylalanine, tyrosine, and tryptophan** are both glucogenic and ketogenic.
- **Aspartatic acid, asparagine, glutamic acid, glutamine, alanine, arginine, histidine, glycine, serine, proline, valine, methionine,** and **cysteine** are glucogenic.

Because each amino acid has a different side chain, each amino acid is catabolized differently to produce energy and waste products. We will not go through all of these reactions for each of the amino acids. The overall fate of the amino acids is shown in Figure 2.11.8.

AMINO ACIDS ARE DEAMINATED TO ENABLE OXIDATION

Many amino acids share a common mechanism for the removal of the amino group to form intermediates in the TCA cycle or glycolytic cascade. This is a **transamination** reaction followed by a **dehydrogenation** reaction, as shown in Figure 2.11.9. The α-ketoglutarate formed in the transamination reaction in the mitochondria can then enter the TCA cycle. The deamination results in the liberation of ammonia. The reaction sequence shown is one of many such involved in the deamination of a variety of amino acids including phenylalanine, tyrosine, aspartate, cysteine, lysine, arginine, alanine, isoleucine, leucine, and valine. The reactions differ only in the α-keto acid formed following the transamination.

UREA IS PRODUCED DURING DEAMINATION AND IS ELIMINATED AS A WASTE PRODUCT

The ammonia released during deamination is removed from the blood almost entirely by conversion into **urea** in the **liver**. This occurs through another metabolic process called the **urea cycle** (see Figure 2.11.10). In this process, the ammonia is combined with bicarbonate ion to form **carbamoyl phosphate**. **The complete operation of the cycle requires continual input of aspartate.** This can be derived from transamination of oxaloacetic acid by glutamic acid, the reverse of the process shown in Figure 2.11.9. Since glutamate is the product of transamination with several amino acids, it can be replenished. Thus one of the amino groups of urea is derived from ammonia and the other is derived from amino groups on various amino acids, transaminated to glutamate.

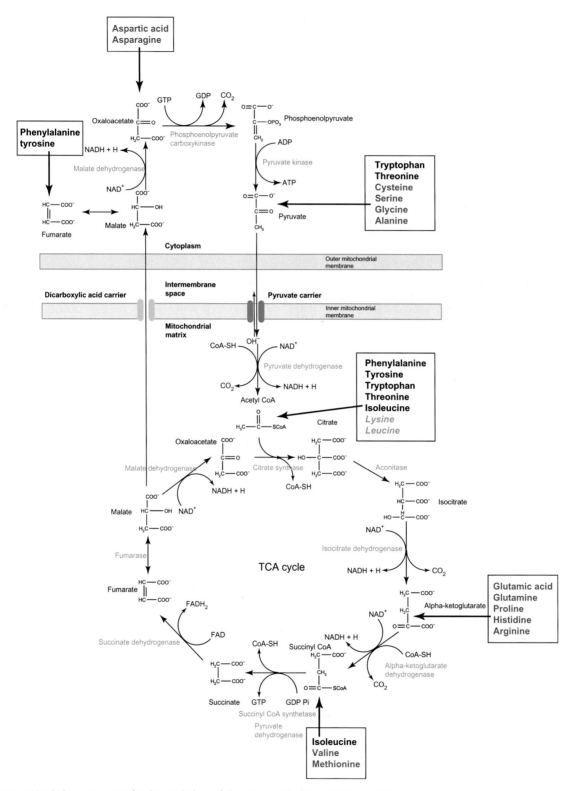

FIGURE 2.11.8 Metabolic entry points for the catabolism of the amino acids. Those amino acids that produce acetyl CoA are called ketogenic. These include leucine, lysine, phenylalanine, tyrosine, tryptophan, and isoleucine. Those amino acids that produce carbohydrate precursors that can be converted to glucose are called glucogenic. These include aspartic acid, asparagine, phenylalanine, tyrosine, tryptophan, alanine, cysteine, serine, threonine, glycine, glutamic acid, glutamine, proline, histidine, arginine, isoleucine, valine, and methionine. Only lysine and leucine are exclusively ketogenic. Exclusively ketogenic amino acids are in light blue italic; exclusively glucogenic are in dark blue; both ketogenic and glucogenic are in black.

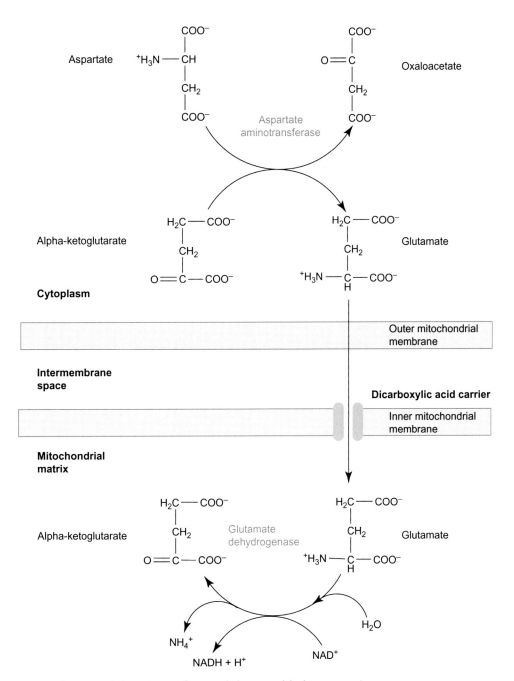

FIGURE 2.11.9 Deamination of amino acids by aminotransferase and glutamate dehydrogenase action.

250 QUANTITATIVE HUMAN PHYSIOLOGY: AN INTRODUCTION

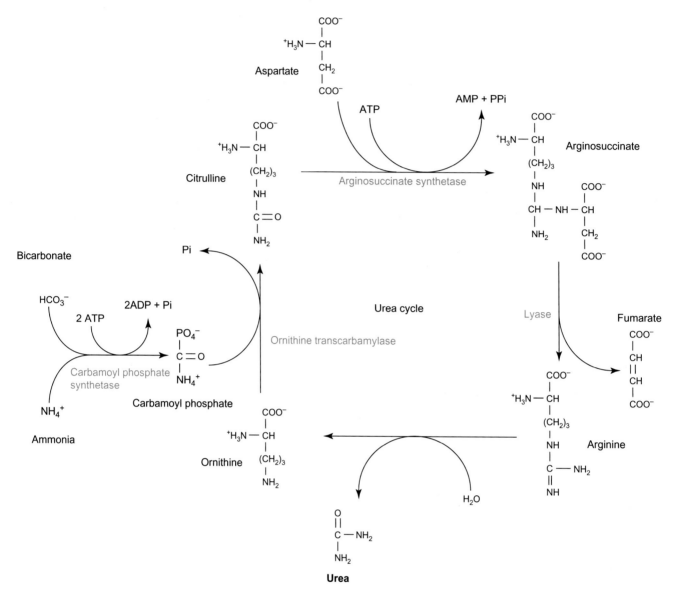

FIGURE 2.11.10 Formation of urea from the urea cycle.

Clinical Applications: Nonshivering Thermogenesis

Immediately after birth, the baby is thrust into a cool and dry environment. It loses heat rapidly by evaporation. Even when its skin dries, the baby continues to lose heat because it has a large surface area relative to its small body mass, it cannot put on warmer clothes, it has poor thermal insulation in the form of fat, and it has limited ability to generate heat by muscular contractions. However, the baby is able to generate heat by nonshivering thermogenesis, also called metabolic thermogenesis.

Babies have specialized fat tissue called brown fat, located mainly in the neck and in the midline of the back. Mitochondria contain cytochromes that contain iron, giving the mitochondria a reddish-brown color. Brown fat contains lots of mitochondria so it takes on a brownish color from mitochondrial pigments. The mitochondria in this fat make ATP primarily from the oxidation of fatty acids that are produced by hydrolysis of stored triglycerides. Under certain circumstances, these mitochondria can become **uncoupled**. Uncoupling of oxidative phosphorylation in brown fat mitochondria generates heat.

Oxidative phosphorylation couples an **exothermic** reaction, one that releases heat, to an **endothermic** reaction, one that requires heat. Pumping of H^+ ions out of the mitochondrial matrix couples the exothermic oxidation reactions to the energy of the H^+ electrochemical gradient. This in turn is coupled to the endothermic synthesis of ATP. The oxidation of foodstuffs (carbohydrates, fats, and amino acids) reduces NAD^+ to NADH, whose oxidation powers the H^+ pumping by the electron transport chain. The ATP synthase couples the energy stored in the electrochemical potential of H^+ to ATP synthesis. This coupling can never be 100% efficient, so that cellular respiration releases some energy as heat. Under certain circumstances a special protein, called the **uncoupling protein**, short-circuits the synthesis of ATP by allowing H^+ ions to cross the inner mitochondrial membrane without making ATP. Then more of the energy stored in the H^+ gradient is dissipated as heat and less is captured by ATP. The exothermic reactions are uncoupled from ATP synthesis and the mitochondria generate heat proportionate to the number of mitochondria and the rate of oxygen consumption.

UCP1, a 32-kDa protein isolated from brown fat mitochondria, was the first uncoupling protein to be described. UCP1 is located mainly in brown fat, whereas UCP2 is expressed in many tissues and UCP3 is found mainly in skeletal muscle. BMCP1 is specific to the brain. UCP1 is activated by fatty acids and inhibited by purine nucleotides (ATP, ADP, GTP, and GDP). Exposure to cold increases UCP1 content in brown fat, and increased caloric content of the diet increases the tissue content of UCP2. Thus these proteins are believed to participate in basal or regulatory thermogenesis, but their exact functions are not yet worked out.

The mechanism of brown fat thermogenesis in response to cold is shown in Figure 2.11.11. Adipose tissue is supplied by sympathetic nerves that release noradrenaline onto the fat cells. This stimulates the breakdown of stored triglycerides to fatty acids, which in turn activate UCP1. The effect depends on the tissue content of UCP1, which is also increased by cold exposure. How UCP1 increases the H^+ flux across the membrane is also not yet known. UCP1 may form a H^+-specific pore or it may transport H^+ ions by cycling an H^+ carrier, most likely fatty acids. Current work favors the H^+ carrier mechanism because some carboxylic acid groups can activate transport without being transported.

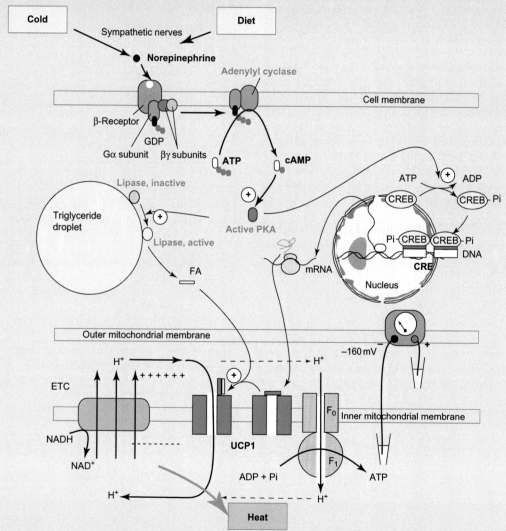

FIGURE 2.11.11 Postulated mechanisms for nonshivering thermogenesis in brown adipose tissue. Noradrenaline released from sympathetic nerve terminals in adipose tissue binds to β receptors on the surface of the adipocytes, leading to increased cytosolic cAMP by activating adenylyl cyclase. cAMP activates protein kinase A (PKA), which phosphorylates several targets, thereby changing their activity. PKA activates hormone-sensitive lipase that increases triglyceride breakdown and increases cytoplasmic fatty acid concentration. The increased fatty acids activate UCP1, short-circuiting the H^+ electrochemical gradient and reducing the synthesis of ATP from ADP and Pi. This uncouples oxidative phosphorylation and the energy of oxidation of fats appears as heat. PKA also phosphorylates a transcription factor, CREB (cyclic AMP response element binding protein), that binds to specific regions of the DNA (CRE—cAMP response element) and activates their transcription into mRNA. This leads to increased numbers of UCP1 protein in the mitochondria, which adapts the body to exposure to cold or some other stress. Diet-induced thermogenesis probably involves increased expression of UCP2.

SUMMARY

The oxidation of fatty acids and amino acids produces ATP through some of the same reactions used to produce ATP by the oxidation of carbohydrates. Fatty acids are released from triglycerides by hormone-sensitive lipase. The fatty acids are bound in the cytoplasm by fatty acid binding protein and carried into the mitochondria by being complexed with carnitine. Inside the mitochondria, fatty acids undergo beta oxidation in which two carbons at a time are cleaved off the carboxyl end and converted into acetyl Coenzyme A. Each turn of the beta oxidation cycle produces one NADH and one $FADH_2$. These feed reducing equivalents into the electron transport chain, which pumps H^+ ions out of the mitochondrial matrix and produces the electrochemical gradient of H^+ that drives ATP synthesis through the F_0F_1ATPase. Beta oxidation also produces acetyl CoA that enters the TCA cycle by combining with oxaloacetate to form citric acid. Each turn of the TCA cycle converts the acetyl CoA into 2 molecules of CO_2, 3 NADH, 1 $FADH_2$, and 1 GTP. Thus palmitic acid (16:0) produces 7 NADH and 7 $FADH_2$ from beta oxidation, and 24 NADH, 8 $FADH_2$, and 8 GTP from the complete oxidation of acetyl CoA. Since NADH produces 2.5 ATP molecules and $FADH_2$ produces 1.5, the ATP count from palmitic acid is 28 from beta oxidation and 80 from the TCA cycle. Two ATP molecules are used to prime the palmitic acid in the thiokinase reaction that converts palmitic acid to palmitoyl CoA.

The liver packages lipid metabolites into ketone bodies, which collectively consist of acetoacetic acid, beta-hydroxybutyric acid, and acetone. Build-up of these ketone bodies during starvation or other metabolic conditions such as diabetes is called ketosis.

Each amino acid has its own metabolic pathway because they all differ chemically. However, they feed into the main metabolic pathways in a limited number of places. Those amino acids that can be used to make glucose are called glucogenic. Those that can be used to make ketone bodies by producing acetyl CoA are called ketogenic. Many amino acids are both glucogenic and ketogenic. Only leucine and lysine are exclusively ketogenic.

Many amino acids are deaminated by a combination of transamination and dehydrogenation. In this reaction, the amino group is transferred from the amino acid to alpha-ketoglutaric acid, forming glutamic acid. The glutamic acid is then converted back to alpha-ketoglutaric acid by glutamate dehydrogenase, resulting in release of ammonia and formation of NADH. The ammonium formed in this way is converted to urea through the urea cycle. The urea cycle begins with the formation of carbamoyl phosphate by condensing ammonium with HCO_3^-. Carbamoyl phosphate then combines with ornithine to form citrulline. Citrulline combines with aspartic acid to form arginosuccinate and arginine in sequence. Arginine gives rise to urea and ornithine to begin the cycle again.

REVIEW QUESTIONS

1. What components make up a triglyceride? Where are triglycerides found in the body? Can they be metabolized as is? What enzyme breaks triglycerides down into components? What activates this enzyme? What inhibits the enzyme?
2. How is glycerol oxidized? How much ATP is produced from glycerol, mole per mole?
3. How are fatty acids carried in the cytosol? Where does oxidation of fatty acids take place? How do the fatty acids get into the mitochondria?
4. What is meant by beta oxidation? What are the main products? How many beta oxidation cycles are there for palmitic acid? Stearic acid? Oleic acid?
5. How much ATP is produced from the complete oxidation of palmitic acid, mole per mole?
6. What are the ketone bodies? Where are they produced?
7. What is a glucogenic amino acid? Which amino acids are glucogenic?
8. What is a ketogenic amino acid? Which amino acids are exclusively ketogenic?
9. What is transamination? What are the major receptors for transamination?
10. What is urea? Where is it produced?

UNIT 3

PHYSIOLOGY OF EXCITABLE CELLS

The Origin of the Resting Membrane Potential 3.1

Learning Objectives

- Write the Nernst equation for any given ion
- Define equilibrium potential
- Be able to calculate the equilibrium potential for any ion
- Recognize the proper form of the Goldman–Hodgkin–Katz equation
- Explain why in the Goldman–Hodgkin–Katz equation the anion concentration in one compartment appears on the opposite side of the argument from the cation concentration in the same compartment
- Define and distinguish between slope conductance and chord conductance
- Write the chord conductance equation
- Explain what happens to membrane potential when the conductance to a particular ion changes

INTRODUCTION

In Chapter 2.6, we analyzed the energetics of membrane transport across resting cardiac muscle cells that have a membrane potential of about -80 mV, negative inside. At rest, the concentration gradient favored Na^+ and Ca^{2+} entry into the cell and K^+ exit. These gradients produced slow leaks of ions that were continually balanced by active transport mechanisms such as the Na,K-ATPase and Na–Ca exchanger so that the ionic composition of the cytosol stayed constant.

The resting membrane potential is extremely important because, first, all cells have a membrane potential (but not the same value!), and modulation of the membrane potential is associated with modulation of cellular activity. Certain cells of the body, called **excitable cells**, can use rapid changes in their membrane potential as a signal. These cells include nerve cells and muscle cells.

Now we ask the question, where did the resting membrane potential come from? To answer this question, we will consider hypothetical membranes. These membranes are not like any biological membrane, but we consider them because they will clarify how the resting membrane potential comes to be what it is. Here we will use concepts of potential and capacitance already covered in Chapter 1.3, and the concept of the electrochemical potential discussed in Chapter 1.7.

THE EQUILIBRIUM POTENTIAL ARISES FROM THE BALANCE BETWEEN ELECTRICAL FORCE AND DIFFUSION

First we consider a **hypothetical membrane that is permeable only to Na^+ ions**. Suppose that $[Na^+]_o$, the outside or extracellular sodium concentration, is 145×10^{-3} M, and $[Na^+]_i = 12 \times 10^{-3}$ M. And suppose that initially there is no membrane potential. What happens? As shown in Figure 3.1.1, the diffusion gradient for Na^+ favors Na^+ entry into the cell, and the initial Na^+ influx carries a charge that builds up on the inside of the cell. This produces a potential (recall that separation of charge produces a potential) across the membrane that impedes further Na^+ ion movement because the positive charges repel the positively charged Na^+ ion. The potential that develops is related to the electric force that now works against further Na^+ movement. Eventually the electric force will get so large that the diffusive flow will be exactly counterbalanced, and net flow will stop. This will occur at the **Na equilibrium potential**. We can calculate what that potential should be in two ways: first by looking at Fick's law and second by analyzing the energetics. We get the same answer either way. The situation for the development of the Na^+ equilibrium potential is shown in Figure 3.1.1.

Fick's law with an electrical force is given as

$$[3.1.1] \qquad J_s = -D\frac{\partial C}{\partial x} - \frac{D}{RT}Cz\Im\frac{\partial \psi}{\partial x}$$

This is Eqn [1.7.19]. Here J_s is the solute flux, in this case the flux of Na^+, D is the diffusion coefficient, C is the concentration of Na^+, R is the gas constant, T is the absolute temperature, z is the charge on the ion ($+1$ for Na^+), and $\Im$ is the faraday, the number of coulombs per mole. At equilibrium, $J_s = 0$, and we equate the diffusive force and the electrical force:

$$[3.1.2] \qquad \frac{RT}{C}\frac{\partial C}{\partial x} = -z\Im\frac{\partial \psi}{\partial x}$$

Integrating both sides from outside to inside of the membrane, we get

$$[3.1.3] \qquad \int_o^i \frac{RT}{C}\frac{\partial C}{\partial x}dx = \int_o^i -z\Im\frac{\partial \psi}{\partial x}dx$$

$$RT\ln\frac{C_i}{C_o} = z\Im(\psi_o - \psi_i)$$

255

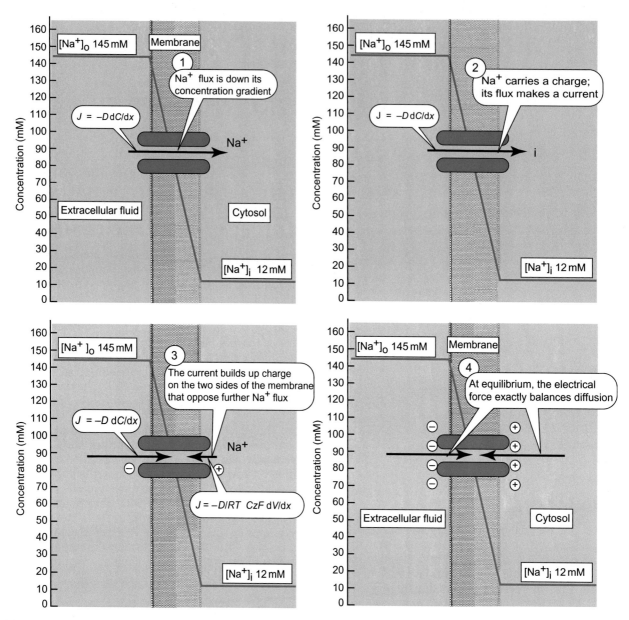

FIGURE 3.1.1 Generation of the Na equilibrium potential across a hypothetical membrane that is permeable only to Na. The [Na]$_o$ in this case is high, about 145 mM, whereas [Na]$_i$ is 12 mM. Thus the diffusion gradient favors Na$^+$ flux from the outside to the inside of the cell. Since Na$^+$ is electrically charged, flux of only Na$^+$ makes a current across the membrane that separates charges and thus produces an electrical potential. The electric field exerts forces on Na$^+$ that retards its movement. When the diffusive force exactly balances the electrical force, flux is zero and the potential is E_{Na}, the sodium equilibrium potential.

This gives the equilibrium membrane potential as

[3.1.4] $$\frac{RT}{z\Im} \ln \frac{C_i}{C_o} = (\psi_o - \psi_i)$$

It is usual for physiologists to take the outside solution as ground ($\psi_o = 0$) because it is potential *differences* that we are concerned with. Using this standard $\psi_o = 0$, we rewrite Eqn [3.1.4] as

[3.1.5] $$\frac{RT}{z\Im} \ln \frac{C_o}{C_i} = (\psi_i - \psi_o)$$

If you always put the outside ion in the numerator of the argument of the logarithm, then the sign of the membrane potential will be correct. This is a famous equation, the **Nernst equation**, named for Walther Nernst (1864–1941), a physical chemist from Berlin, Germany. It calculates the potential at which net flux is zero, which occurs at equilibrium for the ion. **The membrane potential at which the diffusive force is exactly balanced by the electrical force is called the equilibrium potential for that ion.** For sodium, it is usually symbolized as E_{Na}.

The Nernst Equation can also be derived easily from considering the electrochemical potentials. At equilibrium we have

[3.1.6] $$\begin{aligned}\Delta\mu &= \mu_{Na_i} - \mu_{Na_o} \\ 0 &= \mu^o_{Na_i} + RT \ln[Na^+]_i + z\Im\psi_i - \mu^o_{Na_o} \\ &\quad - RT \ln[Na^+]_o - z\Im\psi_o\end{aligned}$$

> **EXAMPLE 3.1.1 Calculate the Equilibrium Potential for Na+**
>
> Inserting values for $[Na^+]_i = 12 \times 10^{-3}$ M and $[Na^+]_o = 145 \times 10^{-3}$ M, into Eqn [3.1.5], we calculate
>
> $$\psi_i - \psi_o = \frac{8.314 \, J \, mol^{-1} \, K^{-1} \times 310 \, K \times \ln\left(\frac{145 \times 10^{-3} \, M}{12 \times 10^{-3} \, M}\right)}{1 \times 9.649 \times 10^4 \, C \, mol^{-1}} = 0.0666 \, V$$
>
> Note that the potential inside is positive, as it should be to impede further influx of a positive ion.

Canceling out the standard free energies, which are equal, we can arrange Eqn [3.1.6] to give

$$[3.1.7] \quad \frac{RT}{z\mathfrak{F}} \ln \frac{[Na^+]_o}{[Na^+]_i} = (\psi_i - \psi_o)$$

The Nernst equation is often written with $\log_{10}$ instead of the natural logarithm. At $37°C = 310$ K, Eqn [3.1.7] can be written as

$$[3.1.8] \quad 0.0615 \log \frac{[Na^+]_o}{[Na^+]_i} = (\psi_i - \psi_o)$$

where the log is now $\log_{10}$. The term 0.0615 is the evaluation of the expression $RT/z\mathfrak{F}$ and conversion of the natural log, ln, to $\log_{10}$. What this means is that every 10-fold gradient in concentration of a singly charged ion gives an equilibrium potential of 61.5 mV at 37°C.

THE EQUILIBRIUM POTENTIAL FOR K+ IS NEGATIVE

Now we **suppose that the membrane is impermeable to Na+ and Cl− ions but it is permeable only to K+**. The [K+] concentrations on the two sides of the membrane are those in a muscle cell, namely, $[K^+]_o = 4 \times 10^{-3}$ M and $[K^+]_i = 155 \times 10^{-3}$ M. Because of its concentration gradient, K+ will diffuse out of the cell, causing an outward current and accumulation of positive charges on the *outside* of the cell. The Nernst equation is

$$[3.1.9] \quad \frac{RT}{z\mathfrak{F}} \ln \frac{[K^+]_o}{[K^+]_i} = (\psi_i - \psi_o)$$

The result of the calculation using $[K^+]_o = 4 \times 10^{-3}$ M and $[K^+]_i = 155 \times 10^{-3}$ M is **EK = −0.0977 V** or **−97.7 mV**. Here the potential is strongly negative inside.

We can repeat this calculation for any ion to obtain its equilibrium potential. We must remember what goes into the equations, however. As an example, consider that **the membrane is not permeable to either K+ or Na+, but is freely permeable only to Cl−**. Suppose further that Cl− is 100×10^{-3} M outside the cell and 5×10^{-3} M inside the cell. The Nernst equation for Cl is

$$[3.1.10] \quad \frac{RT}{z\mathfrak{F}} \ln \frac{[Cl^-]_o}{[Cl^-]_i} = (\psi_i - \psi_o)$$

In this case, $z = -1$ because the Cl− ion has a negative charge. This is equivalent to inverting the argument of the logarithm. Here we get $\psi_i - \psi_o = -0.0615 \log (100/5) = -0.080$ V. Thus $E_{Cl} = $ **−0.080 V = −80 mV**.

Figure 3.1.2 shows the concentration differences for Na+, K+, and Cl− and the equilibrium potential for each of these ions. It is important to remember what the equilibrium potential is. It is the potential at which the electrical force exactly balances diffusion so that the net flux of the ion across the membrane is zero. Since there is no flux, there is no change in the concentrations on the two sides of the membrane and the ion is at equilibrium.

Such hypothetical membranes as the ones we have been considering that are permeable only to Na+, or only to K+ or only to Cl− do not exist. Real membranes have some nonzero permeability to all of these ions. So what is the membrane potential across a membrane that is permeable to all three? The short answer is: it depends on how permeable the membrane is to each of the ions. What we need is some expression that tells us what the magnitude of the membrane potential will be, given the equilibrium potentials and the relative permeabilities of the ions.

INTEGRATION OF THE NERNST–PLANCK ELECTRODIFFUSION EQUATION GIVES THE GOLDMAN–HODGKIN–KATZ EQUATION

When there is a flux of solute, and the solute is charged, there is a current. The relationship between current density and flux is

$$[3.1.11] \quad I_i = z\mathfrak{F} J_i$$

where the subscript i denotes the particular ion that is carrying the current. The total current is the sum of all the ionic currents. Now we substitute in for the flux using Eqn [3.1.1] to get

$$[3.1.12] \quad I_i = z_i \mathfrak{F} J_i = -z_i \mathfrak{F} \left(D_i \frac{\partial C_i}{\partial x} + \frac{D_i}{RT} C_i z_i \mathfrak{F} \frac{\partial \psi}{\partial x} \right)$$

This is the **Nernst–Planck electrodiffusion equation**. We can obtain an expression for I_i by integrating this equation over the thickness of the membrane (from

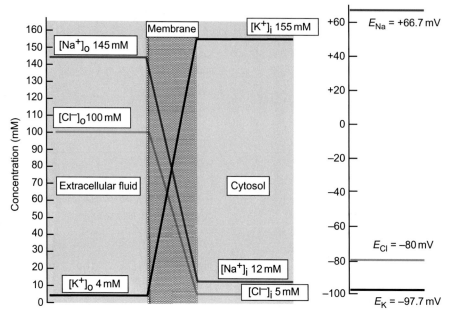

FIGURE 3.1.2 Equilibrium potentials for Na, K, and Cl in a muscle cell. A positive equilibrium potential is needed to prevent Na$^+$ ions from entering the cell from the extracellular fluid; a negative equilibrium potential is needed to prevent Cl$^-$ ions from entering; a negative equilibrium potential is needed to prevent K$^+$ ions from exiting the cell. The solid lines represent ion concentrations.

$x = 0$ to δ, where δ is the thickness of the membrane), but this integration can be accomplished in closed form only if we assume that the potential gradient is linear. This is called the **constant field assumption**. The integration gives

$$[3.1.13] \quad I_i = \frac{z_i^2 \mathfrak{F}^2 D_i}{RT\delta} E_m \left(\frac{C_i e^{(z_i \mathfrak{F} E_m / RT)} - C_o}{e^{(z_i \mathfrak{F} E_m / RT)} - 1} \right)$$

as detailed in Appendix 3.1.A1. This is often written in an alternate form by multiplying both numerator and denominator by $e^{(-z\mathfrak{F}/RT)E_m}$:

$$[3.1.14] \quad I_{\text{ion}} = \frac{(D/\delta)[(z^2 \mathfrak{F}^2 / RT) E_m][C_i - C_o e^{(-z\mathfrak{F}/RT)E_m}]}{[1 - e^{(-z\mathfrak{F}/RT)E_m}]}$$

This is the **Goldman–Hodgkin–Katz (GHK) current equation**. It relates the current carried by each ion to its concentration on both sides of the membrane (C_i and C_o) and to the membrane potential, E_m. Now at steady-state resting membrane potential, the total current across the membrane must be zero, or otherwise E_m would be changing:

$$[3.1.15] \quad I_{\text{total}} = I_{\text{Na}} + I_K + I_{\text{Cl}} = 0$$

By substituting in for the expression for I_{Na}, I_K, and I_{Cl}, we can derive an expression for E_m:

$$[3.1.16] \quad E_m = \frac{RT}{\mathfrak{F}} \ln \left[\frac{P_K[K^+]_o + P_{\text{Na}}[Na^+]_o + P_{\text{Cl}}[Cl^-]_i}{P_K[K^+]_i + P_{\text{Na}}[Na^+]_i + P_{\text{Cl}}[Cl^-]_o} \right]$$

This is the **GHK equation**. It describes the resting membrane potential when only Na$^+$, K$^+$, and Cl$^-$ are permeant, but it can be expanded to include other ions. The inside concentration of Cl$^-$ appears in the numerator with the outside concentrations of Na$^+$ and K$^+$ because z for Cl$^-$ is -1.0 and z for Na$^+$ and K$^+$ is $+1.0$. Additional ions can be added to this equation if they contribute significantly to the currents across the membrane. This equation shows that the resting membrane potential results from the concentration-weighted permeabilities across the membrane because each permeability is multiplied by the concentration of the ion. Appendix 3.1.A1 presents a full derivation of this equation.

SLOPE CONDUCTANCE AND CHORD CONDUCTANCE RELATE ION FLOWS TO THE NET DRIVING FORCE

The GHK current equation (Eqn [3.1.14]) describes the current carried by any given ion in terms of its concentration on both sides of the membrane and the membrane potential. If we assume that the permeability, D/δ, is constant, we can calculate the current carried by each ion as a function of the membrane potential. The currents carried by K$^+$ and Na$^+$ for a muscle cell containing 155 mM [K$^+$]$_i$ and 12 mM [Na$^+$]$_i$ and 4 mM [K$^+$]$_o$ and 145 mM [Na$^+$]$_o$ as predicted from the GHK current equation are shown in Figure 3.1.3. At the equilibrium potential for each ion, there is no current. This equilibrium potential is also called the **reversal potential**, because at this point the current changes from negative (positive charges enter the cell—by convention this is an **inward current**) to positive (positive charges exits the cell—this is an **outward current**).

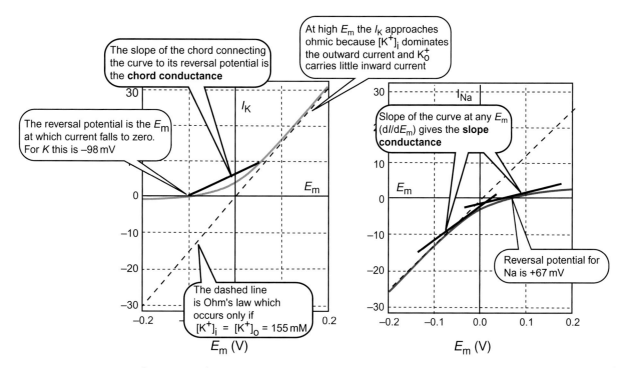

FIGURE 3.1.3 Currents carried by K$^+$ (left) and Na$^+$ (right) as predicted by the GHK current equation if the membrane was permeable only to K$^+$ (left) or to Na$^+$ (right).

The relationship between current and voltage can be described by a **conductance**. Ohm's law states

$$[3.1.17] \quad I_i = \frac{E}{R_i} = g_i E$$

where E is the potential difference that drives current flow (NOT the electric field intensity!), R_i is the **resistance** to the ion i, and g_i is the **conductance**. Resistances have units of ohms. Conductances have units of ohm^{-1}, which is a **siemen**, equal to 1 AV^{-1}. According to Ohm's law, the conductance is defined for a line that passes through the origin. The origin is the point at which there is no net driving force for current flow. For ions that are not uniformly distributed across the membrane, however, the point of no net current flow occurs at the reversal potential. Thus we *define* a **chord conductance**:

$$[3.1.18] \quad I_i = g_i(E_m - E_i)$$

where g_i is the chord conductance at E_m, E_m is the membrane potential at any point in the current–voltage curve, and E_i is the equilibrium potential = reversal potential for a single ion, i. This conductance is the slope of the chord joining the curve to its reversal potential, as shown in Figure 3.1.3. Thus the chord conductance is not constant but varies with membrane potential, E_m.

We can also define a **slope conductance** that relates I to E. This is obtained by differentiating Eqn [3.1.17]:

$$[3.1.19] \quad g = \frac{dI_i}{dE_m}$$

Neither the slope conductance nor the chord conductance is a constant. Even for a straight cylindrical pore that is specific for some ion, the conductance varies with voltage because the current is carried only by one ion and its concentration is not the same on the two sides of the membrane.

THE CHORD CONDUCTANCE EQUATION RELATES MEMBRANE POTENTIAL TO ALL ION FLOWS

There are two ways to answer the problem of finding the resting membrane potential when the membrane is permeable to several ions. One way focuses on the permeabilities as we have already defined them, and a second focuses on the conductances, which are related, but not identical, to the permeabilities. We will use conductances here because it makes it easier to understand other electrical phenomena in cells. We begin with the fact that the total current at the resting membrane potential is zero. This is Eqn [3.1.15]:

$$[3.1.15] \quad I_{total} = I_{Na} + I_K + I_{Cl} = 0$$

Substituting in from Eqn [3.1.18] for the individual currents, we get

$$[3.1.20] \quad I = g_{Na}(E_m - E_{Na}) + g_K(E_m - E_K) + g_{Cl}(E_m - E_{Cl})$$

At rest, $I = 0$, and so we collect terms in E_m on the left-hand side to find

$$[3.1.21] \quad (g_{Na} + g_K + g_{Cl})E_m = g_{Na}E_{Na} + g_K E_K + g_{Cl}E_{Cl}$$

Solving for E_m, we obtain

$$E_m = \frac{g_{Na}}{g_{Na}+g_K+g_{Cl}}E_{Na} + \frac{g_K}{g_{Na}+g_K+g_{Cl}}E_K + \frac{g_{Cl}}{g_{Na}+g_K+g_{Cl}}E_{Cl}$$
[3.1.22]

This is the **chord conductance equation**. What it says is that **the resting membrane potential is the conductance-weighted average of the equilibrium potentials** for the ions that have any conductance across the membrane. We can clearly enlarge Eqn [3.1.22] to include other ions such as Ca^{2+}.

Both the chord conductance equation and the GHK equation indicate that all permeant ions contribute to the resting membrane potential. Those ions with the largest conductances, or the largest permeabilities, are correspondingly greater determinants of the resting membrane potential.

THE CURRENT CONVENTION IS THAT AN OUTWARD CURRENT IS POSITIVE

As mentioned earlier, the reversal potential is the potential at which the current reverses direction: it changes from positive to negative. It is important here to remember the sign convention for these currents. First, **current is taken as the direction of positive charge flow**. Second, a **positive current is taken as an outward current**—it goes from the inside of the cell to the outside. Third, a **negative current is taken as an inward current**. This convention becomes apparent when we consider the individual ionic currents.

According to Eqn [3.1.18], the individual ionic currents are

[3.1.23]
$$I_{Na} = g_{Na}(E_m - E_{Na})$$
$$I_K = g_K(E_m - E_K)$$
$$I_{Cl} = g_{Cl}(E_m - E_{Cl})$$

Recall that g stands for a conductance. **Conductances are always positive**.

We consider the example given above for the concentrations and equilibrium potentials. These values are recapitulated in Table 3.1.1.

Suppose that the resting membrane potential is -85 mV. The Na current could then be calculated as

TABLE 3.1.1 Concentrations of Ions and Their Equilibrium Potentials

n	[Ion]$_{out}$	[Ion]$_{in}$	E_i
Na^+	145×10^{-3} M	12×10^{-3} M	$+66.6$ mV
K^+	4×10^{-3} M	155×10^{-3} M	-97.7 mV
Cl^-	100×10^{-3} M	5×10^{-3} M	-80.0 mV

$$I_{Na} = g_{Na}(-85 \text{ mV} - 66.7 \text{ mV}) = -g_{Na} \times 151.7 \text{ mV}$$

Which way does the Na^+ ion go? Its concentration is higher outside the cell than inside. A potential of $+66.6$ mV *inside* would be necessary to stop the diffusive flow down its concentration gradient, but the resting membrane potential is -85 mV. Thus the negative membrane potential further drives Na^+ influx. Na^+ goes into the cell, but *the current is negative*, as indicated by the sign and the fact that g_{Na} is always positive. The current is negative because that is our convention for determining the direction of current. An inward current (flux of a positive ion) is a negative current. From Eqn [3.1.11], we see that the current is in the same direction as the flux:

[3.1.11] $$I_i = z_i \mathfrak{F} J_i$$

Thus inward Na flux is also a negative flux. The convention for positive flux is also from inside the cell to outside.

In the case of K^+, the current is given as

$$I_K = g_K(-85 \text{ mV} + 97.7 \text{ mV}) = g_{Na} \times 8.7 \text{ mV}$$

which is positive. It takes -97.7 mV to stop K^+ exit from the cell. The resting membrane potential of -85 mV is insufficient to stop K^+ exit, so at rest there is some outward I_K. Because it is an outward current, it is positive. The J_K is similarly positive.

In the case of Cl^-, the current is

$$I_{Cl} = g_{Cl}(-85 \text{ mV} + 80 \text{ mV}) = -g_{Cl} \times 5 \text{ mV}$$

So the current is negative. Here -80 mV is enough to stop Cl entry into the cell. This is the E_{Cl} and you can see that if the membrane potential was -80 mV, I_{Cl} would be zero. But the membrane potential is -85 mV, which is more negative than E_{Cl}; thus the negative inside potential forces Cl^- out of the cell. There is an outward flux of Cl^-, which is a positive flux. But the charge on Cl^-, $z_{Cl} = -1$ and so the current carried by Cl^- is opposite to its flux! (see Eqn [3.1.11]). The negative current is an inward current carried by the outward flux of Cl^-.

This current convention and the convention that membrane potential is defined as $E_m = \psi_i - \psi_o$ are true conventions in that the opposite conventions do not violate any physical law. These conventions are equivalent to the orientation of the x-axis perpendicular to the surface of the cell membrane. The convention is that $x = 0$ is on the inside of the cell and $x = \delta$, where δ is the thickness of the membrane, is on the outside surface. Figure 3.1.4 illustrates this convention and what it means for the gradients in C and ψ.

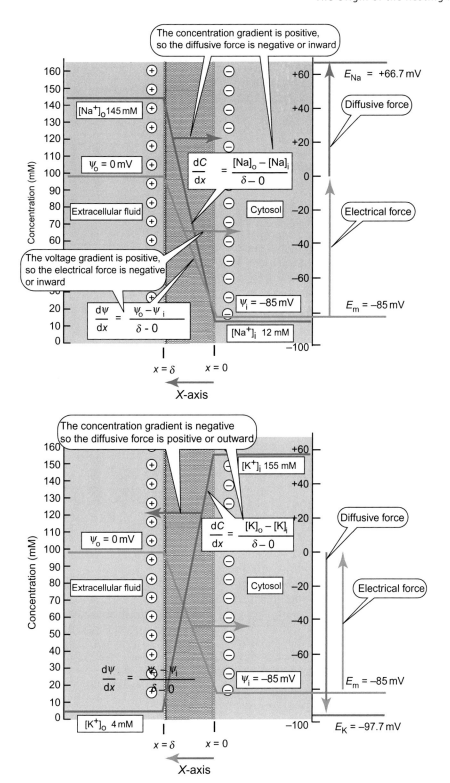

FIGURE 3.1.4 Cartoon of the concentration and voltage gradients and their resulting parts of the ion flux. Top, situation with Na$^+$ ions. The [Na$^+$]$_o$ is about 145 mM, whereas the [Na$^+$]$_i$ is about 12 mM. Thus the concentration gradient is positive (it slopes up with x; the direction of the x-axis is from inside to outside of the cell as indicated at the bottom of the figure). Fick's law gives the diffusive flux as being proportional to the negative of the gradient. Thus the diffusive force favors a negative flux, which is directed inward. The outside potential, ψ_o, is taken as zero. At rest, ψ_i is about −85 mV. Thus the voltage gradient is also positive. The flux produced by the electrical force is also proportional to the negative of the electrical gradient. Thus the positive voltage gradient produces a negative flux, which is inward. Thus for Na$^+$, both the diffusive force and electrical force produce an inward flux and an inward current. The bottom panel illustrates the situation with K$^+$ ions. [K$^+$]$_o$ is just 4 mM and [K$^+$]$_i$ = 155 mM, so the concentration gradient is negative. Thus the diffusive force is positive or directed outward. The negative electrical potential inside makes a positive electrical potential gradient, which favors a negative K$^+$ movement, or an inward current and flux. Thus here the diffusive force and the electrical force oppose each other, but their magnitudes are not equal. Another way of looking at the equilibrium potential is that it is the magnitude of the diffusive force expressed in electrical terms. The net effect on movement is obtained by subtracting the two forces.

SUMMARY

Separation of charges produces an electric field defined as the force experienced by a test positive unit charge. The field at any point is also characterized by a potential, defined as the work necessary to move a positive unit test charge from infinite separation to that point. The potential surrounding positive charge is positive, and that surrounding negative charge is negative. Thus the separation of charges across a membrane produces a potential difference between the two sides of the membrane.

Ions move in response to concentration differences and in response to electrical forces. If a membrane is permeable only to a single ion, and the concentration of the ion is different on the two sides of the membrane, then the ion will diffuse from its high concentration toward its low concentration. Since it is charged, the ion movement makes a current and there is a separation of charge. This separation of charge produces an electrical force that opposes further diffusion. The electric field is the force per unit test charge, and this is the negative spatial derivative of the potential. The membrane potential at which diffusion is exactly balanced by the electrical forces is called the equilibrium potential. It can be calculated by using the Nernst equation:

$$\frac{RT}{z\Im} \ln \frac{C_o}{C_i} = (\psi_i - \psi_o)$$

The equilibrium potentials for K^+, Na^+, and Cl^- depend on their concentrations on both sides of the membrane and they are typically different from each other. In the Nernst equation, put the outside concentration in the numerator to agree with the sign convention.

Real membranes are permeable to a variety of ions, but with different permeabilities. The permeability can be expressed either as a permeability relating flux to driving force in concentration units or as conductance relating current to driving force in voltage units. The resting membrane potential occurs at a net zero membrane current. This occurs at a membrane potential that is the conductance-weighted average of the equilibrium potentials, as described by the chord conductance equation:

$$E_m = \frac{g_{Na}}{(g_{Na} + g_K + g_{Cl})} E_{Na} + \frac{g_K}{(g_{Na} + g_K + g_{Cl})} E_K$$
$$+ \frac{g_{Cl}}{(g_{Na} + g_K + g_{Cl})} E_{Cl}$$

Thus the resting membrane potential is closest to the equilibrium potential of the ion with the highest conductance. The resting membrane potential also can be expressed in terms of the permeabilities: it depends on the concentration-weighted permeabilities as expressed in the GHK equation.

Membrane potential is defined as $E_m = \psi_i - \psi_o$. An outward flow of positive ions is taken as a positive current. The current for any ion is given as

$$I_i = g_i(E_m - E_i)$$

where g_i is the conductance to the particular ion, in siemens. This is a chord conductance which is distinguished from a slope conductance. Here E_i is the equilibrium potential for the ion in question. Conductances are always positive. A negative current means that the current is directed inward.

REVIEW QUESTIONS

1. What is the current of an ion at its equilibrium potential? What would the current look like as a function of membrane potential if the ion had the same concentration on both sides of the membrane?
2. If the conductance of the membrane to K^+ was increased, what would happen to the equilibrium potential for K^+ (E_K)? What would happen to E_m?
3. Why is Cl_o in the denominator of the GHK equation when Na_o and K_o are in the numerator?
4. What potential is calculated by the Nernst equation?
5. What potential is calculated by the GHK equation?
6. What is a siemen?
7. What is meant by an inward current? Outward current? What sign does an inward current have? If Cl^- exits the cell, does it make a positive or negative current?

APPENDIX 3.1.A1 DERIVATION OF THE GHK EQUATION

The flux of an ion through a water-filled channel should be given by the generalized Fick's law given by Eqn [3.1.1]:

[3.1.A1.1] $$J_s = -D\frac{\partial C}{\partial x} - \frac{D}{RT} Cz\Im \frac{\partial \psi}{\partial x}$$

The flux can be obtained by integrating this equation. To begin, we multiply both sides of the equation by an integrating factor, ρ, and we choose ρ so that the right-hand side of the equation becomes an exact differential:

[3.1.A1.2] $$J_s \rho = -D\left[\rho\frac{\partial C}{\partial x} - \frac{Cz\Im \rho}{RT}\frac{\partial \psi}{\partial x}\right]$$

We choose ρ so that the terms in brackets are an exact differential. Thus we want

[3.1.A1.3] $$\rho\frac{\partial C}{\partial x} + \frac{Cz\Im \rho}{RT}\frac{\partial \psi}{\partial x} = \frac{d(\rho C)}{dx} = \rho\frac{\partial C}{\partial x} + C\frac{\partial \rho}{\partial x}$$

From comparing the extreme left of Eqn [3.1.A1.3] to the extreme right, we see that the condition that multiplication by ρ transforms the equation into an exact differential is met if

[3.1.A1.4] $$\frac{\partial \rho}{\partial x} = \frac{z\Im \rho}{RT}\frac{\partial \psi}{\partial x}$$

We rearrange this to get

[3.1.A1.5] $$\frac{\partial \rho}{\rho} = \frac{z\mathfrak{F}}{RT}\partial \psi$$

A solution to Eqn [3.1.A1.5] is

[3.1.A1.6] $$\rho = e^{(z\mathfrak{F}/RT)\psi}$$

We may insert this result back into Eqn [3.1.A1.2] to obtain

[3.1.A1.7] $$J_s e^{(z\mathfrak{F}/RT)\psi} = -D\frac{d(Ce^{(z\mathfrak{F}/RT)\psi})}{dx}$$

which may be rewritten as

[3.1.A1.8] $$J_s e^{(z\mathfrak{F}/RT)\psi} dx = -D d(Ce^{(z\mathfrak{F}/RT)\psi})$$

The flux can be obtained by integrating this equation from $x = 0$ (one side of the membrane) to $x = \delta$, the other side of the membrane for a membrane with thickness δ:

[3.1.A1.9] $$\int_0^\delta J_s e^{(z\mathfrak{F}/RT)\psi} dx = -D \int_0^\delta d(Ce^{(z\mathfrak{F}/RT)\psi})$$

In this case, we limit ourselves to the steady-state condition. In this case, J_s does not vary with distance across the membrane; it is constant. Therefore J_s may be removed from the integral and we get

[3.1.A1.10] $$J_s = \frac{-D \int_0^\delta d(Ce^{(z\mathfrak{F}/RT)\psi})}{\int_0^\delta e^{(z\mathfrak{F}/RT)\psi} dx}$$

The numerator in this equation is the integral of an exact differential and can be immediately evaluated between the boundaries. This gives

[3.1.A1.11] $$J_s = \frac{-D[C(\delta)e^{(z\mathfrak{F}/RT)\psi(\delta)} - C(0)e^{(z\mathfrak{F}/RT)\psi(0)}]}{\int_0^\delta e^{(z\mathfrak{F}/RT)\psi} dx}$$

The denominator in this equation can be evaluated only if $\psi(x)$ is known. However, generally $\psi(x)$ is unknown. We can integrate the denominator if we assume that ψ is a linear function of x. This is true if the electric field between $x = 0$ and δ is constant. Recall that the electric field intensity is the negative derivative of the potential. For this reason, the assumption of a linear potential gradient is called the **constant field** assumption. If the potential is linear, then we can write

[3.1.A1.12] $$\psi = \psi(0) + \frac{\psi(0) - \psi(\delta)}{0 - \delta}x$$

The situation of $C(0)$, $C(\delta)$, $\psi(0)$, and $\psi(\delta)$ is illustrated in Figure 3.1.A1.1. The difference in potential across the membrane, $\psi(0) - \psi(\delta)$, is the the membrane potential E_m. Here $\psi(0)$ is the inside potential and $\psi(\delta)$ is the outside potential. Thus Eqn [3.1.A1.12] becomes

[3.1.A1.13] $$\psi = \psi(0) + \frac{E_m}{\delta}x$$

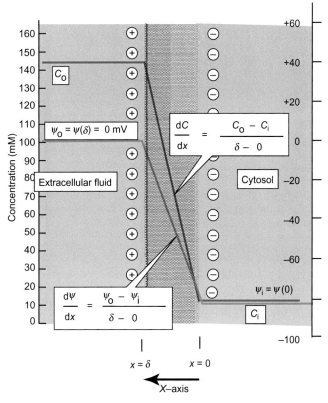

FIGURE 3.1.A1.1 Concentration and potential profile across the membrane along with the convention for the location of the X-axis.

This result can be substituted into the denominator of Eqn [3.1.A1.11] to give

$$\int_0^\delta e^{(z\mathfrak{F}/RT)\psi} dx = \int_0^\delta e^{(z\mathfrak{F}/RT)(\psi(0) - (E_m/\delta)x)} dx$$

$$= e^{(z\mathfrak{F}/RT)\psi(0)} \int_0^\delta e^{-(z\mathfrak{F}/RT)(E_m/\delta)x} dx$$

[3.1.A1.14]

Evaluation of the integral gives

[3.1.A1.15] $$\frac{e^{(z\mathfrak{F}/RT)\psi(0)}}{-(z\mathfrak{F}/RT)(E_m/\delta)}(e^{-(z\mathfrak{F}/RT)E_m} - 1)$$

Inserting the result of Eqn [3.1.A1.15] back into Eqn [3.1.A1.11]

$$J_s = \frac{-D[C(\delta)e^{(z\mathfrak{F}/RT)\psi(\delta)} - C(0)e^{(z\mathfrak{F}/RT)\psi(0)}]}{(e^{(z\mathfrak{F}/RT)\psi(0)})/(-(z\mathfrak{F}/RT)(E_m/\delta))[e^{-(z\mathfrak{F}/RT)E_m} - 1]}$$

[3.1.A1.16]

This can be simplified to

$$J_s = \frac{(D/\delta)[(z\mathfrak{F}/RT)E_m][C(\delta)e^{(z\mathfrak{F}/RT)\psi(\delta)} - C(0)e^{(z\mathfrak{F}/RT)\psi(0)}]e^{-(z\mathfrak{F}/RT)\psi(0)}}{[e^{-(z\mathfrak{F}/RT)E_m} - 1]}$$

[3.1.A1.17]

Multiplying through by the exponent in the numerator, and recalling that $E_m = \psi(0) - \psi(\delta)$, we convert Eqn [3.1.A1.17] into

[3.1.A1.18] $$J_s = \frac{(D/\delta)[(z\mathfrak{F}/RT)E_m][C(\delta)e^{-(z\mathfrak{F}/RT)E_m} - C_0]}{[e^{-(z\mathfrak{F}/RT)E_m} - 1]}$$

This equation is typically rewritten by multiplying numerator and denominator by -1, with no change in meaning:

$$J_s = \frac{(D/\delta)[(z\mathcal{F}/RT)E_m][C(0) - C(\delta)e^{-(z\mathcal{F}/RT)E_m}]}{[1 - e^{-(z\mathcal{F}/RT)E_m}]}$$

[3.1.A1.19]

This flux equation holds true for any ion including Na^+, K^+, and Cl^-. The current carried by each ion is given by Eqn [3.1.19] by multiplying the flux by $z\mathcal{F}$. We identify $C(0)$ with the inside concentration of the ion, and $C(\delta)$ with the outside concentration, as described in Figure 3.1.A1.1 which of course are different for each of the ions. For each ion, we can write an equation for the current that has the form

$$I_{ion} = \frac{(D/\delta)[(z^2\mathcal{F}^2/RT)E_m][C_i - C_o e^{-(z\mathcal{F}/RT)E_m}]}{[1 - e^{-(z\mathcal{F}/RT)E_m}]}$$

[3.1.A1.20]

Thus we can write

$$I_{Na} = \frac{(D/\delta)[(\mathcal{F}^2/RT)E_m][[Na]_i - [Na]_o e^{-(\mathcal{F}/RT)E_m}]}{[1 - e^{-(\mathcal{F}/RT)E_m}]}$$

[3.1.A1.21]

[3.1.A1.22] $$I_K = \frac{(D/\delta)[(\mathcal{F}^2/RT)E_m][[K]_i - [K]_o e^{-(\mathcal{F}/RT)E_m}]}{[1 - e^{-(\mathcal{F}/RT)E_m}]}$$

$$I_{Cl} = \frac{(D/\delta)[(\mathcal{F}^2/RT)E_m][[Cl]_{in} - [Cl]_{out} e^{(\mathcal{F}/RT)E_m}]}{[1 - e^{(\mathcal{F}/RT)E_m}]}$$

[3.1.A1.23]

Note that the exponent in the numerator and denominator of I_{Cl} is positive instead of negative! This is because the z term in the exponent in Eqn [3.1.A1.20] is -1 for Cl. The total current across the membrane is the sum of the currents carried by each ion. At steady state (the resting membrane potential), the current is zero. Thus we can write

[3.1.A1.24] $$I_{total} = I_{Na} + I_K + I_{Cl} = 0$$

We now insert the expressions for each of the individual currents. Here, however, we make an additional substitution that the diffusion coefficient for each ion divided by the thickness of the membrane, D/δ, is equal to the permeability of the membrane to each ion. The result is

$$I_{total} = 0 = \frac{P_{Na}[(\mathcal{F}^2/RT)E_m][[Na]_{in} - [Na]_{out} e^{-(\mathcal{F}/RT)E_m}]}{[1 - e^{-(\mathcal{F}/RT)E_m}]}$$
$$+ \frac{P_K[(\mathcal{F}^2/RT)E_m][[K]_{in} - [K]_{out} e^{-(\mathcal{F}/RT)E_m}]}{[1 - e^{-(\mathcal{F}/RT)E_m}]}$$
$$+ \frac{P_{Cl}[(\mathcal{F}^2/RT)E_m][[Cl]_{in} - [Cl]_{out} e^{(\mathcal{F}/RT)E_m}]}{[1 - e^{(\mathcal{F}/RT)E_m}]}$$

[3.1.A1.25]

Solving this equation, we find

[3.1.A1.26]
$$0 = P_{Na}[[Na]_i - [Na]_o e^{-(\mathcal{F}/RT)E_m}]$$
$$+ P_K[[K]_i - [K]_o e^{-(\mathcal{F}/RT)E_m}]$$
$$+ P_{Cl}[[Cl]_o - [Cl]_i e^{-(\mathcal{F}/RT)E_m}]$$

where we have multiplied the numerator and denominator of the Chloride contribution to the current by $\exp(-\mathcal{F}/RT\, E_m)$ and rearranged the terms. Solving for E_m, we finally obtain

[3.1.A1.27] $$E_m = \frac{RT}{\mathcal{F}} \ln \frac{P_{Na}[Na]_o + P_K[K]_o + P_{Cl}[Cl]_i}{P_{Na}[Na]_i + P_K[K]_i + P_{Cl}[Cl]_o}$$

This is the GHK equation. It tells us that the resting membrane potential (at which point the net current across the membrane is zero) is determined by the concentrations of ions on both sides of the membrane and by their respective permeabilities. The ion with the largest permeability dominates the membrane potential by moving the potential closer to its equilibrium potential.

The Action Potential 3.2

Learning Objectives

- Draw a picture of a motor neuron and identify soma, dendrites, axon, myelin sheath, and terminals
- Describe what is meant by depolarization and hyperpolarization and what currents achieve it
- Explain what is meant by the "all or none" law
- Define latency, absolute refractory period, relative refractory period, and overshoot in an action potential on a nerve
- Define rheobase and chronaxie and how they can be determined from a strength–duration curve
- Recognize the Weiss relation between strength and duration
- Draw a graph showing the conductance changes with time during an action potential on a nerve
- Explain what is meant by the "activation gate" and "inactivation gate" of the Na channel
- Describe the state of the activation gate and inactivation gate during an action potential
- Describe how the inactivation gate is reset
- Distinguish between unitary current and ensemble current
- Using the whole-cell i–v curve, explain the effect of small depolarizations below threshold and of depolarizations above threshold
- Explain why stronger stimuli need shorter durations to achieve an action potential

CELLS USE ACTION POTENTIALS AS FAST SIGNALS

Certain cells in the body are capable of initiating and propagating an **action potential** over their surface. These cells are called **excitable** cells and they include muscle and nerve cells. The action potential is a brief, pulse-like change in the membrane potential. Because it can be **propagated** rapidly over the surface of the cell, it conveys a fast signal from one place to another. We will use a motor neuron as an example of an excitable cell.

THE MOTOR NEURON HAS DENDRITES, A CELL BODY, AND AN AXON

Motor neurons are large cells in the ventral horn of the spinal cord as shown in Figure 3.2.1. They have a number of processes called **dendrites** that bring signals to the motor neuron. The motor neuron also has one large process, the **axon**, that connects the motor neuron on one end with a muscle fiber on the other. Action potentials move along the axon so that activity in the motor neuron alters activity in the muscle.

Axons from neurons can be **myelinated** or **unmyelinated**. Myelin refers to a sheath that covers the axon, but not entirely. In the peripheral nervous system, **Schwann cells** make the myelin by wrapping themselves around the axon, forming a multilayered structure of multiple cell membranes of the Schwann cell. In the central nervous system, **oligodendroglial cells** make the myelin. The sheath is not continuous in either the peripheral or central nervous system. At the end of each Schwann cell, there is a gap in the myelin. This gap is called the **Node of Ranvier** (see Figure 3.2.2).

Like all cells, the motor neuron has a nucleus located in its cell body or **soma**. The soma is also sometimes referred to as the **perikaryon** from the Greek root "peri" meaning "around" or "surrounding" and "karyon" meaning "nut" or "kernel," and referring to the nucleus. There is only one nucleus in the motor neuron and it is the site of mRNA transcription.

PASSING A CURRENT ACROSS THE MEMBRANE CHANGES THE MEMBRANE POTENTIAL

Figure 3.2.3 shows a highly schematic diagram of how the membrane potential can be measured by inserting a microelectrode through the membrane of the axon. The resting membrane potential in these axons is about -60 mV. Recall that the membrane potential is defined as $\psi_i - \psi_o$.

It is possible to pass current across the membrane through the arrangements shown in Figure 3.2.3. In one case, the battery is hooked up so that current will pass into the cell. Recall here that current is defined as the direction of positive charge flow and an outward flow is positive. An inward current, shown in the middle of Figure 3.2.3, will **depolarize** the cell. The cell is already polarized; an inward current would make the cell less polarized and thus it would depolarize it. If the battery was connected with the opposite polarity, the resulting current would be outward and this would make the membrane more polarized. This is a **hyperpolarizing** current.

AN OUTWARD CURRENT HYPERPOLARIZES THE MEMBRANE POTENTIAL

When an outward current is passed across the membrane, the recorded membrane potential is a distorted version of the stimulus. As the magnitude of the current is increased (stimulus intensity is increased), the recorded hyperpolarization grows larger (see Figure 3.2.4). It takes some time for the current to reach a new steady-state membrane potential. In addition, the magnitude of the hyperpolarization depends on the distance from the stimulating electrode to the recording electrode. These phenomena are consequences of the electrical characteristics of the axon, its **cable properties**. In brief, the cable properties define a time constant and a length constant which describe how voltage builds up (or falls off) with time and with distance. The membrane parameters which influence these cable characteristics include the membrane resistance, the membrane capacitance, and the electrical resistance of the axoplasm.

THE RESULT OF DEPOLARIZING STIMULUS OF ADEQUATE SIZE IS A NEW PHENOMENON—THE ACTION POTENTIAL

Depolarization to a small degree produces a change in E_m that mirrors hyperpolarization—the change is a distorted version of the stimulus. With larger depolarization, however, a new phenomenon arises. When E_m exceeds a threshold value, there is an abrupt rise in E_m, reaching positive values near +30 mV. Just as quickly, E_m returns to near normal values. This abrupt change in the membrane potential brought about by depolarization is the **action potential** (see Figure 3.2.5).

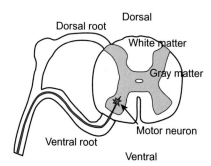

FIGURE 3.2.1 Location of the motor neuron in the spinal cord. The spinal cord is shown in cross-section. The dorsal aspect is toward the back; ventral is toward the front. The dorsal and ventral roots are paired, with one on each side of the cord, but only one side is shown here. The motor neuron is shown in dark blue. The motor neuron's cell body is located in gray matter in the ventral horn, and its long axon leaves the cord via the ventral root and continues on to a muscle where it makes a neuromuscular junction. These cells produce an action potential that propagates along the axon, excites the nerve at the neuromuscular junction, and conveys that excitation to the muscle in order to activate the muscle.

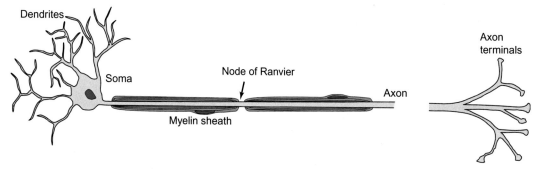

FIGURE 3.2.2 Parts of the motor neuron. Dendrites are multiple processes of the neuron that bring signals to the cell body, or soma. A single long axon exits the cell on one pole and reaches all the way to its target cell, the muscle fiber. The long axon is covered by a myelin sheath made by Schwann cells. The sheath is interrupted at regular intervals at the nodes of Ranvier.

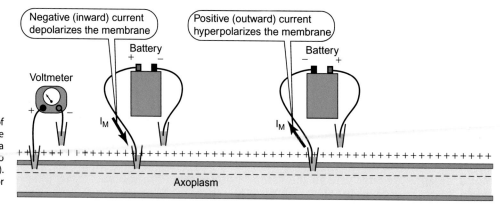

FIGURE 3.2.3 Arrangement of electrodes to record membrane potential (left); to inject a depolarizing current (middle) or to pass a hyperpolarizing current (right). I_M refers to "membrane current," or current across the membrane.

THE ACTION POTENTIAL IS ALL OR NONE

When the strength of depolarization is increased further, the resulting wave form in E_m remains remarkably constant. That is, if the depolarization is enough to trigger an action potential, the resulting action potential is independent of the stimulus. The action potential is "**all or none**." The action potential is not graded, meaning that it does not vary from a minimum to a maximum; rather, a given stimulus will either produce an action potential or it will not, and the resulting shape of the plot of E_m against time is approximately the same.

THE LATENCY DECREASES WITH INCREASING STIMULUS STRENGTH

Once initiated, the action potential moves away from its point of origin, and this is called **propagation**. At some point away from its origin, the action potentials are indistinguishable. However, near the stimulus there is a slight difference, having to do with the time it takes to reach threshold, the **latency**. The membrane potential is not a completely faithful representation of the stimulus: the rise in E_m lags behind the stimulus rise and the fall in E_m lags behind the stimulus fall. The membrane distorts the stimulus. When the stimulus strength is increased above threshold, the time to reach threshold, the latency, decreases. The relationship between latency and stimulus strength is shown in Figure 3.2.6.

THRESHOLD IS THE MEMBRANE POTENTIAL AT WHICH AN ACTION POTENTIAL IS ELICITED 50% OF THE TIME

The **threshold** for a nerve is the membrane potential which must be reached in order to "fire" an action

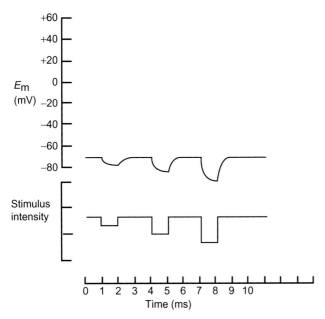

FIGURE 3.2.4 Results of hyperpolarizing stimuli of varying intensity on the axon membrane potential, E_m. The effect of hyperpolarizing stimuli on the membrane potential is a distorted version of the stimulus wave form. The rise time and fall time of E_m are part of the cable properties of the axon.

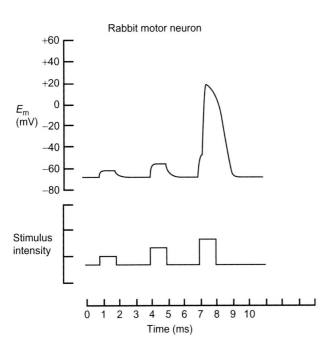

FIGURE 3.2.5 Results of depolarizing stimuli of varying intensity on the axon membrane potential. At low stimulus strength, E_m is a distorted version of the stimulus, as in the case of hyperpolarization. When the depolarization exceeds a threshold value, there is an abrupt rise in E_m, reaching positive values. This abrupt change in the membrane potential brought about by depolarization is the **action potential**.

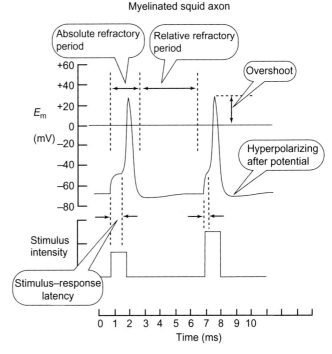

FIGURE 3.2.6 Effect of stimulus strength above threshold on resulting wave forms in a myelinated squid motor neuron axon. Note that the wave form here differs somewhat from that in mammalian motor neurons. It is shorter in the squid and shows a hyperpolarization after potential that is largely lacking in mammalian motor neurons.

potential. Operationally, it is defined as the membrane potential at which the nerve will fire an action potential 50% of the time. However, it is possible to initiate an action potential even when the membrane potential is below threshold, but the probability of this happening is much reduced. Similarly, it is possible to fail to initiate an action potential even when the membrane potential is above threshold, but with a much reduced probability. The probability of initiating an action potential is a steep function of the membrane potential.

THE NERVE CANNOT PRODUCE A SECOND EXCITATION DURING THE ABSOLUTE REFRACTORY PERIOD

While the first action potential is occurring, it is impossible to begin a second action potential, no matter how powerful the second stimulus. The time during which the nerve is **refractory** to a second stimulus is called the **absolute refractory period**. It typically lasts for 1–2 ms. Following the absolute refractory period is a second, **relative refractory period**. This typically lasts some 4 ms or so, and during this time it is possible to stimulate the nerve cell to make another action potential, but it is more difficult to do so than in the resting neuron. That is, the threshold is elevated during the relative refractory period.

THE ACTION POTENTIAL REVERSES TO POSITIVE VALUES, CALLED THE OVERSHOOT

The first recordings of action potentials with intracellular electrodes were accomplished in 1939 and 1940 by A.L. Hodgkin and Andrew Huxley and by K.S. Cole and H.J. Curtis, respectively, using the squid giant axon. They found results similar to those shown in Figure 3.2.6. In particular, they found that the membrane potential not only went to zero, but it took on positive values. This is called the **overshoot**. Given the fact that the equilibrium potentials for all known ions were negative except for Na^+ and Ca^{2+}, these pioneers suggested that the membrane could have a positive potential by becoming selectively permeable to Na^+.

THE STRENGTH–DURATION RELATIONSHIP IS HYPERBOLIC

The relationship between the current necessary to reach threshold and the duration of the current for a compound action potential is shown in Figure 3.2.7. This is related to the strength–latency relationship, as the latency is the time required to elicit an action potential for a given strength of stimulus. The strength–duration relationship shown in Figure 3.2.7 asks the question, for a given strength of stimulus, how long must it continue to produce an action potential? For some strengths of stimulus, no duration is sufficient—you never get an action potential. For a critical strength of stimulus, you get an action potential only if the stimulus lasts infinitely long. This strength of stimulus is

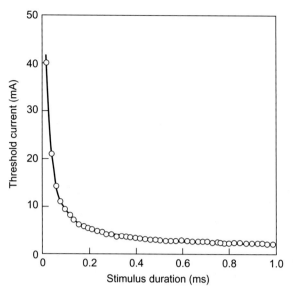

FIGURE. 3.2.7 Strength–duration relationship in human peripheral nerve. The median nerve was stimulated using surface electrodes 1 cm in diameter taped to the skin over the median nerve at the wrist, 4 cm apart oriented along the course of the nerve. Stimulus was a square wave with rise and fall times of 10 μs. The antidromic compound sensory action potential was recorded at the index finger using ring electrodes with 2–3 mm diameter set 2–3 cm apart on the finger. Stimulus intensity was reduced 2–5% until the amplitude of the compound action potential (the aggregate of a bundle of axons) was reduced to 30% of maximum (From Mogyoros, I, Kiernan, MC, and Burke, D. Strength–duration properties of human peripheral nerve. Brain **119**:439–447, 1996).

called the **rheobase** (see Figure 3.2.8). Weiss described this relationship by the equation:

$$[3.2.1] \qquad Q = I_{rh}(t + \tau_{SD})$$

where Q is the charge, I is the rheobase current, t is the time the current is on, and τ_{SD} is a strength–duration time constant, often called the **chronaxie**. Because the total charge is the current times the time, this equation becomes

$$[3.2.2] \qquad I = I_{rh}\frac{(t + \tau_{SD})}{t} = I_{rh} + I_{rh}\frac{\tau_{SD}}{t}$$

This last equation describes a rectangular hyperbole offset from the x-axis by I_{rh}. Also, when $t = \tau_{SD}$, the current $I = 2\ I_{rh}$; the current is twice the rheobase. These relationships are shown diagrammatically in Figure 3.2.8. We will seek to further understand this relationship after we establish how depolarization induces an action potential.

VOLTAGE-DEPENDENT CHANGES IN ION CONDUCTANCE CAUSE THE ACTION POTENTIAL

INCREASE IN ION CONDUCTANCE BEGINS AFTER THE MEMBRANE BEGINS TO DEPOLARIZE

K.S. Cole and H.J. Curtis in 1939 studied the impedance properties of the squid giant axon and discovered a marked decrease in the impedance (equivalently, an increase in the conductance) of the squid axon

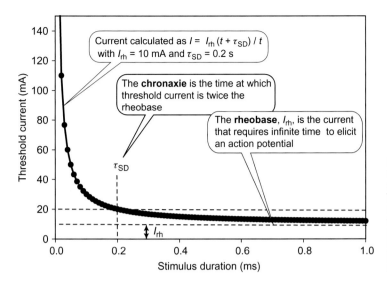

FIGURE 3.2.8 Definition of rheobase and chronaxie. The rheobase current is the current which takes an infinite time to elicit an action potential—it is obtained by extrapolation or curve fitting. The chronaxie is the time at which threshold current is twice the rheobase. These two parameters are parameters of Eqn [3.2.2] that can be obtained by fitting the equation to experimental data. In the figure, the current was calculated according to Eqn [3.2.2] using a rheobase of 10 mA and a chronaxie of 200 ms. Note that the calculated curve resembles the form of the experimental curve shown in Figure 3.2.7.

membrane during the upstroke of the action potential. This increase begins only after the membrane potential rises many millivolts above the resting membrane potential. They argued that the foot of the action potential resulted from the discharging of the membrane from local currents from elsewhere in the cell. At the inflection point on the rising phase of the action potential, the cell generated its own net inward current. Such a current must be carried by an ion, and the most likely candidate is Na^+, because of its high extracellular concentration.

THE ACTION POTENTIAL IS ACCOMPANIED BY NA$^+$ INFLUX

To test the idea that increases in Na^+ conductance might cause the action potential, Alan Hodgkin and Bernard Katz replaced some of the NaCl in seawater with choline chloride. Replacing the Na^+ reduced the upstroke of the action potential and markedly reduced the size of the action potential. Later experiments using radioactive tracers showed that Na^+ influx accompanies the action potential.

THE CHORD CONDUCTANCE EQUATION PREDICTS THAT CHANGES IN CONDUCTANCE WILL CHANGE THE MEMBRANE POTENTIAL

In Chapter 3.1, we developed the chord conductance equation that shows that the resting membrane potential is the conductance-weighted average of the equilibrium potentials for all ions (see Eqn [3.2.3]). At rest, the membrane potential is closer to the K^+ equilibrium potential because conductance K^+ is higher than the conductance of the other ions.

$$[3.2.3] \quad E_m = \frac{g_{Na}}{(g_{Na} + g_K + g_{Cl})}E_{Na} + \frac{g_K}{(g_{Na} + g_K + g_{Cl})}E_K + \frac{g_{Cl}}{(g_{Na} + g_K + g_{Cl})}E_{Cl}$$

During the rising phase of the action potential, the conductance to Na^+ increases, changing E_m from its resting, polarized value toward more positive values. If the conductance to Na^+ becomes large enough, relative to the conductances for K^+ and Cl^-, then E_m will be driven toward the equilibrium potential for Na^+, E_{Na}. Since E_{Na} is positive, E_m will be driven to positive values and will exhibit the overshoot.

G_{NA} INCREASES TRANSIENTLY DURING THE ACTION POTENTIAL; G_K INCREASES LATER AND STAYS ELEVATED LONGER

The results described earlier show that the rapid depolarization and overshoot in the action potential is due to a transient increase in membrane conductance, and this is accompanied by an Na^+ influx. The rapid repolarization of the membrane afterward is due to shutting off the increased Na^+ conductance and increasing the K^+ conductance. These results are consistent with the chord conductance equation. Hodgkin and Huxley succeeded in calculating g_{Na} and g_K during different parts of the action potential in the squid axon, and their results are summarized schematically in Figure 3.2.9. Although these results were determined in the squid, the principles remain the same for action potentials in mammalian excitable cells.

CONDUCTANCE AND EQUILIBRIUM POTENTIALS FOR NA$^+$ AND K$^+$ ACCOUNT FOR ALL OF THE FEATURES OF THE ACTION POTENTIAL

The origin of the action potential can be explained on the basis of the equilibrium potential for Na^+ and K^+ and the time course of their conductances. In the squid axon, the resting potential is on the order of -60 mV, with E_K about -75 mV. When the membrane is depolarized by some means, a conductance pathway for Na^+ begins to open. At this time, Na^+ is far away from its equilibrium

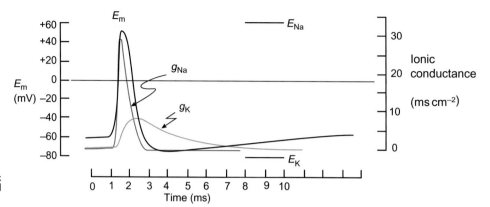

FIGURE 3.2.9 Changes in g_{Na} and g_K during the propagated action potential as calculated by Hodgkin and Huxley.

potential, so the driving force for Na$^+$ is large and the opening of the conductance pathway causes a Na$^+$ influx and a further depolarization of the membrane. This additional depolarization further opens additional conductance pathways. This positive feedback explains the explosive increase in Na$^+$ conductance and the rapid increase in the membrane potential toward E_{Na}.

The pathway for Na$^+$ conductance next undergoes a **time-dependent inactivation**. This by itself reduces the inward Na$^+$ current, except that the inactivation is accompanied by a reduction in E_m (i.e., a repolarization of the membrane) and a subsequent increase in the driving force. The Na$^+$ current is complicated, being the product of the conductance and the net driving force (see Eqn [3.1.18]).

Repolarization of the membrane would occur with a return of Na$^+$ conductance to normal, but it is hastened by a delayed increase in g_K followed by its gradual return to resting levels. The increased g_K increases an outward current carried by K$^+$, and this current more quickly repolarizes the cell. Because g_K remains elevated even when g_{Na} has returned to normal, the cell is hyperpolarized for several ms following the action potential until g_K returns to normal.

G_{NA} IS A FUNCTION OF A NA$^+$-SELECTIVE CHANNEL

The identification of two major components, I_{Na} and I_K, in the ionic currents by Hodgkin and Huxley, has allowed electrophysiologists to characterize the gating properties of the channels that pass these currents. The present view is that the Na$^+$ channel has at least four distinct components:

- **The Selectivity Filter Enables the Na$^+$ Channel to Pass Na$^+$ Preferentially**
  The relative permeabilities of ions in ion channels can be calculated by measuring the reversal potential when the ion composition is changed. The results show that the Na$^+$ channel actually conducts H$^+$ ions much more readily than it does Na$^+$ ions. The current through the Na$^+$ channel, however, is dominated by Na$^+$ because of its much higher concentration. The [H$^+$] is around 10^{-7} M, whereas [Na$^+$] is around 0.1 M, or 10^6-fold higher than [H$^+$]. It is believed that the selectivity relies on a combination of hydrated ion size and the free energy of hydration. Most other ions have a lower permeability through the Na$^+$ channel. Thus the Na$^+$ channel is **selective** for Na$^+$.

- **The Na$^+$ Channel Possesses an Activation Gate**
  At rest the membrane has a low g_{Na} because the Na$^+$ channel is blocked by its **activation gate**. This is a part of the Na$^+$ channel that can be moved to allow Na$^+$ to conduct through the channel. The opening of this gate is voltage dependent: the activation gate begins to open only when the membrane depolarizes. This voltage-dependent opening of the activation gates causes the explosive increase in Na$^+$ conductance in the rising phase of the action potential.

- **The Na$^+$ Channel Possesses an Inactivation Gate**
  Na$^+$ channels also inactivate. This is the function of a separate gate that closes according to time and voltage. When it closes, the Na$^+$ channel is nonconducting and, most importantly, it is **inactivatable**. Conductance through the Na$^+$ channel requires that both the inactivation gate and the activation gate are open. When the inactivation gate is closed, opening of the activation gate alone does not allow the channel to conduct ions.

- **Specific Toxins Bind to the Na$^+$ Channel**
  The puffer fish produces a potent toxin, **tetrodotoxin**, or TTX, that binds to Na$^+$ channels and blocks them. As a result, action potential conduction in nerve and muscle is blocked, with lethal consequences. Another natural toxin, **saxitoxin**, has similar properties. It derives its name from the Alaskan butter clam, *Saxidomus*. Eating a single contaminated shellfish can be fatal. The Na$^+$ channel contains regions that bind these toxins.

THE INACTIVATION GATES MUST BE RESET BEFORE ANOTHER ACTION POTENTIAL CAN BE FIRED

An action potential can occur in a nerve only when the Na$^+$ channels can open. If they are blocked, for example, by tetrodotoxin or saxitoxin, then no action potentials are possible. Similarly, if the channels are in the inactivatable state because the inactivation gate is

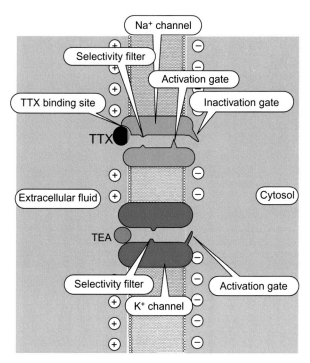

FIGURE 3.2.10 Hypothetical and conceptual model of the voltage-gated Na^+ channel and K^+ channel. The selectivity filter allows Na^+ to pass but not K^+ (top). When TTX (black circle) binds to its site, the channel is blocked. Closure of either the activation gate or inactivation gate will block Na^+ conductance. The K^+ channel also possesses an activation gate and is blocked by tetraethylammonium (TEA).

closed, they also cannot be activated by opening the activation gates. Following a normal action potential, the Na^+ channels are reset when the activation gates close and the inactivation gates open. This takes some time. Part of the refractoriness of the nerve cell immediately following the action potential is due to the lower conductance of the membrane to Na^+ because the inactivation gates are closed.

Figure 3.2.10 shows a conceptual model of the Na^+ and K^+ channels with their activation and inactivation gates and their binding sites for specific blockers. Figure 3.2.11 shows how these channels change during the action potential. Figure 3.2.12 shows the timing of opening of the activation and inactivation gates and that Na^+ entry is possible only when both are open.

CONDUCTANCE DEPENDS ON THE NUMBER AND STATE OF THE CHANNELS

Figure 3.2.11 shows a cartoon of the states of the Na^+ and K^+ channels in axon membranes that give rise to the action potential. Out of necessity, only representative channels are shown in the figure. They are meant to represent what a **population** of channels is doing. The overall current carried through a patch of membrane is given by

[3.2.4] $$I_i = NP_o i$$

where I is the current, N is the number of channels, P_o is the probability of a channel being open, and i is the **unitary current**, the current carried by the open channel under physiological conditions. Cells possess thousands of channels, each of whose behavior is **stochastic**, meaning that their opening and closing are not deterministic but probabilistic. The average behavior of many channels or the behavior of a large number of them is predictable, but the behavior of a single channel appears to be erratic and unpredictable. The sum of the currents of a population of channels is called the **ensemble current**.

PATCH CLAMP EXPERIMENTS MEASURE UNITARY CONDUCTANCES

It is possible to study the behavior of single channels using a patch clamp technique, shown in Figure 3.2.13. In this method, one clamps the voltage across the patch and measures current across it at that voltage. One can also step the potential from some holding value to a new value and measure the currents. Figure 3.2.14 shows the unitary Na^+ and K^+ currents in successive sweeps from patch clamp recordings in neuroblastoma cells (left) and squid giant axon (right). Distinction can be made between I_{Na} and I_K by judicious choice of ionic conditions and use of specific inhibitors. The Na^+ channels open briefly upon depolarization and then do not open later on. Note that the individual channels open and close rather erratically. It is not possible to predict exactly when a particular channel will either open or close. The presence of a large number of channels, however, will smooth out these discontinuities. The ensemble average is obtained by averaging many sweeps. This describes the average behavior of a single channel over many experiments, which is equivalent to the average behavior of a population of channels in a single sweep. The average behavior shows that Na^+ channels open upon depolarization and then close, or inactivate. K^+ channels, on the other hand, open after a slight delay and stay open during depolarization. Repolarization closes the K^+ channels.

THE CURRENT−VOLTAGE RELATIONSHIP FOR THE WHOLE CELL DETERMINES THE THRESHOLD FOR EXCITATION

Now that we know the ionic basis of both the resting membrane potential and the action potential, we are in a better position to understand why there is a critical level of depolarization. The whole-cell current−voltage relationship in a cell is approximated by the curve shown in Figure 3.2.15. This curve results mainly from the $i-v$ relationship shown for the K^+ channels and for the Na^+ channels shown in Figure 3.1.3. These currents were for the open channels, and the whole-cell $i-v$ relationship is not just a sum of these, but also reflects the probability of the channel opening at the particular membrane potential. The resulting curve has negative currents at hyperpolarized membrane potentials, positive current at slight depolarizations, and negative currents at further depolarizations. Recall our convention for current sign: a positive current is an outward current,

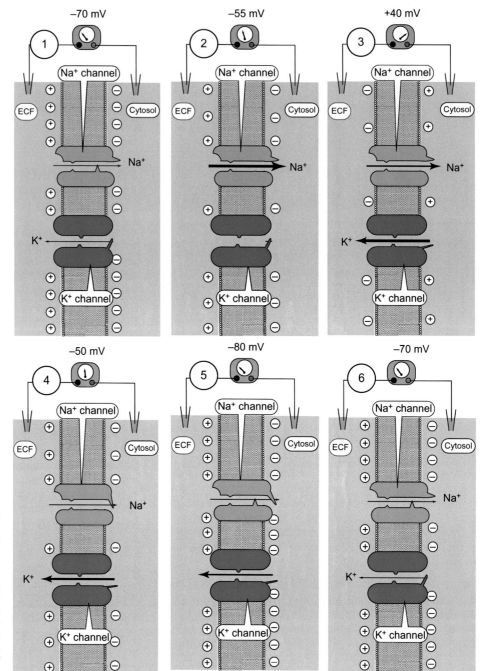

FIGURE 3.2.11 Proposed conceptual changes in the Na$^+$ and K$^+$ channels that give rise to the action potential. At rest (1), fluxes through both Na$^+$ and K$^+$ channels are small and the membrane is polarized. The Na$^+$ activation gate is closed and the inactivation gate is open. The K$^+$ activation gate is closed. During threshold depolarization (2), the Na$^+$ activation gate opens and inward Na$^+$ current further depolarizes the cell, leading to the upstroke of the action potential and the overshoot. Inward Na$^+$ slows as Na$^+$ approaches its equilibrium potential. Meanwhile, the K$^+$ activation gate opens, leading to an outward K$^+$ current (3). This outward K$^+$ current repolarizes the cell and the Na$^+$ inactivation gate closes (4). The continued outward K$^+$ current may lead to a hyperpolarization. The Na$^+$ activation gate closes (5). Upon repolarization, the K$^+$ activation gate closes and the Na$^+$ inactivation gate opens, resetting to the resting condition. The diagram here shows a single channel, but membrane conductance is governed by an ensemble of channels whose states are not necessarily identical.

and current is in the direction of positive ion flow. Thus a positive current removes positive ions from the cell, which hyperpolarizes the cell. Thus at membrane potentials lower than the resting membrane potential there is a negative total current carried by Na$^+$ ions into the cell, the result being a depolarization back toward the resting membrane potential. The total current is zero at the resting membrane potential. At the resting membrane potential, the total $i-v$ curve crosses the x-axis. At slight depolarizations, the current is positive, meaning that positive ions exit the cell and cause a return towards the resting membrane potential. At the uniform threshold, the $i-v$ curve again crosses the x-axis. At this point, the current shifts from positive to negative. A slightly more positive membrane potential shifts the current from positive, which returns the membrane to the resting membrane potential, to negative, caused by influx of Na$^+$, and the membrane depolarizes. This depolarization results in progressively more negative currents. This is the influx of Na$^+$ ions that constitutes the rising phase of the action potential. The point where the $i-v$ curve crosses the x-axis on the high potential side is the "uniform threshold," meaning that if the membrane was uniformly depolarized to this point, an action potential would ensue 50% of the time. It turns out that the membrane is seldom uniformly depolarized: the potential varies with distance from the point of excitation. As a consequence, the actual threshold is generally higher than the uniform threshold.

The Action Potential

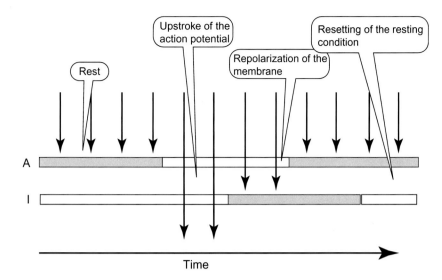

FIGURE 3.2.12 Timing of the opening of the Na^+ channel activation gate (A) and inactivation gate (I). The activation gate is closed at rest while the inactivation gate is open. The activation gate opens briefly during excitation, and Na^+ can cross the membrane because both gates are open. The inactivation gate closes, partly contributing to repolarization of the membrane. This first closes the activation gate, followed by a later resetting of the inactivation gate to the open state. Blue indicates closed state; white indicates open state.

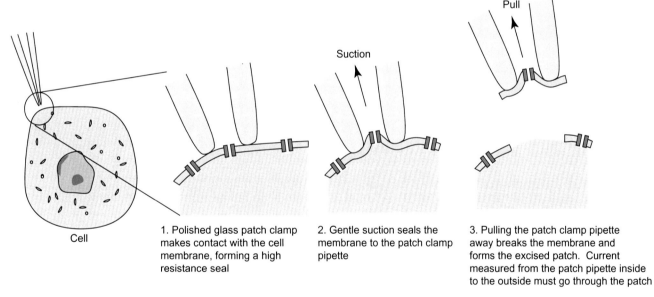

FIGURE 3.2.13 Method for making an excised cell patch. A polished glass microelectrode is brought near to a cell while positive pressure is applied to keep the microelectrode contents uncontaminated by the cell bathing solution. As the microelectrode approaches the cell, the resistance increases, indicating close approach. Positive pressure is turned off and the microelectrode advances to make contact with the membrane. Clean membranes will form a high-resistance seal. Suction is then applied and the microelectrode is reversed to pull off a small patch of membrane. In this configuration, any current that passes across the microelectrode tip must pass through the patch. In this way, single channel currents can be measured. The patch may have none, one, two, or several channels. Experiments can also be performed using cell-attached patches.

THRESHOLD DEPOLARIZATION REQUIRES A THRESHOLD CHARGE MOVEMENT, WHICH EXPLAINS THE STRENGTH–DURATION RELATIONSHIP

Depolarization to threshold requires enough charge to change the voltage across the membrane to the threshold voltage. When stimulation is at a point, the threshold voltage is actually higher because the negative current in the patch of membrane above threshold is partially offset by positive currents in nearby membrane that is below threshold. The condition for threshold is that the total current of the cable is inward. The area of membrane that supplies inward current must be large enough so that outward currents supplied by the rest of the membrane are counterbalanced. The minimum length of fiber that must be depolarized to threshold is called the **liminal length**. This idea is shown diagrammatically in Figure 3.2.16.

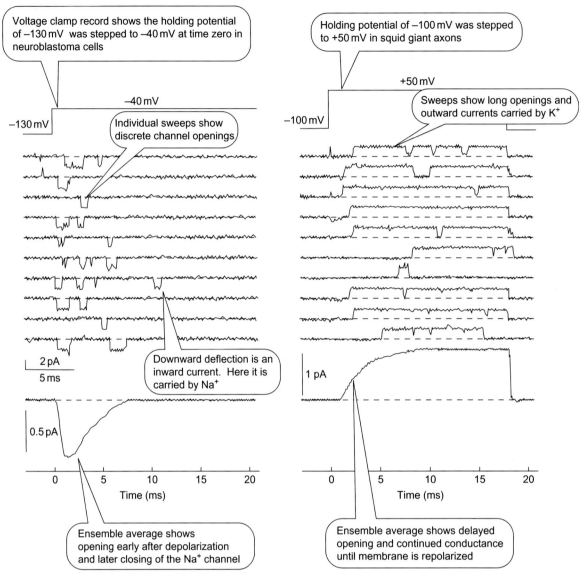

FIGURE 3.2.14 Patch clamp currents of Na$^+$ channels in neuroblastoma cells (left) and K$^+$ channels of squid giant axon (right). (Left) Adapted from data of C.M. Baumgarten, S.C. Dudley, R.B. Rogart and H.A. Fozzard, Unitary conductance of Na$^+$ channel isoforms in cardiac and NB2a neuroblastoma cells. *Am. J. Physiol.* **269**: C1356–C1363, 1995. (Right) Adapted from data of F. Bezanilla and C.K. Augustine cited in B. Hille, *Ionic Channels in Excitable Membranes*, Sinauer, 1992.

THE AMOUNT OF CHARGE NECESSARY TO REACH THRESHOLD EXPLAINS THE STRENGTH–DURATION RELATIONSHIP

Figure 3.2.16 indicates that depolarization to threshold requires movement of sufficient charge according to the capacitance of the membrane: $\Delta V = \Delta q/C$. Thus reaching threshold (a ΔV from the resting potential) requires a defined amount of charge movement. This explains the inverse relationship between stimulus strength (its current) and the duration, according to the Weiss Equation (see Eqn [3.2.1]).

SUMMARY

Resting nerve cells are polarized with a negative resting membrane potential caused by greater K$^+$ conductance in the resting cell. Application of an outward current further polarizes the membrane, and the recording membrane potential is a distorted version of the stimulus. Application of an inward current depolarizes the membrane. If depolarization reaches threshold, nerve cells fire an action potential. The action potential is a brief, pulse-like change in the membrane potential which can move from one area of the cell membrane to another and so it can be used to signal distant parts of the neuron.

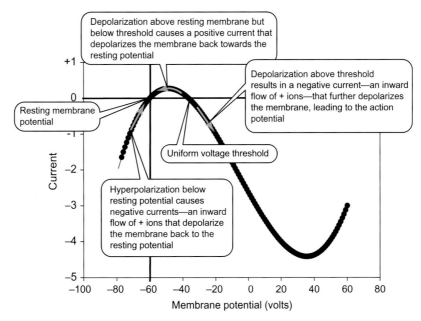

FIGURE 3.2.15 Whole-cell current–voltage relationship. Positive currents above the equilibrium potential for K^+ are mainly due to K^+ exit, mainly due to the increased driving force for K^+. Negative currents are mainly due to Na^+ entry, due to reduction in the driving force for K^+. Hyperpolarization below the resting membrane potential causes a negative current (inward flow of positive ions) that depolarizes the cell back towards the resting membrane potential. Slight depolarizations produce a positive current (outward flow of positive ions) that repolarizes the cell back towards the resting membrane potential. Thus slight hyperpolarization or depolarization returns the cell to rest. Larger depolarizations cause a negative current due to inward flow of Na^+ ions that further depolarize the cell in the action potential.

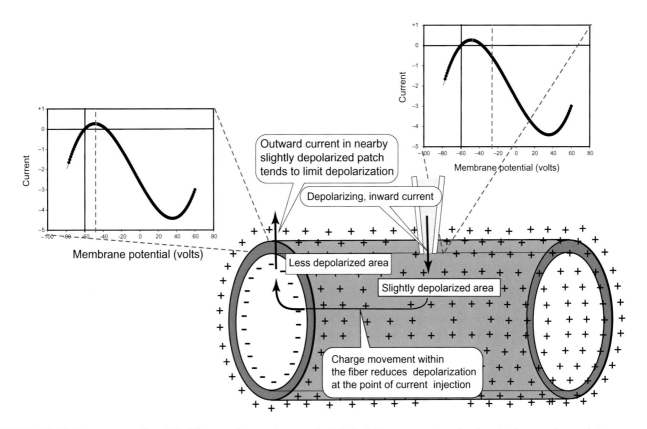

FIGURE 3.2.16 Consequence of spatial differences in membrane potential. Excitation occurs at a location in the membrane that causes a depolarization. This depolarization is sufficiently large to produce a negative or inward current that would further depolarize the cell. However, nearby patches of membrane are depolarized less and they are located on the part of the i–v curve that carries a positive current. This positive current lowers the depolarization at the point of excitation.

Action potentials are all or none: you either get one or you don't. Increasing stimulus strength shortens the delay before the start of the action potential, but it does not alter its peak or its duration. The threshold is the potential at which an action potential is triggered 50% of the time, but this depends on the rate of depolarization. For some period of time after the start of an action potential, nerve cells will not begin another action potential. This period is the absolute refractory period. The relative refractory period follows the absolute refractory period. During the relative refractory period, the threshold for initiating a second action potential is higher.

During the action potential, the membrane potential overshoots zero and becomes positive. Since only Na^+ and Ca^{2+} have positive equilibrium potentials, the membrane potential can become positive only if the permeability to one of these increases. The ionic currents giving rise to the action potential for squid giant axons were described by Alan Hodgkin and Andrew Huxley, who received the Nobel Prize for the work in 1963. They developed a kinetic model that explained the action potential in terms of a single voltage-gated K^+ channel and a Na^+ channel governed by voltage-dependent and time-dependent activation and inactivation gates. During rest, the Na^+ inactivation gate is open and the activation gate is closed. Upon depolarization, the activation gate opens and Na^+ conductance increases markedly. This causes an inward current (carried by Na^+) that further depolarizes the membrane and causes more Na^+ channels to open. The explosive increase in g_{Na} causes the rising phase of the action potential. With time, the Na^+ inactivation gate closes, the K^+ activation gate opens, and the Na^+ activation gate closes. Opening of the K^+ activation gate and closing of the Na^+ inactivation gate cause an outward current that repolarizes the membrane. With further time, the Na^+ inactivation gates reopen and this reestablishes the resting condition.

The squid axon is presented here as an example of how to think about excitable cells. Each excitable cell, and in fact each part of an excitable cell, can have different sets of channels with distinct activation and inactivation properties. The height of the action potential and its duration depend on these characteristics, as well as the electrical characteristics of the cell and the ionic conditions inside and outside the cell.

Channels have discrete states with discrete conductances. Each channel undergoes transitions between conducting and nonconducting or subconducting states. Such discontinuous conductance changes cannot be described by continuum mathematics but rely on probabilistic descriptions. Continuum mathematics can describe the average or ensemble behavior of a population of channels, provided the population is sufficiently large.

Depolarization to threshold requires the movement of sufficient charge to cause the depolarization. This charge can be delivered over a short time at high current or long time at low current. This is the basic nature of the strength–duration relationship. At lower currents there is more time for redistribution of charges within the nerve fiber so that the relationship is not purely reciprocal. The strength–duration relationship is adequately described by the Weiss equation: $I = I_{rh}(t + \tau_{SD})/t$, where I_{rh} is the rheobase and τ_{SD} is the chronaxie.

REVIEW QUESTIONS

1. If a Na-selective channel was to open on the membrane of a motor neuron at rest, which way would current flow? Would this depolarize or hyperpolarize the cell?
2. If a K-selective channel was to open on the membrane of a motor neuron at rest, which way would current flow? Would this depolarize or hyperpolarize the cell?
3. What is an action potential? Why does the membrane potential become positive during the action potential?
4. What do we mean when we say that action potentials are "all or none"? What is the absolute refractory period? Relative refractory period?
5. What is the "activation gate" of the Na^+ channel? When is it open? When does it close? What is the "inactivation gate" of the Na^+ channel? When is it open? When does it close?
6. What is tetrodotoxin? Why does it block action potentials?
7. What is a patch clamp? Why is it useful? What is an ensemble current?
8. How do g_{Na} and g_K change during the action potential? What causes these changes?
9. What is the total current across the membrane at the resting membrane potential? Why does a slight depolarization, below threshold, come back to the resting membrane potential? Why does a slight hyperpolarization return to the resting membrane potential?
10. Why is there an inverse relationship between current and time to reach threshold?

APPENDIX 3.2.
A1 THE HODGKIN–HUXLEY MODEL OF THE ACTION POTENTIAL

ALAN HODGKIN AND ANDREW HUXLEY'S GOAL WAS TO ACCOUNT FOR THE ACTION POTENTIAL BY MOLECULAR MECHANISMS

Hodgkin and Huxley's goal was to explain the ionic fluxes and conductance changes during the action potential in terms of molecular mechanisms. After trying some different mechanisms, they concluded that not enough was known to determine a unique mechanism. Instead, they tried to develop an empirical kinetic description which would allow them to calculate electrical responses and which would correctly predict the shape of the action potential and its conduction velocity. Today this model is referred to as the Hodgkin–Huxley or HH model.

THE HH MODEL DIVIDES THE TOTAL CURRENT INTO SEPARATE NA$^+$, K$^+$, AND LEAK CURRENTS

Hodgkin and Huxley wrote the Na$^+$ and K$^+$ currents in terms of their maximum conductances which are multiplied by coefficients that vary continually between 0 and 1. The overall conductance, then, varies between 0 and the maximum conductance. All of the kinetic properties of the conductances are embedded in the characteristics of the coefficients. The conductances in the model vary with voltage and time but not with concentration of either Na$^+$ or K$^+$.

THE HH MODEL OF THE K$^+$ CONDUCTANCE INCORPORATES FOUR INDEPENDENT "PARTICLES"

On depolarization, the increase in g_K follows an S-shaped curve, whereas g_K decreases exponentially upon repolarization. Hodgkin and Huxley proposed that the gating of the K$^+$ channel could be modeled by four identical membrane "particles." The probability that each channel is in the position to allow K$^+$ conductance is n. All four particles must be correctly situated to allow conductance. The probability that all four are positioned for conductance is n^4. The K$^+$ current is given as

[3.2.A1.1] $$I_K = n^4 g_{K_{max}}(E_m - E_K)$$

Here the probability n depends on time and voltage. In the HH model, values of n are determined by a first-order reaction, written as

[3.2.A1.2] $$1 - n \underset{\beta_n}{\overset{\alpha_n}{\rightleftharpoons}} n$$

where α_n and β_n depend on voltage. The reaction in Eqn [3.2.A1.2] can be written in differential form as

[3.2.A1.3] $$\frac{dn}{dt} = \alpha_n(1-n) - \beta_n n$$

Hodgkin and Huxley determined empirical relationships between the rate constants, α_n and β_n, and the membrane potential. For the squid axon at 6°C, these were:

[3.2.A1.4]
$$\alpha_n = \frac{0.01(10 - (E_m - E_r))}{e^{(10-(E_m-E_r))/10} - 1}$$
$$\beta_n = 0.125 e^{(-(E_m - E_r)/80)}$$

where the membrane potential is in millivolts and the rate constants have units of M s^{-1}. These are not the exact forms originally used by Hodgkin and Huxley, because their sign convention for membrane potential is the reverse of that used today. In addition, Hodgkin and Huxley derived their empirical equations based on voltage clamp experiments in which the degree of variation from resting membrane potential, here symbolized as E_r, was clamped. Equation [3.2.A1.4] is transformed to agree with today's conventions.

An alternative way of expressing Eqn [3.2.A1.3] is as follows:

[3.2.A1.5] $$\frac{dn}{dt} = \frac{(n_\infty - n)}{T_n}$$

where n_∞ is the steady-state value of n at any particular voltage and T_n is a time constant. The values of n_∞ and T_n are given by

[3.2.A1.6]
$$n_\infty = \frac{\alpha_n}{\alpha_n + \beta_n}$$
$$T_n = \frac{1}{\alpha_n + \beta_n}$$

THE HH MODEL OF NA$^+$ CONDUCTANCES USES ACTIVATING AND INACTIVATING PARTICLES

Similar to the case with the K$^+$ conductance model, HH empirically modeled the Na$^+$ conductance with four hypothetical gating particles that make first-order transitions between conductive and nonconductive states. Because the Na$^+$ conductance has two opposing actions, activation and inactivation, Hodgkin and Huxley used two kinds of gating particles, called m and h. Here the probability of an open configuration is m and h, respectively. The probability that all four gates are open is $m^3 h$. In this case, the Na$^+$ current is given as

[3.2.A1.7] $$I_{Na} = m^3 h g_{Na_{max}}(E_m - E_{Na})$$

The same formalism for transitions between open and closed states of the n gates for the K$^+$ channels also governs transitions between the open and closed states of the m and h gates of the Na$^+$ channel:

[3.2.A1.8]
$$1 - m \underset{\beta_m}{\overset{\alpha_m}{\rightleftharpoons}} m$$
$$1 - h \underset{\beta_h}{\overset{\alpha_h}{\rightleftharpoons}} h$$

The rate constants all depend on voltage, according to the following empirical relations:

[3.2.A1.9]
$$\alpha_m = \frac{0.1(25 - (E_m - E_r))}{e^{(25-(E_m-E_r))/10} - 1}$$
$$\beta_m = 4 e^{(-(E_m - E_r)/18)}$$
$$\alpha_h = 0.07 e^{(-(E_m - E_r)/20)}$$
$$\beta_h = \frac{1}{e^{((30-(E_m-E_r))/10)} + 1}$$

Relationships among m_∞, h_∞, T_m, and T_h are given in analogy to Eqn [3.2.A1.6].

CALCULATION OF $G_{NA(T)}$ AND $G_{K(T)}$ FOR A VOLTAGE CLAMP EXPERIMENT

To calculate the time dependence of Na$^+$ and K$^+$ conductances, we need to know the set of rate constants which describe the transitions between states of the conductances in the HH model: (α_n, β_n, α_m, β_m, α_h, β_h). Further, we need to know this set at the two voltages. In the voltage clamp experiment, the nerve is held at some voltage and then is rapidly switched to another voltage. As a result, the conductances go through a transition from one state to another. Because the gates have different kinetics, the overall behavior can be complex. Each of the gates relaxes between the steady-state value before the voltage jump, and the second steady-state value at the new voltage. The equation describing the time course from an initial value of n, m, and h (n_0, m_0, and h_0) and a final value of n, m, and h (n_∞, m_∞, and h_∞) can be derived from integrating Eqn [3.2.A1.3]:

[3.2.A1.10]
$$\frac{dn}{dt} = \alpha_n(1-n) - \beta_n n$$
$$= \left(\frac{\alpha_n}{\alpha_n + \beta_n} - n\right)(\alpha_n + \beta_n)$$
$$\frac{dn}{(\alpha_n/\alpha_n + \beta_n) - n} = (\alpha_n + \beta_n)dt$$
$$\int_{n_0}^{n} \frac{dn}{(\alpha_n/\alpha_n + \beta_n) - n} = \int_0^t (\alpha_n + \beta_n)dt$$

Integrating, we obtain

[3.2.A1.11]
$$\ln\left[\frac{\alpha_n}{\alpha_n + \beta_n} - n\right]_{n_0}^{n} = -\frac{t}{T_n}$$
$$\ln\left[\frac{n_\infty - n}{n_\infty - n_0}\right] = -\frac{t}{T_n}$$
$$n = n_\infty - (n_\infty - n_0)e^{-(t/T_n)}$$

Here we have identified T_n and n_∞ according to Eqn [3.2.A1.6]. It is easy to justify the assignment of n_∞: simply let dn/dt in Eqn [3.2.A1.3] go to zero at infinite time (at which point the steady-state value of n should have been reached). From that constraint, we find $n_\infty = \alpha_n/(\alpha_n + \beta_n)$.

The last equation tells us that at $t=0$, $n = n_0$ and at $t=\infty$, $n = n_\infty$. In between, n relaxes between n_0 and n_∞ with an exponential time course. If we know n_0, n_∞, and T_n, we can calculate the $n(t)$. All of these values depend on the α and β for the given gates. The procedure is entirely analogous for m and h.

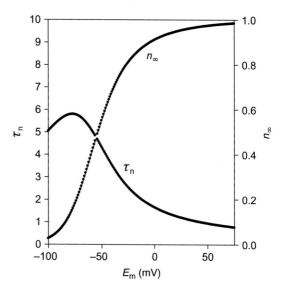

FIGURE 3.2.A1.1 The dependence of τ_n and the steady-state value of n (n_∞) as a function of the membrane potential, E_m. These values of n control the conductance through the K$^+$ channel in the HH formalism.

RESULTS OF THE CALCULATIONS

Figure 3.2.A1.1 shows the voltage dependence of n_∞ and T_n.

The voltage clamp experiments performed by Hodgkin and Huxley involved changing the membrane potential essentially instantaneously from the resting potential (−65 mV) to some set potential and clamping it there. What happens is that the resting values of n, m, and h (called n_0, m_0, and h_0) relax to their new values at the clamped voltage, which we will call n_∞, m_∞, and h_∞. The resulting time course of n, m, and h generates a time course for the conductances, calculated according to Eqns [3.2.A1.1] and [3.2.A1.7]. For a +88-mV clamp from −65 to +23 mV, the relevant values are given in Table 3.2.A1.1 (see Figure 3.2.A1.2).

These values are substituted into equations of the form of Eqn [3.2.A1.11] to derive $n(t)$, $m(t)$, and $h(t)$, from which the instantaneous conductances can be calculated

TABLE 3.2.A1.1 Values of n, m, and h for Voltage Clamp from −65 to +23 mV

K Channel	Na Channel	
$n_0 = 0.3177$	$m_0 = 0.0529$	$h_0 = 0.5961$
$n_\infty = 0.9494$	$m_\infty = 0.9953$	$h_\infty = 0.0009$
$T_n = 1.2028$ ms	$T_m = 0.1577$ ms	$T_h = 1.0022$ ms
$g_{Kmax} = 36$ ms cm^{-2}	$g_{Namax} = 120$ ms cm^{-2}	

Derived from Figures 3.2.A1.1 and 3.2.A1.2.

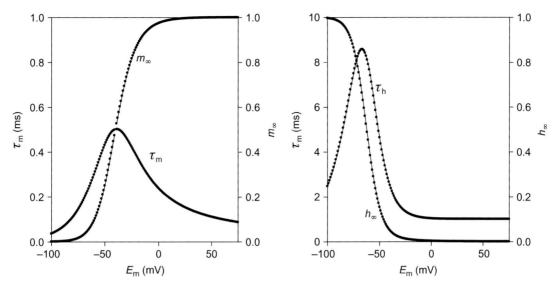

FIGURE 3.2.A1.2 The dependence of τ_m and m_∞ on the membrane potential. These values control the opening of the activation gate of the Na$^+$ channel in the HH fomalism.

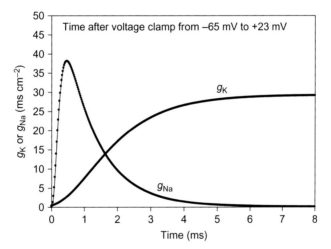

FIGURE 3.2.A1.3 Calculated changes in g_{Na} and g_K during a voltage clamp from -65 mV to $+23$ mV using the HH formalism.

according to Eqns [3.2.A1.1] and [3.2.A1.7]. The results of these calculations for the given voltage clamp are shown in Figure 3.2.A1.3.

Of course, Hodgkin and Huxley had it much more difficult than this, because they had to find the original equations and parameters to fit their voltage clamp results, whereas here we are simply confirming that the equations and parameters they found do, indeed, look like their voltage clamp records. It is important to remember that the Hodgkin–Huxley formalism is an empirical model, designed to fit the data. Although there are deficiencies in the model, it succeeds admirably well in predicting the wave form of the action potential. It is now generally accepted that the basis of the action potential is a rapid switching on of the Na$^+$ conductance, followed by its inactivation and more slowly turning on of the K$^+$ conductance.

3.3 Propagation of the Action Potential

Learning Objectives

- Define propagation of the action potential
- Define conduction velocity
- Describe how conduction velocity varies with axon diameter and with myelination
- Using the formula for a parallel plate capacitor, explain how myelin decreases membrane capacitance
- Using the formula for resistances in parallel, explain how internal resistance of the axoplasm varies with axon diameter
- Define the space constant and time constant
- Describe how the space constant and time constant vary with axon diameter and myelination
- Qualitatively account for how membrane capacitance and axoplasmic resistance explain the dependence of conduction velocity on myelination and axon diameter
- Describe saltatory conduction and explain what "jumps" from node to node

THE ACTION POTENTIAL MOVES ALONG THE AXON

Consider Figure 3.3.1, which shows an axon of a motor neuron that has been impaled at intervals by recording electrodes. If an action potential is begun at the far left by depolarization to threshold, each succeeding recording electrode records an action potential. Note that the successive action potentials have similar waveforms but they are observed at each electrode at successively later times. The action potential is **propagated** along the surface of the nerve. The action potential moves over the surface of the cell, appearing some distance away after some elapsed time.

THE VELOCITY OF NERVE CONDUCTION VARIES DIRECTLY WITH THE AXON DIAMETER

The action potentials shown in Figure 3.3.1 do not have identical waveforms due to the stimulation artifact that dies out with distance along the axon. After this initial stimulation artifact decays away, all subsequent action potentials are essentially identical. The identical waveform of the action potential as it travels over the axon is a variant of the "all-or-none" description of the action potential. As the action potential appears later at longer distances from the point of initiation, we can define a **conduction velocity** of action potential propagation equal to the distance between the recording electrodes divided by the delay in time between action potentials recorded at the two sites. The velocity of action potential conduction has been determined for myelinated and unmyelinated fibers of different sizes (see Table 3.3.1).

Within each category of nerve fiber, myelinated or unmyelinated, the conduction velocity varies with the diameter of the nerve. For myelinated fibers, the conduction velocity varies approximately in proportion to the diameter. In unmyelinated fibers, the conduction velocity varies approximately with the square root of the diameter.

THE ACTION POTENTIAL IS PROPAGATED BY CURRENT MOVING AXIALLY DOWN THE AXON

How is the action potential conducted down the length of the axon? Recall that the action potential is triggered by a depolarization of the membrane to threshold. In order for an action potential to move from one place to another along the axon, the depolarization that triggers the action potential must precede it. Depolarization of the membrane proceeds **electrotonically** or passively. As shown in Figure 3.3.2, the local depolarization of the neuron's axon membrane spreads out from the origin of the depolarization.

THE TIME COURSE AND DISTANCE OF ELECTROTONIC SPREAD DEPEND ON THE *CABLE PROPERTIES* OF THE NERVE

Recall from Figures 3.2.4 and 3.2.5 that a hyperpolarizing or depolarizing stimulus was not faithfully reproduced in the axon: the signal was distorted. This distortion is a consequence of the **cable properties** of the nerve. The cable properties of the nerve refer to the passive or electrotonic properties and not to the active properties that give rise to the action potential. Each length of axon is characterized by an Ohmic resistance to current across the membrane (R_m), a capacitance (C_m), a resistance through the external solution that bathes the membrane (R_o), and a resistance through the

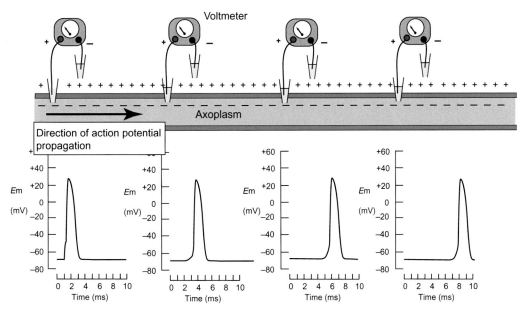

FIGURE 3.3.1 Appearance of action potentials at later times down the axon from the point of stimulation. The output of each voltmeter is shown below it.

TABLE 3.3.1 Velocity of Nerve Impulse Conduction as a Function of Axon Size

Nerve Fiber Type	Diameter (μm)	Conduction Velocity (m s^{-1})	Physiological Function
Aα	12–22	70–120	Somatic motor
Aδ	1–5	12–30	Pain, sharp
C	0.5–1.2	0.2–2	Pain, ache

axoplasm that fills the axon (R_i). A schematic diagram of this electrical model is shown in Figure 3.3.3.

If we pass a constant current across the membrane between nodes A and B, for example, so that some new membrane potential E is established, we should expect that the membrane potential Ex at some point x away from the current source will depend on the distance from the current source and, because of the capacitances, it will also depend on the time since the current was turned on. The cable properties determine this dependence on position x and time t.

CAPACITANCE DEPENDS ON THE AREA, THICKNESS, AND DIELECTRIC CONSTANT

The membrane acts much like a parallel plate capacitor. The expression for the capacitance of a parallel plate capacitor is given as

[3.3.1] $$C = \frac{\kappa \varepsilon_o A}{\delta}$$

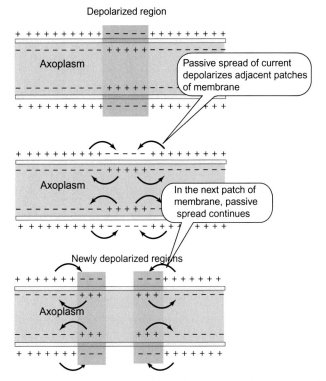

FIGURE 3.3.2 Passive spread of a depolarization to adjacent areas of membrane. Depolarization of a patch of membrane spreads to adjacent areas. If the depolarization reaches threshold in the nearby patch, an action potential will be initiated. If there is no action potential, the spread of depolarization will decay away with time and distance from the original depolarized area.

where C is the capacitance (in F = C V^{-1}), κ is the dielectric constant characteristic of the material between the plates (a dimensionless ratio), ε_0 is the electrical permittivity of the vacuum = 8.85×10^{-12} C^2 J^{-1} m^{-1},

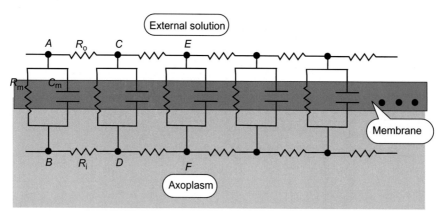

FIGURE 3.3.3 Schematic diagram of the electrical model of an axon.

A is the area (m^2), and δ is the distance between the plates (m). Since 1 J = 1 V C, the units of C come out in C V^{-1}. From this equation, it should be clear that the capacitance depends directly on the area of the membrane. It is common to normalize the capacitance by dividing by the area. The **specific capacitance** $C_m = C/A$ is this normalized capacitance.

The real membrane is *not* a parallel plate capacitor, but a concentric, coaxial capacitor. This does not materially affect the analysis presented here, as detailed in Appendix 3.3.A1.

CHARGE BUILDUP OR DEPLETION FROM A CAPACITOR CONSTITUTES A CAPACITATIVE CURRENT

The relationship among charge, volts, and capacitance is

[3.3.2] $$C = \frac{Q}{V}$$

where C is the capacitance, Q is the charge (in coulombs) on the capacitor, and V is the voltage difference across the capacitor. When the capacitor is charging or discharging, there is a capacitative current given as

[3.3.3] $$i = \frac{dQ}{dt} = C\frac{dV}{dt}$$

The capacitive current is given this special name because current does not pass through the dielectric, but charge builds up on one side and is taken away from the other, so effectively there is a charge movement across the capacitor without a charge flow through the dielectric.

THE TRANSMEMBRANE RESISTANCE DEPENDS ON THE AREA OF THE MEMBRANE

Adding membrane area is like adding resistances in parallel—the overall resistance actually decreases. The conductances add, whereas the inverse of the resistances add. The total conductance of a patch of membrane is the conductance per unit area times the area. Let G be the total conductance of a membrane of area A, and G_M be the specific conductance per unit area. Similarly, let R be the total resistance, which is the inverse of the conductance, and R_M be the specific resistance. Then the resistance of a patch of membrane is given as

[3.3.4] $$G = G_m A$$
$$\frac{1}{G} = \frac{1}{G_M A} = \frac{1}{G_m}\frac{1}{A}$$
$$R = \frac{R_m}{A}$$

Here the specific resistance, R_m, is given in units of Ω cm^2, and R is in Ω. These units may seem peculiar, that the resistance per unit area is given as Ω cm^2, but this is a consequence of Ohm's Law that gives the current as being inversely proportional to the resistance.

THE AXOPLASMIC RESISTANCE DEPENDS ON THE DISTANCE, AREA, AND SPECIFIC RESISTANCE

The resistance of an electrolyte solution such as that in the axon is typically given as its **specific resistance**, ρ. This is the resistance between two faces of a cube 1 cm on a side. Since resistances in series add, the resistance of a length of solution is just the length times the specific resistance. The resistance of an area of electrolyte solution is the specific resistance divided by the area, as in Eqn [3.3.4]. So the equivalent resistance of the axoplasm is given as

[3.3.5] $$R_i = \frac{\rho_i d}{A}$$

where d is the distance and A is the cross-sectional area. Since R_i has the units Ω, ρ_i has the units Ω cm.

THE EXTRACELLULAR RESISTANCE ALSO DEPENDS ON THE DISTANCE, AREA, AND SPECIFIC RESISTANCE

The resistance provided by the extracellular electrolyte solution is entirely analogous to the axoplasmic resistance. However, the area involved here is not precisely known and it is large. Because the area appears in the denominator of Eqn [3.3.5], typically the outside resistance is small compared to the axoplasmic resistance. We will treat R_o as being zero, so that the voltage everywhere along the axon on the outside is zero.

CABLE PROPERTIES DEFINE A *SPACE CONSTANT* AND A *TIME CONSTANT*

Consider part of the schematic diagram of Figure 3.3.3 shown in Figure 3.3.4. Analysis of the currents as a function of distance will allow us to characterize the axon in terms of its **cable properties**. These include a **space constant** and a **time constant**.

One of Kirchoff's circuit laws states that **the sum of all currents out of any node must be zero**. This is just another way of saying that there is a conservation of total charge. Applying this principle to node (x) in Figure 3.3.4, we have

[3.3.6] $\qquad i(x) = i(x + dx) + i_m$

where $i(x)$ is the current passing down the axon into the node at (x), i_m is the current that passes through the membrane at this node, and $i(x + dx)$ is the current that passes down to the next node at $(x + dx)$. The current across the membrane, i_m, has two parts: one part that passes through the resistance and a second part that either charges or discharges the capacitor.

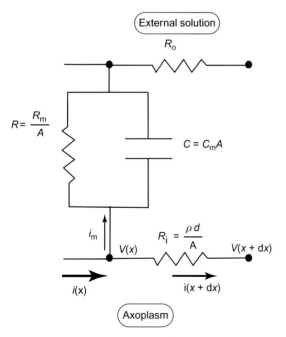

FIGURE 3.3.4 Currents at a patch of membrane area of the axon. C is the capacitance of the membrane; C_m is the specific capacitance; R is the resistance across the axon membrane; and R_m is the resistance per unit area; i_m is the current across the membrane; $i(x)$ is the current down the axon at node x. R_i is the internal resistance of the axoplasm and ρ_i is its specific resistance.

The current i_m can be written as

[3.3.7] $\qquad i_m = \dfrac{V(x) - V_r}{R_m} A + C_m A \dfrac{dV}{dt}$

where $V(x)$ is the membrane potential at position x and V_r is the membrane potential at rest at which the net membrane current is zero. The first part of the right-hand side of this equation is just Ohm's law for the current through the resistance across the membrane. The second part is from Eqn [3.3.3] and describes that part of the current that either charges or discharges the capacitor. Substituting Eqn [3.3.7] into Eqn [3.3.6], we have

[3.3.8] $\qquad i(x) = i(x + dx) + \dfrac{V(x) - V_r}{R_m} A + C_m A \dfrac{dV}{dt}$

This equation can be rearranged, using $A = 2\pi a \, dx$ as the surface area of the membrane element, where a is the radius of the axon. Insertion of the area in Eqn [3.3.8] and rearranging, we obtain

[3.3.9] $\qquad \dfrac{i(x + dx) - i(x)}{dx} = -2\pi a \dfrac{V(x) - V_r}{R_m} A - 2\pi a C_m A \dfrac{dV}{dt}$

In the limit as $dx \to 0$, the left-hand side of this equation is the definition of the derivative. Taking this limit, we derive

[3.3.10] $\qquad \dfrac{di}{dx} = -2\pi a \dfrac{V(x) - V_r}{R_m} - 2\pi a C_m \dfrac{dV}{dt}$

There is another relationship between i and $V(x)$ that we can use here, and that is Ohm's law through R_i. There would be no longitudinal current unless there is a voltage gradient in x. Ohm's law gives the longitudinal current, i, in Eqn [3.3.10] as

[3.3.11]
$$i = \dfrac{V(x) - V(x + dx)}{R_i}$$
$$i = -\dfrac{V(x + dx) - V(x)}{(\rho_i dx / \pi a^2)}$$
$$i = -\dfrac{\pi a^2}{\rho_i} \lim_{dx \to 0} \left[\dfrac{V(x + dx) - V(x)}{dx} \right]$$
$$i = -\dfrac{\pi a^2}{\rho_i} \dfrac{dV}{dx}$$

Substituting this result for i into Eqn [3.3.10], we derive

[3.3.12] $\qquad -\dfrac{\pi a^2}{\rho_i} \dfrac{d^2 V}{dx^2} = -2\pi a \dfrac{V - V_r}{R_m} - 2\pi a C_m \dfrac{dV}{dt}$

which can be rearranged to

[3.3.13] $\qquad V - V_r = \dfrac{R_m \pi a^2}{2\pi a \rho_i} \dfrac{d^2 V}{dx^2} - \dfrac{R_m}{2\pi a} 2\pi a C_m \dfrac{dV}{dt}$

If we let $V' = V - V_r$, this equation has the form

[3.3.14] $\qquad V' = \lambda^2 \dfrac{d^2 V'}{dx^2} - \tau \dfrac{dV'}{dt}$

Here λ is identified as a **space constant** and τ is a **time constant**, so named because they govern the spatial and time derivatives of the voltage when a constant current is injected across the membrane. By comparison with Eqn [3.3.13], their values are given as

$$[3.3.15] \quad \lambda = \sqrt{\frac{R_m \pi a^2}{\rho_i 2\pi a}} = \sqrt{\frac{R_m}{\rho_i} \frac{a}{2}}$$

$$\tau = R_m C_m$$

Note that at steady state, when $dV/dt = 0$, the space constant defines the way in which the voltage varies with distance. This would be the case when a constant current has been passed across the membrane for a sufficiently long time to charge all the capacitors to their steady-state level. In this case, Eqn [3.3.14] becomes

$$[3.3.16] \quad V' = \lambda^2 \frac{d^2 V'}{dx^2}$$

The relevant solution to this differential equation is

$$[3.3.17] \quad V = (V_0 - V_r)e^{(-x/\lambda)} + V_r$$

This equation means that the voltage falls off exponentially from the point of current application to the nerve to its value some large distance away. Here the resting variable V_0 refers to the voltage at $x = 0$ and V_r is the resting voltage some large distance away. The equation looks this way because of the boundary conditions that $V = V_0$ when $x = 0$, and $V = V_r$ when $x = \infty$. These boundary conditions arise because of the way that we set up the situation: the current at a specified point on the axon is constant and it is applied for sufficiently long times that the capacitors are all charged and the capacitative currents go to zero. Under these conditions, at steady state, the voltage at the point of current application will be V_0 and the voltage far away (actually, infinitely far away) will be the resting membrane potential. This condition is called **electrotonus**.

Note that the space constant described in Eqn [3.3.15] consists of two resistances: the resistance to current flow across the membrane and the resistance to current flow down the interior of the axon. If the resistance across the membrane becomes larger, as occurs in myelinated fibers, then the space constant will be larger. If the resistance of the axoplasm is smaller, as occurs when the axon is larger (and therefore has a larger cross-sectional area, which participates in the determination of R_i), then the space constant will be larger. Figure 3.3.5 shows the voltage as a function of distance from the point of application of an inward current sufficient to depolarize the cell to -40 mV without the engagement of the active behavior of the axon, say in a TTX-poisoned nerve. In myelinated fibers, the voltage decreases slowly away from the point of current injection because λ, the space constant, is large. In contrast, the steady-state voltage in unmyelinated fibers decays rapidly away from the point of injection of the current.

From Eqn [3.3.17], the space constant is the distance for the voltage difference, $(V - V_r)$, to fall from $V_0 - V_r$ to within $1/e$ of $V_0 - V_r$, which is $0.367 \times [V_0 - V_r]$. This can be readily seen by letting $x = \lambda$, at which point $V - V_r = [V_0 - V_r] \times 1/e$. The larger space constant in myelinated fibers means that a depolarization maintains a higher voltage at longer distances, because more of the current travel down the axon to change the membrane potential as opposed to going through the membrane. Equation [3.3.15] predicts that the space constant will vary with the square root of the axon diameter. This is the case with unmyelinated fibers. In myelinated fibers, the value of R_m increases proportionate to the radius. So the overall length constant would be proportional to the square root of the square of the radius or approximately proportionate to the radius.

Substituting in for R_m and ρ_i from Eqns [3.3.4] and [3.3.5] into Eqn [3.3.15], and using the area terms $2\pi ad$ for the surface area (relevant to R_m) or πa^2 for the cross-sectional area (relevant to R_i), we obtain

$$[3.3.18] \quad \lambda = \sqrt{\frac{R_m}{\rho_i} \frac{a}{2}}$$

$$= \sqrt{\frac{2\pi a d R}{(\pi a^2/d) R_i} \frac{a}{2}}$$

$$= \sqrt{\frac{dR}{(R_i/d)}}$$

$$= \sqrt{\frac{R}{R_i} d^2}$$

It is important to note that the units of these resistances are different because of the way in which they have been normalized. R_m is the resistance per unit area, and so its units are $\Omega \, cm^2$. The specific resistivity, ρ_i, is in units of resistance per unit area per unit length, so its units are $\Omega \, cm$. Insertion of these units in Eqn [3.3.18] gives the proper units of distance in the calculation of λ. R and R_i are resistances, in Ω. Since these resistances depend on the distance down the axon, the distance used is included in the calculation of R and R_i. Sometimes it is convenient to express the resistances as resistances per unit length. These variables are commonly noted as r_m and r_i, with units of $\Omega \, cm$ and $\Omega \, cm^{-1}$, respectively. This is equivalent to identifying

$$[3.3.19] \quad r_m = dR$$

$$r_i = \frac{R_i}{d}$$

in Eqn [3.3.18]. The utility of this nomenclature is that the space constant is then given simply as

$$[3.3.20] \quad \lambda = \sqrt{\frac{r_m}{r_i}}$$

The **temporal** properties of cables can be made readily apparent by using a **space clamp**. This is accomplished

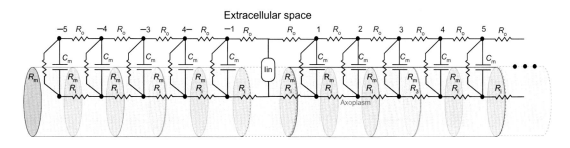
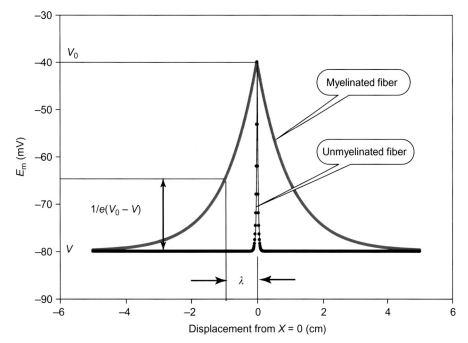

FIGURE 3.3.5 Steady-state voltage as a function of distance from the point of injection of current in myelinated and unmyelinated nerve fibers. In unmyelinated nerves, the voltage drops off quickly with distance, implying a small space constant because R_m is small and R_i is large. In myelinated fibers, the voltage drops off much less with distance, indicating a large space constant.

experimentally by passing a thin wire down the axon, effectively making $R_i = 0$. This removes the spatial dependence of $V(x,t)$ caused by passing a current. Here $dV/dx = 0$ and Eqn [3.3.14] becomes

$$[3.3.21] \qquad V' = -\tau \frac{dV'}{dt}$$

This is a familiar equation for exponential decay. The applicable solution is

$$[3.3.22] \qquad V = (V_0 - V_r)e^{-(t/\tau)} + V_r$$

where the time constant, $\tau = R_m C_m$ does not vary very much from myelinated to unmyelinated fibers because myelination increases R_m according to the thickness of the myelin layer, but it decreases C_m inversely with the thickness of the myelin layer. The time constant describes the characteristic time for the capacitor to charge or discharge from the initial value to the final value. Here $V_0 - V_r$ refers to the difference between the membrane potential initially and the membrane potential at steady state after injection of current. Note that, as with the space constant, the time constant is defined as the time necessary for the voltage difference to decay to $1/e = 0.367$ of the difference to its new value. Each patch of membrane is identical in this way, but the distributed nature of the capacitors and resistances makes the response of each membrane dependent on its distance from the current source. In this way, the patches of membrane nearer the current source charge faster than exponentially, and those further away charge slower than exponentially. This phenomenon is easily modeled using electronic networks.

THE CABLE PROPERTIES EXPLAIN THE VELOCITY OF ACTION POTENTIAL CONDUCTION

As we discussed in the section "The Action Potential Is Propagated by Current Moving Axially Down the Axon," the action potential occurs at a specified location x because the membrane is depolarized to threshold at that location. If the depolarization occurs earlier at location x, then the action potential also occurs earlier at that location and the velocity of nerve impulse conduction is faster. Myelinated nerves have a longer space constant. Thus the depolarization occurring some distance x away

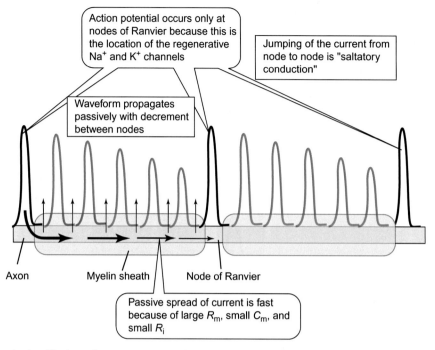

FIGURE 3.3.6 Saltatory conduction. The inward current accompanying an action potential travels down the axon to depolarize adjacent parts of the cell. In myelinated fibers, little of the current is lost across the membrane because the transmembrane resistance is high (myelin insulates the axon) and little is lost to discharging the capacitor of the axon because the capacitance is low. Enough current remains to depolarize the axon at the next node of Ranvier where voltage-gated Na^+ channels reside. In this way, myelin increases conduction velocity. In between nodes, the "action potential" does not involve active currents across the membrane: it is a passive spread of the waveform of the action potential. Its decrement shown here is exaggerated for illustrative purposes.

from the point of stimulation is greater in the myelinated nerve. This is due to the fact that:

- more of the current can travel down the axon, because its resistance is less due to its larger size;
- less of the current crosses the membrane, because its resistance is larger in the myelinated fibers;
- less current is required to depolarize the membrane, because the membrane capacitance is smaller due to the greater thickness of the myelin (see Eqn 3.3.1).

All of this conspires for depolarization to be reached sooner at distances remote from the source of stimulation. When depolarization is reached sooner, the action potential occurs sooner and action potential velocity increases.

SALTATORY CONDUCTION REFERS TO THE "JUMPING" OF THE CURRENT FROM NODE TO NODE

As mentioned earlier in Chapter 3.2, myelinated fibers consist of axons that are surrounded by multiple coverings of the plasma membranes of Schwann cells in the peripheral nervous system and oligodendroglia cells in the central nervous system. These wrappings provide the increased resistance and decreased capacitance that produce a long space constant and allow for rapid nerve impulse conduction. These wrappings are not continuous, however. There are gaps between the Schwann cells, and each gap is called a **node of Ranvier**. The nodes themselves are only a 2–3 μm long, whereas the internodal distance is typically 1–3 mm. These nodes contain abundant voltage-gated Na^+ and K^+ channels. At each node, these channels respond to the depolarization of the membrane by undergoing the conductance changes that we discussed in Chapter 3.2. In between the nodes, the myelin insulates the nerve plasma membrane from the extracellular solution so that the resistance is large and few channels are present. Huxley and Stampfli experimentally established that the current occurs mainly at the nodes. Thus the current appears to "jump" from node to node. This is called **saltatory conduction** (from the Latin "saltatio" meaning "to leap" or "to dance"). In between the nodes, the membrane potential is conducted electrotonically, or passively, without the necessity of regenerating the action potential. The great advantage in speed is provided by the rapid electrotonic spread of depolarization from node to node. In addition, the electrotonic spread does not require the active generation of currents at every patch of membrane, so less inward current is required for action potential propagation and so less energy is used to pump out the Na^+ that enters the nerve with each action potential. Figure 3.3.6 illustrates saltatory conduction.

THE ACTION POTENTIAL IS SPREAD OUT OVER MORE THAN ONE NODE

Each action potential lasts about 2 ms. The conduction velocity in a large myelinated fiber is on the order of 100 m s^{-1}. Thus the action potential is spread out over a distance of $100 \text{ m s}^{-1} \times 0.002 \text{ s} = 0.2 \text{ m}$ or 20 cm. Since each node is separated by about 2 mm, the entire action potential at any one time is spread out over 100 nodes. This calculation should make it clear that the nerve impulse does not "jump" from node to node, as it occupies many nodes. The current that produces the action potential occurs only at the nodes, because that is where the channels are concentrated and where the axon membrane has contact with the extracellular fluid.

SUMMARY

The brief, pulse-like change in membrane potential, the action potential, moves along axons with definite velocity. The action potential propagates with essentially the same waveform all along the axon. The velocity of action potential propagation varies with myelination and axon size. In unmyelinated axons, the velocity varies approximately with the square root of axon diameter. In myelinated fibers, the velocity varies approximately in proportion to the diameter. The velocity of action potential propagation arises from the cable properties of nerves.

Depolarization starts the action potential. The action potential itself is a depolarization that passively spreads to nearby patches of membrane. When nearby patches of membrane reach threshold, they begin an action potential at that point. The rate of passive spread of current depends on the resistance of the membrane, the capacitance of the membrane, and the resistance of the axoplasm. These are quantitatively represented in the time constant and the length constant. The length constant describes the drop-off of membrane potential from a point of current injection at the steady state when we apply a constant inward current. It is an exponential decrease with distance, and the length constant is a factor in the exponential. It is the distance it takes $V_0 - V_r$ to fall to $1/e(V_0 - V_r)$. The time constant describes the time for the membrane capacitor to charge or discharge under conditions of space clamp.

If the resistance of the membrane is increased, more of the depolarizing current passes down the axon and discharges the capacitor on the membrane. Thus insulating the membrane with myelin increases the passive spread of current and increases conduction velocity. Decreasing the membrane capacitance also increases conduction velocity. Adding myelin increases the thickness of the membrane capacitor, which decreases its capacitance. Increasing the diameter of the axoplasm decreases the internal resistance, which facilitates passive spread of current and increases conduction velocity.

Myelinated nerves generate a true action potential only at the nodes of Ranvier. The nodes are breaks in the myelin sheath that occur between neighboring Schwann cells that form the sheath. The nodes are 2–3 μm long and occur about 1–3 mm apart. The spread of the action potential between nodes is electrotonic, meaning that it spreads passively with some decrement between nodes. Current passes only at the nodes. This jumping of current from node to node is called saltatory conduction.

REVIEW QUESTIONS

1. What effect does myelination have on conduction velocity? What does myelin do to transmembrane resistance? Membrane capacitance?
2. What effect does axon diameter have on conduction velocity?
3. What effects does axon size have on the internal resistance to current flow along the axon length?
4. What is the time constant of an axon? Why is it relatively unaffected by myelination? How would you measure the time constant?
5. What is the space constant of an axon? What determines its value? What effect does myelination have on the space constant? How would you measure the space constant?
6. What is meant by "saltatory conduction"? What jumps from node to node?

APPENDIX 3.3.A1 CAPACITANCE OF A COAXIAL CAPACITOR

USE OF GAUSS'S LAW TO DERIVE CAPACITANCE

In Chapter 1.3, we discussed Gauss's law and how it can be used to determine the electric field intensity between the plates of a capacitor, and how that could be used to determine the capacitance. Gauss's law is a direct consequence of Coulomb's law and is written as

[3.3.A1.1] $$\oint \mathbf{E} \cdot d\mathbf{s} = \frac{q}{\varepsilon_0}$$

where the integral is the integral over a closed surface of the dot product of the electric field and the area vector, equal to the area increment and oriented normal to the surface. What this says is that this dot product, evaluated over *any* close surface, is equal to the charge enclosed by the surface divided by ε_0, the electrical permittivity of space. If there is no enclosed charge, the surface integral is zero. This allows us to determine the electric field intensity from a given charge.

The second necessary idea here connects the electric field intensity to the voltage. The electric field is the negative of the spatial derivative of the potential; specifically, the electric field is the negative gradient of the potential (see Chapter 1.3 for a discussion of the gradient). We can write this as

[3.3.A1.2] $$\mathbf{E} = -\nabla V$$

where **E** is the electric field and V is the voltage. Knowing **E** from Gauss's law, and the enclosed charge, q, we can obtain V. This enables us to determine the capacitance as

[3.3.A1.3] $$C = \frac{q}{V}$$

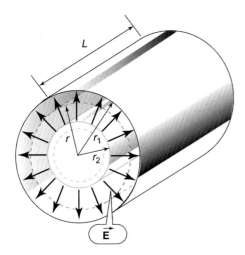

FIGURE 3.3.A1.1 Coaxial capacitor consists of concentric cylinders. The inner cylinder has an outer radius of r_1 (the charges are located on its surface) and the outer cylinder has an inner radius of r_2. Both cylinders are of length L. The electric field is oriented radially outward between the two concentric cylinders.

THE CAPACITANCE OF A COAXIAL CABLE

We now consider the capacitance of a coaxial, cylindrical capacitor as shown in Figure 3.3.A1.1. The inner radius is r_2 and the outer radius is r_1. We consider a length, L, of the capacitor and draw a Gaussian surface as shown so that one side of the closed surface is inside the inner cylinder and the other is in the middle of the capacitor at a radius of r. This is indicated in the figure by a dashed line. The Gaussian surface is of length L and encloses itself at the ends by two circles oriented parallel to the electric field. By symmetry, the electric field is directed radially outward and is normal to the Gaussian surface we have chosen, precisely for that reason. Thus, Gauss's law gives

[3.3.A1.4]
$$\oint \mathbf{E} \cdot d\mathbf{s} = E 2\pi r L = \frac{q}{\varepsilon_0}$$
$$\mathbf{E} = \frac{q}{\varepsilon_0 2\pi r L}$$

From the definition of $\mathbf{E}$ and V, we have

[3.3.A1.5]
$$\mathbf{E} = -\nabla V = -\frac{dV}{dr}$$

because of the radial symmetry of this situation. Here $-dV/dr$ gives the magnitude of $\mathbf{E}$, not its direction. In the same way, Eqn [3.3.A1.4] gives the magnitude of $\mathbf{E}$ and not its direction. We equate these two to evaluate the potential difference across the capacitor:

[3.3.A1.6]
$$-\frac{dV}{dr} = \frac{q}{\varepsilon_0 2\pi r L}$$
$$\int_2^1 -dV = \int_2^1 \frac{q}{\varepsilon_0 2\pi L} \frac{dr}{r}$$
$$V_2 - V_1 = \frac{q}{\varepsilon_0 2\pi L} \ln \frac{r_1}{r_2}$$

The capacitance is then calculated as $q/\Delta V$:

[3.3.A1.7]
$$C = \frac{q}{V_2 - V_1} = \frac{\varepsilon_0 2\pi L}{\ln(r_1/r_2)}$$

This is the final equation for the capacitance for a coaxial capacitor. This reverts to the formula for a parallel plate capacitor if the separation of the plates is small compared to their radii. To see this, we write

[3.3.A1.8]
$$r_1 = r_2 + \delta$$

where δ is the separation of the plates. Inserting this into Eqn [3.3.A1.7], we have

[3.3.A1.9]
$$C = \frac{\varepsilon_0 2\pi L}{\ln((r_2 + \delta)/r_2)}$$
$$= \frac{\varepsilon_0 2\pi L}{\ln(1 + (\delta/r_2))}$$
$$= \frac{\varepsilon_0 2\pi L}{(\delta/r_2)}$$
$$= \frac{\varepsilon_0 2\pi r_2 L}{\delta}$$
$$= \frac{\varepsilon_0 A}{\delta}$$

This last equation is the formula for the capacitance of a parallel plate capacitor. This relies on the approximation $\ln(1 + \alpha) \approx \alpha$ for small values of α. This approximation is only moderately good for axons in which α might be 1.0, where the thickness of the myelin sheath is as great as the radius of the axon itself.

Problem Set
Membrane Potential, Action Potential, and Nerve Conduction 3.1

1. Muscle cells are long, multinucleated cells that are shaped something like cylinders that taper near the ends. Consider a 10-cm length of a muscle cell that is 50 µm in diameter. Assume that the cell is a right circular cylinder and that the capacitance of the muscle cell membrane is 1 µF cm^{-2}. Neglect the ends of the cylinder and edge effects where the membrane bends at the edges.
   A. How many charges have to move across the membrane in order to produce a potential of -80 mV, negative inside?
   B. The total concentration of anions (negatively charged ions) in the cytosol is about 150×10^{-3} M, and the total concentration of cations (positively charged ions) is about the same. About how many cation charges are there in the cytosol of the muscle cell of 10 cm length and 50 µm diameter?
   C. Compare the calculations in part A and part B. The ions that move across the membrane to produce the membrane potential of -80 mV are what fraction of the cytosolic cations?

   Helpful hints: One mole of charge is 9.649×10^4 C. This is the **Faraday**.

2. The chord conductance equation for the membrane potential derived for you Chapter 3.1 is incomplete. This is because at the same time there is passive movement of ions there occurs electrogenic active movement of ions. The Na–K pump carries a current and so contributes to the membrane potential directly. The contribution of the pump to the resting membrane potential can be estimated by noting the effect on the potential after inhibiting the pump with a specific inhibitor, **ouabain**. When ouabain is added, the membrane potential depolarizes by about 2 mV. The pump current can be estimated as

   [3.PS1.1] $$I_p = \frac{\Delta V_p}{R_m}$$

   where I_p is the pump current, ΔV_p is the drop in voltage when the pump is inhibited, and R_m is the membrane resistance.
   A. If $R_m = 1000$ Ω cm^2, what is the pump current per cm^2 of membrane?
   B. Convert this answer to a flux of univalent ions in units of mol cm^{-2} s^{-1}.
   C. The density of Na–K pump sites, estimated by ouabain binding studies, is usually about 1000 µm^{-2}. If the pump has a turnover of about 120 s^{-1}, which is the number of completed reactions per second, and it pumps 3 Na$^+$ ions out for each completed cycle, what is the Na$^+$ flux across the membrane? This answer should be given in units of mol cm^{-2} s^{-1}.
   D. Convert the flux of Na$^+$ in part C to a current in units of amps cm^{-2}.
   E. If the pump transports 2 K$^+$ in for each completed turnover, calculate the K$^+$ flux in units of mol cm^{-2} s^{-1}.
   F. Convert the K$^+$ flux to a current in units of amps cm^{-2}.
   G. What is the net current carried by the pump, calculated both as mol charge cm^{-2} s^{-1} and as amps cm^{-2}?
   H. How does the current and flux carried by the pump, calculated on the basis of the pump density and postulated turnover number, compared to the current and flux calculated by the depolarization caused by inhibition of the pump?

3. Assume that a membrane is permeable only to Na$^+$ and K$^+$. The concentrations of ions are:

   $$[Na^+]_o = 145 \text{ mM} \quad [Na^+]_i = 12 \text{ mM}$$

   $$[K^+]_o = 4 \text{ mM} \quad [K^+]_i = 155 \text{ mM}$$

   A. Calculate the equilibrium potential for Na, E_{Na}.
   B. Calculate the equilibrium potential for K, E_K.
   C. Given the following table for g_K and g_{Na}, calculate the membrane potential, E_m, using the chord conductance equation.

Time (ms)	g_K	g_{Na}
0	0.366751	0.010589
0.4	2.186175	37.48381
0.8	5.478086	31.23947
1.2	9.486194	21.33964
1.6	13.48798	14.36902
2	17.05401	9.674778

   D. Graph the membrane potential against time. Does this vaguely resemble an action potential?

E. The membrane potential during an action potential cannot be calculated solely by the chord conductance equation. Why not? *Hint*: Think of the assumptions used in deriving the chord conductance equation.

4. The action potential on a motor neuron lasts about 1.5 ms. A large motor neuron conducts the action potential at about 100 m s^{-1}.
   A. Calculate the distance between the beginning of the action potential and its end during its conduction down the fiber. (At any time, the action potential is located in a place on the neuron; because the action potential does not occur instantaneously, it must be spread out in space. This question asks how spread out is the action potential.)
   B. The nodes of Ranvier are interruptions in the myelin sheath in myelinated fibers, and typically they are located about 1 mm apart along the length of the axon. The action potential that you calculated in part A is spread out over how many nodes of Ranvier?
   C. In saltatory conduction, what "jumps" from node to node?

5. Pain information from the skin travels to the spinal cord over several different kinds of axons. The $A\delta$ fibers conduct action potentials at up to 30 m s^{-1} whereas the small unmyelinated C fibers conduct more slowly. This information is processed beginning in the spinal cord where interneurons are used to connect the pain information to a motor neuron, which conducts at up to 100 m s^{-1}. Consider a sharp pain that occurs at the end of the fingers, about 1 m away from the spinal cord. Calculate the minimum reaction time based on these speeds of conduction, allowing for two synaptic delays in the cord of 1 ms each.
   Reactions involving higher order nervous function (things you have to think about—such as braking your car) take longer because of the additional numbers of neurons involved in the pathway.

6. Suppose the internal specific resistance of the axoplasm is 50 Ω cm. Calculate the internal resistance of an axon 10 mm long if it has a radius of 5 μm. Repeat the calculation for an axon with an internal radius of 500 μm.

7. Assume that the dielectric constant of myelin is 8 and the thickness of one layer is 7.5 nm. Assume a radius of 5 μm.
   A. Calculate the specific capacitance (the capacitance per unit area) of a single layer assuming it is a flat sheet. Calculate the capacitance if the axon is 10 cm long.
   B. Calculate the specific capacitance if the myelin sheath is 2 μm, assuming it is a flat sheet. Calculate the total capacitance if the axon is 10 cm long.
   C. How do these differences in the capacitances affect the space constant? The time constant?
   D. Recalculate the capacitances in A and B assuming they are a coaxial capacitance. How much difference does it make to assume that the membrane is planar?

8. A cell has the following conditions:

   $[Na^+]_o = 145$ mM $[Na^+]_i = 10$ mM
   $[K^+]_o = 4$ mM $[K^+]_i = 150$ mM
   $[Ca^{+2}]_o = 1.2$ mM $[Ca^{+2}]_i = 0.3$ μM

   A. Calculate E_{Na}, E_K, and E_{Ca}.
   B. If $g_K = 100\, g_{Na}$ and $g_{Na} = 5\, g_{Ca}$, calculate E_m at rest.
   C. Calculate the net driving force for I_{Na}, I_K, and I_{Ca}.
   D. Calculate the relative values of I_{Na}, I_K, and I_{Ca}.

9. From the graph in Figure 3.2.9, estimate the average conductance for Na$^+$ during the action potential by approximating the conductance as a triangle. Estimate the average E_m the same way. Calculate the average flux of Na$^+$ using the E_{Na} in the figure. Estimate the cytosolic content of Na$^+$, assuming the cell is a sphere with a radius of 20 μm. How much does a single action potential change E_{Na}?

10. From the graph in Figure 3.2.14, estimate the unitary *current* carried by the K$^+$ channels in squid axons (right panel). If E_m is -70 mV in these cells, estimate the unitary *conductance* under physiological conditions.

11. Assume an axon has an internal diameter of 1 μm and a myelin sheath 1 μm thick. The internal specific resistance is 100 Ω cm. For the myelin sheath, use a dielectric constant of 8 and a transmembrane resistance of 2×10^5 Ω cm^2.
    A. Calculate the internal resistance per unit length.
    B. Calculate the capacitance per unit length.
    C. Calculate the transmembrane resistance per unit length.
    D. Calculate the time constant.
    E. Calculate the space constant.

12. Assume an unmyelinated axon has an internal diameter of 1 μm and an internal specific resistance of 100 Ω cm. Assume the single plasma membrane is 7.5 nm thick and has a dielectric constant of 8. The transmembrane resistance is 1000 Ω cm^2.
    A. Calculate the internal resistance per unit length.
    B. Calculate the capacitance per unit length.
    C. Calculate the transmembrane resistance per unit length.
    D. Calculate the time constant.
    E. Calculate the space constant.

13. Assume that the extracellular $[Ca^{2+}] = 1.2$ mM and the intracellular $[Ca^{2+}] = 0.1$ μM.
    A. Calculate the equilibrium potential for Ca^{2+}.
    B. If a channel was to open when $E_m = -50$ mV, which way would Ca^{2+} go?
    C. Is this a positive or a negative current?

14. Suppose you depolarize an axon to threshold midway between its initial segment and its terminus.
    A. In which direction would the action potential travel?
    B. What would happen if you simultaneously began an action potential at the initial segment and at a point midway between the initial segment and the terminus?
15. A myelinated nerve is treated with tetrodotoxin in the middle of the axon. What happens to action potentials conducted from the initial segment toward the axon treated with tetrodotoxin?
16. Ouabain is a naturally occurring substance that inhibits the Na,K-ATPase with high affinity. When added to a neuron, the membrane potential depolarizes by about 2 mV. Why does this happen?
17. An excitable cell has the following conditions:

$$[Na^+]_o = 145 \text{ mM} \quad [Na^+]_i = 10 \text{ mM}$$

$$[K^+]_o = 4 \text{ mM} \quad [K^+]_i = 150 \text{ mM}$$

After passing a number of action potentials in a short time, the following changes occurred:

$$[Na^+]_o = 140 \text{ mM} \quad [Na^+]_i = 15 \text{ mM}$$

$$[K^+]_o = 7 \text{ mM} \quad [K^+]_i = 145 \text{ mM}$$

Calculate E_{Na} and E_K for the two conditions. If at rest $g_K = 10\, g_{Na}$, calculate the resting membrane potential at rest for both sets of ionic composition. What effect would this have on action potentials?

18. In the plot of threshold current against stimulus time, the rheobase is difficult to determine because it is the extrapolation of threshold current at infinite time. Using the Weiss equation, suggest a different plot of the equation that would allow determination of I_{rh} by extrapolation of a linear curve, and calculation of τ_{SD} from the slope of the line.
19. Consider an excitable cell with a resting membrane potential of -60 mV and a uniform threshold potential of -40 mV. For the following questions, refer to Figure 3.2.15.
    A. If the membrane is hyperpolarized to -65 mV, what would the sign of the total current be? What would happen as a result of this current?
    B. If the membrane was depolarized to -50 mV, what would the sign of the total current be? What would happen as a result of this current?
    C. If the membrane was depolarized to -35 mV, what would the sign of the total current be? What would happen as a result of this current?
20. Dendrites are not right cylinders but taper as they extend away from the body, or soma, of the neuron. Neurotransmitters cause either a depolarization or hyperpolarization (postsynaptic potentials) that is conveyed electronically through the cytoplasm. Would these postsynaptic potentials decay away differently for thin versus wider dendrites?

3.4 Skeletal Muscle Mechanics

Learning Objectives

- Contrast skeletal, cardiac, and smooth muscles
- Describe an isometric contraction
- Describe a muscle twitch
- Distinguish between slow-twitch and fast-twitch muscle fibers on the basis of their contractile behavior
- Define the motor unit
- Describe how recruitment increases muscle force
- Describe how a muscle responds to increased frequency of stimulation
- Define tetany
- Describe the size principle of motor unit recruitment
- Describe the relative importance of frequency coding versus population coding for grading muscle force
- Define the innervation ratio and explain how the size principle creates proportional control of muscle force
- Define passive and active tension and show how they vary with muscle length
- Define afterload
- Describe an isotonic contraction
- Draw the force–velocity curve for fast-twitch and slow-twitch muscles
- Show how power varies with velocity of contraction
- Distinguish among concentric, isometric, and eccentric contractions
- Describe muscle pinnation and how this contributes to muscle function
- Define muscle fatigue
- Describe Burke's system of muscle classification based on contractile properties

MUSCLES EITHER SHORTEN OR PRODUCE FORCE

The primary action of a muscle is **to contract**. Usually this word means to shorten, but physiologists often use it in a broader sense, meaning **to activate** the muscle. Activation of the muscles can produce force without actually shortening, as you might do when holding something really heavy. As we shall see later, there is a trade-off between shortening and producing force, and both originate from the same mechanism. In everyday motion, some muscle contractions occur while the muscle actually lengthens. These are called **eccentric** contractions.

MUSCLES PERFORM DIVERSE FUNCTIONS

Muscles that are connected to the skeleton are called **skeletal muscles**. Because they are connected to the bones via their tendons, these muscles transfer force to the skeleton and move one bone relative to another. Coordinated activation of muscles allows for the various movements of the joints (flexion and extension; abduction and adduction; protraction and retraction; elevation and depression; pronation and supination; medial rotation and lateral rotation; and circumduction). Skeletal muscles allow us to lift weights and to move ourselves from place to place. These muscles also include the tongue, the muscles that move the eyeballs, and the upper third of the esophagus. A secondary function of these skeletal muscles is the production of body heat. During vigorous exercise, the heat produced by the skeletal muscles raises body temperature and the excess heat must be removed from the body.

Other muscles surround hollow organs. These include the heart, urinary bladder, gallbladder, and uterus. Contraction of the muscles in the walls of these organs develops a tension in the walls and so develops pressure within the organ. Still other muscles are present in the long hollow tubes of the body. Contraction of the muscles in these tubes sets the diameter of a tube and thereby controls flow through it. Examples include the muscular sphincters of the gastrointestinal tract, the arterioles, the ureters, and the airways of the lungs. Contraction of the muscular layers of the intestine also propels material through the intestine.

MUSCLES ARE CLASSIFIED ACCORDING TO FINE STRUCTURE, NEURAL CONTROL, AND ANATOMICAL ARRANGEMENT

STRIATED MUSCLES HAVE STRIPES

Under the microscope, some muscles appear to have stripes running across them. These are said to be **striated** (from the Latin meaning channel or furrow). All of the skeletal muscle and cardiac muscle are striated.

Muscles lacking stripes are called **smooth** muscles (Figure 3.4.1).

SKELETAL MUSCLES ARE USUALLY ACTIVATED VOLUNTARILY

Most skeletal muscles are normally activated by a motor neuron that resides in the anterior or ventral horn of the spinal cord. Commands for the activation of the motor neuron can come from sensory input, as in the spinal reflexes, or from the motor cortex in the cerebrum. These muscles are organized as **motor units**, with nerve fibers providing the signal for activation. In contrast, cardiac and visceral smooth muscles are intrinsically active and do not require neural initiation of activation. However, these muscles are innervated and their activity is modulated by nerves and sometimes by humoral factors. These different types of control are broadly classed as **voluntary** and **involuntary**. Breathing and blinking the eyelids are largely automatic events that occur involuntarily, but we can consciously suppress these activities for some time and we must be able to control breathing voluntarily in order to speak, sing, or swim. This chapter is devoted to voluntary skeletal muscles.

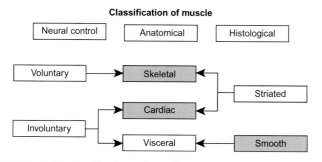

FIGURE 3.4.1 Classification of muscles can be based on control properties, anatomical properties, or the fine structure observed histologically. The preferred classification is that highlighted in the figure, which is a mixture of several classification schemes.

ISOMETRIC FORCE IS MEASURED WHILE KEEPING MUSCLE LENGTH CONSTANT

A muscle can be dissected free from its **origin** and **insertion** (the places where the tendons attach to the bones) and connected to a device to measure the force developed upon activation, as shown in Figure 3.4.2. Here the experimental setup measures force at a nearly constant length, and so these are called **isometric contractions** (from "iso," meaning "the same," and "metric" meaning "distance"). When a stimulus is turned on very briefly (3 ms), a single muscle contraction, called the **muscle twitch**, is recorded. Force rises and falls without muscle shortening. It takes some time for force to rise and generally a longer time for it to fall back to resting levels. When a **gastrocnemius** muscle of a rat is used, the twitch time is about 70 ms. When the **soleus** is used, the twitch time is over 150 ms. Muscles can be distinguished on the basis of their contractile properties, including their twitch times. The gastrocnemius is an example of a fast-twitch muscle; the soleus is a slow-twitch muscle.

The recording in Figure 3.4.2 also shows that the action potential on the muscle lasts only a few ms, whereas the twitch lasts much longer. The action potential is a **trigger** for the muscle contraction.

MUSCLE FORCE DEPENDS ON THE NUMBER OF *MOTOR UNITS* THAT ARE ACTIVATED

The stimulus applied to the nerve can be varied by increasing its voltage or changing its duration, delay, or frequency. When the voltage strength of the stimulus is gradually increased, the force of the twitch also increases until the force reaches a plateau, as shown in Figure 3.4.3. Why should muscle force depend on the voltage of the stimulus?

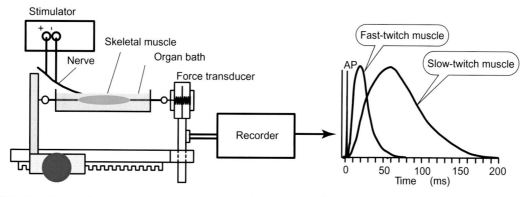

FIGURE 3.4.2 Experimental setup for measuring isometric tension in an isolated muscle. In this preparation, the nerve that usually activates the muscle is severed and action potentials on the motor nerve axon must be initiated by an external stimulator. The muscle is tied at one end to a rigid support and at the other end to a force transducer. When the muscle contracts, it pulls against a stiff spring located in the transducer. The transducer turns its slight movement into an electrical signal that can be recorded. Because the actual shortening of the spring, and its attached muscle, is extremely small, this contraction is called an isometric contraction. A single stimulation results in an action potential in the nerve and subsequently in the muscle that activates the muscle to produce a single contraction called a twitch. Different muscles differ markedly in the time course of force development and relaxation.

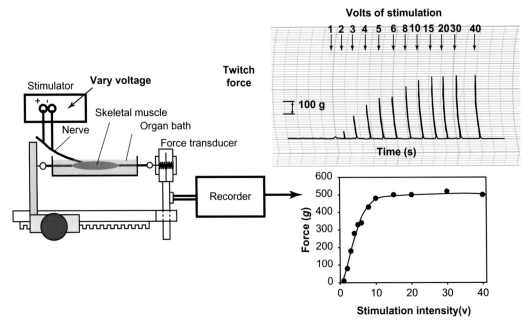

FIGURE 3.4.3 Increase in the muscle twitch with increased recruitment of motor fibers by increasing the strength of the external stimulus. Data obtained from electrical stimulation of the sciatic nerve serving the gastrocnemius muscle of the rat.

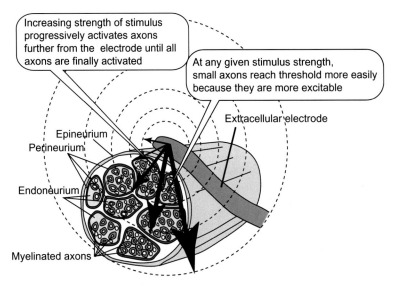

FIGURE 3.4.4 Intact nerve and the extracellular electrode that stimulates it. The whole nerve is a bundle of axons originating in either sensory or motor neurons. As the strength of stimulus increases, more axons are activated until, eventually, all axons in the nerve fire action potentials. Since each axon innervates a set of muscle fibers, called its motor unit, progressive activation of axons causes progressive increases in the number of activated muscle fibers, and progressive increases in the total force produced by the muscle. The smaller axons are more excitable, and they are activated first.

The motor nerve supplying the muscle is actually a bundle of axons from motor neurons carrying excitation to the muscle and sensory nerve fibers that return information to the central nervous system about the state of the muscle. The proportion of motor nerve axons within the bundle varies from 40% to 70% in different muscles, with the remaining being axons carrying sensory information. The motor axons come in two general types: small and large diameter axons (see Figure 3.4.4). The stimulus that we apply is an extracellular stimulus that depolarizes the axons by passing an inward current across their membranes. But the current from the extracellular electrode has many places to go, and most of the current does not go across the axon membrane. Thus higher voltages are necessary to depolarize all of the axons in the bundle. This is why volts are necessary for the external stimulus, when only a few millivolts of membrane depolarization are sufficient to initiate an action potential. Further, the different sized motor neuron axons differ in their rheobase—the amount of current for an infinite time that will bring the axons to threshold (see Figure 3.2.8). Increasing the stimulus

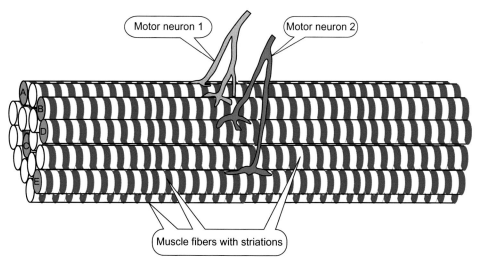

FIGURE 3.4.5 The motor unit consists of the motor neuron and all of the muscle fibers innervated by it. Here motor neuron 1 innervates fibers A, B, and C; motor neuron 2 innervates fibers D and E. When only motor neuron 1 fires an action potential, only fibers A, B, and C contribute to the force. When only motor neuron 2 fires an action potential, only fibers D and E contribute to the force developed by the muscle. When both motor neurons fire an action potential, all of the muscle fibers contribute to the force. Thus the force is greater when more motor units are recruited.

strength (voltage) increases the number of motor neurons that are activated, which increases the number of muscle fibers that are activated, which increases force. Each motor nerve branches and connects to a set of muscle fibers, making a neuromuscular junction with each of them. **The motor neuron and the set of muscle fibers it innervates make up a motor unit.** The entire muscle consists of a large number of muscle fibers that are each typically innervated by a single motor neuron making a junction in the middle of the muscle fiber. Motor neurons typically innervate more than one muscle fiber, but each muscle fiber is innervated by only one motor neuron. The motor unit is designed for 100% fidelity: each time the motor neuron is activated, all of its muscle fibers are subsequently activated. Thus motor units are indivisible quantal elements in all movements (see Figures 3.4.4 and 3.4.5).

The increase in force by activating increasing number of motor units is called **recruitment**. The nearly continuous variation in force from small to large forces means that this force is not all or none; it is **graded**. Although the gradation is fine, it is not really continuous because muscle fibers are activated in a discrete (as opposed to continuous) way. Because there are so many muscle fibers, the force appears to vary nearly continuously. It is important to note that the method of recruitment here, increasing the voltage of stimulation of the nerve, is *not physiological*. Ordinarily motor units are recruited through neuronal connections to the motor neurons, particularly by the activation of command signals originating from the primary motor cortex in the brain, located in the precentral gyrus (see Chapter 4.5). This area of the brain contains "upper" motor neurons that provided excitation to the "lower" motor neurons that are located in the ventral horn of the spinal cord and directly activate muscle fibers. The motor unit specifically refers to a lower motor neuron and the set of muscle fibers that it innervates. Which lower motor neurons that are activated depends on the output of the primary motor cortex, along with modulating influences from other brain areas and from sensory information. The neural control of movement is discussed in Chapter 4.5.

THE *SIZE PRINCIPLE* STATES THAT MOTOR UNITS ARE RECRUITED IN THE ORDER OF THEIR SIZE

Large motor units are innervated by large motor neurons, and smaller motor units are innervated by smaller motor neurons. The small motor neurons are more excitable, so these are recruited first. This corresponds to our everyday experience. When trying to perform delicate movements that require dexterity but little force, control of muscle force must be fine. This is accomplished by recruiting small numbers of muscle fibers. When performing gross motor movements involving a lot of force, the increments of force are large and we recruit successively larger motor units. The recruitment of motor units in order of their sizes is accomplished through other nerves that make connections to the lower motor neurons.

MUSCLE FORCE CAN BE GRADED BY THE FREQUENCY OF MOTOR NEURON FIRING

The action potential on the motor neuron is very short, between 1 and 3 ms. The action potential on the muscle cell membrane is also short, on the order of 3–5 ms. The muscle twitches are long by comparison, some 30–300 ms. This means that it is possible to stimulate the muscle with another action potential before the muscle has relaxed. Indeed, we can stimulate a muscle again before it reaches its peak tension. When a muscle is stimulated before it has completely relaxed,

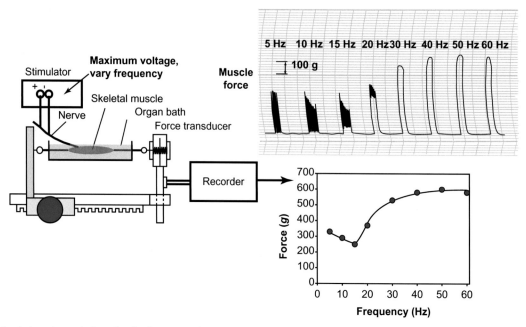

FIGURE 3.4.6 Gradation of muscle force by the frequency of stimulation. Increasing the frequency of stimulation at low frequencies does not increase the force; it merely increases the frequency of the same twitch wave form. When the period of the stimulation frequency is shorter than the period of the twitch, however, force begins to summate. With continued increases in frequency, there is added force until tetany is reached. Data obtained from electrical stimulation of the sciatic nerve serving intact gastrocnemius muscle of the rat.

a new twitch begins force where the first twitch left off (see Figure 3.4.6). The resulting force is greater than with a single twitch. Thus muscle force **summates** with repetitive stimulation. When the frequency of stimulation is great enough, the muscle produces a single forceful contraction with no waviness in the force. This condition is called **tetany**.

The frequency required to reach tetany depends on how fast the muscle contracts and relaxes. If the muscle twitch is 100 ms long, then summation will just begin when the next stimulation arrives at the end of relaxation. Since there are 10 100 ms-intervals in a second, summation for such a fiber should begin at a stimulation frequency of about 10 Hz. (1 Hz is hertz, meaning a cycle per second; 10 Hz is therefore 10 cycles or events per second.) Typically most muscles in the human tetanize between 20 and 100 Hz. **Summation begins when stimulation frequency just exceeds the inverse of the twitch time.**

Although the results shown in Figure 3.4.6 are a laboratory observation involving an isolated nerve and its muscle, the same phenomenon also occurs in living, breathing people. We can grade the force of a muscular contraction, or vary force more or less continuously, by altering the frequency of motor neuron firing. This is one of the principal ways of physiological control of muscle force. Since action potentials are an all-or-none phenomenon, information can be coded only through the frequency of action potentials (frequency code) and in the population of neurons carrying it (population code—this is the same as recruitment). In the case of muscle, high-frequency neuronal activity is converted to high intensity of muscle force.

As you can see from Figure 3.4.6, the tetanic force is much greater than the twitch force. In general, **tetanic force is about five times the twitch force**, but the twitch ratio varies from 2 to 10 in different muscle types.

MUSCLE FORCE DEPENDS ON THE LENGTH OF THE MUSCLE

The device shown in Figures 3.4.2, 3.4.3, and 3.4.6 can be adjusted to vary the length of the muscle. When muscles are relaxed, they exert no force. When the relaxed muscles are stretched passively, without activation by a nerve stimulus, they produce a **passive force**. This is due to elastic properties of the muscle material itself. This passive force increases steeply and nonlinearly with increases in length (see Figure 3.4.7).

When a stretched muscle is stimulated tetanically, it produces a force in addition to the passive force. This increment in force is called the **active tension** because it depends on activation of the muscle by the tetanic stimulation. The additional active force produced by stimulation depends on the length of the muscle. The active force increases to a maximum and then declines with further increases in muscle length. The relationship between active force and muscle length is the **length–tension curve**, originally determined in isolated muscle fibers by Robert Ramsey and Sybil Street. This data became an important observation for explaining the mechanism by which muscles produce force.

The resting length of a muscle is usually designated as L_0. Because muscles are attached to the skeleton, their

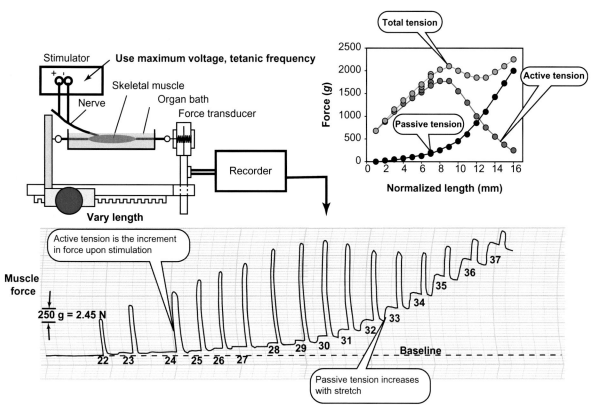

FIGURE 3.4.7 The length–tension relationship in muscle. Increasing muscle length from slack length to relatively long lengths causes a passive force or tension that does not require activation of the muscle. Activation of the muscle at maximum recruitment and tetanic frequency (so that recruitment or frequency response does not confound the results) causes an increment in the force called the active force or tension. Muscle length was increased in increments of 1 mm as indicated by the Vernier scale readings under each tetanic stimulation. Active tension increases biphasically with muscle length. Passive tension increases progressively with muscle length—muscles do not obey Hooke's law, in which force is proportional to length. The total tension is the sum of the passive and active forces. Data obtained by electrical stimulation of the sciatic nerve innervating the gastrocnemius muscle of the rat.

degree of shortening is defined by the movement of the bones and the attachment points of the muscles. Most muscles do not shorten or lengthen by more than about 25% of their rest length. Thus although muscle length can determine muscle force, typically the muscle is physiologically arranged near the top of the length–tension curve, and muscle length changes are relatively unimportant compared to recruitment and frequency of stimulation.

RECRUITMENT PROVIDES THE GREATEST GRADATION OF MUSCLE FORCE

We have described three distinct ways to vary muscle force: (1) recruit muscle fibers by activating larger numbers of motor units, (2) vary the frequency of activation of the activated motor units, and (3) vary the length. As mentioned above, muscle length typically does not vary significantly, and the resting length of muscles is situated at the top of the active length–tension curve, so that varying muscle length is not an important way of varying muscle force in the whole person. Most activation of muscle is not by a single action potential to produce a twitch, but by a train of impulses (see Chapter 3.7) interrupted by rest periods. These series of activations increase force by wave summation. Although this is important, it provides a range of forces that scale according to the tetanus–twitch ratio, which varies from about 2–10, depending on the muscle. Therefore, the frequency of activation provides at most about a 10-fold range of variation in muscle force. Skeletal muscles often have a range of forces that varies 20,000 fold, from the weakest contraction to the most forceful. Thus rate coding or frequency coding provides an important part of muscle force gradation, but population coding (recruitment) provides the greater part of muscle force gradation.

MUSCLE FIBERS DIFFER IN CONTRACTILE, METABOLIC AND PROTEOMIC CHARACTERISTICS

As shown in Figure 3.4.2, muscles differ in their contractile properties. This mechanical difference derives from differences in their constituent motor units. Each motor neuron innervates a number of muscle fibers. The muscle fibers themselves are heterogeneous, and several classification schemes have been devised to describe them. Based on contractile behavior, muscle fibers can be broadly classified as being S, FR, FI and FF: these mean S = slow; FR = fast, fatigue resistant; FI = fast, intermediate in fatigability; FF = fast, fatigable. The differences in the muscle fibers come from their expression of different protein isoforms or from differing quantities of the same proteins or from differing quantities and composition of subcellular organelles. The muscle fibers differ in metabolic ability that shows up in their gross appearance: red muscle contains a lot of myoglobin, a protein that binds oxygen within the cytoplasm of the muscle fiber and aids in oxygen delivery to mitochondria, which are also rich in red muscles. White muscle fibers rely less on oxidative metabolism and use anaerobic pathways more. So muscles can be classified according to their metabolic properties as being SO = slow, oxidative; FOG = fast, oxidative and glycolytic; and FG = fast, glycolytic. Muscles are also differentiated on the basis of myosin isoform expression as type I (equivalent to SO), type IIa (equivalent to FOG), and type IIb (FG). We will return to these different muscle fiber types in Chapters 3.5 and 3.7.

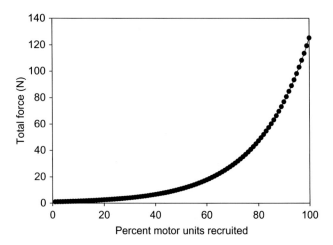

FIGURE 3.4.8 Simulated active force of a muscle as a fraction of motor units that are activated. The muscle was simulated as a population of 100 motor units whose aggregate muscle fiber area was given as $A_{i+1} = 1.05\ A_i$, with a constant specific force of 20 N cm^{-2}.

MOTOR UNITS CONTAIN A SINGLE TYPE OF MUSCLE FIBER

Muscle fibers are heterogeneous, differing in contractile and metabolic properties, but all of those muscle fibers that are innervated by a single motor neuron are of a single type. This fact suggests that there is some kind of communication that passes between the motor neuron and the muscle fibers that directs their expression of protein isoforms. In short, the motor neuron establishes the kind of muscle fibers it innervates and maintains them in that form.

THE INNERVATION RATIO OF MOTOR UNITS PRODUCES A PROPORTIONAL CONTROL OF MUSCLE FORCE

The tetanic force produced by a single motor unit is given as

[3.4.1] $\qquad F_{MU} = NF_{MF} = NAF_S$

where F_{MU} is the force of the motor unit, N is the number of muscle fibers that the motor neuron innervates, A is the area per muscle fiber, and F_S is the specific force or the force per unit area of muscle fiber. **N is the innervation ratio.** Although some reports suggest that there might be differences in the specific force for different muscle fiber types, the effect is not large. Typically the force per unit area for a muscle fiber is about 20 N cm^{-2}. The total force developed by a muscle is given as

[3.4.2] $\qquad F = \sum_{i=1}^{i=j} N_i A_i F_S$

indicating that the forces of the motor units from unit 1 to j sum linearly, and that the innervation ratio and area of the muscle fibers vary with motor unit. Typically the small motor units have low innervation ratios and their muscle fibers have a smaller cross-sectional area. The first recruited motor units typically have Type I muscle fibers (slow twitch, oxidative metabolism) and the last recruited motor units innervate large muscle fibers of Type IIB (fast, glycolytic metabolism). Innervation ratios vary with the type of muscle from about 5 in the lateral rectus muscle that moves the eyeball to over 2000 muscle fibers for a single motor unit in the medial gastrocnemius in the calf. According to the size principle, the small motor units are recruited first, and progressively larger motor units are recruited until all motor units are recruited in maximal voluntary contractions. The theoretical consequence of the dispersion of motor unit innervation ratios and the orderly recruitment of motor units is shown in Figure 3.4.8. The total active force produced by the muscle increases progressively with the recruitment of the motor units in a process referred to as proportional control. In this process, force is added in steps that are proportional to the amount of force already present. During weak contractions, force is adjusted by changing it slightly (e.g., in steps of 5%) whereas strong contractions are adjusted by making large incremental changes that remain 5% of the force already present. The proportional control is accomplished by recruiting motor units in order of increasing strength.

MUSCLE FORCE VARIES INVERSELY WITH MUSCLE VELOCITY

The experimental setups illustrated in Figures 3.4.2, 3.4.3, 3.4.6, and 3.4.7 were all performed using isometric contractions in which muscle length did not change. We now change this situation and hook up one end of the muscle to a pulley system connected to an **afterload** (see Figure 3.4.9). An afterloaded muscle cannot shorten until it produces a force equal to the afterload. When stimulated, the muscle contracts isometrically during the time it takes to develop force equal to the

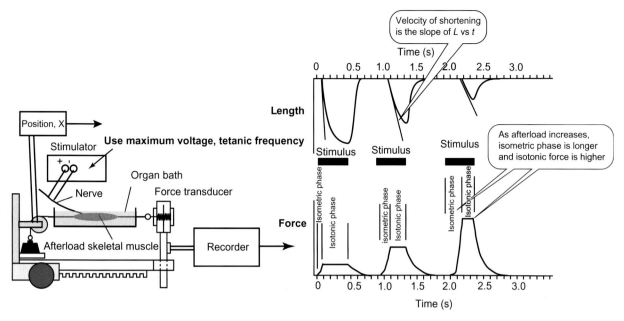

FIGURE 3.4.9 Experimental setup for measuring the force and velocity of isotonic contractions. The afterload is supported by a shelf prior to the activation of the muscle and thus it is not felt by the muscle until all slack in the line are taken up by muscle shortening. When the size of the afterload is increased, the muscle takes longer to develop sufficient force to lift the load, and the velocity of shortening is less. The initial velocity of shortening is inversely related to the afterload or the force developed by the muscle.

afterload. After the muscle develops a force equal to the afterload, it lifts the afterload and continues to shorten. For some part of this contraction, the velocity of shortening is approximately constant. Such a contraction is called an **isotonic contraction**. Thus each contraction consists of an isometric part and an isotonic part, as shown in Figure 3.4.9.

Everyday experience shows that the speed of muscle contraction depends on the load that must be moved. We know that we can move a light load quickly, whereas we move a heavy load slowly. The force–velocity curve is produced by measuring the initial velocity and plotting it against the afterload. The initial velocity can be measured using a device such as that shown in Figure 3.4.9, where the afterload is varied. The initial velocity is used because after the muscle shortens it cannot produce as much force, because it moves off the optimum of the length–tension curve. The force–velocity curve is shown in Figure 3.4.10.

MUSCLE POWER VARIES WITH THE LOAD AND MUSCLE TYPE

Power is *defined* as the rate of energy production or consumption and it has the units of energy per unit time. In mechanical terms, energy is work. Work is defined in mechanics as force × distance. Thus we have the following:

[3.4.3A]   Power = Energy/time

[3.4.3B]   Energy = work

[3.4.3C]   Work = force × distance

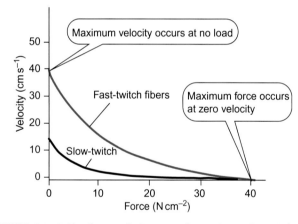

FIGURE 3.4.10 The force–velocity curve for an intact fast-twitch or slow-twitch muscle. Fast contractions can occur only with low loads. Higher loads mean slower velocity of muscle contraction, but more force. Thus there is an inverse relationship between force and velocity.

Inserting Eqn [3.4.3C] into Eqn [3.4.3B] and then into Eqn [3.4.3A], we get

[3.4.4]   Power = force × distance/time

From the definition of velocity as distance/time, we get

[3.4.5]   Power = force × velocity

The force–velocity curve obtained experimentally is displayed in Figure 3.4.10. We can obtain the power–force curve by multiplying force and velocity at every point on this curve. This power is the instantaneous power produced during the initial shortening of the muscle. Power is expressed in units of watts or $N\,m\,s^{-1}$, per unit weight of muscle. The power as a function of muscle force is shown in Figure 3.4.11. This shows

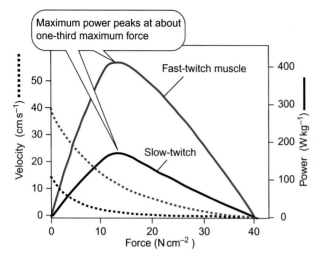

FIGURE 3.4.11 Power as a function of force for slow-twitch and fast-twitch muscles. Velocity is shown as dashes, power as solid lines. Power peaks at about one-third maximum force for both fast-twitch and slow-twitch muscles, but fast-twitch muscles deliver more power due to their faster contractions. These curves are part of the reasons for gears on bicycles, so that a constant power can be delivered to the wheels by keeping the velocity of the muscles and forces near the peak power output when translational velocity of the bike changes going up or down hill.

that the power is about two to three times greater in fast-twitch fibers because of their greater speed of contraction.

ECCENTRIC CONTRACTIONS LENGTHEN THE MUSCLE AND PRODUCE MORE FORCE

According to the way in which we measured velocity, a positive velocity corresponded to a shortening of the muscle. If the support for a *very large* afterload in Figure 3.4.9 is removed, the afterload will cause the muscle to lengthen during contraction. This lengthening is a negative velocity. Contraction of the muscle during a lengthening is called an **eccentric contraction**. Contraction of a muscle that causes a shortening is called a **concentric contraction**. The extension of the force–velocity curve to negative velocities (see Figure 3.4.12) shows that muscles can exert about 40% more force in an eccentric contraction compared to the maximal isometric force measured at zero velocity.

CONCENTRIC, ISOMETRIC, AND ECCENTRIC CONTRACTIONS SERVE DIFFERENT FUNCTIONS

Because concentric contractions shorten, they are useful for the acceleration of one body part relative to another, including parts that are loaded with external objects. Isometric contractions are used to fix joints, usually to produce a platform on which other actions can be made. For example, delicate work by the fingers requires immobilization of the arm and shoulder to hold the hand still while the fingers do the work. Such immobilization is accomplished by simultaneously activating **antagonistic muscles**—those that move joints in opposite directions. Eccentric contractions are used to decelerate body parts, as in activation of the quadriceps muscles in the leg while going downstairs.

Table 3.4.1 shows the three types of contractions, their functions for movement, and the work performed.

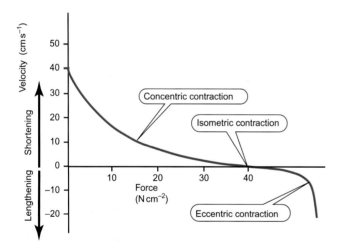

FIGURE 3.4.12 Concentric, isometric, and eccentric contractions. Concentric contractions involve a shortening of the muscle. Eccentric contractions involve a lengthening of the muscle. Isometric contractions occur when the muscle length does not change and occurs at zero velocity. Developed force is greatest for eccentric contractions, next highest for isometric contractions and lowest for concentric contractions.

MUSCLE ARCHITECTURE INFLUENCES FORCE AND VELOCITY OF THE WHOLE MUSCLE

The force that a muscle develops depends on its size: larger muscles produce greater force. Because muscles

TABLE 3.4.1 Types of Contractions and Their Uses

Types of Contractions	Distance Change	Function	Work
Concentric	Shortening (+D)	Acceleration	Positive $W = F \times (+D)$
Isometric	No change (0 D)	Fixation	Zero
Eccentric	Lengthening (−D)	Deceleration	Negative $W = F \times (-D)$

are not right cylinders but take on complicated shapes, determinants of the size present something of a problem. The usual estimate of size is the muscle's cross-sectional area at the belly, or widest part, of the muscle. The maximum force expressed per unit area is typically from 15 to 40 N cm^{-2}. This is enough to lift about 1.5–4 kg cm^{-2}. The variation in this force per unit area derives from the orientation of the muscle fibers within the muscle.

There are three major orientations of the muscle fibers within muscles: **parallel fibers**, **fusiform**, and **pinnate**. The parallel arrangement is present in muscles shaped like a strap or in parts of flat-shaped muscles. In fusiform muscles, the fibers are generally parallel to the longitudinal axis of the muscle, but their number varies with distance from the belly of the muscle. In pinnate muscles, the fibers are oriented at an angle to the tendon or aponeurosis. Because they resemble a feather, these muscles are called pinnate or pennate (see Figure 3.4.13).

The consequence of orienting the muscle fibers at an angle with respect to the tendon is to increase the effective cross-sectional area of the muscle while reducing the distance the muscle can contract along the lines of the tendons. To see this, we consider two muscles, a strap muscle with parallel fibers having a volume of 300 cm^3 and a resting length of 36.8 cm and a second muscle with pinnate architecture with fibers 12 cm long and overall length of 36.8 cm. The pinnate muscle fibers are oriented 15° to the line of action. The geometry of these is shown in Figures 3.4.14 and 3.4.15. The fibers in the strap muscle do not span the distance from tendon to tendon. Instead, long strap muscles are divided into compartments by fibrous bands called **inscriptions**. The sartorius muscle (originating on the lateral hip, wrapping around to the inner thigh, and inserting on the medial tibia) has three inscriptions, giving four compartments; the semitendinosus (part of the hamstring muscles, originating on the medial tuberosity of the ischium and inserting its long tendon on the medial tibia) has three compartments and the biceps femoris (in the back of the thigh, originating in two places and inserting on the lateral side of the head of the fibula) and gracilis (most superficial muscle on the inner thigh, originating on the symphysis pubis and inserting on the medial tibia) have two compartments each.

The conclusion from the calculations in Examples 3.4.1 and 3.4.2 is that the pinnate arrangement of the muscle fibers allows many more muscle fibers to attach to the aponeurosis. The consequence is that the muscle can develop more force, but it does so by reducing its velocity.

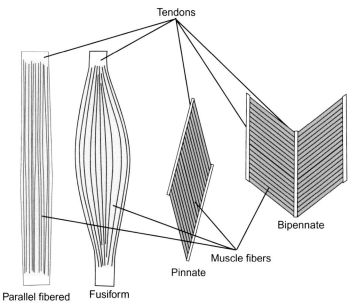

FIGURE 3.4.13 Different arrangement of muscle fibers. Parallel fibers are oriented longitudinally in the direction of the muscle. Fusiform muscles are tapered. Pinnate fibers are parallel but oriented at an angle to the action of the muscle.

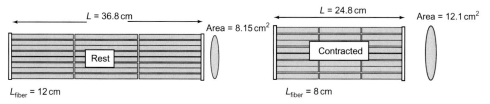

FIGURE 3.4.14 Geometry of a strap muscle with two inscriptions, fiber length = 12 cm and volume = 300 cm^2.

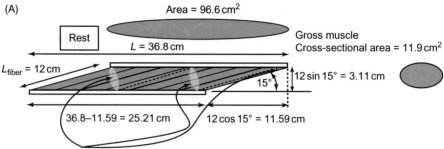

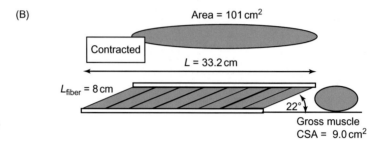

FIGURE 3.4.15 Geometry of a pinnate muscle fiber at rest (A) and contracted (B).

EXAMPLE 3.4.1 Calculate the Isometric Force and Maximal Velocity of a Parallel Muscle

We use the strap muscle shown in Figure 3.4.14 as an example. Its maximal force is its cross-sectional area times the force developed per unit cross-sectional area. Since its volume is 300 cm^3 and its length is 36.8 cm, its cross-sectional area is 300 cm^3/36.8 cm = 8.15 cm^2.

The typical isometric force per unit area is 20 N cm^{-2}. The isometric force of this muscle would be

$F = 8.15$ cm$^2 \times 20$ Ncm$^{-2} =$ **163 N**

Taking the unloaded contraction time as 0.033 s for the muscle fibers to contract from 12 to 8 cm, we get an unloaded velocity = (36.8 cm − 24.8 cm)/0.033 s = **364 cm s^{-1}**.

EXAMPLE 3.4.2 Calculate the Isometric Force and Maximal Velocity of a Unipinnate Muscle

An unipinnate muscle with the same volume and rest length as the strap muscle has fibers oriented at 15° to the direction of force. The fiber rest length is 12 cm; thus the aggregate cross-sectional area of the fibers is 300 cm^3/12 cm = 25 cm^2, which can also be obtained from the area of the aponeurosis × sin 15°. This is not the gross cross-sectional area of the muscle. The gross cross-sectional area can be estimated from the volume and length of the muscle. From the geometry in Figure 3.4.15, the overall length of the muscle is 36.8 cm, but each aponeurosis is 25.21 cm long; the thickness of the muscle is 3.11 cm. From these numbers, the area of the aponeuroses is 96.6 cm^2 and the gross cross-sectional area of the muscle is 11.9 cm^2. Total force generated by the fibers is the cross-sectional area of the fibers × the force per unit area = 25 cm$^2 \times 20$ N cm$^{-2} = 500$ N. This force is not directed along the lines of the aponeurosis; the force transmitted to the tendons will be 500 N × cos 15° = **483 N**.

The apparent isometric force of the muscle, using its gross cross-sectional area, is 483 N/11.9 cm^2 = 40.6 N cm^{-2}.

When the fibers shorten from 12 to 8 cm, the pinnation angle changes. The overall length of the muscle changes from 36.8 to 33.2 cm, or by 3.6 cm, in 0.033 s, giving an unloaded velocity of 3.6 cm/0.033 s = **109 cm s^{-1}**.

MUSCLES DECREASE FORCE UPON REPEATED STIMULATION; THIS IS *FATIGUE*

Everyday experience shows us that maximal effort can be sustained only briefly. The more intense the effort, the faster one fatigues. Intense efforts rely predominantly on fast-twitch fibers. These are generally larger than the slow-twitch fibers and belong to larger motor units, so that these are recruited last. This makes subjective sense, because these large muscle fibers in large motor units increase force in the greatest increments when they are recruited. These fast-twitch fibers are also more easily fatigued than the slow-twitch fibers. Fatigue of slow-twitch muscles takes longer to produce. Therefore, it makes sense that these slow-twitch fibers, which are smaller and belong to smaller motor units, are recruited early on and so are active almost every time the muscle is activated, even for tasks requiring little force. The differences in fatiguability of different muscles led Burke to propose a system of classification of muscles based on four types:

1. S = slow-twitch fibers
2. FR = fast, fatigue resistant
3. FI = fast, intermediate fatigue resistant
4. FF = fast, fatiguable.

Most muscles consist of thousands of muscle fibers. Most muscles contain all of the different fiber types, but they differ in their relative number. The soleus muscle in the human consists predominantly of slow-twitch muscle fibers, whereas the gastrocnemius consists mainly of fast-twitch fibers. However, there is considerable individual variation in the fiber types that are present in individual muscles. Transformation of muscle types appears to be limited and obeys a strict progression. Increased use of a muscle invariably tends to convert muscle fibers into more fatigue-resistant fibers, but conversion of fast fatigable fibers all the way to slow-twitch fibers does not occur.

SUMMARY

Skeletal muscles consist of thousands of muscle fibers which are large multinucleated cells. Each muscle fiber is controlled by a single motor neuron that forms a neuromuscular junction approximately in the middle of the fiber. Motor neurons branch and form junctions with one-to-many muscle fibers. All of the muscle fibers innervated by a single motor neuron constitute the motor unit. Large motor units typically have large motor neurons in the spinal cord, and small motor units have small motor neurons. The small motor neurons are more excitable and are recruited first.

A single action potential on a motor neuron produces a twitch. Brief twitches characterize fast-twitch muscles: they develop force rapidly and relax rapidly. Slow-twitch fibers develop force more slowly and relax more slowly. Muscle force can be graded three ways: (1) by recruiting more motor units; (2) by increasing the frequency of stimulation; and (3) by changing the length of the muscle. Changing the length of the muscle is not so important because muscles are restrained by the position of their origins and insertions on the skeleton. Recruitment follows the size principle: small motor units are recruited first. Increasing frequency of neuronal action potentials excites the muscle again before it has time to relax because the twitch is usually much longer than the action potential. Thus muscle force can summate when excitation frequency exceeds $1/t$, where t is the period of the twitch. The tetanic frequency is the frequency at which all waviness disappears from the force record. Typically tetanic force is about five times the twitch force. Recruitment is usually the most important way of grading muscle force and can be responsible for a 100-fold or greater range of muscle force.

Stretching a muscle produces a passive force that increases nonlinearly with stretch. The active force, the increment of force caused by excitation, changes biphasically with muscle length. It increases with increasing muscle length when the muscle is short, reaches a maximum, and then decreases with further stretch of the muscle.

The velocity of muscle shortening depends on the load. Maximal velocity occurs at zero load, and maximum force develops at zero velocity. The power output of muscle varies biphasically with muscle force, being maximal at about $1/3 F_{max}$. Both fast- and slow-twitch muscles show this behavior. Power output also varies with velocity, being maximal at about $1/3 V_{max}$. During rapid movements, almost all power derive from fast-twitch muscle.

Muscle activation can produce force while the muscle shortens, and this is called a concentric contraction. Isometric contractions refer to the activation of muscle under conditions in which it does not change length. In some cases, muscle activation develops a force while the muscle lengthens. This is an eccentric contraction. Concentric contractions are used to accelerate objects or body parts. Isometric contractions are used to fix joints in some configuration, whereas eccentric contractions are used to decelerate objects or body parts.

Muscle fibers often are oriented at an angle with respect to the direction of muscle action. This is called the pinnation angle. This allows more muscle fibers to fit into the volume occupied by the muscle and increases force but decreases muscle velocity.

Fatigue is the loss of muscle force due to prolonged use. There are two types of fatigue: rapid onset of fatigue brought about by continuous maximal stimulation and slower onset of fatigue brought about by repetitive but submaximal activation of the muscles.

Muscles can be classified on the basis of their contractile activities. Thus there are slow-twitch fibers, fast and fatigue-resistant fibers, fast intermediate (with respect to fatigue) fibers, and fast fatiguable fibers.

REVIEW QUESTIONS

1. What is a muscle twitch? What is tetany?
2. Name three ways that the body varies muscle force. Are all equally important physiologically?
3. What is a motor unit? What is recruitment? How is recruitment usually affected? How much stimulation can be achieved through recruitment?
4. What is tetany? What is the most rapid train of twitches possible without increasing force? By how much can force be increased by increasing the frequency of stimulation?
5. What is active tension? What is passive tension? How do they vary with muscle length? Is this an important way of varying muscle strength?
6. What is the relationship between force and velocity? Is a negative velocity of muscle contraction possible? If so, what do you call such a contraction?
7. What do we mean by "fast-twitch" and "slow-twitch" fibers? Why is muscle power higher in fast-twitch than in slow-twitch fibers?
8. What are concentric, isometric, and eccentric contractions? What are these used for?
9. What advantage does fiber pinnation afford? What is its disadvantage?

Contractile Mechanisms in Skeletal Muscle 3.5

Learning Objectives

- Explain why muscle fibers are multinucleated
- Distinguish among muscle fiber, myofibrils, and myofilaments
- Identify the various cross-striations as evidenced by electron microscopy including the A-band, I-band, Z-line, M-line, and H-zone
- Define the sarcomere
- Describe which filaments are found in which microscopic zone of the sarcomere
- Be able to show the relationship among the sarcoplasmic reticulum, T-tubules, and myofibrils
- Draw the length–tension curve using sarcomere length and explain the origin of its major points
- Describe the structure of the thick filament and explain the origin of the bare zone and its width
- Name the constituents of the thin filament and each of their functions
- Describe the polarity of thin filaments at the Z-disk
- Explain the function of the cross-bridge cycle and write a simplified reaction mechanism
- Describe how myosin isoforms generate muscle heterogeneity
- Define myosin turnover number and explain why it correlates with muscle speed
- Describe costameres and how force is thought to be distributed through the sarcolemma to the extracellular matrix

INTRODUCTION

Chapter 3.4 has shown us how the overall muscle behaves: muscles are heterogeneous with respect to contractile properties. They can be classified according to their twitch times, velocity of shortening, and resistance to fatigue: fibers can be slow, fast fatigue resistant, fast intermediate, and fast fatigable. Muscle force can be graded by the recruitment of motor units, by varying the frequency of motor neuron firing, and by varying the length of the muscle. Of these, recruitment offers control of the greatest range of force, frequency the next greatest, and length the least. There is an inverse relationship between velocity and force of shortening. The power of the muscle peaks at about one-third maximal force and at about one-third maximal velocity. What we seek now is some explanation of these overall behaviors in the subcellular and molecular description of muscle.

MUSCLE FIBERS HAVE A HIGHLY ORGANIZED STRUCTURE

Muscle fibers are typically large cells, some 20–100 μm in diameter and many centimeters long, with the longest fibers being about 12 cm. These cells are **multinucleated**, because they need many nuclei to govern protein synthesis and degradation. The nuclei are typically located near the periphery of the cell and often are more highly concentrated near the myoneural, or neuromuscular, junction. The most striking feature of muscle cells viewed under the light microscope is their banded appearance. The fibers have stripes, or **striations**, that result from the highly organized arrangement of proteins in the muscle fiber. These striations consist of alternating **A-bands** and **I-bands**, named because the I-bands are *isotropic* to polarized light (meaning that they appear the same from all directions) whereas the A-bands are *anisotropic* to polarized light. The cross-striations are perpendicular to the long axis of the muscle fiber. Figure 3.5.1 shows the microscopic appearance of frog skeletal muscle fibers using phase contrast microscopy.

Muscle cells are also striated longitudinally by the organization of contractile proteins into tiny threads called **myofibrils**. These are generally cylinders of material about 1 μm in diameter that also clearly show cross-striations. The myofibrils are kept in register across the entire cell to give rise to the cross-striated appearance. The electron micrograph in Figure 3.5.2 shows how the striations in the myofibrils line up across the cell.

Just as each muscle fiber contains many myofibrils, each myofibril is in turn composed of many filaments. These filaments come in two main varieties: the **thin filament** and the **thick filament**. The major constituent of the thin filament is **actin**; the main component of the thick filament is **myosin**. The microscopic striated appearance of the muscle is due to the way in which the filaments overlap each other.

The thick filaments define the beginning and end of the A-band. The myosin component of the A-band gives rise to the anisotropic behavior under polarized light. Because the thick filaments are 1.6 μm long, the A-band is also 1.6 μm long. Figure 3.5.3 shows a schematic illustration of the structure of the muscle fibers and myofibrils.

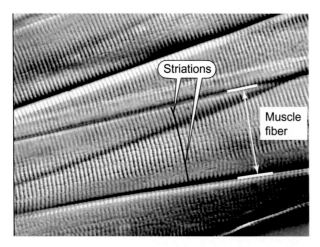

FIGURE 3.5.1 Microscopic appearance of skeletal muscle. A bundle of frog sartorius muscle fibers was teased out and viewed under phase contrast microscopy. Cross-striations are readily apparent in these unstained muscle fibers.

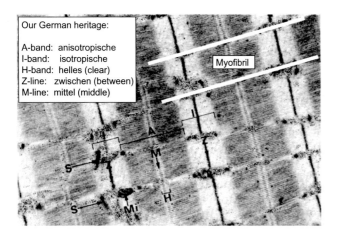

FIGURE 3.5.2 Electron micrograph of muscle. The spaces between myofibrils are filled with membranes of the sarcoplasmic reticulum, mitochondria, and glycogen granules. The myofibrils are bundles of filaments arranged longitudinally parallel to the long axis of the muscle fiber. The various bands are named according to their position, appearance, or by how they rotate the plane of polarized light.

The thin filaments are about 1.0 μm long but their length varies with muscle types and species. In human deltoid muscle biopsies, thin filament length averages 1.19 μm whereas in the pectoralis major it is 1.37 μm. Opposite thin filaments are connected, back to back, at a structure called the **Z-line** (from the German "zwischen" meaning "between"). Because the myofibrils are cylindrical, the Z-line is actually a disk of material and it is also called the **Z-disk**. The thin filaments typically overlap the thick filaments. The **H-zone** is a clearer area in the middle of the thick filaments. Its name derives from the German "helles," meaning "clear." This is the part of the A-band in which the thin filaments do not overlap the thick filaments. The thick filaments are connected in their center by material that forms the **M-line** (from the German "mittel" meaning "middle"; see Figures 3.5.2 and 3.5.3).

Electron micrographs show that the thick filaments form a hexagonal-centered lattice. The thin filaments also form a hexagonal lattice, but it is rotated 30° from the thick filament lattice. Each thick filament is in the center of a hexagon of thin filaments, whereas each thin filament is located equidistant from a triangle of three thick filaments. Thus each thin filament is surrounded by three thick filaments and each thick filament is surrounded by six thin filaments. In some electron micrographs, **cross-bridges** can be seen between the thick and thin filaments. The interaction of the filaments through these cross-bridges produces either shortening or force.

The functional unit of contraction or force production is the **sarcomere**, extending from one Z-disk to the next. The myofibrils consist of thousands of these sarcomeres strung end to end. The length of the sarcomere varies with muscle activation. Typically, the rest length is about 2.0–2.2 μm, depending on the length of the thin filament. This rest length is less than the thick filament (1.6 μm) plus two thin filaments (1.0–1.37 μm each) because of the overlap of the filaments at rest.

At regular intervals, the surface membrane of the muscle fiber, the **sarcolemma**, is invaginated to form a long tubule running perpendicular to the surface and penetrating into the farthest parts of the fiber's interior. These are the transverse tubules or **T-tubules**. The function of these T-tubules is to bring the action potential on the sarcolemma into the interior of the cell. The T-tubules allow for the rapid spread of the excitation to all parts of the sarcoplasm.

Adjacent to the T-tubules and surrounding each myofibril is a membranous network called the **sarcoplasmic reticulum** or **SR**. This organelle surrounds the myofibril like a loosely knit sweater surrounds your arm. It forms an internal compartment, the **lumen**, separate from the cytoplasm of the muscle fiber. The SR is divided into a **longitudinal SR** and **terminal cisternae**. The terminal cisternae are sacs that make contact with the T-tubules, whereas the longitudinal SR are thin tubes of membrane that connect the terminal cisternae from one side of the sarcomere to the other. The longitudinal SR and terminal cisternae are connected and form a single enclosed space. In skeletal muscle, the junction of the T-tubule and the SR forms a **triad**, because it is the junction of one T-tubule and two terminal cisternae. In mammalian skeletal muscle, the triads occur at the junction of the A-band and I-band, so that there are two triads per sarcomere. This arrangement of the T-tubules and SR junction varies with the species and muscle type. Figure 3.5.4 shows the anatomical relation of the SR to the myofibril.

THE SLIDING FILAMENT HYPOTHESIS EXPLAINS THE LENGTH–TENSION CURVE

One of the early observations of muscle contraction was that when a muscle shortens, the Z-disks move closer together but there is no change in the length of the A-band; all of the shortening appears to occur in the

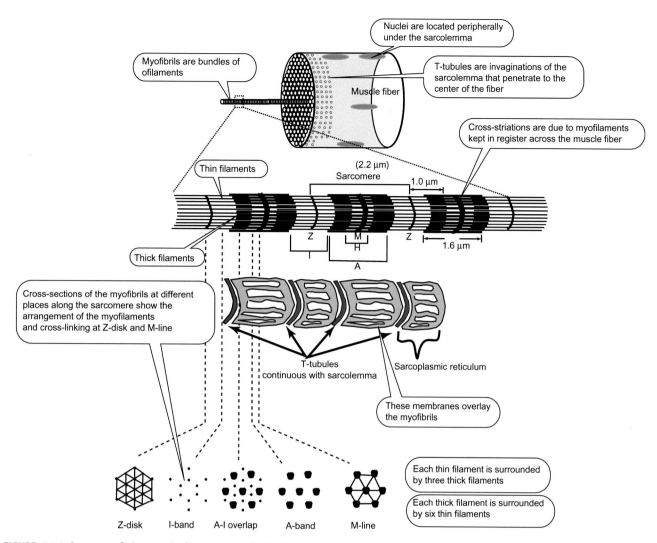

FIGURE 3.5.3 Structure of the muscle fiber and myofibrils. The A-band corresponds to the length of the thick filaments, 1.6 μm. The I-band corresponds to the thin filaments where they do not overlap with the thick filaments. Its width depends on the activation of the muscle. The Z-line or disk is where the thin filaments from opposite sarcomeres are attached. The M-line in the middle of the A-band keeps the thick filaments centered and in register. The clear zone in the middle of the A-band is the region where thin filaments do not overlap thick filaments.

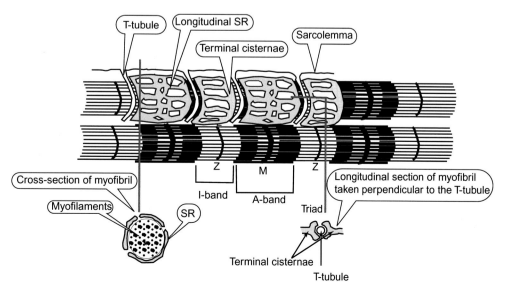

FIGURE 3.5.4 Structure of the SR around the myofibril.

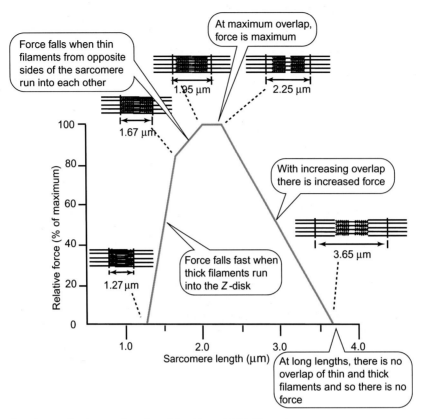

FIGURE 3.5.5 Dependence of tension on the degree of overlap of the thin and thick filaments.

I-bands. The sliding filament theory was first proposed by A.F. Huxley and R. Niedegerke in 1954 to explain these observations and the length–tension curve.

In Chapter 3.4, we showed that the active tension developed by a muscle is low at both extremes of length, and it peaks in the middle. This relationship also holds for individual muscle fibers where it is possible to measure the sarcomere length. The results are shown in Figure 3.5.5 for muscles with thin filaments 1.0 μm long. At long lengths of the muscle, the sarcomeres are proportionately long. At a sarcomere length of 3.65 μm, there is no overlap of the filaments (recall that the A-band is 1.6 μm and the I-band filament is 1.0 μm; thus the length of a sarcomere consisting of two I-band filaments and the A-band is 3.6 μm). At the point where the filaments do not overlap, the muscle cannot generate any force. At shorter lengths, the A-band and I-band filaments overlap increasingly so the force proportionately increases. At 2.25 μm, there is maximum overlap of the force generators and so the muscle generates maximal force. This force does not decrease until the sarcomere shortens to <1.95 μm. The reason for this is that there is still maximal overlap of those portions of the thick filaments that are engaged in force production: the thick filament has a central region that cannot produce force by interacting with the thin filament. When the overlap of the filaments increases, there is no further increase in the area of the thick filament that produces force. From the distances on the graph, we can infer that the bare zone is about 0.3 μm wide. When the sarcomeres shorten still further, the thin filaments begin to run into each other. This occurs at about twice the length of the thin filament or at about 1.95 μm. Although the thin filaments can slide past each other, they can do so only by using some of the force generated by the muscle. Thus the active force falls. At still shorter lengths, the thick filament begins to butt up against the Z-disk. This occurs at about the length of the A-band or 1.6 μm. At this point, further shortening seriously distorts the thick filament and force falls precipitously with further shortening.

FORCE IS PRODUCED BY AN INTERACTION BETWEEN THICK FILAMENT PROTEINS AND THIN FILAMENT PROTEINS

THE THICK FILAMENT CONSISTS PRIMARILY OF MYOSIN

Myosin is a complex of six proteins having a combined molecular weight of about 480,000 Da. It consists of two **heavy chains** of 200,000 Da each, and a total of two pairs of **myosin light chains**, with a molecular weight of about 20,000 and 16,000 Da. The two heavy chains have functionally distinct regions. They consist of a long, rod-shaped **tail** and a globular "**head**" region that contains the "business" end of the molecule. The head is attached to the tail by an arm section. This arm and its heads form the cross-bridges that interact with the thin filament to produce force. The tails of the two heavy chains are twisted together to form a supercoil.

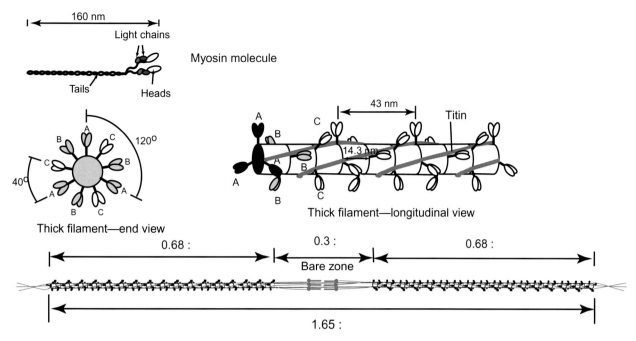

FIGURE 3.5.6 Structure of the thick filament.

Two distinct myosin light chains associate with each of the myosin heads. In smooth muscle, the activity of the myosin is regulated through phosphorylation and dephosphorylation of these light chains. In skeletal muscle, the phosphorylation state of these myosin chains is regulated, and appears to be responsible for "post-tetanic potentiation", in which isometric force increases immediately following a tetanic contraction.

The thick filaments also contain **titin**, a giant protein with a molecular weight of about 3.7 million Da. This protein is also called connectin and is the largest protein known to date. It is the third most abundant protein in skeletal muscle, comprising some 8–10% of the myofibrillar protein. The protein spans the distance from Z-disk to M-line and contains repeat regions for binding of α-actinin, a protein that anchors the thin filaments at the Z-disks. Titin also contains other repeat regions for binding myosin and M-protein, the one responsible for tying together the thick filaments at the M-line and myosin-binding protein C. Although uncertainties remain, the working hypothesis is that titin stabilizes mechanically active sarcomeres at the Z-disk, A-band, and M-line regions, and also serves in the cellular signaling of mechanical force to maintain the sarcomeres.

Myosin binds to titin and to other myosin molecules along its tail region. The myosin heads project out of the thick filaments at regular intervals of 14.3 nm. There appear to be three myosin molecules (six heads) at each plane oriented at right angles to the long axis of the filament. There is an identical repeat every 43 nm. Each thick filament has about 294 myosin dimers projecting from its surface (about 147 facing each of the neighboring Z-disks). Figure 3.5.6 shows myosin's basic structure. Because the myosin molecules lay themselves down along the titin template so that the tail regions point toward the middle of the filament, the middle of the thick filament contains only tails and the heads are located at the ends of the filament. Thus the thick filament is polarized with a "bare zone" in its middle. The titin molecule continues past the end of the myosin to make connections with other proteins in the Z-disk.

THE THIN FILAMENT CONSISTS PRIMARILY OF *ACTIN*

The actin in skeletal muscle is one of several types made by different cells. It can be isolated as a monomeric globular actin or **G-actin**. It has a molecular weight of 41,800 Da and a diameter of about 5.5 nm. Under proper conditions, G-actin will aggregate to form a filament or F-actin. This consists of two strands of actin molecules wound around each other. There are about seven G-actin molecules per half-turn, giving a half-turn distance of about 38.5 nm. Tropomodulin caps the filament at the minus end, which faces the M-line, and cap Z caps the filament at the plus end and anchors it at the Z-disk. The filament grows at the plus end but material can exchange at either end (see Figure 3.5.7).

Nebulin is another giant protein, with a molecular weight between 600,000 and 900,000 Da in different muscles. It makes up some 2–3% of the myofibrillar protein. Nebuin is anchored at its C-terminus at the Z-disk where it binds to cap Z, but it generally extends only 0.9 μm from the Z-line. It does not extend the entire distance of the thin filament. Human nebulin contains a string of 185 tandem repeats of 35 amino acids that are presumed to be actin binding domains. The central 154 of these repeats form 22 "super repeats" of seven units. These are thought to bind tropomyosin and are about 38.5 nm long, corresponding to one-half turn of the actin helix. The 22 repeats of 38.5 nm do not extend to the end of the filament. There are 2 nebulin molecules per thin filament, and these are proposed to occupy the major groove of the filament. The exact location of nebulin is not known. The function of nebulin is not yet settled, but it appears to stabilize the thin filament against depolymerization, but it also interacts with Z-disk proteins.

Thin filaments also contain **tropomyosin**, which consists of two nonidentical polypeptide chains, each with a molecular weight of about 33,000 Da. They are long, rod-shaped molecules, and tropomyosin α and β wrap around each other to form a supercoil. The tropomyosin complex is 38.5 nm long, the same size as a half-turn of the F-actin helix. Along with **troponin**, tropomyosin participates in the regulation of the active state of muscle.

Troponin consists of **troponin T**, a 37,000 Da protein, **troponin C** of 18,000 Da, and **troponin I** of 21,000 Da. These three derive their names from their functions: **TnT** binds to **tropomyosin**; **TnC** binds **calcium** ions (and thereby confers Ca^{2+} sensitivity onto the myofilaments), and **TnI inhibits** the interaction between the thick and thin filaments that cause force development or shortening. These three proteins form a complex at one end of each tropomyosin molecule. The troponin complex is responsible for the final regulation of the contractile state of the myofilaments.

α-ACTININ AT THE Z-DISK JOINS ACTIN FILAMENTS OF ADJACENT SARCOMERES

The Z-disk contains a number of proteins that can bind F-actin or other proteins found at the Z-disk. The precise way that these proteins align the thin filaments is not yet known. One of these proteins, α-actinin, consists of two subunits of 95,000 Da that is located in the Z-disk and anchors the thin filaments there. A proposed arrangement of the Z-disks shown in Figure 3.5.8 illustrates the complexity of these structures. Muscle cells ultimately connect the Z-disk to the surface membrane because the force generated by the myofilaments must eventually be transferred to the outside of the cell, through the sarcolemma.

MYOMESIN JOINS THICK FILAMENTS AT THE M-LINE OR M-BAND

Electron micrographs show a fine structure to the M-line that differs among muscle types. In particular, M-protein contributes to a central line that is present in fast-twitch

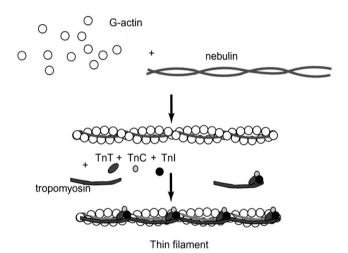

FIGURE 3.5.7 Structure of the thin filament. The actual location of nebulin is not known. The ends of the thin filament are shown. The plus, or barbed end, of the thin filament is anchored at the Z-disk. The minus, or sharp end, is typically in the A-band and is capped by tropomodulin.

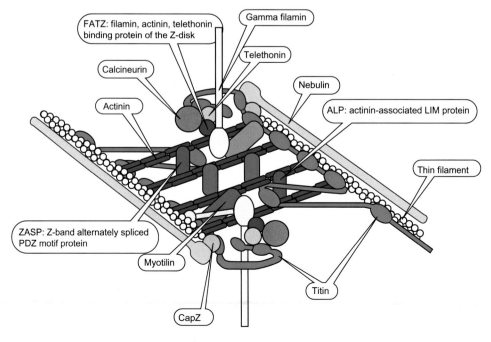

FIGURE 3.5.8 Proposed linkage of thin and thick filaments at the Z-disk. The thin filaments from adjacent sarcomeres have opposite polarity at the Z-disk. The Z-disk contains a variety of proteins that bind to other proteins in the disk. Gamma filamin is a cytoskeletal protein that links the Z-disk to the outside of the cell. Telethonin caps the titin filament. CapZ, a heterodimer, caps the plus or barbed end of the thin filament and also binds titin. Another name for FATZ is calsarcin, which binds telethonin, actinin, and filamin. ZASP is also named cypher and oracle. Actinin is an antiparallel homodimer. *Adapted from Faulkner, G., Lanfranchi, G., and Valle, G. Telethonin and other new proteins of the Z-disk of skeletal muscle. IUBMB Life **51**: 275–282, 2001.*

but not slow-twitch fibers. M-protein and another M-band protein, myomesin, share structural features, both consisting of a head region and 12 modules resembling fibronectin or IgG. Myomesin seems to be expressed in a fixed ratio to myosin in a variety of muscle types. Myomesin binds to myosin, titin, and to itself. A proposed structure for the M-band is shown in Figure 3.5.9.

OVERALL STRUCTURE OF THE SARCOMERE IS COMPLICATED

The overall structure of the sarcomere is shown in simplified form in Figure 3.5.10. The structures of the thin filaments, thick filaments, Z-disks, and M-line have already been reviewed in Figures 3.5.6–3.5.8 and 3.5.9, and their details are omitted in Figure 3.5.10. This figure emphasizes the disposition of the three giant proteins of the sarcomere: titin, nebulin, and obscurin. All of these proteins contain multiple repeats of immunoglobulin G-like and fibronectin-like domains. Titin extends from the M-line to the Z-line, but the portion in the I-band is differently constructed, consisting of a number of repeats that produce a spring-like behavior to the protein. Along with myomesin, it is thought that titin produces a restorative force after contraction and likely also centers the A-band in the sarcomere following force imbalances during contraction.

Nebulin is anchored in the Z-disk to Cap Z and it extends approximately 0.9 μm along the thin filament away from the Z-disk. In most skeletal muscles, this is not enough to reach the end of the thin filament. At the Z-disk, it is proposed that nebulin crosses from one thin filament to the one in the opposite sarcomere, binding only a short distance (0.1 μm or less).

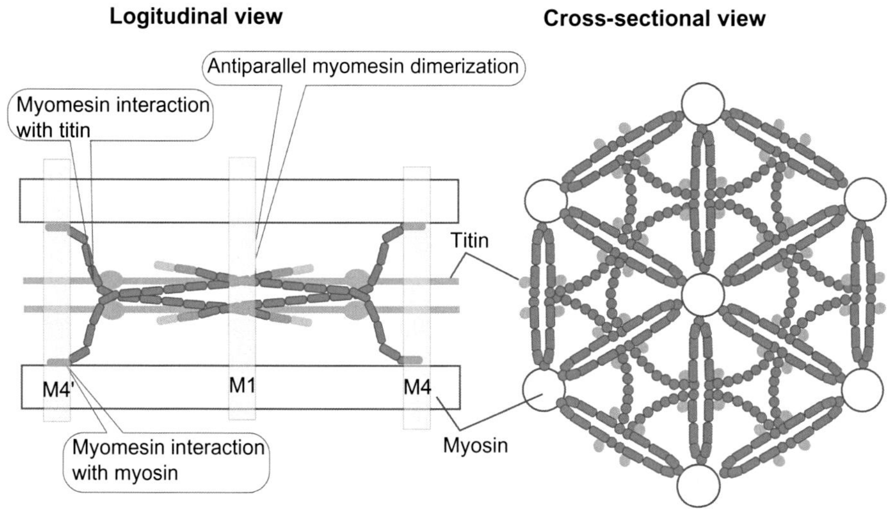

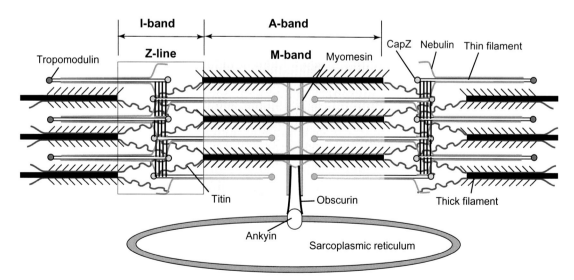

FIGURE 3.5.9 Proposed structure of the M-band. Myomesin binds to myosin, creating increased density on electron micrographs that make the M4 and M4′ lines. The dimerization of myomesin in the center of the A band contributes to the M1 line. Myomesin also binds to titin. Note that the central myosin is surrounded by 12 titin filaments. Six of these originate from the myosin filament on each side of the M-line. Thus, there are six titin filaments per myosin filament in each half sarcomere. Myosin is in white; titin is light blue, and myomesin is dark blue. *Adapted from Lange, S., et al., Dimerisation of myomesin: implications for the structure of the sarcomeric M-band. J. Mol. Biol.* **345**: 289–298, 2005.

FIGURE 3.5.10 Distribution of the giant proteins titin, nebulin, and obscurin in the sarcomere. Titin extends from the center of the M-band to the Z-line. It interacts with the thick filament in the A-band and contains spring-like domains in the I-band. It binds to multiple proteins in the Z-disk. It is thought to provide a restorative force following contraction, and also centers the A-band filaments in the A-band. Nebulin is anchored in the Z-disk and follows the thin filament nearly to its end. It is thought to stabilize the think filament against depolymerization. Obscurin binds to titin at the M-band and to ankyrin on the sarcoplasmic reticulum membrane, thereby helping to stabilize the SR along the outside of the myofibril.

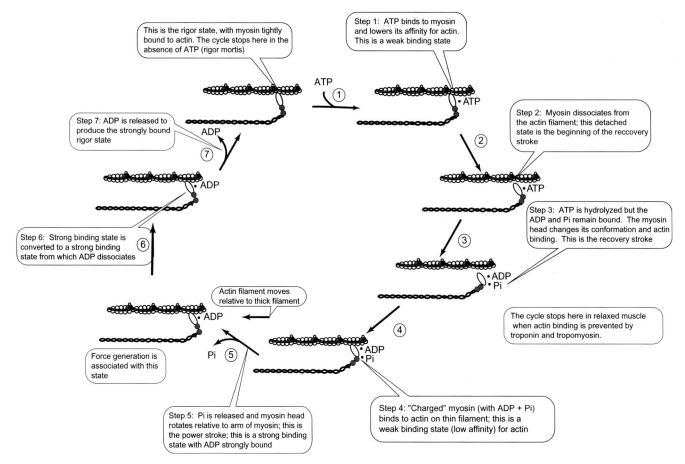

FIGURE 3.5.11 Reaction scheme for the acto-myosin ATPase cycle. Clockwise trace of the cycle shows ATP being converted to ADP + Pi over the entire cycle, steps 1—7. ATP binding to the rigor state in step 1 reduces the affinity of M for A, producing M* that dissociates from actin in step 2. ATP bound to M* is then hydrolyzed and the myosin head engages in a "recovery stroke" that re-establishes its resting position as M**. This form reattaches to A in step 4 and in step 5 it releases Pi and engages in the "power stroke" that moves the actin filament relative to the myosin filament. Here the complex is A•M*•ADP in which myosin has strong affinity for A and for ADP. In step 6, this is converted to A•M•ADP where myosin has strong affinity for A but from which ADP can dissociate. In step 7, ADP dissociates to form the rigor state where the cycle first began.

Obscurin (about 800 kDa) binds to titin and to ankyrin, an integral protein on the surface of the sarcoplasmic reticulum membrane. It is the only protein known to anchor membranes of the SR to the myofibrils. Obscurin is believed to be restricted to the surface of the myofibrils rather than penetrate through them as titin and nebulin do. It is located at the Z-line and M-band of developing muscle but seems to be restricted to the M-band in adult muscle.

CROSS-BRIDGES FROM THE THICK FILAMENT SPLIT ATP AND GENERATE FORCE

The interaction of myosin and actin has two consequences: hydrolysis of ATP to ADP and Pi, and the generation of force and/or shortening. The mechanism by which this happens can be viewed as a limited number of discrete states whose interconversions are governed by rate constants. There are several proposals for the overall mechanism of the acto-myosin ATPase, and a simplified form is shown in Figures 3.5.11 and 3.5.12. The overall reaction is called the **cross-bridge cycle**.

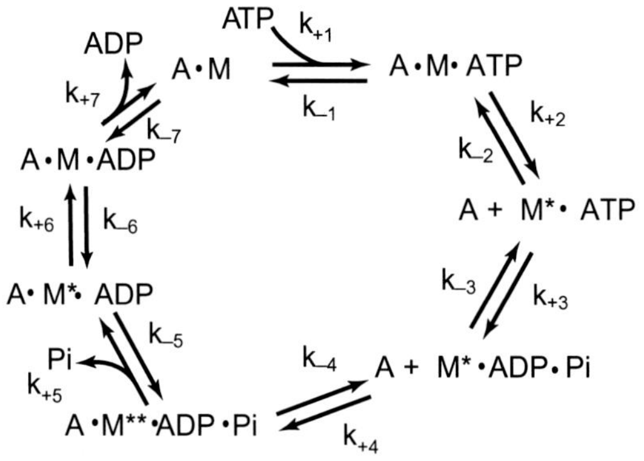

FIGURE 3.5.12 Alternate view of the acto-myosin reaction mechanism. Begin with A•M, the rigor state. Binding of ATP occurs first in step 1, which converts myosin to M* that has reduced affinity for actin and dissociates form A in step 2. ATP is hydrolyzed in step 3, causing strong binding to the actin filament in step 4. In step 5, phosphate (Pi) is released and myosin changes its conformation in the power stroke. In this state, myosin binds actin and ADP strongly. In step 6, this is converted to a myosin state that can dissociate ADP, which occurs in step 7. Each reaction step has forward and reverse rate constants that describe the rates of conversion in the forward and reverse partial reactions. A complete cycle results in the hydrolysis of ATP and the movement of the thin filament relative to the thick filament.

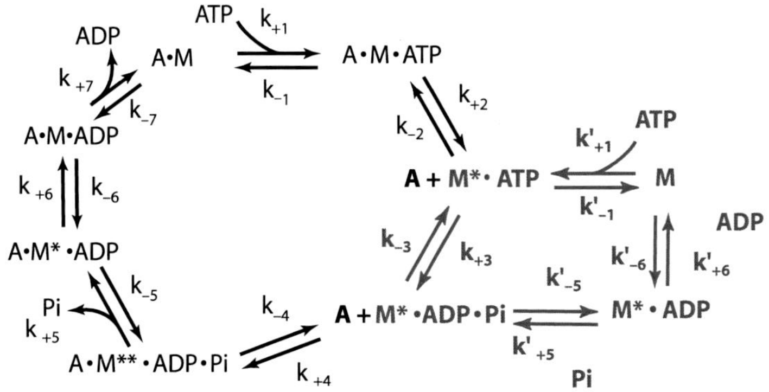

FIGURE 3.5.13 Myosin ATPase activity in the absence of actin. The acto-myosin cross-bridge mechanism is shown in black. Myosin-catalyzed ATPase activity is shown in blue. The rate constants for the transitions of myosin in the absence of actin are smaller than those in the presence of actin, particularly $k'_{+5} \ll k_{+5}$, so that ATPase activity in the absence of actin is much slower than in its presence: myosin is an actin-activated ATPase.

The most important concept of the acto-myosin cross-bridge cycle is that ATP hydrolysis is linked to sequential changes in the conformation, or 3D structure, of the myosin head and lever arm. This allows myosin to "walk" along the actin filament. Since myosin is connected to the thick filament, this movement of the myosin head corresponds to a movement of the thick filament relative to the thin filament.

The reaction scheme shown in Figures 3.5.11 and 3.5.12 is simplified and therefore incomplete. Myosin can split ATP all by itself, without actin, and this reaction is shown added on to the acto-myosin reaction scheme in Figure 3.5.13. However, the rate of ATP hydrolysis without actin is greatly reduced compared to the reaction with myosin in the presence of actin. Myosin is an actin-stimulated ATPase. Myosin binds to actin and this alters the rate of product release so that overall ATP hydrolysis is stimulated some 200- to 300-fold.

MYOSIN HEADS ARE INDEPENDENT BUT MAY COOPERATE THROUGH STRAIN ON THE CROSS-BRIDGE

Each myosin motor spends most of the time in the detached state and takes only one step along the actin filament before detaching again. The fraction of the time spent attached is generally referred to as the "duty cycle," symbolized as f, for the myosin motor. The efficient generation of force or shortening relies on a large number of these motors working together. Most models of muscle force generation assume that the two myosin heads work independently, meaning that the state of one myosin head does not directly affect the kinetics of any step involving a second myosin head. However, some of the reaction rates depend on the stress on the filaments, which in turn produces a strain. Because one myosin motor produces a strain on the thin and thick filaments, such strains can be communicated to nearby myosin heads and alter its kinetics through the effects of the strain.

For example, during eccentric contractions (contractions in which the muscle lengthens), the acto-myosin cross-bridge cycle is not completed. Instead, the myosin cross-bridges are strained and broken without the release of ADP—the myosin head detaches from the actin filament without releasing ADP. This transition is not shown in the reaction scheme of Figure 3.5.12. A consequence of this transition is that ATP hydrolysis is slowed during eccentric contractions. The value of k_{+6} in Figure 3.5.12 is believed to be dependent on the degree of strain, x. For low x, such as occurs in unloaded shortening, k_{+6} is large; for intermediate values of x, k_{+6} has intermediate values; for large x, k_{+6} is believed to be small. The consequence is that under loaded conditions the myosin head remains in the A•M*•ADP state for much longer. The duty ratio, f, could be 0.05 during unloaded shortening but f is much higher during eccentric or isometric contractions.

CROSS-BRIDGE CYCLING RATE EXPLAINS THE FIBER-TYPE DEPENDENCE OF THE FORCE—VELOCITY CURVE

The shortening of a muscle is the result of the shortening of its constituent sarcomeres. Thus if one sarcomere shortens by a distance Δx, and a sarcomere in series with it shortens by the same amount, the total shortening is $2 \times \Delta x$. Since both of these sarcomeres shorten in the same time interval, Δt, the velocity of shortening is $\Delta x / \Delta t$ for one sarcomere and $2 \times \Delta x / \Delta t$ for two sarcomeres in series. Thus the velocity of shortening is directly related to the number of sarcomeres in series. The velocity of shortening of a muscle for any given load should depend on its length.

Each cross-bridge cycle makes a single step estimated from 4 to 11 nm along the thin filament. Rapid turnover of the cross-bridge means that more of these cycles occur per second, and therefore the thin filament slides past the thick filament more quickly. Thus the rate of shortening of each sarcomere, and therefore of the entire muscle, depends on the **turnover rate** of the cross-bridges. The kinetics of the cross-bridges depends in large part on the kinetics inherent in myosin. Muscles make a variety of different **isoforms** of myosin. These are encoded by separate genes. In the adult skeletal muscle, there are two basic varieties: slow myosin and fast myosin that differ in their heavy chains. The slow type I fibers have MHCIβ, fast type IIA fibers have MHCIIa, and type IIB fibers have MHCIIb. Intermediate fibers have MHCIId. The catalytic mechanisms for the two

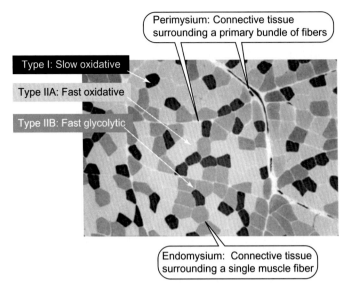

FIGURE 3.5.14 Histological staining of myosin reveals muscle heterogeneity. Myosin staining differentiates among various muscle types. This histological section shows three well-defined classes of staining but several intermediate fibers are also evident.

types of myosin, slow and fast, are similar. However, the turnover number (the number of completed reactions per second) is about $10\,s^{-1}$ for the fast myosin and $3\,s^{-1}$ for slow myosin. The kinetics of the cross-bridge cycling explains the different speeds of contraction of slow-twitch and fast-twitch muscle fibers.

The different myosin isoforms have different catalytic properties even though the basic mechanism is similar. These isoforms can be distinguished by staining for myosin ATPase activity by varying the incubation conditions such as the pH. On this basis, a variety of different staining patterns can be observed. Brooke has classified muscles as type I, type IIA, type IIB, and type IIC on the basis of myosin staining. The type I is the slow type of fiber and type II refers to different fast-twitch fiber types. Two types of myosin staining are available. The histological method uses differences in incubation conditions whereas myosin isoforms can be stained immunologically using antibodies directed against specific regions that vary among the isoforms. Figure 3.5.14 illustrates muscle heterogeneity as evidenced by myosin staining.

The force−velocity curve is also explained by the dependence of the reaction rate constants on strain. As force developed by the muscle increases, strain increases, and the duty cycle, the part of the time myosin stays attached, increases. This results in a decrease in turnover and a slower shortening velocity.

FORCE IS TRANSMITTED OUTSIDE THE CELL THROUGH THE CYTOSKELETON AND SPECIAL TRANSMEMBRANE PROTEINS

So far we have discussed the mechanism of force development or shortening as coming from the interaction of the thin filament protein actin, with the thick filament protein myosin. This force is *inside* the muscle fiber, whereas the force that is measured from muscle is *outside* the fibers. How does the force get outside? The short answer is this: the force of the myofilaments is transmitted laterally to the cytoskeleton elements, which then binds to discrete protein complexes at the **costameres**. Parts of this macromolecular complex also bind extracellular matrix proteins such as collagen and perlecan. Thus the myofilaments make indirect contact with the extracellular matrix through the cytoskeleton and the costameres.

Costameres were named because of their rib-like arrangement, revealed by the staining of the sarcolemma membrane with antibodies directed against specific proteins. These costameres are located at the Z-disk and M-line. In striated muscle, the myofilaments are connected to the costameres using either desmin intermediate filaments or actin microfilaments. There are three main types of complexes within costameres: (1) the **focal adhesion complex** involving transmembrane **integrins** linked to desmin through **talin** and **vinculin**, and linked to extracellular matrix fibers through **laminin**; (2) the **dystrophin−glycoprotein complex (DGC)** involving dystrophin to link γ actin on one end with syntrophin on the other, and a complex involving a number of proteins including α **dystroglycan** and β **dystroglycan** to link syntrophin to laminin, which makes final contact with the extracellular matrix; (3) a **spectrin-based filament network** making contact with membrane proteins through ankyrin. The structures of these costameres are still being worked out, and many more proteins than those listed here participate in it. Notable among these is **dystrophin**, a 427-kDa protein that is absent in **Duchenne muscular dystrophy**. This protein is localized to the periphery of the muscle fibers, on the cytosolic side of the sarcolemma, and binds to cytoskeletal elements. Although it is concentrated at the costameres it is not restricted to this

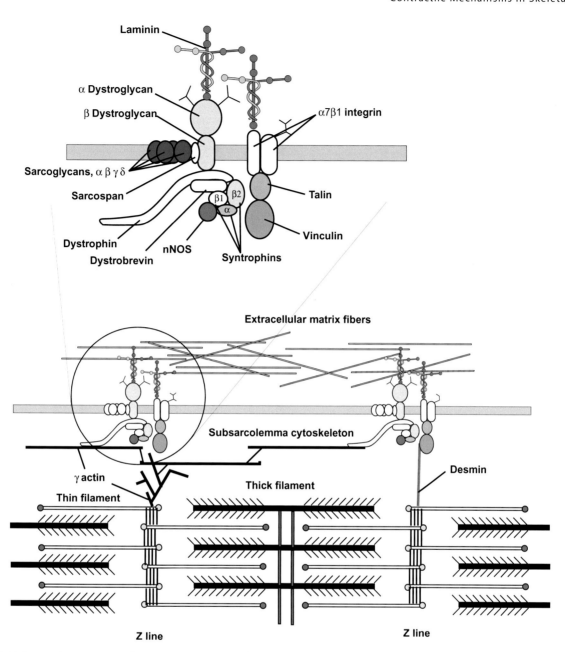

FIGURE 3.5.15 Proposed arrangement of proteins in costameres of skeletal muscle. The dystrophin–glycoprotein complex (DGC) consists of cytosolic proteins, transmembrane proteins, and extracellular proteins that link the myofilaments to the extracellular matrix through cytoskeleton filaments. In the DGC, Z-lines are linked to dystrophin through γ actin cytoskeletal filaments. Dystrophin binds to dystrobrevin and three syntrophins to form a complex with nNOS, neuronal nitric oxide synthetase. This complex thus participates in signaling of force either coming in (outside-in) or out (inside-out). Dystrophin also connects to β dystroglycan, a transmembrane protein that binds sarcospan and a complex of 4 sarcoglycans. The branched addendums to the proteins in the cartoon signify sugar residues added to the proteins. β dystroglycan forms a heterodimer with α dystroglycan, which binds laminin. Laminin is a heterotrimer that forms a cross-shaped protein that binds to collagen and other extracellular matrix proteins. The second major complex is connected to the Z-disks through desmin, an intermediate filament. This binds to vinculin, which binds talin, that binds the integral membrane protein, integrin. Integrin is a heterodimer that can be present in a variety of isoforms. The one in costameres is composed of α7 and β1. The integrin also binds laminin, which anchors the structure in the extracellular matrix. Integrins are associated with FAK (focal adhesion kinase) and ILK (Integrin-linked kinase), not shown.

location. The hypothetical disposition of cytoskeletal elements at the costameres is shown in Figure 3.5.15.

The function of the costameres is not yet established. The possibilities include: (1) transmit force from the contractile filaments to the outside of the cell (inside-out); (2) transmit force from the outside of muscle to the interior, involving signaling mechanisms (outside-in); (3) mechanically support the sarcolemma to protect it against damage, particularly during eccentric contractions; and (4) maintain uniform sarcomeric spacing between resting and active fibers of different motor units.

Clinical Applications: Muscular Dystrophy

The term "muscular dystrophy" includes a wide variety of inherited disorders characterized by progressive muscle weakness and wasting. The primary defect is in the muscle itself rather than in the nerve or neuromuscular junction. Part of this wide diversity of the disorder is summarized in Table 3.5.1.

What this list does is notify the observant student that there is much more to muscle than what we have discussed in the chapter. Some of the proteins in this list are not even mentioned in the chapter because a complete accounting of everything about muscle would occupy far too much space.

A single model to explain all of these muscular dystrophies does not yet exist. Many of the proteins whose defects cause muscular dystrophy are involved in linking the myofilament force generators to the extracellular matrix, and so they may be involved in the transmission of force or in protecting the integrity of the cell when exposed to large forces. They may also be involved in cellular signaling, particularly through nNOS, neuronal nitric oxide synthetase. Other proteins such as myotilin and telethonin may be involved in the integrity of the myofilaments themselves, rather than linking myofilaments to extracellular matrix. How mutations in the nuclear proteins emerin and lamin A/C can cause muscular dystrophy is less well understood. It may be that they participate in a larger karyocytoskeletal network (H.J. Spence, Y.-J. Chen, S.J. Winder, Muscular dystrophies, the cytoskeleton and cell adhesion, *Bio Essays* 24:542–552, 2002).

TABLE 3.5.1 Partial Listing of Muscular Dystrophies, with Their Mode of Inheritance, Gene Product, and the Normal Subcellular Locale of the Gene Product

Name	Mode of Inheritance	Gene Product	Subcellular Localization
Duchenne	XR	Dystrophin	Cytoskeleton
Emery–Dreifuss	XR	Emerin	Nucleus
Emery–Dreifuss	AD	Laminin A/C	Nucleus
LGMD 1A	AD	Myotilin	Myofilaments
LGMD 1B	AD	Lamin A/C	Nucleus
LGMD 1C	AD	Caveolin 3	Sarcolemma
LGMD 1E	AD	Filamin 2	Cytoskeleton
LGMD 2A	AR	Calpain 3	Cytosol
LGMD 2B	AR	Dysferlin	Sarcolemma
LGMD 2C	AR	γ-Sarcoglycan	Sarcolemma
LGMD 2D	AR	α-Sarcoglycan	Sarcolemma
LGMD 2E	AR	β-Sarcoglycan	Sarcolemma
LGMD 2F	AR	δ-Sarcoglycan	Sarcolemma
LGMD 2G	AR	Telethonin	Myofilaments
Congenital MD	AR	Laminin α2	Extracellular matrix

XR, X-linked recessive; AD, autosomal dominant; AR, autosomal recessive; LGMD, limb-girdle muscular dystrophy.

SUMMARY

Muscle fibers are large multinucleated cells whose most obvious histological feature is cross-striations. The cytoplasm contains many myofibrils—tiny cylinders consisting of bundles of myofilaments. The myofilaments include thick filaments, composed mainly of myosin, and thin filaments composed mainly of actin. The thick and thin filaments cause the cross-striations because of their regular overlap that is kept in register all across the diameter of the muscle fiber. The A-band corresponds to the thick filament. The I-band forms where the thin filament does not overlap the thick filament. Z-disks, from the German "zwischen," meaning "between," are centered in the I-band. The functional unit of contraction is the sarcomere, which extends from one Z-disk to the next. The M-line is in the middle of the A-band and corresponds to proteins that keep thick filaments in register and bound at their middle. It is surrounded by the H-zone, which is a clear zone corresponding to the part of the A-band where thin filaments do not overlap. During contraction, the A-band stays the same width whereas the I-band and H-zone both shrink.

The thick filament consists largely of myosin. Six proteins make up myosin: two heavy chains whose tails intertwine to form a supercoil and whose heads contain actin binding sites and a catalytic site for

ATP hydrolysis. Two myosin light chains bind to each head region. These myosin molecules assemble on a titin scaffold with their tails pointing toward the M-line. Because of this, there is a bare zone about 0.3 μm in the center of the thick filament that has no myosin heads.

Globular actin assembles to form a linear chain of filamentous actin. Two chains of F-actin intertwine to form the thin filament. The giant protein nebulin may set the length of the filament. Each half-turn of the filament binds a complex of tropomyosin and troponin. Troponin itself is a complex of: troponin T that binds tropomyosin; troponin I that inhibits actin availability to the thick filament; and troponin C that binds Ca^{2+}. Calcium binding to troponin C regulates the contractile state of the muscle.

Myosin is an actin-activated ATPase. When actin is available, myosin with ADP and Pi bound to it binds to the actin, followed by a release of ADP and Pi and a conformational change in the orientation of the head. Because myosin is bound to the thin filament, the conformational change exerts a force on the thin filament. ATP binding causes the head to dissociate from the actin, and subsequent hydrolysis of ATP to ADP and Pi converts the myosin back to its original form that can bind another actin. This cross-bridge cycle hydrolyzes ATP and develops force or shortens the sarcomere.

Myosin exists in multiple isoforms that can be distinguished by myosin staining. Slow-twitch muscles have myosin I and fast-twitch muscles have isoforms of myosin II. They differ in their turnover numbers. Faster cross-bridge cycling causes faster shortening.

The whole point of muscles is to deliver force to the outside of the muscle. This is accomplished by transferring the force to cytoskeletal elements that connect the internal myofilaments to special proteins that penetrate the membrane. These proteins transfer the force to the extracellular matrix at special regions called costameres. Dystrophin, the protein lacking in some kinds of muscular dystrophy, is concentrated at the costameres.

REVIEW QUESTIONS

1. What causes the cross-striations in skeletal muscle?
2. How large are muscle fibers? Why do they have lots of nuclei? Where are the nuclei located?
3. What is a myofibril? What are thick filaments? Thin filaments? What is the A-band? I-band? H-zone? Z-disk? M-zone? What happens to each during contraction?
4. Why is the top of the length–tension curve flat?
5. What proteins are found on the thick filament? Thin filament? What anchors the filaments?
6. What is the cross-bridge cycle? What hydrolyzes ATP?
7. What determines the speed of contraction? What is responsible for the different speeds of fast- and slow-twitch muscles?
8. How does muscle force, generated by the myofilaments, get out of the muscle fiber?

3.6 The Neuromuscular Junction and Excitation–Contraction Coupling

Learning Objectives

- Identify the major structural features of a motor neuron
- Distinguish between EPSP and IPSP
- Describe how the strength of EPSPs and IPSPs varies with time and distance
- Describe spatial and temporal summations
- Identify the major structural features of the neuromuscular junction
- List in chronological order the events of neuromuscular transmission
- Explain why the end-plate potential is not an action potential
- Define excitation–contraction coupling
- Identify the source of activator Ca^{2+} in E–C coupling
- List the sequential events in E–C coupling including the names of the major proteins
- Describe how the Ca^{2+} transient ends by reuptake of activator Ca^{2+} by SERCA
- Describe how Ca^{2+} binding to TnC activates cross-bridge cycling, include tropomyosin, TnI, and TnT
- Draw the relative time course of the action potential, Ca^{2+} transient, Ca^{2+} binding to TnC, and force
- Explain why force lags behind $[Ca^{2+}]$ and why it persists when $[Ca^{2+}]$ returns to normal
- Define tetany and tetanic frequency
- Explain how repetitive stimulation causes force summation

MOTOR NEURONS ARE THE SOLE PHYSIOLOGICAL ACTIVATORS OF SKELETAL MUSCLES

As described in Chapter 3.2, alpha motor neurons reside in the ventral horn of the spinal cord. They process synaptic inputs leading up to the production of an action potential in the motor neuron. The action potential propagates down the single axon toward the muscle cell where it makes a junction, variously called the neuromuscular junction, motor end plate, or myoneural junction. The process of conversion of the action potential on the motor neuron to an action potential on the muscle fiber is called neuromuscular transmission. The action potential on the muscle fiber then activates the muscle by a process called excitation–contraction coupling. This chapter covers all of the events from the formation of the action potential on the motor neuron to initiation of contraction in the muscle fiber.

THE MOTOR NEURON RECEIVES THOUSANDS OF INPUTS FROM OTHER CELLS

The motor neuron receives literally thousands of inputs from sensory cells, other cells in the spinal cord, or from higher centers in the central nervous system, and all of these inputs are integrated at the level of the motor neuron to determine its activity. These connections all occur through synapses or gaps between the cells. Upon excitation, the presynaptic cell releases a neurotransmitter which then binds to a receptor on the postsynaptic cell to eventually modulate its conductivity to ions. Neurotransmitters that directly alter ion conductances are called ionotropic. Other metabotropic neurotransmitters alter other targets within the postsynaptic cell that indirectly alter membrane conductances. Ionotropic neurotransmitters typically have effects with rapid onset and rapid decay, whereas metabotropic neurotransmitters generally take longer to have their full effect but the effect lasts longer.

POSTSYNAPTIC POTENTIALS CAN BE EXCITATORY OR INHIBITORY

The effect of ionotropic or metabotropic neurotransmitters depends on the receptors in the postsynaptic cell. If binding of neurotransmitter causes an increase in conductance to Na^+ ions, then the postsynaptic cell becomes depolarized. Because depolarization leads to excitation, this depolarization is called an excitatory postsynaptic potential or EPSP. Glutamate that is released from presynaptic terminals can bind to at least four different kinds of receptors: the kainate, AMPA, and NMDA receptors are all ionotropic, and mGluR is metabotropic. Each of these types exists in multiple isoforms. The kainate, AMPA, and NMDA receptors are named for the artificial agonists kainate, κ-amino-3-hydroxy-5-methyl/4-isoxazole propionic acid, and N-methyl D-aspartic acid, respectively. Binding of glutamate to either the kainate or AMPA receptors causes an increase in g_{Na} and g_K, resulting in EPSPs. Activation of the NMDA receptor by glutamate increases g_{Ca}, resulting in an influx of Ca^{2+} ions and depolarizing the cell.

© 2017 Elsevier Inc. All rights reserved.
DOI: http://dx.doi.org/10.1016/B978-0-12-800883-6.00029-X

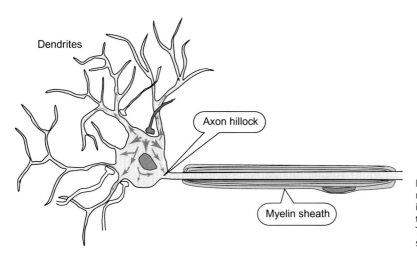

FIGURE 3.6.1 Spread of postsynaptic potential through a motor neuron. The postsynaptic potential intensity is indicated by the size of the arrows. It initiates at a terminal on the soma and spreads through the cytoplasm. The intensity falls off with distance away from the site of stimulation (the synapse) and also decays with time.

Certain other neurotransmitters bind to receptors that selectively increase the conductance of the postsynaptic membrane to Cl^- or K^+. Increased g_K decreases the membrane potential (makes it more negative) toward E_K. This hyperpolarization makes it more difficult to excite the membrane to form an action potential: it is called an **inhibitory postsynaptic potential** or **IPSP**. An example of a typical inhibitory neurotransmitter is **GABA** or gamma-aminobutyric acid. GABA binds to ionotropic $GABA_A$ receptors, which increase g_{Cl} and lead to an IPSP. GABA can also bind to metabotropic $GABA_B$ receptors that indirectly increase g_K to lead to an IPSP. Other examples of neurotransmitters are discussed in Chapter 4.2.

POSTSYNAPTIC POTENTIALS ARE GRADED, SPREAD ELECTROTONICALLY, AND DECAY WITH TIME

Both IPSPs and EPSPs can vary in magnitude, as opposed to the all-or-none character of the action potential. Their sizes depend on the number of neurotransmitter molecules that are released from the presynaptic terminal and on the number and kinds of receptors on the postsynaptic membrane. The postsynaptic potentials spread throughout the cell electrotonically or passively. As they spread, their magnitude decays with distance from the synapse because some of the current travels across the membrane as capacitative current. Because cytoplasmic resistance is little, EPSPs and IPSPs change little in the soma, but decay much more rapidly in the narrow dendrites because of their higher resistance. These EPSPs and IPSPs also decay with time, lasting 15–20 ms. Both are transient changes in the membrane potential (see Figure 3.6.1).

ACTION POTENTIALS ORIGINATE AT THE INITIAL SEGMENT OR AXON HILLOCK

The initial segment of the axon shows a clear area in histological sections and forms the most excitable part of the motor neuron because this area of the cell membrane contains high concentrations of the regenerative Na^+ channels. This part of the cell has the lowest threshold and for this reason it usually originates action potentials. Whenever this part of the neuron reaches threshold, an action potential is fired that propagates backward over the soma and forward down the axon toward the muscle fiber.

MOTOR NEURONS INTEGRATE MULTIPLE SYNAPTIC INPUTS TO INITIATE ACTION POTENTIALS

Usually a single EPSP on the motor neuron is insufficient to excite the trigger zone of the initial segment enough to reach threshold. Simultaneous stimulation of several excitatory connections can summate at the initial segment to produce a suprathreshold potential. The addition of simultaneous excitatory or inhibitory postsynaptic potentials is called spatial summation, because the different inputs originate at different locations on the motor neuron surface. Because EPSPs and IPSPs last some 15–20 ms, repetitive stimulation at a single synapse can cause repetitive EPSPs or IPSPs that can also add. This is called temporal summation, because the postsynaptic potentials are summed over time (see Figure 3.6.2). Spatial and temporal summations can occur together, so that repetitive stimulation of multiple inputs can be summated both temporally and spatially. A spinal motor neuron typically has 10,000 synaptic inputs, about 8000 on the dendrites and 2000 or so on the soma. The motor neuron integrates all of these inputs.

THE ACTION POTENTIAL TRAVELS DOWN THE AXON TOWARD THE NEUROMUSCULAR JUNCTION

Depolarization of the axon membrane in the middle of the axon initiates an action potential that propagates in both directions away from the stimulus. But under normal conditions, the action potential does not begin in the middle of the axon. Under normal physiological conditions, the action potential is unidirectional, beginning at the motor neuron soma and traveling down the axon toward its target, the muscle fiber.

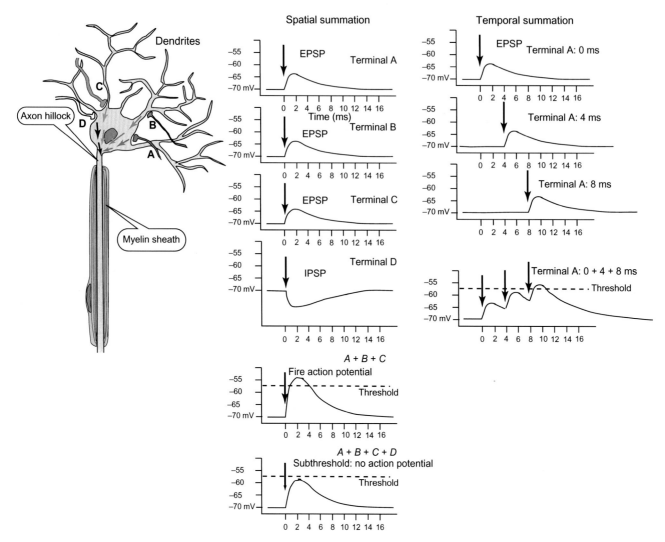

FIGURE 3.6.2 Hypothetical scheme illustrating spatial and temporal summations. Consider a motor neuron with four inputs. Inputs A, B, and C are excitatory inputs that produce EPSPs when excited, and D is an inhibitory input that produces an IPSP. Firing of A, B, or C alone results in EPSPs that are subthreshold. Simultaneous firing of A, B, and C summates to produce a suprathreshold potential that produces an action potential in the motor neuron. Simultaneous firing of A, B, C, and D is below threshold because the IPSP of connection D lowers the postsynaptic potential. This is spatial summation. Even though firing of A is insufficient to reach threshold, repetitive firing of A can summate to produce suprathreshold postsynaptic potentials. Summation of the postsynaptic potentials in time is temporal summation.

THE NEUROMUSCULAR JUNCTION CONSISTS OF MULTIPLE ENLARGEMENTS CONNECTED BY AXON SEGMENTS

As discussed in Chapter 3.4, typically a single motor neuron branches many times and each branch connects to a single muscle fiber. The total number of muscle fibers innervated by a single motor neuron is its **innervation ratio**, which can vary from 5 to 2000. Each motor neuron with its set of muscle fibers makes up a motor unit, whereas usually each muscle fiber typically is innervated by only one motor neuron. The junction of the motor neuron axon with the muscle forms a raised area on the surface of the muscle called a **motor end plate**. The axon breaks up into multiple branches, with enlarged areas called terminal boutons where neurotransmission takes place. Typically there are multiple terminal boutons in each motor end plate (see Figure 3.6.3).

NEUROTRANSMISSION AT THE NEUROMUSCULAR JUNCTION IS UNIDIRECTIONAL

Under normal conditions, an action potential on a motor neuron *always* results in an action potential on the muscle fiber, but an action potential begun on a muscle *never* results in an action potential on the nerve. The neuromuscular junction is designed for 100% fidelity and unidirectionality. This unidirectionality is assured by the highly asymmetric structure of the junction, as shown in Figure 3.6.4. The neuromuscular junction is a chemical synapse in which the action potential on the nerve is converted to a chemical signal, the release of the neurotransmitter acetylcholine. The

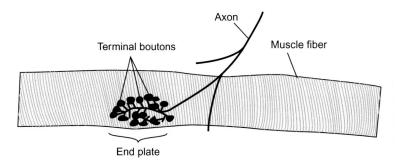

FIGURE 3.6.3 Motor end plate or neuromuscular junction. Motor neuron axons typically branch multiple times, and each branch ends in a neuromuscular junction or motor end plate. The end plate forms a raised area on the surface of the muscle fiber that involves multiple short branches of the axon that end in terminal enlargements of the axon called terminal boutons. These typically occur in a relatively small area of the surface of the fiber.

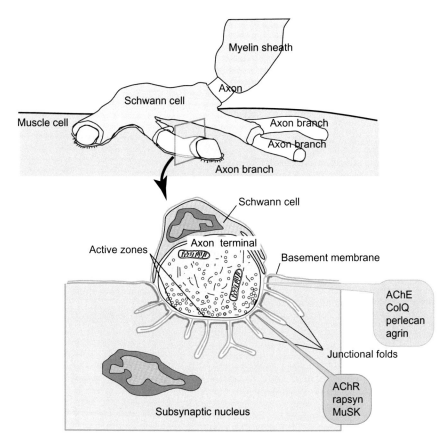

FIGURE 3.6.4 Schematic drawing of a neuromuscular junction. Branches of the motor neuron axon lie within a synaptic trough or gutter on the muscle membrane that has deep junctional folds. The axon terminal is filled with synaptic vesicles that contain neurotransmitter, acetylcholine, and that fuse with the presynaptic membrane only at specific active zones, seen in EM as darker areas on the presynaptic membrane. Particular proteins are located in specific places. AChR, the acetylcholine receptor, is located on the postsynaptic (muscle) membrane at the top of the junctional folds. AchE, acetylcholinesterase, is located in the basement membrane in the gap between nerve and muscle. Other proteins include agrin, which is secreted by motor neurons and aids in the signals that lead to the formation of the neuromuscular junction; rapsyn, which is receptor-associated protein of the synaptic membrane, is also associated with neuromuscular junction formation; MuSK, which is muscle-specific receptor kinase that binds to LRP4 (low-density lipoprotein receptor 4), which in turn is the receptor for agrin. ColQ and perlecan are basement membrane proteins.

chemical signal released into the narrow space between motor neuron and muscle fiber then is converted again into an electrical signal on the muscle. The unidirectionality is assured because the release of neurotransmitter, acetylcholine, occurs only by the presynaptic nerve axon, and receptors for the acetylcholine are localized only on the postsynaptic muscle fiber membrane.

MOTOR NEURONS RELEASE ACETYLCHOLINE TO EXCITE MUSCLES

The action potential on a motor neuron invades the terminal boutons of the neuromuscular junction. The depolarization of the action potential opens voltage-gated Ca^{2+} channels on the presynaptic membrane. Because the extracellular $[Ca^{2+}]$ is about 1.2×10^{-3} M and the intracellular $[Ca^{2+}]$ is less than 1×10^{-6} M, opening of the Ca^{2+} channels causes Ca^{2+} ions to rush into the synaptic terminal. This results in a short-lived, local increase in the cytoplasmic Ca^{2+} concentration.

The motor neuron axon terminal contains about 300,000 small vesicles, about 50 nm in diameter, that each contain between 1000 and 10,000 molecules of acetylcholine. A fraction of these vesicles forms a readily releasable pool (RRP) of neurotransmitter by docking at specific active zones on the presynaptic membrane. When the local $[Ca^{2+}]$ increases during neurotransmission, Ca^{2+} binds to Ca-sensitive proteins (synaptotagmin) that cause the synaptic vesicles to fuse with the presynaptic membrane, releasing their acetylcholine into the cleft between nerve and muscle.

The released acetylcholine diffuses to the postsynaptic membrane where it binds to two α-subunits of the **acetylcholine receptor** (AChR), which is also an ion

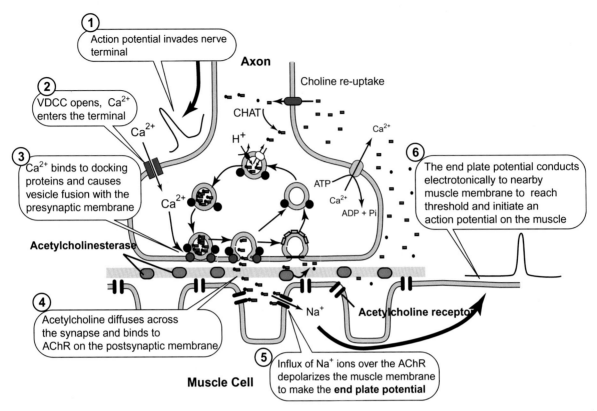

FIGURE 3.6.5 Steps in neuromuscular transmission. The overall process converts an action potential on the motor neuron to an action potential on the muscle fiber membrane. The action potential on the motor neuron propagates along the axon and opens voltage-gated Ca^{2+} channels, leading to an increase in cytoplasmic $[Ca^{2+}]$ in the terminal. The $[Ca^{2+}]$ binds to proteins that dock synaptic vesicles at the active zone, leading to fusion of the vesicles with the presynaptic membrane and release of neurotransmitter, acetylcholine, into the synapse. ACh diffuses across the gap to bind to ACh receptors located on the postsynaptic membrane. ACh binding to this receptor increases the K^+ and Na^+ conductance, which in turn depolarizes the muscle cell membrane. This depolarization is called the end-plate potential. It is conveyed passively to a nearby patch of muscle membrane where, if the end-plate potential is sufficient, it initiates an action potential. Vesicle release is shut off by removal of the synaptic Ca^{2+} by Ca^{2+} pumps or Na^+/Ca^{2+} exchange. The acetylcholine signal is terminated by hydrolysis of acetylcholine in the synapse by acetylcholinesterase. VDCC = voltage-dependent calcium channel.

channel consisting of five subunits. This ligand-gated channel increases its conductance to both Na^+ and K^+, but because Na^+ is so much further away from its equilibrium potential, the main current is an inward current carried by Na^+. This inward current depolarizes the cell, forming the **end-plate potential**. The depolarization spreads passively, or **electrotonically**, to the nearby patch of sarcolemma. This end-plate potential usually drives the muscle cell membrane to threshold, causing an action potential on the muscle membrane.

CA^{2+} EFFLUX MECHANISMS IN THE PRESYNAPTIC CELL SHUT OFF THE CA^{2+} SIGNAL

As long as there is a high $[Ca^{2+}]$ in the region of the vesicles and the active zone of the presynaptic membrane, there will be continued fusion of synaptic vesicles and release of their contents into the gap. The Ca^{2+} signal is shut off in several ways: (1) influx into the terminal is stopped, because the voltage-gated channels close when the membrane repolarizes soon after the action potential passes; (2) Ca^{2+} diffuses away or is bound to intracellular Ca^{2+}-binding proteins; (3) the plasma membrane Ca-ATPase pump and the $Na^+ - Ca^{2+}$ exchanger remove Ca^{2+} that enters the cytoplasm during the brief opening of the Ca channels. The plasma membrane Ca-ATPase is shown in Figure 3.6.5.

ACETYLCHOLINE IS DEGRADED AND THEN RECYCLED

Muscles at rest show spontaneous slight depolarizations that are due to the spontaneous release of a single vesicle full of acetylcholine. These slight depolarizations are all about equally large, each due to the fusion of a single vesicle, and they are called **miniature end-plate potentials**. The size of end-plate potentials is the result of many of these miniature end-plate potentials. In human neuromuscular junctions, some 20–25 vesicles fuse with each motor neuron action potential. At each active zone, the probability of a vesicle fusion event is about 0.7, so that human neuromuscular junctions have about 30–35 active zones. Continuous excitation and subsequent fusion of this number of vesicles with the presynaptic membrane would expand the area of the membrane and deplete it of vesicles. There are two ways to avoid this: (1) prevent full vesicle fusion in the first place and (2) recycle vesicle material that fuses by endocytosis. The first method uses a transient "fusion pore" to connect the vesicle lumen with the extracellular space, and full vesicle fusion with the membrane does not occur. In the second case, an elaborate mechanism involving specialized proteins such as clathrin and dynamin pinches off the fused vesicle membrane area and endocytoses it. Empty vesicles are refilled with acetylcholine through a transport channel that exchanges acetylcholine for H^+ ions. The H^+ ions are accumulated within the vesicles through a vacuolar H-ATPase in the vesicle membrane.

Acetylcholine is also partially recycled. Acetylcholine is degraded to acetate and choline in the synapse through the action of acetylcholinesterase. The acetate is not recycled, but the choline is taken back up into the presynaptic cell through a transporter on the presynaptic cell membrane. In the cell, acetylcholine is resynthesized from acetyl CoA and choline through the enzyme **choline acetyltransferase**. The events occurring in the axon terminal that initiate and shut off neuromuscular transmission are shown in Figure 3.6.5.

THE ACTION POTENTIAL ON THE MUSCLE MEMBRANE PROPAGATES BOTH WAYS ON THE MUSCLE

The end-plate potential is a graded potential (it is not all-or-none) that propagates electrotonically to the neighboring patch of muscle fiber membrane where it initiates an action potential on the muscle much like it does on unmyelinated nerves. Because it begins at the neuromuscular junction, at the belly of the muscle fiber, the action potential propagates in both directions so that all of the muscle is activated within a short interval. The action potential is conducted down the T-tubules to provide nearly simultaneous activation of all parts of the muscle fiber.

Clinical Applications: Diseases of the Neuromuscular Junction

Myasthenia gravis (MG) translates from Greek and Latin to mean, literally, "grave muscle weakness." It is the most common disease associated with the neuromuscular junction but it remains an uncommon disease. The reported prevalence of MG has been increasing with time: in the 1950s the mean prevalence was 22 per million persons and in the 1990s it was 94 per million persons. Reported worldwide prevalence ranges from 40 to 180 per million persons, with an annual incidence of 4—12 per million persons. The increase may be due to improved identification of the disease, prolonged survival of afflicted persons due to improved treatment, and aging of the population at risk. The number of persons with MG in the United States in 1995 was estimated to be 38,000. In 2000, it was estimated to be 59,000. Thus it is a relatively rare disease. It generally affects women younger than 40 and men after turning 60. The ratio of women to men with the disease is about 3:2.

MG is characterized by muscle weakness that worsens with activity and improves after rest. People usually present to their physicians with complaints of specific muscle weakness and not general fatigue. Ocular motor disturbances and ptosis (drooping eyelids) are the initial symptoms in some two-thirds of patients; another one-sixth complain of oropharyngeal muscle weakness, difficulty in chewing, swallowing, or talking; only 10% complain of limb weakness.

The most common cause of MG is an acquired autoimmunity to the AChR in the neuromuscular junction. The antibodies block, alter, or destroy AChRs so that neuromuscular transmission in some muscle fibers fails. Because muscular contractions rely on thousands of muscle fibers, and only some of the junctions fail, the result is weakness rather than paralysis. Treatment aims at removing the antibody attack or boosting the acetylcholine signal to increase the opening time of the remaining active AChRs. Autoimmune suppression using corticosteroids (prednisone) improves or eliminates symptoms in more than 75% of patients. Patients who do not respond to corticosteroids may respond to the immunosuppressant azathioprine. Some patients, particularly those who develop the disease later in life, have a thymoma (tumor of the thymus gland, which is involved in processing of autoimmune antibodies). Removal of the thymus in these cases improves their condition markedly. The acetylcholine signal in the neuromuscular junction can be lengthened by anticholinesterase drugs such as neostigmine bromide or pyridostigmine bromide.

Some people with antiacetylcholine receptor (AChR) antibodies also have antibodies directed against intracellular muscle proteins such as titin and the ryanodine receptor (RyR) (see Chapters 3.6 and 3.7). These patients typically have thymomas and their disease is more severe. The presence of titin and RyR antibodies is a proposed subclassification of MG that correlates with myopathy.

Some 10% of patients with generalized MG do not have anti-AChR antibodies at all. Instead, they have antibodies to a muscle-specific protein kinase (MuSK), low-density lipoprotein receptor 4 (LRP4), or some unidentified protein. The condition is commonly referred to as seronegative MG or SNMG. MuSK is a tyrosine kinase that is involved in the signaling for formation of the neuromuscular junction. Nerve cells release agrin that binds to LRP4 that initiates a signal mechanism that leads to AChR clustering in the neuromuscular junction.

In the Lambert—Eaton Myasthenic Syndrome (LEMS), antibodies are directed against the voltage-gated calcium channels in the presynaptic nerve terminal. There are a variety of such voltage-gated Ca^{2+} channels in the body. Those on the presynaptic nerve terminal are the P/Q-type. Immunoglobulins from LEMS patients bind to P/Q-type Ca^{2+} channels and decrease the Ca^{2+} influx caused by the action potential. Since this Ca^{2+} entry triggers fusion of the vesicles containing acetylcholine, a smaller spike of cytoplasmic $[Ca^{2+}]$ causes fewer vesicles to fuse and a smaller and shorter acetylcholine release. In some cases, neurotransmission fails and muscle weakness results. It is distinguishable from MG because muscle strength increases with repetitive stimulation in LEMS and decreases in MG.

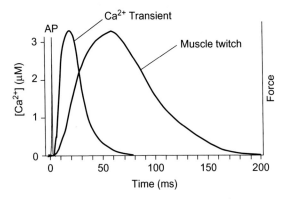

FIGURE 3.6.6 Duration of the action potential (AP), Ca^{2+} transient, and force in a muscle twitch. The action potential occurs first and is the shortest event, lasting some 2–5 ms. The Ca^{2+} transient follows the action potential and force is developed later. Ashley and Ridgway used aequorin, a protein isolated from a luminescent jellyfish, *Aequorea aequorea*, to demonstrate the Ca^{2+} transient. A variety of artificial fluorescent probes can follow the Ca^{2+} transient, and all of these show a Ca^{2+} transient that follows the action potential but precedes force development.

THE MUSCLE FIBER CONVERTS THE ACTION POTENTIAL INTO AN INTRACELLULAR CA^{2+} SIGNAL

The process by which the action potential on the surface of the muscle fiber signals muscle contraction is called excitation–contraction coupling or E–C coupling. The first clue to its mechanism was provided by Christopher Ashley and Ellis Ridgway, who demonstrated that Ca^{2+} transients follow the action potential but precede force in barnacle muscle. This time course is consistent with the action of Ca^{2+} as a second messenger between action potential and contractile activity (force or shortening). Figure 3.6.6 shows the time course.

THE CA^{2+} DURING E–C COUPLING ORIGINATES FROM THE SARCOPLASMIC RETICULUM

Because the rise in myoplasmic $[Ca^{2+}]$ upon depolarization of the sarcolemma in skeletal muscle can occur in the absence of extracellular Ca^{2+}, the intracellular Ca^{2+} must originate from some cellular store. This is the sarcoplasmic reticulum or SR. The SR is a network of a sac-like, membrane-bound structures that surrounds the myofibrils and contains loads of Ca^{2+}. The SR makes close contact with the T-tubules in structures called triads, consisting of one T-tubule and two terminal cisternae. The terminal cisternae are the specialization of the SR that forms a continuous sac near its junction with the T-tubule. The triads are typically found at the junction of the A-band and I-band in skeletal muscles. In electron micrographs, foot structures can be seen connecting the T-tubule with the terminal cisternae. The SR does not form a continuous sleeve but instead it is discontinuous, with "windows" in its surface to allow for transport of metabolites such as ATP, ADP, and Pi into and out of the myofibril. The structure of the SR has been presented in Figure 3.5.4.

CA^{2+} RELEASE FROM THE SR AND REUPTAKE BY THE SR REQUIRES SEVERAL PROTEINS

A protein called the dihydropyridine receptor (DHPR) senses the membrane potential on the T-tubule membrane and relays this to another protein, the Ryanodine receptor (RyR), on the SR membrane inside the fiber. The DHPR consists of five subunits, one of which forms a Ca^{2+} channel. The DHPR is associated with L-type Ca^{2+} channel activity, but Ca^{2+} influx through the DHPR is unnecessary for E–C coupling in skeletal muscle. There is close physical contact between the DHPR and the RyR in the SR terminal cisternae. The RyR provides a channel for rapid Ca^{2+} release from the SR. It thus appears that the DHPR acts as a voltage sensor that in some way triggers Ca^{2+} release from the SR. Although there continues to be some uncertainty as to the mechanism, current results favor a direct interaction of the DHPR and the RyR in E–C coupling.

The RyR is so named because it binds a plant alkaloid, ryanodine, with high affinity. There are several different isoforms. **RyR1** is the type in skeletal muscle; **RyR2** is in the heart and brain; and **RyR3** is present in epithelial cells, in smooth muscle, and also in the brain. RyR1 is a large protein of 565 kDa. It associates as tetramers in the SR membrane, forming a structure of about 2 million Da. The purified RyR1 forms channels in lipid bilayers with large conductances and gating characteristics consistent with its role in E–C coupling. It is believed that part of the DHPR interacts either directly with RyR1 or through another protein to trigger the opening of RyR1 and subsequent release of Ca^{2+} stored within the lumen of the SR. The sequential events in EC coupling and relaxation are shown in Figure 3.6.7.

REUPTAKE OF CA^{2+} BY THE SR ENDS CONTRACTION AND INITIATES RELAXATION

To relax the muscle, activator Ca^{2+} is taken back up into the SR by a calcium pump. This is a membrane protein of 110 kDa that links the hydrolysis of ATP to ADP and Pi with the pumping of 2 Ca^{2+} atoms back into the SR lumen. The SR Ca pump is also referred to as the SR Ca-ATPase. There are several different isoforms in different tissues. The one in fast-twitch muscle is SERCA1a for smooth endoplasmic reticulum calcium ATPase. Slow-twitch muscle has the SERCA2a isoform.

Calsequestrin is a 63-kDa protein in the SR lumen that binds Ca^{2+} with low affinity (dissociation constant, $k_D = 10^{-3}$ M) but high capacity. Calsequestrin allows the SR to store a lot of Ca^{2+} at relatively low concentrations. Thus the SR can release a lot of Ca^{2+} yet it can reaccumulate the Ca^{2+} rapidly to allow relaxation. Without calsequestrin, the same amount of Ca^{2+} in the tiny volume of the SR would reach concentrations so high that the Ca^{2+} pump would reach its thermodynamic limit before all the Ca^{2+} could be reaccumulated. Calsequestrin may also directly participate in E–C coupling by interacting with

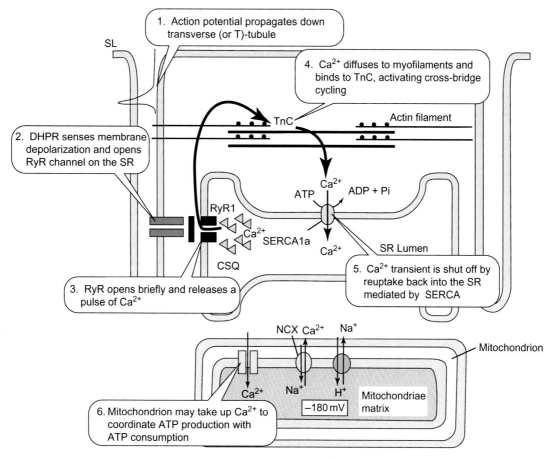

FIGURE 3.6.7 Cartoon schematic of the events that cause the rise and fall of the Ca^{2+} transient in skeletal muscle. The action potential propagates down the T-tubule where its depolarization is sensed by the DHPRs in the T-tubule membrane. These interact with the ryanodine receptors (RyR1) in the membranes of the terminal cisternae of the SR that are in close physical contact with the DHPR. The RyR is the Ca^{2+} release channel that opens briefly, allowing a pulse of Ca^{2+} to leave the SR. This released Ca^{2+} can bind to the myofilaments and activate cross-bridge cycling because the SERCA pumps on the SR membrane do not immediately remove activator Ca^{2+} back into the SR. During relaxation, the SERCA Ca^{2+} pumps remove cytosolic Ca^{2+} from the myofilaments and restore the SR to its resting state of high luminal $[Ca^{2+}]$. Mitochondria can also take up Ca^{2+} but probably do so to regulate their own rate of ATP production commensurate with ATP consumption rather than to regulate the Ca^{2+} transient.

the RyR1. In this way, calsequestrin may be a "Ca^{2+} wire" that feeds Ca^{2+} directly to the RyR1 Ca^{2+} release channel.

CROSS-BRIDGE CYCLING IS CONTROLLED BY MYOPLASMIC [CA^{2+}]

During relaxation, there is some overlap of the thin and thick filaments, but they do not interact because tropomyosin is in the way. Recall the structure of the thin filament (see Figure 3.5.7): it is a double strand of actin molecules with a half-repeat of about seven actins spanning some 38.5 nm. Tropomyosin is a long rod-shaped molecule 38.5 nm long that lies along the filament. At one end is the troponin complex of TnT, TnI, and TnC. At low myoplasmic [Ca^{2+}], the tropomyosin blocks the myosin head from interacting with the actin on the thin filament, and there is no cross-bridge formation. Figure 3.6.8 shows the steric interference of tropomyosin on myosin binding to the thin filament.

The fast-twitch isoform of troponin C has four binding sites for Ca^{2+}. Two of these are called high-affinity or "Ca–Mg" sites. The other two sites have a lower affinity for Ca^{2+} and a much lower affinity for Mg^{2+} and are therefore called "Ca-specific sites." The Ca–Mg sites appear to be always occupied and therefore they do not regulate TnC; Ca^{2+} binding to the Ca-specific sites regulates TnC configuration. TnC binds Ca^{2+} cooperatively, which means that the binding of one Ca^{2+} to an empty Ca-specific site greatly increases the binding to the second empty site. Thus Ca^{2+} tends to bind to both Ca-specific sites or to none at all, and the result is a very steep dependence on [Ca^{2+}] (see Figure 3.6.9). When Ca^{2+} binds to TnC, it changes its conformation. Because TnC connects TnT with TnI, all three shift their position. Because TnT binds tropomyosin, tropomyosin also shifts its position. This shift has the effect of uncovering the binding site for myosin on the actin, allowing cross-bridges to form. Binding to actin activates myosin ATPase activity and cross-bridge cycling, producing force or shortening.

Thus Ca^{2+} binding to TnT has its effect amplified: instead of regulating the interaction of only one actin with the thick filament, it regulates the interaction of one entire half-turn of F-actin. This is more economical of energy. If every actin–myosin interaction required Ca^{2+} binding, it would necessitate a larger Ca^{2+} release

QUANTITATIVE HUMAN PHYSIOLOGY: AN INTRODUCTION

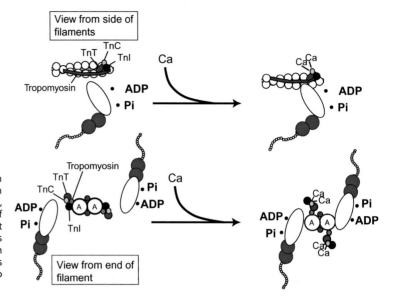

FIGURE 3.6.8 Calcium control of actin–myosin interaction. In relaxed muscle, tropomyosin blocks actin–myosin interaction and is held in place by TnT, which binds tropomyosin, TnC, which binds calcium, and TnI, which inhibits the interaction of actin and myosin. In the presence of micromolar Ca^{2+} brought about by release of Ca^{2+} from the SR, TnC changes its conformation and rearranges TnI and TnT so that tropomyosin moves out of the way of the myosin binding site on actin. Thus actin and myosin can interact, engage in cross-bridge cycling to consume ATP, and shorten or develop force.

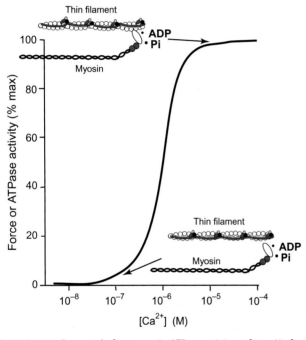

FIGURE 3.6.9 Ca-control of actomyosin ATPase activity or force. No force is generated at resting myoplasmic $[Ca^{2+}]$ of 10^{-7} M, whereas maximum force is generated at peak $[Ca^{2+}]$ of about 10^{-5} M. The steep dependence on $[Ca^{2+}]$ is a consequence of the highly cooperative nature of Ca binding and interaction of the filaments. It makes Ca appear to be a switch between the two states of no force and maximum force.

from the SR, and a greater expenditure of energy to pump the Ca^{2+} back in to relax the muscle.

SEQUENTIAL SR RELEASE AND SUMMATION OF MYOPLASMIC [CA^{2+}] EXPLAINS SUMMATION AND TETANY

Because it takes some time for TnC to bind Ca^{2+}, there is a delay between the rise in the Ca^{2+} transient and TnC binding with Ca^{2+} during a single twitch. TnC also retains some Ca^{2+} even when the Ca^{2+} transient has fallen. The drop in the myoplasmic $[Ca^{2+}]$ is brought

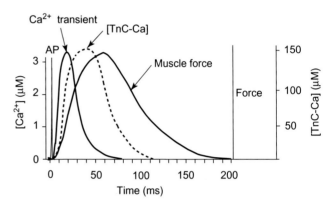

FIGURE 3.6.10 Time course of Ca^{2+} transient, [TnC–Ca], the concentration of TnC with bound Ca^{2+}, and force during the muscle twitch. The presence of elastic elements in series with the force generators is used to explain the discrepancy in the time courses of [TnC–Ca] and force generation.

about by active transport of free Ca^{2+} back into the SR through the operation of the SERCA Ca pump. This pump can transport only that Ca^{2+} that diffuses to it—the free Ca^{2+}. As free $[Ca^{2+}]$ drops, Ca^{2+} dissociates from the TnC–Ca^{2+} complex. This dissociation reaction is much slower than the binding reaction. Therefore, TnC remains bound with Ca^{2+} after the Ca^{2+} transient has peaked and fallen. Thus we would expect force to remain elevated even after the Ca^{2+} transient has fallen. We might expect that force at any time is proportional to the occupancy of TnC with Ca^{2+}. Figure 3.6.10 shows the hypothetical time course of $[Ca^{2+}]$, [TnC–Ca], and force. The force is delayed compared to the occupancy of TnC with Ca and remains elevated when TnC–Ca has returned to baseline.

THE ELASTIC PROPERTIES OF THE MUSCLE ARE RESPONSIBLE FOR THE WAVEFORM OF THE TWITCH

Experiments with rapid stretching of fibers upon initiation of contraction or rapid shortening of fibers during contraction reveal an elastic behavior that appears to be in series with the contractile elements. Rapid stretch produces a much stronger force, and rapid release during contraction reduces the force to zero. The elements of the muscle that produce force appear to be in series with a spring, as shown in Figure 3.6.11, called the series elastic element. Force developed by the contractile elements (the myofilaments engaged in cross-bridge cycling) first stretches the spring. The force transmitted to the outside of the muscle depends on the length of the spring and is determined by its length–tension characteristics. The necessity of stretching the spring to transmit force to the outside of the muscle causes the delay in the force from its origin in the cross-bridges. Similarly, the series elastic element remains stretched when Ca^{2+} dissociates from TnC and the cross-bridge cycling stops. Thus the force transmitted to the outside of the muscle remains elevated for the time required for the series elastic element to return to its normal length.

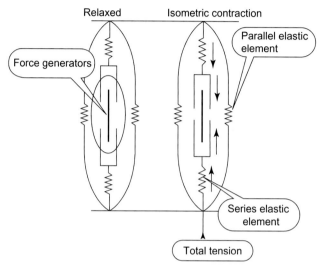

FIGURE 3.6.11 Series and parallel elastic components of muscle contraction. The muscle behaves as if there are elastic components in series with the contractile filaments and other elastic components in parallel with it. Force transmitted to the outside of the muscle lags behind the force generators because they must first stretch out the series elastic elements. After the force generators stop generating force, the total tension remains elevated because the elastic elements remain stretched. What proteins are responsible for these elastic behaviors is not known, although titin is hypothesized to contribute part and some of the series elastic behavior appears to reside in the cross-bridges.

REPETITIVE STIMULATION CAUSES REPETITIVE CA^{2+} RELEASE FROM THE SR AND WAVE SUMMATION

The twitch force lags behind the $[Ca^{2+}]$ and [TnC–Ca] transients because of the delay in stretching the series elastic elements. On repetitive stimulation by its motor neuron, the SR repetitively releases Ca^{2+} and the Ca^{2+} transients meld to form a more continuous high myoplasmic $[Ca^{2+}]$. By simple mass action, the prolonged $[Ca^{2+}]$ prolongs the time that TnC remains bound with Ca^{2+}, and this prolongs the activation of the cross-bridges. This in turn allows more time to fully stretch the series elastic elements. When the series elastic elements are stretched fully, maximum tension is transmitted to the outside of the muscle. Maximum tension is reached at the tetanus. Thus higher force seen in tetanus originates in the summation of the Ca^{2+} transients. Figure 3.6.12 illustrates this schematically.

SUMMARY

Motor neurons receive thousands of synaptic inputs, either on the soma or on the dendrites, from sensory cells, interneurons, and cells higher up in the CNS. Stimulation of excitatory connections depolarizes the

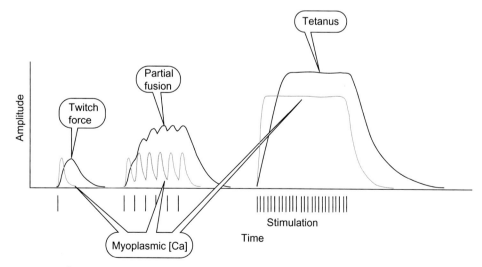

FIGURE 3.6.12 Summation of Ca^{2+} transients and tetanus. The Ca^{2+} transient is shorter than force development in the twitch. Rapid repetitive stimulation can elicit a second release of Ca^{2+} from remaining SR Ca^{2+} stores before the Ca^{2+} transient is finished. Although the amount of released Ca^{2+} in subsequent releases may diminish, because of diminished SR Ca^{2+} remaining in the SR, the total myoplasmic $[Ca^{2+}]$ increases. Rapid repetitive stimulations result in a transfer of Ca^{2+} from the SR to the myoplasm. During the twitch the cross-bridges are fully activated, but there is insufficient time for the force generators to stretch the series elastic elements. During prolonged and rapid repetitive stimulation (tetanus), these force generators have time to stretch these elements and to fully transmit the force to the outside of the muscle. At intermediate stimulation frequencies, muscle force is partially fused.

motor neuron; this is an EPSP. Other connections hyperpolarize the cell (IPSP). Both IPSPs and EPSPs decay with distance and time. If the axon hillock reaches threshold, it fires an action potential back over the soma and down the axon. Multiple simultaneous EPSPs and IPSPs can add by spatial summation. Repetitive EPSPs and IPSPs add by temporal summation.

An action potential in the motor neuron propagates down the axon to the neuromuscular junction. There, axon terminals are filled with thousands of tiny synaptic vesicles that contain acetylcholine (ACh). The action potential activates voltage-gated Ca^{2+} channels in the membrane that open briefly to let in Ca^{2+}, which then binds to special proteins that activate fusion of the synaptic vesicles with the active zone of the nerve terminal. The acetylcholine is released into a gap between muscle and nerve, and diffuses to the muscle membrane where it binds to the AChR, causing it to increase g_{Na} on the muscle membrane. Influx of Na^+ produces an end-plate potential which is conveyed electrotonically to nearby patches of muscle membrane, where an action potential begins and propagates in both directions away from the neuromuscular junction. The ACh signal is shut off by removal of Ca^{2+} from the nerve terminal cytoplasm and by destruction of ACh by acetylcholinesterase.

The muscle membrane action potential propagates along the muscle membrane and penetrates deep into the muscle along transverse tubules (T-tubules). Depolarization of the T-tubule is sensed by DHPRs on the T-tubule membrane. The DHPRs are mechanically linked to RyR located on the SR membrane immediately adjacent to the SR. RyR1 is a large tetramer that forms a gated Ca^{2+} channel across the SR membrane. These RyRs briefly open in response to depolarization of the T-tubules that is sensed by the DHPR. The terminal cisternae contain Ca^{2+} free in solution and bound to calsequestrin. Opening of the RyR1 channel releases the stored Ca^{2+}, which increases cytosolic $[Ca^{2+}]$.

The released Ca^{2+} diffuses throughout the myofibril and binds to troponin C (TnC). Ca^{2+} binding alters the conformation of TnC, which in turn alters the disposition of TnI, TnT, and tropomyosin. At rest, tropomyosin inhibits the interaction of actin and myosin so that cross-bridge cycling cannot occur. Ca^{2+} binding to TnC removes the inhibition of tropomyosin so that cross-bridge cycling occurs. The result is actomyosin ATPase activity and force development or shortening of the muscle. Reuptake of the released Ca^{2+} by the SR Ca pump causes relaxation of the muscle because Ca^{2+} dissociates from TnC by mass action. Removal of Ca^{2+} from TnC causes tropomyosin to move back to its inhibitory position, cross-bridge cycling stops, and the muscle relaxes.

Muscle force on the outside of the muscle fiber lags behind the Ca^{2+} transient. The lag is due to a delay in TnC binding of Ca^{2+} and a delay in the transmission of the force from the cross-bridges to the exterior of the muscle. The muscle behaves as if the force generators have a spring arranged in series with them. Force is transmitted through the spring only to the extent that the spring is stretched. This takes time, so the force transient (the twitch) is both delayed and prolonged compared to the Ca^{2+} transient. At least some of the spring characteristics are in the myofilaments themselves.

Repetitive action potentials can cause additional Ca^{2+} release before the muscle relaxes from the first twitch. The prolonged elevation of cytosolic $[Ca^{2+}]$ gives the myofilaments more time to completely stretch the series elastic elements to fully transmit their force to the exterior of the muscle. Force increases with increasing frequency of stimulation until tetany is reached, in which there is maximum force and no waviness in the force.

REVIEW QUESTIONS

1. Could neuromuscular transmission occur in the absence of extracellular Ca? Why or why not?
2. What is spatial summation? Temporal summation? Can both occur simultaneously?
3. Would you expect some synaptic vesicles to fuse with the presynaptic motor neuron membrane without an action potential? Why or why not?
4. What effect would an inhibitor of acetylcholinesterase have on neuromuscular transmission?
5. Curare blocks the AChR from binding ACh. What effect would this have on neuromuscular transmission?
6. What protein senses the voltage on the T-tubule? What protein releases Ca^{2+} from SR stores? What protein takes Ca^{2+} back up into the SR?
7. What protein confers Ca^{2+} sensitivity on the myofilaments?
8. Why is tetanic force greater than twitch force?
9. Why does rapid stretch of a muscle during its activation increase the force?

Clinical Applications: Malignant Hyperthermia and Central Core Disease

On April 14, 1960, a 21-year-old student sustained compound fractures of his tibia and fibula when he was run over by a car in Melbourne, Australia. In the casualty department, he was more worried about anesthesia than his injuries. He said that 10 of his close relatives had died during anesthesia for minor procedures. Ether had been used in all these anesthesia-associated deaths, so the attending physicians used a new anesthetic, halothane. After 10 minutes of anesthesia, the student's blood pressure fell, his heart rate rose, and he felt very hot. The anesthetic was stopped and the student was rubbed down with ice packs. He recovered. Further clinical examination of the student showed no known abnormality, but study of his family history indicated a previously undescribed inborn error of metabolism inherited as an autosomal dominant trait. The syndrome was characterized by hyperthermia, a rapid and large rise in body temperature, and it was malignant: fully 70% of the early cases were fatal. This first case of malignant hyperthermia was reported in *The Lancet* in 1960 by Michael Denborough.

Persons with malignant hyperthermia are usually clinically inconspicuous. Their condition is brought on by a triggering event, usually exposure to **volatile anesthetics** or to depolarizing muscle relaxants such as **succinylcholine**. The increased metabolism is due to increased cytoplasmic [Ca^{2+}] within muscle cells that hydrolyzes ATP by the activation of cross-bridge cycling and the activation of glycogenolysis, glycolysis, and oxidative metabolism. The muscles become rigid and may undergo **rhabdomyolysis**, in which the muscle membranes lose their integrity and leak myoglobin, creatine kinase (CK), and potassium into the blood. The high plasma [K^+] can cause lethal cardiac arrhythmias. The condition is not uniform: some cases of MH are brought on by exertion, overheating, or infections. Persons suspected of being MH susceptible can be diagnosed using the response of muscle biopsies to halothane or caffeine. Early treatment with **dantrolene sodium** reduces the crisis and reduces the mortality rate from 70% to 10%. Dantrolene inhibits Ca^{2+} release from the SR through the RyR1 Ca^{2+} release channel. This prevents the rapid rise in cytoplasmic [Ca^{2+}] that causes MH.

Central core disease (CCD) derives its name from the histological appearance of type I muscle fibers in affected individuals. In this case, RyR1 exhibits increased Ca^{2+} permeability even in the resting state. In these fibers, the central core of the muscle fibers is damaged, with unstructured and damaged contractile filaments and destruction of the mitochondria, probably due to too much [Ca^{2+}] in the interior of the cell. At the periphery of the cell, the surface membrane presumably pumps out excess Ca^{2+} across the SL membrane, thereby preventing injury to the mitochondria in the peripheral regions of the muscle fiber.

Mutations of the RyR1 make up the most frequent cause of MH and CCD. Over 20 distinct mutations of the RyR1 cause MH. In some other persons, MH is caused by mutations in the DHPR. The DHPR makes close physical contact with RyR1 and during normal excitation–contraction coupling the DHPR provides RyR1 with the signal it needs to release Ca^{2+}.

APPENDIX 3.6.A1 : MOLECULAR MACHINERY OF THE NEUROMUSCULAR JUNCTION

NEUROMUSCULAR JUNCTIONS CONSIST OF A COMPLEX ARRAY OF INTERACTING PROTEINS

As described in Chapter 3.6, neuromuscular transmission involves a series of processes that begin with an action potential on the nerve axon and end with an action potential on the muscle fiber. The main sequence is (1) conversion of the electrical signal, the action potential, to a cytosolic Ca^{2+} signal within the nerve terminal; (2) fusion of docked vesicles with the presynaptic membrane and release of neurotransmitter; (3) diffusion of the neurotransmitter, acetylcholine, to the postsynaptic, muscle fiber membrane; (4) conversion of the chemical signal, acetylcholine, into an electrical signal by opening ligand-gated channels on the muscle membrane; (5) conduction of the resulting end-plate potential to the area of muscle membrane outside of the end plate; (6) initiation of the action potential on the muscle fiber membrane. These events are all mediated by proteins and their spatial arrangement is crucial to the sequence of events. This appendix considers the proteins involved on both the presynaptic and postsynaptic membranes that produce these events. This is a second level of detail that is not generally required for introductory courses. A third level of structural detail involves the parts of the proteins that interact with other proteins' parts, and the detailed structure of these proteins. This third level is the concern of structural biologists and we will not consider it here.

Electron micrographs of the neuromuscular junction shows electron-dense material on the presynaptic membrane near where synaptic vesicles congregate and attach to the presynaptic membrane. These areas of the membrane are called **active zones**, and vesicle fusion occurs only at these areas. The presynaptic membrane at the active zone contains several protein constituents that specifically bind to proteins on the vesicular membrane. These proteins are collectively known as **SNARE** proteins for **SNAP receptor**. These names derive from other factors that were found to be required for synaptic transmission: **NSF** is the *N*-ethyl maleimide sensitive factor and **SNAP** is the **soluble NSF attachment protein**; SNARE is for SNAP receptor. **Synaptobrevin** (also called VAMP, for vesicle-associated membrane protein) is a v-SNARE that binds to synaptic vesicles and to another protein on the vesicles, **synaptotagmin**. Synaptotagmin binds Ca^{2+} and confers Ca^{2+} sensitivity to vesicle fusion. Synaptobrevin interacts with two t-SNARE proteins (t for "target"), **SNAP-25** and **syntaxin** on the presynaptic membrane. The v-SNARE protein synaptobrevin and the t-SNARE proteins SNAP-25 and syntaxin spontaneously coil up in an exergonic reaction that forces the vesicle towards the target membrane. There are a variety of t-SNARE and v-SNARE proteins that interact selectively, which allows for the same general machinery to be used in vesicle trafficking throughout the cell. Neuromuscular transmission is a specific example of it. The priming of the vesicle for fusion is regulated by several other proteins. **SM proteins (Munc-18** is one example) bind to syntaxin and are essential for membrane fusion in the cell. Another protein, complexin, activates the complex but clamps the complex to prevent premature fusion. Lastly, synaptotagmin binds Ca^{2+} when it rises during neurotransmission and reverses the complexin clamp. The location of the voltage-gated calcium channel is crucial here: typically only a few Ca^{2+} channels open at any active zone, and the Ca^{2+} concentration increases locally, limited by binding and diffusion of Ca^{2+} away from the Ca^{2+} channel. Other proteins, such as RIM (Rab-3 interacting molecule), are thought

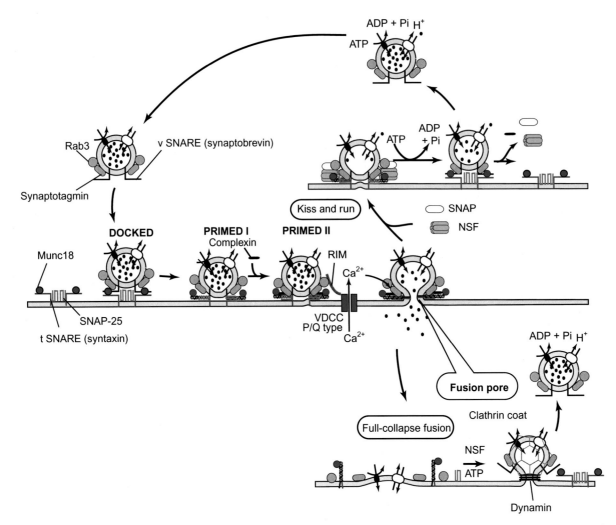

FIGURE 3.6.A1.1 Proteins involved with vesicle fusion and recycling in the neuromuscular junction. Vesicles have a v-SNARE protein (synaptobrevin or VAMP) whereas the presynaptic membrane has t-SNARES, syntaxin, and SNAP-25. These proteins interact and coil up spontaneously in a series of reactions that first dock the vesicle and then prime it in two stages. The reaction requires S/M proteins, in this case Munc-18. The coiling of the SNARES produces a force normal to the membrane that draws the vesicle to the membrane. Premature fusion is prevented by complexin that interacts with the SNARES and prevents the completion of the reaction. RIM, Rab-3 interacting molecule, interacts with Rab-3 on the vesicle and P/Q-type voltage-gated Ca^{2+} channels on the presynaptic membrane to draw these together so that the Ca^{2+} channel is close to the active zone. When the action potential propagates over the presynaptic membrane, the Ca^{2+} channel opens briefly and Ca^{2+} enters the terminal in narrow space, increasing the local $[Ca^{2+}]$. Synaptotagmin binds the Ca^{2+}, and in this form displaces complexin from its inhibitor position, allowing the vesicle to fuse with the presynaptic membrane and empty its content of acetylcholine into the synapse. Fused vesicles may undergo full-collapse fusion, in which case the vesicle membrane proteins are incorporated into the presynaptic membrane, or form a fusion pore that does not expand. This fusion pore is thought to be reversible, so that the vesicle fusion is "kiss and run." Recycling of the membrane component requires the N-ethylmaleimide sensitive factor, NSF, and ATP, and the soluble NSF-associated protein (SNAP, distinguished from SNAP-25).

to position the P/Q-type voltage-gated Ca^{2+} channels near the docked and primed vesicles. This process is illustrated in Figure 3.6.A1.1.

Although neurotransmission at each active zone results in the fusion of at most a few vesicles, transmission at physiological frequencies (50 Hz or more) can result in the fusion of a large number of vesicles. If synaptic vesicles fully fuse with the presynaptic membrane, the readily releasable pool of vesicles will become depleted, the area of the presynaptic membrane will increase, and also the character of the presynaptic membrane will change because of incorporation of vesicle proteins having a different composition than presynaptic membrane proteins.

There are two ways of avoiding these problems. First, the vesicles may not fuse completely with the presynaptic membrane. Instead, the vesicles "kiss and run": they form a transient fusion pore that allows neurotransmitter to escape to the synapse, but the fusion pore does not expand and instead closes off. The vesicle is released by an ATP-requiring reaction that uncoils the t-SNAREs and v-SNARES from each other, mediated by NSF (N-ethyl maleimide sensitive factor), a hexamer that presumably splits 3–6 ATP molecules with each catalytic cycle to uncoil the SNARE complex. Second, complete fusion of vesicles can be reversed by slow endocytosis involving segregation of the vesicle and presynaptic proteins, clathrin

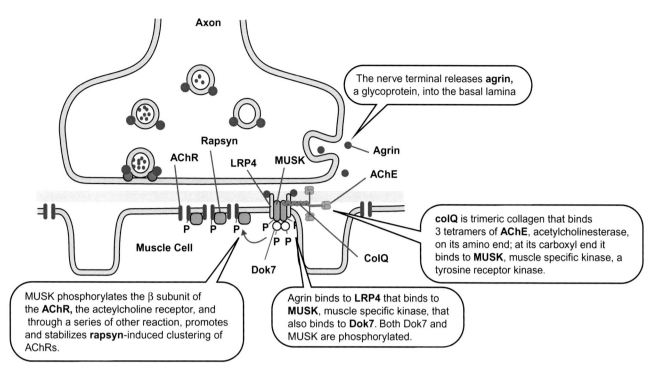

FIGURE 3.6.A1.2 Proteins involved in the formation and maintenance of the neuromuscular junction. Signals pass between the motor neuron and muscle fiber to form and maintain the neuromuscular junction, and only one of these signals is shown here. Neurons release agrin, a glycoprotein, into the extracellular space adjacent to the muscle fiber. Agrin binds to a receptor, LRP4, on the muscle membrane. LRP4 binds to MUSK, muscle-specific kinase, which is a receptor tyrosine kinase. MUSK phosphorylates Dok7, a dimer that binds to MUSK, and also autophosphorylates itself in multiple tyrosine residues. MUSK phosphorylates the β subunit of the acetylcholine receptor, AChR, that promotes its association with Rapsyn, receptor-associated protein of the synapse. Rapsyn and acetylcholine receptor cluster at the neuromuscular junction. This clustering acts on a prepattern of AChR that exists without agrin. Further organization of the neuromuscular junction is provided by colQ, collagen type Q. This consists of a trimer of collagen that binds three tetramers of acetylcholinesterase, AchE, on the amino ends of the collagen. The carboxyl ends bind to MUSK. Thus rapsyn, AChR, MUSK, Dok7, LRP4, and AChE are all colocalized within the neuromuscular junction.

coats of the vesicles and pinching off of the endocytosing vesicles by use of dynamin. These processes are also shown in Figure 3.6.A1.1.

SPECIALIZATIONS OF THE MUSCLE FIBER MEMBRANE PREPARE IT TO RECEIVE THE ACETYLCHOLINE SIGNAL

The key protein on the muscle fiber membrane is the acetylcholine receptor, **AchR**, which consist of 5 subunits: 2 α, and one each of β, γ and ε or γ. Both α subunits must bind acetylcholine in order to open the receptor's high conductance path. Clustering of AchR in the end-plate region is the result of cross-talk between the neuron and the muscle fiber. Nerve cells secrete **agrin** into the extracellular space, and agrin binds to LRP4, low-density lipoprotein receptor 4, which binds to **MUSK** (Muscle-specific kinase) a receptor tyrosine kinase on the surface of the muscle fiber. Activation of MUSK requires **DOK7**, a cytoplasmic protein. DOK7 not only activates MUSK, but is one of its substrates. MUSK phosphorylates the β subunit of the AChR, causing it to bind to another protein, **rapsyn** (receptor associated protein of the synapse). Rapsyn appears to mediate the clustering of the AchR at the neuromuscular junction. The extracellular domain of MUSK interacts with the carboxyl end of collagen Q (colQ) while the amino terminus of colQ interacts with **acetylcholinesterase** (AChE) in the extracellular space. Other signals pass between neuron and muscle fiber to maintain the neuromuscular junction. The main proteins associated with neuromuscular transmission are shown in Figure 3.6.A1.2.

APPENDIX 3.6.A2 : MOLECULAR MACHINERY OF THE CALCIUM RELEASE UNIT

THE CALCIUM RELEASE UNIT CONSISTS OF MULTIPLE PROTEINS ON T-TUBULES AND SARCOPLASMIC RETICULUM MEMBRANES

The Calcium Release Unit of muscle consists of the terminal cisternae of the sarcoplasmic reticulum and its adjacent transverse or T-tubule that is the invaginated part of the surface membrane. The two major proteins are the voltage sensor on the T-tubule, the **DHPR (dihydropyridine receptor)**, and the calcium release channel, **RyR (ryanodine receptor)**, on the sarcoplasmic reticulum. The DHPR is composed of five subunits. The α1 subunit acts as a voltage sensor and changes its conformation when the T-tubule depolarizes during the excitation−contraction coupling. This conformational change is transmitted to the RyR, causing it to open and release Ca^{2+} stored in the lumen of the SR.

Clinical Application: Congenital Myasthenic Syndromes

Myasthenia gravis is an autoimmune disease in which antibodies are directed against proteins of the neuromuscular junction: AchR, MUSK, and LRP4, and some unidentified ones. These antibodies reduce neuromuscular transmission and produce muscle weakness. Antibodies against the voltage-gated Ca^{2+} causes the Lambert–Eaton Myasthenic Syndrome. Mutations in genes encoding the proteins involved in neuromuscular transmission produce **congenital myasthenic syndromes** (CMS), which are a group of rare and heterogeneous disorders.

Genetic disorders of the AChR cause Slow Channel Syndrome, in which the AChR opening is prolonged. This causes desensitization blockade of the receptor and damage to the muscle fiber due to cation overload. Fast Channel CMS is rare, causes severe weakness, and is treated with AChE blockers to help keep channels open longer. Mutations in MUSK produce proximal limb and facial weakness without ocular involvement. Mutations in Dok7 is usually present with limb-girdle myasthenia, accounting for some 20% of CMS cases. Mutations in CHAT (choline acetyltransferase, the enzyme that catalyzes the synthesis of acetylcholine from acetyl CoA and choline) cause a neonatal CMS with life-threatening failure of the respiratory muscles. CMS can also be caused by mutations in LRP4 and ColQ.

Macromolecular complexes such as those involved in neuromuscular function can fail if any of their components fail. Thus it is not surprising that a variety of mutations can produce similar symptoms.

Clinical Application: Botox Treatment

Botox treatment is the most popular cosmetic treatment, with over 6 million treatments in the United States each year in 2013. It removes facial wrinkles by paralysis of the facial muscles and also produces a flat effect—a severe reduction in the external expression of emotions. Typically injection of picogram quantities of the toxin can relax muscles for 2–4 months. It is also used for other muscular disorders, including strabismus (misalignment of the eyes), achalasia (spasm of the lower esophageal sphincter), and blepharospasm (uncontrolled and sustained contractions of the muscles around the eye).

Botulinum toxin (botox) originates from *Clostridium botulinum*, a gram-negative bacterium. Botulism can be fatal, and most often results from ingestion of food contaminated with the toxin. The toxin is inactivated by heating to more than 85°C for more than 5 minutes. Fatal botulism kills by arrest of the respiratory muscles. *Clostridium botulinum* produces seven related toxins, called botulinum toxin types A, B, C1, D, E, F, and G. The main commercial types are A and B. The toxin is synthesized as a protoxin of 150 kDa, which is subsequently cleaved to light (L) and heavy (H) chains that remain linked by a disulfide bond. Nerve terminals have receptors for both the H-chain and L-chain. The L-chain is transported across the nerve terminal membrane by endocytosis, and it becomes incorporated into endosomes. The various L-chains are **metalloproteinases**—proteases that require metal ions for activity. The L-chains bind Zn^{2+}. Their substrates are one of several proteins that make up the **SNARE** complex that is required for neurotransmitter vesicle fusion with the presynaptic nerve membrane. At least three proteins are involved in this process: v-SNARE (synaptobrevin) is associated with neurotransmitter vesicles; t-SNARE (syntaxin) is on the presynaptic membrane; and SNAP-25, another t-SNARE. All of these proteins are required for the docking of vesicles in the active zone of the presynaptic terminal and for subsequent neurotransmission. Each of the botulinum toxins cleaves one of these three proteins and thereby interferes with neurotransmitter release. Synaptobrevin is cleaved, at different loci, by toxins B, D, G, and F; SNAP-25 is cleaved by A, C, and E; syntaxin is cleaved by toxin C. Thus botulinum toxin causes paralysis of the muscles by cutting up the proteins that are essential for the release of acetylcholine at the neuromuscular junction.

The RyR is present in three isoforms: **RyR1**, **RyR2**, and **RyR3**. RyR1 is present in all skeletal muscles. RyR2 is expressed in cardiac muscle and RyR3 is located in a variety of tissues. The RyR consists of a homotetramer, with single copy molecular weight of 565 kDa. The RyR is regulated by phosphorylation/dephosphorylation at several phosphorylation sites. These are phosphorylated by **protein kinase A (PKA), protein kinase G (PKG), and calmodulin-dependent protein kinase (CAM kinase II)** and are dephosphorylated by **protein phosphatases (PP1 and PP2a)**. Each monomer of the tetrameric RyR receptor binds **calmodulin**, either with or without bound Ca^{2+}, and **FKBP-12 (FK506 binding protein)**. FKBP is named for its binding of FK506, an immunosuppressor drug. Binding of FK506 to FKBP causes FKBP to dissociate from the RyR.

The RyR1 also binds **triadin** and **junctin**, two proteins of similar structure that bind each other, RyR and **calsequestrin (CASQ)** within the lumen of the SR. Calsequestrin has two main isoforms: CASQ1 and CASQ2. Fast-twitch muscle expresses only CASQ1. Slow twitch expresses both CASQ1 and CASQ2, and cardiac muscle contains only CASQ2. CASQ binds luminal Ca^{2+} at low affinity ($K_D = 1$ mM) but high capacity. At high $[Ca^{2+}]$, the CASQ polymerizes and forms extended strings within the SR lumen.

Rapid release of Ca^{2+} from the SR entails a movement of positive charges from the lumen to the cytosolic compartment. If there were no compensatory charge movements, this would cause a negative potential inside the SR that would oppose further Ca^{2+} release. This is

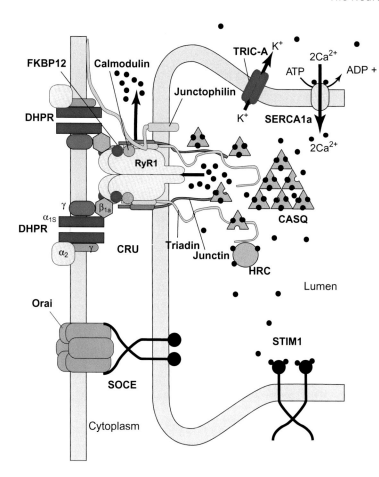

FIGURE 3.6.A2.1 Proteins involved in the release of calcium at the triad in skeletal muscle. Calcium is accumulated into the lumen of the SR by the operation of the SERCA1a calcium pump, a primary active transporter, that pumps in 2 Ca^{2+} atoms for each ATP hydrolyzed. The Ca^{2+} inside the SR is largely bound to calsequestrin (CASQ) and histidine-rich Ca^{2+}-binding protein (HRC). CASQ polymerizes when luminal $[Ca^{2+}]$ is high. Opening of a conductance pathway through the RyR1 releases Ca^{2+} during excitation–contraction coupling. The RyR1 is present as a homotetramer that binds triadin, junction, junctophilin, FKBP, and calmodulin. Triadin and junctin also bind to each other and to CASQ and HRC. These may inform the RyR1 about the level of Ca^{2+} stores in the SR, and they modify the gating behavior of RyR1. Opening of the RyR1 is typically accomplished by conformational changes in the dihydropyridine receptor, DHPR, on the T-tubule membrane. There is a direct mechanical connection between the β_{1a} subunit of the DHPR and RyR1. This requires close apposition of DHPR and RyR1 which is stabilized by junctophilin that binds both SR and T-tubule membranes, DHPR and RyR1. When Ca^{2+} stores are depleted, STIM1 alters its conformation and opens Orai, a store-operated Ca^{2+}-entry channel. Rapid movements of Ca^{2+} across the membrane entail currents that produce a membrane potential that inhibits further Ca^{2+} current. This potential is short-circuited by TRIC-A that allows for counter currents of K^+—outward during Ca^{2+} accumulation by the SERCA1a and inward during Ca^{2+} release by the RyR1.

avoided by **TRIC-A (trimeric intracellular cation-selective channel)** in the SR membrane. These allow K^+ entry into the SR in response to any negative potential that develops.

The Ca^{2+} store in the SR is maintained by the operation of SERCA in the membranes of the SR. This 110-kDa protein directly couples ATP hydrolysis to the transport of 2 Ca^{2+} atoms into the SR. The **SERCA (smooth endoplasmic reticulum Ca-ATPase)** comes in different isoforms. Fast-twitch fibers express SERCA1a, whereas slow-twitch and cardiac muscle express SERCA2a. Slow-twitch and cardiac muscle express a small protein, **phospholamban (PLN)** that associates as pentamers in the membrane and inhibits the SERCA2a pump. Phosphorylation of phospholamban by CAM kinase II or PKA relieves the inhibition of the SERCA2a pump by phospholamban.

The triad junction includes a Store-operated Ca^{2+} entry (SOCE) mechanism. This complex of proteins senses the Ca^{2+} content of the SR and, when it is low, opens a pathway to refill the SR from the extracellular space. The sensor of the SR content is **STIM1 (stromal interacting molecule)** that is present as a homodimer and contains luminal sites for binding Ca^{2+}. Upon depletion, STIM1 undergoes a conformational change and clusters near the triad where it directly interacts with a plasma membrane hexameric channel called **Orai1**. Orai1 opens allowing Ca^{2+} to enter the restricted space between T-tubule and SR.

A cartoon illustrating the main features of the interactions of these proteins in the triad is shown in Figure 3.6.A2.1. The main point of this is to remind the student that these functions are carried out by macromolecular assemblies of proteins and that mutations or deficiencies of any of these participatory proteins could have serious consequences for the operation of the entire assembly.

3.7 Muscle Energetics, Fatigue, and Training

Learning Objectives

- Explain why ATP is the "energy currency" of muscle cells
- List the major uses of ATP during contractions
- Explain the function of the creatine phosphokinase reaction in regenerating ATP
- Describe the myokinase reaction
- Describe the sources of glucose for muscle contraction
- Contrast the rate and capacity of ATP generation through glycolysis and oxidative phosphorylation
- Explain why lactic acid is produced by muscle
- Describe what is meant by the term "lactate shuttle" and describe the Cori cycle
- Explain why lactate release by muscle increases with vigorous exercise
- Describe the "anaerobic threshold" and its likely cause
- Describe the effect of exercise on glucose uptake by muscle
- Distinguish between fatigue at high intensity and low intensity exercises and explain the likely cause of fatigue in these two types of exercises
- Describe the ischemic exercise test and why it reveals defects in glycogenolysis
- Discuss the potential for transformation of muscle types in humans

MUSCULAR ACTIVITY RELIES ON THE FREE ENERGY OF ATP HYDROLYSIS

The cross-bridge cycle links shortening or force development to the hydrolysis of ATP. Through this cycle, chemical energy stored in the γ phosphate bond of ATP is made available for mechanical work. Thermodynamics tells us that the amount of work must be less than the chemical energy released. The chemical energy that is not converted to work is dissipated as heat. The free energy of ATP hydrolysis was given in Chapter 1.7 as

$$[3.7.1] \qquad \Delta G = \Delta G^{0\prime} + RT \ln\left(\frac{[ADP][Pi]}{[ATP]}\right)$$

where $\Delta G^{0\prime}$ refers to the standard free energy change for ATP hydrolysis under the conditions of the cell—typically pH 7.0 and 37°C (310°K), and typical cell concentrations of Mg^{2+} and typical ionic strength. We have used -31.0 kJ mol^{-1} as this value, but the validity of this value for muscle tissue is uncertain. The values of [ATP], [ADP], and [Pi] are also uncertain because of methodological difficulties and the fact that the values can depend on which animal, which muscle, which muscle fiber, and the state of the muscle fiber. These concentrations may also vary with location within the muscle fiber. Some estimated values for human muscle at rest are: [ATP] = 8.2×10^{-3} M; [Pi] = 3.8×10^{-3} M; [ADP] = 10×10^{-6} M. Insertion of these values into Eqn [3.7.1] gives a calculated $\Delta G_{ATP \rightarrow ADP+Pi} = -62.7$ kJ mol^{-1}. During exercise, the concentrations of ATP, ADP, and Pi change and the free energy of ATP hydrolysis becomes less negative.

MUSCULAR ACTIVITY CONSUMES ATP AT HIGH RATES

In an intact muscle, macroscopic force is accompanied by thousands and thousands of cross-bridges cycling at rates that depend on the motor units that are recruited, the frequency with which they are recruited, and the type of muscle fiber (types I, IIA, or IIB). ATP hydrolysis is also used for Ca^{2+} reuptake into the SR and active ion pumping by the Na^+,K^+-ATPase to maintain the Na^+ and K^+ gradients necessary for the muscle action potential. Maintenance chores such as turnover of mRNA, lipids, and proteins also split ATP, but the actomyosin ATPase is the main sink for ATP during activity.

THE AGGREGATE RATE AND AMOUNT OF ATP CONSUMPTION VARIES WITH THE INTENSITY AND DURATION OF EXERCISE

The *rate* of ATP utilization by the aggregate muscles of the body depends on the intensity of the exercise. Intensity of exercise is determined by two variables: the recruitment of motor units and the frequency with which they are recruited. In weight lifting, for example, specific muscles are recruited for specific tasks such as the bench press. During a maximum lift, the primary muscle groups are recruited 100%, and they are recruited with tetanic stimulations. There is no rest phase for the muscles until the lift is completed. There is no more intense exercise than this for the specific muscles involved.

The *total amount* of ATP used during an exercise is its rate of utilization times the duration of the event. High intensity exercise can be maintained only for short periods, whereas moderate intensity exercise can be endured for long times. The relationship between intensity and sustainable effort is not linear. Maximal effort can be sustained for only a few seconds whereas high effort can often be sustained for much longer periods. Table 3.7.1 gives the approximate rate and amount of ATP needed for different track events. The amount needed for rest is not given because its duration is not specified.

IN REPETITIVE EXERCISE, INTENSITY INCREASES FREQUENCY AND REDUCES REST TIME

Figure 3.7.1 shows an electromyogram, or EMG, of leg muscles obtained from a slowly walking rat. The EMG is akin to an electrocardiogram (see Chapter 5.6): it is the trace of *electrical activity* of the muscle on the surface of the body. Larger EMG amplitudes (recorded in volts) indicate greater number of muscle fibers that are firing action potentials. Thus the EMG directly indicates activation and not force, but the expectation is that force would increase when the EMG amplitude increases. Each muscle is activated at appropriate times for definite lengths of time, and this activity is interspersed with periods of rest. In vigorous walking, the frequency of activation increases and the duration of each activation decreases. The **duty cycle**, the part of the time that the muscle is activated, typically increases with intensity because the period of activation decreases less than the period of rest. This describes the situation when we increase the speed of a repetitive action such as walking. In many animals, the speed of locomotion varies through a series of different sets of muscle coordination called gaits. In humans, walking and running form distinctive gaits, but running fast and running slow differ mainly in the speed and recruitment rather than the sequence of activation. Thus high intensity as applied to repetitive exercise means something different than in weight lifting: motor unit recruitment is increased, perhaps to 100%, but the muscles must be controlled by bursts of activation followed by rest periods. Intensity in this case is graded by recruitment and the duty cycle. In weight lifting, intensity is graded entirely by recruitment because the duty cycle is 1.0.

Each muscle fiber, when activated, is activated completely. The control of force for the aggregate muscle is achieved by the temporal recruitment of the fibers: which fibers are being activated and with what frequency and in what sequence, and not by control of the force of each motor unit or muscle fiber. Each activation entails fast rates of ATP hydrolysis as the actomyosin cross-bridges are activated by the Ca^{2+} transient.

TABLE 3.7.1 Rate and Amount of ATP Needed for Different Track Events

Event	Rate of ATP Consumption (mol/min)	Amount of ATP Needed (mol)
Rest	0.07	—
100 m sprint	2.6	0.4
800 m run	2.0	3.4
1500 m run	1.7	6
42,200 m marathon	1.0	150

Adapted from E. Hultman and H. Sjoholm, Biochemical causes of fatigue, in Human Muscle Power, Human Kinetics Publishers, Inc., Champaign, IL, 1986.

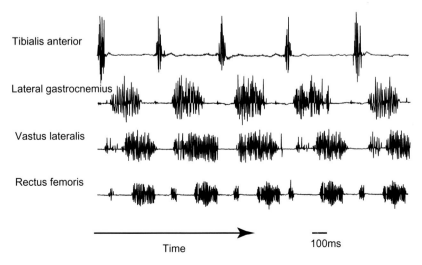

FIGURE 3.7.1 Electromyogram of leg muscles of rats walking slowly on a treadmill at a speed of 1 mph. The tibialis anterior muscle is in the anterior of the lower leg and lifts the foot. It is activated first. The gastrocnemius muscle is in the posterior lower leg, the calf, and extends the foot. The vastus lateralis and rectus femoris muscles are in the anterior thigh and extend the lower leg. *Modified from G.A. Brooks, T.D. Fahey, T.P. White, and K.M. Baldwin,* Exercise Physiology, *3rd ed, McGraw Hill, New York, NY, 1999.*

Increasing the frequency of activation necessarily decreases the time that the muscle is not activated, and this is the time the muscle has for metabolism to oxidize substrates produced during the contractions. Thus every muscle contraction requires rapid ATP consumption and rapid ATP regeneration, but at high intensity exercise there is less time for metabolism to recover the resting state. In the limit of maximum effort, there is no rest phase at all. This difference in maximum effort versus endurance exercise is the origin of the different kinds of fatigue generated by these different types of exercise.

METABOLIC PATHWAYS REGENERATE ATP ON DIFFERENT TIMESCALES AND WITH DIFFERENT CAPACITIES

Because active muscle uses ATP quickly, it must be regenerated quickly. Those systems that regenerate ATP quickly must themselves be regenerated in the rest phases. There are three major systems for the regeneration of ATP that differ in their capacity and rate:

1. Creatine phosphokinase and myokinase regenerate ATP fastest but with low capacity
2. Glycolysis regenerates ATP fast but with low capacity
3. Oxidative phosphorylation regenerates ATP slowest but with highest capacity.

These pathways are illustrated in Figure 3.7.2.

Muscles contain about 15–40 mM creatine phosphate. **Creatine phosphokinase** (CPK) catalyzes the phosphorylation of ADP from creatine phosphate to form ATP. This is an extremely rapid reaction that helps to "buffer" ATP concentrations near the normal 8 mM levels found in the myoplasm of muscle fibers. When creatine phosphate is used to regenerate ATP, myoplasmic [Pi] increases. It is believed that CPK is located very close to ATP-utilizing reactions such as the myosin ATPase and SR Ca-ATPase and may directly transfer ATP to these enzymes. This is called **substrate channeling**. These ATPase enzymes split ATP faster in the presence of CPK and creatine phosphate. **All contractions of muscle require creatine phosphate regeneration of ATP for maximum force.** Another enzyme, **myokinase**, converts two molecules of ADP into ATP and AMP. As ADP builds up in active muscle, AMP also builds up and may be a metabolic indicator of the fuel status of the cell.

Glycolysis breaks glucose into two pyruvate molecules and regenerates ATP from ADP and Pi. Regeneration of ATP requires the incorporation of Pi into 1,3-diphosphoglyceric acid from glyceraldehyde-3P, as shown in Figure 3.7.2. Glycolysis begins with glucose that may arise from muscle glycogen or from the blood. Glycolysis requires 2 ATP molecules and generates 4 ATP, for a net gain of only 2 ATP molecules per molecule of glucose. Additional ATP (3–5 per molecule of glucose) can be generated from the NADH produced by the oxidation of glyceraldehyde-3-phosphate. Glycolysis occurs in all muscle fibers, but the concentration of glycolytic enzymes differs. Fast-twitch glycolytic fibers depend most heavily on glycolysis for ATP regeneration.

The pyruvate formed in the cytoplasm by glycolysis enters the mitochondria to be converted to acetyl CoA by pyruvate dehydrogenase. In this process, 1 CO_2 is released. The acetyl CoA is converted to two more CO_2 molecules through the TCA cycle, which also produces NADH, $FADH_2$, and GTP. In the presence of oxygen, the NADH and $FADH_2$ are oxidized through the electron transport chain (ETC). The ETC pumps H^+ ions out of the matrix, establishing a $[H^+]$ gradient and an electrical potential across the inner mitochondrial membrane. This electrochemical gradient for H^+ is used by the mitochondrial F_0F_1ATPase, the **ATP synthase**, to synthesize ATP from ADP and Pi. Net ATP production from oxidative phosphorylation alone is about 25 ATP molecules per molecule of glucose. Oxygen is needed as the final electron acceptor from the ETC. Without oxygen, the ETC remains reduced and everything backs up. The TCA cycle stops for lack of NAD^+, and beta oxidation of fats stops for the same reason.

THE METABOLIC PATHWAYS USED BY MUSCLE VARIES WITH INTENSITY AND DURATION OF EXERCISE

Muscles can use fats, carbohydrates, and proteins as fuels. Which is used at what rates depends on the type, intensity, and duration of exercise. At rest, muscles use mainly free fatty acids (FFAs). At moderate exercise (<50% maximum O_2 consumption (V_{O_2})), muscles use blood glucose and free fatty acids. At higher intensities of exercise (>50% V_{O_2}), the proportion contributed by glycogen becomes increasingly important so that at 70–80% V_{O_2} aerobic metabolism of glycogen is predominant. During mild exercise of long duration, there is a gradual increase in dependence on FFA over glucose. Table 3.7.2 shows the rates of ATP production and amounts of ATP available from various sources. It is important to remember that every muscle contraction utilizes creatine phosphate and glycogen to lactate, but at low frequency the oxidation of lactate or blood glucose during the rest period provides energy to resynthesize creatine phosphate and glycogen.

AT HIGH INTENSITY OF EXERCISE, GLUCOSE AND GLYCOGEN ARE THE PREFERRED FUEL FOR MUSCLE

Muscle cells burn glucose, but the amount of free glucose is limited and cannot fuel muscle activity by itself. Muscles and liver store carbohydrates as glycogen. Sympathetic nervous activity and circulating epinephrine activate glycogenolysis, the breakdown of glycogen, to provide glucose for muscle activity. The enzyme phosphorylase in muscle and liver breaks down glycogen. It is activated by phosphorylation by phosphorylase kinase, which is in turn controlled by phosphorylation by protein kinase A, PKA. PKA is activated by 3′,5′cyclic AMP that is produced by adenylyl

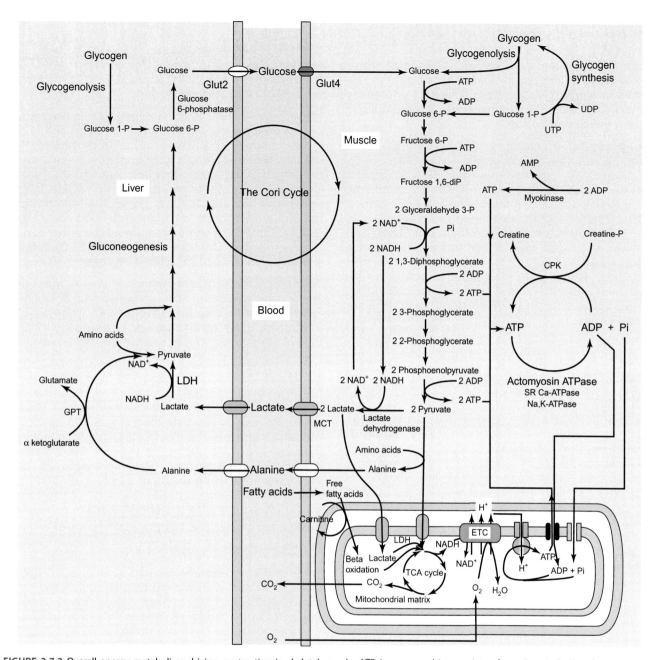

FIGURE 3.7.2 Overall energy metabolism driving contraction in skeletal muscle. ATP is consumed in a variety of reactions including the actomyosin cross-bridges and the SR Ca-ATPase pump. ATP is provided by a variety of routes including creatine phosphokinase (CPK) converting creatine P and ADP to ATP, glycolysis and complete oxidation of carbohydrates through the TCA cycle and electron transport chain (ETC). The source of glucose for glycolysis can be muscle glycogen or plasma glucose. The glucose is imported into the muscle by a glucose transporter, Glut4. Plasma glucose originates from liver and extrahepatic tissues either through glycolysis (liver) or gluconeogenesis (liver, kidneys, intestine). Fatty acids form acetyl CoA through beta oxidation and the acetyl CoA is then completely oxidized, in the presence of adequate oxygen, in the mitochondria. These fatty acids derive principally from adipose stores. When oxygen is insufficient to oxidize myoplasmic NADH, glycolysis continues by the regeneration of NAD^+ by converting pyruvate to lactic acid by lactate dehydrogenase (LDH). Production of lactic acid thereby allows glycolysis to continue. Lactic acid produced in this way is transported into the blood and from there to the liver where it can be converted to glucose again. This cycle of muscle glucose to lactate to liver lactate to glucose is the Cori cycle. Muscle also participates in the alanine–glucose cycle in which amino acids in muscle are converted to alanine and exported to the liver to make glucose by gluconeogenesis.

TABLE 3.7.2 Rate and Amount of ATP Available for Contraction from Various Fuel Sources

Source of Energy	Rate of ATP Production (mol/min)	Amount of ATP Available (mol)
ATP and creatine phosphate	4.4	0.7
Glycogen to lactate	2.4	1.6
Muscle glycogen to CO_2	1.0	84
Liver glycogen to CO_2	0.4	19
Fatty acids to CO_2	0.4	4000

cyclase following the activation of a G_s-coupled receptor. The rapid utilization of ATP in normal contractions appears to require glycogenolysis. Glycogen is partially regenerated during the resting phase of the muscle between trains of impulses.

Glycogen stored in muscle is dedicated to glycolysis because glycogenolysis produces G-1P and then G-6P, and muscle lacks **glucose-6-phosphatase** that converts G-6P to glucose. Glucose can cross the cell membrane over its transporter, but ionically charged G-6P cannot. Because muscle lacks the enzyme to make free glucose, muscle cannot export significant glucose. Liver and other tissues produce glucose that can travel to muscle through the blood. Muscle tissues take up glucose by a transporter, Glut4, that is sensitive to exercise. The Glut4 transporter is recruited to the cell membrane by insulin, but exercise also recruits Glut4 in the absence of insulin.

LACTIC ACID PRODUCED BY ANAEROBIC METABOLISM ALLOWS HIGH GLYCOLYTIC FLUX

As shown in Figure 3.7.2, glycolysis does not require oxygen to produce ATP, so this is called **anaerobic metabolism**, but it does reduce NAD^+ to NADH. The NAD^+ is an obligatory cofactor for the reaction of glyceraldehyde-3-phosphate to 1,3-diphosphoglycerate. If the cell runs out of cytoplasmic NAD^+, glycolysis will halt. Cytoplasmic NADH produced by glycolysis can be oxidized back to NAD^+ by the mitochondria through shuttle systems described in Chapter 2.10. Conversion of NADH to NAD^+ requires an oxidized electron transport chain. During rapid bursts of glycolysis, the mitochondria cannot keep up with the NADH generated and NAD^+ levels could fall. Lactic dehydrogenase converts pyruvic acid to lactic acid, simultaneously converting NADH to NAD^+. This NAD^+ can then be used to allow glycolysis to proceed at the glyceraldehyde-3-phosphate step. Thus **lactic acid production allows glycolysis to proceed during rapid glycolytic bursts of ATP production**.

MUSCLE FIBERS DIFFER IN THEIR METABOLIC PROPERTIES

Muscles can be classified by their mechanical properties (see Chapter 3.4) and by their myosin staining (see Chapter 3.5). Muscle fibers also differ in their metabolic capabilities. Peter and coworkers described three types of fibers: slow oxidative (SO), fast glycolytic (FG), and fast oxidative-glycolytic (FOG). Table 3.7.3 compares the various classification schemes.

The different muscle fiber types originate in the expression of different isoforms of muscle proteins and in the relative amount of different organelles in the cell. Red muscle fibers that contain a lot of myoglobin and mitochondria (giving them their red appearance) have a large oxidative capacity and they are slower and fatigue resistant. White muscle fibers containing little myoglobin and mitochondria are fast-twitch fibers that fatigue rapidly. Table 3.7.4 compares some of the different proteins expressed in different muscle types and the relative abundance of selected subcellular organelles.

Whole muscles contain thousands of muscle fibers that are distributed among the various muscle fiber types. Specific muscles may be predominantly composed of one muscle fiber type or another. The distribution of muscle fiber types within specific muscles varies with the individual.

The classification schemes shown in Table 3.7.3 are useful ways of organizing how we think about muscle, but they do not truly reflect the heterogeneity of muscle fibers. The muscle fibers are more continuous in the distribution of their characteristics, with gradations between them that are lost in the few classes that we recognize. Nevertheless, these schemes help us to think about how muscles do their job.

BLOOD LACTATE LEVELS RISE PROGRESSIVELY WITH INCREASES IN EXERCISE INTENSITY

Early studies of exercise showed a progressive rise in blood lactic acid with increasing intensity of exercise, as shown in Figure 3.7.3. Increased circulating lactic acid originates in the working muscles. Originally lactic acid was thought to be produced during anaerobic metabolism, so that the increased levels in the blood were thought to represent increased reliance on anaerobic metabolism. For this reason, the knee in the curve was called the "anaerobic threshold." The classical view is that lactate is produced only under anaerobic conditions, when muscle P_{O_2} falls below levels that fully energize mitochondria. This view is now thought to be wrong. The main fact that does not fit is that lactate is produced by exercising muscles that are fully oxygenated. Lactic acid is normally part of muscle metabolism, but in high intensity exercise there is a mismatch between lactic acid production and oxidation.

TABLE 3.7.3 Muscle Fiber Type Classification Schemes

Classification Scheme	Muscle Property Used to Classify Types	Fiber Types
Burke	Mechanical properties	S, FR, FI, FF
Brooke	Myosin ATPase staining	I, IIA, IIB, IIC
Peter	Metabolic capacity	SO, FOG, FG

TABLE 3.7.4 Comparison of Selected Protein Isoforms and Organelles in Different Muscle Fiber Types

	Type I Muscle	Type IIa Muscle	Type IIb Muscle	Cardiac Muscle
Twitch	Slow	Fast	Fast	
Fatigue	Resistant	Resistant	Fatiguable	Resistant
Metabolism	Oxidative	Oxidative	Glycolytic	Oxidative
Mitochondria	+++	++++	+	++++
SR volume	++	+++	++++	+
Glycogen	+	+++	++++	++
Myosin heavy chain	MHC-I	MHC-IIa	MHC-IIb, -IIx	MHC-α, MHC-β
Myosin light chain	MLC-1aS, -1bS	MLC-1f, -3f	MLC-1f, -3f	MLC-1v, -1a
SR Ca-ATPase	SERCA2a	SERCA1a	SERCA1a	SERCA2a
Phospholamban	++	−	−	+
Calsequestrin	Fast and cardiac	Fast	Fast	Cardiac
RyR	RyR1	RyR1	RyR1	RyR2
Troponin C	TnC_1	TnC_2	TnC_2	TnC_1
Myoglobin	+++	+++	−	+++
Parvalbumin	−	+	++	−

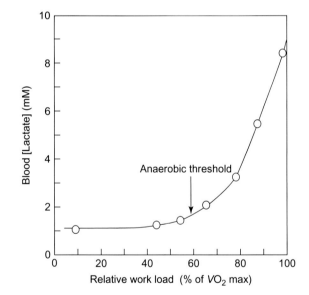

FIGURE 3.7.3 Blood lactate concentration as a function of relative work load. Lactate levels in blood increase only gradually until about 60% of $V_{O_2\ max}$ is reached, and then lactate concentration increases markedly with further increases in exercise intensity.

MITOCHONDRIA IMPORT LACTIC ACID, THEN METABOLIZE IT; THIS FORMS A CARRIER SYSTEM FOR NADH OXIDATION

Mitochondria possess a monocarboxylic acid transporter (MCT1) that allows lactic acid to enter the mitochondria. Mitochondria also possess lactic dehydrogenase, LDH, which converts lactic acid to pyruvate. The pyruvate is then oxidized by the TCA cycle. The ability of lactic acid to enter the mitochondria and be converted back to pyruvate forms a pathway for the oxidation of cytoplasmic NADH, as shown in Figure 3.7.4. Lactic acid is not a shuttle system in that it is consumed by the mitochondria. Thus it is more like a carrier of reducing equivalents.

LACTATE SHUTTLES TO THE MITOCHONDRIA, OXIDATIVE FIBERS, OR LIVER

The fastest muscle fibers are expected to produce lactic acid at the highest rate. The first method of removing lactic acid is through mitochondria in the same cells that produce it. The lactic acid is taken up into the mitochondria by the MCT1 and the lactic acid is converted to pyruvate and oxidized. This intracellular shuttle is shown in Figure 3.7.5. When lactic acid production outstrips the capacity of lactic acid to be metabolized within the fast-twitch muscles, it spills out into the extracellular space. Some of this lactate is taken up by oxidative fibers, which are generally smaller than the large glycolytic fibers. Thus some of the lactic acid produced and released by muscle is metabolized in the muscle itself. This constitutes the cell−cell shuttle, also shown in Figure 3.7.5.

The liver can take up lactate that is released into the blood by the active muscles. The liver either metabolizes the lactate for energy or uses it to make new glucose through gluconeogenesis, and exports the glucose into the blood. Muscles can then take up this glucose and use it again for energy. This cycle of blood glucose to muscle lactate to blood lactate to liver lactate and back to blood glucose is called the Cori cycle, shown in Figure 3.7.5.

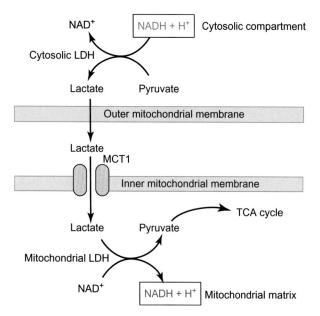

FIGURE 3.7.4 Lactic acid carries reducing equivalents into the mitochondria. Cytosolic NAD^+ is an obligatory requirement for glycolysis. Conversion of pyruvate to lactic acid in the cytoplasm regenerates NAD^+ so that glycolysis can continue. The lactic acid enters the mitochondria over the MCT1 carrier (which also transports pyruvate) and is converted back to pyruvate in the mitochondria by mitochondrial lactate dehydrogenase (LDH), which also converts NAD^+ to NADH in the matrix. The NADH generated in the mitochondria can be oxidized back to NAD^+ by the electron transport chain. The pyruvate can enter the TCA cycle through pyruvate dehydrogenase.

THE "ANAEROBIC THRESHOLD" RESULTS FROM A MISMATCH OF LACTIC ACID PRODUCTION AND OXIDATION

The increase in blood lactate with intensity of exercise results from the release of more lactic acid by the exercising muscles than can be metabolized by the aggregate tissues of the body. This has the appearance of an increase of anaerobic metabolism, and in one sense it is. Every muscle contraction involves "anaerobic" generation of ATP, meaning through the process of glycolysis. However, the tissue is not anaerobic. In less intense exercise, there is sufficient rest time for the lactic acid produced during this period to be oxidized. When exercise intensity increases, three different things happen:

1. The rest period between contractions becomes shorter.
2. The fast glycolytic fibers are increasingly recruited over the oxidative fibers, causing release of more lactic acid.
3. The sympathetic nervous system increases the rate of glycogenolysis, further increasing the supply of pyruvate, and, by mass action, the production of lactate.

Because of these events, lactate release by the active muscles soars with increased exercise intensity and outstrips the ability of the tissues to metabolize the lactate. This occurs without gross anaerobic conditions. The muscles are still fully oxygenated. There is no anaerobiosis, yet there is increased production of lactate by anaerobic pathways.

EXERCISE INCREASES GLUCOSE TRANSPORTERS IN THE MUSCLE SARCOLEMMA

Extracellular glucose enters muscle fibers through specific transporters, Glut4 transporters, in the muscle fiber membrane. The rate of uptake depends on the number of these transporters in the membrane and their activity. Insulin, an important hormone produced by the β cells in the islets of Langerhans in the pancreas, increases the uptake of glucose by the peripheral tissues, especially muscle, by recruiting Glut4 transporters from latent storage in vesicles in the muscle fiber. This mechanism is shown in Figure 3.7.6.

Contractile activity also increases the number of Glut4 transporters, but this effect is additive to the effect of insulin. Thus exercise itself exerts an insulin-like effect and increases glucose uptake without increases in insulin. The mechanism by which exercise increases Glut4 transporters is not yet completely worked out. However, researchers believe that AMPK, a protein kinase stimulated by AMP, and calmodulin-dependent protein kinase, CAMK, may be involved. AMPK is stimulated by AMP, which is produced from ADP by myokinase when ADP concentrations rise during contractions. The AMPK is thought to be a kind of "fuel gauge" that senses low fuel levels and then switches off ATP-consuming reactions and turns on ATP-producing reactions. AMP acts in a negative feedback mechanism to restore ATP levels. AMPK phosphorylates AS160 (Akt substrate molecular weight 160 kDa), a GTPase-activating protein (GAP) that inactivates Rab. The human genome codes for over 60 Rabs, which are involved in vesicular tethering to the cytoskeleton, trafficking and docking and fusion to membranes. Rab is active when it is bound to GTP. Its GTPase activity splits GTP to GDP, and then Rab is inactive. AS160 activates the GTPase of Rab, thereby inactivating Rab. Phosphorylation of AS160 converts it to an inactive state, which thereby activates Rab by removing AS160 inactivation. How contraction causes Glut4 insertion into the plasma membrane is presently unknown, but it probably involves CAMK, NO, and AMPK. CAMK is activated by Ca^{2+} when it rises to activate the myofilaments. If it simultaneously activates Glut4 transporters, it acts as a feed-forward mechanism to begin increasing ATP supply in anticipation of its need to support contraction. PKC, CAMK and nitric oxide, NO, are all increased after activity and are thought to result in increased Glut4 incorporation into the surface membrane, but the mechanisms by which this happens are not yet worked out.

Because exercise has an insulin-like effect on glucose uptake by muscles, diabetic persons who inject insulin must cut back on their insulin shots when they anticipate they will exercise.

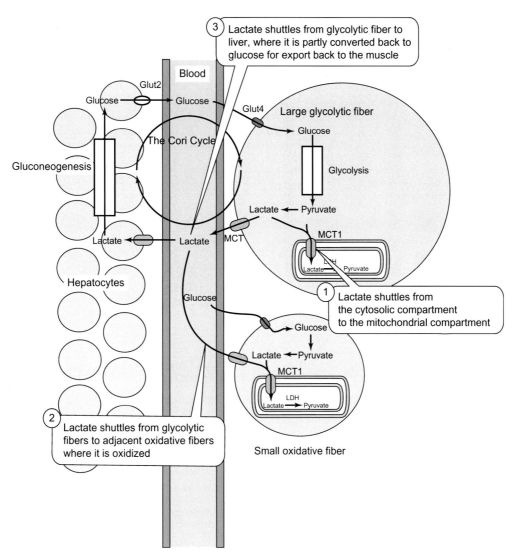

FIGURE 3.7.5 The lactate shuttles. Lactate is produced in the cytoplasm by LDH acting on pyruvate and NADH. Lactate can be shuttled from the cytoplasmic compartment to the mitochondrial compartment by importing the lactate into the mitochondria and linking it to the synthesis of pyruvate and the generation of NADH in the mitochondrial matrix. This is the intracellular lactate shuttle. Secondly, lactate can be exported into the blood where it is taken up by adjacent oxidative muscle fibers and completely oxidized by their mitochondria. This is the cell-to-cell lactate shuttle. Third, lactate released into the blood when lactic acid production is high can be taken up by liver cells (also called hepatocytes). The hepatocytes resynthesize glucose from the lactate and export it back into the blood where it can be taken up by the exercising muscle, for example. This is the Cori cycle.

FATIGUE IS A TRANSIENT LOSS OF WORK CAPACITY RESULTING FROM PRECEDING WORK

There are two types of fatigue. Exercise performed at submaximal effort for many repetitions eventually results in a loss of the ability to generate the submaximal force. This fatigue takes some time to develop and time to recover from. The second type of fatigue occurs at maximal effort. This involves maximal recruitment and a sustained volley of nerve impulses to continuously activate the muscle with no rest phase. Fatigue here is rapid in onset and also recovers rapidly. The definition of a maximum bench press is that you cannot do it twice, so fatigue in this sense occurs within seconds.

Because submaximal effort can be anything from near maximum to very light, fatigue can occur in a continuous spectrum of effort and repetitions. These two archetypes of many repetitions and light load versus a single effort at maximal load have different mechanisms for the origin of the fatigue.

FATIGUE HAS MULTIPLE CAUSES

Anything that happens from the cerebral cortex where motor commands originate, to the lower motor neuron where the signals are integrated, to the neuromuscular junction, EC coupling, and formation of cross-bridges are potential sites for partial failure during fatigue. We might expect that the cause of fatigue will not be the same along the spectrum of effort and repetition.

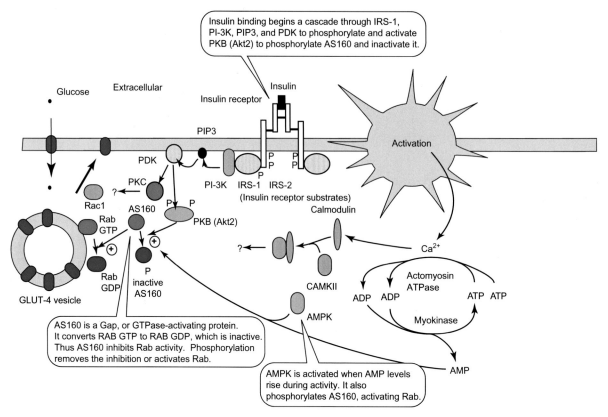

FIGURE 3.7.6 Insulin and exercise increase GLUT4 incorporation into the sarcolemma of muscle fibers by recruiting transporters from a population of latent transporters located in vesicles in the cell. The increased numbers of GLUT4 transporters increase glucose uptake and generation of ATP through glycolysis and oxidation of pyruvate or lactate. Insulin binds to its receptor, a receptor tyrosine kinase, that phosphorylates IRS-1, insulin receptor substrate 1, and transphosphorylates itself. IRS-1 activates PI-3K, phosphatidylinositol 3-kinase, that increases the concentration of PI 3,4,5-P_3 that, in turn, activates PDK, phosphatidylinositol-dependent kinase. PDK phosphorylates PKB (protein kinase B, also called Akt—in this case Akt2). Active PKB phosphorylates AS160, Akt substrate with molecular weight 160 kDa. AS160 in its unphosphorylated state is a GAP, GTPase-activating protein, that stimulates the GTPase activity of Rab proteins on GLUT4 vesicles within the cell. Active Rab is part of the machinery involved in trafficking of the vesicles towards the cell membrane and fusion of the membrane with the sarcolemma. Active Rab has GTP bound. Thus, AS160, by stimulating Rab GTPase activity, converts it from the active to inactive state. Phosphorylation of AS160 makes AS160 inactive and thus preserves active Rab-GTP. In this way, insulin stimulates the incorporation of GLUT4 into the membrane. Activity of the muscle also does this, but through separate paths that converge onto Rab. During activity, AMP levels rise and activate AMPK (AMP Kinase) that also phosphorylates AS160. Calmodulin-dependent protein kinase (CAMK) is also thought to increase GLUT4 incorporation, but its targets are not yet certain.

IN HIGH INTENSITY, SHORT DURATION EXERCISE, FATIGUE IS LIKELY DUE TO INCREASED CYTOSOLIC PI

In high intensity, very short duration exercise such as maximum weight lifting, it is thought that fatigue resides in the muscles alone. Here athletes have extremely strong motivation and most likely recruit all of their muscle fibers. Since the last recruited fibers are generally fast-twitch fibers that fatigue easily, reduction in force with continued effort is likely due to reduced force in those fibers. In fast-twitch fibers activated for short bursts, creatine phosphate regenerates ATP from ADP. The terminal phosphate comes from the creatine phosphate, so that the regeneration of ATP from creatine phosphate results in a large buildup of Pi in the myoplasm. At the same time, the activation of anaerobic glycolysis results in ATP production with a buildup of lactic acid and H^+ ions (due to dissociation of the acidic group on lactate). During exercise, the pH of muscle can fall from pH 7.0 to pH 6.0. Formally it was believed that H^+ interferes with force production, either by directly inhibiting the actomyosin ATPase or by making the myofilaments less sensitive to activator Ca^{2+}. This effect is now thought to be less important, whereas Pi can inhibit force development both by interfering with cross-bridge cycling and by reducing the Ca^{2+} transient. Because creatine phosphate content in fast-twitch fibers can reach as high as 40 mM, regeneration of ATP can increase [Pi] higher than 20 mM. At these concentrations, Pi can enter the SR and precipitate Ca^{2+} within the lumen, making it less available for release during E–C coupling. The resulting reduction in force is perceived as fatigue. **High intensity, short duration exercise causes fatigue mainly through increased cytoplasmic [Pi].**

IN LONG DURATION, REPETITIVE EXERCISE, FATIGUE IS SHARED WITH GLYCOGEN DEPLETION AND CENTRAL FATIGUE

Central fatigue is the term given to the reduction in force related to reduced recruitment by the central nervous system. Peripheral fatigue involves any changes within the muscle that leads to a reduced response to neural excitation. We all have the subjective experience of being fatigued, and that accomplishing a motor task requires an extra dose of mental effort when we are fatigued. How do we obtain the subjective perception of fatigue?

Evidence is mounting that group III (thinly myelinated fibers) and group IV (unmyelinated fibers) carry afferent information from muscle fibers to the dorsal horn of the spinal cord and from there affect both spinal and supraspinal sites. The output of group III and IV fibers increases during exercise and stays increased for the duration of the exercise. Input from these fibers increases blood flow to active muscles and increases ventilatory drive during exercise. This assures oxygen and nutrient supply to the exercising tissues and helps prevent peripheral fatigue in exercising muscles. The output of group III and IV fibers appears to inhibit spinal motor neurons and, presumably, it is the group III and IV supraspinal effects that give rise to the subjective perception of fatigue. It appears that there may be more than one type of receptor involved in these afferent signals. One type, the so-called metaboreceptors or ergoreceptors, respond to normal levels of ATP, lactate, and [H^+] associated with nonfatiguing exercise. A second type, metabonociceptors, respond to higher concentrations of these materials that is associated with fatigue or damage to the tissues.

The performance times at high but submaximal workloads depend on the size of the glycogen stores before exercise. Fatigue appears when glycogen levels fall but before they are zero. This has led to the hypothesis of the "glycogen shuttle" in which glycogenolysis is necessary to maintain ATP during contraction, and it is resynthesized during the rest period between contractions. When glycogen becomes low, it can no longer sustain ATP levels during contraction and force falls, even though glycogen is not completely used up. Thus **low intensity, long duration exercise causes fatigue through glycogen depletion**. This explanation has led to the training regimen of **carbohydrate loading** to postpone fatigue during endurance athletic events. In this regimen, glycogen stores are increased by a combination of exercise and carbohydrate consumption. It is usually accomplished by exhaustive exercise followed within 2 h by a high-carbohydrate meal. Under these conditions, the glycogen stores **supercompensate** and stores larger than normal amounts of glycogen.

GLYCOGENOSES SHOW THE IMPORTANCE OF GLYCOGEN TO EXERCISE

Persons with deficits in the enzymes involved in glycogen metabolism have exercise intolerance (see Clinical Applications: Muscle glycogenoses and Figure 3.7.7). However, the mechanism by which glycogen depletion impairs force development remains unknown. If the only function of glycogen is to regenerate ATP, then the proximate cause of muscle force decline should be reduced ATP concentrations or increased ADP and Pi. But fatigue in endurance exercise appears to be unexplained by gross reductions in ATP or increases in ADP and Pi. The key appears to be in the heterogeneity of glycogen stores. Glycogen is located in three separate compartments in the muscle fiber: (1) subsarcolemmal glycogen; (2) intermyofibrillar glycogen (between the myofibrils); and (3) intramyofibrillar glycogen (within the myofibrils). Glycogen in these compartments is used to fuel specific activities of the cell. The subsarcolemma glycogen fuels the surface membrane Na,K-ATPase, whereas the intramyofibrillar glycogen fuels the SR Ca-ATPase. The intramyofibrillar pool of glycogen is depleted more readily by exercise. The hypothesis is that glycogen depletion in this compartment reduces Ca^{2+} release by the SR, so that the contraction in response to excitation has reduced force, i.e., the muscle fiber is fatigued.

INITIAL TRAINING GAINS ARE NEURAL

During initial training (the first few weeks) of naïve subjects, the maximal voluntary contraction increases whereas the maximal evoked contraction (produced by direct and maximal stimulation of the motor nerve) does not. This suggests that trainees learn to activate their muscles more fully or they improve coordination of the voluntary contraction. Increases in maximal evoked contraction require longer training. Thus early and rapid strength gains come from training the brain.

MUSCLE STRENGTH DEPENDS ON MUSCLE SIZE

The maximum force that can be exerted by a muscle depends on its cross-sectional area and architecture such as pinnation. Strength training employs contractions against large resistances with few repetitions. It is called **resistance training**. Beginning training is associated with **delayed onset muscle soreness** or **DOMS**. Part of the soreness could be due to stretching of sensory neurons either from edema or from mechanical stretch. The afferent information is probably carried by unmyelinated fibers from the muscle. The exercise signals muscle hypertrophy: the diameter of the fibers increases but the number of fibers stays the same. It may be that some limited amount of hyperplasia (increase in cell number) may also occur. Both type I and type II fibers increase in size in response to heavy resistance training. Hypertrophy occurs two ways: muscle fibers make more myofibrils and satellite cells within the muscle fuse with existing muscle fibers to help control the additional cytoplasm. During development, satellite cells are recruited to form myotubes that further differentiate to become muscle fibers. The signals for muscle hypertrophy are summarized in Figure 3.7.8.

ENDURANCE TRAINING USES REPETITIVE MOVEMENTS TO TUNE MUSCLE METABOLISM

Endurance training increases the capillarity of muscles and tunes the muscles' metabolic capabilities. Concentrations of myoglobin and TCA cycle enzymes are increased as well as both the size and number of mitochondria. Muscles of endurance-trained subjects use fats

344 QUANTITATIVE HUMAN PHYSIOLOGY: AN INTRODUCTION

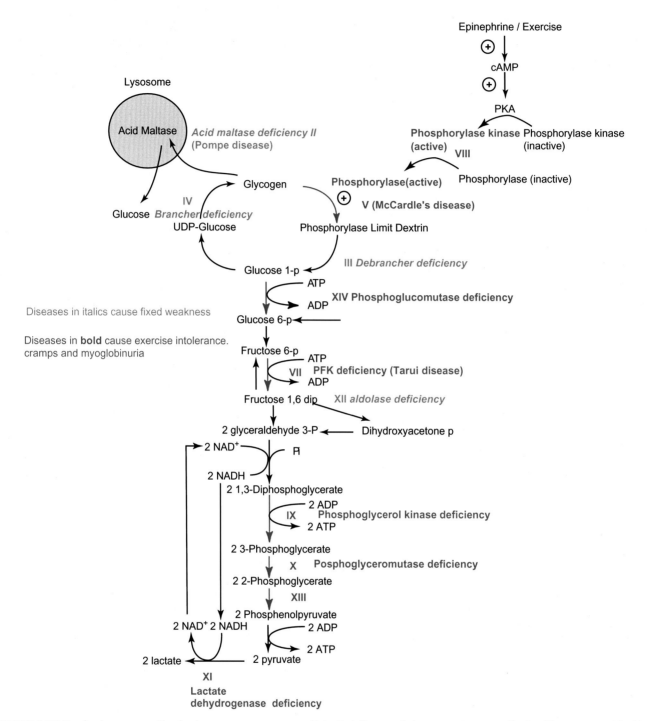

FIGURE 3.7.7 Muscle glycogenoses. Muscle glycogenoses are a group of inherited diseases of glycogen storage or utilization. They present with either exercise intolerance, fatigue, cramps and myoglobinuria, or fixed muscle weakness. Those shown in italics present with fixed muscle weakness. Those shown in the figure in bold present with exercise intolerance, cramps, fatigue, and myoglobinuria. Myoglobinuria typically does not show up unless exercise is strenuous. Many persons afflicted with these diseases learn to exercise within their limits and so maintain fairly normal life styles.

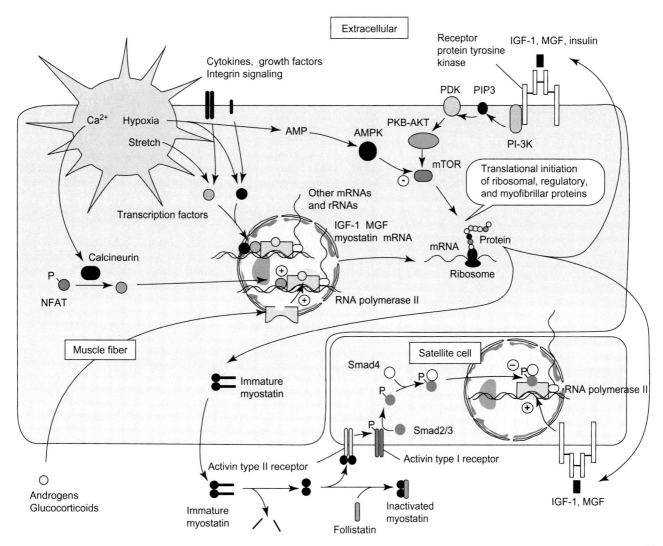

FIGURE 3.7.8 The signaling events in muscle hypertrophy. Muscle grows in response to stretch, increases in the integrated cytoplasmic [Ca^{2+}], androgens, glucocorticoids, and cytokines. Calcium activates calcineurin, a protein phosphatase that dephosphorylates NFAT (nuclear factor of activated T cells) and activates it. Myostatin produced by muscle inhibits satellite cell division and differentiation. Muscle cells also respond to IGF-1 (insulin-like growth factor-1) and MGF (muscle growth factor). Myostatin is also known as growth and differentiation factor 8 (GDF-8). It is a member of the transforming growth factor β family (TGFβ). It is secreted as an inactive peptide that dimerizes and is then cleaved to its active form. Follistatin, another regulator, can bind to myostatin and inhibit it. Myostatin binds to activin receptor type IIB. This phosphorylates activin receptor type I. (These activin receptors are serine–threonine protein kinases; there are currently 5 type II and 7 type I receptors. The type I receptors are also referred to as activin-like kinases, or ALK1, ALK2, and so on.) The activated activin receptor type I then phosphorylates members of a family of proteins called Smads. Eight different Smads have been identified in mammals. These Smads then control gene expression in the muscle fiber and satellite cells.

as the primary fuel for moderate exercise, thereby sparing glycogen for bursts of high intensity activity.

ENDOCRINE AND AUTOCRINE SIGNALS REGULATE MUSCLE SIZE (=STRENGTH)

Growth hormone is secreted by the anterior pituitary. It induces the liver to form insulin-like growth factor (IGF-I) that circulates in the blood. Muscles have receptors for IGF-1 that activate the cascade leading from phosphatidylinositol 3-kinase (PI-3K) to activation of protein kinase B (PKB-AKT) and mammalian target of rapamycin (mTOR) that stimulate muscle growth. This is inhibited by AMP-dependent protein kinase when AMP levels rise during hypoxia, for example. Muscles are also responsive to anabolic steroids and androgens. This is the basis for the generally greater strength and muscle size in men compared to women.

Myostatin is an autocrine and paracrine hormone produced by muscle cells that inhibits muscle differentiation and growth. Knockout mice without myostatin and certain breeds of cattle (Belgian Blue and Piedmontese) that lack effective myostatin are "double muscled." Because myostatin also targets adipocytes, these animals also lack adipogenesis (formation of fat) and so they are lean as well. A report was published about a young boy who had a myostatin mutation that is associated with gross muscle hypertrophy (*New England Journal of*

Clinical Application: Muscle Glycogenoses

Muscle glycogenoses are a group of inherited defects in enzymes that deal with glycogen storage or utilization, or carbohydrate metabolism. Their clinical presentations take two forms. One form presents with exercise intolerance, with cramps, fatigue, and myoglobinuria that results from breakdown of the muscle fibers and spilling of myoglobin into the blood (**rhabdomyolysis**) and from there into the urine. The second form presents with a progressive muscle weakness. There are number of different types of glycogenoses, and some of these are shown in Figure 3.7.7. They differ in the enzymes affected, but many of these defects have subtypes that differ in their specifics. The iconic example of glycogenoses is **McArdle's disease**, a deficiency of muscle phosphorylase. McArdle's disease is also known as glycogen storage disease V (GSD V). Without muscle phosphorylase, muscle glycogen cannot be broken down to glucose-1-phosphate and so muscles cannot use the glycogen for ATP production. The result is an exercise intolerance. In persons with a deficiency of debrancher enzyme, GSDIII, glycogen can be broken down to limit dextrins, and some glucose-1-phosphate can be produced. The result is muscle weakness rather than exercise intolerance.

Glycogenoses can be diagnosed through genome sequencing, but they can also be diagnosed clinically through the ischemic exercise test. In this test, an iv cannula with a three-way valve is inserted into a large vein in the forearm. A blood sample is obtained 2 minutes before exercise as a baseline. A pressure cuff around the upper arm is inflated above systolic pressure to stop blood flow to the forearm. This is the ischemic part. The patient then exercises the forearm by squeezing something—rolled up towel, a ball, or a dynamometer—repetitively for 2 minutes. After each squeeze, the fingers are extended completely. The exercise is valid when the patient can no longer fully extend the fingers. After 2 minutes, the blood pressure cuff is released (mark time zero) and blood samples are obtained at 1, 3, 5 and 10 minutes and analyzed for lactate, ammonia, pyruvate, creatine kinase, and phosphate. In normal persons, lactate levels increase from 1 mM to 3–5 mM, and ammonia levels rise from 40 μM to 100 μM. In persons with muscle phosphorylase deficiency, lactate hardly rises because the muscles can provide no substrate (glucose) without blood flow. This is diagnostic of glycogenosis, particularly McArdle's disease. Lack of increase in ammonia results from myoadenylate deaminase deficiency.

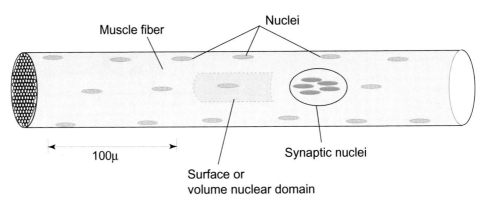

FIGURE 3.7.9 Nuclear and membrane domains in skeletal muscle fibers. Muscle fiber nuclei are located near the periphery of the fiber, nearly aligned in rows with more or less regular spacing with some 35–80 nuclei per mm of fiber. Each nucleus controls protein expression in a volume or surface element called its domain.

Medicine **350**:2682–2688, 2004). Its mode of action is illustrated in Figure 3.7.8.

HUMAN ABILITY TO SWITCH MUSCLE FIBER TYPES IS LIMITED

The overall mechanical and metabolic performance of the muscles is a consequence of the heterogeneous mosaic composition of different muscle fiber types that make up the muscle. One basis of this heterogeneity is the expression of specific myosin isoforms. There are at least 20 structurally distinct classes of myosin heavy chains. Eleven of these are expressed in adult mammalian muscles, but some are specific to one muscle. The most common isoforms are MHCIb, MHCIIa, MHCIIb, and MHCIId.

Immunohistochemistry reveals that some muscle fibers are "pure" types that express only a single myosin heavy chain (MHC) isoform, whereas other muscle fibers contain two or more isoforms and are "hybrids." How can this be? Recall that each muscle fiber contains many nuclei, typically located at the periphery of the cell near the sarcolemma, as shown in Figure 3.7.9. Each nucleus controls a volume of cytoplasm or surface of the fiber called its "**nuclear domain**." A separate population of nuclei congregates near the neuromuscular junction. During transitions between fiber types, it is possible that some of these nuclei receive different signals than others, and therefore transcribe different genes for the expression of myosin. In this way, hybrid muscle fiber types arise. The existence of hybrid muscle fibers allows a more continuous gradation between muscle types, as shown in Table 3.7.5.

Is it possible for humans to convert a type I slow fiber into a type II fast fiber or vice versa? In experimental animals, a fast-twitch muscle removed from its bed and transplanted to a slow-twitch muscle bed converts part way from slow twitch to fast twitch, and vice versa, suggesting that the pattern of neural stimulation determines muscle type. Chronic low-frequency stimulation

Muscle Fiber Type	Myosin Heavy Chain Expression	Muscle Fiber Description
Type I pure fiber	MHCI	Slow
Hybrid	MHCI > MHCIIa	
	MHCIIa > MHCI	
Type IIa pure fiber	MHCIIa	Fast, fatigue resistant
Hybrid	MHCIIa > MHCIIx	
	MHCIIx > MHCIIa	
Type IIx pure fiber	MHCIIx	Fast, fatigable

TABLE 3.7.5 The Muscle Fiber Type Continuum

Hybrid muscle fibers allow transitional forms intermediate between Types I, IIa and IIb. Humans do not make MHCIIb (experimental animals do), so the human form in fast fatiguable muscles is named MHCIIx

of fast-twitch fibers increases the expression of proteins normally expressed only by slow-twitch fibers. Denervation or muscle unloading increases the levels of proteins normally expressed by fast-twitch fibers. Although the evidence is inconclusive, it appears that the stimulation of muscle necessary to transform the fiber types is so severe that no human can train that hard. The consensus seems to be that the transformation of muscle types is limited in part by the original position of the muscle on the muscle fiber-type continuum. The transformation by exercise is always towards a slower type of muscle, but conversion of type IIx fiber to a type I fiber does not occur.

SUMMARY

Muscles produce force or shorten because the actomyosin cross-bridge cycle is activated. This cycle hydrolyzes ATP. Both force and shortening derive their energy from the energy of ATP hydrolysis. Buildup of ADP and Pi slows cross-bridge cycling and reduces the energy available for work. Muscle fibers keep ATP levels high by regenerating ATP through phosphagen buffer systems and metabolism. The first buffer for ATP is creatine phosphate, which rapidly converts ADP to ATP by creatine kinase. Myokinase can convert 2 molecules of ADP to one ATP and one AMP. ATP can also be regenerated from ADP and Pi by glycolysis and by oxidative phosphorylation. Glycolysis makes ATP quickly and anaerobically, but its capacity is low. Oxidative phosphorylation is slower but it has a much larger capacity. During bursts of activity when ATP consumption outstrips oxidative phosphorylation, muscles markedly increase release of lactic acid. Lactic dehydrogenase makes lactic acid from pyruvic acid, simultaneously oxidizing NADH to NAD^+. By regenerating NAD^+, formation of lactic acid allows glycolysis to continue. Lactate is always produced by muscles. The lactate is oxidized within the muscle fibers, or by adjacent oxidative muscle fibers with more mitochondria, or by being exported to the liver where it is used to regenerate glucose. The cycle of blood glucose to muscle glucose to lactic acid to liver and back to blood glucose is called the Cori cycle.

Muscles can burn carbohydrates, fats, and proteins for energy. Which fuel is used depends on the duration and intensity of exercise. Rapid bursts of high activity are generally fueled by glycolysis, whereas slower activity lasting for long periods relies on oxidation of fats. Exercise increases the Glut4 carriers in the muscle to increase their glucose uptake independently of insulin.

Muscles are made up of thousands of muscle fibers that can be grouped into a few kinds of muscle fibers. These have different contractile and metabolic characteristics and express particular isoforms of a variety of muscle proteins. The type I muscle fiber is a slow-twitch fiber that typically contains a slow myosin isoform but contains lots of mitochondria and myoglobin in the cytoplasm. It expresses a slow-twitch SR Ca-ATPase and its TnC has only one regulatory Ca^{2+}-binding site. Type I fibers are oxidative and fatigue resistant. Type II fibers are fast-twitch fibers and there are two main varieties: type IIa fibers are fast oxidative fibers that contain SERCA1a SR pumps and TnC_2, which has two Ca^{2+} regulatory binding sites. Type IIb fibers are fast and glycolytic. Type I and IIa fibers contribute endurance and some speed; type IIb fibers contribute speed and power.

Fatigue is the transient loss in muscle work capacity or strength that results from preceding work. In humans, most fatigue originates in the muscle, as opposed to the brain. There are two types of fatigue: the fatigue that rapidly accompanies the use of maximal force, which can be sustained only for seconds and the fatigue that results from repetitive submaximal contractions such as long-distance running. Generally, fatigue is delayed longer with less strenuous activity. Fatigue has two origins: central fatigue is the reduction in nervous excitation and peripheral fatigue originates in the muscle. The fatigue that accompanies maximum force is due to buildup of Pi that originates from creatine phosphate. Endurance fatigue has both central and peripheral components. The peripheral component is due to depletion of glycogen, most likely that part of glycogen that determines Ca^{2+} release from the SR.

Training can improve fatigue resistance by fine-tuning muscle fiber metabolism and increasing muscle size. Maximal muscle force depends only on muscle size and is increased by resistance training—few repetitions against high resistance. Endurance training increases muscle capillarity, mitochondrial volume, and with proper nutrition, glycogen stores. Although muscle types can be converted in laboratory animals, it is unlikely that humans can substantially alter their muscle types by training. Muscles are heterogeneous with respect to muscle fiber types. They generally contain some population of type I, IIa, and IIb fibers and this distribution varies from person to person.

REVIEW QUESTIONS

1. What happens if ATP is not regenerated during muscle contraction?
2. What reactions consume ATP during muscular activity? Which of these consumes most of the ATP?
3. What does creatine kinase do? What does myokinase do? If ATP is regenerated by creatine kinase, what happens to cytoplasmic [Pi]? If ATP is regenerated by myokinase, what happens to cytoplasmic [AMP]?
4. How do fatty acids produce ATP in muscle? Is oxygen necessary for this process?
5. How does glucose produce ATP in muscle? Is oxygen necessary for this process?
6. Why do muscles make lactic acid? What happens to the lactic acid that muscles make? Why do muscles make more lactic acid during vigorous activity?
7. What is glycogenolysis? What does it produce in muscle? Can muscle contribute to blood glucose? Why or why not?
8. How does the liver help in the energy supply of muscle during exercise?
9. What does exercise do to glucose uptake by muscles? Does this effect require insulin?
10. What causes fatigue at high recruitment, short duration? What causes fatigue at low intensity, long duration?
11. Why does a deficiency in muscle phosphorylase cause exercise intolerance?
12. Describe the characteristics of type I, type IIa, and type IIb muscles. Can these be transformed by training?

Problem Set
Neuromuscular Transmission, Muscle Force, and Energetics   3.2

1. At rest in muscle cells, the E_m is not constant but shows seemingly random fluctuations that appear to be in units of about 0.5 mV. These are called **miniature end-plate potentials** and are due to the fusion of a single vesicle containing acetylcholine with the presynaptic nerve terminal membrane.
   A. If the muscle cell is 50 μm in diameter and 10 cm long, and its specific capacitance is 1 μF cm^{-2}, how much charge must flow for this 0.5 mV change due to a single vesicle fusion event?
   B. If $E_{Na} = +66$ mV and $E_m = -85$ mV, and the MEPP rises in 0.5 ms, what is the average conductance increase in the membrane due to the single fusion event? Assume the current is all carried by Na$^+$.
   C. How many channels do you think open during the MEPP? Use 40 pS for the unitary conductance of the acetylcholine receptor.

2. Synapse A on a motoneuron results in a EPSP with a peak depolarization of 2 mV. Synapse B also results in an EPSP with a peak depolarization of 2 mV. Synapse C results in an IPSP peak of hyperpolarization of 2.5 mV. The resting potential of the nerve is −70 mV.
   A. Simultaneous firing of A and B results in what peak potential? What is this phenomenon called? Do you think it is enough to reach threshold?
   B. Simultaneous firing of A and C results in what peak potential? What is this phenomenon called? Do you think it is enough to reach threshold?
   C. Simultaneous firing of A, B, and C results in what peak potential?

3. The membrane potential of a cell is −70 mV. E_{Na} is +65 mV, $E_K = -95$ mV, $E_{Cl} = -75$ mV.
   A. If a Na channel was to open, in which direction would Na$^+$ go? Is this a negative or positive current?
   B. If a K channel was to open, in which direction would K$^+$ go? Is this a negative or positive current?
   C. If a Cl channel was to open, in which direction would Cl$^-$ go? Is this a negative or positive current?

4. Neuron A makes a synapse on a dendrite on a motor neuron and neurotransmission at this synapse produces a 5 mV depolarization immediately adjacent to the synapse.
   A. Why is the EPSP less when measured further away?
   B. Would you expect the EPSP to be larger in the soma or further away from the synapse up the dendrite? Why?
   C. Repetitive firing of the neuron should result in a larger depolarization. What is this phenomenon called?

5. Consider the sarcomere shown in Figures 3.5.3 and 3.5.4. Assume that the myofibril is about 1 μm thick.
   A. What is the approximate maximum distance between the nearest RyR on the SR and the TnC on the thin filament?
   B. Assume the diffusion coefficient of Ca^{2+} in the cytosol is about 0.4×10^{-5} cm^2 s^{-1}. What is the time of diffusion from the nearest RyR to the furthest TnC?

6. Assume that the sarcomere length is 2.8 μm.
   A. How much overlap is there between the thick and thin filaments?
   B. What fraction of the force generators on the thick filament do not overlap with the thin filament?

7. A.V. Hill derived an empirical equation to describe the force–velocity relationship. He wrote

$$(T + \alpha)(v + \beta) = (T_0 + \alpha)\beta$$

   where T is the tension, or force, v is the velocity, and T_0 is the force at which $v = 0$, the isometric tension.
   A. Derive an expression for the maximum velocity (recall that maximum velocity occurs when $T = 0$) in terms of T_0, α, and β.
   B. Hill's equation can be written in a normalized form, where $v' = v/v_{max}$ and $T' = T/T_0$. Show that this normalization results in

$$v' = (1 - T')/(1 + T'/k) \quad \text{where } k = \alpha/T_0 = \beta/v_{max}$$

   C. Derive an expression for power ($T \times v$) in terms of v, T_0, α, and β.

8. Consider muscle fibers that are 8 cm long and that develop a maximum of 20 N cm^{-2} force. Consider a muscle that has a volume of 20 cm^3 with the fibers aligned with the direction of the tendon ($\theta = 0$).

A. What is the maximum force developed by this muscle?
B. If partial recruitment results in activation of 10% of the muscle fibers (assuming they are all of equal size), how much force could the muscle generate?
C. If the muscle fibers contract 15% of their length in 50 ms under no load, what is the maximum muscle velocity?
D. Suppose the muscle fibers were oriented at 15° relative to the tendon axis, with the same volume of muscle. What is the maximum force delivered to the tendon? What is the maximum velocity?

9. Suppose that the amount of TnC in fast-twitch muscle fibers is about 70 μM. The density of muscle is about 1.06 g cm^{-3}.
   A. From the structure of the thin and thick filaments, estimate the amount of myosin heads. Convert this to nmol per g of tissue.
   B. Suppose that the muscle activation is 75 ms and that the turnover of the heads is 10 s^{-1}. About how many ATP would be hydrolyzed in a single twitch per g tissue? Use a step function for activation: it is on or off with a 75-ms duration. This is not how the muscle works, but reality is a lot more complicated to simulate. We are doing a crude estimate here.
   C. Assume that the TnC binds 2 Ca^{2+}. How much ATP would be split by the SERCA pumps to return activating Ca^{2+} to the SR per g tissue? Suppose now that we consider a slow-twitch muscle fiber which has TnC1 that binds only 1 Ca^{2+} per TnC and has a turnover of 3 s^{-1} and a twitch time of 200 ms.
   D. Estimate the number of ATP hydrolyzed by the cross-bridges during the twitch.
   E. Estimate the ATP hydrolyzed by the SERCA pumps to return activating Ca^{2+} to the SR per g tissue.
   F. Do you think there would be more or less SR necessary in slow-twitch muscle fibers?

Smooth Muscle 3.8

Learning Objectives

- Compare smooth muscle cell morphology to that of skeletal muscle fibers
- Contrast tonic and phasic smooth muscles
- Distinguish electromechanical coupling from pharmacomechanical coupling
- Define unitary smooth muscle and contrast it from multi-unit smooth muscle
- Describe varicosities and their role in smooth muscle stimulation
- List the mechanisms on the sarcolemma for Ca^{2+} entry into the smooth muscle cell
- List the mechanisms on the sarcolemma for Ca^{2+} exit from the smooth muscle cell
- List the mechanisms on the SR for Ca^{2+} exit into the smooth muscle cytoplasm
- List the mechanisms on the SR for Ca^{2+} uptake into the SR from the smooth muscle cytoplasm
- Describe how myosin light chain kinase (MLCK) activates smooth muscle
- Describe how myosin light chain phosphatase (MLCP) inactivates smooth muscle
- Describe what is meant by "Ca^{2+} sensitization"
- Describe how Ca^{2+} activates smooth muscle

SMOOTH MUSCLES SHOW NO CROSS-STRIATIONS

Smooth muscle lines the walls of hollow organs and tubes such as the urinary bladder, gall bladder, uterus, bile ducts, ureters, intestines, and arteries. Contraction of the smooth muscle in tubes either controls the caliber of the tube or propels material through the lumen of the tube. In the hollow organs, contraction of the smooth muscle increases the pressure within the organ, thereby propelling material out of the organ. Smooth muscle cells are small and spindle shaped with no cross-striations. Typically they are only 2–5 μm in diameter and up to about 400 μm long with a single, central nucleus. This contrasts with the very long, multinucleated skeletal muscle fibers.

SMOOTH MUSCLE DEVELOPS TENSION MORE SLOWLY BUT CAN MAINTAIN TENSION FOR A LONG TIME

Skeletal muscle and cardiac muscle develop tension quickly and relax quickly, in accord with their physiological function. Smooth muscle contracts and relaxes much more slowly, but it can sustain contractions for long periods of time without fatiguing, and it can develop as much force as striated muscle. This ability is useful in the **sphincters** of the body. Sphincters are bands of smooth muscle that surround areas that separate adjoining parts of the hollow organs. The sphincters act as valves. Contraction of the band of smooth muscle prevents the movement of material through the sphincter. Examples include the urinary sphincter that controls the movement of urine from the urinary bladder, and the various sphincters in the gastrointestinal tract. These sphincters are usually contracted and open only when necessary.

SMOOTH MUSCLE CAN CONTRACT TONICALLY OR PHASICALLY

Some smooth muscles contract tonically, maintaining force nearly continuously and relaxing only occasionally. Other smooth muscles contract phasically, producing force in phases and alternating between inactive and active states, even in the absence of external stimulation. Still other smooth muscles do not contract at all until they are stimulated. Many smooth muscles are controlled by neurons or other cells that release chemicals that may either stimulate or inhibit the contractile activity of the muscle. Figure 3.8.1 illustrates the force produced by tonically active smooth muscles or phasically active smooth muscles.

SMOOTH MUSCLES EXHIBIT A VARIETY OF ELECTRICAL ACTIVITIES THAT MAY OR MAY NOT BE COUPLED TO FORCE DEVELOPMENT

During relaxation, smooth muscle cells have a resting membrane potential between -40 and -80 mV. The origin of this resting membrane potential is essentially the same as for striated muscle: in the resting state there is a large conductance for K^+, with smaller conductances for Na^+ and Cl^-. Some of these resting membrane potentials are unstable, and the membrane potential undergoes cyclic depolarizations and repolarizations. These are called slow waves. In some muscles, these slow waves trigger bursts of action potentials with associated rhythmic contractions. In other smooth muscles, the developed force tracks the membrane potential but there are no action potentials. In still other smooth muscles, the membrane potential is stable and exogenous agents vary the force. The processes linking changes in membrane potential to force development are called excitation–contraction coupling or electromechanical coupling. Pharmacomechanical coupling refers to the process of force development without changes in membrane potential. Figure 3.8.2 illustrates the variety of electrical and mechanical responses of smooth muscles.

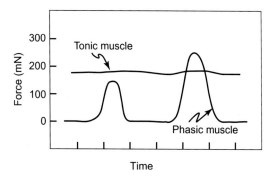

FIGURE 3.8.1 Force developed by tonically active smooth muscle and phasically active smooth muscle. Tonically active muscles develop and maintain force for protracted periods. The degree of tension is their tone. Phasic muscles undergo cycles of contraction and relaxation.

CONTRACTILE FILAMENTS IN SMOOTH MUSCLE CELLS FORM A LATTICE THAT ATTACHES TO THE CELL MEMBRANE

As in skeletal muscle, tension in smooth muscle cells is produced by actin filaments interacting with myosin filaments. Unlike in skeletal muscle, these filaments are not arranged in register but are loosely arranged throughout the cytoplasm. The thin filaments attach to **dense bodies** located throughout the cytoplasm and their ends terminate on myosin filaments in the cytoplasm. These myosin filaments attach to other thin filaments whose ends terminate on the membrane in **attachment plaques**, also sometimes called membrane dense areas. Thus contraction of the myofilaments, in which the actin filaments move along the myosin filaments, draws the dense bodies and attachment plaques closer together. The dense bodies and attachment plaques are believed to be analogous to the Z-disks in skeletal muscle fibers. They are rich in the actin-binding protein, α-actinin, and also contain another protein, vinculin (130,000 Da), which is not found in the Z-disk. Vinculin binds α-actinin and also binds to integral membrane proteins in the attachment plaques. Vinculin appears to anchor the actin filaments to specific membrane adhesion sites (see Figure 3.8.3).

ADJACENT SMOOTH MUSCLE CELLS ARE MECHANICALLY COUPLED AND MAY BE ELECTRICALLY COUPLED

The forces produced by the myofilaments are transmitted to the cell membrane by the attachment plaques on the surface membrane. These forces on the cell membrane are transmitted to adjacent cells through the

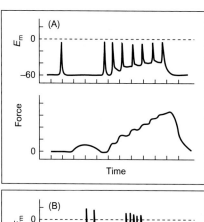
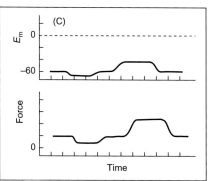
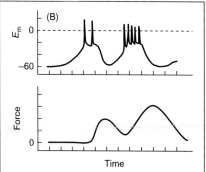
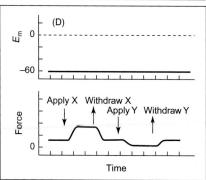

FIGURE 3.8.2 Membrane potential (E_m) and force in different types of smooth muscle. In some smooth muscles (A), pacemaker cells generate action potentials that are transmitted through gap junctions to all of the cells, producing force incremental with the frequency of stimulation. In other smooth muscles (B), slow waves trigger bursts of action potentials with associated force development. In still other muscles (C), the muscle maintains a tone that depends on the membrane potential, but action potentials are not generated. Tissues that show pharmacomechanical coupling (D) have a stable membrane potential that is unaltered by agents that produce force by altering the smooth muscle cytoplasmic $[Ca^{2+}]$ without altering the membrane potential. Adapted from R.M. Berne and M.N. Levy, Principles of Physiology, 3rd Ed., Mosby, St. Louis, 2000.

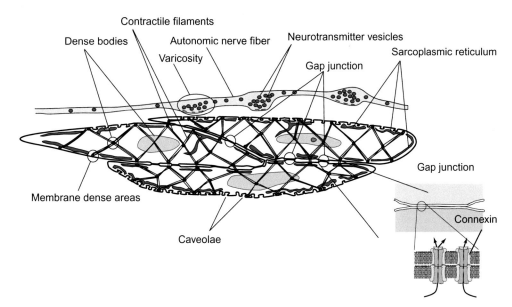

FIGURE 3.8.3 Simplified schematic drawing of the structure of smooth muscle cells. The cells are small with a single, centrally located nucleus. Contractile filaments are anchored in the cytosol at dense bodies, which are the functional analogues of Z-disks in skeletal and cardiac muscles. These form a lattice surrounding the nucleus. The filaments are anchored in the membranes at places called membrane dense areas or attachment plaques. The surface membrane is covered with small invaginations called caveolae. These may serve as local sources of Ca^{2+} for activation of the muscle. The SR provides transient changes in cytosolic $[Ca^{2+}]$, but the extent and activity of the SR in smooth muscle is much less than in the striated muscles. Cells may be coupled with adjacent cells both mechanically through mutual connections to filaments in the extracellular matrix and electrically through gap junctions. At gap junctions, each cell contributes a hemichannel consisting of six connexin molecules in a hexagonal array. Connexins have molecular weights between 25 and 50 kDa and at least 12 different varieties have been identified. Apposition of two hemichannels forms a conducting pathway between the cells so that electrical signals and small-molecular-weight signaling molecules can move between the cells. Smooth muscle is usually innervated but the release of neurotransmitters does not occur in focal regions like the skeletal neuromuscular junction, but instead release occurs from varicosities and the neurotransmitters spread by diffusion throughout the local extracellular fluid.

extracellular matrix. Thus adjacent smooth muscle cells are mechanically coupled. These cells also may be electrically coupled by gap junctions (see Figure 3.8.3). Ions and low-molecular-weight materials pass between the cells, and so cells attached by gap junctions are both electrically and metabolically coupled. Different smooth muscles differ in their degree of coupling. Intestinal smooth muscle cells, for example, are highly coupled so that excitation of one rapidly passes to the others and the cells contract in unison. These muscles are called unitary smooth muscle. Other smooth muscles, such as the vas deferens, lack gap junctions and these muscles are called multiunit smooth muscle, and each cell must be activated individually. Smooth muscles can possess coupling anywhere in the spectrum between these two extremes.

SMOOTH MUSCLE IS CONTROLLED BY INTRINSIC ACTIVITY, NERVES, AND HORMONES

Smooth muscle cells contain a variety of receptors on their surface membrane that respond to chemical signals that are released from nerves or that arrive via the circulation. Many of the chemical signals that arrive via the circulation are hormones. Some smooth muscle cells receive no innervation. Even those that ordinarily receive innervation usually maintain organ function when innervation is interrupted. This is important in transplantation when the vascular tone must be maintained even when the transplanted heart or kidney or liver lacks its usual innervation.

NERVES RELEASE NEUROTRANSMITTERS DIFFUSELY ONTO SMOOTH MUSCLE

There is no neuromuscular junction in smooth muscle as there is in skeletal muscle. Instead, the nerves form varicosities, or bulges in their axons, where vesicles containing neurotransmitter accumulate and release their neurotransmitter in response to action potentials in the axon. The neurotransmitters are released into the extracellular space to diffuse to their receptors on the smooth muscle cell membranes (see Figure 3.8.3).

CONTRACTION IN SMOOTH MUSCLE CELLS IS INITIATED BY INCREASING INTRACELLULAR $[CA^{2+}]$

Phasic and tonic smooth muscle contractions both involve changes in the cytoplasmic free $[Ca^{2+}]$ from resting levels near 80–140 nM to activating concentrations in the range of 500–700 nM. Figure 3.8.4 shows that phasic contractions typically involve a single Ca^{2+} transient that is followed by force development. Repetitive stimulation can cause summation of both the Ca^{2+} transient and force. In tonically contracting smooth muscle, the force is maintained by the

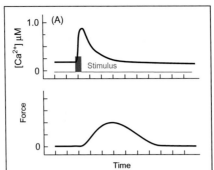
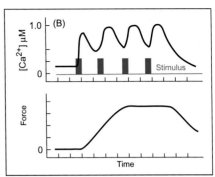

FIGURE 3.8.4 [Ca^{2+}] and force in phasic smooth muscle cells during a single short-lasting stimulation (A) and during repetitive stimulation (B). A single stimulation (blue) results in a transient increase in the cytoplasmic [Ca^{2+}] and subsequent development of force (A). Repetitive stimuli can result in the summation of [Ca^{2+}] transients and greater force (B). *Adapted from R.M. Berne and M.N. Levy,* Principles of Physiology, *3rd Ed, Mosby, St. Louis, 2000.*

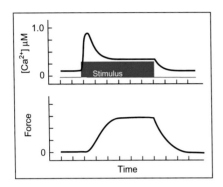

FIGURE 3.8.5 [Ca^{2+}] and force during the stimulation of tonic smooth muscle. The continuous presence of stimulus causes an initial Ca^{2+} spike followed by a continued high cytoplasmic [Ca^{2+}]. The force that is developed depends on the [Ca^{2+}]. However, force can be altered by changing the sensitivity of the force producing system to cytoplasmic [Ca^{2+}]. *Adapted from R.M. Berne and M.N. Levy,* Principles of Physiology, *3rd Ed., Mosby, St. Louis, 2000.*

continued presence of the activating agent. The initial Ca^{2+} transient is followed by a long-lasting increased cytosolic [Ca^{2+}] and long-lasting force (see Figure 3.8.5). When the stimulus is removed, cytosolic [Ca^{2+}] and force return to baseline, unstimulated values.

SMOOTH MUSCLE CYTOSOLIC [CA^{2+}] IS HETEROGENEOUS AND CONTROLLED BY MULTIPLE MECHANISMS

The cytoplasmic [Ca^{2+}] in smooth muscle cells is set by sources and sinks for Ca^{2+} both at the surface membrane and at the sarcoplasmic reticulum (SR). These sources and sinks and their cellular locations are illustrated in Figure 3.8.6. Spatial separation of the sources and sinks allows for spatially distinct regions of cytosolic [Ca^{2+}]. This is further possible because the SR resides immediately underneath the plasma membrane, and so Ca^{2+} influx across the plasma membrane is first conditioned by SR Ca^{2+} transport processes. Either the Ca^{2+} influx is amplified by Ca^{2+}-induced Ca^{2+} release by RyR, or it is reduced by Ca^{2+} sequestration. Ca^{2+} influx through those channels located in the caveolae, on the other hand, passes directly into the deep cytosol.

SMOOTH MUSCLE [CA^{2+}] CAN BE REGULATED BY CHEMICAL SIGNALS

PHOSPHORYLATION OF PHOSPHOLAMBAN REGULATES SR SERCA ACTIVITY

Phospholamban is a 6-kDa protein that associates reversibly into pentamers. Phospholamban tonically inhibits the SERCA pump by increasing its K_m for Ca^{2+}. Phosphorylation of phospholamban relieves this inhibition, speeding up the SR Ca-ATPase at any given [Ca^{2+}] and therefore lowering the cytoplasmic [Ca^{2+}]. Phosphorylation of phospholamban is controlled by protein kinases that are, in turn, controlled by agonists. Epinephrine, for example, binds to β-receptors on the surfaces of smooth muscle cells, which are coupled to a heterotrimeric G-protein (G_s) that activates adenylyl cyclase. The adenylyl cyclase makes more 3′,5′-cyclic AMP from ATP, and the increased [cAMP] in turn activates protein kinase A (PKA). The G_s mechanism is described in Chapter 2.8. In smooth muscle cells, PKA phosphorylates phospholamban, which helps draw down cytoplasmic [Ca^{2+}] and helps relax smooth muscle that possesses β-adrenergic receptors.

[CA^{2+}] IS REGULATED VIA MEMBRANE POTENTIAL BY ALTERING BK_{CA} (K_{CA}, CALCIUM-ACTIVATED K^+ CHANNEL)

Human airway smooth muscle responds to epinephrine through $β_2$ receptors, but human airway smooth muscle is unique in that it lacks phospholamban. Instead, PKA activated in these tissues phosphorylates a large-conductance, Ca^{2+}-activated K^+ channel called BK_{Ca}. Its activation increases K^+ efflux from the cell, causing a hyperpolarization and reduction in Ca^{2+} influx through voltage-gated Ca^{2+} channels. BK_{Ca} channels are activated by PKA and PKG, and inhibited by PKC. Thus, it is expected that PKA and PKG would promote relaxation, whereas PKC would contribute to contraction.

THE IP3R OPENS IN RESPONSE TO RECEPTOR BINDING ON THE SURFACE MEMBRANE

A variety of agonists including acetylcholine, norepinephrine, histamine, endothelin, leukotrienes, and

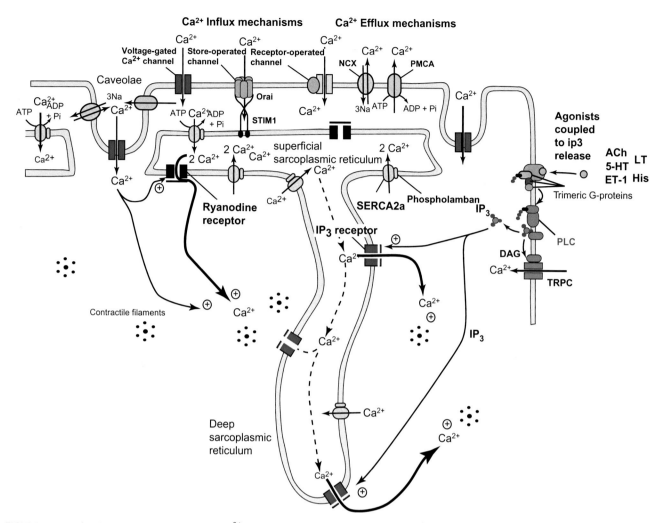

FIGURE 3.8.6 Molecular mechanisms underlying Ca^{2+} homeostasis in smooth muscle cells. Ca^{2+} influx across the plasma membrane is mediated by three separate types of channels: (1) the voltage-gated, L-type Ca^{2+} channel; (2) receptor-operated channels; and (3) store-operated channels that respond to some signal from the SR when it is depleted of Ca^{2+}. Ca^{2+} efflux across the plasma membrane is mediated by (1) the Na^+–Ca^{2+} exchanger (NCX), which exchanges 3 Na^+ for 1 Ca^{2+} and (2) PMCA, which transports 1 Ca^{2+} atom out of the cell per molecule of ATP hydrolyzed. Ca^{2+} influx can occur to a restricted space between the plasma membrane and the SR. In some cases, the Ca^{2+} appears to enter the SR directly without involving active pumping by the SERCA pump on the SR membrane, probably through close apposition of some SR channel with the plasma membrane channels. The store-operated channels use STIM1 to sense SR Ca^{2+} load. With high SR $[Ca^{2+}]$, STIM1 binds Ca^{2+} and disaggregates. In low SR $[Ca^{2+}]$, STIM1 aggregates and interacts with Orai1 to open a Ca^{2+} channel into the restricted space adjacent to the SR. Ca^{2+} that enters the cell can cause additional Ca^{2+} release from ryanodine receptors, RyR, on the SR membrane through a process called Ca^{2+}-induced Ca^{2+}-release. Remaining Ca^{2+} can be taken up into the SR membrane through the SERCA Ca^{2+} pump with a stoichiometry of 2 Ca^{2+} per ATP hydrolyzed. The SERCA pump is regulated by phospholamban, a 6-kDa protein whose tonic inhibition of the pump can be relieved by phosphorylation. The SR is divided into two parts: the part immediately adjacent to the plasma membrane is the superficial SR. Its lumen is continuous with the deep SR. Caveolae are invaginations of the plasma membrane that have concentrations of efflux mechanisms that appear to concentrate Ca^{2+} near the membrane. Ca^{2+} channels there admit Ca^{2+} that can penetrate deeper into the cytosol because there is no SR in the way. IP3 receptors in the deep SR are stimulated by IP3 that is produced by phospholipase C that in turn is activated by agonist binding to a G_q-coupled receptor on the plasma membrane. A variety of agonists can activate Gq-coupled receptors, eventually causing smooth muscle contractions. Examples include acetylcholine (ACh), serotonin (5HT), leukotrienes (LK), histamine (His), and endothelin (ET). The phospholipase C that is activated by these agonists also releases diacylglycerol (DAG) that activates Ca^{2+} influx through a receptor-operated Ca^{2+} channel, TRPC (for transient receptor potential channel). The IP3 receptor releases Ca^{2+} that activates the contractile elements. The spatial separation of sources and sinks of Ca^{2+} produces spatial heterogeneity in the cytosolic $[Ca^{2+}]$.

thromboxane A2 cause release of Ca^{2+} from the SR by indirectly activating inositol triphosphate (IP3) receptors. The IP3 receptors are large-molecular-weight receptors that are related to the ryanodine receptor family. Both consist of tetramers of large molecular weight, both reside in the membrane of the endoplasmic reticulum, and both release Ca^{2+} in their open configuration. The IP3 receptor is gated by binding of IP3. IP3 is a second messenger produced in the cell in response to binding of a first messenger (the agonists listed above) to receptors on the outside of the cell. Agonist binding is coupled to a heterotrimeric G-protein, G_q, which activates yet another enzyme, phospholipase C. Phospholipase C liberates IP3 and diacylglycerol (DAG) from a membrane lipid, phosphatidylinositol 4,5-bisphosphate. This G_q mechanism is described in

Chapter 2.8. A new feature here is the inhibition of BK_{Ca} channels by PKC that tends to depolarize the cell and allow Ca^{2+} entry over the voltage-gated Ca^{2+} channels.

FORCE IN SMOOTH MUSCLE ARISES FROM CA^{2+} ACTIVATION OF ACTIN–MYOSIN INTERACTION

Smooth muscles contain thick and thin filaments, composed predominantly of myosin and actin, respectively. However, their arrangement is quite different from the striated muscles. The filaments are not organized into sarcomeres, and the ratio of thin to thick filaments is closer to 10:1 than it is to the 2:1 ratio found in skeletal and cardiac muscles. The interaction of the actin thin filaments and the myosin in smooth muscle is regulated by cytosolic $[Ca^{2+}]$, but not in the same way as in skeletal and cardiac muscles. Smooth muscle varies its force through recruitment of cross-bridges by phosphorylating the myosin light chains, which is controlled by Ca^{2+}.

MYOSIN LIGHT CHAIN PHOSPHORYLATION REGULATES SMOOTH MUSCLE FORCE

Myosin is a complex of six proteins: two heavy chains of about 200,000 Da each and a total of four myosin light chains. The heavy chains have functionally distinct regions: the long rod-shaped tail imparts rigidity to the myosin; an arm section connects the tail to the globular head region. Each globular head binds actin and contains the site of ATP binding and hydrolysis. The base of each head or neck region binds two light chains. One is called the essential light chain (ELC); the other is the regulatory light chain (RLC or MLC20). As their names imply, regulation of myosin cross-bridge cycling with actin occurs by phosphorylation and dephosphorylation of the RLCs (see Figure 3.8.7).

Myosin light chain kinase (MLCK) phosphorylates serine 19 of the 20,000 Da RLC. This triggers cross-bridge cycling and shortening or force development in smooth muscle. The activity of MLCK depends directly on Ca–**calmodulin**, as shown in Figure 3.8.8. Calmodulin activates MLCK when it binds Ca^{2+}. This is the link between cytosolic $[Ca^{2+}]$ and initiation of force in smooth muscle.

MYOSIN LIGHT CHAIN PHOSPHATASE DEPHOSPHORYLATES THE RLC

Myosin light chain phosphatase (MLCP) catalyzes the dephosphorylation of the RLCs. Dephosphorylating the RLCs of myosin inhibits myosin cross-bridge formation with actin, but dephosphorylating myosin already on actin reduces its off rate, forming the so-called "latch state." The latch state corresponds to the situation where smooth muscle holds tension at low rates of ATP hydrolysis. Although smooth muscle can develop the same force as striated muscle, it does so with much less

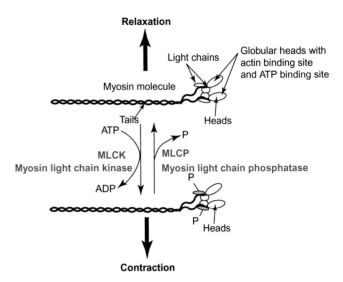

FIGURE 3.8.7 Structure of myosin and regulation of contraction. Each complex consists of two heavy chains and four light chains. The two heavy chains have long tails that form a supercoil. The head groups of the heavy chains contain binding sites for actin and for ATP. This is the "business end" of the heavy chains that is responsible for interacting with actin and hydrolyzing ATP. The RLCs and ELCs bind to the head groups near the neck region. Phosphorylation of the RLCs permits cross-bridge formation and cycling, and thus leads to force development or shortening. The phosphorylation state of the RLCs results from a balance between phosphorylation by MLCK and dephosphorylation by MLCP.

myosin (about one-fifth that of striated muscle) and at much slower rates. The ATP consumption by smooth muscle actomyosin is 1/100–1/500 that of striated muscle, whereas force development is on the order of 1000 times slower in smooth muscle. Figure 3.8.9 shows the effect of MLCP and MLCK on the state of myosin light chain phosphorylation and force development. Thus the state of phosphorylation of the myosin light chains results from a balance between MLCK and MLCP activities (see Figure 3.8.7).

CA^{2+} SENSITIZATION PRODUCES FORCE AT LOWER $[CA^{2+}]$ LEVELS

MLCP activity alters the phosphorylation level independent of MLCK activity. Thus the activation of MLCP would promote relaxation and the inhibition of MLCP should promote contraction. If MLCP is inhibited without altering MLCK activity, there would be increased contraction at cytosolic $[Ca^{2+}]$ that previously did not cause contractions. This is Ca^{2+} sensitization. The mechanism of α_1 adrenergic Ca^{2+} sensitization involves a G_q-coupled receptor which activates phospholipase C in the surface membrane, causing a release of IP3 and DAG. The DAG stimulates protein kinase C (PKC) that phosphorylates another protein target, called CPI-17. The phosphorylated CPI-17 inhibits MLCP. In this way, α_1 adrenergic stimulation of smooth muscle produces a contraction in part by Ca^{2+} sensitization brought about by inhibition of myosin light chain dephosphorylation and in part by activation of MLCK by increasing $[Ca^{2+}]$ through IP_3-coupled Ca^{2+} release from the SR. The regulation of MLCP is shown in Figure 3.8.10.

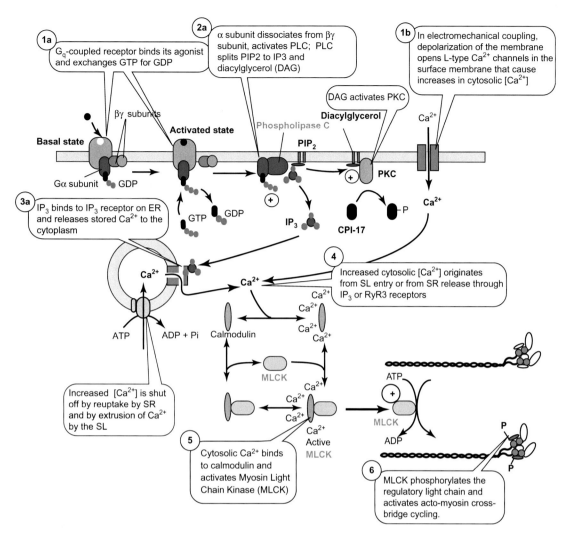

FIGURE 3.8.8 Regulation of cross-bridge cycling by calcium effects on myosin light chain phosphorylation in smooth muscle. When cytoplasmic [Ca^{2+}] rises either through entry from the extracellular fluid over voltage-gated Ca^{2+} channels or activation of G$_q$-coupled receptors acting through IP3-induced release of Ca^{2+} from intracellular stores, it binds to calmodulin and activates MLCK. MLCK then phosphorylates the RCL of myosin and activates cross-bridge formation and force development. The free [Ca^{2+}] is maintained by all those processes shown in Figure 3.8.6 and mostly omitted here. According to this scheme, removal of Ca^{2+} from the cytosol causes relaxation, either by increased efflux from the cytosol or reduced influx into the cytosol.

A second mode of Ca^{2+} sensitization is through phosphorylation of MLCP by **rho kinase**. Several agonists activate the small monomeric G-protein **RhoA** through binding to heterotrimeric G-protein coupled receptors in the cell membrane. RhoA has GTPase activity. It activates RhoA kinase (ROCK) which in turn phosphorylates the myosin-binding subunit (MYPT1) of MLCP. This phosphorylation inactivates the phosphatase, which thus preserves the phosphorylation of the myosin light chains and maintains force development even in the absence of a sustained Ca^{2+} signal. Figure 3.8.10 also shows the regulation of MLCP by rhoA-GTP.

NITRIC OXIDE INDUCES SMOOTH MUSCLE RELAXATION BY STIMULATING GUANYLATE CYCLASE

Nitric oxide synthases (NOS) from endothelial tissue (eNOS), neural tissue (nNOS), and an inducible, Ca^{2+}-independent nitric oxide synthase called iNOS produce nitric oxide, NO, by converting L-arginine and O$_2$ to NO and citrulline. The NO diffuses from its site of production to smooth muscle cells where it activates soluble guanylate cyclase to increase cytosolic cyclic guanylyl monophosphate or cGMP. The increased cGMP then is believed to activate a cGMP-dependent protein kinase, PKG, which relaxes smooth muscle by phosphorylating target proteins. The effects of PKG activation are manifold. PKG phosphorylates phospholamban, which brings about relaxation by increasing Ca^{2+} uptake into the SR. PKG also stimulates MLCP activity, thereby reducing the phosphorylation of the myosin light chains, causing a loss of Ca^{2+} sensitivity. PKG also activates a large-conductance K$^+$ channel (**BK$_{Ca}$**) that hyperpolarizes the cell and reduces Ca^{2+} entry over voltage-gated Ca^{2+} channels in the surface membrane. The effects of NO on smooth muscle are shown in Figure 3.8.11.

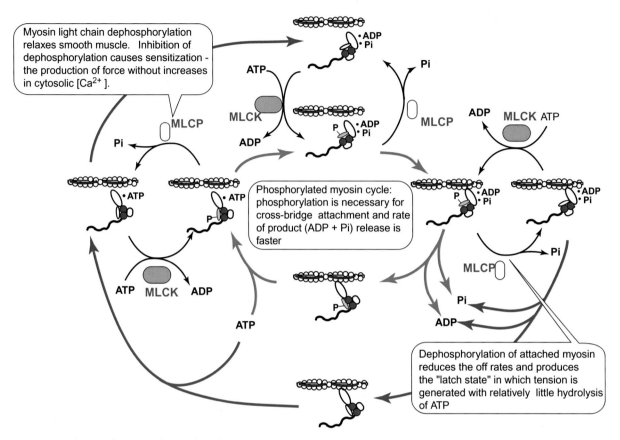

FIGURE 3.8.9 Regulation of cross-bridge cycling by myosin light chain phosphorylation and dephosphorylation in smooth muscle. MLCK phosphorylates the RLCs and activates cross-bridge formation between myosin and actin, initiating the cross-bridge cycling (inner cycle). This cycle is deactivated by dephosphorylation of the light chains by MLCP. MLCP can also dephosphorylate the light chains while myosin is attached to the actin filaments. This produces an alternate cross-bridge cycle (outer cycle) that has slower kinetics. This is the "latch state" that produces force with little expenditure of energy. In effect, it makes the muscle stiff.

ADRENERGIC STIMULATION RELAXES SMOOTH MUSCLES BY REDUCING CYTOSOLIC [CA^{2+}]

Circulating epinephrine or locally released norepinephrine can bind to β-adrenergic, G_S-coupled receptors on some smooth muscle cells. These G_S-coupled receptors activate adenylyl cyclase to increase cytoplasmic concentrations of 3′,5′-cyclic adenosine monophosphate (cAMP). This activates PKA that phosphorylates targets including phospholamban in some smooth muscles. Phosphorylation of phospholamban relieves inhibition of the SERCA2a pump on the SR and increases removal of activator Ca^{2+} from the cytosol, resulting in relaxation. In other tissues, PKA activates BK$_{Ca}$ and relaxes cells by hyperpolarization. These effects are shown in Figure 3.8.11.

SYNOPSIS OF MECHANISMS PROMOTING CONTRACTION OR RELAXATION OF SMOOTH MUSCLE

Figure 3.8.12 illustrates that phosphorylation of the myosin light chains promotes contraction and dephosphorylation promotes relaxation. Contraction is thus controlled by the relative activities of MLCK and MLCP. MLCK is controlled mainly by cytoplasmic [Ca^{2+}], through calmodulin, and the increased [Ca^{2+}] originates either from entry over voltage-dependent Ca^{2+} channels or by release of internal stores. Thus, G_q-coupled receptors cause contraction, and membrane depolarization causes contraction through increases in cytoplasmic [Ca^{2+}], and G_s-coupled receptors and NO cause relaxation by increasing uptake of Ca^{2+} back into the SR through phosphorylation of phospholamban.

Other agonists work by affecting MLCP. $G_{α12}$-coupled receptors activate Rho kinase that phosphorylates MLCP, inactivating it and thereby promoting phosphorylation of the light chains and contraction. G_q-coupled receptors, besides increasing cytoplasmic [Ca^{2+}] through IP3-induced release of Ca^{2+} from the SR, also activate PKC that phosphorylates CPI-17 that inhibits MLCP, thereby promoting contraction. Other agonists can work through modulation of channels that affect the membrane potential, such as BK$_{Ca}$. PKA and PKG both activate BK$_{Ca}$, promoting relaxation, whereas PKC inhibits BK$_{Ca}$, thereby promoting contraction.

This synopsis shows the general effect of a variety of agonist and mechanisms on a variety of tissues. The mechanisms used vary with the tissue and the species.

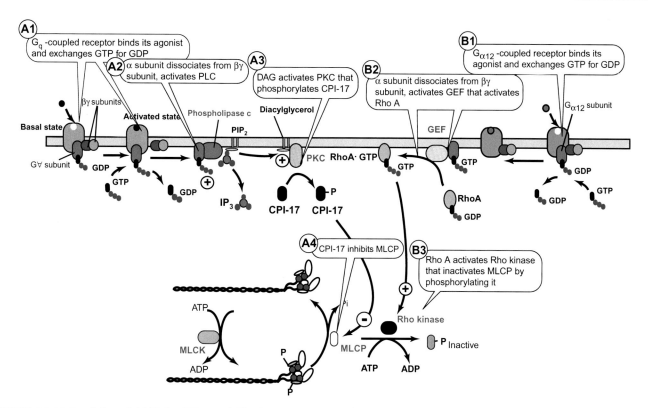

FIGURE 3.8.10 Regulation of MLCP by CPI-17 and Rho kinase. Activation of G_q-coupled receptors (such as adrenergic α_1-receptors) activates phospholipase C which cleaves phosphatidylinositol bisphosphate to form IP3 (which then releases Ca^{2+} from the SR and activates MLCK through calmodulin) and DAG. DAG activates PKC, which phosphorylates target proteins including CPI-17. The phosphorylated form of CPI-17 inhibits myosin light chain phosphates (MLCP), thereby promoting the phosphorylated form of the myosin light chains, which promotes contraction. Other agonists bind to $G_{\alpha 12}$-coupled receptors, whose α-subunit activates a GTP-exchange factor (GEF) to exchange GTP for GDP on a monomeric protein, RhoA. This stimulates Rho kinase, which phosphorylates MLCP, inhibiting it. Thus the activation of the Rho kinase pathway promotes contraction because it inhibits the dephosphorylation of the myosin light chains.

While cAMP-mediated mechanisms may produce relaxation via phosphorylation of phospholamban in one tissue, it may promote relaxation by activation of BK_{Ca} in a different tissue or in the same tissue in a different species. Figure 3.8.12 is meant to illustrate the breadth of the types of effects rather than catalogue the effects in each of the many different types of smooth muscles.

SUMMARY

Smooth muscles line the walls of hollow tubes and organs. They control the caliber of the tubes or move materials by producing a pressure within the hollow organs. They exhibit tremendous variety of physiological responses and mechanisms of actions. Some smooth muscles produce force in association with action potentials, some in response to slow changes in membrane potential, and others produce force completely dissociated from changes in membrane potential. Electromechanical coupling describes the events that link changes in membrane potential to force. Pharmacomechanical coupling refers to the mechanisms that cause force in the absence of changes in membrane potential. Smooth muscles respond to a bewildering list of agonists including acetylcholine, histamine, thromboxane A_2, norepinephrine, endothelin, nitric oxide, and leukotrienes.

Smooth muscle cells are small cells that connect to each other through gap junctions. The actin and myosin filaments form a network within the cell that connects to the cell membrane at attachment plaques. They connect within the cytoplasm at dense bodies. The degree of electrical coupling varies from the extreme of unitary smooth muscles in which all cells are electrically coupled to multiunit smooth muscles in which each cell must be activated separately.

Smooth muscle regulates its force through the phosphorylation state of the RLC of myosin. Ordinarily, smooth muscle contraction begins with increases in cytoplasmic $[Ca^{2+}]$ brought about either by changes in membrane potential or by ligand binding to receptors on the smooth muscle cell membrane. These increase Ca^{2+} influx into the cell by opening voltage-gated Ca^{2+} channels or receptor-operated Ca^{2+} channels on the surface membrane or by releasing Ca^{2+} from the SR. Agonists binding to a G_q protein activate phospholipase C that liberates IP3 from the surface membrane. IP3 then binds to its receptor on the SR and releases Ca^{2+}. Ca^{2+} that enters the cell can also cause Ca^{2+}-induced Ca^{2+}-release from RyR3 receptors on the deep SR.

Increased cytosolic $[Ca^{2+}]$ binds to calmodulin which activates MLCK. This phosphorylates myosin light chains, which activate contraction. Removal of Ca^{2+} by

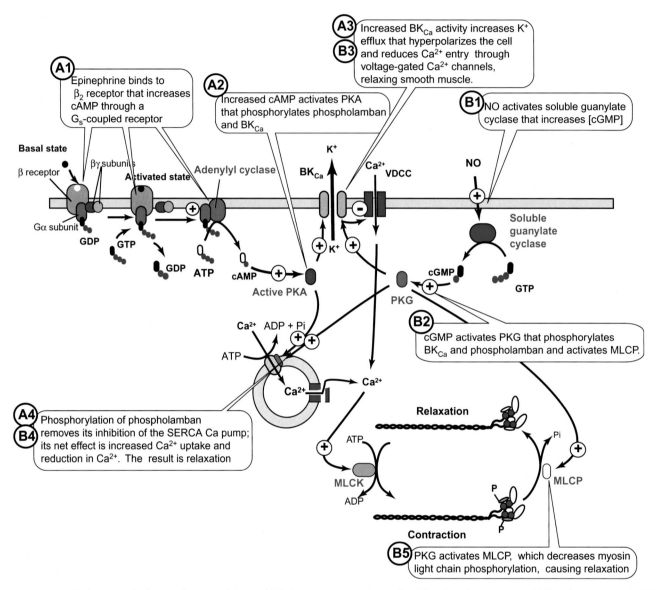

FIGURE 3.8.11 Mechanisms of relaxation by epinephrine and NO. Beta$_2$ agonists activate adenylyl cyclase that increases cAMP and activates PKA. PKA activates BK$_{Ca}$, a large-conductance channel for K$^+$ whose activation hyperpolarizes the cell and reduces Ca^{2+} entry over voltage-gated Ca^{2+} channels. PK also phosphorylates phospholamban that stimulates Ca^{2+} uptake by the SR, thereby promoting relaxation. Nitric oxide, NO, produced by other cells enters the smooth muscle cell and activates a soluble guanylate cyclase, which increases cGMP and activates PKG. PKG activates BK$_{Ca}$ to promote relaxation, phosphorylates phospholamban that aids in relaxation and also activates myosin light chain phosphatase (MLCP) that aids in relaxation by dephosphorylating the myosin regulatory light chain.

reuptake by the SR or by efflux from the cell causes relaxation by deactivating MLCK. Beta adrenergic stimulation relaxes smooth muscle by increasing Ca^{2+} uptake into the SR. It does this through a G-protein coupled increase in cAMP that stimulates PKA that in turn phosphorylates phospholamban (PLB). Phosphorylation of PLB removes its inhibition of the SERCA2a Ca-ATPase on the SR membrane. In other tissues, PKA activates a large conductance for K$^+$ that hyperpolarizes the membrane and reduces cytosolic [Ca^{2+}] by reducing Ca^{2+} influx. Nitric oxide also works through this mechanism except it stimulates a guanylyl cyclase that increases cGMP that activates PKG that phosphorylates PLB and activates BK$_{Ca}$.

The phosphorylation state of the myosin light chains is also controlled by MLCP. Activation of MLCP brings about relaxation. Inhibition of MLCP keeps the light chains phosphorylated. The activity of MLCP is controlled by receptors that activate PLC and by receptors that activate Rho kinase and by nitric oxide, NO. Activation of PLC releases DAG that activates PKC. PKC phosphorylates a protein, CPI-17, which activates its inhibition of MLCP, which prolongs the phosphorylated state of the myosin light chains. Activation of Rho kinase releases a GTP-exchange factor (GEF) that activates a monomeric G-protein, RhoA, which in turn activates RhoA kinase. RhoA kinase phosphorylates MLCP, inactivating it. Thus RhoA also promotes phosphorylation of the myosin light chains. Nitric oxide produced by neighboring cells activates a soluble guanylate cyclase that produces cGMP. This activates PKG which

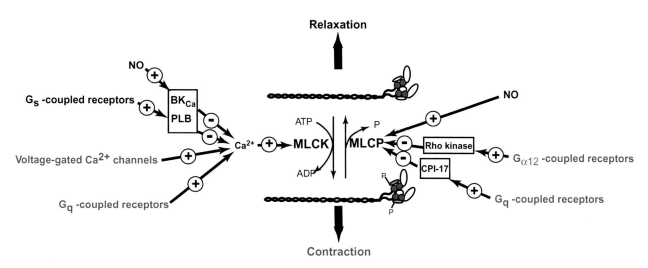

FIGURE 3.8.12 Synopsis of effects on smooth muscle contraction and relaxation. Receptors that promote contraction are shown in light blue; those that promote relaxation are shown in black. Activation of voltage-gated Ca^{2+} channels raises cytoplasmic $[Ca^{2+}]$, activating MLCK and promoting contraction; G_q-coupled receptors increase cytoplasmic $[Ca^{2+}]$ through IP_3-induced release of Ca^{2+} from SR stores, thereby activating MLCK and promoting contraction. G_q also activates PKC that phosphorylates CPI-17 that inhibits MLCP, thereby promoting contraction. $G_{\alpha 12}$ mechanisms activate rho kinase that phosphorylates MLCP, inactivating it and promoting contraction. NO activates a soluble guanylate cyclase that increases cytoplasmic cGMP, activating protein kinase G (PKG) that in some tissues phosphorylates phospholamban (PLB) and increases Ca^{2+} uptake by the SR, reducing cytoplasmic $[Ca^{2+}]$ and promoting relaxation. In other tissues PKG activates BK_{Ca} (a large-conductance channel for K^+) that increases outward K^+ current, making the membrane potential more negative and reducing Ca^{2+} entry through voltage-gated Ca^{2+} channels. In some tissues, G_s-coupled receptors activate adenylyl cyclase that increases cytoplasmic cAMP, activating PKA that similarly phosphorylates phospholamban and promotes relaxation. In other tissues, it activates BK_{Ca} and promotes relaxation. In some tissues, NO activates PKG that activates MLCP, thereby promoting relaxation.

phosphorylates MLCP at a different location from RhoA kinase, and activates it. Thus NO promotes dephosphorylation of the myosin light chains, causing relaxation.

REVIEW QUESTIONS

1. What is a tonic smooth muscle? What is a phasic smooth muscle?
2. What is electromechanical coupling? What is pharmacomechanical coupling?
3. What are slow waves?
4. How are actin filaments anchored at membranes? In the cytosol?
5. What are unitary smooth muscles? Multiunit smooth muscles?
6. How does Ca^{2+} enter the cytosol? How does it exit the cytosol?
7. How does Ca^{2+} initiate smooth muscle contraction? What does MLCK do? What does MLCP do?
8. What effect does phosphorylation of phospholamban have on smooth muscle contraction?
9. What is sensitization? What makes sensitization?
10. What is the latch state?
11. What effect does nitric oxide (NO) have on smooth muscle? What mechanism does it use?
12. How do G_q-coupled receptors promote contraction? How do G_s-coupled receptors promote relaxation?

UNIT 4

THE NERVOUS SYSTEM

Organization of the Nervous System 4.1

Learning Objectives

- Describe the general role of the nervous and endocrine systems in human physiology
- List the three steps in forming a response to an environmental challenge
- Describe what is meant by "afferent" and "efferent" fibers
- List the major components of the central nervous system
- List the major components of the peripheral nervous system
- List the three meninges and describe their relative position in the CNS
- Identify the cerebrum and cerebellum of the brain
- Define gyrus and sulcus
- Identify the longitudinal fissure, central sulcus, and lateral fissure
- Identify the five lobes of the cerebrum
- Describe the composition of the cerebrospinal fluid
- List the ventricles and describe what they are
- Describe what is meant by the blood-brain barrier
- Describe white matter and gray matter
- Define the term "nucleus" when applied to the CNS
- Describe the term "tract"
- Identify the spinal cord, medulla, pons, and midbrain

THE NEUROENDOCRINE SYSTEM CONTROLS PHYSIOLOGICAL SYSTEMS

The organ systems of the body cooperate—work together—to make life possible. Cooperation requires coordination, the proper alignment of the physiological systems so that their cooperation produces homeostasis. The control systems are manifold. **The nervous system provides short-term and rapid control. The endocrine system provides long-term and steady control.** These two systems interact, and physiologists often speak of the neuroendocrine system to reflect the fact that both the nervous system and the endocrine system cooperate to control all other organ systems in the body.

This control system is extraordinarily complex and complicated. The brain itself occupies a volume of ~ 1350 cm^3. Approximately 100 billion (10^{11}) neurons make connections to other neurons. Some neurons make only a few connections, whereas others may make many thousands. Approximately $10-100 \times 10^{12}$ contacts allow these neurons to communicate with each other. Somehow, this extraordinarily large number of contacts produces self-aware human behavior. In addition to all these neurons, about 10 times as many other cells play a supportive role and may also be involved in processing information. Many more neurons reside outside the central nervous system (CNS), in the periphery.

A CENTRAL TENET OF PHYSIOLOGICAL PSYCHOLOGY IS THAT NEURAL PROCESSES *COMPLETELY* EXPLAIN ALL BEHAVIOR

"Behavior" is the way a person or thing acts, especially in response to some outside influence. Because this behavior involves action, it is readily observable. It includes gross and fine body movements intended to accomplish a task, spoken language, and nonverbal modes of communication. In addition, humans exhibit a wide range of **affective behaviors**, relating to emotions, and **cognitive behaviors**, relating to thinking. What causes all of these internal and external behaviors?

Aristotle (384–322 BC) hypothesized that the mind is responsible for behavior. To Aristotle, the mind was not a material entity. The mind perceived, imagined, thought, emoted, desired, felt pain, reasoned, and remembered. The philosophy in which the mind controls behavior is called **mentalism**. Although science denies the controlling presence of a nonmaterial entity, we still use mentalist terms such as perception, imagination, emotion, motivation, memory, attention, and reason. Aristotle's ideas went largely unchallenged until the Renaissance. René Descartes (1596–1650) wrote a book, *Treatise on Man*, in which he tried to explain how the mind might be united to the body. In his view, the interface between mind and body was the brain. Descartes postulated that sensory information traveled from the skin, for example, to the brain. The mind then learned of our sensations through the brain. The mind could command action, but it did so through actions on the brain that were then conveyed to the muscles. Descartes sought to explain everything in terms of a machine analogy. He thought that the operation of all things, however complex, could be explained by some mechanism. He postulated hydraulics analogies to explain sensation: touching an object produces a pressure which squirts fluid along nerves into the brain. The fluid then enters fluid-filled cavities called **ventricles** and is sensed there by the mind. Descartes posited that the mind resided in the **pineal body**, a small bit of tissue located in the center of the brain besides one of the ventricles. On the output side, the pineal gland directed fluid from the ventricles down other nerves to activate muscles. The philosophical doctrine that the body is controlled by two entities, a mind and a brain, is called **dualism**. This left Descartes with a formidable problem: how could a nonmaterial entity interact with a material brain? This has been called the **mind–body problem**.

Besides writing *On The Origin of Species* in 1858, Charles Darwin (1809–1892) wrote another book entitled *On the Expression of the Emotions in Man and Animals*. Darwin's *Origin of Species* sought to explain the origin of the tremendous variety of animals and plants that are now present in the world by **natural selection** working on variations among members of a species already present in the population. Although Darwin had no mechanism to explain natural selection, because genes had not yet been identified, he forcibly argued from an amazing detail of information that evolution produced different species, and that the hallmark of the different species was a different character of body. The brain is part of the body, and so it is reasonable to suppose that evolution also changes the brain. In his second book, *On the Expression of the Emotions in Man and Animals*, Darwin argued that expressions of emotions are like physical attributes of the body: they are subject to evolution. This idea is still controversial today. How much of our behavior is **"hard-wired"** as opposed to **plastic**? Here "plastic" refers to the ability to change behavior based on past experiences or new inputs. Darwin's ideas implied that **the brain causes behavior**. This idea is now so well established that it is a central tenet of neuroscience: behavior arises from neural function *alone*. This idea is called **materialism**.

THE NEW MIND–BODY PROBLEM IS HOW CONSCIOUSNESS ARISES FROM A MATERIAL BRAIN

We all have subjective experiences of consciousness and of self. You are an individual distinct from other human beings, with distinguishable internal feelings and perceptions. What exactly is it in you that sees, hears, or is conscious? Because of overwhelming evidence that specific deficits in brain function produce specific deficits in mental function, we have come to believe that the brain somehow produces this mystical thing that is conscious and self-aware. This thing is not material in the ordinary sense of the word, just like an idea is not a material thing. The new mind–body problem is the inverse of Descartes' mind–body problem: how can a material thing (the brain) produce the nonmaterial thing that we identify as "self"? We can call this the mind and be in reasonable agreement with other notions of the mind. We cannot answer these questions here. These ideas are presented here so that the student can understand the ultimate goal of neuroscience: explain all behavior, including the subjective experience of consciousness and self-awareness. Although these are extremely important questions, most physiologists take a narrower aim of explaining how the nervous system integrates bodily functions, and not mental ones.

EXTERNAL BEHAVIORAL RESPONSES REQUIRE SENSORS, INTERNAL PROCESSES, AND MOTOR RESPONSE

Responding to changes in the external environment has three components:

1. appraisal of environmental conditions;
2. deciding the behavioral response;
3. activating the behavior.

These three components of a behavioral response correspond roughly to three identifiable branches of the nervous system.

THE APPRAISAL OF ENVIRONMENTAL CONDITIONS IS ACHIEVED THROUGH THE USE OF THE SENSES

The senses consist of the five classical senses: vision, hearing, taste, smell, and touch, but they include other senses such as balance, pain, temperature, pressure, and joint position. The "environment" that we sense includes the internal environment as well as the external environment. Thus, we have a sensation of hunger that results directly from our internal environment and indirectly from the external environment. Our internal environment includes the orientation of our body parts in the external environment, which is sensed by a combination of vision, balance, and proprioceptive sensors that apprise us of the location of our joints.

"DECIDING" THE BEHAVIORAL RESPONSE IS ACHIEVED THROUGH INTEGRATIVE CENTERS OF THE CNS

Once the environmental conditions have been sensed, the appropriate behavioral response must be engaged. Higher centers of the CNS determine the appropriate response. **Afferent** fibers bring sensory information from the periphery toward an appropriate processing station. This information is in the form of action potentials and, in general, its **intensity** is encoded by the **number of sensory cells** that are activated and by the **frequency of action potentials** generated by these cells. The **quality** of the sensory information is encoded by the type of sensory cell that fires and from the connections that this sensory cell makes in the brain. The information of the types of sensory cells firing and their intensity is integrated by the higher centers, which then initiate a response.

BEHAVIOR IS BROUGHT ABOUT BY EFFERENT FIBERS ACTIVATING EFFECTOR CELLS

The higher centers initiate a response by activating **efferent** fibers that bring information away from the CNS toward effector cells. In many cases, these effector cells are motoneurons that subsequently activate muscles, causing us to move. Responses to changes in our environment could include increased secretion of hormones as would occur upon perception of danger, or it might include modulation of digestion or any of a host of other responses.

THE NERVOUS SYSTEM IS DIVIDED INTO THE CENTRAL AND PERIPHERAL NERVOUS SYSTEM

The nervous system is broken down into anatomically and functionally discrete elements. The two main divisions are the **central nervous system** and the **peripheral nervous system**.

The **CNS** consists of the **brain** and the **spinal cord**. This part of the nervous system is encased in bone for protection. The brain is surrounded by the bones of the cranium and the spinal cord is surrounded and protected by the vertebrae.

The afferent fibers leading to the CNS and efferent fibers leading away from the CNS comprise the **peripheral nervous system**. The sensory afferent nerve fibers that bring sensory information from the periphery to the CNS constitute the incoming arm of the peripheral nervous system. The outgoing, or efferent arm, is the other half. The peripheral nervous system is further subdivided into:

   A. the somatic sensory system, consisting of sensory afferent fibers;
   B. the somatic motor system, consisting of motor efferent fibers;
   C. the autonomic nervous system, consisting of both afferent and efferent fibers.

The autonomic nervous system is further subdivided into a **sympathetic** and **parasympathetic** system. These divisions are distinguished on the basis of the anatomical location, the types of neurotransmitters, and their effects on the function of their innervated tissues. The autonomic nervous system controls cardiac muscle, smooth muscle, exocrine glands and some endocrine glands, and adipose tissue. The sympathetic nervous system typically prepares the body for emergency action, whereas the parasympathetic system promotes ordinary activity such as digestion. A third division of the autonomic nervous system is the **enteric nervous system**. The enteric nervous system is a network of neurons in the walls of the digestive tract. This system includes sensors in the lining of the tract; **ganglia**, or collections of neurons that integrate the responses; and cells that engage effector cells. Information flows between the enteric nervous system and the CNS, but the enteric nervous system can also function on its own. The basic organization of the nervous system is shown schematically in Figure 4.1.1.

THE BRAIN HAS READILY IDENTIFIABLE SURFACE FEATURES

The brain is covered by tough connective tissue called **meninges**, a three-layered structure that protects the brain from physical damage. The outermost layer is the **dura mater**, which derives from the latin for "hard mother." The dura mater covers the entire CNS in a loose sack. The middle layer is the **arachnoid**, from the Greek meaning "spider-like." This is a delicate sheet that follows the brain's contours and processes a fluid, the **cerebrospinal fluid (CSF)**, that bathes the entire CNS. Immediately apposed to the brain's surface is the **pia mater**, which means "soft mother." The CSF fills the space between the arachnoid and the pia mater. The CSF provides a cushion for the brain and allows the brain to move a little within the cranium.

The most obvious anatomical feature of the brain is its gross bilateral symmetry. A plane drawn perpendicular to the face and passing through the midline reveals a bilateral symmetry to the entire body. Such a plane is called the **midsaggital plane** and any section taken parallel to it is called a **saggital section**.

Looking at the brain from the top or the side, the two largest structures are the **cerebrum** and the **cerebellum**. From the top, the cerebrum is divided into left and right **hemispheres**. These two halves of the brain are separated by a **longitudinal fissure**. This fissure does not penetrate all the way through the brain. The two halves are connected anatomically and functionally through **commissures** that transfer information between the two hemispheres. Much of the information flows between the two hemispheres through a structure called the **corpus callosum**. The commissures are internal to the brain and cannot be seen from its surface.

The cerebellum is a bilaterally symmetrical structure at the back of and underneath the cerebrum. It lies adjacent and posterior to the **brain stem**. The cerebellum helps coordinate complicated motor movements. Both the cerebrum and the cerebellum consist of a fairly thin sheet of tissue, the **cortex**, which is folded so that it can fit into the skull. This allows for an increased area of cortex for each of these important structures. The folds produce bumps and cracks on the surface. A bump is called a **gyrus** and the crack or valley between adjacent bumps is called a **sulcus**. The **central sulcus** divides the front of the brain from its rear. The plural of these structures are **gyri** and **sulci**. Some of the sulci are deep and long and are called fissures. The **longitudinal fissure** divides the cerebrum into right and left hemispheres. The **lateral fissure,** also called the **Sylvian sulcus**, separates some of the lobes of the cerebrum, as shown in Figure 4.1.2 (lateral view).

The cerebrum is the main portion of the brain that is visible from its exterior. It is a convoluted structure that consists of four lobes named for the cranial bones under which they lie and another lobe buried beneath these. The four surface lobes are the **frontal lobe**, at the anterior of the brain; the **temporal lobe**, on the lateral surfaces beneath the temporal bone; the **parietal lobe**,

368 QUANTITATIVE HUMAN PHYSIOLOGY

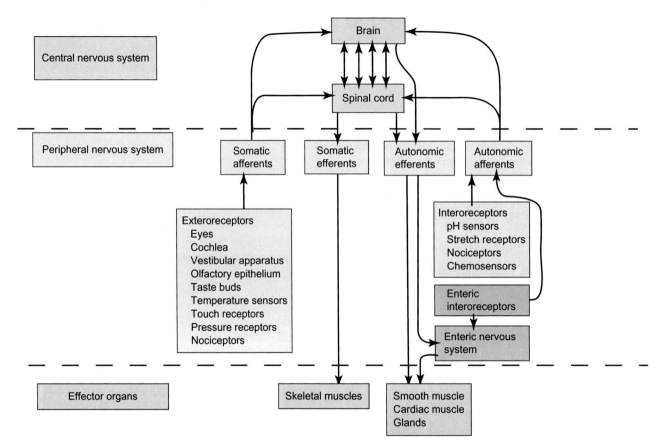

FIGURE 4.1.1 General organizational scheme of the nervous system. A variety of sensors, both internal and external, transmit information to the CNS by somatic or autonomic afferents. The CNS consists of the brain and spinal cord. These process the sensory information and initiate responses that travel to effector organs over somatic or autonomic efferent pathways. The enteric nervous system communicates with the CNS, but it can also function on its own by receiving sensory information, processing it, and initiating responses locally. The peripheral nervous system consists of afferent and efferent fibers and the ganglia that exist outside of the CNS.

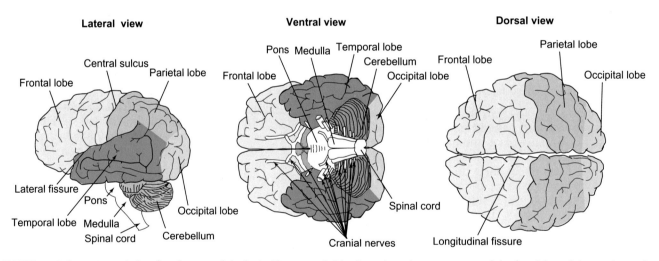

FIGURE 4.1.2 Gross anatomical surface features of the brain. The top and side views show the arrangement of the four lobes of the cerebrum: the frontal, temporal, parietal, and occipital. The cerebellum is a crinkled structure at the back of and underneath the cerebrum, just dorsal to the brain stem. The ventral view shows the cranial nerves I–XII.

at the top of the brain; and the **occipital lobe**, at the posterior surface of the cerebrum. The **insular lobe** can be seen by opening the Sylvian sulcus.

Parts of each of these lobes are dedicated to specific functions. The **frontal lobe** is important in foresight and planning and is important in motor tasks. As discussed in Chapter 4.4, the primary motor cortex is located in the frontal lobe immediately forward of the central sulcus. The premotor and supplementary motor cortexes are located adjacent to the primary cortex in the frontal lobe. These areas are essential for the proper planning of motor tasks. The frontal lobe contains **Broca's area**, in the inferior frontal gyrus, usually in the left hemisphere.

TABLE 4.1.1 Names and Main Function of the Cranial Nerves

Cranial Nerve	Common Name	Function	Cranial Nerve	Common Name	Function
I	Olfactory nerve	Smell	VII	Facial	Motor to face; taste at front of tongue
II	Optic nerve	Vision	VIII	Vestibulocochlear	Sound, rotation, and gravity
III	Oculomotor nerve	Muscles of the eye	IX	Glossopharyngeal nerve	Taste at back of tongue; salivary secretion; blood pressure and gas sense
IV	Trochlear nerve	Superior oblique eye muscle	X	Vagus	Autonomic afferent and efferents
V	Trigeminal nerve	Chewing muscles; sensory from the face	XI	Spinal accessory	Motor to sternocleidomastoid, trapezius
VI	Abducens	Lateral rectus eye muscle	XII	Hypoglossal nerve	Motor for the tongue

Damage to it causes **expressive aphasia**, the inability to form coherent speech.

The **parietal lobe** contains the somatosensory cortex and the sensory association cortex in which a variety of sensory information is processed and integrated to form a coherent model of the body's position in the world in order to plan motor activities. Visual information, balance, and proprioceptive (body sense) information all mingle here. The parietal lobe contains **Wernicke's area**, which is involved in the understanding of speech. Damage to this area causes **receptive aphasia**, in which the person cannot understand speech.

The main function of the **occipital lobe** is to process visual information. Some processing occurs in the retina itself and in the lateral geniculate nucleus, but the visual cortex in the occipital lobe is responsible for generating conscious awareness and identification of objects in the visual field.

If we turn the brain over and look at it from the bottom, we see the bottom parts of the cerebrum—the temporal lobe, frontal lobe, and occipital lobe—and the cerebellum, and we also see a whitish structure in the center, leading toward the posterior or back of the brain. Attached to this structure are a number of what look a little like white wires. These are the **cranial nerves**. The whitish structure itself is the **brain stem**. It has a number of parts. In this view (Figure 4.1.2, ventral view) the most obvious parts of the brain stem are the pons and the medulla. A synopsis of the functions of the cranial nerves is shown in Table 4.1.1.

CSF FILLS THE VENTRICLES AND CUSHIONS THE BRAIN

Transverse slices of the brain reveal hollow spaces called **ventricles**. The ventricles include two **lateral ventricles**, a **third ventricle**, and a **fourth ventricle**. All of these are filled with **CSF**. Channels connect all of the ventricles. The lateral ventricles are connected to the third ventricle by an **interventricular foramen**, and the third ventricle connects to the fourth through the **cerebral aqueduct** (also called the aqueduct of Sylvius). CSF can leave the fourth ventricle through apertures on its lateral and medial surfaces (the **lateral and median apertures**, respectively). CSF leaving through these apertures can circulate through the subarachnoid spaces surrounding the brain and spinal cord. The location of these four ventricles and the circulation of CSF is shown schematically in Figure 4.1.3.

The CSF cushions the brain and spinal cord against mechanical shock. In addition, the extracellular fluid (ECF) of the brain is in direct contact with the CSF, so CSF composition dictates brain ECF composition. Table 4.1.2 shows the composition of the CSF in comparison to that of plasma. The CSF lacks the high protein content of the plasma, and so its osmolarity is balanced by increased concentrations of Na^+ and Cl^- ions. Its density is almost the same as brain tissue, so that the brain nearly floats in this fluid. This gives rise to the phenomenon of **countrecoup**. Blows to the head accelerate the skull which transmits the force to the CSF directly beneath it, which simultaneously transmits force to the brain tissue because of the incompressibility of the fluid. On the opposite side, however, the skull pulls away from the fluid, creating a momentary partial vacuum. When the skull is no longer accelerated by the blow, the vacuum collapses and the brain smashes into the skull on the side opposite to the original blow.

Specialized structures called **choroid plexuses** produce the CSF. These structures consist of specialized blood supplies covered with specialized **ependymal cells** that form a continuous layer separating CSF from the blood. The choroid plexus projects into the temporal horn of each lateral ventricle, the posterior portion of the third ventricle, and the roof of the fourth ventricle. These cells make about 0.35 mL of CSF each minute, or about 500 mL per day. A corresponding amount is reabsorbed daily by **arachnoid villi**, which are specialized finger-like projections of the arachnoid membrane into the venous sinuses. The ventricles contain about 35 mL of CSF, while the brain and spinal cord contain about 100 mL. Thus the CSF turns over 3–4 times per day (500 mL day^{-1}/135 mL). The circulation of the CSF is shown in Figure 4.1.3.

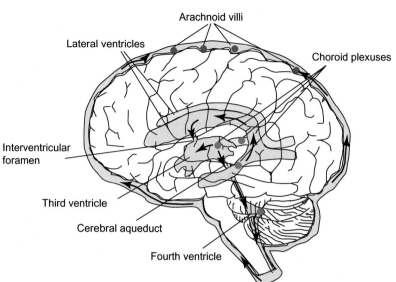

FIGURE 4.1.3 The four ventricles of the brain and circulation of the CSF. Two lateral ventricles have a complicated shape, being continuous with a portion above and another portion below and lateral to the third ventricle, which is centered on the midline. The fourth ventricle is below the third, immediately ventral to the cerebellum. The choroid plexuses, within the ventricles, produce CSF from blood. The CSF circulates between the ventricles through canals. The lateral ventricles communicate with the third ventricle through the interventricular foramen (foramen of Monro); the third and fourth ventricles communicate through the cerebral aqueduct and the fourth ventricle communicates with the spinal fluid in the arachnoid space through two lateral apertures (foramen of Luschka) and a median aperture (foramen of Magendie), not shown.

TABLE 4.1.2 Constituents of CSF Compared to Plasma

Constituent	Lumbar CSF	Plasma
Na^+	148 mM	142 mM
K^+	3 mM	4 mM
Cl^-	125 mM	105 mM
Glucose	50–75 mg/dL	70–100 mg/dL
Protein	15–45 mg/dL	$6-8 \times 10^3$ mg/dL
pH	7.3	7.4

The rate of CSF production is nearly independent of its pressure, whereas the absorption of CSF increases directly with CSF pressure. This results in CSF pressure of about 10–15 mmHg. This is extremely important because the skull presents an incompressible limit to the contents of the cranium. Increases in CSF fluid volume must be paid by decreases in the volume of nervous tissue. Cerebral hemorrhage, for example, causes a rapid rise in intracranial pressure, with subsequent reduction of blood flow, impairment of oxygen delivery, and subsequent rapid deterioration of brain function. For this reason, head trauma patients are monitored carefully for increases in CSF pressure.

THE BLOOD-BRAIN BARRIER PROTECTS THE BRAIN

The endothelial cells of the brain capillaries are joined tightly together, rather than having slit pores or being fenestrated as in the microcirculation of other vascular beds (see Chapter 5.10). This arrangement forms a continuous cellular layer between the blood and the brain ECF, forming a blood-brain barrier. This barrier also exists between the blood and the CSF. The blood-brain barrier is highly permeable to oxygen, CO_2, water, and most lipid-soluble materials such as anesthetics and alcohol. It is slightly permeable to plasma electrolytes but it is impermeable to plasma proteins and a variety of other materials. The cells that line the surfaces of the ventricles, however, are permeable to most materials. Thus, drugs that cannot cross the blood-brain barrier, and thus are ineffective when administered through the blood, can be effective when administered directly into the CSF (Figure 4.1.4).

CROSS SECTIONS OF THE BRAIN AND STAINING REVEAL INTERNAL STRUCTURES

A frontal section of the brain (a section along a plane parallel to the face or front of the body) at about the middle of the brain reveals a number of structures as shown in Figure 4.1.5.

The frontal section shown in Figure 4.1.5 shows readily identifiable differences in the gross appearance of different regions. Some areas appear grayer (shown in dark blue in Figure 4.1.5) because of their greater concentration of cell bodies with their attendant nuclei and synthetic machinery. The white areas (shown in light blue in Figure 4.1.5) owe their appearance to the many neuronal processes, the axons, that are encased with myelin. The myelin consists of many wrappings of the cell membranes of supportive cells and thus contains a large proportion of phospholipids. These produce the white appearance of these areas. These areas are accordingly called **gray matter** and **white matter.** Some areas of the brain have a mottled gray and white appearance, and these areas are called **reticular matter** (from the latin "rete" meaning "network"). As shown in Figure 4.1.5, gray areas have more or less the same locations and extents in different individuals and these conglomerations of cells also serve the same functions. These areas are called **nuclei** and are given specific names. In some cases, functional areas were discovered by their response to ablation (surgical or electrical destruction of the area) or excitation. In these cases the functional areas are also given specific names. There are a large number of named nuclei which serve well-defined functions. These will be introduced during discussions of their functions.

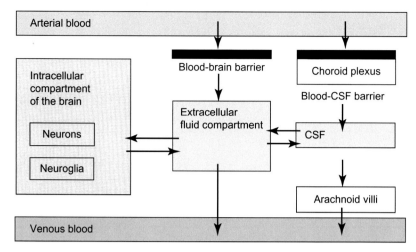

FIGURE 4.1.4 Fluid transfer between blood and brain ECF and CSF. The ECF of the brain is tightly sealed from the blood so that only permeable materials are transferred to the brain's ECF. This excludes most large molecular materials such as plasma proteins. The CSF is similarly insulated from the blood. The blood-brain barrier and the blood-CSF barrier can sometimes determine the efficacy of drugs targeted to the brain. There is ready and easy transfer between brain ECF and CSF so that drugs administered into the CSF have ready access to the target neurons.

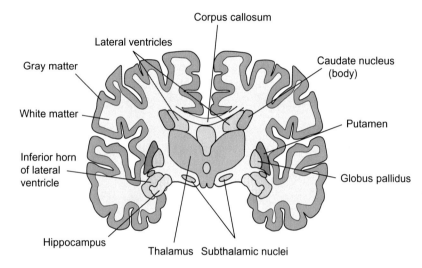

FIGURE 4.1.5 Frontal section of the brain reveals areas distinguishable to the unaided eye. The cortex, or area of the brain next to the surface, appears darker because it contains mostly cell bodies with their darker nuclei. These darker areas are called the "gray matter." Other areas that contain a concentration of cell bodies are called nuclei, and some are shown here. The basal ganglia consist of the caudate nucleus, the putamen, and the globus pallidus. Together with the thalamus, the subthalamic nuclei, and the substantia nigra, the basal ganglia help control movement. The substantia nigra is located in the midbrain. Many other areas can be identified by other sections.

Cell bodies are collected into specialized areas called nuclei and their processes are often collected into **tracts**. These tracts become evident only when many nerve processes (axons) are traveling together toward similar locations. They are evident in the spinal cord, where sensory inputs travel up the cord and efferent fibers travel down the cord, and in the brain itself. The corpus callosum is a large tract that contains some 200 million nerve fibers that transmit information from one hemisphere to the other.

GRAY MATTER IS ORGANIZED INTO LAYERS

The cerebral cortex lies beneath the pia mater and covers the surface of most of the brain, as viewed from the top. It contains many folds, or gyri, as shown in Figure 4.1.5. The cerebral cortex consists of two parts: the **neocortex** and the **limbic cortex**. The neocortex covers most of the surface of the cerebrum as viewed from outside the brain. The limbic cortex covers the **cingulate gyrus**, the fold of the cerebrum immediately adjacent to the longitudinal fissure above the corpus callosum (see Figure 4.1.7). Both of these are organized further into layers. The neocortex is organized into six layers of cells, whereas the limbic cortex is organized into four layers. Each of these is involved in specialized functions such as (a) the input of sensory information, (b) integration of inputs, and (c) outputs to other parts of the brain or efferent pathways. In the same way, neurons within the gray matter of the spinal cord are organized in **laminae**, or layers.

OVERALL FUNCTION OF THE NERVOUS SYSTEM DERIVES FROM ITS COMPONENT CELLS

The function of the nervous system is to command and coordinate responses of the body to change in environmental conditions, either internal or external. It is axiomatic that this response arises somehow from the myriad events occurring in the individual cells within the nervous system. We become apprised of changes in the environmental conditions: first by sensory cells that directly sense these changes; second by the cells that receive inputs from these sensory cells. This information must be integrated along with other signals to select the appropriate response. In the end, effector organs are activated, again by collections of single neurons, but the choice of which neurons and their timing is the result

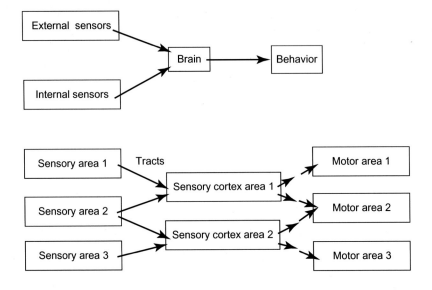
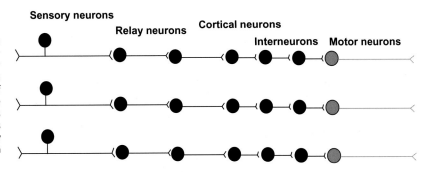

FIGURE 4.1.6 Different levels of organization in the nervous system. The brain mediates our behavior after being apprised of changes in either our internal or external world. This mediation can be viewed from the nuclei level of organization, in which the activity of many neurons is simultaneously modified. All of the activity, of course, also occurs at the cellular level. It is an axiom of neurophysiology that all behavior derives from the activity of neurons. How this activity "summates" to produce conscious behavior is a difficult and still largely unsolved problem.

of central processing. This entire process can be looked at on at least three levels of organization, as shown in Figure 4.1.6. The brain level simply indicates that the brain somehow produces our behavior when information of our external or internal world reaches it. The second level, the nuclei level, indicates that information flows between populations of neurons that comprise relatively large parts of the brain. The third level, the neuron level, indicates that all of these processes occurring on these grand scales arise from the individual events that occur between nerve cells.

OVERVIEW OF THE FUNCTIONS OF SOME MAJOR AREAS OF THE CNS

THE FOREBRAIN CONSISTS OF THE CEREBRAL CORTEX, BASAL GANGLIA, AND THE LIMBIC SYSTEM

The cerebral cortex is the convoluted surface that is visible from the outside of the brain (Figure 4.1.2) and consists of four external lobes: the frontal, temporal, parietal, and occipital lobes. The insular lobe lies beneath these. Each hemisphere has each of these five lobes. The cerebral cortex forms a neural map of the world that integrates external sensors with internal ones so that the body can be placed in this map to plan motor tasks. The external surface of the body maps onto the cortex so that adjacent areas on the surface of the body are processed in adjacent areas of the cortex. This is true of both sensory inputs and motor outputs. In addition, there are topographic projections of "visual space" and "auditory space" onto the cortex. Areas of the cortex are designated as **primary**, **secondary**, or **tertiary** depending on their role in sensory perception or motor control.

The basal ganglia include the **putamen**, **globus pallidus**, and **caudate** nuclei (Figure 4.1.5). These are located deep within the forebrain just below the white matter of the cortex. These structures receive inputs from all lobes of the cerebral cortex and send connections to the prefrontal and premotor cortices by way of the thalamus. The basal ganglia assist in movement and suppress useless and unwanted movements. They also inhibit muscle tone throughout the body.

The limbic system includes the **amygdala**, the **hippocampus**, **cingulate gyrus**, **corpus callosum**, and **fornix**. The cingulate gyrus lies just above the corpus callosum and immediately adjacent to the longitudinal fissure. The amygdala and hippocampus are regions of the cortex buried in the temporal lobe adjacent to the floor of the third ventricle and lateral to it. This complex of forebrain structures plays important roles in the emotions and sociosexual behavior. However, the limbic system has multiple functions and is important in motivation and memory formation as well.

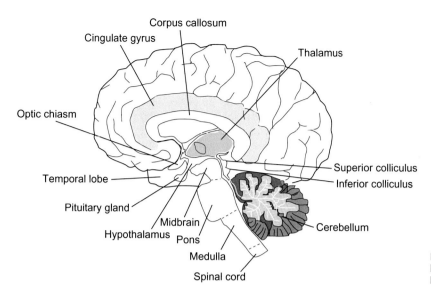

FIGURE 4.1.7 Midsaggital section of the brain and brain stem. The plane of section is parallel to the longitudinal fissure, perpendicular to the face.

THE DIENCEPHALON OR "BETWEEN BRAIN" INCLUDES THE THALAMUS AND HYPOTHALAMUS

The thalamus physically occupies a central position between the brain stem and the forebrain. It is an important relay point for information going both ways. The thalamus contains about 20 nuclei. Many of these relay sensory information to the cerebral cortex. Other regions are engaged in motor functions whereas still others are involved with attention and memory.

The hypothalamus derives its name from its location immediately below the thalamus. It regulates a wide variety of homeostatic functions including (a) thirst; (b) urine output; (c) food intake; (d) body temperature; and (e) hormone secretion. The hypothalamus coordinates the autonomic functions of the body, partly through nerve activity and partly through hormonal activity.

THE BRAIN STEM CONSISTS OF THE MIDBRAIN, PONS, AND MEDULLA OBLONGATA

The midbrain lies above the pons and is divided into two parts: the **tectum** and the **tegmentum**. The tectum receives input from the eyes and the ears to the **superior colliculus** and **inferior colliculus**, which together make up the tectum. These structures allow us to simultaneously process visual and auditory stimuli to search for the source of a sound, for example. The tegmentum lies below the tectum and consists of a number of nuclei that are involved in controlling eye and limb movements.

The pons lies ventral to the cerebellum and superior to the medulla, as shown in Figure 4.1.7. Areas of the pons, together with centers in the medulla, help regulate breathing. The pons also relays information from the cerebral cortex to the cerebellum.

The medulla is the rostral (toward the nose) extension of the spinal cord. It contains centers that regulate blood pressure and breathing and also helps coordinate swallowing, coughing, and vomiting.

THE HINDBRAIN CONSISTS OF THE BRAIN STEM AND THE CEREBELLUM

The cerebellum is the large structure lying beneath the occipital lobe and behind (dorsal to) the pons and medulla. It coordinates complex movements, and helps maintain posture and coordination of head and eye movements. The cerebellum integrates proprioceptive inputs from the spinal cord, motor control input from the cerebral cortex, and inputs about balance from the vestibular apparatus of the inner ear.

THE SPINAL NERVES AND CRANIAL NERVES CONNECT THE CNS TO THE PERIPHERY

The spinal cord extends from the base of the skull down to the sacrum. It is encased in a protective bony case, the spine, to which afferent sensory nerves arrive and efferent effector nerves leave. Tracts of axons travel in both directions up and down the cord to coordinate activities from the CNS to the periphery.

SUMMARY

The neuroendocrine system controls the entire spectrum of human behavior including our fine and gross motor movements, language, affective behavior (relating to the emotions) and cognitive behavior (relating to thinking), and sociosexual function and behavior. The nervous system provides short-term and rapid control whereas the endocrine system provides long-term and steady control. The two systems of control interact extensively, yet they differ in anatomic location and mechanisms of action.

The nervous system is broken down into two major divisions: the central nervous system (CNS) and the peripheral nervous system. The CNS consists of the brain and spinal cord. The peripheral nervous system in turn consists of three parts: the somatic sensory system, the somatic motor system, and the autonomic system. The somatic sensory system is made up of sensors that feed information to the CNS along afferent fibers. These include all of the exteroreceptors, which include sight, touch, taste, hearing, balance, smell, temperature, and pain. The somatic motor system activates skeletal muscles through motor efferent fibers. The autonomic system itself consists of three parts: the sympathetic nervous system, the parasympathetic nervous system, and the enteric nervous system. The autonomic nervous system responds to sensory information about the internal state of the body, for which there are a number of interoreceptors that respond to pH and other chemicals, pain, and stretch. The autonomic nervous system also responds to higher CNS functions such as emotion and mood. The enteric nervous system interfaces with the autonomic nervous system but it can also function independently. Autonomic efferent fibers bring commands to effector organs such as the heart, smooth muscles, and glands.

The fragile parts of the CNS are encased in bone and three membranes, the meninges: the dura mater, arachnoid mater, and pia mater. It is also bathed in the CSF. The CSF fills the four ventricles: two lateral ventricles and the third and fourth ventricle. The CSF is made by the choroid plexuses located in each of the ventricles, and it is absorbed by arachnoid villi. The ventricles are connected by canals called the intraventricular foramina, the cerebral aqueduct, and lateral and medial foramina connecting the fourth ventricle with the spinal fluid. Many blood-borne materials cannot cross the blood-brain barrier because the capillaries are tightly lined with cells.

The brain consists of many parts. From its surface, it appears to be bilaterally symmetrical. A longitudinal fissure separates the left and right hemispheres, but the fissure penetrates only as far as the corpus callosum, a large collection of nerve fibers that communicates information between the two hemispheres. Four lobes of the cerebrum are visible from the surface: the frontal, temporal, parietal, and occipital lobes. The insular lobe is buried and can be seen by opening the lateral fissure, or Sylvian sulcus. The cerebrum folds into bumps (gyri) and valleys (sulci), each of which are devoted to specific functions and which have names. Also visible on the surface of the brain is the cerebellum at the back of and underneath the occipital lobe, part of the brain stem, and the cranial nerves.

Brain tissues consist of gray matter, white matter, and reticular matter. The gray matter appears gray on gross inspection because of a concentration of cell bodies with their nuclei. The white matter contains a preponderance of nerve fibers with high fat content. Reticular matter appears mottled. Collections of cell bodies in the deeper parts of the brain are named and generally serve specific functions. Axons often follow specific tracts in which many axons originate from one nucleus and travel to another. The gray matter itself is organized into layers.

The forebrain consists of the cerebrum, basal ganglia, and limbic system. The "between brain" consists of the thalamus and hypothalamus, and the hindbrain consists of the midbrain, pons medulla, and cerebellum. The spinal cord begins at the end of the medulla.

REVIEW QUESTIONS

1. The central sulcus separates which lobes of the cerebrum? The longitudinal fissure separates which lobes of the cerebrum? Name the five lobes of the cerebrum and identify them in a drawing of the brain.
2. What are the meninges?
3. What is gray matter? White matter? What is a nucleus (in neurophysiological terms)? What is a tract?
4. What are the ventricles? Name them. What is the composition of the CSF? Where is it made? Where is it resorbed? What is normal CSF pressure? Why is this important?
5. What is the blood-brain barrier?
6. What makes up the CNS? Name the major parts of the forebrain, diencephalon, and hindbrain.
7. How is information conveyed between right and left hemispheres?

Cells, Synapses, and Neurotransmitters 4.2

Learning Objectives

- List the four different types of glial cells, recognize their structure, and briefly describe their function
- Contrast the myelin sheath in the CNS and PNS
- Identify soma, dendrites, axon, myelin sheath, node of Ranvier and axon hillock
- Define the synapse and distinguish between electrical and chemical synapses
- List in order the events that occur in chemical synaptic transmission, including the shut-off of the signal
- Describe the major mechanisms for removing neurotransmitters
- Describe the recycling of synaptic vesicles
- Distinguish between ionotropic and metabotropic transmitters
- Define pre-synaptic and post-synaptic membranes or cells
- Describe the conductance changes underlying EPSPs or IPSPs
- Describe fast and slow axonal transport

NERVOUS SYSTEM BEHAVIOR DERIVES FROM CELL BEHAVIOR

An axiom of physiology states that organ physiology derives from cell physiology: the summed behavior of its component cells determines the overall behavior of an organ. How this summation occurs is complex and in many cases our understanding of it is inadequate. The topology of the system, which is the connectedness of the cells and their spatial arrangement, plays an important role in determining how the cells interact to produce organ behavior. Cells in the central nervous system (CNS) are organized in clear spatial patterns. Although it appears that these spatial patterns are extremely important, how they produce their desired effects is not so easily discerned. Regardless of our present inability to show how neuronal function produces higher brain function, most neurophysiologists adhere to the axiom that *all* brain behavior has a cellular basis.

NERVOUS TISSUE IS COMPOSED OF NEURONS AND SUPPORTING CELLS

Neurons are the "functional units" of the nervous system. These cells receive inputs, make a "decision" based on these inputs, and transmit "information" on to other cells. They are characterized by their ability to produce action potentials in response to inputs. The action potentials are useful because they allow one part of the cell to communicate with its extreme edges within a few milliseconds, even though the extreme edges could be as much as a meter away. There is no faster way for a physiological signal to travel from one part of the body to another. There are about 10^{11} of these neurons, distributed among a variety of neuron types. The other category of cells in the nervous system is the supporting cells. These cells actually comprise the major part of the nervous system, outnumbering the neurons 10 to 1. There are four major types of glial cells in the CNS and another type in the peripheral nervous system (PNS). There are collectively called **neuroglia**, which literally means "nerve glue."

GLIAL CELLS PROTECT AND SERVE

In the CNS, there are four distinct types of neuroglial cells. These comprise the following: (A) **ependymal cells**; (B) **astrocytes**; (C) **oligodendroglia**; and (D) **microglial** cells.

Ependymal cells are cuboidal epithelial cells that line the internal cavities of the ventricles of the brain and the central canal of the spinal cord. They contribute to the formation of cerebrospinal fluid. Ependymal cells have cilia that project into the CSF and help circulate it. Neurons in the adult lose their mitotic capabilities so that destruction of neurons results in irreplaceable loss of neurons. However, the glial cells retain mitotic capabilities and so they can replace lost glial cells. Recent studies suggest that ependymal cells may serve as stem cells not only for lost glial cells but also to replace neurons (see Figure 4.2.1).

The word **oligodendroglia** comes from the roots for "few," ("oligo"), "branches" ("dendro"), and "glue" ("glia"). This describes the role of these cells, which is to form the myelin sheaths around a few axonal processes. These cells send out large flat branches of cytoplasm that wrap around axons in the CNS and in the spinal

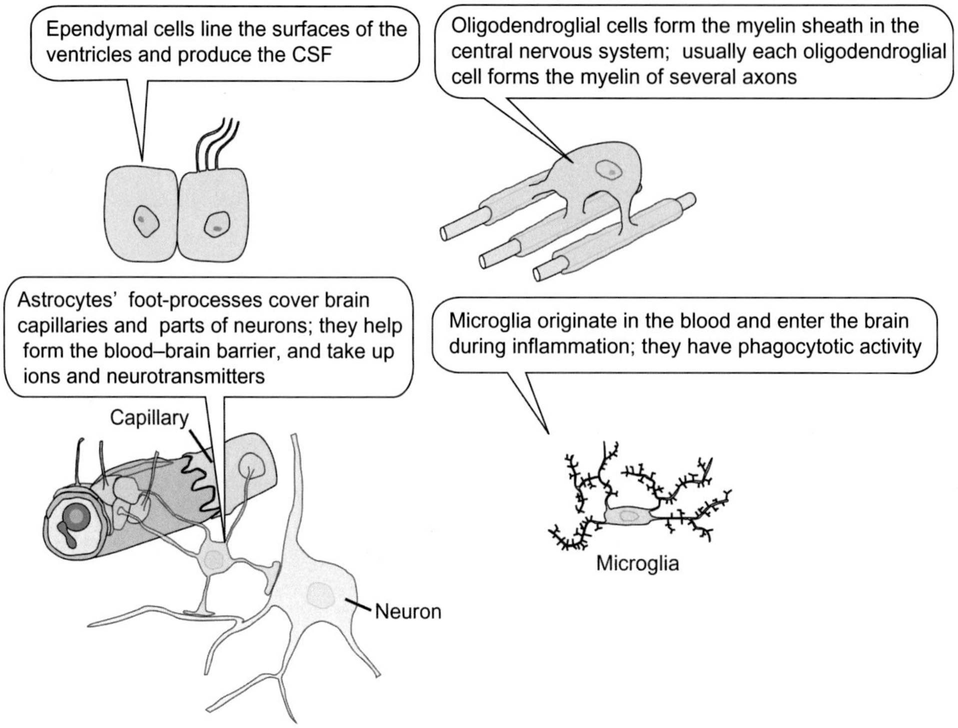

FIGURE 4.2.1 Types of neuroglial cells. Ependymal cells are ovoid or cuboidal cells that line the ventricles and produce the cerebrospinal fluid. Oligodendroglial cells extend their cytoplasm around axons to form an insulating myelin sheath in the brain and spinal cord. In the PNS, Schwann cells form myelin. Astrocytes are star-shaped cells that send out processes to envelope capillaries and neurons. Astrocytes in the white matter are typically fibrous astrocytes; those in gray matter are protoplasmic astrocytes. Microglia do not derive from the same embryonic tissue as neurons, but instead derive from the blood.

Ependymal cells line the surfaces of the ventricles and produce the CSF

Oligodendroglial cells form the myelin sheath in the central nervous system; usually each oligodendroglial cell forms the myelin of several axons

Astrocytes' foot-processes cover brain capillaries and parts of neurons; they help form the blood–brain barrier, and take up ions and neurotransmitters

Microglia originate in the blood and enter the brain during inflammation; they have phagocytotic activity

cord. The glial cells squeeze out the cytoplasm and form an insulating sheath of coiled cell membranes. In cross sections in electron micrographs, these coils appear to be stacks of membrane. The stacked membranes insulate the axon's plasma membrane from the external solution, markedly increasing the resistance between axoplasm and the extracellular fluid and decreasing the electrical capacitance. These effects increase the length constant and the speed of conduction of action potentials (see Chapter 3.3). Small gaps, the **nodes of Ranvier**, separate the cell process of one oligodendroglia from the next. Regenerative currents flow across the **axolemma**, the plasma membrane of the axon, only at the nodes. In the PNS, the cells that form myelin are called **Schwann cells**. Figure 4.2.2 illustrates how the myelin sheath is formed and the nodes of Ranvier in the CNS and PNS.

Astrocytes derive their name from their star shape. There are two basic forms: **fibrous astrocytes** have many filaments in their cytoplasm and they are found in bundles of axons of myelinated fibers in the white matter of the brain; **protoplasmic astrocytes** have fewer filaments and are found in the gray matter. Astrocytes perform a number of functions:

1. Astrocytes provide a kind of scaffold for the proper generation of spatial relationships of neurons during development and for the maintenance of this relationship during adulthood.
2. Astrocytes interact with the capillaries within the CNS to form tight junctions, which make up the blood–brain barrier. The astrocytes send out foot processes to cover the capillaries, reinforcing the tight junctions.
3. Astrocytes help shut off neurotransmission by taking up released neurotransmitters such as **GABA** (gamma amino butyric acid) and **glutamate**. They degrade the neurotransmitters into materials that can be used to resynthesize the neurotransmitters in the neurons.
4. Astrocytes take up K^+ released from the neuron during repolarization after the action potential. Most of the time the Na^+–K^+ ATPase returns the Na^+ that entered the neuron and the K^+ that left it during the action potential. During high activity the Na^+–K^+ ATPase can be outpaced, so K^+ builds up in the Extracellular fluid (ECF). Removal of the excess K^+ helps keep the resting membrane potentials at the proper level so that neuronal activity can continue.
5. After injury to areas of the brain, **reactive astrocytes** appear that **phagocytose** cellular debris. Phagocytosis refers to "cell eating" and describes the clean up of cellular debris. Together with fibroblasts, the astrocytes form glial scars.
6. Astrocytes contain the enzyme **carbonic anhydrase**, which speeds up the equilibrium between dissolved CO_2 and carbonic acid, H_2CO_3, which promptly dissociates into HCO_3^- and H^+. By speeding up the reaction, astrocytes help regulate the pH of the interstitial fluid.

Unlike the other glial cells, **microglia** originate from the blood and migrate into the brain when the brain is damaged. They remove cellular debris by phagocytosis.

NEURONS DIFFER IN SHAPES AND SIZE

The common features of all neurons include the following: (a) the cell body, or **soma**; (b) processes that gather information either from a sensory device or from other neurons; and (c) one or more processes that convey excitation either towards other neurons or towards

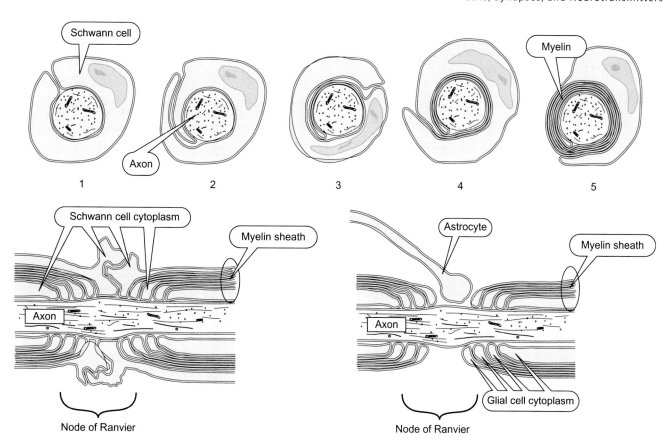

FIGURE 4.2.2 Stages in the formation of myelin by Schwann cells. Schwann cells wrap around axons in the PNS, forming myelin by the compact stacking of their cell membranes. Although the myelin is actually a spiral stack of Schwann cell membranes, cross sections appear as stacks. In the CNS oligodendroglial cells form myelin around several axons. The lower illustrations show longitudinal sections of myelinated nerves in the PNS and CNS. At the node of Ranvier, the axon has access to ions in the extracellular fluid.

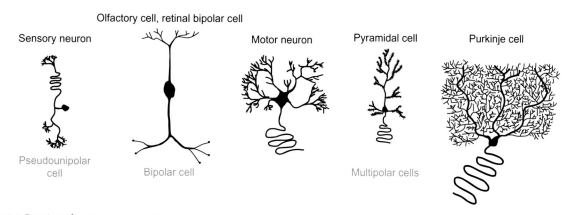

FIGURE 4.2.3 Drawing of various neurons. The sensory neuron is typical of somatosensory neurons in the skin. It is a pseudounipolar neuron because one process exits the cell body, but it divides into a sensory branch that reaches the periphery and a central branch that makes connections to neurons in the CNS. Cells in the olfactory epithelium, in the eye, and in the ear are bipolar neurons and make connection between sensory stimuli (odorants, light, and sound, respectively) and sensory areas of the brain along cranial nerves I, II, and VIII, respectively. Examples of multipolar neurons include the motor neuron, an efferent neuron found in the ventral horn of the spinal cord; the pyramidal cell, an interneuron in the cerebral cortex; and a Purkinje cell, an interneuron found in the cerebellum.

effector cells in muscles or glands. The soma surrounds the cell nucleus, or **karyon**, and so sometimes the soma is referred to as the **perikaryon**. Nerve cells come in a variety of shapes and sizes that can be categorized according to their processes (**unipolar, bipolar,** or **multipolar**) or according to their function (**sensory, interneurons,** and **efferent neurons**). A variety of names are given to the many varieties of specific cell types found in specific locations in the brain. Examples of these are the Purkinje cells of the cerebellum and pyramidal cells of the cerebral cortex, which are interneurons, and the lower motor neurons in the ventral horn of the spinal cord, which are efferent neurons. Figure 4.2.3 shows examples of some different neurons.

INPUT INFORMATION TYPICALLY CONVERGES ON THE CELL AND OUTPUT INFORMATION DIVERGES

A "typical" nerve cell (Figure 4.2.4) has multiple dendrites and a single axon that branches to form collaterals near the point of termination. This typical nerve cell forms junctions with other neurons by closely apposing its axonal end feet to the membrane of the other cell, forming junctions called **synapses**. Typically, information transfer across synapses is unidirectional. The cell sending the message is the **pre-synaptic cell**, and the cell receiving it is the **post-synaptic cell**. The dendrites receive inputs from a staggeringly large number of sources and its output diverges onto the set of cells with which it forms synapses. Integration of signals on the input side makes use of **temporal summation** and **spatial summation**, as discussed previously in Chapter 3.6. Spatial summation depends heavily on the location of synapses with respect to the site of initiation

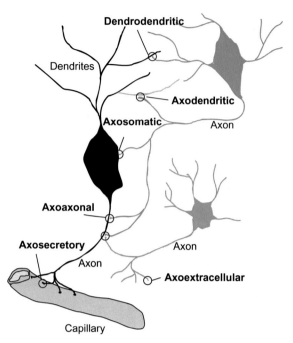

FIGURE 4.2.5 Locations of synapses. Synapses can occur between two dendrites on different cells (dendrodendritric); between an axon and a dendrite (axodendritic); between an axon and a soma (axosomatic); between an axon and another axon (axoaxonal); between an axon and another axon near a synapse onto a third axon (a pre-synaptic axoaxonal connection); between an axon and the extracellular fluid (axoextracellular); and at an axon terminus on a blood vessel (axosecretory).

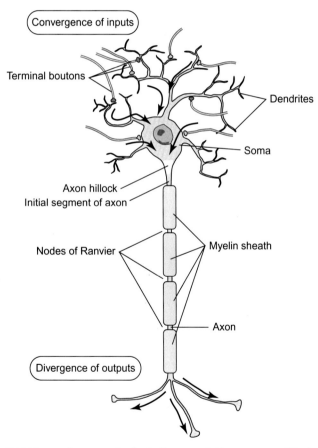

FIGURE 4.2.4 Structure of a "typical" nerve cell. The soma, or cell body, contains the organelles associated with protein synthesis including the nucleus, nucleolus, nuclear membrane, rough and smooth endoplasmic reticulum, and Golgi apparatus. The dendrites are long extensions of the soma that generally receive inputs, which are then integrated in the soma. Action potentials are initiated in the axon hillock, a clear, cone-shaped region of the cytoplasm where the initial segment of the axon leaves the soma. The axon may branch many times and produce hundreds of terminals that end either on other neurons or on effector cells. Thus, the neuron converges inputs into a single effect—the firing of action potentials along the axon, and it diverges its output onto a number of effector cells.

of the action potential. Figure 4.2.5 shows that synapses occur on almost any neuronal surface. Synapses are named for the anatomical parts that participate in their formation. **Axosomatic** synapses occur between a pre-synaptic axon and a post-synaptic soma; **axodendritic** synapses between an axon and a dendrite; **axoaxonic** synapses between an axon and another axon; **dendrodendritic** synapses occur between two dendrites; **axoextracellular** synapses have no post-synaptic membrane—the axon dumps its transmitter into the extracellular fluid; and in **axosecretory** connections the axon dumps its transmitter into the blood. Sometimes axoaxonal synapses occur close to the point at which an axon makes another axoaxonal connection. These are called **pre-synaptic connections**.

CHEMICAL SYNAPSES ARE OVERWHELMINGLY MORE COMMON

The root word for "synapse" means "connection." There are two basic kinds of synapses: **chemical synapses** and **electrical synapses**. In electrical synapses, the membranes make very close approach and in electron micrographs appear to be fused, and there is a direct electrical connection between the two cells. Electrical synapses form when a **connexin** hexamer on one cell membrane joins up with a second connexin hexamer on the other cell membrane. The connexin hexamer forms a pore that allows low-molecular-weight materials such as ions (Na^+, K^+, and Ca^{2+}) and signaling molecules such as cAMP and cGMP to pass from one cell to another. Electrical events such as depolarization, hyperpolarization, or an action potential are readily transferred from one cell to another through the channels formed by the aligned connexins. The close juxtaposition of these membranes is actually accomplished through multiple connexins, and the structure is called a **gap junction**. These synapses are bidirectional.

In chemical synapses, there remains some separation between the pre-synaptic and post-synaptic membrane, typically on the order of 50 nm, but it can be as large as 150 nm. The transmission of excitation across the gap in chemical synapses occurs by the fusion of vesicles containing neurotransmitter with the pre-synaptic membrane. The chemical released into the gap diffuses to the post-synaptic membrane where it binds to specific receptors in that membrane. The effect of the neurotransmitter on the post-synaptic cell depends on the neurotransmitter and on the receptor. Chemical synapses are by far more numerous than electrical synapses. Typically, **transmission across chemical synapses is unidirectional** because the vesicles containing neurotransmitter must be present on the pre-synaptic side. By this criterion, electron micrographs reveal that some synapses appear to be bidirectional. They in fact comprise two adjacent unidirectional synapses. Figure 4.2.6 illustrates the types of synapses.

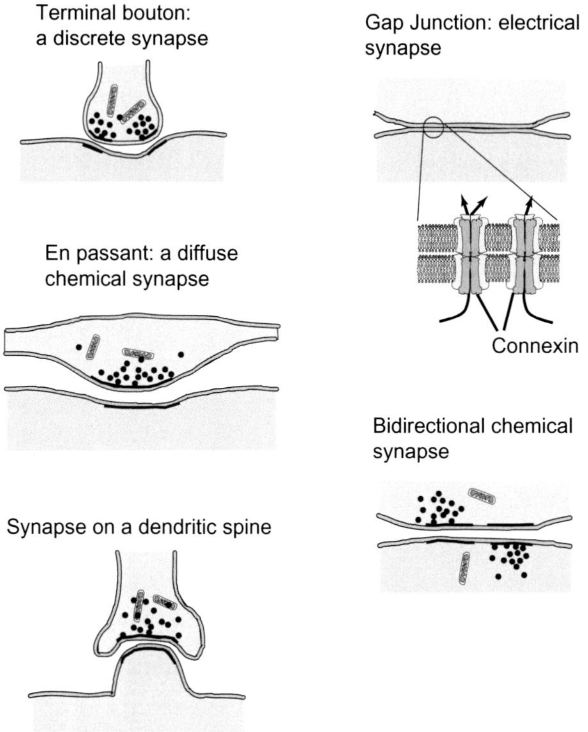

Synaptic transmission requires modifications of the pre-synaptic membrane. The pre-synaptic cell has an accumulation of vesicles containing neurotransmitter in the vicinity of the synapse and has an accumulation of proteins necessary for the fusion of these vesicles with the pre-synaptic membrane. This produces an electron-dense area on the pre-synaptic membrane called the **active zone**. The post-synaptic cell is also modified: it has an accumulation of receptors for the neurotransmitter.

Ca^{2+} SIGNALS INITIATE CHEMICAL NEUROTRANSMISSION

There are a variety of neurotransmitters, and their effects on the post-synaptic cell depend on the receptors there. Synaptic transmission follows the same general plan for all of the neurotransmitters. In the pre-synaptic terminal, tiny membrane-bound spheroids sequester the transmitter into an internal compartment, separate from the cytosol in the pre-synaptic terminal. These are **synaptic vesicles**. When an action potential invades the terminus, it triggers a series of events, beginning with increasing intracellular $[Ca^{2+}]$ from depolarization opening voltage-dependent Ca^{2+} channels, which causes the synaptic vesicle to fuse with the pre-synaptic cell membrane, releasing the neurotransmitter into the gap between the cells (see Figure 4.2.7).

VESICLE FUSION USES THE SAME MOLECULAR MACHINERY THAT REGULATES OTHER VESICLE TRAFFIC

Synaptic vesicles appear to come in two major types: small synaptic vesicles (SSVs) and large dense core vesicles (LDCVs). The differences between SSVs and LDCVs are noted in Table 4.2.1. The kinetics of release vary markedly between the two but nonetheless they both appear to use the same machinery that is used for trafficking of vesicles throughout the cell. The mechanism involves an estimated 25 different proteins in the cytoplasm, vesicle membrane, and pre-synaptic membrane and is called the **SNARE hypothesis**. **NSF** (N-ethylmaleimide-sensitive factor) is an ATPase that is attached to vesicles by **SNAP** (soluble NSF attachment protein). **Synaptobrevin** (a member of the v-SNARE family; v for vesicle membrane, SNARE for SNAP receptor) binds to vesicles and also binds to **synaptotagmin** on the vesicles. Synaptotagmin binds Ca^{2+} and confers Ca^{2+} sensitivity to vesicle fusion. A trimer of **syntaxin** (a member of a family of t-SNARE; t for target) binds to the pre-synaptic membrane along with **SNAP-25**. The t-SNARE and v-SNARE spontaneously coil up in an exergonic reaction that forces the vesicle towards the target. v-SNARE and t-SNARE proteins interact selectively, which allows for the same general machinery to be used in vesicle trafficking throughout the cell. The

FIGURE 4.2.6 Different types of synapses. Synapses can be either chemical or electrical. Chemical synapses transmit excitation across the gap between cells by the fusion of vesicles with the pre-synaptic membrane and diffusion of transmitter to the post-synaptic cell membrane. Electrical synapses occur where the pre-synaptic cell membrane fuses with the post-synaptic cell membrane, so that diffusable ions and other small molecular weight materials can pass from one cell to the other. The fusion of these membranes requires lining up of connexin hexamers in the two membranes, and these connexins form the pore through which low-molecular-weight materials diffuse. Chemical synapses can be discrete, meaning that the release and reception of neurotransmitter is highly localized or diffuse. Diffuse chemical synapses usually have a wider gap between the cells and the receptors for the neurotransmitters are more spread out.

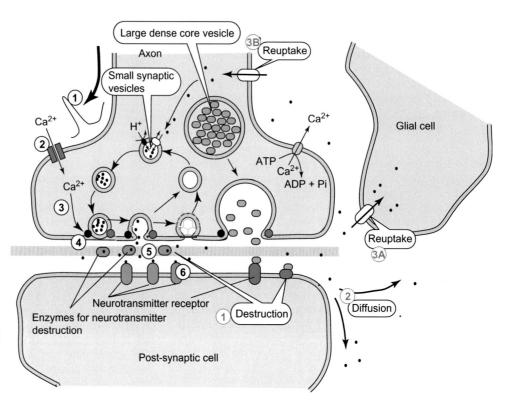

FIGURE 4.2.7 General mechanisms for the origin and shut off of neurotransmission. Neurotransmitters are stored in two types of vesicles: small synaptic vesicles and large dense core vesicles. Both appear to use the same machinery for release of their contents, though the kinetics differs. An action potential that is conducted into the pre-synaptic terminal (1) opens voltage-gated Ca^{2+} channels. (2) This lets Ca^{2+} into the cell and increased cytosolic $[Ca^{2+}]$ (3) binds to synaptotagmin, which causes fusion of vesicles with the pre-synaptic membrane. (4) The neurotransmitter diffuses across the gap (5) and binds to receptors on the post-synaptic cell membrane. (6) The effect of the neurotransmitter depends on the receptor. The neurotransmitter signal is shut off by enzymatic destruction of neurotransmitter (blue 1), diffusion away from the receptors (blue 2), or by reuptake by glial cells or the pre-synaptic cell (blue 3A and 3B).

TABLE 4.2.1 Comparison of Small Synaptic Vesicles (SSVs) with Large Dense Core Vesicles (LDCVs)

Characteristic	SSVs	LDCVs
Size (nm)	50	100–300
EM morphology	Clear particle	Dense core
Neurotransmitters	Acetylcholine, glutamate, glycine, GABA, ATP	Amines (catecholamines, serotonin, histamine), peptides
Release kinetics	0.2 ms after single action potential	>50 ms after repetitive stimulation
Distance from Ca^{2+} channel (nm)	20	300
Vesicle recycling	Local recycling	Slow endocytosis
Location	Made in nerve terminals	Homogeneously distributed

spontaneous fusion is regulated by three other proteins: SM proteins bind to the t-SNAREs, and then complexin activates but clamps the proteins to prevent premature fusion. Lastly, synaptotagmin binds Ca^{2+}, when it rises to signal neurotransmission, and reverses the clamping action of complexin. The coiling of the synaptobrevin and syntaxin is reversed by NSF, which uses the energy of ATP hydrolysis to uncoil the proteins. See Figure 4.2.8.

Ca^{2+} EFFLUX MECHANISMS IN THE PRE-SYNAPTIC CELL SHUT OFF THE Ca^{2+} SIGNAL

Increased intracellular $[Ca^{2+}]$ signals neurotransmitter release. The plasma membrane Ca^{2+}-ATPase (PMCA) and a plasma membrane Na^+–Ca^{2+} exchanger (NCX) expel the Ca^{2+} that entered over the voltage-dependent Ca^{2+} channel, returning cytoplasmic $[Ca^{2+}]$ to normal and shutting off further neurotransmission.

REMOVAL OR DESTRUCTION OF THE NEUROTRANSMITTER SHUTS OFF THE NEUROTRANSMITTER SIGNAL

Neurotransmitters bind to their receptor by mass action. This principle states that the rate of binding is proportional to the concentration of free ligand (neurotransmitter) and free receptor, and the rate of unbinding or desorption is proportional to the concentration of bound ligand. This is stated succinctly in the equations

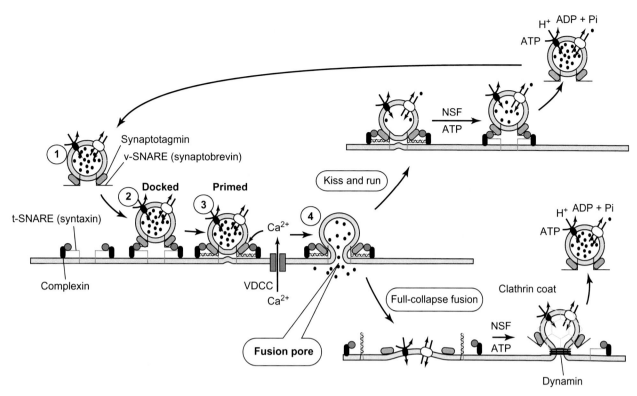

FIGURE 4.2.8 Molecular mechanism for vesicle fusion and recycling. Vesicle membranes have a different composition from the plasma membrane. Synaptobrevin, a v-SNARE (1), interacts with syntaxin, a t-SNARE (2) and coils up, bringing the vesicle close to the plasma membrane (3). This constitutes docking of vesicles and then priming for fusion. Complete fusion is prevented by SM protein Munc18-1 and complexin. Synaptotagmin, bound to the vesicle and interacting with SNAP-25 and syntaxin, binds Ca^{2+} and upon doing so releases the complexin clamp and the vesicles form a fusion pore (4). Recycling of membranes appears to have a fast mode and a slow mode. The fast mode reverses the fusion pore without a full collapse and fusion of the vesicle with the pre-synaptic membrane. When vesicles fully fuse, vesicle membrane is retrieved by endocytotic mechanisms, involving the formation of a clathrin coat and using dynamin to pinch off the endocytotic vesicle. Vesicles reload with neurotransmitter by secondary active transport powered by the vacuolar H^+-ATPase.

[4.2.1]
$$L + P \xrightarrow{k_{on}} L \cdot P$$
$$L \cdot P \xrightarrow{k_{off}} L + P$$

Thus, the occupancy of the receptor P with the neurotransmitter L will decrease only when the free ligand concentration falls. Lowering the concentration of free neurotransmitter in the synaptic gap, therefore, will shut off the continued effect on the post-synaptic cell. As shown in Figure 4.2.7, there are three general ways to achieve this end: (1) destruction of the neurotransmitter by degradative enzymes; (2) diffusion of the neurotransmitter away from the post-synaptic receptors; and (3) reuptake of the neurotransmitter either by the pre-synaptic terminal or by other cells.

THE PRE-SYNAPTIC TERMINAL RECYCLES NEUROTRANSMITTER VESICLES

Each action potential results in the fusion of a large number of synaptic vesicles. If the vesicles fully fuse with the pre-synaptic membrane, the vesicle pool becomes partially depleted, the area of the pre-synaptic membrane is increased, and the character of the pre-synaptic membrane changes because the vesicles have a different composition.

It appears there are two ways to avoid this problem. First, the vesicles may not fuse completely with the pre-synaptic membrane, but instead "kiss and run": they form a fusion pore that allows neurotransmitter to escape, but the fusion pore is not expanded and instead closes off and the vesicle is released to refill with neurotransmitter. Second, complete fusion of vesicles can be reversed by slow endocytosis involving clathrin coats and dynamin pinching off the endocytosing vesicles. These processes are shown schematically in Figure 4.2.8.

IONOTROPIC RECEPTORS ARE LIGAND-GATED CHANNELS; *METABOTROPIC RECEPTORS* ARE GPCR

The effects of neurotransmitters on the post-synaptic cell are mediated by the receptors and not by the neurotransmitters themselves. The neurotransmitters *activate* the receptors to exert their effects. There are two general classes of receptors. The ionotropic receptors are ligand-gated ion channels, as described in Chapter 2.8, whereas the metabotropic receptors are G-protein-coupled receptors (GPCRs). The ionotropic receptors operate very rapidly, causing membrane potential change within 0.1–2 ms. The metabotropic receptors take longer to activate, but their effects can remain activated far longer because metabotropic receptors can turn on phosphorylation of channels that remains until they are dephosphorylated.

ACETYLCHOLINE BINDS TO *NICOTINIC RECEPTORS* OR *MUSCARINIC RECEPTORS*

A classic example of receptors that we have already discussed is the **nicotinic receptor** at the neuromuscular junction (see Chapter 3.6). The nicotinic receptor derives its name from its stimulation by **nicotine**. This receptor is a complex of five subunits with the composition $\alpha_2\beta\gamma\delta$. Each of the two α subunits contains a binding site for acetylcholine, the ligand that binds to this receptor. When both α subunits bind acetylcholine, the receptor changes its conformation, opening a large conductance pathway for cations. Because Na^+ is furthest away from its equilibrium potential and its concentration is higher, most of the current when the nicotinic receptor opens is carried by Na^+ influx. The result is a depolarization of the post-synaptic membrane.

Acetylcholine also binds to a class of receptors that are coupled to heterotrimeric G-proteins whose general mechanism was discussed in Chapter 2.8. These are **muscarinic receptors** because **muscarine** is an agonist for this receptor. There are at least five different varieties of muscarinic receptors, labeled M1, M2, M3, M4, and M5. The M1, M3, and M5 receptors are coupled to G-proteins that activate phospholipase C (they use G_q mechanisms), whereas M4 couples to G_i, which inactivates adenylyl cyclase and M2 couples to a K^+ channel and to G_i (see Figure 4.2.9).

CATECHOLAMINES: DOPAMINE, NOREPINEPHRINE, AND EPINEPHRINE DERIVE FROM TYROSINE

Dopamine, norepinephrine, and epinephrine are all catecholamines, so-named because they consist of catechol, a phenyl group with two adjacent hydroxyl groups and an ethylamine side chain. The synthesis of dopamine (dihydroxyphenylethylamine) begins with the conversion of tyrosine to L-DOPA (dihydroxyphenylalanine) by the enzyme tyrosine hydroxylase. L-DOPA is then converted to dopamine by DOPA decarboxylase, as shown in Figure 4.2.10. Norepinephrine and epinephrine are synthesized from dopamine. Terminals that synthesize norepinephrine must first synthesize dopamine, which is placed in vesicles just like it is in dopamine-secreting terminals. The vesicles contain dopamine β hydroxylase, whereas tyrosine hyroxylase and L-DOPA decarboxylase and PNMT (phenylethanolamine *N*-methyltransferase) are all in the cytoplasm.

DOPAMINE COUPLES TO G_S AND G_I-COUPLED RECEPTORS THROUGH D_1 AND D_2 RECEPTORS

Dopamine released by pre-synaptic terminals crosses the synapse and binds to D_1 or D_2 receptors, both

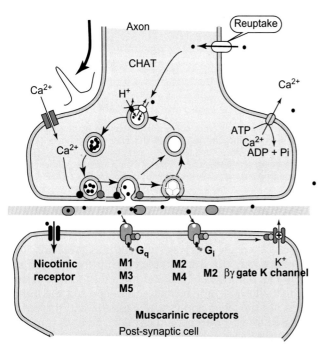

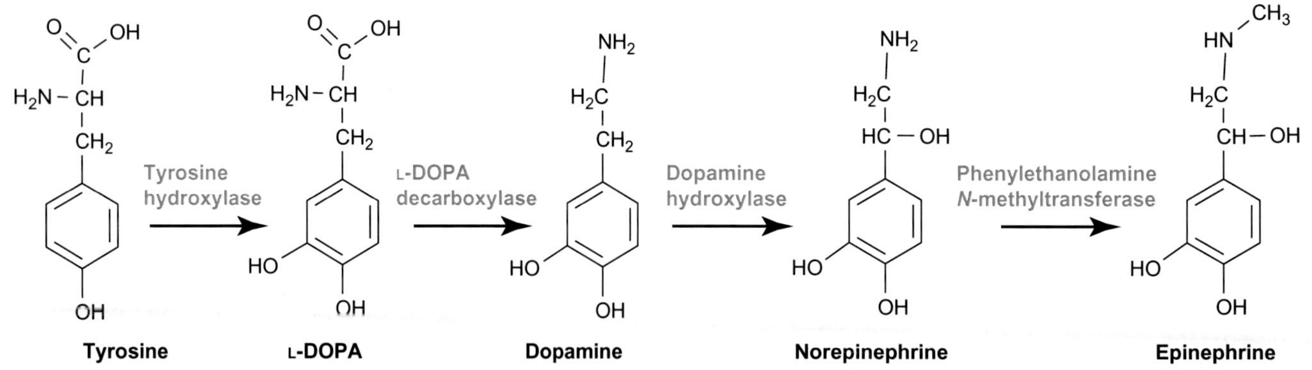

FIGURE 4.2.9 Receptors for acetylcholine. Post-synaptic cells can contain nicotinic receptors or one of five different muscarinic receptors. Nicotinic receptors, of which there are different types, are pentameric ionotropic receptors. All of the M receptors are metabotropic. Acetylcholine is degraded by acetylcholinesterase.

FIGURE 4.2.10 Synthesis of dopamine from tyrosine, and norepinephrine and epinephrine from dopamine.

GPCR; D_1 couples to a G_s mechanism, whereas D_2 couples to a G_i mechanism. Activation of D_2 receptors increases g_K, thereby hyperpolarizing the post-synaptic cell and causing an IPSP. This metabotropic effect develops more slowly and lasts longer than a single IPSP or EPSP. Dopamine can have opposite effects on the post-synaptic cell, depending on which GPCR it expresses (see Figure 4.2.11).

ADRENERGIC RECEPTORS ARE CLASSIFIED ACCORDING TO THEIR PHARMACOLOGY

Epinephrine is a hormone released from the adrenal medulla in response to stress, mediated by sympathetic fibers. The word *epinephrine* derives from *epi*, meaning *above*, and *nephros*, the root word for *kidney*, because the gland sits atop the kidney. Epinephrine is also called **adrenaline**, derived from the name of its gland. For this reason, receptors for both epinephrine and norepinephrine are called **adrenergic receptors**. Ahlquist in 1948 classified the adrenergic receptors as α or β, based on their response to epinephrine, norepinephrine, and isoproterenol, an adrenergic agonist. The β receptors respond to much lower concentrations of epinephrine or norepinephrine than the α receptors. β_1 receptors respond to epinephrine and norepinephrine about equally, whereas β_2 receptors are more sensitive to epinephrine. Propanolol blocks the β receptors; phenoxybenzamine blocks α receptors. All of the β receptors are linked to G_s; α_1 receptors are linked to G_q and α_2 receptors are linked to G_i (see Figure 4.2.12).

GLUTAMATE AND ASPARTATE ARE EXCITATORY NEUROTRANSMITTERS

Transamination reactions in neurons, as well as most other cells, readily interconvert glutamic acid into aspartic acid, and vice versa. Both of these amino acids stimulate the same receptors, and because of the difficulty in distinguishing neurons that use glutamate from those that use aspartate, we classify neurons that use either as a group, the **glutamatergic neurons**. The glutamate in these neurons derives from α-ketoglutarate in the Kreb's Cycle, or from ingested food. The neurons store it in synaptic vesicles and release it into the gap when an action potential invades the nerve terminus. All of the receptors on the post-synaptic cell are ionotropic (see Figure 4.2.13).

The **NMDA receptor**, named for an artificial agonist, *N*-methyl-D-aspartate, increases g_{Ca}, the conductance to Ca^{2+}, in response to glutamate binding. The increased g_{Ca} produces an inward current that depolarizes the post-synaptic membrane and therefore makes an EPSP. Prolonged exposure to glutamate leads to pathological increases in neuronal cell $[Ca^{2+}]$ that can kill the post-synaptic cell. This phenomenon is called **excitotoxicity**.

Binding of glutamate to the **AMPA** or **kainate receptors** leads to an increase in g_{Na} and g_K. These receptors are named for their artificial agonists: AMPA is α-amino-3-hydroxy-5-methylisoxazole-4-propionate. Occupancy of these receptors produces an EPSP.

GABA INHIBITS NEURONS

GABA, or **gamma amino butyric acid**, is synthesized from glutamic acid by glutamic acid decarboxylase in

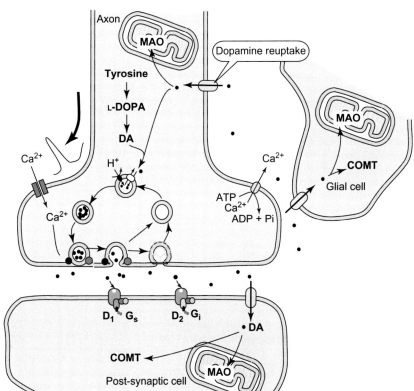

FIGURE 4.2.11 Fate of dopamine in pre- and post-synaptic terminals. Dopamine is synthesized from tyrosine in the pre-synaptic terminal and packaged into vesicles. The action potential opens voltage-gated Ca^{2+} channels on the pre-synaptic membrane that trigger fusion of dopamine-loaded vesicles. Dopamine diffuses across the gap and binds to post-synaptic receptors, which may be either D_1 receptors linked to a G_s protein or D_2 receptors linked to G_i. Dopamine is taken back up by the pre-synaptic cell, post-synaptic cell, or glial cell. It may be recycled by the pre-synaptic cell or degraded by MAO (monoamine oxidase) or COMT (catechol-O-methyl transferase). COMT places a methyl group on the 3 OH of the catechol ring; MAO removes the amine group. MAO is mitochondrial; COMT is in the cytosol.

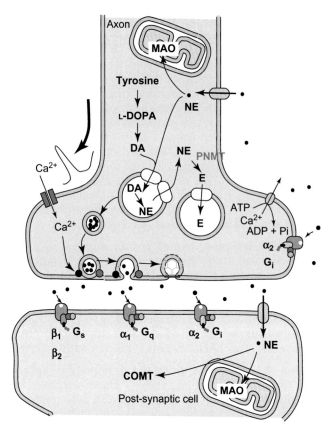

FIGURE 4.2.12 Processing of adrenergic receptors. Norepinephrine is made from tyrosine with dopamine as an intermediate step. It can be released from adrenergic autonomic nerve terminals. Epinephrine is synthesized from norepineprhine and released mainly from the adrenal medulla. Binding to β receptors on the post-synaptic membrane is coupled to a G_s mechanism. There are several different types of β receptors. Binding to α_1 receptors couples to a G_q mechanism that increases IP3 and leads to increased cytosolic [Ca^{2+}]. Activation of α_2 receptors is linked to a G_i mechanism. Thus, adrenergic stimulation can be stimulatory (G_s or G_q) in those tissues with the appropriate receptor, or inhibitory (G_i) in cells having those receptors. Like dopamine, norepinephrine is degraded by MAO or COMT.

the pre-synaptic terminal and stored in synaptic vesicles until released. Released GABA binds to $GABA_A$ or $GABA_B$ receptors on the post-synaptic membrane, both producing IPSPs. The $GABA_A$ receptor is an ionotropic receptor and produces a fast IPSP by increasing g_{Cl}. This increases flux of Cl^- into the cell, because $E_m > E_{Cl}$, and this flux makes an outward current (because Cl has a negative charge, current is opposite to Cl flux) that polarizes the cell. Occupancy of the $GABA_B$ receptor produces a slower IPSP because it activates a GPCR that activates a K^+ channel. The increased g_K increases K^+ efflux, a positive current that polarizes the cell away from threshold (see Figure 4.2.14).

SEROTONIN EXERTS MULTIPLE EFFECTS IN THE PNS AND CNS

Serotonin was identified in the serum of mammals in 1946 as a material that had tonic effects on the vasculature, hence its name. It is also known as 5-hydroxy-tryptamine, or 5-HT. It has profound effects on sleep, circadian rhythms, appetite, mood, cognition, reproductive behavior, thermoregulation, and endocrine, cardiovascular, and respiratory function. Serotonin derives from tryptophan. Tryptophan hydroxylase converts tryptophan to 5-hydroxy-tryptophan, and then L-amino acid decarboxylase converts 5-hydroxy-tryptophan to 5-hydroxy-tryptamine, as shown in Figure 4.2.15 and then the cell transports it into synaptic vesicles.

Like all other neurotransmitters, 5-HT works through its receptors, which have been classified into seven families (5-HT_1–5-HT_7) and at least 14 diffferent subtypes. Most of these work through metabotropic receptors, except for 5-HT_3, which is an ionotropic receptor. Figure 4.2.16 illustrates these receptors.

NEUROPEPTIDES ARE SYNTHESIZED IN THE SOMA AND TRANSPORTED VIA AXONAL TRANSPORT

A large variety of neuropeptides are used as neurotransmitters in **peptidergic synapses**. A partial listing is given in Table 4.2.2. These differ from the low-molecular-weight neurotransmitters in that they are often synthesized as larger precursors that are proteolytically cleaved after transport into LDCVs, and they are synthesized in the soma, not in the terminus. They typically bind their receptors with high affinity, and shut-off of the signal is achieved by degradation rather than by uptake by glial cells or pre-synaptic cells. For many years, it was believed that each neuron used a single neurotransmitter, but now it is known that some neurons secrete a low-molecular-weight neurotransmitter and one high-molecular-weight neurotransmitter. The known pairs are listed in Table 4.2.3.

The transport of materials along the axon is called **axonal transport** and it occurs in both directions. Movement from the soma to the axon is called **anterograde transport**, and this is associated with growth of the axon and renewal of synaptic vesicles. **Fast axonal transport** moves materials at 200–400 mm/day; **slow axonal transport** moves materials at 1–2 mm/day. **Retrograde transport** brings worn out parts of the cell back to the soma for destruction, disposal, or recycling. These movements require cytoskeletal elements, **microtubules**, and **microfilaments**, whose structure is described in Chapter 2.1. Materials are carried along microtubule tracts by motor proteins that bind to the tract and "walk" along it, using ATP hydrolysis to power the movement. **Kinesin** is a large protein of 380 kDa that resembles myosin in that it is a dimer with two subunits that each bind to microtubules, a long tail, and an associated light chain. Synaptic vesicles bind to kinesin and are carried when kinesin walks along the tract. Most members of the kinesin family are + directed, meaning that they walk towards the + end of the microtubule (its growing end away from the soma). K1FA kinesin transports synaptic vesicles; K1FB transports mitochondria. **Dynein** is another large protein containing a number of subunits. It carries cargo retrograde, from the + end to the − end of microtubules.

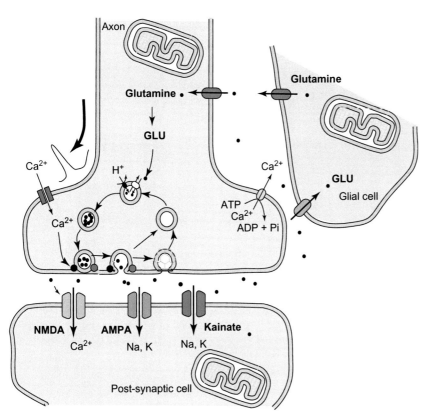

FIGURE 4.2.13 Handling of glutamate at glutamatergic synapses. Glutamate is synthesized from glutamine by glutaminase in the pre-synaptic terminal. The glutamic acid is stored in synaptic vesicles and released in response to an action potential on the pre-synaptic membrane. The released glutamic acid binds to one of its receptors, either the NMDA, AMPA, or kainate variety. Each of these incorporates ion channels and binding of glutamate increases the conductance to some cation, producing an inward current and an EPSP. The released glutamic acid is taken up by glial cells and resynthesized into glutamine by glutamine synthetase.

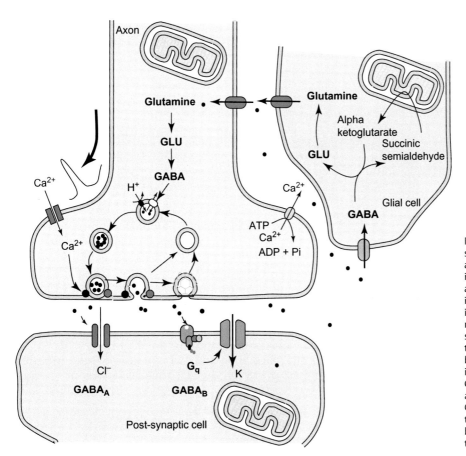

FIGURE 4.2.14 Handling of GABA at GABAergic synapses. Cells synthesize GABA from glutamic acid by glutamic acid decarboxylase and then import the GABA into synaptic vesicles. An action potential in the pre-synaptic terminal increases the local $[Ca^{2+}]$ by increasing Ca^{2+} influx through a voltage-gated channel. This results in fusion of the vesicles with the pre-synaptic membrane and release of GABA into the synapse. GABA diffuses across the gap and binds to GABA$_A$ receptors that increase g_{Cl}, inhibiting the post-synaptic cell. Binding to GABA$_B$ receptors increases g_K indirectly, which also inhibits the post-synaptic cell. The released GABA is taken up by glial cells which convert the GABA to glutamine, which can be taken up by the pre-synaptic terminal to convert it back to GABA.

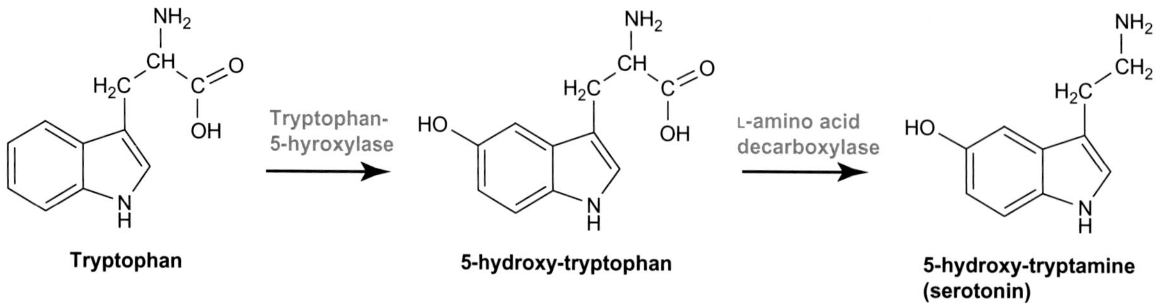

FIGURE 4.2.15 Synthesis of serotonin from tryptophan.

Clinical Applications: Multiple Sclerosis

Multiple sclerosis (MS) is a progressively debilitating disease of the nervous system. Its hallmark is demyelination, the loss of myelin and destruction of the oligodendroglial cells that produce it. The brains of affected persons develop plaques or lesions in seemingly random areas of the white matter in which the myelin is lost. Because the actual areas that are demyelinated vary from person to person, the presenting symptoms also vary considerably. Each affected person develops their own set of symptoms and progresses along individual paths. Patients may first present themselves to their physician with symptoms of numbness, paresthesias (pins and needles), muscle weakness, muscle spasms, spasticity, cramps, pain, blurred vision, slurred speech, loss of balance, nausea, fatigue, depression, incontinence, constipation, inability to swallow, loss of sexual function—the list is nearly endless.

MS is not contagious, but its etiology remains elusive. A favored hypothesis is autoimmune destruction of myelin. What triggers the autoimmune response is also unknown. The progress of the disease is characterized by four patterns: relapsing-remitting MS; secondary progressive MS; progressive relapsing MS; and primary progressive MS. These forms differ in the presence of relapses and the degree of recovery during remission. During periods of MS activity, the brain appears to be involved in an inflammatory response. Leukocytes invade the brain and myelin is stripped from the axons. Oligodendrocytes are killed. However, it is unclear which causes which: does inflammation kill the oligodendrocytes or do dead oligodendrocytes trigger the inflammation? The resulting demyelination alters the propagation properties of the axons, and some axons are also damaged by the inflammation. During this period the affected neurons cannot effectively transmit action potentials. Scar tissue replaces the myelin, and this scarification is the origin of the name, multiple sclerosis. "Multiple" means "many", referring to the many plaques that form; "sclerosis" means "scar forming."

Remission may be accompanied by reduction in inflammation. In the early stages, neurons that have not been damaged can resume their normal functions. Demyelinated nerves may become re-myelinated or the brain may develop new connections to circumvent the damage. In this way normal function can be partially restored. Demyelinated nerves adapt by increasing the number of regenerative Na^+ channels on their surface membrane. This cannot speed action potential conduction velocity, but it can prevent it from failing. If the Na^+ channels were present only at spacings equivalent to the Nodes of Ranvier, and the axon were demyelinated, there might not be enough depolarization left to open the Na^+ channels because of loss of current across the membrane. If the Na^+ channels are closer together, the action potential can still be conducted. Thus the nerve can remain active even if conduction velocity is slowed. For some functions, such as control of movement that requires rapid decision making, reduced conduction velocity greatly and adversely affects function.

SUMMARY

Brain behavior derives from the behavior of brain cells. The set of brain cells consists of neurons and the neuroglia. Neuroglia support, nurture, and protect the neurons, and make up some 90% of the total number of cells in the brain. The neuroglia include ependymal cells, oligodendroglia, astrocytes, and microglia. The ependymal cells make cerebrospinal fluid. The oligodendroglia form the myelin sheaths around axons in the white matter of the brain. The astrocytes help form the blood–brain barrier, take up used neurotransmitters and K^+ ions, and regulate the pH of the extracellular fluid in the brain. The microglia clean up damaged and infected tissue by phagocytosis. In the PNS, Schwann cells make the myelin sheath. Myelin consists of multiple layers of the cell membranes of oligodendroglia or Schwann cells. It insulates the axon. This insulation increases the membrane resistance and decreases the capacitance, both of which speed conduction velocity.

Neurons themselves vary considerably. Somatic sensory cells are pseudounipolar; some sensory cells in olfaction, vision, and hearing are bipolar. The majority of neurons are multipolar. The dendritic fields, size of the soma, and axon branching are highly variable. Most neurons have thousands of synapses. These multiple inputs converge on the cell whereas its output diverges onto other cells.

The synapse is the connection between neurons. There are axodendritic, axoaxonal, axosomatic, dendrodendritic, axoextracellular, and axosecretory synapses. Two

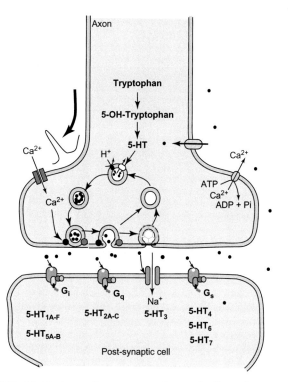

FIGURE 4.2.16 Serotonin receptors. Seven families of serotonin receptors have been identified, with several subtypes of some of these families. Most are metabotropic, working through heterotrimeric GPCRs either through G_i mechanisms (5-HT$_1$ and 5-HT$_5$ receptor families), G_q mechanisms (5-HT$_2$), or G_s mechanisms (5-HT$_4$, 5-HT$_6$, and 5-HT$_7$ families). The 5-HT$_3$ receptors are ionotropic, increasing conductance to cations nonselectively. In this case, the largest flux should be carried by the Na$^+$ ion because it is furthest from equilibrium and has one of the highest concentrations. The result is a depolarization of the post-synaptic cell.

main types of synapses are recognized: the electrical synapse and the chemical synapse. The electrical synapse is a gap junction consisting of a field of connexin pores that pass ions and signaling molecules directly from one cell to another without passing through the extracellular fluid. The pores form by aligning one connexin hexamer in one cell with another connexin hexamer in another cell. Chemical synapses have a gap between the cells. Vesicles containing neurotransmitters accumulate on the pre-synaptic side of the synapse, the side from which the signal originates. When an action potential propagates into the terminal of the pre-synaptic cell, voltage-gated Ca^{2+} channels open and Ca^{2+} enters the terminal. The increased local [Ca^{2+}] binds to proteins that attach the vesicles to the pre-synaptic membrane. This triggers fusion of the vesicle with the pre-synaptic membrane, and the vesicles release neurotransmitter into the gap. The neurotransmitter diffuses to the post-synaptic membrane where it binds to its receptors. The Ca^{2+} signal for neurotransmitter release is shut off by pumping Ca^{2+} out of the terminal by the PMCA and the NCX. The neurotransmitter signal to the post-synaptic membrane is shut off by three mechanisms: (1) reuptake of the neurotransmitter by the pre-synaptic membrane or glial cells; (2) destruction of the neurotransmitter by the post-synaptic cell; (3) diffusion away from the synapse. Vesicles that fuse with the pre-synaptic membrane are recycled and loaded back up with neurotransmitter.

The action of the neurotransmitter on the post-synaptic cell depends on the receptor. Ionotropic receptors directly alter the conductance properties of the post-synaptic membrane that produce currents that either

TABLE 4.2.2 Partial Listing of Neurotransmitters with Receptor Types

Neurotransmitter	Receptor	Peptide Neurotransmitter
Acetylcholine	Nicotinic (ionotropic)	**Opioids**
	Muscarinic (metabotropic)	β Lipotropin
		α MSH
Biogenic amines		α Endorphin
Epinephrine	Adrenergic β (metabotropic)	β Endorphin
Norepinephrine	Adrenergic α,β	Met enkephalin
Dopamine	D_1, D_2, D_4, D_5	Leu enkephalin
Serotonin	5-HT$_{1-7}$	Dynorphin A
Histamine	H_{1-3}	Dynorphin B
Amino acids		**Gastrointestinal peptides**
Glutamate/Aspartate	NMDA, AMPA, kainate	Cholecystokinin
GABA	GABA$_A$, GABA$_B$	Secretin
Glycine		Substance P
		Vasoactive intestinal polypeptide
Purines		
Adenosine		
ATP		
Gases		**Hypothalamic peptides**
CO		LHRH
NO		Oxytocin
		TRH
		Somatostatin
		Vasopressin
		Corticotrophin

TABLE 4.2.3 Examples of Co-Localization of Low-Molecular Weight Neurotransmitters and Neuropeptides

Low-Molecular-Weight Neurotransmitter	Neuropeptide
Acetylcholine	Vasoactive intestinal polypeptide
Dopamine	Enkephalin
	Cholecystokinin
Norepinephrine	Enkephalin
	Somatostatin
Serotonin	Substance P
	TRH (thryotropin-releasing hormone)

depolarize (EPSP) or hyperpolarize (IPSP) the post-synaptic cell. Metabotropic receptors are linked to longer-lasting changes in membrane conductance through signaling pathways such as cyclic AMP and protein kinase A, or PLC and IP3. Metabotropic effects take longer to produce and last longer than ionotropic mechanisms. Each neurotransmitter has its own set of receptors and receptor mechanisms.

Materials move from the soma to the terminus, and back again to the soma, by fast and slow axonal transport. These occur over microtubule and microfilament tracts using motor proteins that attach cargo. Kinesins are generally + directed motor proteins, traveling to the + end of microtubules away from the soma. Dyneins generally transport material from the + end to the − end.

REVIEW QUESTIONS

1. Name the four types of glial cells. What do ependymal cells do? What do astrocytes do? What do oligodendroglial cells do? What do microglia do?
2. What is a myelin sheath? What cells make it in the CNS? In the PNS? What happens if myelinated cells become demyelinated?
3. What is a pseudounipolar neuron? A bipolar neuron? A multipolar neuron? What is meant by "convergence of inputs"? What is meant by "divergence of output"?
4. What is a synpase? What is the pre-synaptic cell? What is the post-synaptic cell? Are synapses bidirectional? What is an electrical synapse? Is it bidirectional?
5. How does a small synaptic vesicle differ from a large, dense core vesicle? Are they present simultaneously in a single synapse?
6. What is meant by "kiss and run" synaptic vesicle fusion? Could synaptic vesicle fusion occur in the absence of extracellular Ca^{2+}?
7. What is an ionotropic receptor? What is a metabotropic receptor?
8. Acetylcholine binds to what kinds of receptors? Which are ionotropic? Which are metabotropic? How do M1, M3, and M5 receptors work? What does M2 do? M4?
9. How can dopamine have different effects on post-synaptic cells? What mechanism do D_1 receptors use? D_2?
10. What amino acids is used to make dopamine? Serotonin? Norepinephrine? Epinephrine? GABA?
11. Is glutamate generally inhibitory or excitatory? Why? How about GABA? Why?
12. Where are neuropeptides synthesized? How do they get to the nerve terminals?
13. What is anterograde axoplasmic transport? What is retrograde axoplasmic transport? What motor proteins are generally used in each?

Cutaneous Sensory Systems 4.3

Learning Objectives

- List examples of exteroreceptors and interoreceptors
- Define adequate stimulus, modality, and perception
- Distinguish between long receptors and short receptors
- Describe how the nervous system codes quality of sensation and intensity of sensation
- Define receptive field
- Describe adaptation
- List the five mechanoreceptors in the skin and distinguish them on the basis of their frequency response and rate of adaptation
- Describe hot and cold sensors
- Distinguish between first and second pain
- Draw the dorsal column pathway for most mechanoreceptors
- Explain somatotropic representation and how it originates
- Define dermatome, myotome, and sclerotome
- Describe the parts of the body surface projecting to the nucleus gracilis and nucleus cuneatus
- Draw the anterolateral tract for pain and temperature
- Describe the gate control theory of pain and explain how rubbing the body surface reduces the subjective experience of pain
- Indicate the somatosensory cortex in the brain and its somatotopic representation
- Describe the typical receptive field of somatosensory cortical neurons
- Explain how lateral inhibition sharpens spatial discrimination

SENSORS PROVIDE A WINDOW ONTO OUR WORLD

Sensory systems are the link between the central nervous system (CNS) and events that occur outside of it. They inform the CNS of what is happening in both the external world and the internal world. Sensors that convey information about the external environment are called **exteroreceptors**; sensors reporting on the internal environment are **interoreceptors**.

EXTERORECEPTORS INCLUDE THE FIVE CLASSICAL SENSES AND THE CUTANEOUS SENSES

The exteroreceptors include all those sensory systems that apprise the CNS of conditions in the external environment. These include the following:

- Eyes (vision)
- Cochlea (hearing)
- Vestibular apparatus (balance, rotation and linear acceleration, gravity)
- Olfactory epithelium (smell)
- Taste buds (taste)
- Touch receptors (touch)
- Temperature sensors
- Pressure receptors
- Nociceptors (pain)
- Skin stretch receptors.

INTERORECEPTORS REPORT ON THE CHEMICAL AND PHYSICAL STATE OF THE INTERIOR OF THE BODY

The interoreceptors also include a wide variety of receptors. They report on a variety of variables ranging from the stretch of skeletal muscles to the pH of the blood. They include the following:

- Stretch receptors (arteries, veins, atria, intestines, bladder, etc.)
- Chemosensors (CO_2, O_2, glucose)
- pH sensors (blood, intestinal lumen)
- Nociceptors (damage, pain)
- Muscle length sensors
- Muscle tension sensors
- Proprioceptors
- Temperature sensors
- Osmoreceptors.

SENSORY SYSTEMS CONSIST OF THE SENSE ORGAN, THE SENSORY RECEPTORS, AND THE PATHWAYS TO THE CNS

Sensory systems include **receptors** that respond to what is called an **adequate stimulus**, which is the *kind* of stimulus to which receptors respond preferentially. A **sensory modality** is an identifiable class of sensation. Sensory receptors respond to the adequate stimulus with the lowest **threshold**, referring to the lowest stimulus intensity that elicits a response from the sensor. For example, vision is a sensory modality whose adequate stimulus is light within the narrow band of wavelengths that we can see. The receptors for vision are the rods and cones in the retina that actually respond to light. However, rods and cones will also respond to pressure on the eyeballs, causing us to "see" light, called **phosphenes**. The proper sensing of light requires not just the rods and cones but all of the accessory structures that enables us to see. Table 4.3.1 shows the various modalities, their receptor cells, and their sense organs.

PERCEPTION REFERS TO OUR AWARENESS OF A STIMULUS

Sensory systems bring information about the external or internal world into the CNS where it is processed to bring about an awareness or consciousness of the excitation. This awareness is called **perception**, and the process by which we become aware is called **sensation**. Both of these are distinct from what receptor cells do, which is to **transduce** an adequate stimulus into action potentials.

LONG AND SHORT RECEPTORS DIFFER IN THEIR PRODUCTION OF ACTION POTENTIALS

The body has two basic plans for sensory **transduction**: the long and short receptors as shown in Figure 4.3.1. **Long receptors** link sensory energy to action potentials. Those in skin and muscle have nerve endings at the periphery and their cell bodies lie in the dorsal root of the sensory nerve (see Figure 4.3.5). The single receptor cell transduces an adequate stimulus, such as mechanical deformation, into a **receptor potential**. The receptor potentials are **graded** responses whose intensity depends on the intensity of the stimulus. They are conducted electrotonically and so die out with distance away from the stimulus as well as with time after the stimulus. Such graded responses would die out before they could reach the end of the cell's long processes. Therefore, the sensory cell converts these receptor potentials into action potentials which can be conducted along the length of the receptor without decrement. The next cell to receive this information is a **second-order neuron**, located some distance away, in the spinal cord.

Short receptor cells have a short distance between the detection of the adequate stimulus and its transmission to the next cell, which is typically a **primary afferent neuron**. Here the sensory cells produce a receptor potential in response to the adequate stimulus. This receptor potential is linked to the transmission of excitation through the use of neurotransmitters, usually at the opposite end of the sensory cell. Fusion of neurotransmitters at the opposite end of the sensory cell requires a receptor potential but does not require an action potential. The released neurotransmitter then causes a **generator potential** in the primary afferent neuron. When the

TABLE 4.3.1 Sensory Modalities, Their Receptors, and Their Organs

Sensory Modality	Receptor Cells	Sense Organ
Vision	Rods or cones	Eye
Hearing	Hair cells	Ear (organ of Corti)
Rotational acceleration	Hair cells	Ear (semicircular canal)
Linear acceleration	Hair cells	Ear (utricle and saccule)
Smell	Olfactory neurons	Olfactory membrane
Taste	Taste receptor cells	Taste buds
Touch/pressure	Nerve endings	Skin
Heat	Nerve endings	Skin
Cold	Nerve endings	Skin
Pain	Nerve endings	Skin
Proprioception	Nerve endings	Joints, capsules, muscles
Muscle length	Nerve endings	Muscle spindles
Muscle tension	Nerve endings	Golgi tendon organ

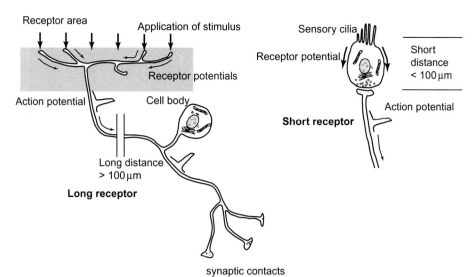

FIGURE 4.3.1 Long and short receptors. Long receptors initiate a receptor potential near the stimulus. If sufficiently strong, the receptor potential causes the cell to fire an action potential, which travels over a long fiber to reach a second-order neuron. Short receptors initiate a graded receptor potential in response to stimulation that causes release of neurotransmitters without an action potential. The graded release causes a graded generator potential which, if sufficiently strong, initiates an action potential in the sensory neuron.

generator potential reaches threshold, the primary afferent neuron fires an action potential.

ANATOMICAL CONNECTION DETERMINES THE QUALITY OF A SENSORY STIMULUS

How do we know that a stimulus is visual, or tactile, or pressure, or pain, or whatever modality it is? Each sensory modality conveys sensory information to the brain over distinct neural pathways with distinct connections in the brain. This method of encoding sensory information is called the **labeled line** of stimulus coding. Each sensory modality has its own pathway and its own destinations within the CNS, so action potentials carried along these lines have the "label" of that modality. Some people have the bizarre ability to experience one modality as another. This capacity is called **synesthesia**, which literally means "feeling together". About 1 person in 25,000 has some form of synesthesia. The most common form is colored hearing. People with this synesthesia say that they "see" music and speech in color. Curiously, the reverse is not true: they do not hear sounds when they see something. Why this happens is unknown, but it has some genetic component. Perhaps some auditory input lost its way in development and innervated the visual cortex. Other types of synesthesia also exist. Individuals have reported experiencing words that "taste"; others report experiencing shapes with different tastes.

THE INTENSITY OF SENSORY STIMULI IS ENCODED BY THE *FREQUENCY* OF SENSORY RECEPTOR FIRING AND THE *POPULATION* OF ACTIVE RECEPTORS

The nervous system encodes the quality of the sensory modality largely through labeled lines, but the intensity is encoded by the population of cells that respond and by the frequency of their response. Intense stimuli cause larger or longer lasting receptor potentials, which are converted into a train of action potentials on the sensory neuron. Thus, the first method of encoding the stimulus strength is through the frequency of action potentials on the sensory neuron. Repetitive firing of sensory neurons excites CNS interneurons by temporal summation. This is called **frequency coding**. Intense stimuli also excite more sensory neurons than do less intense stimuli. Thus, more intense stimulation recruits additional sensory input. The inputs of these sensory neurons travel to overlapping sets of interneurons in the CNS, so that these interneurons are excited by spatial summation. Thus, intensity can also be coded by the number of sensory cells that participate. This is **population coding**.

FREQUENCY CODING IS THE BASIS OF THE WEBER–FECHNER LAW OF PSYCHOLOGY

In 1846, Weber found that blindfolded subjects could discriminate between small increments of weights placed in their hands, but the sensitivity depended on how much weight was already there. As weight was added, a proportionately larger weight had to be added in order to discriminate between two weights. He found that the minimal detectable difference was about 3% of an object's weight. Fechner realized that this empirical observation implied a logarithmic relationship between stimulus and response.

Consider that we have two weights, W_1 and W_2, each of which gives rise to a sensation S_1 and S_2. We say that we can discriminate between the weights when we can detect their difference, $\Delta W = W_2 - W_1$; but what we really detect is the difference in their associated sensations, $\Delta S = S_2 - S_1$. Weber's observations suggest that the relationship between sensation and stimulus is approximately given as

[4.3.1] $$\Delta S = k \frac{\Delta W}{W}$$

Writing this in differential form and then integrating, we would expect the relationship between sensation and stimulus intensity to have the form

[4.3.2] $$S = k_1 \log W + k_2$$

The receptor potential is approximately related to the logarithm of stimulus intensity, and it appears that this receptor mechanism may be responsible for the Weber–Fechner psychological "law". However, the relationship fails at the extremes of ranges of sensory sensitivity. The exact relationship between stimulus intensity and sensation differs among the various modalities. A more general power law holds over a wider range of intensities and is given as

[4.3.3] $$\Psi = k\Phi^n$$

where Ψ is the psychological sensation, Φ is the stimulus intensity, and n is the exponent. The exponent n varies from 0.33 to 3.5 depending on the modality. This is called **Stevens' Power Law** and was first proposed in 1957. The logarithmic relationship between stimulus and response allows us to code a wide range of stimulus intensities. A three-fold change in frequency of firing on a logarithmic scale corresponds to a 1000-fold change in stimulus intensity.

ADAPTATION TO A STIMULUS ALLOWS SENSORY NEURONS TO SIGNAL POSITION, VELOCITY, AND ACCELERATION

Adaptation refers to the decrease in sensation that occurs upon continued stimulation. It results from processes within the receptors themselves, although central mechanisms can also play a part. The rate of adaptation depends on the sensory modality. Touch receptors adapt rapidly whereas sensors in the muscle spindle adapt slower and sensors for blood pressure do not adapt at all. Different rates of adaptation allow sensory systems to signal position, velocity, or acceleration as shown in Figure 4.3.2.

RECEPTIVE FIELDS REFER TO THE PHYSICAL AREAS AT WHICH A STIMULUS WILL EXCITE A RECEPTOR

For cutaneous sensory receptors, **receptive fields** are the areas of the body surface that, when stimulated, excite the sensory neuron. Stimulation within only a small area of the finger tips excites touch receptors there, and they have small receptive fields. Stimulation of a person's back over a much larger area excites cutaneous receptors there. Because of the small receptive fields on the finger tips, we can easily distinguish between two closely placed stimuli. But on the back we cannot distinguish between two closely placed stimuli. The ability to distinguish between two stimuli on the body surface is called **two-point discrimination**. The receptive field can be defined for the sensory neurons or for higher order interneurons that receive inputs from a number of

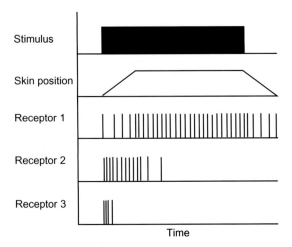

FIGURE 4.3.2 How adaptation can be used to signal position, velocity, or acceleration. Application of a square wave pressure pulse to the skin (top) results in a displacement of the skin (skin position, second from top). Sensors that are sensitive to position (stretch of the receptors) respond with a train of action potentials, the frequency of which is related to position. Lack of adaptation of these sensors allows their action potentials to signal skin position (Receptor 1). Rapidly adapting receptors respond only when there is movement (Receptor 2) and so they signal the velocity of skin movement. Still more rapidly adapting sensors (bottom, Receptor 3) inform the CNS of the acceleration of the skin.

sensory neurons or other interneurons. More integrative neurons involved in processing of sensory information have more complicated receptive fields.

CUTANEOUS RECEPTORS INCLUDE MECHANORECEPTORS, THERMORECEPTORS, AND NOCICEPTORS

THE SURFACE OF THE BODY CONTAINS A VARIETY OF MECHANORECEPTORS

The skin contains a variety of mechanoreceptors including Pacinian corpuscles, Meissner's corpuscles, Ruffini's corpuscles, Merkel's disks, and free nerve endings in the skin and surrounding hair follicles. All of these are long receptors, in which the receptor or generator potential is developed within the sensory cell and it fires action potentials based on this receptor potential without an additional connection to a sensory cell. Their axons are all myelinated, so that the CNS is informed quickly whenever any of these receptors detects a stimulus (see Figure 4.3.3).

The **Pacinian corpuscles** consist of a free nerve ending enclosed by a layered capsule, much like an onion. They lie in the subcutaneous layers of the skin, are rapidly adapting, and respond best to vibration. This response is due to the mechanical properties of the capsule, as the frequency response is removed when the capsule is removed.

Meissner's corpuscles reside in the dermis just below the epidermis. These also rapidly adapt and are thought to respond to fluttering types of stimuli. That is, Meissner's corpuscles also respond to vibration but at a lower frequency than Pacinian corpuscles.

Cutaneous Sensory Systems

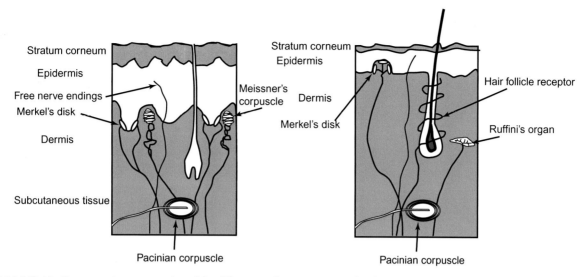

FIGURE 4.3.3 Highly diagrammatic representation of the different mechanoreceptors in the skin. The glabrous or nonhairy skin, shown at left, contains a different set of mechanoreceptors from the hairy skin, as shown at the right. Both contain a variety of receptors.

Merkel's disks are located in the dermis and have small receptive fields. They are slowly adapting and respond to steady touch-pressure on the skin.

Ruffini's corpuscles are located in the dermis and are slowly adapting. They have much larger receptive fields and so they may participate both in touch-pressure sense and in proprioceptive sense by detecting the push or pull of skin from one segment of the body on another.

All of the encapsulated nerve endings listed above are present in both the hairy and nonhairy (**glabrous**) skin. **Free nerve endings** are present in the epidermis in glabrous skin. They contribute to the tactile sense. They also wrap around hair follicles in the hairy skin, where they detect movement of the hair.

THERMORECEPTORS CONSIST OF COLD RECEPTORS AND WARM RECEPTORS

Cold receptors are free nerve endings with thin myelinated fibers, whereas the warm receptors are free nerve endings with unmyelinated axons with low conduction speeds. They differ from the mechanoreceptors in that they exhibit tonic level of activity at most temperatures. They respond to temperature changes with a phasic component followed by a tonic component that depends on the temperature. This is why when you get into a hot bath it feels hot for a while and then it feels warm. The phasic component indicates the change in temperature upon immersion in the hot water and the tonic component indicates that it is still warm. Much of this perception is due to central processing of the peripheral thermoreceptor input (see Figure 4.3.4).

NOCICEPTORS PRODUCE FIRST AND SECOND PAIN

Free nerve endings called **nociceptors** in the skin have a high threshold for mechanical, chemical, or thermal

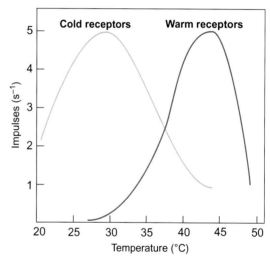

FIGURE 4.3.4 Response of thermoreceptors to skin temperature. The skin temperature was held constant at the indicated temperature, while the frequency of action potentials was recorded from fibers representing each of the thermoreceptor types.

stimuli and respond only when the intensity of these stimuli is high enough to damage tissue. We perceive the input from these receptors as pain. Superficial pain that arises from the skin has two components. The onset of an intense stimuli is sensed by an immediate, sharp, and highly localized pain called **first pain** or **initial pain**. After a delay of about 1 s or so, we are aware of a more diffuse, dull, and aching sensation which is **second pain** or **delayed pain**. First pain is carried over Aδ fibers, which are myelinated fibers that conduct action potentials relatively fast. Second pain is carried over small unmyelinated C fibers that conduct more slowly. This explains the difference in the onset of the sensations, but their subjective experience must be explained by their CNS connections, according to the labeled line theory of sensation (Table 4.3.2).

TABLE 4.3.2 Summary of Cutaneous Receptors

Type of Receptor	Location	Sensation	Fiber Type	Conduction Velocity (m s^{-1})	Adaptation
Pacinian corpuscle	Subcutaneous	Vibration	Aβ, large myelinated	30–70	Rapid
Meissner's corpuscle	Dermis of nonhairy skin	Flutter, tapping	Aβ, large myelinated	30–70	Rapid
Merkel's disks	Dermis	Touch, pressure	Aδ, small myelinated	12–30	Slow
Ruffini's corpuscles	Dermis	Touch, pressure, propioception	Aβ, large myelinated	30–70	Slow
Nerve ending	Hair follicle	Touch	Aβ, large myelinated	30–70	Rapid
Cold receptor	Dermis	Cold	Aδ, small myelinated	12–30	Phasic and tonic components
Warm receptor	Dermis	Warmth	C, small unmyelinated	0.5–2	Phasic and tonic components
Nociceptors	Epidermis	First pain	Aδ, small myelinated	12–30	Slow
Nociceptors	Epidermis	Second pain	C, small unmyelinated	0.5–2	Slow

SOMATOSENSORY INFORMATION IS TRANSMITTED TO THE BRAIN THROUGH THE *DORSAL COLUMN PATHWAY*

All of the cutaneous receptors we have discussed so far have a nerve ending in or near the skin and a cell body that resides in the dorsal root of the afferent or sensory nerve leading to the spinal cord (see Figure 4.3.5). The primary afferent neuron is a first-order neuron, being the first neuron to be affected by environmental stimuli. In many cases, the axon from the sensory neuron enters the spinal cord and turns upward and travels to the brainstem in tracts of axons located in the dorsal part of the spinal cord. Accordingly, these tracts form part of the **dorsal columns**, which consist of the **fasciculus cuneatus** and the **fasciculus gracilis**. These parts derive from their destination: the **nucleus cuneatus** and the **nucleus gracilis**, both located in the **medulla**. Neurons in the nucleus gracilis serve the lower parts of the body, whereas neurons in the nucleus cuneatus serve upper parts of the body. These neurons are second-order neurons, being the second stage in the communication between cutaneous sensory receptors and the sensory cortex.

The second-order neurons located in the nucleus gracilis and nucleus cuneatus send axons to the opposite side of the medulla and upward toward the thalamus. They send collaterals to an area of the brainstem called the **reticular activating system (RAS)** that functions in alertness. Using this pathway, somatic stimuli can wake us from a deep sleep. Neurons in the ventral posterolateral thalamus form third-order neurons that relay sensory information up to the cerebral cortex. The fourth-order neurons reside in the **somatosensory cortex**.

Our basic body plan, an evolutionary legacy from our ancient segmented ancestors, is based on body segments which contain a skin element (**dermatome**), a muscle element (**myotome**), and a bone element (**sclerotome**). The vertebrae in each segment comprise part of the sclerotome. Each spinal segment corresponds to a body segment and the sensory nerves that enter the cord serve the corresponding dermatome. The efferent motor nerves at each segment serve the corresponding myotome. When the sensory nerves enter the cord, their input is added to the previous input, from parts lower down the cord, in a layered fashion. Thus, the nerves traveling up the cord are well-organized into the two main tracts and within the tracts preserve a topology. Their mapping is preserved all the way to the cortex, where there is a **somatotopic representation** of the body on the cortex. "Somatotopic" means that adjacent areas of the body have sensory inputs into adjacent areas of the cortex, so that the body surface maps onto the neural surface.

The sensory pathways for cutaneous senses cross over from the left side of the body onto the right side of the brain and from the right side of the body to the left brain. This crossover occurs in the medulla, as shown in Figure 4.3.5, in the dorsal column pathway, but it is a recurrent theme in neurophysiology. Sensory input from the right side of the body travels to the left hemisphere of the brain, and the left side of the body is felt in the right hemisphere of the brain. En route, however, the sensory inputs branch to a variety of areas for

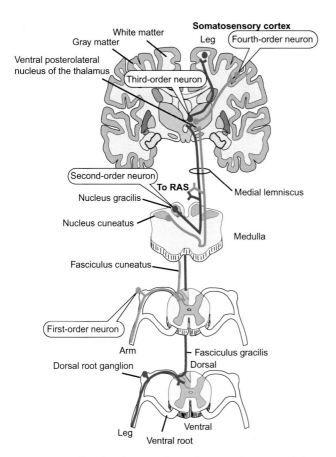

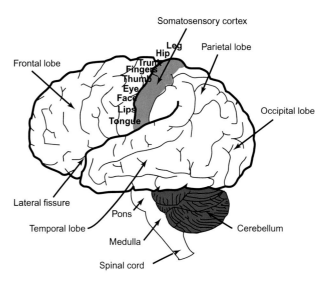

FIGURE 4.3.5 Dorsal column pathway for somatic senses. Primary afferent sensory neurons have cell bodies in the dorsal root ganglia and long axons that terminate in receptors at the skin. Graded receptor potentials are produced in the skin approximately in proportion to the logarithm of the stimulus intensity; those that reach threshold produce an action potential that is conducted toward the cord. The primary afferent fibers make synapses in the cord but also send a long fiber up the cord toward the brainstem. The sensory fibers travel in tracts, the fasciculus gracilis and the fasciculus cuneatus, in the dorsal part of the spinal cord. Because sensory inputs are arranged according to body segments, or dermatomes, there is a regular arrangement of fibers traveling in these fasciculi. These fibers make synapses with second-order neurons in the nucleus gracilis and nucleus fasciculatus. The axons from these second-order neurons cross over to the opposite side of the medulla, sending collaterals to the RAS and to the ventral posterolateral thalamus. These relay sensory information onto the cerebral cortex. The area of the cerebral cortex that receives sensory input is called the primary somatosensory cortex.

FIGURE 4.3.6 The somatosensory cortex. Sensory inputs reach the postcentral gyrus after having been relayed there by the ventral posterolateral thalamus. The projections of sensory neurons form a kind of neural map of the body, with adjacent areas of the cortex receiving sensory input from adjacent areas of the body. The inputs from the foot and toes are on the postcentral gyrus adjacent to the longitudinal fissure, out of view in this diagram.

defined reasons. Sensory information entering the cord branches immediately to join the dorsal column tracts and to make synapses with interneurons within the spinal cord. These synapses allow for the spinal reflexes that we will discuss in Chapter 4.4. In addition, collaterals synapse onto the RAS to alert us.

THE CUTANEOUS SENSES MAP ONTO THE SENSORY CORTEX

The primary somatosensory cortex is called S1. This area of the cerebral cortex receives sensory information from the somatic senses, plus proprioceptive senses and some visceral senses. It is located on the postcentral gyrus of the parietal lobe, as shown in Figure 4.3.6. The topological arrangement of the somatic senses is preserved as they enter the spinal cord, travel up the dorsal column tracts, to the nucleus gracilis or nucleus cuneatus, and is preserved through the thalamus to eventually map onto the cortex. Thus the surface of the body maps onto the surface of the brain.

PAIN AND TEMPERATURE INFORMATION TRAVEL IN THE ANTEROLATERAL TRACT

Nociceptors and thermoreceptors do not send axons up the dorsal columns. Instead, these receptors synapse on interneurons within the spinal cord that immediately send axons across the cord to the opposite side, where they ascend in the **anterolateral tract**, or the **ventrolateral tract**. These are also called the **ventral spinothalamic tract** and the **lateral spinothalamic tract**. The neurons making these tracts are second-order neurons whose processes ascend to the thalamus. There, third-order neurons project to the cerebral cortex. Cells in the ventral spinothalamic tract give off collateral branches in the medulla, whereas some cells terminate there (see Figure 4.3.7).

DISORDERS OF SENSATION CAN PINPOINT DAMAGE

Because pain and temperature signals cross the spinal cord at their level of entry and the other cutaneous senses cross in the brainstem, a relatively small lesion affecting only one side of the spinal cord could affect the sensation of pain and temperature on the **contralateral** side (the opposite site of the lesion) while it affects

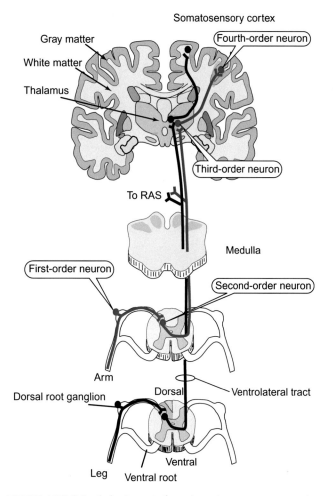

FIGURE 4.3.7 Spinothalamic tracts for pain and temperature sensation. Primary afferent sensory neurons synapse on interneurons within the spinal cord at the level of the primary afferent. These second-order neurons send an axon across the midline that ascends in the antero or ventral spinothalamic tract or the lateral spinothalamic tract. These fibers connect to third-order neurons in the thalamus, which then project to the cerebral cortex. Some cells in the ventral spinothalamic tract give off collaterals in the medulla and some terminate in the reticular formation as part of the spinoreticular tract.

other sensations on the **ipsilateral** side (the same side of the lesion). Both senses project to the contralateral cerebral cortex. In addition, the vertical location of sensory damage can be assessed from the part of the skin that shows sensory loss because each dermatome enters the spinal cord at known places.

PAIN SENSATION CAN BE REDUCED BY SOMATOSENSORY INPUT

Our subjective experience shows that perceived pain can be reduced by gently stroking the skin over the affected area. The explanation for this forms the **gate control theory** of pain. Large diameter somatosensory Aβ fibers make excitatory connections with interneurons in the dorsal horn of the spinal cord. These interneurons also receive inhibitory inputs from small diameter C afferents. When these interneurons are excited, they inhibit type C primary afferents by presynaptic inhibition, thereby acting as a gate to control transmission from type C primary afferents. The activity of the interneuron is determined by the balance between activity on the small nociceptive fibers and the larger somatic cutaneous receptors. When activity on C fibers dominates, the interneuron is inhibited and the gate is opened; when the skin is stroked, Aβ activity may dominate and the interneuron is activated, closing the gate (see Figure 4.3.8).

THE RECEPTIVE FIELD OF SOMATOSENSORY CORTICAL NEURONS IS OFTEN ON-CENTER, OFF-SURROUND

The receptive field of primary cutaneous sensory neurons is the area of the skin where stimulation produces excitation. The receptive field of neurons in the somatosensory cortex is more complex than this because the output of many different cells converges on these neurons. Many cortical neurons exhibit an **inhibitory**

FIGURE 4.3.8 The gate theory of pain modulation by sensory fibers. Pain information enters the spinal cord over small unmyelinated C fibers. Normally these sensory fibers excite a second-order neuron that crosses over to the ventrolateral or spinothalamic tracts and ascends to the medulla. Sensory Aβ fibers send processes up the dorsal columns to synapse with second-order neurons in the medulla. They also synapse with interneurons in the dorsal horn of the gray matter of the spinal cord. These neurons are excited by somatosensory input, and they have inhibitory connections to the second-order neurons of the pain pathway. Thus, stimulation of the somatic sensory neurons inhibits the activity of neurons that signal pain.

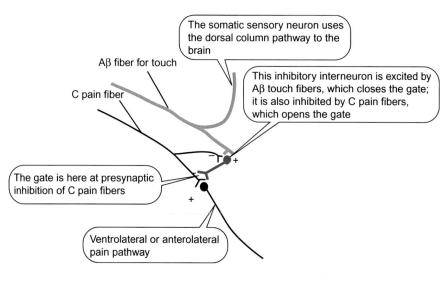

surround. Stimulation of a central region on the skin stimulates the neuron, whereas stimulation of the area around that central region inhibits the neuron (see Figure 4.3.9). This phenomenon has its origins in **lateral inhibition**, which refers to the inhibition of a higher order neuron by primary sensory afferents serving the areas around an excitatory area. Figure 4.3.10 shows a hypothetical wiring diagram that would produce lateral inhibition.

SUMMARY

Sensory systems apprise the body of the conditions of the external environment (exteroreceptors) and the internal environment (interoreceptors). Long receptors have sensory processes in the periphery, but their cell bodies are in the dorsal root of the sensory nerve just outside the spinal cord. These receptors must convert sensory energy into action potentials. Short receptors typically do not fire action potentials; they generate receptor potentials in response to stimuli which causes them to release neurotransmitters onto primary afferent neurons. This release of neurotransmitter is graded

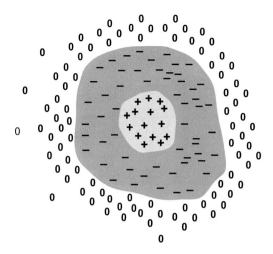

FIGURE 4.3.9 Receptive field of a somatosensory cortical neuron. Stimulation of the skin in a certain area causes increased frequency of firing of the cortical cell as indicated by "+" in the figure. Stimulation of the surrounding area decreases the excitation of the cell, as indicated by a "−" in the figure. Stimulation of an area still further removed results in no effect on the cell, as indicated by a "0" in the figure. The receptive fields are not necessarily circular but may take on complex shapes.

Clinical Applications: Neuropathic Pain and Transcutaneous Electrical Neural Stimulation

Subjective experience tells us that pain differs from other sensations such as cold, warmth, touch, and sound. Pain is a complex experience derived from nociception, somewhat like vision is a complex experience derived from light sensation. But pain carries with it distressful affect (from the latin *affectus*, meaning "state of mind") that the other modalities ordinarily lack. Under some circumstances pain becomes emotionally distressful. The mechanism by which this occurs is not fully understood.

Clinicians recognize two broad classes of pain: nociceptive or physiological pain, and neuropathic or intractable pain. Physiological pain arises from stimulation of nociceptors and is carried by $A\delta$ and C fibers to the CNS. Stimulation of $A\delta$ fibers causes a sharp and highly localized pain that is usually short lived. Stimulation of C fibers produces a dull, aching, or burning pain that persists even when the stimulation is removed. These two types of pain are often discriminated by the adjectives **epicritic** (from the Greek *epi*, meaning "on" and *krino* meaning "separate" or "judge"), referring to the sharp pain that is easily localized, and **protopathic** (from the Greek *protos* meaning "first" and *pathos* meaning "suffering") that describes the suffering of second pain. Physiological pain can be increased by peripheral sensitization or central sensitization. Peripheral sensitization results from changes in the nociceptors brought about by inflammation. Central sensitization is caused by changes in neuronal activity in the spinal cord, usually interneurons. Many of these neurons have NMDA receptors that respond to glutamic acid. These receptors at rest typically bind Mg^{2+} ions, and the bound Mg^{2+} inhibits depolarizing Ca^{2+} influx. When the cells are depolarized further, Mg^{2+} dissociates from the receptor and the cell becomes hypersensitive. Although both peripheral and central sensitization can last for minutes to hours after removal of the stimulus, it does not cause permanent changes in the CNS.

Neuropathic pain, on the other hand, results from injury to the CNS that permanently changes CNS connections. For example, injury can break axons, and the somas will regenerate new axons. But recovery is incomplete, and small fibers are less likely to regenerate than larger ones. Injury to a nerve will lose C and $A\delta$ fibers more than the large $A\beta$ fibers. In attempts to make new synaptic connections, the $A\beta$ fibers can make new synapses on neurons abandoned by the C fibers. Thus, previously innocuous stimuli become severely painful. This change from innocuous to noxious perception is called **allodynia**. Persons with amputations sometimes report this type of pain that appears to them to originate from the removed limb. Such pain is called **phantom pain**.

Neuropathic or intractable pain is severely distressing and affected people seek relief from the misery. Surgical remedies include sectioning of the dorsal root (**rhizotomy**) or of the anterolateral tract (**tractotomy**). These procedures often provide immediate relief, but the pain almost always returns when the severed axons once again form inappropriate synapses.

Transcutaneous electrical neural stimulation (TENS) sometimes provides relief from intractable pain. This less invasive procedure uses gate control theory. Electrodes are placed on the skin over a peripheral nerve. The large fibers have a lower threshold for stimulation and are preferentially activated by TENS. Preferential activation of the large fibers closes the "gate" by which pain enters the CNS. TENS devices successfully reduce some types of intractable pain.

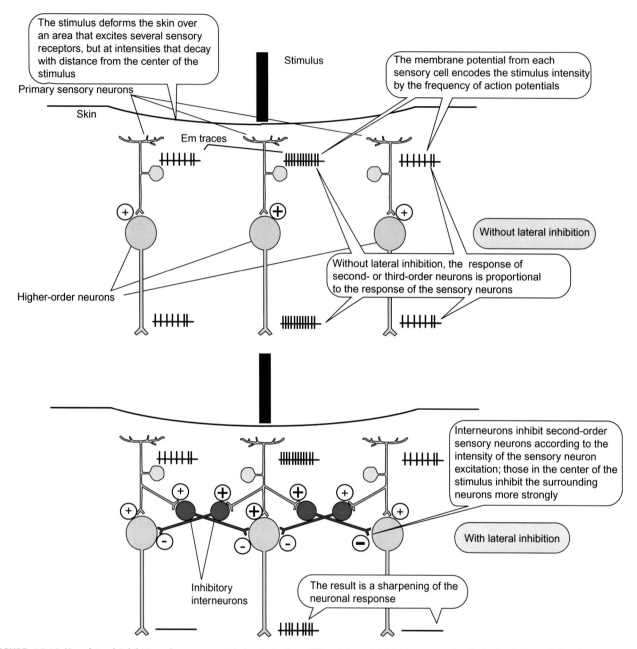

FIGURE 4.3.10 How lateral inhibition sharpens spatial discrimination. Without lateral inhibition, a mechanical stimulation of the skin produces a response from three sensory neurons which is passed forward up to the somatosensory cortex. Lateral inhibition by interneurons inhibits the cells responding to the periphery of the stimulus more than at the center. This sharpens the spatial discrimination of the stimulus and also produces the on-center and off-surround receptive field. That is, a stimulus localized to the periphery inhibits the response at the center. The sharpening of the response enhances two-point discrimination for the cutaneous senses. The effects shown here are exaggerated and are the net effect of events at several layers in the sensory pathway and not at the level of the primary sensory afferents.

according to the strength of the stimulus, and the primary afferent neuron responds with a train of action potentials.

The quality of the stimulus (light, touch, sound, etc.) is conveyed to the CNS by labeled lines: the connectivity of the neurons determines its interpreted quality. The intensity is encoded by the frequency of sensory neuron action potentials (the frequency code) and by the number of sensory neurons that are excited (the population code). The frequency of action potential firing is roughly proportional to the logarithm of the stimulus intensity, so that a three-fold change in frequency of action potentials encodes a 1000-fold change in stimulus intensity.

The dynamics of sensory receptor response allows sensation of position, velocity, and acceleration. The physical location of the receptor means that it will respond only to stimuli within its receptive field. Receptive fields can also be defined for higher order neurons within the CNS. Regions of high spatial sensitivity are characterized by neurons with small receptive fields.

Cutaneous sensory cells bring information into the CNS by the spinal cord. Somatic sensory neurons for touch and pressure enter the cord through the dorsal root and send fibers up to two dorsal columns, the fasciculus gracilis and the fasciculus cuneatus. The fasciculus gracilis serves the lower body and the fasciculus cuneatus serves the upper torso. The axons are laid down at each level of the cord so that there is a topological representation of the body surface within the cord. This topology is preserved all the way to the cortex, where the somatic senses map onto the somatosensory cortex immediately posterior to the central sulcus. The area of the body surface served by each level of the spinal cord is called a dermatome. The dorsal columns make contact with second-order neurons in the medulla, whose axons cross over and project onto third-order neurons in the thalamus. These third-order cells then send axons up to the somatosensory cortex. Thus, sensory information crosses over at the level of the medulla. Somatic senses of pain and temperature, however, make a synapse in the spinal cord at the level of the spinal root. The second-order neuron crosses over at that level and sends fibers up the spinothalamic tracts. Both somatic sensory neurons and pain and temperature sensors send collaterals to the reticular activating system in the medulla that will awake you if you are sleeping, and otherwise contributes to alertness.

Pain reception can be reduced by somatic senses due to interneurons that are excited by touch but inhibit pain fibers. This is the basis by which pain is lessened by movement or by mechanical stimulation of the skin.

REVIEW QUESTIONS

1. What is a sensory modality? What is the adequate stimulus? How is quality of sensory information encoded?
2. How is intensity of sensory information encoded?
3. What is one way to distinguish between position, velocity, and acceleration?
4. What is lateral inhibition and how does it sharpen spatial discrimination? How does this give rise to an on-center, off-surround receptive field? What is meant by "receptive field"?
5. What is the dorsal column pathway? What sensory modalities use it? What is the anterolateral pathway, and what sensory modalities use it?
6. What is the somatosensory cortex? What is somatotopic mapping? Why does it occur?
7. What is a fasciculus? What is the fasciculus gracilis, and what part of the body does it serve? What is the fasciculus cuneatus, and what part of the body does it serve?

4.4 Spinal Reflexes

Learning Objectives

- Define "reflex"
- List four spinal reflexes
- Describe the withdrawal reflex and draw a "wiring diagram" that explains it
- Explain why two different interneurons are required for the withdrawal reflex
- Describe the crossed-extensor reflex
- Draw a "wiring diagram" to explain the crossed-extensor reflex
- Describe the myotatic reflex
- Distinguish between extrafusal and intrafusal muscle fibers
- Distinguish between primary and secondary afferent information arising from the muscle spindle
- Draw a "wiring" diagram to explain the myotatic reflex
- Explain how the gamma efferent system maintains muscle spindle sensitivity during contraction
- Describe the function of the Golgi tendon organs and the inverse myotatic reflex
- Describe what is meant by the "final common pathway" in motor control
- List the major tracts of axons from higher centers that modulate spinal reflexes and control lower motor neurons

A *REFLEX* IS A STEREOTYPED MUSCULAR RESPONSE TO A SPECIFIC SENSORY STIMULUS

The word "reflex" derives from "reflection" meaning a mirror image of an event or stimulus. Thus the spinal reflexes are processes that cause sensory stimuli to be "reflected" onto the musculature without the involvement of conscious deliberation. The "stereotype" means that the reflexes always occur in the same manner and no training is required. These reflexes result from hardwired connections that are the same in all humans and in most mammals. We will discuss four basic reflexes:

1. The withdrawal reflex
2. The crossed-extensor reflex
3. The myotatic reflex
4. The inverse myotatic reflex.

THE WITHDRAWAL REFLEX PROTECTS US FROM PAINFUL STIMULI

A painful stimulus in a limb extremity causes us to reflexly withdraw the limb from the stimulus, protecting it from further harm. Withdrawal is accomplished by activation of the **flexor muscles** and inactivation of the **extensor muscles**, and the coordination of these is accomplished by the spinal cord.

Sensory neurons entering the spinal cord synapse on interneurons in the dorsal horn of the spinal cord, which subsequently make polysynaptic connections to the large α motor neurons that reside in the ventral horn of the cord and control the flexor muscles. At the same time, interneurons also synapse on inhibitory interneurons that innervate the motor neurons for the extensor muscles. In this way, painful sensory input activates the flexor muscle and relaxes the extensors. The wiring diagram for this reflex is shown in Figure 4.4.1. A minimum of two interneurons are required to transform the excitatory input of the interneuron into an inhibitory one. The α motor neurons all have the same set of receptors for neurotransmitters. Since the interneuron produces only one set of neurotransmitters, it cannot simultaneously excite the flexor's motor neuron and inhibit the extensor's. Instead, it must activate another interneuron that uses a different neurotransmitter to inhibit the extensor motor neuron.

THE CROSSED-EXTENSOR REFLEX USUALLY OCCURS IN ASSOCIATION WITH THE WITHDRAWAL REFLEX

If we step on a sharp piece of glass, withdrawal of the affected leg requires the cooperation of the opposite leg to maintain balance. If the affected limb is an arm, the arm on the contralateral side to the pain will extend rather than flex, in an effort to move the body away from the source of the pain. The flexion of the limb on the side of the pain is the withdrawal reflex. **The reflex extension of the limb on the opposite side is called the crossed-extensor reflex**. The mechanism is similar to that of the withdrawal reflex: interneurons form excitatory and inhibitory connections to the appropriate motor neurons. A circuit diagram for the crossed-extensor reflex is shown in Figure 4.4.2. The net effect of any pathway can be determined by multiplying the signs of

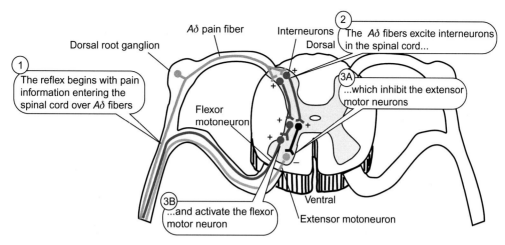

FIGURE 4.4.1 Wiring diagram for the withdrawal reflex. Fast pain fibers make connections with an interneuron in the spinal cord. There is always at least one and there may be more interneurons between the primary afferent fiber and the motor neurons. Excitation of the interneuron passes on to the flexor motor neurons, while other interneurons inhibit the extensor muscles.

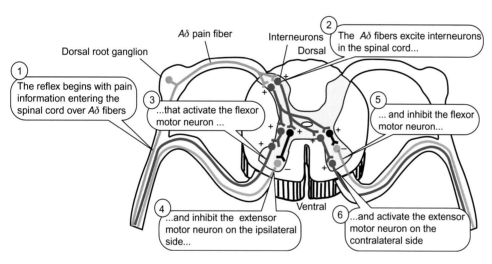

FIGURE 4.4.2 The crossed-extensor reflex. A strong nociceptive stimulus elicits the withdrawal reflex on the ipsilateral side. At the same time, fibers cross over the cord to synapse onto excitatory interneurons that activate the extensor motor neuron and inhibitory interneurons for the flexor muscles. Thus, activation of the sensory afferents causes a withdrawal on the ipsilateral side and an extension on the contralateral side.

activation (+) or inhibition (−). Thus, pain activates the ipsilateral (meaning the same side) flexion (+ × + × + = +) and inhibits extensors (+ × + × − = −); it inhibits contralateral (the opposite side) flexion (+ × + × − = −) and activates contralateral extension (+ × + × + = +).

THE MYOTATIC REFLEX INVOLVES A MUSCLE LENGTH SENSOR, THE MUSCLE SPINDLE

The **myotatic reflex** is the **muscle stretch reflex**. A classical example of this is the knee-jerk reflex. Striking the patellar ligament (connecting the patella to the tibial tuberosity) on a flexed, but relaxed knee joint with a rubber mallet causes the quadriceps to stretch. The stretch is sensed and the information carried to the cord, resulting in activation of the muscle, causing the knee to extend. The process begins with the sensor for muscle length, the muscle spindle.

THE MUSCLE SPINDLE IS A SPECIALIZED MUSCLE FIBER

The regular muscle fibers are called **extrafusal** fibers, and the activation and contraction of these fibers are discussed in Chapters 3.4–3.7. Interspersed among these fibers are small encapsulated sensory receptors that have a fusiform or spindle shape. The entire apparatus is called the **muscle spindle**, and its job is to inform the central nervous system (CNS) of the contractile state of the muscle by sending afferent impulses to the spinal cord when the muscle spindle is stretched. The muscle spindle consists of a group of fine muscle fibers, called **intrafusal muscle fibers**, 4–10 mm long, whose central portions are not contractile.

Typically the connective tissue **capsule** encloses two different types of intrafusal fibers. The **nuclear chain** intrafusal fibers have a set of aligned nuclei in the center. The **nuclear bag** fibers have a clump of nuclei randomly arrayed in a bag-like structure in the center of the intrafusal fiber. Typically a spindle has two to three nuclear bag fibers and about five nuclear chain fibers. There are also two distinct types of afferent sensory fiber endings. Large myelinated nerves termed **Ia** or **primary afferents** surrounds the central portion of all of the intrafusal muscle fibers. This nerve ending forms a coiled structure called the **annulospiral ring**. Stretching this nerve ending activates **stretch-activated channels** in its surface that depolarize the neuron and therefore increase its firing rate. Lengthening the muscles stretches these receptors and increases the firing rate. Shortening of the muscle alleviates stretch of these receptors and their firing rate decreases. Thus, the firing rate of the intrafusal sensory nerve is related to the stretch of the muscle relative to the muscle spindle. The nuclear bags actually come in two varieties: "static bags" and "dynamic bags". The static nuclear bag and nuclear chain fibers receive a second kind of innervation, classified as **II afferents**. These innervate the juxtaequatorial regions of the intrafusal fibers. The II afferents consist of medium myelinated fibers that adapt slowly. Their tonic activity carries information about the static muscle length. In addition to these sensory receptors, each intrafusal fiber is innervated with a motor neuron, the **gamma motor efferent**, that controls the length of the intrafusal fiber by activating its contractile mechanism just as an ordinary motor neuron would. This has the effect of maintaining the sensitivity of the muscle spindle when muscles contract. The arrangement is shown schematically in Figure 4.4.3.

THE MYOTATIC REFLEX IS A MONOSYNAPTIC REFLEX BETWEEN Ia AFFERENTS AND THE α MOTOR NEURON

The myotatic reflex is the "knee-jerk" reflex in which a muscle contracts in direct response to its stretch. It is typically elicited by tapping on the tendon of a muscle, which deforms the tendon and stretches the muscle. Ia afferents of the muscle spindles sense the stretch and make direct connections to the α motor neurons that innervate the muscle. The increased rate of firing upon stretch causes excitation of the α motor neuron and a contraction of the muscle. Because the stretch is short-lived, so is the excitation of the Ia afferent and so is the excitation of the α motor neuron, and a short twitch of the muscle is observed. The wiring of this reflex is shown in Figure 4.4.4.

The purpose of the myotatic reflex is to resist changes in muscle length. This is useful when one is trying to maintain the position of a limb during purposeful work or in maintaining posture, but it is counter-productive during purposeful movements when the muscles must contract while their antagonist member relaxes. Thus, this system is over-ridden during purposeful movements. At other times, the reflex can be enhanced. Its strength can be facilitated or depressed by inputs from the motor cortex.

The myotatic reflex relies on excitation of the α motor neuron by input from the Ia afferents. Their ability to depolarize the motor neuron to produce action potentials depends on the resting membrane potential of the motor neuron, which in turn can be influenced by other

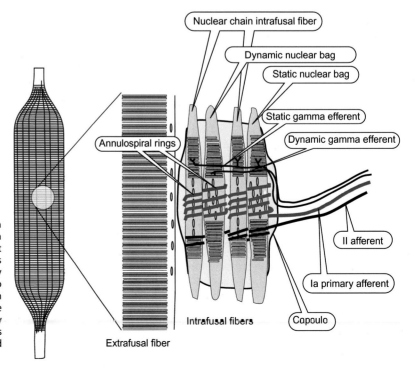

FIGURE 4.4.3 The muscle spindle. The muscle spindle is a group of smaller and specialized muscle fibers within a muscle. Typical spindles contain two bag fibers and about five chain fibers. The nuclear bag fibers are classified as static or dynamic. The intrafusal fibers are innervated by motor nerves (the gamma motor neurons) and two different types of sensory fibers: the type Ia stretch receptors and the type II afferent sensory receptors. The annulospiral receptors sense stretch and are rapidly adapting. Thus they sense the rate at which the muscle is stretched. The type II afferents are slowly adapting and inform the CNS about the static stretch of the muscle.

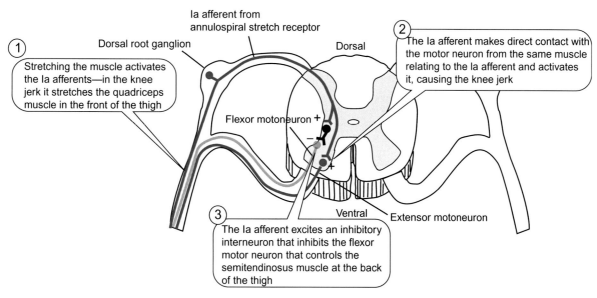

FIGURE 4.4.4 The circuitry of the myotatic reflex. Increased firing of the annulospiral stretch receptor of the Ia muscle spindle afferents activates the muscle, here denoted as the extensor muscle. In the knee-jerk reflex this corresponds to the quadriceps muscle. Simultaneously, the stretch receptor synapses on an interneuron that inhibits the antagonistic muscle, which in this case is a flexor muscle, the semitendinosus. In this way, stretch of the muscle is accompanied by contraction of the muscle and relaxation of its antagonist.

connections in the cord and from higher centers. Usually it is possible to enhance the knee-jerk reflex by **Jendrassik's maneuver**. In this maneuver, the subject interlocks the fingers of the two hands and tries to pulls the hands apart while resisting the movement. This isometric exercise increases the general outflow from the cerebral motor cortex, depolarizing a variety of motor neurons and thereby facilitating the myotatic reflex.

THE GAMMA MOTOR SYSTEM MAINTAINS TENSION ON THE INTRAFUSAL FIBERS DURING MUSCLE CONTRACTION

The primary and secondary sensory receptors of the muscle spindle are stretch receptors. The rapidly adapting annulospiral sensors encode the rate of stretch, whereas the slowly adapting secondary sensors encode the static length of the muscle. During contractions, the extrafusal fibers shorten and so the tension on the intrafusal fibers disappears. To maintain the tension on the stretch receptors, the γ **efferent system** simultaneously activates the intrafusal fibers with the extrafusal fibers so that the stretch receptors are tight enough to remain able to report stretch. Thus, the γ efferent system maintains the sensitivity of the muscle spindle stretch receptors over the range of muscle lengths that occur during contraction and relaxation. The cell bodies of the γ motoneurons are located in the ventral horn of spinal cord, like those of the α motor neurons. The γ efferent motor system is shown in Figure 4.4.5.

Coactivation of the γ efferents also helps activate the muscles. While under tension, the Ia and II fibers return action potentials to the cord to facilitate the α motor neurons. When the muscle contracts, this facilitation would disappear because the Ia and II fibers would decrease their firing rates. Coactivation of the intrafusal fibers solves this problem.

THE INVERSE MYOTATIC REFLEX INVOLVES SENSORS OF MUSCLE FORCE IN THE TENDON

Stretch receptors called **Golgi tendon organs** are found within the collagen fibers of tendons and within joint capsules. They are generally located in series with the muscle rather than the parallel arrangement of the intrafusal muscle fibers. Therefore, the stretch of the tendons reflects the force on the tendon that is developed by all of the muscle, and the firing rate of the Golgi tendon organ encodes muscle force rather than stretch, even though it actually senses stretch. The action potentials are carried to the spinal cord over **type Ib afferent fibers**, which are large myelinated fibers. The sensory neurons synapse in the cord on interneurons which then make connections to α motor neurons. In this way, high force development that could injure muscles can be prevented by relaxation of the muscle. This reflex, which connects high force in the Golgi tendon organs with relaxation, is the opposite of the myotatic reflex, the stretch reflex, in which stretch elicits a reflex contraction. Thus, this reflex is called the **inverse myotatic reflex** even though the muscle is not necessarily stretched. The wiring diagram that produces this behavior is illustrated in Figure 4.4.6.

Table 4.4.1 summarizes the afferent and efferent inputs relating to control of muscle function.

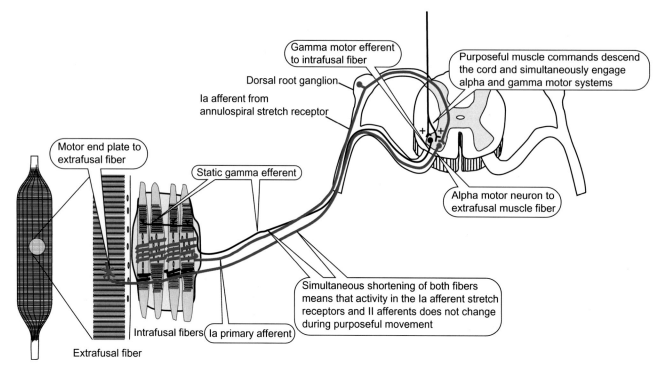

FIGURE 4.4.5 Coactivation of the γ efferent system with the α motor neurons. Command signals from the motor cortex activate α motor neurons. Simultaneously, γ efferents are activated so that the intrafusal fibers contract along with the muscle, thereby preserving sensitivity of the stretch receptors during contraction. In this way, the CNS can remain informed of the rate of muscle shortening through sensory inputs from the Ia stretch receptors on the muscle spindles, and the stretch receptors do not inhibit purposeful movement. Some connections are only partially shown for clarity.

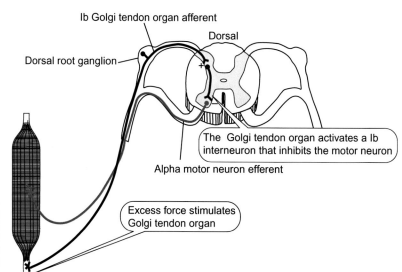

FIGURE 4.4.6 Wiring of the inverse myotatic reflex. Sensors in the tendon, the Golgi tendon organ, are activated upon stretch of the tendon, which requires considerable force. These sensors synapse on interneurons in the spinal cord that inhibit further activity of the motor neurons innervating the muscle. This relaxation of the muscle prevents damage from excess force.

THE SPINAL CORD POSSESSES OTHER REFLEXES AND INCLUDES *LOCOMOTOR PATTERN GENERATORS*

The spinal reflexes discussed above are only part of the story. There are a variety of other reflexes such as the **placing reflex**, which helps maintain posture and support. The placing reflex refers to the reaction to tactile stimuli applied to the back of the paw of lower animals. The reaction is to flex the limb and then swing it forward and extend it. For example, a twig touching the dorsum of the foot during its swing phase results in the foot being lifted over the twig. In humans, this is the reaction upon stubbing one's toe: the affected limb is flexed and then rapidly swung forward and extended to catch the falling body. This is a far more complicated reflex than what we have already discussed. It involves a complex sequence of motor neuron activation and inhibition.

The spinal cord also possesses **central pattern generators** that are the basis of gait. Lower animals, because of their quadripedal locomotion, show distinct differences

TABLE 4.4.1 Summary of the Afferent and Efferent Fibers to Skeletal Muscle Involved in the Spinal Reflexes and Control of Muscle Function

Afferents

Receptor	Axon	Axon Diameter (μm)	Conduction Velocity (m s^{-1})	Function
Primary muscle spindle	Ia	12–20	70–120	Rate of length change
Secondary muscle spindle	II	4–12	20–70	Static length
Golgi tendon organ	Ib	12–18	70–110	Muscle tension

Efferents

Motor neuron	Target	Axon Diameter (μm)	Conduction Velocity (m s^{-1})	
α Motor neuron	Extrafusal muscle	8–13	40–80	Voluntary control
γ Motor neuron	Intrafusal muscle	3–8	20–50	Adjust spindle sensitivity

in gaits from walking, trotting, pacing, and galloping. Humans also show differences in gait but, because of our bipedal locomotion, the differences center on speed of movement rather than differences in the sequence of flexion and extension of four legs. In the gallop, for example, the two hind legs are simultaneously or nearly simultaneously extended to push the animal forward. Out of phase with the extension of the hind legs is the extension and then flexion of the forelegs to pull the animal forward. In bipedal motion this would become a hop, which is extraordinarily inefficient locomotion. Humans pay for their bipedal locomotion by being one of the slowest animals around. The advantage is that it frees the hands to throw a sharp rock or a spear at much higher velocities than locomotion.

Experiments with lower animals show that part of locomotion control resides in pattern generators located in the spinal cord. When the thoracic spinal cord of a cat is cut, severing all connections to the lower motor neurons from higher centers in the brain, and the cat is suspended in a sling above a treadmill, the animal can still raise and place the hind legs, and this motion will keep pace as the treadmill is accelerated. Thus, the basic circuitry for movement of the hind legs resides in the spinal cord.

THE SPINAL CORD CONTAINS DESCENDING TRACTS THAT CONTROL LOWER MOTOR NEURONS

The cell bodies of the **lower motor neurons** reside in the ventral horn of the spinal cord and provide the only direct neural control of skeletal muscle. All of the reflexes, and all voluntary control of muscle, act through their effects on these relatively small number of motor neurons. We have already discussed the spinal reflexes. In addition, voluntary and unconscious higher control of muscle descend in bundles from the cerebral cortex and brainstem down the spinal cord to make synapses onto interneurons and motor neurons within the spinal cord. Sherrington coined the term "final common path" to describe the motor neurons that innervate the muscles because all of the variety of reflex activities and all complex behavior involving skeletal muscles must eventually be directed to these lower motor neurons.

The descending tracts are named first from their point of origin and secondly from their termination. In addition, tracts are named for their specific anatomical location. Thus the descending tracts controlling the lower motor neurons are as follows:

LATERAL CORTICOSPINAL TRACT

This tract originates in the cerebral cortex and descends to the lower motor neurons in the spine. Hence it is a corticospinal tract. In addition, it descends the spinal cord in the lateral aspect. As we will see in Chapter 4.5, the motor and premotor areas of the cerebral cortex in front of the central sulcus forms this tract, which crosses the midline (**decussates**) to descend on the lateral aspect of the contralateral spinal cord. The axons make synapses with interneurons and motor neurons within the cord. The fibers in this tract control motor neurons for the muscles of the distal extremities. Thus, damage to this pathway results in the loss of fine motor skills.

RUBROSPINAL TRACT

The **red nucleus** is a structure in the midbrain from which the rubrospinal tract originates. It receives inputs from the motor cortex and from the cerebellum. The fibers cross over the midline and descend in the rubrospinal tract, terminating primarily on interneurons within the spinal cord.

VENTRAL CORTICOSPINAL TRACT

This tract also originates from the premotor and motor cortex and descends the ventral contralateral spinal cord. These fibers control motor neurons for axial and proximal muscles and so are involved in posture and gross motor movements.

LATERAL VESTIBULOSPINAL TRACT

As its name implies, the lateral vestibulospinal tract originates in the **lateral vestibular nucleus.** These fibers receive inputs from the ipsilateral vestibular apparatus (see Chapter 4.5), which informs the brain about balance. The tract descends without crossing to make connections to medial motor neurons that control postural muscles on the axis of the body. The tract is involved in the reaction of the muscles to gravity and thus it helps maintain posture.

MEDIAL VESTIBULOSPINAL TRACT

The medial vestibulospinal tract originates from the **medial vestibular nucleus**, which in turn receives inputs from the vestibular apparatus and from stretch receptors in the muscles of the head and neck. The tract descends ipsilaterally to make contact with interneurons and motor neurons within the thoracic spinal cord to help position the torso relative to gravity and the head during movements.

TECTOSPINAL TRACT

The tectospinal tract originates in the **superior colliculus**, which is part of the **tectum**, or roof of the midbrain. The tract crosses the midline before descending to the upper cervical spinal cord. It receives inputs from the eyes and ears and serves to coordinate the positioning of the head and eyes in relation to auditory and visual targets.

PONTINE RETICULOSPINAL TRACT

The pontine reticulospinal tract begins in part of the **reticular formation** located in the pons. It descends ipsilaterally and terminates on interneurons that control medial extensor motor neurons. This tract is largely excitatory.

MEDULLARY RETICULOSPINAL TRACT

This tract is similar to the pontine reticulospinal tract except that it originates in the part of the reticular formation located in the medulla. It is largely inhibitory, and it serves to balance the excitatory drive of the pontine reticulospinal tract.

The anatomical pathways for these tracts are shown in Figure 4.4.7.

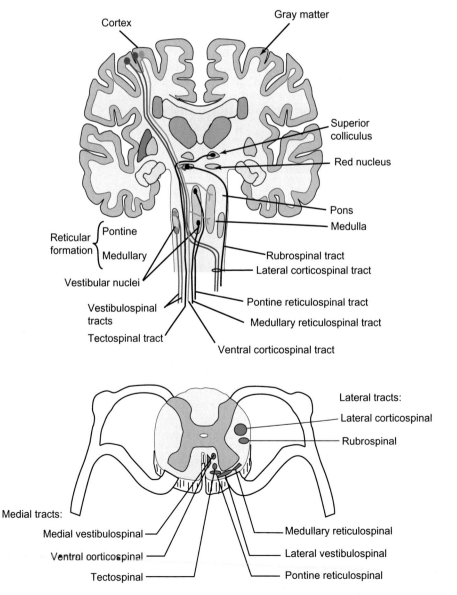

FIGURE 4.4.7 Major descending motor pathways. The lateral tracts supply motor neurons located in the lateral portion of the ventral horn of the spinal cord that control the distal muscles of the limbs. The medial tracts supply motor neurons located in the medial portion of the ventral horn of the spinal cord. These control the axial and proximal muscles. Cells in the primary motor cortex send axons to the contralateral spinal cord in the lateral corticospinal tract. These send collaterals to the red nucleus located in the midbrain. Cells in the red nucleus also send fibers to the contralateral spinal cord in the rubrospinal tract. Some primary motor neurons do not cross over; they form the ventral corticospinal tract. These neurons in the ventral corticospinal tract send collaterals to the contralateral reticular formation, from which axons descend ipsilaterally, forming the pontine reticulospinal tract and the medullary reticulospinal tract. The vestibular nuclei send axons down the spinal cord ipsilaterally, forming the lateral vestibulospinal tract and the medial vestibulospinal tract. The vestibular nuclei receive inputs from the vestibular apparatus, which senses gravity, and so these tracts are involved in postural responses to gravity. The superior colliculus is part of the tectum, or roof of the midbrain, and it receives inputs from both the eyes and the ears. It sends axons down the contralateral spinal cord in the tectospinal tract. It is involved in moving the head and eyes in search of auditory or visual targets. The lower part of the figure indicates the approximate location of these tracts on one side of the spinal cord. These tracts are symmetrically placed on both sides of the spinal cord, but for clarity only one side is shown.

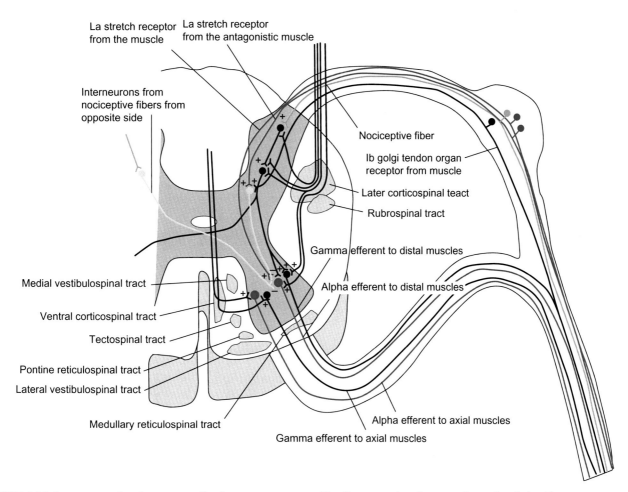

FIGURE 4.4.8 Convergence of pathways controlling lower motor neurons. **The flexor muscles alone are shown for clarity.** The α motor neurons receive inputs from the stretch receptors (Ia fibers) located in the intrafusal fibers of the muscle. These directly activate the α motor neurons in a monosynaptic pathway. The α motor neurons also receive inputs from the Ia stretch receptors of the antagonistic muscle, which synapse on an interneuron which then inhibits the motor neuron. Afferent Ib fibers from the Golgi tendon organs make a polysynaptic, inhibitory connection to the α motor neuron. Nociceptive fibers from the ipsilateral side make a polysynaptic excitatory connection to α motor neurons for the flexors and an inhibitory connection to motor neurons of the extensors, which are not shown. The nociceptors from the contralateral side synapse onto interneurons within the cord. These send fibers across the cord to have opposite effects on the contralateral α motor neurons. Inputs from the higher centers in the nervous system descend in the tracts indicated. Only those inputs from the corticospinal tracts are shown. The lateral corticospinal tract descends from the primary motor cortex and engages generator patterns and motor neurons directly. The lateral corticospinal tract predominantly activates muscles in the limb extremities. The ventral corticospinal tract provides input to α motor neurons that control the axial muscles.

ALL OF THE INPUTS TO THE LOWER MOTOR NEURONS FORM INTEGRATED RESPONSES

All of the reflexes that we have discussed, and all of the descending tracts with their attendant activity, are simultaneously active to varying degrees. Thus the response of the motor neurons depends on the relative intensity of the inputs from all of these different sources. A simplified version of all of these combined inputs is shown diagrammatically in Figure 4.4.8.

SUMMARY

Strong stimuli applied to the body are *reflected* back in a muscular response. These reflexes are stereotypical and hardwired. The withdrawal reflex refers to the withdrawing of a limb upon painful stimulation of the skin of the limb. The pain information is carried by Aδ pain fibers that synapse onto interneurons within the spinal cord. These in turn make synapses on the motor neurons of both the flexors and extensors. Pain activates the flexors and inhibits the extensors.

The crossed-extensor reflex generally occurs with the withdrawal reflex and refers to the extension of the opposite limb. Here interneuron connections cross over to the other side of the spinal cord, but now contralateral pain activates extensors and inhibits flexors. The withdrawal reflex and crossed-extensor reflex are polysynaptic but occur entirely within the spinal cord. Input from higher centers may facilitate or inhibit reflexes but they cannot eliminate them.

Muscles have receptors for stretch and force. Extrafusal muscle fibers comprise the bulk of muscle and forms the major force-generating structure. Intrafusal muscle fibers are buried in the muscle and they contain afferent receptors for stretch, but they also contain contractile elements. Nuclear bag and nuclear chain fibers both contribute to the afferents. Primary Ia fibers report on dynamic stretch; they innervate the dynamic nuclear bag fibers. Group II afferents are receptors that report on the static stretch of the muscle; these innervate static nuclear bag fibers and all of the nuclear chain fibers. Rapid stretch of the intrafusal fibers sends action potentials to the spinal cord, where Ia fibers make a monosynaptic activation of α motor neurons that innervate the same muscle as the Ia afferents. The stretch receptors also inhibit the antagonist muscles. The stretch reflex is called the myotatic reflex.

Gamma motor neurons innervate intrafusal fibers and control their contraction. Contraction of a muscle relieves stretch on the parallel intrafusal fibers and this defacilitates muscle contraction and also removes the effectiveness of the receptors as stretch receptors. Simultaneous coactivation of the α motor neurons and γ motor neurons during purposeful movements preserves the sensitivity of the stretch receptors during contraction.

The Golgi tendon organs are arranged in series with the muscle, and they inform the CNS of the force developed by the muscle. When force is excessive, afferent activity on the Ib Golgi tendon organ afferents inhibits further α motor neuron activity. This reflex is called the inverse myotatic reflex.

Spinal reflexes form the basis of movement, but more complicated locomotion circuits are present in the spinal cord. Although these locomotion pattern generators play an important role in movement, purposeful movement requires motor commands from higher centers in the CNS. The α motor neurons are the sole conduit for excitation of muscles, and all commands pass through this final common pathway. These are the lower motor neurons. Purposeful control of the lower motor neurons is provided by upper motor neurons in the cortex. Many tracts of axons descend the cord to control the lower motor neurons and the interneurons in the cord. These tracts are named for their location in the spinal cord and for the origin and terminus of the axons. The lateral corticospinal tract, for example, is located in the lateral aspect of the cord and connects neurons in the cortex with those in the spinal cord. Other tracts provide input from vision and balance sensory systems.

REVIEW QUESTIONS

1. What is a reflex?
2. What sensory information elicits a withdrawal reflex?
3. Why are interneurons necessary for the withdrawal reflex?
4. What is the crossed-extensor reflex? Why is it necessary to inhibit antagonistic muscles during operation of the reflexes?
5. Which reflex is the only monosynaptic reflex? Would you expect this reflex to be faster or slower?
6. What is a muscle spindle? What afferents carry what information to the spinal cord? Are you conscious of this information? Why is it necessary for muscle spindles to have contractile ability?
7. Which spinal tracts do not cross over from their origin?
8. What is the purpose of the inverse myotatic reflex?
9. What is a locomotor pattern generator?

Balance and Control of Movement 4.5

Learning Objectives

- Describe the population code and frequency code for controlling muscle force
- Describe the somatotopic organization of lower motor neurons
- Define what is meant by the motor neuron pool in the spinal cord and its organization by myotome
- Indicate the location of the primary motor cortex, supplementary motor cortex, premotor cortex, and sensory association area
- List the components of the basal ganglia
- Describe the role of the direct pathway in enhancing desired movement
- Describe the role of the indirect pathway in suppressing unwanted movement
- Describe the role of the substantia nigra in balancing the direct and indirect pathways
- List the major inputs into the cerebellum and its function
- Identify the semicircular canals and distinguish them from the utricle and saccule
- Distinguish the perilymph from the endolymph in terms of location and composition
- Identify the information provided by the semicircular canals and the utricle and saccule
- Identify the major features of hair cells and how they respond to deformation of the kinocilium

THE NERVOUS SYSTEM USES A POPULATION CODE AND FREQUENCY CODE TO CONTROL CONTRACTILE FORCE

Large multinucleated cells called muscle fibers make up a skeletal muscle. The overwhelming majority of them are extrafusal fibers that produce force to move or rigidly fix body parts, when antagonistic pairs of muscles are simultaneously activated. As described in Chapter 3.4, the intrafusal fibers report to the CNS on muscle length and its rate of change. The extrafusal fibers differ in size, relative contributions of various subcellular organelles, expression of protein and enzyme isoforms, metabolic capabilities, and contractile properties. Muscle fibers are broadly classified as slow, fast glycolytic or fast oxidative-glycolytic (see Chapter 3.7). Each extrafusal fiber is innervated by a single α motor neuron, but each α motor neuron typically innervates a number of muscle fibers. The motor neuron and the set of muscle fibers innervated by it constitute a **motor unit**. Activation of the α motor neuron activates all of the muscle fibers in its motor unit.

Motor neurons that innervate a large number of muscle fibers are larger than those that innervate a small number of muscle fibers. The smaller motor neurons are more easily excitable, probably because they have smaller distances to convey EPSPs to their axon hillocks. During the activation of muscle, the smaller motor units are activated first, and progressively larger motor units are activated until all motor units are **recruited**. This is the **size principle**. The size principle makes subjective sense because delicate movements that require dexterity but little force demand increasing force in the smallest possible increments by recruiting small numbers of muscle fibers. When performing gross motor movements involving lots of force, larger increments of force can be accomplished by recruiting successively larger motor units and more of them. Thus, the contractile force of a muscle can be **graded** (varied more or less continuously from low to high levels), through the recruitment of successively larger proportions of the population of motor neurons. This is the **population code** for encoding the strength of the muscle force. Activation of a larger population of motor neurons activates more muscle fibers and produces more force.

Muscles develop more force upon rapid repetitive stimulation. This wave summation is due to the release of additional activator Ca^{2+} from the sarcoplasmic reticulum before the force from the previous activation has returned to baseline. Thus, repetitive stimulation by the α motor neurons results in a larger force produced over a longer time. For typical movements, muscles must be activated by bursts of activity of their α motor neurons. This is the **frequency code** for varying muscle force.

CONTROL OF MOVEMENT ENTAILS CONTROL OF α MOTONEURON ACTIVITY

Muscular activity is controlled entirely by the activity of the α motor neurons that innervate the muscle. These

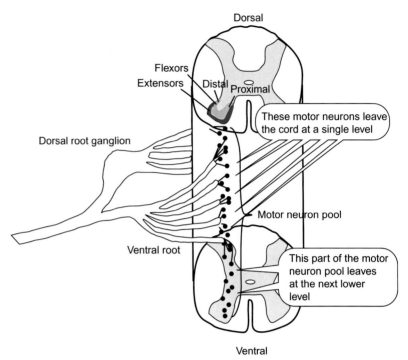

FIGURE 4.5.1 Spatial arrangement of α motor neurons in the spinal cord. The cell bodies of α motor neurons innervating flexor muscles are located in the dorsal part of the ventral horn; those innervating extensor muscles are found more ventrally. Similarly, the cell bodies of α motor neurons innervating the muscles of the distal limbs are located laterally, whereas those motor neurons for the proximal muscles are located medially. The motor neurons are not located merely in one spinal segment but are distributed up and down the cord. Groups of these form connections to the ventral root, which joins the dorsal root distal to the dorsal root ganglion. Motor neurons serving specific muscles are grouped together in adjacent spinal segments. These form a **motor neuron pool**. Most motor neurons for a particular muscle leave the cord at a single spinal segment.

cells are called the **lower motor neurons**, to distinguish them from the neurons in the cortex, the **upper motor neurons** that provide the initial drive for the α motor neurons. Some of the control of the α motor neurons resides in the spinal cord, some resides in the motor cortex, and some resides in other structures such as the basal ganglia and in the cerebellum. Most movements are ballistic: muscles are activated in a particular sequence, for short bursts of action potentials, through learned sequences to achieve the desired end. Such coordinated trains of activity are evident in the EMG of muscle, as shown in Figure 3.7.1.

THE CIRCUITRY OF THE SPINAL CORD PROVIDES THE FIRST LAYER IN A HIERARCHY OF MUSCLE CONTROL

The motor neurons in the spinal cord are topographically organized. The cell bodies or somas of the motor neurons lie in the ventral horn of the gray matter of the spinal cord. Their position there is not haphazard. Instead, there is a topographical organization to their arrangement within the cord (see Figure 4.5.1).

THE MOTOR NERVES ARE ORGANIZED BY MYOTOMES

There are thirty pairs of spinal nerves that carry sensory afferents and motor efferents to their targets: There are 8 cervical, 12 thoracic, 5 lumbar, and 5 sacral spinal nerves. Each spinal nerve supplies an embryologically defined **somite** or **body segment**, which is characterized by a **dermatome**, **myotome**, and **sclerotome**. The somatic sensory nerves report on sensory information obtained from a dermatome, the segmental unit of the skin. Each motor nerve innervates a designated area of muscle, called a myotome. These motor nerves exit the spinal cord via the ventral roots, which join the dorsal roots after a short distance, and travel to their targets alongside incoming sensory nerves. Despite this segmental organization, both sensory and motor organization is not absolutely limited to single spinal segments. Instead, motor and sensory innervation often overlaps one or two segments above and below a given spinal segment. Thus, the motor neurons innervating a single muscle are grouped together vertically in adjacent spinal segments, forming a **motor neuron pool**. The spatial arrangement of these is shown in Figure 4.5.1.

SPINAL REFLEXES FORM THE BASIS OF MOTOR CONTROL

The basic wiring of the spinal cord produces reciprocal actions of **synergistic** muscles and **antagonistic** muscles. Synergistic muscles are muscles whose coordinated activity produces a desired result. Flexion of the elbow, for example, requires fixation of the shoulder as well as contraction of the biceps brachii, brachialis, and radiobrachialis. The antagonists are those muscles that work against the desired movement. Typically antagonistic pairs of muscles are reciprocally activated, except when rigid fixation of a joint is desired. Thus, the spinal cord provides circuits which can activate synergists while inhibiting their antagonists, while allowing for coactivation of antagonistic pairs of muscles upon demand by higher centers.

The spinal cord also possesses **locomotor pattern generators**, which are neural circuits that produce or

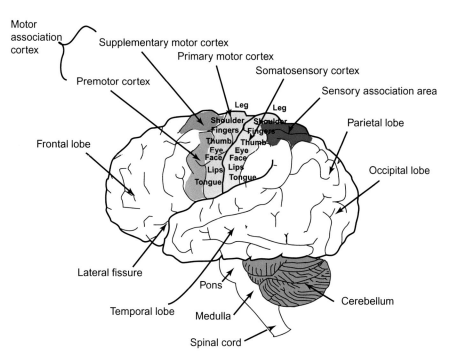

FIGURE 4.5.2 Areas of the cortex involved in motor control. The primary motor cortex has a somatotopic representation that mirrors the somatotopic representation in the somatosensory cortex. The other areas noted (sensory association area; premotor cortex; supplementary motor cortex) have roles in the selection of appropriate muscle programs.

control a particular set of muscles in a defined sequence. Usually the muscles activated by the pattern generator form rhythmic behaviors such as locomotion or breathing. Upon removal of higher inputs by transection of the spinal cord, many animals will exhibit locomotive behaviors upon stimulation of afferent somatosensory nerves. It is not clear how strong the locomotor pattern generators are in humans, but the basic circuitry allowing this behavior remains.

PURPOSEFUL MOVEMENTS ORIGINATE IN THE CEREBRAL CORTEX

The reflexes in the spinal cord form the basic motor plan by which complicated motor movements can be coordinated. However, these circuits are reflexes. As complicated as they might be, they cannot form purposeful movements. Instead, purposeful movements originate in higher brain centers, areas of the cerebral cortex.

THE PRIMARY MOTOR CORTEX HAS A SOMATOTOPIC ORGANIZATION

The **primary motor cortex** is a strip of cortical tissue in the frontal lobe immediately anterior to the central sulcus. It is called the "primary" motor cortex because it is required for the initiation of purposeful movements, and stimulation of areas of this cortex has the lowest threshold for eliciting a motor response. The motor responses that are elicited upon stimulation of the primary motor cortex are organized **somatotopically**: there is a spatial arrangement of motor responses in which adjacent muscles are controlled by adjacent regions of the primary cortex. This somatotopic mapping mirrors that of the somatosensory cortex. In fact, the somatosensory cortex is just on the other side of the central sulcus. These two areas are actually adjacent, being connected by the cortical tissue that follows the contours of the sulcus (the **paracentral lobule**).

The motor cortex is a layered structure, with cells of different overall shape and structure occupying different layers. Layer 5 of the cortex contains giant pyramidal or Betz cells. These cells send processes out over three separate tracts.

A. The **corticobulbar tract** supplies motor commands to cranial nerve motor nuclei of the head and neck
B. The **ventral corticospinal tract** innervates lower motor neurons of the axial and proximal muscles
C. The **lateral corticospinal tract** supplies inputs to lower motor neurons controlling the muscles of the distal extremities. As these muscles control fine movements, this tract is essential for delicate and precise actions of the fingers and hands.

These three tracts also contain fibers from other regions of the cerebrum, including the **motor association cortex**, the **somatosensory cortex**, and the **sensory association area**. These areas, with the somatotopic map of motor function, are illustrated in Figure 4.5.2.

MOTOR ACTIVITY ORIGINATES FROM SENSORY AREAS TOGETHER WITH PREMOTOR AREAS

Although the motor cortex provides an important role in driving the lower motor neurons, which motor neurons are activated and when they are activated is determined elsewhere. Selection of appropriate motor actions depends critically on knowledge of where the body is in space, where it should go, and how it should get there. Thus, sensory information from the somatosensory cortex and other sensory areas (such as the vestibular apparatus) must be coordinated. One of the major regions for the planning of motor programs is the **motor association cortex**, consisting of the **premotor cortex** and the **supplementary motor cortex**. The supplementary motor cortex is required for bilaterally coordinated movements. Lesions of this area cause **apraxia**, in which simple movements are retained but the affected individual is unable to perform coordinated movements involving both hands, such as tying a shoe or buttoning a shirt.

Sensory information ascends the spinal cord and is relayed in the thalamus to the somatosensory cortex. This information is sent to other areas of the cerebral cortex such as the **sensory association area** in the lateral aspect of the parietal lobe. This area integrates many different sensory modalities to form a kind of neural map of body, visual, and auditory space. Lesions within the sensory association area produce a variety of **neglect syndromes** in which the affected individual appears to lose awareness of body regions or portions of visual or auditory space that are served by the damaged area. Complex movement involving muscles shows specific deficits having to do with only one side of the body or involving near space or distant space, depending on the location of the lesion. These deficits result from incomplete formation of the neural map that forms the target for movement. The movement of the body in real space is planned by its representation in neural space. The sensory association area mingles sensory information to form a neural map of space with the position of the body superimposed on it. Planning of movements requires this mingling of sensory information.

MOTOR CONTROL IS HIERARCHICAL AND SERIAL

The individual connections of neurons controlling motor programs are enormous and cannot be diagrammed or understood in the particulars. Therefore we will approach the overall control systems in terms of block diagrams as shown in Figure 4.5.3. As this figure shows, cortical cells in the sensory association cortex and in the prefrontal region initiate movement by (1) focusing attention and (2) forming a working memory for a motor task. This *working memory* allows us to evaluate the consequences of future actions and to update these plans depending on new sensory information. Outputs from these areas feed into the supplementary motor cortex (see Figure 4.5.2) and the premotor cortex. Somatotopic mapping is present in

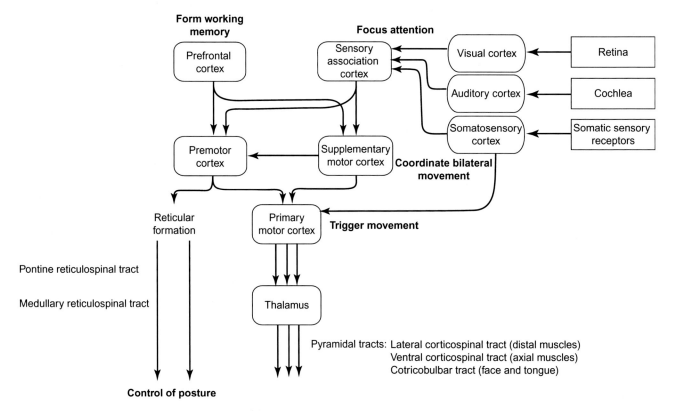

FIGURE 4.5.3 Block diagram for neuronal processing of motor programs. Although lower motor neurons are controlled by cells in the primary motor cortex, voluntary movement is initiated in the prefrontal cortex, which forms a "working memory" that anticipates possible future movements, and in the sensory association cortex that integrates sensory information from a variety of senses and focuses attention on the sensory information relevant to the motor task. This information is forwarded to the next step in the process, which is the formation of a coordinated motor plan involving the premotor cortex and the supplementary motor cortex. When the proper sequence of events is established, the primary motor cortex, containing the upper motor neurons, actually triggers the movement by exciting the proper lower motor neurons through tracts that descend the spinal cord.

the primary motor cortex and in the premotor cortex and the supplementary motor area. The supplementary motor cortex coordinates bilateral movement. Output from the premotor cortex goes to the primary motor cortex and to the reticular formation, whose output is used to orient the body for the motor task at hand. Cells in the primary motor cortex trigger the movement through the pyramidal tracts leading from the motor cortex to the lower motor neurons. Thus information involving motor programs is hierarchical: there is a definite sequence of events from accumulating different sensory information about future actions (where is the baseball?), to thinking about the action, and engaging the appropriate motor program. The goal of training is to shorten the times required for these processes and to make them as automatic as possible. The faster you can identify the curve or the slider, the faster you can adjust the swing of the bat, and the higher the likelihood that you get a piece of that pitch. If you have to think about it, you will strike out.

THE *BASAL GANGLIA* AND *CEREBELLUM* PLAY IMPORTANT ROLES IN MOVEMENT

The basal ganglia and cerebellum compose the major subcortical motor control systems and are critically important for the coordination of movement. The basal ganglia consist of five nuclei (**caudate nucleus, putamen, globus pallidus, subthalamic nucleus,** and **substantia nigra**). They receive widespread inputs from the entire cortex and project back to the cortex over the thalamus. They are important not only for the fine-tuning of motor systems but also have important effects on cognition, mood, and behavior. The **cerebellum**, on the other hand, receives cortical inputs only from the sensory and motor areas but projects to spinal motor areas that do not receive direct basal ganglia connections.

The motor cortex has direct connections to motor neurons in tracts collectively called the **pyramidal system**, named for the pyramid-like structure these tracts form as they pass through the medulla. The basal ganglia constitute part of the **extrapyramidal motor system**, which consists of the pathways from noncortical regions that lead ultimately to the lower motor neurons. The extrapyramidal system also includes output from the cerebellum.

Figure 4.5.4 shows the location of the basal ganglia between the motor cortex and targets in the brain stem and spinal cord. Bundles of fibers that connect sensory information to the cortex (spinocortical tracts and bulbar cortical tracts) and that connect the cortex to lower targets (corticofugal tracts—including the lateral corticospinal tract, the ventral corticospinal tract, and the corticobulbar tract) pass through this region. These bundles of myelinated fibers collectively form a broad band called the **internal capsule**. It penetrates the three basal ganglia (caudate, putamen, and globus pallidus) and the contrast between the myelinated fibers and the ganglia gives the area a striated or striped appearance. For this reason, the caudate, putamen, and globus pallidus are collectively called the **corpus striatum** (literally, *striped body*). The caudate and putamen have similar structures and embryologic origins and are collectively called the **neostriatum**; the globus pallidus consists of two regions, the **internal globus pallidus** and the **external globus pallidus**.

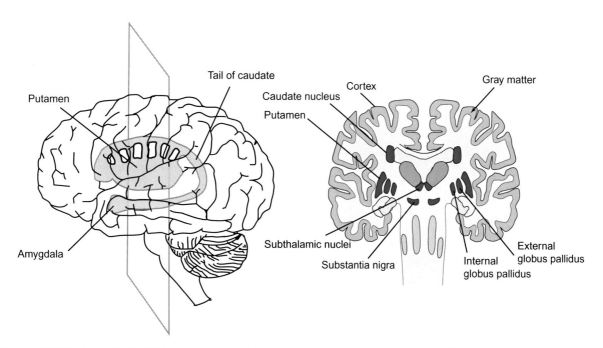

FIGURE 4.5.4 Anatomic arrangement of the basal ganglia. Left, the shape of the caudate and outer aspect of the putamen as seen from the left side. The structures lie internal to the brain and are not visible from its surface. The caudate nucleus follows the line of the lateral ventricles. It is long and thin and its shape resembles a ram's horn. The putamen and globus pallidus are shaped likes lenses and are collectively called the lenticular nucleus.

The basal ganglia scale motor programs for the task at hand and enable automatic performance of practiced motor acts. For example, you can write your name large on a blackboard or you can write your name small on a check. The motor program originates in the cortex but is conditioned by inputs from the basal ganglia and the cerebellum. Part of this conditioning is the scaling factor that determines the size of the movements and their strength. The fine tuning and updating of the motor program is performed by the cerebellum. Thus, the basal ganglia are a side-loop pathway that fine tunes motor instructions as they are being formulated in the premotor area.

The functional connections of the basal ganglia are shown in Figure 4.5.5. The putamen receives inputs from a variety of cortical areas including the somatosensory cortex, sensory association cortex, and prefrontal cortex. The caudate receives input primarily from the prefrontal cortex. Both the putamen and caudate nucleus send fibers to the internal globus pallidus in what is called the **direct pathway**. These connections use GABA as a neurotransmitter and they inhibit activity of the globus pallidus neurons. These neurons in the internal globus pallidus send inhibitory connections to the **ventroanterior** and **ventrolateral nuclei** of the thalamus. These neurons from the globus pallidus are tonically active and therefore they continuously inhibit the excitation of the supplementary motor cortex by the thalamus. Inhibition of the internal globus pallidus therefore removes inhibition of this excitatory pathway. This is referred to as **disinhibition**. Thus activation of the direct pathway is equivalent to exciting the supplementary motor cortex and providing additional neuronal drive for movement.

The **indirect pathway** inhibits the excitatory output of the thalamus to the supplementary association cortex. In this case the cortical input to the neostriatum is sent to the external globus pallidus and then to the subthalamic nucleus, finally sending processes to the internal globus pallidus. As shown in Figure 4.5.5, the connections from the striatum to the external globus pallidus are inhibitory; the connections from the external globus pallidus to the subthalamic nucleus are also inhibitory, whereas the connections from the subthalamic nucleus to the internal globus pallidus are excitatory. Thus activation of this indirect pathway causes activation of the internal globus pallidus, which inhibits the excitatory flow from thalamus to supplementary motor cortex. Activation of the indirect pathway, then, finally inhibits the supplementary motor cortex. **The indirect pathway is inhibitory whereas the direct path is excitatory**.

Inhibition or activation is simply kept track of by multiplying the signs. Activation is given a positive sign, $+$; inhibition is given a negative sign, $-$. Inhibition of an inhibitory pathway is $- \times - = +$; thus, inhibition of inhibition, or disinhibition, activates. Following the direct pathway we have $-$ for the connection between neostriatum and internal globus pallidus, $-$ for the connection of internal globus pallidus to thalamus, and $+$ from thalamus to supplementary motor cortex; the whole is $- \times - \times + = +$, or activation. Similarly, the

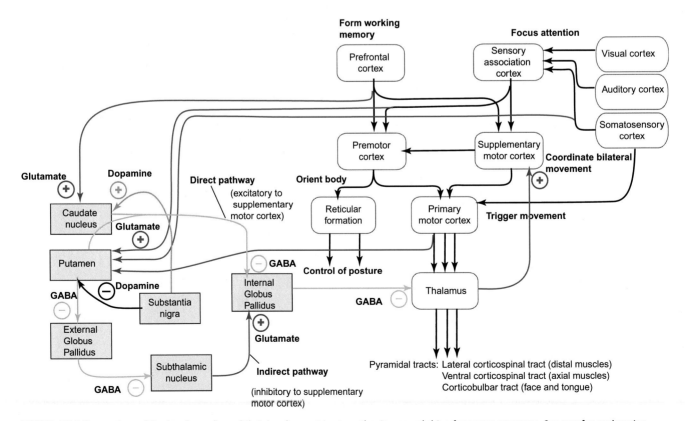

FIGURE 4.5.5 Connections of the basal ganglia and their involvement in strengthening neural drive for motor programs. See text for explanation.

indirect pathway involves − for the connection from neostriatum to external globus pallidus; − for the connection from external globus pallidus to the subthalamic nucleus; + for the connection between subthalamic nucleus and the internal globus pallidus; − for the connection from internal globus pallidus to thalamus; and + for the connection between thalamus and supplementary motor cortex. The final effect is $-\times-\times+\times-\times+=-$. Thus the indirect pathway inhibits excitation of the supplementary motor cortex.

THE *SUBSTANTIA NIGRA* SETS THE BALANCE BETWEEN THE DIRECT AND INDIRECT PATHWAYS

Dopaminergic connections from the **substantia nigra** form the nigrostriatal pathway that connects the substantia nigra to the caudate and putamen. These connections excite the neurons running over the direct, stimulatory pathway, and they inhibit those neurons running over the indirect, inhibitory pathway. Activity of the substantia nigra therefore alters the balance between excitatory (direct pathway) and inhibitory (indirect pathway) motor drive. Activation and inhibition by a single neurotransmitter is possible because of the receptors on the post-synaptic cell (see Chapter 4.2).

THE *CEREBELLUM* MAINTAINS MOVEMENT ACCURACY

Persons whose cerebellum is damaged exhibit abnormal movements marked by a loss of motor coordination, and lack of smooth sequences of motor programs. The lack of coordination is called **cerebellar ataxia**. Part of this coordination includes the integration of the sense of balance, which originates from receptors in the **vestibular apparatus**. The clumsiness, loss of equilibrium and inability to target motor tasks in inebriated persons derives largely from depression of cerebellar function. The cerebellum also is critically important in learning motor skills. The point of practicing motor skills, from athletics to playing a musical instrument or typing, is to make the sequence of motions automatic. During these activities we voluntarily engage an entire muscle program without having to think about each action. If we had to think about each part of the complicated program separately, we could not perform it rapidly or smoothly. The block diagram of some of the cerebellar connections is shown in Figure 4.5.6.

When we engage a motor program, commands pass from the primary motor cortex down the pyramidal tracts to the lower motor neurons in the spinal cord. At the same time, a duplicate of these commands is sent to the cerebellum through relay nuclei in the pons. These nuclei send projections called **mossy fibers** to the cerebellar cortex where they synapse on **granule cells** (see Figures 4.5.7 and 4.5.8). Excitatory input to these granule cells is directed to **parallel fibers** that make contact with **Purkinje cells**, which form the major output of the cerebellar cortex. Purkinje cells receive as many as 200,000 synapses from parallel fibers. The small EPSPs generated by these connections summate temporally and spatially to produce Purkinje cell action potentials, but because of their large number of contacts the granule cells drive Purkinje cells with a tonic level of 50 to 100 action potentials per second. The excitatory drive of the granule cells is modulated by interneurons, **stellate cells** and **basket cells**, which are excited by parallel fibers but inhibit the Purkinje cells. **Golgi cells** are other interneurons in the cerebellar cortex that receive input from the parallel fibers and feed back to inhibit the granule cells.

Sensory information from the spinal cord, brain stem, and cerebral cortex passes through the **inferior olivary complex** in the medulla which sends processes that synapse directly on the primary output cells of the cerebellar cortex, the **Purkinje cells**. Because of their vine-like appearance, these processes are called **climbing fibers**. The climbing fibers form strong connections that make large EPSPs on the Purkinje cells.

The mossy fibers from the pontine nuclei bring in a duplicate copy of the motor program. At the same time, sensory information about the movement that was actually made enters the cerebellum through the climbing fibers. Both are processed in the cerebellar cortex. This mingling of the motor programs and sensory information about what the body actually did allows us to make adjustments to the motor program so that when we activate it, the body will do what we want.

The Purkinje cells form the major output of the cerebellar cortex. Fibers connect to the **dentate nucleus**, which relays output to the red nucleus and the thalamus. The red nucleus in turn projects to spinal motor neurons through the rubrospinal tract, and the thalamus projects back to the premotor and primary motor cortex. Purkinje cells also send projections to two deep cerebellar nuclei, the **fastigial nucleus** and the **interposed nucleus**, which send information to the spinal cord through three paths:

A. through the reticular formation and the reticulospinal tract;
B. indirectly through the red nucleus and the rubrospinal tract;
C. directly to spinal motor neurons.

THE SENSE OF BALANCE ORIGINATES IN *HAIR CELLS* IN THE *VESTIBULAR APPARATUS*

Each inner ear contains a **vestibular apparatus**, also called the **labyrinth**, that consists of three mutually orthogonal semicircular canals. The canals are encased in the **bony labyrinth** of the temporal bone. The three semicircular canals are the **posterior, anterior**, and **horizontal canals**. The canals are filled with **endolymph**, a special fluid that is extracellular but nevertheless is high in K^+. They are surrounded by **perilymph**, a fluid similar in composition to the cerebrospinal fluid, or CSF. The posterior and anterior canals are oriented

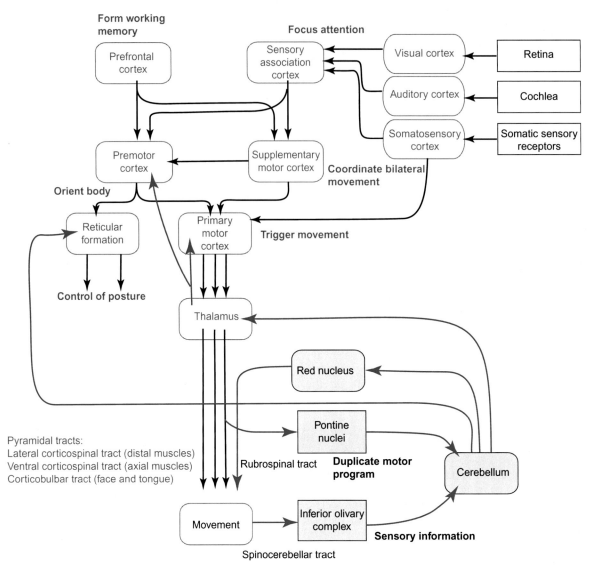

FIGURE 4.5.6 Block diagram of connections between the cerebellum and other centers involved with the coordination of muscle movement.

vertically when the head is held upright. In addition, each vestibular apparatus contains two chambers, the **utricle** and **saccule**, that contain a field of **hair cells** that provide information about the position of the head relative to gravity. The utricle is horizontal whereas the saccule is oriented vertically (see Figure 4.5.9). The three canals are continuous with each other and with the utricle and saccule, as well as the **cochlea**, that part of the inner ear that is devoted to hearing (see Chapter 4.6).

Each semicircular canal forms an enlarged region, the **ampulla**, near its junction with the utricle. Each ampulla contains a sensory epithelium that consists of a ridge of columnar hair cells. The hair cells are embedded in a gelatinous mass called the **cupola**. The cupola extends to the opposite wall of the ampulla. The entire structure is called the **ampullary crest**. The "hairs" on these cells are not true hairs, nor are they true **cilia** because they do not contain the 9 + 2 arrangement of microtubules that characterize true cilia, nor do they possess propulsive power. Because of their cilia-like appearance, the "hairs" of these mechanosensitive cells are called **stereocilia**. The hair cells have some 50 to 100 sterocilia and one **kinocilium**. The kinocilium is the longest "hair" and it defines a polarity of the hair cell. The stereocilia become progressively shorter as their distance from the kinocilium increases. The axis of the hair cell is defined by the line of sterocilia that point toward the kinocilium.

Electron micrographs reveal thin connections, called **tip links**, between the tips of neighboring stereocilia. These are believed to connect to K^+ channels on the surfaces of the stereocilia. Bending of the stereocilia modulates the conductance to K^+ ions and alters the membrane potential. The surface of the hair cells faces the endolymph which contains higher $[K^+]$ than the intracellular fluid, so opening a K^+ channel causes influx of K^+, rather than the efflux that would occur with most cells that face the low $[K^+]$ of the extracellular fluid. Movement of the kinocilium away from the stereocilia

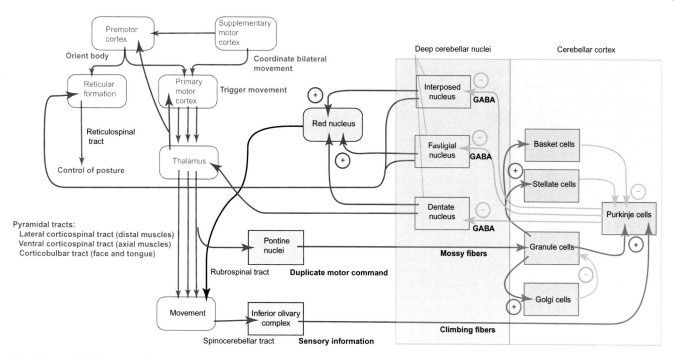

FIGURE 4.5.7 Block diagram of connections within the cortex and deep nuclei of the cerebellum. Duplicate motor commands enter the cerebellum on mossy fibers from the pontine nuclei. These synapse on granule cells that form parallel fibers that make excitatory synapses on Purkinje cells. The parallel fibers also synapse on stellate cells and basket cells that inhibit Purkinje cells. This inhibition sharpens the excitatory drive by lateral inhibition. Granule cells also synapse on Golgi cells that feed back to inhibit the granule cells. Purkinje cells project to the dentate, fastigial, and interposed nuclei. The dentate nucleus projects to the thalamus and to the red nucleus. The fastigial and interposed nuclei project to the reticular formation and to the red nucleus. The red nucleus sends activating processes down the cord on the rubrospinal tract. The reticular formation sends fibers to control muscles of posture on the reticulospinal tract. The thalamus sends processes to the premotor cortex and the primary motor cortex to adjust for inaccuracies in the desired motor act. Sensory information enters the cerebellum from the inferior olivary complex over climbing fibers that have strong connections to the Purkinje cells. In this way, motor information over the mossy fibers can be compared to sensory information over the climbing fibers, and motor tasks can be refined and conditioned.

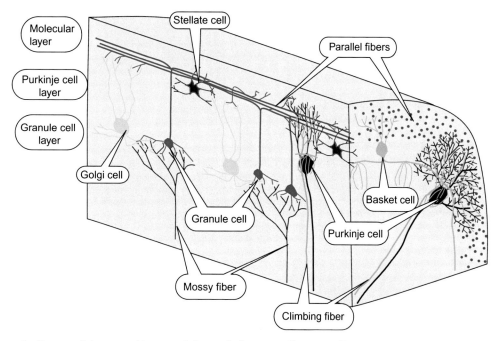

FIGURE 4.5.8 Schematic diagram of the cytoarchitecture of the cerebellar cortex. The mossy fibers synapse on granule cells whose axons divide to form two parallel fibers that synapse on Golgi cells that feed back onto granule cells, inhibiting them. They also make many synapses on Purkinje cells, whose dendritic fields form a fan whose long axis is perpendicular to the parallel fibers. Basket cells are excited by parallel fibers and inhibit Purkinje cells. Their processes are perpendicular to the parallel fibers. Stellate cells provide lateral inhibition of Purkinje cells. Climbing fibers bring in sensory information and form strong excitatory connections with just a few Purkinje cells. (Source: Modified from L.R. Johnson (ed.) "Essential Medical Physiology", Lipincott-Raven, Philadelphia (1998).)

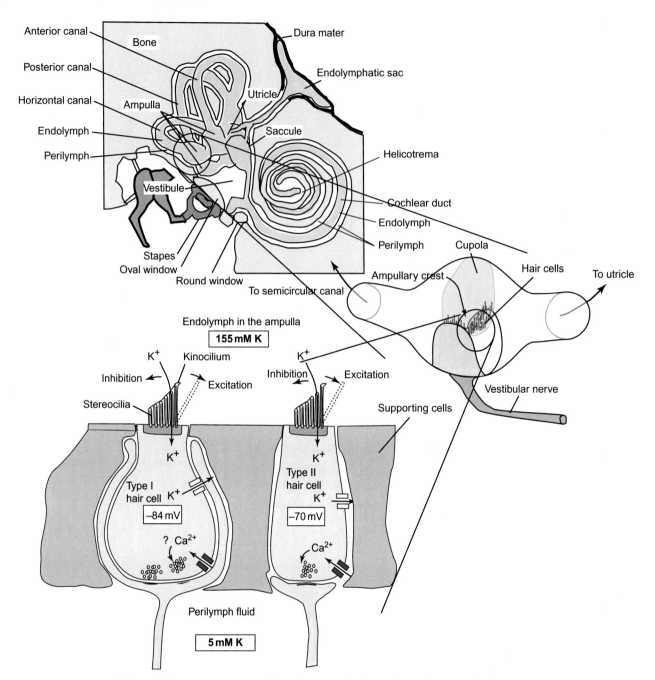

FIGURE 4.5.9 Structure of the vestibular apparatus. Hair cells in three semicircular canals provide dynamic sensation of movement of the head. The hair cells reside in an enlarged part of the canal called the ampulla. Three mutually perpendicular canals provide information that allows computation of the final vector of movement. The semicircular canals form part of the labyrinth that is surrounded by temporal bone (osseous labyrinth). The fluid within the labyrinth is endolymph, a special fluid that contains higher [K^+] than the intracellular fluid, so that opening of K^+ channels on the surface membrane produces an inward K^+ flux and depolarization of the cell. The hair cells are located on a ridge called the ampullary crest within the ampulla. A gelatinous mass called the cupola extends from the base of the hair cells to the opposite wall of the ampulla. Movement of the endolymph relative to the hair cells results in deformation of the "hairs" and changes in the membrane potential that results in neurotransmission to afferent sensory nerves that closely appose the hair cells' membrane. Static response to gravity originates from hair cells in the utricle and saccule.

produces excitation (it stretches the tip link) and movement toward the stereocilia inhibits the cell. There are two types of hair cells. Type I has a resting E_m of −84 mV and type II has a resting E_m of −70 mV. In type I cells the afferent sensory nerve nearly surrounds the hair cell and leaves a tiny microdomain between the two cells. The response of the afferent nerve cells may be related to ionic changes in this microdomain. Type I cells respond to stereocilia deformation with large receptor potentials, whereas the response of type II cells is smaller and shows **rectification**: the response is much greater to deformation toward the kinocilium than it is to deformation away from it. In type II cells, depolarization increases the conductance to Ca^{2+} along its basal-lateral membrane. The local [Ca^{2+}] in this region causes vesicles containing neurotransmitter to fuse with the basal-lateral membrane, which in turn activates the sensory afferent neuron that innervates the hair cells (Figure 4.5.9).

ROTATION OF THE HEAD GIVES OPPOSITE SIGNALS FROM THE TWO VESTIBULAR APPARATUSES

Because the semicircular canals and the hair cells are rigidly attached to the head, they rotate with the head and initially sweep through the stationary endolymph. The endolymph does not immediately move, because of its inertia and lack of rigid attachment to the skeleton. Thus, rotation to the right entails a relative motion of the endolymph to the left. The labyrinth in the right ear is a mirror image of the one in the left, and the axes of the hair cells are also mirror images. The direction of endolymph movement will be to the left in both ears. Thus, the relative motion of the endolymph will be in the direction of the hair cell axis in one ear, and opposite in the other. Movement of the endolymph in the direction of the axis (toward the kinocilium) excites the cell, whereas movement opposite to the axis (away from the kinocilium) inhibits the cell. The hair cells in one ear respond with increased firing frequency upon rotation of the head, while the hair cells in the other ear respond with decreased firing frequency. This situation is shown schematically in Figure 4.5.10. The firing of hair cells in the three orthogonal semicircular canals informs the central nervous system about rotational acceleration in three dimensions.

THE UTRICLES AND SACCULES CONTAIN HAIR CELLS THAT RESPOND TO STATIC FORCES OF GRAVITY

The bilateral utricles and saccules contain specialized epithelia that is analogous to the ampullary crest. Each of these regions is called a **macula**. The macula in the utricle and saccule contains an array of hair cells whose stereocilia project into the **otolithic membrane**, a gelatinous mass that contains tiny crystals of calcium carbonate, called **otoliths** (literally, "ear stones"). When the head is held erect, the macula in the utricles is oriented in the horizontal plane whereas the macula in the saccule is oriented vertically. Gravity pulls on the dense otoliths, which deform the gelatinous mass and subsequently press on the stereocilia and influence the firing rate of the hair cells. Since gravity pulls constantly, the hair cells in the utricle and saccule provide tonic information about the orientation of the head.

THE AFFERENT SENSORY NEURONS FROM THE VESTIBULAR APPARATUS PROJECT TO THE VESTIBULAR NUCLEI IN THE MEDULLA

Sensory neurons from the vestibular apparatus joins the auditory sensory afferents to form **cranial nerve VIII**,

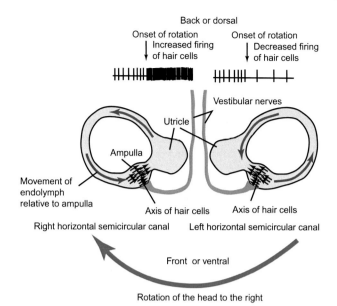

FIGURE 4.5.10 Response of hair cells in the horizontal semicircular canal to rotation of the erect head. Cross-sections of the horizontal semicircular canal are shown, as viewed from above. The nose faces the bottom of the page. Rotation of the head to the right, indicated by the arrow at the bottom of the figure, sweeps the ampullary crest through endolymph which resists the movement because of the fluid's inertia. Thus there is a relative movement of the endolymph to the left across the ampullary crest in both right and left inner ear. The hair cells, however, have mirror image orientations in the two inner ears, so that in the right ear the endolymph deforms the stereocilia toward the kinocilium whereas in the left ear (to the right in the figure) the deformation of the stereocilia is away from the kinocilium. Since movement toward the kinocilium is excitatory, the right ampulla increases its firing rate while the left ampulla decreases its firing rate.

the vestibular and auditory nerve. This nerve enters the central nervous system near the junction of the pons and the medulla. The vestibular sensory afferents synapse on neurons found in the vestibular nuclei in the lateral medulla. Four recognized areas make up the vestibular nuclei: the lateral vestibular nuclei have an inferior and a superior nucleus, and the medial vestibular nuclei also are divided into superior and inferior portions. Information in these nuclei travel down the cord in the **lateral vestibulospinal tract** to provide excitation for maintaining body posture. The medial and superior vestibular nuclei receive inputs primarily from the semicircular canals. They send processes onto three pairs of nuclei that control eye movement. The motor outputs of these three pairs of nuclei travel in cranial nerves **III (oculomotor)**, **IV (trochlear)**, and **VI (abducens)**. This pathway enables the reflex movement of the eyes to the left when the head rotates to the right. Because of this reflex, rapid movement of the head back and forth does not produce as blurred an image of the hand compared to rapid movement of the hand back and forth.

SUMMARY

The lower motor neurons provide the final common pathway by which muscles are activated. The sequence

Clinical Applications: Parkinsonism

In 1817 James Parkinson, an English physician, wrote a book entitled "An Essay on the Shaking Palsy" in which he described the disorder that is now known as **Parkinson's disease**. People with Parkinsonism exhibit rhythmic tremor at rest with lessened muscular power, rigidity, difficulty in initiating movements (**akinesia**) and slowness in executing movements (**bradykinesia**). More complex motor programs show the largest deficits. Patients with this disease cannot execute simultaneous or sequential motor tasks, indicating that the deficit lies in the activation of motor plans for complex motor tasks. Patients also cannot execute normal rapid movements. When thrown off balance, such a patient falls before righting reflexes can be activated.

Parkinson's disease is caused by damage to the projections of the substantia nigra to the neostriatum. These projections use dopamine as a neurotransmitter. The loss of these projections removes activation of the direct pathway that stimulates the supplementary motor cortex, and removes inhibition of the indirect pathway that inhibits the supplementary motor cortex through its action on the thalamus. This lack of dopaminergic stimulation removes the excitatory boost that is necessary for a critical level of activity in the supplementary motor cortex, and the result is tentative movement and reduced muscle recruitment.

Oral administration of L-DOPA (L-dihydroxy phenyl alanine, a precursor for dopamine synthesis; see Chapter 4.2) increases dopamine synthesis in the remaining substantia nigra cells. This temporarily relieves symptoms, but it does not postpone or mitigate further loss of dopaminergic cells in the nigrostriatal pathway. Parkinson's disease affects about 1% of the population over age 50.

---

of activation of particular lower motor neurons, their frequency of firing and duration of the train of action potentials determines the movement of the torso, limb or digits. Purposeful movement requires coordinated activation and inhibition of synergistic muscles and antagonistic muscles. Activation of the population of motor neurons that make up the pool supplying any given muscle determines the strength of contraction. Much of the basic pattern of these movements is provided by spinal reflexes, but purposeful movement itself originates in the cortex and is coordinated and modulated by the basal ganglia and cerebellum.

The primary motor cortex is a strip of cortex lying rostral to the central sulcus. Stimulation of this region has the lowest threshold for eliciting a motor response. Areas of the musculature map onto this primary motor cortex so that the leg is controlled by areas lying beside the longitudinal fissure, the arm is controlled by areas lateral to the fissure, and the facial muscles are controlled by areas further lateral. This somatotopic mapping mirrors the somatotopic sensory projection of the body. The somatosensory map and somatomotor map are connected through cortex lining the central sulcus. Although neurons in this primary motor cortex drive lower motor neurons most strongly, the primary motor cortex does not initiate motor commands. Rather, they execute them. Motor commands begin by comparing one's position in the world to the position one wants to be. The sensory association cortex receives sensory input from a variety of sources and integrates the information to form a neural map of the real world and the body's position in it. Motor plans originate in the prefrontal cortex, which forms its plans based on the output of the sensory association area. The prefrontal cortex forwards a "working memory" of the movement to the premotor cortex and supplementary motor cortex which prepares the body for movement by positioning the axis of the body and engaging the primary motor cortex.

The basal ganglia consists of a set of structures, the caudate nucleus, putamen, external and internal globus pallidi, the subthalamic nucleus, and the substantia nigra. These modulate movement by providing an excitatory boost to desired movement and inhibiting undesired movements. The direct pathway excites the supplementary motor cortex and the indirect pathway inhibits it. The substantia nigra sets the balance between the indirect and direct pathways.

The cerebellum receives a duplicate of the motor commands over the mossy fibers arising from pontine nuclei. Simultaneously, sensory fibers from the inferior olivary nucleus reach the cerebellum over climbing fibers. This mingling of motor commands and sensory information allows the cerebellum to condition movement to match its intended target.

Hair cells in special structures of the inner ear detect the static pull of gravity and the acceleration of the head. These hair cells have stereocilia that respond to movement. Hair cells in the utricle and saccule detect static gravity while hair cells in the ampulla of the three semicircular canals detect acceleration. These cells are bathed in endolymph, which is distinctive because of its high $[K^+]$. Opening of K^+ channels on the apical membrane of the hair cells depolarizes the cells, causing them to release neurotransmitters to afferent sensory neurons adjacent to the hair cells. This occurs when the endolymph moves relative to the hair cells during acceleration. This sensory information is carried over cranial nerve VIII to the vestibular nuclei in the medulla, which sends information down the cord to maintain posture and to centers that control eye movement.

REVIEW QUESTIONS

1. What neurons directly control muscle contraction? Where are the cell bodies of these neurons?
2. What is a motor unit? How is muscle contraction graded?
3. What is the primary motor cortex? Where is the sensory association area? Where is the premotor cortex and the supplementary motor cortex? What is the somatotopic mapping of motor control?
4. What role does the spinal cord play in control of movement?
5. What happens when there is damage to the sensory association area?
6. Name the basal ganglia. What is the major input to the caudate nucleus? To the putamen?
7. What is the direct pathway? Does it promote or inhibit wanted effort? How does it do this?
8. What is the indirect pathway? Does it promote or inhibit wanted effort?
9. What role does the substantia nigra play in the control of movement by the basal ganglia? What disease is association with dysfunction of the substantia nigra?
10. What information is carried by the mossy fibers? The climbing fibers? What happens when cerebellar function is compromised?
11. What is the endolymph? What is the perilymph? How do hair cells detect acceleration? Where is the information processed in the CNS?

4.1 Problem Set
Nerve Conduction

1. Renshaw (*J. Neurophys.* 3:373–387, 1940) provided an ingenious method for estimating the time of synaptic delay in the spinal cord. He stimulated the intermediate gray of the spinal cord with successively larger stimuli and recorded the output of the ventral root. A schematic of his experimental set-up and results are shown in Figure 4.PS1.1. *Recordings were performed using extracellular electrodes.* A postulated schematic of the neuronal "wiring diagram" is provided to help you think about this. With successively larger stimuli, more and more of the cord interneurons, and eventually the motoneurons, were excited, as indicated by the dashed lines in the neuronal wiring diagram.

    A. Explain the origins of peaks S, T, and M in the record.
    B. Why does S decline as M increase?
    C. Estimate the synaptic delay from the traces.

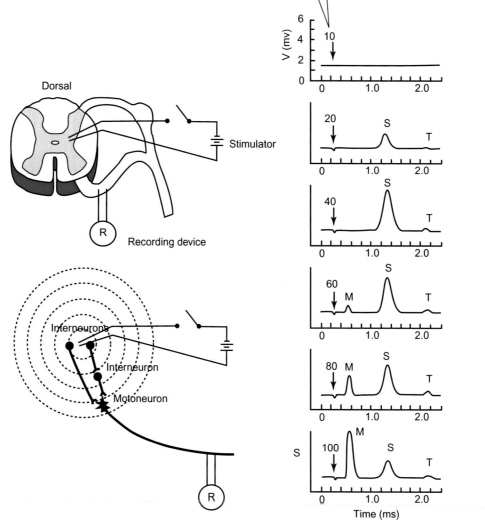

FIGURE 4.PS1.1

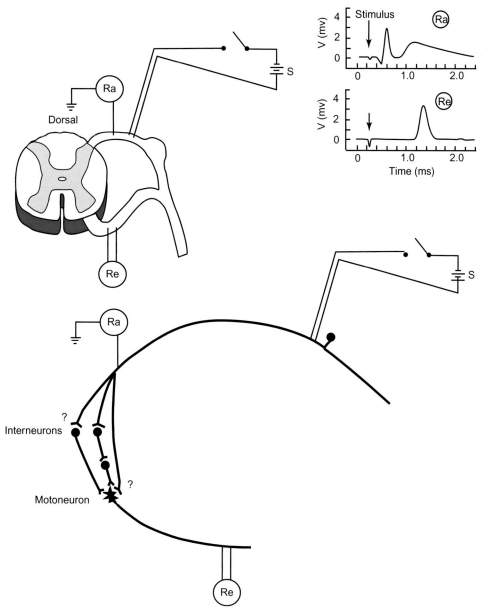

FIGURE 4.PS1.2

2. Renshaw also set up the experiment shown in Figure 4.PS1.2, in which the dorsal root was stimulated weakly (which stimulates the Ia afferents) and the resulting electrical activity was recorded in the dorsal root entry zone and in the ventral root. The results are shown at the top right of the figure, where Ra represents the afferent root (sensory) and Re indicates the efferent root (motor). Explain how this result, coupled with that of problem 1, allowed Renshaw to conclude that the Ia afferent reflex was monosynaptic.

3. The **Hoffman reflex** refers to the reflex contraction of a muscle caused by stimulation of the nerve. The experimental set-up is shown in Figure 4.PS1.3. The muscle contraction is monitored by electromyography (EMG). An example of the EMG response is given. The Ia fibers in the nerve have a lower threshold for activation than do the motor axons, and so the EMG recording depends on the magnitude of the stimulus. At low stimulus strengths a pure H-reflex is recorded, with no M-wave. As the strength of stimulus increases, motor axons supplying the muscle are excited and two distinct responses are recorded, the H-reflex and the M-wave. As the stimulus strength increases still further, the H-reflex progressively is extinguished. From the "wiring diagram", deduce why the H-wave disappears at high stimulus strength. Hint: stimulation of a nerve in its middle conducts **orthodromically** (toward its normal terminal) *and* **antidromically**—in the opposite direction toward the cell body.

4. The International Olympic Committee has approached you for help in making sprint races fair. In these races, sprinters are started by three commands: "On your marks," "Get set," and

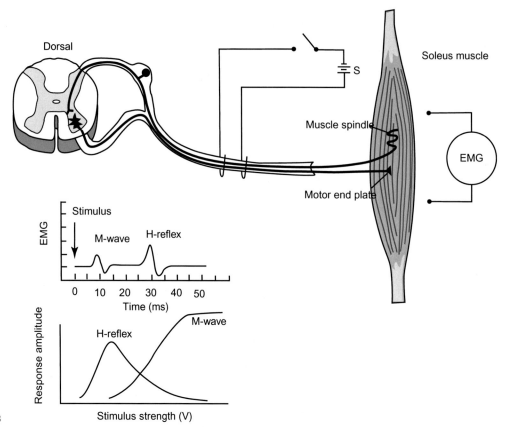

FIGURE 4.PS1.3

"Blam!" A blank pistol report starts the runners. Winners of these races are often determined by a scant hundredth of a second, or 10 ms, over the course of 100 m. The slightest advantage by a sprinter can mean victory. If a sprinter anticipates the gun, just slightly, he can win the race. In an unfair race, when a runner starts before the gun, the race is stopped by a second blank pistol report. The field is allowed one false start, and the next false start results in the disqualification of the runner who first begins. The IOC wants an automatic determination of when a false start occurs, and their question to you is: what is the minimum reaction time for a sprinter to come off the blocks *after* hearing the pistol report? How would you determine this reaction time? Knowing what you do about the pathways involved, synaptic delays, and propagation velocities, estimate the minimum reaction time.

5. Estimate the time of diffusion for a neurotransmitter across a 100-nm synaptic gap using the equations developed in Chapter 1.6.
6. A typical standard for military hikes is about 4 miles per hour with 30 in. strides. This is *not* double-time, but it is a rigorous pace. It may not seem like much, but it is harder than it sounds.
   A. Estimate the frequency of foot strikes.
   B. The definition of walking as opposed to running is that there is always contact with the ground. Estimate the duty cycle (the fraction of time that is active) for the gastrocnemius muscle (the calf muscle) during the hike.
   C. Based on your answer to 6A and 6B, what do you suppose the action potential train to the gastrocnemius muscle looks like?
7. Suppose that a vesicle 100 nm in external diameter, with 10 nm wall thickness, contains 10,000 molecules of neurotransmitter. It dumps its neurotransmitter into a cleft 50 nm wide and 500 nm in diameter.
   A. What is the concentration, in molar, of neurotransmitter inside the synaptic vesicle?
   B. What is the concentration, in molar, of neurotransmitter in the synaptic cleft assuming that it becomes evenly distributed and there is no degradation?
8. When you roll down a long hill, and then try to stand immediately upon reaching the bottom, you discover that you are dizzy. Why?
9. You are seated on a rotating chair, and your pals quickly rotate you to your right for about 20 s, and then stop you. They get a chuckle out of watching your eyes move slowly to one side and then being rapidly reset, a condition called **nystagmus** (from the Greek "nod" because it resembles the nod when one falls asleep, the slow phase as the head drops and a quick phase as it is snapped back to an erect position). The slow phase is part of the vestibulo-ocular reflex and brain stem circuits generate the quick phase. What causes this slow phase, and in what direction does it occur?
10. A diagram of the dermatomes is shown in Figure 4.PS1.4. These are areas of the skin that

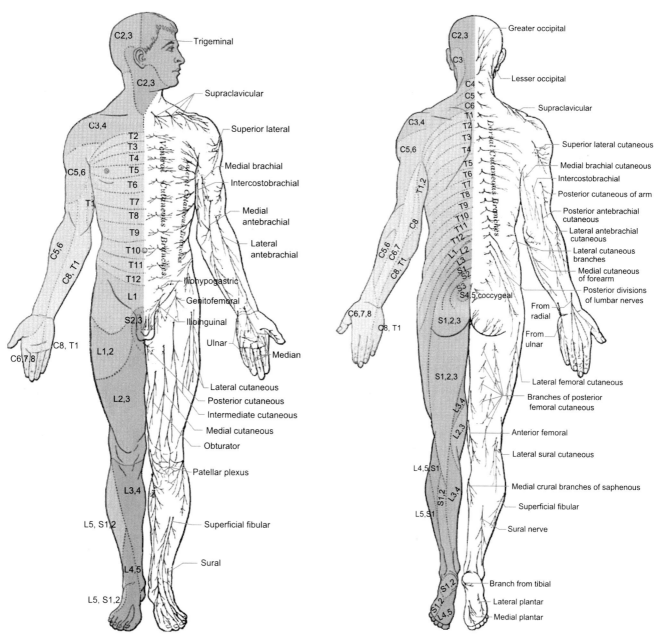

FIGURE 4.PS1.4 Dermatomes.

are innervated by the indicated spinal nerves. The myotomes are muscles that generally lie under the corresponding dermatomes.
  A. Your friend took a nasty spill playing intramural soccer and now complains of loss of sensation in his left shoulder. What might cause this?
  B. Another friend noticed the other day that his left foot sort of flaps along when he walks and he has trouble lifting the foot and cannot walk on his heels, but only on his left foot. What might cause this problem?
11. The Weber–Fechner Law uses what are called "just noticeable differences", or jnd. Persons were asked to determine the jnd for various sensory modalities. The just noticeable difference was the one that was identified 50% of the time. In this sense, it describes a threshold for sensory perception. The Stevens power law, on the other hand, asked people to rate the intensity of a their perception relative to some standard. The standard stimulus was assigned a number, the modulus, and all perceptions were given a number in proportion to that modulus. Thus, if a subject reported that the stimulus was twice the standard, its assigned number was twice the modulus.
  A. For weight lifting, Stevens reports an exponent of 1.45 in his power law. Given an initial weight of 1 kg, at what mass would the perceived stimulus double?

B. At what weight would it double again (be four times as great as the 1 kg mass)?
   C. What additional information is needed to compare the Weber–Fechner law to the Stevens power law?
12. Nerve conduction studies are performed to distinguish among several causes of sensory or motor deficits. These can take multiple forms. Motor nerve conduction studies stimulate motor nerves and record from the muscle innervated by the nerve. Sensory nerve conduction studies stimulate purely sensory portions of a nerve and record from proximal sections of the nerve.

    In the F-wave study, supramaximal stimulation of a nerve sends action potentials antidromically towards the ventral horn where a small population of motor neurons backfire towards the muscle, and the second small peak is measured at the site of stimulation. The distance is measured from the site of stimulation to the spinal cord and doubled because the nerve is conducted both ways, and this is divided by the latency, with 1 ms subtracted.
    A. The measured distance between the site of stimulation on the arm and the spinal cord was 85 cm, and the latency in the F-wave study was 30 ms. Calculate the nerve conduction velocity. Is this normal?
    B. The measured distance between the site of stimulation on the calf and the spinal cord was 145 cm, and the latency in the F-wave study was 50 ms. Calculate the nerve conduction velocity. Is this normal?
13. Botulinum toxin is produced by *Clostridium botulinum*, a gram-negative bacterium. The toxin actually consists of seven related toxins, types A, B, C1, D, E, F, and G. The toxin is synthesized as a protoxin of 150 kDa, which is proteolytically cleaved to a light (L) and a heavy (H) chain that remain linked by a disulfide bond. Nerve terminals have receptors for both L and H chains. The L-chain is transported across the nerve terminal membrane by endocytosis. These L chains are metalloproteinases—proteases that require metal ions for activity. The L-chain binds Zn^{2+}, and it proteolytically degrades several proteins that make up the SNARE complex.

    What do you suppose happens when botulinum toxin is injected near neuromuscular junctions?

    From this conclusion, what do you suppose is the basis for botox injections?
14. The two-point discrimination test generally gives a discrimination of 2–4 mm on the lips and finger pads, 8–15 mm on the palms, and 30–40 mm on the shins or on the back. What are the approximate relative densities of cutaneous touch receptors in these regions?
15. A muscle strain is a tear in the muscle fiber or associated tendons. The tear or damage to ligaments is called a sprain. It is usually caused by stretching of the muscle while it is contracted through the antagonistic muscles.
    A. Would you expect the Golgi tendon organ to prevent muscle strains?
    B. Given that the distance from the hamstring neuromuscular junction to the spinal cord is 50 cm, what is the expected delay in response of the Golgi tendon organ?

The Chemical Senses 4.6

Learning Objectives

- Indicate the location of the olfactory epithelium
- List four classes of cells in the olfactory epithelium and briefly describe their function
- Describe the mechanism of olfactory sensory transduction (G_s mechanism linked to increased Cl^- conductance)
- Explain how increased Cl^- conductance leads to depolarization in olfactory sensory cells
- Describe the postulated function of the vomeronasal organ
- Explain how some "smells" are actually irritants of the trigeminal nerve
- List three types of papillae on the tongue
- Indicate the parts of the tongue and pharynx innervated by the glossopharyngeal nerve (CN IX), facial nerve (CN VII), and vagus nerve (CN X)
- List the five major taste sensations
- Describe how the "hot" taste of peppers is distinguished from the other taste qualities

THE CHEMICAL SENSES INCLUDE TASTE AND SMELL

The adequate stimuli for the chemical senses are environmental chemicals originating outside the body. The chemical senses include taste and smell. Sensing of these environmental chemicals conveys information vital to the survival of the individual, allowing detection of prey or predators in the case of smell or nourishing or poisonous foods in the case of taste. Smell often carries social and sexual signals as well.

TASTE AND OLFACTORY RECEPTORS TURN OVER REGULARLY

Taste and olfactory receptors line epithelia that are regularly exposed to potentially noxious materials. Because of this exposure, most epithelial cells are regularly sloughed off and replaced with new cells. The same is true for the taste and olfactory receptors. Olfactory receptors turn over every 4–8 weeks. Since the taste receptors are modified epithelial cells, this ability is not surprising. Olfactory cells, on the other hand, are true neurons whose cell bodies are located in the olfactory mucosa and which project axons directly to the olfactory bulb in the brain. Nevertheless, these olfactory neurons are continually replaced throughout life. Basal cells in the olfactory epithelium undergo mitosis to produce new olfactory receptor cells that must grow new axons into the olfactory bulb and make new connections there.

THE OLFACTORY EPITHELIUM RESIDES IN THE ROOF OF THE NASAL CAVITIES

The olfactory epithelium consists of two patches, each with areas of about 5 cm^2, located in the roof of the nasal cavities. The epithelium's surface is defined by a thin perforated bony plate, the cribriform plate, that separates the nasal cavities from the brain. The plate extends horizontally in a plane just below the eyes. A secondary area of olfaction, the vomeronasal organ, lines the turbinates in the nasal cavity. It appears to be sensitive to pheromones and may be involved in sexual function. Figure 4.6.1 shows the location of the olfactory epithelium.

OLFACTORY RECEPTOR CELLS SEND AXONS THROUGH THE CRIBRIFORM PLATE

Figure 4.6.2 illustrates the cells of the olfactory epithelium. They include olfactory receptor cells, supporting cells, basal cells, and secreting cells. The olfactory receptor cells are those that directly respond to odorants, the volatilized chemicals present in the inhaled air that passes by the olfactory epithelium. The molecular shape of the odorants is the adequate stimulus for the modality of olfaction. The olfactory receptor cells are bipolar nerve cells that extend a single dendrite to the surface of the epithelium where it enlarges to form a knob. From this knob, some 5–20 cilia protrude into the layer of mucus that coats the epithelium. The axon extends through the cribriform plate and travels on to make contact with secondary neurons in the main olfactory bulb, which is a specialized region below the frontal lobe, but not part of the cerebral cortex. Thus these olfactory receptor cells are neurons and are called **olfactory sensory neurons** (OSN).

Supporting cells fill in the gaps between the olfactory receptor cells and form a continuous sheet of epithelial cells. They are often called **sustentacular cells**, which literally means "sustaining cells." **Basal cells** are pluripotent stem cells that produce new olfactory receptor

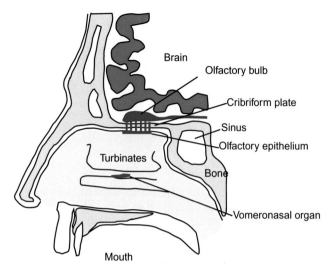

FIGURE 4.6.1 Anatomical location of the olfactory epithelium. The sensors that detect odors lie in the nasal epithelium. They send processes through a perforated bony structure, the cribriform plate, to make connections within the olfactory bulb. The olfactory bulb then sends nerve fibers to the brain for the conscious perception of odors.

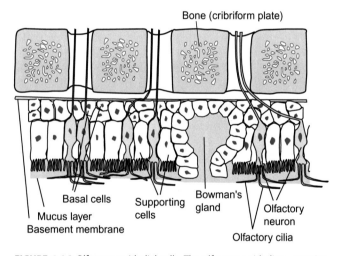

FIGURE 4.6.2 Olfactory epithelial cells. The olfactory epithelium contains olfactory neurons which send an axon through the cribriform plate to make contact with cells in the olfactory bulb. The olfactory neurons are supported by other cells called supporting cells or sustentacular cells. Interspersed among the neurons and supporting cells are secretory glands organized into Bowman's glands that produce a layer of mucus. The olfactory epithelial cells have olfactory cilia that project into the mucus layer. These cilia have the 9 + 2 arrangement of microtubule doublets but they lack the dynein arms, and so they are not motile. Basal cells along the basement membrane are stem cells that differentiate to form new olfactory neurons.

cells, secretory cells, or supporting cells to replace cells that die. Pluripotent cells are already partially differentiated but remain capable of becoming a variety of cell types. Totipotent cells occur early in embryogenesis and are capable of becoming any cell in the body.

A thin layer of mucus covers the entire olfactory epithelium. This material consists of mucopolysaccharides, antibodies, electrolytes, and a variety of soluble **odorant binding proteins**. **Bowman's glands**, consisting of groups of cells interspersed throughout the olfactory epithelium, produce this mucus. The entire mucus content of the olfactory epithelium is replaced about every 10 min. The mucus layer with its associated odorant binding proteins may concentrate odorants and facilitate their interaction with receptors on the olfactory receptor cells, or aid in the disposal of odorants.

HUMANS RECOGNIZE A WIDE VARIETY OF ODORS BUT ARE OFTEN UNTRAINED IN THEIR IDENTIFICATION

Humans readily recognize a variety of images and assign to them their proper names. Many of us can identify and distinguish among a host of sounds: an oboe is easily distinguished from a clarinet or a violin. We easily recognize voices. However, we often have difficulty identifying an odor even though we know we can smell it and we know that there is something vaguely familiar about it. Dogs, on the other hand, recognize the odors of particular people much like we recognize voices. Our inability to identify odors is probably partly due to a lack of training. When presented with a variety of odorants, people often cannot identify them unless provided with a list of descriptors for those odors. The list prompts correct identifications, suggesting that we are untrained in the association of the odors with their descriptors. Typically humans can recognize 10,000 distinct odors, most being disagreeable.

THE RESPONSE TO SPECIFIC ODORANTS IS MEDIATED BY SPECIFIC ODORANT BINDING PROTEINS

Odorant molecules comprise a diverse set of chemicals. To recognize them, the olfactory system uses a large family of receptors called odorant binding proteins that are G-protein-coupled receptors, as described below. The odorant binding proteins are synthesized by individual olfactory sensory neurons, the receptor cells, and are expressed on their cilia where they are exposed to the odorants. The mouse genome contains some 1300 different genes that code for odorant binding proteins, but each olfactory sensory neuron expresses only one of them. Humans have about 350 intact genes for these olfactory receptors (ORs). Each of these responds to a limited spatial arrangement on a chemical, called its molecular receptive range. A second class of chemosensory receptors, first reported in 2001, are the **trace amine-associated receptors (TAARs)**, which originally were thought to recognize volatile trace amines such as trimethylamine, present in mouse urine. Humans have six distinct TAARs. These reside in cells that express the same G-protein expressed in OR olfactory sensory neurons, so the mechanism of signal transduction is likely to be the same as for the olfactory receptors.

The binding of odorant proteins to their receptor on the olfactory receptor cell cilia activates a G-protein called **Golf**. The α subunit of G_{olf} stimulates adenylyl cyclase, which forms cAMP from ATP. The cAMP binds directly to cation-specific channels in the olfactory receptor cell membrane, increasing conductance to Ca^{2+} or Na^+. These are called cyclic nucleotide-gated channels or CNC. The resulting Ca^{2+} influx raises the local $[Ca^{2+}]$ which, in turn, leads to activation of Cl^- channels, TMEM 16B. In most cells, the activation of a Cl^- channel, resulting in an increased Cl^- conductance, would lead to a hyperpolarization or to a decreased ability to move away from the Cl^- equilibrium potential. However, olfactory neurons have an unusually high $[Cl^-]$, so that the activation of Cl^- channels results in a Cl^- efflux. Thus, the combined effect of cation entry and anion exit is a depolarization of the cell. Figure 4.6.3 illustrates these events.

The olfactory neuron returns to its baseline state first by removal of the odorant, probably aided by rapid replacement of the mucus that lines the olfactory epithelium. Soluble odorant binding proteins in the mucus may aid in this disposal. This stops the stimulation of G_{olf}. The α subunit spontaneously hydrolyzes its bound GTP to GDP and reassociates with the βγ subunit, deactivating adenylyl cyclase. Cyclic AMP phosphodiesterase converts cAMP to AMP, thereby deactivating the cation ion channel. The increased cytosolic $[Ca^{2+}]$ is removed by transport to the extracellular space or by uptake by the endoplasmic reticulum (ER).

THE OLFACTORY RECEPTOR CELLS SEND AXONS TO SECOND-ORDER NEURONS IN THE OLFACTORY BULB

Axons leaving olfactory receptor cells cross the cribriform plate and make contact with second-order olfactory neurons in the main olfactory bulb, which is a specialized region below the frontal lobe, but not part of the cerebral cortex. The second-order neurons are mitral cells and tufted cells. The contacts of the second-order neurons and the primary olfactory receptors form glomeruli, which consist of the grouped axonal processes of a large number of olfactory receptors (some 25,000 per glomerulus) and the apical dendrites of some 100 or so second-order neurons; about one-third of these are mitral cells and two-thirds are tufted cells. Each olfactory receptor contacts several second-order neurons, and each second-order neuron receives several thousand inputs from olfactory receptors. The olfactory bulb includes periglomerular cells and granule cells. Information in the form of nerve impulses travels from the olfactory bulb to the brain and back from the brain. A simplified schematic diagram of these connections is shown in Figure 4.6.4.

EACH GLOMERULUS CORRESPONDS TO ONE RECEPTOR THAT RESPONDS TO ITS MOLECULAR RECEPTIVE RANGE

The olfactory neurons expressing a single odorant binding protein send axons through the cribriform plate to the olfactory bulb where they converge on one or at most a few glomeruli. In at least some cases, the synapses in a single glomerulus contain only processes from a single type of olfactory receptor cell. Because mitral cells and tufted cells have a single primary dendrite that projects to a single glomerulus, the response of the mitral cells reflects the response of the glomerulus they innervate. Figure 4.6.5 shows the response of two mitral cells when the olfactory epithelium was exposed to a variety of disubstituted benzene odorants. Mitral cell A is tuned selectively to benzenes substituted in the para position, whereas mitral cell B responds to disubstituted benzenes with short side chains. Note that both mitral cells respond to para xylene. The molecular receptive range of individual mitral cells corresponds to a set of characteristic structural features that may be shared by a range of odorants.

OLFACTION REQUIRES PATTERN RECOGNITION OVER ABOUT 350 INPUT CHANNELS

As described above, each OSN expresses only one type of OR and the axons of those OSN that express the same ORs converge typically on one or two glomeruli located in the olfactory bulb. Each odorant excites a set of OR that depends on the interaction between the OR and the odorant. Thus each odorant activates an odorant-specific pattern of glomeruli. Identification of an odor becomes the problem of recognizing the pattern of activation of the two-dimensional sheet of glomeruli. This recognition occurs in higher centers.

OLFACTORY OUTPUT CONNECTS DIRECTLY TO THE CORTEX IN THE TEMPORAL LOBE

Mitral cells and tufted cells send their processes to several areas including the anterior olfactory nucleus (AON), the piriform cortex (PC), the cortical amygdala, and the entorhinal complex. These areas traditionally constitute the primary olfactory cortex. However, some of these areas do not contain pyramidal neurons that form distinct layers, and so they are not cerebral cortex. These connections are unique in that they are the only sensory connections that do not travel through the thalamus before making connections in the cortex. These connections in the temporal lobe overlay those of the hippocampus and the amygdala, components of the limbic system, which is important in setting mood and emotional behavior. These are the areas that are responsible for associating emotional response with odors. Presumably these areas are also responsible for the highly evocative experience of memory upon odor sensation. These are the basis of the "involuntary memories" as described by Marcel Proust in "A la Recherce du Temps Perdu" (traditionally translated as "Remembrances of Things Past," but more recently translated as "In Search of Lost Time"). Proust describes eating madeleine cakes dipped in tea and the odors and taste evoke childhood memories of eating such cakes with his aunt, and these involuntary evoked memories

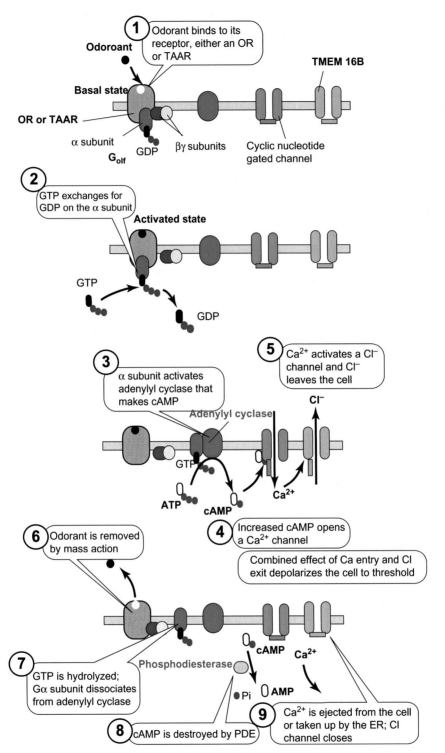

FIGURE 4.6.3 Mechanism of olfactory transduction. Odorants bind to receptor proteins located on the olfactory cilia. These receptor proteins are either an olfactory receptor of a trace amine-associated receptor. The human genome encodes about 350 ORs and only 6 TAARs. Each OSN (olfactory sensory neuron) expresses only one kind of receptor (1). Reception triggers an exchange of GTP for GDP on G_{olf}, a G-protein located in these cells (2). GTP–GDP exchange results in the dissociation of the G_{olf}-α subunit from the βγ subunit. The G_{olf}-α subunit activates adenylyl cyclase to produce cAMP from ATP (3). The cAMP directly activates a cation channel that admits Ca^{2+} into the cell (4). The Ca^{2+} activates a Cl^- conductance pathway which, because of the high $[Cl^-]$ in these cells, produces a Cl^- efflux (5). The cation entry and anion exit produce a depolarization to threshold and the initiation of an action potential on the basal side of the cell toward the cribriform plate. The system is reset by removal of the odorant (by washout and diffusion) (6), inactivation of the Gα subunit by GTP hydrolysis (7), removal of cAMP (by phosphodiesterase) (8), and removal of Ca^{2+} by surface membrane active transport and ER uptake (9).

The Chemical Senses

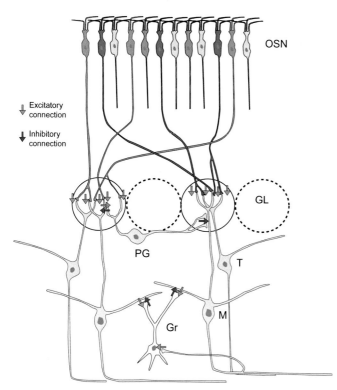

FIGURE 4.6.4 Connections between the olfactory epithelium and the olfactory bulb. Olfactory sensory neurons (OSN) make contact with the dendritic trees of tufted cells (T) and mitral cells (M) in a concentrated area of nerve process called a glomerulus (GL). The individual OSNs each make a single odorant binding protein, and the OSNs making these are randomly distributed within zones of the olfactory epithelium. OSNs making the same kind of odorant binding protein are shown here by the same color. Their processes converge on at most a few glomeruli. Therefore, the response of the mitral cells is tuned to specific odorants. Other cells, such as the periglomerular cells (PG), engage in a kind of lateral inhibition to sharpen the response of the mitral cells. The granule cells (Gr) produce a negative feedback that limits the response of the mitral cells, which form the major output of the olfactory bulb to the olfactory cortex. Light arrows signify excitation; dark arrows signify inhibition. *From K. Mori, H. Nagao, and Y. Yoshihara, The Olfactory bulb: coding and processing of odor molecule information.* Science **286**: 711–715, 1999.

bring back additional memories of the town during his forgotten childhood years.

A SECOND OLFACTORY OUTPUT IS THROUGH THE THALAMUS TO THE ORBITOFRONTAL CORTEX

Olfactory output is also funneled through the thalamus to the orbitofrontal cortex. These connections are shown schematically in Figure 4.6.6. Lesions of this area prevent the conscious perception of odors, so this area may be more important than the primary olfactory cortex for this purpose. The primary olfactory cortex may be more important in odor-related memories and emotional responses. The orbitofrontal cortex is adjacent to the primary taste cortex, so that the overall perception of flavor in a food may be produced by overlapping inputs from these two regions. This is in accord with the subjective experience of the tastelessness of food when you have a cold and cannot smell.

THE DETECTION LIMITS FOR ODORS CAN BE LOW

Only volatile chemicals can be smelled because only volatile materials can enter the nostrils along with the inhaled air. In addition, we can smell only those chemicals that are reasonably water soluble because they must penetrate the mucus layer that covers the olfactory receptor cells. These requirements can be overwhelmed by mass action. The volatility of a chemical describes the relationship between the concentration of chemical in the gas phase in equilibrium with the liquid phase. The higher the concentration of chemical in the liquid phase, the higher its concentration in the gas phase. Similarly, water solubility of a volatile material describes the concentration in the water phase in relation to the concentration in the gaseous phase. Thus increasing the concentration in the gas phase would increase the concentration in the water phase. These physical requirements for odorants explain in part their different thresholds for smell.

Methyl mercaptan, CH_3SH, can be detected in inhaled air at concentrations of 10^{-12} M. Because of its low threshold, methyl mercaptan is added to natural gas, which has no natural odor, so that people can smell it to detect gas leaks.

ADAPTATION TO ODORS INVOLVES THE CENTRAL NERVOUS SYSTEM

Subjective experience tells us that sensations of smell nearly disappear upon continued exposure to odorants. Part of this adaptation occurs at the level of the olfactory receptor cells, whose firing rates decrease 50% after the first few seconds of stimulation. This is partly explained by depolarization block—cells that do not repolarize cannot reset their Na^+ channels from the inactivatable state in order to initiate action potentials. However, the subjective experience is greater than this, implying that at least some of the adaptation is due to central nervous system (CNS) processing. The mechanism for this is postulated to occur through granule cells, another nerve cell type within the olfactory bulb (see Figure 4.6.4). These cells receive input from efferent fibers coming back from the brain in the olfactory nerve. These cells release gamma-amino butyric acid on synapses with tufted cells and mitral cells, inhibiting the output of the these latter two cell types.

SOME "SMELLS" STIMULATE THE TRIGEMINAL NERVE AND NOT THE OLFACTORY NERVE

Some irritants produce a strong sensation that is subjectively akin to smell but operates over separate nerve paths. Ammonia, for example, elicits a strong reaction that is carried by the trigeminal nerve, cranial nerve V. The olfactory nerve is cranial nerve I. The trigeminal nerve also carries motor control for mastication and

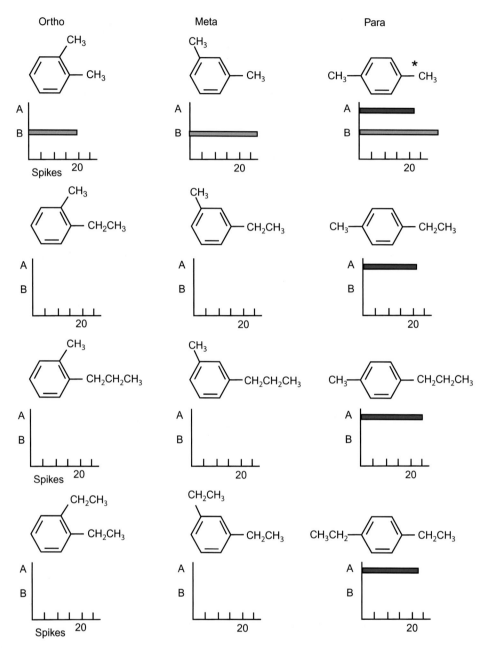

FIGURE 4.6.5 Different mitral cells respond to different molecular features. Responses of mitral cells in the rabbit main olfactory bulb were recorded as the average number of spikes per inhalation cycle for each of the indicated odorants. Two mitral cells were recorded, indicative of the response of two glomerular modules. The molecular structure of the odorants is shown above each graph. Mitral cell A responds to para isomers of disubstituted benzenes (dark bars), whereas mitral cell B responds to disubstituted benzenes with short side chains (light bars). Both cell types respond to para xylene, indicated by the asterisk. Thus mitral cells respond to a range of molecules that incorporate specific structural features. *Modified from K. Mori, H. Nagao, and Y. Yoshihara, The olfactory bulb: coding and processing of odor molecule information. Science* **286**: *711–715, 1999.*

facial sensory input. Other sensations carried over the trigeminal nerve include those elicited by chlorine, peppermint, and menthol.

HUMANS DISTINGUISH AMONG FIVE PRIMARY TYPES OF TASTE SENSATIONS

Taste performs an essential role in physiology by ensuring the consumption of nourishing foods and the aversion of potentially harmful or noxious foods. Humans discriminate among five primary stimuli: sweet, bitter, sour, salty, and umami. This last is translated as "meaty" or "savory" and is the taste associated with monosodium glutamate (MSG). The entire hedonistic experience of food incorporates additional senses to these five primary modalities, including the hot taste of peppers, coolness of peppermint, texture, weight, and temperature. These additional features are sensed and transmitted separately from the classical sense of taste but become integrated with taste in the hedonistic appreciation of food.

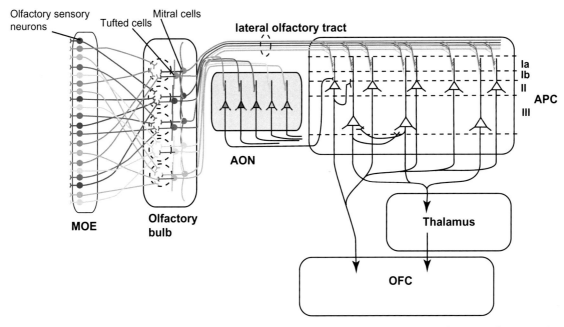

FIGURE 4.6.6 Part of the central connections of the olfactory system. Olfactory sensory neurons (OSN) in the main olfactory epithelium (MOE) make connections in the olfactory bulb in which like OSN converge on one or two glomeruli. These excite tufted cells and mitral cells according to the sensitivity of the OSNs. Mitral and tufted cells send axons to the anterior olfactory nucleus (AON) and to the anterior pyriform cortex (APC) where sensory information is processed further. The APC is cortical, consisting of pyramidal neurons in layers. Pyramidal cells in layer II send axons directly to the orbital frontal cortex (OFC) whereas those in layer III send axons first to the thalamus (submedius nucleus and mediodorsal nucleus, not shown). Cells in the thalamus then send axons to areas of the orbitofrontal cortex.

THE TASTE BUDS ARE GROUPS OF TASTE RECEPTORS ARRANGED ON TASTE PAPILLAE

Taste receptors are modified epithelial cells that are concentrated in specialized structures called taste buds (see Figure 4.6.7). Taste buds are found on the tongue and also distributed throughout the rest of the oral cavity, epiglottis, and esophagus. The taste buds also contain sustentacular or supporting cells and basal cells that divide and differentiate into new taste receptors. Taste receptors die and are replaced after a life span of about 2 weeks. Each taste bud consists of a group of some 50–100 taste receptor cells (TRCs) together with their supporting and basal cells. The cells are arranged like the slices of an orange with a central pore that opens onto the surface of the tongue. Microvilli from the TRCs protrude into the central pore where they sense dissolved chemicals. The taste buds reside on taste papillae. These are larger structures with three distinct varieties that differ with respect to their structure and their location on the tongue (see Figures 4.6.8 and 4.6.9).

Circumvallate papillae are located in the posterior of the tongue and respond to bitter substances such as plant alkaloids, which are often poisonous. As their name implies, these are large, round structures surrounded by a depression or valley. They are innervated by the **glossopharyngeal nerve** or **cranial nerve IX**. There are typically only 10–12 of these in the tongue.

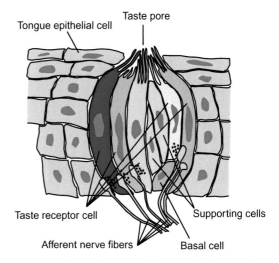

FIGURE 4.6.7 Histology of a taste bud. Taste buds are collections of 50–100 TRCs, basal or stem cells, and supporting cells, arranged something like the slices in an orange. The receptor cells contain synaptic vesicles on their basal sides where the terminus of an afferent sensory neuron makes a synapse. The receptor cells depolarize in response to a tastant, resulting in a cascade of events that eventually cause an action potential in the afferent nerve fiber. Each TRC makes receptors for a single class of tastants, so individual TRCs are tuned to one of each of the primary taste modalities. These are signified by the different colors of each TRC.

The taste buds are located on the sides of the papillae, as many as 250 per papilla.

Several hundred **fungiform papillae** are located in the anterior two-thirds of the tongue and respond best

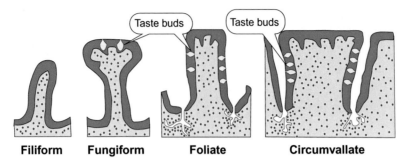

FIGURE 4.6.8 Types of papilla. The filiform papillae are found all over the tongue and contain no taste buds. The fungiform papillae are located anteriorly, have one to five taste buds and are innervated by the facial nerve. The foliate papillae are lateral and posterior, and are innervated by the glossopharyngeal nerve. The few circumvallate papillae are located at the base of the tongue.

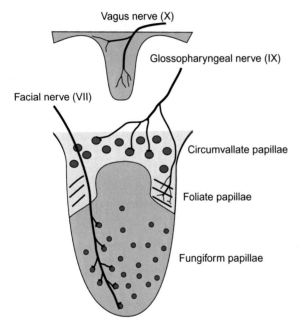

FIGURE 4.6.9 Innervation of the tongue and pharynx. The posterior part of the tongue, near the base, contains 10–12 circumvallate papillae and is innervated by the glossopharyngeal nerve, cranial nerve IX. The foliate papillae are located laterally and posteriorly, and are also innervated by the glossopharyngeal nerve. The anterior two-thirds of the tongue contain mainly fungiform papillae that are innervated by the chorda tympani branch of the facial nerve, cranial nerve VII. The pharynx and epiglottis also have scattered taste buds that are innervated by the vagus nerve, cranial nerve X.

to sweet and salty substances, but also respond to sour. As their name implies, fungiform papillae are shaped like mushrooms. Each papilla has 1-5 taste buds. They send sensory information over the **chorda tympani** branch of the **facial nerve** or **cranial nerve VII**.

Foliate papillae are located in the lateral and posterior edges of the tongue and respond best to sour substances. Once it was believed that particular taste modalities were concentrated in particular areas of the tongue. Although it may be true that some parts of the tongue are more sensitive to particular modalities than other parts of the tongue, all parts of the tongue appear to respond to all five classes of tastants. The lateral and posterior edges are innervated by the glossopharyngeal nerve.

A fourth type of papilla, the filiform papilla, about 2–3 mm long, is located all over the surface of the tongue and is the most abundant type of papilla. These have no taste buds.

The **vagus nerve** innervates taste buds in the posterior pharynx and epiglottis.

TRCs RESPOND TO SINGLE MODALITIES

Recent evidence points to a labeled-line model of sensory reception for taste. In this model, individual TRCs are tuned to respond to a single taste modality to the exclusion of the others. They are also innervated by similarly tuned nerve fibers, so that the excitation of a single TRC induces the firing of its nerve fiber that carries the signal of reception of that modality. The activity is labeled by the kind of cell that produces it.

SALTY TASTE HAS TWO MECHANISMS DISTINGUISHED BY THEIR AMILORIDE SENSITIVITY

Some TRCs possess a highly selective Na^+ channel, the epithelial Na^+ channel or ENaC, on their apical membrane, the one facing the oral cavity. This channel is blocked by amiloride. Because this channel is highly selective for Na^+ over K^+, TRCs possessing this mechanism are also Na^+ selective. Entry of Na^+ into the cell depends on the concentration of Na^+ in the mucosal fluid bathing the cells. It makes an inward current that depolarizes the cell. The depolarization is transmitted electrotonically to the base of the cell where voltage-gated Ca^{2+} channels open, causing Ca^{2+} influx from the extracellular fluid that increases cytosolic $[Ca^{2+}]$ near synaptic vesicles. This causes vesicle fusion with the TRC cell plasma membrane, releasing neurotransmitter. The neurotransmitters in turn change the ionic conductance in the sensory afferent neuron, producing an excitatory postsynaptic potential (EPSP). If the resulting EPSP reaches threshold, the sensory neuron initiates an action potential. Because fibers innervating TRCs with this mechanism fire selectively to an Na^+ stimulus, they are called N fibers.

The second mechanism for salt sensation also involves an inward current carried by a channel that responds to Na^+, K^+, and NH_4^+. This channel is blocked by another inhibitor, **cetylpyridinium chloride (CPC)**, but is insensitive to amiloride. Together, both channels appear to explain all of the response to NaCl. Humans have mostly amiloride-insensitive salt taste. Because the amiloride-insensitive component does not distinguish highly between Na^+ and K^+, we can use KCl on the table as a substitute for NaCl. Fibers that innervate TRCs with this mechanism are called **H fibers** (see Figure 4.6.10).

SOUR TASTE DEPENDS ON TRC CYTOSOLIC pH

The apical membrane of sour TRCs has an H^+ channel that allows entry of H^+ ions. This channel is not yet identified. Exposure of these cells to an acid solution causes the cytosolic pH to decrease (because $[H^+]$ increases) and this is presumably coupled to Ca^{2+} entry near the synapse. This Ca^{2+} entry may be linked to depolarization caused by H^+ entry or by its activation of Na^+ channels, or inhibition of K^+ channels. Weak acids cause excitation of sour taste receptors by a different mechanism: they cross the apical membrane in the neutral, nonionized form which then forms H^+ inside the cell. Thus the increased $[H^+]$ itself is the initial signal. The precise cascade of events is not yet clear. These TRCs have a Na^+-H^+ exchanger located on the basolateral surface that is activated by cytosolic Ca^{2+}. This can explain the adaptation to sour substances: acidic solutions presented to the apical membrane increase the cytosolic $[H^+]$ which then increases cytosolic $[Ca^{2+}]$. The increased $[Ca^{2+}]$ increases the rate of H^+ exchange out of the cell, bringing the cytosolic pH back toward normal and reducing the activity of the receptor. Thus the system naturally adapts. Figure 4.6.11 illustrates the sour taste mechanism.

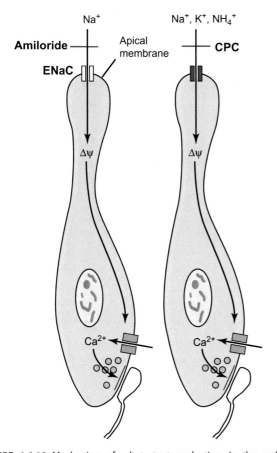

FIGURE 4.6.10 Mechanism of salt taste transduction. In the sodium-selective mechanism, Na^+ enters the cell through an apical Na^+-selective channel, the ENaC. This entry depends directly on the sodium concentration in the fluid bathing the apical surface. This is blocked by amiloride. The resulting depolarization spreads electrotonically to the basal–lateral membrane of the cell where voltage-gated Ca^{2+} channels admit Ca^{2+} depending on the depolarization. The resulting increase in cytosolic $[Ca^{2+}]$ causes fusion of synaptic vesicles with the basal–lateral membrane of the TRC, causing release of neurotransmitter. The neurotransmitter then excites the primary sensory afferent fiber. A second mechanism uses a less selective cation channel which is insensitive to amiloride but is blocked by CPC. The resulting depolarization causes the same events as Na^+ influx through ENaC.

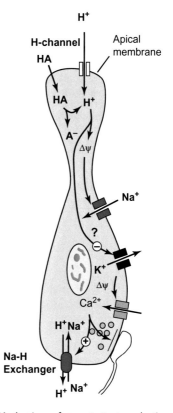

FIGURE 4.6.11 Mechanism of sour taste transduction. TRCs for sour taste respond to the intracellular pH. Decreases in pH result in fusion of vesicles and release of neurotransmitter to the primary sensory afferent neurons. Increasing $[H^+]$ in the apical solution or exposure to high concentrations of weak acid, HA, both lead to an increased cytosolic $[H^+]$, which somehow is linked to Ca^{2+} entry into the cell and fusion of synaptic vesicles with the TRC's plasma membrane. Adaptation to the sour taste occurs by the activation of efflux pathways for H^+. Activation of the Na-H exchange protein is shown.

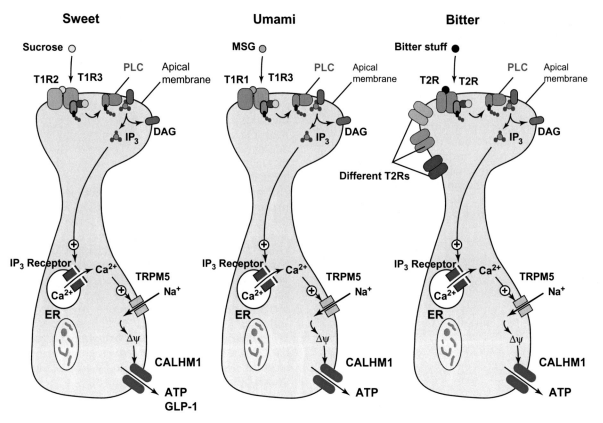

FIGURE 4.6.12 Mechanism of sweet, umami, and bitter tastes. All three of these work through G-protein-coupled receptors. In the case of sweet, a heterodimer composed of T1R2 and T1R3 senses sweet materials in the solution on the surface of the tongue. This is coupled to a G-protein that activates phospholipase C to hydrolyze phosphatidyl inositol 4,5-bisphosphate to DAG and IP3 which binds to an IP3-receptor on the ER, releasing Ca^{2+} ions into the cytoplasm. The increased $[Ca^{2+}]$ activates a TRPM5 which increases conductance for Na^+, producing a depolarization. The depolarization results in the release of glucagon-like peptide 1 (GLP-1) and ATP in the case of sweet taste and probably ATP in the case of umami and bitter taste. Release of these neurotransmitters seems to not involve vesicle fusion but may involve a pore provided by CALMH1, for calcium homeostasis modulator. This basic signaling cascade also occurs for umami and bitter tastes, except the surface taste receptors are different. In the case of umami, the receptors on the surface of the cell, on the microvilli at the central pore, are heterodimers of T1R1 and T1R3. In the case of bitter taste, about 25 different T2R genes are expressed, many if not all on a single TRC. All of the T2R receptors eventually produce the same subjective taste of "bitter."

SWEET, BITTER, AND UMAMI TASTE ARE TRANSDUCED BY THREE SETS OF G-PROTEIN-COUPLED RECEPTORS

Sweet, bitter, and umami tastes share the common feature of activating a GPCR that uses a G_q mechanism to activate a channel called TRPM5 for transient receptor potential type M5. This downstream signaling pathway is shared by all three modalities. What distinguishes them is the receptors that initiate the cascade. There are three T1 receptors, T1R1, T1R2, and T1R3, and a family of about 25 T2R receptors. T1R3 forms a heterodimer with T1R2 to form the sweet taste receptor, and with T1R1 to form the umami taste receptor. Mice with knockouts of the T1R3 gene do not respond to either umami or sweet tastants; knockouts of T1R2 alone cannot respond to sweet but do respond to umami tastants; knockouts of T1R1 alone cannot respond to umami but respond to sweet. Bitter taste, on the other hand, is enabled by the expression of a family of about 25 different receptors in the T2R family. These form homodimers that respond to bitter tastants. TRCs typically express many and perhaps all of these different receptors in individual cells. Thus many different tastants of considerably different chemical structure produce the same subjective bitter taste. Figure 4.6.12 shows the mechanisms for sweet, bitter, and umami tastes.

THE "HOT" TASTE OF JALAPENO PEPPERS IS SENSED THROUGH PAIN RECEPTORS

The "hot" taste of jalapeno peppers is due to a chemical, capsaicin, that is present in the chili peppers. This compound activates the vanilloid receptor (VR1), a nonselective cation channel found on small unmyelinated fibers (type C fibers and Aδ fibers). These fibers are also activated by pain and heat. Some of these fibers enter the CNS over the trigeminal nerve. This explains the "hot" sensation of chilis that can border on painful and its subjective pungent quality. Other sensations that are carried over the trigeminal nerve include those elicited by peppermint, chlorine, and menthol.

Clinical Applications: Ageusia

The ability to taste things in food or drinks depends on specific chemicals in the food binding to receptors on the taste cells. Many of these receptors are metabotropic, coupled to G-proteins that are linked to intracellular signaling mechanisms that eventually cause the release of neurotransmitters at the basal pole of the taste cells. These receptors are proteins whose amino acid sequence is directed by DNA. Individuals with mutations in the DNA can produce receptors that have reduced affinity for the tastants or the receptors may be absent altogether. Such persons can experience taste deficits. Ageusia means the inability to taste. Specific ageusia refers to the inability to taste a specific chemical. Hypogeusia means that the person has an elevated threshold for detection of tastants—they are less sensitive than normal.

The ability to taste phenylthiocarbamide, PTC, is genetic, and the human population is heterogeneous with respect to this ability. The PTC receptor is a member of the T2R family of bitter taste receptors. Three single nucleotide polymorphisms have been identified, with three single amino acid substitutions. These polymorphisms are pro/ala at amino acid 49; ala/val at amino acid 262; and val/ile at amino acid 296. The ancestral and normal form of the protein is pro—ala—val, and persons with this receptor can taste PTC. Persons homozygous for the ala—val—ile form of the receptor cannot taste PTC or do so at much reduced sensitivity.

Other specific ageusias exist. In a sample of 109 people, some 83% were classified as "normal" tasters for MSG, whereas the remaining 17% were "hypotasters." Because psychophysical studies are difficult to do, multiple testing regimens are employed and classification as a "taster," "hypotaster," or "nontaster" depends on the regimen. Further testing showed a smaller proportion of people who could be regarded as "nontasters."

Ageusia for PTC and MSG are currently the only known representatives of taste blindness in humans.

Clinical Applications: Foreign Chemicals and Endogenous Ligands

The vanilloid receptors are widespread throughout the animal kingdom, but those of birds differ from those of mammals; the birds' VR1 does not respond to capsaicin. It has been postulated that this is an example of coevolution, where one species hones its chemicals to fit the environment partly provided by other species. Plants need to disperse their seeds, and they engage in a variety of mechanisms for doing so. One mechanism is to attract animals who will eat the plants and disperse the seeds in their feces. The seeds must resist digestion themselves. Experiments have shown that chili seeds will not germinate after ingestion by rodents, but they will germinate after being eaten by birds. The "hot" taste of chilis discourages herbivores from feeding on the plant because it is painful, but it does not deter the birds. It is difficult to say which came first: the chili plant that exploits a difference in the VR1 receptors or birds that exploited a difference in their VR1 receptor to harvest a food source unsavory to the rodents.

This prompts the question: why do we have receptors that are sensitive to capsaicin? This question extends to a host of foreign chemicals that have very specific effects on the body. Why do naturally occurring foreign compounds have such specific effects? The final answer to such a question is elusive, of course, but there are two potential answers: (1) plants evolved the synthetic machinery to make these compounds to target our receptors for some effect—usually bad; (2) the effects are entirely accidental and result from the similarity in the shape of the chemicals to some endogenous ligands that we use as part of our normal signaling pathways.

An outstanding example of this is opium. Opium is an extract of the opium poppy, *Papaver somniferum*. It has been used for thousands of years to produce euphoria, analgesia, and sleep, and for relief of coughs. Friedrich Sertürner synthesized codeine and morphine from the poppy plant in 1805. Morphine is named after Morpheus, the Greek god of dreams, and is a potent analgesic (pain reliever). Heroin is synthesized from morphine. The basis of these effects was first illuminated by Candice Pert and Solomon Snyder when they demonstrated that opiates have specific receptors. This brings us back to the question as to why we would have receptors for poppy plant extracts. John Hughes and Hans Kosterlitz found that these receptors responded to two naturally occurring endogenous peptides, named **endorphins** for *endo*genous *morphine*-like substances. Thus the pharmacological properties of morphine are probably entirely accidental, resulting from its ability to bind to receptors at the same site as our endogenous ligands.

Is there an endogenous ligand for the capsaicin receptor? Recent investigations suggest that a chemical called **anandamide** may be an endogenous ligand for the VR1 receptor. This compound is derived from arachidonic acid. Anandamide was originally proposed to be the endogenous ligand for the cannabinoid receptors, CB_1 and CB_2, that respond to the active chemical from marijuana (*Cannabis sativa*), Δ^9-tetrahydrocannabinol. But anandamide stimulates only the CB_1 receptors and another endogenous compound, 2-arachidonoyl-glycerol, fully excites both CB_1 and CB_2 receptors. Anandamide excites the VR1 receptors and may be its endogenous ligand.

TASTE RECEPTORS PROJECT TO THE CORTEX THROUGH THE SOLITARY NUCLEUS AND THE THALAMUS

The TRCs are not neurons, even though they release neurotransmitters, because they have no dendritic of axonal processes. They make synapses onto dendritic processes of primary afferent sensory neurons whose cell bodies reside in three cranial nerve ganglia. The anterior two-thirds of the tongue and the palate are innervated by the facial nerve or cranial nerve VII. The cell bodies for the taste fibers are in the geniculate ganglia. The posterior third of the tongue is supplied by the glossopharyngeal nerve or cranial nerve IX. The cell bodies for this sensory nerve are located in the inferior glossopharyngeal ganglia. The vagus nerve, cranial nerve X, supplies the scattered taste receptors in the throat regions, including the glottis, epiglottis, and pharynx. The cell bodies of the vagus reside in the inferior vagal ganglia.

Sensory fibers from all three of these cranial nerves enter the lateral medulla and make synapses on cells in the gustatory division of the **solitary nucleus** in the medulla. The second-order neurons in the solitary nucleus send fibers up to the **ventral posterior medial nucleus** of the **thalamus**, where they make synapses on third-order neurons. These thalamic neurons then project to the **primary gustatory cortex** located in the insular and orbitofrontal regions.

FLAVOR IS IN THE BRAIN

The overall sensation of flavor is a mixture of sensations. Mechanical stimuli ("mouth feel"), odors, taste and temperature all figure into the overall perception of flavors. These diverse nervous signals are first integrated in the anterior insular cortex. This core flavor percept is then conveyed to regions of the amygdala, orbitofrontal cortex, and anterior cingulate cortex to produce the final flavor perception. Exactly how this is accomplished is not yet understood.

SUMMARY

The chemical senses include the sense of taste and smell. Both detect foreign chemicals when they interact with receptors at surfaces of the body. Smell detects volatile chemicals in the inhaled air; taste detects dissolved chemicals in food and drink. Because these receptors are on the surfaces that are exposed to noxious chemicals, both the olfactory and the taste epithelium are replaced regularly. The olfactory cells are true neurons, with an axon exiting the base of the cell, crossing the cribriform plate and making synapses in the olfactory bulb. The taste cells are receptor cells that do not produce action potentials, but instead release neurotransmitter onto a primary afferent sensory cell.

Olfactory receptor cells have long cilia that protrude into a mucus layer that is secreted by Bowman's glands in the olfactory epithelium. Odorant binding proteins on the cilia membrane bind odorants and couple to a G_s protein that stimulates adenylyl cyclase and increases cAMP in the olfactory cells. The cAMP activates protein kinase A that phosphorylates a Ca^{2+} channel, activating it. The increased cytoplasmic $[Ca^{2+}]$ activates a Cl^- channel. Because these cells have higher than normal cytoplasmic $[Cl^-]$, opening the channel causes Cl^- to leave, which makes an inward current that depolarizes the cell. The olfactory cell action potentials are further processed in the olfactory bulb and the signal is relayed to the primary olfactory cortex and the orbitofrontal cortex through the olfactory nerve, cranial nerve I. Some odors are actually irritants whose signals pass to the CNS over the trigeminal nerve, cranial nerve V. Examples include ammonium, chlorine, peppermint, and menthol.

Taste receptors are located in taste buds, conglomerates of about 50–100 cells, on taste papillae on the tongue and elsewhere in the pharynx and palate. The facial nerve (cranial nerve VII) innervates the front of the tongue, the glossopharyngeal (cranial nerve IX) innervates the back of the tongue, and the vagus nerve (cranial nerve X) innervates the palate, pharynx, and epiglottis. There are four types of papilla: circumvallate, fungiform, foliate, and filiform. The filiform papillae have no taste buds. There are five primary taste modalities: salty, sour, sweet, bitter, and umami.

Salty taste has two basic mechanisms: Na^+ entry through the amiloride-sensitive ENaC depolarizes the cell. This is highly selective for Na^+. The second mechanism uses a CPC-sensitive channel that lets in Na^+, K^+, and NH_4^+, which depolarizes the cell. This does not distinguish well between Na^+ and K^+. Humans probably have both mechanisms but do not discriminate well between Na^+ and K^+, so K^+ tastes salty.

Sour taste detects the receptor cell's intracellular $[H^+]$. This is coupled to Ca^{2+} entry that releases neurotransmitter onto the afferent sensory nerve. Increased cytoplasmic $[Ca^{2+}]$ activates H^+ efflux mechanisms, so that the tongue adapts to sour taste.

Sweet, bitter, and umami tastes all use G-protein-coupled receptors. There are three T1 receptors: T1R1, T1R2, and T1R3. These form heterodimers that are responsible for the sweet and umami sensing. TRCs expressing T1R2 and T1R3 on their surface respond to sweet tastants; removal of either protein ablates the ability to taste sweet materials. TRCs expressing T1R1 and T1R3 are responsible for the umami taste. TRCs that sense bitter compounds express several members of the T2R receptor family, all of which subjectively appear to be bitter. Each TRC responds to only one taste modality, and thus the sense of taste is a labeled-line mechanism.

The intracellular signal cascade for sweet, bitter, and umami modalities are similar. Binding of tastant to the receptor dissociates the heterotrimeric G-protein, activating phospholipase C that splits phosphatidyl inositol 4,5-bisphosphate to diacylglycerol (DAG) and inositol trisphosphate (IP3). The released IP3 binds to IP3-receptors on the ER, which then form a Ca^{2+} conducting pathway and release Ca^{2+} that is stored in the ER. The increase in cytosolic $[Ca^{2+}]$ activates a transient receptor protein (TRPM5) which increases Na^+ influx, thereby depolarizing the TRC. The depolarization releases neurotransmitter at the base of the TRC, activating the primary afferent neuron.

The "hot" taste of chili peppers is due to capsaicin, which stimulates small unmyelinated fibers in the trigeminal nerve. Capsaicin excites the vanilloid receptor, VR1.

REVIEW QUESTIONS

1. Where is the olfactory epithelium? What is the cribriform plate? Describe the shape of the olfactory sensory neuron. What is unique about it?
2. How do olfactory sensory neurons get depolarized when odorants bind to their olfactory cilia? How specific is the response?
3. Where do the axons of the olfactory sensory neurons go? Is there a specific glomerulus at which they form synapses? What cells form the output of the olfactory bulb?
4. How many different odorants can humans detect? How many odorant binding proteins do we make?
5. What are the five taste modalities? Are these sensed in particular areas of the tongue?
6. What is a taste bud? Where are they located?
7. What are the four different types of tongue papillae? Do they all have taste buds?
8. Which part of the tongue is innervated by the facial nerve? Glossopharyngeal nerve? Vagus nerve?
9. How is salty taste transduced? Why do both NaCl and KCl taste salty to humans?
10. How is sour taste transduced? Why does it adapt?
11. What is the general mechanism for sweet, bitter, and umami tastes? Why are there multiple receptors for bitter taste but not sweet taste?

4.7 Hearing

Learning Objectives

- Distinguish among the qualities of sound perception including pitch, timbre, and loudness
- Define decibels and be able to calculate loudness in decibels given either pressure or intensity
- Describe the function of the outer ear, middle ear, and inner ear
- Describe the anatomic location of the eustachian tube and describe its function
- Briefly explain why the tympanic membrane is much larger than the oval window
- Indicate the following on an anatomic drawing of the cochlea: modiolus, perilymph, endolymph, Reissner's membrane, scala vestibuli, scala media, scala tympani, tectorial membrane, outer hair cells, inner hair cells
- Explain why "stereocilia" are not true cilia
- Describe the afferent and efferent innervation of outer and inner hair cells. Explain why this means that the inner hair cells form the sensory response to sound
- Identify the cochlea, helicotrema, oval window, round window. Explain what is meant by "tonotopic mapping"
- Define "evoked otoacoustic emissions" and identify their source
- Trace auditory sensory information from spiral ganglia to dorsal and ventral cochlear nuclei to inferior colliculus to medial geniculate nucleus to auditory cortex
- Identify the auditory cortex and indicate its tonotopic representation
- Identify Broca's area and Wernicke's area and associate with each deficits in oral communication
- Describe the basic operation of cochlear implants
- Define ABR and BAER and describe how they can be used to test hearing in newborns

THE HUMAN AUDITORY SYSTEM DISCRIMINATES AMONG TONE, TIMBRE, AND INTENSITY

We know from everyday experience that we can discriminate among different aspects of sound. We can discriminate the pitch, or the perceived highness or lowness (as opposed to loudness or softness) of a sound. Sound waves of a single frequency are called **tones**. **Pitch** is the perceived counterpart of tones. The greater the frequency, the higher the pitch. The human ear is most sensitive to tones between 500 and 5000 Hz, but can detect tones from 10 to 20,000 Hz with reduced sensitivity.

Sounds in general are a composite of many different frequencies. Pleasing sounds consist of mixtures of tones with frequencies in particular ratios. For example, two tones with frequencies in the ratio of 2:1 are said to be separated by an **octave**. Similar ratios form the basis of **intervals** in music, as described in Table 4.7.1. The presence of a mixture of frequencies is the **timbre** of a sound. This is due to the presence of overtones, which are frequencies imposed on a tone. A tuning fork produces a pure tone. If you were to sing a "C" or play it on a clarinet, the sound wave will have overtones that allow us to identify the source as voice or clarinet. Further, the overtones in voices are produced by the characteristics of vocal cords, pharynx, and nasal passages so that we are able to identify voices based on their characteristic overtones. The human ear is exquisitely sensitive to differences in frequency, being able to detect differences between separate tones of as little as 2 Hz. Musicians use this ability to tune their instruments. Simultaneous sounds of frequencies f_1 and f_2 produce an interference pattern with a beat frequency of $|f_1 - f_2|$. Musicians detect these beats and tune their instruments until the beats disappear.

Humans also discriminate sounds on the basis of their intensity or loudness. In physical terms, the intensity of sound is the amount of energy transmitted per second through a unit area perpendicular to the direction of propagation of the sound. The usual units are $W\,cm^{-2}$ or $W\,m^{-2}$. We use a logarithmic transform of the intensity to represent the entire range of audible sounds. This is the **decibel**, abbreviated as dB. It is defined as

[4.7.1] $$\text{Loudness (decibels)} = 10 \log \frac{I_{\text{sound}}}{I_{\text{ref}}}$$

Here the intensity is in units of $W\,cm^{-2}$. The intensity of a sound requires a reference sound. Usually this is taken as the threshold of hearing, which varies with the frequency. For this reason, there are two decibel scales: one for **pressure sound level** and another for **hearing sound level**. They differ in using a physical intensity for a reference or threshold of hearing. The pressure sound level uses a reference of $10^{-16}\,W\,cm^{-2} = 10^{-12}\,W\,m^{-2}$. Because the intensity is proportional to the square of

TABLE 4.7.1 Frequency Ratios in Musical Intervals

Interval	Frequency Ratio	Examples
Octave	2:1	Five octaves of C: 1024, 512, 256, 128, 64 Hz
Third	5:4	320 and 256 Hz (middle C)
Fourth	4:3	342 and 256 Hz
Fifth	3:2	384 and 256 Hz

the pressure amplitude (see Appendix 4.7.A1), the loudness can be given as

$$[4.7.2] \quad \text{Loudness (dB)} = 20 \log \frac{\Delta P_{\text{sound}}}{\Delta P_{\text{ref}}}$$

Here ΔP is the increment in pressure over ambient pressure: the extra pressure that is propagated in the sound wave. The factor of 2 appears in Eqn [4.7.2] compared to Eqn [4.7.1] because of the proportionality of the intensity to the square of the pressure amplitude. The auditory threshold at 2000 Hz is about 2×10^{-5} Pa (1 Pa = 1 N m^{-2}). This corresponds to an intensity of about 0.5×10^{-12} W m^{-2}. Older texts sometimes use units of dyne cm^{-2} for units of pressure. Since 1 dyne = 10^{-5} N and 1 cm^{-2} = 10^4 m^{-2}, the conversion is 1 dyne cm^{-2} = 0.1 N m^{-2}. The threshold for hearing as a function of sound intensity (expressed in dB, ΔP, and I) is shown in Figure 4.7.1. This threshold is not absolute: background noise raises the hearing threshold through a process called **masking**. Masking is a signal-to-noise ratio problem that explains why it is difficult to understand conversation in a crowded, noisy room.

THE AUDITORY SYSTEM CAN LOCATE SOURCES OF NOISE USING TIME DELAYS AND INTENSITY DIFFERENCES

In addition to the sensations of pitch, timbre, and loudness, we can also locate the sources of sound using a combination of the time delay in the sounds striking the right and left ear, and the sound shadow produced by the head.

THE EAR CONSISTS OF THREE PARTS: THE *OUTER EAR*, *MIDDLE EAR*, AND *INNER EAR*: EACH HAS A DEFINITE FUNCTION

THE OUTER EAR COLLECTS AND CHANNELS SOUND TO THE MIDDLE EAR

The outer or external ear consists of the **pinna**, also called the **auricle**, and the **external auditory meatus**, the ear canal. The ear canal ends, and the middle ear begins, at the **tympanic membrane**, the eardrum. The shape of the auricle helps focus sound into the ear. In some animals, the auricle is funnel shaped and can be moved independently to search for sounds. Humans' auricle lacks this ability. The external auditory meatus contains **sebaceous glands** and **ceruminous glands** that secrete oils and waxes, respectively. The wax is the ester of a fatty acid and a long chain alcohol. These sticky secretions trap dust, dirt, insects, and bacteria that enter the auditory canal. The combination of these secretions and sloughed-off epithelial cells from the canal forms ear wax, or cerumen (from the Latin "cera" meaning "wax"). Impacted ear wax can cause pain or deafness. It can be removed by dissolution with a warm bicarbonate-rich solution or by mechanical removal (Figure 4.7.2).

THE MIDDLE EAR TRANSFORMS AIR PRESSURE WAVES TO FLUID PRESSURE WAVES

The middle ear is an air-filled cavity between the tympanic membrane on one side and the promontory of the temporal bone on the other. The middle ear contains three **ossicles**, tiny bones that connect the tympanic membrane to the membrane covering the **oval window** on the inner ear. The oval window leads to a fluid-filled chamber which coils around in a structure called the **cochlea**. The three bones are the **malleus (hammer)**, **incus (anvil)**, and **stapes (stirrup)**. They transfer

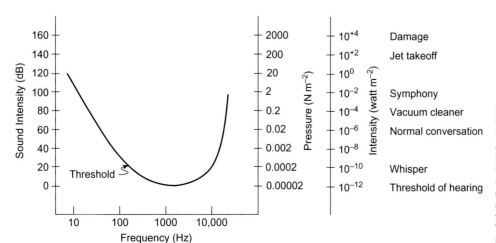

FIGURE 4.7.1 Threshold of hearing as a function of frequency. The ear is most sensitive to frequencies between 500 and 5000 Hz. The sound intensity can be described in decibels, or dB, as indicated on the scale to the left; as the maximum of the pressure amplitude, as given on the scale on the near right; and as energy per unit area per unit time, given on the scale to the far right. Typical sources of each of these sound intensities are indicated at the far right.

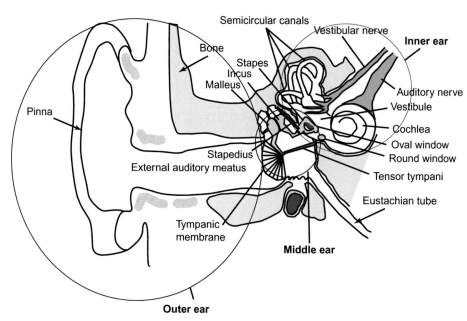

FIGURE 4.7.2 Structure of the ear. The outer ear includes the pinna and the external auditory meatus, the ear canal. These focus sound onto the tympanic membrane, which begins the middle ear. The air-filled middle ear contains three tiny bones, the ossicles (malleus, incus, and stapes), that transmit vibrations of the tympanic membrane to the inner ear. The stapes abuts the inner ear at the oval window. The inner ear consists of the cochlea for transduction of sound, and the semicircular canals for balance and detection of motion. The eustachian tube connects the middle ear to the atmosphere through the oral cavity. The auditory nerve transmits sound information to the CNS; the vestibular nerve transmits information about balance and movement. The two nerves combined make up cranial nerve VIII.

vibration of the tympanic membrane to vibration of the fluid in the cochlea.

The fluid in the external ear canal is air. Compressional waves in air cause the tympanic membrane to vibrate. This vibration must be transferred to the fluid in the cochlea of the inner ear. The cochlear fluid is a watery solution. Compressional waves in water require a completely different pressure because of the different inertia (the density) and different elastic properties (the bulk modulus: see Appendix 4.7.A1). Because of these differences, sound usually reflects off the air/water interface rather than being conducted into the water. The tympanic membrane and bones of the middle ear provide an impedance matching device to transfer sound energy to the cochlear fluid. The effective area of the tympanic membrane is about 0.4–0.6 cm^2; the area of the oval window is 0.03 cm^2. If there is no dissipation of energy in the bones themselves, then the sound energy can be concentrated some 15-fold where the stapes contacts the oval window.

The position of the tympanic membrane sets the position of the ossicles and therefore sets the transfer of vibration from air to cochlear fluid. If the pressure within the middle ear is higher than that in the external ear canal, the tympanic membrane will bulge outward and sensitivity to sound will decrease. If the pressure within the middle ear is lower than that in the external ear canal, the tympanic membrane will bulge inward and sound sensitivity will also decrease. This explains our loss of hearing sensitivity when we change altitude quickly, either ascending or descending. The **eustachian tube** connects the inner ear to the oral cavity and equilibrates the pressure on the two sides of the eardrum. Opening this tube by yawning wide or chewing equilibrates the pressure on the two sides of the membrane and pops the tympanic membrane back into place. This is sometimes accompanied by a popping or crackling auditory sensation and return of normal hearing sensitivity.

Thread-like ligaments and two muscles, the **tensor tympani** and the **stapedius**, keep the tiny ossicles in place. The two muscles contract reflexly with loud noises, body movements, stimulation of the ear canal, chewing, and contraction of facial muscles during vocalization. Contraction of these muscles reduces hearing sensitivity by making the ossicles more rigid. The latency of the reflex contraction to loud noises is about 40–60 ms, too long to prevent damage from brief intense sounds like explosions. However, these muscles may protect against prolonged, low-frequency sounds.

THE INNER EAR TRANSDUCES FLUID PRESSURE WAVES INTO NERVE IMPULSES THAT ARE TRANSMITTED TO THE BRAIN

The inner ear lies in a bony cavity in the temporal bone medial to the middle ear. The bony part of the inner ear is called the **bony labyrinth**, and it contains a variety of membranous ducts and sacs which are collectively called the **membranous labyrinth**. The membranous labyrinth floats in a fluid, the **perilymph**, that occupies the space between the membranous labyrinth and the bony labyrinth. The perilymph has low [K$^+$] and high [Na$^+$]. The part of the membranous labyrinth that makes contact with the stapes is the vestibule. The cochlea forms a spiral around a bony core called the **modiolus**. Two membranes, **Reissner's membrane** and the **basilar membrane**, divide the cochlea into three compartments. The outer two compartments contain perilymph and are continuous with the vestibule. The inner compartment contains **endolymph**, which has high [K$^+$] and low [Na$^+$]. The structures of the inner ear and details of the cochlea are shown in Figure 4.7.3.

The cochlea lies in a negative space—a hollow region of bone—making about 2 and 3/4 turns around a central core, the modiolus, that is shaped like a cone, with edges hollowed out to accommodate the cochlea. The nerves that receive input from the auditory receptor cells

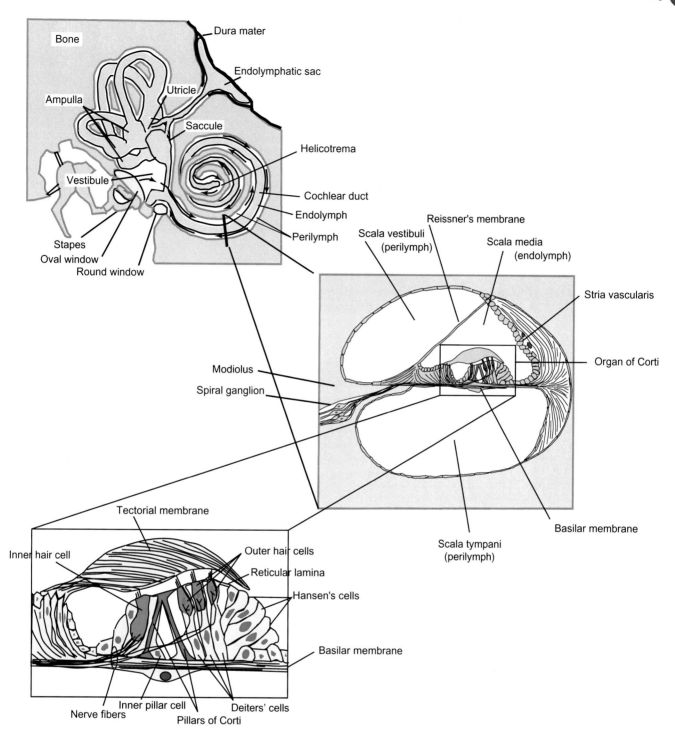

FIGURE 4.7.3 Structure of the inner ear. For explanation of cell types, see the text.

have cell bodies that reside in **spiral ganglia** within the modiolus. Their fibers collect to form the auditory nerve, which is part of cranial nerve VIII. Figure 4.7.3 shows cross sections of the cochlea.

The fluid in the scala vestibuli and scala tympani is perilymph. These two compartments are connected at the **helicotrema**, which is at the apex of the cochlea. At the base, the scala vestibuli is continuous with the vestibule. The foot of the stapes connects to the vestibule at the oval window and imparts pressure waves to the vestibule perilymph. These pressure waves are conducted all the way up the cochlea in the scala vestibuli, through the helicotrema and back down the scala tympani. Pressure waves at the far end of the scala tympani cause the round window to vibrate.

The **organ of Corti** is the specialized part of the cochlea that senses the vibrations of the basilar membrane and transduces these vibrations into electrical impulses. The vibrations are sensed by hair cells. There are two types of hair cells, as shown in Figure 4.7.3. There are three

Clinical Applications: Ear Infections and Tubes

Large numbers of children suffer from recurrent or chronic middle ear infections, called **otitis media**. This is an inflammation of the middle ear caused by viruses or bacteria, which invade the middle ear, usually by way of the eustachian tube, which connects the middle ear to the oral cavity. The resulting infection can be accompanied by fluid buildup in the middle ear that pushes on the eardrum. This causes a temporary loss of hearing sensitivity, and it hurts. Ordinarily the eustachian tube drains fluids into the oral cavity, but it may not open sufficiently due to congestion or a variety of other reasons. Most children are treated with antibiotics for bacterial infections. Children (or adults) with chronic or recurrent otitis media with effusions may require tubes to be inserted in their ears. The procedure of inserting tubes in the eardrum is called **tympanostomy**.

There are over 50 different designs for the tubes inserted into the tympanic membrane. The tubes can be made of rigid plastic or metal or silastic. Some are simple tubes whereas others have flanges to help secure them in the eardrum. The procedure requires making an incision in the membrane and inserting the tube in a place on the eardrum that will not interfere with the motion of the malleus. Because the surgery is delicate, the young patient is anesthetized with a general anesthetic. The procedure is very common. The National Institute for Child Health and Development estimates that in the United States in 1996, 280,000 children under the age of 3 years underwent tympanostomy.

The tubes provide an effective drain for the middle ear to the external ear canal. This extra drain allows the inner ear to become ventilated and reduce its fluid buildup. This reduces the temporary hearing loss and pain of acute otitis media, and reduces its recurrence. However, the tubes do not guarantee that ear infections will not recur. Much of the justification for the procedure relies on the argument that loss of hearing in young children harms their verbal and social development, but recent studies suggest that tympanostomy has no effect on children's scores on tests of expressive language or general cognition.

---

outer hair cells for every **inner hair cell**. The inner hair cells are innervated by about 20 unbranched afferent nerve fibers. Afferents from the outer hair cells form branches, with each neuron connecting to about 10 outer hair cells and each outer hair cell connecting to about 4 afferents. The hair cells also receive efferent nerve fibers, and most of these (80%) innervate the outer hair cells. The hair cells have tiny **stereocilia** that project from a structure called the **cuticular plate** on the apical aspect of the hair cells. The hair cells are overlaid by a gelatinous, acellular structure called the **tectorial membrane**. Shear between the tectorial membrane and stereocilia of the hair cells produces deformation of the stereocilia that eventually activates afferent nerve fibers. The inner hair cells generate afferent input to the auditory system, whereas the outer hair cells change their length to "tune" the inner hair cells' response.

The chief structural elements of the organ of Corti are provided by inner and outer **pillar cells**. These pillar cells contain trunks of dense, interlaced microfibrils, and microtubules that lock together at the apical side and contact the basilar membrane at widely separated points. They form a cell-free space, **the tunnel of Corti**, beneath the **reticular lamina**, a sheet of cells through which the apical surfaces of the hair cells penetrate. The thin reticular lamina is composed of **phalangeal processes** arising in part from **Deiters' cells**. These Deiters' cells support the outer hair cells only on their basal and apical surfaces, leaving the sides of the outer hair cells free of cellular contact.

HAIR CELLS OF THE COCHLEA RESPOND TO DEFORMATION OF STEREOCILIA TOUCHING THE TECTORIAL MEMBRANE

The inner and outer hair cells share the common feature of having their apical surfaces immersed in endolymph, the fluid of the scala media. In addition, their apical surfaces have a **cuticular plate** in which are embedded rows of **stereocilia**. The two types of hair cells differ substantially. The inner hair cell is flask shaped and is innervated mostly by afferent nerves. The outer hair cells are rod shaped and are innervated predominantly by efferent nerves. The inner hair cells form a spiral row running from oval window to apex of the cochlea; the outer hair cells form three parallel spiral rows over the same surface. Figure 4.7.4 shows these cells. The stereocilia of the inner hair cells are longer than those of the outer hair cells. Both sets of stereocilia form several rows, arranged either as a "V" shape of "W" shape, with the points of the "V" pointing away from the modiolus around which the cochlea spirals. The stereocilia on the side away from the modiolus are longer. Only the longest stereocilia contact the tectorial membrane that overlays the hair cells. These structures are called stereocilia because they resemble cilia without having the same basic structure of 9 double microtubules surrounding a central pair. Electron micrographs reveal thin filamentous connections between the tips of the short stereocilia and the taller ones. These **tip links** connect to a K^+ channel. Bending of the stereocilia away from the center of the modiolus excites the hair cell by stretching the tip links, opening a K^+ channel and depolarizing the cell.

As shown in Figure 4.7.4, the stereocilia protrude through a cellular layer called the **reticular lamina.** This structure divides the fluid in the scala media, which is filled with endolymph, from the scala tympani, which is filled with perilymph. The endolymph is the most unusual extracellular fluid in the body. This extracellular fluid has [K^+] of about 150 mM and [Na^+] of about 1 mM. In addition, it is maintained at a high positive potential, +50 to +100 mV. This highly unusual fluid is maintained by the **stria vascularis** (see Figure 4.7.3). The positive endolymph potential is maintained by active processes within the stria vascularis, and it collapses within minutes after interrupting the oxygen supply. The basolateral surfaces

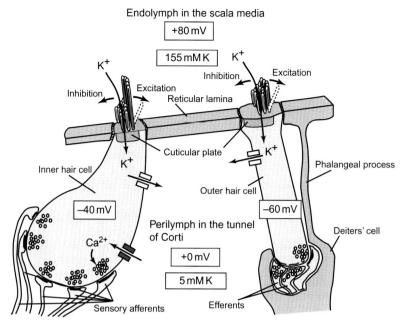

FIGURE 4.7.4 Diagram of the inner and outer hair cell arrangement. The inner hair cells have stereocilia whose first row is much longer than the others. The endolymph has a high $[K^+]$ and a positive potential that is actively maintained by the stria vascularis. Stretching of the tip links between the stereocilia open K^+ channels on the stereocilia membranes that results in K^+ inward flow and depolarization. These channels at rest have some conductance so that relaxation of the tip links produces a reduction in the K^+ current with resulting hyperpolarization and inhibition of the nervous activity on the sensory afferent nerves. (Source: Modified from Geisler, From Sound to Synapse, Oxford University Press, 1998.)

of both the inner and outer hair cells lie beneath the reticular lamina. The basilar membrane is freely permeable to small ions, whereas the reticular lamina seals the fluid in the scala media (endolymph) from that in the tunnel of Corti (perilymph). Thus, the fluid bathing the basolateral surfaces of the hair cells is perilymph, with high $[Na^+]$ and low $[K^+]$. The low $[K^+]$ and high $[Na^+]$ is necessary for the propagation of the action potential along the afferent sensory axon.

When the long stereocilia tug on the shorter ones through the tip links, K^+ channels are opened, causing K^+ ions to enter the hair cell, because the potential in the scala media is $+80$ mV whereas that in the hair cell is some -40 to -60 mV, and $[K^+]$ is comparable between the scala media and the hair cell. This flow depolarizes the cell, leading to opening of voltage-gated Ca^{2+} channels on the basolateral aspect of the hair cells. The depolarization of the hair cells is its **transduction potential**. Because there is a potential difference between the scala media, containing endolymph, and the scala tympani, containing perilymph, there is always some current passing through the reticular lamina, and movement of the stereocilia towards the modiolus reduces the conductance to K^+ and reduces the current. Thus, movement of the stereocilia causes either excitation or inhibition.

OUTER HAIR CELLS MOVE IN RESPONSE TO EFFERENT STIMULATION AND THEREBY TUNE THE INNER HAIR CELLS

Outer hair cells change their axial dimensions in response to electrical stimulation. Hyperpolarization lengthens the cells, and depolarizations shorten them. This "**electromotility**" depends on membrane potential, and it occurs extremely rapidly. Outer hair cells can follow sinusoidal electrical commands up to 24 kHz. The outer hair cells can shorten or lengthen only a few percent of their rest length of 30 μm, but this is about the same magnitude as sound evoked vibrations of the basilar membrane. The mechanism of electromotility is not established but requires the protein **prestin**. Prestin knockouts in mice result in 20–40 dB hearing loss.

Instead of sound vibration being dampened by the absorption of sound energy by the lamina, the electromotility response of the outer hair cells adds energy, something like negative damping. This phenomenon has been termed the **cochlear amplifier**. This allows for the exquisite response of the inner hair cells. Loss of outer hair cells produces a loss of sensitivity and frequency discrimination.

In 1978, Kemp reported that brief clicks directed to the ear resulted in faint sounds coming out of the ear. These are called **evoked otoacoustic emissions**. Later studies found that otoacoustic emissions could emanate spontaneously from unstimulated cochleas. Otoacoustic emissions from one ear can be modulated by sound presented to the contralateral ear, suggesting that otoacoustic emissions probably originate from the outer hair cells.

THE COCHLEA PRODUCES A TONOTOPIC MAPPING OF SOUND FREQUENCY

As described above, hair cells sense sound through deformation of the stereocilia on their apical membranes. How is this deformation produced? George von Békésy first described vibration of the basilar membrane in a series of experiments performed on cochleas from cadavers during the 1940s (G. von Békésy, *Experiments on Hearing*, McGraw Hill, 1960). This remarkable achievement earned a Nobel Prize for von Békésy in 1961. He found that pressure waves applied to the oval window resulted in a "traveling wave" of displacement of the basilar membrane. The wave is conveyed from the base of the cochlea, near the oval window, toward the apex of the cochlea near the helicotrema. The amplitude of the displacement varies with the distance along the cochlea

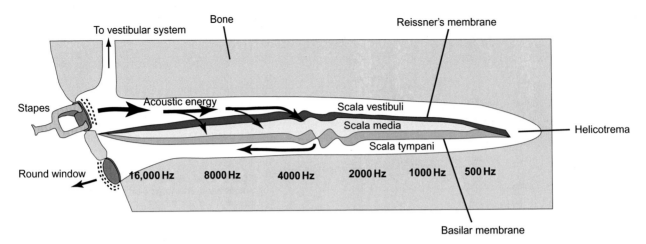

FIGURE 4.7.5 Highly schematic diagram of the cochlea, unwound from its spiral as if it were straight. The cochlear partition between scala vestibuli and scala tympani consists of Reissner's membrane and the basilar membrane, including the organ of Corti. The fluid bathing the scala vestibuli and scala tympani is perilymph, whereas the scala media is filled with endolymph. Acoustic energy produced from a loud tone is conveyed to the perilymph in the scala vestibuli by the vibration of the oval window by the stapes. Most of the energy is absorbed by an area of the cochlear partition that depends on the frequency of the sound, and this produces tuned vibrations of the cochlear partition. Some of the sound energy remains to displace the round window, resulting in sound emission from the cochlea back into the middle ear. *(Source: Modified from Geisler, From Sound to Synapse, Oxford University Press, 1998.)*

and with the frequency of the sound. For a particular tone, the maximum amplitude is located within a fairly narrow region of the cochlea. In Von Békésy's experiments, the displacement was broadly distributed along the cochlea because he measured traveling waves at high sound levels and he used dead cochleas. Later, more sensitive techniques with live cochleas showed a much more sharply tuned displacement of the basilar membrane.

At the base of the cochlea, near the oval window, the basilar membrane is narrow. As it winds around the modiolus, the basilar membrane becomes progressively wider, reaching its widest dimensions at the apex of the cochlea. Because of this, high-frequency sounds produce a maximum displacement of the basilar membrane near the base of the cochlea, whereas low-frequency sounds maximally displace the basilar membrane near the apex. The result is that inner hair cells located near the oval window, near the base of the cochlea, respond to high-frequency sounds and inner hair cells located near the apex respond to low-frequency sounds. Thus the cochlea is organized **tonotopically**: there is a relationship between frequency response and location on the cochlea. This idea is illustrated in Figure 4.7.5.

AUDITORY INFORMATION PASSES THROUGH THE BRAIN STEM TO THE AUDITORY CORTEX

The inner hair cells make synapses on the processes of **bipolar cells** whose cell bodies are located in the **spiral ganglion**, buried in the bone of the modiolus. There are about 30,000 of these cells in the human spiral ganglion, and the vast majority of these make contact with a single inner hair cell, and each inner hair cell contacts between 10 and 20 primary afferent auditory nerve fibers. The fibers leave the spiral ganglion and are collected in the **auditory nerve** that joins the vestibular nerve to form **cranial nerve VIII** (Figure 4.7.6).

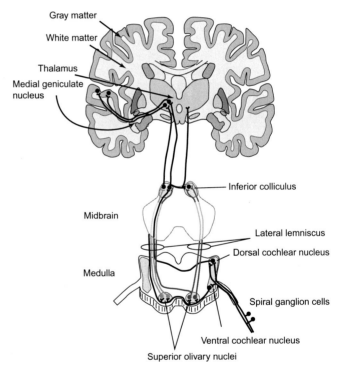

FIGURE 4.7.6 Schematic diagram of central auditory pathways. Spiral ganglion cells receive afferent signals directly from inner hair cells. The primary afferent processes enter in the medulla and make synapses on cells in the dorsal and ventral cochlear nuclei in the medulla. Second-order fibers ascend in the contralateral lateral lemniscus to make contact with cells in the inferior colliculus. Neurons in the ventral cochlear nucleus send collaterals to the superior olivary nucleus. Third-order cells in the olivary nuclei send ascending fibers to the inferior colliculus. Cells in the inferior colliculus in turn send fibers to the medial geniculate nucleus of the thalamus, which relays the information to the auditory region of the temporal lobe in the cerebral cortex. *(Source: Modified from Nicholls, Martin, Wallace, and Fuchs, From Neuron to Brain, Sinauer Associates, 2001.)*

Primary auditory nerve fibers in cranial nerve VIII make synapses on secondary afferent neurons in the **dorsal and ventral cochlear nuclei** which are located in the brain stem. The secondary afferent fibers originating in the

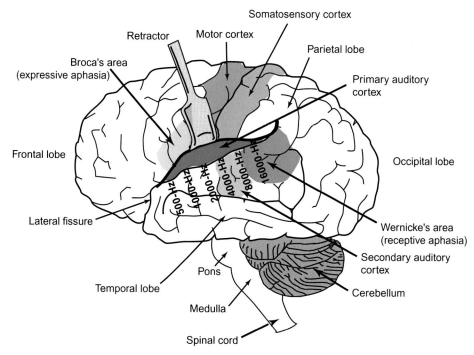

FIGURE 4.7.7 The primary auditory cortex (A1) and areas for producing and understanding speech. Afferents from the apex of the cochlea feed into anterior areas of A1, and these respond to low sound frequencies. Afferents from the base of the cochlea feed into posterior regions of A1, and these process high-frequency sounds. Broca's area lies in the frontal lobe adjacent to the areas of the motor cortex responsible for motor control of the face, lips, jaw, tongue, pharynx, and larynx. Damage to this area causes expressive aphasia. Damage to Wernicke's area, in the posterior part of the temporal lobe adjacent to the secondary auditory cortex, causes receptive aphasia. (Source: Modified from Kolb and Whishaw, An Introduction to Brain and Behavior, Worth Publishers, 2000.)

cochlear nuclei may take a variety of paths. Some synapse on neurons located in the **superior olivary complex**, both ipsilateral and contralateral. The superior olivary complex is a collection of nuclei located in the medulla. Fibers leaving the superior olive then synapse on neurons in the **inferior colliculus**. The inferior colliculus also receives input from the eyes, so these connections make possible complex behavior such as seeking the source of a sound.

Fibers from the inferior colliculus project to the **medial geniculate nucleus** (MGN) of the **thalamus** and also cross over to the contralateral side. These make contact with another set of neurons that send fibers directly to the **primary auditory cortex**, located laterally near the Sylvian sulcus. The primary auditory cortex is also called A1 and is located within Heschl's gyrus, the transverse temporal gyrus. The perception of pitch is mapped onto A1 **tonotopically**. The neurons in the cochlea code for frequency based on their location in the cochlea: cells at the base respond to high frequencies and cells at the apex of the cochlea respond to low frequencies. This spatial arrangement of sensitivity to tone is preserved throughout the neural connections so that there is a mapping of location on the A1 cortex and pitch: cells in the anterior part of A1 respond to low frequencies and cells in the posterior part respond to high frequencies (see Figure 4.7.7).

LANGUAGE IS PROCESSED IN AREAS NEAR THE PRIMARY AUDITORY CORTEX IN THE LEFT HEMISPHERE, BUT MUSIC IS PROCESSED IN THE RIGHT HEMISPHERE

Based on "natural experiments," Paul Broca (1824–1880) identified an area immediately in front of the central fissure and superior to the lateral fissure, now called **Broca's area**. Damage to this area results in **expressive aphasia**, the inability to speak despite being able to understand spoken language and despite having normal vocal apparatus. Karl Wernicke later identified a separate area, **Wernicke's area**, located in the posterior of the left temporal lobe. Damage to this area results in **receptive aphasia**, which is the inability to understand the spoken word. People with receptive aphasia can speak, but their speech is garbled and unintelligible.

Human experience is replete with anecdotes of persons whose brains have been damaged and have lost the ability to speak, and yet they can sing. This is possible because the processing of music is located in separate areas in the temporal lobe of the right hemisphere. Despite this, cooperation among many brain areas is needed for full function, as damage to the left hemisphere can interfere with the ability to read or write music.

PERCEPTION OF PITCH IS ACCOMPLISHED BY A COMBINATION OF TUNING AND PHASE LOCKING

As described earlier, the primary afferents for sound detection arise from the inner hair cells that respond to deformation of the stereocilia in response to movement of the tectorial and basilar membranes. Pressure waves transmitted to the oval window through the action of the ossicles produce a wave of basilar membrane deformation that has a maximum displacement that depends on the frequency, but it also depends on the magnitude of the pressure waves. The physical response of these membranes, along with the electrical properties of the inner hair cells, produces a "tuning curve" for the primary afferent fibers. These tuning curves are the threshold intensities for impulse generation as a function of

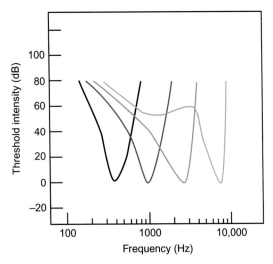

FIGURE 4.7.8 Tuning curves for primary afferents at different locations on the cochlea. Afferents from inner hair cells near the apex show a maximum sensitivity (lowest threshold) to sounds with low frequencies (black curve). Primary afferents from inner hair cells near the base show maximum sensitivity to higher frequencies (curve on far right). Cells in between show intermediate behavior (middle two curves). The sensitivity falls off sharply at higher frequencies, but lower frequencies can still excite the afferents off their maximum sensitivity. *(Source: Modified from Lieberman and Mulroy, in Hamernik, Henderson, and Davis, eds., New Perspectives on Noise-Induced Hearing Loss, Raven Press, 1982.)*

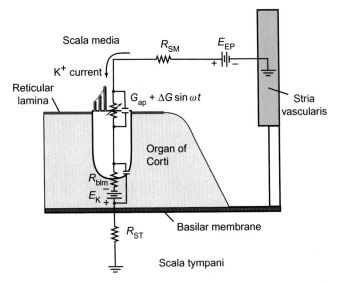

FIGURE 4.7.9 Schematic diagram of the resistances and potentials across the basilar membrane. Ground is taken as the vascular system. The stria vascularis produces an endolymphatic potential, E_{EP}, which is approximately +80 mV with respect to the scala tympani (ST). The scala media (SM) and scala tympani present resistances to current flow. Current flows across the cell membrane in accordance with a steady-state conductance G, which varies when K^+ channels are opened on the apical membrane by movement of the stereocilia. A sinusoidal variation in the apical conductance results in a sinusoidal variation in the K^+ current and a sinusoidal variation in the membrane potential. *(Source: Modified from Geisler, From Sound to Synapse, Oxford University Press, 1998.)*

frequency. Typical tuning curves for primary afferents of the auditory nerve are shown in Figure 4.7.8.

The tuning curves show a fairly sharp maximum sensitivity (lowest threshold) at a **characteristic frequency**. The sensitivity falls off sharply at higher frequencies but generally shows a "tail" at lower frequencies. This is because the traveling wave for high frequencies travels only so far up the cochlea, whereas all low-frequency sounds must pass through the inner hair cells' location to reach their place of maximum basilar membrane displacement. Louder sounds produce larger deformations of the basilar membrane, so that a louder sound might be confused with one of higher pitch. There is psychophysical evidence for a shift in the perceived pitch with sound intensity. The auditory system deals with this problem by using a phase-lock coding of sound frequency in addition to the tuning curves of the primary afferents.

Because a pure tone transmitted to the perilymph induces a sinusoidal displacement of the basilar membrane, the primary afferent neurons tend to fire in phase with the sound wave. Thus, the neurons fire at the same part of the sound wave, usually its maximum pressure. Thus, their firing rate encodes the original frequency of the sound. However, each primary afferent has a limited firing rate of something less than 200 Hz; they cannot keep up with tones having a higher frequency than this. Because each inner hair cell is innervated by 10 or more primary afferents, the ensemble does indeed track the stimulus waveform. The convergence of these primary afferent neurons on targets within the cochlear nucleus suggests that summation within the cochlear nucleus allows a frequency coding of sound pitch as well as a place coding (the tonotopic map). This requirement probably explains the curious multiple innervation of the inner hair cells. These combined codes allow for simultaneous discrimination of sound intensity and pitch.

THE *COCHLEAR MICROPHONIC* SHOWS THAT THE INNER HAIR CELLS HAVE AN AC RESPONSE THAT CAN KEEP UP WITH MODERATE FREQUENCY VIBRATIONS

In a rather bizarre experiment, Weaver and Bray in 1930 reported that the electrical activity recorded from a cat's auditory nerve could be amplified and hooked up to speakers. If you then spoke into the cat's ear, you could simultaneously hear yourself over the speaker! In 1931, Adrian found that the source of the AC signal was not the nerve, but the cochlea, and he coined the term **cochlear microphonic** to describe it. Figure 4.7.9 shows an equivalent electrical circuit for a hair cell in which a sinusoidal perturbation of the hair cells' stereocilia, at the frequency of the sound, would produce a similar sinusoidal variation in the current across the organ of Corti and a sinusoidal variation in the membrane potential.

Note that all of the current that passes through the hair cell in Figure 4.7.9 also passes through the resistances R_{SM} and R_{ST}, and gives rise to a transduction voltages in the scala media and scala tympani. As currents from all hair cells share these pathways, the total current is the sum of all the individual hair cell currents. The voltage

Clinical Applications: Cochlear Implants

The first cochlear implant was developed by Graeme Clark at the University of Melbourne, Australia, and implanted in 1978. Cochlear implants are now commercially available. The devices are not hearing aids, which merely amplify sound. Instead, the devices themselves receive sound and distribute the power into a series of bandwiths and stimulate the hearing nerve tonotopically by bypassing the ossicles and the oval window. It can restore partial hearing to the deaf.

Cochlear implants consist of both internal and external parts. The internal parts are surgically implanted, whereas the external parts can be detached at any time. The external parts consist of a microphone, a speech processor, and a transmitter. The microphone picks up sounds and converts them into electronic signals. These are sent over a thin wire to the speech processor, which divides the sound into frequency bands and sends the coded signal to a transmitter worn outside the skull. The transmitter sends the coded signal across the skin by radio waves.

The internal parts consist of a receiver and an electrode array. The receiver converts the radio waves into electrical signals that it sends down an electrode array implanted in the cochlea. Because the cochlea detects sound frequency according to the location of the inner hair cells along the cochlea, with high frequency at the base of the cochlea and low frequency at the apex, the electrode array can stimulate local regions of the cochlea and elicit specific perceptions of sound frequency. The Nucleus® 24 divides sound into a maximum of 22 channels. By dividing sound into 22 frequency bands and stimulating 22 different regions of the cochlea, the implant approximates normal hearing.

Fourier analysis divides any function into an infinite series of sines:

$$f(x) = a_0 + \sum_{n=1}^{\infty} a_n \sin \omega_n x$$

In the case of sound, $f(x)$ is the pressure as a function of time, which can be approximated by a series of 20 terms of different angular frequencies, ω. Ideally the power of received sound could be divided up according to a Fourier analysis, but this takes too much time to be done on the fly, in real time, by a small device. The speech processors divide the interesting part of the sound spectrum into 20 different frequency bands by using a system of bandpass filters. It then stimulates parts of the cochlea according to the power of the sound signal in each of these bands. A variety of speech coding strategies can be used.

Sound perception with cochlear implants is not normal. Wearers detect a "metallic" sound to their perception and it takes time and motivation to learn to hear with the device. The benefits vary with the user and their motivation to learn. Over half of the recipients can distinguish speech with no visual cues.

Clinical Applications: Hearing Tests for Newborns

Until recently, parents became aware of their child's deafness only when the child failed to respond to noises or failed to make sounds or develop speech. Modern instruments allow testing of hearing in infants as soon as 6 h after birth.

Two major methods are available: otoacoustic emissions and auditory brain stem response (ABR), also called brain stem auditory evoked response (BAER).

As mentioned in the text, the ear not only hears sounds but emits sounds. The otoacoustic emissions tests the ear for the sounds it emits as a consequence of hearing sounds. Clicks presented to the ear generate transient evoked otoacoustic emissions from a variety of regions of the cochlea corresponding to different frequencies of emitted sounds. Analysis of the spectrum of otoacoustic emissions can diagnose problems in hearing.

The ABR is a kind of evoked electroencephalogram. Each time a brief click is presented to the ear, neurons are activated sequentially in the pathway from ear to brain. The activity of these neurons causes minute potential changes that can be detected on the surface of the skin overlying the brain stem. The signals from these neurons are usually lost in the sea of noise produced by all of their neighboring neurons. The signal-to-noise ratio is improved by averaging many sweeps. Each sweep is triggered on a signal when a click is presented into the ear of the newborn. Since each response of the auditory system follows sequentially, signals from the auditory system occur at definite intervals following the click, but ongoing brain activity is not synchronized in this way. By taking many sweeps and averaging them, the random background noise of the brain cancels out, whereas the tiny signals from the auditory system do not cancel out. Movement of the head produces large signals (electromyograms) which typically do not happen often enough to cancel out during signal averaging, so the patient must be still. The test is usually performed during sleep.

Figure 4.7.10 shows a typical trace during a BAER. Each peak is attributed to a part of the neuronal chain linking ear to brain.

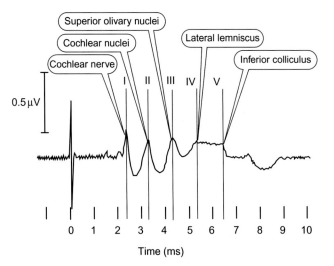

FIGURE 4.7.10 Tracing of ABR. Wave I is attributed to activity along the auditory nerve, cranial nerve VIII. This activates neurons in the cochlear nuclei (wave II) and then those in the superior olivary nuclei (wave III). The activity then passes through the lateral lemniscus (wave IV) to the inferior colliculus (wave V).

registered in the scala tympani is the sum of contributions from many cells. The time-varying portions, the AC response, is the cochlear microphonic. There is also a DC component that occurs at higher amplitude with greater frequency of sound stimulus.

SUMMARY

The sense of hearing informs us of the loudness, tone, and timbre of sounds. Loudness is measured in decibels, the logarithmic ratio of intensity or pressure relative to a reference. Each 20 decibel unit is a 10-fold increase in pressure level and a 100-fold increase in sound intensity. The ear is most sensitive to sounds between 500 and 5000 Hz.

The outer ear or pinna collects sound and funnels it into the external auditory meatus. The tympanic membrane at the end of the ear canal divides the outer ear from the inner ear. Sound makes the tympanic membrane vibrate, and this vibration transfers to three tiny bones—the malleus, incus, and stapes (hammer, anvil and stirrup)—that focuses the vibration onto the oval window, which marks the beginning of the inner ear. The pressure on the two sides of the tympanic membrane equilibrates through the eustachian tube that connects the inner ear with the oral cavity. Yawning or chewing can open the tube and pressure can be equilibrated.

The relevant part of the inner ear for hearing is the cochlea. The entire inner ear is surrounded by a bony labyrinth. The cochlea forms a spiral around a bony core called the modiolus. The cochlea itself consist of three compartments: the scala vestibuli, the scala media, and the scala tympani. The scala vestibuli and scala tympani are connected at the apex of the cochlea at the helicotrema, and both are filled with perilymph. The scala media is filled with endolymph, which has a high $[K^+]$ and a high voltage ($+80$ mV) relative to the perilymph. The stria vascularis in the walls of the scala media maintains this unusual fluid. Reissner's membrane separates the scala vestibuli from the scala media, and the basilar membrane separates the scala tympani from the scala media. Vibration of the oval window is transferred to the scala vestibuli and eventually returns via the scala tympani to the round window. The receptor cells that respond to sound vibrations reside in the organ of Corti, which lies on the basilar membrane. This complicated device consists of a single row of inner hair cells and three rows of outer hair cells, supported and arranged by supporting structures, all overlaid by a gelatinous tectorial membrane. The inner hair cells are richly innervated with afferent nerves whose cell bodies reside in the spiral ganglia in the modiolus. The outer hair cells receive a rich supply of efferent nerves. The outer hair cells move in response to nervous stimulation. It is believed that the inner hair cells respond to sound, while the outer hair cells tune the response of the inner hair cells.

The hair cells owe their name to a field of stereocilia on their apical surface. This field has a polarity, with stereocilia becoming progressively larger as one moves away from the center of the spiral. The long sterocilia connect to shorter ones by tip links. Movement of the stereocilia away from the modiolus tugs on the tip links that open K^+ channels in the apical membrane, which depolarizes the cells and activates the spiral ganglion cells.

Sensory information from the cochlea joins the vestibular nerve in cranial nerve VIII and enters the brain stem to make synapses with secondary afferent neurons in the dorsal and ventral cochlear nuclei. These neurons project to neurons in the superior olivary nucleus both ipsilaterally and contralaterally, and cross over to ascend to the inferior colliculus. Neurons in the inferior colliculus project to the medial geniculate nucleus in the thalamus, which then projects to the primary auditory cortex. The auditory cortex is found beneath the lateral fissure. Sound tone is represented tonotopically: the base of the cochlea responds to high-frequency sound and the apex responds to low frequencies. This mapping is preserved through the auditory pathways so that the anterior region of the auditory cortex senses low frequencies and the posterior region senses high frequencies.

REVIEW QUESTIONS

1. How are decibels related to sound intensity? What is the relationship between sound intensity and pressure amplitude? What is timbre?
2. Where is the eustachian tube and what does it do? What bone connects to the eardrum? What bone connects to the oval window?
3. What is the function of the tensor tympani and stapedius muscles? Why do you suppose you can sometimes hear your heartbeat?
4. What cells form the major afferent component of hearing? What do the outer hair cells do?
5. Where are the scala tympani and scala vestibuli? Where is the scala media? What fluid does each contain? What is so unusual about the endolymph?

6. Where are the cell bodies of auditory nerve primary afferents? Where do their axons project? Where is the primary auditory cortex?
7. What is meant by tonotopic mapping? What regions of the cochlea respond to high frequencies? Low frequencies? What areas of A1 respond to high frequencies? Low frequencies?
8. Where is Broca's area? What deficits result from its damage. Where is Wernicke's area? What deficits result from its damage?

APPENDIX 4.7.A1 THE PHYSICS OF SOUND

THE SPEED OF SOUND IS RELATED TO AIR DENSITY AND BULK MODULUS

Consider a right cylinder with cross-sectional area, A, that is filled with a gas of density ρ. This gas has two characteristics that are required for the transmission of sound: it possesses inertia and it is elastically deformable. Gases are characterized by their **bulk modulus**, defined as

[4.7.A1.1] $$B = \frac{\Delta P}{-\Delta V/V}$$

where ΔP is the change in pressure, V is the volume, and ΔV is the change in volume. The bulk modulus is always taken as being positive. What this equation says, then, is that a positive pressure increment results in a negative volume increment. That is, applying a pressure to a gas causes it to compress; reducing the pressure causes it to expand.

Suppose now that the cylinder of gas is fitted with a piston by which we can deliver a rapid force to the right. When this occurs, the piston will compress the gas immediately adjacent to it. This increased pressure will cause the gas molecules to move to the right, transmitting the pressure to the next volume element. This repeated process generates a compressional wave pulse that moves to the right (see Figure 4.7.A1.1).

For simplicity we assume that the piston moves with constant velocity v, and this causes all particles in the pressure pulse to move to the right at velocity v. If the piston moves for the time Δt, the trailing edge of the pressure pulse will have moved a distance $v\Delta t$. During this interval, the leading edge of the wave pulse will have advanced a distance $c\Delta t$, where c is the **wave speed**.

The longitudinal displacement of the particles is taken as y, and their original location is designated as x. Thus, after the interval Δt, the longitudinal displacement y will vary with x, as shown in Figure 4.7.A1.1.

The mass of gas originally at rest in the volume $V = c\Delta tA$ is just the undeformed density, ρ, times this volume, or $m = \rho c\Delta tA$. The change of momentum of the gas molecules provided by the piston, and present in the wave pulse, is

[4.7.A1.2] $$\Delta(mv) = \rho c\Delta tAv$$

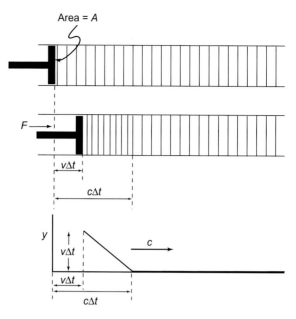

FIGURE 4.7.A1.1 Cartoon of a right circular cylinder fitted with a piston that moves with velocity v for time Δt. This produces a change in volume $\Delta V = -v\Delta tA$; this compression pulse travels at velocity c, so that the effective volume element which is being compressed is $V = c\Delta tA$.

The average force, F, times the time applied gives the momentum change. This is the impulse–momentum theorem, which is a direct result of Newton's third law. Insertion of this result into Eqn [4.7.A1.2] gives

$$F\Delta t = \Delta(mv) = \rho c\Delta tAv$$
$$F = \rho cAv$$

[4.7.A1.3] $$\frac{F}{A} = \rho cv$$

$$\Delta P_0 = \rho cv$$

The force divided by the area, A, gives the longitudinal stress, which is equal to the pressure increment, ΔP_0, which is the maximal pressure that is produced at the end of the piston's excursion at $t = \Delta t$. This is the increase in pressure over the pressure that was present before the pulse. We will insert this into the definition of the bulk modulus, but first we want to solve for v in terms of volume and c. We recognize that the undisturbed volume was $V = Ac\Delta t$, and the change in volume was $\Delta V = -Av\Delta t$. The minus sign indicates that the volume decreased upon application of the positive longitudinal stress. This allows us to solve for v:

$$V = cA\Delta t$$
$$\Delta V = -vA\Delta t$$

[4.7.A1.4] $$v = -\frac{\Delta V}{V}c$$

Inserting this last result into Eqn [4.7.A1.3] gives

[4.7.A1.5] $$\Delta P_0 = \rho c^2 \left(-\frac{\Delta V}{V}\right)$$

We can solve for c using the definition of the bulk modulus, to obtain

$$[4.7.A1.6] \qquad c = \sqrt{\frac{B}{\rho}}$$

THE SPEED OF SOUND DEPENDS ON TEMPERATURE AND THE MOLECULAR WEIGHT OF THE AIR MOLECULES

The bulk modulus, B, in Eqn [4.7.A1.6] is that defined in Eqn [4.7.A1.1]. It contains terms of dP, V, and dV. We have a relation that links these variables, and that is the ideal gas law: $PV = nRT$. However, we do not know the relation between P and V unless we know the temperature. The propagation of sound waves involves compression and expansion of gas. Compression of gas heats it and expansion of gas cools the gas. This is the basis for refrigeration and air-conditioning. Newton solved the problem of determining c by assuming that the temperature was constant. He reasoned that the propagation was so rapid that the temperature could not rise or fall. His results give the isothermal speed of sound, and it is wrong. Laplace made the argument that the propagation of sound occurred **adiabatically**, or without heat flow from or to the surroundings. What we need is the adiabatic relation between P and V.

The Molar Specific Heat at Constant V Is Not the Same as the Molar Specific Heat at Constant P

The specific heat of a substance is defined as the amount of heat energy required to raise the temperature of 1 g of the substance by 1 K. For gases, this is generally expressed in terms of moles rather than grams, and so it is called the **molar specific heat**. There are two common ways of determining the molar specific heat of gases: either by keeping the volume constant or by keeping the pressure constant. The molar specific heat at constant volume is denoted as C_V and the molar specific heat at constant pressure is denoted as C_P. These molar specific heats have the units of joules $\text{mol}^{-1}\,\text{K}^{-1}$.

The first law of thermodynamics is equivalent to the conservation of energy. We can write it as

$$[4.7.A1.7] \qquad dQ = dU + dW$$

where dQ is the increment of heat energy that enters a system from the surroundings, dU is the change in the internal energy of the system, and dW is work done *by* the system. In the case of a gas, the work it will do will be $dW = \int F dx = \int F/A\; A dx = \int P dV$. Expansion involves positive work; compression involves negative work (work is done *on* the system by the surroundings). Consider the case where volume is constant. Under these conditions, $dV = 0$ and so $dW_V = 0$. By the definition of C_V, we have

$$[4.7.A1.8] \qquad C_V = \frac{dQ_V}{n\,dT}$$

where dQ_V is the increment of heat energy at constant V, n is the number of moles of gas being heated isovolumetrically, and dT is the temperature increment. Solving for dQ_V and inserting into the conservation of energy, Eqn [4.7.A1.7], we get

$$[4.7.A1.9] \qquad \begin{aligned} dQ_V &= dU_V + dW_V \\ dQ_V &= nC_V\,dT \\ dW_V &= 0 \quad \text{(condition of isovolumetric heating)} \\ dU_V &= nC_V\,dT \end{aligned}$$

Now let us suppose that we heat the same amount of gas by the same increment, dT, but this time we do it at constant pressure rather than constant volume. Here the work increment, dW, will not be zero. As we heat the gas it will expand. Here the heat absorbed is given from the definition of the molar specific heat:

$$[4.7.A1.10] \qquad C_P = \frac{dQ_P}{n\,dT}$$

We insert this and the work term into the conservation of energy:

$$[4.7.A1.11] \qquad \begin{aligned} dQ_P &= dU_P + dW_P \\ dQ_P &= nC_P\,dT \\ dW_P &= P\,dV_P \\ dU_P &= nC_P\,dT - P\,dV_P \end{aligned}$$

Now the internal energy of a gas depends on the temperature alone and not on P or V singly. Thus for the same temperature increment, the internal energy increment is the same. Therefore, we can equate the last line in Eqns [4.7.A1.9] and [4.7.A1.11] to obtain:

$$[4.7.A1.12] \qquad nC_V\,dT = nC_P\,dT - P\,dV_P$$

Here we make use of the Ideal Gas Law, $PV = nRT$, which at constant pressure gives $P\,dV_P = nR\,dT$.

Substitution into Eqn [4.7.A1.12] gives

$$[4.7.A1.13] \qquad nC_V\,dT = nC_P\,dT - nR\,dT$$

which finally gives us

$$[4.7.A1.14] \qquad C_P - C_V = R$$

The value of the gas constant here should be in heat units: 8.314 J $\text{mol}^{-1}\,\text{K}^{-1}$ or 1.99 kcal $\text{mol}^{-1}\,\text{K}^{-1}$. The reason Eqn [3.8.A1.14] is true is that all of the heat energy entering a gas at constant volume goes into raising the gas's temperature. When the gas is allowed to expand, some of this energy also goes into the work of expansion.

For Adiabatic Processes, PV^γ = constant, Where $\gamma = C_P/C_V$

We next consider an adiabatic exansion of a gas. Here **adiabatic** means that there is no heat transfer during the process. The gas performs positive work during expansion, which is $dW = \int P\,dV$. The energy for this work comes from the internal energy. Thus, the internal energy decreases by the amount $dU = nC_V\,dT$. We use C_V here because we are writing only the internal energy change; we use C_P to include both the work term and the internal energy term. This should be clear from inspecting Eqns [4.7.A1.11] and [4.7.A1.12], from which we see that $dU_P = nC_V\,dT$. We insert this result into the First Law of Thermodynamics, Eqn [4.7.A1.7], to find

[4.7.A1.15]
$$dQ = dU + dW$$
$$0 = nC_V\,dT + P\,dV$$

We can insert the Ideal Gas Law, $PV = nRT$, into this equation, separate the variables, and integrate:

$$0 = nC_V\,dT + \frac{nRT}{V}dV$$

$$nC_V dT = -\frac{nRT}{V}dV$$

[4.7.A1.16]
$$\int \frac{dT}{T} = -\frac{R}{C_V}\int \frac{dV}{V}$$

$$\ln T = -\frac{R}{C_V}\ln V + K$$

$$TV^{\frac{R}{C_V}} = K_1$$

Since PV is proportional to T for an ideal gas, and $R = C_P - C_V$, we can rewrite the last equation as

[4.7.A1.17]
$$PVV^{\frac{R}{C_V}} = K$$
$$PVV^{\frac{C_P - C_V}{C_V}} = K$$
$$PVV^{\frac{C_P}{C_V}}V^{-1} = K$$

The final relation is written as

[4.7.A1.18]
$$PV^{\frac{C_P}{C_V}} = PV^{\gamma} = K$$

Inserting the Adiabatic Relation $PV^{\gamma} = K$ Gives the Speed of Sound as a Function of Temperature

Our three key equations are outlined above: they are the definition of the bulk modulus, B; the dependence of the speed of sound on the square root of B/ρ; and the relation between P and V during adiabatic expansion (and compression). These can be united by taking the total differential of Eqn [4.7.A1.18] and incorporating the result into Eqns [4.7.A1.1] and [4.7.A1.6]. The total differential of Eqn [4.7.A1.18] can be rearranged to obtain

[4.7.A1.19]
$$\Delta PV^{\gamma} + P\gamma V^{(\gamma-1)}\Delta V = 0$$
$$\Delta P = -P\gamma V^{-1}\Delta V$$
$$\frac{\Delta P}{-\Delta V/V} = \gamma P$$

This last equation gives an expression for B, the bulk modulus, in terms of the heat capacities and the pressure. Inserting it into our equation for the velocity of sound, we find

[4.7.A1.20]
$$C = \sqrt{\frac{\gamma P}{\rho}}$$

Here we can substitute in again for P from the ideal gas law, $PV = nRT$. Recall that ρ is the density of the gas. This is its mass, m, divided by its volume, V. The mass in any volume is the number of moles, n, times the molecular weight, M, in g mol^{-1}. Substituting in $P = nRT/V$ and $\rho = nM/V$ into Eqn [4.7.A1.20], we get

[4.7.A1.21]
$$C = \sqrt{\frac{\gamma nRT/V}{nM/V}}$$

[4.7.A1.22]
$$C = \sqrt{\frac{\gamma RT}{M}}$$

This final equation shows that the speed of sound varies with the square root of the temperature and inversely with the molecular weight of the gas. For air at $0°C$, $T = 273$ K, $M = 28.8$ g mol^{-1}, $R = 8.314$ J mol^{-1} K^{-1}, $C_P = 29.72$ J mol^{-1} K^{-1}, and $C_V = 21.41$ J mol^{-1} K^{-1}. Plugging in these values, we calculate a speed of sound to be 331 m s^{-1}. This is equal to the measured speed of sound in dry air at $0°C$.

THE SPEED OF SOUND IS RELATED TO THE AVERAGE SPEED OF AIR MOLECULES

The equipartition theorem of thermodynamics states that each degree of freedom of movement has an equilibrium energy of $kT/2$, where k is Boltzmann's constant (the gas constant per molecule) and T is the temperature in K. In a diatomic molecule such as N_2, which is the major constituent of air, there are three translational degrees of freedom (the x-, y-, and z-directions) so that the average kinetic energy is $3kT/2$. In addition, the molecule can rotate about its center of mass with three axes of rotation. One of these, however, passes directly through the nuclei of the N atoms and so the energy associated with this rotation is negligibly small. Thus the remaining two degrees of rotation have a total energy of kT. The total energy is thus $5kT/2$. This is the equilibrium energy per molecule in these modes of movement. For a mole of material, the internal energy content is $U = 5RT/2$. The molar heat capacity for these diatomic molecules at constant volume (where all the heat energy is converted to internal energy) is $C_V = 5R/2$. Equation [4.7.A1.14] tells us that $C_P = C_V + R = 7R/2$. This simple calculation suggests that the molar heat capacity for air should be approximately $3.5 \times R = 3.5 \times 8.314$ J mol^{-1} K^{-1} = 29.10 J mol^{-1} K^{-1}.

The observed value, 29.72 J mol^{-1} K^{-1}, is close to this theoretical value.

In addition, if $C_V = 5R/2$ and $C_P = 7R/2$, then their ratio $\gamma = C_P/C_V = 1.40$. This is close to the 1.39 observed for air.

According to the equipartition theorem, then, the average kinetic energy of gas molecules should be given by

[4.7.A1.23] $$\frac{1}{2}m\mathbf{v}^2 = \frac{3}{2}kT$$

Multiplying through by Avogadro's number, $N_0 = 6.02 \times 10^{23}$ particles mol^{-1}, we convert m, the mass per molecule, to the molecular weight and we convert k, Boltzmann's constant, to R:

[4.7.A1.24] $$M\mathbf{v}^2 = 3RT$$

We can substitute in for RT in Equation [4.7.A1.22] to obtain

[4.7.A1.25] $$c = \sqrt{\frac{\gamma(M\mathbf{v}^2/3)}{M}}$$
$$c = \sqrt{\frac{\gamma}{3}}\mathbf{v}$$

The value of γ is about 1.4. This equation predicts that the speed of sound is on the same order of magnitude as the average velocity of air molecules. This makes intuitive sense, because it is the motion of the air molecules that propagates the sound.

THE INTENSITY OF SOUND IS PROPORTIONAL TO THE SQUARE OF THE PRESSURE

The intensity of sound can be derived in a variety of ways, usually by writing a wave equation for the propagating sound wave and summing the energy of the oscillating particles making up that wave, treating them all as harmonic oscillators. We can derive it perhaps more easily from the energy delivered by the piston that began the wave, realizing that, in this model with no dissipation of energy, the power in must be equal to the power out.

The intensity of sound is defined as the energy per unit area per unit time, or the power per unit area. It is usually given in units of W m^{-2} = J s^{-1} m^{-2}. The piston that produces the sound moved at a constant velocity, v, for the time interval Δt. The intensity of the resulting sound wave, transmitting the power of the piston without loss, is just the total energy delivered by the piston divided by the time and area. The total energy delivered is the force times the distance traveled in that interval, which is $v\Delta t$. The piston moves at a constant velocity but does not deliver a constant force because the gas is compressed more at the end than at the beginning. The energy transferred from piston to gas is

[4.7.A1.26] $$E = \int \Delta P A \, dx$$

where ΔP is the excess pressure, A is the area, and dx is the distance increment. This is just the force × distance work. The intensity, I, is given as

[4.7.A1.27] $$I = \frac{E}{A\Delta t} = \int \frac{\Delta PA \, dx}{A\Delta t}$$

The piston advances at a constant velocity, v, so that we can transform dx into dt:

[4.7.A1.28] $$x = vt$$
$$dx = v \, dt$$

Similarly, the longitudinal displacement, y, is linearly related to t. We can substitute in $v \, dt$ for dx in Eqn [4.7.A1.28]. In addition, we can substitute in for ΔP from Eqn [4.7.A1.5] to get

[4.7.A1.29] $$I = \int_0^{\Delta t} \frac{\rho c^2 (-\Delta V/V) V \, dt}{\Delta t}$$

Lastly, we substitute in for $V = cA\Delta t$ and $\Delta V = -vAt$. We use t here instead of Δt because the longitudinal displacement is varying with time as the piston moves. This gives

[4.7.A1.30] $$I = \int_0^{\Delta t} \frac{\rho c^2 (Avt/Ac\Delta t)v \, dt}{\Delta t}$$
$$I = \int_0^{\Delta t} \frac{(\rho c/\Delta t)v^2 t \, dt}{\Delta t}$$

Evaluation of the integral gives

[4.7.A1.31] $$I = \frac{\rho c v^2}{2}$$

From our initial analysis of the pressure and velocity, from Eqn [4.7.A1.3], we have $v = \Delta P_0/\rho c$. Inserting this value for v, we get our final equation for the intensity:

[4.7.A1.32] $$I = \frac{\Delta P_0^2}{2\rho c}$$

Thus the intensity of sound depends on the square of the pressure amplitude. The intensity has units of J s^{-1} m^{-2} = W m^{-2}.

PROPAGATION OF SOUND CAN BE WRITTEN AS A WAVE EQUATION FOR WHICH THE INTENSITY IS PROPORTIONAL TO THE SQUARE OF THE PRESSURE AMPLITUDE

The transmission of a force to the air produces a longitudinal displacement of the gas molecules, which propagates through the air. We can write a wave equation for a traveling sinusoidal wave through a compressible medium like air as

[4.7.A1.33] $$y = A_0 \sin(\omega t - kx)$$

Here A_0 is the amplitude, ω is the angular frequency, in rad s^{-1}, and k is the wave number, in units of rad cm^{-1}. Thus $\omega = 2\pi f = 2\pi/T$, where f is the frequency in cycles s^{-1} and T is the period of the cycle, in s cycle^{-1}. We can see from these relationships that each addition of T to the time t in the argument adds 2π to the argument of the sin, making a complete cycle. Similarly, the addition of a wavelength, λ, to the value of x in the argument also adds 2π to the argument of the sin; therefore, $k = 2\pi/\lambda$. Since the velocity of the wave is the frequency times the wavelength, we have

[4.7.A1.34]
$$c = \lambda f = \frac{2\pi}{k}\frac{\omega}{2\pi}$$
$$c = \frac{\omega}{k}$$

From the definition of the bulk modulus (Eqns [4.7.A1.1] and [4.7.A1.4]) we have

[4.7.A1.35] $$\Delta P = B\frac{v}{c}$$

The ratio of v/c can be discerned from Figure [4.7.A1.1]. The slope of y, the longitudinal displacement, as a function of x, the location, is given as

[4.7.A1.36] $$\frac{\partial y}{\partial x} = \frac{v\Delta t}{c\Delta t - v\Delta t} \approx \frac{v}{c}$$

since $v \ll c$. When v approaches or exceeds c, a new phenomenon results, the shock wave. The equations for sound are valid only when $v \ll c$. Thus Equation [4.7.A1.35] becomes

[4.7.A1.37] $$\Delta P = B\frac{\partial y}{\partial x}$$

Taking the partial derivative of Eqn [4.7.A1.33], we obtain

[4.7.A1.38] $$\frac{\partial y}{\partial x} = -kA_0 \cos(\omega t - kx)$$

where A_0 is the amplitude of the oscillation. From Eqn [4.7.A1.6], we write

[4.7.A1.39] $$B = c^2\rho$$

Inserting both of these last two equations into Eqn [4.7.A1.37], we arrive at

[4.7.A1.40] $$\Delta P = -kA_0 c^2 \rho \cos(\omega t - kx)$$

This equation can be rewritten as

[4.7.A1.41] $$\Delta P = -\Delta P_0 \cos(\omega t - kx)$$

where

[4.7.A1.42] $$\Delta P_0 = kA_0 c^2 \rho$$

Equations [4.7.A1.41] and [4.7.A1.33] show that the pressure wave and the longitudinal displacement are 90° out of phase: the pressure is greatest (and least) when the displacement is zero. We can use the relation $ck = \omega$ from Eqn [4.7.A1.34] to write

[4.7.A1.43] $$\Delta P_0 = \omega A_0 c\rho$$

The intensity of a sound wave is the energy per unit area per unit time. The unit area is oriented at a right angle to the direction of propagation of the wave. We can determine this by determining the energy in a single wavelength and dividing by the time it takes for that wavelength to be completed. This time is $T = 1/f$. Each particle in the sound wave can be viewed as undergoing simple harmonic oscillation at the angular frequency of the sound. It undergoes this oscillation in response to a mechanical stimulus which is related to the pressure. The energy of a simple harmonic oscillator is given as

[4.7.A1.44] $$E = \frac{1}{2}k'A_0^2$$

where k' is an equivalent force constant. For mechanical systems, $k' = \omega^2 m$, where ω is the angular velocity in rad s^{-1} and m is the mass. In this case, all points on a single wavelength have the same total energy, and so the energy in one wavelength is the energy per unit mass times the total mass. The total mass of air that is moving is the density of the air times its volume. The volume is λA, where λ is the wavelength and A is the area. Thus, the total energy in a wavelength is given as

[4.7.A1.45] $$E = -\frac{1}{2}\omega^2 \rho \lambda A A_0^2$$

Now the intensity is the energy per unit area per unit time. Thus we divide E by A and by T to obtain the intensity:

[4.7.A1.46] $$I = \frac{E}{AT} = \frac{1}{2}\omega^2 \rho \frac{\lambda}{T} A_0^2$$

Recognizing that $\lambda/T = c$, we write

[4.7.A1.47] $$I = -\frac{1}{2}\frac{\omega^2 \rho^2 c^2 A_0^2}{\rho c}$$

We write it in this way to better recognize the substitution of Eqn [4.7.A1.43] into Eqn [4.7.A1.47]. The result is the same as what we obtained before:

[4.7.A1.48] $$I = \frac{\Delta P_0^2}{2\rho c}$$

Note that the intensity of sound is independent of its frequency. This result is less apparent when the derivation was performed from the integration of energy delivered to the air, as opposed to the analysis of the resulting sound wave.

4.8 Vision

Learning Objectives

- List the three fundamental steps in vision
- List the three layers of the outer structure of the eyeball
- Describe the function of the iris and ciliary body
- Identify the fluid chambers of the eyeball and describe control of intraocular pressure
- Identify parts of the retina including optic disk, macula lutea, fovea
- Define "near point," "accommodation," "convergence," "refractive index"
- Write Snell's law of refraction and the thin lens formula
- Be able to calculate the refractive power of the lens
- Compare the major features of rods and cones
- List the events in phototransduction
- List the functions of the retinal pigmented epithelium
- Describe the response of bipolar cells to light in their receptive field
- Identify the cells in the retina that produce action potentials
- Distinguish between the magnocellular and parvocellular pathway
- Trace the optic nerve input from retina to cortex and identify the visual cortex

OVERVIEW OF THE VISUAL SYSTEM

Vision can be broken down into three sequential processes. These are:

A. Focus light on the retina
B. Transduce light to a nervous signal
C. Process the nervous signals to form conscious perception of objects.

The structure of the eye provides the means for focusing light onto the retina. The retina contains specialized photoreceptors that absorb light and transduce the light energy into neural signals. The processing of these neural signals begins at the retina and continues along pathways to the occipital lobe of the brain where the visual cortex processes the signals. Since we do not yet understand consciousness, we do not understand how conscious perception occurs. However, we understand something of how the signals are processed to form bits and pieces of the final **percept**.

THE STRUCTURE OF THE EYE ENABLES FOCUSING OF LIGHT ON THE RETINA

The outer structure of the eyeball contains three layers: the **sclera**, the **uvea**, and the **retina** (Figure 4.8.1). The sclera is a tough layer of connective tissue that forms the outermost layer of the eye bulb. Its anterior parts are visible as the "whites" of the eye. It is continuous with the **cornea**, which is a specialized transparent covering of the anterior one-sixth of the eye. The junction of the sclera and the cornea is called the **limbus**. At this transition, the cornea gives rise to the sclera and a delicate membranous tissue, the **conjunctiva**. This structure is reflected back to form the layer of the eyelid that rides over the eyeball.

The **uvea** is the middle layer of the eye that contains the **choroid**, the **ciliary body**, and the **iris**. The iris is shaped like a squashed doughnut. It has a central aperture, the **pupil**, whose size varies from 1 to 8 mm depending on the light conditions. The iris contains a **dilator** and a **sphincter muscle**. Sympathetic nerves control the dilator muscle and parasympathetic nerves activate the sphincter (see Chapter 4.9). Bright illumination of one eye causes constriction of the pupil in a reflex called the **direct pupillary light reflex**. The contralateral pupil also constricts, even if that eye is not illuminated, in the **consensual pupillary light reflex**.

The ciliary body contains the **ciliary muscle**, which controls the shape of the **lens**. The ciliary muscle is a smooth muscle with fibers oriented in various directions. This muscle connects to the lens through **suspensory ligaments**. The ciliary muscle forms folds arranged radially around the lens, and the suspensory ligaments arise from the valleys of these folds in zones (hence the term "**zonules of Zinn**"). Contraction of the ciliary muscle makes its diameter smaller, which loosens the tension on the suspensory ligaments, and the lens becomes thicker.

The ciliary body produces a fluid called the **aqueous humor** with a composition similar to plasma except that its protein content is much lower, $5-15$ mg dL^{-1} compared to about 7 g dL^{-1} in plasma. This fluid provides nutrients to the cornea and lens which do not have a blood supply. It flows from the **posterior chamber** between the zonules and the iris, out of the pupil and into the **anterior chamber** between the iris and the cornea (see Figure 4.8.2).

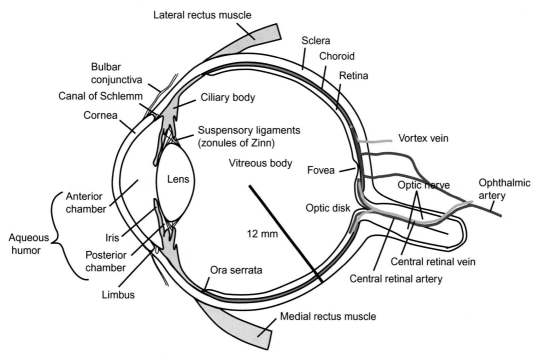

FIGURE 4.8.1 Structure of the eye. The diagram represents a horizontal section taken through the right eye and viewed from above. Individual parts of the eye are described in the text.

Clinical Applications: Glaucoma

Continual production of the aqueous humor ensures adequate intraocular pressure for maintaining the structural integrity of the eyeball. Healthy values of the intraocular pressure range from 13 to 18 mmHg. Because the ciliary body produces a constant flow of aqueous humor of about 2.5 µL min^{-1}, the intraocular pressure is determined largely by the outflow through Schlemm's canal. Large increases in the intraocular pressure damage the eye and cause blindness. This condition is called **glaucoma**, but there is overlap of the intraocular pressures among normal persons and persons with glaucoma. Glaucoma is a heterogeneous class of conditions for which increase in intraocular pressure is a major risk factor.

Intraocular pressures are measured with **tonometers**. Indentation tonometers measure the amount of indentation of the cornea by a known force. Applanation tonometers estimate the intraocular pressure by measuring the force required to flatten a known area of the eyeball. The intraocular pressure is approximately equal to the force required divided by the area that is flattened (the Imbert–Fick law). Most applanation tonometers make contact with the cornea with a plunger, and this contact creates the danger of the transmission of pathogens including hepatitis B and human immunodeficiency virus (HIV). The air-puff tonometer deforms the cornea by a jet of air whose force increases linearly with time. Flattening of the cornea is detected by light reflected off the cornea. The air-puff tonometer avoids corneal abrasion, reaction to the local anesthetics used in the plunger-type applanation tonometers, and reduces the transmission of infectious diseases.

The choroid extends from the **ora serrata** (the junction of the choroid and ciliary body, see Figure 4.8.1) and forms a middle layer over the rest of the eyeball. It consists primarily of blood vessels and connective tissue. It provides nutrients to the outer retina and removes metabolites from the retina.

The retina is the innermost layer of the eye ball, lying between the choroid layer and the **vitreous body**. It extends from the circular edge of the **optic disk**, where the optic nerve leaves the eye, to the ora serrata. The retina transduces incident light into nervous signals. The retina consists of several distinct layers. The outermost layer (the layer closest to the choroid) is the **retinal pigmented epithelium** consisting of pigmented columnar cells arranged in hexagonal close packing. Photoreceptor cells lie adjacent to the retinal pigmented epithelium. Two kinds of photoreceptor cells make up the photoreceptor layer. These are the **rods** and the **cones**. Rods are extremely sensitive and form the basis for **scotopic vision**, vision in dim light. Cones are less sensitive but allow for color vision in **photopic vision**, or vision in bright light. Cones vary in their spectral sensitivity. Red, green, or blue cones respond best to those colors of light. The other layers of the retina consist of nerve cells that process visual information even before it leaves the eye.

The retina possesses two general regions: a peripheral region in which rods predominate and a central region

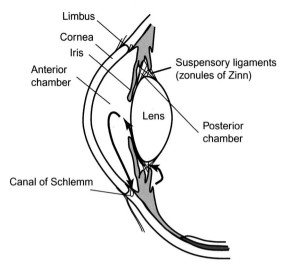

FIGURE 4.8.2 Flow of the aqueous humor. The ciliary body produces about 2.5 μL of aqueous humor each minute. This flows through the zonule fibers, into the posterior chamber between the lens and the iris. From there the fluid flows out of the pupil and into the anterior chamber. There it drains through a trabecular meshwork that feeds into the canal of Schlemm. From there the fluid drains into the veins.

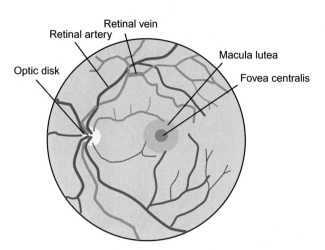

FIGURE 4.8.3 Schematic drawing of a fundus, the inner lining of the eye visible through the pupil. The optic disk, 1.5 mm in diameter, is where nerve and blood vessels enter or exit the eyeball. Branches of the ophthalmic artery form the posterior ciliary arteries that supply the choroid. Anterior ciliary arteries originate from blood vessels supplying the rectus muscles. The inner retina (toward the center of the eye) is supplied by the central retinal artery that branches off the ophthalmic artery and enters the eye along with the optic nerve. Four or five vortex veins drain the choroid, about one vortex vein in each quadrant of the eye. The retinal arteries and veins supply the inner retina with oxygen and nutrients. The macula lutea is a darker area of pigmentation about 5–6 mm in diameter with a central fovea, a depression in the retina in which cones predominate.

dominated by the cones. The peripheral region detects gross form and motion, whereas the central region possesses high resolution. Dim light involving the peripheral retina allows for the identification of objects, but the level of fine detail is poor and color vision is absent. In bright light, the eyes focus light onto the central region of the retina called the **macula lutea**. The macula lutea is about 5–6 mm in diameter. Its most central region is called the **fovea**. This is a central depression in the retina caused by a displacement of the neural cells in the retinal layers, so that the fovea contains only the photoreceptors. There are no photoreceptors in the optic disk, where the nerve fibers from the retina come together and exit the eyeball. Thus light that falls on the optic disk remains undetected. The visual field, the points in visual space that can can be seen, includes a **blind spot** that corresponds to the projection of visual space onto the optic disk. The retina can be viewed through an **ophthalmoscope**, which simultaneously illuminates the retina and projects the reflected light back to the observer. This fundus view provides the only easy observation of the circulatory system. The pattern of blood vessels in the fundus is one possible basis for physical identification of persons. "Red eye" in photographs results from light reflected back to the camera from the fundus (Figure 4.8.3).

THE VITREOUS BODY MAINTAINS EYE SHAPE

The vitreous chamber is filled with a gel-like vitreous body that consists of a solution of salts, soluble proteins, and a mucopolysaccharide called **hyaluronic acid**. A fine meshwork of collagen fibers courses through the vitreous body and ties it to various eye structures. The strongest attachments tie the vitreous body to the ora serrata. Other attachments occur at the posterior lens, optic disk, macula, and retinal blood vessels. Because of its viscoelastic properties, the vitreous body protects the eye by absorbing shocks to it.

THE EYE FOCUSES LIGHT ON THE RETINA BY REFRACTION

Light is focused when all rays emanating from each point of an object are bent to reach a point on the image. The path of light through transparent surfaces such as the cornea and lens depends on the curvature of the surfaces and the relative refractive indices for each material. The equation governing the light path is **Snell's law of refraction** (see Appendix 4.8.A1). This law states that

[4.8.1] $$n_1 \sin \theta_1 = n_2 \sin \theta_2$$

where n_1 is the refractive index of medium 1, n_2 is the refractive index of medium 2, θ_1 is the angle formed by the incident light ray and a line perpendicular to the surface of medium 1 (its **normal**) and θ_2 is the angle formed by the ray in medium 2 and its normal. When light travels from a region of low n to a region of high n, the light is bent toward a line normal to the surface. Thus the ability to bend light depends on the difference in refractive index at the interface and the curvature of the surface. Light entering the eye sequentially encounters a series of transparent surfaces with different refractive indices:

- Air, $n = 1.00029$
- Thin film of tears, $n = 1.33$
- Cornea, $n = 1.376$
- Aqueous humor, $n = 1.336$
- Lens, $n = 1.386$
- Vitreous body, $n = 1.336$.

> **EXAMPLE 4.8.1 Calculate the Refractive Power of the Eye at Rest**
>
> At rest, the eyes are focused on objects far away, and we can approximate $1/O = 0$. The image forms on the retina about 1.6 cm = 0.016 m away from the nodal plane. Then we can calculate
>
> $$\frac{1}{f} = 0 + \frac{1}{0.016 \text{ m}} = \mathbf{63 \text{ D}}$$
>
> This is the refractive power when the lens is most relaxed.

> **EXAMPLE 4.8.2 Refractive Power at the Near Point**
>
> The near point of a young person is about 10 cm. Calculate the refractive power.
>
> According to the thin lens formula, the refractive power is given as
>
> $$\frac{1}{f} = \frac{1}{O} + \frac{1}{I}$$
>
> So calculating the refractive power means identifying O and I. The distance to the object is just the near point: 10 cm = 0.1 m. The distance to the image depends on the size of the eye. We will use 1.6 cm = 0.016 m. Then we have
>
> $$\frac{1}{f} = \frac{1}{0.1 \text{ m}} + \frac{1}{0.016 \text{ m}} = \mathbf{73.5 \text{ D}}$$

Bending light more means a reduction in the focal length. The focal length of the eye is the distance from the eye's nodal point to the focused image. In a spherical lens, the nodal point lies in the center of the lens along an equatorial plane. Because there are multiple refractive surfaces in the eye, its nodal point can be considered to lie 1.5–1.7 cm in front of the retina. The **refractive power** in **diopters** is the inverse of the focal length:

[4.8.2] $$\text{Diopters (D)} = \frac{1}{f}$$

where f is the focal length in m. For any optical system, there is a point beyond which all objects can be considered to be at infinity. For the eye, this distance is about 6 m or 20 ft from the eye. Light rays emanating from such an object can be considered to be parallel to each other and to the optical axis of the eye. The distances from the object and the image obey the thin lens formula (see Appendix 4.8.A1):

[4.8.3] $$\frac{1}{O} + \frac{1}{I} = \frac{1}{f}$$

where O is the distance from the eye's nodal point to the object, I is the distance from the nodal point to the image, and f is the focal length.

When the lens is relaxed, about two-thirds of the refractive power, 43 D, is contributed by the cornea and about 20 D is contributed by the lens. The reason the cornea provides so much refraction is that its interface with air entails a large difference in refractive index (air has $n = 1.000$, whereas for the cornea $n = 1.376$). Thus, the refractive power results from the interface of a surface with its neighboring structures as well as from its own structure.

THE LENS CHANGES SHAPE TO FOCUS NEAR OBJECTS

Light rays from objects that are close to the eye are not parallel. Therefore, the eye must bend them more in order to focus the rays on the retina. To accomplish this, the lens change shape. Constriction of the ciliary muscle causes its diameter to decrease, which releases tension on the suspensory ligaments or zonule fibers connecting the ciliary muscle to the lens. Because the lens is inherently elastic, the release of tension causes the lens to become rounder, which increases its refractive power (see Figure 4.8.4). The increased curvature of the lens increases its refractive power. This process is called **accommodation**. Accommodation usually also involves **convergence**, when both eyes turn toward the midline to focus on close objects. The increased refractive power allows us to see near objects. The nearest an object can be and still be clearly seen is called the **near point**.

The difference in the refractive power of the relaxed eye and the maximally accommodated eye is the **power of accommodation**. In young persons, this is 10–12 D. The power of accommodation decreases with age as the elasticity of the lens decreases. This causes a recession of the near point with age, or **presbyopia**, as shown in Table 4.8.1. Close inspection of the thin lens formula shows that the power of accommodation is equal to the inverse of the near point when the near point is expressed in m.

NEAR-SIGHTEDNESS AND FAR-SIGHTEDNESS ARE PROBLEMS IN FOCUSING THE IMAGE ON THE RETINA

In the normal resting eye, parallel light rays are focused on the retina. This condition is called **emmetropia**, from the Greek *en* meaning *in* and *metron* meaning *measure*. In some cases, the resting eye does not focus light on the retina. In **myopia**, or nearsightedness, the image is formed in front of the retina. Usually an abnormally long eyeball causes nearsightedness but sometimes abnormally powerful refractive power causes myopia. Lenses that diverge light (indicated by a − sign) correct myopia. In **hypermetropia**, or farsightedness, the image forms behind the retina because the eyeball is

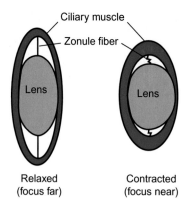

FIGURE 4.8.4 Accommodation of the eye. Far objects are focused on the retina by a combination of the refraction of the cornea and lens. Focusing near objects requires an increase in the refractive power of the eye, which is accomplished by changing the shape of the lens. The lens shape is controlled by the ciliary muscles. Contraction of the muscles removes tension on the zonule fibers that suspend the lens. Removal of tension allows the lens to recoil to a rounder shape.

too short. Converging lenses (+ lenses) correct hypermetropia.

PHOTORECEPTOR CELLS IN THE RETINA TRANSDUCE LIGHT SIGNALS

Rod-shaped cells and cone-shaped cells in the retina detect light. Their structures are shown schematically in Figure 4.8.5. The rod-shaped cells are more highly sensitive to light because they contain larger numbers of visual pigments and their effects are more efficiently transduced to changes in rod cell membrane potential. The overall mechanism for transduction in rod cells is shown in Figure 4.8.6. Neither rods nor cones produce action potentials. Instead, they alter the membrane potential of another kind of retinal cell, the bipolar cell, which also does not produce action potentials.

The outer segments of the rods contain discrete disks of membranes that do not contact the surface membrane. Each rod has some 700–2000 of these disks, and each contains some 50,000 molecules of **rhodopsin**, consisting of the protein **opsin** and its chromophore, **11-*cis*-retinal**, a derivative of vitamin A. Rhodopsin maximally absorbs light at a wavelength of 507 nm. The visual pigments in the cones reside in folds of the surface membrane of the cone outer segment. There are three types of cones that differ in the protein that binds 11-*cis* retinal. A single cone contains only one of three different visual pigments. Cones are maximally sensitive to red, green, or blue light, and they maximally absorb light at 555, 530, and 426 nm, respectively.

THE RETINA CONSISTS OF SEVERAL LAYERS AND BEGINS PROCESSING OF VISUAL SIGNALS

The outermost layer of the retina is the **retinal pigmented epithelium**. This cell layer consists of cuboidal cells with tentacle-like processes at the apical surface that extend

TABLE 4.8.1 The Effect of Age on the Near Point and Power of Accommodation

Age (years)	Near Point (m)	Power of Accommodation (D)
10	0.07	14
20	0.10	10
30	0.13	7.7
40	0.19	5.3
50	0.54	1.9
60	0.83	1.2

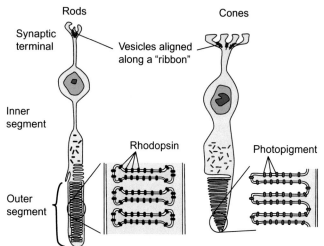

Long outer segment	Shorter outer segment
Large synaptic terminal	Larger synaptic terminal
High sensitivity	Low sensitivity
Scotopic vision	Photopic vision
Saturates in daylight	Saturates only in intense light
Lower temporal resolution (20 Hz)	High temporal resolution (70 Hz)
Rhodopsin	Erythrolabe Chlorolabe Cyanolabe
Low spatial acuity	High spatial acuity
Color blind	Color sensitive

FIGURE 4.8.5 Schematic diagram of the structures of rods and cones in the human retina. Both rods and cones consist of an outer segment that contains the light-sensitive chemicals and an inner segment that contains the nucleus and an abundance of mitochondria. Rods and cones differ in their structure, sensitivity to light intensity, sensitivity to the wavelength of light, and saturability. Rods are concentrated outside the fovea, whereas cones are concentrated within the fovea. Cones give rise to color vision in bright light (photopic vision), whereas rods allow vision in dim light (scotopic vision).

between photoreceptor cells. These cells contain **melanin granules**, a high molecular weight, insoluble polymer of oxidized tyrosine molecules. These absorb stray light, preventing light scatter between photoreceptor cells. The photoreceptor cells do not divide, but they continually renew themselves by making new layers in the outer

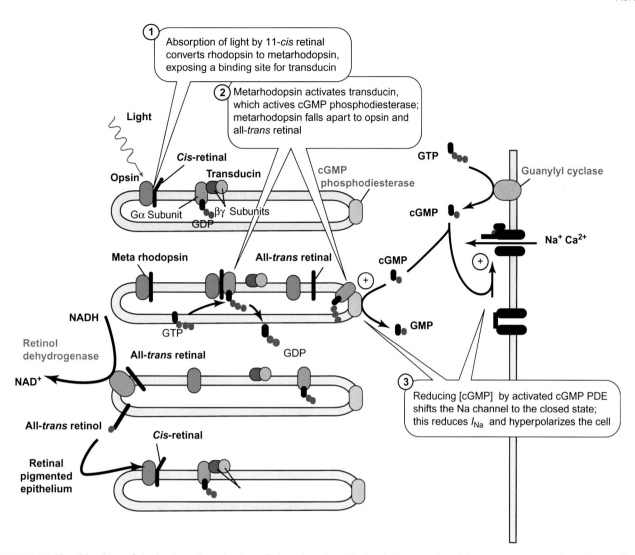

FIGURE 4.8.6 Photobleaching of rhodopsin and mechanism of photodetection. Rhodopsin is a complex of the protein, opsin, with its chromophore, 11-*cis* retinal. Absorption of a photon of light causes 11-*cis* retinal to isomerize to all-*trans* retinal. The complex of opsin with all-*trans* retinal is converted rapidly to metarhodopsin II. This metarhodopsin interacts with a heterotrimeric G-protein, transducin, whose α subunit activates a cGMP phosphodiesterase. The cGMP phosphodiesterase lowers the concentration of cGMP in these cells. cGMP sets the equilibrium between the open and closed state of a channel on the surface membrane that conducts both Na^+ and Ca^{2+}. Reducing [cGMP] favors the closed state, which reduces the inward I_{Na}, hyperpolarizing the cell. Meanwhile, the metarhodopsin dissociates to form opsin and all-*trans* retinal. Regeneration of rhodopsin begins with conversion of all-*trans* retinal to all-*trans* retinol inside the rods. The all-*trans* retinol is exported to the retinal pigmented epithelium where it is converted back to 11-*cis* retinal, which recombines with opsin to form rhodopsin again.

segments and shedding the old layers. The retinal epithelial cells phagocytose the shed outer segments. Photoreception produces **all-*trans* retinal** from **11-*cis* retinal** (see Figure 4.8.6), and the photoreceptor cells convert this to all-*trans* retinol. The retinal pigmented epithelial cells take up the all-*trans* retinol and convert it to 11-*cis* retinal which is taken back up by the photoreceptor cells. The retinal pigmented epithelium also synthesizes proteins that form the interphotoreceptor matrix and aid in the exchange of nutrients and metabolites between the photoreceptor cells and blood (Figure 4.8.7).

The **photoreceptor layer** contains the parts of the photoreceptor cells that absorb light and begin transduction between light absorption and neural signal. The retina contains some 80–110 million rod cells and some 4–5 million cones. The number of rod cells per unit area of retina is greatest in the retinal area immediately surrounding the fovea and decreases with distance toward the ora serrata (the line where the retina ends). The fovea contains a high density of tightly packed cones, with 200,000 cones mm^{-2}.

The **outer nuclear layer** contains the nuclei of the photoreceptor cells.

The **outer plexiform layer** contains synaptic contacts between photoreceptor cells, **bipolar cells**, and **horizontal cells**. The photoreceptor cells end in a **synaptic terminal** that makes contact with processes from **bipolar cells**. The bipolar cells derive their name from the fact that their cell bodies have two major processes. One extends to the outer plexiform layer and the other to the inner plexiform layer. Each bipolar cell receives inputs from a set of rod photoreceptor cells or from a set of cones, but not from both. **Rod bipolar cells** collect signals from 15 to 20 rod cells; **midget**

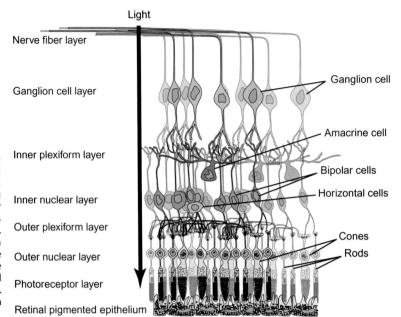

FIGURE 4.8.7 Schematic diagram of the retina. The outermost part of the retina is the retinal pigmented epithelium that absorbs scattered light and participates in the maintenance of the photoreceptors and in recycling bleached photopigments. This is the layer furthest away from incoming illumination. The photoreceptor layer is immediately adjacent to the retinal pigmented epithelium. The signal arising from the photoreceptors converge on several different bipolar cells. The bipolar cells, like the photoreceptor cells, exhibit graded changes in membrane potential but do not produce action potentials. A second line of signal processing occurs at the inner plexiform layer. Ganglion cells produce action potentials that travel down their axon. These collected axons make up the optic nerve.

bipolar cells receive inputs from a single cone; **flat bipolar cells** have inputs from 5 to 20 cones. As mentioned above, there are some 80–110 million rod cells and about 4–5 million cones. There are about 36 million bipolar cells. Thus there is convergence of inputs from the photoreceptors onto the bipolar cells. The **horizontal cells** are interneurons that receive inputs from many photoreceptor cells and make synapses back onto other photoreceptor cells and with bipolar cells.

The **inner nuclear layer** contains the nuclei of retinal interneurons including the bipolar cells, horizontal cells, and **amacrine cells**. The **inner plexiform layer** contains synaptic contacts between bipolar cells, amacrine cells, and **ganglion cells**.

The **ganglion cell layer** forms the output of the retina. The retina contains about 1–2 million ganglion cells. They form nerve fibers that make up the major parts of the **optic nerve**. The ganglion cells are the only retinal cells that produce action potentials. The photodetector cells continually secrete glutamate neurotransmitters in the dark. Illumination closes Na$^+$ channels in the outer segments, through transducin activation of cGMP phosphodiesterase, which hyperpolarizes the photoreceptor cell and diminishes release of glutamate neurotransmitter. Thus the photoreceptor cells have a graded membrane potential but do not produce action potentials. Horizontal, bipolar cells and amacrine cells also do not produce action potentials. Their response is a graded potential. Ganglion cells, however, produce action potentials that travel along their axons down the optic nerve.

The **nerve fiber layer** contains the axons of ganglion cells.

BIPOLAR CELLS ARE OFF-CENTER OR ON-CENTER

Bipolar cells receive inputs from a set of photoreceptor cells that define the bipolar cell's **receptive field**. **The neurotransmitter released from all photoreceptor cells is glutamate**. Because glutamate release is *decreased* upon exposure to light, a bipolar cell that responds to glutamate by excitation will be excited when the light is off. These are called **off-center bipolar cells** because they **are active when the light is off** in the center of their receptive field (Figure 4.8.8).

Some bipolar cells respond to glutamate by hyperpolarization, or inhibition. In the dark, the photoreceptors maximally release glutamate and these bipolar cells are maximally hyperpolarized. In the light, the photoreceptors are hyperpolarized and they release less glutamate. These bipolar cells are called **on-center bipolar cells** because they **are active when the light is on**.

THE OUTPUT OF BIPOLAR CELLS CONVERGE ONTO ON-CENTER AND OFF-CENTER GANGLION CELLS

As information flows centrally, the number of cells carrying the information decreases. Thus, there are many more photoreceptor cells (80–110 million rods, 4–5 million cones) than bipolar cells (36 million), and there are many more bipolar cells than ganglion cells (1–2 million ganglion cells per eye).

GANGLION CELLS IN THE PARVOCELLULAR PATHWAY RECEIVE INPUT FROM A FEW BIPOLAR CELLS

Some ganglion cells receive inputs that are equivalent to a single or just a few bipolar cells. This small receptive field allows the ganglion cell to encode fine details of the visual field. The axons from these ganglion cells segregate in the optic nerve and the segregation remains intact all the way up the neural pathways to the visual cortex. These ganglion cells are small, and therefore they are named **parvocellular** or P cells from the root meaning "small." These cells project to the four dorsal layers of the **lateral geniculate nucleus** (two ipsilateral

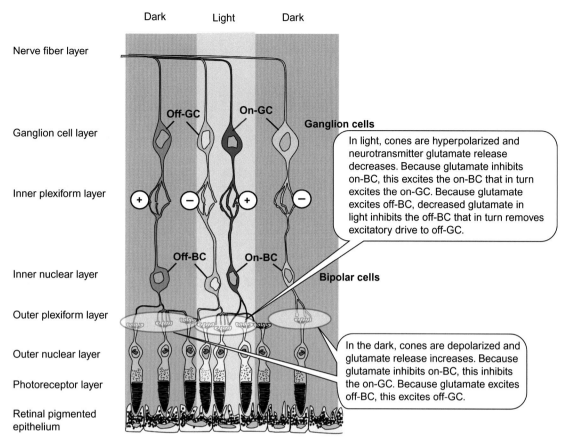

FIGURE 4.8.8 On-center and off-center bipolar cells and ganglion cells. In the dark, photoreceptors are depolarized (dark gray color) and increase their release of glutamate neurotransmitter. Light causes these photodetectors to hyperpolarize and decrease their glutamate release (light blue color). **Glutamate inhibits on-BCs and excites off-BCs**. In the light, decreased glutamate relieves the inhibition of on-BCs. Thus the on-BCs in the light are depolarized (dark blue) which increases their release of neurotransmitter. This activates on-GCs (dark blue). Thus on-GCs are activated when the light is on. By similar arguments, the off-GCs are activated when the light is off (dark gray).

and two contralateral) forming the parvocellular division. These cells carry detailed visual information.

GANGLION CELLS IN THE MAGNOCELLULAR PATHWAY RECEIVE INPUT FROM HUNDREDS OF BIPOLAR CELLS

Another type of ganglion cell are much larger and are called magnocellular or M cells. These cells receive inputs from many bipolar cells and so produce less visual acuity. Their segregation, like that of the parvocellular division, remains intact through the visual pathway. They project to the two ventral layers of the lateral geniculate nucleus, one from the ipsilateral retina and one from the contralateral retina.

SURROUND EFFECTS ON GANGLION CELLS ARE DUE TO CONVERGENCE OF BIPOLAR CELLS AND LATERAL PROCESSING BY HORIZONTAL CELLS AND AMACRINE CELLS

The response of ganglion cells to light patterns on the retina is more complicated than a simple "on-center" or "off-center." Illumination or lack of illumination of the surrounding areas of the retina also influences ganglion cell response. This effect arises from convergence of input from bipolar cells onto ganglion cells and is partly produced by integrative functions of the horizontal cells. The rod bipolar cells receive inputs from up to 50 photodetectors, and they also have on-center, off-surround types of receptive fields. Thus the ganglion cells do not create a receptive field pattern; they relay a receptive field organization that is already present at the level of the bipolar cells. The response to light of ganglion cells with on- or off-centers is shown in Figure 4.8.9.

SIGNALS FROM THE TWO EYES CROSS OVER DURING THE CENTRAL VISUAL PATHWAYS

Each retina possesses some 1–2 million ganglion cells that are distributed throughout the retina. The retina itself is divided into four quadrants: the superior and inferior halves are divided into nasal and temporal quadrants. The image of the external world on the retina is inverted and reversed. Thus, illumination of the left peripheral visual field projects to the nasal retina of the left eye, and the right peripheral visual field projects to the temporal retina of the left eye. Accordingly, the output of the ganglion cells represents particular areas

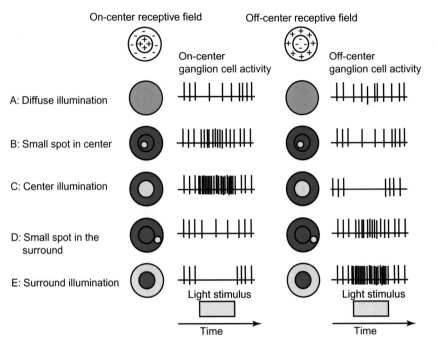

FIGURE 4.8.9 Effect of the size and position of a light stimulus on the response of on-center ganglion cells and off-center ganglion cells. Diffuse illumination (stimulus A) produces conflicting inputs to on-center ganglion cells: the center excites the ganglion cell while the surround inhibits it. Because the surround has a larger area, the result is mild inhibition. Off-center ganglion cells, on the other hand, are inhibited by light in the center and excited by light in the surround. On-center ganglion cells respond most vigorously to light that fills the center and dark in the surround (stimulus C). Off-center ganglion cells respond most vigorously to dark in the center and light that fills the surround (stimulus E). (Source: *Adapted from J. Beatty, The Human Brain, Sage Publications, Inc. 2001.*)

of the visual field. Each of these travels along the optic nerve. To produce a single, coherent percept of the visual world, the outputs of these ganglion cells must be merged.

Nerve fibers from ganglion cells from all parts of the retina come together at the **optic chiasm** and from there travel on to the **lateral geniculate nucleus** or the **superior colliculus**. The destination of the optic nerve fibers depends upon their location in the eye. The parts in the left retina of both eyes pass on to the left brain, and those on the right retinas go to the right brain. Thus, the ganglion cells that correspond to the same parts of the visual field mingle together (Figure 4.8.10).

The magnocellular pathway includes fibers from large ganglion cells that receive convergent inputs from many bipolar cells. These fibers from the nasal quadrants cross over at the optic chiasm and project to the **ventral lateral geniculate nucleus** in the thalamus. The parvocellular ganglion cells receive inputs from fewer bipolar cells and therefore these cells are smaller. They project to the **dorsal lateral geniculate nucleus**. These fibers synapse on cell bodies that form 6 layers in the lateral geniculate nucleus, as shown in Figure 4.8.10. These layers are numbered 1–6 from the most ventral to the most dorsal. Layers 3–6 constitute the parvocellular input, and the inputs to these layers alternate between the contralateral and the ipsilateral eye. Layers 1 and 2 form the magnocellular pathway, and these have also ipsilateral and contralateral input.

SOME GANGLION CELLS PROJECT TO OTHER AREAS OF THE BRAIN

Some ganglion cells in the magnocellular pathway do not project to the lateral geniculate nucleus, but project instead to the **superior colliculus**, part of the tectum or roof of the midbrain. The superior colliculus makes no contribution to the conscious visual percept but provides information for visual orientation, focusing of the image, and control of **saccades**, the rapid movement of the eyes used in tracking objects in visual space.

ADDITIONAL PROCESSING OF VISUAL IMAGES OCCURS IN THE VISUAL CORTEX

The cells in the lateral geniculate nucleus project to the visual cortex through fibers called **optic radiations** that spread out like a fan traveling laterally and inferiorly around the horn of the lateral ventricle. These make connections in the visual cortex at the back of the brain in the occipital lobe. This region of the brain is often called the **striate cortex** because a layer of myelinated fibers is found there. The processes from the lateral geniculate nucleus connect primarily to layer 4 of the visual cortex. Like much of the cortex, the visual cortex is organized into 6 layers. The visual cortex has additional levels of organization. Alternate columns of cortex receive inputs primarily from one eye. These alternating bands form **ocular dominance columns**.

The responses of specific cells in the visual cortex were first described by Hubel and Wiesel, who earned a Nobel Prize in 1981. They described **simple cortical cells** that have receptive fields that are basically elongated on-center, off-surround receptive fields. These simple cortical cells respond when bars of light stimulate the on-center and the off-surround is dark. These cells respond preferentially to bars of light with a particular orientation, and the same stimulus at 90° elicits a poor response. These cells respond to bars only in the correct position; movement of a bar into the off-surround results in inhibition rather than stimulation. Hubel and Wiesel also found **complex cortical cells**

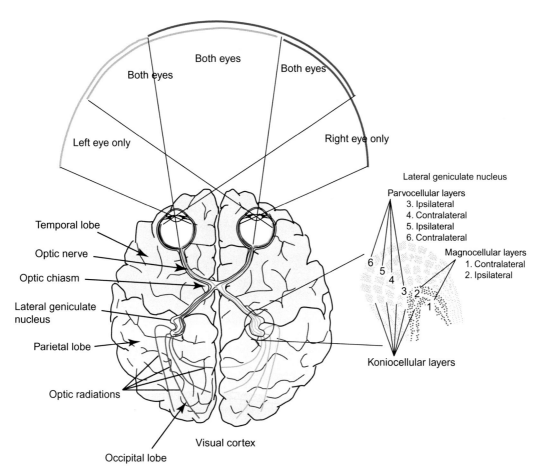

FIGURE 4.8.10 The visual pathway. Areas of the external world constitute the visual field. The optics of the eye focus light from these areas on areas of the retina and the light alters the firing patterns of retinal ganglion cells. The axons of these cells travel over the optic nerve to the optic chiasm, where the fibers from the nasal quadrants cross over to the opposite side of the brain. The fibers from all parvocellular ganglion cells and most magnocellular ganglion cells travel to the lateral geniculate nucleus, where the signals are relayed to the cerebral cortex that lines the occipital lobe. There the visual cortex further processes the visual image. Thus, visual signals from the temporal retina are ipsilateral and those from the nasal retina are contralateral.

that respond to bars of light with the proper orientation anywhere in their receptive field. These remain sensitive to orientation but not position. Another type of cell in the visual cortex responds to bars that have ends to them. These "end-stopped cells" may be corner detectors.

Edge detectors can be constructed by selectively connecting some on-center, off-surround ganglion cells in an array. The retina produces some million of these kinds of receptive fields. A bar along the on-centers of a number of ganglion cells excites them all, so that a cortical cell having excitatory connections to this set of ganglion cells will be excited optimally when the bar is aligned with their centers.

THE VISUAL CORTEX SENDS OUTPUT TO THE TEMPORAL AND PARIETAL LOBES

The visual cortex sends fibers along two distinct output pathways: the **dorsal stream** and the **ventral stream**. The dorsal stream feeds information into the parietal lobe which appears to be necessary for using visual information in forming movements. The ventral stream feeds visual information into the temporal lobe where details of the objects are identified. The ventral stream is the "what" pathway, whereas the dorsal stream is the "how" pathway.

WE STILL DO NOT KNOW HOW WE PERCEIVE VISUAL IMAGES

Despite the remarkable progress in identifying all of those processes involved in focusing light, transducing the signal, and interpreting the signal at the cellular level, we still do not understand exactly what it is that perceives and how that perception is accomplished. It is clear that the visual system is hierarchical; each level builds upon processing accomplished at a lower level. Is it likely that there is a "Grandmother cell" whose firing tells us, whoever the "us" is, that we are seeing our Grandmother? Or is it more likely that there is some delocalized set of neurons, which by their nature are a fleeting occurrence, that leads to the interpretation of "Grandmother" in our visual field? In the final analysis, visual perception remains mysterious.

Clinical Applications: Cataracts

The transparency of the lens depends on lens proteins called **crystallins**. These water-soluble proteins comprise 90% of the total lens protein. The remaining 10% consist of water-insoluble proteins including membrane proteins and cytoskeletal proteins. The lens contains about 66% water and 33% protein, double the protein content of most other tissues. This high protein content causes the high refractive index of the lens. These proteins last longer than any other proteins in the body. Most lens cells lose the ability to make new proteins and so their lens proteins last a lifetime. Because cells are added to the outside layers of the lens, the crystallin concentration varies from 15% in the cortex to 70% in the interior of the lens. This concentration gradient of lens proteins produces a gradient in refractive index.

An opaque area in the lens is called a **cataract**. The opacity interferes with light transmission to the retina and produces glare during night driving, blurred or distorted vision, reduction in visual acuity, and double vision in the affected eye. The opacity may occur in different zones of the eye, and the cataract is classified accordingly. Cataracts may result from trauma, medications, radiation, inflammation, metabolic diseases, and simple aging.

Poor vision caused by cataracts can be improved by surgical removal of the lens. Assessment of the potential for improved vision requires the use of a potential acuity meter (PAM) or laser interferometer. The PAM projects a miniature Snellen eye chart (the chart used to measure visual acuity in the typical physicians's office) on the retina through a 0.1 mm pinhole aperture through a clear area of the lens, thus bypassing the cataract. The interferometers project an interference pattern on the retina. Removal of the lens may be followed by insertion of an intraocular lens implant. Intracapsular cataract extraction removes the entire lens; extracapsular cataract extraction leaves the posterior lens capsule to support an implant.

SUMMARY

The visual system has three problems: focus light on the retina, transduce the light signal to neural signals, and process these neural signals to form a conscious perception. Light passing from the external environment to the retina sequentially passes through a thin film of tears, the cornea, the aqueous humor, the lens, and the vitreous humor. Light is refracted at each interface according to Snell's law:

$$n_1 \sin \theta_1 = n_2 \sin \theta_2$$

where n_1 and n_2 are the refractive indices of media 1 and media 2 and θ_1 and θ_2 are the angles between the incident light ray and surface normal. The purpose of the refraction is to bend light so as to focus the light on the retina. The refractive power of the eye is given by the thin lens formula:

$$\frac{1}{O} + \frac{1}{I} = \frac{1}{f}$$

where O is the distance from the nodal point to the object, I is the distance to the image, and f is the focal length. The refractive power is $1/f$. At rest, the refractive power is about 63 D, most of which is provided by the cornea. The refractive power of the eye depends on the shape of the lens, which is controlled by the ciliary muscle. Contraction of the muscle releases tension on suspensory ligaments of the lens, causing the lens to become rounder because of its inherent elasticity. Accommodation is the ability to change refractive power to focus on nearby objects. This is generally 10–12 D in young people and decreases with age. The gradual loss of accommodation with age is presbyopia.

Retinal photoreceptors are mainly rods and cones. Both release glutamate onto bipolar cells and reduce the amount of release when exposed to light. Rods are highly sensitive but saturate easily, have lower spatial acuity, and do not detect color. Cones are responsible for color vision and vision in bright light.

The photoreceptors contain a chromophore, 11-*cis*-retinal, complexed to proteins that confer sensitivity to different wavelengths of light. Absorption of light converts 11-*cis*-retinal to all-*trans* retinal and uncovers a binding site for a heterotrimeric G-protein called transducin. After excitation, transducin exchanges GTP for GDP and activates a cGMP phosphodiesterase. This reduces [cGMP] which closes a Na^+ channel, leading to hyperpolarization and less release of glutamate.

Glutamate excites off-bipolar cells, so these off-BC cells are excited when the light is off. Glutamate inhibits on bipolar cells; they are excited with the light is on. These connect and converge onto ganglion cells that produce action potentials that are relayed to the lateral genicular nucleus and then to the visual cortex in the occipital lobe. Fibers from the nasal quadrants cross over to the opposite side at the optic chiasm; fibers from the temporal quadrants remain ipsilateral. The magnocellular ganglion cells convey general information about shape; parvocellular cells convey detailed visual information.

The visual cortex has 6 layers and is organized into ocular dominance columns that have information from one eye. Specific cells in the cortex respond to higher-order patterns in the visual field. Simple cells respond to a bar of light with the proper position and orientation. Complex cells respond to bars of light of the proper orientation but regardless of position. Other cells respond to more complex features in the visual field.

REVIEW QUESTIONS

1. What part of the eye causes the most refraction? What part of the eye is most responsible for accommodation? What muscles are responsible for it?

2. What is the near point? At what point is accommodation maximal? What is the name for loss of accommodation with age?
3. What produces the aqueous humor? Where does it drain? What happens if intraocular pressure gets too high?
4. Where are the cones concentrated? Where are the rods? What differences explain their relative functions?
5. How do photoreceptor cells transduce light into a neural signal? What transmitter is used?
6. What neurons connect directly to the photoreceptors? Do they make action potentials? What is an "on-bipolar cell" and an "off-bipolar cell"?
7. How do "on-ganglion cells" respond to light within their receptive fields? How do "off-ganglion cells" respond? Do ganglion cells make action potentials?
8. What is the blind spot? Where do nerve fibers go after they leave the eyeball in the optic nerve?
9. What is the difference between parvocellular and magnocellular ganglion cells?
10. What is the lateral geniculate nucleus? Where is the visual cortex? What do simple cortical cells do? What do complex cortical cells do?

APPENDIX 4.8.A1 REFRACTION OF LIGHT AND THE THIN LENS FORMULA

LIGHT INTERACTS WITH MATTER AS IT PASSES THROUGH

Electromagnetic Radiation Induces Oscillation of Electrically Charged Particles in Matter

Light is an oscillating electromagnetic field. An oscillating, electrically charged particle generates an oscillating electromagnetic field that propagates away in all directions at right angles to the direction of oscillation. Similarly, the oscillating electromagnetic wave of light interacts with charged particles in matter, causing them to oscillate at the frequency of the incident light and to reradiate electromagnetic waves at right angles to their motion. This is equivalent to Huygens' principle, which states that every point on an advancing wave at speed v is the origin of a secondary wavelet that propagates at the same speed. This phenomenon of reradiation of incident light is called **scatter**. When the oscillating charged particles have appropriate spatial separation, the reradiated electromagnetic waves undergo destructive interference in all directions except in the direction of the original incident light. This is the basis of **transparency**. In transparent substances, only the forward scatter survives the destructive interference among the secondary waves produced by the moving charged particles within the material. The transparency of the eye lens depends on the dense packing of scattering elements within the lens, and this is maintained by its relatively dry state. Cataracts are a disruption of this structure so that the lens scatters light in all directions.

The Oscillating Particles Re-emit Light After a Delay, Slowing the Speed of Light Within Matter

The speed of light in a vacuum, c, is 3×10^9 m s^{-1}. When light passes through a transparent medium, it passes between the atoms at the speed c. However, each interaction with charged particles in the medium delays the phase of the light, so that its speed within the medium, v, is reduced. The ratio of the speed of light in vacuum to the speed of light in the medium is called the **refractive index** of the medium, denoted by n:

[4.8.A1.1] $$n = \frac{c}{v}$$

THE DIFFERENT SPEED OF LIGHT IN TWO MEDIA CAUSES REFRACTION, THE BENDING OF LIGHT AT THEIR INTERFACE

According to Huygens' principle, every point on a wave front traveling at speed v is the origin of a secondary wavelet that also propagates at speed v. Consider Figure 4.8.A1.1, which shows the interface between air and a liquid transparent medium. The refractive index of air is 1.00029, so the speed of light in air is very near to its speed in a vacuum. The refractive index of water, however, is 1.33, so that light is significantly slowed when it enters the water. We assume here that the incident light approaches the interface obliquely at some angle, θ_i, the angle of incidence. As the wave front enters the water, it is slowed compared to the remaining wave front that still travels through the air. The result is that the wave front's direction is changed to θ_r, the angle of refraction. During some time t, the wave front advances $v_i t$ in the incident medium, where v_i is its speed in this medium. During that same time, the wave front advances a shorter distance, $v_r t$, in the second medium. We can choose t so that the geometry in Figure 4.8.A1.1 occurs. The arrows in the figure are at 90° to the wave fronts, so that $\angle$ACD and $\angle$ABD are both at right angles. The arrows correspond to the rays that are at right angles to the wave fronts. Thus we can write

[4.8.A1.2] $$\sin \theta_i = \frac{v_i t}{AD}$$
$$\sin \theta_r = \frac{v_r t}{AD}$$

These two equations can be combined, by taking their ratios and canceling the like factors, to obtain

[4.8.A1.3] $$\frac{\sin \theta_i}{\sin \theta_r} = \frac{v_i}{v_r}$$

Since $v_i = c/n_i$ and $v_r = c/n_r$, by Eqn [4.8.A1.1], we rewrite Eqn [4.8.A1.3] as

[4.8.A1.4] $$n_i \sin \theta_i = n_r \sin \theta_r$$

This equation is **Snell's law of refraction**.

It is often more convenient to deal with refraction in terms of **light rays** rather than the wave fronts. Light

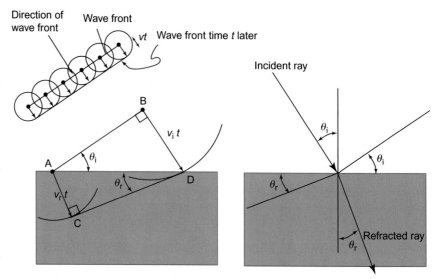

FIGURE 4.8.A1.1 Refraction of light at an interface between two media. The wave front can be imagined as proceeding by the propagation of wavelets at each point along the wave front. Only the leading edge of the wavelets contain energy. When the wave front obliquely enters a surface with a higher refractive index, the wave front is slowed while the wave front remaining in the incident medium is not slowed. The consequence of this slowing is a change in the angle of propagation. The right-hand side of the figure shows the light ray representation of the light, with the same conclusion: movement of light from a region of lower to higher refractive index bends the light toward a line normal to the surface. (Source: *Adapted from J.T. McIlwain, Introduction to the Biology of Vision, Cambridge University Press, 1996.*)

rays are normal to the wave fronts and give the direction of propagation of the light. The geometry of the light rays is such that Snell's law still holds, except that the angles are defined as the angles made between the rays and a line drawn perpendicular to the surface of the interface, rather than between the wave front and the surface itself.

REFRACTION DEPENDS ON THE ANGLE OF INCIDENCE AND THE REFRACTIVE INDICES IN THE TWO MEDIA

According to Snell's law, the angle of refraction, θ_r, depends on the angle of incidence, θ_i, and the ratio of the refractive indices of the two media that make up the interface. If the media have identical refractive indices, there is no refraction and the angle $\theta_r = \theta_i$. Similarly, light that enters the interface normal to its surface is not deviated, because $\theta_i = 0°$ and $\sin \theta_i = 0$ and therefore $\theta_r = 0°$ by Eqn [4.8.A1.4].

CONVEX LENSES CONVERGE LIGHT; CONCAVE LENSES DIVERGE IT

Consider a piece of glass with surfaces consisting of three planes that intersect a bundle of parallel light rays, as shown in Figure 4.8.A1.2. The refractive index of the glass is higher than that of the surrounding air, and so incident light is bent toward the normal at all surfaces. However, each surface presents a different angle to the incident light, so that light is bent toward the middle of the glass in its convex arrangement and away from the midline in its concave arrangement. This is the basis for the formation of lenses and the focusing of light.

THE THIN LENS FORMULA LINKS REFRACTIVE POWER AND IMAGE FORMATION

Lenses Can Refract Light to Intersect at a Single Point Called the Focal Point

Figure 4.8.A1.2 indicates that a crude convex lens causes light rays to converge. It is possible to construct a lens in which all of the parallel incident rays intersect at a single point. Here it is clear that the incident rays must deviate more strongly the further from the axis of the lens, and this increased deviation must occur gradually rather than abruptly as in the crude lens. That is, the surfaces of the lens must smoothly curve from the center toward the edge of the lens. This is accomplished practically by making spherical lenses in which the two curved surfaces subtend a solid angle of a sphere, as shown in Figure 4.8.A1.3. These spherical lenses refract parallel incident light so that, for practical purposes, they intersect at a single point. This single point is called the focal point. If the radii of curvature of the two surfaces are large compared to the thickness of the lens, then the lens may be considered to be a "thin lens" for which the following equations are valid.

The Focal Length Is a Measure of the Refractive Power of a Lens

The focal length is the distance from the focal point to the lens. Specifically, the focal length is the distance from the focal point to the plane that passes through the equator of the lens, as shown in Figure 4.8.A1.3. In physiological optics, the strength of the lens is expressed in **diopters**. This is the reciprocal of the focal length, f. We write

[4.8.A1.5] $$\text{Diopters (D)} = \frac{1}{f}$$

A lens has a single focal length, regardless of whether parallel incident light strikes it from the right or from the left. This is illustrated in Figure 4.8.A1.3.

Spherical Lenses Do Not Exactly Focus at a Single Point

As mentioned above, the refraction of light depends on the angle of incidence and the refractive indices of the two media. Because of its spherical curvature, light passing through the extreme edges of a spherical lens does not focus at exactly the same point as light that passes through the more central parts of the lens. This phenomenon is called **spherical aberration**. The refractive

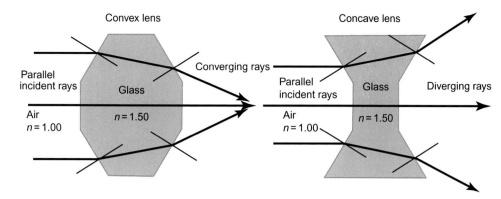

FIGURE 4.8.A1.2 Convergence of light rays by a convex lens (left) and divergence of light rays by a concave lens (right). The refraction of light rays depends not only on the relative refractive indices of the two media but on the angle of incidence.

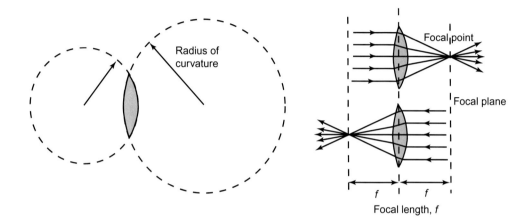

FIGURE 4.8.A1.3 Spherical lenses focus parallel incident light at the focal point. The focal length, f, is the same for light approaching the lens from either direction.

index also varies with the wavelength of light, a phenomenon having to do with how light is retarded by its interaction with charged particles within the medium. Blue light is bent more than yellow light, which is bent more than red light. This dependence of refractive index on the wavelength of light causes the separation of white light into its component colors by a prism. The result of this variation in refractive index is that white light produces separate images for each color that makes up the white light. This separation of images is called **chromatic aberration**. Spherical aberration and chromatic aberration are shown schematically in Figure 4.8.A1.4. The eye possesses some degree of both spherical aberration and chromatic aberration.

A SPHERICAL LENS FORMS AN IMAGE OF AN OBJECT THAT IS RELATED TO THE DISTANCE FROM THE OBJECT AND THE FOCAL LENGTH OF THE LENS

The image formed by a thin lens can be located by drawing three rays: (1) a ray which passes through the center of the lens is unchanged because the lens faces are parallel here and the lens is assumed to be thin. This ray is normal to the surface of the lens; (2) a ray parallel to the lens axis is refracted to pass through the focal point on the opposite side; (3) a ray which passes through the focal point on the side of the object emerges from the lens parallel to the lens axis. These three rays are shown in Figure 4.8.A1.5. The image is inverted and found on the far side of the focal point away from the lens. The distance from the object to the lens' nodal plane is O and the distance from the image to the lens' nodal plane is I. The height of the real object is h_O and the height of its image is h_I. The object is oriented at a right angle to the lens axis, and so is the image. The geometry of the situation allows us to identify two sets of similar triangles:

[4.8.A1.6]
$$\Delta AJD \sim \Delta HGD$$
$$\Delta BDF \sim \Delta HGF$$

These triangles are similar by the angle–angle–angle theorem of geometry. Because the ratio of corresponding parts of similar triangles are equal, we write

[4.8.A1.7]
$$\frac{h_O}{h_I} = \frac{O}{I}$$

From the second set of similar triangles, we observe that

[4.8.A1.8]
$$\frac{\overline{DF}}{\overline{FG}} = \frac{\overline{BD}}{\overline{HG}}$$

From the geometry, we can identify these line segments as

[4.8.A1.9]
$$\overline{DF} = f$$
$$\overline{FG} = I - f$$
$$\overline{BD} = h_O$$
$$\overline{HG} = h_I$$

470 QUANTITATIVE HUMAN PHYSIOLOGY

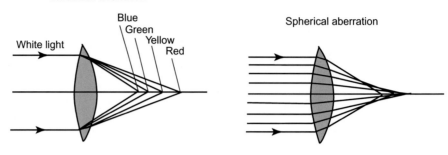

FIGURE 4.8.A1.4 Schematic diagram of the light paths taken and the resulting abnormal focus in lenses with chromatic aberration (left) and spherical aberration (right). In chromatic aberration, light of short wavelengths is bent more than light with long wavelengths, and the result is a different focus for light of different colors. In spherical aberration, light at the periphery of the lens is bent more, resulting in a different focus for light near the optical axis compared to light at the edge of the lens.

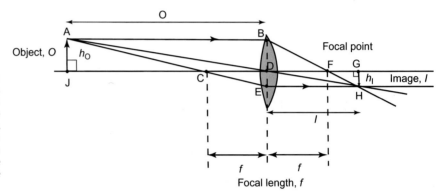

FIGURE 4.8.A1.5 Formation of an image by a spherical thin lens. The position of the image is determined by the three rays shown. One is a construction between the object and the image that passes roughly through the center of the lens. Parallel light from the object is refracted through the focus of the lens. A light ray emanating from the image and parallel to the optical axis of the lens intersects the object at its extreme. From the geometry, the thin lens equation may be derived.

Inserting these values into Eqn [4.8.A1.8], we find

[4.8.A1.10] $$\frac{f}{I-f} = \frac{h_O}{h_I} = \frac{O}{I}$$

Equation [4.8.A1.10] can be rearranged to give:

[4.8.A1.11]
$$\frac{f}{O} = \frac{I-f}{I}$$
$$\frac{f}{O} = 1 - \frac{f}{I}$$
$$\frac{1}{O} = \frac{1}{f} - \frac{1}{I}$$

This last equation can be rearranged to give the **thin lens formula**:

[4.8.A1.12] $$\frac{1}{O} + \frac{1}{I} = \frac{1}{f}$$

Once again, here O is the distance from the nodal plane of the lens to the object, I is the distance from the nodal plane of the lens to the image, and f is the focal length. This equation clearly meets the condition when $O \to \infty$; under these conditions, the light entering from the object is parallel and the image is found at the focal point ($I = f$ in Eqn [4.8.A1.12]).

Problem Set
Sensory Transduction 4.2

1. The reference intensity for sound is 10^{-12} W m^{-2}. The density of dry air at 273 K is 1.293 g L^{-1} and the velocity of sound is 331 m s^{-1}. This reference intensity corresponds to what maximum pressure amplitude? (Hint: see Eqn [4.7.A1.32]).
2. My old vacuum cleaner makes a noise of about 80 dB. What is the sound intensity? What is the maximum pressure amplitude?
3. Increasing the loudness of a sound from 10 to 20 dB increases the intensity by what factor? It increases the pressure amplitude by what factor?
4. You are subjected to a pure pitch with an amplitude of 40 dB. What is its pressure amplitude? Given that the area of the tympanic membrane is 0.5 cm^2 and that the area of the oval window is 0.03 cm^2, what is the pressure amplitude, assuming no loss during transmission by the ossicles, at the oval window?
5. Assume that the distance between your ears is about 20 cm. Assume that sound waves originate directly to your right. What is the delay in the arrival of the sound between the right and left ear? Do you think that your nervous system could locate the source of sound based on this difference? Why or why not?
6. A. You are looking at an object 1 m tall and 20 m away from you. Assuming the eye is completely relaxed and that the nodal point of the eye is 1.6 cm in front of the retina, calculate the height of the image of the object on the retina, assuming that it is in focus. What is the focal length and the refractive power of the eye under these conditions?
   B. You have shifted your focus to a nearby object 10 cm tall located a mere 0.5 m away. In doing so, your eye accommodates to bring its image into focus on the retina. What is the refractive power of the eye in this case? How tall is the image on the retina?
7. The fovea is responsible for vision of highest acuity, and you move your eyes so as to focus light there. The fovea is about 0.5 mm in diameter. How large an area of attention does this make at 0.5 m distant? At 5 m? 10 m?
8. The resolution of the eye is $1/x$ where x is the spatial separation of a line pair. Typical physiological resolution of the human eye is about 1.7 in good light by this criterion, which corresponds to about 0.6 arc minutes per line pair, or about 0.01. Consider that you are viewing a grating of lines 25 cm from the eye.
   A. If the lines can just be resolved, how large is the distance between line centers?
   B. What is the spacing between the lines in the image on the retina?
   C. In the fovea the light-sensitive areas of the cones are about 1.5×10^{-6} m in diameter and about 2.5×10^{-6} m separates their centers in close hexagonal packing. Does your answer in part B make sense with respect to this spacing of the cones?
9. The energy of a photon of light is given by $E = h\nu$, where h is Planck's constant = 6.625×10^{-34} J s and ν is the frequency of light in s^{-1}. Assume that the average wavelength of light emitted from a normal incandescent bulb is 500 nm = 5000 Å. Assume further that a 100 W bulb's energy is converted 100% to light energy of this wavelength.
   A. How many photons does the bulb emit per second?
   B. Consider that your pupil is constricted to a diameter of 4 mm and that you are 1 m away from the bulb which for this purpose we will consider to be a point source. How many photons enter each of your eyes each second from such a source?
10. Your friend puts on 0.5 mL of a special alluring scent at a concentration of 10 mM of odorant. Assume that all of the odorant is volatilized. What is its final concentration in a room with dimension 11 feet × 15 feet × 8 feet tall? Would you expect to be able to smell this odorant at these concentrations?
11. Two objects are drawn on a sheet of white paper, 4 cm apart, in the horizontal plane. You are instructed to close the right eye and focus on the object to the right, starting with the paper about 30 cm from your eye. As you draw the paper toward you, you notice that at about 13 cm the left object disappears from view! As you draw the paper still closer, it reappears!

Why did it disappear, and then reappear? From the information provided, calculate the angular displacement of the optic nerve from the fovea.

12. To sense any signal, cells must distinguish between the signal and the background noise. For the chemical senses, the signal is the binding of a chemical to a receptor on the sensing cells. At the molecular level, background noise corresponds to thermal fluctuations in the movement of molecules. The equipartition theorem of thermodynamics states that each mode of motion at equilibrium has the energy $\frac{1}{2}kT$, where k is Boltzmann's constant and T is the temperature in Kelvin. $k = 1.38 \times 10^{-23}$ J K^{-1} mol^{-1}.
    A. Calculate the thermal background energy fluctuations, or noise.
    B. Noncovalent binding energies of ligands to their receptors varies with the number of chemical groups on the two surfaces that interact. A typical value is around 10–20 kJ mol^{-1}. How much is this per molecule?
    C. Do you think it is possible to sense the binding of a single ligand?

13. A young adult's eye has a nodal point 1.6 cm in front of the fovea. His near point is 7 cm.
    A. What is his maximum power of refraction?
    B. What is his power of refraction for far objects?
    C. What is his power of accommodation?

14. After increased thirst, increased urination, and increased hunger, the next most common complaint of diabetic persons is blurred vision. Although the cause is still debated, most think that the high plasma glucose causes glucose to enter the lens causing it to swell. What should this swelling do to the refractive power of the lens? What kind of lens is necessary to correct it?

15. Age-related macular degeneration results from atrophy of the retinal pigmented epithelium which causes vision loss through loss of photoreceptor cells. The peripheral retina is unaffected.
    A. What would happen to the ability to read in a person affected by macular degeneration?
    B. What would happen to the ability of a person to walk along a city sidewalk for a person affected by macular degeneration?

Autonomic Nervous System   4.9

Learning Objectives

- List the three components of the autonomic nervous system
- Compare the pathways of somatic motor neurons to sympathetic efferents
- Describe the sympathetic chain, identifying the cervical ganglia, celiac ganglion, and superior and inferior mesenteric ganglia
- Distinguish between paravertebral ganglia and prevertebral ganglia
- Distinguish between preganglionic and postganglionic sympathetic fibers
- Describe the anatomic location of the parasympathetic nervous system
- Compare preganglionic and postganglionic fibers in the sympathetic nervous system to those in the parasympathetic nervous system
- Identify the organs innervated by the parasympathetic nervous system through cranial nerves III, VII, IX, and X and the pelvic nerve

THE AUTONOMIC NERVOUS SYSTEM SERVES A HOMEOSTATIC FUNCTION AND AN ADAPTIVE FUNCTION

The autonomic nervous system (ANS) is divided into three divisions: the **sympathetic**, **parasympathetic**, and **enteric**. The sympathetic division regulates the use of metabolic resources and coordinating the emergency response of the body to potentially life-threatening situations (**"fight or flight"**). The parasympathetic division usually presides over the restoration of metabolic reserves and the elimination of wastes (**"rest and digest"**). The enteric nervous system is usually considered separately because of its restricted location to the intestines and related organs (see Figure 4.9.1).

The ANS controls the day-to-day operations of most of the internal organs including

- the cardiovascular system, including heart and blood vessels;
- digestive system;
- respiratory tract;
- kidney and urinary bladder.

Certain involuntary responses to external stimuli are also mediated by the ANS. These include

- constriction of the pupil in bright light and dilation of the pupil in dim light;
- vasodilation of the skin and sweating in response to high core temperature, and vasoconstriction, "goose bumps," and increased fat metabolism in response to low core temperature;
- the "flight or fight" response to threatening stimuli.

Most organs receive both parasympathetic and sympathetic efferents that often have opposite effects. For example, parasympathetic stimulation of the heart slows the heart rate and reduces the strength of contraction; sympathetic stimulation accelerates the heart rate and increases the strength of contraction. However, the **skin and blood vessels receive only sympathetic efferents**. Responses for these tissues occur by adjusting the level of stimulation above or below a tonic level. Most organs have some basal level of stimulation called parasympathetic tone, or sympathetic tone, that establish basal levels of function that can be changed either by increasing or decreasing the tonic frequency of firing of efferent fibers. Thus, removal of parasympathetic output to the heart removes part of its brake on heart rate, and the heart rate increases. The same response would occur if sympathetic output to the heart were increased.

AUTONOMIC REFLEXES ARE FAST

One of the advantages of neural control of organ function over hormonal control is its rapid action and equally rapid return to normal levels. Through nervous stimulation, the heart rate can double within 3–5 s. Goose bumps, sweating, or change of blood pressure can occur within seconds of stimulation. These rapid responses are **reflexes**, stereotypical responses in which specific groups of neurons respond to specific sensory inputs. Generally, these actions occur subconsciously and we are hardly aware of either the sensory information or the reflex responses to them. Some autonomic functions, such as defecation and urination, reach consciousness because they require contraction or relaxation of voluntary skeletal muscle.

THE EMOTIONAL STATE GREATLY AFFECTS AUTONOMIC EFFERENT FUNCTION

Emotions such as excitement, euphoria, fear, anxiety, or anger all influence the level of autonomic activity. These

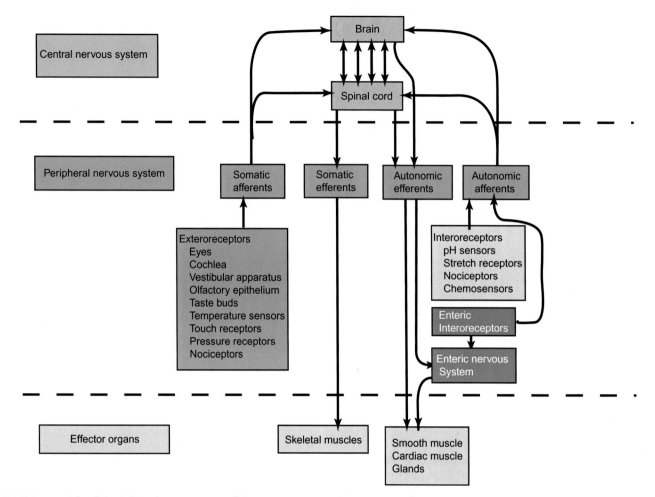

FIGURE 4.9.1 Role of the ANS in the organization of the nervous system. The autonomic efferent systems control the involuntary functions of the body, and they are driven by the integration of signals arriving from these organs along autonomic afferents and from information about the external world that enters the CNS through the exteroreceptors.

emotions stimulate the ANS to cause a variety of responses such as cold sweats, indigestion, excess stomach acid production, cardiac palpitations, high blood pressure, or low blood pressure. These responses are influenced by a variety of psychosocial factors and are conditioned by our previous experiences and individual variation. We live in a complex world of subtle social cues whose effect on us is filtered through the cerebral cortex and limbic system (the seat of the emotions), often subconsciously, into the autonomic system. Upheaval in our social world can literally make us sick, but only by our own permission. We in fact make ourselves sick by our reaction.

AUTONOMIC EFFERENT NERVES HAVE TWO NEURONS

As we saw in Chapter 4.4, motor nerves to skeletal muscles follow an uninterrupted path from the cell body located in the ventral horn of the spinal cord to the muscle. In the ANS the efferent path consists of two neurons: **a preganglionic neuron** and a **postganglionic neuron**. A comparison between the two systems is shown in Figure 4.9.2. Both the sympathetic and the parasympathetic system use these two-neuron paths to the effector cells, but their anatomic arrangement is different. Figure 4.9.2 illustrates the sympathetic division only.

PREGANGLIONIC NEURONS ARE IN THE CNS

The cell body of a preganglionic neuron lies in an appropriate nucleus of the brain or in the gray matter of the spinal cord (**intermediolateral column**). Its axon is thin (2–4 μm) and myelinated. It usually terminates on a postganglionic neuron within a ganglion. A ganglion is a collection of nerve cell bodies outside of the CNS.

POSTGANGLIONIC NEURON CELL BODIES ARE IN GANGLIA

The postganglionic neuron cell body lies in the ganglion, but its typically thin and unmyelinated axon (1.5 μm) leaves the ganglion to make contact with the target organ. Thus, the name "postganglionic" is used to describe its axon. It receives its input from preganglionic neurons that lie within the CNS.

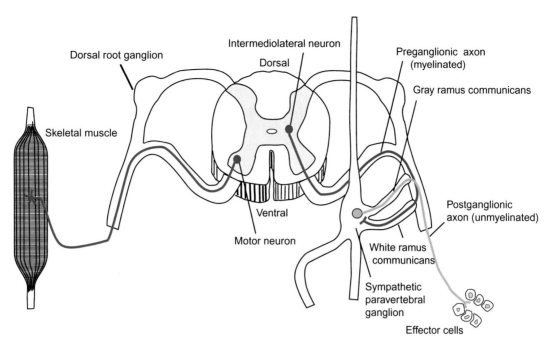

FIGURE 4.9.2 Schematic comparison of the two-neuron efferent pathway of the sympathetic nervous system with the single-neuron efferent pathway for the somatic motor system. The somatic motor system is driven by motor neurons with cell bodies in the ventral horn of the spinal cord, with direct connections to the muscle. The sympathetic neurons have cell bodies in the intermediate gray of the spinal cord and make a synapse on a ganglion cell, which then makes connections to effector cells.

THE SYMPATHETIC NERVOUS SYSTEM ORIGINATES IN THE *THORACOLUMBAR SPINAL CORD*

THE CELL BODIES OF PREGANGLIONIC SYMPATHETIC FIBERS LIE IN T1–T12 AND L1–L2

The preganglionic sympathetic fibers originate from cell bodies in the intermediolateral area of the gray matter of the spinal cord in thoracic segments T1–T12 and lumbar segments L1–L2. The axons leave the cord through the ventral roots and then leave the ventral roots through short branches called **white ramus communicans**. The axons in the white ramus enter a **paravertebral ganglion** at the same level of the cord.

THE PARAVERTEBRAL GANGLIA FORM A SYMPATHETIC CHAIN

Each spinal cord segment has two ventral roots that connect by a white ramus to a spinal sympathetic ganglion. These ganglia communicate with each other up and down the spinal cord, forming two sympathetic chains, one on each side of the vertebral column. Only those pairs of spinal sympathetic ganglia from T1 to L2 receive inputs via the white rami. Those above T1 receive inputs from fibers from the thoracic segments that climb the sympathetic chain. Those ganglia below L2 receive inputs from preganglionic fibers that descend the chain from the lower thoracic and lumbar ganglia (see Figure 4.9.3). Each pair of spinal sympathetic ganglia send out an efferent branch to effector organs. These fibers travel from the ganglia to the spinal nerve by a **gray ramus communicans**. It is gray because the axons it contains are unmyelinated postganglionic fibers. Thus, only some of the sympathetic ganglia have white rami, but all have gray rami.

PREVERTEBRAL GANGLIA ARE UNPAIRED AND LIE IN THE ABDOMEN

Postganglionic fibers that supply the visceral organs originate in three ganglia found within the abdominal cavity. These unpaired ganglia receive input from segments T5–L2 that travel through the sympathetic ganglia without making a synapse, to synapse on neurons within these ganglia. These ganglia are the:

- celiac ganglion
- superior mesenteric ganglion
- inferior mesenteric ganglion.

Fibers from these ganglia affect smooth muscles of the intestinal walls, smooth muscles of the blood vessels in the viscera, intestinal glands, and the enteric nervous system.

THE ADRENAL MEDULLA IS A MODIFIED GANGLION

Preganglionic fibers travel through the prevertebral ganglia and the celiac ganglion to synapse with **chromaffin cells** in the **adrenal medulla**. These cells derive their name from the colored reaction they produce in fixed tissue. These cells are modified postganglionic cells. They release their neurotransmitter, **epinephrine**, directly into the blood instead of at nerve terminals. They also release norepinephrine. Epinephrine is commonly known as **adrenaline**, and its circulating form prepares the body for emergency action. The adrenal medulla is discussed more thoroughly in Chapter 9.6.

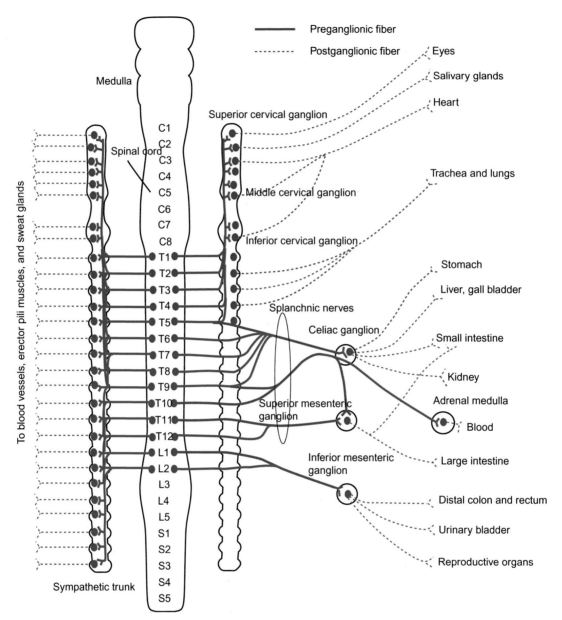

FIGURE 4.9.3 Schematic diagram of the connections of the sympathetic nervous system. Each segment of the spinal cord corresponds to a pair of sympathetic ganglia (paravertebral ganglia), one on each side of the column. These communicate up and down by interganglionic branches, so the set of ganglia resembles beads on a string and is called the sympathetic chain. The preganglionic cells originating in the cord synapse in the ganglia with postganglionic cells. These postganglionic cells send efferents to the skin, sweat glands, and blood vessels of the body, as shown at the left. The efferent output to the viscera, shown to the right, can occur via postganglionic cells in the paravertebral ganglia or via prevertebral ganglia. The prevertebral ganglia include the celiac, superior mesenteric, and inferior mesenteric ganglia. They receive inputs from preganglionic fibers that travel through the chain. All of the visceral organs receive sympathetic efferents as shown. Although the targets of efferent output are shown separately here, in reality both sympathetic chains give rise to both types of efferents. The adrenal medulla receives preganglionic input directly; it acts like a ganglion.

PREGANGLIONIC FIBERS HAVE THREE FATES

According to the description given above, preganglionic fibers with cell bodies in the interomedial column of the spinal cord have one of three fates (shown schematically in Figure 4.9.4):

- synapse on a postganglionic neuron within the paravertebral ganglion;
- travel through the paravertebral ganglion and synapse on a postganglionic cell in the prevertebral ganglia (celiac, superior mesenteric, inferior mesenteric, adrenal medulla);
- travel up or down the sympathetic trunk to synapse on a postganglionic cell in other segments.

CERVICAL GANGLIA *ARE FORMED FROM ASCENDING AXONS FROM T1 TO T5*

Preganglionic axons that inform the cervical ganglia arise from thoracic segments T1–T5. The three cervical ganglia are the **superior**, **middle**, and **inferior** cervical ganglia.

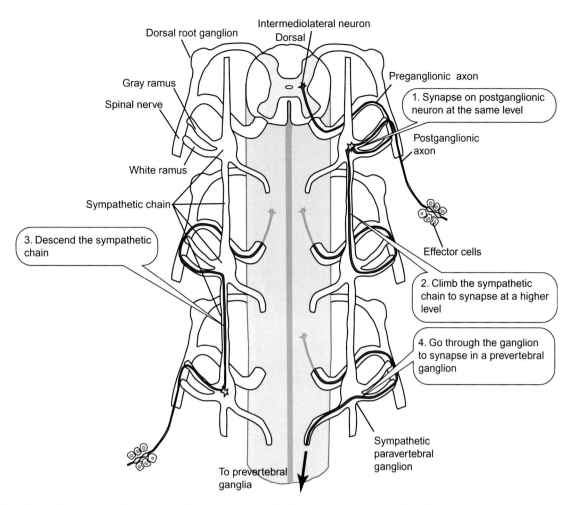

FIGURE 4.9.4 Schematic diagram of the wiring of the sympathetic chain. Preganglionic neurons with cell bodies in the interomedial column of the spinal cord exit the ventral root and their axons travel over the white ramus communicans to the sympathetic ganglia. There the axon can travel up (2) or down (3) the sympathetic chain to neighboring ganglia. Alternatively, the axon can make synapses with a neuron within the ganglia at the same level as the spinal nerve (1) or the axon can pass through the ganglia to make a synapse in a prevertebral ganglion located in the mesentery (4).

PREGANGLIONIC AXONS BRANCH TO ENABLE WIDESPREAD RESPONSES

The number of sympathetic postganglionic axons exceeds the number of preganglionic axons by a factor of about 100. Thus, each preganglionic axon branches many times and thereby spreads its influence over many postganglionic cells. This is an example of **divergence** (see Chapter 4.2). Divergence in the sympathetic nervous system allows widespread activation of many effectors when necessary. Preganglionic axons also **converge** on individual postganglionic neurons. The activity of the postganglionic neurons thereby reflects an integrated response of many preganglionic neurons.

THE PARASYMPATHETIC NERVOUS SYSTEM ORIGINATES IN CRANIAL AND SACRAL NERVES

Preganglionic parasympathetic nerve fibers originate in specific nuclei in the brain or in segments S2–S4 of the sacral spinal cord. These preganglionic axons (2–4 μm in diameter) are myelinated and make synapses with postganglionic cells that reside close to or within the effector organ. Thus, the parasympathetic nervous system has long preganglionic fibers and short postganglionic fibers. This is in stark contrast to the sympathetic nervous system in which the preganglionic fibers are short and the postganglionic fibers are long. In contrast to the sympathetic nervous system, the parasympathetic nervous system is more localized and less diffuse. The preganglionic fibers branch far less; the ratio of preganglionic neurons to postganglionic neurons is near 1:1 or 1:2 (Figure 4.9.5).

Parasympathetic efferents flow out through cranial nerves III, VII, IX, and X. **Cranial nerve III** is the **oculomotor nerve**, originating in the tectum of the midbrain where inputs from the optic nerve provide input for ocular reflexes. **Parasympathetic stimulation constricts the pupil and contracts the ciliary muscle**.

Parasympathetic output of the **facial nerve**, **cranial nerve VII**, originates in the superior salivary nucleus in the rostral medulla. Some fibers synapse on postganglionic neurons in the pterygopalatine ganglion, which innervates the **lachrymal glands** and the nasal and palatine mucosa. Parasympathetic stimulation enhances secretion of tears. Other fibers in the facial nerve travel in

478 QUANTITATIVE HUMAN PHYSIOLOGY

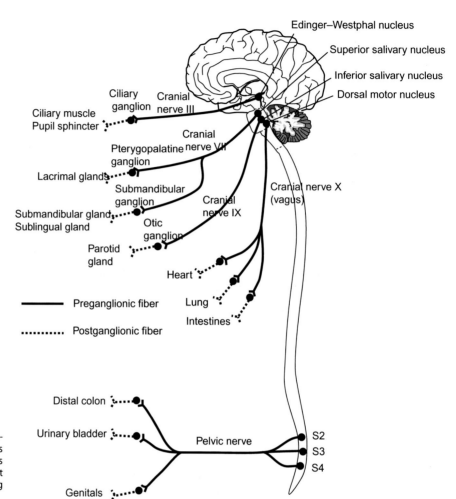

FIGURE 4.9.5 Connections of the parasympathetic nervous system. This system originates from cranial and sacral centers and is characterized by long preganglionic fibers that synapse in ganglia near or in the organ being regulated.

the chorda tympani, a division of the facial nerve, to synapse with cells in the submandibular ganglion. These innervate the submandibular and sublingual glands; **parasympathetic stimulation increases salivary secretion of fluid**.

Preganglionic parasympathetic neurons reside in the inferior salivary nucleus in the medulla and travel over the **glossopharyngeal nerve**, or cranial nerve IX. These synapse on cells in the **otic ganglion**, where the postganglionic fibers join cranial nerve V to travel to the parotid gland. **Parasympathetic stimulation of the parotid gland increases its rate of saliva secretion**. The glossopharyngeal nerve also brings sensory information into the medulla. The carotid bodies sense the arterial P_{O_2} and P_{CO_2} and relay this information to the respiratory centers of the medulla. Baroreceptors in the carotid sinus sense arterial blood pressure and relay that information to the **tractus solitarius** in the medulla.

The **vagus nerve**, cranial nerve X, is the major parasympathetic nerve. The **nucleus ambiguus** and the **dorsal motor nuclei** in the **medulla** provide efferent output to the vagus nerve that supplies a variety of internal organs including the heart, lungs, kidney, liver, spleen, pancreas, and the gastrointestinal tract. Long preganglionic fibers travel over the vagus nerves to ganglia located in the target tissues. The right vagus nerve supplies the **sinoatrial node** (SA node) of the heart whereas the left vagus nerve supplies the **atrioventricular node** (AV node) (see Chapter 5.5). **Vagal stimulation of the heart slows its rate and reduces its strength of contraction**. Vagal efferents to the lung control the caliber of the bronchioles through control of the smooth muscles in the bronchiole walls. **Vagal stimulation constricts the bronchioles and also regulates secretory activity**. Vagal inputs to the esophagus and stomach make synapses with enteric ganglia. Innervation by the vagus regulates gastrointestinal motility and secretion.

Preganglionic parasympathetic nerves originate in the interomedial gray of segments S2, S3, and S4 of the spinal cord. Their long presynaptic axons reach enteric ganglia in the lower portion of the gastrointestinal tract (the descending colon, sigmoid colon, rectum, and internal anal sphincter), the urinary bladder, and the genitalia. **Parasympathetic stimulation causes urination, defecation, and erection**, but not simultaneously.

AUTONOMIC REFLEXES LINK SENSORY INPUT TO MOTOR EFFERENTS

Traditionally, the ANS is viewed as an efferent system. However, efferent activity is brought about in response to sensory input, which travels over the same nerves as the efferents. The vagus nerve, for example, carries 10 times as many afferent fibers as efferent fibers: the afferents are a necessary part of the ANS as much as the cutaneous and muscle sensors are necessary to the somatic motor system. Sensory information from the visceral organs, blood vessels, and skin forms the afferent limb of the autonomic reflex arc. This sensory input modulates the activity of the sympathetic neurons in the interomedial lateral column of the spinal cord from T1 to L2, or of parasympathetic neurons in the parasympathetic nuclei of the medulla or in the sacral cord. The sensory input that drives autonomic reflexes may or may not reach consciousness. Thus some of the inputs climb the cord to reach higher centers. Sensations perceived from the viscera may be vaguely localized or may be felt in a somatic structure other than the organ. Thus, damage to the heart can be referred to the left arm. This is called **referred pain**.

Just as sensory input from the internal organs can reach higher centers to produce a conscious perception, output from higher centers can influence autonomic function. Thus, the traffic of information up the cord is accompanied by modulatory inputs from higher centers down the cord. These "higher centers" are autonomic ganglia in the brain stem that receive input from the amygdala and the hypothalamus, but have no direct input from the somatic motor cortex. Thus there is no conscious control of autonomic function. A schematic of these events is shown in Figure 4.9.6 for sympathetic reflexes.

THE MAJOR AUTONOMIC NEUROTRANSMITTERS ARE *ACETYLCHOLINE* AND *NOREPINEPHRINE*

Both sympathetic and parasympathetic preganglionic neurons release acetylcholine at their terminals. The postsynaptic membrane on the postganglionic cell has nicotinic receptors for acetylcholine, so named because nicotine is an agonist. This receptor is similar to the nicotinic receptor at the neuromuscular junction, but the two receptors are not identical. Unlike the nicotinic receptor at the neuromuscular junction, the ANS receptor is not blocked by **curare**, but it is blocked by **hexamethonium**. The nicotinic receptor is ionotropic and binding of acetylcholine opens a channel for cations that causes a depolarization of the postsynaptic cell membrane. Similar to the neuromuscular junction, acetylcholine is rapidly degraded by acetylcholinesterase, which shuts off the signal.

Postganglionic fibers of the sympathetic division mainly release norepinephrine whereas postganglionic parasympathetic fibers release acetylcholine. An exception to this rule is the postganglionic sympathetic fibers to sweat glands, which release acetylcholine. Figure 4.9.7 shows the neurotransmitters released by preganglionic and postganglionic fibers in both the sympathetic and parasympathetic divisions of the ANS.

PARASYMPATHETIC RELEASE OF ACETYLCHOLINE WORKS ON *MUSCARINIC* RECEPTORS

Postganglionic neurons in parasympathetic the division generally are located within the target tissue and have short postganglionic axons. They release acetylcholine onto **muscarinic receptors** located on the target cells, so named because the plant alkaloid **muscarine** is an agonist for these receptors. The muscarinic receptors are all metabotropic. There are multiple isoforms of the muscarinic receptors designated M1, M2, M3, M4, and M5. Muscarinic receptors are blocked by **atropine**.

As described in Chapter 4.2, M1, M3, and M5 receptors are linked to G_q-**coupled receptors**. Recall that these heterotrimeric G-proteins consist of α, β, and γ subunit

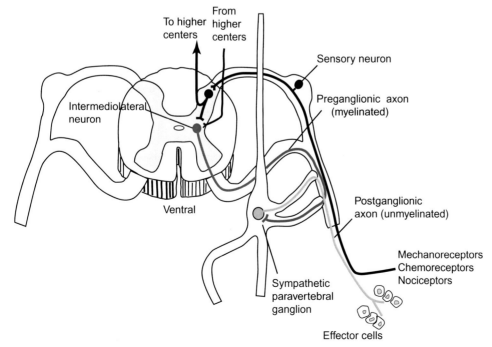

FIGURE 4.9.6 Hypothetical example of an autonomic reflex. Sensory information due to mechanoreceptors, chemoreceptors, or nociceptors enters the spinal cord along the dorsal root. This information is conveyed to an interneuron that conveys it to autonomic preganglionic neurons located in the intermediolateral column of the spinal cord. Some of these sensory fibers send sensory information up the cord to the brain stem, enabling conscious perception of visceral events. Higher centers return input to the intermediolateral neurons, thereby providing a mechanism for the influence of autonomic function by the emotional state. Most of these reflex activities occur subconsciously, but visceral pain can quickly draw one's attention.

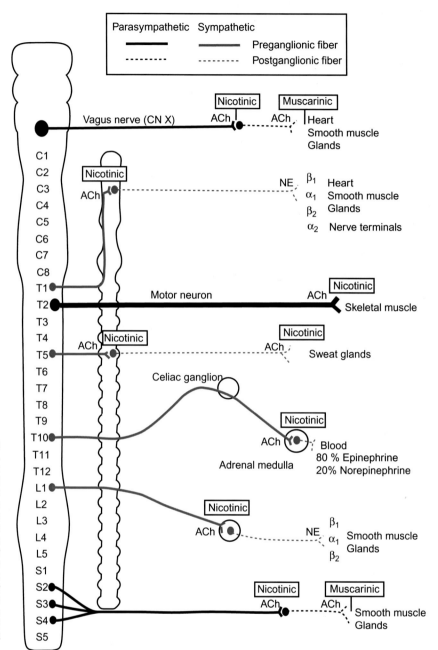

FIGURE 4.9.7 Neurotransmitters used by the parasympathetic and sympathetic nervous systems. Preganglionic sympathetic and parasympathetic fibers both use acetylcholine as neurotransmitter and the postganglionic cells have nicotinic cholinergic receptors. The postganglionic parasympathetic nerves also release acetylcholine, and the postsynaptic target cells have muscarinic cholinergic receptors. In contrast, postganglionic sympathetic fibers release norepinephrine onto cells that have α_1-, α_2-, β_1-, or β_2-receptors. An exception to this rule is postganglionic fibers that release acetylcholine onto muscrinic receptors in the sweat glands. For comparison, a somatic α-motor neuron that controls skeletal muscle is also shown. Preganglionic fibers are solid lines; postganglionic fibers are dashed lines. Red color signifies sympathetic nervous system; black indicates parasympathetic connections.

exchanges GTP for GDP upon ligand binding and dissociates from the $\beta\gamma$ subunits. The α subunit for M1, M3, and M5 receptors activates **phospholipase C** to release **inositol tris phosphate (IP3)** and **diacylglycerol (DAG)** from phosphatidylinositol 4,5-bisphosphate. The released IP3 releases Ca^{2+} from intracellular stores, and this activates **CAM kinase**, a protein kinase that phosphorylates target proteins within the cell. The DAG activates **protein kinase C (PKC)** of which there are multiple subtypes. This phosphorylation brings about the final response to the neurotransmitter.

The M4 receptor is coupled to a **G$_i$ receptor**: binding of acetylcholine to these receptors inhibits the enzyme **adenylyl cyclase**, which produces 3',5'-**cyclic AMP** from ATP. Cyclic AMP stimulates **protein kinase A (PKA)**, which phosphorylates its target proteins. Thus, binding of acetylcholine to cells with M2 and M4 receptors results in less phosphorylation of target proteins. The M2 receptor is also coupled to a G$_i$ mechanism, but the $\beta\gamma$ subunit also directly activates a K^+ channel, which typically causes a hyperpolarization in target cells.

NOREPINEPHRINE RELEASED BY POSTGANGLIONIC SYMPATHETIC NEURONS ACTS THROUGH α- AND β-RECEPTORS

Epinephrine is a hormone released from the adrenal medulla in response to stress through activation of sympathetic preganglionic efferents. "Epinephrine" derives its name from "epi," meaning "above" and "nephros," for kidney, because the adrenal gland is located on top of the kidney. Its other name "adrenaline" derives from "renal," another root for kidney. Receptors for both norepinephrine and epinephrine are called **adrenergic receptors**. There are four main types of these receptors, which were first described by their different responses to pharmacological agents. The β-receptors respond to much lower concentrations of epinephrine or norepinephrine than do the α-receptors. $β_1$-receptors respond to epinephrine and norepinephrine about equally whereas $β_2$-receptors are much more sensitive to epinephrine than to norepinephrine. These receptors were described previously in Chapter 4.2.

Both $β_1$- and $β_2$-receptors couple to a G_s protein that increases [cAMP] and activates PKA to phosphorylate target proteins. The β-receptors are blocked by **propranolol**. The $α_1$-receptors are coupled to G_q proteins that activate phospholipase C to increase cytosolic concentrations of DAG and IP3 as detailed above for the M1, M3, and M5 receptors. The $α_2$-receptors are linked to a G_i protein that inhibits adenylyl cyclase, decreasing [cAMP]. The α-receptors are blocked by **phenoxybenzamine**.

Norepinephrine is typically released by postganglionic sympathetic fibers, but those that innervate the general sweat glands of the skin release acetylcholine. However,

TABLE 4.9.1 Effect of Parasympathetic and Sympathetic Stimulation on Target Tissues

Target Tissue	Parasympathetic Stimulation	Sympathetic Stimulation
Eye		
Pupil	Contraction (miosis)	Dilation ($α_1$)—mydriasis
Ciliary muscle	Contraction	Relaxation ($β_2$)
Mueller's muscle	—	Contraction ($α_1$)
Lachrymal glands	Increased secretion	—
Salivary glands	Increased secretion (M3)	Increased protein secretion ($β_1$)
Nasal glands	Increased secretion	Decreased secretion ($α_1$)
Skin		
Sweat glands	—	Increased secretion (M)
Palms	—	Emotional sweating ($α_1$) contraction ($α_1$)
Erector pili	—	
Heart		
Rate	Decreased (M2)	Increased ($β_1$)
Contractility	Decreased (M2)	Increased ($β_1$)
Lungs		
Bronchioles	Constriction	Dilation ($β_2$)
Secretion	Increased secretion	—
Blood vessels		
Skin	—	Constriction ($α_1$)
Skeletal muscle	—	Dilation ($β_2$) constriction ($α_1$)
Viscera	—	Constriction ($α_1$)
Gastrointestinal tract		
Motility	Increased	Decreased ($α_2$, $β_2$)
Sphincters	Relaxation	Contraction ($α_1$)
Glands	Increased secretion (M3 in pancreas)	Decreased secretion
Gall bladder	Contraction	Relaxation
Liver	—	Glycogenolysis and gluconeogenesis ($β_2$, $α_1$)
Urinary system		
Detrusor muscle	Contraction	Relaxation ($β_2$)
Ureter	Relaxation	Contraction ($α_1$)
Sphincter	Relaxation	Contraction ($α_1$)
Fat tissue	—	Lipolysis (β)

fibers to sweat glands of the palms of the hands release norepinephrine.

AUTONOMIC NERVE TERMINALS ALSO RELEASE OTHER NEUROTRANSMITTERS

Sympathetic nerves that cause vasoconstriction release both norepinephrine and **neuropeptide Y**. **Vasoactive intestinal peptide (VIP)** is released along with acetylcholine by sympathetic nerves to the sweat glands and also by parasympathetic nerves to the enteric nervous system. Some postganglionic parasympathetic neurons release **nitric oxide (NO)**. These neurons are important in the intestine and in erectile function and are the basis of action of Viagra. These neurons are called nonadrenergic noncholinergic (NANC).

EFFECTS OF AUTONOMIC STIMULATION DEPEND ON THE RECEPTOR ON THE TARGET CELL

The specificity of autonomic effects derives from the use of a variety of neurotransmitters from the presynaptic cell and the use of a variety of receptors on the postsynaptic cell. Postganglionic parasympathetic fibers releasing acetylcholine will have no effect on target cells that have only α- or β-adrenergic receptors. These same fibers will have different effects on target cells expressing M1 on their surface compared to those expressing M4. Typically cells with M1, M3, and M5 receptors are activated by parasympathetic stimulation, whereas those with M2 and M4 are inhibited. Table 4.9.1 shows the effects of parasympathetic or sympathetic stimulation on a variety of target tissues.

THE PUPILLARY LIGHT REFLEX REGULATES LIGHT INTENSITY FALLING ON THE RETINA: A PARASYMPATHETIC REFLEX

The size of the pupil is determined by two thin layers of muscle within the **iris**, as described in Chapter 4.8. Muscle fibers in the **dilator** muscle orient radially and are controlled by sympathetic nerves (α_1-receptors). Sympathetic stimulation dilates the pupils, but only in unusually stressful situations. The fibers in the **sphincter** muscle orient circumferentially, as shown in Figure 4.9.8. Parasympathetic nerves contract this muscle and constrict the pupil. The **direct pupillary reflex** refers to the constriction of the pupil upon exposure to bright light. The sphincter muscle receives tonic parasympathetic stimulation, so dilation of the pupil in dim light normally results from reduction in parasympathetic tone rather than from sympathetic stimulation of the dilator muscle. Special retinal ganglion cells (RGCs) containing a photodetector called **melanopsin** detect the level of illumination without contributing to image formation. These RGCs project fibers to the midbrain on the ipsilateral side. These areas of the midbrain project to the **Edinger–Westphal nucleus** near the midline of the medulla, which then sends efferent motor fibers to the ciliary ganglia and from there to the pupil sphincter muscle and the **ciliary body**, which controls tension on the lens. The effect is to constrict the pupil and release tension on the lens, causing it to increase its refractory power. Simultaneously, information about the bright illumination passes over to the other side of the brain through the posterior commissure, and project to the Edinger–Westphal nucleus on the contralateral side. This causes constriction of the pupil in the other

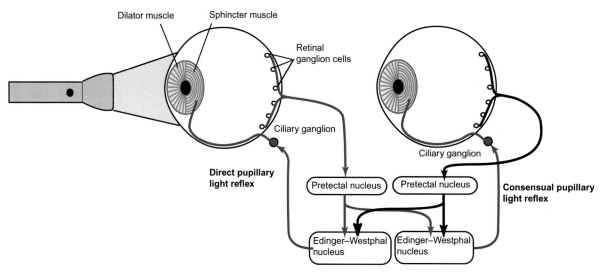

FIGURE 4.9.8 Direct and consensual pupillary reflex. Special retinal ganglion cells that contain melanopsin, now called "intrinsically photosensitive retinal ganglion cells," or ipRGCs, begin a four-neuron reflex pathway. The first neurons, the ipRGCs, project to the ipsilateral pretectal nucleus at the level of the superior colliculus. A second cell there projects to each Edinger–Westphal nucleus. A third cell in the Edinger–Westphal nucleus connects to the ciliary ganglion, where the final postganglionic parasympathetic fiber innervates the sphincter muscle. In the direct pupillary reflex, exposing one eye to light causes constriction of the pupil in that eye. This is accomplished by the reflex in the figure. Activated axons are shown in blue, whereas quiescent fibers are in black. Because the pretectal fibers project to both Edinger–Westphal nuclei, light incident on one eye also causes pupil constriction in the other eye, even when it is kept dark. Light exposure causing constriction of the pupil in the contralateral eye is the consensual pupillary light reflex. Constriction of the pupils to a pin point is called **miosis**. A separate sympathetic innervation of the dilator muscle occurs through the sympathetic system from nerve cells in the superior cervical ganglion. Widening of the pupil is called **mydriasis**. Dilation of the pupils in dim light is normally due to removal of parasympathetic tone.

eye, even though that eye is not exposed to bright light. This is the **consensual pupillary light reflex**. These reflexes have the effect of limiting light exposure to the retina so that good visual discrimination is maintained over wide ranges of illumination.

MICTURITION INVOLVES AUTONOMIC REFLEXES AND VOLITIONAL NERVOUS ACTIVITY

The urinary bladder is a muscular sac that stores between 150 and 250 mL of urine produced by the kidneys until it can be voided. The ability to hold urine in the bladder is called **continence** and emptying the bladder is called **micturition**. Micturition can occur automatically by a spinal reflex, but it can be overridden by higher control. Reflex contraction of the **detrusor muscle** in the bladder is controlled by afferent stretch receptors in the walls of the bladder that travel over the **pelvic nerve** to the sacral spinal cord (S2–S4) where they engage efferent fibers. Information from these stretch receptors ascends the cord to make us consciously aware of the need to urinate. Simultaneously, parasympathetic efferents relax the internal sphincter muscle on the urethra.

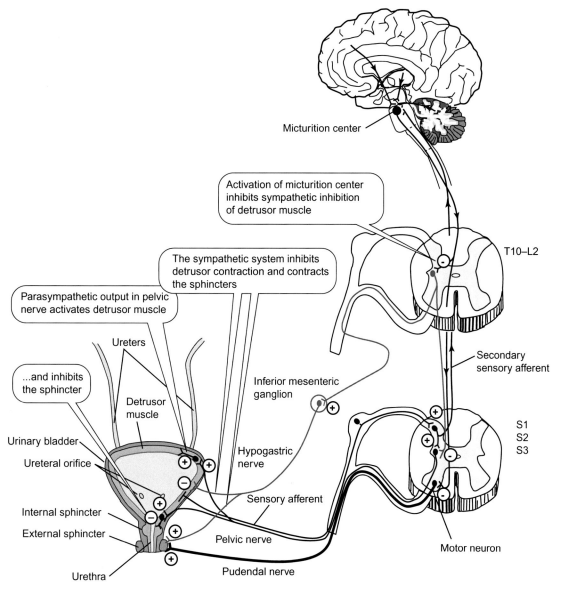

FIGURE 4.9.9 Parasympathetic and sympathetic control of micturition. Stretch receptors in the wall of the bladder activate a parasympathetic urination reflex in the sacral spinal cord. The parasympathetic afferents and efferents both travel over the pelvic nerve. The efferent parasympathetics contract the detrusor muscle that empties the bladder and relaxes the internal sphincter. Sympathetic efferents originate in the intermediolateral gray column of T10–L2 and travel to the inferior mesenteric ganglion. The hypogastric nerve carries the sympathetic efferents to the bladder where they promote relaxation of the detrusor and contraction of the internal sphincter. An intraspinal tract from the intermediolateral sympathetics also inhibits the sacral micturition reflex. Micturition can be overridden by conscious control of somatic α-motor neurons in the ventral horn of the sacral spinal cord. Efferent signals from the motor neuron travel over the pudendal nerve to contract the external sphincter. The micturition center located in the rostral pons receives inputs from the cerebrum, cerebellum, and other superpontine centers. It coordinates the parasympathetic, sympathetic, and somatic motor control of micturition. The sympathetic fibers are shown in blue; parasympathetics are in black; somatic motor neuron in thick black. The effects shown are for promotion of micturition; a positive sign indicates excitation, a negative sign indicates inhibition. The overall effect of any pathway can be determined by multiplying the signs. Thus the parasympathetic micturition reflex is stretch receptor (+) × interneuron (+) × preganglionic fiber (+) × postganglionic fiber on detrusor (+) = detrusor contraction (+).

Sympathetic fibers originating in the intermediolateral horn in the distal thoracic and lumbar spinal cord travel to the inferior mesenteric ganglion where they synapse on postganglionic sympathetic fibers. These fibers travel in the hypogastric nerve to the bladder where they inhibit the detrusor muscle through β_2-receptors and contract the internal sphincter muscle through α_1-receptors. The sympathetic fibers also inhibit the stretch reflex through intraspinal connections. The sympathetic system inhibits bladder emptying whereas the parasympathetic system promotes it.

Nearly everyone has at one time or another found themselves with a full bladder with no socially acceptable way of relieving themselves. The only recourse is to "hold it" until a proper place is found. This is the function of the **external sphincter** muscles that are controlled by somatic α motor neurons in the sacral spinal cord that reach the external sphincter muscle over the **pudendal nerve**. These motor neurons are under conscious control through higher centers and through spinal tracts.

A **micturition center** in the rostral pons receives input from the cortex, cerebellum, and other centers above the pons and controls micturition through the **reticulospinal tract**. This center can be either activated, to void the bladder, or suppressed to inhibit micturition. Figure 4.9.9 shows the effects of nervous activity that promotes micturition. To promote micturition, the micturition center inhibits sympathetic output to remove inhibition of detrusor contraction and contraction of the internal sphincter. Simultaneously, it strengthens the spinal stretch reflex and inhibits somatic motor neuron contraction of the external sphincter. All of these actions result in contraction of the detrusor muscle and relaxation of the sphincters.

The control of defecation follows the same basic principles as the control of micturition.

Clinical Applications: Nerve Gases

Nerve gases bind to and inhibit acetylcholinesterase. As a result, all cholinergic synapses are chronically stimulated because the neurotransmitter, acetylcholine, is not properly hydrolyzed following excitation. The resulting symptoms depend somewhat on the dose of poison. All cholinergic parasympathetic effects are activated: salivation, lacrimation, urination, defecation, emesis, (SLUDE), bronchial secretion, bronchioconstriction, and bradycardia. Other cholinergic synapses are also activated, producing sweating, muscle spasm, and effective paralysis. The agents affect both the respiratory center and the respiratory muscles so that death typically occurs by suffocation. At high doses, victims may lose consciousness and convulse, and death may occur so quickly that the parasympathetic effects may not have time to develop. Thus, these nerve gases kill by jamming the off switch to cholinergic neurotransmission.

The most important nerve gases are **Sarin** and **VX**, whose chemical structures are shown below. Sarin was first produced by the Germans in 1938 and named for the four scientists who discovered it: **S**chrader, **A**mbros, **R**udinger, and L**in**de. By the end of the war the German military had 30 tons of it. Allied military planners assumed the gas would be used, and Britain had stockpiled 30 million gas masks in preparation for the event. Fortunately, Germany never used Sarin during World War II, probably in fear of like retaliation and because its dispersal is not easily controlled. Its only known use was by the Japanese cult Aum Shinrikyo, who dropped plastic bags wrapped in newspaper on five separate subway trains in Tokyo on March 20, 1995. The bags were punctured by sharpened umbrella tips in a coordinated attack. The liquid leaked out and vaporized, poisoning several thousand commuters and killing 12.

British chemists made VX in 1952 at the Porton Down Weapons Research Center in Wiltshire. They traded information about its methods to the United States in return for secrets on thermonuclear weapons. This nasty chemical has the advantage of being stable and so can contaminate areas for longer times. The British abandoned nerve agents in favor of thermonuclear weapons. The LD_{50} (the dose that kills 50% of those exposed) for VX is 10 mg.

The term "nerve gas" is something of a misnomer because the agents are liquids at ambient temperatures and pressures. However, they can be dispersed as gas or aerosols, and the liquid form is readily absorbed through the skin.

One antidote for these nerve agents is **atropine**. This does not reverse the effect of the agents, but blocks the effects of excess acetylcholine neurotransmitters. The dose is about 1 mg. A second agent, pralidoxime chloride, is part of the Mark I Kit issued to the U.S. military and contains 600 mg. Pralidoxime rescues the acetylcholinesterase from nerve agent inhibition.

Who has nerve agents, in what quantities, and with what delivery capabilities, is a question for Intelligence Agencies. The United States and Russia have admitted to having stockpiles of VX, and both are committed to destroying them. Although there is no proof, some claim that Iraq used nerve agents in attacks on the Kurds in Halabja in Northern Iraq in 1988.

Sarin

VX

Clinical Applications: Muscarine Poisoning

Eating mushrooms can be a tricky business because some are poisonous. The mushroom *Amanita muscaris* makes an alkaloid poison called **muscarine**, whose structure is shown below. Note that it has some structural similarities to acetylcholine, shown next to it. Because of its structural similarity to acetylcholine, muscarine can bind to the acetylcholine receptor and activate it. Thus, muscarine is a cholinergic **agonist**. Muscarine acts on the muscarinic receptors, but not on the nicotinic receptors. The chemical structure of nicotine is shown for comparison. However, muscarine does not have the acetyl ester group that is hydrolyzed by acetylcholinesterase, so its stimulation cannot be shut off by hydrolysis. Instead, muscarine is slowly cleared from the body by urinary excretion.

Muscarine poisoning causes a chronic stimulation of the muscarinic receptors. These are largely on target cells of parasympathetic stimulation. Their effects include stimulation of glandular secretions including tearing, salivation, sweating, and nasal and bronchial secretions. Smooth muscle of the intestinal tract is stimulated, causing both defecation and emesis. The urinary bladder contracts and the sphincter relaxes, causing urination. Constriction of the bronchioles makes it difficult to breathe, and the heart rate slows (bradycardia). It is altogether a nasty business.

Fortunately, there is help. Another plant, the deadly nightshade, *Atropa belladona*, produces another poison called **atropine**, whose chemical structure is shown above. This chemical is a **muscarinic receptor antagonist**, and it blocks the effects of muscarine. One effect of parasympathetic stimulation is constriction of the pupillary sphincter muscle, which constricts the pupil. Ordinarily, this occurs in response to bright light to limit the light intensity on the retina. Blocking this effect with atropine therefore dilates the pupils. Ladies in Europe discovered this effect and administered it to themselves to appear more attractive. Hence the plant was named "belladona," meaning "beautiful lady." Although atropine can block the effects of muscarine, it is a poison in its own right.

SUMMARY

The sympathetic nervous system consists of nuclei in the brain stem that connect to preganglionic neurons in the interomedial gray in the spinal cord. The preganglionic fibers exiting the cord through the ventral root reach the paravertebral ganglia over the white rami communicans. The paravertebral ganglia form a sympathetic chain on each side of the cord. Preganglionic fibers synapse with neurons in the paravertebral ganglion at the same level of the cord, ascend the chain to make synapses above, descend the chain to make synapses below, or travel through the ganglion to prevertebral ganglia: the celiac ganglion, superior mesenteric ganglion, the inferior mesenteric ganglion, or the adrenal gland. The cervical part of the sympathetic chain form the inferior, middle, and superior cervical ganglia. Sympathetic preganglionic fibers originate in T1–L2 and are relatively short. The postganglionic fibers are long. The preganglionic synapse uses acetylcholine and the postganglionic synapse uses norepinephrine, but sympathetic postganglionic fibers use acetylcholine on sweat glands. The sympathetic nervous system prepares the body for action. It increases heart rate and cardiac contractility, dilates bronchioles, redistributes blood flow to the muscles, inhibits gastrointestinal motility and secretion, and mobilizes metabolic fuels.

The targets of sympathetic stimulation respond through two major classes of surface receptors: the α- and β-receptors. The α_1-receptors act through a G_q mechanism which raises intracellular $[Ca^{2+}]$ by IP3-induced release of Ca^{2+} stores; the α_2-receptors act through G_i which inhibits adenylyl cyclase and reduces phosphorylation of target proteins within the cell. Both β_1- and β_2-receptors activate a G_s that stimulates adenylyl cyclase and increases phosphorylation of target proteins. Activation of β-receptors on smooth muscle relaxes them; activation of α_1-receptors causes constriction.

The parasympathetic nervous system originates in the cranium and sacral cord, has long preganglionic fibers and short postganglionic ones, and uses acetylcholine at both preganglionic and postganglionic synapses.

It diverges less than the sympathetic nervous system. It constricts the pupil and controls accommodation through the oculomotor nerve, cranial nerve III. Parasympathetic innervation also controls lacrimation (cranial nerve VII) and salivation (cranial nerve VII and IX). The heart, lungs, and intestines are all controlled through the vagus nerve, cranial nerve X. The shutting off of the parasympathetic stimulation requires destruction of acetylcholine by acetylcholinesterase. Inhibition of acetylcholinesterase by nerve gas causes SLUDE: salivation, lacrimation, urination, defecation, and emesis. Other effects include bradycardia, bronchoconstriction, and bronchial secretion.

The receptors at the preganglionic parasympathetic synapse are nicotinic; those at the postganglionic synapse are muscarinic. The muscarinic receptor subtypes M1, M3, and M5 use a G_q mechanism (activation of phospholipase C, releasing IP3 and diacylglycerol (DAG); IP3 releases intracellular stores of Ca^{2+} while DAG activates PKC); M2 and M4 receptors use a G_i mechanism, but M2 receptors also modulate a K^+ channel through the $\beta\gamma$ subunit of the G protein.

The pupillary light reflex involves special retinal cells that sense light but do not contribute to image formation. These intrinsically photosensitive retinal ganglion cells use a different chromophore, melanopsin, to detect light. Their output is to the pretectal nucleus, which then connects to cells in the Edinger–Westphal nucleus in the medulla. These cells then project to a fourth neuron in the ciliary ganglion, which in turn innervates the sphincter muscle in the iris that constricts the pupil.

REVIEW QUESTIONS

1. In general, what does the sympathetic nervous system do? What does the parasympathetic nervous system do?
2. Where are the preganglionic sympathetic neurons? Where do their axons go? Are these myelinated? Where are the ganglionic sympathetic neurons? What is a paravertebral ganglion? What is meant by "sympathetic chain"? Where do the postganglionic fibers go? Are they myelinated? What is the white rami communicans? Why is it white? What is the gray rami communicans? Why is it gray?
3. Are sympathetic preganglionic fibers long or short? Are sympathetic postganglionic fibers long or short?
4. What neurotransmitters are used at the preganglionic sympathetic nerve terminal? Postganglionic? What is the exception to this generality?
5. What effect does sympathetic stimulation have on sweating? Blood vessel caliber? Bronchiole caliber? Heart rate?
6. Name the two major classes of adrenergic receptors. What signaling mechanisms do they use?
7. What are prevertebral ganglia? Name three. What does the adrenal medulla secrete? What stimulates its secretion?
8. Where are the preganglionic parasympathetic neurons? Where are the ganglionic neurons? Are the preganglionic fibers long or short? Are the postganglionic fibers long or short?
9. What neurotransmitters are used at the preganglionic parasympathetic nerve terminal? Postganglionic? What types of receptors are present at the two locations?
10. What effect does parasympathetic stimulation have on sweating? Blood vessel caliber? Bronchiole caliber? Intestinal motility? Lacrimation? What is SLUDE?

UNIT 5

THE CARDIOVASCULAR SYSTEM

Overview of the Cardiovascular System and the Blood 5.1

Learning Objectives

- Calculate the time of diffusion as a function of distance
- Describe the components of the cardiovascular system and their function
- List the classes of important materials carried by the blood
- Describe the location and function of the interstitial fluid
- Identify the major arteries and veins from their anatomic location
- Distinguish between the systemic and pulmonary circulation
- Describe what is meant by a portal circulation
- Explain why the output of the heart is almost always exactly equal to its input
- Define vascular resistance
- Define vascular compliance
- Define the hematocrit
- Describe hemostasis in general terms
- Describe the role of platelets in blood clotting
- Distinguish between the intrinsic and extrinsic pathway in blood clotting
- Describe the chemical reaction that directly forms a clot
- Describe the function of plasmin and how it is activated

THE CIRCULATORY SYSTEM IS A TRANSPORT SYSTEM

The central organizational theme of large, multicellular creatures such as ourselves is that the cells, in general, are not directly connected to the external environment. They are directly connected to the internal environment, which is the interstitial fluid, and indirectly connected to the environment through a medium of exchange, and that medium is the blood. The blood allows cells to have indirect contact with the external environment through the filters of the lungs, intestines, kidneys, and the skin. The internal environment that bathes the cells must be supplied with nutrients for the cells, must have waste products removed, and must have the heat that the cells generate dissipated to the external environment. This overall organization is shown schematically in Figure 5.1.1

THE CIRCULATORY SYSTEM CONSISTS OF THE *HEART, BLOOD VESSELS*, AND *BLOOD*

Figure 5.1.1 shows the three main components of the circulatory system: the heart, the blood vessels, and the blood. Each performs an irreducible function that cannot be provided by the other components.

The **heart** is a bag of muscle that encloses a part of the blood and that contracts rhythmically to move blood from the veins, at low pressure, to the arteries at high pressure. It is actually four bags of muscles (see Figure 5.1.1) arranged in series, but their contraction is nearly simultaneous. The heart has valves that make flow unidirectional. The heart contracts about once a second (70 beats per minute) and has its own rhythm generator, the pacemaker. Specialized excitable cells form an electrical system that coordinates the heart beat between its four chambers.

The blood vessels are hollow tubes that are conduits for the blood. They are collectively called the **vascular system**. The blood vessels are of five major types: **large arteries**, **arterioles**, **capillaries**, **venules**, and **major veins**. **Arteries always carry blood away from the heart** and withstand high pressures.

Arterioles are smaller versions of the arteries that are important in the regulation of blood flow. The capillaries are tiny tubes that are often about the same size as the red blood cells. They allow close but indirect contact of the blood with the tissues. Here, water, gas, and solute exchange between the blood and the tissues. **The venules and veins carry blood toward the heart** and generally have a low pressure.

The vascular system has four main functions:

1. Transform pulsatile flow from the heart beat into more continuous flow.
2. Distribute the blood to various organs.
3. Exchange materials in the tissues.
4. Veins serve as a volume reservoir.

THE CIRCULATORY SYSTEM CARRIES NUTRIENTS, WASTES, CHEMICAL SIGNALS, AND HEAT

The whole point of the circulatory system is to carry something in the blood. Blood contains a variety of

QUANTITATIVE HUMAN PHYSIOLOGY

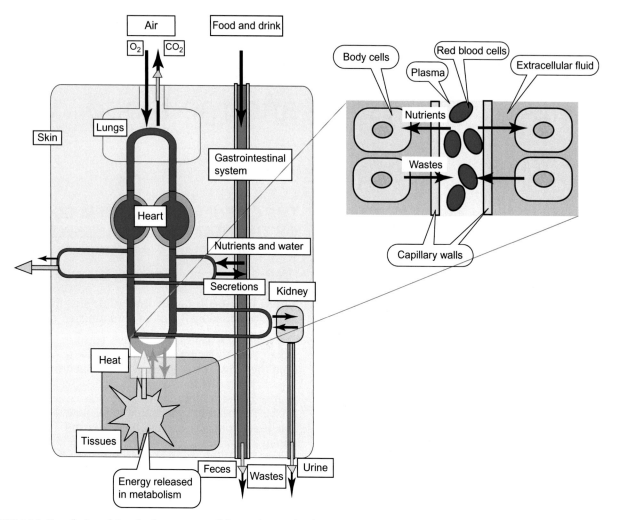

FIGURE 5.1.1 Overall plan of the circulatory system. Cells require supply of nutrients and removal of waste material and heat. Because they are far from body surfaces, nutrients are delivered and wastes removed through the medium of the blood, contained within vessels, and moved through the system by pressure generated by the beating heart. The final purpose of the cardiovascular system is to transport materials to the tissues (inset) where materials can exchange into the **interstitial fluid** that immediate surrounds the cells. Black arrows indicate mass exchange; light arrows indicate energy exchange.

materials, including water, salts, dissolved proteins, nutrients, and a variety of different cell types. Table 5.1.1 contains a partial listing of the huge variety of materials carried by the blood. These come under four main categories: nutrients, wastes, chemical signals, and thermal energy. Although not commonly thought of as a nutrient, the primary material being transported to the cells is oxygen. Because it occupies a central place in metabolism, oxygen levels in the tissues is homeostatically regulated. Many of these items are transported from a **source** to a **sink**. In addition, the blood contains a large variety of things that are being carried for the purpose of being part of the blood. These include the plasma proteins and the formed elements of the blood, the red and white blood cells.

THE CIRCULATION IS NECESSARY BECAUSE DIFFUSION FROM AND TO THE ENVIRONMENT IS TOO SLOW

Single-celled, free-living organisms do not have closed circulatory systems because they are small enough that simple diffusion between the cell and the outside environment adequately exchanges gases such as oxygen and carbon dioxide and nutrients such as glucose and amino acids. The time of one-dimensional diffusion is related to the diffusion distance according to

$$[5.1.1] \qquad \bar{x}^2 = 2D\,\Delta t$$

where $\bar{x}^2$ is the average square displacement from $x = 0$, the starting point of diffusion, D is the diffusion coefficient, and Δt is the elapsed time. Consider the diffusion of oxygen, with a diffusion coefficient of $1.8 \times 10^{-5}\,\text{cm}^2\,\text{s}^{-1}$. Table 5.1.2 shows that diffusion is extremely rapid over short distances, 0–1 μm. At cellular distances, oxygen diffusion is also rapid, requiring a few seconds to travel 50 μm. Over longer distances, diffusion becomes quite slow. As organisms increased in size, they encountered diffusion limitations. The solution to increase delivery of materials in large, multicellular animals is to increase the flow of materials by **convection**. The one-dimensional convection–diffusion equation was given earlier as Eqn [1.6.36]:

TABLE 5.1.1 Partial Listing of Materials Carried in the Blood

Nutrients	Waste Products	Chemical Signals	Thermal Energy
Water	CO_2	Growth hormone	Metabolic heat
Oxygen	Urea	ACTH, LH, FSH	
Glucose	Ammonia	Prolactin, TSH	
Amino acids	Uric acid	ADH, oxytocin	
Fatty acids	Creatinine	Thyroxine	
Vitamins	Lactic acid	PTH	
Lipids as lipoproteins	Acetone	Insulin, glucagon	
Na, K, Ca, Mg	Acetoacetic acid	Epinephrine	
Fe, Zn, Cu, Mn	β-hydroxybutryic acid	Erythropoietin	
Cl, I, PO_4, SO_4	Enzymes: SGOT, LDH	GI hormones	
		Gastrin, CCK	
		Motilin, secretin	
		Somatostatin	

TABLE 5.1.2 The Calculated Time for One-Dimensional Diffusion for O_2 in Water for Various Distances

Physiological Distance	Distance	Time for O_2 Diffusion
Cell membrane	10 nm	28×10^{-9} s
Mitochondrion	1 μm	0.28×10^{-3} s
Nucleus	10 μm	28×10^{-3} s
Large muscle cell	100 μm	2.8 s
Small muscle length	1 cm	7.8 h
Larger muscle	10 cm	32 days

Note that diffusion is extremely fast over short distances but is very slow for longer distances.

$$[5.1.2] \quad J_s = -D \frac{\partial C(x,t)}{\partial x} + J_v C(x,t)$$

Convection, the movement of materials by flow, adds the term, $J_v C$, to the diffusional flux. This convectional term is the part of the flux or transport of solute that occurs when solute is carried along by the bulk flux of fluid. This one-dimensional equation is a simplification of the equation that considers the flux as a three-dimensional vector. In three dimensions, the spatial partial derivative of $C(x, t)$ in Eqn [5.1.2] is replaced with the **gradient** of C, a vector, and J_v is the velocity of fluid that carries the solute, which is also a vector. The resulting flux of solute is also a vector that gives the magnitude of the flux and its direction. In large animals the convection term dominates the total flux, because the distances are large and the gradients of C are small.

THE CIRCULATORY SYSTEM CONSISTS OF THE *PULMONARY CIRCULATION* AND *SYSTEMIC CIRCULATION*

The anatomic arrangement of the major blood vessels is shown in Figure 5.1.2. Only the major vessels are shown because otherwise the entire diagram would be occupied by vessels. Every cell of the body is within a few hundred microns of a vessel.

The heart is actually a dual pump. The right heart pumps blood into the **pulmonary artery** to the lungs where blood is replenished with oxygen and waste CO_2 is removed. The blood collects in the **pulmonary veins** and returns to the heart. The flow of blood from the right heart to the lungs and back to the left heart is called the **pulmonary circulation**. The pulmonary

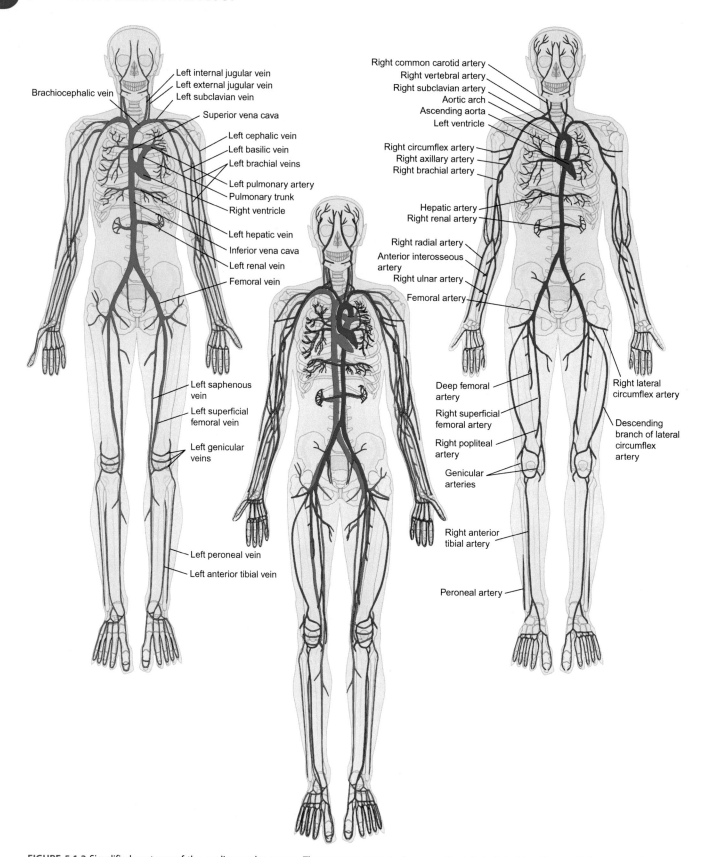

FIGURE 5.1.2 Simplified anatomy of the cardiovascular system. The venous system is shown on the left, in light blue. The arterial system is shown on the right, in darker blue. The combined circulatory system is shown in the middle. Only the major arteries and veins are shown and the vessels supplying the intestines have been omitted for clarity. Vessels have been labeled for only one side of the body. As can be seen, the veins and arteries typically are paired so that every artery has an associated vein.

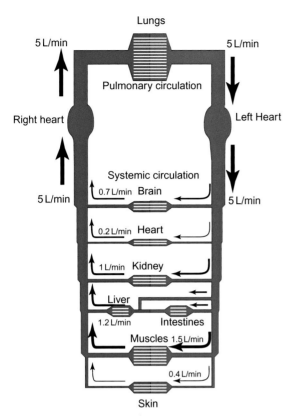

FIGURE 5.1.3 General layout of the circulation and typical flows at rest. The pulmonary circulation is in series with the systemic circulation, and therefore the flow through the pulmonary circulation is equal to the flow through the systemic circulation.

circulation is in series with the **systemic circulation**, the circulation from the left heart into the aorta and then to the rest of the body. Although the dual pump is in series, the pumping is simultaneous, not sequential. The vessels of the systemic circulation are generally arranged in parallel. This arrangement allows for independent regulation of blood flow to the various organs.

Because the circulatory system is closed, the flow at steady state into the right heart must be equal to the flow out of the right heart, which must be equal to the flow into the left heart and the flow out of the left heart. Because the vessels are slightly elastic, temporary differences in the output of the right heart and left heart cause shifts of fluid from the venous to the arterial side or vice versa.

MOST CIRCULATORY BEDS ARE ARRANGED IN *PARALLEL*

Most of the major arteries supplying the organs are arranged in parallel, as shown in Figure 5.1.3. The flow of blood encounters a hydrodynamic resistance in the vessels that perfuse each organ. By adjusting the resistance (by vasodilation or vasoconstriction), flow through each organ can be regulated more or less independently from the flow through other organs.

There are some exceptions to the parallel arrangement of flow. In the gastrointestinal tract, blood first perfuses the intestine and then perfuses the liver in the portal circulation. In this case, blood travels through two capillary circulations in series. A second example occurs in the kidneys. The first set of capillaries is a high-pressure circulation that forms the glomerular filtrate (see Chapter 6.2) as the first step in making urine. Blood that drains from these capillaries then flows through the efferent arteriole and then breaks up into a second set of capillaries, the peritubular capillaries.

PRESSURE DRIVES BLOOD FLOW THROUGH THE VASCULAR SYSTEM

Newtonian mechanics tells us that materials at rest remain at rest unless acted upon by an outside force and that materials in motion remain in motion unless acted upon by an outside force. After the heart accelerates a volume of blood, its energy would be gradually dissipated by distension of the arterial walls and friction with the walls of the vessels and within the fluid itself, unless a second heart beat, and then a third and fourth, continually provided force to keep the fluid moving. To maintain a flow through a resistance, there must be a continuous application of a force. When this force is normalized to the area over which it operates, it is the pressure. The first law of the cardiovascular system is written as:

$$[5.1.3] \qquad Q = \frac{\Delta P}{R}$$

where Q is the flow (in units of volume per unit time), ΔP is the pressure difference separated by the distance along which flow occurs, and R is the resistance of the tube through which flow occurs. This is a hydrodynamic analogue of Ohm's law for the flow of current. This law is valid only for laminar flow.

Laminar flow is streamlined flow that occurs, for example, when water is gently poured from a pitcher. Laminar flow through rigid, straight tubes can be described by Poiseuille's law (Eqn [1.2.17]), which has a steep dependence on the caliber of the tube. Poiseuille's law applies only to long, straight tubes under conditions of laminar flow. It is not valid in a network as complicated and varying as the cardiovascular system. Turbulent flow is chaotic, like the flow of a river around an obstacle at flood stage: the velocity of water flow at different points in the stream can point almost anywhere, including back upstream.

Despite the limitations to application of Poiseuille's law to the vasculature, resistance of the vessels can increase by vasoconstriction and decrease by vasodilation. The rules of addition of resistances in series and parallel are exactly analogous to those in electrical circuits.

VESSELS ARE CHARACTERIZED BY A *COMPLIANCE*

The cardiovascular system is a closed system, but it is *elastic*. What this means is that it does not have a defined volume. The volume can increase or decrease

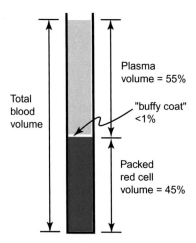

FIGURE 5.1.4 Whole blood after centrifugation in a cylindrical tube. The blood separates into a fluid phase, the plasma, that contains electrolytes, proteins and lipoproteins, and a packed cell layer containing primarily erythrocytes. At the interface between the plasma and the packed red cells is a "buffy coat" of white blood cells and platelets. The **hematocrit ratio** is the ratio of the packed cell volume to the total volume. The hematocrit is the hematocrit ratio $\times$ 100. Normal values for the hematocrit are 42–52% for men and 37–47% for women.

and the system responds by expanding or shrinking. The insertion of extra volume and the accompanying expansion of the vessels is accompanied by a change in pressure. The relationship between a volume change and a pressure change is

[5.1.4] $$\Delta P = \frac{\Delta V}{C}$$

where ΔP is the pressure change, ΔV is the volume change, and C is the **compliance**. The meaning of compliance here corresponds to its nontechnical usage: a compliant system is one that expands easily. In the case of a very compliant system, an expansion by ΔV produces little pressure change. Rigid or noncompliant vessels do not expand easily and expansion by a small ΔV produces a large pressure change.

BLOOD CONSISTS OF CELLS SUSPENDED IN PLASMA

Blood makes up about 6–8% of the body mass by weight. Its density ranges from 1.05 to 1.06 g cm^{-3} and its viscosity ranges from 2.5×10^{-3} to 3.9×10^{-3} Pa s, about 3.5–5.5 times that of water. Blood is a non-Newtonian fluid, meaning that its viscosity is not independent of the applied shear rate (dv/dr, the gradient of velocity with radius, in units of s^{-1}), or of the diameter of the tube. The blood consists of two main components: **plasma** and **red blood cells**. The **white blood cells** make up a third, much smaller, component. Centrifugation of *unclotted* blood results in the separation of the cells from the plasma as shown in Figure 5.1.4. The plasma is a straw-colored fluid that contains electrolytes, proteins, metabolites, hormones, nutrients, and wastes.

Besides carrying innumerable materials, the plasma also defines the electrolytic composition and the osmolarity of the extracellular fluid and contains proteins that form and dissolve blood **clots**. The red blood cells carry primarily oxygen, but also buffer blood pH. The white blood cells are responsible for inflammation and the response to invasion by parasites and microbes.

HEMOSTASIS DEFENDS THE INTEGRITY OF THE VASCULAR VOLUME

Hemostasis is the arrest of bleeding when blood vessel integrity is breached. The complete healing of an injury to a vessel includes:

- vasoconstriction of the vessel;
- formation of a platelet plug;
- coagulation of the blood;
- clot retraction;
- clot dissolution;
- formation of connective tissue throughout the wound to form a permanent seal.

VASOCONSTRICTION AND BACK PRESSURE REDUCES BLEEDING

Disruption of a blood vessel allows blood to escape either to the outside of the body or into the extravascular space. Bleeding into the extravascular space builds up pressure in the tissue that partly restrains further bleeding. How fast the back pressure builds up depends on the compliance of the tissues. Surface wounds shed blood out of the body and no back pressure develops. Bleeding in this case is limited by vasoconstriction. Damaged vascular cells contract reflexly by a local myogenic mechanism elicited by vascular damage. Platelets release **thromboxane A$_2$ (TXA$_2$)** and **serotonin**, both potent vasoconstrictors. **Thrombin**, a protein that is activated during the coagulation cascade, elicits TXA$_2$ and serotonin release from platelets.

THE PLATELET PLUG CAN SEAL SMALL VASCULAR HOLES

Injury to blood vessels exposes collagen that binds platelets, causing them to **degranulate**, releasing adenosine diphosphate (ADP), TXA$_2$, serotonin, and lipoproteins. ADP and TXA$_2$ recruit platelets from the blood. Lipoproteins help initiate coagulation. Aggregation of platelets is accompanied by activation of myosin light chain kinase within the platelets, activating cytoskeletal elements to change the shape of the platelets. Aggregation of the platelets at the site of injury forms a plug that can seal small vessels temporarily.

BLOOD COAGULATION SEALS THE LEAK

Blood plasma contains **fibrinogen**, a large molecular weight (330 kDa), soluble protein. Conversion of fibrinogen into insoluble fibrin, and then cross-linking the fibrin, produces a tangled meshwork of filaments that comprises the blood clot. Plasma itself can clot, because it possesses fibrinogen. Serum is the fluid left after plasma has clotted, and so it can no longer clot. The clot is called a **thrombus**. If it contains only platelets, it

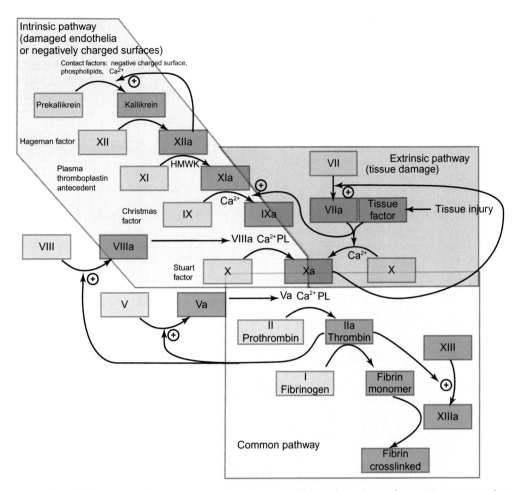

FIGURE 5.1.5 Summary of the clotting cascade. Roman numerals denote zymogen forms of the clotting factors. These inactive forms are enclosed in light blue boxes. The activated, proteolytic forms of the clotting factors are denoted by the suffix "a" and are enclosed in dark blue boxes. Blood typically clots *in vivo* by activation of the extrinsic pathway. Tissue factor, also called tissue thromboplastin, is a nonenzymatic lipoprotein expressed on the surface of cells that normally are not in contact with the blood. Tissue damage exposes blood to tissue factor, which binds factor VIIa and together they activate factor X to Xa. Factor Xa converts prothrombin to thrombin. Thrombin then converts fibrinogen to fibrin, which is the key step in clot formation. The clot is polymerized by factor XIIIa to form the dense network of fibrin strands. Although factor VII is activated to VIIa by the latter's product, factor Xa, it appears that blood contains trace amounts of VIIa sufficient to begin the clotting cascade. Blood coagulation by the intrinsic pathway occurs when blood is exposed to a negatively charged surface. HMWK stands for high-molecular-weight kininogen. Factors V, VIII, tissue factor, and HMWK are all nonenzymatic protein cofactors that increase the proteolytic activity of thrombin and factors Xa, VIIa, and XIIa, respectively.

is a **white thrombus**; a **red thrombus** entraps red blood cells as well. The formation of the clot and its cross-linking are controlled by **thrombin**, an enzyme formed from its plasma precursor, **prothrombin**. Activation of prothrombin to thrombin is the culmination of a series of events, shown diagrammatically in Figure 5.1.5.

CLOT RETRACTION DRAWS THE EDGES OF THE WOUND TOGETHER

During the first few hours after a clot forms, it retracts and extrudes serum. Clot retraction draws the wound surfaces together and also reopens any vessels that may be occluded by the clot. Close apposition of the wound surfaces enhances wound healing. Retraction is caused by the platelets.

PLASMIN DISSOLVES CLOTS

Endothelial cells release inactive **tissue plasminogen activator (tPA)** following injury. Fibrin activates tPA, which in turn activates plasmin by proteolytic cleavage of an inactive β globulin, **plasminogen**. Plasminogen binds to both fibrin and fibrinogen, so it is incorporated into the clot. Plasmin then proteolyzes fibrin. This mechanism accounts for most of the **fibrinolysis**, but other activators of the conversion of plasminogen to plasmin are present in plasma and urine (**urokinase**). Circulating baseline levels of plasmin are inhibited by α2-antiplasmin. tPA is also inhibited by specific inhibitors called **plasminogen activator inhibitor.** Four distinct inhibitors have been identified; PAI-1 and PAI-2 seem to be most important.

BLOOD COAGULATION SITS ON A KNIFE EDGE OF ACTIVATION AND INHIBITION

Coagulation normally occurs only when there is trauma to the blood vessels. What stops the clot from progressing throughout the entire vasculature? The answer is that the blood contains a number of anticoagulants whose job is to prevent the clotting mechanism from getting out of hand. These are:

- thrombin inhibitors;
- antithromboplastin, also called tissue factor pathway inhibitor;
- heparin.

THERE ARE FOUR DISTINCT THROMBIN INHIBITORS

The thrombin inhibitors include **antithrombin III**, **α_2 macroglobulin**, **heparin cofactor II**, and **α_1-antitrypsin**. All of these are serine protease inhibitors. Antithrombin III is the most important because it inhibits the activities of thrombin and factors IXa, Xa, XIa, and XIIa. Antithrombin activity is stimulated by binding of **heparin**, a sulfated polysaccharide that is produced by basophils and mast cells and is present on the surfaces of endothelial cells. Heparin increases the activity of antithrombin III some 100- to 1000-fold, effectively preventing clot formation. The inhibition of factors IXa, Xa, XIa, and XIIa effectively stops the intrinsic pathway of blood coagulation.

ANTITHROMBOPLASTIN STOPS THE EXTRINSIC PATHWAY

Tissue factor pathway inhibitor (TFPI) is a 34 kDa protein that is found in plasma lipoproteins and bound to the vascular endothelium. It binds to and inhibits factor Xa. The complex Xa-TFPI then interacts with the complex of tissue factor VIIa and inhibits its activation of factors X and IX.

Clinical Applications: Bleeding Disorders

Hemophilia refers to the inability to clot blood. As you can see from Figure 5.1.5, there are a number of proteins involved in clotting and dysfunction of any of them could, in principle, lead to problems in clotting. Hemophilia A is the classical hemophilia due to an X-linked disorder that causes a deficiency in factor VIII. Factor VIIIa has no enzymatic activity in itself; it is a cofactor that increases the activity of IXa to convert factor X to Xa. To date, some 150 different point mutations in the factor VIIIa gene have been identified. Hemophilia A occurs in from 1:5000 to 1:10,000 males in all populations. Persons with factor VIIIa deficiency suffer from joint and muscle hemorrhage, easy bruising, and prolonged bleeding. The severity depends on the level of factor VIII activity. Treatment involves infusion with factor VIII isolated from human plasma or prepared by recombinant DNA technology.

Hemophilia B results from a deficiency in factor IX. Its prevalence is about 1/10th that of hemophilia A and, as with hemophilia A, the severity ranges from mild to severe depending on the factor IX activity in plasma. Some 300 unique factor IX mutations have been identified.

The most common inherited bleeding disorder is **von Willebrand disease**, due to the inherited deficiency in von Willebrand factor (**vWF**). vWF has two roles in clotting. First, the adhesion of platelets to collagen exposed on endothelial cell surfaces is mediated by vWF. vWF acts as a bridge between a specific glycoprotein on the surface of the platelets (GPIb/IX) and collagen fibrils. vWF also binds to and stabilizes factor VIII in the plasma. Persons deficient in vWF will also have a concomitant decrease in plasma factor VIII. Clinically significant von Willebrand disease occurs in about 125 persons per million, about twice the frequency of hemophilia A.

Clinical Applications: Anticoagulant Therapy

Inappropriate clot formation is a clinical problem as common as hemorrhage. Several agents can be used to prevent clotting. Figure 5.1.5 shows that several of the proteolytic reactions in the clotting cascade require Ca^{2+} ions. These Ca^{2+} ions bind to special γ-carboxyglutamic acid residues on factors II (prothrombin), VII, IX, and X. Chelation of plasma Ca^{2+} by adding Na citrate or Na oxalate prevents these reactions and stops the coagulation cascade. Ca^{2+} chelators can be used to stop coagulation of the blood outside the body, but they cannot be used *in vivo* because the heart and nerves depend on normal plasma $[Ca^{2+}]$ for their operation. The strength of the heart beat decreases directly with decreases in plasma $[Ca^{2+}]$, and ceases altogether when plasma $[Ca^{2+}]$ falls below 0.2 mM.

As mentioned in the text, heparin is produced by mast cells and is present on the surface of endothelial cells. It prevents blood clotting by activating antithrombin III some 100- to 1000-fold. During tissue damage, the activation of thrombin outstrips the natural inhibition of it and clot formation occurs. Antithrombin III activity can be stimulated by injecting heparin, which effectively inhibits clot formation.

The liver synthesizes clotting factors II (prothrombin), VII, IX, and X. All of these proteins contain γ-carboxyglutamic acid residues, also called gla, near their amino terminus. This amino acid contains two exposed carboxyl groups that bind Ca^{2+}. The γ-carboxyglutamic acid is formed by a posttranslational modification of the proteins that requires vitamin K as a cofactor. Vitamin K was discovered in 1929 by Henrik Dam in Copenhagen when he fed chicks fat-free diets and noted that they developed hemorrhages under the skin and in the muscles. He subsequently discovered that the antihemorrhagic factor was fat soluble but none of the known fat-soluble vitamins could prevent the hemorrhagic disease. Dam named the new vitamin "K" for "koagulation." In 1939 vitamin K was isolated from alfalfa. In 1941, Campbell and Link discovered that bishydroxycoumarin (dicumarol) was the active agent in spoiled clover that caused a hemorrhagic disease of cattle. This compound has structural similarities to vitamin K, and prevents the synthesis of vitamin

K-dependent gla proteins by interfering with the vitamin K cycle, shown below. A second compound, warfarin (named for the Wisconsin Alumni Research Foundation, and known as Coumadin®), also inhibits vitamin K recycling and thereby prevents formation of the vitamin K-dependent clotting factors.

SUMMARY

The cardiovascular system transports nutrients and chemical signals to the tissues and removes waste materials and heat. This transport system is necessary mainly because diffusion is too slow to exchange materials over the distances between our tissues and the environment. The medium of transport is the blood, which normally remains contained within blood vessels. The final exchange of materials occurs across the walls of the capillaries from the blood to the interstitial fluid that bathes all cells of the body. To assure rapid exchange, the blood reaches all parts of the body, coming within short distances of all cells. This is accomplished by an extensive network of vessels. The arteries take blood away from the heart whereas the veins return blood to the heart.

The heart provides the motive force for movement of the blood. It consists of a dual pump. The right heart pumps blood through the pulmonary circulation whereas the left heart pumps it through the systemic circulation. These two circulations are in series, whereas all other organs supplied by the systemic circulation are perfused in parallel. This arrangement allows for the separate regulation of blood flow to the organs to meet demands.

Because the circulatory system is closed, at steady state the output of the right heart must match its input, which in turn must match the output of the left heart. Transient deviations from this balance are accompanied by shifts in blood to or from the systemic circulation into the pulmonary circulation.

Flow through the cardiovascular system is driven by pressure which originates with the beating of the heart. The vascular component of the cardiovascular system can be characterized by two major characteristics: its resistance to flow and its compliance. The resistance relates total flow to the pressure difference that drives the flow, whereas the compliance relates the volume within the vessels to the pressure.

Blood consists of a fluid, the plasma, in which cells are suspended. The fraction of the blood volume occupied by the red blood cells is called the hematocrit.

Blood clotting usually begins with tissue injury that exposes platelets to collagen, which causes the platelets to adhere together, and release signaling molecules including serotonin, ADP, TXA_2, and a lipoprotein factor that hastens coagulation. The key reaction in coagulation is the conversion of fibrinogen to fibrin and its cross-linking into a dense network of fibers. Thrombin is a serine protease that converts fibrinogen to fibrin, and its activation from prothrombin involves a complicated cascade of events. The body contains anticoagulants to prevent accidental clotting. Once formed, clots retract and then are dissolved by plasmin, which is activated from plasminogen by proteolysis.

REVIEW QUESTIONS

1. What is convection? Why is it important?
2. Define vascular resistance. How, in general, does it depend on vessel caliber?
3. How is the pulmonary circulation distinct from the systemic circulation?
4. Define compliance. Why is compliance higher in the veins than in the arteries?
5. What enzyme converts fibrinogen to fibrin? How is it activated?
6. What enzyme dissolves clots? How is it activated?

5.2 Plasma and Red Blood Cells

Learning Objectives

- Write the concentrations of the major electrolytes of plasma
- List in order the major proteins of plasma with their functions
- Describe the structure of immunoglobulins
- Describe the shape of erythrocytes
- Calculate mean cell volume and mean corpuscular hemoglobin concentration
- Describe the structure of hemoglobin and list the major variants
- Recognize the structure of heme and the central role of Fe
- Write the negative feedback regulation of erythrocyte formation by erythropoietin
- Describe the destruction of erythrocytes with the life span and metabolic fate of heme
- Describe the handling of Fe in the body
- Explain the basis of blood typing, identifying universal donor and recipient types

PLASMA CONSISTS MAINLY OF WATER, ELECTROLYTES, AND PROTEINS

All of the components of blood are either suspended or dissolved in a watery phase. The dissolved materials include all of the electrolytes, a variety of proteins, nutrients, gases, wastes, and hormones. Some of these are not readily soluble in water and so they are helped along by little "flotation devices"—typically proteins that coat the materials and allow them to be carried in the water phase. Because some of the volume of the plasma is occupied by these materials, **water makes up only 92% of the mass of plasma**. Because these materials occupy volume, water also makes up about 92% of the volume of plasma. Thus, concentrations of dissolved substances when expressed per L of plasma are lower than those expressed per L of plasma water.

Electrolytes are dissolved substances that dissociate into ions and therefore confer electrical conductivity onto the solution. The electrolytes in plasma include Na^+, K^+, Cl^-, HCO_3^-, Ca^{2+}, H^+, Mg^{2+}, and $H_2PO_4^-$, and a variety of other ions. These ions contribute to the overall osmotic pressure of plasma and also determine the concentrations of these ions in the **interstitial fluid**.

The equilibrium potential for any ion across the cell membrane depends on the concentrations of the ion in the interstitial fluid and in the cell (see Chapter 3.1). These equilibrium potentials and the conductances to the ions determine the resting potential and the excitability of the cells to external stimuli. The concentrations of ions in normal plasma and in the interstitial fluid are given in Table 5.2.1.

Plasma contains a number of different proteins that can be classified on the basis of their solubility and coagulability, among other characteristics, and each of these classes contain a variety of members. Figure 5.1.1 illustrates the components that make up the plasma proteins.

The serum **albumin** concentration varies from 2.8 to 4.5 g%, where "g%" refers to the g of albumin per 100 mL of plasma. It makes up more than 50% of the total plasma protein but, because it is smaller than other plasma proteins, it makes up a larger fraction of the total molar concentration of protein. Thus, albumin makes up a larger fraction of the plasma osmolarity due to proteins. Albumin is made in the liver and secreted into the blood. Its regions of high hydrophobicity bind a variety of hydrophobic materials including fatty acids, bilirubin, and steroid hormones. This binding allows albumin to transport these materials in the blood.

The **globulins** are the second most abundant class of plasma proteins, accounting for about 3.1 g% of the plasma protein. About 80% of the globulins are synthesized by the liver and secreted into the blood. The rest are synthesized in various parts of the body. The γ-globulins are synthesized in the **lymph nodes** and in the **reticuloendothelial cells** that are distributed widely throughout the body, especially in the spleen. The globulins are divided into subclasses based on their separation by electrophoresis (see Figure 5.2.1).

The α_1-globulins include the **high-density lipoproteins** or **HDL**, which are complexes of proteins and lipids that are used to transport lipids through the blood. In this case the lipids are noncovalently bound in large macromolecular assemblies that are coated with proteins called **apolipoproteins**. The apolipoproteins, also called apoproteins, maintain the solubility of the lipoproteins and provide recognition sites for the tissues that metabolize the lipids found in these lipoproteins.

The major α_2-globulin, **macroglobulin**, is present at about 0.2 g%. It has a molecular weight of 842 kDa and

TABLE 5.2.1 Concentration of Ions in Plasma and Interstitial Fluid

Electrolyte	Plasma C (mEq/L)	ISF C (mEq/L)
Na^+	142	144
K^+	4	4
Ca^{2+}	5	2.4
Mg^{2+}	3	1.5
H^+	3.9×10^{-5}	3.7×10^{-5}
Cl^-	105	114
HCO_3^-	24	25
Phosphates	2.5	2.5
Sulfate	1	1
Organic acids	5	5
Protein	16	4
Total osmolarity	296	296

The concentration is given in units of milliequivalents/L. This unit is related to molarity through the charge per ion, z. The concentration in milliequivalents/L is related to the concentration in millimolar by $C_{mEq} = |z C_{mM}|$. The concentrations given are the total concentration and not the free concentrations. In the case of Ca^{2+}, about half of the 5 mEq/L are bound to plasma proteins, so that the *free* plasma $[Ca^{2+}]$ is closer to 2.5 mEq/L or 1.25 mM. The actual concentrations in individuals vary within a range of normal, and typical values are given. Plasma proteins occupy a significant volume, so that the water content of plasma is reduced. The concentrations shown here are expressed per L of plasma, not per L of plasma water.

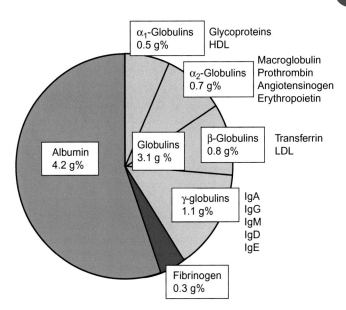

FIGURE 5.2.1 Composition of plasma proteins. "Albumin" refers to a class of proteins that are readily soluble and that coagulate upon heating. Examples are egg ovalbumin and serum albumin. Serum albumin (molecular weight of 66.5 kDa) is the smallest and most abundant of the major plasma proteins. "Globulin" refers to proteins that are insoluble or sparingly soluble in water, but whose solubility is greatly increased by adding salts such as NaCl. These proteins coagulate when heated. The names of the various classes of the globulins derive from how they separate during electrophoresis. The numbers indicate typical values within normal ranges of 2.8–4.5 g% for albumin; 0.4–0.5 g% for α_1-globulins; 0.4–0.9 g% for α_2-globulins; 0.6–1.1 g% for β-globulins.

its major function appears to be inhibition of proteolytic activity. Other α_2-globulins include **prothrombin**, a 62.7 kDa protein that is essential in blood clotting; **angiotensinogen**, a precursor for **angiotensin**, a potent vasoconstrictor; **erythropoietin**, a 34 kDa glycoprotein synthesized and secreted by the kidney that controls the formation of red blood cells.

The major β-globulins are **transferrin**, an 80 kDa protein that binds and transports iron, and the β-lipoproteins. These lipoproteins include the **low-density lipoproteins (LDL)** and the **very low-density lipoproteins** or **VLDL**. As with the HDL described above, these are macromolecular assemblies of insoluble lipid and its apolipoprotein coat that keeps the lipoprotein in suspension and helps tissues bind and absorb the lipid for metabolism.

Specialized cells produce and release **antibodies** into the plasma. These are all members of the γ-globulin class of proteins, of which there are several types. Because they form part of the body's immune defense, these globulins are also called **immunoglobulins**. Antibodies are proteins composed of multiples of heavy and light protein chains that are linked together by disulfide bonds, as shown in Figure 5.2.2.

About 80% of the immunoglobulins in blood are IgG. IgG and IgM provide the major defense against bacteria. The IgM class is polymerized to give $(LH)_{2n}$, where L is the light chain, H is the heavy chain, and $n = 5-6$. Its molecular weight is about 750 kDa. This

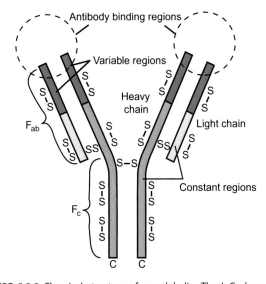

FIGURE 5.2.2 Chemical structure of a γ-globulin. The IgG class shown here has a molecular weight of 150 kDa. It consists of four polypeptide chains. Two heavy chains and two light chains are connected by disulfide bonds and noncovalent interactions. Both heavy and light chains consist of regions that are constant and other regions that vary. These variable regions allow the immunoglobulins to bind to specific **antigens**. The fragment of the immunoglobulin containing the **a**ntigen **b**inding region is called F_{ab}; the fragment containing the **c**onstant region is called F_c. Binding of the antigen by the antibody allows the body to recognize the antigen and "tag" it for destruction.

immunoglobulin is a receptor on special cells of the immune system called **B-lymphocytes**. It is also secreted into the plasma by **plasma cells**, B-lymphocytes to which an antigen has bound and activated the production and secretion of antibodies. IgA is found in secretions of the gastrointestinal tract, respiratory tract, genitourinary system, milk, and tears. IgE appears to help defend against parasitic worms and be involved in allergic responses such as hay fever, asthma, and hives.

Human plasma contains 0.3 g% of fibrinogen. This large protein (330 kDa) is a precursor to **fibrin**, which forms the blood clot.

PLASMA PROTEINS AND IONS BUFFER CHANGES IN PLASMA pH

Proteins and phosphate groups in plasma provide the first defense against changes in the plasma H^+ by binding or releasing H^+ according to **Le Chatelier's Principle**. Le Chatelier's Principle states that a reaction at equilibrium will react to a disturbance of the equilibrium in the opposite direction to the disturbance. The reactions are:

[5.2.1]
$$HPr \rightleftharpoons Pr^- + H^+$$
$$H_2PO_4^- \rightleftharpoons HPO_4^{-2} + H^+$$

where Pr indicates plasma protein. When the $[H^+]$ increases, the available Pr^- and HPO_4^{-2} binds it and *buffers* the change in $[H^+]$; conversely, when H^+ ions are being consumed by some process, Pr and $H_2PO_4^{-1}$ can release more H^+ and *buffer* decreases in $[H^+]$. These chemical buffers for $[H^+]$ are the first and most immediate response to changes in plasma $[H^+]$ but they do not actively regulate pH. Instead, both the lungs and kidneys participate in the regulation of body $[H^+]$ (see Chapters 6.5 and 7.7). The **buffer capacity** is defined as Δ **acid**/Δ**pH** and for plasma it is about 7.3 mEq L^{-1} per pH unit.

THE *ONCOTIC PRESSURE* OF PLASMA PROTEINS RETAINS CIRCULATORY VOLUME

The total osmotic pressure of plasma is given by the van't Hoff equation:

[5.2.2]
$$\pi = RT \sum \phi_S C_S$$

where π is the osmotic pressure, R is the gas constant (= 0.082 L atm mol^{-1} K^{-1}), T is the temperature (in K), C_s is the molar concentration of species s, and φ_s is the osmotic coefficient of the species s. The contributions of the solutes can be divided into two components: (1) the **crystalloids** and (2) the proteins (**colloids**). Because the crystalloids cross the capillary membranes easily, whereas the plasma proteins do not, the osmotic pressure difference between the interstitial fluid and the plasma is due almost entirely to the plasma proteins alone. Therefore, the osmotic pressure due to the proteins alone is given a separate name, the **colloid osmotic pressure** or the **oncotic** pressure. Because the plasma protein concentration exceeds that in the interstitial fluid, there is a net osmotic pressure favoring fluid movement from the interstitial space to the plasma. This movement is necessary to counteract the pressure-driven filtration of fluid out of the capillaries because of the higher hydrostatic pressure within the capillaries. Thus, reduction in plasma protein concentration is associated with **edema**, or swelling of the tissues (see "Clinical Applications: Edema" in Chapter 5.10).

THE *ERYTHROCYTE* IS THE MOST ABUNDANT CELL IN THE BLOOD

Erythrocytes have a distinctive biconcave disk shape that distinguishes them from all other cells. The cells are about 7–8 μm in diameter and about 2 μm thick at the widest, with a depression in the middle. A micrograph of a blood smear is shown in Figure 5.2.3. A cytoskeleton holds the cell in this shape and allows the cell to deform in order to fit through capillaries that are often smaller than the diameter of the red blood cell. Thus, the cell bends to fit through. A cartoon illustrating the possible cytoskeletal attachments of red blood cells is shown in Figure 5.2.4. Each mm^3 of blood contains some $4.5-5.5 \times 10^6$ of these cells. The main function of erythrocytes is to carry oxygen bound to hemoglobin.

ERYTHROCYTES CONTAIN A LOT OF *HEMOGLOBIN*

Hemoglobin is the protein component of blood that gives it its red color. It is named for its two components: a protein or **globin** component, and a **heme** group that contains iron and is responsible for hemoglobin's ability to bind oxygen. The normal hemoglobin concentration in plasma is 14–17 g% for adult males and 12–16 g% for females. This is g of hemoglobin per deciliter (= 0.1 L) of whole blood.

HEMOGLOBIN CONSISTS OF FOUR POLYPEPTIDE CHAINS, EACH WITH A *HEME* GROUP

Hemoglobin is made up of four polypeptide chains. Two of these are α-globin molecules, each containing

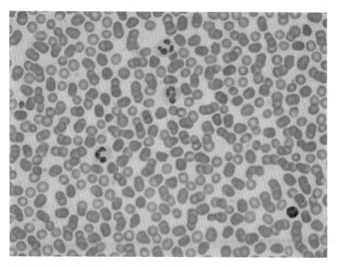

FIGURE 5.2.3 Wright's stain of whole blood. Most of the cells visible in this micrograph are red blood cells, which have no nuclei. White blood cells have nuclei that stain purple with this stain and therefore are easily seen.

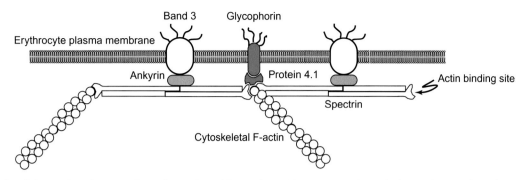

FIGURE 5.2.4 Cytoskeleton of erythrocytes. **Spectrin** consists of four polypeptides, two α and two β chains of 240 and 220 kDa, respectively. The β-chain contains an **actin** binding site at one end, so the assembled tetramer has an actin binding site at each end. The actin–spectrin complex is attached to the membrane through two proteins: **ankyrin** binds to spectrin and to **band 3**, a protein that is inserted in the plasma membrane of the red blood cell. Additional links between the cytoskeleton and the membrane may be provided by other proteins and transmembrane proteins such as **glycophorin**.

141 amino acids, and the other two are globins of another type (β, γ, δ, or ε), each with 146 amino acids. About 98% of adult human hemoglobin is HbA$_1$, consisting of two α and two β chains, and so it is designated α$_2$β$_2$. Some 2% or so is type HbA$_2$, with the composition α$_2$δ$_2$. The type of hemoglobin changes during embryonic development. Early on during embryogenesis hemoglobin has the composition α$_2$ε$_2$, which then switches to the fetal form with the composition α$_2$γ$_2$. These forms are necessary because the fetus must be able to "steal" oxygen from the maternal circulation, and therefore it must bind oxygen more avidly than the maternal hemoglobin. A cartoon of the structure of hemoglobin is shown in Figure 5.2.5.

In addition to its variation during embryogenesis and its heterogeneity in the adult, hemoglobin differs among individuals. The human population makes over 500 different kinds of hemoglobins due to mutations at one or more loci. Most of the changes from the predominant form do not alter hemoglobin's ability to transport oxygen and otherwise do not adversely affect the carrier of the abnormal gene. On the other hand, the abnormal hemoglobin in **sickle cell anemia**, a hereditary disease, arises from a single amino acid substitution (see Clinical Applications Box "Sickle Cell Disease and Malaria").

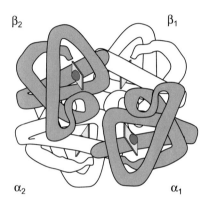

FIGURE 5.2.5 Highly schematic drawing of the structure of hemoglobin. The drawing shows hemoglobin A$_1$ with the composition α$_2$β$_2$. Each globin protein forms a complex with a heme moiety, which in turn consists of porphyrin complexed with an iron atom. (*Source: Adapted from Dickerson and Geis*, The Structure and Action of Proteins, *Harper and Row, New York, NY, 1969.*)

Each of the four polypeptide chains in hemoglobin contains a heme group that consists of a **porphyrin skeleton** and a central **iron** atom. Its chemical structure is shown in Figure 5.2.6.

EXAMPLE 5.2.1 Mean Cell Volume

The typical hematocrit (Hct) is about 45. If the density of cells is 5×10^6 mm^{-3}, what is the RBC mean cell volume?

As shown in Figure 5.1.4, centrifugation separates plasma from the cellular elements of blood. The red blood cells make up almost all of the cellular elements. Hct is defined as:

Hct = packed cell volume/total blood volume × 100

The hematocrit can be reported as the **hematocrit ratio**, which is the ratio without multiplying it by 100 to convert it to a percentage. Normal values of Hct are 42–52% for adult males and 37–47% for adult females. **Anemia** occurs when Hct < 25%.

If Hct > 75%, the condition is called **polycythemia**, but it can be caused transiently by dehydration.

The mean cell volume is calculated as MCV = Hct ratio/density of cells. For the normal values given above, we calculate

$$\text{MCV} = 0.45/5 \times 10^{12} \text{ cells L}^{-1} = .09 \times 10^{-12} \text{ L cell}^{-1}$$
$$= 90 \times 10^{-15} \text{ L cell}^{-1} = 90 \text{ fL cell}^{-1}$$

The mean cell volume can be used to distinguish among different types of anemias. **Microcytic** and **macrocytic** anemias are those in which erythrocytes are abnormally small or large, respectively.

> **EXAMPLE 5.2.2 Mean Corpuscular Hemoglobin Concentration**
>
> The hematocrit ratio is 0.45 and the hemoglobin concentration is 15 g%. What is the mean corpuscular hemoglobin concentration?
>
> The mean corpuscular hemoglobin concentration is found by dividing the aggregate hemoglobin content of the red blood cells by its aggregate volume. Since both are determined in whole blood, the mean corpuscular hemoglobin concentration can be found by
>
> MCHC = hemoglobin concentration/hematocrit ratio
> = 150 g L^{-1}/0.45 = 333.3 g L^{-1}
>
> This can be converted to molarity by dividing by the gram molecular weight of hemoglobin, 64,500 g mol^{-1}:
>
> 333 g L^{-1}/64,500 g mol^{-1} = 5.17 × 10^{-3} M
>
> These calculations are useful in describing different kinds of anemias. **Hypochromic** and **normochromic** anemias describe conditions in which the hemoglobin concentration is reduced or normal, respectively.

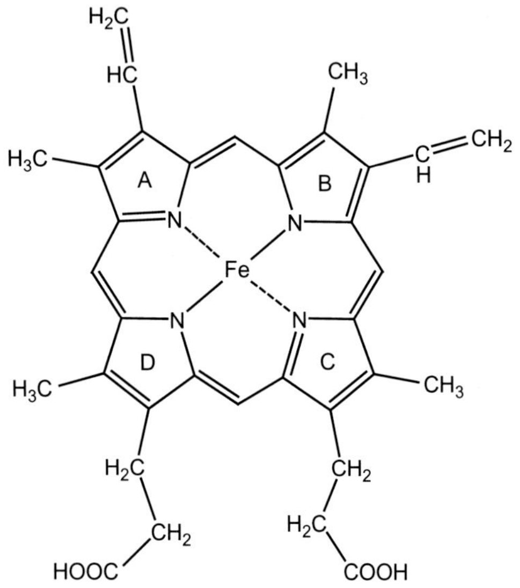

FIGURE 5.2.6 Chemical structure of heme. The porphyrin skeleton consists of the four rings (A–D) and their associated side chains. The Fe atom in the center forms a coordination complex with oxygen. Heme binds oxygen only when iron is in the ferrous (Fe^{2+}) oxidation state: oxidation of Fe to the ferric state (Fe^{3+}) prohibits oxygen binding. Hemoglobin with Fe^{3+} is called **methemoglobin**. Free heme without the protein also does not bind oxygen because some of the coordination bonds of the Fe atom with oxygen are provided by the protein component of hemoglobin.

ERYTHROPOIETIN CONTROLS FORMATION OF ERYTHROCYTES FROM *PLURIPOTENT STEM CELLS* IN BONE MARROW

Erythrocytes and every kind of white blood cell derive from **pluripotent stem cells** that reside in the bone marrow. "Potency" is used by embryologists to define what kinds of cells a precursor cell may become by **differentiation**. A "totipotent" cell, which occurs early during embryogenesis, is capable of becoming any cell in the body. A "pluripotent" cell is already partially differentiated but remains capable of forming a variety of cells. The pluripotent stem cells in the marrow cannot become nerve or muscle cells, but they can give rise to any of the blood cells. Pluripotent stem cells differentiate into **committed stem cells** (those with a set fate) under the influence of **cytokines**. Cytokines are chemical factors released from one cell that affects the growth or activity of another cell. The most important cytokine for **erythropoiesis** (from the Greek "poiesis," meaning "a making") is **erythropoietin**. This glycoprotein hormone is made in the kidney and secreted into the blood in response to **hypoxia**, or low oxygen levels in the tissue. The hypoxia generally results from low oxygen content in the perfusing blood or from low perfusion of the tissue. This system forms a negative feedback loop, in which low oxygen levels stimulate erythropoietin release, which increases erythropoiesis, which in turn raises oxygen delivery to the tissue, which raises oxygen levels, as shown schematically in Figure 5.2.7.

PHAGOCYTES IN THE RETICULOENDOTHELIAL SYSTEM DESTROY WORN ERYTHROCYTES

The number of erythrocytes in the circulation results from a balance between their continual synthesis and destruction. Oxidative damage of red blood cell membranes with age makes them increasingly fragile. **Macrophages** in the spleen, liver, and bone marrow consume the worn out red blood cells by engulfing them and then destroying them. These macrophages make up part of the **reticuloendothelial system**, a combination of monocytes, mobile macrophages and fixed tissue macrophages, and a few specialized endothelial cells in the bone marrow, spleen, liver, and lymph nodes. All of these cells are phagocytotic and they form a system for the elimination of foreign materials or worn out body parts. Isotopic experiments show that **the average life span of an erythrocyte is about 126 ± 7 days**.

Macrophages engulf red blood cells and break down all of its components: membrane, cytosolic enzymes, and hemoglobin. The macrophages remove heme from hemoglobin and break the globin part down to its constituent amino acids. The iron is removed from heme and reused to make new heme, and the porphyrin skeleton is broken down first into **biliverdin** and then into **bilirubin**. Bilirubin is sparingly soluble. It is solubilized in the liver by covalently attaching glucuronic acid. The conjugated bilirubin is then excreted into the intestinal tract through

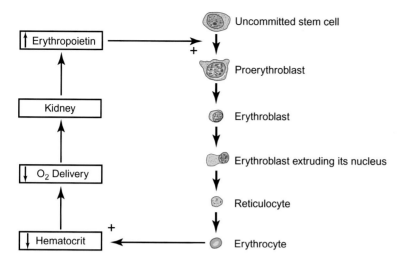

FIGURE 5.2.7 Control of erythropoiesis by erythropoietin. Decreased hematocrit or decreased blood volume will decrease oxygen delivery to the kidney, which decreases oxygen within this tissue, increasing release of erythropoietin, which travels through the blood to the bone marrow where it stimulates differentiation of uncommitted stem cells to begin forming erythrocytes. The formation of erythrocytes takes place in stages, some of which are shown here. Increased formation of erythrocytes completes a negative feedback loop, in which the original signal, a decreased hematocrit or blood volume, is corrected by increased formation of erythrocytes.

the bile (see Chapter 8.4). In the intestine, bacteria deconjugate bilirubin diglucuronide and further convert bilirubin to urobilinogen and stercobilinogen. These are oxidized to **urobilin** and **stercobilin**, respectively, which are excreted in the urine and feces, respectively. These colored materials make significant contributions to the color of urine and feces. **Destruction of 1 g of hemoglobin produces about 35 mg of bilirubin.** Poor liver function leads to a buildup of bilirubin levels in the blood and **jaundice**, the yellow appearance of the skin due to accumulation of bilirubin in the tissues.

IRON RECYCLES INTO NEW HEME

Each day about 20–30 mg of iron is recycled from destroyed red blood cells, which is much more than the daily intake or absorption of iron from the diet. Iron released from hemoglobin is transferred to **transferrin**, a plasma protein of 80 kDa with two iron binding sites per molecule. Its normal plasma concentration is 0.22–0.35 g%, but clinicians monitor its physiological function by measuring the **total iron binding capacity of serum (TIBC)**, which ranges from 45 to 80 μM (also reported as 250–450 μg dL^{-1}). The serum iron concentration reports the amount of iron that is actually bound to transferrin; its normal values are about 10–30 μM. Thus iron typically occupies some 30% of the available binding sites in serum.

Most cells need iron to form heme groups that are incorporated into proteins of the electron transport chain of mitochondria. Cells import the iron from the blood by binding the Fe–transferrin complex to a **transferrin receptor** located on the cell's plasma membrane. Cells then endocytose the receptor-diferric transferrin complex. The iron is released and incorporated into heme or stored as **ferritin**. Ferritin is the iron complex of a protein, apoferritin, and a ferric hydroxide phosphate. The apoferritin monomer has a molecular weight of about 20 kDa; some 24 monomers associate to form a protein shell with a cavity that holds crystals of FeOOH with some phosphate. Each ferritin complex has the capacity for 4300 iron atoms, but typically they contain about 2000 iron atoms. Figure 5.2.8 shows typical elemental Fe processing per day.

Iron differs from most nutrients in that **there is no excretory pathway for excess iron**. Instead, it accumulates as an insoluble complex of ferritin and iron called **hemosiderin**. The liver can accumulate enough hemosiderin to destroy the organ in a disease called **hemosiderosis**.

HUMAN BLOOD CAN BE CLASSIFIED INTO A SMALL NUMBER OF BLOOD TYPES

The surface of red cell membranes carries two related antigens, called type A and type B. Whether an individual human has these antigens or not depends on their genotype: an individual may have neither of the antigens, or they may have either one, or they may have both. In addition to these antigens, individuals almost always have plasma antibodies that will react with whatever antigen is *not* on their red blood cell membranes. These antibodies are not inherited but develop after birth due to exposure to small quantities of the antigens in the food and air. These antibodies will bind to the antigens and cause the red cells to clump together, or **agglutinate**. Thus, the type A and type B antigens are called **agglutinogens** and the antibodies that react to them are called **agglutinins**. Agglutination occurs because the antibodies have multiple binding sites for the antigens. The IgG class of antibodies has two sites whereas the IgM class has 10 sites per molecule. A single antibody molecule can bind to antigen molecules on two or more different cells, thereby linking them. If this agglutination occurs within the circulatory system, the individual is at great risk of cardiovascular collapse and death.

The blood groups are produced by three alleles at a single location on each of two chromosomes. The three alleles are type O, type A, and type B. These three different alleles allow for six different genotypes, as shown in Table 5.2.2. A single allele of type A or type B is sufficient to express antigen A or B on the surface of the red cell. The blood types, agglutinogens, and agglutinins are also shown in Table 5.2.2.

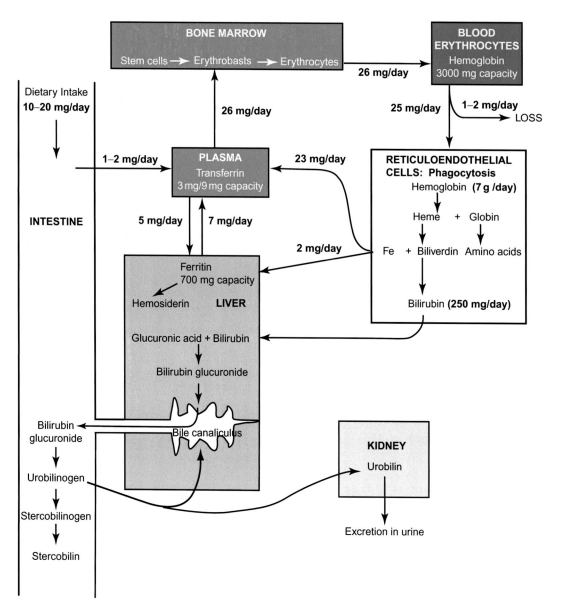

FIGURE 5.2.8 Iron balance and heme degradation pathways. The numbers indicate the amounts of elemental iron that are processed each day. For simplicity, some minor pathways have been omitted. The two numbers in parentheses for hemoglobin breakdown and bilirubin formation refer to the amounts of these materials processed each day.

TABLE 5.2.2 Genotypes, Agglutinogens, and Agglutinins for the Major Blood Groups

Genotypes	Blood Group	Agglutinogens	Agglutinins
OO	O	—	anti-A and anti-B
OA and AA	A	A	anti-B
OB and BB	B	B	anti-A
AB	AB	A and B	—

Cells from a person with Type O blood will not agglutinate when placed in the serum of persons with Type A, Type B, or Type AB blood, because the cells lack the agglutinogen. On the other hand, infusion of Type O blood does not clump the recipient's cells because the agglutinins present in the Type O serum are so diluted that there are insufficient numbers of antibodies to clump the cells. Thus, **Type O blood is a universal donor type**. Persons with Type AB blood do not agglutinate blood originating from Type O, Type A, or Type B blood, because they lack the agglutinins. So people with **Type AB blood are called universal recipients**.

Clinical Applications: Erythroblastosis Fetalis

In addition to the four major blood groups, there are a variety of other factors, including the Rh factor, that influence compatibility of blood. In all cases, blood must be matched to avoid transfusion reactions. The distinction between the Rh factor and the other major blood groups is that agglutinins to the Rh factor do not develop spontaneously as do agglutinins to the Type A or Type B agglutinogens. Instead, people who are Rh− must be exposed directly to Rh+ blood in order to develop the antigens. This occurs when an Rh− mother carries an Rh+ baby. In the first pregnancy, there is usually little contact between the baby's blood and the mother's so that the mother does not develop antibodies to the Rh agglutinogens. During birth, there usually is some contact between placental fetal blood and maternal blood that begins the production of anti-Rh agglutinins in the mother. In the next pregnancy, the anti-Rh agglutinins can penetrate the placenta and cause red cell agglutination in the fetus. The condition is called **erythroblastosis fetalis**, in which the fetal or newborn red cells agglutinate and are subsequently phagocytosed and degraded. The probability of developing the disease increases with each successive pregnancy involving an Rh− mother and Rh+ baby. The baby is at risk due to severe anemia and also due to precipitation of bilirubin in neuronal cells that may result in mental retardation and motor impairment.

SUMMARY

Blood consists of plasma, red blood cells, and white blood cells. The fraction of the blood volume occupied by red blood cells is called the hematocrit ratio, and is typically about 0.37–0.52. It is usually higher for males than females because of the monthly blood loss in the menstrual cycle. The red blood cells are packed with hemoglobin, which binds most of the oxygen that is carried in blood. Hemoglobin consists of four polypeptide chains associated noncovalently. Each chain binds one heme group, which consists of a porphyrin ring and a complexed iron atom. Erythrocytes are made from pluripotent stem cells in the bone marrow in a process called erythropoiesis. Erythropoiesis is stimulated by a hormone, erythropoietin, that is made in the kidney in response to poor oxygenation.

The erythrocyte has a distinctive biconcave disk shape that is maintained by a cytoskeleton. This shape allows it to travel through narrow capillaries. Cells become damaged with age and are degraded by macrophages that comprise the reticuloendothelial system. The average life span of an erythrocyte is about 120 days. The iron in the erythrocyte hemoglobin is recycled, and the porphyrin pigments are degraded and excreted in urine and feces.

Erythrocytes also contain agglutinogen proteins on their surfaces. Persons with type A blood carry the A agglutinogen; persons with type B carry the B agglutinogen; persons with AB contain both; and persons with type O blood carry neither. The plasma contains antibodies directed against the agglutinogens that are not present in the blood. Thus, persons with type O blood have antibodies to both A and B agglutinogens. These agglutinogens form the basis of the blood types.

Plasma contains water and electrolytes, a host of nutrients, and a variety of proteins. The major protein classes include fibrinogen, which is necessary for blood clotting; albumin, which forms a large part of the plasma oncotic pressure and carries many insoluble materials; and the globulins, which are involved in lipid transport and immunological reactions. Plasma is distinct from serum in that serum lacks the clotting proteins.

REVIEW QUESTIONS

1. How do you determine the hematocrit?
2. From the hematocrit and number of red blood cells (obtained by hemacytometry), calculate MCV.
3. What is the oncotic pressure? What protein is mainly responsible for it?
4. What are the three most highly concentrated cations in plasma?
5. What are the three most highly concentrated anions in plasma?
6. What is an antibody? What proteins are antibodies?
7. What protein clots the blood?
8. What controls red blood cell formation? How is it regulated?
9. What major proteins maintain the shape of the erythrocyte?
10. What is hemoglobin? What does it do? Where is it? What is its structure?
11. How long do red blood cells live? Where are they broken down? What happens to the heme?
12. How is Fe from broken heme recycled? What does transferrin do? What does ferritin do? What is hemosiderosis?
13. What causes blood types? What are the major types? Which is the universal donor? Universal recipient?

Clinical Applications: Anemia

Anemia literally means "no blood." It refers to any condition in which the oxygen carrying capacity of blood is diminished due to a reduction in the number or volume of red blood cells, or a reduction in the hemoglobin content of blood. Anemia therefore refers to the concentration of oxygen carrying material and not its total amount in the body. There are a variety of causes of anemia that can broadly be classified according to increased loss of oxygen carrier or decreased production.

(Continued)

Clinical Applications: Anemia (Continued)

Increased loss of blood through the stools or internal bleeding can reduce the number of red blood cells per unit volume of blood, while the cells maintain their normal size and hemoglobin content. Such abnormal blood loss can be the result of drugs such as aspirin and other NSAIDs (nonsteroidal anti-inflammatory drugs) that interfere with blood clotting.

Increased destruction of red blood cells can be accelerated when the red cells show an abnormal fragility. Persons with mutations in the red blood cytoskeleton, for example, make spherical red blood cells rather than flattened biconcave disks. These cells rupture more easily when they pass through the circulation and so the lifetime of these cells is shortened. This condition is called **hereditary spherocytosis**. Abnormal hemoglobin, such as occurs in **sickle cell disease**, also shortens the life span of the erythrocytes. Parasitic infections such as malaria rupture red blood cells, thereby increasing the loss of red blood cells.

Decreased ability to form red blood cells can be due to deficiency in the materials needed to make the cells. Thus, insufficient iron leads to an **iron-deficiency anemia** that is characterized by normal numbers of red cells, but with reduced hemoglobin content. These cells are usually smaller than normal (**microcytic**) and they contain less hemoglobin (the cells are **hypochromic**). Formation of red blood cells requires two vitamins: **vitamin B_{12}** (also called cobalamin) and **folic acid**. Both of these are required to make the materials for DNA replication. Proliferation of the stem cells and maturation of the erythrocytes requires DNA production and so deficiencies in these nutrients results in a reduction in the number of red blood cells being produced. Dietary deficiency of vitamin B_{12} is rare. The lack of the vitamin is often associated with a deficiency in the production of **intrinsic factor**, a substance secreted by cells lining the stomach. Intrinsic factor is necessary for the proper absorption of vitamin B_{12} from the ingested food. Lack of intrinsic factor thus translates to a lack of vitamin B_{12}. This condition is called **pernicious anemia**. In addition to anemia, there are central nervous consequences of vitamin B_{12} deficiency that are independent of the anemia.

Aplastic anemia refers to the situation in which the bone marrow fails to produce adequate numbers of red blood cells. Radiation, chemotherapy, or exposure to toxic chemicals can interfere with the erythropoietic capability of the marrow. Failure to produce enough red blood cells can also result from insufficient circulating levels of erythropoietin. This can occur in renal anemia, in which the kidneys no longer produce sufficient erythropoietin.

Clinical Applications: Sickle Cell Disease and Malaria

Sickle cell disease results from a single amino acid substitution in which the amino acid valine replaces glutamic acid at position 6 in the β-chain of hemoglobin. This changes the physical characteristics of hemoglobin so that it polymerizes when it becomes deoxygenated. The polymerized hemoglobin deforms the red blood cells, causing them to resemble a sickle shape, or a crescent moon. The misshapened cells do not bend easily when they pass through the narrow capillaries, causing poor circulation and reducing the life span of the cells. Because erythrocytes are more easily destroyed, sickle cell anemia results.

Darwin's theory of evolution predicts that maladaptive mutations such as sickle cell anemia should be removed from the population because such traits reduce the ability to produce offspring to carry the trait forward. So why does such an inherited disorder persist in the population? The short answer is that although sickle cell disease produces a potentially debilitating anemia, it also provides protection against malarial infections.

Malaria literally means "mala aria" or "bad air." This name reflects the early idea that malaria was spread by bad air. The Italian scientist Giovanni Batista Grassi established in 1898 that this disease results from the parasitic infection of a protozoan of the genus *Plasmodium*, which is transmitted to people through the bite of a mosquito. The life cycle of *Plasmodium falciparum* has three distinct stages. The sporogenic phase occurs in the *Anopheles* mosquito, in which the protozoan migrates to the salivary gland and forms sporozoites. The female Anopheles mosquito injects the sporozoites into the human host when she obtains her blood meal. The second stage occurs in the liver of the human host. The sporozoites penetrate liver cells, reproducing into hundreds of merozoites which are released when the liver cell ruptures. The third stage occurs in the blood when the merozoites penetrate red blood cells to produce hundreds of microgametocytes. About 3 days after initial penetration by the merozoite, the red blood cells rupture to release microgametocytes and more merozoites. These can reinfect other red blood cells or be taken up when another mosquito bites the infected person. The rupture of the red blood cells is accompanied by fever, from which malaria derives its second name, "swamp fever." The disease is characterized by recurrent fevers corresponding to the cycle of microgametocyte and merozoite formation and rupture of red cells.

Malaria for some reasons speeds the sickling process, so that infected red blood cells are rapidly deformed and then destroyed by the spleen rather than producing more merozoites. Thus the infection is curtailed. People who are heterozygous for the sickle trait are partly protected from malaria while the sickle cell anemia is also mild. People who are homozygous for the sickle trait may suffer from sickle cell anemia of variable severity. Because of its advantage against malaria, the sickle cell trait has reached a kind of equilibrium in populations indigenous to areas of natural malaria infestation. In areas of malaria absence, however, the sickle cell trait confers a selective disadvantage.

Malaria most likely originated in Africa and spread to the New World by European slave traders. In the early 1600s, Jesuit missionaries in South America learned that extracts of the bark of the Cinchona tree could cure malaria. Its active ingredient is a toxic plant alkaloid called **quinine**. Quinine has now been synthesized and the synthetic analogue is called mefloquine. Other drugs include chloroquine, a synthetic drug, and Qinghaosu, derived from the sweet wormwood plant in China. There are a variety of types of malaria owing to different species of *Plasmodium* and their sensitivities to these drugs, and their ability to infect man and various animals also varies.

White Blood Cells and Inflammation 5.3

Learning Objectives

- List the five types of leukocytes in order of their abundance
- List the polymorphonuclear granulocytes and the agranular leukocytes
- Describe the origin of the platelets and the role of thrombopoietin in their regulation
- List the function of each of the white blood cells
- Describe the difference between the innate and specific or adaptive immune responses
- Describe the reticuloendothelial system
- List the four classical signs of inflammation
- Distinguish between innate and adaptive immunity
- List the events in order in the recruitment of neutrophils and monocytes to the extravascular space
- Describe the function of the complement system

THE WHITE BLOOD CELLS INCLUDE NEUTROPHILS, EOSINOPHILS, BASOPHILS, MONOCYTES, LYMPHOCYTES, AND PLATELETS

The white blood cells are called **leukocytes** (from the Greek "leukos" meaning "white" and "kytos," meaning "cell"). The granular leukocytes (**eosinophils, neutrophils, and basophils**) are named for the granules in their cytoplasm; the agranular leukocytes (**monocytes and lymphocytes**) lack cytoplasmic granules. All of the granular leukocytes have **polymorphic** nuclei that appear as two to five ovate-shaped lobes that are connected by strands. Histological stains help differentiate among the different white blood cells. The microscopic appearance of these cells is shown in Figure 5.3.1.

Each mm^3 of blood contains some $5-5.5 \times 10^6$ red blood cells but only about 4000–11,000 white blood cells. Of these, the neutrophils comprise 50–70%, the lymphocytes 20–40%, the monocytes 2–8%, the eosinophils 1–4%, and the basophils <1%. **Platelets** are cell fragments of **megakaryocytes**. The platelets typically number between 0.2 and 0.5×10^6 particles per mm^3 of blood. The relative fraction of white blood cells contributed by each type is shown in Figure 5.3.2.

WHITE BLOOD CELLS ORIGINATE FROM PLURIPOTENT STEM CELLS

THE PLURIPOTENT STEM CELLS FORM TWO DISTINCT LINEAGES

Pluripotent stem cells in the bone marrow produce **myeloid** and **lymphoid** progenitors. The myeloid progenitor differentiates further into a granulocyte/macrophage progenitor that further differentiates into the granulocytes and the monocytes. **All of these cells are capable of phagocytosis, but it is primarily the neutrophils and the monocytes that defend against bacterial and viral infection by phagocytosis.** The neutrophil is a mature cell that is capable of phagocytosis, whereas the phagocytic abilities of circulating monocytes are small. When the monocytes enter the tissues, they enlarge, develop large numbers of lysosomes, and are then called **macrophages** (see Figure 5.3.3).

The myeloid progenitors also give rise in the marrow to **megakaryocytes** and **erythrocytes**. The megakaryocytes disintegrate in the bone marrow to produce fragments called **platelets**. Lymphocytes originate from various lymphogenous organs, including the lymph nodes or glands, the thymus, tonsils, the bone marrow, and the gut. The lymphoid stem cell further differentiates into the lymphocyte.

HEMATOPOIETIC GROWTH FACTORS STIMULATE FORMATION OF WHITE BLOOD CELLS

Thrombopoietin (TPO) is a 332 amino acid glycoprotein made primarily in the liver that stimulates the formation of megakaryocytes from CFU-Meg (colony forming unit, megakaryocyte). TPO binds to receptors that dimerize and activate cytoplasmic kinases. These kinases phosphorylate STAT-1, 3, and 5 and CREB. (See Chapter 2.8—STAT stands for **s**ignal **t**ransduction and **a**ctivator of **t**ranscription; CREB stands for **c**yclic AMP **r**esponse **e**lement **b**inding protein.) Hematopoietic stem cells, megakaryocyte progenitors, megakaryocytes, and platelets have receptors for TPO (Figure 5.3.4). The liver makes TPO at a constant rate, but platelets degrade TPO. Thus, lower platelets results in higher [TPO], which stimulates the differentiation of megakaryocytes into platelets. This regulation of TPO differs significantly from the feedback mechanisms involving erythropoietin and formation of erythrocytes. Other growth factors for differentiation of white blood cells are indicated in Figure 5.3.3.

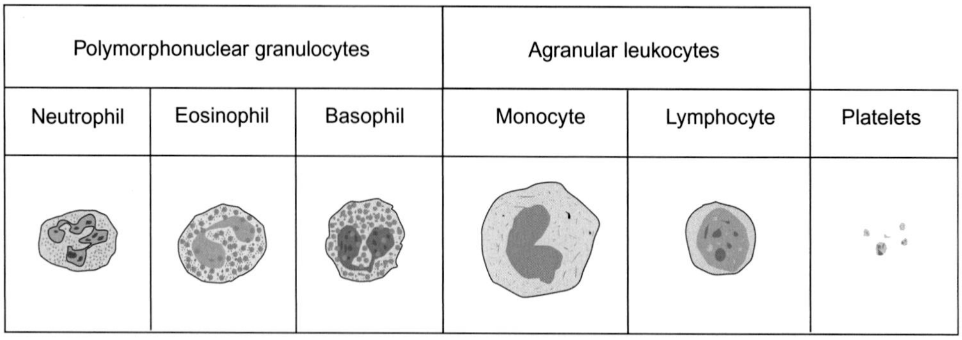

FIGURE 5.3.1 The various types of white blood cells. The polymorphonuclear granulocytes possess lobed nuclei and cytoplasmic granules. Cytoplasmic granules in eosinophils take up eosin, a histological dye; cytoplasmic granules of basophils stain with a basic, blue dye. Neutrophils have granules that do not stain with either eosin or basic dyes. The monocytes are larger than the polymorphonuclear granulocytes. They have a single oval or horseshoe-shaped nucleus and relatively few cytoplasmic granules. Lymphocytes contain a single, oval-shaped nucleus and small amounts of cytoplasm. Platelets are cell fragments.

Polymorphonuclear granulocytes			Agranular leukocytes		
Neutrophil	Eosinophil	Basophil	Monocyte	Lymphocyte	Platelets

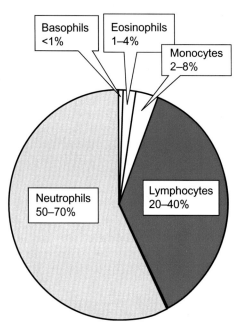

FIGURE 5.3.2 Fraction of leukocytes contributed by each class. The major class of white blood cells is the neutrophil, followed in order by the lymphocytes, monocytes, eosinophils, and basophils.

Basophils <1%
Eosinophils 1–4%
Monocytes 2–8%
Neutrophils 50–70%
Lymphocytes 20–40%

response to bacterial invasion. They form part of the nonspecific or innate defense system. An elevated white cell count is often used clinically as an indicator of bacterial versus viral infections. Neutrophils contain a variety of cytoplasmic granules that contain proteins and peptides that kill and digest microbes. Activation of these proteins requires production of **reactive oxygen species (ROS)** including **superoxide**, **hydrogen peroxide**, **hydroxyl radical**, and **hypochlorous acid**. These ROS are produced by the enzyme **NADPH oxidase** on the plasma membrane that invaginates to form the phagosome, and **myeloperoxidase** that is present in the cytoplasmic granules.

MONOCYTES LEAVE THE CIRCULATORY SYSTEM TO BECOME *TISSUE MACROPHAGES*

Monocytes are the largest of the leukocytes. They are released into the blood from the marrow in an immature form with little phagocytic ability. They circulate in the blood until they find a suitable home in the tissues, where they greatly enlarge to become tissue macrophages. Some of these large phagocytic cells remain mobile, while others attach themselves to tissue components and become **fixed macrophages**. Their life span varies from months to years, depending on their activity.

BASOPHILS RESEMBLE *MAST CELLS*

Basophils are the least numerous of the leukocytes. Structurally and functionally they resemble mast cells. However, basophils originate in the marrow whereas mast cells originate from precursor cells in the connective tissue. Both basophils and mast cells contain secretory granules that store **histamine** and **heparin**, among other chemicals. Histamine release occurs in allergic reactions and in response to inflammation. Histamine profoundly increases capillary permeability and in this way contributes to the inflammatory response by causing tissue **edema**, or swelling. Heparin activates lipoprotein lipase that degrades triglycerides in the blood from dietary sources. Heparin also prevents blood clotting.

THE CIRCULATING WHITE BLOOD CELLS ARE IN TRANSIT FROM PRODUCTION SITE TO THE TISSUES

White blood cells are produced in the marrow, but they function in the tissues. Their presence in blood represents transport from the site of production to the site of use. Normally, the marrow stores about three times as many white cells as are present in the blood. Once released into the blood, a typical white cell spends some 4–8 h circulating and about another 4–5 days in the tissues.

NEUTROPHILS ARE *PHAGOCYTES*

The neutrophils are the most numerous of the leukocytes. They leave the circulatory system in response to signals from injured tissues early in the inflammatory

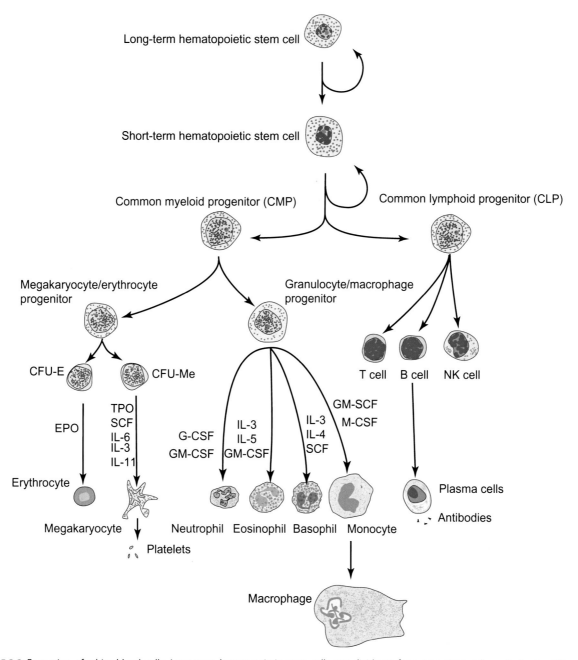

FIGURE 5.3.3 Formation of white blood cells. Long-term hematopoietic stem cells can divide to form new stem cells or to begin differentiation. Differentiating long-term hematopoietic stem cells become short-term hematopoietic stem cells that can also self-renew or begin differentiation by producing multipotent progenitors of two main lines: the lymphoid line and the myeloid line. The lymphoid line gives rise to the T-lymphocytes and B-lymphocytes and natural killer cells (NK cells). The myeloid line gives rise to two further progenitor lines: the megakaryocyte/erythroid progenitor and the granulocyte/macrophage progenitor. These give rise to the erythrocytes, platelets, and the white blood cells. EPO = erythropoietin; TPO = thrombopoietin; IL-3, IL-4, IL-6, IL-11 = interleukins; SCF = stem cell factor (Steel factor); GM-CSF = granulocyte/megakaryocyte colony stimulating factor; G-CSF = granulocyte colony stimulating factor; M-CSF = monocyte colony stimulating factor.

Heparin released from the basophils probably acts primarily as an anticoagulant.

EOSINOPHILS ARE INVOLVED IN DEFENSE OF PARASITIC INFECTIONS AND ALLERGIES

Eosinophils comprise only a few percent of the total number of leukocytes, but are readily identifiable in blood smears because their cytoplasmic granules take on an orange-red to bright yellow color when stained with eosin. Their numbers increase in persons with allergic conditions such as hay fever and asthma and in persons with parasitic infections. Parasites are typically much larger than eosinophils and therefore eosinophils cannot engulf them to destroy them. Eosinophils attach themselves to the juvenile form of parasites and secrete materials that injure or kill the parasite. These materials

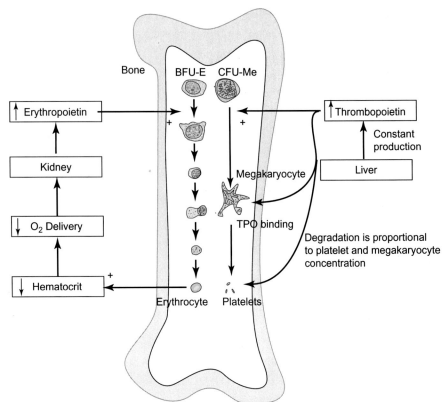

FIGURE 5.3.4 Comparison of EPO and TPO regulation of erythropoiesis and thrombopoiesis. EPO is synthesized mainly in the kidney in response to low O_2 delivery caused by a decreased hematocrit. The increased circulating [EPO] stimulates differentiation of burst-forming units for erythropoiesis BFU-E, so that the production of red blood cells is increased. In the absence of other events, this raises the hematocrit and increases kidney P_{O_2}, decreasing the initial signal for EPO. This control system forms a negative feedback loop. TPO is synthesized at a constant rate by the liver, but it is bound and then degraded by megakaryocytes and thrombocytes (platelets). Thus, when platelets are high, TPO is low and thrombopoiesis is depressed. When platelet counts are low, TPO is high and thrombopoiesis proceeds rapidly.

include hydrolytic enzymes that probably originate from lysosomes, ROS, and a polypeptide that has larvicidal activity.

LYMPHOCYTES FORM A SPECIFIC DEFENSE SYSTEM

The neutrophils, basophils, eosinophils, and monocytes are all part of a *non-specific* defense system. This means that the cells are capable of destroying a large variety of bacterial or viral invaders or cellular debris resulting from tissue damage, without specific recognition of which bacterium or viral invader. They accomplish this by recognizing common patterns of binding sites on the surfaces of these invaders. Lymphocytes, on the other hand, provide defense against *specific* foreign objects. Lymphocytes come in a variety of types, broadly classified as **T-lymphocytes**, **B-lymphocytes**, and **natural killer cells**. These descriptors derive from the processing of the lymphocytes during their differentiation. All three cell types derive from a single lymphoid stem cell. The thymus modifies the cells that become T-lymphocytes and these cells promote cell-mediated immunity. The bone marrow influences the cells that become B cells, and these cells differentiate to form **plasma cells** that produce circulating antibodies that comprise humoral immunity. The processing of T-lymphocytes occurs during the perinatal period. Thus, removal of the thymus during late fetal life can seriously interfere with cell-mediated immunity, whereas later ablation of the thymus has little or no effect. The natural killer cells arise in the bone marrow and destroy virus-infected cells and cancer cells. The T-lymphocytes constitute a heterogeneous group of cells that is further broken down into two main subsets: **cytotoxic T cells** and **helper T cells**. Cytotoxic T cells, as their name implies, kill cells. They bind to antigens on the surfaces of target cells and kill them by secreting toxic chemicals. This interaction does not require antibodies. Helper T cells, as their name also implies, do not kill cells on their own. Instead, they help activate both B cells and cytotoxic T cells by secreting **cytokines**. "Cytokine" literally means "causing the cell to move"; in this case it refers to chemicals secreted by one cell that activate other cells. Cytotoxic T cells and helper T cells are distinguishable based on their surface content of specific proteins. Cytotoxic T cells express a protein called CD8 on their surfaces; helper T cells express CD4. Thus, these cells are sometimes referred to as CD8 cells or CD4 cells. Another type of T cell is called the **suppressor T cell** or **regulatory T cell**. These T cells actively suppress other T cells that can produce tissue damage.

TISSUE MACROPHAGES, MONOCYTES, AND SPECIALIZED ENDOTHELIAL CELLS FORM THE *RETICULOENDOTHELIAL* SYSTEM

Monocytes leave the blood and enter tissues where they provide a mobile defense against foreign materials, including bacteria and viruses. The monocytes have little ability to destroy foreign material until they enter the tissues and transform into macrophages. Sometimes tissue macrophages attach to the tissues and remain fixed and inactive for long periods, months to years, until they become activated. The combination of the circulating monocytes, mobile macrophages within tissues, fixed macrophages, and some specialized cells in the spleen, lymph nodes, and bone marrow constitute the **reticuloendothelial system**. The cells in this system form a network for preventing bacterial or viral invasion of the blood. In addition, almost all of these cells derive from monocytic stem cells and for this reason they are categorized together. This "system" is a phagocytotic defense system that exists in all tissues, but is especially concentrated in those tissues where bacterial invasion is most likely. These are the tissues that interface with the environment:

- **Skin:** Although the skin is typically impervious to bacteria or viruses, breaks in the skin allow their entry. Macrophages in the subcutaneous connective tissue, called **histiocytes**, phagocytose bacteria, viruses, and necrotic tissue.
- **Lymph:** Except during the brief period of bleeding following injury, bacteria and viruses cannot enter the blood because the capillary walls exclude them. However, they can enter the lymphatics that drain the tissues. The lymph vessels drain into lymph nodes located intermittently throughout the lymphatic system. The nodes consist of a meshwork of cavities or **sinuses** that are lined with epithelial cells and large numbers of macrophages. Foreign particles become entrapped in the meshwork and then are engulfed by macrophages.
- **Lung:** Each lobe of the lung contains many millions of **alveoli**, tiny thin-walled sacs in which the blood is separated from the air by a thin layer of endothelial cells and alveolar cells. This short separation is necessary for rapid gas exchange, but it also presents an opportunity for airborne organisms or other materials to enter the blood. To combat this, the alveolar walls also contain large numbers of tissue macrophages that phagocytose materials that stick to the alveoli. If the particles are not digestible by the macrophages, the cells form a giant cell capsule around an offensive particle.
- **Liver:** Bacteria and other materials constantly enter the venous blood that drains the intestine, eventually entering the portal circulation. The portal blood passes through the liver, where macrophages called **Kupffer cells** line the sinuses. These cells effectively scrub the portal blood so that bacteria and other foreign particles that enter the blood through the gut do not pass into the general circulation.
- **Spleen and bone marrow:** If the peripheral defense systems fail and bacteria succeed in entering the general circulation (a condition called **septicemia**), the spleen and bone marrow contain macrophages that can clear the blood. The spleen differs from all other organs in that the red cells leave the capillaries and course through a meshwork of fibers and return to the circulation through the walls of the venous sinuses. The red pulp of the spleen, through which the red cells squeeze, is lined with macrophages that engulf worn and abnormal red blood cells as well as any foreign matter.

INFLAMMATION IS THE NET RESPONSE OF THE BODY TO *TISSUE INJURY*

In the first century, Aulus Celsus in *De Medicina* first described four of the five **cardinal signs of inflammation**. These were

- rubor (redness)
- calor (heat)
- tumor (swelling)
- dolor (pain).

Later, Galen added a fifth sign: impaired function. Tissue injury can be caused by bacteria, trauma, toxic or caustic chemicals, and heat. Regardless of the cause, injured tissue releases a variety of substances that evoke a complex series of changes within the tissue. This entire complex of changes is called **inflammation**. These processes include:

A. **Vasodilation** of the local blood vessels with consequent increased blood flow.
B. **Increased permeability of the capillaries** with consequent leakage of fluid into the interstitial space.
C. **Clotting of the fluid in the interstitial space** due to leakage of clotting proteins from the plasma into the interstitial fluid.

MIGRATION OF GRANULOCYTES AND MONOCYTES FROM THE BLOOD INTO THE TISSUES

The redness and warmth arises from the increased local blood flow. The edema or swelling arises from the increased capillary permeability and subsequent **exudation** of fluid into the tissues. The swelling stretches free nerve endings that can cause pain. The clotting of the fluid in the interstitial space effectively walls off the area so that pathogens cannot easily spread to the rest of the body. The migration of granulocytes and monocytes into the tissues is called **extravasation**. These actions limit the spread of infection and eventually remove it from the body.

INFLAMMATION BEGINS WITH THE RELEASE OF SIGNALING MOLECULES FROM THE DAMAGED TISSUE

Tissue injury caused by trauma, bacterial invasion, chemicals, heat, or other factors releases a set of signaling molecules that initiates the inflammatory response. The nature of the inflammatory response depends on the set of signaling molecules that are released, and this set is determined in part by the kind of injury. A partial list of the signaling molecules and their sources is given in Table 5.3.1.

THE INNATE IMMUNE RESPONSE REQUIRES NO PRIOR EXPOSURE—SPECIFICITY OF THE RESPONSE IS INHERITED IN THE GENOME

The **innate response** to infection requires no prior exposure to the infectious agent. The **adaptive immune** response occurs only when the innate response is overwhelmed, generally because the pathogens evade recognition by the cells that mediate the innate response. In the adaptive immune response, cells arise that recognize specific parts of the infectious agent. These cells generate antigen-specific cells that target the pathogen, and memory cells that prevent reinfection by the same pathogen.

Microorganisms typically have a cell membrane that is covered on the outside by a cell wall. The cell membrane is a lipid structure similar to the mammalian biological membranes (see Chapter 2.4). The cell wall imparts some structural rigidity to the cell while remaining porous. This cell wall differs among various bacteria, and Gram's stain is used to differentiate many species. **Gram-positive** bacteria have little lipid in their cell walls, whereas **Gram-negative** bacteria are rich in lipid. These walls consist of covalently linked polysaccharide and peptide chains. The wall typically presents repeating patterns of sugar molecules. Phagocytic macrophages within the tissues have surface receptors that recognize common constituents of bacterial surfaces. These receptors include:

- **mannose receptor**, a transmembrane receptor with multiple binding sites for mannose, recognizes the pattern of mannose on the microbes' cell wall.
- **scavenger receptor**, a structurally heterogeneous set of molecules, recognizes anionic polymers and acetylated low-density lipoproteins. These recognize structures on red blood cells that are masked by sialic acid. Old cells that lose their sialic acid are scavenged. Cells with these receptors take up lipoproteins from the blood and are involved in atherogenesis.
- **LPS receptor (CD14).** Gram-negative bacteria produce a component of their cell walls called **lipopolysaccharide** or LPS. The plasma contains a protein that binds LPS, called LPS-binding protein or **LBP**. The complex of LPS and LBP binds to another receptor on the macrophage, CD14.
- **β-glucan receptor** recognizes glucan structures in yeast.

When constituents of bacterial cell walls bind to the receptors, they trigger the macrophage to engulf and destroy the invaders, and simultaneously the macrophage releases **cytokines** and **chemokines**. The cytokines activate cells that have receptors for them; the chemokines attract cells with receptors for them. The cytokines include interleukin-1 (IL-1), interleukin-6 (IL-6), interleukin-12 (IL-12), and tumor necrosis factor-α (TNF-α). The chemokine interleukin-8 attracts neutrophils, basophils, and T cells to the site of infection. The process of macrophage recognition, engulfment, and release of cytokines and chemokines is illustrated in Figure 5.3.5.

TABLE 5.3.1 Sources of Selected Inflammatory Mediators Produced During Inflammation

Mast Cells	Platelets	Macrophages	Eosinophils	Complement	Kinins	Nerves
Histamine	5-HT	TNF-α	PAF	C3a	Bradykinin	ATP
ATP	ADP	IL-1		C5a		SP
Prostaglandins		IL-6				NKA
Leukotrienes		IL-8				CGRP
TXA$_2$		IL-12				

TXA$_2$, thromboxane A$_2$; 5-HT, 5-hydroxy tryptamine; TNF-α, tumor necrosis factor-α; IL-1, interleukin 1; IL-6, interleukin 6; IL-8, interleukin 8; IL-12, interleukin 12; PAF, platelet activating factor; SP, substance P; NKA, neurokinin A; CGRP, calcitonin-gene related peptide.
Source: Adapted from J. Linden, Blood flow regulation in inflammation, in K. Ley, ed., Physiology of Inflammation, Oxford University Press, New York, NY, 2001.

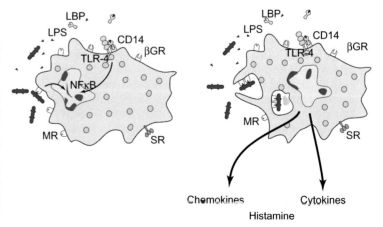

FIGURE 5.3.5 Macrophage phagocytosis and secretion of chemokines and cytokines. Macrophages have a variety of surface receptors that recognize bacterial invaders without prior exposure. Examples include a mannose receptor (MR), β-glucan receptor (βGR), and scavenger receptor (SR). Bacterial LPS binds to a plasma LPS binding protein (LPB) that then binds to a surface receptor, CD14, that is linked to toll-like receptor 4 (TLR-4). This binding initiates a cascade of events that activate a series of protein kinases that eventually activates NF-κB by phosphorylating its inhibitor. The NF-κB then activates the transcription of genes that promote the inflammatory response, the release of chemokines such as interleukin-8, and cytokines such as IL-1, IL-6, IL-12, and TNF-α. These trigger part of the inflammatory response and begin recruitment of leukocytes to the site of infection.

NEUTROPHILS AND MONOCYTES LEAVE THE CIRCULATORY SYSTEM BY *DIAPEDESIS* IN RESPONSE TO CHEMOTAXIC COMPOUNDS

Tissues react to bacterial or viral invasion by producing a variety of chemotaxic compounds that attract circulating neutrophils and monocytes. These include chemicals produced by the bacteria or by damaged tissue or by the reaction of cells to the foreign invaders.

Chemotaxis refers to the migration of cells in response to a chemical signal (see Figure 5.3.6).

- **Leukocytes are "captured"**: Leukocytes first move toward the outer edge of the circulation ("marginate"), allowing the leukocytes to closely approach the endothelial surface. The endothelial cells expose a protein called **P-selectin** in response to inflammatory mediators such as histamine. P-selectin makes the surface of the capillary sticky to leukocytes. It captures the leukocytes by binding to PSGL-1

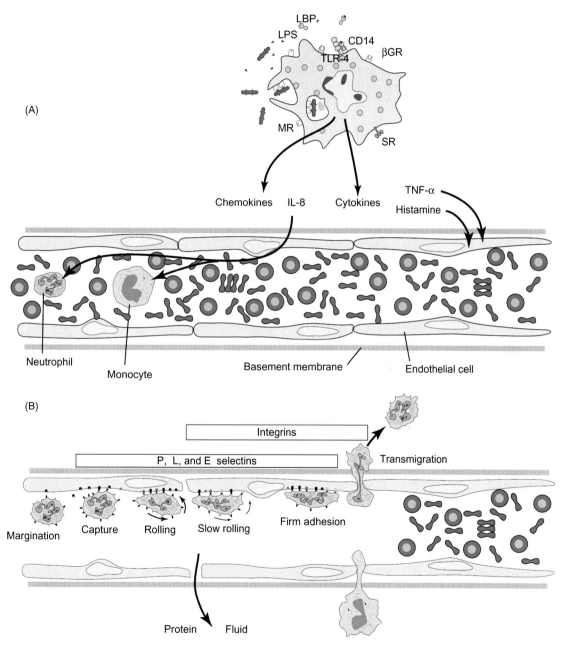

FIGURE 5.3.6 Recruitment of leukocytes to the site of infection. Macrophages release chemokines and cytokines in response to phagocytosis of foreign organisms. The chemokines attract the leukocytes. Cytokines released by the macrophages elicit proteins on the surface of the endothelial cells in two phases: a rapid phase involving proteins already present, and a delayed phase corresponding to new synthesis of membrane proteins. P-selectins bind to PSGL-1 on the leukocytes, capturing them. These cells first rapidly roll along the endothelial surface, mediated by binding to P- and L-selectin on the endothelial cells. Binding to E-selectin on the endothelium slows the rolling. Eventually, the leukocytes adhere firmly to the endothelium and begin transmigration across the capillary wall.

(**P**-selectin **g**lyco**p**rotein **l**igand-1) on the surface of the leukocytes. P-selectin exposure is rapid and short-lived, reaching a peak within 10 min after stimulation by cytokines.

- **Leukocytes roll along the endothelial surface:** The transient association of P-selectin with its ligand on the leukocyte allows the leukocyte to roll along the endothelial cell surface. Other proteins involved in the rolling include **E-selectin** and **L-selectin**. E-selectin is on the surface of the endothelium and L-selectin is on the leukocyte.
- **The rolling slows:** All three selectins will support leukocyte rolling, but at different velocities. E-selectin slows the rolling to $5-10\ \mu m\ s^{-1}$. The slowing is due in part to increased densities of E-selectin, which are increased 1–2 h after the onset of inflammation.
- **The leukocytes firmly adhere:** The binding of E-selectin is believed to begin the firm attachment of leukocytes to the endothelial surface. Other proteins involved include **ICAM-1** (**i**ntercellular **a**dhesion **m**olecule) on the endothelial cell that binds to an integrin, **LFA-1** (**l**eukocyte **f**unctional **a**ntigen) on the leukocyte.
- **The leukocytes transmigrate by ameboid motion through the capillary wall:** After adhering, leukocytes further interact with other proteins on the endothelial surface, particularly **PECAM-1** (**p**latelet-**e**ndothelial **c**ell **a**dhesion **m**olecule) and other leukocyte integrins. Both neutrophils and monocytes cross through the capillary wall by **ameboid motion**. This process consists of the extension of **pseudopodia** (literally, "false feet") by the cell, by deformation of the cell's cytoskeleton. Since the cytoskeleton is attached to the cell membrane, its deformation deforms the entire cell. Retraction of the opposite end of the cell results in a net forward movement of the cell. This **transmigration** is also called **diapedesis**.

THE COMPLEMENT SYSTEM DESTROYS MICROBES THAT HAVE ATTACHED ANTIBODIES

The complement system consists of about 12 soluble plasma proteins that are made mainly in the liver and circulate in the blood. They are inactive until activated by an infection. They are called "complement" because of their ability to amplify and complement the activity of antibodies. The complement proteins form a proteolytic cascade in which successive members cleave the next member, forming an active serine protease and releasing peptide fragments that can be used in recruiting phagocytes and lymphocytes. The final result of complement activation is the formation of a large transmembrane channel from aggregation of several units of C9.

IgG OR IgM ACTIVATES COMPLEMENT IN THE CLASSICAL PATHWAY

IgG or IgM antibodies bound to the surface of a microbe activates C1, a six-headed molecule that binds to a polymeric antigen–antibody aggregate. Binding of C1 activates a protease in a subunit of C1, which then activates a serine protease in another subunit of C1. The activated C1 then sequentially activates C4 to C4b, which activates C2 to C2b. The complex C4b2b cleaves C3.

A MANNAN-BINDING PROTEIN ACTIVATES C3

In a separate pathway, a plasma protein called mannan-binding lectin (MBL) forms a cluster of six mannose-binding heads around a central stalk. This protein specifically binds to mannose and fucose residues on the surface of bacterial cell walls that have the proper spacing and orientation. This complex is associated with a MBL-associated serine protease or MASP. These cleave C4 and C2 as in the classical pathway, resulting in the proteolytic cleavage of C3.

PROTEOLYTIC CLEAVAGE OF C3 IS THE CENTRAL EVENT

Proteolytic cleavage of C3 releases a large fragment, C3b, and a smaller fragment, C3a. The smaller fragment as well as fragments of C4 and C5 recruit phagocytes from the circulation. C3b binds covalently to the surface of the pathogen where it activates the remainder of the complement cascade, and where it can be recognized by phagocytic cells to enhance phagocytosis.

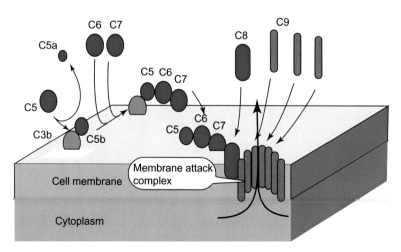

FIGURE 5.3.7 Mechanism of cell lysis by complement. See text for detailed explanation. (Source: *Modified from Alberts et al.,* Molecular Biology of the Cell, *Garland Science, New York, NY, 2002.*)

C9 UNITS AGGREGATE TO FORM A PORE THAT LYSES THE CELL

The membrane-immobilized C3b triggers a cascade of events beginning with the cleavage of C5 to C5a and C5b. C5b remains loosely bound to C3b and rapidly binds C6 and C7 to form C567. This firmly attaches to the pathogen membrane via C7. A molecule of C8 then binds to form C5678. This complex then binds C9, which subsequently undergoes a conformational change that exposes a hydrophobic region on C9. It inserts into the membrane and binds other units of C9, forming a large aqueous channel. This is called the **membrane attack complex**. This large channel causes the cytoplasm of the pathogen to leak out, killing the cell. The process is illustrated in Figure 5.3.7.

SUMMARY

The white blood cells include the polymorphonuclear granulocytes (neutrophils, eosinophils, and basophils) and the agranular leukocytes (monocytes and lymphocytes). Platelets are particles of megakaryocytes. The white cells are formed in bone marrow from pluripotent stem cells that form two lines: the myeloid line and the lymphoid line. The neutrophils are the most numerous white cell that phagocytose bacteria and other materials. Monocytes leave the circulatory system to become tissue macrophages that may live for long times in the tissues and have phagocytotic capability. Basophils resemble mast cells except that basophils originate in the marrow. They secrete histamine and heparin. Eosinophils are believed to defend against parasitic infections. Neutrophils, basophils, and eosinophils form a nonspecific defense system. The lymphocytes form a specific defense by recognizing a specific antigen. Lymphocytes come in a variety of types including T-lymphocytes, B-lymphocytes, and natural killer cells. The T-lymphocytes are further classified into cytotoxic T cells, helper T cells, and suppressor T cells. T cells provide cell-mediated immunity whereas B cells confer humoral immunity. The phagocytic cells occupy all regions of the body that may provide a route for bacterial infection. Together these phagocytic cells comprise the reticuloendothelial system.

Inflammation is the response of the body to tissue injury. Its five cardinal signs are: redness, increased temperature, edema, pain, and loss of function. Inflammation begins with the release of one or more of a host of signaling molecules by cells in the damaged area. Tissue macrophages, for example, have receptors that recognize bacterial cell wall constituents or LPS released from gram-negative bacteria. Binding of these ligands causes the macrophage to release cytokines and chemokines to both activate and attract white blood cells to the site of damage. Neutrophils and monocytes can cross the capillary wall by diapedesis to enter the infected or damaged area. Recruitment of neutrophils requires the expression of endothelial surface proteins called P-, L-, and E-selectins, and intercellular adhesion molecules. Receptors for these proteins reside on the leukocyte cell membrane.

Complement destroys pathogens by punching holes in their cell membranes. In the classic pathway, complement is activated by immunoglobulins IgG or IgM bound to the pathogen. This begins a cascade of proteolytic events and recruitment of complement factors until, at last, a series of complement proteins called C9 aggregate to form a water-filled channel across the membrane of the cell.

REVIEW QUESTIONS

1. What is the most common leukocyte? The least common?
2. How does TPO regulate the number of platelets?
3. What do platelets do?
4. How does the innate immune system differ from the adaptive immune system?
5. What cells constitute the reticuloendothelial system?
6. What cells do monocytes become in the tissues?
7. What are the steps involved in recruitment of neutrophils outside of the vascular system?
8. How does complement kill bacteria?

5.4 The Heart as a Pump

> **Learning Objectives**
>
> - Describe the anatomical location of the heart and identify its major external features
> - Define diastole and systole
> - Trace the flow of blood from the venae cavae to the aorta
> - Name the four valves of the heart and describe their location
> - Describe the origin of the first and second heart sound
> - Be able to determine whether flow is turbulent or laminar
> - Write Laplace's law for a sphere
> - Define stroke volume and ejection fraction
> - Describe the general function of the SA node
> - Describe the general structure of the electrical conduction system of the heart

THE HEART IS LOCATED IN THE CENTER OF THE THORACIC CAVITY

The heart is located in the middle of the thoracic cavity, oriented obliquely, with the apex of the heart pointing down and to the left, as shown in Figures 5.4.1 and 5.4.2. It is suspended within a tough fibrous sac, the **pericardium**, by its connections to the great vessels: the superior and inferior venae cavae, the pulmonary artery and veins, and the aorta. The pericardium is fused to the diaphragm, and so downward movement of the diaphragm during inspiration pulls the heart into a more vertical orientation. The heart sits atop the diaphragm and its **apex** is close to the anterior surface of the thoracic cavity. With every beat, the heart twists forward and the apex taps against the chest wall, producing the **apex beat**. This can be felt in the fifth left intercostal space. The heart itself is built around a ring of fibrous tissue called the **annulus fibrosus** that separates the atria from the ventricles and contains the four heart valves, as shown in Figure 5.4.6. The right and left atria are two thin-walled sacs that collect blood either from the veins (right atria) or from the lungs (left atria), whereas the ventricles are thick-walled muscular sacs that propel blood though the arteries. We introduce the function of the heart here by tracing the flow of blood through it.

THE HEART IS A MUSCLE

The heart consists of three types of muscle tissue: atrial muscle, ventricular muscle, and specialized muscle tissue that coordinates electrical signals through the heart. Like other muscles, the main function of the atrial and ventricular muscle fibers is to contract and produce force. These muscle fibers surround hollow cavities that are filled with blood. Therefore, their contraction produces a pressure within the cavity, and this pressure drives flow. The period of contraction of the heart muscle is called **systole**; its period of relaxation is called **diastole**. The mechanism of contraction is similar to that of skeletal muscles that we have already discussed (Chapters 3.4–3.7), except that the duration of cardiac contractions is longer. As we will discuss later, entirely different mechanisms regulate cardiac force than those that regulate the force of skeletal muscle.

The conductive fibers contract only weakly because they contain few contractile elements. Instead, they are specialized to conduct an electrical signal through the heart so as to coordinate cardiac contraction. Cardiac muscle fibers also conduct electrical signals, but they coordinate contraction locally rather than coordinating the activity of the entire heart. The heart is actually two pumps in series, both of which consist of an atrium and a ventricle. Uncoordinated contractions would have the pumps operating against each other. Therefore, coordination of mechanical activity of the chambers is essential for efficient pumping.

CONTRACTION OF CARDIAC MUSCLE PRODUCES A PRESSURE WITHIN THE CHAMBER

Contraction of cardiac muscle produces a tension within its walls that generates a pressure within the chamber. The relationship between tension and pressure in the heart is complicated because the geometry is not simple and because the wall tension is not uniform. Nevertheless, some approximations can be made. We consider first a thin-walled sphere.

LAW OF LAPLACE FOR THIN-WALLED SPHERES

Consider a sphere as shown in Figure 5.4.3. We draw an imaginary plane that bisects the sphere. The sphere is in mechanical equilibrium, meaning that the total forces on any part of the sphere sum to zero. The pressure difference between the inside and outside of the sphere is P. Really there are two pressures, but the outside pressure is taken to be zero gauge pressure (i.e., equal to the ambient pressure). The net pressure force on the upper

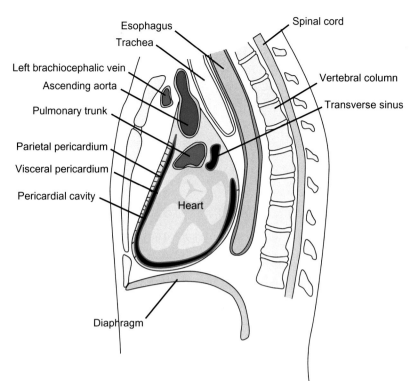

FIGURE 5.4.1 Location of the heart in the center of the chest. The figure shows a transverse section. The left ventricle actually lies behind and to the left of the right ventricle. (Source: *Redrawn from C. Rosse and P. Gaddum-Rosse*, Textbook of Anatomy, *Lippincott Raven, New York, NY, 1997.*)

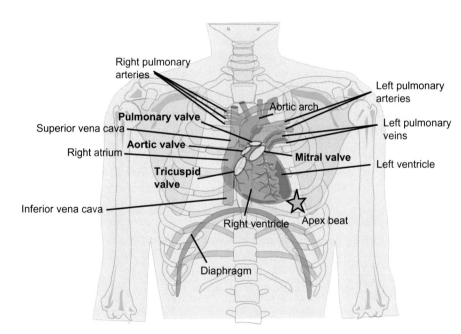

FIGURE 5.4.2 Location of the heart in the center of the chest. The figure shows a frontal, surface view of the heart. Its major axis lies obliquely to the plane of symmetry. The connective tissue ring containing the four heart valves (shown schematically as light blue rings) serves as a structural base for the heart. The tricuspid, mitral, aortic, and pulmonary valves are all grouped in this connective tissue ring set in an oblique plane beneath the sternum, at right angles to the major axis of the heart. The apex of the heart taps against the chest wall, causing the apex beat in the fifth left intercostal space.

hemisphere is thus the pressure, P, times the area of the hemisphere exposed to the pressure: $A = \pi r^2$. This pressure is balanced by the tension in the walls, which is the force per unit distance. The tension per unit length, T, acts all the way around the equator shown in the figure, so that the total tension is $F = T \times 2\pi r$. At mechanical equilibrium these two forces, the pressure force and the tension, are equal:

[5.4.1]
$$P\pi r^2 = T\, 2\pi r$$
$$P = \frac{2T}{r}$$

This last equation is the Law of Laplace for thin-walled spheres.

THE LAW OF LAPLACE FOR THICK-WALLED SPHERES

The model here is a hollow sphere with an inner radius r_1 and outer radius r_2, and wall thickness $w = r_2 - r_1$. We define here the **wall stress**, which is the force per unit area in the wall. Recall that the wall tension in the thin-wall model was the force per unit length. Thus we have

QUANTITATIVE HUMAN PHYSIOLOGY

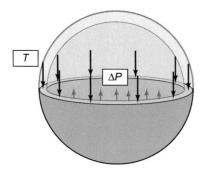

FIGURE 5.4.3 Schematic diagram of the component forces experienced by one hemisphere of a thin-walled sphere. The force on the upper hemisphere by the walls of the lower hemisphere is the tension times the circumference and is directed downward. The force of the contents of the sphere on the upper hemisphere is the pressure times the area and is directed upward. At mechanical equilibrium the total force on the upper hemisphere is zero.

[5.4.2]
$$\sigma = \frac{F}{A}$$
$$T = \frac{F}{l}$$
$$A = w\, l$$
$$\sigma = \frac{T}{w}$$

where σ is the wall stress, F is the force, A is the area, T is the tension, l is the length, and w is the width or, in this case, the wall thickness.

The condition of mechanical equilibrium demands that the internal pressure times the area must balance the wall force:

[5.4.3]
$$P_1\, \pi r_1^2 = \int_{r_1}^{r_2} \sigma(r)\, 2\pi r\, dr$$

This cannot be integrated unless we know the form of $\sigma(r)$. In general, $\sigma(r)$ is not constant with radius or with time and consists of two components: an elastic component having to do with the passive resistance of the heart muscle to stretch and an active component that produces the high pressures in the chamber to drive blood flow. If we assume $\sigma(r)$ is constant with r, we obtain:

[5.4.4]
$$P_1\, \pi r_1^2 = \sigma\, \pi\, [r_2^2 - r_1^2]$$

Substituting in from Eqn [5.4.2] for $\sigma = T/w = T/(r_2 - r_1)$, we get

[5.4.5]
$$P_1 = \frac{T}{r_2 - r_1} \frac{\pi (r_2^2 - r_1^2)}{\pi r_1^2} = T\frac{r_2 + r_1}{r_1^2}$$

This form of Laplace's law reverts to the thin-walled case when $r_2 = r_1$.

The point of these formulas is to emphasize that development of tension within the walls of the heart produces a pressure within the chamber. Because the heart is not a sphere but is more like a prolate spheroid, and the fact that the wall stress is not constant with r, the equations derived here are only approximately valid.

An alternate form of the Law of Laplace keeps σ in the equation. Taking Eqn [5.4.4], we have

[5.4.6]
$$P_1 = \frac{\sigma\, \pi\, [r_2^2 - r_1^2]}{\pi r_1^2}$$
$$P_1 = \frac{\sigma\, (r_2 - r_1)\, (r_2 + r_1)}{r_1^2}$$
$$P_1 \approx \frac{2\, \sigma\, w}{r_1}$$

This has a practical consequence for dilated hearts. When hearts enlarge, as in dilated cardiomyopathy, to achieve the same pressure the wall stress must also increase. To reduce the wall stress, the thickness of the heart wall also increases. Thus, dilated hearts are also hypertrophied: they have thicker walls.

BLOOD IS PUMPED THROUGH FOUR CHAMBERS

A surface view of the heart is shown in Figure 5.4.4; Figure 5.4.5 shows a cross-section of the heart with the direction of blood flow. Blood flow is unidirectional, from the veins to the arteries. Blood comes to the heart by collection into successively larger veins, finally flowing into the **superior vena cava**, which drains the head, neck, and arms, and the **inferior vena cava**, which drains the abdominal organs and the lower limbs. Blood from the venae cavae and the **coronary sinus** (the main drain of blood that supplies the heart) enters the **right atrium**. The right atrium is one of four chambers within the heart that collect blood and, by its contraction, propels blood forward. The right atrium is a thin-walled chamber that serves mainly as a blood reservoir and gateway to the right ventricle. Its contraction only modestly enhances flow into the ventricle.

When the heart is relaxed, blood that enters the right atrium continues to flow into the **right ventricle**. The **tricuspid valve** between the right atrium and right ventricle ensures that blood flow is unidirectional. This valve derives its name from its three cusps. The valve consists of a ring of connective tissue in the wall between atrium and ventricle, to which thin flaps of connective tissue are attached. Upon contraction of the right ventricle, the tricuspid valve closes and prevents blood from flowing back into the right atrium. The valve is kept from inverting by a series of cords, the **chordae tendineae**, that connect the valve to the **papillary muscles** on the opposite ventricular wall. These papillary muscles also contract, preventing **prolapse** of the valve (its inversion into the right atrium). Thus, the tricuspid valve is closed by increasing pressure within the right ventricle, and it opens when pressure in the right atrium exceeds that in the right ventricle. This valve, like the other three cardiac valves, is pressure operated.

The right ventricle in human adults is about 0.5 cm thick. It resembles a pocket sewn onto the larger left

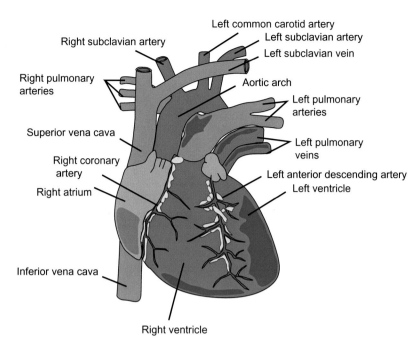

FIGURE 5.4.4 Surface view of the heart and its associated veins and arteries. Blood exits the heart by the pulmonary arteries and the aorta. The coronary arteries originate at the base of the aorta. The blood enters the heart through the large veins, the superior and inferior venae cavae. The large veins and pulmonary arteries contain blood with lower oxygen content because it is returning from the tissues where oxygen has been extracted. The pulmonary veins and aorta contain blood that has recently passed through the lungs without going to the tissues, and so it has a higher oxygen content. Vessels containing lower oxygen content are shown in lighter colour; vessels with higher oxygen content are shown in darker colour.

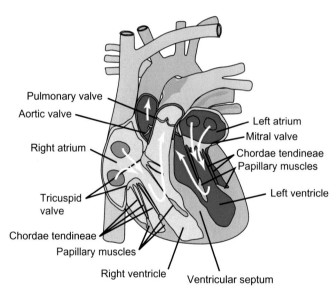

FIGURE 5.4.5 Sectional view of the heart showing the direction of blood flow. This is a schematic construct because no section of the heart simultaneously shows the ventricles and all four valves. During diastole, blood enters the right atrium and flows into the right ventricle through the tricuspid valve. During systole, the right ventricle contracts, closing the tricuspid valve and sending blood through the pulmonary valve to the pulmonary circulation. Blood returning from the lungs enters the left atrium and flows through the mitral valve into the left ventricle. During systole, the mitral valve also closes and contraction of the left ventricle sends blood through the aortic valve into the systemic circulation. Thus, the valves produce unidirectional flow through the heart. The chordae tendineae anchor the valve flaps to the papillary muscles in the ventricles, preventing valve prolapse.

ventricle. The part of the left ventricle between the two ventricles is called the **septum**. Contraction of the free anterior wall of the right ventricle squeezes the blood between it and the septum, forcing the blood out through the **pulmonary artery to the lungs**. The pulmonary artery has a valve, the **pulmonary valve**, that prevents blood from running back into the right ventricle during diastole. The pulmonary valve consists of three baggy cusps. It opens when pressure within the right ventricle exceeds that in the pulmonary artery, and it snaps shut when the pressure in the pulmonary artery exceeds that in the right ventricle.

The blood in the right heart comes from the peripheral tissues and so it is depleted of oxygen and carries extra waste CO_2 that was produced by the tissues. In the lungs, the blood loses its extra CO_2 and replenishes its O_2 content. The reoxygenated blood returns to the heart through the **pulmonary veins**, flowing into the **left atrium**. Blood that enters the left atrium flows directly into the **left ventricle** during diastole. Left atrial contraction precedes left ventricular contraction, and this helps load the ventricle with blood prior to its contraction. The flow is kept unidirectional here by the **mitral valve**. As with the other atrioventricular valve, the margins of the cusps are tethered by chordae tendineae to two papillary muscles on the wall of the ventricle.

Contraction of the left ventricle decreases its chamber's length and diameter. Contraction begins at the apex of the heart and proceeds up the ventricle. In this way, the ventricle pushes blood out into the systemic circulation through the **aorta**. Because the pressure in the systemic circulation is much higher than in the pulmonary circulation, the left ventricle must provide more pressure to push blood into the high-pressure system. Thus, the left ventricular walls are about three times thicker than those of the right ventricle. The aorta contains another three-cusp valve, the **aortic valve**. The aortic valve and the pulmonary valves are called the **semi-lunar valves** because of their shape. When pressure within the left ventricle exceeds that in the aorta, the valve opens and blood rushes forward. As the pressure within the ventricle falls upon relaxation (diastole), the aortic valve snaps shut, preventing backflow from the aorta to the left ventricle.

Although blood flow through the heart is sequential, the contractions of the chambers are not: both atria contract nearly simultaneously, followed shortly thereafter by the simultaneous contractions of the ventricles. At any time, the blood pumped out by the left ventricle is blood that was pumped out by the right ventricle a couple of beats ago, but both contract at the same time.

THE FOUR VALVES ARE NEARLY COPLANAR

A fibrous connective tissue ring called the **annulus fibrosus** contains all four of the heart valves. Figure 5.4.5 does not adequately show this because otherwise it becomes nearly impossible to show the valves and their connectivity with clarity. Figure 5.4.6 shows a cross-section of the heart at the level of the annulus fibrosus.

CLOSURE OF THE VALVES PRODUCES THE *HEART SOUNDS*

Valves are pressure operated. Greater pressure on the outflow side of a heart valve causes it to close. The cusps snap back as they check the movement of the refluxing blood. The sudden movement of the cusps produces a brief turbulence that transmits a vibration to the chest wall. These vibrations are audible to the ear pressed against the chest and are better heard through a **stethoscope**. The procedure of carefully monitoring sounds within the body is called **auscultation** (from the Latin "auscultare", meaning "to listen"). The sounds can be recorded by a microphone and displayed graphically in a tracing called a **phonocardiogram**. There are a total of four heart sounds, but two are normally clearly audible. The **first heart sound** corresponds to the closure of the tricuspid and mitral valves. It has a frequency of about 100 Hz. The second heart sound, at a slightly higher frequency, corresponds to closure of the aortic and pulmonary valves. Figure 5.4.7 shows a phonocardiogram overlaying a trace of ventricular volume and pressure during the cardiac cycle.

ADDITIONAL TURBULENCE CAUSES *HEART MURMURS*

Laminar flow is streamlined flow, and it is silent. Chaotic flow is also called turbulent flow, and it is noisy. The occurrence of turbulent flow is often estimated from the Reynolds number, named after Osborne Reynolds, who studied the patterns of flows in tubes by injecting a thin stream of visible dye into the moving fluid. The Reynolds number is given as

$$[5.4.7] \qquad Re = \frac{2a <V> \rho}{\eta}$$

where Re is the Reynolds number, a is the radius of the tube, $<V>$ is the average velocity, ρ is the density, and η is the viscosity. Thus, the Reynolds number is a dimensionless number that is a ratio of the inertial forces to the viscous forces. Turbulence normally occurs when $Re \sim 2000$. Abnormal movement of fluid can produce turbulence in the heart, causing additional sounds that can be heard by auscultation. These are **heart murmurs**. Examples include **aortic regurgitation**. In this case, the aortic valve incompletely seals the left ventricle from the aorta during ventricular relaxation. Blood squirts back through the leaky valve into the ventricle, driven by the higher pressure in the aorta. The squirting produces a noise, or **bruit**.

SUMMARY OF THE CONTRACTILE EVENTS IN THE CARDIAC CYCLE

VENTRICULAR FILLING

The ventricles fill over a period of about 0.45 s. Note in Figure 5.4.7 that while the left ventricle fills, its pressure decreases. This occurs because the ventricles are still recoiling from systole. During ventricular filling, both the tricuspid and mitral valves are open, whereas the

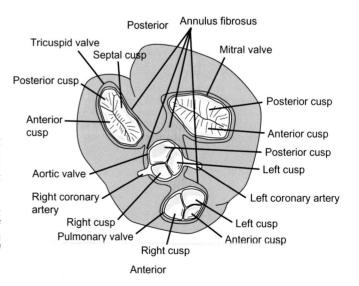

FIGURE 5.4.6 Surface view of the top of the heart with the atria dissected away. The annulus fibrosus, the connective tissue ring that separates the two atria from the two ventricles, is shown in gray. This ring approximates a plane in which all four valves (shown in white) are embedded. The tricuspid valve separates the right atria from the right ventricle. It has three cusps: an anterior cusp, a septal cusp, and a posterior cusp. The mitral valve separates the left atria and ventricle, and it has an anterior and posterior cusp. The pulmonary valve and aortic valve are located in the pulmonary artery and aorta, respectively. Each of these have three cusps. The valves open in a definite sequence during the pumping action of the heart. (Source: Adapted from C. Rosse and P. Gaddum-Rosse, Textbook of Anatomy, Lippincott Raven, New York, NY, 1997.)

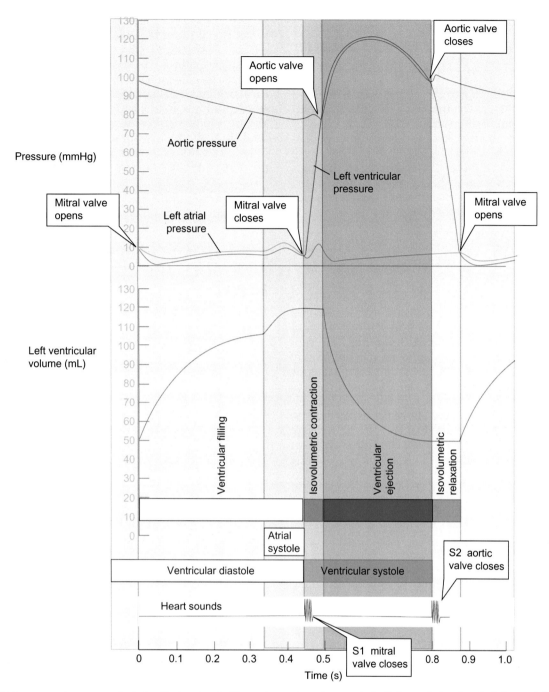

FIGURE 5.4.7 Sequence of some events during the heart cycle. The cycle consists of one period of left ventricular diastole (relaxation) followed by a phase of left ventricular systole (contraction). Because there is a lag between contraction and development of pressure, these periods do not precisely correspond to the periods of left ventricular filling and ejection, as shown. The mitral valve opens when left atrial pressure exceeds left ventricular pressure and closes when contraction of the ventricle raises left ventricular pressure to equal, and then exceed, left atrial pressure. Closure of the mitral valve contributes to the first heart sound, S1. Similarly, the aortic valve opens when contraction of the heart increases pressure to levels above aortic diastolic pressure. Continued contraction causes ejection of blood from the left ventricle. On relaxing, the pressure in the left ventricle falls. The first phase is isovolumetric relaxation. When left ventricular pressure falls below aortic pressures, the aortic valves snap shut, contributing to the second heart sound, S2.

aortic and pulmonary valves are closed. Ventricular filling is completed by contraction of the atria during the last 0.12 s. The contraction of the atria is not really necessary, but without it blood can stagnate in the atria, leading to clots, which can be released into the circulation and cause strokes or heart attacks. The volume of blood in the left ventricle at the end of ventricular filling is called the **end-diastolic volume (EDV)**, which is about 120 mL in the adult human. The corresponding pressure, the end-diastolic pressure (EDP), is about 4–7 mmHg.

- ventricles fill for about 0.45 s;
- mitral and tricuspid valves open;

- aortic and pulmonary valves close;
- End-diastolic volume ≈120 mL;
- End-diastolic pressure ≈4–7 mmHg.

ISOVOLUMETRIC CONTRACTION

Ventricular contraction occurs over about 0.35 s and consists of two phases. The first phase is a brief isovolumetric contraction, lasting about 0.05 s and a longer ejection phase of about 0.30 s. The isovolumetric contraction causes left ventricular pressure to rise above atrial pressure, which closes the mitral valve and produces the first heart sound. The aortic valve opens at the end of isovolumetric contraction when left ventricular pressure exceeds aortic pressure.

- ventricular contraction for about 0.05 s;
- mitral and tricuspid valves closed;
- aortic and pulmonary valves closed.

EJECTION

When ventricular pressure rises further to exceed aortic pressure or pulmonary artery pressure, the aortic and pulmonary valves open and blood flows from the ventricles into the systemic circulation or the pulmonary circulation. The ejection phase of the cardiac cycle ends when the aortic valve snaps shut, producing the heart's second sound. At the end of the ejection phase the volume of blood remaining in the left ventricle, its end-systolic volume (ESV), is about 50 mL. Thus, the **stroke volume**, the volume of blood ejected with each heart beat, is the difference between the end-diastolic volume and the end-systolic volume = 120 mL − 50 mL = about 70 mL. The **ejection fraction** is the fraction of the EDV that is ejected. The typical value for the ejection fraction is 70/120 = 0.58.

- ejection takes about 0.30 s;
- mitral and tricuspid valves closed;
- aortic and pulmonary valves open;
- peak pressure of about 25 mmHg (pulmonary circulation) or 120 mmHg (systemic circulation).

ISOVOLUMETRIC RELAXATION

When the aortic and pulmonary valves close, the heart relaxes isovolumetrically because both the outflow valves and inflow valves are closed, and so no fluid moves across them. This phase lasts until the intraventricular pressure falls below the pressure in the atria, at which time the mitral and tricuspid valves open again. The isovolumetric relaxation lasts about 0.08 s. The volume throughout this phase is the ESV, approximately 50 mL. The atrioventricular (AV) valves open at an atrial pressure of about 7 mmHg.

- isovolumetric relaxation lasts about 0.08 s;
- aortic and pulmonary valves closed;
- mitral and tricuspid valves closed.

AN AUTOMATIC ELECTRICAL SYSTEM CONTROLS THE CONTRACTION OF THE HEART

THE HEART BEAT ORIGINATES IN THE SA NODE, THE HEART'S PACEMAKER

The heart contains its own rhythm generator or **pacemaker** and will beat on its own when isolated from the body and perfused appropriately, or when it is transplanted from a donor to a recipient. The normal pacemaker activity is located in a specialized group of cells that comprise the **sinoatrial node** or **SA node**. The SA node consists of a small strip of modified muscle tissue, about 20 mm × 4 mm, on the posterior wall of the atrium near the superior vena cava. Cells in the SA node have an unstable membrane potential that spontaneously depolarizes to produce an action potential about once per second. This action potential is conducted through the atrial tissue to the AV node, as shown in Figure 5.4.8. The wave of electrical activity that spreads outward from the SA node is called the **cardiac impulse**. Other parts of the heart also have intrinsic rhythm, but the SA node has the higher intrinsic frequency, so it entrains all parts of the heart to its rhythm.

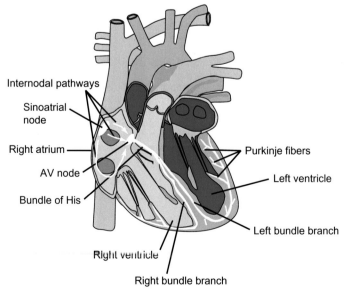

FIGURE 5.4.8 Conduction system of the heart. The heart beat originates in the SA node at the junction of the venae cavae with the right atria. Excitation spreads through the atrial muscle to the AV node, which connects to the bundle of His. This conducts the excitation across the annulus fibrosus, which insulates the ventricles from the atria. The bundle of His separates into a left bundle branch that activates the left ventricle and a right bundle branch. The conducting fibers terminate in wide conducting fibers called Purkinje fibers.

SPECIAL PATHWAYS CONDUCT THE ELECTRICAL SIGNAL FOR CONTRACTION

Effective pumping requires coordinated contraction of the heart's four chambers. This is accomplished through special conducting fibers. Fibers called **Bachmann's bundle** relay the impulse to the left atria. This coordinates right and left atrial contraction. The velocity of conduction in Bachmann's bundle is about 1 m s^{-1}.

THE AV NODE DELAYS THE TRANSMISSION OF THE CARDIAC IMPULSE

The cardiac impulse travels to the AV node through the atrial tissue. The AV node is a small group of cells and connective tissue at the posterior and lower area of the right atrium. The annulus fibrosus forms an insulating band between the atria and the ventricles, and the AV node begins the only electrical pathway across it. Transmission of the cardiac impulse is delayed here for about 0.1 s. This **nodal delay** allows for contraction of the atria and complete filling of the ventricles before beginning contraction of the ventricles. AV nodal cells conduct the impulse at 0.01–0.05 m s^{-1}. AV nodal cells also have intrinsic rhythm and will take over if transmission from the SA node fails. Because AV nodal cells have a lower frequency, failure of the SA node will cause lower frequency of heart beats.

THE BUNDLE OF HIS AND ITS BRANCHES CONDUCT THE IMPULSE TO THE APEX OF THE HEART

The bundle of His consists of wide, fast-conducting muscle fibers that carry the cardiac impulse through the insulating annulus fibrosus into the fibrous upper part of the ventricular septum. There, it becomes the **left bundle branch** and the **right bundle branch**. The right bundle branch travels along the right side of the septum and supplies excitation to the right ventricle. The left bundle branch is really two sets of fibers, one anterior and another posterior. These course down the left side of the septum and supply the left ventricle. The bundle fibers terminate in an extensive network of large fibers called **Purkinje fibers**. These fibers conduct the impulse at high velocity, 3–5 m s^{-1}. They distribute the impulse to the subendocardium, the layer of cells beneath the endocardium. The endocardium lines the cavities of the heart.

HEART CELLS CONDUCT ELECTRICAL SIGNALS TO NEIGHBORING CELLS THROUGH INTERCALATED DISKS

Contractile cells in the heart generally have one or two nuclei and they form a branching network, as shown schematically in Figure 5.4.9. The junction between adjacent contractile cardiac cells is called the **intercalated disk**. It is easily distinguished in both light and electron microscopy because it forms a dark and wide band running across the width of the cell. Here the membranes of two adjacent cells interdigitate and are linked by cytoskeletal elements. In addition, the intercalated disks form a conductive pathway between two adjacent cells—the cells are electrically coupled and excitation of one cell readily passes to the second. All cardiomyocytes have electrical connections to at least two and often more cells. The branches assure that the cells form a **functional syncytium** (from "syn", meaning "together" and "kyto", meaning "cell"). Thus electrical activity is conducted along the contractile cells of the heart and local neighbors contract in sequence, one shortly after the other. The ventricular muscle conducts the impulse at about 0.5–1 m s^{-1}. This local connectivity of the cardiomyocytes coordinates local contraction but not global contraction. Global coordination is provided by the distribution network.

Clinical Applications: Tetralogy of Fallot

The tetralogy of Fallot is a congenital heart defect that consists of four components:

1. Pulmonary stenosis—a narrowing of the outflow of the right ventricle.
2. Overriding aorta—the aorta attaches to both right and left ventricle.
3. Ventricular septal defect—a hole in the septum connects deoxygenated with oxygenated blood.
4. Right ventricular hypertrophy—this is secondary to the defects listed above.

Because of the mixing of deoxygenated with oxygenated blood, due to the ventricular septal defect, and preferential flow of blood from both ventricles through the aorta due to obstructed flow through the pulmonary artery, the blood in the systemic circulation has low oxygenation. Babies with the condition are called "blue babies" because of the resulting cyanosis.

Surgical treatment originally consisted of constructing a shunt between the subclavian artery and the pulmonary artery, redirecting a large proportion of partially oxygenated blood to the lungs and increasing flow through the pulmonary circulation and greatly relieving symptoms. The procedure was invented by the team of Alfred Blalock, Helen Taussig, and Vivien Thomas at Johns Hopkins in 1944 and is portrayed in the movie *Something the Lord Made*. Although palliative, the Blalock-Thomas-Taussig shunt was not curative. Total surgical repair of tetralogy of Fallot was first successful in 1954, and since then survival rates have increased markedly.

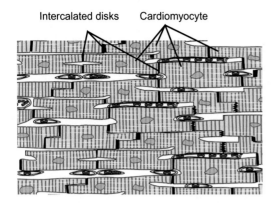

FIGURE 5.4.9 Schematic diagram of the functional syncytium of the heart. Contractile cells have cross striations and connect to each other at intercalated disks, special structures that allow passage of electrical activity from one cell to the next.

SUMMARY

The heart is a four-chambered pump. Blood returns to the heart from the tissues through the superior and inferior venae cavae and enters the right atrium. During ventricular filling, blood drains from the right atrium through the tricuspid valve into the right ventricle. The ventricles are given an extra boost of filling by atrial contraction. When the right ventricle contracts, the increased pressure closes the tricuspid valve and, at higher pressure, the pulmonary valve opens and blood flows into the pulmonary artery to the lungs. Blood returns to the heart through the pulmonary veins, which empty into the left atrium. During ventricular filling, blood leaves the left atrium through the mitral valve into the left ventricle. Contraction of the left ventricle increases the pressure within the ventricle, which closes the mitral valve and opens the aortic valve. Blood is then ejected into the aorta to supply the rest of the body. Thus, the four pressure-operated valves provide for unidirectional flow of blood sequentially through the pulmonary and systemic circulations. The four valves are situated within a fibrous ring, the annulus fibrosus, that approximates a plane that divides and electrically isolates the atria from the ventricles. Simultaneous closure of the tricuspid and mitral valves produces the first heart sound; closure of the aortic and pulmonary valves causes the second heart sound.

Coordination of the contraction of the heart is provided by a special conduction system that consists of modified muscle tissues rather than nerve. The rhythm of the heart is set by the sinoatrial (SA) node at the posterior of the right atrium near the vena cava. The SA node is the pacemaker because cells in the SA node have the fastest spontaneous rate of depolarization, but other parts of the heart also have spontaneous rhythm. The wave of electrical activity, the cardiac impulse, can be conducted by the contractile cells of the heart because they are electrically coupled at their junctions, the intercalated disks. The cardiac impulse is conducted to the AV node, which delays the impulse and then sends it through the annulus fibrosus by the bundle of His. The bundle of His consists of larger, specialized muscle cells that conduct the impulse more rapidly. The right and left bundle branches derive from the bundle of His and separately conduct the impulse to the right and left ventricles. In this way, the conducting system distributes excitation to the heart muscle to coordinate its contraction from the apex upwards.

REVIEW QUESTIONS

1. Where does blood come from that fills the right atrium? Left atrium? Where does blood go when it is pumped from the right ventricle? Left ventricle?
2. What valve separates right atrium from right ventricle? What valve separates left atrium from left ventricle?
3. Why is the right ventricular wall thinner than the left ventricular wall?
4. What valve prevents backflow in the pulmonary artery? In the aorta?
5. What controls opening and closing of the valves in the heart?
6. What causes the first heart sound? The second? Why are there not more heart sounds?
7. What is a bruit? What causes it?
8. What is the pacemaker of the heart? Where is it? What is the AV node and where is it?

Problem Set 5.1
Blood

1. The value of L_p for the red blood cell is about $1.8 \times 10^{-11}\,cm^3\,dyne^{-1}\,s^{-1}$. Its surface area is about $1.35 \times 10^{-6}\,cm^2$ (Solomon, *Methods Enzymol.* 192–222, 1989).
   - A. What is the initial osmotic flow if the osmolarity inside is initially 300 mOsm and the osmolarity outside is 275 mOsm? (Assume $\sigma = 1.0$ for all solutes).
   - B. If the volume of the cell is $100 \times 10^{-12}\,cm^3$, how long would it take to double its volume provided that the osmotic pressure and area of the membrane and L_p did not change?
   - C. In the case described, how much water would be required to enter the cell to equilibrate the osmotic pressure between inside and outside? Assume that the outside bath is essential infinite so that its osmotic pressure is kept constant.

2.
   - A. Assume that water movement is fast compared to solute movement when red blood cells are placed in contact with solutions of different tonicity. Derive an equation that relates the volume of the cell, V_c, after equilibrating with the outside solution, in terms of the original volume, V_0; the osmotic pressure of the solution under isotonic conditions, $\pi_{isotonic}$; the osmotic pressure of the new solution, π.
   Hints: In this situation where water movement is fast compared to solute movement, the total amount of osmotically active solutes inside the cell will be constant. Write this in terms of the original volume and the total osmotic pressure. After equilibration with a new solution, the volume changes but the total number of osmotically active solutes inside the cell does not. At this new volume, V_c, the cell's osmotic pressure equilibrates with the osmotic pressure of the outside solution. From this you should be able to derive an expression for a perfect osmometer.
   - B. Because some of the cell's volume is occupied by structures, it appears to have an "osmotically inactive" volume, which we will denote as V_b. How would this volume alter your equation in Part A?

3.
   - A. The hematocrit in most healthy adult males is about 45%. If the number of red blood cells in whole blood is about 5.2×10^6 per mm^3, calculate the apparent volume per cell. How does this compare to the volume given in problem 1?
   - B. The red blood cell is a biconcave disk about 7.5 μm in diameter and about 1.9 μm at its thickest point. *Estimate* the surface area from these dimensions. How does this calculation compare to the surface area given in problem 1?

4.
   - A. Assume that the mass of a person is 70 kg and that 8% of the body mass is blood. If the density of blood is 1.055 g/mL, what is the blood volume?
   - B. Assume that the hematocrit ratio is 0.45. What is the plasma volume in this person?
   - C. Assume that the average life span of the red blood cell is 120 days. What volume of packed red blood cells is being destroyed each day? This corresponds to what volume of whole blood? On average, what volume of whole blood is being made each day?
   - D. Assume that the hemoglobin concentration in the whole blood is 16 g%. How much hemoglobin is degraded each day?
   - E. The molecular weight of hemoglobin is $66{,}600\,g\,mol^{-1}$ and each hemoglobin binds four Fe atoms. How much iron is liberated from hemoglobin each day? Give the answer in mg/day and mol/day.
   - F. The molecular weight of bilirubin is $584.7\,g\,mol^{-1}$. How much bilirubin is liberated from hemoglobin each day in this person, in mg/day?
   - G. From Part A and Part D, calculate the total amount of hemoglobin in the blood. From this calculate the total amount of iron in the blood.

5. A patient's blood was centrifuged in a capillary tube. The total height of the column of blood was 3.5 cm and the column of packed cells was 1.8 cm. Analysis of whole blood gave a hemoglobin concentration of 12 g%, and counting of the RBCs in a hemacytometer gave a count of 4×10^6 cells per mm^3.
   - A. What is the hematocrit?
   - B. What is the MCV?
   - C. What is the MCHC?

6. Assume that the average density of hemoglobin is $1.34\,g\,cm^{-3}$. If the hemoglobin concentration of whole blood is 15 g% and the hematocrit is

45%, what fraction of the red blood cell volume is occupied by hemoglobin?

7. Assume that the shear rate in the aorta is about $120\ s^{-1}$. Assume further that the aorta has a diameter of 1 inch and that the stroke volume is 70 mL and that the viscosity of whole blood is 3×10^{-3} Pa s.
   A. What is the drag force of the vessel wall on each stroke volume?
   B. If the density of whole blood is $1.05\ g\ cm^{-3}$ and its average velocity in the aorta is $0.2\ m\ s^{-1}$, what is the initial deceleration of the blood flow in the aorta?

8. Normal dimensions of the human heart for live persons can be determined noninvasively using magnetic resonance imaging (MRI) or two-dimensional echocardiography (2DE). (S. Kaul, et al., Measurement of normal left heart dimensions using optimally oriented MR images, *Am. J. Radiol.* 146:75–79, 1986.) During systole the left ventricular cavity has a diameter of about 34 mm and the wall thickness is about 16 mm. If systolic pressure is 130 mmHg, estimate the wall tension assuming that the wall stress is constant and that the ventricle is a sphere of the dimensions given. Estimate the wall stress.

9. A. If the left ventricle contains 120 mL at end diastole, and we assume it is a sphere, what should its cavity radius be?
   B. If the left ventricle contains 50 mL at end systole, and we assume it is a sphere, what should its cavity radius be?
   C. How do these answers compare to measurements of cavity diameter of normal persons of about 46 mm at end diastole and about 34 mm at end systole? What causes the discrepancy?

10. You sliced your leg accidentally with an axe and lost 1.5 L of blood before you stopped the bleeding. Assume that you had 15 mg% hemoglobin, 45% hematocrit, and 5 L of blood before your accident.
    A. The blood volume is generally made up quickly by fluid entering plasma from the ISF and from fluid intake (persons who lose blood get thirsty). What would the new hemoglobin concentration be after you have recovered a blood volume of 5 L?
    B. What would the new hematocrit be?
    C. How much iron did you lose in the accident?
    D. The rate of formation of red blood cells depends in part on the recycling of iron from destroyed red blood cells, but can be increased by erythropoietin and mobilization of Fe from ferritin stores in the liver. Suppose that the life span of the red blood cells is a constant 120 days. Suppose further that the rate of incorporation of Fe into Hb is a constant 21 mg per day. Write an equation for the recovery of the hemoglobin concentration after the loss of blood. How long does it take to recover one-half of the lost Hb?

11. The red blood cell count is typically $5 \times 10^6\ mm^{-3}$ and the blood volume is 5 L, and Hct is 45%.
    A. How many red blood cells are there in the total circulation?
    B. If the average life span is 126 days, how many red blood cells are destroyed each day?
    C. How many red blood cells are made each day?
    D. If the white blood cell count is $5 \times 10^3\ mm^{-3}$ and the volume of the white blood cell is the same as that of the red blood cell, what is the total number of white blood cells in the circulation?
    E. What is the total volume of white blood cells in the circulation?

12. The end systolic volume in an athlete's heart is 50 mL and his end diastolic volume is 140 mL.
    A. What is his stroke volume?
    B. Is this larger or smaller than normal?
    C. What explains the difference?
    D. What is his ejection fracton?
    E. Is the ejection fraction larger or smaller than normal?
    F. What explains the difference?

13. The last time you gave blood your Hct was 40% and your hemoglobin was 14 g%. That was about 3 months ago. In the blood bank they determined you are type A−. Today you have been in an accident and your doctors suspect internal bleeding. They took a sample of your blood and found a Hct of 25%.
    A. What would you expect your hemoglobin to be?
    B. Can you tell how much blood you have lost? If you cannot, why not?
    C. Can you determine the fraction of blood that you have lost?
    D. What types of blood can you receive without a transfusion reaction?

14. During a hot August, a football player weighing 100 kg lost 5 L of hypotonic sweat. His Hct before the workout was 45% and RBC count was $5 \times 10^6\ mm^{-3}$. During the workout he drank 1.5 L of water which had been absorbed into the blood from the GI tract. Assume that the blood mass is 8% of his body mass and that the density of blood is $1.05\ g\ cm^{-3}$. Further assume that the fluids lost from his body were distributed proportionately from the extracellular volume and intracellular volume, which make up 20% and 40% of the body mass, respectively (assume the density of the fluids is $1.0\ g\ cm^{-3}$).
    A. How much volume did he lose?
    B. How much of the lost volume came from the extracellular fluid compartment? How much from the intracellular volume?
    C. What was the MCV before the workout?

D. What is the MCV after the workout? (The red blood cell has an intracellular volume)
E. What is the hematocrit after the workout?

15. A backpacker weighs 220 pounds. Assume his blood weight is about 7% of his body weight. His hematocrit at rest is 40% and total plasma proteins is 6 g%. After hiking on a hot day, the backpacker loses a net 4 L of body fluid by sweating and drinking, where the drinking is insufficient to match his sweating. Assume that the plasma makes up 6% of his total body water and that the fluid lost from each compartment is in proportion to the size of each fluid compartment.
    A. Assuming no loss of blood, what is his hematocrit after his hot hike?
    B. What is his total plasma protein concentration after hiking?
    C. What effect would these changes have on his blood viscosity?

16. A person has a total Hb concentration of 14 g%, a hematocrit of 40%, and a RBC count of 5×10^6 cells mm^{-3}.
    A. What is the MCV?
    B. What is the MCHC?
    C. Calculate the Hb content of an average RBC.
    D. Calculate the number of molecules of Hb in an average RBC.
    E. If the density of Hb is 1.34 g cm^{-3}, what volume of the average RBC is occupied by Hb?

5.5 The Cardiac Action Potential

Learning Objectives

- Identify an action potential as originating from SA nodal cells, atrial cells, or ventricular cells
- Distinguish between T-type and L-type calcium channels
- Describe what is meant by "pacemaker potential"
- Describe the general effects of sympathetic and parasympathetic stimulation of the SA node
- Identify the signaling pathways used by sympathetic and parasympathetic nervous systems for the electrical system of the heart
- Identify phases 0, 1, 2, 3, and 4 of the ventricular action potential
- Describe the major ion fluxes in phases 0, 1, 2, 3, and 4
- Define "chronotropic" and "inotropic" effects
- Describe the effect of epinephrine on the ventricular myocyte action potential

DIFFERENT CARDIAC CELLS DIFFER IN THEIR RESTING MEMBRANE POTENTIAL AND ACTION POTENTIAL

As described in Chapter 5.4, the heart consists of three main muscle types: atrial muscle, ventricular muscle, and specialized muscle that coordinates electrical signals through the heart. These cells are all excitable, meaning that they can produce a brief, pulse-like change in their membrane potential, and this **action potential** can be conducted over the surface of the cell to activate all parts of the cell nearly simultaneously.

The different cell types have different sets of ion channels on their membranes. The different ion channels have different conductances and these conductances respond differently over time and to changes in membrane potential. These differences cause differences in the resting membrane potentials and the action potentials. Figure 5.5.1 shows the action potentials of sino atrial (SA) nodal cells, atrial muscle cells, and ventricular muscle cells. The resting membrane potentials of the contractile cells are stable and stay at −80 to −90 mV until stimulated by depolarization. The SA nodal cells, on the other hand, have unstable resting membrane potentials that begin at about −60 mV and gradually depolarize and reach threshold.

SA NODAL CELLS SPONTANEOUSLY GENERATE ACTION POTENTIALS WHEREAS CONTRACTILE CELLS HAVE STABLE RESTING MEMBRANE POTENTIALS

The SA nodal cells have a resting membrane potential of about −60 mV. This potential is due to a larger conductance to K^+ ions compared to Na^+ or Ca^{2+} ions. Recall that the resting membrane potential is the chord conductance-weighted average of the equilibrium potentials for all diffusable ions:

$$[5.5.1] \quad E_m = \frac{g_{Na}}{(g_{Na} + g_K + g_{Ca})} E_{Na} + \frac{g_K}{(g_{Na} + g_K + g_{Ca})} E_K + \frac{g_{Ca}}{(g_{Na} + g_K + g_{Ca})} E_{Ca}$$

where E_m is the membrane potential; E_i is the equilibrium potential for the ith ion; g_i is the conductance for the ith ion. Recall here that the chord conductance is a coefficient that relates the current carried by a particular ion to its driving force:

$$[5.5.2] \quad I_i = g_i(E_m - E_i)$$

where I_i is the current carried by the ith ion, g_i is its conductance, E_m is the membrane potential, and E_i is the equilibrium potential for the ith ion. This is a variant of Ohm's law for ions moving across a membrane. The net driving force for the ion is $E_m - E_i$. **Current flow is always taken as the direction of positive ion flow, outward flow being positive.**

Like most excitable cells, the SA nodal cell at rest has a larger g_K than g_{Na} or g_{Ca}, but the channel responsible for g_K is different from that in nerve or in the contractile cells of the ventricle. The stable E_m in contractile cardiac cells is produced by the **inward rectifying K^+ channel** (carrying the current I_{K1}) which SA nodal cells lack. Instead, SA nodal cells have a **delayed rectifying K^+ channel** (carrying the current I_K) that opens slowly upon depolarization and deactivates with time. The initial negative membrane potential depolarizes slowly with time toward a threshold of about −40 to −55 mV. This slow depolarization is called the **pacemaker potential**. Potentials of about −60 mV activate an inward Na^+ current called I_f—for **"funny" current**. It earns this name because it has the peculiar property of being activated by hyperpolarization

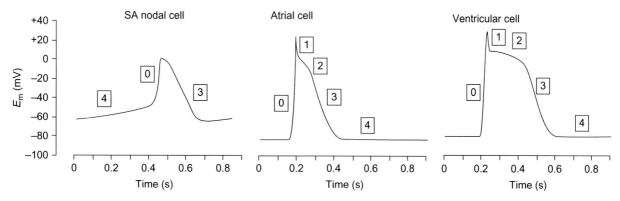

FIGURE 5.5.1 Action potentials in SA nodal cells, atrial muscle cells, and ventricular muscle cells. The SA nodal cell resting membrane potential initially is about −60 mV. It is unstable, gradually spontaneously depolarizing toward threshold at about −50 mV. The gradual increase is called the pacemaker potential. Each of the action potentials occurs in phases numbered 0–4.

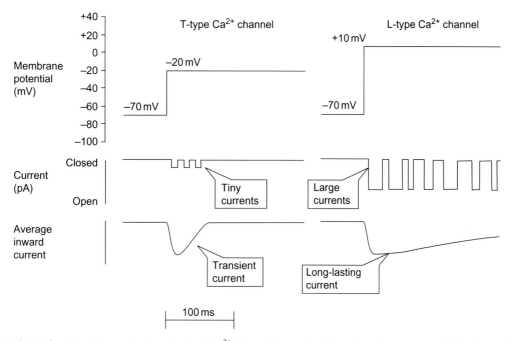

FIGURE 5.5.2 Recordings of the currents passing through single Ca^{2+} channels in small patches of membranes as studied by the patch clamp method. When the membrane is depolarized above about −55 mV, a single Ca^{2+} channel opens sporadically, admitting tiny currents (left, middle). The average response of many of these channels produces the ensemble response of a larger patch of membrane (left, bottom). These T-type channels open transiently. When the membrane is depolarized further, a single L-type Ca^{2+} channel opens. It also flickers between open and closed states, but its open state current is larger (right, middle). Also, the average of many channels produces a longer lasting current (right, bottom), although it also eventually inactivates. Thus the T-type channel's name derives from its tiny current and transient opening, and the L-type channel's name derives from its larger current and its long-lasting openings.

rather than by depolarization, whereas most channels in excitable tissues are activated by depolarization. The membrane potential gradually depolarizes from a combination of the inward Na^+ current (I_f) and from a decay in the outward K^+ current (I_K). When the potential reaches about −55 mV, voltage-gated Ca^{2+} channels contribute to an inward current that depolarizes the membrane potential. Two types of Ca^{2+} channels work here: **T-type Ca^{2+} channels** have tiny conductances that transiently open; **L-type Ca^{2+} channels** have large conductances with long-lasting openings. The T-type Ca^{2+} channels contribute to the last third of the pacemaker potential, whereas the inward Ca^{2+} current from L-type Ca^{2+} channels largely forms the upstroke of the action potential. Traces showing the differences between T-type and L-type channels are shown in Figure 5.5.2. The approximate time course of changes in membrane currents that produce the SA node action potential are shown in Figure 5.5.3.

AUTONOMIC NERVES ALTER THE HEART RATE BY AFFECTING THE PACEMAKER POTENTIAL

SYMPATHETIC STIMULATION ACCELERATES THE HEART BY INCREASING THE SLOPE OF THE PACEMAKER POTENTIAL

As described in Chapter 4.9, sympathetic nerves originating in spinal segments T1–T5 travel to the sympathetic chain where they ascend to the superior cervical

QUANTITATIVE HUMAN PHYSIOLOGY

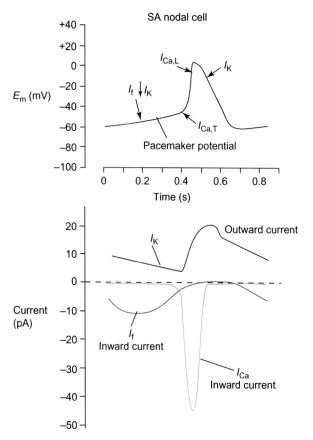

FIGURE 5.5.3 SA nodal currents that produce the SA node action potential. Recall that a positive current is an outward current. Thus, a negative current in the figure denotes an inward flow of positive charge. The pacemaker potential is caused by a decreased outward K^+ current as I_K falls and by an increased inward current of Na^+ as the "funny" current, I_F, increases. As the membrane reaches about −55 mV, T-type and then L-type Ca^{2+} channels open, carrying an inward Ca^{2+} current that contributes to the last third of the pacemaker potential and which produces the rapid upstroke of the SA nodal action potential. Then the delayed rectifying K channel opens and increases outward K^+ current at the same time that I_f and I_{ca} are decreasing. Thus, I_K returns the membrane potential toward its original −60 mV. (Source: Redrawn from J.R. Levick, An Introduction to Cardiovascular Physiology, Arnold, London, 2000.)

ganglia where they synapse with ganglion cells. These ganglion cells send postganglionic fibers along the great vessels to the heart where they release **norepinephrine**. Norepinephrine released from these terminals, and circulating epinephrine, bind to $β_1$ **receptors** on SA nodal cells, on AV nodal cells, and on contractile cardiomyocytes. As described in Chapter 2.8, the $β_1$ receptors are linked to a G_s mechanism. Occupancy of the $β_1$ receptor causes exchange of GTP for GDP on the α subunit of the heterotrimeric G_s protein. The α subunit dissociates from the β and γ subunits and binds to adenylyl cyclase, activating it. This increases the production of **3′,5′-cyclic AMP** from ATP. The increased concentration of cAMP activates **protein kinase A**, which phosphorylates a number of protein targets within nodal or contractile cells. This increases L-type I_{Ca} and the hyperpolarizing I_K. Cyclic AMP directly activates I_f. This increases the slope of the pacemaker potential, thereby reducing the time required for the SA nodal cells to reach threshold. Increasing the hyperpolarizing I_K current also shortens the time for repolarization. All of these effects increase the frequency of action potentials produced from the SA node (see Figure 5.5.4) and forms the basis for the **chronotropic effects** of the sympathetic stimulation, referring to its ability to accelerate the heart rate. Sympathetic stimulation has a positive chronotropic effect because it accelerates the heart rate. Sympathetic stimulation has other chronotropic effects: it decreases AV nodal delay and shortens the action potential on the contractile cells. If sympathetic stimulation did not shorten the contractile cells' action potential, the long duration of the action potential could limit the heart rate.

PARASYMPATHETIC STIMULATION SLOWS THE HEART RATE BY DECREASING THE SLOPE OF THE PACEMAKER POTENTIAL

Parasympathetic nerves to the heart originate from the vagal motor nuclei in the brainstem and travel over the **vagus nerve** (cranial nerve X) to the heart. The right vagus nerve supplies the SA node and slows its pacemaker; the left vagus innervates the AV node and slows its conduction of the cardiac impulse to the bundle of His. The vagus fibers are preganglionic; they make synapses with parasympathetic neurons within the heart. These ganglionic fibers send short postganglionic fibers to the nodal and muscle tissue. These terminals release **acetylcholine**, the main neurotransmitter for postganglionic parasympathetic nerves. Acetylcholine can bind to a variety of receptors. In the heart, its main receptor is the **M2 receptor**. Binding to the M2 receptor has two main effects:

1. It activates a G_i mechanism that inhibits adenylyl cyclase, in opposition to the $β_1$ mechanism that norepinephrine activates.
2. Second, the βγ subunits released by acetylcholine binding to G_i activates an acetylcholine-sensitive K^+ channel that carries a current called I_{K-ACh}.

These two effects slow the heart rate by:

- hyperpolarizing the SA nodal cells, thereby increasing the time required to depolarize to threshold;
- decreasing the slope of the pacemaker potential by decreasing cAMP and decreasing phosphorylation of target proteins.

The overall effects of sympathetic and parasympathetic stimulation on the SA node action potential are shown in Figure 5.5.4. The subcellular basis for these effects is shown in Figure 5.5.5.

THE IONIC BASIS OF THE VENTRICULAR CARDIOMYOCYTE ACTION POTENTIAL

THE RESTING MEMBRANE POTENTIAL, PHASE 4, IS SET BY E_K AND LARGE G_K

The resting membrane potential of ventricular contractile cells is determined by the conductance-weighted average of the equilibrium potentials for all of the diffusable

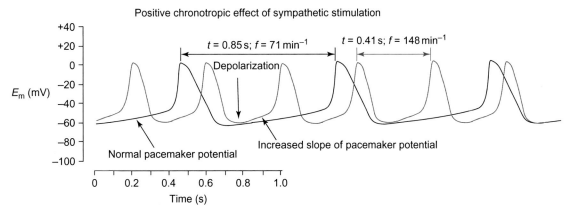

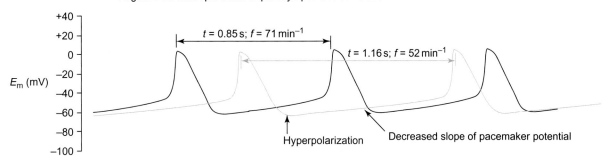

FIGURE 5.5.4 Effects of sympathetic (top) and parasympathetic (bottom) stimulation on the SA nodal action potential. Since the SA node is the pacemaker of the heart, the heart rate depends directly on the rate of action potential production by the SA node. Sympathetic stimulation causes phosphorylation of membrane channels that alters their conductance. The result is a depolarization, an increase in the slope of the pacemaker potential, a shorter time to reach threshold resulting in an increase in the rate of firing of action potentials, and consequently a higher heart rate. Parasympathetic stimulation decreases phosphorylation and independently stimulates a K$^+$ channel that carries I_{K-ACh}. These effects cause a hyperpolarization, a decrease in the slope of the pacemaker potential, a decrease in the rate of firing of action potentials, and a decreased heart rate.

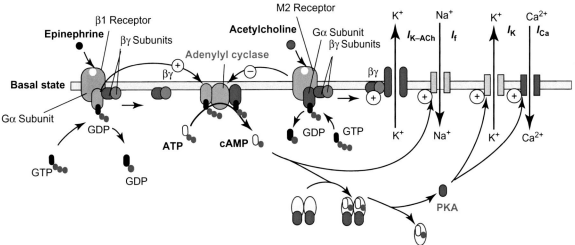

FIGURE 5.5.5 Molecular mechanism of action of sympathetic and parasympathetic stimulation of SA nodal cells. See text for details.

ions, as described in Eqn [5.5.1]. The equilibrium potential for Na$^+$ is given by the Nernst Equation, whose derivation was described in Chapter 3.1:

$$[5.5.3] \quad E_{Na} = \frac{RT}{z\Im} \ln \frac{[Na^+]_o}{[Na^+]_i}$$

where E_{Na} is the equilibrium potential for Na$^+$, R is the gas constant, which in electrical units is 8.314 J mol^{-1} K^{-1}, T is the temperature in K, z is the valence on the Na$^+$ ion, which is 1.0, and $\Im$ is the faraday = 9.649×10^4 C mol^{-1}. The logarithm is the natural log. Expressions such as that in Eqn [5.5.3] can

QUANTITATIVE HUMAN PHYSIOLOGY

TABLE 5.5.1 Concentrations of Selected Ions in the Extracellular Fluid and Intracellular Fluid and Their Calculated Equilibrium Potentials for Cardiomyocytes

Ion, X	$[X_o]$ (M)	$[X_i]$ (M)	E_x (V)
K^+	4×10^{-3}	140×10^{-3}	-0.094
Na^+	145×10^{-3}	10×10^{-3}	$+0.071$
Ca^{2+}	1.2×10^{-3}	1×10^{-7}	$+0.125$
Cl^-	114×10^{-3}	30×10^{-3}	-0.036

be used for each ion, and the resulting equilibrium potentials are listed in Table 5.5.1 for the ionic conditions in the cardiomyocytes.

The resting membrane potential in the ventricular cells is -0.080 to -0.090 V. This is because at rest in these cells, $g_K \gg g_{Na}, g_{Ca},$ and g_{Cl}, so by Eqn [5.5.1] the resting membrane potential is closer to E_K than to $E_{Na}, E_{Ca},$ or E_{Cl}. Two specific channels account for the large resting g_K. These channels carry the **delayed rectifying K^+ current** (I_K) and the **inwardly rectifying K^+ current** (I_{K1}). The resting E_m is never as negative as E_K, however, because there is residual conductance to Na^+ that carries the **background current**, I_b. This is the stable situation that pertains to the resting cell and accounts for phase 4 of the action potential for ventricular cells shown in Figure 5.5.1.

THE UPSTROKE OF THE ACTION POTENTIAL, PHASE 0, IS DUE TO INWARD Na^+ CURRENT

Ventricular muscle cells contain Na^+ channels like those in nerve tissue (Chapter 3.2). These voltage-gated channels open upon depolarization, causing an inward Na^+ current (because E_m at rest is -85 mV and E_{Na} is $+71$ mV, the driving force for Na^+ is: $E_m - E_{Na} = -156$ mV; the negative sign means the Na^+ current enters the cell). This I_{Na} causes a further depolarization, opening still more Na^+ channels and increasing I_{Na} still further. This causes an explosive depolarization of the cell, reaching about $+50$ mV in less than 2 ms. Like the regenerative Na^+ channels in nerve, the Na^+ channels close spontaneously. The model for the Na^+ channel is the Hodgkin–Huxley model, in which the Na^+ channel acts as if it had two gates, called m and h. The m gate is the activation gate, and the h gate is the inactivation gate. At rest, the h gate is open and the m gate is closed. Depolarization activates the channel by opening the m gate. Since both m and h gates are open, the channel conducts Na^+ into the cell, depolarizing it further. Depolarization also causes the delayed inactivation of the channel by closing the h gate. Thus, I_{Na} is transient. This Na^+ channel is blocked by **tetrodotoxin**. The channel returns to its resting configuration first by closing the m gate and then by opening the h gate. In this state it is closed but activatable. According to this scheme, there is a time when the channel has a closed inactivation gate. During this time, it cannot be opened and so the membrane is **refractory** to excitation: it cannot be excited to form an action potential when the Na^+ channel is in the inactivated state. We can identify an **absolute refractory period** (ARP) during which no stimulus, no matter how large, can induce the cell to fire an action potential. During the **relative refractory period**, the cell can fire an action potential but a larger than normal stimulus is required to excite the cell. The states of the channel and refractory periods are shown in Figure 5.5.6.

PHASE 1 REPOLARIZATION IS CAUSED BY K^+ OUTWARD CURRENT (I_{to}) WHILE THE Na^+ CHANNEL INACTIVATES

Repolarization occurs when the outward current exceeds the inward current. At the membrane potential at the end of phase 0, the driving force for Na^+ is inward, but not so strong because E_m is closer to E_{na}, and the driving force for K^+ entry is large because $E_m - E_K$ is large. The falling Na^+ conductance alone would tend to repolarize the cell. However, the conductance of the channel carrying I_{K1} is **inwardly rectified**. Inwardly rectifier K^+ channels derive their name from the fact that they pass much larger inward currents at negative E_m than outward currents at positive E_m, even in symmetrical $[K^+]$ solutions. Thus, at positive E_m the outward current I_{K1} is suppressed by reducing the conductance of its channel. This prevents the immediate repolarization of the action potential and allows the rest of the action potential to develop.

Another channel, the transient outward K^+ channel, causes a current called I_{to} that contributes to the repolarization of phase 1. Depolarization rapidly but transiently activates this channel, which inactivates over about 100 ms, thereby contributing to net membrane current over phases 1 and 2 of the action potential.

Ca^{2+} INWARD CURRENT MAINTAINS THE PLATEAU OF PHASE 2

The L-Type Ca^{2+} Channels Conduct Inward Ca^{2+} Current During the Early Plateau

Up to now, the action potential that we have described is similar to that in nerve: the resting membrane potential is set by a large g_K compared to g_{Na} or g_{Ca}; the upstroke is due to rapidly but transiently increasing g_{Na}, followed by a repolarization due to the decreased g_{Na}

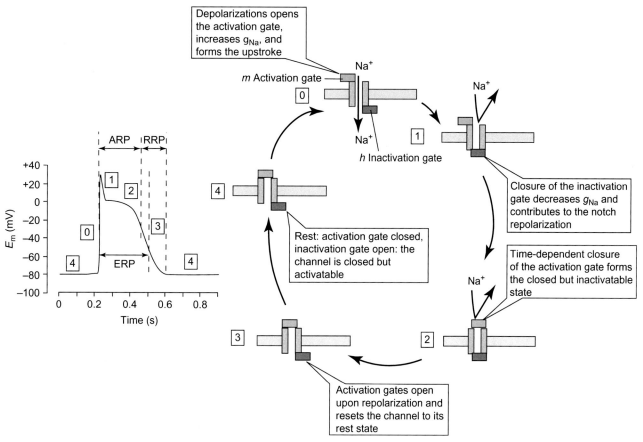

FIGURE 5.5.6 States of the fast Na$^+$ channel during the action potential. During phase 4, the channel is closed but activatable, meaning that the activation gate, m, is closed, whereas the inactivation gate, h, is open. Upon depolarization, the activation gate opens regeneratively, causing a rapid increase in g_{Na} that produces the upstroke of the action potential, phase 0. During phase 1, the Na$^+$ channels close their inactivation gates. This is followed by closure of the activation gates. During repolarization in phase 3 the inactivation gates open in a progressively larger fraction of the channels, leading to the resting state with activation gate closed but activatable. When the inactivation gate is closed, opening of the activation gate still does not produce an open channel. Thus, when the inactivation gate is closed, the Na$^+$ channel cannot be activated, and the membrane containing Na$^+$ channels in this state cannot elicit an action potential. This state of the cell membrane is called a refractory state. The absolute refractory period (ARP) corresponds to the period in which no action potential can be elicited, regardless of the depolarization. The relative refractory period (RRP) indicates that an action potential can be elicited but at higher thresholds. The effective refractory period (ERP) corresponds to the time during which an action potential will not have sufficient current to depolarize adjacent cells.

and brief increase in I_{to}, carried by K$^+$. In ventricular cells, however, repolarization is delayed. There is a plateau phase that keeps the membrane potential elevated for 200–400 ms and constitutes most of the **action potential duration** (APD). The inward current during the plateau phase is primarily due to L-type Ca^{2+} channels. These channels open upon depolarization but more slowly than the Na$^+$ channels that produce the upstroke of the action potential. The L-type Ca^{2+} channels inactivate slowly during the plateau.

$I_{Ca,L}$ Is Balanced in Part by the Delayed Rectifier K$^+$ Channel

The plateau phase is characterized by a nearly steady membrane potential between about 0 and −20 mV. This slowly changing membrane potential means that the inward $I_{Ca,L}$ must be balanced by some outward current. The outward current is a K$^+$ current carried by the delayed rectifier (I_K), so-named because they activate slowly upon depolarization. The amplitude and time course of $I_{Ca,L}$ and I_K set the time course of phases 2 and 3.

CLOSURE OF THE L-TYPE CHANNEL AND INCREASED g_K CAUSE REPOLARIZATION IN PHASE 3

During the plateau phase, $I_{Ca,L}$ decays and I_K increases, leading to an acceleration of the rate of repolarization. As the membrane repolarizes, I_{K1}, the inwardly rectifying K$^+$ current, increases. This further increases g_K and further accelerates repolarization. During this time, the Na$^+$ and Ca^{2+} channels reset themselves. These channels need repolarization to recover from the closed and inactivated state to the closed but activatable state.

THE Na$^+$−Ca^{2+} EXCHANGER PASSES A CURRENT

The Ca^{2+} carried by I_{Ca} that enters the cardiac myocytes with each contraction must be pumped out again during some other part of the cycle or else Ca^{2+}

would accumulate indefinitely. There are two mechanisms on the surface membrane of the cells that pump out Ca^{2+} ions: a plasma membrane Ca-ATPase (PMCA) and a Na^+–Ca^{2+} exchanger (NCX). The PMCA couples uphill Ca^{2+} transport to the chemical energy in ATP hydrolysis, and thus it is a primary active transport mechanism. The NCX was discussed in Chapter 2.6 as an example of secondary active transport, which derives its energy for Ca^{2+} extrusion from the energy of the Na^+ gradient. It transports three Na^+ ions for each Ca^{2+}, so each turnover is accompanied by a net movement of one positive charge. The NCX is reversible, depending on the electrochemical potentials for each ion on both sides of the membrane. During the plateau phase, NCX transports Ca^{2+} in and Na^+ out, and thereby it contributes an outward current. This is the NCX reverse mode. During the latter stages of the plateau, $I_{Ca,L}$ gradually inactivates and the NCX pumps one Ca^{2+} out and three Na^+ in, producing an inward current that delays repolarization.

Figure 5.5.7 reprises the ventricular action potential, the conductance changes, and the currents that produce the action potential.

EPINEPHRINE ENHANCES THE L-TYPE Ca^{2+} CHANNELS, WHICH ELEVATES THE ACTION POTENTIAL PLATEAU

β_1 Adrenergic stimulation of the heart activates adenylyl cyclase through a G_s-coupled system, as illustrated in Figure 5.5.5. Increased adenylyl cyclase activity increases cytoplasmic [cAMP], which in turn activates PKA that phosphorylates target proteins. One of these target proteins is the L-type Ca^{2+} channel. Its phosphorylation increases $I_{Ca,L}$ during the plateau phase, causing a dome-shaped action potential (see Figure 5.5.8). There are multiple effects of β_1 adrenergic agonists on the heart. We have already discussed the chronotropic effects that increase the heart rate through effects on the SA nodal cells. β_1 Adrenergic agonists also exert a **positive inotropic effect**, which refers to their ability to increase the strength of myocardial contraction. Part of this derives from the extra Ca^{2+} that enters during the plateau phase of the action potential. Like skeletal muscle, cardiac muscle is activated by a transient increase in cytoplasmic [Ca^{2+}]. In skeletal muscle, all of this cytoplasmic Ca^{2+} derives from sarcoplasmic reticulum (SR) stores. The skeletal muscle system is geared for total or near total activation of the actomyosin cross bridges by having large Ca^{2+} transients from an abundant SR. Fine control of the force of skeletal muscle is provided by neural control. In cardiac muscle, the strength of contraction is regulated in large part by the size of the Ca^{2+} transient, and it has two sources: the intracellular stores in the SR and influx into the cell across the cell membrane, the sarcolemma. When Ca^{2+} entry increases, it increases the Ca^{2+} transient and increases the number of actomyosin cross bridges, thereby increasing the force of cardiac muscle contraction. Thus, increasing $I_{Ca,L}$ by β_1 adrenergic stimulation increases the force of cardiac muscle contraction.

THE ACTION POTENTIAL IS CONDUCTED TO NEIGHBORING CELLS THROUGH GAP JUNCTIONS IN THE INTERCALATED DISKS

The space constant in cardiac tissue is some 20 times the length of an individual cell in the longitudinal direction and many times the cell diameter in the transverse direction. The cells are electrically coupled through **gap junctions** where they join at the intercalated disks. As in the nervous system, gap junctions in cardiac tissue form when **connexin** hexamers on one cell membrane join up with connexin hexamers on an adjacent membrane. Each hexamer forms a hemi-channel, and together they form a complete channel that allows diffusion of ions and small molecular weight signaling molecules between cells. There are at least six different isoforms of connexin, and four of these are expressed in heart. The gap junctions allow the action potential to spread passively from one cell to another without using neurotransmitters and without using large numbers of conducting fibers. A schematic diagram of a gap junction is shown in Figure 5.5.9.

SUMMARY

The cardiac action potential takes a different form in different cardiac cells, which include SA nodal cells, atrial muscle cells, AV nodal cells, Purkinje fibers, and ventricular muscle cells. We consider here the action potential of SA nodal cells and ventricular muscle cells. The SA node contains the most excitable cells in the heart, and so it sets the pace of the heart. The SA nodal cells have an unstable resting membrane potential that spontaneously depolarizes due to a pacemaker potential. This is caused by the "funny" Na^+ current and a decrease in the conductance of the inward rectifier K^+ channel. On reaching a threshold of about -55 mV, an action potential begins by the progressive opening of T-type and L-type Ca^{2+} channels. The spike returns to baseline because of an increase in the delayed rectifier K^+ current. The slope of the pacemaker potential and the resting membrane potential determine the time necessary to reach threshold: the sooner threshold is reached, the sooner an action potential is fired and the faster the heart rate. Sympathetic stimulation increases the slope of the pacemaker potential and depolarizes the resting membrane potential. Both of these help increase the heart rate. Sympathetic stimulation releases norepinephrine that acts on the SA node through β_1 receptors that are coupled to a G_s protein. This increases [cAMP] within the cell, which activates protein kinase A that subsequently phosphorylates a number of target proteins. Parasympathetic stimulation is coupled to M2 receptors that decrease [cAMP] and therefore opposes the positive chronotropic effect of sympathetic stimulation. Parasympathetic stimulation releases acetylcholine at terminals of the vagus nerve in the heart, which lowers the pacemaker potential and hyperpolarizes the cell. This reduces the frequency of action potentials originating at the SA node.

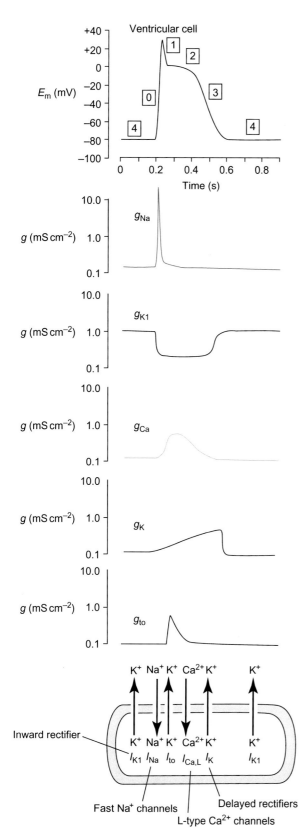

FIGURE 5.5.7 Conductance changes and currents responsible for the major features of the ventricular myocyte action potential. The action potential consists of five phases; phase 4 is the resting phase, produced largely by inward rectifying K^+ channels that keep g_K greater than g_{Na} or g_{Ca} at rest. These channels carry I_{K1}. Phase 0 is the upstroke of the action potential, caused by the opening of fast Na^+ channels that carry I_{Na}. These open transiently, and a notch in the action potential (phase 1) is formed when the fast Na^+ channels close and L-type Ca^{2+} channels open and I_{to} contributes to phase 1. The plateau phase (phase 2) is maintained by a combination of $I_{Ca,L}$, the delayed rectifier (I_K) and the current carried by the NCX I_{NaCa}. Repolarization in phase 3 is brought about by inactivation of $I_{Ca,L}$ and increases in I_K (the delayed rectifier current) and I_{K1} (the inward rectifier current). The time courses of the conductance changes for each of the main currents involved in the ventricular myocyte action potential are shown individually. The main currents that produce the action potential are shown in the bottom of the figure.

The action potential in ventricular cardiomyocytes begins with a rapid upstroke (phase 0) caused by the regenerative opening of fast Na^+ channels. This is followed by a partial repolarization (phase 1) caused by closing of the fast channels, opening of a transient outward K^+ channel, and reduced conductance of the inward rectifier K^+ channel. The membrane potential enters the plateau phase (phase 2) due to inward Ca^{2+} currents carried by

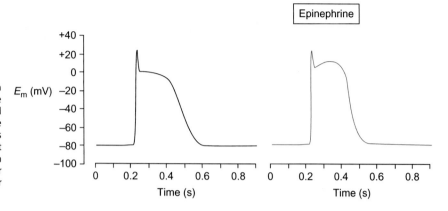

FIGURE 5.5.8 Effect of epinephrine, which stimulates β receptors, on the cardiomyocyte action potential. Epinephrine begins a signal cascade that results in the phosphorylation of the L-type Ca^{2+} channel, among other targets. Its effects include an increase in inward Ca^{2+} current during the plateau phase (phase 2) that results in an elevation of the plateau. Left, normal ventricular cardiomyocyte action potential; right, after exposure to epinephrine.

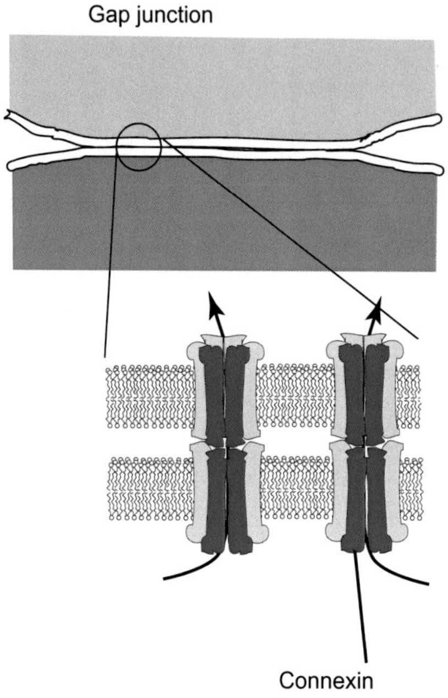

FIGURE 5.5.9 The gap junction. A small conductive, aqueous pathway is available between adjoining heart cells by the juxtapositioning of two hemi-channels consisting of connexin hexamers. Two hemi-channels from adjacent membranes can align, forming a direct aqueous channel between the two cells.

the L-type Ca^{2+} channel. The repolarization phase, phase 3, occurs because these Ca^{2+} channels gradually inactivate and delayed rectifier K^+ channels contribute to an outward K^+ current. During repolarization, the fast Na^+ channels revert to their resting state of closed but activatable.

The action potential spreads passively throughout regions of ventricular cardiomyocytes through the gap junction connections between cells.

REVIEW QUESTIONS

1. Draw and label the phases of the action potentials for SA nodal cells, atrial cells, and ventricular cells.
2. What is responsible for the pacemaker potential? How does sympathetic stimulation alter it? How does parasympathetic stimulation alter it? What does sympathetic stimulation do the resting membrane potential in SA nodal cells? What does parasympathetic stimulation do? How do these stimulations alter heart rate?
3. What currents are responsible for the upstroke of the action potential in SA nodal cells and in ventricular muscle cells?
4. What is phase 1 of the ventricular myocyte action potential, and what causes it?
5. What causes the plateau of the action potential?
6. What currents are responsible for repolarization of the ventricular myocyte (phase 3)?
7. During what part of the action potential do the Na channels inactivation gates close and activation gates open?

The Electrocardiogram 5.6

Learning Objectives

- Define what is meant by ECG and describe its utility in medicine and physiology
- Define Einthoven's Triangle and write the convention for leads I, II, and III
- Explain why Kirchhoff's law results in lead I + lead III = lead II
- Identify the main features of the ECG including the P, QRS complex, and T wave
- Be able to determine the heart's electric dipole from the value of leads I and III
- Identify the PR interval, PR segment, ST segment, and QT interval from an ECG trace
- Describe the events in the heart that produce the P wave
- Describe the events in the heart that produce the QRS complex
- Describe the events in the heart that produce the T wave
- Describe the general idea of augmented limb leads
- Define mean electrical axis, left axis deviation, and right axis deviation
- Describe why Einthoven's triangle is an idealized abstraction that is useful but not strictly valid

THE ECG IS THE PROJECTION OF CARDIAC ELECTRICAL ACTIVITY ONTO THE BODY SURFACE

The electrocardiogram (ECG) is well known as an integral part of the examination of patients, but its physical basis and the source of its signals are less well known. **The ECG is a record of the electrical activity of the heart that is projected onto the surface of the body where it is measured by surface electrodes.** Because this record represents the heart's electrical events, it aids diagnosis of electrical abnormalities. Its usefulness resides entirely in its clinical application. The cardiologist uses the ECG to understand the electrical character of the heart including (a) excitation of the pacemakers; (b) spread of activation from one region of the heart to another; (c) the pathways by which the wave of activation spreads; and (d) the basis of the action potential itself. The ECG will not illuminate the mechanical effectiveness of the heart: the ECG can be perfectly normal in mechanically ineffective hearts, whereas abnormal ECGs can be recorded in a mechanically sound heart. Clinical abnormalities in the ECG are legion, and many years of experience have produced systematic analyses of these abnormalities. Here we focus on how the normal ECG derives from the normal heart beat.

THE HEART MUSCLE FIBERS ACT AS ELECTRIC DIPOLES

As excitation of the cardiac muscle is conveyed through the myocardium, the electrical state of the muscle fibers differs from place to place. At some time during the conduction of the action potential, some parts of the muscle are depolarized and other parts are not. This produces the equivalent of an electrical dipole, as shown in Figure 5.6.1. The electric dipole is discussed in Chapter 1.4 (See Appendix 1.4.A1). The **dipole moment** is given as

$$[5.6.1] \qquad \vec{p} = q_+ \vec{d}$$

where d is a vector pointing from q_- to q_+, and q_+ is the magnitude of charge separated by distance, d. A potential surrounds the dipole, which is just the sum of the potentials due to each of the charges. It is given (Appendix 1.4.A1) approximately as

$$[5.6.2] \qquad V = \frac{p \, \cos \theta}{4\pi\varepsilon_0 \, r^2}$$

The electric dipole of the heart produces an electrical potential throughout the thorax and projected onto the skin. Under ideal conditions, this electric potential can be calculated or graphically constructed by the projection laws of dipole vectors, but these simple calculations or projections are valid only if the body is a spherical, homogeneous conductor, with the dipole lying at the center of the sphere. The body is not such an ideal spherical conductor.

EINTHOVEN IDEALIZED THE THORAX AS A TRIANGLE

Although not the first to record electrical activity from a heart, Einthoven (1860–1927) devised a string galvanometer to record the small potentials accurately, and he provided a simplified analysis of the ECG that is still used today. Einthoven was awarded the 1924 Nobel Prize in Medicine and Physiology for his contributions to the measurement and interpretation of the ECG.

Einthoven imagined that the electrical state of the heart at any time could be represented by a single vector, representing the electric dipole moment, located in the center of the thorax. Electrodes attached to the right

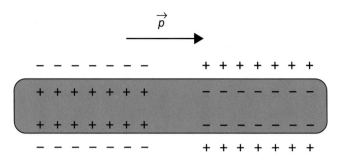

FIGURE 5.6.1 The equivalent dipole of the heart muscle during partial activation. The muscle at rest is polarized, negative inside, as shown to the right in the muscle. Activation of the muscle begins by its depolarization, with positive potential inside. This situation produces a dipole moment as shown. As activation proceeds, the dipole moment changes and so does the voltage projected to the surface of the skin.

arm (RA), left arm (LA), and left leg (LL) ideally measure voltages at the apices of an equilateral triangle that has the heart at its center (see Figure 5.6.2) The voltage differences between the different electrodes can be measured in three combinations:

$$\text{Left arm} - \text{right arm} = \text{LEAD I}$$
$$\text{Left leg} - \text{right arm} = \text{LEAD II}$$
$$\text{Left leg} - \text{left arm} = \text{LEAD III}$$

These leads are bipolar leads, consisting of the potential difference between two sites. The voltages recorded at these bipolar leads are not independent. Because of Kirchoff's voltage law, which states that the voltage drop around any closed circuit is zero, we have

[5.6.3] $\qquad$ I + III = II

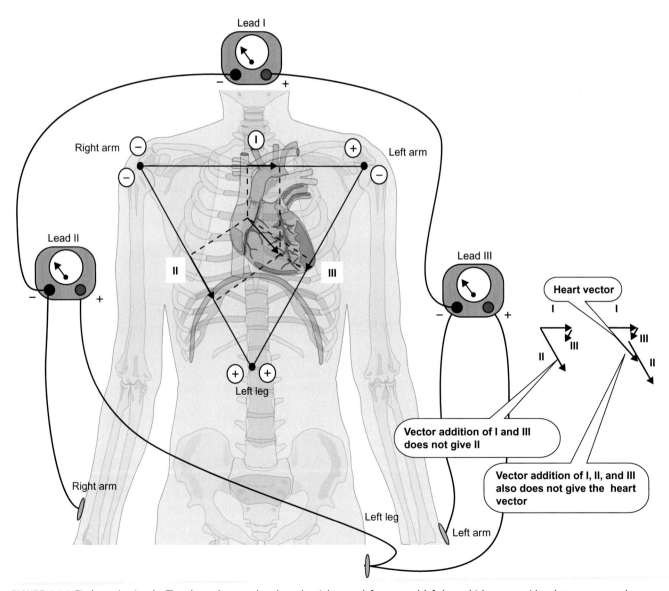

FIGURE 5.6.2 Einthoven's triangle. The electrodes are placed on the right arm, left arm, and left leg, which are considered to measure voltages as indicated on the torso. The three electrodes ideally form an equilateral triangle with the heart's electric dipole at its center. The voltage measured along any axis is the projection of the dipole moment onto that axis, as shown. The dipole vectors do not add like normal vectors, head to tail, because the triangular coordinate system is not orthogonal.

so that the voltage in the third bipolar lead can always be calculated from the other two. This peculiar form of Kirchoff's voltage law is due to the fact that lead II is reported in the counterclockwise direction in Figure 5.6.2 (left leg—right arm) rather than in the clockwise direction that is used for leads I and III. This convention causes the voltages to add as in Eqn [5.6.3]. Einthoven selected this convention because he wanted positive voltages corresponding to the cardiac impulse in all leads in normal individuals.

The voltages recorded in each of the bipolar leads at any instant is equal to the projection of the heart's electric dipole at that instant onto the bipolar lead. This projection is shown in Figure 5.6.2.

THE HEART'S ELECTRIC DIPOLE MOMENT VARIES WITH TIME—AND SO DOES ITS RECORDING ON LEADS I, II, AND III

At rest, the heart fibers are uniformly polarized and so there is no net dipole moment and therefore no voltage recorded on leads I, II, or III. The heart muscle depolarizes in a coordinated matter, with the atria depolarizing first, and the ventricles depolarizing later and eventually repolarizing. This sequence of the cardiac impulse within the muscle fibers causes the electric dipole of the entire heart to change magnitude and direction with time. The resulting projections of the electric dipole moment onto the limb leads produces a record of voltage that varies with time—the **scalar electrocardiogram**, or **ECG**. The ECG is typically produced at a chart speed of 25 mm s^{-1} on a scale of 1 mV cm^{-1}. A typical example of an ECG is shown in Figure 5.6.3.

THE VALUES OF LEADS I AND III CAN BE USED TO CALCULATE THE ELECTRIC DIPOLE MOMENT OF THE HEART

The description above indicates that the values recorded on leads I, II, and III at any time are the result of the electric dipole moment of the heart at that time. However, what we measure are the voltages on these leads while the electric dipole moment remains unknown. The direction and magnitude of the electric dipole moment can be approximately reconstructed from the measured voltages. Reconstruction of the heart's electrical axis cannot be done by ordinary vector addition, however. Figure 5.6.2 shows that addition of vectors I and III does not produce vector II, nor does addition of I, II, and III produce the electrical axis of the heart. The reason for this is that the vectors I, II, and III are projections of the heart's dipole moment onto axes that are not **orthogonal**.

In orthogonal coordinate systems, a vector parallel to one coordinate has no components in the direction of any other coordinate. In the triangular coordinate system, a vector along one axis *does* have components along both of the other axes. What this means is that the electric dipole can be reconstructed only by reversing the projection onto the axes, as described in the legend to Figure 5.6.4.

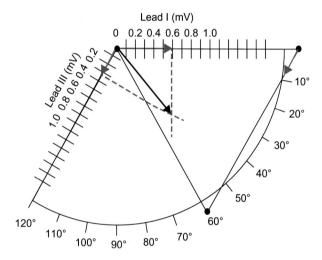

FIGURE 5.6.4 Reconstruction of the heart's electrical dipole from leads I and III. A vector whose magnitude is the voltage of lead I is drawn with its starting point at the origin and its end on the lead I axis. A second vector equal to the voltage of lead III is drawn the same way but oriented on an axis parallel to the lead III axis and starting at the origin. The electric dipole extends from the origin to the intersection of the two lines drawn perpendicular to the axes at the ends of the lead I and lead III vectors. The electric dipole of the heart obtained this way pertains to the time leads I and III were measured. This dipole has a magnitude (its length in the same units as for the two lead axes) and a direction. The direction is usually given in degrees using the convention shown.

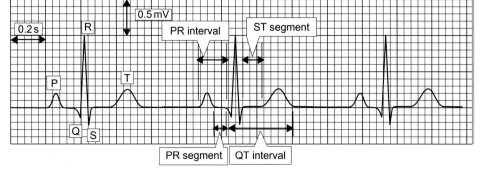

FIGURE 5.6.3 Typical appearance of the ECG on lead II ($V_{LL} - V_{RA}$). The ECG consists of several named electrical events, beginning with the P wave. The P wave is the first electrical event of the heart cycle and corresponds to the depolarization of the atria. The QRS complex corresponds to depolarization of the ventricles. Buried within it is the electrical signature of the repolarization of the atria. The T wave corresponds to repolarization of the ventricles. It is upright because of the sequence of repolarization of different parts of the ventricles, as described in the text.

ATRIAL DEPOLARIZATION CAUSES THE P WAVE

The size and magnitude of the cardiac electric dipole changes continuously as excitation spreads through the heart. The stepwise state of the heart and its corresponding electric dipole throughout the cardiac cycle is shown in Figures 5.6.5 and 5.6.6. The heart beat begins with SA node depolarization to threshold and initiation of the cardiac impulse in the SA node. This impulse travels outward through the right atrium and, via Bachmann's bundle, to the left atrium. This produces an electric dipole with positive contributions along the lead II axis, and so it is recorded as a positive voltage on this lead. When the atria are completely depolarized, there is no remaining dipole except over the small amount of tissue that makes up the AV node. Because of its small size, this tissue does not produce much voltage at the leads.

SEQUENTIAL DEPOLARIZATION OF THE VENTRICLES PRODUCES THE QRS COMPLEX

As described in Chapter 5.4, conduction of the cardiac impulse to the ventricles is delayed by the AV node. This allows the atria to contract and fully load the ventricles prior to their contraction and ejection of blood into the pulmonary and systemic circulations. This delay shows up as a lag between the P wave and the QRS complex, the part of the ECG that derives from ventricular depolarization. The nodal delay is typically about 0.1 s. Depolarization of the ventricles then occurs in a sequential fashion beginning with the left ventricular septum. At the very beginning of the QRS complex, Q is usually slightly negative. This is due to the fact that the cardiac dipole at the beginning of activation points slightly away from the lead II axis, and therefore its projection is in the opposite direction. Part of the Q wave in the QRS complex also results from atrial repolarization (see Figure 5.6.6).

THE SUBEPICARDIUM REPOLARIZES BEFORE THE SUBENDOCARDIUM, CAUSING AN UPRIGHT T WAVE

The **epicardium** is the layer of cells on the outer surface of the heart, facing the pericardial fluid. The **endocardium** is the layer of cells lining the inner surface of the heart, facing the blood. Cardiomyocytes adjacent to these layers are called the **subepicardium** and **subendocardium**, respectively. Cells in the middle layer of the ventricle form the **midmyocardium**. These cells are somewhat heterogeneous in that their action potentials have different durations. Cells in the subendocardium depolarize before cells in the subepicardium. However, these cells repolarize roughly in

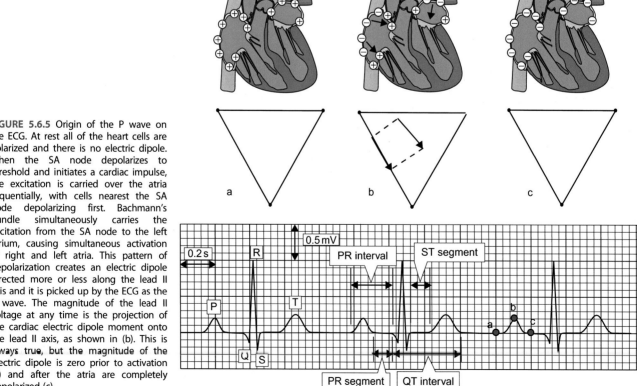

FIGURE 5.6.5 Origin of the P wave on the ECG. At rest all of the heart cells are polarized and there is no electric dipole. When the SA node depolarizes to threshold and initiates a cardiac impulse, the excitation is carried over the atria sequentially, with cells nearest the SA node depolarizing first. Bachmann's bundle simultaneously carries the excitation from the SA node to the left atrium, causing simultaneous activation of right and left atria. This pattern of depolarization creates an electric dipole directed more or less along the lead II axis and it is picked up by the ECG as the P wave. The magnitude of the lead II voltage at any time is the projection of the cardiac electric dipole moment onto the lead II axis, as shown in (b). This is always true, but the magnitude of the electric dipole is zero prior to activation (a) and after the atria are completely depolarized (c).

the inverse order of their depolarization: the last cells to depolarize (the subepicardium) are the first to repolarize. This is due to the length of their action potentials. Cells in the subepicardium have shorter action potentials than cells in the subendocardium. The consequence is that the cardiac dipole points along the lead II axis during repolarization, and the T wave, the electrical event corresponding to ventricular repolarization, is positive (see Figure 5.6.6).

THE CARDIAC DIPOLE TRACES A CLOSED CURVE DURING EACH HEART BEAT

Figure 5.6.6 d, e, and f show that the cardiac electric dipole moment changes its direction and magnitude through the QRS complex. A more detailed analysis shows that the cardiac dipole changes continuously so that the end of the vector traces a closed curve every time the heart beats. The entire cycle lasts the duration of the QRS complex, which is approximately 80 ms. An approximation of the dipole and its location at various times after the beginning of the QRS complex is shown in Figure 5.6.8.

THE LARGEST DEPOLARIZATION DEFINES THE *MEAN ELECTRICAL AXIS*

The cycle shown in Figure 5.6.8 is elongated toward the vector produced upon depolarization of the thick left ventricular wall. This elongation defines an **electrical axis** of the heart, which normally should relate to the heart's anatomical position within the thorax. The electrical axis is most easily defined as the cardiac electric dipole at the

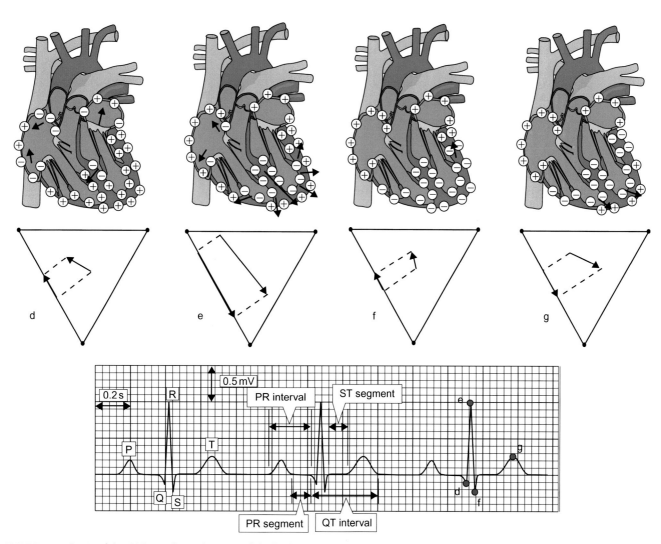

FIGURE 5.6.6 Origin of the QRS complex and T wave of the lead II scalar ECG. The spread of activation during ventricular activation creates a cardiac dipole that varies in magnitude and direction because of the sequence of activation produced by the left and right bundle branches and the Purkinje fibers. The Q wave is negative because the depolarization of the ventricles begins with the septum and the resulting dipole points away from the lead II axis (d). The largest vector arises from depolarization of the inner (subendocardium) apex of the heart, while the outer (subepicardium) apex remains polarized. This forms the R wave (e). The S wave is also negative and arises from the last depolarization of the free left ventricular wall near the left atrium. At this time the dipole vector points away from the lead II axis (f). Repolarization of the atria is buried in the QRS complex. Repolarization of the T wave is upright because the subepicardium has a shorter action potential and repolarizes before the subendocardium, creating a dipole pointing along the lead II axis (g).

542 QUANTITATIVE HUMAN PHYSIOLOGY

EXAMPLE 5.6.1 Calculate the Heart Vector

At the top of the R wave, lead I reads 0.35 mV and lead III reads 0.75 mV. Calculate the heart vector at this time.

Refer to Figure 5.6.7. What we desire is the length (magnitude) of vector V_H and the angle it makes with respect to the lead I (horizontal) axis.

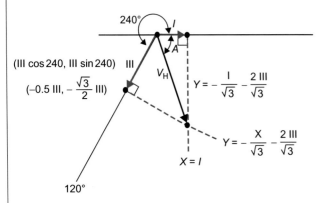

FIGURE 5.6.7 Calculation of the heart vector from lead I and lead III. The vector is the line from the origin to the point of intersection of two perpendiculars drawn from the end point of lead I and lead III. It is given as the angle with respect to lead I and its magnitude, in mV.

This can be calculated in various ways. Using analytical geometry, we can write the equation for the two lines perpendicular to the lead I and lead III axes at the end of the lead I and lead III vectors. The line perpendicular to the lead I is just $X = I$, where I is the magnitude of lead I. The line perpendicular to lead III can be calculated by knowing that its slope is $-1/\tan 240 = -1/\sqrt{3}$ and it passes through the point (X,Y) at the end of the vector, which is given as (III cos 240, III sin 240) = $(-0.5\ III, -\sqrt{3}/2\ III)$, where III is the magnitude of lead III. This equation is given in the point-slope form as

$(Y + \sqrt{3}/2\ III)/(X + 0.5\ III) = -1/\sqrt{3}$ which can be rearranged to

$Y = -X/\sqrt{3} - 0.5\ III/\sqrt{3} - \sqrt{3}\ III/2$ or $[-X\ 2\sqrt{3} - III\ 4\sqrt{3}]/6$

The intersection of the two lines is given as $(I, [-I\ 2\sqrt{3} - III\ 4\sqrt{3}]/6)$

$$\text{Angle } A = \tan^{-1}(Y/I) = \tan^{-1}(-1/\sqrt{3} - 2/\sqrt{3}\ III/I)$$

For $I = 0.35$ and III $= 0.75$, this is $A = \tan^{-1}(-0.5774 - 2.4744) = \tan^{-1}(-3.05176) = -71.86°$. Convention takes this as positive **71.86°**.

The magnitude of the heart vector can then be calculated as $V_H = I/\cos A = 0.35/\cos(-71.86) = $ **1.12 mV**.

---

peak of the curve traced out in Figure 5.6.8. Because of its tilt, however, the projection of the vector tracing this curve peaks at different times on the lead I and lead III axes. The electrical axis can be determined from lead I and lead III ECGs using the voltage of the R wave and determining the resulting cardiac dipole as described in Figure 5.6.4.

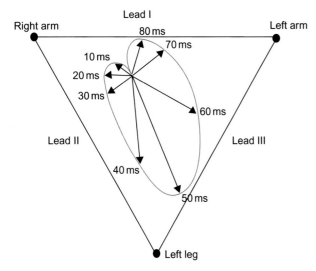

FIGURE 5.6.8 Approximate changes in the direction and magnitude of the cardiac dipole during ventricular excitation. The projection of the dipole on the different leads produces the ECG and accounts for different shapes of the QRS complex recorded in leads I, II, and III.

Typically the electrical axis is oriented about 60° below the horizontal, but it varies widely from −30° to 110°. The electrical axis depends in part on the anatomical orientation of the heart. A tall, thin person generally would have a more vertically oriented electrical axis compared to a short, stout person. As described in Chapter 5.4, the heart lies within a tough fibrous sac, the pericardium, which in turn is fused to the diaphragm. During inspiration, the diaphragm moves downward to expand the thoracic cavity. This movement pulls the heart into a more vertical orientation. Thus, the electrical axis normally becomes more vertical during inspiration and more horizontal during expiration.

The electrical axis also depends on the relative contributions of the right and left ventricles to the ECG. Hypertrophy of the left ventricle shifts the axis, causing **left axis deviation**. Hypertrophy of the right ventricle occurs in response to pulmonary hypertension and is associated with **right axis deviation**.

UNIPOLAR LEADS RECORD THE DIFFERENCE BETWEEN AN ELECTRODE AND A ZERO ELECTRODE

Every system for recording voltages consists of two electrodes whose potential difference is fed into the input of an amplifier. If the two electrodes are placed at two different places in an electric field, the system records $V_1 - V_2$. The two electrodes are called **bipolar electrodes**. If one of the electrodes is placed in a "zero" area of the field, the potential difference recorded between the zero, or "indifferent" electrode and the "different" electrode is called a **unipolar potential**. The zero area surrounding an electric dipole lies along the line perpendicular to the dipole moment, as shown in Figure 1.4.A1.2. Using unipolar electrodes for recording ECGs requires knowing where to place the zero electrode. In practice, we never know the area of zero potential, and its location changes as the heart's electric dipole changes during the cardiac cycle. Using greatly simplified assumptions, we can construct a zero electrode by combining the three Einthoven leads through resistances >5 kΩ, as shown in Figure 5.6.9. The connection produces a **central terminal (CT)**. Its definition in this manner is equivalent to Kirchhoff's voltage law. By adding a large resistance, the effect of small differences in the output resistances for each of the three leads is minimized.

AUGMENTED UNIPOLAR LIMB LEADS USE COMBINATION OF ONLY TWO ELECTRODES FOR THE INDIFFERENT ELECTRODE

The three standard limb leads have magnitudes defined by:

[5.6.4]
$$\begin{aligned} I &= V_{LA} - V_{RA} \\ II &= V_{LL} - V_{RA} \\ III &= V_{LL} - V_{LA} \end{aligned}$$

where V is the voltage at an electrode, LL signifies the left leg electrode, RA is the right arm, and LA is the left arm. The **augmented limb leads** use each of the three Einthoven electrode positions but combine each with the average of the opposite two. Their magnitude is given as:

[5.6.5]
$$\begin{aligned} aVR &= V_{RA} - \frac{[V_{LA} + V_{LL}]}{2} \\ aVL &= V_{LA} - \frac{[V_{RA} + V_{LL}]}{2} \\ aVF &= V_{LL} - \frac{[V_{RA} + V_{LA}]}{2} \end{aligned}$$

where aVR stands for the augmented voltage for the right arm. In an ideal field where the heart sits at the center of an equilateral triangle, the average of two electrodes is the voltage found midway between them. Thus, voltages measured using these augmented limb leads correspond to the projection of the cardiac dipole onto the lead axes shown in Figure 5.6.10, oriented at 30°, 90°, and 150°.

THE EINTHOVEN TRIANGLE IS ONLY APPROXIMATELY VALID

Einthoven's triangle involves several assumptions. These are as follows:

1. The heart's electrical activity can be represented as a single electric dipole.
2. The heart is small compared to the field so that the heart can be considered to be located as a point in its center.
3. The thorax is a homogeneous conductor.
4. The thorax is a sphere.

The individual heart cells are tiny and may be regarded as an electric dipole. It is not obvious that the superpositioning of all simultaneously generated fields produced by the many muscle fibers can be regarded as being produced by a single dipole. This assumption is

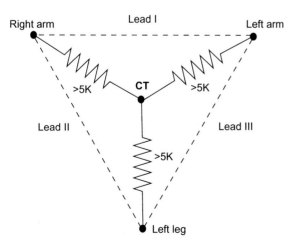

FIGURE 5.6.9 Construction of a zero terminal. Under simplified assumptions, the voltage drop around a loop is zero. Under this assumption, connection of all three leads to a CT should produce a zero reference point. The 5 kΩ resistances are added to balance small output resistances at each electrode.

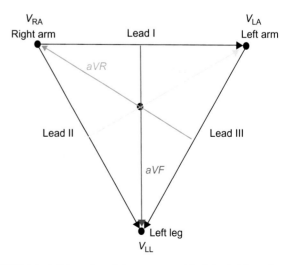

FIGURE 5.6.10 Orientation of the augmented limb leads. The voltage is measured between one of the Einthoven electrodes (right arm for aVR, left arm for aVL, and left leg for aVF) and the average of the two opposite leads. This average approximates the voltage at a position midway between the two electrodes. Thus, the voltages on the augmented limb leads are equivalent to the projection of the cardiac dipole onto the three lead axes as shown.

closely related to the second that the heart is small compared to the field so that it can be considered to be a point located in its center.

There is no question that the leads are not truly equidistant and that the thorax is not a homogeneous conductor. Each individual has somewhat different distances and different distributions of conductive media. A strict analysis of the projection would need to account for the exact shape and conductivity within the thorax. If the distribution of conductivity were known in detail, it is theoretically possible to predict the surface voltage for every electrical dipole of the heart. However, such a distribution is not known for any individual.

Despite the fact that Einthoven's triangle is strictly invalid, a large base of clinically relevant information has been assembled using its assumptions and diagnosis is possible within that frame of experience. In a sense, then, correcting for the errors in the concept amounts to a redefinition of the normal and doesn't produce clinical benefits worth the effort. A number of other electrode placement systems have been tried and evaluated.

THE CARDIAC CYCLE, REVISITED

Figure 5.4.7 shows the mechanical events that occur during the cardiac cycle, but it does not include the electrical events. Figure 5.6.11 shows these mechanical events along with the electrical events, as recorded by the ECG. Note in this figure that the electrical events precede the mechanical events. The mechanical events are caused by contraction of the heart muscle, and the electrical activation of the muscle triggers this contraction. The electrical event is not the same as contraction in the same sense that the electrical activation of a skeletal muscle is not the same as its contraction. The events that connect electrical excitation to contraction comprise excitation–contraction coupling. This topic will be covered in the next chapter.

Clinical Applications: Arrhythmias

Irregular rhythms of the heart vary from the benign and normal to the pathological but harmless palpitation to serious and life-threatening disorders. They can be diagnosed through the use of the ECG and are broadly classified as follows.

- **Sinus arrhythmia:** the SA node is the pacemaker and increased firing frequency produces **sinus tachycardia**, whereas decreased firing frequency causes **sinus bradycardia**. Normal heart rate varies with respiration, increasing during inspiration, and decreasing during expiration.
- **AV conduction blocks:** the excitation initiated by the SA is passed on to the AV node, which then relays it via the bundle of His and then the left and right bundle branches. A variety of ailments can interfere with the conduction here. **First degree heart block** shows a lengthening of the PR interval due to slowing of conduction from the AV node. A PR interval greater than 0.2 s is abnormal. **Second degree heart block** occurs when P waves intermittently fail to pass from the atria to the ventricles. Often the ratio between the number of P waves and QRS complexes is the ratio of two small integers, such as 2:1, 3:1, or 3:2. **Third degree heart block** occurs when no P waves are conducted through the AV node. In this case the ventricles initiate their own rhythm, typically at much slower rates. Thus, in first degree block the PR interval is extended; in second degree block conduction sometimes fails; in third degree block it fails all the time.
- **Premature depolarization:** abnormal myocytes occasionally can spontaneously reach threshold and begin an action potential before the normal excitation arrives from the SA node. Such an event can trigger a premature beat called an **ectopic beat** or **extrasystole**. The region that originates the premature beat is called an **ectopic focus**. An ectopic focus in the ventricles typically causes a broad QRS complex and a poorly coordinated contraction that fails to eject blood because the excitation is not properly distributed through the His–Purkinje conducting fibers. When the normal excitation resulting from the SA node appears after an extrasystole, the ventricles are refractory and no contraction results. This results in a long delay between the extrasystole and the next heart beat. The delay is called the **compensatory pause**.
- **Reentry or circus arrhythmia:** abnormal conduction pathways can develop that cause excitation to spread in a circuit. Cardiomyocytes that have just contracted and emerge from their refractory state are stimulated once again by the return of the excitation. This process is called re-entry. Re-entry in the atria or AV node can lead to tachycardia. This often begins and ends abruptly and is called **paroxysmal supraventricular tachycardia**, or **PSVT**, meaning that it is episodic (paroxysmal), originates from tissues above the ventricles (supraventricular), and causes a rapid heart rate (tachycardia). The QRS complexes are normal because excitation in the ventricles follows the normal pathways.
- **Fibrillation:** Fibrillation is an uncoordinated, repetitive excitation of the heart muscle that causes a writhing motion of the muscle but no effective ejection of blood. It probably results from multiple reentry circuits. **Atrial fibrillation** is not immediately life-threatening because atrial contraction is not essential to ventricular filling. However, regions of stagnant blood can form in the atria, predisposing to thrombus formation. The clot can dislodge and travel on to clog up part of the pulmonary or systemic vascular beds. **Ventricular fibrillation** is usually fatal unless the rhythm reverts to normal, which rarely occurs unaided. It causes loss of consciousness within seconds due to loss of cerebral perfusion and death ensues within minutes unless resuscitation efforts maintain perfusion.

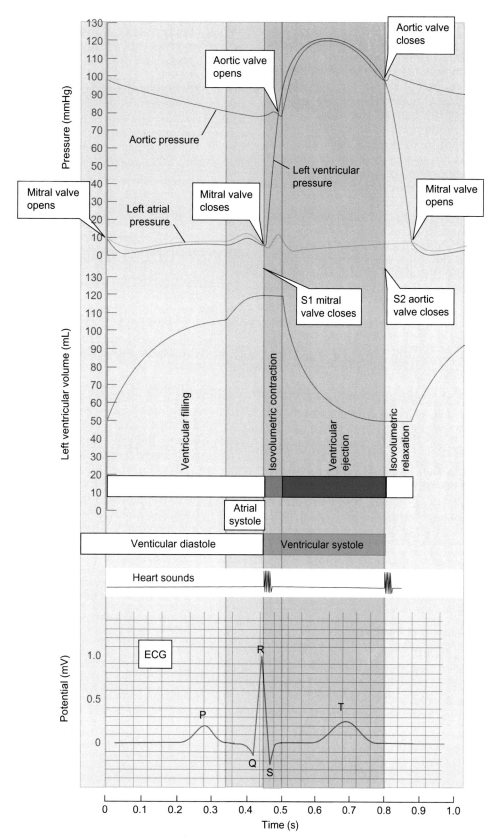

FIGURE 5.6.11 Mechanical and electrical events in the cardiac cycle.

SUMMARY

During diastole, all of the heart cells are polarized. Cells in the SA node spontaneously depolarize first and spread their depolarization through the atria. During the spread of excitation, part of the atria is depolarized and part is not, creating a cardiac electric dipole. This dipole can be detected by surface electrodes and its magnitude is equal to the projection of the heart's dipole moment, a vector, onto the axis connecting the bipolar electrodes. Einthoven invented a recording system that uses three electrodes: one on the right arm, one on the left arm, and a third on the left leg. The three leads are *defined* as: lead $I = V_{LA} - V_{RA}$; lead $III = V_{LL} - V_{LA}$; lead $II = V_{LL} - V_{RA}$. The analysis is simplified by assuming that the three bipolar leads from these three electrodes form an equilateral triangle. At any time the cardiac dipole projects differently onto the three leads. Because the leads are not orthogonal, the voltages in the three leads are not independent. The sequence and the mass of tissue being depolarized cause the cardiac electric dipole to change magnitude and direction throughout activation. This causes the voltages on the three leads to vary with time. The recording of this voltage is the scalar electrocardiogram, or ECG.

The normal ECG has defined parts. Atrial depolarization produces the first part, the P wave, which precedes atrial contraction or systole. The AV node conducts the excitation through the annulus fibrosus but delays it so that the atria have enough time to fully fill the ventricles. This delay appears as a short, flat region of the ECG called the PR segment. The depolarization of the ventricles produces the QRS complex, which lasts about 80 ms. The Q wave may be negative, depending on the recording lead, because the cardiac dipole moment at that time points up and toward the right arm. The R wave has the largest amplitude and corresponds to ventricular depolarization. During the S wave the main part of the ventricles has already depolarized and only the base of the heart undergoes depolarization. During the ST segment the ventricles are depolarized fully and the cardiac dipole disappears. It reappears during ventricular repolarization to form the T wave. The T wave is upright because the last cells to depolarize, in the subepicardium, are the first to repolarize. Thus the cardiac dipole during repolarization has the same direction as that during depolarization. Atrial repolarization occurs during the QRS wave.

The angle of the largest dipole is called the electrical axis of the heart.

REVIEW QUESTIONS

1. In general terms, define the ECG. What is an electric dipole?
2. What are the three primary limb leads? How are they connected? What is the relationship betwen leads I and III and II?
3. Why do the three main leads not add as vectors? How can you reconstruct the heart vector from I and III?
4. What is the P wave? What electrical event in the heart does it correspond to? What is the PR interval? What electrical event does it correspond to?
5. What is the QRS complex? What electrical event in the heart does it correspond to? Where is atrial repolarization in the ECG? What is the ST segment? What is the T wave? Why is it upright?
6. What is the mean electrical axis of the heart? How does it change standing up vs lying down?
7. How would you create a zero terminal?

The Cellular Basis of Cardiac Contractility   5.7

Learning Objectives

- Describe the general ultrastructural features of cardiomyocytes including intercalated disks, myofibrils, mitochondria, sarcoplasmic reticulum, sarcolemma, nuclei, and T-tubules
- Compare the structure of cardiomyocytes to skeletal muscle fibers
- Describe the components of the thick filaments
- Describe the components of the thin filaments
- Identify A band, I band, H zone, Z line, M line, and sarcomere
- Trace the events in excitation–contraction coupling from the action potential on the SL to activation of the contractile elements
- Describe calcium-induced calcium release
- List the cellular mechanisms responsible for cytosolic calcium homeostasis
- Identify the mechanisms by which cardiac strength of contraction can be varied
- Describe what is meant by the positive staircase and describe its mechanism
- Describe the basis of the inotropic effect of sympathetic stimulation
- Describe how cardiac contractile force is modulated by stretch

CARDIAC MUSCLE SHARES MANY STRUCTURAL FEATURES WITH SKELETAL MUSCLE

The main structural features of cardiac muscle reprise those of skeletal muscle, but there are notable differences (see Figure 5.7.1). Cardiomyocytes are typically 10–20 μm across and about 50–100 μm long, whereas skeletal muscle fibers are larger in diameter (20–100 μm) and many centimeters long. The cardiomyocytes typically have one or two nuclei, whereas skeletal muscle fibers have many hundreds of nuclei. Muscle fibers themselves do not branch, whereas cardiomyocytes often branch. Because of their different structures, the surface membrane of skeletal muscle fibers generally fuses with tendons, whereas cardiomyocytes form junctions with neighboring cells. These junctions are the **intercalated disks**. Because the heart must contract rhythmically throughout all of life, it is a highly aerobic organ. Accordingly, mitochondria occupy 30–35% of the cellular volume, a greater fraction than any skeletal muscle.

INTERCALATED DISKS ELECTRICALLY COUPLE CARDIOMYOCYTES

The intercalated disks appear dense in histological specimens or under the electron microscope. They contain **gap junctions**, **adherens junction**, and **desmosomes**. Gap junctions (see Chapter 4.2) consist of fields of connexin hexamers that form one-half of a channel linking adjacent cells. The full channel forms when one hemi-channel in the membrane of one cell joins up with a second hemi-channel in the adjacent cell. The two hemi-channels form an aqueous pathway between the two cells. These gap junctions allow passage of current so that adjacent cells are **electrically coupled**, allowing nearly simultaneous electrical activation of all of the cardiomyocytes. The desmosomes rivet cells together like miniature spot welds. They consist of plaques of cadherin molecules, a transmembrane glycoprotein that spans the gap between cells and anchors the desmosome to desmin filaments in the cytoplasm. Thus, the desmosome joins intermediate filaments in one cell to those in its neighbor. In the adherens junction, the extracellular part of cadherin molecules binds those from the adjacent cell. The cadherin molecules link to actin filaments in the cytoplasm through a variety of anchoring proteins.

THE STRENGTH OF CARDIAC MUSCLE CONTRACTION IS NOT REGULATED BY RECRUITMENT OR BY SUMMATION

As in skeletal muscle, electrical activation of cardiomyocytes begins with an action potential that is propagated along the surface of the cell. As discussed in Chapter 5.5, the cardiac action potential differs markedly from the skeletal muscle action potential both in its overall character and in its duration. The cardiac action potential lasts nearly as long as the cardiac muscle contraction, whereas the skeletal muscle action potential is very brief, on the order of a few milliseconds. Because of this, cardiac muscle cannot summate. The force of skeletal muscle force is varied by recruitment of motor units and summation of force by repetitive stimulation. In cardiac muscle, all of the cardiomyocytes are activated for each heart beat because the cells are electrically coupled: there is 100% recruitment all of the time. Because the action potential lasts so long, it is not possible to stimulate the cardiomyocytes for a second time during the heart beat, so

© 2012 Elsevier Inc. All rights reserved.
DOI: http://dx.doi.org/10.1016/B978-0-12-800883-6.00051-3

QUANTITATIVE HUMAN PHYSIOLOGY

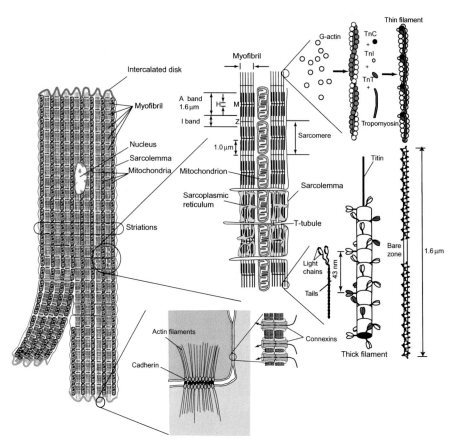

FIGURE 5.7.1 Gross and fine structure of a ventricular cardiomyocyte. The cell is surrounded by a sarcolemma (SL) which invaginates at regular intervals to form transverse or T-tubules. These bring the excitation of the SL into the interior of the cell. The cell typically has one or two nuclei. The cytoplasm is filled with myofibrils and mitochondria. The myofibrils consist of bundles of contractile filaments. The thin filaments are bundles of filamentous actin that are centered on the Z line. These filaments also contain tropomyosin, and the troponin complex of TnC, TnI, and TnT. The tropomyosin and troponin complex regulate contraction. The thick filaments are centered on the M line and consist of a titin scaffold on which myosin molecules assemble. The thick and thin filaments interdigitate, forming a series of cross-bands as shown. The A band corresponds to the thick filaments, but the H zone in its middle is the part where thin filaments do not overlap the thick filaments. The I band corresponds to the part of the thin filaments that do not overlap the thick filaments. Contraction is controlled by cytoplasmic [Ca^{2+}]. The SR is a membranous network that surrounds the myofibril. The SR releases Ca^{2+} during excitation to trigger contraction. The SR ends contraction by transporting activator Ca^{2+} back into the SR lumen.

cardiac muscle does not respond to repetitive stimulation. So what regulates cardiac muscle contractility? We will first discuss cardiac muscle structure and excitation–contraction coupling before we attempt an answer.

CARDIAC MYOFIBRILS HAVE THICK AND THIN FILAMENTS AND FORM THE CROSS-STRIATIONS

Both cardiac and skeletal muscles are **striated**. The striations arise from the regular arrangement of the **myofilaments** which are kept in register all the way across the diameter of the cell. These myofilaments consist of repeating units called **sarcomeres**. Each sarcomere is 1.8–2.0 μm long. The bands in each sarcomere are named for their microscopic appearance. The "A" band resides in the middle of the sarcomere and derives its name from the German "anisotropische," referring to the anisotropic way the proteins in this band handle polarized light. The "I" bands lie adjacent to the A bands on each side. Their name means "isotropische," referring to their isotropic handling of polarized light. The Z line is a thin partition in the middle of the I band. The sarcomere extends from Z line to Z line. The middle of the A band typically has an "M" line, meaning "mittel" or middle. The M line is in the middle of the "H" zone, from "helles," meaning "clear." The contractile proteins in cardiac muscle, like skeletal muscle, are arranged in longitudinal bundles called **myofibrils**. In cardiac muscle, these myofibrils are adjacent to rows of mitochondria. The myofibrils are typically 1 μm across and span the distance from one end of the cell to the other. The myofibrils in turn are made up of contractile proteins that are contained in smaller myofilaments.

The A band contains **thick filaments**, about 15 nm in diameter, composed mainly of **myosin** and **titin**. The I band contains **thin filaments**, 6 nm in diameter. The A band is always about 1.6 μm long because the thick filaments are this long. The giant protein titin extends from the M line to the Z line. It contains many binding sites for myosin, which assemble on the titin. Each thick filament contains about 300 myosin molecules. Myosin molecules are about 160 nm long and consist of six proteins. Two myosin heavy chain proteins of about 200,000 Da wind around each other to make the supercoiled tail, and each heavy chain contributes one of the two globular heads. These project away from the coiled tail and enable the myosin molecule to interact with **actin** in the thin filament. Two pairs of **myosin light chains** bind to the globular heads. The myosin assembles on the titin scaffold so that the tails face the M line and the heads are oriented toward the Z line. Thus, the thick filament is polarized. Because of this arrangement, the center of the A band contains a short region, about 0.3 μm long, that contains only tails and no heads. This is the bare zone.

The thin filaments consist of **filamentous** or **F-actin**, constructed by linking **globular**, or **G-actin**, into two intertwined chains that are about 1.0 μm long. G-actin has a molecular weight of about 42,000 Da. About seven of these make up each half-turn of the F-actin filament, which has a repeat distance of 38.5 nm. The filaments bind to **actinin** at the Z line and extend into the A band in both directions. The I band is only about 0.25 μm wide because the major part of the thin filament interdigitates between the thick filaments in the A band. **Tropomyosin** is a heterodimer made up of α and β subunits that coil around each other to form a rod-like structure that lays in the groove of the F-actin. Each tropomyosin extends one half-turn, 38.5 nm. A troponin complex consisting of three separate proteins resides on the end of each tropomyosin. **Troponin I** inhibits actin–myosin interaction; **troponin C** binds Ca^{2+} ions; and **Troponin T** binds to tropomyosin. The interaction of thick and thin filament is controlled by Ca^{2+} binding to TnC as it is in skeletal muscle, except the cardiac TnC is a different isoform from skeletal muscle TnC.

ACTIN-ACTIVATED MYOSIN ATPase ACTIVITY PRODUCES FORCE AND SHORTENING

The globular head of the myosin heavy chains is an actin-activated ATPase. During rest, the myosin heads cannot interact with the actin filaments because tropomyosin is in the way. When the muscle is activated, tropomyosin shifts out of the way and the globular heads engage the actin filaments, forming cross-bridges between the thick and thin filaments. These cross-bridges cycle between states with high and low affinity for the thin filaments. The cycling requires ATP hydrolysis and results in movement or force development. An abbreviated version of the overall ATPase cycle is shown in Figure 5.7.2.

CYTOPLASMIC [Ca^{2+}] CONTROLS ACTOMYOSIN CROSS-BRIDGE CYCLING

Cytoplasmic [Ca^{2+}] controls the inhibitory position of tropomyosin on the thin filament. During diastole, cytoplasmic [Ca^{2+}] is low and tropomyosin inhibits interaction between the thin and thick filaments. During excitation, the sarcoplasmic reticulum (SR) releases Ca^{2+} from its stores and the Ca^{2+} binds to TnC, causing a series of conformational changes that remove tropomyosin from its inhibitory position. Cross-bridge cycling then follows and the cardiomyocyte shortens and develops tension. The relation between force and [Ca^{2+}] is steep, as shown in Figure 5.7.3.

CALCIUM-INDUCED CALCIUM RELEASE COUPLES EXCITATION TO CONTRACTION IN CARDIAC MUSCLE

THE SR RELEASES STORED Ca^{2+} TO ACTIVATE CONTRACTION

In cardiac muscle, the **transverse** or **(T)-tubules** penetrate the muscle cell interior at the level of the Z line, so that cardiac muscle has only one T-tubule per sarcomere. The **sarcoplasmic reticulum** or **SR** membrane is a membranous network that surrounds the myofibrils and makes contact with the T-tubule at junctions called **dyads**. So-called "feet" structures join the junctional SR to the T-tubules. These consist of **ryanodine receptors (RyR)** which form a Ca^{2+} channel across the SR membrane. Cardiac muscle also has **peripheral junctional SR** that make contact directly with the sarcolemma (SL), and **extended junctional SR** (also called corbular SR) that have feet structures while making no direct contact with either SL or T-tubules. Three different ryanodine receptor isoforms exist: **RyR1**, **RyR2**, and **RyR3**. Skeletal muscles express mainly RyR1; heart and brain express RyR2; epithelial tissues, smooth muscle, and brain also express RyR3. At the T-tubule, the RyR closely approaches **dihydropyridine receptors (DHPRs)**. A schematic diagram of the approximate anatomic arrangement of the SR in cardiac muscle is shown in Figure 5.7.4.

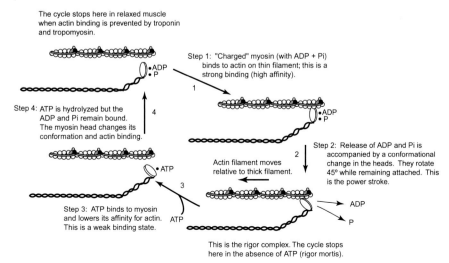

FIGURE 5.7.2 The actomyosin cross-bridge cycle. At rest, ATP binds to myosin (M) forming M−ATP. This lowers its affinity for the thin filament actin (A) (3). The ATP is spontaneously hydrolyzed, forming M−ADP−Pi, but the product ADP and Pi remains bound (4). This form of myosin has a high affinity for A. At rest, it cannot bind to A because tropomyosin (Tm) prevents the binding. During activation, Tm shifts out of the way and M−ADP−Pi binds to A to form A−M−ADP−Pi. This complex releases first Pi and then ADP, and the myosin head rotates with the attached A filament (A−M) (2). This is the power stroke that moves the thin filament some 5−10 nm past the thick filament. The A−M state is the rigor complex. It forms a stiff attachment of the actin filament to the myosin filament. Binding of ATP to M dissociates the complex so that the cycle can begin again (3). Thus, cycling through the reaction scheme hydrolyzes ATP and causes shortening or force development by the myofibrils.

THE DHPR SENSES T-TUBULE VOLTAGE AND FORMS A Ca^{2+} CHANNEL

The dihydropyridines are a class of drugs that include nitrendipine, nisoldipine, nifedipine, and BayK 8644. All of these affect the L-type Ca^{2+} channel current that is present on ventricular cardiomyocytes. BayK 8644 activates it; the other three inhibit it. The skeletal muscle DHPR consists of five subunits (α_1, α_2, β, γ, and δ), one of which (α_1) forms the Ca^{2+} channel and has the DHPR binding site. The cardiac DHPR appears to lack the γ subunit. This difference probably explains the different mechanisms of excitation–contraction coupling in skeletal and cardiac muscle. The DHPR senses depolarization of the T-tubule membrane and opens its Ca^{2+} channel, allowing Ca^{2+} to enter the cell.

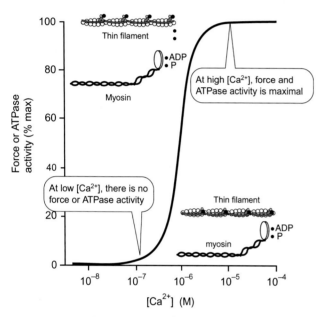

FIGURE 5.7.3 Control of actomyosin cycling and force development by cytoplasmic $[Ca^{2+}]$. No force is developed when cytoplasmic $[Ca^{2+}]$ is less than about 10^{-7} M, and force is maximal when $[Ca^{2+}]$ is about 10^{-6} to 10^{-5} M. The steep dependence of force on $[Ca^{2+}]$ is due to cooperativity. Thus, the cytoplasmic $[Ca^{2+}]$ is a switch that controls force development and actomyosin cross-bridge cycling.

Ca^{2+} ENTRY THROUGH THE DHPR TRIGGERS SR Ca^{2+} RELEASE

Release of Ca^{2+} from the SR in cardiac muscle involves a process called **calcium-induced calcium release**, or **CICR**. Depolarization of the SL membrane opens voltage-dependent Ca^{2+} channels in the DHPR that let Ca^{2+} into the cell near the RyR2 receptors on the SR. This Ca^{2+} is insufficient by itself to activate contraction, but it is amplified by causing a further and greater Ca^{2+} release from the SR. This process is shown schematically in Figure 5.7.5.

THE SIZE OF THE SL Ca^{2+} TRIGGER AND THE AMOUNT OF Ca^{2+} STORED IN THE SR DETERMINE THE SIZE OF THE Ca^{2+} TRANSIENT

The amount and rate of Ca^{2+} that enters the cardiomyocyte upon electrical activation controls the number of RyR2 channels that open. Thus, increasing SL Ca^{2+} entry generally increases Ca^{2+} release from the SR. Increasing the store of Ca^{2+} inside the SR also increases the amount of Ca^{2+} that is released.

REUPTAKE OF Ca^{2+} BY THE SR AND SL EXTRUSION OF Ca^{2+} CAUSE RELAXATION

Binding of Ca^{2+} to TnC initiates contraction, and removal of Ca^{2+} from TnC brings about relaxation. Removal of activator Ca^{2+} is accomplished in cardiac muscle by two membranes: the SL and the SR. A **Ca-ATPase pump** on the SR membrane couples movement of two Ca^{2+} atoms from the cytoplasm into the SR lumen to the hydrolysis of one ATP molecule. This active transport pump is an isoform of the **SERCA** pump family, for smooth endoplasmic reticulum Ca-ATPase. The isoform in cardiac SR is SERCA2a. This is regulated by phosphorylation of **phospholamban**, a 5-kDa protein in the SR membrane.

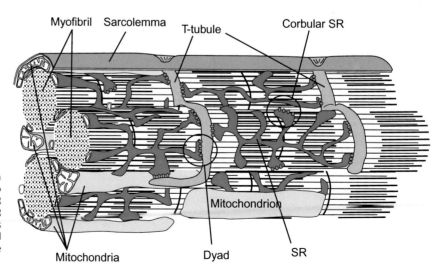

FIGURE 5.7.4 Arrangement of T-tubules and SR in ventricular cardiomyocytes. The SL invaginates to form T-tubules at the level of the Z lines. The SR forms dyad junctions with the T-tubule. Cardiac SR also forms junctions with the peripheral SL and has regions called corbular SR that contain junctional proteins without forming a junction with T-tubule or SL.

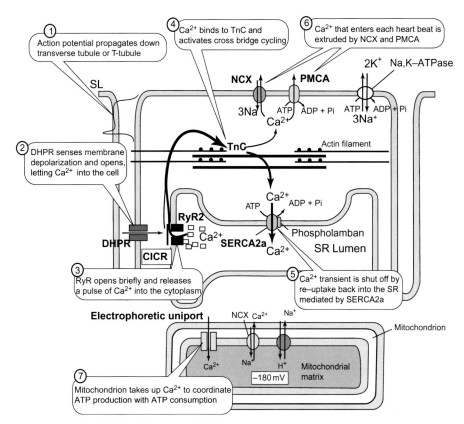

FIGURE 5.7.5 Overall processes involved in Ca^{2+} transport in the ventricular cardiomyocyte. Excitation begins with the cardiac action potential (1) that propagates over the surface of the cell and down into the T-tubules. Both the surface of the cell and T-tubules possess L-type Ca^{2+} channels that open in response to depolarization (2). This lets in a small amount of Ca^{2+} in the vicinity of the junction of the SR with the T-tubule or SL membrane (peripheral junctional SR and extended junctional SR). This trigger Ca^{2+} binds to ryanodine receptors (RyR2) on the SR membrane and briefly opens a Ca^{2+} channel in the RyR2 (3). This releases Ca^{2+} from the SR in a process called "calcium-induced calcium release" (CICR). The rapid release of Ca^{2+} transiently raises the cytoplasmic $[Ca^{2+}]$, causing Ca^{2+} to bind to troponin C on the thin filament (4). This binding moves tropomyosin out of its inhibitory position, thereby allowing the thin filament to interact with the thick filament, resulting in force development and actomyosin ATPase activity. The contraction is shut off by the removal of activator Ca^{2+} from the cytoplasmic compartment mainly by reuptake into the SR mediated by the SERCA2a pump in the SR (5), and secondarily by extrusion into the extracellular space by a NCX (6) that couples the influx of three Na^+ to the exit of one Ca^{2+} ion. A relatively minor component of Ca^{2+} exit is through PMCA. The mitochondria can also take up Ca^{2+} through the electrophoretic uniport, and release it through Na^+–Ca^{2+} exchange (7). The largest source of activator Ca^{2+} is the SR and its greatest sink during relaxation is also the SR.

Ca^{2+} enters each cardiomyocyte with every heart beat. At steady state, the Ca^{2+} that enters with each heart beat must also exit each beat, or else the heart will continue to accumulate Ca^{2+}. The entry of Ca^{2+} is mediated mainly by the L-type Ca^{2+} channels. Most of the efflux of Ca^{2+} from the cell is carried by the Na^+-Ca^{2+} exchanger (NCX) that brings in three Na^+ ions for every Ca^{2+} ion that is extruded from the cell. This is an example of secondary active transport. Because of the changes in $[Ca^{2+}]$ and membrane potential during the action potential, NCX can mediate either Ca^{2+} influx or efflux from the cell. A second efflux mechanism is through the **p**lasma **m**embrane **C**a-ATPase pump, or **PMCA**. This protein has a molecular weight of 138 kDa and is distinct from the SR Ca^{2+} pump. It pumps one Ca^{2+} out of the cell for each ATP hydrolyzed, which differs from the 2:1 stoichiometry of the SR Ca pump. PMCA is another example of primary active transport. It is regulated by phosphorylation by 3′,5′-cAMP-activated protein kinase (PKA), and calmodulin-dependent protein kinase (CAM-PK).

MITOCHONDRIA CAN TAKE UP Ca^{2+}

Mitochondria can take up Ca^{2+} through a mechanism called the **electrophoretic uniport** that uses the negative transmembrane potential difference across the inner mitochondrial membrane to drive Ca^{2+} movement down its electrochemical gradient. Mitochondria probably take up Ca^{2+} mainly to regulate ATP production rather than to participate in beat-to-beat regulation of cytoplasmic $[Ca^{2+}]$.

CALSEQUESTRIN AUGMENTS SR Ca^{2+} UPTAKE AND RELEASE

Pumping Ca^{2+} from the cytoplasm into the SR lumen increases the luminal $[Ca^{2+}]$. The enclosed SR volume is a tiny fraction of the entire cell volume, and so holding all of the activator Ca^{2+} would require high intraluminal $[Ca^{2+}]$ which would slow down pumping because the pump would have to work against a large gradient. **Calsequestrin**, a Ca^{2+}-binding protein inside the SR, keeps the luminal [Ca^{2+}] low without sacrificing Ca^{2+} content. Calsequestrin has many acidic amino acids that bind Ca^{2+} with low affinity ($K_D \approx 10^{-3}$ M). This low affinity allows the SR to store plenty of Ca^{2+} while still allowing a gradient that favors rapid release of Ca^{2+} when the RyR2 channels open. Calsequestrin also interacts with RyR2 to promote Ca^{2+} release.

WHAT REGULATES CARDIAC CONTRACTILITY?

Skeletal muscle controls force through recruitment of motor units and increased frequency of motor neuron activation. The heart can use neither of these mechanisms because the heart is fully recruited every heart beat and it cannot tetanize. So what regulates cardiac contractility? Muscle force depends on the number of actomyosin cross-bridges. Skeletal muscle SR normally releases enough Ca^{2+} to fully saturate all of the TnC in the fiber. Cardiac SR typically does *not* raise cytoplasmic [Ca^{2+}] high enough to saturate TnC. Thus more cross-bridges can be activated by increasing the size of the [Ca^{2+}] transient or by changing the affinity of the myofilaments for Ca^{2+}. Accordingly, cardiac contractility can be altered by one of two means:

1. Change the size of the [Ca^{2+}] transient
2. Change the sensitivity of the myofilaments to a given Ca^{2+} transient.

The heart uses both of these entirely different mechanisms to vary cardiac contractility.

THE FORCE GENERALLY INCREASES WITH THE FREQUENCY OF THE HEART BEAT: THE *FORCE−FREQUENCY RELATION*

In 1871 HG Bowditch observed that increasing the frequency of heart beats causes a gradual increase in the force of contraction. When the frequency is increased from low values, the first beat is usually weaker, but the force of subsequent beats gradually increases until a new steady state is reached. The record of force resembles a staircase, and so the phenomenon is called a **positive staircase** or **Treppe**. A typical record is shown in Figure 5.7.6.

The size of the Ca^{2+} transients explains the positive staircase or Bowditch effect. Increasing the frequency increases the total influx of Ca^{2+} across the SL because there are more action potentials. The result is that the SR becomes more loaded with Ca^{2+} and releases more Ca^{2+}. The SR gradually loads more because it is exposed to a higher average [Ca^{2+}] and therefore pumps in more Ca^{2+}. The increased [Ca^{2+}] transient recruits more actomyosin cross-bridges to make more force.

The first beat after increasing the frequency is usually weaker because there has been insufficient time for the SR RyR2 channels to recover from the previous excitation. By the next beat, the SR Ca^{2+} store has increased and so SR Ca^{2+} release increases.

Increasing force with increasing frequency has a limit: if the heart rate becomes too fast, force decreases again, for two reasons: (1) ventricular filling cannot keep pace with the higher contraction frequency, and so there is less volume in the ventricle and less pressure; (2) at high frequencies the action potential duration is shorter and the time for Ca^{2+} influx shortens. Thus, at very high frequencies the Ca^{2+} transient becomes smaller again.

SYMPATHETIC STIMULATION INCREASES FORCE BY INCREASING THE Ca^{2+} TRANSIENT

Noradrenaline released from sympathetic nerves in the heart binds mainly to β_1 receptors that are coupled to a heterotrimeric G_s protein. The α subunit of this G protein activates adenylyl cyclase, raising the cytosolic concentration of 3′,5′-cyclic AMP, which in turn activates protein kinase A (PKA). PKA phosphorylates a number of proteins within the cell. In the SA node, sympathetic stimulation increases the heart rate (**positive chronotropy**), which by itself also increases the force of contraction through the force−frequency relation. Ventricular myocytes also have β_1 receptors that activate PKA to phosphorylate a number of targets, including:

- L-type Ca^{2+} channel
- Phospholamban (PLB)
- RyR2, the SR Ca^{2+} release channel
- TnI.

Phosphorylation of the L-type Ca^{2+} channel increases I_{Ca}, leading to greater SL Ca^{2+} influx that contributes

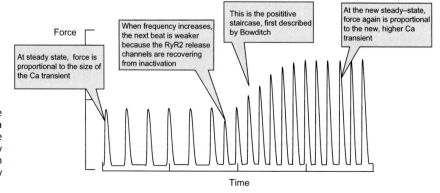

FIGURE 5.7.6 The positive staircase, or Treppe. The curves show pressure developed by the heart at a low frequency and at higher frequencies. When the frequency is increased, the first beat is generally weaker and subsequent beats gradually increase in force, forming a staircase response until a new steady state is reached.

directly to a larger Ca^{2+} transient and increased SR Ca^{2+} load. It also increases the size of the Ca^{2+} trigger to release Ca^{2+} from the SR by CICR. All of these effects lead to an increased Ca^{2+} transient.

PLB is a 5-kDa protein that forms pentamers in the SR membrane. The dephosphorylated PLB inhibits the SERCA2a Ca-ATPase by reducing its affinity for Ca^{2+}. Phosphorylation by PKA removes this inhibition, leading to increased Ca^{2+} pumping at submaximal $[Ca^{2+}]$. The result is increased loading of the SR and increased release of Ca^{2+} upon excitation. This also increases the rate of relaxation, a phenomenon called the **lusitropic effect**.

There are multiple phosphorylation sites on the RyR2 which are believed to increase the size of the Ca^{2+} transient.

TnI is a 32-kDa protein that forms part of the control mechanism for EC coupling. Phosphorylation of TnI by PKA increases the rate of Ca^{2+} dissociation from TnC. Since the affinity of TnC is the ratio of the on and off rate constants, increasing k_{off} decreases the affinity. This should reduce force development unless the Ca^{2+} transient increases enough to overcome the reduced sensitivity of the myofilaments. TnI phosphorylation aids the rate of removal of Ca^{2+}, thereby increasing the rate of relaxation and contributing to the lusitropic effect of sympathetic stimulation.

Figure 5.7.7 illustrates the effect of sympathetic stimulation on parts of the cardiomyocyte that cause its inotropic and lusitropic effects. The chronotropic effects are due to sympathetic stimulation on the SA node as detailed in Chapter 5.5.

PARASYMPATHETIC STIMULATION OPPOSES SYMPATHETIC EFFECTS (SEE FIGURE 5.7.7)

Parasympathetic stimulation of the heart releases acetylcholine onto M2 receptors which are coupled to a G_i protein that inhibits adenylyl cyclase. Since almost all of the sympathetic effect on the heart is mediated by β receptors linked to G_s, parasympathetic stimulation should oppose all of the effects of sympathetic stimulation.

The SA node and atria receive tonic vagal sympathetic stimulation, so **removal of vagal tone causes a rapid increase in heart rate**. Sympathetic stimulation of heart rate takes longer to produce than removing vagal tone. The negative chronotropic effects of parasympathetic stimulation were explained in Chapter 5.5. They are due to occupancy of M2 receptors and activation of I_{K-ACH} that causes a hyperpolarization and a delayed rise in the pacemaker potential.

M2 receptors in the ventricles are less dense than in the atria. During normal activity there is little sympathetic tone to the ventricles. Thus, cAMP levels during normal conditions are low and vagal stimulation or vagal withdrawal does not change this low basal [cAMP]. Vagal stimulation in the presence of sympathetic stimulation would reduce sympathetic inotropy, but otherwise stimulation of M2 receptors has no direct negative inotropic effect.

CARDIAC GLYCOSIDES INCREASE CARDIAC CONTRACTILITY BY INCREASING THE Ca^{2+} Transient

The cardiac glycosides are a class of drugs that inhibit the sarcolemmal Na,K-ATPase. Drugs in this class include **ouabain** and **digoxin**. Digoxin is a natural product of a plant called **foxglove**. It has been used for over two centuries to increase the force of contraction of failing hearts. The SL Na,K-ATPase pumps three Na^+ ions out of the heart for every two K^+ that it pumps in, at the expense of ATP hydrolysis. The cardiac glycosides inhibit this pump, which increases the cytoplasmic $[Na^+]$, which decreases the electrochemical gradient for Na^+ across the SL membrane. This, in turn, slows down the Ca^{2+} exit across the SL that relies mainly on the energy of the Na^+

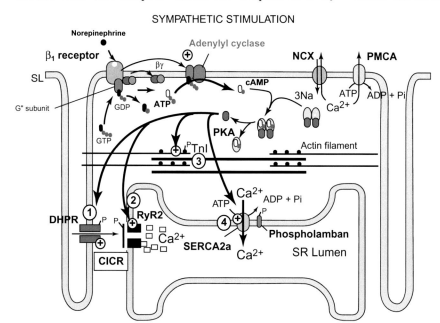

FIGURE 5.7.7 Autonomic effects on cardiac contractility. Sympathetic stimulation activates adenylyl cyclase that increases the [cAMP] in the cell and activates protein kinase A (PKA). PKA phosphorylates many target proteins, including (1) the L-type Ca^{2+} channel; (2) RyR2; (3) TnI; and (4) phospholamban. These effects all increase the Ca^{2+} transient either by increasing influx across the SL or by increasing the SR load or by directly increasing SR Ca^{2+} release. Phosphorylation of TnI reduces myofilament Ca^{2+} sensitivity by increasing the rate of Ca^{2+} dissociation. This increases the rate of relaxation of the cardiomyocyte. Parasympathetic stimulation opposes sympathetic effects but does not in itself produce negative inotropy.

gradient to move Ca^{2+} out of the cell. As a result, efflux of Ca^{2+} across the SL is reduced and over several beats the SR competes more effectively for Ca^{2+} that enters the cell. The SR loads to higher levels and therefore releases more Ca^{2+} during excitation. In this way, the $[Ca^{2+}]$ transients increase and force development increases. Drugs that increase the force of contraction of the heart have a **positive inotropic effect** and are said to exhibit positive inotropy. Thus, the cardiac glycosides are positively inotropic (see Figure 5.7.8).

CARDIAC CONTRACTILE FORCE IS POWERFULLY MODULATED BY STRETCH

The force developed by skeletal muscle has two components: a passive force caused by elastic elements in the muscle and an active force that is the additional force a muscle develops when it is stimulated. The active force in skeletal muscle has a biphasic dependence on muscle length that corresponds to the degree of overlap of the thick and thin filaments. The reduction in developed

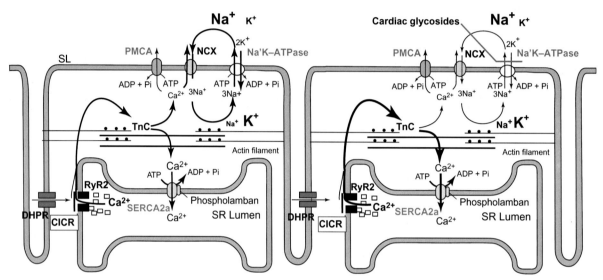

FIGURE 5.7.8 Basis of positive inotropy of cardiac glycosides. Left panel: Ca^{2+} that enters every heart beat is removed mainly by the NCX that uses the energy of the Na^+ gradient to extrude Ca^{2+}. This energy gradient is maintained by the Na^+,K^+-ATPase. Right panel: Cardiac glycosides inhibit this pump, which alters the Na^+ gradient so that less Ca^{2+} is removed by the NCX. Instead, the Ca^{2+} is taken up by the SR, which increases its load of Ca^{2+}, which increases the Ca^{2+} released upon excitation, creating a larger Ca^{2+} transient and more force.

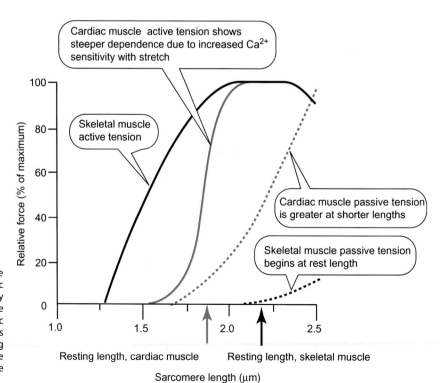

FIGURE 5.7.9 Comparison of part of the length–tension curve for skeletal muscle and cardiac muscle. As skeletal muscle is stretched from very short lengths, its tension increases because excessive overlap of myofilaments is removed. As cardiac muscle is stretched from very short lengths, its tension increases from removing interfering myofilament overlap and from increasing the number of active cross-bridges by increasing the sensitivity of the myofilaments to Ca^{2+}.

force at short muscle lengths when overlap is maximal is caused by interference of the filaments when sarcomere length is less than twice the length of the thin filament.

In skeletal muscle, the length–tension curve is not terribly important because the rest length of the muscle lies near the optimum of the length–tension curve and the fractional shortening or lengthening of skeletal muscle does not deviate too far from this optimum. The active force of cardiac muscle also depends on the overlap of thick and thin filaments, but it must avoid going over the top of the length–tension curve. **If the heart overfills, the active force would decrease and there would be no way to empty the heart!** This is accomplished by a steep rise in the passive force when stretched beyond 2.2–2.3 μm. Normally the heart muscle is never stretched to these long sarcomere lengths because the pressure during ventricular filling is not great enough to stretch it that far. Thus, **the heart normally operates only on the ascending limb of the length–tension curve** (see Figure 5.7.9).

The length–tension curve in skeletal muscle is typically observed under normal conditions of 100% recruitment and 100% saturation of TnC. Cardiac muscle does not normally saturate its TnC with its smaller Ca^{2+} transient. Stretching the muscle at short lengths increases its sensitivity to Ca^{2+} so that **stretching increases the fraction of active cross-bridges at any given $[Ca^{2+}]$**. In cardiac muscle, the length–tension curve at a given $[Ca^{2+}]$ has two components: the alteration of force due to overlap of filaments and the alteration of force due to changes in the number of active cross-bridges.

SUMMARY

Unlike skeletal muscle, all of the cardiomyocytes are electrically activated every heart beat, and so there is maximum recruitment at every beat. The action potential is nearly as long as the contraction, and therefore there is also no tetany or summation in cardiac muscle. Cardiac muscle regulates its strength of contraction two ways: (1) by regulating the size of the Ca^{2+} transient; (2) by regulating the sensitivity of the myofilaments.

The size of the Ca^{2+} transient can be experimentally altered in a variety of ways, including: (1) changing extracellular $[Ca^{2+}]$; (2) altering Na^+–Ca^{2+} exchange across the SL by varying the Na^+ electrochemical potential difference by changing extracellular $[Na^+]$ or by inhibiting the Na^+,K^+-ATPase with cardiac glycosides; (3) by increasing the heart rate, which increases the time-averaged Ca^{2+} entry; (4) by regulatory phosphorylation of a number of transport proteins dealing with the Ca^{2+} transient, including the L-type Ca^{2+} channel, phospholamban, SERCA2a, and RyR2. In physiological conditions, alterations of the heart rate and sympathetic/parasympathetic stimulation are the main mechanisms for the cellular regulation of cardiac contractility.

Cardiac muscle normally operates on the ascending limb of the length–tension curve because there is significant passive forces at short muscle length in the heart, and the forces that fill the ventricles are not large enough to stretch the heart beyond the ascending limb. Stretch of cardiac muscle at short sarcomere lengths markedly increases force by removing interfering overlap of the myofilaments and by increasing the Ca^{2+} sensitivity of the myofilaments. Thus, the degree of filling of the heart strongly influences its force of contraction.

REVIEW QUESTIONS

1. How are cardiomyocytes electrically coupled?
2. How does excitation–contraction coupling differ in cardiac muscle compared to skeletal muscle?
3. Force generation in cardiac muscle depends on the interaction of the thin filament with the thick filaments. What is the main constituent of the thin filament? Thick filament? What prevents their interaction at rest? What controls activation of the interaction of thick and thin filaments?
4. What is the main source of activator Ca^{2+}? How is it released? What determines how much Ca^{2+} is released?
5. How is the activator Ca^{2+} removed to bring about relaxation? Name three transport mechanisms for Ca^{2+} removal from the cardiomyocyte cytoplasm. Which is the main mechanism?
6. What would happen to the Ca^{2+} transient following sympathetic stimulation? How are the transport mechanisms affected?
7. What would happen to the Ca^{2+} transient following inhibition of the Na,K-ATPase? What mechanisms are affected?
8. What is the effect of increasing frequency on the force of heart muscle contraction? Why does this happen?
9. What does stretch do to the force of cardiac contractility?
10. Why does the heart operate only on the rising part of the length–tension curve?

5.8 The Cardiac Function Curve

Learning Objectives

- Define the stroke volume
- Calculate the cardiac output from the stroke volume and heart rate
- List the determinants of the stroke volume
- On a PV diagram, identify: ventricular filling, isovolumetric contraction, ejection, isovolumetric relaxation
- List three components of cardiac work
- Describe the Frank–Starling Law of the Heart
- Define central venous pressure, preload, and afterload
- Draw a normal, resting cardiac function curve
- Describe what happens to stroke volume when preload is increased at constant afterload
- Describe what happens to stroke volume when afterload is increased at constant preload
- Describe Fick's method for estimating cardiac output
- Describe the indicator dilution method for estimating cardiac output
- Describe the effect of positive inotropic agents on the cardiac function curve

CARDIAC OUTPUT IS THE FLOW PRODUCED BY THE HEART

As described in Chapter 5.4, each beat of the heart ejects a volume of blood, the **stroke volume**, equal to the difference between the end-diastolic volume of the left ventricle and the end-systolic volume:

$$[5.8.1] \quad SV = EDV - ESV$$

where SV is the **stroke volume**, EDV is the **end-diastolic volume**, and ESV is the **end-systolic volume**. The ejected volume then travels through the **systemic circulation** that perfuses all of the tissues except the lungs. The **cardiac output** is the average flow into the aorta, calculated as:

$$[5.8.2] \quad CO = SV \times HR$$

where CO is the **cardiac output**, SV is the stroke volume, and HR is the **heart rate**, in beats per minute. The units of CO are $L\,min^{-1}$. Typical values at rest are $4-6\,L\,min^{-1}$, but this can increase as much as 5-fold during strenuous exercise. According to Eqn [5.8.2], CO can be varied by changing SV or changing HR or both.

STROKE VOLUME IS DETERMINED BY *PRELOAD, AFTERLOAD*, AND *CONTRACTILITY*

The amount of blood ejected by the heart each beat depends on the force of the heart's contraction and by how much blood is in the heart. These can be altered by two basic mechanisms: the degree of stretch of the ventricle prior to systole, and the heart's inherent ability to produce tension, or its contractility. The degree of stretch of the ventricle depends on the pressure in the veins that passively fill the heart, or the **central venous pressure**. This is also called the **preload** because it is the pressure load prior to contraction of the heart. Stretching the heart up to a point shifts the contractility up the ascending limb of the length–tension curve, and more force is generated. **The heart's contractility is its ability to produce force at any given stretch**. Sympathetic stimulation increases the heart's contractility and cardiac disease reduces it.

The heart ejects blood into the high-pressure arteries. The pressure in the arteries is called the **afterload** because the heart muscle feels this pressure after contraction has begun and intraventricular pressure rises to equal or exceed aortic pressure. If arterial pressure is raised, it takes longer for the heart to generate pressure in its isovolumetric contraction phase, and more of the contractile energy is consumed in raising the pressure as opposed to ejecting the blood. This makes subjective sense: the heart is pumping blood against the arterial pressure, much like a weight that you might lift. Increasing the arterial pressure reduces the rate of pumping blood just like increasing the weight reduces your rate of lifting it.

THE INTEGRAL OF THE PRESSURE–VOLUME LOOP IS THE PV WORK

The Wiggers diagram (Figure 5.6.11) shows the time course of pressure development and the volume of the left ventricle with time during a single cardiac cycle. At each time, we can plot the pressure in the left ventricle against its volume. The result of such a plot is the **pressure–volume loop** shown in Figure 5.8.1.

During each cycle, the pressure–volume relationship traces a counter-clockwise loop. We begin with the opening of the mitral valve (point A in Figure 5.8.1)

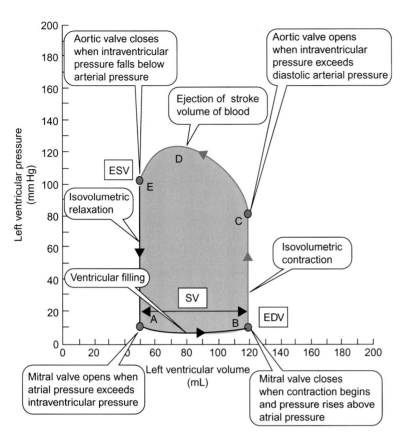

FIGURE 5.8.1 Pressure–volume loop of the cardiac cycle. Blood enters the left ventricle during diastole after the mitral valve opens (A) and continues to enter until it reaches the end-diastolic volume, about 120 mL. Isovolumetric contraction raises pressure to the level of the arterial pressure, which causes the aortic valve to open and the heart ejects about 70 mL of blood. When ventricular pressure is less than arterial pressure, the aortic valve closes (E) and the heart relaxes isovolumetrically. At this point the left ventricle contains the ESV, about 50 mL. The indicated area is the pressure–volume work done by the heart during each heart beat.

when left atrial pressure exceeds that of the left ventricle. Blood flows from the left atria into the left ventricle, and pressure in the left ventricle actually decreases because the heart is continuing its relaxation and expanding slightly faster than it is filling. At the end of diastole the heart is filled and pressure has again risen to about 7 mmHg. At this point, the **EDV** is about 120 mL. The heart begins to contract (point B in Figure 5.8.1) and enters the isovolumetric phase of contraction in which pressure develops until left ventricular pressure exceeds that in the aorta. At the point at which the aortic valve opens, arterial pressure is at its lowest point, the **diastolic pressure**. This is typically about 80 mmHg (point C in Figure 5.8.1). Once the aortic valve opens, the heart ejects blood into the aorta, causing its pressure to rise and causing passive stretch of the elastic aorta. The pressure momentarily rises and then falls again as the ejected blood travels down the arterial system and as the heart begins to relax (around point D in Figure 5.8.1). When intraventricular pressure falls below arterial pressure, the aortic valve snaps shut (point E in Figure 5.8.1) and the heart undergoes isovolumetric relaxation. The volume at this point is the **ESV**. The stroke volume is the volume of blood ejected in the heart cycle, and it corresponds to EDV − ESV = 120 mL − 50 mL = 70 mL. This is a typical value for the stroke volume.

The work increment, dW, is $F \times dx$, where dx is the distance increment; simultaneously multiplying and dividing by the area, we have

[5.8.3] $$dW = \frac{F}{A} A\, dx = P\, dV$$

Thus the net pressure–volume work done *by* the heart is the integral of the pressure–volume curve minus the work done *on* the heart *by* the blood. The work done on the heart by the blood is the integral of the pressure–volume curve from points A to B in Figure 5.8.1. The net work is the area shown in Figure 5.8.1.

TOTAL WORK OF THE HEART INCLUDES PRESSURE, KINETIC, AND GRAVITATIONAL TERMS

The heart does not just raise the blood from the low end-diastolic pressure to its high systolic pressure, but it also accelerates it. The blood in the aorta has a velocity and therefore it has a kinetic energy. In addition, in the standing position, the heart lifts the blood against the force of gravity. The total work is thus given as

[5.8.4] $$W = W_P + W_K + W_G = \frac{PV + 1/2\, \Delta V\, v^2 + \Delta g\, h\, V}{2\, \rho\, V\, v^2 + \rho\, g\, h\, V}$$

where W_P is the pressure term, W_K is the kinetic energy term, and W_G is the gravitational energy term. V is the volume of blood, ρ is the density, g is the acceleration due to gravity, and h is the height to which the blood is raised. Ordinarily the kinetic energy is relatively small, but it can be significant in situations in which the kinetic energy is converted to pressure or *vice versa*. Each of these energy terms can be divided by the volume to derive the energy per unit volume, and are referred to as "equivalent pressures," all of which have the units of force per unit area. The net equivalent pressure, P', is the sum of the equivalent pressures:

[5.8.5] $\qquad P' = P + 1/2\ \rho\ v^2 + \rho\ g\ h$

The total energy difference, per unit volume, between any two points in the cardiovascular system can be expressed in terms of equivalent pressure:

$$P'_1 - P'_2 = (P_1 - P_2) + 1/2\ \rho(v_1^2 - v_2^2) + \rho g(h_1 - h_2)$$
[5.8.6]

STRETCH OF THE HEART DETERMINES THE STROKE VOLUME: THE FRANK–STARLING LAW OF THE HEART

In 1895, the German physiologist Otto Frank ligated the aorta of a frog heart so that the contractions of the heart were purely isovolumetric, and he then measured the pressure generated when the heart was stretched by increasing the diastolic volume. He found that **the isovolumetric pressure increased with stretch**. These experiments were extended by the English physiologist, Ernst Starling, who used a more physiological preparation of the isolated heart and lung of a dog. The heart was filled with warm oxygenated blood from a reservoir whose height controlled the **central venous pressure**, the pressure at the entrance to the right atrium. The right heart fills during diastole until the pressure in the right atrium and right ventricle is equal to this central venous pressure. Thus, the degree of distension of the right ventricle at the end of diastole is determined largely by this central venous pressure. Similarly, the pressure in the pulmonary vein determines the degree of left ventricular distension at the end of diastole. These two pressures, the central venous pressure and the pulmonary vein pressure, are called **filling pressure**. When the central venous pressure is increased, the right atrial end-diastolic pressure also increases and the right ventricle stretches. According to what Otto Frank found, this increases the force of contraction of the right ventricle, and more blood is ejected. This ejected blood flows through the lungs into the pulmonary vein and, because there is more blood filling the veins, increases the pressure in the pulmonary veins. Thus, the increased output of the right ventricle increases the pressure in the pulmonary veins, which in turn increases the end-diastolic stretch of the left ventricle. This increases the force of contraction of the left ventricle, ejecting more blood. Thus, increases in central venous pressure increases the output of both ventricles. This is the Frank–Starling Law of the Heart: **increasing right atrial pressure increases the stroke volume of both ventricles** (Figure 5.8.2).

THE VENTRICULAR FUNCTION CURVE PLOTS CARDIAC FUNCTION AGAINST RIGHT ATRIAL PRESSURE

Any curve that plots a measure of the energy of cardiac contraction (stroke volume, cardiac output) against some measure of cardiac fiber length is a ventricular function curve. Here we plot cardiac output against the right atrial pressure, as shown in Figure 5.8.3. This plot represents the **intrinsic regulation** of the heart to its inputs that occurs independently of nervous regulation of the heart. This ventricular function curve can also be regulated by the autonomic nervous system, called **extrinsic regulation**.

INCREASING PRELOAD INCREASES THE STROKE VOLUME, INCREASING AFTERLOAD DECREASES IT

The **afterload** for the heart is the arterial pressure into which the heart ejects its stroke volume. Because the heart is really two pumps in series, there are two afterload pressures. These are the pulmonary arterial pressure in the pulmonary circulation and the arterial pressure in the systemic circulation. As mentioned earlier, increasing afterload reduces the stroke volume

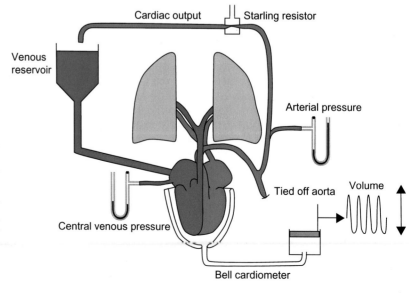

FIGURE 5.8.2 The Starling experiment. Starling perfused an isolated heart–lung preparation from dogs. The pressure of the blood returning to the right atrium was adjusted by raising or lowering the height of the venous reservoir. Both central venous pressure and arterial pressure were measured by mercury manometers. Arterial pressure were held constant by using a "Starling resistor" that mimicked the resistance of the systemic circulation, the total peripheral resistance. Heart volume was measured by using a bell cardiometer, an inverted glass bell that is attached to the atrioventricular groove by a rubber diaphragm. Ventricular volumes were recorded on a rotating drum.

because it takes longer for the heart to develop enough pressure to force open the aortic valve and more of its energy is taken to increase the pressure rather than eject the stroke volume. The PV loops that result from increased preload and increased afterload are shown in Figure 5.8.4. These PV loops are bounded by the passive and active tension curves of the heart, which are shown in Figure 5.8.4 as a function of volume, not length. When preload increases, the end-diastolic volume increases along the passive tension curve. Contraction of the heart begins its isovolumetric phase which raises the pressure to the diastolic arterial pressure. When it reaches this pressure, the aortic valve opens and the heart ejects blood into the aorta. Thus, if the afterload is kept constant while the preload increases, the stroke volume increases. If the afterload (diastolic arterial pressure) is also elevated while the preload is kept constant, it takes longer for the heart to develop pressure and it ejects blood for a consequently shorter period. Thus, the stroke volume of blood ejected at higher afterload is less and cardiac output is correspondingly less.

POSITIVE INOTROPIC AGENTS SHIFT THE CARDIAC FUNCTION CURVE UP AND TO THE LEFT

As described in Chapter 5.7, positive inotropic agents increase the force of cardiac contraction. Examples of these include the cardiac glycosides, norepinephrine that is released from sympathetic terminals, and epinephrine that is released from the adrenal gland and reaches the heart through the blood. Sympathetic stimulation increases the rate at which intraventricular pressure rises, which in turn decreases the time at which the ventricle begins ejecting blood into the aorta. Because it begins ejecting sooner, the heart empties more completely. The effect of sympathetic stimulation on left ventricular pressure is shown in Figure 5.8.5.

Alternate views of the effects of sympathetic stimulation are provided by the PV loop diagram (Figure 5.8.6) and the cardiac function curve (Figure 5.8.7). In the PV loop diagram, sympathetic stimulation increases cardiac contractility by raising the isovolumetric systolic curve and shifting it to the left. Higher pressure is developed at lower left ventricular volumes. This curve, along with the passive, diastolic curve, sets the limits of the PV loop. Increasing cardiac contractility also increases cardiac output at any given right atrial pressure, as shown in Figure 5.8.7.

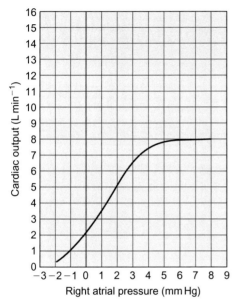

FIGURE 5.8.3 Ventricular function curve. The graph shows the cardiac output of the heart when it is pumping against a constant arterial resistance and heart rate when right atrial pressure is varied.

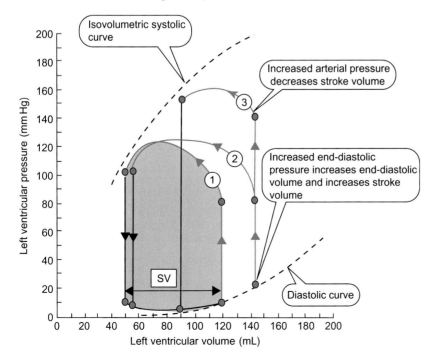

FIGURE 5.8.4 PV loops of the human ventricle. Loop 1 shows the control situation, which is identical to that shown in Figure 5.8.1. When the preload is increased while the arterial pressure is held constant (Loop 2), the end-diastolic volume increases along the passive tension curve (diastolic curve). This passive stretch of the ventricle causes an increase in stroke volume and cardiac output. Loop 3 shows the effect of increasing both the preload and the afterload. Increasing the arterial pressure reduces the stroke volume compared to the situation without increasing the afterload.

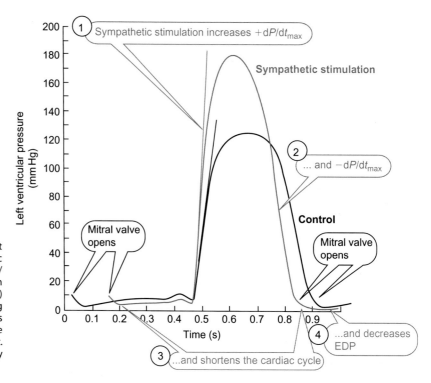

FIGURE 5.8.5 Effect of sympathetic stimulation on left ventricular pressures in the cardiac cycle. Sympathetic stimulation increases the rate of pressure rise ($+dP/dt_{max}$) and the rate of pressure fall upon relaxation ($-dP/dt_{max}$), decreases the end-diastolic pressure (EDP) and shortens the cardiac cycle. All other things being equal, this increases cardiac output. As it turns out, this effect is limited because of the venous return—the heart cannot pump more blood than is delivered to it. Thus, sympathetic stimulation of the heart alone, by itself, only marginally increases cardiac output.

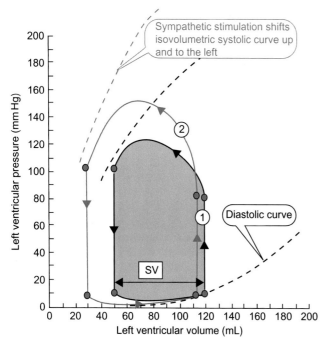

FIGURE 5.8.6 Effect of sympathetic stimulation on the PV loop of the left ventricle. Control loop (1); sympathetic stimulation (2). Sympathetic stimulation shifts the isovolumetric systolic curve up and to the left. The curve shows the effect of sympathetic stimulation on the heart alone, independent of effects on the vasculature. Thus, the diastolic and systolic pressure are shown unchanged.

FICK'S PRINCIPLE ESTIMATES CARDIAC OUTPUT FROM O_2 CONSUMPTION

In 1870, Adolf Fick pointed out that the oxygen absorbed by the lungs must be carried away by the blood that perfuses the lungs. This statement results in a simple mass balance equation:

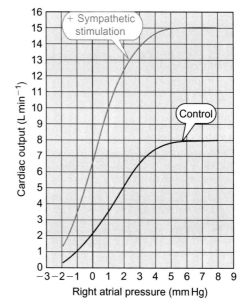

FIGURE 5.8.7 Effect of sympathetic stimulation on the cardiac function curve. Sympathetic stimulation shifts the cardiac function curve up and to the left.

$$[5.8.7] \qquad Q_a\,[O_2]_a + Q_{O_2} = Q_v\,[O_2]_v$$

where Q_a is the blood flow in the pulmonary arteries (in L min^{-1}), $[O_2]_a$ is the total oxygen concentration in the pulmonary arterial blood (in units such as mL O_2 per liter of blood), Q_{O_2} is the oxygen taken up by the lungs (in mL O_2 min^{-1}), Q_v is the blood flow in the pulmonary veins, and $[O_2]_v$ is the total oxygen concentration in the pulmonary venous blood. Thus the left hand side of Eqn [5.8.7] is the oxygen input to the lungs, and the right hand side is the oxygen output. Figure 5.8.8 illustrates the situation.

EXAMPLE 5.8.1 Calculate the Cardiac Output from O_2 Consumption and A–V O_2

At rest, Q_{O_2} (oxygen consumption) is typically 250 mL O_2 min^{-1} (volumes of gas are given in volumes at STPD—standard temperature and pressure, dry: see Chapter 6.3). The arterial blood obtained from the radial, brachial, or femoral artery, $[O_2]_a$, is 19.5 mL% (volume at STPD per 100 mL of whole blood). The venous blood (obtained from the right ventricle outflow tract through a cardiac catheter inserted through the antecubital (elbow) vein had $[O_2]_v = 14.5$ mL%. What is the cardiac output?

We can use Eqn [5.8.7] to solve for the cardiac output when we recognize that Q_a and Q_v are equal to each other and to the cardiac output. Then we have

$$Q_a ([O_2]_a - [O_2]_v) = Q_{O_2}$$

$$Q_a = \frac{Q_{O_2}}{([O_2]_a - [O_2]_v)} = \frac{250 \text{ mL min}^{-1}}{(195 \text{ mL L}^{-1} - 145 \text{ mL L}^{-1})} = 5 \text{ L min}^{-1}$$

CARDIAC OUTPUT CAN BE DETERMINED BY THE *INDICATOR DILUTION METHOD*

Suppose that we inject someone with n moles of an indicator. Ideally, this indicator should be confined to the bloodstream and be relatively easy to measure. Examples include dyes that bind to plasma albumin or plasma albumin that has itself been tagged with radioactive iodine. Let the mass (or, equivalently, the number of moles) of the injected indicator be m. It becomes distributed in some volume of blood that passes through the heart and into the systemic circulation. In some time t, all of the volume of blood possessing indicator will pass each point in the circulation. If we know the volume and the time, we can calculate the cardiac output. How do we get the volume? The short answer is: by measuring the concentration of indicator over time. We assume that cardiac output, Q_a, is constant. Then in each time increment the aggregate volume of blood passing through the systemic circulation is

[5.8.8] $$dV = Q_a \, dt$$

At any time, the concentration of indicator is C_m, given as

[5.8.9] $$C_m = \frac{dm}{dV}$$

where dm is the amount of mass (or moles) of indicator in the volume element dV. The total mass of indicator, m, is determined by integrating Eqn [5.8.9]:

[5.8.10] $$m = \int_0^t dm = \int_0^t C_m \, dV$$

From Eqn [5.8.8] we substitute in for dV to obtain:

[5.8.11] $$m = \int_0^t Q_a \, C_m \, dt$$
$$m = Q_a \int_0^t C_m \, dt$$

And, therefore, cardiac output can be calculated as

[5.8.12] $$Q_a = \frac{m}{\int_0^t C_m \, dt}$$

If we know the quantity m and measure C_m with time, we can determine Q_a by dividing m by the area under the C_m versus t curve. This procedure is illustrated in Figure 5.8.9.

The plot of $C(t)$ against t is complicated. Typically, $C(t)$ rises quickly to a peak and then decays exponentially because the ventricle ejects only about two thirds of its end-diastolic volume with each heart beat. The indicator that remains in the heart at the end of systole is then diluted with fresh blood that enters the heart, and the residual indicator is then diluted a second time, and so on. The concentration of the indicator after n heart beats is given as

[5.8.13] $$C_m(n) = \left[\frac{EDV - SV}{EDV}\right]^n C_m(0)$$

Taking the logarithm of both sides, we find

[5.8.14] $$\ln C_m(n) = n \ln \left[\frac{EDV - SV}{EDV}\right] + \ln C_m(0)$$

Because n is proportional to t through the heart rate ($n = HR \times t$), the plot of $\ln C_m$ in the decay phase of $C(m)$ versus t is linear with t. Equation [5.8.14] can be rewritten as

[5.8.15] $$\ln C_m(n) = \ln C_m(0) + \left(HR \ln \left[1 - \frac{SV}{EDV}\right]\right) t$$

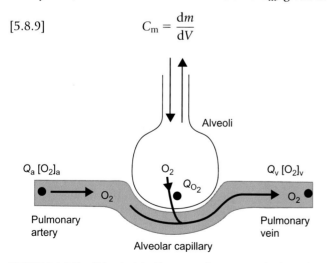

FIGURE 5.8.8 The Fick principle. The input of oxygen to the lungs from the blood is $Q_a[O_2]_a$, where Q_a is the blood flow through the pulmonary arteries, and $[O_2]_a$ is the total concentration of O_2 in this blood. This input is added to by absorption of O_2 from the alveolar air, which at steady state is the same as the rate of O_2 consumption, Q_{O_2}. This total input of O_2 is carried off by the pulmonary vein blood at the rate of $Q_v[O_2]_v$.

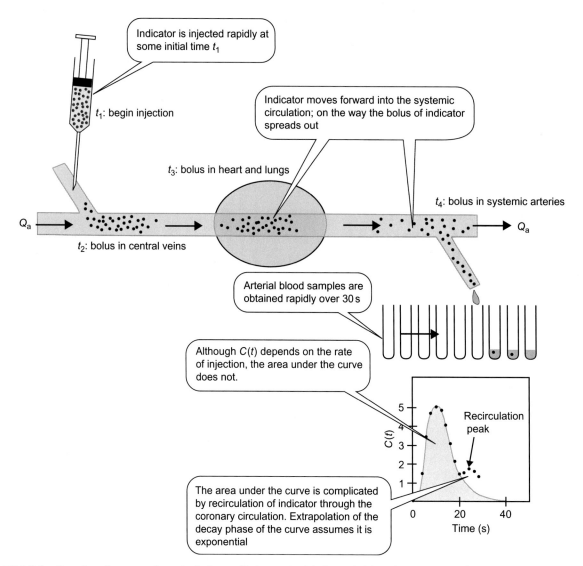

FIGURE 5.8.9 Estimation of cardiac output from the indicator dilution method. Indicator is injected into a vein and arterial concentration is measured with time. The shape of the concentration curve depends on the rate of injection, but the integral of the $C(t)$ versus t curve depends only on the amount of indicator and the cardiac output. Since the amount of indicator injected is known, the cardiac output, Q_q, can be calculated from Eqn [5.8.12]. However, the area under the curve can be overestimated if recirculation of indicator occurs. Estimation of the area can be made by extrapolating the decay curve, assuming that it decays exponentially, and integrating this extrapolated curve.

where **we identify SV/EDV as the ejection fraction**, the fraction of the end-diastolic volume that is ejected into the systemic circulation with each heart beat. Typically SV = 70 mL and EDV = 120 mL, so that the ejection fraction is typically about 0.58. Thus the slope of the decay curve (ln $C(t)$ against t) should typically be about 70 min^{-1} × ln (1 − 0.58) = −60.7 min^{-1} = −1.01 s^{-1}.

This analysis is valid for the single left ventricle, but for the entire heart it is incomplete because both the right and left heart participate in this extension of the concentration profile in time. The blood entering the left heart is only gradually depleted of indicator by the right heart. If we restrict ourselves to the time before indicator can recirculate to the venous side of the circulation, our analysis does apply to the right heart. Because the blood from the right heart feeds into the left heart, the final concentration coming out of the left heart would be

$$C_m(n) = \left[\frac{EDV_l - SV_l}{EDV_l} + \frac{EDV_r - SV_r}{EDV_r}\frac{SV}{EDV_l}\right]^n C_m(0)$$

[5.8.16]

This equation describes the situation in which the initial concentrations in both right and left ventricles are $C_m(0)$ and the blood entering the right heart has no indicator. The subscript "l" indicates the left heart and "r" indicates the right heart. This situation would pertain exactly only if the indicator were injected over a particular time course. If we assume equal ejection fractions for right and left ventricles, the slope becomes 70 min^{-1} × ln [(1 − 0.58) + (1 − 0.58) × 70/120] = −28.6 min^{-1} = −0.48 s^{-1}. Thus serial dilution by the right heart and left heart more slowly distributes indicator than either chamber alone. In practice, the injection is usually slower and subsequently the decay of indicator is also slower, with smaller slopes on the curve of ln C_m against t.

The plot of $\ln C(t)$ against t is needed because in the middle of the decay curve the concentration usually shows a recirculation hump that corresponds to indicator reentering the venous blood after completing one circuit of the circulation. Because the coronary circulation is the shortest circuit, this blood contributes first to the recirculation hump. The concentration due only to the initially injected indicator can be estimated by extrapolation of $\ln C(t)$ against t from the decay part of the curve that corresponds to indicator dilution in the heart.

THE THERMAL DILUTION METHOD

In most clinical departments, the indicator dilution method has been replaced by thermal dilution methods. In this case, a known mass of cold saline is injected rapidly into the right atrium and a thermistor placed in the distal pulmonary artery records temperature with time. The principle is the same as the indicator dilution method except that the quantity being injected is an amount of "cold", calculated as $m \times C_p \times (T_{blood} - T_{saline})$, where m is the mass of saline, C_p is its specific heat (the amount of energy necessary to raise its temperature by $1°C$) and T_{blood} and T_{saline} are the temperatures of the blood and of the saline, respectively. The cardiac output in this case is calculated by dividing the amount of "cold" by the integral of the temperature–time plot.

The thermal dilution method avoids the recirculation problem because blood returning to the atrium has been warmed to tissue temperatures before it recirculates. However, heat transfer from the tissue to blood will overestimate the cardiac output by reducing the integral. Corrections can be made for this heat transfer.

SUMMARY

The function of the heart is to pump blood. A quantitative measure of this function is the cardiac output, which is the rate of blood flow out of the left ventricle. It can be determined experimentally by Fick's principle using oxygen consumption and oxygen content of mixed venous blood and arterial blood or by using indicator dilution or thermal dilution. Typical values of cardiac output at rest are about $5 \, L \, min^{-1}$. The cardiac output can be calculated as the stroke volume times the heart rate.

The total work of the heart includes three terms: pressure–volume work, kinetic energy work, and gravitational work. By far the largest of these is the pressure–volume term. The pressure–volume loops of the heart are set between the passive stretch of the heart (the diastolic curve) and the isovolumetric systolic curve that the heart can develop against a closed aortic valve.

The cardiac output is determined by the preload, afterload, and cardiac contractility. The preload refers to the pressure in the large veins that feed into the atria. When preload increases, the ventricles are stretched and subsequently provide a greater force. This is the Frank–Starling Law of the Heart: increasing right atrial pressure increases the stroke volume of both ventricles. The ventricular function curve is a plot of the cardiac output against right atrial pressure.

The afterload refers to the pressure in the large arteries and derives its name from the fact that the heart does not "feel" the afterload until pressure rises enough to open the aortic valve. The afterload is the pressure

Clinical Applications: Heart Failure

Heart failure, generally called congestive heart failure, is the inability of the heart to pump enough blood to meet the needs of the body. It is a constellation of disease states generally characterized by exercise intolerance of varying degrees. In left heart failure, the left ventricle fails to pump blood out of the pulmonary circulation, leading to pulmonary congestion (edema) with attendant symptoms such as tachypnea (increased breathing frequency) and increased work of breathing. This may or may not cause right heart failure due to pulmonary hypertension. Right heart failure fails to pump blood out of the great veins, leading to systemic congestion and peripheral edema. Right and left heart failure can occur independently or together. Causes of heart failure include:

- **Coronary artery disease and heart attack.** Fat deposits in the coronary arteries build up over time, a condition called atherosclerosis, and reduce the lumen of the arteries supplying the heart muscle, leading to weak contractions due to hypoxia of the muscle. In some cases, the vessels occlude entirely, leading to death of parts of the muscle (heart attack) and complete dysfunction of part of the heart.
- **Hypertension.** This is increased pressure that can occur either in the pulmonary or systemic circulations. It increases the afterload of the heart. Chronic hypertension causes the heart to hypertrophy, or enlarge, and this requires stronger contractions to develop the same pressure, according to the Law of Laplace. Eventually the heart becomes unable to effectively pump blood.
- **Faulty valves.** Faulty valves can be due to congenital defects, or cardiac infections can lead to **stenosis** (a narrowing of the orifice through which blood exits the chamber) or leaks. Either causes an additional work load on the heart that can cause hypertrophy and eventual failure. **Aortic regurgitation**, for example, can cause left heart failure.
- **Cardiomyopathy.** This is damage to the heart muscle directly, which can be congenital or acquired through infections, alcoholism, drug abuse, or thyrotoxicosis.
- **Arrhythmias.** Tachycardia in the absence of exercise can lead to heart failure.

Treatment of heart failure generally focuses either on removal of the load on the heart or on increasing cardiac contractility. Reversible causes of heart failure need to be addressed. Front line medications include ACE (angiotensin-converting enzyme) inhibitors such as captopril, loop diuretics to inhibit salt reabsorption in the loop of Henlé (Chapter 7.5), and beta blockers. Lifestyle changes include alcohol limitation, smoking cessation, weight loss, exercise, and salt and fluid restriction.

against which the heart ejects blood. Thus, increasing the afterload increases the pressure against which the heart ejects blood. Increasing the afterload decreases the stroke volume and thereby the cardiac output.

Cardiac contractility refers to the ability of the heart to produce force at any given stretch. Positive inotropic agents increase cardiac contractility by shifting the isovolumetric systolic curve up and to the left, producing more pressure at a given ventricular volume. This increases the stroke volume at any specified preload. Positive inotropic agents also shift the ventricular function curve up and to the left, giving a greater cardiac output for any given right atrial pressure.

Cardiac output can be measured by Fick's dilution principle, the indicator dilution principle, and thermal dilution.

REVIEW QUESTIONS

1. What determines the end-diastolic volume?
2. What is the stroke volume? How do preload and afterload affect the stroke volume?
3. What is the cardiac output? How does it depend on heart rate and stroke volume?
4. Draw a pressure–volume curve for the left ventricle. Label the region of ventricular filling during diastole, the period of isovolumetric contraction, the period of ejection of blood, and the period of isovolumetric relaxation.
5. Identify components of cardiac work.
6. What is the ventricular function curve? What is meant by intrinsic regulation?
7. What effect do positive inotropic agents have on the ventricular function curve? What do these agents do to the pressure–volume curve? What do they do to stroke volume? Cardiac output?
8. Why is the output of the right heart the same as the output of the left heart?
9. How can you determine cardiac output from oxygen consumption?
10. How can you determine cardiac output from indicator or thermal dilution?

Problem Set 5.2
Cardiac Work

1. The work done by the heart in each beat contains at least two components. The first is the amount of work done to generate and maintain the pressure in the ventricle above that in the arterial outlet port. This is defined as **pressure–volume work**. As blood is ejected, it acquires a velocity. Hence there is **kinetic work** done to propel the blood forward.

   The pressure–volume work can be found by integrating the cardiac pressure–volume loop. This requires detailed knowledge of the form of the loop. However, we can *estimate* this area easily if we use some simple assumptions. Instead of working against a variable pressure, for example, we can assume that all of the blood is ejected against a constant *average* pressure of 93 mmHg. What counts is not the absolute pressure, of course, but the difference in pressure. Assume that the end diastolic pressure is 5 mmHg and the end systolic pressure is the same. This estimation is supported by the mean value theorem of integrals:

   [5.P2.1]
   $$\frac{\int_A^B f(x)dx}{B-A} = \overline{f(\varsigma)}$$
   $$A \leq \varsigma \leq B$$

   Here the overline denotes the average value of the function. Thus, the average value of a function is *defined* as the value that, when multiplied by the interval, gives the definite integral. If you substitute $dx = dV$, the volume increment, and $f(x) = P$, you can see what is meant by "average" pressure.

   The kinetic energy of the ejected blood is calculated as

   $$\text{Kinetic work} = \frac{1}{2}mv^2$$

   where m = stroke volume × density, v = flow/area; stroke volume = end diastolic volume (EDV) − end systolic volume (ESV); flow = stroke volume/duration of systole.

   Numbers you may need to know: duration of ventricular systole = QT interval on ECG = 0.4 s. You should recognize this as the time over which the blood is ejected. Aortic radius = 1 cm; assume a heart rate of 70 bpm; assume an EDV = 150 mL and an ESV = 75 mL. Density of blood is approximately 1.055 g cm^{-3}.
   Calculate the work done by the heart and compare the two forms of work, i.e., pressure–volume versus kinetic. Which is greater?

2. Using the results from the problem above, calculate the mechanical efficiency of the heart using its O_2 consumption as a measure of energy input. O_2 consumption at rest (which is the situation in Problem #1) is 0.09 mL O_2 (at STPD) per g of tissue per min. Assume a left ventricular weight of 250 g. The average metabolic energy derived from oxidation of substrates by oxygen is 20.3 J mL^{-1} of O_2 (at STPD).

3. A. Calculate the approximate length of time for a molecule of oxygen to diffuse from your lungs to your calf muscles. Assume that $D_{O_2} = 1.78 \times 10^{-5}$ cm^2 s^{-1}, the value of the O_2 diffusion coefficient in water.

   B. Compare the time of oxygen diffusion to the time required to move blood from the heart to the muscles. Estimate this from the velocity of blood as it leaves the heart (as calculated in Problem #1) and the distance from the heart to the calf muscle.

4. The radius of the aorta in a test subject determined by MRI is 1.2 cm. At rest, his end diastolic volume is 140 mL and his end systolic volume is 55 mL. His diastolic pressure is 70 mmHg and his systolic pressure is 115 mmHg. If we start the cardiac cycle at the opening of the mitral valve, the aortic valve in this person opens at 0.55 s and closes at 0.9 s. The entire cycle lasts 1.1 s in this person. The density of blood is 1.055 g mL^{-1} and its viscosity is 3.0×10^{-3} Pa s. A pascal is one N m^{-2}. Using this information, answer the following questions:
   A. What is the stroke volume?
   B. What is the heart rate?
   C. What is the cardiac output?
   D. What is the period of ejection?
   E. What is the *average* velocity of blood in the aorta?
   F. What is the Reynolds number for blood during ejection?
   G. Is blood flow in the aorta laminar or turbulent?
   H. What is the kinetic energy of the ejected blood?

I. What pressure–volume work is done during ejection?
5. For the test person in Problem #4, the total oxygen content of blood at the root of the aorta where the coronary arteries come off is 20 mL%, or 20 mL of O_2 per dL of whole blood. Most of this is carried by hemoglobin inside erythrocytes and a small amount is dissolved. Cardiac venous blood contains 10 mL% of O_2. The heart weighed 250 g and blood flow at rest is 275 mL min^{-1}. Assume that two-thirds of the heart mass forms the left ventricle.
   A. What is the resting oxygen consumption of the heart, in mL O_2 per g heart per min?
   B. The average metabolic energy derived from oxygen consumption is 20.3 J mL^{-1} of O_2. Assuming that all parts of the heart consume oxygen equally, and using the answer to parts H and I of Problem #4, what is the mechanical efficiency of the left ventricle?
6. A person has a mass of 50 kg and 7.5% of her body weight is blood. The density of her blood is 1.053 g mL^{-1}, and her hematocrit is 40%. Her hemoglobin concentration is 14 g%.
   A. What is her blood volume?
   B. What is her plasma volume?
   C. Assume that the average life span of the red blood cells is 120 days. What volume of blood is being destroyed each day? On average, what volume of blood is being made each day?
   D. How much hemoglobin is degraded each day?
   E. The molecular weight of hemoglobin is 64,500 g mol^{-1} and each hemoglobin binds 4 Fe atoms. How much iron is liberated from hemoglobin each day? Give the answer in mg/day and mol/day.
   F. The molecular weight of bilirubin is 584.7 g mol^{-1}. How much bilirubin is liberated from hemoglobin each day in this person, in mg/day?
   G. How much total hemoglobin is in her blood? How much iron?
7. Injection of KCl raises its concentration in the plasma to 20 mM. What happens to E_K? What do you expect would happen to the heart?
8. At the top of the R wave, lead I reads 0.35 mV and lead III reads 0.75 mV. What is the electrical axis of the heart?
9. The end diastolic volume in a patient was estimated by echocardiography to be 170 mL and the end systolic volume was 120 mL. Calculate the ejection fraction. Is this normal?
10. During exercise, an athlete consumed 1.0 L of O_2 per min. Arterial content of O_2 was 20.5 mL% and mixed venous blood had 12.5 mL%. What is the athlete's cardiac output during exercise?
11. Estimate the distance from the AV node to the apex of the heart. Compare the time it would take for conduction of an action potential to travel from the AV node to the apex if it were conducted only by nodal cells, atrial cells, ventricular cells, or Purkinje fibers respectively.
12. Assume that the diameter of a biceps muscle is 4 cm.
    A. How long would it take O_2 to diffuse from the surface of the muscle to its interior if its diffusion coefficient was equal to that in water, $D = 1.78 \times 10^{-5}$ cm^2 s^{-1}?
    B. Each muscle fiber generally contacts several capillaries. If the muscle fiber is 50 μm in diameter, how long would it take O_2 to diffuse to its interior, given the same diffusion coefficient as in part A?
13. At steady state A, blood flow through a muscle is 0.025 mL min^{-1} g^{-1}. Arteriolar O_2 content is 20 mL% and venous O_2 content is 14 mL%. This question deals with the relationship between perfusion and extracellular concentration of O_2.
    A. What is the rate of O_2 consumption of the muscle?
    B. Suppose that blood flow increases from 0.025 to 0.1 mL min^{-1} g^{-1}. If arteriolar and venous O_2 content is unchanged, what happens to O_2 consumption?
    C. Under the new condition of higher flow, what is the venous O_2 content if O_2 consumption does *not* change?
14. The end diastolic volume of a heart is 140 mL. Assume that it is a sphere. At end diastole, the intraventricular pressure is 7 mmHg. The wall thickness at this time is 1.1 cm. At the end of isovolumetric contraction, the intraventricular pressure is 80 mmHg.
    A. What is the wall tension at end diastole?
    B. What is the wall tension at the end of the isovolumetric contraction?
    C. At the end of systole, the intraventricular volume is 65 mL, the pressure is 100 mmHg, and its wall thickness is 1.65 cm. What is the wall tension at this time?
    D. The wall stress is related to tension by $\sigma = T/w$, where σ is the wall stress, T is the tension, and w is the wall thickness. Calculate the wall stress from A, B, and C.
15. At the top of the R wave, lead I reads 0.55 mV and lead III reads 0.70 mV. What is the electrical axis of the heart?
16. The concentration of TnC in heart muscle cells is estimated at about 70 nmol g^{-1} wet weight of tissue. O_2 consumption is estimated to be about 0.09 mL O_2 per min per g wet weight of tissue. Assume the TnC is 60% saturated during normal, resting contractions. Assume that 80% of the activating Ca^{2+} originates from SR stores. Assume a heart rate of 70 min^{-1}. Assume free energy of ATP hydrolysis is 57 kJ mol^{-1} and that metabolism obtains 20.3 kJ L^{-1} of O_2 consumed.
    A. How much Ca^{2+} is released and taken back up per heart beat, per g of tissue?

B. How much energy is consumed to take the Ca^{2+} back up into the SR, per heart beat?
C. What fraction of the cell's energy consumption is used for this purpose?
D. How much Ca^{2+} enters and leaves the cell across the SL membrane?
E. Assuming that all of the Ca^{2+} exits the SL over the NCX, how much Na^+ enters the cell over the NCX?
F. How many ATP are consumed to remove the Na^+ that enters over the NCX?
G. What fraction of the cell's energy consumption is used for this purpose?

5.9 Vascular Function: Hemodynamics

Learning Objectives

- List the four distinct functions of the vasculature
- Write the equations for the four principles of the circulation
- Distinguish between lateral pressure and end pressure
- Describe the relative compliance of the veins vs the arteries
- Contrast the velocity of the pressure pulse with the bulk fluid velocity
- Identify the parts of the aortic pressure trace including systolic pressure, diastolic pressure, dicrotic notch, pulse pressure, and average pressure
- Describe the relationship between stroke volume, compliance, and pulse pressure
- Calculate the mean arterial pressure from systolic and diastolic pressure
- Describe the determination of blood pressure by sphygmomanometry
- Identify the Korotkoff sounds and their origin
- Define artery, arteriole, capillary, venule, vein
- Write Poiseuille's law for laminar flow through long narrow tubes
- List the conditions required for Poiseuille's law to be valid
- Describe how hydrodynamic resistances add in series and in parallel

THE VASCULAR SYSTEM DISTRIBUTES CARDIAC OUTPUT TO THE TISSUES

As described in Chapter 5.1, the vascular system serves four distinct functions:

1. Transforms the pulsatile flow from the heart to more continuous flow (**arteries**).
2. Distributes the cardiac output to the tissues (**arterioles**).
3. Exchanges materials with the tissue (**capillaries**).
4. Provides a volume reservoir (**veins**).

THE CIRCULATORY SYSTEM USES FOUR MAJOR PHYSICAL PRINCIPLES

The physical principles of flow through the circulatory system are incorporated into four statements:

[5.9.1]
$$(1) \quad Q_V = \frac{\Delta P}{R}$$
$$(2) \quad \Delta P = \frac{\Delta V}{C}$$
$$(3) \quad <V> = \frac{Q_V}{A}$$
$$(4) \quad R = \frac{8\eta l}{\pi a^4}$$

The first principle listed above describes steady-state flow in terms of an Ohm's law analogue. Here Q_V is the volume flow in L min^{-1}, ΔP is the difference in pressure that drives flow, and R is the resistance. We will discuss further the validity of this law and the meaning of ΔP.

The second principle is a statement of the elasticity of the vessels, where ΔP is the pressure increment produced when a volume ΔV is inserted into the vessel. The pressure increment is related to the volume increment through the **compliance** of the vessel, C.

The third principle describes the relation between average velocity in the vessels, $<v>$, the flow, and the cross-sectional area, A. As we will see, in laminar flow the velocity across the vessel is not uniform.

The fourth principle is Poiseuille's law, which describes steady-state flow in tubes. We have seen this equation before (Chapter 1.2). Here we write the resistance of a long tube under laminar flow conditions where R is the resistance, η is the viscosity of the flowing fluid, l is the length of the tube, and a is its radius.

FLOW IS DRIVEN BY A PRESSURE DIFFERENCE

It is not just the static pressure that drives flow. Flowing fluids have a kinetic energy that can be converted to pressure, and vice versa, and they also possess gravitational potential energy. The total mechanical energy was given in Chapter 5.8 as

[5.9.2] $\quad E = E_P + E_K + E_G = PV + 1/2\ \rho V v^2 + \rho g h V$

where E, the total mechanical energy, is the sum of the pressure–volume energy, the kinetic energy, and the gravitational energy. Here ρ is the density of the blood, v is its velocity, g is the acceleration due to gravity, and h is the height of the fluid above a reference height.

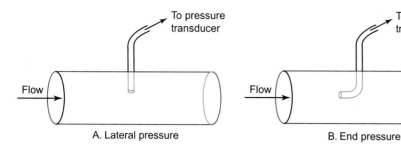

FIGURE 5.9.1 "Lateral pressure", A, and "end pressure", B, measured in a tube filled with a flowing fluid. Blood that flows into the end of the catheter loses some or all of its kinetic energy, which is converted to pressure. This measures the end pressure, which is slightly higher than the lateral pressure.

Dividing by the volume, we derive the total equivalent pressure, P', as

$$[5.9.3] \qquad P' = P + 1/2\ \rho v^2 + \rho g h$$

This equivalent pressure drives steady-state flow through the circulatory system. This is the basis of Bernoulli's law and also allows us to determine the difference between **lateral pressure** and **end pressure**.

In a static fluid the pressure is the same in all directions. In a moving fluid the kinetic energy contributes to the pressure only in the direction of its velocity. Thus, a catheter inserted into a blood vessel will record slightly different pressures if the tip of the catheter is pointing perpendicular to the fluid flow, into the fluid flow, or away from the fluid flow, as illustrated in Figure 5.9.1. Similarly, the pressure measured in a tube when the fluid is flowing depends on the velocity of the flow.

EXAMPLE 5.9.1 Calculate the difference in lateral and end pressure in the aorta

Assume that the thoracic aorta in man has a radius of 1.1 cm. Assume a steady flow of 5 L min^{-1}. What is the difference between lateral and end pressure?

We estimate the velocity of blood from $\langle v \rangle = Q_v/A$ where $\langle v \rangle$ is the average velocity, Q_v is the volume flow, and A is the cross-sectional area. The area is $\pi(d/2)^2 = 3.8$ cm^2 and the flow is

$$\langle v \rangle = [5000\ \text{cm}^3/60\ \text{s}]/3.8\ \text{cm}^2 = 22\ \text{cm s}^{-1}$$

The pressure corresponding to this velocity is $1/2\ \rho v^2$. The density of blood is about 1.05 g cm^{-3} (Chapter 5.1). This is

$1/2 \times 1.05$ g cm$^{-3} \times [22$ cm s$^{-1}]^2 = 254$ g cm s^{-2} cm^{-2} or 254 dyn cm^{-2}. We convert to Pa by 254 dyn $\times 10^{-5}$ N dyn^{-1} cm$^{-2} \times 10^4$ cm^2 m^{-2} = **25.4 Pa**. Since 1 mmHg = 133.3 Pa (Chapter 1.2), this is 25.4/133.3 = **0.2 mmHg**.

This calculation depends on the velocity of the blood. In the aorta, the flow is pulsatile and not constant. If flow were 100 cm s^{-1}, the end pressure would be 4 mmHg higher than lateral pressure.

As an example, consider that a blood vessel has a partial stenosis—a narrowing of its caliber—for part of its length. The steady-state flow through all sections of the tube must be the same because at steady state there is no net storage or loss of volume. Thus Q_v is the same through every cross-section of the vessel. According to principle (3), this means that the average velocity in the narrow portions must increase inversely with the cross-sectional area. The fluid accelerates, converting pressure energy into kinetic energy. On the other side of the stenosis the fluid slows again. The lateral pressure varies inversely with the fluid velocity, which is Bernoulli's law. The lateral pressure, equivalent pressure due to kinetic energy, and total pressure for the vessel with a partial stenosis is shown in Figure 5.9.2.

COMPLIANCE DESCRIBES THE RELATION BETWEEN PRESSURE AND VOLUME IN THE VESSELS

The elasticity of the vessels is described empirically by the compliance, as described in Eqn [5.9.1] and reproduced here:

$$[5.9.4] \qquad \Delta P = \frac{\Delta V}{C}$$

We may rearrange this to define compliance as

$$[5.9.5] \qquad C = \frac{dV}{dP}$$

This definition fits our subjective sense of what we mean by compliance. A compliant system is one that stretches. According to Eqn [5.9.5], a high compliance means that you can fit in more volume per unit change in pressure—it stretches easily. Conversely, a low compliance means that adding small volumes markedly increases the pressure. The compliance depends on the makeup of the vessel and varies according to location within the circulatory system. The large arteries have thick walls and are not as compliant as the more distensible veins. In particular, in humans

$$[5.9.6] \qquad \frac{C_V}{C_A} \approx 19$$

where C_V is the compliance of the veins and C_A is the compliance of the arteries. The compliance makes the ejected blood functionally compressible even though the fluid itself is not compressible—the expansion of

QUANTITATIVE HUMAN PHYSIOLOGY

the vessels in response to the ejected blood achieves the same result as if the blood were compressible. After the heart stops ejecting blood, the elastic arteries recoil and continue to push the blood forward. The fluid movement is driven by the blood pressure, whose time dependence results from the size and time course of the ejected volume of blood by the heart, the viscosity of the blood, and the size and compliance of the vascular tree leading away from the heart. The pressure pulse that results propagates down the arterial tree at some $5-8 \text{ m s}^{-1}$. The pressure pulse is transmitted faster than fluid flow, which is only about 0.2 m s^{-1}.

THE HEART'S EJECTION OF BLOOD INTO THE ARTERIAL TREE CAUSES THE ARTERIAL PRESSURE PULSE

The pressure pulse obtained from a catheter in the subclavian artery is shown in Figure 5.9.3. The lowest pressure, the **diastolic pressure**, about **80 mmHg** in healthy adults, occurs immediately prior to the opening of the aortic valve. The heart ejects blood faster than it can run off through the arteries, so part of the ejected volume (67–80%) distends the elastic arteries and raises the pressure toward its maximum value, the **systolic pressure**, typically about **120 mmHg**. When ejection slows, runoff of blood catches up and exceeds ejection. The volume in the elastic arteries falls and pressure falls with it. Intraventricular pressure falls precipitously, causing the aortic valve to snap shut. This causes a brief interruption in the fall of pressure, called the **incisura** or **dicrotic notch**.

THE PULSE PRESSURE DEPENDS ON THE STROKE VOLUME AND COMPLIANCE OF THE ARTERIES

The rise of pressure from diastolic to systolic levels is called the **pulse pressure**, defined as

[5.9.7]
$$\Delta P_{\text{pulse}} = P_{\text{systolic}} - P_{\text{diastolic}}$$

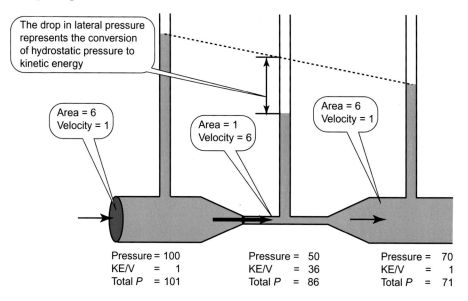

FIGURE 5.9.2 The effect of tube caliber on steady-state pressures. A single tube is shown with a narrow portion in its middle. During steady-state flow, the fluid in the narrow portion flows more quickly, and in this section the lateral pressure is reduced because the pressure–volume energy is converted to kinetic energy. When the tube widens again, the kinetic energy is converted back to pressure–volume energy. Fluid flows from the narrow part to the wide part in opposition to the pressure gradient but in accord with the total mechanical energy of the fluid. (Source: *Adapted from J.R. Levick, Cardiovascular Physiology, Arnold, New York, NY, 2003.*)

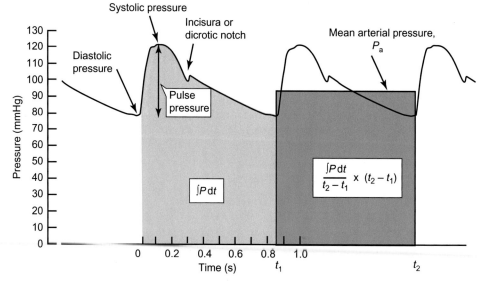

FIGURE 5.9.3 Arterial pressure pulse measured in the subclavian artery. The pressure rises upon ejection of blood during systole, peaking at the systolic pressure. When ejection slows, blood flows down the artery, volume within the elastic artery decreases, and pressure falls. The incisura or dicrotic notch occurs when the aortic valve closes. Pressure then falls to the diastolic pressure. The mean arterial pressure is the pressure that, when multiplied by the period, equals the area under the arterial pressure trace.

The compliance of the arteries is nonlinear, being stiffer at higher volumes. Thus, the pulse pressure for a given stroke volume is greater at higher pressures (see Figure 5.9.4). When the compliance of the arteries is decreased, as occurs in general in older individuals due to **arteriosclerosis** (hardening of the arteries) and **atherosclerosis** (build up of fatty deposits in the arterial walls), the pulse pressure also increases (see Figure 5.9.5).

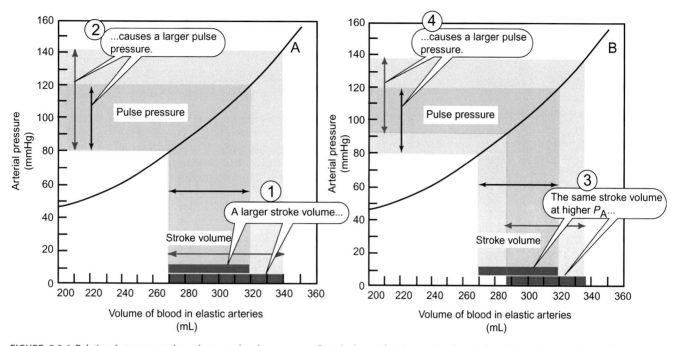

FIGURE 5.9.4 Relation between stroke volume and pulse pressure. Part A shows that increasing the stroke volume increases the pulse pressure. Because the compliance is nonlinear, the increase in pulse pressure is even greater than the increase in stroke volume. Part B shows that the same stroke volume at a higher mean blood pressure results in a larger pulse pressure. Note that the elevation of the pulse pressure is due to a greater increase in the systolic pressure than in the diastolic pressure.

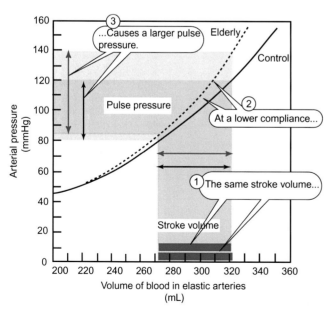

FIGURE 5.9.5 Effect of decreased compliance on the pulse pressure. Elderly persons often exhibit decreased compliance of their arteries, which manifests itself on these plots as an increased slope because $C = dV/dP$ and the plot is P against V. The decreased compliance causes an increase in pulse pressure, with a greater effect on increased systolic pressure.

DIASTOLIC PRESSURE PLUS ONE-THIRD PULSE PRESSURE ESTIMATES THE MEAN ARTERIAL PRESSURE

The arterial pressure has a complicated time course, as shown in Figure 5.9.3. The mean arterial pressure is the constant pressure that would produce the same flow as the arterial pressure. It is defined mathematically as

$$[5.9.8] \qquad P_A = \frac{\int_{t_1}^{t_2} P \, dt}{t_2 - t_1}$$

This means that the mean arterial pressure multiplied by the time of the cycle gives the integral of the pressure–time trace. Thus, the area under the pressure–time trace is equal to the area of a box one cycle long with a constant pressure of P_A. The relationship between the pressure trace and P_A is shown in Figure 5.9.3.

The mean arterial pressure can be approximated by a rule-of-thumb calculation:

$$[5.9.9] \qquad \begin{aligned} P_A &= P_{\text{diastolic}} + \frac{\Delta P_{\text{pulse}}}{3} \\ &= P_{\text{diastolic}} + \frac{(P_{\text{pulse}} - P_{\text{diastolic}})}{3} \end{aligned}$$

Thus, for a person with a brachial artery systolic pressure of 120 mmHg and diastolic pressure of 80 mmHg, the mean arterial pressure is approximately 93 mmHg.

PRESSURE AND FLOW WAVES PROPAGATE DOWN THE ARTERIAL TREE

The ejection of blood by the heart generates pressure waves in the aorta and pulmonary artery that propagate down the arterial tree at some $5-8\,\mathrm{m\,s^{-1}}$, whereas the flow is only about $0.2\,\mathrm{m\,s^{-1}}$. This should present no conceptual difficulties because pressure waves in air (sound) propagate in dry air at about $331\,\mathrm{m\,s^{-1}}$ in the absence of any flow whatever! Blood that is ejected by the heart is not compressible, and therefore it has to make room for itself in the aorta. It does this partly by distending the aorta (and thereby increasing the pressure there) and partly by pushing the blood already in the aorta into the next section of artery. This displaced blood moves forward and distends the artery in the next section of the circulation, increasing the pressure there, and also pushing some of the blood forward. This sequence of events is repeated for each segment of the arterial tree causing the pressure wave to propagate down the arterial tree. Because the vessels change size and compliance as one progresses from aorta forward, the shape of the pressure wave also changes. Figure 5.9.6 shows representative pressure pulses obtained by a catheter as it was gradually withdrawn from the subclavian artery to the radial artery. Several changes occur in the waveform as it travels peripherally:

- The dicrotic notch is damped out and eventually disappears.
- The mean pressure falls by about 2 mmHg from the subclavian artery to the radial artery.
- The systolic wave steepens and increases in magnitude.
- A diastolic wave appears due to reflection of the pressure pulse from the periphery.

CLINICIANS USE A *SPHYGMOMANOMETER* TO MEASURE BLOOD PRESSURE

In 1773, Stephen Hales made the first measurement of blood pressure when he connected a 3-m glass tube to the carotid artery of a horse via a goose trachea, and he noted the height to which the blood rose in the tube. Noninterventional clinical measurements were made possible by the invention of the mercury manometer by Poiseuille, which has evolved into the **sphygmomanometer**.

In this procedure, an inflatable cuff is placed around the upper arm. The lower end of the cuff should be at heart level to prevent gravitational pressures being added or subtracted from the measured pressure. The cuff is inflated to stop all flow through the brachial artery. The clinician places a stethoscope over the antecubital space (the inner surface of the elbow joint) over the brachial artery. Slow release of the pressure cuff will lessen the occlusion until some blood squirts through

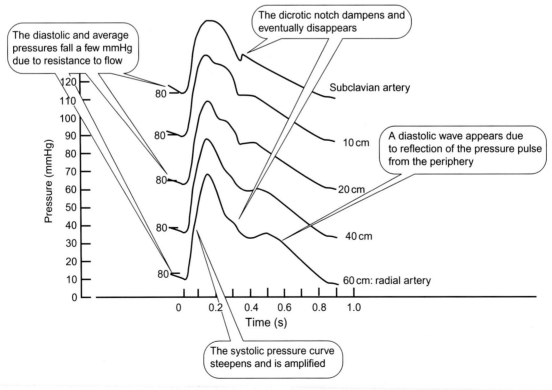

FIGURE 5.9.6 The pressure pulse obtained sequentially by withdrawal of a catheter from the subclavian artery to the radial artery. The distances by each trace refer to the distance withdrawn from its central location in the subclavian artery. Each trace is offset vertically and realigned so that the times of initiation of the pressure pulses coincide.

when the cuff pressure is just below systolic pressure, creating the **first Korotkoff sound**. The cuff pressure at the first Korotkoff sound is the **systolic pressure**. As the cuff pressure is further lowered, more blood spurts through the partially occluded brachial artery and the Korotkoff sounds become louder. When the cuff pressure is further lowered, near the diastolic pressure, the artery remains open almost all of the time. The Korotkoff sounds suddenly muffle and then, as cuff pressure decreases further, the sounds disappear. The pressure at which the sound first disappears is the diastolic pressure. Figure 5.9.7 illustrates this method of obtaining blood pressure.

BLOOD VESSELS BRANCH EXTENSIVELY, REDUCING THEIR DIAMETER BUT INCREASING THE OVERALL AREA

A single ascending blood vessel, the aorta, leaves the left ventricle, and it branches immediately to form the coronary arteries at the base of the aorta. The aorta branches to form the major arteries, which branch again to form small arteries and then smaller arteries. At each branching, the subsequent vessels become smaller in diameter but larger in number. The number of branches increases

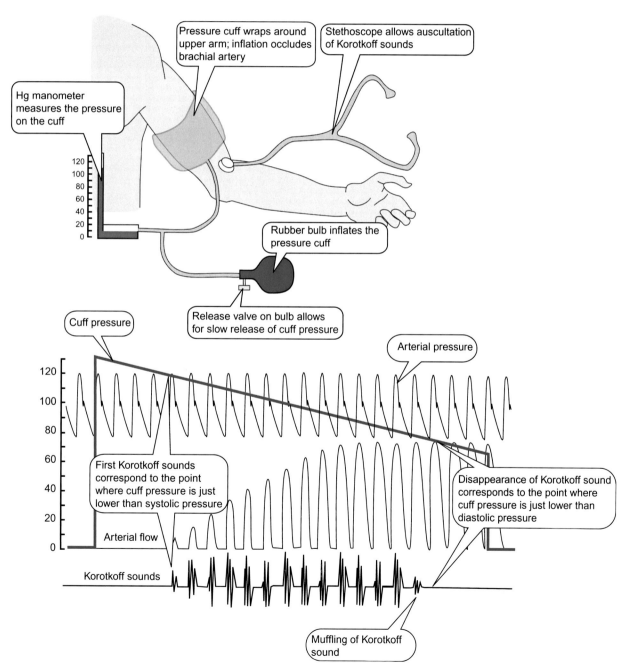

FIGURE 5.9.7 Estimation of arterial blood pressure using a sphygmomanometer. The clinician inflates the pressure cuff to a pressure higher than the anticipated systolic pressure and slowly releases the pressure while watching the manometer and listening for Korotkoff sounds. (See text for details).

faster than the decrease in area, so that the aggregate area increases with each branching. Because the flow through each level of the circulatory system must be the same, the velocity of flow becomes slower with each branching because $<v> = Q_v/A$, where A is the total area. Arteries with radius $a < 50\ \mu m$ are called **arterioles**. These branch several times, forming smaller arterioles until at last the vessels form the **capillaries**, small structures with radii between 1 and 3 μm. Exchange of nutrients in the tissues occurs at the small arterioles, capillaries, and small venules (see Chapter 5.10). The blood from the capillaries then collects into venules, small veins, and the larger veins, in which the cross-sectional area of the vessels progressively increases and their number decreases, the opposite of what occurs in the arteries. The key parameters that vary with branching are given in Table 5.9.1.

THE MAJOR PRESSURE DROP IN THE ARTERIAL CIRCULATION OCCURS IN THE ARTERIOLES

The pressure drops in the circulatory system indicate hydrodynamic resistances, just like voltage drops in a network of electrical resistances indicate the magnitude of the electrical resistances. There is little change in the average pressure going from the subclavian artery to the radial artery (Figure 5.9.6) indicating little resistance to flow along that length of artery. Figure 5.9.8 shows how the pressure drops along the systemic and pulmonary circulations, showing that the major pressure drop occurs in the arterioles.

POISEUILLE'S LAW ONLY APPROXIMATELY DESCRIBES FLOW IN THE VASCULATURE

In 1817, Jean Leonard Marie Poiseuille (1799–1869) introduced the mercury manometer to measure blood pressure. His early investigation using this device led him to believe that the blood pressure remained constant as blood traveled through the large arteries because the actual small decrease in the average pressure was not resolved with his method. He knew that the blood pressure within the veins was low and that therefore there must be a drop in pressure in the small vessels <2 mm in diameter. This led him to study flow in very small tubes. He discovered that the relation between flow, Q_v, and the drop in pressure, ΔP, along a tube of length l and inner diameter D was

$$[5.9.10] \qquad Q_v = \frac{K D^4 \Delta P}{l}$$

where K is an empirical coefficient that Poiseuille found was independent of tube caliber, length or flow, but fell with decreasing temperature. Wiedemann in 1856 and Hagenbach in 1860 independently derived a theoretical solution to this problem:

$$[5.9.11] \qquad Q_v = \frac{\pi a^4 \Delta P}{8 \eta l}$$

where a is the radius and η is the viscosity. Thus, Poiseuille's experimental result identifies his coefficient as $K = \pi/128\ \eta$. The derivation of Poiseuille's law is presented in Chapter 1.2.A1. The theoretical derivation of Poiseuille's law requires several assumptions:

- The fluid is a Newtonian fluid (viscosity is independent of the shear rate, dv/dr).
- Flow is laminar.
- There is no "slippage" at the vascular wall (zero velocity of the fluid in contact with the wall).
- Flow is steady.
- The tube is cylindrical with circular cross section and parallel walls.
- The walls of the tube are rigid.

TABLE 5.9.1 Key Parameters Dealing with the Branching of the Systemic Circulatory System[a]

Vessel	Number	Radius (cm)	Area per Unit: $A_i = \pi a^2$ (cm^2)	Aggregate Cross-sectional Area $A_{total} = n\ \pi a^2$ (cm^2)	$<v> = Q_v/A_{total}$ (cm s^{-1})	Single Unit Flow $q_v = Q_v/n$ (cm^3 s^{-1})
Aorta	1	1.1	4	4	21	83
Arteries[b]	8000	5×10^{-2}	8×10^{-3}	64	1.3	1×10^{-2}
Arterioles[b]	1×10^7	25×10^{-4}	2×10^{-5}	200	0.4	8.3×10^{-6}
Capillaries[b]	1×10^{10}	4×10^{-4}	5.0×10^{-7}	5000	1.7×10^{-2}	8.3×10^{-9}
Venules[b]	4×10^7	5×10^{-3}	7.9×10^{-5}	3160	2.6×10^{-2}	2.1×10^{-6}
Veins[b]	8000	0.09	1×10^{-1}	800	0.1	1×10^{-2}
Vena Cava	2	1.4	6	12	6.9	41

[a]The area per unit is the cross-sectional area calculated from the radius as $A_i = \pi a^2$ where a is the radius. The aggregate cross-sectional area was calculated by multiplying the cross-sectional area per vessel times the number of vessels of that generation or order: $A_{total} = n\ \pi a^2$ where n is the number of vessels at that generation. The average velocity of blood flow within the vessel was calculated as $<v> = Q_v/A_{total}$ where A_{total} is the aggregate cross-sectional area. The single unit flow is the average blood flow through a single vessel of the size described.
[b]For each category of arteries, arterioles, capillaries, venules, and veins there are several orders of vessels. The calculations shown are for a representative generation of the approximate size indicated.

The assumption of Newtonian behavior is reasonably good. This is surprising, given that blood is not even a fluid but a suspension of cells. Deviations from Newtonian behavior occur in vessels with radius less than 0.025 cm.

"Laminar" flow refers to streamlined flow. Its counterpart is turbulent flow. The point at which turbulence begins is estimated from the **Reynolds number**, a dimensionless constant that is the ratio of the inertial forces to the viscous forces:

$$[5.9.12] \qquad Re = \frac{2a <V> \rho}{\eta}$$

where Re is the Reynolds number, a is the radius of the tube, $<V>$ is the average velocity, ρ is the density, and η is the viscosity. Flow becomes turbulent at $Re \sim 2000$, but in the circulatory system Re is significantly lower than 2000. Turbulent flow sometimes occurs in the aortic root and around irregularities in the arterial surface such as atheromatous plaques. Turbulence produces a **bruit** ("noise") that is audible through a stethoscope.

The assumption of no slippage forms part of the boundary conditions that allows us to solve the differential equations leading to Poiseuille's law. Hydrodynamic experiments in general confirm it.

The assumption of steady flow is not met in the circulation because flow and pressure are pulsatile until the blood meets very small vessels. The derivation of Poiseuille's law shows that the velocity profile within steady-state flow is parabolic:

$$[5.9.13] \qquad V = V_{max}\left[1 - \frac{r_2}{a_2}\right]$$

where r is the distance from the center of the vessel and a is its radius. Thus the velocity is maximal in the center and zero when $r = a$, and it varies parabolically (see Appendix 1.2.A1 for the derivation of this velocity profile). This velocity profile develops as fluid flows down a tube. At the entrance, the velocity profile is blunt and slowly develops as the viscous drag on the vessel wall and subsequent layers of the fluid gradually communicates through the entire moving fluid. This is shown schematically in Figure 5.9.9. The distance necessary to establish the parabolic profile is called the **entrance length**. This idea pertains only to vessels where $a > 50\ \mu m$. Thus Poiseuille's law is valid only for tubes that are long compared to the entrance length.

Arteries typically have circular cross-sections, while veins typically have elliptical cross-sections. Further, arteries typically taper toward their periphery so that their walls are not parallel. The arteries also branch extensively and typically branches occur every 3–4 cm. These branches alter the geometry of the tubes and introduce complexities into the analysis of the relation between flow and pressure.

Blood vessels are distensible and therefore their diameter depends on the transmural pressure (the pressure difference between the inside and outside of the tube). Because pressure falls along the length of the vessel, tubes of constant dimensions and composition must taper along their length. Thus, even under conditions of constant initial pressure the vessels would not meet the

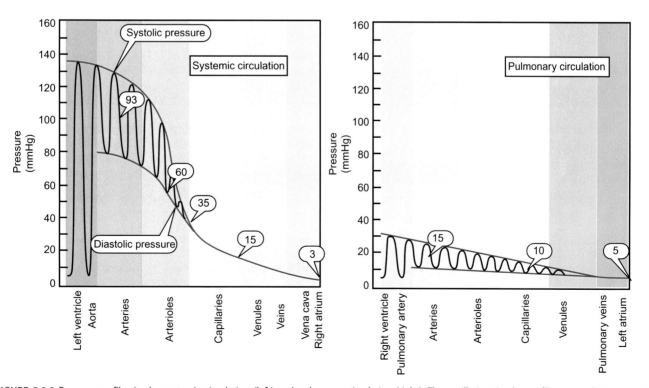

FIGURE 5.9.8 Pressure profiles in the systemic circulation (left) and pulmonary circulation (right). The oscillations in the profile recapitulate pressure variation in time, not distance.

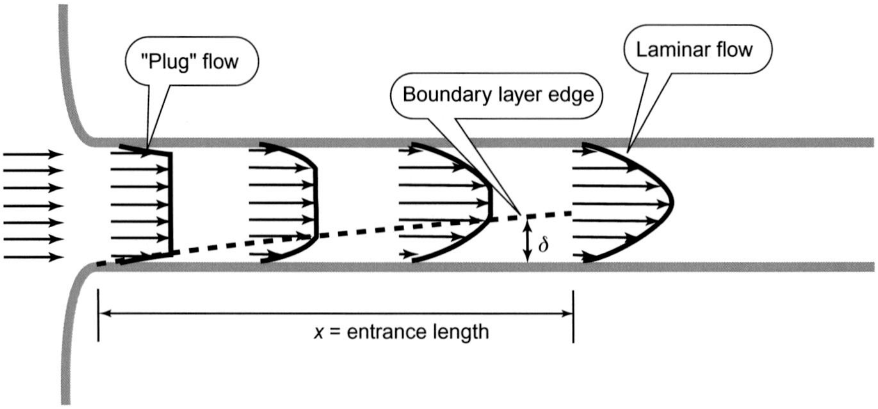

FIGURE 5.9.9 Development of the parabolic velocity profile at the entrance of a blood vessel. At the entrance of the vessel nearly all laminae, or layers, advance with a common velocity, but those laminae near the vessel edge are first slowed by viscous drag. The blunt velocity profile in the inner core becomes smaller as the fluid moves further along. The parabolic velocity profile is achieved at an entrance length, x.

assumption of Poiseuille flow. The generation of pressure pulses by the heart further complicates matters.

The circulatory system satisfactorily meets the assumptions of Newtonian fluid, laminar flow, and no slippage at the vascular wall, but it does not meet the assumptions of steady flow, cylindrical shape of the vessels, and rigid walls. Pulsatile flow, the complicated geometry of the vascular tree, and its distensibility all seriously limit the validity of Eqn [5.9.11] in quantifying blood flow.

THE RATIO OF ΔP TO Q_V DEFINES THE VASCULAR *RESISTANCE*

Despite the fact that Poiseuille's law only approximates the flow through blood vessels, flow through the vasculature is still linear with the pressure difference (actually, the total mechanical energy) between the beginning and end of the vessel. Thus we may write

$$[5.9.14] \qquad Q_v = \frac{\Delta P}{R}$$

where R is the vascular resistance. This is the same as Poiseuille's law if we identify

$$[5.9.15] \qquad R = \frac{8\eta l}{\pi a^4}$$

which is Principle (4) listed at the beginning of this chapter. Equation [5.9.14] is the hydrodynamic analogue of Ohm's law

$$[5.9.16] \qquad I = \frac{\Delta V}{R}$$

where I is the current, ΔV is the voltage drop, and R is the electrical resistance. The resistance of a network of vessels can therefore be modeled as a network of electrical resistances, and we find that vascular resistances in series add like electrical resistances, and vascular resistances in parallel add like electrical resistances in parallel. Thus for vascular resistance in series the resistances add:

$$[5.9.17] \qquad R_{total} = R_1 + R_2 + R_3 + \cdots$$

and for vascular resistances in parallel, the conductances add:

$$[5.9.18] \qquad \frac{1}{R_{total}} = \frac{1}{R_1} + \frac{1}{R_2} + \frac{1}{R_3} + \cdots$$

$$R_{total} = \frac{1}{\frac{1}{R_1} + \frac{1}{R_2} + \frac{1}{R_3} + \cdots}$$

SUMMARY

The total mechanical energy of blood includes its pressure energy, kinetic energy, and gravitational potential energy. Blood flows down its total mechanical energy gradient, but the most important measured variable is the pressure.

Insertion of extra volume into any part of the cardiovascular system raises the pressure. The relationship between the extra volume and the pressure increment is the compliance. The veins are some 19 times more compliant than the arteries. When the heart ejects its stroke volume into the aorta, the resulting increased volume produces a pressure pulse that propagates down the arterial tree. Its velocity is about $5-8$ m s^{-1}, but this increases as arterial compliance decreases with age. The actual flow of blood is about 0.2 m s^{-1}, so the pressure pulse travels much faster than the blood itself. The pressure pulse displays a maximum at the systolic pressure and a minimum at the diastolic pressure. Closure of the aortic valve produces a brief interruption, the dicrotic notch, in the diastolic fall in pressure. As it moves toward the periphery the pressure pulse loses its dicrotic notch but gains a diastolic wave due to pressure wave reflection as it travels down the arteries.

Blood pressure can be measured with a sphygmomanometer. The pulse pressure is the difference between systolic and diastolic pressure. The mean arterial pressure can be estimated as the diastolic pressure plus one-third of the pulse pressure. The size of the pulse pressure depends on the compliance of the arteries and the volume of blood ejected by the heart, the stroke volume. Increasing the stroke volume proportionately increases the pulse pressure. Decreasing the compliance increases the pulse pressure.

Blood pressure falls only a few mmHg over the large arteries but falls more quickly traveling over the arterioles. This indicates that most of the vascular resistance resides in the arterioles. Poiseuille studied flow through small tubes and found that the flow is proportional to the fourth power of the radius. Although this relationship cannot be rigorously applied to the complicated geometry of the vascular system, its basic conclusion is correct: vascular resistance shows a steep dependence on vascular radius.

Flow in the vascular system is linearly related to the pressure difference that drives it. The ratio of ΔP to flow defines the vascular resistance, which acts like electrical resistances in that vascular resistance of a series arrangement is the sum of the individual resistances, and in a parallel arrangement the conductances add to give the total equivalent conductance.

REVIEW QUESTIONS

1. What is meant by the total mechanical energy of a fluid? Why does kinetic energy contribute to pressure only in the direction of velocity? Why is end pressure greater than lateral pressure? Is this a significant problem in cardiovascular physiology?
2. Where is compliance greatest in the cardiovascular system?
3. Draw an arterial pressure trace. Label its components. How does one determine systolic and diastolic pressure clinically?
4. What is meant by "average arterial pressure"? How is it typically estimated?
5. What is the pulse pressure? What produces it? How does compliance affect it? Is it generally greater or smaller in older persons? Why?
6. What happens to the pressure waveform as it travels down the arteries toward the periphery?
7. What is the typical velocity of the pressure wave? What is the typical average velocity of fluid flow?
8. Define vascular resistance. Where is most of the resistance in the circulatory system?
9. Does Poiseuille's law describe flow in the vascular system? Why or why not?

5.10 The Microcirculation and Solute Exchange

Learning Objectives

- List three different types of capillaries and where they are typically found
- Describe the various routes of transfer of materials across the capillary wall
- List those determinants of transfer that increase passive transport
- Describe what is meant by flow-limited and diffusion-limited transport
- Define solute extraction
- Describe vasomotion
- Write the Starling equation that describes volume transfer across the capillaries
- Define the filtration coefficient
- Define colloid osmotic pressure and estimate it from the protein concentration
- Describe the function of the lymph

THE EXCHANGE VESSELS INCLUDE CAPILLARIES, TERMINAL ARTERIOLES, AND VENULES

Direct exchange of materials between the cells and the external environment is impractical because the diffusion distance is too large. The diffusion distance is effectively shortened by convection, the movement of materials by bulk transport, through the circulatory system. The final perfusion of tissues is through the **microcirculation**, which closely apposes nearly all cells of the body. The arteries branch extensively, and each generation of branches increases the number of vessels, decreases their caliber, but increases the overall cross-sectional area, so that flow slows with each succeeding branch. The last arteries branch to form **first-order arterioles**, which finally branch to form **terminal arterioles**, which then form **capillaries**. The capillaries are tiny vessels 4–8 μm wide that extend for 0.5–1 mm (500–1000 μm). These are invisible to the naked eye. In 1628, Harvey postulated their existence to explain the circulation of the blood, and Malpighi first observed them in frog lungs in 1661. These small channels coalesce to form **venules**, vessels some 15 μm wide that contain pericytes but no smooth muscle. These smallest venules unite to form progressively larger venules and eventually small and then larger veins. Smooth muscles are present in the walls of the veins when their width exceeds 30–50 μm. A schematic of this branching is shown in Figure 5.10.1. Exchange occurs across vessels on both sides of the capillary bed—some oxygen diffuses across the arteriolar walls and some fluid flows across the walls of pericytic venules, but the **major site of material exchange occurs across the capillary walls**.

ULTRASTRUCTURAL STUDIES REVEAL THREE DISTINCT TYPES OF CAPILLARIES

There are three distinct types of capillaries:

A. Continuous capillaries
B. Fenestrated capillaries
C. Discontinuous capillaries.

The overall ultrastructural appearance of these capillaries is shown schematically in Figure 5.10.2.

Continuous capillaries are the most abundant, being present in muscle, skin, lung, fat, connective tissue, and nervous tissue. This type of capillary consists of a

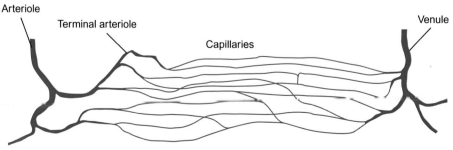

FIGURE 5.10.1 Schematic drawing of an arteriole giving rise to a capillary bed perfusing a tissue (cells not shown). The capillaries may branch extensively within the tissue. In some tissues, particularly the skin, arteriovenous anastomoses may directly link the arterioles to the venules, allowing blood flow to bypass the capillary bed.

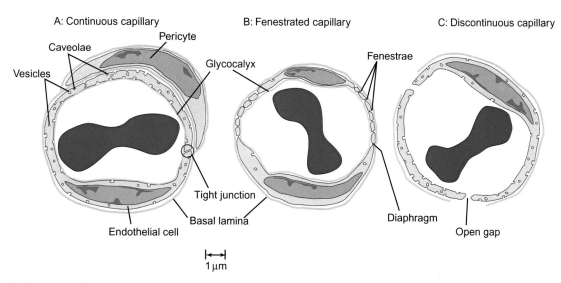

FIGURE 5.10.2 Types of capillaries. See text for their description.

continuous barrier of 1–3 endothelial cells with a continuous basement membrane. Its wall is only one cell thick—the endothelial cell—the distance separating plasma and interstitial fluid is only about 0.5 μm. **Pericytes** partially cover the outside of the capillary and may help direct the formation and proper structure of capillaries. A glycocalyx (literally "sweet husk") covers the inner surface of the endothelial cells. The negative charge on the glycocalyx repels the negatively charged plasma proteins and helps retain these in the plasma. Neighboring endothelial cells join at **tight junctions**. Despite their name, fluid can pass through these so-called tight junctions where **junctional strands** incompletely seal the junction. Both the inner and outer surface of these capillaries bind fluid-filled vesicles about 70 nm in diameter. At the surfaces, the vesicles fuse with the endothelial cell membrane, forming a narrower neck about 20–30 nm across. Thus, at the surface, these vesicles form flask-shaped vesicles called **caveolae**. The protein **caveolin** stabilizes the cytoplasmic face of these vesicles, and the vesicular membrane itself contains receptors for a variety of plasma materials such as albumin, insulin, transferrin, and ceruloplasmin. Electron micrographs show free-floating vesicles that are closely apposed to these caveolae. Serial sections show that these vesicles connect with others. These vesicles allow transport of macromolecules into the cell (**endocytosis**) or across it (**transcytosis**).

Fenestrated capillaries have "windows" for solute and fluid exchange. These capillaries are found in tissues that specialize in fluid exchange—such as the kidney, intestinal mucosa, exocrine glands, choroid plexus, and ciliary body of the eye. The **fenestrae** or gaps are about 50–60 nm wide. In most tissues, a thin diaphragm bridges the gap.

Discontinuous capillaries are also called **sinusoidal** capillaries. They are present in organs that require transfer of proteins or cells across the capillary wall.

Discontinuous capillaries possess large gaps in the endothelial cell lining—over 100 nm—and gaps in the basement membrane. Both large proteins and cells can leave the capillaries through these gaps. These types of capillaries are found in: the liver, where many plasma proteins are synthesized and placed into the plasma; the spleen, which destroys worn erythrocytes; and the bone marrow, which makes erythrocytes and therefore must be able to get the erythrocytes into the capillaries.

CAPILLARY EXCHANGE USES PASSIVE MECHANISMS

Capillary exchange occurs through three mechanisms:

A. Passive diffusion
B. Bulk flow of fluid
C. Transcytosis.

The routes used for these mechanisms are shown schematically in Figure 5.10.3. The capillary presents a series of permeability barriers beginning with the innermost glycocalyx and continuing with the inner plasma membrane of the endothelial cell, its cytoplasm, outer plasma membrane, and basement membrane. Shortcuts to these barriers exist. Cells meet at intercellular junctions whose clefts remain permeable to water and low molecular weight, water-soluble solutes. At some places, the endothelial cell has "windows" or fenestrae that are more permeable to water and low molecular weight solutes. In other places, the continuity of the endothelial cells is interrupted by open gaps that allow nearly all materials to exchange with the extravascular fluid. More specific transport systems recognize specific proteins and carry them across the cell within vesicles. Water channels called aquaporins allow water to cross the membrane. There are multiple isoforms of aquaporin; AQP1 is the one present in endothelial cells.

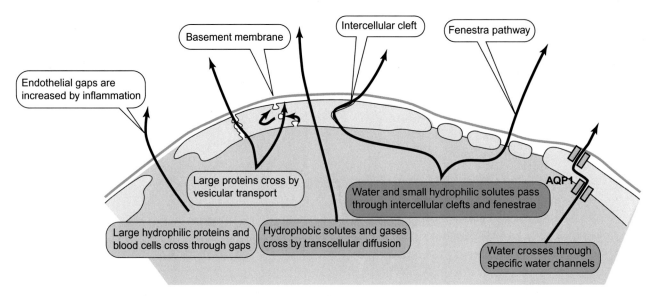

FIGURE 5.10.3 Routes of solute transfer across capillaries. (Source: Adapted from J.R. Levick, *Cardiovascular Physiology*, Arnold, New York, NY, 2003.)

PASSIVE DIFFUSION OBEYS FICK'S LAW OF DIFFUSION ACROSS MULTIPLE BARRIERS

In Chapter 2.5, we discussed diffusion through narrow pores and diffusion across a lipid membrane. In both cases the steady-state rate of diffusion was given by an equation of the form

[5.10.1] $$J_s = p\Delta C$$

where p is the **permeability** of the barrier, with units of cm s^{-1} when flux is in mol cm^{-2} s^{-1} and C is the concentration of solute in mol cm^{-3}. The value of the permeability depends on the microscopic character of the route of transfer. The permeability includes the diffusion coefficient of the solute, the partition coefficient of the solute if it permeates through the lipid membranes, the thickness of the membrane, and the number and size of channels or pores if the solute permeates through these. The flow is just the flux times the area:

[5.10.2] $$Q_s = Ap\Delta C$$

where Q_s is the flow of solute, in mol s^{-1}, and A is the area in cm^2. The driving force for diffusion is the concentration difference. This fundamental fact is modified when the diffusing material is a gas because the actual concentration of the gas depends on **the solubility** of the gas in different phases. With gases, the **partial pressure** of the gas drives diffusion. The partial pressure of the gas in solution is the partial pressure of gas that would be in equilibrium with its concentration. The solubility is a related concept: it is the coefficient that relates concentration to partial pressure. Thus, for gases we write

[5.10.3] $$[A] = \alpha_A P_A$$

where $[A]$ is the concentration of gas A, α_A is the solubility of gas A, and P_A is its partial pressure. The value of α depends on the units used for $[A]$ and P_A. This equation is discussed in more detail in Chapter 6.3. As a consequence of these facts, the diffusion of a gas across a single lipid barrier, for example, is given as

[5.10.4] $$Q_s = \frac{A k_s \alpha_s D_s}{\delta} \Delta P_s$$

where Q_s is the diffusive flow of gas s, A is the area through which it flows, k_s is the partition coefficient in the lipid, α_s is the solubility in the aqueous phases, D_s is the diffusion coefficient through the lipid membrane, δ is the thickness of the membrane, and ΔP_s is the partial pressure difference between one side of the membrane and the other. This equation is also Eqn [6.3.14]. For a complex barrier consisting of several individual barriers, Eqn [5.10.4] must be modified, yet the result has a similar form. For oxygen, we write

[5.10.5] $$Q_{O_2} = \frac{A k_{O_2} D_{O_2}}{\delta_{O_2}} \left[P_{O_2\,\text{capillary}} - P_{O_2\,\text{cell}} \right]$$

where k_{O_2} is a constant that incorporates both the partition coefficient and the solubility from Eqn [5.10.4]. In this equation, δ_{O_2} is the average diffusion distance for oxygen. Thus the flow of oxygen to the cells is enhanced by:

- increased area through which transfer occurs;
- decreased diffusion distance;
- increased oxygen gradient (ΔP_{O_2}).

The other variables in Eqn [5.10.5] that influence oxygen delivery, such as k_{O_2} and D_{O_2}, are characteristic of oxygen as a molecule and thus are not subject to physiological variation.

EITHER FLOW OR DIFFUSION CAN LIMIT DELIVERY OF MATERIALS TO CELLS

The relation between blood flow and delivery of materials to cells by diffusion from the blood can be illustrated for a simple system with the following assumptions:

- Flow is constant with the value Q_V. Blood velocity is also constant: $Q_V = AJ_V$.
- The concentration of material in the interstitial fluid that bathes the cells is constant.
- The system is at steady state: delivery of material to the cells exactly balances metabolism.
- Exchange occurs only across the capillary wall and the permeability is constant.
- The capillary is a right circular cylinder with radius r and length l.

The situation is shown in Figure 5.10.4.

Let the concentration of solute entering from the arteriolar side at the left be C_a, and the concentration leaving the capillary at the venule end be C_v. We investigate what happens to the material in a slab of the cylinder that is dx thick between x and $x + dx$. Its volume is $V = \pi r^2 dx$ and its surface area $S = 2\pi r\, dx$; the cross-sectional area is $A = \pi r^2$. There is a flow of solute from the left which is the flow of the blood times the concentration of solute in the blood at x: $Q(x) = AJ_V C(x)$, where $C(x)$ is the concentration of solute at point x. The solute in the volume element leaves the element two ways: some diffuses out of the cylinder into the tissue and the remainder is carried forward to the next volume element in the cylinder. The amount that diffuses out of the cylinder per unit time, normal to the surface of the cylinder, is $Q_D = 2\pi r\, dx\, p [C(x) - C_i]$, where $C(x)$ is the concentration of material in the volume element and C_i is the constant concentration in the interstitial fluid. The amount carried forward to the next volume element is $Q(x + dx) = AJ_V C(x + dx)$. The conservation of material dictates that the input of material must be equal to the output:

[5.10.6]
$$Q(x) = Q_D + Q(x + dx)$$
$$AJ_V C(x) = 2\pi r p\, dx[C(x) - C_i] + AJ_V C(x + dx)$$

The second line of this equation can be rearranged to give

[5.10.7]
$$\frac{[AJ_V C(x) - AJ_V C(x + dx)]}{dx} = 2\pi r p [C(x) - C_i]$$
$$-\frac{dC(x)}{dx} = \frac{2\pi r p}{AJ_V}[C(x) - C_i]$$

where we have taken the limit as $dx \to 0$ and used the definition of the derivative. We may separate variables in Eqn [5.10.7] and integrate between $x = 0$ at the arteriolar end of the capillary to position x somewhere along the capillary:

[5.10.8]
$$\int_0^x \frac{dC(x)}{(C(x) - C_i)} = -\int_0^x \frac{2\pi r p}{AJ_V} dx$$
$$\ln\frac{(C(x) - C_i)}{(C_a - C_i)} = -\frac{2\pi r p}{AJ_V} x$$

Taking the exponent of both sides and rearranging gives C as a function of distance x along the capillary:

[5.10.9]
$$C(x) = C_i + (C_a - C_i)e^{-(2\pi r p/AJ_V)x}$$

If we let $x = l$, then $C(l) = C_v$, the concentration of solute as it exits the capillary at the venule end:

[5.10.10]
$$C_v = C_i + (C_a - C_i)e^{-(2\pi r l p/AJ_V)}$$
$$C_v = C_i + (C_a - C_i)e^{(-Sp/Q_V)}$$

where S is the surface area of the capillary, p is its permeability, and Q_V is the flow through the capillary. This equation describes the rate of equilibration between the capillary contents and the interstitial fluid. The equation shows that the difference between C_a and C_i decays exponentially with distance along the capillary, and that the rate of this decay is directly related to the surface area of the capillary, directly related to its permeability and inversely related to its rate of flow. This all makes subjective sense: if the capillary is highly permeable to solute, it ought to equilibrate rapidly between blood and interstitial fluid. If the surface area is large, it also ought to equilibrate rapidly. If the flow is large, then the material washes out before it has time to equilibrate. The concentration profiles for $C(x)$ for various values of S, p, and Q_V are shown in Figure 5.10.5.

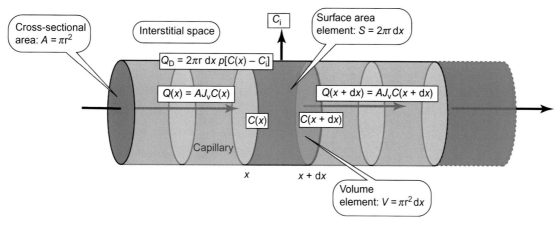

FIGURE 5.10.4 Movement of material across an idealized capillary. We use these definitions of flows to derive an expression for $C(x)$.

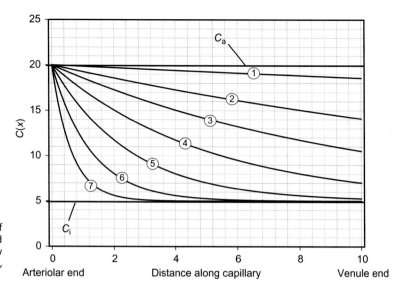

FIGURE 5.10.5 Concentration profile of the concentration of solute within the capillary when the interstitial fluid concentration is kept constant. For each simulation, the flow is steady. Values of Sp/Q_V were 0.1 (curve 1), 0.5 (curve 2), 1 (curve 3), 2 (curve 4), 4 (curve 5), 8 (curve 6), and 16 (curve 7).

One of the purposes of calculating the concentration profile as we have done is to allow calculation of the total diffusional transfer of solute over the length of the capillary. This can be done most easily by recognizing that the total diffusional transfer is just the input into the arteriolar end of the capillary minus the output at the venule end:

[5.10.11] $$Q_D = Q_V C_a - Q_V C_v$$

here Q_D is the total material flow from capillary to interstitial fluid, not the increment in flow as used in Eqn [5.10.6]. This is easily calculated because we know what C_v is by Eqn [5.10.10]. Insertion of Eqn [5.10.10] into Eqn [5.10.11] allows us to obtain

[5.10.12] $$Q_D = Q_V[C_a - C_i]\left[1 - e^{-(Sp/Q_V)}\right]$$

The graph of total diffusional transfer, Q_D, against blood flow, Q_V, is shown in Figure 5.10.6. This graph shows that at low flow the diffusional transfer of solute is **flow limited** because increases in flow directly increase the transfer of solute to the tissues. This makes sense with Figure 5.10.5 because, given the dimensions of the vessel and its permeability characteristics, diffusional flow can be increased only by increasing the gradient. The gradient in Figure 5.10.5 is greatest at high flows. In this case, transfer of materials becomes **diffusion limited**. These descriptions refer to general domains of the graph, and there is no sharp demarcation between them.

THE INTERSTITIAL FLUID CONCENTRATION IS SET BY THE BALANCE BETWEEN CONSUMPTION AND DELIVERY

Equations [5.10.9] and [5.10.12] were derived with the assumption that the interstitial fluid concentration was constant, but actually it depends on the activity of the tissue in contact with it. In the case of oxygen, carbon

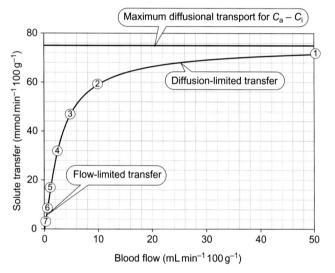

FIGURE 5.10.6 Diffusion and flow-limited diffusional transport across the capillary wall. For a given arterial and interstitial concentration of a solute, surface area, and permeability, flow can determine the transfer of material. Slow flow limits transfer to the amount of material the flowing fluid contains, fast flow maintains the highest diffusion gradient and the highest material transfer, so that increasing flow further only marginally increases transfer. Points 1–7 refer to the same values of Sp/Q_V as in Figure 5.10.5, obtained for constant values of S and p by varying Q_V.

dioxide, and nutrients, there are wide swings in the rate at which tissues consume or produce these materials. Let us consider a material like oxygen that is consumed during metabolism. In the steady state (another assumption!), the rate of metabolism equals the rate of transfer of material to the interstitial fluid. Let Q_m indicate the rate of metabolism. In the case of O_2 or glucose, Q_m is negative because these materials disappear from the interstitial fluid. Then Eqn [5.10.12] becomes

[5.10.13] $$-Q_m = Q_D = Q_V[C_a - C_i]\left[1 - e^{-(Sp/Q_V)}\right]$$

Here the expression

$$[5.10.14] \qquad \left[1 - e^{-(Sp/Q_V)}\right] = E$$

is called the **solute extraction**. It is equal to the fraction of arteriolar solute that would be transferred across the capillary wall if C_i were zero. It is a dimensionless number with range 0–1.0. We can solve Eqn [5.10.13] for C_i:

$$[5.10.15] \qquad \begin{aligned} C_i &= \frac{Q_V C_a E + Q_m}{Q_V E} \\ C_i &= \frac{C_a + Q_m}{Q_V E} \end{aligned}$$

In the last equation, a negative Q_m causes C_i to be less than C_a. How much less depends on the flow, the extraction of the solute, and the rate of consumption. If a material is not consumed by the tissues, $Q_m = 0$ and Eqn [5.10.15] shows that at steady state its concentration will equal that in the arterial blood. Production of a solute such as CO_2 or lactic acid entails a positive Q_m so that the concentration of such a solute will be higher in the interstitial fluid than in the blood, and its concentration will also obey Eqn [5.10.15].

REGULATION OF PERFUSION REGULATES SOLUTE TRANSFER

The equation for diffusional transfer of solute that we derived from simplified assumptions is reproduced below:

$$[5.10.16] \qquad Q_D = Q_V[C_a - C_i]\left[1 - e^{-(Sp/Q_V)}\right]$$

To increase diffusional transfer, physiological systems can:

A. increase the blood flow (Q_V in Eqn [5.10.16]);
B. increase the gradient for diffusion (($C_a - C_i$) in Eqn [5.10.16]);
C. increase the effective area for diffusion (S in Eqn [5.10.16]), or increase the apparent permeability, p, by decreasing the effective diffusion distance or by changing the permeability barriers of the capillaries.

Diffusional transfer is increased by increasing S, p, and Q_V by increasing the number of open capillaries. Figure 5.10.7 shows a schematic of a muscle consisting of a population of muscle fibers. Each set of capillaries

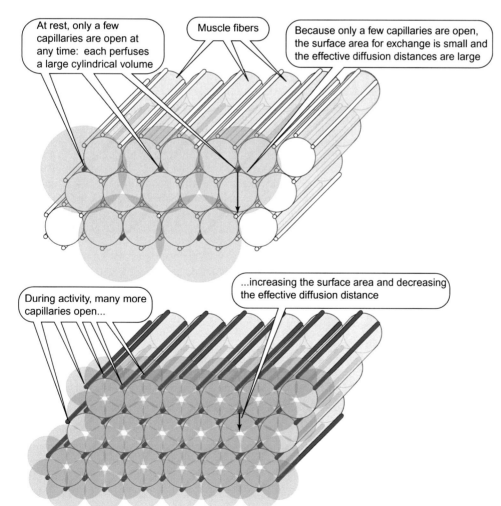

FIGURE 5.10.7 Regulation of solute delivery to the tissues. Top: at rest, tissues have low metabolic rates and low perfusion delivers sufficient nutrients and removes waste metabolites. Only a fraction of the capillaries are open, and these cycle between closed and open states. Bottom: during strenuous activity, metabolism increases, and the capillaries progressively open, increasing the effective surface area for exchange, and decreasing the effective diffusion distance, which both increase diffusional delivery of nutrients and removal of wastes.

arises from a separate terminal arteriole, which is invested with smooth muscle. At rest, many of these terminal arterioles close, so that each capillary perfuses a large volume. At rest, these arterioles cycle between open and closed states, a phenomenon called **vasomotion**. The capillaries that bear the perfusion burden change with time, and the various tissue parts are sometimes nearby and sometimes further away from open capillaries. At rest, tissue metabolism is slow and this lower perfusion is sufficient to supply nutrients and remove metabolites at the lower metabolic rate. Equation (5.10.15) shows that, for a constant extraction, the ratio of metabolism to perfusion dictates the concentration of solutes in the interstitial fluid. During strenuous activity, the metabolism of the muscle rises many fold and many more arterioles open up to perfuse the tissue, effectively increasing the area for diffusion, S, and the apparent permeability, p, by decreasing the effective diffusion distance.

SOME MACROMOLECULES CROSS THE CAPILLARY WALL BY *TRANSCYTOSIS*

Large molecules have low permeabilities through most of the diffusion barriers presented by the glycocalyx, endothelial cells, and basement membrane. These materials can still cross the capillary, albeit at low rates, by uptake into the endothelial cells and thence into **transcytotic vesicles**. The endothelial cell transports these vesicles across their narrow cytoplasm and they fuse with the opposite membrane on the tissue side of the capillary. Particular proteins, such as ferritin, may have specific recognition sites for uptake into these vesicles, whereas others may be taken up as part of **fluid phase endocytosis**. Transport in this way does not obey diffusion kinetics because the mechanism is not diffusion.

STARLING FIRST DESCRIBED THE FORCES THAT DRIVE BULK FLUID MOVEMENT ACROSS CAPILLARIES

The major pathway for water and fluid transport is not diffusion: it is pressure-driven flow. In Chapter 2.7, we derived an expression for the bulk flow of fluid across a membrane:

[5.10.17] $$J_V = L_p[(P_L - P_R) - \sigma(\pi_L - \pi_R)]$$

where J_V is the volume flux, in cm s^{-1}, L_p is the **hydraulic conductivity**, P_L and P_R are the pressures on the left and right of the membrane, σ is the **reflection coefficient**, and π_L and π_R are the osmotic pressures on the left and right sides of the membrane. Here the hydraulic conductivity results from the combination of flows across all pathways provided by the set of AQP1 channels, gaps, fenestrae, and intercellular clefts. Similarly, the reflection coefficient is an aggregate character of the capillary wall. The net filtration or reabsorption of fluid from the capillaries to the interstitial fluid or from interstitial fluid to the capillary blood depends on the **net filtration pressure**, $\Delta P - \sigma \Delta \pi$. Ernest Starling first recognized the origin of the forces that contribute to the net filtration pressure in 1896. Strictly speaking, Eqn [5.10.17] is written as

$$J_V = L_p \left[(P_L - P_R) - \left(\sum_x \sigma_x \pi_{L,x} - \sum_x \sigma_x \pi_{R,x}\right)\right]$$

[5.10.18]

where the subscript x denotes each of the chemical species in solution on each side of the membrane. When applied to the capillary, the notation is

$$J_V = L_p \left[(P_C - P_i) - \left(\sum_x \sigma_x \pi_{C,x} - \sum_x \sigma_x \pi_{i,x}\right)\right]$$

[5.10.19]

where the subscript "C" denotes capillary and "i" denotes interstitial fluid. Thus, each dissolved constituent of either the capillary fluid or interstitial fluid contributes to the overall driving force. Many of the solutes in plasma are also present in the interstitial fluid, and their concentrations are nearly identical. In Eqn [5.10.19], their osmotic effects cancel each other because their osmotic pressure subtracts from both the capillary side and the interstitial fluid side. Because solutes having the same concentration on both sides of the capillary make no *net* osmotic force, we *define* a new osmotic pressure called the **colloid osmotic pressure**.

Large proteins, the **colloids**, cannot easily cross the capillary wall, whereas the **crystalloids**, the small highly soluble salts, easily cross. Because these proteins are negatively charged, their unequal distribution sets up a Donnan effect in which the concentration of positive ions is slightly greater on the side with greater concentration of impermeable negative ions. The combination of the osmotic effects of the protein and its attendant Donnan effect is called the **colloid osmotic pressure**. A synonym for it is the **oncotic pressure**. We rewrite Eqn [5.10.19] as

[5.10.20] $$Q_V = A k_f [(P_C - P_i) - \sigma(\pi_C - \pi_i)]$$

where Q_V is the volume flow, not flux, A is the available surface area through which bulk fluid flow occurs, k_f is the **filtration coefficient** that is characteristic of the capillary, and which is identical to L_p, and π_C and π_i now denote the **oncotic pressure** in the capillary and in the interstitial fluid, respectively.

IN MOST ORGANS, NET FILTRATION PRESSURE DRIVES FLUID OUT OF THE CAPILLARIES AT THE ARTERIOLAR END

The oncotic pressure of plasma, 25 mmHg, normally does not vary significantly with location in the body because the plasma is regulated and it mixes rapidly. The oncotic pressure of the interstitial fluid varies depending on the leakiness of the capillaries to plasma proteins. Direct measurement of interstitial fluid oncotic pressure is difficult because samples cannot be obtained easily. Because **lymph** drains the interstitial fluid, the oncotic pressure of lymph approximates the interstitial fluid oncotic pressure, and this differs with location. Sinusoidal or discontinuous capillaries such as those in the liver result in lymph with 4–6 g protein per dL; lymph from the leg contains 1–3 g dL^{-1}; in the intestine it is 3–4 g dL^{-1} and in the lung it is 4–5 g dL^{-1}. The oncotic pressure is influenced by which proteins are present as well as their concentration by weight, because the oncotic pressure is produced by the number of dissolved particles. These proteins do not obey van't Hoff's law, in which osmotic pressure is linearly related to concentration, because these proteins are not ideal. The empirical relation between protein concentration and osmotic pressure is

[5.10.21]
$$\pi_{\text{albumin}} = 2.8C + 0.18C^2 + 0.012C^3$$
$$\pi_{\text{globulin}} = 1.6C + 0.15C^2 + 0.006C^3$$

where π is in mmHg and C, the concentration of albumin or globulin, is in g dL^{-1}. The interstitial fluid oncotic pressure varies from nearly the same as plasma to values less than one-third that of plasma.

The capillary hydrostatic pressure, P_C, also varies with location in the body. The renal corpuscle is specialized for rapid filtration, and its capillary pressure at the arteriolar end of the capillary is as high as 60 mmHg, whereas those in the lung are low, nearly 10 mmHg. Average values at the arteriolar end are 32–36 mmHg. Because the fluid encounters resistance along the capillary, its pressure falls from the arteriolar end to an average of 12–25 mmHg at the venule end.

The interstitial fluid hydrostatic pressure is generally thought to be small, but it depends on the location. Subatmospheric interstitial fluid pressures have been recorded in the lung, skeletal muscle, and subcutaneous tissues, whereas small positive pressures have been recorded in the kidney, liver, and intestine. Values of P_i range from -2 mmHg in the lungs to $+10$ mmHg in the kidney.

Table 5.10.1 lists typical values for the Starling forces at the arteriolar and venule end of an idealized capillary, and Figure 5.10.8 shows approximately how these forces vary with distance along the capillary. The end result is that fluid is filtered out of the capillaries at the arteriolar end and reabsorbed back into the capillaries at the venule end. Note that this is not the case in all capillary beds.

THE LYMPHATICS DRAIN THE FLUID FILTERED THROUGH THE CAPILLARIES BACK INTO THE BLOOD

As shown in Figure 5.10.8 and Table 5.10.1, many capillaries filter fluid at their arteriolar end and reabsorb it at their venule end. Generally more fluid is filtered out than reabsorbed. The extra fluid is collected into specialized vessels that collectively make up the **lymphatic system**. The lymphatic system's functions include:

- preservation of the circulatory volume;
- absorption of nutrients;
- defense against bacterial and viral invasion.

Each day the aggregate capillaries of the body filter some 2–4 L more than is reabsorbed. The lymph vessels return the capillary filtrate to the general circulation by draining into the large veins near the neck. This completes the extravascular circulation of fluid and protein as shown in Figure 5.10.9. If lymph flow is blocked, the tissues develop **lymphedema**, a severe accumulation of massive amounts of protein-rich fluid in the tissues.

Lymph vessels in the intestine transport fat as **chylomicrons** (see Chapter 8.5). These are tiny globules of fat that are coated with proteins to help carry them in the blood. Because these tiny globules scatter light, they make their suspension appear white. For this reason, the lymph vessels that drain the intestine are called **lacteals**, because they appear milky white after the ingestion of a fatty meal. These lymph vessels coalesce and eventually form the **thoracic duct**, which empties into the left subclavian vein.

As lymph drains the tissues, it carries bacteria, viruses, and other foreign matter to the lymph nodes that contain cells that comprise part of the reticuloendothelial system. These macrophages engulf these foreign particles, thereby scrubbing the lymph clean. The foreign particles in the lymph can mobilize white blood cells from the lymph nodes, thereby helping prevent bacterial invasion of the bloodstream, **septicemia**.

MUSCLE ACTIVITY HELPS PUMP LYMPH THROUGH THE LYMPHATICS

Lymph must be pumped along the lymphatics because the pressure at the venous outlet is higher than in the initial lymphatics. Lymph is pumped along by

TABLE 5.10.1 Starling Forces at the Arteriolar and Venule Ends of Idealized Subcutaneous Capillaries in the Legs and Lungs

Location		P_C (mmHg)	P_i (mmHg)	π_C (mmHg)	π_i (mmHg)	Net Pressure (mmHg)
Legs, subcutaneous	Arteriolar	35	-1	25	3	14
	Venule	15	-1	25	3	-6
Lungs	Arteriolar	10	-2	25	18	5
	Venule	8	-2	25	18	3

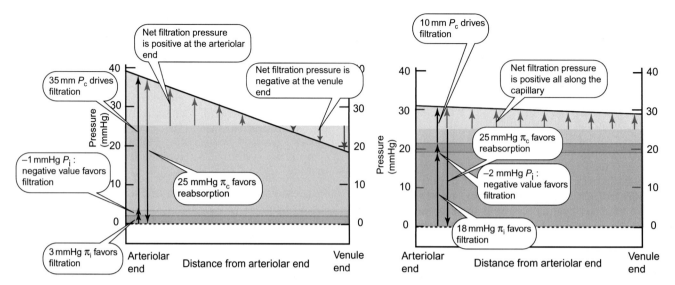

FIGURE 5.10.8 Balance of forces for net filtration across capillaries in the legs (left) and lungs (right) as a function of distance along the capillary. At the arteriolar end in the legs the capillary hydrostatic pressure is high, about 35 mmHg, and it gradually diminishes as the blood encounters resistance through the capillary, to reach 15 mmHg at the venule end. This filtration pressure is augmented by the low interstitial oncotic pressure and the negative interstitial fluid pressure. It is opposed by the interstitial fluid oncotic pressure. The sum of all forces favors filtration at the arteriolar end and reabsorption at the venule end. In other capillary beds, such as the lung, the high interstitial oncotic pressure drives a net filtration along the entire length of the capillary.

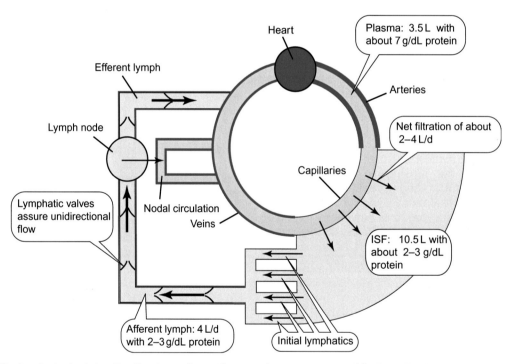

FIGURE 5.10.9 The lymphatic circulation. Net filtration at the capillaries produces some 2–4 L of fluid per day that is added to the interstitial fluid. This is continually returned to the circulation via the lymphatics. Some of this fluid returns by nodal circulation, the remainder through the thoracic duct. Valves in the lymph vessels assure unidirectional flow toward the circulation.

two main mechanisms: **extrinsic propulsion** caused by tissue movements—both passive and active—and **intrinsic propulsion** caused by rhythmic contractions of lymphatic smooth muscle. The lymph vessels possess valves that assure unidirectional flow from the tissues to the circulation. Muscle activity contributes to the extrinsic propulsion.

SUMMARY

The central purpose of the cardiovascular system is to exchange materials and heat between the blood and the tissues, and all of this exchange occurs in the small vessels—the terminal arterioles, capillaries, and small venules. The capillaries are located between the

Clinical Applications: Edema

Edema is defined as an excess of interstitial fluid. Edema can occur locally. Excess subcutaneous interstitial fluid is called **peripheral edema**. In the lungs it is **pulmonary edema** and in the abdominal cavity it is called **ascites**. Whenever capillary filtration rate exceeds lymphatic drainage, fluid accumulates and produces edema. So edema can be caused by either excessive capillary filtration or reduced lymphatic drainage.

Increased capillary filtration can be caused by elevated capillary hydrostatic pressure (P_C in Eqn [5.10.20]), decreased plasma oncotic pressure (π_C in Eqn [5.10.20]), or increased capillary permeability (k_f in Eqn [5.10.20]).

Capillary pressures rise when venous pressure rises, and this occurs when the heart fails to pump blood out of the veins back into the arteries. Thus, right ventricular failure leads to systemic edema and left ventricular failure leads to pulmonary edema.

Decreased oncotic pressure results from insufficient formation of plasma proteins or excessive loss. Decreased formation occurs in protein malnutrition. The big bellies of small children who suffer from malnutrition results from ascites when these children are displaced from their mother's breast by the next baby. Their diet of mother's milk is replaced by a high-carbohydrate, protein-poor diet that does not support synthesis of adequate plasma protein. Africans have named this condition **kwashiorkor**, the disease of the displaced child. Plasma albumin, the most abundant single class of plasma proteins, is made in the liver, along with fibrinogen and α and β globulins. **Liver failure** reduces the synthesis of these and lowers plasma oncotic pressure, leading to edema. In the **nephrotic syndrome**, the kidney loses protein in the urine, sometimes more than 20 g per day. This excessive loss cannot be balanced by increased synthesis, and plasma oncotic pressure falls. Excessive protein loss can also result from intestinal disease.

Inflammation releases signaling molecules that vasodilate local blood vessels and increase capillary permeability. These cause the redness and swelling associated with the inflammatory response. Increased capillary permeability results in increased formation of a protein-rich interstitial fluid.

Anything that interferes with lymphatic drainage will lead to edema because the lymph is the only route for the return of escaped proteins to the circulation. Although surgery and radiation treatment for cancer can cause this, the most common cause worldwide is **lymphatic filariasis**, a nematode infection transmitted by mosquitoes. In chronic cases **elephantiasis** occurs, so called because the grossly swollen extremities resemble the limbs of elephants. The disease is caused by parasitic worms, including *Wuchereria bancrofti*, *Brugia malayi*, and *B. timori*. These are all carried to their human host by the bite of infected female mosquitoes. Injected microfilariae reproduce and spread throughout the bloodstream. As they accumulate, they block lymphatic drainage and can produce gross edema in the arms, legs, genitalia, and breasts.

---

arterioles and venules and fall into three classes: continuous, fenestrated, and discontinuous. Continuous capillaries are the tightest, fenestrated capillaries are leakier, and discontinuous capillaries allow large molecular weight proteins and blood cells to pass between blood and tissue. Continuous capillaries are the most abundant.

Capillary exchange occurs through three mechanisms: diffusion, transcytosis, and bulk flow of fluid. Many soluble materials cross the capillary by simple diffusion. Hydrophilic materials pass through intercellular clefts and fenestrae, and lipid-soluble materials can pass through the endothelial cell membranes. Increasing the perfusion of a tissue increases the delivery of materials by several means: (1) increasing the perfusion typically occurs by opening more capillaries. This increases the surface area through which diffusion occurs; (2) because opening more capillaries reduces the distance between capillaries, it also reduces the diffusion distance to the tissue, thereby increasing diffusive flux. Increasing the concentration gradient also increases the diffusive flux into the tissue. This can be accomplished by decreasing the interstitial fluid concentration or keeping the capillary concentration high by increasing the flow in the capillary.

Total material transfer can be flow limited or diffusion limited. These descriptions define different domains of the plot of material transfer against blood flow. In the flow-limited domain, increasing blood flow increases solute transfer proportionately. In this domain, the blood flowing through the tissues nearly equilibrates with the interstitial fluid. In this case, the extraction of the solute (the fraction of solute which is taken up by the tissues) is near 1.0. Additional flow in this case provides additional solute. In the diffusion-limited domain, increasing flow only marginally increases solute transfer because the concentration gradient is already maximal. In this case, the extraction of the solute is nearly 0 and only increases in the permeability of the capillary or surface area can increase solute transfer.

Bulk movement of fluid occurs through the intercellular clefts, fenestrae, and gaps in the capillaries. The forces that govern fluid movement include the hydrostatic pressures inside and outside of the capillaries, and the oncotic pressures inside and outside of the capillaries. The oncotic pressure of a solution is the osmotic pressure due to the impermeant proteins and the colloids, and their attendant affect on the distribution of ions (the Donnan effect). The oncotic pressure is also called the colloid osmotic pressure. In most tissues, the major driving force for filtration of fluid across the capillary wall is the capillary hydrostatic pressure. This is augmented by the interstitial fluid oncotic pressure, whose value depends on the tissue. It is high in liver and lung and low in most subcutaneous tissues and skeletal muscle. The plasma oncotic pressure provides the major opposing force for filtration. It is relatively constant at about 25 mmHg. Positive interstitial fluid pressure also opposes filtration. The interstitial fluid pressure varies with location. It is generally small and sometimes negative. The balance of these forces causes filtration at the arteriolar end of some capillaries and reabsorption at the venule end. In other capillary beds, there is filtration throughout the length of the capillary. Capillaries produce a net fluid filtration that is collected by the lymphatics and drained back into the blood.

REVIEW QUESTIONS

1. What are the three different types of capillaries? Where are they found in the body?
2. How do things get across capillary walls? What kinds of things are retained by the capillaries?
3. What does flow-limited mean? What does diffusion-limited mean?
4. How can material exchange be increased?
5. What is vasomotion?
6. What is solute extraction?
7. What forces favor fluid extrusion into the interstitial fluid? What forces favor fluid reabsorption? What is colloid osmotic pressure? What is its typical value?
8. What happens to fluid that is filtered by the capillaries but not reabsorbed?
9. What would happen if lymph flow were blocked?

Regulation of Perfusion 5.11

Learning Objectives

- Write an equation that expresses capillary pressure in terms of arteriolar resistance and venule resistance
- Define vasoconstriction and venoconstriction
- Describe how MLCK and MLCP regulate the phosphorylation of myosin light chains in vascular smooth muscle
- Distinguish between electromechanical coupling and pharmacomechanical coupling
- Describe Ca^{2+} sensitization
- Describe the mechanism of sympathetic vasoconstriction
- List four classes of intrinsic control of arteriolar caliber
- give two classes of extrinsic control of arteriolar caliber
- Describe the myogenic response
- Describe the effects of endothelial secretions on arteriolar caliber
- Describe metabolic control of arteriolar caliber
- Describe the effects of hormones epinephrine, vasopressin, angiotensin, and natriuretic peptide on arteriolar caliber

FOR ANY GIVEN INPUT PRESSURE, THE CALIBER OF THE ARTERIOLES CONTROLS PERFUSION OF A TISSUE

In Chapter 5.9, we saw that flow through a vessel is driven by the pressure gradient, steeply depends on the radius of the vessel, and that the major pressure drops occur in the resistance vessels, the small arteries and arterioles. In Chapter 5.10, we saw that terminal arterioles supply a set of capillaries that exchange nutrients, waste, heat, and signals with the tissues. Increasing the proportion of open capillaries by increasing the number of open arterioles enhances flow-limited transfer of materials. The caliber of the arterioles thus determines tissue perfusion. The arterioles are invested with vascular smooth muscle whose constriction or dilation regulates perfusion. This chapter is concerned with how constriction or dilation of the arterioles occurs and their consequence for perfusion and capillary pressure.

VASOCONSTRICTION DECREASES CAPILLARY PRESSURE

Figure 5.11.1 shows terminal arterioles branching into capillaries, which then collect again to form venules. What determines the pressure in the capillaries? We can approximate this answer by lumping resistance terms together. Capillary pressure decreases slightly along the capillary so there is no single capillary pressure. But we can choose some point in the capillary and ask what the pressure is in that place. Let the total flow through the arterioles be Q_{VA}. Because net filtration of plasma is small, Q_{VA} is nearly the same as the aggregate flow through the capillaries, which is nearly the same as the aggregate flow through the veins, Q_{VV}. This is the hydraulic analogue of Kirchhoff's current law. As we have taken the aggregate flows, we also take aggregate or equivalent resistances. The arterioles obey the hydraulic analogue of Ohm's law:

$$[5.11.1] \qquad Q_{VA} = \frac{\Delta P_A}{R_A}$$

where the subscript "A" denotes arterioles. The pressure difference that drives flow through the arterioles is just $P_A - P_C$, where the subscripts denote arteriole and capillary, respectively.

The flow through the venules that drain this capillary bed is the same as the flow through the arterioles. We write the relation between flow through the venules, driving force, and resistance as

$$[5.11.2] \qquad Q_{V,V} = \frac{P_C - P_V}{R_V}$$

Since these flows in Eqns [5.11.1] and [5.11.2] are equal, we can combine them to solve for P_C:

$$\frac{P_A - P_C}{R_A} = \frac{P_C - P_V}{R_V}$$

$$[5.11.3] \qquad P_A - P_C = \frac{R_A}{R_V} P_C - \frac{R_A}{R_V} P_V$$

$$P_C \left[1 + \frac{R_A}{R_V} \right] = P_A + \frac{R_A}{R_V} P_V$$

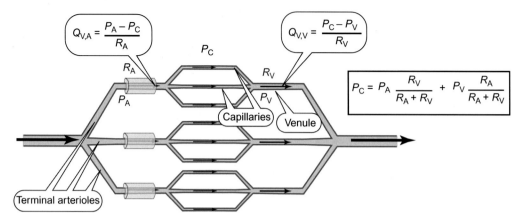

FIGURE 5.11.1 Capillary pressure is determined by the relative values of precapillary resistance, R_A and postcapillary resistance, R_V.

which leads to

$$[5.11.4] \qquad P_C = \frac{P_A + \frac{R_A}{R_V} P_V}{1 + \frac{R_A}{R_V}}$$

Although this is the form of the equation as used by cardiovascular physiologists, because it expresses the resistances as a ratio, it is perhaps easier to understand if rewritten (by multiplying through by R_V) as

$$[5.11.5] \qquad P_C = P_A \left[\frac{R_V}{R_A + R_V}\right] + P_V \left[\frac{R_A}{R_A + R_V}\right]$$

This hydraulic circuit is analogous to a voltage divider. Equation [5.11.5] shows that the capillary pressure must be intermediate between the arterial and venous pressures, and that its value depends on the relative resistances of the precapillary resistance, R_A, and the postcapillary resistance, R_V. **Increasing R_A lowers the capillary pressure toward P_V, and decreasing R_A raises it toward P_A.** Capillary pressure favors filtration of fluid into the interstitial space, so increasing P_C leads to edema and lowering it favors reabsorption of fluid from the interstitial space into the blood. Thus, the body reacts to **hemorrhage**, the loss of blood due to external or internal bleeding, by **vasoconstriction** mediated by the sympathetic nerves. Vasoconstriction refers to constriction of the arteries and arterioles, whereas **venoconstriction** refers to constriction of the veins. After hemorrhage, vasoconstriction raises R_A/R_V, thereby reducing P_C and favoring movement of fluid from the interstitial fluid back into the circulation.

VASCULAR SMOOTH MUSCLE CONTRACTS BY ACTIVATION OF MYOSIN LIGHT CHAIN KINASE

Most arteries have three major layers. Outermost is the **tunica adventitia**, which consists mainly of loose connective tissue, nerves, and blood vessels that supply the arterial wall. The middle layer is the **tunica media**, which contains elastic tissue and layers of vascular smooth muscle cells. Innermost is the **tunic intima**, which contains the endothelial cells that line the inner surface of the vessel and also contains elastic tissue (see

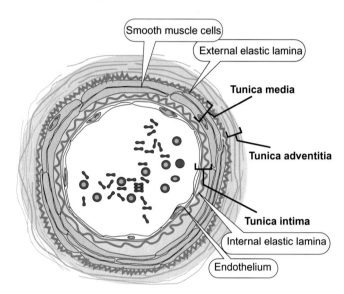

FIGURE 5.11.2 Structure of a small artery. The outermost layer, the tunica adventitia, is largely connective tissue. Inward is the middle layer, the tunica media, containing an outer layer of elastic fibers and a layer of smooth muscle cells. The innermost layer is the tunic intima, consisting of the internal elastic lamina and the endothelial cell layer that lines the lumen of the vessel.

Figure 5.11.2). Vascular smooth muscle cells in the tunica media control the caliber of the vessels. In vascular smooth muscle cells, force is regulated through phosphorylation of the myosin light chains, as shown diagrammatically in Figure 5.11.3.

MULTIPLE SIGNALS REGULATE THE ACTIVITY OF MLCK AND MLCP

Unlike both cardiac and skeletal muscle, there is no troponin C in smooth muscle to regulate acto-myosin interaction directly. Instead, the contractile elements themselves must be activated either by **electromechanical coupling** or by **pharmacomechanical coupling**. In electromechanical coupling, depolarization of the smooth muscle cell opens voltage-gated Ca^{2+} channels on the surface membrane (**sarcolemma**) that allows Ca^{2+} to enter the cell and raise cytoplasmic $[Ca^{2+}]$. The increased $[Ca^{2+}]$ binds to calmodulin, which then activates myosin

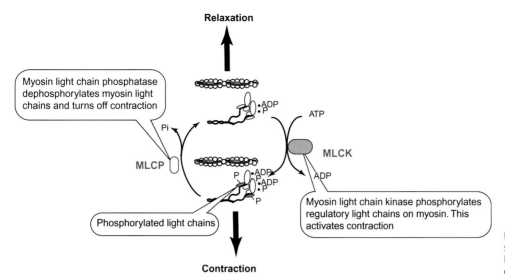

FIGURE 5.11.3 Regulation of vascular smooth muscle contraction by myosin light chain phosphorylation and dephosphorylation.

light chain kinase (**MLCK**). In pharmacomechanical coupling, chemical signals such as norepinephrine released from postganglionic sympathetic nerves bind to receptors on the sarcolemma and activate signaling systems that eventually raise cytoplasmic [Ca^{2+}]. Norepinephrine released from sympathetic terminals binds to **α₁ receptors** on vascular smooth muscle, which is linked to a **G$_q$** mechanism (see Chapter 2.8). Mechanisms of activation of contraction are shown in Figure 5.11.4.

Inhibition of MLCP has the same effect as activation of MLCK in that it increases phosphorylation of the myosin light chains, but this effect does not depend on Ca^{2+}. Increasing myosin phosphorylation by inhibition of MLCP is called **Ca^{2+} sensitization**, because it causes more force at any given cytoplasmic [Ca^{2+}]. There are several ways to inhibit MLCP, as shown in Figure 5.11.5. One is by phosphorylation of **CPI-17** by protein kinase C (PKC) that is activated by diacylglycerol (DAG) liberated when phospholipase C splits phosphatidylinositol bisphosphate. This is part of the G$_q$-coupled mechanism elicited by norepinephrine acting on α₁ receptors. A second pathway is through activation of G$_{12/13}$-coupled receptors to activate a GTP-exchange factor that converts rhoA-GDP to membrane-bound rhoA-GTP. **RhoA** is a monomeric G-protein that binds to and hydrolyzes GTP. RhoA-GTP activates rhoA kinase, a protein kinase that phosphorylates MLCP. Phosphorylation of MLCP inactivates it, thereby preserving the phosphorylation state of the myosin light chain. This mechanism is used by **endothelin-1** acting on **ET-A** receptors on vascular smooth muscle.

MULTIPLE MECHANISMS CAUSE VASODILATION

GS MECHANISMS RELAX VASCULAR SMOOTH MUSCLE

Circulating epinephrine causes vasodilation of some vascular beds, due to β₂ receptors on the vascular smooth muscle cells. These beds include skeletal muscle and myocardium. Both β₁ and β₂ receptors are linked to heterotrimeric G-proteins that are G$_s$-proteins which activate **adenylyl cyclase**. Adenylyl cyclase converts ATP to 3′,5′-**cyclic AMP**, usually denoted as cAMP, that activates **protein kinase A**, or **PKA**. PKA phosphorylates several proteins in the vascular smooth muscle cell including **phospholamban** and **SL K⁺ channels**. These mechanisms are shown in Figure 5.11.6.

Phospholamban is a 5 kDa protein in the SR that associates as pentamers and contains multiple phosphorylation sites. Phosphorylation of phospholamban increases the affinity of the Ca^{2+} pump on the SR so that the SR takes up Ca^{2+} more quickly, thereby relaxing the muscle by removing activation of MLCK.

PKA phosphorylates K$_{ATP}$ channels and K$_{Ca}$ channels in the sarcolemma. K$_{ATP}$ channels are inhibited by ATP. Thus, they open when cells are metabolically challenged, causing hyperpolarization and no electromechanical activity. K$_{Ca}$ channels are activated by elevated cytoplasmic [Ca^{2+}], causing vasodilation. Phosphorylation of the K⁺ channels increases their outward currents, reducing the excitability of the cells.

NITRIC OXIDE RELAXES VASCULAR SMOOTH MUSCLE BY STIMULATING GUANYLATE CYCLASE

A variety of agonists and physical forces cause endothelial cells to form **nitric oxide**, **NO**. As described in Chapter 3.8, three different enzymes convert L-arginine and O₂ to NO and citrulline. These include the endothelial nitric oxide synthase (**eNOS**), the neuronal NOS (**nNOS**), and an inducible NOS (**iNOS**). NO produced by the endothelial cells diffuses to the vascular smooth muscle where it relaxes the cells. The early name for NO was endothelium derived relaxing factor, or EDRF.

NO activates a **soluble guanylate cyclase** that increases cGMP in the cells. This in turn activates **protein kinase G (PKG)** that phosphorylates target proteins. PKG phosphorylates phospholamban, the protein that regulates

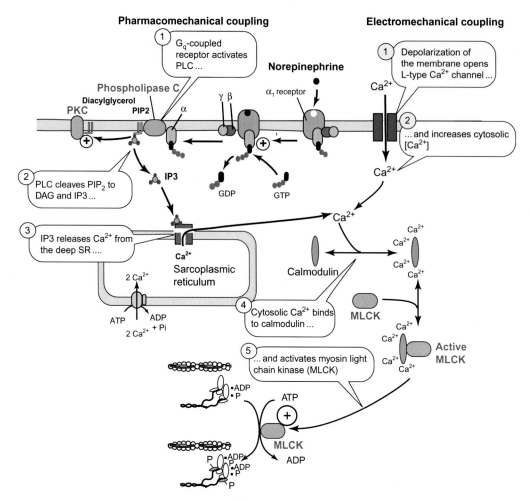

FIGURE 5.11.4 Electromechanical coupling and pharmacomechanical coupling in vascular smooth muscle cells. In electromechanical coupling, cytoplasmic [Ca^{2+}] rises because voltage-gated Ca^{2+} channels open in response to membrane depolarization; the cytoplasmic Ca^{2+} binds to calmodulin and activates MLCK. MLCK phosphorylates the myosin light chains and activates actin–myosin interaction and contraction. In pharmacomechanical coupling, cytoplasmic [Ca^{2+}] rises by IP3-induced release of Ca^{2+} from SR stores. IP3 arises from G$_q$-coupled activation of phospholipase C.

SERCA2a in the SR membrane, thereby increasing Ca^{2+} uptake and bringing about relaxation by removing activation of MLCK. PKG also stimulates MLCP activity, thereby reducing the phosphorylation of the regulatory light chains and decreasing Ca^{2+} sensitivity. The effects of NO are integrated into Figure 5.11.6.

CONTROL OF BLOOD VESSEL CALIBER IS LOCAL (INTRINSIC) AND SYSTEMIC (EXTRINSIC)

Many different physiological agents control blood vessel constriction or dilation. These are broadly classified as **intrinsic** and **extrinsic**. Intrinsic control refers to control that originates within the tissue being perfused. Extrinsic control refers to control that is imposed on the vascular bed by control systems that reside outside the organ being perfused. Each of these broad classifications have several classes of control.

INTRINSIC CONTROL

- myogenic response;
- endothelial secretions (NO, prostacyclin, endothelin);
- metabolic control (adenosine, acid pH, K$^+$);
- local paracrine secretions (histamine, bradykinin, prostaglandins, thromboxane, serotonin).

EXTRINSIC CONTROL

- nerves;
- hormones (epinephrine, vasopressin, angiotensin, atrial natriuretic peptide (ANP)).

THE MYOGENIC RESPONSE ARISES FROM THE CONTRACTILE RESPONSE TO STRETCH

William Bayliss discovered the myogenic response in 1902 when he found that distension of blood vessels by increased blood pressure caused their contraction. Thus, the myogenic response results from stretch-induced

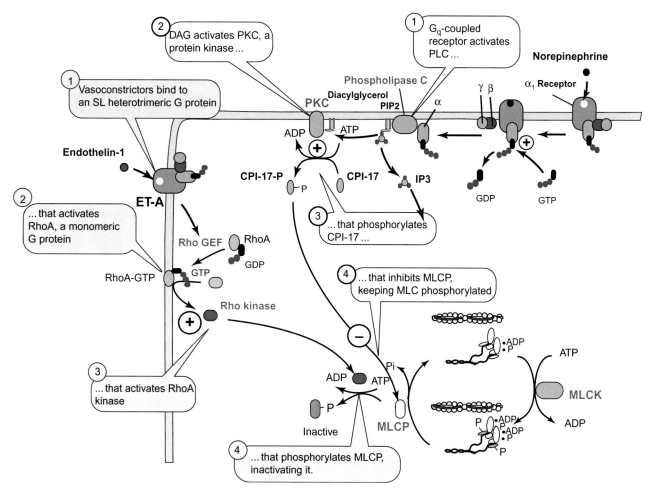

FIGURE 5.11.5 Mechanisms of Ca^{2+} sensitization. See text for description.

contraction. Its mechanism is not thoroughly known, but probably involves stretch-activated cation channels that depolarize the cells, thereby opening voltage-gated Ca^{2+} channels. Volume-regulated Cl^- channels may also be involved. Maintenance of stretch-induced vasoconstriction for long periods appears to involve a vasoconstrictor, 20-HETE (20 hydroxy eicosa-tetra-enoic acid), a hydroxylated derivative of arachidonic acid. This vasoconstrictor helps prevent repolarization of the cell, thereby maintaining its contraction.

ENDOTHELIAL SECRETIONS DILATE ARTERIOLES

Endothelial cells produce at least three substances that affect blood vessels: **NO**, **prostacyclin (PG_{I2})**, and **endothelin**. NO and prostacyclin are vasodilators, whereas endothelin is a vasoconstrictor. All three substances are secreted as they are synthesized—they are not stored for later release.

NO is synthesized by **eNOS** converting L-arginine + O_2 to citrulline + NO. Agonists such as thrombin, bradykinin, substance P, and acetylcholine stimulate NO production by endothelial cells. Blood flowing past the surface of the endothelial cells exerts a shear stress on the cells that is proportional to the gradient of velocity normal to the surface (dv/dr is the **shear rate**). This shear stress produces a continuous, basal rate of NO production. During exercise, the increased blood flow increases the shear rate and induces vasodilation. This is called **flow-induced vasodilation**. NO vasodilates vessels by activating soluble guanylate cyclase, as shown in Figure 5.11.6.

The enzyme **cyclooxygenase** (COX) produces **prostacyclin** by acting on arachidonic acid in the cell membrane. Arachidonic acid is a 20:4 ω-6 fatty acid found in cell membranes. It is released from its phospholipid by **phospholipase A_2**. It is converted to a variety of active substances including thromboxane, prostacyclin, and 20 HETE, as described above for the myogenic response. The chemical structures of these compounds are shown in Figure 5.11.7.

Endothelin is a 21 amino acid peptide secreted by endothelial cells in response to hypoxia, angiotensin II, vasopressin, and thrombin. There are three isoforms: ET-1, ET-2, and ET-3. They mediate their effects through endothelin receptors, ET_A, ET_{B1}, ET_{B2}, and ET_C.

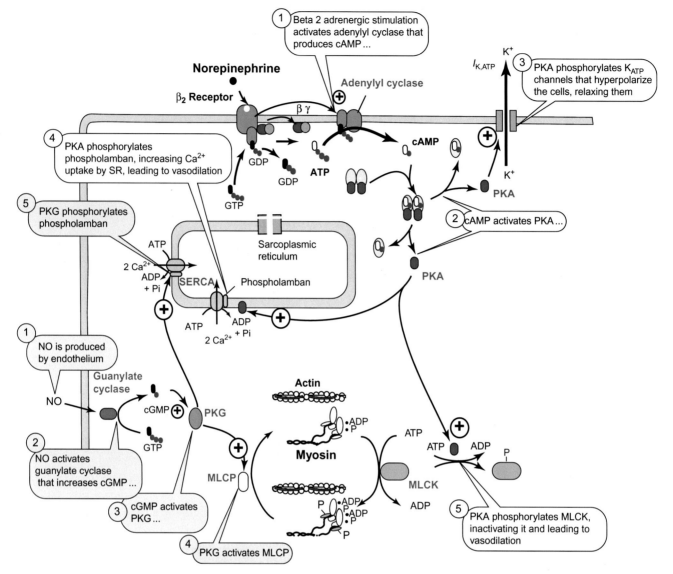

FIGURE 5.11.6 Mechanism of vasodilation. Reduction in the level of phosphorylation of the myosin regulatory light chain brings about relaxation. Activation of β_2 receptors activates adenylyl cyclase that increases cytoplasmic cAMP levels. This activates PKA, which phosphorylates K_{ATP} channels on the surface membrane, which hyperpolarizes the cell and reduces Ca^{2+} influx into the cell. Phosphorylation of phospholamban increases Ca^{2+} uptake into the SR and reduces cytoplasmic $[Ca^{2+}]$, thereby relaxing the cell. Phosphorylation of MLCK inactivates MLCK activity, thereby reducing the phosphorylation state of the myosin light chains. Nitric oxide (NO) produced in endothelial cells activates guanylate cyclase, which increases cGMP, which in turn activates PKG. PKG phosphorylates phospholamban with similar results as with β_2 stimulation. PKG also activates MLCP, thereby reducing the phosphorylation of the light chains.

METABOLIC PRODUCTS GENERALLY VASODILATE

Increased metabolism of skeletal muscle, cardiac muscle, or regions of the brain causes increased blood flow to that region. This is the basis of **functional magnetic resonance imaging** of the brain (fMRI), which actually monitors blood flow to regions of the brain when they alter their activity. The response is called **metabolic hyperemia** or **functional hyperemia**. Vasodilators released from metabolizing tissue diffuse out of the tissue to the tunica media of the vessels and cause vasodilation.

These vasodilators include **adenosine**, CO_2, and K^+. The response is extremely sensitive, so that usually oxygen delivery is directly proportional to oxygen demand (Figure 5.11.8).

During high rates of energy consumption, cells rapidly break ATP down to ADP and Pi. The ATP is regenerated in a variety of ways: muscle cells use creatine phosphate and creatine phosphokinase to regenerate ATP quickly. Myokinase generates ATP from 2 molecules of ADP in the reaction $2ADP \rightarrow ATP + AMP$. Some of this AMP leaks out of the cells where AMP 5′ nucleotidase breaks it down to

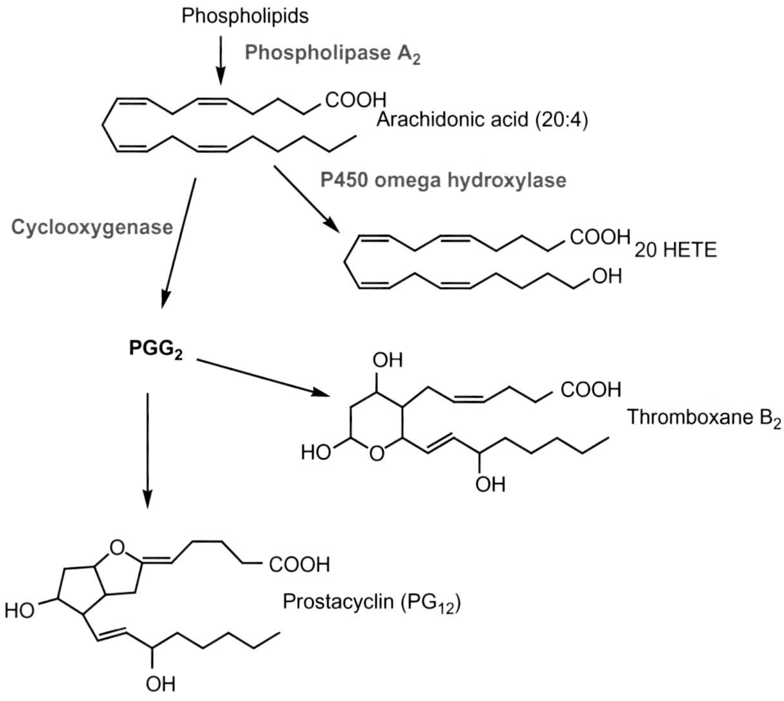

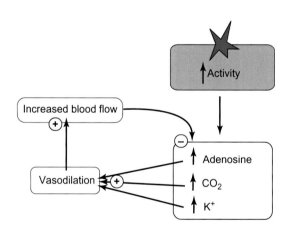

FIGURE 5.11.7 Chemical structures of arachidonic acid, 20-HETE, thromboxane B_2 and prostacyclin.

FIGURE 5.11.8 Metabolic vasodilators. Increased activity results in increased levels of adenosine, CO_2, and K^+. These all cause vasodilation, which increases blood flow and washes out the metabolites and in some cases reduces their production by increased O_2 supply.

adenosine. This extracellular adenosine binds to receptors that ultimately cause vasodilation. Adenosine causes only part of the metabolic hyperemia, because about half of the effect persists when adenosine receptors are blocked.

To regenerate ATP through oxidative phosphorylation, metabolism produces CO_2. This combines with water to form carbonic acid, a process that is catalyzed by the enzyme carbonic anhydrase. The carbonic acid dissociates to form H^+ and HCO_3^-. The increased $[H^+]$ causes vasodilation, particularly in cerebral blood vessels. The effect is less pronounced in cardiac and skeletal muscle.

Increased activity of muscle or nerves means increased frequency of action potentials, and each action potential entails an inward Na^+ current to depolarize the cell and an outward K^+ current to repolarize it. During rest, the Na^+–K^+-ATPase restores the Na^+ and K^+ to their proper compartments, but during activity it cannot keep up. The $[K^+]$ outside the cell increases. Ordinarily you would expect this to raise E_K and depolarize the cell, but in the range of increased interstitial fluid $[K^+]$ (4–9 mM), an inwardly rectifying K^+ channel is activated, causing hyperpolarization of vascular smooth muscle cells and vasodilation.

PARACRINE SECRETIONS AFFECT VASCULAR CALIBER

Local paracrine secretions that affect vascular smooth muscle include **histamine, bradykinin, thromboxane, prostaglandins,** and **serotonin**.

Histamine is a major mediator of **inflammation** (see Chapter 5.2). Mast cells and basophilic leukocytes synthesize and store histamine, and release it in response to trauma and allergic reactions such as **urticaria** (hives), **asthma**, and **anaphylaxis**. Histamine dilates arterioles, constricts veins, and increases capillary permeability. The different effects on arterioles and venules are mediated by different receptors. H1 receptors link to a G_q mechanism that raises cytoplasmic $[Ca^{2+}]$ and ultimately causes venoconstriction and increased vascular permeability. H2 receptors are linked to a G_s mechanism that increases cAMP in smooth muscle cells and ultimately causes relaxation and vasodilation.

During inflammation, the enzyme kallikrein is activated by proteolytic fragments of clotting factor XII. Kallikrein then cleaves kininogen to form a 9 amino acid fragment, **bradykinin**. Bradykinin contributes to the vasodilation and tissue swelling and it produces pain.

Thromboxane A_2 and **prostacyclin** are both produced from arachidonic acid in a cascade involving **cyclooxygenase**. This enzyme is the site of action of the **nonsteroidal antiinflammatory drugs (NSAIDs)** such as aspirin, ibuprofen, and indomethacin. Thromboxane A_2 is released by platelets upon contact with collagen. Its potent vasoconstriction helps stop bleeding and also helps aggregate the platelets to form the platelet plug. Prostacyclin, on the other hand, is a vasodilator.

Platelets also release serotonin as part of the normal clotting reaction. Its vasoconstriction assists in the arrest of bleeding.

THE SYMPATHETIC NERVOUS SYSTEM PREDOMINANTLY CONTROLS THE VASCULAR SYSTEM

The sympathetic nervous system provides the main innervation of vascular smooth muscle. It releases norepinephrine at varicosities near the smooth muscle cells. The effects of the norepinephrine depend on the receptors on the vascular smooth muscle. Those having α_1 receptors activate a G_q-protein that ultimately constricts the smooth muscle by increasing cytoplasmic $[Ca^{2+}]$. Those having α_2 activate a G_i-protein that also constricts the smooth muscle by decreasing [cAMP] within the cells. Other smooth muscle cells have β_2 receptors that cause vasodilation by increasing [cAMP]. The α and β receptors derive their description from the relative potency of norepinephrine, released from sympathetic nerve terminals, and epinephrine, released from the adrenal medulla. Vascular smooth muscle in skeletal muscle contains mainly β_2 receptors that dilate upon exposure to epinephrine or norepinephrine. Vascular smooth muscle in the intestines and skin have mainly α_1 receptors and constrict upon sympathetic stimulation.

In most vessels, the sympathetic nervous system provides tonic activity. For tissues such as the skin and intestine, which constrict in response to sympathetic stimulation, vasodilation is accomplished by removal of tonic sympathetic activity.

In a few tissues, the vascular smooth muscle is innervated by parasympathetic nerves that dilate the vessels. These tissues include the salivary glands, some gastrointestinal glands and the erectile tissue of the external genitalia. The postganglionic parasympathetic fibers to these tissues release acetylcholine onto endothelial cells which are activated by it to produce NO

CIRCULATING HORMONES THAT AFFECT VESSEL CALIBER INCLUDE EPINEPHRINE, ANGIOTENSIN, ANP, AND VASOPRESSIN

EPINEPHRINE ORIGINATES MAINLY FROM THE ADRENAL MEDULLA

Preganglionic sympathetic fibers originating in the thoracic spinal cord travel through the paravertebral ganglia and celiac ganglia to reach the adrenal medulla. There they synapse on specialized cells, the **chromaffin cells**, in the adrenal medulla. These cells are ganglionic sympathetic neurons. They have nicotinic acetylcholine receptors on their membranes and respond to sympathetic stimulation by secreting epinephrine and norepinephrine into the blood, in a ratio of about 3 epinephrine:1 norepinephrine. The effects of these on the target tissues depends on the receptors. In general, α receptors respond best to norepinephrine and β receptors respond better to epinephrine. The main stimuli for adrenal medulla secretion are exercise, hypoglycemia, hypotension, and the alerting response (the "fight or flight" reaction).

VASOPRESSIN CONSTRICTS BLOOD VESSELS

Vasopressin is also known as **antidiuretic hormone, ADH**. These names describe its major actions, which is to constrict the blood vessels and reduce urine output, respectively. Its synthesis and secretion is described in detail in Chapter 9.2 and its renal mechanism is detailed in Chapter 7.6. Briefly, ADH is a 9 amino acid peptide synthesized in cells in the supraoptic nucleus and paraventricular nucleus of the hypothalamus. It travels down axons to be released in the posterior pituitary in response to **hypovolemia** or increased **plasma osmolarity**. Its actions are to limit urinary excretion of water so as to help counteract hypovolemia by not losing more fluid. This action retains water but not salt, so that the end result of vasopressin is the excretion of a highly concentrated urine. This compensates for increased plasma osmolarity by discarding salt in the urine in excess of water.

The actions of vasopressin are mediated by V1a, V1b, and V2 receptors. V1 receptors on blood vessels use a G_q-coupled mechanism that ultimately raises cytoplasmic $[Ca^{2+}]$ to induce contraction. In the kidney, vasopressin activates V2 receptors that couples a G_s receptor mechanism to insertion of aquaporin channels into membranes of the renal tubule cells. This increases water reabsorption from the incipient urine.

ANGIOTENSIN II CONSTRICTS BLOOD VESSELS, INDUCES THIRST, AND RELEASES ADH AND ALDOSTERONE

The control of circulating levels of angiotensin begins with secretion of **renin** by the afferent arterioles in the glomeruli of the kidney. It constitutes part of the **renin–angiotensin–aldosterone (RAA)** system and is discussed in detail in Chapter 7.6. Renin is secreted in response to three inputs:

1. Renal sympathetic nerves stimulate renin release.
2. Decreased distal tubule load of Na^+ increases renin release.
3. Decreased afferent arteriolar pressure increases renin release.

The released renin cleaves a protein in the plasma, **angiotensinogen**, to form angiotensin I, a 10 amino acid peptide fragment. **Angiotensin converting enzyme, ACE**, cleaves off 2 more amino acids to form angiotensin II. Angiotensin II exerts several effects, including:

- vasoconstriction;
- release of aldosterone from the adrenal cortex;
- increasing the sensation of thirst;
- release of ADH.

All of these effects are designed to counteract the original stimulation for renin release, and multiple negative feedback loops are found in the system (see Chapter 7.6).

ANP OPPOSES THE RAA SYSTEM

ANP is a 28 amino acid hormone that is released by the atrium and induces a **natriuresis**. Natriuresis is increased urinary excretion of Na^+. Stretch of the atrial wall caused by high cardiac filling pressures induces ANP release. ANP stimulates renal salt and water excretion to rid the body of this excess volume. In addition, ANP exerts moderate vasodilatory effects. The consequence of this is to lower blood pressure but also to increase capillary pressure. This increases the flow of fluid out of the capillaries into the interstitial compartment. This temporarily unloads the excess vascular volume by shifting it to the interstitial fluid, and more permanently by increasing urine volume.

SUMMARY

"Perfusion of tissues" means the blood flow through the tissue. Perfusion is determined by the pressure difference between the input arterial pressure and the output venous pressure, and the resistance. The resistance in turn is determined by the caliber of the vessels, approximately in accord with Poiseuille's law. Constriction of the arterioles decreases perfusion and dilation increases it. Venoconstriction can also influence flow and capillary pressure.

Vascular smooth muscle must be activated to contract. Smooth muscle controls the interaction between myosin and actin by the phosphorylation state of regulatory light chains on the myosin, instead of by Ca^{2+} binding to troponin C as is done in skeletal muscle and cardiac muscle. MLCK phosphorylates the regulatory light chains, and myosin light chain phosphatase (MLCP) dephosphorylates them. The balance between these two reactions sets the phosphorylation state and the contractile force of the muscle.

Contraction is initiated by increased cytoplasmic $[Ca^{2+}]$ either by increasing influx into the cell through voltage-gated Ca^{2+} channels by depolarizing the cell membrane (electromechanical coupling), or by increasing Ca^{2+} influx into the cytoplasm through receptor-operated channels or by Ca^{2+} release from internal stores in the sarcoplasmic reticulum (pharmacomechanical coupling). The increased cytoplasmic $[Ca^{2+}]$ binds to calmodulin, and the Ca_4CAM complex activates MLCK.

Physiological regulation of MLCP regulates the Ca^{2+} sensitivity of smooth muscle contraction, so that high force can be elicited at low cytoplasmic $[Ca^{2+}]$ by inhibiting MLCP. Alpha$_1$ adrenergic stimulation, usually by norepinephrine, increases Ca^{2+} sensitivity by activating an inhibitor of MLCP through PKC. Other vasoconstrictors activate a monomeric G-protein called RhoA. This in turn activates RhoA kinase that phosphorylates MLCP, inactivating it.

Vasodilators exert their effects through multiple paths. Beta$_2$ adrenergic receptors, which respond to epinephrine more than norepinephrine, activate adenylyl cyclase through a G_s-protein, increasing [cAMP]. The increased [cAMP] activates PKA that phosphorylates phospholamban, a regulatory protein on the SR membrane. Phosphorylation of phospholamban removes its inhibition of the SR Ca^{2+} pump; cytoplasmic $[Ca^{2+}]$ falls and the muscle cell relaxes. Nitric oxide (NO) produced by endothelial cells in response to shear stress or agonists activates a guanylate cyclase in smooth muscle cells that ultimately activates PKG that activates MLCP and inactivates MLCK.

Vascular smooth muscle is regulated by factors within the tissue (intrinsic regulation) and factors external to it (extrinsic regulation). Intrinsic factors include the myogenic response, in which stretch induces contraction. Endothelial secretions include NO, endothelin, and prostacyclin. Metabolic factors include adenosine, acidosis, and increased extracellular $[K^+]$. Other paracrine secretions include bradykinin, thromboxane, histamine, and prostaglandins.

Most vascular beds contain only sympathetic innervation, but there are some exceptions. The sympathetic system activates differentially, causing constriction here and vasodilation there depending on the degree of activation of the nerves and on the receptors on the vessels. Many vessels contain both α receptors that contract the vessels and β receptors that dilate them. Circulating epinephrine originates largely from the adrenal medulla and has metabolic effects as well as vasoconstrictor and vasodilator activity. Vasopressin, also called ADH, is a vasoconstrictor. It is secreted by the posterior pituitary in response to hypovolemia and plasma hyperosmolarity. Angiotensin II also vasoconstricts. ANP, secreted by atrial cells in response to stretch, vasodilates.

REVIEW QUESTIONS

1. If arterioles constrict, what happens to capillary pressure? Venous pressure? What happens to the resistance of the vessel? What happens to blood flow?
2. What kind of muscle is in the vasculature? How is its contraction regulated?
3. What enzyme activates smooth muscle? How is this enzyme activated? Is the phosphorylation state of myosin light chains regulated any other way? What inhibits this other pathway?
4. What brings about vascular smooth muscle relaxation? What effect does this have on vessel caliber?
5. What is the myogenic response?
6. What materials do endothelial cells make to influence arteriolar caliber? What does endothelin do? What is NO? What does it do? What is prostacyclin? What enzyme starts the production of prostacyclin?
7. What effects do each of the following have on vascular smooth muscle: adenosine, CO_2 and K.
8. What effects do histamine, prostaglandins, bradykinin, thromboxane, and serotonin have on vascular smooth muscle?
9. What effect does sympathetic stimulation have on the vasculature?
10. What effects do angiotensin II, epinephrine, vasopressin, and ANP have on the vasculature?

Integration of Cardiac Output and Venous Return 5.12

Learning Objectives

- Explain why cardiac output generally equals venous return
- Define total peripheral resistance, TPR
- Define mean systemic pressure
- Define stressed and unstressed volume
- Draw the curve relating mean systemic pressure to blood volume
- Draw the vascular function curve as cardiac output versus right atrial pressure
- Explain the plateau, knee, slope, and intersection on the horizontal axis of the vascular function curve
- Explain the effects of vasoconstriction and vasodilation on the vascular function curve
- Determine the operating point of the cardiovascular system by the intersection of the cardiac function curve and vascular function curve
- Explain the effects of hemorrhage and transfusion on the vascular function curve
- Explain the effects of cardiac disease and sympathetic stimulation on the cardiac function curve and the operating point of the cardiovascular system
- List the components of the sympathetic tetralogy
- Describe the effects of exercise on the cardiac and vascular function curves

THE CARDIOVASCULAR SYSTEM IS CLOSED

The overall flow for the cardiovascular system is recapitulated in Figure 5.12.1. In the long term, the cardiovascular system is open because it receives input of materials and fluids from the intestine and it discards materials and fluids through the skin, lungs, intestines, and kidneys. However, the rates of fluid transfer from these inputs and outputs are slow compared to the flow through the system. Fluid exchanges vary greatly; Figure 5.12.1 shows typical values for a comfortable environment. The flow of blood around the circulation is about 5 L min^{-1} or 7200 L day^{-1}. On average, loss and gain from the environment are about equal and the blood volume does not change. Imbalances in the inputs and outputs are extremely important because they determine the circulatory volume, but they occur on a different timescale than circulation. Thus, for the minute-to-minute regulation of the circulation, the circulatory volume is very nearly constant and the circulatory system approximates a **closed system**. Inflow to any part of the system must be the same as outflow through that part, or volume would build up or be depleted wherever the imbalance occurs. The consequence is that the input to the right heart from the veins, the **venous return**, is equal to the output of the left heart, the **cardiac output**.

THE CARDIOVASCULAR SYSTEM CAN BE SIMPLIFIED FOR ANALYSIS

In Chapter 5.8, we learned about the Frank–Starling Law of the Heart: increased filling pressure stretches the heart and increases its force of contraction. Increasing the force of contraction expels more blood from the left ventricle, so that cardiac output increases when the **preload** increases. This preload is generally expressed as the **right atrial pressure**, the pressure which drives filling of the heart. The **afterload** also affects cardiac output. Typically the cardiac function curve is obtained at constant afterload. The normal cardiac function curve is reproduced in Figure 5.12.2.

The cardiac function curve was determined in isolated heart–lung preparations, which contains both sides of the heart and the intervening pulmonary circulation. In these experiments, the outflow pressure was held constant while the input pressure was varied. Thus, the cardiac function curve describes the heart without the attached systemic circulation. In this isolated heart–lung preparation, flow out of the right heart must match flow into the left heart, or else fluid would accumulate or be drawn out of the lungs. The Frank–Starling law of the heart indicates that the increased filling pressure of the right heart results in increased cardiac output. Any increase in output of the right heart is quickly communicated to the left heart as an increased filling pressure. Thus, increased output of the *right* heart is matched to increased output of the *left* heart. Because of this tight coupling, we can collapse the heart and lungs to a single equivalent pumping mechanism. Similarly, we can

QUANTITATIVE HUMAN PHYSIOLOGY

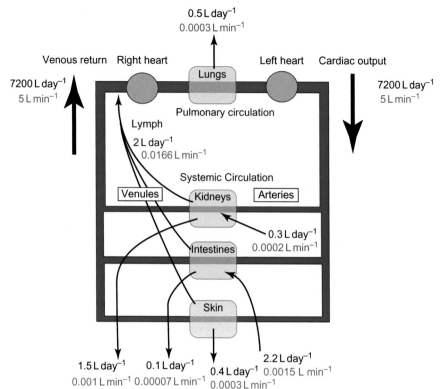

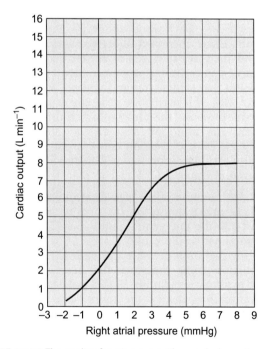

FIGURE 5.12.1 Balance of fluids that traverse the cardiovascular system. The bulk of the flow is retained inside the vessels of the cardiovascular system. Although there are exchanges of material into the blood from intestine and other tissues, and from the blood into many tissues, the flow of fluid that leaves the cardiovascular system is much smaller than the flow around the circulation. Thus, on short timescales, the circulatory system approximates a closed system.

FIGURE 5.12.2 The cardiac function curve. The graph shows the output of the left ventricle when it is pumping against a constant arterial pressure and when right atrial pressure is varied.

lump the systemic circulation into a single equivalent set of vessels in series. This lumping follows the rules of combination of parallel or series resistances. Figure 5.12.3 shows this situation.

THE OPERATING POINT OF THE CARDIOVASCULAR SYSTEM MATCHES CARDIAC FUNCTION TO VASCULAR FUNCTION

Figure 5.12.3 clarifies the role of the heart and the vasculature in determining the cardiac output. The heart pumps an increment of blood, the stroke volume, from the venous side of the circulation to the arterial side. The arteries, arterioles, capillaries, venules, and veins return this blood to the heart. The analysis which follows breaks the cardiovascular system into two components: the heart and the vasculature. The heart is characterized by its output as a function of its input and output pressures: P_{RA} and P_A. P_{RA} is the pressure at the right atria, the **central venous pressure**, or the **right atrial pressure** (the preload); P_A is the arterial pressure against which the heart pumps (the afterload). For every P_A, there is a cardiac function curve.

The venous return also has a relationship between P_A and P_{RA}. It is

$$[5.12.1] \qquad Q_{veins} = \frac{P_A - P_{RA}}{R_{A \to RA}}$$

where Q_{veins} is the flow through the systemic circulation (in the intact circulatory system this is equal to the cardiac output) and $R_{A \to RA}$ is the resistance to flow from the left heart side, or arterial side, of the systemic circulation to the right heart side, or venous side. This resistance lumps together all of the resistances of arteries, arterioles, capillaries, venules, and veins that supply the peripheral

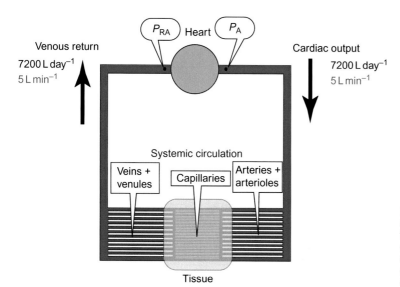

FIGURE 5.12.3 Simplified cardiovascular system. Because right and left heart outputs are tightly coupled, we collapse these two sides of the heart and its intervening pulmonary circulation into a single pump. Also, the arteries and arterioles, consisting of both series and parallel arrangements, can be combined to a single aggregate. Similarly, capillaries and veins can be combined to form their own equivalent aggregates.

tissues of the body. Hence, it is called the **total peripheral resistance**, or **TPR**. The equation is rewritten as

$$[5.12.2] \qquad Q_{veins} = \frac{P_A - P_{RA}}{TPR}$$

Although Eqn [5.12.2] is true and can be used to estimate TPR, it does not help us understand what determines the flow through the system, because TPR, P_A, and P_{RA} are not independent. What we desire is a relationship between Q and P_{RA} for the vasculature, at the same P_A used for the cardiac function curve, because the cardiac function curve is such a relation. The operating point of the combined system can then be found by the simultaneous solution of both functions. The simultaneous solutions match the cardiac output to venous return and the boundary pressures (both curves are obtained at the same P_A and P_{RA}). To derive the **vascular function curve**, we introduce the idea of mean systemic pressure.

THE MEAN SYSTEMIC PRESSURE NORMALLY EQUALS THE MEAN CIRCULATORY PRESSURE

The **mean circulatory pressure** is the pressure that would be measured at all points in the *circulatory system* when the heart is stopped and blood is instantaneously redistributed (meaning before any reflexive changes in the vasculature) so that pressure is the same everywhere. The **mean systemic pressure** is the pressure that would be measured at all points in the *systemic circulation* if all inputs and outputs to the systemic circulation were stopped and blood were distributed so that pressure is the same everywhere. The mean systemic pressure and pulmonary pressure are measured by clamping both the pulmonary artery and aorta while measuring the pressure within the two circulations. The results of the measurement show that the mean pulmonary pressure is normally about the same as the mean systemic pressure, about 7 mmHg. The mean circulatory pressure is the average of both systemic and pulmonary circulations but is heavily weighted to the systemic circulation because its compliance is some seven times greater than that of the pulmonary circulation. In normal individuals, the mean circulatory pressure is equal to the mean systemic pressure. **In man, the mean systemic pressure is normally 7 mmHg.**

FILLING THE EMPTY CIRCULATORY SYSTEM REVEALS STRESSED AND UNSTRESSED VOLUMES

Suppose that we perform a "thought experiment" in which we *completely* drain the blood from the cardiovascular system. In this experiment we stop the heart and we imagine that the vessels retain all of their characteristics that they had in the intact system before we drained the blood, with no change in caliber or compliance. Next we begin filling the system back up with blood, noting the volume of the blood and the pressure in the system. Because there is no flow, except transiently when we inject volume, the pressure is the same everywhere in the system. We find that it takes some volume to fill up the circulatory system at no pressure because many of the vessels do not collapse when blood is removed. After the circulatory system is filled to zero pressure, additional volume causes significant pressure rise, and this rise is linear with volume. Figure 5.12.4 illustrates the results of such an experiment. The graph is extrapolated from experiments in dogs in which only relatively small volumes of blood were added or removed from the circulation.

The slope of the plot is related to the compliance of the entire system, defined as

$$[5.12.3] \qquad C_S = \frac{\Delta V_S}{\Delta P_S}$$

where C_S is the **compliance** of the system, ΔV_S is the total volume of blood in the system, and ΔP_S is the pressure in the system. The entire system is comprised of

QUANTITATIVE HUMAN PHYSIOLOGY

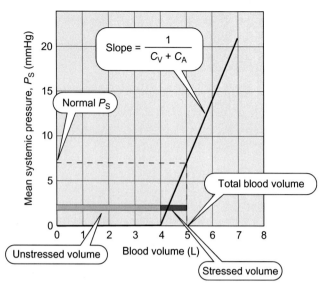

FIGURE 5.12.4 Relation between blood volume and mean systemic pressure. The graph is an extrapolation from experiments performed in dogs in which mean circulatory pressure (which approximates mean systemic pressure) was measured after withdrawal or addition of blood.

two major parts, as shown in Fig. [5.12.3]: the arteries and arterioles and the venules and veins. Each can be described by its own compliances:

[5.12.4]
$$C_V = \frac{\Delta V_V}{\Delta P_V}$$
$$C_A = \frac{\Delta V_A}{\Delta P_A}$$

where C_V is the compliance of the venous side and C_A is the compliance of the arterial side of the vasculature. In this case, the two are connected with no flow so that $\Delta P_V = \Delta P_A = \Delta P_S$ and the total additional volume is the sum of the additional volumes on both sides of the vasculature: $\Delta V_S = \Delta V_A + \Delta V_V$

Thus, we can write

[5.12.5]
$$\begin{aligned} C_S &= \frac{\Delta V_S}{\Delta P_S} \\ &= \frac{\Delta V_A + \Delta V_V}{\Delta P_S} \\ &= \frac{\Delta V_A}{\Delta P_S} + \frac{\Delta V_V}{\Delta P_S} \\ &= \frac{\Delta V_A}{\Delta P_A} + \frac{\Delta V_V}{\Delta P_V} \\ &= C_V + C_A \end{aligned}$$

Thus, the compliance of the entire system is just the sum of the compliance of its components. The slope of the line in Figure 5.12.4 is thus $1/C_S = 1/(C_V + C_A)$.

The volume of blood that just fills the circulatory system with no pressure is called the **unstressed volume**. The added amount of blood necessary to bring the pressure from zero to the mean systemic pressure is the **stressed volume**. Thus, some physiologists refer to the mean systemic pressure as the **mean systemic filling pressure**. The values for the unstressed volume and stressed volume are not constant: constriction of the veins (venoconstriction) lowers the unstressed volume and increases the stressed volume. Thus, **venoconstriction increases the mean systemic pressure**. Similarly, changes in blood volume alone will change the stressed volume without changing the unstressed volume. We will return to these effects later.

THE VASCULAR FUNCTION CURVE CAN BE DERIVED FROM ARTERIAL AND VENOUS COMPLIANCES AND TPR

We now proceed to derive a relationship between flow through the veins, Q_{veins}, and the right atrial pressure, P_{RA}. This derivation is a modification of one provided by Levy (M.N. Levy, The cardiac and vascular factors that determine systemic blood flow, *Circulation Research* 44:739–747, 1979).

When the heart is stopped and no autonomic reflexes are activated, the blood redistributes itself so that pressure is equal everywhere in the circulation and there is no flow. At this point, the pressure is the mean systemic pressure, P_{MS}. If we start up the heart again, it begins pumping blood from the venous side of the circulation to the arterial side. As it does so, the added volume in the arteries expands the arteries and increases its pressure. Similarly, removal of volume from the venous side decreases its pressure. The increments or decrements in pressure are described by their respective compliances, as dictated by Eqn [5.12.4]:

[5.12.6]
$$\Delta P_A = \frac{\Delta V}{\Delta C_A}$$
$$\Delta P_{RA} = \frac{-\Delta V}{C_V}$$

where ΔP_A is the *increment* in pressure caused by adding the volume ΔV, ΔP_{RA} is the *increment* in pressure on the venous side caused by removing the volume ΔV, C_A and C_V are the compliances for the arterial and venous part of the circulation, respectively. Note that the increment of pressure for the venous side is negative (it is really a *decrement*) because volume is being withdrawn. ΔV itself is always taken as positive. We use P_{RA}, the pressure at the level of the right atrium, as a measure of the venous compartment pressure.

The volume being added to the arterial side is equal to the volume being withdrawn from the venous side. Thus, the two equations in Eqn [5.12.6] can be combined. The increment in arterial pressure is then given as

[5.12.7]
$$\Delta P_A = -\Delta P_{RA}\frac{C_V}{C_A}$$

This is the *increment* in pressure that occurs on top of the pressure initially in the arteries, which was P_{MS}. The pressures are thus given as

[5.12.8]
$$P_A = P_{MS} + \Delta P_A$$
$$P_{RA} = P_{MS} + \Delta P_{RA}$$

Note that ΔP_{RA} is negative in Eqn [5.12.6] so that the pressure in the veins is lowered by virtue of the heart pumping a volume out of it. We can rewrite the lower equation in Eqn [5.12.8] as

[5.12.9] $$\Delta P_{RA} = P_{RA} - P_{MS}$$

This result can in turn be substituted into Eqn [5.12.7] to obtain

[5.12.10] $$\Delta P_A = -[P_{RA} - P_{MS}]\frac{C_V}{C_A}$$

We can now substitute this result into the top equation of Eqn [5.12.8] to find

[5.12.11] $$P_A = P_{MS} - [P_{RA} - P_{MS}]\frac{C_V}{C_A}$$
$$= P_{MS}\left[1 + \frac{C_V}{C_A}\right] - P_{RA}\frac{C_V}{C_A}$$

The total flow through the veins is given in Eqn [5.12.2]. We can substitute in from Eqn [5.12.11] to obtain

$$Q_{veins} = \frac{P_A - P_{RA}}{TPR}$$
$$= \frac{P_{MS}[1 + (C_V/C_A)] - P_{RA}(C_V/C_A) - P_{RA}}{TPR}$$
$$= \frac{P_{MS}[1 + (C_V/C_A)] - P_{RA}[1 + (C_V/C_A)]}{TPR}$$
$$= -\frac{[1 + (C_V/C_A)]}{TPR}[P_{RA} - P_{MS}]$$

[5.12.12]

The last equation is what we need, and is rewritten below:

[5.12.13] $$Q_{veins} = -\frac{[1 + (C_V/C_A)]}{TPR}[P_{RA} - P_{MS}]$$

Equation [5.12.13] describes the **vascular function curve**. Q_{veins} is the flow through the veins, P_{RA} is the pressure at the right atrium, and P_{MS} is the mean systemic pressure. This equation only approximates the actual relationship between Q_{veins} and P_{RA} because we have calculated the pressures using the compliances, and this assumes no flow. We then inserted these static pressures into a formula for flow!

THE EXPERIMENTALLY DETERMINED VASCULAR FUNCTION CURVE FOLLOWS THE THEORETICAL RESULT ONLY FOR POSITIVE RIGHT ATRIAL PRESSURES

The vascular function curve was determined experimentally in dogs by replacing the heart with a pump so that the flow could be set and right atrial pressure measured. The design of this experiment should make it clear that the right atrial pressure is not an independent variable but depends on the pumping action of the heart. Indeed, according to our definition, mean systemic pressure is the pressure measured when the heart is stopped. The right atrial pressure differs from the mean systemic pressure only when the heart is pumping. We can rewrite Eqn [5.12.13] as

[5.12.14] $$P_{RA} = -\frac{TPR}{\left[1 + \frac{C_V}{C_A}\right]}Q_{veins} + P_{MS}$$

The dependence of P_{RA} on Q_{veins} determined experimentally is shown in Figure 5.12.5. Here we plot P_{RA} against Q_{veins} specifically to indicate how this was accomplished: the pump rate was controlled (it is the independent variable) and the resulting right atrial pressure (P_{RA}) was measured. Here Q_{veins} is labeled "cardiac output" to indicate that in the intact system the heart beat draws down the right atrial pressure. When it pumps faster, right atrial pressure drops further.

The curve shown in Figure 5.12.5 generally agrees with the theoretical analysis that results in Eqn [5.12.14]. At zero flow, the right atrial pressure equals the mean systemic pressure. This fits the definition of the mean systemic pressure as the pressure throughout the systemic circulation at zero flow. As flow increases, the right atrial pressure becomes smaller until it reaches 0 mmHg. The negative slope agrees with the equation. At about 0 mmHg right atrial pressure, the curve bends, and this bend is *not* predicted by Eqn [5.12.14]. Here a new phenomenon occurs that we have not considered—the partial collapse of veins exposed to small or negative transmural pressures. The negative luminal pressure flattens the veins, changing their cross-sectional profile.

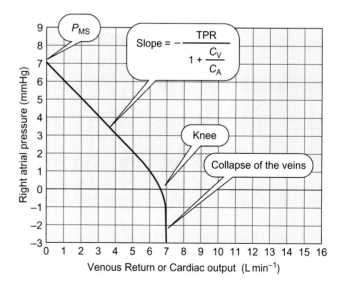

FIGURE 5.12.5 Dependence of right atrial pressure, P_{RA}, on the flow through the vasculature, which is equal to the cardiac output. The slope is related to the compliances of the venous and arterial sides of the circulation and to the TPR. At zero flow, the right atrial pressure is the mean systemic pressure. At low P_{RA}, near zero and lower, the flow levels off because the veins collapse partly due to their lower luminal pressure transmitted to the veins from the atrium. Flow remains high, but its increase is limited by reduced aggregate cross-sectional area of the veins.

Flow still continues at a high rate but cannot increase further due to the flutter of the veins.

SIMULTANEOUS SOLUTION OF THE CARDIAC FUNCTION CURVE AND VASCULAR FUNCTION CURVE DEFINES THE STEADY-STATE OPERATING POINT OF THE CARDIOVASCULAR SYSTEM

The right atrial pressure determines the stretch of the right ventricle, which in turn determines the output of the right heart, which in turn determines the output of the left heart. The output of the left heart is the cardiac output. The plot of cardiac output as a function of right atrial pressure is the cardiac function curve. The vascular function curve describes how the flow through the veins sets the right atrial pressure. In the intact cardiovascular system operating at steady state, the cardiac output is the same as the flow through the veins. Thus, the right atrial pressure sets the cardiac output, which then sets the right atrial pressure through its interaction with the vasculature. This forms a compact negative feedback system that naturally finds its steady state. The steady-state operating point of the combined cardiovascular system can be found by simultaneously solving the cardiac function curve and the vascular function curve. This can be done graphically by plotting the two curves on the same graph. To do this, we need to exchange the axes of Figure 5.12.5. This curve with right atrial pressure plotted on the abscissa (the horizontal axis) is called the **vascular function curve**. The cardiac function curve and vascular function curve are shown overlaid in Figure 5.12.6. The steady-state operating point is the intersection of the two curves: in normal resting conditions this is $5\,L\,min^{-1}$ at a right atrial pressure of 2 mmHg.

CHANGING ARTERIOLAR RESISTANCE ROTATES THE VASCULAR FUNCTION AROUND P_{MS}

The TPR is approximated as the sum of the aggregate arterial and venous resistances. Most of that resistance is in the arterial side of the circulation and specifically in the arterioles. **Vasodilation** refers to dilation of these resistance vessels and reduction in their resistance; **vasoconstriction** refers to a reduction in their caliber and an increase in their resistance. Equation [5.12.13] describes the vascular function curve. TPR is in the denominator of the slope of flow (cardiac output) against right atrial pressure. Increasing TPR thus decreases the slope, while leaving the intercept at P_{MS}, the mean systemic pressure. P_{MS} is unaffected by changes in TPR. Changes in TPR also do not affect the compliances. Thus, increasing TPR rotates the vascular function curve downward, and decreasing TPR rotates the curve upward, as shown in Figure 5.12.7.

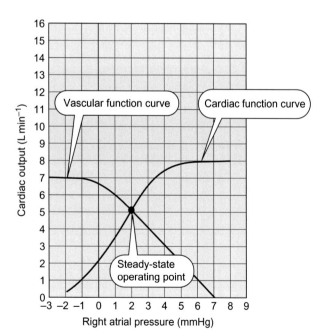

FIGURE 5.12.6 Graphical solution of the steady-state operation of the cardiovascular system. The cardiac function curve describes the dependence of cardiac output on the right atrial pressure. The vascular function curve describes the dependence of right atrial pressure on cardiac output. At steady state, both functions must be true because there is only one steady-state flow (the cardiac output) and one right atrial pressure. The steady-state operating point corresponds to the intersection of the two curves.

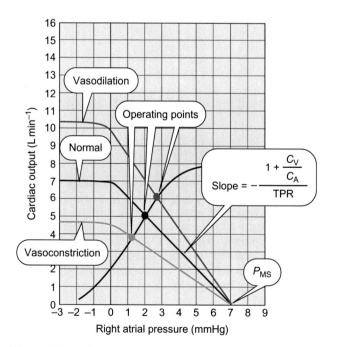

FIGURE 5.12.7 Changing arteriolar resistance rotates the vascular function curve and establishes new steady-state operating points. The normal vascular function curve and cardiac function curves are shown in black. Vasoconstriction increases the TPR which decreases the slope of the vascular function curve but does not alter its intercept at P_{MS} (the light blue curve). Thus, it rotates the curve downward around P_{MS}. The result is a decrease in the cardiac output at the steady-state operating point. Vasodilation decreases the TPR and rotates the vascular function curve upward around P_{MS} (the dark blue curve) Vasodilation increases the steady-state cardiac output.

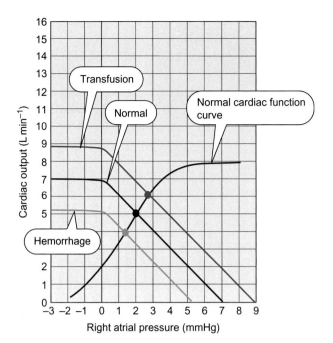

FIGURE 5.12.8 Changes in blood volume shift the vascular function curve vertically. The normal vascular function curve is shown in black. Hemorrhage causes loss of blood from the circulation, which reduces the stressed volume and lowers the mean systemic pressure. This shifts the vascular function curve downward (the light blue curve). Transfusion of extra blood increases the stressed volume and raises the mean systemic pressure, shifting the vascular function curve upward (the dark blue curve). The steady-state operating points shift accordingly to the new intersection of the vascular function curve with the normal cardiac function curve.

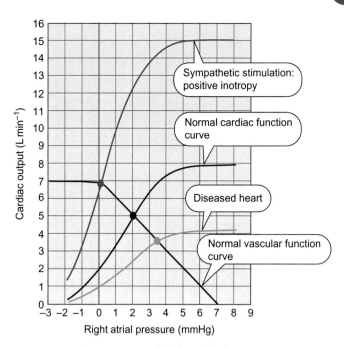

FIGURE 5.12.9 Inotropic agents shift the cardiac function curve. Positive inotropic agents such as sympathetic stimulation rotate the cardiac function curve counterclockwise, increasing the cardiac output at its steady-state operating point (dark blue curve). The normal cardiac function curve and the vascular function curve are shown in black. Depressed contractility such as in disease states reduces the cardiac output at the steady-state operating point (light blue curve).

CHANGES IN BLOOD VOLUME SHIFT THE VASCULAR FUNCTION CURVE VERTICALLY

Hemorrhage refers to the loss of blood from the cardiovascular system. Inspection of Figure 5.12.4 shows that reduction in blood volume will reduce the stressed volume, causing a decrease in the mean systemic pressure. This has the effect of shifting the vascular function curve downward. Increase in the blood volume by transfusion has the opposite effect: it increases the stressed volume and raises the mean systemic pressure. It shifts the vascular function curve upward. These shifts and their effects on the steady-state operating point are shown in Figure 5.12.8.

CHANGES IN THE CARDIAC FUNCTION CURVE CHANGE THE STEADY-STATE OPERATING POINT

As described in Chapter 5.8, the cardiac function curve can be rotated counterclockwise by positive inotropic agents such as cardiac glycosides or by sympathetic stimulation. Conversely, the cardiac function curve can rotate clockwise in pathological conditions or by negative inotropic agents. These effects are due to changes in the heart's contractility. Figure 5.12.9 illustrates the effect of these changes on the steady-state operating point of the cardiovascular system.

STRENUOUS EXERCISE ALTERS MULTIPLE PARTS OF THE CARDIOVASCULAR SYSTEM

Strenuous exercise has four major effects on the cardiovascular system. These are:

1. Increased heart rate
2. Increased cardiac contractility
3. Vasoconstriction and vasodilation
4. Venoconstriction.

Vasoconstriction occurs in some beds that are inessential to the immediate problem posed by an emergency. Vasodilation occurs in the muscle beds that are being used. The net result is a pronounced reduction in the TPR, which increases the slope of the vascular function curve. Venoconstriction has the effect of shifting the point at which the circulatory system fills with blood. By reducing the caliber of the veins, it effectively increases the stressed volume and thereby increases the mean systemic pressure. This shifts the vascular function curve upward. The increased heart rate and increased cardiac contractility combine to markedly shift the cardiac function curve upward and to the left. The net result is that the steady-state operating point shifts to much higher cardiac output. These effects are displayed in Figure 5.12.10.

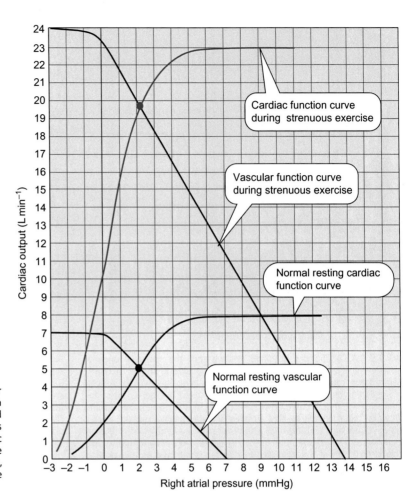

FIGURE 5.12.10 Operating point of the cardiovascular system during strenuous exercise. The cardiac function curve is greatly enhanced by increased heart rate and increased contractility. The vascular function curve is affected by venoconstriction, which raises mean systemic pressure; by vasodilation, which increases the slope of the linear portion of the curve; and by the muscle pumps, which also increase the slope and shift the vascular curve upward.

Muscular activity markedly shifts both the mean systemic pressure and the slope of the vascular function curve. It does this by providing "muscle pumps" in the venous circulation that have the effect of negative resistances. Contraction of limb muscles squeezes the vessels inside the muscle. Because veins have valves that assure unidirectional flow, this squeeze propels blood toward the heart. When the muscle relaxes during rhythmic exercise, such as running, the veins fill up again from blood that drains the muscle. In this way, blood flow into and out of the muscle is pulsatile, deriving its frequency from the frequency of activation of the muscle. Thus, each active muscle acts as a miniature auxiliary pump that assists the heart in circulating the blood. During strenuous exercise, fully one-half of the total energy of circulation can be provided by skeletal muscles.

SUMMARY

During the short term, the cardiovascular system can be regarded as being closed. Because of this, at steady state, the cardiac output must match the venous return. Mismatches between cardiac output and venous return are quickly resolved by the transfer of blood either from the venous side of the circulation to the arteries or the other way around. This transfer of volume alters the pressure in each side so that flows become equal.

The mean systemic pressure is the pressure throughout the systemic circulation when all of its inputs and outputs are simultaneously blocked. This is similar to the mean circulatory pressure, which is the pressure throughout all of the circulatory system (systemic and pulmonary systems) when the heart is stopped and fluid redistributed before any cardiovascular reflexes occur. The mean systemic pressure depends on the volume of the vessels, the compliance of the venous and arterial side of the circulation, and the volume of blood. The volume of blood required to fill the vessels to 0 mmHg pressure is the unstressed volume. The remainder of the blood volume is the stressed volume. Mean systemic pressure increases linearly with the stressed volume according to the compliance of the whole system. Transfusion of extra blood and venoconstriction raises the mean systemic pressure.

The cardiac function curve plots the cardiac output, in units of $L\,min^{-1}$, against the right atrial pressure. The right atrial pressure is the preload of the heart that determines its degree of stretch and hence its force of contraction. The cumulative volumes ejected by the heart into the arterial side of the circulation produce a pressure gradient leading back to the heart and establish the right atrial pressure. Thus there is a relation between flow through the veins and the right atrial pressure. This relationship forms the vascular function curve, which is plotted as flow against right atrial pressure. The steady-state operating point of the system is the simultaneous solution of the cardiac function curve and the vascular function curve. It is solved graphically by the intersection of the two curves.

The vascular function curve intercepts the abscissa (right atrial pressure axis) at the mean systemic pressure. The slope of the linear portion of the vascular function curve is inversely related to the TPR. Increasing arteriolar resistance (vasoconstriction) therefore decreases the slope of the vascular function curve and lowers the cardiac output. Vasodilation has the opposite effect: the slope increases and cardiac output increases.

Hemorrhage displaces the vascular function curve downward, decreasing mean systemic pressure but it does not significantly change the slope of the vascular function curve. Transfusion of extra blood has the opposite effects: mean systemic pressure is increased and the vascular function curve is displaced upward.

Changing the contractility of the heart alters the operating point of the system by shifting the cardiac function curve. This allows for large increases in cardiac output. During strenuous exercise, the activity of the muscles helps propel blood around the circulatory system, greatly reducing the work of the heart and allowing for large cardiac output.

REVIEW QUESTIONS

1. What is the mean systemic pressure? Is this the same as mean circulatory pressure?
2. How does right atrial pressure determine cardiac output from the *left* heart?
3. What is the unstressed volume? What is the stressed volume? What is the slope of the curve of volume against pressure with the heart stopped?
4. Draw the vascular function curve (flow versus right atrial pressure). Why does flow go down when pressure rises?
5. What is the slope of the vascular function curve (flow versus right atrial pressure)? What is the relative capacitance of the venous and arterial systems? What effect does vasodilation have on the vascular function curve? What effects does vasoconstriction have?
6. What effect does hypovolemia have on the vascular function curve? What is the effect of transfusion on the vascular function curve?
7. Why is the operating point of the cardiovascular system defined by the intersection of the vascular function curve and cardiac function curve?
8. What effect does exercise have on the operating point of the cardiovascular system?

5.13 Regulation of Arterial Pressure

Learning Objectives

- Explain why arterial pressure both drives and derives from flow
- Describe what is meant by the term "baroreceptor"
- Describe the anatomic location of the cardiac baroreceptors
- Describe the response of the baroreceptors to arterial pressure
- Explain the reflex effect of raising blood pressure on parasympathetic and sympathetic input to the heart and vasculature
- Explain the reflex effect of lowering blood pressure on parsympathetic and sympathetic input to the heart and vasculature
- Identify the main regions of the brain involved in the baroreflex
- Describe the respiratory sinus arrhythmia and explain its origin
- Recognize and describe the individual block components of the renal–cardiovascular systems diagram
- List the major hormonal influences on long-term regulation of blood pressure
- Describe the effect of ADH, RAA, and ADH on the renal function curve

ARTERIAL PRESSURE *DRIVES* FLOW BUT ARTERIAL PRESSURE ALSO *ARISES* FROM FLOW

The first law of the cardiovascular system, which we have used repeatedly, is written as

[5.13.1]
$$Q_V = \frac{\Delta P}{R}$$
$$CO = \frac{P_A - P_{RA}}{TPR}$$

where Q_V is the flow around the entire circulatory loop, equal to the cardiac output, CO, P_A is the pressure in the arteries, specifically in the aorta, P_{RA} is the pressure in the right atrium, and TPR is the total peripheral resistance. Generally the pressure in the right atrium is very small compared to the arterial pressure, and this equation can be approximated as

[5.13.2] $\qquad P_A = CO \times TPR$

This is an empirical relationship, not a causal one. For example, this equation indicates that decreasing the TPR by a factor of 2 will halve the arterial pressure *if* CO is maintained constant. Although this is true, decreasing the TPR by a factor of 2 generally will increase cardiac output (see Chapter 5.12).

The arterial pressure is pulsatile. The proper pressure to use in Eqns [5.13.1] and [5.13.2] is the **mean arterial pressure**, which is typically approximated as the diastolic pressure plus one-third of the pulse pressure (Eqn [5.9.9]). The actual mean arterial pressure depends on the systole duty cycle, the fraction of time the heart spends in systole, which varies with heart rate. A more recent study (Razminia et al., Validation of a new formula for mean arterial pressure calculation. Catheterization and Cardiac Interventions, 63:419–425, 2004) suggests that we use the equation

[5.13.3] $\quad P_A = P_{diastolic} + (0.33 + 0.0012 \times HR)\Delta P_{pulse}$

which corrects for the changes in the duty cycle of the heart with the heart rate. Here P_A is the mean arterial pressure, $P_{diastolic}$ is the diastolic pressure, HR is the heart rate, and ΔP_{pulse} is the pulse pressure.

Although Eqn [5.13.1] is consistent with our idea that pressure drives flow, pressure within the arteries is also generated by flow. Consider the experiment in Chapter 5.12 in which we imagined that we drained the circulatory system of blood completely, and then began to fill it up again. When we add enough blood to make the pressure everywhere equal to zero, we have added the **unstressed volume**. On adding more volume, the pressure increases linearly with the added volume. The volume above the unstressed volume is the **stressed volume**. The stressed volume of blood distributes itself between the venous and arterial compartments. Since the venous compliance is much larger than the arterial compliance, most of the stressed volume is in the veins. We can calculate the fraction of volume in each compartment. We define three compliances: one for the venous side, C_V, one for the arterial side, C_A, and one equivalent compliance for the entire system, C_S:

[5.13.4a] $\qquad C_S = \dfrac{\Delta V_S}{\Delta P_S}$

[5.13.4b] $\qquad C_V = \dfrac{\Delta V_V}{\Delta P_V}$

$$[5.13.4c] \quad C_A = \frac{\Delta V_A}{\Delta P_A}$$

Since the pressure is the same everywhere, we can write

$$[5.13.5] \quad \Delta P_V = \Delta P_A = \Delta P_S = \frac{\Delta V_S}{C_S}$$

Inserting this back into Eqns [5.13.4b] and [5.13.4c] and rearranging for the volume increments, we get

$$[5.13.6] \quad \Delta V_V = \frac{C_V}{C_A + C_V} \Delta V_S$$
$$\Delta V_A = \frac{C_A}{C_A + C_V} \Delta V_S$$

This makes use of Eqn [5.12.5], which states that the compliance of the system is the sum of the compliances. Thus, the volumes in the venous and arterial side distribute themselves according to their relative compliances. The ratio of the venous compliance to the arterial compliance has been estimated in experimental animals to be about 18:1, so that $C_V/(C_V + C_A) \approx 18/19$ and $C_A/(C_V + C_A) \approx 1/19$. Thus, most of the stressed volume is in the venous side of the circulation. We can estimate the individual compliances from the slope of the line in Figure 5.12.4. When withdrawing or adding blood, mean systemic pressure increases according to Eqns [5.13.4a,b,c] with the compliance of the entire system. The slope is about 7 mmHg L^{-1}. This is the inverse of C_S. Thus we have two equations in two unknowns:

$$[5.13.7] \quad C_S = C_V + C_A \approx 0.142 \text{ L mmHg}^{-1}$$
$$C_V \approx 18 \times C_A$$

Solving for C_A and C_V, we find $C_A \approx 0.007$ L mmHg^{-1} and $C_V \approx 0.135$ L mmHg^{-1}.

Suppose now that we have added back the original blood volume. The pressure in the system, with the heart stopped, is the **mean systemic pressure**. Most of the blood will be in the venous system because it has the larger volume and the larger compliance. Now suppose we start up the heart. As in our analysis in Chapter 5.12, removing a volume ΔV_{SV} from the venous compartment to the arterial compartment will raise the arterial pressure and lower the venous pressure. For a typical stroke volume of about 70 mL, the pressure increases by $\Delta V_{SV}/C_A$ in the arteries and decreases by $\Delta V_{SV}/C_V$ in the veins. Using the approximate values for C_A and C_V that we calculated above, this increases P_A by 10 mmHg and decreases P_V by 0.5 mmHg. A single heart beat does not move enough blood to set the normal arterial pressure. Repetitive heart beats continue to move blood from the venous to the arterial side of the circulation. If compliance is independent of volume, then a mean arterial pressure would obtain when about 650 mL of blood was removed from the venous side to the arterial side (93 mmHg × 0.007 mL mmHg^{-1} = 651 mL). This would lower the venous pressure from 7 mm Hg mean systemic pressure to about 2.2 mmHg. Thus, the operating pressures in the arterial side and venous side require about nine stroke volumes. The heart actually pumps more than that to achieve the shift in volume, because part of the transferred volume returns to the venous side while the heart makes the transfer. So if we were to start the heart up again when the blood is evenly distributed so that the pressure is the mean systemic pressure everywhere, it would take more than nine heart beats to reach normal pressure.

This purpose of this calculation is to emphasize that **the heart produces arterial pressure**. The pulsatile pressure of the left ventricle is translated to the sustained pressure of the vasculature by the elasticity or compliance of the vessels.

REGULATION OF ARTERIAL PRESSURE OCCURS ON THREE SEPARATE TIMESCALES INVOLVING THREE DISTINCT TYPES OF MECHANISMS

As pointed out above, the heart causes the high arterial pressure by pumping blood from the low-pressure, venous side of the circulation to the high-pressure, arterial side. The pressure that results could be increased by increasing the stroke volume of the heart, by changing the compliance of the arterial system or its resistance to run off of the stroke volume placed in it by the heart, or by changing the volume of blood in the vasculature. These three distinct mechanisms of regulating the arterial pressure are controlled by three separate and distinct parts of the cardiovascular system. Thus arterial pressure is regulated by controlling three fundamental and independent variables:

1. **Cardiac contractility** (control of the heart's strength of contraction)
2. **Vascular smooth muscle contractility** (caliber and compliance of the vessels)
3. **Blood volume** (regulation of renal function).

The mechanisms involve:

- **Neurogenic mechanisms**: fast response (seconds)
- **Hormonal mechanism**: intermediate response (minutes to hours)
- **Intrinsic mechanisms**: slow response of blood volume (days to weeks).

BAROUR CEPTORS IN THE CAROTID SINUS AND AORTIC ARCH SENSE BLOOD PRESSURE

Sensors for the arterial pressure are called **baroreceptors**, but these actually sense stretch. Baroreceptors are found in the **carotid sinus** and the **aortic arch** (see Figure 5.13.1). The carotid sinus is a thin-walled dilatation located at the proximal end of the internal carotid artery just above the bifurcation of the common carotid artery into the internal and external carotid artery. Increasing pressure in the internal carotid stretches the walls of the carotid sinus, which excites the baroreceptors and increases their firing rate.

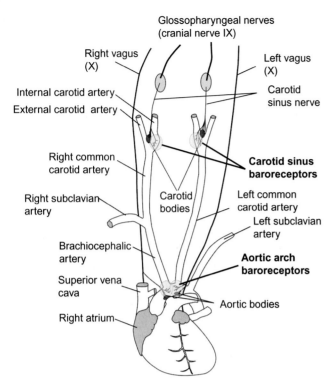

FIGURE 5.13.1 Location of the arterial baroreceptors. Stretch receptors that sense arterial pressure are located in the carotid sinus, near the bifurcation of the common carotid artery form that is the internal and external carotid arteries, and in the aortic arch. Stretch of these vessels excites baroreceptor afferents. The afferents from the carotid bodies travel to the central nervous system over the glossopharyngeal nerve. Afferents from the aortic arch reach the CNS over the vagus nerve.

Large diameter, myelinated **baroreceptor A fibers** activate at blood pressure between 30 and 90 mmHg. Therefore, they are active at normal blood pressure levels but become saturated with higher blood pressure and therefore cannot inform the CNS about high blood pressure.

Unmyelinated **baroreceptor C fibers** are more numerous than the A fibers. They have lower conduction velocities, but higher thresholds, in the range of 70–140 mmHg. At normal blood pressure, only about a quarter of the C fibers are activated. At progressively higher pressures, progressively more C fibers are recruited (see Figure 5.13.2).

The rate of firing of the baroreceptors responds not only to stretch but the rate of stretch. Thus, a rapid rise in blood pressure elicits a burst of impulses that subsequently slows. The rapid burst of impulses is called the **dynamic** response. The slower, steady response to static stretch corresponds to the **static** response. This phenomenon is illustrated in Figure 5.13.3.

As blood pressure increases in the range of normal, the rate of firing of individual baroreceptor A fibers increases, but additional baroreceptor C fibers begin firing as the pressure exceeds their threshold. This is called recruitment. Both the carotid sinus nerve and aortic nerves contain a large number of fibers. Because of the different thresholds, the aggregate response of the

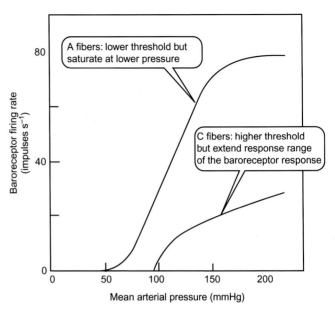

FIGURE 5.13.2 Heterogeneity of baroreceptor responses. Type A fibers are myelinated fibers that respond to low pressures but saturate first at pressures around 150 mmHg. C fibers have higher thresholds, between 70 and 140 mmHg, but continue to respond at higher pressures. (Source: Adapted from H.M. Coleridge, J.C.G. Coleridge, and H.D. Schultz, Journal of Physiology **394**:2910–313, 1987.)

afferent nerves has a greater usable range than the response of any one fiber or of any one type of fiber.

As P_A increases, the aggregate afferent impulses from the baroreceptors will increase, with greater proportions from the C fibers as the pressure increases. When the pulse pressure increases, the incremental burst of activity during systole is roughly proportional to the pulse pressure. The ability to sense the size of the pulse is important whenever the stroke volume decreases, such as in **orthostasis** (adoption of an upright posture) and in moderate **hemorrhage**, when the stroke volume falls without much change in mean arterial pressure (see Figure 5.13.4).

THE BAROREFLEX REGULATES HEART AND VASCULATURE TO STABILIZE BLOOD PRESSURE

Increases in blood pressure increase the baroreceptor input to the brain stem through the glossopharyngeal and vagus nerves. The brain stem integrates this information, subsequently reducing sympathetic outflow to the cardiovascular system and increasing parasympathetic outflow over the vagi to the heart. The effects include:

- Decreased heart rate due to increased parasympathetic stimulation of the SA node. Together with reduced sympathetic activity, the result is **bradycardia** and **reduced myocardial contractility**. These reduce cardiac output.
- Reduced sympathetic output to the vasculature causes vasodilation and therefore a **fall in total peripheral resistance (TPR)**.

Since Eqn [5.13.2] tells us that the average arterial pressure is CO × TPR, and the baroreflex reduces both

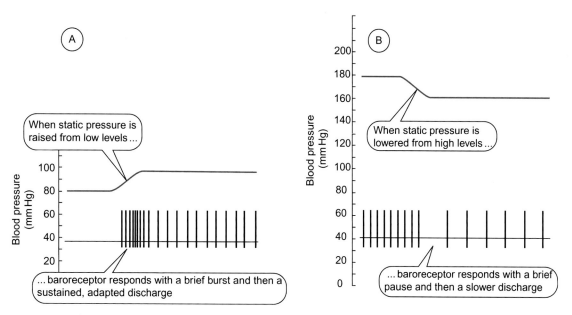

FIGURE 5.13.3 Response of the baroreceptors to changes in static pressure. When pressure is raised from low levels, there is a brief, rapid burst of impulses followed by a sustained but slower discharge rate. Thus, the receptors show a partial adaptation. Conversely, when pressure is lowered from a high level, there is a brief pause followed by sustained, slower rate of discharge. The responses are shown from single myelinated baroreceptor afferents from cat carotid sinus. (Source: *Adapted from S. Landgren*, Acta Physiologica Scandinavica **26**:1–34, 1952.)

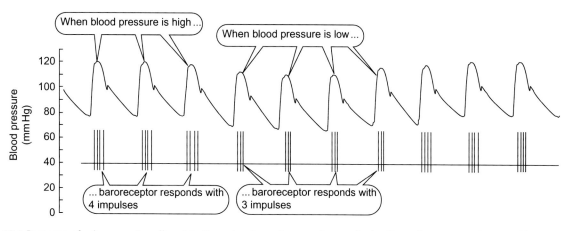

FIGURE 5.13.4 Response of a baroreceptor afferent to the pulse. Recordings are from a single afferent baroreceptor from a rabbit aorta. When the arterial pressure is high, the baroreceptor responds with 4 impulses. When the pressure is slightly lower, it responds with 3. Summation of responses from many fibers allows for a graded response of the average pressure and of the pulse pressure. (Source: *Modified from S.E. Downing*, Journal of Physiology **150**:210–213, 1960.)

CO and TPR, the consequence of the baroreflex is to reduce arterial pressure toward normal.

Reduced arterial blood pressure engages the opposite response of the baroreflex to increased blood pressure. The firing frequency of the baroreceptors decreases and engages the brain stem to cause:

1. Increased heart rates due to decreased parasympathetic stimulation of the SA node along with increased sympathetic activity. The **tachycardia** is accompanied by **increased myocardial contractility**.
2. Increased sympathetic output causes vasoconstriction and a **rise in TPR**. Both of these raise the blood pressure toward normal.

The baroreflex is a negative feedback mechanism that corrects acute changes in blood pressure. It acts rapidly.

The delay between an increase in high blood pressure and bradycardia is about 0.5 s, and the delay between high blood pressure and vasodilation is about 1.5 s.

THE BAROREFLEX MEDIATES PARASYMPATHETIC AND SYMPATHETIC OUTPUT FROM CENTERS LOCATED IN THE MEDULLA

SENSORY AFFERENTS TERMINATE IN THE NUCLEUS TRACTUS SOLITARIUS

The dorsal–medial medulla contains an elongated nucleus of cells called the **nucleus tractus solitarius (NTS)**, which is the terminus of nearly all of the cardiovascular afferents including the baroreceptors, cardiopulmonary afferents, arterial chemoreceptors, and pulmonary stretch receptors. Sensory integration of these various signals occurs here, and the NTS sends its output to the **nucleus ambiguus (NA)**, the **caudal ventrolateral medulla (CVLM)**, and the **hypothalamus**.

NTS TONICALLY INHIBITS HEART RATE THROUGH VAGAL EFFERENTS

Vagal efferents originating mainly in the **nucleus ambiguus**, but some in the **dorsal motor nucleus** of the vagus in the medulla, inhibit the heart rate by reducing the slope of the pacemaker potential in SA nodal cells. Baroreceptors tonically activate cells in the NTS, and this tonic excitation is relayed to the nucleus ambiguus where vagal efferents are tonically active. Thus **in humans there is a constant vagal tone** that continuously inhibits the heart rate. Reductions in blood pressure relieve this tonic inhibition, thereby accelerating the heart rate by removing parasympathetic tone. This completes a negative feedback loop: the accelerated heart rate increases the cardiac output and tends to restore the blood pressure. When blood pressure rises, the reverse occurs: NTS stimulates the NA more and there is greater vagal inhibition of heart rate (see Figure 5.13.5).

SYMPATHETIC STIMULATION OF THE HEART IS THROUGH THE ROSTRAL VENTROLATERAL MEDULLA

The output of the NTS also goes to the CVLM, where it stimulates inhibitory connections to the rostral ventrolateral medulla, RVLM. The RVLM connects to sympathetic preganglionic neurons in the interomedial gray of the thoracic spinal cord. Thus, decreased blood pressure that is sensed by the baroreceptors reduces the output of the NTS to the CVLM, which relieves its inhibition of the RVLM, thereby increasing sympathetic stimulation of the heart and the vessels. This accelerates the heart rate and vasoconstricts, both of which tend to restore the decrease in blood pressure. When blood pressure rises, the opposite occurs: increased activation of the inhibition of CVLM on RVLM causes reduces sympathetic outflow to the heart and vessels.

PARASYMPATHETIC WITHDRAWAL ACCELERATES THE HEART FASTER THAN SYMPATHETIC STIMULATION

In humans, there is continual parasympathetic tone. Its effect is much faster than the sympathetic effects, so that acceleration of heart rate in the short-term regulation of blood pressure is typically first accomplished by either removal of parasympathetic tone (when blood pressure falls) or adding to the parasympathetic tone when blood pressure rises.

INSPIRATION INFLUENCES HEART RATE—THE *RESPIRATORY SINUS ARRHYTHMIA*

If one continuously monitors heart rate and blood pressure in a resting person, one quickly observes a regular oscillation in both heart rate and blood pressure due to breathing. This is called the **respiratory sinus arrhythmia**. Fourier analysis of the frequency of the R–R interval, as an immediate indicator of heart rate, shows that the main power in the frequency spectrum is at the heart rate, around 70 bpm, and the second source of power is in the respiratory rhythm, around 12 bpm. The **heart**

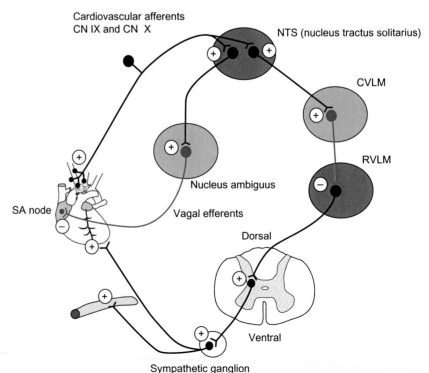

FIGURE 5.13.5 Central pathways for control of heart rate and blood pressure. Cardiac and baroreceptor afferents terminate on cells in the NTS in the medulla, entering over the glossopharyngeal nerve (CN IX) and the vagus (X). These cells send excitatory connections to the nucleus ambiguus which, along with the dorsal motor nucleus of the vagus, sends inhibitory connections over the right vagus to the SA node. Cells in the NTS also excite cells in the CVLM, which inhibit cells in the RVLM, which reduces excitation of sympathetic preganglionic cells in the thoracic spinal cord.

rate accelerates during inspiration and slows during expiration. This is due to connections to the vagal nuclei from the inspiratory centers in the medulla. These connections hyperpolarize the cardiac motor neurons in the NTS, inhibiting their response to the baroreceptors. This causes the acceleration of the heart rate during inspiration. This effect is shown diagrammatically in Figure 5.13.6.

HIGHER CENTERS INFLUENCE BLOOD PRESSURE AND HEART RATE

Everyday experience shows that emotional stress and fear markedly affect both heart rate and blood pressure. The alerting response is a stereotypical response to a dangerous or threatening environmental stimulus. It typically involves

- tachycardia, with attendant increased cardiac output;
- vasodilation of skeletal muscle;
- vasoconstriction of visceral perfusion (skin, splanchnic, and renal circulation);
- increased blood pressure and reduced sensitivity to the baroreceptor reflex.

The brain regions involved in this effect are distributed along the central long axis of the brain, being found in the amygdala, hypothalamus, and periaqueductal gray matter.

LONG-TERM REGULATION OF BLOOD VOLUME DETERMINES LONG-TERM REGULATION OF BLOOD PRESSURE

The main idea for long-term regulation of blood pressure is this: if you put more blood into the system, the pressure goes up; if you remove blood from the system, the pressure goes down. This is true only if the caliber of the blood vessels does not change. In the short term, the caliber of the vessels does change, which is the basis for sympathetic regulation of the blood pressure. In the short term, the heart rate (and therefore cardiac output) can change, and this is the basis of parasympathetic regulation of blood pressure. If we change the blood volume, we can change the baseline on which these regulations occur. The long-term regulation of the blood volume involves hormonal regulation of renal function.

A block diagram illustrating the components of long-term regulation is shown in Figure 5.13.7. The diagram is explained sequentially in terms of its blocks here.

Block I: The Blood Volume–Mean Systemic Pressure Relation

The mean systemic pressure is the pressure when the heart is stopped and all flow stops. It is the pressure that the system would have if it were filled to the indicated blood volume and under the indicated conditions. This curve can be shifted by changes in the caliber of the arteries or veins, such as occurs during sympathetic venoconstriction.

Block II: The Mean Systemic Pressure–Cardiac Output Relation

The mean systemic pressure forms the *x*-intercept of the vascular function curve. Along with the TPR and relative compliances, this determines the vascular function curve. Its point of intersection with the cardiac function curve determines the operating point of the entire system, which is the cardiac output (CO), equal to the venous return.

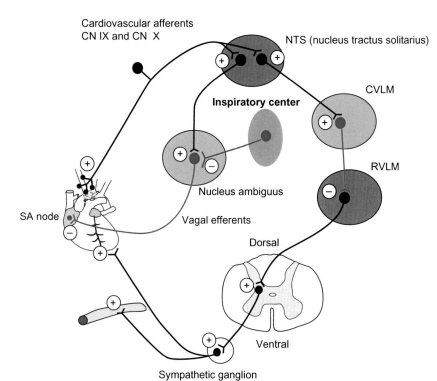

FIGURE 5.13.6 Origin of the respiratory sinus arrhythmia. During inspiration, the heart rate accelerates. This is thought to be due to inhibitory connections from the inspiratory center to cardiac motor neurons in the nucleus ambiguus. These connections momentarily desensitize the cells to the baroreceptor input, so that there is less brake on the heart, and therefore an acceleration of the heart during inspiration.

QUANTITATIVE HUMAN PHYSIOLOGY

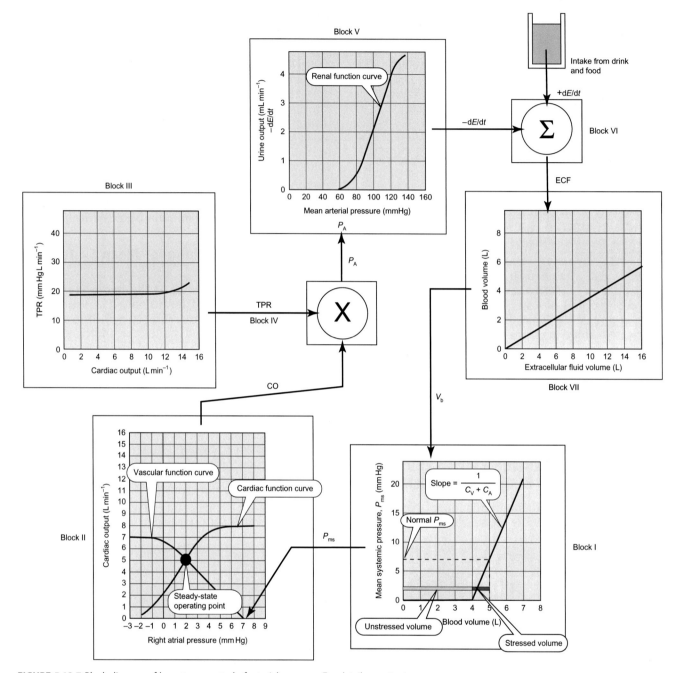

FIGURE 5.13.7 Block diagram of long-term control of arterial pressure. For details, see text.

Block III: The Cardiac Output–TPR Relationship

The TPR converts cardiac output, CO, into the mean arterial pressure. The main determinant of TPR is the caliber of the resistance vessels, the arterioles. What affects the caliber of these vessels was discussed in Chapter 5.12. TPR is independent of CO until CO gets large.

Block IV: $P_A = CO \times TPR$

The mean arterial pressure, P_A, can be calculated as $P_A = CO \times TPR$. This is an empirical relationship, not a causal one, because the variables are not independent.

Block V: The Renal Function Curve

The renal function curve relates urinary output to mean arterial pressure. The curve shown is the approximate normal curve. This is not a causal relationship but reflects the fact that both P_A and urinary excretion increase with circulatory volume. Increasing volume increases both P_A and the rate of urinary excretion, but there is a threshold between 60 and 80 mmHg below which urinary output ceases. Fluid lost from the kidney directly derives from the extracellular fluid (ECF), but this fluid also exchanges with the intracellular fluid. Thus, urinary excretion represents an equivalent volume of the ECF, with Na as its most important component, because Na content determines the ECF. The renal function curve is influenced by several hormones including **antidiuretic hormone** (ADH), the **renin–angiotensin–aldosterone** (RAA) system, and **atrial natriuretic peptide** (ANP).

Block VI: ECF Volume Is the Integral of Excretion and Intake

While some ECF volume is being lost by excretion in the urine, additional fluid is taken into the body in food and drink. The ECF is the sum of fluid gains and losses.

Block VII: The Blood Volume–Extracellular Fluid Volume Relationship

The ECF volume is the sum of the plasma volume and interstitial fluid volume. Plasma makes up about 58% of the blood volume. Blood volume is typically about 5 L, whereas ECF volume is about 14 L. Thus blood comprises about 36% of the ECF volume. Expansion of the ECF generally means expansion of the blood volume, and contraction of the ECF volume generally means contraction of the blood volume.

All of these blocks interact to produce the observed cardiac output, arterial pressure, and renal output. The systems will interact to find the steady state in which the net change to the blood volume will be zero. We will consider several perturbations of this system to illustrate its operation.

SYMPATHETIC TETRALOGY

General sympathetic stimulation exerts several cardiovascular effects. These include:

- increased heart rate;
- increased cardiac contractility;
- vasoconstriction;
- venoconstriction.

These effects are indicated in the systems diagram as shown in Figure 5.13.8.

- **Venoconstriction increases P_{ms}.** As shown in Block I, venoconstriction increases the stressed volume—it shifts the curve relating P_{ms} to blood volume to the left.
- **Vasoconstriction changes the slope of the vascular function curve.** Venoconstriction increases P_{ms}, and this shifts the vascular function curve upward. At the same time, vasoconstriction increases TPR, which decreases the slope of the vascular function curve (Block II).
- **Sympathetic stimulation rotates the cardiac function curve counterclockwise.** Sympathetic stimulation increases the force of cardiac contraction and also increases the heart rate. Both of these effects markedly increase the cardiac output for a given input. This raises the cardiac function curve and rotates it counterclockwise. (Block II).
- **Sympathetic stimulation shifts fluid from interstitial fluid to blood.** Constriction of the arterioles reduces the pressure downstream in the capillaries. This reduces the net driving force for filtration of fluid out of the capillaries and increases the movement of fluid from the interstitial fluid into the blood. This is a relatively small effect (Block VII). In addition, sympathetic constriction of renal arterioles reduces urinary excretion, preventing loss of circulatory volume and maintaining blood pressure.

HORMONAL REGULATION OF BLOOD PRESSURE

Figure 5.13.7 should clarify a central truth relating to the arterial blood pressure: it depends on the state of contraction of the arteries and how much blood is in them. How much blood is in them is the blood volume. If you reduce the blood volume, you reduce the mean systemic pressure, which lowers the vascular function curve, which lowers CO and lowers P_A. Similarly, raising the blood volume raises P_{ms} and raises the vascular function curve, which raises CO and raises P_A. The blood volume is controlled by several hormones working in concert. These are:

- RAA system hormones
- ADH
- ANP.

RAA SYSTEM

As discussed in Chapter 5.11, the RAA system begins with the secretion of the enzyme **renin** by granule cells in the afferent arteriole of the glomeruli of the kidney. Renin is secreted in response to three stimuli:

1. Renal sympathetic nerves stimulate renin release.
2. Decreased distal tubule [Na^+] stimulates renin release.
3. Decreased afferent arteriolar pressure stimulates renin release.

Renin cleaves an α_2 macroglobulin in blood called **angiotensinogen**, which is made by the liver, to form **angiotensin I**, a 10 amino acid peptide. **Angiotensin converting enzyme**, ACE, cleaves off 2 more amino acids to make **angiotensin II**. Angiotensin II has four major effects:

1. vasoconstriction;
2. release of aldosterone from the adrenal cortex;
3. increasing the sensation of thirst;
4. Release of ADH.

All of these effects counteract the cause of the initial stimulus for renin release, which was reduced afferent arteriolar pressure or reduced perfusion of the kidney. The angiotensin II causes vasoconstriction, which tends to raise blood pressure back toward normal. More importantly, the aldosterone that is released from the adrenal cortex increases Na^+ reabsorption from the kidney, and thereby reduces renal excretion of Na^+. Since Na^+ is the major extracellular ion, retention of Na^+

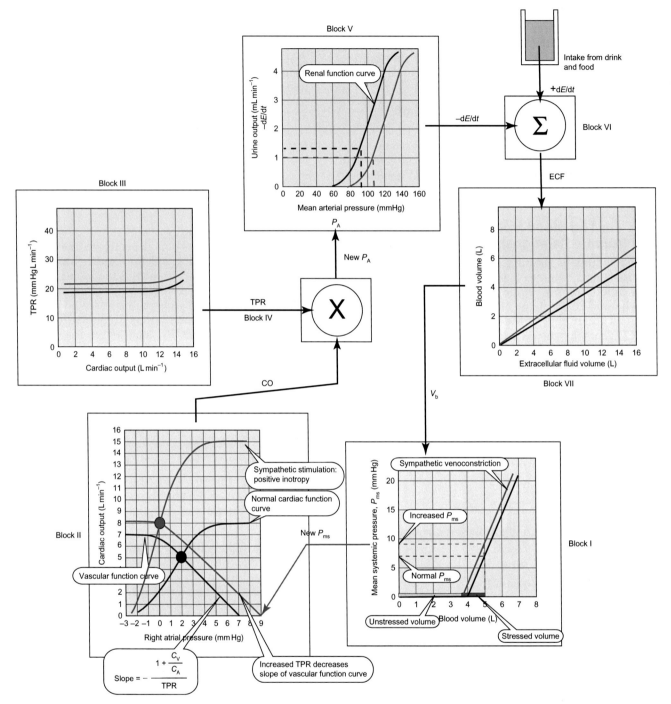

FIGURE 5.13.8 Effects of sympathetic stimulation on the cardiovascular–renal system. Venoconstriction increases P_{ms} according to Block I. The increased P_{ms} causes an upward shift in the vascular function curve, whose slope is simultaneously decreased by the increase in TPR caused by sympathetic vasoconstriction. Sympathetic stimulation increases the contractility of the heart and also its rate, rotating the cardiac function curve up and to the left (Block II). These effects on cardiac and vascular function curves cause an increase in cardiac output. The increased output along with the increased TPR (Block III) causes an increase in the arterial pressure. The increased arterial pressure increases urinary excretion according to the normal renal function curve (Block V), but sympathetic stimulation shifts the curve to the right, reducing urinary loss for a given pressure. Constriction of the arterioles causes a reduction in the capillary pressure that favors filtration out of the capillaries and into the interstitial space. This favors more blood volume per unit of ECF volume (Block VII).

retains ECF. This raises the blood volume and restores blood pressure and perfusion. **The overall effect of the RAA system is to shift the renal function curve to the right.** This effect is illustrated in Figure 5.13.9.

ANTIDIURETIC HORMONE

ADH is also known as vasopressin. ADH constricts blood vessels and reduces urine output. ADH is a 9 amino acid peptide synthesized in cells in the supraoptic nucleus and paraventricular nucleus in the hypothalamus. It travels

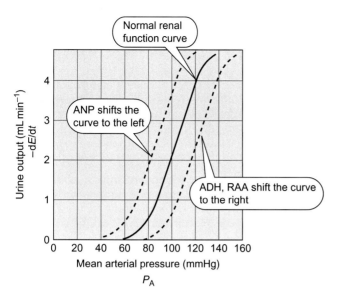

FIGURE 5.13.9 Effects of hormones on the renal function curve. ADH increases the reabsorption of water by the kidneys so that the kidney excretes a small volume of highly concentrated urine. This has the effect of reducing the extracellular volume excreted at any given pressure: this equates to a shift in the renal function curve to the right. The RAA system reduces the excretion of Na^+, which means it reduces the volume of ECF excreted by the kidney. ANP promotes the excretion of water and salt, so it shifts the renal function curve to the left.

down the axons of these cells and is secreted in the posterior pituitary in response to **hypovolemia** and to **plasma hyperosmolarity**. Thus, it helps prevent further hypovolemia by preventing, as much as possible, further loss of fluid through the urine. It retains water but not salt, so that aldosterone and ADH both tend to increase the ECF volume by limiting urinary excretion, but they do so through entirely different mechanisms that allow differential excretion of water or salt. **The overall effect of ADH is also a shift of the renal function curve to the right.** This effect is illustrated in Figure 5.13.9.

ATRIAL NATRIURETIC PEPTIDE

ANP is a 28 amino acid peptide that is secreted by atrial cells in response to stretch. The atria stretch more when right atrial pressure is high, which occurs when blood volume expands and P_{ms} increases. This shifts the vascular function curve upward and raises P_{RA}. ANP stimulates salt and water excretion to rid the body of the excess fluid. Thus, **ANP shifts the renal function curve to the left.**

Clinical Applications: Hypertension

Hypertension is defined as an abnormally elevated blood pressure. Since blood pressure is elevated during activity, its definition centers on resting blood pressure. The diagnosis of hypertension requires three successive measurement of high blood pressure on three separate visits to the physician's office, separated each by at least a week. Often hypertension is asymptomatic, but its control is vitally important for a number of chronic diseases including heart disease, myocardial infarction and heart failure, stroke, and kidney failure. The classification of hypertension is shown in Table 5.13.1.

Essential hypertension has no readily identifiable cause, whereas secondary hypertension, by definition, results from an identifiable cause such as Cushing's syndrome (overproduction of glucocorticoids, see Chapter 9.5), hyperthyroidism, pheochromocytoma (overproduction of epinephrine by the adrenal gland), and pre-eclampsia. Although there is no identifiable cause in essential hypertension, known risk factors include sedentary lifestyle, smoking, stress, obesity, salt intake, alcohol intake, and genetics.

The general treatment of hypertension is two fold: modification of lifestyle and pharmacological. Lifestyle modification include loss of weight, reduction of salt intake, exercise, stress reduction, limitation of alcohol intake and smoking.

Pharmacological intervention aims to reduce the stressed volume by reducing overall ECF volume, thereby reducing mean systemic pressure and lowering the vascular function curve. This will lower cardiac output at the same TPR. Other drugs aim to reduce TPR. Drugs used to treat hypertension include:

1. **ACE inhibitors.** These prevent the formation of angiotensin II from angiotensin I and thereby remove a potent vasoconstrictor and also reduce aldosterone secretion. This tends to waste Na^+ and reduce the ECF volume. Other drugs block the angiotensin II receptor (ARB, such as losartan) that prevents the vasoconstrictor effects of angiotensin II.
2. **Diuretics.** Hydrochlorothiazide inhibits salt reabsorption in the distal nephron (see Chapter 7.5); loop diuretics such as furosemide also waste Na^+ and excrete more of the ECF.
3. **Beta blockers** such as labetolol antagonize the effects of circulating epinephrine on beta receptors, reduce heart rate and cardiac contractility to reduce cardiac output, and thereby reduce mean arterial pressure.
4. **Calcium channel blockers** such as amlodipine block voltage-gated Ca^{2+} entry into smooth muscle cells, thereby causing vasodilation and reduction in TPR, so causing reduced mean arterial pressure.

TABLE 5.13.1 Classification of Hypertension

	Normal	Prehypertensive	Stage 1	Stage 2
Systolic (mmHg)	90–119	120–139	150–159	>160
Diastolic (mmHg)	60–79	80–89	90–99	>100

SUMMARY

The average arterial pressure originates from the heart pumping blood from the venous side of the circulation to the arterial side. The pulse pressure derives from the ejection of a volume of blood, along with the compliance of the arteries and the TPR.

Regulation of the average arterial blood pressure is accomplished in the short run through the baroreceptor reflex. In the long run it is accomplished by regulation of the volume of ECF by the kidneys.

The baroreceptors are stretch receptors located in the carotid body at the bifurcation of the internal and external carotid arteries, and in the aortic arch. Different receptors respond at different pressure levels, so that response is assured over a wide range of pressure. These receptors typically respond to the systolic pressure and are silent during the lower, diastolic pressure wave.

Sensory afferents from the carotid body and aortic bodies travel over the glossopharyngeal nerve, CN IX, and the vagus nerve, CN X, respectively, to make connections within the NTS in the medulla. Their firing rate is approximately proportional to the pressure. Cells in the NTS send fibers to the nucleus ambiguus, which in turn forms the efferent limb of the vagus nerve. These send parasympathetic fibers to the SA node through the right vagus nerve. Thus increased blood pressure activates cells in the NTS which in turn activate cells in the nucleus ambiguus, which reflexly slows the heart rate through parasympathetic efferents. When blood pressure falls, tonic parasympathetic stimulation is removed, relieving the heart of its brake, and therefore accelerating the heart rate. This increases cardiac output, which raises arterial pressure back toward normal.

The baroreceptor reflex also influences sympathetic outflow. The NTS makes contact with cells in the CVLM which sends inhibitory connections to the RVLM which activates sympathetic preganglionic interneurons in the spinal cord. Thus, increases in arterial pressure cause an increase in the firing rate of the baroreceptors which increases inhibition of the RVLM by CVLM, which inhibits sympathetic outflow to the heart and vasculature. This contributes to bradycardia and vasoconstriction, both of which tend to restore the blood pressure toward normal.

The inspiratory center is active during inspiration, and it sends collateral fibers to inhibit cells in the nucleus ambiguus that drive parasympathetic efferents to the heart. Inspiration therefore inhibits parasympathetic tone, which accelerates the heart rate during inspiration. This effect is called the **respiratory sinus arrhythmia**.

Long-term regulation of blood pressure is achieved by regulating the volume of the ECF and its distribution between veins and arteries. Regulation of the ECF volume is controlled by hormones including ADH, the RAA system, and ANP.

REVIEW QUESTIONS

1. Where are the baroreceptors located? How do they respond to arterial pressure? Why are there two types of baroreceptors?
2. What is the baroreflex? What is the cardiovascular response to high blood pressure? Low blood pressure? What mediates bradycardia? What mediates tachycardia?
3. What is respiratory sinus arrhythmia? Is the heart rate faster during inspiration or expiration? Why?
4. What regulates blood pressure in the long term?
5. Diuretics increase urinary excretion at any arterial pressure. Why do these aid in controlling blood pressure?
6. What is the sympathetic tetralogy? Does this regulate blood pressure or is it used to keep pressure high? Why would you want high blood pressure?
7. What is renin? What causes its release? What does it do?
8. What is aldosterone? What causes its release? What does it do?
9. What is ADH? What causes its release? What does it do?

Problem Set 5.3
Hemodynamics and Microcirculation

1. A normal individual experiences a deep cut that severs the radial artery near the elbow. Ignoring air resistance, approximately how high will the blood spurt? (Hint: the specific gravity of blood is 1.050 g cm^{-3} and the specific gravity of mercury is 13.6 g cm^{-3}.)

2. At the bifurcation of the femoral artery, the pressure is 100 mmHg. The pressure in the veins draining either the left or right femoral artery is 10 mmHg. The resistance in the right femoral artery is 1.67 times the resistance through the left femoral artery. The total flow at the bifurcation is 800 mL min^{-1}. How does the flow divide into the left and right femoral arteries? (Hint: what is the conservation of flow? Use this to solve one of the resistance values.)

3. A subject has normal values of 125/80 mmHg blood pressure, heart rate of 65 bpm, and a cardiac output of 5 L min^{-1}.
   A. If the subject's blood pressure rises to 140/80, what is the new stroke volume? (Hint: compliance of the aorta does not change.)
   B. What is the new mean arterial pressure?
   C. If the heart rate does not change, what is the new peripheral resistance?
   D. Is this an increase or a decrease?

4. An arteriole has a radius of 25 μm and it is 1000 μm long. The viscosity of blood is 3×10^{-3} Pa s and its density is 1.055 g cm^{-3}. Assume the arteriole is a right circular cylinder.
   A. Assuming laminar flow, what is the resistance of this arteriole?
   B. If it constricts to 20 μm, what is its resistance?
   C. What is the resistance of 20 of these arterioles (as in (A)) arranged in parallel?
   D. If the input pressure to these arterioles is 100 mmHg and the output pressure at the arteriolar end of the capillaries is 35 mmHg, what is the flow?
   E. What is the average velocity of blood through the arteriole?
   F. How much time does the blood spend in the arteriole?
   G. What is the Reynolds number for flow through the arteriole under these conditions?

5. An arteriole with a radius of 30 μm and a length of 1000 μm feeds into a set of 200 capillaries each with a radius of 4 μm and a length of 500 μm. The viscosity of blood is 3×10^{-3} Pa s and its density is 1.055 g cm^{-3}. The input pressure to the arteriole is 100 mmHg and the output pressure at the end of the capillary is 15 mmHg. Assume that the capillaries all come off the arteriole at the same place.
   A. What is the resistance of the arteriole?
   B. What is the resistance of a single capillary?
   C. What is the resistance of the set of capillaries?
   D. What is the pressure at the arteriolar end of the capillary?
   E. What is the total resistance through arterioles and capillary?
   F. What is the flow through the system?
   G. What is the velocity of flow through the arteriole?
   H. What is the velocity of flow through the capillaries?
   I. How much time does blood spend in the capillary?
   J. What is the Reynolds number for flow through the arteriole?
   K. Do the values for pressure and flow and resistance seem reasonable to you?

6. The net filtration across relaxed skeletal muscle capillaries is about 0.005 mL min^{-1} per 100 g of tissue. Assume the following values: the density of muscle is 1.08 g cm^{-3}; the length of the capillaries is 500 μm and their cross-sectional density is 250 capillaries mm^{-2}; their average radius is 4 μm. Pressure at the arteriolar end of the capillary is 40 mm Hg and it decays linearly to 15 mmHg at the venule end. The oncotic pressure of plasma is 25 mmHg. Interstitial fluid pressure in relaxed muscle is −1 mmHg and intersitial fluid oncotic pressure is 5 mmHg.
   A. Calculate the surface area of the capillaries per 100 g of muscle. (Hint: calculate the mass of muscle 1 mm^2 in cross section and 500 μm long.)
   B. Does the capillary filter fluid along its entire length, or does it reabsorb fluid at the venule end?
   C. Calculate the L_p for the capillary, assuming $\sigma = 1.0$. You will need to integrate the rate of filtration along the length of the capillary.

7. About 20% of the cardiac output at rest goes to the kidneys. What fraction of the TPR is due to the kidneys?

8. The arterial glucose concentration is usually between 80 and 120 mg%. Assume it is 90 mg% (this is mg per dL (0.1 L) of blood) in this problem.
   A. The blood flow to a muscle at rest is 0.025 mL min^{-1} g^{-1}. Its venous glucose concentration in the blood draining the muscle is 80 mg%. What is the glucose consumption rate, in mol min^{-1} g^{-1}? The molecular weight of glucose is 180.2 g mol^{-1}.
   B. The interstitial fluid glucose concentration is about 75 mg%. What is the solute extraction at rest?
   C. What is the diffusing capacity, equal to the surface area times the permeability?
   D. Suppose there are 250 capillaries open per mm^2 cross section of the muscle, and that each capillary is 500 μm long. Assume each capillary has a radius of 4 μm. If the density of muscle is 1.08 g cm^3, how much area of capillaries is available for glucose exchange?
   E. Using the results to C and D, calculate the permeability of the capillaries to glucose.
   F. Net filtration of fluid across the capillaries in muscle is about 0.005 mL min^{-1} per 100 g of tissue. If glucose is dragged across with this bulk flow, what percent of glucose transport across the capillary is due to diffusion and what per cent to bulk flow?
   During exercise, glucose consumption increases to 60 μmol min^{-1} per 100 g. Assume arterial [glucose] does not change. Blood flow increases to 0.60 mL min^{-1} g^{-1}.
   G. What is the venous concentration of glucose during exercise?
   H. The number of open capillaries increases to 1000 mm^{-2} cross section. How much area is now available for glucose exchange?
   I. Assuming that p is unchanged, what is the new solute extraction during exercise?
   J. What is the interstitial fluid concentration of glucose during exercise?

9. The radial artery in a person has an inner diameter of 4 mm and it is 25 cm long. The viscosity of blood is 3×10^{-3} Pa s and its density is 1.055 g cm^{-3}.
   A. Assuming laminar flow, what is the resistance of the radial artery?
   B. The average blood velocity in the radial artery is about 20 cm s^{-1}. What is the total flow?
   C. What is the Reynolds number for flow in this radial artery?
   D. From A and B, calculate the pressure difference between the beginning and end of the artery.

10. A defensive back for the Ohio State University football team weighs 89 kg. His blood is 7.5% of his body mass. Its density is 1.06 g cm^{-3}, the hematocrit ratio is 0.45 and hemoglobin is 15 g %. His plasma albumin is 4.5 g%, and his globulins are 3.1 g%.
    A. Calculate the contribution to the osmotic pressure of plasma made by albumin and globulin according to Eqn [5.10.17].
    B. During a heavy work-out in August preseason camp, he loses 3 L of protein-free fluid as sweat. If all of this fluid came from his plasma, calculate his new hematocrit, new albumin, and new globulin concentration.
    C. From part B, calculate the contribution to the osmotic pressure this albumin and globulin concentration would make to plasma.
    D. What effect would this have on the balance of fluid across tissue capillaries?

11. The Law of Laplace for a cylinder differs from that in a sphere. For a sphere, $P = 2T/r$; for a cylinder, $P = T/r$.
    A. Consider the abdominal aorta with $r = 0.8$ cm. What is the tension at systolic pressure if $P = 120$ mmHg?
    B. What is the tension in the radial artery if $r = 2$ mm and systolic pressure is 120 mmHg?

12. The brachial artery has an inner diameter of 5 mm. The viscosity of blood is 3×10^{-3} Pa s and its density is 1.055 g cm^{-3}.
    A. Assuming laminar flow, what is the resistance of the vessel per unit length?
    B. If the brachial artery is 20 cm long, what is its resistance?
    C. If it constricts to 4 mm, what is its resistance?

13. The diastolic pressure is 80 mmHg and the systolic pressure is 120 mmHg. The cardiac output is 5 L min^{-1}. The mean systemic pressure is 7 mmHg with a stressed volume of 1 L and an unstressed volume of 4 L. The right atrial pressure is 2 mmHg.
    A. Calculate the TPR.
    B. If $C_V = 18 C_A$, calculate C_V and C_A from the stressed and unstressed volumes.
    C. Calculate the slope of the vascular function curve. Compare it to Figure 5.12.6. Is it comparable?
    D. Estimate C_A from the stroke volume and pressure pulse. How does it compare to C_A calculated in part B?

14. Assume the following numbers for the arterioles: There are 1×10^7 of them; diameter is 50×10^{-4} cm; length is 1 cm; viscosity of blood is 3×10^{-3} Pa s. Calculate the aggregate resistance of the arterioles assuming Poiseuille flow.

15. An individual was studied by echocardiography and it was determined that his end diastolic volume was 125 mL, end systolic volume 65 mL and his heart rate at this time 75 bpm.
    A. What is his stroke volume?
    B. What is his ejection fraction?
    C. What is his cardiac output?

UNIT 6

RESPIRATORY PHYSIOLOGY

The Mechanics of Breathing 6.1

Learning Objectives

- Describe the overall function of the respiratory system
- Identify the force that moves air into and out of the lungs
- List the conditioning effects of the nasal passageways
- Explain what is meant by "mucus escalator"
- Identify the muscle responsible for quiet inspiration
- Identify the cause of air movement during quiet expiration
- Identify on a drawing: trachea, visceral pleural, parietal pleura, pleural space
- Describe the location and function of the external and internal intercostal muscles
- Identify the components of elastic recoil of the lungs/chest system
- Write the Law of Laplace and explain its significance for lung expansion
- Describe the chemical nature of surfactant and list its functions
- Explain why intrapleural pressure is nearly always negative (subatmospheric)
- Explain why a pneumothorax collapses the lungs
- Identify the FRC on the relaxation volume–pressure curve

THE RESPIRATORY SYSTEM SUPPLIES O_2 AND REMOVES WASTE CO_2

As described in Chapters 2.9–2.11, all cells of the body derive their energy from the slow and controlled oxidation of foodstuffs. The overall oxidation of glucose is described by the chemical equation:

$$C_6H_{12}O_6 + 6O_2 \Rightarrow 6CO_2 + 6H_2O + \text{heat}$$

and the complete oxidation of a triglyceride such as tripalmitin is described by

$$C_{51}H_{98}O_6 + 72.5O_2 \Rightarrow 51CO_2 + 49H_2O + \text{heat}$$

Both of these oxidations have large negative changes in free energy, ΔG, shown here as heat energy, which can be harnessed to produce work. Collectively, these reactions make up **cellular respiration**. In the body, the oxidation reactions occur slowly and the released chemical energy is captured in the terminal phosphate bond of ATP. The energy captured in ATP is used to perform mechanical work, concentration work, synthetic work, and electrical work. As ATP is consumed, metabolic pathways regenerate it so that the cellular concentration of ATP remains relatively constant while ATP continually turns over. The maintenance of the constancy of the body's systems, homeostasis, depends on a continual throughput of energy, which is possible only by continually supplying O_2 and removing CO_2. This is the function of the respiratory system: to exchange CO_2 and O_2 with the atmosphere so that the circulatory system can supply cells with adequate amounts of O_2 and remove the waste CO_2.

FOUR CORE ASPECTS OF RESPIRATORY PHYSIOLOGY

The four main parts of respiratory physiology are:

1. **Pulmonary ventilation**: The process in which a volume of gas is added to or removed from the lungs.
2. **Gas exchange**: The process in the lung by which blood is recharged with O_2 and dumps its waste CO_2.
3. **Transport of O_2 and CO_2** between the tissues and the lungs.
4. **Regulation of ventilation**: The process by which bodily needs are translated into more rapid or slower ventilation.

We will discuss these four topics in order in this unit. We begin by discussing pulmonary ventilation.

AIR FLOWS THROUGH AN EXTENSIVE AIRWAY SYSTEM THAT *FILTERS*, *WARMS*, AND *HUMIDIFIES* THE AIR

Inspired air enters through the nose or mouth, or both, where it is **conditioned**: it is filtered, warmed, and humidified. Many hairs and sticky mucus trap large dust particles and thus filter the inspired air. Outcroppings of tissue in the nasal cavity called **turbinates** expose the air to a large surface area and mix the air within the nasal passages, humidifying and warming the air to 37°C.

After leaving the nasal passages, air travels through the throat, or pharynx, and then through the larynx or voice box. It then enters the **trachea**, which subsequently branches many times. The pharynx, larynx, and early generations of the airways leading to the lungs do not participate in gas exchange. The airways are lined with **goblet cells** that produce mucus and a **ciliated epithelium** that constantly moves mucus toward the mouth. The mucus traps dirt and dust. The movement of the mucus is like a "mucus escalator" that constantly cleans the lungs. The mucus brought to the mouth is typically **expectorated** (spat out) or swallowed. Persons with **cystic fibrosis** produce thick mucus that cannot be easily removed. Part of their treatment is a vigorous thumping of the chest to clear out the lungs (see Clinical Applications: Cystic Fibrosis box).

The airways beyond the trachea constitute the **tracheobronchial tree**. Each branching of the airways produces the next generation, like a family tree. The first few generations conduct the air toward the gas exchanging regions, but they do not themselves exchange gases. The larger branches are the **bronchi** and **bronchioles**. These conductive airways become progressively smaller and more numerous as they branch, leading eventually to **terminal bronchioles, respiratory bronchioles**, and **alveolar ducts**, ending in dead-end sacs called **alveoli** (see Figure 6.1.1). The alveoli are small sacs, about 0.3 mm in diameter, where tiny distances separate the blood from the alveolar gases. The lungs contain some 300×10^6 alveoli. Thus the airways from nose to alveoli consist of a **conducting zone** of generations 0–16 and a **respiratory zone** in series with the conducting zone, with the respiratory zone consisting of generations 17–23.

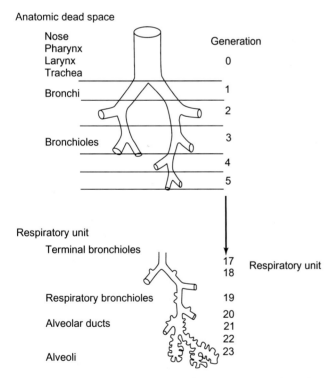

FIGURE 6.1.1 Highly schematic diagram of the extensive branching of the tracheobronchial tree to form the gas exchanging regions of the lung. Inspired air is conditioned in the anatomic dead space, so named because this region does not participate in gas exchange. The conditioning includes cleansing by nasal hairs and the mucus escalator, humidification, and warming. The airways branch successively to form bronchi, bronchioles, terminal bronchioles, respiratory bronchioles, alveolar ducts, and alveoli. They are on the order of 23 generations of branches, which generate 2^{24} airways. The last airways contain progressively more alveoli until the airways end in a blind sac covered with multiple alveoli. Gas exchange occurs in the respiratory unit from generations 17 to 23, consisting of terminal bronchioles, respiratory bronchioles, alveolar ducts, and alveoli.

Clinical Applications: Cystic Fibrosis

Cystic Fibrosis is a disease caused by mutations in the gene encoding for a protein called **CFTR**, which stands for **c**ystic **f**ibrosis **t**ransmembrane conductance **r**egulator. This protein belongs to a large superfamily of proteins called ABC transport proteins, which stands for **A**TP-**b**inding **c**assette transport proteins. CFTR consists of a single amino acid chain of 1480 amino acids that has five distinct functional regions. Two transmembrane domains are involved in forming a Cl^- channel, two domains bind nucleotides, and another regulatory domain is phosphorylated by protein kinase A, which is controlled by the second messenger, cAMP. The full function of CFTR is not yet known. It is a voltage-independent Cl^- channel, but it may also regulate the conductance of other apical channels. There are a variety of mutations of CFTR that produce symptoms. The most common mutation (about 70% of cases) is the deletion of phenylalanine at amino acid 508.

CFTR is present in the apical membrane of a variety of epithelial cells including those of the intestine, airways, secretory glands, and bile ducts. It appears to be necessary for the proper secretion of watery solutions by these epithelia. Insufficient watery secretion by tissues possessing defective CFTR results in insufficient dilution of the mucus secreted by the epithelium. The tissues become covered with sticky mucus that is difficult to remove. Persons afflicted with defective CFTR suffer from increased bacterial infections because of the failure of the mucus escalator.

In 1970, the median life expectancy of persons with defective CFTR was 8 years. Because of advances in treatment, the median life expectancy today is near 30 years. New treatments include an enzyme that breaks down the mucus to help get it moving and antibiotics. About 40,000 persons in the United States suffer from cystic fibrosis, and about 2500 new cases arrive every year.

GAS FLOWS IN RESPONSE TO PRESSURE DIFFERENCES

Gas is a fluid that obeys the same general principles of other fluids: it flows from a region of high pressure to a region of low pressure. In laminar flow through straight cylindrical pipes, in which the flow is streamlined, flow is described by Poiseuille's law:

$$[6.1.1] \qquad Q_V = \frac{\pi a^4}{8\eta} \frac{\Delta P}{\Delta x}$$

where $P/\Delta x$ is the pressure gradient, Q_V is the flow, a is the radius of a right cylindrical tube through which flow occurs, and η is the viscosity of the flowing medium. According to this equation, the flow of gas is directly proportional to the pressure difference. In a straight cylindrical pipe of radius a and length L, the equation becomes a hydraulic equivalent of Ohm's law:

$$[6.1.2] \qquad Q_V = \frac{\Delta P}{R}$$

where R is the equivalent resistance to gas flow through the airways. These equations are identical to those we use in cardiovascular physiology (see Unit 5).

CHANGES IN LUNG VOLUMES PRODUCE THE PRESSURE DIFFERENCES THAT DRIVE AIR MOVEMENT

The Ideal Gas Law describes the relation between pressure and volume in an ideal gas:

$$[6.1.3] \qquad PV = nRT$$

where P is the pressure measured in atmospheres or mmHg or Pa ($=N\ m^{-2}$), or some other appropriate unit, V is the volume, n is the number of moles, R is the gas constant, and T is the temperature in kelvin. When P is in atmospheres and V is in L, $R = 0.082\ L\ atm\ mol^{-1}\ K^{-1}$. The inverse relation between pressure and volume is the basic principle responsible for pulmonary ventilation: increasing the volume of the thoracic cavity, with the enclosed lungs, decreases the pressure of the gas in the lungs and so air rushes in from the outside. Conversely, the reduction of the volume of the thoracic cavity increases the pressure of the gas in the lungs and so air moves from the lungs back out into the ambient air.

SKELETAL MUSCLES POWER INSPIRATION

THE DIAPHRAGM IS THE MAJOR MUSCLE DRIVING NORMAL INSPIRATION

Inspiration refers to taking air into the lungs. **Expiration** refers to the removal of air from the lungs. Alternative terms are **inhalation** and **exhalation**. Normal resting inspiration begins with contraction of the diaphragm, a dome-shaped muscle that separates the **thoracic cavity** containing the heart and lungs from the **abdominal cavity** containing the liver, stomach, and intestines. The **phrenic nerve** innervates the diaphragm. We must consciously control the diaphragm in order to eat or speak or swim or sing. Because breathing is essential to life, back-up systems exist so that breathing becomes automatic when we sleep or when we lose consciousness (see Chapter 6.6).

When the diaphragm contracts, it expands the thoracic cavity at the expense of the abdominal cavity. The increase in volume of the thoracic cavity decreases the intrathoracic pressure so that, if the **glottis** is open, air will flow from the atmosphere into the lungs (see Figure 6.1.2). The glottis consists of the two vocal folds in the larynx and the slit between them.

THE EXTERNAL INTERCOSTAL MUSCLES EXPAND THE THORACIC CAGE BY ELEVATING AND EXTENDING THE STERNUM

The external intercostal muscles originate on the inferior surfaces of the proximal parts of the ribs and insert on the superior and distal parts of the next lower rib. These

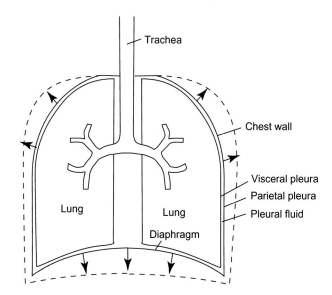

FIGURE 6.1.2 Expansion of the thoracic cavity during inspiration. The lungs are shown as "floating" within the thoracic cavity, being anchored by the bronchi. The surface of the lungs is covered by a membrane, the **visceral pleura**, which is continuous with the pleura lining the inside of the chest wall, the **parietal pleura**. The intrapleural space is very thin (about 10–20 μm) and filled with pleural fluid. Contraction of the diaphragm causes expansion of the lungs, which lowers the pressure of its enclosed air. The inspiration is aided by external intercostal muscles and accessory muscles connected to the bones of the rib cage which raise the rib cage and increase its posterior–anterior dimensions.

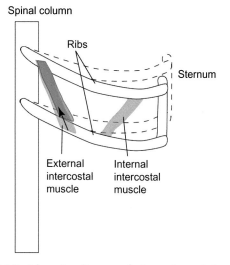

FIGURE 6.1.3 Schematic diagram of the action of the external intercostal muscles. The external intercostal muscles connect the ribs in such a way that the contraction of the muscles lifts the ribs and rib cage and expands the anterior–posterior dimensions of the rib cage. Contraction of the internal intercostals, which are oriented opposite of the external intercostals, produces the opposite effect: lowering of the ribs and reduction in the anterior–posterior dimension.

are innervated by intercostal nerves originating in thoracic segments of the spinal cord. Contraction of these muscles raises the rib cage upward and outward, thereby contributing to the expansion of the thoracic cavity and reduction of pressure within the lung, causing air to move from the air into the lungs (see Figure 6.1.3). Contraction of the internal intercostal muscles has the opposite effects and aids in expiration.

ACCESSORY MUSCLES AID IN FORCEFUL INSPIRATION

The sternocleidomastoid muscles connect the sternum to the mastoid process behind the ear. Contraction of this muscle helps lift the sternum along with the rest of the rib cage. The **scalene** muscles lift the first two ribs. These muscles generally contribute little to expansion of the rib cage during rest but are recruited at high levels of ventilation during strenuous exercise, for example.

RESTING EXPIRATION IS PASSIVE; ABDOMINAL MUSCLES AID IN FORCEFUL EXPIRATION

By expanding the lungs and chest wall, the muscles of inspiration store mechanical energy in the elastic lungs and chest wall system. When these inspiratory muscles relax, the lungs and chest wall recoil, producing a contraction of the thoracic cavity and subsequent rise in the intrathoracic pressure. This forces air out of the lungs, but the force is passive.

Contraction of the abdominal muscles compresses the contents of the abdominal cavity, thereby reducing the volume of the thoracic cavity, and pulls down on the rib cage. Both actions increase the flow of gas during expiration, allowing for more rapid pulmonary ventilation. These abdominal muscles include the rectus abdominus, transverse abdominus, external oblique, and internal oblique. The internal intercostal muscles are oriented opposite to the external intercostals: they lower the rib cage and decrease its anterior dimensions. These muscles assist expiration only during strenuous breathing.

THE PLEURA AND THE PLEURAL FLUID JOIN THE LUNGS TO THE CHEST WALL

According to Figure 6.1.2, there is no rigid structural connection between the lungs and the chest wall and the lungs mostly "float" in the thoracic cavity. So how can the lungs follow the chest wall as if they were attached? The tensile properties of the thin layer of pleural fluid effectively join the lungs to the chest wall. Consider two flat, horizontal glass plates separated by a thin layer of water, as shown in Figure 6.1.4. The layer of water allows the two plates to slip easily past each other, in the horizontal plane, but it is extremely difficult to separate the plates vertically because the water effectively binds them together. Application of a large force normal to the surface produces a large negative pressure within the fluid between the plates. This is analogous to the situation with the lung and the chest wall.

COMPLIANCE MEASURES THE EASE OF EXPANDING THE LUNGS

The compliance describes the ease of expansion of a volume. It is *defined* as

[6.1.4] $$C = \frac{\Delta V}{\Delta P}$$

where C is the compliance, ΔV is the change in volume, and ΔP is the change in pressure. A large compliance means that the volume is easily distended, because a large volume change results from a small pressure change. The compliance of the lungs can be obtained from the slope of the volume—pressure curve. Figure 6.1.5 shows the volume—pressure curves when excised lungs are filled first with air and then with saline. These results illustrate several key concepts:

1. Filling the lungs with air shows **hysteresis**—the curves are different for inflation and deflation.
2. The curves are nonlinear and so the compliance depends on the volume at which it is measured.
3. The compliance is greatly different for inflation with air and with saline. This shows that a large component of the compliance is the **surface tension** between the lungs and air.

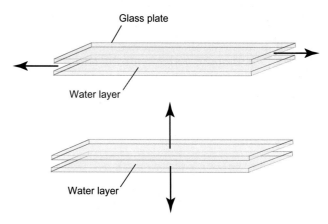

FIGURE 6.1.4 Lubricating and tensile effects of a thin water layer. A thin layer of water separates two flat glass plates. Application of a force in the plane of the glass causes the glass plates to slip past each other. The water layer reduces the friction between the glass plates and facilitates movement in the plane of the plates. Application of a force normal to the plane of the plates causes a slight expansion of the water layer and production of a large negative pressure within that layer. Separation of the plates in this direction requires much larger forces compared to movement of the plates along their planes.

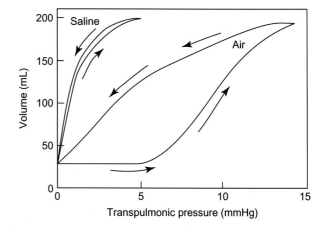

FIGURE 6.1.5 Compliance of the isolated lung. The isolated lung expands when positive pressure is applied to fluids entering the lung. The slope of the volume—pressure curve is the compliance at that volume. The compliance demonstrates hysteresis or markedly different behavior upon inflation or deflation. Also, the compliance is nonlinear: it requires more incremental pressure to begin inflation than to continue it. Lastly, the compliance is greatly increased when saline instead of air is used to inflate the lung.

THE COMPLIANCE AND RECOIL TENDENCY OF THE LUNG IS PRODUCED BY *ELASTIC FIBERS* AND BY *SURFACE TENSION*

As described above, it takes effort to expand the lungs but the lungs recoil on their own accord. The lung's elasticity is due to elastic fibers in the lung and airways and to the surface tension. The elastic fibers include **elastin** and **collagen** fibers in the extracellular space surrounding the alveoli, bronchioles, and pulmonary capillaries. They account for about one-third of the recoil tendency.

As discussed in Chapter 2.4, surface tension arises from the unequal forces applied to molecules at the interface between two phases; in this case the phases are air and the thin layer of fluid that lines the alveoli. Surface tension produces a force along a plane surface that is measured in units of force per unit length. It takes effort to expand the surface, and this effort is expressed as the increment in surface energy, as described in Eqn [2.4.1]:

[6.1.5] $$dG = \gamma dA$$

where dG is the increment in the surface free energy, γ is the **surface tension**, and dA is the increment in area.

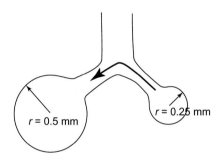

FIGURE 6.1.6 Schematic diagram of the predicted instability of alveoli. According to the Law of Laplace, the pressure within alveoli at mechanical equilibrium would be given by $P = 2\gamma/r$. If the surface tension is the same in both alveoli, but r is smaller on the right, then the pressure on the right will be larger and air will move from the smaller alveolus to the larger one. This does **not** happen, in fact, because the surface tension is highly dependent on r and because the alveoli are not independent, individual spheres. The alveoli are connected to other alveoli because they share walls and as one collapses it is supported or kept open by its neighbors.

In a sphere at equilibrium, the surface tension is related to the pressure and the size of the sphere according to the **Law of Laplace**:

[6.1.6] $$P = \frac{2\gamma}{r}$$

where P is the pressure within the sphere, γ is the surface tension, and r is the radius of the sphere. The Law of Laplace can be derived from energetic principles (see PS2.1) or mechanical principles (see Chapter 5.4). One way of looking at this is to say that the surface tension produces a pressure on the gas within an elastic sphere that opposes its expansion. Another way of looking at it is that the surface tension promotes the recoil of the sphere. About two-thirds of the recoil tendency of the lung originates in the aggregate surface tension in the millions of alveoli. When the lungs are filled with saline, there is no air–water interface and therefore there is no surface tension. Under these conditions, the compliance of the lungs increases (see Figure 6.1.5).

THE LAW OF LAPLACE PREDICTS ALVEOLAR INSTABILITY

In the lungs the alveoli approximate spheres, but instead of being closed they are open to the airways and are connected to each other through the gas within the lungs. According to the Law of Laplace, small alveoli would have larger pressures than the larger alveoli, and thus the smaller alveoli should collapse and the larger ones grow larger (see Figure 6.1.6).

PULMONARY SURFACTANT LOWERS THE SURFACE TENSION IN THE ALVEOLI

Alveolar type II cells secrete a lipoprotein material called **surfactant**, whose primary function is to **reduce the surface tension** in the alveoli. Surfactant is a lipoprotein that consists mainly of dipalmitoylphosphatidylcholine and some glycoprotein components. This material spreads out over the surface of the alveoli and reduces its surface tension by interaction of the hydrophilic parts of the surfactant molecule with the water layer next to the alveolar cells and by the interaction of

---

EXAMPLE 6.1.1 Pressure Within Alveoli of Different Sizes

Consider the two alveoli in Figure 6.1.6. Suppose that the surface tension was constant at 50 dyne cm^{-1} (1 dyne is 1 g cm s^{-2}, the unit of force in the CGS system, which is not part of the SI system of units). What is the pressure in each of the alveoli?

Pressure is given by the Law of Laplace, which is only an approximation to the situation here because of the fact that the alveoli are not spheres and there are other forces besides the surface tension involved. Nevertheless, we use it here as an approximation. The pressure on the left is given as $P = 2 \times 50$ dyne cm^{-1}/0.5 × 10^{-1} cm = **2000 dyne cm^{-2}**. The pressure on the right is $P = 2 \times 50$ dyne cm^{-1}/0.25 × 10^{-1} cm = **4000 dyne cm^{-2}**.

These values can be converted to Pa, atm, or mmHg using the conversion 1 dyne = 1 g cm s^{-1} × 10^{-3} kg g^{-1} × 10^{-2} m cm^{-1} = 10^{-5} N; 1 dyne cm^{-2} = 10^{-5} N cm^{-2} × (10^2 cm m^{-1})^2 = 10^{-1} N m^{-2} = 10^{-1} Pa.

The pressure on the left is 2000 dyne cm^{-2} × 10^{-1} Pa (dyne cm^{-2})$^{-1}$ = **200 Pa** × 9.87 × 10^{-6} atm Pa^{-1} = **0.00197 atm** × 760 mmHg atm^{-1} = **1.5 mmHg**.

the hydrophobic parts of the surfactant with the air. Surfactant has multiple functions:

- Surfactant causes the hysteresis in the lung volume—pressure curve.
- Surfactant stabilizes alveolar size.
- Surfactant reduces the work of breathing.
- Surfactant keeps alveoli dry.

Surfactant can be extracted from the lung and spread out over water as a monolayer. The surface tension measured on the water film shows hysteresis, just like that of the isolated lung. Thus it is likely that surfactant is responsible for the hysteresis in the volume—pressure curves shown in Figure 6.1.5.

The ability of surfactant to lower the surface tension depends on its surface concentration. When an alveolus shrinks, the concentration of surfactant increases and the surface tension is reduced. The pressure inside the alveolus still obeys the Law of Laplace, but the surface tension in the Law of Laplace depends on the size of the alveolus. Higher concentration of surfactant lowers the alveolar pressure by lowering γ in Eqn [6.1.6] while shrinking of the alveolus raises the pressure by making r smaller. The final result is that the surfactant stabilizes alveolar size.

Surface tension contributes about two-thirds of the recoil tendency of the lungs. Inspiration requires work to expand the lungs against this recoil tendency. Because surfactant lowers that recoil tendency, surfactant reduces the work of breathing. Persons with insufficient surfactant indeed have difficulty breathing (see Clinical Applications: Respiratory Distress Syndrome).

The surface tension that tends to collapse alveoli (thus raising the pressure within the enclosed space) tends to draw the surface away from the interstitial fluid, thus *reducing* the pressure within the interstitial space. Low pressure in the interstitial space favors fluid filtration out of the pulmonary capillaries, according to the Starling—Landis equation (see Chapter 5.9). Excess surface tension thus leads to pulmonary edema, and the reduction of the surface tension by surfactant helps prevent edema.

THE LUNGS AND CHEST WALL INTERACT TO PRODUCE THE PRESSURES THAT DRIVE VENTILATION

STATIC PRESSURES ARE MEASURED WHEN AIRFLOW IS ZERO

The static lung volumes refer to the lung volumes when there is no flow of air through the airways. As mentioned earlier, both the lungs and chest wall are elastic structures whose volume depends on the pressure difference between the inside and outside, like a balloon. Because the muscles pull on the chest wall, part of the pressure is generated by the muscles. The balance of pressures in the static lung during no-flow conditions can be written as

[6.1.7] $\qquad P_A - P_B = P_{muscles} + P_{lung} + P_{chest\ wall}$

where P_B is the ambient, barometric pressure, P_A is the pressure within the alveoli, $P_{muscles}$ is the pressure due to muscular compression or expansion of the chest wall, P_{lung} is the pressure due to elastic properties of the lung, and $P_{chest\ wall}$ is the pressure due to the properties of the chest wall. These components can be separated in humans by having a person inhale or exhale to achieve a given volume, and then relaxing the muscles against a closed airway, while at the same time measuring $P_B - P_A$. Because the airway is closed, the airflow is zero; because the muscles are relaxed, $P_{muscles} = 0$ and Eqn [6.1.7] becomes

[6.1.8] $\qquad P_A - P_B = P_{lung} + P_{chest\ wall} = P_{rs}$

where P_{rs} is the pressure of the entire respiratory system consisting of lung and chest wall.

THE PRESSURE DROP ACROSS THE LUNG AND CHEST WALL CAN BE ESTIMATED USING P_{PL}

The contribution of the chest wall and the lung to the total pressure, P_{rs}, can be determined by recognizing that the **intrapleural space** divides the total pressures into two components, as shown in Figure 6.1.7. The difference between the alveolar pressure and the barometric pressure is P_{rs}.

THE RELAXATION VOLUME—PRESSURE CURVES SHOW THAT THE LUNGS AND CHEST WALL USUALLY PULL IN OPPOSITE DIRECTIONS, CAUSING A NEGATIVE INTRAPLEURAL PRESSURE

The relaxation volume—pressure curves for the lungs, chest wall, and total respiratory system can be obtained as described in Figure 6.1.7, by measuring P_A, P_B, and P_{pl}, and the results are shown in Figure 6.1.8. The chest acts like a spring; at low volumes (below point A in Figure 6.1.8) it recoils back to higher volumes, creating a

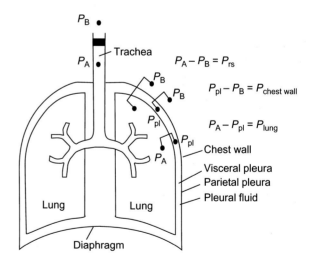

FIGURE 6.1.7 The intrapleural pressure is the pressure within the narrow space between the lung and the chest wall. Since the pressure drop across the entire respiratory system (P_{rs}) consists of the sum of the pressure drops across the lung and chest wall, each of the drops can be determined if the intrapleural pressure is known. The intrapleural pressure can be measured by insertion of a needle into the intrapleural space, but may be more easily approximated by the esophageal pressure. This pressure divides P_{rs} into a pressure drop across the chest wall and a pressure drop across the lung. The pressure drop across the chest wall is given as $P_{chest\ wall} = P_{pl} - P_B$, where P_{pl} is the pressure within the intrapleural space; the pressure drop across the lungs is $P_{lung} = P_A - P_{pl}$. The relaxation volume—pressure curves are obtained by relaxing the respiratory muscles when the airways are closed, as shown.

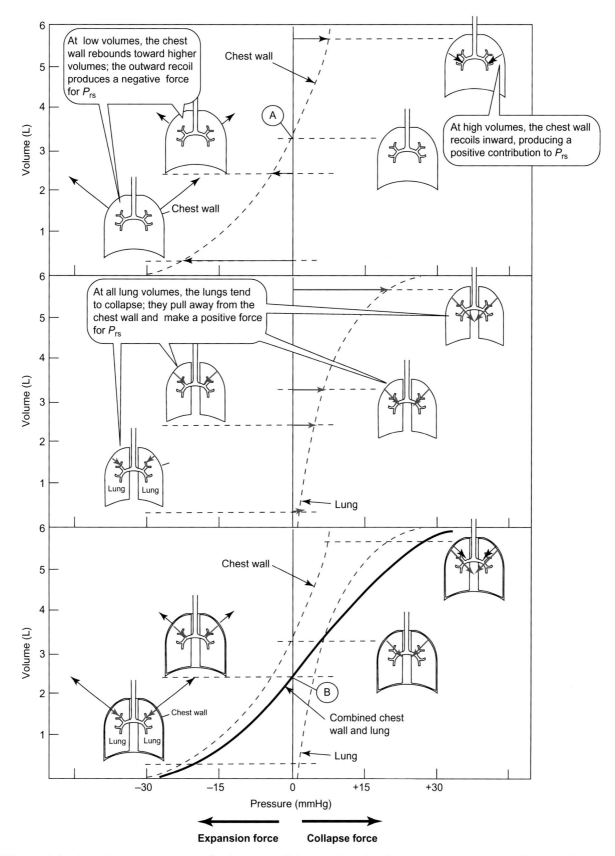

FIGURE 6.1.8 Relaxation volume–pressure curves for the chest wall, lungs, and combined respiratory system, so named because the muscles are relaxed when pressures are measured. At a single volume, the contribution to the total pressure is the sum of the pressures contributed by the chest wall and by the lungs (see Eqn [6.1.8]). Thus the pressures add as vectors along the dashed horizontal lines to form the combined chest and wall curve (solid line). Throughout most of the volume range, the chest wall tends to recoil toward a larger volume. At large volumes, the chest wall tends to recoil inward. Such large volumes can be accomplished by the action of respiratory muscles. The lungs always tend to collapse, and the tendency becomes larger with increasing volume. Except at high volumes, then, the lungs and the chest wall act in opposite directions: the lung tends to shrink whereas the chest wall tends to expand. *Modified from E. Agostoni and J. Mead, Statics of the respiratory system, in W.O. Fenn and H. Rahn, eds.,* Handbook of Physiology, *American Physiological Society, Washington, DC, 1964.*

negative pressure across the respiratory system. At high volumes it recoils back to lower volumes, creating a positive pressure across the respiratory system. Thus negative static pressures means there is a tendency to expand (which can be balanced only by negative pressures) and positive pressures means there is a tendency to collapse (which can be balanced only by positive pressures). Because these are static volumes, the negative pressure is necessary to stop expansion, and positive pressures are necessary to stop the collapse. These tendencies are illustrated by arrows in Figure 6.1.6, with an arrow to the left (negative pressure) showing a force for expansion and an arrow to the right showing a force for collapse. The lungs at all volumes tend to recoil toward a smaller volume, so they always contribute a positive pressure to the respiratory system. Because the lung is plastered to the chest wall by the pleural fluid, the combination creates either a negative or a positive pressure. The balance point is point B in the lower panel—here no muscle force is necessary to set the volume of the system.

THE LOW INTRAPLEURAL PRESSURE IS CAUSED BY THE RECOIL TENDENCY OF THE LUNG

The recoil of the lung away from the chest wall causes a relatively lower pressure in the intrapleural space than in the alveolar space. As a consequence, the lung always experiences a positive pressure that keeps it inflated against its tendency to recoil. Although the net pressure on the lung is positive, the pressure is actually *caused* by the recoil tendency of the lung. That is, the lung recoils away from the chest wall until it develops a negative intrapleural pressure (positive $P_A - P_{pl}$) large enough to keep the lung inflated.

BREAKING THE SEAL ON THE INTRAPLEURAL SPACE COLLAPSES THE LUNGS

A puncture of the chest wall allows air to enter the intrapleural space and equilibrate its pressure with the ambient barometric pressure. In this case, the pressure inflating the lung becomes zero or negative and is not enough to counteract the natural recoil of the lung: the lungs collapse. This condition is called a **pneumothorax** (literally, air in the thorax), as illustrated in Figure 6.1.9.

> ### Clinical Applications: Respiratory Distress Syndrome
>
> The first breath of air after being born poses its own set of problems. *In utero*, the baby can "inhale" amniotic fluid, which has a high degree of viscous resistance but has no air–water interface. In addition, the baby does not obtain gases from this fluid, as gas exchange is accomplished through the placental circulation. After birth, the baby must expand its lungs for the first time in the presence of the surface tension associated with air breathing. Accordingly, the concentration of surfactant in full-term infants is actually higher than that of adults, presumably to ease the transition from fluid to air breathing upon birth. Surfactant production by alveolar type II cells requires maturation of these cells, and this is not fully accomplished until about 36 weeks of gestation. Premature infants may lack sufficient surfactant, and the extra recoil tendency of the lungs produces difficulty in breathing. The condition is called **infant respiratory distress syndrome** or RDS. The condition is also called **hyaline membrane disease**, because the improper secretion of surfactant results in coagulation of cellular debris and plasma proteins in structures called hyaline membranes that line the alveoli.
>
> RDS occurs in about 1% of newborns, but is most common in premature infants, affecting 75% of infants that are delivered after only 26 weeks of gestation. Affected infants take rapid and shallow breaths, have low oxygen tensions in the blood (hypoxemia), and are acidotic because they cannot get rid of excess CO_2 (see Chapter 6.5). These babies rapidly perish from the condition if they are left untreated. Treatment includes instillation of exogenous surfactant into the trachea, from which it spreads throughout the lungs during positive pressure artificial respiration. Most premature infants can be rescued from RDS by these means.

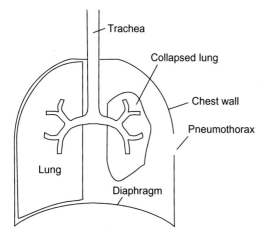

FIGURE 6.1.9 Schematic diagram of lung collapse upon the formation of a pneumothorax. Under normal conditions, a lung's recoil tendency produces a lower pressure within the intrapleural space. This lower pressure produces a net positive pressure that holds the lung against the chest wall. When air is introduced, the intrapleural pressure equilibrates with the ambient barometric pressure and the lung's natural recoil tendency causes it to collapse.

AIRWAY RESISTANCE PARTLY DETERMINES DYNAMIC PRESSURES

The relaxation volume–pressure curves shown in Figure 6.1.8 are obtained from the nonphysiological case when the muscles are relaxed and airflow is zero. This does not pertain during breathing, yet the curves help us to understand the origin of the forces. During dynamic breathing, air actually flows into or out of the lungs and respiratory muscles are active. In this case, the alveolar pressure and intrapleural pressure reflect not only the elastic properties of the lungs and chest wall but also the airway resistance.

The alveolar and intrapleural pressures during quiet breathing are shown in Figure 6.1.10. We begin at (1) in the figure, which corresponds to the situation at the end of normal expiration. At this point, the volume of

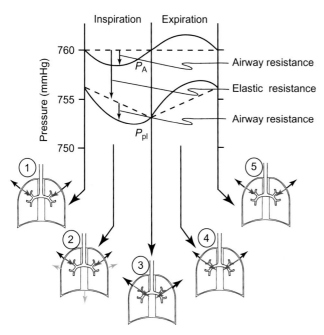

FIGURE 6.1.10 Alveolar and intrapleural pressures during dynamic breathing. If there were no airway resistance, the pressures would be nearly the same in the lungs as outside (the dashed lines). The alveolar pressure becomes lower when the inspiratory muscles pull the chest wall away from the lungs, making the intrapleural pressure lower, causing the lung to expand (because P_{lung} is more positive; see Figure 6.1.8) and alveolar pressure to drop. The expansion of the chest wall in normal breathing occurs when the chest wall forces are near zero, so that normal expansion occurs near the volume at which chest wall forces switch from expansion to recoil. At this larger volume, the expansion force of the chest wall is reduced so that removal of the inspiratory muscle forces causes the combined chest wall and lung system to recoil. In this case the intrapleural pressure is higher, because the muscles pulling on the chest wall are now relaxed. This in turn raises the alveolar pressure and air leaves the lung.

air in the lungs is called the **functional residual capacity** or **FRC**. Here the inspiratory and expiratory muscles are relaxed and alveolar pressure, P_A, is equal to the ambient barometric pressure, P_B. Thus the pressure across the entire respiratory system, P_{rs}, is zero according to Eqn [6.1.8]. This occurs when the recoil force of the lungs exactly balances the expansion force of the chest wall. At this point, the intrapleural pressure is lower than P_A or P_B. The lung and chest wall are pulling in opposite directions. This corresponds to point B in the lower panel of Figure 6.1.8.

At the beginning of a normal inspiration, the inspiratory muscles begin to pull the chest wall away from the lungs (light blue arrows at (2) in Figure 6.1.10), causing the intrapleural pressure to become lower. This causes the lungs to expand and alveolar pressure to drop, driving air into the lungs. If there were no airway resistance, then gas movement would be nearly instantaneous and the pressure differences driving air movement would be tiny. The airway resistance requires greater pressure drops to drive airflow. Normal inspiration of some 500 mL is enough to reduce the expansion tendencies of the chest wall to zero (point A in the top panel of Figure 6.1.8) and possibly large enough to enter the region of the curve where the chest wall contributes to overall respiratory system recoil. The intrapleural pressure falls even more than is necessary only to compensate for airway resistance. This additional fall overcomes elastic resistance.

At the beginning of expiration ((3) in Figure 6.1.10), the inspiratory muscles relax. This removes one of the expansion forces, and so at this point the recoil forces of the lungs dominate the balance of forces. The lungs recoil and intrapleural and alveolar pressures rise. Because the lungs are always tending to pull away from the chest wall, the intrapleural pressure remains subatmospheric throughout normal inspiration and expiration. The increased alveolar pressure overcomes airway resistance. The pressure increase in the intrapleural pressure overcomes both the airway resistance and the elastic resistance.

SUMMARY

The purpose of the respiratory system is to provide O_2 for metabolism and to remove waste CO_2. To do this, the respiratory system brings in new air and expires air that is depleted of O_2 and enriched in CO_2. Movement of air responds to pressure differences: inspiration is brought about by lowering the pressure in the lung by expanding the thoracic cavity. Expiration is brought about by increasing the pressure in the lungs. Inspiration is active. Contraction of the diaphragm expands the thoracic cavity against the recoil tendency of the lungs. In forceful breathing, the external intercostals may aid inspiration. Expiration is usually passive but may be aided by the abdominal muscles and internal intercostal muscles. The recoil tendency of the lungs arises from elastic fibers and surface tension in the millions of alveoli, little sacs of lung tissue that closely appose blood capillaries for the exchange of gases. The surface tension is reduced by surfactant, a lipoprotein produced by alveolar type II cells. This helps reduce the recoil tendency of the lung, which in turn reduces the work of breathing. Surfactant also helps keep alveoli the same size because it reduces the surface tension when alveolar size decreases. Surfactant also keeps the alveoli dry (by reducing negative pressure within the interstitial spaces) and causes hysteresis in the pressure–volume curves of the lungs.

Air passes along mucus membranes that line the nasal cavity, where it is conditioned by removal of dust particles, humidification, and warming. Air then passes through the tracheobronchial tree, which consists of progressively smaller and more numerous airways that lead eventually to the alveoli. The lung is suspended within the thoracic cavity. It is covered by a membrane, the visceral pleura, which is continuous with the parietal pleura that lines the chest wall. The pleural fluid is a thin layer of fluid that allows the lung to slide along the chest wall while remaining attached through the tension in the fluid. The intrapleural pressure is typically subatmospheric, caused by the recoil tendency of the lungs and the tendency of the chest wall to expand. Because the intrapleural pressure is less than alveolar pressure, the pressure across the lung itself is positive, keeping it expanded. During inspiration, the chest wall moves outward, making the intrapleural pressure more negative and thereby expanding the lung. This reduces the pressure within the lung, causing air to move from the atmosphere into the lung. The reverse occurs during expiration.

REVIEW QUESTIONS

1. What functions of the tracheobronchial tree disappear during mouth breathing?
2. At the end of a normal expiration, the glottis is open but no air moves into or out of the lung. What is the alveolar pressure at this time with respect to barometric pressure? Find this point on the relaxation volume–pressure curve (see Figure 6.1.8)
3. Explain why surfactant (a) lowers surface tension, (b) reduces the work of breathing, (c) helps stabilize alveoli size, and (d) prevents low interstitial fluid pressures.
4. The relaxation volume–pressure curve shows that large lung volumes must be supported by large positive pressure (solid line in Figure 6.1.8). In dynamic breathing, alveolar pressure during inspiration is never positive! How do the lungs get to that position?
5. At rest, the pulmonary ventilation is about $6 \, L \, min^{-1}$. During exercise, it can increase markedly, depending on the size of the person and their physical fitness, to highs near $120 \, L \, min^{-1}$. What happens to alveolar pressure, P_A, during inspiration and expiration during exercise, with respect to its value at rest?

Lung Volumes and Airway Resistance 6.2

Learning Objectives

- Identify on a spirometer trace and be able to calculate TV, IRV, ERV, and RV
- Identify on a spirometer trace and be able to calculate IC, VC, FRC, and TLC
- Define the respiratory minute volume and pulmonary ventilation
- Describe the procedure followed in obtaining the forced vital capacity
- Indicate how a spirometer can measure airway resistance
- Distinguish between laminar and turbulent flows
- Describe how resistance varies as a function of flow in the airways
- Write the equation for the Reynolds number and be able to predict when flow is turbulent
- Define dynamic compression
- Explain how airway resistance is modified by tracheo-bronchiole smooth muscle
- Explain how pursed-lips breathing can increase airflow

SPIROMETERS MEASURE LUNG VOLUMES AND ALLOW IDENTIFICATION OF SEVERAL LUNG VOLUMES AND LUNG CAPACITIES

The volume of air that can be inhaled or exhaled can be estimated using **spirometry**. Several different kinds of spirometers exist. Here we describe a volume–displacement Collins respirometer in which a person breathes through a small tube that is connected to an air space located within a lightweight bell jar that is isolated from the ambient air by a layer of water. When air enters the lungs, it leaves the bell jar and reduces the volume of air in the jar by the same volume of air that is inhaled, and this can be measured continuously by linking movement of the bell jar to a recording pen. The setup is illustrated in Figure 6.2.1.

The spirometer aids in identifying several useful lung volumes and capacities. These are:

1. The **tidal volume, TV**, is the amount of air breathed in and out during normal, restful breathing. Its typical value, as shown in Figure 6.2.1, is about 500 mL = 0.5 L. The magnitude of the TV depends on the size of the individual and their metabolic state. The spirometer recording shown in Figure 6.2.1 is idealized: the TV usually varies from breath to breath.
2. The **inspiratory reserve volume, IRV**, is the additional volume of air that can be inspired at the end of a normal or tidal inspiration. The typical value for a young adult male of normal size is about 3000 mL.
3. The **expiratory reserve volume, ERV**, is the additional volume of air that can be expired after a normal or tidal expiration. A typical value is about 1100 mL for a young adult male.
4. The **residual volume, RV**, is the volume remaining in the lung after a maximum expiration. Even a maximum effort cannot void the lungs of all air. This volume of air cannot be measured by spirometry but it can be calculated by measuring the functional residual capacity by two other techniques: **gas dilution** and **body plethysmography**. The value of the RV is typically about 1200 mL.

LUNG CAPACITIES ARE COMBINATIONS OF TWO OR MORE LUNG VOLUMES

We have identified four lung volumes, three of which are easily measured by spirometry. We also identify four lung capacities. These are:

1. The **functional residual capacity, FRC**, is the volume of air remaining in the lungs after a normal or tidal expiration. It is the sum of two lung volumes: the RV and the ERV. A typical value for a young adult male is FRC = RV + ERV = 1.2 L + 1.1 L = 2.3 L. The FRC is that point during the respiratory cycle when the pressure across the entire respiratory system, P_{rs}, is zero. At this point, the recoil force of the lungs exactly balances the expansive force of the chest wall.
2. The **inspiratory capacity, IC**, is the volume of air that can be inspired after a normal or tidal expiration. The inspiratory capacity thus is the volume of air that can be inspired beginning at the FRC. According to Figure 6.2.1, the IC is the sum of the IRV and the TV. Thus, IC = IRV + TV = 3.0 L + 0.5 L = 3.5 L.
3. The **vital capacity**, **VC**, is the sum of all lung volumes above the RV: the ERV, the TV, and the

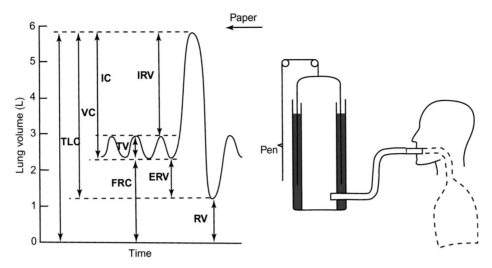

FIGURE 6.2.1 Typical setup for a volume–displacement spirometer. The person places a clip on the nose so that all air passes through the mouth. The person then breathes in air that is connected to a closed space in a light cylindrical bell jar that is sealed from the outside by a layer of water retained between two concentric cylinders. The movement of the bell jar is converted to the movement of a pen that writes on moving paper. The pen movement is inverted from the original bell jar. Inspiration results in downward movement of the bell jar and upward movement of the pen. The person is instructed to breathe normally for a few breaths, then to inspire as deeply as possible, and then expire as deeply as possible. The trace allows identification and measurement of several lung volumes and capacities, as described in the text.

TABLE 6.2.1 Ratios of VC, RV, and TLC to the Cube of the Height in Meters

Age (years)	VC/h³ (L m⁻³)	RV/h³ (L m⁻³)	TLC/h³ (L m⁻³)
18–19	0.990	0.24	1.23
20–29	1.025	0.275	1.30
30–34	1.020	0.30	1.30
35–39	1.010	0.31	1.32
40–44	1.000	0.32	1.32
45–49	0.990	0.33	1.32
50–54	0.970	0.35	1.32
55–59	0.950	0.37	1.32
60–64	0.930	0.39	1.32

The values were obtained from European men of different ages in the upright posture.
Data from E. Agostoni and J. Mead, Statics of the respiratory system, in W.O. Fenn and H. Rahn, eds., *Handbook of Physiology*, 1964.

IRV, as shown in Figure 6.2.1. Its volume is given as VC = ERV + TV + IRV = 1.1 L + 0.5 L + 3.0 L = 4.6 L. The VC is the largest volume of air that can be expired from a maximal inspiration.

4. The **total lung capacity, TLC**, is the maximum volume of air that the respiratory system can hold. It is the sum of all four lung volumes. Thus TLC = VC + RV. Because this includes the RV, it cannot be determined using spirometry. Its value is given as TLC = IRV + TV + ERV + RV = 3.0 L + 0.5 L + 1.1 L + 1.2 L = 5.8 L. Again, this value is a "typical" value for a young adult male. Such a person is largely fictitious in that most people are not normal young adult males. Particular values should be interpreted by considering the age, sex, and body size of the particular person.

LUNG VOLUMES AND CAPACITIES VARY MAINLY WITH BODY SIZE

Generally speaking, larger people have larger lung volumes and capacities compared to their smaller counterparts. But what do we mean by "larger"? Suitable measures of size include height, weight, and surface area, and these are not independent. Taller people often weigh more and have a larger surface area—but not always. Extensive studies of European males in the 1950s showed a fairly constant relation between the cube of the height and VC, RV, and TLC. These results (see Table 6.2.1) also show that the relationship between lung volumes and height changes with age. The values for women are typically about 10% less than those for men of the same age and height.

PULMONARY VENTILATION IS THE PRODUCT OF RESPIRATORY RATE AND TIDAL VOLUME

The purpose of the respiratory system is to exchange gases between the blood and the inspired air in order to supply oxygen to metabolizing tissues and to remove waste CO_2 to the air. When the tissues metabolize more quickly, as during exercise, the exchange of gases must increase to match demand. This is not a static problem, but a dynamic one, and it is the rate of ventilation, rate of gas exchange, and rate of O_2 consumption and CO_2 production, that matter. The lung volumes and lung capacities are all measured in liters, whereas rates of ventilation, gas exchange, and consumption are expressed in $L\ min^{-1}$.

The pulmonary ventilation is the rate at which air moves out of the lungs. Suppose that 500 mL leaves the lungs with each tidal respiration at rest and that the resting respiratory rate (RR) is 12 breaths per minute. Then the pulmonary ventilation, $Q_V = TV \times RR = 0.5\ L\ breath^{-1} \times 12$ breaths $min^{-1} = 6\ L\ min^{-1}$. Because usually more O_2 is consumed than CO_2 is produced, the volume of dry air that enters the lung is a tiny bit larger than the volume of dry gas that is expired. The resting pulmonary ventilation is often called the **respiratory minute volume**.

DURING EXERCISE, PULMONARY VENTILATION INCREASES DUE TO INCREASED RR AND TV

Typically, exercise increases both the RR and the TV. The increased TV usually uses only part of the IRV and part of the ERV. During exercise the pulmonary ventilation can reach $80-100\ L\ min^{-1}$ in healthy young adults. This is about a 15-fold increase from the resting respiratory minute volume.

THE MAXIMUM VOLUNTARY VENTILATION EXCEEDS PULMONARY VENTILATION DURING EXERCISE

The **maximum voluntary ventilation** refers to the maximum rate of pulmonary ventilation. This can be measured using a spirometer, but the period of maximal breathing is limited to 15 s because the individual can become alkalotic and he can fatigue. Figure 6.2.2 shows a spirometer trace of an individual moving as much air as possible. In this case, the TV during the maximum voluntary ventilation was 2.8 L (1.7 L of IRV + 0.5 L V + 0.6 L of ERV), and the RR was 15 breaths in the 15 s period, giving 60 breaths min^{-1}. This gives a maximum voluntary ventilation of 60 breaths $min^{-1} \times 2.8\ L\ breath^{-1} = 168\ L\ min^{-1}$. This is an impressive maximum voluntary ventilation which typically can be reached only in very fit young adults. Normal values depend on size, age, and sex. The maximal voluntary ventilation exceeds pulmonary ventilation during exercise because the person focuses only on moving as much air as possible for a very short time. It is not possible to continue such maximal ventilation for protracted times.

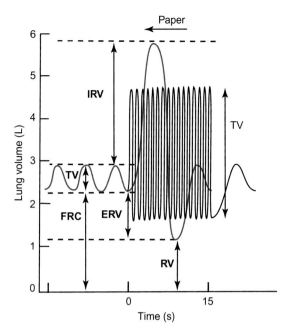

FIGURE 6.2.2 Spirometer trace of maximum voluntary ventilation. The blue trace represents maximum inspiration and expiration, as described in Figure 6.2.1. Superimposed on this is 15 s of maximum ventilation in which the person increases both the rate and depth of ventilation. In this case, some of the IRV and some of the ERV were used to increase the TV from the normal 0.5 to 2.8 L. The RR increased from about 12 to 60 min^{-1}.

SPIROMETRY ALSO PROVIDES A CLINICALLY USEFUL MEASURE OF AIRWAY RESISTANCE

The major difference between the static behavior of the lungs and its dynamic behavior is the resistance that accompanies airflow and partially limits the ability to ventilate the lungs. A clinically useful measure of the ability to move air quickly is the **forced vital capacity (FVC)**. In this procedure, a person connected to a spirometer is instructed to breathe in as deeply as possible and then to expire as *completely* and *rapidly* as possible. A hypothetical spirometer tracing of this procedure is shown in Figure 6.2.3. The volume of air that can be expired this way is the FVC. The timescale is expanded to resolve the time course of the FVC, and the volumes expired at 1, 2, and 3 s are measured. These are referred to as FEV_1, FEV_2, and FEV_3 for forced expiratory volume at 1, 2, and 3 s. The absolute values of FEV_1, FEV_2, and FEV_3 vary with FVC, and so these values are normalized by dividing by FVC.

The clinical presentation of the FVC is often turned around and presented as shown in Figure 6.2.4. This spirogram provides four test results: **FVC, FEV1, FEV1/VC, and FEF25−75**, also called the **maximal midexpiratory flow rate**. This is the averaged forced expiratory flow rate from 25% of FVC to 75% of FVC.

A more modern approach is the **flow−volume loop**, obtained from clinical spirometers (see Figure 6.2.5). The main results from this presentation of the data are the **FVC** and the **peak expiratory flow rate, PEFR**, the maximum flow during forceful expiration. For healthy

EXAMPLE 6.2.1 Lung Volumes and Airway Resistance

The lung capabilities in an older individual were measured by spirometry, and the following data were obtained:

TV = 0.4 L; IRV = 2.3 L; ERV = 1.2 L; FEV_1 = 2.0 L

Calculate the VC and FRC. Are these normal? Is the airway resistance normal?

The VC is calculated as VC = ERV + TV + ERV = 1.2 L + 0.4 L + 2.3 L = **3.9 L**. We cannot calculate the FRC because this requires knowledge of the RV, which we cannot obtain from spirometric data. You cannot determine whether this VC is normal or not without information about the size and age of the person.

We calculate an indicator of airway resistance as FEV_1/VC = 2.0 L/3.9 L = **0.51**. This number is definitely below the normal value of 0.80–0.90.

---

adult males, the PEFR is in the range of 9–11 L s^{-1}; for adult females, it is about 6–7 L s^{-1}.

AIRWAY RESISTANCE DEPENDS ON WHETHER AIRFLOW IS LAMINAR OR TURBULENT

STREAMLINE OR LAMINAR FLOW ENTAILS A RESISTANCE THAT IS INDEPENDENT OF FLOW

The relation between laminar flow and pressure in a right cylindrical tube was covered in Chapters 1.2 and 5.8. It is described by Poiseuille's law:

$$[6.2.1] \qquad Q_V = \frac{\pi a^4}{8\eta l}\Delta P$$

where Q_V is the flow, in volume per unit time, a is the radius, η is the viscosity of the flowing medium, I is the length of the cylinder, and ΔP is the pressure difference between the beginning and the end of the cylinder. This is the hydraulic analogy of Ohm's law:

$$[6.2.2] \qquad Q_V = \frac{\Delta P}{R}$$

where R is the **resistance** of the airway to flow. This equation can be rewritten as

$$[6.2.3] \qquad \Delta P = R Q_V$$

Equation [6.2.3] is presented this way to emphasize a different point: the pressure difference necessary to maintain a steady flow of Q_V is linearly related to Q_V, and the coefficient is the resistance. Here the resistance to flow

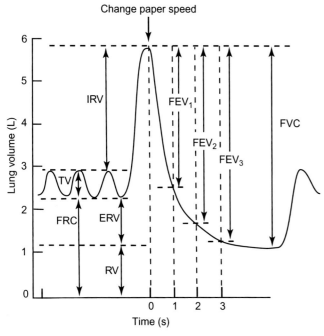

FIGURE 6.2.3 Spirometer tracing of a FVC. The person connected to the spirometer is instructed to breathe normally for a few breaths, then to inspire as deeply as possible, and then to expire as rapidly and completely as possible. Near the peak of inspiration, the chart speed is changed to resolve the time course of the FVC. The volumes expired at 1, 2, and 3 s are recorded as the forced expiratory volume (FEV_1, FEV_2, and FEV_3). The ratio of FEV_1, FEV_2, and FEV_3 to the FVC provides clues to the airway resistance. The FVC should be equal to the VC, but it is typically expired much faster than for the measurement of the VC. FEV_1/FVC is typically greater than 0.80; a value significantly lower than this indicates abnormally high airway resistance. FEV_2/FVC should exceed 0.90; FEV_3/FVC should exceed 0.95.

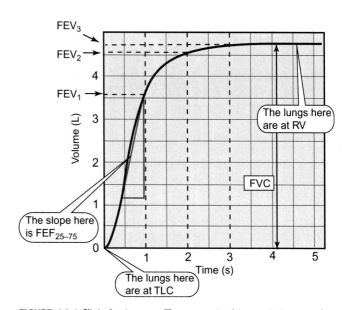

FIGURE 6.2.4 Clinical spirogram. The person in this case is instructed to inspire as deeply as possible, and then to exhale as rapidly and completely as possible. The expired volume is plotted as a function of time after beginning expiration, so the plot is the reverse of the positive-displacement spirometer. The clinically useful test results are the **FVC**, equal to the total volume of gas expired following maximal inspiration; FEV_1, the forced expiratory volume after 1 s; FEV_1/FVC; and the **maximal midexpiratory flow rate** or FEF_{25-75}, which is the slope of the volume–time curve between 25% of FVC and 75% of FVC.

Lung Volumes and Airway Resistance

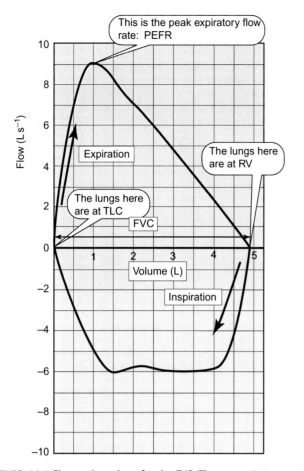

FIGURE 6.2.5 Flow–volume loop for the FVC. The person is instructed to inspire as deeply as possible and then to exhale as rapidly and completely as possible. The expiratory phase is shown above the abscissa, and volume refers to the volume of exhaled air. The zero volume begins with the lungs as full as possible; their volume at this time is the TLC. After exhaling as much as possible, the volume remaining in the lungs is the RV. The volume of air expired in the maneuver is the FVC. The maximum flow during expiration is the PEFR. The maximum flow rate during inspiration can be nearly as great as that during maximal expiration. In diseased lungs, maximum inspiratory flow can be greater than the maximal expiratory flow due to **dynamic compression** of the airways during expiration.

depends only on the viscosity of the air, the radius of the airway, and its length. It does not depend on the rate of airflow and we would expect that the plot of Q_V against ΔP would be linear with slope $= 1/R$. Equation [6.2.3] can be rearranged to emphasize a different point: **the resistance is defined as the ratio of $\Delta P/Q_V$**.

TURBULENT FLOW PRODUCES A NONLINEAR RELATION BETWEEN FLOW AND PRESSURE DIFFERENCE

Turbulent flow differs from laminar flow in that the pressure difference needed to sustain a constant turbulent flow is proportional to the square of the flow:

[6.2.4] $$\Delta P = K_2 Q_V^2$$

where K_2 is a coefficient for the equation but it is not the resistance because the relationship is not Ohmic. This equation describes the relation between ΔP and Q_V only over the range of turbulent Q_V.

THE REYNOLDS NUMBER INDICATES THE TENDENCY TO TURBULENT FLOW

Whether the flow of air is laminar or turbulent depends on several physical factors including the velocity of flow, the density of the air, the viscosity of the air, and the size and physical characteristics such as the smoothness or roughness of the wall of the tubes through which flow occurs. The Reynolds number incorporates some of these factors (see Chapter 5.8). The Reynolds number is a dimensionless ratio of the inertial forces to the viscous forces of the fluid. It is given as

[6.2.5] $$Re = \frac{2a\langle V \rangle \rho}{\eta}$$

where Re is the Reynolds number, a is the radius of the airway, $\langle V \rangle$ is the average velocity of airflow, ρ is the density of air, and η is its viscosity. When $Re > 2500$, turbulent flow dominates; at lower values of Re, laminar flow is more likely. Equation [6.2.5] shows that Re is large when the airway radius is large and flow velocities are high. The density of air (at 1 atm) at 37°C is 1.13 g L^{-1} and its viscosity is 19.1×10^{-6} Pa s, where 1 Pa = 1 N m^{-2} = 10 dyne cm^{-2}. Q_V is the flow in units of volume time^{-1} and J_V, the flux, is the flow per unit area, in units of length time^{-1} (volume area^{-1} time^{-1}). J_V is the velocity of fluid flow. For streamline flow, J_V is not constant across the tube but varies from zero at the sides of the tube to a maximum in the center of the tube. We calculate the average velocity as

[6.2.6] $$\langle V \rangle = \frac{Q_V}{A}$$

where A is the area through which the flow occurs. Insertion of $\langle V \rangle$ from Eqn [6.2.6] into Eqn [6.2.5] gives

[6.2.7] $$Re = \frac{2aQ_V\rho}{A\eta}$$

Here A is the total cross-sectional area of airways of radius a. Thus $A = n\pi a^2$, where n is the number of airways of the same generation. Using this equation and the results shown in Figure 6.2.6 for the morphometry of the lungs, we can calculate the Reynolds number for each generation of airway. It is clear from Eqn [6.2.7] that the Reynolds number increases directly with Q_V and inversely with total cross-sectional area. The results show that for normal TV, $Re < 2000$ everywhere in the tracheobronchial tree, suggesting that turbulent flow most likely does not occur during resting respiration. At higher flows, turbulence becomes apparent in the upper airways mainly because the total cross-sectional area is small.

THE POISEUILLE EQUATION DERIVED FOR RIGHT CYLINDERS DOES NOT MODEL THE COMPLICATED AIRWAYS

A complicated system of tubes such as the tracheobronchial tree with its branches, irregular wall surfaces, and changes in caliber is not amenable to direct application of equations that have been derived to describe flow through right cylinders. Calculation of airway resistance,

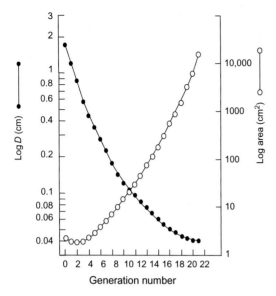

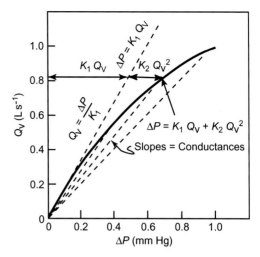

FIGURE 6.2.6 Morphometry of the lungs. The morphometry of human lungs was estimated from plastic casts for the larger airways and from histological sections for the smaller airways. The diameter of the airways is plotted against the generation number (*closed circles*). The generation number refers to the number of times the airways have branched. The diameter decreases nearly exponentially with the generation number during the first few generations and then decreases more slowly thereafter. Even though the airways are markedly smaller, there is a tremendous number of smaller airways, so that the total cross-sectional area increases markedly with generation number. Because the cross-sectional area increases, the velocity of airflow through the airways diminishes as the airways divide toward the alveoli. *Data from W.R. Weibel, Morphometry of the Human Lung, Springer, 1963.*

FIGURE 6.2.7 Relationship between flow and pressure difference in the lungs. The slope of the chord connecting the line with origin is the airway conductance and its inverse is the resistance (K_1). The resistance is not constant: it depends on the airflow. This is because as airflow increases, a greater fraction of the tracheobronchial tree exhibits turbulent flow with its higher resistance.

however, suggests that both laminar flow and turbulent flow occur within the airways.

AIRWAY RESISTANCE IS THE SLOPE BETWEEN ΔP AND Q_V

Equation [6.2.3] describing laminar flow predicts a linear relation between the pressure and the flow. On the other hand, Eqn [6.2.4], which describes turbulent flow predicts a parabolic relationship between ΔP and Q_V. Which describes the lungs? Figure 6.2.7 illustrates that the relationship between ΔP and Q_V is neither linear nor parabolic but is a sum of both processes.

TURBULENT AND LAMINAR FLOWS RESULT IN RESISTANCES THAT OCCUR IN SERIES AND ADD

The tracheobronchial tree consists of a series of airways that branch. The airways become progressively smaller, as shown in Figure 6.2.6, and at the same time the number of parallel airways becomes larger. Thus the airways consist of both series and parallel components. The collection of parallel airways at each generation can be lumped to form a single equivalent resistance, and these lumped resistances then are arranged in series and the equivalent resistances add. At some levels of the generations of airways, the flow is laminar and resistance is given by Eqn [6.2.3] as

[6.2.8] $$R = \frac{\Delta P}{Q_V} = K_1$$

For those airways in which flow is turbulent, the resistance is given by Eqn [6.2.4] as

[6.2.9] $$R = \frac{\Delta P}{Q_V} = K_2 Q_V$$

Since the airways in which laminar flow occurs are in series with those in which turbulent flow occurs, we might expect the resistances to add. Thus we would expect

[6.2.10] $$R_{\text{total}} = \frac{\Delta P}{Q_V} = K_1 + K_2 Q_V$$
$$\Delta P = K_1 Q_V + K_2 Q_V^2$$

These equations form an adequate approximate description of overall airway resistance. As predicted from Eqn [6.2.10], the airway resistance, calculated as $\Delta P/Q_V$, is linearly related to Q_V with intercept K_1 and slope K_2. Thus the resistance is not constant but depends on the rate of flow.

AIRWAY RESISTANCE DEPENDS ON LUNG VOLUME

As the lungs expand, their recoil tendency increases and so they pull harder on the chest wall, resulting in a more negative intrapleural pressure. Lung tissue is also anchored to the airways, and so the expansion of the lungs also causes an expansion of the airways by pulling on them from the tissue side of the airways. The bronchial caliber is set in part by the radial traction of the surrounding lung tissue. The relationship is nearly hyperbolic, described by

[6.2.11] $$R_{\text{total}} = \frac{K}{V}$$

EXAMPLE 6.2.2 Calculate the Reynolds Number

The diameter of the trachea in one individual was 1.8 cm. The person's PEFR was measured to be 10 L s^{-1}. During normal resting respiration, the flow was about 0.3 L s^{-1}. Calculate the Reynolds number for the two circumstances.

The Reynolds number is $2a\langle V\rangle \rho/\eta$, where a is the radius, $\langle V\rangle$ is the average velocity ($=Q_V/A$), ρ is the density, and η is the viscosity. We use $\rho = 1.13$ g L^{-1} and $\eta = 19.1 \times 10^{-6}$ Pa s = 191×10^{-6} dyne cm^{-2} s. These conversions use the cgs system, which is often more convenient.

First, we calculate $\langle V\rangle = 10$ L s$^{-1}/\pi \times (0.9$ cm$)^2 = 3.93$ L s^{-1} cm$^{-2} \times$ 1000 cm^3 L$^{-1} = 3.93 \times 10^3$ cm s^{-1}.

During peak expiration, the Reynolds number is

$$1.8 \text{ cm} \times 3.93 \times 10^3 \text{ cm s}^{-1} \times 1.13 \text{ g L}^{-1}$$
$$\times 10^{-3} \text{ L cm}^{-3}/191 \times 10^{-6} \text{ g cm s}^{-2} \text{ cm}^{-2} \text{ s} = \mathbf{41{,}851}$$

This is clearly turbulent. At the lower flow, $\langle V\rangle = 0.3$ L s$^{-1} \times$ 1000 cm^3 L$^{-1}/2.54$ cm^2 = 118 cm s^{-1}.

$$R_e = 1.8 \text{ cm} \times 118 \text{ cm s}^{-1} \times 1.13 \text{ g L}^{-1}$$
$$\times 10^{-3} \text{ L cm}^{-3}/191 \times 10^{-6} \text{ g cm s}^{-2} \text{ cm}^{-2} \text{ s} = \mathbf{1256}$$

which is not turbulent.

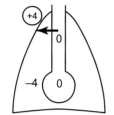 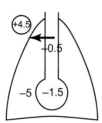 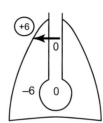 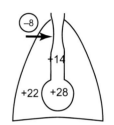

At FRC, before inspiration | During inspiration | At end inspiration | During forced expiration

FIGURE 6.2.8 Schematic diagram of the pressures acting on the airways during various stages of breathing. Approximate intrapleural, alveolar, and airway pressures are given in mm Hg. The net pressure acting on the airways helps keep them open except in the case of a forceful expiration, in which the pressure within the airways becomes less than that in the intrapleural space. The pressure drops in the airways due to the resistance to flow at high volumes. The result is the dynamic compression of the airways and limitation of maximal expiration rates. *Modified from J.B. West, Respiratory Physiology, Williams and Wilkins, Baltimore, 1979.*

where V is the lung volume and K is some constant. The inverse of the resistance, R_{total}, is the **conductance**. According to Eqn [6.2.11], the airway conductance is linearly related to the lung volume.

DYNAMIC COMPRESSION OF THE AIRWAYS DURING FORCEFUL EXPIRATION LIMITS AIRFLOW

The pressure difference between the intrapleural space and the airways is part of the balance of forces keeping them open. During forceful expirations, the pressure outside the airways is higher than it is inside. This transmural pressure difference diminishes the caliber of the airways, thereby increasing their resistance. Figure 6.2.8 illustrates the net pressures acting on the airways.

AIRWAY RESISTANCE IS MODIFIED BY SMOOTH MUSCLE CONTRACTION OF THE AIRWAYS

The airways are invested with a coat of smooth muscle that is innervated by parasympathetic fibers. Contraction of these smooth muscles causes **bronchoconstriction**, and relaxation causes **bronchodilation**. Parasympathetic excitation with its accompanying release of acetylcholine brings about bronchoconstriction. In addition, local factors can cause bronchoconstriction. The lungs sit at the interface between the body and the environment and provide a convenient route for the entry of all manner of foreign organisms and objects. As a result, the lung is provided with potent immune mechanisms and cells. Release of histamine from these cells upon presentation with an antigen contracts the smooth muscle and constricts the bronchioles. In severe cases, the bronchoconstriction can be fatal. This is the case in severe asthma attacks that claim the lives of approximately 2500 persons in the United States each year.

The smooth muscles of the airways possess β_2 receptors. Their stimulation increases cAMP in the cells and causes relaxation (see Chapter 3.8). Thus the rescue treatment for a person with bronchoconstriction is injection of epinephrine to relax the muscles and relieve the constriction. Sympathetic stimulation causes bronchodilation.

PURSED LIPS INCREASE AIRFLOW IN CASES OF INCREASED AIRWAY RESISTANCE

Persons with decreased lung recoil exhibit increased airway resistance because the lungs do not pull on the airways enough to keep them open. The effect is to increase the FRC because the balance between lung

Clinical Applications: COPD

COPD stands for **c**hronic **o**bstructive **p**ulmonary **d**isease. This is a set of different diseases that differ in their causes but have a common physiological effect: increased airway resistance. It includes **emphysema**, **chronic bronchitis**, and **bronchiectasis**. Although asthma is also a chronic obstructive condition, it is not considered to be a COPD because the term is reserved for irreversible conditions. Unlike emphysema, chronic bronchitis, and bronchiectasis, the airway obstruction in asthma is usually reversible. The increased airway resistance can be due to the loss of elastic recoil due to changes in the lung tissue, as seen in **emphysema**, or to airway inflammation as seen in **chronic bronchitis**.

Emphysema claimed 106,000 lives in the United States in 1998, making it the fourth leading cause of death nationwide. It entails abnormal enlargement of the air spaces distal to the terminal bronchioles with destruction of the tissue without fibrosis. Smoking tobacco is the primary cause of COPD, accounting for 80–90% of the risk of developing COPD. However, only 15% of smokers actually develop the disease. The mechanism by which cigarette smoke causes lung tissue destruction is controversial. For over 30 years, the favored explanation was that cigarette smoking caused an imbalance in the ratio of elastase/antielastase. Smoking causes neutrophils to collect in the lungs and these neutrophils contain elastase, an enzyme that degrades elastin, an important component of the extracellular matrix that allows the lungs to recoil. A serum protein called α_1-antitrypsin is made in the liver but is found mainly in the lungs. It inactivates neutrophil elastase. Persons with inherited deficits in α_1-antitrypsin who smoke have a much higher incidence of COPD.

This theory has come into question because some persons with deficiency of α_1-antitrypsin who smoke do not develop the disease. Other causative elements include free radicals produced by neutrophils that are activated by cigarette smoke. The free radicals are postulated to oxidize a methionine residue of α_1-antitrypsin, thereby inactivating its inhibitory effects on elastase.

Whatever its cause, destruction of the elastin and collagen fibers of the lung results in the destruction of the alveoli. The air spaces become larger and the surface area of gas exchange becomes much smaller. The reduced recoil tendency no longer holds open the bronchioles and so these tend to collapse, producing the increased airway resistance that characterizes COPD.

Bronchiectasis results in dilated bronchi that easily collapse, causing the obstruction, due to the destruction of their muscle layers and supportive elastic tissue. It is characterized by persistent cough and production of large volumes of yellow or green sputum. Although multiple conditions lead to it, it usually is associated with bacterial infections. Persons with acquired immunodeficiency syndrome (AIDS) are susceptible to opportunistic infections, and AIDS is a leading cause of bronchiectasis. Tuberculosis is another leading cause of acquired bronchiectasis. Congenital causes include cystic fibrosis (which accounts for half of the cases of bronchiectasis in children and young adults) and primary ciliary dyskinesia, in which the cilia in the respiratory epithelium are defective, and the mucus escalator fails. This leads to respiratory infections and predisposes to bronchiectasis. Treatment includes surgery to remove localized bronchiectasis, controlling infections and removing obstruction and bronchial secretions.

Clinical Applications: Positive Pressure Breathing

In normal breathing, the pressures within the alveoli immediately prior to inspiration (at the end of a tidal expiration) and after inspiration (at the end of tidal inspiration) are equal to the atmospheric pressure. Fibrosis decreases the FRC because it is more difficult to expand the lungs, and movement of air is reduced because the work of breathing is greater, even though the pressure differences exceed normal values. This can be helped by increasing the pressure of the incoming air, artificially expanding the lung with air pressure instead of muscle action. This enables the system to operate at higher volumes and restores volume exchange. This is accomplished by ventilators.

Common mechanical ventilators use positive pressure to force air into the lungs, causing inflation of the lungs. The ventilator can be set to introduce a known tidal volume and then stop pushing more air, and allow the natural recoil of the respiratory system to produce expiration. Although expiration is passive, the ventilatory can be altered to allow expiration only when the expiratory pressure within the lungs exceeds a small positive value, the **PEEP** (positive end-expiratory pressure). The purpose of the PEEP is analogous to the actions of pursed-lips breathing: it keeps the airways open. The mechanical ventilator in this case allows the respiratory system to operate at a higher volume, leading to better gas exchange.

Negative-pressure ventilators, such as the iron lung used in the Polio epidemics in the United States during the 1950s, apply negative pressure to the thorax, thereby causing expansion of the lungs. Expiration is accomplished by positive pressure to the thorax.

Persons with obstructive sleep apnea typically have issues with intermittent closure of the airway due to physical obstruction. Commonly, extra tissue in obese persons applies weight to the airways, requiring abnormally high transmural pressures to keep them open. This obstruction occurs while laying down to sleep and produces periods of no breathing—apnea—better known as **sleep apnea**. Other anatomic structures may also obstruct the airways, also causing sleep apnea. Inflating the respiratory system at all points of the breathing cycle physically ensures that such structures no longer obstruct the airways. This requires continuous positive airway pressure (CPAP) that aids breathing by opening airways internally by inflation. CPAP machines are becoming more common and more portable. By wearing a tight-fitting mask while sleeping, patients with sleep apnea can use CPAP to remove or reduce airway obstructions, leading to the reduction of episodes of sleep apnea and reducing snoring is those so afflicted. About 12 million Americans have been diagnosed with sleep apnea but there are likely many more because persons so affected are often asymptomatic.

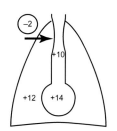
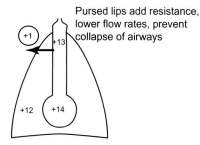

FIGURE 6.2.9 Pursed-lip breathing. For persons with obstructive airway disease, elastic tissue in the lungs is reduced, lowering the pull on the airways and causing them to collapse more easily. Forced expiration, by applying a positive intrapleural pressure rather than the negative one typically produced by lung recoil, causes the airways to collapse. Paradoxically, the airways can be kept open by pursing the lips and transferring the resistance to the lips rather than the airways.

recoil and chest expansion occurs at a higher lung volume. The patient has a difficult time getting air out of the lungs because of the reduced recoil force of the lungs. One way to increase airflow, somewhat counterintuitively, is to increase outflow resistance at the lips. That is, to breathe through pursed lips. This produces an increased resistance at the outflow point of the air, which increases its pressure in the airways, helping to keep the airways open. The patient is instructed to breathe slowly through pursed lips. Effectively, more of the resistance to airflow is transferred from the airways to the lips. This is shown diagrammatically in Figure 6.2.9.

SUMMARY

Spirometry can be used to measure three lung volumes and two capacities: the TV, IRV, ERV, IC and VC. The RV cannot be measured by spirometry and therefore the FRC and TLC also cannot be determined by spirometry. Combinations of these four volumes define the four lung capacities: the inspiratory capacity (IC = IRV + TV), vital capacity (VC = IRV + TV + ERV), the functional residual capacity (FRC = ERV + RV), and the total lung capacity (TLC = IRV + TV + ERV + RV). The lung volumes and capacities vary with body size, age, and sex.

Pulmonary ventilation is the product of the TV and RR. The maximum voluntary ventilation can also be measured using a spirometer. It usually exceeds the maximum ventilation during exercise. Spirometers also can provide a measure of airway resistance from the fraction of the FVC that can be expired in 1, 2, and 3 s.

At low flow rates, laminar or streamline airflow is described by an ohmic relationship between flow and the net pressure difference (ΔP) that drives flow. At high flow rates, the relationship between flow and ΔP is not linear. The presence of turbulence in the airways depends on the velocity of airflow, the diameter of the airways, and the density and viscosity of the air. The velocity of the air and diameter of the airways vary considerably, whereas density and viscosity of the air are nearly constant. Turbulence is likely when the Reynolds number exceeds 2500. The Reynolds number is calculated from

$$Re = \frac{2a \langle V \rangle \rho}{\eta}$$

The diameter of the airways decreases nearly exponentially with generation number, whereas the total cross-sectional area of the airways increases because the number of airways increases with generation number. Because the total cross-sectional area increases so much, the average velocity of airflow decreases with generation number. Thus the airways can be approximated by laminar flow and turbulent flow in series. At low pulmonary ventilation, there is little turbulent flow. At high flow rates, dynamic compression of the airways limits expiratory airflow. Airway resistance is also modified by smooth muscle contraction of smooth muscle surrounding the bronchioles. Parasympathetic stimulation constricts the bronchioles, whereas sympathetic stimulation dilates them.

REVIEW QUESTIONS

1. Draw a spirometer trace and label TV, IRV, ERV, RV, IC, FRC, VC, and TLC. Indicate which cannot be measured by spirometry alone.
2. Describe the procedure for measuring the FVC and indicate what results from the procedure indicate airway resistance.
3. Describe how you could determine pulmonary ventilation and maximum voluntary ventilation.
4. Write the equation for the Reynolds number. Do you think the transition between laminar flow and turbulent flow should be a narrow range of values for the Reynolds number or a fairly broad range?
5. Would you expect turbulent flow in the larger airways or in the smaller ones? Why?
6. What endogenous materials alter airway resistance, and how do they do it?

6.3 Gas Exchange in the Lungs

Learning Objectives

- Distinguish between respiratory quotient and respiratory exchange ratio
- List the three steps in the uptake of oxygen by the blood
- Write the Ideal Gas Equation
- Define partial pressure of a gas
- Define vapor pressure
- Give the vapor pressure of water at body temperature
- Write Henry's law for dissolution of gas in aqueous solutions
- Define what is meant by STPD
- Be able to convert partial pressures and volumes at STPD to BTPS
- List the sequential barriers to diffusion that are present between air and blood in the lungs
- List the factors that make gas exchange in the lungs rapid
- Describe the driving force for gas movement
- Define the diffusing capacity of a gas
- Define the anatomic dead space
- Distinguish between alveolar ventilation and pulmonary ventilation
- Write the alveolar ventilation equation
- Recognize the alveolar gas equation and identify its variables
- Explain why alveolar gas composition is relatively constant
- Describe the time course of gas equilibration across the alveoli

THE RESPIRATORY SYSTEM EXCHANGES BLOOD GASES WITH ATMOSPHERIC GASES

As pointed out in Chapter 6.1, cellular metabolism of carbohydrates and fats entails their complete oxidation. The overall reactions are written as

$$C_6H_{12}O_6 + 6O_2 \Rightarrow 6CO_2 + 6H_2O + \text{heat}$$

$$C_{51}H_{98}O_6 + 72.5O_2 \Rightarrow 51CO_2 + 49H_2O + \text{heat}$$

During constant metabolic conditions, there is a constant production of CO_2 and constant consumption of O_2. The ratio of these two is given a special name, the **respiratory quotient**. It is defined as

$$[6.3.1] \qquad R = \frac{Q_{CO_2}}{Q_{O_2}}$$

where Q_{CO_2} is the rate of production of CO_2, in moles per unit time or proportional units, and Q_{O_2} is the rate of consumption of O_2, in the same units as Q_{CO_2}. The value of R depends on the type of materials being oxidized for energy. When carbohydrate is the only fuel being oxidized for energy, the respiratory quotient is 1.0: there is a molecule of CO_2 produced for every molecule of oxygen consumed. When fats alone are being oxidized, the respiratory quotient is $51/72.5 = 0.70$. Protein oxidation produces a respiratory quotient intermediate between these extremes. Normally the body consumes a mixture of carbohydrates, fats, and proteins, and a typical value for R is 0.80. Its range is $0.7 < R < 1.0$.

The **respiratory exchange ratio** is the ratio of the rate of CO_2 to O_2 exchanged between the body and the atmosphere. At steady state, this is the same as the ratio of gas produced and consumed by the tissues, the respiratory quotient. However, the transient state between one metabolic state and another involves storage or depletion of gases in the tissues and blood and for short times the respiratory exchange ratio can be different from the respiratory quotient.

An understanding of gas exchange requires understanding the behavior of gases both in the gas phase and dissolved in the aqueous phase. We begin with a review of these principles.

THE PARTIAL PRESSURE OF A GAS IS ITS MOLE FRACTION TIMES THE TOTAL PRESSURE

A gas contains an assortment of molecules that expands to fill any container. The gas molecules move rapidly in random directions until they collide with each other or with the walls of the container. When they hit the sides of the wall they rebound, thereby undergoing a change in momentum. This change in momentum corresponds to force exerted on the molecule *by* the wall, which is equal but opposite in direction to the force of the molecule *on* the wall. When many molecules collide with the

TABLE 6.3.1 Components of Air and Partial Pressure at STPD (1 atm pressure, 0°C with 0% Humidity)

Component of Air	Content in Volume Percent	Content in Mole Fraction	Partial Pressure (mmHg)
N_2	78.08	0.7808	593.5
O_2	20.95	0.2095	159.2
Ar	0.93	0.0093	7.1
CO_2	0.03	0.0003	0.2
H_2O	0.00	0.00	0.0
Total	100.00	1.0000	760.0

wall, they produce an average force distributed over an area. This is the pressure, the force per unit area. The velocity, and hence the momentum, of the molecules increases with the temperature. These ideas are subsumed into the **Ideal Gas Equation**:

$$[6.3.2] \qquad PV = nRT$$

Here P is the pressure in atmospheres or mmHg or pascals (1 Pa = 1 N m^{-2}), V is the volume in L, n is the number of moles of gas, R is the gas constant (0.082 L atm mol^{-1} K^{-1}; 1.986 cal mol^{-1} K^{-1}; 8.314 J mol^{-1} K^{-1}), and T is the temperature in K. This can be rewritten easily as

$$[6.3.3] \qquad P = \frac{n}{V} RT$$

where n/V is the concentration of the gas. The **Ideal Gas Equation** given here is approximately true for a variety of gases, including O_2, N_2, He, CO_2, and many other gases. These gases are not ideal and the corrections for nonideality depend on the kind of gas.

If a gas consists of only one kind of molecule, adding successive increments of gas adds incremental pressure according to the number of moles in the increment. Thus if pressure P_1 is produced by n_1 moles of the gas, then adding n_2 moles to the same volume adds pressure P_2:

$$[6.3.4] \qquad \begin{aligned} P &= \frac{n}{V} RT \\ &= \frac{n_1 + n_2}{V} RT \\ &= \frac{n_1}{V} RT + \frac{n_2}{V} RT \\ &= P_1 + P_2 \end{aligned}$$

Thus we can partition the total pressure in a gas according to the number of moles of gas that contributes to the total pressure. In a mixture of gases, we ought to expect the same additive properties of the pressure. Suppose now that we have a mixture of n_A moles of gas A and n_C moles of gas C. According to Eqn [6.3.4], we would have a total pressure given as

$$[6.3.5] \qquad \begin{aligned} P &= \frac{n}{V} RT \\ &= \frac{n_A}{V} RT + \frac{n_C}{V} RT \\ &= P_A + P_C \\ &= \frac{n_A}{n}\frac{n}{V} RT + \frac{n_C}{n}\frac{n}{V} RT \end{aligned}$$

Here the pressure contributed by each gas is called the **partial pressure**. Each gas in a mixture contributes to the total pressure according to its concentration. Because the concentrations of all gases at the same P and T are proportional to the number of molecules, within the limits of ideal behavior, the concentrations in a mixture are proportional to the **mole fractions**. This is evident from the last expression in Eqn [6.3.5], from which we identify the partial pressure of A as

$$[6.3.6] \qquad P_A = \frac{n_A}{n} P_B = f_A P_B$$

where f_A is the **mole fraction** of gas A, defined as the ratio of the number of moles of A to the total number of moles in the mixture, and P_B is the total or barometric pressure. Thus from Eqn [6.3.5], we see that the total pressure is the sum of the partial pressures which, by Eqn [6.3.6], is proportional to the mole fractions. The mole fraction and partial pressure of each kind of gas in the atmosphere at sea level (atmospheric pressure = 1 atm = 760 mmHg) are given in Table 6.3.1.

Because N_2 gas is inert with respect to body metabolism, the partial pressures of all gases other than H_2O vapor, O_2, and CO_2 are typically included in the contribution made by N_2. Although air contains Ar and a variety of other gases including He, H_2, Ne, Kr, CH_4, and NO_2, we will lump the partial pressures of all these gases into that made by N_2. Thus we write

$$[6.3.7] \qquad P_{total} = P_{CO_2} + P_{O_2} + P_{N_2} + P_{H_2O}$$

THE VAPOR PRESSURE OF WATER IS THE PARTIAL PRESSURE OF WATER IN THE GAS PHASE THAT IS IN EQUILIBRIUM WITH LIQUID WATER

Water molecules in the gas phase exert a pressure just like all other molecules. If a gas is placed in contact with liquid water in a closed container, water will evaporate until an equilibrium is reached in which the rate of water evaporation is equal to its rate of condensation. The gas phase at this equilibrium will be **saturated** with water vapor. The partial pressure of water in the gas phase under these conditions is *defined* as the vapor pressure (see Figure 6.3.1). Water can evaporate into a gas phase that is not in a closed container. In this case, the gas phase may or may not be saturated with water vapor. If there is any water in the gas phase, its partial pressure will be determined by its mole fraction as described in Eqn [6.3.6].

THE VAPOR PRESSURE OF WATER AT BODY TEMPERATURE IS 47 MMHG

At $37°C = 310$ K, the vapor pressure of water is 47 mmHg. If we inspire dry air, the mucus membranes lining the nasal passages and airways will moisten the air by adding water. Generally, the air that reaches the alveoli has already become saturated with water vapor at body temperature. The added water expands the gas phase and dilutes all other gaseous components. If we assume that the inspired air is completely dry, then the partial pressures of gases in moist tracheal air will be given by

$$[6.3.8] \qquad P_A = f_A(P_B - P_{H_2O})$$

where P_B is the barometric pressure. It is important to recognize that this equation converts the mole fraction in dry air, f_A, to the partial pressure in water-saturated air, P_A, in this equation. Thus humidifying the inspired air reduces the partial pressures of all of the gas components. The mole fractions and partial pressures of moist tracheal air are shown in Table 6.3.2 along with these values for dry atmospheric air. The values for P_{N_2} differ from those in Table 6.3.1 because it also includes the partial pressure of argon and other rare atmospheric gases.

The **relative humidity** is a common measure of the amount of water in the gaseous phase. It is defined as the ratio of the measured partial pressure of water to the vapor pressure of water.

HENRY'S LAW DESCRIBES THE DISSOLUTION OF GASES IN WATER

Henry's law, discovered in 1803, relates the mole fraction of gas that is dissolved in the liquid, watery phase to the partial pressure:

$$[6.3.9] \qquad x_A = \beta_A P_A$$

where x_A is the mole fraction of gas A in the aqueous phase, β_A is its solubility, and P_A is the partial pressure of the gas in the gaseous phase in equilibrium with the aqueous phase, in atmospheres. Physiologists typically convert partial pressures to units of mmHg. The mole fraction is simply the ratio of the moles of gas to the total number of moles in the mixture. For dilute solutions of gases, when n_A, the number of moles of gas, is very much less than n_W, the number of moles of water, the mole fraction is linearly related to the concentration of gas:

$$[6.3.10] \qquad \begin{aligned} x_A &= \frac{n_A}{n_A + n_W} \approx \frac{n_A}{n_W} \\ &= \frac{n_A}{V/\overline{V}_W} = \overline{V}_W \frac{n_A}{V} \end{aligned}$$

where $\overline{V}_W$ is the volume of water per mole (its partial molar volume) and n_A and n_W are the moles of gas A and water, respectively, in the mixture. Inserting this result into Eqn [6.3.9], we get

$$[6.3.11] \qquad \begin{aligned} \overline{V}_W \frac{n_A}{V} &= \beta_A P_A \\ [A] &= \frac{\beta_A}{\overline{V}_W} P_A \\ [A] &= \alpha_A P_A \end{aligned}$$

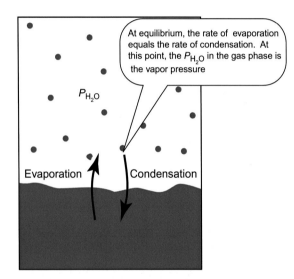

FIGURE 6.3.1 Water vapor pressure. Liquid water in contact with the gas phase will evaporate (form water vapor) until there is a dynamic equilibrium between the rate of evaporation and the rate of condensation. At this point of equilibrium, the partial pressure of water in the gas phase is equal to the vapor pressure. Heating the water increases the rate of evaporation and so raises the vapor pressure. The boiling point of water is the temperature at which the vapor pressure is equal to the atmospheric pressure.

TABLE 6.3.2 Composition and Partial Pressure of Gases in Dry Air and Moist Tracheal Air

Gas	Dry Inspired Atmospheric Air		Moist Tracheal Air	
	Mole Fraction	Partial Pressure (mmHg)	Mole Fraction	Partial Pressure (mmHg)
N_2 (with Ar, etc.)	0.7901	600.6	0.7412	563.3
O_2	0.2095	159.2	0.1966	149.4
CO_2	0.0003	0.2	0.0003	0.2
H_2O	0.0	0	0.0618	47.0
Total	1.0	760.0	1.0	760.0

TABLE 6.3.3 Solubilities of Gases of Physiological Interest at 37°C and STPD

Gas	β (mole fraction atm^{-1})	α (mL gas mL water^{-1} atm^{-1})	α (mL gas dL water^{-1} mmHg^{-1})
N_2	1.026×10^{-5}	0.0127	0.00169
O_2	1.932×10^{-5}	0.0241	0.00317
CO_2	4.560×10^{-4}	0.567	0.0747
CO	1.524×10^{-5}	0.01897	0.00250

Solubilities were calculated by extrapolation of β versus temperature or from formulas of $\beta(T)$ given in *Handbook of Chemistry and Physics*, 73rd ed., CRC Press Inc., Baton Rouge, 1992.

From Eqn [6.3.11], we can relate the molar concentration of gas to its partial pressure. However, physiologists use a particular unit of concentration which is mL of gas per unit volume of solution, and the volume of solution is usually expressed in deciliters (dL = 100 mL = 0.1 L). Because the volume of a gas depends on the pressure and temperature and the water content, the volumes of gases used in respiratory physiology are often expressed as the volume the gas would occupy under standard conditions. These standard conditions are referred to as **STPD**, meaning **s**tandard **t**emperature and **p**ressure, **d**ry. STPD refers to the conditions of 0°C = 273.16 K, 1 atm pressure = 760 mmHg and zero water vapor. Under these conditions, the molar volume of the ideal gas is 22.4 L mol^{-1}. Because of these choices of units, the final solubility factors that we will use are in units of mL dL^{-1} mmHg^{-1}. The solubility of gases depends strongly on the temperature. The values for the solubilities given in Table 6.3.3 are for water at 37°C = 310.2 K, with gas volumes at STPD.

CONVERSION OF PARTIAL PRESSURES AND VOLUMES AT STPD TO THOSE AT BTPS

As noted above, the rates of O_2 consumption and CO_2 production are expressed in units of volume at STPD or standard temperature and pressure, dry. The lung volumes and rates of movement of lung volumes, on the other hand, are measured at BTPS or body temperature and pressure, saturated. In addition, sometimes measurements are made at ATPS—ambient temperature and pressure, saturated. This is the case for the measurements made with the spirometer. Conversion of these volumes follows the principles described in Appendix 6.3.A2.

GASES DIFFUSE ACROSS THE ALVEOLAR MEMBRANE PASSIVELY

There are no carriers or pumps for gases. Until recently, the model of membrane permeation by gases was by dissolution within the membrane and diffusion through the lipid core. However, recent discoveries suggest that cholesterol reduces CO_2 permeability of lipid bilayers and suggest that proteins may mediate CO_2 diffusion. These proteins likely act as gas channels. AQP1 is one likely candidate for a CO_2 channel. At the present time there is no evidence for channels for O_2. Regardless of the path, the passive flux can be described by

[6.3.12] $$J_s = p\Delta C$$

where p is the permeability, J_s is the flux of solute s, which in this case is a gas such as O_2 or CO_2, and ΔC is the concentration difference across the alveolar membrane. This equation was developed to describe passive transport across a microporous membrane or a lipid bilayer membrane (see Eqn [2.5.9]). It describes diffusion kinetics: the overall flux depends linearly on the concentration difference and shows no saturation. This equation applies to diffusion of gas through any single phase, but it must be modified when diffusion occurs across phases due to the different solubility of gases within those phases. The result (see Eqn [6.3.14]) shows that **the partial pressure of gas drives diffusion**.

Eqn [6.3.11] states that the concentration of gas in the aqueous phase is linearly related to the partial pressure of gas in equilibrium with it. Thus the ΔC in Eqn [6.3.12] is linearly related to ΔP_s, the difference in partial pressure across the alveolar membrane.

In traveling from the alveolar air to the blood, O_2 encounters a variety of barriers. These are shown schematically in Figure 6.3.2. These barriers include:

- the alveolar lining, extracellular to the alveolar cell;
- alveolar cell, consisting of two plasma membranes and intervening cytosol;
- interstitial fluid;
- endothelial cell, consisting of two plasma membranes and intervening cytosol;
- plasma;
- red blood cell membrane.

The barriers include the watery lining of the alveoli that contains surfactant, the alveolar cell, the interstitial fluid, the endothelial cell, the plasma, and the red blood cell membrane.

Figure 6.3.2 illustrates how the diffusional barrier between air and blood is made up of many sequential layers. Each of these layers is characterized by an equation like Eqn [6.3.12]. The permeabilities of each layer depend on a variety of factors. We may model diffusion of gas across each layer similarly to the penetration of lipid soluble solutes through lipid bilayer membranes

QUANTITATIVE HUMAN PHYSIOLOGY: AN INTRODUCTION

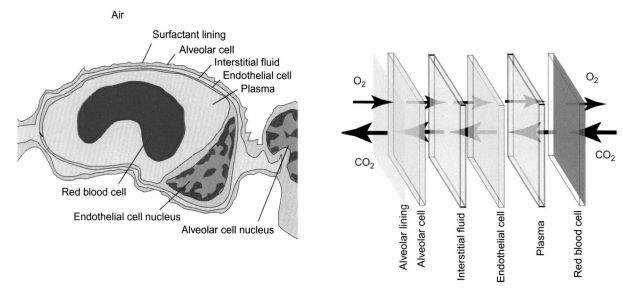

FIGURE 6.3.2 Alveolar membrane with successive barriers to diffusional transfer of gases from air to blood.

(see Chapter 2.5). In this case, the permeability is given by

[6.3.13] $$p = \frac{k_s D_s}{\delta}$$

where k_s is the partition coefficient, D_s is the diffusion coefficient, and δ is the thickness of the barrier. In the case of gas penetration through the alveolar membrane, the partition coefficient is related to the solubility. The diffusion coefficient that we use in Eqn [6.3.13] must be the one that pertains through the barrier. In the case of the respiratory gases, it is the diffusion coefficient either through the membrane lipids or through the watery phase making up the interstitial fluid, or the cytosol of the alveolar cell, and so on. It is not the diffusion coefficient in the gaseous phase. The permeability is enhanced through each layer by:

- *High solubility.* Gases that dissolve rapidly and to high concentrations (have large k_s) produce larger concentrations of materials that enhance the flux.
- *Large diffusion coefficients*, D_s. These are fixed by physical characteristics of the gases and are not physiological variables.
- *Short diffusion distances*, δ. The overall separation of the air from the blood is amazingly tiny. Typically the entire alveolar membrane is only 0.5×10^{-4} cm = 0.5 μm thick.

The flux of gas across a single barrier is given by Eqn [6.3.12]. This flux is the amount of gas that diffuses per unit time per unit area of membrane. This is an intensive property that does not depend on how large the surface is. The flow of gas, on the other hand, is an extensive property given by

[6.3.14] $$Q_s = \frac{A k_s \alpha_s D_s}{\delta} \Delta P_s$$

Here the terms $k_s \alpha_s$ enable us to express the flow in terms of the observable partial pressure of the gas. Each of the terms in Eqn [6.3.14] affects the total flow. The system is adapted for rapid flow by the magnitude of these factors:

- Because of the large number of alveoli, the total surface area for gas exchange is about 70 m². **This large surface area increases the flow proportionate to the area.**
- **The diffusion distance**, δ, **is small**. As mentioned for the flux across each layer, short diffusion distances increase the flux. They do the same here for the flow.
- **The physical behavior of gases and interaction with alveolar components sets the solubility and partition coefficients.** Because CO_2 has a higher solubility than O_2 (see Table 6.3.3), the flow of CO_2 for a given partial pressure difference is 20 times faster than the flow of O_2.
- **The pressure gradients drive flow**. Even though CO_2 can diffuse more rapidly per unit pressure difference, larger pressure differences of O_2 can assure the same rate of flow. Because the rate of O_2 consumption is nearly equal to the rate of CO_2 production (the respiratory quotient, R, 0.8), the exchange of O_2 is in fact about the same as that of CO_2.

THE *DIFFUSING CAPACITY* IS THE FLOW PER UNIT PARTIAL PRESSURE

The permeability of a complex sandwich of barriers is an algebraic function of all of the individual permeabilities. The permeabilities act like conductances. In a series of barriers such as occurs in the alveolar membranes, the resistances add; thus the total permeability for a series of six barriers that comprise the alveolar membrane becomes a function of all of the permeabilities. It is not practical to measure all of the terms in Eqn [6.3.14] that help determine Q_s. Instead, these terms can be lumped into a single term, the diffusing capacity. It is given by

[6.3.15] $$Q_s = D_L \Delta P$$

where D_L is the diffusing capacity. Its units are in mL of gas $\min^{-1}$ mmHg^{-1}, and it is calculated as $Q_s/\Delta P$. Each gas has its own diffusing capacity because the solubility and diffusion coefficients for each gas are different. The diffusing capacity is not constant because the effective surface area for gas exchange can be altered by increasing the blood perfusion of underperfused areas of the lung. At rest the diffusing capacity of O_2 is about 21 mL $\min^{-1}$ mmHg^{-1} and that of CO_2 is about 20 times higher. The greater diffusing capacity of CO_2 derives from its much higher solubility. During exercise, $D_{L_{O_2}}$ increases to as much as 65 mL $\min^{-1}$ mmHg^{-1}.

THE *ANATOMIC DEAD SPACE* REDUCES THE VOLUME OF INSPIRED AIR THAT EXCHANGES WITH THE BLOOD

As described in Chapter 6.2, each breath at rest moves a tidal volume of air into the lungs. Some of this volume replaces air within the trachea and upper parts of the tracheobronchial tree that do not exchange with the blood. The volume of gas that does not exchange is said to occupy the **anatomic dead space**. The **alveolar ventilation** can be calculated as

[6.3.16] $\qquad Q_A = \nu_R(V_T - V_D)$

where Q_A is the alveolar ventilation, in units of L $\min^{-1}$, ν_R is the respiration rate, V_T is the tidal volume, and V_D is the volume of the anatomic dead space. The anatomic dead space is typically about 150 mL. Typical alveolar ventilation at 12 breaths $\min^{-1}$ with $V_T = 500$ mL and $V_D = 150$ mL is 12 $\min^{-1}$ (0.5−0.15 L) = 4.2 L $\min^{-1}$. The alveolar ventilation differs from the pulmonary ventilation, which is $\nu_R V_T$, by an amount equal to the dead space ventilation.

Recall from Chapter 6.2 that the functional residual capacity is the volume of air left in the lungs after a normal expiration. The FRC is the sum of the expiratory reserve volume (ERV) and the residual volume (RV). Although these volumes vary individually, typical values are 1.1 L for ERV and 1.2 L for RV, and the FRC is about 2.3 L. Since $V_T = 0.5$ L and $V_D = 0.15$ L, each breath adds 0.35 L of new atmospheric air to 2.3 L of alveolar air. Because each breath renews only a small fraction of alveolar air, the composition of alveolar air remains fairly constant. This idea is presented schematically in Figure 6.3.3.

The consequence of the functional residual capacity is that expired air is not devoid of O_2 nor is the CO_2 levels in expired air as high as that in alveolar air. As a result, rescue breathing ("mouth-to-mouth resuscitation") is possible. These facts have led to revision of the standards for CPR, cardiopulmonary resuscitation. Since alveolar air is only gradually altered by ventilation, it is more important in CPR to focus on the cardiac part than the pulmonary part, and emphasis is placed on external stimulation of circulation with fewer rescue breaths. The current recommendation is 30 chest pumps to 2 rescue breaths.

PHYSIOLOGIC DEAD SPACE IS LARGER THAN THE ANATOMIC DEAD SPACE

The rate of CO_2 elimination in the expired air is the rate of CO_2 expiration minus the rate of CO_2 inspired. Since the expired air originates in part from the alveolar air, we can write

[6.3.17] $\qquad Q_E f_{E_{CO_2}} - Q_I f_{I_{CO_2}} = Q_A f_{A_{CO_2}} - Q_A^* f_{I_{CO_2}}$

where Q indicates flow of gas, in L $\min^{-1}$, and f indicates a mole fraction that is proportional to the partial pressure of gas. Here Q_E is the flow of expired gas, Q_I is the flow of inspired gas, Q_A is the flow of *expired* air that exchanges with blood, Q_A^* is the flow of *inspired* air that exchanges, and the various values of f correspond. Since $f_{I_{CO_2}}$ is near zero, we can eliminate its terms from Eqn [6.3.17]. The flow of expired air is just the respiratory rate times the volume of expired air per breath: $Q_E = \nu_R V_T$, where V_T is the tidal volume. The alveolar ventilation, Q_A, is similarly given in Eqn [6.3.16] as $\nu_R(V_T - V_D)$. Thus Eqn [6.3.17] becomes

[6.3.18] $\qquad V_T f_{E_{CO_2}} = (V_T - V_D) f_{A_{CO_2}}$

We can rearrange this equation to solve for V_D, the **physiologic dead space**:

[6.3.19] $\qquad V_D = V_T \dfrac{(f_{A_{CO_2}} - f_{E_{CO_2}})}{f_{A_{CO_2}}} = V_T \left[1 - \dfrac{f_{E_{CO_2}}}{f_{A_{CO_2}}}\right]$

In the practical determination of V_D, the mole fractions are replaced by the partial pressures of CO_2 in their respective volumes. The physiologic dead space defined by Eqn [6.3.19] is always larger than the anatomic dead space and reflects areas of the lungs that are underperfused with blood. Gas exchange is complete in these areas, but additional blood flow would result in additional gas exchange, lower $P_{A_{O_2}}$ and higher $P_{A_{CO_2}}$. The "wasted volume" is the difference between the physiological dead space and the anatomic dead space. Thus the anatomic dead space is the volume of ventilation

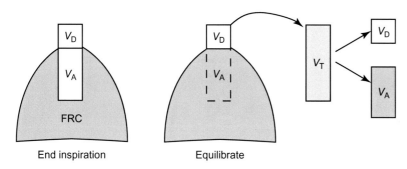

FIGURE 6.3.3 Tidal volume and the anatomic dead space. At the end of a normal inspiration, about 2.3 L of air remains in the lungs. About 0.5 L of tidal air is inspired, but only about 0.35 L reaches the alveoli because about 0.15 L is in the anatomic dead space that does not exchange gas with the blood. The expired air contains a mixture of dead space air and alveolar air. Thus the mole fraction of CO_2 in the expired air is the volume-weighted average of the mole fractions in alveolar air and inspired air.

where gas exchange cannot occur (because the diffusion distances to the blood are too great and anatomic structures intervene); the **physiological dead space** is the virtual or equivalent volume of ventilation where gas exchange does not occur; and the **wasted volume = physiological dead space − anatomic dead space** is the volume where gas exchange could occur but does not because of underperfusion with blood.

THE RATE OF CO₂ PRODUCTION ALLOWS CALCULATION OF ALVEOLAR VENTILATION

At steady state, the net rate of CO_2 production by the body is equal to the net rate of CO_2 elimination through the lungs. This is simply expressed as the difference between the rate of CO_2 input and CO_2 output:

$$[6.3.20] \quad Q_{CO_2} = Q_A f_{A_{CO_2}} - Q_A^* f_{I_{CO_2}}$$

where Q_{CO_2} is the rate of CO_2 elimination, in mL min⁻¹, Q_A is the flow of expired air that exchanges gas, Q_A^* is the flow of inspired air that exchanges, and f refers to the mole fraction of gas in the expired or inspired alveolar gas. Since the mole fraction of CO_2 in inspired air is very low, we can neglect this and rearrange Eqn [6.3.20] to obtain the **alveolar ventilation equation**:

$$[6.3.21] \quad Q_A = \frac{Q_{CO_2}}{P_{A_{CO_2}}}(P_B - 47)$$

where P_B is in mmHg and 47 is the vapor pressure of water at body temperature, in mmHg.

A full derivation of this equation is given in Appendix 6.3.A1. It is important to remember that **all of the flows are measured in volumes of gases at STPD**. The alveolar ventilation equation illustrates the inverse relationship between alveolar ventilation and $P_{A_{CO_2}}$, the partial pressure of CO_2 in alveolar air. If Q_{CO_2}, the production of CO_2 by the body, is constant, then **increases in alveolar ventilation will decrease** $P_{A_{CO_2}}$. Conversely, **decreases in Q_A will increase** $P_{A_{CO_2}}$. This inverse relation between alveolar ventilation and $P_{A_{CO_2}}$ holds for the steady state at any given metabolic rate. The alveolar ventilation equation can be rearranged to solve for $P_{A_{CO_2}}$:

$$[6.3.22] \quad P_{A_{CO_2}} = \frac{Q_{CO_2}}{Q_A}(P_B - 47)$$

THE ALVEOLAR GAS EQUATION ALLOWS CALCULATION OF $P_{A_{O_2}}$

By considering that the net consumption of O_2 at steady state is the difference between inspired O_2 content and expired O_2 content, it is possible to derive an equation that relates $P_{A_{O_2}}$ to the inspired $P_{O_2}, P_{A_{CO_2}}$ and the respiratory quotient, R. The alveolar gas equation is derived in Appendix 6.3.1 and is given as

$$[6.3.23] \quad P_{A_{O_2}} = P_{I_{O_2}} - \frac{1}{R}P_{A_{O_2}} + f_{I_{O_2}}\left(\frac{1-R}{R}\right)P_{A_{O_2}}$$

Because the last term usually gives a slight correction, this is often approximated as

$$[6.3.24] \quad P_{A_{O_2}} = f_{I_{O_2}}(P_B - 47) - \frac{1}{R}P_{A_{O_2}}$$

The value of $P_{A_{O_2}}$ at rest is approximately 100 mmHg.

BLOOD IS IN THE LUNGS FOR LESS THAN A SECOND—BUT THAT IS LONG ENOUGH TO EQUILIBRATE THE GASES

The total amount of blood in the lung capillaries is normally about 70 mL, approximately one stroke volume of the heart. At a heart rate of 72 bpm, the blood is in the lungs only 60 s min⁻¹/72 min⁻¹ = 0.83 s. During this short time, venous blood entering the lungs equilibrates completely with the alveolar air so that the exiting blood has the same P_{CO_2} and P_{O_2} as alveolar air. Blood actually equilibrates faster than its dwell time, indicating that there is some reserve in the diffusing capacity to accommodate the increased cardiac output and higher needs for gas exchange during exercise. Figure 6.3.4 illustrates the time course of blood P_{O_2} and P_{CO_2} during its brief course through the alveolar capillaries. The relative roles of diffusion and perfusion in capillary exchange are discussed in Chapter 5.10. Since, during normal conditions, P_{CO2} and P_{O2} equilibrate before the end of the capillary, this exchange is perfusion limited: if more blood flowed through, the transit time would be shorter yet the blood would still equilibrate, meaning more gas would be transferred into the blood. The transfer is perfusion limited.

EXAMPLE 6.3.1 Calculate $P_{A_{CO_2}}$ at Rest

At rest, the rate of CO_2 production is about 200 mL min⁻¹, STPD; tidal volume is 0.51 L, anatomic dead space is about 150 mL, and respiratory rate is about 12 min⁻¹.

The alveolar ventilation, Q_A, is calculated as 12 min⁻¹ × (510 − 150 mL) = 4320 mL min⁻¹. This volume is measured at body temperature, saturated. It must be converted to STPD. The conversion (see Appendix 6.3.A2) is given as

$Q_{A_{STPD}} = Q_{A_{BTPS}}/1.2104 = 4320$ mL min⁻¹/1.2104 = 3570 mL min⁻¹

We can then calculate $P_{A_{CO_2}}$ = 200 mL min⁻¹/3570 mL min⁻¹ × (760 mmHg − 47 mmHg)

$$P_{A_{CO_2}} = 40 \text{ mmHg}$$

BLOOD FLOW TO THE LUNG VARIES WITH POSITION WITH RESPECT TO GRAVITY

As described in Chapter 5.8, the pulmonary circulation is a low-pressure system that is influenced by gravity much more strongly than the systemic circulation. The pulmonary arterial pressure averages about 15 mmHg and in the left atrium it averages about 5 mmHg. The total pulmonary vascular resistance (PVR) can be calculated to be about 2 mmHg L^{-1} min, which is about a tenth of the systemic total peripheral resistance. The relatively small driving force for blood flow is affected by gravity: a 1 cm vertical ascent in the lungs reduces the pressure by 0.77 mmHg. Thus the pulmonary arterial pressure 10 cm above the heart will be reduced by 7.7 mmHg. Traditionally, the lung has been divided into three zones, each about one-third of the lungs. In the upper third, blood flow is reduced because the arteriolar blood pressure, P_a, is less than the alveolar gas pressure, P_A, and so blood flow is limited. In the middle third, $P_a > P_A$, but $P_A > P_v$, the blood pressure on the venous side. This partially collapses the capillaries and dams up the blood in the circulation. In the lower third, $P_a > P_A$ and $P_v > P_A$ and blood flow is maximal. This situation is illustrated in Figure 6.3.5.

The result of decreased blood flow to Zone 1 because of gravity in an upright person is a mismatch between ventilation and perfusion. The ventilation/perfusion ratio, Q_A/Q_V, is about 3 in this region. At the base of the lung, the Q_A/Q_V ratio is about 0.6.

REGULATION OF THE PULMONARY CIRCULATION HELPS RESTORE THE VENTILATION/PERFUSION RATIO

Pulmonary circulation depends on the pulmonary vascular resistance (PVR), gravity, alveolar pressure, and the hydrostatic pressure gradient provided by the right heart. The PVR, in turn, is influenced mainly by two factors: the inflation of the lung and the reaction of the arterioles to the partial pressure of oxygen in the blood.

The PVR, which determines blood flow, is affected by three separate variables: the alveolar gas pressure that compresses the capillaries, the resistance of alveolar vessels (mainly the capillaries), and the resistance of extraalveolar vessels. Unlike the systemic circulation, the capillaries in the lungs accounts for about 40% of the PVR. At lung volumes greater than the FRC, capillaries are stretched and compressed, and the resistance of the alveolar vessels increases. At lower lung volumes, the extraalveolar vessels are not held open by their tethers to the alveolar tissues, and their resistance increases.

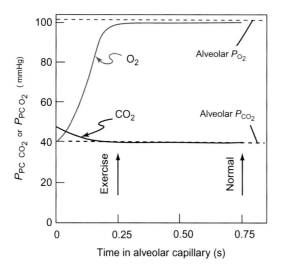

FIGURE 6.3.4 Time course of gas equilibration during the transit of blood through the pulmonary capillaries. Blood remains in the lung capillaries about 0.75 s. Venous blood has a $P_{V_{O_2}}$ of about 40 mmHg and a $P_{V_{CO_2}}$ of about 46 mmHg. The alveolar gas has a relatively constant composition of $P_{A_{O_2}}$ of about 102 mmHg and $P_{A_{CO_2}}$ of about 40 mmHg. Blood coursing through the capillaries equilibrates $P_{PC\,CO_2}$ and $P_{PC\,O_2}$ (pulmonary capillary plasma P_{CO_2} and P_{O_2}, respectively) with the alveolar gas within about 0.25 s, leaving considerable reserve time for gas exchange. During exercise, the cardiac output is increased and the dwell time of blood in the lung capillaries is reduced. Even in this case, the gases have enough time to equilibrate between alveolar gas and blood.

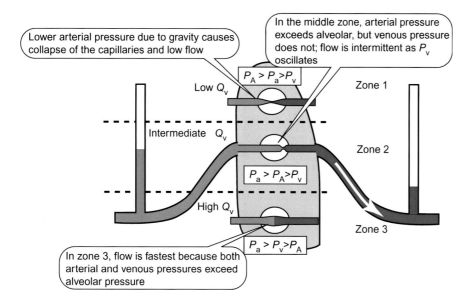

FIGURE 6.3.5 Effect of gravity on blood flow in the lung. The pulmonary arterial blood is shown in lighter blue; pulmonary venous blood, returning to the left heart, is shown in darker blue. The pressures are low, averaging around 15 mmHg in the pulmonary artery and about 8.5 mmHg in the venous circulation. Because gravity reduces the pressure, the perfusion pressures are lower at the top of the lung and therefore this area has less perfusion. In the middle zone, arterial pressure exceeds alveolar pressure, but venous pressure is sometimes insufficient, and the vessels close intermittently. In the lower zone, there is sufficient pressure to allow perfusion.

Clinical Applications: Asthma

Asthma is an obstructive disease marked by increased airway resistance. Although asthma is a chronic condition, it is differentiated from the other chronic obstructive pulmonary diseases by its reversibility. A sudden, acute exacerbation of the condition, with marked interference in gas flow, is called an asthma attack. The severity of the disease varies from rare and intermittent attacks to persistent symptoms requiring frequent use of short-term β_2 agonists.

The cause of asthma remains puzzling. At least two types of asthma exist: an allergic form, also called extrinsic or atopic, and a nonallergic, intrinsic nonatopic form. In the allergic form, asthma attacks are brought about by exposure to specific allergens. Exposure activates secretion of histamine from mast cells in the lung, inducing bronchospasm that is relieved by β_2 agonists such as salbutamol. About half of asthma cases, however, have no connection with allergic reactions.

Epidemiological studies show that the incidence of asthma is markedly increasing worldwide. Currently some 300 million people are affected, and in 2009 it caused an estimated 250,000 deaths globally. In 1989, David Strachan, a British epidemiologist, first proposed the **hygiene hypothesis** to explain the rapid rise in asthma. He observed that people living in unhygienic circumstances were less likely to have asthma than those living in cleaner environments. In this hypothesis, dirt, bacteria, and viruses in early life train the immune system to produce cells geared to defend against these problems, keeping the cells that respond to allergens in check. Unfortunately, more studies have shown that the hygiene hypothesis often does not explain all of the epidemiology. Children living on farms where cows and pigs are raised, drinking unpasteurized milk, almost never have asthma of any sort. There is something about Western, urban, or suburban culture that helps produce asthma, perhaps a change in the bacterial fauna both within the respiratory and GI tracts, but the full story is not yet known.

Other types of asthma also exist. Exercise-induced asthma, or exercise-induced bronchospasm (EIB), is thought to be related to dehydration of the airways when increased airflow requires increased humidification. This increases the osmolarity within cells in the airways, leading to release of histamine or leukotrienes, bringing on the bronchospasm.

These effects contribute to a biphasic relationship between PVR and lung volume. The PVR is minimum around the functional residual capacity, FRC.

When the ventilation/perfusion ratio is low, blood flow into the region carries away all of the excess oxygen until the alveolar gas becomes depleted of it. The resulting relative hypoxia causes a vasoconstriction called **hypoxic vasoconstriction**. This increases the local resistance to blood flow and shifts perfusion to underperfused regions of the lungs with high Q_A/Q_V ratios. Similar hypoxic vasoconstriction can occur in underventilated areas of the lungs. In these areas, neither ventilation nor perfusion is normal, but the hypoxic vasoconstriction helps to normalize their ratios.

SUMMARY

The total pressure within a gas mixture can be partitioned into the contributions made by each gas. These are their partial pressures, which are proportional to the mole fractions in the mixture. Water vapor also has a partial pressure, which is 47 mmHg at 37°C. The relation between partial pressure and mole fraction is written as

$$P_A = f_A(P_B - 47)$$

where P_A is the partial pressure in alveolar air (in mmHg), f_A is the mole fraction in dry air, P_B is the barometric pressure (mmHg), and 47 is the vapor pressure of water at body temperature.

The mole fraction of gases in liquid phases is also proportional to the partial pressure of gas in equilibrium with that phase (Henry's law). The solubility is the proportionality constant between mole fraction of gas in the aqueous phase and partial pressure; this solubility depends on the temperature. Gas content is always expressed as mL of gas per deciliter under STPD conditions: standard temperature and pressure, dry. Henry's law is written as

$$X_A = \beta_A P_A \quad [A] = \alpha_A P_A$$

Transport of gases into the blood traverses a thin sandwich of several layers including the alveolar lining, the alveolar cell, interstitial fluid, endothelial cell, plasma, and red cell membrane. Gas exchanges across these barriers occur passively. The rate of gas transport is linearly related to the partial pressure difference. The lung area is large (70 m^2) and thin (0.5 μm) so that gas exchange is rapid, reaching equilibrium within 0.25 s, whereas blood is in the exchanging pulmonary capillaries perhaps 0.8 s during quiet sitting. The overall transport of gas can be described by the diffusing capacity, the coefficient that relates flow of gas in mL min^{-1} to the partial pressure difference in mmHg. The diffusing capacity for CO_2 is about 20 times higher than that of O_2 because CO_2 has a higher solubility.

Each breath ventilates a dead space that does not exchange gas as well as regions of the lungs that do exchange. The anatomic dead space corresponds to areas of the tracheobronchial tree that do not exchange. The physiologic dead space includes areas that are underperfused. The alveolar P_{CO_2} can be calculated from the alveolar ventilation equation: it increases with increase in CO_2 production (as occurs in exercise) and decreases with increase in ventilation. The alveolar P_{CO_2} at rest is typically 40 mmHg. The alveolar P_{O_2} can be calculated from the alveolar gas equation and is typically about 100 mmHg. Because gas exchange is complete, these are also the typical values of P_{CO_2} and P_{O_2} in the blood that leaves the lungs.

REVIEW QUESTIONS

1. Why is the expired gas generally not the same volume as inspired gas?
2. Why is it necessary to express Q_{O_2} and Q_{CO_2} in volumes at STPD?
3. What is pulmonary ventilation? If pulmonary ventilation increases, what happens to alveolar ventilation? How do the two differ? If alveolar ventilation increases, what happens to $P_{A_{CO_2}}$ if there is no change in Q_{CO_2}? What happens to $P_{A_{CO_2}}$ if pulmonary ventilation decreases?
4. What is the vapor pressure of water? Does it vary with P_B? Does it vary with temperature? What is its value at body temperature? Why do we have to be concerned about it?
5. Distinguish between anatomic dead space and physiological dead space. Which is larger and why?
6. Would alveolar air have a constant composition if V_T were much larger, as in exercise?
7. What is the relative solubility of CO_2 and O_2?
8. Why do we say that "pressure gradients drive gas flow" instead of concentration gradients?

APPENDIX 6.3.A1 DERIVATION OF THE STEADY-STATE GAS EXCHANGE EQUATIONS

ALVEOLAR VENTILATION CONSISTS OF THE PART OF PULMONARY VENTILATION THAT PARTICIPATES IN GAS EXCHANGE

As discussed in Chapter 6.2, the pulmonary ventilation is the total flow of air into and out of the lungs. This is given by the respiratory rate times the tidal volume:

[6.3.A1.1] $$Q_P = \nu_R \langle V_T \rangle$$

where Q_P is the pulmonary ventilation, ν_R is the respiratory frequency, and $\langle V_T \rangle$ is the average tidal volume. This equation is ambiguous because the inspired part of the tidal volume is not the same as the expired part. Three things happen to change the expired volume relative to the inspired volume: first, the air is equilibrated to body temperature. In temperate climates, this generally means that the expired air has a larger volume because the ambient temperature is usually below body temperature. Second, the inspired air is humidified. This also generally expands the volume of air. Both of these effects can be corrected by converting the volumes of gas to some standard such as STPD. Third, the volume of expired air is altered by the consumption of O_2 and the excretion of waste CO_2. If the consumption of O_2 is a mole-to-mole match for CO_2 excretion, then the expired volume will match the inspired volume. But in general the consumption of O_2 is not one-for-one with CO_2 production. We will return to these issues.

Part of each tidal volume renews air in regions of the lung where gas exchanges with the blood, whereas dead space ventilation does not exchange gas. The alveolar ventilation is given as

[6.3.A1.2] $$\nu_R V_A = \nu_R V_E - \nu_R V_D$$
$$Q_A = Q_E - Q_D$$

where V_A is the part of the tidal volume that exchanges gas, V_E is the expired volume, V_D is the volume of gas in the dead space, Q_A is the alveolar ventilation, in L min^{-1}, Q_E is the flow of expired air, and Q_D is the dead space ventilation.

INSPIRED AIR DIFFERS IN VOLUME FROM EXPIRED AIR TO THE EXTENT OF DIFFERENCES BETWEEN Q_{O_2} AND Q_{CO_2}

Consider a single tidal volume. The volume of inspired air that exchanges gas is V_A^* and the volume of expired air that originates from the exchanging regions is V_A. The volume of this expired air is equal to the volume of inspired air less the volume of O_2 taken up plus the volume of CO_2 excreted:

[6.3.A1.3] $$V_A = V_A^* + V_{CO_2} - V_{O_2}$$

We can multiply both sides of this equation by ν_R to obtain the flow of gas instead of the volumes:

[6.3.A1.4] $$Q_A = Q_A^* + Q_{CO_2} - Q_{O_2}$$

This can be rewritten in terms of Q_A^*:

[6.3.A1.5] $$Q_A^* = Q_A - Q_{CO_2} + Q_{O_2}$$

Since there is no gas exchange in the dead-space volumes, we write the conservation of mass equations for CO_2 and O_2 as

[6.3.A1.6A] $$Q_{CO_2} = f_{A_{CO_2}} Q_A - f_{I_{CO_2}} Q_A^*$$

[6.3.A1.6B] $$Q_{O_2} = f_{I_{O_2}} Q_A^* - f_{A_{O_2}} Q_A$$

Substituting in for Q_A^* from Eqn [6.3.A1.5] into Eqn [6.3.A1.6A], we obtain

[6.3.A1.7] $$Q_{CO_2} = f_{A_{CO_2}} Q_A - f_{I_{CO_2}} (Q_A + Q_{O_2} - Q_{CO_2})$$

This can be solved for $f_{A_{CO_2}}$:

[6.3.A1.8] $$f_{A_{CO_2}} = \frac{Q_{CO_2}}{Q_A} + f_{I_{CO_2}} + f_{I_{CO_2}} \frac{(Q_{O_2} - Q_{CO_2})}{Q_A}$$

The mole fractions can be converted to partial pressures by multiplying by $(P_B - 47)$, because the mole fractions of gas are always expressed in terms of STPD. The results give

[6.3.A1.9] $$P_{A_{CO_2}} = \frac{Q_{CO_2}}{Q_A}(P_B - 47) + P_{I_{CO_2}} + P_{I_{CO_2}} \frac{(Q_{O_2} - Q_{CO_2})}{Q_A}$$

When there is little CO_2 in the inspired air, the last two terms on the right-hand side of the equation can be ignored. The resulting approximation gives

[6.3.A1.10] $$P_{A_{CO_2}} = \frac{Q_{CO_2}}{Q_A}(P_B - 47)$$

This is the alveolar ventilation equation. It illustrates the inverse relationship between $P_{A_{CO_2}}$ and Q_A. When alveolar ventilation increases, for the same metabolic condition, $P_{A_{CO_2}}$ decreases and vice versa. Similarly, when CO_2 production increases, $P_{A_{CO_2}}$ increases in parallel unless Q_A also increases.

THE ALVEOLAR GAS EQUATION GIVES $P_{A_{O_2}}$

In the same manner as our derivation of $P_{A_{CO_2}}$, we can find an expression for $P_{A_{O_2}}$ by inserting the result for Q_A^* from Eqn [6.3.A1.5] into Eqn [6.3.A1.6B]. The result gives

[6.3.A1.11] $\quad Q_{O_2} = f_{I_{O_2}}(Q_A + Q_{O_2} - Q_{CO_2}) - f_{A_{O_2}} Q_A$

Solving for $f_{A_{O_2}}$, we find

[6.3.A1.12] $\quad f_{A_{O_2}} = f_{I_{O_2}} + f_{I_{O_2}}\left(\dfrac{Q_{O_2}}{Q_A} - \dfrac{Q_{CO_2}}{Q_A}\right) - \dfrac{Q_{O_2}}{Q_A}$

here we make use of another definition. We define the **respiratory quotient**, R, to be the ratio of Q_{CO_2}/Q_{O_2}. Then we substitute in for $Q_{O_2} = Q_{CO_2}/R$. Then Eqn [6.3.A1.12] becomes

$$f_{A_{O_2}} = f_{I_{O_2}} + f_{I_{O_2}}\left(\dfrac{1}{R}\dfrac{Q_{CO_2}}{Q_A} - \dfrac{Q_{CO_2}}{Q_A}\right) - \dfrac{1}{R}\dfrac{Q_{CO_2}}{Q_A}$$

[6.3.A1.13]

which is rearranged to

[6.3.A1.14] $\quad f_{A_{O_2}} = f_{I_{O_2}} + f_{I_{O_2}}\left(\dfrac{(1-R)}{R}\right)\dfrac{Q_{CO_2}}{Q_A} - \dfrac{1}{R}\dfrac{Q_{CO_2}}{Q_A}$

Now we may substitute in for Q_{CO_2}/Q_A from Eqn [6.3.A1.10] to obtain

$$f_{A_{O_2}} = f_{I_{O_2}} + f_{I_{O_2}}\left(\dfrac{(1-R)}{R}\right)\dfrac{P_{A_{CO_2}}}{(P_B - 47)} - \dfrac{1}{R}\dfrac{P_{A_{CO_2}}}{(P_B - 47)}$$

[6.3.A1.15]

Multiplying both sides of Eqn [6.3.A1.15] by $P_B - 47$, we obtain the partial pressure of O_2 in the alveolar air:

[6.3.A1.15] $\quad P_{A_{O_2}} = P_{I_{O_2}} - \dfrac{1}{R}P_{A_{CO_2}} + f_{I_{O_2}}\left(\dfrac{(1-R)}{R}\right)P_{A_{CO_2}}$

This last equation is the **alveolar gas equation**. To obtain this equation, we used Eqn [6.3.A1.10], which is an approximation for the case when the inspired CO_2 is negligible. Thus this equation is strictly true only when the inspired concentration of CO_2 is as low as it is in atmospheric air. In addition, this equation describes alveolar P_{O_2} during the steady state at constant R. At any time that the metabolic condition rapidly changes, such as the onset of or cool down in exercise, this equation will be only approximately true.

APPENDIX 6.3.A2 CONVERSION OF PARTIAL PRESSURES AND VOLUMES BETWEEN STPD AND BTPS

Respiratory physiologists report $P_{A_{O_2}}$ and $P_{A_{CO_2}}$, Q_{O_2} and Q_{CO_2} in terms of mmHg or mL min^{-1} STPD, or standard pressure and temperature, dry. The standard temperature is $0°C = 273.16$ K, standard pressure is 1 atm = 760 mmHg = 1.01×10^5 Pa, and dry means $P_{H_2O} = 0$. Many measurements are not performed under these conditions, however. BTPS refers to body temperature ($37°C = 310.16$ K), pressure of 1 atm and saturated with water vapor so that $P_{H_2O} = 47$ mmHg. Sometimes measurements are also performed at ATPS. The relevant temperature then is the ambient temperature. If the measurements are performed at some elevation, the ambient pressure may be less than 1 atm, and P_{H_2O} depends on the temperature. Between 21°C and 37°C, the vapor pressure decreases by about 2 mmHg per °C change in temperature.

What we need is some way to convert partial pressures and volumes between conditions, particularly between STPD and BTPS. The equation we use is the Ideal Gas Equation

[6.3.A2.1] $\quad PV = nRT$

If we consider two conditions, labeled by "1" and "2," this equation holds for both:

[6.3.A2.2] $\quad \begin{array}{l} P_1 V_1 = nRT_1 \\ P_2 V_2 = nRT_2 \end{array}$

If the number of moles of gas is constant ($n_1 = n_2$), then Eqn [6.3.A2.2] is easily rewritten as

[6.3.A2.3] $\quad V_2 = \dfrac{T_2}{T_1}\dfrac{P_1}{P_2}V_1$

Suppose that state "1" represents STPD and state "2" represents BTPS. Now what complicates matters is that $n_1 \neq n_2$ because the total number of moles in the gas is increased by equilibration with water vapor at 37°C. However, the dry gases still obey this equation. What we need to use in Eqn [6.3.A2.3] is the pressure of the dry gases alone. This is $P_B - P_{H_2O}$, the difference between the barometric pressure and the vapor pressure of saturated air. For STPD conditions, $P_{H_2O} = 0$. For state 1, representing STPD, we have $T_1 = 273$ K, $P_1 = 760$ mmHg; for state 2, represented by BTPS, we have $T_2 = 310$ K ($=273$ K $+ 37$ K) and $P_2 = 760 - 47$. Conversion of $V_2 = V_{BTPS}$ from $V_1 = V_{STPD}$ is thus given as

[6.3.A2.3] $\quad V_{BTPS} = \dfrac{310}{273}\dfrac{760}{(760-47)}V_{STPD}$

$\qquad\qquad = 1.2104\, V_{STPD}$

Problem Set 6.1
Airway Resistance and Alveolar Gas Exchange

1. The density of Hg is 13.6 g cm^3 and the acceleration due to gravity is 981 cm s^{-2}. Convert the pressure exerted by a column of mercury 1 mm high to pascals, N m^{-2}. If the atmospheric pressure on top of a 3000 m peak is 527 mmHg, what is it in pascals?
2. A. The typical diameter of an alveolus is about 0.3 mm. If there are 300×10^6 alveoli in the lungs, calculate the approximate area for gas exchange, assuming that the alveoli are spheres.
   B. Based on this number and size of alveoli, calculate the total volume of gas in the alveoli. How does this compare to typical values for FRC? Differences between these two ought to be due to errors or rounding in either the size or number of alveoli. If the number of alveoli is correct, what would the diameter have to be to make this volume equal to the FRC? If the diameter is correct, what would the number of alveoli have to be to make this volume equal to the FRC?
3. A. The typical diameter of an alveolus is 0.3 mm. The surface tension of water is about 70 dyne cm^{-1}. The dyne is the unit of force in the cgs system: 1 g cm s^{-2}. It is converted to N by multiplying by 10^{-3} kg g^{-1} × 10^{-2} m cm^{-1} = 10^{-5} kg g^{-1} m cm^{-1}. What is the pressure within this alveolus assuming no surfactant at mechanical equilibrium? Give the answer in pascals (N m^{-2}) and in mmHg.
   B. Surfactant lowers the surface tension to about 28 dyne cm^{-1}. What would the pressure be in an alveolus with a diameter of 0.3 mm?
   C. Expanding the lungs increases the size of the alveoli and increases the surface area. How much work is necessary to expand the alveoli in a normal tidal volume of 0.5 L when the FRC is 2.3 L and assuming there is surfactant. *Hint:* Calculate work either as PdV or γdA; use the change in volume given to calculate dV or dA given that the rest diameter of alveoli is 0.3 mm and that there are N of them (see Problem 2B).
   D. During normal tidal breathing, the airflow is about 0.5 L s^{-1} and the pressure driving airflow is about 1 mmHg. Calculate the overall airway resistance. Estimate the energy required to overcome this airway resistance for a tidal volume of 0.5 L.
   E. The value calculated in 3C is the work required per breath just to expand the alveoli. The work of breathing includes the work to overcome airway resistance and to expand the chest wall. Assume that the work of expanding the chest wall is about one-half of the work of expanding the lungs. If the resting metabolism is 5200 J min^{-1}, what fraction of resting metabolism is used in the work of breathing? If breathing is 8% efficient, about what fraction of resting metabolism is used for breathing?
4. The table below lists the airway diameter and aggregate area (the number of airways times the airway area per airway) as a function of generation number.

Generation No.	Diameter (cm)	Area (cm^2)	Reynold's No.
0	1.8	2.54	
1	1.22	2.33	
2	0.83	2.13	
3	0.56	2	
4	0.45	2.48	
5	0.35	3.11	
6	0.28	3.96	
7	0.23	5.1	
8	0.186	6.95	
9	0.154	9.56	
10	0.130	13.4	
11	0.109	19.6	
12	0.095	28.8	
13	0.082	44.5	
14	0.074	69.4	
15	0.066	113	
16	0.060	180	
17	0.054	300	
18	0.050	534	
19	0.047	944	
20	0.045	1600	
21	0.043	3220	
22	0.041	5880	
23	0.041	11,800	

A. Calculate the Reynolds number for generation 0, 2, and 10 for airflow during normal resting respiration. The maximum flow rate during these conditions is about 0.5 L s^{-1}.

What generalizations can you make about laminar and turbulent flows during rest?

B. Calculate the Reynolds number for generation 0, 2, and 10 for airflow during high ventilation that occurs during exercise, for example. Assume that the flow rate during these conditions is $5\,L\,s^{-1}$. What conclusions can you make regarding laminar and turbulent flows during high flow conditions?

5. The surface area of the lungs is about 70 m² and the thickness of the alveolar diffusion layer is about 0.5 μm. The $P_{A_{O_2}}$ is about 100 mmHg, whereas the $P_{v_{O_2}}$, the partial pressure of O_2 in venous blood, is about 40 mmHg. The diffusion coefficient of O_2 in water is about $1.5 \times 10^{-5}\,cm^2\,s^{-1}$. The solubility of O_2 is given by Henry's law with a coefficient α given in Table 6.3.3 (with gas volumes at STPD!). This gives

$$[O_2] = 0.024 P_{O_2}$$

where $[O_2]$ is expressed in mL O_2 per mL of water and P_{O_2} is in atmospheres. Only the dissolved O_2 diffuses. What is the initial rate of O_2 diffusion from the aggregate alveoli to the blood? Assume that the diffusion of O_2 across the lipid bilayers in the alveolar membrane is essentially instantaneous because the partition coefficient of O_2 is high and the diffusion distance, the thickness of the membrane, is very small (~10 nm). Thus to O_2 the alveolar membrane looks like a thin aqueous layer. The initial rate means clamp the venous P_{O_2} at 40 mmHg. Give the answer in mL O_2 (STPD) per min and in mol per min using the Ideal Gas Law. How does this compare to Q_{O_2} at rest? Explain the difference.

6. Resting O_2 consumption is about 250 mL min^{-1} while resting CO_2 production is about 200 mL min^{-1}. Both these volumes refer to volumes at STPD. How much mass is lost per day through the lungs, exclusive of water loss, due to resting metabolism?

7. Assume that the inspired air is dry air. The vapor pressure of 47 mmHg at 37°C corresponds to a volume and mass of liquid water. If the tidal volume is 500 mL and respiration rate is 12 min^{-1}, how much water is lost from the body per day? Would you expect this to increase or decrease with activity?

8. A. At rest V_T at BTPS is 510 mL. If the dead space is 150 mL, what is V_A?
B. If the respiratory rate is 12 min^{-1}, what is Q_A at BTPS?
C. Convert Q_A from part B to STPD.
D. Calculate $P_{A_{CO_2}}$ if $Q_{A_{CO_2}}$ is 200 mL min^{-1} and $P_B = 760$ mmHg.

9. A. The mole fraction of O_2 in dry atmospheric air is 0.21. If $P_B = 760$ mmHg, what is P_{O_2}?
B. What is $P_{I_{O_2}}$ in moist tracheal air?
C. If $R = 0.8$ and $P_{A_{CO_2}} = 40$ mmHg, calculate $P_{A_{O_2}}$.

10. A number of empirical formulas have been developed to predict lung volumes for people as a function of body size and age. These are used to provide spirometric standards to determine what is normal and what is abnormal. Two such formulas are given below (C.J. Gore et al., Spirometric standards for healthy adult lifetime nonsmokers in Australia. *European Respiratory Journal* **8**:773–782, 1995):

$$FVC\ (female) = -3.598 - 0.0002525A^2 + 4.680H$$
$$FVC\ (male) = 12.675 - 0.0002764A^2$$
$$+ 10.736H^2 + 4.790H^3$$

Another empirical equation (J.F. Morris et al., Spirometric standards for healthy nonsmoking adults. *American Review of Respiratory Disease* **103**:57–67, 1971) gives

$$FVC\ (female) = -2.852 - 0.024A + 4.528H$$
$$FVC\ (male) = -4.241 - 0.025A + 5.827H$$

where A is the age in years and H is the height in m.
A. For a male 5'10" tall and 20 years of age, compare these two predictions for FVC.
B. For a female 5'4" tall and 20 years old, compare these two predictions for FVC.
C. How could you judge which of the many available predictive equations was the best to use?

11. A common clinical unit in respiratory physiology is cmH$_2$O instead of mmHg. Convert the two units. Assume the density of water is $1.00\,g\,cm^{-3}$.

12. The FRC can be measured using a whole body plethysmograph. This consists of a refrigerator-sized, air-tight chamber in which the patient can sit comfortably and which is usually transparent so that the patient can see the operator and vice versa. The patient breathes through a tube connected to a pneumotach, which can measure airflow and pressure at the mouthpiece and which is fitted with a shutter. After a normal expiration, the shutter closes and the patient pants against the closed shutter, while simultaneously holding the cheeks in with the hands. The pressure at the mouthpiece is recorded. The pressure goes up and down cyclically due to the panting effort. When the pressure goes up, the gas remaining in the lung is compressed—its volume decreases. This decrease in volume of the thoracic cavity is accompanied by an increase in volume outside the body, in the chamber. If the volume change measured by the body plethysmograph is 71 mL and the pressure change measured at the mouthpiece was 20 mmHg, estimate the FRC. (Hint: Let V be the FRC prior to expiratory effort, and $V - \Delta V$ be its volume after; the pressure prior to expiratory effort is $P_B = 760$ mmHg, and the pressure after is $P_B + \Delta P$. An important point to remember is that the vapor pressure of water does *not* change!)

13. Normal values of $Q_{CO_2} = 200$ mL min^{-1} (STPD), $V_T = 510$ mL (BTPS), $V_D = 150$ mL (BTPS), and $\nu_R = 12$ min^{-1}. During exercise, Q_{CO_2} increased to 1000 mL min^{-1} (STPD), and ν_R increased to 20 min^{-1}.
    A. What would V_T be if $P_{A_{CO_2}}$ did not change?
    B. If $R = 0.8$, calculate $P_{A_{O_2}}$ (assuming that $P_{A_{CO_2}} = 40$ mmHg).
14. The solubility of CO_2 in water at 37°C is given in Table 6.3.3 as 0.0747 mL dL^{-1} mmHg^{-1}. Blood is not water. Assume that CO_2 is soluble only in water, and that 93% of plasma is water by volume. Assume that the hematocrit is 0.45 and that in the erythrocytes only 72% of their volume is water. Calculate the amount of dissolved CO_2 in blood at 37°C at 40 mmHg in mL dL^{-1}.
15. The diffusion coefficient of O_2 is about 1.5×10^{-5} cm^2 s^{-1}. Its solubility given in Table 6.3.3 is 0.00317 mL dL^{-1} mmHg^{-1}.

    A. Given that plasma is 93% water by volume, the hematocrit is 0.45 and that in the erythrocytes only 72% of the volume is water, calculate the concentration of [O_2] in the blood if $P_{O_2} = 40$ mmHg.
    B. Estimate the concentration of [O_2] in the alveolar lining if its water volume is 0.80 of the tissue volume, and $P_{A_{O_2}} = 1000$ mmHg.
    C. The diffusing capacity of O_2 is 21 mL min^{-1} mmHg^{-1}. If $P_{A_{O_2}} = 100$ mmHg and $P_{V_{O_2}} = 40$ mmHg, what is the *initial* flow of O_2 from alveolar air to blood? (Initial means assume that the gradient of partial pressures remains constant.)
    D. What area of water surface would give the same diffusive flow from D_{O_2} and [O_2] gradient if the diffusion distance is 0.5 μm?

6.4 Oxygen and Carbon Dioxide Transport

Learning Objectives

- Be able to calculate the dissolved oxygen content of blood from its solubility
- Be able to calculate the bound oxygen content of blood from its hemoglobin concentration and oxygen saturation
- Write the Hill equation for oxygen binding to hemoglobin
- Describe the consequence of positive cooperativity on the oxygen saturation curve of hemoglobin
- Be able to calculate oxygen delivery to tissue based on blood flow and arteriovenous differences in oxygen content
- Describe the gradient of oxygen partial pressure from capillaries to tissue
- Indicate the site of oxygen consumption within cells
- Describe the function of myoglobin in muscle cells
- Describe the effect on the oxygen saturation curve of the following: temperature, [H$^+$], P_{CO_2}, and 2,3-DPG
- List three ways CO_2 is carried in the blood
- Distinguish between CO_2 content of the blood and CO_2 transport by the blood
- Identify the quantitatively largest component of CO_2 transport by the blood
- Describe the chloride shift

DISSOLVED OXYGEN CONTENT OF BLOOD IS SMALL

The solubility of oxygen in water at 37°C is given in Table 6.3.3 as 0.003 mL of O_2 at STPD per 100 mL of water per mmHg. The partial pressure of oxygen in arterial blood (P_{aO_2}) is about 100 mmHg, so the dissolved [O_2] is about 0.3 mL dL^{-1}. It is actually slightly less than this because blood is not 100% water.

MOST OF THE OXYGEN IN BLOOD IS BOUND TO *HEMOGLOBIN*

The aggregate red blood cells make up about 40–45% of the volume of blood (the **hematocrit**). Each red blood cell is packed with hemoglobin, a protein consisting of four polypeptide chains ($\alpha_2\beta_2$) with an aggregate molecular weight of 64,500 Da. The cell is so full that only about 0.72 of the red cell volume is water. Each of the four polypeptide chains contains a **heme** group consisting of a porphyrin backbone that complexes an Fe^{2+} atom (see Chapter 5.2). Each of the heme groups can bind oxygen noncovalently.

The rate of oxygen binding to hemoglobin depends on the concentration of dissolved O_2 and the concentration of vacant hemoglobin sites. Similarly, the rate of O_2 dissociation depends on the concentration of oxygen bound to hemoglobin [Hb·O_2]. Equilibrium is reached when the two rates are equal. Since the concentration of dissolved O_2 is proportional to P_{aO_2}, by Henry's law, the binding equilibrium depends on P_{aO_2}. Hemoglobin saturation is defined as the percent of binding capacity that is occupied by oxygen. The relationship between hemoglobin saturation and P_{aO_2} is shown in Figure 6.4.1. This curve is called the **oxygen dissociation curve** or the **oxygen saturation curve**. The oxygen saturation curve does *not* obey simple saturation kinetics as described by the Langmuir adsorption isotherm:

$$[6.4.1] \quad [\text{Hb} \cdot O_2] = \frac{[O_2]}{K + [O_2]} [\text{Hb} \cdot O_2]_{max}$$

This equation describes a dissociation curve that is hyperbolic, as shown in Figure 6.4.1. The best fit of this

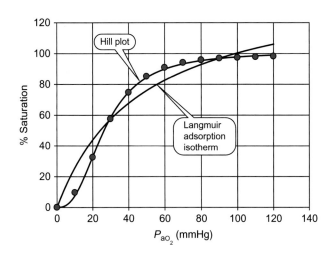

FIGURE 6.4.1 Oxygen dissociation curve. Oxygen binding to hemoglobin is a sigmoidal, or S-shaped, function of the arterial P_{O_2}. The binding is not satisfactorily fit by a simple curve describing saturation kinetics (see Eqn [6.4.1]). The data are much better described when the oxygen tension or partial pressure is raised to a power, h, called the Hill coefficient. *Data from F.J.W. Roughton, Transport of oxygen and carbon dioxide, in W.O. Fenn and H. Rahn, eds.,* Handbook of Physiology, *1964.*

equation to the oxygen binding data is not satisfactory. Binding is better described by the **Hill equation**:

$$[6.4.2] \quad [Hb \cdot O_2] = \frac{[O_2]^h}{K + [O_2]^h}[Hb \cdot O_2]_{max}$$

where h is the **Hill coefficient**. For particular choices of K and h, this equation fits the observed binding isotherm much better, as seen in Figure 6.4.1. The Hill coefficient is a measure of the **cooperativity** of binding. There are four sites for O_2 binding to Hb. If binding of the first O_2 increases the affinity of the next binding site for O_2, the binding is said to be **positively cooperative** and $h > 1.0$. This cooperativity arises from interaction between the hemoglobin subunits that depends on their state of oxygen binding. In Figure 6.4.1, the best fit to the Hill plot gives $h = 2.56 \pm 0.07$ indicating that oxygen binding to hemoglobin is highly positively cooperative. See Appendix 9.1.A1 for a discussion of the Langmuir adsorption isotherm and Hill plots.

OXYGEN CONSUMPTION CAN BE CALCULATED BY BLOOD FLOW TIMES THE A−V DIFFERENCE IN OXYGEN

The partial pressures of the respiratory gases in alveolar air and blood are shown in Figure 6.4.2. We can use the P_{O_2} from the arterial and venous sides of the circulation, along with the oxygen dissociation curve in Figure 6.4.1, to calculate the oxygen consumption of the tissues. The balance between input and output is as follows:

$$[6.4.3] \quad Q_a[O_2]_a = Q_v[O_2]_v + Q_{O_2}$$

where Q_a is the total arterial blood flow, in mL min^{-1}, $[O_2]_a$ is the arterial blood total concentration of O_2, Q_v is the venous blood flow, $[O_2]_v$ is the venous blood total $[O_2]$, and Q_{O_2} is the aggregate rate of oxygen consumption.

The concentrations of O_2 are expressed as milliliters gas per unit volume of blood, and the volume of gas is expressed under STPD conditions. This equation is simply the conservation of mass. Since at steady state, the arterial blood flow is nearly equal to the venous drainage, we can rewrite the equation where we identify Q_a with the cardiac output when Q_{O_2} is the total body oxygen consumption in units of milliliters O_2 at STPD per minute:

$$[6.4.4] \quad Q_a([O_2]_a - [O_2]_v) = Q_{O_2}$$

See Example 6.4.2 for an application of this equation.

EXAMPLE 6.4.1 Blood Carrying Capacity for Oxygen

The normal hemoglobin concentration in whole blood is 15 g%. How much O_2 can this amount of hemoglobin hold?

Each of the four heme groups in hemoglobin can bind oxygen noncovalently. Thus the carrying capacity of blood is four times the concentration of hemoglobin. The molar concentration of hemoglobin is

$$(15 \text{ g dL}^{-1})/(64,500 \text{ g mol}^{-1}) = 2.32 \times 10^{-4} \text{ mol dL}^{-1}$$
$$= 2.32 \times 10^{-3} \text{ M}$$

Since each molecule of hemoglobin can bind four molecules of O_2, the oxygen binding capacity is

9.30×10^{-4} mol dL^{-1}

At STPD the molar volume of O_2 is 22.4 L mol^{-1}, and so this capacity corresponds to

9.30×10^{-4} mol dL^{-1} × 22.4 L mol^{-1} = **20.8 mL O_2 dL^{-1}**

Because the hemoglobin concentration varies between individuals, it is convenient to describe the O_2 binding capacity per unit hemoglobin:

20.8 mL O_2 dL^{-1} / 15 g dL^{-1} = **1.39 mL O_2 g hemoglobin^{-1}**

Gas concentrations are typically expressed in milliliters of gas at STPD per deciliter.

EXAMPLE 6.4.2 Calculation of O_2 Consumption from A−V [O_2] Difference and Q_a

The cardiac output is typically about 5 L min^{-1}. The P_{aO_2} of arterial blood is 95 mmHg. According to Figure 6.4.1, Hb at this P_{aO_2} is about 98% saturated. We assume a typical [Hb] = 15 g%, which has a carrying capacity of 20.8 mL dL^{-1}, from Example 6.4.1. Therefore, the total [O_2] content of the arterial blood is the dissolved quantity + the amount bound to Hb:

$[O_2]_a = \alpha \times 95$ mmHg $+ 0.98 \times 20.8$ mL dL^{-1}, where $\alpha = 0.003$ is the solubility given in Table 6.3.3:

$[O_2]_a = 0.3$ mL dL^{-1} + 19.8 mL dL^{-1} = 20.7 mL dL^{-1}

The venous P_{O_2}, P_{vO_2}, drops to about 40 mmHg. According to Figure 6.4.1, this P_{O_2} is in equilibrium with blood that is about 75% saturated. The $[O_2]_v$ is calculated as

$[O_2]_a = 0.003$ mL dL^{-1} mmHg^{-1} × 40 mmHg $+ 0.75 \times 20.8$ mL dL^{-1}
≈ 15.72 mL dL^{-1}.

Inserting this into Eqn [6.4.4], we get

$Q_{O_2} = 5$ L min^{-1} × (20.7 mL dL^{-1} − 15.72 mL dL^{-1}) = **249 mL min^{-1}**

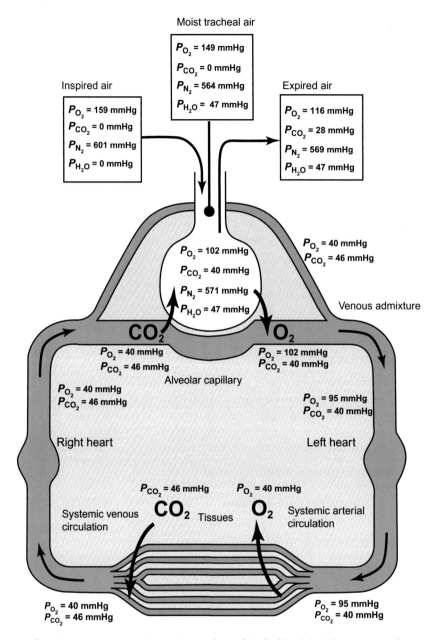

FIGURE 6.4.2 Partial pressures of respiratory gases in respiratory air samples and in the blood. Partial pressures in dry inspired air are proportional to the mole fraction of the gas in atmospheric air. Humidifying the moist tracheal air dilutes all respiratory gases because water is added to the inspired air. Alveolar air has stable partial pressures because the volume of air entering the alveoli with each breath is a small fraction of the air remaining in the lung. The expired air consists of a mixture of the moist tracheal air that does not exchange gas (the dead volume) and expired alveolar air. Venous blood entering the alveolar capillaries is partially depleted of O_2 and enriched in CO_2 from exchange with the tissues. The alveolar capillaries have a large surface area and small diffusion distances, so the blood in the alveolar capillaries equilibrates with alveolar air. Some blood that perfuses poorly ventilated regions of the lungs or from anatomic shunts contributes volume but less O_2 to the arterial blood. This **venous admixture** reduces the P_{aO_2} from 102 mmHg to about 95 mmHg. The gradients for O_2 diffusion from alveolar air to blood and from blood to tissue are much greater than the gradients for CO_2 diffusion because the higher solubility of CO_2 enables faster diffusion.

OXYGEN CONSUMPTION CAN BE CALCULATED FROM THE DIFFERENCE BETWEEN O₂ INSPIRED AND O₂ EXPIRED

At steady state, the oxygen consumption calculated by the cardiac output times the arterial–venous difference in [O_2] should match the oxygen that is taken up by the respiratory system. Oxygen uptake by the respiratory system is governed by its own conservation relation, written as

[6.4.5] $$Q_T^* f_{IO_2} = Q_T f_{EO_2} + Q_{O_2}$$

where Q_T^* is the flow of inspired air, Q_T is the flow of expired air, f_{IO_2} and f_{EO_2} are the mole fractions of O_2 in the inspired and expired tidal volumes, respectively. Because Q_{O_2} is expressed in terms of volumes of O_2 gas at STPD, both Q_T^* and Q_T must also be expressed in

EXAMPLE 6.4.3 Calculation of O_2 Consumption from Inspired and Expired Volumes and P_{O_2}

The *expired* volume at BTPS is about 500 mL per breath, and the respiratory rate is about 12 min^{-1}. Using the values for the gases in Figure 6.4.1, and respiratory quotient $R = 0.8$, estimate the oxygen consumption.

The flow of *expired* air, $Q_T = 500$ mL $\times$ 12 min^{-1}/1.2104 = 4.96 L min^{-1}. The value of 1.2104 is used to convert Q_T at BTPS to STPD, according to Eqn [6.3.A2.3].

$f_{I_{O_2}}$ is the mole fraction of O_2 in inspired air, which is 159/760 = 0.209.

$f_{E_{O_2}}$ can be calculated from $P_{E_{O_2}}$ shown in Figure 6.4.2 and Eqn (6.3.8): $f_{E_{O_2}} = P_{E_{O_2}}/(P_B - P_{H_2O})$:

$$f_{E_{O_2}} = 116 \text{ mm Hg}/713 \text{ mm Hg} = 0.163$$

The flow of inspired air, Q_T^* at STPD can be calculated from $Q_T + Q_{O_2} - Q_{CO_2}$. If $R = 1.0$, then $Q_T^* = Q_T$.

If we use $R = 0.8$ and estimate $Q_{O_2} = 250$ mL min^{-1}/12 min^{-1} = 21 mL breath^{-1} and $Q_{CO_2} = 200$ mL min^{-1}/12 min^{-1} = 17 mL breath, then $V_T^* = V_T + 4$ mL = 504 mL; converting this to STPD per minute, we have $Q_T^* = 504$ mL $\times$ 12 min^{-1}/1.2104 = 5.00 L min^{-1}.

Then we calculate $Q_{O_2} = 5.00$ L min^{-1} $\times$ 0.209 – 4.96 L min^{-1} $\times$ 0.163 = **237 mL min^{-1}**.

This is in reasonable agreement with Q_{O_2} calculated in Example 6.4.2. The discrepancy in the values is due to rounding and in the rounded value of V_T. If we use just a slightly larger value for V_T the discrepancy disappears.

STPD, and $f_{I_{O_2}}$ and $f_{E_{O_2}}$ are the mole fractions of O_2 in dry inspired or expired airs, respectively.

O_2 DIFFUSES FROM BLOOD TO THE INTERSTITIAL FLUIDS AND THEN TO THE CELLS

Oxygen diffuses down its partial pressure gradient from blood to the interstitial fluid (ISF). The systemic arterial blood has a $P_{a_{O_2}}$ of about 95 mmHg. The average ISF P_{O_2} is variable, depending on the metabolism of the tissue, but it is generally near 40 mmHg. The higher P_{O_2} in the blood produces a net diffusion of O_2 out of the blood and into the ISF. Only the dissolved O_2 diffuses. Because of Henry's law, the dissolved $[O_2]$ is directly proportional to P_{O_2}, so the gradient of P_{O_2} is proportional to the gradient in $[O_2]$ in any single phase. The large area and small diffusion distance between blood and the ISF lead to rapid exchange between blood and the ISF.

The low P_{O_2} in the ISF is caused by diffusion of O_2 into the cells, where the P_{O_2} is even lower because the mitochondria consume the O_2. Mitochondria typically require 3–5 mmHg P_{O_2}, which is considerably lower than the P_{O_2} of the interstitial fluid. The cytosolic $[O_2]$ is intermediate between that of the interstitial fluid and the intracellular fluid, with a P_{O_2} of nearly 20 mmHg. Thus there is a continuous gradient of P_{O_2} from about 95 mmHg in the arterial blood to 40 mmHg in the interstitial fluid to 20 mmHg in the cytosol to 5 mmHg in the mitochondria. Figure 6.4.3 illustrates the P_{O_2} gradients that drive diffusion from the blood to its place of consumption, the mitochondria.

HEMOGLOBIN DELIVERS OXYGEN TO THE TISSUES

Only dissolved O_2 can diffuse across membranes. When O_2 diffuses out of the capillaries into the interstitial fluid, blood P_{O_2} decreases. According to the law of mass action,

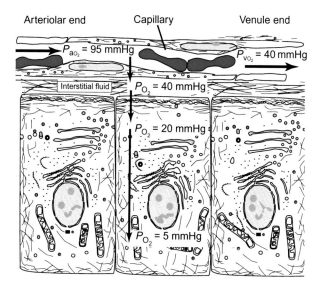

FIGURE 6.4.3 Diffusion of O_2 from blood to the mitochondria. The arteriolar blood has a $P_{a_{O_2}}$ of about 95 mmHg. The interstitial fluid P_{O_2} is about 40 mmHg. Thus there is a gradient of dissolved O_2 from blood to ISF. As O_2 diffuses, other O_2 molecules bound to Hb dissociate. The cytosolic P_{O_2} is lower still, being least near the mitochondria where the O_2 is being consumed. Thus O_2 diffuses down its P_{O_2} gradient from its source, the blood, to its sink, the mitochondria. By the time blood leaves the tissue, it has equilibrated with the ISF P_{O_2} of about 40 mmHg.

O_2 dissociates from hemoglobin and thereby resists the fall in P_{O_2}. In the process, the amount of bound O_2 decreases. When the blood finally leaves the capillaries, its P_{O_2} has equilibrated with the interstitial fluid P_{O_2} and the percent saturation of blood hemoglobin decreases according to the oxygen dissociation curve.

MYOGLOBIN STORES O_2 IN OXIDATIVE MUSCLE AND MAY ENHANCE DIFFUSION

Myoglobin is a low-molecular weight protein of 16,000 Da that contains one heme and binds one molecule of O_2 per molecule of protein. Tissue content of myoglobin depends on the tissue and the species. Highly oxidative muscle fibers contain a lot of myoglobin. Because it consists of a single polypeptide chain, myoglobin does not have subunits that can interact to produce cooperative binding. Instead, oxygen binding to myoglobin obeys a simple saturation equation with half-maximal saturation at about 5 mmHg P_{O_2} (see Figure 6.4.4). It has two functions in muscle: it stores oxygen for use during heavy exercise, and it enhances diffusion through the cytosol by carrying the oxygen. By binding O_2, myoglobin (Mb) provides a second diffusive pathway for O_2 through the cell cytosol. The total diffusion is the sum of the diffusion of dissolved O_2 and the diffusion of oxygen carried by Mb, Mb·O_2. The diffusion of oxygen through aqueous solutions in the presence of myoglobin is described in Eqn [6.4.6].

[6.4.6]
$$J_{total} = -D_{O_2}\frac{d[O_2]}{dx} - D_{Mb}\frac{d[Mb-O_2]}{dx}$$
$$= -D_{O_2}\frac{d[O_2]}{dx} - D_{Mb}[Mb]_{total}\frac{d\nu}{dx}$$

where ν is the fraction of myoglobin that is bound with O_2. The diffusion coefficient of O_2 in water is about 1.5×10^{-5} cm^2 s^{-1}, whereas the diffusion coefficient of myoglobin is about 0.1×10^{-5} cm^2 s^{-1}, about 15-fold lower than that of oxygen. Even though the diffusion coefficient of myoglobin is small compared to that of dissolved O_2, its concentration gradient can be much higher and so myoglobin can contribute to the overall diffusion of O_2 through the tissue.

SHIFT OF THE O_2 DISSOCIATION CURVE TO THE RIGHT HELPS DELIVER O_2 TO EXERCISING MUSCLES

Increased body temperature, increased [H^+] in the red blood cell, increased P_{CO_2} in the blood, and increased concentrations of 2,3-diphosphoglycerate (DPG) all shift the O_2 dissociation curve to the right (see Figures 6.4.5–6.4.8). These shifts to the right mean that higher P_{O_2} is required to saturate hemoglobin; alternatively, at any given P_{O_2}, hemoglobin gives up more of its bound O_2. Thus, increased temperature, increased [H^+], increased P_{CO_2}, and increased 2,3-DPG all increase the availability of O_2 to the tissues.

Consider the situation with increased temperature. At rest, blood enters the muscles at 37°C with P_{aO_2} = 95 mmHg. According to our earlier calculation (see Example 6.4.1), blood contains about 20 mL O_2 dL^{-1} at P_{aO_2} = 95 mmHg. When it equilibrates with ISF at 40 mmHg and 37°C, the saturation is 75%, and the blood's O_2 content is 15 mL dL^{-1}: the muscles extract 5 mL dL^{-1}. When the muscles are exercising, their temperature increases. Arterial blood still contains 20 mL dL^{-1} but the venous blood at 40 mmHg P_{vO_2} and 43°C is only about 50% saturated: it contains 10 mL dL^{-1}. Thus the exercising muscles at the increased temperature extract 10 mL dL^{-1} if the ISF is kept at 40 mmHg.

The O_2 dissociation curve also is sensitive to blood pH (see Figure 6.4.6). Increasing [H^+] (lowering the pH) shifts the curve to the right, and decreasing [H^+] (raising the pH) shifts the curve to the left. Because exercising muscles also produce acid, this effect helps unload O_2 to active muscles.

Decreasing the P_{CO_2} from the normal value of 40 mmHg shifts the curve to the left; increasing P_{CO_2} shifts the curve to the right. This is called the **Bohr effect** after its discoverer. This effect also helps unload

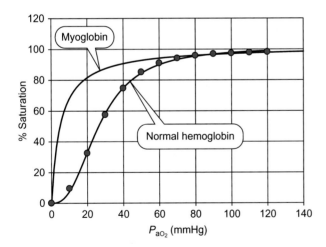

FIGURE 6.4.4 Comparison of the oxygen dissociation curve for hemoglobin and myoglobin. Hemoglobin consists of four subunits that can interact, giving rise to cooperative behavior, and causing deviations from simple saturation kinetics. Myoglobin consists of a single polypeptide chain that cannot interact with other subunits, and thus its dissociation curve shows simple saturation behavior. Half-saturation of myoglobin occurs at P_{O_2} of about 5 mmHg, intermediate between the P_{O_2} of the ISF and the mitochondria.

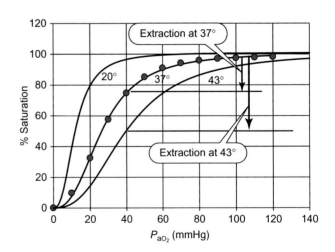

FIGURE 6.4.5 Effect of temperature on the oxygen dissociation curve. Reduced temperature shifts the curve to the left and elevated temperature shifts the curve to the right. The reduced affinity for O_2 at higher temperatures helps dissociate O_2 from Hb during exercise. Therefore, at the same P_{O_2}, increased temperature results in more O_2 being delivered to the hot, exercising muscles.

Oxygen and Carbon Dioxide Transport

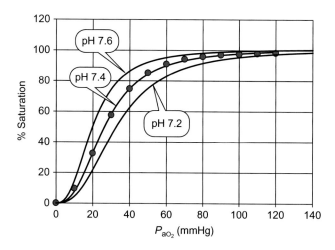

FIGURE 6.4.6 Effect of pH on the oxygen dissociation curves. Alkalinization shifts the curve to the left; acidification shifts the curve to the right. Shifting the curve to the right means that Hb can hold less oxygen at the same P_{O_2}, and thus the blood unloads more O_2 when the tissues are acidic, as happens when muscles are active.

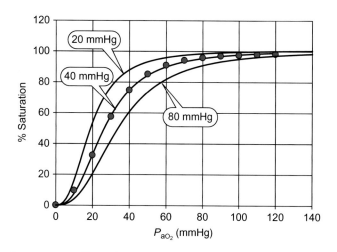

FIGURE 6.4.7 Effect of CO_2 on the oxygen dissociation curve. Half-maximal saturation of Hb is shifted to lower P_{O_2} by reducing P_{CO_2} and is shifted to higher P_{O_2} by raising P_{CO_2}.

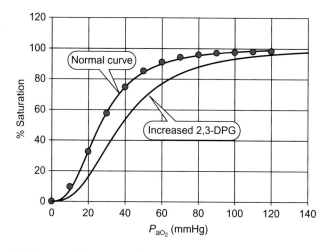

FIGURE 6.4.8 Shift of the O_2 dissociation curve to the right by 2,3-DPG.

O_2 during the transit of blood through active tissue. When the tissues are actively metabolizing, P_{CO_2} rises and more O_2 dissociates from hemoglobin. Figure 6.4.7 illustrates this effect.

Since mature red blood cells have no mitochondria, they derive their energy by anaerobic metabolism or glycolysis. The high concentrations of glycolytic intermediates create high concentrations of 2,3-DPG from a side reaction. The concentration of 2,3-DPG increases during hypoxia (low P_{O_2}) and alkalosis; it decreases with increased blood [H^+]. By shifting the O_2 dissociation curve to the right, the increased 2,3-DPG helps deliver O_2 to the tissues despite the hypoxia (see Figure 6.4.8).

INCREASED O_2 DELIVERY IN EXERCISE IS CAUSED BY INCREASED BLOOD FLOW AND SHORTER DIFFUSION

Although pH, P_{CO_2}, temperature, and 2,3-DPG all shift the O_2 dissociation curve for hemoglobin, **the main mechanism for increasing O_2 delivery to active tissues is by increasing blood flow**. Dilation of arterioles increases the number of open capillaries, increases the pressure that drives flow, increases the cross-sectional area through which blood moves, decreases the effective diffusion distance, and increases the effective diffusion surface area. All of these factors increase O_2 delivery.

DISSOLVED CO_2 ACCOUNTS FOR A SMALL FRACTION OF BLOOD CO_2 TRANSPORT

The dissolved [CO_2] is given by Henry's law as $\alpha \times P_{CO_2}$. The value of α at 37°C and in water is given in Table 6.3.3 as 0.0747 mL dL^{-1} mmHg^{-1}. However, only 93% of plasma is water, and hemoglobin in red blood cells displaces blood water still further (about 0.72 of the red cell volume is water) so this solubility coefficient must be adjusted. In normal blood, the overall solubility is about 0.062 mL dL^{-1} mmHg^{-1}. Thus arterial blood with $P_{a_{CO_2}} = 40$ mmHg contains about 2.5 mL CO_2 dL^{-1}; venous blood with $P_{v_{CO_2}} = 46$ mmHg contains about 2.9 mL CO_2 dL^{-1}. The net transport of dissolved CO_2 from tissues to lungs is thus $Q_a \alpha (P_{v_{CO_2}} - P_{a_{CO_2}}) = 5$ L min^{-1} (2.9 − 2.5 mL dL^{-1}) ≈ 20 mL min^{-1}. Since the overall CO_2 production is about 200 mL min^{-1}, only about 0.10 of the total is transported as dissolved CO_2.

MOST CO_2 IS CARRIED IN THE BLOOD AS HCO_3^-

Dissolved CO_2 reacts with water to form carbonic acid, H_2CO_3, which then dissociates into H^+ and bicarbonate, HCO_3^-. The sequential reactions are written as

[6.4.7]
$$CO_2 + H_2O \rightleftharpoons H_2CO_3$$
$$H_2CO_3 \rightleftharpoons H^+ + HCO_3^-$$

Clinical Applications: Carbon Monoxide Poisoning

Carbon monoxide is a lipophilic gas that binds to hemoglobin at the same site as oxygen, forming **carboxyhemoglobin**. Its binding, however, is about 220 times stronger than that of oxygen. Hemoglobin saturates with CO at a partial pressure of about 0.5 mmHg. CO binding to hemoglobin shifts the oxygen dissociation curve to the left and converts it from a sigmoidal to a more hyperbolic shape. This interferes with the ability of Hb to dissociate O_2 at low P_{O_2}, and therefore CO severely hampers the extraction of O_2 by the tissues.

CO occupies some 1–2% of the O_2 binding sites on Hb in people living in urban environments. Heavy smokers may have 10% of their Hb binding capacity occupied by Hb·CO. Because the binding reaction is reversible, high Hb·CO levels can be brought down by simply breathing air devoid of CO. The high affinity of Hb for CO means that the off rate constant is slow, and therefore the reversal of CO binding to Hb is slow. The half-time of the reverse reaction is about 4 h.

Persons with acute CO poisoning present a cherry-red appearance. Their main difficulty lies in the ability to extract O_2 in the tissues. Treatment consists of providing the victim with high concentrations of O_2 to breathe. The high P_{O_2} increases the dissolved [O_2] and competes better with the CO for Hb binding sites. The high P_{O_2} also speeds up CO elimination by competitive interference with the rebinding of CO. **Hyperbaric** O_2, in which P_{O_2} exceeds atmospheric pressures, speeds up CO elimination further.

Unaided, the formation of H_2CO_3 is slow, with a half-time greater than 5 s. CO_2 readily enters the red blood cell where **carbonic anhydrase**, a Zn-containing enzyme of 30,000 Da, converts CO_2 and H_2O to HCO_3^- and H^+ directly. These can then combine readily to form H_2CO_3. This enzyme completes the equilibration of CO_2, H_2CO_3, HCO_3^-, and H^+ within milliseconds. The dissociation of carbonic acid is rapid, and most of the CO_2 is converted to HCO_3^-. Most of the HCO_3^- formed in the red blood cell exchanges for Cl^- in the plasma, so that the HCO_3^- is largely carried in the plasma instead of the red blood cells. The exchange of HCO_3^- for Cl^- is called the **chloride shift** (see Figure 6.4.9). Because the P_{CO_2} is different in venous and arterial blood, the [HCO_3^-] is also different. Venous blood typically contains 20.7 mM HCO_3^-, whereas arterial blood contains about 19.1 mM HCO_3^-. The net transport of HCO_3^- is therefore Q_a (20.66 − 19.14 mM) = 5 L min^{-1} × 1.52 mM = 7.6 mmol min^{-1}. This converts to about 171 mL min^{-1} of CO_2 at STPD. **Thus, the fraction of CO_2 transported as HCO_3^- is 171 mL min^{-1}/200 mL min^{-1} = 0.85.**

CARBAMINOHEMOGLOBIN ACCOUNTS FOR A SMALL FRACTION OF TRANSPORTED CO_2

CO_2 reacts with free NH_2 terminal groups on both the α and β chains of hemoglobin to form a new compound, **carbaminohemoglobin** (see Figure 6.4.9). This reaction can also occur with plasma proteins. The combination of CO_2 with NH_2 groups is called a **carbamate**. Carbamate formation is reversible and influenced by P_{O_2}, pH, and [2,3-DPG]. When P_{O_2} increases, as it does when the blood enters the alveoli and exchanges O_2 with alveolar air, carbaminohemoglobin dissociates to CO_2 and Hb−NH_2. The reduction in CO_2 content of the hemoglobin by increased P_{O_2} is called the **Haldane effect** (see Figure 6.4.10). It is the converse of the Bohr effect, in which O_2 binding by Hb is reduced by increased P_{CO_2}. Typically arterial blood contains about 0.75 mM carbaminohemoglobin, whereas venous blood contains 0.84 mM. **Thus carbaminohemoglobin contributes Q_a (0.84 mM − 0.75 mM) = 5 L min^{-1} × 0.09 mM = 0.45 mmol min^{-1} = 10 mL min^{-1} or about 0.05 of the total CO_2 transport.**

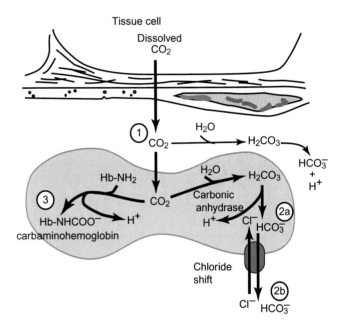

FIGURE 6.4.9 Transport of CO_2 in blood. CO_2 produced in the tissues diffuses through the interstitial fluid and capillaries in the dissolved state. It diffuses through the red blood cell membrane and encounters carbonic anhydrase, which catalyzes the hydration of CO_2 to form bicarbonate, HCO_3^-, and H^+. Hemoglobin within the red blood cell is a potent buffer for the released H^+. The HCO_3^- exchanges for Cl^- across the red blood cell membrane, a movement called the chloride shift. CO_2 reversibly combines with NH_2 groups on the hemoglobin to form carbaminohemoglobin. Thus CO_2 is carried in three ways: (1) dissolved in plasma and the red blood cell cytoplasm; (2) as HCO_3^- in the red blood cell cytoplasm (2a) and plasma (2b); and (3) as carbaminohemoglobin.

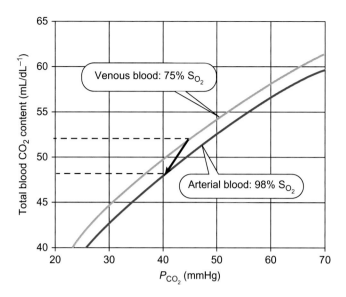

FIGURE 6.4.10 Effect of oxygenation on the total CO_2 content of blood. The reduction in the CO_2 content at the same P_{CO_2} by oxygenation is called the Haldane effect. In this way, oxygenation in the lungs aids in the removal of CO_2 from the venous blood. Without the Haldane effect, the change from $P_{vCO_2} = 46$ mmHg to $P_{aCO_2} = 40$ mmHg would follow the venous curve to release about 2.2 mL dL^{-1}. With the additional change from 75% O_2 saturation to 98% O_2 saturation, the CO_2 content follows the arrow to jump to the arterial curve, releasing a total of about 4 mL dL^{-1}. *Adapted from N.C. Staub, Basic Respiratory Physiology, Churchill Livingstone, New York, NY, 1991.*

SUMMARY

Oxygen is carried in two ways: dissolved in the plasma and bound to hemoglobin within red blood cells. The arterial blood contains about 20 mL dL^{-1} O_2, and about 98% of this is bound to hemoglobin. Hemoglobin can carry about 1.35 mL O_2 per gram, and blood normally contains about 15 g Hb dL^{-1}. Venous blood normally contains about 15 mL O_2 dL^{-1}, so the tissues extract at rest about 25% of the arterial O_2. With a cardiac output of 5 L min^{-1}, the net transport of O_2 is therefore about

$$5\,L\,min^{-1} \times (20\,mL\,dL^{-1} - 15\,mL\,dL^{-1}) = 250\,mL\,O_2\,min^{-1}$$

This O_2 transport matches the metabolic consumption of O_2 and the amount calculated from the flow of respiratory air and the difference between inspired air and expired air O_2 content.

Hemoglobin displays marked cooperativity in O_2 binding, so that its O_2 dissociation curve is steepest at physiological P_{O_2} levels. There is a continuous gradient of P_{O_2} from about 100 mmHg in the blood to 40 mmHg in the interstitial fluid, 20 mmHg in the cell, and about 5 mmHg in the mitochondria. The Hb dissociation curve shifts to the right with increased temperature, increased P_{CO_2}, decreased pH, and increased DPG. The Bohr effect describes the decreased affinity for O_2

Clinical Applications: Blood Substitutes

Trauma at disaster sites, automobile accidents, and on the battlefield often entail loss of blood and consequent hypotension and hypovolemic shock that can be fatal. The best treatment is to replace the lost blood. Transfusion with other people's blood poses numerous problems. Human blood requires donors, exacting storage conditions in order to prolong clinical effectiveness and reduce risk of infections, and an entire infrastructure of collection and storage centers. Human blood comes in a variety of types that are not compatible and so each recipient must be cross-matched with the potential transfused blood. Lastly, human blood transmits communicable diseases such as the human immunodeficiency virus (HIV) and hepatitis C virus (HCV). Each problem has been overcome. The infrastructure is in place, cross-matching is routinely performed, and screening of donors and testing for contaminants make the donor supply increasingly safe. All of this effort comes at significant cost. Donor blood shortages and all of the problems listed here have given impetus to developing safe and economical blood and plasma substitutes.

Blood substitutes must: (1) carry oxygen in the circulation; (2) deliver oxygen to the tissues; (3) require no cross-matching or compatibility testing; (4) have a long shelf-life; (5) survive in the circulation for suitable times before being cleared; (6) have no side effects; (7) have no pathogens; (8) not significantly alter blood viscosity. Two general types of blood substitutes that are currently being developed are broadly classed as hemoglobin-based oxygen carriers (HBOCs) and perfluorocarbon emulsions.

The best blood substitute would mimic hemoglobin's O_2 dissociation curve. A cell-free hemoglobin solution retains its ability to bind oxygen, and it does not possess the surface proteins responsible for transfusion reactions, so cross-matching is not required. However, unaltered hemoglobin has unacceptably short survival times in the circulation, an abnormally high O_2 affinity, and its clearance by the kidneys gums up the works. The attempted solution to these problems has been to polymerize the hemoglobin. Three such polymerized HBOCs are currently in advanced clinical trials.

Perfluorocarbons are biochemically inert liquids that carry O_2 as dissolved gas. Their O_2 content is linearly related to P_{O_2}. Perfluorocarbons are not miscible with watery solutions and can be used only as an emulsified preparation. The second generation of fluorocarbon preparations uses egg yolk phospholipids as emulsifiers. The droplets must be a specific size (about 0.17 μm) in order to be tolerated. The droplets are taken up by cells of the reticuloendothelial system and the perfluorocarbons are eventually excreted by exhalation through the lungs. (J.E. Squires, Artificial blood, *Science* **295**:1002–1005, 2002; R. Winslow, Blood substitutes, *Adv. Drug Del. Rev.* **40**:131–142, 2000; D.R. Spahn, Current status of artificial oxygen carriers, *Adv. Drug. Del. Rev.* **40**:143–151, 2000.)

caused by CO_2. These shifts help hemoglobin unload O_2 to the tissues when they become more active.

Blood transports CO_2 from the tissues to the lungs in three ways: dissolved as CO_2, as HCO_3^-, and bound to hemoglobin as carbaminohemoglobin. Dissolved CO_2 accounts for about 10% of the total CO_2 transport; about 85% of the CO_2 transport is carried as HCO_3^-, and only about 5% is transported as carbaminohemoglobin. In the lungs, increased P_{O_2} helps unload CO_2 so that each 100 mL of blood transports about 4 mL of CO_2 to the atmosphere. Overall CO_2 production is about 200 mL min^{-1}.

REVIEW QUESTIONS

1. Use Henry's law to calculate the dissolved oxygen concentration at P_{aO_2} = 100 and 40 mmHg.
2. What is the utility of an S-shaped oxygen dissociation curve? What equation describes this curve? What parameter of the curve indicates cooperativity? What does cooperativity mean?
3. What causes a rightward shift in the oxygen dissociation curve? Does a rightward shift mean oxygen dissociates more or less easily from Hb?
4. Where is hemoglobin in the body? Where is myoglobin? How do these differ in their oxygen binding? What are their functions within the body?
5. Where would you expect to find the lowest P_{O_2} within a cell?
6. In what forms is CO_2 carried in the blood? Which form accounts for the greatest transport?
7. Distinguish between transport of gas (O_2 or CO_2) and content of gas in the blood.
8. Would you expect fetal hemoglobin to have a higher or lower affinity for O_2 than adult hemoglobin? Why?

Acid−Base Physiology I: The Bicarbonate Buffer System and Respiratory Compensation 6.5

Learning Objectives

- Define pH
- Explain why plasma pH is regulated within narrow limits
- Define alkalosis and acidosis
- List the three major systems for regulating plasma pH
- Define the association and dissociation constants for a chemical buffer
- List the major chemical buffers in plasma
- Describe the function of the chemical buffers
- Explain what is meant by the isohydric principle
- Describe the role of carbonic anhydrase in the bicarbonate buffer system
- Write the Henderson−Hasselbalch equation for the bicarbonate buffer system using P_{CO_2} for carbonic acid and with the appropriate constants
- Describe the consequences of hyper- or hypoventilation as a primary defect
- Describe the response of the respiratory system to metabolic acidosis or alkalosis
- Explain why respiratory compensation for pH disturbances cannot be complete

pH IS A MONOTONICALLY DECREASING FUNCTION OF [H$^+$]

Because the [H$^+$] in solution can vary over many orders of magnitude, Sorenson devised the pH scale, originally named for the "potentz" (meaning "power" in German) of the [H$^+$]. The pH is *defined* as

[6.5.1] $$pH = -\log a_{H^+}$$

where P_{CO_2} is the **activity** of the H$^+$ ion. In Eqn [6.5.1], the logarithm is to the base 10. The activity is related to concentration by

[6.5.2] $$a_{H^+} = \gamma [H^+]$$

where γ is the **activity coefficient**. For dilute ideal solutions, $\gamma = 1$. In plasma, where [H$^+$] is about 10^{-7} M, the assumption that $\gamma = 1.0$ is justified. In this case, we write:

[6.5.3] $$pH = -\log [H^+]$$

The [H$^+$] in bodily fluids varies from a high of about 0.1 M in gastric juice to a low of about 2×10^{-9} M in the most alkaline pancreatic juice. The corresponding pH values are about pH 2 for the gastric juice and pH 8.8 in pancreatic juice.

PLASMA pH IS MAINTAINED WITHIN NARROW LIMITS

Failure to regulate [H$^+$] within fairly narrow limits causes death. Below pH 6.9, a person slips into a coma and death follows. Above 7.8, death is accompanied by tetany and convulsions. These are the extremes of pH compatible with life. Normally arterial plasma pH is maintained within more narrow limits, between about pH 7.35 and 7.44. These normal limits (7.35 and 7.44) correspond to [H$^+$] of 45×10^{-9} and 36×10^{-9} M. Regulation of plasma [H$^+$] within this narrow range of low concentrations illustrates its importance. **Acidemia** refers to [H$^+$] in the plasma below pH 7.35, the condition is called **acidosis**. Similarly, **alkalemia** refers to [H$^+$] above pH 7.44, the condition is **alkalosis**.

The extreme sensitivity to [H$^+$] is due to the binding (e.g., absorption and association) or unbinding (e.g., desorption and dissociation) of H$^+$ to or from ionizable groups on amino acids that make up proteins. These groups affect the local charge on the protein that affects the 3D structure of the proteins and therefore affects their function. Most enzymes, for example, have well-defined optimum pH and their activity falls off when pH is away from this optimum.

THE BODY USES *CHEMICAL BUFFERS*, THE *RESPIRATORY SYSTEM*, AND THE *RENAL SYSTEM* TO REGULATE pH

Chemical buffers respond rapidly and are the first line of defense in acid−base imbalances. This system resists changes in blood pH but cannot, in itself, restore acid or base excess. The respiratory system works by adjusting plasma P_{CO_2}. Increasing P_{CO_2} lowers the pH and decreasing P_{CO_2} raises it. The renal system works by adjusting plasma [HCO$_3^-$]. Increasing [HCO$_3^-$] raises the pH and decreasing [HCO$_3^-$] lowers the pH. How this works is described in more detail in Chapter 7.7.

CHEMICAL BUFFERS ABSORB OR DESORB H$^+$ ACCORDING TO THE LAW OF MASS ACTION

The Bronsted−Lowry theory defines an acid as a proton donor, whereas a base is a proton acceptor; according to Lewis' definition, an acid is an electron acceptor, whereas a base is an electron donor. For our purposes, an acid is a chemical that dissociates in water to produce a hydrogen ion, H^+, and its conjugate base:

[6.5.4] $$HA \rightarrow H^+ + A^-$$

A base is any chemical that can remove H^+ from solution. In Eqn [6.5.4], the anion A^- is a base because the reverse reaction

[6.5.5] $$A^- + H^+ \rightarrow HA$$

removes H^+ from solution. The dissociation reaction (see Eqn [6.5.4]) and the association reaction (see Eqn [6.5.5]) occur simultaneously with forward and reverse rate constants, k_f and k_r. The reactions rates are given by

[6.5.6A] $$J_f = k_f[HA]$$

[6.5.6B] $$J_r = k_r[H^+][A^-]$$

where J_f is the rate of the forward reaction and J_r is the rate of the reverse reaction. Equilibrium occurs when the two rates are equal. When $J_f = J_r$, we can take the ratio of Eqn [6.5.6B] to Eqn [6.5.6A] to get

[6.5.7] $$\frac{k_f}{k_r} = K_D = \frac{[A^-][H^+]}{[HA]}$$

where K_D is the **dissociation constant**. In Eqn [6.5.7], the concentrations are not just any concentrations. The equation is true only for equilibrium concentrations. Although these can vary widely, setting any two determines the third uniquely. The equilibrium also can be written as the inverse:

[6.5.8] $$\frac{k_r}{k_f} = K_A = \frac{[HA]}{[A^-][H^+]}$$

where the equilibrium is equivalently described in terms of the **association constant**, $K_A = 1/K_D$. The units of K_D are M, and the units of K_A are M^{-1}. This nomenclature is sometimes confusing because chemists also refer to the dissociation constant as K_a, meaning the acid equilibrium constant.

Strong acids have large dissociation constants. When these compounds dissolve in water, almost all of the acid dissociates to H^+ and A^- and the equilibrium [HA] is small. Similarly, strong bases have small dissociation constants or large association constants. In water, strong bases bind nearly all of the available H^+ until all of the base is present as HA and the equilibrium $[A^-]$ is small. Physiological buffers are weak acids or weak bases, with a K_D close to physiological $[H^+]$. Any buffer has the property of releasing or absorbing H^+. Upon dissociation, the acid becomes a base. The acid and base of a buffer exist in a pair, and so we speak of an acid and its **conjugate** base, where the term "conjugate" means that they occur in pairs.

In chemistry, a pH buffer is a chemical compound that resists change in the pH of a solution when acid or base is added to it. Adding acid or base to pure water results in large pH changes. Addition of a buffer before adding the acid or base results in smaller pH changes, as shown in

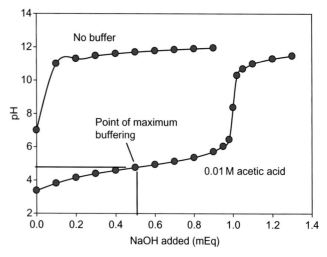

FIGURE 6.5.1 Adding a small amount (0.1 mEq) of NaOH to 100 mL of water at pH = 7.0 results in a large pH change. Adding NaOH to 100 mL of 0.01 M acetic acid produces a solution containing both undissociated acid and the Na acetate. The change in pH is buffered by combination with acetic acid. The pH change per unit NaOH is least at the point at which pH = pK (in this case, pK = 4.76). This is the point of highest buffer capacity.

Figure 6.5.1. The buffer capacity of a solution is defined as the number of moles of strong acid or base that must be added per unit change in pH: $\Delta n/\Delta pH$. Large buffer capacities require more added strong acid or base for the same change in pH. The buffer capacity depends on the concentrations of [HA] and $[A^-]$; the greatest buffering capacity of a solution occurs when the $[H^+] = K_D$.

Eqn [6.5.7] can be written in terms of $[H^+]$ as

[6.5.9] $$[H^+] = K_D \frac{[HA]}{[A^-]}$$

Taking the logarithm of both sides gives us

[6.5.10] $$\log[H^+] = \log K_D + \log \frac{[HA]}{[A^-]}$$

Multiplying both sides of the equation by -1, we get

[6.5.11] $$-\log[H^+] = -\log K_D + -\log \frac{[HA]}{[A^-]}$$

Recognizing that $-\log [H^+] = pH$ and identifying $-\log K_D = pK$, we write Eqn [6.5.11] as

[6.5.12] $$pH = pK + \log \frac{[A^-]}{[HA]}$$

This is the **Henderson–Hasselbalch equation**, which becomes very useful in dealing with problems of acid–base physiology.

There are a variety of buffers in plasma and within cells that absorb or desorb H^+ ions. Examples include protein buffers, phosphate buffer, and the bicarbonate buffer system. Proteins have ionizable groups such as $-COOH$ that can release H^+ and have pKs in the range of 3.5–5.5; they also possess $-NH_2$ groups that can accept H^+ and have pKs in the alkaline range, 8–9.

Imidazole groups from histidine also buffer pH changes. For each ionizable group we can write

[6.5.13] $\quad H_n PROT^{-r} \rightleftharpoons H_{n-1} PROT^{-(r+1)} + H^+$

and we can write a corresponding Henderson–Hasselbalch equation for each dissociation reaction. Because proteins usually contain a number of acidic and basic residues, proteins in general have a number of different pKs. This allows proteins to buffer pH changes over a wide range of pH.

Phosphate undergoes three ionizations:

[6.5.14] $\quad \begin{aligned} &H_3PO_4 \rightleftharpoons H_2PO_4^- + H^+: K_D = 7.5 \times 10^{-3} \text{ M} \\ &H_2PO_4^- \rightleftharpoons HPO_4^{-2} + H^+: K_D = 6.2 \times 10^{-8} \text{ M} \\ &HPO_4^{-2} \rightleftharpoons PO_4^{-3} + H^+: K_D = 4.8 \times 10^{-13} \text{ M} \end{aligned}$

According to the Henderson–Hasselbalch equation, at physiological pH of around 7.4, the first H^+ is completely dissociated and essentially no H_3PO_4 remains. Similarly, the last H^+ does not dissociate until the pH becomes far more alkaline. The only dissociation that contributes meaningfully to the chemical buffers at normal plasma pH is the second dissociation reaction with a pK = 7.21. Phosphate buffer amounts to about 1 mM in plasma, but there are multiple forms of organic phosphates (ATP, ADP, pyrophosphate, creatine phosphate, etc.) inside cells that contribute to the intracellular buffer capacity.

By far the most important buffer system in the body is the HCO_3^- buffer system. It is written as

[6.5.15] $\quad CO_2 + H_2O \rightleftharpoons H_2CO_3 \rightleftharpoons HCO_3^- + H^+$

The hydration of CO_2 to form **carbonic acid**, H_2CO_3, occurs slowly. Many tissues contain the enzyme **carbonic anhydrase (CA)** a Zn-containing enzyme that converts CO_2 and H_2O to H^+ and HCO_3^-. This bypasses the carbonic acid. However, H^+ and HCO_3^- can rapidly equilibrate with H_2CO_3, so that effectively the enzyme equilibrates all components of the carbonic acid–bicarbonate system. There are multiple isoforms of carbonic anhydrase in human tissues, and all have large turnover numbers on the order of $10^3 - 10^6$ completed reactions s^{-1}.

THE *ISOHYDRIC PRINCIPLE* STATES THAT ALL BUFFERS IN A SOLUTION ARE IN EQUILIBRIUM WITH THE SAME [H^+]

A Henderson–Hasselbalch equation can be written for each of the three buffer systems listed above. The three equations for the protein, phosphate, and bicarbonate buffers are as follows:

[6.5.16] $\quad \begin{aligned} pH &= pK_{PROT} + \log\frac{[H_{n-1}PROT^{-(r+1)}]}{[H_n PROT^{-r}]} \\ pH &= pK_{Pi} + \log\frac{[HPO_4^{-2}]}{[H_2PO_4^-]} \\ pH &= pK + \log\frac{[HCO_3^-]}{[H_2CO_3]} \end{aligned}$

where the pK for each reaction is different because the chemical environment for each H^+ binding site is different. All three of these reactions occur simultaneously in plasma, but there is only one [H^+] and one pH in plasma. Thus all three of these chemical buffers are in equilibrium simultaneously with a single [H^+]. This is the **isohydric principle**, which states simply that all buffers in a solution are simultaneously in equilibrium with one [H^+]. All buffer systems are connected by virtue of their shared dependence on [H^+]. The consequence of this is that **adjustments of a single buffer system will adjust them all through changes in [H^+]**. This is what makes the bicarbonate buffer system so important: it is the one that is physiologically adjusted.

EXPRESSING [H_2CO_3] IN TERMS OF P_{CO_2} MAKES THE HENDERSON–HASSELBALCH EQUATION MORE USEFUL

The Henderson–Hasselbalch equation for the dissociation of carbonic acid is given as

[6.5.17] $\quad pH = pK + \log\frac{[HCO_3^-]}{[H_2CO_3]}$

The K_D for this reaction at 37°C is 3.38×10^{-7} M. This value for K_D can be used only if we know the concentrations of HCO_3^- and H_2CO_3. However, the H_2CO_3 concentration in plasma cannot be measured easily. Because of the action of carbonic anhydrase, [H_2CO_3] is in equilibrium with the dissolved [CO_2] and is proportional to it; the [H_2CO_3] is 0.235% of the dissolved [CO_2]. From the definition of K_D [see Eqn [6.5.7]), we write for the reaction forming carbonic acid:

[6.5.18] $\quad \begin{aligned} K_D &= \frac{[HCO_3^-][H^+]}{[H_2CO_3]} \\ &= \frac{[HCO_3^-][H^+]}{0.00235 [CO_2]} \end{aligned}$

We can define a new K'_D by substituting [CO_2] for [H_2CO_3]:

[6.5.19] $\quad K'_D = 0.00235 \times K_D = 7.94 \times 10^{-10} \text{ M} = \frac{[HCO_3^-][H^+]}{[CO_2]}$

The Henderson–Hasselbalch equation then must be rewritten in terms of [CO_2] instead of [H_2CO_3]. Eqn [6.5.17] becomes

[6.5.20] $\quad pH = 9.10 + \log\frac{[HCO_3^-]}{[CO_2]}$

where the $9.10 = pK'_D = -\log K'_D$ when K'_D is in units of M. **When the [HCO_3^-] is expressed in mM, then $pK'_D = 6.10$ at 37°C.** This equation still needs work because typically it is the P_{CO_2} that is measured and not the concentration of dissolved CO_2. The conversion between the two is the solubility. According to Henry's Law, we write:

[6.5.21] $$[CO_2] = \alpha P_{CO_2}$$

When $[CO_2]$ is expressed in mL per dL of water, and P_{CO_2} is expressed in mmHg, the solubility coefficient, α, is 0.0747 mL dL^{-1} mmHg^{-1} (see Table 6.3.3). This can be converted to the solubility in plasma by multiplying by 0.93, the percent of the plasma volume that is water. Thus the solubility in plasma is 0.0695 mL dL^{-1} mmHg^{-1}. The volume of gas dissolved is expressed as the volume at STPD. It is converted to molarity by dividing by the molar volume, which is 22.4 L mol^{-1} = 22,400 mL mol^{-1}. This gives a solubility coefficient of 3.08×10^{-6} when $[CO_2]$ is expressed in mol dL^{-1} and P_{CO_2} is in mmHg. Because $[HCO_3^-]$ is expressed in mM = 10^{-3} mol L^{-1}, the equation requires $[CO_2]$ in mM. Conversion of the solubility coefficient to give $[CO_2]$ in mM gives 3.08×10^{-6} mol dL^{-1} mmHg^{-1} $\times 10^3$ mmol mol^{-1} $\times$ 10 dL L^{-1} = 0.0308 mmol L^{-1} mmHg^{-1}. The final relation is

[6.5.22] $$[CO_2] = 0.0308 P_{CO_2}$$

when $[CO]_2$ is expressed in units of mM and P_{CO_2} is in mmHg. Inserting this result into Eqn [6.5.20] and using the value of pK appropriate for the units and the variables, we arrive at the useful form of the Henderson–Hasselbalch equation for the HCO_3^- buffer system:

[6.5.23] $$pH = 6.10 + \log \frac{[HCO_3^-]}{0.0308\, P_{CO_2}}$$

THE RESPIRATORY SYSTEM REGULATES pH BY ADJUSTING PLASMA P_{CO_2}

The HCO_3^- buffer system can be described by its chemical reaction and by the Henderson–Hasselbalch equation that derives from this reaction. These descriptions are reproduced below.

[6.5.15] $$CO_2 + H_2O \rightleftharpoons H_2CO_3 \rightleftharpoons HCO_3^- + H^+$$

[6.5.23] $$pH = 6.10 + \log \frac{[HCO_3^-]}{0.0308\, P_{CO_2}}$$

To adjust plasma pH, the respiratory system has only one response: it can change pulmonary ventilation, which in turn alters alveolar ventilation. Recall the alveolar ventilation equation that we derived in Chapter 6.3:

[6.3.22] $$P_{A_{CO_2}} = \frac{Q_{CO_2}}{Q_A}(P_B - 47)$$

where $P_{A_{CO_2}}$ is the alveolar P_{CO_2}, Q_{CO_2} is the rate of CO_2 production by the body, Q_A is the alveolar ventilation, and P_B is the ambient, barometric pressure. If the respiratory system lowers alveolar ventilation (given a constant rate of CO_2 production, Q_{CO_2}), alveolar P_{CO_2} will increase, and so will arterial P_{CO_2} which is brought into equilibrium with alveolar P_{CO_2} when the blood passes through the lungs. The body will reach a new steady state in which the elimination of CO_2 through the lungs will equal its production, but it will be accomplished with less ventilation and a higher P_{CO_2} in the expired air. The increased P_{CO_2} will increase dissolved $[CO_2]$ which will, by the law of mass action, increase $[H_2CO_3]$ and increase $[HCO_3^-]$ and $[H^+]$. The increased $[H^+]$ lowers the pH of the plasma and acidosis results. All of this can be seen by an application of mass action to Eqn [6.5.15]. It can also be seen by applying Eqn [6.5.23]. The logarithm is a monotonically increasing function of its argument. The argument is the ratio $[HCO_3^-]/0.03 P_{CO_2}$. If P_{CO_2} increases, the argument decreases and so does the logarithm of the argument. Thus the pH also decreases because it is the sum of a constant and the logarithm. The decreased pH is called an **acidosis**. Thus **hypoventilation causes respiratory acidosis**.

By the same line of reasoning, in reverse, **hyperventilation causes respiratory alkalosis**.

HYPOVENTILATION IN RESPONSE TO ALKALOSIS IS CALLED *RESPIRATORY COMPENSATION OF ALKALOSIS*

Alkalosis can be produced by a variety of conditions including **vomiting**, in which stomach acid is lost from the body, or by **alkali treatment** for peptic ulcers or indigestion. When plasma pH rises from some problem other than the lungs, the respiratory system can compensate by hypoventilating. This raises plasma P_{CO_2} and adjusts plasma pH back toward normal. The depression of ventilation results from chemosensors for pH that help control ventilatory drive (see Chapter 6.6). The respiratory compensation that results is rapid but incomplete; the pH remains only partly compensated. Full compensation requires the kidneys to adjust $[HCO_3^-]$.

HYPERVENTILATION IN RESPONSE TO ACIDOSIS IS CALLED *RESPIRATORY COMPENSATION OF ACIDOSIS*

Acidosis can be produced by a variety of conditions in which H^+ production is increased or in which H^+ is not excreted. These conditions include **diarrhea**, in which buffer HCO_3^- is lost; **diabetes mellitus**, in which metabolic acid production is increased; and **renal tubular acidosis**, in which the kidneys fail to excrete sufficient acid. When plasma pH falls from some problem other than with the lungs, the respiratory system can compensate by hyperventilating. This lowers plasma P_{CO_2} and adjusts plasma pH back toward normal. The increased ventilatory drive results from the stimulation of chemosensors for pH that help control ventilatory drive (see Chapter 6.6). The respiratory compensation that results is rapid but incomplete; the pH remains only partly compensated. Once again, full compensation requires the kidneys to adjust $[HCO_3^-]$.

RESPIRATORY ACIDOSIS AND RESPIRATORY ALKALOSIS

Hypoventilation in the absence of alkalosis to inhibit ventilatory drive is a primary respiratory problem. This can result from **CNS depression** caused by trauma or by some drugs that inhibit respiratory drive such as barbiturates. Hypoventilation increases P_{CO_2} and leads to acidosis. There can be no respiratory compensation for a primary respiratory problem. In addition, anything that interferes with gas exchange can also cause respiratory acidosis. These conditions include **emphysema** or **asthma**. Interference with airflow as in asthma or reduction of the area of exchange as in emphysema can both increase P_{CO_2} and cause respiratory acidosis.

In the absence of acidosis to stimulate ventilatory drive, hyperventilation is a primary respiratory problem. This can result from excess CNS stimulation caused by anxiety or other psychogenic phenomena or by voluntary hyperventilation prior to holding one's breath under water, for example. Hyperventilation decreases P_{CO_2} and leads to alkalosis. There can be no respiratory compensation for a primary respiratory problem.

THE pH – HCO_3^- DIAGRAM DEPICTS ACID–BASE BALANCE GRAPHICALLY

There are a variety of graphical techniques that have been developed to simplify the analysis of acid–base physiology. Two common graphs are the pH – HCO_3^- diagram, sometimes also called the Davenport diagram, and the pH–log P_{CO_2} diagram, also called the Siggaard–Anderson nomogram. We will discuss only the pH – HCO_3^- diagram here.

The Henderson–Hasselbalch equation describes the relationship between P_{CO_2} and HCO_3^- at a given pH:

$$[6.5.23] \quad pH = 6.10 + \log \frac{[HCO_3^-]}{0.0308 \, P_{CO_2}}$$

In the pH – HCO_3^- diagram, the pH is plotted on the abscissa and [HCO_3^-] is plotted on the ordinate. Because the values of pH, [HCO_3^-], and P_{CO_2} are related according to Eqn [6.5.23], any two values determine the third. If we set P_{CO_2} = 40 mmHg, as an example, we can determine the set of all points whose values of pH and [HCO_3^-] satisfy the Henderson–Hasselbalch equation. All points in the set define a P_{CO_2} isobar, meaning "same pressure." P_{CO_2} isobars at 20, 40, 60, and 80 mmHg are shown in Figure 6.5.2.

The P_{CO_2} isobars describe how pH and [HCO_3^-] change at constant P_{CO_2}. The **plasma buffer line** describes how pH and [HCO_3^-] change when the blood is equilibrated with various P_{CO_2}. It is called the buffer line because adding or removing CO_2 is equivalent to adding or removing acid as would occur in a titration. Because the red blood cells contain hemoglobin, which is a powerful pH buffer, the buffering power of plasma depends on both plasma proteins and hemoglobin, and therefore the slope of the buffer

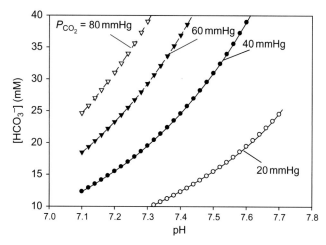

FIGURE 6.5.2 The pH – HCO_3^- diagram showing the variation of pH and [HCO_3^-] at constant P_{CO_2}. The lines of equal P_{CO_2} are called isobars. *Adapted from H.W. Davenport, The ABC of Acid–Base Chemistry, 6th edition, University of Chicago Press, 1974.*

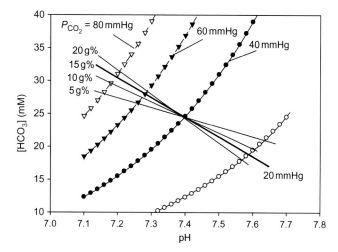

FIGURE 6.5.3 The pH – HCO_3^- diagram showing the buffer lines of blood that contains 5, 10, 15, and 20 g hemoglobin per dL whole blood. Increasing [Hb] increases the buffer capacity of blood. The buffer line at the normal hemoglobin concentration (15 g%) is the normal buffer line and is indicated in the figure by the darker line.

line depends on the hemoglobin concentration of blood. Figure 6.5.3 shows the plasma buffer line as a function of [Hb].

Disturbances of acid–base balance can be depicted graphically on the pH – HCO_3^- diagram as illustrated in Figure 6.5.4. The isobars at 20, 40, 60, and 80 mmHg P_{CO_2} are shown here along with the normal buffer line. The normal situation occurs at point A in the figure, where the pH is 7.4, [HCO_3^-] is 24 mM, and P_{CO_2} is 40 mmHg. Pure respiratory alkalosis corresponds to a decrease in P_{CO_2} with no change other than by mass action to changes in HCO_3^-; this occurs at point B. Respiratory acidosis moves the point describing acid–base in the opposite direction: P_{CO_2} increases and the point describing the acid–base balance moves along the normal buffer line toward the higher P_{CO_2} isobar (point C).

EXAMPLE 6.5.1 Normal Acid–Base Conditions

The normal pH of arterial blood is 7.4. From previous discussion (see Chapter 6.4), the arterial P_{aCO_2} is 40 mmHg. What is the normal [HCO_3^-]?

Here we use the Henderson–Hasselbalch equation:

$$pH = 6.1 + \log[HCO_3^-]/0.0308\, P_{aCO_2}$$

where pH = 7.4 and P_{aCO_2} = 40 mmHg. The constant, 0.0308, is designed for [HCO_3^-] in mM and P_{aCO_2} in mmHg. Therefore, we have

$$7.4 - 6.1 = \log[HCO_3^-]/1.232; \quad 1.3 = \log[HCO_3^-]/1.232$$

The log here is log base 10. Thus we have

$$10^{1.3} \times 1.232 = [HCO_3^-] = \mathbf{24.6\ mM}$$

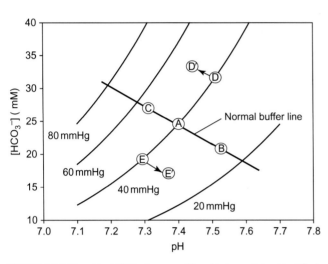

FIGURE 6.5.4 The pH − HCO_3^- diagram with acid–base status identified. The normal situation is at point A. P_{CO_2} is normal, pH is normal, and [HCO_3^-] is normal. Respiratory alkalosis occurs at point B, where P_{CO_2} is reduced by hyperventilation. Point C corresponds to respiratory acidosis, in which P_{CO_2} is elevated as the primary problem. Uncompensated metabolic alkalosis occurs at point D; respiratory compensation rapidly moves the body from point D to D', returning pH to near normal but with elevated [HCO_3^-]. Point E is uncompensated metabolic acidosis; respiratory compensation moves the body from E to E' by hyperventilation.

Changes in acid–base status parallel to the normal P_{CO_2} isobar correspond to **uncompensated metabolic alkalosis** or **uncompensated metabolic acidosis**. These states correspond to points D and E in Figure 6.5.4. Respiratory compensation for metabolic alkalosis consists of hypoventilation to raise P_{CO_2} to help bring pH toward normal. This corresponds to position D' in Figure 6.5.4. Note that the respiratory compensation here cannot be complete because there must be some residual alkalosis to continue to inhibit respiratory drive to maintain the hypoventilation. Also, respiratory compensation for metabolic alkalosis does not return us to point A, the normal acid–base situation. This must be accomplished by lowering [HCO_3^-]. This job is performed by the kidneys.

On the other hand, respiratory compensation for metabolic acidosis consists of hyperventilation and reduction in P_{CO_2}. This moves the acid–base status from point E to E' in Figure 6.5.4. Here again respiratory compensation cannot be complete because there must be a continual stimulation of hyperventilation to maintain the lower P_{CO_2}. Respiratory compensation cannot return the acid–base status to normal at point A. This requires addition of [HCO_3^-]. This final task of adjusting plasma pH is accomplished by the kidneys.

SUMMARY

The plasma pH is defined as $-\log[H^+]$; when [H^+] increases, the pH decreases. The condition of high plasma pH is called alkalosis; low plasma pH is acidosis. The body has three lines of defense against departures from normal plasma pH: the chemical buffers, the respiratory system, and the renal system. The chemical buffers passively resist changes in pH by absorbing excess H^+ when pH falls or by releasing H^+ ions when pH rises. Chemical buffers include proteins, phosphate, and bicarbonate buffers. All of these equilibrate with a single [H^+], and so the buffer systems are linked. This is the isohydric principle. Because of this link, adjustment of the bicarbonate buffer system controls all buffer systems. The bicarbonate buffer system has two components: plasma [CO_2] and [HCO_3^-]. The respiratory system controls plasma pH by adjusting the [CO_2]; the renal system adjusts [HCO_3^-].

Each buffer system can be described by its Henderson–Hasselbalch equation:

$$pH = pK + \log[A]/[HA]$$

where the $pK = -\log K_D$ and K_D is the dissociation constant that varies with the chemical nature of each buffer. The acid in the bicarbonate buffer system is H_2CO_3. The equilibrium between dissolved CO_2 and H_2CO_3 is accelerated by carbonic anhydrase. Because this equilibrium is established so quickly, we can transform the Henderson–Hasselbalch equation for the bicarbonate buffer system to

$$pH = 6.1 + \log[HCO_3]/0.0308 \times P_{CO_2}$$

where [HCO_3^-] is in mM and P_{CO_2} is in units of mmHg. The respiratory system can increase P_{CO_2} by hypoventilation. This increases [H^+] by adding acid as H_2CO_3 and decreases the pH. Respiratory acidosis is caused by hypoventilation as the primary disturbance. Hypoventilation also forms the respiratory response to metabolic alkalosis. The respiratory system can also decrease P_{CO_2} by hyperventilating. This decreases [H^+]

and causes an alkalosis. Respiratory alkalosis results from hyperventilation as the primary disturbance. Hyperventilation also forms the respiratory compensation of metabolic acidosis. Respiratory compensation for either acidosis or alkalosis is incomplete because some residual pH disturbance must remain to maintain the hyperventilation or hypoventilation. Complete compensation of pH disturbances requires the kidney to change plasma [HCO_3^-].

REVIEW QUESTIONS

1. Why is the bicarbonate buffer system so important when its pK is not very close to the normal pH of blood?
2. In the Henderson–Hasselbalch equation, what units are necessary for [HCO_3^-]? For P_{CO_2}? If different units were used, would the value for pK remain the same?
3. Draw the pH – HCO_3^- diagram. Label the axes. Draw the isobar for the normal P_{CO_2} of 40 mmHg. Draw the normal buffer line. Indicate the normal position of acid–base balance.
4. Describe the three systems for acid–base regulation, their relative speeds of operation, and their ability to compensate for imbalances.
5. Write the alveolar ventilation equation and predict the effect of changes in Q_A on $P_{a_{CO_2}}$ and $P_{a_{O_2}}$.

6.6 Control of Ventilation

Learning Objectives

- List the muscles of respiration
- Identify the origin of voluntary and involuntary controls of ventilation
- Define PRG and indicate its location and its effect on ventilation
- Define DRG and indicate its location and its influence on ventilation
- Define VRG and indicate its location and its influence on ventilation
- Describe the location and sensitivity of the peripheral chemosensors
- Describe the afferent nerves for the carotid bodies and aortic bodies
- Describe how pH, P_{CO_2}, and P_{O_2} affect the firing rates of chemosensors in the carotid and aortic bodies
- Describe the Hering–Breuer inflation reflex
- Describe the location of the central chemosensors
- Explain how the central chemosensors respond to plasma pH and P_{CO_2}
- Name the only chemosensor that responds to low P_{O_2}
- Explain how increased ventilation occurs during exercise

NERVES REGULATE BREATHING

The **phrenic nerve** is actually a pair of nerves, the right and left phrenic nerves, that activate contraction of the diaphragm that expands the thoracic cavity. Because the lungs are stuck to the thoracic cavity, this expands the lungs and thereby draws air into them. The cell bodies of the motor neurons that make up the phrenic nerve reside in a longitudinally oriented column in segments C3–C5 of the spinal cord. Initiation of action potentials in the **external intercostals** elevates and expands the rib cage, aiding inspiration. The cell bodies of the motor neurons that control the external intercostals form a column that extends the entire length of the thoracic spinal cord. Similarly, motor neurons that control the **internal intercostals**, which aid expiration, form a separate column. The **abdominal muscles** also aid expiration, and the cell bodies of their motor neurons are found in the lower thoracic and upper lumbar spinal cord segments. The locations of these are shown schematically in Figure 6.6.1.

These observations permit an obvious conclusion: ventilation requires the activation of muscles which is accomplished through motor neurons. Thus the generation of the basic rhythm of breathing and its regulation during activity such as swimming or singing or talking or eating or sleeping and during exercise must ultimately involve nerve centers that control these motor neurons.

CONTROL OF BREATHING INVOLVES VOLUNTARY AND INVOLUNTARY COMPONENTS

Everyday experience tells us that we can regulate our breathing to accomplish tasks such as swimming, singing, talking, and eating. This ability constitutes a voluntary component to control of ventilation. During sleep we continue to breathe without voluntary control. **Voluntary control arises from the cerebral cortex**, whereas **involuntary control arises from centers in the medulla and pons**. Both areas project to the same final pathway: the spinal motor neurons that control respiratory muscles. However, these dual control systems project to these motor neurons via different pathways (see Figure 6.6.2). Cutting through the brain stem above the pons in experimental animals removes all voluntary control and only automatic mechanisms in the brain stem drive ventilation.

THE BRAIN STEM CONTAINS A *PONTINE RESPIRATORY GROUP* IN THE PONS, AN *APNEUSTIC CENTER* IN THE LOWER PONS, AND *DORSAL AND VENTRAL RESPIRATORY GROUPS* IN THE MEDULLA

Part of the evidence for the contemporary model of ventilatory control arose from experiments in animals in which the brain stem was transected at different places (see Figure 6.6.3). These experiments indicated the existence of a **pontine respiratory group (PRG)** that switches off inspiration; an **apneustic center** in the middle pons that prevents the switch off of inspiration, and two more areas in the medulla, the **dorsal respiratory group (DRG)** and **ventral respiratory group (VRG)**. Further investigations have continued to use experimental animals as direct information from humans is scarce.

Control of Ventilation

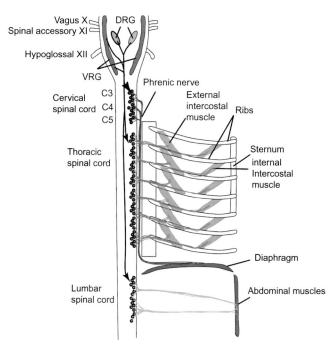

FIGURE 6.6.1 Location of the spinal motor neurons that control respiratory muscles. The cell bodies of motor neurons that activate the diaphragm are located in C3–C5. Their axons collect in the phrenic nerve. Cell bodies of motor neurons controlling the intercostal muscles are located in the thoracic spinal cord. The abdominal muscles are controlled by motor neurons whose cell bodies are in the lower thoracic and upper lumbar spinal cord. All of these motor neurons receive inputs from controlling centers in the medulla. The DRG is the dorsal respiratory group in the medulla; the VRG is the ventral respiratory group. *Adapted from M.P. Hlastala and A.J. Berger, Physiology of Respiration, Oxford University Press, 2001.*

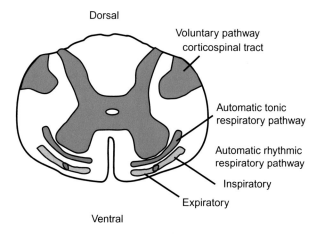

FIGURE 6.6.2 Location of descending pathways for the control of spinal respiratory motor neurons. The diagram shows a midcervical section of the spinal cord. The corticospinal tract in the dorsolateral cord contains axons of cells in higher centers. Descending automatic respiratory drive is carried in ventrolateral columns. Axons controlling expiratory motor neurons are situated medially, whereas those controlling inspiration are situated more laterally. The tonic involuntary pathway originates in the medial reticular formation of the medulla. *Redrawn from M.P. Hlastala and A.J. Berger, Physiology of Respiration, Oxford University Press, 2001.*

THE DRG RECEIVES A VARIETY OF INPUTS AND EXCITES INSPIRATORY MOTOR NEURONS

The DRG, located in the dorsal medial medulla, receives afferent signals over the **glossopharyngeal** and **vagus** nerves. These provide the DRG with sensory information from chemoreceptors in the carotid body and aortic body, as well as slowly and rapidly adapting stretch receptors in the lungs, and from C fiber endings in the lung. The DRG contains mainly neurons that directly excite motor neurons whose axons exit the spinal cord in the phrenic nerve to activate the diaphragm. These DRG neurons are called **I neurons**, for neurons that drive inspiration. These I neuron fibers cross the midline in the medulla. This is consistent with the general plan of the nervous system in which much sensory and motor information cross over to the contralateral side (see Figure 6.6.4). The DRG projects to the PRG and to the VRG. Thus the DRG tells the PRG and VRG that inspiration has begun at the same time that it begins.

THE VRG CONTAINS BOTH I AND E NEURONS

The VRG is located in the ventrolateral region of the medulla as shown in Figure 6.6.3. It has both inspiratory or **I neurons** that fire action potentials during the inspiratory phase of the respiratory cycle, and expiratory or **E neurons** that fire action potentials during the expiratory phase. I neurons predominate in the middle of the VRG, whereas E neurons are more abundant in both the rostral (toward the nose) and caudal (toward the tail) parts of the VRG. The I and E neurons are segregated in the spinal cord as well (see Figure 6.6.2). Both I and E neurons cross over the midline to project to the contralateral spinal motor neurons. Axons from I neurons cross rostrally, whereas the axons from E neurons cross more caudally. E neurons in the caudal VRG make monosynaptic contacts with expiratory internal intercostal motor neurons (see Figure 6.6.4).

I neurons from the intermediate VRG project contralaterally to descend the cord and make contact with spinal motor neurons. Axons from this part of the VRG also project to the contralateral VRG, making contact with both I and E areas.

The most rostral part of the VRG contains E neurons and is postulated to be the site of central pattern generation for the respiratory system. It sends inhibitory processes to the contralateral DRG and others to the ipsilateral caudal VRG, which contains other E neurons. The exact connectivity of these neurons is unknown. A specialized area, called the **pre-Botzinger complex (PBC)**, is located in the rostral part of the VRG and appears to be the "kernel" of respiratory rhythm generation.

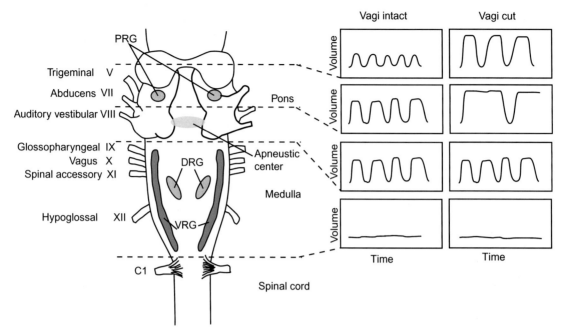

FIGURE 6.6.3 Effect of transections of the brain stem on respiratory patterns. The drawing shows a dorsal view of the brain stem after removal of the cerebellum. Transection above the pons removes voluntary control but respiration has a normal depth and rhythm. If the two vagus nerves are cut, inspiration is deeper than usual and the breathing frequency decreases. Thus the vagi supply sensory information that switches inspiration to expiration. Transection in the mid pons results in deeper inspiration with decreased frequency, suggesting that the transection removes a center that switches from inspiration to expiration. This center is the **PRG**, the **pontine respiratory group**. Section of the vagi with transection at mid pons results in apneusis, a breathing pattern consisting of long and deep inspirations punctuated with rapid expirations. Transection between medulla and pons with the vagi cut does not result in apneusis, suggesting there is a center in the lower pons that inhibits the switch from inspiration to expiration. This is the **apneustic center**, whose normal function remains unclear. Transection below the medulla removes all rhythmic breathing, indicating that the major sources of normal respiratory rhythm are located in the medulla. These are the **dorsal respiratory group (DRG)** and **ventral respiratory group (VRG)** of respiratory neurons.

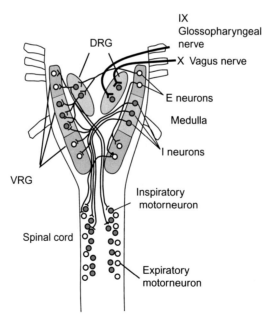

FIGURE 6.6.4 Connections between the DRG and VRG and their projections to the spinal motor neurons. Neurons that fire during inspiration (I neurons) are shown in dark; neurons that fire during expiration (E neurons) are light. The DRG contains predominantly I neurons that cross over and project onto spinal neurons that control inspiratory muscles. The DRG neurons also project to the ipsilateral VRG to both I and E neurons. The VRG has three sections. The most caudal and rostral parts of the VRG contain mainly E neurons, whereas the middle part contains I neurons. E neurons from the rostral VRG form inhibitory synapses on the contralateral DRG neurons and also form excitatory synapses on the ipsilateral caudal VRG E neurons. I neurons in the central VRG cross the midline and innervate spinal motor neurons for inspiratory muscles on the contralateral side. They also synapse on the contralateral I and E neurons of the VRG. Only one side of each connection is shown in the diagram. These connections are only a fraction of the total connections that occur between these structures.

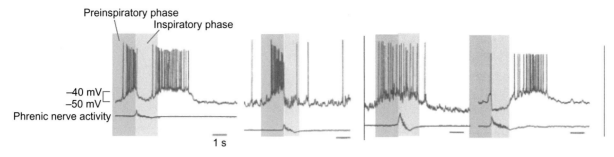

FIGURE 6.6.5 Responses of neurons classified as "pre-I" neurons. These neurons typically produce action potentials during the preinspiratory phase of respiration. They are generally inactive during the respiratory phase, but become variably active again during the postrespiratory phase. *From H. Onimura, A. Arata, and I. Homma, Neuronal mechanisms of respiratory rhythm generation: an approach using an in vitro preparation. Jap. J. Physiol. 47:385–403 (1997).*

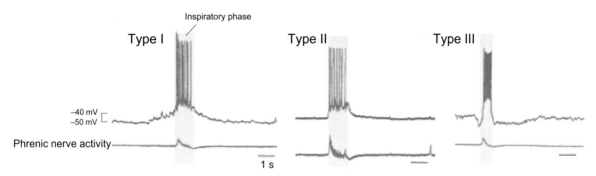

FIGURE 6.6.6 Responses of neurons classified as inspiratory or Insp neurons. These typically are active during the inspiratory phase of respiration and either show depolarizations around inspiration (Type I) or no postsynaptic potential (Type II) or hyperpolarizations (Type III). *From H. Onimura, A. Arata, and I. Homma, Neuronal mechanisms of respiratory rhythm generation: an approach using an in vitro preparation. Jap. J. Physiol. 47:385–403 (1997).*

NEURONS IN THE VRG HAVE A MORE VARIED ACTIVITY THAN "EXPIRATORY" OR "INSPIRATORY"

Respiratory neurons have a more complicated electrical pattern than simply "inspiratory" or "expiratory." Electrical recordings reveal neurons that depolarize to form bursts of action potentials prior to the inspiratory drive along the phrenic nerve. These have been called "preinspiratory" (Pre-I) neurons. These Pre-I neurons are usually hyperpolarized during the inspiratory phase and initiate action potentials again after inspiratory drive is over. Thus these Pre-I neurons are biphasic. This biphasic pattern is variable, with about 20% of the Pre-I neurons showing no inhibition during the inspiratory phase (see Figure 6.6.5). Other neurons fire mainly during the inspiratory phase, called Inspiratory (Insp) neurons, and these can be divided into three subtypes based on their differing postsynaptic potentials during the pre- and postinspiratory phases (see Figure 6.6.6). Type I Insp neurons have gradual depolarization during the preinspiratory phase, and similar diminishing depolarizations during the postinspiratory phase, probably due to synaptic connections to pre-I cells. Type II Insp neurons lack such connections, and therefore lack the depolarizations in either the pre- or postinspiratory phase. Type III Insp neurons show hyperpolarizations during the pre- and postinspiratory phases.

DESPITE PROGRESS, THE NEURAL MECHANISM OF THE RESPIRATORY PATTERN REMAINS UNKNOWN

The whole point of determining the nerve centers that control respiration and their inhibitory and excitatory connections to other nerve centers is to produce a model of how the nervous system generates and modulates the respiratory rhythm. Unfortunately, the neural basis for the respiratory rhythm remains elusive despite the identification of the PRG, DRG, VRG, PBC, and some of their main connections (see Figure 6.6.4). Part of the difficulty lies in the complexity of multiple connections and inputs.

There appear to be two major contenders for the rhythm generator itself. The first are cellular pacemakers. These are neurons with voltage- and time-dependent channels that produce rhythmic behavior at the cellular level. Indeed, some neurons with promising characteristics have been identified in the PBC in the rostral VRG.

The second possibility is that rhythm generation is a system property that arises from interactions between neurons due to patterns of inhibition and excitation along with appropriate sensory input. It is clear, for example, that sensory input from the vagus nerves has profound effects on the respiratory patterns generated by the brain stem. In the mature animal, experiments suggest that the respiratory rhythm does not originate from pacemaker cells, but it is a network property requiring interaction among neurons.

Despite this lack of a comprehensive neural model, there is more known about control of ventilation from peripheral sensors, which we will now discuss.

PERIPHERAL CHEMOSENSORS MODULATE RESPIRATION IN RESPONSE TO CHANGES IN P_{aO_2}, P_{aCO_2}, AND pH

The peripheral sensory organs that detect changes in arterial P_{O_2}, P_{CO_2}, and pH are located in specialized areas called the **carotid bodies** and the **aortic bodies**. The carotid bodies dominate the signal—the aortic bodies are of secondary importance. The ventilatory response to lowering P_{aO_2} is almost exclusively determined by the carotid bodies. Their anatomic location is shown diagrammatically in Figure 6.6.7.

PERIPHERAL ARTERIAL CHEMOSENSORS INCREASE FIRING RATES WITH INCREASED P_{aCO_2}, DECREASED PH, AND DECREASED P_{aO_2}

The carotid chemosensors respond to changes in blood pH and blood gases by altering the frequency of action potentials transmitted by the glossopharyngeal nerve (cranial nerve IX) to the brain stem. At any pH, increasing P_{aCO_2} causes an increase in the firing rates of these chemosensors (see Figure 6.6.8). The results of Figure 6.6.8 also show that, at constant P_{aCO_2}, increasing [H$^+$] increases the rate of firing of the carotid chemosensors.

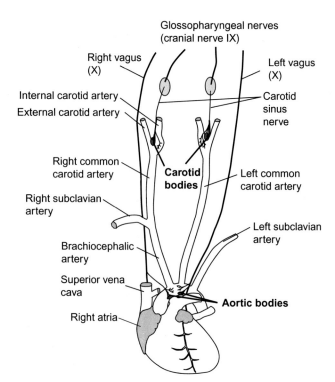

FIGURE 6.6.7 Anatomic location of the carotid and aortic bodies. The carotid bodies are located at the bifurcation of the common carotid to form the external and internal carotid arteries. These structures are about 5 mm long in the adult human and they are extremely well perfused with blood, some 20 mL per gram of tissue per minute. This is far above the metabolic needs of the tissue, so there is very little extraction of O$_2$ from the arterial blood. Thus the venous blood draining the carotid body has almost as high a P_{O_2} as the arterial blood. This high flow enables the carotid bodies to sense *arterial* gases and pH. The carotid body is innervated by the carotid sinus nerve, and sensory information travels to the brain stem through the **glossopharyngeal nerve**, cranial nerve IX. The aortic bodies are located in the aortic arch. They are innervated by the **vagus nerve**, cranial nerve X.

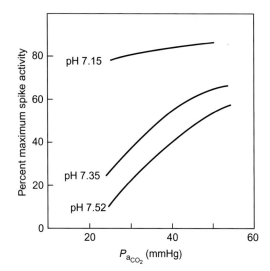 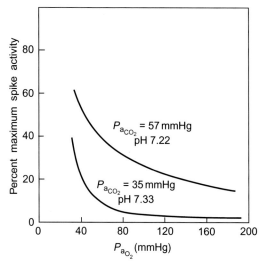

FIGURE 6.6.8 Response of the carotid chemoreceptors to pH, P_{aCO_2}, and P_{aO_2}. The response is measured as the frequency of action potentials. Increasing P_{aCO_2} increases the firing activity (left) at a constant pH, and decreasing pH increases firing activity at constant P_{aCO_2}. Decreasing P_{aO_2} increases the firing rate (right) at constant pH and P_{aCO_2}, but increasing P_{aCO_2} (with consequent acidemia) augments the response. Hypoxia (low P_{aO_2}) does not stimulate the chemoreceptors until P_{aO_2} falls below about 60 mmHg. *Adapted from M.P. Hlastala and A.J. Berger,* Physiology of Respiration, *Oxford University Press, 2001.*

Decreasing $P_{a_{O_2}}$ below about 60 mmHg also stimulates the firing of carotid chemosensors. Note that this is arterial P_{O_2}, which normally is about 100 mmHg. Thus there must be a fairly drastic drop in blood $P_{a_{O_2}}$ before the peripheral chemoreceptors become activated. The data of Figure 6.6.8 indicate that these receptors are more sensitive to pH and $P_{a_{CO_2}}$ than they are to $P_{a_{O_2}}$.

PERIPHERAL CHEMOSENSORS FOR $P_{a_{O_2}}$ ARE MORE IMPORTANT THAN THOSE FOR $P_{a_{CO_2}}$

Changes in ventilation caused by **hypercapnia** (increased $P_{a_{CO_2}}$) are only slightly reduced when the sensory input from the carotid bodies is cut, indicating that there are other sensors for $P_{a_{CO_2}}$. On the other hand, **in humans the carotid bodies are the only sensors for** $P_{a_{O_2}}$ and the hyperventilation that accompanies low $P_{a_{O_2}}$ is entirely due to input from the carotid bodies.

THE VENTILATORY RESPONSE TO INCREASED CHEMORECEPTOR FIRING RATE IS INCREASED VENTILATION

The chemoreceptor input travels over the glossopharyngeal nerve and the vagus nerve to the brain stem where it makes synapses with neurons in the DRG. These neurons are I neurons, and so increased rates of firing from the chemoreceptors cause increased inspiratory activity and increased pulmonary ventilation. This system makes a complete negative-feedback loop: hypercapnia increases chemoreceptor firing rate that increases inspiratory neuronal activity that increases ventilation that tends to reduce $P_{a_{CO_2}}$ by removing more CO_2 to the atmosphere, thereby reducing the original signal of increased $P_{a_{CO_2}}$. Similarly, acidemia increases chemoreceptor firing rate, leading to increased ventilation and, by lowering $P_{a_{CO_2}}$, the pH tends to normal by respiratory compensation (see Chapter 6.5). When $P_{a_{O_2}}$ falls, the chemoreceptors increase their firing rate, which increases ventilation which tends to restore $P_{a_{O_2}}$ toward normal. These negative-feedback loops are shown schematically in Figure 6.6.9.

CENTRAL CHEMOSENSORS PROVIDE THE MAJOR RESPONSE TO CHANGES IN $P_{a_{CO_2}}$

Although the peripheral chemoreceptors respond to acidemia and hypercapnia by increasing their firing rate, surgical removal reduces the ventilatory response only by about 10%. Thus, 90% of the response is due to some other process. This process resides in chemosensors located in the ventral medulla. These chemosensory zones have been localized by a variety of techniques including lesioning, electrical stimulation, and focal application of chemicals. Three zones are located in the ventral medulla just beneath its surface, as shown in Figure 6.6.10.

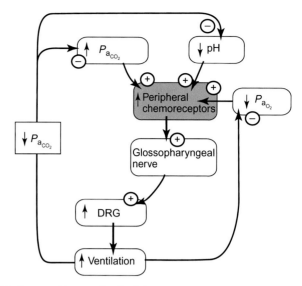

FIGURE 6.6.9 Negative-feedback loops in the control of ventilation by peripheral chemosensors. Chemosensors in the carotid bodies and aortic bodies respond to increased $P_{a_{CO_2}}$ by increasing their firing rate. This signal is carried to the DRG via the glossopharyngeal nerve and vagus nerve. These connect to inspiratory neurons in the DRG that increase the depth and frequency of breathing to increase pulmonary ventilation. This lowers the alveolar P_{CO_2} and hence arterial $P_{a_{CO_2}}$. Thus the loop acts to lower the original perturbation of hypercapnia. Similarly, decreased pH stimulates ventilation, which lowers $P_{a_{CO_2}}$ and therefore raises the pH toward normal. Decreased $P_{a_{O_2}}$ also stimulates the peripheral chemosensors, leading to increased ventilation and increased $P_{a_{O_2}}$.

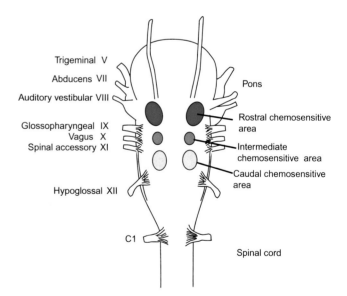

FIGURE 6.6.10 Location of the chemosensitive areas of the brain stem. The figure shows a ventral view of the brain stem. There appear to be three separate areas, one located rostrally, another caudally, and another intermediate between the two.

In experimental situations, the medulla can be perfused with solutions that vary the pH and P_{CO_2} separately by changing the $[HCO_3^-]$ according to the Henderson–Hasselbalch equation. In these cases, the hyperventilatory response (measured by the integrated response of the phrenic nerve) occurs when the pH is

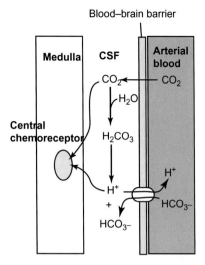

FIGURE 6.6.11 Connection between blood P_{CO_2} and CSF pH. CO_2 readily penetrates the blood–brain barrier whereas H^+ does not. During sustained acidosis, the CSF pH is initially depressed and then gradually returns to normal, suggesting the existence of some mechanism for regulating CSF pH independent of the pH of blood. Active pumping of HCO_3^- into the CSF (or of H^+ ions from CSF to blood) has been postulated as one possible mechanism for this regulation.

unchanged but P_{CO_2} is increased. It also occurs when the P_{CO_2} is unchanged and the pH is lowered. Thus it appears that the central chemoreceptors respond to either P_{CO_2} or the pH as independent stimuli. However, these experiments do not identify the stimulus at the surfaces of the central chemosensors.

The central chemoreceptors are functionally and structurally separated from blood by the **blood–brain barrier** (see Chapter 4.1). The interstitial fluid bathing the cells is separated from blood by a layer of cells which is composed of tightly joined **endothelial cells** and **astrocytes**. The **cerebrospinal fluid (CSF)** that fills the major brain cavities and the spaces between the meninges is made by **ependymal** cells that line these cavities but are especially prominent in the **choroid plexuses** within the ventricles of the brain. Ions such as H^+ and HCO_3^- do not easily cross the blood–brain barrier, whereas lipophilic materials such as CO_2 easily cross. Thus CO_2 in the blood crosses the blood–brain barrier and equilibrates with CO_2 in the CSF. There it hydrates to form carbonic acid, H_2CO_3, which then dissociates to form H^+ and HCO_3^-. Increasing blood P_{aCO_2} increases CSF P_{CO_2} which in turn increases CSF $[H^+]$ (see Figure 6.6.11). The chemoreceptors sense this decreased pH in the CSF and increase their firing rate, which in turn increases respiration. The cells that respond to changes in CSF composition have not yet been identified, and it is not yet clear whether they respond directly to local $[H^+]$, P_{CO_2}, or both.

THE BRAIN ADJUSTS THE $[HCO_3^-]$ OF THE CSF

Hypercapnia that causes a decreased pH of the CSF is followed by a gradual return of the CSF pH toward normal. Thus the brain possesses regulatory mechanisms that locally adjust disturbances in the CSF's pH. Nobody knows exactly how this is accomplished and several competing theories exist. The events depicted in Figure 6.6.11 are one way by which this regulation could occur. Changes in CSF $[HCO_3^-]$ occur slowly and the return to normal pH may take many hours or days. Ventilation may reflect differences in the sensory information provided by central and peripheral chemoreceptors.

VENTILATORY DRIVE INCREASES BY INTEGRATED RESPONSE TO ELEVATED P_{aCO_2}, METABOLIC ACIDOSIS, OR HYPOXIA

HYPERCAPNIA STIMULATES VENTILATORY DRIVE

Acute increase in P_{aCO_2} decreases arterial blood pH. Both elevated P_{aCO_2} and decreased blood pH stimulate the peripheral chemoreceptors to increase ventilatory drive. The elevated CO_2 crosses the blood–brain barrier, acidifies the CSF, and excites the central chemoreceptors to increase ventilatory drive. So in the acute phase of hypercapnia, increased ventilatory drive derives from both peripheral and central chemoreceptors. The hypercapnia cannot remain in the face of increased ventilation unless there is some problem with gas exchange such as occurs in emphysema.

In the chronic phase of hypercapnia, blood pH returns toward normal and so only the elevated P_{aCO_2} remains to stimulate the peripheral chemoreceptors. Similarly, the CSF regulates its pH so that this component of central chemoreceptor stimulation is reduced. The result is a continued increased ventilatory drive but with greatly reduced magnitude.

METABOLIC ACIDOSIS INCREASES VENTILATORY DRIVE

Acute metabolic acidosis decreases the pH of the arterial blood and strongly stimulates the peripheral chemoreceptors to increase ventilatory drive. The increased ventilatory drive results in decreased P_{aCO_2} and subsequent rise in plasma pH. This respiratory compensation of the metabolic acidosis occurs relatively fast. Since H^+ penetrates the blood–brain barrier very slowly, the reduced P_{aCO_2} produces a paradoxical alkalinization of the CSF with a *decrease* in the stimulation of central chemoreceptors. Thus the overall ventilatory drive provided by the acidosis stimulating the peripheral chemoreceptors is blunted by the reduction in central chemoreceptor drive.

In the chronic phase of metabolic acidosis, there is a gradual return of the CSF from its initially alkalemic state to a normal pH. Thus the CSF has a normal pH with a reduced P_{aCO_2} caused by the increased ventilatory drive provided by peripheral chemoreceptors. Here the ventilatory drive provided by the central chemoreceptors returns toward normal (see Figure 6.6.12).

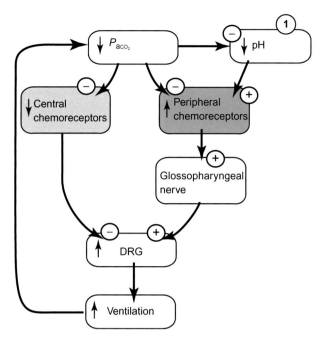

FIGURE 6.6.12 Feedback loops in ventilatory adjustment to metabolic acidosis. Start at 1 with the decreased pH. This stimulates the peripheral chemoreceptors, whose increased firing rates stimulate the DRG to increase ventilation. The increased ventilation decreases P_{aCO_2}, which decreases its tonic stimulation of peripheral chemoreceptors. The decreased P_{aCO_2} also raises the pH because the excreted CO_2 originates from H^+ combining with HCO_3^-. The decreased P_{aCO_2} alkalinizes the CSF, thereby reducing central ventilatory drive. Thus the increased ventilatory drive provided by the decreased pH is suppressed by the resulting decreased P_{aCO_2}.

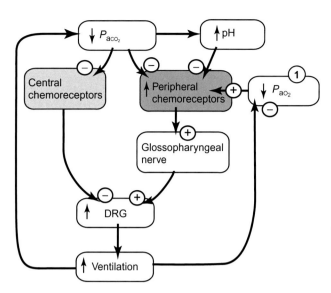

FIGURE 6.6.13 Feedback loops in ventilatory adjustment to hypoxia. Start at 1 with decreased arterial P_{O_2}. This stimulates only peripheral chemosensors to increase ventilatory drive. This reduces P_{aCO_2}, which also increases plasma pH. Both the reduced P_{aCO_2} and alkaline blood reduce peripheral ventilatory drive. In addition, the reduced P_{aCO_2} reduces central ventilatory drive.

HYPOXIA INCREASES VENTILATORY DRIVE

Recall that the peripheral chemoreceptors are the only sensors for P_{aO_2}. When P_{aO_2} falls below about 60 mmHg, the peripheral chemoreceptors are stimulated and ventilatory drive is increased. The increased ventilation causes P_{aCO_2} to fall, with a resulting alkalemia. The reduced P_{aCO_2} also alkalinizes the CSF, which reduces the ventilatory drive provided by the central chemoreceptors. The result is an attenuation of the ventilation stimulated by the peripheral chemoreceptors. This type of response would be seen by persons ascending to an altitude of about 3000 m (10,000 ft) where the P_{IO_2} is about 100 mmHg and P_{aO_2} is about 60 mmHg (see Figure 6.6.13).

After a few days at altitude, ventilation increases further. This is called **ventilatory acclimatization**. A progressive return of the CSF from alkaline to normal pH could explain this acclimatization, but the evidence does not support this mechanism. Because ventilatory acclimatization is ablated by removal of the peripheral chemoreceptors, it may be due to time-dependent changes in the sensitivity of the peripheral chemoreceptors.

Additional adaptations occur in persons who stay at altitude. Hypoxia stimulates secretion of **erythropoietin**, a glycoprotein hormone made in the kidneys. Erythropoietin stimulates the differentiation of uncommitted stem cells to begin forming erythrocytes. Thus hypoxia leads eventually to an increase in the number of circulating red blood cells, which increases the hematocrit and the hemoglobin concentration in blood, leading to increased O_2 content at any P_{aO_2}. This condition is called **polycythemia**. Other adaptations to altitude are increased levels of 2,3-DPG in the blood and a persistent hypocapnic alkalosis. The increased level of 2,3-DPG shifts the O_2 dissociation curve to the right, while the hypocapnic alkalosis shifts the curve to the left. Up to an altitude of about 4250 m (14,000 ft) the result is an O_2 dissociation curve close to that at sea level. At extreme altitudes, the left shift due to alkalemia wins out.

AIRWAY AND LUNG MECHANORECEPTORS ALTER BREATHING PATTERNS

Sensory receptors are found in a variety of places throughout the respiratory system and are responsible for a wide variety of behaviors. These include the following:

- **Sneezing**: Mechanical stimulation of the nasal passage results in the sneeze reflex, consisting of a strong inspiration followed by a rapid expiration with partially closed airways to increase airflow velocity. The reflex clears foreign material from the passages.
- **The diving reflex**: Water in the nose elicits a complex series of respiratory and cardiovascular responses including cessation of respiration, closure of the larynx, bronchoconstriction, bradycardia, and vasoconstriction of many vascular beds except those of the brain and heart.

- **The aspiration reflex**: Mechanoreceptors in the epipharynx initiates a series of strong, brief inspirations that dislodge materials from the epipharynx into the pharynx where they can be coughed up or swallowed.
- **The swallowing reflex**: Adult humans cannot breathe and swallow at the same time. Receptors in the pharynx cause cessation of respiration, closure of the larynx, and coordinated muscular contractions that move material from the oral cavity into the esophagus.
- **Cough reflexes**: Rapidly adapting receptors in the airway epithelia respond both to mechanical deformation and chemical irritation. These receptors initiate reflexes that include coughing, mucus secretion, and bronchoconstriction. All of these actions have the common theme of removing the offending irritant. Coughing creates a high velocity of airflow that helps eliminate foreign matter. Increasing mucus secretion helps trap the foreign matter. Bronchoconstriction decreases the diameter of the airways and therefore increases the velocity of airflow produced by the cough, so that foreign matter can be more easily expelled.
- **The Hering–Breuer inflation reflex**: The lower airways contain slowly adapting stretch receptors. Afferent sensory information from these stretch receptors travels over the vagus nerve to the brain stem where it inhibits inspiration by stimulating the neurons in the PRG. This is the Hering–Breuer inflation reflex. Inflation of the lungs stimulates the stretch receptors, which reflexly inhibit further inflation.
- **The pulmonary chemoreflex**: In addition to stretch receptors, the vagus nerve carries sensory inputs from C fibers, which are unmyelinated nerves with slow conduction velocities on the order of 2.5 m s^{-1}. C fibers in the bronchi appear to respond mainly to stretch whereas those fibers near the capillaries respond to exogenous and endogenous chemicals including **capsaicin**, the irritant in red pepper, **histamine**, **bradykinin**, **serotonin**, and **prostaglandins**. These chemicals elicit the pulmonary chemoreflex that includes apnea, bradycardia, and hypotension that is immediately followed by rapid, shallow breathing (tachypnea).

INCREASED RESPIRATION DURING EXERCISE MAY BE NEURAL AND MAY INVOLVE LEARNING

Everyday experience tells us that exertion is accompanied by increased pulmonary ventilation. What regulates ventilation during exercise? Exercise increases O_2 consumption, CO_2 production, and H^+ generation from metabolic acids such as lactic acid, acetoacetic acid, and β-hydroxybutyric acid. These changes would reduce $P_{a_{O_2}}$, raise $P_{a_{CO_2}}$, and produce acidemia. All of these would lead to increased ventilatory drive. However, none of these are altered during moderate exercise. Respiratory adjustments are so precise that $P_{a_{O_2}}$, $P_{a_{CO_2}}$, and blood pH are perfectly normal. At higher work rates, pH and $[HCO_3^-]$ may actually fall.

Passive movements of the limbs of experimental animals increase ventilation; direct stimulation of muscles through their nerves also stimulates ventilation; blockade of the sensory afferents from the muscles blocks the effect. Muscle mechanoreceptors and nociceptors make the largest contribution to the hyperpnea of exercise. These results suggest that muscles and joints notify the respiratory centers about their activity through sensory afferents, thereby stimulating ventilation.

Stimulation of motor systems in paralyzed animals can elicit ventilatory drive similarly to exercise. This suggests that the motor systems that drive muscles send collaterals to the respiratory system to simultaneously drive ventilation. This is a **feedforward** mechanism in which the motor system anticipates the respiratory demands of its commands and adjusts ventilation in advance of those demands rather than waiting for discrepancies of respiratory gases or pH to stimulate ventilation.

Another possible explanation for exercise hyperpnea is altered sensitivity of the central controllers to input from the peripheral and central chemoreceptors. Feedforward input from motor systems or from muscle afferents could alter the response of the brain stem respiratory neurons so that the same input from the chemoreceptors produces a larger ventilatory drive.

The respiratory response to exercise may be in part conditioned by previous bouts of exercise. The observation is that exercise involves an error-free adjustment of ventilation so that blood gases and pH are normal during exercise of varying intensities. A negative-feedback mechanism can achieve zero error only by having an infinite gain. Real negative-feedback systems ought to have residual error that drives the increased ventilation. In principle, the brain could learn to titrate respiration according to muscle afferent information or output from the motor cortex so as to prevent blood gas abnormalities. In this hypothesis, each bout of exercise teaches the nervous system how much ventilation is necessary to prevent disorders of blood gases or pH with each increment in intensity of exercise.

Clinical Applications: Sleep Apnea

Absence of airflow for 10 s or longer is **apnea**; reduction in tidal volume between 50% and 75% for at least 10 s is hypopnea. Both apnea and hypopnea that occur during sleep can be classified as **central**, **obstructive**, or **mixed**. Central apnea is caused by reduced respiratory effort, whereas obstructive apnea involves persistent effort during the period of the apnea. Mixed apneas begin with a central component that is followed by an obstructive component. Sleep apnea is defined as cessation of airflow for 10 s or longer with a frequency of at least 15 episodes per hour. Normal young adults typically have up to five apneas per hour during sleep.

Obstructive sleep apnea, OSA, is predominantly a male problem, affecting 10 times as many men as women, afflicting some 2–4% of the general population. Nearly all persons with OSA snore loudly and most complain of daytime sleepiness because their apnea wakes them repetitively during the night. The cause is generally airway obstruction in the oropharynx related in part to relaxation of the tongue and pharynx during sleep. The majority of persons with OSA are overweight. Increasing body weight positively correlates with pharyngeal resistance and weight reduction generally improves the condition.

Weight reduction in persons with OSA often causes significant clinical improvement. Since most people with OSA experience sleep apneas in the supine position, sleep posture modification often improves the condition. Antidepressant drugs such as the tricyclic antidepressants reduce REM sleep, which is the stage of sleep most often associated with OSA. A tongue-retaining orthodontic appliance moves the tongue forward, thereby improving the opening of the pharyngeal airway. Supplemental oxygen or nocturnal ventilation by tracheostomy tube or face mask eliminates nocturnal hypoxemia and hypercapnia. For patients who have tried these or other methods and have dangerous OSA, surgical remedies may help. These include tracheostomy, uvulopalatopharyngioplasty, in which soft tissues of the soft palate, uvula, and posterior lateral pharyngeal wall is removed, and removal of the tonsils or adenoids.

SUMMARY

Ventilation of the lungs is controlled by the central nervous system. Voluntary control arises from the cerebral cortex, and involuntary control arises from the medulla and pons. These send inspiratory signals out to the diaphragm over the phrenic nerve, arising from C3–C5 spinal segments, and to the internal intercostals over thoracic segments. The pons contains a PRG that switches off inspiration and an apneustic center in the lower pons that sends signals to prevent switching off of inspiration. Sensory information coming into the medulla along the vagus nerves also helps switch off inspiration. The medulla contains DRG and VRG. The DRG contains primarily inspiratory neurons that fire action potentials during inspiration. The DRG receives sensory information from chemoreceptors in the carotid body over the glossopharyngeal nerve (CN IX) and from the aortic bodies over the vagus nerve (CN X). The VRG contains both inspiratory and expiratory neurons. A region in the rostral VRG may be the origin of the respiratory rhythm. There are multiple connections among the DRG, VRG, and spinal motor neurons that drive the respiratory muscles.

The peripheral chemoreceptors respond to $P_{a_{O_2}}$, $P_{a_{CO_2}}$, and pH. Increasing $P_{a_{CO_2}}$ or decreasing pH increases the frequency of action potentials of the peripheral chemoreceptors, which in turn increases ventilation by stimulating the DRG. The peripheral chemoreceptors are the only sensors for $P_{a_{O_2}}$. The ventral medulla contains central chemoreceptors that respond to the pH of the CSF. Metabolic acids and H^+ itself cannot easily cross the blood–brain barrier, whereas CO_2 easily crosses. Thus elevated $P_{a_{CO_2}}$ acidifies the CSF, causing increased ventilation. The CSF possesses mechanisms to regulate its pH independently of blood. In chronic conditions, the ventilatory drive provided by CSF acidosis gradually disappears, so that only the peripheral chemoreceptors drive ventilation. Increased ventilation caused by hypoxemia, resulting from ascent to altitude, e.g., decreases $P_{a_{CO_2}}$, which alkalinizes the CSF. This reduces the central drive for ventilation. After a few days at altitude the ventilatory drives increase further. In metabolic acidosis, the decreased pH increases ventilation due to excitation of the peripheral chemoreceptors. The increased ventilation lowers $P_{a_{CO_2}}$, which alkalinizes the CSF and reduces central chemoreceptor-driven ventilation.

REVIEW QUESTIONS

1. The actor Christopher Reeve suffered a high cervical spinal cord injury after falling while horseback riding. What is the consequence of this injury with respect to the control of breathing?
2. In metabolic acidosis, trace the signal from the peripheral chemoreceptor to its final effect on ventilation. Using the alveolar ventilation equation, predict in what direction this alters alveolar P_{CO_2}. In what direction does this alter arterial P_{CO_2}? Using the Henderson–Hasselbalch equation, predict the effect of this change on plasma pH. How does the change in arterial P_{CO_2} affect CSF pH? What effect does this have on central ventilatory drive?
3. Repeat review question 2 for metabolic alkalosis.
4. Why does the acute ventilatory response to metabolic acidosis differ from the chronic response?
5. For involuntary control of ventilation, cutting the two vagus nerves produces deeper ventilation at lower frequency (see Figure 6.6.3, top right trace). Why?
6. What provides increased ventilatory drive during exercise?

6.2 Problem Set
Gas Transport and pH Disturbances

1. A. An elderly woman has a hemoglobin concentration of 10 g dL^{-1}. Her O$_2$ dissociation curve is normal (when expressed as S_{O_2}). Assume that her resting O$_2$ consumption (Q_{O_2}) is within normal limits, 225 mL min^{-1}, and that Q_a, the cardiac output, is 4.5 L min^{-1}; P_{aO_2} = 95 mmHg is normal.
   - A.1 What is the total oxygen content (in mL dL^{-1}) of her arterial blood?
   - A.2 What is the total oxygen content of her mixed venous blood?
   - A.3 What is the S_{O_2} of venous blood?
   B. Assume that the S_{O_2} of venous blood in this woman is the normal 75%.
   - B.1 What would be the P_{vO_2}?
   - B.2 What would the oxygen content of her mixed venous blood be?
   C. If arterial blood has P_{aO_2} = 95 mmHg as in part A and mixed venous blood has S_{O_2} = 75% (part B), how much O$_2$ would be extracted by the tissues per L of blood?
   D. Given the [O$_2$]$_a$ − [O$_2$]$_v$ from part C, what must Q_a be for tissue extraction to match her Q_{O_2} of 225 mL min^{-1}?
   E. According to your calculations, the response to anemia in principle could be to keep cardiac output, Q_a, constant and let P_{vO_2} fall, or to raise Q_a to keep P_{vO_2} constant. With your new knowledge of respiratory and circulatory physiology, which do you think happens?

2. The circulating blood volume typically is 7% of body weight. Assume that a person weighs 80 kg and that the hemoglobin concentration is 14.5 g dL^{-1}.
   A. What is the total amount of hemoglobin in the circulation?
   B. What is the maximum amount of O$_2$ that could be in the blood if *all* the blood has a P_{O_2} = 100 mmHg?
   C. If Q_{O_2} = 275 mL min^{-1}, how long would this O$_2$ last if the person stopped breathing? (Assume that Q_{O_2} is constant. In reality, it would decrease as P_{aO_2} falls.)
   D. If all the blood had a P_{CO_2} = 46 mmHg, estimate how much total CO$_2$ it would contain. (Include dissolved, carbamino Hb, and HCO$_3^-$) Hint: Use the numbers in Table 6.PS2.1.
   E. If R = 0.8, how much would the total CO$_2$ change during the time calculated in Part C?

3. Table 6.PS2.2 gives the [Hb·O$_2$] in blood containing 15 g Hb dL^{-1}. Perform a Hill plot on these data to obtain h and the P_{O_2} at 50% saturation. This is called the P_{50}, the oxygen partial pressure at one-half maximal saturation. The theoretical basis for the Hill plot is described below. To do this, define a saturation fraction, v, given as

   [6.PS2.1] $$v = \frac{[\text{Hb}\cdot\text{O}_2]}{[\text{Hb}\cdot\text{O}_2]_{max}}$$

   From Eqn (6.4.2), this equation is

   [6.PS2.2] $$v = \frac{[\text{O}_2]^h}{K + [\text{O}_2]^h}$$

   We can write from this equation that

   $$1 - v = 1 - \frac{[\text{O}_2]^h}{K + [\text{O}_2]^h}$$

   [6.PS2.3] $$= \frac{K + [\text{O}_2]^h}{K + [\text{O}_2]^h} - \frac{[\text{O}_2]^h}{K + [\text{O}_2]^h}$$

   $$= \frac{K}{K + [\text{O}_2]^h}$$

   Taking the ratio of Eqn (6.PS2.2) to Eqn (6.PS2.3), we get

   [6.PS2.4] $$\frac{v}{1-v} = \frac{[\text{O}_2]^h}{K}$$

   Taking the logarithm on both sides, we get

   [6.PS2.5] $$\log\frac{v}{1-v} = -\log K + h \log [\text{O}_2]$$

   This last equation forms the basis of the Hill plot with K as a dissociation constant rather than as an association constant. Plotting log $(v/1-v)$ against log [O$_2$] allows calculation of h as the slope. Since [O$_2$] = αP_{O_2}, we can plot log P_{O_2} and get the same h. When $v = ½$, log $(v/1-v) = 0$ and at this point $P_{O_2} = P_{50}$.

TABLE 6.PS2.1 Transport of O_2 and CO_2 to and from the Tissues

Variable	Arterial Blood	Mixed Venous Blood	A − V Difference	Transport = $Q_a \times$ (A − V)
S_{O_2} (%)	98	74	24	
P_{O_2} (mmHg)	95	40	55	
Plasma $[O_2]$[a]	0.16 mL dL^{-1}	0.07 mL dL^{-1}	0.09 mL dL^{-1}	5 mL min^{-1}
RBC $[O_2]$[a]				
Dissolved	0.13 mL dL^{-1}	0.05 mL dL^{-1}	0.08 mL dL^{-1}	4 mL min^{-1}
Bound	19.8 mL dL^{-1}	15.0 mL dL^{-1}	4.8 mL dL^{-1}	240 mL min^{-1}
Total $[O_2]$	20.1 mL dL^{-1}	15.1 mL dL^{-1}	5.0 mL dL^{-1}	250 mL min^{-1}
pH	7.4	7.37		
P_{CO_2} (mmHg)	40	46		
Plasma $[CO_2]$[a]				
Dissolved	1.53 mL dL^{-1}	1.76 mL dL^{-1}	−0.23 mL dL^{-1}	−11 mL min^{-1}
HCO_3^-	30.28 mL dL^{-1}	32.50 mL dL^{-1}	−2.22 mL dL^{-1}	−111 mL min^{-1}
RBC $[CO_2]$[a]				
Dissolved	0.97 mL dL^{-1}	1.11 mL dL^{-1}	−0.14 mL dL^{-1}	−7 mL min^{-1}
HCO_3^-	12.6 mL dL^{-1}	13.8 mL dL^{-1}	−1.20 mL dL^{-1}	−60 mL min^{-1}
Carbamino Hb	1.68 mL dL^{-1}	1.89 mL dL^{-1}	−0.21 mL dL^{-1}	−11 mL min^{-1}
Total $[CO_2]$	47.1 mL dL^{-1}	51.1 mL dL^{-1}	−4.0 mL dL^{-1}	200 mL min^{-1}

The various forms of dissolved and bound O_2 and CO_2 are partitioned into the parts contributed by plasma and the red blood cells for each form of gas. Normal values assumed were as follows: 15 g Hb dL^{-1}; cardiac output (Q_a) of 5 L min^{-1}; temperature of 37°C; hematocrit of 0.45; respiratory quotient, $R = 0.8$; resting O_2 consumption of 250 mL min^{-1}.

[a]Values are given in terms of dL of blood, not the fluid indicated. For example, the plasma $[O_2]$ is *not* the concentration per dL of plasma; it is the contribution that plasma makes to the total $[O_2]$ per dL of *blood*. The RBC $[O_2]$ is not the concentration of O_2 per dL of RBCs; it is the contribution that the RBCs make to the total $[O_2]$ per dL of blood.

Adapted from R. Klocke, Carbon dioxide transport, in A.P. Fishman, ed., Handbook of Physiology, Section 3, Volume IV, American Physiological Society, Washington, DC, 1986; C. Geers and G. Gros, Carbon dioxide transport and carbonic anhydrase in blood and muscle, Physiological Review **80**:681–715, 2000.

TABLE 6.PS2.2 Oxygen Saturation of Hemoglobin

P_{O_2} (mmHg)	[Hb · O_2] (mL dL^{-1})	Fractional Saturation, v	Log [v/(1 − v)]	Log P_{O_2}
10	1.93			
20	6.51			
30	11.51			
40	15.01			
50	17.10			
60	18.27			
70	18.91			
80	19.28			
90	19.40			
100	19.58			
110	19.66			
120	19.74			

4. In the table, calculate the missing values using the Henderson–Hasselbalch equation and identify the acid–base condition.

Part	pH	P_{aCO_2} (mmHg)	$[HCO_3^-]$ (mM)	Condition
A	7.40	40		
B		35	24.13	
C	7.14		33.0	
D	7.60	40		
E	7.31		20.0	

5. Consider the 80-kg person in Problem no. 2. Assume the person has a hematocrit of 0.40. If we infuse this person with 100 mL of 150 mM $NaHCO_3^-$, assuming instant equilibration with the plasma, exchange across the red blood cells is slow, and before respiratory or renal compensation can occur, what would happen to blood pH? If we could somehow wave a magic wand and remove the equivalent amount of HCO_3^-, what would happen to plasma pH?

6. Five samples of whole blood were equilibrated at 37°C with gas mixtures containing various percentages of carbon dioxide. The P_{CO_2} in blood was calculated by assuming that it was equal to the P_{CO_2} in the gas mixture. Plasma was removed from the erythrocytes anaerobically (why?) And the plasma was analyzed for the total carbon dioxide. Given the [total CO_2] and P_{CO_2} in the table below, calculate $[CO_2]$ and $[HCO_3^-]$. Hint: [Total CO_2] is the sum of all forms of CO_2 in the sample; for free $[CO_2]$ use the solubility and pH. Graph the results on the pH – HCO_3^- diagram as the **normal buffer line**.

Sample	P_{CO_2} (mmHg)	[Total CO_2] (mM)	$[CO_2]$ (mM)	$[HCO_3^-]$ (mM)	pH
1	85.1	32.3			
2	46.5	26.4			
3	40.0	25.2			
4	33.3	24.0			
5	23.3	21.5			

7. During exercise, both Q_{O_2} and Q_{CO_2} increase as well as cardiac output, Q_a. Suppose that during exercise Q_{O_2} increases to 2000 mL min^{-1}, and the respiratory quotient remains at 0.8, and that Q_a increases to 18 L min^{-1}. Assume that P_{aO_2} remains at 95 mmHg and P_{aCO_2} is 40 mmHg and that blood [Hb] = 15 g%.
   A. What is the total arterial content of O_2?
   B. What is the venous content of O_2?
   C. What is the P_{vO_2}?
   D. What is the Q_{CO_2}?
   E. Assuming an arterial pH of 7.4, what is the total CO_2 content of arterial blood? (Hint: See the table in Problem no. 4.)
   F. Can you determine the total CO_2 content of venous blood?
   G. Can you determine the P_{vCO_2}?
   H. Determine the new alveolar ventilation from the alveolar ventilation equation.
   I. Calculate the predicted P_{aO_2} from the alveolar gas equation.

8. In the Hill equation, $K = 4550$ mmHg$^{2.62}$ and $h = 2.62$ (the exponent of mmHg is 2.62 because it adds with P_{O_2} in the Hill equation. If P_{O_2} is in units of mmHg, $P_{O_2}^{2.62}$ is in units of mmHg$^{2.62}$). Note that $K = 4550$ is $24.9^{2.62}$. Calculate the P_{O_2} for the following saturation fractions. Hint: See the equations in Problem no. 3. $S_x = [Hb \cdot O_2]x/[Hb \cdot O_2]_{max}$
   A. S_{25}
   B. S_{50}
   C. S_{75}
   D. S_{98}
   E. Repeat the calculations for A–D if h were equal to 1. In this case, $K = 24.9$ mmHg (why?)

9. For the phosphate buffer system, calculate the concentrations of $[H_2PO_4^{-1}]$ and $[HPO_4^{-2}]$ at pH 7.4 if the total phosphate concentration is 1 mM. Hint: Use the Henderson–Hasselbalch equation and the K_D given in Eqn (6.5.14).

10. Using the Henderson–Hasselbalch equation, plot the normal isobar on the pH – HCO_3^- diagram for $P_{aCO_2} = 40$ mmHg.

11. Venous blood has a higher P_{CO_2} than arterial blood. As a consequence, it is slightly more acidic.
    A. If **arterial** plasma has a $P_{CO_2} = 40$ mmHg and a pH = 7.4, what is the $[HCO_3^-]$? Convert this to mL of CO_2 per dL of plasma (STPD).
    B. If **venous** plasma has a $P_{CO_2} = 46$ mmHg and a pH = 7.37, what is its $[HCO_3^-]$? Convert this to mL of CO_2 per dL of plasma (STPD).
    C. From the results in A and B, calculate the separate contributions of venous and arterial plasma HCO_3^- to the total blood CO_2 (sum of dissolved CO_2, HCO_3^-, and carbamino Hb), **in mL CO_2 per dL of blood**, using a hematocrit of 0.45.
    D. The intracellular pH of the red blood cell in the **arteries** is 7.25. Calculate the $[HCO_3^-]$ in the red blood cell cytoplasm assuming equilibration of 40 mmHg P_{CO_2} across the RBC membrane. Convert this to mL of CO_2 per dL of RBC ICF (STPD).
    E. The intracellular pH of the red blood cell in the veins is 7.23. Calculate the $[HCO_3^-]$ in the red blood cell cytoplasm assuming equilibration of 46 mmHg P_{CO_2} across the RBC membrane. Convert this to mL of CO_2 per dL of RBC ICF (STPD).
    F. From the results of D and E, calculate the separate contributions of venous and arterial blood RBC HCO_3^- to the total blood CO_2 (sum of dissolved CO_2, HCO_3^-, and carbamino Hb), **in mL CO_2 per dL of blood**, using a hematocrit of 0.45 and RBC water content of 72% of its volume.

G. From the results of C and F, calculate the total CO_2 as HCO_3^- in arterial and venous blood, in mL dL^{-1}. Assuming a cardiac output, $Q_a = 5$ L min^{-1}, calculate the delivery of CO_2 from tissues to lungs in the form of HCO_3^-.

12. The solubility of CO_2 in water at 37°C is 0.0747 mL dL^{-1} mmHg^{-1} (see Table 6.3.3). Given a hematocrit of 0.45, a RBC water content of 72% of its volume, and a water content of 93% of the plasma volume:
   A. What is the dissolved CO_2 content of arterial and venous blood? ($P_{aCO_2} = 40$ mmHg and $P_{vCO_2} = 46$ mmHg). This should be in units of mL CO_2 per dL.
   B. If $Q_a = 5$ L min^{-1}, calculate the delivery of CO_2 from tissues to lungs in the form of dissolved CO_2.

13. Assume that the [carbamino Hb] is 0.00123 [Hb][P_{CO_2}] where [carbamino Hb] is in units of mM, [Hb] is in g dL^{-1} **of red blood cell volume**, and P_{CO_2} is in mmHg. The Hct = 0.45.
   A. Calculate the concentration of carbamino Hb in arterial and venous blood given $P_{aCO_2} = 40$ mmHg and $P_{vCO_2} = 46$ mmHg, and [Hb] = 15 g dL^{-1} **of whole blood**; give this in mL dL^{-1}.
   B. If $Q_a = 5$ L min^{-1}, calculate the delivery of CO_2 from tissues to lungs in the form of carbamino Hb.

14. Add up the delivery of CO_2 from tissues to lungs in Problems 11–13. Is this reasonable compared to Q_{CO_2}?

15. You are approaching your final exam in Quantitative Physiology and you are nervous and find yourself hyperventilating. Your tidal volume is now 0.6 L and your breathing frequency is 20 min^{-1}. Assume your dead space volume is 150 mL. Normally your tidal volume is 0.5 L, breathing frequency is 12 min^{-1}, and P_{aCO_2} is 40 mmHg.
   A. What is your pulmonary ventilation?
   B. What is your alveolar ventilation?
   C. If your Q_{CO_2} has not changed from 200 mL min^{-1}, what is your new P_{aCO_2}?
   D. Based on the normal buffer line in Figure 6.5.3, estimate your plasma pH.
   E. Name your acid–base status.

UNIT 7

RENAL PHYSIOLOGY

Body Fluid Compartments 7.1

Learning Objectives

- Calculate the volume of distribution of a material based on the dilution principle
- Define the term "marker" for volumes of distribution
- List one marker for plasma, extracellular fluid, and total body water
- Calculate the ICF volume from total body water and ECF
- Calculate ISF volume from ECF and plasma volume
- Define "lean body mass"
- Give approximate values for plasma, ISF, and ICF volumes as a percent of LBM
- Describe how ISF and plasma exchange materials
- Describe how ECF and ICF equilibrate osmotic pressure
- Using Darrow–Yannet diagrams, predict the consequences of adding or subtracting water, salt, or saline from the ECF on fluid compartment size and osmolarity
- Describe how changes in plasma can alter all body fluid compartments' volume and composition

FICK'S DILUTION PRINCIPLE ALLOWS DETERMINATION OF BODY FLUID COMPARTMENTS

The physiological regulation of the amount and distribution of water in the body is critically important. We have powerful mechanisms for insuring its proper regulation including thirst and other behavioral responses and the operation of our renal systems. Before we discuss this in detail, we need to know something about how much water is there, and where it is.

The amount of water in any space can be measured by **Fick's dilution principle**. In this method, a known amount of a measurable material is injected into a person and, after the material has had time to become evenly distributed, the concentration of the material in an aliquot of the plasma is measured. The volume of distribution is calculated from the simple formula:

[7.1.1] $$\text{Volume} = \frac{\text{amount}}{\text{concentration}}$$

The volume calculated by this method is the **volume of distribution** of the material that was injected. Different materials have access to different compartments. **Deuterium oxide**, for example, is a chemical nearly identical to water and it distributes itself according to the **total body water (TBW)**. **Inulin** is a polymer of fructose that readily crosses capillary walls but cannot enter cells. Thus it distributes itself through, and is a marker for, the **extracellular volume**. Evans' blue dye is substance that binds to plasma proteins and is restricted to the space occupied by these proteins. On the time scale of these measurements, Evans' blue dye is restricted to the plasma because the plasma proteins largely do not leave the vasculature. Thus Evans' blue dye marks the **plasma volume**.

The calculation of the volume of distribution given in Eqn (7.1.1) is valid only if the amount of marker material in the volume of distribution is known at the time at which the concentration is measured, and only if the marker is homogeneously distributed in that volume. In some cases, some of the marker may be lost to the urine, gastrointestinal tract, skin, or lungs during the time during which equilibration takes place. These losses must be accounted for in order for the volume of

---

EXAMPLE 7.1.1 Determine the TBW from the Distribution of D_2O

Seven grams of D_2O was injected into a healthy 70-kg adult (100 mg kg^{-1} body weight). After a 2-hour period of equilibration, the concentration of D_2O in a sample of plasma was measured by using the isotope ratio ($^2H/^1H$) determined in a mass spectrometer. Calculations showed the [D_2O] in the plasma was

$$[D_2O] = 0.0166 \text{ g dL}^{-1}$$

During the 2 hours, some of the D_2O was lost in the urine, lungs, and skin. Normally these losses are not large, averaging about 0.4% over a 2-hour period. If we assume the losses are 0.4%, we can calculate the volume of distribution of D_2O at the time of measurement as

Volume = 7 g (1 − 0.004)/0.0166 g dL^{-1}
= 6.972 g/0.0166 g dL^{-1} = 420 dL = **42.0 L**

This is the volume of distribution of D_2O, taken as an estimate of the **TBW**.

distribution to be accurately determined. Example 7.1.1 illustrates this for the determination of TBW.

INULIN MARKS THE EXTRACELLULAR FLUID; EVANS' BLUE DYE MARKS PLASMA

If we inject our test person with a solution of **inulin** instead of D_2O, we find that the plasma concentration of inulin rapidly falls from its initial value, losing about 64% of its initial value within 2 hours. The inulin is rapidly excreted in the urine and the plasma inulin concentration does not reach a stable value with a single injection. The calculation of the volume of distribution of inulin by Eqn (7.1.1) is valid only if the amount of inulin in the volume of distribution is known at the time the concentration is determined, and only if the distribution of inulin in that volume is homogeneous. We can get around this problem by infusing a solution of inulin at a constant rate to reach a steady-state plasma concentration at which the rate of inulin infusion exactly balances the rate of inulin excretion in the urine (see Figure 7.1.1). When this steady state is reached, the infusion is stopped and the bladder is emptied. From then on, the plasma concentration of inulin falls as inulin is excreted in the urine. We determine the total amount of inulin in the body at the time of steady-state inulin concentration by collecting all of the urine during the next 4–6 hours and measuring its volume and the urinary concentration of inulin. Eqn (7.1.1) can then be used to calculate the volume of distribution of inulin. In this case, the amount of inulin is experimentally determined as the total amount excreted after stopping the infusion, and its concentration is the steady-state concentration. When this procedure is performed, the volume of distribution of inulin is found to be about 14 L or about one-third of the TBW.

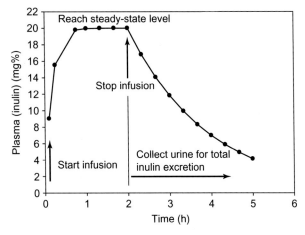

FIGURE 7.1.1 Determination of the volume of distribution of inulin by infusing an inulin solution continuously until a steady-state plasma concentration is reached. At this point, the bladder is emptied and the infusion is stopped. All urine is then collected for 4–6 hours. During this time all of the inulin in the body is excreted. This can be determined by measuring the volume of the urine and its concentration of inulin. Then Eqn (7.1.1) can be used to determine the volume of distribution.

Now suppose that we inject a solution of Evans' blue dye. In this case, the clearance of Evans' blue dye from the plasma is much slower than that of inulin, and some 4–6% of the material is lost from the plasma each hour. If we measure the plasma concentration of the dye 10 minutes after injection (using spectrophotometry and the absorbance of plasma at 627 nm, subtracting the absorbancy at 627 nm of a plasma sample taken before injection of dye), there will be sufficient time for equilibration of the dye within its volume of distribution without large losses of material. When we calculate the volume of distribution under these circumstances, we find that Evans' blue dye distributes in a volume of about 3.5 L or about 5% of the body weight.

THE MAIN FLUID COMPARTMENTS ARE THE *INTRACELLULAR* COMPARTMENT, THE *INTERSTITIAL* COMPARTMENT, AND THE *PLASMA*

Table 7.1.1 shows the three markers we have discussed here and their volume of distribution. The volumes of distribution for D_2O, inulin, and Evans' blue are different because only certain fluid compartments are accessible to these materials. D_2O distributes itself like water and it is a *marker* for the total body water or TBW. Inulin is a polymer of fructose that can cross the capillary wall but cannot penetrate cells. Thus the intracellular volume is inaccessible to inulin, but inulin can get just about everywhere else. Inulin is a *marker* for the extracellular fluid compartment or ECF. Evans' blue is a dye that binds to plasma proteins that are confined to the intravascular volume, the volume contained in the closed circulatory system. Thus Evans' blue dye is a *marker* for the volume of the plasma, the fluid contained within the circulatory system that is not within the cells of the blood. These relationships are shown in Figure 7.1.2. One of the main fluid compartments is not directly measured using these markers. This is the **interstitial fluid (ISF)** that lies between the cells and the vasculature.

Clearly, the volume of fluid that is part of the TBW but that is not extracellular is the intracellular fluid (ICF). From Figure 7.1.2, it is clear that we can calculate the ICF volume as

TABLE 7.1.1 Volumes of Distribution of D_2O, Inulin, and Evans' Blue Dye

Marker	Volume of Distribution (L)	Percent of Body Weight	Fluid Compartment
D_2O	42	60	Total body water
Inulin	14	20	Extracellular fluid
Evans' blue dye	3.5	5	Plasma

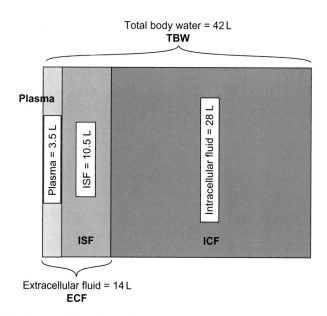

FIGURE 7.1.2 The relative volumes of distribution of D_2O, which marks the total body water, inulin, which marks the extracellular fluid, and Evans' blue, which marks the plasma.

$$[7.1.2] \qquad ICF = TBW - ECF$$

where ICF is the intracellular fluid volume, TBW is the total body water, and ECF is the extracellular fluid volume. In this case, the ICF is $42 - 14$ L $= 28$ L. Most of the body's water is inside the cells that make up the body. Figure 7.1.2 also makes it clear that there is some fluid that is outside the cells but is not contained within the vasculature. This fluid is called the **interstitial fluid**. From Figure 7.1.2, we can see that we can calculate the ISF volume as

$$[7.1.3] \qquad ISF = ECF - plasma$$

In the case that we consider here, the ISF is $14 - 3.5$ L $= 10.5$ L.

The three fluid compartments shown in Figure 7.1.2, the **plasma**, the **ISF**, and the **ICF**, comprise the three major fluid compartments of the body. There are a variety of smaller compartments such as the **intraocular fluid**, **cerebrospinal fluid**, **synovial fluid**, and **gastrointestinal secretions**. These smaller compartments are sometimes called **transcellular fluid compartments** because they are separated from the plasma by a cellular layer other than the capillaries. Although these compartments usually are smaller than the major fluid compartments, sometimes their volume can become much larger, and there are pathological conditions that can be associated with each of them. Examples of these include **glaucoma**, **hydrocephaly**, and **diarrhea**. Despite their importance, we will not discuss them further.

THE TBW VARIES WITH BODY COMPOSITION

Most of the soft tissues are 70–80% water, but bone and fat contain little water (see Table 7.1.2). Many different tissues contribute to overall body composition and the mixture of these tissues leads to the normal

TABLE 7.1.2 Percent Water Composition by Weight of Different Tissues

Tissue	Percent Water by Weight
Muscle	76
Blood	83
Brain	75
Liver	68
Bone	22
Fat	10

body composition of 60% water by weight. This number is a generalization and clearly does not pertain to all people because people differ considerably in their bodily composition.

In the early 1940s, motivated in part by standards used for admission to the US armed forces, Albert Behnke proposed that the body could be partitioned into two parts: (1) **lean body mass, LBM**, and (2) excess fat. The LBM included "essential fat" that was necessary for health and included the fat in the brain, spinal cord, bone marrow, and internal organs. Originally the essential fat was set at 10% of body weight, but it has since been revised to about 3%. The LBM is typically 73% water by weight. This percentage is a consequence of the definition of LBM. Thus the LBM of any individual can be calculated from the TBW as

$$[7.1.4] \qquad \text{Lean body mass} = \frac{TBW}{0.73}$$

The "excess body fat" is defined as the body weight in excess of the LBM. We calculate it as

$$[7.1.5] \qquad \text{Excess body weight} = \text{body weight} - LBM$$

The criterion for obesity is somewhat arbitrary. Where does one draw the line between obese and "overfat" in a continuous distribution? A common **definition of obesity is body fat in excess of 20% of total body weight**.

WATER COMPOSITION OF THE LBM VARIES WITH AGE AND SEX

Aging is a gradual process of desiccation. This begins soon after birth. The LBM of the neonate is 81% water; that of the adult is 73% water, as described above. There appears to be little change in the composition of the LBM from early adulthood to middle age. The change in relative water content is additive to the gradual increase in body fat seen in many people with age, so that the TBW gradually decreases with age when expressed as a percentage of body weight.

Generally speaking, females have less TBW when expressed as a percentage of body weight. This is generally attributed to the greater contribution of body fat to

	Cell	Interstitial fluid	Plasma
Na^+	10	142	142
K^+	145	4	4
Ca^{2+}	0.0002	2.8	5
Mg^{2+}	1	1.4	1.5
Total cations	156	150.2	152.5
HCO_3^-	10	27	24
Proteins	60	4	18.5
Phosphates	50	1.2	2
Cl^-	5	111	100.5
Other anions	31	7.0	7.5
Total anions	156	150.2	152.5

FIGURE 7.1.3 Composition of the plasma, interstitial fluid, and typical cell fluid. All values are in milliequivalents per liter. The equivalent is a chemical unit meaning the amount of something required to remove or supply 1 mol of H^+ ions, or supply or remove 1 mol of electrons in a redox reaction. The concentration unit of $mEq\ L^{-1}$ is related to $mmol\ L^{-1}$ by the charge on an ion: $mEq\ L^{-1} = |z| \times mM$, where z is the valence (+/− integer) for the ion. Here $z = mEq\ mmol^{-1}$. Thus 1 mM $Ca^{2+} = 2\ mEq\ L^{-1}$.

the total body weight rather than any difference in the composition of the LBM.

THE FLUID COMPARTMENTS CORRESPOND TO ANATOMIC COMPARTMENTS

The three major fluid compartments exist in well-defined anatomical regions that have barriers to the direct transfer of material. The plasma is separated from the ISF by the layer of cells that forms the capillary walls. The ISF is separated from the cells by the aggregate of the membranes of the cells of the body. These barriers are not impervious, however, and exchange of both solutes and water across these barriers is required for the continued health of the cells. As an example, the capillary wall effectively restricts transfer of plasma proteins from the plasma to the interstitial space, whereas most small molecular weight solutes and water can cross the capillary wall fairly easily. The result is that the ISF is low in protein compared to the plasma, but otherwise it is nearly identical to plasma. However, not all capillary beds are identical (see Chapter 5.10) and the interstitial fluid composition varies among the tissues. The approximate composition of the fluid compartments is shown in Figure 7.1.3.

BODY FLUIDS OBEY THE PRINCIPLE OF MACROSCOPIC ELECTRONEUTRALITY

Because electrical forces are so large, separation of charges within an electrolytic solution is not possible because the ions would quickly move to neutralize charge separation. The practical consequence of this is that the sum of all positive charges in a solution must be equal to the sum of all the negative charges:

$$[7.1.6] \qquad \sum_i C_i^+ = \sum_i C_i^-$$

where i indicates each of the chemical species in the solution. This principle is called **macroscopic electroneutrality**. Its application is evident in Figure 7.1.3 in which the total cation concentration is the same as the total anion concentration. It is possible for there to be local separation of charge, but the fraction of charge separated in this way is very small and not noticeable on the macroscopic scale. Separation of charge across a membrane produces the membrane potential, and this is extremely important, but the imbalance in charge on the macroscopic scale is tiny.

THE GIBBS–DONNAN EQUILIBRIUM ARISES FROM UNEQUAL DISTRIBUTION OF IMPERMEANT IONS

The major difference in composition between plasma and ISF (see Figure 7.1.3) is that the plasma contains much higher concentrations of large, negatively charged proteins. This difference also produces differences in the concentration of small diffusible ions. To analyze this situation, we begin with a hypothetical example shown in Figure 7.1.4.

We can write the free energy change for Na^+ transfer across the membrane

$$[7.1.7] \qquad \begin{aligned} \Delta \mu_{Na_o \Rightarrow Na_i} &= \mu_{Na_i} - \mu_{Na_o} \\ &= \mu^0 + RT \ln[Na^+]_i + z\Im \psi_i \\ &\quad - \mu^0 - RT \ln[Na^+]_o - z\Im \psi_o \\ &= RT \ln \frac{[Na^+]_i}{[Na^+]_o} + z\Im(\psi_i - \psi_o) \end{aligned}$$

and we can write the free energy for Cl^- transfer similarly as

$$[7.1.8] \qquad \begin{aligned} \Delta \mu_{Cl_o \Rightarrow Cl_i} &= \mu_{Cl_i} - \mu_{Cl_o} \\ &= \mu^0 + RT \ln[Cl^+]_i + z\Im \psi_i - \mu^0 \\ &\quad - RT \ln[Cl^+]_o - z\Im \psi_o \\ &= RT \ln \frac{[Cl^+]_i}{[Cl^+]_o} + z\Im(\psi_i - \psi_o) \end{aligned}$$

At equilibrium, $\Delta \mu = 0$ for any process. Setting both Eqns (7.1.7) and (7.1.8) to zero, we get

$$[7.1.9] \qquad \begin{aligned} -\frac{RT}{z\Im} \ln \frac{[Na^+]_i}{[Na^+]_o} &= (\psi_i - \psi_o) \\ -\frac{RT}{z\Im} \ln \frac{[Cl^+]_i}{[Cl^+]_o} &= (\psi_i - \psi_o) \end{aligned}$$

Recalling that $z = 1$ for Na^+ and $z = -1$ for Cl^{-1} and that $-\ln a = \ln 1/a$, we rewrite Eqn (7.1.9) as

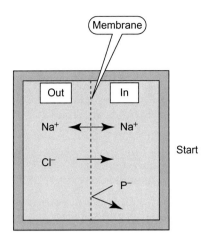

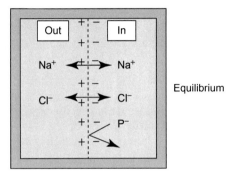

FIGURE 7.1.4 Establishment of the Gibbs–Donnan equilibrium. Initially, a membrane separates a solution of NaCl from NaP, where P is an impermeant anion. Initially, [Cl⁻] is higher on the left than on the right, so it diffuses to the right. Movement of Cl⁻ carries a charge so that a negative charge builds up on the right. This negative charge attracts the Na⁺ ions, which follow the Cl⁻. Eventually a concentration difference in both Cl⁻ and Na⁺ is established across the membrane and their gradients are in equilibrium with the membrane potential so that the net flux of both ions goes to zero. The ratio of the ions at equilibrium is the Gibbs–Donnan ratio.

[7.1.10]
$$\frac{RT}{\Im} \ln \frac{[Na^+]_o}{[Na^+]_i} = (\psi_i - \psi_o)$$
$$\frac{RT}{\Im} \ln \frac{[Cl^-]_i}{[Cl^-]_o} = (\psi_i - \psi_o)$$

These equations are variants of the Nernst equation. The right-hand side of both of these is just the membrane potential. Thus we can equate the two left-hand sides of the equations in Eqn (7.1.10) and we have

[7.1.11]
$$\frac{RT}{\Im} \ln \frac{[Na^+]_o}{[Na^+]_i} = \frac{RT}{\Im} \ln \frac{[Cl^-]_i}{[Cl^-]_o}$$

which gives

[7.1.12]
$$\frac{[Na^+]_o}{[Na^+]_i} = \frac{[Cl^-]_i}{[Cl^-]_o} = r$$

The ratio of the ions in the two compartments is a constant, r, called the **Gibbs–Donnan ratio**. **The Gibbs–Donnan ratio between plasma and the ISF in most capillary beds is 0.95**, when "in" refers to the plasma compartment and "out" refers to the ISF compartment.

CHANGES OF PLASMA VOLUME AND COMPOSITION TRANSFER TO ALL FLUID COMPARTMENTS

As discussed in Chapter 5.10, fluid in the plasma exchanges rapidly with fluid in the interstitial space. In addition, many of the solutes of plasma also readily exchange with solutes in the interstitial space. Although the ISF generally is not identical to the plasma, it is very similar and differs mainly in the lower protein concentration in the ISF. The capillary wall keeps the fluids separate,

---

EXAMPLE 7.1.2 Calculate the Gibbs–Donnan Potential Across the Capillaries

The potential is given by the Nernst equation (see Eqn 7.1.9):

$E_m = \psi_i - \psi_o = RT/\Im \ln[Na^+]_o/[Na^+]_i = 8.314 \text{ J mol}^{-1} \text{ K}^{-1}$
$\times 310 \text{ K}/96,500 \text{ C mol}^{-1} \times \ln 0.95$
$= 0.0267 \text{ V} \times (-0.05) = -\mathbf{1.37 \text{ mV}}$

The concentrations of ions listed in Figure 7.1.3 do not conform to the Gibbs–Donnan ratio of 0.95. The reason for this is that the values in Figure 7.1.3 are expressed as mEq per liter of solution and not per liter of solvent water. Plasma proteins, at about 7 g%, occupy a significant fraction of the volume of plasma, and only about 93% of the plasma volume is water. Using a figure of 98% for ISF, we can calculate the concentration per liter of water for plasma and ISF and get the values in Table 7.1.3.

The capillaries in different vascular beds differ in their permeabilities, so that the interstitial fluid protein content also differs, which subsequently slightly alters the interstitial fluid ionic composition as a consequence of the Gibbs–Donnan equilibrium (see Chapter 5.10).

TABLE 7.1.3 Concentration of Na+ and Cl− in Plasma and ISF

Ion	ISF (mEq L⁻¹)	Plasma (mEq L⁻¹)	ISF (mEq L⁻¹ water)	Plasma (mEq L⁻¹ water)	Gibbs–Donnan Ratio
Na⁺	142	142	144.9	152.7	0.95
Cl⁻	111	100.5	113.3	108.0	0.95

so that vascular flow can be maintained, while allowing this exchange.

The aggregate plasma membranes of the body cells separate the ICF from the ISF. These membranes prevent the free movement of solutes, as must be the case in order for the ICF to maintain a composition vastly different from the ISF. Maintenance of the composition of the ICF is the job of the plasma membrane, and this requires the input of energy and the operation of both passive and active transport mechanisms—the ion pumps and secondary active transporters. Despite this, alterations of the composition of the ISF have consequences for the composition of the ICF. In almost all cells, **aquaporins allow water transport that equilibrates the osmotic pressure across the plasma membrane.** Altering the osmolarity of the ISF nearly immediately alters the osmolarity of the ICF. Thus changes in the plasma are reflected in changes in the ISF, which in turn are reflected in changes in the ICF.

DARROW–YANNET DIAGRAMS DEPICT FLUID COMPARTMENT COMPOSITION AND VOLUME

Darrow–Yannet diagrams divide the body fluids into the ICF and the extracellular fluid (ECF), represented by rectangles. The height of the rectangles represents the concentration of osmotically active particles or the osmolarity; the width represents the volume of the compartment. The area of the rectangles is the osmolarity times the volume which is the amount of osmotically active particles, in osmoles, in the compartment.

The principles used in thinking about the volume and composition of the body fluids are simple:

- First, water equilibrates the osmotic pressure throughout all body fluids.
- Second, Na^+ stays extracellular.
- Third, intracellular materials stay intracellular.

Although these generalizations are not strictly true, they are close enough so that we can see what happens when various alterations in body fluid composition or volume occur. We consider here three changes in the overall composition of body volume and composition.

INGESTING OF WATER EXPANDS BOTH ECF AND ICF VOLUMES

What would happen if a "typical" individual were to drink 1 L of water? Before drinking the water, the individual has an ICF compartment of 28 L and an ECF compartment of 14 L, and both have an osmotic pressure of about 300 mOsM. This situation is depicted in Figure 7.1.5. After 1 L of water is consumed, water is

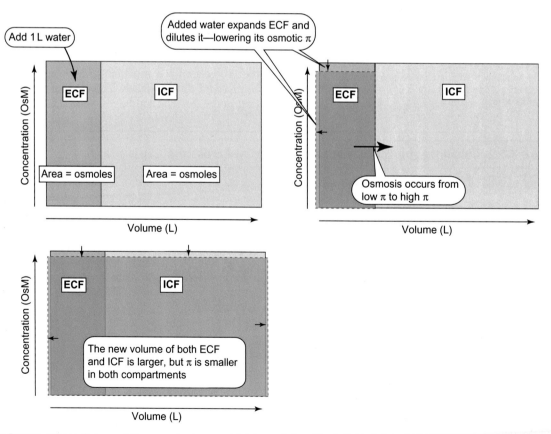

FIGURE 7.1.5 Darrow–Yannet diagram of the normal volume and total osmolarity of body fluids (left top). The height of the rectangles indicates the osmolarity and the width of the rectangles indicates the volume of the body fluid. After ingestion and absorption of 1 L of water, the extracellular fluid is diluted and expanded. This causes osmotic flow of water into the intracellular compartment, diluting and expanding it (top right). At steady state, both the ECF and ICF are diluted to about 293 mOsM. Both the extracellular and intracellular fluid compartments are expanded by a total of 1 L; 0.33 L of this is in the extracellular fluid compartment and 0.67 L in the intracellular fluid compartment (bottom).

absorbed from the intestines into the plasma, diluting it. This dilution sets up osmosis of water from the region of low to high osmotic pressure: from the plasma to the ISF. This movement occurs rapidly and the ISF is diluted. Thus all of the ECF is diluted. This in turn sets up an osmosis of water from the ECF to the ICF: the ICF becomes diluted. At the end, the osmotic pressures in all three compartments are equilibrated and both are diluted to the same degree. This can happen only when both extracellular and intracellular compartments are expanded by the same degree.

INGESTING NACL EXPANDS THE ECF BUT SHRINKS THE ICF VOLUME

What happens after eating and absorbing 10 g of NaCl? In this case, we assume that no water is ingested at the same time. The absorbed NaCl will be placed into plasma, causing an increase in the concentration of osmotically active solutes. We assume here that the NaCl remains entirely extracellular. The increased osmolarity of the ECF will draw fluid out of the intracellular compartment, shrinking it and increasing its osmolarity. These events are shown diagrammatically in Figure 7.1.6.

EXAMPLE 7.1.3 Osmolarity and Volume of Fluid Compartments After Drinking Water

The TBW in a person is 42 L and the extracellular volume is 14 L. Assuming the initial osmolarity of the fluids is 300 mOsM, what would be the volume and osmolarity of the body fluids after drinking 1 L of water?

The ICF volume is given by ICF = TBW − ECF = 42 − 14 L = 28 L.

The total osmotically active solute is 42 L × 300 mOsM L^{-1} = 12,600 mOsM.

The final osmolarity will be the total osmoles divided by the total volume:

12.6 Osmol/43 L = 293 mOsM

The volume of the ECF will be given by V = amount/concentration = 14 L × 300 mOsM/293 mOsM = **14.33 L**.

The volume of the ICF will be given by the similar relation: 28 × 300 mOsM/293 mOsM = **28.67 L**.

Thus drinking water expands both ECF and ICF and dilutes both compartments.

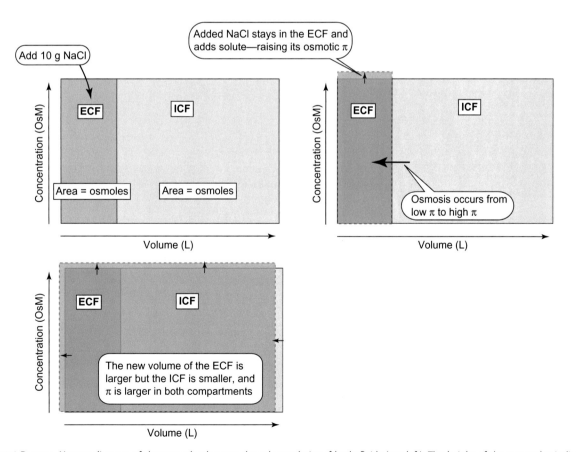

FIGURE 7.1.6 Darrow–Yannet diagram of the normal volume and total osmolarity of body fluids (top left). The height of the rectangles indicates the osmolarity and the width of the rectangles indicates the volume of the body fluid. After ingestion and absorption of 10 g NaCl, the extracellular fluid becomes more concentrated because there are more osmotically active solutes in it. Water moves from the intracellular fluid into the extracellular fluid because the osmotic pressure is higher in the ECF. Because Na remains extracellular, the extracellular body fluid compartment expands.

QUANTITATIVE HUMAN PHYSIOLOGY: AN INTRODUCTION

INFUSION OF ISOTONIC SALINE EXPANDS THE ECF VOLUME BUT DOES NOT CHANGE THE ICF VOLUME

Isotonic saline is *defined* as a solution of NaCl that does not cause movement of fluid when placed in contact with body cells. Thus we would expect no movement of solutes and presumably negligible movement of body cell solutes, when isotonic saline is infused. If we were to infuse 1 L of a solution of NaCl, we would expect all of it to expand only the ECF compartment. Thus the final volume of the ECF would be 15 L with an osmolarity of 300 mOsM, and the ICF volume and osmolarity would remain unchanged. This situation is shown in Darrow–Yannet diagrams in Figure 7.1.7.

THE KIDNEYS REGULATE BODY FLUID VOLUME AND COMPOSITION BY ACTING ON THE PLASMA

From the above discussion, we see that alterations in the plasma volume or concentration become transferred to the ISF and from there to the ICF. This establishes a fundamental concept of the regulation of body fluid volume and composition: it is accomplished by acting directly on only the plasma. Thus the kidneys and, indeed, other organ systems participate in the regulation of the body fluid volume and composition by acting solely on the plasma. Figure 7.1.8 illustrates this concept schematically.

EXAMPLE 7.1.4 Osmolarity and Fluid Compartment Volumes After Ingesting Salt

After ingesting and absorbing 10 g of NaCl, without water, what would the volume and osmolarity be of the ECF and ICF?

The total osmotically active solute before ingesting the NaCl was 42 L × 300 mOsM = 12.6 Osmol.

10 grams of NaCl adds 2 Osmol mol^{-1} × 10 g NaCl/58.4 g NaCl mol^{-1} = 342 mOsmol.

The total osmolarity after ingesting 10 g NaCl = total osmoles/total volume = 12.942 Osmol/42 L = **308.1 mOsM**.

The volume of the ECF is given by volume = total osmotic solutes/osmolarity. The total osmotic solute is what was present before the salt, plus the salt:

ECF = [14 L × 300 mOsM + 342 mOsM]/308.1 mOsM = **14.74 L**

The volume of the ICF is calculated similarly as: ICF = 28 L × 300 mOsM/308.1 mOsM = **27.26 L**.

Thus ingesting salt alone concentrates both fluid compartments, expands the ECF, and shrinks the ICF.

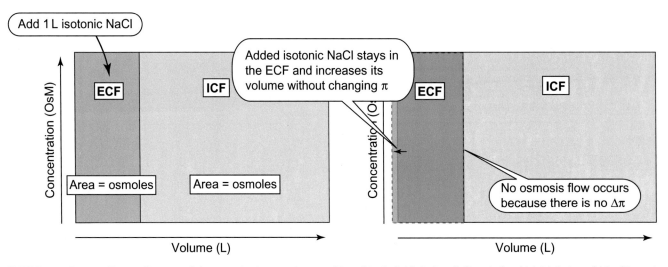

FIGURE 7.1.7 Darrow–Yannet diagrams of the normal volume and composition of body fluids before (left) and after (right) infusion of 1 L of isotonic saline (0.9% NaCl). In this case, all of the volume expands the extracellular fluid compartment with no change in the intracellular fluid compartment volume or osmolarity.

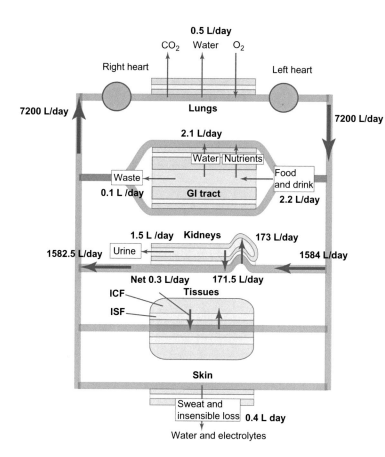

FIGURE 7.1.8 Regulation of the body fluid compartments by the kidneys, lungs, and gastrointestinal systems. The common connection between all organ systems regulating fluid volume and composition is their operation on the plasma. The kidneys have the primary function of homeostatic regulation of the body fluids by processing a large fraction of the cardiac output. The average cardiac output is 7200 L day^{-1} of which about 20%—about 1584 L day^{-1}—flows through the kidneys. Of this, about 173 L day^{-1} is filtered by the kidneys and mostly reabsorbed back into the circulation (171.5 L day^{-1}). The numbers given are approximate and average values that vary widely depending on conditions. Most variable is water loss through sweat and compensatory intake through the gastrointestinal system. Overall body balance of water and electrolytes is determined mainly by the gastrointestinal system and the kidneys, with loss through the lungs and the skin. Typical intake is 2.5 L day^{-1} = 2.2 L day^{-1} via food and drink and 0.3 L day^{-1} from metabolism; losses also total 2.5 L day^{-1} = 0.1 L day^{-1} from feces + 0.5 L day^{-1} from lungs + 1.5 L day^{-1} from urine, and 0.4 L day^{-1} from sweat and insensible loss from the skin. Changes in composition of the plasma are relayed to the tissues through the interstitial fluid.

SUMMARY

The overall function of the renal system is to regulate the volume and composition of body fluids. The body fluids consist of three major compartments: the ICF, the ISF, and the plasma. In a normal young adult male, these comprise about 40%, 15%, and 5% of body weight, respectively. These volumes can be calculated from the volume of distribution of marker substances that distribute themselves according to the TBW, ECF, or the plasma. The intracellular volume is the TBW minus the ECF; the ISF volume is calculated as the ECF minus the plasma. The volumes of the various fluid compartments vary with age, gender, and body composition. Fat tissue has little water, so excess body fat contributes excess body weight without much added water. Therefore, using the idea of an LBM containing 73% water, it is possible to estimate excess body fat from body weight and TBW.

The presence of anionic proteins in the plasma which cannot equilibrate across the capillary wall sets up a slightly unequal concentration of diffusible ions across the capillary. This is due to the Gibbs–Donnan effect, which produces a negative potential on the side of the impermeant anion. The potential can be calculated from the Nernst equation.

The body fluids reside in well-defined anatomic compartments, and there are barriers and forces that determine exchange between the compartments. The distribution of electrolytes and water can be visualized using Darrow–Yannet diagrams. In these calculations, we assume that water equilibrates its osmotic pressure across all boundaries, Na$^+$ remains extracellular, and K$^+$ remains intracellular. From this analysis, we can see that ingestion of water expands ICF and ECF compartments while diluting their osmolarity; ingesting NaCl alone expands the ECF and concentrates both ECF and ICF solutes; ingestion or infusion of isotonic saline expands only the ECF with no changes in ICF.

The renal system accomplishes its task of regulating body fluid compartment size and composition by operating solely on the plasma. Changes in plasma fluid volume and composition are then transferred to the ISF and ICF. Other organ systems also participate in this regulation.

REVIEW QUESTIONS

1. Describe the dilution principle method for determining volumes of fluid compartments.
2. What characteristics make for good "markers" for any fluid compartment?
3. What three main principles are the basis for approximate calculations of fluid compartment changes with changes in water or electrolytes?
4. Why is Na$^+$ present mostly in the ECF?
5. Why does water equilibrate osmolarity between intracellular fluid and extracellular fluid?
6. Why does the area of a Darrow–Yannet diagram indicate total osmotically active solutes?
7. What quantities are conserved in calculations involving volume and osmolarity of fluid compartments?
8. Based on your knowledge of Darrow–Yannet diagrams, what do you suppose would be the renal response to decreased ECF volume?

7.2 Functional Anatomy of the Kidneys and Overview of Kidney Function

Learning Objectives

- Identify the gross anatomical structures of the renal system including renal artery and vein, ureter, urinary bladder, and urethra
- Identify: renal medulla, renal cortex, renal capsule
- Distinguish between cortical and juxtamedullary nephrons
- Identify: afferent arteriole, efferent arteriole, glomerulus, Bowman's capsule, proximal convoluted tubule, proximal straight tubule, loop of Henle, distal tubule, collecting duct, peritubular capillaries, vasa recta, juxtaglomerular apparatus, macula densa
- Describe the signals that turn on renin secretion
- Describe the action of renin on angiotensinogen
- List the main functions of angiotensin II
- List the functions of the kidney unrelated to the formation of urine

FUNCTION FOLLOWS FORM IN FUNCTIONAL UNITS CALLED *NEPHRONS*

As discussed in Chapter 7.1, the kidneys regulate the volume and composition of the body fluids and it accomplishes this task by acting solely on the plasma. The effects of kidneys transfer to the interstitial fluid and to the intracellular fluid by altering water flow and material transport across the barriers between these compartments. The remainder of renal physiology seeks to answer the question: how does the renal system alter plasma volume and composition? The full answer is complicated, but a partial answer is that all functions of the kidney occur in miniature in small functional units called **nephrons**. Their mechanisms in turn are entirely dependent on transport properties of nephron parts and the spatial arrangement of these parts within the kidneys. The function of the nephrons follows their form and understanding how the kidneys accomplish their task requires an understanding of the structure of the component nephrons and their spatial arrangement within the kidney.

THE PAIRED KIDNEYS HAVE AN ENORMOUS BLOOD SUPPLY AND DRAIN URINE INTO THE *BLADDER* THROUGH THE *URETERS*

The kidneys are bean-shaped organs that lie against the back of the abdominal wall beneath the **peritoneum**, the membranous connective tissue sheet that lines the abdominal cavity and reflects back over all of the intestines. The two kidneys together weigh about 300 g or about 0.4% of the body mass. The blood supply is provided by the **renal artery** that comes directly off the **abdominal aorta** and is drained by the **renal vein** that drains into the **inferior vena cava**. These large vessels deliver about 20% of the cardiac output, or about 1 L min^{-1}, to the kidneys. This is an enormous fraction of the cardiac output for the size of the kidneys and it provides the first clue to their operation: they must do something to the blood other than extract oxygen from it. The kidneys clear the blood of waste products and adjust plasma volume and composition. To do this, the blood has to run through the kidneys at a high rate, far more than is required for the metabolism of the kidneys (see Figure 7.2.1).

The waste products and excess fluid and electrolytes are disposed of in the urine. The kidneys collect the urine from their many nephrons into the **ureter**. The ureter is a tube that contains smooth muscle. The ureters can actively propel the urine into the bladder by **peristalsis**, a wave of muscular contraction that begins in the kidney and continues to the urinary bladder, which also has a layer of smooth muscle. The urinary bladder stores the urine and voids it to the outside of the body through the **urethra**. Urination requires the active contraction of the bladder and relaxation of the urinary sphincters.

INTERMEDIATE LEVEL OF KIDNEY STRUCTURE REVEALS FUNCTIONAL AREAS

A longitudinal cross-section of the kidney shows two easily distinguished parts (see Figure 7.2.2). The **cortex** lies beneath the tough **renal capsule** and has a reddish-brown

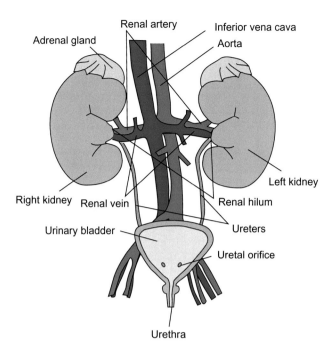

FIGURE 7.2.1 Diagram of the anatomic arrangement of the urinary system. The kidneys are paired structures located behind the peritoneum, a connective tissue sheet that lines the abdominal cavity. The hilum is a longitudinal slit in the medial aspect of the kidney through which the renal artery enters and the renal vein and ureters exit. The kidneys are also supplied by afferent and efferent nerves.

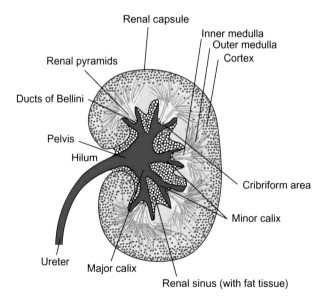

FIGURE 7.2.2 Longitudinal cross-section of the kidney. The reddish-brown, granular cortex lies immediately below the tough connective tissue capsule. The medulla lies more interiorly. Urine is formed inside tiny tubules that eventually coalesce to form ever-larger tubules, eventually forming the ducts of Bellini. These penetrate through the cribriform area of the renal pyramids and drain their fluid, the urine, into the minor calices. The urine eventually drains into the ureters and from there into the urinary bladder from which it is eliminated from the body.

and granular appearance. The cortex contains many **glomeruli** (singular is **glomerulus**), which are little balls of capillaries that are all crammed together and surrounded by a thin epithelial capsule called **Bowman's capsule**. The granular appearance is due to the many glomeruli in the cortex. The **medulla** comprises the inner part of the kidney. The medulla in turn consists of an outer part and an inner part. The outer part has a striated appearance that is due to the many tubules running from the cortex down into the medulla. These tubules are part of the nephrons, the functional units of the kidneys, and the tubules collect into structures called **renal pyramids**. Each kidney typically has 8–10 of these renal pyramids. The tubules coalesce into progressively larger tubules, the **collecting ducts**. These collecting ducts in turn fuse to form **ducts of Bellini**, which pierce through the apex of the renal pyramids in a series of 18–24 tiny holes in each pyramid. This flattened area of the renal pyramid that is pierced by the ducts of Bellini is called the **cribriform area**. Urine in its final stages passes through the tiny orifices in the cribriform area to reach a minor calix (from the Greek "kalyx," meaning "cup"). Two or more minor calices fuse to form a major calix. The urine eventually collects into the **ureters**, which drain into the urinary bladder.

THE RENAL ARTERIES ARISE FROM THE ABDOMINAL AORTA

The single renal artery enters the hilum and then branches to form the **interlobar arteries**, so-named because they pass between the lobes of the kidney. At the junction of the cortex and medulla, the interlobar arteries bend over to form incomplete arches. Accordingly, this section of the blood supply is called the **arcuate artery**. Many small **radial arteries** branch off at right angles from the arcuate artery, carrying blood toward the cortex. These radial arteries give rise to short lateral branches called **afferent arterioles** that supply blood to the glomeruli.

The venous drainage of the kidney follows the arterial supply. **Radial veins** arise from stellate ("star shaped") sinuses in the superficial cortex that drain capillaries from the cortex. These penetrate the cortex and join the **arcuate veins** that arch over the renal pyramids. The arcuate veins join to form the **interlobar veins**, which in turn join to form the single **renal vein**. The anatomy of the kidney's vasculature is shown in Figure 7.2.3.

THE FUNCTIONAL UNIT OF THE KIDNEY, THE NEPHRON, PARTICIPATES IN ALL ELEMENTARY RENAL PROCESSES

The kidneys use three elementary processes to produce urine and to regulate the volume and composition of the plasma. These are as follows:

1. Ultrafiltration
2. Secretion
3. Reabsorption.

All of these elementary processes occur in each nephron and so these nephrons are functional units of renal function. Each kidney contains some 1–1.3 million of these nephrons.

THE NEPHRON IS A TUBULE WITH FUNCTIONALLY AND MICROSCOPICALLY DISTINCT REGIONS

The nephrons are all long tubes with parts that are oriented approximately parallel to the renal capsule, or perpendicular to it. Each portion of the tube has its own set of transport properties. There are two basic kinds of nephrons: **cortical nephrons** and **juxtamedullary nephrons**. These distinctions have to do with the location of the glomerulus, the tiny ball of capillary network, and the penetration into the medulla by the loops of the nephron tubule. Cortical nephrons have a glomerulus located nearer to the outer parts of the cortex and their **loops of Henle** are short. Juxtamedullary nephrons have a glomerulus near the junction of the cortex and medulla and their loops of Henle penetrate deep into the medulla. The relative number of cortical and juxtamedullary nephrons and the lengths of their loops of Henle determine the ability of the kidney to concentrate the urine. In humans, about 85% of the nephrons are cortical nephrons and about 15% are juxtamedullary nephrons. Kangaroo rats, which live in extremely arid environments, never need to drink water. Their only source of water is from metabolism of food. They achieve this feat by having extraordinarily long loops of Henle (see Figure 7.2.4 for a diagrammatic description of nephron structure).

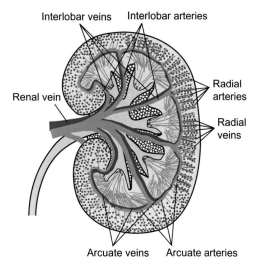

FIGURE 7.2.3 Diagram of the vasculature of the kidney. Blood enters the kidney through the single renal artery that breaks up into several interlobar arteries. These penetrate the renal columns where, at the junction of the cortex and medulla, they bend back over the bases of the renal pyramids, forming the arcuate arteries. The radial arteries come off the arcuate arteries at right angles and these supply blood to the cortex. The afferent arterioles that supply blood to the glomeruli are short lateral branches of the radial arteries. The venous drainage more or less follows the arterial supply except that the arcuate veins form complete arches over the renal pyramids whereas the arcuate arteries form incomplete arches.

Although the blood supply is technically not part of the nephron, its anatomical arrangement is crucial to renal function. We have already traced the macroscopic blood flow: blood enters the kidney by the single **renal artery** that divides into the **interlobar arteries**. These curve back over the base of the renal pyramids to form the **arcuate arteries**. **Radial arteries** arise at right angles to the arcuate arteries and rise through the cortex. Short lateral branches of the radial arteries are called the **afferent arterioles**. The afferent arterioles supply blood to the **glomerulus** and regulate its hydrostatic pressure.

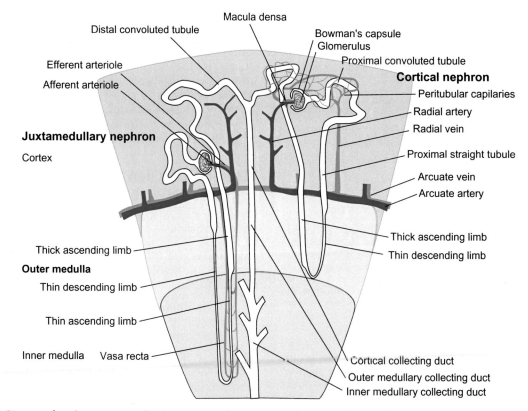

FIGURE 7.2.4 Diagram of nephron structure. Structures are not drawn to scale. Two general kinds of nephrons are shown: a cortical nephron and a juxtamedullary nephron. Blood supply is partially shown with most capillaries omitted. See text for explanation.

The glomerulus is a group of anastomosing capillaries that, together with its closely apposed **Bowman's capsule**, forms an ultrafiltrate of plasma. This ultrafiltration is the first step in the formation of urine.

After leaving the glomerulus, blood enters the **efferent arteriole**. Thus the capillary network of the glomerulus is highly unusual in that it sits between two arterioles rather than between an arteriole and a venule. It is a high-pressure capillary bed. The efferent arteriole helps regulate the fraction of plasma that is filtered by the glomerulus. The fate of blood in the efferent arterioles depends on the location of the glomerulus. For cortical nephrons, the efferent arterioles break up into an anastomosing network of capillaries called the **peritubular capillaries**. These supply oxygen to cells in the nephron to energize transport processes and they provide materials for secretion and a conduit to remove materials that are reabsorbed from the ultrafiltrate. These capillaries eventually drain into sinuses that drain into the **radial veins**, which drain into the **arcuate veins**, into the **interlobar veins** and, finally, into the single **renal vein**.

In juxtamedullary nephrons, the efferent arterioles give rise to the **vasa recta**, which plunges down into the renal papilla to supply blood to that tissue. The **descending vasa recta** and **ascending vasa recta** make a loop. The vasa recta forms a capillary network within the renal medulla. It acts as a **countercurrent exchanger** which exchanges material with the medullary interstitium without destroying the osmotic gradient that is established there. These ideas will be discussed in depth in Chapter 7.5.

Now let us trace the path of urine formation. Urine formation begins in **Bowman's space**. As mentioned earlier, the glomerulus together with the closely apposed **Bowman's capsule** forms an ultrafiltrate of plasma, and this is the first step in the formation of urine. This ultrafiltrate flows through the tube of the nephron, traveling first through the **proximal convoluted tubule** where many of the filtered substances are reabsorbed from the ultrafiltrate including water, salt, amino acids, glucose, and bicarbonate. Secretion of some materials into the incipient urine also occurs here. Fluid then enters the **loop of Henle**. The loop of Henle can be considered to be a loop of nephron tubule oriented perpendicular to the surface of the capsule of the kidney: it descends through the cortex into the medulla, makes a hairpin turn, and returns again to the cortex. This arrangement is crucial to its function as a **countercurrent multiplier** to establish an osmotic gradient that runs from 300 mOsM in the cortex to as high as 1200 mOsM in the inner medullary interstitium. This osmotic gradient is essential for the kidney to concentrate urine. How this is accomplished will be discussed in Chapter 7.5. For now, you should learn to associate functional descriptors with the various parts of the nephron: the glomerulus is an ultrafilter; the proximal convoluted tubule reabsorbs and secretes; the vasa recta is a countercurrent exchanger; the loop of Henle is a countercurrent multiplier.

The loop of Henle consists of all of the tubule parts between the proximal convoluted tubule and the distal convoluted tubule. In cortical nephrons, the loop of Henle is not so long, penetrating down through the outer medulla to the edge of the inner medulla. It consists of a proximal straight tubule, a thin descending limb, and a thick ascending limb. Each thick ascending limb returns to its own glomerulus where it lies closely apposed to the afferent and efferent arterioles of its own glomerulus. The complex of the thick ascending limb, afferent arteriole, and efferent arteriole is the **juxtaglomerular apparatus**. The thick ascending limb ends at this point and the **distal convoluted tubule** begins. Here additional reabsorption of water or salt occurs and the distal tubule connects to a **collecting duct**. The collecting duct may be considered as part of the nephron, but many distal convoluted tubules drain into a single collecting duct. Thus the collecting ducts individually belong to several nephrons. The final concentration or dilution of the urine occurs in the collecting ducts as the fluid flows successively through the **cortical collecting duct**, **outer medullary collecting duct**, and **inner medullary collecting duct**.

The structure of **juxtamedullary nephrons** differs from that of cortical nephrons. In juxtamedullary nephrons, the loop of Henle penetrates further into the inner medulla. It consists of the proximal straight tubule, thin descending limb, thin ascending limb, and thick ascending limb. Here the proximal straight tubule becomes the thin descending limb at about the same depth as that of the cortical nephrons. In juxtamedullary nephrons, the thin descending limb descends all the way to the renal papilla. It returns as a thin ascending limb to the junction of the inner and outer medulla, becoming the thick ascending limb. This thick limb returns to its own glomerulus, forming its own juxtaglomerular apparatus with it.

THE JUXTAGLOMERULAR APPARATUS PRODUCES *RENIN*

RENIN IS AN ENZYME RELEASED BY GRANULE CELLS IN THE AFFERENT ARTERIOLE

As mentioned above, the thick ascending limb of the loop of Henle rises out of the outer medulla and travels toward the glomerulus from which it arose. It makes close contact with the afferent arteriole and efferent arteriole at the point at which they enter and exit the glomerulus. At this point, the cortical thick ascending limb ends and the distal convoluted tubule begins. The region of close apposition of the afferent arteriole, efferent arteriole, and cortical thick ascending limb is called the **juxtaglomerular apparatus** because it is next to the glomerulus. This structure contains a number of specialized cells. The cells in the thick ascending limb adjacent to the arterioles form the **macula densa**. The volume of these cells responds to the [NaCl] in the tubular fluid. These macula densa cells send signals through **mesangial cells** to affect cells in the afferent arteriole. The afferent arteriole contains vascular smooth muscle cells and specialized **granule cells** that contain granules of

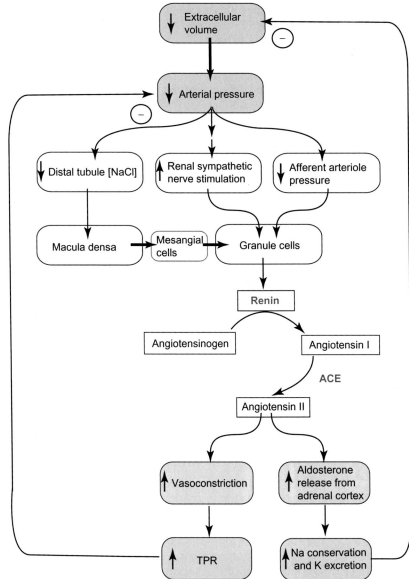

FIGURE 7.2.5 The renin–angiotensin–aldosterone system. Renin is an enzyme secreted by granule cells in the afferent arteriole in response to lower afferent arteriolar pressure, renal sympathetic nervous stimulation, or signals arriving from the macula densa and probably relayed through mesangial cells. Renin converts a precursor plasma protein, made in the liver, to angiotensin I, a decapeptide. Angiotensin I is further converted to angiotensin II by ACE, predominantly located in the lung. Angiotensin II has two major effects: it is a potent vasoconstrictor and it releases aldosterone from the adrenal cortex. The vasoconstrictor effects of angiotensin II increase total peripheral resistance, thereby tending to return the blood pressure toward normal. Thus the vasoconstrictor effects of angiotensin II complete a negative feedback loop in which the original perturbation, a decrease in blood pressure, is corrected. The increased circulating aldosterone has the effect of lessening Na^+ excretion and increasing K^+ excretion. The result is a tendency to preserve the extracellular volume, thereby defending against decrease in blood pressure. Thus the system forms a second negative feedback loop.

renin. These granule cells are also called juxtaglomerular or JG cells. Activation of the macula densa cells leads to afferent arteriolar constriction and release of renin into the circulation. The events leading to renin secretion and the consequences of renin secretion are shown in Figure 7.2.5.

RENIN CLIPS ANGIOTENSIN I OFF A PLASMA PRECURSOR, WHICH IS THEN CONVERTED TO ANGIOTENSIN II

Renin is an enzyme that converts an inactive plasma protein, **angiotensinogen**, to **angiotensin I**. This is the first step in the renin–angiotensin–aldosterone system of blood pressure and volume regulation that we will discuss in detail in Chapter 7.6. Angiotensin I is a polypeptide 10 amino acids long. It is converted to **angiotensin II** when **angiotensin converting enzyme (ACE)** clips off two amino acids. ACE is mainly located in the lungs, but the kidney also has some ACE activity.

ANGIOTENSIN II VASOCONSTRICTS AND RELEASES ALDOSTERONE

Angiotensin II binds to AT_1 or AT_2 receptors, exerting two major effects: it is a potent vasoconstrictor of the resistance vessels of the body, thereby producing an increase in total peripheral resistance and a rise in blood pressure, and it releases **aldosterone** from the **adrenal cortex**. Aldosterone is a steroid hormone that conserves salt by decreasing salt excretion from the kidney. Although aldosterone cannot add NaCl to the extracellular fluid of its own accord, it can retain the NaCl that enters the body through the food. Since NaCl is largely extracellular, retaining salt increases the extracellular fluid volume, which tends to expand the blood volume. Thus the retention of salt caused by aldosterone indirectly tends to elevate blood pressure and cardiac output.

OUR EVOLVING KNOWLEDGE OF THE RAAS HAS ADDITIONAL COMPONENTS

The above description applies to the classical RAAS (renin–angiotensin–aldosterone system). Newer results show that another enzyme, ACE2, can cleave angiotensin II to produce angiotensin (1–7) that interacts with a different receptor, the Mas receptor, and that aminopeptidase A and aminopeptidase N cleave angiotensin II to angiotensin IV that interacts with the AT_4 receptor. In addition, granule cells release prorenin, a precursor form of renin, that binds to a prorenin receptor. The physiological importance of these accessory pathways remains to be clarified.

NONEXCRETORY FUNCTIONS OF THE KIDNEY

THE KIDNEYS PRODUCE THE HORMONE ERYTHROPOIETIN

In addition to its primary function of regulating the volume and composition of the plasma, and hence contributing to homeostasis of all body fluids, the kidney performs other essential functions. These include the secretion of **erythropoietin** in response to hypoxia. Erythropoietin is a glycoprotein hormone that stimulates the rate of formation of red blood cells or erythropoiesis. This forms another negative feedback loop: poor supply of oxygen to the kidneys stimulates the secretion of erythropoietin, which stimulates the formation of red blood cells that in turn increase the supply of oxygen to the kidney (see Chapter 5.2).

THE KIDNEYS ACTIVATE VITAMIN D

Vitamin D is synthesized in the skin and then subjected to two hydroxylation reactions. The first occurs in the liver and produces 25-OH-cholecalciferol. The second hydroxylation reaction occurs in mitochondria of renal proximal tubule cells and produces $1,25\text{-}(OH)_2$ cholecalciferol, the active circulating form of vitamin D. The hydroxylation reaction in the kidney is stimulated by **parathyroid hormone**, **PTH**, and by low plasma [Pi]. $1,25\text{-}(OH)_2$ cholecalciferol stimulates Ca^{2+} and Pi absorption from the intestine and helps maintain plasma $[Ca^{2+}]$ and [Pi] by actions on bone and kidney. The kidney hydroxylation reaction is crucial to the activity of vitamin D and forms part of the negative feedback loop that controls homeostasis of both plasma Ca^{2+} and phosphate. See Chapter 9.8 for detailed descriptions of these negative feedback loops.

KIDNEYS PRODUCE GLUCOSE BY GLUCONEOGENESIS

The kidneys help maintain blood glucose during fasting by converting amino acids into glucose. This metabolic process is called **gluconeogenesis**, which literally means "new glucose origination."

THE KIDNEYS DEGRADE MANY HORMONES

Hormones generally are secreted in response to some stimulus in order to affect some system that then counteracts the original stimulus. This is the essence of negative feedback loops and their role in maintaining homeostasis. To have effective control, the hormone signals must be shut off. Theoretically, this could be accomplished several ways. One way is to regulate negative feedback by a dual system in which hormone release and degradation are both regulated. A second, simpler solution is to regulate release of hormone and degrade the hormones according to first-order kinetics. In this case, the rate of degradation is proportional to hormone concentration. This is the typical manner of regulation of circulating hormone levels. The rate of release depends on the stimulus, whereas degradation is accomplished by mass action. The kidney participates in the degradation of a number of polypeptide hormones including angiotensin II, parathyroid hormone, insulin, and glucagon.

Clinical Applications: Kidney Failure

The best illustration of the importance of the kidney is what happens when it fails. Complete kidney failure, also called renal failure, universally leads to death unless the blood is cleansed by dialysis.

Such death is caused by a buildup of waste products and water and electrolyte imbalances.

Kidney failure can result from acute kidney injury, which is a rapid and progressive loss of renal function characterized by oliguria (low urine output, less than 400 mL per day for an adult) and fluid and electrolyte imbalance. There are a variety of causes, usually having to do with reduced blood supply to the kidney or when the kidneys are overwhelmed by toxins. Drug overdoses and crush syndrome are two examples. In the crush syndrome, breakdown of muscle tissue (rhabdomyolysis) releases myoglobin, potassium, and phosphate into the blood stream. The myoglobin clogs up the kidney and reduces its filtration, the first step in the formation of urine.

Kidney failure can also result from chronic conditions that progressively reduce kidney function. The most common cause is diabetes mellitus, which produces a nephropathy having to do with dysfunction of renal capillaries. Hypertension also slowly and progressively damages the kidney. Polycystic kidney disease is a third well-known cause of chronic kidney disease. Polycystic kidney disease has a known genetic component.

Clinicians use the glomerular filtration rate (see Chapter 7.3) as an indicator of the stages of renal failure. Without filtration, the kidney cannot rid the body of specific wastes such as creatinine and urea, and these substances build up in the blood. Accumulation of nitrogenous wastes in the blood is called **azotemia**. An older term for this is uremia, referring, literally, to urine in the blood. The term "uremia" is now used to describe the illness accompanying kidney failure. In the early stages of kidney failure, excretion of urea and creatinine may be near normal, but the normal excretion requires higher than normal plasma levels. At progressive stages of failure, urea and creatinine levels soar and the patient needs dialysis or kidney transplantation to remove these wastes.

(Continued)

Clinical Applications: Kidney Failure (Continued)

Symptoms of renal failure include the following:

- high levels of urea in the blood that can lead to nausea and weight loss;
- high levels of phosphate in the blood, leading to bone damage or failure to heal fractures and muscle cramping due to hypocalcemia;
- increased plasma [K^+], leading to cardiac arrhythmias;
- swelling of the legs and ankles and shortness of breath;
- anemia due to reduced erythropoietin secretion;
- protein in the urine.

SUMMARY

The kidney has a well-defined microscopic anatomy that is integral to its function of regulating the volume and composition of body fluids by performing three fundamental operations on the plasma: ultrafiltration, secretion, and reabsorption. All three of these fundamental operations are performed by the functional unit of the kidney, the nephron.

Blood enters the kidney by the renal artery, which subdivides into the interlobar arteries, arcuate arteries, radial arteries, and finally the afferent arterioles. The afferent arterioles supply two major categories of nephrons: cortical nephrons and juxtamedullary nephrons, named for the location of the glomerulus within the kidney. The glomerulus is a group of capillaries located between the afferent arteriole and the efferent arteriole. Together with its closely apposed Bowman's capsule, the glomerulus produces an ultrafiltrate of plasma in the first step of forming the urine. The efferent arteriole in cortical nephrons forms the peritubular capillaries that carry off reabsorbed materials and supply blood for secretion by the first tubular part of the nephron, the proximal tubule. The efferent arterioles of the juxtamedullary nephrons make a hairpin loop to the inner medulla of the kidney, forming vessels called the vasa recta. Venous blood collects into radial veins, leading to arcuate veins and interlobar veins and eventually into the renal veins.

The nephron consists of the glomerulus, proximal tubule, loop of Henle, distal tubule, and collecting ducts. The collecting ducts are formed from several nephrons, so they actually belong to a group of nephrons. Specific modification of the tubular fluid occurs along each segment of the nephron. The loop of Henle differs between the cortical nephrons, which have short loops, and the juxtamedullary nephrons, which have long loops. These loops form an osmotic gradient within the kidney interstitium that is crucial to the ability to concentrate urine.

The juxtaglomerular apparatus refers to the junction of the afferent arteriole, efferent arterial, and the thick ascending limb of the loop of Henle from the same nephron. Specialized granule cells in the afferent arteriole secrete renin, an enzyme that cleaves a plasma protein, angiotensinogen, to form angiotensin I. Angiotensin I is further cleaved by ACE to form angiotensin II. Angiotensin II has several effects, including vasoconstriction and release of a steroid hormone, aldosterone, from the adrenal cortex. The effects of angiotensin II are to restore blood pressure and increase the ECF volume. The major stimuli for renin release are as follows: (1) decreased afferent arteriolar pressure; (2) increased activity of renal sympathetic nerves; and (3) decrease in distal tubule [NaCl].

REVIEW QUESTIONS

1. Trace the blood flow in the kidneys from the abdominal aorta through the kidneys and back through the veins to the inferior vena cava. How does the blood flow differ in cortical and juxtamedullary nephrons?
2. Trace the flow of fluid in the nephron from Bowman's capsule to the collecting duct. How does this differ in cortical and juxtamedullary nephrons?
3. What is the main function of the glomerulus together with its Bowman's capsule (the renal corpuscle)? Loop of Henle? Proximal convoluted tubule? Vasa recta?
4. The kidney is involved in the formation/activation of several circulating substances. Name three of these and describe their function.
5. What is renin? Where does it originate? What does it do? What is the juxtaglomerular apparatus?

Glomerular Filtration 7.3

Learning Objectives

- Summarize the evidence that the renal corpuscle forms an ultrafiltrate
- List the three elementary processes found in the nephrons
- Define the clearance for any substance
- Calculate the GFR as the clearance of inulin and explain why the clearance of inulin is the GFR
- Calculate effective renal plasma flow from the clearance of PAH
- Describe what is meant by filtration fraction and give its normal range
- Describe the layers of the glomerular filtration barrier and how they provide a sequential size filter
- Define the sieving coefficient
- Write the Starling equation for the forces that produce the glomerular ultrafiltrate
- Provide approximate estimates of the components of the driving forces for ultrafiltration

MORPHOLOGICAL STUDIES FIRST LED TO THE IDEA OF GLOMERULAR FILTRATION

Unlike the description of the nephron given in Chapter 7.2, the cortex is actually a compact mass of coiled tubules that resembles a bag of worms. The blood vessels and tubules are jumbled together without the spaces between them as shown in Figure 7.2.4. Despite this confusion of tubules, early microscopists such as Malpighi, who published his findings in 1666, and Bowman, in 1842, recognized that each **renal corpuscle** (the glomerulus together with its Bowman's capsule) is connected to a single tubule. Bowman described the glomerular capillaries and found that their outer covering was continuous with the outer layer of the capsule and with the tubule. Thus any fluid formed in Bowman's space would drain down into the tubule. Bowman thought that the glomerulus secreted water that flushed away solutes secreted into the urine by the tubule cells. Bowman's contemporary, Ludwig, proposed that the glomerulus produced a protein-free ultrafiltrate. According to this view, reabsorption of water from the ultrafiltrate concentrated the waste products and reduced the volume of the urine. Although Ludwig's ideas substantially agree with the modern view, it is one thing to propose a mechanism and quite another to provide experimental evidence for it.

MICROPUNCTURE STUDIES SHOWED THAT THE FLUID IN BOWMAN'S SPACE IS AN ULTRAFILTRATE

A.N. Richards succeeded in removing samples of fluid from Bowman's space by using micropuncture techniques. Applying microchemical analysis to these samples, Richards and colleagues showed that the fluid was free of protein, but the concentrations of a variety of small molecular weight solutes such as creatinine, Cl^-, glucose, Pi, K^+, urea, and uric acid were the same in the Bowman's space fluid as they were in plasma. A filter removes suspended particles from a mixture. An ultrafilter removes not only the particulate matter but also removes colloidal materials such as proteins. Small molecular weight solutes, however, are not removed by an ultrafilter. Thus the fluid in Bowman's space had the composition of an ultrafiltrate. Richards accomplished this remarkable demonstration in the early 1920s. A diagram of the micropuncture technique along with the major features of the renal corpuscle is shown in Figure 7.3.1.

TUBULAR REABSORPTION EXPLAINS THE LACK OF NUTRIENTS IN THE FINAL URINE

The glomerular ultrafiltrate contains amino acids and glucose and a host of other materials in concentrations nearly identical to those in plasma. But substances such as glucose and amino acids are completely absent from the final urine. These two observations establish that some materials must be removed from the tubular fluid after formation of the ultrafiltrate and while the fluid travels down the nephron. The removal of materials from the tubular fluid is called **tubular reabsorption**. Reabsorbed materials move from the tubular fluid into the peritubular capillaries.

TUBULAR SECRETION ADDS MATERIAL TO THE ULTRAFILTRATE

Tubular secretion refers to a direction of movement of materials from the blood to the tubule lumen. Tubular reabsorption refers to the direction of movement in

QUANTITATIVE HUMAN PHYSIOLOGY: AN INTRODUCTION

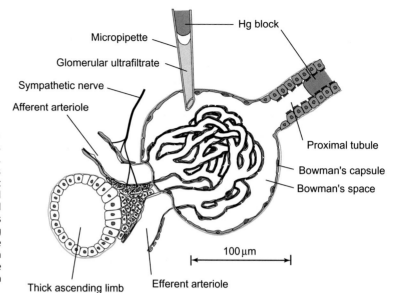

FIGURE 7.3.1 Micropuncture of the renal corpuscle. The renal corpuscle consists of the glomerulus, a tuft of capillaries between the afferent arteriole and the efferent arteriole, and its closely apposed Bowman's capsule. The outer fibrous layer of Bowman's capsule is continuous with the basement membrane of the proximal tubule. The flattened epithelial layer of Bowman's capsule is also continuous with the cuboidal epithelium of the proximal tubule. Micropuncture was accomplished by A.N. Richards in the 1920s by inserting a micropipette into Bowman's space and blocking the movement of fluid down the proximal tubule by inserting a plug of mercury. A second plug of Hg in the micropipette sealed the sample from air and thereby prevented evaporation of the tiny sample.

the opposite direction, from lumen to blood. Tubular secretion was demonstrated by showing that some materials show up in the urine in excess of what can be explained by filtration. This is a quantitative argument that depends on the rates of excretion and not the concentration in the urine because the concentrations of excreted materials depend on the reabsorption of water relative to the solute.

THE THREE ELEMENTARY NEPHRON PROCESSES ARE ULTRAFILTRATION, REABSORPTION, AND SECRETION

As outlined above, the first step in the formation of urine is the formation of an ultrafiltrate of plasma by the renal corpuscle. The fluid then flows down the nephron beginning with the proximal convoluted tubule, where water and solutes are either removed from the tubular fluid and returned to the blood (reabsorbed), or water and solutes are added to the tubular fluid from the blood (secretion). **Ultrafiltration, reabsorption,** and **secretion** constitute the three elementary functions of the nephron. The remainder of renal physiology is how these processes produce the final urine and how they are regulated to meet the demands of homeostasis.

THE CLEARANCE OF INULIN PROVIDES AN ESTIMATE OF THE GLOMERULAR FILTRATION RATE

In order to understand how the body regulates ultrafiltration, reabsorption, and secretion, we need to be able to quantify these processes. The clearance of inulin is an experimental measure of the rate of glomerular filtration. To see this, consider the simplified nephron shown in Figure 7.3.2. The nephron consists of a tube running from the renal corpuscle down into the duct of Bellini. At steady state, there is a constant throughput

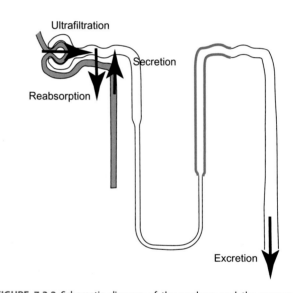

FIGURE 7.3.2 Schematic diagram of the nephron and the sources and sinks of materials across the tubule. At steady state for a material which is neither synthesized nor degraded by the nephron, the amount entering the tubule per unit time is the amount filtered + the amount secreted. During the same time, the amount leaving the tubule per unit time is the amount reabsorbed + the amount excreted.

of materials from the blood and into the urine, and the nephron is the agent of this throughput. However, the nephron itself, and the volume and contents of the tubular fluid, is constant. That is, the input of all materials into the tubular fluid is equal to the output of all materials at steady state:

[7.3.1] $$\text{Input} = \text{output}$$

In this equation, we assume that the material is neither synthesized nor degraded by the tissue. The flow of material into the tubular fluid may occur at a number of sites in the nephron. These include glomerular filtration and secretion, denoted by the arrows leading into the tubule lumen in Figure 7.3.2. The output is the sum of all the material excreted in the final urine and the material that is reabsorbed from the tubular fluid back into the blood. These are denoted by the arrows leading away from the tubule lumen in Figure 7.3.2. So we can write

[7.3.2] Amount filtered + amount secreted = amount excreted + amount reabsorbed

What this simple equation means is that if something shows up in the urine (the amount excreted), it had to come from someplace and that someplace is either the filtered plasma or secretion from plasma, or both. Conversely, if a substance is filtered but does not show up in the urine, or shows up less in the urine than in the filtrate, then it must have been reabsorbed. It is possible to deduce which of these events is occurring by looking at the rate of excretion of a substance as a function of its plasma concentration.

The amounts filtered, secreted, excreted, or reabsorbed in Eqn (7.3.2) refer to the amounts filtered, secreted, excreted, or reabsorbed *in some definite time interval*. We can divide both sides of Eqn (7.3.2) by the time interval over which the amounts are measured, to obtain

[7.3.3] Rate of filtration + rate of secretion = rate of excretion + rate of reabsorption

Inulin is a polymer of fructose derived from plants. It is small enough that it is freely filtered by the glomerulus, meaning that the concentration of inulin in the ultrafiltrate is the same as its concentration in the plasma. It is also neither secreted nor reabsorbed by the kidney tubules. Thus, in Eqn (7.3.3) the rate of secretion = 0 and the rate of reabsorption = 0, and Eqn (7.3.3) is rewritten as

[7.3.4] Rate of inulin filtration = rate of inulin excretion

The amount of inulin that is filtered per unit time is given as GFR × UF_{inulin}, where the GFR refers to the **glomerular filtration rate**, in units of volume per unit time, and UF_{inulin} refers to the concentration of inulin in the ultrafiltrate. Because inulin is freely filtered, the ultrafiltrate concentration is the same as its plasma concentration, denoted as P_{inulin}, so that the rate of filtration is GFR × P_{inulin}. The amount of inulin excreted per unit time is the urinary inulin concentration times the rate of urine excretion, denoted here as Q_u. Thus Eqn (7.3.4) becomes

[7.3.5] $$GFR\, P_{inulin} = Q_u U_{inulin}$$

This equation is easily rearranged to solve for GFR:

[7.3.6] $$GFR = \frac{Q_u U_{inulin}}{P_{inulin}}$$

A typical value for the GFR for the two kidneys is 120 mL min^{-1}. This is the aggregate rate of formation of glomerular filtrate from all functional nephrons. Since there are about 1.3×10^6 nephrons in each kidney, the aggregate GFR is produced by a single nephron GFR that averages about 120 mL min^{-1}/2.6×10^6 nephrons $\approx 45 \times 10^{-9}$ L min^{-1} per nephron. This is an average value for the SNGFR, the single nephron GFR. Some nephrons may have greater SNGFR and some may have smaller SNGFR.

THE CLEARANCE OF PARA AMINO HIPPURIC ACID ALLOWS ESTIMATION OF RENAL PLASMA FLOW

Para amino hippuric acid, or PAH, is avidly secreted by the renal tubules so that nearly all of the blood that enters the kidneys is "cleared" of PAH and renal venous blood contains very little PAH. The amount of PAH appearing in the urine per unit time, then, is equal to the amount contained in the blood that perfuses the kidney per unit time. This conservation relation is written as

[7.3.7] $$ERPF\, P_{PAH} = Q_u U_{PAH}$$

where ERPF stands for the **effective renal plasma flow**, P_{PAH} is the plasma concentration of PAH, Q_u is the rate of urine formation, and U_{PAH} is the concentration of PAH in the urine. The left-hand side of this equation is the amount of PAH that enters the kidney per unit time, and the right-hand side is the amount of PAH that is excreted per unit time. This equation can be rewritten as

[7.3.8] $$ERPF = \frac{Q_u U_{PAH}}{P_{PAH}}$$

As in Eqn (7.3.6), the units of ERPF are in mL min^{-1}. A typical value for ERPF calculated from PAH excretion is about 600 mL min^{-1}. This calculation gives the ERPF because part of the blood plasma that supplies the kidneys goes to regions of the kidney that do not filter the blood or secrete materials from it into the nephrons. These regions include the adipose tissue and the renal capsule and the renal medulla. The true renal plasma flow (RPF) can be estimated by redefining the boundaries through which the flows of material occur, as shown in Figure 7.3.3. Here the input of PAH to the kidneys is the RPF times the renal arterial plasma [PAH]. The output includes two sinks: the urine and the venous blood. The conservation principle equates the input and the output to give

[7.3.9] $$RPF\, RA_{PAH} = RPF\, RV_{PAH} + Q_u U_{PAH}$$

where RA indicates the renal arterial plasma and RV denotes the renal venous plasma. Solving for RPF, we obtain:

[7.3.10] $$RPF = \frac{Q_u U_{PAH}}{RA_{PAH} - RV_{PAH}}$$

Equation (7.3.10) gives the true RPF. It differs from Eqn (7.3.8) in that the equation for ERPF requires that the renal venous plasma [PAH] is zero. Typically about 10% of the renal plasma supplies regions of the kidneys that do not actively secrete PAH. Thus the renal plasma flow is about 10% higher than the ERPF. RPF averages about 660 mL min^{-1}.

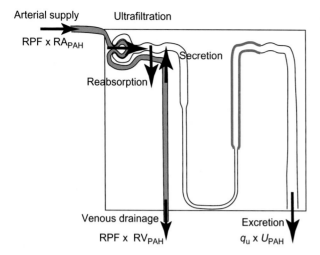

FIGURE 7.3.3 Estimation of the renal plasma flow from the renal handling of PAH. The boundaries of the system are expanded to include the vasculature. The balance of input and output refers to inputs and outputs across the boundaries of the system, and movement of materials inside the "box" do not enter into the conservation equation.

The plasma flow is related to the blood flow. The **hematocrit** is the fraction of blood volume occupied by the blood cells, mostly the erythrocytes (see Chapter 5.1). Typically this is about 0.45, meaning that 45% of the blood volume consists of the blood cells. This means that the plasma volume is the remaining volume or 55% of the blood volume. The plasma volume is thus $(1 - \text{Hct}) \times$ blood volume. Thus the effective renal blood flow and renal blood flow can be calculated from the effective renal plasma flow and renal plasma flow as

[7.3.11]
$$\text{ERBF} = \frac{\text{ERPF}}{1 - \text{Hct}}$$
$$\text{RBF} = \frac{\text{RPF}}{1 - \text{Hct}}$$

Using Hct = 0.45, the typically ERBF would be 600 mL min^{-1}/(1 − 0.45) = 1090 mL min^{-1}. The total renal blood flow would be given as RBF = 660 mL min^{-1}/(1−0.45) = 1200 mL min^{-1}. The typical cardiac output is about 5 L min^{-1}, so that the kidneys receive 1.2 L min^{-1}/5.0 L min^{-1} = 0.24. Thus nearly one-quarter of the cardiac output at rest is directed to the kidneys.

THE CLEARANCE OF A SUBSTANCE DEPENDS ON HOW IT IS HANDLED BY THE KIDNEY

Equations (7.3.6) and (7.3.8) both have the form

[7.3.12]
$$C_x = \frac{Q_u U_x}{P_x}$$

where C_x is called the **clearance** of substance x. Q_u is the flow of urine, in mL min^{-1}; U_x is the concentration of substance x in the excreted urine; and P_x is the plasma concentration of substance x. The clearance of any substance x is a flow measured in mL min^{-1} but it represents a virtual flow, not a real flow. For example, the clearance of inulin is 120 mL min^{-1}, meaning that the amount of inulin that appears in the urine each minute is equal to the amount of inulin contained in 120 mL of plasma, and therefore 120 mL of plasma has been "cleared" of inulin in each minute. Similarly, when the clearance of PAH is 600 mL min^{-1}, it means that the amount of PAH in the urine is equal to the amount contained in 600 mL of plasma, and thus 600 mL of plasma has been "cleared" of PAH each minute. The interpretation given to the clearance of a substance depends on the way in which the nephron processes it. In the case of inulin, the clearance is equal to the GFR because inulin is freely filtered by the glomerulus, but it is neither secreted nor reabsorbed. The clearance of PAH equals the ERPF because blood that perfuses the tubules is completely cleared of PAH by secretion.

GLOMERULAR FILTRATION IS LIKE A LEAKY HOSE; ABOUT 20% OF THE PLASMA CONSTITUENTS END UP IN THE FILTRATE

The GFR is typically 120 mL min^{-1} and the ERPF is about 600 mL min^{-1}. Thus the kidney takes 600 mL min^{-1} of plasma and makes 120 mL min^{-1} of ultrafiltrate. The plasma flow through the aggregate efferent arterioles, then, should be about 480 mL min^{-1}. Thus the kidneys do not "clear" inulin from 120 mL of plasma per minute. Instead they incompletely clear the inulin from 600 mL min^{-1}. The amount of inulin that is filtered, and ultimately excreted, per minute is equal to that contained in 120 mL of plasma. In this sense, ultrafiltration "clears" 120 mL of plasma per minute. Thus glomerular filtration is like a leaky hose: most of the fluid passing through the afferent arteriole continues on through the efferent arteriole, with a relatively small fraction passing into the tubule lumen as an ultrafiltrate. This idea is expressed in the **filtration fraction**, defined as

[7.3.13]
$$\text{Filtration fraction} = \frac{\text{GFR}}{\text{RPF}}$$

Since GFR ≈ 120 mL min^{-1} and RPF ≈ 660 mL min^{-1}, the filtration fraction is normally about 120/660 = 0.18.

MULTIPLE STRUCTURES CONTRIBUTE TO THE SELECTIVITY OF THE GLOMERULAR FILTRATE

The glomerular filtration barrier consists of a sandwich of three layers: the **endothelial cell layer** of the glomerular capillaries, a **basement membrane** consisting of a meshwork of fibers of the extracellular matrix, and a second cell layer provided by specialized cells called **podocytes**. Each of these layers contributes to the filtration barrier. Figure 7.3.4 shows these layers.

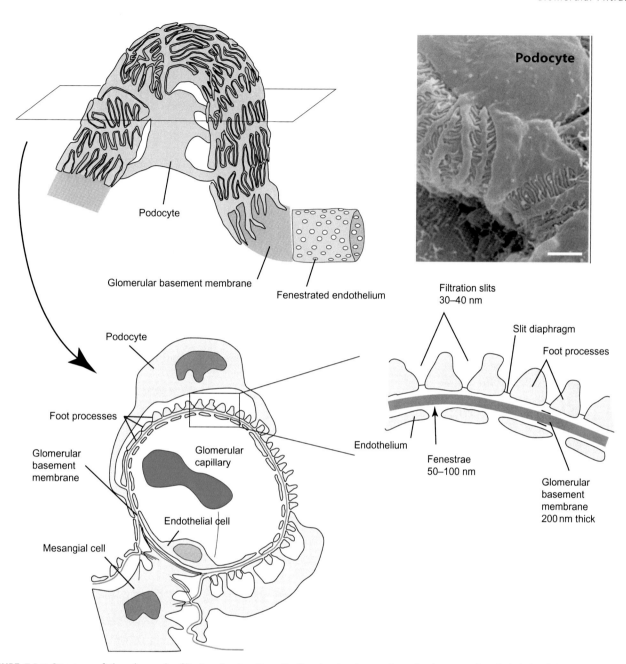

FIGURE 7.3.4 Structure of the glomerular filtration barrier. The ultrafiltration barrier consists of a fenestrated endothelial cell layer, a glomerular basement membrane (GBM), and a layer of glomerular podocytes. The endothelial cell is penetrated by a large number of fenestrae, or windows, about 50–100 nm in diameter. The endothelial cells lie along the glomerular basement membrane, which consists of a meshwork of fibers. This basement membrane presents part of the filtration barrier. Glomerular podocytes line the glomerular basement membrane on the side of the ultrafiltrate, in Bowman's space. These cells send out foot processes that interdigitate with other foot processes, thereby forming filtration slits, about 30–40 nm across. The glomerular epithelium slit diaphragms (GESD) link foot processes together. The slit diaphragms also provide a crucial filtration barrier to proteins 70 kDa and larger. At the right is a scanning EM in false color of a podocyte with interdigitating foot processes that effectively cover a glomerular capillary. *From R.P. Scott and S.E. Quaggin, The cell biology of renal filtration, J. Cell Biol.* **209**:199–210 (2015).

THE ENDOTHELIAL CELL LAYER RETAINS THE CELLULAR ELEMENTS OF BLOOD

The endothelial cells that line the glomerular capillaries are specialized for their function. These capillaries are **fenestrated**. The root word for fenestra means "window." These capillaries are full of holes or windows that provide nearly no resistance to fluid flow but retain all of the blood cells and platelets. The fenestrae are typically about 50–100 nm in diameter. This distance is much larger than the width of most plasma proteins, suggesting that the fenestrae exclude the cells but all of the plasma solutes, including the plasma proteins, easily pass through this endothelial layer. The endothelial layer is specialized for filtration. Electron microscopy of rat glomeruli indicates that the fenestrae cover about 20% of the endothelial surface. This is a far greater area than the fenestrae that occur in other capillary beds, such as muscle.

Although the fenestrae seem too large to exclude proteins, evidence suggests that proteins are at least partially retained by the glomerular endothelial layer, due to the presence of the **glycocalyx**, a fibrous network of negatively charged glycoproteins on the surface of the endothelial cells. "Glycocalyx" derives from roots meaning "sweet husk." The glycocalyx in the glomerular capillaries is expanded to form the endothelial surface layer, ESL, more than 200 nm thick, that may form a restrictive barrier to protein filtration. The role of the ESL in protein restriction is controversial.

THE BASEMENT MEMBRANE EXCLUDES SOME PROTEINS

THE GLOMERULAR BASEMENT MEMBRANE IS A MESHWORK OF FIBERS INCLUDING TYPE IV COLLAGEN, LAMININ, NIDOGEN, AND PROTEOGLYCANS

The glomerular basement membrane derives from a fusion of the basement membranes of the glomerular endothelial cells and the podocytes. It consists of a meshwork of fibers. The human glomerular extracellular matrix contains some 144 different proteins. The most abundant of these includes **Type IV collagen**, **laminin**, nidogen, and proteoglycans, and is illustrated in Figure 7.3.5. Type IV collagen forms the basement membrane backbone. It consists of a protomer fiber made up of three α chains, each about 180 kDa. Six different genes encode for various α subunits of Type IV collagen. The Type IV collagen protomers self-associate to form a network that supports the basement membrane.

Laminin is the second most abundant basement membrane component. These large molecules (600–800 kDa) are heterotrimers of α, β, and γ chains. Currently recognized isoforms include five distinct α chains (α1–5), three isoforms of the β chains (β1–3), and two types of γ chains (γ1 and γ2). There are 30 possible combinations of these isoforms, but only 11 have been identified. They are named according to their α, β and γ components. In the adult, LM-521 predominates. The three subunits arrange themselves to produce a long arm and three short arms. In the presence of Ca^{2+}, laminin polymerizes into a hexagonal lattice by interactions between the short arms.

Nidogen, also known as entactin, is about 150 kDa and is shaped like a dumbbell, consisting of three globular domains connected by linker segments. Nidogen binds the γ1 chain of laminin, and it also binds type IV collagen and to another component of basement membranes, perlecan.

Proteoglycans have a protein core with attached chains of glycosaminoglycans. Heparan sulfate proteoglycans are the most abundant class in basement membranes. The negative charges of the heparan sulfate confer a net negative surface charge to the basement membrane. These proteoglycans include perlecan (see above) and agrin.

THE GLOMERULAR BASEMENT MEMBRANE FILTERS OUT SOME PROTEINS

Whether the glomerular basement membrane (GBM) or the glomerular slit diaphragms determine the selectivity of the glomerular filtration membrane has been debated for more than 30 years. Several kidney diseases that result in **glomerulonephritis** and consequent leakage of protein into the urine are caused by defects in basement membrane proteins. Mutations in COL4A3 and COL4A4 are examples. Mice deficient in the β2 component of laminin die shortly after birth due to massive **proteinuria**, loss of proteins in the urine.

THE SLIT MEMBRANE RETAINS PROTEINS 70 kDa OR LARGER

The slit diaphragms are specialized junctions between neighboring podocyte foot processes. They contain a variety of proteins including nephrin, podocin, CD2AP, α-actinin-4, and other key proteins associated with adherens junctions and tight junctions including ZO-1 (zona occludens-1), JAM-1 (junctional adhesion molecule 1), catenins, occludin. The exact disposition of these is not yet known, but a proposed structure is shown in Figure 7.3.5. Early experiments showed that proteins like horse radish peroxidase, with a molecular weight of 40 kDa, penetrate through the endothelial layer, basement membrane, and the slit pores to enter Bowman's space. Larger proteins such as myeloperoxidase, about 160 kDa, cross the endothelium and the basement membrane but pile up at the slit diaphragms. Other experiments with knock-outs of slit diaphragm proteins in mice suggest that the glomerular epithelial slit diaphragm is the crucial barrier to plasma proteins. According to this view, mutations of the glomerular basement proteins cause proteinuria by secondarily changing the slit diaphragms.

THE SIEVING COEFFICIENT DEPENDS MAINLY ON THE SLIT DIAPHRAGM

The local sieving coefficient of proteins or of any solute is defined as

$$[7.3.14] \qquad \text{Sieving coefficient} = \Theta = \frac{C_B}{C_P}$$

where C_B is the concentration of protein in Bowman's space and C_P is its concentration in the glomerular capillary plasma. Theoretical calculations of the sieving coefficient for membranes of different composition indicate that the slit diaphragm provides the filtration barrier to proteins (see Figure 7.3.6).

THE GLOMERULUS SELECTIVELY EXCLUDES PROTEINS BASED ON SIZE AND CHARGE

The negative charges on the proteins of the GBM and the slit diaphragm selectively restrain negatively charged proteins more than positively charged proteins.

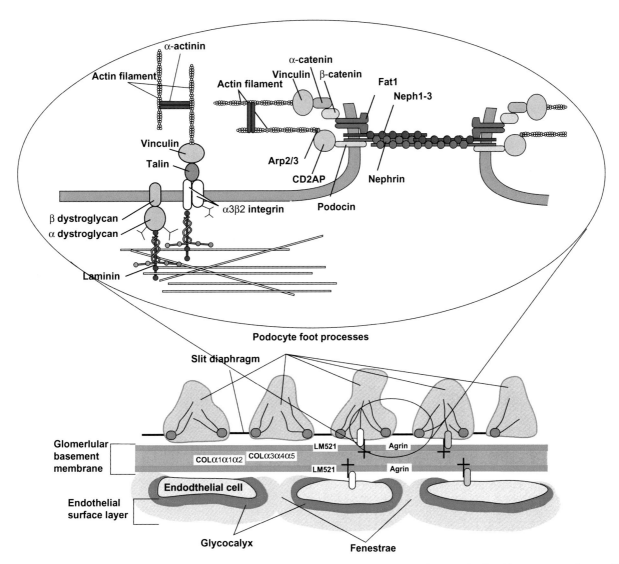

FIGURE 7.3.5 Molecular structure of the glomerular filtration barrier. The glomerular basement membrane is the fused basement membrane of both the endothelial cell layer and the podocyte cell layer. The endothelial cells have a glycocalyx coat that is extended to form the endothelial surface layer, over 200 nm thick. The charges in this layer may be important in restricting plasma protein filtration. The basement membrane is ordered, with a collagen core lined with layers relatively enriched in agrin, perlecan, and laminin 521. The collagen is mostly α3α4α5 type IV collagen, but the less abundant α1α1α2 type IV collagen is distributed more to the endothelium side. Both endothelial cells and foot processes anchor to the basement membrane through integrins and dystroglycans that bind laminin in the basement membrane. The slit diaphragms themselves contain specialized proteins including Nephrin and Neph1−3 and the cadherin Fat1. Podocin within the foot processes binds nephrin and CD2AP that links the slit diaphragm to the actin cytoskeleton. The slit diaphragms likely consist of mainly of cross-bound nephrin and neph1−3. Additional proteins, such as occludin, junctional adhesion molecule, and zona occludens-1, not shown, are also present in the foot processes.

The combined effects of protein size and charge on their sieving coefficient are shown in Figure 7.3.7.

THE STARLING FORCES DRIVE ULTRAFILTRATION

The forces that govern the formation of the glomerular ultrafiltrate are identical in form to those that govern exchange of material across any capillary in the body. The equation that describes the relationship between flow and forces across a porous membrane is

$$[7.3.15] \quad J_v = L_P \left[(P_L - P_R) - \left(\sum_i \sigma_i \pi_{i\,L} - \sum_i \sigma_i \pi_{i\,R} \right) \right]$$

where J_v is the volume flux; L_P is the hydraulic permeability; P_L and P_R are the hydrostatic pressures on the left and right sides of the membrane, respectively; σ_i is the reflection coefficient of the ith species; and π_{iL} and π_{iR} are the osmotic pressures contributed by the ith species on the left and right sides of the membrane, respectively. If we multiply both sides by the aggregate area of the glomeruli, we obtain

$$\text{GFR} = Q_v = A\, L_v$$
$$= A\, L_P \left[(P_L - P_R) - \left(\sum_i \sigma_i \pi_{i\,L} - \sum_i \sigma_i \pi_{i\,R} \right) \right]$$

[7.3.16]

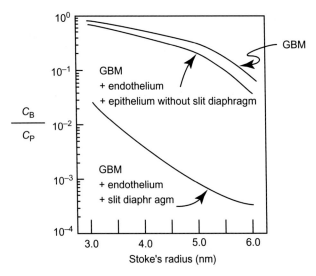

FIGURE 7.3.6 Results of theoretical calculations of the sieving coefficient as a function of molecular size for hypothetical membranes consisting of glomerular basement membrane alone (GBM); glomerular basement membrane plus the attached endothelial cell layer and podocyte epithelial cell layer, but without the slit diaphragms; the entire capillary wall consisting of endothelium, glomerular basement membrane, and podocytes with intact slit diaphragms. *Modified from A. Edwards, B.S. Daniels, and W.M. Deen, Ultrastructural model for size selectivity in glomerular filtration, Am. J. Physiol. 276:F892–F902, 1999.*

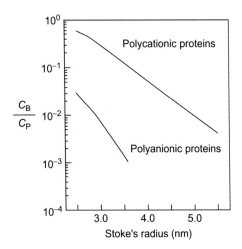

FIGURE 7.3.7 Effect of charge and size on the sieving coefficient of renal corpuscles *in vivo*. Values of the sieving coefficient were obtained by direct micropuncture or by urinary clearance corrected for tubular reabsorption. Note that cationic proteins (those with + net charge) are filtered far more easily than anionic proteins of the same size. The filterability of proteins diminishes precipitously with larger size. For reference, the Stokes radius for serum albumin is about 3.6 nm. *Modified from B.D. Myers, Determinants of the glomerular filtration of macromolecules, in S.G. Massry and R.J. Glassock, eds., Massry & Glassock's Textbook of Nephrology, 4th edition, Lippincott, Williams and Wilkins, Philadelphia, PA, 2001.*

We further simplify the equation by assuming that materials are either freely filtered or they are not filtered at all. If they are freely filtered, then $\sigma = 0$; if they are not filtered at all, then $\sigma = 1.0$. We identify the "left" side of the membrane with the capillary, and the "right" side with Bowman's space. Then Eqn (7.3.16) becomes

[7.3.17] $\quad \text{GFR} = K_f[(P_{GC} - P_{BS}) - (\pi_{GC} - \pi_{BS})]$

where K_f is the filtration coefficient; P_{GC} is the hydrostatic pressure within the glomerular capillaries; P_{BS} is the hydrostatic pressure within Bowman's space; and π stands for the total osmotic pressure contributed only by the proteins that are not filtered by the glomerulus. Recall that the osmotic pressure is the sum of all contributions to the osmotic pressure and that it is a property of the solution itself. Here we have partitioned the total osmotic pressure into the part contributed by the freely permeable solutes and the part contributed by the impermeant proteins. We have discarded the part contributed by freely permeable solutes because it makes no contribution to the *effective* osmotic pressure since $\sigma = 0$ for these solutes. The remaining part of the osmotic pressure, that part contributed by the proteins, is called the **colloid osmotic pressure** or the **oncotic pressure** (see Chapter 5.2). Because the protein content of the ultrafiltrate is near zero, $\pi_{BS} \approx 0$ and Eqn (7.3.17) is simplified further to

[7.3.18] $\quad \text{GFR} = K_f[P_{GC} - P_{BS} - \pi_{GC}]$

Thus the driving force favoring glomerular filtration is the glomerular capillary hydrostatic pressure. The glomerular capillary is a high-pressure capillary, with hydrostatic pressures near 60 mmHg over its length. The formation of the glomerular filtrate is opposed by the pressure within Bowman's space, which is a fairly constant 20 mmHg. The oncotic pressure of the glomerular capillaries is not constant over their length because the blood is concentrated by removal of about 20% of its volume (the filtration fraction) as a protein-free filtrate. Thus the protein left behind is concentrated. The oncotic pressure is not linear with protein concentration but can be fit by a higher order polynomial (see Problem 15 of PS2.2). The oncotic pressure at the beginning of the glomerular capillary is about 25 mmHg. As the blood passes through the capillaries and protein-free filtrate is removed, the oncotic pressure climbs to 35 mmHg. The degree to which the blood is concentrated depends in part on its flow. Faster flowing blood has a lower filtration fraction because it is in the glomerular capillaries a shorter time. If the flow is sufficiently sluggish, **filtration equilibrium** can be reached. At this point, the net force for filtration becomes zero and no further filtration occurs. The balance of Starling forces is shown schematically in Figure 7.3.8.

Average values for the Starling forces are as follows:

$$P_{GC} = 60 \text{ mmHg}$$
$$P_{BS} = 20 \text{ mmHg}$$
$$\pi_{GC} = 30 \text{ mmHg}$$

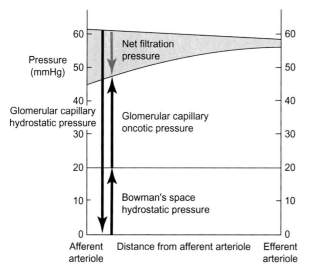

FIGURE 7.3.8 Starling forces as a function of distance from the afferent arteriole to the efferent arteriole. Glomerular capillary hydrostatic pressure is about 60 mmHg and declines only slightly through the glomerular capillary. Similarly, the hydrostatic pressure within Bowman's space is about 20 mmHg and is not different along the length of the capillary. The oncotic pressure of the glomerular capillary begins at about 25 mmHg and increases as protein-free filtrate is removed from the blood. At the efferent arteriole end of the glomerular capillary, the oncotic pressure increases to about 35 mmHg. These profiles in pressures depend on the flow of blood through the capillary. If flow is slower, the glomerular capillary oncotic pressure may rise further and the net filtration pressure will shrink to zero. At this point, the glomerular capillary reaches filtration equilibrium.

Insertion of these into Eqn (7.3.18) shows that the net driving force averages about 10 mmHg in favor of filtration.

WHY DOES THE GLOMERULAR FILTRATION BARRIER NOT CLOG?

How the glomerular filtration barrier provides both a size and charge selectivity, and the respective roles of the glomerular endothelial cell, the basement membrane, and the slit diaphragms have been intensely investigated and debated for over 50 years, and the issues are still not resolved. There are two major questions that need answering: (1) why doesn't the glomerular filtration barrier clog? And (2) why does the sieving coefficient for proteins increase when the glomerular filtration rate decreases? Hydraulic-driven filters invariably clog, and the solution is replacement of the filter or back-flushing the filter to remove the clogging material. The kidney cannot use either of these solutions. The most likely case is that the kidneys do not become clogged because the retained proteins diffuse back into the blood. If the major pathway for albumin across the glomerular filtration barrier is by diffusion, then reduction of filtration to zero would lead to diffusion equilibration across the barrier, or an increase in the sieving coefficient.

Clinical Applications: Nephrotic Syndrome

The nephrotic syndrome is a set of symptoms that include the following:

- protein in the urine;
- low blood protein levels;
- swelling or edema.

It may also include elevated levels of serum lipids, anemia, and vitamin D deficiency, all because of loss of plasma proteins into the urine. This can have multiple causes, but all involve defects in the glomerular barrier to proteins so that excess proteins are filtered and thereby excreted in the final urine. The three barriers were discussed in the text: the fenestrated endothelial cell layer, the GBM, and the podocyte and slit diaphragm.

Nephrotic syndrome can be primary or secondary. Primary causes are described by their histological changes: minimal change disease, focal segmented glomerulosclerosis, and membranous nephropathy. Secondary causes are described by their underlying cause, which include diabetes mellitus, sarcoidosis, hepatitis B, hepatitis C, bacterial infections, parasitic infections, and more.

All of the diseases are characterized by protein in the urine, at least 3.5 g per 24 h. The loss of protein can cause hypoalbuminemia, with resulting edema that may show as puffiness around the eyes, pitting edema in the legs, and pleural effusion. Loss of proteins stimulates liver synthesis, including lipoproteins. Because lipoprotein lipase levels fall, lipoprotein levels increase. Loss of vitamin D binding protein can lead to vitamin D deficiency diseases, with calcium malabsorption and bone disease.

Mutations of nephrin, a protein of the filtration slit, cause nephrotic syndrome. Mutations of podocin also cause nephrotic syndrome that is insensitive to steroid treatment. Podocin is an integral protein of the podocyte cell membrane that segregates into lipid rafts and is required to recruit nephrin into those rafts. Current thought is that podocin and nephrin form a signaling complex that activates protein kinases involved in glomerular structural integrity. These mutations cause minimal change diseases in which structural changes are evident only at the electron microscope level and not at the histological level. Until recently, these were part of the set of nephrotic syndrome called idiopathic nephrotic syndrome.

Membranous glomerulonephritis is one of the more common causes of nephrotic syndrome in adults. It is an inflammatory disease, believed to be caused by binding of antibodies to antigens in the GBM that triggers the formation of a membrane attack complex from complement (see Chapter 5.3). This triggers release of proteases and oxidants that damage the capillary walls, causing them to become leaky. Histology reveals thickened basement membranes.

Treatment depends on etiology. For all nephrotic syndromes, monitoring and maintaining normal fluid levels and distribution among the body compartments are the goal. This could include restriction of fluid intake, restriction of salt intake, regular monitoring of blood pressure and urine output, and the use of diuretics. Inflammatory causes of nephrotic syndrome are treated with immunosuppressants such as prednisolone and dietary modificaton.

SUMMARY

Glomerular filtration is the first step in the formation of urine. It was originally posited on morphological grounds and was established by micropuncture and microanalysis studies in the 1920s. The GFR can be estimated from the inulin clearance because inulin is neither secreted nor reabsorbed by the nephron. The ERPF can be estimated from the clearance of para amino hippuric acid because PAH is so avidly secreted by the nephrons. The clearance of any substance is calculated as

$$C_x = \frac{Q_u \times U_x}{P_x}$$

where C_x is the clearance, Q_u is the urine flow rate, in mL min^{-1}, U_x is the urinary concentration of x, and P_x is the plasma concentration of x. Clearance has the units of mL min^{-1} and its interpretation depends on the way in which the nephrons handle substance x. The RPF is about 660 mL min^{-1}, whereas the GFR is about 120 mL min^{-1}. The filtration fraction, GFR/RPF, is about 0.2.

The rapid filtration of large volumes of fluid is accomplished by the specialized structures of the glomerulus and Bowman's capsule. The filtration barrier consists of the capillary endothelial cell, basement membrane, and podocytes of Bowman's capsule. The endothelial cells are highly fenestrated. These openings increase the area available for filtration but retain the cellular elements of blood. The basement membrane restricts filtration of larger molecular weight proteins. Slit diaphragms appear in the openings between podocyte foot processes. These may form the final filtration barrier to proteins like plasma albumin. The sieving coefficient is given as

$$\Theta = \frac{C_B}{C_P}$$

the ratio of concentration of material in Bowman's space to its concentration in plasma. For most small molecular weight materials, $\Theta = 1.0$.

The forces producing the ultrafiltrate are the same as those that govern exchange across any capillaries. The net force is the balance between hydrostatic and osmotic pressures within the glomerular capillary and the fluid in Bowman's space. The glomerular capillary is specialized in that its filtration coefficient is large and its major driving force, the hydrostatic pressure within the glomerular capillary, is high.

REVIEW QUESTIONS

1. What is the sieving coefficient for small molecular weight solutes such as amino acids, glucose, and electrolytes?
2. What characteristics of molecules determine their sieving coefficient?
3. Why is the clearance of inulin equal to the GFR, whereas the clearance of PAH is equal to the ERPF?
4. What is the main force favoring formation of the glomerular ultrafiltrate?
5. What are the main forces opposing the formation of the glomerular ultrafiltrate?
6. What layers constitute the glomerular filtration apparatus and what specializations allow them to produce a rapid ultrafiltrate that excludes most plasma proteins?

Problem Set 7.1
Fluid Volumes, Glomerular Filtration, and Clearance

1. The time course of decay of plasma [inulin] shown in Figure 7.1.1 can be simultaneously used to determine the ECF volume and the GFR. The disappearance of inulin is given by the first-order equation

   [7.PS1.1] $\quad \dfrac{dN_{inulin}}{dt} = -\text{GFR}\, P_{inulin}$

   where N_{inulin} is the number of moles of inulin in the body and P_{inulin} is its plasma concentration. The plasma concentration is the total number of moles of inulin in the body divided by its volume of distribution, the ECF:

   [7.PS1.1A] $\quad P_{inulin} = \dfrac{N_{inulin}}{\text{ECF}}$

   Combining these two equations, we get

   [7.PS1.1B] $\quad \dfrac{dN_{inulin}}{dt} = -\dfrac{\text{GFR}}{\text{ECF}} N_{inulin}$

   This equation describes a first-order decay curve.
   A. Separate variables and integrate the first-order decay equation between the definite limits of time $t = 0$ and t, corresponding to $N_{inulin} = N_0$ and N_t.
   B. Two hours after establishing steady state in a 70-kg person and then stopping infusion, [inulin] falls from the steady-state value of 20 mg% at steady state to 7.2 mg%. Calculate GFR/ECF from the first-order decay equation.
   C. The total urine collected during 8 hours was 500 mL, and the average inulin concentration was 560 mg%. Calculate the ECF.
   D. From B and C, calculate the GFR.

2. A woman weighing 60 kg is given 10 mg of Evan's Blue dye intravenously. Ten minutes later a blood sample was obtained from another vein and colorimetric analysis of the plasma showed 0.4 mg% of the dye. Assume that the administered dye was evenly distributed in the plasma compartment by the end of the 10 minutes and that no dye was lost from the plasma during this period.
   A. Calculate the woman's plasma volume.
   B. If the woman's hematocrit is 0.40, what is her total blood volume?

3. A person has TBW = 42 L and ECF = 14 L and a plasma osmolarity of 300 mOsM. After losing 4 g of NaCl and 2 L in sweat, what is the new ECF and ICF volume and osmolarity?

4. A person has TBW = 50 L and ECF = 16 L and an original osmolarity of 295 mOsM. The person eats a meal that contains 6 g of NaCl and 1 g of KCl and 0.5 L of water. Assume it is all absorbed and distributed. Calculate the estimated new ICF and ECF volume and osmolarity.

5. Suppose you injected someone with deuterated water, ^2H_2O. At some time, the marker water would be evenly distributed among all exchangeable fluid compartments. The ^2H_2O is filtered through the glomerulus, and then reabsorbed. Assume that the total body water is 42 L and ECF is 14 L. The GFR is 120 mL min^{-1} and urine flow is a constant 1 mL min^{-1}. Assume that ingested fluids replenish the urinary losses on a continuous basis.
   A. How long would the ^2H_2O remain in the body? Derive an equation for plasma $[^2H_2O]$ as a function of time.
   B. What would the half-life of ^2H_2O be in the plasma?
   Suppose you inject the same person with inulin and that a peak [inulin] was obtained some minutes later.
   C. Derive an equation for plasma [inulin] as a function of time.
   D. What would the half-life of inulin be in the plasma?
   E. How does this compare to B?

6. A person weighs 105 kg. His total body water is 48 L.
   A. Calculate his lean body mass.
   B. Calculate the percent of body fat in excess of the lean body mass.
   C. Would you say that this person is obese? Why or why not?

7. The empirical fit to the oncotic pressure due to albumin is given by Landis and Pappenheimer (Handbook of Physiology, vol. 2, Section 2, pp. 961–1034, 1963) as

   [7.PS1.1C] $\quad \pi_{albumin} = 2.8C + 0.18C^2 + 0.012C^3$

   where $\pi_{albumin}$ is in units of mmHg and C is in units of g% (gram per 100 mL of plasma).
   A. Calculate albumin's contribution to plasma oncotic pressure at the afferent arteriole when [albumin] = 4 g%.

TABLE 7.PS1.1 Diffusion Coefficients and M_r for a Variety of Proteins

Protein	Molecular Weight	$D \times 10^7$ (cm² s⁻¹)	Stokes Radius (a_s)
Milk lipase	6600	14.5	
Metallothionein	9700	12.4	
Cytochrome C	12,000	12.9	
Ribonuclease	12,600	13.1	
Myoglobin	16,890	11.3	
Chymotrypsinogen	23,200	9.5	
Carbonic anhydrase	30,600	10.0	
Peroxidase II	44,050	6.8	
Albumin	68,500	6.1	
Lactoperoxidase	92,620	6.0	
Aldolase	149,100	4.6	

B. Assume the filtration fraction is 0.2 and that the sieving coefficient for albumin = 0. What is the concentration of albumin at the venule end of the glomerular capillary?

C. Calculate albumin's contribution to the plasma oncotic pressure at the efferent arteriolar end of the glomerulus.

8. Table 7.PS1.1 lists the diffusion coefficients of a variety of proteins. These were determined in water at 25°C at a viscosity of 1×10^{-3} Pa s, where Pa is pascal = 1 N m⁻².

A. Calculate the Stokes radii for the proteins using the Stokes Einstein equation:

$$[7.PS1.2] \qquad D = \frac{kT}{6\pi\eta a_s}$$

where D is the diffusion coefficient, k is Boltzmann's constant ($=1.38 \times 10^{-23}$ J mol⁻¹ K⁻¹), η is the viscosity, and a_s is the Stokes radius.

B. Based on these calculations and Figure 7.3.6, what is the approximate molecular weight cutoff of the kidney glomerulus? (This is the molecular weight of a substance corresponding to 50% retention by the membrane.)

9. Given that the average glomerular capillary pressure is about 55 mmHg, plasma oncotic pressure is about 28 mmHg, the hydrostatic pressure within Bowman's space is 20 mmHg, ultrafiltrate oncotic pressure is 0 mmHg, the GFR is 120 mL min⁻¹, and each kidney weighs 125 g, define a filtration coefficient and find its magnitude.

10. The following test results were obtained over a 24-hour period:

Urine volume = 1.4 L
Urine [inulin] = 100 mg%
Urine [urea] = 220 mmol L⁻¹
Urine [PAH] = 70 mg mL⁻¹
Plasma [inulin] = 1 mg%
Plasma [urea] = 5 mmol L⁻¹
Plasma [PAH] = 0.2 mg mL⁻¹
Hematocrit = 0.40

Calculate:
A. $C_{inulin}, C_{urea}, C_{PAH}$
B. ERPF
C. The rate of urea filtration
D. The rate of urea excretion
E. The rate of urea reabsorption
F. The rate of PAH filtration
G. The rate of PAH excretion
H. The rate of PAH secretion (assuming no PAH is reabsorbed).

11. The following data were obtained from a research animal:

Urine flow rate   1.5 mL min⁻¹
Urine [inulin]   150 mg%
Plasma [inulin]   1.5 mg%
Urine [A]   18 mg%
Plasma [A]   1.8 mg%
Urine [B]   200 mg%
Plasma [B]   1.0 mg%
Urine [C]   20 mg%
Plasma [C]   0.2 mg%

A. Based on this information, postulate how the kidney handles substances A, B, and C and justify your answer.
B. Propose experiment(s) to test your postulates.

12. One model of the glomerular membrane is a microporous membrane in which right cylindrical pores penetrate all the way through the membrane. Assume that the pores have a length of 50 nm and a radius of 3.5 nm. The viscosity of plasma is 0.002 Pa s. The average hydrostatic pressure in the glomerulus is 60 mmHg, hydrostatic pressure in Bowman's space is 20 mmHg, and the average oncotic pressure of glomerular capillary blood is 28 mmHg.

A. Calculate the flow through a single pore assuming laminar flow (use the Poiseuille flow equation).
B. How many pores would there have to be to produce a normal GFR?
C. If the total aggregate area of the kidneys for filtration is 1.5 m², what is the density of the pores (the number of pores per unit area)?
D. What fraction of the area is present as pores?

13. Suppose that the aggregate area of the ultrafiltration surface in the two kidneys combined is 15,000 cm² (1.5 m²). Suppose further that the aggregate area of pores in the glomerular barrier is

5% of the total area. The viscosity of plasma is 0.002 Pa s and the thickness of the barrier is 50 nm. The pressure driving the flow is the balance between the average glomerular pressure at 60 mmHg, the pressure within Bowman's space at 20 mmHg, and the oncotic pressure of plasma at average of 28 mmHg. Assume the rate of filtration is 120 cm^3 min^{-1}.
   A. Calculate the average velocity of flow across the membrane.
   B. Assuming Poiseuille flow, equate this to the average velocity of flow within the pores, and calculate the equivalent radius of the pores (the radius they would have if they were right cylindrical pores or uniform size).
14. Assume that myoglobin is a spherical protein with a Stokes radius of about 1.9 nm and that the pores in the renal corpuscle have a radius of 3.5 nm.
   A. Give an estimate of σ, the reflection coefficient, for myoglobin. Recall here that if the protein hits the rim, it is assumed to be reflected back into the plasma side of the membrane.
   B. Calculate the sieving coefficient for myoglobin, assuming that the only restriction for myoglobin filtration is at the pore entrance.
   C. In rhabdomyolysis, muscle membranes leak myoglobin into the blood. What do you think happens to that myoglobin, based on your answer to A and B?
15. The following test results were obtained over a 24-hour period:

    Urine volume   = 1.2 L
    Urine [inulin]   = 110 mg%
    Urine [creatinine]   = 170 mg%
    Plasma [inulin]   = 0.8 mg%
    Plasma [creatinine] = 1.2 mg%
    Hematocrit   = 0.40

   A. Calculate the clearance of inulin.
   B. Calculate the clearance of creatinine.
   C. What is the GFR?
   D. Creatinine is an endogenous by-product of muscle metabolism, originating from creatine. Assuming a steady state, estimate the daily production of creatinine.
   E. If the filtration fraction is 0.18, estimate the effective renal plasma flow.
16. The following test results were obtained over a 24-hour period:

    Urine volume   = 1.3 L
    Urine [inulin]   = 133 mg%
    Urine [glucose]   = 0 mg%
    Plasma [inulin]   = 0.8 mg%
    Plasma [glucose]   = 90 mg%
    Hematocrit   = 0.40

   A. Determine the GFR.
   B. Calculate the clearance of glucose.
   C. Estimate the daily filtered load of glucose. This is how much glucose is filtered per day.
   D. How much glucose is excreted per day?
   E. How much glucose is reabsorbed from the ultrafiltrate, per day?
17. The Stokes radius for water is 0.1 nm; urea's radius is 0.16 nm, and the radii for glucose and sucrose are 0.36 and 0.44 nm, respectively. The radius for inulin, a fructose polymer with a molecular weight of 5.5 kDa, is about 1.6 nm. Assume the radius of the pores in the glomerulus is 3.5 nm. *Hint*: The area available to the solvent is NOT the dimensions of the pore. The reflection coefficient is a parameter of area available to solute relative to water.
   A. Calculate the reflection coefficient for each of the solutes given above, on the assumption that the available area for the solutes is $\pi(a - a_s)^2$, where a is the radius of the pore and a_s is the radius of the solute.
   B. Based on this reasoning, estimate the sieving coefficient, Θ, for these small molecules.
   C. Does your calculation support the idea that inulin is freely filtered? Why or why not?
18. A 60-year-old man is admitted at 5 PM to the hospital after hiking all day and complaining of dizziness. He says he ran out of water the last 5 miles of his hike and the temperature was near 100°F. The patient reports that he last urinated at noon, and his urine was deep yellow. The results of lab tests were:

Na	143 mM
K	5.2 mM
pH	7.38
P_{O_2}	38 mmHg
Cl$^-$	102 mM
HCO$_3^-$	23 mM
P_{CO_2}	97 mmHg
Hemoglobin	15.1 g%
Creatinine	3.1 mg%
Systolic blood pressure	120 mmHg
Diastolic blood pressure	75 mmHg

   A. Which of these variables are abnormal?
   B. Using Eqn (7.4.8), calculate the GFR for this individual. Use $K = 1$. Is this value normal?
   C. Do you think the GFR is low or lower when the patient was admitted?
19. Suppose a person has a GFR of 120 mL min^{-1} and produces creatinine at 1.5 mg min^{-1}. Assume the ECF is 14 L and the rate of secretion is a constant 0.1 mg min^{-1}.
   A. Calculate the steady-state creatinine concentration in plasma.
   B. Suppose that one renal artery experienced a complete blockage and that its contribution to the GFR went to zero, and the total GFR of the body was then 60 mL min^{-1}. Assume also that the secretion of creatinine was halved to 0.05 mg min^{-1}. What would the new steady-state creatinine concentration be?
   C. If the GFR instantaneously changes from 120 mL min^{-1} to 60 mL min^{-1}, and renal secretion is constant but undergoes a step changes from 0.1 to 0.05 mg min^{-1} at the same time the GFR drops, and creatinine

synthesis does not change, derive an equation that describes the time course of the creatinine concentration from its initial steady state to the second steady state. Can this be characterized by a half-life?

20. Alternate pathways for macromolecular ultrafiltration have been considered. Albumin or other substances can get into Bowman's space in several ways: (1) they can diffuse into that compartment across the glomerular filtration barrier; (2) they can be dragged into that compartment by the flow of fluid across the barrier; (3) they can be electrophoresed into the compartment, responding to potential differences between the two compartments. The total flow of albumin is given as

[7.PS1.1] $$Q_A = Q_V(1-\sigma)\overline{C} - D_A A \frac{dC}{dx} - \mu_e \overline{C} A \frac{d\psi}{dx}$$

where the first term is the flow resulting from solvent drag (where $\overline{C}$ is the average concentration), the second is the flow derived from diffusion, and the third is the flow caused by electrophoresis, where μ_e is the electrophoretic mobility, which depends on the magnitude of charge on the particle, its shape, and the viscosity of the medium. The equation can be written solely in terms of driving forces if we replace

[7.PS1.2] $$Q_V = -L_p A \frac{dP}{dx}$$

where L_p is the hydraulic permeability, A is the area, and P is the pressure that drives fluid flow. Let us suppose that diffusion is the *only* mechanism working to transport albumin across the glomerulus. In the steady state, the input into the aggregate glomerular capillaries must balance the output. The input of albumin is the flow carried by the aggregate afferent arterioles and the output is carried by the efferent arterioles and the ultrafiltrate. This is given as

[7.PS1.3] $$Q_{va} C_p = Q_{ve} C_g + Q_{vf} C_f$$

Here Q_{va} represents the volume flow into the glomerulus from the afferent arterioles, C_p is the concentration of albumin in the incoming plasma, Q_{ve} is the volume flow of plasma out the efferent arterioles, C_g is the average concentration of albumin in the glomerulus, Q_{vf} is the volume flow into the ultrafiltrate, equal to the GFR, and C_f is the concentration of albumin in the ultrafiltrate. Here we view the aggregate of the fluid in the glomerular capillaries as being well mixed, with a single concentration of albumin. Although this is probably not true, it simplifies the problem. The presence of multiple diffusional barriers within the glomerulus makes the diffusion complex. We *define* a permeability, related to the diffusion coefficient within the barriers and the thickness of the barriers, for the entire complex, as

[7.PS1.4] $$p_f = \frac{J_f}{\Delta C} = \frac{Q_{vf} C_f}{A(C_g - C_f)}$$

where C_g is the concentration of albumin within the glomerular capillary. At steady state, the gradient in albumin must be linear.
Given that $Q_{va} = 600$ mL min^{-1}, $C_p = 3.7$ g% = 3.7 g/100 mL; $Q_{ve} = 480$ mL min^{-1}; apparent $\theta = 0.00062$; area of diffusion = 670 cm^2 and thickness of the barrier = 300 nm.

A. Calculate the concentration of protein in the ultrafiltrate from C_p and θ.
B. Given the values for flows and concentrations, calculate the concentration of protein in the glomerular capillaries, C_g.
C. Suppose that the protein that enters the ultrafiltrate does so solely by diffusion and not by solvent drag or electrophoresis. The flow would then be given as

[7.PS1.4] $$Q_A = Q_{Vf} C_f = \frac{D_m A}{\Delta x}(C_g - C_f)$$

Calculate D_m the diffusion coefficient in the glomerular filtration membrane that would produce the same flow of albumin as is observed.

D. The free diffusion coefficient for albumin in water is 6×10^{-7} cm^2 s^{-1}. Calculate the **hindrance factor**, given as

[7.PS1.5] $$D_m = H_D D_W$$

where D_m is the diffusion coefficient in the membrane, H_D is the diffusional hindrance factor (which could differ from the hindrance factor due to filtration), and D_W is the diffusion coefficient in water.

Tubular Reabsorption and Secretion 7.4

Learning Objectives

- Calculate the filtered load from the sieving coefficient, plasma concentration, and GFR
- From the filtered load and rate of excretion, calculate the rate of secretion or reabsorption
- From a renal titration curve, identify T_m, renal threshold, and splay
- List the causes of splay in renal titration curves
- From renal titration curves, identify substances that are reabsorbed or those that are secreted
- Explain why endogenous creatinine clearance is a good estimate of the GFR
- Describe the mechanism of glucose reabsorption in the proximal tubule
- Explain why glucose appears in the urine of diabetic persons
- Describe the mechanism of amino acid and phosphate reabsorption in the proximal tubule
- Explain how the double ratio of $(TF/P)_x/(TF/P)_{inulin}$ illuminates the site of reabsorption
- Estimate the fraction of water and electrolytes that are absorbed in the proximal tubule
- Describe the mechanism of protein reabsorption

THE *FILTERED LOAD* OF WATER AND VALUABLE NUTRIENTS IS ENORMOUS

For many small molecular weight solutes, such as amino acids, glucose, and electrolytes, the sieving coefficient, Θ, at the glomerulus is about 1.0—the concentration of the materials in the ultrafiltrate is the same as in plasma. The rate of filtration of any plasma material can be calculated as

[7.4.1] $\qquad$ Filtered load$_x$ = GFR $\Theta_x P_x$

where GFR is the glomerular filtration rate, normally expressed in mL min^{-1}, Θ_x is the sieving coefficient for substance x, and P_x is the plasma concentration of substance x. The normal GFR is about 120 mL min^{-1} or about 170 L day^{-1}. For a 70-kg adult male, the total body water is only 42 L. Thus every day the kidneys filter an equivalent of about four times the entire body water. Since the total urine output is only 1–2 L day^{-1}, most of the filtered load of water must be returned to plasma and not excreted in the urine. The kidneys reabsorb nearly the entire filtered load of water.

THE RENAL TITRATION CURVE OF INULIN IS LINEAR

We can find insights into the overall handling of materials by the kidneys by examining the rate of excretion as a function of the plasma concentration. In this process, we vary plasma concentration, typically by infusing increasing amounts of the material being studied, and we measure the urine flow rate (Q_u), urinary concentration (U_x), and plasma concentration (P_x). We then calculate the rate of excretion as $Q_u U_x$ and plot this against P_x. The result of such an experiment using inulin as a test substance is shown in Figure 7.4.1. We can analyze these curves using the equations that we presented in Chapter 7.3. During the steady state, the overall renal handling of inulin is described by the conservation relation, assuming that inulin is neither destroyed nor produced metabolically.

[7.4.2] $\qquad$ Input = output

Considering the origin of fluxes into and out of the tubule, we write

[7.4.3] $\qquad$ Rate of filtration + rate of secretion = rate of excretion + rate of reabsorption

which is identical to Eqn (7.3.3). In the case of inulin, the rate of excretion is proportional to the plasma [inulin] and is linearly related to the rate of filtration. This linear relationship must hold true if inulin is freely filtered but is neither reabsorbed nor secreted, but the linear relationship by itself is not proof that inulin is handled this way by the kidney. Other experiments have shown that inulin is indeed neither secreted nor reabsorbed. If a single nephron is blocked just below the glomerulus and inulin is perfused into the nephron just below the block, the inulin is quantitatively recovered in the distal loop of the same nephron, indicating that none is reabsorbed. If inulin is also perfused systemically, the recovery of inulin is unchanged. This indicates that inulin is not secreted. The way inulin is handled by the kidney allows us to use its clearance to estimate the GFR accurately.

EXAMPLE 7.4.1 Calculate the Daily Filtered Load of Glucose and Na+

The normal plasma glucose concentration is about 80 mg dL^{-1}. Since $\Theta_{glucose} = 1.0$, we can calculate the daily filtered load as

GFR $\Theta_{glucose} P_{glucose}$ = 170 L day^{-1} × 1.0 × 80 mg dL^{-1} × 10 dL L^{-1}
= **136 g glucose day^{-1}**

The normal plasma Na$^+$ is about 142 mmol L^{-1}, and it is readily filtered because it is small: $\Theta_{Na} = 1.0$.

The daily filtered load is

GFR $\Theta_{Na} P_{Na}$ = 170 L day^{-1} × 1.0 × 142 mmol L^{-1}
= 24,140 mmol day^{-1}

This is 24.14 mol × 22.99 g mol^{-1} = **555 g day^{-1}**.

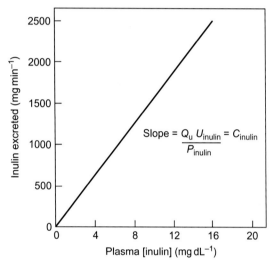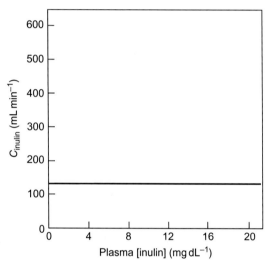

FIGURE 7.4.1 Renal titration of the kidneys with inulin. The rate of excretion of inulin is linearly related to its plasma concentration (left). The ratio of the rate of excretion ($Q_u U_{inulin}$) to the plasma [inulin] (P_{inulin}) is the clearance of inulin in units of mL min^{-1}, which is the slope of the line of rate of excretion versus P_{inulin}. A typical value for an adult human male is about 120 mL min^{-1}. In the case of inulin, the clearance is independent of its plasma concentration (right).

In Example 7.4.1, we calculated that the daily filtered load of glucose is 136 g. Normally there is zero glucose in the urine. Thus all of this filtered glucose is reabsorbed by the kidney. The filtered load of Na$^+$ is calculated to be about 555 g Na$^+$ day^{-1}. Typically the urinary excretion of Na$^+$ is about 4 g day^{-1}. Thus the kidneys reabsorb some 551 g of Na$^+$ from the ultrafiltrate every day. These numbers show that the filtered load of many substances is enormous and that a major job of the kidney is reabsorbing all of these necessary materials from the ultrafiltrate.

THE RENAL TITRATION OF GLUCOSE SHOWS REABSORPTION AND SATURATION KINETICS

We can perform a renal titration of glucose by infusing glucose at constant rates to establish different steady-state plasma [glucose]. The rate of glucose filtration will vary linearly with the $P_{glucose}$ according to Eqn (7.4.1). The rate of excretion is $Q_u \times U_{glucose}$, meaning the urinary flow times the urinary [glucose]. We have

Rate of filtration = GFR$\Theta_{glucose} P_{glucose}$
Rate of secretion = 0
Rate of excretion = $Q_u U_{glucose}$
Rate of reabsorption = GFR$\Theta_{glucose} P_{glucose} - Q_u U_{glucose}$
[7.4.4]

The amount filtered per unit time is a linear function of the plasma glucose concentration, as described in Eqn (7.4.4), with slope = GFR × $\Theta_{glucose}$ when plotted against $P_{glucose}$ where the sieving coefficient for glucose = 1. The amount excreted per unit time is experimentally determined. The experimental observation is that urinary excretion of glucose is zero until plasma [glucose] concentrations exceed a **renal threshold**. This is the plasma glucose concentration at which glucose first appears in the urine. After a short nonlinear portion called **splay**, the amount of glucose excreted per unit time increases linearly with further increases in plasma glucose. From these two quantities, the rate of filtration and the rate of excretion, we can calculate the rate of glucose reabsorbed. The results are shown in Figure 7.4.2.

The titration curve for glucose (see Figure 7.4.2) shows that the rate of reabsorption of glucose increases

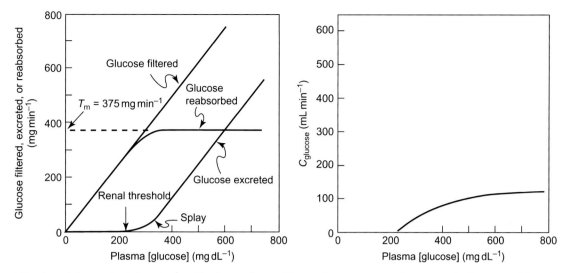

FIGURE 7.4.2 Titration of the renal tubules of man with plasma glucose. Plasma glucose concentration was gradually increased by infusing glucose. The rate of glucose excretion was measured simultaneously with the clearance of inulin and plasma glucose concentration. The amount filtered per unit time was calculated as $C_{inulin} \times P_{glucose}$, and the amount reabsorbed per unit time was calculated as the amount filtered per unit time minus the amount excreted per unit time. The amount reabsorbed shows saturation kinetics with a well-defined transport maximum, $T_{m\ glucose}$. Redrawn from R.F. Pitts, *Physiology of the Kidney and Body Fluids*, Year Book Medical Publishers, Chicago, IL, 1974.

with the filtered load until a maximum rate is reached and then reabsorption does not increase further. The transport mechanism displays **saturation kinetics** and the kidney has a distinct transport maximum or T_m. The transport maximum for glucose, $T_{m\ glucose}$, is about 375 mg min^{-1}. With a GFR of 120 mL min^{-1}, the filtered load just equals the T_m when the plasma glucose concentration is about 312 mg dL^{-1}. However, glucose begins appearing in the urine before this value of plasma glucose concentration is reached.

SATURATION KINETICS AND NEPHRON HETEROGENEITY CAUSE SPLAY

Gradual saturation is typical of carriers that bind transported substrate. The transport rate initially increases nearly linearly with the concentration of transported material and then gradually tapers off. The curve of rate versus concentration of transported material does not show sharp breaks (see Chapter 2.5). The second explanation for splay is the heterogeneity of nephrons. Nephrons are not uniformly long and their glomeruli are not all the same size. Their relative rate of ultrafiltration and reabsorption can vary. This is **morphologic heterogeneity**. Thus not all nephrons have the same transport maximum, and the overall observed T_m is a weighted average of all the individual nephron T_m.

HIGH PLASMA GLUCOSE IN *DIABETES MELLITUS* CAUSES GLUCOSE EXCRETION

The normal plasma glucose concentration is about 80–120 mg dL^{-1}, and therefore the normal filtered load is about 100–150 mg min^{-1}. This is far below $T_{m\ glucose}$. Usually the urine contains no glucose. Its clearance, defined as

[7.4.5] $$C_G = \frac{Q_u U_G}{P_G}$$

is normally zero. Here C_G is the clearance of glucose, Q_u is the urine flow rate, in mL min^{-1}; U_G is the urine concentration of glucose, which is generally zero; and P_G is the plasma [glucose]. C_G is zero at low plasma [glucose] but rises with higher plasma [glucose], as shown in Figure 7.4.2, to approach the clearance of inulin at very high plasma [glucose]. The clearance at any P_G is the slope of the chord in Figure 7.4.2 connecting the origin to the curve for the rate of glucose excretion.

Because plasma [glucose] is usually regulated at levels below the renal threshold, **the kidneys normally do not participate in the homeostatic regulation of plasma [glucose]**. Instead, plasma glucose homeostasis is controlled by hormones including insulin, glucagon, epinephrine, and glucocorticoids. Insulin is secreted by the β cells of the islets of Langerhans in the pancreas (see Chapter 9.4) and stimulates glucose uptake by peripheral tissues. If insulin secretion is inadequate, or if the target tissues become insensitive to insulin, then plasma [glucose] rises. If uncorrected, this hyperglycemia can exceed the renal threshold and glucose appears in the urine. This condition is called **glucosuria** (or glycosuria) and the disease is called **diabetes mellitus**. The excreted glucose exerts an osmotic effect that markedly increases urine flow, Q_u. This explains the name of the disease: "diabetes" means "to siphon off" and "mellitus" means "sweet." This describes the sweet taste of the copious and dilute urine produced by persons with uncontrolled diabetes mellitus. The uncontrolled disease is characterized by frequent drinking (**polydipsia**) to replenish fluids lost by frequent urination (**polyuria**) and weight loss due to loss of calories in the form of glucose.

THE KIDNEYS HELP REGULATE PLASMA PHOSPHATE

The renal titration for phosphate PO_4^- is shown in Figure 7.4.3. At physiological pH, plasma phosphate exists as HPO_4^{-2} and $H_2PO_4^{-1}$, and so plasma [phosphate] is expressed as their sum, in mM, or in terms of mg phosphorus dL^{-1}. Normal plasma [phosphate] is about 0.9–1.5 mM. The filtered load shown in Figure 7.4.3 is a linear function of the plasma [phosphate], with the slope of the line being the GFR. The equation for the line is given by Eqn (7.4.1). The renal threshold for phosphate occurs at about 1 mM. Increasing plasma [phosphate] above the renal threshold increases the amount of phosphate that is excreted. As with glucose, the relationship between excreted phosphate and plasma [phosphate] above the renal threshold is linear, with a slope equal to the GFR. The amount of phosphate absorbed shows a distinct $T_{m\ phosphate}$ of about 0.125 mmol min^{-1}. Thus the overall handling of phosphate by the kidneys is similar in form to that of glucose. However, the renal threshold for phosphate is close to the physiologically normal plasma [phosphate], so that the urine nearly always contains some amount of phosphate. Because of this, **the kidneys participate in the normal regulation of plasma [phosphate]**. If plasma [phosphate] increases, the filtered load increases. Because the filtered load already slightly exceeds the renal threshold, more filtered load means that the $T_{m\ phosphate}$ is further exceeded and more phosphate is excreted in the urine. Rises in plasma [phosphate] increase phosphate excretion so that plasma [phosphate] returns towards normal.

The $T_{m\ phosphate}$ is influenced by circulating hormones that control plasma [phosphate]. Parathyroid hormone (PTH) is secreted by the parathyroid glands in response to hypocalcemia (low plasma $[Ca^{2+}]$). The increased PTH increases bone resorption that liberates both Ca^{2+} and phosphate from the bone and releases them into plasma. PTH also decreases $T_{m\ phosphate}$ so that phosphate reabsorption is decreased and more phosphate is excreted. The net effect of PTH is to raise plasma $[Ca^{2+}]$ without increasing plasma [phosphate].

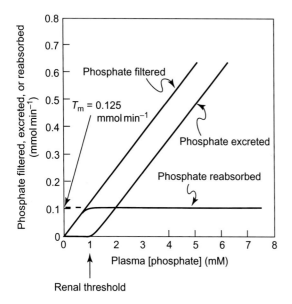

FIGURE 7.4.3 Renal titration of plasma phosphate. The filtered load increases linearly with the plasma [phosphate]. Above about 1 mM, increases in excreted phosphate are linear with the increases in plasma phosphate, with a slope equal to the GFR. The rate of phosphate reabsorption can be calculated from the difference between the filtered load and the rate of phosphate excretion. The resulting curve shows saturation kinetics, with a well-defined transport maximum, $T_{m\ phosphate}$ that is about 0.125 mmol min^{-1}. Since $T_{m\ phosphate}$ is nearly the same as the normal filtered load, the kidney acts like a spillway for phosphate: it keeps phosphate at the level of the spillway and any excess drains off into the urine. *Redrawn from R.F. Pitts,* Physiology of the Kidney and Body Fluids, *Year Book Medical Publishers, Chicago, IL, 1974.*

RENAL TITRATION CURVE OF PAH SHOWS SECRETION

The renal titration of *para*-amino hippuric acid (PAH) is shown in Figure 7.4.4. This titration curve differs from the renal titration curves of glucose or phosphate

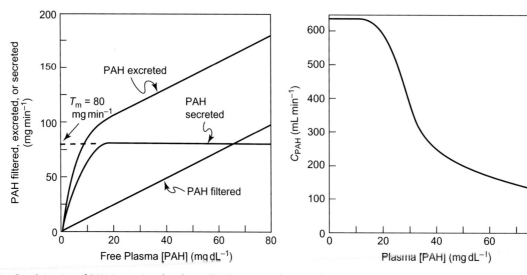

FIGURE 7.4.4 Renal titration of PAH. Increasing the plasma [PAH] increases the rate of PAH excretion. At low plasma [PAH], $T_{m\ PAH}$ is not exceeded and all blood perfusing the secreting portions of the nephrons is cleared of PAH. In this region of plasma [PAH], the clearance is equal to the effective renal plasma flow, ERPF. At higher plasma [PAH] the clearance falls because the transport mechanism is saturated and some plasma PAH escapes in the renal venous blood.

because the rate of excretion of PAH exceeds its rate of filtration. The rate of filtration is calculated as GFR $\times P_{PAH}$ ($\Theta_{PAH} = 1.0$). The rate of excretion can exceed the rate of filtration only if the material is secreted. The rate of secretion is calculated by using Eqn (7.4.3):

Rate of secretion = rate of excretion − rate of filtration

[7.4.6]

The principle of secretion was first established for injected dyes whose concentration could be easily measured and whose route in the kidney could be traced microscopically. The rate of secretion of PAH obeys saturation kinetics with a well-defined $T_{m\ PAH}$.

The clearance of PAH is constant and high at low plasma [PAH], below the $T_{m\ PAH}$. Under these circumstances the blood that perfuses the secreting portions of the nephron is nearly completely cleared of PAH. Under these conditions, the clearance of PAH approximates the effective renal plasma flow, ERPF. At high P_{PAH} that exceeds the renal threshold, the clearance no longer approximates the ERPF. The clearance in Figure 7.4.4 is the slope of the chord connecting the origin to the curve for rate of PAH excretion.

THE MEANING OF THE CLEARANCE DEPENDS ON THE RENAL HANDLING

Figures 7.4.1, 7.4.2, and 7.4.4 show the dependence of the clearance on the plasma concentration of three different solutes: inulin, glucose, and PAH. The clearance for inulin is independent of plasma [inulin]. The clearance of glucose is zero at low plasma [glucose] and climbs asymptotically toward the inulin clearance when plasma [glucose] increasingly exceeds its T_m. The clearance of PAH is highest at lowest plasma [PAH] and it asymptotically approaches the inulin clearance as plasma [PAH] increasingly exceeds its secretory T_m. Figure 7.4.5 shows all of these curves together plus an additional curve for the clearance of creatinine. These curves illustrate that the clearance of a solute depends on its concentration and on how the kidney handles it. For inulin, the clearance is equal to the GFR; for glucose, the clearance is normally zero, meaning that it is entirely reabsorbed; for PAH, the clearance at plasma [PAH] below $T_{m\ PAH}$ is a measure of ERPF.

ENDOGENOUS CREATININE CLEARANCE APPROXIMATES THE GFR

Creatinine is a by-product of muscle metabolism in which creatine in the muscle is converted nonenzymatically to creatinine. Because the total body content of creatine is fairly constant, there is a continual production of creatinine and a continual excretion of it in the urine. The typical 70-kg adult man produces about 2 g of creatinine per day. As evident from Figure 7.4.5, creatinine is slightly secreted by the kidneys so that at low plasma [creatinine] the clearance of creatinine is about 5–10% greater than the inulin clearance. However, creatinine is already present in the blood at steady-state

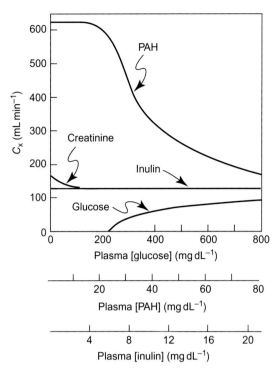

FIGURE 7.4.5 Comparison of the clearances of inulin, glucose, PAH, and creatinine as a function of their plasma concentrations. The clearance of inulin is independent of its concentration and is interpreted as being equal to the GFR because inulin is freely filtered and not reabsorbed or secreted. The clearance of glucose is zero at physiological plasma [glucose] because glucose is avidly reabsorbed by the renal tubules. The clearance of PAH is high at low plasma [PAH] because the kidneys avidly secrete it into the tubular fluid. The clearance of creatinine is about 10% higher than the inulin clearance at physiological plasma [creatine]. The clearance of endogenous creatinine can be used to approximate the GFR.

levels, so the **endogenous creatinine clearance** can be used without the necessity of having to infuse inulin. All that is necessary is a timed urine sample and a plasma sample. Measuring the [creatinine] in both the urine and plasma sample allows the calculation of the endogenous creatinine clearance that provides an approximate measure of the GFR.

PLASMA CREATININE CONCENTRATION ALONE INDICATES THE GFR

At steady state, the rate of creatinine production by the body will be equal to its rate of excretion. The rate of excretion is $Q_u U_{creatinine}$. The major source of the excreted creatinine is from ultrafiltration, with a small fraction from secretion. Thus we can write

[7.4.7] $\qquad Q_u U_{creatinine} = GFR P_{creatinine} + T_s$

where T_s, the rate of secretion of creatinine, is generally small compared to the rate of filtration. The rate of excretion is relatively constant. Dividing both sides of Eqn (7.4.7) by GFR, we obtain

[7.4.8] $\qquad \dfrac{Q_u U_{creatinine} - T_s}{GFR} = P_{creatinine}$

Since T_s is generally small compared to the other term in the numerator, there is an approximate inverse relationship between the GFR and plasma [creatinine]. The normal production of creatinine is about 2 g day^{-1} or about 1.39 mg min^{-1}, which is equal to $Q_u U_{creatine}$ in the numerator. Because T_s is small, the total rate of filtration, the amount in the numerator in Eqn (7.4.8), is about 1.3 mg min^{-1}. If the GFR is 120 mL min^{-1}, the plasma [creatinine] should be about 1.3 mg min^{-1}/120 mL min$^{-1} \approx 1.1$ mg dL^{-1}. Average values of plasma [creatinine] range from 0.5 to 1.5 mg dL^{-1}, with an average of 1.2 mg dL^{-1}.

Clinicians have developed two widely used equations that estimate the GFR from serum creatinine measurements. These are the Modification of Diet in Renal Disease (MDRD) Study equation and the Chronic Kidney Disease Epidemiology Collaboration (CKD-EPI) equation. These use measurements of serum creatinine from the isotope dilution mass spectrometer method. The MDRD equation is

$$[7.4.9] \quad \text{GFR}(\text{mL min}^{-1}/1.73\,\text{m}^2) = 175 \times [\text{Cr}]^{-1.154} \times \text{Age}^{-0.203} \times K$$

the constant, K, is adjusted for different groups. It is 0.742 if the patient is a female, and 1.212 if the patient is African American. The CKD–EPI equation is actually a set of eight equations with slightly different constants depending on race, gender, and values of the measured serum creatinine. These all take the form

$$[7.4.10] \quad \text{GFR} = K \times \left(\frac{[\text{Cr}]}{\alpha}\right)^{-\beta} \times 0.993^{\text{Age}}$$

where the constants K, α, and β differ depending on the range of the creatine measurement or if the patient is male or female or Caucasian or black. These equations have the advantage of avoiding errors in the 24-hour urine collection necessary for the creatinine clearance measurement, but cannot be used for persons with abnormal creatinine production. Abnormal creatinine production may occur in persons with extreme body size or muscle mass, such as amputees, paraplegics, morbidly obese persons or persons eating vegetarian diets or taking creatine supplements.

(TF/P)$_{\text{INULIN}}$ MARKS WATER REABSORPTION

The GFR is about 120 mL min^{-1}, whereas urine flow is about 1 mL min^{-1}. Thus most of the ultrafiltrate fluid is reabsorbed. Where in the nephron is it absorbed? We can use the ratio of concentrations TF/P for inulin to get an answer. Because the sieving coefficient is 1.0 for inulin, its concentration in Bowman's space is the same as it is in plasma: the ratio (TF/P)$_{\text{inulin}} = 1.0$ at this point in the nephron. Once the inulin is filtered, it is neither reabsorbed nor excreted, but water reabsorption from the tubular fluid will concentrate the remaining inulin and (TF/P)$_{\text{inulin}}$ will increase. Let V_{TF} be the volume of the tubular fluid that remains in the tubule and passes through a particular cross-section of the nephron, per

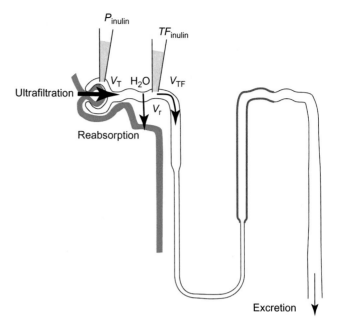

FIGURE 7.4.6 Simplified nephron showing the sampling of tubular fluid at Bowman's capsule and further down the nephron. The volume of filtrate is V_T, which is equal to the volume of fluid remaining in the tubule at the place of sampling, V_{TF}, plus the volume of fluid that is reabsorbed, V_R. The concentration of inulin in the ultrafiltrate is equal to its plasma concentration, P_{inulin}, and its concentration at the point of sampling of tubular fluid is designated as TF_{inulin}.

second, and let V_R be the volume of fluid that has been reabsorbed by the nephron from the glomerulus up to that point (see Figure 7.4.6). The total volume of these two is just the volume of ultrafiltrate formed, V_T, per second:

$$[7.4.11] \quad V_{TF} + V_R = V_T$$

At steady state, the total amount of inulin that remains in the tubule and passes by any point in the nephron per second must be the same everywhere or else there will be a buildup or depletion of the amount of inulin:

$$[7.4.12] \quad V_{TF} TF_{\text{inulin}} = V_T P_{\text{inulin}}$$

We use the plasma [inulin], P_{inulin}, on the right-hand side of the equation because the total volume is the volume of ultrafiltrate and the ultrafiltrate concentration of inulin equals its plasma concentration. The fraction of water reabsorbed is V_R/V_T. This is obtained by dividing both sides of Eqn (7.4.11) by V_T and rearranging:

$$[7.4.13] \quad \frac{V_R}{V_T} = 1 - \frac{V_{TF}}{V_T}$$

From Eqn (7.4.12), we see that

$$[7.4.14] \quad \frac{V_{TF}}{V_T} = \frac{1}{(TF_{\text{inulin}}/P_{\text{inulin}})}$$

Substituting this result into Eqn (7.4.13), we find

$$[7.4.15] \quad \frac{V_R}{V_T} = 1 - \frac{1}{(TF/P)_{\text{inulin}}}$$

Thus the fraction of water reabsorbed can be estimated by measuring plasma [inulin] and tubular fluid [inulin] alone. At the end of the proximal tubule $(TF/P)_{inulin} = 3$, so that the fraction of water reabsorbed in the proximal tubule is about $1 - 1/3 = 67\%$.

THE DOUBLE RATIO $(TF/P)_x/(TF/P)_{INULIN}$ IS THE FRACTION OF THE FILTERED LOAD OF X REMAINING

The TF/P ratio of a substance x is ambiguous. Samples of tubular fluid at the end of the proximal tubule show that $(TF/P)_{Na^+} = 1.0$. What does this mean? It could mean that neither Na^+ nor water was reabsorbed in this section of the nephron, or it could mean that they were reabsorbed in the same ratio so that the concentration of Na^+ did not change. The relative movement of substance x and water can be determined by taking the double ratio, $(TF/P)_x/(TF/P)_{inulin}$. The conservation equation is as follows:

$$[7.4.16] \qquad V_{TF}TF_x + R_x = V_T P_x$$

where TF_x is the concentration of substance x in the tubular fluid of volume V_{TF}, R_x is the amount of substance x that is reabsorbed, V_T is the total volume of ultrafiltrate, and P_x is the plasma concentration. The fraction of substance remaining in the tubule is given as

$$[7.4.17] \qquad \text{Fraction remaining} = \frac{V_{TF}TF_x}{V_T P_x}$$

We can insert the result for V_{TF}/V_T from Eqn (7.4.14) to obtain

$$[7.4.18] \qquad \text{Fraction remaining} = \frac{(TF/P)_x}{(TF/P)_{inulin}}$$

MICROPUNCTURE STUDIES SHOW THAT THE PROXIMAL TUBULE REABSORBS TWO-THIRDS OF THE ULTRAFILTRATE

Sampling of the tubular fluid at the end of the proximal tubule shows that **all of the glucose and amino acids are reabsorbed in the proximal tubule and that the remaining fluid is isosmotic with plasma**. This means that the fluid that is reabsorbed in the proximal tubule is also isosmotic. The $(TF/P)_{inulin}$ is about 3 at the end of the proximal tubule. By Eqn (7.4.13), this means that two-thirds of the ultrafiltrate volume has been reabsorbed. The ratio $(TF/P)_{Na^+}$, however, remains about 1.0 at the end of the proximal tubule, so that the double ratio is about 1/3. This is the fraction of filtered Na^+ that remains in the tubule. Thus two-thirds of the filtered Na^+ is also reabsorbed in the proximal tubule. The proximal tubule also absorbs about two-thirds of the filtered load for a variety of other materials including Cl^-, HCO_3^-, and K^+.

THE *PROXIMAL CONVOLUTED TUBULE* CONTAINS MANY TRANSPORT MECHANISMS

GLUCOSE ENTERS BY Na^+-LINKED COTRANSPORT AND LEAVES BY FACILITATED DIFFUSION

Nearly all filtered glucose is reabsorbed in the early proximal tubule. A sodium–glucose linked cotransporter (**SGLT**) resides on the microvillus membrane facing the tubular fluid. There are five isoforms of these Na^+–glucose cotransporters. SGLT1 and SGLT2 reside in the proximal tubule, SGLT2 in the proximal convoluted tubule and SGLT1 in the proximal straight tubule. The SGLT1 couples two Na^+ atoms per glucose imported into the cell; with SGLT2 the stoichiometry is 1 Na^+:1 glucose. The energy cost for glucose entry is paid by Na^+ running from a high electrochemical potential to a lower one. The low electrochemical potential of Na^+ inside the cell is maintained by the **Na,K-ATPase** located on the basolateral membrane of the proximal tubule cells. Glucose that enters the cell leaves it by facilitated diffusion on carriers called GLUT. The early proximal tubule has GLUT2 carriers; the proximal straight tubule has GLUT1 (see Figure 7.4.7).

AMINO ACIDS TRANSPORTERS ARE ON BOTH APICAL AND BASOLATERAL MEMBRANES

Amino acid transport across the renal tubule epithelium reprises that found in the intestines, with some differences. A set of at least five distinct amino acid transporters on the apical membrane bring filtered amino acids into the proximal tubule cell. This set of amino acid transporters are necessary because of the heterogeneity of the amino acids that must be reabsorbed. One of these carries neutral amino acids; a second carries the anionic amino acids; another transports cationic amino acids and cystine; another carries proline and hydroxyproline; another carries proline, glycine, and alanine. Some of these carriers require Na^+ as a cotransported ion (**secondary active transport**), and some do not (**facilitated diffusion**). Another set of carriers carry the amino acids across the basolateral membrane. Some of these carriers are identical to those on the apical membrane, and some are distinct (see Figure 7.4.8).

Na^+ COTRANSPORTERS CARRY ORGANIC ANIONS

The apical membrane contains a number of other transporters that bring organic anions into the proximal tubule cell by secondary active transport with Na^+ energetics driving the inward movement of the anion. Examples include phosphate, lactate, citrate, succinate, and acetate. As shown by the renal titration curve, the urine generally contains some phosphate, so not all of the phosphate is reabsorbed. Lactate, on the other hand, is completely reabsorbed from the urine. Specific carrier proteins carry lactate and phosphate out of the cell and into the blood flowing

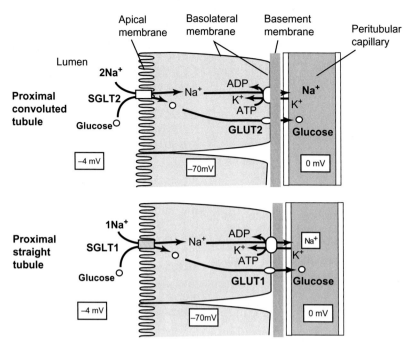

FIGURE 7.4.7 Mechanism of glucose absorption in the proximal convoluted tubule and proximal straight tubule. See text for details.

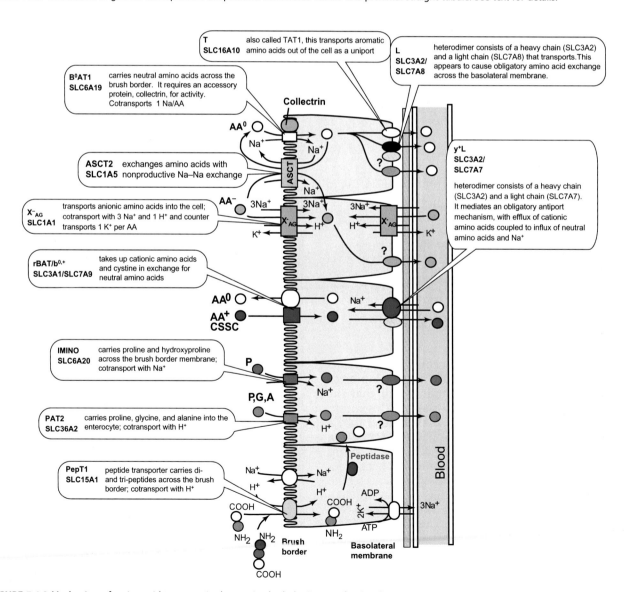

FIGURE 7.4.8 Mechanism of amino acid transport in the proximal tubule. See text for details.

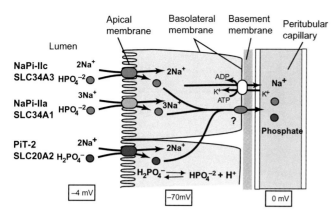

FIGURE 7.4.9 Mechanism of phosphate absorption in proximal tubule cells. Three distinct phosphate carriers on the brush border membrane have been identified. All use Na$^+$ co-transport and thus are powered by the Na$^+$ gradient established by the basolateral Na-K-ATPase. The NaPi-IIc and PiT-2 transporters both use a stoichiometry of 2 Na$^+$ per phosphate. The Napi-IIa uses 3 Na$^+$ per phosphate. All three are located in the proximal convoluted tubule.

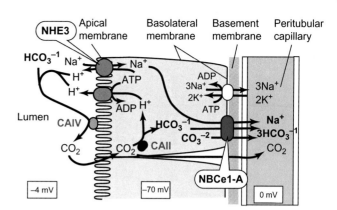

FIGURE 7.4.10 Mechanism of HCO$_3^-$ absorption in the proximal tubule. Filtered HCO$_3^-$ combines with H$^+$ secreted into the tubule by Na–H exchange over **NHE3**, or, less importantly, by an H$^+$ pump. This is converted to CO$_2$ by an apical membrane-bound carbonic anhydrase (CAIV). CO$_2$ from the luminal fluid or peritubular blood is converted to HCO$_3^-$ and H$^+$ in the cytosol by the aid of a soluble carbonic anhydrase (CAII). A Na–HCO$_3^-$ cotransporter, **NBCe1-A**, transports one Na$^+$ along with the equivalent of three HCO$_3^-$ (one HCO$_3^{-1}$ and one CO$_3^{-2}$). In this way, Na$^+$ and HCO$_3^-$ are removed from the filtrate and Na$^+$ and HCO$_3^{-1}$ appear in the blood.

through the peritubular capillaries. These carriers employ facilitated diffusion (see Figure 7.4.9).

BICARBONATE ABSORPTION IS LINKED TO ACID SECRETION

The absorption of HCO$_3^-$ begins with H$^+$ secretion into the lumen through two mechanisms: **Na$^+$–H$^+$ exchange** and a **H$^+$-ATPase pump**. The secreted H$^+$ along with filtered HCO$_3^-$ is converted to H$_2$O and CO$_2$ by the action of **carbonic anhydrase** on the apical membrane (**CAIV**). The CO$_2$ is transported into the cell and equilibrates with peritubular capillary CO$_2$. In the cell, CO$_2$ is converted to H$^+$ and HCO$_3^-$ by a cytosolic carbonic anhydrase (**CAII**). HCO$_3^-$ can dissociate to form H$^+$ and CO$_3^{-2}$. The sodium–bicarbonate exchanger (NBCe1-B) transports HCO$_3^-$ and CO$_3^{-2}$ out of the cell with a cotransport of Na$^+$. The H$^+$ formed from the hydration reaction of CAII feeds into the Na$^+$–H$^+$ exchanger and the H$^+$-ATPase pump. The result of the entire operation is that Na$^+$ and HCO$_3^-$ disappear from the lumen and Na$^+$ and HCO$_3^-$ appear in the peritubular blood. The HCO$_3^-$ that appears in the blood is not the same as that which disappeared from the lumen, but the net effect is the same as if it were (see Figure 7.4.10).

THE PROXIMAL CONVOLUTED TUBULE PASSIVELY REABSORBS UREA

Urea is a small nitrogenous compound (molecular weight is 60) that is the main end product of protein catabolism in mammals (see Chapter 2.11). It is made predominantly in the liver from ammonia and bicarbonate and is one of the main components of urine. The rate of synthesis varies from 300 to 600 mmol/day depending on the protein intake. All of this urea eventually finds its way into the urine. Because urea makes up a large part of the obligatory solute excretion, its osmotic pressure requires significant volumes of water to carry the urea. Urea passively crosses biological membranes, but its permeability is low because of its low solubility in the lipid bilayer. Some cells speed up this process through **urea transporters**, which move urea by facilitated diffusion. Urea is passively reabsorbed in the proximal tubule, but its route of transport is not clear. Urea transporters have not yet been identified for the proximal tubule. SGLT1 can transport urea via Na–urea transport.

WATER FOLLOWS THE OSMOTIC PRESSURE GRADIENT THROUGH WATER CHANNELS

Water reabsorption is by osmosis through water channels in the membrane. These water channels consist of a family of proteins called **aquaporin**. At least seven different aquaporin isoforms are expressed in the kidney. The proximal tubule has abundant **AQP1** on the apical and basolateral membranes and **AQP7** on the apical membrane of the late proximal tubule.

The blood that flows through the peritubular capillaries that surround the proximal tubules originates from the efferent arterioles of cortical nephrons. This blood has passed through the glomerulus and has had a protein-free filtrate abstracted from it. Thus the protein concentration in the efferent arterioles is increased by removal of 20% of the plasma volume while leaving the proteins behind. Therefore, the peritubular capillaries contain plasma with a higher oncotic pressure. As Na$^+$ is reabsorbed with other solutes, the concentration of osmolytes in the spaces between the cells increases, causing a local increase in the osmotic pressure in this space. Water moves in response to the high oncotic pressure of the peritubular capillaries and the slight hyperosmolarity of the lateral intracellular space, so that water flows across the basolateral membrane into the lateral

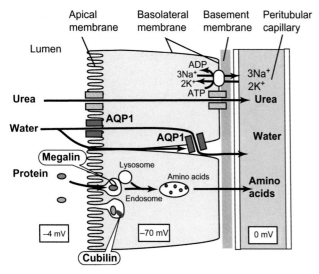

FIGURE 7.4.11 Mechanism of urea, water, and protein reabsorption in the proximal tubule. See text for details.

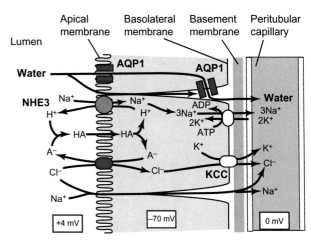

FIGURE 7.4.12 Mechanism of water and NaCl reabsorption in the late proximal tubule. By the time the tubular fluid gets to the late proximal tubule, a large fraction of the filtered water and Na^+ has already been reabsorbed. In the early proximal tubule, Na^+ reabsorption is preferentially accompanied by HCO_3^- reabsorption. This concentrates the Cl^- that is left behind so that its concentration exceeds that of the peritubular blood. Transport of Cl^- occurs through a cellular route and a paracellular route. In the cellular route, Cl^- enters the cell in exchange for an anion, which could be OH^-, formate (HCO_2^-), oxalate, HCO_3^-, or sulfate. This process occurs in parallel with a Na^+-H^+ exchanger so that the net effect is entry of NaCl into the cell. Chloride exits the cell passively because the interior of the cell is sufficiently negative to overcome the unfavorable difference in Cl^- concentration between the cell and the peritubular fluid. Some of the chloride is believed to be carried by K–Cl cotransport channels, KCC, KCC1, KCC3 and KCC4 are expressed in the kidney. The Na^+,K^+-ATPase pumps the Na^+ out of the cell at the basolateral membrane. NaCl also travels through the junctions between the cells in the paracellular route. This draws water from the tubular fluid by osmosis, and some of this water goes through the paracellular path. The osmotic pressure difference between the peritubular fluid and the tubular fluid is small, on the order of 2–6 mOsM. This is sufficient to drive water reabsorption.

intracellular space and into the interstitial space surrounding the capillaries, and from there into the peritubular capillaries. As water moves from the cell, it concentrates the cell contents so that the osmotic gradient is transferred to the apical membrane. Water moves from the tubular fluid into the cell in response to this gradient. The net effect is water reabsorption from the tubular fluid into the peritubular capillaries, caused by the increased oncotic pressure of the capillary blood and the active reabsorption of Na^+ and other solutes (see Figure 7.4.11).

THE PROXIMAL TUBULE REABSORBS FILTERED PROTEINS BY ENDOCYTOSIS

The concentration of proteins in the ultrafiltrate is small but not zero because the sieving coefficient is not exactly zero but is some small value that depends on the size of the protein. Normal plasma albumin concentrations are $3.5-5.0$ g dL^{-1}, whereas the concentration in the ultrafiltrate is about $1-3$ mg dL^{-1}, giving a sieving coefficient of about 0.0003–0.0008. Thus about 1.8 g of albumin are filtered per day. In addition, a variety of low molecular weight proteins and polypeptide hormones are filtered by the kidney and subsequently destroyed by it. These filtered proteins and polypeptides are taken up by an extensive network of coated pits in the proximal tubule. Two broad-spectrum binding proteins, megalin and cubilin, recognize and bind filtered proteins and direct them into the coated pits. Kidney epithelial cells take up the proteins by endocytosis, and the resulting endosome fuses with a lysosome, with its content of hydrolytic enzymes, and the proteins are hydrolyzed to their constituent amino acids. The amino acids are then reabsorbed into the peritubular capillary blood (see Figure 7.4.11)

ABSORPTION OF WATER AND SALT ACROSS THE LATE PROXIMAL TUBULE

The transport of Na^+, Cl^-, and water is different in the late proximal tubule than it is in the early proximal tubule. In the early proximal tubule, Na^+ is reabsorbed preferentially with HCO_3^-. In the late proximal tubule, Na^+ is reabsorbed with Cl^-. As Na^+ gets absorbed along with the amino acids and glucose and HCO_3^- in the early proximal tubule, the $[Cl^-]$ in the tubular fluid increases. This higher $[Cl^-]$ drives Cl^- diffusion through the so-called tight junctions between the cells at the apical aspect. This creates a lumen positive potential that favors Na^+ reabsorption. Chloride enters the tubule cells in part in exchange for secreted anions. The mechanism is illustrated in Figure 7.4.12.

SUMMARY

The filtered load of water and many nutrients is enormous. Most of the filtered material does not show up in the final urine because it is reabsorbed back into the peritubular capillaries that perfuse the nephron. The overall handling of materials is made evident from renal titration curves. Because inulin is freely filtered but not reabsorbed or secreted, its renal excretion rate is linear with its plasma concentration. This means that the clearance of inulin is independent of its plasma concentration. In contrast, no glucose appears in the urine until the plasma concentration exceeds its renal threshold, which is substantially higher than normal plasma glucose concentrations. Thus normally the kidney does not participate in regulating plasma glucose concentration. Normally the glucose clearance is zero because it is avidly reabsorbed in the proximal tubule. The renal titration curve for phosphate is similar to that of glucose except that the renal threshold is near the normal plasma concentration for phosphate. Normally there is some phosphate excretion, and if plasma phosphate levels rise, more phosphate spills over into the urine. The renal titration curve for para-amino hippuric acid (PAH) shows that it is secreted by the kidney. Because it is so avidly secreted, the clearance of PAH approximates the renal plasma flow. Creatinine is a waste product of muscle metabolism whose clearance approximates the GFR.

The ratio of tubular fluid concentration of inulin to plasma concentration of inulin can be used to estimate absorption of water. The double ratio $[TF_x/P_x]/[TF_{inulin}/P_{inulin}]$ can be used to estimate the relative handling of water and material.

About two-thirds of filtered water is absorbed in the proximal tubule, along with all of the filtered glucose and amino acids and about two-thirds of the filtered Na^+, Cl^-, and HCO_3^-. Glucose and amino acids are absorbed by secondary active transport at the apical membrane of proximal tubule cells and by facilitated transport across the basolateral membrane into the peritubular capillary blood. The secondary active transport requires Na^+ cotransport down its electrochemical gradient. The electrochemical gradient for Na^+ is maintained by a basolateral Na,K-ATPase pump. Urea and water are reabsorbed passively by specific carriers. HCO_3^- reabsorption is linked to H^+ secretion powered by $Na^+–H^+$ exchange and an apical H^+-ATPase pump. The small amount of filtered protein is reabsorbed by endocytosis.

REVIEW QUESTIONS

1. Why is the clearance of inulin independent of its plasma concentration?
2. Under what conditions is the clearance of PAH equal to the effective renal plasma flow, ERPF?
3. The renal threshold is usually less than $T_m/(GFR \times \Theta)$. Why?
4. How do you calculate the rate of reabsorption or rate of secretion?
5. What does the ratio TF_{inulin}/P_{inulin} signify? What is its value at the end of the proximal tubule?
6. What is the significance of $(TF/P)_x/(TF/P)_{inulin}$? What is this double ratio for Na at the end of the proximal tubule? What is it for glucose?
7. How are amino acids reabsorbed from the proximal tubule? Where do they go?
8. How is glucose reabsorbed from the proximal tubule?
9. How is HCO_3^- reabsorbed from the proximal tubule?

7.5 Mechanism of Concentration and Dilution of Urine

Learning Objectives

- Describe how the proximal tubule reabsorbs an isosmotic fluid
- Summarize the transport characteristics of the various parts of the loop of Henle and the distal tubule, focusing on its treatment of salt, water, and urea
- Describe how ADH varies the permeability of the distal nephron to water and urea
- Explain what is meant by countercurrent multiplier
- Explain how these transport characteristics produce an osmotic gradient from low in the cortex to high in the inner medulla
- Draw on a graph the relative contributions of NaCl and urea to the osmotic gradient in cortex and medulla
- Explain the function of the vasa recta as countercurrent exchangers
- Predict the concentration and urine flow in low and high ADH states
- Explain the origin of osmotic diuresis in diabetes mellitus, for example

LIFE ON DRY LAND STRUGGLES AGAINST DESICCATION

The end product of protein metabolism is urea. The end product of purine metabolism is uric acid. These and other materials must be excreted every day. The total obligatory excretion depends on the dietary intake of protein and electrolytes, but typically the body must excrete about 600 mOsmol of solutes every day through urine. If we can excrete urine that is only isosmotic with plasma, about 300 mOsM, the obligatory solute excretion of 600 mOsmol would require about 2 L of water. In fact, the kidney can excrete urine that is about 4 times as concentrated as plasma or 1200 mOsM. The obligatory solute load could then be excreted in 0.5 L of water, saving us the trouble of having to find and drink 1.5 L of water every day. This may not seem like a big deal to people who obtain their water from a tap, but it is a big deal to primitive people who drank from streams or ponds and had no other source of water and no way to carry it. Our kidneys evolved to handle the natural environment, not the modern, artificial one. The natural environment on land is generally an arid one in which the kidneys must conserve water and excrete excess water when necessary. Water is conserved by using as little as possible to excrete the obligatory solute load in a small volume of concentrated urine. Excess water is eliminated by excreting a large volume of dilute urine. The subject of this chapter is how the kidneys achieve this feat.

CONTROL OF URINE CONCENTRATION USES AN OSMOTIC GRADIENT AND REGULATED WATER PERMEABILITY

Here, we consider the overall plan of controlling urine concentration. We consider two extreme cases: (1) the kidney maximally concentrates the urine and (2) the kidney maximally dilutes the urine. The ability of the kidney to concentrate the urine is controlled by a hormone, antidiuretic hormone (ADH) also called arginine vasopressin (AVP). This hormone has this name because it decreases the volume of urine. Thus it has an antidiuretic effect. This hormone also causes vasoconstriction; hence its other name, vasopressin. The overall mechanism for concentrating or diluting the urine is shown diagrammatically in Figure 7.5.1.

In the first case, where the kidneys maximally concentrate urine, the aggregate activity of the nephrons with their associated vasculature establishes an osmotic gradient running from about 300 mOsM in the cortex to about 1200 mOsM in the inner medulla. It takes energy to establish this osmotic gradient, and the energy is derived from kidney metabolism. The gradient is established by pumping NaCl out of the tubule into the interstitial fluid and by the countercurrent arrangement of the loop of Henle. How this happens is described in detail later. Because the ascending limb of the loop of Henle pumps out NaCl, the fluid presented to the distal tubule is hypoosmotic. In the presence of ADH, the late distal tubule is permeable to water. When the tubule fluid enters the collecting duct it has equilibrated with the osmotic pressure of the cortex, about 300 mOsM. ADH increases the water permeability of the outer and inner medullary collecting ducts, so that when the fluid travels along the collecting duct from cortex to outer medulla to inner medulla, through areas of increasing osmotic pressure, water moves from the tubule into the interstitium. Thus water is removed from the tubular fluid and the tubular fluid becomes increasingly concentrated. In this case, the kidneys excrete a small volume of concentrated urine.

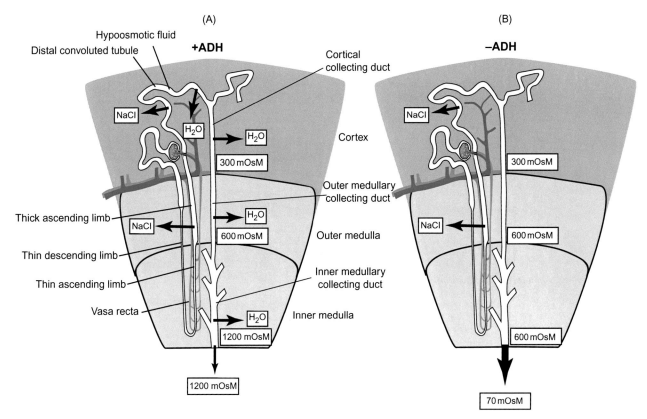

FIGURE 7.5.1 Overview of concentration and dilution of urine. In the presence of ADH (also known as AVP), the distal nephron (late distal tubule and collecting duct) is permeable to water (A). Therefore, the osmotic pressure of the tubular fluid equilibrates with the osmotic pressure within the kidney interstitium. This runs from 300 mOsM in the cortex to about 1200 mOsM in the inner medulla. Water moves from the hypoosmotic tubule contents to the hyperosmotic interstitium. As a result, water is reabsorbed and tubular fluid becomes concentrated. In the absence of ADH (B), the distal nephron remains impermeable to water. The tubular fluid flowing from cortex to inner medulla cannot equilibrate its osmotic pressure with the hyperosmolar interstitium. As a result, the kidneys excrete a large volume of dilute urine.

In the second case, the kidneys maximally dilute the urine. As before, the fluid presented to the distal tubule is hypoosmotic to plasma because the thick ascending limb of the loop of Henle pumps out NaCl but it is impermeable to water. Now, however, the distal tubule and collecting ducts remain impermeable to water because ADH secretion is low. The consequence is that the tubular fluid cannot equilibrate its osmotic pressure with the surrounding interstitium because water movement is slow. The result is that water is not reabsorbed in the collecting duct and the final urine has a large volume and a low osmolarity. The low ADH levels also alter the osmotic gradient in the interstitium. The increased osmolarity seen in the inner medulla disappears in the absence of ADH.

The above description provides an overview of how the kidney produces concentrated or dilute urine. How does the kidney establish the osmotic gradient in the first place?

TUBULAR TRANSPORT MECHANISMS DIFFER ALONG THE LENGTH OF THE NEPHRON

THE PROXIMAL TUBULE REABSORBS AN ISOSMOTIC SOLUTION

The reabsorption of nutrients, water, and salt from the proximal tubule was described in Chapter 7.4 and is summarized again in Figure 7.5.2. The Na^+, K^+-ATPase provides the motive force for all of the cotransport processes by establishing a favorable electrochemical gradient for Na^+ entry into the cell at the apical membrane. This favorable gradient powers the movement of a large number of solutes including glucose, amino acids, phosphate, lactate, sulfate, and indirectly through Na^+–H^+ exchange, HCO_3^-. Water and urea reabsorption passively follow the movement of osmotically active solutes so that the fluid that remains at the end of the proximal tubule is isosmotic with plasma. At this point, all of the nutrients are reabsorbed but the concentration of some secreted materials is higher. This fluid is presented to the loop of Henle.

THE DESCENDING LIMB OF THE LOOP OF HENLE IS PERMEABLE TO WATER AND UREA

The descending limb of the loop of Henle is permeable to both water and urea. This is due to the presence of **aquaporin** channels (**AQP1**) and **urea transporters** (**UT-A2**). As we shall see later, the effect of these is to allow osmotic equilibration between the tubular fluid and the kidney interstitium (see Figure 7.5.3).

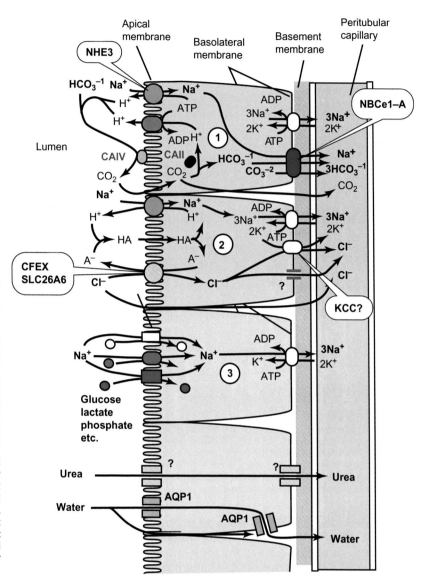

FIGURE 7.5.2 Synopsis of the mechanism of water, urea, Na, Cl, K, and HCO_3^- reabsorption in the proximal convoluted tubule. Na^+ reabsorption occurs in three ways: (1) by entry into the cell by the Na–H exchanger (NHE3) that is coupled to the reabsorption of HCO_3^- on the basolateral membrane by the Na–bicarbonate exchanger, electrogenic (NBCe1-A); (2) by entry into the cell by the NHE3 again that is pumped out of the cell by the Na,K-ATPase on the basolateral membrane, accompanied by Cl^- that is mostly reabsorbed passively and paracellularly, but also enters the cell over the chloride-organic exchanger (CFEX) and exits the cell by the K-Chloride channel (KCC1, KCC3 and KCC4 are expressed in the kidney) or possibly by a chloride channel; (3) Na^+ entry into the cell by secondary active transport mechanisms that couple Na^+ entry to entry of other substrates such as glucose, amino acids or anionic acids such as phosphate, citrate or lactate, followed by pumping out across the basolateral membrane. The anions carried by the CFEX can be HCO_3^-, formic acid, or oxalic acid, with oxalic acid being most important. Water and urea are passively reabsorbed through aquaporins and unidentified pathways for urea.

THE THIN ASCENDING LIMB OF THE LOOP OF HENLE IS PERMEABLE TO NaCl BUT NOT WATER

The thin ascending limb of the loop of Henle brings tubular fluid around from the descending limb to the thick ascending limb. It is impermeable to water and urea but may be permeable to Cl. The basolateral membrane of the thin ascending limb cells has a chloride channel, **ClC-Ka**, that functions only when associated with an accessory protein called **barttin**. Deletion of ClC-Ka homologue in rats causes diabetes insipidus— an inability to concentrate the urine, and in humans defects in **ClC-Kb** are linked to **Bartter syndrome**, characterized by impaired ability to concentrate the urine. ClC-Kb is found in the thick ascending limb of the Loop of Henle (see Figures 7.5.4 and 7.5.5).

THE THICK ASCENDING LIMB PUMPS OUT SALT BUT IS IMPERMEABLE TO WATER

The mechanism of NaCl transport in the cells of the thick ascending limb of the loop of Henle is shown schematically in Figure 7.5.5. These cells possess very active basolateral Na^+,K^+-ATPase activity that powers the overall transport. Transported ions enter the apical membrane by a **Na:K:2Cl cotransporter (NKCC2 = SLC12A2)**. This carrier is inhibited by the "loop diuretics," **furosemide** and **bumetanide**. The apical membrane contains a K^+ channel (renal outer medullary K = ROMK1) that recycles K^+ across the apical membrane. Cells of the thick ascending limb also have chloride channels (ClC-Kb) and a K-Cl electroneutral cotransporter (KCC4) on the basolateral membrane.

THE DISTAL TUBULE CAN PUMP OUT SALT

The distal tubule consists of two functionally distinct regions. The early distal tubule is impermeable to water and urea and its permeability is not regulated by ADH. The late distal tubule has water channels and its permeability is regulated by ADH. The late distal tubule is also called the **connecting duct** because it connects the rest of the nephron to the collecting duct. As in most parts of the nephron, the Na^+,K^+-ATPase powers Na^+ reabsorption in the early distal tubule. Na^+ enters the cell

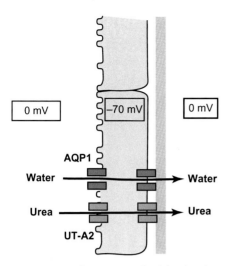

FIGURE 7.5.3 Transport mechanisms in cells of the thin descending limb of Henle in long loops of Henle. Short loops of Henle, originating from the cortical nephrons, have no AQP1.

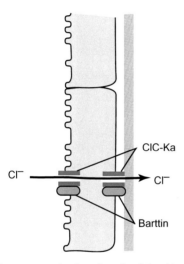

FIGURE 7.5.4 Transport mechanisms in cells of the thin ascending limb of the loop of Henle. Chloride channels on the basolateral membrane here and in the thick ascending limb of the loop of Henle are necessary for the ability to concentrate the urine. The ClC-Ka and ClC-Kb channels both require an accessory protein, barttin

by a Na^+–Cl^- cotransporter on the apical membrane, and the cotransported Cl^- leaves the cell passively through a chloride channel on the basolateral membrane. The apical Na^+–Cl^- cotransporter is inhibited by a class of diuretics called **thiazides**. Thus the early distal tubule can reabsorb NaCl, whereas water and urea are not reabsorbed. The transport processes in the early distal tubule are shown diagrammatically in Figure 7.5.6.

ADH REGULATES THE WATER PERMEABILITY OF THE LATE DISTAL TUBULE AND COLLECTING DUCT

The late distal tubule and collecting duct are heterogeneous, containing several cell types. The main types are the **principal cells** and the **intercalated cells**. The principal cells are responsible for regulating ion and water transport, whereas the intercalated cells are involved in excretion of acid; they are discussed in Chapter 7.7. The principal cells contain aquaporins on both the apical and the basolateral membranes. AQP2 is located on the apical membrane, whereas AQP3 and AQP4 are on the basolateral membrane. The number of AQP2 molecules on the apical membrane is influenced by ADH. Thus the permeability of the entire distal nephron is controlled by ADH (see Chapter 7.6). These cells also can reabsorb NaCl. Here, Na^+ enters the cell through **EnaC**, the epithelial Na channel. This channel is blocked by **amiloride**. Once again, the Na^+,K^+-ATPase provides the gradient for passive Na^+ entry at the apical membrane. The principal cells also play an important role in the regulation of K^+ excretion, as will be discussed in Chapter 7.6. The transport mechanisms of the principal cells of the late distal tubule (also known as the connecting tubule) and collecting duct are shown in Figure 7.5.7.

ADH REGULATES UREA TRANSPORT IN THE INNER MEDULLARY COLLECTING DUCT

The inner medullary collecting duct differs from the outer medullary and cortical collecting ducts in that it is permeable to urea when stimulated by ADH. The UT-A1 type of urea transporter is on the apical membrane of the principal cells of the inner medullary collecting duct and the UT-A3 is on the basolateral membrane (see Figure 7.5.8).

THE NEPHRON PRODUCES AN OSMOTIC GRADIENT WITHIN THE KIDNEY INTERSTITIUM

Experimental results for the concentrations of urea and NaCl in slices of kidneys from water-deprived dogs are shown in Figure 7.5.9. These results show that the osmotic gradient in the outer medulla is produced by NaCl, and then the NaCl gradient levels off. In the inner medulla, the osmotic gradient rises because the urea concentration rises. The explanation for these contributions rests with the countercurrent arrangement of the loop of Henle and in the transport mechanisms present in the different segments of the nephron. First, we will consider the contribution of NaCl.

The characteristics of the loop of Henle dealing with NaCl and water transport are as follows:

- The thin descending limb of the loop of Henle is permeable to water but not NaCl.
- The thin ascending limb is impermeable to water but is permeable to NaCl.
- The thick ascending limb actively pumps out NaCl but is impermeable to water (see Figure 7.5.3).
- The descending limb and ascending limb are arranged with countercurrent flow.

These characteristics allow the formation of an osmotic gradient. The descending limb and ascending limb are closely apposed, with a small intervening interstitial volume, so that flow down into the medulla within the descending limb is opposite to the flow up and out

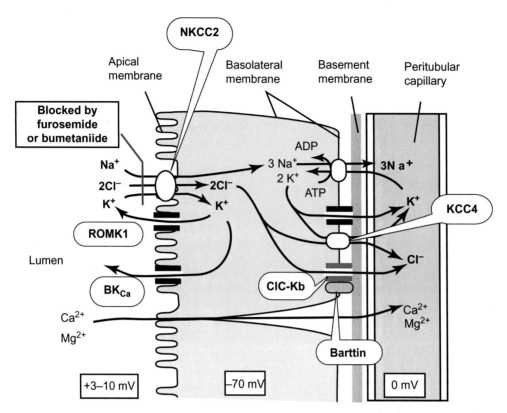

FIGURE 7.5.5 Mechanism of transport in cells of the thick ascending limb of the loop of Henle. Na$^+$, K$^+$, and Cl$^-$ all enter the cell at the apical membrane through the operation of a Na:K:2Cl cotransporter. The protein is referred to as NKCC2 and its gene is referred to as SLC12A2. The K$^+$ that enters is recycled through an apical K$^+$ channel, ROMK1. Na$^+$ and Cl$^-$ that enter are removed from the cell by the operation of the Na$^+$,K$^+$-ATPase and by two efflux routes for Cl$^-$: a K:Cl symport (or cotransporter, KCC4) and by a chloride channel (ClC-Kb). K$^+$ that exits the cell through the K:Cl symport and through a basolateral K$^+$ channel is recycled back into the cell through the Na$^+$,K$^+$-ATPase. Na$^+$ can also be absorbed through the paracellular route in these cells.

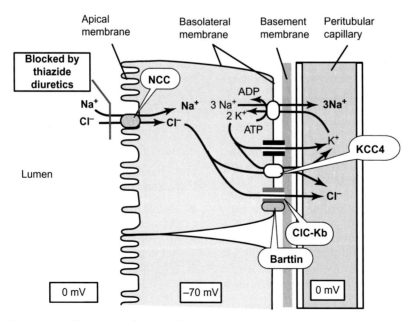

FIGURE 7.5.6 Mechanisms of transport in the early distal tubule. The early distal tubule is impermeable to both water and urea but it can pump modest amounts of NaCl. It does this by electroneutral uptake of Na$^+$ and Cl$^-$ on the apical membrane through the Na–Chloride cotransporter (NCC). The Na,K-ATPase pumps the Na$^+$ out of the cell on the basolateral membrane whereas Cl$^-$ exits the cell over the K Cl channel (KCC4) and a chloride channel (ClC-Kb). K$^+$ that exits the cell over KCC4 is recycled into the cell by the Na,K-ATPase. The distal nephron epithelium is "tight," whereas the proximal tubule epithelium is "leaky." Thus the distal nephron can support large differences in concentrations of ions between tubular fluid and interstitium, whereas the proximal tubule cannot.

Mechanism of Concentration and Dilution of Urine

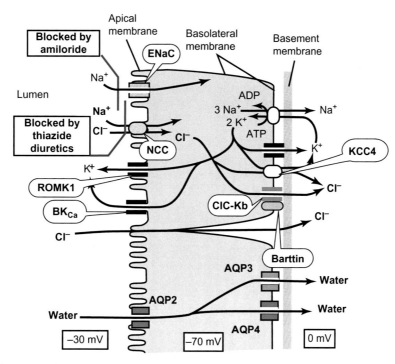

FIGURE 7.5.7 Transport mechanisms in the principal cells of the cortical and outer medullary collecting ducts. These cells alone among the kidney tubule possess an amiloride-sensitive Na^+ entry mechanism, mediated by **ENaC**, the epithelial Na^+ channel, which is inhibited by amiloride. These cells can reabsorb NaCl. These cells are also important in K^+ secretion. They have water channels whose insertion into the apical membrane is controlled by circulating levels of ADH, also known as AVP.

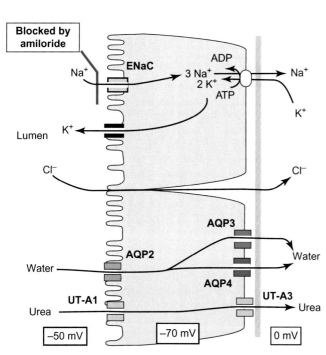

FIGURE 7.5.8 Major transport pathways in inner medullary collecting duct principal cells. The apical membrane has urea transporter UT-A1 and aquaporin 2 (AQP2); the basolateral membrane has UT-A3 and AQP3 and AQP4. These cells remain capable of absorbing NaCl mediated by ENaC on the apical membrane and the Na,K-ATPase on the basolateral membrane.

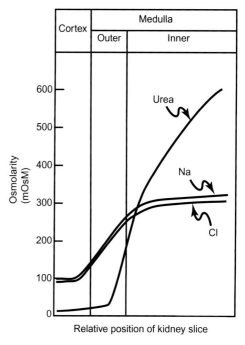

FIGURE 7.5.9 Contributions of Na, Cl, and urea to the interstitial osmotic gradient that runs from the kidney cortex to the inner medulla. In the cortex, the osmotic pressure is about 300 mOsM with about half contributed by Na and half contributed by Cl, with little contribution by urea. In the outer medulla, the interstitial osmolarity climbs to about 700 mOsM at the border of the inner and outer medulla, with about 100 mOsM contributed by urea and 600 mOsM by NaCl. In the inner medulla, the osmotic pressure gradually increases with distance to reach a maximum of about 1200 mOsM, with about 600 mOsM due to NaCl and another 600 mOsM due to urea. These contributions are due to the spatial separation of transporters within the nephron and its countercurrent flow.

of the medulla within the ascending limb. Suppose that the thick ascending limb is capable of creating a transepithelial concentration gradient of NaCl, called the **single effect**, which corresponds to the difference between the osmolarity of the interstitial fluid and the tubular fluid in the thick ascending limb. Since the thick ascending limb pumps out NaCl but is impermeable to water, the interstitial fluid becomes hyperosmotic. Because the thin descending limb is permeable to water, water leaves the descending limb and equilibrates with the interstitial fluid. Thus the fluid in the descending limb is also concentrated because water leaves the tubule without accompanying NaCl. This more highly concentrated fluid then flows around the bend in the loop of Henle and presents itself to the thick ascending limb. So now the thick ascending limb has a higher starting concentration of NaCl. When it produces its single effect, it does so from a higher starting concentration. Repetition of the concentration, equilibration, and countercurrent flow produces a standing gradient running from 300 mOsM at the juncture of cortex and medulla to 600 mOsM at the juncture of the inner and outer medulla. The loop of Henle is described as a **countercurrent multiplier** because its single effect is **multiplied by the countercurrent flow**.

THE VASA RECTA ARE COUNTERCURRENT EXCHANGERS

The experimental observation is that the fluid presented to the distal tubule is hypoosmotic to plasma. The active pumping of NaCl out of the thick ascending limb, and the inability of water to follow the NaCl, partially explains this observation. However, the **vasa recta** are necessary to carry off the water reabsorbed from the descending limb and to carry off the extra NaCl reabsorbed from the thick ascending limb.

Unlike the thick ascending limb, the vasa recta passively exchange salts and water along their course from the cortex down into the medulla and back again to the cortex. Equilibration of osmotic pressure, however, lags behind the osmotic gradient because the vasa recta blood is flowing. The consequence is that residual salt and water remain in the vasa recta from its course through the medulla, so that vasa recta blood returns from the medulla with more salt than water, and hence it is hypertonic. Without this return flow, the thick ascending limb could not dilute its tubular contents.

UREA CONTRIBUTES TO THE OSMOTIC GRADIENT IN THE INNER MEDULLA

Figure 7.5.9 shows that urea makes the major contribution to the osmotic gradient in the inner medulla. This is also a consequence of the spatial separation of transport characteristics of the loop of Henle and the collecting duct. Urea transporters are located in the descending limb of the loop of Henle (UT-A2), in the inner medullary collecting duct (UT-A1 and UT-A3) and in the descending vasa recta (UT-B1). We consider the case where ADH secretion is high so that the water permeability of the distal nephron is high. The tubular fluid entering the late distal tubule has some urea, and it is hypoosmotic due to active pumping out of NaCl in the thick ascending limb of the loop of Henle. Water is reabsorbed in the late distal tubule so when the fluid reaches the collecting duct it is isosmotic once again. This water reabsorption occurs without urea reabsorption, so urea becomes slightly concentrated in the late distal tubule. As the fluid moves down the cortical and outer medullary collecting ducts, water continues to be reabsorbed but the nephron remains impermeable to urea. Thus the urea becomes progressively more concentrated. When the fluid enters the inner medulla, urea permeability is high so that urea passively leaves the collecting duct and equilibrates with the inner medullary interstitial fluid. The descending limb of the loop of Henle possesses UT-A2 urea transporters, so that the fluid in the inner medullary loop of Henle also equilibrates with this high urea concentration. This tubular fluid moves up to the thick ascending limb, now carrying more urea. But here the urea permeability is low: all of this urea survives to the distal tubule, where once again it is concentrated by removal of water. In this way, urea produces a cycle that establishes its contribution to the inner medullary osmotic gradient. Urea concentration in the urine is set by its highest concentration in the kidney interstitium.

Figure 7.5.10 illustrates the role of NaCl, urea, and various nephron segments in the establishment of the osmotic gradient within the kidney interstitium.

TRANSPORT BY THE VASA RECTA IS ESSENTIAL TO THE OPERATION OF THE LOOP OF HENLE

In the absence of the vasa recta, all of the NaCl pumped into the interstitial fluid by the thick ascending limb would eventually have to return to the cortex through the kidney tubule. Likewise, in the absence of the vasa recta, all of the water reabsorbed into the kidney interstitial fluid would also have to return to the cortex through the kidney tubule. Thus the loop of Henle could produce a concentration gradient, but the fluid leaving the tubule at the top of the thick ascending limb would have the same constituents at steady state as what entered the tubule at the end of the proximal tubule. In short, nothing could be accomplished by the loop of Henle in the absence of the vasa recta. The vasa recta carries away all of the water and salt reabsorbed by the loop. It does this by passive exchange, but the magnitude of the flow with respect to the vasa recta's permeability is crucial: the vasa recta blood does not completely equilibrate with the interstitial fluid because its flow is too fast relative to its permeability. The result is that the vasa recta blood contains more reabsorbed salt than reabsorbed water, and it is hypertonic. The numbers in boxes in Figure 7.5.10 indicate the approximate flows and osmolarities of either vasa recta blood or tubular fluid at various points along the nephron.

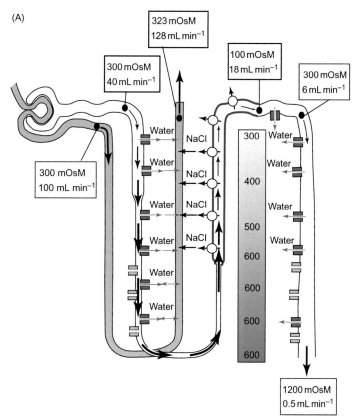
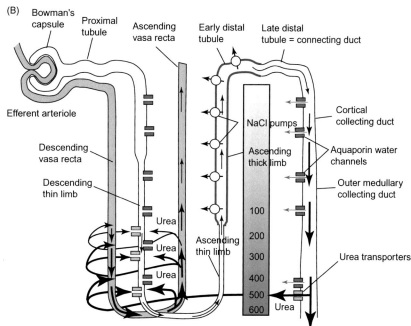

FIGURE 7.5.10 Contribution of NaCl (A) and urea (B) to the osmotic gradient in the kidney interstitium. The thick ascending limb can actively transport NaCl out of the tubule and into the interstitial fluid, but it is impermeable to water. Therefore, the thick ascending limb concentrates the interstitial fluid and dilutes the tubular fluid. The increased osmotic pressure draws water out of the descending thin limb, which is permeable to water but not NaCl. The result is that the fluid in the descending limb becomes as concentrated as the interstitial fluid. This fluid then flows around the bend in the loop of Henle, providing yet more salt for the thick ascending limb. As a result, NaCl builds up in the interstitial fluid and a gradient of osmotic pressure is set up in the outer medulla. The increase in osmolarity of the inner medulla is caused by urea recycling, as shown in (B). The thin ascending limb, thick ascending limb, and distal nephron up to the inner medulla collecting duct are impermeable to urea. The inner medulla, however, has urea transporters. The urea in the collecting duct is progressively concentrated when water leaves the duct in response to the progressively higher interstitial osmotic pressure. In the inner medulla, the high urea concentration favors massive urea movement into the interstitial fluid, which then enters both the descending thin limb and the descending vasa recta. This urea is then recycled in two ways: in the vasa recta and in the tubule. The descending vasa recta turns and becomes the ascending vasa recta. Although the ascending vasa recta has no urea transporters, it is highly fenestrated and urea probably leaves the ascending vasa recta as it flows up toward the cortex. This produces a gradient in urea concentration, high nearest the innermost medulla and progressively dilute going from the inner to outer medulla. Urea that enters the descending limb of the loop of Henle cannot escape anywhere along the ascending thin or thick limb, distal nephron, or cortical and outer medullary collecting duct. All of the urea present in the tubular fluid at the tip of the loop of Henle will come around again to be concentrated by water removal and then recycled again. In this way, the urea concentration builds up in the interstitial fluid. About one-half of the 1200 mOsM osmolarity at the tip of the renal pyramid is due to urea (see Figure 7.5.9).

At steady state, there must be material balance for the entire medulla. The mass balance equation is given as

Input = output
$Q_{vr\ in} + Q_{pct\ out} + Q_{dt\ out} = Q_{vr\ out} + Q_{tal\ out} + Q_{urine\ out}$
$M_{vr\ in} + M_{pct\ out} + M_{dt\ out} = M_{vr\ out} + M_{tal\ out} + M_{urine\ out}$
[7.5.1]

where Q is the flow, in mL min^{-1}, and M is the moles of osmotically active solutes, in OsMoles. The subscript "vr" means "vasa recta," "pct" means "proximal convoluted tubule," "dt" means "distal tubule," "tal" means "thick ascending limb," and "urine" refers to the final fluid that exits the collecting duct. The vasa recta input of fluid and osmoles respectively in Figure 7.5.10 is 100 mL min^{-1} and 100 mL min^{-1} × 300 mOsM = 30.0 mOsmol min^{-1}. Similarly, the output of the proximal tubule (input to the medulla) is 40 mL min^{-1} and 12.0 mOsmol min^{-1}. The flow of fluid and osmoles at the end of the distal tubule is 6 mL min^{-1} and 1.8 mOsmol min^{-1}, respectively. Thus the sum of the fluids inputs is 100 + 40 + 6 = 146 mL min^{-1} and osmole inputs sum as 30.0 + 12.0 + 1.8 = 43.8 mOsmol min^{-1}. On the output side, the vasa recta output is 128 mL min^{-1} and 128 mL min^{-1} × 323 = 41.3 mOsmol min^{-1}. The output of the thick ascending limb is 18 mL min^{-1} and 1.8 mOsmol min^{-1}, and urinary output is 0.5 mL min^{-1} and 0.6 mOsmol min^{-1}. The sums of the outputs are 128 + 18 + 0.5 = 146 mL min^{-1}, and 41.3 + 1.8 + 0.6 = 43.7 mOsmol min^{-1}, which match the inputs for fluid and for total osmoles. This exercise should make it clear that, without the vasa recta removing reabsorbed water and salt, the composition of the fluid leaving the thick ascending limb would of necessity be the same as the fluid leaving the proximal tubule. Although the thick ascending limb does all the work of reabsorption, all parts of the loop of Henle are involved, including the vasa recta.

INCREASED SOLUTE LOADS IN THE DISTAL NEPHRON PRODUCE AN OSMOTIC DIURESIS

The proximal convoluted tubule avidly reabsorbs filtered glucose into the peritubular capillaries so that it is all reabsorbed by the end of the proximal tubule. The mechanism for glucose reabsorption was described in Chapter 7.4. The proximal tubule is the only site for glucose reabsorption. If the filtered load of glucose overwhelms the proximal tubule transport mechanisms, glucose escapes to the loop of Henle. There is no reabsorption of glucose beyond the proximal tubule, and the glucose becomes progressively more concentrated as the nephron reabsorbs water and salt. The glucose exerts an osmotic pressure and produces an **osmotic diuresis**, the severity being directly proportional to the amount of excreted glucose. This is the origin of the **polyuria** of persons with uncontrolled diabetes mellitus in which the plasma concentration of glucose exceeds its renal threshold. Any osmotically active material in the distal nephron will have this effect. Mannitol is freely filtered by the kidney but neither secreted nor reabsorbed. Injection of mannitol will produce an osmotic diuresis that is directly proportional to the amount of mannitol injected.

ADH CONTROLS DISTAL NEPHRON PERMEABILITY

ADH increases the water permeability of the late distal tubule (or connecting duct) and all parts of the collecting duct. It also increases the urea permeability of the inner medullary collecting duct. When the distal nephron permeability to water and urea is high, all of the mechanisms for concentrating the urine operate, and the kidneys excrete a small volume of highly concentrated urine. In the absence of ADH, the distal nephron is not permeable to either water or urea. Fluid that enters the distal tubule is hypoosmotic, about 100 mOsM. Low water permeability prevents water from leaving the hypoosmotic fluid to equilibrate with the interstitial fluid. Thus the tubular fluid stays hypoosmotic all the way through the collecting duct. Because the late distal tubule and collecting duct can pump out some Na$^+$, the final concentration of the urine is actually lower than 100 mOsM. The kidneys can maximally dilute urine to about 70 mOsM. The low urea permeability in the absence of ADH shuts off the urea recycling, so that the concentration profile in the kidney interstitium is changed. Figure 7.5.11 shows approximate values for tubular fluid concentration along the nephron in the presence and absence of ADH.

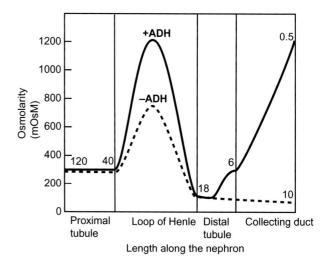

FIGURE 7.5.11 Concentration profile of the tubular fluid along the nephron ± ADH. In either the presence or absence of ADH, the glomerulus filters an isosmotic solution that is reabsorbed in the proximal tubule as an isosmotic solution. Fluid traveling down into the medulla is progressively concentrated as water leaves the fluid through aquaporin channels to equilibrate with the hyperosmolar interstitial fluid. This occurs in either the presence or absence of ADH, but the interstitial osmotic pressure gradient is maximal when the kidney is stimulated by ADH, and minimal when ADH is absent. This is due to the removal of the component of osmolarity contributed by urea when ADH is absent, because urea recycling is broken. In the absence of ADH, the urine reaches its minimum concentration due to continued ion pumping in the distal nephron and low water and urea permeability in the distal nephron.

Clinical Applications: Diabetes Insipidus and SIADH

"Diabetes insipidus" derives its name from "diabetes," meaning "to siphon off" and "insipidus," meaning "tasteless." This describes the urinary output from persons afflicted with the condition: they excrete a large volume of highly dilute, and hence tasteless, urine. The urine production can be enormous. This symptom can be caused by any of four separate defects:

1. Hypothalamic diabetes insipidus—inability to synthesize or secrete ADH.
2. Nephrogenic diabetes insipidus—inability of the kidney to respond to normal ADH.
3. Transient diabetes insipidus of pregnancy—accelerated metabolism of ADH.
4. Primary polydipsia—excess ingestion of fluids rather than decreased ADH.

Hypothalamic diabetes insipidus can be caused by:

- hereditary defects,
- tumors,
- head trauma or surgery.

People with hereditary defects in ADH often do not present with symptoms at birth, but develop them in childhood. This contrasts with babies with congenital nephrogenic diabetes insipidus, who have the disease during the first weeks of life. Most hereditary diabetes insipidus involves defects in neurophysin.

A working hypothesis is that the neurophysin somehow folds badly and progressively gums up ADH synthesis or secretion.

Hereditary nephrogenic diabetes insipidus involves genetic defects either in the V2 receptor or in its target, AQP2. More than 90% of persons afflicted with nephrogenic diabetes insipidus are males whose X chromosome has one of over 100 different types of defective V2 receptors. A female with nephrogenic diabetes insipidus most likely has a defect in AQP2. Acquired nephrogenic diabetes insipidus can be caused by chronic kidney disease.

In pregnant women, the placenta produces an enzyme called **cysteine aminopeptidase** that is released into the plasma and degrades both oxytocin and ADH. This accelerates degradation of ADH, decreases concentrations of circulating ADH, and temporarily produces large volumes of dilute urine.

The syndrome of **inappropriate antidiuretic hormone secretion**, or **SIADH**, occurs when plasma ADH levels are high when ordinarily they would be low. The inappropriate ADH secretion retains water even when plasma osmolarity is low. Thus the hallmark of SIADH is plasma hypoosmolarity. Because Na^+ is the major extracellular cation, plasma hypoosmolarity almost always means hyponatremia, low blood $[Na^+]$. Although SIADH has a number of causes, its most common is an ADH-secreting tumor.

SUMMARY

One of the chief jobs of the renal system is excreting an obligatory solute load while simultaneously either retaining water or getting rid of excess water. Thus it is designed to excrete a concentrated urine under conditions of water deprivation or a dilute urine under conditions of water excess. This is regulated through ADH.

The transport characteristics of the loop of Henle and vasa recta, together with their spatial arrangement as parallel hair-pin loops, allow the loop of Henle to establish an osmotic gradient within the kidney interstitium, running from low osmolarity in the cortex to high osmolarity within the inner medulla. Once this gradient is established, production of a concentrated urine consists of allowing the incipient urine to equilibrate with this hyperosmotic interstitium. The gradient is most concentrated in the presence of ADH, when the gradient reaches a high of about 1200 mOsM. About half of this is due to NaCl concentrated by the thick ascending limb of the loop of Henle, whereas the remaining half is due to recirculation of urea. In the absence of ADH, the contribution of urea to the osmotic gradient is washed out.

The establishment of the osmotic gradient is a complicated business which has not yet been thoroughly understood. In principle, the loop of Henle functions as a countercurrent multiplier. It produces a single effect, the creation of an osmotic gradient across its epithelium, and this effect is multiplied by the countercurrent arrangement of flow. Both descending and ascending limbs of the loop of Henle, plus the vasa recta, are required components. The fluid presented to the distal tubule is hypoosmotic, and this can be sustained only if the vasa recta contains hyperosmotic fluid. The vasa recta functions as countercurrent exchangers. About half of the final osmotic gradient is produced by urea recycling within the kidney inner medulla. This is produced by the fact that urea is permeable only to the inner medullary collecting duct.

REVIEW QUESTIONS

1. How are filtered fluid and solutes reabsorbed in the proximal tubule? What is the main transport mechanism in the thick ascending limb of Henle? How is fluid and salt absorbed in the early distal tubule? Late distal tubule?
2. How does NaCl contribute to the osmolarity of the medullary interstitium? What transport features of the nephron allow this to happen?
3. How does urea contribute to the osmolarity of the medullary interstitium? What transport features of the nephron allow this to happen?
4. What hormone controls the final dilution and concentration of urine?
5. What is the limit of concentration of human urine? What limits it?
6. What is the limit of dilution of human urine? What limits it?
7. What is the role of the vasa recta in concentration or dilution of urine?

7.6 Regulation of Fluid and Electrolyte Balance

Learning Objectives

- Describe how afferent and efferent arteriolar contractions affect the GFR
- Define the autoregulation of RBF and GFR
- Describe the myogenic mechanism and tubuloglomerular feedback as mechanisms of autoregulation
- Describe what is meant by glomerulotubular balance and distinguish it from tubuloglomerular feedback
- Describe the chemical nature of ADH
- Describe the glandular origin of ADH
- Describe the mechanism by which ADH alters distal tubule water permeability
- List the signals leading to increased ADH secretion
- Describe the chemical nature and origin of renin
- List the signals leading to increased renin release
- Describe the cascade of events leading to angiotensin II
- List the actions of angiotensin II
- Recognize the chemical structure of aldosterone and identify its glandular origin
- List the signals leading to increased aldosterone release
- List the actions of aldosterone

REGULATION OF GLOMERULAR FILTRATION RATE AFFECTS URINE OUTPUT

Pressures produce the glomerular ultrafiltrate according to

[7.6.1] $$GFR = K_f[P_{GC} - P_{BS} - \pi_{GC}]$$

Theoretically, the regulation of the glomerular filtration rate (GFR) can be accomplished by regulating K_f, the filtration coefficient; P_{GC}, the hydrostatic pressure within the glomerular capillary; P_{BS}, the hydrostatic pressure within Bowman's space; or π_{GC}, the oncotic pressure of the glomerular capillary blood.

P_{BS} IS NORMALLY CONSTANT AND NOT SUBJECT TO PHYSIOLOGICAL REGULATION

The hydrostatic pressure within Bowman's space is a consequence of the volume within that structure and its compliance. Typically the pressure in Bowman's space is a modest 20 mmHg, and it is not subject to large variation. If the kidney's urinary path becomes blocked, pressure within the urinary conduits will rise until filtration equilibration is reached. This is a pathological condition. Although there is no evidence that P_{BS} is physiologically regulated, it seems that increased GFR ought to increase P_{BS} as a direct effect of compliance. Increased P_{BS} should then increase flow down the tubule because the driving force is increased. Similarly, reductions in GFR ought to decrease P_{BS} because there would be less fluid distending the cavity. Thus the P_{BS} may not be independently regulated, but it may vary directly with the GFR. The consequence is that the resistance of the tubule to flow naturally opposes changes in GFR. There is little information about the relative importance of this resistance.

PLASMA ONCOTIC PRESSURE INCREASES WITH DEHYDRATION AND REDUCES GFR

The plasma oncotic pressure is that part of the total osmotic pressure of the plasma that is due to impermeant proteins. During prolonged water restriction or after water loss due to sweat, the blood becomes more concentrated and its oncotic pressure increases. Since this pressure opposes the GFR, dehydration also reduces the GFR due to reduction in the net driving force. This produces a negative feedback loop: dehydration increases π_{GC} which decreases the GFR, thereby lessening further loss of fluid through the urine. However, the kidneys are merely conservatory organs. They cannot alone set aright the dehydration. Drinking water is necessary. Dehydration engages the mechanism of **thirst** to motivate us to seek water.

CONTRACTION OF MESANGIAL CELLS ALTERS K_F

Mesangial cells surround the glomerular capillaries and have contractile properties. In response to a variety of vasoactive agents, these cells can alter the effective surface area for filtration, perhaps also altering the dimensions of glomerular capillary fenestrae and podocyte slit pore size. The result is that K_f is reduced by vasoconstrictors and increased by vasodilators. The physiological importance of this mechanism is unknown.

THE AFFERENT ARTERIOLE AND EFFERENT ARTERIOLE EXERT SEPARATE EFFECTS ON P_{GC}

The glomerular capillary hydrostatic pressure is the driving force for formation of the glomerular ultrafiltrate.

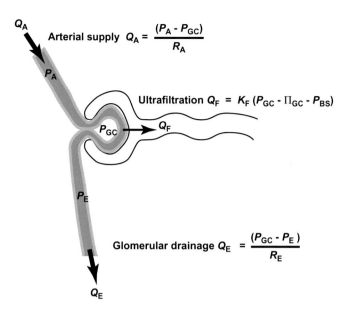

FIGURE 7.6.1 Flows into and out of the glomerulus. Input is the flow through the afferent arteriole, which is determined by the resistance of the afferent arteriole and the pressure difference that drives flow. Output consists of the ultrafiltrate and flow through the efferent arteriole. The afferent and efferent arteriolar resistances largely determine the glomerular capillary hydrostatic pressure.

Consider the arrangement of flows shown in Figure 7.6.1. The conservation of fluid volume at steady state requires that the flow into the glomerular capillaries through the afferent arterioles equals the flow out of the glomerular capillaries through the efferent arterioles and the glomerular ultrafiltrate. We write this as

$$[7.6.2] \quad Q_A = Q_E + Q_F$$

where Q_A is the blood flow through the afferent arteriole, Q_E is the flow through the efferent arteriole, and Q_F is the flow of the ultrafiltrate, or the GFR. The flow through each is determined by the pressure difference and the resistance. We write these as

$$[7.6.3] \quad \frac{(P_A - P_{GC})}{R_A} = \frac{(P_{GC} - P_E)}{R_E} + K_f[P_{GC} - P_{BS} - \pi_{GC}]$$

where P is the pressure in the respective spaces, R_A and R_E are the resistances of the afferent and efferent arterioles, respectively, and K_f is the filtration coefficient for the glomerular filtration, as written in Eqn (7.6.1). We can manipulate Eqn (7.6.3) algebraically to solve for P_{GC}, the pressure in the glomerular capillary. We find

$$P_{GC} = P_A \frac{R_E}{R_E + R_A + K_f R_E R_A} + P_E \frac{R_A}{R_E + R_A + K_f R_E R_A}$$
$$+ (\pi_{GC} + P_{BS}) \frac{K_f R_E R_A}{R_E + R_A + K_f R_E R_A}$$

[7.6.4]

The utility of this equation is reduced by the fact that π_{GC} and, to a lesser extent, P_{BS} are not independent of P_{GC}. Because P_{GC} drives the ultrafiltration, the filtration fraction generally increases with increases in P_{GC}—but not always—and so π_{GC} also increases with P_{GC} and glomerular filtration is opposed. Therefore, the last term on the right-hand side of the equation does not have a clear interpretation. Nevertheless, the relative normal values of R_E, R_A, and K_f can illuminate the relative importance of the three terms in determining P_{GC}. If we use $P_A \approx 90$ mmHg and $P_{GC} \approx 60$ mmHg, and the normal value for $Q_A \approx 1.1$ L min^{-1}, we can calculate $R_A \approx 27.3$ mmHg L^{-1} min. Similarly, we can estimate R_E for the normal condition as $(60$ mmHg $- 10$ mmHg$)/0.98$ L min^{-1} $= 51.0$ mmHg L^{-1} min. K_f under normal conditions can be estimated from the average values of the GFR and driving forces as $K_f = 0.12$ L min$^{-1}/10$ mmHg $= 0.012$ L min^{-1} mmHg^{-1}. Inserting these values into Eqn (7.6.4) gives

$$[7.6.5] \quad P_{GC} = 0.537 P_A + 0.287 P_E + 0.176(\pi_{GC} + P_{BS})$$

This equation is NOT causative, but it is descriptive: although true for the given state of the glomerulus and afferent and efferent arterioles, the coefficients will change with other states. If we double R_E without changing any other variable, e.g., the three coefficients become 0.627, 0.167, and 0.205 for P_A, P_E, and $(\pi_{GC} + P_{BS})$, respectively, and P_{GC} increases from 60 to 67.3 mmHg. The equation allows us to see that the factors multiplying the afferent arteriolar pressure dominate the calculation. Because P_A is very much greater than P_E, and the coefficient multiplying $(\pi_{GC} + P_{BS})$ is so small, the coefficient multiplying P_A has the greatest importance. This coefficient is given as

$$[7.6.6] \quad \frac{R_E}{R_E + R_A + K_f R_E R_A}$$

so the resistance of the efferent arteriole is the most important factor in determining P_{GC} and, therefore, the GFR. Thus the constriction of the efferent arteriole causing an increase in R_E results in an increase in P_{GC} and an increase in GFR. This conclusion has its limits. If R_E increases too much, the flow through the glomerular capillaries is reduced and GFR similarly falls because pressure equilibrium will be reached and there will be less blood flow to support the GFR. On the other hand, increasing R_A by vasoconstriction should increase the denominator in the P_A term, reducing the pressure in the glomerular capillaries and reducing the GFR. The effect of dilation or constriction of afferent and efferent arterioles on GFR and filtration fraction is shown in Figure 7.6.2.

RBF AND GFR EXHIBIT AUTOREGULATION

AUTOREGULATION MAINTAINS A RELATIVELY CONSTANT RBF AND GFR

According to Eqn (7.6.1), we should expect that decreased arterial blood pressure would decrease RBF and decrease the GFR and that increased arterial blood pressure would increase both RBF and GFR. The experimental observation in dogs is that RBF and GFR remain

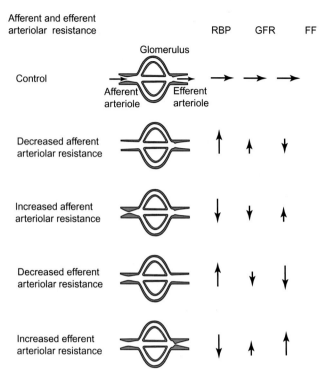

FIGURE 7.6.2 Effect of afferent arteriolar and efferent arteriolar constrictions and dilation on renal blood flow (RBF), glomerular filtration rate (GFR), and filtration fraction (FF). Dilating the afferent arteriole increases RBP because the overall resistance to flow is decreased. The GFR is also increased because of the increase in glomerular capillary pressure, but this increase is slight. The result is a decrease in the filtration fraction. Constricting the afferent arteriole reduces RBP because the vascular resistance is increased. This constriction also reduces the pressure downstream from the constriction, which reduces the GFR approximately proportionately. However, reducing RBF gives more time for filtration to reach equilibrium, so that the filtration fraction will rise slightly upon afferent arteriolar constriction. Constriction of the efferent arteriole alone also reduces RBF but with an increase in glomerular capillary pressure. This favors a relative increase in the GFR over the RBF, so that the filtration fraction is increased.

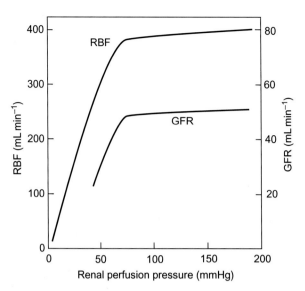

FIGURE 7.6.3 Autoregulation of renal blood flow (RBF) and glomerular filtration rate (GFR) as a function of renal perfusion pressure in dog kidneys. Assuming that the hematocrit was 0.45 in this animal, the filtration fraction would be approximately 50 mL min^{-1}/(400 mL min^{-1} × 0.55) = 0.23. *Adapted from L.G. Navar, Renal autoregulation: perspectives from whole kidney and single nephron studies. Am. J. Physiol.* **234**:F357, 1978.

nearly constant over a wide range of arterial blood pressure, as shown in Figure 7.6.3. This phenomenon is called **autoregulation**. Because both RBF and GFR are maintained fairly constant between 80 and 180 mmHg perfusion pressure, we conclude that afferent arteriolar constriction causes autoregulation. If the efferent arteriole was involved, we would expect changes in the GFR.

THE MYOGENIC MECHANISM AND TUBULOGLOMERULAR FEEDBACK ACCOUNTS FOR AUTOREGULATION

The dynamic response of autoregulation reveals a fast component and a slow component. The fast component is due to **myogenic vasoconstriction**, whereas the slower component is mediated by feedback from the early distal tubule, called **tubuloglomerular feedback**. Tubuloglomerular feedback requires changes in GFR to affect the composition or volume of fluid delivered to the distal tubule, and this takes time to work its way down from the glomerulus to the distal tubule.

Alterations of blood pressure that stretch the afferent arteriole, however, occur rapidly.

THE MYOGENIC MECHANISM AUTOREGULATES GLOMERULAR BLOOD FLOW AND PROTECTS AGAINST RENAL DAMAGE FROM OVERPRESSURE

The myogenic response is the reflex response of the afferent arterioles to changes in blood pressure. Increased blood pressure increases the tension in the vascular wall, and the vascular smooth muscle contracts. Similarly, decreased blood pressure decreases the tension and the smooth muscle relaxes. These responses originate in the smooth muscle cells themselves, triggered by stretch. Although the myogenic mechanism provides a nearly constant P_{GC}, thereby regulating the GFR, the myogenic mechanism also protects the glomerulus from damage by high pressure. Studies of the frequency response of autoregulation show that oscillating pressure causes sustained contractions. The myogenic mechanism responds to the systolic pressure and not the mean pressure. The mean pressure drives the GFR, whereas the myogenic mechanism protects the glomerulus from the systolic pressure. Since systolic and mean pressure generally change in concert, protection against systolic pressure autoregulates the GFR as a by-product.

The mechanism of the myogenic response connects mechanical stretch of the renal vasculature with the contractile machinery of vascular smooth muscle cells. The general mechanism of myogenic responses has been discussed in Chapter 5.11. The myogenic response involves three stages: a membrane sensor for mechanical stretch, transduction of that stretch response to a cytosolic signal, and then, lastly, the contractile response to that signal. The membrane response involves detection of stretch by integrins, membrane-bound dimers of one α and one β subunit, whose deformation activates tyrosine protein kinases, mitogen-activated protein kinases (MAPKs), or extracellular receptor kinases (ERKs). These modulate surface membrane voltage-operated calcium channels (VOCC) and ryanodine receptors (RyR) on the sarcoplasmic reticulum to alter cytosolic $[Ca^{2+}]$, thereby altering the activity of myosin light chain kinase (MLCK) and the phosphorylation state of the myosin light chains, altering the contractile state of the vascular smooth muscle. A variety of other mechanosensitive sensors and pathways have been proposed to explain the myogenic response.

TUBULOGLOMERULAR FEEDBACK REGULATES THE SINGLE NEPHRON GFR

Tubuloglomerular feedback refers to the feedback regulation of the GFR in a single nephron based on sensory information about the distal tubule fluid. This feedback regulation involves the **juxtaglomerular apparatus (JGA)**, a collection of specialized cells where the thick ascending limb contacts the afferent and efferent arterioles. The JGA includes the **macula densa** cells that line the thick ascending limb at its junction with the distal tubule, **granule cells** in the afferent arteriole wall that release the enzyme **renin** into the circulation, and **mesangial cells** that lie between these structures and which may relay signals between them (see Figure 7.6.4). Increased distal tubular Na^+ concentration causes the macula densa cells to swell by activation of the NKCC transporter. These cells release ATP that either directly activates arteriolar cells through purinergic receptors or is converted to adenosine by extracellular nucleotidases. The resulting adenosine can activate adenosine A1 receptors that activate afferent arteriolar contraction. Swelling of the macula densa cells signals constriction of the afferent arteriole of its own nephron so that the glomerular filtration of the same nephron is decreased. This forms a negative feedback loop in which the increased distal tubular load of Na^+ is decreased by reducing the GFR and subsequently allowing more time for Na^+ reabsorption because flow is slower. This regulation of the GFR occurs locally in a single nephron. The thick ascending limb of the nephron regulates its own single nephron GFR (SNGFR).

THE NEPHRON ADJUSTS REABSORPTION OF WATER AND SALT TO MATCH CHANGES IN THE GFR

Autoregulation of the GFR by either the myogenic mechanism or tubuloglomerular feedback does not mean that the GFR does not change. It can change drastically due to sympathetic nervous stimulation of the renal arterioles, for example. When the GFR changes as a primary event, the amount of salt and water reabsorbed by the proximal tubule also changes, so that a constant fraction of the filtered load, about 0.67, is reabsorbed. This phenomenon is called **glomerulotubular balance**. It should not be confused with tubuloglomerular feedback. Glomerulotubular balance is partially explained by two mechanisms: peritubular oncotic pressure and Na^+-cotransport mechanisms. If the GFR increases without a concomitant change in RBF, as might occur if the efferent arterioles constrict while the afferent arterioles relax, then the filtration fraction will increase. The blood perfusing the peritubular capillaries then would have increased oncotic pressure because a greater fraction of protein-free ultrafiltrate would concentrate the remaining proteins more. This effect is not linear, so that the oncotic pressure would increase more than the concentration. The increased oncotic pressure would favor reabsorption of fluid from the tubule lumen. In addition, an increased GFR without a change in the RBF implies that a greater pressure drop occurs across

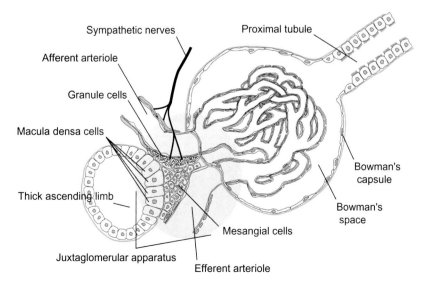

FIGURE 7.6.4 Structures of the juxtaglomerular apparatus. The macula densa cells of the thick ascending limb of the loop of Henle sense the distal tubular load of Na^+ by swelling in response to high Na^+ loads. Signals from the macula densa cells are believed to be relayed by mesangial cells to smooth muscle and granule cells in the wall of the afferent arteriole.

the efferent arteriole than before, so that the hydrostatic pressure within the peritubular capillaries will also be less. Thus both the decrease in hydrostatic pressure and the increased oncotic pressure of the peritubular blood will favor fluid reabsorption along the peritubular capillaries.

Increasing the GFR increases the filtered load of all solutes, including glucose and amino acids and other solutes that are absorbed by Na^+ cotransport. Since these solutes are reabsorbed completely in the proximal tubule, increasing their load increases the amount of Na^+ reabsorption proportionately.

WATER BALANCE IN THE BODY IS MEDIATED BY ANTIDIURETIC HORMONE

The typical water exchanges in moderate climates are shown in Figure 7.6.5. Water inputs include the water content of drink and food plus water made in the body during oxidation. Water outputs include the **insensible loss** through the skin and lungs, even in the absence of sweating, plus losses in urine and feces. These numbers are highly variable. High ambient temperatures or elevated body temperature due to exercise can cause losses by sweating as much as 4 L of water per hour. Sweat is a hypoosmotic fluid that contains NaCl and KCl, and excessive sweating also leads to excessive electrolyte loss. Large losses of water can occur through vomiting and diarrhea. Urinary water excretion is a major daily loss of water. After prolonged and strenuous exercise, urine volume is small and its color is deep, indicating concentration of its solutes. After drinking excessively, the urine volume is large and it is nearly colorless because it is dilute. This everyday experience shows us that the urinary volume varies inversely with the amount of sweat and directly with fluid intake, whereas urinary concentration typically varies inversely with urinary volume. Thus the regulation of urinary volume and concentration is a major component in the regulation of total body water. However, the kidneys must always excrete an obligatory solute load of about 600 mOsmol per day, which requires at least 500 mL of urine per day. The kidneys cannot make water. Loss of body water can be corrected only by drinking water. Thus the behavioral mechanism of thirst is also crucial for maintenance of total body water. The regulation of urinary water excretion is mediated by **ADH**, **antidiuretic hormone**.

ADH SECRETION BY THE BRAIN IS INCREASED BY HYPOVOLEMIA AND HYPEROSMOLARITY

As its name implies, ADH has an "antidiuresis" effect; it inhibits urine flow. It is also called vasopressin because it also causes vasoconstriction. This hormone is a peptide consisting of nine amino acids whose chemical structure is shown in Figure 7.6.6. The hormone is synthesized in the brain in cells in the **supraoptic nucleus** and **paraventricular nucleus** as a precursor protein called propressophysin. It is carried by axoplasmic transport down to the posterior pituitary gland where the precursor is cleaved by proteolysis to its final products: ADH, neurophysin II, and a glycoprotein. Upon stimulation of the cell bodies in the supraoptic nucleus and paraventricular nucleus, ADH and its carrier neurophysin are released from their nerve terminals by exocytosis. ADH dissociates from neurophysin and circulates to its target tissues (see Chapter 9.2).

Specialized cells in the anterior hypothalamus, close to the supraoptic nucleus, sense the osmolarity of blood. Increases in the concentration of impermeant solutes cause these **osmoreceptor cells** to shrink, and this activates them to signal cells in the supraoptic nucleus and paraventricular nucleus to release ADH. Permeable solutes like urea and glucose do not activate the osmoreceptors, so these solutes could increase plasma osmolarity without evoking an increased ADH release. **Changes in blood osmolarity are the strongest factor in regulating ADH release** (see Figure 7.6.7).

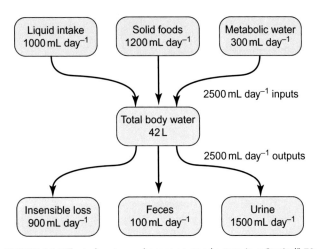

FIGURE 7.6.5 Typical water exchanges at steady state in a "typical" 70-kg adult male. Total body water is about 42 L. Daily loss is about 2.5 L day^{-1}, with most through the urine (1.5 L day^{-1}) and through insensible loss through the skin and lungs (0.9 L day^{-1}). Almost all of this loss is replenished through food (1.2 L day^{-1}) and drink (1 L day^{-1}) with a relatively small component through metabolic water (0.3 L day^{-1}). These numbers vary greatly depending on environmental conditions and activity level.

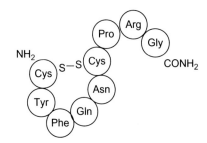

FIGURE 7.6.6 Chemical structure of antidiuretic hormone. The cyclic peptide is released from nerve cell terminals in the posterior pituitary to circulate to target tissues. The cell bodies, located in the hypothalamus, make the hormone as a precursor that is carried by axoplasmic transport along axons that terminate in the posterior pituitary. ADH is released in response to low blood pressure, low circulating blood volume, and blood hyperosmolarity.

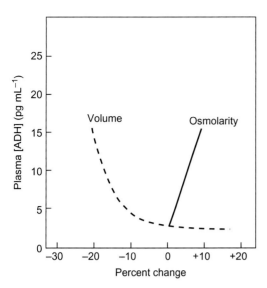

FIGURE 7.6.7 Relative sensitivity of ADH secretion to changes in plasma osmolarity and blood volume. Hyperosmolarity markedly increases ADH secretion as evidenced by higher circulatory [ADH]. Hypovolemia also increases circulating levels of ADH, but larger changes are required compared to changes in plasma osmolarity. The steep curve for osmolarity indicates that ADH secretion is most sensitive to changes in plasma osmolarity.

Stretch receptors in the low-pressure side of the circulation increase their action potential frequency with stretch of the walls of the left atria and intrathoracic veins. The afferent input of these stretch receptors tonically inhibits ADH release. Thus increased stretch caused by excess venous volume reduces ADH release, and decreased stretch increases ADH release. Similarly, immersion in water or exposure to cold tends to send blood back to the central veins, causing a reflex diuresis mediated through reduced ADH release by increased activity of the central stretch receptors (see Figure 7.6.7).

Other factors that influence ADH release include pain, trauma, emotional distress, nausea, ethanol ingestion, angiotensin II, catecholamines, and atrial natriuretic factor. Stimulation of α adrenergic sympathetics is diuretic, whereas β adrenergic stimulation is antidiuretic. Alcohol inhibits ADH release, causing a diuresis.

ADH INCREASES WATER PERMEABILITY OF THE DISTAL NEPHRON

Like most peptide hormones, ADH exerts its effect on cells through receptors on the surface of its target cells. The vasopressin receptors include V1a, V1b, and V2. The V1 receptor subtype is linked to a G_q mechanism involving the activation of phospholipase C and the liberation of inositol trisphosphate (IP3) and diacylglycerol (DAG). (See Chapter 2.8 for a description of G_q-linked receptors.) The V1 receptor subtypes are responsible for the vasoconstrictor effects of ADH. The kidney possesses V2 receptors, which are linked to a G_s mechanism. Occupancy of the V2 receptor liberates the α subunit of the heterotrimeic G_s protein, and this subunit then activates adenylyl cyclase to produce cAMP from ATP. The increased cytosolic concentration of cAMP then activates protein kinase A (PKA). PKA phosphorylates subunits of **aquaporin**, the water channel protein. In the kidney, **AQP2** is phosphorylated at multiple amino acids. Phosphorylation of AQP2 signals the insertion of AQP2 into the apical membrane of the cell, thereby rendering the entire cell permeable to water. The AQP2 channels continuously cycle between the apical membrane and endocytotic vesicles. Channels that are insufficiently phosphorylated are removed from the membrane and phosphorylated channels are inserted into the membrane. **Protein phosphatase** assures that in the unstimulated state the channels become endocytosed and the water permeability of the cell is low. Endocytosis is facilitated by ubiquitinylation of AQP2. Thus the continued presence of ADH is necessary to maintain high water permeability. This mechanism of ADH action is shown in Figure 7.6.8. Although many parts of the nephron possess aquaporins, only the late distal tubule and collecting duct have AQP2 on the apical membrane. AQP1 is located in the proximal tubule and descending thin limb of the loop of Henle, and AQP3 and AQP4 are on the basolateral membrane of the collecting duct principal cells. AQP1, AQP3, and AQP4 are not sensitive to ADH. ADH also increases the total AQP2 content of the tubule cells by increasing transcription of the gene coding for AQP2 (see Figure 7.6.8).

ADH also increases the urea permeability of the inner medullary collecting duct. The urea transporter on the apical membrane has been identified as UT-A1, and ADH increases the insertion of these transporters in the apical membrane, increasing the urea permeability, and producing a higher osmotic gradient along the cortical–medullary axis due to urea recycling (see Chapter 7.5). The action of ADH of increasing the urea permeability of the inner medullary collecting duct occurs through similar mechanisms as the insertion of AQP2 into the apical membranes (see Figure 7.6.9).

THE ADH-RENAL SYSTEM FORMS NEGATIVE FEEDBACK LOOPS

Increasing the osmolarity of blood increases ADH release, which increases the water permeability of the terminal nephron. This causes the excretion of a small volume of concentrated urine. Since solutes are excreted in excess of water in a concentrated urine, this has the effect of removing salt from the extracellular fluid. Thus this system would correct the hyperosmolarity and return body fluids to normal.

Increased blood volume increases stretch that in turn increases inhibition of ADH release. The lower circulating [ADH] results in low water and urea permeability of the terminal nephron and excretion of a large volume of dilute urine. This tends to deplete the blood volume, correcting the perturbation in blood volume. Similarly, lowering the blood volume engages the system so that water is lost less rapidly and any additional fluids

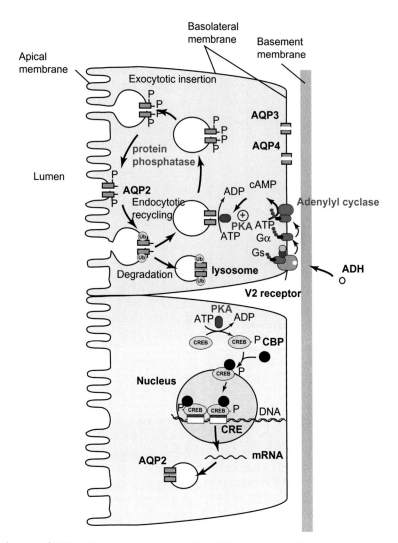

FIGURE 7.6.8 Simplified mechanism of ADH action on water permeability of distal tubule and collecting duct principal cells. AQP2 channels are sequestered in subapical vesicles in the nonstimulated state. ADH binds to V2 receptors on the basolateral membrane, activating a G_s mechanism. The α subunit of a heterotrimeric G-protein exchanges GTP for GDP and dissociates from the βγ subunits. The Gα subunit activates adenylyl cyclase, converting ATP to 3′,5′-cAMP. The cytosolic cAMP then activates protein kinase A (PKA), which phosphorylates AQP2 channels in vesicles. The AQP2 channels are then inserted into the apical membrane, thereby increasing the cell's permeability to water. AQP2 is dephosphorylated by protein phosphatases and is also ubiquitinylated. These signal for endocytosis of the AQP2, which may then be recycled or degraded by lysosomal enzymes. ADH also increases the total AQP2 in the cell over a longer time by PKA-mediated phosphorylation of the cyclic AMP response element binding protein (CREB), which then binds the CREB-binding protein (CBP) and then activates the AQP2 promoter to increase expression of AQP2. The long-term effect of ADH is mediated by another protein, EPAC (exchange protein activated directly by cAMP).

that are consumed are retained by the kidneys. These feedback loops are illustrated in Figure 7.6.10.

THE *FREE WATER CLEARANCE* QUANTIFIES THE OVERALL CONCENTRATION OR DILUTION OF URINE

The free water clearance is given by

$$[7.6.7] \qquad C_{H_2O} = Q_U \left(1 - \frac{U_{osm}}{P_{osm}}\right)$$

where C_{H_2O} is the free water clearance, Q_U is the urine flow, U_{osm} is the total osmolarity of urine, and P_{osm} is the total osmolarity of plasma. Note that this equation is not the same as the typical clearance, $C_x = Q_U U_x / P_x$, because the concentration of water in a solution is inversely related to the osmolarity: higher osmolarity implies less water per unit volume of urine. You can see that this equation makes intuitive sense. If the urine osmolarity is the same as plasma, the urine is isosmolar and there is no gain or loss of free water. There is still loss of water, equal to Q_U, but the body fluids are not being concentrated or diluted. If the free water clearance is positive, it means that $U_{osm} < P_{osm}$ and free water is being excreted or cleared. In this case the urine is hyposmolar and body fluids are being concentrated by removal of water. If the free water clearance is negative, it means $U_{osm} > P_{osm}$ and the urine is hyperosmolar: free water is being retained in the body while excess salt

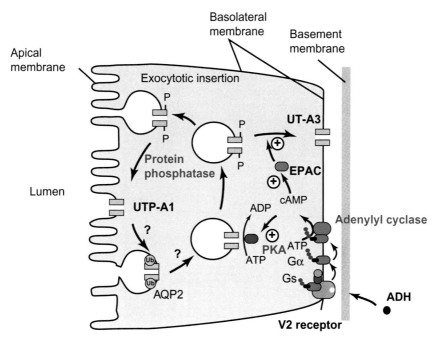

FIGURE 7.6.9 Simplified mechanism of action of ADH on urea permeability of the inner medullary collecting duct cells. ADH (=vasopressin) binds to a GPCR (G-protein-coupled receptor) on the basolateral membrane of the cells, causing the exchange of GTP for GDP on the α subunit of the heterotrimeric G-protein. The G_α subunit activates adenylyl cyclase to increase cytosolic concentrations of 3',5' cyclic AMP. This cAMP activates PKA (protein kinase A) and EPAC (exchange protein activated by cAMP) that phosphorylates UT-A1 channels, signaling their incorporation into the apical membrane. cAMP also activates EPAC that also mediates UT-A1 and UT-A3 function.

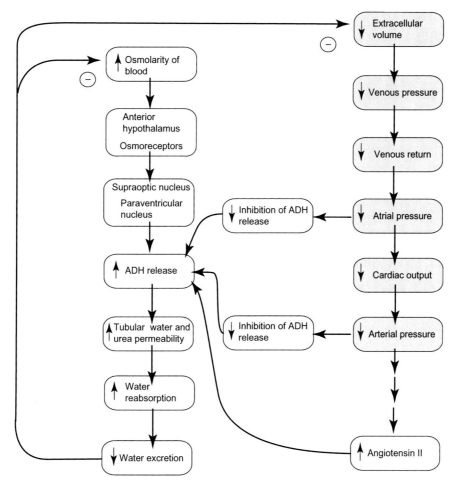

FIGURE 7.6.10 Negative feedback loops involving ADH. The main determinants of ADH release are increased plasma osmolarity and decreased blood volume. ADH released in response to these perturbations increases the water and urea permeability of the distal nephron resulting in more water reabsorption and excretion of a small volume of concentrated urine. This can correct the hyperosmolarity but cannot correct the reduced blood volume. However, it sets the stage for retention of water when we drink.

and other osmolytes are being excreted. The free water clearance thus measures the volume of plasma that is being completed cleared of water. Unlike other clearances, the free water clearance can be positive, zero, or negative, and reflects the relative treatment of osmolytes—primarily NaCl—and water.

REGULATION OF Na$^+$ BALANCE INVOLVES THE *RENIN–ANGIOTENSIN–ALDOSTERONE* SYSTEM

RENIN *BEGINS A CASCADE OF EVENTS LEADING TO THE PRODUCTION OF CIRCULATING* ANGIOTENSIN II

Specialized granule cells called juxtaglomerular cells or JG cells in the afferent arteriole release **renin** into the circulation. Renin is a proteolytic enzyme that converts an inactive plasma protein, an α_2 globulin, called **angiotensinogen**, into **angiotensin I**. Angiotensin I is a polypeptide 10 amino acids in length. **Angiotensin converting enzyme (ACE)** cleaves off two additional amino acids to produce **angiotensin II**. ACE is located predominantly in the lung, but the kidney also has some ACE activity. Angiotensin II exerts multiple effects that are discussed below. A third enzyme, **ACE2**, converts angiotensin II to **angiotensin (1–7)** that binds to a GPCR called Mas whose activation opposes many of the functions of angiotensin II.

JG cells synthesize prorenin which is converted to renin by proteases. JG cells release prorenin at a constant and slow rate that is not affected by stimuli for renin release. Prorenin is not converted to renin in the circulation.

MULTIPLE SIGNALS CAUSE RENIN RELEASE FROM THE GRANULE CELLS OF THE AFFERENT ARTERIOLE

Three main signals stimulate renin release:

1. Decreased afferent arteriolar pressure increases renin release.
2. Renal sympathetic nerve stimulation increases renin release.
3. Decreased distal tubule [NaCl] increases renin release.

Decreased afferent arteriolar pressure is sensed locally within the arteriole and a signal passes to the granule cells directly. The renal sensor for arteriolar pressure is called the "intrarenal baroreceptor." The response of the system shows a threshold arterial pressure of about 85 mmHg that separates an almost constant level of renin secretion at higher pressures from a steep increase at lower pressures (see Figure 7.6.11). Renal sympathetic nervous stimulation can occur when there is a drop in systemic arterial pressure or during strenuous exercise or in the "fight or flight" reaction in which the renal arteries are vasoconstricted. Sympathetic stimulation of renin release occurs at low levels of sympathetic renal nerve activity that precedes the renal vasoconstriction that occurs at higher sympathetic output to the kidneys. This mechanism involves β_2 adrenergic receptors

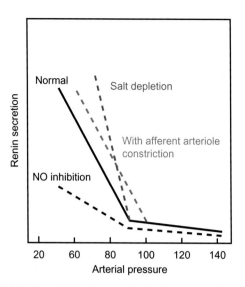

FIGURE 7.6.11 Control of renin secretion. Renin secretion has a low and nearly constant level at high arterial pressure. At a threshold of about 85 mmHg, renin secretion rises steeply with decrease in pressure. This effect is blocked by the inhibition of NO production. The threshold is increased by afferent arteriolar constriction caused by high levels of renal sympathetic nerve stimulation or by pathological conditions such as renal stenosis. Salt depletion increases renin release at any given pressure, indicating the multifactorial nature of renin secretion.

on the granule cells. However, the vasoconstriction of the afferent arteriole through α-adrenergic receptors also lowers the pressure in the glomerulus and sensitizes the intrarenal baroreceptor, causing increased renin secretion for a given reduction in arterial pressure.

Decreased distal tubule [NaCl] is sensed by cells in the **macula densa**, which swell in response to increased [NaCl] in the distal tubule. A diffusible paracrine signal passes from the macula densa cells to the granule cells, probably being relayed by the intervening mesangial cells. Three candidates for this paracrine signal include adenosine, nitric oxide (NO), and prostaglandins. Isolated juxtaglomerular apparatus (JGA) produces PGE$_3$ that can stimulate renin secretion through EP2 or EP4 receptors. In isolated perfused JGA, stimulation of renin secretion by low [NaCl] in the lumen is blocked by COX-2 (cyclooxygenase, an enzyme necessary for prostaglandin synthesis) blockers. Adenosine inhibits renin release and may be involved in the suppression of renin release by high [NaCl] or increased arterial pressure, but it appears to be unimportant in the increased secretion caused by low [NaCl] in the tubule fluid at the macula densa. Macula densa cells expresses the neuronal nitric oxide synthase (NOS-I) that produces NO that stimulates renin release in response to low luminal [NaCl].

Renin release is also feed back inhibited by angiotensin II.

ANGIOTENSIN II EXERTS MULTIPLE EFFECTS

Angiotensin II exerts five major effects:

1. AII causes vasoconstriction.
2. AII promptly releases **aldosterone** from the **zona glomerulosa** of the adrenal gland.
3. AII induces the sensation of thirst.

4. AII releases ADH.
5. AII increases reabsorption of Na^+ and HCO_3^- in the proximal tubule.

All of these effects work to increase the extracellular fluid volume and to maintain the integrity of the circulation. The vasoconstriction helps to bring the systemic circulatory pressure back up by increasing total peripheral resistance, TPR. Aldosterone helps reabsorb filtered Na^+. This helps expand, or prevents further loss of, extracellular fluid volume. The sensation of thirst drives us to drink fluids to replenish the ECF volume. Release of ADH helps recapture filtered water, which prevents further loss of ECF. Increasing reabsorption of Na^+ reduces Na^+ loss in the urine, further helping to preserve the ECF volume. These effects of ANG II are mediated by angiotensin receptors (AT_1 and AT_2) that are G-protein-coupled receptors (GPCR) residing on the cell membrane of target cells.

ALDOSTERONE IS A MINERALOCORTICOID THAT INCREASES Na^+ RETENTION AND K^+ EXCRETION

Aldosterone is a steroid hormone produced in the outermost layer of the adrenal gland, the **zona glomerulosa**. Its secretion is controlled predominantly by angiotensin II and plasma $[K^+]$. Increasing plasma $[K^+]$ increases aldosterone secretion. Other stimuli include decreased plasma $[Na^+]$ and **adrenocorticotrophic hormone (ACTH)**. Aldosterone's chemical structure is shown in Figure 7.6.12.

Aldosterone acts according to the classical steroid hormone model in which the hormone binds to cytosolic receptors and is transferred to the nucleus where it binds to nuclear receptors that stimulate the transcription of specific genes. Aldosterone affects Na^+ and water absorption and K^+ secretion in the gut, salivary glands, sweat glands, and kidneys. In the kidneys, aldosterone increases Na^+ reabsorption and increases K^+ secretion in cells in the cortical collecting duct. It increases the activity of the basolateral Na^+,K^+-ATPase, ENaC, and the apical K^+ channel, ROMK (see Figure 7.5.7). Its major effect is to increase ENaC. The mechanism of action of aldosterone is discussed in Chapter 9.5.

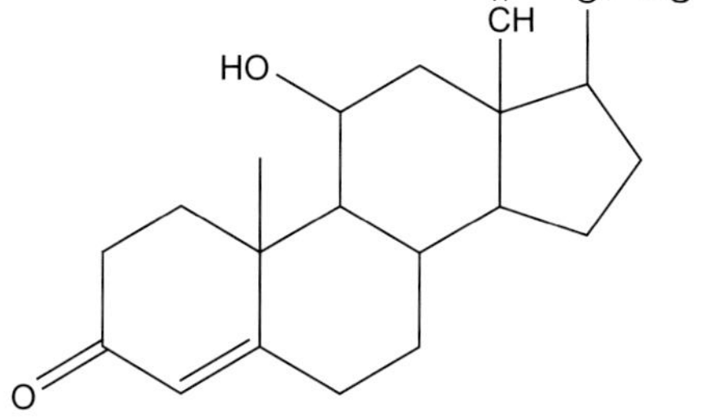

FIGURE 7.6.12 Chemical structure of aldosterone.

ATRIAL NATRIURETIC PEPTIDE AND ENDOGENOUS DIGITALIS-LIKE SUBSTANCE INCREASE Na^+ EXCRETION IN HYPERVOLEMIA

Atrial natriuretic peptide (ANP) originates in the atria and B-type or brain natriuretic peptide (BNP) originates in the ventricles. A third form, CNP, is made in many organs including the brain, kidney, bone, and vasculature. Both ANP and BNP are made as pro-hormones that are converted to their active forms by a membrane-bound protease called corin. Stretch of the atrial wall and other stimuli such as endothelins cause release of atrial natriuretic peptide (ANP) into the blood. Pressure or volume overload that stretches the ventricular wall causes BNP release. Both ANP and BNP bind to natriuretic peptide receptor type A (NPR-A) which is a quanylyl cyclase. ANP decreases TPR and increases renal Na^+ and water excretion. ANP appears to increase Na^+ excretion by inhibiting apical ENaC and basolateral Na, K-ATPase in the medullary collecting duct.

The adrenal cortex synthesizes a steroid hormone that appears to inhibit the Na^+,K^+-ATPase. This substance may be structurally similar to a class of compounds called cardiac glycosides because they increase the contractile force of the heart. The compound **digitalis** was first isolated from the Foxglove plant. Related drugs are **ouabain** and **digoxin**. These compounds all inhibit the Na^+,K^+-ATPase. For this reason, the compound is called endogenous digitalis-like substance or EDLS. The role of EDLS in Na^+ balance is not established.

THE INTEGRATED RESPONSE TO DECREASED BLOOD VOLUME

Figure 7.6.13 illustrates the main events in the response to decreased blood volume. The kidney engages the renin–angiotensin–aldosterone (RAA) system in response to decreased afferent arteriolar pressure, increased renal sympathetic nervous stimulation, and decreased NaCl delivery to the macula densa. Angiotensin II that results from increased renin release causes vasoconstriction, increased aldosterone release, increased ADH release, increased Na^+ reabsorption, and increased thirst. All of these actions tend to raise the ECF volume or to prevent its further loss. The kidneys cannot make new volume. Loss of water or salt through sweat or diarrhea or vomiting or hemorrhage can be reversed only by the consumption of water and electrolytes.

INTEGRATED RESPONSE TO INCREASED Na^+ LOAD OR VOLUME EXPANSION

Excess ingestion of NaCl, followed by its absorption from the gastrointestinal tract, leads to an increased osmolarity of the blood that stimulates thirst and ADH release. Thus there is an increase in water intake and

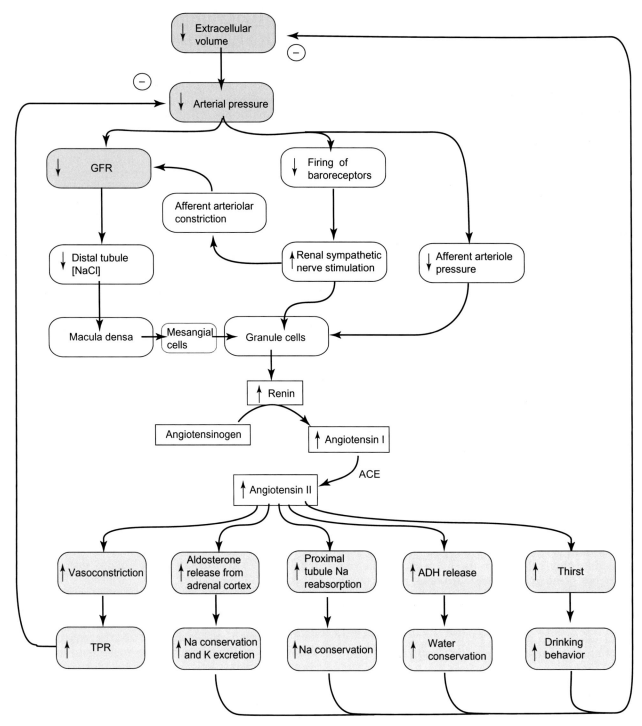

FIGURE 7.6.13 Feedback mechanisms for the response to decrease in blood volume.

conservation until the osmolarity of the blood returns to normal. This can be achieved only by an expansion of the ECF volume, as described in Chapter 7.1. These events occur rapidly and then the body is presented with a normal blood osmolarity but an expanded blood volume. All of the systems shown in Figures 7.6.10 and 7.6.13 are then brought to bear to correct the situation, but this time they all work in reverse. Thus there is a reduction in ADH levels and in aldosterone levels, with the resulting excretion of a more dilute urine, approximately isoosmolar with plasma, so that the ECF volume returns to normal while maintaining normal osmolarity.

SUMMARY

The renal system regulates the volume and osmolarity of the plasma and, by extension, of the ISF and ICF, by excreting either a concentrated urine or a dilute urine. This is controlled principally by ADH, but other factors are involved including the RAA system, stretch receptors in the left atria and intrathoracic veins, and the thirst mechanism. Increased osmolarity increases ADH release, which results in excretion of a highly concentrated urine. This eliminates salt in excess of water, so that the blood osmolarity returns toward normal. Stretch of the left atria inhibits ADH release, and less stretch during plasma depletion relieves the inhibition, causing increased ADH release. This in itself cannot restore plasma volume, but it helps retain water when it is ingested.

ADH exerts its effects on the kidney through V2 receptors linked to a G_s mechanism. Binding of ADH to V2 receptors increases cytosolic [cAMP] that activates PKA to phosphorylate aquaporin channels (AQP2) that in turn signal the cell to transport the channels to the apical membrane. This increases the water permeability of the late distal tubule and the collecting duct. In the absence of ADH, protein phosphatases gradually dephosphorylate the AQP2 channels and they are removed from the apical membrane and stored in endocytotic vesicles.

The RAA system is activated by: (1) decreased afferent arteriolar pressure; (2) increased renal sympathetic nervous stimulation; and (3) decreased distal tubule [NaCl]. These all signal the granule cells in the afferent arteriole to release renin, an enzyme that breaks down plasma angiotensinogen to angiotensin I, which is further converted to the active angiotensin II by ACE. Angiotensin II has multiple effects, including: (1) vasoconstriction; (2) release of aldosterone from the adrenal cortex; (3) release of ADH; (4) increased thirst; and (5) increased absorption of Na^+ and HCO_3^- from the proximal tubule. All of these actions defend against reduced blood pressure and blood volume. The vasoconstriction helps return blood pressure to normal, aldosterone retains Na^+ by preventing renal loss, ADH minimizes renal water loss, increased thirst motivates replacement of lost fluids, increasing Na^+ reabsorption in the proximal tubule complements actions of aldosterone in the distal nephron.

REVIEW QUESTIONS

1. What effect would increased blood pressure have on the GFR? On the filtration fraction? What effect does autoregulation have on this relationship?
2. What mechanisms are responsible for autoregulation of renal blood flow and GFR? Which operates more quickly?
3. What effect does dehydration have on GFR?
4. What is ADH? Where does it originate? What are the stimuli for its release? What are its renal actions? What are its actions on the vasculature?
5. What is renin? What cells secrete it? What are the stimuli for its release? What does it do?
6. What is angiotensin? How is it produced? What does it do?
7. What is aldosterone? What stimuli release it? What does it do?
8. What are the sensors for hypovolemia? How does hypovolemia get corrected?
9. What are the sensors for hyperosmolarity? How does hyperosmolarity get corrected?

7.7 Renal Component of Acid—Base Balance

Learning Objectives

- List the three body systems for regulating plasma pH
- Define titratable acid
- Describe how the kidney can make new HCO_3^- linked to urinary titratable acid and ammonium
- Describe the metabolic origin of urinary ammonium
- Explain why ammonium does not show up as a titratable acid
- Describe the renal response to respiratory acidosis
- Describe the renal response to respiratory alkalosis
- Describe the renal response to metabolic acidosis
- Describe the renal response to metabolic alkalosis

THE KIDNEYS ELIMINATE THE ACID PRODUCED FROM METABOLISM

Addition of CO_2 to blood raises the H^+ concentration of the blood, decreasing its pH. This is caused by the hydration of CO_2 with the formation of carbonic acid (H_2CO_3) and subsequent dissociation of (H_2CO_3) to bicarbonate and H^+:

$$[7.7.1] \quad CO_2 + H_2O \rightleftharpoons H_2CO_3 \rightleftharpoons HCO_3^- + H^+$$

This reaction occurs spontaneously in physiological fluids, but it is slow. The enzyme **carbonic anhydrase** catalyzes the hydration reaction but without the release of H_2CO_3. Carbonic anhydrase catalyzes the reaction:

$$[7.7.2] \quad CO_2 + H_2O \rightleftharpoons HCO_3^- + H^+$$

The H^+ and HCO_3^- can then combine to form carbonic acid. Because carbonic acid is in equilibrium with the volatile CO_2, H_2CO_3 is called a **volatile acid**. At rest, the body produces approximately 200 mL of CO_2 per minute (at STPD). This is 200 mL min^{-1} × 1440 min day^{-1}/ 22.4 L mol^{-1} ≈ 13 mol of CO_2 per day under resting conditions. The actual production averages about 20 mol of CO_2 per day because activity increases O_2 consumption and CO_2 production. Because the hydration reaction that occurs in the venous blood that collects CO_2 is reversed in the lungs where the CO_2 is eliminated to the atmosphere, there is no net H^+ produced from CO_2.

In addition to CO_2, the body produces about 0.3–1 mmol of acid per kilogram of body weight per day or about 20–70 mmol of acid per day. This acid originates from the oxidation of sulfur-containing amino acids as in the oxidation of methionine:

$$[7.7.3] \quad \begin{array}{l} 2C_5O_{11}NO_2S + 15H_2O \Rightarrow 4H^+ + 2SO_4^{2-} \\ + CO(NH_2)_2 + 7H_2O + 9CO_2 \end{array}$$

Additional acid originates from oxidation of phospholipids, producing H_3PO_4, and from partial metabolism of carbohydrates and fats that produce organic acids (lactic acid, acetoacetic acid, and β-hydroxybutyric acid). In rapid muscular activity or hypoxia in which anaerobic metabolism dominates, the production of these nonvolatile acids is markedly enhanced. Thus the normal acid—base situation in blood is the defense of the alkaline blood against a constant assault by acid. The kidneys excrete the excess acid that is produced daily.

THE BODY USES *CHEMICAL BUFFERS*, THE *RESPIRATORY SYSTEM*, AND THE *RENAL SYSTEM* TO REGULATE pH

THE CHEMICAL BUFFERS RESPOND RAPIDLY AND ARE THE FIRST DEFENSE

The chemical buffers resist change in plasma pH by binding H^+ when it is present in excess and releasing H^+ when there is a deficit. There are a variety of buffers, and their affinity for H^+ is described by their K_A, the acid association constant (see Chapter 6.5). The ability to buffer changes in pH is maximum when the pH = log K_A = −log K_D = pK. All of these buffers are linked by their binding of the H^+ ion (the **isohydric principle**), so that adjustment of any one buffer system adjusts them all through the adjustment of [H^+]. The body adjusts the **bicarbonate buffer system** because CO_2 can be regulated by adjusting CO_2 elimination to the atmosphere.

THE RESPIRATORY SYSTEM RESPONDS RAPIDLY BY ADJUSTING P_{aCO_2}

Recall the Henderson—Hasselbalch equation for the bicarbonate buffer system that we derived in Chapter 6.5:

$$[7.7.4] \quad pH = 6.1 + \log \frac{[HCO_3^-]}{0.0308\, P_{CO_2}}$$

Here 6.10 is the pK when [HCO_3^-] is given in mmol L^{-1} and arterial P_{CO_2} is in mmHg. The log is the logarithm base 10, not the natural logarithm. Increasing P_{CO_2} decreases the argument of the logarithm, which decreases the value of the logarithm, which decreases pH. The alveolar ventilation equation (Eqns 6.3.21 and 6.3.22) shows that the alveolar P_{aCO_2}, which equilibrates with the arterial P_{CO_2}, is set by the ratio Q_{CO_2}/Q_A, the rate of CO_2 production divided by the rate of alveolar ventilation. For a given rate of metabolism (Q_{CO_2}), the P_{aCO_2} is inversely proportional to the alveolar ventilation, Q_A. Thus increasing the breathing rate and depth increases Q_A and decreases P_{aCO_2}. According to Eqn (7.7.3), increasing alveolar ventilation will lower P_{CO_2}, increase the argument of the logarithm, increase its value, and thereby increase plasma pH. Similarly, reducing Q_A raises the plasma P_{CO_2} and lowers pH. Alveolar ventilation is stimulated in part by pH sensors located in the central nervous system (CNS) in the medulla and peripherally in the carotid bodies. Thus acidosis stimulates respiratory drive by stimulating the peripheral chemoreceptors, and this increases Q_A, blowing off CO_2 and raising the pH of the blood back toward normal. The opposite events occur in alkalosis. This response of the respiratory system is rapid but incomplete: it cannot by itself return the blood to a normal pH and P_{CO_2}.

The Henderson–Hasselbalch equation is a different way of representing Eqn (7.7.1), the buffer reaction of carbonic acid. Acid–base balance can also be understood by a straightforward application of **Le Chatelier's principle** to the bicarbonate buffer reaction (Eqn 7.7.1), except it is more difficult to be quantitative about it. Le Chatelier's principle states that any reaction will respond to a perturbation in its constituents in such a way as to minimize the perturbation. The bicarbonate buffer reaction is as follows:

[7.7.5] $$CO_2 + H_2O \rightleftharpoons H_2CO_3 \rightleftharpoons HCO_3^- + H^+$$

Thus increasing P_{CO_2} increases [CO_2], which forces the reaction to the right: the reaction consumes CO_2 to produce H_2CO_3, HCO_3^-, and H^+ in order to minimize the perturbation of increased CO_2. In this way, reduction in ventilation increases P_{CO_2}, which subsequently increases [H^+] and decreases the pH. Similarly, reducing P_{CO_2} forces the reaction in the opposite direction, consuming H^+ and raising the pH.

THE RENAL SYSTEM RESPONDS SLOWLY BUT COMPLETELY BY ADJUSTING [HCO_3^-]

The kidney constitutes the third line of defense against pH disturbances. Although it acts much more slowly than the chemical buffers and the respiratory system, it completely compensates for the addition of acid or base to the blood. It does this either by acidifying the blood by alkalinizing the urine or by alkalinizing the blood by acidifying the urine. The mediator of both of these processes is the bicarbonate ion, HCO_3^-. These relationships can be seen by inspection of the Henderson–Hasselbalch equation or by applying Le Chatelier's principle to the bicarbonate buffer reaction. In acidosis, log ([HCO_3^-]/0.0308 P_{CO_2}) is too low. It can be corrected by lowering P_{CO_2}, which is the respiratory component of compensation, or by raising [HCO_3^-], which is the renal component of compensation. In the bicarbonate buffer reaction, increased acidity can be corrected by causing the reaction to go to the left, consuming H^+. This is accomplished by lowering [CO_2] by hyperventilation, or by raising [HCO_3^-], which binds the extra H^+ ions. Similar arguments apply to the respiratory and renal components of compensation for alkalosis, with the direction of the reactions reversed compared to acidosis.

ACID EXCRETION BY THE TUBULE ADJUSTS BLOOD pH

Regulation of plasma pH by the kidney is accomplished by secreting more or less acid into the tubule fluid *relative to the filtered load of HCO_3^-*. If the plasma is acidic, secretion of more acid into the tubule fluid than the filtered HCO_3^- acidifies the tubule fluid and alkalinizes the blood. As we shall see, acid secretion is linked to HCO_3^- transfer to the blood. Therefore, acid secretion in excess of the filtered HCO_3^- alkalinizes the blood by increasing its [HCO_3^-], thereby returning the blood pH toward normal. Similarly, if the blood is too alkaline, less acid is secreted into the tubule than the filtered HCO_3^-, and HCO_3^- is lost in an alkaline urine. The HCO_3^- that is lost originated from the blood by glomerular filtration. Thus less acid secretion is linked to reduction in plasma [HCO_3^-], which acidifies the blood. The key to this process is the link between acid secretion and HCO_3^- appearance in the plasma.

THE KIDNEY LINKS ACID SECRETION TO HCO_3^- APPEARANCE IN PLASMA

The basic mechanism for the reabsorption of filtered HCO_3^- in **proximal tubule cells** is shown in Figure 7.7.1. Secretion of acid begins with plasma CO_2. CO_2 from the plasma can enter the tubule cells where it is hydrated to form HCO_3^- and H^+. The hydration reaction is catalyzed by **carbonic anhydrase**. Two mechanisms on the apical membrane pump out the H^+ across the apical membrane into the lumen. These are as follows: (1) an active **H^+-ATPase** that is a primary active transporter and (2) an **Na^+–H^+ exchanger (NHE3)** that operates by secondary active transport. In other parts of the nephron, a third secretory mechanism, a H^+,K^+-ATPase, can pump H^+ from the cell in exchange for K^+.

In the tubule lumen, carbonic anhydrase IV on the luminal membrane rapidly combines the secreted H^+ with HCO_3^- that was filtered from the plasma. These form CO_2 and H_2O. The CO_2 can then diffuse through the cell and into the plasma. Thus acid secretion results in the disappearance of HCO_3^- from the tubule fluid.

HCO_3^- in the tubule cell is transported across the basolateral membrane into the peritubular fluid by a specific

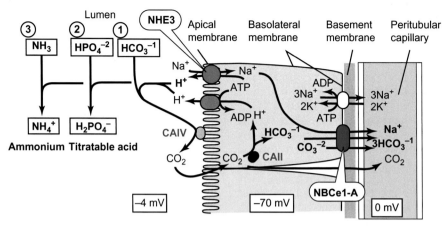

FIGURE 7.7.1 Mechanism of reabsorption of HCO_3^- in proximal tubule cells. H^+ ions within the tubule cell are actively secreted into the lumen across the apical membrane by the operation of primary active transport (H^+-ATPase) or secondary active transport (Na^+–H^+ exchanger). The secreted H^+ combines with HCO_3^- to form carbonic acid, H_2CO_3. Carbonic anhydrase converts this to CO_2 and H_2O. The H_2O joins the rest of the water in the tubule lumen and may be either reabsorbed or excreted. The CO_2 diffuses into the cell where it may be hydrated to form H_2CO_3 again, catalyzed by cytosolic carbonic anhydrase. The carbonic acid dissociates to form H^+ and HCO_3^-. The H^+ replenishes the cytosolic H^+ initially pumped out of the cell. The HCO_3^- is transported out of the cell into the peritubular fluid by a secondary active transporter, using Na^+ as the cotransported ion. The net result is a disappearance of $Na^+ HCO_3^-$ from the lumen and appearance of $Na^+ HCO_3^-$ in the blood. Secreted H^+ ions can also combine with other filtered buffers. The figure shows binding of secreted H^+ with Na_2HPO_4, which has a $pK = 7.21$. According to the Henderson–Hasselbalch equation, an acid is half neutralized when the pH = pK, and this is the point of maximal buffering. Addition of strong base will neutralize this acid, and so it forms part of the titratable acid. Its excretion is linked to the formation of new HCO_3^- in plasma. Secreted H^+ ions can also combine with NH_3, ammonia, to produce NH_4^+, the ammonium ion. Excretion of NH_4^+ is also linked to the formation of new HCO_3^- in the plasma.

carrier. This **Na^+–HCO_3^- cotransporter type NBCe1-A** appears to transport one HCO_3^- and one CO_3^{2-} for each Na^+, which is the equivalent of three HCO_3^- for each Na^+. The NBCe1-A is electrogenic. Since it carries three negative charges for each positive charge, it carries a net current.

In the overall process shown in Figure 7.7.1, a HCO_3^- ion is returned to the blood for every H^+ ion that is secreted. It is important to note that the HCO_3^- that appears in the blood is not the same as that which disappeared from the tubule lumen. The one that disappeared from the tubule lumen became H_2O and CO_2, whereas the HCO_3^- that is placed in the blood originates from plasma or tubule fluid CO_2. However, the net effect of the system's operation is the transfer of filtered HCO_3^- to the blood. If H^+ secretion is insufficient to reclaim all of the filtered HCO_3^-, then HCO_3^- will be lost in the urine. If H^+ secretion is sufficiently rapid, all of the filtered HCO_3^- can be reclaimed. If H^+ secretion is more rapid yet, new HCO_3^- can be added to the blood.

SECRETED ACID RECLAIMS HCO_3^- OR COMBINES WITH TITRATABLE ACID OR AMMONIUM

Figure 7.7.1 illustrates three possible fates of secreted H^+. The first is binding to HCO_3^-, which is equivalent to reabsorption of filtered HCO_3^-. Two other fates are possible. Secreted H^+ can bind to **ammonia (NH_3)** to produce **ammonium (NH_4^+)**, or it can bind to **titratable acids**. An example of titratable acids is $H_2PO_4^-$.

Titratable acid is defined as the amount of strong base needed to titrate the pH of the urine back to pH 7.4. This is the normal pH of the glomerular ultrafiltrate, so the amount of strong base that is used to titrate the urine back is equal to the amount of titratable acid that was excreted into the tubular fluid to produce the urine. As can be seen in Figure 7.7.1, excretion of titratable acid is linked to the formation of new HCO_3^-; for every H^+ ion that is secreted into the tubule fluid and combines with a titratable acid, a new HCO_3^- ion is placed into the plasma. Thus the venous blood that drains the kidney can have a net gain in plasma $[HCO_3^-]$. The increased HCO_3^- binds plasma H^+, thereby reducing the $[H^+]$ and raising the pH.

Figure 7.7.1 also shows that the excretion of NH_4^+ is linked to the formation of new HCO_3^-. For every H^+ that is secreted and combines with NH_3, a new HCO_3^- ion is placed in the plasma. The mechanism is not exactly as shown in Figure 7.7.1 because the binding of H^+ usually occurs within the tubule cell rather than in the lumen, but the net effect is the same: excretion of NH_4^+ increases the plasma $[HCO_3^-]$.

AMMONIA DOES NOT SHOW UP AS TITRATABLE ACID BECAUSE ITS pK IS TOO HIGH

The acid dissociation of NH_4^+ is written as

[7.7.6] $\quad NH_4^+ \rightleftharpoons NH_3 + H^+: K_D = 6.3 \times 10^{-10}$ M

The pK for this reaction is 9.2. The Henderson–Hasselbalch equation for this reaction is as follows:

[7.7.7] $$pH = 9.2 + \log\frac{[NH_3]}{[NH_4^+]}$$

where NH_4^+ is the acid and NH_3 is the base. What this equation means is that at ordinary pH of 7.4, the equilibrium $[NH_3]$ will be much smaller than $[NH_4^+]$. At pH 7.4, in cells and in the tubular fluid, most of the buffer will be present as NH_4^+. The NH_4^+ will not react with base and be converted to NH_3 until the pH approaches the pK. Thus the titration of the urine to pH 7.4 does not reveal its content of NH_4^+.

NEW HCO_3^- FORMED IS THE SUM OF TITRATABLE ACID AND NH_4^+ MINUS THE EXCRETED HCO_3^-

For each H^+ excreted as titratable acid, there is one new HCO_3^- transferred to blood. For each H^+ excreted as NH_4^+, there is also one new HCO_3^- transferred to blood. Urine typically contains some HCO_3^- that has not been reabsorbed. The net gain of HCO_3^- to the body is the balance of these sources and sinks:

$$\text{Net } HCO_3^- \text{ gain} = \text{titratable acid} + NH_4^+ - HCO_3^-$$
[7.7.8]

where the titratable acid, NH_4^+, and HCO_3^- refer to the amounts present in the urine.

AMMONIUM ORIGINATES FROM AMINO ACIDS IN PROXIMAL TUBULE CELLS

Proximal tubule cells deaminate **glutamine** to glutamic acid and then form α-ketoglutaric acid, liberating one NH_3 group at each step. These reactions are shown in Figure 7.7.2. Glutamine enters the proximal tubule cell two ways: it is reabsorbed from filtered glutamine

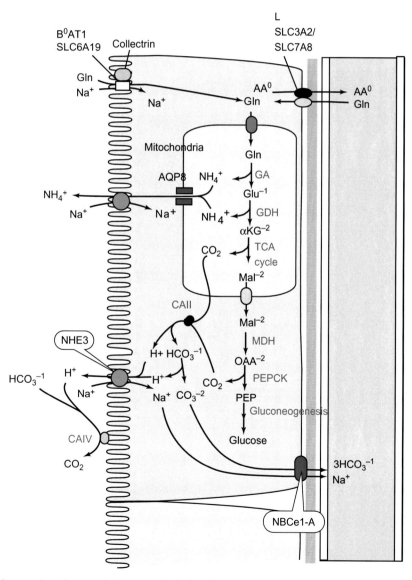

FIGURE 7.7.2 Mechanism of generation of ammonia in proximal tubule cells. Deamination of glutamine occurs in two steps: conversion to glutamic acid and conversion to α-ketoglutaric acid. Ammonia generated in this way quickly is protonated to ammonium, which can be transported across the apical cell membrane by the Na^+–H^+ exchanger. Every excreted NH_4^+ is equivalent to the secretion of H^+ that is linked to the formation of a new HCO_3^- that appears in the blood. In this way, the kidney can replenish blood HCO_3^- that has been depleted by acidosis.

through the neutral amino acid transporter, B^0AT1, or it is taken up from the peritubular capillaries by exchange with neutral amino acids. During normal acid–base balance, filtered glutamine is taken up by these cells and largely returned to the blood. During chronic acidosis, the activity of several enzymes and transporters are increased to increase the rate of deamination of glutamine. The NH_3 liberated from these reactions nearly instantly equilibrates with the cells' $[H^+]$, with only a small fraction remaining free as NH_3. The NH_4^+ crosses the apical membrane, probably carried by the $Na^+–H^+$ exchanger. Some free NH_3 may also diffuse across the apical membrane because it is uncharged and lipophilic. In the tubule fluid, NH_3 binds to H^+ ions secreted by the cells. Once it binds H^+ to form ammonium, it can no longer easily penetrate the apical membrane. This is called **diffusion trapping** because NH_3 diffuses across the membrane and then becomes trapped in the acid environment as NH_4^+. For every NH_4^+ excreted, there is the equivalent of one H^+ excreted that is linked to transfer of new HCO_3^- to the plasma. Other amino acids also can generate ammonium but most of it derives from glutamine.

THE THICK ASCENDING LIMB SECRETES ACID, REABSORBS BICARBONATE AND AMMONIUM

About 80–90% of the filtered HCO_3^- is reabsorbed in the proximal tubule. Distal tubule fluid has about the same $[HCO_3^-]$ as the late proximal tubule. Because water is reabsorbed in the loop of Henle, this means that significant HCO_3^- reabsorption must occur in the loop, about 10–15% of the filtered load. In addition, much of the NH_4^+ in the proximal tubule does not continue on to the distal tubule: it also must have been reabsorbed. These transport activities occur in the thick ascending limb, through mechanisms outlined in Figure 7.7.3. As in the proximal tubule, the majority of

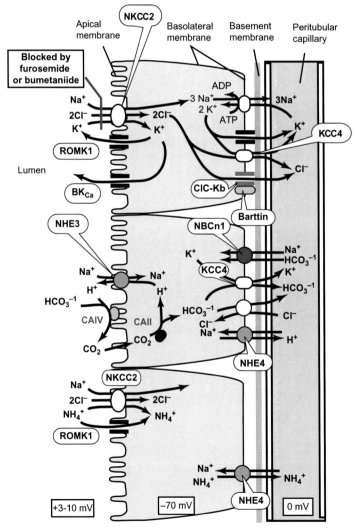

FIGURE 7.7.3 Mechanisms of H^+ secretion, HCO_3^- reabsorption, and NH_4^+ reabsorption in thick ascending limb. Top, mechanisms of Na^+, K^+, and Cl^- transport. Middle, mechanisms of HCO_3^- reabsorption. Bottom, mechanisms of NH_4^+ transport. The mechanism of H^+ secretion is similar to that in the proximal tubule. Most of the H^+ is secreted through secondary active transport powered by the Na^+ gradient through an $Na^+–H^+$ exchanger. The thick ascending limb also possesses a H^+-ATPase. HCO_3^- in the tubule fluid is neutralized by the secreted H^+, while a new HCO_3^- ion made in the cell is transported out of the cell by multiple mechanisms that include $Cl^-–HCO_3^-$ exchange, and $K^+–HCO_3^-$ cotransport, probably over KCC4. NH_4^+ in the fluid may enter the cell through the Na:K:2Cl cotransporter or through K^+ channels. NH_4^+ exits the cell over the $Na^+–H^+$ exchanger, NHE4.

acid secretion occurs through the Na^+-H^+ exchanger (NHE3). The secreted acid combines with HCO_3^- in the lumen to become CO_2. HCO_3^- exit from the cell is accomplished by multiple mechanisms including a $Cl^--HCO_3^-$ exchange and $K^+-HCO_3^-$ cotransport probably mediated by KCC4. Cells in the thick ascending limb express NBCn1, a $Na^+-HCO_3^-$ cotransporter on the basolateral membranes, but this is thought to mediate an HCO_3^- entry into the cells. NH_4^+ reabsorption is favored by a positive lumen potential. NH_4^+ enters the thick ascending limb cell either through the Na:K:2KCl transporter or through an apical K^+ channel (ROMK). Most NH_4^+ exits the cells by the NHE4, a Na^+-H^+ exchanger that can carry NH_4^+. The net result of these transport processes is to reabsorb $NaHCO_3^-$ and to add NH_3/NH_4^+ to the kidney interstitium. This sets up another countercurrent multiplier that concentrates NH_3/NH_4^+ in the medullary interstitium. NH_4^+ is secreted by the proximal tubules, reabsorbed into the interstitium by cells in the thick ascending limb, and then NH_3/NH_4^+ enters the collecting duct by diffusion trapping driven by the acid pH of the collecting duct fluid.

Increasing $[K^+]$ in the lumen of the thick ascending limb interferes with NH_4^+ reabsorption, presumably by competition between K^+ and NH_4^+ for transport via NKCC2. Hyperkalemia thus induces an acidosis by preventing the gradient in NH_4^+ between the vasa recta and collecting duct.

DIFFERENT CELL TYPES IN THE DISTAL NEPHRON AND COLLECTING DUCT HANDLE ACID AND BASE DIFFERENTLY

The late distal tubule and collecting duct are heterogeneous structures that contain several cell types. The distal nephron and cortical collecting duct can secrete H^+ or HCO_3^-, depending on the chronic acid–base status of the body. Distinct cell types mediate these functions. Cells called α-intercalated cells secrete acid and β-intercalated cells secrete base. These cells are also called **A-intercalated** and **B-intercalated** cells, for acid secreting and base secreting. The mechanisms are shown diagrammatically in Figure 7.7.4. The distal nephron also has other types of cells besides these A-intercalated and B-intercalated types.

The α-intercalated cells secrete H^+ ions across their apical membrane by two primary active transporters: the H^+-ATPase and the H^+,K^+-ATPase. This secretion of H^+ is linked to the formation of new HCO_3^- within the cell, which exits the basolateral membrane over a $Cl^--HCO_3^-$ exchanger, AE1. The $Cl^--HCO_3^-$ exchange is a product of alternate gene splicing of the same gene that produces the $Cl^--HCO_3^-$ exchanger of the red blood cells. HCO_3^- can also exit the cell over a $Cl^--HCO_3^-$ exchanger, SLC26A7.

The ion transport properties of β-intercalated cells look largely like a mirror image of those of the α-intercalated cell. They have H^+-ATPase primary active transport on the basolateral membrane that pumps H^+ into the peritubular fluid rather than into the lumen as in the case

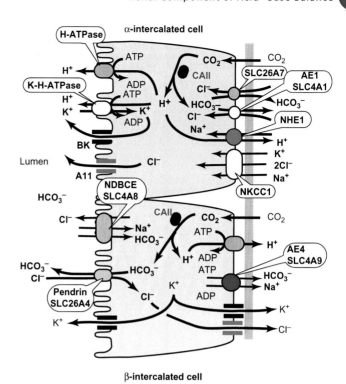

FIGURE 7.7.4 Mechanism of acid and base secretion by α-intercalated cells (top) and β-intercalated cells (bottom) in the distal nephron. The α-intercalated cells secrete acid whereas the β-intercalated cells secrete base. α-intercalated cells secrete H^+ on the apical membrane by primary active transport catalyzed by the H-ATPase and by the K-H-driven ATPase. HCO_3^- linked to this acid secretion is placed into blood by carriers (AE1 and SLC26A7) on the basolateral membrane. In B cells, acid is secreted into the blood and HCO_3^- is secreted into the tubule by Pendrin.

of the α-intercalated cells. Similarly, HCO_3^- generated in the cell exits on an apical membrane $Cl^--HCO_3^-$ exchanger, Pendrin (SLC4A8). HCO_3^- in the lumen can be taken up by a Na-driven bicarbonate exchanger, NDBCE (SLC4A8). This device uses the energy of the Na^+ gradient to drive $Cl^--HCO_3^-$ exchange. HCO_3^- in B-intercalated cells can exit on the basolateral side via AE4, a Na^+ and HCO_3^- cotransporter.

SYNOPSIS OF ACID–BASE HANDLING BY THE NEPHRON

Figures 7.7.1–7.7.4 show a plethora of transport mechanisms that become an unwieldy obstacle to our understanding of acid–base balance. However, it is almost certainly true that all of these transport mechanisms play a role under some conditions. Figure 7.7.5 shows a simplified version of these transport mechanisms, with the identifying names of the transporters.

TUBULAR pH AND CELLULAR P_{CO_2} REGULATE HCO_3^- REABSORPTION AND H^+ SECRETION

THE LIMITING pH OF THE URINE IS ABOUT 4.4

The active transport pumps present in the apical membrane of acid secreting parts of the nephron can

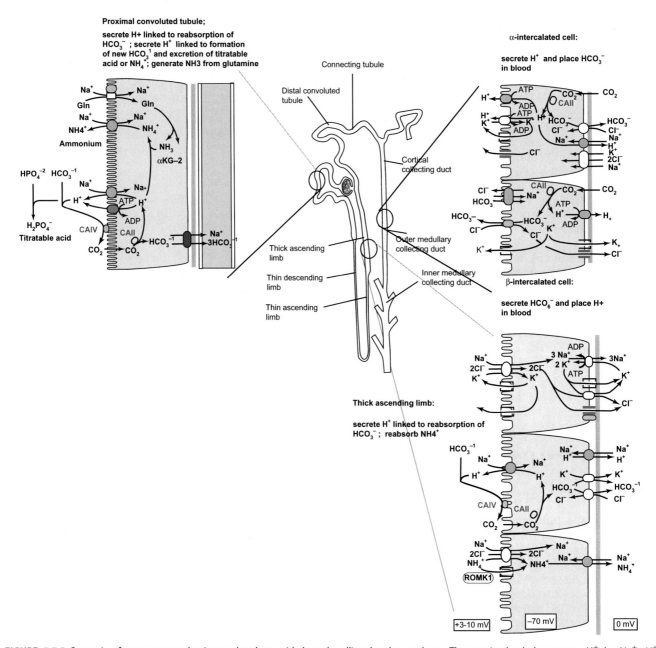

FIGURE 7.7.5 Synopsis of transport mechanisms related to acid–base handling by the nephron. The proximal tubule secretes H^+ by Na^+–H^+ exchange, powered by the basolateral Na,K-ATPase that creates a gradient for Na^+ entry into the cell. The secreted H^+ can combine with filtered HCO_3^- to form CO_2 via a luminal, membrane bound carbonic anhydrase IV. The secreted H^+ can also combine with filtered acids, such as HPO_4^{-2} to form $H_2PO_4^{-1}$, which is a titratable acid. The third fate of secreted H^+ is to combine with NH_3 produced by the deamination of glutamine. The formation of titratable acid and ammonium is linked to the formation of new HCO_3^- that is placed in the blood. Thus the acidification of the urine by the proximal tubules is accompanied by alkalinization of the blood by addition of HCO_3^-. Cells in the thick ascending limb can also secrete H^+ through the Na^+–H^+ exchanger and reabsorb remaining filtered HCO_3^-. These cells can also reabsorb filtered NH_4^+ by entry across the apical membrane on the NKCC1 transporter, and exit across the basolateral membrane via the NHE4. Cells in the distal nephron are heterogeneous. A-intercalated cells secrete acid and alkalinize the blood by adding HCO_3^- to it. B-intercalated cells secrete base (HCO_3^-) and acidify the blood by secreting H^+.

produce a limited gradient of [H^+] because this gradient requires energy. As the gradient gets large, the pump is inhibited. These pumps can maximally concentrate [H^+] about 1000-fold, from pH 7.4 to about pH 4.4. As the pH of the luminal contents falls, the ability of the pumps to continue secreting H^+ also falls. By binding secreted H^+, buffers in the tubule fluid enhance H^+ secretion.

HCO_3^- REABSORPTION DISPLAYS GLOMERULOTUBULAR BALANCE

Because HCO_3^- binds H^+, increased filtered load of HCO_3^- prevents the fall in tubular pH that otherwise would accompany H^+ secretion. This effect allows H^+ secretion to continue, and this H^+ secretion is the first step in the mechanism for HCO_3^- reabsorption.

Thus increased filtered load of HCO_3^- increases the rate of HCO_3^- reabsorption. This constitutes part of glomerulotubular balance, in which increased filtration at the glomerulus increases reabsorption by the tubules.

P_{CO_2} DRIVES ACID SECRETION AND HCO_3^- REABSORPTION

Figure 7.7.1 illustrates how the H^+ ions that drive H^+ secretion from the tubule cells are replenished ultimately from dissolved CO_2. The dissolved CO_2, in turn, derives from interstitial fluid CO_2 which derives from arterial CO_2. Experimental control of P_{aCO_2} levels in the dog reveals increasing HCO_3^- reabsorption (which is approximately equal to acid secretion) with increasing P_{aCO_2}. Thus hypoventilation that increases P_{aCO_2} and causes respiratory acidosis is met with increasing renal H^+ secretion. Figure 7.7.6 illustrates the relationship between P_{aCO_2} and HCO_3^- reabsorption.

RENAL COMPENSATION FOR ACID–BASE DISTURBANCES

Acid–base balance is affected by multiple processes within the body, and dysfunction of the compensatory mechanisms leads to acid–base imbalance. The four major disorders for which the kidneys compensate include:

- respiratory acidosis
- respiratory alkalosis
- metabolic acidosis
- metabolic alkalosis.

We will consider the renal response to each of these and then consider special circumstances such as K^+ excess or deficit, and volume contraction or expansion.

THE KIDNEYS COMPENSATE FOR RESPIRATORY ACIDOSIS BY INCREASING [HCO_3^-]

In respiratory acidosis, the P_{aCO_2} is increased because either something interferes with ventilation (e.g., CNS depression) or something interferes with gas exchange (e.g., emphysema). The result is increased P_{aCO_2} driving an increased plasma [H^+] and acidosis. There is also an increased [HCO_3^-], as is expected from Le Chatelier's principle applied to the bicarbonate buffer system. The increased P_{CO_2} drives acid secretion, so acid secretion is increased. The secreted acid indirectly causes HCO_3^- reabsorption, but there is also more HCO_3^- filtered. Which is increased more, acid secretion by increased P_{CO_2} or filtered load? Here the increased acid secretion is more than the filtered load, so there is reabsorption of most of the filtered HCO_3^- plus formation of additional HCO_3^- linked to the excretion of titratable acid and NH_4^+. This is what you would expect from the Henderson–Hasselbalch equation, based on teleological arguments: the kidney can restore the acidic plasma pH by raising [HCO_3^-]. This situation is shown in the pH–HCO_3^- diagram (see Figure 7.7.7). Note that the acid–base condition does not return to normal, even if the pH can normalize, because of the continued presence of respiratory acidosis. The normal condition can be attained only by resolving the initial disturbance. Thus plasma pH can be normal even though P_{CO_2} and [HCO_3^-] is not normal.

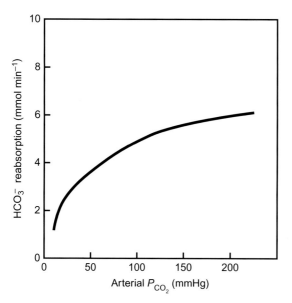

FIGURE 7.7.6 Relationship between arterial P_{CO_2} and the rate of HCO_3^- reabsorption in the dog. Since HCO_3^- reabsorption requires H^+ secretion, this illustrates how H^+ secretion is driven by H^+ supply to the tubule cells through hydration of CO_2. From F.C. Rector, D.W. Seldin, A.D. Roberts, and J.S. Smith, The role of plasma CO_2 tension and carbonic anhydrase activity in the renal reabsorption of bicarbonate, J. Clin. Invest. **39**:1706, 1960.

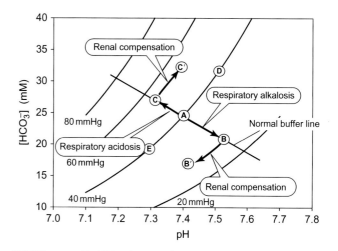

FIGURE 7.7.7 pH–HCO_3^- diagram showing the renal compensation of respiratory acidosis and alkalosis. The normal pH situation is indicated as A. Respiratory acidosis entails an increase in P_{CO_2} and [HCO_3^-] along the normal buffer line, indicated as point C. Renal compensation is to increase the plasma [HCO_3^-] along the abnormally high isobar. This can return the pH to normal levels (point C′) in the continued presence of the respiratory condition. Respiratory alkalosis involves a decrease in P_{CO_2} and [HCO_3^-] along the normal buffer line (point B). Renal compensation consists of lowering [HCO_3^-] by excreting HCO_3^- in an alkaline urine. Compensated respiratory alkalosis is indicated by point B′.

THE KIDNEYS COMPENSATE FOR RESPIRATORY ALKALOSIS BY DECREASING [HCO$_3^-$]

In respiratory alkalosis, everything happens in reverse to respiratory acidosis. The alkalosis is caused by increased, inappropriate ventilation, usually caused by CNS stimulation. The alveolar P_{CO_2} is decreased, and therefore the arterial P_{CO_2} is decreased, and [HCO$_3^-$] is also decreased due to Le Chatelier's principle. The change in a secondary parameter—one that derives from the primary one—is always less than the change in the primary parameter. The decrease in [HCO$_3^-$] is less than the decrease in P_{CO_2}. Thus there is insufficient P_{CO_2} to drive acid secretion to reabsorb all of the filtered [HCO$_3^-$], and HCO$_3^-$ is lost in an alkaline urine. The pH disturbances of respiratory alkalosis and subsequent renal compensation are shown graphically in Figure 7.7.7.

These changes can be predicted by looking carefully at the Henderson–Hasselbalch equation for the bicarbonate buffer system (see Eqn 7.7.4)

[7.7.4] $$pH = 6.1 + \log \frac{[HCO_3^-]}{0.0308\, P_{CO_2}}$$

In respiratory acidosis, pH is low because P_{CO_2} is high, making the argument of the log ($=[HCO_3^-]/0.0308\, P_{CO_2}$) smaller and therefore the value of log is smaller and the pH is less: it is acidic. Renal compensation consists of raising [HCO$_3^-$] to increase the argument of the log, thereby increasing its value and increasing the pH. That is, the pH returns towards normal, but the P_{CO_2} and [HCO$_3^-$] remain abnormal as long as the respiratory problem persists. The rate of new HCO$_3^-$ formation is equal to the rate of titratable acid and ammonium excretion.

Using similar arguments but from the opposite direction, respiratory alkalosis occurs when alveolar ventilation increases and plasma P_{CO_2} decreases. Here the argument of the log is too high because of the reduced P_{CO_2}. Renal compensation consists of lowering plasma [HCO$_3^-$] so that the argument of the logarithm returns to normal values, though neither numerator nor denominator are normal. The kidneys lower plasma [HCO$_3^-$] by failing to reabsorb all of the filtered [HCO$_3^-$], because the P_{CO_2} that drives acid secretion is lowered more than the filtered HCO$_3^-$.

THE OVERALL RESPONSE TO METABOLIC ACIDOSIS INVOLVES BOTH LUNGS AND KIDNEYS

THE LUNGS RAISE PH BY LOWERING P_{CO_2} BY HYPERVENTILATION

In acidosis, the ratio [HCO$_3^-$]/P_{CO_2} is lower than normal. Return to normal pH requires either raising [HCO$_3^-$] or lowering P_{CO_2}. The kidneys raise [HCO$_3^-$] while the lungs lower P_{CO_2}. Decreased plasma pH by itself excites chemoreceptors in the medulla and carotid body to increase ventilatory drive. Increased ventilation increases Q_A, which decreases the P_{CO_2}. According to the Henderson–Hasselbalch equation, this raises the pH. The situation is shown graphically in Figure 7.7.8.

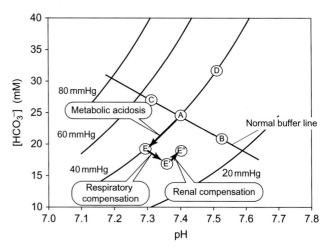

FIGURE 7.7.8 Respiratory and renal compensation of metabolic acidosis. The normal pH status is indicated by point A. Metabolic acidosis is indicated by point E. Acid stimulates ventilation to cause an excursion parallel to the normal buffer line, to point E'. This respiratory compensation occurs rapidly. Renal compensation consists of an excursion parallel to the isobars and results in increasing plasma [HCO$_3^-$], as indicated in the path from E' to E''. Hyperventilation requires increased stimulatory drive; removal of this stimulation by renal compensation would return the acid–base situation toward the normal isobar.

THE KIDNEYS RAISE [HCO$_3^-$] BY EXCRETING AN ACID URINE

In response to acidosis, the kidneys raise plasma [HCO$_3^-$]. Both P_{CO_2} and HCO$_3^-$ are reduced, but P_{CO_2} is reduced less because the HCO$_3^-$ has been removed by combining with plasma H$^+$ produced by metabolism. Thus there is reabsorption of most of the filtered HCO$_3^-$ and secretion of H$^+$ by the kidney is linked to formation of new HCO$_3^-$. The new HCO$_3^-$ is equal to the excretion of **titratable acid** and **ammonia** (see Figure 7.7.8).

THE OVERALL RESPONSE TO METABOLIC ALKALOSIS INVOLVES BOTH LUNGS AND KIDNEYS

THE LUNGS LOWER PH BY RAISING P_{CO_2} BY HYPOVENTILATION

In alkalosis, the ratio [HCO$_3^-$]/P_{CO_2} is elevated. Return to normal pH requires either lowering [HCO$_3^-$] or raising P_{CO_2}. The kidneys lower [HCO$_3^-$] while the lungs raise P_{CO_2}. The alkaline pH decreases the respiratory drive tonically provided by the central and peripheral chemoreceptors, reducing Q_A, alveolar ventilation, and raising P_{CO_2}. This lowers the ratio [HCO$_3^-$]/P_{CO_2}, returning the pH toward normal. The lungs cannot return to normal pH because then there would be no reduced ventilatory drive to keep the P_{CO_2} elevated. The result is a partial compensation of alkalosis. The situation is shown graphically in Figure 7.7.9.

THE KIDNEYS LOWER [HCO$_3^-$] BY EXCRETING AN ALKALINE URINE

Following respiratory compensation for metabolic alkalosis, P_{CO_2} and HCO$_3^-$ are both elevated, but P_{CO_2} is elevated less because the HCO$_3^-$ is increased by more dissociation from H$_2$CO$_3$ due to lower [H$^+$] and more H$_2$CO$_3$ made from the increased P_{CO_2}. Thus there is insufficient acid secretion to reabsorb is all of the filtered HCO$_3^-$ and HCO$_3^-$ lost in an alkaline urine (see Figure 7.7.9).

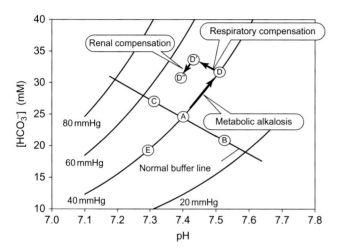

FIGURE 7.7.9 Respiratory and renal compensation of metabolic alkalosis. The normal pH status is indicated by point A. Metabolic alkalosis is indicated by point D. Alkalosis inhibits ventilation to cause an excursion parallel to the normal buffer line, to point D'. This respiratory compensation occurs rapidly. Renal compensation consists of an excursion parallel to the isobars and results in decreasing plasma [HCO$_3^-$], as indicated in the path from D' to D". Hypoventilation requires decreased stimulatory drive; removal of this inhibition by renal compensation would return the acid–base situation toward the normal isobar.

HCO$_3^-$ INCREASES ACID SECRETION BY A-INTERCALATED CELLS THROUGH A SOLUBLE ADENYLYL CYCLASE

The apical membrane H-ATPase in type A-intercalated cells is recruited from a population of pumps on endosomes. Translocation to the surface is activated by **PKA** (protein kinase A) and is inhibited by **AMPK** (AMP-activated protein kinase, a metabolic sensor that responds to metabolic stress when ATP levels fall). These two protein kinases phosphorylate the H-ATPase at distinct sites. PKA responds to the acid–base condition through the actions of a **soluble adenylyl cyclase**, sAC, which is activated by elevated HCO$_3^-$ levels within the cell, which is, in turn, rapidly equilibrated with the P_{CO_2} in the cells through carbonic anhydrase (see Figure 7.7.10). Elevated HCO$_3^-$ activates sAC to increase cAMP levels within the cell, which then activates PKA and recruits more of the H-ATPase to the apical membrane. This increases the acid secretion of the A-intercalated cells and increases HCO$_3^-$ reabsorption or the formation of new HCO$_3^-$ linked to the excretion of titratable acid or ammonium. This seems paradoxical: why would these cells increase acid secretion when cellular HCO$_3^-$ is high, only to further increase HCO$_3^-$? This is exactly what happens, however, and may be the mechanistic explanation behind the observation of Figure 7.7.6 that shows the monotonic relationship between P_{CO_2} and the rate of acid secretion by the kidney. In the pH-HCO$_3^-$ diagrams, all cases in which HCO$_3^-$ is elevated is accompanied by increased H$^+$ secretion. In some cases, such as respiratory acidosis, HCO$_3^-$ increases and the renal compensation is the reabsorption of filtered HCO$_3^-$ and formation of new HCO$_3^-$. In other cases, such as metabolic alkalosis with respiratory compensation, the kidneys secrete more acid, but the filtered load of HCO$_3^-$ is increased more, so that the kidney actually wastes HCO$_3^-$ despite increasing its acid secretion.

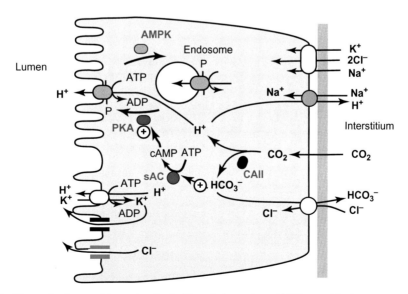

FIGURE 7.7.10 Regulation of acid secretion in type A-intercalated cells. The apical membrane H-ATPase resides in reservoir endosomal vesicles. Movement to the apical membrane is stimulated by phosphorylation by PKA and inhibited by phosphorylation at a different site by AMPK. PKA is activated by increased concentrations of cAMP that is produced by a soluble adenylyl cyclase that is activated by HCO$_2^-$. Thus increased P_{CO_2} results in increased HCO$_3^-$ through the action of carbonic anhydrase, CA, that results in increased secretion of acid. Similar mechanisms pertain to the proximal tubule.

CHRONIC ACIDOSIS INCREASES EXCRETION OF NH_4^+

During normal acid–base balance, the kidneys filter about 20% of the plasma glutamine, but the arterial-venous difference is only about 3%. This means that most of the filtered glutamine is returned to blood and is not metabolized by the kidneys. Acute metabolic acidosis alters this metabolism. Within 3 hours of the onset of acute metabolic acidosis, the net extraction of glutamine by the kidneys climbs to 35%, exceeding the filtered load. Thus the kidneys import glutamine from the blood across the basolateral membrane and metabolize it to produce NH_4^+. In addition, plasma glutamine levels rise due to release from muscle. In chronic metabolic acidosis, plasma glutamine levels normalize, but glutamine extraction remains high. Chronic metabolic acidosis causes an upregulation of several kidney enzymes that metabolize glutamine including glutaminase (GA), glutamic acid dehydrogenase (GDH), and phosphoenol pyruvate carboxykinase (PEPCK, see Figure 7.7.2). Thus chronic metabolic acidosis increases the production of ammonia. Figure 7.7.11 illustrates that the urinary excretion of NH_4^+ increases with decreasing urinary pH, but, at any given pH, it is increased by chronic acidosis.

POTASSIUM AND ACID–BASE BALANCE INTERACT

Potassium concentrations and H^+ concentrations in cells are generally inversely related: when K is depleted, H^+ is increased and therefore acid secretion by renal tubule cells is increased. Thus **hypokalemia is often associated** with a metabolic alkalosis and, conversely, **hyperkalemia is often associated with metabolic acidosis.**

VOLUME CONTRACTION INCREASES ACID SECRETION

Volume contraction, or loss of fluid volume, activates the renin–angiotensin–aldosterone system through the mechanisms described in Chapter 7.6. In this case, both ANG II and aldosterone are elevated. These increase ENaC activity in the principal cells of the distal nephron, which increase Na^+ uptake and K^+ secretion, and they also activate H^+ secretion from A-intercalated cells. Thus volume contraction tends to produce metabolic alkalosis. Volume expansion has opposite effects.

THE OVERALL PICTURE

Normal metabolism produces excess acid that must be excreted in the urine. Accordingly, the urine is typically acidic, with a pH normally around 6.0. This is accompanied by modest amounts of NH_4^+ and titratable acid, so that normally the kidney returns more HCO_3^- to the body than is filtered. This combines with plasma H^+ and is excreted to the air as CO_2. Thus the acid that is xproduced in the tissues is removed by its being linked to the final excretion of the same amount of acid in the urine. The total amount of extra base delivered to the blood as HCO_3^- is equal to the titratable acid + ammonia − excreted HCO_3^-, as described in Eqn (7.7.8). In the steady state, the excretion of acid must match its metabolic production. In acidosis, the kidneys must excrete more H^+ and produce more HCO_3^- linked to the excretion of titratable acid and ammonium. In alkalosis, HCO_3^- is lost from the body by not reabsorbing all of the filtered load of HCO_3^-. Table 7.7.1 lists typical values for excretion of titratable acid, ammonium, and HCO_3^- in various conditions of acid–base balance. The overall response of the kidneys to acid–base disturbances is shown in Figure 7.7.12.

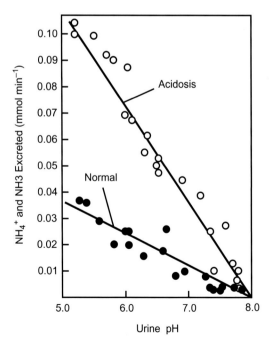

FIGURE 7.7.11 Excretion of $NH_4^+ + NH_3$ when urine pH is varied by infusion of $NaHCO_3$ under normal conditions and during chronic acidosis produced by feeding NH_4Cl. Data were obtained in a single dog. *From R.F. Pitts, The renal excretion of acid, Fed. Proc. 7:418, 1948.*

TABLE 7.7.1 Typical Approximate Values for the Excretion of Titratable Acid, NH_4^+, and HCO_3^- During Normal Acid–Base Balance, Alkalosis, and Acidosis

Kidney Acid/Base Excretion	Acid–Base Status		
	Normal	Alkalosis	Acidosis
Titratable acid (mmol day^{-1})	20	0	40
NH_4^+ (mmol day^{-1})	40	0	160
HCO_3^- excreted (mmol day^{-1})	5	80	0
Net HCO_3^- gain (mmol day^{-1})	+55	−80	+200
Urinary pH	6.0	8.0	4.6

In the normal situation, the kidney produces new HCO_3^- to balance the net acid production from metabolism. During alkalosis, acid–base status is returned to normal by excreting base as HCO_3^-. During acidosis, there is an increase in new HCO_3^- formation that is linked mainly to increased excretion of NH_4^+.

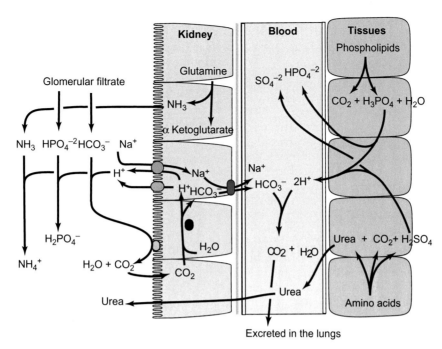

FIGURE 7.7.12 Overall function of the kidneys in acid–base balance. The tissues generally produce acid at a rate that depends largely on protein intake. The acid is buffered in the blood by HCO_3^-, which forms CO_2 and H_2O. The lungs eliminate the CO_2 from the body. This buffering action depletes the blood of HCO_3^-. The exact amount of HCO_3^- can be replenished by linking H^+ excretion in the kidneys to new HCO_3^- production. Thus, under steady-state circumstances, the kidney excretes the same amount of acid that is produced by the tissues, except that the acid does not derive directly from the tissues.

Clinical Applications: Renal Tubular Acidosis

Continual production of acid by metabolism necessitates excretion of an acid urine. Failure to secrete acid by the kidney will cause an acidosis that is classified as metabolic acidosis but is due to kidney failure. The failure can have multiple causes. The term "renal tubular acidosis" is reserved for cases in which insufficient urinary acidification occurs in otherwise adequate renal function.

There are two major categories of renal tubular acidosis: Type 1 is distal renal tubular acidosis, caused by failure of the α-intercalated cells to secrete enough acid. The urine cannot be acidified below pH 5.3. This can lead to: (1) acidemia; (2) hypokalemia; (3) urinary stone formation; (4) nephrocalcinosis (calcification of the kidney tissue); and (5) bone demineralization.

Causes of distal renal tubular acidosis include autoimmune diseases, hereditary mutations of the basolateral HCO_3^-–Cl^- exchanger or the apical membrane H^+-ATPase, and chronic urinary obstruction, among others.

Type 2 is proximal renal tubular acidosis. It is due to a failure of the proximal tubule cells to adequately reabsorb sufficient quantities of the filtered bicarbonate, leading to wasting of bicarbonate and acidemia. Because the distal tubule intercalated cells are generally normal, the urine can still be acidified and the acidemia is less severe than in distal renal tubular acidosis.

Proximal renal tubular acidosis may sometimes present as a solitary defect, but more usually it is associated with a general failure of proximal tubule transport called Fanconi's syndrome, where there is phosphaturia, glucosuria, aminoaciduria, uricosuria, and proteinuria.

Treatment of distal renal tubular acidosis requires correction of the acidemia by oral bicarbonate. Hypokalemia that presents is addressed with potassium citrate, which also prevents nephrocalcinosis by complexing calcium. Treatment of proximal tubular acidosis also uses bicarbonate to reverse the acidosis.

SUMMARY

Defense against acid–base imbalance is accomplished through three interacting systems: the chemical buffers of the blood, the respiratory system, and the renal system.

The chemical buffers resist changes in plasma pH by binding H^+ ions when they are in excess and dissociating to form H^+ ions when the $[H^+]$ falls. The most important chemical buffer is the HCO_3^- buffer because it can be adjusted by both the respiratory and renal systems.

The respiratory system adjusts plasma pH by adjusting $P_{a_{CO_2}}$. During acidosis, respiration is stimulated, resulting in decreased $P_{a_{CO_2}}$. According to the Henderson–Hasselbalch equation:

$$pH = 6.1 + \log([HCO_3^-]/0.0308\, P_{a_{CO_2}})$$

decreasing P_{aCO_2} increases the pH. Thus acidosis evokes a compensatory respiratory alkalosis. Similarly, alkalosis inhibits respiration so that P_{aCO_2} rises, and the acidosis is countered by a respiratory compensation.

The respiratory compensation is never complete because there must be residual pH imbalance to maintain the respiratory response that elevates or reduces P_{aCO_2}. The renal compensation, however, is complete. The renal system adjusts pH by either excreting HCO_3^- by failing to reabsorb all of the filtered HCO_3^- or by forming new HCO_3^- linked to the excretion of titratable acid or NH_4^+. The main activity of the kidneys is to secrete H^+ through either an H^+-ATPase or an Na^+–H^+ exchanger located on the apical membrane. Secreted H^+ then either combines with filtered HCO_3^-, or with filtered buffers such as Na_2HPO_4, or with NH_3. Every secreted H^+ causes HCO_3^- to appear in the plasma. Thus when secreted H^+ ions combine with filtered HCO_3^-, the overall process is equivalent to reabsorption of filtered HCO_3^-; when H^+ combines with titratable acid or NH_3, the kidney effectively places new HCO_3 in plasma.

Thus the kidney can excrete HCO_3^- in an alkaline urine, thereby reducing the $[HCO_3^-]$ in blood. By the Henderson–Hasselbalch equation, this would lower blood pH. Alternatively, the kidney can fight acidosis by adding HCO_3^- to blood, alkalinizing the blood while excreting an acid urine.

REVIEW QUESTIONS

1. How are filtered HCO_3^- ions reabsorbed?
2. What happens if the amount of acid secreted exceeds the amount of filtered HCO_3^-?
3. In metabolic acidosis, what is the respiratory response? What is the renal response? Do the kidneys excrete more or less acid after respiratory compensation for metabolic acidosis compared to normal? Do they excrete more or less acid after respiratory compensation for metabolic acidosis, compared to normal?
4. What is the renal response to respiratory acidosis?
5. What drives H^+ secretion?
6. Why does hyperkalemia predispose to acidosis?

Problem Set 7.2
Fluid and Electrolyte Balance and Acid–Base Balance

1. The following were measured in a 24-hour urine collection:

   Urine volume = 1.2 L
   Titratable acid = 33 mmol L^{-1}
   Urine [HCO$_3^-$] = 4 mmol L^{-1}
   Urine [NH$_4^+$] = 90 mmol L^{-1}

   Calculate the amount of HCO$_3^-$ gained or lost from the body through renal mechanisms.

2. A patient with a history of bronchopulmonary disease was admitted to the hospital and the following laboratory data were obtained for arterial blood:

Na	143 mM
K	4.8 mM
pH	7.34
P_{O_2}	47 mmHg
Cl	78 mM
HCO$_3^-$	38 mM
P_{CO_2}	73 mmHg
Hemoglobin	17.1 g%

   Classify the acid–base status of this patient in terms of acidosis/alkalosis, respiratory or metabolic, compensated or uncompensated.

3. The concept of **free water clearance** is sometimes used to assess the kidneys' ability to handle a water load. The free water clearance is *defined* as follows:

   $$C_{H_2O} = Q_U - \left[\frac{(Q_U U_{osmol})}{P_{osmol}}\right]$$

   Here C_{H_2O} is the free water clearance, Q_U is the flow of urine, in mL min^{-1}, U_{osmol} is the total concentration of materials in the urine, in osmol^{-1} L^{-1}, and P_{osmol} is the osmolarity of plasma. The term in brackets is the clearance of osmolarity, and it represents the volume of plasma that contains the total osmolarity that is excreted. Calculate the free water clearance for the following conditions:

Condition	Q_U (mL min^{-1})	U_{osmol} (mOsM)	P_{osmol} (mOsM)	C_{H_2O}
A	0.8	900	300	
B	1.5	300	300	
C	6.0	70	300	

   What does a negative free water clearance mean?

4. The acid NH$_4^+$ dissociates to its conjugate base (NH$_3$) and H$^+$, with a dissociation constant $K_D = 6.3 \times 10^{-10}$ M:

   $$NH_4^+ \rightleftharpoons NH_3 + H^+$$

   Suppose that NH$_4^+$ is originally present in two compartments separated by a cellular membrane. The membrane is freely permeable to NH$_3$, but it is impermeable to NH$_4^+$. The pH on one side is 9.0, and on the other side it is 7.4. Let T_1 and T_2 be the total concentration of NH$_3$ + NH$_4^+$ on sides 1 and 2, respectively. Calculate T_1/T_2.

5. The acid HA dissociates at physiological pH into its conjugate base (A$^-$) and H$^+$ with a dissociation constant K_D:

   $$HA \rightleftharpoons A^- + H^+$$

   Suppose that HA is originally present in two compartments separated by a membrane. The membrane is freely permeable to HA, but it is impermeable to A$^-$. The pH on one side of the membrane is pH$_1$, and on the other side it is pH$_2$. Derive an expression for the ratio of the total concentration of material ([HA] + [A$^-$]) on the two sides of the membrane in terms of its K_D and the two pHs.

6. Given the data below on the handling of urea by the kidneys of an adult man at varying rates of urine flow, calculate (a) the GFR (in mL min^{-1}), (b) filtered load of urea (in mmol min^{-1}), (c) excreted urea (in mmol min^{-1} and as % of filtered load), and (d) reabsorbed urea (in mmol min^{-1} and as % of filtered load).

Data	Q_U, Urine Flow (mL min^{-1})	U_{inulin} (mg mL^{-1})	U_{urea} (mM)	P_{inulin} (mg mL^{-1})	P_{urea} (mM)
A	0.4	144	300	0.5	5
B	0.8	75	263	0.5	5
C	1.0	60	240	0.5	5
D	3.1	20	119	0.5	5
E	10.2	5.8	37	0.5	5

7. Suppose that the rate of reabsorption from the aggregate tubules can be described by the equation:

$$T = \frac{T_m P_x}{(k_t + P_x)}$$

where P_x is the plasma concentration of substance x, T is the rate of reabsorption, T_m is the maximum transport rate, and k_t is the transport constant, equal to the concentration at half-maximal transport.
   A. Write an equation for the rate of substance x excreted, $Q_U U_x$, as a function of plasma concentration, P_x, assuming that $\Theta_x = 1.0$.
   B. Show that for these assumptions the excretion rate can be zero only when $P_x = 0$.
   C. What interpretation would you give for negative values of the rate of excretion as predicted by the equation in part A?
   D. For values of GFR = 120 mL min^{-1}, $R_m = 312.5$ mg mL^{-1}, $k_t = 132$ mg min^{-1}, plot the rate of excretion versus P_x. Can you identify a renal threshold? Is splay evident?

8. A 60-year-old woman was admitted to the hospital with a diagnosis of pneumonia. She had been on a thiazide diuretic for 6 months previously for treatment of heart failure. Her lab results were as follows:

K$^+$	2.1 mM
pH	7.64
P_{CO_2}	32 mmHg
P_{O_2}	75 mmHg
HCO$_3^-$	33 mM

   Classify her acid–base disturbance. *Hint*: Graph it out on the pH–HCO$_3^-$ diagram.

9. Creatinine production in a particular well-muscled individual is 3 g day^{-1}. His plasma creatinine is 1.5 mg dL^{-1}. Estimate the GFR.

10. The renal blood flow into the kidney is 1200 L min^{-1} with a concentration of total solutes of 296 mOsM. Urinary output is 0.5 mL min^{-1} of 1150 mOsM.
    A. What is the flow and osmolarity of the renal venous blood?
    B. According to Figure 7.5.10, the fluid coming out of the vasa recta is hyperosmotic. How does the final renal venous blood become hyposmotic during antidiuresis?

11. During diuresis, inflow into the distal tubule is about 18 mL min^{-1} at 100 mOsM. Outflow from the distal tubule into the collecting duct is about 6 mL min^{-1} and 300 mOsM.
    A. How much fluid (per unit time) is reabsorbed in the distal tubule?
    B. How much solute (in osmoles per unit time) is reabsorbed in the distal tubule?
    C. Where does this fluid and solutes go?
    D. What does this do to the osmolarity of the venous blood that drains the kidney?

12. Using the figures for volume flow given in Figure 7.5.10, and given that plasma inulin is maintained at 10 mg dL^{-1}, indicate the inulin concentration during diuresis.
    A. At the end of the proximal tubule
    B. At the end of the thick ascending limb/early distal tubule
    C. At the late distal tubule
    D. In the final urine

13. Mannitol diuresis is generally provided by injecting a 10% or 20% solution of mannitol at a dose of 0.5–1.5 g kg^{-1} body weight. A patient weighs 180 lbs and we wish to give a dose of 1.0 g kg^{-1}. Plasma creatinine is 1.1 mg%. The ECF is 15 L in this person. The formula weight of mannitol is 182.17 g mol^{-1}.
    A. How much 20% mannitol should we infuse?
    B. Urine flow before administration of mannitol was 1.0 mL min^{-1} with a creatine concentration of 1.09 mg mL^{-1}. What is the GFR?
    C. Suppose that the person is at maximum antidiuresis, with a urine flow of 0.5 mL min^{-1} prior to mannitol injection, with a urine osmolarity of 1200 mOsM. What is the increase in urinary flow rate caused by mannitol infusion?
    D. How does mannitol increase urine flow?

14. A woman is admitted into the hospital with the following findings:

    pH = 7.25
    P_{CO_2} = 30 mmHg
    Plasma [creatinine] = 4 mg%

    A. Classify her acid–base status.
    B. Would you expect her renal acid secretion to be increased, normal, or decreased, and why?

15. An athlete weighs 80 kg. Total body water is 60% of his body mass and ECF is 20% of body mass. His plasma osmolarity before working out is 295 mOsM. He works out and loses 3 L of fluids containing 100 mOsM solutes: assume it is all NaCl. Assume ECF [Na] = 140 mM.
    A. What is his plasma osmolarity after the workout?
    B. How much urine at 1200 mOsM must he excrete to bring his plasma osmolarity back to normal?

16. A person has the following blood values:

    pH = 7.50
    P_{CO_2} = 50 mmHg
    [HCO$_3^-$] = 38.7 mmHg

    A. Classify this person's acid–base status.
    B. Is renal acid secretion increased or decreased in this person?
    C. Is the urine acidic or alkalotic? Why?

17. The following test results were obtained in a person who was infused with inulin to achieve a constant plasma concentration:

    Urine volume in 4 hours = 480 mL
    Urine [inulin]   = 30 mg%
    Plasma [inulin]   = 0.5 mg%
    Urine [glucose]   = 200 mg%
    Plasma [glucose]   = 400 mg%

    A. What is the GFR?
    B. What is the filtered load of glucose in mg min^{-1}?
    C. What is the rate of glucose reabsorption?
    D. Do you think this rate of glucose reabsorption is about equal to the T_m for glucose? Why or why not?

18. The following blood values were measured in a person:

    $[HCO_3^-] = 38$ mM
    pH = 7.15

    A. Calculate the P_{aCO_2} using the Henderson–Hasselbalch equation.
    B. Describe the persons acid–base condition.
    C. What is the respiratory compensation for this condition?
    D. What is the renal compensation for this condition?
    E. What happens to ammonium excretion?
    F. What happens to titratable acid excretion?

UNIT 8

GASTROINTESTINAL PHYSIOLOGY

Mouth and Esophagus 8.1

Learning Objectives

- State the overall function of the gastrointestinal system
- Describe the role of sphincters in the gastrointestinal tract
- Name the major parts of the gastrointestinal tract from mouth to anus, in order
- Describe the overall function of liver, gallbladder, and pancreas in gastrointestinal function
- List the major purposes of chewing food
- Distinguish between serous and mucus saliva
- List the major purposes of saliva
- Describe the three major components of salivary glands (acinus, intercalated duct, and striated duct)
- Explain why salivary composition varies with flow rate
- Describe CNS control of salivation and its major sources of variation
- Describe the pharyngeal phase of swallowing
- Describe peristalsis
- Distinguish between primary peristalsis and secondary peristalsis
- Describe the overall layers of the gastrointestinal tract
- Describe the location and function of the intestinal nerve plexuses
- Define achalasia and GERD

THE GASTROINTESTINAL SYSTEM SECURES NUTRIENTS FOR MAINTENANCE, MOVEMENT, AND GROWTH

In thermodynamics, entropy S is a state variable (meaning that it depends solely on the state and not on the path to that state) that is related to the degree of disorder. Boltzmann's tomb in the central cemetery in Vienna has inscribed on it his famous equation:

[8.1.1] $$S = k \ln \Omega$$

where Ω is the number of indistinguishable ways of obtaining a state. This is related to the probability of obtaining a state. Living beings are highly ordered and the probability of assembling one spontaneously from scratch is pretty low. According to Boltzmann's formulation, living things are characterized by low entropy. All natural processes require that the entropy of the universe increases—they proceed from less probable to more probable arrangements. Local decreases in entropy (such as occurs in the assembly of living things) are possible only by increasing the entropy of the surroundings to compensate for the local decrease. **Living things maintain their highly ordered structure by taking in energy and degrading it (producing entropy)**. Our assembly and continued existence as a low entropy state far from equilibrium depends on a throughput of energy that we degrade continuously to increase the entropy of the universe. We secure this energy and the chemical building blocks of our structure through digestion and absorption of the food we eat. This is accomplished by the gastrointestinal (GI) system. **The function of the GI tract is to extract the nutrients we need from ingested food.** These nutrients maintain our highly ordered structure far away from equilibrium, provide energy for our activity, and provide the raw building materials needed for growth from birth to adulthood. The GI tract also provides another route for the excretion of metabolic wastes.

NUTRIENTS ARE NECESSARY MATERIALS THAT MUST BE SUPPLIED BY FOOD

"Nutrient" is defined as any substance that is used by the body that must be supplied in adequate amounts from consumed foods. There are six classes of nutrients:

1. water
2. fats
3. proteins
4. carbohydrates
5. vitamins
6. minerals.

Probably the single most important function of the gastrointestinal tract is to absorb sufficient water to replace losses from the lungs, skin, urine, and feces. We produce water from metabolism, but this is very much less than what is needed to replenish losses. Fats, proteins, and carbohydrates are macronutrients, because we require them in large quantities, both as a source of energy and as building blocks for the tissues. The macronutrients are first broken down into simple constituents by digestion, and these simple constituents are then absorbed into the blood stream for use by peripheral tissues. Some of these building blocks can be synthesized, but the ones that we can't make are called "essential," meaning that they must be obtained directly

TABLE 8.1.1 Daily Requirements for Nutrients for Humans

Water	Fats	Proteins	Carbohydrates	Vitamins	Minerals
3.7 L	20–35% of total energy *Essential fatty acids:* Linoleic acid (18:2 $\omega-6$) **17 g** α-Linolenic acid (18:3 $\omega-3$) **1.6 g**	**56 g** *Essential amino acids:* Histidine Isoleucine Leucine Lysine Methionine Phenylalanine Threonine Tryptophan Valine	**130 g**	*Fat-soluble vitamins:* Vitamin A **900** μg Vitamin D **15** μg Vitamin E **15 mg** Vitamin K **120** μg *Water-soluble vitamins:* Vitamin C **90 mg** Thiamine (B1) **1.2 mg** Riboflavin (B2) **1.3 mg** Niacin (B3) **16 mg** Pantothenic acid (B5) **5 mg** Vitamin B6 **1.3 mg** Biotin (B7) **30** μg Folic acid (B9) **400** μg Vitamin B12 **2.4** μg Choline **550 mg**	Sodium **2.3 g** Potassium **4.7 g** Chloride **2.3 g** Calcium **1000 mg** Phosphorus **700 mg** Fluoride **4 mg** Iodine **120** μg Iron **8 mg** Copper **900** μg Selenium **55** μg Magnesium **400 mg** Manganese **2.3 mg** Molybdenum **45** μg Zinc **11 mg** Chromium **35** μg

The values are given per day for a healthy male between 19 and 30 years old. Value for overall fat is given as a percent of total dietary calories. For all other nutrients, values given are the **Recommended Dietary Allowances** or **RDA** (in bold black), which is the average daily intake that is sufficient to meet the nutrient requirements of at least 97.5% of all healthy individuals within the group. RDAs are calculated from the **EAR**, the **estimated average requirement**, plus twice the standard deviation for the EAR. If there is insufficient data to establish an EAR, an **AI** or **adequate intake** is developed and given for some nutrients (bold blue).

from the diet. The subsets of the six classes of nutrients are given in Table 8.1.1. The **RDA** (recommended dietary allowance) or **AI** (adequate intake) is given for each. The RDA varies with age, body size, gender, and reproductive status (pregnant or lactating). The figures given are for a 20-year-old 70-kg male.

THE GASTROINTESTINAL SYSTEM IS A TUBE RUNNING FROM MOUTH TO ANUS

Our basic body plan is that of a hollow cylinder with thick walls. The outside is the **integument** or skin; the thick wall encases the bones, muscles, nerves, and all other organs, and the inner layer is the GI system. As shown in Figure 8.1.1, the GI system is a long tube, with various shapes and bulges, that runs from the mouth to the anus. Topologically, the **lumen**, the space enclosed by the tube, and all food in the lumen, is outside of the body. The GI system moves ingested food along the tube, breaks it down into usable materials, and absorbs these building blocks into the blood. Unabsorbable materials and other wastes are excreted in the feces.

THE GASTROINTESTINAL SYSTEM PROPELS MATERIAL BETWEEN DEFINED AREAS DEMARCATED BY *SPHINCTERS*

Smooth muscles embedded in the walls of the GI tract squeeze the food and move it along. Neurons in the gut wall control and coordinate the smooth muscle contractions. Movement of material between regions is controlled in part by a series of **sphincters**, which are bands of smooth muscle that generally are tonically contracted, narrowing the lumen. Food in the mouth is chewed and lubricated and then swallowed. The swallowed **bolus** of food enters the esophagus through the **upper esophageal sphincter**. The **esophagus** transports materials to the **stomach**. The **lower esophageal sphincter (LES)**, at the junction of esophagus and stomach, controls entry to the stomach and reflux of stomach contents into the esophagus. Digestion, the breakdown of food into its constituent parts, begins in the mouth and continues in the stomach. The stomach holds food and releases it gradually into the **small intestine**. The **pyloric sphincter**, at the junction of the stomach and the first segment of the small intestine, controls stomach emptying. The small intestine provides the major part of digestion and absorption of nutrients. It consists of three segments: the **duodenum**, **jejunum**, and **ileum**. The ileum dumps its contents into the **large intestine**, which consists of the **cecum, ascending colon, transverse colon, descending colon**, and **sigmoid colon**. The **ileocecal sphincter** controls the movement of material from the ileum into the large intestine and prevents the backward movement of material from the colon into the ileum. Eventually the material that is not absorbed moves into the **rectum** from where it is expelled from the body during **defecation**. Defecation is controlled by two sphincters, the **internal anal sphincter** and the **external anal sphincter**. Thus the entire GI tract consists of a series of specialized areas of a tube that is invested with sphincters that control the movement of materials between the different regions. The upper esophageal sphincter and the external anal sphincter are both composed of skeletal muscle (even though they do not connect to bone). The external anal sphincter is controlled voluntarily. The movement of materials through the GI tract is largely unconscious

Mouth and Esophagus

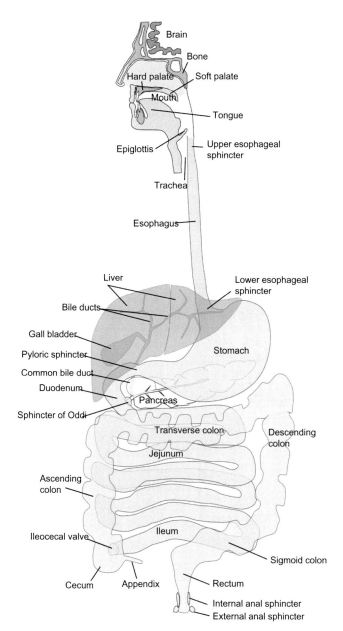

FIGURE 8.1.1 Overall view of the GI system. Food enters the mouth where it is chewed and lubricated and then swallowed. The food passes through the upper esophageal sphincter into the esophagus, which propels it through the LES into the stomach. The stomach in turn moves material through the pyloric sphincter into the first part of the small intestine, the duodenum. The duodenum leads into the jejunum and ileum, with no sphincters separating these regions of the small intestine. The ileum drains into the cecum of the large intestine at the ileocecal sphincter or valve. Material sequentially traverses the ascending colon, transverse colon, descending colon, and sigmoid colon before it enters the rectum prior to defecation. Defecation is controlled in part by two sphincters, the internal anal sphincter and the external anal sphincter. The liver contributes to GI function by secreting bile. Bile flows down the hepatic bile ducts to the gallbladder where it is stored between meals. During meals the gallbladder empties into the duodenum through the common bile duct. The pancreas secretes a variety of digestive enzymes that enter the duodenum through the pancreatic duct, which joins the common bile duct just before it enters the duodenum. The sphincter of Oddi controls bile and pancreatic juice entry into the duodenum.

and involuntary, whereas we begin and end the process (by eating and defecation) through voluntary control.

THE LIVER AND PANCREAS SECRETE MATERIALS INTO THE INTESTINE TO AID DIGESTION AND ABSORPTION

The **liver** and the **pancreas** secrete materials into the intestine to aid digestion or absorption of nutrients. The pancreas lies in the angle between the stomach and the duodenum, and it secretes a watery fluid that contains a host of pancreatic enzymes that break down food into its parts. The secretion flows along the pancreatic duct, which enters the duodenum. Another sphincter, the **sphincter of Oddi**, controls entry of pancreatic juice into the duodenum.

Liver cells make **bile** that flows down **hepatic bile ducts** to be stored in the **gallbladder**. After a meal, the gallbladder contracts and squeezes out the stored bile that travels down the **common bile duct** where it joins the pancreatic duct just before it enters the duodenum. The bile helps digest and absorb fats and also serves as an excretory route for a variety of materials.

THE TRIGEMINAL NUCLEUS IN THE BRAINSTEM SETS THE RHYTHM OF MASTICATION

The muscles of the jaw, lips, tongue, and cheek all participate in chewing. Most of these muscles are innervated by the **trigeminal nerve (cranial nerve V)**. The **masseter** muscle originates on the inner surface and anterior two-thirds of the zygomatic bone (the bone that forms the cheekbone) and inserts onto the ramus of the mandible (the back and lower part of the lower jaw). The **temporalis** muscle is a large fan-shaped muscle that originates on the temporal bone of the skull and inserts on the coronoid process of the mandible. The **medial pterygoideus** is synergistic with the masseter in closing the jaw. It originates on the pterygoid plate and inserts on the inner side of the ramus of the mandible. Contraction of the temporalis muscle abducts the mandible to the side, whereas contraction of the medial pterygoideus pulls the mandible medially. These muscles are paired, and so contraction of the left medial pterygoideus pulls the mandible to right, but only if the right medial pterygoideus is relaxed. Proper synchronization of these muscles produces the grinding motion of the teeth. The **digastric muscle** is a paired muscle that inserts on the mastoid process, part of the temporal bone behind the ear, and the suture that joins the two halves of the lower jaw. It is mainly responsible, with accessory muscles, for opening the jaw. It has two bellies that are connected to an intermediate tendon that forms a connective tissue sheet that also connects to the hyoid bone. When the hyoid bone is pulled from below, contraction of the digastric muscle opens the jaw. The trigeminal nerve innervates the anterior part of the digastric, whereas the **facial nerve, cranial nerve VII**, innervates the posterior belly.

Coordination of muscles for chewing is accomplished by a neural center in the brainstem, with parts in the midbrain, pons, and medulla. The final pathway to the masseter and digastric muscles is provided by two motor sections of the nucleus of the trigeminal nerve, NV_{mot}. Output of these motor neurons is controlled by a variety of inputs from other areas of the nucleus of the trigeminal nerve (NV) that relay sensory signals from the muscles, teeth, and commands from the **cortical masticatory area (CMA)** in the primary motor cortex. Part of the NV, NV_{spnr} contains a **central pattern generator (CPG)** that sets the rhythm for chewing but is modulated by inputs from sensory afferents such as the stretch receptors in the respective muscles. The overall control of mastication is shown in Figure 8.1.2. Other accessory muscles are also involved. The tongue positions food under the teeth while keeping out of the way of the teeth, and cheek muscle (buccinator) helps keep the food there.

The origin of the masticatory rhythm remains an active field of investigation. Current thought is that some cells in the NV_{snpr} have intrinsic bursting abilities that make it act like a pacemaker. The ionic mechanisms that produce this bursting behavior are not yet clear. How the neurons act in concert to transmit this rhythm to sets of antagonistic muscles is also not yet known. Neuronal networks allow emergence of new network properties because the intrinsic rhythm can be modulated by synaptic contacts. A working hypothesis of the CPG for mastication is shown in Figure 8.1.3.

CHEWING HAS MULTIPLE PURPOSES

- **Chewing grinds food**: The jaw muscles can produce a force of up to 700 N on the molars and about 350 N on the incisors. This is enough to crack a hazelnut. These forces are concentrated in the small surface area of occlusive contact of the teeth.
- **Grinding increases the surface area**: Because digestion occurs at the surfaces of the ingested food, increasing the surface area increases the speed of digestion. However, this makes little difference to the completeness of digestion. Chunks of meat that are swallowed whole are as completely digested as chunks that are thoroughly chewed, but chewing reduces the time for digestion.
- **Chewing reduces food size so that it can be swallowed**: Wild predators generally rip meat off their kills with their canine teeth and swallow large pieces. We tend to choke on such pieces. Chewing breaks

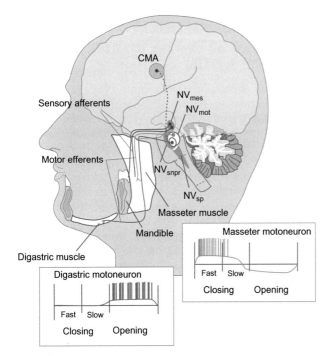

FIGURE 8.1.2 Nervous control of mastication. The masseter muscle helps close the jaw. The digastric muscle helps open it. The trigeminal nerve (cranial nerve V) innervates both muscles. Motoneurons in the motor part of the nucleus of the trigeminal nerve (NV_{mot}) drive the muscles. NV_{mot} receives inputs from NV_{mes} that contains primary sensory afferents from stretch receptors within the muscles and from the teeth and NV_{spnr} that contains the central pattern generator. The CPG receives input from the cortical masticatory area (CMA) in M1 and feedback through sensors. Motoneurons for the masseter muscle produce the trace shown in the figure: a depolarization followed by a train of action potentials occurs during the closing phase and these gradually decrease in frequency at the beginning of the slow phase of jaw closing. During the opening phase the cells are hyperpolarized. The horizontal line indicates the resting membrane potential. Motoneurons for the digastric muscle show little change in the membrane potential during jaw closing, followed by a depolarization and train of action potentials during the opening phase. These eventually fatigue and the motoneuron repolarizes. *From P. Morquette et al., Generation of the masticatory central pattern and its modulation by sensory feedback. Progr. Neurobiol.* **96**:340–355, 2012.

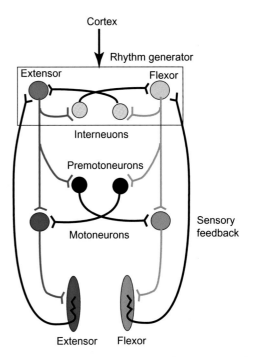

FIGURE 8.1.3 Model of the nervous control of mastication. Cells in the NV_{snpr} have intrinsic rhythm or pacemaker behavior and make up the rhythm generator. All colored connections are excitatory. All black connections are inhibitory. Mutual inhibition through interneurons assures alternating activity between extensors and flexors. Fatigue of the interneurons switches between the two centers. The first level of connections generates the rhythm, the second level composed of premotoneurons and motoneurons determines the envelope of the bursting pattern of the motoneurons. *From P. Morquette et al., Generation of the masticatory central pattern and its modulation by sensory feedback.* Progr. Neurobiol. **96**:340–355, 2012.

Clinical Applications: Temporomandibular Joint Syndrome

Temporomandibular joint syndrome (TMJ), also called temporomandibular disorder (TMD), is the most common source of pain after toothache. It can be classified broadly as: (1) secondary to myofacial pain and dysfunction (MPD) and (2) secondary to a primary articular disease. It is estimated that TMD affects some 10 million people in the United States alone. It affects women, 4:1, over men. The highest incidence is among young adults, especially women between 20 and 40 years old.

The symptoms of TMD include the following: (1) pain, usually around the ears, that is associated with chewing; the pain may radiate to the head, but it is not like a typical headache; (2) clicking, popping, or snapping sounds associated with pain and movement of the temporomandibular joint; (3) episodes of limited jaw opening or locking; the locking may be in the open or closed position and can be very painful; and (4) associated symptoms of neck pain or stiffness, shoulder pain, dizziness, and ear pain.

MPD comprises the majority of cases of TMD. X-rays of the jaw show no destructive changes of the temporomandibular joint. It has a variety of causes. Frequently, MPD is associated with **bruxism** (grinding of the teeth) and jaw clenching due to stress or anxiety. The pain, tenderness, and spasm of the mastication muscles are due to muscular hyperactivity and dysfunction due to malocclusion. Psychological factors increase the risk of MPD. Persons with MPD tend to exhibit obsessive–compulsive behavior and are more likely to deny problems in their life.

TMD caused by true articular disease is most often due to disk displacement. The temporomandibular joint is a gliding joint formed by the condyle of the mandible and the squamous (flat) part of temporal bone. The articular surface of the temporal bone forms a convex articular eminence anteriorly and a concave articular fossa posteriorly. The articular surface of the mandible is the top of the condyle. An articular disk separates the mandible from the temporal bone and divides the joint cavity into two small spaces. Degenerative joint disease, rheumatoid arthritis, dislocations, infections, and neoplasias (cancers) are all possible causes of individual cases of TMD due to articular disease.

---

the food into pieces small enough to pass into the pharynx.
- **Chewing breaks cell walls of vegetative matter**: The cellulose cell walls of vegetative matter resist digestion because we lack the enzymes to break down cellulose. The crushing force of the teeth breaks down these cell walls and releases the cell contents for digestion.
- **Chewing takes the edges off food particles**: Breaking the food into smaller pieces and crushing it between the teeth helps prevent food particles from excoriating the gut lining.
- **Chewing mixes the food with the saliva**: The saliva moistens the food, lubricates it in preparation for swallowing, and allows us to taste the food.
- **Chewing moves food around so that the taste buds can sample it**: The taste buds sense only dissolved substances. Chewing mixes the saliva with the food and moves the food around in the mouth. Both of these actions present material to the taste buds so that they can be sampled. In addition, mechanoreceptors on the palate, tongue, and cheek walls create a "mouth feel" for the food that forms part of the hedonistic enjoyment of food and informs the brain when the food has the proper consistency for swallowing.

SALIVA MOISTENS, LUBRICATES, DIGESTS, AND PROTECTS

There are two general types of salivary glands: **serous glands** secrete mainly a watery fluid; **mucus glands** secrete more viscous saliva that contains **mucin**. Mucin is a class of high molecular weight glycoproteins that are expressed by epithelial tissues. They consist of a variable number of tandem repeat sequences that are rich in serine and threonine that provide hydroxyl groups to which oligosaccharide chains are O-linked.

The **parotid gland** is the largest of the salivary glands and is located at the back of the jaw between the angle of the jaw and the ear. The parotid gland secretes up to 50% of the saliva volume and secretes only serous saliva. These serous salivary glands also secrete α-amylase, an enzyme that hydrolyzes α-1,4-glycosidic linkages in starch.

The **submandibular glands** lie under the mandible and consist of a mixture of serous- and mucin-secreting cells. The **sublingual glands** lie under the tongue and are also a mixture of serous- and mucin-secreting cells. In humans the submandibular gland appears to cosecrete mucin and TFF3, one of three kinds of TFF-peptides. Its physiological role is not yet established; it seems to alter the **rheological** properties of saliva (its ability to flow), and it may also help repair injured parts of the GI tract.

The three major salivary glands (parotid, submandibular, and sublingual) account for about 90% of salivary secretion. A number of minor glands contribute the remaining 10%. Although serous lingual glands secrete small amounts of salivary **lipase**, it has little functional importance in humans. The location of the major salivary glands is shown in Figure 8.1.4.

The major functions of saliva are as follows:

- **Saliva lubricates food**: The watery secretion of saliva moistens food, making it easier to swallow. In addition, mucins make the food slippery, making it easier to swallow.
- **Saliva begins digestion**: Salivary amylase and lipase begin digestion even before the food leaves the mouth.
- **Saliva aids in speech by keeping the mouth moist**: By keeping the moving parts moist, saliva aids in the movement of lips and tongue both in the action of chewing and in speech.

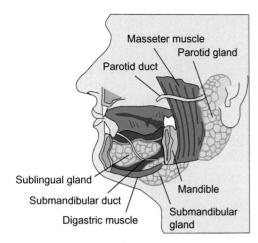

FIGURE 8.1.4 Location of the main salivary glands. The parotid gland is the largest, is located at the back of the jaw between the angle of the jaw and ear, and secretes only serous saliva. It drains into the roof of the mouth through the parotid duct (Stenson's duct). The submandibular gland lies under the mandible and is a mixture of serous and mucin-secreting cells. It drains into the mouth under the tongue through the submandibular duct (Wharton's duct). The sublingual glands lie under the tongue and also are a mixture of serous and mucin-secreting cells.

- **Saliva buffers mouth pH**: Saliva pH ranges from 6 to 8, with more alkaline pH at high salivary secretion rates. It contains HCO_3^-, phosphate, and proteins that protect the hard and soft tissues of the mouth against large pH changes by binding excess H^+ ions.
- **Saliva keeps the mouth clean**: The continuous production of saliva washes the mouth. This helps flush away food particles and reduces dental caries.
- **Saliva helps prevent infections**: Saliva contains a variety of antibacterial and antifungal components. These include **lysozyme**, **lactoferrin**, **myeloperoxidase**, **salivary peroxidase**, **immunoglobulin A (IgA)**, histatins, **statherin**, and **alpha and beta defensins**. Dry mouth, or **xerostomia**, is associated with infections of the buccal mucosa and with dental caries. The lysozyme destroys bacterial cell walls. Lactoferrin reduces bacterial growth by complexing essential iron. The peroxidases produce hydrogen peroxide (H_2O_2) and hypochlorus acid (HOCl) that are antibacterial. Histatin 5, in particular, inhibits the growth of *Candida albicans*, an oral fungus normally present whose overgrowth causes **thrush**. Statherins inhibit precipitation of calcium phosphate in saliva, and it also inhibit growth of anaerobic bacteria. Alpha defensin originates from neutrophils and beta defensins from epithelial cells. Both have antibacterial and antiviral effects. Secretory IgA has activity against bacteria but its significance is unknown as persons with hereditary lack of IgA show no increase in oral disease.
- **Saliva may help heal mouth and esophageal damage**: The mouth repeatedly experiences wounds that range from cheek, tongue, and lip biting to tooth extraction, but these wounds heal much more quickly than skin wounds with less scarring and infrequent infections. First, saliva contains tissue factor that aids in hemostasis that limits bleeding and exposure of the circulation to infectious agents in the mouth. Second, in addition to the electrolytes and mucins, saliva also contains a variety of **growth factors**. Growth factors are polypeptides that help tissues repair themselves following injury. Saliva contains **epidermal growth factor (EGF)**, **fibroblast growth factor (FGF)**, **nerve-derived growth factor (NGF)**, vascular endothelial growth factor (VEGF), **transforming growth factor α (TGF-α)**, and **trefoil factor 3 (TFF3)**. All of these are thought to enhance healing in the mouth. Their function in protecting the esophagus from damage either by excoriation (scratching) by food particles, by acid reflux from the stomach, or by bacterial or viral agents is not yet established. The instinctive licking of wounds by humans and many other animals may be beneficial because of the healing factors found in saliva.

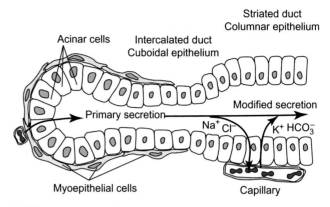

FIGURE 8.1.5 Schematic diagram of a salivary gland unit and the mechanisms of saliva secretion and modification. Saliva produced by acinar cells is isotonic with plasma. The lumen of the acinus leads to intercalated ducts that direct the saliva through striated salivary ducts where the fluid is modified by reabsorption of Na^+ and Cl^- and secretion of K^+ and HCO_3^-. The resulting saliva is hypotonic to plasma.

SALIVARY GLANDS PRODUCE AN ISOTONIC FLUID THAT IS SUBSEQUENTLY MODIFIED

The salivary glands are tubes that end in blind sacs (see Figure 8.1.5). Some of these sacs are shaped like grapes, from which they derive their name **acinus** (meaning "grape shaped"). Others end in a smaller cap of a tubular end piece. Cells in the acinus secrete a fluid that is isotonic with plasma and whose composition initially is independent of the flow rate. The acinar cells are connected by apical **tight junctions** that are permeable to cations. The acini lead into **intercalated ducts** that connect the acini to **striated ducts**. The intercalated ducts consist of cuboidal epithelial cells that may serve as a reservoir of stem cells for both acini and striated ducts. The striated ducts consist of columnar epithelial cells that possess deep infoldings of their basal membrane. These deep infoldings produce their striated appearance from which the striated ducts derive their name. These ducts modify the composition of the plasma by reabsorbing some components (Na^+ and Cl^-) back into the blood and secreting other components (K^+) into the saliva. The striated duct cells are impermeable to water. Reabsorption of Na^+ and Cl^- is not accompanied by the osmotic transport of water, resulting in hypotonic saliva.

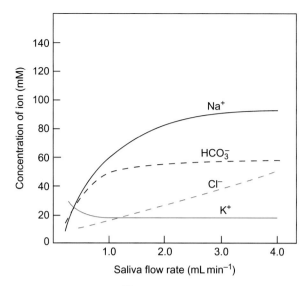

FIGURE 8.1.6 Composition of human parotid saliva as a function of flow rate. As flow increases, Na^+ and Cl^- concentrations increase toward plasma levels, whereas HCO_3^- concentrations level off and K^+ concentrations are less than at low flow rates. *Redrawn from J.H. Thaysen, N.A. Thorn, and I.L. Schwartz, Excretion of sodium potassium, chloride and carbon dioxide in human parotid saliva. Am. J. Physiol. 178:155, 1954.*

Myoepithelial cells cover the acini and line the outside of the intercalated ducts. Both sympathetic and parasympathetic fibers innervate them. They support the salivary gland structures and promote secretion by their contraction.

SALIVA COMPOSITION DEPENDS ON THE FLOW RATE

Because saliva is formed in two stages, production of an isotonic secretion and then modification during passage through the ducts, the composition of saliva varies with its rate of flow. Saliva flow varies from $0.1\ mL\ min^{-1}$ during sleep to $4-6\ mL\ min^{-1}$ during eating. Unstimulated saliva production during the day is about $0.3-0.4\ mL\ min^{-1}$ and shows a circadian rhythm, peaking generally in mid-afternoon. Total salivary secretion has been estimated to be about $0.6\ L\ day^{-1}$. The composition of human parotid saliva as a function of saliva flow rate is shown in Figure 8.1.6.

THE SALIVARY NUCLEI OF THE MEDULLA CONTROL SALIVATION

Direct control of salivation occurs through the superior and inferior salivary nuclei located in the pons and medulla (see Figure 8.1.7). Parasympathetic

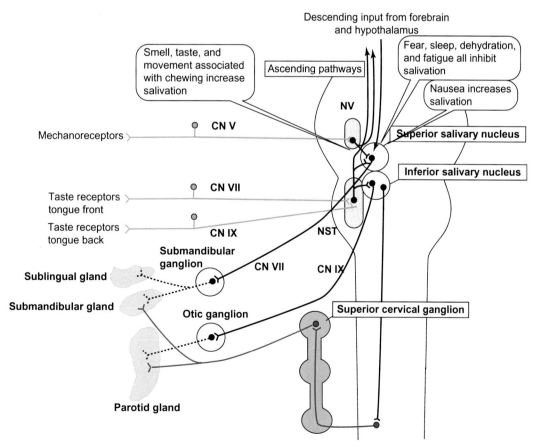

FIGURE 8.1.7 Nervous control of salivation. The superior and inferior salivary nuclei located in the pons and medulla directly control salivation. Preganglionic parasympathetic fibers from the superior salivary nucleus travel over the facial nerve, CN VII, to the submandibular ganglion and then to secretory cells in the submandibular and sublingual glands. Preganglionic fibers from the inferior salivary nucleus travel to the otic ganglion and stimulate parotid gland secretion. Sympathetic fibers exit the cord at T1, T2, and T3, travel up the sympathetic chain to the superior cervical ganglion, and then to the parotid and submandibular glands. Taste and mechanoreceptor afferents enter the nucleus of the solitary tract (NST) and NV, the nucleus of the trigeminal nerve, and modulate salivary secretion. Inputs from higher centers also modulate salivary secretion.

preganglionic fibers from the superior salivary nucleus travel over a branch of the facial nerve (CN VII) to the submandibular ganglion where they synapse onto postganglionic fibers that stimulate serous secretion of the sublingual and submandibular salivary glands. Preganglionic fibers from the inferior salivary nucleus travel over the glossopharyngeal nerve (CN IX) to synapse on postganglionic fibers in the otic ganglion, and these stimulate serous secretion of the parotid glands. The salivary nuclei are influenced by sensory afferents from the taste buds, olfactory receptors, and mechanoreceptors. Olfactory information enters via the olfactory nerve (CN I). The facial nerve (CN VII) transmits taste information from the anterior tongue; the glossopharyngeal nerve (CN IX) carries taste sensation from the posterior tongue. Both synapse on cells in the nucleus of the solitary tract (NST) which then connects to the salivary nuclei. The trigeminal nerve (CN V) carries mechanoreceptor signals to the trigeminal nucleus (NV), which then also connects to the salivary nucleus. Cells in the inferior salivary nucleus also stimulate sympathetic motor neurons in T1, T2, and T3 of the thoracic spinal cord. Their axons travel to the sympathetic chain and ascend to the superior cervical ganglion to synapse onto postganglionic fibers that stimulate the parotid and submandibular glands.

Higher centers of the brain, such as the forebrain and hypothalamus, can also influence salivation through inputs to the salivary nuclei. Fatigue, dehydration, sleep, fear, and mental depression all depress saliva flow. Nausea and appetizing odors stimulate saliva production. These higher centers receive sensory information from a variety of sources including the NST and NV in order to correctly instruct the salivary nuclei.

PARASYMPATHETIC STIMULATION RESULTS IN HIGH-VOLUME, WATERY SALIVA

Salivary glands are innervated by both the sympathetic and parasympathetic divisions of the autonomic nervous system and both stimulate secretion. Parasympathetic stimulation mainly increases the watery secretion, whereas sympathetic stimulation mainly increases protein secretion (see Figure 8.1.8).

Parasympathetic nerves contain a variety of neurotransmitters including acetylcholine, vasoactive intestinal polypeptide (VIP), and substance P. These nerves are associated with salivary glands and the vasculature that supplies the glands. Substance P and acetylcholine stimulate a watery secretion by activating a G_q mechanism that activates phospholipase C to split phosphatidylinositol-4,5-bisphosphate to diacylglycerol and inositol-1,4,5 triphosphate or IP_3. The released IP_3 binds to IP_3 receptors (IP_3R) on the endoplasmic reticulum and causes the IP_3R to release Ca^{2+} to the cytoplasm. The increased $[Ca^{2+}]$ is the final trigger that stimulates a variety of ionic transport mechanisms that result in increased secretion of primary saliva. In addition, parasympathetic stimulation releases **vasoactive intestinal peptide (VIP)** which is a potent vasodilator and **increases blood flow to the salivary glands**. Since the saliva originates from the blood, the increased

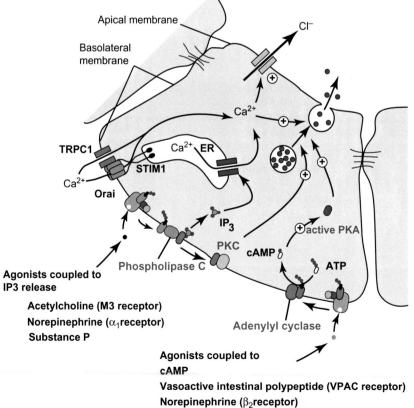

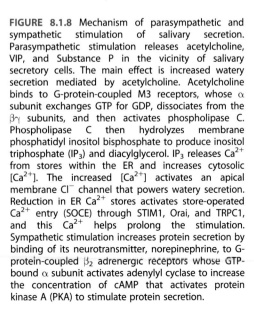

FIGURE 8.1.8 Mechanism of parasympathetic and sympathetic stimulation of salivary secretion. Parasympathetic stimulation releases acetylcholine, VIP, and Substance P in the vicinity of salivary secretory cells. The main effect is increased watery secretion mediated by acetylcholine. Acetylcholine binds to G-protein-coupled M3 receptors, whose α subunit exchanges GTP for GDP, dissociates from the $\beta\gamma$ subunits, and then activates phospholipase C. Phospholipase C then hydrolyzes membrane phosphatidyl inositol bisphosphate to produce inositol triphosphate (IP_3) and diacylglycerol. IP_3 releases Ca^{2+} from stores within the ER and increases cytosolic $[Ca^{2+}]$. The increased $[Ca^{2+}]$ activates an apical membrane Cl^- channel that powers watery secretion. Reduction in ER Ca^{2+} stores activates store-operated Ca^{2+} entry (SOCE) through STIM1, Orai, and TRPC1, and this Ca^{2+} helps prolong the stimulation. Sympathetic stimulation increases protein secretion by binding of its neurotransmitter, norepinephrine, to G-protein-coupled β_2 adrenergic receptors whose GTP-bound α subunit activates adenylyl cyclase to increase the concentration of cAMP that activates protein kinase A (PKA) to stimulate protein secretion.

blood flow allows increased salivary secretion. The net effect of decreased blood flow is increased protein content of a more viscous saliva. VIP binds to a G-protein-coupled receptor (VPAC) that activates adenylyl cyclase and increases 3′,5′ cyclic AMP within the cells. This activates protein kinase A (PKA) that stimulates protein secretion. In addition, cholinergic stimulation can increase protein secretion by means independent of cAMP, by increasing [Ca^{2+}], and activating PKC. Thus parasympathetic stimulation increases both the watery and protein secretion.

Sympathetic stimulation releases norepinephrine near the basal membranes of acinar and ductal cells. These cells are heterogeneous. Some cells possess α_1 receptors, which act through a G_q mechanism, in much the same way as M3 acetylcholine receptors, to increase the rate of production of a watery saliva. Other cells possess β_1 receptors, which act through a G_s protein to stimulate adenylyl cyclase to increase the concentration of cAMP within the cells. This results in increased protein and glycoprotein secretion. VIP also can increase cAMP. Thus the simultaneous activation of both parasympathetic and sympathetic nerves, as occurs during reflex secretion, leads to augmented secretion of salivary proteins.

IN THE FIRST STAGE OF SALIVA PRODUCTION, ACINAR CELLS SECRETE A FLUID ISOTONIC IN NaCl

Figure 8.1.9 illustrates the mechanism of fluid secretion by salivary acinar cells. The ultimate energy for the process is derived from the Na,K-ATPase located on the basolateral membranes of these cells. By pumping Na^+ out at the basolateral membrane, the pump maintains

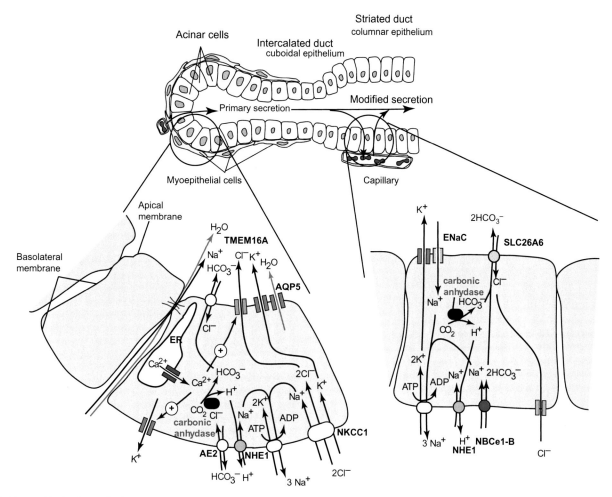

FIGURE 8.1.9 Mechanisms of salivary secretion and reabsorption. Salivary secretion takes place in two stages: the production of the primary secretion isotonic with plasma, containing mostly Na and Cl; and reabsorption of Na and Cl with secretion of K^+ and HCO_3^-. Secretion is turned on by increasing cytosolic [Ca^{2+}] through a cascade of events shown in Figure 8.1.8. The main effect of increased cytosolic [Ca^{2+}] is activation of K^+ channels at the **basolateral membrane** (the membrane lining the lateral and basal aspects of the cell, facing the blood) and Cl^- channels (**TMEM16A**) at the **apical membrane** (the membrane at the apex of the cell, facing the lumen). Cl^- efflux into the lumen and K^+ efflux into the interstitial space at the basolateral membrane create a transepithelial potential that draws Na^+ through the tight junctions. Water follows the osmotic pressure of these ions both through the tight junctions and through AQP5 in the cell. The Na,K-ATPase maintains the membrane potential and the driving forces for Na^+ entry into the cell and K^+ exit. The removal of ions shrinks the cell, turning on a Na–K–2Cl cotransporter (**NKCC1**) that further supplies ions for transport into the lumen. Because water follows passively, the saliva initially formed is isotonic with plasma. The salivary duct cells reabsorb Na^+ and Cl^- and secrete K^+ and HCO_3^-. Na^+ is reabsorbed by the **EnaC** and the Na,K-ATPase. Cl^- is reabsorbed over the apical Cl^-–$2HCO_3^-$ exchanger (**SLC26A6**) and the basolateral Cl^- channel. HCO_3^- is secreted by entry into the cell over the Na^+–$2HCO_3$ cotransporter (**NBCe1-B**) or from hydration of CO_2 by carbonic anhydrase, and then efflux from the cell over SLC26A6. The final modified secretion is hypotonic because water does not follow electrolyte reabsorption in the striated duct cells. Other transport mechanisms are present in these cells that are of lesser importance.

an inwardly directed electrochemical gradient for Na^+ and an outwardly directed electrochemical gradient for K^+. The K^+ that accumulates exits the cell by Ca^{2+}-activated channels on the basolateral membranes. The $Na^+-K^+-2Cl^-$ (**NKCC1**) cotransporter on the basolateral membrane mediates Cl^-, Na^+, and K^+ entry into the cell. The Na^+ that enters is pumped out by the Na,K-ATPase and by Na^+-H^+ exchange (**NHE1**) on the basolateral membrane. The K^+ that enters exits over Ca^{2+} activated K^+ channels. These processes lead to markedly increased $[Cl^-]$ in the cell. A second mechanism of accumulating Cl^- relies on the paired NHE1 and $Cl^- - HCO_3^-$ exchanger (**AE2**). The operation of these two brings Na^+ and Cl^- into the cell along with the net outward movement of CO_2 as $HCO_3^- + H^+$. Ca^{2+}-activated opening of Cl^- channels (**ANO1 = TMEM16A**) on the apical membrane causes a Cl^- efflux into the lumen of the acinus. Chloride entry into the lumen and K^+ exit on the basolateral side produce a transepithelial potential difference. Acinar cells have relatively leaky tight junctions that allow Na^+ to enter the lumen, along with water that moves in response to the osmotic pressure of the transported ions. Water appears to travel by both transcellular and paracellular routes. The transcellular route is through an aquaporin (**AQP5**). Thus the primary acinar secretion contains mainly NaCl.

IN THE SECOND STAGE OF SALIVA PRODUCTION, DUCT CELLS REABSORB NaCl AND SECRETE A HYPOTONIC KHCO$_3$

The second stage of saliva secretion occurs in the ducts (see Figure 8.1.9). At the apical membrane, Na^+ enters the ductal cell through **ENaC**, the amiloride-sensitive epithelial Na^+ channel. Na^+ that enters is pumped out of the cell by the basolateral Na,K-ATPase. Conversely, K^+ that enters the cell is secreted over an apical K^+ channel. The apical membrane contains a $Cl^- - 2HCO_3^-$ exchanger (**Slc26a6**) that secretes 2 HCO_3^- and reabsorbs 1 Cl^-. HCO_3^- enters the cell by $Na^+-HCO_3^-$ cotransport (**NBCe1-B**) on the basolateral membrane. This minimal model can explain Na^+ and Cl^- absorption from the primary salivary secretion, and K^+ and HCO_3^- secretion into the incipient saliva.

A SWALLOWING CENTER IN THE MEDULLA ORCHESTRATES SWALLOWING

Swallowing is a tricky business because it involves passing a **bolus** of food through the pharynx into the esophagus past the trachea. One wrong move and the bolus enters the trachea and the person chokes. Further, leakage of small amounts of food into the lungs (**aspiration**) can cause potentially fatal lung problems. It is vitally important that swallowing follows a programmed series of events every time we swallow. The sequence of events is controlled by a **swallowing center** located in the **medulla** in or near the nucleus of the solitary tract (NST). This center receives sensory input from the mouth and pharynx and delivers efferent motor commands via the trigeminal (CN V), facial (CN VII), and hypoglossal (CN IX) nerves.

SWALLOWING IS A COMPLEX SEQUENCE OF EVENTS

Swallowing is initiated voluntarily by actions of the tongue on material in the mouth. Sensors within the mucosa of the mouth inform the brain of the size and physical characteristics of the food. Unlike other animals we generally cannot swallow large chunks of food. When the food is sufficiently broken down, the tongue separates the food into the part to be swallowed by pressing the tip of the tongue against the **soft palate**. The tongue then elevates and retracts against the palate, pushing the bolus of food back toward the **pharynx**, the topmost part of the throat. The sequence of events in swallowing is illustrated in Figure 8.1.10.

Although the swallowing reflex can begin voluntarily, it cannot be sustained without sensory input from fluid or matter on sensory receptors that line the pharynx. Afferent sensory information travels over the trigeminal (CN V), vagus (CN X), and glossopharyngeal (CN IX) nerves to the swallowing center, which then coordinates swallowing through efferent nerves. This programmed sequence, once initiated, cannot be easily stopped. Blockade of sensory information by local anesthesia, for example, interferes with the proper coordination of swallowing and may result in aspiration. Thus the proper coordination of swallowing requires sensory feedback from mechanosensors distributed throughout the pharynx and larynx.

SWALLOWING CONSISTS OF A *PHARYNGEAL PHASE* AND AN *ESOPHAGEAL PHASE*

The process of swallowing described in Figure 8.1.10 refers to the **pharyngeal phase** of swallowing that encompasses movement of food or liquids from the mouth past the upper esophageal sphincter. In the **esophageal phase**, food is propelled down the esophagus by smooth muscle contraction. The wave of muscular contraction that immediately follows the pharyngeal phase is called **primary peristalsis**. It is a programmed event that is controlled by neural efferents from the swallowing center. The esophagus is highly unusual in that the uppermost 5–10% of the esophagus consists of striated skeletal muscle. The lower 50% is entirely smooth muscle, whereas the middle 30–40% is a mixture of both striated and smooth muscles. Thus the vagus efferent from the swallowing center controls a skeletal muscle in the upper esophagus. The vagus sequentially activates the skeletal muscle portion and the smooth muscle portion in a seamless, coordinated manner to produce smooth movement of the food to the stomach.

Generally primary peristalsis successfully clears the esophagus of the bolus and the food or liquid enters the stomach. Food remaining in the esophagus following primary peristalsis stretches the esophagus, and mechanoreceptors in the esophagus sense this stretch.

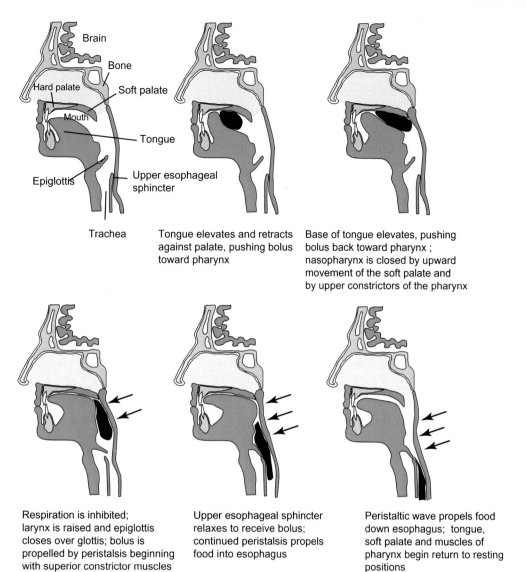

FIGURE 8.1.10 **Sequential events in swallowing.** Swallowing is initiated by voluntarily pushing a bolus of food toward the pharynx by elevating the tongue against the soft palate. The nasopharynx closes by apposition of the soft palate with the upper constrictor muscles of the pharynx. The tongue continues to push the food back into the pharynx. This initiates a sequential constriction of the muscles of the pharynx in a wave of contraction called peristalsis. This moves the food down into the larynx. Respiration is inhibited, the larynx is elevated, and the **epiglottis** closes over the glottis to prevent aspiration of food or fluid into the lungs. The peristaltic wave pushes the food down toward the esophagus, while the **upper esophageal sphincter** opens to accept it.

The stretch receptors directly affect smooth muscle contraction by actions on interneurons contained within the esophagus itself and by afferent input to the swallowing center to activate **secondary peristalsis** (see Figure 8.1.11). Secondary peristalsis is a **vagovagal reflex** because both the sensory and motor travels over the vagus nerve.

THE ESOPHAGUS CONTAINS AN INNER CIRCULAR SMOOTH MUSCLE LAYER AND AN OUTER LONGITUDINAL SMOOTH MUSCLE LAYER

Epithelial cells make up the **mucosa** that lines the lumen of the esophagus. Immediately below the mucosa is a layer of smooth muscle called the **muscularis mucosae**. Below the muscularis mucosa is a layer of **submucosa** containing blood vessels, connective tissue, and immunological cells such as macrophages. Below the submucosa (toward the outside of the esophagus) are two other layers of muscle, generally thicker and more developed than the thin muscularis mucosa. The layer closest to the lumen consists of cells oriented circumferentially. This is the **circular muscle** layer. Contraction of these cells constricts the tube. Outside the circular layer is another layer of smooth muscle oriented longitudinally, the **longitudinal muscle**. This controls the length of the tube. This layered organization of the esophagus is similar to that of the entire GI tract (see Figure 8.1.12).

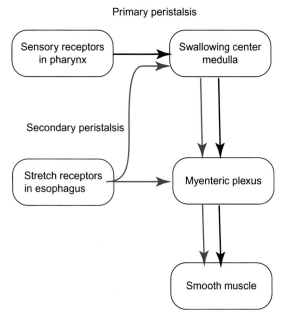

FIGURE 8.1.11 Block diagram of the control of primary and secondary peristalsis. Primary peristalsis arises as part of the programmed sequence of events initiated by the swallowing center. Secondary peristalsis arises from sensory input from stretch receptors within the esophagus that communicates both with the myenteric plexus in the esophagus and with the swallowing center. Both the sensory arm and the efferent arm of secondary peristalsis travel over the vagus nerve, and so this reflex is a **vagovagal reflex**.

THE GUT CONTAINS TWO GANGLIONIC PLEXUSES OF NERVE CELLS, THE BODY'S "LITTLE BRAIN"

In the intestine, nerve cells are collected into small groups in between the outer longitudinal muscle layer and the inner circular muscle layer. These bunches of nerve cells collectively form the **myenteric plexus**. These nerve cells receive local sensory input from the GI tract and respond to central efferent commands. Each group of neurons communicates with others up and down the GI tract. The arrangement of muscle layers and nerves and their control is illustrated in Figure 8.1.12. Underneath the circular muscular layer is a second nerve plexus, the **submucosal plexus**. It contains sensory neurons that detect stimuli from the lumen of the GI tract, and secretomotor cells that control secretion and contraction of the inner circular layer and the **muscularis mucosa**.

THE *LES* MUST RELAX FOR FOOD TO ENTER THE STOMACH

The lower esophageal sphincter (LES) tonically constricts the opening between the stomach and the esophagus. This prevents reflux of acidic stomach contents into the esophagus, which is poorly equipped to deal with it. Peristalsis of the esophagus is generally followed by relaxation of the LES and entry of food into the stomach. The vagus nerve relaxes the LES through the release of **VIP** and **nitric oxide (NO)**.

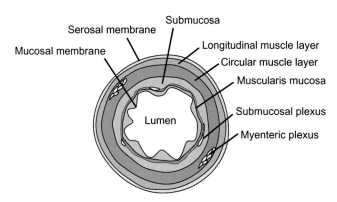

FIGURE 8.1.12 Cross-section of the esophagus showing the innermost lining of the lumen, the mucosal membrane, and its underlying structures. The muscularis mucosa is a thin layer of muscle whose contraction folds the mucosa to form ridges and valleys. Below the muscularis mucosa is the submucosa containing connective tissue, blood vessels, and nerves. The nerve complex here is the submucosal plexus. Further outward is an inner circular layer and an outer longitudinal layer. Between them is another interconnected nerve network called the myenteric plexus. These nerve plexuses control GI motility and secretion.

SUMMARY

The function of the GI tract is to extract nutrients from the ingested food to provide building blocks for our bodies and energy to power its movement and metabolism. The GI tract is basically a tube that runs from mouth to anus. The tube has specialized areas that are separated by muscular sphincters that control the movement of material from one part to the next. The esophagus brings food to the stomach, which stores it and reduces its particle size. The esophagus is marked by an upper esophageal sphincter and a lower esophageal sphincter. The pyloric sphincter controls stomach emptying into the small intestine, which consists of the duodenum, jejunum, and ileum. The ileocecal sphincter controls movement of the luminal contents into the large intestine, which consists of the cecum, ascending colon, transverse colon, descending colon, and sigmoid colon. The sigmoid colon connects to the rectum. Defecation is controlled by two sphincters, the internal and external anal sphincters.

Digestion begins in the mouth where food is chewed and mixed with saliva. Chewing decreases particle size, increases the surface area, breaks down vegetable cell walls, mixes the food with saliva and other foods, and moves it around so it may be sampled.

Saliva moistens and lubricates the food, buffers the mouth pH, helps ward off infections, and helps heal the esophagus. In addition, it facilitates speech by keeping the mouth moist. Saliva is formed in acinar (grape shaped) glands in the parotid, submandibular, and sublingual glands. The parotid gland produces only fluid, whereas the submandibular and sublingual glands add mucin, glycoproteins that help lubricate food, and other proteins. The fluid flows down an intercalated duct to reach a striated duct, which modifies the saliva by reabsorbing ions. Saliva as formed is isotonic with plasma,

Clinical Applications: Achalasia

Achalasia is a disorder of esophageal motility which interferes with food passing from the esophagus to the stomach. It is characterized by increased LES pressure, diminished peristalsis in the distal third of the esophagus, and lack of coordinated relaxation of the LES in response to primary or secondary peristalsis in the esophagus. Sir Thomas Willis first described the condition in 1674, and he successfully treated his patient by dilating the LES with a cork-tipped whalebone. The condition is relatively rare, with an incidence of four to six cases per million persons and a prevalence of eight cases per million. Its etiology is unknown. Current theories of causality include autoimmune disorders and infection. Commonly, the myenteric plexus (Auerbach's plexus) degenerates, with death of inhibitory neurons that secrete **VIP** and **NO**. Both of these neurotransmitters inhibit GI motility and cause relaxation of sphincters (see Chapter 8.3).

Failure to relax the LES causes entrapment of food in the esophagus with subsequent enlargement of the esophagus (megaesophagus). This inflames and irritates the esophagus (**esophagitis**). The trapped food causes regurgitation, which in turn can cause nocturnal coughing and aspiration. Aspiration of esophageal contents leads to pneumonia and lung abscesses.

Achalasia can be diagnosed by **video esophagraphy** in which the patient swallows a radio-opaque solution (barium) during fluoroscopy. The barium absorbs X-rays and therefore it shows the enlarged end of the esophagus and a narrow tapering at the sphincter. In persons with achalasia, the barium solution empties into the stomach more slowly. This test must be confirmed by endoscopic examination and esophageal manometry. **Esophageal manometry** is the standard criterion for diagnosis of the disease. Esophageal manometry measures the pressures in the esophagus by passing a thin tube into the esophagus. In persons with achalasia, there is no pressure wave in the distal esophagus following a swallow, and the pressure within the contracted sphincter does not decrease with the swallow.

There are several treatment options for achalasia, including oral medications, dilation of the sphincter by a balloon placed within the esophagus, surgical cutting of the muscle (myotomy), and botox injection. The oral medications consist of four distinct classes, all of which aim to relax smooth muscle. The goal of LES dilation and surgery is to tear or cut the muscle fibers to reduce the restriction of food entering the stomach.

Injection of botulinum toxin (botox) into the LES relaxes the sphincter by a complicated mechanism. *Clostridium botulinum* is a gram-negative bacterium that produces seven related toxins, called botulinum toxin types A, B, C1, D, E, F, and G. The toxin is synthesized as a protoxin of 150 kDa, which is subsequently cleaved to light (L) and heavy (H) chains that remain linked by a disulfide bond. Nerve terminals have receptors for both the H- and L-chains. The L-chain is transported across the nerve terminal membrane by endocytosis, and it becomes incorporated into endosomes. The various L-chains are **metalloproteinases**—proteases that require metal ions for activity. The L-chains bind Zn^{2+}. Their substrates are one of several proteins that make up the **SNARE** complex that is required for neurotransmitter vesicle fusion with the presynaptic nerve membrane. At least three proteins are involved in this process: v-SNARE (synaptobrevin) is associated with neurotransmitter vesicles; t-SNARE (syntaxin) is on the presynaptic membrane; and SNAP-25, another t-SNARE. All of these proteins are required for the docking of vesicles in the active zone of the presynaptic terminal and for subsequent neurotransmission. Each of the botulinum toxins cleaves one of these three proteins and thereby interferes with neurotransmitter release. Thus botulinum toxin causes paralysis of the muscles. Injection of botulinum toxin into the LES thus causes its relaxation and improves emptying of food into the stomach.

Clinical Applications: GERD and Hiatal Hernia

The Clinical Applications box about achalasia details the condition, complications, diagnosis, and treatment for the situation when the LES does not relax coordinately with esophageal peristalsis. Gastroesophageal reflux disease, or GERD, is its opposite: the LES does not constrict enough to adequately retain stomach contents within the stomach. The result is that the stomach contents **reflux** into the esophagus. The acidic stomach contents causes a burning sensation in the throat or chest which is colloquially called **heartburn**. However, gastroesophageal reflux may be asymptomatic. Persistent heartburn more than twice weekly may be considered to indicate GERD.

The cause of the condition is unknown. Anyone can have GERD, including infants and children. Obesity is a risk factor and losing weight is one of the lifestyle changes recommended for persons with GERD. Hiatal hernia is also a risk factor. However, the majority of persons with hiatal hernias are asymptomatic, and many persons with GERD do not have hiatal hernia.

The esophagus is approximately 30 cm long, passing from pharynx to stomach. The diaphragm forms a dome that separates the thoracic cavity from the abdominal cavity. Contraction of the diaphragm expands the thoracic cavity, reducing the pressure of the gas in the lungs. This causes air to move into the lungs, and so contraction of the diaphragm drives inhalation. To do this, the diaphragm must seal off the two cavities. To reach the stomach, the esophagus must pass through the diaphragm. Thus there is a hole in the diaphragm to allow the esophagus to pass through. This is the **hiatus**. The diaphragm is sealed around the esophagus by the phrenoesophageal ligament. This fibrous layer of connective tissue not only seals the hiatus but anchors the esophagus so that the LES lies within the abdominal cavity and not the thorax. Sometimes the LES anchor gives way and the LES moves upward through the diaphragm into the thorax. This is a hiatal hernia. The lower pressure within the thorax is transmitted to the esophagus, which thereby loses some of its effectiveness as a sphincter.

The esophagus lacks protection against luminal acid, and so it may be damaged by acid reflux. The damage may cause bleeding. Esophageal ulcers and scars that develop may narrow the esophagus, making swallowing difficult. Some people develop **Barrett's esophagus**, in which the cells lining the esophagus become more like intestinal cells. It is generally regarded as being a precancerous condition.

but reabsorption of ions produces a hypotonic saliva. Slow flow allows the striated duct cells more time to reabsorb ions, so the osmolarity decreases as the flow decreases. Both parasympathetic and sympathetic nerves influence saliva production. Parasympathetic stimulation increases serous saliva (without protein), whereas sympathetic stimulation increases protein secretion. The salivary nuclei in the medulla control salivation. Smell, taste, and feel of food in the mouth, as well as just thinking about food, all promote salivation. Fatigue, dehydration, fear, and sleep all depress salivation.

Swallowing is a tricky business because you have to avoid getting material in the trachea. The mechanism for swallowing is hard wired before birth and does not require learning. The swallowing center in the medulla uses sensory information provided by the trigeminal (cranial nerve V), glossopharyngeal (cranial nerve IX), and vagus (cranial nerve X) to determine when food is ready to be swallowed. The swallowing program is initiated voluntarily but cannot be interrupted voluntarily. The swallowing program sequentially pushes the food backward toward the pharynx, closes the nasopharynx, constricts the pharynx, lifts the larynx and blocks the glottis with the epiglottis, and relaxes the upper esophageal sphincter. Primary peristalsis is a wave of contraction of the esophagus that immediately follows swallowing. It usually sweeps the swallowed food bolus into the stomach. If food remains in the esophagus, it stretches the esophagus and begins a secondary peristalsis.

The upper part of the esophagus is striated muscle like skeletal muscle though it is not connected to bone. The middle part is a mixture of striated and smooth muscles, whereas the part near the stomach is smooth muscle. The basic structure of the esophagus is like that of the rest of the GI tract: the lumen of the intestine is lined with an epithelial mucosa. The submucosa contains the muscularis mucosa that folds the mucosa, the submucosal plexus, blood vessels, and glands. Further outward is the circular muscle layer and the longitudinal muscle layer. The circular muscle determines the caliber of the tube and the longitudinal muscle controls its length. Between them is the myenteric plexus of nerves that responds to local sensory stimulation and to commands from the autonomic nervous system.

REVIEW QUESTIONS

1. What is a sphincter? Name the GI sphincters, in order, from mouth to anus.
2. What is mucus? What does it do? What salivary glands produce it? What stimulates its production?
3. How does parasympathetic stimulation stimulate serous saliva production?
4. Why is swallowing so complicated? What controls swallowing? What would happen if that area is damaged by a stroke, for example?
5. What is peristalsis? In the esophagus, what causes primary peristalsis? What causes secondary peristalsis? What is a vagovagal reflex?
6. What muscle layers are responsible for peristalsis? What controls these muscle layers?

The Stomach 8.2

Learning Objectives

- Identify the parts of the stomach including fundus, body, antrum, and pyloric sphincter
- Distinguish between receptive relaxation and adaptive relaxation
- Identify the location of the pacemaker zone and describe its function
- Describe interstitial cells of Cajal and their function
- Distinguish among propulsion, grinding, and retropulsion
- Define chyme
- List the gastric factors that regulate gastric emptying
- List the duodenal factors that inhibit gastric emptying
- Define the MMC and describe its function
- List the stomach secretions and their cells of origin
- Describe the function of intrinsic factor
- List the three phases of acid secretion
- Describe the role of GRP, SST, and histamine in acid secretion
- List the major factors that regulate acid secretion
- List the apical transport mechanisms for acid secretion by parietal cells

THE STOMACH STORES FOOD AND RELEASES IT GRADUALLY TO THE SMALL INTESTINE

The purpose of the gastrointestinal (GI) system is to break down food into its component parts and absorb the released nutrients into the blood for use by the rest of the body. Digesting the food into its parts requires enzymatic reactions that take time. The trick is to keep the food in the GI tract long enough to break it down and absorb it, but not so long that we carry around useless extra weight. The stomach stores food and releases it to the intestines at a rate that the intestines can properly process it. It also secretes materials, begins digestion, and absorbs some materials.

THE STOMACH HAS DISTINCT REGIONS

The stomach is a bulge in the alimentary tract between the **lower esophageal sphincter**, between the stomach and the esophagus, and the **pyloric sphincter** between the stomach and the **duodenum**, the first segment of the small intestine. The stomach has several named parts that have no clear anatomical demarcations (see Figure 8.2.1). The stomach can be divided into two regions based on their different patterns of motility. The term "orad" means "toward the mouth," and the term "caudad" means, literally, "toward the tail." The orad part of the stomach has thinner muscle layers and serves to receive food from the esophagus. The caudad part has thicker walls that generate more force to mix and grind stomach contents and to propel the material into the duodenum.

Most parts of the GI tract have two layers of muscle that contribute to the movement of luminal contents. The stomach has three muscle layers that move food: an outer longitudinal layer, a middle circular layer, and an inner oblique layer that is unique to the stomach. The longitudinal layer is thickest over the curvatures of the stomach and does not cover the anterior and posterior surfaces. The oblique muscle consists of two bands that radiate from the lower esophageal sphincter to fuse with the circular layer in the caudad stomach. The thickness of the longitudinal and circular layers increases from the orad to the caudad stomach.

GASTRIC MOTILITY IS FUNDAMENTALLY INTRINSIC, BUT IT IS MODULATED BY NERVES AND HORMONES

Intrinsic myogenic contraction forms the basic of gastric motility, and this fundamental motility occurs in the absence of any other influences. The resting membrane potential of stomach smooth muscle cells exhibits a gradient from about -48 mV in cells in the orad stomach to about -71 mV in cells from the antrum. The threshold for contractile activity in all of these cells is about -50 mV. Gastric peristalsis occurs primarily in the distal stomach and is associated with **gastric slow waves**, rhythmic depolarizations of the resting membrane potential, consisting of a rapid depolarization followed by a plateau phase (see Figure 8.2.2). Although all sites of the caudad stomach generate these membrane potential oscillations, cells along the greater curvature of the stomach show the highest frequency of slow waves, 3–5 $\min^{-1}$, and these entrain the rest of the stomach to their high frequency. The gastric slow waves are paced by **interstitial cells of Cajal (ICC)**, stellate

786 QUANTITATIVE HUMAN PHYSIOLOGY: AN INTRODUCTION

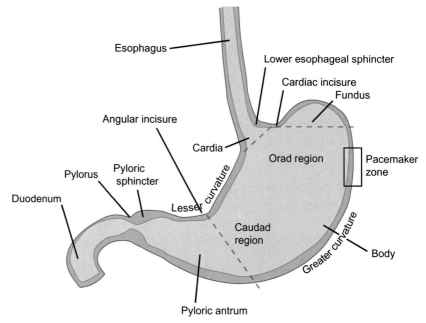

FIGURE 8.2.1 Structure of the stomach. The esophagus empties into the stomach at the cardia where there is a functional sphincter, the lower esophageal sphincter, also sometimes called the cardiac sphincter. The left margin of the esophagus makes a sharp angle with the stomach called the cardiac incisure. The **fundus** is the part of the stomach superior to this incisure. The **body** and the **antrum** make up the remainder of the stomach. The **pyloric sphincter** regulates the movement of material between the stomach and the duodenum. The stomach can also be divided into an orad and a caudad region on the basis of its motility.

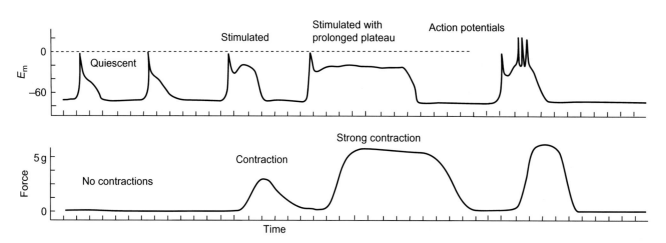

FIGURE 8.2.2 Electrical and mechanical activities of stomach smooth muscle cells. Rhythmic depolarization of the smooth muscle cell membrane potential, called slow waves, originates from most cells in the distal stomach. Cells in the pacemaker zone have a higher frequency and establish the **basic electrical rhythm** of the stomach. The cyclic depolarizations consist of a rapid upstroke and a plateau phase. During the interdigestive period, the plateau is usually below threshold for contraction. Stimulation by parasympathetic nerves raises and prolongs the plateau so that contraction occurs because activator Ca^{2+} enters the cells during the plateau. Action potentials can be produced which cause even stronger contractions.

cells that lay along the greater curvature, receive inputs from the autonomic nerves or the enteric plexuses, and relay this to muscle cells. These interstitial cells of Cajal are accordingly called **pacemaker cells**, and they set the **basic electrical rhythm** for the stomach.

EXTRINSIC AND INTRINSIC NERVES CONTROL GASTRIC MOTILITY

Afferent extrinsic nerves, mainly traveling over the vagus and splanchnic nerves, convey sensory information to the central nervous system (CNS), and efferent signals control the smooth muscle. The intrinsic innervation consists of the myenteric plexus, sandwiched between the outer longitudinal and inner circular layers of smooth muscle, and a submucosal plexus.

The myenteric plexus is most prominent of these two systems of intrinsic innervation. Parasympathetic extrinsic innervation originates from the **vagus** and sympathetic from the **celiac ganglion**. The vagus nerve primarily innervates cells in the myenteric plexus. Preganglionic parasympathetic fibers release acetylcholine to activate nicotinic receptors in the enteric ganglia, which excites gastric motility. Postganglionic sympathetic fibers release norepinephrine which inhibits excitatory neurotransmission within the enteric ganglia. Some vagal efferents activate inhibitory myenteric neurons that release vasoactive intestinal peptide (VIP) or nitric oxide (NO). These neurons are important in **receptive relaxation**, discussed below. The net effect of vagotomy is an increased gastric tone, suggesting that normal vagal activity provides a net inhibition of gastric tone.

A NUMBER OF HORMONES INFLUENCE GASTRIC MOTILITY

Gastrin, ghrelin, and motilin all stimulate gastric motility and decrease the time of emptying of the stomach. These are all hormones that originate from gastrointestinal sources in response to gastrointestinal signals, and details of their origin, secretion, and action will be addressed later. All are protein hormones. Gastrin and ghrelin originate mainly from the stomach, and motilin is believed to initiate the fasting pattern of motility. Cholecystokinin, glucagon, glucagon-like peptide 1 (GLP-1), peptide YY, and somatostatin all decrease stomach motility and delay gastric emptying. These protein hormones originate from the small intestine.

THE ORAD STOMACH RELAXES TO ACCOMMODATE LARGE MEALS

One of the main functions of the stomach is to act as a reservoir of ingested food that can be released to the intestines gradually so as to spread absorption of nutrients over a longer period of time. To do this, the stomach must stretch to accommodate meals. The stomach can hold about 1.5 L of material. Because the fundus maintains a constant tension, or **tone**, the stomach muscles must relax in order for the stomach to expand to receive more food. This gastric accommodation has two parts: **receptive relaxation** and **adaptive relaxation**. Receptive relaxation is the reduction in gastric tone as part of the swallowing program. Receptive relaxation occurs with a dry swallow or with mechanical stimulation of the pharynx or esophagus, and does not require movement of the bolus of food into the stomach. The vagus nerve mediates receptive relaxation by releasing vasoactive intestinal polypeptide (VIP) or nitric oxide (NO). On the other hand, adaptive relaxation begins with activation of stretch receptors upon distension of the stomach. The afferents travel along the vagus nerve and initiate a reflex in which efferent signals return over the vagus nerve to relax the stomach. Because both afferent and efferent pathways travel over the vagus nerve, this is called a **vagovagal reflex**.

AFTER A MEAL, GASTRIC CONTRACTIONS RESULT IN *PROPULSION*, *GRINDING*, OR *RETROPULSION* OF STOMACH CONTENTS

The slow waves paced by ICC propagate along the smooth muscle layers through electrical connections between smooth muscle cells. The slow waves travel faster circumferentially than longitudinally, so that the resulting contraction travels as a ring around the stomach from the body towards the antrum. In the interdigestive period, the slow waves are smaller and the plateau phase is below threshold. After a meal (the **postprandial** period), stimulation by the parasympathetic nervous system increases the amplitude and duration of the plateau phase and induces action potentials. These electrical events cause more intense contractions. Sympathetic stimulation reduces the amplitude and duration of the plateau phase, thereby decreasing the frequency of action potentials and the strength of contractions.

Cinefluoroscopic studies (X-ray movies) of gastric motility reveal a small ring of contraction that begins in the body of the stomach and propagates toward the antrum. This minor ring of contraction reaches the pylorus and causes it to contract to retain stomach contents. A few seconds after the minor contractile ring begins, a second contractile ring follows it. This more forceful ring propels food toward the pylorus and grinds it against itself. The increased pressure within the antrum causes the stomach contents to squirt back into the body (see Figure 8.2.3). This **retropulsion** mixes the

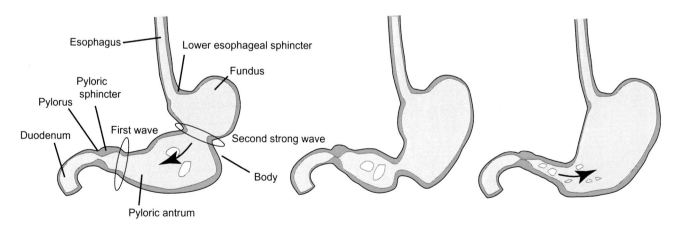

FIGURE 8.2.3 Contractions of the stomach that propel the stomach contents toward the antrum, grind it, and cause retropulsion back into the body of the stomach. These contractions break up the food into a fine dispersion called chyme. The first weak wave causes the pyloric sphincter to close. The second, stronger wave propels food into the antrum where it is ground and then dispersed back into the body of the stomach through the narrow orifice produced by the encircling surfaces of the stomach walls.

stomach contents and helps reduce particle size. This motility of the stomach produces **chyme**, a finely dispersed mixture of food and stomach secretions.

STOMACH AND DUODENAL CONTENTS REGULATE STOMACH EMPTYING

After a meal, the load of food and gastric secretions must be transferred into the small intestine where it can be further digested and absorbed. However, homeostasis of the nutrient content of blood is best served when there is a continuous supply of nutrients. This is achieved by slowing the passage of food so that nutrients are absorbed over a period of hours. Typically the stomach takes about 3 hours to empty. Usually the stomach contains a mixture of liquids, digestible solids, fats, and indigestible solids. The particle size, the amount, and the proportion of each of these constituents influence the rate of gastric emptying. Figure 8.2.4 shows that inert liquids such as water or isotonic saline empty quickest. The rate of gastric emptying of these liquids obeys first-order kinetics:

$$[8.2.1] \quad \frac{dV}{dt} = -kV$$

where V is the volume in the stomach and dV/dt is its rate of removal from the stomach to the duodenum. First-order kinetics describe processes that are proportional to the first power of the starting material. The negative sign in the equation indicates that the volume in the stomach decreases with time. The rate is generally described in terms of its half-life or the time required for one-half of the material to disappear. Integration of Eqn [8.2.1] gives

$$[8.2.2] \quad V(t) = V_0 e^{-kt}$$

where V_0 is the starting volume at $t = 0$. The half-time of exponential decay occurs when the volume is reduced to one-half of its starting value: $V(t_{1/2}) = \tfrac{1}{2}V_0$. Inserting this into Eqn [8.2.2], we find that the half-life is related to k, the rate constant, by

$$[8.2.3] \quad t_{1/2} = \frac{\ln 2}{k} = \frac{0.693}{k}$$

The half-life for gastric emptying of water is about 8–18 minutes.

Gastric emptying of digestible solids is quite different from emptying of liquids: it shows a lag phase of up to an hour in which particle size is continuously reduced. During that time there is little gastric emptying. Following the lag, stomach contents empty with zero-order kinetics, being independent of the volume remaining in the stomach (see Figure 8.2.4). Consumption of homogenized food in which particle size is less than 0.25 mm abolishes the lag time, suggesting that it is entirely due to inhibition of gastric emptying by large particles that are sensed by mechanoreceptors in the stomach itself. During the processing of a meal, particles that pass into the duodenum are usually less than 0.5 mm.

A variety of stimuli in the duodenum produces reflex inhibition of gastric emptying. These include the following:

- distension of the duodenum;
- irritation of the duodenum;
- acid pH in the duodenum;
- increased osmolarity of the duodenal contents;
- amino acids in the duodenum;
- fat digestion products (e.g., fatty acids, monoglycerides, and diglycerides) in the duodenum.

These factors slow gastric emptying to the rate at which the intestine can process the incoming load of nutrients. The inhibition of gastric emptying by the intestine is called the **enterogastric inhibitory reflex**. This reflex seems to present a constant rate of caloric load to the intestine: fatty substances possess the highest caloric density and inhibit gastric emptying the most.

NERVES AND GI HORMONES ALTER STOMACH EMPTYING

The vagus nerve is essential to normal gastric emptying because vagotomy causes rapid delivery of liquids regardless of content, probably related to the failure of the stomach to accommodate incoming material. The slowing of gastric emptying by particle size is a function of the stomach alone.

Fat digestion products in the duodenum cause the secretion of **cholecystokinin**, or **CCK**, from the duodenum. CCK is a polypeptide hormone secreted from endocrine I cells scattered throughout the proximal two-thirds of the small intestine. In the blood, CCK has a variety of forms that differ in their number of amino acids. CCK-83 has 83 amino acids, CCK-58 has 58, and so on for CCK-39, CCK-33, and CCK-8. The smaller CCKs are formed by proteolytic cleavage of the larger ones. All of these are active because they all share the same N-terminal sequence of Asp—Tyr—Met—Gly—Trp—Met—Asp—Phe. CCK activates a vagovagal reflex that releases VIP and NO onto the stomach muscle cells, which inhibit the muscle

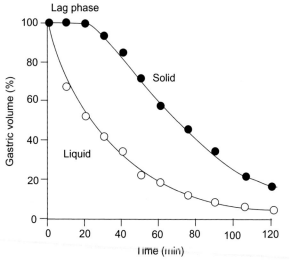

FIGURE 8.2.4 Gastric emptying of liquid versus solid materials. Solid foods exhibit a lag phase while particle size is being reduced. Liquids begin emptying immediately and show first-order kinetics.

and thereby slow gastric emptying. Slowing of gastric emptying by CCK makes physiological sense. It takes time to digest and absorb fats. Fat in the duodenum causes CCK release which slows gastric emptying, thereby providing time for the duodenum to process the fats that are delivered to it.

Glucagon, glucagon-like peptide-1 (GLP-1), peptide YY, and somatostatin also delay gastric emptying. Enteroendocrine cells in the small intestine secrete GLP-1 and peptide YY in response to luminal nutrients. GLP-1 is the C-amidated 7-36 amino acid peptide derived from proteolysis of proglucagon. GLP-1 is an incretin, meaning that it increases the secretion of insulin by the β cells of the islets of Langerhans in the pancreas, but it also markedly decreases gastric emptying. This has the effect of delaying increases in plasma glucose that typically follows a meal. It is believed that GLP-1 decreases stomach emptying by activating vagal afferents.

Enteroendocrine cells, D-cells, in the stomach secrete **somatostatin** that mainly inhibits gastric, biliary, and pancreatic secretions. It has two forms: SST-14 and SST-28 for the number of amino acids in the polypeptide. Some cells in the CNS also use somatostatin as a neurotransmitter. It decreases gastric emptying, which reduces the acid input into the duodenum. SST also has a variety of other effects, including inhibition of the release of a number of protein hormones.

Gastrin, ghrelin, and motilin all increase the rate of gastric emptying. All three are gastrointestinal protein hormones. Gastrin is secreted by G-cells in the pyloric glands in the antrum of the stomach and in the duodenum. It circulates in two forms: Gastrin-17 and Gastrin-34, referring to the number of amino acids in the peptides. The circulating half-life of Gastrin-17 is just 7 minutes, whereas gastrin-34 has a half-life of 30 minutes. Both forms are equally potent in stimulating acid secretion by the parietal cells. Gastrin has structural similarities to CCK and, in fact, exerts its effects through the CCK-2 receptor.

Ghrelin is a 28 amino acid peptide secreted by the oxyntic glands of the stomach. It is made as preproghrelin, which is then cleaved to form ghrelin and **obestatin**. Ghrelin is highly unusual in that it is octanoylated at serine 3 by the enzyme **GOAT (ghrelin O-acyl transferase)**. Ghrelin has a multitude of effects: it stimulates growth hormone release, increases appetite and feeding behavior, increases gastric emptying and gastric and intestinal motility.

M-cells in the duodenum and jejunum secrete motilin, a 22 amino acid peptide that has structural similarities to ghrelin. Motilin is temporally associated with phase III of the interdigestive pattern of motility, the migratory myoelectric complex, and is believed to initiate the MMC in humans.

THE *MIGRATING MOTILITY COMPLEX* CLEARS THE STOMACH AND INTESTINE DURING FASTING

The **migrating motility complex (MMC)**, sometimes also called the **migrating motor complex** or **migrating myoelectric complex**, and abbreviated as **MMC**, describes the pattern of stomach motility during the interdigestive period. This interdigestive motility is quite different from the motility after eating a meal. The MMC consists of three phases that empty the stomach of undigested material. The frequency of MMC varies with the time of day, being slowest in sleep ($0.25\ h^{-1}$) and more active during the day ($0.64\ h^{-1}$). After eating, the fed pattern of motility that we have already discussed replaces the MMC. The MMC is not a property of the stomach alone. It describes the interdigestive motility pattern of the whole GI tract. This pattern is said to perform a "housekeeping" function for the GI tract because it cleans it of all materials.

The MMC consists of three phases that together last about 90 minutes. In phase I, the stomach is quiescent for about 50 minutes, some 40–60% of the cycle length. Phase II, lasting another 25–30 minutes, is characterized by increasingly strong but irregular contractions. Phase III lasts only 5–10 minutes, but it includes intense contractions that completely occlude the stomach lumen, sweeping before it all undigested particles that still remain in the stomach.

Motilin is a polypeptide 22 amino acids long that is released in pulses every 90 minutes or so during fasting by special M-cells in the duodenum. Spikes of motilin concentration in the blood temporally correlate with the MMC. Myenteric neurons in the stomach possess receptors for motilin, and **atropine** (an antagonist of acetylcholine's action on muscarinic receptors) blocks motilin's effects. This suggests that motilin may activate the stomach myenteric plexus which then induces the MMC through cholinergic pathways.

Clinical Applications: Gastroparesis

Gastroparesis literally means "stomach paralysis." It refers to a condition of abnormal stomach motility, generally the lack of the normal postprandial motility pattern and delayed gastric emptying. Complete stomach paralysis is fatal in that it is equivalent to a gastrointestinal blockage. Generally gastroparesis results in failure to produce finely divided chyme or failure to propel the food into the small intestine properly.

The cause of gastroparesis often cannot be identified, and is called idiopathic, meaning having no identified cause. Common causes are diabetes, infections, hypothyroidism, scleroderma, autoimmune diseases, psychological disorders, radiation treatment in the area, and surgery to the area. Surgery or other damage to the vagus nerve interferes with stomach emptying, and the effect can occur either immediately or years after the surgery. Intestinal blockage can mimic the symptoms of gastroparesis.

Symptoms of gastroparesis included bloating, eructation (burping or belching) early satiety, heartburn, and epigastric pain. Symptoms can depend on the type of food consumed: solid foods, foods high in fat, and carbonated drinks can all trigger symptoms. Nausea and vomiting are also common. Vomiting can occur hours after eating the last meal. Diagnosis can be accomplished by endoscopy (visual

(Continued)

Clinical Applications: Gastroparesis (Continued)

inspection of the stomach and intestine via an inserted tube), gastric manometry, scintigraphic analysis of gastric emptying and gastric accommodation (using Tc^{99m}), x-rays and undigestible wireless capsule monitoring using the Smartpill that records pH, temperature and pressure along the GI tract.

There is no cure for gastroparesis. The main approach is to alter the diet to reduce the necessity of the stomach to grind the food. Fatty foods and fibrous foods take longer to empty and should be avoided. Fibrous foods can be cooked longer to soften them. Poorly digested food can collect into a mass called a **bezoar** that can form a blockage, causing nausea and pain. Breaking up bezoars may require endoscopic tools. Even with significant impairment of gastric emptying, thick and thin liquids such as puddings and smoothies can be tolerated well. Severe gastroparesis may require insertion of a feeding tube past the stomach into the jejunum.

Additional treatments include attempts to normalize motility either medically or electrically. Erythromycin, for example, stimulates motilin receptors and can increase gastric emptying, but often patients develop a tolerance to the drug. Emerging treatment options include gastric pacing or gastric electrical stimulation (GES). Gastric pacing uses a set of pacing wires attached to the stomach and an external power source that delivers high energy excitation at a frequency of about 3 per minute. GES uses an implantable device to deliver low energy stimulation at a frequency of about 12 per minute. Newer designs envision series of electrodes that could provide sequential stimulation of the muscle so as to produce propulsive contractions. These GES devices remain experimental.

THE STOMACH SECRETES HCl, PEPSINOGEN, MUCUS, GASTRIC LIPASE, AND INTRINSIC FACTOR

The lining of the stomach at rest is thrown into thick, velvety folds called **rugae**. These contain microscopic invaginations, called **gastric pits**, that each open into four or five gastric glands. Gastric glands come in two varieties. The **oxyntic glands** (see Figure 8.2.5) in the orad stomach make up 75% of the total number of glands. The remaining 25% are **pyloric glands** in the antrum and pylorus (see Figure 8.2.6). Specialized **parietal cells** secrete **HCl** and **intrinsic factor**. The HCl activates the proteolytic enzyme, **pepsinogen**, to form **pepsin**, and sets the acid environment for peptic proteolysis of dietary protein. The acid environment is also bacteriostatic but not bacteriocidal—the stomach has a thriving microbiota, but the acid prevents its overgrowth. Intrinsic factor is a 45-kDa glycoprotein that binds **vitamin B_{12} (cobalamin)**. Intrinsic factor is necessary for absorption of vitamin B_{12} in the terminal ileum by receptor-mediated endocytosis. **Intrinsic factor is the only essential secretion of the stomach**, and human life is not sustainable without it. Persons with total gastrectomies (removal of the stomach) must inject themselves with vitamin B_{12}.

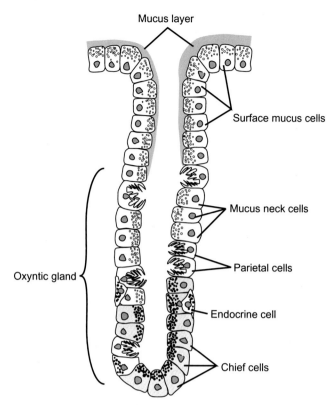

FIGURE 8.2.5 Structure of the oxyntic gland. Gastric pits open into a channel leading into a cluster of oxyntic glands, one of which is shown here. The surface mucus cells secrete mucus to line the stomach and protect it from its acid environment. The mucus contains mucin and HCO_3^- to neutralize stomach acid. The acid is secreted by parietal cells, whose appearance changes drastically from the quiescent to the stimulated states. These cells also secrete intrinsic factor, which is necessary for the absorption of vitamin B_{12} in the terminal ileum. Mucus is also secreted by cells in the neck of the gland. Pepsinogen is a heterogeneous mixture of isozymes that are secreted by the chief cells. Endocrine cells of various types lie interspersed among the secretory cells.

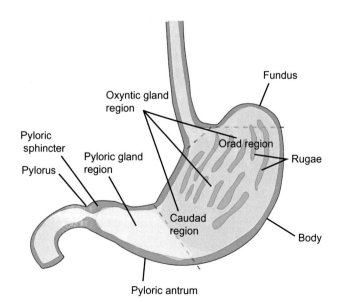

FIGURE 8.2.6 Location of the pyloric glands and oxyntic glands in the stomach. The oxyntic glands, comprising 75% of the gastric glands, secrete HCl, intrinsic factor, mucus, and pepsinogen. The pyloric glands cover the gastric antrum and pylorus and contain G-cells that secrete **gastrin** into the circulation. The **rugae** are folds in the stomach lining.

Surface epithelial cells, specialized mucus cells of the neck, and mucus cells in the glands also secrete mucin, a high molecular weight glycoprotein. The mucin monomers are cross-linked by disulfide bonds to form a hydrated gel that lines the stomach. It protects the epithelium from the corrosive effects of the acid and from the physical abrasion of stomach contents, and it lubricates stomach contents. The mucus gel traps HCO_3^- that neutralizes stomach acid, so there is a gradient in pH from acidic in the lumen to near neutral pH adjacent to the cells that are covered with mucus.

Chief cells synthesize **pepsinogen** as a 43-kDa inactive precursor and package it into granules that are stored in the apex of the cells. Stimulation of the cells in response to food in the stomach causes exocytosis of the granule. Contact of pepsinogen with acid converts it to its active form, pepsin, with a molecular weight of 35 kDa. Pepsinogen actually consists of seven different isozymes. Although pepsin is a potent protease, it is unnecessary for the complete digestion of dietary protein because pancreatic enzymes can handle the job in the absence of pepsin.

ACID SECRETION IS REGULATED IN FOUR PHASES: THE *BASAL PHASE*, *CEPHALIC PHASE*, *GASTRIC PHASE*, AND *INTESTINAL PHASE*

THE CEPHALIC PHASE IS MEDIATED BY THE VAGUS NERVE THROUGH ACETYLCHOLINE AND GASTRIN

The cephalic phase of gastric secretion is mediated entirely through the vagus nerve. A variety of sensory stimuli including the sight, smell, and taste of food elicits acid secretion in the stomach. It contributes about 30–50% to the total postprandial acid production. The vagus nerve is the sole neural link between the brain's higher functions and gastric secretion. It exerts its effects through two separate pathways: direct stimulation by acetylcholine and indirect through gastrin.

Parietal cells possess M3 cholinergic receptors that turn on acid secretion. Stimulation of the vagus excites postganglionic parasympathetic neurons in the stomach, which then release acetylcholine onto parietal cells to stimulate acid secretion.

Cephalic efferents in the vagus nerve release gastrin-releasing peptide (**GRP**) onto **G-cells** in the pyloric glands and these release **gastrin**. Atropine, a muscarinic cholinergic antagonist, blocks the direct effects of vagal stimulation on parietal cells, but it does not block the release of gastrin. Figure 8.2.7 illustrates the cephalic phase of gastric acid secretion.

Gastrin is a polypeptide hormone secreted by G-cells in pyloric glands in the antrum of the stomach and in the duodenum. It circulates in two forms: Gastrin-17 and Gastrin-34, referring to the number of amino acids in their sequences. The circulating half-life of Gastrin-17 is just 7 minutes, whereas Gastrin-34 has a half-life of 30 minutes. Both forms are equally potent in stimulating acid secretion by the parietal cells. Gastrin has structural similarities to CCK and, in fact, exerts its effects through the CCK-2 receptor.

Gastrin released from the G-cells during the cephalic phase stimulates parietal cells directly and indirectly through **histamine** released from enterochromaffin-like cells (**ECL cells**) in the lamina propria nearby the parietal cells. Thus histamine is a **paracrine** hormone that affects nearby cells rather than traveling through the blood. Mast cells in the stomach also release histamine, but they are not closely apposed to the parietal cells. The histamine diffuses through the extracellular fluid to the parietal cells where it binds to H_2 receptors. These receptors are blocked by **cimetidine**.

STRETCH RECEPTORS AND CHEMORECEPTORS IN THE STOMACH REGULATE H^+ SECRETION IN THE GASTRIC PHASE

The gastric phase accounts for about 60% of the secretory response to a meal. Stretch receptors in the walls of the stomach increase their firing rate upon distension of the stomach following a meal. Afferent stretch receptor sensory information travels along the vagus nerve to the CNS where efferent vagal connections begin. These complete a **long reflex arc** that increases parietal cell HCl secretion. This effect of stretching the fundus and body of the stomach is nearly abolished by proximal vagotomy. Stretch of the antrum also elicits a long vago-vagal reflex that stimulates parietal cell acid secretion. Stretch of the antrum, in addition, releases gastrin that stimulates acid secretion (see Figure 8.2.8).

Antral pH below 3.0 inhibits gastrin release evoked by antral distension, by nutrients in the stomach, or by cephalic stimulation. This forms a negative feedback loop: gastrin stimulates acid secretion which in turn shuts off gastrin secretion. The effect is mediated by acid stimulation of D-cells, enteroendocrine cells that secrete **somatostatin**, which mainly inhibits gastric, biliary, and pancreatic secretion. It has two forms: SST-14 and SST-28 for the number of amino acids in the polypeptide.

Amino acids produced by the digestion of dietary protein potently stimulate acid secretion by increasing gastrin release. Intact proteins are poor secretagogues, whereas proteolytic digests of these proteins are effective. Phenylalanine and tryptophan have the greatest ability to release gastrin. This effect is blocked by stomach pH below 3.0, but it is not blocked by atropine or vagotomy, suggesting that the amino acids directly affect the G-cells.

Caffeine, **alcohol**, and **calcium** also stimulate gastric acid secretion. Pure ethanol only modestly stimulates acid secretion, whereas red wine and beer potently increase gastrin release and acid secretion. This suggests that amines or amino acids in these drinks are the actual secretagogues and that the effect is mediated by gastrin. Caffeine stimulates acid secretion, but so does decaffeinated coffee. Oral calcium also stimulates gastrin release and acid secretion. The effect is independent of its buffering action and is probably due to direct effects of the Ca^{2+} ion.

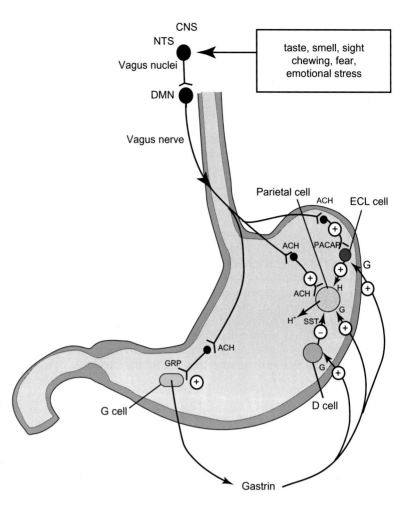

FIGURE 8.2.7 The cephalic phase of gastric acid secretion. Sensory stimuli such as the sight, smell, and taste of food produce efferent activity from the brain (the nucleus tractus solitarius and dorsal motor nucleus of the vagus) to stimulate parietal cells to secrete acid. The mediator in this pathway is acetylcholine (ACh), which exerts its effects on the parietal cells through a muscarinic (M3) cholinergic receptor. In a second pathway, vagal stimulation releases gastrin from G-cells in the antrum, which then stimulates acid secretion. Gastrin stimulation of the parietal cells is both direct and indirect. In the direct pathway, parietal cells respond to gastrin through gastrin receptors on their membranes. In the indirect pathway, gastrin stimulates histamine (H) release from **enterochromaffin-like (ECL) cells**. Vagal stimulation also relieves inhibition of acid secretion by **somatostatin, SST**, secreted by D-cells in the body and antrum. Parietal cells thus have receptors for acetylcholine (ACh), gastrin (G), histamine (H), and somatostatin (SST).

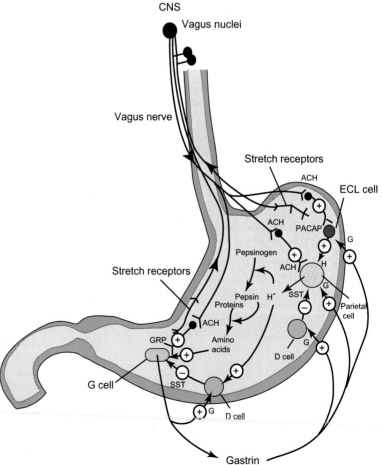

FIGURE 8.2.8 The gastric phase of acid secretion. Acid secretion by the parietal cell is stimulated by acetylcholine, gastrin (G), and histamine (H). Acetylcholine (ACh) is released by vagal efferents in response to stretch receptors located in the fundus. These exert their effects over a vagovagal reflex. Enterochromaffin-like cells (ECL) release histamine in response to gastrin levels, and gastrin directly stimulates parietal cell acid secretion. Gastrin is secreted by G-cells in the antrum. Gastrin release is stimulated by stretch of the antrum, mediated by a long vagovagal reflex. Amino acids in the stomach lumen also stimulate gastrin release. Gastrin release is feedback inhibited by acid through stimulation of D-cells that secrete somatostatin (SST). Somatostatin release is stimulated by stomach acid and by gastrin, which forms a short negative feedback loop.

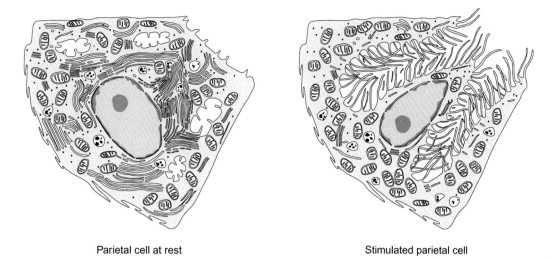

Parietal cell at rest   Stimulated parietal cell

FIGURE 8.2.9 Resting and stimulated parietal cells. The resting cell contains a large number of tubulovesicles that contain the H^+,K^+-ATPase pump protein. When the cell is stimulated, the tubulovesicles fuse with the plasma membrane, forming a plasma membrane with a well-developed system of microvilli from which the cell secretes HCl.

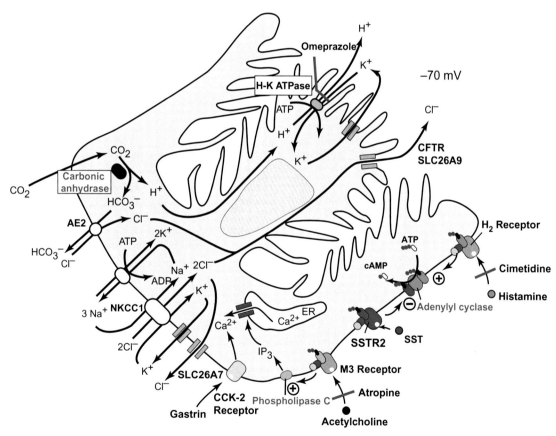

FIGURE 8.2.10 The mechanism of HCl secretion by parietal cells. M3 receptors bind acetylcholine that is released from vagus efferent nerves. These M3 receptors are coupled to a G_q protein whose α subunit activates phospholipase C that releases IP3 from phosphatidylinositol bisphosphate. The IP3 releases Ca^{2+} from intracellular stores. This effect of acetylcholine is blocked by **atropine**. H_2 receptors bind histamine released by ECL (enterochromaffin-like cells) near the parietal cell in response to circulating gastrin. These H_2 receptors are blocked by **cimetidine**. Histamine binding to H_2 receptors activates adenylyl cyclase and formation of cAMP and activation of protein kinase A, PKA. The signaling mechanism of gastrin is not established, but it appears to involve CCK-2 receptors that increase cytosolic Ca^{2+} but by a mechanism distinct from acetylcholine. SST binds to a receptor that activates a G_i mechanism. Activation of these pathways results in a variety of effects, including fusion of tubulovesicles in the cytosol to the apical membrane, which results in insertion of the H^+,K^+-ATPase, K^+ channel, and Cl channel into the apical cell membrane. All three of these proteins are required for HCl secretion. Inhibition of the H^+,K^+-ATPase by proton pump inhibitors, such as **omeprazole**, inhibits acid secretion. The basolateral membrane contains a HCO_3^-–Cl^- exchanger (anion exchanger 2 (AE2) = SLC4A2) that helps supply Cl^- for apical secretion. The HCO_3^- that is exchanged for Cl^- is produced by the hydration of CO_2 catalyzed by **carbonic anhydrase**. The basolateral membrane also contains the Na^+,K^+-ATPase, which transports three Na^+ out of the cell and two K^+ ions into the cell for each ATP that is hydrolyzed. A Na^+–H^+ exchanger is also present on the basolateral membrane. Secretion of acid into the stomach is thus accompanied by secretion of base HCO_3^- into the blood. The parietal cell acidifies the stomach contents while it alkalinizes the blood.

Clinical Applications: Peptic Ulcers

Gastric and duodenal ulcers form when the protective linings of the stomach or duodenum are destroyed by acid and pepsin, causing **peptic ulcers**. These are an erosion of the lining of the stomach that can penetrate to the blood vessels, causing a bleeding ulcer. They can penetrate all the way through the wall, causing a perforated ulcer in which stomach contents leak into the peritoneal cavity.

Normally, peptic ulcers are prevented by the combination of the protective layer of mucus, secretion of acid-neutralizing HCO_3^- into the mucus, and replacement of epithelial cells by new cells produced from stem cells within the epithelium. Ulcers form when these protective measures fail, either because of overproduction of acid or weakened protective barriers. It was once believed that excess acid production was the primary cause of peptic ulcers. Now it is accepted that a bacterium, *Helicobacter pylori*, weakens the protective barrier and so contributes to the etiology of peptic ulcers. Some of the remaining cases are due to **NSAIDs: n**onsteroidal **a**nti-**i**nflammatory **d**rugs such as aspirin and a host of artificial drugs such as ibuprofen, indomethacin, and naproxen. These drugs inhibit an enzyme called COX-2 or **cyclooxygenase**. As a side effect, these NSAIDs reduce HCO_3^- secretion by the stomach mucosa.

H. pylori was first associated with gastritis (an inflammation of the stomach) in 1983 by Warren and Marshall. These bacteria are curved, gram-negative bacteria about 0.5 μm × 3 μm, with four to seven sheathed flagella at one pole of the cell. These bacteria survive in the inhospitable stomach environment because they possess large amounts of **urease**. This enzyme produces NH_4^+ from urea, and the ammonium neutralizes the acid in the local environment of the bacterium. Warren and Marshall earned the 2005 Nobel Prize in Physiology or Medicine.

About 80% of persons with duodenal ulcers test positive for *H. pylori*, and cure of their infection cures the ulcer. How the bacterium causes ulcers is not established. However, at least some people develop ulcers without *H. pylori*'s help. Initially the scientific community was incredulous at the prospect that peptic ulcers could be an infectious disease. The pendulum has swung to near universal acceptance of the central importance of *H. pylori*, and now the pendulum is swinging again to a more qualified view of its importance.

NUTRIENTS IN THE INTESTINE AFFECT STOMACH ACID SECRETION

In 1929, Feng and coworkers observed that infusion of fat into the duodenum inhibited gastric acid secretion by releasing an inhibitory hormone from the small intestinal mucosa. They named this hormone enterogastrone. It is likely that enterogastrone is not one hormone but a family of them, all of which feed back onto the stomach to limit its acid secretion. Candidate hormones include CCK, secretin, somatostatin, and neurotensin. Release of these materials and their control of gastric acid secretion constitute the **intestinal phase** of gastric secretion.

THE SURFACE MEMBRANE H^+,K^+-ATPASE, OR PROTON PUMP, ACTIVELY SECRETES HCL

At rest, parietal cells contain a number of small intracellular **canaliculi** (small canals) and a number of **tubulovesicles**. These are intracellular, membrane-bound structures that appear as long tubes within the cytoplasm. These tubulovesicles contain the H^+,K^+-ATPase pump. Stimulation of the parietal cell causes fusion of the tubulovesicles with the apical cell membrane. Upon its translocation to the apical membrane, the H^+,K^+-ATPase pump begins pumping H^+ ions into the lumen in exchange for K^+ ions. The changes in parietal cell structure upon stimulation of acid secretion are illustrated in Figure 8.2.9. How parietal cells secrete an acidic solution while maintaining a neutral cytosolic pH is not known. The scheme shown in Figure 8.2.10 is a reasonable hypothesis that incorporates the known ion transport mechanisms in the parietal cells.

SUMMARY

The main function of the stomach is to store food and to release it to the intestines at a rate whereby the intestines can process it. The stomach mixes the food and grinds it into a finely divided chyme that increases the surface area of the food in preparation for digestion. The stomach also secretes mucin, water, HCl, pepsinogen, and intrinsic factor. Pepsinogen begins protein digestion and intrinsic factor is necessary for the absorption of vitamin B_{12} (cobalamin) in the terminal ileum.

The lower esophageal sphincter marks the beginning of the stomach, and the pyloric sphincter marks its caudal end. The stomach consists of the fundus, the body, and the antrum. The stomach expands to accommodate a meal. Receptive relaxation refers to the relaxation of the stomach in response to swallowing, and it is part of the swallowing program. Stretch receptors in the stomach also relax it through a vagovagal reflex. Stomach motility is controlled locally through the myenteric plexus, which is influenced by vagal inputs from the parasympathetic nervous system and by sympathetic inputs from the celiac ganglion. ICCs in the greater curvature have the highest intrinsic frequency of slow waves, which are rhythmic membrane potential depolarizations, and therefore these cells are pacemakers for the stomach. These cells set the frequency of stomach motility. The stomach grinds the food to produce a finely dispersed mixture of food and secretions.

Sensors in the stomach and in the duodenum regulate stomach emptying. Large particles excite mechanosensors in the stomach that delay stomach emptying. Distension, irritation, acid, increased osmolarity, amino acids, and fat digestion products in the duodenum all inhibit gastric emptying. All of these regulate the rate of stomach emptying to be commensurate with intestinal processing. This is the enterogastric inhibitory reflex. It is caused by the vagus nerve and by a variety of GI hormones including CCK and secretin.

During fasting, motility shifts to the MMC (migrating motility complex, migrating myoelectric complex, migrating motor complex), which lasts about 90 minutes. It includes a phase of strong propulsive contractions that moves all luminal contents forward, clearing the gut. Increase in motilin, a 22 amino acid hormone produced in the duodenum, is temporally associated with the MMC.

Gastric glands in the stomach secrete mucin, acid, intrinsic factor, and pepsinogen. The parietal cells secrete acid and intrinsic factor. Chief cells secrete pepsinogen that is activated to pepsin by stomach acid.

Acid secretion requires an apical proton pump (H^+,K^+-ATPase), K^+ channel, and Cl^- channel. In the unstimulated state, the proton pump resides in tubulovesicles beneath the apical membrane of the parietal cells. Upon stimulation, the tubulovesicles fuse with the apical membrane and the proton pump begins pumping H^+ ions into the lumen. The parietal cell has receptors for gastrin, acetylcholine, histamine, and somatostatin. Acid secretion has cephalic, gastric, and intestinal phases. The cephalic phase is initiated by sensations associated with food intake: smell, sight, and taste, or even thinking about food. These stimulate acid secretion directly through the vagus nerve and release of acetylcholine, and indirectly through the vagus nerve by increasing gastrin secretion by G-cells in the antrum.

Vagal stimulation also causes histamine release from ECL cells in the vicinity of the parietal cells. Gastrin also activates the ECL cells. Somatostatin, SST, is released by D-cells and inhibits acid secretion. Stretch, acid, and protein digestion products regulate acid secretion in the gastric phase. Stretch of the antrum increases gastrin release through a long vagovagal reflex, and gastrin then stimulates acid secretion. Amino acids stimulate gastrin release. High stomach acidity activates D-cells in the antrum that release SST to inhibit gastrin release to stop acid secretion.

REVIEW QUESTIONS

1. What is gastric accommodation and how does it differ from receptive relaxation? What causes each?
2. Why do gastric contractions occur in two waves, one stronger than the other?
3. How does gastric motility differ after a meal and during the interdigestive period?
4. What is the enterogastric inhibitory reflex? What is its purpose?
5. What stimulates acid secretion? What inhibits it?
6. The Zollinger–Ellison syndrome often results from a carcinoma of the duodenum or pancreas that produces gastrin. What do you expect happens to the stomach in these cases?
7. What is the only essential secretion of the stomach?
8. What cells secrete HCl? What cells secrete pepsinogen? What cells secrete gastrin?

8.3 Intestinal and Colonic Chemoreception and Motility

Learning Objectives

- List the three sections of the small intestine and six anatomic parts of the large intestine
- List those variables that comprise the enterogastric inhibitory reflex
- List three components of the intrinsic innervation of the intestine
- Describe the anatomic location of the myenteric plexus
- Describe the anatomic location of the submucosal plexus
- Define enteroendocrine cell
- Describe two pathways for the secretion of GLP-1 as an example of an EEC
- List four secretions of EECs and what stimulates their secretion, and the effect of those secretions
- Describe the origin of sympathetic and parasympathetic innervations of the intestine
- Distinguish between the motility patterns of segmentation, peristalsis, and migrating motor complex
- Describe what is meant by ascending contraction and descending relaxation during peristalsis
- Describe each of the following gut reflexes: receptive relaxation, enterogastric inhibitory reflex, gastrocolic reflex, gastroileal reflex, ileal brake, and rectoanal inhibitory reflex
- Describe the events that occur during defecation
- List the various causes of vomiting
- Describe what is meant by "chemical trigger zone"

THE SMALL INTESTINE CONSISTS OF *DUODENUM, JEJUNUM,* AND *ILEUM*

The small intestine extends from the pylorus at the end of the stomach to the beginning of the colon. It is about 6 m long, but its length varies with the degree of longitudinal smooth muscle contraction. The first segment of the small intestine, the **duodenum**, derives its name from anatomists who defined its length as being "12 finger breadths," about 25 cm. This shortest and widest part of the small intestine extends from the stomach to the **ligament of Treitz**. The ligament of Treitz is not really a ligament, but it is a suspensory muscle that attaches the duodenum to the posterior abdominal wall and causes the acute angle the small intestine makes at the end of the duodenum. Pancreatic and biliary secretions enter the intestine in the duodenum.

By definition, the **jejunum** consists of the upper two-fifths (40%) of the small intestine from the ligament of Treitz to the first part of the colon, the **cecum**. The distal three-fifths (60%) comprises the **ileum**. Although there are distinct histological differences between the upper jejunum and lower ileum, the tissues gradually make the transition so there is no clear demarcation between the two.

INTRINSIC NERVES, EXTRINSIC NERVES, PARACRINE AND ENDOCRINE HORMONES REGULATE INTESTINAL AND COLONIC MOTILITY

If your intestinal motility stops, you die. This function is so important that there are multiple levels of control. In Chapter 8.1, we described that a swallowing center in the medulla directs peristalsis in the esophagus as part of the swallowing program, mediated by the vagus nerve. If that nerve is cut, intrinsic nerves take over to produce propulsive action. If these are blocked, the smooth muscles of the esophagus can direct similar motor patterns. This redundancy of levels of control produces a robust system, one that is resistant to failure, because failure carries a high price. Smooth muscle contraction constitutes the final step in intestinal motility. These smooth muscle cells are regulated by specialized cells, called **interstitial cells of Cajal (ICC)**, and also by intrinsic nerves within the intestinal wall. These are all modulated by extrinsic efferent nerves originating in the central nervous system, and both paracrine and endocrine secretions. The extrinsic nerves respond to sensory information that derives from afferent sensors in the intestine. Thus extrinsic nerves carry both sensory and motor information, with sensory afferents far outnumbering the motor efferents.

INTRINSIC INNERVATION OF THE INTESTINE CONSISTS OF THE *MYENTERIC PLEXUS, SUBMUCOSAL PLEXUS,* AND THE *INTERSTITIAL CELLS OF CAJAL*

THE INTERSTITIAL CELLS OF CAJAL GENERATE INTRINSIC INTESTINAL RHYTHM

The **interstitial cells of Cajal (ICCs)** were first described histologically by Ramon y Cajal nearly a century ago, but their function has only recently become

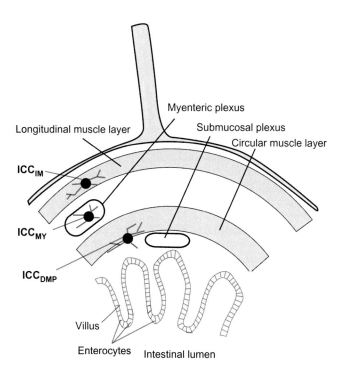

FIGURE 8.3.1 Schematic diagram of the location of interstitial cells of Cajal (ICCs) in a cross-section of small intestine. ICCs are stellate cells that may be found in the myenteric plexus (ICC$_{MY}$), in the muscle layers (ICC$_{IM}$), or in the inner surface of the circular muscle (ICC$_{DMP}$). These cells make connections to muscle cells and interneurons in the plexuses. The ICCs are shown in black.

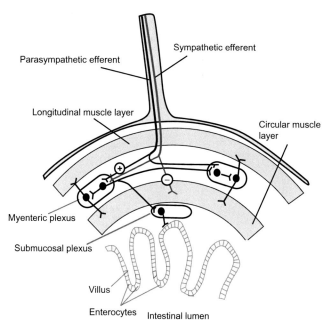

FIGURE 8.3.2 Schematic diagram of the myenteric plexus and submucosal plexus. These plexuses are collections of neurons and interstitial cells of Cajal located between the longitudinal and circular layers (myenteric plexus) or inside the circular layer (submucosal plexus). They receive innervation from the sympathetic and parasympathetic systems and project to muscle and secretory cells. They send connections up and down the intestine to connect with other ganglia.

known. These stellate cells are specialized smooth muscle cells that receive inputs from nerve cells within the myenteric and submucosal plexuses and also communicate electrically with smooth muscle cells in the longitudinal and circular muscle layers. There are several separate functional classes of ICC. A network of ICC$_{MY}$ cells lies within the myenteric plexus between the longitudinal and circular muscle layers. These cells set the rhythm and trigger slow waves. **The smooth muscle cells themselves do not originate slow waves**. A second network of ICCs, called intramuscular ICC or ICC$_{IM}$, is dispersed among the smooth muscle cells. In the small intestine, the ICCs are concentrated at the inner surface of the circular muscle layer to form ICC$_{DMP}$, denoting deep muscle plexus. Figure 8.3.1 highlights the ICCs.

THE MYENTERIC PLEXUS AND SUBMUCOSAL PLEXUS MAKE UP THE ENTERIC NERVOUS SYSTEM

As in most of the gastrointestinal tract, the small intestine contains an inner circular muscle layer that determines the caliber of the lumen and an outer longitudinal smooth muscle layer that sets the length. The myenteric plexus is a network of ganglia, connected by fiber tracts, that is sandwiched between these two layers. It receives sensory information from stretch receptors in the outer muscle layers and free sensory endings in the mucosa that respond to mechanical and chemical stimuli. The submucosal plexus is located between the circular smooth muscle and the muscularis mucosae. The function of the muscularis mucosa is poorly understood; it may help fold the epithelium into finger-like projections or villi (the singular form is villus). The enteric plexuses are highlighted in Figure 8.3.2.

INTRINSIC PRIMARY AFFERENT NEURONS REGULATE LOCAL RESPONSES

In response to local signals, enterochromaffin cells (EC cells) release 5-hydroxytryptamine (5-HT) and ATP to stimulate intrinsic primary afferent neurons (IPANs) in the submucosal plexus and myenteric plexus. These connect to interneurons that control secretion and motor activity of the mucosa and to interneurons within the myenteric plexus to control local contractile reflexes. The second component carries mechanosensory stimuli. Physiological information passes over the vagus nerve. Spinal afferents carry nociceptive (pain) information over the splanchnic nerves. The IPANs and sensory afferents are shown in Figure 8.3.3.

SOME ENTEROENDOCRINE CELLS ARE CHEMORECEPTORS THAT RESPOND TO A VARIETY OF STIMULI

Fifteen distinct types of enteroendocrine cells (EECs) lie amid the absorptive cells of the intestinal epithelium. These are either "open," meaning that they have

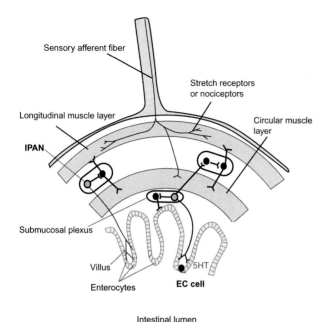

FIGURE 8.3.3 Intrinsic primary afferent neurons and sensory afferents. Enterochromaffin cells in the villus lining release 5-hydroxytryptamine or ATP onto terminals of intrinsic primary afferent neurons (light blue color) whose cell bodies lie in the submucosal and myenteric plexes. These control local contractile or secretory responses. Other sensory components sense either stretch or pain and are carried to the CNS over splanchnic nerves.

microvilli that extend into the lumen or "closed" because they do not have direct contact with luminal contents (see Figure 8.3.4). The EECs collectively sense a wide variety of materials in the intestinal lumen by various mechanisms including G-protein-coupled receptors (GPCR family) on the surface membrane of the EEC or through solute carriers (SLC family). The mechanisms for several luminal materials are identical to those found in the taste receptors of the tongue. Although these receptors do not contribute to the subjective perception of taste, in some sense the intestine "tastes" these chemicals to determine the nature of the materials being absorbed into the blood so as to regulate the gastrointestinal tract and adjust metabolism. The EEC and enterochromaffin cells release materials that act on afferent nerves or circulate in the blood to produce either local (paracrine) or systemic (endocrine) effects.

L-type enteroendocrine cells are heterogeneous. The mechanism of signal transduction for an L-type cell in the jejunum and ileum is shown in Figure 8.3.5. These cells release glucagon-like polypeptide-1 (GLP-1) in response to glucose in the lumen. There are two separate pathways by which these cells sense glucose as described in the figure legend. A summary of the enteroendocrine cells is given in Table 8.3.1.

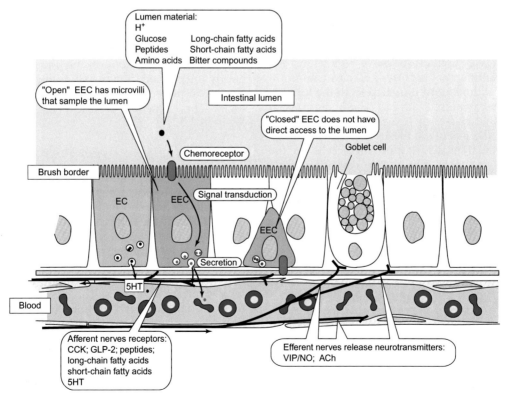

FIGURE 8.3.4 Chemoreception by enteroendocrine cells (EEC). EEC may be "open" if they have direct access to intestinal lumen contents, or "closed" if they do not. EEC cells come in many varieties. They sense materials in the lumen or blood and generally release polypeptide hormones in response to these materials. The materials that are actively sensed include H^+ (indirectly), glucose, peptides, amino acids, long-chain fatty acids, short-chain fatty acids, and bitter compounds. The chemoreceptors are coupled to signaling pathways that link chemoreception to secretion of polypeptide hormones. The released hormones include secretin, CCK, GLP-1 and GLP-2, PYY, SST. EECs are also in the stomach and release gastrin, SST, and ghrelin. These released polypeptides excite afferent nerves or travel in the circulation as hormones to have autocrine, paracrine, or endocrine effects. Efferent nerves can regulate secretion or blood flow. Enterochromaffin cells (EC) release 5-hydroxytryptamine (5-HT) that acts on nerves and circulates in the blood.

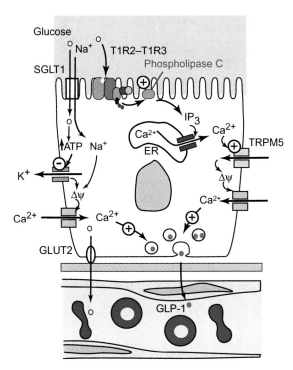

FIGURE 8.3.5 Mechanism of glucose sensing in enteroendocrine L cells to release GLP-1. There are two basic pathways proposed to explain glucose sensing, developed from localization of transport elements to these cells through knock-out models, immunological localization of proteins, and pharmacological effects on GLP-1 secretion. One pathway is identical to the taste receptors on the tongue, using T1R2–T1R3 that is coupled to the G protein, α gustducin, a G_q mechanism, to increase cytosolic $[Ca^{2+}]$ to increase secretion of GLP-1. The second uses glucose entry into the cell coupled to Na^+ entry over the Na–glucose cotransporter, SGLT1. Na^+ entry depolarizes the cell, leading to increased $[Ca^{2+}]$ and also increases cytosolic [ATP] that shuts down an ATP-sensitive K^+ channel that also leads to depolarization and increased influx of Ca^{2+}. The L cells in the ileum may differ from those in the colon. For example, glibencamide, a drug that inhibits the K_{ATP} channel, did not affect GLP-1 secretion from ileum explants, but did so from colonic explants.

EXTRINSIC INNERVATION OF THE GUT ARISES FROM PARASYMPATHETIC AND SYMPATHETIC NERVES

POSTGANGLIONIC SYMPATHETIC NERVES ORIGINATE FROM THE CELIAC GANGLION, SUPERIOR MESENTERIC GANGLION, AND INFERIOR MESENTERIC GANGLION

Efferent preganglionic sympathetic neurons project from the spinal cord to the prevertebral ganglia, where they form nicotinic cholinergic synapses. The long postganglionic sympathetic fibers project to the myenteric plexus, where they release norepinephrine primarily onto $α_2$ receptors that inhibit excitatory cholinergic neurotransmission. Some sympathetic fibers directly innervate the sphincters, causing contraction, consistent with the inhibitory role of the sympathetic nervous system on GI function. Other sympathetic fibers make connections directly on smooth muscle cells to inhibit them. These smooth muscle cells contain $β_2$ receptors that increase cAMP in the muscle cells that relaxes the smooth muscle (see Chapter 3.8). The extrinsic innervation is highlighted in Figure 8.3.6.

THE VAGUS SUPPLIES PARASYMPATHETIC INNERVATION TO THE GUT

Parasympathetic preganglionic cholinergic fibers excite excitatory postganglionic neurons, which use a variety of neurotransmitters including **substance P, neurokinin A**, and **acetylcholine** as a neurotransmitter. The parasympathetic preganglionic fibers also excite inhibitory postganglionic neurons which use **vasoactive intestinal polypeptide (VIP)** and **nitric oxide (NO)** as neurotransmitters, among others.

AFFERENT FIBERS TRAVEL OVER SPLANCHNIC NERVES AND THE VAGUS

The vagus nerve contains both afferent and efferent fibers, with the afferent fibers outnumbering the efferents by a factor of 10. Vagus afferents travel to the nucleus solitarius in the brain stem and from there to the dorsal motor nucleus of the vagus and the nucleus ambiguus. These carry physiological information. In addition, sensory information outnumbers efferents 3:1 in the splanchnic nerves. Afferent information passes by way of the splanchnic nerves to the dorsal horn of the spinal cord. The afferent neurons have cell bodies in the dorsal root ganglia. Second-order neurons in the cord project to the brain stem and cerebral cortex. Most nociceptive information travels over the splanchnic nerves.

THE ENTERIC NERVOUS SYSTEM CAN FUNCTION AUTONOMOUSLY

The enteric nervous system contains some 10^8 neurons, whereas the vagus nerve has only about 2000 efferent fibers. There are as many neurons in the enteric nervous system as there are in the spinal cord. Although intestinal motility is modulated by central nervous system efferents, it contains all the elements necessary for reflex activity including sensory neurons, interneurons, and motor neurons and can regulate GI function autonomously using local sensory information.

SLOW WAVE ACTIVITY FORMS THE BASIS OF INTESTINAL SMOOTH MUSCLE CONTRACTION

Intestinal smooth muscle cells have resting membrane potentials between -40 and -80 mV. These membrane potentials rhythmically oscillate, a phenomenon called slow waves. The slow waves consist of a rapid depolarization followed by a partial repolarization and then a prolonged plateau phase of depolarization, ending with complete repolarization (see Figure 8.2.2). In the stomach the frequency of the slow waves is set by the pacemaker cells at about $3-5$ min^{-1}. In the duodenum, the **basic electrical rhythm** is about $10-12$ min^{-1}. The frequency decreases from $10-12$ min^{-1} in the duodenum to $7-8$ min^{-1} in the distal ileum. The basic electrical rhythm is set by the ICCs.

TABLE 8.3.1 Enteroendocrine Cells (EEC) in the Gastrointestinal Tract

EEC	Location	Secretion	Stimulus	Receptors	Main Function
G-cell	Antrum	Gastrin	GRP + peptides + amino acids + SST−	CaSR, GPRC6A, SSTR	Increases acid secretion by parietal cells
P-D1-cell	Fundus	Ghrelin obestatin	Sugars bitter	T1R3 T2R	Hunger control and release of GH
D-cell	Stomach antrum (open) stomach body (closed)	SST SST	H+, gastrin, CCK		Inhibits gastrin secretion Inhibits acid secretion
S-cell	Duodenum	Secretin	SRP		Stimulates pancreatic secretion Delays gastric emptying
I-cell	Duodenum	CCK	Fat products + peptides + amino acids	GPR120; FFAR1 GPR92 CaSR; T1R1-T1R3	Stimulate vagus nerves to increase pancreatic zymogen secretion and gallbladder contraction
K-cell	Duodenum−jejunum	GIP	Sugars LCFA amino acids	SGLT1 GPR120; FFAR1	
L-cell	Proximal s.i.	GLP-1	Sugars LCFA AA bitter	T1R2-T1R3; SGLT1 GPR120; FFAR1 GPRC6A T2R	Delays gastric emptying; increases insulin secretion
	distal s.i., colon	PYY, GLP-1	SCFA	FFAR2; FFAR3	Inhibits gastric and intestinal motility Inhibits gastric and pancreatic secretion
M-cell	Small intestine	Motilin			Initiates phase III of MMC
N-cell	Ileum	Neurotensin	Fatty acids		Inhibits gastric secretion and delays emptying; stimulates pancreatic and intestinal secretion

GRP = gastrin releasing peptide; SST = somatostatin; CaSR = extracellular Ca^{2+}-sensing receptor; GPR = G-protein-coupled receptor, many of which have trivial names (e.g., FFAR1 = GPR40); T1R1, T1R3, T2R = taste receptors; T1R1–T1R3 combination detects "sweet"; various T2Rs detect "bitter"; CCK = cholecystokinin; SRP = secretin releasing peptide; GIP = glucose-dependent insulinotropic polypeptide; LCFA = long-chain fatty acid; SGLT1 = sodium, glucose cotransporter type 1; FFAR1 = free fatty acid receptor type 1; AA = amino acids; PYY = protein YY; SCFA = short-chain fatty acid.

INTESTINAL MOTILITY HAS SEVERAL DIFFERENT PATTERNS: SEGMENTATIONS, PERISTALSIS, MIGRATING MOTOR COMPLEX OR MIGRATING MYOELECTRIC COMPLEX, AND REVERSE PERISTALSIS

SEGMENTATION CONTRACTIONS MIX INTESTINAL CONTENTS

Coordinated constriction or relaxation of the outer longitudinal muscle and inner circular muscle gives rise to stereotypical patterns of motility. Segmentation contractions involve local regions of the intestine in which the circular muscle contracts in one region and relaxes in adjacent regions. This moves materials in both directions, orad and caudad. Thus segmental contractions mix intestinal contents with digestive juices. This is the main pattern of motility after a meal (see Figure 8.3.7).

PERISTALSIS PROPELS THE CHYME

Peristalsis consists of a wave of circular muscle contraction that propagates caudally. Both circular and longitudinal muscle layers participate: caudad to the peristaltic wave the circular muscle relaxes and the longitudinal muscle contracts. Contraction of the longitudinal layer shortens the distance the chyme must move and helps increase the caliber of the intestine. At the wave front the circular layer contracts and the longitudinal muscle relaxes. Peristaltic waves generally do not propagate along the entire intestine but instead propel the chyme only a few tens of centimeters.

The peristaltic reflex is a neurally mediated reflex in the small intestine and colon that causes caudad propulsion of chyme. Mechanical or chemical stimulation of the mucosa causes EC cells to release 5-HT (serotonin) onto local receptors of IPANs. These neurons project to the myenteric plexus and submucosal plexus where they activate myenteric ganglion cells to send impulses up

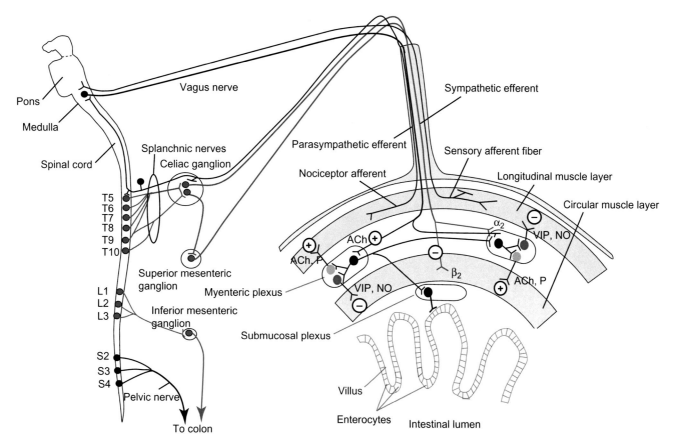

FIGURE 8.3.6 Extrinsic and intrinsic innervations of the intestine. A cross-section of the small intestine is shown. The vagus nerve carries parasympathetic innervation of the upper small intestine. Parasympathetic input to the colon originates in the sacral segments of the spinal cord (S2–S4) and travels over the pelvic nerve. Sympathetic efferents originate in T5–T10 and travel to the small intestine over the splanchnic nerves to the celiac and superior mesenteric ganglia. Sympathetic efferents to the colon originate in L1–L3 to the inferior mesenteric ganglia. The sympathetic postganglionic efferents mainly inhibit release of acetylcholine at parasympathetic preganglionic cholinergic fibers, mediated by α_2 receptors on the parasympathetic preganglionic terminals. Parasympathetic preganglionic fibers mainly release acetylcholine onto nicotinic cholinergic receptors in the myenteric plexus. One type of postganglionic cell (about 45–50% of myenteric neurons) uses substance P, acetylcholine, and neurokinin A as transmitter and these excite smooth muscle activity. A second kind, about 20–30% of myenteric neurons, uses VIP and NO as neurotransmitters, and these inhibit smooth muscle activity. The myenteric neurons mainly synapse on other myenteric neurons, interstitial cells of Cajal, and on circular smooth muscle. A smaller fraction connects to longitudinal muscle, which receives mainly excitatory input, and to the submucosal ganglia. Afferent sensory information flows over the vagus, pelvic, and splanchnic nerves. This includes stretch and chemosensory information that passes through the prevertebral ganglia where a second level of integration occurs. Pain stimuli are carried mainly over the splanchnic nerves.

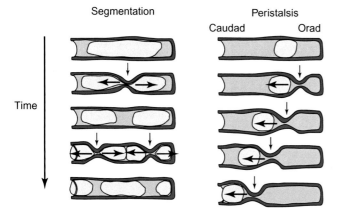

FIGURE 8.3.7 Segmentation and peristaltic motility in the small intestine. Segmentation (left) occurs when an area of circular muscle contracts. Material is propelled in both directions, orad and caudad. Segmental contractions do not propagate along the intestine and occur randomly along the intestine. The bidirectional movement of luminal contents mixes the chyme with digestive juices. Peristaltic contractions (right) consist of a propagating wave of contraction that generally moves from the orad to the caudad direction, sweeping luminal contents before it. It consists of a wave of contraction of the circular muscle, with longitudinal muscle contracting out of phase.

and down the intestinal tract through adjacent myenteric ganglia. Impulses traveling orad activate the **ascending contraction**. The neurotransmitters here are **acetylcholine**, **substance P**, and **neurokinin A**. The impulses traveling caudad activate a **descending relaxation** using **VIP**, **NO**, and **PACAP (pituitary adenylyl cyclase activating polypeptide)**. Figure 8.3.8 illustrates the enteric nervous system control of peristalsis.

THE MIGRATING MOTOR COMPLEX OCCURS IN THE FASTING STATE

The migrating motor complex (MMC), also called the migrating myoelectric complex, describes the pattern of motility during the interdigestive period. As discussed in Chapter 8.2, this stereotypical pattern consists of three phases over a period of 85–115 minutes. It is the "intestinal housekeeper" that propels undigested matter and everything else into the colon. Phase I of the MMC is a quiescent period lasting 40–60% of the cycle length. Phase II, lasting another 20–30% of the cycle length, consists of

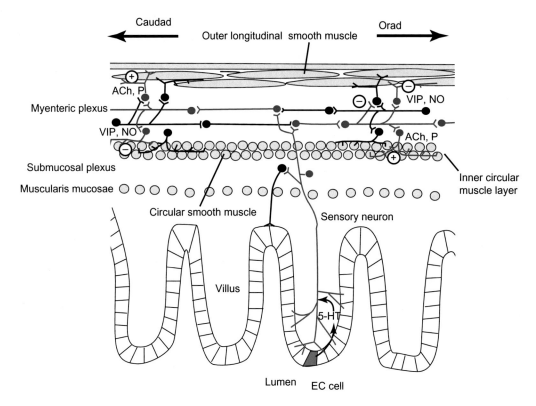

FIGURE 8.3.8 Nerve activity during peristalsis. A longitudinal section of the intestinal wall is shown. The activated cells are shown in blue, quiescent cells in black. Peristalsis generally begins with sensory afferent stimulation resulting in the release of 5-HT from EC cells in the mucosa lining. These sensory neurons make synapses on interneurons within the myenteric ganglia. The interneurons send impulses in the orad direction that activate excitatory myenteric motor neurons for the circular muscle and inhibit excitatory motor neurons for the longitudinal layer. The excitatory motor neurons release acetylcholine, substance P, and neurokinin A. The inhibitory neurons release VIP, NO, and others. These form the ascending contraction of the peristaltic reflex. At the same time, the sensory neurons activate interneurons that send signals in the caudad direction. These activate inhibitory myenteric motor neurons for the circular layer and excitatory myenteric neurons for the longitudinal layer. The result is a descending relaxation that precedes the contractile wave. In the figure, the blue axons are activated while the black lines are quiescent. Thus the longitudinal muscle in the orad direction relaxes while the circular muscle contracts; conversely, the longitudinal muscle in the caudad direction contracts and the circular muscle relaxes.

increasingly strong but irregular contractions. Phase III lasts 5–10 minutes but consists of strong contractions that propagate caudally and that completely occlude the lumen.

THE *ILEOCECAL SPHINCTER* PREVENTS REFLUX OF COLONIC CONTENTS INTO THE ILEUM

The junction between the terminal ileum and the colon produces a region of high pressure that resists movement of material from the colon into the ileum. However, the slow electrical waves and MMC migrate through the ileocecal junction and evoke colonic contractions. Movement of materials from the ileum into the colon is characterized by bolus movements separated by periods of no flow. Distension of the colon produces reflex relaxation of the ileum and contraction of the ileocecal sphincter. This effect is not blocked by transection of either the vagus or pelvic nerves, but it is blocked by transection of the splanchnic nerves. This suggests that the extrinsic sympathetic nerves are involved in this reflex. Distension of the ileum promotes contraction of the ileum and propulsion of material into the cecum.

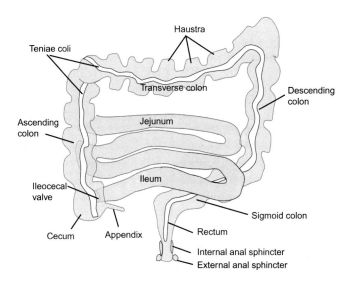

FIGURE 8.3.9 The gross anatomy of the large intestine.

THE LARGE INTESTINE OR COLON HAS SEVERAL ANATOMIC REGIONS

The colon begins at the ileocecal sphincter and consists of the **cecum**, the **ascending colon**, the **transverse colon**, and the **descending colon**. This leads to the **sigmoid colon** and the **rectum** (see Figure 8.3.9). As in the rest of the gastrointestinal system, the colon contains an inner layer of circular muscle that is continuous from cecum to rectum. In the rectum, this layer thickens to form the **internal anal sphincter**. Contraction of the circular muscle divides the colon into segments called **haustra**, giving the appearance of a chain of small sacs. The haustra disappear during contractions of a segment and their locations shift. The outer longitudinal muscle layers are not continuous in the large intestine but instead form three bands called **teniae coli** that run the length of the large intestine. They fan out and become continuous at the rectum. The external anal sphincter consists of striated muscle that overlaps the internal anal sphincter. This muscle provides voluntary control over defecation.

COLONIC MOTILITY SHOWS SEVERAL DISTINCT PATTERNS

Short duration contractions mix colonic contents to aid in water extraction. Each day some 7–10 L of ingested or secreted fluid enters the GI tract, and about 0.6 L of this reaches the colon. The colon absorbs an additional 0.5 L and compacts the feces for excretion. These short duration contractions do not propagate. They last less than 15 seconds and have a frequency of 2–13 min^{-1}.

Long duration contractions last 40–60 seconds. These contractions propagate either orally or caudally. They both mix and propel colonic contents. Both long and short duration contractions may occur simultaneously, giving a mixed pattern of contraction.

Giant migrating contractions are high-amplitude contractions that propagate caudally over extended distances in the colon. They produce mass movement of fecal material caudally. At their onset, haustra disappear. After the contents are propelled caudally for some distance, the haustra reappear and short and long duration contractions occur again. These mass movements of feces precede the conscious need to defecate by propelling fecal matter into the sigmoid colon and rectum.

LOCAL AND EXTRINSIC NERVOUS INNERVATIONS CONTROL ILEAL AND COLONIC MOTILITY

COLONIC STRETCH RECEPTORS INITIATE A LOCAL CONTRACTION REFLEX

Colonic distension or sensing of fecal material mediated by short-chain fatty acids activates receptors that in turn activate circular muscle contraction through local reflexes mediated by the myenteric plexus. As mentioned above, distension of the ileum promotes contraction of the ileum and movement of material into the colon. Part of this response is the reflex relaxation of the colon to accept material from the ileum. These responses are probably mediated by the local myenteric plexus.

DISTENSION OF THE STOMACH EXCITES ILEAL AND COLONIC MOTILITY AND RELAXES THE ILEOCECAL SPHINCTER

Centrally mediated pathways account for longer reflexes. The proximal large intestine is supplied by sympathetic nerves arising from the superior mesenteric ganglion. Sympathetic nerves arising from the inferior mesenteric ganglion supply the descending and sigmoid colon and rectum. The large intestine receives parasympathetic nerves from the pelvic nerve arising from spinal segments S2–S4. Ingesting a meal increases ileal contraction and relaxes the ileocecal sphincter. This is the **gastroileal reflex**, which appears to be mediated by extrinsic nerves. Similarly, eating a meal increases colonic activity simultaneously throughout all of the colon. This is the **gastrocolic reflex**. The gastrocolic reflex can produce mass movement of feces in the colon and stimulate defecation. The sensory afferents involved in the reflex include stretch receptors in the stomach and chemoreceptors in the small intestine. Figure 8.3.8 summarizes the excitatory and inhibitory feedbacks of motility between different regions of the gastrointestinal tract.

DEFECATION INVOLVES VOLUNTARY AND INVOLUNTARY MUSCLES

The combined action of the internal anal sphincter and the external anal sphincter retains fecal matter in the rectum. The ability to do this is called **continence**, and the loss of this ability is called **incontinence**. As in the rest of the GI tract, nervous control of colonic and rectal motility is set by a balance between parasympathetic stimulation of peristalsis and inhibition of the sphincters and sympathetic inhibition of peristalsis and excitation of the sphincters. The parasympathetics originate in the sacral cord and sympathetics in the thoraco-lumbar cord. Involuntary parasympathetic nerves traveling over the **pelvic nerve** control the tone of the internal anal sphincter. The external anal sphincter consists of striated muscle that is controlled voluntarily by somatic motor neurons that exit the spinal cord at S2–S4 and travel over the pudendal nerve. The internal anal sphincter makes the largest contribution to continence; about 85% of the resting anal tone of 40–80 mmHg comes from the internal anal sphincter.

Epithelial nerve endings in the rectum and anus inform the CNS of the nature of rectal contents (solid, liquid, or gas). This information allows a conscious decision about how to evacuate the rectal contents. The anal passage of gas is technically called **flatulation** and the gas is referred to as the **flatus**.

Stretch receptors in the sigmoid colon and rectum send afferent signals to the spinal cord along the pelvic nerve. These afferents excite preganglionic parasympathetic neurons that send efferent connections to postganglionic parasympathetic neurons in the smooth muscle of the large intestine. This initiates contractions that further move fecal material toward the rectum. At the same time, the afferent sensory information ascends the spinal cord to the cerebrum to create the conscious urge to defecate. Voluntary contraction of the abdominal muscles can help move fecal matter into the distal large intestine, thereby bringing on defecation. Distension of the rectum reflexly relaxes the internal anal sphincter and contracts the external anal sphincter. This switches control of defecation from involuntary retention by the internal anal sphincter to voluntary retention by the external anal sphincter. Voluntary contraction of the external anal sphincter can increase pressure within the anal sphincter to 150 mmHg. The inhibition of internal anal sphincter tone by rectal distension is called the **rectoanal inhibitory reflex**. It is mediated by intrinsic neural pathways in the wall of the rectum and anus, mediated by VIP and NO.

Distension of the rectum provides the initial sensory stimulation for defecation. Actual defecation involves parasympathetic relaxation of the internal anal sphincter, opening of the angle between the rectum and the anus, voluntary relaxation of the external anal sphincter, and increasing the pressure within the sigmoid colon and rectum to propel the feces out through the anus. The increase in pressure is caused by a peristaltic wave in the sigmoid colon and rectum and is assisted by increasing intra-abdominal pressure by voluntarily contracting the abdominal muscles and diaphragm. The nerves involved in defecation are shown in Figure 8.3.10. These pathways are essentially the same as those used for micturition, and there is cross-talk between the defecation and micturition pathways.

SUMMARY OF REGULATORY CONNECTIONS WITHIN THE GI TRACT

The inhibition of gastric emptying by stimuli in the small intestine is called the **enterogastric inhibitory reflex**. The sensory stimuli in the duodenum that inhibits gastric emptying include:

> distension of the duodenum
> irritation of the duodenum
> acid pH in the duodenum
> increased osmolarity of duodenal content
> amino acids (particularly tryptophan) in the duodenum
> fat digestion products (fatty acids, monoglycerides, and diglycerides) in the duodenum.

Osmoreceptors are limited to the duodenum, whereas receptors for acid, glucose, or oleic acid are present within the first 1.5 m of the small intestine. Perfusion of the ileum with glucose or lipids delays gastric emptying. This last phenomenon is called the **ileogastric reflex** or the **ileal brake**.

Similarly, distension of the stomach causes contraction of the ileum and colon, referred to as the **gastroileal reflex** and **gastrocolic reflex**. Distension of the ileum causes relaxation of the ileocecal valve, whereas distension of the ascending colon constricts it. Distension of the rectum inhibits internal anal sphincter tone, the **rectoanal inhibitory reflex**. Figure 8.3.11 summarizes the excitatory and inhibitory feedbacks of motility between different regions of the gastrointestinal tract.

VOMITING REMOVES POTENTIALLY DANGEROUS MATERIAL FROM THE GUT

A variety of stimuli in the GI tract results in emesis or vomiting. These include specific irritants instilled in the stomach, and distension of the stomach or the intestine. Some stimuli, such as hypertonic saline and copper sulfate, excite mucosal sensory receptors, whereas emesis caused by distension begins with smooth muscle stretch receptors. These sensory afferents travel to a vomiting center, located in the medulla, over the vagus nerve. Stimulation of sensory receptors over the splanchnic nerves causes pain and not emesis. Emetic agents such as **syrup of ipecac** cause emesis through activation of stomach afferents.

Other stimuli also can cause emesis. These include mechanical stimulation of the back of the pharynx. This is the **gag reflex** that normally prevents us from swallowing large pieces of material.

The dorsal surface of the medulla at the caudal aspect of the fourth ventricle of the brain responds to a wide variety of circulating chemicals by initiating emesis. This region is called the **chemical trigger zone** or **CTZ**. Administration of a number of compounds directly to this area causes emesis and surgical removal of the area prevents the effect. The vomiting that accompanies certain cancer medications and radiation therapy may act through the CTZ. However, some radiation therapy and other cancer medications may also act through gastrointestinal receptors because vagotomy removes their effect. Infusion of epinephrine causes emesis and may explain emesis upon strenuous exertion.

Motion sickness and a variety of diseases of the inner ear cause emesis by activating the brain stem vestibular nuclei. Anticholinergic drugs, such as **dramamine**, bind to M1 receptors and prevent motion sickness by increasing the habituation to motion stimuli. Figure 8.3.12 shows the stimuli that can lead to vomiting.

Exceedingly unpleasant sensory stimulation, including noxious odors, taste, visual stimulation, or pain, can result in emesis through cerebral cortical afferents. The smell of vomit itself is a powerful stimulant for vomiting.

VOMITING IS A COMPLICATED PROGRAMMED EVENT

Vomiting, like swallowing, entails complicated reflexes that are controlled by a vomiting center in the medulla. This center coordinates the gastrointestinal muscles and

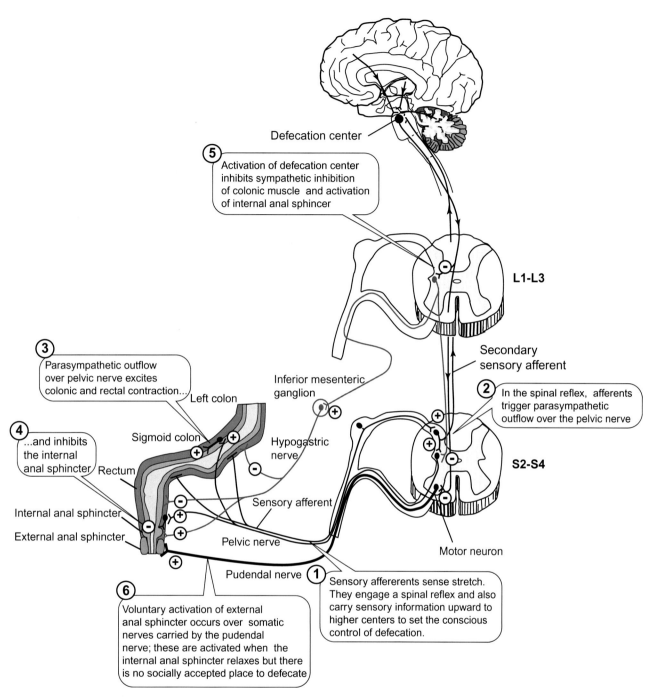

FIGURE 8.3.10 Nervous control of defecation. When material or fluid is presented to the distal colon and rectum, it stretches the tissues and activates stretch-activated sensors in the walls of the gut. These send afferent fibers to the sacral cord over the pelvic nerve, and this begins a reflex but also sends fibers up the cord to higher centers in the brain. This upward flow of sensory input informs the brain of the status of the colon and creates the urge to defecate. In the cord, parasympathetic outflow over the pelvic nerve causes contraction of the left colon, sigmoid colon and rectum, and relaxation of the internal anal sphincter. At the same time, outflow from the brain inhibits sympathetic inhibition of colonic and rectal motility and activation of internal anal sphincter contraction. These combined activities relax the internal anal sphincter and contract colonic and rectal muscles. These actions can be overridden by conscious contraction of the external anal sphincter, mediated by voluntary motor nerves exiting the cord in the sacrum, and traveling over spinal roots that form the pudendal nerve.

somatic motor systems to expel potentially noxious luminal contents from the gastrointestinal tract. It is often preceded by **nausea**, the subjective feeling of being about to vomit. Vomiting is generally also preceded by **reverse peristalsis**, which is controlled by extrinsic nerves and is entirely abolished by vagotomy. Intraluminal pressures generated by reverse peristalsis are nearly twice as great as those produced by normal

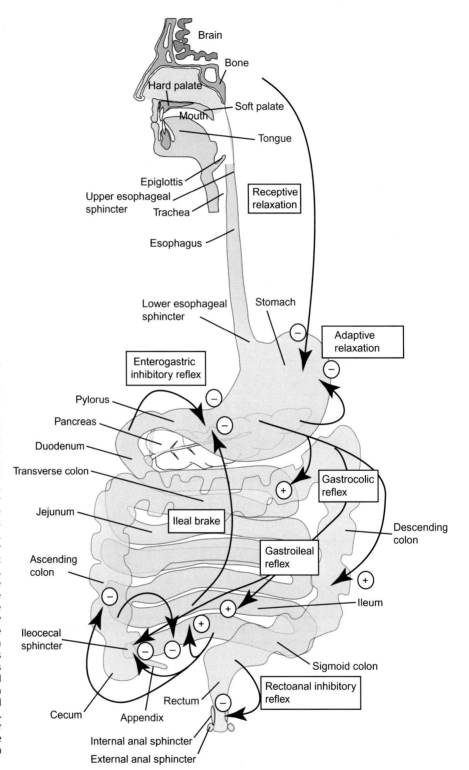

FIGURE 8.3.11 Regulation of gastrointestinal motility by remote parts of the gastrointestinal tract. Each arrow represents the effect of stretch or other stimulation on motility in another part of the GI tract. Receptive relaxation refers to the relaxation of stomach muscle caused by the swallowing center in the medulla. Distension of the stomach causes a further relaxation of the orad stomach as part of adaptive relaxation. Emptying of the stomach is inhibited by a variety of reflexes arising from the duodenum and proximal jejunum. This includes acid pH in the lumen, distension of the duodenum, hyperosmolarity of duodenal contents, irritation of the duodenum, amino acids, and lipolytic products within the duodenum and proximal jejunum. This comprises the enterogastric inhibitory reflex. Nutrients in the ileum also inhibit gastric emptying; this is the ileal brake. However, normally these nutrients are absorbed by the time the material reaches the ileum, so that the ileal brake is an abnormal pattern. Food in the stomach and in the duodenum excites motility throughout the large intestine in the gastrocolic reflex. This is mediated by stretch and by chemosensors in the duodenum. This same information promotes ileal contractions and relaxes the ileocecal valve in the gastroileal reflex. Stretch of the ileum promotes its own contraction, relaxes the ileocecal valve, and relaxes the cecum and ascending colon. Distension of the ascending colon, on the other hand, relaxes the ileum. Distension of the rectum relaxes the internal anal sphincter in the rectoanal inhibitory reflex.

phasic intestinal contractions. They last several times longer and are propagated over longer distances at greater speeds. Reverse peristalsis brings intestinal contents into the stomach from which it can be expelled during vomiting. Contraction of the abdominal muscles produces a large positive intra-abdominal pressure that forces gastric contents past the lower esophageal sphincter and into the esophagus. Reverse peristalsis also occurs in the esophagus, and the propulsion of the vomitus through the mouth is aided by moving the hyoid bone and larynx upward and forward. Ventilation is suppressed during vomiting, and the glottis is closed. Parasympathetic stimulation increases saliva secretion to help protect the teeth when exposed to the acid vomit.

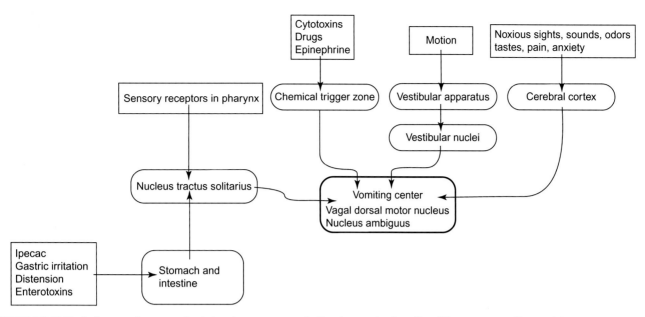

FIGURE 8.3.12 Block diagram of sensory stimulation that causes emesis. Emesis can arise from five different sensory afferents: (1) sensory stimulation of the pharynx; (2) distension or irritation of the intestine or stomach; (3) motion stimulation of the vestibular apparatus; (4) chemical stimulation of the CTZ; and (5) sensory processing through the cerebral cortex.

Clinical Application: Opioid-Induced Constipation

Physicians prescribe opioids as an analgesic to treat persons with a variety of pain conditions ranging from chronic pain resulting degenerative diseases of the back, knees, and hips, other musculoskeletal pain, neuropathic pain, and cancer pain, to acute pain resulting from sprains, strains, fractures, dental work, and surgery. Opioids act centrally on μ-receptors, κ-receptors, δ-receptors, and the opioid-receptor like ORL-1 to alleviate pain. However, they also act on μ-receptors, κ-receptors, δ-receptors in the GI tract, resulting in opioid-induced bowel dysfunction, OIBD, commonly producing opioid-induced constipation, OIC. Approximately 5% of the adult population in the United States is treated with opioids for chronic pain and many develop OIBD as a complication. The effects on the gut appear to be mediated mainly through the μ-receptor.

In the human gut, μ-receptors mainly are found in the myenteric and submucosal plexuses, as shown in Figure 8.3.13. In normal physiology, these receptors bind endogenous ligands such as enkephalins, endorphins, and dynorphins. Analgesic drugs also bind to these receptors and inhibit neurotransmitter release through a G_i mechanism. The result is reduced coordination of motility.

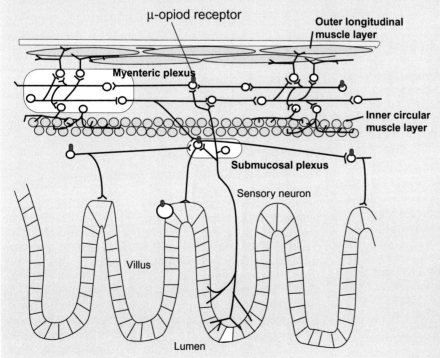

FIGURE 8.3.13 Opioid receptors in the enteric nervous system. From Brock, C. et al., Opioid-induced bowel dysfunction. Drugs 72:1847–1865 (2012).

Clinical Applications: Diverticular Disease of the Colon

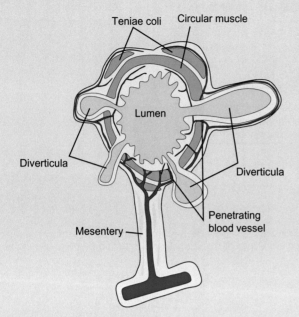

FIGURE 8.3.14 Cross-section of the colon showing diverticula that form as pouches of mucosa and submucosa that bulge outward through the muscle layers of the colon. *Redrawn from C.M. Friel and J.B. Matthews, Diverticular disease of the colon, Clin. Perspect. Gastroenterol. 3:187–197, 2000.*

Diverticulosis refers to the presence of small saccules in the colon that form when the wall of the intestine bulges out through weaknesses in the muscle layers. There are two types of diverticula: true diverticula have all of the normal components of the gut wall, whereas false diverticula contain only mucosa and submucosa. True diverticula are uncommon and probably congenital. False diverticula are acquired as a consequence of chronically increased luminal pressure. A diagram of diverticula is shown in Figure 8.3.14.

The prevalence of diverticulosis in undeveloped countries in Asia and Africa is less than 10%, whereas in developed countries it increases with age from about 10% in persons over 40 years old to 80% in persons over 85 years of age. Dennis Burkitt first championed the idea that many diseases of the colon, including diverticulosis, ulcerative colitis, and colon cancer, were related to the lack of dietary fiber in the purified and enriched diets of developed countries. Dietary fiber expands the colon, increasing its radius, r. According to the Law of Laplace for a cylinder, $P = T/r$, where P is the pressure, T is the wall tension, and r is the radius. When r decreases, P increases for the same wall tension. Thus the reduced caliber of the colon when ingesting a fiber-poor diet causes an increase in the pressure. Whether this hypothesis adequately explains the pathophysiology of diverticular disease is unknown. The decreased stool size might initiate other changes, such as thickening of the muscle bands, that exaggerate the colonic pressure response to food and other agents that increase colonic pressure.

Inflammation of the pouches causes diverticulitis. Although many people have asymptomatic diverticulosis, the condition can lead to serious complications including abscess formation, perforation leading to peritonitis (an infectious inflammation of the **peritoneum**, the membrane lining the abdominal cavity), stricture of the colon, and bleeding.

SUMMARY

The small intestine begins at the end of the pyloric sphincter and continues to the ileocecal sphincter. The duodenum makes up the first short segment, followed by the jejunum, and then the ileum. The large intestine begins at the ileocecal sphincter and consists of the cecum, followed by the ascending, transverse, descending, and sigmoid colon. The colon leads to the rectum and the anus.

Intestinal function is regulated by extrinsic nerves arising from the parasympathetic and sympathetic nervous systems and an intrinsic enteric nervous system consisting of the myenteric plexus, the submucosal plexuses, and the ICCs. The ICCs come in several varieties. These cells set the basic electrical rhythm of the gut. The myenteric plexus lies between the outer longitudinal muscle layer and the inner circular muscle layer. The submucosal plexuses lie between the muscularis mucosa and the inner circular muscle layer.

Sensory information comes to the enteric nervous system and CNS through mechanosensitive enterochromaffin cells (EC) and chemosensitive and mechanosensitive enteroendocrine cells (EEC). Multiple types of EEC use sensory mechanisms similar to those of the taste buds to monitor types and amounts of nutrients in the ingested food. Enteroendocrine cells detect a variety of materials including glucose, amino acids, peptides, long-chain fatty acids, short-chain fatty acids, and stretch. They release a large number of endocrine peptides including secretin, gastrin, cholecystokinin, somatostatin, protein YY, glucagon-like polypeptide-1 (GLP-1), ghrelin, and motilin.

The parasympathetic system promotes motility and secretion. Parasympathetic efferents arrive over the vagus nerve and release acetylcholine on nicotinic receptors on postganglionic parasympathetic neurons. These excite motility by releasing substance P, neurokinin, or acetylcholine. The parasympathetic nervous system also may stimulate inhibitory postganglionic neurons that use VIP or NO as neurotransmitters. Parasympathetic innervation of the colon originates in spinal segments S2–S4 and travels over the pelvic nerve. Sympathetic nerves arrive from the prevertebral ganglia: the celiac and superior mesenteric ganglia are supplied from sympathetics from T1 to T10 and supply the proximal gut; the distal gut is supplied by the inferior mesenteric ganglion from L1 to L3. Afferent sensory information travels over the vagus nerve and pain information travels over the splanchnic nerve. Sensory information begins with EC cells that release serotonin or ATP onto IPANs whose cell bodies lie within the enteric plexuses. The intestine has as many neurons as the spinal cord and can function autonomously without extrinsic innervation.

Gastrointestinal motility has three basic patterns: segmentation contractions that mix the contents, peristalsis that moves the contents a short distance, and the MMC that propels the contents forward during the fasting state. Segmentations and peristalsis result from local stimulation of the mechanosensors in the intestinal epithelium. These project to the myenteric plexus where they cause an ascending contraction and a descending relaxation. The ascending contraction occurs above a bolus of food. It involves contraction of the inner circular layer and relaxation of the outer longitudinal layer. The descending relaxation involves relaxation of the inner circular layer and contraction of the outer longitudinal layer. Peristalsis occurs when the areas of ascending contraction and descending relaxation move along the GI tract.

Colonic motility also shows three patterns. Short duration contractions mix the contents and produce the haustra, segments of the colon divided by regions of circular muscle contraction. Longer duration contractions propel material either orally or caudally. Giant migrating contractions propel material into the sigmoid colon or rectum.

Regions of the gut influence other regions through long reflexes. Distension and other stimuli in the duodenum inhibit gastric emptying in the enterogastric inhibitory reflex. Distension of the stomach promotes motility in the ileum and relaxation of the ileocecal sphincter in the gastroileal reflex. Distension of the stomach promotes colonic motility in the gastrocolic reflex. Distension of the ileum promotes its contraction, whereas distension of the colon inhibits ileal contraction. Distension of the rectum inhibits the internal anal sphincter in the rectoanal inhibitory reflex and contracts the external anal sphincter, switching control of defecation from the involuntary internal anal sphincter to the voluntary external anal sphincter.

REVIEW QUESTIONS

1. Where is the myenteric plexus? The submucosal plexus?
2. What are interstitial cells of Cajal? Why are they important?
3. What are intrinsic primary afferent neurons? To what to they respond? What cells do they activate?
4. Where do sympathetic nerves to the intestine originate? What do they do? What neurotransmitter do they use?
5. Where do parasympathetic nerves to the intestine originate? What do they do? What neurotransmitters do the preganglionic and postganglionic fibers use?
6. What are enterochromaffin cells?
7. What are enteroendocrine cells? What do they sense? What do they secrete?
8. What is peristalsis? What is meant by "ascending contraction" and "descending relaxation"?
9. What are receptive relaxation, gastric accommodation, enterogastric inhibitor reflex, ileal brake, and gastrocolic reflex?
10. Where is the vomiting center? Why is it necessary for vomiting to be a programmed event? What is the chemical trigger zone?

8.4 Pancreatic and Biliary Secretion

Learning Objectives

- List the endocrine secretions of the pancreas
- List the four major classes of exocrine secretions of the pancreas
- Identify in a drawing: the acinar cells, intralobular duct, interlobular duct, and main pancreatic duct
- List the proteolytic secretions of the pancreas
- Explain why the pancreas does not digest itself with its proteolytic secretions
- List the amylolytic secretions of the pancreas
- List the lipolytic secretions of the pancreas
- Describe the composition of the watery secretion of the pancreas, especially noting its pH
- Define what is meant by CFTR; note its subcellular location and function
- List three phases of regulation of pancreatic secretion
- Describe CCK: include its chemical nature, site of secretion, secretagogues, and actions
- Describe secretin: include its chemical nature, site of secretion, secretagogues, and actions
- Identify in a drawing: hepatic bile duct, cystic duct, gallbladder, common bile duct, and sphincter of Oddi
- List the major components of bile
- Name the primary and secondary bile acids and identify their source for synthesis
- Describe the enterohepatic circulation of bile acids
- Describe the function of the gallbladder

THE EXOCRINE PANCREAS SECRETES DIGESTIVE ENZYMES AND HCO_3^-

The pancreas has both **exocrine** and **endocrine** functions. Exocrine glands are glands with ducts that secrete materials onto some surface—generally the skin, the gastrointestinal tract, or respiratory epithelium. Endocrine glands are ductless and secrete hormones into the blood. The pancreas is both of these. Clusters of cells called the **islets of Langerhans**, distributed throughout the pancreas, secrete the hormones **insulin, glucagon,** and **somatostatin**, which regulate metabolism and the fate of absorbed nutrients. These endocrine functions are discussed in Chapter 9.4. The exocrine pancreas consists of clusters of **acini**, hollow spheroids of some 20–50 pyramidal cells arranged around a central lumen. The acini form lobules that are separated by loose connective tissue. Each acinus is drained by an intralobular duct which joins other ducts to form interlobular ducts and then progressively larger ducts until they reach the main pancreatic duct. The **acinar cells secrete protein enzymes that digest food**. The **ductal cells secrete a watery, alkaline solution** that neutralizes stomach acid and carries the enzymes forward (see Figure 8.4.1).

THE PANCREAS SECRETES FOUR CLASSES OF ENZYMES

The protein content of human pancreatic juice varies between 0.7% and 10% because of varying rates of secretion of protein and the watery solution that carries the proteins into the intestine. Digestive enzymes make up the major part of the protein secretion. Trypsin inhibitor, plasma proteins, and mucin make up the remainder. The pancreas secretes four classes of enzymes: **proteolytic enzymes, amylolytic enzymes, lipolytic enzymes,** and **nucleolytic enzymes**.

THE PANCREAS SECRETES INACTIVE FORMS OF THE PROTEOLYTIC ENZYMES

The pancreas secretes a set of proteolytic enzymes in an inactive form. These inactive forms include the following:

- Trypsinogen
- Chymotrypsinogen
- Proelastase
- Procarboxypeptidase A
- Procarboxypeptidase B.

Enterokinase, an enzyme secreted by the duodenal mucosa, converts trypsinogen to **trypsin**, which then autocatalytically converts more trypsinogen to trypsin. Trypsin then converts chymotrypsinogen to **chymotrypsin**, proelastase to **elastase**, and procarboxypeptidase A and B to **carboxypeptidase A and B**. Trypsin, chymotrypsin, and elastase are all endopeptidases: they hydrolyze peptide bonds on the interior of a protein. Carboxypeptidases A and B are exopeptidases: they cleave amino acids off the end of proteins. Carboxypeptidases split one amino acid at a time off the carboxyl end of a polypeptide chain. Carboxypeptidase A releases neutral amino acids and carboxypeptidase B releases cationic amino acids.

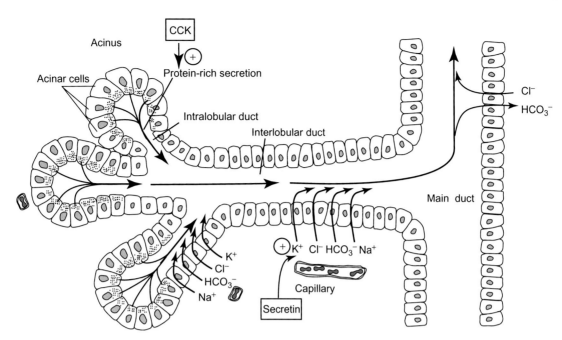

FIGURE 8.4.1 Schematic cartoon of the exocrine pancreas. The acinar cells store protein secretions in **zymogen granules**. Fusion of the granules with the apical membrane releases the enclosed proteins into the lumen of the acinus. **CCK** stimulates enzyme secretion. The interlobular duct cells' secretion is isoosmotic with plasma. **Secretin** stimulates secretion by the duct cells. As the fluid flows down the main pancreatic ducts, HCO_3^- exchanges for Cl^-.

The pancreas secretes small amounts of **trypsin inhibitor**. Between pH 3 and 7, this material forms a 1:1 complex with trypsin that inactivates its proteolytic activity. This protects the pancreas from being digested by small amounts of active trypsin, and it also prevents the premature activation of all of the other proteolytic enzymes. This inhibitor does not prevent activity of trypsin in the intestinal lumen because it is secreted in small amounts that are overwhelmed when trypsin becomes fully activated.

PANCREATIC AMYLASE BREAKS DOWN STARCHES

Most of the digestion of carbohydrates occurs on the brush border membrane of the **enterocytes**, the absorptive cells that line the villi. However, the salivary glands and the pancreas secrete distinct forms of **amylase**. This enzyme hydrolyzes the α-1,4-glycosidic bond in starch, but cannot hydrolyze the α-1,6-glycosidic bonds at which the starch chain branches, nor can it hydrolyze the β-1,4 linkages present in cellulose. See Chapter 8.5 for details about carbohydrate digestion and absorption. Amylase is secreted in its active form.

THE PANCREAS SECRETES A SET OF LIPOLYTIC ENZYMES

The pancreas secretes a variety of lipolytic enzymes, including:

- Pancreatic lipase
- Pancreatic colipase
- Phospholipase A_2
- Cholesterol esterase.

The role of these enzymes in lipid digestion and absorption is discussed in Chapter 8.5. The optimal activity of these enzymes depends upon the bile acids delivered to the intestine through the bile.

THE PANCREAS SECRETES NUCLEOLYTIC ENZYMES

The pancreas also secretes pancreatic **ribonuclease (RNAase)** and pancreatic **deoxyribonuclease (DNAase)**. These degrade nucleic acids in the ingested food.

PANCREATIC DUCT CELLS SECRETE AN ALKALINE SOLUTION IN TWO STAGES

The pancreas daily secretes 1–2 L of pancreatic juice that consists mainly of water, electrolytes, and pancreatic enzymes. The secretion occurs in two stages. The first phase consists of acinar secretion that makes up little volume but contains most of the pancreatic enzymes. Most of the fluid in the final pancreatic juice consists of ductal secretion that itself is different from secretion of the small and main pancreatic ducts. The acinar secretion is isoosmotic with plasma, and the major cations, Na^+ and K^+, are present in concentrations close to those in plasma, independent of the rate of pancreatic secretion. The major anions are Cl^- and HCO_3^-. In the main pancreatic duct, HCO_3^- is reabsorbed in exchange for Cl^-. Thus at high flow rates the $[HCO_3^-]$ is high because there is not enough time to exchange much HCO_3^- for Cl^-. At low flow rates, the $[HCO_3^-]$ and $[Cl^-]$ vary reciprocally (see Figure 8.4.2).

Figure 8.4.3 shows the main ion transport mechanisms involved with acinar secretion. The mechanisms of secretion here are far less investigated that those in salivary glands, but, to the extent that they are known, the pancreatic acinar seem to use the same mechanisms as the salivary acini, with the noted absence of AQP5 from the apical membrane of pancreatic acini.

Pancreatic duct secretion differs from the acinar secretion in that the primary secretion is HCO_3^-. The secreted HCO_3^- derives from CO_2 in the circulation or from pancreatic cell metabolism, which is converted to HCO_3^- within the cells by the enzyme **carbonic anhydrase**. In the intralobular and small interlobular ducts, cells have a cAMP-regulated chloride channel in the apical membranes (those facing the lumen of the duct). This is the site of the defect in persons afflicted with **cystic fibrosis** and is called the **CFTR** for **cystic fibrosis transmembrane conductance regulator**. HCO_3^- secretion across the apical membrane depends on the Cl^- concentration in the lumen, which in turn depends on the activity of CFTR. The activity of CFTR is increased by cAMP, which is increased upon stimulation of the duct cells with **secretin**. Na^+ and K^+ in the ductal fluid arrive through **paracellular** pathways—between the cells and through the junctions at their apical ends rather than through the cells themselves. The pathway through the cells is called the **transcellular** pathway. The main idea here is that the pancreatic ducts secrete a highly alkaline fluid, and they accomplish this by secreting HCO_3^- and acidifying the blood. Figure 8.4.4 illustrates pancreatic duct secretion.

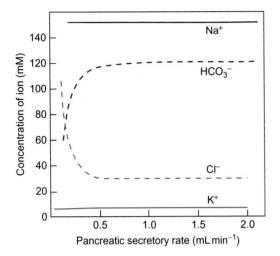

FIGURE 8.4.2 Concentration of pancreatic juice electrolytes as a function of secretory rate. The composition of the initial secretion in the acini is similar to plasma with high [Na^+] and low [K^+], but instead of high [Cl^-] there is a high [HCO_3^-]. At high flow rates, the final pancreatic secretion retains these characteristics. At low flow rates, the HCO_3^- exchanges with Cl^- in the larger pancreatic ducts, and so the [HCO_3^-] falls and [Cl^-] rises.

POSTPRANDIAL PANCREATIC ENZYME SECRETION IS REGULATED IN *CEPHALIC, GASTRIC,* AND *INTESTINAL* PHASES

CEPHALIC INPUTS ACCOUNT FOR UP TO 50% OF MAXIMAL PANCREATIC ENZYME SECRETION

The sight, smell, and taste of food stimulate pancreatic secretion rich in enzymes. In humans, this effect can produce rates of secretion up to 50% of the maximal rates produced by infusion of secretin and

FIGURE 8.4.3 Mechanism of pancreatic acinar secretion. Secretion is stimulated mainly by parasympathetic stimulation mediated by cholinergic M3 receptors that operate via a G_q mechanism. CCK stimulates zymogen secretion mainly by activating vagal afferents that eventually release acetylcholine onto the acinar cells. Release of Ca^{2+} from the ER stimulates further influx of Ca^{2+} from the extracellular fluid through store-operated calcium entry (SOCE) involving STIM1 as a sensor of ER Ca^{2+} depletion and Orai and TRPC1 as routes of Ca^{2+} entry. Increased Ca^{2+} stimulates zymogen secretion and also activates an apical Cl^- channel and basolateral K^+ channels. Cl^- for secretion is provided by the Na–K–2Cl transporter (NKCC1) on the basolateral membrane, whose influx is made possible by the Na^+ gradient established by the Na,K-ATPase pump, also located on the basolateral membrane. The enzyme carbonic anhydrase provides HCO_3^- that is secreted in exchange for Cl^- at the apical membrane. The H^+ produced by the carbonic anhydrase is removed from the cell in exchange for Na^+ over the Na–H exchanger (NHE1) on the basolateral membrane. Water and Na^+ follow the transepithelial electrical potential and osmotic pressure of secreted solutes. The anion exchanger AE2 is present on the basolateral membrane and may either provide Cl^- or HCO_3^- for secretion into the lumen.

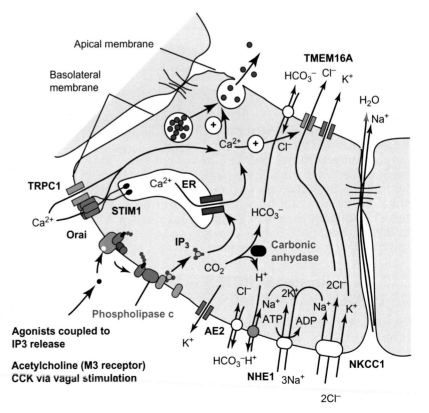

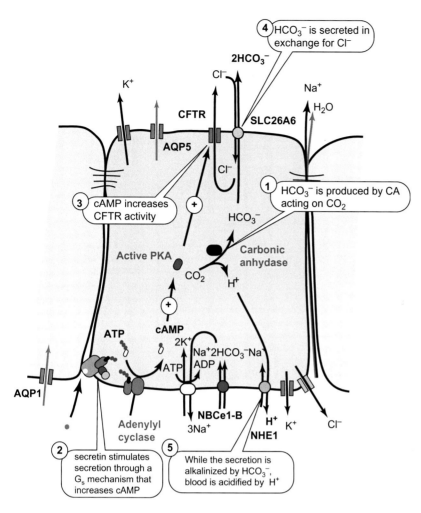

FIGURE 8.4.4 Mechanism of pancreatic duct cell secretion of electrolytes and water. CO_2 in the blood diffuses into the cell where it is hydrated by carbonic anhydrase (CA) to form H^+ and HCO_3^-. The HCO_3^- formed feeds into a HCO_3^-–Cl^- exchanger located in the apical cell membrane. This HCO_3^-–Cl^- exchanger is coupled to an apical membrane Cl^- channel that recirculates the Cl^-. This apical Cl^- channel is the cystic fibrosis transmembrane conductance regulator or CFTR. In its absence, the duct cells produce limited quantities of secretory fluid in a variety of epithelia including the pancreas and the respiratory epithelium, and the result is a thick, viscous, and sticky secretion. Na^+ and K^+ reach the lumen through paracellular pathways. The H^+ produced from the carbonic anhydrase acting on CO_2 exits the cell by exchange with Na^+, which in turn is pumped out of the cell by the Na^+, K^+-ATPase. The K^+ entering the cell by the Na^+, K^+-ATPase exits the cell through K^+ channels on the basolateral membrane Secretory HCO_3^- is also provided by a basolateral Na^+-HCO_3^- cotransporter, NBCe1-B. The co-transported Na^+ is removed by the basolateral Na^+-K^+-ATPase. Secretin stimulates ductal cell secretion by increasing the conductance of the apical CFTR. The mechanism involves stimulation of adenylyl cyclase through a G_s protein that increases cytoplasmic [cAMP].

cholecystokinin (CCK). The cephalic phase is mediated by vagus efferents to the pancreas. Acetylcholine stimulates pancreatic secretion through an M3 receptor, which activates a G_q protein that activates phospholipase C to release IP3 and DAG (diacylglycerol). The IP3 raises the cytosolic [Ca^{2+}] by releasing Ca^{2+} from stores in the ER through the IP3 receptor (see Figure 8.4.3).

A VAGOVAGAL REFLEX MEDIATES THE GASTRIC PHASE OF PANCREATIC ENZYME SECRETION

Distension of the orad stomach, but not the antrum, stimulates pancreatic enzyme secretion. This effect is mediated by the vagus nerves and is abolished by vagotomy or atropine, an acetylcholine antagonist. This effect accounts for perhaps 5–10% of the postprandial pancreatic enzyme secretion in humans.

DELIVERY OF NUTRIENTS TO THE INTESTINE POWERFULLY STIMULATES PANCREATIC ENZYME SECRETION

Cholecystokinin or **CCK** is a peptide hormone secreted by endocrine I cells in the duodenal mucosa in response to amino acids and lipid digestion products in the duodenum. Recall from Chapter 8.2 that circulating CCK is a heterogeneous mixture of forms up to 83 amino acids in length. After a meal, fasting levels of CCK increase from 1 to 6–8 pM within 10–30 minutes. The CCK stimulates pancreatic zymogen secretion by activating vagal afferents in the duodenum. These afferents travel to the CNS and engage vagal efferents whose postganglionic fibers stimulate the acinar cells. Phenylalanine, valine, and methionine stimulate pancreatic enzyme secretion the most. Undigested fats do not release CCK, whereas fatty acids and monoglycerides effectively release CCK. In addition, sensory afferents detecting duodenal hyperosmolarity and stretch receptors initiate vagovagal reflexes that increase pancreatic enzyme secretion.

SECRETIN PRIMARILY REGULATES PANCREATIC DUCT SECRETION

Secretin is a linear polypeptide of 27 amino acids. S-cells located in the duodenum and proximal jejunum secrete secretin in response to increased [H^+] in the lumen, mediated through the release of secretin releasing peptides from the duodenum. Secretin increases HCO_3^- secretion from the bile ducts and pancreatic ducts and from Brunner's glands in the intestine and delays gastric emptying and therefore the delivery of acid into the duodenum. This makes a neat feedback loop: the acid that stimulates secretin release is

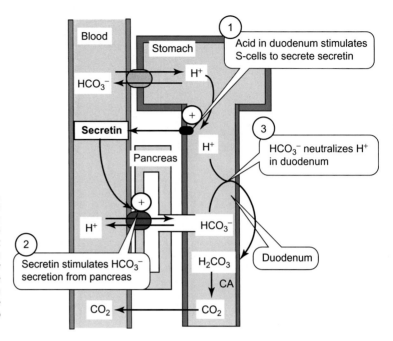

FIGURE 8.4.5 Negative feedback loop for H^+ in the duodenum. Parietal cells in the stomach secrete H^+ into the stomach while alkalinizing the blood with HCO_3^-. The acid in the stomach enters the duodenum where it stimulates S-cells to secrete secretin into the blood. Secretin stimulates pancreatic duct secretion of HCO_3^- by stimulating the CFTR through increasing cAMP. The secretion of HCO_3^- into the ducts is accompanied by acidification of the blood with H^+. The HCO_3^- secreted into the duodenum combines with, and neutralizes, the H^+ which entered from the stomach. In addition, the H^+ added to the blood by the pancreas is neutralized by the HCO_3^- added to the blood by the stomach.

neutralized by the HCO_3^- secreted in response to secretin. Secretin is linked to a G_s protein that increases cytosolic cAMP in its target cells by increasing the activity of adenylyl cyclase. The negative feedback of acid in the duodenum is shown in Figure 8.4.5.

THE LIVER PRODUCES BILE AND STORES IT IN THE GALLBLADDER IN THE INTERDIGESTIVE PERIOD

Bile is produced by the liver, which collects the bile first in **bile canaliculi** that drain just a few **hepatocytes**, or liver cells. These bile canaliculi drain into progressively larger bile ducts, eventually forming right and left hepatic bile ducts. These fuse into the **common hepatic duct**. The **cystic duct** comes off the common hepatic duct and leads to the **gallbladder**. The gallbladder is a small sac invested with smooth muscle that resides in a small indentation of the liver just above the duodenum. The continuation of the bile duct below the divergence of the cystic duct is called the **common bile duct**. It enters the duodenum alongside the pancreatic duct at the **ampulla of Vater**. The sphincter around the ampulla is the **sphincter of Oddi**. Figure 8.4.6 shows the anatomic arrangement of the **biliary tree**.

HEPATOCYTES ARE POLARIZED CELLS WITH SPECIAL ACCESS TO PLASMA

Hepatocytes are typically arranged in sheets of single cells that fan out from a **central vein** to form an **hepatic lobule**. The hepatocytes are polarized with a basolateral membrane facing the blood and an apical membrane facing the bile canaliculus. The canaliculus is defined by two adjoining cells, which are connected by junctional complexes. The liver is perfused with blood from the **hepatic portal vein**, which drains the intestines, and the **hepatic artery**. Venous blood drains into a central vein. In the liver, endothelial cells lack basement membranes, and this facilitates transfer of materials into the **space of Disse** between endothelial cells and hepatocytes and allows the hepatocytes to contribute to blood proteins and lipoproteins. Figure 8.4.7 illustrates the histological arrangement of the hepatocytes and the mechanism of bile formation. This figure illustrates how blood flow from the portal vein is countercurrent to the flow of bile. This arrangement is critical to the recirculation of bile acids during a meal.

BILE CONSISTS OF *BILE ACIDS, PHOSPHOLIPIDS, CHOLESTEROL, BILE PIGMENTS, MUCIN, XENOCHEMICALS,* AND *ELECTROLYTES*

The bile acids are actively secreted into the bile and function as emulsifying agents for the digestion of fats in the intestine (see Chapter 8.5). Bile also contains phospholipids and cholesterol that bind the bile acids and protect the cell from the cytotoxic detergent effects of bile acids. Mucin in the bile helps protect the epithelium lining the bile ducts from the damaging effects of the bile acids. The electrolytes in the bile exert an osmotic pressure that helps produce the flow of bile. The osmotic activity of the bile acids, phospholipids, and cholesterol is diminished because they aggregate to form large structures called **micelles**. The average composition of hepatic bile is described in Table 8.4.1.

THE LIVER MAKES AND RECYCLES BILE ACIDS AS AN INTEGRAL PART OF BILIARY SECRETION

Primary bile acids (cholic acid and chenodeoxycholic acid) are synthesized from cholesterol in hepatocytes.

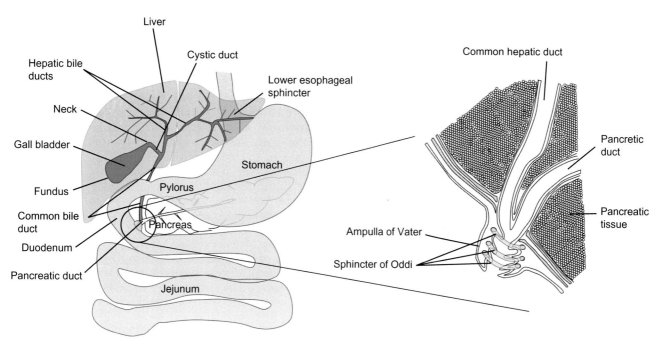

FIGURE 8.4.6 Anatomical structures of the biliary tree, with a close-up view of the hepatopancreatic junction with the duodenum.

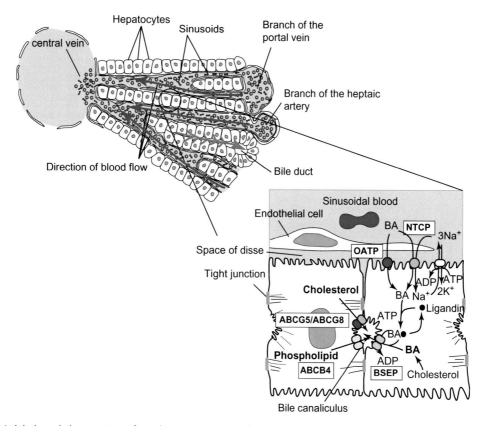

FIGURE 8.4.7 Hepatic lobule and close-up view of two hepatocytes. Bile is formed by the liver cell and placed into a bile canaliculus. These canaliculi aggregate to form progressively larger bile ductules, culminating in the right and left hepatic ducts. Bile acids (BA) are reclaimed from sinusoidal blood by a Na^+-dependent cotransporter (NTCP, Na^+-taurocholate cotransporting polypeptide, also known an SLC10A1) and Na^+-independent mechanisms (OATP, organic anion transporting protein = SCLO1B3). A family of cytosolic proteins called ligandins bind bile acids. The function of the ligandins in bile transport is unknown. They may protect the cell from the solubilizing effects of the bile acids or they may help in cytoplasmic transport of the bile acids. Bile acids derive mostly from uptake from the sinusoidal blood, but some is synthesized anew each day from cholesterol. The bile acids are actively secreted into the bile canaliculus via the bile salt export pump (BSEP) which is an ATP-binding casette transporter, known as ABCB11. Phospholipids are secreted by the multidrug resistant receptor 4 (MDR4), also known as ABCB4. Cholesterol is also secreted by an ATP-binding casette transporter. These ATP-binding casettes are all dimers that bind two ATP, one for each subunit, and hydrolyze them during the transport cycle. The BSEP and MDR4 are both homodimers; the cholesterol transporter is a heterodimer consisting of ABCG5 and ABCG8. The secreted phospholipids, cholesterol, and bile acids aggregate to form micelles within the bile.

The activity of a microsomal cholesterol 7 hydroxylase determines the rate of formation of the primary bile acids, and this enzyme is inhibited by bile acids. Thus the synthesis of the bile acids is under negative feedback control. This degradation of cholesterol is the largest metabolic sink for cholesterol in the body. The primary bile acids are secreted into the bile by a primary active transport mechanism, the bile salt export pump (BSEP = ABCB11) as described in Figure 8.4.7.

In the intestine, these primary bile acids are altered by intestinal bacteria to produce **secondary bile acids**: **lithocholic acid**, **deoxycholic acid**, and **ursodeoxycholic acid**. The bile acids are reabsorbed into the portal blood by the terminal ileum. This portal blood returns to the liver and the bile acids are taken up from the sinusoidal blood by a Na^+-dependent cotransporter that links bile acid uptake to Na^+ entry (NTCP = SLC10A1, see Figure 8.4.7) and by a second mechanism on the basolateral membrane that does not require Na^+ (OATP = SLCO1B3, see Figure 8.4.7). This recycling of bile acids from liver to intestine and back to liver is called the **enterohepatic circulation** (see Figure 8.4.8). The entire bile acid pool of the body may turn over three to five times during a single meal.

The liver conjugates the —COOH group of the bile acids by covalently linking it with glycine or taurine. This occurs after the synthesis of the primary bile acids and

TABLE 8.4.1 Average Composition of Hepatic Bile

Components	Concentration (mM)
Na^+	140–165
K^+	3–7
Cl^-	77–117
HCO_3^-	12–55
Bile salts	3–45
Bilirubin	1–2
Phospholipids	140–810 mg%
Cholesterol	100–320 mg%

From R.H. Mosely, Bile Secretion, in T. Yamada, et al., eds., Textbook of Gastroenterology, vol. I, Lippincott Williams and Wilkins, Philadelphia, PA, 1999.

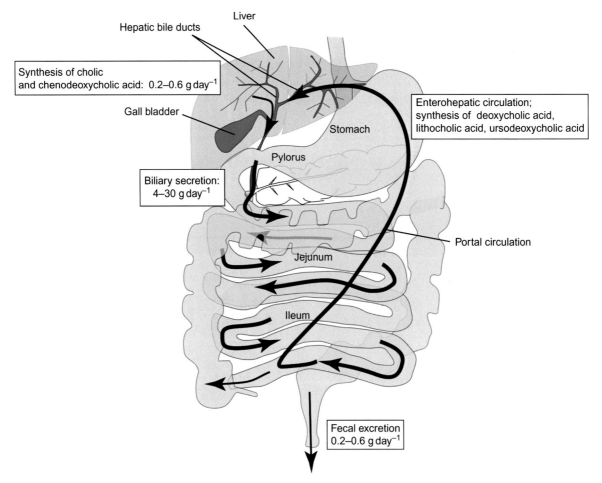

FIGURE 8.4.8 Enterohepatic circulation of the bile acids. Bile acids are synthesized in the liver, secreted into the bile, and stored temporarily in the gallbladder. After a meal, the gallbladder contracts and bile enters the duodenum at the sphincter of Oddi. Intestinal flora modifies the bile acids to form secondary bile acids. Both primary and secondary bile acids are absorbed into the portal blood at the terminal ileum. The liver takes up the bile acids from the portal blood and recycles them back into the bile. Numbers in boxes give the average bile acid amounts per day.

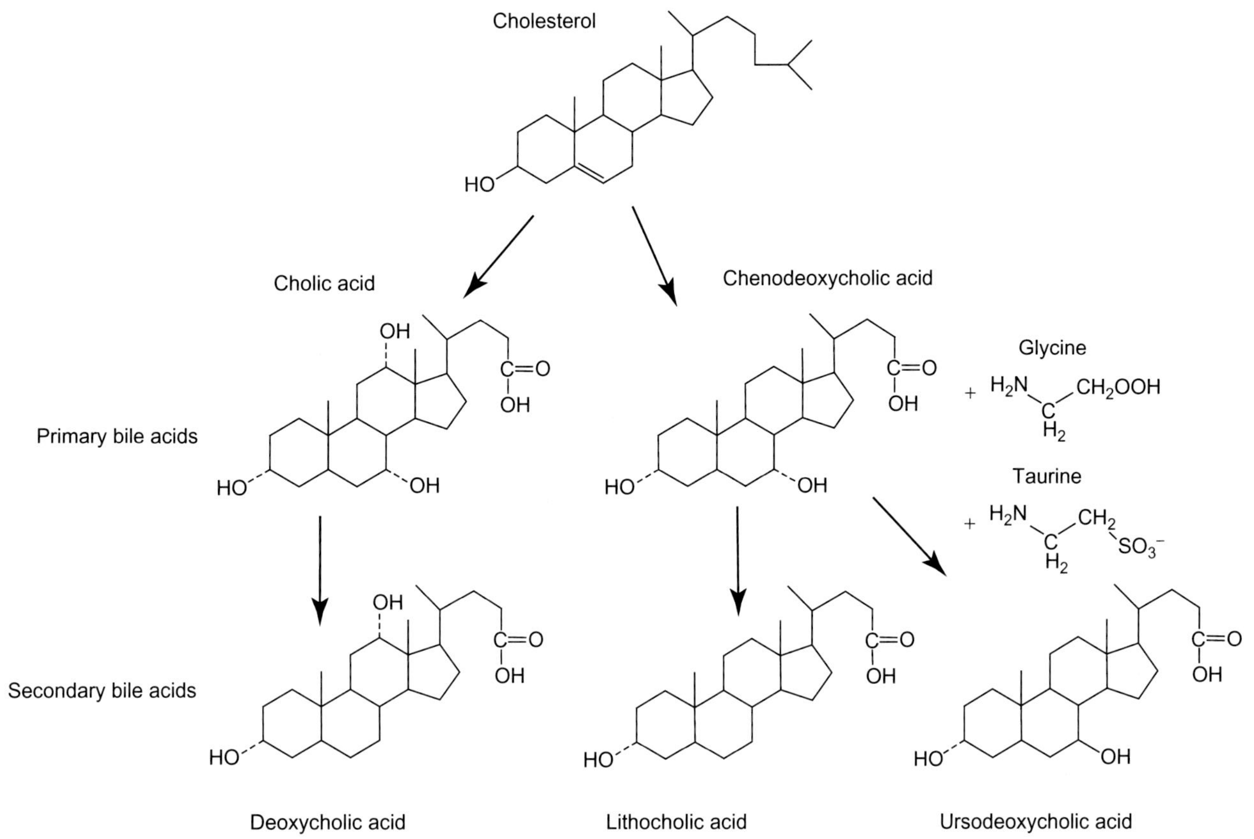

FIGURE 8.4.9 Structures of the primary and secondary bile acids. The primary bile acids are synthesized in the liver cells from cholesterol and are excreted into the bile as such. Intestinal bacteria convert the primary bile acids into secondary bile acids, which are reabsorbed along with unchanged primary bile acids in the terminal ileum.

during the hepatic phase of the enterohepatic circulation. Thus the bile contains cholic acid, glycocholic acid, taurocholic acid, and the corresponding free and conjugated forms of chenodeoxycholic acid, lithocholic acid, deoxycholic acid, and ursodeoxycholic acid. Figure 8.4.9 shows the chemical structures of the primary and secondary bile acids.

Bile acids have potent **surface activity** and can solubilize membrane proteins and lipids. They typically exhibit a **critical micellar concentration** or **CMC**. At low concentrations, below the CMC, the bile acids are monomers in solution. At higher concentrations, above the CMC, the bile acids aggregate to form micelles, small aggregates in which the hydrophilic hydroxyls of the bile acids face the water phase and the hydrophobic organic backbone of the steroid nucleus faces the interior of the micelle (see Chapter 8.5 for illustrations of these structures). In this form, the bile acids can absorb hydrophobic materials such as cholesterol and phospholipids to form **mixed micelles**. To prevent the bile acids from solubilizing the liver cells, the cytoplasmic bile acid concentration must be kept low. This is achieved by **ligandins**, a family of proteins that bind bile acids with high affinity. These ligandins have catalytic activity; they are glutathione S-transferases. But they also bind bile acids and bilirubin with high affinity. Their function in this regard is not established, but they probably protect the liver cell from the toxic effect of bile acids, while simultaneously carrying the bile acids to the bile canaliculus where the bile acids are excreted into the bile.

THE LIVER EXCRETES *XENOBIOTICS* (FOREIGN BIOLOGICALLY ACTIVE CHEMICALS)

The bile provides an excretory pathway for organic lipophilic xenobiotics such as drugs because these are not easily eliminated by the kidneys through the urine. Such compounds are poorly filtered by the glomerulus and are not readily secreted by the kidney tubules (see Chapter 7.2). The liver contains multi-specific enzymes called **monooxygenases** of the **cytochrome P450** family. Their general reaction is to insert hydroxyl groups into foreign compounds as shown in Figure 8.4.10.

Drugs or other foreign compounds such as pesticides, plant alkaloids, and organic pollutants can be recognized by nonspecific transporters called **MDR**, for **multidrug resistance-associated protein**. This protein is a member of a family of **ABC transporters**, so-named because each member of the family possesses an ATP-binding cassette. These ABC transporters split ATP and transport a variety of materials across membranes. The MDR2 protein, also known as ABCB4, secretes phospholipids across the canalicular membrane.

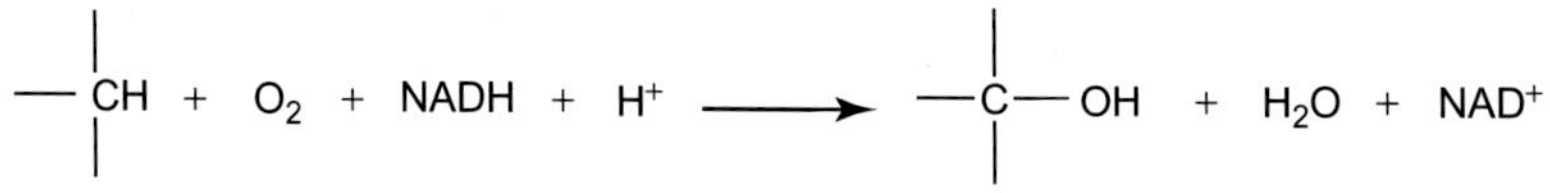

$$-\overset{|}{\underset{|}{C}}H + O_2 + NADH + H^+ \longrightarrow -\overset{|}{\underset{|}{C}}-OH + H_2O + NAD^+$$

FIGURE 8.4.10 Generalized reaction of liver monooxygenases. These enzymes use NADH and molecular oxygen to hydroxylate a variety of lipophilic xenobiotics to make them more water soluble and more easily excreted either in the bile or in the urine.

ABCG5 AND ABCG8 SECRETE CHOLESTEROL INTO THE BILE

Cholesterol enters liver cell metabolism two ways: by synthesis from acetyl CoA or from import from blood-borne lipoproteins. Lipoproteins are complexes of lipid and proteins called apolipoproteins. Liver cells take up a variety of lipoproteins including chylomicron remnants, HDL (high-density lipoproteins), and LDL (low-density lipoproteins). Both newly synthesized and imported cholesterol can be exported from the cell into the bile across the bile canaliculus. Two ABC transporters, ABCG5 and ABCG8, form a heterodimer that transports cholesterol into the bile.

THE *GALLBLADDER* STORES AND CONCENTRATES BILE AND RELEASES IT DURING DIGESTION

The liver makes 0.5–1.0 L of bile per day. During the interdigestive periods, much of this is directed along the cystic duct to fill the gallbladder, which stores the bile and concentrates it. The gallbladder itself can hold only some 15–60 mL of fluid. The gallbladder absorbs water by actively pumping Na^+ out of the fluid. Cl^- and HCO_3^- are also absorbed, and water is drawn out of the bile by the osmotic pressure of the absorbed electrolytes. In this way, the gallbladder can concentrate the bile some 10-fold.

CCK is released from I cells in the mucosa of the duodenum in response to amino acids and fat digestion products. As described earlier, it increases the enzyme secretion of the pancreatic acinar cells. In addition, CCK stimulates contraction of the gallbladder and relaxation of the sphincter of Oddi. The effects of CCK on gallbladder contraction in the intact biliary tree are mediated by its stimulation of cholinergic nerves. Since CCK is released upon the entry of chyme into the duodenum, this action serves to deliver bile to the intestine when it is needed.

THE BILE DUCT CELLS SECRETE A HCO_3^--RICH SOLUTION MUCH LIKE PANCREATIC DUCT CELLS

The bile ducts secrete a watery fluid using the same mechanisms as the pancreatic ducts. As with the pancreatic ducts, secretin stimulates bile duct secretion through increasing cAMP in the ductal cells. Secretin is secreted from the duodenal mucosa S-cells primarily in response to duodenal acidification. The overall flow of bile consists of a **bile acid-dependent bile flow** and a second component that is independent of bile acid

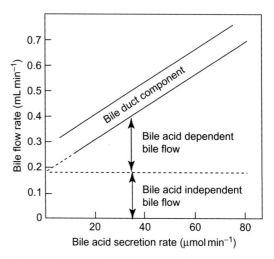

FIGURE 8.4.11 Components of bile flow. The actual relationship between biliary flow rate and bile acid secretion rate is probably curvilinear, but the diagram helps illustrate the division of bile flow into a bile acid-dependent flow and a bile acid-independent flow. On top of this flow, the bile ducts contribute fluid to augment the hepatic bile flow.

secretion and is attributed to the secretion of the electrolytes. This **bile acid-independent bile flow** is defined as the Y-intercept of the graph of bile flow against bile acid secretion, as shown in Figure 8.4.11.

Clinical Applications: Gallstones

Gallstones come in several varieties, among which are cholesterol gallstones and pigment gallstones. Cholesterol gallstones contain 50–75% cholesterol and most contain a pigmented center that suggests that a pigment stone first formed and acted as a center for crystallization (or **nidus**) for cholesterol. Cholesterol by itself is nearly insoluble in water. It is transported in blood and in bile solely by its association with other materials that are soluble. Thus the solubility of cholesterol in bile depends entirely on the concentrations of bile acids and phospholipids that hold cholesterol in solution. The condition of having gallstones has the technical name **cholelithiasis**, which derives from the roots **chole** (bile) and **lith** (stone). Bile that is supersaturated with cholesterol is called **lithogenic** bile. In persons who form gallstones, the liver makes lithogenic bile, but how this unstable solution can be formed is not yet known. Persons susceptible to cholesterol gallstones generally (but not always!) fit a pattern of the four Fs: **f**at, **f**air, **f**orty, and **f**ertile.

Often cholesterol gallstones are asymptomatic and they are discovered accidentally during tests for other reasons. Asymptomatic or "silent stones" need no treatment. Gallstones may also provoke a gallbladder "attack" which may include

(Continued)

> **Clinical Applications: Gallstones (Continued)**
>
> steady pain in the upper abdomen, in the back between the shoulder blades, or under the right shoulder; nausea or vomiting; abdominal bloating; excessive belching; indigestion; and colic (cramping of gastrointestinal sections). Stones can sometimes lodge in the cystic duct, causing stagnant bile and **acute cholecystitis**. Blockade of the common bile duct by a stone can produce **cholangitis** and **jaundice**. Yellow pigments arise when blood is broken down, and normally the liver excretes these in the bile. If the bile cannot be excreted, or if the liver fails, these pigments accumulate and give a yellow cast to skin and eyes. This condition is called jaundice. In addition, stones blocking the lower end of the bile duct where it enters the intestine can obstruct pancreatic secretions leading to **acute pancreatitis**. Since bilirubin in the bile secretions produces the color of the stools, clay-colored stools indicate lack of biliary secretions into the gastrointestinal tract. Sweating, chills, or low-grade fever and jaundice in conjunction with the other symptoms require immediate medical attention.
>
> Most symptomatic stones result in surgical removal of the gallbladder or **cholecystectomy**. This is one of the most common surgical procedures, with some 500,000 performed annually in the United States. Laparascopic cholecystectomy is most common and is used when there are no complications. This ordinarily does not seriously impair fat digestion because bile is still continuously produced by the liver. In some cases, the ability to digest dietary fat may be reduced. Other treatments include **oral dissolution therapy** in which cholesterol gallstones are dissolved by increasing the bile acid content of the bile by orally ingesting ursodeoxycholic acid (Actigall) or chenodeoxycholic acid (Chenix). Such therapy may take months or years to dissolve stones, and the stones will recur unless medication is continued for life. Small stones can be broken into tiny parts by **extracorporeal shockwave lithotripsy**, and the small stones can be passed into the gastrointestinal tract through the bile ducts. Shock waves are produced by an underwater high voltage discharge and the concentration of shock waves is achieved by reflection. Oral bile acid therapy prevents the recurrence of gallstones following lithotripsy. Extracorporeal shockwave lithotripsy is only sparingly used. Presently, the overwhelmingly favored treatment of symptomatic gallstones is cholecystectomy.

SUMMARY

The pancreas has both endocrine and exocrine functions. It produces the hormones insulin, glucagon, and somatostatin and also an exocrine secretion of a watery solution containing digestive enzymes. The exocrine acini produce and secrete proteolytic enzymes including trypsinogen, chymotrypsinogen, proelastase, and procarboxypeptidases A and B. These forms are all inactive and are activated in the intestine by a cascade begun by enterokinase that activates conversion of trypsinogen to trypsin. The pancreas also produces a trypsin inhibitor to prevent premature activation of trypsinogen within the pancreas. The pancreas also produces pancreatic amylase to digest starch, pancreatic lipase and co-lipase, phospholipase A_2, cholesterol esterase, RNAase, and DNAase.

The pancreas produces a watery solution that is rich in HCO_3^-. Secretin is released from the duodenum in response to acid, and secretin increases the activity of the CFTR (cystic fibrosis transmembrane conductance regulator) in the apical membranes of acinar cells. The CFTR secretes Cl^- which then drives HCO_3^- secretion through an exchange mechanism. Thus secretin increases the pancreatic secretion of a HCO_3^--rich solution. As the fluid flows down the pancreatic ducts, the HCO_3^- is exchanged for Cl^-, so that the composition of the pancreatic secretion depends on its flow rate: at high rates the HCO_3^- remains high because there is insufficient time for complete exchange for Cl^-.

Vagal efferents control pancreatic enzyme secretion that results from the sight, smell, and taste of food. CCK increases pancreatic secretion that results from food in the duodenum. Fat digestion products and amino acids potently stimulate CCK release from the duodenum, which excites vagal afferents to begin a vagovagal reflex to stimulate pancreatic enzyme secretion.

Bile consists of a HCO_3^--rich solution that contains bile pigments derived from blood breakdown products, mucin, bile acids, cholesterol, phospholipids, electrolytes, and detoxified xenochemicals. Bile is produced in the liver by the hepatocytes and collected first into tiny canaliculi, which progressively fuse to form the hepatic bile duct. The gallbladder stores and concentrates bile during interdigestive periods and releases it during meals. The cystic duct branches off the hepatic duct to fill the gallbladder. The common bile duct enters the duodenum at the ampulla of Vater. The sphincter of Oddi controls the opening of this duct. After a meal, the duodenum releases CCK that stimulates gallbladder contraction and relaxation of the sphincter of Oddi. The bile acids in the bile help digest and absorb fats.

The liver makes primary bile acids, cholic acid, and chenodeoxycholic acid, from cholesterol. In the intestine, bacteria convert these to secondary bile acids, litho-cholic acid, deoxycholic acid, and ursodeoxycholic acid. These are absorbed into the blood at the terminal ileum and come back to the liver in the portal circulation. Thus the primary and secondary bile acids are recirculated. The recycling of the bile acids is called the enterohepatic circulation. The entire bile acid pool may turn over several times during a single meal.

REVIEW QUESTIONS

1. What exocrine secretions does the pancreas make? Where do these things go?
2. What are the three phases of pancreatic secretion? What mediates them? What are the initial sensory stimuli in each case?
3. Where is bile formed? Trace its flow from liver to duodenum.
4. What is meant by "primary" bile acids and "secondary" bile acids? Which bile acids are primary

and which are secondary? How do the secondary bile acids get into the bile?
5. What is the "enterohepatic circulation"? Where are bile acids absorbed in the gastrointestinal tract?
6. What else is in bile besides the bile acids?
7. Some amount of circulating cholesterol is absorbed from dietary sources. This absorption requires the bile acids to solubilize the cholesterol in the gut so it can be absorbed. Cholestyramine is an ion-exchange resin that binds bile acids and therefore interferes with the enterohepatic circulation. Since bile acids are necessary for dietary cholesterol absorption and also are the main sink for cholesterol in the body, ingesting cholestyramine should lower blood cholesterol. Do you anticipate any negative side effects of cholestyramine ingestion?
8. What does the gallbladder do? What happens if you remove it (cholecystectomy) due to gallstones?
9. What is the main stimulus for gallbladder contraction?

Digestion and Absorption of the Macronutrients 8.5

Learning Objectives

- List three ways the intestine increases its surface area for digestion and absorption of nutrients
- Name the major enzyme involved in the gastric phase of protein digestion
- Describe the activation of pancreatic proteolytic enzymes and their major actions in the luminal phase of protein digestion
- Describe the enzymes involved in the brush border phase of protein digestion
- Describe in general terms how amino acids are transported from lumen to blood
- Explain why humans can digest starch but not cellulose
- List the products of amylase digestion of starch
- List the brush border enzymes involved in the final digestion of carbohydrates
- Describe lactose intolerance and explain the origin of its symptoms
- Name three major monosaccharides and describe how they enter the blood
- Describe what is meant by emulsification of lipids and how this is accomplished in the gut
- List the major lipolytic enzymes
- Define what is meant by "lipoprotein"
- Distinguish among chylomicron, VLDL, LDL, and HDL
- Describe the pathway for lipid digestion and absorption
- Describe the location and mechanism of bile acid absorption

THE INTESTINE INCREASES ITS SURFACE AREA BY FOLDS UPON FOLDS

Digestion involves breaking down the macronutrients (fats, carbohydrates, and proteins) into their constituent parts (fatty acids, glycerol, monosaccharides, and amino acids). This occurs in the intestine, and then the intestine absorbs these nutrients into the blood for use by the rest of the body. To do this, the intestine presents a large surface area to the ingested food because the absorption of nutrients occurs through this surface. The surface area is increased by three levels of folding (see Figure 8.5.1).

The first level of folding is the **folds of Kerckring** that increase the area about threefold. The folds of Kerckring are folds of the entire epithelium along with its vasculature and nerve supply. Lining the walls of the folds are the next level of folding, the **villi**, that increase the area another 10-fold. The villi are finger-like projections of the intestinal epithelium that extend outward from the surface of the folds about 0.5–1.5 mm. The villi in the duodenum are typically longer than those in the distal small intestine. These villi are covered with a layer of columnar epithelial cells, the **enterocytes**. The apical membrane of the enterocyte, facing the lumen, forms a forest of small projections called **microvilli** that increase the surface area another 20-fold. These structures are maintained by cytoskeletal elements that are anchored in the tips of the microvilli and in a web of cytoskeletal elements just below the microvilli, called the **terminal web**. Because of their appearance in the light microscope, the forest of microvilli is also called the **brush border**.

THE INTESTINAL LINING CONTINUOUSLY RENEWS ITSELF

Interspersed among the enterocytes are **goblet cells** that secrete **mucus** to line the luminal surface. Between the villi are pits called the **crypts of Lieberkuhn**. Cells in the crypts are continuous with the villus layer. Some crypt cells secrete water and electrolytes, whereas other cells, **enterochromaffin cells**, secrete 5-hydroxytrypamine upon mechanical stimulation to excite **intrinsic primary afferent neurons**. Still other cells in the crypts, epithelial stem cells, divide to renew the villus lining. The daughter cells migrate up the villus and differentiate to form goblet cells or absorptive cells. As they migrate, the cells express different proteins. At the tips of the villus, the cells are extruded into the lumen. Thus there is a continuous parade of cells from the crypts toward the lumen. The entire epithelium renews itself every 4–5 days.

The villi also surround a central lymph vessel, the **central lacteal**, and arteriolar and venule capillaries. The close approach of these vessels to the absorptive epithelium reduces the diffusion distance from the absorptive cells into the blood and the lymph.

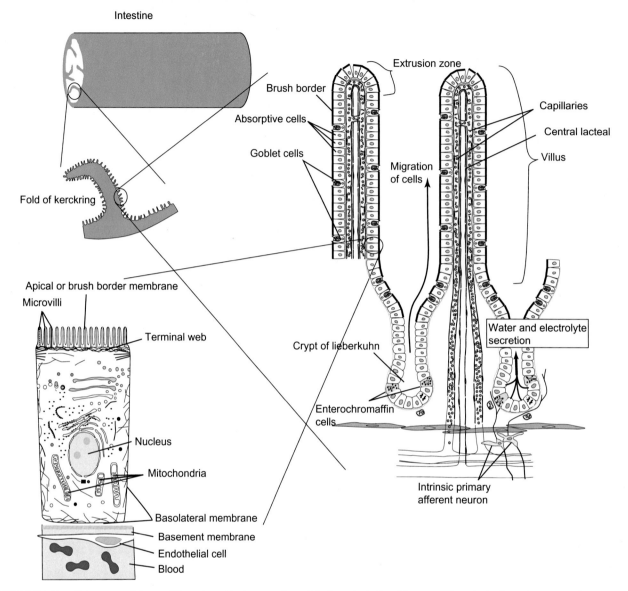

FIGURE 8.5.1 Overall structure of the small intestine with close-up view of the intestinal villus. The overall intestine's dimensions approximate a simple cylinder about 4 cm in diameter and about 2.8 m in length. The folds of Kerckring increase the surface area about threefold. On these folds are the villi, finger-like projections of the intestinal lining, which further increase the surface area about 10-fold. The villi are lined with enterocytes, whose apical membrane is covered by microvilli that increase its absorptive surface an additional 20-fold. The intestinal villi arise from pits called crypts of Lieberkuhn where cells divide and then migrate up the villus, where eventually they are extruded at the tips. As they migrate up the villus the cells differentiate and mature into goblet cells, which secrete mucus, and absorptive cells that take up the digested food and transfer it to the blood. The total area of the intestine is thus increased some 600-fold from about 0.3 m^2 to about 200 m^2.

PROTEIN DIGESTION OCCURS IN A GASTRIC PHASE AND AN INTESTINAL PHASE

PROTEIN DIGESTION BEGINS IN THE STOMACH

The acid in the stomach **denatures** many proteins—it unfolds them and exposes internal peptide bonds for proteolytic attack. The acid also activates pepsinogens to pepsins. The pepsins are a class of **endopeptidases** that cleave peptide bonds in the middle of the polypeptide chain. The pepsinogens are secreted by the chief cells in the gastric glands. Pepsin is active only at pH < 4. Upon mixing with the alkaline duodenal contents (provided by HCO_3^- in the pancreatic and biliary secretions), the pepsins become inactive.

The digestion products of pepsin cause release of gastrin from G-cells in the antrum, thereby stimulating acid secretion (see Chapter 8.2). They are also potent secretagogues for CCK release from the duodenum, thereby indirectly stimulating pancreatic enzyme secretion and gallbladder contraction. Despite these effects, people with total gastrectomies are capable of fully digesting dietary protein. The gastric phase is not essential to protein digestion.

THE INTESTINAL PHASE CONSISTS OF A LUMINAL PHASE, BRUSH BORDER PHASE, AND INTRACELLULAR PHASE

The exocrine pancreas secretes three endopeptidases (**trypsin**, **chymotrypsin**, and **elastase**) and two

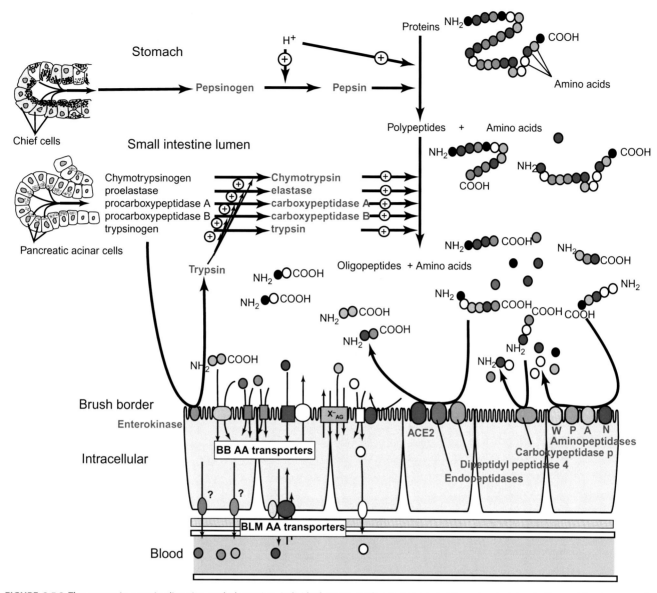

FIGURE 8.5.2 The events in protein digestion and absorption. Individual amino acids are represented by colored circles. In the gastric phase, pepsins break proteins down into polypeptides and some amino acids. In the intestinal phase, pancreatic proteolytic enzymes are activated by enterokinase and further break down proteins to smaller oligopeptides and amino acids. A number of proteases in the brush border complete digestion and feed the released amino acids into transport mechanisms that take the amino acids and some di- and tri-peptides up into the enterocytes. See text for details.

exopeptidases (**carboxypeptidase A** and **carboxypeptidase B**) in inactive forms. **Enterokinase** on the brush border begins a cascade of activation of the pancreatic enzymes by converting trypsinogen into trypsin. The activated trypsin converts more trypsinogen to trypsin and activates all of the remaining pancreatic proteases. The cascade of events is shown in Figure 8.5.2. The endopeptidases cleave proteins in the middle of their chains with specificity. Trypsin, for example, cleaves the peptide bonds in which basic amino acids (lysine and arginine) contribute the carboxyl group. Chymotrypsin cleaves those peptide bonds in which aromatic amino acids (tyrosine, phenylalanine, and tryptophan) contribute the carboxyl group. The carboxypeptidases cleave single amino acids off the free carboxyl ends of proteins. Carboxypeptidase A cleaves off aromatic or branched chain amino acids; carboxypeptidase B cleaves off basic amino acids. The end result of pancreatic proteolysis is some free amino acids and a mixture of oligopeptides.

The brush border contains a variety of peptidases to complete protein digestion. **Aminopeptidase N** cleaves off one neutral amino acid and **aminopeptidase A** cleaves off one anionic amino acid (glutamate, aspartate) from the amino terminus. **Aminopeptidase P** cleaves off one amino acid which is bound to a penultimate proline, P. **Aminopeptidase W** cleaves off one amino acid which is bound to a penultimate tryptophan, W. **Carboxypeptidase P** cleaves a single amino acid preferentially bound to a penultimate proline, P. **Dipeptidylcarboxypeptidases** are like carboxypeptidase except they cleave off dipeptide fragments from the carboxyl end. There are three types of dipeptidylcarboxypeptidases: dipeptyl peptidase 4 and **angiotensin-converting enzyme** (ACE) 1 and 2. Two endopeptidases also reside on the brush border.

Inside the enterocytes, intracellular proteases split small peptides into their components. There are several of these. An aminotripeptidase cleaves the amino acids off the NH$_2$ terminus of tripeptides. The enterocytes contain a variety of dipeptidases that are specific for dipeptides containing specific amino acids. Some peptides are resistant to proteolysis and are not degraded.

SPECIFIC CARRIERS MOVE AMINO ACIDS ACROSS THE BRUSH BORDER AND BASOLATERAL MEMBRANES

Gastric, pancreatic, and brush border proteases reduce ingested protein to a mixture of amino acids and small peptides of from two to six amino acids. Six major transport systems carry these amino acids and small peptides across the brush border membrane, as listed below and shown in Figure 8.5.3.

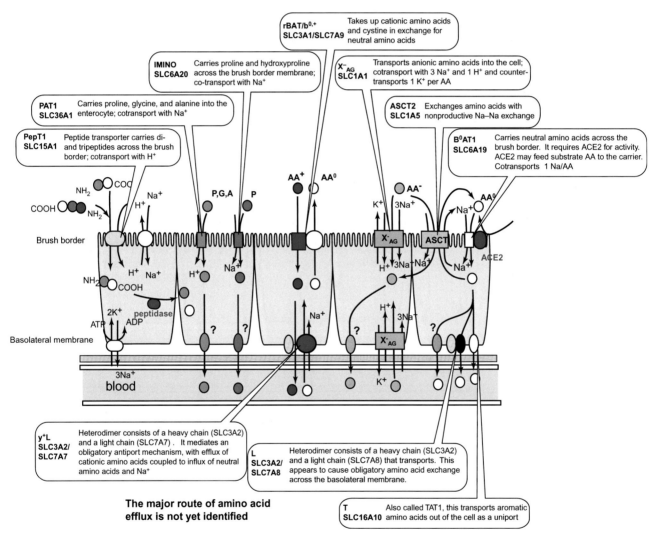

FIGURE 8.5.3 The cellular mechanisms responsible for amino acid absorption from the intestinal lumen into the blood. Proteolytic digestion produces free amino acids and small peptides in the lumen of the intestine. Some of these small peptides are digested on the surface of the enterocytes by aminopeptidase, carboxypeptidases, and dipeptidase that cleave amino acids or dipeptides off small peptides. The free amino acids are taken up into the cell by a set of transporters. Some of these require Na$^+$ as a cotransported ion and some do not. There are six identified transporters in the brush border that carry neutral amino acids (AA^0), anionic amino acids (AA$^-$), cationic amino acids (AA$^+$), imino amino acids, and P (proline), G (glycine), and A (alanine). The anionic amino acid transporter, X$^-_{AG}$, also cotransports 3 Na$^+$ and 1 H$^+$ and countertransports 1 K$^+$. The ASCT on the brush border produces no net amino acid uptake, but the neutral amino acid carrier, B^{0}AT1 (=SLC6A19), can supply neutral amino acids so that ASCT can take up a variety of amino acids in exchange for the neutral amino acids. The neutral amino acid carrier requires ACE2 (angiotensin-converting enzyme, a protease) for activity, and evidence suggests it also binds aminopeptidase N. This may form a complex that feeds amino acids directly to the transporter. The cationic amino acid transporter (rBAT/b$^{0,+}$ (=SLC3A1/SLC7A9)) also mediates cationic amino acid influx coupled to neutral amino acid efflux. SLC6A20 carries proline (P) specifically and is a cotransporter. Proline, glycine, and alanine (P,G,A) are carried by PAT1 (=SLC36A1) with energy derived from H$^+$ influx into the cell. Absorbed amino acids are transported out across the basolateral membrane by a different set of transporters. The major efflux pathway has not yet been identified. Dipeptides are also taken up by the cell, using H$^+$ as a cotransported ion. The dipeptide carrier exhibits a broad specificity. Some dipeptides and tripeptides are split by proteases within the enterocytes, and the resulting amino acids are transported across the basolateral membrane by the free amino acid transporters. Some dipeptides and tripeptides are absorbed into the blood intact.

Neutral amino acid system: B⁰AT1 (=SLC6A19) carries neutral amino acids (M,L,I,V,Q,N,F,C,A,S,G,Y,T,H,P,W,K). This carrier is a symport, using the Na^+ gradient to drive the inward flux of amino acids. The stoichiometry is 1 Na^+:1 amino acid. Its activity requires ACE2, a dipeptidylcarboxypeptidase that apparently releases amino acids that are substrates for B⁰AT1. This carrier also associates with aminopeptidase N.

Anionic amino acid system: X^-_{AG} (=SLC1A1) carries anionic amino acids, aspartic acid, and glutamic acid (E,D). This cotransports 3 Na^+ and 1 H^+ for each amino acid, and return of the carrier is facilitated by K^+.

Cationic amino acid system: rBAT/b⁰,⁺ (=SLC3A1/SLC7A9) is a heterodimer that carries the cationic amino acids (R,K, Cystine). This works through an obligatory exchange mechanism so that there is no net amino acid transport.

Proline transport system: IMINO (=SLC6A20) carries proline and hydroxyproline, and will not carry glycine. Its transport is coupled to the Na^+ electrochemical gradient.

Proline and glycine transport system: PAT1 (=SLC36A1) carries proline or glycine (P,G,A) coupled to H^+ entry.

Di- and tri-peptide transport systems: (PepT1 = SLC15A1) carries di- and tri-peptides across the brush border membrane, coupled to H^+ cotransport. Most di- and tripeptides are hydrolyzed but some di- and tripeptides exit the cell to be absorbed intact into the circulation.

DISTINCT CARRIERS TRANSPORT AMINO ACIDS ACROSS THE BASOLATERAL MEMBRANE

Amino acid absorption is completed by the transport of amino acids across the basolateral membrane (BLM) of the enterocyte where they can enter the blood. The main mechanism for efflux of amino acids at the BLM has not yet been identified. However, several mechanisms that partially explain amino acid efflux are known. These are listed below and illustrated in Figure 8.5.3.

Aromatic amino acids: The T system, also known as TAT1 (=SLC16A10), is thought to mediate the transport of F, Y, and W, across the basolateral membrane.

L system: This system consists of a heterodimer, with SLC3A2 making up a heavy chain that appears to direct the transporter to the BLM. The light chain is SLC7A8. It appears to mediate an obligatory exchange of amino acids.

y⁺L system: This system also is a heterodimer, with SLC3A2 and SLC7A7. It also mediates an obligatory exchange of amino acids, with the influx of amino acids linked to Na^+ cotransport.

CARBOHYDRATES ARE MAINLY DIGESTED IN THE SMALL INTESTINE

In Western cultures, 200–300 g of dietary carbohydrates provide 40–50% of the daily caloric intake. Another 30–40% is provided by fats, with the remainder, about 15%, deriving from protein. These carbohydrates are mostly plant products. Lactose (milk sugar) and small amounts of glycogen in meat are the only animal carbohydrates in the diet.

Carbohydrates, or **saccharides**, are organic compounds containing one or more aldehyde or ketone groups and multiple hydroxyl groups. **Monosaccharides** such as **glucose**, **galactose**, and **fructose** have a single aldehyde or ketone group. **Disaccharides** such as **sucrose**, **lactose**, or **maltose** contain just two monosaccharide units. Sucrose consists of glucose and fructose; lactose is made of glucose and galactose; and maltose is composed of two glucose molecules. Oligosaccharides consist of 2–10 monosaccharide units, and polysaccharides are longer yet. About 50% of the calories derived from carbohydrates come from starch. The two main varieties of starch, α-amylose and **amylopectin**, are found in grains such as wheat, barley, and rice and also in many vegetables. Amylose is a linear polymer of glucose units linked by α(1,4)-glycosidic bonds. Amylopectin is a branched chain of glucose units, with α(1,6)-glycosidic linkages occurring at branch points every 15–20 glucose units. These starches form gels upon hydration, which gives certain foods part of their characteristic "mouth feel." The structure is shown in Figure 8.5.4.

Starch digestion has a luminal phase and a membrane phase. The luminal phase is due to α-amylase, an enzyme secreted by the salivary glands and by the pancreas. The α-amylase secreted by the parotid gland originates from the AMY1 gene, whereas the pancreatic product derives from the AMY2 gene, and their products show 94% sequence homology. Salivary amylase is of minor importance because the acid pH in the stomach inactivates the enzyme. Amylase digestion produces a mixture of **maltose** (glucose–glucose linked by α(1,4)-glycosidic bond), **maltotriose** (three glucose molecules linked by α(1,4)-bonds), and structures called **limit dextrins** (see Figure 8.5.4).

INDIGESTIBLE CARBOHYDRATES MAKE UP PART OF DIETARY FIBER

The digestive enzymes in humans can hydrolyze only some carbohydrates bonds. The undigestible carbohydrates make up part of **dietary fiber**. Indigestible fiber can be insoluble (e.g., cellulose) and soluble (e.g., hemicellulose, pectin, and gums). Fibers containing β(1,4) or β(1,3) linkages are indigestible because these bonds are immune to amylase attack.

THE BRUSH BORDER COMPLETES STARCH DIGESTION

The brush border contains several key enzymes that digest the products of luminal digestion to produce monosaccharides. These enzymes are **sucrase-isomaltase**, **lactase**, **maltase-glucoamylase**, and **trehalase**. Sucrase-isomaltase is a single gene product that

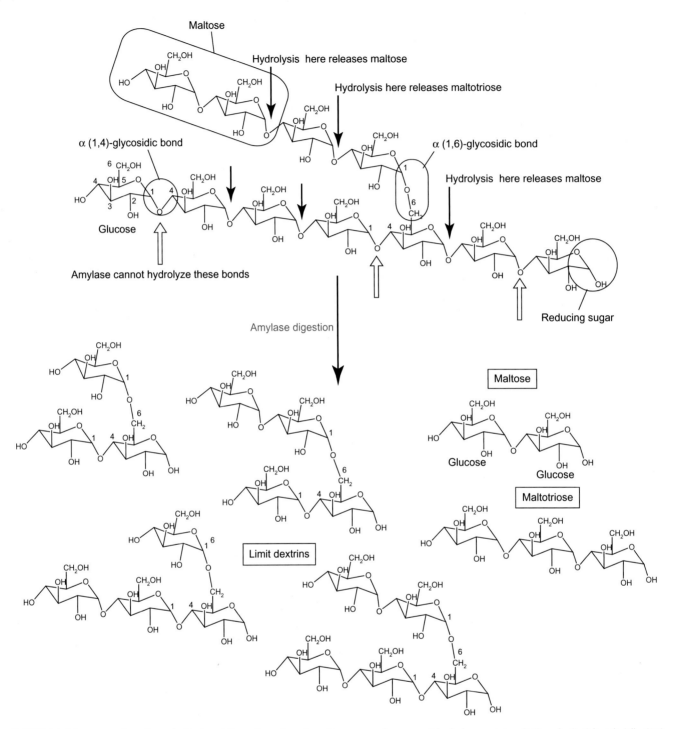

FIGURE 8.5.4 Structure of amylopectin (top) and the action of amylase. Because amylase cannot hydrolyze certain α(1,6) and α(1,4) bonds (all α(1,6) bonds and those α(1,4) bonds at the end of the chains or adjacent to α(1,6) bonds), its digestion of amylopectin is incomplete. Amylase's complete digestion of starch produces maltose (a glucose–glucose disaccharide), maltotriose, and a set of limit dextrins. The set of limit dextrins shown would be produced from the substrate shown.

has two active sites. The **sucrase** site splits sucrose into glucose and fructose. These two monosaccharides can then be absorbed by brush border transporters. The **isomaltase** active site cleaves maltose at its α(1,4) bond and it cleaves limit dextrins at their α(1,6) bond. This enzyme site is also called α-dextrinase. Lactase on the brush border membrane splits dietary lactose, obtained solely from human milk or dairy products, into galactose and glucose, which are both absorbed. Maltase-glucoamylase has two subunits that cleave the α(1,4)-glycosidic bond at the nonreducing end of oligosaccharides, thereby releasing the terminal glucose. Trehalose is an α(1,1)-linked dimer of glucose that is present in some natural foods. Trehalose breaks this bond to form glucose monosaccharides that can be absorbed.

GLUCOSE, FRUCTOSE, AND GALACTOSE ABSORPTION IS CARRIER MEDIATED

Complete digestion of carbohydrates produces three monosaccharides: **glucose**, **galactose**, and **fructose**. These are absorbed into the absorptive cell by two carriers. Galactose and glucose are both carried into the cell by secondary active transport, using Na^+ as the cotransported ion. The carrier is called **SGLT1** (=SLC5A11) for **s**odium **g**lucose-**l**inked **t**ransporter. Fructose is carried into the cell by facilitated diffusion by **GLUT5** (=SLC2A5), which, despite its name, appears to be specific for fructose in humans.

The final step in absorption of monosaccharides requires their transfer to the blood. The monosaccharides penetrate the basolateral membranes by carrier mechanisms. Glucose and fructose exit the cell by the **GLUT2** (=SLC2A2) carrier, which uses a facilitated diffusion mechanism. The process of carbohydrate digestion and absorption is illustrated in Figure 8.5.5.

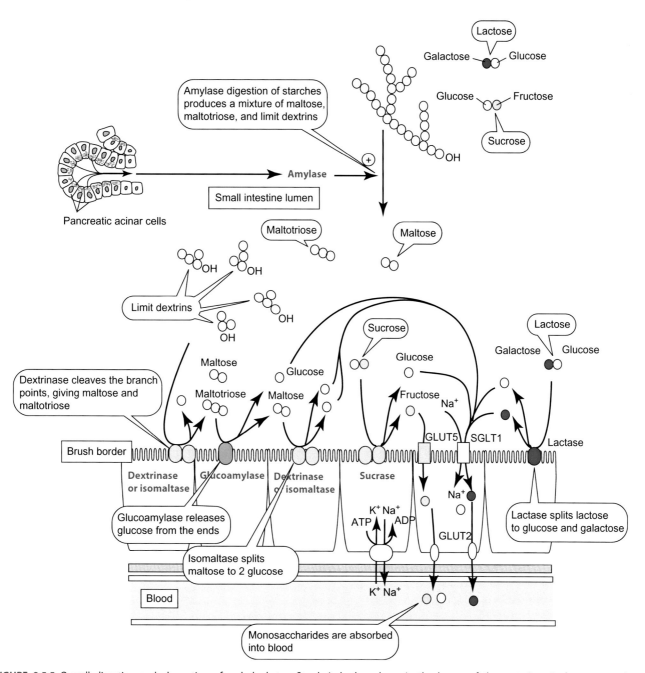

FIGURE 8.5.5 Overall digestion and absorption of carbohydrates. Starch is broken down in the lumen of the gastrointestinal tract to maltose, maltotriose, and limit dextrins. Dextrinase, which is the same enzyme as isomaltase, breaks the limit dextrins into maltotriose and maltose. Glucoamylase breaks maltotriose to glucose and maltose and isomaltase completes the digestion of maltose to two glucose molecules. Lactase cleaves lactose to galactose + glucose. Ordinary table sugar, sucrose, is broken down to glucose and fructose by sucrase, which is the same enzyme as dextrinase and isomaltase. Glucose and galactose are taken up into the cell by SGLT1; fructose is taken up by GLUT5. All of the monosaccharides leave the cell via GLUT2 on the basolateral membrane.

LIPID DIGESTION BEGINS WITH EMULSIFICATION

The defining characteristic of lipids is their insolubility in water and their solubility in organic solvents. Digestion of dietary lipids first requires access to the lipids, which is achieved by their dispersion into a stable form with a large surface area. This dispersion of lipid in an aqueous environment is called **emulsification**. Emulsification begins in the stomach with mechanical mixing and absorption of oligopeptides onto the lipid surface. In this crude emulsion, most of the lipid occupies the core of the lipid droplets with little on the surface where digestion occurs. In the small intestine, the bile salts and lipolytic digestion products form finer dispersions of lipid in which the surface area is greatly increased.

MOST LIPOLYTIC ACTIVITY OCCURS IN THE SMALL INTESTINE

Many experimental animals produce a **lingual lipase** and a **gastric lipase**. In humans, lingual lipase makes little or no contribution to the preduodenal lipase activity. Cells in the fundus of the stomach in humans secrete a 43-kDa lipase. This enzyme has a low pH optimum and preferentially cleaves the fatty acyl ester bond at position 3, producing fatty acid and diglyceride. The bulk of lipase activity comes from the pancreas.

The high $[H^+]$ in the stomach binds to the $-COO^-$ group at the ends of fatty acids, keeping them in the uncharged $-COOH$ form. This uncharged form stays dissolved in the hydrophobic core of lipid droplets. In the duodenum, the HCO_3^- neutralizes the acid from the stomach, and the fatty acids are ionized between pH 6 and 7.5. These charged fatty acids line the surface of the droplets and aid in their dispersion. The pancreas secretes a 50-kDa lipase and a 10-kDa colipase. The **pancreatic colipase** binds to the triglyceride surface with the help of the **bile acids** and anchors **pancreatic lipase**. In the absence of colipase, bile acids remove pancreatic lipase from the lipid surface and thereby inhibit lipolysis. The pancreatic enzymes **phospholipase A2** (MW 13.6 kDa) and **cholesterol esterase** also absorb to this surface. Pancreatic lipase hydrolyzes the positions 1 and 3 of triglycerides, releasing fatty acids and 2-monoglyceride. This reaction is shown schematically in Figure 8.5.6. The surface

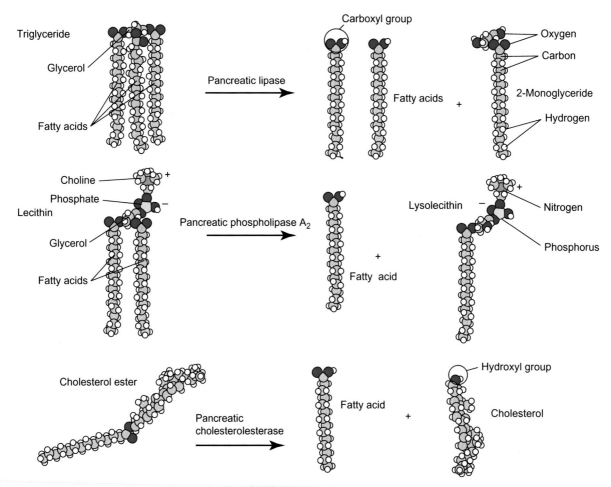

FIGURE 8.5.6 Lipolytic digestion by pancreatic enzymes in the lumen of the small intestine. Pancreatic lipase, anchored and activated by pancreatic colipase, breaks triglycerides down into fatty acids and monoglycerides. Pancreatic phospholipase cleaves phospholipids into fatty acids and lysophospholipids. Cholesterol esterase cleaves the fatty acid off cholesterol esters.

activity of these products is higher than that of the starting triglyceride, and they aid in the dispersion of lipids in the lumen. Pancreatic phospholipase A_2 hydrolyzes the fatty acyl ester bond at position 2 in phospholipids, releasing a free fatty acid and lysolecithin, or **lysophosphatidylcholine**. Like free fatty acids, lysolecithin helps disperse the lipid droplets (see Figure 8.5.6). Cholesterol esters are cleaved by pancreatic cholesterol esterase and absorbed into enterocytes as free cholesterol. The pancreatic cholesterol esterase requires bile acids for activity.

All of the lipid hydrolysis products are removed from the large lipid droplets to form complexes with bile acids and phospholipids called **mixed micelles**. Bile acids stabilize mixed micelles because of their unique topology (see Figure 8.5.7). The mixed micelles form a tiny disk-shaped aggregate (see Figure 8.5.8).

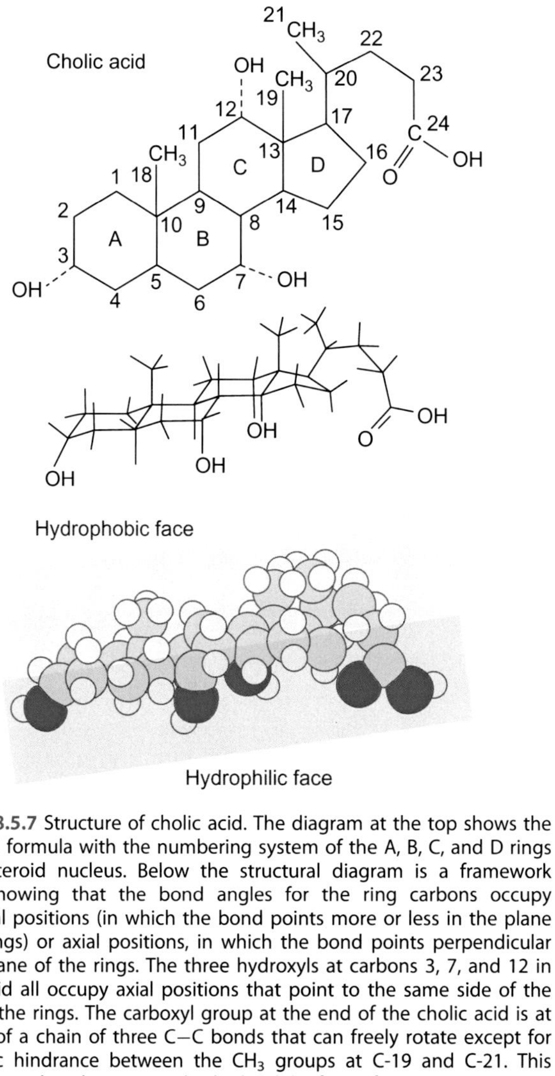

FIGURE 8.5.7 Structure of cholic acid. The diagram at the top shows the structural formula with the numbering system of the A, B, C, and D rings of the steroid nucleus. Below the structural diagram is a framework model showing that the bond angles for the ring carbons occupy equatorial positions (in which the bond points more or less in the plane of the rings) or axial positions, in which the bond points perpendicular to the plane of the rings. The three hydroxyls at carbons 3, 7, and 12 in cholic acid all occupy axial positions that point to the same side of the plane of the rings. The carboxyl group at the end of the cholic acid is at the end of a chain of three C–C bonds that can freely rotate except for the steric hindrance between the CH_3 groups at C-19 and C-21. This forces the carboxyl group to the hydrophilic face of the molecule. Thus the plane of the steroid ring structure divides the molecule into a hydrophilic half and a hydrophobic half. The bottom diagram shows a space-filling model in which the hydrophilic groups (in blue) are clearly separated from the hydrophobic groups.

HYDROLYSIS PRODUCTS OF LIPIDS ARE ABSORBED AND THEN REPACKAGED INTO *LIPOPROTEINS*

The mixed micelles deliver lipolytic products to the microvilli membrane. The mechanism of transport across the brush border membrane is not yet established, but it may involve a number of lipid-binding proteins including FABP pm (fatty acid binding protein, plasma membrane), FATP4 (fatty acid transport protein), CD36 (cluster of differentiation 36), cytosolic FABP (fatty acid binding protein), and other proteins such as Cav-1 (caveolin-1) and clathrin. Similarly, the mechanism for cholesterol absorption is not established, but it involves NPC1-L1 (Nieman = Pick C1-like 1) protein and it is specifically inhibited by some drugs (**ezetimibe**).

The current hypothesis is that at least some fatty acids, such as the short-chain fatty acids, can enter the enterocyte by simple diffusion. Long-chain fatty acids and 2-monoglycerides may enter either by diffusion or by endocytosis mediated through caveolin to form **CEV—caveolae endocytotic vesicles**. These vesicles then fuse with the smooth endoplasmic reticulum and the lipids are reesterified to form triglycerides through the sequential action of monoacylglycerol acyl transferase (MGAT) and diacylglycerol acyl transferase (DGAT). Cholesterol is also esterified by acyl cholesterol acyl transferase (ACAT). The reesterified triglycerides, phospholipid, and cholesterol are packaged into **lipoproteins** in the Golgi apparatus of the enterocytes by **microsomal triglyceride transfer protein (MTP)**. Lipoproteins are a class of particles found in the lymph and the blood that contain both lipid and protein coats called **apolipoproteins**. The protein coats help solubilize the lipids and also allow cells to latch onto the lipoproteins. The intestine makes a variety of lipoproteins including **chylomicrons**, very low density lipoprotein (**VLDL**), and high density lipoprotein (**HDL**). These are secreted at the basolateral membrane and reach the lymph and blood vessels that perfuse the intestine. The chylomicrons are transported solely in the lymph. The lymph ducts coalesce to drain into the **thoracic duct** that empties into the subclavian vein and thence into the systemic circulation. After a fatty meal, the lymph vessels appear white because of the light scattering of the many chylomicrons. This white color of the lymphatics gives the **lacteals** their name because they look milky white. The lymph transports about 50% of the absorbed lipids, whereas the portal vein carries the other 50%. The shorter the fatty acyl chain, the larger the percentage of absorbed lipids that is carried in the blood instead of the lymph. Chylomicrons are about 80–500 nm in diameter, are coated with apolipoproteins A-I, A-II, and B48, and are 95% lipid by mass, mostly triglyceride. In the circulation, they pick up apolipoproteins E and C. VLDLs are much smaller than chylomicrons, some 30–80 nm in diameter, are coated with apolipoproteins A-IV and B48, and consist of about 30% triglycerides and 40%

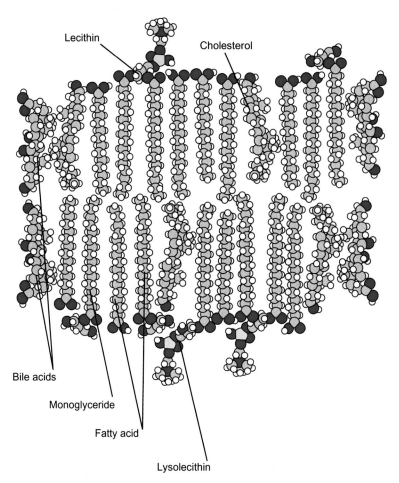

FIGURE 8.5.8 Cross-section through a mixed micelle. The core of the mixed micelles in the intestinal lumen contains lipolytic products such as monoglycerides, free fatty acids, phospholipids, lysophospholipids, and cholesterol. The hydrophobic domains of these lipids face the hydrophobic core of the micelle while their polar ends face the aqueous phase. The sides of the hydrophobic core are coated with bile acids that absorb to it. Their hydrophobic face covers the core of the micelle and their hydrophilic face interacts with water. In this way the bile acids stabilize the structure. The entire structure is actually disk shaped, and the bile acids coat a cylindrical surface.

Clinical Applications: Lactose Intolerance

The defining characteristic of mammals is that new born animals are fed milk by their mothers. Milk contains lactose as its major carbohydrate, and so it is crucial that the young animal be able to digest and absorb this disaccharide of galactose and glucose. **Lactase** is the only enzyme that can digest lactose. After weaning, many persons lose much of their ability to digest lactose. They therefore become **lactose intolerant**. Ingestion of too much lactose produces a variety of symptoms including nausea, cramps, bloating, and diarrhea. The lactose passes undigested into the colon where bacteria have a feast on the free sugar. They produce hydrogen gas which can be detected in the exhaled air as part of the diagnosis of lactose intolerance. (More than 20 ppm H_2 in exhaled air after ingestion of 1 g lactose per kg body weight.) The undigested lactose also exerts an osmotic pressure that retains water in the intestine. These two consequences, osmotic diarrhea and excess gas production, can produce an "explosive" diarrhea. The degree of diarrhea and gas depends on the amount of lactose ingested. Even "lactose intolerant" people generally can tolerate small amounts of lactose. Conversely, persons who normally have no problems digesting and absorbing lactose can be overwhelmed by excessive intake.

Lactose increases calcium absorption from the intestine, but how it accomplishes this feat is unknown. Milk has a lot of calcium, and its consumption aids the young animal in its early rapid growth. Lactose aids this growth. Calcium absorption from the intestine requires the hormone $1,25(OH)_2$ cholecalciferol, which is derived from vitamin D (cholecalciferol) by two successive hydroxylation reactions. Hydroxylation at C-25 occurs in the liver and hydroxylation at C-1 occurs in the kidney. The skin makes vitamin D when it is exposed to ultraviolet radiation from the sun, and so it is sometimes called "the sunshine vitamin." People with adequate exposure to the sun have no dietary requirement for vitamin D. This and its mechanism of action, similar to that of the classic steroids, argue that vitamin D is not a vitamin, but a hormone.

In primary lactose malabsorption, lactase activity is high at birth and decreases in childhood to remain low in the adult. This is the "normal" condition for humans. Typically 70–100% of adults worldwide are lactose intolerant, with the exception of the population of Northern and Central Europe and their descendants in America and Australia. In North America, about 15% of the white population has lactose malabsorption, whereas about 80% of the black population and 53% of the Hispanic population have lactase insufficiency. The cause of this ethnic distribution is unknown, but one interesting hypothesis is that it arose from the adaptive response to diminished exposure to sunlight in northern climates, both as a result of the tilt of the earth and from covering the skin with insulating clothes. Reduced ultraviolet illumination of the skin reduces synthesis of vitamin D in the skin, which is the only hormone that regulates intestinal absorption of dietary calcium. Over time, these indigenous people shed dark skin pigmentation to favor vitamin D synthesis, but many retained the tanning reaction to protect the skin from the harmful effects of UV radiation. They also retained intestinal lactase into adulthood to enable them to consume dairy foods throughout life. Dairy foods contain the most calcium in a favorable ratio to phosphorus. Thus light skin pigmentation, lactose tolerance, domestication of cattle, and dietary consumption of dairy products were physical and cultural adaptations that coevolved to improve nutrition of neolithic humans who migrated north from Africa. There is no proof for this hypothesis, but it is supported by the observation that even in Europe the indigenous people show a gradation of skin color and lactose tolerance, with the whitest and most tolerant people at the northernmost reaches.

Persons with lactose intolerance can find relief by eating **yogurt**. Yogurt contains live cultures of the bacterium *Lactobacillus acidophilus* that contain β-galactosidase activity that splits lactose into glucose and galactose. The lactase activity survives passage through the acid stomach if it is associated with bacterial cell walls that protect it. Yeast lactase (**lactaid**) can be ingested simultaneously with dairy products to improve lactose tolerance.

Clinical Applications: The Benefits of Dietary Fiber

The modern diet in Western cultures consists of high caloric density foods that have been highly enriched. Much of the nonnutritive components of food have been stripped away by milling and refining. In the last 30 years, we have come to realize that these indigestible components of food are an indispensible part of a healthy diet and that they are needed for normal gastrointestinal function.

Chemical definition of dietary fiber is difficult because of its diversity. Fiber includes cellulose, hemicellulose, pectins, gums, β-glucans, and lignins. Lignin is a mixture of phenolic compounds and is not a polysaccharide.

Much of the current interest in dietary fiber stems from the work of Denis Burkitt and Hugh Trowell, physicians who practiced medicine in Africa after World War II. They postulated that the "roughage" in the African diet afforded them some protection from a variety of ailments that afflicted persons in Western countries. These included cardiovascular disease; colonic dysfunction including diverticulitis, ulcerative colitis, constipation, and colonic cancer; cholelithiasis; appendicitis; hemorrhoids; diabetes; and obesity. The role of dietary fiber in many of these conditions remains to be established, but the effects appear to be consistent with what we know of fiber and gastrointestinal function.

Dietary fiber affects gastrointestinal motility. Depending on the fiber type and particle size, fiber can reduce gastric emptying while enhancing intestinal motility, so that the overall effect on time from mouth to anus can be either an increase or a decrease. Most fiber increases stool bulk, which decreases the intestinal **transit time**. The increased stool bulk dilutes potential colon carcinogens in addition to decreasing the time these carcinogens are in the colon. These effects explain the effect of fiber on colonic cancer and constipation. Dietary fiber, particularly lignin, binds bile acids. Some of these bile acids are believed to be transformed into carcinogens in the gut. Thus formation of bile salts reduces carcinogen formation. In addition, binding the bile acids increases excretion of cholesterol by removing some of the negative feedback on bile acid synthesis in the liver. This may explain part of the beneficial effect of fiber in preventing cardiovascular disease. Dietary fiber expands the bile acid pool and increases the proportion of chenodeoxycholic acid over deoxycholic acid. This lowers the **lithogenic index** of bile and therefore reduces the chances of having gallstones.

Diverticula are little pouches or blind sacs that develop in the lining of the large intestine. They are potentially painful and can become inflamed, causing **diverticulitis**. The condition of having diverticuli is **diverticulosis**. Increasing the stool volume decreases the intraluminal pressure by the Law of Laplace, $P = T/r$. Thus increasing r decreases P at the same wall tension. Lastly, fiber may reduce hemorrhoids by reducing straining during defecation.

Although the cause and effect relationships between dietary fiber and chronic diseases are not fully established, there is reason to think that increasing dietary fiber is good for you. The current recommendation is to consume 18 g of nonstarch polysaccharides per day. This can be accomplished by consuming whole-grain foods, cereal products, fruits, and vegetables that contain natural fiber.

phospholipid. In the circulation, the VLDLs are coated with apolipoproteins B100, C, and E. The intestine secretes discoid-shaped HDL particles. These are 6–13 nm across and are 70% protein, 30% lipid. Half of the lipid is phospholipid. The major apolipoprotein in the intestinal HDL is A-I and A-IV. The overall pathway of lipid digestion and absorption is shown diagrammatically in Figure 8.5.9.

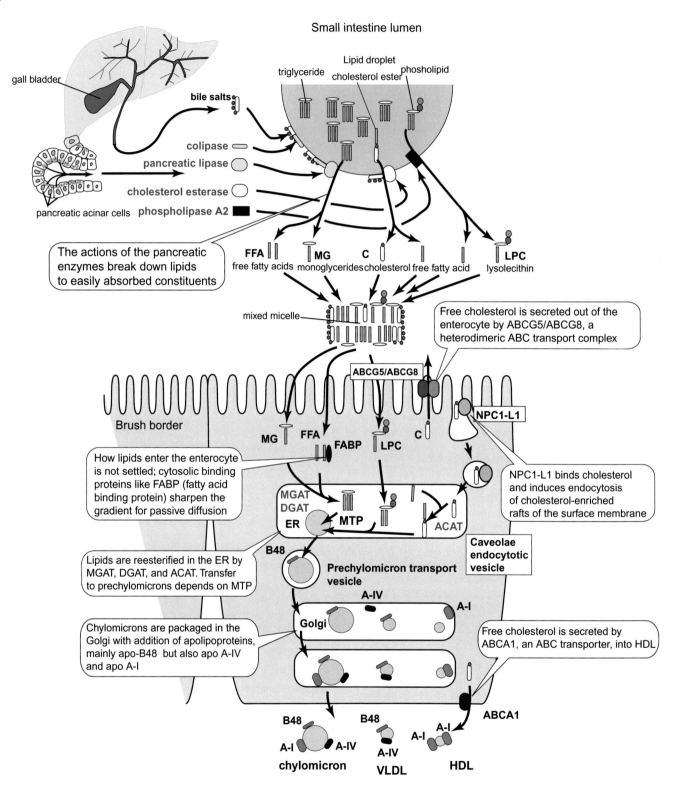

FIGURE 8.5.9 Overall processing of lipids by the intestine. Some lipid digestion occurs in the stomach, but most occurs in the intestine by pancreatic lipase, acting in concert with bile acids and colipase. This breaks down triglycerides into 2-monoglycerides and fatty acids. Pancreatic phospholipase degrades lecithin to fatty acids and lysolecithin, and cholesterol esters are hydrolyzed by pancreatic cholesterol esterase. All of the products are abstracted from lipid droplets to form mixed micelles. These transfer the hydrolysis products to the brush border where they are absorbed and then reesterified in the smooth endoplasmic reticulum of the enterocytes. The lipids are then packaged into lipoproteins that are released to travel in the lymph or in the portal blood. The lipoproteins consist of lipid interiors covered by apolipoproteins (B48, A-I, and A-IV). VLDLs are very low density proteins; HDLs are high-density lipoproteins.

MG, monoacylglycerides; FFA, free fatty acids; C, cholesterol; LPC, lysophosphatidylcholine; FABP, fatty acid binding protein; NPC1-L1, Nieman Pick C1-like-l protein; MGAT, monoacylglycerol acyl transferase; DGAT, diacylglycerol acyl transferase; ACAT, acyl cholesterol acyl transferase; ABCG5 and ABCG8, ATP-binding cassette transport proteins G5 and G8 that form a heterodimer for export of cholesterol from the enterocyte; ABCA1, ATP-binding cassette for cholesterol export along the basolateral membrane.

BILE ACIDS ARE ABSORBED IN THE TERMINAL ILEUM

As described in Chapter 8.4, the bile acids that are secreted by the liver are absorbed by the terminal ileum back into the portal blood and taken back up again by the liver. Passive absorption of bile acids occurs throughout the small intestine, but the active transporters are present only in the ileum. Absorption is mediated by secondary active transport, using Na^+ influx to drive transport of the bile acids into the enterocyte.

SUMMARY

In order to efficiently absorb nutrients, the intestine presents a huge surface area for absorption. The folds of Kerckring multiply the area by 3; the villi add another factor of 10; the microvilli on the enterocytes multiply this by 20. The villus cells continuously parade up the villus from their origins in the crypts of Lieberkuhn to their extrusion at the villus tips. The entire villus surface renews itself every 4–5 days.

Protein digestion begins in the stomach where chief cells secrete pepsinogen that is activated to pepsin by stomach acid. Pepsins are endopeptidases. The intestinal phase begins with proteolytic attack by the pancreatic enzymes trypsin, chymotrypsin, elastase, and carboxypeptidases A and B. These are secreted by the pancreas as inactive precursors. Enterokinase in the duodenum cleaves trypsin from trypsinogen, which then activates all the other proteolytic enzymes. These degrade dietary proteins into amino acids and oligopeptides. The brush border contains aminopeptidases and carboxypeptidases that cleave off single amino acids from the amino or carboxyl end of oligopeptides. Dipeptidylcarboxypeptidase cleaves off two amino acids from the carboxyl end of oligopeptides. Amino acids are transported into the blood using six distinct amino acid transporters and one dipeptide transporter. Some of these use secondary active transport with Na^+ as a cotransporter, but others use countertransport of other amino acids. Basolateral efflux of amino acids is less well known.

Starch digestion begins in the mouth with salivary amylase, but its action is quickly quenched by stomach acid. It resumes with the secretion of pancreatic amylase, which degrades starch into maltose, maltotriose, and limit dextrins. A number of enzymes on the brush border complete the digestion of starch. Dextrinase debranches the limit dextrins. Glucoamylase on the brush border cleaves maltotriose, producing maltose and glucose. Isomaltose breaks maltose down to two glucose molecules. Sucrase, another activity of the dextrinase/isomaltase enzyme, cleaves sucrose to glucose and fructose. Lactase breaks down lactose to galactose and glucose. These monosaccharides are reabsorbed by brush border carriers, either by facilitated diffusion (GLUT5) or by secondary active transport using Na^+ cotransport (SGLT1). The monosaccharides exit the enterocyte into the blood by facilitated diffusion via GLUT2.

Lipids are first broken down in order to gain entry into the enterocyte, and then they are reassembled into lipoproteins. Gastric lipase digests some of the ingested fat, but most of it is digested by pancreatic enzymes. Bile acids secreted by the liver and stored in the gallbladder are released by CCK stimulation of gallbladder contraction and relaxation of the sphincter of Oddi. The bile acids help secure pancreatic colipase on the surface of lipid droplets in the intestine. Pancreatic colipase then anchors pancreatic lipase, which then sets about hydrolyzing dietary triglycerides into fatty acids and 2-monoglyceride. Pancreatic phospholipase A_2 and cholesterol esterase also absorb to the lipid surface and hydrolyze phospholipids and cholesterol esterase. The released fatty acids, monoglycerides, lysolecithin, and cholesterol form mixed micelles with the bile acids. The mixed micelles deliver the lipids to the brush border where the lipids are taken up into the enterocytes. The enterocytes then reesterify the absorbed lipolytic products and reassemble them into lipoproteins. The intestines produce chylomicrons, VLDLs, and HDLs. These are distinguished by their lipid composition and presence of specific apolipoprotein coats. The chylomicrons are absorbed into the lymph. About 50% of dietary lipid travels through the lymph, the rest through the portal circulation.

REVIEW QUESTIONS

1. How much of an increase in area is caused by the folds of Kerckring? By the villi? By the microvilli?
2. How fast do cells in the intestine turn over? Where do new ones come from? Where do the old ones go?
3. What does acid do to proteins in the stomach? What enzyme breaks down proteins in the stomach? What cells secrete it? How is it activated?
4. What proteolytic enzymes does the pancreas make? How are they activated? Why doesn't trypsin inhibitor inhibit trypsin in the lumen of the gut? What products result from luminal proteolysis?
5. What proteolytic enzymes are on the brush border?
6. How are amino acids absorbed into the enterocyte? How do they get into the blood? Are all proteins broken down to single amino acid for absorption?
7. What does pancreatic amylase do? What is a limit dextrin?
8. What enzymes for carbohydrate digestion are on the brush border? What sugars get absorbed? How are they absorbed?
9. What is lactose intolerance?
10. What is emulsification? Why does pancreatic lipase need pancreatic colipase and bile acids? What is a mixed micelle? What happens to fatty acids, glycerol, and other lipolytic products in the enterocyte?
11. Where and how are bile acids absorbed?

8.6 Energy Balance and Regulation of Food Intake

Learning Objectives

- Describe a whole-body calorimeter
- Describe a bomb calorimeter
- List the Atwater factors
- Be able to use the Atwater factors to calculate the energy content of food
- Describe indirect calorimetry
- Use the volume of O_2 consumed, CO_2 produced, and urinary nitrogen to calculate the proteins, carbohydrates, and fats that are burned
- Use allometric formulas to estimate BMR
- Give subjective evidence that body weight is homeostatically regulated
- Distinguish between satiety signals and adiposity signals in regulation of food intake
- Identify the satiety center and the "feeding center" and why they are so designated
- List short-term signals that regulate food intake
- List long-term signals that regulate food intake
- Identify the following abbreviations for neurotransmitters involved in food regulation: POM, CART, AgRP, CCK, NPY
- Describe what is meant by a "glucose stat" and distinguish between glucose-sensitive and glucose-responsive neurons
- Indicate the part of the brain where food intake signals are integrated with hormonal signals

EARLY STUDIES ON ENERGY BALANCE USED CALORIMETERS

The scientific foundations of nutrition began in the 1780s when Antoine Lavoisier and Simon Pierre de Laplace measured oxygen consumption and carbon dioxide production in humans and found that both increased after a meal and during exercise, even though the temperature of the subject did not change. They constructed a small calorimeter for guinea pigs and showed a direct relationship between heat given off by the animal and the respiratory exchange. These early experiments were improved upon by Atwater and colleagues, who constructed a calorimeter large enough for humans (see Figure 8.6.1). A summary of one of his experiments is shown in Table 8.6.1.

These results showed, within experimental error, that man was not an exception to the laws of thermodynamics. Man obeyed the conservation of energy law:

[8.6.1] Energy input = energy output

This equation can be expanded to give

Energy content of food = work output
  + net heat output + net storage

[8.6.2]

This equation clarifies what is necessary to cause a negative storage of energy (a loss of body stores): either reduce the energy content of the food consumed or increase the work or heat output. Translated into clinical practice, this means that if you want to lose weight you must eat less or move more. Although this is conceptually easy, it is complicated by several facts: net heat output is related to the energy content of the food and the work output; and heat output is regulated by fat stores, so that heat output declines when one tries to lose weight.

"THE ENERGY CONTENT OF FOOD" IS ITS HEAT OF COMBUSTION

Atwater and others had to determine the energy content of the food consumed by their experimental subjects. They did this using a bomb calorimeter, shown in Figure 8.6.2. Passing a small electric current ignites the food sample in the presence of high oxygen tension, which causes the organic material to burn. The chemical reactions are similar to those occurring in the body, except in the body the reactions occur at lower temperatures through enzyme-catalyzed reactions. The human body is capable of completely oxidizing carbohydrates and fats, but we only partially oxidize proteins, excreting urea as a waste product. Therefore, the energy of combustion of proteins in the bomb calorimeter exceeds that in the body by the amount of chemical energy stored in the waste urea. Atwater used these facts to derive the **Atwater factors**, which are still widely used today along with tables of food composition to calculate the energy content of specific foods (see Table 8.6.2). These factors make allowance for the energy lost in feces and in urine. They represent physiological approximations based on experiments with a limited number of subjects. In food tables, the protein content is usually obtained from the nitrogen content by multiplying by 6.25 (because most proteins

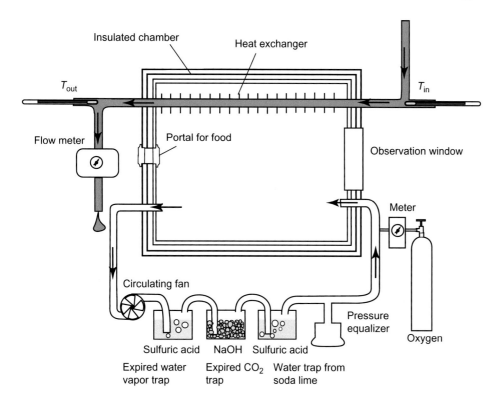

FIGURE 8.6.1 The Atwater–Benedict respiration calorimeter. The chamber was thermally insulated from its surroundings. The heat produced by the person contained in the calorimeter was estimated from the temperature of the incoming water (T_{in}), the temperature of the water exiting the chamber (T_{out}), and the rate of water flow. Air was circulated by a blower. The air that left the chamber was scrubbed of water by passing it through sulfuric acid. A solution of NaOH or soda lime pellets absorbed the CO_2 produced. Known quantities of O_2 could be added to the air before it reentered the chamber. The amount of CO_2 produced could be estimated from the gain in weight of the soda lime.

TABLE 8.6.1 Results from an Experiment by Atwater and Benedict 1899, in MJ

Source of Energy	Total 4 Days	Average per Day
(a) Heat of combustion of food eaten	41.22	10.31
(b) Heat of combustion of feces	1.26	0.32
(c) Heat of combustion of urine	2.25	0.56
(d) Heat of combustion of alcohol	0.35	0.09
(e) Heat of protein gained "+" or lost "−"	−1.16	−0.29
(f) Heat of fat gained "+" or lost "−"	−2.26	−0.56
(g) Energy of food oxidized: $a - (b+c+d+e+f)$	40.78	10.19
(h) Heat determined by calorimetry	40.06	10.02
(i) Difference more "+" or less "−" than g	−0.68	−0.17
(j) Difference (%)	−1.6	−1.6

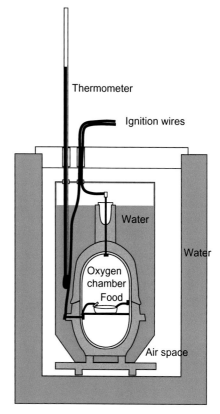

FIGURE 8.6.2 A bomb calorimeter. A sample of food is placed in a crucible within a steel-walled vessel containing a high oxygen tension. Passing a current ignites the food, combining it with the oxygen and producing heat from the oxidation reaction. The heat produced is determined by measuring the temperature increase in the water surrounding the calorimeter. The calorimeter is calibrated in advance to obtain a "calorimeter constant" that reflects the heat capacity of both the water and the metal of the calorimeter. Energy added by the electric current is calculated and subtracted from the heat produced.

are typically 16% nitrogen by mass). However, this factor is too high for cereals and too low for milk (5.7 is better for cereals, 6.4 for milk).

The units for the Atwater factors are $kcal\ g^{-1}$. The **calorie** is the amount of heat energy necessary to warm 1 g of water from 14.5°C to 15.5°C. The **kcal** is often written as **Calorie**, which is a source of endless confusion. J. Joule (1818–1889) showed the equivalence of mechanical, electrical, and heat energy. The joule is a N m ($= kg\ m^2\ s^{-2}$) or a volt-coulomb. The conversion factors are as follows:

$$1\ cal = 4.184\ J \quad 1\ J = 0.239\ cal$$

TABLE 8.6.2 The Atwater Factors and Their Derivation for Different Macronutrients

Macronutrients	Bomb Calorimetry (kcal g^{-1})	Urinary Loss (kcal g^{-1})	Digestibility (%)	Atwater Factor (kcal g^{-1})
Proteins	5.65	1.25	92	4.0
Carbohydrates	4.1	0	99	4.0
Fats	9.4	0	95	9.0
Alcohol	7.1	Trace	100	7.0

TABLE 8.6.3 Energy Yields from Oxidation of Substrates

Substrate	O$_2$ Consumed L g^{-1}	CO$_2$ Produced L g^{-1}	RQ	Heat kJ	Heat kcal	Energy Equivalents in Energy per Liter of O$_2$ kJ	Energy Equivalents in Energy per Liter of O$_2$ kcal	Energy Equivalents in Energy per Liter of CO$_2$ kJ	Energy Equivalents in Energy per Liter of CO$_2$ kcal
Starch	0.829	0.829	1.00	17.6	4.20	21.2	5.06	21.2	5.06
Sucrose	0.786	0.786	1.00	16.6	3.96	21.1	5.04	21.1	5.04
Glucose	0.746	0.746	1.00	15.6	3.74	21.0	5.01	21.0	5.01
Lipid	2.019	1.427	0.71	39.6	9.46	19.6	4.69	27.7	6.63
Protein	1.010	0.844	0.83	19.7	4.70	19.0	4.66	23.3	5.58
Lactic acid	0.746	0.746	1.00	15.1	3.62	20.3	4.85	20.3	4.85

Data from G. Livesey and E. Marinos, Am. J. Clin. Nutr. **47**:608–626, 1988.

MEASUREMENT OF ENERGY EXPENDITURE BY INDIRECT CALORIMETRY

The measurement of energy expenditure by measuring the heat dissipated by the body is called **direct calorimetry**. Direct calorimetry is technically difficult and tedious and subject to additive errors. Atwater found that the total energy expenditure is quantitatively related to the total oxygen consumption. Thus an indirect measure of energy expenditure can be accomplished by measuring gas exchange.

The **energy equivalence** of O$_2$ consumption for the three macronutrients was determined calorimetrically in 1897 by Zuntz in Switzerland, and much more recently as well. The recent results offer no substantial changes from the original data of Zuntz. The recent data are given in Table 8.6.3.

The theoretical energy equivalence of proteins is much harder to assess because of the mixture of amino acids that make up proteins—proteins are heterogeneous. However, the energy equivalence can be determined experimentally, as shown in Table 8.6.3. According to this table, the energy equivalence of oxygen is about 5.0 kcal L^{-1} for carbohydrates, 4.7 kcal L^{-1} for fats, and about 4.6 kcal L^{-1} for proteins. Since we burn a mixture of these macronutrients, we can take an average of about 4.85 kcal L^{-1} and estimate the total metabolic rate or energy consumption as

[8.6.3]
$$M = 4.85 \text{ kcal L}^{-1} \times Q_{O_2}$$
$$= 20.3 \text{ kJ L}^{-1} \times Q_{O_2}$$

EXAMPLE 8.6.1 Calculate the Energy Equivalence of Carbohydrates and Fats

The complete oxidation of glucose is written as

[8.6.4] $C_6H_{12}O_6 + 6O_2 \Rightarrow 6H_2O + 6CO_2 + \text{heat}$

For every mole of glucose, 6 moles of oxygen are consumed and 6 moles of carbon dioxide are produced. Bomb calorimetry gives 3.73 kcal g^{-1} of heat liberated or 671 kcal mol^{-1} = 2807 kJ mol^{-1}. The 6 moles of O$_2$ at STPD (standard temperature, 0°C, and pressure, 1 atm = 760 mmHg, and dry) occupies approximately 134.4 L. Thus the energy equivalence of O$_2$ when glucose is burned is 671 kcal mol^{-1}/134.4 L mol^{-1} = **4.99 kcal L^{-1}**.

The complete oxidation of tripalmitin is written as

[8.6.5] $C_{51}H_{98}O_2 + 72.5O_2 \Rightarrow 49H_2O + 51CO_2 + \text{heat}$

The heat of combustion for tripalmitin is 7415 kcal mol^{-1}, corresponding to the consumption of 72.5 moles of oxygen, for an energy equivalence of 7415 kcal mol^{-1}/1624 L mol^{-1} = **4.56 kcal L^{-1}**.

where Q_{O_2} is the rate of oxygen consumption in Liters at STPD per unit time and M is the metabolic rate.

INDIRECT CALORIMETRY AND URINARY NITROGEN ALLOW ESTIMATION OF CATABOLISM OF MACRONUTRIENTS

Suppose an individual oxidizes c grams of glucose, f grams of fat, and p grams of proteins. The p grams of protein will translate to 0.16 g of urinary nitrogen, n, because the only source of urinary nitrogen is protein catabolism, and proteins are nearly uniformly 16% nitrogen by mass. Thus the amount of protein being catabolized can be calculated directly from urinary nitrogen excretion as

[8.6.6] $$p = 6.25n$$

where p is the grams of protein and n is the grams of urinary nitrogen. According to the coefficients in Table 8.6.3, we can write the oxygen consumption as

[8.6.7] $$\begin{aligned} V_{O_2} &= 0.746c + 2.02f + 1.01p \\ V_{CO_2} &= 0.746c + 1.43f + 0.844p \end{aligned}$$

where V_{O_2} and V_{CO_2} are the volumes of O_2 consumed and CO_2 produced, respectively, at STPD. If we measure urinary nitrogen, we can calculate p according to Eqn (8.6.6). If we measure the O_2 consumption and CO_2 production, we can calculate c and f by solving the simultaneous equations of Eqn (8.6.7).

ENERGY EXPENDITURE CONSISTS OF BASAL METABOLISM PLUS ACTIVITY INCREMENT

The **basal metabolic rate**, BMR, is *defined* as **the metabolic rate during rest but while the person is awake**. The person should be in a postabsorptive state, not having eaten within the last 12 hours. The person should also not have strenuously exercised within the previous 12 hours. The air in the room should be comfortable with all sources of excitement removed. The BMR is usually determined by indirect calorimetry by measuring Q_{O_2}, the rate of oxygen consumption. The resting energy expenditure (REE) differs from the BMR in that the determination of the REE does not require fasting for 12 hours.

Body size, composition, age, and gender have marked effects on the BMR. The overall volume of the body increases approximately according to the cube of the linear dimensions, whereas the surface area increases according to the square. Thus larger people have a smaller surface area to volume ratio. Since body heat must be shed on the surface, this means that larger people must produce less heat per unit body dimension, or they will get too hot too easily. Max Rubner (1854–1932) showed in 1883 that mouse, dog, man, and horse had greatly different BMR when expressed per unit body weight, but they were all very similar when compared per unit surface area. Based on the geometric argument above, Rubner proposed that BMR = $KM^{2/3}$, where M is the mass and K is a constant. The exponent of 0.67 in Rubner's equation is the subject of some debate. Max Kleiber reevaluated the effect of body size on metabolism and found an exponent of 3/4 (actually 0.754). Although the absolute differences in these exponents (0.67 vs 0.75) does not seem large, much has been made of a 2/3 or 3/4 power law because it was thought to be one of the few unifying principles of biology that applied equally well to microorganisms as to elephants. It is likely that no single process determines either the preexponential or exponential factor in the allometric formula:

[8.6.8] $$BMR = aM^b$$

It is likely that herbivores have a different relationship between BMR and M because of their extensive microbial activity and their nearly continuous state of feeding. Inclusion of herbivores in regressions of BMR against M skews the curve.

EMPIRICAL FORMULAS FOR BMR

Measurements of human BMR have been made as a function of age and body composition (see Figure 8.6.3). This gives the BMR in terms of kcal per square meter of surface per hour. By carefully measuring the surface area of a few persons (by covering them with tiny squares), researchers have derived an empirical equation for surface area. The DuBois–Meeh formula for surface area is

[8.6.9] $$BSA(m^2) = 0.007184 \times W(kg)^{0.245} \times H(cm)^{0.725}$$

where BSA is the **body surface area**, in m^2, W is the weight, in kg, and H is the height, in cm. These are empirical data obtained using only nine subjects and its

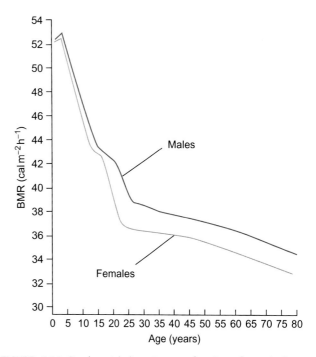

FIGURE 8.6.3 Basal metabolic rate as a function of age in human females and males. The BMR is given here as the kilocalories of energy expenditure per square meter of body surface area, per hour.

use is debated. Harris and Benedict in 1919 developed an equation for predicting the BMR for men and women based on their weight, W, in kg, their height, H, in cm and their age, A, in years. The equation gives the BMR in **kcal per day**:

$$\text{Men BMR} = 66.5 + (13.7 \times W) + (5.0 \times H) - (6.8 \times A)$$
$$\text{Women BMR} = 655.1 + (9.56 \times W) + (1.85 \times H) - (4.7 \times A)$$

[8.6.10]

A more recent equation published by Mifflin and St. Jeor in 1990 gives the REE, in kcal per day, as

$$\text{Men REE} = 5 + (10 \times W) + (6.25 \times H) - (5 \times A)$$
$$\text{Women REE} = -161 + (10 \times W) + (6.25 \times H) - (5 \times A)$$

[8.6.11]

EATING FOOD INCREASES METABOLISM

The metabolic response to the ingestion of food has been variously called "**diet-induced thermogenesis**," "**thermic effect of food**," "**specific dynamic action**," and "**heat increment**." All of these refer to the increased heat production which follows ingestion of food. Its magnitude varies with the diet and individual and can range from 5% to 20% of the ingested calories. Usually the thermic effect is greater when the meal is larger and when the individual has a greater fat-free mass. In addition, protein in the diet exerts a much larger thermic effect than starches or fats. Measurement of the thermic effect of food is illustrated in Figure 8.6.4.

ACTIVITY ADDS THE GREATEST INCREMENT TO METABOLISM

The single largest source of variation in energy expenditure is the level of activity. There are two main considerations when estimating the effect of physical activity: the **duration** and the **intensity**. Table 8.6.4 details the energy cost associated with physical activity, expressed in terms of milliliters of O_2 consumed per kilogram of body weight per minute because activity becomes energetically more costly as body size increases. It is also expressed in METS (metabolic equivalents). One MET is the average O_2 consumption at rest, equal to 3.5 mL of O_2 per kg per minute.

THE BODY HOMEOSTATICALLY REGULATES ITS WEIGHT

Over the course of a year, most adults consume between 400 and 700 kg of food, yet their weight fluctuates at most a few kilograms. A mismatch of just 5% between food intake and energy expenditure would result in a weight gain or loss of about 5 kg per year—but most people gain or lose far less than that. **Body weight regulation is extraordinarily precise**. Major changes in body weight and composition can be forced by under- or overfeeding, or by forced exertion, but body weight returns when *ad libitum* feeding resumes.

THE CENTRAL NERVOUS SYSTEM REGULATES FEEDING BEHAVIOR

The act of choosing what to eat, how much to eat, and when to eat it is a conscious decision that involves higher order neural processing. In the human psychosocial environment, food intake is associated with other activities. We entertain and socialize around food and attach culturally dependent status to certain foods. There are traditional breakfast foods, lunch foods, snack foods, and supper foods. Consumption is dictated by the time food is prepared rather than the time we are hungry. Food becomes much more than mere nourishment. Its presentation and social setting become as important as the food itself. On the other hand, noxious sights,

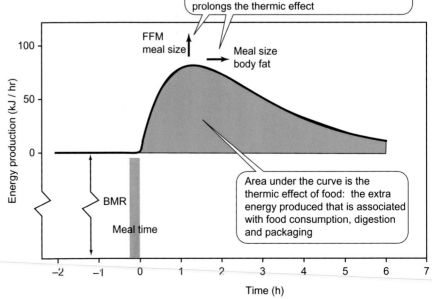

FIGURE 8.6.4 The thermic effect of food. During rest, energy production is more or less constant. Following ingestion of a meal, heat production increases to a maximum and then tapers off back to baseline levels. The peak is affected by the size of the meal and the free fat mass (FFM). Increased meal size also prolongs the thermic effect. Increased body fat typically delays the peak of the thermic effect. The extra heat production is the area under the curve of heat production above the BMR. The curve can be fit to the equation $A + Bte^{-t/C}$ where the maximum of the curve occurs at $t = C$ at an increment of $B C e^{-1}$. For the curve shown, $B = 175.9$ kJ h^{-2} and $C = 1.3$ h.

TABLE 8.6.4 Energy Expenditure for Various Activity Levels for Men and for Women

Exercise Intensity	Energy Expenditure			
	kcal min^{-1}	L O$_2$ min^{-1}	mL O$_2$ kg^{-1} min^{-1}	METs
Men				
Light	2.0–5.0	0.4–1.0	6–15	1.6–4
Moderate	5.0–7.5	1.0–1.5	15–22.5	4–6
Heavy	7.5–10.0	1.5–2.0	22.5–30	6–8
Very heavy	10.0–12.5	2.0–2.5	30–37.5	8–10
Women				
Light	1.5–3.5	0.3–0.7	5.5–12.5	1.5–3.5
Moderate	3.5–7.5	0.7–1.1	12.5–20	3.5–5.5
Heavy	7.5–10.0	1.1–1.5	20–27.5	5.5–7.5
Very heavy	10.0–12.5	1.5–1.9	27.5–35	7.5–9.5

Modified from W.D. McCardle, F.I. Katch, and V.L. Katch, Exercise Physiology, Lea and Febiger, Philadelphia, PA, 1991

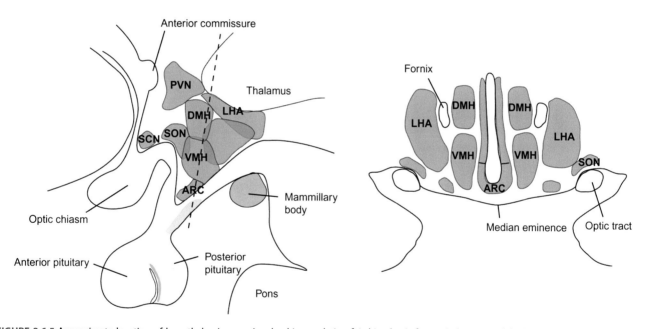

FIGURE 8.6.5 Approximate location of hypothalamic areas involved in regulating food intake. Left, saggital section of the brain (plane perpendicular to the face). Right, coronal section (plane parallel to the face). Labeled structures include the arcuate nucleus (ARC), the paraventricular nucleus (PVN), the lateral hypothalamic area (LHA), the dorsal medial hypothalamus (DMH), and the ventral medial hypothalamus (VMH). Other structures not directly involved in food intake regulation but shown for reference include the suprachiasmatic nucleus (SCN) and the supraoptic nuclei (SON).

odors, pain, or distressful psychological events cause us to lose our appetite. Underneath all of the cultural and hedonistic aspects of consuming food are the physiological signals that instruct us how much to eat.

EARLY STUDIES SHOWED THAT THE HYPOTHALAMUS DRIVES FEEDING BEHAVIOR

Electrolytic lesioning of the **ventromedial hypothalamus (VMH)** causes hyperphagia (excess food consumption) in rats. Conversely, electrical stimulation of this area inhibits feeding. This ventromedial hypothalamus is classically known as a **"satiety center,"** as stimulation shuts down feeding and destruction of it causes overeating and weight gain (see Figure 8.6.5).

Electrical stimulation of the **lateral hypothalamic area (LHA)** causes feeding behavior, whereas destruction by electrolytic lesioning attenuates feeding and causes weight loss. Thus this area seems to turn on feeding and so it was classically called the **"feeding center."**

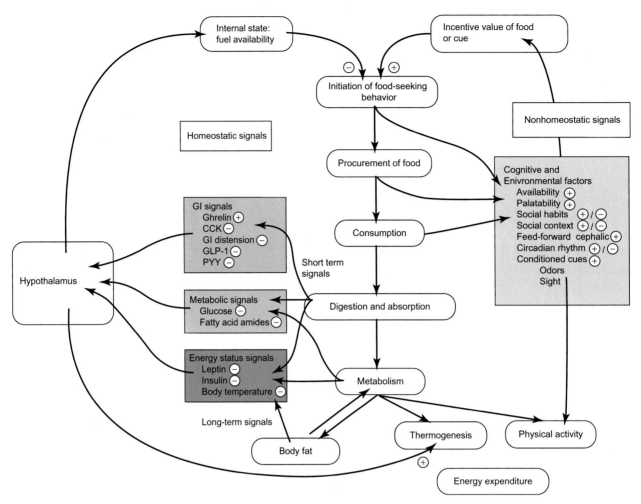

FIGURE 8.6.6 Overall scheme for the regulation of food intake. Homeostatic signals are shown on the left of the figure; nonhomeostatic signals are on the right. Nonhomeostatic signals include all those sensory modalities that make food attractive or unattractive, including odor, sight, taste, mouth feel (all of which constitute part of palatability), and in addition the availability and social context of the food. Physical exertion is also nonhomeostatic. Homeostatic signals include short-term signals that inform the CNS about the kinds and amounts of food consumed in a meal and long-term signals that inform the CNS about the body's adiposity. Most signals from the GI tract and fuel stores shut down food intake, with the exception of ghrelin. The − and + signs refer to the effect of the signal on the indicated output when the signal increases.

THE SIMPLISTIC EARLY VIEW IS SUPPLANTED BY A PICTURE OF MULTIPLE CENTERS AND MULTIPLE SIGNALS

Since the early studies identified "satiety center" and "feeding center," investigators have discovered a number of materials that affect food intake and have shown that multiple centers are involved. In addition to these, odors, sights, and higher order psychological events input into our subjective feeling of being hungry. Most of this work has been performed in experimental animals and so the relevance to human physiology is often not yet established. Although there is as yet no complete "wiring diagram" that describes in detail how feeding behavior is governed, much of the puzzle has been uncovered. Figure 8.6.6 shows the general scheme of how food intake and energy expenditure is regulated.

SHORT-TERM SIGNALS LIMIT THE SIZE OF MEALS: THEY ARE SATIETY SIGNALS

Short-term signals originate from the periphery and depend on how much and what kind of food is consumed in a meal, as opposed to the body's adiposity and energy status, and limit the size of individual meals. Short-term signals that originate from the intestine or metabolism are summarized in Table 8.6.5.

All parts of the gastrointestinal tract are invested with **mechanoreceptors** that sense stretch or irritation and **chemosensors** that sense nutrients and noxious chemicals. These receptors send afferent information over the vagus nerve to the **nucleus tractus solitarius (NTS)** in the brain stem and from there to the dorsal motor nucleus of the vagus for vagovagal reflexes. The NTS also receives inputs from the sense of taste from cranial nerves VII (facial nerve) and IX (glossopharyngeal nerve).

TABLE 8.6.5 Synopsis of Short-Term Homeostatic Mechanisms Regulating Food Intake

Signal	Source and Secretagogues	Chemical Nature	Targets	Primary Action
Gut distension	Stretch receptors in stomach and small intestine	Vagal afferent nerve activity	Nucleus tractus solitarius (NTS) to parabrachial nucleus to hypothalamus	Inhibits feeding; also involved in reflexes for gut motility and secretion
GI chemoreceptors	Sugars, amino acids, fatty acids, and peptides in small intestine	Vagal afferent nerve activity	Nucleus tractus solitarius (NTS) to parabrachial nucleus to hypothalamus	Inhibits feeding; also involved in reflexes for gut motility and secretion
Cholecystokinin (CCK)	Endocrine I cells in proximal third of intestine	83-aa chain + fragments	CCK1 receptors on vagal afferents; interacts with GI distension signals	Inhibits feeding; also increases pancreatic exocrine secretion and gallbladder contraction
Glucose	Digested and absorbed carbohydrates	Glucose	Glucose-responsive and glucose-sensitive neurons in hypothalamus	Inhibits feeding; increases insulin secretion and decreases glucagon secretion
GLP-1	Endocrine L cells in jejunum, ileum, and colon in response to intestinal nutrients	30-aa chain	Hypothalamus pancreatic islet cells stomach	Inhibits feeding; stimulates insulin secretion; delays gastric emptying
Ghrelin	Endocrine X/A cells in the stomach	28-aa chain: acylated at ser3		Stimulates feeding; binds to GHS-R to stimulate GHRH release and by directly stimulating somatotrophs
PYY(3-36)	Endocrine cells in distal intestine in response to digestion products	34-aa chain	Y2 receptors in the arcuate nucleus that inhibit NPY release	Inhibits feeding but probably at the next meal

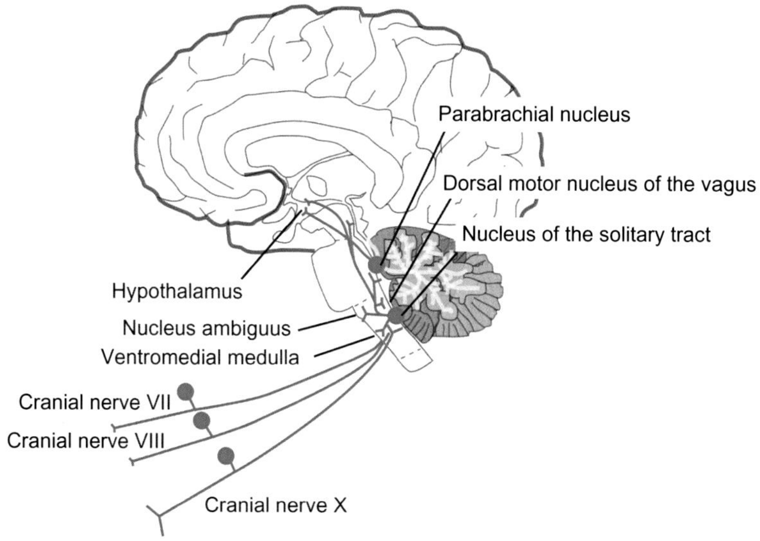

FIGURE 8.6.7 Ascending pathways for visceral information relating to a meal. Gustatory mechanoreceptors, and chemoreceptors from the facial, glossopharyngeal, and vagus nerves all feed into the nucleus of the solitary tract (NTS) and from there ascend to the parabrachial nucleus and from there to the hypothalamus, amygdala, and cortex.

Information passes up to the hypothalamus directly and through the parabrachial nucleus. Afferent information also travels along the sympathetic nerves in the splanchnic nerves. Figure 8.6.7 shows these ascending pathways for visceral information.

As discussed in Chapter 8.2, **cholecystokinin**, or **CCK**, is secreted by endocrine I cells in the proximal intestine in response to amino acids and fat digestion products. It stimulates CCK1 receptors on afferent fibers of the vagus. These fibers may be the same as those that respond to distension, so some vagal fibers may integrate different kinds of signals related to ingestion of food. The vagal signal proceeds to the NTS which then stimulates pancreatic acinar cells to increase zymogen output. CCK is also synthesized in the hypothalamus and is used as a neurotransmitter there. The central effects of CCK remain unclear. Vagotomy only partially reduces CCK's inhibition of food intake, suggesting that both peripheral and central effects occur.

The **glucostatic hypothesis**, first proposed by Jean Mayer in 1953, holds that sensors in the brain detect hypoglycemia and then initiate subjective feelings of hunger to motivate the person to seek food. Glucose levels increase after eating a meal, and sensors detect this rise and stop eating behavior. Although this makes subjective sense, typically eating stops before

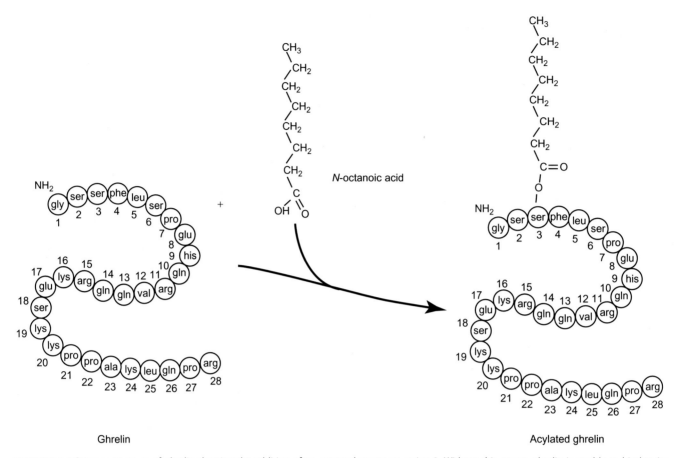

FIGURE 8.6.8 Primary structure of ghrelin showing the addition of an octanoyl group on serine 3. Without this group, ghrelin is unable to bind to its receptors.

postprandial glucose levels rise. Nevertheless, hypoglycemia or inhibition of glucose metabolism with 2-deoxyglucose increases feeding behavior. Some neuronal cells are **glucose responsive**: they **increase their firing rate when glucose levels rise**. Others are **glucose sensitive**: they **increase their firing rate when glucose levels fall**. These neurons are alternatively described as glucose excited (GE) or glucose inhibited (GI). These glucose-sensing neurons reside in all of the hypothalamic areas important to the regulation of food intake and energy expenditure: the arcuate nucleus (ARC), dorsal medial hypothalamus (DMH), paraventricular hypothalamus (PVN), ventral medial hypothalamus (VMH), and lateral hypothalamic area (LHA). The NTS also contains glucose-sensing neurons. Glucose-responsive or glucose-excited neurons respond by depolarization by mechanisms like those in β cells of the pancreatic islets. These GE neurons also respond to other metabolites and other stimuli, suggesting that they integrate signals for feeding behavior.

Glucagon-like peptide 1 is secreted by L-type endocrine cells that are located in the jejunum, ileum, and colon. It stimulates insulin secretion and inhibits gastric emptying. This would promote gastric distension that in turn would increase afferent vagal stimulation to inhibit feeding. The effect of GLP-1 on feeding behavior is abolished by vagotomy, but its effect on stomach emptying is not. However, GLP-1 peaks too late to effectively signal satiety to the central nervous system (CNS). GLP-1 is also synthesized in the brain, and GLP-1 receptors have been located in the supraoptic nuclei (SON), PVN, and ARC of the hypothalamus.

Ghrelin is a 28-aa peptide secreted primarily by endocrine cells in the stomach during fasting. Its main effects include release of growth hormone (GH) and increased feeding. Ghrelin is named because of its **gr**owth **h**ormone **rel**easing action. Ghrelin binds to growth hormone secretagogue receptors (GHS-R) on cells in the ARC of the hypothalamus and on somatotrophs in the anterior pituitary. Ghrelin is unusual in that its activity requires an octanoyl group added to the hydroxyl of serine 3 (see Figure 8.6.8). Only the acylated form binds to its GHS-R, but the major part of the circulating ghrelin is unacylated. The main stimulus for ghrelin secretion is fasting, but hypoglycemia and insulin also cause ghrelin secretion. After each meal, circulating ghrelin levels fall due to inhibition of ghrelin secretion by nutrients in the stomach. Thus ghrelin may be a peripheral signal for satiety by removal of its feeding stimulation.

Ghrelin stimulates neurons in the ARC that use **neuropeptide Y (NPY)** and **agouti related protein (AgRP)**. These neurons project to neurons in the PVN, LHA, and DMH that help regulate feeding. Both NPY and AgRP

TABLE 8.6.6 Synopsis of Long-Term Regulators of Food Intake

Signal	Source and Secretagogues	Chemical Nature	Targets	Primary Action
Insulin	β Cells of pancreatic islets	30 + 21-aa chain	Liver, muscle, fat, and brain	Regulates glucose uptake and lipolysis; inhibits feeding
Amylin	β Cells of pancreatic islets	37-aa chain	Neurons in hypothalamus	Inhibits feeding
Leptin	Adipocytes and endocrine cells in the stomach	146-aa chain	Neurons in the arcuate nucleus; vagal afferents	Inhibits feeding

stimulate feeding and are called **orexigenic**. Other neurons in the ARC use α **melanocyte stimulating hormone** (αMSH) and **cocaine–amphetamine regulated transcript** (CART) as neurotransmitters. The αMSH derives from **POMC** or **proopiomelanocortin** and so the neurons are called POMC/CART neurons. These neurons ordinarily not only inhibit feeding but also regulate energy expenditure through sympathetic stimulation of adipose and muscle tissues (see Chapter 2.11, Clinical Applications: Nonshivering Thermogenesis). These POMC/CART neurons are inhibited by the NPY/AgRP neurons. Because ghrelin stimulates the orexigenic NPY/AgRP neurons, it indirectly inhibits the POMC/CART anorexigenic neurons.

Protein YY(3-36) is secreted from the distal intestine in response to digestion products in the lumen. In contrast to CCK, PYY(3-36) secretion occurs postprandially rather than during a meal. It exerts an **anorexigenic** effect (it inhibits food intake) through peripheral and central mechanisms. It is proposed to activate presynaptic Y2 receptors. **NPY** is released by cells in the ARC that exert potent orexigenic control. They act on Y receptors, of which there are at least six different varieties: Y1–Y6. The Y1 receptor activates feeding. Y2 receptors are on NPY cells and are autoreceptors. By activating the Y2 receptor, PYY may inhibit NPY cells in the ARC. This inhibition indirectly activates neurons that contain POMC/CART. These POMC-containing cells inhibit food intake and increase energy expenditure. Thus PYY appears to rise too late to influence meal size while it is being eaten but may reduce the size of the subsequent meal by inhibiting cells that stimulate food intake and indirectly activating cells in the ARC that inhibit food intake.

LONG-TERM SIGNALS MAINTAIN BODY COMPOSITION: THEY ARE ADIPOSITY SIGNALS

In 1953, Kennedy formulated the **lipostatic hypothesis** for the regulation of body weight, in which fat tissue produces a signal that informs the CNS of the level of body adiposity, and the CNS accordingly activates eating or fasting to keep adipose stores constant. Hervey surgically connected the circulatory systems of two rats and then made one rat obese by destroying its satiety center, the VMH. The other rat became hypophagic and thin in response to the obesity of its partner. We now know that fat tissue secretes blood-borne substances that are detected by the CNS. In the parabiotic rats, the thin rat detected this material from its obese partner and reduced food intake, thinking that it was too fat. Thus adipose tissue makes substances that inform the CNS of the adiposity of the body. The regulatory mechanisms appear to be much better at correcting deficits of adipose tissue than excesses. The long-term signals that originate from adipose and pancreatic tissues are listed in Table 8.6.6.

Adipose tissue secretes **leptin** (from the Greek "leptos," meaning "thin") proportional to the amount of fat, so that circulating levels of leptin indicate adiposity. Leptin secretion is increased by insulin and adipocyte glucose metabolism. High fat and high fructose diets, both of which more poorly stimulate insulin secretion and blood glucose levels, are associated with lower leptin levels. This may partly explain the epidemic of obesity in Western cultures where people consume high fat diets laden with high fructose corn syrup. Leptin binds to two kinds of receptors, a "long" form, LRb, and a short form, LRa. It is transported across the blood–brain barrier by a mechanism involving LRa. After transport into the CSF, leptin binds to LRb on specific cells in the hypothalamus. This works through a JAK/STAT pathway, ending in the phosphorylation of STAT3 (see Figure 8.6.9). **Leptin suppresses appetite and increases energy expenditure.** It binds to NPY/AgRP neurons and POMC/CART neurons in the ARC. It decreases production of the orexigenic NPY and AgRP and increases production of anorexigenic αMSH and CART. In addition, leptin increases energy expenditure through activation of the sympathetic nervous system. Leptin also has receptors in the LHA and PVN.

Although insulin is secreted by β cells in the pancreatic islets of Langerhans rather than by adipocytes, the circulatory levels of insulin are also proportional to the body adiposity: fatter people secrete proportionately more insulin to a given increase in blood glucose than do leaner people. Insulin also crosses the blood–brain barrier by specific transport mechanisms, and neurons in the ARC of the hypothalamus have receptors for insulin.

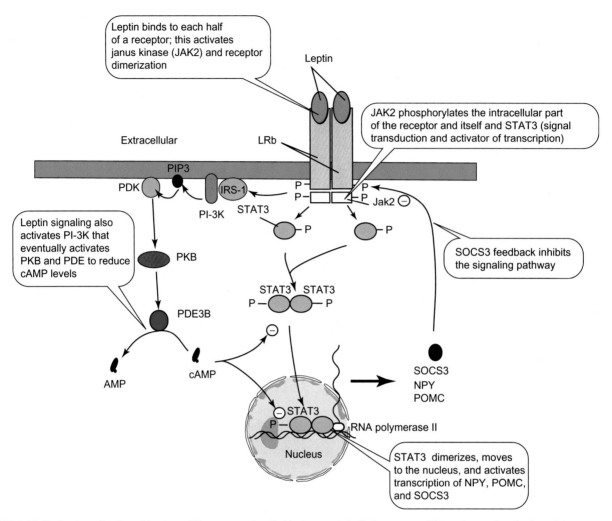

FIGURE 8.6.9 Mechanism of action of leptin on LRb receptors. Leptin binds to each half of a receptor. This activates dimerization of the receptor and activation of its associated janus kinase (JAK2). JAK2 phosphorylates itself and the intracellular part of the receptor and other targets. One of these is STAT3, for **s**ignal **t**ransduction and **a**ctivation of **t**ranscription. The activated STAT3 dimerizes and moves to the nucleus where it enhances transcription of target genes including those coding for NPY, proopiomelanocortin (POMC), and suppressor of cytokine signaling (SOCS). Leptin also activates the phosphorylation of insulin receptor substrates (IRS-1) that bind phosphatidyl inositol 3 kinase (PI-3K) which makes PIP3 in the membrane. PIP3 in turn activates a phosphoinositide-dependent protein kinase (PDK) that in turn activates protein kinase B (PKB) which activates phosphodiesterase (PDE3B) that degrades cAMP. This removes inhibition of STAT3 transcription. SOCS3 feeds back on the receptor to inhibit its further activity.

Insulin levels rise with feeding, fall with fasting, and rise in obesity, similarly to leptin. Unlike leptin, insulin secretion is stimulated acutely in response to meals. Insulin administered into the brain suppresses food intake.

Amylin, a polypeptide containing 37 amino acids, is cosecreted with insulin in response to glucose and other nutrients. Its secretion is proportional to body adiposity. It delays gastric emptying and in experimental animals amylin reduces food intake. Like insulin, amylin is transported across the blood–brain barrier by a carrier. Low doses administered directly into the ventricles of the brain also reduce food intake, suggesting that its action lies in the brain itself rather than indirectly through inhibition of gastric emptying. Blockade of central amylin receptors leads to increased food intake. Although the data supporting amylin's role are less extensive, the evidence suggests that amylin and insulin both inform the CNS of body adiposity.

INTEGRATED MECHANISM OF FOOD INTAKE REGULATION

The signals for food intake regulation are integrated in the hypothalamus, involving the ARC, LHA, VMH, DMH, and PVN. Neurons in these structures use a variety of neurotransmitters. Those that are orexigenic (promoting feeding behavior) include NPY/AgRP, Orexin A and B (also known as the hypocretins), and melanin concentrating hormone (MCH). Anorexigenic neurotransmitters include POMC/CART. A complete "wiring" diagram that describes the neural mechanism for regulation of food intake is hopelessly complicated for presentation here. A simplified version is shown in Figure 8.6.10.

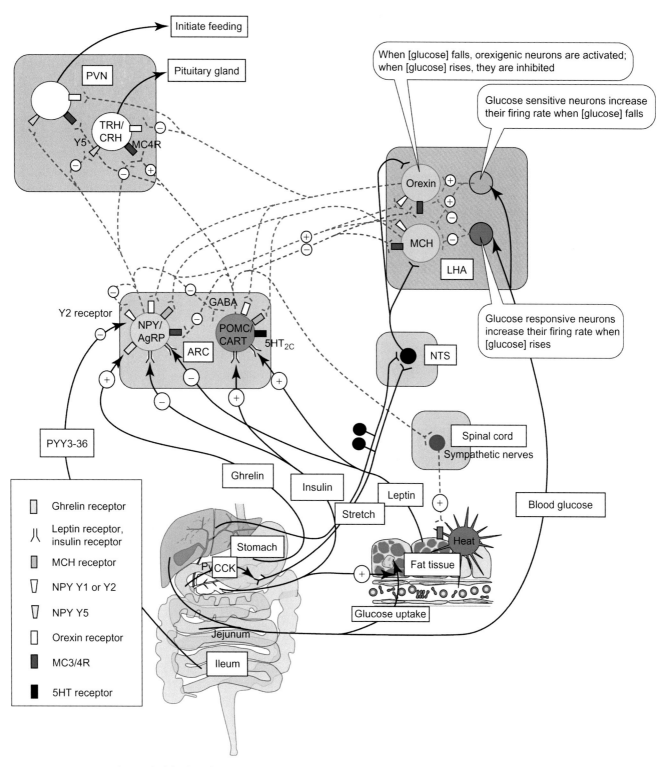

FIGURE 8.6.10 Integrated control of food intake. Some of the connections are shown for NPY/AgRP and POMC/CART neurons in the arcuate nucleus (ARC) and for orexin and MCH neurons in the lateral hypothalamic area (LHA). Information entering the hypothalamus from the periphery is in black. Final output is also shown in black. NPY/AgRP cells in the ARC are orexigenic. They are stimulated by ghrelin, inhibited by leptin, insulin, and PYY3-36, and are sensitive to glucose (not shown). They project to cells in the PVN, LHA, and DMH (not shown) and inhibit POMC/CART cells in ARC. POMC/CART cells use αMSH as a neurotransmitter onto MC3/4R receptors. αMSH and CART are anorexigenic. These cells are stimulated by leptin and insulin and activated indirectly by PYY3-36 through its inhibition of NPY/AgRP input onto the POMC/CART cells. Orexin A- and B-containing neurons are in the LHA. They are turned on by low blood glucose and by the balance of inputs from NPY/AgRP and POMC/CART cells from the ARC. They also receive input from the NTS, the nucleus of the solitary tract.

SUMMARY

Energy balance in humans obeys the laws of thermodynamics. The energy available from ingested food is its heat of combustion from initial foodstuff to final waste products. Glucose and fats are potentially oxidized completely, but proteins are only partially oxidized. The Atwater factors are approximations that give the available metabolic energy for fats, carbohydrates, proteins, and alcohol, and take into account digestibility and losses in urine and feces.

Energy production is nearly perfectly predicted by oxygen consumption. The amount of energy derived from fats, lipids, and carbohydrates differs markedly when expressed per gram of starting material, but it is essentially the same when expressed in terms of the oxygen consumed. Measurement of V_{O_2} and V_{CO_2} allows estimation of energy production. If urinary protein is also measured, the amount of fats, carbohydrates, and proteins consumed can be estimated.

The largest component of energy expenditure is the basal metabolic rate, but the greatest variable is the duration and intensity of activity. In highly active people, activity can consume more energy than maintenance. The BMR varies with age and gender and can be estimated by empirical formulas.

The hypothalamus regulates food intake by integrating a variety of signals. These include short-term satiety signals from the gastrointestinal tract and long-term adiposity signals from adipose tissue and metabolism. On top of this are the psychosocial cues including the availability, palatability, and social environment of eating.

The short-term signals nearly uniformly inhibit food intake. These include the following: GI distension; chemosensor detection of nutrients in the GI tract; CCK; blood glucose; GLP-1; and PYY3-36. All of these increase after a meal and all inhibit feeding. PYY3-36 is released from the ileum and therefore increases too late to influence food consumption at the meal. It probably influences food consumption at the next meal. Ghrelin is a 28-aa peptide produced by the stomach during fasting that increases appetite.

Blood glucose appears to regulate feeding behavior through glucose-responsive neurons that increase their firing when glucose levels rise and glucose-sensitive neurons that increase their firing rate when glucose levels fall.

The two most important long-term peripheral regulators of food intake are insulin and leptin. Adipose tissue secretes leptin in proportion to the amount of adipose tissue and insulin levels are also proportional to body adiposity. Several hypothalamic areas have receptors for both insulin and leptin. Leptin not only turns off feeding behavior but it also turns on thermogenesis through sympathetic stimulation of adipose tissue and muscle.

The central regulators of feeding behavior are classified as orexigenic (promoting feeding) and anorexigenic (inhibiting feeding). The orexigenic transmitters include NPY and agouti-related peptide (AgRP), both expressed by a population of cells in the ARC; and orexin A and B, expressed by cells in the LHA and MCH, produced by separate cells in the LHA. The main anorexigenic transmitters include αMSH, derived from POMC in cells in the ARC; and cocaine and amphetamine regulated transcript, or CART, coexpressed in POMC cells. In addition to its anorexigenic effect, POMC/CART cells activate sympathetic stimulation of the tissues, leading to increased energy expenditure.

REVIEW QUESTIONS

1. What are the Atwater factors for protein, carbohydrate, and fats? Are they simply the results of bomb calorimetry? What is a bomb calorimeter, anyway?
2. What is indirect calorimetry? What is meant by "energy equivalent" of oxygen for proteins, fats, and carbohydrates? Why do you think the energy equivalent of oxygen is so close for all three macronutrients?
3. What is the BMR? Why is this important? What accounts for the largest part of variation in energy requirement or expenditure?
4. What part of the brain is involved in homeostatic regulation of body adiposity?
5. What short-term signals turn on feeding behavior? What short-term signals turn off feeding behavior? What is GLP-1? Where is it made? What increases its secretion? What is ghrelin? Where is it secreted? What stimulates its secretion?
6. What are glucose-responsive neurons? What are glucose-sensitive neurons? How would these influence feeding behavior?
7. What candidates are there for long-term regulation of feeding behavior? What is leptin? Where is it made? What does it do?

Problem Set 8.1
Energy Balance

Conversion Factors: 2.54 cm = 1 in.; 1 lb = 453.6 g; 16 oz = 1 lb; 1 J = 0.239 cal; 1 W = 1 J s^{-1}; °F = 9/5 × °C + 32; 1 fluid oz = 0.029574 L; 1 L = 1.0566 qt

1. Using the Atwater Factors, calculate the energy content of 100 g (3/4 cup) of vanilla ice cream of composition 4% protein, 13% fat, and 21% carbohydrate. Percent composition is by weight, i.e., 4% is 4 g per 100 g ice cream.
2. Raw, long-grain brown rice has the following composition (http://www.ars.usda.gov):

Nutrients	Grams per 100 g
Water	10.37
Protein	7.94
Fat	2.92
Carbohydrate	77.24
Ash	1.53

   Using the Atwater Factors, calculate the total energy per 100 g of dry rice, in both kJ and kcal.
3. According to NHANES 2011–2012 (National Health and Nutrition Examination Survey), the average diets of 20–29-year-old people in the United States has the following composition of macronutrients (Table PS 8.1.2):
   Complete the table by calculating how much energy is supplied by each macronutrient and what fraction this constitutes of the whole, using the Atwater factors. Why do you suppose that females, on average, consume fewer calories than males?
4. According to the USDA, the following data were obtained for whole wheat flour (Table PS 8.1.3): What is the apparent Atwater Factor for these macronutrients based on these numbers? Why are they not identical to the Atwater Factors of 4, 4, and 9 kcal per g? See http://www.nutribase.com/449.html.
5. Consider a 22-year-old male who is 5′10″ tall and weighs 175 lb.
   A. Estimate his body surface area using the Dubois equation.
   B. Using Figure 8.6.3, estimate the BMR in energy per day.
   C. Compare the BMR in part B to that obtained from the Harris–Benedict equation.
   D. Add 25% for activity to the BMR to estimate normal energy requirements. This varies widely depending on the activity level. This is the estimated energy requirement and estimated energy consumption.
   E. The RDA (recommended dietary allowances) for protein is 0.8 g per kg of body weight per day. Estimate the RDA for protein for this individual.
   F. According to the USDA (http://www.ars.usda.gov), 4 oz of lean ground beef (10% fat) contains 22.60 g of protein. How many ounces of hamburger would supply this person with the RDA for protein?
   G. Using the Atwater Factor for protein, what fraction of the dietary consumption is made up from the recommended level of protein?
6. Adipose tissue varies in its composition, both in terms of water content, fat, and protein content, and distribution of fatty acyl chains in the lipids. Omental fat biopsies obtained from patients undergoing surgery were analyzed and found to contain about 10% water, 87% fat, and 2% protein (L.W. Thomas, *Exp. Physiol.* 47:179–188, 1962). Calculate the energy equivalent of 1 lb of adipose tissue in the body. This would be the excess energy expenditure necessary to lose 1 lb of fat.
7. In obese persons, biopsies of fat show lower water content than biopsies of fat from leaner persons. Give a reasonable explanation of why this should be the case.
8. Based on your answer to PS 8.1 #6, about how much extra oxygen must you consume to lose 1 lb of adipose tissue?
9. Although the rate of oxygen consumption varies with body size, composition, and age, an average figure is about 250 mL min^{-1}. Based on this oxygen consumption, and using an average value for the energy equivalent of O_2, what is a typical BMR in kcal per day? What is this BMR in watts? Based on this figure, how much cooling would you need to keep an auditorium with 500 people at a constant temperature?
10. Consider a male person 20 years of age who is 5′8″ tall and weighs 160 lb.
    A. If the basal oxygen consumption is 3.5 mL O_2 per minute per kg of body weight (=1 MET), what is the resting energy expenditure?
    B. Calculate their BMR from the Harris–Benedict formula. How do the values from A and B compare?

TABLE PS 8.1.2

Macronutrients	Males	Females	Calories (kCal or kJ)		% of energy	
			Males	Females	Males	Females
Protein (g)	102.9	72.1				
Carbohydrate (g)	332	255				
Fat (g)	102.3	75.5				
Total (g)						

TABLE PS 8.1.3

Macronutrients	g/120 g	kcal/120 g	Atwater Factor
Protein	16.9	59.8	
Carbohydrate	87.1	329	
Total fat	2.2	18.8	

11. If n is known, then p is known from Eqn (8.6.6). If p is known, then Eqn (8.6.7) are two equations in two unknowns. Solve Eqns (8.6.6) and (8.6.7) to derive expressions for c and f in terms of V_{CO_2}, V_{O_2}, and n.

12. Solve PS 8.1 #11 first. A middle-aged person walks for 2 hours on a treadmill at 4 mph. He voided his bladder at the beginning of the exercise and again at the end. The second urine sample was 130 mL, and its nitrogen concentration was 11.1 mg mL^{-1}. The total oxygen consumed during the 2 hours was 125.1 L, and the CO_2 produced was 97.0 L.
    A. Calculate the total urinary nitrogen and estimate g of protein utilized during the period.
    B. Calculate the grams of carbohydrates and fats consumed during the period.

13. Lavoisier determined metabolic rate of experimental animals by surrounding them with insulated ice and measuring the rate at which the ice melted. The heat of fusion of ice is 80 cal g^{-1}.
    A. If your oxygen consumption is 250 mL min^{-1}, at what rate can you melt ice?
    B. Assume you weigh 150 lb. The overall heat capacity of the human body is about 3500 J kg^{-1} °C^{-1}. You cool down 8 oz of coke to 0°C and in it is still floating 50 g of ice. How much will drinking that coke at 0°C alter your body temperature, assuming no compensatory mechanisms and instantaneous thermal equilibrium?

14. Assume you weigh 170 lb and your heat capacity is 3500 J kg^{-1} °C^{-1}; your oxygen consumption is 1 MET = 3.5 mL kg^{-1} min^{-1} and your resting temperature is 37°C. Your "friends" try to play a joke on you by covering you with a layer of metallic film that prevents all routes of energy transfer between you and the environment: no radiative, conductive, convective, or evaporative loss of energy can occur. Assuming your metabolic rate does not change with temperature (an invalid assumption), how long will it take before your temperature rises from 37°C to 39°C?

15. Assume you consume a typical American diet containing 20% of calories from protein, 40% from fat, and 40% from carbohydrates. Assume further that you caloric intake is 2800 kcal day^{-1} and that you maintain a steady state of both energy balance and protein balance. Estimate
    A. urinary nitrogen excretion, per day;
    B. O_2 consumed, per day;
    C. CO_2 produced, per day;
    D. the average respiratory quotient.

16. Flying Dog Brewery makes a beer called Doggie Style that is 4.7% ethanol by volume and contains 11.4 g of carbohydrate per 12 oz bottle. Using the Atwater Factors, calculate the calories in a 12-oz bottle of this beer. The density of alcohol is 0.785 g cm^{-3}.

17. In Chapter 2.11, we calculated that the ATP yield from glucose was 32 moles of ATP per mole of glucose and the ATP yield from tripalmitin was 336.5 moles of ATP per mole of tripalmitin. Compare these yields to the Atwater Factors. Are the ATP yields proportional to the energy of combustion? The molecular weight of glucose is 180 g mol^{-1}. The molecular weight of tripalmitin is 807.3 g mol^{-1}.

18. The heat of combustion for palmitic acid is 9977.9 kJ mol^{-1} for the crystalline form, 10,031.3 kJ mol^{-1} for liquid form, and 10,132.3 kJ mol^{-1} for the gas form. The heat of combustion of glycerol is 1655.4 kJ mol^{-1} for the liquid.
    A. Write a balanced chemical reaction for the complete oxidation of palmitic acid ($C_{16}H_{32}O_2$).
    B. Write a balanced chemical reaction for the complete oxidation of glycerol ($C_3H_8O_3$).
    C. Why is the energy of combustion different for crystalline, liquid, and gas forms of palmitic acid? Which should be used in biochemical calculations?
    D. What is the energy equivalence of O_2 when palmitic acid is oxidized?
    E. What is the energy equivalence of O_2 when glycerol is oxidized?

F. What is the predicted Atwater Factor for palmitic acid?
G. What is the predicted Atwater Factor for glycerol?

19. An indication of adiposity is the **body mass index**, **BMI**. It is defined as the mass in kilograms divided by the height in meters squared: BMI = kg m^{-2}. Although this obviously does not directly measure body fat, it was found that this variable best predicts chronic diseases associated with obesity. BMI > 30 indicates obesity; BMI > 40 indicates morbid obesity; BMI < 16 indicates dangerous underweight.
    A. For a person 5′8″ and 180 lb, calculate the BMI. Is this person obese?
    B. For a person 5′2″ tall and 100 lb, calculate the BMI. Is this person seriously underweight?
    C. Do you think BMI is a useful measure of obesity? Why or why not?

20. Suppose a person has a BMR of 1770 kcal day^{-1}. The person consumes a meal containing 700 kcal. The thermic effect of the food is given as $179.5\, t e^{-t/1.3}$ where t is time in hours, and the units of the constant, 179.5, are kJ h^{-2}. This constant is a fitted parameter that varies with individuals—it is NOT a biological invariant.
    A. Calculate the thermic effect by integrating the equation from 0 to 8 hours (negligible thermic effect remains after 6 hours but we will do it for 8 to be certain).
    Hint:
    $$\int A\, x\, e^{\frac{x}{C}}\, dx = -AC^2\left(\frac{x}{C} + 1\right) e^{\frac{x}{C}}$$
    B. What is the fraction of calories of the meal expended in the thermic effect?
    C. Prove the hint in part A by taking the derivative of both sides of the equation.

UNIT 9

ENDOCRINE PHYSIOLOGY

General Principles of Endocrinology 9.1

Learning Objectives

- Distinguish among autocrine, endocrine, paracrine, and neurocrine secretions
- List the classical endocrine glands
- Define endocrine gland and hormone
- Describe how the neuroendocrine system affects physiological control
- List those hormones that are polypeptides including
  A. anterior pituitary releasing hormones
  B. anterior pituitary hormones
  C. posterior pituitary hormones
  D. five gastrointestinal hormones
  E. hormones that originate from pancreas, kidney, heart, and parathyroid glands
- List those hormones that are derivatives of amino acids
- List those hormones that are derived from steroid metabolism
- Describe the process of making, processing, storing, and releasing polypeptide hormones
- Describe the process of making, processing, storing, and releasing steroid hormones
- Draw the Scatchard plot, titration curve, and Hill plot. Label the axes and indicate what parameters can be obtained from each plot
- Identify threshold, maximum response, and ED_{50} from titration curves
- Define upregulation and downregulation
- Describe desensitization and distinguish heterologous from homologous desensitization
- Define metabolic clearance rate and how it relates to the first-order rate constant for hormone degradation

ENDOCRINE GLANDS RELEASE SIGNALING MOLECULES INTO THE BLOOD

The classical definition of an **endocrine gland** is: "A comparatively small, circumscribed organ, devoid of ducts, and having access to a rich blood supply. It releases its products (hormones) directly into the blood stream." The classical endocrine glands include the

- Pituitary
- Hypothalamus
- Thyroid gland
- Parathyroid glands
- Adrenal glands
- Pancreatic islets
- Gastrointestinal system
- Gonads (testes, ovaries)
- Placenta.

The classical definition of a hormone is an organic substance that is released into the blood at low concentrations by a ductless gland, and that travels through the circulation to a remote target tissue where it evokes systemic adjustments by engaging specific responses. It is now recognized that a number of cells release hormone-like signaling molecules into the blood but do not fit into the classical definition of comprising circumscribed organs, but are more diffusely located. These cells include neurons, renal cells, cardiac atrial cells, adipocytes, osteocytes, and platelets.

MODERN DEFINITIONS OF HORMONE INCLUDE LOCAL AND DISTANT EFFECTS AND INTEGRATION OF THE ENDOCRINE AND NEURAL SYSTEMS OF CONTROL

It is now recognized that the complicated control systems of the classical endocrine glands probably evolved from more "primitive" systems that control local events in small multicellular animals without circulatory systems. These local control systems are discussed in Chapter 2.8 and are partly recapitulated in Figure 9.1.1. **Paracrine** hormones influence cells that are near the source of the paracrine hormone. **Autocrine** hormones affect the cell that secretes it. If the hormone is released into the blood to travel to its target, it fits the classical definition of an **endocrine** hormone. Neurons release materials that affect only those neighboring cells that have receptors for the neurotransmitters. Thus neurotransmitters are a kind of paracrine secretion, but the axon extends the distance between the cells. In neurocrine systems, neurons release neurotransmitters into the blood, and they exert their effects on distant target tissues that are reached through the blood.

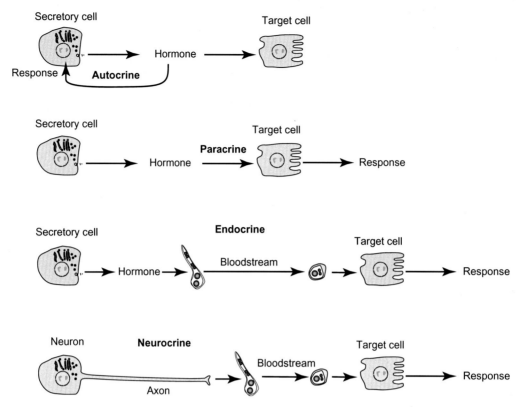

FIGURE 9.1.1 Various classes of hormones. Autocrine hormones affect the secretory cell; this is an autocrine function. Paracrine hormones diffuse through the neighboring extracellular matrix to affect nearby cells. Endocrine hormones are released into the blood to be transported to target cells, where they engage physiological responses. Neurocrine hormones are released into the blood after having been transported down axonal processes and stored in terminals near the point of release. These are released in response to action potentials that invade the nerve terminal, and so nerve cell activity precedes release of hormone.

THE NEURAL SYSTEM PROVIDES FAST, SHORT-LIVED CONTROL; ENDOCRINE CONTROL IS SLOWER AND LONGER LASTING

Both the neural and endocrine systems have components that

- secrete materials by exocytosis;
- generate membrane potentials and can be depolarized;
- affect their target cells by binding to specific receptors on target cells.

In addition,

- Some neurotransmitters are also used as hormones.

The main difference between the neural system of control and the endocrine system is the use of the axon. The axon extends the cell body and allows it to release its neurotransmitter close to its intended target in response to action potential propagation into the axon terminus. The endocrine system relies on the blood to transport its signaling molecules great distances. Both systems derive their specificity from the interaction of the signaling molecules with their receptors. Because of the axon and the extraordinarily fast conduction velocity of nerve impulses, the neural system provides fast control with rapid onset and nearly equally rapid shut-off. Endocrine systems typically produce a signal that is slower in onset but persists much longer. Overlap in the neural and endocrine control systems is illustrated in Figure 9.1.2. The endocrine system is controlled mainly through **negative feedback loops**, in which the response of the target cells restores the internal conditions towards normal, thereby lessening stimulation of the endocrine and neural systems and turning off the activating signal.

HORMONES CAN BE CLASSIFIED BY THEIR CHEMICAL STRUCTURE AND SOURCE

Hormones are chemically diverse but they can be grouped according to their chemical composition. The major hormones are listed in Tables 9.1.1A and 9.1.1B. The three major classes of hormones are:

1. **Modified amino acids**
2. **Polypeptides**
3. **Steroids**.

Hormones that are modified amino acids include **epinephrine**, also called **adrenaline**. It is synthesized from the amino acid, **tyrosine**, and secreted by the adrenal medulla in response to preganglionic sympathetic nervous stimulation. **Thyroxine** is another hormone derived by iodination of tyrosine molecules.

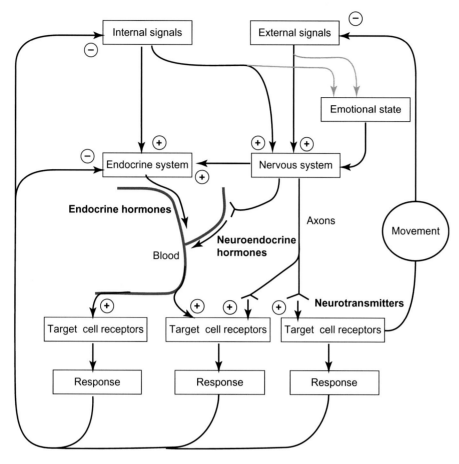

FIGURE 9.1.2 Overall control of body systems by the neuroendocrine systems. Interoreceptors sense the internal environment, and exteroreceptors sense events in the surrounding world. Interoreceptors can activate the endocrine branch directly, or they can activate the neural branch that then activates other nervous components or parts of the endocrine system. Exteroreceptors work exclusively through activation of the nervous system, which may then activate other neural, endocrine, or neuroendocrine responses. External and internal signals also influence the psychological state of the individual, which has profound, neurally mediated effects on the endocrine system. The two systems work together to provide integrated responses to challenges of homeostasis.

TABLE 9.1.1A Summary of the Major Human Hormones

Hormone Name	Source	Chemical Nature	Target Cells	Primary Action
Releasing hormones	Hypothalamus			
TRH (thyrotrophin releasing hormone)		3 aa chain	Thyrotrophs	Stimulates TSH release
CRH (corticotrophin releasing hormone)		41 aa chain	Corticotrophs	Stimulates ACTH release
GHRH (growth hormone releasing hormone)		40 or 44 aa chain	Somatotrophs	Stimulates GH release
GnRH (gonadotrophin releasing hormone)		10 aa chain	Gonadotrophs	Stimulates FSH and LH release
Somatostatin		14 aa chain	Somatotrophs	Inhibits GH release
Anterior pituitary hormones	Anterior pituitary			
TSH (thyroid stimulating hormone)	Thyrotroph cells	92 aa + 112 aa chains	Thyroid follicle cell	Stimulates thyroid hormone production + release
ACTH (adrenocorticotrophic hormone)	Corticotroph cells	39 aa chain	Adrenal cortex	Stimulates glucocorticoid, androgen production + release
GH (growth hormone)	Somatotroph cells	191 aa chain	Bone, liver, muscle	Stimulates growth
LH (luteinizing hormone)	Gonadotroph cells	92 aa + 112 aa chains	Ovarian follicle cells	Triggers ovulation
FSH (follicle stimulating hormone)	Gonadotroph cells	92 aa + 119 aa chains	Sertoli cells, ovarian cells	Stimulates growth and development
PRL (prolactin)	Lactotroph cells	199 aa chain	Alveolar cells of the breasts	Stimulates milk production

(Continued)

TABLE 9.1.1A (Continued)

Hormone Name	Source	Chemical Nature	Target Cells	Primary Action
Posterior pituitary hormones	Synthesized in hypothalamus, released in posterior pituitary			
ADH (antidiuretic hormone)		9 aa chain	Renal tubule and blood vessels	Increases water retention; vasoconstriction
Oxytocin		9 aa chain	Uterine muscle and breasts	Stimulates uterine contraction and milk "let-down"
Gastrointestinal hormones				
Gastrin	G-cells of stomach	17 aa or 34 aa chain	Parietal, ECL cells	Stimulates acid secretion
Cholecystokinin	Duodenal I cells	83 aa chain + fragments	Smooth muscle, neurons	Contracts gall bladder, slows gastric emptying
Secretin	S-cells in intestine	27 aa chain	Duct cells	Increases HCO_3^--rich secretion
Motilin	M-cells of duodenum	22 aa chain	Myenteric neurons	Coordinates migrating motor complex
Ghrelin	oxyntic glands	28 aa chain	hypothalamic neurons	stimulates appetite and release growth hormone
Somatostatin	D-cells of stomach	14 or 28 aa chain	Secretory cells	Inhibits gastric, biliary, and pancreatic secretion
Miscellaneous Polypeptide Hormones				
Insulin	β-cells of pancreas	30 aa + 21 aa chain	Liver, muscle, fat	Regulates glucose uptake and lipolysis
Glucagon	α-cells of pancreas	29 aa chain	Liver	Regulates glucose metabolism
Somatostatin	δ-cells of pancreas	14 aa chain	Pancreatic islets	Inhibits insulin and glucagon secretion
Parathyroid hormone (PTH)	Parathyroid gland	84 aa chain	Bone, kidney, lymphocytes	Regulates blood $[Ca^{2+}]$ and $[H_2PO_4^-]$
Calcitonin (CT)	C-cells of thyroid	32 aa chain	Bone	Inhibits bone resorption of Ca^{2+} and $H_2PO_4^-$
Atrial natriuretic peptide (ANP)	Cardiac atrial cells	28 aa chain	Renal tubule cells	Stimulates Na^+ excretion
Erythropoietin	Renal cells		Bone marrow stem cells	Stimulates erythropoiesis

TABLE 9.1.1B Summary of the Major Human Hormones

Hormone Name	Source	Chemical Nature	Target Cells	Primary Action
Thyroid hormone	Thyroid follicle	Modified tyrosine	Nearly all cells	Regulates energy metabolism
Epinephrine (adrenaline)	Adrenal medulla	Modified tyrosine	Circulatory system, liver	Engages "fight or flight" reaction
1,25-$(OH)_2$ cholecalciferol (vitamin D)	Skin, activated in liver and kidney	Secosteroid	Bone, intestine, kidney tubules	Mineralizes the skeleton by keeping plasma $[Ca^{2+}]$ high
Glucocorticoids	Adrenal cortex	Steroid	Fat, muscle, liver	Regulates metabolism
Aldosterone	Adrenal cortex	Steroid	Kidney tubule cells	Decreases Na^+ excretion and increases K^+ excretion
Androgens	Testes and adrenal	Steroid	Reproductive tract	Stimulate growth and development
Estrogens	Ovaries and placenta	Steroid	Reproductive tract	Stimulate growth and development
Progesterone	Ovaries and placenta	Steroid	Uterus and breasts	Stimulate growth and development

A great many of the endocrine hormones are polypeptide hormones, and many of these are glycosylated. These range in length all the way from **thyroid releasing hormone (TRH)**, which is a tripeptide, to **human chorionic gonadotrophin** (hCG), which is about 34 kDa. Tables 9.1.1A and 9.1.1B lists the major hormones along with their source, target, and major physiological action. The polypeptide hormones can be classified in a variety of ways. These hormones include the following:

ANTERIOR PITUITARY RELEASING OR INHIBITING HORMONES

Cells in the hypothalamus release these hormones into a local circulation that delivers them to the anterior pituitary where they control the release of yet another set of hormones. They include:

- **Thyroid releasing hormone (TRH)**; 3 amino acids
- **Corticotrophin releasing hormone (CRH)**; 41 amino acid chain
- **Growth hormone releasing hormone (GHRH)**; 44 or 40 amino acids
- **Gonadotrophin releasing hormone (GnRH)**; 10 amino acids
- **Somatostatin (SST)**; 14 amino acids.

ANTERIOR PITUITARY HORMONES

The anterior pituitary secretes a set of trophic polypeptide hormones that influence other endocrine glands. The anterior pituitary hormones include:

- **Thyroid stimulating hormone (TSH)**; two chains; glycoprotein
- **Adrenocorticotrophic hormone (ACTH)**; 39 amino acids
- **Growth hormone (GH)**; 191 amino acids
- **Luteinizing hormone (LH)**; two chains; glycoprotein
- **Follicle stimulating hormone (FSH)**; two chains; glycoprotein
- **Prolactin (PRL)**; 199 amino acids.

For most of these, the names are self-explanatory. TSH stimulates the thyroid gland. ACTH stimulates the adrenal cortex. GH stimulates whole body growth and also has particular effects on the skeleton. LH, FSH, and PRL are all hormones involved in sexual reproduction. Each of these hormones will be discussed in later chapters.

POSTERIOR PITUITARY HORMONES

Hypothalamic neurons project to the posterior pituitary where they release neuroendocrine hormones into the general circulation. These include:

- **Antidiuretic hormone (ADH)**, also known as vasopressin; 9 amino acids
- **Oxytocin**; 9 amino acids.

ADH regulates the excretion of water by the kidney. Oxytocin contracts the uterus during **parturition** and also stimulates the **milk ejection reflex** during breastfeeding of the infant.

GASTROINTESTINAL HORMONES

The gastrointestinal system releases its own set of hormones that mainly regulate gastrointestinal function. The function of these hormones was discussed in Unit 8. These polypeptide hormones include:

- **Gastrin**; 17 or 34 amino acid chain
- **Cholecystokinin**; 83 amino acids
- **Secretin**; 27 amino acids
- **Motilin**; 22 amino acids
- **Somatostatin**; 14 or 28 amino acid chain
- **Ghrelin**; 28 amino acids.

OTHER POLYPEPTIDE HORMONES ARE RELEASED BY INDIVIDUAL GLANDS

A number of other polypeptide hormones are released by a variety of endocrine glands.

- **Insulin**, released by β-cells in the islets of Langerhans in the pancreas
- **Glucagon**, released by α-cells in the pancreas
- **Somatostatin**, released by δ-cells in the pancreas
- **Parathyroid hormone (PTH)**, released by the parathyroid glands
- **Calcitonin (CT)**, released by C-cells in the thyroid
- **Atrial natriuretic peptide (ANP)**, released by cardiac atrial cells
- **Erythropoietin**, released by the kidney
- **FGF-23, fibroblast growth factor**, released by osteocytes
- **IGF-I and IGF-II**, released by the liver.

STEROID HORMONES ARE SYNTHESIZED FROM CHOLESTEROL

A variety of steroid hormones are produced in the adrenal cortex, gonads, and placenta. These include:

- **Androgens**, including **testosterone** and **dehydroepiandrosterone (DHEA)**
- **Estrogens**, primarily **estradiol**
- **Glucocorticoids**, including **cortisol** and **corticosterone**
- **Mineralocorticoids**, including **aldosterone**.

The skin produces

- **Cholecalciferol**, precursor to its most active form, **1,25-dihydroxycholecalciferol.**

POLYPEPTIDE HORMONES ARE TYPICALLY SYNTHESIZED AS LARGER PRECURSORS

Cells synthesize polypeptide hormones using the same processes for the synthesis of other secretory products such as digestive enzymes and mucin that we have already described (see Chapters 2.3 and 2.4). The process is summarized in Figure 9.1.3.

Briefly, RNA polymerase II transcribes specific regions of the DNA that carries the instruction for synthesizing the hormone. After processing, mRNA leaves the nucleus to attach to ribosomes on the rough endoplasmic reticulum (ER) where it is translated into protein. Generally, the mRNA first directs the synthesis of a larger precursor called

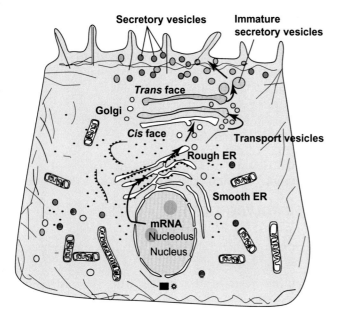

FIGURE 9.1.3 Overall processing of peptide hormones. The mRNA originates in the nucleus and is transported to the rough ER where it specifies the synthesis of a preprohormone. The preprohormone is elongated and transferred to the lumen of the ER. It is cleaved to form a prohormone. From there it travels to the Golgi stack by transport vesicles that bud off the ER membrane. In the Golgi stack, the hormone is further processed by proteolysis and possibly by glycosylation. The Golgi apparatus packages the polypeptide into immature secretory vesicles which concentrate the hormone into secretory vesicles. These vesicles form a storage pool of hormone that can be released rapidly upon stimulation of the secretory cell.

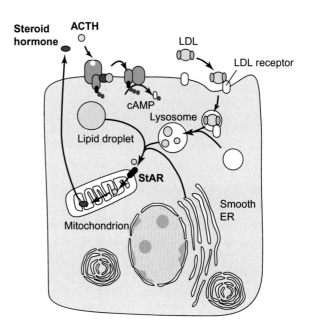

FIGURE 9.1.4 Process of synthesis and secretion of steroid hormones in cells of the adrenal cortex. Synthesis is stimulated by ACTH, or corticotropin, which is coupled to increased cytoplasmic concentrations of 3′,5′cAMP through a heterotrimeric G-protein. The synthesis of the steroid hormones occurs in the inner mitochondria and is controlled by the activity of StAR, a 30-kDa protein that transfers cholesterol from the outer mitochondrial membrane to the inner membrane. The steroid hormones leave the cell by passive diffusion.

a **preprohormone** that has a signal sequence at the N-terminus of the growing peptide chain. This signal sequence is recognized by a **signal recognition particle (SRP)** (see Chapter 2.4), which stops elongation and transfers the peptide chain to a **translocon**, a protein channel in the ER membrane. After SRP dissociates from the ribosome, chain elongation resumes and the growing peptide chain passes into the lumen of the ER via the translocon.

The portion of the ER containing enclosed polypeptide buds off the ER membrane and travels to the *cis*, or forming, face of the **Golgi stack**. The opposite pole of the Golgi stack facing the plasma membrane is the *trans* or maturing face. Vesicles are able to transport materials through the Golgi stack in either direction: from *cis* face to *trans* face or from *trans* face to *cis* face. During transit, proteins are processed in a variety of ways, including proteolytic cleavage, glycosylation, and association of different subunits. The finished product buds off the *trans* face to form an immature secretory vesicle, which matures by concentration of its enclosed contents, probably by removal of excess membrane back to the Golgi. These **secretory vesicles** form a reservoir of hormone that can be mobilized when the secretory cell is stimulated.

STEROID HORMONES ARE METABOLIZED FROM CHOLESTEROL AND ARE NOT STORED

Unlike cells that secrete polypeptides, steroid producing cells do not make secretory vesicles with stores of ready-made hormone that await the signal for secretion. Instead, steroid hormone synthesis and secretion is all part of a single process. The steroid hormones are synthesized from cholesterol that originates either from circulating low-density lipoproteins that are taken up by the secreting cells, from cholesterol stores in lipid droplets within these cells, or from cholesterol that is newly synthesized from acetyl CoA on the smooth endoplasmic reticulum of these cells. The rate-limiting step appears to be the transfer of cholesterol from the outer mitochondrial membrane to the inner mitochondrial membrane, where the enzymes that synthesize the hormones are located. A protein called **steroid acute regulatory protein (StAR)** carries the cholesterol across. This 30-kDa protein is regulated by cytosolic levels of cAMP. ACTH increases cAMP by binding to its G_s-coupled receptor on the surface of these cells. The overall process of steroid hormone synthesis and secretion is shown schematically in Figure 9.1.4.

BLOOD CARRIES HORMONES IN EITHER FREE OR BOUND FORMS

The polypeptide hormones and epinephrine are readily soluble in water, and so the blood carries them in the dissolved state. These hormones do not readily penetrate cell membranes. Instead, they bind to receptors on the outside surface of the target cells. How this binding is used to affect the target cell has been discussed in general terms in Chapter 2.8. A variety of other hormones are not readily soluble in water. Large carrier

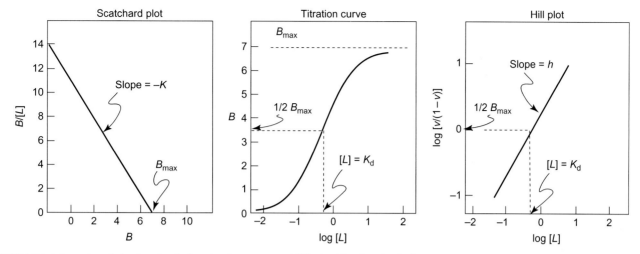

FIGURE 9.1.5 Examples of the Scatchard plot, titration curve, and Hill plot. The Scatchard plot is generally used to determine the affinity of the receptor for its ligand and the number of binding sites; the titration curve best shows how the affinity is determined by points above and below K_d, and shows the whole range of response; the Hill Plot is generally used to determine the cooperativity of the ligand–receptor interaction. See Appendix 9.1.A1 for a full discussion.

proteins bind to these materials and carry them in the plasma. These binding proteins include:

- Thyroxine binding globulin (TBG)
- Corticosteroid binding globulin (CBG or transcortin)
- Sex hormone binding globulin (SHBG)
- Vitamin D binding protein
- Albumin.

These proteins are made by the liver and secreted into the circulation. Most of these are about 50–60 kDa plasma glycoproteins. They bind the hormones with affinities that depend on the hormone. TBG binds mainly triiodothyronine (T3) and tetraiodothyronine (T4). Corticosteroid binding protein binds both cortisol and progesterone, whereas the SHBG carries testosterone and estrogen. Albumin binds many hydrophobic materials nonspecifically. All of these carrier proteins are in dynamic equilibrium with free hormone. Typically, only 1–10% of the total blood hormone is free in solution; the remaining 90–99% remains bound to the carrier. All of the materials carried by plasma binding proteins are believed to exert their effects by penetrating cells directly and binding to receptors within the target cells. Thus only the free material is biologically active, because only it can penetrate cells to exert biological effects. Most of these hydrophobic hormones alter genomic expression of their target cells, although nongenomic effects also occur.

ONLY TARGET CELLS WITH RECEPTORS TO THE HORMONE RESPOND TO THE HORMONE

Hormones are very much like neurotransmitters in that **only cells that have receptors to the hormone can respond**. Peptide hormones bind to the plasma membrane of their target cells, initiating a signaling cascade to change target cell behavior. This requires receptors for the hormone on the surface membrane. The collection of receptors on the cell's membrane defines its responsiveness to the hormones. The hydrophobic hormones bind to intracellular receptors and only those cells with the appropriate intracellular receptors can respond. Specificity of hormone action derives from the specificity of the molecular interaction of hormone and receptor.

DOSE–RESPONSE CURVES DERIVE FROM THE KINETICS OF HORMONE BINDING AND POST-RECEPTOR EVENTS

The kinetics of ligand binding to receptors can be graphically represented in a variety of ways including the **Scatchard plot**, the **titration curve**, or the **Hill plot**. The mathematical basis for these plots is discussed in Appendix 9.1.A1. The three types of plots are shown in Figure 9.1.5.

THE SCATCHARD PLOT PLOTS BOUND/FREE VERSUS BOUND

The Scatchard analysis plots the total amount of bound ligand divided by its free concentration ($B/[L]$) against the total bound ligand, B. For a single kind of receptor, the slope is equal to $-K$, the association constant for the binding of ligand to receptor, and the x-intercept is equal to the number of binding sites for the ligand, B_{max}.

THE TITRATION CURVE PLOTS THE AMOUNT OF BOUND LIGAND AGAINST THE LOG OF THE FREE LIGAND CONCENTRATION

The titration curve plots the amount of bound ligand against the logarithm of the free ligand concentration. Its utility is in identifying the concentration of ligand at which one-half of the receptors are occupied. This occurs when the $[L] = K_d$, the dissociation constant for ligand binding.

QUANTITATIVE HUMAN PHYSIOLOGY: AN INTRODUCTION

HILL PLOTS ARE USED WHEN SCATCHARD PLOTS ARE NOT LINEAR

Hill plots define a saturation fraction, ν, which is the amount of bound ligand divided by the number of binding sites: $\nu = B/B_{max}$. The value of $\log(\nu/(1-\nu))$ is plotted against $\log[L]$. The slope of the line between $0.1 < \nu < 0.9$ is the Hill coefficient, h, and is an index of cooperativity. Values of $h < 1.0$ indicate negative cooperativity. This means that binding of ligand makes further binding of ligand more difficult; positive cooperativity occurs with values of $h > 1.0$, meaning that binding of ligand makes further binding easier.

DOSE–RESPONSE CURVES ARE MOST EASILY VISUALIZED WITH TITRATION CURVES

The dose–response curve for a target tissue usually uses some unit of biological response instead of occupancy of the receptors, on the assumption that biological response relates linearly to receptor occupancy.

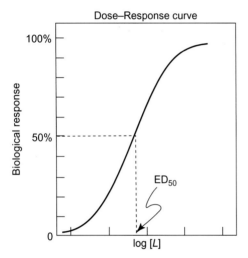

FIGURE 9.1.6 Typical dose–response curve for the effect of a hormone on a target tissue. Generally the dose–response curve has a sigmoid shape. The concentration of hormone at one-half maximal target cell activity is called the ED_{50}, for effective dose at 50% of maximal response.

Typically, targets exhibit some **basal activity** that occurs in the absence of the hormone. The response to increasing hormone passes a **threshold** level, which is the smallest discernible increase in biological response, to eventually reach a **maximal response**. The concentration of hormone required to produce a response halfway between the basal level and the maximal level is called the **median effective dose** or ED_{50} (the effective dose at 50% response). The overall dose–response curve depends on the kinetics of hormone binding to its receptor and to the kinetics of postreceptor events (see Figure 9.1.6).

DOSE–RESPONSE CURVES CAN BE "UPREGULATED" OR "DOWNREGULATED"

The sensitivity of target tissues to hormones can change with conditions. Two major ways in which the dose–response curve can change is in the maximal response and in the ED_{50}. These are shown diagrammatically in Figure 9.1.7.

Theoretically, the maximal response of a target tissue can be altered by changing

- the number of active cells in the tissue
- the number of receptor molecules in the tissue
- the effectiveness of each receptor molecule.

UPREGULATION AND DOWNREGULATION ALTER THE NUMBER OF ACTIVE RECEPTORS

Cells continually synthesize and degrade receptor molecules. Exposure to their ligand almost always causes **downregulation**, a decrease in the number of available or responsive receptors. This can occur rapidly by moving receptors from the surface membrane into intracellular sites where the latent sites cannot react to extracellular hormone. Although less common, **upregulation** also occurs. FSH and estradiol, for example, upregulate the number of their receptors in order to amplify their action.

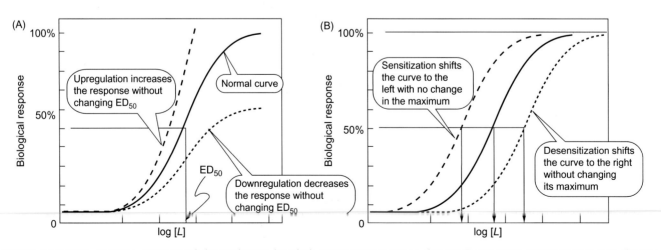

FIGURE 9.1.7 Dose–response curves reveal the mechanism by which target tissue response changes. Target tissue responsiveness may change, indicated by a change in the maximal response with no change in the ED_{50} (A). Target tissue sensitivity may also change (B) as indicated by a change in the ED_{50} without a change in the maximal response.

DESENSITIZATION *CHANGES THE CONCENTRATION DEPENDENCE OF TARGET RESPONSE*

Changes in the sensitivity of the target tissue shift the dose–response curve right or left without changing its maximum (see Figure 9.1.7). Desensitization refers to a shift to the right, so that higher concentrations of hormone are necessary to elicit the same response; **sensitization** indicates a shift to the left. If exposure to a hormone desensitizes the target to itself, the process is **homologous desensitization**. If desensitization results from exposure to some other agent, the process is **heterologous desensitization**.

THE *HALF-LIFE* AND *METABOLIC CLEARANCE RATE* QUANTITATIVELY DESCRIBE HORMONE METABOLISM

Hormones disappear from the circulation by

- uptake by target cells
- metabolism
- excretion into the urine or bile.

The overall process can be described quantitatively by the **metabolic clearance rate (MCR)**, *defined* as:

$$[9.1.1] \quad \text{MCR} = \frac{\text{amount of hormone removed per unit time (in units of mg min}^{-1})}{\text{plasma[hormone] (in units of mg mL}^{-1})}$$

The units of MCR are mL min^{-1}, which are the same units used to express renal clearance. MCR corresponds to the volume of plasma that contains all of the hormone that is cleared in 1 minute. This is usually determined by injecting someone with a small amount of radioactive hormone and determining the rate of disappearance of the labeled hormone from serial blood samples. The rate of disappearance of labeled hormone generally obeys **first-order kinetics**: the rate of hormone loss is proportional to how much hormone is present. This is expressed quantitatively as

$$[9.1.2] \quad \frac{dH}{dt} = -kH$$

where H is the amount of labeled hormone and k is the aggregate rate constant. The description "first-order kinetics" derives from this rate law being proportional to the first power of the substrate. This equation can be easily solved for $H(t)$ by separation of variables and integrating to give

$$[9.1.3] \quad H = H_0 \, e^{-kt}$$

First-order plots are plots of ln H against t. If the process is first order, then plots of ln H against t should be linear with slope $= -k$. The **half-life** is defined as the time required for H to decay to $H_0/2$. Inserting $H = H_0/2$ into Eqn (9.1.3) we find

$$[9.1.4] \quad t_{1/2} = \frac{\ln 2}{k} = \frac{0.693}{k}$$

Thus the half-life $t_{1/2}$ is inversely related to the rate constant for disappearance of material. If k is large, then the half-life is short and the material disappears rapidly.

THE MCR IS INVERSELY RELATED TO THE HALF-LIFE

The MCR tells us what volume of plasma is cleared of hormone per minute. If this is large, it means that the hormone is rapidly cleared, which means that the half-life is short. Thus the half-life is inversely proportional to the MCR. The exact relationship between MCR and $t_{1/2}$ is apparent when we recognize from its definition that

$$[9.1.5] \quad \text{MCR} = \frac{-dH/dt}{H/V_d}$$

where dH/dt is an amount per unit time; since it is disappearing, dH/dt is negative and $-dH/dt$ is positive; H is the amount of hormone, and V_d is its volume of distribution; H/V_d is its concentration. We substitute in from Eqn (9.1.2) to obtain

$$[9.1.6] \quad \text{MCR} = \frac{kH}{H/V_d} = kV_d$$

Stated another way, we can write

$$[9.1.7] \quad k = \frac{\text{MCR}}{V_d} = \frac{\text{MCR}[H]}{V_d[H]}$$

Note that MCR[H] is the size of the pool that turns over every minute, and $V_d[H]$ should be equal to the size of the pool of hormone (in units of amount) that is distributed in the volume V_d. The ratio of these two is the **fractional turnover rate**: the fraction of the entire hormone pool that turns over per minute. Thus Eqn (9.1.7) tells us that **the first-order rate constant for disappearance of hormone is equal to the fractional turnover rate**.

Half-life of hormones must be established using radioactive labels or some other technique because the total amount of hormone in the body does not obey Eqn (9.1.2). It obeys the equation:

$$[9.1.8] \quad \frac{dH}{dt} = -kH + S$$

where kH is the rate of degradation and S is the rate of secretion. The rate of secretion can be highly variable and episodic. If S is constant, then the amount of hormone in the body, and therefore its concentration, would be constant, set so $dH/dt = 0$ and $H = S/k$. Normally this is not the case. The rate of secretion is generally episodic and dependent on other variables that change with physiological condition, and the hormone levels fluctuate within some normal limits.

A VARIETY OF TECHNIQUES CAN MEASURE HORMONE LEVELS

THE EARLIEST HORMONE ASSAYS WERE BIOASSAYS

Many hormones were initially discovered by their effects on live animals or people, often through "experiments of nature" in which the circulating hormone concentrations were either excessive or insufficient. Because of this, many of the early hormone assays were biological assays, or **bioassays**, in which a biological response was noted after injection of a sample. Biological assays have the disadvantages of lack of specificity, lack of sensitivity, slow response time, and the need to use many animals. They have been replaced by physical and immunological techniques.

CHROMATOGRAPHY SEPARATES MATERIALS BASED ON PARTITIONING BETWEEN PHASES

Chromatographic assays for hormones involves **high performance liquid chromatography (HPLC)** systems. These generally involve a solid phase column of material that interacts with dissolved materials as they flow past in a liquid phase. There are two general kinds: **normal phase** and **reverse phase** HPLC. The column packing in normal phase HPLC contains polar groups such as amino or nitrile ions. In reverse phase HPLC, the solid phase presents a nonpolar stationary stage. The liquid phase can be constant (**isocratic elution**), changed stepwise (**step elution**), or changed gradually in **gradient elution**. The trick is to choose the columns and elution solvents that separate the compounds of interest. The compounds can be detected in the eluate by a variety of techniques including light absorption, fluorescence, electrochemical properties, and mass spectroscopy. The advantage of HPLC is that it can allow the simultaneous detection of several related materials, such as the steroid hormones, in a single sample.

MASS SPECTROMETRY

Mass spectrometry detects materials on the basis of their charge to mass ratio. After cleaning up the samples by HPLC, for example, the sample is fragmented into charged ions. This fragmentation occurs in specific ways, depending on the chemical nature of the material, and so the resulting spectrum of fragments with mass/charge ratios allows qualitative and quantitative identification of the material.

RADIOIMMUNOASSAYS

Radioimmunoassays measure hormone concentration by competing with radioactively labeled material for

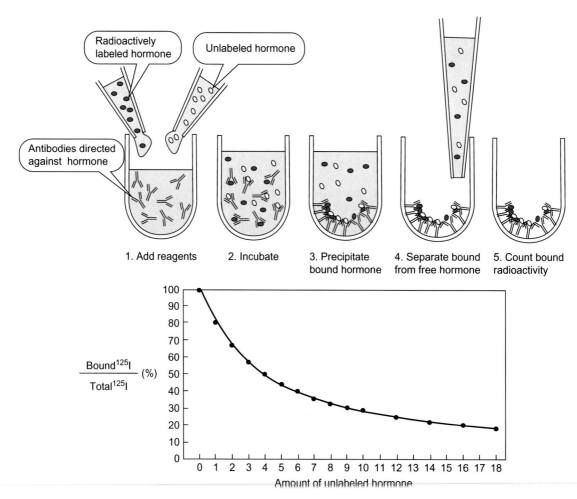

FIGURE 9.1.8 Basic principles of radioimmunoassay. Antibodies against the hormone are incubated in a series of tubes containing a constant amount of radioactively labeled hormone and various amounts of standard, unlabeled hormone or samples containing unknown amounts of hormone. The labeled and unlabeled hormones compete for binding sites on the antibodies. When more unlabeled hormone is present, less radioactivity binds to the antibody. After equilibration, the antibody–hormone complex is precipitated and collected into a pellet by centrifugation. The free unlabeled and radioactive hormone is removed and the bound radioactivity is measured using a radiation counter. Plots of the percent of radioactivity bound against unlabeled hormone concentration are highly nonlinear. The amount of hormone in the unknowns is determined by interpolation.

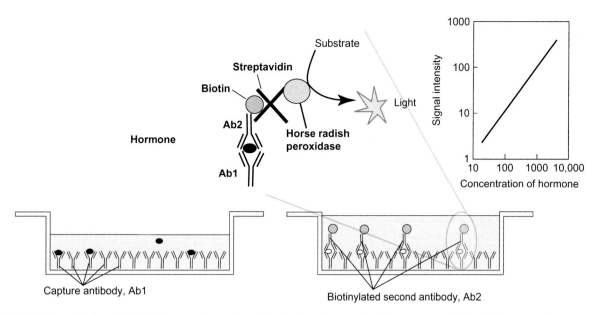

FIGURE 9.1.9 Basic principles of ELISA. The assay is conducted in plastic wells coated with a capture antibody, Ab1, present in great excess of hormone. Hormone binding to Ab1 is therefore proportional to its concentration. Incubation of a second antibody (Ab2) covalently linked with biotin makes an antibody–hormone–antibody sandwich. The biotin is a convenient handle to link Ab2 to an enzyme, HRP, by incubating the sandwich with streptavidin conjugated with HRP. Streptavidin has a very high affinity for biotin. Incubation of the resulting complex with appropriate substrates causes a chemiluminescent reaction. The intensity of the emitted light is proportional to the concentration of hormone over three orders of magnitude.

binding sites on antibodies that bind both labeled and unlabeled hormones. If there is little "cold" hormone, then most of the radioactivity will be bound. A lot of "cold" hormone will compete for the hormone binding site and less radioactivity will be bound. Standard curves are prepared by using a series of known amounts of cold hormone in separate tubes (see Figure 9.1.8).

ENZYME-LINKED IMMUNOSORBENT ASSAY

This assay is sometimes called the sandwich assay because the hormone is sandwiched between two antibodies. The assay is usually conducted in a 3 × 5 inch plastic plate containing 96 wells that are coated with a **capture antibody** (Ab1) that extracts hormone from the standards containing various known amounts of the hormone or the samples. The capture antibody is used in large excess over the hormone to be detected. The hormone–Ab1 complex is then reacted with a second antibody (Ab2) that can be linked to a detection system. One method of doing this is to complex Ab2 with **biotin**. Biotin is a water-soluble vitamin that is required as a cofactor for carboxylase reactions in the body. Its utility is that it can be easily covalently linked to protein molecules, and it is bound extremely tightly by **streptavidin**, a 70-kDa protein. The biotin on Ab2 serves as a link to streptavidin conjugated to **horseradish peroxidase** (HRP). Addition of substrates for HRP produces a signal that can be detected photoelectrically. This entire process is illustrated in Figure 9.1.9.

SUMMARY

Endocrine hormones are a class of signaling molecules that originate from a comparatively small and localized part of the body. They travel through the blood stream and reach distant target tissues where they exert specific effects. These classical endocrine hormones have been supplemented by a variety of signaling molecules that affect only nearby cells (paracrine hormones) or affect the cells that made them (autocrine hormones). In addition, nerve cells release neurocrine hormones. These signaling molecules work together with the nervous system to regulate normal bodily function.

Hormones can be grouped according to their chemical structure as (1) modified amino acids; (2) polypeptides; and (3) steroid derivatives. Each group contains a large number of examples.

Amino acid derivatives include epinephrine and thyroxine.

Polypeptide hormones include hypothalamic releasing factors, including TRH, CRH, GHRH, GnRH, and SST. The anterior pituitary hormones, all polypeptides, include TSH, ACTH, GH, LH, FSH, and PRL. The posterior pituitary hormones include ADH and oxytocin. The name of a polypeptide protein often describes its action.

Gastrointestinal polypeptide hormones include gastrin, cholecystokinin, secretin, motilin, ghrelin and SST. Other polypeptide hormones, secreted by individual glands, include insulin, glucagon, SST, parathyroid hormone, CT, ANP, and erythropoietin.

The steroid hormones include androgens, estrogens, glucocorticoids, mineralocorticoids, and cholecalciferol.

Peptide hormones are typically synthesized as larger prohormones and stored in secretory vesicles awaiting signals for their release. Steroid hormones are typically secreted as they are synthesized.

The response of target cells to circulating hormones depends on the hormone receptors. Lipid soluble hormones have intracellular receptors; polypeptide hormones bind to receptors on the cell surface. Binding of hormone to its receptor can be analyzed using Scatchard plots, titration curves, or Hill plots. The titration curve is often most useful because the full range of ligand concentration is easily visualized.

Secretion of most hormones is regulated whereas degradation follows simple first-order kinetics. The MCR is the volume of plasma completely cleared of hormone per minute. It is equal to the fractional turnover rate times the volume of distribution of hormone.

Hormone assays include radioimmunoassay, in which radioactively labeled hormone competes with "cold" hormone for binding sites on antibodies directed against the hormone. Enzyme-linked immunosorbent assay (ELISA) involves sandwiching the hormone between two antibodies, one of which is linked through biotin to streptavidin-linked HRP. Reaction of the HRP produces light that is proportional to the amount of captured hormone.

REVIEW QUESTIONS

1. What is a hormone? How is it distinguished from a neurotransmitter?
2. Name three major chemical classes of hormones. What is different about their synthesis and storage?
3. What are the two major types of hormone receptors?
4. What is a Scatchard plot? What is a titration curve? What is a Hill plot? What information can you derive from each?
5. What is a dose–response curve? What is downregulation? Upregulation? Desensitization?
6. What is the metabolic clearance rate? How is it related to fractional turnover?

APPENDIX 9.1.A1 ANALYSIS OF LIGAND BINDING

THE LANGMUIR ADSORPTION ISOTHERM

It was observed long ago that a finely divided solid exposed to a dye solution or a gas at low pressure caused a reduction in the dye concentration or a reduction in the gas pressure. In these cases the dye or gas was **adsorbed** onto the surface of the solid. The fine division of the solid ensured that there was a great total surface area of varying shapes and sizes to accommodate the dye or gas. This adsorption can be represented by a chemical equilibrium

[9.1.A1.1] $$A + S \rightleftharpoons A \cdot S$$

where A is the material being absorbed and S is the vacant site on the surface of the solid. We cannot write the equilibrium in the usual way by considering the concentration of A and S, because the concentration of the vacant sites is poorly defined: S is distributed on an area, not throughout a volume. The concentration of S would have different units than the concentration of A. However, these different units will actually cancel in the definition of the equilibrium. Let Θ be the **fraction** of sites occupied by A. Then $1 - \Theta$ will be the fraction of sites which remain vacant. Then we can write

[9.1.A1.2] $$K = \frac{\Theta}{(1-\Theta)A}$$

to define the equilibrium in the heterogeneous phase of liquid solution and solid absorbant. Here K is an association constant. This can be rearranged to give

[9.1.A1.3] $$\Theta = \frac{KA}{1+KA}$$

THE SCATCHARD PLOT

Consider a protein which has a single site for binding a ligand. The word "ligand" originates from the Latin "ligo" meaning "to tie." The protein may be soluble or itself bound to a membrane fraction. We define the **intrinsic association constant** to be

[9.1.A1.4] $$K = \frac{[L \cdot P]}{[L][P]}$$

where $[L]$ is the free ligand concentration in equilibrium with the binding site, $[P]$ is the concentration of the protein with the site vacant, and $[L \cdot P]$ is the concentration of protein with ligand bound. In strict analogy to the Langmuir Adsorption Isotherm, we define the **saturation fraction**, ν, as the proportion of the site which is occupied by L:

[9.1.A1.5] $$\nu = \frac{[L \cdot P]}{[L \cdot P] + [P]}$$

By Eqn (9.1.A1.4),

[9.1.A1.6] $$[L \cdot P] = K[L][P]$$

Substitution into Eqn (9.1.A1.5) and rearranging, we obtain:

[9.1.A1.7] $$\nu = \frac{K[L][P]}{K[L][P] + [P]} = \frac{K[L]}{1 + K[L]}$$

This may be rearranged to give

[9.1.A1.8] $$\nu + \nu K[L] = K[L]$$
$$\frac{\nu}{[L]} = K(1-\nu)$$

This equation is the foundation of the Scatchard plot. In this plot, the values of $\nu/[L]$ are plotted against the values of ν. According to Eqn (9.1.A1.8), this plot should result in a straight line intersecting the vertical, $\nu/[L]$ axis at K and intersecting the horizontal, ν axis at 1.0, the slope of the line being $-K$.

The real experimental situation is hardly ever so clear as this theoretical analysis. First, the binding reaction being investigated may be occurring in a heterogeneous mixture consisting of more than one protein. The total number of binding proteins, and the number of binding sites on each protein, may be unknown. According to the definition of the saturation fraction, it is limited to the range $0 \leq \nu \leq 1.0$. In the experiment, some quantity of protein containing the binding site is incubated with a known concentration of radioactively labeled ligand. After incubation for sufficient time to ensure equilibration, as determined by previous experiments, the free ligand is separated from the bound ligand, and the amount of bound ligand is determined. The binding is then expressed as the amount of bound ligand per unit of protein, typically as pmol/mg or some similar unit. This is the amount bound, B. In the Scatchard plot, the axes are $B/[L]$ against B; the experimental value can be treated as ν, even though it is clear that it is no longer restricted to values between 0 and 1.0. In the Scatchard analysis using the dimensionless values of ν, the slope is $-K$ in units of M^{-1}. Using the experimental values of B gives the identical slope, but the intercept on the B axis is now the B_{max}, the maximum binding at saturation.

The above description holds for a mixture of proteins, each of which binds at a single site and which are independent of each other. Suppose now that we introduce the notion of multiple sites. We define the intrinsic association constant of the ith site to be

[9.1.A1.9] $$\nu_i = \frac{K_i[L][P]}{K_i[L][P] + [P]} = \frac{K_i[L]}{1 + K_i[L]}$$

Suppose now that there are more than one sites on the protein, and that the ligand binding can be modeled as a set of j independent sites, each with the identical intrinsic association constant, K_i. Now we define the **gross saturation fraction** as

[9.1.A1.10] $$\nu_i = \sum_{i=1}^{j} \frac{K_i[L][P]}{K_i[L][P] + [P]}$$

If all of these j independent sites are characterized by the same intrinsic association constant, then Eqn (9.1.A1.10) becomes

[9.1.A1.11] $$\nu_i = \frac{nK[L]}{1 + K[L]}$$

where n is the number of binding sites. This equation may be rearranged into the Scatchard form:

[9.1.A1.12] $$\frac{\nu}{[L]} = K(n - \nu)$$

In using the experimental values for the ligand binding, this is

[9.1.A1.13] $$\frac{B}{[L]} = K(B_{max} - B)$$

The result of the plot of $B/[L]$ against B, then, is a slope equal to $-K$, and an intercept on the B axis equal to B_{max}, and an intercept on the $B/[L]$ axis equal to $B_{max}K$.

The discussion above pertains to the situation in which there is a single set of noninteracting sites with identical association constants. Consider now the case in which there are two sets of ligand-binding sites. Consider for now that the sites do not interact and each set of sites is characterized by one association constant, but the two sets have different K_s. Each set of sites is described by its own saturation fraction, which we express here in terms of the experimentally observable B, the amount of ligand bound per unit total protein:

[9.1.A1.14] $$B_1 = \frac{B_{max\ 1}K_1[L]}{1 + K_1[L]}$$
$$B_2 = \frac{B_{max\ 2}K_2[L]}{1 + K_2[L]}$$

and each of these relations can be transformed to the Scatchard equation:

[9.1.A1.15] $$\frac{B_1}{[L]} = K_1(B_{max\ 1} - B_1)$$
$$\frac{B_2}{[L]} = K_2(B_{max\ 2} - B_2)$$

Now the binding of ligand to these two sets of sites occurs in a single reaction medium which has a single free ligand concentration $[L]$. The total amount of ligand bound is just the sum of the amounts bound to the different sets of sites, as given by Eqn (9.1.A1.10). We may divide the total amount bound by the free ligand concentration to obtain the relations:

[9.1.A1.16] $$B = B_1 + B_2$$
$$\frac{B}{[L]} = \frac{B_1}{[L]} + \frac{B_2}{[L]}$$
$$\frac{B}{[L]} = K_1(B_{max\ 1} - B_1) + K_2(B_{max\ 2} - B_2)$$

According to Eqn (9.1.A1.15), each set of sites independently plotted as $B/[L]$ against B gives an intercept on the $B/[L]$ axis of K_1B_{max1} or K_2B_{max2}, and an intercept on the B axis of B_{max1} or B_{max2}, with a slope of $-K_1$ or $-K_2$. For a protein (or mixture of proteins) containing both sets of sites, the resulting Scatchard plot is the vector sum of the individual plots. Figure 9.1.A1.1 shows a Scatchard plot for two sets of sites determined separately. The resulting Scatchard plot when the two sets of sites are present in a single mixture can be constructed from the individual graphs by realizing that all points on a line passing through the origin are characterized by the same free ligand concentration. If the two sets of sites are present in the mixture and are at equilibrium, they must be simultaneously in equilibrium with a single $[L]$. Along a line passing through the origin, $B = B_1 + B_2$ and $B/[L] = B_1/[L] + B_2/[L]$. The composite B and $B/[L]$ for the two sets of sites is the sum of the individual values along the lines of equal $[L]$. Figure 9.1.A1.1 shows the way in which the composite Scatchard results from the vector addition of the two component sets of sites.

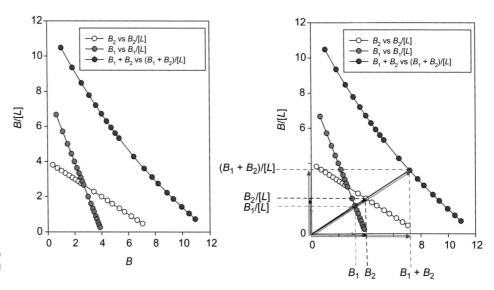

FIGURE 9.1.A1.1 Scatchard plot for two independent sites and their resulting composite plot.

This theoretical analysis shows how the binding constants and number of binding sites produce a composite Scatchard plot. The experimental problem, however, is this: how does one determine the affinities (association constants) and number of binding sites given an experimentally derived Scatchard plot? This is generally called the **inverse problem**. Predicting the behavior of a complex system given its characteristics is generally much easier than uniquely determining what those characteristics are, given the behavior of the system. Taking the theoretical results of Figure 9.1.A1.1, for example, the resulting Scatchard plot for two sets of sites is nearly linear. Given experimental error, it would be a common error to assume there is only one set of binding sites because the two sites do not differ enough in their association constants to produce a sufficiently curved composite plot. Even if there is only one set of binding sites, the analysis of the Scatchard plot can be complicated. Application of linear regression to such a plot, for example, is inappropriate because linear regression assumes no error in determination of the abscissa value, and all the error is in the ordinate value. In the case of the Scatchard plot, it is generally assumed that the free ligand concentration is known exactly, and that the error is entirely in the determination of B. In this case, the error bars would lie on a line passing through the origin. In practice, however, there is error in both B and $[L]$. Computer programs such as LIGAND are available which peel the Scatchard plot into its component parts, while minimizing the global error in fitting the binding parameters to the data. There are a number of cautions which must be exercised in this process. Basically, the cautions boil down to one piece of advice: do not waste powerful mathematics on bad data. It is better to spend the time to get good binding data.

THE TITRATION CURVE

The Scatchard analysis given above used the association constant, K, as one descriptor of the binding reaction. As noted above, the association constant has the units of M^{-1}.

The dissociation constant, K_d, is the inverse of the association constant, with units of M. For the case of a single set of noninteracting sites with identical K, we write:

[9.1.A1.17] $$B = \frac{B_{max}K[L]}{1 + K[L]}$$

The **titration curve** is a plot of B against log $[L]$. It is analogous to the alkaline titration of acid, in which the amount of added strong alkali (NaOH) is plotted against the pH. The curve is S-shaped. Its maximum slope occurs when $[L] = K_d$. To see this, we differentiate Eqn (9.1.A1.17) with respect to log $[L]$:

[9.1.A1.18]
$$\frac{dB}{d\log[L]} = [L]\frac{dB}{d[L]}$$
$$= [L]\left[\frac{B_{max}K}{1+K[L]} - \frac{B_{max}K^2[L]}{1+K[L]^2}\right]$$
$$= B_{max}\left[\frac{K[L]}{1+K[L]} - \left(\frac{K[L]}{1+K[L]}\right)^2\right]$$
$$= B_{max}[\nu - \nu^2]$$

where ν is the saturation fraction defined by Eqn (9.1.A1.7), whose range is restricted to values between 0 and 1.0. The slope of the titration curve can be seen from this relation to be zero when $\nu = 0$ and when $\nu = 1.0$, and to have a maximum when $\nu = 0.5$. The maximum slope occurs at an inflection point in the S-shaped curve. The inflection point at $\nu = 0.5$ can be derived by setting the second derivative of B against log$[L]$ equal to zero. When $\nu = 0.5$, from the definition of ν, $[L] = 1/K = K_d$; this is the point at which the free ligand concentration is equal to the dissociation constant for the sites. An example of the titration curve is shown in Figure 9.1.A1.2.

The utility of the titration curve is that it allows all of the binding data to be represented in a single plot and it visually allows judgment of an important point: do the binding data encompass values of B both above and

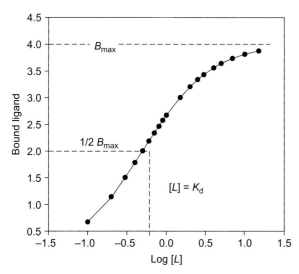

FIGURE 9.1.A1.2 Titration binding curve with $B_{max} = 4$ and $K_d = 0.5$.

with an association constant given by

$$[9.1.A1.20] \quad K = \frac{[P \cdot L_h]}{[P][L]^h}$$

Here the fractional saturation is given by

$$[9.1.A1.21] \quad \nu = \frac{[P \cdot L_h]}{[P] + [P \cdot L_h]} = \frac{K[L]^h}{1 + K[L]^h}$$

Hill proposed this equation purely on an empirical basis, and the equation is now known as the **Hill equation**, with the variable Y usually being used instead of ν. This equation can be rearranged to give

$$\frac{\nu}{1-\nu} = K[L]^h$$

$$[9.1.A1.22] \quad \log\left[\frac{\nu}{1-\nu}\right] = \log K + h \log[L]$$

Thus a plot of $\log(\nu/(1-\nu))$ against $\log[L]$ should give a straight line with slope h. This plot is the **Hill plot**, and it allows evaluation of K and h. This plot may also be applied to enzyme reactions, where ν is usually calculated as v/V_{max}, where v is the enzyme velocity and V_{max} is the maximum enzyme velocity.

The exponent h is referred to as the **Hill coefficient**. It is an index of the cooperativity as the larger h, the greater the degree of cooperativity. Its interpretation will become clearer after we consider the Adair equation.

Adair considered that the binding of oxygen to hemoglobin could be described as a four-step process (because the hemoglobin molecule was a tetramer), rather than a single concerted step as required by the Hill equation. In this case, we write:

$$[9.1.A1.23] \quad \begin{aligned} P + L &\rightleftharpoons P \cdot L \\ P \cdot L + L &\rightleftharpoons P \cdot L_2 \\ P \cdot L_2 + L &\rightleftharpoons P \cdot L_3 \\ P \cdot L_3 + L &\rightleftharpoons P \cdot L_4 \end{aligned}$$

We suppose that the binding sites are identical, but they may be influenced identically by prior occupation of adjacent sites. Thus we define an **intrinsic or site association constant**, K_1, K_2, K_3, and K_4 for the binding to the first, second, third, and fourth sites, respectively. The rate of binding to the protein molecule, however, is affected by the number of vacancies. To convert the site constants to **molecular constants**, we must multiply by the number of sites involved in the reaction per molecule. Equilibrium is attained when the forward rate is equal to the reverse rate. So for the binding of ligand to the first site, we write:

$$[9.1.A1.24] \quad \begin{aligned} Q_{on\ 1} &= 4\,K_{on\ 1}[P][L] \\ Q_{off\ 1} &= K_{off\ 1}[P \cdot L] \end{aligned}$$

where Q_{on1} and Q_{off1} are the rates of the on reaction and off reaction at the first site, respectively. The molecular on rate constant is $4\,K_{on1}$ because there are four

below $0.5\ B_{max}$? This is important quality control in determining if the binding data are up to the task of characterizing K_d and B_{max}. Unless the inflection point is well defined, how is it possible in the titration curve to predict where the binding will level off to B_{max}? Both K_d and B_{max} are best defined when there are B, $[L]$ data taken above and below $\nu = 0.5$. (See I.M. Klotz, Number of receptor sites from Scatchard graphs: facts and fantasies. *Science* **217**:1247–1249, 1982.)

COOPERATIVITY AND THE HILL PLOT

According to the above discussion, there are three plots one can make to represent ligand-binding data: (1) the adsorption isotherm in which the degree of binding, B (or, alternatively, the saturation fraction, ν) is plotted against the free ligand concentration, $[L]$; (2) the titration curve, in which the degree of binding is plotted against the logarithm of $[L]$; and (3) the Scatchard plot, in which the ratio of the degree of binding to $[L]$ is plotted against the degree of binding. According to the analysis presented so far, the Langmuir adsorption curve should show a hyperbolic curve, the titration curve should be S-shaped, and the Scatchard plot should be linear. The analysis so far, however, has been limited to identical sets of **noninteracting** sites. Sometimes, there is interaction between the sites so that the binding at one site affects the binding at another. This interaction is referred to as **cooperativity**. If binding ligand at one site increases the association constant at a vacant site, then the cooperativity is positive; if binding the first ligand decreases the association constant at the vacant site, the cooperativity is negative. A quantitative index of cooperativity is the **Hill coefficient**.

Hill attempted to explain the binding of oxygen to hemoglobin by postulating empirically that h molecules of oxygen were bound in a single step:

$$[9.1.A1.19] \quad P + hL \rightleftharpoons P \cdot L_h$$

empty sites on each protein molecule. At equilibrium, the on and off rates are equal, so we define:

[9.1.A1.25]
$$Q_{on\ 1} = Q_{off\ 1}$$
$$4K_{on\ 1}[P][L] = K_{off\ 1}[P \cdot L]$$
$$\frac{[P \cdot L]}{[P][L]} = \frac{4K_{on\ 1}}{K_{off\ 1}} = 4K_1$$

In analogy, we write the relations as

[9.1.A1.26]
$$\frac{[P \cdot L_2]}{[P \cdot L][L]} = \frac{3K_{on\ 2}}{2K_{off\ 2}} = \frac{3}{2}K_2$$
$$\frac{[P \cdot L_3]}{[P \cdot L_2][L]} = \frac{2K_{on\ 3}}{3K_{off\ 3}} = \frac{2}{3}K_3$$
$$\frac{[P \cdot L_4]}{[P \cdot L_3][L]} = \frac{K_{on\ 4}}{4K_{off\ 4}} = \frac{1}{4}K_4$$

In this formalism, all four constants would be equal ($K_1 = K_2 = K_3 = K_4$) if the four sites were identical and acted independently.

In Adair's model, the concentration of the various bound forms is given by Eqns (9.1.A1.25) and (9.1.A1.26) as

[9.1.A1.27]
$$[P \cdot L] = 4K_1[P][L]$$
$$[P \cdot L_2] = \frac{3}{2}K_2[P \cdot L][L] = 6K_1K_2[P][L]^2$$
$$[P \cdot L_3] = \frac{2}{3}K_3[P \cdot L_2][L] = 4K_1K_2K_3[P][L]^3$$
$$[P \cdot L_4] = \frac{1}{4}K_4[P \cdot L_3][L] = K_1K_2K_3K_4[P][L]^4$$

We now use these results to calculate the saturation fraction as

[9.1.A1.28]
$$\nu = \frac{[P \cdot L] + 2[P \cdot L_2] + 3[P \cdot L_3] + 4[P \cdot L_4]}{4([P] + [P \cdot L] + [P \cdot L_2] + [P \cdot L_3] + [P \cdot L_4])}$$
$$= \frac{4K_1[P][L] + 12K_1K_2[P][L]^2 + 12K_1K_2K_3[P][L]^3 + 4K_1K_2K_3K_4[P][L]^4}{4[P] + 16K_1[P][L] + 24K_1K_2[P][L]^2 + 16K_1K_2K_3[P][L]^3 + 4K_1K_2K_3K_4[P][L]^4}$$

Factoring out $4[P]$ from the numerator and denominator, we have

[9.1.A1.29]
$$\nu = \frac{K_1[L] + 3K_1K_2[L]^2 + 3K_1K_2K_3[L]^3 + K_1K_2K_3K_4[L]^4}{1 + 4K_1[L] + 6K_1K_2[L]^2 + 4K_1K_2K_3[L]^3 + K_1K_2K_3K_4[L]^4}$$

This last equation is known as the **Adair equation** for four sites. If all four intrinsic association constants are equal, then this equation simplifies as

[9.1.A1.30]
$$\nu = \frac{K[L](1 + K[L])^3}{(1 + K[L])^4} = \frac{K[L]}{1 + K[L]}$$

which is the expression derived for noninteracting sites. If all four association constants are equal, the adsorption isotherm will be a hyperbola, and the Scatchard plot will be linear. Conversely, deviations from a hyperbola on the adsorption isotherm or from linearity on the Scatchard plot mean that the association constants cannot all be equal. In particular, if $K_4 >>> K_1$, K_2, and K_3, then Eqn (9.1.A1.29) becomes

[9.1.A1.31]
$$\nu = \frac{K_1K_2K_3K_4[L]^4}{1 + K_1K_2K_3K_4[L]^4}$$

which is the Hill equation (Eqn 9.1.A1.21) with $K = K_1K_2K_3K_4$ and $h = 4$. There is no way to simplify Eqn (9.1.A1.29) to give a Hill coefficient greater than 4. Also, when [L] is very small, [L] dominates the terms in higher orders of [L], and the observed Hill coefficient tends to 1.0. In general, the Hill coefficient approaches unity at both extremes of [L] and cannot exceed the number of binding sites in the linear region of the curve from $0.1 \leq \nu \leq 0.9$. Thus the interpretation of the Hill coefficient, h, is that it is a minimum estimate of the number of binding sites.

COMPETITIVE INHIBITION OF BINDING

Here we consider that a second ligand, an inhibitor, I, is able to bind to the same binding site, and in so doing blocks the binding of the first. The reactions are written as

[9.1.A1.32]
$$P + L \rightleftharpoons P \cdot L$$
$$P + I \rightleftharpoons P \cdot I$$

In competition experiments, it is customary to use the dissociation constants rather than the association constants. The dissociation constants are defined as

[9.1.A1.33]
$$K_d = \frac{[P][L]}{[P \cdot L]}$$
$$K_I = \frac{[P][I]}{[P \cdot I]}$$

where both K_d and K_I are in units of M. We define the saturation fraction as before:

[9.1.A1.34]
$$V = \frac{B}{B_{max}} = \frac{[P \cdot L]}{[P] + [P \cdot L] + [P \cdot I]}$$

substituting in for $[P \cdot L]$ and $[P \cdot I]$ from Eqn (9.1.A1.33), we obtain

[9.1.A1.35]
$$\nu = \frac{[P][L]/K_d}{[P] + ([P][L]/K_d) + ([P][I]/K_I)}$$
$$= \frac{[L]/K_d}{1 + ([L]/K_d) + ([I]/K_I)}$$
$$= \frac{[L]}{K_d(1 + ([I]/K_I)) + [L]}$$

There are several possible ways to evaluate K_I. One is to choose a single concentration of free ligand, [L], and determine the inhibitor concentration at which the binding is inhibited 50%. This concentration of inhibitor is the IC_{50}, the inhibitor concentration at 50% inhibition. Taking the ratio of ν in the absence and presence of inhibitor at the IC_{50}, we have

[9.1.A1.36]
$$\frac{B \text{ with}[I] = 0}{B \text{ with}[I] = IC_{50}} = 2 = \frac{[L]/(K_d + [L])}{[L]/(K_d(1 + ([IC_{50}]/K_I)) + [L])}$$

Canceling out the identical free ligand concentration in the two experiments and rearranging, we obtain

[9.1.A1.37]
$$2 = \frac{K_d(1 + (IC_{50}/K_I)) + [L]}{K_d + [L]}$$

$$K_d + \frac{IC_{50}}{K_I} + [L] = 2K_d + 2[L]$$

$$K_I = \frac{IC_{50}}{1 + ([L]/K_d)}$$

This last equation enables us to determine K_I for the inhibitor from the free ligand concentration used to determine IC_{50} and from the K_d for the ligand binding alone.

THE DIXON PLOT

A second method for determining the K_I for an inhibitor does not require prior knowledge of K_d. Equation (9.1.A1.35) can be rewritten as

[9.1.A1.38]
$$B = \frac{B_{max}[L]}{K_d(1 + ([I]/K_I)) + [L]}$$

This can be inverted and rearranged to give

[9.1.A1.39]
$$\frac{1}{B} = \frac{K_d(1 + ([I]/K_I)) + [L]}{B_{max}[L]}$$

$$\frac{1}{B} = \frac{K_d + [L]}{B_{max}[L]} + \frac{K_d/K_I}{B_{max}[L]}[I]$$

According to this equation, plots of $1/B$ against $[I]$ at a constant $[L]$ should be linear with an intercept on the $1/B$ axis of $(K_d + [L])/(B_{max}[L])$ and a slope of $(K_d/K_I)/(B_{max}[L])$. This plot of $1/B$ against $[I]$ is the **Dixon plot**. If we experimentally determine two lines at differing $[L]$, $[L]_1$, and $[L]_2$, then the two curves will intersect at $1/B_1 = 1/B_2$. It can be seen from Eqn (9.1.A1.39) that this condition is met when $[I] = -K_I$. That is, the point of intersection of the lines in a Dixon plot is $(-K_I, 1/B_{max})$.

9.2 Hypothalamus and Pituitary Gland

> **Learning Objectives**
>
> - Identify the anatomical location of the pituitary gland, hypothalamus, and parts of the gland including the adenohypophysis and neurohypophysis
> - List the hormones secreted by the posterior lobe of the pituitary and state their chemical nature
> - Trace the route of posterior pituitary hormones from their point of synthesis to point of secretion
> - List the signals leading to oxytocin release
> - List the major actions of oxytocin
> - List the signals leading to ADH release
> - List the major actions of ADH
> - List the hormones released from the anterior pituitary
> - Describe in general how the hypothalamus controls release of anterior pituitary hormones
> - Describe in particular the signals for control of GH release by somatotrophs including GHRH, SST, ghrelin, IGF, and GH
> - Summarize the effects of GH
> - Name the diseases associated with GH excess or deficit

THE *PITUITARY GLAND* LIES BELOW THE BRAIN AND CONNECTS TO THE HYPOTHALAMUS BY A NARROW STALK

The pituitary gland, also called the **hypophysis**, sits below the brain in a depression of the sphenoid bone called the **sella turcica** or "Turkish saddle" (see Figure 9.2.1). Its name derives from the Greek "ptuo" and the Latin "pituita," meaning phlegm. It has two basic parts, the **adenohypophysis** and the **neurohypophysis**. The adenohypophysis derives from an evagination of the pharyngeal cavity, whereas the neurohypophysis derives from an invagination of the brain tissue. The two coalesce to form the pituitary gland. They remain connected to the hypothalamus through a narrow stalk called the **hypophyseal stalk**.

The adenohypophysis is further divided into three parts: the **pars distalis** is the largest and is also called the **anterior lobe of the pituitary**. The **pars tuberalis** forms the outer covering of the hypophyseal stalk. The third part is the **pars intermedia**, a specialized group of cells lying between the adenohypophysis and the neurohypophysis. In humans, the pars intermedia is small and consists of a thin, diffuse region of cells.

The neurohypophysis consists of the **median eminence**, the bottom part of the hypothalamus, the **infundibular stem** that forms the core of the hypophyseal stalk, and the **infundibular process**, also called the **posterior lobe**. The word "infundibulum" describes "funnel" from which these structures derive their names (see Figure 9.2.1).

CELLS IN THE HYPOTHALAMUS SYNTHESIZE ADH AND OXYTOCIN AND SECRETE THEM IN THE POSTERIOR PITUITARY

The two hormones secreted by the posterior pituitary are **antidiuretic hormone**, or **ADH**, also known as **vasopressin**, and **oxytocin**. Their structures are shown in Figure 9.2.2. Both hormones are synthesized in large cells in the hypothalamus, transported into nerve terminals in the posterior pituitary, and released in response to stimulation of the cell bodies in the hypothalamus. These magnocellular neurons are located in the **supraoptic nucleus** just above the optic chiasm and in the **paraventricular nucleus** on the side of the hypothalamus. The cells that synthesize oxytocin are distinct from those that secrete ADH.

OXYTOCIN AND ADH ARE CHAINS OF NINE AMINO ACIDS

Cells in the hypothalamus synthesize ADH as a large precursor, propressophysin, that contains four separate functional regions: a signal peptide that directs the protein into ER, a 9 amino acid section that becomes the active hormone, a 93–95 amino acid polypeptide that is cleaved off to form **neurophysin**, and a 39 amino acid glycopeptide. ADH is stored in secretory granules bound to neurophysin, which is a 10-kDa protein that binds to other neurophysin molecules and to ADH. The complex of ADH, neurophysin, and glycoprotein is stored within secretory vesicles, and transported down the axon to its terminal in the posterior pituitary. On stimulation, all three are released and then ADH dissociates from the neurophysin in the blood (see Figures 9.2.2 and 9.2.3). Oxytocin is synthesized similarly to vasopressin, except the neurophysin analog is different and there is no glycopeptide.

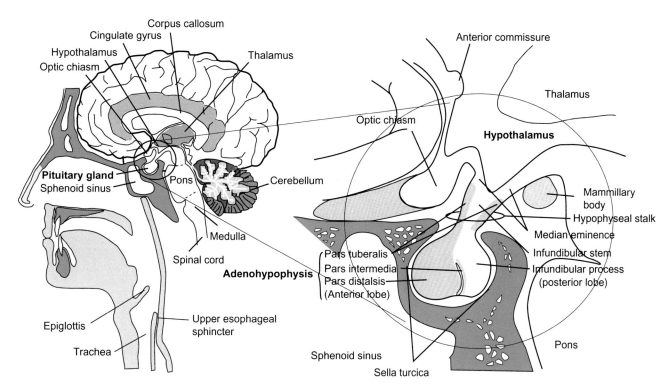

FIGURE 9.2.1 Location and identification of parts of the pituitary gland. The gland lies below the hypothalamus and is connected to it by a hypophyseal stalk. Its mass is about 0.5 g in adult males and is slightly larger in females. It is about 1.2–1.5 cm from side to side, about 1 cm long, and about 0.5 cm thick. The pituitary consists of the adenohypophysis and neurohypophysis. The adenohypophysis in turn consists of the pars tuberalis that covers the hypophyseal stalk, the pars distalis (also called the anterior lobe) and the pars intermedia, which in humans is small. The neurohypophysis consists of the median eminence at the base of the hypothalamus, the infundibular stem in the core of the hypophyseal stalk, and the infundibular process, also called the posterior pituitary. The sphenoid bone surrounds the pituitary with a structure called the sella turcica.

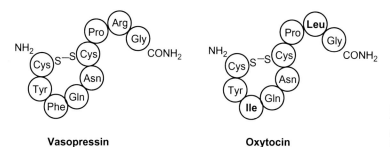

FIGURE 9.2.2 Chemical structures of vasopressin and oxytocin. The two 9 amino acid polypeptides differ only in positions 3 and 8. The terminal carboxyl group is amidated in both hormones.

OXYTOCIN CONTRACTS THE UTERUS AND MYOEPITHELIAL CELLS OF ALVEOLI CELLS IN THE BREAST

Oxytocin has no known humoral function in human males. In females, oxytocin causes the uterus to contract and it causes the "milk ejection reflex," and it may be involved in some maternal behaviors. Stretch or dilation of the cervix and vagina are strong stimuli for oxytocin secretion, mediated by neural pathways called the **Ferguson reflex**. During birth, oxytocin in the mother is released explosively in pulses, causing contraction of the uterus and aiding delivery of the baby. The density of oxytocin receptors increases as much as 200-fold as parturition approaches, which markedly increases the sensitivity of the uterus to oxytocin. Oxytocin binds to oxytocin receptor, OXTR, that is a G-protein-coupled receptor, working through a G_q mechanism. Occupancy of OXTR activates phospholipase C, which releases IP_3, which in turn increases plasma $[Ca^{2+}]$ by release of Ca^{2+} through IP_3 receptors on the ER, which then activates MLKC within smooth muscle cells to activate contraction.

Suckling of an infant at the breast excites sensory afferents that cross in the medulla and eventually connect to the oxytocinergic magnocellular neurons in the supraoptic nucleus and paraventricular nucleus. This sensory information causes a synchronous pulsatile release of oxytocin that pumps milk out of the breast by stimulating the glandular cells of the breast that produce the milk and contraction of myoepithelial cells that empty the alveoli that contain the milk. This effect is called the **milk ejection reflex** or the **milk let-down reflex**. In humans, the release of oxytocin can be elicited by psychological stimuli such as preparing for nursing or hearing the baby cry. Suckling causes pulsatile release of oxytocin, whereas breast massage causes a more continuous oxytocin release (see Figure 9.2.4).

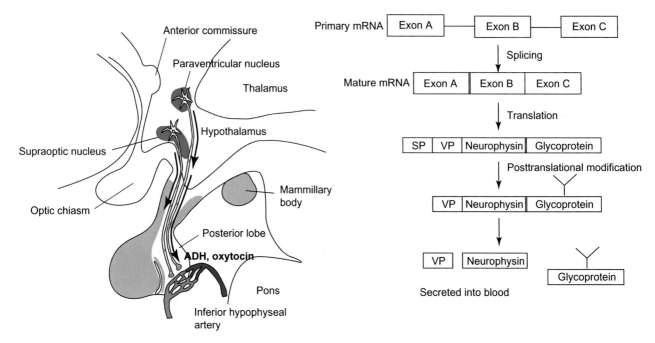

FIGURE 9.2.3 Processing of ADH. Chromosome 20 has three exons that code for ADH in magnocellular neurons located in the supraoptic nucleus and paraventricular nucleus of the hypothalamus. Transcription of the DNA results in a primary mRNA transcript which is then spliced to form a mature mRNA. Translation of the mRNA produces a single peptide strand that contains four functional regions: a signal peptide, ADH (also known as vasopressin, VP), neurophysin, and glycoprotein. The signal sequence is removed and the peptide is cleaved into ADH, neurophysin, and glycoprotein. The neurophysin binds ADH in vesicles that are transported from the soma to its axon terminals in the posterior pituitary where it is released on excitation. Neuroendocrine cells in the supraoptic nucleus and paraventricular nucleus make either ADH or oxytocin, but not both. The same basic mechanism as shown here also applies to oxytocin, except there is no glycoprotein.

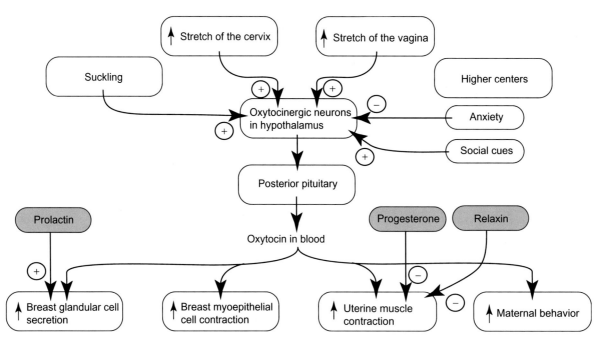

FIGURE 9.2.4 Control and effects of oxytocin. Oxytocin is synthesized in the hypothalamus but released in the posterior pituitary. Its release is stimulated by afferent sensation from the breast caused by suckling, by stretch of the cervix or vagina, and by higher centers. Its main effects are on uterine contraction and lactation. Uterine contraction prior to parturition is inhibited by progesterone and relaxin.

OXYTOCIN HAS BECOME KNOWN AS THE "TRUST HORMONE" OR "LOVE HORMONE"

Oxytocin that is produced by cells in the hypothalamus and released into the blood is a hormone. Oxytocin that is released at nerve terminals elsewhere in the brain is a neurotransmitter, and this release results in detectable increases in plasma levels of oxytocin. Increased plasma levels of oxytocin have been detected in prosocial environments, and administration of oxytocin by nasal sprays promotes a variety of prosocial behaviors. Oxytocin inhibits fear responses in the amygdala, and this reduction in fear thereby promotes prosocial interactions. For this reason, the hormone has become known as the "trust hormone."

INCREASED PLASMA OSMOLARITY AND DECREASED BLOOD VOLUME STIMULATE ADH RELEASE

Increasing plasma osmolarity and decreasing blood volume independently increase ADH release (see Figure 9.2.5). Osmoreceptors in the anterior hypothalamus tonically stimulate magnocellular neurons to secrete ADH. Lowering the osmolarity reduces ADH secretion and increasing plasma osmolarity increases it. This forms a negative feedback loop, as ADH retains water by action on the kidneys, as described in Chapter 7.6. Briefly, ADH engages a G_s mechanism through V2 receptors that increase the water and urea permeability of principal cells of the collecting duct, by recruiting latent aquaporin-2 water channels to the apical membrane. High ADH increases reabsorption of water and produces a low volume of highly concentrated urine; low ADH is associated with a high volume of highly dilute urine. Lowered osmolarity decreases ADH secretion, causing loss of water over salt in the kidney and the blood osmolarity returns toward normal. Increased osmolarity increases ADH secretion, leading to reabsorption of water. Salt can be excreted in excess of water, leading to a return toward normal plasma osmolarity.

Reduction in blood volume and pressure also stimulates ADH release, but not as strongly as increased osmolarity. High-pressure receptors in the carotid sinus and aortic arch, and low-pressure receptors in the atria and pulmonary veins, inform the central nervous system of the state of the circulation. The afferents travel over cranial nerves IX (glossopharyngeal nerve) and X (vagus nerve) to the medulla. These inputs tonically inhibit ADH release. Reduction in blood volume reduces the firing rate of the stretch receptors, thereby reducing the tonic inhibition and increasing ADH release, causing water retention by the kidney. This cannot raise blood volume by itself, but it helps conserve water that is consumed. ADH also binds to V1 receptors on the blood vessels, causing vasoconstriction through a G_q mechanism and raising the pressure toward normal.

Although the adjustment of water and salt excretion can adjust plasma osmolarity and correct for excess plasma volume, conservation of water alone cannot correct reduced plasma volume. This requires drinking fluids and absorbing the fluid into the blood. **Thirst** is stimulated by the same sensory afferents that control ADH release: high-pressure and low-pressure receptors, and osmoreceptors in the anterior hypothalamus.

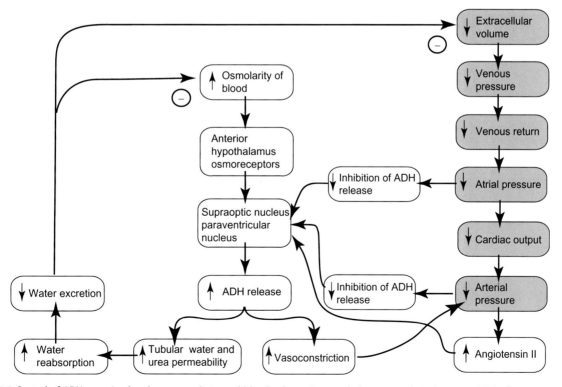

FIGURE 9.2.5 Control of ADH secretion by plasma osmolarity and blood volume. Increased plasma osmolarity increases ADH release. Decreased blood volume, sensed by stretch receptors in the great veins and atria, also increases ADH release. ADH increases water and urea permeability of the distal nephron, leading to excretion of a small volume of concentrated urine, thereby minimizing further loss of blood volume and decreasing the osmolarity of the plasma back toward normal.

THE HYPOTHALAMUS CONTROLS RELEASE OF HORMONES FROM THE ANTERIOR PITUITARY

The anterior pituitary releases hormones in response to stimulation by cells in the hypothalamus. These hormones include:

- thyroid stimulating hormone (TSH)
- adrenocorticotropic hormone (ACTH)
- growth hormone (GH)
- luteinizing hormone (LH)
- follicle stimulating hormone (FSH)
- prolactin (PRL).

These hormones are secreted in the anterior hypothalamus by five distinct cell types (see Figure 9.2.6). These are **trophic** cells because their hormones affect the growth of their target tissues. **Thyrotrophs** target the thyroid gland; **gonadotrophs** secrete FSH and LH that target cells in the gonads; **somatotrophs** secrete GH that influences overall body (soma) growth; **corticotrophs** secrete ACTH that stimulates secretions and size of the adrenal cortex; and **lactotrophs** secrete PRL that targets the mammary gland.

The superior hypophyseal artery supplies blood to the median eminence. It breaks up into a series of capillaries, forming long and short capillary loops. Neurons in the hypothalamus secrete small molecular weight factors that enter these capillaries. The blood vessels course down into the anterior pituitary, where they once again form an anastomosing network of capillaries. This is a **portal** circulation: a second set of capillaries between artery and vein. These capillaries are fenestrated and lie outside of the blood–brain barrier. The fenestrations allow relatively unrestricted diffusion of materials into the extracellular space. The releasing factors thus travel down to the anterior pituitary undiluted by the general circulation. In the anterior pituitary, these factors bind to receptors on the trophic cells, controlling their secretion of hormones. The anatomic arrangement of these cells and the portal circulation is illustrated in Figure 9.2.7. A summary of the releasing factors is given in Table 9.2.1.

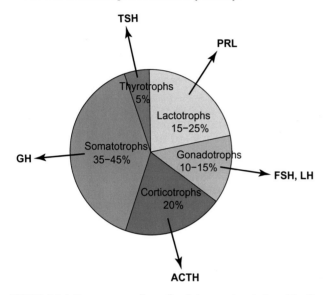

FIGURE 9.2.6 Hormone secreting cells of the anterior pituitary. The five distinct types of cells each secrete definite sets of hormones. The size of the slices reflects the approximate percentage of the cell type in the overall cell population of the anterior pituitary. The locations of the cells are also not arbitrary, being collected in regions. The thyrotrophs, for example, congregate in the anteromedial areas of the gland, whereas somatotrophs are predominantly located in the lateral wings of the gland.

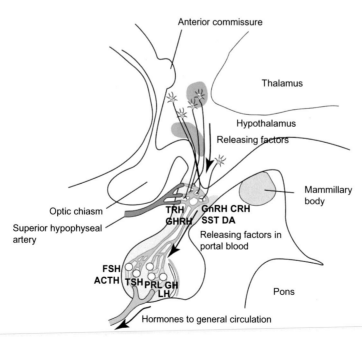

FIGURE 9.2.7 The blood supply to the anterior pituitary and control of anterior pituitary by hypothalamic factors. The superior hypophyseal artery enters the median eminence where it forms a capillary network. This drains into sinuses that form a portal circulation to the anterior pituitary. Neurons in the hypothalamus project axons to the capillary network in the median eminence, and release factors into the blood there. These factors travel down the portal circulation to the anterior pituitary, where they bind to the surface membrane of the trophic cells. The factors may either stimulate or inhibit hormone secretion by these cells. Hormones released in the anterior pituitary eventually travel into the systemic circulation.

TABLE 9.2.1 Hypothalamic Releasing Factors

Releasing Factor	Chemical Nature	Target Cells	Primary Action
TRH (thyrotropin releasing hormone)	3 aa chain	Thyrotrophs	Stimulates TSH release
CRH (corticotropin releasing hormone)	41 aa chain	Corticotrophs	Stimulates ACTH release
GHRH (growth hormone releasing hormone)	40 or 44 aa chain	Somatotrophs	Stimulates GH release
SST (somatostatin)	28 aa chain or 1–12, 14–28 fragments	Somatotrophs	Inhibits GH release
GnRH (gonadotropin releasing hormone)	10 aa chain	Gonadotrophs	Stimulates FSH and LH release
DA (Dopamine)	Neurotransmitter	Lactotrophs	Inhibits PRL release

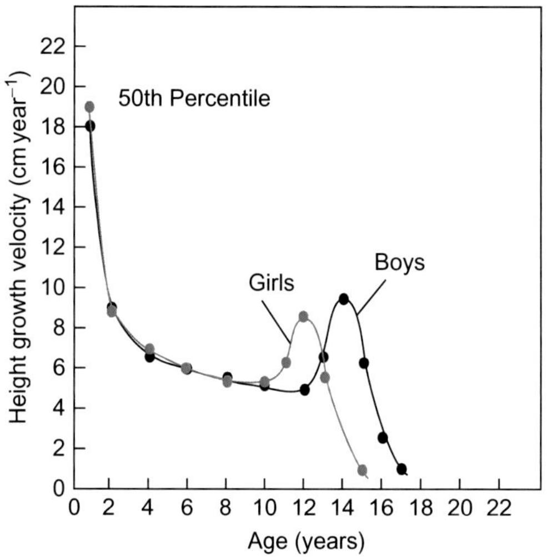

FIGURE 9.2.8 Growth velocity curves for humans as a function of age in years. The curves show the growth rate of height for British boys and girls who were followed longitudinally. Note that the growth spurt for girls occurs earlier and is smaller. This accounts for the difference in average heights between adult men and women. The lines show the 50th percentile. *Adapted from E.O. Reiter and R.G. Rosenfeld, "Normal and Aberrant Growth", in* Williams Textbook of Endocrinology, *10th ed., Saunders, 2003.*

MULTIPLE SIGNALS PRODUCE PULSATILE RELEASE OF GH

GH is a 191 amino acid polypeptide (22 kDa) that is synthesized and secreted by somatotrophs located in the lateral aspects of the anterior pituitary. Circulating GH levels show profound diurnal variation, with peak GH secretion during the night. This is true regardless of the onset of sleep, though GH secretion is further stimulated during slow wave sleep. GH secretion also varies with age. The velocity of growth is shown graphically in Figure 9.2.8, which represents data from British youth followed longitudinally. The pubertal growth spurt corresponds to a peak in circulating GH levels. Total secretion varies from about 2 mg day^{-1} during puberty to about 1/100th of that, 0.02 mg day^{-1}, in elderly or obese adults. The increased secretion during puberty is due to increased secretion at each pulse, rather than a change in the frequency of pulses.

THE COMPLICATED GH SECRETION PATTERN IS PRODUCED BY COMPLICATED NEURONAL CIRCUITS

Pulsatile GH secretion that varies with the time of day and with development requires multiple inputs and complicated circuitry. These secretion patterns largely reflect the interplay between two hypothalamic controls of GH: **GHRH, growth hormone releasing hormone**, secreted by cells in the arcuate nucleus, and **somatostatin (SST)** secreted by cells in the periventricular nucleus. SST is also referred to as **somatotropin release inhibiting factor (SRIF)**. Although a lot of the "wiring" has been worked out, it is unclear how much more remains undescribed (see Figure 9.2.9).

The somatotroph cells that secrete GH respond to multiple signals, including:

- **GHRH**, the main stimulator of GH synthesis and release;
- **SST**, the major inhibitor of GH synthesis and release;
- **GH**, which forms a negative feedback inhibition of its own secretion;
- IGF (**insulin-like growth factor**, formerly called **somatomedin**);
- Ghrelin, a polypeptide made by the stomach that promotes GH secretion.

GHRH TRAVELS TO SOMATOTROPHS THROUGH THE HYPOPHYSEAL PORTAL CIRCULATION

Cells in the arcuate nucleus in the hypothalamus produce GHRH and release it into the hypophyseal portal circulation. These hypothalamic cells receive a variety of inputs, as shown in Figure 9.2.9. Physiological stimuli for GHRH secretion include:

- Episodic, spontaneous release (neuronal patterns)
- Exercise
- Stress (physical or psychological)
- Slow wave sleep
- Fasting
- Postprandial glucose decline
- Gonadal steroids.

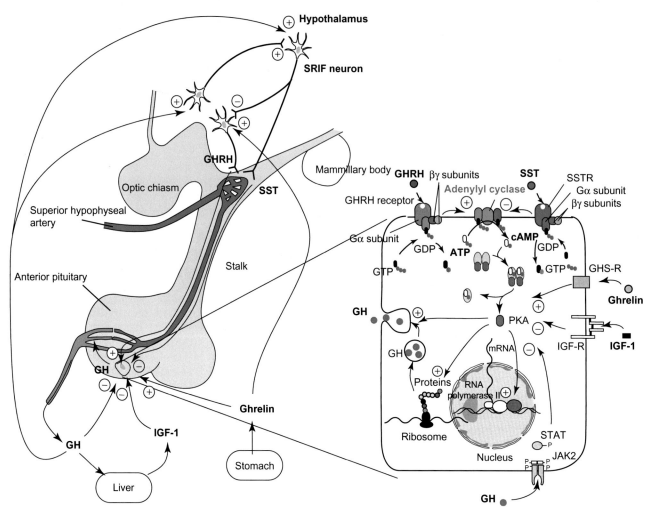

FIGURE 9.2.9 Control of GH secretion. GHRH secreted by cells in the hypothalamus is released into the portal circulation and stimulates GH synthesis and secretion by somatotrophs in the anterior pituitary. SST, also secreted by cells in the hypothalamus, inhibits GH secretion. Ghrelin is secreted by the stomach and promotes GH secretion by direct action in the anterior pituitary and by increasing GHRH secretion from the hypothalamus. GH causes negative feedback inhibition of GH secretion both by itself and through its stimulation of IGF-I and IGF-II secretion into the blood by the liver. All five of these stimuli appear to work through separate receptors on the surface of the somatotroph. GHRH secretion is inhibited by SST and activated by ghrelin and other inputs. Stimulation of SRIF neurons by GH increases SST secretion, which inhibits GH secretion.

GHRH binds to receptors on the somatotrophs' plasma membrane that is coupled to a G_s protein. This stimulates adenylyl cyclase activity and increases cytoplasmic cAMP levels, and activates **protein kinase A (PKA)** that phosphorylates target proteins, causing release of preformed GH, increased GH mRNA transcription, and increased GH synthesis.

SST INHIBITS GH RELEASE

SRIF cells in the periventricular nucleus secrete SST into the hypophyseal portal circulation. SST was found unexpectedly in early efforts to isolate GHRH from pituitary extracts, and the term "SST" was originally applied to a 14 amino acid peptide, now referred to as SST-14. SST-14 is a fragment of SST-28. There are five SST receptors identified so far, and all are present in brain, stomach, and the islets of Langerhans in the pancreas. In somatotrophs, SST is coupled to a G_i protein that inhibits GH synthesis and release.

GH FORMS A SHORT NEGATIVE FEEDBACK LOOP ON GH SECRETION

Somatotrophs have receptors for GH itself. Thus GH released by the somatotrophs feeds back directly on the secretory cells, inhibiting further GH secretion. This forms a short negative feedback loop, so-called because there are no intermediate steps in the loop. The overall control of GH secretion is shown diagrammatically in Figure 9.2.10.

IGF-I INHIBITS GH SECRETION

IGF stands for **insulin-like growth factor**. There are two major types, IGF-I and IGF-II. IGF-I contains 70 amino acids, whereas IGF-II has 67, but the two are separate gene products. These were originally called **somatomedins**, whose properties include:

- serum concentration dependent on GH;
- insulin-like activity;
- stimulation of DNA synthesis and cell multiplication;
- stimulation of sulfate incorporation into cartilage.

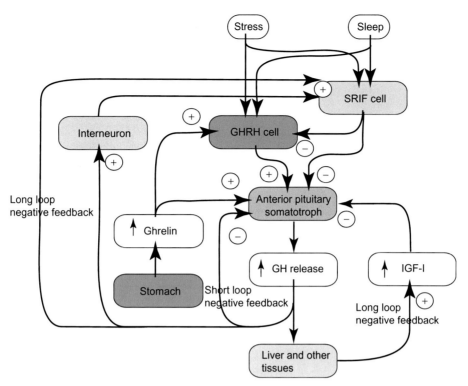

FIGURE 9.2.10 Overall control of GH secretion. The ultimate effect of any pathway can be determined by multiplying the signs of their effects: negative signs denote a decrease or inhibition, and positive signs denote stimulation or increase. Thus GH causes a negative feedback inhibition on its own secretion in a variety of ways: by direct feedback in a short loop ($- = -$); by a long loop through the SRIF cell ($+ \times - = -$); and by a long loop through hypothalamic interneurons ($+ \times + \times - = -$). Light blue boxes denote sources of inhibition of GH secretion, dark blue boxes denote sources of stimulation. The carriers of the signals are in white boxes.

Because of their structural similarity to proinsulin, the somatomedins were renamed as IGF in 1978. GH stimulates IGF-I production by the liver and other tissues. The somatotrophs have receptors for IGF-I and IGF-II, and their occupancy inhibits GH secretion, forming a **long negative feedback loop**.

GHRELIN STIMULATES GH SECRETION

Ghrelin is a 28 amino acid polypeptide secreted by endocrine cells of the stomach, in response to fasting. Somatotrophs have another receptor, called GHS receptor, for **g**rowth **h**ormone **s**ecretagogue. Ghrelin binds to this receptor in both the pituitary and in the hypothalamus.

GH MEDIATES SOME OF ITS EFFECTS THROUGH INCREASED IGF-I

Many of the effects of GH are mediated through somatomedin, or IGF, IGF-I, and IGF-II. IGF-I and II are about the same size but are produced by entirely different genes. GH is the primary regulator of IGF-I gene expression through the JAK-STAT pathway. Binding of one GH molecule to two of its receptors activates JAK2 (Janus kinase). The activated tyrosine kinase activity then phosphorylates a number of signaling molecules including STAT1, STAT3, and STAT5 (STAT refers to **s**ignal **t**ransduction and **a**ctivators of **t**ranscription). The phosphorylated STATs travel to the nucleus where they bind to DNA and elicit specific gene transcription.

Binding of GH to its receptor also activates the MAP/ERK pathway and the mTOR pathway in some tissues. These pathways are shown in Figure 9.2.11.

The somatomedin hypothesis posits that *all* effects of GH are mediated by IGFs. However, IGF administration cannot duplicate all of the effects of GH, and some effects of GH, such as raising blood glucose, are opposite to the effects of IGF. Thus it appears that some effects of GH are independent of IGF.

The overall physiological effects of GH are listed in Table 9.2.2.

SKELETAL GROWTH OCCURS MAINLY AT THE GROWTH PLATES

Most bones are formed by a process called endochondral ossification. This is a highly regulated process shown in Figure 9.2.12. It begins with condensation of mesenchymal cells that differentiate into chondrocytes. These undergo a regular pattern of proliferation, hypertrophy and calcification and cell death (see Figure 9.2.13) that leaves a mineralized cartilage that is destroyed and replaced with bone tissue by osteogenesis. A large number of signals regulate this process, including insulin-like growth factor (IGF), fibroblast growth factors (FGFs), bone morphogenic proteins (BMPs), and vascular endothelial growth factors (VEGFs). Linear growth of the long bones, which largely determine the height of a person, occurs by proliferation of cells at the epiphyseal plate or the **growth plate**. This is influenced by a number of hormones including GH, IGF-1, thyroid hormone, glucocorticoids, androgens, and estrogens. When the cartilage cells stop proliferation and all become calcified, the growth plate closes and no more linear growth is possible. This is referred

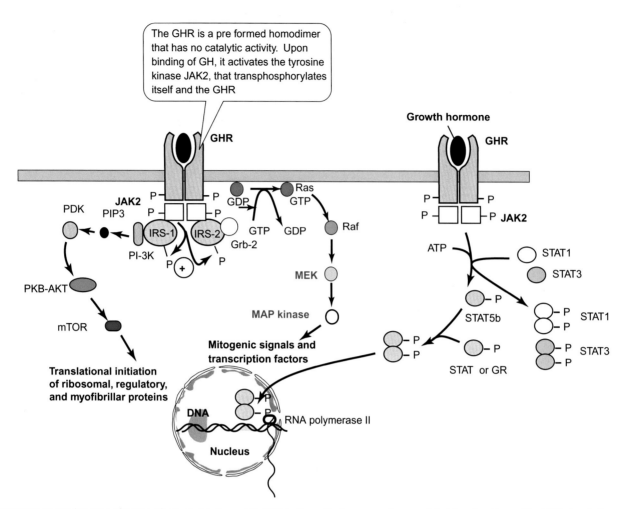

FIGURE 9.2.11 Mechanisms of action of growth hormone, GH. GH binding activates a receptor-associated tyrosine kinase. The GHR already is present in target tissues as a homodimer. Binding of GH activates Janus Kinase (JAK2) that phosphorylates the GHR and itself and phosphorylates interesting proteins including STAT 1, 3, 5a and 5b for **s**ignal **t**ransduction and **a**ctivation of **t**ranscription. These then dimerize and modulate gene expression in the target cell. Most of the effect is mediated by STAT 5b. The GHR-associated tyrosine kinase also phosphorylates IRS (insulin receptor substrates) that activates PI-3K (phosphatidylinositol 3kinase) that produces PIP3, phosphatidylinositol tri phosphate, that activates PDK (protein kinase D) that activates PKB, also known as AKT (activin), which further activates mTOR (mammalian target of rapamycin). mTOR turns on particular genes. In yet a third pathway, MAP kinase (mitogen activated protein kinase) is turned on through a cascade of events.

TABLE 9.2.2 Overview of the Physiological Actions of Growth Hormone

1. **The skeleton**
    1.1. Increases formation of cartilage and extracellular matrix
    1.2. Increases sulfate incorporation into cartilage (chondroitin sulfate)
    1.3. Increases thickness of epiphyseal cartilage of long bones
    1.4. Stimulates epiphyseal growth so that long bones are longer
    1.5. Stimulates osteoclast differentiation and osteoblast activity, and thereby increases bone mass
2. **Cell proliferation and size**
    2.1. Increases the number of cells in most organs of the body
    2.2. Increases the size of cells in most organs of the body
3. **Protein metabolism**
    3.1. Increases uptake of amino acids into muscle, heart, liver
    3.2. Increases total body nitrogen retention
    3.3. Increases lean body mass
    3.4. Stimulates protein synthesis
4. **Carbohydrate metabolism**
    4.1. Increases plasma glucose (the "diabetogenic" effect of GH)
    4.2. Mobilizes liver glycogen and increases liver gluconeogenesis
    4.3. Increases insulin release
    4.4. Inhibits glucose uptake by muscle and adipose tissue
5. **Fat metabolism**
    5.1. Increases breakdown of triglycerides to glycerol and fatty acids (lipolysis)
    5.2. Increases plasma free fatty acids
    5.3. Decreases glucose uptake into adipose tissue
    5.4. Decreases body fat
6. **Synergistic actions with other hormones**
    6.1. Many hormones (ACTH, TSH, FSH, LH) are more effective with GH present
    6.2. Full response to GH requires T3, insulin, sex steroids

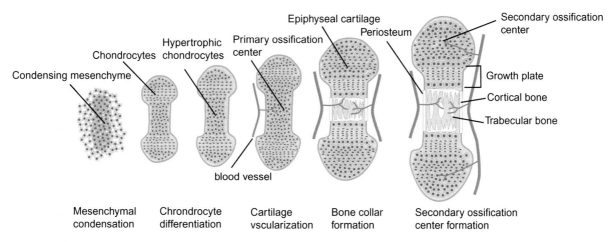

FIGURE 9.2.12 Endochondral bone formation. Endochondral bone formation begins with condensation of mesenchymal cells that then differentiate into chondrocytes that form the extracellular matrix of cartilage. The cells undergo defined steps of proliferation, hypertrophy, and calcification. The primary ossification center begins in the center of the long bones and migrates to each side of the long bones at the epiphyseal growth plates. Here proliferation of chondrocytes continues until closure of the growth plate during early adulthood. The process is regulated by a number of endocrine and paracrine secretions. *Adapted from Y. Xie, S. Zhou, H. Chen, X. Du and L. Chen, Recent research on the growth plate. Advances in fibroblast growth factor signaling in growth plate development and disorders.* J. Mol. Endocrinol. **53**: T11-34, 2014.

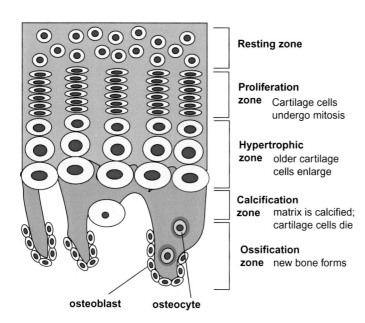

FIGURE 9.2.13 Cartoon of chondrocyte involvement in the formation of bone. Chondrocyte shows distinct stages that are reflected in spatial zones within the growth plate. The resting zone has small round cells that are adjacent to the articular surface. These cells undergo mitosis to form flat chondrocytes that are arranged in proliferative columns. Resting cells and proliferative chondrocytes secrete collagen II, aggrecan, and other matrix proteins to form the cartilage matrix. The proliferative chondrocytes form prehypertrophic and then hypertrophic condrocytes that secrete collagen X. These hypertrophic chondrocytes remodel the cartilage and then die. The matrix is calcified in the calcification zone. The calcified matrix is then ossified by resorption of the mineralized cartilage, vascularization, and formation of bone through osteoblasts. A mnemonic for these zones is "Real People Have Career Options", for Resting, Proliferation, Hypertrophy, Calcification and Ossification.

to as closure of the growth plate. Thus the adult height is determined by the regulation of proliferation and its final ending at the growth plate, and this process is controlled by multiple genes.

BOTH STARVATION AND ESTROGEN REDUCE ADULT HEIGHT

Children who suffer from protein-calorie malnutrition exhibit stunted growth. It is possible that this effect is mediated by **fibroblast growth factor 21 (FGF21)**. In rodents, FGF21 is secreted by mainly by liver and adipose tissue in response to energy deprivation, mediated by PPARα (peroxisome proliferator-activated receptor) which is in turn activated by increased free fatty acids in plasma that occur following lipolysis stimulated by starvation. FGF21 binds to its receptor, β-Klotho, and initiates signals that interferes with JAK/STAT signaling by GH. FGF21 acts on both the liver and the chondrocytes in the growth plate to inhibit GH actions. Osteoblasts secrete IGF-1 in response to GH. This IGF-1 acts as a paracrine hormone, and its effects are attenuated by FGF21. This effect is shown schematically in Figure 9.2.14. Whether this explains growth inhibition in malnourished children is less clear, as FGF21 in humans is less responsive to fasting, and circulating FGF21 level is elevated in obese humans.

Other hormones, such as estrogen, have biphasic effects on growth. At low doses, estradiol stimulates GH effects on the liver, apparently explaining the stimulation of growth in early puberty. High doses of estradiol inhibit longitudinal growth, explaining the reduced stature of women. In the liver, estrogen attenuates GH effects by stimulating SOCS2 (suppressor of cytokine signaling 2).

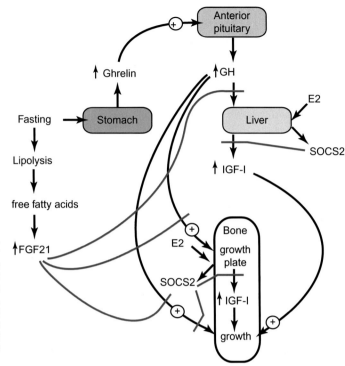

FIGURE 9.2.14 Interactions of growth hormone with other modulators. The major effects of GH are mediated by IGF-1 released by the liver systemically and by IGF-1 released within the tissues as a paracrine messenger. Some effects of GH are independent of IGF-1. The effects of GH are inhibited by fibroblast growth factor 21 (FGF21) that is released mostly by liver and adipose tissue in response to fasting. FGF21 interferes with GH signaling in liver and bone, thereby inhibiting linear growth during energy deprivation. Estadiol (E2) at low levels stimulates growth and in higher levels inhibits growth. Inhibition likely occurs by increased expression of SOCS2 (suppressor of cytokine signaling 2).

The family of SOCS proteins use a variety of mechanisms to disrupt signaling including masking of binding sites on the GHR/JAK2 complex, inhibition of JAK2 activity, and stimulation of the ubiquitinylation of the GHR, leading to its degradation. It is likely that this is part of the inhibition of growth by estrogen that is responsible for the shorter height of human females, as the estrogen produced later on during puberty leads to an earlier closure of the growth plate, and therefore less linear growth.

SUMMARY

The pituitary gland consists of two major components: the adenohypophysis and the neurohypophysis. Part of the adenohypophysis is the pars distalis, also referred to as the anterior lobe. The infundibular process of the neurohypophysis forms the posterior pituitary. The gland sits below the hypothalamus and is connected to it by the hypophyseal stalk. This stalk contains neuronal processes and blood vessels that collect neurotransmitters released by nerve cells in the hypothalamus. In this way, the endocrine system is connected to and controlled by the nervous system.

The hormones released by the posterior pituitary include ADH and oxytocin. These are synthesized by cells in the paraventricular and supraoptic nucleus of the hypothalamus and are transported down into the posterior pituitary by axoplasmic transport, bound within vesicles to neurophysin. Stimulation of the cells in the hypothalamus causes fusion of the vesicles and release of the hormones into the blood. Both ADH and oxytocin contain nine amino acids and show structural similarity. ADH release is increased by increased plasma osmolarity and decreased blood volume. The actions of ADH are twofold: it increases the water and urea permeability of the distal nephron and thereby causes the kidneys to excrete a low volume of highly concentrated urine. This effect on water and urea permeability is due to cAMP and PKA-mediated phosphorylation of aquaporin channels which are then inserted into the apical membrane of kidney tubule cells. The second action of ADH is vasoconstriction, from which ADH derives its other name: vasopressin.

Stimulation of oxytocin is provided by stretch of the uterus and suckling or associated psychosocial cues. The hormone causes uterine contraction and is essential in parturition. It also causes the milk "let-down" reflex or the milk ejection reflex: it stimulates contraction of myoepithelial cells in the breast and milk secretion by the breast.

The anterior pituitary releases a number of "master hormones" including TSH, LH, FSH, PRL, GH, and ACTH. These are released from specific cells in the anterior pituitary in response to releasing factors produced by neurons in the hypothalamus. These neurons package the releasing factors in neurotransmitter vesicles that dump their contents into the hypophyseal portal circulation that carries the factors to the anterior pituitary without dilution.

GH is synthesized and released by specific cells called somatotrophs in the anterior pituitary. These cells integrate at least five separate signals: GHRH, SST, ghrelin, IGF-1, and GH itself. GHRH is released from cells in the arcuate nucleus of the hypothalamus and stimulates GH release by somatotrophs by a G_s mechanism. SST is released from cells in the periventricular nucleus and inhibits GH release by a G_i mechanism. Ghrelin is released from the stomach during fasting and stimulates somatotrophs directly and also stimulates release of GHRH by hypothalamic cells. GH negatively feeds back onto somatotrophs to reduce GH secretion.

Clinical Applications: GH Excess or Deficiency

Excess GH secretion from childhood is **gigantism** and results in abnormally tall persons. Probably the documented record for the tallest human being was Robert Wadlow (February 22, 1918—July 15, 1940) who reached the height of 8'11" (2.72 m) but had not stopped growing at the time of his death at the age of 22. The average height in the United States for men is 5'10" or 1.72 m. His weight at the time of death was 485 lb (220 kg) with size 37AA shoes. Wadlow is sometimes known as the Alton Giant, named for his home town of Alton, Illinois (see Figure 9.2.15).

Excess GH has its consequences. Wadlow suffered from muscle weakness and tendency toward infections. He required braces to walk, and one of these irritated his ankle, causing a blister and subsequent infection. This probably progressed to septicemia and he died in his sleep at age 22.

Excess GH secretion in the adult causes **acromegaly**, first described by Pierre Marie in 1886 as disordered somatic growth and proportion. Pituitary adenomas cause 95% of the cases of acromegaly. The adenomas typically are slow-growing tumors that appear in the third to fifth decade of life. The symptoms of acromegaly include glucose intolerance, widening of bones leading to coarser facial features and enlarged hands and feet, enlarged heart, liver, and kidneys, and thickened skin and enlarged muscles.

GH deficiency in childhood produces short adults with a general tendency toward obesity. Any malfunction in the cascade from GHRH secretion to target cell responsiveness could account for dwarfism. These people have delayed skeletal growth and sexual maturation, but they are otherwise healthy with normal mental capacity. However, persons with GH deficiency have reduced life expectancy due to cardiovascular and cerebrovascular diseases. Because the window of opportunity closes early in life, diagnosis of GH deficiency must be made early. Children 3 standard deviations (SD) below average or with growth deceleration (2 SD below average for 1 year), or with combinations of these, should be evaluated for the cause of poor growth. A variety of conditions can cause secondary growth disorders: malnutrition, chronic diseases such as malabsorption and GI diseases, chronic liver disease, cardiovascular disease, or renal disease. Primary causes of GH deficiency lie in pituitary or hypothalamic dysfunction, and IGF deficiency due to GH insensitivity.

FIGURE 9.2.15 Robert Wadlow compared to his father, Franklin Wadlow, at 5'11".

Probably the most famous dwarf was Charles Stratton (January 4, 1838—July 15, 1883) who achieved fame through association with P.T. Barnum. He was born in Bridgeport, CT, weighing 4.3 kg at birth. He stopped growing at 6 months of age and 25" (64 cm). At 9 years of age he began to grow again, reaching 82.6 cm at age 18. He toured the United States and Europe as an entertainer, with the stage name General Tom Thumb, earning a fortune. He died in 1883 from a stroke at the age of 45.

The hypothalamic cells respond to a variety of stimuli. The final GH release is episodic and pulsatile.

GH has multiple effects on multiple systems. Excess produces gigantism in youth and acromegaly in adults. Deficits in childhood cause dwarfism. It increases the growth of long bones and increases the uptake of amino acids, increases blood glucose and mobilizes glycogen, and increases lipolysis. Linear growth stops upon closure of the epiphyseal growth plate in the long bones. Effect of GH on the growth plates is inhibited by fasting through elevation of fibroblast growth factor 21 (FGF21) and by high levels of estradiol, probably mediated by increased expression of suppressor of cytokine signaling 2 (SOCS2).

REVIEW QUESTIONS

1. Where is the pituitary gland? What is the adenohypophysis? What is the neurohypophysis?
2. What hormones are secreted by the posterior pituitary? What are their chemical natures?
3. What stimulates ADH release? What are the main effects of ADH? What signaling mechanism is involved in the renal effects of ADH? What signaling mechanism is involved in the vascular effects of ADH?
4. What stimulates oxytocin release? What are the main effects of oxytocin?

5. What hormones are secreted by the anterior pituitary? What controls their release?
6. What cells secrete GH? What hypothalamic factors increase secretion? What hypothalamic factors decrease secretion?
7. Describe the anatomic relationship between hypothalamic cells and control of GH release.
8. What other factors control GH release?
9. What signaling mechanism does GH use? What does GH do directly? List the effects of GH.
10. What happens when GH is present in excess in youth? In adulthood? What happens when it is deficient?

The Thyroid Gland 9.3

Learning Objectives

- Indicate the anatomic location of the thyroid gland
- Identify in a histological drawing: thyroid follicle, colloid, parafollicular cell, and identify the state of the gland (normal, stimulated, unstimulated)
- Recognize the chemical structures of T3 and T4
- Describe the mechanism of iodine uptake by the thyroid gland
- List the steps in the synthesis of T4
- Define monoiodotyrosine (MIT) and diiodotyrosine (DIT)
- List the major signals for release of TRH from the hypothalamus
- List the major signals for release of TSH from thyrotrope cells in the anterior pituitary
- Draw appropriate feedback loops for thyroid gland status for the following conditions: normal, central hypothyroidism, iodine deficiency, and thyroid defect hypothyroidism
- Describe T4 and T3 transport in blood
- Describe the major mechanism of T3 action
- Describe the consequence of hypothyroidism in children
- Describe the consequence of hypothyroidism in adults
- Discuss the basis for goiter formation
- Describe the suspected cause of Graves' disease—hyperthyroidism in adults

THE THYROID GLAND IS ONE OF THE LARGEST ENDOCRINE GLANDS

The **thyroid gland** is located in the neck just below the cricoid cartilage (Figure 9.3.1). It consists of two lobes separated by a narrow **isthmus** and is covered by two layers of connective tissue that form its capsule. Normal thyroid glands weigh 25–40 g, but size varies with age, reproductive state, and diet. Its rich blood supply, 4–6 mL min^{-1} g^{-1}, is one of the highest in the body. Two pairs of parathyroid glands are embedded within the thyroid. These glands secrete parathyroid hormone (PTH) that helps regulate plasma [Ca^{2+}] and [$H_2PO_4^-$]. PTH will be discussed separately.

THE THYROID GLAND CONSISTS OF THOUSANDS OF *FOLLICLES* THAT STORE *THYROGLOBULIN*

The thyroid gland is made up of thousands of spherical or ovate **follicles**, as shown in Figure 9.3.2. Secretory epithelial cells line the follicles, and the follicle contains a homogeneous, gelatinous **colloid**. The epithelial cells' appearance correlates with their function: inactive thyroid consists of low cuboidal or squamous epithelium; active thyroid consists of columnar epithelial cells with basilar infoldings and numerous apical microvilli (see Figure 9.3.3). Interspersed among the follicles are **parafollicular cells** or **C cells**. These cells secrete another hormone, **calcitonin**, that helps regulate bone resorption and plasma [Ca^{2+}] homeostasis. This hormone is unrelated to the principal hormones of the follicular cells.

THE THYROID FOLLICLE SECRETES *THYROXINE* AND *TRIIODOTHYRONINE*

Upon stimulation, the thyroid follicle secretes thyroxine and triiodothyronine. The chemical structures of these hormones are shown in Figure 9.3.4. Both are iodinated derivatives of the amino acid tyrosine. Because thyroxine contains four iodine atoms per molecule, it is referred to as T4. Triiodothyronine has only three iodine atoms, lacking the second iodine at the 5′ position on the phenyl ring, and it is abbreviated as T3. These hormones are incorporated as part of **thyroglobulin**, the main constituent of the colloid.

FOLLICULAR CELLS SECRETE THYROGLOBULIN PRECURSOR INTO THE FOLLICLE

Follicular cells secrete a glycosylated, 300 kDa thyroglobulin precursor into the follicle that forms a 660-kDa dimer linked by disulfide bonds. Either during its secretion into the follicle or soon after, thyroglobulin is iodinated to form **monoiodotyrosine** (MIT) and **diiodotyrosine** (DIT). Figure 9.3.5 illustrates the cellular processes of synthesis, storage, and secretion of thyroxine.

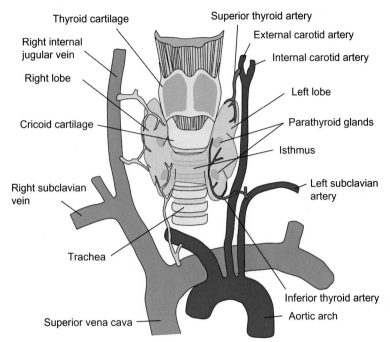

FIGURE 9.3.1 Anatomical position of the thyroid gland. The gland consists of two lobes that are closely affixed to the lateral and anterior aspects of the trachea near the cricoid cartilage. The gland receives rich blood flow through the superior thyroid artery originating from the external carotid artery, and the inferior thyroid artery originating from the left subclavian artery. Blood drains the gland into the superior, middle, and inferior thyroid veins that drain into the internal jugular and innominate veins. The parathyroid glands constitute a separate endocrine function and are considered separately. The parathyroid glands consist of a pair of glands embedded in each lobe of the thyroid gland.

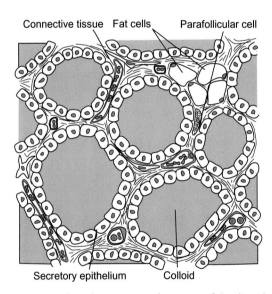

FIGURE 9.3.2 Histological appearance of a section of the thyroid gland. The gland is filled with follicles consisting of an internal colloid surrounded by a single layer of secretory epithelium. Interspersed among the follicles are connective tissue, including fibroblasts and fat cells, and special cells called parafollicular cells that secrete another hormone, calcitonin, that is involved in plasma $[Ca^{2+}]$ regulation. Calcitonin is unrelated to the hormones produced by the follicular cells.

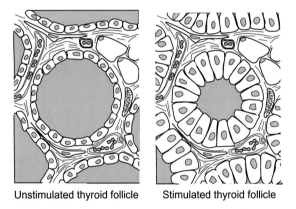

FIGURE 9.3.3 Histological appearance of the thyroid under different physiological states. In the unstimulated state, the cells appear as low cuboidal epithelial cells with abundant colloid. Highly stimulated thyroid follicles have columnar epithelium and their colloid becomes depleted.

SYNTHESIS OF THYROXINE REQUIRES FOUR STEPS

1. Accumuation of iodine
2. Oxidation of I^- to I^0
3. Organification
4. Coupling.

Thyroid secretory cells actively pump I into the cytoplasm through a **2Na:I symport (NIS)** located on the basolateral membrane. The uphill transport of I^- derives its energy from the Na^+ gradient that is maintained by the Na^+–K^+-ATPase. The human NIS gene codes for a protein of 643 amino acids. NIS can concentrate I^- some 25-fold over that in plasma. A second carrier, **pendrin**, carries I^- across the apical membrane.

Thyroid peroxidase on the apical membrane of follicular cells **oxidizes I^- to I^0**. Iodine in this state spontaneously displaces H on the phenyl group of tyrosine residues. The iodination reaction itself is not catalyzed. The process of adding iodine to thyroglobulin is called **organification**. The products of the reaction, MIT and DIT, remain attached to the peptide chain of thyroglobulin.

The final stage in synthesis of the storage form of T4 and T3 is the **coupling** of two molecules of DIT to form

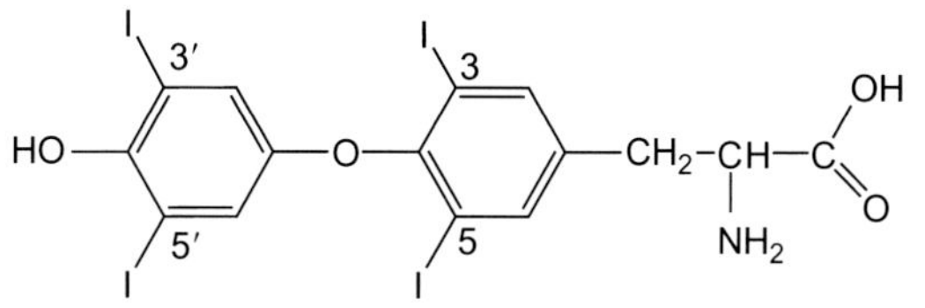

FIGURE 9.3.4 Chemical structures of thyroxine (T4) and triiodothyronine (T3).

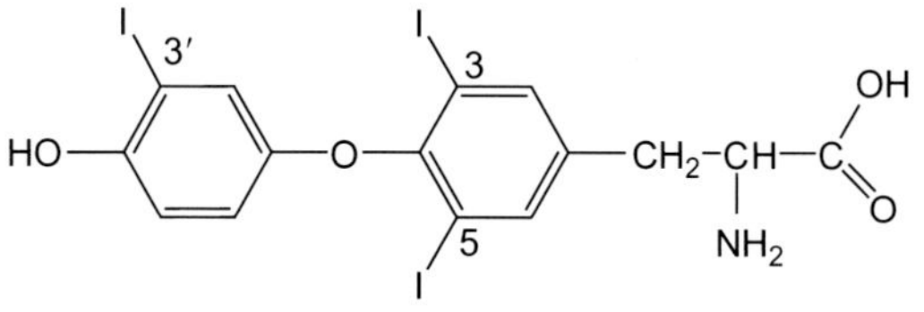

T4 or the coupling of one molecule of MIT to DIT to form T3. The exact chemical mechanism by which coupling occurs is not known, but it involves thyroid peroxidase. After coupling, the T3 or T4 remain attached to thyroglobulin's peptide chain. Some of the MIT and DIT remain as MIT or DIT attached to thyroglobulin.

FOLLICULAR CELLS PROTEOLYZE THYROGLOBULIN TO RELEASE T4 AND T3

Thyroid stimulating hormone, or **TSH**, is secreted from the anterior pituitary and acutely stimulates the thyroid follicular cells to extend long strands called **pseudopodia** to surround chunks of colloid and take them into the cell in endocytotic vesicles. These fuse with lysosomes, whose enzymes proteolyze the thyroglobulin and release T4 and T3 that were bound to thyroglobulin's peptide backbone. MIT and DIT are also released. Only T4 and T3 are released into the bloodstream, in the ratio of about 20:1 of T4 and T3, respectively.

The thyroid follicular cells have no mechanism for making T3 or T4 from MIT and DIT in its cytoplasm. **Thyroid deiodinase** located on the endoplasmic reticulum removes iodine from MIT and DIT. The iodine released in this process is recycled into thyroglobulin. Two-thirds of the iodine that gets incorporated into thyroglobulin originates from iodine recycled from MIT and DIT rather than freshly accumulated iodine that enters the cell through the sodium–iodine symport (NIS).

TSH REGULATES STATE OF THE THYROID GLAND

TSH is an N-linked glycoprotein of 28 kDa that is synthesized, stored, and released in basophilic cells, called thyrotrophs, of the anterior pituitary. TSH is also called **thyrotropin** because it is a "trophic" hormone, derived from the Greek "trophos," meaning "to nourish." Anterior pituitary trophic hormones increase the size of their target tissue. Like the other pituitary glycoproteins, FSH, LH, and HCG (human chorionic gonadotrophin), TSH consists of an α and β chain that are not covalently linked. The α subunits of FSH, LH, HCG, and TSH are identical, whereas the β chains confer biological specificity.

TSH binds to specific G_s-coupled receptors on the basolateral membrane of thyroid follicle secretory cells. These receptors link to phosphorylation of cell proteins that regulate thyroid cell metabolism and T3 and T4 synthesis. The short-term effect of TSH on the thyroid is the release of T3 and T4 from already synthesized colloid material. Long-term exposure to TSH increases the size of the thyroid gland by increasing the number of follicular secretory cells (**hyperplasia**) and increasing the size of the cells (**hypertrophy**).

THE HYPOTHALAMUS PARTLY CONTROLS TSH RELEASE

Neurons in the **paraventricular nucleus** of the **hypothalamus** secrete **thyrotropin releasing hormone (TRH)**, a tripeptide, into the portal blood that travels from the hypothalamus down to the anterior pituitary. In the anterior pituitary, TRH binds to G_q-coupled receptors on the surfaces of thyrotrophs to increase release of TSH. The paraventricular nucleus that produces TRH receives inputs from a variety of sources within the brain (see Figure 9.3.6).

T4 AND T3 INHIBIT SECRETION OF TSH

T3 and T4 inhibit TSH secretion and synthesis and decrease the sensitivity of the thyrotrophs to TRH. The effect of T4 and T3 is mediated by T3, as it is in all target tissues, through effects on gene transcription. T3 binds to a thyroid hormone receptor (TR) which then binds to the thyroid hormone response element on the DNA (TRE). These modify genetic expression in the thyrotrophs. T3 decreases the expression of the genes for TSH and for the TRH receptor on the cell membrane. In this way, T3 decreases the release of TSH, completing a negative feedback loop.

ALMOST ALL CIRCULATING T4 AND T3 ARE BOUND TO PLASMA PROTEINS

T4 and T3 are lipophilic and bind to hydrophobic or lipophilic domains of circulating proteins:

- eighty percent to thyroxine-binding globulin (TBG);
- fifteen percent to thyroxine-binding prealbumin;
- five percent to albumin.

These proteins are large enough that they are not filtered by the kidney, and they cannot escape the capillaries to enter target cells. Only 0.03% of the T4 and 0.3% of T3 are free to cross the capillaries and enter cells. This small fraction of the total circulating thyroxine is the biologically active fraction.

TBG is a glycoprotein that is made in the liver and binds one T4 or T3 molecule. Pregnancy and estrogen therapy increase plasma levels of TBG; chronic liver disease such as **cirrhosis** decreases TBG because the rate of

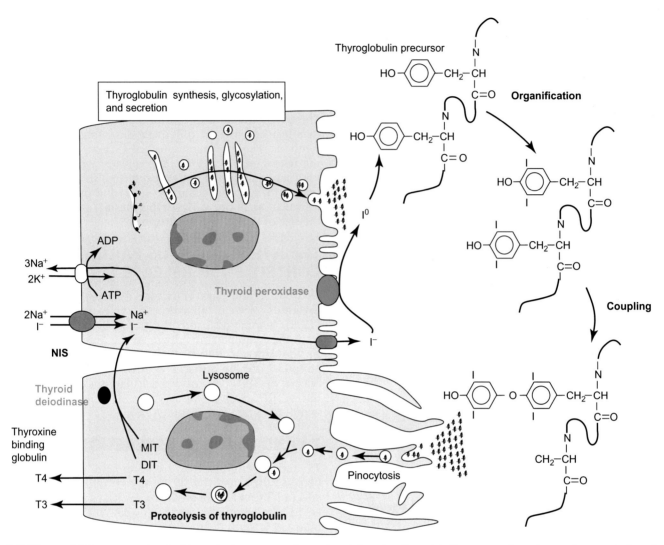

FIGURE 9.3.5 Cellular processes involved in synthesis, storage, and secretion of thyroxine. Thyroglobulin precursor is made on polysomes in the cell, enters the ER, and is transferred to the Golgi apparatus where it is glycosylated and packaged into secretory vesicles that carry the precursor to the apical membrane and secrete it into the follicle. During or soon after exocytosis into the follicle, it is iodinated. The iodine originates from dietary iodine that is absorbed by the gastrointestinal tract. It is actively pumped into the secretory cells by secondary active transport using a Na^+-I^- symport (NIS) on the basolateral membrane of the cell. Iodine is then transported into the follicle, where thyroid peroxidase on the apical membrane oxidizes the iodine. The iodine chemically incorporates itself into tyrosine residues on the thyroglobulin precursor, forming MIT and DIT. In an incompletely understood reaction, thyroid peroxidase helps couple two DITs to form T4 coupled to thyroglobulin, or one MIT and one DIT to form T3. When the cell is stimulated, pseudopodia engulf chunks of the follicle contents, which migrate in endocytotic vesicles toward the base of the cell where they fuse with lysosomes migrating to meet them. The proteases within the lysosomes completely degrade the thyroglobulin, releasing T4 and T3 that are subsequently secreted from the cell into the blood. Thyroid deiodinase liberates iodine from MIT and DIT produced from the proteolysis of thyroglobulin, and the iodine is recycled back into thyroglobulin.

synthesis is decreased; chronic kidney disease such as the **nephrotic syndrome** decreases TBG because it is lost in the urine.

THE TISSUES METABOLIZE T4 TO T3 AND rT3; T3 IS THE ACTIVE METABOLITE

LIVER AND KIDNEY POSSESS 5′ DEIODINASE TYPE I THAT CONVERTS T4 TO T3 AND rT3

The thyroid gland secretes about 80–100 μg of T4 each day. This is the main circulating form of the hormone, with plasma concentrations of about 8 μg dL^{-1}. Injected radioactively labeled T4 has a half-life of about 6 days. T4 is deiodinated by the enzyme **deiodinase Type I**, which is particularly rich in liver and kidney. This enzyme removes iodine at either the outer ring 5′ position to form T3 or at the inner ring 5 position to form reverse T3 (rT3) (Figure 9.3.7).

Brain, anterior pituitary, brown adipose tissue, and placenta have 5′ deiodinase type II that is similar to Type I 5′ deiodinase in that it incorporates a rare amino acid, **selenocysteine** (Sec), and it is an integral membrane protein. It has only outer ring deiodinase activity. Thus, it converts T4 to T3 and rT3 to 3,3′-T2. Most of the T3 in these tissues derives from Type II 5′ deiodinase acting on circulating T4.

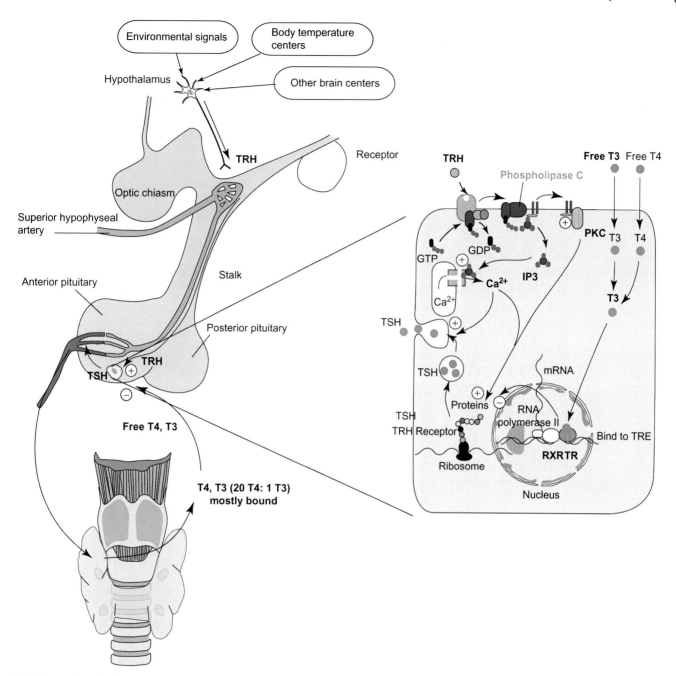

FIGURE 9.3.6 Control of T4 and T3 secretion. T3 and T4 secretion is stimulated solely by TSH, or thyrotropin, which in turn is secreted by the anterior pituitary solely in response to TRH, thyrotropin-releasing hormone. TRH reaches the anterior pituitary through the hypophyseal portal circulation. TRH is synthesized in neurohormonal cells residing in the paraventricular nucleus of the hypothalamus and is secreted in response to environmental stimuli, body temperature, and other inputs from the brain. TRH activates secretion of TSH in thyrotrophs through activation of G_q-coupled receptors. TSH is released into the blood wherein it travels to the thyroid gland where it causes release of T3 and T4 in the short term and causes enlargement of the thyroid and increased synthesis of thyroglobulin in the long term. Over-secretion of T3 and T4 is prevented by a negative feedback of T3 on the anterior pituitary. T3 binds to its nuclear receptor (TR) and modulates gene expression in the thyrotrophs. T3 reduces the number of receptors for TRH, thereby reducing the sensitivity of the thyrotrophs to TRH levels, and reduces the synthesis of TSH within the thyrotrophs.

A type III deiodinase is present in most tissues. It also contains a selenocysteine residue. It removes I from the 5 position on the inner ring. Thus, it converts T4 to rT3 and T3 to 3,3′-T2.

A fraction of the circulating T4 is deaminated and decarboxylated to form tetraiodoacetic acid, or **tetrac**. This compound circulates at about $0.06\ \mu g\ dL^{-1}$. Thyroid sulphotransferases, mainly in the liver, conjugate T4, T3, and rT3 with sulfate. The liver also has iodothyronine UDP-glucuronyltransferase activity, which catalyzes the conjugation of T4 and T3 with glucuronic acid. The liver excretes T4 sulfate and T4 glucuronate in the bile. The normal plasma concentrations, metabolic clearance rate,

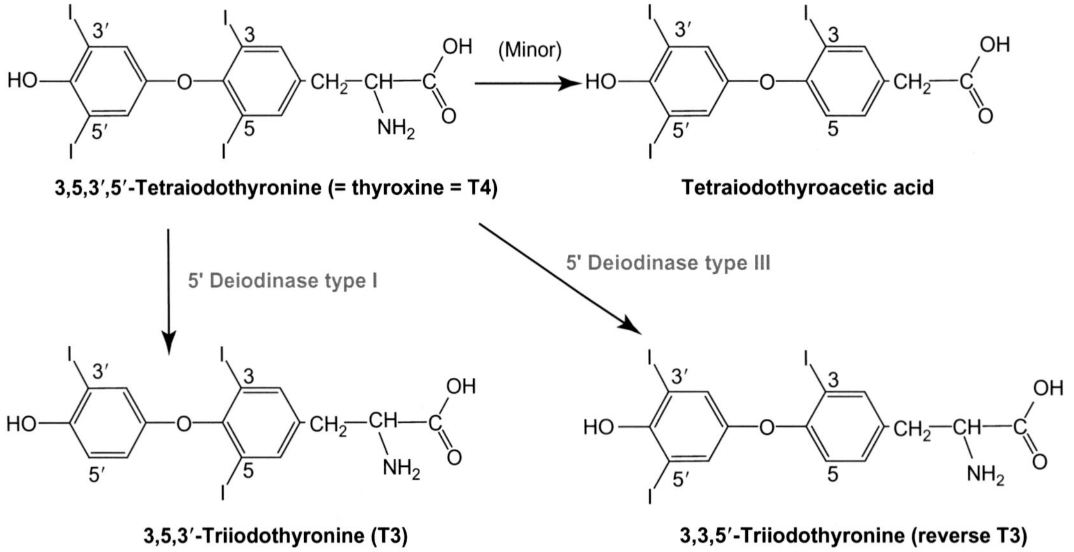

FIGURE 9.3.7 Metabolism of T4. The major circulating hormone is T4. It is deiodinated in peripheral tissues to T3 and rT3, primarily in the kidney and liver. T3 is the active metabolite in tissues. Deiodination to rT3 is the beginning of a degradative pathway.

TABLE 9.3.1 Approximate Values for the Normal Plasma Concentration, Metabolic Clearance Rate, and Production Rate of the Major Thyroid Hormone Derivatives in Normal Adult Humans

Compound	Plasma Total [ ] (μg dL^{-1})	Metabolic Clearance Rate (L day^{-1}) per 70 kg Body Weight	Production Rate (μg day^{-1}) per 70 kg Body Weight
T4	8.6	1.2	100
T3	0.14	24	31
rT3	0.04	111	39

and production rates for the major thyroid hormone derivatives are summarized in Table 9.3.1.

The estimates shown here are the average mean values of multiple studies. Values can be converted to nmol by using the formula weights: T4, 1 nmol = 777 ng; T3 and rT3, 1 nmol = 651 ng. (Data from I.J. Chopra and L. Sabatino, Nature and sources of circulating thyroid hormones, in L.E. Braverman and R.D. Utger, eds., *The Thyroid, A Fundamental and Clinical Text*, 8th edition, Lippincott, Williams and Wilkins, Philadelphia, PA, 2000.)

T3 ALTERS GENE EXPRESSION

Most tissues contain TR in the nucleus of their cells. These 50–55 kDa receptors structurally resemble the nuclear receptors for steroid hormones and vitamin D. Tissues often have all three types: TR-α1, TR-β1, and TR-β2; however, their proportion varies with the tissue.

The TRs bind to specific sequences in the DNA called thyroid responsive elements, or **TREs**. Each of these regulates the transcription of mRNA that codes for a protein. Binding of the TR to **positive TREs** increases expression of the mRNA and the protein. Binding of TRs to **negative TREs** decreases the mRNA and protein expression. All positive TREs require dimers to activate transcription, but some negative TREs contain only a single binding site for TR. It is currently thought that heterodimer formation with **RXR**, the **9-cis-retinoic acid receptor**, forms the most effective configuration of TR for regulation of transcription (see Figure 9.3.8).

TR proteins bind to the TRE in the presence or absence of T3. In the absence of T3, TRs bind to positive TREs and inhibit or repress the rate of transcription. Binding of T3 to the TR on these positive TREs relieves this inhibition. The repression of gene expression is due to **TR corepressors**, of which there are several varieties. These proteins bind to the TR when the T3 binding site is vacant, and dissociate from the TR when the receptor binds T3.

THYROID HORMONE PLAYS A CRUCIAL ROLE IN GROWTH AND DEVELOPMENT AND IN GENERAL METABOLISM

Thyroid hormone has multiple effects on every tissue in the body. During particular times during development thyroid hormone plays a crucial role and its insufficiency has devastating consequences. An overview of the effects is shown in Table 9.3.2.

TABLE 9.3.2 Overview of the Physiological Actions of Thyroid Hormone

1. CNS development
   1.1 Normal T3 is required for proper development and function of the CNS
   1.2 Low levels of T3 during development cause mental retardation
   1.3 High levels of T3 increase irritability and excitability
   1.4 T3 inhibits nerve cell replication
   1.5 T3 stimulates neuron cell body growth and branching of dendrites
   1.6 T3 stimulates axon myelination

2. Body growth
   2.1 T3 stimulates growth hormone synthesis in somatotrophs in the anterior pituitary. Thus hypothyroidism is associated with growth retardation
   2.2 T3 stimulates synthesis of structural proteins in skeletal muscle, heart, liver, etc.
   2.3 T3 stimulates calcification of the growth plates of the long bones, limiting linear growth

3. Basal energy expenditure
   3.1 T3 increases the basal metabolic rate (BMR). This is called the **thermogenic effect**
   3.2 T3 increases O_2 consumption, energy production, and heat production
   3.3 T3 promotes mitochondrial growth and replication
   3.4 T3 increases expression of a variety of respiratory enzymes

4. Intermediary metabolism
   4.1 T3 promotes protein synthesis in a variety of tissues
   4.2 T3 potentiates the effects of epinephrine in the liver and increases liver glycogenolysis and gluconeogenesis; thus, it raises blood glucose and decreases liver glycogen
   4.3 T3 potentiates the effects of insulin on skeletal muscle; it increases uptake, utilization, and storage of glucose
   4.4 T3 potentiates the effects of insulin on adipose tissue; it increases lipolysis and increases circulating levels of free fatty acids and decreases plasma cholesterol

5. Cardiovascular system
   5.1 The heart has receptors for T3 which increase transcription of specific genes. Among these are the genes coding for SERCA and for myosin heavy chain α and β
   5.2 T3 increases cardiac output by increasing both stroke volume and heart rate
   5.3 T3 increases sensitivity to β-adrenergic stimulation
   5.4 T3 increases respiratory ventilation secondary to increased CO_2 production

6. TSH secretion
   6.1 The TSH gene is negatively regulated by T3; T3 decreases the synthesis of mRNAs coding for TSH α and β subunits, thereby decreasing synthesis of TSH

HYPOTHYROIDISM REFERS TO REDUCED CIRCULATING LEVELS OF T4 AND T3

Figures 9.3.5 and 9.3.6 show numerous steps in the synthesis, storage, and secretion of T4 and T3. Defects in any one of these steps can interfere with the secretion of sufficient thyroid hormone. Defects in the thyroid gland itself are classified as **primary hypothyroidism**. Insufficient T4 and T3 production because of lack of stimulation by TSH is called **central hypothyroidism**.

The most common cause of hypothyroidism worldwide is **iodine deficiency**. The iodine required to make T4 and T3 comes from the diet. Insufficient dietary iodine results in primary hypothyroidism. The low circulating levels of T4 and T3 no longer inhibit TSH synthesis and secretion, and TSH levels rise. The increased TSH levels exert a trophic effect on the thyroid, causing hyperplasia and hypertrophy of the cells (see Figure 9.3.3), resulting in an enlarged thyroid gland, called a **goiter**. Goiters are common in primary hypothyroidism (Figure 9.3.9).

There are many causes of hypothyroidism. Defects are known to occur in:

- iodine availability in the diet;
- NIS, the sodium–iodine symport, that traps iodine within the secretory cell;
- pendrin, the protein that transports iodine into the follicle across the apical membrane;
- thyroid peroxidase, the enzyme that oxidizes iodine for organification;
- thyroglobulin synthesis;
- iodine recycling in the thyroid secretory cells;
- the TSH receptor that signals the follicular secretory cells to secrete T4 and T3;
- the $G_{\alpha s}$ subunit that couples the TSH receptor to adenylyl cyclase to turn on secretion;
- the development of the thyroid gland.

These all cause primary hypothyroidism. Central or secondary hypothyroidism can be caused by defects in:

- development of the hypothalamus or pituitary;
- the TRH receptor;
- regulation of TSH synthesis or secretion.

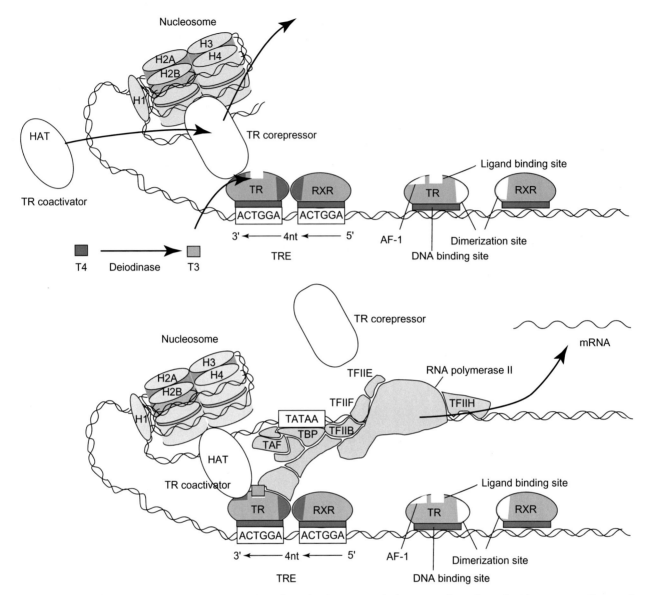

FIGURE 9.3.8 Action of T3 on gene expression at positive TREs. The TR binds to a TRE, which consists of specific nucleotide sequences. The one shown is called a **direct repeat**, consisting of two sets of six nucleotides separated by four nucleotides of variable composition. The TR binds to one half-site in the TRE on its DNA-binding domain. The second half-site is occupied by a retinoid X receptor (RXR), which forms a heterodimer with the TR. This binding occurs in the absence of T3. In this case, the TR binds one of a variety of TR corepressors that inhibits transcription, probably through histone deacetylase activity. In the hypoacetylated state, DNA is compact and provides poor access for transcription. When T3 binds to the TR, a TR coactivator displaces the corepressor. There are several varieties of coactivator. Many of these activators have histone acetyl transferase activity (HAT). The unraveling of the DNA and exposure of the TATA box recruits a variety of proteins, culminating in the stabilization of the preinitiation complex and activation of RNA polymerase II to transcribe mRNA from the DNA template.

Hypothyroidism can also be caused by:

- goitrogens, environmental compounds that interfere with iodine transport;
- inflammation of the thyroid gland (**thyroiditis**);
- autoimmune destruction of the thyroid gland, such as in **Hashimoto's thyroiditis**.

Many of these disorders are exceedingly rare (see Clinical Applications: Pendred Syndrome), whereas primary hypothyroidism due to iodine deficiency is quite common. The prevalence of hypothyroidism increases with age. For most of those patients, the cause is unidentified because the diagnosis and treatment, exogenous replacement therapy with T4, does not require detailed knowledge of its etiology. The regulatory mechanisms leading to goiter are summarized diagrammatically in Figure 9.3.9.

THE CLINICAL SYMPTOMS OF HYPOTHYROIDISM ARE MANIFOLD

Table 9.3.2 shows that thyroid hormone exerts many physiological effects. The sensitivity of these effects to

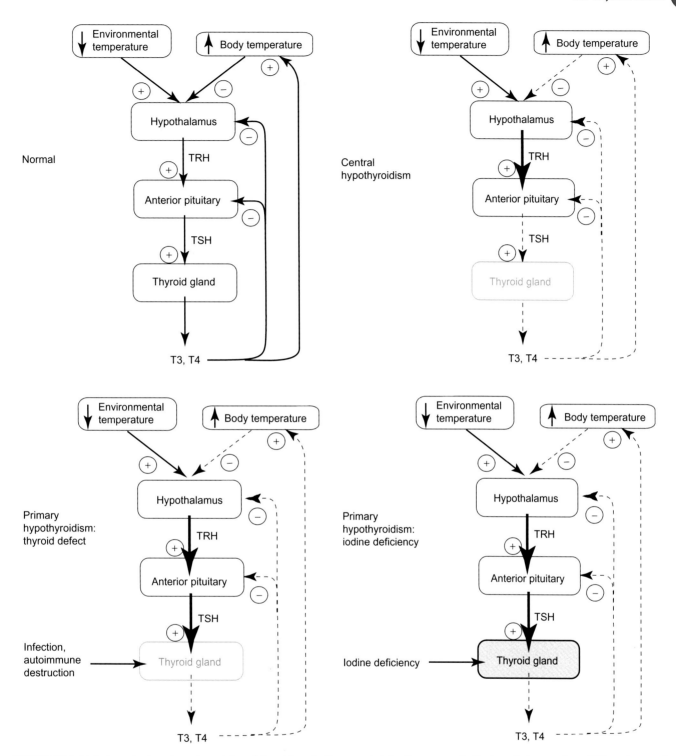

FIGURE 9.3.9 Control mechanisms in various states of thyroidism. In the normal state, the neurons in the hypothalamus release TRH to the portal circulation leading to the anterior pituitary gland in response to integrated signals from the environment and the body, mostly dealing with temperature and energy production. Decreases in environmental temperature stimulate TRH release, whereas rises in body temperature or energy production decrease it. TRH releases TSH from the anterior pituitary, which then stimulates T3 and T4 release from the thyroid. T3 and T4 feed back directly on TSH synthesis and secretion and indirectly through body energy production on the release of TRH. In persons with thyroid defects, TSH is ineffective in increasing T3 and T4 secretion because the thyroid cannot respond. The thyroid may or may not enlarge because it is simultaneously being stimulated by TSH and destroyed by autoimmune or inflammatory processes. In central hypothyroidism, TRH may fail to elicit TSH secretion. Thus, TRH remains high and T3 and T4 levels do not rise. In iodine deficiency, the thyroid also cannot make sufficient T3 and T4. There is no feedback inhibition of TRH or TSH secretion and so both remain high. The high TSH levels stimulate thyroid growth, producing a goiter.

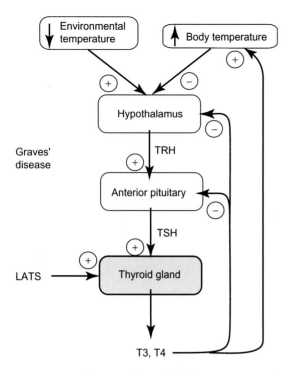

FIGURE 9.3.10 Mechanism of hyperthyroidism in Graves' disease. Persons with Graves' disease produce antibodies against their TSH receptors on the thyroid gland. These antibodies bind to the receptor and stimulate the thyroid gland inappropriately and regardless of the thyroid hormone status of the blood. The result is a hyperactive thyroid and high circulating T3 and T4 levels.

hypothyroidism varies according to the age of onset of the hypothyroidism. Hypothyroidism in utero and in early postnatal life causes **cretinism**, marked by severe mental retardation, deaf-mutism, spastic gait and lack of motor control, and poor growth. Early recognition and treatment of this condition is crucial because the untreated condition leads to permanent damage that cannot be reversed by later return to a euthyroid state.

Hypothyroidism in the juvenile leads to delayed growth and sexual maturation with symptoms intermediate between those of the infant and the adult. Hypothyroidism in the adult is characterized by **myxedema**. Accumulation of hyaluronic acid in the dermis absorbs water and it swells to one-thousand times its weight, resulting in a puffy thickening of the skin. Part of the edema is due to albumin escape from the capillaries. Other symptoms in the adult include the following:

- Sluggish mental functions including memory defects, lethargy, and somnolence
- Slow myotatic reflexes
- Decreased BMR and appetite
- Cold intolerance
- Decreased cardiac output due to bradycardia and decreased stroke volume
- Pale and cool skin due to cutaneous vasoconstriction
- Dry and coarse skin due to reduced sweat and oil secretions
- Myxedematous skin.

Clinical Applications: Goitrogens

Some plants produce natural **goitrogens**, compounds that produce goiters in humans in the absence of dietary iodine deficiency. Broccoli, brussel sprouts, cauliflower, and cabbage are all varieties of the species *Brassica oleracea*. These plants all produce thioglucosides which, after digestion, produce thiocyanate and isothiocyanate. Both of these compounds interfere with the NIS in the thyroid gland. In sufficient amounts, they can cause primary hypothyroidism, increased TSH secretion, and hypertrophy of the thyroid gland, or goiter.

Cassava or **manioc**, is a basic food staple in certain areas of the tropics. It is a starchy root that people grind up to make flour or meal. This plant contains cyanoglucosides that are converted to cyanide, which is then detoxified to thiocyanate. The **thiocyanate is a potent goitrogen**. These cyanoglucosides are also found in bamboo shoots, sweet potatoes, and lima beans. The determining factor in the development of goiter is the ratio of dietary iodine to thiocyanate. Goiter develops when the urinary iodine/thiocyanate ratio drops below about 3 µg iodine per mg of thiocyanate.

Clinical Application: Pendred Syndrome

In 1896, Vaughan Pendred (1869–1946) described a syndrome of deafness associated with goiter, an enlargement of the thyroid gland, in two members of a large family. Its etiology was further illuminated in 1958 when it was found that that Pendred syndrome is accompanied by a positive perchlorate discharge test. In this test, the thyroid is first loaded with radiolabeled iodine by i.v. injection. Two hours later, further uptake is inhibited by injecting 100 mg sodium perchlorate, which blocks the iodine uptake mechanism. If the radiolabeled iodine is already incorporated into thyroglobulin, it will not be released. In Pendred syndrome, a larger fraction of the radiolabeled iodine is released because it has not been transferred to the follicle. It is now known that the Pendred gene, PDS, codes for a protein called Pendrin that is probably involved in transporting iodine into the follicle across the apical membrane. The prevalence of Pendred syndrome is between 1:15,000 and 1:100,000. Some 35 different mutations of pendrin have been identified. The function of pendrin in the inner ear is not yet understood.

THE MOST IMPORTANT CLINICAL ABNORMALITY OF HYPERTHYROIDISM IS *GRAVES' DISEASE*

In 1835, R.J. Graves described women who had goiter and palpitations and included one with **exophthalamos**, a protrusion of the eyeballs. Graves thought that the disease derived from defects in the heart, and it was not until the 1890s that surgeons discovered the thyroid origin of the disease. The etiology of the disease was clarified in 1956 when it was found that serum from persons with Graves' disease could cause release of radiolabeled thyroid hormone when injected into guinea pigs, but its action was more prolonged than that of TSH. (The circulating half-life of TSH is 1 h.) The active substance was termed **long-acting thyroid stimulator**, or **LATS**. This substance is an immunoglobulin directed against the TSH receptor. Some of these antibodies stimulate the TSH receptor, whereas others bind and block TSH stimulation. Graves' hyperthyroidism is caused by autoantibodies that stimulate the TSH receptor but this is not subject to the normal negative feedback loops for control of blood T3 and T4 levels. The etiology of Graves' disease is illustrated in Figure 9.3.10.

Clinical Applications: Iodine Deficiency Disorders

One of the four major public health problems relating to nutrition is iodine deficiency. The thyroid requires iodine to make thyroid hormone, and this iodine is obtained from the diet. Historically, the major natural source of iodine is from seafood. Sea water contains about 50–60 µg of iodide per liter. Iodide is oxidized by sunlight to elemental iodine, which is volatile. Every year some 400,000 tons of iodine escapes from the sea. Air contains about 0.7 µg m^{-3}. The iodine returns to the soil in rain, but this is insufficient to replenish the leaching of iodine from deficient soils. Thus, plants raised in iodine-deficient areas, particularly mountainous regions that have experienced glaciation, and inland areas, are low in iodine. People indigenous to these regions are at risk of developing iodine deficiency disorders (IDDs).

The clinical manifestations of IDDs are **hypothyroidism, goiter, dwarfism, impaired neurological development, myxedema**, and **cretinism**. Goiter is an enlargement of the thyroid gland, for whatever reason. In iodine deficiency, the deficit in T4 and T3 removes negative feedback on TSH synthesis and secretion in the thyrotrophs of the anterior pituitary. As a result, TSH levels rise and exert trophic effects on the thyroid gland, causing its enlargement. Goiter in areas of iodine deficiency is called **endemic goiter**. The dwarfism arises from the lack of stimulation by T3 of GH synthesis and secretion in somatotrophs in the anterior pituitary. Hypothyroidism leads to impaired neurological function in the adult. Hypothyroidism in the fetus and early postnatal life leads to cretinism. Cretinism is a polymorphous collection of abnormalities of intellectual and physical development. **Neurological cretins** are extremely mentally retarded, deaf-mutes with spastic gaits and impaired motor abilities. **Myxedematous cretins** are less severely retarded but have retarded growth and sexual development, myxedema, retarded maturation of body proportions, dry skin, and sparse development of nails and hair. These two forms comprise the extreme limits of a spectrum of abnormalities caused by fetal iodine deficiency. The evidence that cretinism is an IDD is largely epidemiological: its frequency correlates with the degree of iodine deficiency; iodine treatment of a population reduces its incidence; it appears together with iodine deficiency of recent onset. The prevalence of goiters in the persons with neurological cretinism is about the same as in the noncretin population, whereas goiters in persons with myxedematous cretinism are relatively rare. Nevertheless, myxedematous cretins have hypothyroidism with high TSH levels. Neurological cretinism is the more prevalent form.

These IDDs are a huge public health problem. In 1986, the International Council for the Control of Iodine Deficiency Disorders (ICCIDD) was formed at a meeting in Kathmandu, Nepal. This organization consists of a global interdisciplinary network of experts who work closely with the WHO and UNICEF and a number of governments in efforts to eradicate IDDs. According to the ICCIDD (www.ICCIDD.org), as much as 30% of the world's population is at risk of IDDs; 750 million people suffer from some degree of goiter, 43 million have IDD-related brain damage, and some 5.7 million are afflicted with cretinism.

Prevention of IDDs in populations can be as simple as iodination of ordinary table salt. The dietary requirements for iodine scale with age: 40 µg day^{-1} from 0 to 6 months; 50 µg day^{-1} from 6 to 12 months; 60–120 µg day^{-1} from 1 to 10 years; and 120–150 µg day^{-1} from 11 years onward. The US standard for iodized salt is 0.006% by weight as KI, which is equivalent to 45 µg iodine g^{-1}. One teaspoon of salt weighs 6 g, and so this degree of iodination provides 270 µg of iodine, more than enough for the needs of most people. Iodized salt has a definite shelf-life because the iodine becomes volatilized. Heating the salt in solution removes the iodine.

Despite our collective scientific knowledge, IDDs continue to be a veritable scourge, both medically and socially. Eradication is hampered by lack of education, lack of infrastructure, and cultural resistance to new dietary products.

Hyperthyroidism in Graves' disease causes the following:

- Diffuse goiter from extended stimulation of the gland through its TSH receptors
- Exophthalamos, protrusion of the eyeballs
- Increased CNS excitability, nervousness, emotional lability, hyperkinesia, tremor, insomnia
- Increased BMR
- Increased appetite
- Heat intolerance

- Increased cardiac output caused by tachycardia and increased stroke volume
- Increased sensitivity to adrenergic stimulation
- Peripheral vasodilation
- Cutaneous vasodilation and excessive sweating.

SUMMARY

The thyroid gland is found in the neck surrounding the trachea. It secretes thyroxine and triiodothyronine when stimulated by TSH, a glycoprotein hormone secreted by the anterior pituitary. The gland itself consists of spherical or oblate follicles lined by an epithelium and containing a colloid made up of thyroglobulin. Tyrosine residues in the thyroglobulin are iodinated by the gland. Iodine is taken up by the gland through secondary active transport at the basolateral membrane (facing the blood) through a Na–I symport (NIS). The I^- is carried across the apical membrane by another carrier, probably pendrin. The I^- in the follicle is converted to I^0 by thyroid peroxidase. The I^0 forms covalent bonds with tyrosine residues in a process called organification. MIT and DIT residues are coupled to form thyroxine (T4—four I atoms on two phenyl groups) or triiodothyronine (T3), still bound to thyroglobulin. Release of T4 or T3 occurs when the gland is stimulated by TSH. Thyroid cells endocytose the thyroglobulin and release T3 or T4 after proteolytic digestion of the colloid.

TSH is a glycoprotein hormone secreted by thyrotrophs in the anterior pituitary, when these cells are stimulated by TRH. There are two basic controls of these cells: TRH released from nerve cells in the paraventricular nucleus of the hypothalamus and negative feedback inhibition by T3 and T4. The signals for secretion of TRH are neuronal and derive from environmental temperature, body core temperature, and other brain centers. TRH stimulates TSH secretion through a G_q mechanism.

Circulating T3 and T4 are largely carried by proteins that cannot leave the circulation. Only a small fraction of free T3 is active. It enters cells and binds to multiple nuclear receptors called TR. These bind to TREs. Positive TREs increase mRNA expression in response to T3; negative TREs decrease mRNA synthesis and expression. TR alone binds to TRE and inhibits mRNA expression; binding of T3 to the TR relieves this inhibition. The circulating T3 and T4 are degraded by deiodination followed by deamination and decarboxylation.

Hypothyroidism in infancy and childhood produces devastating dysfunction. Cretinism is marked by severe mental retardation, short stature, deaf-mutism, and lack of motor control. Hypothyroidism in the adult causes myxedema, caused by buildup of dermal hyaluronic acid and its swelling with fluid. This is accompanied by sluggish mental function, decreased metabolic rate, and cold intolerance. Hypothyroidism has many causes. The most common is iodine deficiency, which is readily treated with iodized salt.

Goiters are hyperplastic and hypertrophied thyroid glands, for whatever reason. In iodine deficiency, lack of T3 and T4 output by the gland relieves negative feedback inhibition of TSH secretion so that TSH remains high and the thyroid gland is stimulated. Goiters also occur in Graves' disease, which is a hyperthyroid condition caused by antibodies to the TSH receptor. These persons make a long-acting thyroid stimulator that stimulates the thyroid gland inappropriately. In this case, TSH secretion is inhibited by the high circulating T3 and T4 levels.

REVIEW QUESTIONS

1. Describe the hypothalamic control of TSH release. What is TRF? What stimulates its release? How does it cause TSH release?
2. What is TSH? Where is it released? What stimulates its release? What inhibits it? What does it do? What mechanism does it use to stimulate thyroid secretion?
3. What is thyroid hormone? List the four steps involved in iodination of thyroglobulin. Where does the iodine come from? How does it get into the follicle? How does it combine with tyrosine? How are T3 and T4 formed? How are they released? Which is the predominant secreted form? Which is most active biologically?
4. How are T3 and T4 carried in blood? What physiological effects do they have? What diseases result from hypothyroidism? What is the most common cause of hypothyroidism? How can you cure it?
5. How are T4 and T3 metabolized? Where is T3 formed? What is the mode of action of T3?
6. What diseases result from hyperthyroidism? What is the most common cause of hyperthyroidism?
7. What is a goiter? How can the goiter be enlarged in both hypothyroid or hyperthyroid states?

The Endocrine Pancreas and Control of Blood Glucose 9.4

Learning Objectives

- List three hormones and their cell types that secrete them in the islets of Langerhans of the pancreas
- Describe the chemical structure of insulin in terms of its polypeptide chains
- Explain how increased plasma [glucose] causes insulin release
- Explain the mechanism of oral hypoglycemic agents on insulin release
- Describe the mechanism by which insulin increases peripheral glucose uptake
- Summarize the effects of insulin on carbohydrate, fat, and protein metabolism
- Indicate the major physiological stimulus for glucagon release
- Describe the mechanism of glucagon stimulation of liver glycogenolysis
- Describe the mechanism of glucagon stimulation of liver gluconeogenesis
- Explain why diabetics must cut back on their insulin when they exercise
- Describe the levels of glucagon and insulin after a meal and during fasting

THE PANCREAS HAS BOTH EXOCRINE AND ENDOCRINE FUNCTIONS

The pancreas lies between the greater curvature of the stomach and the duodenum. It consists mostly of **acinar glands** that secrete pancreatic juice that is carried by ducts into the duodenum where the exocrine pancreatic secretion neutralizes stomach acid and provides enzymes for digestion. About 1% of the pancreatic mass makes up the **islets of Langerhans**, which are endocrine glands. There are about a million islets distributed throughout the acinar tissue, set off from the acinar pancreas by a connective tissue sheath (see Figure 9.4.1). The islets contain four distinct cells types. The β cells make up about 60% of the islet population and secrete **insulin**. This hormone is the most important hormone in regulating carbohydrate and lipid metabolism. The α cells make up 25% of the islet population and secrete **glucagon**, which increases blood glucose by increasing its formation by gluconeogenesis and glycogenolysis. Insulin is secreted during times of nutrient abundance, and it promotes metabolic fuel storage. Glucagon is secreted during times of nutrient deficit, and it mobilizes metabolic fuel stores.

About 10% of the islet cells are δ cells that secrete **somatostatin**. Among its other actions as a growth hormone antagonist, somatostatin suppresses insulin release. The remaining islet cells, F cells, secrete **pancreatic polypeptide**.

β CELLS SYNTHESIZE INSULIN AS A PROHORMONE AND SECRETE INSULIN AND C PEPTIDE 1:1

Like many secreted proteins, insulin is synthesized as a preproinsulin, 110 amino acids long. The signal sequence is cleaved to form proinsulin, 86 amino acids long, which is further processed to form the A and B chains of insulin, and C peptide. The B chain corresponds to amino acids 1–30 of proinsulin; the A chain corresponds to amino acids 64–86, and the peptide connecting the two gives rise to the C peptide after two pairs of basic amino acids are cleaved off. The A and B chains are held together by two disulfide links. The relationship between the A, B, and C chains is shown in Figure 9.4.2.

Both mature insulin and the C peptide are stored in secretory granules and released into the portal blood upon stimulation of the β cells. The portal blood travels to the liver prior to entering the systemic circulation. The liver removes 50–60% of the insulin before it reaches the systemic circulation, so systemic levels of insulin do not indicate the rate of insulin secretion. However, C peptide is not cleared by the liver but instead is excreted by the kidney with a plasma half-life of about 30 min. The half-life of circulating insulin is only about 5 min. Because of these differences, circulating C peptide is a better indicator of β cell function, but it does not reflect rapidly changing rates of insulin secretion.

HIGH PLASMA GLUCOSE STIMULATES INSULIN SECRETION

Glucose has a molecular weight of 180 and cannot easily penetrate lipid bilayers. It enters β cells by facilitated

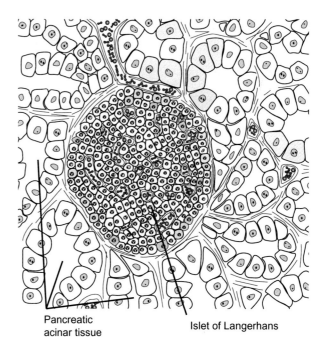

FIGURE 9.4.1 Diagrammatic representation of a histological section of pancreatic tissue showing exocrine and endocrine cells. The islets of Langerhans are separated from the exocrine tissue by a connective tissue sheath. The islets contain several distinct cell types that secrete distinct hormones.

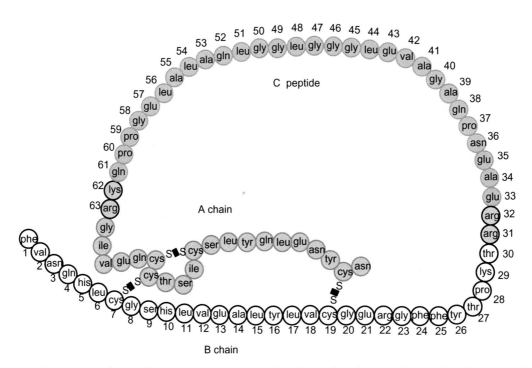

FIGURE 9.4.2 Amino acid sequence of proinsulin, showing the cleavage points that produce the A and B chain. The B chain contains amino acids 1–30 (white circles); the A chain contains amino acids 64–84 (gray circles). The A and B chain are linked by two disulfide bonds, and there is an additional disulfide bond within the A chain. The C peptide is produced by cleaving two pairs of basic amino acids from both ends of the connecting peptide (blue circles with dark outlines). The C peptide is secreted in a 1:1 molar ratio with insulin upon stimulation of the β cells (blue circles).

diffusion using the GLUT2 carrier. **Glucokinase** converts cytoplasmic glucose to glucose 6-phosphate. This is the rate-limiting step in glycolysis and therefore the enzyme serves as a **glucose sensor** for the β cell. As plasma glucose increases, cytoplasmic glucose increases and the rate of glycolysis increases, producing a local increase in cytoplasmic ATP concentrations. ATP inhibits an ATP-sensitive K^+ channel (K_{ATP}), so that increasing ATP blocks the channel, reducing K^+ efflux and reducing an outward current. This depolarizes the cell.

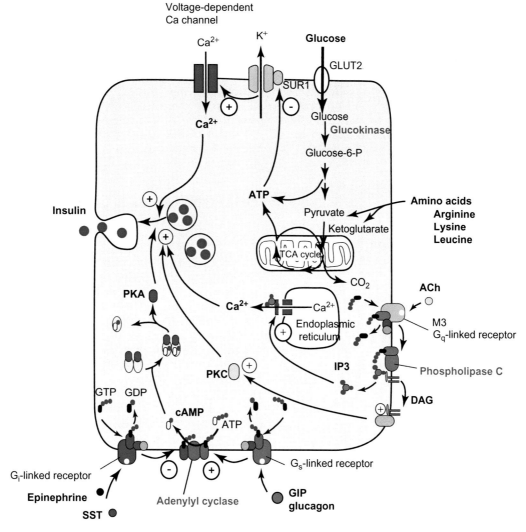

FIGURE 9.4.3 Mechanisms of regulation of insulin secretion. Exocytosis of insulin-containing granules is dependent on increases in cytoplasmic [Ca^{2+}]. This is achieved by depolarization of the cell that opens a voltage-dependent Ca^{2+} channel. Depolarization of the cell is generally achieved by closing an ATP-dependent K^+ channel (K_{ATP}). However, acetylcholine may increase a Na^+ conductance pathway to effect depolarization. Secretion of insulin is modulated by a variety of hormones and neurotransmitters. Epinephrine and norepinephrine inhibit insulin secretion through inhibition of adenylyl cyclase. Glucagon-like peptide (GLP-1) and glucose-dependent insulinotropic polypeptide (GIP) both stimulate insulin secretion by activating adenylyl cyclase, increasing cAMP and stimulating protein kinase A (PKA). Acetylcholine increases insulin secretion through an M3 receptor that activates PLC through a G_q mechanism. PLC releases diacylglycerol (DAG) that activates PKC, and inositol trisphosphate (IP3) that releases Ca^{2+} from intracellular stores in the endoplasmic reticulum. The released Ca^{2+} contributes to increased cytoplasmic [Ca^{2+}] that causes exocytosis of insulin-containing granules. Acetylcholine also depolarizes the cell, probably by activating a Na^+ channel.

Voltage-dependent Ca^{2+} channels in the membrane open in response to the depolarization, causing an influx of Ca^{2+} and an increase in cytoplasmic [Ca^{2+}]. The increased cytoplasmic [Ca^{2+}] induces vesicle fusion and insulin secretion (see Figures 9.4.3 and 9.4.4).

GLP-1 AND GIP STIMULATE INSULIN SECRETION; *SOMATOSTATIN* INHIBITS IT

Endocrine cells in the small intestine release **glucagon-like peptide 1 (GLP-1) and glucose-dependent insulinotropic peptide (GIP)**. In general, these hormones are not secretagogues in themselves, but they increase the sensitivity of the islet β cells for glucose. GLP-1 and GIP both increase cAMP through G_s-coupled receptors. Activation of G_q-coupled receptors also potentiates insulin release. Somatostatin, SST, inhibits insulin release, probably by a paracrine mechanism. These effects explain the fact that insulin secretion is greater with an oral dose of glucose than with an infusion—the **incretin** effect.

PARASYMPATHETIC STIMULATION INCREASES INSULIN SECRETION; SYMPATHETIC STIMULATION INHIBITS IT

Parasympathetic stimulation increases insulin secretion through M3, muscarinic receptor type 3, which is a G_q-coupled receptor. Sympathetic inhibition appears to be mediated by α_2 receptors that are linked to G_i proteins that inhibit adenylyl cyclase. Acetylcholine also appears to depolarize β cells by activating a Na^+ conductance on the surface of the cells. This depolarization would open voltage-dependent Ca^{2+} channels and stimulate insulin secretion. Despite these effects, innervation cannot be considered essential to islet function because the islets continue to secrete insulin and maintain blood sugar levels when they are transplanted to another location in the body where reinnervation is impossible.

AMINO ACIDS STIMULATE INSULIN SECRETION

Amino acids such as arginine, lysine, and leucine stimulate insulin secretion. The mechanism by which this occurs is not established. The working hypothesis is that these amino acids increase ATP supply by their oxidation as fuels, thereby stimulating insulin release. The enzyme glutamine dehydrogenase (GDH) converts glutamic acid to α-ketoglutaric acid, which feeds into the TCA cycle (see Chapter 2.11). Leucine stimulates this enzyme, and the stimulation disappears with high ATP/ADP ratio. Thus, leucine can raise the ATP/ADP ratio at low glucose concentrations, thereby stimulating insulin release, but it cannot stimulate insulin release at high glucose levels. This hypothesis is strengthened by the occurrence of hyperinsulinism in persons with mutations in the regulatory site of GDH.

SULFONYLUREAS CLOSE THE K_{ATP} CHANNEL AND THEREBY INCREASE INSULIN SECRETION

The K_{ATP} channel consists of a **sulfonylurea receptor (SUR1)** and a K^+ channel, either KIR6.1 or KIR6.2 (referring to inwardly rectifying K^+ channel). In the pancreatic β cell, SUR1 and KIR6.2 form a pair, and four pairs form an octameric K_{ATP} channel. The SUR is sensitive to a class of drugs called **sulfonylureas** that includes oral hypoglycemic agents such as **tolbutamide** and **glyburide**. Other insulin secretagogues also bind to the SUR1 but at distinct sites from the sulfonylureas. Closing the K_{ATP} channel depolarizes the cell, and opens the voltage-sensitive Ca^{2+} channels, leading to exocytosis of granules containing insulin and C peptide.

INSULIN RELEASE IS PULSATILE

About 50% of the total amount of insulin secreted each day is due to basal stimulation. The remaining 50% is secreted in response to meals. The average insulin secretion rate over a 24-h period for normal persons is shown in Figure 9.4.5A. The average secretion rates tend to blunt random spikes in the individual records because these are not synchronized. Individual records show that insulin is secreted in multiple bursts (Figure 9.4.5B).

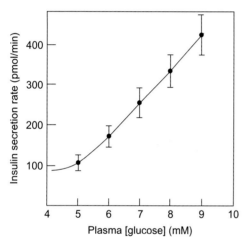

FIGURE 9.4.4 Rate of insulin secretion as a function of plasma glucose concentration. Over a considerable range, insulin secretion is quasi-linear with plasma glucose concentration. (Source: From C.N. Jones, D. Pei, and P. Staris, Alterations in the glucose-stimulated insulin secretory response curve and insulin clearance in nondiabetic insulin-resistant individuals, J. Clin. Endocrinol. Metab. **82**:1834–1838, 1997.)

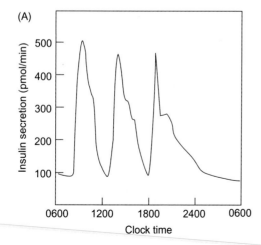
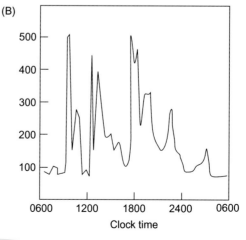

FIGURE 9.4.5 Pulsatile pattern of insulin secretion. Basal secretion amounts to about 100 pmol min^{-1}. This is augmented tremendously by feeding, as shown in the three large peaks from averaged data shown at left **(A)**. In the averaged data, pulsatile patterns of release are averaged out. Individual records of insulin secretion (right, **B**) show multiple peaks of insulin secretion after each meal. (Source: Redrawn from K.S. Polonsky, B.D. Given, and E. van Cauter, Twenty four hour profiles and pulsatile patterns of insulin secretion in normal and obese subjects, J. Clin. Invest. **81**:442–448, 1988.)

This pattern of pulsatile release is also seen in isolated islets of Langerhans from experimental animals. Stepwise increases in the glucose concentration cause a brief, pulse-like insulin release, followed by a longer, more sustained period of lesser release (see Figure 9.4.6). Analysis of repetitive stimulations suggests that the β cells contain about 10,000 vesicles full of insulin, but only a small fraction, some 0.5%, is in a **readily releasable pool** (RRP) that is adjacent to the voltage-dependent Ca^{2+} channel. Another 1000 or so vesicles are morphologically docked with the membrane but are not readily releasable. They constitute the **immediately releasable pool** (IRP). These undergo "priming" to convert the vesicles into readily releasable vesicles.

INSULIN PHOSPHORYLATES INSULIN RECEPTOR SUBSTRATES VIA A TYROSINE KINASE

Insulin binds to the extracellular surface of its receptor, which is a $\alpha_2\beta_2$ dimer, even in the absence of insulin. Insulin can bind to both halves of this dimer, activating its intrinsic tyrosine kinase activity. The activated tyrosine kinase phosphorylates the intracellular parts of the β chain of the receptor and a set of proteins called **insulin receptor substrates**, or **IRS**, consisting of IRS-1, IRS-2, IRS-3, and IRS-4.

Phosphorylation of these IRS proteins exposes a docking surface for other proteins, including **phosphatidyl inositol-3-kinase**, which makes phosphatidyl inositol-3,4,5-trisphosphate, or PIP3. The PIP3 activates **phosphoinositide-dependent protein kinase**, PDK1 and PDK2. This activates **protein kinase B** and isoforms of **protein kinase C** (PKC) that go on to phosphorylate target proteins. Some of these targets are involved with apoptosis, or programmed cell death. Phosphorylation of these by PKB inactivates the "death" proteins and activates death-inhibitory proteins. Insulin activates another pathway through **Grb-2** leading to a cascade resulting in the activation of **MAP kinase**, **m**itogen **a**ctivated **p**rotein **k**inase. MAP kinase phosphorylates components of a gene regulatory complex that alters gene expression. Thus, insulin promotes survival and growth of cells (see Figure 9.4.7). The overall physiological actions of insulin are summarized in Table 9.4.1.

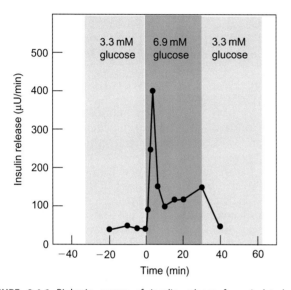

FIGURE 9.4.6 Biphasic course of insulin release from isolated rat pancreatic islets in response to a step change in glucose concentration. Insulin release shows a prompt phase lasting less than 10 min, followed by a slower and more sustained phase. *(Source: From R. Nesher and E. Cerasi, Modeling phasic insulin release. Immediate and time-dependent effects of glucose,* Diabetes **51**: S53–S59, 2002.)

LOW GLUCOSE STIMULATES GLUCAGON RELEASE FROM α CELLS IN THE ISLETS OF LANGERHANS

The α cells make up 25% of the islets of Langerhans and secrete glucagon, a 29-amino-acid peptide, in response to hypoglycemia. The α cells synthesize proglucagon which is then cleaved by a proprotein convertase to

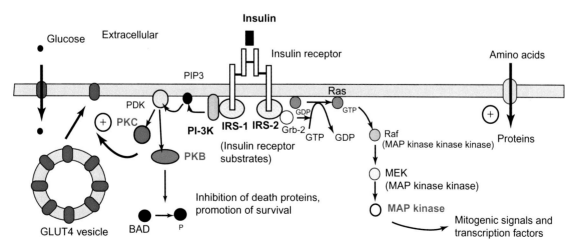

FIGURE 9.4.7 Mechanism of action of insulin on target cells. Insulin binds to both halves of a homodimer with intrinsic tyrosine kinase activity, activating phosphorylation of insulin receptor substrates. These activate PI-3K and Grb-2. PI-3K forms PIP3 in the membranes, which activates PDK1 and PDK2, which in turn activate PKC and PKB. PKB phosphorylates another protein, BAD, that inhibits BAD's ill effects. A second signaling pathway occurs through Grb-2, growth factor receptor binding protein. Grb-2 catalyzes the exchange of GTP for GDP on Ras, which then initiates a phosphorylation cascade through Raf and MEK, finally activating MAP kinase. MAP kinase signals the nucleus to promote growth and survival.

TABLE 9.4.1 Overview of the Physiological Actions of Insulin

1 Effects on Carbohydrate Metabolism
1.1 Insulin recruits the GLUT4 transporter to the surface membrane
1.2 Insulin increases glucose uptake by adipose and muscle tissue
1.3 By increasing peripheral uptake, insulin reduces plasma [glucose]
1.4 Insulin increases glucokinase activity in the liver, so [glucose-6-P] increases
1.5 Insulin stimulates glycogen synthase so that glycogenesis in liver and muscle increases
1.6 Insulin decreases glycogenolysis in liver and muscle
1.7 Insulin decreases gluconeogenesis in liver

2 Effects on protein metabolism
2.1 Insulin increases amino acid transport into liver, muscle, and adipose cells
2.2 Insulin increases protein synthesis and decreases protein breakdown

3 Effects on fat metabolism
3.1 Insulin increases uptake of free fatty acids (FFAs) by adipose tissue by stimulating lipoprotein lipase
3.2 Insulin increases lipogenesis
3.3 Insulin decreases lipolysis by inhibiting hormone-sensitive adipose lipase
3.4 Insulin decreases β oxidation of FFA in the liver
3.5 Insulin decreases plasma [FFA]

4 Effects on ions
4.1 Insulin increases cellular uptake of K^+, PO_4^{-3}, and Mg^{2+}
4.2 Insulin stimulates Na^+–K^+-ATPase

5 Overall effects
5.1 Insulin stimulates anabolic reactions of carbohydrates, lipids, and proteins
5.2 All tissues of the body are affected either directly or indirectly

---

form glucagon. Glucagon restores blood glucose levels by mobilizing glucose primarily from the liver. Glucagon also affects adipose tissue.

In addition to low glucose, ingestion of proteins stimulates glucagon secretion and insulin inhibits it. Free fatty acids (FFAs) inhibit glucagon release, and both parasympathetic and sympathetic stimulation increase it. These effects are relatively unimportant compared to the profound stimulation of glucagon release caused by low blood glucose.

GLUCAGON STIMULATES LIVER GLYCOGENOLYSIS THROUGH A G_s AND G_q MECHANISM

Glucagon binds to G-protein-coupled receptors on the surface of hepatocytes. The G_s-coupled receptor activates **adenylyl cyclase**. The G_q-coupled receptor activates **phospholipase C (PLC)**. Both of these actions result in the activation of protein kinases. Increasing cytoplasmic [cAMP] activates **protein kinase A**, and increasing cytoplasmic [Ca^{2+}] through IP3-induced release of Ca^{2+} from the endoplasmic reticulum activates calmodulin-dependent protein kinase (**CAM kinase**). Activation of protein kinase A begins a cascade of events that results in the activation of **phosphorylase**, the enzyme that breaks down glycogen and inhibits glycogen synthase. At the same time, specific enzymes are activated that decrease glycolysis and enhance gluconeogenesis, the formation of glucose from amino acids. The final result is as follows:

- Decreased glycolysis
- Decreased glycogen synthesis
- Increased glycogenolysis
- Increased gluconeogenesis.

The combined effects of glucagon on liver cells are shown diagrammatically in Figure 9.4.8.

BLOOD GLUCOSE IS MAINTAINED BETWEEN 70 AND 110 mg% IN THE FACE OF CONSTANT DEPLETION

The fasting blood glucose level is typically between 70 and 110 mg% (4.0 to 6.0 mM) with an average of about 90 mg% (5 mM). This level is maintained by a balance between influx of glucose into the circulation and efflux out of it. The brain needs a continuous supply of glucose because it requires glucose and cannot make glucose or store more than a few minutes' worth as glycogen. The rate of glucose uptake by the brain depends entirely on the plasma glucose concentration. At rest, the brain accounts for about 60% of the basal glucose utilization. During exercise, muscle glucose consumption increases many fold, and the fraction of glucose used by the brain is greatly reduced. Increased glucose consumption during exercise must be met with increased glucose supply, or blood glucose levels would fall precipitously and cause **hypoglycemia**.

PLASMA GLUCOSE CONCENTRATIONS ARE MAINTAINED BY *ABSORPTION, GLYCOGENOLYSIS,* AND *GLUCONEOGENESIS*

Plasma glucose is derived from three sources:

1. **Absorption** of glucose from the gastrointestinal tract following ingestion of carbohydrates.
2. **Glycogenolysis**, the breakdown of the polymerized storage form of glucose.
3. **Gluoneogenesis**, the formation of glucose from other precursors including lactate, pyruvate, amino acids, and glycerol.

After a meal, carbohydrates in the food are digested to monosaccharides and absorbed into the blood (see Chapter 8.5). The rate of glucose absorption into the circulation can be more than twice the rate of fasting endogenous glucose production. If glucose utilization is not increased, glucose absorption causes blood glucose to rise. The postprandial hyperglycemia increases insulin secretion, which blunts the hyperglycemia due to insulin

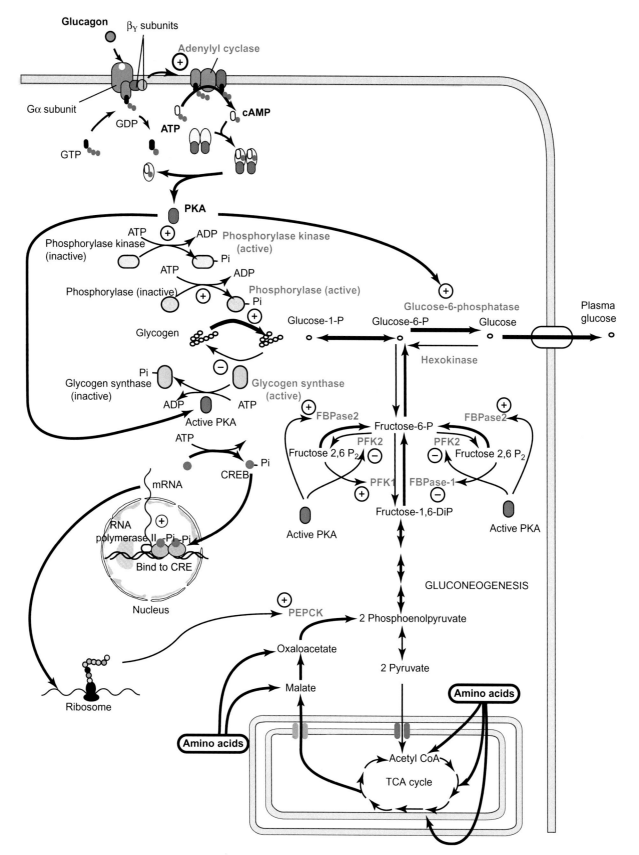

FIGURE 9.4.8 Mechanism of action of glucagon on liver cells to put glucose into the blood. Binding of glucagon to receptors on the surface of the liver cell is coupled through a heterotrimeric G_s protein to the activation of **adenylyl cyclase**, which increases the production of 3′,5′ cyclic AMP. Increased cytosolic [cAMP] activates **PKA**, which phosphorylates **phosphorylase kinase**, thereby activating it. The activated phosphorylase kinase then phosphorylates **phosphorylase**, converting it from its inactive to active form. The activated phosphorylase then phosphorylates **glycogen**, which then can be broken down to release **glucose-1-phosphate**. Phosphoglucomutase converts glucose-1-phosphate to **glucose-6-phosphate**.

stimulation of peripheral glucose uptake by liver, muscle, and fat. In persons with a deficit in insulin secretion, the postprandial glucose peak is larger and longer lasting. This is the basis of the **oral glucose tolerance test** for prediabetic persons, in which 1.75 g of glucose per kilogram of body weight is given orally (or 75 g for an adult) and blood glucose is measured with time.

Most tissues can make glycogen and break it down. The **breakdown of glycogen produces glucose-1-phosphate**, which is isomerized to glucose-6-phosphate by phosphoglucomutase. Thus, glycogenolysis never produces free glucose. Glycolysis of cytoplasmic glucose also begins by forming glucose-6-phosphate from glucose and ATP. **Gluconeogenesis is a reversal of glycolysis to form glucose-6-phosphate**. Thus, glycogenolysis, gluconeogenesis, and glycolysis all funnel through glucose-6-phosphate (see Figure 9.4.8). Tissues that lack glucose-6-phosphatase cannot form free glucose from either glycogen or amino acid precursors. Only the liver, kidney, and intestine possess **glucose-6-phosphatase**, so only these tissues can produce cytoplasmic glucose. Glucose-6-phosphate cannot exit the cell, so only these tissues can contribute glucose to the blood, either from glycogenolysis or from gluconeogenesis.

MULTIPLE HORMONES AND NERVES CONTROL GLUCOSE FLUX

Systemic glucose levels are controlled by two competing systems that regulate the influx of glucose into the circulation and the efflux of glucose out of the circulation. Insulin primarily lowers plasma glucose by controlling efflux of glucose from the blood—it controls the GLUT4 receptors on skeletal muscle and fat to increase their uptake of glucose. Glucagon and epinephrine raise plasma glucose by increasing glycogenolysis and gluconeogenesis. Because glucagon is released into the portal circulation, it is believed to affect only the liver under physiological conditions. Epinephrine, on the other hand, affects multiple organ systems. These two control systems prevent excessive hyperglycemia following feeding and hypoglycemia during fasting. Other hormones are also involved in the regulation of metabolism, including growth hormone and cortisol. Both of these defend against hypoglycemia. The response of insulin and glucagon and their effects on liver, muscle, and adipose tissue is shown schematically in Figure 9.4.9.

EXERCISE HAS AN INSULIN-LIKE EFFECT

Glucose carriers can be inserted into the muscle cell membrane independently of insulin when the muscle is repetitively activated. Thus, exercise has an insulin-like effect and muscle activity promotes peripheral glucose uptake at lower levels of insulin. For this reason, diabetic persons must titrate their injected insulin according to their daily carbohydrate intake and taking into account their projected level of exercise. Strenuous or long-lasting exercise requires less insulin.

In the fasting state, glucose influx into the circulation from the intestines stops because there is no more food in the intestines. Constant drains on blood glucose by the brain and other metabolizing tissues tend to reduce blood glucose concentration. This shuts off insulin secretion and turns on glucagon secretion to reduce plasma glucose efflux from the circulation and to increase glucose influx. Muscles use their glycogen stores and plasma fatty acids instead of plasma glucose, and adipose tissue reduces glucose uptake. Neither of these tissues can provide glucose for the other tissues of the body because they lack glucose-6-phosphatase activity. Glucagon stimulates the liver to form new glucose from amino acids, glycerol, and lactic acid produced by active muscle. Complete regulation of carbohydrate, amino acid, and fat metabolism is considerably more complicated, involving additional actions by epinephrine, growth hormone, and cortisol and involving more metabolic pathways than described here.

SUMMARY

β Cells in the islets of Langerhans in the pancreas secrete insulin in response to hyperglycemia. Insulin is synthesized as a single polypeptide chain that is cleaved repeatedly to form the final insulin and a C peptide. All derive from the single proinsulin. The secreted insulin consists of two chains, A and B, linked by two disulfide bonds. Because the C peptide is degraded more slowly than insulin itself, its level can be used as an indicator of insulin secretion.

Insulin exerts its effects by binding to a homodimer insulin receptor with intrinsic tyrosine kinase activity. The receptor phosphorylates a set of proteins called insulin receptor substrates. Among other actions, these activate a phosphatidyl inositol-3-kinase which makes phosphatidyl inositol-

---

FIGURE 9.4.8 (*Continued*) Glucose-6-phosphatase removes the phosphate from glucose-6-phosphate to produce glucose. This glucose can be released into the bloodstream. Simultaneous with this activation of **glycogenolysis**, the increased PKA also phosphorylates glycogen synthase, inactivating it. This decreases glycogen synthesis. PKA also phosphorylates **CREB**, the **cyclic AMP responsive element binding protein**. This activates its binding to the CRE, cAMP responsive element. Activation of CRE increases the transcription of another transcriptional activator that then turns on the synthesis of **PEPCK (phosphoenolpyruvate carboxy kinase)**. This enzyme converts **oxaloacetate** to **phosphoenolpyruvate**. The oxaloacetate is a common carbohydrate intermediate formed from the glucogenic amino acids. PKA also indirectly regulates a key controlling enzyme in glycolysis: **phosphofructokinase 1/fructose biphosphatase 1 (PFK1/FBPase1)**. PFK1 converts fructose-6-phosphate to fructose-1,6-biphosphate; FBPase1 converts fructose-1,6-biphosphate to fructose-6-phosphate. The FBPase1 activity and PFK1 activities are regulated by cytosolic levels of **fructose-2,6-biphosphate (FBP)**. Fructose-2,6-biphosphate stimulates PFK activity and inhibits FBPase activity. Fructose-2,6-biphosphate levels are determined by the activity of **phosphofructo kinase 2** and **fructose-2,6-biphosphatase (FBPase2)** that convert fructose-6-phosphate to fructose-2,6-biphosphate. The activities of PFK2/FBPase2 reside on a single polypeptide chain. PKA phosphorylates PFK2/FBPase2, stimulating the FBPase2 activity and inhibiting the PFK2 activity. This reduces the level of fructose-2,6-biphosphate, which subsequently removes activation of PFK1 and removes inhibition of FBPase1. The net result is an inhibition of PFK1, which thereby slows glycolysis, and activation of FBPase1, which increases gluconeogenesis.

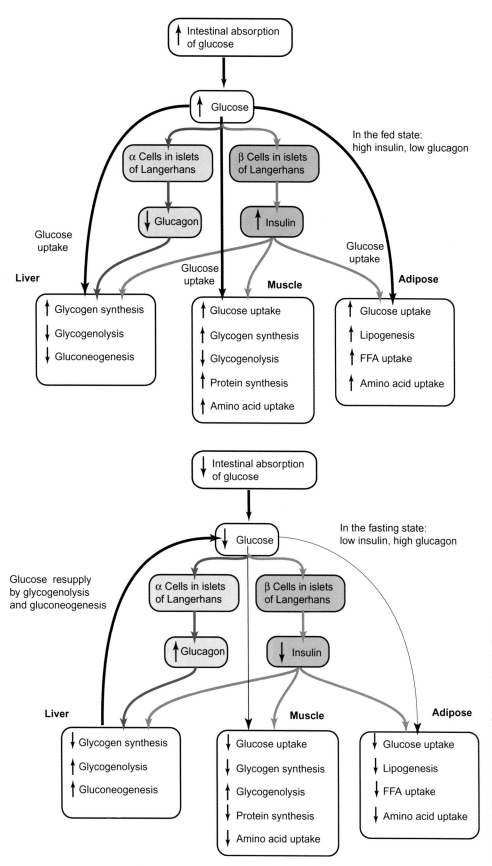

FIGURE 9.4.9 Regulation of plasma glucose concentration during the fed state (top) and the fasting state (bottom). During the fed state, absorption of glucose from the intestine and metabolism of amino acids to glucose produce a prolonged elevation of plasma glucose concentration. This suppresses glucagon secretion and stimulates insulin secretion. Glucagon mainly affects the liver. The withdrawal of glucagon and stimulation by insulin increases liver glycogen synthesis and protein synthesis and reduces glycogenolysis and gluconeogenesis. Insulin increases glucose uptake in both muscle and adipose tissue and the glucose taken up is stored either as glycogen in muscle or triglycerides in adipose tissue.

Clinical Applications: Diabetes Mellitus

The descriptive term "diabetes mellitus" derives from the Greek meaning "to siphon" ("diabetes") and "sweet" ("mellitus"). These terms describe the large volume of urine produce by persons with diabetes mellitus and the sweet taste of the urine. The hallmark of the disease is high plasma glucose concentrations that exceed the plasma threshold of the kidney. The filtered load of glucose is too great for the kidney to reabsorb it all, and the glucose that exceeds the transport maximum is carried through to the final urine. The glucose in the distal nephron exerts an osmotic pressure that draws water in from the surrounding kidney interstitium, thereby causing a large volume of urine with high concentrations of glucose. All of this wasted glucose represents a metabolic drain, and even well-hydrated persons gradually waste away.

There are two major types of diabetes mellitus. Both are characterized by high plasma glucose concentrations. The elevated plasma glucose results from a relative or absolute deficiency in insulin action. This is due to either (1) deficient secretion of insulin by the β cells of the islets of Langerhans or (2) a deficient response of the target cells to adequate insulin levels. Deficient secretion of insulin is called **insulin-dependent diabetes mellitus (IDDM)** or **Type 1 diabetes**. Reduced sensitivity to insulin is called **noninsulin-dependent diabetes mellitus (NIDDM)** or **Type 2 diabetes**.

Type I diabetes arises from destruction of the β cells in the pancreas, usually from autoimmune processes. The etiology of the disease is not entirely clear. Some individuals appear to have a genetic predisposition to the disease, which is brought about, generally in adolescence, by environmental components such as viral infections. The affected individuals are diagnosed from their polyuria and hyperglycemia. Because they lack insulin, persons with Type I diabetes cannot take up adequate glucose into muscle and adipose tissue, and so these tissues experience starvation in the midst of plenty. As a result, these tissues rely on fat metabolism and produce large quantities of **ketone bodies: acetoacetic acid, β-hydroxybutyric acid,** and **acetone**. These ketone bodies are acids, and the resulting condition is called **ketoacidosis**. These ketone bodies can exceed their renal threshold and spill over into the urine. Because they are acids, they draw with them cations such as Na^+ and K^+. If left untreated, the resulting acidosis and electrolyte imbalance eventually causes coma and death.

As little as 80 years ago Type I diabetes was uniformly fatal. Now afflicted persons can be kept healthy by injecting insulin. However, this form of diabetes makes up just 5% of the total cases in the United States. Some 95% of diabetics have Type II diabetes. In this disease, the afflicted person has normal or higher than normal concentrations of insulin, but the target tissues do not respond normally. This condition is called **insulin resistance**. The condition is related to obesity; losing excess body fat often reduces the severity of the disease and can return the person to normal.

Most diabetics are managed through a program of modified diet, insulin injections for IDDM or oral hypoglycemic agents for NIDDM, and exercise. Exercise increases peripheral uptake of glucose, an insulin-like effect. Thus, diabetics taking insulin injections adjust their dose depending on the amount of exercise they have taken.

The degree of control of a diabetic is measured by the long-term average glucose concentrations. Persons with poor control have widely varying levels of plasma glucose that generally average higher than normal. Glucose forms a glycosylated hemoglobin, HbA_{1C}, whose value reflects the long-term average glucose concentration. Its upper limit of normal is 6%. Long-term control of blood glucose levels is important because diabetes is associated with a number of long-term changes in the basement membranes of capillaries. Microvasculature lesions can lead to **diabetic retinopathy** and **diabetic nephropathy**, in which the microvasculature nourishing the retina degrades, causing blindness, or the microvasculature in the kidney fails to nourish the kidney. Long-standing diabetics also develop circulatory problems in the extremities that can result in amputation.

3,4,5-triphosphate, or PIP3, which subsequently activates a phosphoinositide-dependent protein kinase, PDK. This activates PKB and PKC. One of the consequences of the phosphorylation of their targets is the insertion into the plasma membrane of more copies of the glucose transport protein GLUT4. This increases glucose uptake by tissues that have insulin receptors, which applies to most peripheral tissues. Insulin also has profound effects on lipid metabolism by activating lipoprotein lipase and inhibiting the hormone-sensitive adipose lipase. The result is decreased lipolysis and increased lipogenesis.

α Cells in the islets of Langerhans secrete glucagon in response to hypoglycemia. Glucagon is a single polypeptide chain of 29 amino acids. It raises blood glucose by increasing glycogenolysis and gluconeogenesis. In hepatocytes, glucagon acts through a G_s and G_q mechanism. PKA phosphorylates a crucial enzyme, phosphofructokinase 2/fructose-2,6-biphosphatase. Both of these activities reside on a single polypeptide chain. Phosphorylation inhibits the PFK activity and activates the FBPase2 activity. The consequence is a reduction in fructose-2,6-biphosphate. Fructose-2,6-biphosphate stimulates PFK activity and inhibits FBPase activity, so its reduction lowers the PFK activity (necessary for glycolysis) and increases FBPase (necessary for gluconeogenesis). In this way, glucagon enhances gluconeogenesis.

In the fed state, glucose and other nutrients enter the bloodstream and increase plasma glucose concentration. Consequently, postprandial levels of glucagon are low and insulin levels are high. Peripheral tissues take up glucose, and make glycogen, proteins, and lipids. In the fasting state, plasma glucose levels fall and glucagon rises while insulin falls. This situation is marked by glycogenolysis, gluconeogenesis, and lipolysis.

REVIEW QUESTIONS

1. What is insulin? Where is it produced? What cells secrete it? What stimulates its release?
2. How is insulin secretion stimulated? How do oral hypoglycemic agents work?
3. What effects does insulin have on muscle? On liver? Adipose tissue? Can you consider insulin to be mainly a hormone of carbohydrate metabolism?
4. What is insulin's mechanism of action?
5. What is glucagon? Where is it produced? What cells secrete it? What stimulates its release?
6. What effects does glucagon have on liver? Does it affect muscle or adipose tissue?
7. What is glucagon's mechanism of action?

9.5 The Adrenal Cortex

Learning Objectives

- Describe the anatomical location of the adrenal glands
- List the three major histological divisions of the adrenal cortex and the major hormones secreted by each part
- Draw the steroid nucleus
- Distinguish between the structures of DHEA, cortisol, and cortisone
- Identify ACTH by name and indicate the location of cells that secrete it
- Define CRH and indicate the location of cells that secrete it
- Define POMC and name the peptide hormones that derive from it
- Describe the mechanism by which ACTH increases cortical steroid secretion
- Describe the mechanism of the anti-inflammatory effect of cortisol
- List the major secretagogues for aldosterone
- Describe the negative feedback loops involved in the renin–angiotensin–aldosterone system
- Describe the major features of aldosterone action on the distal nephron

THE ADRENAL GLANDS LIE ATOP THE KIDNEYS, ARE RICHLY VASCULARIZED, AND SECRETE MANY HORMONES

In 1856, Brown-Sequard performed adrenalectomies in dogs, cats, and guinea pigs and showed that the procedure was uniformly fatal. He declared that the adrenals were "organs essential for life." Each of two adrenal glands sits on top of the kidneys along their posteromedial aspect. The glands are roughly pyramidal, weighing about 4 g apiece and having a rich blood supply. Each gland consists of two main divisions, the **cortex** and the **medulla**, which together secrete four main classes of hormones. The cortex produces:

- glucocorticoids (cortisol, corticosterone);
- mineralocorticoids (aldosterone, deoxycorticosterone);
- sex hormones (androgens);

whereas the medulla produces

- catecholamines (epinephrine and norepinephrine).

These hormones are essential for the regulation of metabolism including blood glucose, protein turnover and fat metabolism, regulation of blood volume and pressure through Na^+ balance, tissue response to infection or injury, and the whole body response to stress.

The whole gland is covered by a connective tissue **capsule**. The outer zone, the **cortex**, comprises 80–90% of the weight of the gland. The innermost zone, the **medulla**, makes up the remainder. The cortex consists of an outermost area, the **zona glomerulosa**, so named because the cells form into tiny balls. This layer is only a few cells thick. The **zona fasciculata** lies below the zona glomerulosa; it consists of long cords of polyhedral cells running radially out toward the zona glomerulosa. The innermost layer of the cortex is the **zona reticularis**. The zona reticularis forms a more branching network so that the radial arrangement of the cords is less obvious (see Figure 9.5.1).

The zona glomerulosa produces the **mineralocorticoids** because only this part of the adrenal gland expresses the enzymes required for their synthesis. It secretes some 100–150 μg/day. Similarly, the zona fasciculata produces some 10–20 mg of glucocorticoids each day. The androgen steroid **dehydroepiandrosterone (DHEA)** is sulfated to form DHEAs only in the zona reticularis. The adult adrenal secretes more than 20 mg/day of DHEA and DHEAs (dehydroepiandrosterone sulfate).

STEROID HORMONES DERIVE FROM CHOLESTEROL

All adrenal steroidogenesis begins with cholesterol. The cholesterol has two origins: (1) uptake from low-density lipoproteins (LDL) by specific LDL receptors on the surfaces of adrenal gland cells and (2) *de novo* synthesis of cholesterol within the adrenal cortex from acetyl CoA. The structure of cholesterol is shown in Figure 9.5.2. Cholesterol has a four ring structure called the **steroid nucleus** that is common to all steroid hormones. These steroid hormones possess a wide variety of activities that are due to surprisingly small variations in their structure. The classification of the steroid hormones is based not only on their activities but also on their structure, as shown in Figure 9.5.2.

The synthesis of the adrenal cortex steroid hormones begins with the import of cholesterol into the mitochondria. This transport is mediated by StAR, for

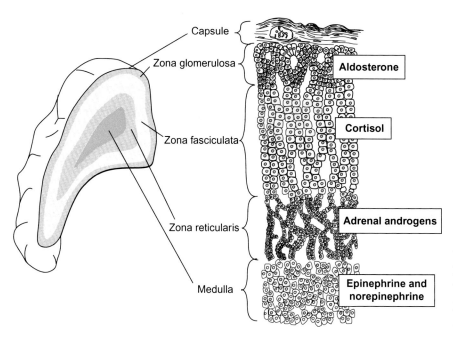

FIGURE 9.5.1 Appearance of the adrenal gland. A cross section of the adrenal gland reveals several macroscopically distinct regions: the capsule, cortex, and medulla. The cortex is further differentiated into three layers: the zona glomerulosa, the zona fasciculata, and the zona reticularis, from outer to inner cortex. The zona glomerulosa secretes predominantly aldosterone, a mineralocorticoid. The zonae fasciculata and reticularis secrete glucocorticoids and androgens. The medulla secretes epinephrine.

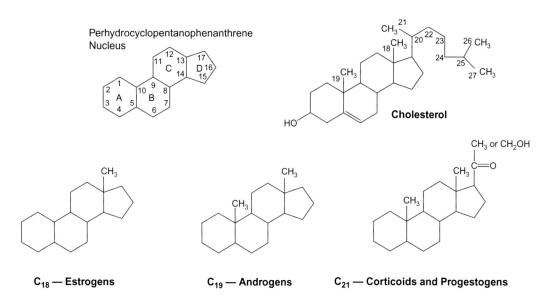

FIGURE 9.5.2 Chemical structures of cholesterol and the three major structural groups of steroid hormones. All of these have the classic steroid nucleus consisting of three six-carbon rings (perhydrophenanthrene) and another five-carbon ring (cyclopentano).

steroidogenic acute regulatory protein, a 30 kDa protein. The activity of StAR is regulated by cAMP levels that are controlled in these cells by **ACTH, adrenocorticotrophic hormone**. Steroidogenesis begins in the mitochondria, but some crucial enzymes lie in the endoplasmic reticulum. The relevant biochemical pathways are shown in Figure 9.5.3.

THE PITUITARY–HYPOTHALAMUS AXIS CONTROLS ADRENAL FUNCTION THROUGH ACTH

As described in Chapter 9.2, corticotrophs in the anterior pituitary secrete ACTH in response to corticotropin-releasing hormone (CRH) secreted by cells in the paraventricular nucleus of the hypothalamus (see Figure 9.5.4). CRH is a 41-amino-acid peptide that is secreted into the hypophyseal portal circulation in response to a variety of stimulati and inputs including: circadian rhythms, hypoglycemia, surgery, fever, and injury. CRH stimulates the release of ACTH from corticotrophs through a G_s-coupled mechanism.

Corticotrophs make proopiomelanocortin, POMC, as their principal secretory protein. This protein has 241 amino acids and contains within it several other smaller peptide hormones. Tissue-specific proteolytic cleavage releases these other peptides. POMC is cleaved within secretory granules and therefore stimulation of secretion releases ACTH and POMC's other cleaved product, β **lipotropin** (βLPH). βLPH has effects on lipid metabolism,

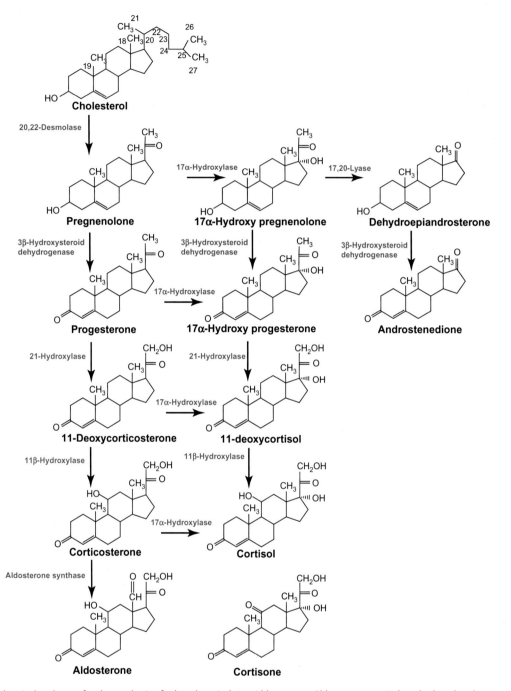

FIGURE 9.5.3 Biochemical pathways for the synthesis of adrenal cortical steroid hormones. Aldosterone, cortisol, and adrenal androgens are synthesized in specific zones of the cortex in a coordinated series of reactions. Each zone of the cortex contains its own set of steroidogenic enzymes.

but its effects in humans are not established. Figure 9.5.5 shows the origin of POMC and the peptide hormones it generates after proteolytic cleavage.

Human skin contains melanocytes that produce melanin, which determines the color of the skin. These cells respond to melanophore stimulating hormone, MSH. In lower animals, the pars intermedia or the pituitary gland secretes MSH. The pars intermedia is the intermediate lobe of the pituitary gland, between the anterior and posterior pituitary. In humans, the pars intermedia is vestigial and produces insignificant amounts of MSH. Persons with Addison's disease (primary adrenal insufficiency) have high levels of ACTH and they develop increased skin pigmentation that is thought to arise from ACTH stimulation of the melanocortin receptor on melanocytes. Peptide sequences homologous to MSH can be found in ACTH and βLPH.

ACTH INCREASES ADRENAL CORTICAL STEROID SECRETION

ACTH binds to a G_s-coupled melanocortin-2 receptor on adrenocortical cells. Binding is followed principally by increases in cytoplasmic [cAMP], although intracellular $[Ca^{2+}]$ also plays a role. ACTH stimulation results in short-

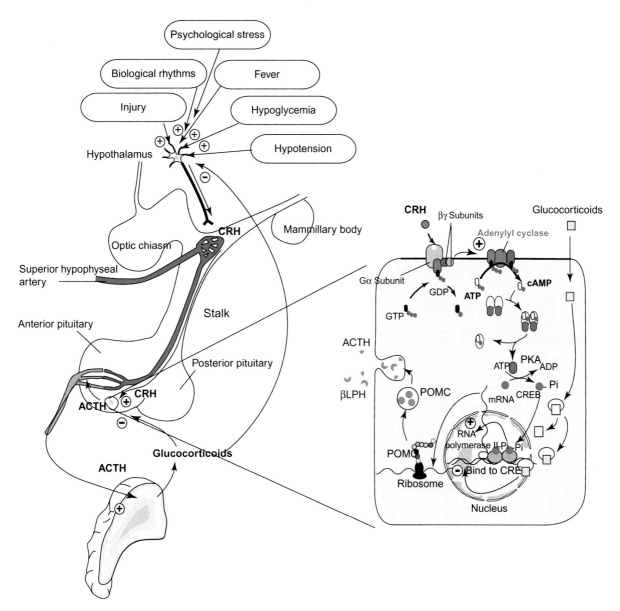

FIGURE 9.5.4 Control of adrenal cortical hormone secretion. Neurons in the paraventricular nucleus of the hypothalamus release CRH into the hypophyseal portal circulation in response to a variety of stimuli. CRH activates G_s-coupled receptors on corticotrophs in the anterior pituitary. The increased [cAMP] stimulates synthesis of POMC (proopiomelanocortin, a precursor of ACTH) probably through a CREB (cyclic AMP response element binding protein) mechanism. POMC is cleaved within granules and both ACTH and βLPH (lipotropin) are released. Glucocorticoids provide negative feedback by inhibition of CRH secretion in the hypothalamus and by inhibition of ACTH secretion.

term and long-term effects. Acutely, **StAR (steroidogenesis acute regulatory protein)** increases cholesterol delivery to the cholesterol side chain cleavage enzyme located in the inner mitochondria. Increasing this rate-limiting step increases steroid hormone synthesis. Prolonged exposure to ACTH increases the transcription of genes that produce all of the enzymes for steroid hormone synthesis: cholesterol side chain cleavage enzyme (also known as CYP11A1), 17α-hydroxylase (CYP17), 21-hydroxylase (CYP21A2), and 11β-hydroxylase (CYP11B1). ACTH also increases the number of surface receptors for LDL. In addition to these effects, ACTH also exerts a general trophic effect on the adrenal gland, increasing its weight due to both **hyperplasia** and **hypertrophy**. These effects are shown diagrammatically in Figure 9.5.6.

CORTISOL BINDING PROTEIN CARRIES GLUCOCORTICOIDS IN BLOOD

An $α_2$ globulin called **cortisol binding globulin (CBG)** binds about 90% of the total circulating cortisol. CBG is synthesized and secreted by the liver and binds cortisol with high affinity. Levels of CBG are increased by estrogen and reduced by glucocorticoids. Because of this, pregnant women may have elevated total cortisol levels even though the free cortisol concentration remains unchanged. Only the free hormone can diffuse into the tissues to bind to target cells and alter their behavior. As free hormone is taken up by peripheral tissues, the bound hormone dissociates from its carrier to take the place of the hormone that has left.

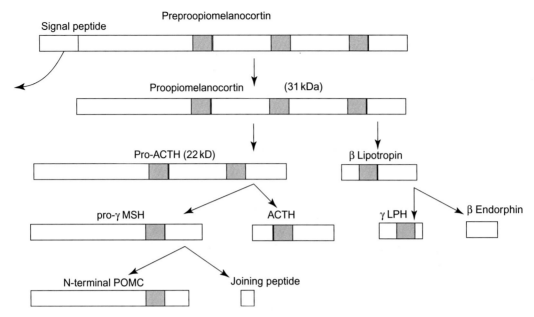

FIGURE 9.5.5 Relationship between POMC and its various cleavage products. Corticotrophs in the anterior pituitary synthesize prePOMC. This contains a signal sequence to get the protein into the secretory granules. Cleavage of the signal sequence produces POMC, proopiomelanocortin. Further cleavage results in the formation of ACTH and β lipotropin, which are the principal secretory products of the corticotrophs. Further cleavage can produce γMSH and β endorphin. β Endorphins are endogenous ligands for opioid receptors, from which POMC partly derives its name. Shaded area represents regions of MSH structural units.

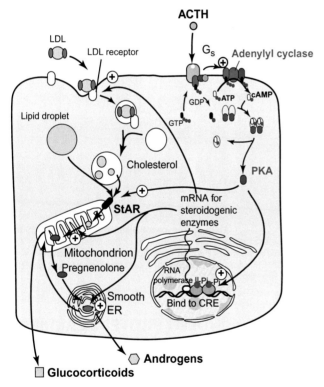

FIGURE 9.5.6 Stimulation of steroidogenesis by ACTH in adrenocortical cells. ACTH acutely increases the activity of StAR, a 30-kDa protein that imports cholesterol into mitochondria. The cholesterol originates from uptake of cholesteryl esters from low-density lipoproteins (LDL) from the blood, or from cholesterol in lipid stores, or from cholesterol newly synthesized on the smooth endoplasmic reticulum. Cholesterol is converted to pregnenolone within the mitochondria by cholesterol side chain cleavage enzyme (CYP11A1). Steroid conversions continue in the smooth endoplasmic reticulum to generate glucocorticoids and androgens. ACTH stimulates steroidogenesis by increasing the number of LDL receptors on the surface of the cell, by increasing cholesterol transport into the inner mitochondria, and by increasing the amounts of several enzymes in the steroid synthesis pathway, including CYP11A1, CYP17, CYP21A2, and CYP11B1. ACTH also stimulates the growth and division of adrenocortical cells.

The circulating half-life of cortisol is about 70–120 min. Corticoid hormones are excreted by the kidney after structural modification destroys their hormone activity and increases their water solubility.

CORTISOL AFFECTS TARGET CELLS THROUGH REGULATION OF TRANSCRIPTION

Glucocorticoids exert their effects on target cells by binding to receptors in the cytosol, called glucocorticoid receptors, or GR. In the unstimulated state, the GR associate with **heat shock proteins**, **HSP70** or **HSP90**. Upon binding of steroid hormone to GR, HSP70 or HSP90 dissociates from the complex, and the GR–steroid complex migrates to the nucleus. The GR–steroid complex dimerizes and then binds to specific sequences of nucleotides in the promoter regions of target genes. These specific nucleotide sequences that bind the GR–steroid complex are **glucocorticoid-response elements** (GREs). The complex begins transactivation by recruiting proteins with histone acetyl transferase (HAT) activity. Acetylation of the histones facilitates their unwinding and exposure of regions of the DNA for transcription. The complex stabilizes RNA polymerase II, and transcription is initiated. The target genes differ among the many cell types that are affected by the glucocorticoids. For example, glucocorticoids reduce the expression of TNF-α in macrophages but decrease the expression of osteocalcin in bone cells.

In some cells, glucocorticoids exert a negative effect. For example, cortisol possesses potent anti-inflammatory activity. The likely mechanism for this effect in macrophages is described in Figure 9.5.7. Glucocorticoids also exhibit anti-inflammatory effects in other cell types. For example, glucocorticoids inhibit

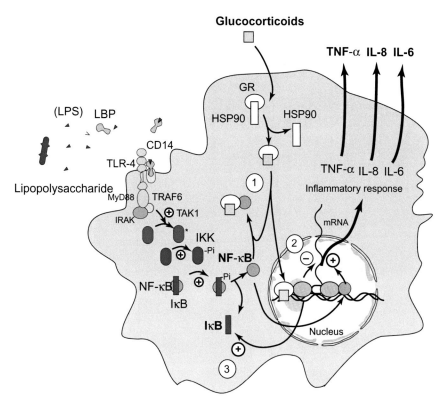

FIGURE 9.5.7 Mechanism of the anti-inflammatory action of the glucocorticoids. Inflammation in this macrophage is brought about by the binding of a bacterial product, lipopolysaccharide or LPS, to plasma LPS binding protein (LBP). The complex of LPS–LBP then binds to a membrane receptor on the surface of the macrophage, CD14, which is linked to another protein called TLR-4, for toll-like receptor 4. This is linked by another complex of proteins to a MAP kinase kinase kinase called TAK1 that phosphorylates and activates an IκB kinase (IKK) that phosphorylates IκB. IκB in its unphosphorylated state inactivates nuclear factor κB (NF-κB), and the phosphorylation removes this inactivation. NF-κB travels to the nucleus where it activates the genes that cause the release of inflammatory cytokines and chemokines, TNF-α, IL-8, IL-6. Glucocorticoids enter the cell passively by diffusion (they are lipid soluble and sparingly water soluble) and bind to a cytoplasmic receptor (glucocorticoid receptor, GR) that is complexed with heat shock protein 90 (HSP90). Binding of glucocorticoid displaces HSP90. The GR–glucocorticoid complex then interferes with inflammation in three separate ways. First, the GR–glucocorticoid complex binds NF-κB so that it cannot exert its stimulatory effect in the nucleus (1). Second, the GR–glucocorticoid complex binds to its own DNA sequences and competes with NF-κB for available coactivators (2). Third, GR–glucocorticoids induce the synthesis of IκB that binds and inactivates NF-κB (3).

transmigration of leukocytes. Glucocorticoids induce the expression of **annexin 1** (formerly called lipocortin), a protein that binds Ca^{2+} and phospholipids and translocates from cytoplasm to cell surface in neutrophils when they adhere to the endothelial cell wall. Annexin 1 causes detachment of the neutrophils, and thereby it inhibits neutrophil transmigration.

CORTISOL AFFECTS MANY BODY FUNCTIONS

The glucocorticoids have many effects on multiple organ systems, and the multitude of these effects is summarized briefly in Table 9.5.1. Cortisol affects carbohydrate, protein, and lipid metabolism and also muscle, skeleton, kidney, and the immune system.

THE ZONA GLOMERULOSA MAKES ALDOSTERONE IN RESPONSE TO ANGIOTENSIN II, ACTH AND K^+

Aldosterone is a mineralocorticoid synthesized and secreted by cells in the zona glomerulosa. Its structure is shown in Figures 9.5.3 and 7.6.9. It is called a mineralocorticoid because its principal actions are on the excretion of minerals, particularly Na^+ and K^+, by various epithelial cells, including the **distal nephron**, **colon**, **salivary glands**, and **sweat glands**. Its synthesis and secretion are controlled mainly by angiotensin II, but plasma $[K^+]$ and ACTH also directly stimulate aldosterone levels.

Granule cells that line the afferent arteriole in the juxtaglomerular apparatus of the kidney secrete **renin** into the blood in response to:

- decreased afferent arteriolar pressure;
- renal sympathetic nerve stimulation;
- decreased distal tubule load of NaCl.

Renin is an enzyme that cleaves circulating **angiotensinogen** to form **angiotensin I**, a 10-amino-acid-long peptide. The liver makes angiotensinogen. Another enzyme, **angiotensin converting enzyme**, or **ACE**, converts angiotensin I to **angiotensin II**, an eight-amino-acid peptide. The lungs contain most ACE activity, but it is also found in the kidney.

ANGIOTENSIN II EXERTS MULTIPLE EFFECTS

Angiotensin II formed by renin action on angiotensinogen and ACE conversion, acts on several tissues. It causes the following:

- Vasoconstriction
- The sensation of thirst
- Release of antidiuretic hormone (ADH)
- Release of aldosterone
- Increased reabsorption of Na^+ and HCO_3^- in the proximal tubule.

All of these effects of angiotensin II work to increase the circulatory volume and maintain circulatory pressure. It motivates us to drink more water while it simultaneously increases the retention of extracellular ions, mainly Na^+.

TABLE 9.5.1 Overview of the Physiological Actions of the Glucocorticoids

1	**Effects on carbohydrate metabolism**	
	1.1	Glucocorticoids increase liver gluconeogenesis
	1.2	Glucocorticoids promote protein and fat catabolism, providing amino acids and glycerol for gluconeogenesis
	1.3	Glucocorticoids inhibit glucose utilization by muscle and adipose tissue
	1.4	Glucocorticoids enhance glycogen stores in liver
	1.5	Glucocorticoids raise blood glucose levels
2	**Effects on protein metabolism**	
	2.1	Glucocorticoids accelerate protein catabolism
	2.2	Glucocorticoids increase amino acid uptake and protein synthesis in the liver
	2.3	Long-term effects of glucocorticoids are "protein wasting"
3	**Effects on fat metabolism**	
	3.1	Glucocorticoids increase lipolysis
	3.2	Glucocorticoids increase plasma levels of free fatty acids
4	**Effects on muscle tissue**	
	4.1	Basal levels of glucocorticoids are essential for normal muscle contractility and performance
	4.2	Excessive glucocorticoids cause muscle wasting and atrophy
5	**Effects on the skeleton**	
	5.1	Glucocorticoids decrease Ca^{2+} absorption from the intestines
	5.2	Glucocorticoids decrease bone formation
	5.3	Glucocorticoids increase bone resorption
6	**Effects on the kidneys**	
	6.1	Glucocorticoids increase the GFR
	6.2	Glucocorticoids increase the free water clearance
	6.3	Glucocorticoids inhibit ADH secretion and action
7	**Effects on the immune system**	
	7.1	Glucocorticoids induce annexin I that inhibits neutrophill transmigration
	7.2	Glucocorticoids inhibit NF-κB
	7.3	Glucocorticoids decrease production of cytokines (IL-1, IL-6, and TNF-α)

The overall negative feedback loops are shown diagrammatically in Figure 9.5.8.

Angiotensin II binds to AT_1 or AT_2 receptors on the surface of its target cells. Most of its effects are mediated by AT_{1A} receptors except in the adrenal gland where increase of aldosterone secretion appears to be mediated by AT_{1B} receptors. The AT_{1B} receptors couple to G_q-coupled receptors. Activation of CAM kinase (through increased cytosolic [Ca^{2+}] through IP3-induced release of ER stores) appears to be responsible for the early increase in aldosterone secretion in response to AII. DAG released by phospholipase C activates protein kinase C and appears to be responsible for more sustained secretion of aldosterone.

ALDOSTERONE INCREASES Na^+ REABSORPTION AND K^+ SECRETION BY GENOMIC AND NONGENOMIC MECHANISMS

The primary renal target of aldosterone is the **principal cells in the cortical collecting duct**. The mechanism of Na^+ reabsorption and K^+ in these cells is shown diagrammatically in Figure 9.5.9. Aldosterone increases Na^+ reabsorption by increasing the activity of apical membrane **epithelial Na channel (ENaC)**, and an **apical K^+ channel**, which increases Na^+ reabsorption and K^+ secretion. The effects of aldosterone on these activities are indirect. Aldosterone exerts these effects by altering genomic expression of another set of proteins that include **SGK1** (for serum and glucocorticoid-inducible kinase) and **K-RasA**. The increased expression of these proteins is accomplished by classical genomic mechanisms for the steroid hormones.

Aldosterone first binds to its cytosolic mineralocorticoid receptor, MR. This receptor has similar affinities for aldosterone and cortisol, but the circulating levels of cortisol are 100–1000 times those of aldosterone. Why are not the mineralocorticoid receptors continuously activated by cortisol? Those tissues that express MR also express 11β-hydroxysteroid dehydrogenase, which has been proposed to protect the MR from being occupied by glucocorticoids by metabolizing the glucocorticoids to products with much lower affinity for the MR.

SGK1 increases ENaC activity by reducing its interaction with another protein, **Nedd4-2**, which appears to tag ENaC with ubiquitin, a 76-amino-acid protein that is used to tag proteins for degradation. SGK1 phosphorylates Nedd4-2, blocking its ubiquitination of ENaC. By decreasing its degradation, SGK1 increases the membrane density of this channel. In addition to increasing the density of epithelial Na^+ channels, aldosterone may alter the activity of these channels indirectly through increasing expression of K-RasA. Aldosterone-induced methylation activates ENaC activity, but the effect may be mediated by other proteins in a cascade of activation. This complicated pathway is under active investigation.

SUMMARY

An adrenal gland sits atop each kidney. These small glands are encapsulated and contain a cortex and a medulla. The cortex secretes steroid hormones, and the medulla secretes primarily epinephrine. The cortex consists of three regions: the zona glomerulosa is the outermost and it secretes primarily aldosterone; the next layer is the zona fasciculata and it secretes glucocorticoids; the innermost layer of the cortex, the zona reticularis, secretes adrenal androgens.

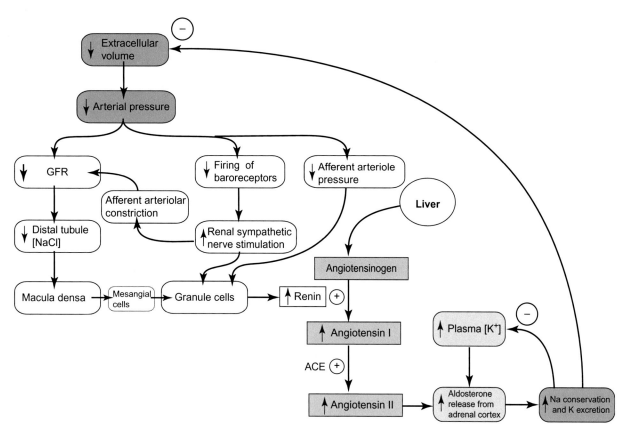

FIGURE 9.5.8 Overall control of aldosterone secretion and its effects on mineral balance. Aldosterone secretion is primarily controlled by angiotensin II levels and plasma [K^+]. Aldosterone's primary effect is retention of Na^+ and excretion of K^+ by the kidney. Since aldosterone secretion is stimulated by increased plasma [K^+], its effect completes a simple negative feedback loop. Similarly, aldosterone secretion is increased by angiotensin II, whose circulating levels are controlled indirectly by the rate of secretion of renin from granule cells in the juxtaglomerular apparatus. The stimuli for renin release include decreased afferent arteriolar pressure, increased renal sympathetic stimulation, and decreased distal tubule [NaCl]. The aldosterone increases Na^+ reabsorption and K^+ secretion. The result is a tendency to retain the extracellular fluid volume. This tends to raise arteriolar pressure toward normal, decrease renal sympathetic nervous stimulation, and return distal tubule NaCl loads back toward normal, completing another set of negative feedback loops.

Clinical Applications: Cushing's Disease and Cushing's Syndrome

Cushing's *syndrome* is caused by an excess of glucocorticoids, regardless of source or cause. Cushing's *disease* refers to those cases that are caused by inappropriately increased pituitary production of ACTH. Both originate in clinical observations of Harvey Cushing, who in 1912 first described a woman with obesity, hirsuitism and amenorrhea, and who in 1932 recognized the syndrome as a primary pituitary abnormality causing adrenal hyperplasia. Causes include exogenous corticosteroid therapy, adrenal tumors, ectopic ACTH or CRH production, and pituitary tumors.

The clinical symptoms of Cushing's syndrome involve multiple organ systems because glucocorticoids regulate metabolism all over. Afflicted persons often have a distinct pattern of centripetal obesity while legs and arms may be normal. Patients develop adipose deposits over the thoracocervical spine, the so-called buffalo hump. Fat buildup over the cheeks and temporal regions of the face produce the rounded, "moon" face. The skin is fragile and thin, and easily bruised. The thin skin is stretched by the underlying fat and produces "purple striae," or stripes of livid red-purple usually found on the abdomen but also occurring on the arms and thighs. Stretch marks caused by birth or rapid weight loss are less pigmented. Skin pigmentation is rare in Cushing's disease but is common in ectopic ACTH syndrome. Cushing's syndrome also impairs glucose intolerance. Overt diabetes mellitus presents in a third of cases. Persons with Cushing's syndrome have high blood pressure.

Cushing's syndrome is also associated with muscle wasting and weakness, osteoporosis, hirsuitism, and reproductive dysfunction. The organic matrix of bone is depleted with particular susceptibility of the vertebral column. Patients may lose height due to vertebral collapse. Hirsuitism is due to overproduction of adrenal androgens. The high circulating glucocorticoids inhibit gonadotropin-releasing hormone pulsatility and subsequent LH and FSH secretion, causing hypogonadism. These effects are reversed upon correction of the hypercortisolism.

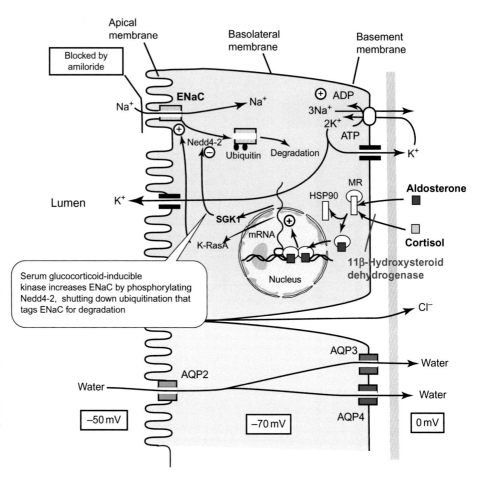

FIGURE 9.5.9 Mechanism of action of mineralocorticoids on principal cells of the cortical collecting duct. Aldosterone binds to mineralocorticoid receptors in the cytosol of the cells. The overall effect of aldosterone is to increase the activity of the epithelial Na channel, ENaC, as well as the apical K channel and basolateral Na,K-ATPase. The signal pathways by which these actions occur are still being actively investigated. The current hypothesis is that ENaC levels are increased by inhibition of degradation by phosphorylation of Nedd4-2, which inhibits its ubiquitination of ENaC, a process that "tags" the protein for degradation by the cell. Synthesis of K-RasA appears to regulate the function of ENaC by activating it by as yet unknown mechanisms. Activity of K-RasA requires methylation.

Clinical Applications: Addison's Disease

Thomas Addison first described **primary hypoadrenalism** in a classical monograph published in 1855. It is a rare condition with a prevalence of some 4–11 affected persons per 100,000. The majority of these cases (about 70%) are caused by autoimmune adrenalitis. The adrenal glands atrophy, with complete loss of the cortex while the medulla remains intact. In primary adrenal failure, the patient usually has glucocorticoid and mineralocorticoid deficiency. Secondary hypoadrenalism is caused by a lack of ACTH, and afflicted persons have glucocorticoid deficiency but generally have an intact renin–angiotensin–aldosterone system. This presence of adequate aldosterone in secondary hypoadrenalism accounts for the difference in clinical presentation between primary and secondary hypoadrenalism.

The most visible distinction between primary and secondary hypoadrenalism is the presence of excess skin pigmentation in primary hypoadrenalism due to high ACTH secretion because there is little negative feedback from blood levels of glucocorticoids. Increased stimulation of melanocortin-2 receptors by the high circulating ACTH levels causes the pigmentation. In persons with secondary hypoadrenalism, low ACTH causes their adrenal insufficiency.

The clinical features of adrenal insufficiency usually develop slowly and diagnosis is made when the patient is stressed with some other illness. Patients may present themselves with an acute adrenal crisis, with dehydration, hypotension, and imminent circulatory collapse. They have unexplained hypoglycemia, hyponatremia, hyperkalemia, nausea, vomiting, diarrhea, and sometimes abdominal pain. This adrenal crisis needs prompt attention. Patients that do not present with an adrenal crisis show postural hypotension. In normal individuals, blood levels of aldosterone respond rapidly to shifts in posture, which suggests that some actions of aldosterone must be nongenomic: a fast-responding hormone would, from a teleologically perspective, require a fast-acting mechanism of response.

All of the steroid hormones derive from cholesterol. They are not stored in secretory granules like the peptide hormones but are synthesized as needed from cholesterol that is imported from the blood or from cholesterol within the adrenal cells either in lipid droplets or freshly synthesized. The cholesterol is first imported into the mitochondria where its side chain is cleaved. Cholesterol is carried into the mitochondria by StAR, or steroidogenesis acute regulatory protein. Its activity is controlled by ACTH, which is secreted by the corticotrophs in the anterior pituitary in response to CRH release from neurons in the paraventricular nucleus of the hypothalamus. CRH release, in turn, is stimulated by a variety of stresses sensed by the central nervous system, including circadian rhythms, hypoglycemia, fever, and injury. Secreted glucocorticoids provide negative feedback to both CRH and ACTH release.

ACTH is synthesized by corticotrophs as prePOMC—pre proopiomelanocortin. Its cleavage produces POMC, and further cleavage makes ACTH and lipotropin. Further cleavage can produce γMSH and β endorphin. Some of these parts of POMC stimulate melanocortin receptors that cause increased pigmentation of the skin. For this reason, excess POMC changes skin pigmentation. ACTH binding to the melanocortin-2 receptor is coupled to a G_s mechanism in the adrenal secretory cells.

Glucocorticoids have metabolic effects on all organ systems. They promote gluconeogenesis in liver and raise blood glucose levels. They cause muscle wasting and thinning of the skin and promote adipose deposition around the viscera. They decrease osteoblast activity and therefore ultimately contribute to bone loss and osteoporosis. They are anti-inflammatory and suppress the immune system. These effects seem counterproductive, but the glucocorticoids help the body respond to stress by mobilizing fuel resources.

The glucocorticoids classically act by binding to a cytosolic receptor which is bound with heat shock protein. Binding of glucocorticoids sheds the heat shock protein and allows the receptor to migrate to the nucleus where it binds to GREs on the DNA. The result is either a stimulation or an inhibition of transcription of DNA into mRNA.

Aldosterone is a mineralocorticoid and is involved in regulating the extracellular volume. Its synthesis and secretion is controlled mainly by angiotensin II and plasma [K]. Angiotensin II in turn is produced by a cascade of proteolytic cleavage of angiotensinogen, a precursor molecule made in the liver and circulating in the blood. When renal arteriolar blood pressure falls, renal sympathetic stimulation increases, or the distal tubule [Na] falls, granule cells in the afferent arteriole are stimulated to secrete renin. This enzyme cleaves angiotensinogen to angiotensin I, containing 10 amino acids. Angiotensin converting enzyme, or ACE, cleaves angiotensin I to angiotensin II, containing eight amino acids. Angiotensin II then exerts multiple effects, including vasoconstriction (to increase blood pressure) and stimulating secretion of aldosterone and ADH. Aldosterone increases Na reabsorption and K excretion by altering the activity of ENaC on the apical membrane of distal tubule cells.

REVIEW QUESTIONS

1. What part of the adrenal gland makes glucocorticoids? What stimulates it?
2. What controls ACTH secretion? Where is CRH produced? What is CRH?
3. Why does excess ACTH production increase skin pigmentation? What is POMC?
4. What effects does cortisol have on blood glucose, liver, muscle, bone, and fat?
5. What effects does cortisol have on the immune system?
6. How does cortisol exert its effects?
7. What is aldosterone and where is it made? What causes its secretion?
8. What are aldosterone's main effects?
9. What is aldosterone's mechanism of action?

9.6 The Adrenal Medulla and Integration of Metabolic Control

Learning Objectives

- Describe the relationship between the adrenal medulla and the sympathetic nervous system
- List the biochemical precursors of epinephrine in sequence of their synthesis
- Distinguish between the local effects of the sympathetic nervous systems and systemic effects
- Identify the signal transduction pathway for β adrenergic effects
- Identify the signal transduction pathways for α_1 and α_2 adrenergic effects
- List two enzymes involved in catecholamine degradation
- Describe the overall effect of epinephrine on carbohydrate metabolism
- Describe the overall effect of epinephrine on fat metabolism
- List the hyperglycemic hormones
- List the hypoglycemic hormones
- Describe in general terms the effects of hormones on intermediary metabolism, fat and protein stores in the body

THE ADRENAL MEDULLA IS PART OF THE SYMPATHETIC NERVOUS SYSTEM

The gross and histological appearance of the adrenal gland was described in Chapter 9.5, Figure 9.5.1. The innermost part of this gland is the medulla, which secretes the catecholamine **epinephrine** into the systemic circulation. It also secretes small amounts of norepinephrine, which is typically released at postganglionic nerve endings in the sympathetic nervous system and acts as a local neurotransmitter. Epinephrine released into the circulation therefore is a neuroendocrine hormone and has diverse effects on multiple organs. The medulla is actually an enlarged and specialized ganglion of the sympathetic nervous system, as recapitulated in Figure 9.6.1. Here, preganglionic sympathetic nerves exit the spinal cord at the lower thoracic level and release acetylcholine onto **chromaffin cells** in the adrenal medulla that synthesize and release epinephrine into the general circulation.

EPINEPHRINE DERIVES FROM TYROSINE

Epinephrine, dopamine, and norepinephrine are all catecholamines. Catechol is dihydroxy benzene, shown in Figure 9.6.2 along with the biosynthetic pathway for epinephrine. All of the enzymes involved in the synthesis are located in the cytoplasm, except for dopamine β hydroxylase that converts dopamine to norepinephrine. This occurs within vesicles in the chromaffin cells. Thus, dopamine is imported into the vesicles to be converted to norepinephrine and is exported back out of the vesicles to be converted into epinephrine, and then transported back into the vesicles for secretion. Uptake into the vesicles is mediated by the vesicular H-ATPase pump and carrier proteins called vesicular monoamine transporters (VMATs). The granulated vesicles contain dopamine β hydroxylase, ascorbic acid, ATP, and chromogranin A. The rate limiting step for synthesis of epinephrine is tyrosine hydroxylase, the first step that forms L-DOPA. Several of the enzymes are influenced by circulating concentrations of glucocorticoids, whereas epinephrine inhibits phenyl N-methyl transferase (PNMT) activity.

CATECHOLAMINES ARE RELEASED IN RESPONSE TO SYMPATHETIC STIMULATION

Preganglionic sympathetic fibers release acetylcholine onto the chromaffin cells in the adrenal medulla, and this activates nicotinic cholinergic receptors. The resulting depolarization, in turn, activates voltage-gated Ca^{2+} channels that raise intracellular $[Ca^{2+}]$. The increased $[Ca^{2+}]$ causes exocytosis of granules containing epinephrine and norepinephrine. Epinephrine makes up about 80% of the catecholamines released from normal adrenal medulla and norepinephrine, about 20%.

Secretion of epinephrine from the adrenal medulla is an integral part of the "**fight or flight**" response of the body to perceived emergency situations. Stimulation of adrenal secretion is induced by a number of conditions, such as:

- trauma
- pain

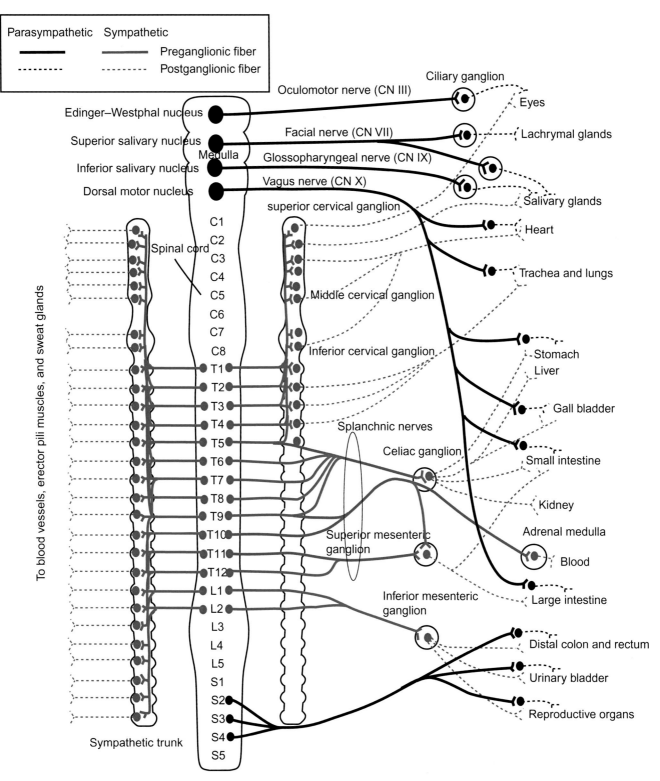

FIGURE 9.6.1 Connections of the sympathetic and parasympathetic nervous system. The sympathetic nervous system connections are shown in blue and the parasympathetic nervous system connections are shown in black.

- hypoglycemia
- hypotension
- anxiety
- temperature extremes
- severe exercise
- anoxia
- hypovolemia.

Figure 9.6.3 illustrates the different uses of epinephrine and norepinephrine by the sympathetic nervous system.

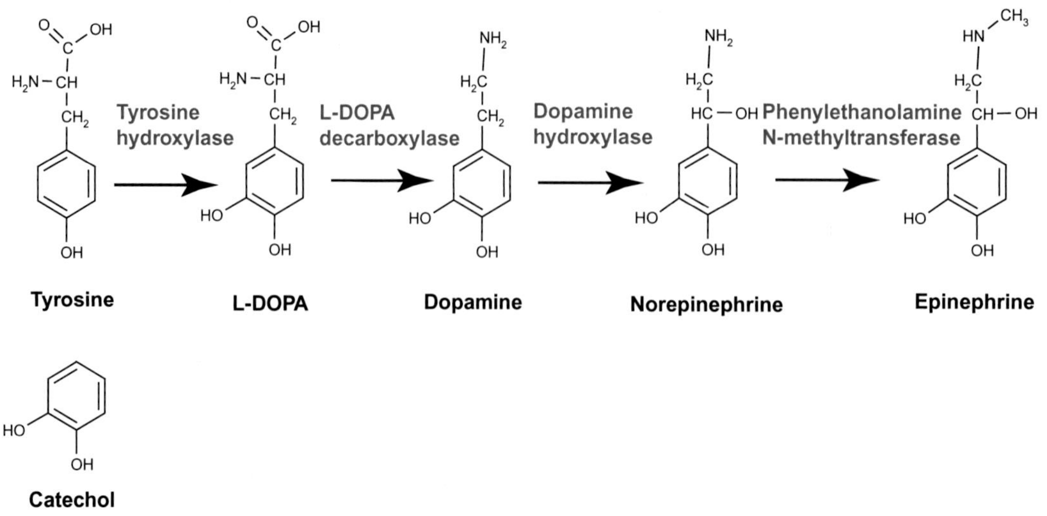

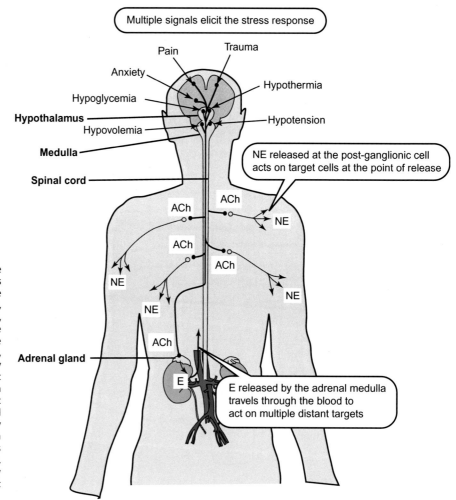

FIGURE 9.6.2 Synthesis of norepinephrine and epinephrine in the adrenal medulla.

FIGURE 9.6.3 Different function of epinephrine (E) and norepinephrine (NE) in the stress response. A variety of signals can initiate the stress response including hypovolemia, hypotension, hypoglycemia, pain, trauma, anxiety, and hypothermia. These all activate the sympathetic nervous system. Some of the responses are mediated by norepinephrine, which is typically secreted in a widely dispersed network through preganglionic sympathetic fibers that use acetylcholine as a neurotransmitter to activate postganglionic fibers that release norepinephrine onto local target cells. Other responses are mediated by circulating epinephrine that is released from the adrenal medulla by sympathetic fibers carried in the preganglionic splanchnic nerves. These increase epinephrine release by depolarizing the chromaffin cells (those that secrete epinephrine) with acetylcholine.

CATECHOLAMINES ARE DEGRADED RAPIDLY

The basal circulating plasma epinephrine concentration ranges from 25 to 50 pg mL^{-1} (= about $1.5 - 3 \times 10^{-10}$ M) with an estimated daily secretion of 150 μg. Nearly all of the epinephrine in the circulation derives from the adrenal medulla, whereas most of the norepinephrine derives from sympathetic nerve terminals in the peripheral tissues and the brain. Most of the neurotransmitter is retaken up by either the pre- or postsynaptic cell, but some escapes immediate uptake and leaks into the general circulation. Both epinephrine and norepinephrine remain in the circulation only a very short time: their half-lives are about 1–3 min and their metabolic clearances are from 2 to 6 L min^{-1}. Circulating epinephrine is degraded mostly in the liver and kidney. The enzymes largely responsible for this degradation are catecholamine O-methyl transferase (COMT) and a combination of mono amine oxidase (MAO) and aldehyde oxidase (AO). The pathways for metabolism of the catecholamines are shown in Figure 9.6.4. Oxidation and O-methylation can occur in random order.

ACTIONS OF CATECHOLAMINES ARE MEDIATED BY ADRENERGIC RECEPTOR TYPES

All actions of epinephrine and norepinephrine are mediated through receptors on the surfaces of their target cells. As discussed in Chapters 4.2 and 4.9, these receptors are the **adrenergic receptors**. Ahlquist in 1948 classified them as α or β based on their response to epinephrine, norepinephrine, and isoproterenol. We now identify several subclasses of receptors: α_1 and α_2 and β_1, β_2, and β_3. These receptors differ in their relative responses to epinephrine and norepinephrine and also in their response to specific pharmacological agents. Some of these differences are summarized in Table 9.6.1. The β_1, β_2, and β_3 receptors are all G-coupled receptors whose occupancy stimulates adenylyl cyclase and so increases cAMP concentrations in the target cell: they are all G_s mechanisms because they stimulate adenylyl cyclase. The α_1 mechanism is a G_q mechanism. It is coupled to activation of phospholipase C in the surface membrane, which cleaves phosphatidyl inositol bisphosphate to release IP3 and diacylglycerol (DAG). The IP3 in turn binds to IP3 receptors on the endoplasmic reticulum membrane, causing them to release stored Ca^{2+}. The increase cytoplasmic [Ca^{2+}] then activates specific cell responses. The DAG activates protein kinase C, which exerts its effects by phosphorylating target proteins. The α_2 receptors are coupled to a G_i mechanism, in which binding to the receptor inhibits adenylyl cyclase and therefore reduces cytoplasmic levels of cAMP.

The multitude of responses of target cells to epinephrine and norepinephrine derive from the distribution of these receptor types among the different tissues. For example, heart and liver tissue have β_1 receptors that are responsible for cardioacceleration and positive inotropy and glycogenolysis. Smooth muscles of the

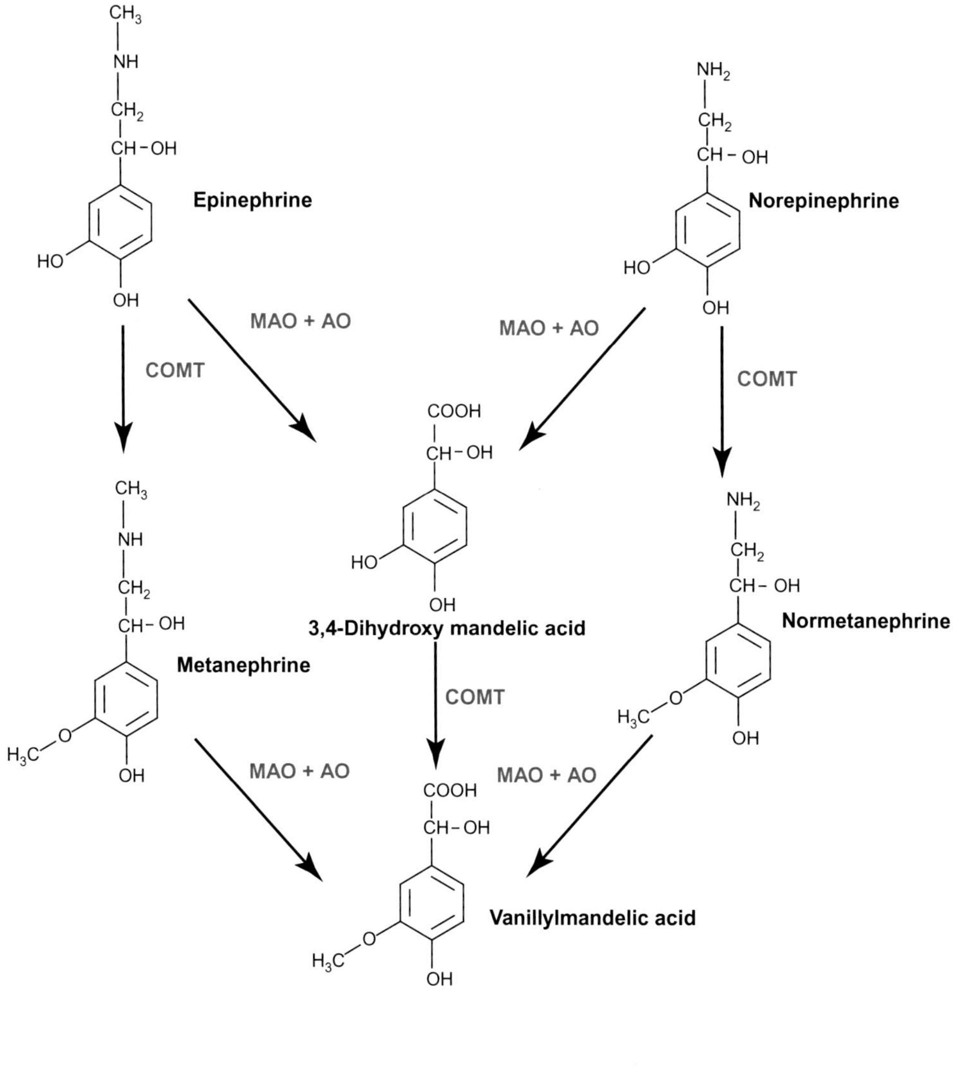

FIGURE 9.6.4 Catabolism of epinephrine and norepinephrine. Both catecholamines are degraded by a combination of catechol-O-methyl transferase (COMT), which incorporates a methyl group onto the meta-hydroxy position on the ring, and monoamine oxidase (MAO) and aldehyde oxidase (AO), which oxidize the amino group and then cleave it off, and then convert the aldehyde to a carboxyl group.

TABLE 9.6.1 Comparison of α- and β Adrenergic Receptors

Repeptor Type	Agonist Potency	Action of Agonist	Mechanism	Agonists	Antagonists
α_1	E>NE≫ISO	Smooth muscle contraction	G_q	Phenylephrine	Phentolamine prazosin phenoxybenzamine
α_2	E>NE≫ISO	Nerve terminal inhibition; smooth muscle contraction	G_i	Clonidine	Yohimbine
β_1	ISO>E = NE	Increased heart rate	G_s	Isoprenaline dobutamine	Atenolol metaprolol
β_2	ISO>E≫NE	Smooth muscle relaxation	G_s	Albuterol	Propranolol
β_3	ISO = E>NE	Increased lipolysis	G_s		

E, epinephrine; NE, norepinephrine; ISO, isoproterenol.

TABLE 9.6.2 Overview of the Physiological Actions of the Catecholamines

1. **Effects on carbohydrate metabolism**
   1.1 Overall effect is to increase plasma [glucose]
   1.1 Catecholamines increase liver glycogenolysis (β_1 and α_1)
   1.2 Catecholamines increase gluconeogenesis (β_2 and α_1)
   1.3 Epinephrine increases muscle glycogenolysis
   1.4 Norepinephrine decreases insulin secretion (α_2)
   1.5 Epinephrine increases insulin secretion (β_2)
   1.6 Catecholamines stimulate glucagon secretion (α)
2. **Effects on fat metabolism**
   2.1 Catecholamines increase lipolysis (β_3 and β_1)
   2.2 Catecholamines increase plasma levels of free fatty acids
3. **Effects on overall metabolism**
   3.1 Catecholamines increase BMR
   3.2 Catecholamines increase Q_{O_2}
   3.3 Catecholamines increase calorigenesis (β_1)
4. **Effects on the cardiovascular system**
   4.1 Catecholamines increase heart rate (β_1)
   4.2 Catecholamines increase cardiac contractility (β_1)
   4.3 Catecholamines increase conduction velocity (β_1)
   4.4 Catecholamines constrict arterioles in skin, kidney, GI tract, genitalia, spleen (α_1)
   4.5 Catecholamines dilate arterioles in skeletal and cardiac muscle, liver, and lungs (β_2)
   4.6 Net effect is to increase cardiac output and increase systolic (but generally not diastolic) pressure and to divert blood from splanchnic circulation to muscle
5. **Effects on the GI tract and sphnicters**
   5.1 Catecholamines relax smooth muscle in GI tract, urinary tract, and bronchioles (β_2)
   5.2 Catecholamines constrict GI and urinary sphincters (α_1)
6. **Other effects**
   6.1 Catecholamines cause pupillary dilation (α_1)
   6.2 Catecholamines cause emotional sweating
   6.3 Catecholamines increase platelet aggregation (α_2)

vasculature, intestine, uterus, and bronchi have β_2 receptors that are responsible for smooth muscle relaxation that causes vasodilation and bronchodilation. Vascular smooth muscle, pupillary dilator muscle, liver, and heart also have α_1 receptors that cause vasoconstriction, dilation of the pupil, glycogenolysis, and positive inotropy. Adipose tissue has β_3 receptors that cause lipolysis.

THE EFFECTS OF CATECHOLAMINES ARE TO PREPARE THE BODY FOR "FIGHT OR FLIGHT"

The concerted actions of the catecholamines have the net effect of preparing the body for emergency action. These actions are various and widely spread over the body systems. Table 9.6.2 summaries the wide variety of physiological effects of the catecholamines. The heart rate accelerates due to an increase in the slope of the sinoatrial node pacemaker potential, caused by β_1 stimulation of the SA nodal cells with increases in I_f and L-type I_{Ca} (see Chapter 5.5). Stimulation of β_1 receptors on cardiomyocytes also phosphorylates a number of proteins in these cells that lead to increased inotropy (Chapter 5.7). This leads to an increased cardiac output and increased systolic pressure and mean arterial pressure. Simultaneously, the catecholamines constrict the blood vessels that supply the blood to organs that are inessential to emergency action: the GI tract, kidneys, and reproductive organs. Receptors on the bronchioles cause dilation of the airways to ready the respiratory system for increase air flow. The catecholamines mobilize liver and muscle glycogen, with only the liver glycogen contributing to the increase in blood levels of glucose. They also stimulate gluconeogenesis to further increase blood glucose and inhibit insulin release and stimulate glucagon release. The catecholamines also mobilize fat stores by increasing lipolysis and circulating levels of fatty acids. Dilation of the pupil, caused by sympathetic stimulation, enables better vision of distant objects. All of these actions prime the body for emergency action and also help rescue the body from trauma or hypoglycemia.

INTEGRATION OF METABOLIC CONTROL

Over the last few chapters, we have discussed several hormones that affect intermediary metabolism. These include: growth hormone, thyroxine, insulin, glucagon, cortisol, and the catecholamines. The purpose of this section is to attempt to integrate their effects on intermediary metabolism. First, we list here the general effects of each hormone on intermediary metabolism, and then we focus on the integrated effects of all of them on intermediary metabolism.

GROWTH HORMONE
1. Carbohydrate metabolism
   - GH increases blood glucose (the "diabetogenic" effect of GH);
   - GH mobilizes liver glycogen;
   - GH increases insulin release;
   - GH inhibits glucose uptake by muscle and adipose tissue.
2. Fat metabolism
   - GH increases plasma free fatty acids;
   - GH increases lipolysis;
   - GH decreases glucose uptake into fat tissue;
   - GH decreases body fat.
3. Protein metabolism
   - GH increases total body nitrogen retention;
   - GH increases lean body mass;
   - GH increases uptake of amino acids into muscle, heart, and liver;
   - GH stimulates protein synthesis.

THYROID HORMONE
1. Carbohydrate metabolism
   - T3 potentiates the effect of epinephrine in the liver; increases liver glycogenolysis and gluconeogenesis;
   - T3 therefore decreases liver glycogen and raises blood glucose;
   - T3 potentiates the effect of insulin on muscle; it increases glucose uptake, utilization and storage of glucose.
2. Fat metabolism
   - T3 potentiates the effect of insulin on adipose tissue;
   - T3 increases plasma free fatty acids;
   - T3 increases lipolysis.
3. Protein metabolism
   - T3 promotes protein synthesis in a number of tissues.

INSULIN
1. Carbohydrate metabolism
   - insulin increases glucose uptake by muscle and adipose tissue;
   - increased uptake decreases plasma glucose concentration;
   - insulin decreases glycogenolysis in liver and muscle;
   - insulin decreases gluconeogenesis in liver and muscle.
2. Fat metabolism
   - insulin increases lipogenesis;
   - insulin decreases lipolysis;
   - insulin decreases plasma free fatty acids.
3. Protein metabolism
   - insulin increases amino acid uptake by liver and muscle;
   - insulin increases protein synthesis and decreases protein breakdown.

GLUCAGON
1. Carbohydrate metabolism
   - glucagon increases liver glycogenolysis and gluconeogenesis;
   - glucagon increases plasma glucose concentration;
   - glucagon does not affect glucose uptake or utilization by peripheral tissues.
2. Fat metabolism
   - glucagon activates hormone-sensitive lipase, so increases lipolysis.
3. Protein metabolism
   - glucagon promotes gluconeogenesis mainly in the liver.

GLUCOCORTICOIDS
1. Carbohydrate metabolism
   - glucocorticoids raise plasma glucose concentrations;
   - glucocorticoids increase gluconeogenesis;
   - glucocorticoids raise liver glycogen stores;

TABLE 9.6.3 General Effects of the Various Hormones on Intermediary Metabolism and Fuel Stores

Hormone	Plasma [glucose]	Glycogen Stores		Gluconeo-genesis	Glucose Uptake, Utilization	Fat Stores in Adipose Tissue	Protein Content of Tissues
		Liver	Muscle				
Insulin	↓	↑	↑	↓	↑	↑	↑
Glucagon	↑	↓	→	↑	→	↓	→
GH	↑	↓	→	→	↓	↓	↑
Cortisol	↑	↑	→	↑	↓	↓	↓
Epinephrine	↑	↓	↓	↑	↓	↓	→
Thyroxine	↑	↓	↑	↑	↑	↓	→

- glucocorticoids reduce muscle and adipose tissue glucose utilization.
2. Fat metabolism
   - glucocorticoids increase lipolysis;
   - glucocorticoids increase plasma free fatty acids.
3. Protein metabolism
   - glucocorticoids accelerate protein catabolism;
   - glucocorticoids are "protein wasting" in the long term.

EPINEPHRINE

1. Carbohydrate metabolism
   - epinephrine increases blood glucose concentration;
   - epinephrine increases liver glycogenolysis;
   - epinephrine increases muscle glycogenolysis;
   - epinephrine increases gluconeogenesis.
2. Fat metabolism
   - epinephrine increases lipolysis in adipose tissue;
   - epinephrine increases plasma free fatty acid concentration.
3. Protein metabolism
   - epinephrine increases gluconeogenesis.

These effects of hormones on intermediary metabolism are summarized in Table 9.6.3.

This table is not so formidable as it first appears. First, all of the listed hormones save insulin raise blood glucose. Because of the liver's importance in raising blood glucose, their effect on the liver is just the opposite: insulin raises liver glycogen and all others lower it. The one exception is cortisol because it raises blood glucose mainly through gluconeogenesis. Similarly, the effects on plasma glucose are opposite to the effects on fat stores: if blood glucose goes up, fat stores go down, and vice versa, with one exception: epinephrine mobilizes both carbohydrate and lipid stores.

Clinical application: Pheochromocytoma

Pheochromocytoma is an excess secretion of epinephrine and norepinephrine usually caused by a tumor of the adrenal medulla but sometimes due to extraadrenal production. Symptoms include elevated heart rate, palpitations, elevated blood pressure, anxiety resembling a panic attack, headache, excessive sweating, elevated blood glucose, and pallor. This is a rare condition, with about 1000 cases diagnosed in the United States each year, with a prevalence of 2–8 per million persons. It typically strikes in young to mid-adult persons. The disease is often associated with intermittent rather than continuous secretion of epinephrine, so that persons suffer sporadic attacks with a classic triad of severe headaches, palpitations, and diaphoresis (excessive sweating).

Diagnosis relies on high plasma levels of metanephrine or 24 h urine collections tested for epinephrine, total metanephrines, and creatinine. The purpose of measuring creatinine is to normalize the values for undercollection of urine. Plasma levels of epinephrine are unreliable because of the sporadic nature of the disease and the short half-life of circulating catecholamines.

Treatment is generally surgical. Preoperative care is essential as the victims are generally volume depleted. Patients are typically volume expanded with a high Na^+ diet and then treated with α blockers, followed by β blockers prior to surgery.

SUMMARY

The adrenal medulla is essentially a ganglion of the sympathetic nervous system that releases neurotransmitter into the blood instead of near local targets. Preganglionic sympathetic nervous fibers originating mainly in the thoracic spinal cord reach the adrenal medulla through the splanchnic nerves, and release acetylcholine onto chromaffin cells in the adrenal gland, causing release of epinephrine into the blood; the epinephrine is then transported to distant targets. Epinephrine is synthesized from tyrosine in the sequence tyrosine–dihydroxyphenylalanine–dopamine–norepinephrine–epinephrine. Circulating epinephrine and norepinephrine are degraded by catechol-O-methyl transferase (COMT) and monoamine oxidase (MAO). A variety of stimuli increase epinephrine secretion including hypoglycemia, hypovolemia, hypotension, fear and anxiety, pain, and trauma.

The effects of epinephrine and norepinephrine are mediated through adrenergic receptors, of which there are at least five types. The α_1 receptors work through a G_q mechanism that activates smooth muscle contraction mainly in the arterioles of skin, GI system and kidney, and the urethral sphincter. Adrenergic α_2 receptors activate a G_i mechanism. All of the β adrenergic receptors exert their effects through a G_s mechanism.

The overall effect of adrenergic stimulation is to prepare the body for emergency action. The pupils dilate (α_1 receptors), blood pressure increases, bronchioles dilate (β_2 receptors), blood flow to inessential organs is reduced, heart rate and contractility are increased, and stored metabolic fuels, glycogen and triglycerides, are mobilized to increase plasma glucose and free fatty acids for metabolism.

REVIEW QUESTIONS

1. Where is the adrenal gland? Where is the medulla? What does it secrete? What stimulates its secretion?
2. What is the proportion of epinephrine and norepinephrine typically secreted by the adrenal gland?
3. From what amino acid is epinephrine synthesized?
4. What are adrenergic receptors? What mechanism is used by α_1 receptors, α_2 receptors, and β receptors?
5. Where are α_1 and β receptors located?
6. What effect does epinephrine have on heart rate, contractility, blood pressure, glucose levels, airway caliber, lipid mobilization, and liver glycogen stores? What is the mechanism in each of these?
7. Which hormones lower blood glucose? Which hormones raise it?

9.7 Calcium and Phosphorus Homeostasis I: The Calcitropic Hormones

Learning Objectives

- Describe the ways calcium is carried in plasma and identify the form that is homeostatically regulated
- Identify the major sources and sinks of plasma calcium
- Explain why calcium and phosphate homeostasis are linked
- Identify the major sources and sinks of plasma phosphate
- List the three hormones involved in calcium and phosphate homeostasis and their sites of origin
- List the three targets of these hormones
- Describe the chemical nature of PTH and its stimulus for release
- List three major actions of PTH
- Describe the chemical nature of CT and its stimulus for release
- List two major actions of CT
- Describe the chemical nature of vitamin D and explain why it isn't a vitamin
- Describe the control of vitamin D activation
- Describe the chemical nature of FGF23 and its stimulus for release
- List the major actions of FGF23

CALCIUM HOMEOSTASIS IS REQUIRED FOR HEALTH

Calcium homeostasis refers to the maintenance of a constant concentration of calcium ions in the extracellular fluid. It includes all of the processes that contribute to maintaining calcium at its "set point." Because plasma $[Ca^{2+}]$ rapidly equilibrates with the extracellular fluid, ECF $[Ca^{2+}]$ is kept constant by keeping the plasma $[Ca^{2+}]$ constant. Maintaining a constant plasma $[Ca^{2+}]$ is important for:

- nerve transmission
- nerve conduction
- muscle contraction
- cardiac contractility
- blood clotting
- bone formation
- excitation−secretion coupling
- cell-to-cell adhesion
- cell-to-cell communication.

Some of these (bone formation, blood clotting, and cell adhesion) depend directly on the extracellular $[Ca^{2+}]$; others depend directly on intracellular $[Ca^{2+}]$. But since ICF $[Ca^{2+}]$ depends indirectly on plasma $[Ca^{2+}]$, all are linked to plasma $[Ca^{2+}]$. Calcium homeostasis can be viewed as having two components: a microcomponent dealing with the intracellular environment and a macrocomponent dealing with the extracellular environment. This chapter concerns regulation of $[Ca^{2+}]$ and phosphate concentrations in the extracellular fluid.

ABOUT HALF OF PLASMA CALCIUM IS FREE: THE OTHER HALF IS COMPLEXED OR BOUND TO PLASMA PROTEINS

Total plasma $[Ca^{2+}]$ is tightly regulated at about 10 mg% (mg per deciliter of blood) or 2.5×10^{-3} M. This total plasma is partitioned among three forms: free Ca^{2+} comprises about half of the total, or 1.2×10^{-3} M; another 40% is bound to plasma proteins, particularly albumin; and the last 10% of the total plasma Ca^{2+} is bound to low-molecular-weight anions such as lactate, citrate, and bicarbonate. The principal plasma proteins are albumin (4 g%), globulin (3 g%), and fibrinogen (0.3 g%). For every g% change in plasma albumin, the total plasma $[Ca^{2+}]$ changes about 0.8 mg%. Thus the total plasma $[Ca^{2+}]$ varies with the plasma protein concentration. The endocrine system regulates the free $[Ca^{2+}]$.

Most of the binding sites for Ca^{2+} on plasma proteins are free carboxyl groups of the side chains of the amino acids, glutamic acid and aspartic acid. When ionized, the negative charge attracts the Ca^{2+} ion and loosely binds it. When two or more of these groups are in close proximity, the binding site has a higher affinity because the negative charges more strongly attract the Ca^{2+} ion. When a hydrogen ion binds to a carboxyl group, it neutralizes its negative charge and so it no longer binds Ca^{2+}. Thus plasma $[H^+]$ competes for Ca^{2+} binding sites. Acidosis (increased plasma $[H^+]$) lowers the protein-bound Ca^{2+} and increases the free $[Ca^{2+}]$. Lowering the $[H^+]$, as in alkalosis, shifts free Ca^{2+} to bound $[Ca^{2+}]$.

FAILURE TO REGULATE PLASMA $[Ca^{2+}]$ CAUSES SYSTEM MALFUNCTION

When plasma $[Ca^{2+}]$ falls to 7 mg% or less, the permeability of neuron membranes is increased so that they become more excitable. As a result, **paresthesias** ("pins and needles") may be felt and **tetany** may develop due to increased neuronal excitability both centrally and peripherally. The tetany that develops can be fatal. This is called **hypocalcemic tetany**. Plasma $[Ca^{2+}]$ >12 mg% leads to depression of the central nervous system and sluggish reactions. Prolonged increased plasma $[Ca^{2+}]$ can lead to ectopic calcification of the soft tissues with loss of function.

PLASMA Ca^{2+} HOMEOSTASIS RESULTS FROM A BALANCE OF SOURCES AND SINKS

Sources of Ca^{2+} include absorption from the intestinal tract, resorption from bone, and reabsorption from the tubular fluid in the kidney. The major sinks occur in the same organs: secretion into the intestine, accretion onto bone, and ultrafiltration in the kidney. Figure 9.7.1 illustrates these sources and sinks for calcium, along with estimates of the daily magnitude of these sources and sinks. They show a person in **calcium balance**, in which body Ca^{2+} content does not change: ingested Ca^{2+} exactly matches the amount in the feces, sweat, and urine. This is not usually the case. The growing infant or teen excretes less Ca^{2+} than the amount ingested and stores the accumulated Ca^{2+} in bone. These growing youth are in **positive calcium balance**. Ultimately, all of the Ca^{2+} in bone comes either from maternal sources *in utero* or from food. When dietary Ca^{2+} is insufficient to keep up with excretion in feces, urine, and sweat, plasma $[Ca^{2+}]$ falls and bone Ca^{2+} is mobilized to return plasma $[Ca^{2+}]$ to normal. This produces a **negative calcium balance** in which the lost Ca^{2+} derives from bone. Positive and negative calcium balance refer to the long-term averages of intake minus loss. On a day-to-day basis, a given individual can be in either negative or positive calcium balance. What determines the health of the bone is the long-term balance. Figure 9.7.1 shows that **the vast majority, more than 99%, of body Ca^{2+}, resides in bone**. A small, nearly undetectable loss of 40 mg day^{-1} over 30 years loses 440,000 mg, or about 37% of the skeletal Ca^{2+} content. But loss of 40 mg day^{-1} is only 5% of the dietary intake. This small loss can result in **osteoporosis**, a thinning and subsequent weakening of bone.

PLASMA Ca^{2+} HOMEOSTASIS IS LINKED TO P HOMEOSTASIS

Maintenance of cytoplasmic $[Ca^{2+}]$ at very low levels is an axiom of cell biology. Inorganic phosphate, Pi, is an integral part of a variety of important organic constituents of cells including DNA, RNA, ATP, phospholipids, and many intermediates of metabolic pathways. One postulate for the origin of low cytoplasmic $[Ca^{2+}]$ is calcium's insolubility with Pi: Ca^{2+} and Pi solutions produce calcium phosphate crystals. This reaction is governed by a solubility product

$$(9.7.1) \qquad [Ca^{2+}][Pi] = K_{sp}$$

According to this postulate, cells evolved Ca^{2+} pumps to prevent Ca–Pi crystallization.

Plasma Ca^{2+} homeostasis is linked to *plasma* Pi homeostasis because the bone mineral is a Ca–Pi salt that is similar to the mineral **hydroxyapatite**, with the empirical formula of $Ca_{10}(PO_4)_6(OH)_2$. When bone is resorbed, both Ca^{2+} and Pi are liberated. When new bone is formed, both Ca^{2+} and Pi are removed from plasma to form the new mineral. Plasma $[Ca^{2+}]$ tends to obey Eqn (9.7.1): rising plasma

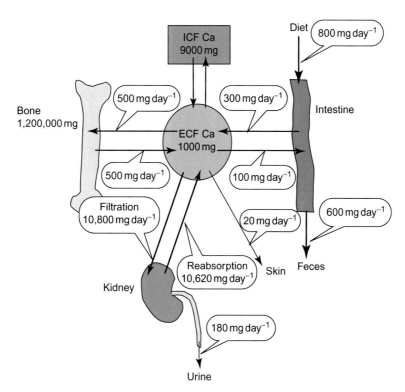

FIGURE 9.7.1 Sources and sinks of plasma Ca^{2+}. Flows given are in mg elemental Ca^{2+} per day, whereas pool sizes are in mg. The numbers in this figure are estimated averages that vary daily and individually with age, gender, and diet.

[Ca^{2+}] tends to lower plasma [Pi], and rising plasma [Pi] tends to lower plasma [Ca^{2+}]. However, plasma [Ca^{2+}] and [Pi] do not obey a solubility product rule, as this would imply that it is in equilibrium with a Ca^{2+}-Pi precipitate, and there is no such equilibrium being obeyed.

PLASMA PI IS PRESENT IN MULTIPLE IONIZED FORMS

Plasma inorganic phosphate participates in a number of reactions with H$^+$:

(9.7.2)
$$H_3PO_4 \rightleftharpoons H_2PO_4^- + H^+ \; K_D = 7.5 \times 10^{-3} \text{ M}$$
$$H_2PO_4^- \rightleftharpoons H_2PO_4^{-2} + H^+ \; K_D = 6.2 \times 10^{-8} \text{ M}$$
$$HPO_4^{-2} \rightleftharpoons PO_4^{-3} + H^+ \; K_D = 4.8 \times 10^{-13} \text{ M}$$

Each of these dissociation reactions is characterized by the given equilibrium constant. These dissociation constants correspond to the [H$^+$] at which the reaction is half complete, when [A$^-$] = [HA]. Thus the first dissociation is half complete at pH 2.1, the second is half complete at pH 7.2, and the third at pH 12.3. Because of these reactions, the form of Pi in the plasma depends on the pH. Because of this ambiguity, plasma concentration of phosphate is expressed in terms of the mg% *phosphorus*. In adults, the normal value is 3.5–4 mg%; in the growing child it is 4–5 mg%.

PLASMA [Pi] IS SET BY A BALANCE BETWEEN SOURCES AND SINKS

Like Ca^{2+} balance, Pi balance results from the flows between the intestine, kidney, bone, and intracellular compartment. Estimates of the flows in a typical adult are shown in Figure 9.7.2. Note that the magnitude of the numbers is near those for Ca^{2+} balance.

The difference is that bone contains more Ca^{2+} than P on a weight basis, and P is higher in the intracellular compartment compared to Ca^{2+}.

OVERALL Ca^{2+} AND Pi HOMEOSTASIS IS CONTROLLED BY FOUR HORMONES ACTING ON THREE TARGET TISSUES

Plasma [Ca^{2+}] is controlled by four hormones:

1. parathyroid hormone (PTH)
2. calcitonin (CT)
3. vitamin D (cholecalciferol)
4. FGF23.

These hormones are collectively called the **calciotropic** hormones. They accomplish Ca^{2+} and Pi homeostasis by acting on three separate targets:

1. intestine
2. bone
3. kidney.

These hormones also exert effects on other tissues that are not directly involved with Ca^{2+} and Pi homeostasis.

HYPOCALCEMIA STIMULATES PTH SECRETION

One pair of the four parathyroid glands is embedded in each half of the thyroid gland. The glands consist of chief cells and oxyphile cells. The chief cells synthesize a 115 amino acid precursor called preproPTH. The N-terminal signal sequence of 25 amino acids is clipped within the ER to produce a 90 amino acid polypeptide called proPTH. The N-terminal 6 amino acids are clipped again to form an 84 amino acid polypeptide,

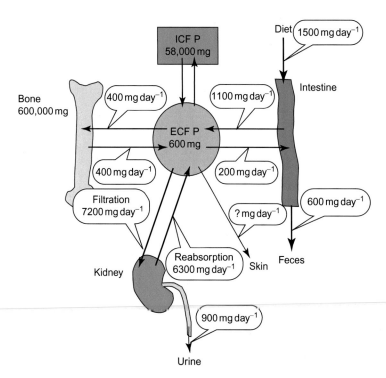

FIGURE 9.7.2 Pi balance in a typical adult. Flows are in units of mg *phosphorus* per day; pool sizes are in units of mg. These numbers differ individually and daily.

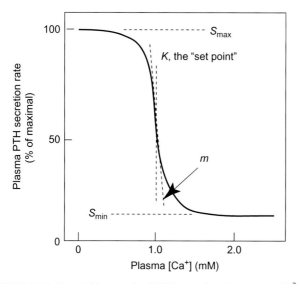

FIGURE 9.7.3 Sigmoidal control of PTH secretion by plasma [Ca^{2+}]. Decreases in plasma [Ca^{2+}] below the normal value cause a steep rise in PTH secretion. Conversely, rises in plasma [Ca^{2+}] decrease in PTH secretion. The curve can be approximately described by the empirical equation:

$$S = [S_{max} - S_{min}]/[1 + ([Ca]/K)^m] + S_{min}$$

where S is the rate of secretion; S_{max} is its maximum rate of secretion and S_{min} is the minimum rate; K is the "set point" at which S is half way between S_{max} and S_{min}; m is the slope of the response at K.

PTH, that is stored in secretory granules within the gland, and secreted in response to hypocalcemia.

The parathyroid cells express an extracellular Ca^{2+} sensor (CaSR), a homodimeric G-protein-coupled receptor, that interacts with a variety of G proteins in various tissues. In the parathyroid gland, CaSR activates phospholipase C and inhibits adenylate cyclase. When plasma [Ca^{2+}] rises, phospholipase C activity in parathyroid cells increases, thereby increasing the hydrolysis of phosphatidylinositol bisphosphate, in turn increasing the cytoplasmic concentration of IP$_3$ that releases Ca^{2+} from stores in the endoplasmic reticulum. Thus cytoplasmic [Ca^{2+}] rises at the same time that cAMP levels decrease, and PTH secretion also decreases. When plasma [Ca^{2+}] falls, the reverse occurs: cell [Ca^{2+}] falls, cAMP rises, and PTH secretion is increased. Thus parathyroid cells are an exception to the rule that rises in cytoplasmic [Ca^{2+}] cause secretion. The overall result (see Figure 9.7.3) shows an impressive, sigmoidal response to plasma [Ca^{2+}] that ranges from maximum to minimum over a range from 0.8 to 1.5 mM. The plasma [Ca^{2+}] that results in 50% of the maximum plasma [PTH] is called the "set point." It is near the normal plasma [Ca^{2+}] levels.

PTH SECRETION IS AN EXAMPLE OF DERIVATIVE CONTROL

At least three different functional relationships between plasma [Ca^{2+}] and the rate of PTH secretion have been considered. In **direct control**, the rate is directly proportional to the plasma [Ca^{2+}]. In **derivative control**, the rate of PTH secretion is proportional to d[Ca^{2+}]/dt. In this case, the rate of change of plasma [Ca^{2+}] influences the rate of PTH secretion. In **integrative control**, the rate of PTH secretion is proportional to ∫[Ca^{2+}]dt. In this case, the gland responds to some average plasma [Ca^{2+}] and therefore it would be sensitive to slow but prolonged changes in [Ca^{2+}]. Experiments suggest that PTH is under derivative control. When plasma [Ca^{2+}] falls, there is a higher rate of PTH secretion when the rate of fall is faster and the rate of secretion is measured at the same plasma [Ca^{2+}]. At the same plasma [Ca^{2+}], PTH secretion is higher when d[Ca^{2+}]/dt is negative compared to when it is positive.

PTH IS DESTROYED RAPIDLY AFTER SECRETION

The normal circulating levels of PTH, about 10–65 ng/L, are a heterogeneous mixture of intact PTH and PTH fragments. Only the N-terminal 34 amino acids are biologically active. A variety of PTH fragments are produced by proteolysis, particularly in the kidney. The half-life of circulating PTH is extremely short, about 2 minutes. Thus there must be a continuous secretion of PTH to maintain plasma [Ca^{2+}], and the circulating levels reflect the rate of secretion by the parathyroid glands. Because of this rapid turnover, only a tiny fraction of circulating PTH ever binds to its receptors on target tissues.

PTH DEFENDS AGAINST HYPOCALCEMIA BY ACTIONS ON BONE AND KIDNEY

The mechanism by which PTH alters Ca^{2+} and Pi handling by bone, intestine, and kidney will be discussed in Chapter 9.8. Here, we summarize its actions:

- PTH causes resorption of bone mineral. This releases both Ca^{2+} and Pi from bone into plasma. By itself, this action would tend to raise both plasma [Ca^{2+}] and [Pi].
- PTH increases the reabsorption of Ca^{2+} from the renal tubular fluid, whereas it decreases the reabsorption of Pi. This helps retain the Ca^{2+} that is removed from bone and discards the Pi. This would tend to raise the plasma [Ca^{2+}] alone while decreasing plasma [Pi].
- PTH activates vitamin D. Vitamin D exerts its major hormonal influence on intestinal Ca^{2+} and Pi absorption, but it must be activated to exert these effects. By contributing to the control of vitamin D activation, PTH *indirectly* stimulates intestinal Ca^{2+} absorption.
- PTH increases release of FGF23 from osteoblasts and osteocytes.

CT IS SECRETED IN RESPONSE TO HYPERCALCEMIA AND GASTROINTESTINAL HORMONES

C-cells of the thyroid gland, also called parafollicular cells because they reside in between the thyroid follicles, synthesize a glycosylated precursor with an M_r of about 17 kDa. These cells secrete **CT**, an polypeptide hormone 32 amino acids long with an M_r of 3.4 kD, in response to hypercalcemia. Because of its origin in the thyroid gland, CT is often called **thyrocalcitonin**. Figure 9.7.4 shows the relationship between CT secretion and plasma $[Ca^{2+}]$. The sensor for extracellular Ca^{2+} in the C-cells is the same CaSR that is present in chief cells of the parathyroid gland, but in this case increases in plasma $[Ca^{2+}]$ result in an increase in secretion rather than an inhibition. Thus CaSR in different tissues are connected to the cellular mechanisms for secretion in different ways, producing inhibition in one case (PTH) and stimulation in another (CT). In addition to hypercalcemia, injections of the gastrointestinal hormones **gastrin** or **cholecystokinin** powerfully stimulate CT release. C-cells have CCK2R (cholecystokinin 2 receptor) on their surfaces. The normal circulating levels of CT are 5–40 ng/L. Circulating CT is degraded by proteolysis mainly in the kidney and liver, with a half-life less than 10 minutes.

CT TENDS TO LOWER PLASMA $[Ca^{2+}]$

CT was discovered relatively recently, in 1961, partly because there are no clinical symptoms resulting from CT overproduction or underproduction. Thyroidectomy does not produce hypercalcemia, and massive overproduction of CT in medullary thyroid tumors does not cause hypocalcemia. Nevertheless, injection of CT promptly lowers plasma $[Ca^{2+}]$ in experimental animals. Lack of pathology associated with abnormal CT levels does not mean that CT is unimportant in human physiology. The body might adjust for these abnormal levels. For example, lack of symptoms in medullary thyroid carcinoma, when CT levels are high, could be due to "CT escape." Downregulation of the CT receptors could result in tissues that no longer respond to CT, whereas they would remain responsive if CT levels were normal. The main effects of CT are to:

- inhibit bone resorption;
- increase Ca^{2+} and Pi excretion in the urine.

"VITAMIN D" IS A HORMONE, NOT A VITAMIN, SYNTHESIZED IN THE SKIN

The naturally occurring vitamin D is called **cholecalciferol** and is derived from ultraviolet radiation of **7-dehydrocholesterol** in the skin. The 7-dehydrocholesterol is an intermediate in the synthesis of cholesterol from lanosterol. The conjugated diene group of 7-dehydrocholesterol absorbs UV radiation in the 250–310 nm range, which causes the B ring of the steroid nucleus to open up to form previtamin D. This then isomerizes in a nonenzymatic reaction to form cholecalciferol, also known as **vitamin D3** (see Figure 9.7.5).

A deficiency of cholecalciferol causes **rickets** in the young and **osteomalacia** in the adult. These diseases are characterized by the **failure to mineralize bone**. This results in a weakened bone that is pliable, and hence children afflicted with rickets often have painful, bowed legs. The noun is "rickets" and the adjective for the disease is "rachitic."

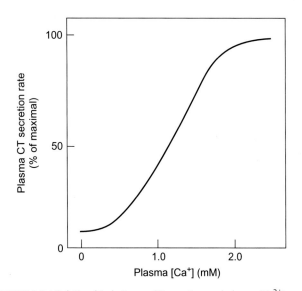

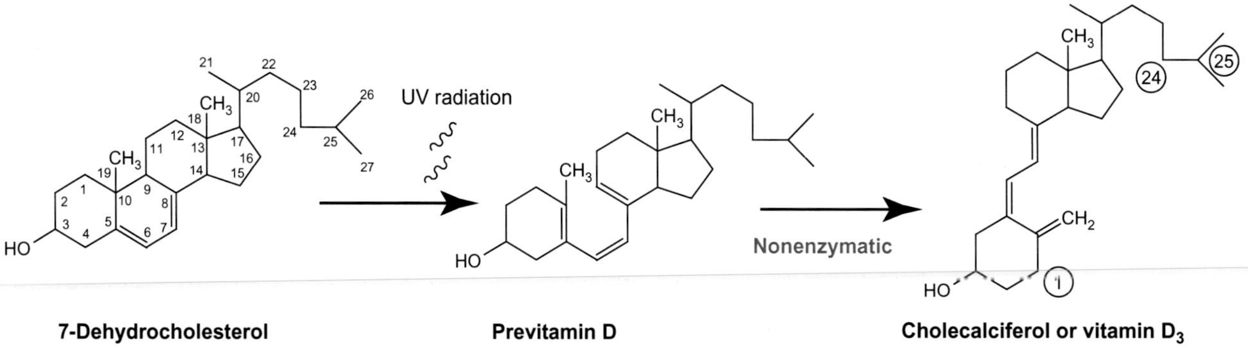

FIGURE 9.7.4 Relationship between CT secretion and plasma $[Ca^{2+}]$.

FIGURE 9.7.5 Conversion of 7-dehydrocholesterol in the skin to cholecalciferol. The numbering system at the far right of the figure indicates three important positions for activation and deactivation of vitamin D at the 1, 24, and 25 positions.

The distinction between vitamin D as a hormone or as a vitamin goes back to research performed at the beginning of the 20th century. In 1919, Huldschinsky in Vienna exposed one arm of a rachitic child to the UV light from a mercury vapor lamp and demonstrated by X-rays that mineralization began in both irradiated and nonirradiated arms. This was clear proof that something traveled (presumably via the blood) from the irradiated area to the nonirradiated area. In that same year, Sir Edward Mellanby produced rickets in puppies and showed that cod liver oil cured them. At that time it was known that cod liver oil contained vitamin A, so Mellanby attributed the cure to vitamin A. In 1922, McCollum, the discoverer of vitamin A, showed that vitamin A in cod liver oil was destroyed by bubbling oxygen through the oil, but the antirachitic factor was still present. McCollum named the new antirachitic factor "vitamin D." At that time there was enormous interest in the micronutrients and the "vitamine" concept. The name "vitamin D" stuck. Obviously, if vitamin D cured rickets, it must be a vitamin deficiency disease! It is only now in hindsight that we can set the historical record straight.

VITAMIN D DOES NOT FIT STANDARD DEFINITIONS OF EITHER VITAMIN OR HORMONE

A vitamin is defined as an "organic substance, present in minute amounts in natural foodstuffs, which is essential to normal function and lack of which in the diet causes deficiency diseases." "Vitamin D" fails to meet many of these requirements. First, it is not naturally present in most foodstuffs. Milk contains 400 USP units of vitamin D per quart, but this is added as a public health measure and is not naturally present in milk. Second, humans exposed to sufficient sunlight (1 h at noon on the back of the hand is sufficient) have no *dietary* requirement for vitamin D.

Hormones are secreted into the blood by endocrine glands in response to some stimulus and travel by the blood to some distant target where they have specific effects. Vitamin D meets some of these criteria. As we will see later, vitamin D affects intestinal absorptive cells by a mechanism entirely analogous to that used by the steroid hormones. Yet vitamin D is not secreted by a gland, circumscribed or not. Instead, its precursor is formed in the deep layers of the skin and then absorbed into the blood where it binds to a carrier protein.

Making the distinction between vitamin D as a hormone or a vitamin is equivalent to defining a natural environment for humans. Primitive peoples, who live out of doors and regulate their activities by the sun and moon, have no requirement for vitamin D and are seldom afflicted by rickets even though their diet may contain gross deficiencies. Civilized people, who live indoors and regulate their activities by clocks and electric lighting, require dietary vitamin D even though their diet may otherwise contain no errors.

THE LIVER ACTIVATES VITAMIN D BY 25-HYDROXYLATION: THE KIDNEY ACTIVATES IT BY 1-HYDROXYLATION

Vitamin D synthesized in the skin is transported in blood by a 52-kDa globulin also called **transcalciferin**. Vitamin D may also be absorbed efficiently (80%) from the diet and then transported to the liver in chylomicrons. In the liver, vitamin D is converted to 25(OH) vitamin D by a microsomal enzyme. This reaction is inhibited by its product, 25(OH) vitamin D.

25(OH) vitamin D is further hydroxylated in the kidney proximal tubule by a mitochondrial enzyme, 25(OH) 1α hydroxylase. The reaction produces **1,25 (OH)$_2$vitamin D**. This is the most active form of vitamin D with regard to Ca^{2+} and Pi homeostasis. It is referred to as **calcitriol** because it has three hydroxyl groups: one at the 3 position, originally present in 7-dehydrocholesterol, and the ones at the 25 position and 1 position added by hydroxylation in the liver and kidney, respectively. The 1α hydroxylase enzyme is regulated by PTH, plasma [Ca^{2+}] and [Pi] and by fibroblast growth factor 23 (FGF23). Figure 9.7.6 summarizes vitamin D metabolism.

VITAMIN D INACTIVATION BEGINS WITH 24-HYDROXYLATION

Kidney mitochondria also contain another hydroxylase that converts 25(OH) vitamin D to 24,25(OH)$_2$ vitamin D. The product, another triol, may have some biological activity or it may represent the beginning of degradation. This reaction, and that of the 1α hydroxylase, is regulated.

PTH CONTROLS METABOLISM OF VITAMIN D

Figure 9.7.7 shows that conversion of 25(OH) vitamin D to 1,25(OH)$_2$ vitamin D or 24,25(OH)$_2$ vitamin D depends critically on the plasma [Ca^{2+}]. At just about the normal plasma [Ca^{2+}], production of 1,25(OH)$_2$ vitamin D turns on when plasma [Ca^{2+}] falls, and the falling [Ca^{2+}] turns off production of 24,25(OH)$_2$. This metabolic switch makes sure the active form of vitamin D helps raise plasma [Ca^{2+}] back toward normal. These effects are mediated through changes in the amounts of 1α hydroxylase present in the kidney.

PTH and low plasma [Pi] stimulate the 1α hydroxylation reaction that converts 25(OH)D to 1,25(OH)$_2$D. The product 1,25(OH)$_2$D weakly inhibits further formation of 1.25(OH)$_2$D$_3$. Plasma [Ca^{2+}] also directly controls the 1α hydroxylase enzyme, probably through CaSR present in the kidney. Thus the 1α hydroxylase appears to be under the control of plasma [Ca^{2+}] both directly and indirectly through PTH: when plasma [Ca^{2+}] falls near the normal plasma [Ca^{2+}], it activates PTH secretion (see Figure 9.7.3). PTH then stimulates the 1α hydroxylase enzyme, forming the active form of vitamin D, 1,25(OH)$_2$D. Because PTH

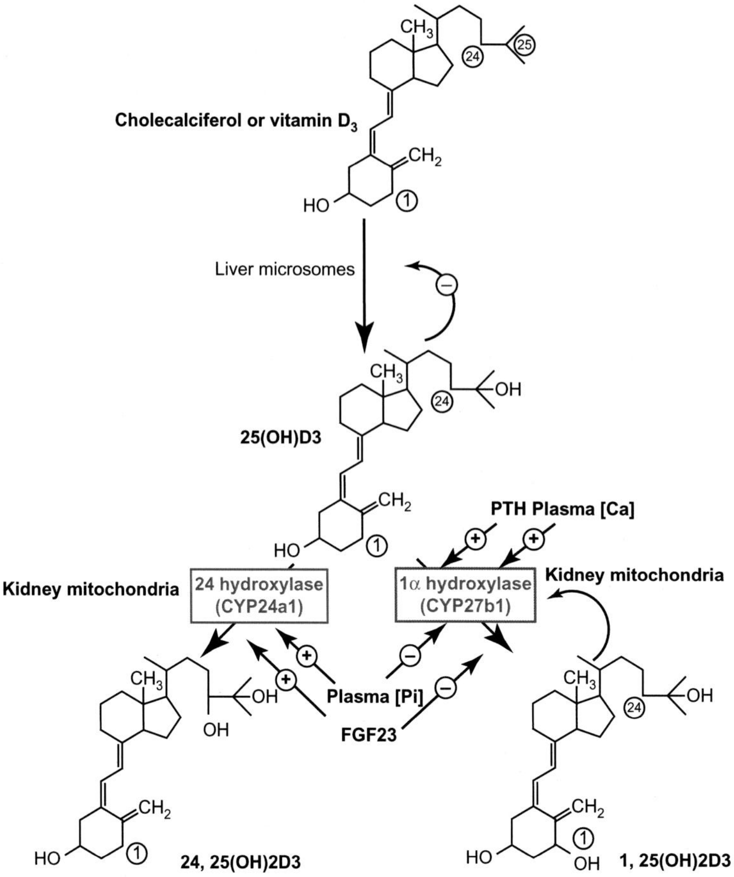

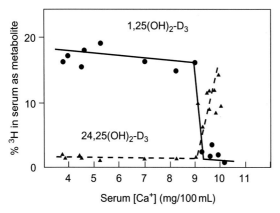

FIGURE 9.7.6 Summary of the control of vitamin D metabolism. Cholecalciferol that is made in the skin (see Figure 9.7.5) travels through the blood bound to a carrier protein. It is hydroxylated first in the liver at the 25 position to form 25(OH)D$_3$. The kidneys then add another hydroxyl group at the 1 position to make 1,25(OH)$_2$D$_3$, the most potent form of the vitamin. The kidney 1α hydroxylase enzyme is regulated by plasma [Pi], [Ca^{2+}] and FGF23. Inactivation of vitamin D also begins in the kidney by hydroxylation at the 24 position.

FIGURE 9.7.7 Switch of vitamin D metabolism from 1,25(OH)$_2$D to 24,25(OH)$_2$D with increasing plasma [Ca^{2+}] above normal.

is the major stimulant of the 1α hydroxylase enzyme, it is the major determinant of circulating levels of 1,25 (OH)$_2$D$_3$. FGF23 inhibits synthesis of the 1α hydroxylase enzyme.

The 24-hydroxylase enzyme is stimulated by normal plasma Pi levels and by 1,25(OH)$_2$D.

VITAMIN D HAS TWO FORMS OF EQUAL POTENCY IN HUMANS

Cholecalciferol is known as vitamin D$_3$ for historical reasons. The chemical structure of vitamin D$_2$ was worked out earlier than that of D$_3$ because of the availability of ergosterol, a sterol derived from fungus. UV light converts ergosterol to **ergocalciferol**, an analogue of cholecalciferol. It is known as vitamin D$_2$ and equals vitamin D$_3$ in preventing rickets. One international unit of vitamin D = 0.025 μg; 400 IU or 10 μg of irradiated ergosterol are typically added per quart of milk. The US RDA (recommended dietary allowance) for vitamin D is 400 IU day^{-1}. The structures of ergosterol and ergocalciferol are shown in Figure 9.7.8. Ergocalciferol differs from cholecalciferol by a double bond and an extra methyl group.

25(OH)$_2$D IS THE MAJOR CIRCULATING FORM OF VITAMIN D

Table 9.7.1 summarizes the half-lives and circulating levels of vitamin D, 25(OH)D and 1,25(OH)$_2$D. The circulating levels of vitamin D depend on exposure to

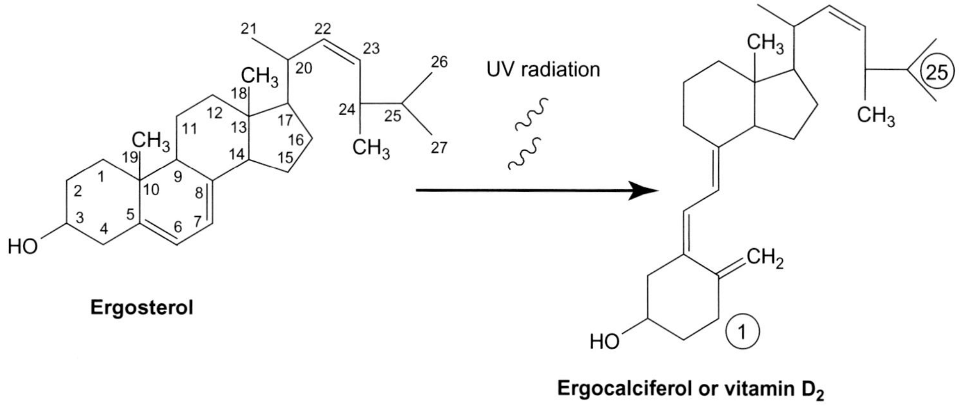

FIGURE 9.7.8 Structure of ergosterol and vitamin D_2, ergocalciferol.

TABLE 9.7.1 Half-Life and Circulating Levels of Vitamin D Metabolites

Metabolite	Half-Life	Plasma Concentration
Vitamin D	~1 day	0–220 μg/L = 0–310 nM
25(OH)D	~3 weeks	8–60 μg/L = 20–150 nM
1,25(OH)$_2$D	~5 h	16–60 ng/L = 40–150 pM

sunlight and dietary content of vitamin D; the circulating levels of 25(OH)D integrate the circulating vitamin D levels and so it serves as a long-term index of vitamin D status. The levels of 1,25(OH)$_2$D are adjusted more quickly and so it reflects shorter term regulation of Ca^{2+} and Pi homeostasis.

VITAMIN D MAINTAINS CONDITIONS FOR BONE MINERALIZATION

The mechanism of action of vitamin D on its target tissues will be discussed in Chapter 9.8. In summary:

- Vitamin D increases Ca^{2+} and Pi absorption from the intestine.
- Vitamin D increases Ca^{2+} and Pi reabsorption from the renal tubules.
- Vitamin D promotes bone resorption.
- Vitamin D increases FGF23 release from bone cells.

The "goal" of vitamin D differs from that of PTH. PTH responds to hypocalcemia and its effects correct the condition by resorbing bone and discarding the Pi generated from the resorption. Vitamin D also resorbs bone, but it also promotes bone mineralization by raising both plasma $[Ca^{2+}]$ and [Pi]. The actions of vitamin D on both intestine and kidney are to raise both $[Ca^{2+}]$ and [Pi].

BONE CELLS RELEASE FGF23 IN RESPONSE TO A VARIETY OF SIGNALS

FGF23, fibroblast growth factor 23, primarily regulates phosphate metabolism, but in the course of doing so also regulates calcium balance. Bone cells, particularly osteocytes, mainly release FGF23 in response to PTH, 1,25(OH)$_2$D$_3$, P$_i$ and sympathetic nervous stimulation. Osteocytes synthesize FGF23 as a 251-amino acid protein with a M_r of 32 kDa due to several sites of O-glycosylation. The N-terminal 24 amino acids comprise the signal peptide that directs the protein to the endoplasmic reticulum. FGF23 interacts with **α-Klotho (KL)** on the surface membrane of target tissues. However, some KL is also secreted into the circulation, forming soluble KL or sKL. Klotho is a coreceptor that interacts with fibroblast growth factor receptor to bind FGF23. In some tissues, FGF23 may have effects independent of Klotho.

The main effects of FGF23 are:

- reduce synthesis of 1α vitamin D hydroxylase in the kidney and increase 24 hydroxylase
- increase phosphate excretion by the kidney
- inhibit PTH secretion.

SUMMARY

The body regulates free plasma calcium concentration within narrow limits. Total plasma $[Ca^{2+}]$ is about 10 mg%, but only about half of this is free in the plasma with the remainder being bound primarily by albumin. Plasma $[Ca^{2+}]$ is linked to plasma [phosphate] because when bone is resorbed both Ca^{2+} and phosphate are released into the plasma, and mineralization of the organic matrix of bone removes both from plasma. All body Ca^{2+} and Pi derives ultimately from maternal stores or from the diet. Youngsters exhibit a positive calcium balance in which calcium is stored in the bone. Adults often make small but regular withdrawals from this store of mineral in order to stabilize blood levels of Ca^{2+} and Pi. The resulting long term but small negative calcium balance over many years can lead to **osteoporosis**.

Four hormones regulate calcium and Pi balance: **PTH, CT, vitamin D** and FGF23. These affect mineral transport in three organs: the **intestine**, the **bone**, and the **kidney**.

The parathyroid glands secrete an 84 amino acid protein hormone, PTH, in response to low plasma $[Ca^{2+}]$. The relationship between secretion and plasma $[Ca^{2+}]$ is steep and depends on the rate of fall of plasma $[Ca^{2+}]$ as well as the level of plasma $[Ca^{2+}]$. PTH increases the resorption of mineral from bone, increases renal excretion of Pi and retention of Ca^{2+}, and activates vitamin D. These actions highlight the "goals" of PTH, which is to keep plasma $[Ca^{2+}]$ within its normal limits.

Parafollicular cells in the thyroid gland secrete a 32 amino acid polypeptide, CT, in response to hypercalcemia. CT stops bone resorption by direct action on bone cells called osteoclasts. Its importance in adult humans is not established, because there are no diseases associated with over- or underproduction of CT.

Vitamin D is inappropriately named because free-living humans in a natural, outdoor environment have no dietary requirement for it. UV radiation from the sun converts 7-dehydrocholesterol in the skin to pre-vitamin D, which nonenzymatically converts to cholecalciferol, or vitamin D_3. Cholecalciferol undergoes two hydroxylation reactions: the liver converts it to $25(OH)D_3$ and the kidney converts $25(OH)_2D_3$ to $1,25(OH)_2D_3$. PTH hormone stimulates the 1α hydroxylase enzyme in the kidney, whereas high plasma [Pi] inhibits the reaction. The active form of vitamin D, $1,25(OH)_2D_3$ increases Ca and Pi absorption from the intestine and kidney and has multiple effects on bone.

FSG23 is a protein hormone produced by osteocytes in response to PTH, increased plasma Pi, and $1,25(OH)_2D_3$. Its major effects are to inhibit PTH secretion, inhibit production of $1,25(OH)_2D_3$, and increase loss of phosphate in the urine.

REVIEW QUESTIONS

1. What is meant by positive calcium balance? Negative calcium balance? Where is most of the body's calcium found?
2. Why is total plasma $[Ca^{2+}]$ dependent on plasma proteins? How does plasma pH change free plasma $[Ca^{2+}]$? What is regulated?
3. Why is Ca^{2+} homeostasis linked to Pi homeostasis? What is the normal plasma free $[Ca^{2+}]$? What is normal total plasma $[Ca^{2+}]$? What is normal plasma P? Why is the normal plasma [Pi] given as phosphorus instead of phosphate?
4. What is PTH? Where is it secreted? What cells secrete it? What is the stimulation for secretion? What does PTH do (in general terms)?
5. What is CT? Where is it secreted? What cells secrete it? What is the stimulation for secretion? What does CT do?
6. What is vitamin D? Where is it secreted? What is the stimulation for secretion? How is it activated? Where does activation occur? What stimulates activation? What inhibits it?

Calcium and Phosphorus Homeostasis II: Target Tissues and Integrated Control 9.8

Learning Objectives

- Identify the parts of bone including endosteum, periosteum, diaphysis, epiphysis, and epiphyseal plate
- Identify the three major cell types in bone: osteoblast, osteocyte, and osteoclast
- Describe the organic matrix of bone and its mineralization
- Describe the location and function of osteoblasts
- Describe the recruitment of osteoclasts from hematopoietic stem cells
- Describe the mechanism of osteoclastic resorption of bone
- Describe remodeling of bone
- List the sequential steps in Ca and Pi absorption from the intestine
- Describe how vitamin D increases Ca and Pi absorption from the intestine
- Describe how the intestine adapts to diets containing differing amounts of Ca and Pi
- Describe the actions of PTH, vitamin D, and FGF23 on Ca and Pi reabsorption by the kidney tubule
- Compare the overall effect of PTH on Ca and Pi homeostasis to that of vitamin D and FGF23
- Trace the negative feedback loops involved in the homeostatic response to low plasma Ca

THE SKELETON GIVES US FORM AND SUPPORT

The skeleton is one of the largest organ system of the body, consisting of 206 bones that vary greatly in size and shape. The larger bones of the limbs are called **long bones**. Smaller bones of the wrist and ankle are called **short bones**. The bones of the skull are classified as **flat bones** and the remainder, such as the vertebrae, are **irregular bones**. All bones consist of a compact **cortex** of bone that surrounds a meshwork of **trabeculae**. Cortical bone is dense or compact bone that comprises about 80% of the mass of the skeleton. The gross appearance of trabecular bone resembles a sponge, and so it is also called **cancellous** bone or **spongy bone**. The interstices between the trabeculae of bone are filled with the red marrow that makes red blood cells. Bones consist mostly of mineralized organic matrix and a small but active cellular component. Figure 9.8.1 shows the general structure of bone.

The articular surfaces of bones are covered with articular cartilage. Elsewhere the surface is covered by a connective tissue membrane, the **periosteum**. The outer fibrous layers of the periosteum contain periosteal blood vessels. The inner layers of the periosteum contain mesenchymal stem cells that can proliferate and differentiate into **osteoblasts**. Osteoblasts lay down the organic matrix of bone and aid in its mineralization. Therefore, these osteoblasts make bone, either in response to injury or due to remodeling of existing bone. The **endosteum** covers the surfaces of the bone that face the marrow. The endosteum is thinner than the periosteum, but it also contains cells that can make bone (see Figure 9.8.1).

The shaft of the long bones is called the **diaphysis**, whereas the end of the bone is the **epiphysis**. Mineralization of the diaphysis and epiphysis begins independently. The region between contains cartilage, the **epiphyseal plate** responsible for the growth of the long bones. When the epiphyseal plate is mineralized, no more growth occurs. This is referred to as **closure** of the epiphyseal plate.

OSTEOBLASTS ARE SURFACE CELLS THAT LAY DOWN THE ORGANIC MATRIX OF BONE

Osteoblasts form a continuous layer over most of the bone. These cells synthesize and secrete a variety of proteins which together make up the **organic matrix** of bone, called the **osteoid**, which consists primarily of **type I collagen**. Noncollagen proteins make up about 10% of the organic matrix and play important roles in the regulation of mineralization. The collagen fibers line up in regular arrays and form nucleation sites along which Ca^{2+} and Pi salts can crystallize. Osteoblasts are recruited from mesenchymal or fibroblast-type cells that form osteoprogenitor cells and then mature osteoblasts. The normal mineralization of the osteoid depends critically on the concentration of Ca^{2+} and Pi in the plasma.

OSTEOCYTES ARE EMBEDDED DEEP WITHIN BONE

Osteoblasts that become completely surrounded by the organic matrix become embedded in the bone and become **osteocytes**. They reside in little pockets called

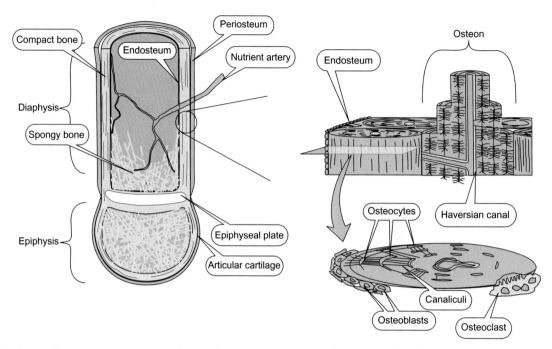

FIGURE 9.8.1 Gross and microscopic components of bone. Bone surfaces are completely covered with cells. On the outside, the periosteum is thick and consists of an outer fibrous layer and an inner layer of cells that can proliferate to form osteoblasts. The inner endosteum is thinner. The shaft of mature bones consists of compact bone organized into osteons. An Haversian canal is central to each osteon, which consists of several layers or lamellae of osteocytes, bone cells that are deeply embedded in the bone. These communicate through thin cytoplasmic processes that course through tiny channels in the bone, the canaliculi. Osteoclasts are large, multinucleated cells that resorb bone.

lacunae that communicate through tiny **canaliculi**, which connect with other osteocytes and with surface osteoblasts. These canaliculi provide an enormous surface area over which Ca^{2+} and Pi can exchange between plasma and bone.

OSTEOCLASTS DESTROY THE ORGANIC MATRIX OF BONE AND RELEASE BOTH Ca^{2+} AND Pi

Osteoclasts are giant, multinucleated cells that are recruited from a pool of circulating stem cells that also give rise to tissue macrophages. These circulating cells fuse together and attach to the surface of bone. The osteoclast forms a "ruffled border" adjacent to the bone where enzymes and H^+ ions are secreted onto the bone to dissolve the mineral and digest the organic matrix. Thus osteoclastic activity forms a resorption cavity and completely destroys the bone that is resorbed.

BONE IS CONSTANTLY BEING REMODELED

Osteoclasts form a resorption cavity and migrate along, leaving a resorption canal in their wake. Osteoblasts invade the resorbed area and begin laying down new bone, eventually becoming entrapped in the bone as new osteocytes. This process of bone remodeling occurs continuously throughout life. During maturation of bone, the bone forms in a sequence of **woven bone** and **lamellar bone** (see Figure 9.8.2). The balance of bone is set by the rates of resorption versus rebuilding. If the osteoblasts cannot keep up with the osteoclasts, there will be a net loss of bone and eventually a loss of bone strength.

OSTEOBLASTS MAKE OSTEOID AND SIGNAL BONE RESORPTION

As described earlier and shown in Figure 9.8.2, osteoblasts originate from mesenchymal stem cells which form **osteoprogenitor cells** or **preosteoblasts**. Osteoblasts vary in size and function; they form woven bone in early development and lamellar bone later on. The osteoblasts secrete proteins that form the organic matrix of bone. Table 9.8.1 summarizes some of these proteins and their postulated function. **Osteocalcin** is unusual, in that this 6-kDa protein has three γ carboxyglutamic acid residues. These amino acids are produced by posttranslational modification involving vitamin K-dependent enzymes (see Chapters 2.3 and 5.1). These residues bind Ca^{2+} avidly. $1,25(OH)_2D_3$ stimulates the synthesis of this protein, and it is postulated to have some role in mineralization. Osteocalcin also feeds back on osteoblasts, inhibiting further bone formation and osteoblast function.

In addition, cells of the osteoblast lineage are important to the initiation of bone resorption. Most of the hormonal factors that stimulate bone resorption act directly on **osteoblasts** or their precursors. Parathyroid hormone (PTH) and $1,25(OH)_2D_3$ stimulate these cells to release **macrophage colony stimulating factor (M-CSF)**

Calcium and Phosphorus Homeostasis II: Target Tissues and Integrated Control

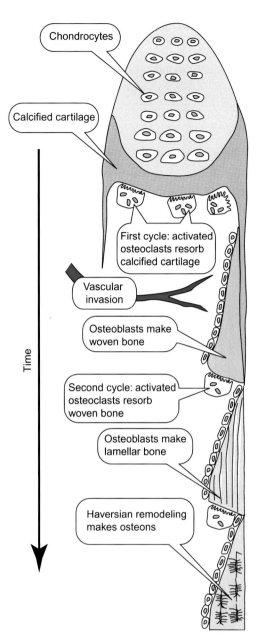

FIGURE 9.8.2 Maturation of endochondral bone. Chondrocytes lay down the cartilage that eventually forms bone. The calcified cartilage is resorbed by osteoclasts and blood vessels invade the structure. Bone resorption then reverses to form bone by osteoblasts. This first bone is woven bone. A second cycle of bone resorption and formation makes lamellar bone. Cortical bone in the adult undergoes further osteoclastic resorption to form resorption canals, which, when filled in by the osteoblasts, form the osteons.

and express **receptor activator of nuclear factor** κB **ligand (RANKL)** on the surface of the osteoblasts.

M-CSF is a factor in the hematopoietic generation of macrophages (see Figure 5.3.3). Osteoblasts make a membrane-bound and soluble form of M-CSF in response to stimulators of bone resorption (1,25 $(OH)_2D_3$, PTH, IL-1, IL-6, TNF, prostaglandins). Hematopoietic cells stimulated by GM-CSF (granulocyte-macrophage colony stimulating factor) and M-CSF differentiate into either monocyte macrophages or preosteoclasts. The preosteoclasts fuse to form multinucleated osteoclasts.

Osteoblasts also make a membrane-bound RANKL in response to stimulators of bone resorption. Its action is opposed by another osteoblast product, **osteoprotegerin (OPG)**. OPG is a soluble receptor for RANKL that prevents RANKL from interacting with its receptor, RANK, on osteoclast precursors and osteoclasts. OPG thus opposes bone resorption, and the balance between RANKL and OPG sets the activity of the osteoclasts. For many stimulators of bone resorption (PTH, 1,25 $(OH)_2D_3$, PGE_2), inhibition of OPG secretion accompanies stimulation of RANKL production, so there is reciprocal action on RANKL and OPG that activates osteoclast genesis and bone resorption (see Figure 9.8.3).

OSTEOCLASTS RESORB BONE

Osteoclasts are giant cells containing between 10 and 20 nuclei. They closely attach to the bone matrix by binding its surface integrins to a bone protein called vitronectin. This close apposition seals off an area of the bone beneath the osteoclast and allows the osteoclast to form a microenvironment that resorbs bone. The area of the osteoclast next to bone forms a "ruffled border" consisting of multiple infoldings of the osteoclast cell membrane. It secretes acid and proteases across the ruffled border, and these dissolve the mineral of bone and destroy the organic matrix (see Figure 9.8.4).

CALCITONIN SHUTS OFF OSTEOCLAST RESORPTION

According to the previous sections, both PTH and 1,25 $(OH)_2D_3$ promote bone resorption, but they do this without directly affecting the osteoclast or its precursors. Indeed, osteoclasts have no receptors for PTH. Instead, PTH and 1,25$(OH)_2D_3$ increase bone resorption by acting on osteoblasts to recruit stem cells to form osteoclasts. Osteoclasts do have receptors for calcitonin (CT). CT binds to a G_s protein on the surface of the osteoclasts. This increases cAMP in these cells and shuts off bone resorption. This effect may be of little importance in adults as there are no diseases associated with CT undersecretion or oversecretion.

SUMMARY OF HORMONE EFFECTS ON BONE

PTH INCREASES OSTEOCYTIC OSTEOLYSIS

Osteocytes have receptors that respond to PTH by increasing cAMP. This activates them to resorb bone by a rapid **osteocytic osteolysis**. This removes exchangeable Ca^{2+} from bone mineral but does not destroy bone.

PTH INCREASES OSTEOCLASTIC OSTEOLYSIS

PTH stimulates the formation of new osteoclasts from circulating stem cells and stimulates bone resorption from existing osteoclasts. However, osteoclasts do not have receptors for PTH, so these effects are mediated by

TABLE 9.8.1 Some Proteins Synthesized and Secreted by Osteoblasts

Protein Name	Chemical Nature	Possible Function
Type I collagen	Two α1 chains + one α2 chain; forms triple helix that is stabilized by hydroxylation of lysine and proline	Provides tensile strength to bone; makes nidus for nucleation of Ca and P salts
Osteocalcin	6 kDa; has three γ carboxyglutamic acid residues	May regulate mineralization; negative regulator of osteoblasts
Matrix GLA protein	Contains γ carboxyglutamic acid residues	Inhibits mineralization
Vitronectin		Attaches cells; binds collagen and heparin
Osteopontin		Binds cells; mediates effects of mechanical stress on osteoblasts and osteoclasts
Alkaline phosphatase		Hydrolyzes inhibitors of mineralization
Osteonectin		May regulate mineralization

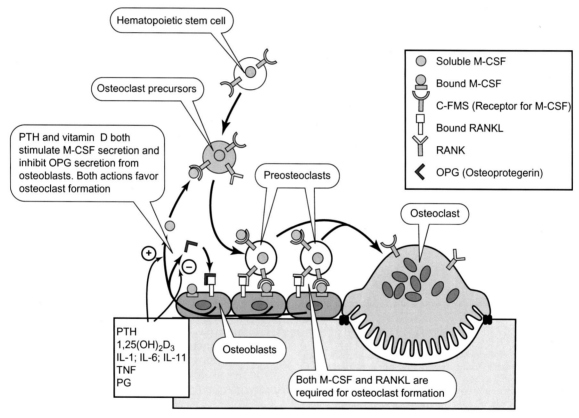

FIGURE 9.8.3 Formation of osteoclasts from hematopoietic precursors. Cells in the granulocyte/macrophage progenitor line have receptors for M-CSF, macrophage colony stimulating factor. When stimulated by factors that favor bone resorption, osteoblasts secrete soluble M-CSF and express it on their surfaces. The hematopoietic stem cells differentiate into monocyte macrophages (not shown) or preosteoclasts. Osteoclasts and their immediate precursors express RANK (receptor activator for nuclear factor κB, NF-κB). Osteoblasts also express surface RANK ligand (RANKL) that binds RANK on the preosteoclasts. These mononuclear preosteoclasts then fuse to form the giant, multinucleated osteoclasts. Osteoblasts also secrete a soluble protein called osteoprotegerin (OPG) that binds to RANKL, preventing it from activating RANK. Both M-CSF and RANKL are essential for formation of osteoclasts and animals lacking either M-CSF or RANKL develop osteopetrosis (bone completely filling the medullary cavity). PTH and vitamin D stimulate synthesis of both soluble and membrane-bound M-CSF and RANKL. PTH and vitamin D also inhibit synthesis of OPG, which removes inhibition of osteoclast formation.

paracrine signals from osteoblasts. The result is that both Ca^{2+} and Pi are moved from bone to plasma. As could be expected, recruitment of additional osteoclasts takes some time, so **osteoclastic osteolysis** is a later response of PTH stimulation.

The result of osteocytic osteolysis is to remove the mineral from the osteoid without destroying the osteoid, whereas osteoclastic osteolysis both removes the mineral and destroys the osteoid. In both cases, Ca^{2+} and Pi are moved from bone to plasma.

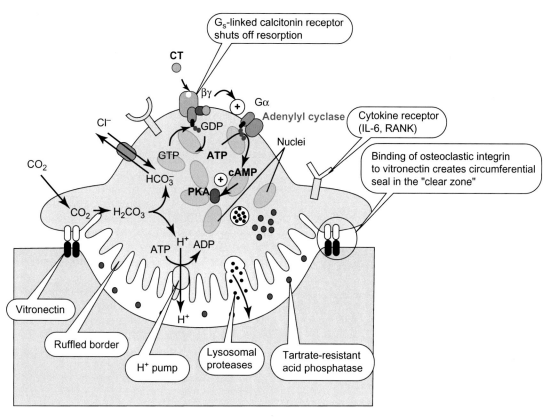

FIGURE 9.8.4 Mechanism of osteoclastic resorption of bone. Osteoclasts form a seal around a resorption area by binding their integrins with a bone protein, vitronectin. The ruffled border incorporates a vacuolar-type H^+ pump that acidifies the extracellular area beneath the osteoclast. Lysosomal proteases and acid phosphatases are released by the osteoclast to break down the organic matrix. Osteoclastic activity is stimulated by cytokines such as IL-6 and RANK and inhibited by calcitonin.

CT DECREASES BONE RESORPTION

The major effect of CT on bone is to inhibit osteoclastic osteolysis. CT acts through surface membrane receptors that activate adenylyl cyclase and increase cAMP levels in osteoclasts.

1,25(OH)$_2$D STIMULATES BONE RESORPTION

1,25(OH)$_2$D is the active form of vitamin D (cholecalciferol). PTH controls its production, and PTH levels, in turn, respond to plasma [Ca^{2+}]. 1,25(OH)$_2$D stimulates bone resorption by itself, but it is most potent in stimulating PTH-induced bone resorption. It also induces osteocalcin production and inhibits collagen production by osteoblasts.

MANY OTHER FACTORS AFFECT THE SKELETON

Factors that alter skeletal metabolism include the following:

- thyroid hormone (hypothyroidism leads to stunted growth);
- glucocorticoids inhibit bone formation (hypercortisolism is associated with osteoporosis);
- gonadal hormones are critical for growth and maintenance;
- insulin is necessary for proper growth of the skeleton;
- paracrines: IL-1, Il-6, TNF-α;
- in youth, growth hormone and FGF-21 regulate the growth of the bones (see Chapter 9.2).

ONLY 1,25(OH)$_2$D DIRECTLY AFFECTS INTESTINAL Ca^{2+} AND Pi ABSORPTION

Ca^{2+} absorption from the intestine entails the movement of Ca^{2+} from the lumen across a sheet of columnar epithelial cells, the **enterocytes**, and into the blood. These epithelial cells are welded together by a ring of connections that goes all the way around the cells near the luminal border. These "welds" have been called the "tight junctions" because in electron micrographs they appeared to hold the cells tightly together. However, some materials can pass between the cells, in what is called the **paracellular** pathway. Alternatively, materials can move from the lumen to the blood by going through the cell, the **transcellular** pathway. The overall Ca^{2+} absorption appears to consist of a saturable, active transport pathway and a nonsaturable passive pathway (see Figure 9.8.5). Ca^{2+} enters the cells through apical Ca^{2+} channels, TRPV5 and TRPV6 (for transient receptor potential vanilloid type 5 and 6). Both of these are expressed in both intestine and kidney, but TRPV6 is the major isoform in the intestine and TRPV5 is the major isoform in the kidney. Both TRPV5 and TRPV6 form heterotetrameric channel complexes with different properties. Both are inactivated by intracellular Ca^{2+}. Intestine also possesses an L-type Ca^{2+} channel, $Ca_v1.3$, that may contribute to apical uptake in the transcellular active Ca^{2+} transport.

In the transcellular route, Ca^{2+} is then transported across the cell and is pumped across the basolateral membranes by the **plasma membrane Ca-ATPase (PMCA)** and probably by Na^+–Ca^{2+} exchange as well.

The active form of vitamin D, $1,25(OH)_2D$, increases Ca^{2+} absorption from the intestines by classical steroid mechanisms that are described in Chapter 2.8. The hormone is carried in blood by a 52-kDa globulin, the vitamin D binding protein. The free hormone penetrates the basolateral membranes of the enterocyte and binds to a nuclear receptor that alters its conformation and binds to **vitamin D responsive elements** on the DNA, resulting in transcription of mRNA that codes for specific proteins. Vitamin D stimulates the transcription of mRNA that codes for **calbindin**, a soluble, low-molecular-weight protein (9 kDa) that binds two Ca^{2+} atoms per molecule with high affinity. Calbindin is found in highest concentration in tissues that actively transport Ca^{2+}, such as the intestine and kidney, but it is also found in other cells such as the Purkinje cells of the cerebellum. The function of calbindin is not yet established, but two functions seem likely: it buffers Ca^{2+} during its transport across the cell, and it enhances the diffusion of Ca^{2+} through the cytosol by carrying it. The total diffusion through the cytosol consists of free Ca^{2+} diffusion and diffusion of Ca^{2+} bound to calbindin. The diffusive flux is dominated by the slower bound Ca^{2+} because its total concentration is much higher than that of the free $[Ca^{2+}]$. Thus the cytosol acts as a permeability barrier because the free $[Ca^{2+}]$ is maintained so low, and calbindin acts much like a carrier in facilitated diffusion. The buffering action of calbindin relieves intracellular inhibition of Ca^{2+} entry by lowering the $[Ca^{2+}]$ immediately adjacent to the apical membrane, and increases transport at the basolateral membrane by increasing $[Ca^{2+}]$ to the Ca^{2+}-starved PMCA (see Figure 9.8.6). $1,25(OH)_2D$ also increases the transcription of genes coding for TRPV6 and possibly TRPV5, and for the exit mechanisms on the basolateral membrane, PMCA and NCX. Evidence is accruing suggesting that $1,25(OH)_2D$ also regulates flux through the paracellular pathway by regulating the expression of claudin proteins Cldn-2 and Cldn-12.

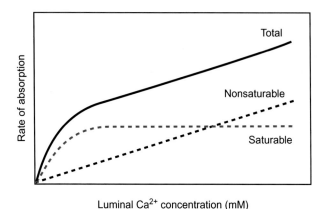

FIGURE 9.8.5 Kinetics of Ca^{2+} absorption from the gastrointestinal tract as a function of luminal $[Ca^{2+}]$. In most parts of the intestine there are two processes, a saturable process that is due to active transport and is transcellular, and a paracellular pathway that obeys passive transport kinetics. The overall transport is a sum of the two processes. The relative importance of each varies with location within the intestine.

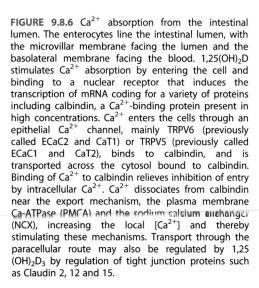

FIGURE 9.8.6 Ca^{2+} absorption from the intestinal lumen. The enterocytes line the intestinal lumen, with the microvillar membrane facing the lumen and the basolateral membrane facing the blood. $1,25(OH)_2D$ stimulates Ca^{2+} absorption by entering the cell and binding to a nuclear receptor that induces the transcription of mRNA coding for a variety of proteins including calbindin, a Ca^{2+}-binding protein present in high concentrations. Ca^{2+} enters the cells through an epithelial Ca^{2+} channel, mainly TRPV6 (previously called ECaC2 and CaT1) or TRPV5 (previously called ECaC1 and CaT2), binds to calbindin, and is transported across the cytosol bound to calbindin. Binding of Ca^{2+} to calbindin relieves inhibition of entry by intracellular Ca^{2+}. Ca^{2+} dissociates from calbindin near the export mechanism, the plasma membrane Ca-ATPase (PMCA) and the sodium calcium exchanger (NCX), increasing the local $[Ca^{2+}]$ and thereby stimulating these mechanisms. Transport through the paracellular route may also be regulated by $1,25(OH)_2D_3$ by regulation of tight junction proteins such as Claudin 2, 12 and 15.

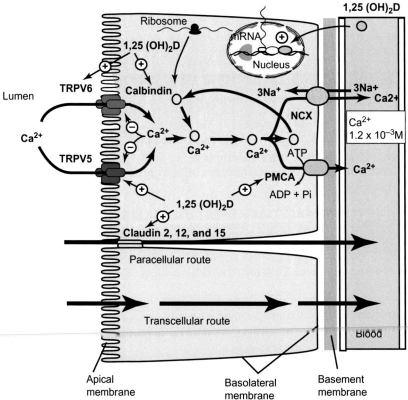

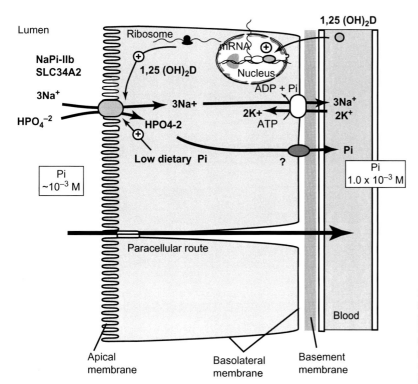

FIGURE 9.8.7 Absorption of Pi from the intestinal lumen. Pi absorption follows a transcellular route and a paracellular route. In the transcellular route, the NaPi-IIb transporter brings Pi into the cell along with 3 Na^+ ions. Pi exits the cell presumably by facilitated diffusion over a carrier that is not yet identified. 1.25$(OH)_2$D stimulates the transcription of the gene coding for the NaPi-IIb transporter. Low dietary phosphate also increases Pi absorption.

Like Ca^{2+}, Pi absorption from the intestine appears to have two components: a Na^+-dependent active component and a Na^+-independent passive component that probably occurs through the paracellular pathway. The active pathway begins with the cotransport of Pi and Na^+ at the apical or luminal surface of the cell, mediated mainly by NaPi-IIb (SLC34A2). The intestine expresses other NaPi transporters, PiT1 (SLC20A1) and PiT2 (SLC20A2), but their contribution appears to be minor. Pi concentrations in the cell are higher than plasma concentrations, so the exit of Pi at the basolateral cell does not require energy. It is believed that the basolateral membrane contains a carrier for Pi exit into the extracellular fluid adjacent to the cell, from which Pi can then diffuse into the blood (see Figure 9.8.7).

1,25$(OH)_2$D stimulates intestinal Pi absorption by increasing the transcription of the gene coding for NaPi-IIb. Low dietary phosphate stimulates Pi absorption by posttranslational modification of the NaPi-IIb. **Alkaline phosphatase**, an enzyme with relatively nonspecific ester phosphatase activity, is located on the intestinal brush-border or microvillar membrane, and 1,25$(OH)_2$D increases the activity of this enzyme. Its role in Pi or Ca^{2+} absorption is not known.

THE INTESTINE ADAPTS TO DIETS CONTAINING DIFFERING AMOUNTS OF Ca^{2+} AND Pi

The body adapts to diets containing differing amounts of Ca^{2+} and Pi by adjusting its ability to absorb these minerals from the food. The absorptive capability of the intestine can be expressed as the **fractional absorption**, meaning the fraction of dietary Ca^{2+} that is absorbed.

When the diet is low in Ca^{2+}, the fractional absorption increases. Thus the fractional absorption measures the affinity by which the intestine absorbs Ca^{2+}, but it does not reflect the actual amount of Ca^{2+} that is absorbed. All other things being equal, increasing dietary Ca^{2+} always results in increased amounts of Ca^{2+} that is absorbed. Adaptation blunts the effects of changes in dietary Ca^{2+} by increasing the fractional absorption when dietary Ca^{2+} is low and by decreasing it when dietary Ca^{2+} is high.

The intestinal absorption of Ca^{2+} also adapts to the dietary content of Pi. When dietary Pi is low, the absorption of Ca^{2+} remains high and decreases little when dietary Ca^{2+} changes. When dietary Pi is high, there is a steep response of absorption with dietary Ca^{2+}. Thus low dietary Pi stimulates absorption of Ca^{2+} by itself, regardless of Ca^{2+}, whereas in the presence of adequate Pi, low dietary Ca^{2+} can stimulate absorption.

Experiments in animals have established that adaptation to the mineral content of the diet requires functioning parathyroid glands. The intestinal absorption of Ca^{2+} is controlled solely by 1,25$(OH)_2D_3$ whose synthesis in the kidney is stimulated by low Pi and high PTH and blocked by high FGF23. Low dietary Pi leads to a low plasma Pi regardless of the plasma $[Ca^{2+}]$. This stimulates formation of 1,25$(OH)_2D_3$ and intestinal Ca^{2+} absorption remains high, even if the dietary content of Ca^{2+} is high. If dietary Pi is adequate, then reducing the dietary content of Ca^{2+} stimulates PTH secretion, which stimulates formation of 1,25$(OH)_2$D and increases intestinal absorption. High plasma phosphate increases FGF23 that inhibits formation of 1,25$(OH)_2$D. See Figure 9.8.8 for a schematic representation of these effects.

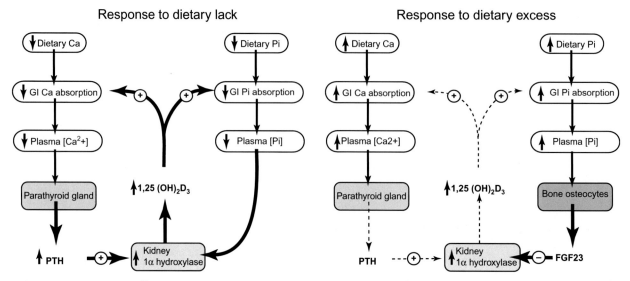

FIGURE 9.8.8 Adaptation of GI Ca^{2+} and Pi absorption to changes in the diet. Low dietary Ca^{2+} increases the ability of the gastrointestinal tract to absorb Ca^{2+} by PTH-induced stimulation of the kidney vitamin D 1α hydroxylase, to increase the circulating levels of $1,25(OH)_2D_3$. Low dietary Pi also increases the 1α hydroxylase enzyme, but not through PTH. High dietary Ca^{2+} reduces the ability to absorb Ca^{2+} by removing stimulation of the 1α hydroxylase by lower levels of PTH. High dietary Pi causes the release of fibroblast growth factor 23 (FGF23) from bone, which inhibits the kidney 1α hydroxylase. PTH, $1,25(OH)_2D_3$, and FGF23 also exert effects on kidney handling of Ca^{2+} and Pi.

REGULATION OF URINARY EXCRETION OF Ca^{2+} AND Pi IS ACHIEVED IN THE DISTAL NEPHRON

The free Ca^{2+} and Ca^{2+} bound by low-molecular-weight solutes are freely filtered by the glomerulus, whereas Ca^{2+} that is bound to plasma proteins such as albumin, globulin, and fibrinogen is retained in the blood. About 60% of the total plasma Ca^{2+} is present as free and complexed forms or about 6 mg%. The filtered load of Ca^{2+} can be readily calculated as GFR × [total Ca^{2+}]$_{ultrafiltrate}$. If the GFR is 125 mL min^{-1}, the filtered load is about 7.5 mg min^{-1}. Of this filtered load, only 1–2% remains in the final urine. The rest is absorbed in various segments of the nephron, the majority being absorbed in the proximal tubule. Significant amounts are absorbed in the thick ascending limb of the loop of Henle and in the distal convoluted tubule. Figure 9.8.9 shows the fraction of filtered Ca^{2+} that is reabsorbed in each nephron segment.

The proximal tubule actively reabsorbs Na^+, with water and a variety of substances following passively through the paracellular route. About 80% of the Ca^{2+} that is reabsorbed in the proximal tubule is absorbed paracellularly, with another 20% being reabsorbed through transcellular route. **PTH binds to a G_s-coupled receptor** on the surface of proximal tubule cells. The resulting increased cAMP has a number of effects, one of which is to **decrease the reabsorption of Na^+, water, HCO_3^-, Ca^{2+}, and Pi.** The mechanism of Pi reabsorption in the proximal tubule is shown in Figure 9.8.10. PTH inhibits Pi reabsorption, but this effect is modulated by Ca^{2+} in the lumen of the tubule. The proximal tubule contains the extracellular Ca sensor, CaSR, on its apical surface. Increases in Ca^{2+} in the lumen inhibit the effects of

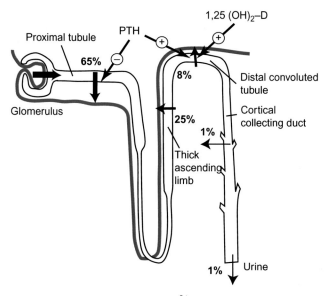

FIGURE 9.8.9 Fraction of filtered Ca^{2+} that is reabsorbed at each nephron segment. Note that PTH *decreases* Ca^{2+} reabsorption in the proximal tubule, whereas it *increases* Ca^{2+} reabsorption in the distal tubule. $1,25(OH)_2D$ increases Ca^{2+} absorption in the distal nephron.

PTH on both the Pi transporters and its stimulation of the vitamin D 1α hydroxylase enzyme that activates vitamin D by producing $1,25(OH)_2D_3$. FGF23 inhibits the transcription of the genes that code for NaPi-IIa and NaPi-IIc and for the 1α hydroxylase. Thus FGF23 inhibits both Pi transport in the proximal tubule and synthesis of $1,25(OH)_2D_3$.

The thick ascending limb of the loop of Henle reabsorbs some 25% of the filtered load mainly through

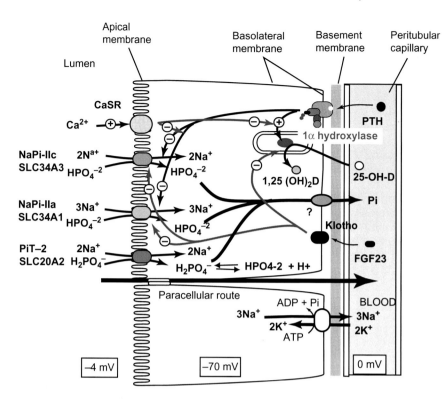

FIGURE 9.8.10 Regulation of Pi transport in the proximal tubule. NaPi-IIa and NaPi-IIb transport Pi into the proximal tubule cell with some contribution by PiT2. Pi exit across the basolateral membrane is not identified. PTH inhibits these transport mechanisms through a G_s mechanism. PTH also activates the vitamin D 1α hydroxylase enzyme that converts circulating $25(OH)D_3$ to its active form, $1,25(OH)_2D_3$. Ca^{2+} in the lumen inhibits these actions of PTH via the extracellular Ca^{2+} sensor, CaSR, on the apical membrane of these cells. FSG23 inhibits the transcription of genes for the Pi transporters on the apical membrane, NaPi-IIa and NaPi-IIc, and the gene for the 1α hydroxylase. FSG23 interacts with αKlotho on the basolateral membrane.

the paracellular route, aided by the lumen positive potential in this segment of the nephron. This reabsorption is passive and is linked to the reabsorption of Na^+, but it does not occur by solvent drag. The extracellular Ca^{2+} sensor is located on the basolateral membrane of the thick ascending limb cells and inhibits Ca^{2+} reabsorption by inhibiting the apical transporters (NKCC2 and ROMK1; see Figure 7.5.5), thereby reducing the transepithelial potential that drives Ca^{2+} reabsorption.

The distal tubule consists of a heterogeneous assortment of cells. Some of these cells contain calbindin, produced in these cells like it is in enterocytes. These cells also contain the TRPV5 that is thought to be involved in transcellular Ca^{2+} transport. **Both PTH and 1,25 $(OH)_2D$ stimulate Ca^{2+} absorption in the distal tubule.**

In the distal tubule, PTH increases the reabsorption of Ca^{2+} but not Pi. Thus PTH causes an increased excretion of Pi or a **hyperphosphaturia**. In the proximal tubule, $1,25(OH)_2D$ increases the reabsorption of Pi by increasing expression of NaPi-IIa and NaPi-IIc.

THE "GOALS" OF PTH, CT, AND VITAMIN D ARE DISTINCT

PTH, CT, and vitamin D, strictly speaking, do not have "goals" because they are inanimate objects without thoughts and anticipatory behavior. The entire physiological system, however, takes on aspects of behavior that are not present in any one part of the system. The systems appear to have goals and if we speak about them as if they do, their behavior makes sense. This way of speaking is called "**teleological**." This means to refer to the final causes of behavior or their purpose.

THE GOAL FOR PTH IS TO RAISE FREE PLASMA [Ca^{2+}]

Hypocalcemia stimulates PTH secretion. PTH then stimulates the bone to release both Ca^{2+} and Pi. Pi released from bone complexes Ca^{2+} and thus prevents the free plasma [Ca^{2+}] from rising as much as it would in the absence of the increased Pi. PTH causes the kidney to retain more of the filtered Ca^{2+} and dispose of the Pi. Thus there is a hyperphosphaturia and more of the Ca^{2+} that is resorbed from bone is retained in the blood. Because PTH can raise the filtered load of Ca^{2+}, the effect of PTH can be either a hypocalciuria or hypercalciuria, depending on the balance of PTH effects on bone resorption and kidney reabsorption of Ca^{2+}. Further, PTH stimulates the kidney 1α hydroxylase to increase production of $1,25(OH)_2D$, which stimulates Ca^{2+} and Pi absorption from the intestine. The net effect of PTH secreted in response to hypocalcemia is generally not hypophosphatemia because it increases Pi input into the plasma at the same time that it increases Pi output from the plasma.

THE GOAL OF CT IS TO STOP BONE RESORPTION—PARTICULARLY IN THE YOUNG

The main physiological effect of CT is the rapid cessation of bone resorption. CT exerts no effect on intestinal Ca^{2+} and Pi handling and the effect of CT on the kidney is of minor physiological importance. There are no known abnormalities in Ca^{2+} handling associated with known CT undersecretion (in people with thyroidectomies, for example) or hypersecretion (in medullary thyroid carcinomas). On the other hand, it is known that the sensitivity to CT changes with age; young people are much more sensitive than older individuals. Further, the gut hormones are potent secretagogues for CT. This has led to the hypothesis that CT prevents postprandial hypercalcemia in the suckled young. Young mammals get a lot of Ca^{2+} in the milk, and this Ca^{2+} could cause a transient hypercalcemia when it is absorbed from the intestine. The gut hormones (e.g., gastrin, cholecystokinin, and secretin) increase after eating but before Ca^{2+} has a chance to be absorbed. CT assures that young mammals deposit dietary Ca^{2+} directly into bone by shutting down bone resorption even before Ca^{2+} is absorbed.

THE GOAL OF VITAMIN D IS TO RAISE BOTH [Ca^{2+}] AND [Pi] TO MINERALIZE THE BONES

The "goal" of vitamin D differs from that of PTH because it acts on bone, kidney, and intestine to increase both plasma [Ca^{2+}] and [Pi]. Its goal is to assure adequate mineralization of the skeleton, and this mineralization requires both Ca^{2+} and Pi. Thus vitamin D activation to $1,25(OH)_2D$ is stimulated not only by low plasma [Ca^{2+}] but also by low plasma [Pi].

THE GOAL OF FGF23 IS TO PREVENT HYPERPHOSPHATEMIA

FGF23 secretion is stimulated by high plasma Pi and by $1,25(OH)_2D_3$. Its main actions are to inhibit Pi reabsorption from the kidney and to decrease formation of $1,25(OH)_2D_3$. Thus FGF23 and $1,25(OH)_2D_3$ exhibit a negative feedback loop within the other homeostatic feedback loops for plasma [Ca^{2+}] and [Pi] regulation. The main effect here appears to be avoidance of hyperphosphatemia when bone is resorbed.

INTEGRATED CONTROL OF PLASMA [Ca^{2+}] AND Pi INVOLVES MULTIPLE NEGATIVE FEEDBACK LOOPS

The integrated control of plasma [Ca^{2+}] and [Pi] is shown in Figure 9.8.11. Here the negative feedback loops are indicated in bold arrows. As an example, consider the physiological response to a lowering of the free plasma [Ca^{2+}]. Several events occur. We trace the negative feedback loops, beginning with:

1. Plasma [Ca^{2+}] falls.
2. The low plasma [Ca^{2+}] stimulates PTH secretion.
3. PTH has several effects:
   - PTH induces a prompt hyperphosphaturia that, by itself, tends to raise [Ca^{2+}] slightly by reducing Ca^{2+} complexation in the plasma. This effect is small.
   - PTH induces a rapid osteocytic osteolysis that resorbs both Ca^{2+} and Pi into plasma.
   - The plasma Ca^{2+} originating from bone is retained to a greater degree by the kidneys because PTH stimulates Ca^{2+} reabsorption. This tends to raise the plasma [Ca^{2+}] back toward normal.
   - PTH stimulates the kidney 1α hydroxylase enzyme to increase $1,25(OH)_2D$.
4. $1,25(OH)_2D_3$ levels increased by PTH stimulation of the kidney 1 hydroxylase increase more slowly and have several effects:
   - $1,25(OH)_2D_3$ stimulates intestinal absorption of both Ca^{2+} and Pi. The increased Ca^{2+} absorption tends to restore plasma [Ca^{2+}c] toward normal.
   - $1,25(OH)_2D$ is synergistic with PTH in resorbing bone. This helps raise plasma [Ca^{2+}] toward normal.
   - $1,25(OH)_2$ helps retain Ca^{2+} and Pi in the kidney, which tends to raise plasma [Ca^{2+}] toward normal.
   - $1,25(OH)_2D_3$ increases the secretion of FGF23 from bone cells.
5. The increased bone resorption, increased intestinal absorption, and mixed effects on the kidney can lead to increased plasma [Pi]. This increases the secretion of FGF23.
6. Increased FGF23 has several effects:
   - FGF23 inhibits Pi reabsorption from the kidney
   - FGF23 inhibits $1,25(OH)_2D_3$ production by the proximal tubule of the kidney.
7. All of these effects work to increase the plasma [Ca^{2+}] from its low value: the negative feedback loop is complete.

The effects of PTH on the kidney and on osteocytic osteolysis are rapid, whereas the effects acting through $1,25(OH)_2D$ or osteoclastic osteolysis take much longer to develop and last longer. It takes considerable time to change the circulating levels of $1,25(OH)_2D$ and for this hormone to regulate the genetic expression of intestinal absorptive cells.

"Normal" plasma [Ca^{2+}] can be maintained at identical levels but at different physiological states. For example, the plasma [Ca^{2+}] can be within normal ranges when a person consumes a diet that is high in readily available Ca^{2+}. Under these conditions, we would expect the PTH levels to be relatively low and $1,25(OH)_2D$ levels would also be low. The plasma [Ca^{2+}] is being maintained primarily through intestinal absorption of Ca^{2+} and the person is likely to be in positive Ca^{2+} balance. A person who consumes a low Ca^{2+} diet might also have a normal plasma [Ca^{2+}], but the PTH levels and $1,25(OH)_2D$ levels would be high because it takes these high values to achieve a normal plasma [Ca^{2+}]. In this case, plasma [Ca^{2+}] would be maintained by resorption of bone Ca^{2+}, and the person would be likely to be in negative Ca^{2+} balance.

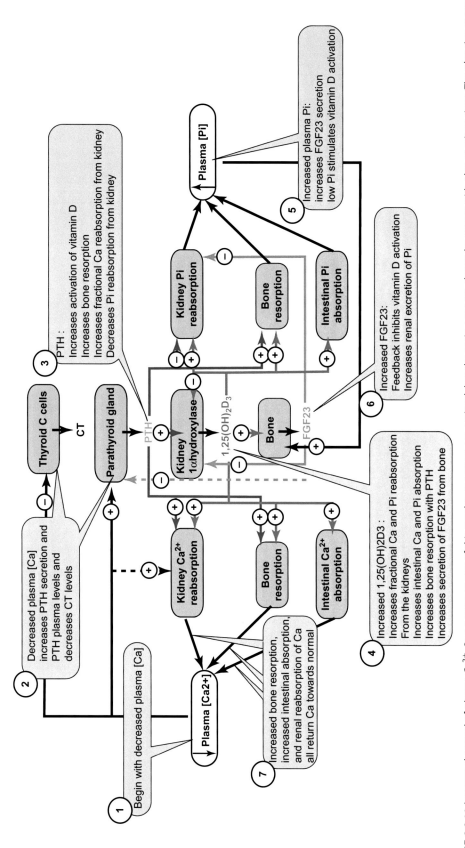

FIGURE 9.8.11 Integrated control of plasma Ca^{2+}. Some components of this regulatory system are omitted, but the main ones are shown by arrows with plus or minus signs. The plus sign means that when the event at the beginning of the arrow occurs, such as decrease in plasma $[Ca^{2+}]$, the effect is an increase in the event at the end of the arrow. The minus sign on an arrow means that when the event at the beginning of the arrow occurs, there is a decrease in the event at the end of the arrow.

Clinical Applications: Dual Energy X-ray Absorbimetry and Bone Density

Diagnosis of some bone diseases, such as osteoporosis, requires reliable and accurate estimations of bone density. Dual energy X-ray absorbimetry provides one such method. Known as DEXA and now as DXA this method uses high- and low-energy X-rays to provide estimates of both the soft tissue and bone mineral composition.

The equation describing absorption of X-rays by materials follows an analogue of the absorption of visible light, the Beer–Lambert Law. The equation is

(9.7.1) $$I = I_0 \, e^{-\mu x}$$

where I is the transmitted intensity of the X-ray beam, I_0 is the incident intensity (prior to absorption), μ is a constant called the **linear attenuation coefficient**, and x is the distance through which the beam passes. Absorption of X- and gamma-rays by materials occurs by photoelectric mechanisms and Compton scattering, in which orbital electrons interact with the radiation and are ejected from the atom, while the radiation loses some energy and changes its direction. These effects are much larger as the Z-number of the absorber increases. Thus, μ is much greater for calcium than it is for hydrogen, oxygen, and carbon. The linear attenuation coefficient also depends on the density of the atoms within the radiation beam: more atoms per unit cross-sectional area mean more opportunities to absorb the radiation. The linear attenuation coefficients also vary with the energy of the radiation, becoming smaller with increasing energy. When layers of soft tissue and bone are both present, the absorption of radiation is given by

(9.7.2) $$I = I_0 \, e^{-(\mu_t x_t + \mu_b x_b)}$$

where the subscript t denotes tissue and b denotes bone. These attenuation coefficients for standard tissue and bone can be determined and tabulated. For a single radiation beam, there is one equation and two unknowns: x_t and x_b. If we use two energies, however, he have two equations:

(9.7.3) $$I_1 = I_{0,1} \, e^{-(\mu_{t1} x_t + \mu_{b1} x_b)}$$
$$I_2 = I_{0,2} \, e^{-(\mu_{t2} x_t + \mu_{b2} x_b)}$$

where the subscripts 1 and 2 refer to the low- and high-energy beams. If the sets of coefficients are known, measurement of $I_1/I_{0,1}$ and $I_2/I_{0,2}$ allows calculation of x_t and x_b. However, the linear attenuation coefficients can be expressed as **mass attenuation coefficients** by dividing by the density of the materials. In this case, the equations become

(9.7.4) $$I_1 = I_{0,1} \, e^{-\left(\frac{\mu_{t1}}{\rho_{t1}} m_t + \frac{\mu_{b1}}{\rho_{b1}} m_b\right)}$$
$$I_2 = I_{0,2} \, e^{-\left(\frac{\mu_{t2}}{\rho_{t2}} m_t + \frac{\mu_{b2}}{\rho_{b2}} x_b\right)}$$

where m_t is the mass of tissue and m_b is the mass of bone. Solution of these equations gives the mass of either bone or tissue within the radiation beam. This allows calculation of the **areal density**: g of bone mineral per cm^2. These areal densities are reported in T-scores, the number of standard deviations from the normal bone density of a 30-year-old person of your gender and race. A T-score of −2.5 (2.5 standard deviations below normal) is considered to be diagnostic of osteoporosis.

SUMMARY

The calciotropic hormones affect Ca and Pi handling in three main target organs: the bone, the intestine, and the kidneys.

Osteoblasts line the surfaces of the bone and secrete the organic matrix that gives bone its form and serves as nucleation sites for mineralization. When these cells become embedded in the matrix they become osteocytes. Osteocytes remain in communication through cytoplasmic processes that travel through tiny canaliculi that link osteocytes to osteoblasts or other osteocytes. Osteoclasts are giant, multinucleated cells that resorb both mineral and organic matrix of bone. Regulation of the number and activity of these bone cells determines the fate of mineral deposition or resorption from bone. Osteoblasts have receptors for PTH and $1,25(OH)_2D_3$ and respond by increasing secretion of M-CSF, a factor that induces differentiation of hematopoietic stem cells to form mononuclear preosteoclasts. PTH and $1,25(OH)_2D_3$ also stimulate osteoblasts to make RANKL—a ligand that binds to *receptor activator of nuclear factor κB* (RANK) expressed on the surface of preosteoclasts and osteoclasts. The two factors M-CSF and RANKL are essential for formation of osteoclasts. $1,25(OH)_2D_3$ also increases osteocalcin formation, inhibits collagen synthesis, and inhibits secretion of OPG.

Osteoclasts resorb bone by sealing off an area of the surface and pumping acid and enzymes into the restricted space. The acid dissolves the mineral away from its nucleation sites on collagen, and the enzymes destroy the organic matrix. Osteoclasts make a resorption cavity and release both Ca and Pi into the blood.

Osteoclasts have no receptors for $1,25(OH)_2D_3$ or PTH, but they do possess receptors for CT. CT binds to a G_s-linked receptor that increases cAMP in these cells. The result is cessation of bone resorption. This does not appear to be important in the adult but may aid the young growing mammal in making strong bones during youth.

The intestine responds to $1,25(OH)_2D_3$ but not to either PTH or CT. $1,25(OH)_2D_3$ binds to a nuclear receptor in enterocytes and increases the transcription of genes coding for calbindin, TRPV6 and NaPi-IIb. Calbindin is a small molecular weight protein that binds Ca^{2+} ions with high affinity and likely either buffers Ca^{2+} during its transcellular transport or increases its diffusion through the cytoplasm of the enterocytes.

1,25(OH)$_2$D$_3$ increases both Ca and Pi transport across the intestine.

Osteocytes and osteoblasts release FGF23 in response to high Pi or 1,25(OH)$_2$D$_3$. FGF23 signals through αklotho to inhibit Pi reabsorption from the kidney tubule by inhibiting Pi entry.

The kidney responds to PTH, 1,25(OH)$_2$D$_3$, and CT, but the effect of CT appears to be unimportant. PTH acts through a G$_s$ mechanism in the kidney. PTH decreases absorption of both Ca and Pi in the proximal tubule but increases Ca absorption in the distal tubule. The result is an increased fraction of reabsorption of Ca and increased Pi excretion. 1,25(OH)$_2$D$_3$ increases reabsorption of both Ca and Pi in the distal nephron. FGF23 inhibits Pi reabsorption by decreasing NaPi-IIa and NaPi-IIc.

These actions of PTH, 1,25(OH)$_2$D$_3$, and CT show different "goals." PTH's goal is to maintain plasma [Ca^{2+}], and it achieves this by increasing bone resorption to place both Ca and Pi into the blood. It then discards Pi at the kidney and retains Ca. 1,25(OH)$_2$D$_3$, on the other hand, increases intestinal absorption of both Ca and Pi, resorbs both Ca and Pi from bone, and attempts to retain both at the kidney. Thus 1,25(OH)$_2$D$_3$ keeps both Ca and Pi levels high to promote bone mineralization. Although CT has definite effects, there are no disease states from known underproduction or overproduction of CT, suggesting that the body can adjust well regardless of CT levels.

REVIEW QUESTIONS

1. What are the endosteum, periosteum, diaphysis, epiphysis? What is the difference between compact bone and cancellous bone?
2. What are osteocytes, osteoblasts, and osteoclasts?
3. Both PTH and CT increase cAMP in bone, but PTH causes bone resorption and CT causes cessation of bone resorption. Explain how this can be.
4. Describe the origin of osteoclasts and the role PTH and 1,25(OH)$_2$D$_3$ play in their formation.
5. What is bone made of? What forms bone? What degrades bone? How is bone resorbed? The continual resorption and reformation of bone is called what?
6. How does Ca get absorbed from the ingested food? What regulates intestinal Ca and Pi absorption?
7. What happens to intestinal Ca absorption in persons who consume a low Ca diet? What is the mechanism that brings this about?
8. What does PTH do to Ca and Pi reabsorption in the proximal tubule? In the distal tubule? What are the overall effects of PTH? Would urinary Ca excretion go up or down? Would urinary Pi excretion go up or down?
9. What does 1,25(OH)$_2$D$_3$ do to Ca and Pi handling in the kidney?
10. What effect does CT have on gut and kidney Ca and Pi handling?

9.9 Female Reproductive Physiology

> **Learning Objectives**
>
> - Identify vagina, cervix, uterus, fallopian tube, fimbria, infundibulum, and ovary and summarize their functions
> - Describe when oogonia arise and when the first and second meiotic divisions occur
> - Draw on a graph the approximate relative levels of LH, FSH, estrogen, and progesterone during the menstrual cycle and indicate when ovulation occurs
> - List the three phases of the menstrual cycle
> - Describe the general changes in the endometrium during the menstrual cycle
> - Name the enzyme that converts testosterone and androstenedione to estradiol and estrone
> - Describe the requirements for positive feedback of estrogen on gonadotrophs
> - Define GnRH and indicate its site of origin
> - Explain why the LH surge is greater than the FSH surge immediately prior to ovulation
> - Distinguish among primary follicle, Graafian follicle, and corpus luteum
> - Describe the main function of the corpus luteum

SEXUAL REPRODUCTION COSTS A LOT BUT IT IS WORTH THE PRICE

Most animal and plant species on this planet engage in sexual reproduction. The whole process is quite an elaborate dance of biochemistry, physiology, and ecology, and in many cases entails risk to both parents and offspring. In humans, especially, death by childbirth historically has been a major cause of death of women in their childbearing years. Similarly, early childhood death historically has been one of the more potent selection processes driving our biological and cultural evolution. One is naturally tempted to ask the question: Why do we reproduce in this way? The short answer is that it confers a benefit on the species. The benefit derives from the scrambling of genes from one generation to the next. Certain gene combinations are deleterious to the survival of the species under specific ecological conditions, whereas other combinations are advantageous. If the environment changes in some way, recombination of already existing genes can produce successful offspring much faster than mutation of an already existing but relatively static genome. In a similar vein, isolated but small breeding populations may not have enough genetic diversity to get them through difficult times. The population as a whole becomes more vulnerable to extinction. This scrambling of the genes occurs not only by mixing the genes from two separate organisms but also by scrambling the genes provided by each of those organisms. The resulting potential diversity of the offspring is staggering. The proof of this is the differences between siblings in a single family. Such sibling groups often share enough genetic material that they can be identified as siblings, yet it is clear that each is a unique individual. The source of this variation—the cell divisions that give rise to the reproductive cells, or **gametes**—is discussed in Chapter 9.10.

THE ANATOMY OF THE FEMALE REPRODUCTIVE TRACT

The internal components of the female reproductive tract consist of the **ovaries**, **fallopian tubes**, **uterus**, and **vagina**. The external genitalia consist of the **labia majora**, **labia minora**, **clitoris**, and **vaginal opening**. The labia are folds of skin that cover fat and smooth muscle and meet along the midline above the vaginal opening. They cover and protect the urinary and vaginal openings as well as the clitoris. The clitoris is invested with numerous nerve endings so that it is highly sensitive to touch, pressure, and temperature. It has an exclusively sexual function to encourage the woman to engage in sexual intercourse and have babies as a consequence. The internal components of the female reproductive system are shown in Figure 9.9.1.

OVERIEW OF FEMALE REPRODUCTIVE FUNCTION

Each of the anatomical features of the female reproductive track has defined functions that are sometimes analogous to the functions of their counterparts in the male reproductive system. They differ markedly from the male in their periodicity. Males produce gametes and are normally fertile continuously. Females produce gametes periodically and have a relatively short-lasting window of fertility. The approximately 28-day cycle of female reproductive function is called the **menstrual cycle**. It is outwardly apparent from **menstruation**, the

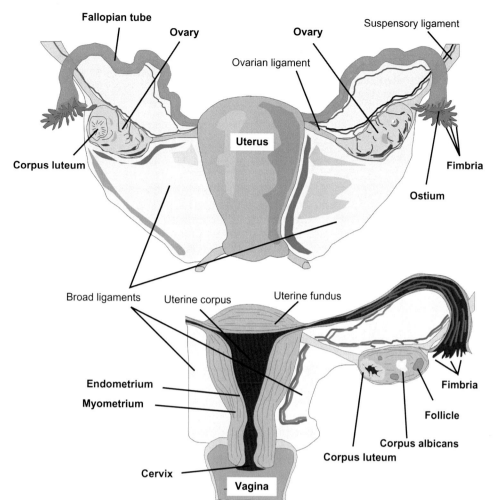

FIGURE 9.9.1 The female reproductive apparatus. See text for an explanation of the structures.

shedding of the uterine lining when fertilization has not occurred and there is no baby on board.

The gonad in the female is the ovary. The two ovaries are arranged bilaterally, one on each side of the pelvic abdomen. Each ovary in the adult weighs about 15 g and is attached to the pelvic wall and to the side of the uterus by the ovarian ligament. The ovarian artery, vein, lymphatics, and nerve supply run along the ovarian ligament. As in the male, these gonads produce gametes and reproductive hormones and are sensitive to the gonadotropins, LH (luteinizing hormone) and FSH (follicle stimulating hormone). In fact, these hormones derive their names from their effects in the female in whom their roles were first discovered and clarified. Each month the ovaries generally produce one ovum or at most a few **ova**, large haploid germ cells (containing half of the genetic information of the mother) that can be fertilized by the male's spermatozoa and subsequently begin development into a baby. This release of the ovum from the ovary is called **ovulation**. The ovum is released from an ovarian **follicle**. It is actually released into an open space in the abdominal cavity but immediately adjacent to a funnel-shaped portion of the **fallopian tube** called the **infundibulum**. The fallopian tube is also called the **oviduct**. Little finger-like processes called **fimbriae** extend from the infundibulum to closely appose the lateral portions of the ovary. The fimbriae undulate so as to draw the ovum into a fallopian tube, which projects about 10 cm from the uterus toward the ovary on each side of the body. The fallopian tube consists of an epithelial lining with secretory and ciliated cells, and a smooth muscular layer capable of peristaltic-like contractions. The number and activity of the cilia and secretory cells change throughout the female's period. Their actions and the peristaltic contractions of the smooth muscle layers of the fallopian tubes propel the ovum from the ovary, where it is released, toward the uterus. The fallopian tubes also provide a conduit for the movement of the sperm toward the ovum.

The **uterus** is colloquially known as the womb. A thick smooth muscle wall encloses a cavity that is lined with a mucous epithelium called the **endometrium**. The composition and thickness of the endometrium varies cyclically with the periodic variation in reproductive hormones that accompanies ovulation. The shedding of the lining when fertilization does not occur makes up the monthly menstrual flow in adult females. It is accompanied by contraction of the uterus which is generally referred to as "cramping" and which can sometimes be very painful. The function of the uterus is to protect and nourish the developing fetus from its time of conception to birth, and to provide the motive force for delivery of the baby during **parturition**, or birth.

Once captured by the fallopian tubes, the ova secrete chemoattractant compounds that attract the sperm. An ovum is viable for 12–24 h. Sperm deposited in the female remain viable for about 48 h. Thus, timing of sexual intercourse surrounding ovulation is crucial for fertilization of the ovum. Fertilization of the ovum typically occurs in the fallopian tube but may occur in rare cases in the abdominal cavity. Implantation of the fertilized ovum outside of the uterus is called **ectopic pregnancy** and is a medical emergency. Normally, the fertilized ovum begins to divide at once and gradually moves down the fallopian tubes to reach the uterus and completes the trip in about 2 or 3 days. After another 2 or 3 days in the uterus, the developing zygote, now typically in the blastocyst stage, initiates implantation. This is the process by which the developing zygote attaches itself to the uterine lining and invades it to eventually form the **placenta**. The placenta has two main functions: to exchange materials for the developing fetus with the mother and to secrete hormones that affect both mother and fetus. The placenta derives nutrients and oxygen from the mother and gives back to her waste products and carbon dioxide.

The cervix is the most inferior part of the uterus that projects out into the vagina. The mucus secreted by the cervix also varies with the cycle to form channels for the sperm to travel into the uterus. The geometry of the cervix and its relatively narrow opening help protect the uterus from foreign objects and thereby protects against injury and infection.

The vagina is essentially a tube leading from its external opening at the surface of the body to its end at the uterus. It is invested with smooth muscle that can contract or expand and a stratified epithelium that secretes a lubricating mucus during sexual arousal that aids in sexual intercourse. The vagina itself has few sensory nerves and is relatively insensitive to touch and pressure.

OOGENESIS BEGINS IN THE FETUS

Oogenesis is the formation of the ova. It begins in the early fetus when primordial germ cells migrate from the yolk sac of the embryo to the genital ridge at 5–6 weeks of gestation. In the developing ovary, these germ cells divide by mitosis to form a population of primary **oogonia**, which are analogous to the male's primary spermatogonia. By 20–24 weeks of gestation, the number of oogonia has peaked at about 7 million. Beginning at 8 or 9 weeks of gestation, some of these oogonia enter into the prophase of meiosis I to become primary oocytes and they arrest at this stage due to inhibitory hormones. Eventually all of the oogonia convert to primary oocytes or degenerate. At birth, the female ovaries contain in excess of 2 million primary oocytes which enlarge gradually to become the single largest uninuclear cell in the body at 100–120 μm in diameter. After birth, the number of primary oocytes decreases because there is no further mitosis and some of the primary oocytes are lost by **apoptosis**, or programmed cell death. At the time of puberty, there are typically 400,000 primary oocytes remaining. Over the reproductive span, typically a woman will ovulate only 400–500 of these oocytes. The remaining oocytes disappear by a process called **atresia**.

PUBERTY INITIATES OVULATION AND DEVELOPMENT OF SECONDARY SEX CHARACTERISTICS

Puberty is the transition from the childhood, nonreproductive state to the adult reproductive state. It requires maturation of the entire hypothalamic–pituitary–gonadal axis. Prior to puberty, LH and FSH secretions are low despite the fact that both gonadal steroids and **inhibin**, which both negatively regulate LH and FSH secretion in adults, are low. This points to other mechanisms, most likely the neuronal circuits that give rise to GnRH release by neurons in the hypothalamus, as the cause of puberty. The process by which puberty begins remains unclear, partly because there is no single precipitating event. Instead, there are gradual changes in a variety of signals each of which, by themselves, does not constitute puberty.

The beginning of menses, or the monthly blood flow representing the sloughing of the uterine lining, is called **menarche**. This is accompanied by full reproductive capacity in the female. However, changes in secondary sex characteristics precede menarche. These include the following:

- Accelerated linear growth associated with the puberty growth spurt
- Development of breasts
- Maturation of the genitalia
- Appearance of pubic and axillary hair
- Widening of the hips
- Enlargement of the uterus, fallopian tubes, and vagina.

Testosterone and estradiol levels in the blood rise during puberty, and these increase the amplitude of the GH pulses in response to GHRH. Although GH is as high or higher in girls than in boys, the estradiol interferes with the effect of GH on the long bones, whereas testosterone, and lack of estradiol, increases the effect in boys. The result is that the boys grow larger. Since girls typically enter puberty earlier, there is a period in which the girls are briefly taller than the boys, but the boys rapidly catch up and surpass the girls. Most of the effects listed above are mediated by estrogen.

The time course of events in puberty is illustrated in Figure 9.9.2.

LH AND FSH DRIVE THE MENSTRUAL CYCLE

The menstrual cycle refers to all of the events that occur during the approximately month-long reproductive cycle of the adult female. The menstrual period is the

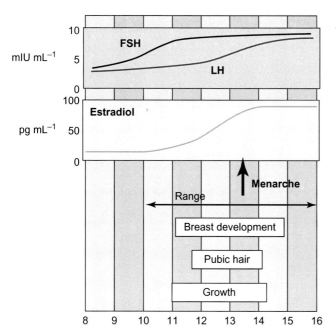

FIGURE 9.9.2 Chronology of events in female puberty. Although the average age of menarche is about 12 years in the United States, regular cycles do not begin until about age 14. There is great variability in the age of menarche. Menstruation can begin as early as 10 years or as late as 16 years of age.

time of the cycle in which the uterine lining sloughs off and is discarded. A summary of some of the events that occur during the cycle is shown in Figure 9.9.3. The entire cycle typically lasts 28 days, but it can last from 25 to 30 days. It consists of three phases:

THE FOLLICULAR PHASE

The follicular phase begins at the first day of menstrual bleeding and lasts until ovulation. It is more variable and accounts for the fluctuations in cycle length. It averages 15 days.

THE OVULATORY PHASE

The ovulatory phase lasts 1–3 days and ends with ovulation. It is characterized by a sharp spike in LH and FSH levels, with LH increasing much more than FSH.

THE LUTEAL PHASE

After ovulation, the dominate follicle becomes the **copus luteum**, or "yellow body," and its secretions dominate the rest of the cycle.

OVERVIEW OF FOLLICULAR DEVELOPMENT

The development of ovarian follicles can be divided into three stages as shown in Figure 9.9.4.

A. *Stage 1*: Development of the follicle prior to ovulation. This stage is usually not less than 13 years long but may be as long as 50 years. The follicle develops in this stage to a primordial follicle (Figure 9.9.4) and then falls quiescent until puberty.

B. *Stage 2*: Activation of the follicle to compete for ovulation. This stage lasts 70–85 days. A small group of the many primary follicles is recruited for the next phase. The primary, secondary, and tertiary follicles can grow in the absence of the pituitary, whereas full development of the **Graafian follicle** requires FSH from the anterior pituitary. Details of the structures of the follicles are shown in Figure 9.9.5.

C. *Stage 3*: Some 5–7 days after the beginning of the menstrual flow, one of the 20 or so sister Graafian follicles becomes the **dominant follicle** of that cycle. This is the one that will ovulate. The ones that will not ovulate, involute and undergo **atresia** through apoptosis, or programmed cell death. After the LH and FSH surges, the follicle ruptures and releases a single ovum into the peritoneal cavity. The first meiotic division is complete at this time. The second meiotic division does not occur until after the sperm penetrates the ovum. Thus, the released ovum is a secondary oocyte.

D. The remnants of the ruptured follicle form a new structure, the **corpus luteum**. This structure produces hormones that optimize the conditions for implantation of the fertilized ovum and maintain the zygote until the placenta can assume this role. The normal life of the corpus luteum is about 14 days. If there is no pregnancy, the corpus luteum spontaneously regresses, forming an avascular scar called the **corpus albicans**, or "white body."

It is important to emphasize that the stages in follicular development described here and shown in Figure 9.9.4 do not happen in a single ovarian cycle. Rather, the stages leading up to ovulation take approximately 85 days, beginning its growth in the late luteal phase of the third cycle preceding ovulation. This point is made clear in Figure 9.9.6, in which the size of the follicle and the time line over several cycles are shown.

CELLULAR ASPECTS OF FOLLICULAR DEVELOPMENT

THE PRIMORDIAL FOLLICLE IS NOT GROWING

Primordial follicles consist of the ovum and a number of spindle-shaped granulosa cells that surround the ovum but lie within the basement membrane (see Figure 9.9.5). Because of this arrangement, the granulosa cells are avascular and do not have ready access to the circulation. These follicles constitute the majority of follicles, making up 90–95% of all follicles, and they do not grow.

THE PRIMARY FOLLICLE HAS BEEN RECRUITED

In the late luteal phase of a cycle, several primordial follicles begin to grow. The granulosa cells form a

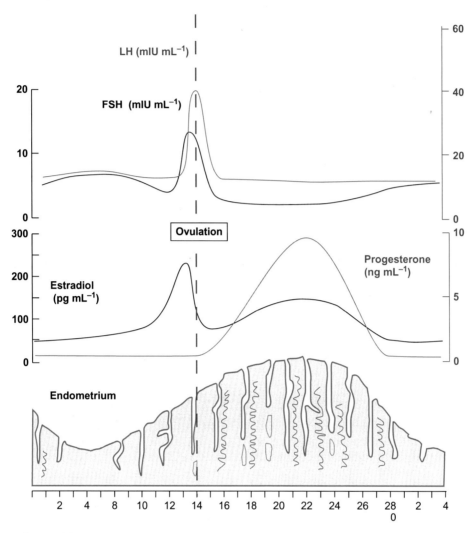

FIGURE 9.9.3 Summary of some of the events in the adult human female menstrual cycle. Estradiol secretion by a maturing follicle gradually increases and exerts a positive feedback on LH and FSH release by gonadotrophs in the anterior pituitary. The surge in LH and FSH causes ovulation or release of the ovum from the dominant follicle. Increasing estradiol levels in the follicular phase of the cycle increase the thickness and vascularity of the uterine lining. After ovulation, progesterone levels rise and alter the vascular tortuosity of the uterine lining and promote glycogen storage within it. The abrupt loss of estradiol and progesterone, if pregnancy does not occur, causes spasmodic contractions of the uterine blood vessels and uterine muscles. The resulting ischemia produces necrosis (cell death) and the lining condenses and degenerates.

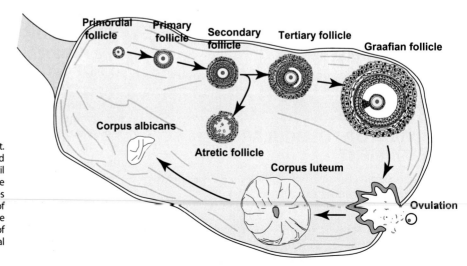

FIGURE 9.9.4 Stages in follicular development. During gestation, some follicles are advanced to secondary follicles where they arrest until puberty. During adult reproductive years, some secondary follicles become Graafian follicles with the appearance of the antrum. One of these in one ovary is promoted to the dominant follicle, which ruptures at the time of ovulation, releasing its ovum into the peritoneal cavity.

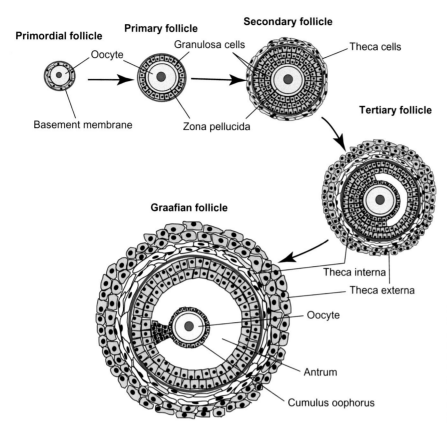

FIGURE 9.9.5 Progression of follicles. Primordial follicles reside in the ovaries and do not grow. After recruitment in the late luteal phase, their spindle-shaped granulosa cells become cuboidal and the oocyte enlarges, forming the primary follicle. This grows progressively into the secondary and tertiary follicle without gonadotropins. The Graafian follicle forms upon selection during the early days of the current cycle and matures into a preovolatory follicle.

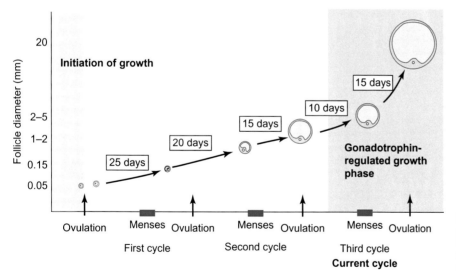

FIGURE 9.9.6 Time course of follicular development from primordial follicle to Graafian follicle. The entire development takes approximately 85 days, which spans several menstrual cycles.

single layer of cuboidal cells, and the oocyte enlarges. The oocyte forms the **zona pellucida**, a clear region surrounding the oocyte and separating it from the granulosa cells. The zona pellucida is a glycoprotein, mucoid substance.

THE SECONDARY FOLLICLE HAS THE BEGINNINGS OF THE THECAL LAYER

The secondary follicle has several layers of cuboidal granulosa cells by attainment of full oocyte size of about 120 μm (= 0.12 mm); it is just visible to the naked eye. The secondary follicle recruits a population of cells that reside immediately adjacent to the basement membrane on the side opposite the granulosa cells. These constitute the **theca**. The secondary follicle acquires its own blood supply, with one or two arterioles breaking up into a capillary network immediately outside the basement membrane. The granulosa cells and oocyte remain avascular. The thecal cells have receptors for LH and make androgens during the preovulatory phases. The granulosa cells have FSH

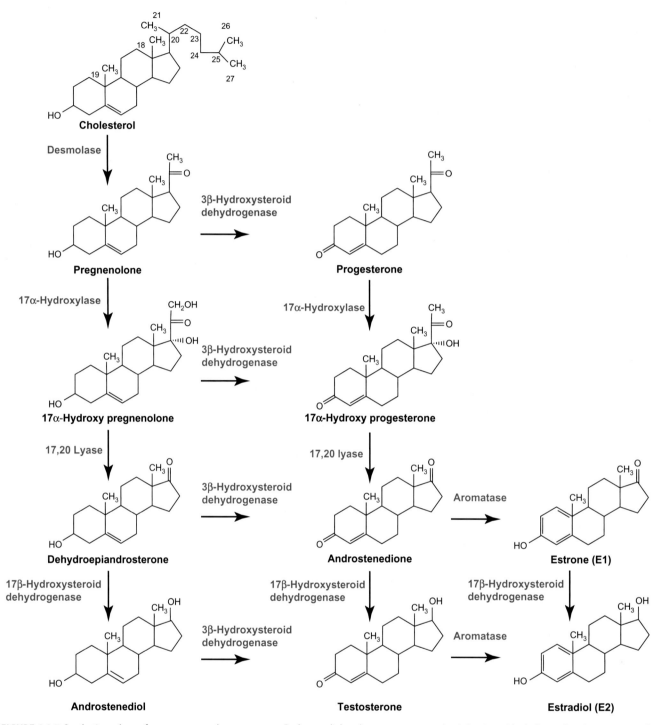

FIGURE 9.9.7 Synthetic pathway for estrogens and progesterone. Both estradiol and progesterone synthesis begins with cholesterol and require specific enzymes. Only some cells have each of the required enzymes, so that synthesis of estradiol requires two cell types, thecal cells and granulosa cells.

receptors and make estrogens from the androgens supplied by the thecal cells (see Figure 9.9.5).

THE TERTIARY FOLLICLE HAS AN ANTRUM

The tertiary follicle is marked by the appearance of a fluid-filled cavity, the **antrum**, adjacent to the oocyte and further development of the theca into **theca interna**—the cells adjacent to the basement membrane and circumferentially organized; and the **theca externa**—less well-organized cells that merge with the stroma of the ovary. The fluid in the antrum is a plasma transudate.

GRAAFIAN FOLLICLE GROWS UNDER THE INFLUENCE OF GONADOTROPINS

The Graafian follicle grows under the influence of gonadotropins. The antrum enlarges and the oocyte, surrounded by a layer of granulosa cells, forms the **cumulus oophorus**, which occupies a polar and eccentric position within the follicle. One of the Graafian follicles becomes dominant and all others undergo atresia. The granulosa cells acquire LH receptors as well as FSH receptors. The follicle becomes preovulatory.

OVARIAN STEROIDOGENESIS REQUIRES TWO CELL TYPES AND TWO HORMONES

The principal biologically active ovarian steroids are estrogen and progesterone, but the ovary also secretes dihydroepiandrosterone (DHEA), androstenedione, and testosterone. The preovulatory follicle secretes estrogen during the follicular phase of the menstrual cycle, whereas the corpus luteum produces both estrogen and progesterone during the luteal phase. The production of these steroids is controlled in a cell-specific manner by LH and FSH.

The pathways for estrogen and progesterone synthesis are shown in Figure 9.9.7. Thecal cells have the enzymes necessary for production of androstenedione, but **thecal cells lack aromatase that is necessary for making estradiol**. Granulosa cells have the aromatase, but **granulosa cells lack 17α hydroxylase and 17,20 lyase** (also called desmolase) necessary for the syntheses of the androstenedione precursor of estradiol. Also, granulosa cells are not near a source of cholesterol for synthesis of steroids, namely LDL cholesterol, whereas the thecal cells have access to LDL cholesterol. In the preovulatory follicle, thecal cells synthesize androstenedione under the influence of LH, and granulosa cells use the androstenedione to make estradiol, under the influence of FSH. The arrangement in the preovulatory follicle is shown in Figure 9.9.8.

After ovulation, granulosa cells become vascularized and gain access to cholesterol through low-density lipoprotein (LDL) uptake. These granulosa cells also acquire LH receptors, which stimulates them to take up LDL cholesterol as they do in thecal cells. As a result, the corpus luteum produces progesterone and estradiol. These are the most active steroidogenic cells in the body, producing up to 40 mg of progesterone per day.

In these cells, LH and FSH both activate SF-1, a transcription factor called steroidogenesis factor-1, which regulates the expression of the genes that code for StAR, desmolase, 17α dehydrogenase, 17,20 lyase, 3β hydroxysteroid dehydrogenase, and aromatase.

CENTRAL HORMONAL CONTROL OF THE MENSTRUAL CYCLE

Hormonal control of the menstrual cycle is complicated by the fact that at any one time there are several types of follicles in various stages of development. The follicle that becomes dominant and ovulates in any one cycle,

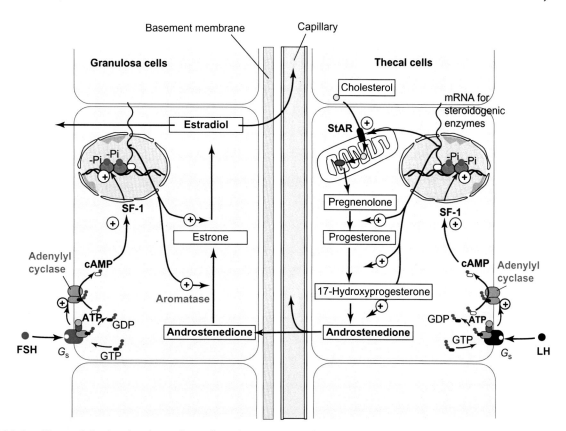

FIGURE 9.9.8 Steroidogenesis by thecal and granulosa cells in the preovulatory follicle. The basement membrane separates the thecal cells from the granulosa cells and granulosa cells have no direct access to the circulation. Thecal cells respond to LH through a G_s mechanism that activates SF-1, the steroidogenesis factor transcription factor that turns on the synthesis of several steroidogenic enzymes. These cells produce androstenedione but cannot produce estrone because they lack aromatase. The androstenedione is transferred to granulosa cells that convert it to estradiol under the control of FSH, also working through a G_s mechanism in the granulosa cells.

for example, typically begins its run 2.5 cycles previously. In the early luteal phase of a cycle, a small group of primary follicles begins further development. Their development depends on the relatively lower concentrations of FSH and LH during the luteal phase. The cohort develops over the next 60–70 days. During this time, FSH increases the growth of the **granulosa cells** that surround the ovum and line the **antrum** (see Figure 9.9.5). FSH increases the enzyme **aromatase** that converts androstenedione to estrone and testosterone to estradiol (see Figure 9.9.7). The increase in granulosa cell mass and increase in aromatase increases estrogen synthesis during the follicular phase of the cycle. The increased estradiol also has a local effect on the granulosa cells to increase their receptors for FSH, which further boost estrogen levels. These effects are illustrated in Figure 9.9.9.

Simultaneously, FSH along with estradiol induces LH receptors on the granulosa cells. Both FSH and LH work through a G_s mechanism. As the follicle approaches ovulation, the increased estrogen production becomes more influenced by LH than by FSH, and this prepares the follicle for its later function as an LH-regulated corpus luteum following ovulation.

The continuously rising estrogen levels also sensitize the anterior pituitary gonadotrophs to GnRH, so that similar amounts of GnRH now cause greater LH and FSH release, which then further stimulate estrogen release from the follicle. This positive feedback requires a critical estrogen level of at least 200 pg mL^{-1} for at least 2 days and relatively low progesterone levels, conditions that pertain only during the late follicular phase of the cycle. This results in the LH and FSH surge that induces ovulation.

Estradiol also increases the number of LH receptors on theca cells, the cells that surround the entire follicle on its outer borders. These theca cells synthesize androstenedione and testosterone, which then diffuse to the granulosa cells that possess aromatase activity to convert these androgens to estrogens.

LH AND FSH SURGE INDUCES OVULATION

As stated above, the positive feedback provided by high estrogen for protracted periods with low progesterone induces ovulation, the rupture of the Graafian follicle, and release of the ovum into the peritoneal cavity. LH receptors on the granulosa cells neutralize factors within the ovum that inhibit the completion of the first meiotic division. Removal of the inhibition allows this first meiotic division to be completed. LH and FSH activate a number of enzymes that induce a pseudo-inflammatory response that leads to follicular rupture.

AFTER OVULATION, THE FOLLICLE FORMS THE *CORPUS LUTEUM*

As its name implies, LH is instrumental in inducing changes in the follicle that produce the corpus luteum. LH levels drop following the LH surge immediately prior to ovulation, but partly because the pattern of LH release changes to a low-frequency, high-amplitude

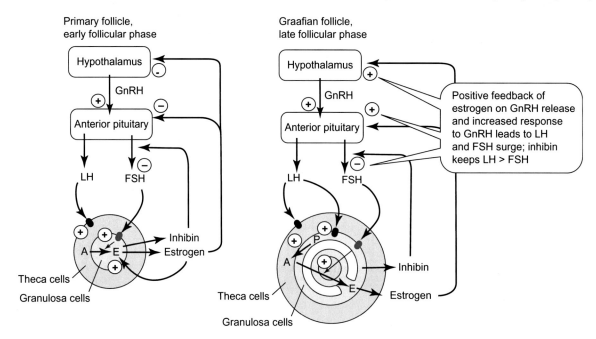

FIGURE 9.9.9 Control of LH and FSH in the early and late follicular phases of the menstrual cycle. In the early follicular phase, FSH stimulates estrogen (E) production by the granulosa cells and LH stimulates androgen (A) production by theca cells in primary follicles. Estrogen feeds back to both the hypothalamus and anterior pituitary to limit GnRH secretion and LH and FSH secretion. FSH induces growth of the follicle and so does estrogen. The two together induce LH receptors on the granulosa cells. After the follicle has grown, LH induces progesterone secretion which begins to increase prior to ovulation, which can be used to produce androgens in the thecal cells to subsequently produce estrogens in the granulosa cells. The continuous increased estrogen levels now exert a positive feedback on the hypothalamus and anterior pituitary to bring about the LH and FSH surge. The LH surge is much greater than the FSH surge because inhibin selectively inhibits FSH synthesis and release by the anterior pituitary. The resulting LH and FSH surge, with progesterone, induces ovulation of the Graafian follicle.

pulsatile release. This pattern produces the increased progesterone and estradiol levels in the luteal phase of the cycle. Because the corpus luteum also produces **inhibin**, LH levels gradually fall and the corpus luteum will regress and undergo atresia through apoptosis. If the ovum is fertilized, it releases hCG, human chorionic gonadotropin. Recall that hCG, LH, FSH, and TSH are all glycoproteins that share structural similarities with identical α subunits and different, but similar, β subunits. hCG is the first hormone to increase in pregnancy, and it is responsible for the early increase in progesterone prior to the establishment of the placenta.

SUMMARY

The female gonads are the ovaries, paired structures that secrete estrogens, androgens, and progestins at various stages in the menstrual cycle, and that release, generally, a single ovum in the middle of the cycle. The ovum is released from a mature Graafian follicle that takes about 85 days to complete its development. The early phases of this development are independent of LH or FSH but maturation depends on both.

The menstrual cycle begins on the first day of menses, or sloughing of the uterine lining when there is no pregnancy. The cycle has three phases: the follicular phase, marked by the growth of the Graafian follicle, and characterized by a peak of estrogen exceeding 200 pg mL^{-1} in plasma around day 13, that originates from the granulosa cells of the Graafian follicle. This is followed by the brief (1–3 days) ovulatory phase, marked by peaks in FSH and LH originating from the gonadotroph cells in the anterior pituitary, and controlled by the GnRH-releasing cells of the hypothalamus in response to secretions of the follicle. Ovulation occurs, and the remnant of the Graafian follicle becomes the corpus luteum in the luteal phase of the cycle. It secretes both estrogen and prodigious quantities of progesterone. This sustains the endometrial lining until the fertilized ovum produces hCG.

Production of estrogen by the follicle during the follicular phase is explained by the two-cell, two-gonadotropin hypothesis. Thecal cells have receptors for LH that are linked to a G_s mechanism. The increased cAMP activates steroidogenesis factor-1 (SF-1) that acts as a switch to turn on a program of steroidogenic enzymes in the thecal cells, but these cells do not express aromatase to convert androstenedione to estrone, which is then converted to estradiol. The thecal cells make androstenedione which is then supplied to the granulosa cells, which are on the other side of the basement membrane of the follicle. Granulosa cells have receptors for FSH, which also works through a G_s mechanism, to convert thecal cell androstenedione to estrone and then to estradiol. This is the origin of circulating estradiol during the follicular phase of the cycle. After ovulation, the basement membrane is no longer an impediment to uptake of LDL cholesterol by the granulosa cells, which also acquire receptors for LH and then make large quantities of progesterone and estrogen during the luteal phase of the cycle.

Ovulation itself is a dance between central secretion of LH and FSH and follicular secretion of estradiol and inhibin. The estradiol peak during the follicular phase informs the gonadotroph cells that the follicle is mature and ready for ovulation, and these cells respond with a peak of LH and FSH secretion that induces ovulation.

REVIEW QUESTIONS

1. What is the fallopian tube? Why does it have a ciliated lining? Why does it have smooth muscle? What do the fimbriae do?
2. What is the start of the menstrual cycle? What hormone increases during the follicular phase? Where is it made? What hormones activate its synthesis?
3. When in the cycle is the LH surge? Where does LH come from? Why does it spike? What happens after the spike?
4. What happens to the ovum after ovulation? What happens to the follicle after ovulation?
5. What hormones are secreted by the corpus luteum? What sustains it? What happens if this sustenance is removed?

9.10 Male Reproductive Physiology

Learning Objectives

- List the phases of mitosis in order with brief descriptions of the events in each phase
- Explain how meiosis differs from mitosis
- Explain the differences in chromosome handling in meiosis I and meiosis II divisions
- Explain the origin of the genetic differences among germ cells
- Trace the path of sperm from testes to urethra during ejaculation
- Identify the major parts of a spermatozoa
- Identify the major sex accessory glands in the human
- Define GnRH and indicate its site of origin
- Describe the chemical nature of LH and FSH and indicate their cells of origin
- Identify the major targets of LH and FSH in males
- Describe the nervous control of erection in males
- Describe the nervous control of ejaculation

SOMATIC CELLS DIVIDE BY *MITOSIS*; GERM CELLS DIVIDE BY *MEIOSIS*

Humans have in general two types of cells: the **somatic** cells make up the vast majority of the cells and are the body cells. These cells grow and divide by **mitosis**. The second type of cell is the **germ cell**, which is used solely for reproduction and derives from somatic cells by a special kind of cell division called **meiosis**.

Long threads of double-stranded DNA make up the genetic material of all cells, and this material is organized into compact units called **chromosomes**. "Chromosome" literally means "colored body" and refers to the ability of these structures to stain with particular dyes that enables visualization under the microscope. Each chromosome consists of a single long piece of double-stranded DNA that is complexed with a variety of proteins that organize its structure and regulate the transcription of the DNA. The organization of the chromosomes varies depending on the stage of the cell's cycle. Chromosomes are most visible and most compact during cell division.

Each chromosome in the somatic cell is part of a homologous pair. These are two versions of the chromosome that have similar but generally not identical DNA sequences. The double-stranded DNA in each chromosome consists of a number of genes. **The gene is *defined* as the unit of inheritance**. Each gene consists of a sequence of nucleotides that carries the instructions for the synthesis of a protein that has some function. Because the chromosomes are paired, nearly every gene also comes in pairs, one on each chromosome. However, the two genes often differ in tiny details. These alternate forms of the genes in a single person are called **alleles**. The complete set of alleles of a particular person is that person's **genotype**. The set of proteins and other materials that the person actually makes, and which determines their outward appearance and behavior, is called the **phenotype**. Humans have 23 pairs of chromosomes. Twenty-two of these pairs are called **autosomes** which are common to all persons. The last pair of chromosomes are the sex chromosomes that are distributed differently in men and women. The sex chromosomes are grossly unequal, and some of the genes on the sex chromosomes are not paired.

MITOSIS PRODUCES TWO DAUGHTER CELLS WITH THE SAME DNA CONTENT AS THE ORIGINAL CELL

Because each somatic cell has two genes for nearly every trait, the condition is called **diploid** (2N). During development, cells grow and divide by mitosis so that each daughter cell also contains the full genetic complement of paired alleles. This process of cell division is summarized in Figure 9.10.1. The entire process is the cell cycle, which is under complex control. During **interphase**, the cell acquires more material and enlarges. It duplicates its DNA and the **centrosome**. The duplicate DNA condense during **prophase** forming **sister chromatids** that remain attached to each other. At this stage, there are four copies of each gene, two of each allele. However, the genes are not active at this time; in the condensed state, they are not being used to make mRNA which codes for the synthesis of new proteins. The centrosome consists of two centrioles oriented at right angles and a centrosome matrix that contains many copies of γTuRC (γ tubulin ring complex) which provides nucleation sites for microtubule assembly. During prophase, the **mitotic spindle**

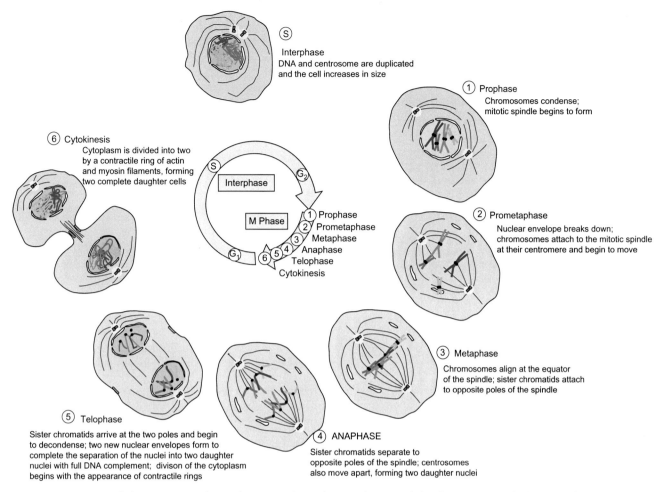

FIGURE 9.10.1 Mitotic cell division. During the synthetic part of interphase, S, the DNA of the chromosomes is duplicated to form identical sister chromatids for each chromosome. The centrosomes are also duplicated. In prophase, the chromosomes condense and the mitotic spindle begins to form. In prometaphase, the nuclear envelope disappears and the chromosomes attach themselves to the mitotic spindle at their kinetochore and begin to move. At metaphase, all of the chromosomes line up at the equator of the mitotic spindle. In anaphase, the sister chromatids separate to opposite poles of the mitotic spindle and the centrosomes move apart. During telophase, the nuclear envelope forms again, making two distinct daughter nuclei. The cell division is completed by separation of the cytoplasm into two daughter cells by constriction of the surface membrane by a contractile ring of cytoskeletal filaments. For clarity, only two homologous pairs of chromosomes are shown.

begins to form. The two centrosomes migrate toward opposite ends of the cell and form a complex arrangement of microtubules that provide tracks along which the chromosomes can move. This movement separates the chromosomes into two groups, but it occurs later, in anaphase.

In the next step, **prometaphase**, the nuclear envelope breaks down and the chromosomes attach to the microtubules of the mitotic spindle. In **metaphase**, the chromosomes align themselves along the equator of the spindle midway between the two centrosomes. They attach to the microtubules at the **kinetochore**. The kinetochore consists of a number of proteins that bind to a central part of the DNA called the centromere. Figure 9.10.2 illustrates the gross structure of a mitotic chromosome.

In **anaphase**, the sister chromatids separate to the opposite poles of the mitotic spindle. Cells have a mechanism that detects unattached kinetochores, so that anaphase does not occur until all of the chromosomes

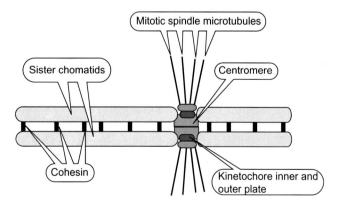

FIGURE 9.10.2 Simplified structure of the mitotic chromosome. The DNA of the chromosome has been duplicated and then condensed to make two sister chromatids that are identical in nearly all respects. They are linked along their length by the protein cohesin. A central region of DNA, the centromere, binds a set of proteins that form the kinetochore, which in turn anchors the microtubules of the mitotic spindle. During anaphase, the sister chromatids are separated by proteolytic cleavage of the cohesin and movement of the sister chromatids in opposite directions along the microtubule tracks provided by the spindle.

are attached and located at the equator of the spindle. Anaphase then begins with the activation of an anaphase promoting complex that has several functions. One is to activate a protease that clips the protein cohesin that glues the two sister chromatids together. When this is accomplished, the two sister chromatids can separate into the opposite poles of the mitotic spindle.

In **telophase**, a new nuclear envelope forms around the newly separated sister chromatids, which now form full chromosomes. They begin to decondense. This ends nuclear division, which is called mitosis or karyokinesis. The division of the cytoplasm, to separate the organelles of the cell into the component daughter cells, is called **cytokinesis**. It begins with a contractile ring that encircles the cell approximately midway between the two centrosomes. The contractile ring consists of actin and myosin filaments and its contraction squeezes the cell so that the cytoplasm is divided into the two daughter cells.

MEIOSIS DIVIDES THE PARENT GENOME IN HALF—BUT CROSSING-OVER DIVERSIFIES THE RESULT

Sexual reproduction in humans produces a single fertilized cell that grows and divides by mitosis. Eventually, through complex developmental events, the single fertilized cell becomes a baby, and eventually an adult. The fertilized cell is diploid, containing 23 pairs of chromosomes. Therefore, all of the cells it produces from mitosis also contain 23 pairs of chromosomes. The fertilized cell arises from fertilization of the **ovum**, produced by the mother, and a **spermatozoon** produced by the father. The 23 pairs of chromosomes results from the fusion of 23 chromosomes in the ovum with 23 chromosomes in the spermatozoon. The ova and sperm cells are collectively called **gametes**, or **germ cells**. They contain only half of the genetic complement of each parent and are called **haploid** (N). The process that produces these haploid cells is called **meiosis**.

Meiosis begins with the duplication of the full DNA complement of the cell to produce two sister chromatids for each chromosome, just as occurs in mitosis. In the first meiotic prophase, homologous chromosomes pair up to form a structure called a **bivalent**. This first meiotic prophase takes days and sometimes years. An extremely important event occurs in this bivalent, i.e., **crossing-over** of genetic material. This scrambles the genome and allows us to confidently assert that every human, save identical twins, is genetically distinct.

Division I of meiosis separates the homologous chromosomes, not the sister chromatids, along the spindle, so that each daughter cell receives one of each homologous pair of chromosomes. This produces two haploid daughter cells that still have two copies of each gene. Thus, these cells have a haploid number of chromosomes but a diploid amount of DNA. The sister chromatids are identical except where genetic recombination, crossing-over, has occurred. The daughter cells produced by meiosis I then undergo a second meiotic division to separate the sister chromatids. The process of meiosis for the generation of spermatozoa is illustrated in Figure 9.10.3.

REASSORTMENT OF GENETIC MATERIAL ARISES FROM TWO SOURCES: INDEPENDENCE OF HOMOLOGOUS CHROMOSOME SORTING AND CROSSING-OVER

As shown in Figure 9.10.3, each somatic cell contains pairs of homologous chromosomes, one-half contributed by the father and the other half contributed by the mother. In the meiotic divisions that produce the gametes, the first meiotic division separates the homologous pairs from each other. These homologous pairs are separated randomly. Since there are two ways to sort each member of a single pair, and the sorting is independent, the number of possible combinations of the maternal and paternal homologs in the final gamete is 2^N, where N is the number of chromosomes in the haploid state, or the number of pairs of chromosomes in the diploid state. Since in humans $N = 23$, the number of possible combinations of maternal and paternal homologous chromosomes is $2^{23} = 8.4 \times 10^6$.

The actual number of unique combinations of genes in the germ cells is actually much greater than this because there is a second source of reassortment of genes in the spermatogonia called crossing-over. In the long prophase of meiosis I, in which the two homologous chromosomes are in close contact, parts of the homologous chromosomes exchange with each other. On average, two or three crossover events occur in each chromosome during this stage. How this produces a large number of unique germ cells is illustrated in Figure 9.10.4. Briefly, each crossover in a given primary spermatocyte will produce four possible arrangements of the sister chromatids. But the number of possible crossovers is much greater. Thus, in one primary spermatocyte the crossover may occur between an extreme end of the chromosome and produce four unique sister chromatids; in another the crossover may occur in a different place. Thus, the number of unique combinations of genetic information from one parent is large.

The production of gametes or germ cells occurs in both males and females, but there are significant differences between them. The above discussion describes both processes in general but its emphasis is on the production of sperm, as shown in Figure 9.10.3. The purpose of the sperm, of course, is to combine genetic information with that of the ovum during the process of fertilization. All of the structures associated with the male reproductive tract are evolved with the purpose of delivering sperm to the female so that the probability of fertilization is enhanced. A section of the male reproductive system is shown in Figure 9.10.5.

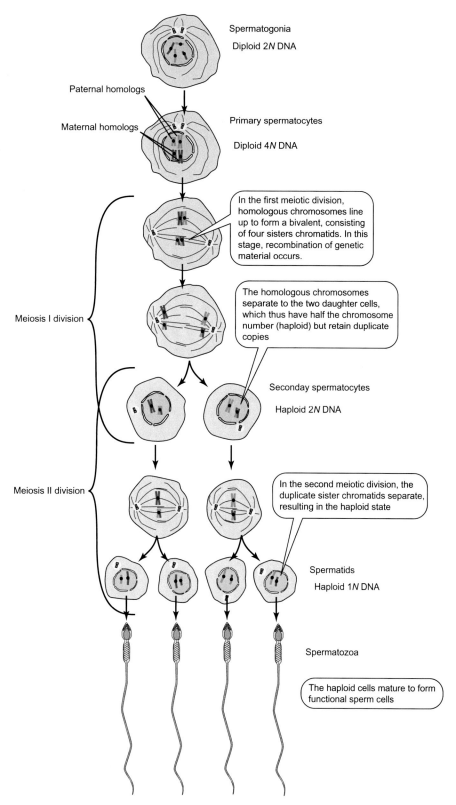

FIGURE 9.10.3 Meiosis in the production of spermatozoa. These cells continue to divide mitotically throughout adulthood. Some of the daughter cells become primary spermatocytes and undergo meiotic division, shown here. These primary spermatocytes are diploid cells with duplicated DNA, so they have 2N chromosomes and 4N DNA. In the first division, the duplicated homologous chromosomes are condensed and then aligned on the equator of the spindle in the form of a bivalent, consisting of four sister chromatids, two from each homologous pair of chromosomes. The homologous pairs separate randomly, one of each going to each daughter cell, which are called secondary spermatocytes. The secondary spermatocytes thus are haploid, with N chromosomes and 2N DNA. The chromosomes in the secondary spermatocyte are then aligned again on the spindle in the second meiotic division, and the sister chromatids are separated into the spermatids, which are haploid, with N chromosomes and 1N DNA. These spermatids then mature into spermatozoa.

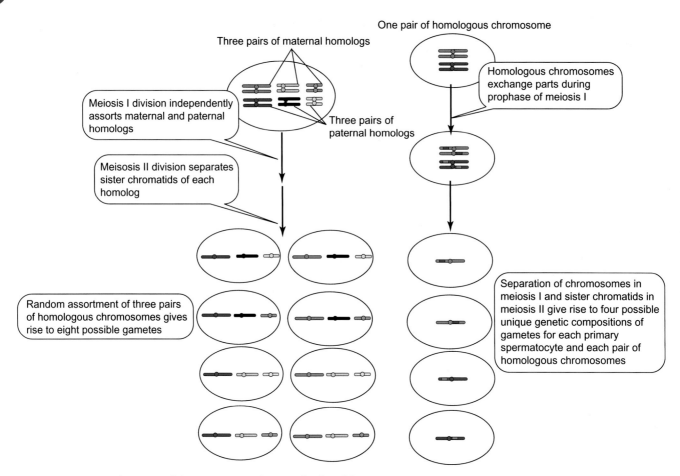

FIGURE 9.10.4 Origin of variation of the DNA content of germ cells. The adult somatic cells have paired chromosomes that contain different genetic information, some resulting from the mother (the set of maternal homologous chromosomes) and some resulting from the father (the set of paternal homologous chromosomes). This is the diploid number, 2N. During interphase, the chromosomes are duplicated to form sister chromatids and a diploid cell with 4N DNA. In meiosis I division, the homologous chromosomes are separated, creating haploid cells with 2N DNA. The sister chromatids are then separated in meiosis II, producing haploid cells with 1N DNA. Because meiosis I division randomly separates maternal and paternal homologs, there are multiple possible combinations of the final haploid germ cells. For N chromosomes, there are 2^N possible combinations. This genetic diversity is greatly enhanced by crossover between the maternal and paternal homologs in prophase of meiosis I. In this stage, the two homologous chromosomes with their sister chromatids are in close proximity, and parts of the maternal homolog on one sister chromatid may cross over to the paternal homolog and vice versa. This produces four nonidentical chromatids. For every crossover, there are four possible unique combinations of genes in the final haploid cells. Crossover can occur at a variety of points in the chromosome, so that the final number of possible germ cells is $2^N \times 4 \times$ the number of unique crossovers possible. (Source: Adapted from B. Alberts et al., The Molecular Biology of the Cell, 2002.)

TESTICLES PRODUCE SPERM AND TESTOSTERONE

TESTICLES REQUIRE LOWER TEMPERATURE FOR SPERM PRODUCTION

In man and most other mammals, the testicles reside in a pouch of skin called the **scrotum** that hangs outside of the body. This position lowers the temperature within the testicle some 1°C or 2°C. This lower temperature facilitates sperm production. A thin muscle called the cremaster muscle can retract the testicles toward the abdomen to increase their temperature when environmental temperatures are low and relax to extend the testicles when environmental temperatures are higher. Research suggests that the number of sperm produced by men today is less than that produced by men of former times, perhaps due to the wearing of tighter clothes and subsequent higher temperatures within the testicles.

TESTES PRODUCE SPERMATOZOA

Spermatozoa are produced within structures called the **seminiferous tubules**. These tubules contain stem cells called **spermatogonia** (singular is spermatogonium) that derive from primordial germ cells that enter the testis during embryogenesis. The spermatogonia reside near the basement membrane of the stratified epithelium of the seminiferous tubules, as shown in Figure 9.10.6. After the onset of puberty, and continuing throughout adult life, these spermatogonia divide mitotically and some of them enter the first meiotic division to form **primary spermatocytes**. These are diploid, having 2N chromosomes, with duplicate sister chromatids for 4N DNA. These primary spermatocytes separate the pairs of homologous chromosomes in the first meiotic division to become **secondary spermatocytes**, which are haploid (1N) with 2N DNA. The spermatocytes then undergo meiosis

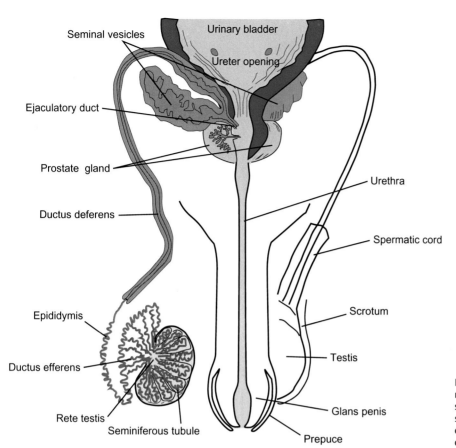

FIGURE 9.10.5 Anatomy of the male reproductive system. Sperm is produced in the seminiferous tubules, then traversing in succession the rete testis, ductuli efferentes, epididymis, vas (ductus) deferens, ejaculatory duct, and the urethra.

II in which the sister chromatids separate into the spermatid cells. Each primary spermatocyte gives rise to two secondary spermatocytes, and each of the two secondary spermatocytes gives rise to two **spermatids**.

Spermatids transform into spermatozoa in a process called **spermiogenesis**. This involves reduction of the volume of cytoplasm and formation of the specialized parts of the spermatozoa. Figure 9.10.7 shows a mature spermatozoon, also called a sperm cell.

The **Sertoli cells** (see Figure 9.10.6) provide nutrients and support for the formation and maturation of spermatids. These cells are large, polyhedral-shaped cells that extend from the basement membrane of the seminiferous tubules to the lumen. Extensions of the Sertoli cell cytoplasm surround the spermatids during the early stages of spermiogenesis. Eventually the maturing spermatids are released from the Sertoli cells in a process called spermiation.

Spermatogenesis is not random. Groups of spermatogonia are often in step and groups of them will initiate a cycle of development about every 16 days. Throughout the testes, these groups are not synchronized so that overall there is a continuous production of spermatozoa. The whole of spermatogenesis from spermatogonia to spermatozoa takes approximately 74 days (Figure 9.10.8).

THE TESTES MAKE TESTOSTERONE

The second main function of the testes is to make testosterone. Testosterone is made by the **Leydig cells**, which are interstitial cells interspersed among the seminiferous tubules. The biochemical pathways leading to the synthesis of various androgens is shown in Figure 9.10.9. Some of these reactions are not predominant in the Leydig cells themselves, but occur in target tissues. Notable among these is the conversion of testosterone to dihydrotestosterone by the **5α reductase** enzyme. This reaction typically occurs in target tissues of testosterone, not in the Leydig cells.

THE HYPOTHALAMUS AND ANTERIOR PITUITARY CONTROL TESTICULAR FUNCTION

CELLS IN THE HYPOTHALAMUS SECRETE GONADOTROPIN-RELEASING HORMONE

Neurons that synthesize and release GnRH, gonadotropin-releasing hormone, are dispersed throughout the hypothalamus but are believed to be concentrated in the arcuate nucleus and preoptic area. These cells synthesize GnRH as a precursor preprohormone containing 92 amino acids. Cleavage of the signal sequence releases a 69-amino-acid prohormone which is again cleaved to form a 10-amino-acid fragment which is the active GnRH. When stimulated, these neurons release GnRH into the hypothalamic portal circulation, which then travels to the anterior pituitary gland where the GnRH stimulates gonadotrophs to release follicle stimulating hormone (FSH) and luteinizing hormone (LH) into the general circulation. GnRH increases LH and FSH release by activating a

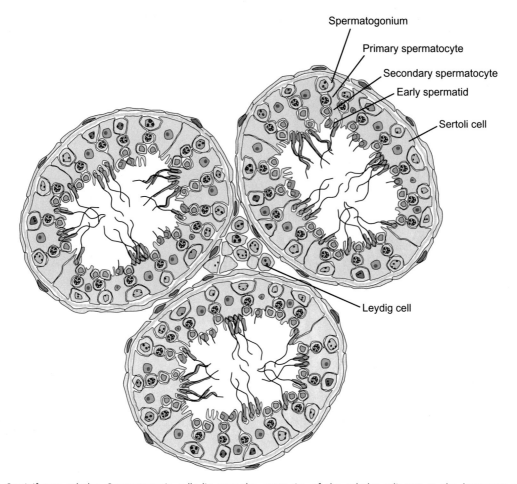

FIGURE 9.10.6 Seminiferous tubules. Spermatogonia cells lie near the outer rim of the tubules adjacent to the basement membrane. They continuously divide by mitosis to form additional spermatogonia. Some spermatogonia become primary spermatocytes, which undergo crossing-over in the prophase to the meiosis I division. After meiosis I division, the daughter cells are called secondary spermatocytes and these cells are haploid. The secondary spermatocytes undergo a second meiotic division to form the spermatids. These then undergo spermiogenesis to transform into spermatozoa.

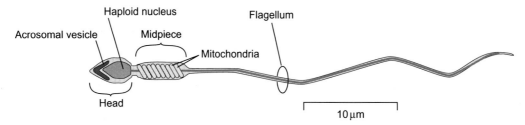

FIGURE 9.10.7 Mature spermatozoon. The genetic information is packaged in a tight bundle in the haploid nucleus in the head region. Surrounding it is an acrosomal vesicle that helps the sperm penetrate the ovum. The spermatozoa's motility derives from its flagellum, which is powered by a group of mitochondria in the midpiece.

G_q mechanism in which the α subunit of the heterotrimeric G protein activates phospholipase C, which hydrolyzes membrane phosphatidyl inositol bisphosphate to liberate IP3 and diacylglycerol (DAG). The released IP3 binds to IP3 receptors on the endoplasmic reticulum membrane and causes Ca^{2+} release from the ER stores; the released DAG activates protein kinase C (PKC). The net effect of GnRH binding to the gonadotrophs is increased synthesis and release of FSH and LH.

CONTROL OF GnRH SECRETION IS STILL INCOMPLETELY UNDERSTOOD

The exact mechanisms responsible for GnRH release are not known in detail. It is known that GnRH release is pulsatile, and therefore the release of LH and FSH from the anterior pituitary is also pulsatile. This pulsatile release is absolutely essential: continuous release of GnRH produces a downregulation of their receptors

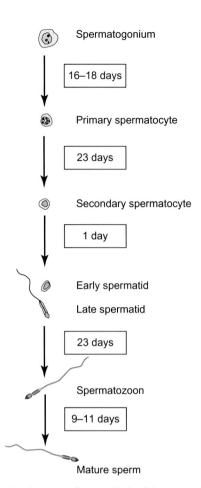

FIGURE 9.10.8 Development of sperm. Each of the stages that make up spermatogenesis has its own specific duration; the spermatogonia take 16–18 days to form the primary spermatocytes; the primary spermatocytes take 23 days to make the secondary spermatocytes; the secondary spermatocytes take only one day to become spermatids, and the spermatids spend another 23 days maturing into spermatozoa. This does not add up to 74 days because after release of the spermatozoa into the lumen of the seminiferous tubules, the spermatozoa still require maturation. The released spermatozoa are not yet motile, and they are transferred to the rete testis and epididymis by movement of the luminal fluid, not by motility of the sperm. In the epididymis, the spermatozoon undergoes maturation processes that progressively enable it to bind to the zona pellucida of the ovum and to fertilize the ovum.

and subsequent reduction in LH and FSH release. Secretion of GnRH is inhibited by circulating levels of testosterone.

BOTH LH AND FSH CONTROL TESTICULAR FUNCTION

LH AND FSH ARE GLYCOPROTEINS THAT BELONG TO THE SAME FAMILY AS TSH AND HUMAN CHORIONIC GONADOTROPIN

Human chorionic gonadotropin (hCG), thyroid-stimulating hormone (TSH), FSH, and LH are all glycoproteins composed of an α and a β chain. All four of these hormones have identical α chains consisting of a 92-amino-acid sequence. Thus, the β chain confers specificity of action.

LH STIMULATES LEYDIG CELLS TO PRODUCE TESTOSTERONE

Although LH derives its name (luteinizing hormone) from its effects on the female reproductive system, it also has important effects in the male. Leydig cells have receptors for LH on their surface membranes that operate through a G_s mechanism. The increased cAMP levels result in increased synthesis and subsequent activity of a number of enzymes involved in the synthesis and secretion of testosterone. The mechanisms for this activation are similar to those of ACTH on adrenal steroid synthesis (Figure 9.5.6). Some 95% of the circulating testosterone originates from the Leydig cells of the testes.

TESTOSTERONE HAS LOCAL AND SYSTEMIC EFFECTS

Testosterone released from the Leydig cells appears in the general circulation and also has local effects on the Sertoli cells that reside just a few hundred microns away from the Leydig cells. Sertoli cells contain **androgen binding protein** (ABP) that binds testosterone and is secreted into the seminiferous tubules. Presumably this presents high concentrations of testosterone to the developing spermatozoa so as to maintain spermatogenesis. Testosterone exerts multiple effects including:

- Maintenance of spermatogenesis in seminiferous tubules and maturation of spermatozoa in the epididymis.
- Maintenance of size and function of the accessory sexual organs.
- Development of secondary sex characteristics at puberty and maintenance postpuberty:
  - size of external genitalia;
  - laryngeal enlargement leading to deepening of the voice;
  - secretion of sebaceous glands, tending toward acne;
  - masculine pattern of muscle mass and anabolism;
  - male hair distribution.
- Behavioral effects: increased libido and aggressiveness.
- Pubertal growth spurt.

FSH STIMULATES SERTOLI CELLS TO PRODUCE A NUMBER OF PROTEINS

Sertoli cells have receptors for FSH on their surface membranes that are linked to a G_s mechanism. The resulting increase in cAMP alters the expression of some 100 proteins. Important among these are the following:

- Increasing aromatase: this enzyme converts testosterone to estradiol or androstenedione to estrone.
- Increasing **ABP**.
- Increasing **inhibin**, a protein that feeds back onto the anterior hypothalamus to shut off gonadotroph secretion.
- Increasing growth factors that support spermatogenesis.

The feedback regulation of GnRH, LH, and FSH secretion is shown in Figure 9.10.10.

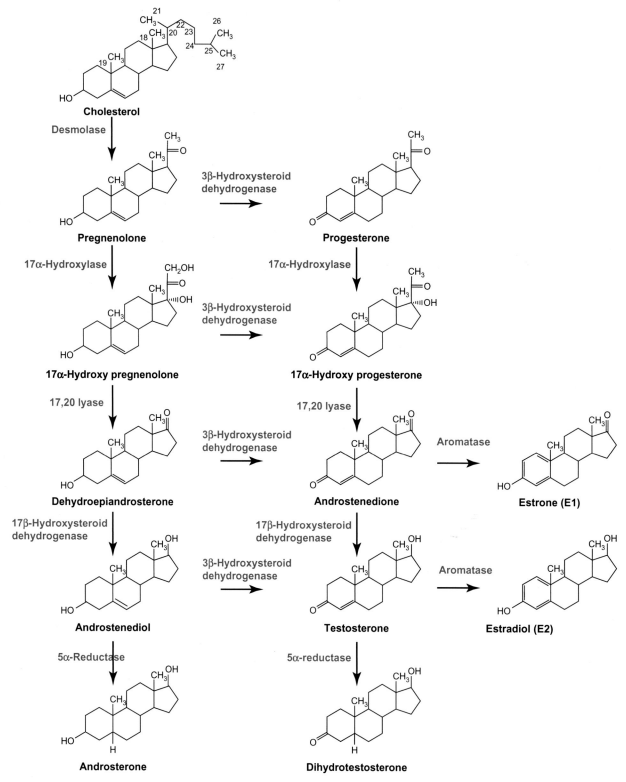

FIGURE 9.10.9 Pathway of androgen synthesis in Leydig cells and androgen metabolism in target cells. Testosterone is synthesized and secreted by the Leydig cells and conversion to dihydrotestosterone generally occurs in the target tissues.

THE MALE SEXUAL RESPONSE

The sexual response consists of four phases:

1. Arousal or excitement phase
2. Plateau phase
3. Orgasmic phase
4. Resolution phase or refractory period.

All of these phases involve the autonomic nervous system, integrated with higher order functions. The most important sexual organ is the brain. Mechanical

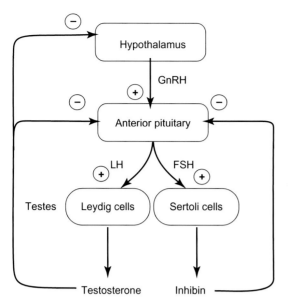

FIGURE 9.10.10 Feedback regulation of GnRH, LH and FSH secretion in the male.

stimulation of the penis or expectation of sexual interactions produces the excitement phase, which in the male is the **erection** of the penis in preparation for sexual intercourse. Penile erection results from a complex interaction of psychological, neural, vascular, and endocrine factors and is usually, but not exclusively, associated with sexual arousal. Penile erection can also occur in response to a full urinary bladder or during rapid eye movement sleep.

Penile erection results from engorgement of the **corpus cavernosa** with blood. The corpus cavernosa are two tubular structures running the length of the penis. Thus, it is ultimately the blood pressure that enables the hardening and enlargement of the penis during erection. Parasympathetic nerves supplying the vessels release NO, nitric oxide, from nerve endings and endothelial cells. The NO activates guanylate cylcase to increase levels of cGMP in the tissue. This relaxes the vascular smooth muscle, as described in Chapter 3.8. Drugs such as sildenafil (viagara) act by inhibiting cGMP phosphodiesterase, preventing cGMP breakdown. These drugs cannot act without parasympathetic activation of erection.

Ejaculation, the emission of semen from the penis, is mediated by sympathetic nerves. The normal volume of the ejaculate is 3–4 mL consisting of spermatozoa and the seminal plasma made up of the secretions of the testes and accessory sex organs including seminal vesicles, prostate gland, and bulbourethral gland. The seminal vesicles contribute 65–75% of the semen volume; the prostate contributes 25–30%, and about 5% is contributed by the testes. The semen is a white opalescent fluid that coagulates after ejaculation. The seminal vesicles add a protein, **semenogelin**, to the ejaculate and this protein causes the coagulation reaction after mixing with prostate fluid. Semenogelin is the major protein in the seminal plasma. After 10–20 min, the semen liquefies again due to proteolysis of the semenogelin by **prostate specific antigen (PSA)**, which is a protease added to the semen by the prostate gland. The seminal vesicle also adds **fructose** to the semen, in excess of 100 mg%, which is the predominant energy source available to the sperm. The ejaculate typically contains 1×10^8 sperm mL^{-1}. Values below 20×10^6 sperm mL^{-1} are considered insufficient for fertilization.

SUMMARY

Cell division in the somatic, or body, cells is called mitosis and consists of an elaborate mechanism for duplication and separation of the entire genome. Duplication occurs during interphase. In prophase, the chromosomes condense and the mitotic spindle forms. In metaphase, the nuclear envelope breaks down and chromosomes align along the equator, bound to the spindle at their kinetochore. In anaphase, the sister chromatids separate to opposite poles. In telophase, division of the cell cytoplasm occurs, the nuclear membrane reforms, and the chromosomes decondense.

Formation of the gametes, or sex cells, requires meiosis. In this case, the pairs or chromosomes duplicate, producing four sequences for each gene: the original and a copy of each of two alleles or alternate forms of a gene found on homologous chromosomes. In meiosis I, homologous chromosomes separate randomly, so that many different combinations of genes are possible. In meiosis II, the sister chromatids separate, producing the haploid (N) gametes. Parts of homologous chromosomes can also cross over or exchange material between homologous chromosomes. This process greatly increases the number of possible rearrangements of the genes in the germ cells.

The male sexual organs consist of the testicles, penis, and accessory organs consisting of a pair of seminal vesicles, a single prostate gland, and a bulbourethral gland. The paired testes produce sperm and testosterone. These testes are suspended outside the body in the scrotum, as the sperm require a slightly lower temperature for maturation. The penis is a delivery system for the sperm to the female's vagina. Spermatogenesis occurs in the seminiferous tubules, requiring 74 days for completion to mature, motile spermatozoa, with final maturation in the epididymis. This process is regulated by FSH and testosterone. Leydig cells in between the seminiferous tubules synthesize testosterone under stimulation by LH. FSH increases synthesis of ABP and inhibin by the Sertoli cells that nourish the developing sperm. Testosterone inhibits LH release from the anterior pituitary; inhibin inhibits FSH release.

Circulating testosterone promotes development of secondary sex characteristics, masculine body type and behavior, and development and maintenance of the genitalia and accessory sex glands.

Vasodilation of vessels supplying the corpus cavernosa, and restriction of the outlet, produces penile erection. This is accomplished through parasympathetic release of NO. Ejaculation is controlled by the sympathetic nervous system. The ejaculate consists of 3–5 mL of fluid originating from the testes (5%), seminal vesicles (70%), and prostate (25%). The ejaculate gels quickly due to semenogelin and then liquefies after 10–20 min from the action of PSA. This probably enables the sperm to stick to the vaginal walls.

REVIEW QUESTIONS

1. List the phases of mitosis in order. At which phase is DNA duplicated? When do the chromosomes condense? When do they line up at the equator? When do they separate? What part of the genotype separates in this phase of mitosis?
2. What is separated in meiosis I? How many ways can this occur? What is the composition of the separated chromosomes? What is separated in meiosis II?
3. What keeps sister chromatids together? What separates them? What is the centromere? What is the kinetochore?
4. Where are sperm made? Where do they mature? What hormones are necessary for their development?
5. What causes penile erection? What causes ejaculation?
6. Name the accessory sex glands. Which contributes the highest proportion of volume to the ejaculate? What contributes the least? What energy source is added to the ejaculate to support the sperm? What causes coagulation of the semen? What cause liquefaction of the semen? How many sperm does semen contain?
7. Where is GnRH synthesized? What cells release LH and FSH? What inhibits their release?
8. What cells does LH stimulate? What do they make in response? What cells does FSH stimulate? What do these cells do in response to FSH stimulation?

Problem Set 9.1
Ligand Binding

1. The specific ryanodine binding to isolated cardiac sarcoplasmic reticulum was determined and gave the following results:

Bound Ryanodine (pmol mg^{-1})	Free Ryanodine (nM)
0.31	0.19
0.73	0.52
1.23	1.27
1.67	2.08
2.01	2.99
2.53	7.47
2.82	17.18
3.12	26.88

   What is the B_{max} and K_d for ryanodine binding? Is the binding cooperative? Be sure to include any graphs that you might want to make. Mathematics without the graphics is a bad mistake because the human mind is a much better filter than mathematical constructs like the correlation coefficient.

2. The following table shows the inhibition of ryanodine binding by ruthenium red. Is ruthenium red a competitive or noncompetitive inhibitor of ryanodine binding? Calculate the K_i for ruthenium red.

Ryanodine (nM)	Ruthenium Red (μM)	Bound Ryanodine (pmol mg^{-1})
5	0	3.14
5	0.043	1.76
5	0.118	0.99
5	0.370	0.39
10	0	4.28
10	0.043	2.43
10	0.118	1.46
10	0.370	0.71
30	0	5.33
30	0.043	4.33
30	0.118	2.99
30	0.370	1.77

3. The following data were obtained for the Ca^{2+} dependence of ryanodine binding to rat heart homogenates:

Bound Ryanodine (pmol mg protein^{-1})	pCa
0.025	7.5
0.048	6.3
0.087	6.0
0.185	5.8
0.329	5.6
0.480	5.4
0.547	5.2
0.552	4.7
0.568	4.3
0.543	3.4

   Here pCa $= -\log [Ca^{2+}]$. The nonspecific binding of ryanodine, measured with excess cold ryanodine, averaged 0.025 pmol mg protein^{-1} at pCa 7.5, 5.4, and 3.4. That is, the nonspecific binding did not depend on pCa. What is the Hill coefficient and K_{Ca} for this binding? (Hint: first plot the saturation curve. Determine B_{max} for use in the Hill plot. Subtract $B_{nonspecific}$ from all values to obtain the specific binding and then perform the Hill analysis. Use for the analysis only those points on the Hill plot that are between 10% and 90% saturation.)

4. The following data were obtained for the velocity of the SERCA2a Ca^{2+}-ATPase of cardiac sarcoplasmic reticulum as a function of the free $[Ca^{2+}]$:

$[Ca^{2+}]$ (μM)	ATPase Rate (nmol min^{-1} mg SR protein^{-1})
61.6	519
25.8	519
6.9	82
4.3	395
3.07	311
2.37	271
1.92	195
1.09	115
0.66	61
0.04	19

What is the Hill coefficient and K for the Hill equation? (Hint: The value of the ATPase rate at very low $[Ca^{2+}]$ is due to contaminating Mg^{2+}-ATPase of unknown origin. Its value must be subtracted from all ATPase values prior to the Hill analysis. Plot v against pCa to convince yourself of this.)

5. The velocities of the pyruvate kinase reaction obtained at a number of concentrations of phosphoenolpyruvate, at constant [ADP] and in the absence of modifiers, are given below:

[PEP] (mM)	Rate (a.u.)
0.0195	0.006
0.0325	0.012
0.065	0.031
0.195	0.094
0.325	0.114
Saturating	0.130

Determine the Hill coefficient and K for the Hill equation.

6. EGTA complexes Ca^{2+} and thereby allows for experimental buffering of solutions containing Ca^{2+}. Assume the association constant for 1:1 complex of Ca^{2+} with EGTA is $0.5 \times 10^{-6}\, M^{-1}$. If the total Ca^{2+} content (the sum of free and bound) is 100 μM, and the total [EGTA] is 400 μM, what is the free $[Ca^{2+}]$? What is $pCa = -\log[Ca^{2+}]$?

7. Suppose that PTH is destroyed only by the kidneys in a single pass. Suppose that the blood volume is 5 L. How much blood would have to flow through the kidneys to achieve a half-life for PTH of 2 min? Is it reasonable to suppose that only the kidney destroys PTH?

8. The half-life of insulin is about 5 min. What is its fractional turnover? If the blood volume is 5 L and the GFR is 120 mL min^{-1}, and assuming a sieving coefficient of 1.0, is loss through glomerular filtration sufficient to explain the half-life of insulin?

9. Thyroxine binding globulin (TBG) has a M_r of 54 kDa. Its metabolic clearance rate is 800 mL day^{-1} and its volume of distribution is 7 L. The plasma [TBG] = 2 mg dL^{-1}.
   A. What is TBG's half-life?
   B. If TBG binds one molecule of T4, what is its binding capacity, expressed as μg T4 dL^{-1}?

10. The following data are normal for humans:

Total T4	100 nM
Free T4	20 pM
Total T3	1.8 nM
Free T3	5 pM

The concentration of TBG is about 1.5 mg%. Its molecular weight is 54 kDa. Assume that 80% of the total T4 and T3 are bound to TBG. Estimate the association constant of TBG for T3 and T4.

11. The daily production rate of cortisol is about 15 mg day^{-1} (10–20). Its circulating half-life is about 95 min (70–120 min). If its volume of distribution is 7 L, what is the average total concentration in blood?

12. The peak of progesterone during the luteal phase of the menstrual cycle is about 10 ng mL^{-1}.
   A. If the molecular weight of progesterone is 314.5 g mol^{-1}, what is the peak concentration in nM?
   B. Suppose that the daily production of progesterone is 20 mg day^{-1} during the luteal phase. What is the metabolic clearance rate?
   C. If the volume of distribution of progesterone is 14 L, what is the fractional turnover?

Printed in the United States
By Bookmasters